Congratulations! You have purchased a 12-month, prepaid subscription to the Special Edition of *The Biology Place* for *BIOLOGY: Concepts and Connections*, Third Edition, by Campbell, Mitchell, and Reece, a Web learning environment for general biology that includes learning activities, study and testing aids, and a wide range of content to help you succeed in your biology course.

To activate your prepaid subscription:

1. Point your Web browser to **www.biology.com/cmr**
2. Select The Biology Place Special Edition Login
3. Enter your preassigned activation ID and password exactly as they appear below in the user ID and password fields:

Activation ID BICMST12257566

Password micron

4. Select Login
5. Complete the online form to establish your personal ID and password.
6. Once your personal ID and password are confirmed, go back to **www.biology.com/cmr/login** to enter the site with your new ID and password.

Your activation ID and password can be used only once to establish your subscription, which is not transferable.

If you did not purchase this book new and in a shrink-wrapped package, this activation ID and password may not be valid. However, if your instructor is recommending or requiring use of *The Biology Place*, you can find more information on purchasing a subscription directly online at **www.biology.com/cmr**, or you can purchase a subscription at your local college bookstore if your professor has specifically requested it.

BIOLOGY

CONCEPTS & CONNECTIONS

THIRD EDITION

Neil A. Campbell
Lawrence G. Mitchell
Jane B. Reece

An imprint of Addison Wesley Longman, Inc.

San Francisco • Reading, Massachusetts • New York • Harlow, England
Don Mills, Ontario • Sydney • Mexico City • Madrid • Amsterdam

Executive Editor: Erin Mulligan
Senior Developmental Editor: Shelley Parlante
Senior Developmental Editor for Media: Pat Burner
Managing Editor: Laura Kenney
Senior Production Editor: Angela Mann
Art and Design Manager: Don Kesner
Senior Art Supervisor: Donna Kalal
Project Editor: Kirsten Watrud
Associate Producer: Maureen Kennedy
Senior Multimedia Artist: Russell Chun
Principal Artists for the Third Edition: Karl Miyajima, Precision Graphics, Carla Simmons, Mary Ann Tenorio
Photo Researcher: Stephen Forsling
Permissions Editor: Ariane dePree-Kajfez
Text Designer: William Seabright and Associates
Layout Artist: Pat Coleman
Cover Designer: Yvo Riezebos
Copyeditor: Chad Colburn
Proofreader: Carol Lombardi
Indexer: Shane-Armstrong Information Systems
Prepress Supervisor: Vivian McDougal
Compositor: Thompson Type
Prepress: H&S Graphics, Inc.
Marketing Manager: Gay Meixel

On the cover:
Gray tree frog, © Dwight R. Kuhn.

Copyright © 1994, 1997, 2000 by Benjamin/Cummings, an imprint of Addison Wesley Longman, Inc.

For permission to use copyrighted material, grateful acknowledgment is made to the copyright holders on pp. 804 to 809, which are hereby made part of this copyright page.

All rights reserved. No part of this publication may be reproduced, stored in a retrieval system, or transmitted in any form or by any means, electronic, mechanical, photocopying, recording, or any other media embodiments now known or hereafter to become known, without the prior written permission of the publisher. Manufactured in the United States of America. Published simultaneously in Canada.

Library of Congress Cataloging-in-Publication Data
Campbell, Neil A., 1946–
Biology : concepts & connections / Neil A. Campbell, Lawrence G. Mitchell, Jane B. Reece. — 3rd ed.
p. cm.
ISBN 0-8053-6585-0
1. Biology. I. Mitchell, Lawrence G. II. Reece, Jane B. III. Title.
QH308.2.C34 1999
570—dc21 99-23620
CIP

3 4 5 6 7 8 9 10—RNT—03 02 01 00

Benjamin/Cummings, an imprint of Addison Wesley Longman, Inc.
1301 Sansome Street
San Francisco, CA 94111

To Rochelle and Allison, with love

N.A.C.

To Mary, with love

L.G.M.

In memory of my parents

J.B.R.

Of Related Interest from the Benjamin/Cummings Series in the Life Sciences

General Biology

N. A. Campbell, J. B. Reece, and L. G. Mitchell
Biology, Fifth Edition (1999)

R. A. Desharnais and J. R. Bell
Biology Labs On-Line (2000)

J. Dickey
Laboratory Investigations for Biology (1995)

J. B. Hagen, D. Allchin, and F. Singer
Doing Biology (1996)

R. Keck and R. Patterson
Biomath: Problem Solving for Biology Students (2000)

A. E. Lawson and B. D. Smith
Studying for Biology, First Edition (1995)

J. Lee
The Scientific Endeavor (2000)

J. G. Morgan and M. E. B. Carter
Investigating Biology, Third Edition (1999)

J. A. Pechenik
A Short Guide to Writing about Biology, Third Edition (1997)

G. I. Sackheim
An Introduction to Chemistry for Biology Students, Sixth Edition (1999)

R. M. Thornton
The Chemistry of Life CD-ROM (1998)

Biotechnology

D. Bourgaize, T. Jewell, and R. Buiser
Biotechnology: Demystifying the Concepts (2000)

Cell Biology

W. M. Becker, L. J. Kleinsmith, and J. Hardin
The World of the Cell, Fourth Edition (2000)

Biochemistry

R. F. Boyer
Modern Experimental Biochemistry, Third Edition (2000)

C. K. Mathews, K. E. van Holde, and K. G. Ahern
Biochemistry, Third Edition (2000)

Genetics

R. J. Brooker
Genetics: Analysis and Principles (1999)

J. P. Chinnici and D. J. Matthes
Genetics: Practice Problems and Solutions (1999)

R. P. Nickerson
Genetics: A Guide to Basic Concepts and Problem Solving (1990)

P. J. Russell
Fundamentals of Genetics, Second Edition (2000)

P. J. Russell
Genetics, Fifth Edition (1998)

Molecular Biology

M. V. Bloom, G. A. Freyer, and D. A. Micklos
Laboratory DNA Science (1996)

J. D. Watson, N. H. Hopkins, J. W. Roberts, J. A. Steitz, and A. M. Weiner
Molecular Biology of the Gene, Fourth Edition (1987)

Microbiology

G. J. Tortora, B. R. Funke, and C. L. Case
Microbiology: An Introduction, Sixth Edition (1998)

Anatomy and Physiology

E. N. Marieb
Essentials of Human Anatomy and Physiology, Sixth Edition (2000)

E. N. Marieb
Human Anatomy and Physiology, Fourth Edition (1998)

E. N. Marieb and J. Mallatt
Human Anatomy, Second Edition (1997)

Ecology and Evolution

C. J. Krebs
Ecological Methodology, Second Edition (1999)

C. J. Krebs
Ecology: The Experimental Analysis of Distribution and Abundance, Fourth Edition (1994)

J. W. Nybakken
Marine Biology: An Ecological Approach, Fourth Edition (1997)

E. R. Pianka
Evolutionary Ecology, Sixth Edition (2000)

R. L. Smith
Ecology and Field Biology, Fifth Edition (1996)

R. L. Smith and T. M. Smith
Elements of Ecology, Fourth Edition Update (2000)

Plant Ecology

M. G. Barbour, J. H. Burk, W. D. Pitts, F. S. Gilliam, and M. W. Schwartz
Terrestrial Plant Ecology, Third Edition (1999)

Zoology

C. L. Harris
Concepts in Zoology, Second Edition (1996)

About the Authors

Left to right, authors Campbell, Reece, and Mitchell

Neil A. Campbell

has taught general biology for 30 years and, with Drs. Reece and Mitchell, has coauthored *Biology,* Fifth Edition, the most widely used text for biology majors. His enthusiasm for sharing the fun of science with his students stems from his own undergraduate experience. He began at Long Beach State College as a history major, but switched to zoology after general education requirements "forced" him to take a science course. Following a B.S. from Long Beach, he earned an M.A. in Zoology from UCLA and a Ph.D. in Plant Biology from the University of California, Riverside. He has published numerous research articles on how certain desert plants survive in salty soil and how the sensitive plant (*Mimosa*) and other legumes move their leaves. His diverse teaching experiences include courses for nonbiology majors at Cornell University, Pomona College, and San Bernardino Valley College, where he received the college's first Outstanding Professor Award in 1986. Dr. Campbell is currently a visiting scholar in the Department of Botany and Plant Sciences at UC Riverside.

Lawrence G. Mitchell

has 21 years of experience teaching a broad range of life science courses at both undergraduate and graduate levels. He holds a B.S. in zoology from Pennsylvania State University and a Ph.D. in zoology and microbiology from the University of Montana. Following postdoctoral research with the National Institute of Allergy and Infectious Diseases, Dr. Mitchell joined the biology faculty at Iowa State University in 1971. He received the Outstanding Teacher Award at Iowa State in 1982. In addition to numerous research publications in aquatic parasitology, Dr. Mitchell has coauthored the textbooks *Zoology* and *Biology,* Fifth Edition. He has also developed television courses in general biology and has written, produced, and narrated programs on wildlife biology for public television. Dr. Mitchell is currently an affiliate professor in the Division of Biological Sciences at the University of Montana. He devotes most of his time to writing and conservation issues.

Jane B. Reece

has worked in biology publishing since 1978, when she joined the editorial staff of Benjamin/Cummings. Her education includes an A.B. from Harvard University, an M.S. from Rutgers University, and a Ph.D. in bacteriology from the University of California, Berkeley. At UC Berkeley and later as a postdoctoral fellow in genetics at Stanford University, her research focused on genetic recombination in bacteria. She has taught biology at Middlesex County College (New Jersey) and Queensborough Community College (New York). During her 12 years as an editor at Benjamin/Cummings, Dr. Reece played major roles in a number of successful textbooks, including *Microbiology: An Introduction,* by G. J. Tortora, B. R. Funke, and C. L. Case; and *Molecular Biology of the Gene,* Fourth Edition, by J. D. Watson et al. She was a coauthor of *The World of the Cell,* Third Edition, with W. M. Becker and M. F. Poenie, and of *Biology,* Fifth Edition.

Preface

As we head into the new millennium, understanding the concepts of biology and their connections to our lives is more important than ever. Whether we're concerned with our own health or the health of our planet, a familiarity with biology is essential. This basic knowledge and an appreciation for how science works have become elements of good citizenship in an era when informed evaluations of health issues, environmental problems, and applications of new technology are critical.

The "connections" to which the title of this book refers go beyond the practical applications of biology. Biology has important connections with the other natural sciences and with the humanities and social sciences as well. And the study of life has no coherence without an understanding of the connections among the different areas of biology and an appreciation of the grand unifying theme of evolution. From its first edition, the hallmarks of this book have included an emphasis on connections within biology and between biology and other fields, and a focus on engaging students from a wide variety of majors.

We could not have hoped to meet our ambitious goals for this book without extensive discussions with teaching colleagues throughout the world and feedback from many of the hundreds of instructors and hundreds of thousands of students who have used our earlier editions. We have been gratified by their enthusiastic responses and have paid close attention to their thoughtful suggestions for improvement. For this third edition, we set out to create a book that would be an even more effective tool for learning biology. In addition, we worked to ensure that the new book would integrate smoothly with an exciting new program of supporting materials in electronic media.

TEACHING STRATEGIES OF *BIOLOGY: CONCEPTS & CONNECTIONS,* 3RD EDITION

How can we help students learn—and enjoy—biology? How can we help instructors teach biology? Our responses to these questions are reflected in the teaching strategies we bring to the book. In this new edition, we build on the approach that has been so successful in earlier editions. Below we describe our five main teaching strategies as they are embodied in the third edition of *Biology: Concepts & Connections.*

This book brings the main concepts of biology into sharp focus—and now makes their organizing framework more explicit

Biology is a vast subject that gets bigger every year, but a general biology course is still only one or two terms long. In that brief time, we explore all of life, from molecules to ecosystems, while also trying to share the excitement of important research breakthroughs. For beginning biology students confronting this avalanche of information, it can seem as important to memorize all the scientific terms and facts as it is to master and apply the major ideas. This situation changes, however, when students can acquire a framework of key biological concepts into which they can fit the many new things they learn. It is this framework of concepts that will serve them long after they have forgotten specific facts and terms.

Biology: Concepts & Connections was the first introductory biology textbook to use concept modules to help students recognize and focus on the main ideas of each chapter. The heading of each module is not simply a topic but a carefully crafted statement of a biological concept. Printed in large type, each concept serves as a focal point for a module, and all of the module's text and illustrations converge on that concept with explanation and, often, evidence. For example, "Sensory receptor cells convert stimuli into electrical energy" announces a key concept in Chapter 29 (Module 29.2). The text and illustrations introduce the general principles of sensory reception and transduction, using the human sense of taste as an example. In this and other modules, we integrate the words and pictures to an unprecedented degree: The text "walks" the student through the illustrations, just as an instructor would do in class. In teaching a sequential process, such as the functioning of the receptor cells of the taste buds (Figure 29.2A), we number the steps in the text to correspond to numbered steps in the figure. The synergy between a module's verbal and graphic components transforms the concept heading into an idea with meaning to the student.

In this new edition, we make the framework that connects a chapter's concepts more explicit by grouping the concept modules under prominent general headings. For example, the concept modules in Chapter 4 that describe the structures and functions of the nuclear envelope, endoplasmic reticulum, Golgi apparatus, lysosomes, and vacuoles are grouped under the general heading "Organelles of the Endomembrane System." Students first see the framework of general headings and subordinated concepts in the outline that now opens each chapter, and the general headings are displayed on maroon bars where they fall within the chapter. They are also used to organize the Chapter Summary at the end of the chapter.

Reinforcing the message of every module in this edition is a checkpoint question. Located at the end of each module, these questions encourage students to test themselves repeatedly as they proceed through a chapter. The questions vary in approach. Some questions directly test the student's understanding of the main concept or part of it; others test understanding of the supporting evidence or ask the student to connect the concept to another concept in the book; still others require the student to carry out a calculation using information in the module. Feedback is provided on the spot: The answer is printed upside down beneath the question. These questions are intended to make students *think* about the material they are studying and to build confidence; opportunities for more comprehensive self-testing are provided at the end of the chapter and in the materials that accompany the book.

Application modules relate biological concepts to students' lives

Students are more motivated to study biological concepts that they can connect to their own lives and interests—concepts associated with health issues, economic problems, environmental quality, ethical controversies, and social responsibility. In this edition, heading statements in green type with a red shading in the background distinguish the numerous application modules that go beyond the core biological concepts. We have also reorganized Unit II, Cellular Reproduction and Genetics, placing more modules dealing with human biology and medicine closer to related core concepts. At the end of each chapter in the book, "Science, Technology, and Society" questions encourage students to apply the concepts they have learned to various social and environmental issues.

Organisms and biological diversity remain center stage throughout the book

Students are more familiar with animals, plants, and other organisms than they are with molecules, cells, or even ecosystems. *Biology: Concepts & Connections* builds on that interest by keeping the spotlight on organisms in every chapter. The illustrated essay that begins each chapter usually portrays a natural scene featuring a biological species and how it is adapted to its environment. The biology of that organism and others is woven through the chapter, even when the chapter takes students inside organisms for a look at cells or stands back for a view of populations, communities, or ecosystems. Our hope is that the chapter-opening essays will help nurture students', and thus society's, appreciation for biological diversity.

This edition ends with a new, capstone chapter on conservation biology, Chapter 38, which further highlights the importance of biological diversity and scientific efforts to sustain it.

The many connections to whole organisms also contribute to the book's versatility in serving a variety of syllabi. Whether a course begins with biological chemistry, with ecology, or somewhere in between, instructors can customize the chapter sequence without having students lose their orientation; whatever the topic, students can connect the concepts to whole organisms.

The theme of evolution unifies the book

Organisms are this book's stars, and evolution provides its story line. The history of life on Earth goes back more than three and a half billion years, and this past is the key to the present diversity of organisms. As the unifying theme of this textbook, evolution elevates biology from a bewildering collection of facts to a coherent study of changing life on a changing planet. In *Biology: Concepts & Connections,* students will study the structure, function, and behavior of organisms in an evolutionary context. And throughout the book, students will view the unity and diversity of life—the similarities and differences among organisms—as the dual consequences of descent with modification. Our updated and enhanced coverage of evolution in Chapter 1 and Unit III, including the new Chapter 19 on human evolution, strengthen the unifying theme of the book.

The process of science continues to be prominent in this new edition

A biology course should make students familiar with the scientific process of posing and testing hypotheses. With an improved introduction to the process of science in Chapter 1, students will be better equipped to appreciate the many examples throughout the book of how scientific concepts emerge from observations and experimental evidence. The book also puts human faces on science with "Talking About Science" modules. A number of these profiles of interesting scientists are new to this edition.

In the review material that concludes each chapter, questions called "Thinking as a Scientist" give students some personal practice with science as a process. Other ways of connecting to the process of science are provided to users of this book via electronic media. Activities on the Student Interactive Study Partner CD-ROM that comes with each new book, as well as activities available at the book's web site, provide further exposure to the scientific process. In fact, this book will work best for students who participate actively in learning about biological concepts and their applications.

NEW CONCEPTS AND CONNECTIONS IN THIS EDITION

Chapter 1, "Introduction: The Scientific Study of Life," includes a modern introduction to the diversity of organisms and their classification, focusing on the three domains—Bacteria, Archaea, and Eukarya. The chapter now features ongoing studies of rainforest epiphytes as an extended example of how the process of science works. Module 1.5 presents life's fundamental features more explicitly than in previous editions.

In **Unit I, The Life of the Cell,** our introduction to the basic chemistry, structure, and energetics of cells, we have fine-tuned a number of modules to help clarify topics that students find difficult. Topics that have been updated include the functions of the cytoskeleton (Module 4.17) and the extracellular matrix (Module 4.19). Among the improved figures is Figure 7.6B, showing the interaction of light with chloroplasts; the same module has new information about the functions of carotenoids in plant cells. A new Talking-About-Science module (7.14) features Nobel laureate Mario Molina, who discusses his research on the Earth's ozone layer.

In reorganizing **Unit II, Cellular Reproduction and Genetics,** we have integrated topics in human genetics that were formerly in the last chapter of the unit into earlier chapters to support the core genetic concepts where they are first presented. The result is a unit that will be more accessible and appealing to students. Other highlights of the revision of Unit II include clearer diagrams of mitosis and meiosis (Modules 8.6 and 8.14), a new emphasis on human pedigrees in Module 9.8, updated coverage of animal cloning (Module 11.3), and a more focused account of the regulation of gene expression in eukaryotes (see especially Module 11.6). To our already strong coverage of the genetic basis of cancer (Modules 11.13–11.15), we have added an interview with leading breast-cancer researcher Mary-Claire King. The final chapter in the unit, Chapter 12, is now titled "DNA Technology and the Human Genome," highlighting the book's updated description of human DNA and the progress of the Human Genome Project (Modules 12.13–12.15). The verbal and graphic explanations of the major methods of DNA technology and their applications to law, industry, medicine, and agriculture have been improved. Two modules on the risks and ethical questions arising from the use of the technology conclude the chapter and the unit.

Major new features in this edition's **Unit III, Concepts of Evolution,** include a discussion of alternative species concepts (Module 14.2) and a new interview with popular author and educator Richard Dawkins (Module 14.9). Also, in Chapter 15, we have added new material on evolutionary trends in horses (Module 15.8) and have extensively revised our coverage of DNA and RNA sequence analysis in evolutionary research (Module 15.12). The new Module 15.14 provides a step-by-step illustration of cladistic analysis.

Unit IV, The Evolution of Biological Diversity, includes a modern treatment of archaea (Modules 16.8 and 16.12) and recent hypotheses about the origins of metabolism (Module 16.11). In Module 16.15, on pathogenic bacteria, we explain how certain strains of *E. coli* can cause life-threatening illness. Module 17.9 now includes a discussion of the loss of biological diversity in harvested forests and includes some thoughts on forest management by Michael Dombeck, Chief of the U.S. Forest Service. Several new photos enhance our coverage of animal diversity in Chapter 18. We also have added a section on incomplete and complete metamorphosis (Module 18.13), and updated our discussion of the evolutionary relationships between annelids and arthropods (Module 18.12), lungfishes and amphibians (Module 18.19), and dinosaurs and birds (Module 18.21). Module 18.23 includes a discussion of the current controversy about the origins of segmentation. Opening with an illustrated essay on the Tyrolean Ice Man, Chapter 19, Human Evolution, now completes this unit.

Our revision of **Unit V, Animals: Form and Function,** includes new information on the hormonal control of fat metabolism (Module 21.15), more attention to the health hazards of tobacco and smoking (Module 22.7), updated applications of monoclonal antibodies (Module 24.12), and a clearer account of the roles of T cells in immunity (Module 24.13). The introduction to Chapter 26 has an updated discussion of the hormone melatonin. Reflecting the rapid pace of research in neurobiology, our revision of Chapter 28, Nervous Systems, was the most extensive in this unit. New art and added text enhance our survey of nervous-system structure in Modules 28.10 and 28.11. In Modules 28.14 and 28.15, a new figure and a table illustrate human brain development and adult structure. The chapter ends with a new module on cellular changes in long-term potentiation and long-term depression.

Reflecting exciting new discoveries in plant biology, the most extensively revised chapter of **Unit VI, Plants: Form and Function,** is Chapter 33, Control Systems in Plants. New to this chapter are an updated discussion of biological clocks in plants (33.10), an interview with leading *Arabidopsis* researcher Gloria Coruzzi (33.13), and an up-to-date account of the defenses against herbivores and pathogens that have evolved in plants (33.14).

A significant new feature of **Unit VII, Ecology,** is a greater emphasis on disturbance and environmental change, both natural and human-caused. We make the connection between habitat alteration and the current high rate of species loss early in the unit (Module 34.9). Module 36.6 elaborates on the pivotal roles of disturbance in communities, providing the necessary background for a detailed look at ecosystem alteration and its effects in our new conservation biology chapter (Chapter 38). New topics in this capstone chapter include the value of biodiversity, human impacts on habitats in developed and developing nations, biodiversity hotspots and endemic species, population fragmentation, source and sink habitats, population viability analysis, landscape ecology, gap analysis, movement corridors, and restoration ecology. Other highlights of the revision of our ecology unit include the addition of temperate rainforests in our biome survey (34.10), an updated section on cognition and cognitive maps (37.11), a heavily revised and updated module on age structures of human populations in developed and developing nations (35.9), and a timely interview with climate-change researcher Camille Parmesan of the University of Texas at Austin.

* * *

Introductory biology is the only science course that many students will take during their college years. Long after today's students have forgotten most of the specific content of their biology course, they will be left with general impressions and attitudes about science and scientists. We hope this new edition of *Biology: Concepts & Connections* helps make those impressions positive and supports the instructor's goals for sharing the fun of biology. To help us produce an even better text in the next edition, please send your comments and suggestions to any member of the author team.

Neil Campbell
Department of Botany and Plant Sciences
University of California
Riverside, CA 92521

Larry Mitchell
540 Lower River Road
Heron, MT 59844

Jane Reece
Benjamin/Cummings Publishing
1301 Sansome Street
San Francisco, CA 94111

Acknowledgments

Biology: Concepts & Connections, Third Edition, results from the combined efforts of many people, and the authors wish to extend heartfelt thanks to all those who contributed to this and previous editions. Our work on this edition began with input from the biologists acknowledged in the Third Edition reviewer list on p. xi, who shared with us their experiences teaching introductory biology and provided specific suggestions for improving both the text and art of this book. Camille Parmesan of the University of Texas at Austin helped us update information on global warming and provided the interview in Chapter 35. Frank Gilliam of Marshall University and Michael Dombeck and Ariel Lugo of the U.S. Forest Service graciously provided comments and photographs that appear in our ecology unit. The unsolicited comments and suggestions we received from other biologists and from biology students were also very helpful. In addition, this book has benefited in countless ways from the stimulating contacts we had with numerous biologists during the recent preparation of our larger text, *Biology*, Fifth Edition. We are fortunate to be part of a global community dedicated to excellence in biology education.

For creating the ancillary materials that support this book, special thanks go to Richard Liebaert, who is the author of the outstanding *Student Study Guide* and the Interactive Study Partner CD-ROM. (He also originated many of the chapter review materials in the text itself.) We also much appreciate all the contributions of Eugene Fenster, who revised the *Instructor's Guide* and the *Test Bank;* the other authors of the *Test Bank* (Deborah Langsam, Linda Simpson, Lisa Shimeld, Marshall Sundberg, and David Tauck); and Iain Miller, the author of the helpful new *Instructor's Guide to Media.*

The superlative publishing team for this edition was headed up by executive editor Erin Mulligan and senior editor Adam Ray. We appreciate their hard work, creativity, and commitment to excellence in biology education and to our biology group. Senior developmental editors Shelley Parlante and Pat Burner were indispensable; they deserve much credit for the high quality of this book and the enclosed CD-ROM, respectively. We are also grateful to project editor Kirsten Watrud, who handled the printed supplements for the book, and publishing assistant Maureen Kennedy, who was in charge of the electronic supplements; we expect great things of these two talented individuals.

Production professionals were involved in this edition from its inception. The components of the teaching package are both attractive and pedagogically effective in large part because of their hard work and creativity. We wish to express our gratitude to senior production editor Angela Mann, production managing editor Laura Kenney, senior art supervisor Donna Kalal, senior production editor Larry Olsen (who handled the supplements), and art and design manager Don Kesner (who personally created some outstanding new pieces of art). Senior multimedia artist Russell Chun made important contributions to the electronic components of our learning package. Freelancers played a number of key roles in book production, and we'd like to give special thanks to Kelly Murphy (who helped organize the art program), copyeditor Chad Colburn, layout artist Pat Coleman, proofreader Carol Lombardi, permissions expert Ariane DePree-Kajfez, and indexer Charlotte Shane. It was a pleasure to work with photoresearcher Stephen Forsling, and we love the cover designed by Yvo Riezebos.

We are indebted to a number of artists for the book's beautiful and instructive illustrations. These include the artists for previous editions of this book and several artists who contributed to our larger text for life science majors, *Biology.* We'd especially like to acknowledge the major influence of artist Carla Simmons on the entire art program. The revised art for this edition of the book was mainly the work of the artists of Precision Graphics and freelancer Karl Miyajima. The book itself would not have come into existence without the efforts of prepress manager Lillian Hom and prepress supervisor Vivian McDougal.

The members of the Addison Wesley Longman sales group and the Benjamin/Cummings marketing department have continued to help us connect with biology instructors and their teaching needs. We thank them for all their hard work and enthusiastic support.

Finally, we are again deeply grateful to our families and friends for their support, encouragement, and patience.

Neil Campbell, Larry Mitchell, and Jane Reece

Reviewers

Third Edition Reviewers

Jane Aloi-Horlings, *Saddleback College*
Loren Ammerman, *University of Texas at Arlington*
Jane Beiswenger, *University of Wyoming*
William Bradshaw, *Brigham Young University*
Becky Brown-Watson, *Santa Rosa Junior College*
Paul Boyer, *Univerity of Wisconsin-Parkside*
Michael Bucher, *College of San Mateo*
George Cain, *University of Iowa*
M. Carabelli, *Broward Community College*
Mary Colavito-Shepanski, *Santa Monica College*
Lewis Deaton, *University of SW Louisiana*
Susan Dunford, *University of Cincinnati*
Thomas Emmel, *University of Florida*
Bernard Frye, *University of Texas at Arlington*
Gail Gasparich, *Towson University*
Grant Gerrish, *University of Hawaii*
Sylvia Greer, *City University of New York*
Maggie Haag, *University of Alberta*
Jean Hegelsen, *Colin County Community College*
Carl Hoagstrom, *Ohio Northern University*
Mark Ikeda, *San Bernardino Valley College*
Charles Jacobs, *Henry Ford Community College*
Fred James, *Presbyterian College*
Tom Kantz, *California State University, Sacramento*
Peter Kish, *Southwestern Oklahoma State University*
Richard Mortensen, *Albion College*
Bette Nybakken, *Hartnell College*
Debra Pearce, *Northern Kentucky University*
James Pru, *Belleville Area College*
Charles Pumpuni, *Northern Virginia Community College*
Bruce Reid, *Kean College*
Laurel Roberts, *University of Pittsburgh*
Duane Rohlfing, *University of South Carolina*
Robert Schoch, *Boston University*
Brian Scholtens, *College of Charleston*
Ross Strayer, *Washtenaw Community College*
Marshall Sundberg, *Emporia State University*
Hilda Taylor, *Acadia University*
Donald Trisel, *Fairmont State College*
Ann Vernon, *St. Charles County Community College*
Sandra Winicur, *Indiana University South Bend*
Tony Yates, *Seminole State College*
Uko Zylstra, *Calvin College*

Scientists Featured in *Talking About Science*

the late Linus Pauling
Mario Molina, *Massachusetts Institute of Technology*
David Satcher, *Centers for Disease Control and Prevention*
Mary-Claire King, *University of Washington*
the late Barbara McClintock
Richard Dawkins, *Oxford University*
Stanley Miller, *University of California, San Diego*
Roger Sperry, *California Institute of Technology*
the late Katherine Esau
Stephen Moore, *Moore Ranch*
Gloria Coruzzi, *New York University*
Eloy Rodriguez, *Cornell University*
Ariel Lugo, *U.S. Forest Service, Puerto Rico*
Camille Parmesan, *University of Texas at Austin*
Frank Gilliam, *Marshall University*
Jane Goodall, *Jane Goodall Institute*
Edward O. Wilson, *Harvard University*

Second Edition Reviewers

Michael Abbott, *Westminster College*
Tania Beliz, *College of San Mateo*
Rudi Berkelhamer, *University of California, Irvine*
Linda Butler, *University of Texas, Austin*
Carolee Caffrey, *University of California, Los Angeles*
Van Christman, *Ricks College*
Don Cox, *Miami University*
Norma Criley, *Illinois Wesleyan University*
Jean DeSaix, *University of North Carolina at Chapel Hill*
Alfred Diboll, *Macon College*
Susan Dunford, *University of Cincinnati*
Norman Ellstrand, *University of California, Riverside*
David Essar, *Winona State University*
Laurie Faber, *Grand Rapids Community College*
Eugene Fenster, *Longview Community College*
Robert Galbraith, *Crafton Hills College*

George Garcia, *University of Texas, Austin*
Shelley Gaudia, *Lane Community College*
Douglas Gayou, *University of Missouri–Columbia*
Rebecca German, *University of Cincinnati*
Frank Gilliam, *Marshall University*
David Glenn-Lewin, *Wichita State University*
Robert Grammer, *Belmont University*
Leah Haimo, *University of California, Riverside*
James Hampton, *Salt Lake Community College*
Ira Herskowitz, *University of California, San Francisco*
Jean Higgins-Fonda, *Prince George's Community College*
John Holt, *Michigan State University*
Michael Hudecki, *State University of New York, Buffalo*
Russell Johnson, *Ricks College*
Tracy Kahn, *University of California, Riverside*
Mahlon Kelly, *University of Virginia*
Robert Koch, *California State University, Fullerton*
Deborah Langsam, *University of North Carolina, Charlotte*
Kevin Lien, *Portland Community College*
David Magrane, *Morehead State University*
Presley Martin, *Drexel University*
Mitchell McGinnis, *North Seattle Community College*
Iain Miller, *University of Cincinnati*
Christopher Murphy, *James Madison University*
Peter Nordloh, *Southeastern Community College–Burlington*
Stephen Novak, *Boise State University*
Michael O'Donnell, *Trinity College*
Karen Olmstead, *University of South Dakota*
Steven O'Neal, *Southwestern Oklahoma State University*
Margaret Peterson, *Concordia Lutheran College*
Rebecca Pyles, *East Tennessee State University*
Douglas Reynolds, *Central Washington University*
Jim Rosowski, *University of Nebraska*
Lisa Shimeld, *Crafton Hills College*
Linda Simpson, *University of North Carolina, Charlotte*
John Stolz, *Duquesne University*
Donald Streuble, *Idaho State University*
Leslie Vander Molen, *Humboldt State University*
Patrick Woolley, *East Central College*

First Edition Reviewers

Daryl Adams, *Mankato State University*
Dawn Adrian Adams, *Baylor University*
Olushola Adeyeye, *Duquesne University*
Dan Alex, *Chabot College*
Sylvester Allred, *Northern Arizona University*
Jane Aloi, *Saddleback College*
Marjay Anderson, *Howard University*
Chris Barnhart, *University of San Diego*
William Barstow, *University of Georgia*
Ernest Benfield, *Virginia Polytechnic Institute*
Harry Bernheim, *Tufts University*
Richard Bliss, *Yuba College*
Lawrence Blumer, *Morehouse College*
William Bowen, *University of Arkansas, Little Rock*
Chris Brinegar, *San Jose State University*
Virginia Buckner, *Johnson County Community College*
Jerry Button, *Portland Community College*
James Cappuccino, *Rockland Community College*
Cathryn Cates, *Tyler Junior College*
Vic Chow, *San Francisco City College*
Mary Colavito-Shepanski, *Santa Monica College*
Robert Creek, *Western Kentucky University*
Judy Daniels, *Monroe Community College*
Lawrence DeFilippi, *Lurleen B. Wallace College*
James Dekloe, *Solano Community College*
Loren Denney, *Southwest Missouri State University*
Jean DeSaix, *University of North Carolina, Chapel Hill*
Stephen Dina, *St. Louis University*
Jean Dickey, *Clemson University*
Gary Donnermeyer, *Iowa Central Community College*
Charles Duggins, *University of South Carolina*
Betty Eidemiller, *Lamar University*
Cindy Erwin, *City College of San Francisco*
Nancy Eyster-Smith, *Bentley College*
Terence Farrell, *Stetson University*
Jerry Feldman, *University of California, Santa Cruz*
Dino Fiabane, *Community College of Philadelphia*
Kathleen Fisher, *San Diego State University*
Robert Frankis, *College of Charleston*
James French, *Rutgers University*
Shelley Gaudia, *Lane Community College*
Robert Gendron, *Indiana University of Pennsylvania*
Dana Griffin, *University of Florida*
Richard Haas, *California State University, Fresno*
Martin Hahn, *William Paterson College*
Leah Haimo, *University of California, Riverside*
Laszlo Hanzely, *Northern Illinois University*
Jim Harris, *Utah Valley Community College*
Mary Harris, *Louisiana State University*
Jean Helgeson, *Collin County Community College*
Ira Herskowitz, *University of California, San Francisco*
Paul Hertz, *Barnard College*
Margeret Hicks, *David Lipscomb University*
Laura Hoopes, *Occidental College*
Robert Howe, *Suffolk University*
George Hudock, *Indiana University*
Kris Hueftle, *Pensacola Junior College*
Charles Ide, *Tulane University*

Ursula Jander, *Washburn University*
Alan Jaworski, *University of Georgia*
Florence Juillerat, *Indiana University–Purdue University at Indianapolis*
Lee Kirkpatrick, *Glendale Community College*
Mary Rose Lamb, *University of Puget Sound*
Carmine Lanciani, *University of Florida*
Deborah Langsam, *University of North Carolina, Charlotte*
Steven Lebsack, *Linn-Benton Community College*
Richard Liebaert, *Linn-Benton Community College*
Ivo Lindauer, *University of Northern Colorado*
William Lindsay, *Monterey Peninsula College*
Melanie Loo, *California State University, Sacramento*
Joseph Marshall, *West Virginia University*
Presley Martin, *Drexel University*
William McComas, *University of Iowa*
Steven McCullagh, *Kennesaw State College*
James McGivern, *Gannon University*
Henry Mulcahy, *Suffolk University*
James Nivison, *Mid Michigan Community College*
Lowell Orr, *Kent State University*
William Outlaw, *Florida State University*
Kay Pauling, *Foothill College*
Patricia Pearson, *Western Kentucky University*
Paula Piehl, *Potomac State College*
Ben Pierce, *Baylor University*
Barbara Pleasants, *Iowa State University*
Kathryn Podwall, *Nassau Community College*
Judith Pottmeyer, *Columbia Basin College*
Donald Potts, *University of California, Santa Cruz*
Brian Reeder, *Morehead State University*
Fred Rhoades, *Western Washington University*
Stephen Rothstein, *University of California, Santa Barbara*
Donald Roush, *University of North Alabama*
Douglas Schamel, *University of Alaska, Fairbanks*
Robert Schoch, *Boston University*
Thomas Shellberg, *Henry Ford Community College*
Jane Shoup, *Purdue University*
Gary Smith, *Tarrant County Junior College*
Gerald Summers, *University of Missouri, Columbia*
Marshall Sundberg, *Louisiana State University*
David Tauck, *Santa Clara University*
Kathy Thompson, *Louisiana State University*
Robert Wallace, *Ripon College*
James Wee, *Loyola University*
Larry Williams, *University of Houston*
Uko Zylstra, *Calvin College*

Supplements

FOR THE STUDENT

Interactive Study Partner CD-ROM

Richard Liebaert, Linn-Benton Community College

Packaged free with the text, the Study Partner contains 126 interactive exercises, animations, and lab simulations, and 20 test questions per chapter with feedback and text references. Icons appear throughout the text to guide students to applicable activities on the CD-ROM.

Special Edition of *The Biology Place*™ for *Biology: Concepts and Connections,* Third Edition

This web-based learning environment is specifically keyed to the text chapters and includes interactive tutorials, investigative learning activities, lab simulations, more than 3,000 quiz questions, and relevant news articles that help students apply concepts of biology to everyday life. A 12-month subscription is now available at no extra cost to all college students who purchase a text. See www.biology.com/cmr for more details.

Study Guide

Richard Liebaert, Linn-Benton Community College

Written by the author of many *Biology: Concepts and Connections* Chapter Review questions, this popular study guide offers a variety of interactive exercises. Icons throughout the study guide refer students to applicable activities on the Interactive Study Partner CD-ROM.

Laboratory Investigations for Biology

Jean L. Dickey, Clemson University

The order of topics in this lab manual parallels the organization of *Biology: Concepts and Connections.* Each of the 20 labs is designed with options to allow instructors to customize both the topics covered and the extent of student-guided inquiry.

Biology Labs Online

The product of a unique collaboration between the California State University System and Addison Wesley Longman, these web-based labs allow students to actively learn about biological principles not normally covered in class. **http://biologylab.awlonline.com**

Related Titles

An Introduction to Chemistry for Biology Students, Sixth Edition by George I. Sackheim

The Chemistry of Life CD-ROM by Robert M. Thornton

A Short Guide to Writing about Biology, Third Edition by Jan A. Pechenik

Doing Biology by Joel B. Hagen, Douglas Allchin, and Fred Singer

Studying for Biology by Anton E. Lawson and Brenda D. Smith

Biomath: Problem Solving for Biology Students by Robert Keck and Richard Patterson

FOR THE INSTRUCTOR

Transparency Acetates

Contains virtually all the art from *Biology: Concepts and Connections* with clear, easy-to-read labels. (Available to qualified college adopters.)

Instructor's Presentation CD-ROM

Animations, tables from the text, and all transparency art from *Biology: Concepts and Connections,* Third Edition are included. A presentation program enables instructors to design a customized slide show of images, import illustrations and photos from other sources, export figures into other presentation software programs, and edit them.

Instructor's Guide to Media

Iain Miller, University of Cincinnati

Provides guidance on how to use the electronic tools available with the text and how to integrate them into your course.

Instructor's Guide

Eugene J. Fenster, Longview Community College, and Fred Rhoades, Western Washington University

The *Instructor's Guide* contains lecture outlines and instructional activities. It also includes course outlines for instructors who use alternative syllabi and references to resources on the World Wide Web. Available in printed form, on disk, and on-line.

Test Bank

Contributions from Eugene J. Fenster, Longview Community College, Deborah Langsam and Linda Simpson, University of North Carolina–Charlotte, Lisa Shimeld, Crafton Hills College, David Tauck, Santa Clara University, and Marshall Sundberg, Emporia State University

The test bank includes questions at three levels: factual, conceptual, and applications-based. Available in print form and on Macintosh, DOS, and Windows test-generating programs. (Available to qualified college adopters.)

Brief Contents

Contents

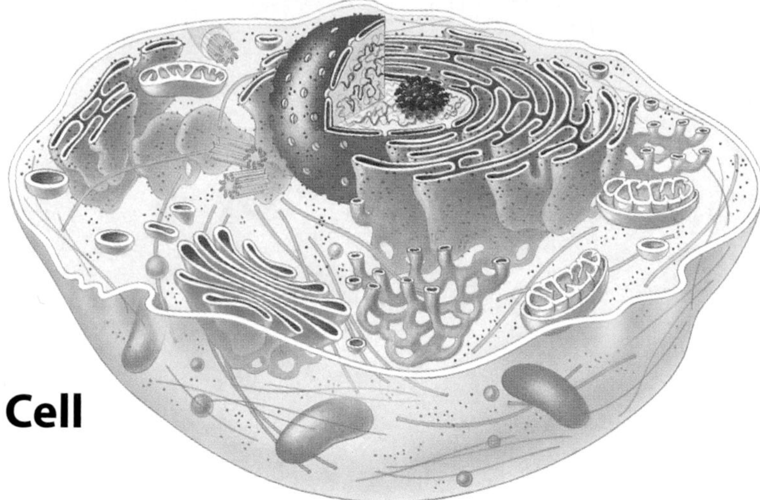

UNIT I The Life of the Cell

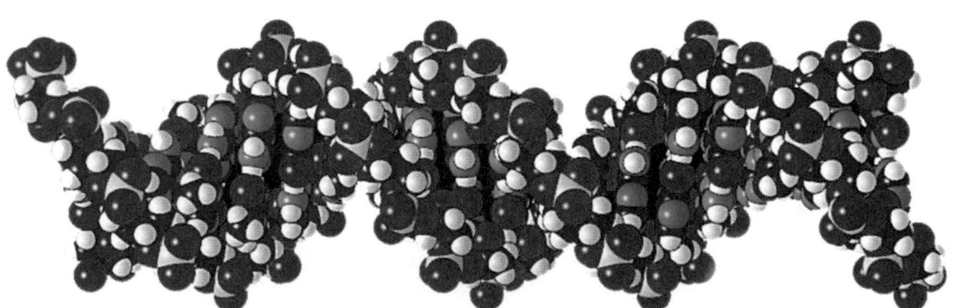

UNIT II Cellular Reproduction and Genetics

11 The Control of Gene Expression 208

12 DNA Technology and the Human Genome 230

UNIT III Concepts of Evolution

13 How Populations Evolve 256

14 The Origin of Species 280

15 Tracing Evolutionary History 294

UNIT IV The Evolution of Biological Diversity

16 The Origin and Evolution of Microbial Life: Prokaryotes and Protists 316

17 Plants, Fungi, and the Colonization of Land 344

UNIT V Animals: Form and Function

21 Nutrition and Digestion 428

22 Respiration: The Exchange of Gases 450

23 Circulation 464

24 The Immune System 482

UNIT VI Plants: Form and Function

31 Plant Structure, Reproduction, and Development 618

32 Plant Nutrition and Transport 640

33 Control Systems in Plants 658

UNIT VII Ecology

34 The Biosphere: An Introduction to Earth's Diverse Environments 678

35 Population Dynamics 698

36 Communities and Ecosystems 714

37 Behavioral Adaptations to the Environment 736

38 Conservation Biology 762

How to Use This Book

To the Student

Like the world it describes, *Biology: Concepts and Connections* is a system of connected and complementary elements. This section shows you how to use the text and its media partners to make studying easier.

After reviewing this guide, read Chapter One, "Introduction: The Scientific Study of Life." Chapter One gives you a framework for the process of science and the scope of biology as a science. Attention to the concepts and connections in this chapter will pay dividends throughout your course.

Chapter Openers Hold a Key to the Chapter

- Most chapters begin with a story about an organism—its way of life, its unique characteristics, its interactions with its environment, and how biologists study it. This story highlights the chapter's main subject and often relates the process of science to the topics at hand.
- An outline, showing the chapter's main topics and their underlying concepts, offers a preview to the chapter.

1 Introduction: The Scientific Study of Life

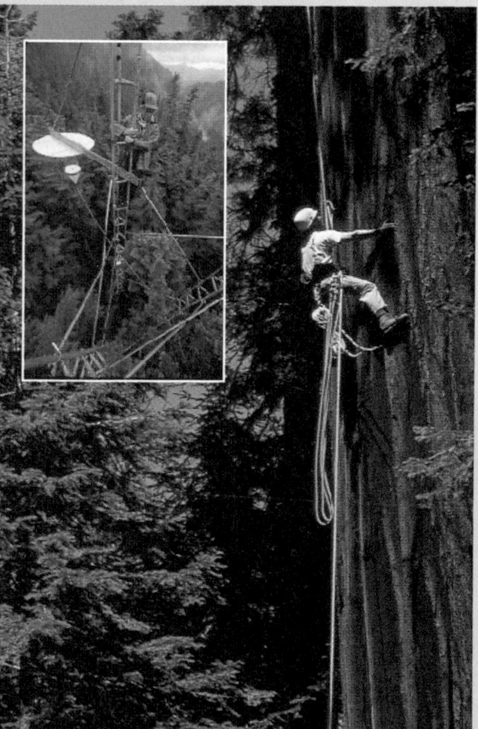

Outline

The Scope of Biology
1.1 Life's levels of organization define the scope of biology

The Process of Science
1.2 Scientists pose and test hypotheses to answer questions about nature
1.3 The process of science: A case study in a tropical rain forest

Evolution, Unity, and Diversity
1.4 The diversity of life can be arranged into three domains
1.5 Unity in diversity: All forms of life have common features
1.6 Evolution explains the unity and diversity of life
1.7 Living organisms and their environments form interconnecting webs

Connections to Everyday Life
1.8 Biology is connected to our lives in many ways

The canopies, or treetops, of forests are one of nature's great showplaces of different kinds of living organisms. Biologists estimate that fully half of the 5 million to 50 million species on Earth spend all or much of their time here, yet canopies were virtually unexplored until about 15 years ago. The canopies of tropical rain forests, areas near the equator where yearly rainfall often exceeds 200 centimeters (80 inches), harbor the greatest number of species. Equally wet forests in the coastal regions of western Canada and the northwestern United States also are home to remarkable numbers. The dominant trees in many northern rain forests are centuries-old firs, hemlocks, and spruces. Extending into northern California are remnants of truly ancient rain forests dominated by redwoods and giant sequoia trees that are thousands of years old.

Research on forest canopies blossomed in the 1970s when scientists began using mountain climbers' roping techniques to reach the treetops. In the large photograph at the left, Humboldt State University researcher Steve Sillett is roping his way into the 90-meter-high canopy of a 2000-year-old giant sequoia tree in northern California.

Roping techniques allow researchers to move from one area of forest to another, but they are not the only way to get into a canopy. For projects that can be done in a limited area, some researchers prefer large construction cranes with horizontal booms that rotate over the canopy. More permanent are platforms, walkways, and towers built into the treetops. In the small photograph on the facing page, a researcher from the University of Washington stands on a tower set up to collect samples of mist, part of a long-term study of how water circulates between the forest and the atmosphere.

Canopy research is not for everyone. Researchers often spend a week or more in treetops, sleeping in hammocks strung between branches higher than a 20-story skyscraper. Their rewards include sights, sounds, and smells found nowhere else. Pioneer forest researcher Dr. Jerry Franklin expresses his enthusiasm for canopy research this way: "Trees are one of Earth's main points of contact with the air. This is our chance to learn about the dialogue between trees and the atmosphere."

A summer evening is a good time to experience a canopy's rich array of life. In a rain forest in the northwestern U.S., brown and greenish fungi, lichens, and earthy mats of moss cling to the high branches. Your eye might catch the split-second shadow of a flying squirrel across a bright moon. Rarely seen because they are active only at night, these foot-long rodents speed-glide between the big trees. Insects are especially abundant in the canopy twilight, and bats flit about, often eating close to their own body weight in insects in a single night. Owls, another group of nocturnal hunters, are silent on the wing, but each species has its own distinctive hoot.

A flying squirrel

The enormous variety of living organisms in forest canopies and the science of canopy research set the stage for this chapter's introduction to biology, the scientific study of life.

1

Chapters help you understand concepts

- Each module features a central concept, which is announced in the heading. You can grasp each concept fully before going on to the next.
- Modules are numbered for easy reference and are never longer than two facing pages—you never have to turn the page to see the figures.
- Each module ends with a self-test question. Feedback is provided on the spot.
- A few main headings organize each chapter into groups of concept modules. The main headings appear in maroon bars across the page.

Illustrations and text work together

- Use both as you study. The illustrations reinforce the message of the text and vice versa. They are both integral to understanding the concepts.
- Some figures take you through a sequence of steps labeled with red-circled numbers. These numbers also appear in the text, keyed to the explanation of the steps in the process.

Islands are living laboratories of speciation 14.5

When oceanic islands are far enough apart to permit populations to evolve in isolation, but close enough to allow occasional dispersions to occur, they are effectively outdoor laboratories of evolution. Located west of Ecuador, the Galápagos Archipelago is one of the world's great showcases of evolution. Each island was born naked from underwater volcanoes and was gradually clothed by plants, animals, and microorganisms derived from strays that rode the ocean currents and winds from other islands and continents. Organisms can also be carried to islands by other organisms, such as sea birds that travel long distances with seeds clinging to their feathers (Figure 14.5A).

The species on the Galápagos Islands today, most of which occur nowhere else, descended from stragglers that floated, flew, or were blown over the sea from the South American mainland. For instance, the Galápagos island chain has a total of 13 species of closely related birds called Galápagos finches (also known as Darwin's finches because Darwin wrote about them). These birds have many similarities but differ in their feeding habits and their beak type, which is correlated with what they eat. The three examples shown in Figure 14.5B illustrate the range of beak shape and size. (Also see Figure 15.9.)

Evidence accumulated since Darwin's time indicates that all 13 finch species evolved from a single small population of ancestral birds that colonized one of the islands. Figure 14.5C illustrates how this might have happened. Completely isolated on the island after ① migrating from the mainland, the founder population (species A in Figure 14.5C) may have undergone significant changes in its gene pool and become a new species, which we'll call species B. ② Later, a few individuals of species B may have been blown by storms to a neighboring island. Isolated on this second island, the second founder population could have evolved into a second new species, species C. ③ Species C could later recolonize the island from which its founding population emigrated, and it might coexist there with its ancestral species if reproductive barriers kept the species distinct. In the case of the Galápagos finches, the two species came to rely on different foods, besides being unable to interbreed. In this simplified model, after species C evolved on the second island, ④ it colonized a third island, where it evolved into species D. ⑤ Species D then dispersed to the two islands of its ancestors. Actually, the Galápagos finches colonized and speciated repeatedly on many separate islands of the Galápagos.

Today, each of the Galápagos Islands has multiple species of finches, with as many as ten on some islands. In contrast, the island of Cocos, about 700 km (420 mi) north of the Galápagos, has only one finch species, found nowhere else. This single species was apparently derived from an ancestral founder population in the same way as the Galápagos birds, but Cocos Island is so isolated that there apparently has been no opportunity for its finch species to colonize other islands, or for other finches to become established on Cocos.

The emergence of numerous species from a common ancestor introduced to new and diverse environments is called **adaptive radiation.** The effects of adaptive radiation of Darwin's finches are evident in the many types of beaks, specialized for different foods.

Figure 14.5B Examples of differences in beak shape and size in Darwin's finches

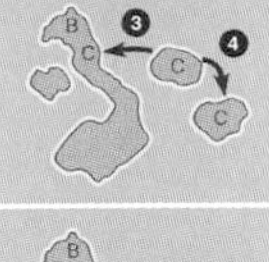
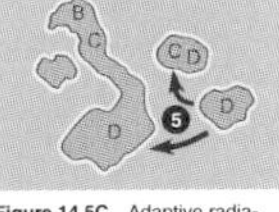

Figure 14.5C Adaptive radiation on an island chain

Figure 14.5A Long-distance dispersal of plant seeds (black spots) on the feathers of a tern

? Why would allopatric speciation be less common on an island close to a mainland than on a more isolated island of the same size?

Continued gene flow between mainland populations and those on nearby islands reduces the chance of enough genetic divergence for speciation.

Photosynthesis, Solar Radiation, and Earth's Atmosphere

Photosynthesis moderates the greenhouse effect; deforestation can intensify it 7.13

...ees in the photograph (Figure 7.13A) form part of an ...owth forest in the Pacific Northwest. Such a forest is ...ient one that has never been seriously disturbed by ...s. Many of the trees—mainly Douglas fir here—are ...nds of years old. The forest they dominate is one of a ...naining undisturbed areas that contain harvestable in the United States.

...growth forests are the focus of a long-standing con... between the timber industry and conservationists. ...ke these contain a lot of marketable lumber, and eco... arguments can be made for harvesting them. On the ...side of the controversy, conservationists argue that ...owth forests are home to many species that can sur... where else, and that we should save these remnants ancient forests for future generations.

...photosynthesis carried out by old-growth forests ...ect bearing on the controversy. At center stage is ...e gas that plants use to make sugars in photosyn... and that all organisms give off as waste from cellular ...tion.

...on dioxide normally makes up about 0.03% of the ...breathe. This amount of CO_2 in the atmosphere pro... plants with plenty of carbon. It also helps moderate climates, because CO_2 retains heat from the sun that otherwise radiate from Earth back into space. Warm... ...duced by CO_2 is called the **greenhouse effect** be... atmospheric CO_2 traps heat and warms the air just as ...lass does in a greenhouse (Figure 7.13B). The green... effect keeps the average temperature on Earth some ...armer than it would be otherwise. ...ically, planet Earth may now be overheating from ...eenhouse effect. The amount of atmospheric CO_2 has been on the rise in the past century, mainly bec... worldwide industrialization and the increased us... gas, coal, and wood as fuels. When these substan... burned, the carbon in them is released as CO_2. The... increase in atmospheric CO_2 seems to be contribut... rise in global temperatures. Photosynthesis consum... tending to counteract the greenhouse effect, ... global rate of photosynthesis may decline as we cle... tracts of forest for farming and urban expansion. Ex... deforestation continues, especially in the tropics, th... western and southeastern United States, Canad... Siberia.

The greenhouse effect has come up on both side... old-growth forest controversy. Those who favor sa... old-growth trees argue that these large photosynt... remove a lot of potentially harmful CO_2 from th... sphere. Those favoring harvesting argue that replac... old trees with seedlings would actually increase ph... thesis and reduce atmospheric CO_2.

Relative to their size and weight, young, rapidly g... trees do, in fact, take up CO_2 at a faster rate than o... However, when old trees are harvested, much less t... their bulk becomes lumber. All the roots and... branches are left behind to decompose. Most of th... itself is turned into paper, sawdust, or fuel—produ... usually decompose or are burned within a few ye... composition and burning turn the carbon compou... were in the tree into CO_2, and this puts much more ... the atmosphere than young trees can take up.

In the next module, we'll hear more about har... fects of human activities on Earth's atmosphere.

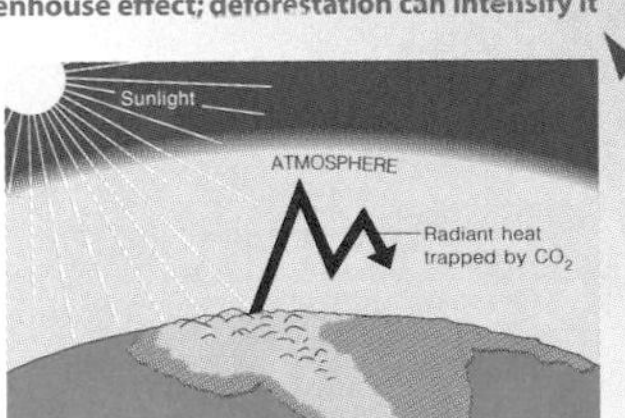

Figure 7.13B The greenhouse effect of CO_2 in the atmosphere

7.13A An old-growth forest in the Pacific Northwest

? Explain the greenhouse effect.

Talking About Science

7.14 Mario Molina talks about Earth's protective ozone layer

As photosynthesis reduces the CO_2 in Earth's atmosphere, it also produces the O_2 on which plants, animals, and most other organisms depend for cellular respiration. This O_2 benefits organisms in another way: High in the atmosphere it is converted to ozone, a form of oxygen that plays an important protective role for life on Earth. Among the scientists who study the ozone layer is Mario Molina of MIT, who in 1995 shared a Nobel Prize for his research on how certain pollutants are damaging that layer. In a recent interview, Dr. Molina explained why the ozone layer is important.

The ozone layer shields the Earth's surface from powerful ultraviolet radiation that comes from the sun. This UV radiation is harmful to organisms, including humans. For example, UV radiation causes sunburn, and skin cancer can be a cumulative result of exposure. There is also evidence that UV can damage crops. Certain developing animals, such as the larvae of fish, seem to be particularly sensitive.

He described how the ozone forms and how it is destroyed.

The ozone forms when high-energy solar radiation breaks apart O_2 molecules and frees oxygen atoms. These then react with unbroken O_2 molecules. The result is ozone, which has *three* oxygen atoms (O_3). So ozone is continuously forming by the action of sunlight on the atmosphere. This is balanced by continuous destruction of the ozone molecules when they react with other chemical compounds that are naturally present in the atmosphere. Humans have disrupted that balance by releasing certain industrial chemicals that hasten this destruction.

The research that won Dr. Molina the Nobel Prize dealt with the destruction of ozone by one particular class of chemicals, called chlorofluorocarbons, or CFCs. In 1974, when Dr. Molina and his colleagues first published their CFC-ozone depletion theory, CFCs were used in large amounts as refrigerants, as propellants in spray cans, as solvents, and in the process for making plastic foams.

We predicted that the continuous release of CFCs would damage the protective ozone layer. CFCs are very stable compounds, and this stability allows them to make it up to the ozone layer, which is about 15 miles above the surface of the Earth. There solar radiation converts them to very reactive chemicals called free radicals, which then destroy the ozone.

How did scientists and the general public react to the alarming possibility that the ozone layer was thinning?

At first, there was very little reaction because not many people were aware of the importance of invisible things like the ozone layer and UV radiation. Experts in our field ...

Figure 7.14 Mario Molina

more people, including politicians, became concerned about ozone depletion. There was also growing awareness about the role of UV radiation in skin cancer and other biological damage. Then, in 1985, scientists documented a ...

... CFCs ... fuels, ... sophi... devel... dard of living?

That's an issue, but there is another important element that I've observed from my own experience. When I was going to the university in Mexico, no one cared about pollution. Dealing with pollution seemed like a luxury. Today, everybody in Mexico City cares about pollution, so solving this problem no longer seems like something that industrialized countries want to impose upon them. The costs of pollution are enormous in Mexico City, where I was born. If only there had been insight about what would happen to the city, pollution could have been reduced at a much lower cost. It's much more difficult to repair the environment now, but I think there is a strong incentive to do that in developing countries. It's not easy because of the complications of social and economic issues. Science and technology are very important components, but not the only ones, so scientists will have to be humble and realize their limits.

The connections between science, technology, and society, so clearly exemplified by the work of Mario Molina, are a major theme of this book. This theme will come up again in the next unit, on cellular reproduction and genetics.

Red shading identifies modules that connect biology to human issues

- "Talking About Science" modules profile men and women working in the world of science. Most of these modules feature interviews.
- Other modules relate biological concepts to health, social, or environmental issues.

10.15 Review: The flow of genetic information in the cell is DNA → RNA → protein

10E

This figure summarizes the main stages in the flow of genetic information from DNA to protein. In transcription (DNA → RNA), the RNA is synthesized on a DNA template (stage ①). In eukaryotic cells, transcription occurs in the nucleus, and the messenger RNA must travel from the nucleus to the cytoplasm.

Translation (RNA → protein) can be divided into four stages (②–⑤), all of which occur in the cytoplasm. When the polypeptide is complete, the two ribosomal subunits come apart, and the tRNA and mRNA are released (not shown in this figure). Translation is rapid; a single ribosome can make an average-sized polypeptide in less than a minute. Typically, an mRNA molecule is translated simultaneously by a number of ribosomes. Once the start codon emerges from the first ribosome, a second ribosome can attach to it, and thus several ribosomes may trail along on the same mRNA molecule.

Each polypeptide coils and folds, assuming a three-dimensional shape, its tertiary structure. Several polypeptides may come together, forming a protein with quaternary structure (see Module 3.18).

What is the overall significance of transcription and translation? These are the processes whereby genes control the structures and activities of cells, or, more broadly, the way the genotype produces the phenotype. The chain of command originates with the information in a gene, a specific linear sequence of nucleotides in DNA. The gene serves as a template, dictating transcription of a complementary sequence of nucleotides in mRNA. In turn, mRNA dictates the linear sequence in which amino acids appear in a specific polypeptide. Finally, the proteins that form from the polypeptides determine the appearance and the capabilities of the cell and organism.

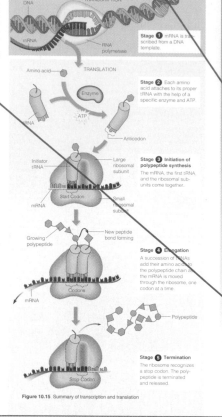

Which of the following molecules or structures does not participate directly in translation: ribosomes, transfer RNA, messenger RNA, DNA, ATP, enzymes?

TRANSCRIPTION

DNA

mRNA

RNA polymerase

Stage ① mRNA is transcribed from a DNA template.

TRANSLATION

Amino acid

Enzyme

ATP

tRNA

Anticodon

Stage ② Each amino acid attaches to its proper tRNA with the help of a specific enzyme and ATP.

Initiator tRNA

Large ribosomal subunit

Start Codon

Small ribosomal subunit

mRNA

Stage ③ Initiation of polypeptide synthesis The mRNA, the first tRNA, and the ribosomal subunits come together.

Growing polypeptide

New peptide bond forming

Codons

mRNA

Stage ④ Elongation A succession of tRNAs add their amino acids to the polypeptide chain as the mRNA is moved through the ribosome, one codon at a time.

Polypeptide

Stop Codon

Stage ⑤ Termination The ribosome recognizes a stop codon. The polypeptide is terminated and released.

Figure 10.15 Summary of transcription and translation

200 UNIT II

OVERVIEW AND REVIEW MODULES PRESENT OVERARCHING CONCEPTS

- Overview modules usually include an overview figure. In later sections, the overarching concept is broken down into several subordinate concepts, each with a more detailed illustration.
- Some chapters also include review modules. Use them with the overview modules to remind yourself how all the details fit together.

CONSISTENT SYMBOLS AND COLORS HELP YOU UNDERSTAND CONCEPTS

- A number of symbols are used consistently in figures throughout the book–the ATP starburst and the shapes and colors of DNA, RNA, and proteins are some examples.
- These symbols provide an immediate visual clue to how the figure relates to the other figures you've seen.
- Many sections of the book connect to additional activities on the Student Interactive Study Partner CD-ROM. Look for this icon.

CHAPTER REVIEW

Chapter Summary

The Biodiversity Crisis: An Overview (Introduction–38.4) Modern human culture and the rapidly growing global human population have created a biodiversity crisis, a rapid decrease in the variety of species on Earth. Extinction rates are unusually high. The goal of conservation biology is to find ways to counter the rapid loss of species **(Introduction)**. Habitat destruction poses the single greatest threat to biodiversity. Competition with exotic species (those that have been transferred to an area where they did not occur naturally) also threatens many species in their native habitats. Overexploitation of wildlife often compounds problems caused by habitat loss and exotic species **(38.1)**. Biodiversity is vital to human welfare. We evolved in Earth's ecosystems, and large-scale changes in them threaten us as well as other species **(38.2)**. The explosive growth of the human population and of technology continues today. Populations of developing nations are growing the fastest, but the technology and resource consumption of the less-populous developed nations put a much greater strain on the biosphere. Oil spills, acid rain, ozone depletion, and chemical pesticides, which are concentrated in food chains by biological magnification, affect the entire world **(38.3)**. Burning of fossil fuels is increasing the amount of CO_2 and other greenhouse gases in the air, which may warm Earth enough to change climate patterns, melt polar ice, and flood coastal regions. Global warming may increase the rate of species loss **(38.4)**.

The Geographic Distribution of Biodiversity (38.5) Environments are patchy and as a result species are not evenly distributed. Some areas are unusually rich in biodiversity. Biodiversity hot spots are relatively small areas with a large concentration of species. Many organisms in these hot spots are endemic, found nowhere else. Endemic species are especially prone to extinction. Many migratory species, both terrestrial and aquatic, require international protection.

Conservation of Species Populations (38.6–38.8) Habitat degradation has seriously fragmented many species populations. Conservation efforts often center on sustaining fragmented populations. A source habitat for a portion (subpopulation) of a fragmented population is where reproductive success exceeds the death rate. In a sink habitat, the subpopulation's death rate exceeds reproduction. Protecting source habitats is vital in species conservation **(38.6)**. Identifying critical habitat factors in source habitats is a central goal of conservation biology **(38.7)**. Increased fragmentation threatens many species, including those whose populations were historically highly fragmented. Conservation biologists often use computer simulations in a population viability analysis (PVA) to assess a species' chances for survival **(38.8)**.

Managing and Sustaining Ecosyste[...] 38.12) Conservation efforts are incre[...] aimed at learning how to sustain wh[...] tems and landscapes. Landscape eco[...] employs ecological principles to stud[...] patterns and make species conservat[...] functional part of the patterns. Lands[...] researchers often use a computerized technique called gap analysis **(38.9)**. (boundaries) between ecosystems, su[...] forests and grasslands, have their ow[...] features and assemblages of species. increased frequency and abruptness caused by human activities can incre[...] species losses. Movement corridors, s[...] clumps of quality habitat connecting isolated habitat patches, may be help[...] harmful to fragmented populations ([...] Restoration of degraded habitats is ir[...] important. Two strategies are biorem[...] the use of living organisms to detoxif[...] areas, and augmentation of ecosyster[...] processes, resupplying an area with k[...] that were removed **(38.11)**. We are th[...] environmentally manipulative of all a[...] species. The attitudes and environme[...] awareness of people are of utmost im[...] in the search for solutions to the biod[...] crisis and other problems facing hum[...] Understanding the biosphere's limits vulnerability and our own linkages to natural world may help us make decis[...] lead to a sustainable future **(38.12)**.

Testing Your Knowledge

Multiple Choice

1. ______________ poses the single greatest threat to biodiversity.
   a. introduction of exotic species
   b. overhunting
   c. movement corridors
   d. habitat loss
   e. global warming
2. Which of the following is characteristic of endemic species?
   a. They are often found in biodiversity hot spots.
   b. They are distributed widely in the biosphere .
   c. They all require edges between ecosystems.
   d. Their trophic position makes them very susceptible to the effects of biological magnification.
   e. Researchers have completed PVAs for most of them.
3. Which of the following is least likely to be a sound conservation practice?
   a. identifying critical habitat factors
   b. developing a population viability analysis
   c. identifying and protecting biodiversity hot spots
   d. establishing movement corridors
   e. reducing the effects of population fragmentation
4. Landscape ecology is an indication of what major trend in co[...] biology?
   a. moving away from a focus on air and water pollution to a fo[...] species conservation
   b. switching from an ecosystem approach to a population appr[...]
   c. a trend toward a central focus on fragmented populations
   d. a trend toward development of more PVAs
   e. a trend toward sustaining biodiversity of multiple ecosystem[...]
5. Ospreys and other top predators in food chains are most severe[...] by pesticides such as DDT because
   a. their systems are especially sensitive to chemicals
   b. of their rapid reproductive rates
   c. the pesticides become concentrated in their prey
   d. they cannot store the pesticides in their tissues
   e. they are directly exposed to pesticides in the air
6. Some biologists think that half the species on Earth may be ext[...] in the next century. Mass extinctions have happened before. How [...] one be different?
   a. This time species are threatened by a change in climate.
   b. This extinction is happening at a much faster rate.
   c. A much larger fraction of Earth's species are likely to be affe[...]
   d. All life on Earth may eventually be extinguished.
   e. This extinction is much more gradual and difficult to notice.

780 UNIT VII ECOLOGY

THE CHAPTER SUMMARY CAN HELP YOU IN SEVERAL WAYS

- Use the Chapter Summary to review the chapter.
- Under each main heading, the key concepts are rephrased. Interspersed module numbers refer you back to the places in the chapter where you can review further if you need to.
- Another way to review is simply to page through the chapter, rereading the concept module headings and looking over all the illustrations.

QUESTION, EXPLORE, AND DISCOVER BIOLOGY

- Each book comes with a free Interactive Study Partner CD-ROM. Interactive exercises, animations, videos, and simulations help you understand dynamic processes. The Interactive Study Partner also contains a glossary with recorded pronunciations, along with chapter quizzes.
- You can also use a special edition of The Biology Place™ web site keyed specifically to *Biology: Concepts and Connections*. This site features interactive learning activities, news articles, quizzes, web links, and additional activities published throughout the course of your class.
- ➔ Icons in the text identify topics with corresponding activities on the Interactive Study Partner.

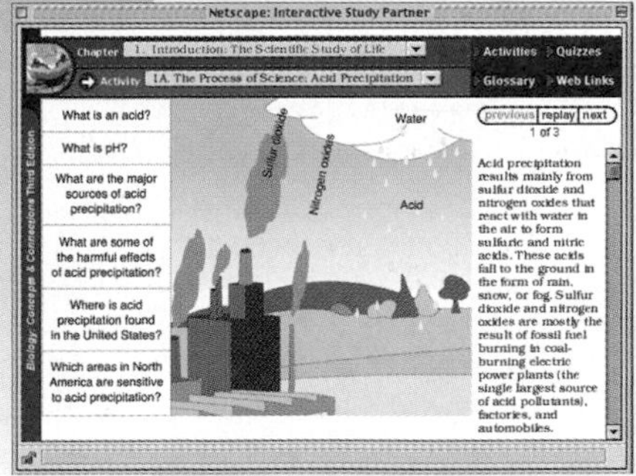

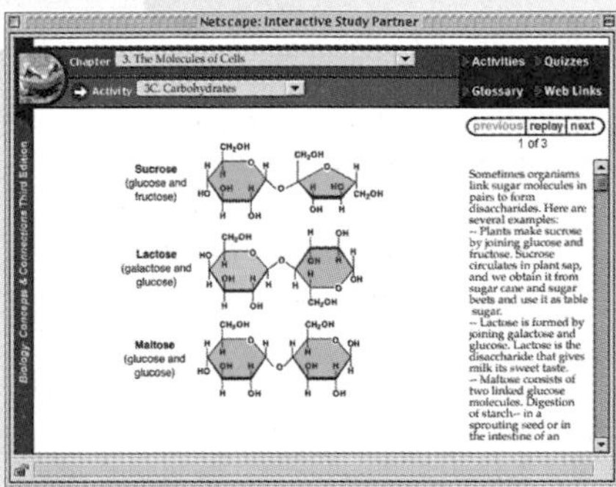

of the following consumes the most energy each day? *(Explain your*

United States
ıth America
ina
lia
ica

g, Comparing, and Explaining

s the greenhouse effect? How is it important to life on Earth?

are the possible causes and consequences of global warming? Why is onal cooperation necessary if we are to solve this problem?

ng as a Scientist

tanker runs aground in a rocky bay, and a local environmental nts to assess the damage that the spilled oil caused to the bay's ma- No previous survey of the organisms in this area has ever been done. nbers of the environmental group collect intertidal organisms in oil-affected bay and a neighboring bay roughly similar to it. The en hires a biologist to classify and count the organisms from both report states that several important mollusks have significantly oulations in the oil-affected bay than in the other bay. Concluding oil spill killed off the mollusks, the environmental group hires a d sues the oil company for criminal damage to the local environ- udge throws the case out for lack of solid evidence. Which of the arguments do you suppose the oil company's attorney used? Ex- r choice.

effort was made to determine why the animals were damaged by spill, and a proposed effect without a proposed mechanism has no ific credibility.
amount of data can prove a hypothesis. Instead of trying to do so, oup or the biologist should have designed the experiment to falsify pothesis that the oil was having an effect.
e oil-free bay was not a true control.
e group's conclusion was hasty because it was based on only some species present.

ears after the oil spill discussed above, most of the oil has disap- om the intertidal area, but the environmental group rehires the biol- o some more work on the bay. The new piece of evidence that would ngthen their conclusion that the oil had depressed the mollusk pop- would be for the biologist to find that

two bays now have nearly identical populations of all intertidal ls, including mollusks
mollusk populations in the oil-affected bay are even lower than after the spill
mollusks in the unpolluted bay have changed in species composi- o they now share few species with the oil-affected bay
oratory tests show that mollusks in both bays can tolerate far oil concentrations than those experienced during the spill

ceans are a poorly understood factor in global warming. Some re- are concerned that warming of the oceans might cause the large of CO_2 dissolved in seawater to come out of solution, the way air orm in a glass of tap water when it warms up. Why might this con- n? On the other hand, how might the oceans help reduce global

gists in the United States are concerned that populations of many y songbirds, such as warblers, are declining. Evidence suggests that hese birds might be victims of pesticides. Most of the pesticides im- n songbird mortality have not been used in the U.S. since the 1970s. hypothesis to explain the current decline in songbird numbers.

, Technology, and Society

ay have heard that human activities cause the extinction of one ery hour. Such estimates vary widely, because we do not know how cies exist, nor how fast their habitat is being destroyed. You can make estimate of the rate of extinction. Start with the number of species thought to exist. To keep things simple, ignore extinction in the temperate latitudes and focus on the 80% of plants and animals that live in the tropical rain forest. Assume that destruction of the forest continues at a rate of 1% per year, so the forest will be gone in 100 years. Assume (optimistically) that half the rain forest species will survive in preserves, forest remnants, and zoos. How many species will disappear in the next century? How many species is that per year? Per day? Recent studies of the rain forest canopy have led some experts, such as Terry Erwin of the Smithsonian Institution, to suggest that there might be as many as 30 million species on Earth. How does starting with this figure change your estimates?

2. One of the reasons the developed countries consume so much energy is that the price of energy does not reflect its real costs. For example, the monetary cost of the 1991 Gulf War would have doubled the price of every barrel of oil imported into the United States that year. What kinds of hidden environmental costs are not reflected in the prices of fossil fuels? How are these costs paid, and by whom? Do you think these costs could or should be figured into the price of oil? How might that be done?

3. Many atmospheric scientists say that global warming is under way and we need to take action now to avoid drastic environmental change. Some say it is too soon to tell and we should gather more data before we act. What are the advantages and disadvantages of doing something now to slow global warming? What are the advantages and disadvantages of waiting until more data are available? Is there any way to know for sure whether there is a link between CO_2 and global warming?

4. Until recently, response to environmental problems has been fragmented—an antipollution law here, incentives for recycling there. Meanwhile, the problems of the gap between the rich and poor nations, diminishing resources, and pollution continue to grow. Now people and governments are starting to envision a sustainable society. The Worldwatch Institute, a respected environmental monitoring organization, estimates that we must reach sustainability by the year 2030 to avoid economic and environmental disaster. To get there, we must begin shaping a sustainable society during this decade. In what ways is our present system not sustainable? What might a sustainable society be like? Do you think a sustainable society is an achievable goal? Why or why not? What is the alternative? What might we do to work toward sustainability? What are the major roadblocks in the way of achieving sustainability? How would your life be different in a sustainable society?

Media Resources

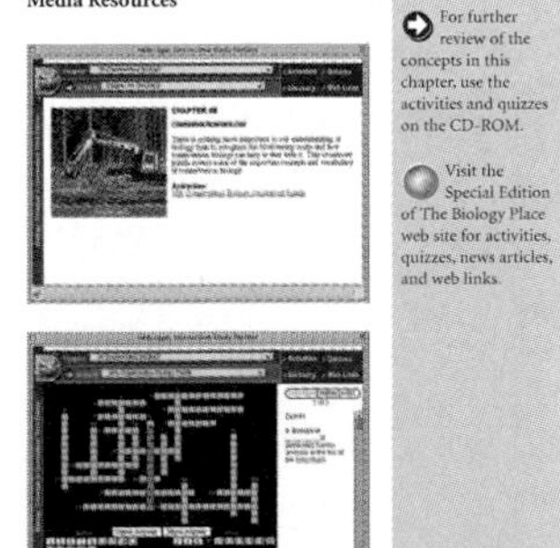

➔ For further review of the concepts in this chapter, use the activities and quizzes on the CD-ROM.

Visit the Special Edition of The Biology Place web site for activities, quizzes, news articles, and web links.

Chapter 38 Conservation Biology **781**

CHAPTER REVIEW QUESTIONS HELP YOU MAKE CONNECTIONS—AND REVIEW FOR TESTS

- "Testing Your Knowledge" questions let you check what you've learned through multiple-choice questions and "Describing, Comparing, and Explaining" questions, which ask you to restate concepts and explanations in your own words.
- "Thinking as a Scientist" questions give you a chance to analyze and interpret data.
- "Science, Technology, and Society" questions involve you in thinking about the ways biological concepts connect to your own life and to society in general.
- The Chapter Review Answers section provides answers and explanations for the Chapter Review questions. You can find it in Appendix 4 at the back of the book.
- Chapter quizzes on the Interactive Study Partner CD-ROM offer another way to test your knowledge. The quizzes provide feedback statements that link directly to pages in your book.
- Go to TestFlight at The Special Edition of The Biology Place™ web site for more practice exams keyed directly to the content of the book.

1 Introduction: The Scientific Study of Life

Outline

The canopies, or treetops, of forests are one of nature's great showplaces of different kinds of living organisms. Biologists estimate that fully half of the 5 million to 50 million species on Earth spend all or much of their time here, yet canopies were virtually unexplored until about 15 years ago. The canopies of tropical rain forests, areas near the equator where yearly rainfall often exceeds 200 centimeters (80 inches), harbor the greatest number of species. Equally wet forests in the coastal regions of western Canada and the northwestern United States also are home to remarkable numbers. The dominant trees in many northern rain forests are centuries-old firs, hemlocks, and spruces. Extending into northern California are remnants of truly ancient rain forests dominated by redwoods and giant sequoia trees that are thousands of years old.

Research on forest canopies blossomed in the 1970s when scientists began using mountain climbers' roping techniques to reach the treetops. In the large photograph at the left, Humboldt State University researcher Steve Sillett is roping his way into the 90-meter-high canopy of a 2000-year-old giant

Evolution, Unity, and Diversity

Connections to Everyday Life

sequoia tree in northern California.

Roping techniques allow researchers to move from one area of forest to another, but they are not the only way to get into a canopy. For projects that can be done in a limited area, some researchers prefer large construction cranes with horizontal booms that rotate over the canopy. More permanent are platforms, walkways, and towers built into the treetops. In the small photograph on the facing page, a researcher from the University of Washington stands on a tower set up to collect samples of mist, part of a long-term study of how water circulates between the forest and the atmosphere.

Canopy research is not for everyone. Researchers often spend a week or more in treetops, sleeping in hammocks strung between branches higher than a 20-story skyscraper. Their rewards include sights, sounds, and smells found nowhere else. Pioneer forest researcher Dr. Jerry Franklin expresses his enthusiasm for canopy research this way: "Trees are one of Earth's main points of contact with the air. This is our chance to learn about the dialogue between trees and the atmosphere."

A summer evening is a good time to experience a canopy's rich array of life. In a rain forest in the northwestern U.S., brown and greenish fungi, lichens, and earthy mats of moss cling to the high branches. Your eye might catch the split-second shadow of a flying squirrel across a bright moon. Rarely seen because they are active only at night, these foot-long rodents speed-glide between the big trees. Insects are especially abundant in the canopy twilight, and bats flit about, often eating close to their own body weight in insects in a single night. Owls, another group of nocturnal hunters, are silent on the wing, but each species has its own distinctive hoot.

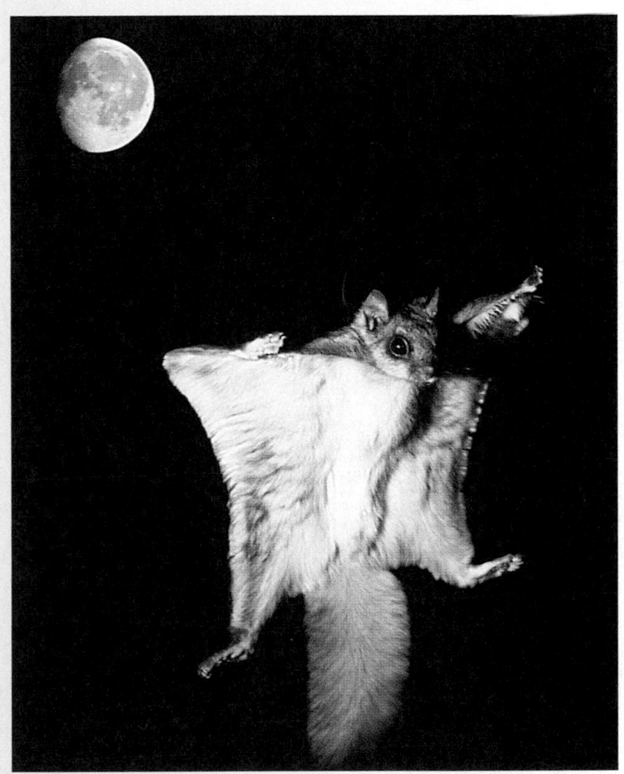

A flying squirrel

The enormous variety of living organisms in forest canopies and the science of canopy research set the stage for this chapter's introduction to biology, the scientific study of life.

The Scope of Biology

1.1 Life's levels of organization define the scope of biology

A forest researcher mentions the dialogue between trees and the atmosphere. What does this mean? The word "dialogue" refers to interactions between living organisms and nonliving matter, the gases in the atmosphere. Such interactions are a fundamental property of ecosystems, the highest of several structural levels into which life is organized. An **ecosystem** (for example, a rain forest) consists of all the organisms living in a particular area, as well as all the nonliving, physical components of the environment that affect the organisms, such as air, soil, and sunlight. As indicated in Figure 1.1 at the right, the ecosystem and the structural levels below it form a hierarchy, with each level building on the ones below it. Below the ecosystem level, all the organisms in a rain forest are collectively called a **community.** Below the community, an interacting group of individuals of one species, flying squirrels in our example, is called a **population.** Below population in the hierarchy is the **organism,** an individual living thing.

Below the organism level, life's hierarchy unfolds within the individual organism. The flying squirrel's body consists of several **organ systems,** such as a circulatory system, an excretory system, and a nervous system, shown here. Each organ system consists of **organs.** For instance, the main organs of the nervous system are the brain, the spinal cord, and the nerves, which transmit messages between the spinal cord and other parts of the body.

As we continue downward through the hierarchy, each organ is made up of several different **tissues,** each of which consists of a group of similar cells. A **cell** is a unit of living matter separated from its environment by a boundary called a membrane. Each tissue has a specific function, which is performed by the cells that compose it. The nervous tissue that makes up most of the brain, for example, consists of nerve cells. The nervous tissue in the squirrel's brain has millions of microscopic nerve cells organized into a communication network of spectacular complexity. The nerve cells transmit signals that coordinate the squirrel's body parts, such as the muscles that stretch out its legs during a glide.

Finally, we reach the molecular level in the hierarchy. We show as our example DNA (deoxyribonucleic acid). DNA molecules provide the blueprints for constructing the organism's other important molecules and transmit this information, as genes, from parents to offspring. A **molecule** is a cluster of atoms, the smallest particles of ordinary matter. In the computer graphic at the right, which illustrates only a tiny segment of one DNA molecule, each of the spheres represents an atom. Each DNA molecule is a very long double helix, two chains coiled around each other. (We discuss DNA further in Module 1.5.)

As we discuss in later chapters, life's hierarchy builds from molecules to ecosystems. It takes many molecules to

Figure 1.1 Life's hierarchy of organization

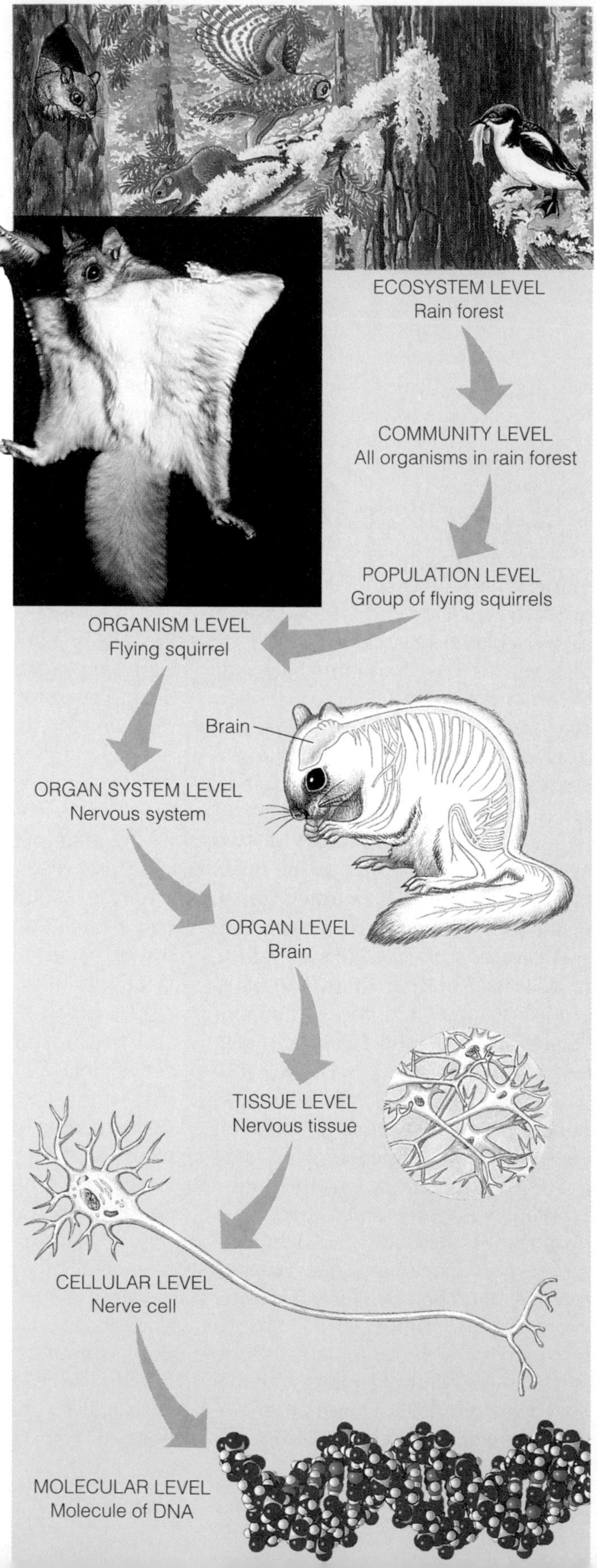

make a cell, many cells to make a tissue, several kinds of tissues to make an organ, and so on. Most biologists specialize in the study of life at a particular level. For instance, a researcher analyzing the body postures of a gliding squirrel focuses on the organism level. However, understanding gliding posture may require studying, at the organ-system level, the interaction between muscles and bones, so the same researcher often works at more than one level. The full spectrum of life's hierarchy, from molecules to ecosystems, encompasses the scope of biology. With this in mind, let's see how biological scientists go about their work. Although we focus on examples of outdoor research in forest ecosystems, the same scientific approach is used in all types of biological research. 

? Which of the following levels of biological organization includes all others in the list? Cell, molecule, organ, tissue

Organ

The Process of Science

Scientists pose and test hypotheses to answer questions about nature 1.2

There is no single, rigid method that scientists use in studying the natural world, but all scientists use a similar investigative process. The key elements of this **process of science** are the following:

1. Observations
2. Questions
3. Hypotheses
4. Predictions
5. Tests

In a scientific study, the initial *observations* take a variety of forms. They often come from previous studies. For example, an observation that has often been made about flying squirrels is that, when they land on a tree, they quickly scramble to the other side. This observation raises a *question:* Assuming this behavior benefits the squirrels' survival, how does it do so? A scientist who wishes to answer this question generates a testable **hypothesis**, a tentative explanation. One hypothesis is that by moving quickly away from a landing site, a squirrel prevents an owl or other predator from catching it when it lands. This reasonable hypothesis is often cited in popular articles about flying squirrels, but let's make no assumptions about its accuracy. Flying squirrels are nocturnal and glide very fast, making them hard to observe without special equipment. Also possibly relevant: Flying squirrels are temporarily blinded by bright light. Do these observations make you wonder how often flying squirrels have actually been seen landing? What if most observations have been made of squirrels landing on the bright sides of trees on moonlit nights? Critical thinking like this leads to *alternative hypotheses;* for example, maybe scrambling around the tree enables flying squirrels to avoid bright moonlight.

With observations and hypotheses in mind, scientists use *deductive reasoning* to predict the results of new observations or experiments. A *prediction* states the results they expect if the hypothesis is correct. Deductive reasoning is the use of "*If . . . then*" logic. In this case, *if* a hypothesis is correct, *and* we perform a test of the hypothesis, *then* we can expect a particular outcome. For example, *if* flying squirrels scramble around a tree simply to avoid bright light (hypothesis), *and* we devise a way to observe them landing in both shadow and light (test), *then* we would expect that squirrels landing on the dark sides of trees would not immediately move to the other side (prediction).

Figure 1.2 What explains the landing behavior of a flying squirrel?

Suppose you make this hypothesis and prediction and then observe numerous landings of flying squirrels, perhaps by taking infrared photos while monitoring light conditions. You find that, in all light conditions, the squirrels scramble to the other side of the tree after landing. This result does not match your prediction and therefore indicates that your hypothesis may be false. However, your result neither supports nor falsifies the alternative hypothesis that the squirrels' behavior enables them to avoid predators. (Can you think of a way to test this hypothesis?)

Our discussion of flying squirrels has brought up several features of the scientific process, in addition to the roles of observations, questions, hypotheses, predictions, and tests. We have seen that science involves critical thinking at every step. Furthermore, science is cumulative, with the results of earlier studies often serving as the basis for new studies. We have also seen that the scientific process can falsify or support hypotheses. Note the key phrase "support hypotheses." Science does not pretend to prove hypotheses with absolute certainty, because it is not possible to accumulate enough observations or experiments to be absolutely certain results will always be the same. We look at additional features of the scientific process next. 

? What is the relationship of hypothesis and prediction in the process of science?

If the hypothesis is correct, then we predict the following outcome from our experiment.

1.3 The process of science: A case study in a tropical rain forest

1A
See Activity 1A on the CD-ROM.

In Figure 1.3A, rain forest canopy researcher Nalini Nadkarni begins a climb into the treetops using modern roping techniques. In the forefront of canopy research, Dr. Nadkarni is a biology professor at Evergreen State College in Olympia, Washington. She has a special interest in epiphytes, plants that grow on other plants. Epiphytes that grow on high tree limbs make up a large part of the plant life in many rain forests. We examine some of her studies as real-world examples of the thought processes that connect observations, questions, hypotheses, and deductive reasoning in scientific work. You might find it helpful to take notes and outline these studies and processes as we describe them.

Figure 1.3A Nalini Nadkarni begins a climb

Scientific background leads to focused questions Before beginning work, scientists obtain as much background information as possible. Preliminary observations (those made before starting new research) help a scientist pose questions that can lead to important discoveries. The scientific literature and other scientists are the main sources of this background information.

Suppose you had a chance to join Dr. Nadkarni's research team, and you decided to spend a summer with the team in a rain forest in Costa Rica. Dr. Nadkarni assigns you to a group studying how epiphytes obtain the resources they need to survive and grow. Some background you would need to know is that all plants require chemicals, such as nitrogen and phosphorus, to construct the molecules that make up their bodies; biologists call these materials essential nutrients. The roots of most plants absorb essential nutrients from the soil, but epiphytes' roots are not anchored in the ground and do not absorb resources from it. Also, epiphytes that have been studied are not parasitic, meaning they do not obtain nutrients or energy from the trees on which they live.

These observations lead to the question: *How do epiphytes obtain the essential chemical elements that most plants absorb through their roots from the soil?* This has been a key question in rain forest research, because epiphytes are so common and are a major food source for other forest organisms. Figure 1.3B shows Dr. Nadkarni next to a mat of epiphytes that she has peeled back from a canopy tree limb. Most of the epiphyte roots are very thin. The tree also has canopy roots, and Nadkarni is gripping one of these in the photograph. Together, the roots of the host and the epiphytes are tightly bound to brownish material that scientists call canopy soil. Nadkarni and others have found that

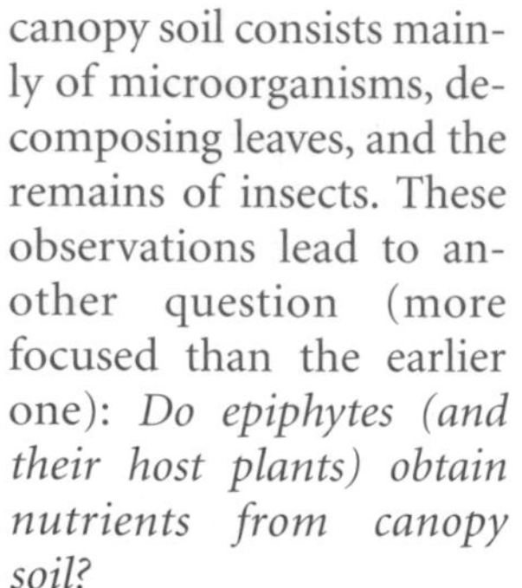

canopy soil consists mainly of microorganisms, decomposing leaves, and the remains of insects. These observations lead to another question (more focused than the earlier one): *Do epiphytes (and their host plants) obtain nutrients from canopy soil?*

Figure 1.3B Nalini Nadkarni examining epiphyte roots

In general, the most sharply focused research questions are most likely to lead to important new discoveries. In your case, choosing a single species of epiphyte to study would be a good way to sharpen focus. Suppose the leader of your research team wants you to study the nutrient relationships of an epiphytic species of bromeliad, a member of the pineapple family, found in the tropics. Dr. Nadkarni informs you that this species is new to science. Your research question now becomes: *Does this particular species of epiphyte obtain nutrients from canopy soil?*

Previous results, observations, and analysis lead to testable hypotheses, predictions, and tests With a sharply focused research question in mind, the next steps in the scientific process are geared toward posing a testable hypothesis.

Researchers often formulate hypotheses by looking at how previous research has answered questions similar to the new one. Since your bromeliad species has not been studied yet, you cannot assume that it is in any way the same as others. However, it is appropriate to use observations made on the roots of other species in planning your research. For instance, you may draw on what others have found out about other epiphytes' canopy roots and the contents of the soil. You may also analyze your own observations to formulate a hypothesis. For example, you could reason that the close association of epiphyte roots and host-plant roots with canopy soil and the known contents of canopy soil make the soil a likely source of nutrients for both the host plants and the epiphytes. You are now close to posing a testable hypothesis, but there are some other factors to consider. Most of Nalini Nadkarni's research is carried out in the field—in actual rain forest canopies—not in a laboratory setting where conditions can be closely controlled. The setting, the research equipment that is available, and timing are practical matters that she (and you) must consider. With all these factors in mind, a reasonable hypothesis that seems testable in a rain forest canopy is: *The bromeliad depends on canopy soil.* Your next step is to think about testing this hypothesis and to make predictions of what the results of tests will be if your hypothesis is correct. Predictions help you evaluate whether your tests are likely to give you useful information about your hypothesis.

One way to test your hypothesis in the canopy would be to locate tree limbs where your bromeliad is the prevalent epiphyte, remove all the epiphytes and soil from a specific area, and monitor the rate at which epiphytes and canopy soil reappear. If your hypothesis is true, you would predict that your bromeliads would reappear after some canopy soil accumulated on the limbs.

Suppose you go ahead and perform a test; you find some accessible tree limbs that are covered with bromeliads, remove all epiphytes and soil from them, and build a temporary work platform to record their reappearance each week for three months. The graph in Figure 1.3C summarizes the results of your test; more or less consistent with your prediction and supporting your hypothesis, the graph shows a general increase in the number of bromeliads (red) as canopy soil accumulates. However, your results do not agree entirely with your prediction (notice on the left side of the graph that some bromeliads reappeared before any soil accumulated on the limbs). This raises an additional question: *Can the bromeliads obtain nutrients from a source other than the canopy soil?* Scientific tests often lead to new questions, which in turn stimulate more research.

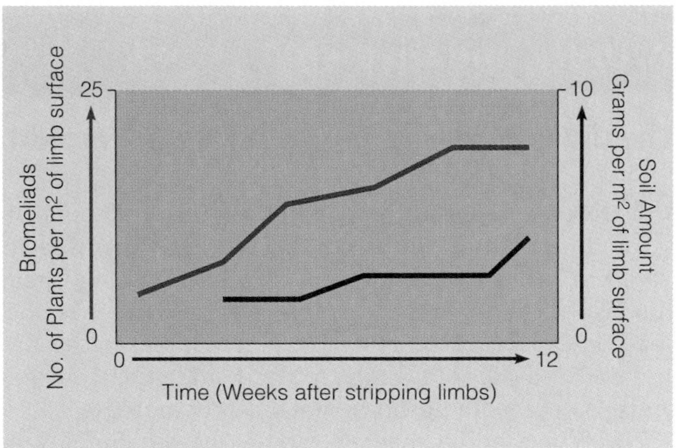

Figure 1.3C Results of a test on regrowth of bromeliads (epiphytes)

Repeated testing and controlled experiments validate results Repetition of tests is an essential part of the process of science. Only after several tests yield results similar to those in Figure 1.3C would you be ready to say that your hypothesis of bromeliad dependence on canopy soil was supported. It would also be premature to pursue other questions raised by your results before repeated testing.

Another component of the scientific process is the **controlled experiment.** In performing a controlled experiment, a scientist actually carries out two parallel tests, one called the experimental, the other the control. Ideally, the experimental test differs from the control test by only a single factor, called the *variable.*

For example, you might wonder whether the bromeliads that appear to grow before canopy soil appears are actually nourished by traces of soil you have been unable to remove because of treetop conditions. To find out, you could perform an experiment in the laboratory where you would attempt to grow bromeliads on two limbs, one that has canopy soil (the experimental test), and the other scrubbed completely clean of soil and isolated from wind or other conditions that might deposit soil (the control test). In this case the presence of canopy soil would be the single variable that is present in the experimental test, but absent in the control. If bromeliads grew only in the experimental test, you would suspect that the few bromeliads that appeared to grow in the treetops without canopy soil were actually nourished by traces of contaminating soil. Since variables are difficult to control in the field, researchers often perform controlled experiments in the regulated environment of a laboratory. However, such experiments may yield results that are not reliable under field conditions.

Many biological studies have to be performed according to the conditions that researchers encounter. Science is an adaptable process that accommodates many less-than-ideal situations.

Scientific knowledge remains open to question In science, hypotheses are not proved, and all knowledge is open to challenge, reinterpretation, and refinement. Nalini Nadkarni and others have amassed evidence supporting the hypothesis that epiphytes and their host trees obtain nutrients from canopy soil. However, studies also show that the roots of epiphytes and their hosts obtain some nutrients from mist in the air. Thus, our canopy example demonstrates that research involves a succession of scientists working on the same or similar problems building on what has been learned earlier.

Science is a powerful means of gaining information and understanding. Scientists share information through publications, meetings, and personal communication. They also critique one another's work and often check one another's results and conclusions by repeating observations and experiments. Science deals only with hypotheses that are testable and subject to modification as knowledge and methods of testing advance. Hypotheses that are not testable, such as those involving supernatural causes, lie outside the realm of science. Also unapproachable by science are questions such as, What is the spiritual meaning of life? The quest for answers to questions like this lie in the realms of philosophy and religion.

? Why is it difficult to draw a conclusion from an experiment that is not controlled?

Without a control, you don't know if the experimental outcome is due to the variable you are trying to test or to some other variable.

1.4 The diversity of life can be arranged into three domains

Rain forests abound with the sights, sounds, and scents of living things. Ants, mosquitoes, beetles, and other insects are literally everywhere—flying, crawling, jumping—and you hear them day and night. In a tropical rain forest, the sweet fragrance of showy orchids often hangs in the air, and the loud calls of parrots, toucans, and other colorful birds compete with the hoots and howls of monkeys.

The richness of life in a rain forest—the vast diversity of species—can be almost overwhelming. To make diversity somewhat more comprehensible, scientists have devised ways of grouping (classifying) organisms. Today, biologists generally favor classification schemes with at least eight kingdoms, which are themselves classified into three higher groups called **domains.**

The organisms in these six photographs representing the three domains could all be found in one small area of a tropical rain forest. Figures 1.4A and 1.4B show microscopic organisms called prokaryotes. Found literally everywhere there is life, from rain forests and polar oceans to your own skin and intestines, prokaryotes are the most widespread of all living organisms. Prokaryotes are distinguished from all other forms of life by their structure. Every living being is composed of cells, but only prokaryotes have cells without a nucleus, a discrete internal structure that controls cellular activities. There are two very different groups of prokaryotes, which make up two of the three domains: **Bacteria** and **Archaea** (Figures 1.4A and 1.4B). All organisms except prokaryotes are members of a third domain, the **Eukarya,** organisms made of cells with a nucleus and discrete internal structures called organelles.

Most pools of water containing prokaryotes would also support members of Domain Eukarya commonly called protists. One group of protists, commonly called algae, make their own food molecules by the process of photosynthesis. Another group, commonly called protozoans, are single-celled and are animal-like in that they eat other organisms, including algae and prokaryotes. Figure 1.4C shows a number of protists in a drop of water viewed with a microscope. The large, irregular, bluish cell on the right is an amoeba (a protozoan), and the smaller cells are mostly single-celled algae. Also present are multicellular algae (the long rodlike filaments), which are considered protists because of their similarities to single-celled algae. Until recently, protists were classified in a single kingdom, but evidence from molecular studies now indicates that they are more diverse than any other group of eukaryotes. As we will see when we study them in more detail, protists comprise several kingdoms within the domain Eukarya.

Organisms in the remaining three groups (kingdoms) of the Eukarya are all multicellular. Kingdom Plantae, the plants, are photosynthetic and consist of cells with strong walls made of cellulose. The plant in Figure 1.4D is a tropical bromeliad, closely related to the epiphytes we discussed in Module 1.3.

Kingdom Fungi, represented in Figure 1.4E by the fly mushroom, is a diverse group that includes the molds,

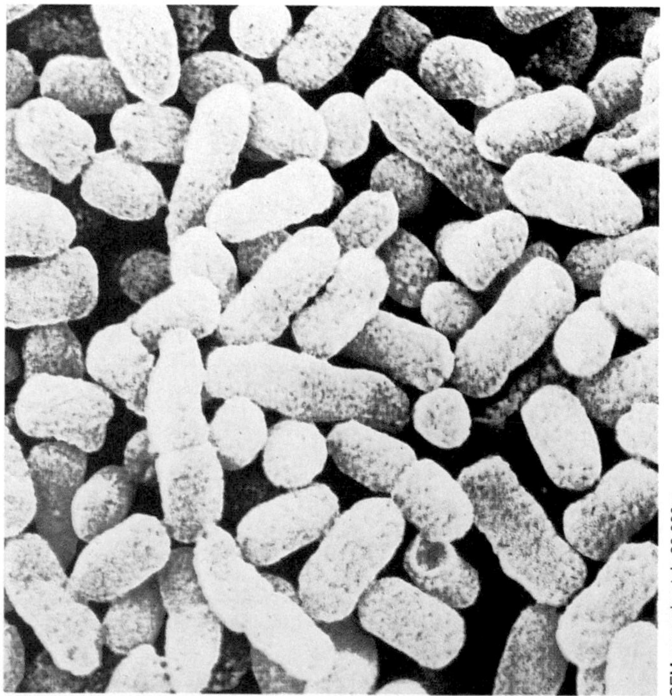

Figure 1.4A Domain Bacteria (prokaryotes)

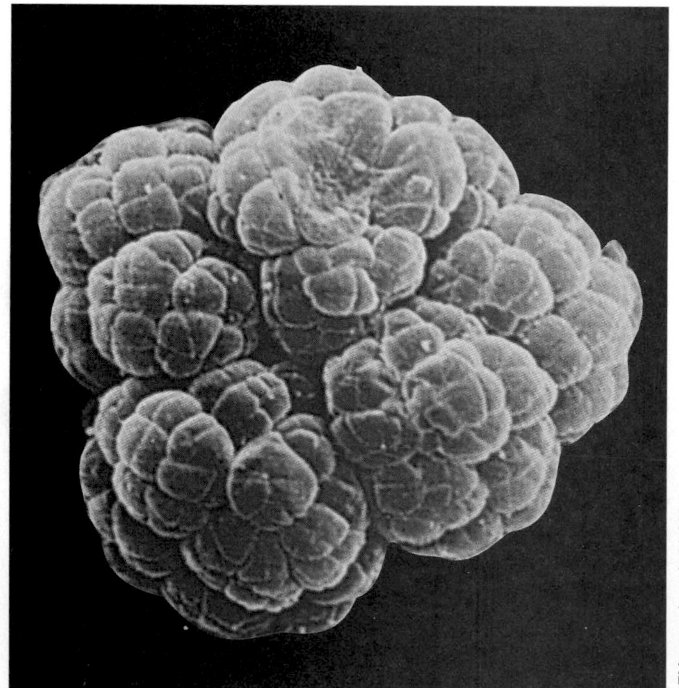

Figure 1.4B Domain Archaea (prokaryotes)

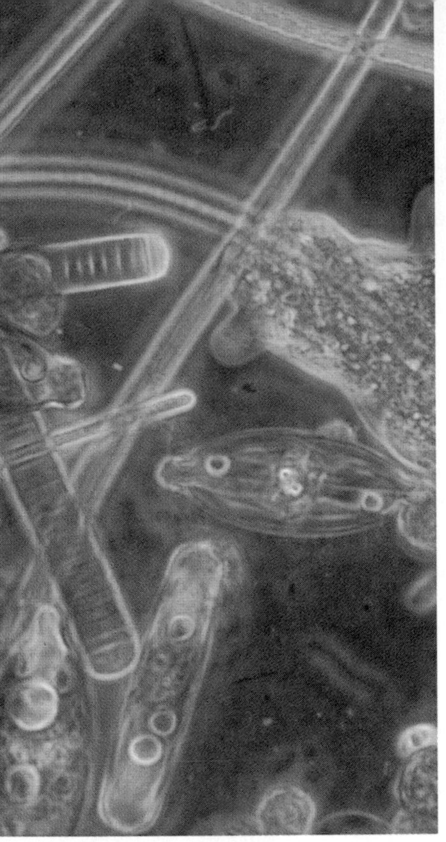

Figure 1.4C Domain Eukarya (protists)

Figure 1.4D Domain Eukarya, Kingdom Plantae

Figure 1.4E Domain Eukarya, Kingdom Fungi

yeasts, and mushrooms. Fungi decompose the remains of dead organisms and absorb nutrients from the leftovers.

Representing the kingdom Animalia (animals), the sloth in Figure 1.4F resides in tropical rain forest canopies. Animals eat other organisms and are made of cells that lack rigid walls. Most animals are motile. The sloth is a slow-moving animal that spends most of its time hanging upside down eating leaves.

There are actually members of three major groups in Figure 1.4F. The sloth is clinging to one of the tall trees (Kingdom Plantae) dominating the rain forest. And the greenish tinge in the animal's hair is a luxuriant growth of photosynthetic prokaryotes (Domain Bacteria). This photograph emphasizes a theme reflected in our book's title: connections among living things. The sloth depends on trees for food and shelter, the prokaryotes gain access to the sunlight necessary for photosynthesis by living on the sloth, and the tree uses some chemical nutrients supplied mainly by prokaryotes.

Life's diversity and its interconnectedness are evident almost everywhere. You can find representatives of the major groups on many city streets. There the most obvious examples of Animalia are likely to be people, with trees, shrubs, and grass representing Plantae. With the help of a microscope, you can find prokaryotes, fungi, and protists in any puddle of water or patch of moist soil. Less obvious, but just as significant, are signs of the basic similarities shared by all organisms. Life's great paradox is the unity in its diversity—the fact that the millions of species of organisms are all variations on a relatively small set of basic features. We discuss the nature and significance of these similar features in the next module.

Figure 1.4F Domain Eukarya, Kingdom Animalia (with Plantae and Domain Bacteria)

To which of the three domains of life do we belong?

Eukarya

1.5 Unity in diversity: All forms of life have common features

Figure 1.5A Three species of orchids from a rain forest

The three species of tropical rain forest plants in Figure 1.5A look quite different. However, they are all members of the orchid family, united by a set of features common to all. For instance, the showy, highly varied flowers of orchids all have a large liplike petal that attracts pollinating insects. Orchids are, in fact, variations on a common theme.

The basis of life's variety and common themes is the genetic information in DNA molecules. All cells have DNA, and the genetic information in DNA is coded in the same way in all organisms. DNA's helical chains are made up of four kinds of chemical building blocks, called nucleotides (indicated by the four colors in Figure 1.5B). All DNA has the same four nucleotides, and the blueprint for every species (and for every individual) is encoded in the particular sequence of nucleotides along the length of its DNA. For instance, the order of the colors in Figure 1.5B represents the nucleotide sequence of a segment of DNA of one species. A different species might have a different order of nucleotides in the corresponding segment of its DNA. Life's vast diversity basically comes from variations in DNA sequences—in other words, from variations on the common theme of storing genetic information in DNA.

Let's now list some properties that are common to all organisms. Taken together, these properties distinguish life from nonlife. The genetic information in DNA underlies all of these properties. (1) *Order.* All organisms exhibit complex organization. (2) *Regulation.* The environment outside an organism may change markedly, but regulatory mechanisms maintain an organism's internal environment within limits that sustain life. (3) *Growth and development.* Each organism has a pattern of growth and development characteristic of its species. (4) *Energy utilization.* Organisms take in energy and transform it to perform all of life's activities. (5) *Response to the environment.* All organisms respond to environmental stimuli. (6) *Reproduction.* DNA lends itself to precise replication, and all organisms reproduce their own kind. (7) *Evolution.* Reproduction underlies the capacity of species to change (evolve) over time. Evolutionary change has been a central, unifying feature of life since it arose about 4 billion years ago.

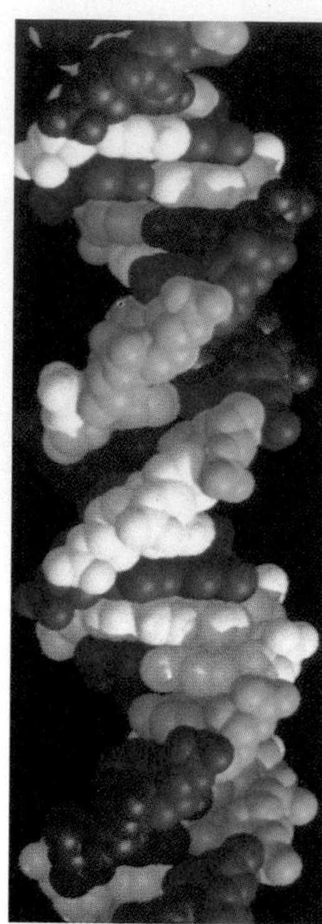

Figure 1.5B DNA, with nucleotides in four different colors

What is the chemical basis for all of life's kinship?

DNA as the genetic material

1.6 Evolution explains the unity and diversity of life

In November 1859, British biologist Charles Robert Darwin published one of the most important and controversial books ever written. Entitled *On the Origin of Species by Means of Natural Selection,* Darwin's book was an immediate best-seller and soon made his name almost synonymous with the concept of evolution.

Darwin stands out in history with people like Newton and Einstein, scientists who synthesized ideas with great explanatory power. Such comprehensive ideas, much broader in scope than hypotheses (which may explain only a single set of observations), are called **theories.** Darwin showed how evolution could explain the common threads underlying life's diversity. According to the evolutionary view of life, members of the orchid family, for instance, are fundamentally similar because they have evolved from a common ancestor. Darwin perceived that the fossil record chronicles the evolution of spe-

Figure 1.6A Charles Darwin in 1859

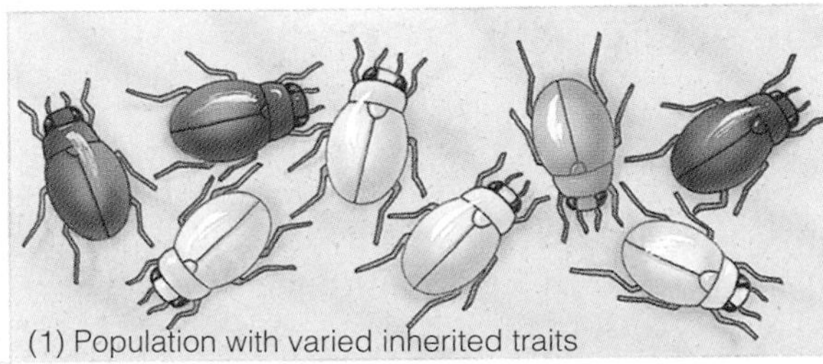

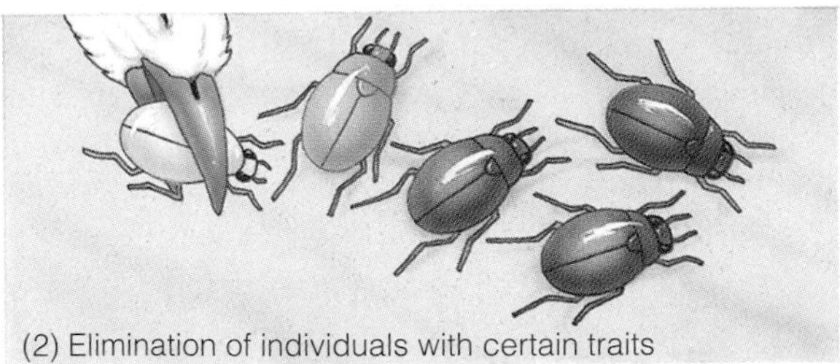

Figure 1.6B Natural selection

cies. In the case of the horse, for example, the fossil record reveals that the species living today was preceded by a number of ancestors. In Darwin's words, species arise through a process of "descent with modification."

Most importantly, Darwin's *Origin of Species* proposed the theory of **natural selection** to explain how evolution occurs. Figure 1.6B uses a simple example to show how natural selection works. In part 1, we see a group of imaginary beetles of a single species. The group is a population; the individuals breed among themselves. The individuals exhibit varied traits that are inherited—in this case, three different body colors. As Darwin realized, heritable variation must be present in the population for natural selection to operate.

In part 2, a predatory bird eats the beetles it sees most easily, the yellow ones. The yellow beetles tend to be eliminated before they have a chance to reproduce and pass on the gene for yellow color. In part 3, the surviving, less colorful beetles have reproduced. The group is now quite different from the original one; natural selection has produced a change in the proportions of the colors in the population. Here we see that natural selection is not a creative process, but an editing mechanism. *Natural selection occurs as heritable variations are exposed to environmental factors that favor the reproductive success of some individuals over others.* In our example, birds are the relevant environmental factor.

Darwin realized that numerous small changes in populations caused by natural selection could eventually result in big changes that alter species. He proposed that the evolution of new species results from an accumulation of minute changes coming from natural selection over time. As we will see in Chapters 13 and 14, there have been some refinements of Darwin's ideas, but his theory of natural selection has stood the test of time.

We see the exquisite results of natural selection in every kind of organism. Each species has its own special set of **adaptations,** features that evolved by means of natural selection. On the left in Figure 1.6C, we see the pangolin, a mammal that lives in East African rain forests. One of its main adaptations, its tough body armor of overlapping scales, protects it from most predators. The pangolin also has an unusually long tongue, which it uses to prod termites and ants out of their nests. Another mammal, the killer whale (orca), is adapted for life at sea (Figure 1.6C, right). It breathes air through nostrils on the top of its head and keeps in touch with its companions by emitting clicking sounds into the water. Orcas use sound echoes to detect obstacles and to locate schools of fish or other prey, such as seals and sea lions.

The pangolin's armor and the orca's echolocating ability did not result from changes in individuals during their lifetime. These traits arose over many, many generations, as individuals with heritable traits that made them best adapted to their environment had the greatest reproductive success.

Understanding how adaptations evolve by natural selection is key to the study of life. Indeed, evolution is biology's core theme—the one idea that makes sense of all we know about life. 

? How does natural selection enable a population of organisms to adapt to their environment?

On average, those individuals with heritable traits best suited to the local environment produce the greatest number of offspring that survive and reproduce. This increases the frequency of those traits in the population.

Figure 1.6C Adaptations to the environment

Pangolin

Killer whale

1.7 Living organisms and their environments form interconnecting webs

Darwin's theory of evolution by natural selection focuses on the response of organisms to interactions with the living and nonliving components of their environment. At the highest level in nature's hierarchy—the ecosystem—these interactions make up a complex web of relationships connecting all the organisms and the components of their environment. The painting in Figure 1.7A (below) provides a simplified view of some of the relationships among organisms in an African rain forest ecosystem. The arrows indicate the directions in which energy and nutrients pass.

Plants dominate the scene depicted in Figure 1.7A, and they provide much of the food that supports the ecosystem. Plants, as well as certain prokaryotes and some protists (not shown), trap energy from sunlight and use carbon dioxide (CO_2) from the air, along with water (H_2O), to make food molecules by photosynthesis. Plants also absorb mineral nutrients from the soil.

Animals like the colorful sunbird (upper right in Figure 1.7A; Figure 1.7B, p. 11), the gorilla, and many insects eat plants, plant parts, or plant products, such as nectar.

Sun
CO_2
H_2O
Light
PLANTS make food
ANIMALS eat plants and other animals
DECOMPOSERS in soil break down organisms
Minerals in soil
Dead organisms

Figure 1.7A A web of interactions in a rain forest ecosystem

Mice and parrots eat mainly plant material but also some insects. Other animals, such as the leopard, many kinds of snakes, and meat-eating insects, prey on animals. In all cases, however, plants and photosynthetic prokaryotes and protists are the ultimate sources of food.

Comprising another vital part of the ecosystem are the prokaryotes, fungi, and small animals in the soil that decompose the remains of dead organisms. These decomposers act as recyclers, changing the complex dead matter into simple mineral nutrients that plants can use. Thus, the web of relationships among plants, animals, microorganisms, and the physical environment gives an ecosystem its structure. The weblike pattern of arrows in an ecosystem diagram indicates this basic structure in a visual way.

Figure 1.7B An African sunbird and a mistletoe plant, whose nectar is food for the bird

Each level in nature's hierarchy has a unique structure (see Module 1.1), and a set of functional properties results from that structure. For example, at the organism level, the thin, curved shape of a sunbird's beak makes it an effective probe for reaching nectar inside flowers. In contrast to a bird's body parts, which are physical entities, the structural web of an ecosystem is abstract. Nonetheless, function comes from structure in an ecosystem, just as it does in an organism. In an ecosystem, the strands of the web represent interactions among living organisms, and between organisms and their nonliving surroundings. The interactions account for the passage of chemical nutrients and energy throughout the ecosystem.

Figure 1.7C highlights a major difference between the flow of chemical nutrients and the flow of energy within an ecosystem. The most basic chemicals that are necessary for life—the atoms making up carbon dioxide, oxygen, water, various minerals, and the more complicated chemicals of organisms—flow from the air and soil to plants, to animals, and back to the air and soil. In other words, chemical nutrients cycle more or less continuously within an ecosystem's structural web. By contrast, an ecosystem gains and loses energy constantly. Energy flows into the ecosystem when plants and other photosynthesizers trap light energy from the sun and convert it to chemical energy by using it in the manufacture of molecules. Chemical energy, in the form of molecules, is then shuttled through the ecosystem's web, powering each organism in turn. In the process of energy shuttling, some of the energy is converted to heat, which eventually leaves the ecosystem. All life on Earth depends on these abilities of ecosystems to cycle chemical nutrients and shuttle energy. 

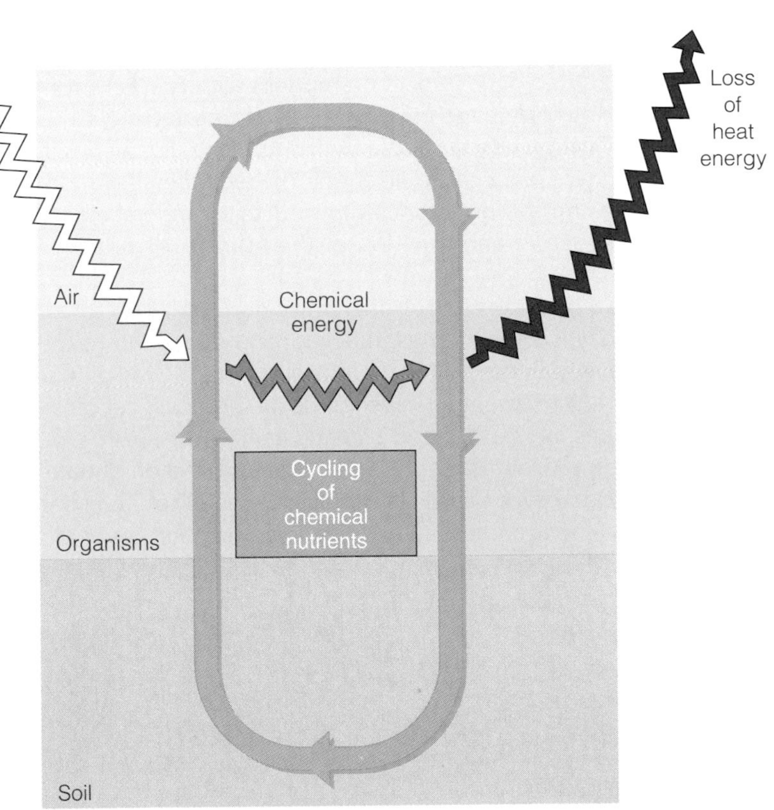

Figure 1.7C Nutrient cycling and energy flow in an ecosystem

? Explain how the photosynthesis of plants functions in both the cycling of carbon and the flow of energy in an ecosystem.

Photosynthesis uses light to convert atmospheric CO_2 to energy-rich food, making it the pathway by which both carbon and energy become available to most organisms.

1.8 Biology is connected to our lives in many ways

Global warming, air and water pollution, endangered species, genetic engineering, test-tube babies, nutrition, aerobic exercise and weight control, medical advances, AIDS and the immune system—is there ever a day that we don't see several of these issues featured in the news? These topics and many more have biological underpinnings. Biology, the science of life, has an enormous impact on our everyday life, and it is impossible to take an informed stand on many important issues without a basic understanding of life science.

A Move to Limit Antibiotic Use In Animal Feed

By DENISE GRADY

Faced with mounting evidence that the routine use of antibiotics in livestock may diminish the drugs' power to cure infections in people, the Food and Drug Administration has begun a major revision of its guidelines for approving new antibiotics for animals and for monitoring the effects of old ones.

Figure 1.8A Biology in the news

Much of biology's impact on modern society stems from its contributions to technology and medicine. Technology is the application of scientific knowledge. Many discoveries in biology have practical applications. The technology of modern birth control, for instance, grew out of an understanding of the structure and function of the human reproductive system.

Perhaps the most important application of biology to our lives today is in helping us understand and respond to the environmental problems we currently face. One of our biggest environmental challenges is the possibility of global changes in weather and climate. Rain forests, which we have featured in this chapter, have a major effect on climate. In this capacity, tropical rain forests are vital to life as far away as Siberia and Antarctica. Consider the photographs in Figure 1.8B, which contrast a tropical rain forest area that is being burned to prepare it for farming and an intact rain forest area. Every year, as human demands for wood, food, and minerals increase, vast areas of tropical rain forest are destroyed. (At least 85% of North America's rain forests have been heavily logged.) Destroying rain forests kills off untold numbers of species. It also produces large amounts of carbon dioxide (CO_2) gas. The CO_2 traps heat from sunlight and can warm the atmosphere. Many scientists contend that the destruction of rain forests at the current rate, coupled with CO_2 (and other gases) increases from sources such as industrial pollution, is raising global temperatures. Higher temperatures might melt glacial and polar ice, cause worldwide flooding, and alter the world's climates even more drastically.

Evaluating news reports on problems of this magnitude requires critical thinking and familiarity with many aspects of biology. For instance, in considering the possible effects of rain forest destruction, it is useful to know something about whole plants, cells, and molecules, as these subjects relate to photosynthesis and other kinds of energy transformation. It is also useful to know about carbon and water cycling in ecosystems, the growth patterns of the human population, and the effect of climate and soil conditions on the distribution of life on Earth.

Biology—from the molecular level to the ecosystem level—is directly connected to our everyday lives. It may also help us find solutions to the many environmental problems that confront us. Biology offers us a deeper understanding of ourselves and our planet, and a chance to more fully appreciate life in all its diversity. 

? What is technology?

It is the application of scientific knowledge.

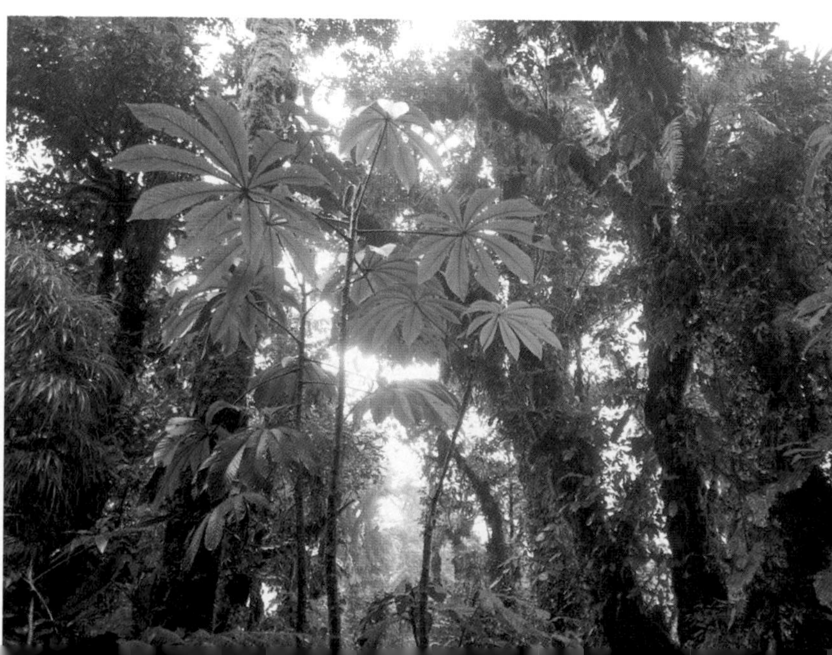

Figure 1.8B A tropical rain forest: intact (right) and being burned for farmland (above)

Chapter Summary

The Scope of Biology (Introduction–1.1) Millions of different kinds (species) of living organisms populate Earth. Many species, especially those living in rain forests, are yet to be discovered by scientists. A structural hierarchy of life, from molecules to ecosystems, defines the scope of biology, the scientific study of life. At the top of life's hierarchy, ecosystems comprise all the organisms living in a particular area, as well as the nonliving environmental components with which they interact. Below the ecosystem, the hierarchy unfolds as follows: All the living organisms in an ecosystem make up a community, a group of interbreeding individuals of a species make up a population, an individual living entity is an organism, an organism is composed of organ systems, organ systems of organs, organs of tissues, tissues of cells, and cells of molecules. A cell is a basic unit of living matter separated from its surroundings by a membrane. Molecules of DNA contain the genetic information for constructing the other molecules that make up cells and organisms.

The Process of Science (1.2–1.3) Key elements of the process of science are observations, questions, hypotheses, predictions, and tests. Observations raise questions, and scientists pose testable hypotheses as tentative answers to the questions. Scientists use deductive reasoning in testing hypotheses: *If* a hypothesis is correct, *and* we test it (by performing an experiment or making observations), *then* we can expect (predict) a particular outcome (1.2).

In generating testable hypotheses, a scientist uses previous results, observations, and analysis. In experiments designed to test hypotheses, control tests must accompany the experimental tests for the meaning of the results to be clear. Firm conclusions can be reached only after repeated tests yield consistent results. The scientific process is adaptable to conditions as different as those in a strictly controlled laboratory and in a highly varying outdoor setting. Science is cumulative, self-correcting, and dependent on the sharing of information. All scientific knowledge is considered subject to challenge (1.3).

Evolution, Unity, and Diversity (1.4–1.7) Grouping organisms by fundamental features, such as cellular structure and mode of nutrition, helps make the vast diversity of life manageable for study. A popular classification scheme groups organisms into three domains; domains Bacteria and Archaea are prokaryotes; the domain Eukarya includes protists (protozoans and algae), Kingdom Fungi (mushrooms, molds, and yeasts), Kingdom Plantae (plants), and Kingdom Animalia (animals) (1.4).

All organisms share a set of common features. For instance, all are made of cells, and all have DNA as their genetic blueprint. Such similarities are signs of unity in life's vast diversity, indicating that all organisms are variations on a common theme. DNA is made of chemical units called nucleotides, and genetic information is coded in the sequences of nucleotides making up DNA's two helically coiled chains. Each species of organism has its own nucleotide sequences. The genetic information in DNA underlies features that together distinguish life from nonlife: growth and development, the use of energy from the environment, response to environmental stimuli, and the ability to reproduce and evolve. Evolutionary change has always been a key feature of life (1.5). Charles Darwin is a central figure in biology because he synthesized the theory of evolution by natural selection. A theory in science is a comprehensive idea with broad explanatory power, in contrast to a hypothesis, which may explain only a single set of observations. For example, the theory of natural selection explains the main mechanism whereby all species of organisms change, or evolve. Natural selection is an editing mechanism. It occurs when populations of organisms having inherited variations are exposed to environmental factors that favor the reproductive success of some individuals over others. For instance, a predator (one kind of environmental factor) may selectively eat individuals whose body color makes them relatively easy to see. As a result, individuals of other colors may survive and produce greater numbers of offspring. All organisms have features, called adaptations, that have evolved by means of natural selection. Evolution is the core theme of biology, its single great explanatory concept (1.6). Focusing on the responses of organisms to their interactions with the living and nonliving components of their environment, the theory of natural selection applies at all levels in life's hierarchy. In an ecosystem, these interactions make up a complex web of relationships. Another feature of life's hierarchy is the emergence of function from structure. The functional attributes of an ecosystem, the passage of chemical nutrients and energy among its living and nonliving components, come from the structure of the ecosystem's web. Whereas chemical nutrients cycle within an ecosystem's web, energy flows in and out continually (1.7).

Connections to Everyday Life (1.8) Biology is connected to a great number of important issues, such as environmental problems and solutions, genetic engineering, and medicine. Many technological advances stem from the scientific study of life. Evaluating everyday reports in the popular press about a large range of subjects requires critical thinking and some familiarity with many areas of biology.

Testing Your Knowledge

Multiple Choice

1. Which of the following best describes the logic of the scientific process?
   a. If I generate a testable hypothesis, tests and observations will support it.
   b. If my prediction is correct, it will lead to a testable hypothesis.
   c. If my observations are accurate, they will support my hypothesis.
   d. If my hypothesis is correct, I can expect certain test results.
   e. If my tests are set up right, they will lead to a testable hypothesis.

2. Amoebas and bacteria are grouped into different domains because
   a. amoebas eat bacteria
   b. bacteria are not made of cells
   c. bacterial cells lack a nucleus
   d. bacteria decompose amoebas
   e. amoebas are photosynthetic

3. A biologist studying interactions among the protists in an ecosystem is *not* working at which level in life's hierarchy? *(Explain your answer.)*
   a. the population level
   b. the molecular level
   c. the community level
   d. the organism level
   e. the organ level

4. Which of the following questions is outside the realm of science?
   a. Which organisms play the most important role in energy input to a rain forest canopy?
   b. What percentage of music majors take a biology course?
   c. What is the physical nature of the universe?
   d. What is the nature of the supernatural?
   e. What is the historical basis for the division of Earth's human population into ethnic groups?

5. Which of the following statements best distinguishes hypotheses from theories in science?

a. Theories are hypotheses that have been proven.
b. Hypotheses are tentative guesses; theories are correct answers to questions about nature.
c. Hypotheses usually are narrow in scope; theories have broad explanatory power.
d. Hypotheses and theories are different terms for essentially the same thing in science.
e. Theories are proven to be true in all cases; hypotheses are usually falsified by tests.

6. Most of the organisms in which group are photosynthetic?

a. Animalia
b. Plantae
c. Domain Archaea
d. Protists
e. Fungi

7. Which of the following represents the correct sequence of levels in life's hierarchy, proceeding downward from an individual animal?

a. brain, spinal cord, organ system, nerve cell, nervous tissue
b. organ system, population of cells, nervous tissue, brain
c. organism, organ system, tissue, cell, organ
d. nervous system, brain, nervous tissue, nerve cell
e. organ system, tissue, molecule, cell

8. The organisms in your backyard include trees, shrubs, grass, ants, mushrooms, birds, spiders, beetles, flies, and bacteria. Together, all these organisms make up

a. an ecosystem
b. a community
c. a population
d. an experimental group
e. both a and b

9. A controlled experiment is one that

a. proceeds slowly enough that a scientist can make careful records of the results
b. may include experimental groups and control groups tested in parallel
c. is repeated many times to make sure the results are accurate
d. proceeds slowly enough that a scientist can test predictions
e. proceeds at a rate controlled by the experimenter

10. The core idea that makes sense of all we know about life is

a. the process of science
b. the idea that ecosystem function comes from its structural web
c. deductive reasoning
d. evolution
e. unity in diversity

Describing, Comparing, and Explaining

1. Explain what is meant by this statement: Natural selection is an editing mechanism rather than a creative process.

2. Explain the role of heritable variations in Darwin's theory of natural selection.

3. Explain what is meant by this statement: The scientific process is not a rigid method.

4. Compare science and religion as means of pursuing answers to questions. How do religious answers differ from those obtained by the scientific process?

5. Contrast technology with science. Give two examples to illustrate the difference.

Thinking as a Scientist

1. The scientific literature on flying squirrels states that these rodents were once common in virtually every forested area of North America, but that they are now no longer present in many areas, especially where housing developments and farms have replaced forests. Suppose you live near a wooded area and are skeptical of reports that flying squirrels are no longer found there. For one thing, you thought you saw a small moving shadow flash between two trees while you were walking by the woods last night. And recently you read an article about the work of biologist Margaret Lowman reporting her discovery that flying squirrels near Williamstown, Massachusetts, spend virtually all their time in the tops of tall oak trees. The article mentioned that, before her study, flying squirrels had been assumed to be completely absent from the area. Dr. Lowman and a student documented the presence of the squirrels by live-trapping them in the treetops. Assuming this is impractical for you, what kinds of evidence, other than actually seeing the squirrels, might you gather that would indicate their presence in your woods?

2. A newspaper headline reads, "Scientific Study Shows That Coffee Can Cut Risk of Suicide." The article states that in a 10-year study of 86,626 female nurses, there were 10 suicides among individuals who drank two to three cups of caffeinated coffee per day, compared with 21 suicides among individuals who almost never drank coffee. The researcher notes that his results were consistent with those of a previous study of 128,934 men and women done virtually the same way. The article also includes the following: (a) mention of a previous study indicating that the amount of caffeine in 2–3 cups of coffee tends to increase the drinker's general energy level, sense of well-being, energy, and motivation to work; (b) criticism by another scientist, who found the study flawed because it failed to address how many nurses used antidepressant drugs; (c) mention that the researcher recorded the use of alcohol and tobacco by the nurses but did not record the number of nurses who were depressed who were told by their physicians not to drink coffee. Do you think the conclusion in the headline is justified? Why or why not? How would you reword the headline to reflect more accurately the study as described? What key elements of the scientific process did the researcher use? Which ones were not used?

Science, Technology, and Society

1. A distinguished scientist at the Marine Biological Laboratory in Woods Hole, Massachusetts, Dr. Shinya Inoué, expressed a concern of many scientists: "I worry about science being learned as a collection of facts and theories. One needs to have a certain body of knowledge, but one also needs to understand how the knowledge is acquired—that really is at the heart of science." What have your prior experiences with science been like? What benefit do you see in nonscientists learning about the process of science?

2. The news media and popular magazines frequently report stories that are connected to biology. In the next 24 hours, record all the ones you hear or read in three different sources, and briefly describe the biological connections you perceive in each.

Media Resources

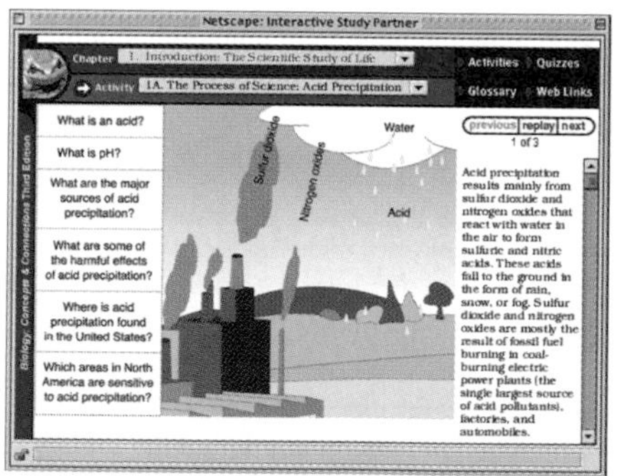

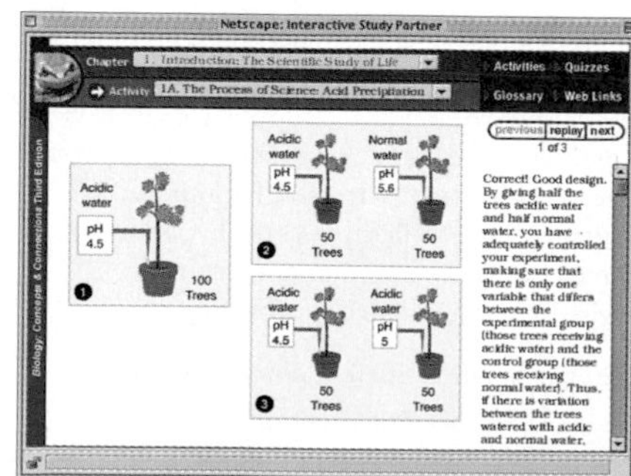

For further review of the concepts in this chapter, use the activities and quizzes on the CD-ROM.

Visit the Special Edition of The Biology Place web site for activities, quizzes, news articles, and web links.

UNIT I
The Life of the Cell

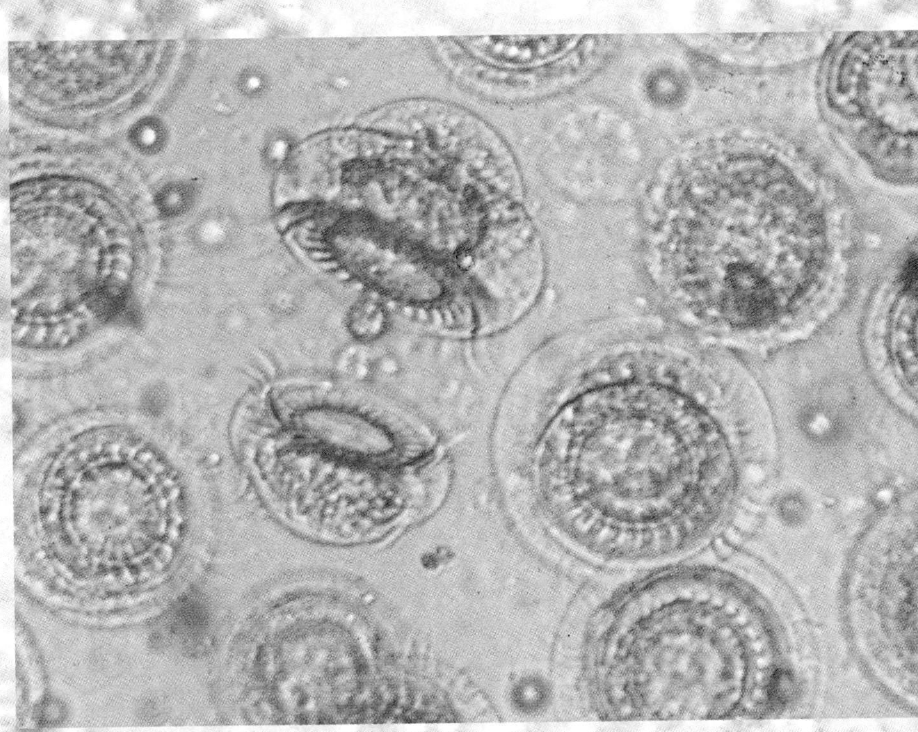

2 The Chemical Basis of Life

Outline

In the autumn, groves of quaking aspen trees turn Colorado mountainsides a rich gold. Up close you can hear the soft rustling of the leaves. It's almost constant, and in a gust of wind, whole trees seem to tremble, or quake. The trembling is an unusual property of aspen leaves, one that results from the unique combination of the parts of the leaves (Figure A). The stalk of each aspen leaf is flattened and acts like a hinge, letting the small, light leaf blade flutter back and forth in the slightest breeze. The fluttering seems to help the leaves breathe—that is, take up carbon dioxide and give off oxygen gas to the air. Plants use carbon dioxide and solar energy to make food (sugar molecules) by photosynthesis. They give off oxygen as a by-product of photosynthesis.

As we saw in Chapter 1, life exhibits unique properties at each level in its structural hierarchy. Underlying the property of fluttering aspen leaves, for instance, is a particular combination and arrangement of cells making up the stalks and leaves. Another property of aspen leaves, their brilliant gold color in autumn, arises from molecules of yellow and orange pigments, colored substances contained in cells near the leaf surface. These pigments are present in the leaf even when it is green during the summer months. The yellow and orange are unmasked when leaf cells die and lose their green pigment.

The green pigment in aspens and other plants is chlorophyll, a substance that absorbs energy from sunlight and uses it to drive photosynthesis. The computer model in Figure B shows a single molecule of chlorophyll. The

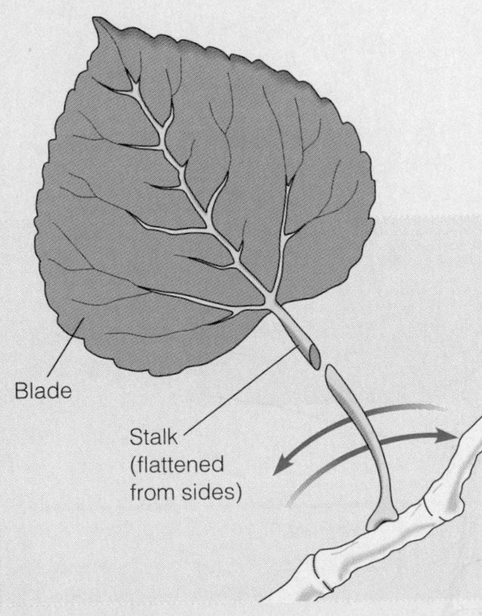

A. Aspen leaf

spheres represent atoms, the tiny units of matter that make up all molecules. The function of a molecule such as chlorophyll depends on its atoms: their kind, number, and arrangement. Ultimately, all the properties of aspens—everything the plants do, from quaking in a breeze to absorbing solar energy and manufacturing food—depends on how atoms are arranged in molecules and how molecules are arranged in the plant's leaves, stems, and roots.

Much of what we know about the functions of plants and other organisms comes from what is called a reductionist approach—a scientific approach based on the idea that the whole can best be understood by studying its parts. In fact, only by taking plants apart and studying the structure and function of their cells and molecules have biologists been able to determine how plants trap and use sunlight, take up carbon dioxide from the air, and absorb water from the soil. Nevertheless, a full understanding of any organism requires that the reductionist approach be combined with the study of the functioning whole. An aspen tree clearly has properties beyond those of its individual leaves,

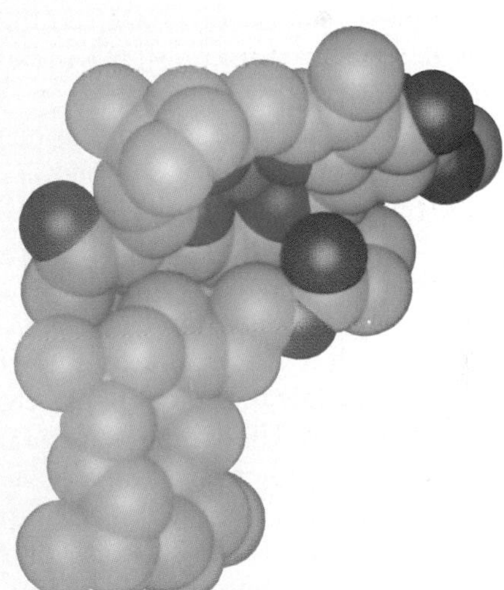
B. Chlorophyll molecule

roots, trunk, and molecules. And at the chemical level, an intact chlorophyll molecule has properties lacking in an unorganized pile of its atoms, such as the property of absorbing light that can be used for photosynthesis.

In this unit of chapters, we take a reductionist approach and focus on the chemical components of life. Keep in mind, however, that the properties of life emerge from the organization of these chemical parts into cells and organisms that we will study throughout the book.

Atoms and Molecules

2.1 Biological function starts at the chemical level

Taking any biological system apart, we eventually end up at the chemical level. Let's now *begin* at the chemical level, with the chlorophyll molecule, and travel to the level of the plant leaf. Throughout this hierarchy, you will see the relationship between structure and the function of photosynthesis, and the emergence of new properties.

Figure 2.1A is a chemical diagram of chlorophyll. It shows every atom making up the molecule. There are five kinds of atoms in chlorophyll: C = a carbon atom, H = hydrogen, O = oxygen, N = nitrogen, and Mg = magnesium. Both this diagram and the computer model on the previous page give information about the specific arrangement of atoms in the chlorophyll molecule that enables it to absorb light.

Chlorophyll by itself performs only the very first step in photosynthesis. The whole process requires an array of many molecules working together with chlorophyll. These molecules are all housed in chloroplasts, tiny structures inside plant cells. Figure 2.1B is a micrograph—a photograph taken with a microscope—of a chloroplast. The ability of a chloroplast to carry out photosynthesis results from the specific arrangement of all the molecules in the chloroplast.

Illustrating the next level in this hierarchy of properties, the micrograph in Figure 2.1C shows about 10 cells from a plant leaf. Each of the multisided structures is a plant cell, and within the cells you can see the chloroplasts, which look like small spheres. Chloroplasts and other structures that perform specific functions inside cells are called organelles. Each of these plant cells functions as a living unit because of cooperation between its chloroplasts and many other organelles. There is also cooperation among the cells themselves. Together, they function as a tissue, a group of similar cells performing a specific function. The main function of this particular tissue is photosynthesis.

Finally, we come to the leaves (Figure 2.1D), which are plant organs. A leaf is composed of several different tissues, including photosynthetic tissue, epidermal tissue forming the leaf's protective covering, and vascular tissue. Among other functions, vascular tissue transports water needed for photosynthesis from the roots to the leaf. The leaf cannot carry out photosynthesis for long without help from the roots and stems, which are also organs. All the organs work together as parts of the whole plant. We could extend this example to the level of the whole organism (the aspen tree) and beyond. In fact, we could continue all the way to the involvement of photosynthesis at the ecosystem level, where, for example, beavers and many other animals use the aspens for food.

We will now look more closely at life's most fundamental level of organization, beginning with the chemical elements.

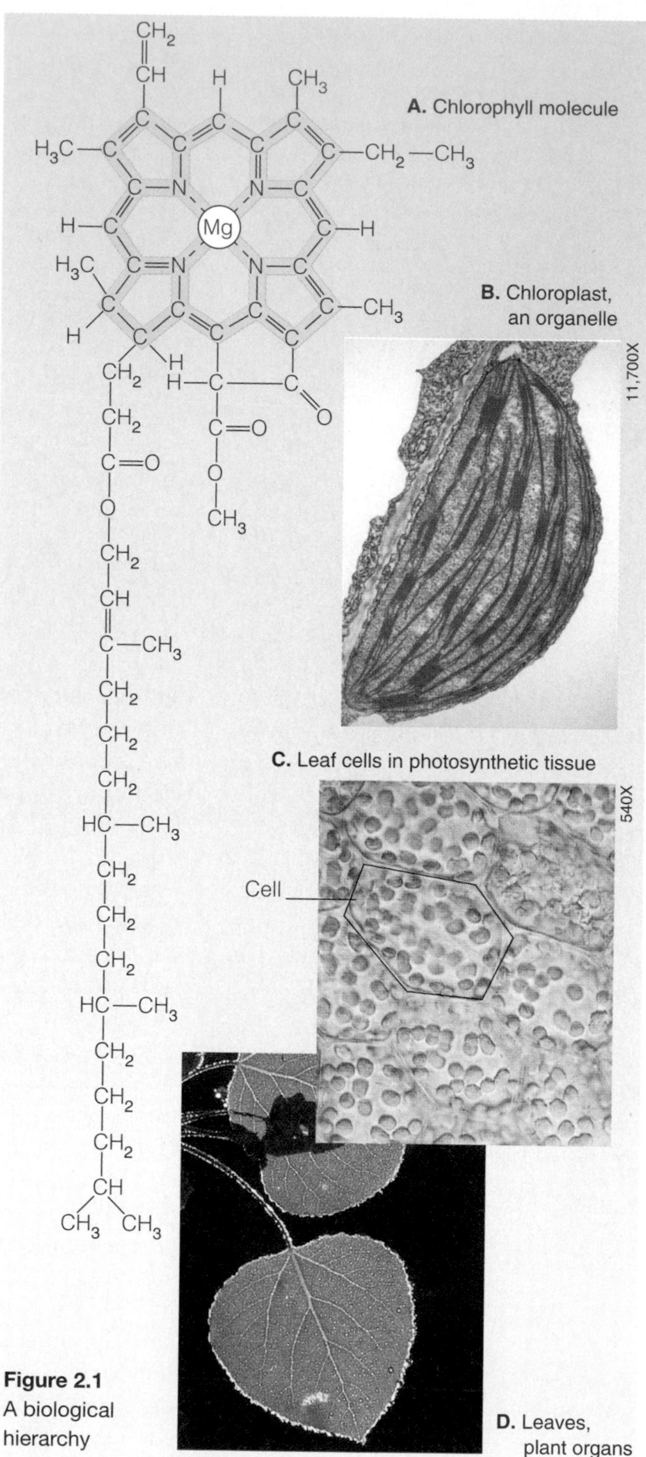

Figure 2.1
A biological hierarchy

Why can't an illuminated sample of pure chlorophyll perform photosynthesis?

Photosynthesis depends on a specific arrangement of chlorophyll and other molecules within the chloroplast.

2.2 Life requires about 25 chemical elements

Life is composed of **matter,** which is anything that occupies space and has mass. Matter, in forms as diverse as rock, wood, metal, and air, is composed of chemical elements. A **chemical element** is a substance that cannot be broken down to other substances by ordinary chemical means. Today, chemists recognize 92 elements occurring in nature; gold, copper, carbon, and oxygen are some examples. About a dozen more elements have been made in the laboratory. Each element has a symbol, the first letter or two of its English, Latin, or German name. For instance, the symbol for gold, Au, is from the Latin word *aurum;* the symbol O stands for the English word *oxygen.*

About 25 of the 92 natural elements are essential to life. As you can see in the table below, four of these—oxygen (O), carbon (C), hydrogen (H), and nitrogen (N)—make up about 96% of the human body, which is typical of living matter. Calcium (Ca), phosphorus (P), potassium (K), sulfur (S), and a few other elements account for most of the remaining 4%. The **trace elements** listed at the bottom of the table are essential to at least some organisms, but only in minute quantities. Some trace elements, such as iron, are needed by all forms of life. Others are required only by certain species. The average human, for example, needs about 0.15 milligram (mg) of the trace element iodine each day. A deficiency of iodine prevents normal functioning of the thyroid gland and results in an abnormal enlargement called a goiter. Too much iodine can also cause a goiter.

Naturally Occurring Elements in the Human Body

Symbol	Element	Wet Weight Percentage*	
O	Oxygen	65.0	96.3
C	Carbon	18.5	
H	Hydrogen	9.5	
N	Nitrogen	3.3	
Ca	Calcium	1.5	
P	Phosphorus	1.0	
K	Potassium	0.4	
S	Sulfur	0.3	
Na	Sodium	0.2	
Cl	Chlorine	0.2	
Mg	Magnesium	0.1	

Trace elements (less than 0.01%): boron (B), chromium (Cr), cobalt (Co), copper (Cu), fluorine (F), iodine (I), iron (Fe), manganese (Mn), molybdenum (Mo), selenium (Se), silicon (Si), tin (Sn), vanadium (V), and zinc (Zn).

*Includes water.

Figure 2.2 Magnesium requirement by a plant

Figure 2.2 above demonstrates the importance of one essential element in plants. The plant on the left is a normal, well-nourished tobacco plant. The other is a tobacco plant grown in soil deficient in magnesium (Mg). Forming the core of the chlorophyll molecule (see Figure 2.1A), magnesium is required for photosynthesis. A plant that does not obtain an adequate supply of this essential element cannot produce enough food to survive in nature.

? What four chemical elements are most abundant in living matter?

Carbon, oxygen, hydrogen, and nitrogen

2.3 Elements can combine to form compounds

Chlorophyll is a **compound,** a substance containing two or more elements in a fixed ratio. Compounds are much more common than pure elements. In fact, few elements exist in a pure state in nature. Many compounds consist of only two elements—for instance, table salt (sodium chloride, NaCl) has equal parts of the elements sodium (Na) and chlorine (Cl). In contrast, most of the compounds in living organisms contain at least three or four different elements, mainly carbon, hydrogen, oxygen, and nitrogen. Proteins, for example, are all formed mainly of these four elements, as is chlorophyll. Different arrangements of the atoms determine unique properties for each compound.

? Why are chlorophyll and proteins both classified as compounds?

They both consist of multiple elements, mainly carbon, oxygen, hydrogen, and nitrogen.

2.4 Atoms consist of protons, neutrons, and electrons

2A
See Activity 2A on the CD-ROM.

Each element consists of one kind of atom, which is different from the atoms of other elements. An **atom,** named from a Greek word meaning indivisible, is the smallest unit of matter that still retains the properties of an element. It would take about a million atoms to stretch across the period printed at the end of this sentence.

Physicists have split the atom into more than a hundred types of subatomic particles. However, to understand basic atomic structure, we only have to consider three. A **proton** is a subatomic particle with a single positive electrical charge (+). An **electron** is a subatomic particle with a single negative electrical charge (−). The positive charge of a proton is equal in magnitude to the negative charge of an electron. Opposite charges (+ and −) attract each other. A third type of subatomic particle, the **neutron,** is electrically neutral (has no electrical charge).

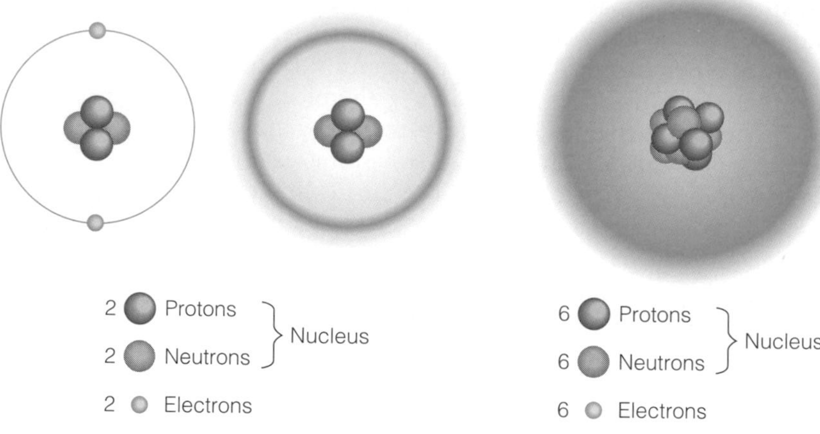

Figure 2.4A Helium atom

Figure 2.4B Carbon atom

Figure 2.4A shows two models of an atom of the element helium (He). (A chemical symbol such as He stands for an atom of an element as well as for the element in general.) Both models in Figure 2.4A are highly diagrammatic, and we use them only to show the basic properties of atoms. Notice that two neutrons and two protons are tightly packed in the atom's central core, or **nucleus.** Two electrons, much smaller than the nuclear particles, orbit the nucleus at nearly the speed of light. The attraction between the negatively charged electrons and the positively charged protons keeps the electrons near the nucleus. The left-hand model shows the number of electrons in the atom. The right-hand model, slightly more realistic, shows the electrons as a spherical cloud of negative charge surrounding the nucleus. Neither model is drawn to scale. In real atoms, the electron cloud is much bigger compared to the nucleus.

Elements differ in the number of subatomic particles in their atoms. All atoms of a particular element have the same unique number of protons. This is the element's **atomic number.** Thus, an atom of helium, with 2 protons, has an atomic number of 2. Carbon, with 6 protons, has an atomic number of 6 (Figure 2.4B). Note that in these atoms, the atomic number is also the number of electrons. When an atom has an equal number of protons and electrons, its net electrical charge is 0 (zero).

In contrast to its atomic number, an atom's **mass number** is the sum of the numbers of protons and neutrons in its nucleus. For helium, the mass number is 4; for the carbon atom shown here, it is 12. Mass is a measure of the amount of matter in an object. A proton and a neutron are almost identical in mass, and for convenience physicists use the mass of a single proton or neutron as a unit of mass. An electron has very little mass—only about 1/2000 the mass of a proton. So an atom's mass is essentially equal to its mass number. An element's atomic mass is commonly referred to as its **atomic weight,** and this is given as a whole number—for example, 4 for helium.

Some elements have variant forms called **isotopes.** The different isotopes of an element have the same numbers of protons and electrons but different numbers of neutrons. The table below shows the numbers of subatomic particles in the three isotopes of carbon. Carbon-12 (usually written ^{12}C), with 6 neutrons, makes up about 99% of all naturally occurring carbon. Most of the other 1% consists of ^{13}C, with 7 neutrons. A third isotope, ^{14}C, with 8 neutrons, occurs in minute quantities. Notice that all three isotopes have 6 protons—otherwise, they would not be carbon. Both ^{12}C and ^{13}C are stable isotopes, meaning their nuclei remain permanently intact. The isotope ^{14}C, on the other hand, is unstable, or radioactive. A **radioactive isotope** is one in which the nucleus decays spontaneously, giving off particles and energy. Radioactive isotopes pose serious risks to living organisms, but they also have many uses in biological research and medicine, as we see next.

Isotopes of Carbon

	Carbon-12	Carbon-13	Carbon-14
Protons	6	6	6
Neutrons	6	7	8
Electrons	6	6	6

A nitrogen atom has 7 protons, and the most common isotope of nitrogen has 7 neutrons. A radioactive isotope of nitrogen has 8 neutrons. What is the atomic number and mass number of this radioactive nitrogen?

Atomic number = 7; mass number = 15

2.5 Radioactive isotopes can help or harm us

Living cells cannot distinguish radioactive isotopes from nonradioactive atoms of the same element. Consequently, organisms take up and use compounds containing radioactive isotopes in the usual way. Because radioactivity is easily detected, radioactive isotopes are useful as tracers—biological spies, in effect—for monitoring the fate of atoms in living organisms. To detect radioactivity, scientists use photographic film or instruments such as Geiger counters.

Biologists often use radioactive tracers to follow molecules as they undergo chemical changes in an organism. For example, plant researchers have used them to study photosynthesis. Knowing that plants take in carbon dioxide (CO_2) from the atmosphere and use it to make sugar molecules, researchers used the radioactive isotope ^{14}C to find out what particular molecules plants make in the process. When researchers provided plants with $^{14}CO_2$, the radioactive carbon appeared first in a compound called 3-PGA and then in several other compounds before showing up in the sugar glucose. These results showed the route by which plants make glucose from CO_2.

Medical diagnosis also makes use of radioactive tracers. Certain kidney disorders, for example, can be diagnosed by injecting tiny amounts of radioactive isotopes into a patient's blood and then measuring the amount of radioactive tracer passed in the urine. In this and most other diagnostic uses of radioactive tracers, the patient receives only a tiny amount of an isotope that decays completely in minutes or hours.

In recent years, radioactive tracers have been used for diagnosis in combination with sophisticated imaging instruments. Figure 2.5A shows a patient being examined by a powerful technique called PET (positron-emission tomography), which can monitor chemical processes as they actually occur in the body. The patient is first injected with an isotope that emits subatomic particles called positrons. A PET scanner then detects energy released by the positrons as the isotope moves through the body. Linked to the scanner is a computer that translates the energy data into an anatomical image. The color of the image varies with the amount of the isotope present in an area.

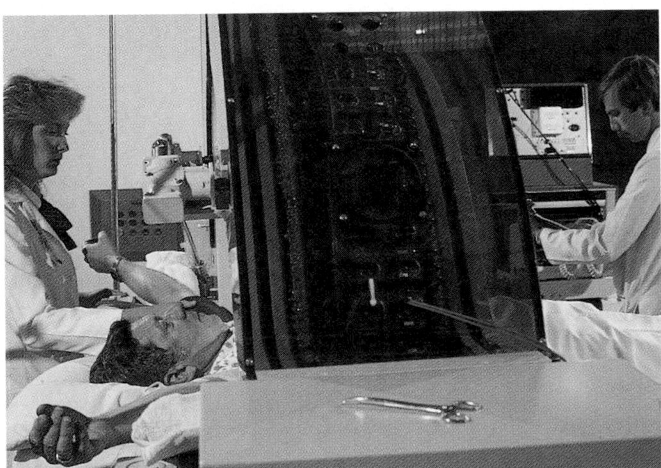

Figure 2.5A PET scanner

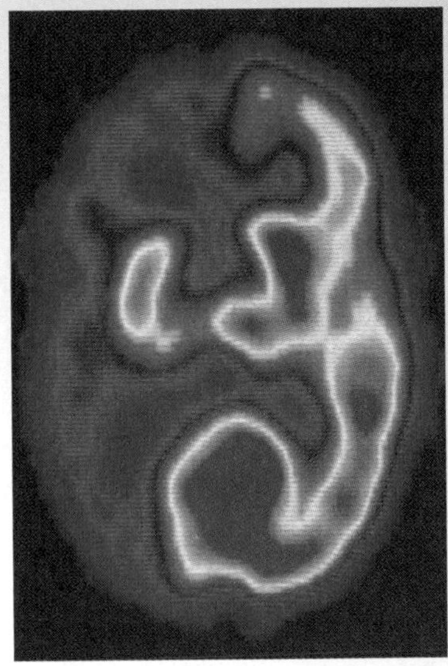

Figure 2.5B PET image of a human brain

Figure 2.5B shows a PET image of the brain (top view) of a 2-year-old boy who had been suffering from seizures. The image shows normal activity in the brain's right half (red, yellow, and green colors) but very little activity in the left half (blue). With this information, surgeons removed the part of the left half that was producing seizures and saved the boy's life. PET is also useful for diagnosing certain heart disorders and cancers.

Though radioactive isotopes have many beneficial uses, uncontrolled exposure to them can harm living organisms by damaging molecules, especially DNA, in cells. The particles and energy thrown off by radioactive atoms can break apart the atoms of molecules and may cause abnormal connections to form between atoms. The severity of the damage depends on the type and amount of radiation an organism absorbs. The explosion of a nuclear reactor at Chernobyl, Ukraine, in 1986 released large amounts of radioactive isotopes into the environment, killing 30 people within a few weeks. The survivors have suffered increased rates of thyroid cancer and of birth defects in their children, and thousands may develop cancers in the future. Many survivors of the atomic bombs exploded over Japan in World War II are now, more than 50 years later, fighting radiation-induced cancers.

Natural sources of radiation can also pose a threat. Radon, a radioactive gas, may be a cause of lung cancer. Radon is a common household contaminant in regions underlain by uranium-bearing rocks, and houses sealed up to prevent heat loss in winter are the ones most likely to contain potentially hazardous levels of radon. Homeowners can buy a radon detector or hire a company to test their homes (state departments of environmental protection will provide references), in order to ensure that radon concentrations are at safe levels. 

? Why are radioactive isotopes useful as tracers in research on the chemistry of life?

Organisms incorporate radioactive isotopes of an element into their molecules just as they do the nonradioactive isotopes, and researchers can detect the presence of the radioactive isotopes.

2.6 Electron arrangement determines the chemical properties of an atom

2B
2C

Of the subatomic particles we have discussed, it is mainly electrons that determine how an atom behaves when it encounters other atoms. Electrons vary in the amount of energy they possess. The farther an electron is from the nucleus, the greater its energy. Electrons in an atom occur only at certain energy levels, called **electron shells.** Depending on their atomic number, atoms may have one, two, or more electron shells, with electrons in the outermost shell having the highest energy. Each shell can accommodate up to a specific number of electrons. The innermost shell is full with only 2 electrons. So in atoms with more than 2 electrons, the remainder are found in shells farther from the nucleus. The outermost (highest-energy) shell can hold up to 8 electrons. It is the number of electrons in the outermost shell that determines the chemical properties of an atom. Atoms whose outer shells are not full tend to interact with other atoms—that is, to participate in chemical reactions.

Figure 2.6 shows the electron shells of four biologically important elements. Small dashed circles represent the unfilled "spaces" in the outer electron shells. Because the outer shells of all four atoms are incomplete, all these atoms react readily with other atoms. The hydrogen atom is highly reactive because it has only 1 electron in its single electron shell, which can accommodate 2 electrons. Atoms of carbon, nitrogen, and oxygen are also highly reactive because their outer shells, which can hold 8 electrons, are incomplete. In contrast, the helium atom in Figure 2.4A has a single, first-level shell that is full with 2 electrons. As a result, helium is chemically inert (unreactive).

How does a chemical reaction enable an atom to fill its outer electron shell? When two atoms with incomplete outer shells react, each atom gives up or acquires electrons so that both partners end up with completed outer shells. Atoms do this by either sharing or transferring outer electrons. These interactions usually result in atoms staying close together, held by attractions called **chemical bonds.** In the next two modules, we look at the two strongest types of chemical bonds.

Sodium has an atomic number of 11. How many electron shells does a sodium atom have, and how many electrons are in the outermost shell?

Three electron shells; one electron in the outermost shell

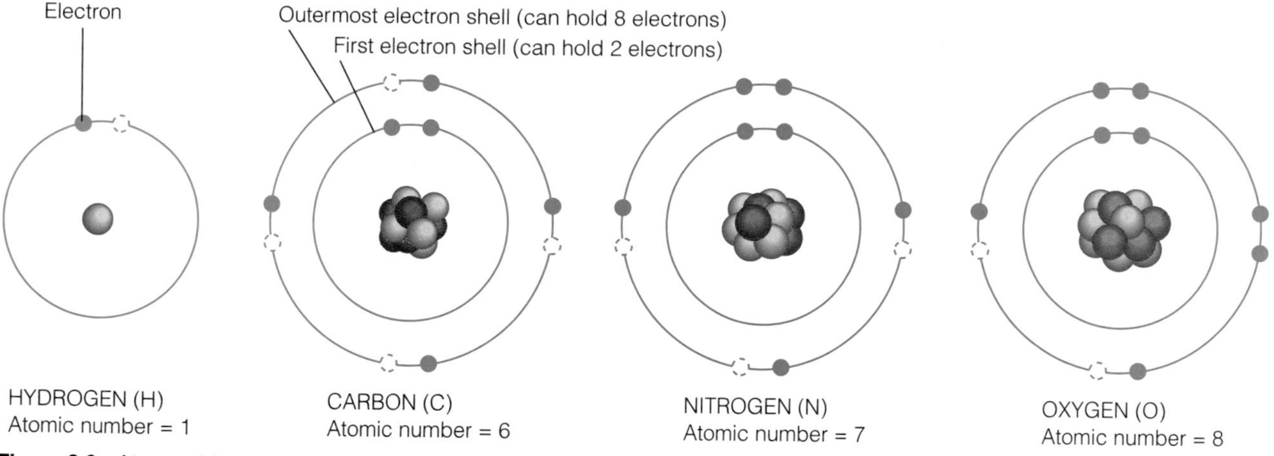

Figure 2.6 Atoms of four elements that are important for life

2.7 Ionic bonds are attractions between ions of opposite charge

2D

At the top of the next page, Figure 2.7A shows how a sodium atom and a chlorine atom form the compound sodium chloride (NaCl). Notice that sodium has only 1 electron in its outer shell, whereas chlorine has 7. When these atoms collide, the chlorine atom strips sodium's outer electron away. In doing so, chlorine fills its outer shell with 8 electrons. Sodium, in losing 1 electron, ends up with only two shells, the outer shell having a full set of 8 electrons.

Since electrons are negatively charged particles, the electron transfer between the two atoms moves one unit of negative charge from sodium to chlorine. Sodium, with 11 protons but now only 10 electrons, has acquired a net electrical charge of $+1$. Chlorine, having gained an extra electron, now has a net electrical charge of -1. In each case, an atom has become what is called an ion. An **ion** is an atom or a molecule with an electrical charge resulting from a gain or loss of one or more electrons. As you can see in Figure 2.7A, the ion formed from chlorine is called a chloride ion. Two ions with opposite charges attract each other; when the attraction holds them together, it is called an **ionic bond.** The resulting compound, in this case NaCl, is electrically neutral.

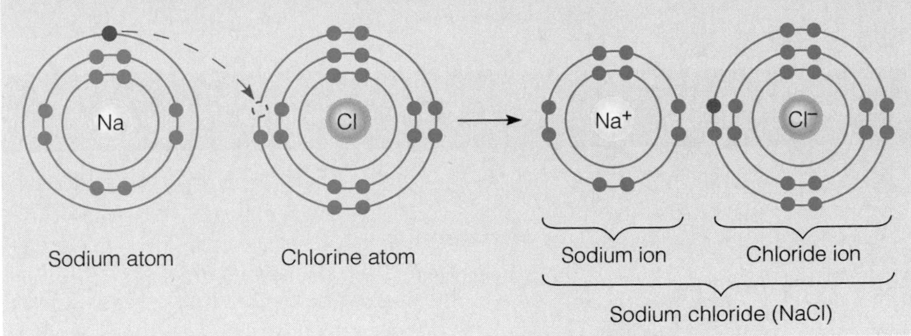

Figure 2.7A Formation of an ionic bond, producing sodium chloride

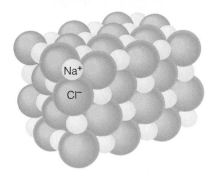

Figure 2.7B A crystal of sodium chloride

Sodium chloride is a type of salt. As a group, salts are ionic compounds that often exist as crystals in nature. Figure 2.7B shows the atoms in a piece of crystalline sodium chloride. A crystal of NaCl can be of any size (there is no fixed number of ions), but sodium and chloride ions are always present in a 1:1 ratio. The ratio may be different in crystals of other salts.

Explain what holds together the atoms in a crystal of table salt (NaCl).

Opposite charges attract. The positively charged sodium ions (Na^+) and the negatively charged chloride ions (Cl^-) are held together by ionic bonds, the attraction between oppositely charged ions.

2.8 Covalent bonds, the sharing of electrons, join atoms into molecules

2E

The second kind of strong chemical bond is the **covalent bond,** in which two atoms share one or more pairs of outer-shell electrons. Two or more atoms held together by covalent bonds form a **molecule.** For example, a covalent bond connects the two hydrogen atoms in the molecule H_2, a common gas in the atmosphere. The figure at the right shows three ways to represent this molecule. The symbol H_2, called the molecular formula, merely tells you that the molecule consists of two atoms of hydrogen. The middle diagram shows that the atoms share 2 electrons, and in doing so, both fill their outer (only) shells. At the right, you see a structural formula. The line between the hydrogen atoms stands for the single covalent bond formed by the sharing of one pair of electrons.

The number of single covalent bonds an atom can form is equal to the number of additional electrons needed to fill its outer shell. Looking back at Figure 2.6, we can see that H can form one covalent bond; O can form two; N, three; and C, four. However, as you can see in the figure here, in an O_2 molecule, each O atom does not form two single bonds. Instead, the two O atoms share two pairs of electrons, forming a **double bond.** In the structural formula, the double bond is indicated by a pair of lines between the O atoms.

H_2 and O_2 are molecules, but because they are composed of only one element, they are not compounds. An example of a molecule that is a compound is methane (CH_4), a common gas produced by certain bacteria. As you can see in the figure, each of the four hydrogen atoms in this molecule shares one pair of electrons with the single carbon atom. The same type of bonding occurs in molecules of water, a compound so important to life that we devote the next six modules to it.

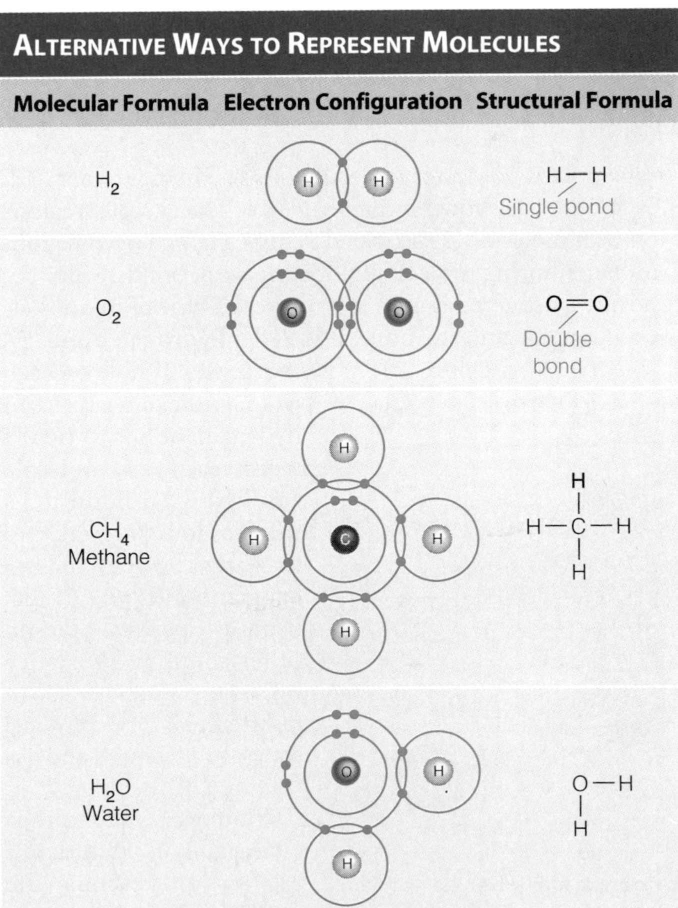

Alternative Ways to Represent Molecules

Molecular Formula	Electron Configuration	Structural Formula
H_2		H—H (Single bond)
O_2		O═O (Double bond)
CH_4 Methane		H—C—H with H above and below C
H_2O Water		O—H with H below O

What is chemically nonsensical about this structure?

H—C═C—H

Each carbon atom has only 3 covalent bonds instead of the required 4.

2.9 Water is a polar molecule

2F

A water molecule (H_2O), as shown in Figure 2.9, consists of two hydrogen atoms covalently bonded to a single oxygen atom. Atoms in a covalently bonded molecule are in a constant tug-of-war for the electrons of their covalent bonds. An atom's attraction for the shared electrons of the bond is called its **electronegativity.** The more electronegative an atom, the more strongly it pulls shared electrons toward itself. In molecules formed of only one element, such as O_2 and H_2, the two identical atoms exert an equal pull on the electrons. The covalent bonds in such molecules are said to be **nonpolar** because the electrons are shared equally between the atoms. Some compounds—for instance, methane (CH_4)—also have nonpolar bonds. The atoms in such compounds are not substantially different in electronegativity.

In contrast to O_2, H_2, and CH_4, water is composed of atoms with very different electronegativities. Oxygen is one of the most electronegative of the elements. As indicated by the arrows in the figure, O attracts the shared electrons in H_2O much more strongly than does H, so that the shared electrons are actually closer to the O atom than to the H atoms. This unequal sharing of electrons produces what is called a polar covalent bond between the oxygen atom and each of the hydrogen atoms in the molecule. A **polar covalent bond** is a chemical bond in which shared electrons are pulled closer to the more electronegative atom, making it partially negative and the other atom partially positive. Thus, in H_2O, the O atom actually has a slight negative charge and each H atom a slight positive charge, even though H_2O as a whole is neutral. Because of its polar covalent bonds, water is a **polar molecule**—that is, it has a slightly negative pole and two slightly positive ones.

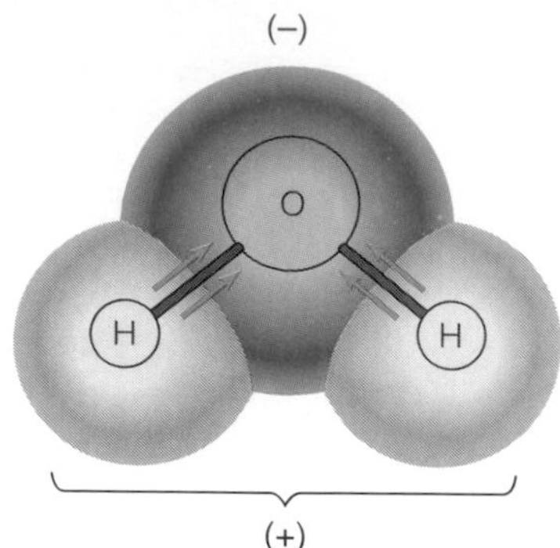

Figure 2.9 A water molecule

Why is it unlikely that two neighboring water molecules would be arranged like this?

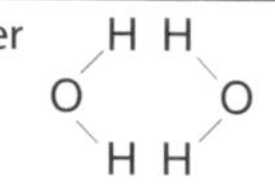

The hydrogen atoms of one molecule, with their partial positive charge, would repel the hydrogen atoms of the adjacent molecule.

2.10 Overview: Water's polarity leads to hydrogen bonding and other unusual properties

2G

The polarity of water molecules makes them interact with each other, as shown in Figure 2.10A. The charged regions on each molecule are attracted to oppositely charged regions on neighboring molecules, forming weak bonds. Since the positively charged region in this special type of bond is always an H atom, the bond is called a **hydrogen bond.** As you can see in Figure 2.10A, the negative (O) pole on each water molecule can form hydrogen bonds to two H atoms; each H_2O molecule can hydrogen-bond to as many as four partners.

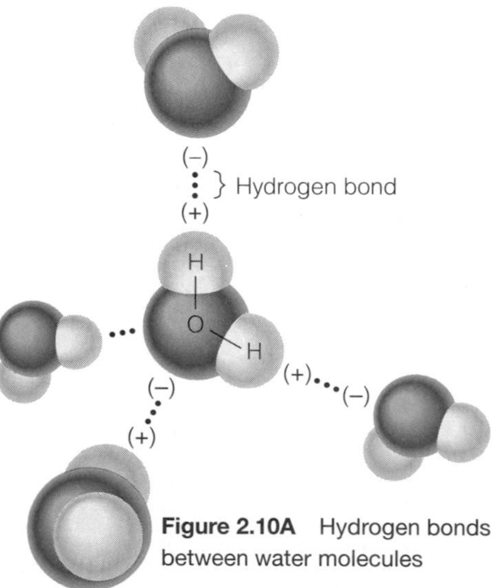

Figure 2.10A Hydrogen bonds between water molecules

Water's polarity and hydrogen bonds give it unusual properties. Like no other common substance on Earth, water exists in nature in all three physical states: solid, liquid, and gas (Figure 2.10B).

Moreover, water's properties and its abundance are major reasons life thrives on Earth.

What enables neighboring water molecules to hydrogen-bond to one another?

The molecules are polar, with the negative end (oxygen end) of one molecule attracted to the positive end (hydrogen end) of its neighbor.

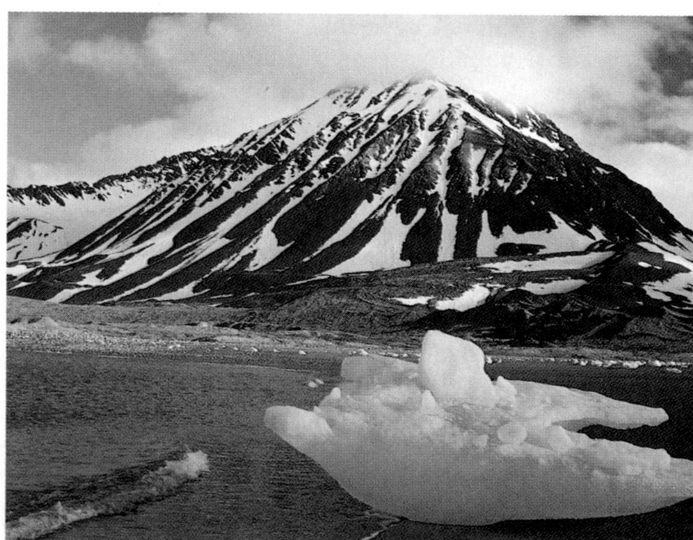
Figure 2.10B The three states of water: ice, liquid water, and vapor

2.11 Hydrogen bonds make liquid water cohesive

2H

Hydrogen bonds between molecules of liquid water last for only a few trillionths of a second, yet at any instant, most of the molecules are hydrogen-bonded to others. This tendency of molecules to stick together, called **cohesion,** is much stronger for water than for most other liquids. The cohesion of water is important in the living world. Trees, for example, depend on cohesion to help transport water from their roots to their leaves. The evaporation of water from a leaf exerts a pulling force on water within the veins of the leaf. Because of cohesion, the force is relayed through the veins all the way down to the roots. As a result, water rises against the force of gravity.

Figure 2.11 Surface tension of water exploited by a water strider

Related to cohesion is **surface tension,** a measure of how difficult it is to stretch or break the surface of a liquid. Hydrogen bonds give water unusually high surface tension, making it behave as though it were coated with an invisible film. The insect in Figure 2.11, called a water strider, takes advantage of the high surface tension of water. The insect is denser than water, yet it walks on ponds without breaking the surface. 

? In a tall tree, water in thin tubes within the trunk is pulled upward by evaporation from the leaves. What keeps the water molecules at the bottom of the tree moving?

Hydrogen bonds hold neighboring water molecules together in the liquid water; this cohesion helps the molecules move against the downward pull of gravity.

2.12 Water's hydrogen bonds moderate temperature

If you have ever burned your finger on a metal pot while waiting for the water in it to boil, you know that water heats up much more slowly than metal. In fact, because of hydrogen bonding, water has a better ability to resist temperature change than most other substances. Because of this property, Earth's giant water supply moderates temperatures, keeping them within limits that permit life.

Temperature and heat are related, but different. A swimmer crossing San Francisco Bay has a higher temperature than the water, but the bay contains far more heat because of its immense volume. **Heat** is the amount of energy associated with the movement of the atoms and molecules in a body of matter. **Temperature** measures the intensity of heat—that is, the *average* speed of molecules rather than the *total* amount of heat energy in a body of matter.

When water is heated, the heat energy first disrupts hydrogen bonds and then makes water molecules move faster. Because heat is absorbed as the bonds break, water absorbs and stores a large amount of heat while warming up only a few degrees. Conversely, when water is cooled, more hydrogen bonds form. Heat energy is released when the bonds form, slowing the cooling process.

A large body of water can store a huge amount of heat from the sun during warm periods. At cooler times, heat given off from the gradually cooling water can warm the air. That's why coastal areas generally have milder climates than inland regions. Water's resistance to temperature change also stabilizes ocean temperatures, creating a favorable environment for marine life.

Hydrogen bonds also decrease water's tendency to evaporate, or vaporize. Liquids vaporize when some of their molecules move fast enough to overcome the attractions that keep the molecules close together. Heating a liquid increases vaporization by increasing the energy of the molecules. Water must absorb an unusually large amount of heat in order to vaporize because its hydrogen bonds tend to hold the molecules in place. Thus, hydrogen bonds slow vaporization and give water a high boiling point (100°C). Liquid water is abundant on Earth largely because temperatures near its boiling point are rare on the planet's surface.

Another way water moderates temperatures is by evaporative cooling. When a substance evaporates, the surface of the liquid remaining behind cools down. This occurs because the molecules with the greatest energy (the "hottest" ones) tend to vaporize first. It's as if the 100 best runners at your college left school, lowering the average speed of the remaining students. Evaporative cooling prevents some land-dwelling organisms from overheating; sweating performs that function, for example (Figure 2.12). On a much larger scale, surface evaporation cools tropical oceans. 

Figure 2.12 Evaporative cooling

? Explain the popular adage "It's not the heat, it's the humidity."

High humidity hampers cooling by resisting the evaporation of sweat.

2.13 Ice is less dense than liquid water

In contrast to the hydrogen bonds in liquid water, the hydrogen bonds in ice are stable, with each molecule bonded to four neighbors, forming a three-dimensional crystal. In Figure 2.13, compare the spaciously arranged molecules in the ice crystal to the tightly packed molecules in the liquid water. The ice crystal has fewer molecules than an equal volume of liquid water. In other words, ice is less dense than liquid water. Therefore ice floats, rather than sinking to the bottom of a pond or lake. Being less dense as a solid than as a liquid is one of water's unusual properties that help maintain life as we know it. If ice sank, it would seldom have a chance to thaw, and eventually all ponds, lakes, and even oceans would freeze solid. 

? Explain how the freezing of water can crack boulders.

Water expands as it freezes, because the water molecules become spaced farther apart in forming ice crystals. When there is water in a crevice of a boulder, expansion of the water due to freezing may crack the rock.

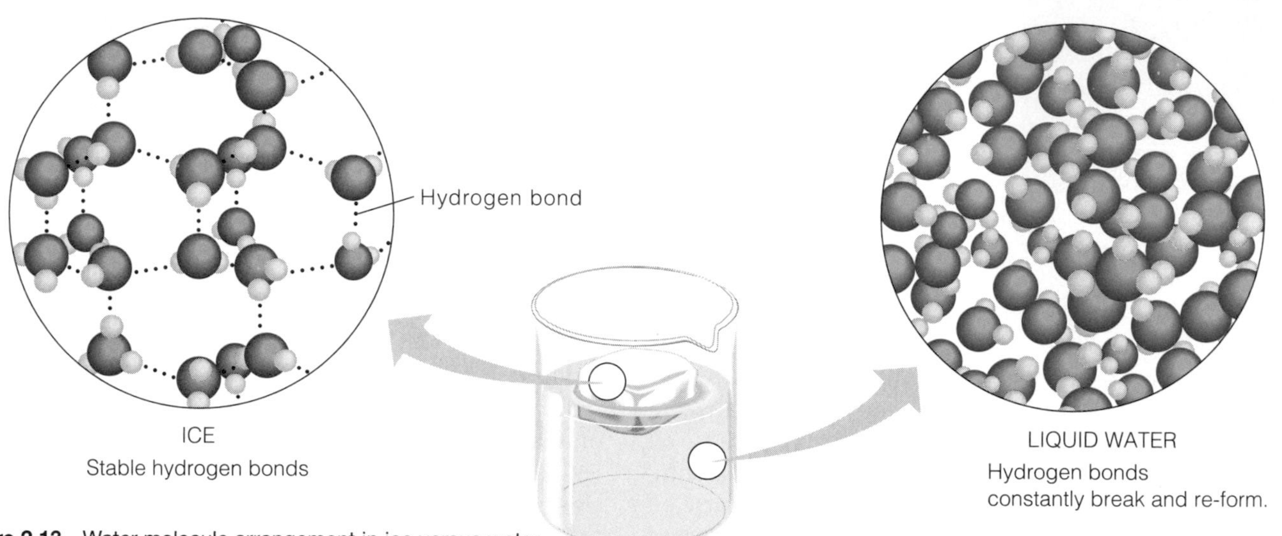

Figure 2.13 Water molecule arrangement in ice versus water

2.14 Water is a versatile solvent

A **solution** is a liquid consisting of a homogeneous mixture of two or more substances. The dissolving agent is called the **solvent,** and a substance that is dissolved is a **solute.** When water is the solvent, the result is called an aqueous solution (Latin *aqua,* water). As the solvent inside all cells, in blood, and in plant sap, water dissolves an enormous variety of solutes necessary for life.

Water's versatility as a solvent results from the polarity of its molecules. Figure 2.14 shows how a crystal of table salt dissolves in water. The sodium and chloride ions at the surface of the crystal have affinities for different parts of the water molecules. The positive Na^+ ions (yellow) attract the electronegative regions (oxygen, red) of the water molecules. The negative Cl^- ions (green) attract the positively charged hydrogen regions (gray). As a result, H_2O molecules surround and separate individual Na^+ and Cl^- ions, dissolving the crystal in the process. 

? Why are blood and most other biological fluids classified as aqueous solutions?

Because the solvent is water.

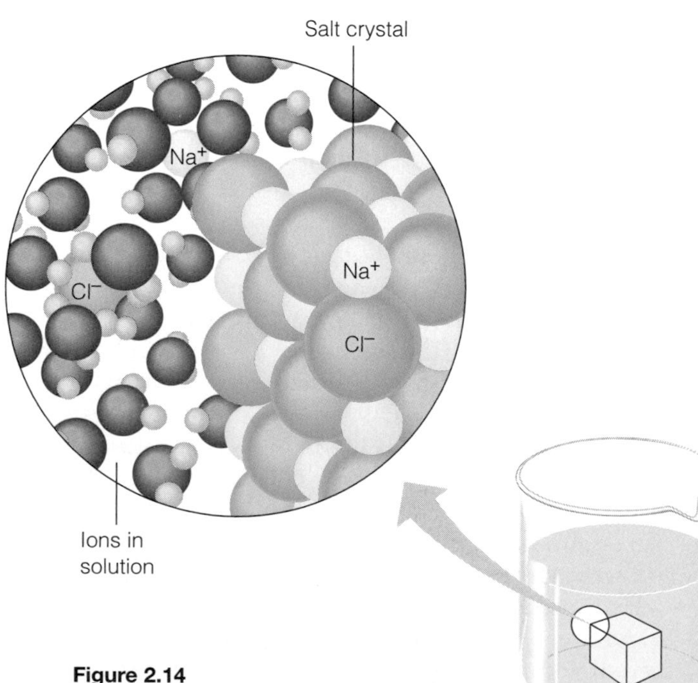

Figure 2.14
A crystal of salt (NaCl) dissolving in water

The chemistry of life is sensitive to acidic and basic conditions

In the aqueous solutions within organisms, most of the water molecules are intact. However, some of the water molecules actually break apart (dissociate) into ions. The ions formed are called hydrogen ions (H^+) and hydroxide ions (OH^-). For the proper functioning of chemical processes within organisms, the right balance of H^+ ions and negatively charged ions, such as OH^-, is critical.

A chemical compound that donates H^+ ions to solutions is called an **acid.** One example of a strong acid is hydrochloric acid (HCl), the acid in your stomach. In solution, HCl dissociates completely into H^+ and Cl^- ions. The more acidic a solution, the higher its concentration of H^+ ions.

A **base** (or alkali) is a compound that accepts H^+ ions and removes them from solution. Some bases, such as sodium hydroxide (NaOH), do this by donating OH^- ions; these combine with H^+ to form H_2O. The more basic (alkaline) a solution, the lower its H^+ concentration (and, for bases like NaOH, the higher its OH^- concentration).

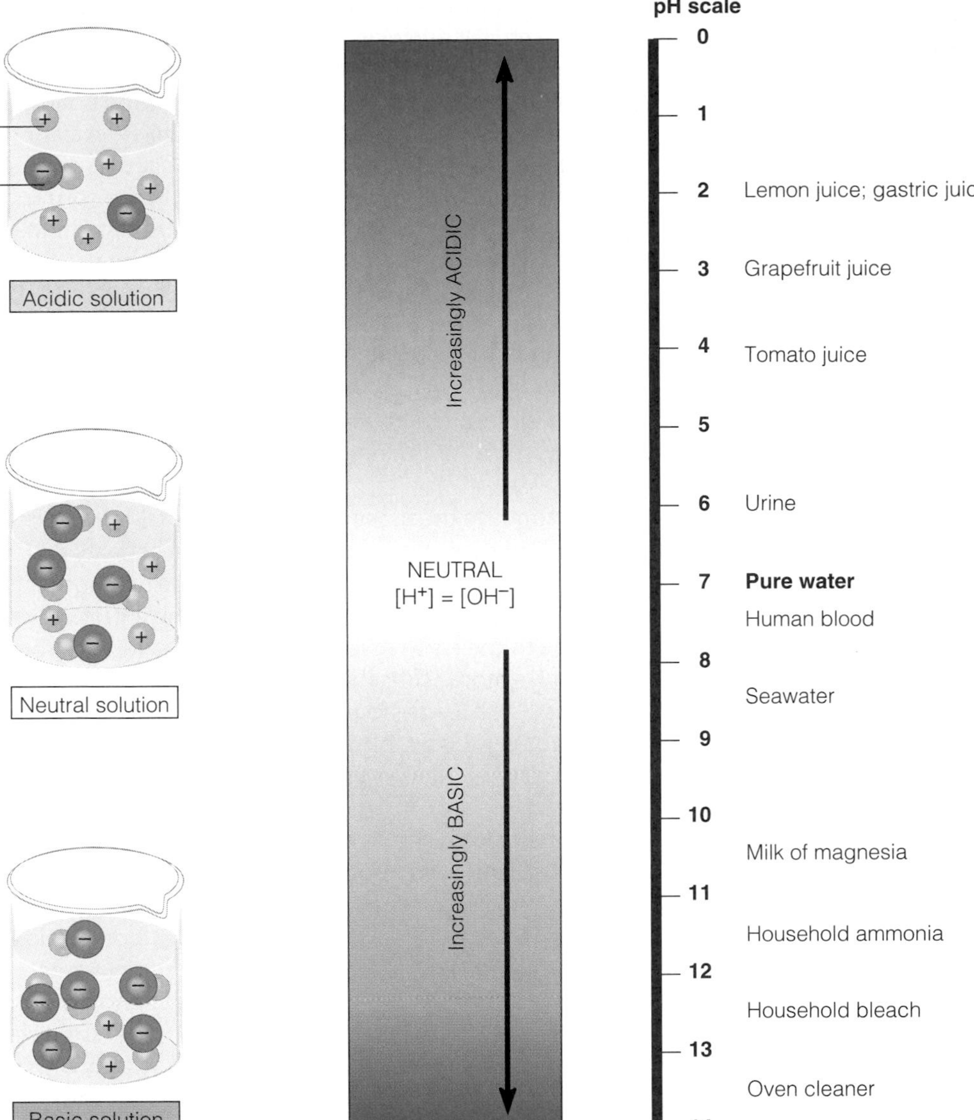

Figure 2.15 The pH scale

To describe the acidity of a solution, we use the **pH scale** (pH stands for potential hydrogen). As shown in Figure 2.15, the scale ranges from 0 (most acidic) to 14 (most basic). Each pH unit represents a tenfold change in the concentration of H^+. For example, lemon juice at pH 2 has 100 times more H^+ than an equal amount of tomato juice at pH 4.

Pure water and aqueous solutions that are neither acidic nor basic are said to be neutral; they have a pH of 7. They do contain some H^+ and OH^- ions, but the concentrations of the two kinds of ions are equal. The pH of the solution inside most living cells is close to 7. Even a slight change in pH can be harmful, because the molecules in cells are very sensitive to H^+ and OH^- concentrations. However, biological fluids contain **buffers,** substances that resist changes in pH by accepting H^+ ions when they are in excess and donating H^+ ions when they are depleted. Buffers resist some changes in pH, but they are not foolproof. As we discuss in the next module, increasing acidity is a current environmental threat.

? Compared to a basic solution at pH 9, the same volume of an acidic solution at pH 4 has ___× more hydrogen ions (H^+).

100,000

2.16 Acid precipitation threatens the environment

2J

Imagine arriving for a long-awaited vacation at a mountain lake only to discover that, since your last visit a few years ago, all fish and other forms of life in the lake have perished because of increased acidity of the water. Over the past two decades, thousands of lakes in North America, Europe, and Asia have suffered that fate, primarily as a result of **acid precipitation,** usually defined as rain or snow with a pH below 5.6. About 4% of the lakes in the U.S. are now dangerously acidic, with the number close to 10% in the eastern part of the country.

Effects of acid precipitation are also felt on land. Figure 2.16 shows dead spruce and fir trees on Mount Mitchell in North Carolina, where acid precipitation and acid fog have greatly reduced the numbers of these high-mountain trees. In cities, acid in the air eats away the surfaces of buildings and contributes to smog.

Acid precipitation results mainly from the presence in the air of sulfur oxides and nitrogen oxides, air-polluting compounds composed of oxygen combined with sulfur or nitrogen. These oxides react with water vapor in the air to form sulfuric and nitric acids, which fall to the earth in rain or snow. Rain with a pH between 2 and 3—more acidic than vinegar—has been recorded in the eastern U.S. Acid fog of pH 1.7, approaching that of the digestive juices in the human stomach, has been recorded downwind from Los Angeles.

Sulfur and nitrogen oxides arise mostly from the burning of fossil fuels (coal, oil, and gas) in factories and automobiles. Electrical power plants that burn coal produce more of these pollutants than any other single source. Ironically, the tall smokestacks built to reduce local pollution by dispersing factory exhaust help spread airborne acids. Winds carry the pollutants away, and acid rain may fall thousands of miles away from industrial centers.

The effect of acid in lakes and streams is most pronounced in the spring, as snow begins to melt. The surface snow melts first, drains down, and sends much of the acid that has accumulated over the winter into lakes and streams all at once. Early meltwater often has a pH as low as 3, and this acid surge hits when fish and other forms of aquatic life are producing eggs and young, which are especially vulnerable to acidic conditions. Strong acidity can break down the molecules of living organisms. And even if the molecules remain intact, they may not be able to carry out the essential chemical processes of life at very low pH.

While acid precipitation can clearly damage lakes and streams, its effects on forests and other land life are controversial. The damage to the forest on Mount Mitchell, on one hand, almost certainly results from acid fog and precipitation. The acid apparently causes changes in the soil that lead to mineral imbalances, lowered tolerance to cold, and general weakness in the trees. On the other hand, careful studies over the past decade seem to show that the vast majority of North American forests are not suffering substantially from acid precipitation.

Many questions remain. We do not know for sure what the long-term effects of acid precipitation may be on plants and soils. Nor do we know much about the effects of airborne acids on terrestrial animals, including humans. Perhaps most importantly, we do not know how much we must reduce fossil-fuel emissions in order to prevent more damage.

As with most environmental issues, there are no easy solutions to the acid precipitation problem. There is some hopeful news, however. In the United States, Canada, and Europe, emissions of sulfur oxides have declined significantly in recent decades, causing a decrease in acid precipitation. Laws that require reductions in emissions are thus already helping to alleviate the problem. But just as important is energy conservation. We all need to realize that unless we decrease our consumption of electricity and our dependence on gasoline-powered automobiles, we will continue to contribute to acid precipitation and other threats to the environment. 

Figure 2.16 Spruce and fir trees on the right were killed by acid precipitation

? What is the relationship between fossil-fuel consumption and acid precipitation?

Compounds in the exhaust react with water vapor in the atmosphere to form sulfuric and nitric acids.

2.17 Chemical reactions rearrange matter

The basic chemistry of life has an overriding theme: The structure of atoms and molecules determines the way they behave. As we have seen, the chemical properties of an atom are determined by the number and arrangement of its subatomic particles, particularly its electrons. Other properties emerge when atoms combine to form molecules and when molecules combine to form more complex substances, such as liquid water. Water is a good example, because its unusual properties form the foundation of life.

Water can be made from hydrogen and oxygen:

$$2\,H_2 + O_2 \longrightarrow 2\,H_2O$$

This is a **chemical reaction,** a process leading to chemical changes in matter. In this particular case, two molecules of hydrogen (2 H_2) react with one molecule of oxygen (O_2) to give two molecules of water (2 H_2O). The arrow indicates the conversion of the starting materials, called **reactants** (H_2 and O_2), to the resulting **product** (H_2O). Notice that the same *numbers* of hydrogen and oxygen atoms appear on the right and left sides of the arrow, although they are grouped differently. Chemical reactions do not create or destroy matter; they only rearrange it in various ways. Chemical reactions involve the making and breaking of chemical bonds. In the example above, the bonds holding hydrogen atoms together in H_2 and those holding oxygen atoms together in O_2 are broken, and new bonds are formed to yield the H_2O product molecules (Figure 2.17A).

Organisms cannot make water from H_2 and O_2, but they do carry out a great number of chemical reactions that rearrange matter in significant ways. Let's examine one that relates to the chapter's opening essay. Much of the yellow color in autumn leaves is from molecules of pigments called carotenoids. These organic molecules are important in photosynthesis, absorbing solar energy as chlorophyll does. One of the most common carotenoids is beta-carotene. Shown in Figure 2.17B, beta-carotene is also abundant in carrots and many green vegetables and is sold in concentrated form as a food supplement. This substance can be important in our diet as a source of vitamin A, which is essential for normal vision and healthy skin. Our cells can make vitamin A from beta-carotene in the following way:

$$\underset{\text{Beta-carotene}}{C_{40}H_{56}} + O_2 + 4\,H \longrightarrow \underset{\text{Vitamin A}}{2\,C_{20}H_{30}O}$$

Although there are many atoms in beta-carotene and vitamin A, the chemical reaction that converts one to the other is essentially a simple one. Two molecules of vitamin A are made from each beta-carotene molecule by splitting the beta-carotene molecule in half. Notice that beta-carotene has 40 carbon (C) atoms, whereas each vitamin A molecule has 20 carbons. The red arrow in Figure 2.17B shows where beta-carotene is split. The other reactants are a molecule of O_2 and 4 H atoms contributed by other molecules in the cell. If you count up all the atoms, you will see that the same number of each type appears on each side of the reaction.

The conversion of beta-carotene to vitamin A is only one example of the thousands of chemical reactions routinely carried out in living cells. Like most of these reactions, our example involved compounds of the element carbon. We look at the carbon compounds of cells in more detail in Chapter 3.

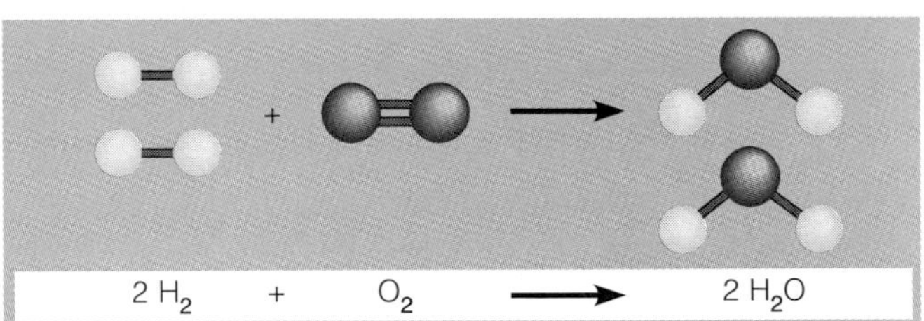

Figure 2.17A Breaking and making of bonds in a chemical reaction

Fill in the blanks with the correct numbers in the following chemical process:

$$C_6H_{12}O_6 + \_\_O_2 \longrightarrow \_\_CO_2 + \_\_H_2O$$

$C_6H_{12}O_6 + 6\,O_2 \longrightarrow 6\,CO_2 + 6\,H_2O$

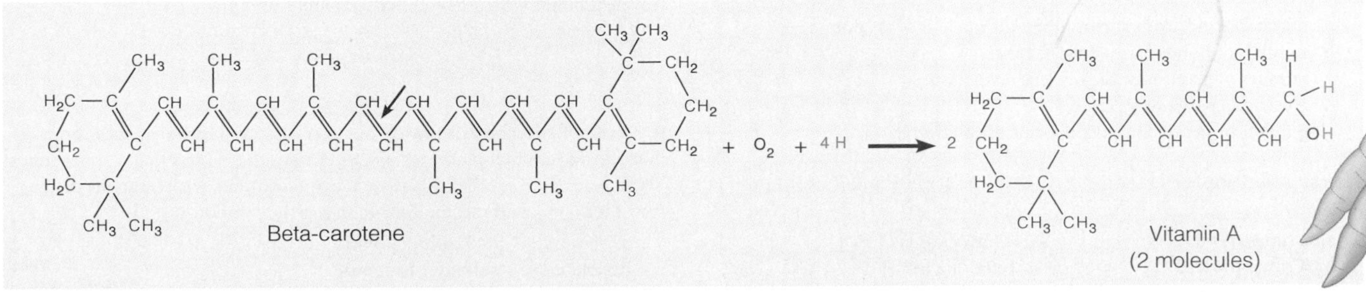

Beta-carotene $\quad + O_2 + 4\,H \longrightarrow 2\quad$ Vitamin A (2 molecules)

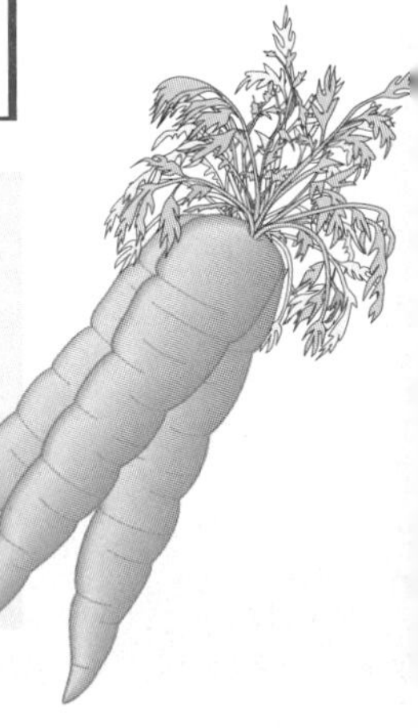

Figure 2.17B Chemical reaction converting beta-carotene to vitamin A

CHAPTER REVIEW

Chapter Summary

Atoms and Molecules (Introduction–2.8) Everything an organism is and does depends on chemistry, which in turn depends on how atoms are arranged in molecules. Much of what we know about life has come from taking living things apart, all the way down to atoms. Studying the parts in order to understand the whole is called the reductionist approach. But this approach must be combined with study of the functioning whole **(Introduction).** At each level of organization—molecule, cell, tissue, organ, or organism—unique properties emerge from structure **(2.1).**

About 25 chemical elements are essential to life. Carbon, hydrogen, oxygen, and nitrogen make up the bulk of living matter **(2.2).** Chemical elements combine in fixed ratios to form compounds, such as sodium chloride and chlorophyll **(2.3).** The smallest particle of an element is an atom. An atom consists of protons and neutrons in a central nucleus, surrounded by electrons. Each atom is held together by attraction between its positively charged protons and negatively charged electrons; neutrons are neutral. Atoms of each element are characterized by a certain number of protons; the number of neutrons may vary **(2.4).** Variant forms of an atom are called isotopes. Some isotopes are radioactive. Though their radioactivity is sometimes harmful to life, isotopes can be useful tracers for studying biological processes **(2.5).**

Electrons are arranged in electron shells. The outermost shell determines the chemical properties of an atom. In most atoms, a complete outer shell holds 8 electrons. Atoms whose shells are not full tend to interact with other atoms and gain, lose, or share electrons **(2.6).**

Electron gain and loss create charged atoms, called ions, whose electrical attraction results in ionic bonds between them. Sodium loses an electron and chlorine gains one to become the ions that form sodium chloride, NaCl **(2.7).** Other atoms share pairs of electrons, binding the atoms together with covalent bonds. Atoms joined by covalent bonds form molecules, such as H_2O (water), O_2 (oxygen), and $C_{20}H_{30}O$ (vitamin A) **(2.8).**

The Properties of Water (2.9–2.16) Atoms in a covalently bonded molecule may share electrons equally, in which case the molecule is said to be nonpolar. In a water molecule, the oxygen atom exerts a stronger pull on the electrons than hydrogen does, making the oxygen slightly negative and the hydrogens slightly positive. Water is therefore a polar molecule **(2.9).** The charged regions on water molecules are attracted to oppositely charged regions on nearby molecules, forming weak bonds called hydrogen bonds. Polarity and hydrogen bonds give water some unique properties. Like no other common substance, it exists in nature in solid, liquid, and gaseous states **(2.10).** Hydrogen bonds make water molecules cohesive, allowing for the movement of water from plant roots to leaves. Insects can walk on water because of surface tension created by cohesive water molecules **(2.11).** Water's ability to store heat moderates body temperature and climate. It takes a lot of energy to disrupt hydrogen bonds, so water is able to absorb a great deal of heat energy without a large increase in temperature. As water cools, a slight drop in temperature releases a large amount of heat. A water molecule takes much energy with it when it breaks away from its neighbors and evaporates, making evaporative cooling possible **(2.12).** Finally, molecules in ice are farther apart than in liquid water. Because ice is less dense than water, it floats, and lakes do not freeze solid **(2.13).** Solutes whose charges or polarity allow them to stick to water molecules will dissolve in water, forming aqueous solutions **(2.14).**

A compound that releases H^+ ions in solution is an acid, and one that accepts H^+ ions is a base. Acidity is measured on the pH scale, from 0 (most acidic) to 14 (most basic). Pure water and solutions that are neither acidic nor basic are neutral, having a pH of 7. Cells are close to pH 7 and kept that way by buffers, substances that resist pH change **(2.15).**

Some ecosystems are threatened by acid precipitation, which is formed when air pollutants from burning fossil fuels combine with water vapor in the air to form sulfuric and nitric acids. These acids can kill fish and damage buildings and may injure trees. Regulations, new technology, and energy conservation may help us reduce acid precipitation **(2.16).**

Rearrangements of Atoms (2.17) In a chemical reaction, reactants interact, atoms rearrange, and products result. Living cells carry out thousands of chemical reactions that rearrange matter in significant ways.

Testing Your Knowledge

Multiple Choice

1. Your body contains the smallest amount of which of the following?
   a. nitrogen
   b. phosphorus
   c. carbon
   d. oxygen
   e. hydrogen

2. Changing the ____ would change it into an atom of a different element.
   a. number of electrons surrounding the nucleus of an atom
   b. number of bonds formed by an atom
   c. number of protons in the nucleus of an atom
   d. electrical charge of an atom
   e. number of neutrons in the nucleus of an atom

3. A solution at pH 6 contains ____ than the same amount of solution at pH 8.
   a. 2 times more H^+
   b. 4 times more H^+
   c. 100 times more H^+
   d. 4 times less H^+
   e. 100 times less H^+

4. Most of the unique properties of water result from the fact that water molecules
   a. are very small
   b. are held together by covalent bonds
   c. easily separate from one another
   d. are constantly in motion
   e. tend to stick together

5. A sulfur atom has 6 electrons in its outer shell. As a result, it forms ____ covalent bonds with other atoms. *(Explain your answer.)*
   a. 2
   b. 3
   c. 4
   d. 6
   e. 8

6. A can of cola consists mostly of sugar dissolved in water, with some carbon dioxide gas that makes it fizzy and makes the pH less than 7. In chemical terms, you could say that cola is an aqueous solution, where water is the ____, sugar is a ____, and carbon dioxide makes the solution ____.
   a. solvent . . . solute . . . basic
   b. solute . . . solvent . . . basic
   c. solvent . . . solute . . . acidic

d. solute . . . solvent . . . acidic
e. not enough information to say

7. Which of the following is *not* a chemical reaction?
   a. Sugar and oxygen gas combine to form carbon dioxide and water.
   b. Sodium metal and chlorine gas unite to form sodium chloride.
   c. Hydrogen gas combines with oxygen gas to form liquid water.
   d. Solid ice melts to form liquid water.
   e. Sulfur dioxide and water vapor join to form sulfuric acid.

True/False *(Change false statements to make them true.)*

1. Table salt, water, and carbon are compounds.

2. Atoms in a water molecule are held together by the sharing of electrons.

3. A bathtub full of lukewarm water may hold more heat than a teakettle full of boiling water.

4. If the atoms in a molecule share electrons equally, the molecule is said to be nonpolar.

5. Ice floats because water molecules in ice are more tightly packed than in liquid water.

6. The smallest particle of an element is a molecule.

7. The pH scale is set up so that pure water has a pH of 1.

8. Most acid precipitation results from the presence of pollutants from aerosol cans and air conditioners.

9. An atom that has gained or lost electrons is called an ion.

10. Reactants are the substances produced by a chemical reaction.

Describing, Comparing, and Explaining

1. Make a sketch that shows how water molecules hydrogen-bond with one another. Why do water molecules form hydrogen bonds? What unique properties of water result from the tendency of water molecules to form hydrogen bonds?

2. Describe two ways in which the water in your body helps stabilize your body temperature.

3. Compare covalent and ionic bonds.

4. What is an acid? A base? How is the acidity of a solution described?

Thinking as a Scientist

1. The diagram below shows the arrangement of electrons around the nucleus of a fluorine atom (left) and a potassium atom (right). Predict what would happen if a fluorine atom and a potassium atom came into contact. What kind of bond do you think they would form?

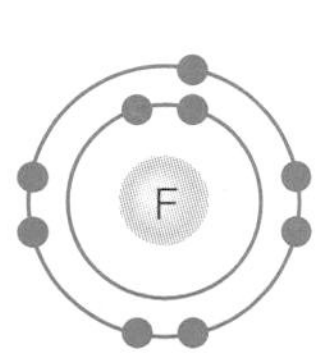

Fluorine atom

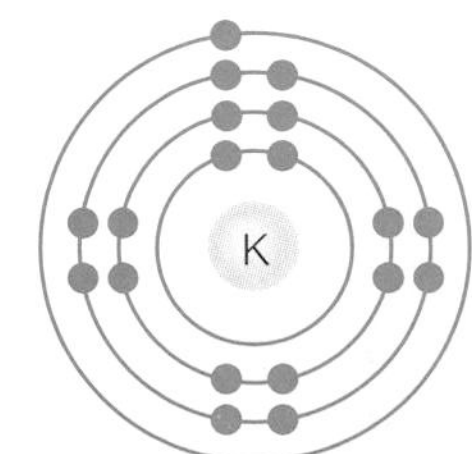

Potassium atom

2. Animals obtain energy through a series of chemical reactions in which sugar ($C_6H_{12}O_6$) and oxygen gas (O_2) are reactants. This process produces water (H_2O) and carbon dioxide (CO_2) as waste products. Suppose you wanted to find out whether the oxygen in CO_2 comes from sugar or oxygen gas. How might you use a radioactive isotope to find out?

3. A newspaper headline reads, "Acid Precipitation Not a Threat." The article states that the city's air pollution control board has been monitoring sulfur oxides and nitrogen oxides in the air and recording the pH of precipitation for the past 5 years. Air samples have been collected daily at two sites; at the busiest downtown intersection and at the entrance to a privately owned electrical power generating station. On most days, the power station is downwind from the downtown area. Over the period studied, sulfur oxides and nitrogen oxides have increased at both sites, with the power station always having higher levels than the downtown site. However, at no time in the past 5 years have any of the pollutant levels exceeded the nationally acceptable standards. Moreover, the pH of precipitation has remained essentially the same (slightly acidic) at both monitoring sites throughout the period. City officials have concluded that there is no immediate reason for alarm, that acid precipitation has not been a problem, but that monitoring should continue.

The news article, however, goes on to point out that not everyone agrees with the city officials. One resident living 10 miles outside the city is quoted: "The trees on our streets and in our yards are dying from acid precipitation. Five years ago all our trees were healthy, but we've lost a dozen in this neighborhood in the past 2 years. We're directly downwind from that power plant, and I know acid rain killed my trees. My neighbors and I are going to sue the power company for losses."

Do you think these residents have the scientific evidence needed to win their lawsuit? Why or why not? Would they have a clear-cut case if a monitoring site had been set up in their neighborhood, and data collected there showed an increase in acid precipitation during the 5-year period? If the residents hired you as a scientific consultant, what advice would you give them?

Science, Technology, and Society

1. One solution to the problem of acid precipitation is to use nuclear energy to produce electricity. Development of nuclear power in the United States virtually stopped after the accident at the Three Mile Island power plant in Pennsylvania in 1979. But proponents of nuclear power contend that accelerated development of nuclear energy is the answer to U.S. energy needs and tightened pollution standards. Besides reducing acid precipitation, what are other potential benefits of nuclear power? What are its possible costs and dangers? Do you think we ought to pursue the development of nuclear power? Why or why not? If a new power plant were to be built near where you live, would you prefer it to be a nuclear or coal-fired plant? Why?

2. While waiting at an airport, one of the authors of this book once overheard this claim: "It's paranoid and ignorant to worry about industry or agriculture contaminating the environment with their chemical wastes. After all, this stuff is just made of the same atoms that were already present in our environment anyway." How might you counter this argument?

Media Resources

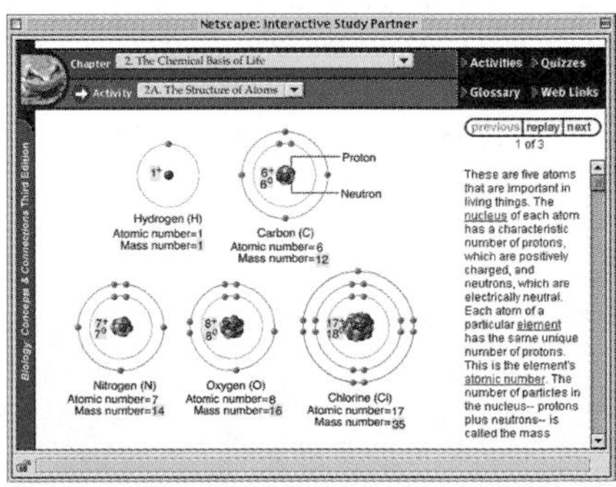

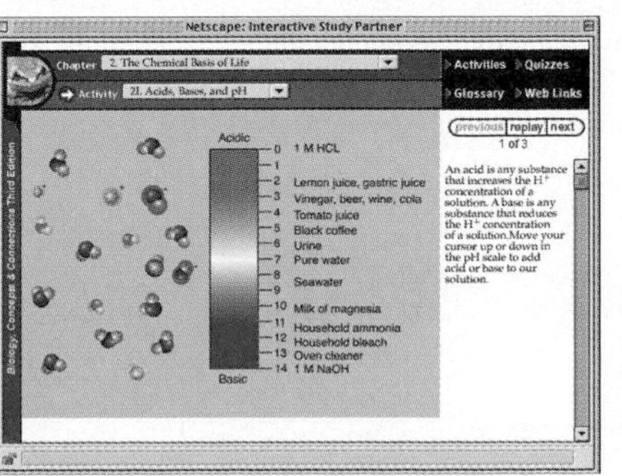

For further review of the concepts in this chapter, use the activities and quizzes on the CD-ROM.

Visit the Special Edition of The Biology Place web site for activities, quizzes, news articles, and web links.

3 The Molecules of Cells

Outline

Ever since ancient times, people have been fascinated with spiders and the webs they spin. An old Navajo legend tells of a Spider Woman who invented weaving and taught all the textile arts to the Native Americans of the Southwest. According to Greek mythology, spiders originated when the maiden Arachne was transformed into one, as a punishment for boasting of her weaving skills. Our technical name for spider is arachnid, which is derived from Arachne's name.

The ancient legends allude to special qualities of spiders and their webs—qualities that arise from unique properties of molecules. Web builders, like the spider in the photograph at the left, spin their webs with remarkable speed and agility. This web probably took no longer than an hour to construct, yet its symmetry and efficacy in capturing insects are without parallel. The spider's web-building skill depends on its genetic programming, built into the DNA molecules it inherited from its parents. The special qualities of the web, which is made of silk produced by glands in the spider's abdomen, result from unique properties of silk proteins associated with a mix of other molecules.

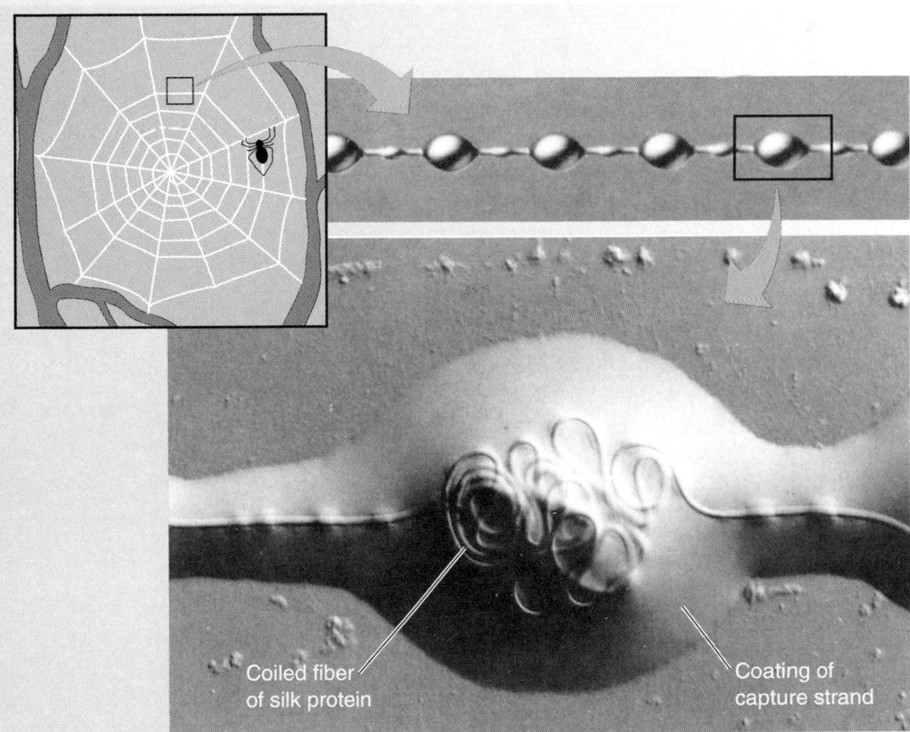

Structure of a spider's capture strand

Silk proteins make a spiderweb remarkably strong and resilient, able to withstand a struggling insect's attempts to escape. The type of web in the photograph on the facing page is known as an orb web for its roughly circular shape. The strands extending straight out from the center of the orb—the radial strands—are composed of dry, relatively inelastic proteins; they maintain the web's position and overall shape. In contrast, the orb's spiraling strand, which actually captures insects, is wet, sticky, and highly elastic. In fact, it can stretch up to four times its original length and then recover with little sag. How does a capture strand extend and contract in response to thrashing insects, to the weight of rainwater and dew, or to the tearing force of a strong wind? The web's elasticity apparently results from the coiling and uncoiling of silk fibers. The photographs on this page show part of a capture strand at two magnifications. The strand contains a single coiled silk fiber (lower photo), which can unwind as needed and then recoil rapidly. Droplets of a sticky fluid coat the coiled thread. The fluid helps trap prey, and when the prey no longer presses against the strand, the fluid's surface tension—its tendency to "bead"—rewinds the fiber.

Spiders and their webs illustrate life's molecules in action. A spider's DNA and the proteins in its silk represent two major classes of molecules in living organisms. Life's diversity results from a seemingly endless variety of these and other molecules. However, a relatively small number of basic structural patterns underlies this diversity and helps us make sense of it. In this chapter, we describe the major molecules of living cells and how they are put together.

Introduction to Organic Compounds and Their Polymers

3.1 Life's molecular diversity is based on the properties of carbon

3A
See Activity 3A on the CD-ROM.

Almost all the molecules a cell makes are composed of carbon atoms bonded to one another and to atoms of other elements. Carbon is unparalleled in its ability to form large, diverse molecules. Next to water, compounds containing carbon are the most common substances in living organisms.

Compounds synthesized by cells and containing carbon are known as **organic compounds.** Well over two million organic compounds are known, and chemists identify more each day. As we discussed in Chapter 2, an element's chemical properties are determined by the electrons in the outermost shell of its atoms. Thus, the enormous variety of carbon-based molecules derives from the tendency of carbon's outer-shell electrons to form chemical bonds, specifically covalent bonds (see Module 2.8). A carbon atom has 4 outer electrons in a shell that holds 8. Consequently, it has a strong tendency to complete its outer shell by sharing electrons with other atoms in four covalent bonds.

At the top of Figure 3.1 are three illustrations of methane, one of the simplest organic molecules. The structural formula simply shows that covalent bonds link four hydrogen atoms to the carbon atom. Each of the four lines in the formula represents a pair of shared electrons. The two models show that methane is three-dimensional, with the space-filling version portraying its overall shape more accurately. The ball-and-stick model indicates the angles of the molecule's covalent bonds (the white "sticks"). The four hydrogen atoms of methane are at the corners of an imaginary tetrahedron (an object with four triangular sides); the same tetrahedral shape occurs wherever a carbon atom has four single bonds. Different bond angles occur where carbon atoms form double bonds. The shape of a molecule, which depends partly on its bond angles, usually helps determine its function.

Methane and other compounds composed of only carbon and hydrogen are called **hydrocarbons.** The figure illustrates some variations in hydrocarbon structure. Carbon atoms, with attached hydrogens, can bond together in chains of various lengths to form compounds such as ethane or propane, a gas used as household fuel. The chain of carbon atoms in organic molecules like these is called a **carbon skeleton.** Carbon skeletons can be unbranched, as in butane, or branched, as in isobutane. Carbon skeletons may also include double bonds, as in 1-butene and 2-butene. These two compounds have the same molecular formula, C_4H_8, but differ in the position of their double bond. Compounds like these with the same molecular formula but different structure are called **isomers.** (Butane and isobutane are also isomers.) Each isomer has unique properties. Cyclohexane and benzene, the only hydrocarbons in the figure that are liquids rather than gases, are examples of carbon skeletons arranged in rings.

? Draw a structural formula for C_2H_4.

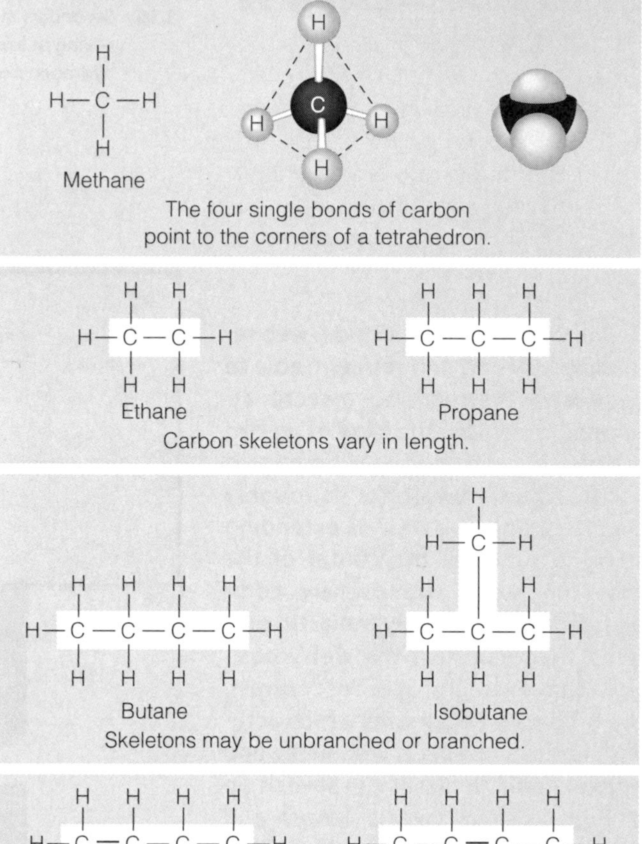

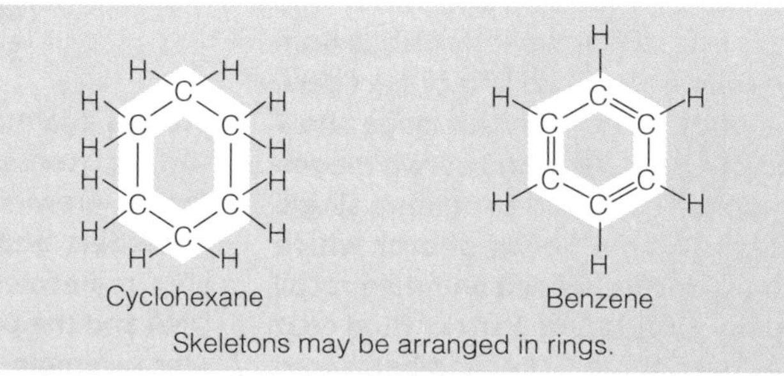

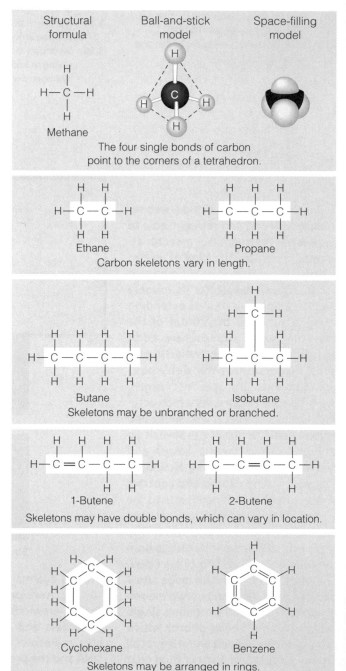

Figure 3.1 Variations in carbon skeletons

Functional groups help determine the properties of organic compounds

3.2

3B

The unique properties of an organic compound depend not only on its carbon skeleton but also on the atoms that are attached to the skeleton. In an organic molecule, the groups of atoms that usually participate in chemical reactions are called **functional groups.** Figure 3.2 below shows four functional groups important in the chemistry of life. Notice that two of the groups (hydroxyl and amino) are appendages to a molecule's carbon skeleton; the other two (carbonyl and carboxyl) include a carbon atom of the skeleton. All of these functional groups are polar, because their oxygen or nitrogen atoms exert a strong pull on shared electrons. The polarity tends to make compounds containing these groups **hydrophilic** (water-loving) and therefore soluble in water—a necessary condition for their roles in water-based life.

A **hydroxyl group** consists of a hydrogen atom bonded to an oxygen atom. The oxygen is bonded to the carbon skeleton of a molecule. Ethanol and other organic compounds containing hydroxyl groups are called alcohols.

A **carbonyl group** is a carbon atom linked by a double bond to an oxygen atom. If the carbon atom of the carbonyl group is at the end of a carbon skeleton, the compound is called an aldehyde. Compounds whose carbonyl groups are within a carbon chain are called ketones.

A **carboxyl group** consists of an oxygen atom double-bonded to a carbon that is also bonded to a hydroxyl group. The carboxyl group acts as an acid by contributing an H^+ ion to a solution (see Module 2.15), and compounds containing carboxyl groups are called **carboxylic acids.** Acetic acid, shown in the figure, gives vinegar its sour taste.

An **amino group** is composed of a nitrogen atom bonded to two hydrogen atoms. It acts as a base by picking up an H^+ ion from a solution. Organic compounds with an amino group are called **amines.**

Though the examples here each contain only one functional group, many biological molecules have two or more. For example, sugars have both hydroxyl and carbonyl groups and thus can be classified as both alcohols and aldehydes (or ketones). Compounds called amino acids have carboxyl as well as amino groups. Amino acids are the building blocks of proteins, as we will see later.

Which of the following functional groups do *not* contain carbon: carboxyl, hydroxyl, amino, and carbonyl?

The amino and hydroxyl groups

Figure 3.2 Some common functional groups

Functional Group	General Formula	Name of Compounds	Example	Where Else Found
Hydroxyl —OH (or HO—)	—O—H	Alcohols	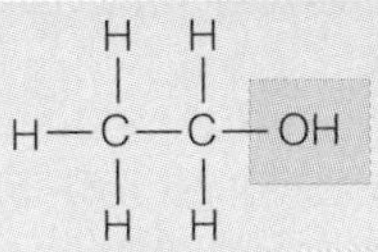 Ethanol	Sugars; water-soluble vitamins
Carbonyl  >CO	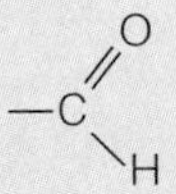 —C(=O)—H	Aldehydes	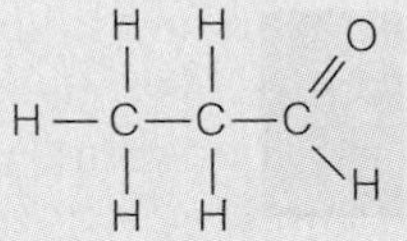 Propanal	Some sugars; formaldehyde (a preservative)
	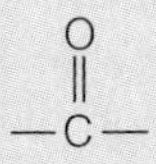 —C(=O)—	Ketones	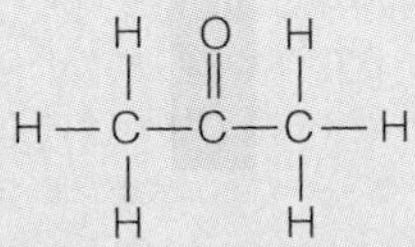 Acetone	Some sugars; "ketone bodies" in urine (from fat breakdown)
Carboxyl —COOH	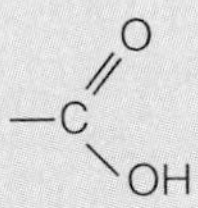 —C(=O)—OH	Carboxylic acids	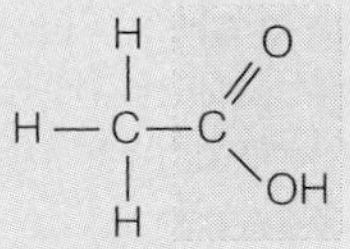 Acetic acid	Amino acids; proteins; some vitamins; fatty acids
Amino $—NH_2$ (or $H_2N—$)	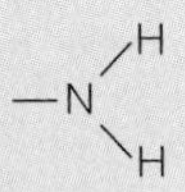 —N(—H)—H	Amines	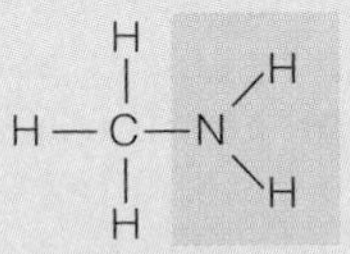 Methylamine	Amino acids; proteins; urea in urine (from protein breakdown)

3.3 Cells make a huge number of large molecules from a small set of small molecules

On a molecular scale, many of life's molecules are gigantic; in fact, biologists call them **macromolecules.** Proteins, one class of macromolecules, may consist of thousands of covalently connected atoms. A second class is nucleic acids, one of which is DNA. A third class of macromolecules is made up of carbohydrates—specifically, a group of large carbohydrates called polysaccharides. Lipids, including the fats, comprise a diverse fourth group of large organic molecules important in cells.

Cells make most of their large molecules by joining smaller organic molecules into chains called **polymers** (from the Greek *polys,* many, and *meros,* part). A polymer is a large molecule consisting of many identical or similar molecular units strung together, much as a train consists of many individual cars. The units that serve as the building blocks of polymers are called **monomers.**

Living cells make a vast number of different polymers. For proteins alone, there are about a trillion different kinds in nature, and the variety is potentially endless. One of life's most remarkable features is that a cell makes all of its diverse macromolecules from a small list of ingredients—about 40 to 50 common monomers and a few others that are rare. Proteins, for example, are built from only 20 kinds of amino acids, arranged in chains typically several hundred amino acids long. DNA is built from just four kinds of monomers called nucleotides. As we will see, the key to the great diversity of protein and DNA molecules is arrangement—variation in the sequence in which the monomers are strung together.

The variety in polymers accounts for the uniqueness of each organism. The monomers used to make polymers, however, are essentially universal. Your proteins and those of a tree or spider are assembled from the same 20 amino acids; the amino acids are just arranged in different sequences. Life has a simple yet elegant molecular logic: Small molecules common to all organisms are ordered into macromolecules, which vary from species to species and even from individual to individual.

Cells link monomers together to form polymers by a process called **dehydration synthesis** (Figure 3.3A). All unlinked monomers have hydrogen atoms (H) and also hydroxyl groups (—OH). For each monomer added to a chain, a water molecule (H_2O) is removed. Notice that two monomers contribute to the H_2O molecule, one monomer (the one at the right end of the short polymer in this example) losing a hydrogen atom, and the other monomer losing a hydroxyl group. As this occurs, a new covalent bond forms, linking the two monomers. The same process occurs regardless of the specific monomers.

Cells not only make macromolecules, they also have to break them down. For example, food that an organism ingests is often in the form of macromolecules. To digest these substances and make their monomers usable, a cell carries out **hydrolysis,** essentially the reverse of dehydration synthesis. Hydrolysis means to break (lyse) with water

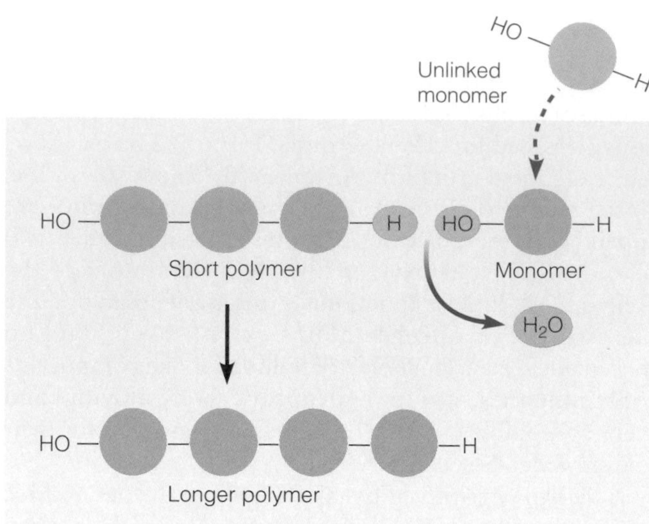

Figure 3.3A Dehydration synthesis of a polymer

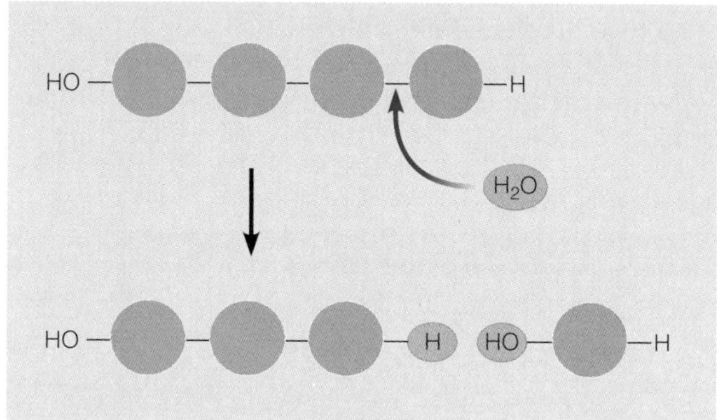

Figure 3.3B Hydrolysis of a polymer

(hydro-), and cells break bonds between monomers by adding water to them, as Figure 3.3B shows. In the process, a hydroxyl group from a water molecule joins to one monomer, and a hydrogen joins to the adjacent monomer.

In the remainder of the chapter, we examine each of the four classes of large molecules in more detail. We will see that polysaccharides, proteins, and nucleic acids are all polymers assembled by dehydration synthesis from their respective monomers: polysaccharides from monosaccharides, proteins from amino acids, and nucleic acids from nucleotides. Lipids are generally smaller than polysaccharides, proteins, and nucleic acids, and they are not truly polymers. However, some of the largest lipids, the fats, are formed by dehydration synthesis from several smaller molecules. 

? How many molecules of water are needed to completely hydrolyze a polymer that is 100 monomers long?

Ninety-nine

3.4 Monosaccharides are the simplest carbohydrates

3C

The name **carbohydrate** refers to a class of molecules ranging from small sugar molecules to large polysaccharides, which are long polymers of sugar monomers. The carbohydrate monomers (single-unit sugars) are **monosaccharides,** from the Greek *mono-*, single, and *sacchar,* sugar. The honey shown in Figure 3.4A consists mainly of monosaccharides called glucose and fructose. These and other single-unit sugars can be hooked together by dehydration synthesis to form more complex sugars and polysaccharides.

Monosaccharides generally have molecular formulas that are some multiple of CH_2O. For example, the formula for glucose, a common monosaccharide of central importance in the chemistry of life, is $C_6H_{12}O_6$. Its molecular structure, shown in Figure 3.4B, has the two trademarks of a sugar: a number of hydroxyl groups (—OH) and a carbonyl group (>C=O). The hydroxyl groups make a sugar an alcohol, and the carbonyl group, depending on its location, makes it either an aldehyde or a ketone as well. Glucose is an aldehyde, and fructose is a ketone.

If you count the numbers of different atoms in the fructose molecule in Figure 3.4B, you will find that its molecular formula is $C_6H_{12}O_6$, identical to that of glucose. Thus, glucose and fructose are isomers; they differ only in the arrangement of their atoms (in this case, the positions of the carbonyl groups, highlighted in white). Seemingly minor differences like this give isomers different properties, such as the ability to react with other molecules. In this case, the differences also make fructose taste considerably sweeter than glucose.

While the carbon skeletons of both glucose and fructose are six carbon atoms long, other monosaccharides have three to seven carbon atoms. The five-carbon sugars, called pentoses, and the six-carbon sugars, called hexoses, are among the most common. Note that many names for sugars have the suffix "-ose."

It is convenient to draw sugars as if their carbon skeletons were linear, but this does not always create an accurate picture. In aqueous solutions, many monosaccharides form rings, as shown for glucose in Figure 3.4C. (The small identifying numbers on the carbon atoms and the red and blue type should help you follow along.) The arrows indicate that glucose switches back and forth between the linear and ring forms when in solution. As shown on the right, the ring diagram of glucose and other sugars may be abbreviated by not showing the carbon atoms at the corners of the ring. Also, as shown in the abbreviated structure, the bonds in the ring are often drawn with varied thickness, indicating that the ring is a relatively flat structure with atoms and functional groups, such as —OH, extending above and below it.

Monosaccharides, particularly glucose, are the main fuel molecules for cellular work. Cells also use the carbon skeletons of monosaccharides as raw material for manufacturing other kinds of organic molecules, including amino acids. Monosaccharides that cells do not use immediately are usually incorporated as monomers into disaccharides and polysaccharides, as described next.

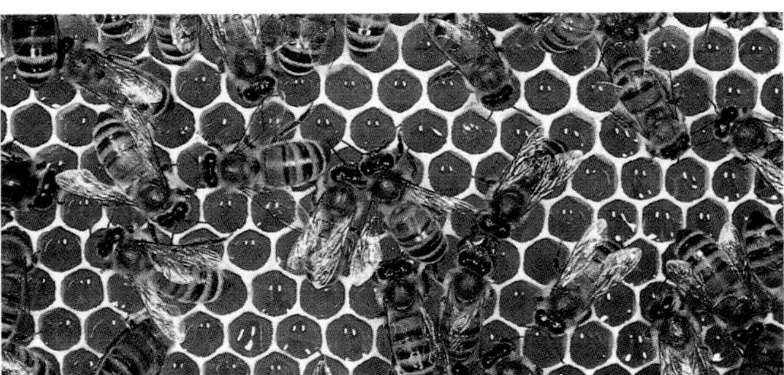

Figure 3.4A Bees with honey, a mixture of two monosaccharides

Write the formula for a monosaccharide that has three carbons.

$C_3H_6O_3$

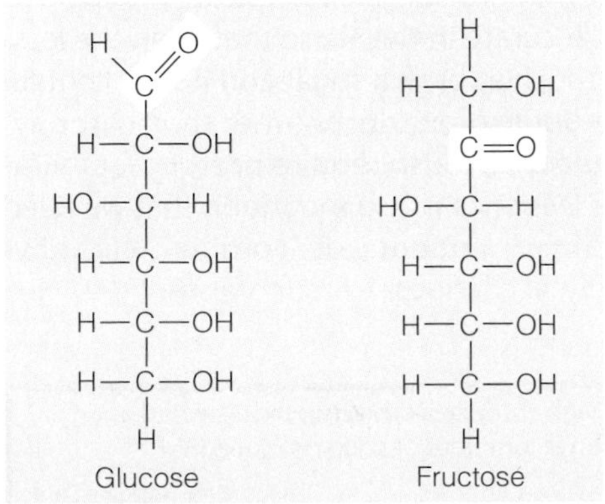

Glucose

Fructose

Figure 3.4B Structures of glucose and fructose

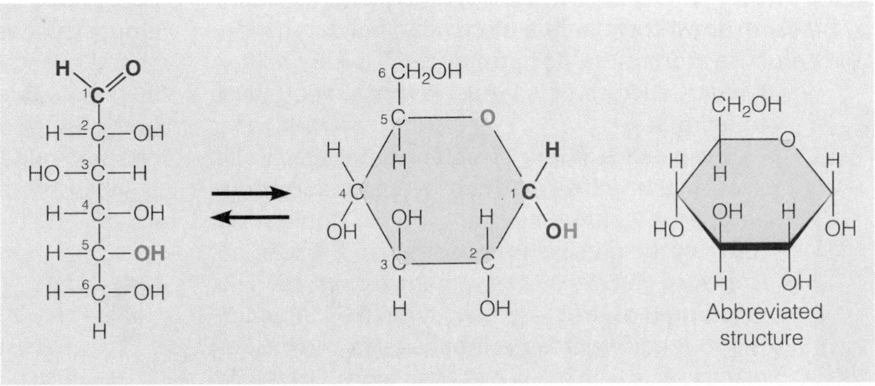

Abbreviated structure

Figure 3.4C Linear and ring forms of glucose

3.5 Cells link single sugars to form disaccharides

3C

Cells construct a **disaccharide,** or double sugar, from two monosaccharides by dehydration synthesis. The figure here shows how maltose, also called malt sugar, forms from two glucose monomers. A linkage forms when one monomer gives up a hydrogen atom from a hydroxyl group and the other gives up an entire hydroxyl group. As H_2O forms, an oxygen atom is left between two covalent bonds, linking the two monomers. Maltose, which is common in germinating seeds, is used in making beer.

The most common disaccharide is sucrose, which is made of a glucose monomer linked to a fructose monomer in a reaction just like the one for maltose. The main carbohydrate in plant sap, sucrose nourishes all the parts of the plant. We extract it from the stems of sugarcane or the roots of sugar beets to use as table sugar.

> **?** Dehydration synthesis joins two glucose molecules to form maltose. Glucose is $C_6H_{12}O_6$. What is the formula for maltose?
>
> $C_{12}H_{22}O_{11}$

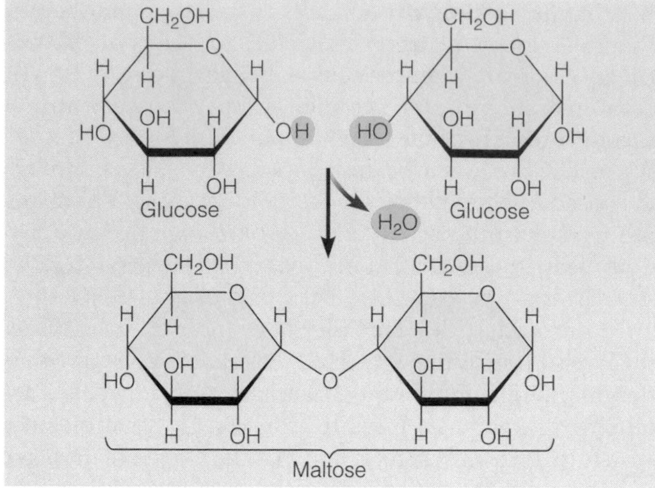

Figure 3.5 Disaccharide formation

3.6 "How sweet it is . . ."

We are born preferring sweet-tasting foods, and although this preference lessens after childhood, it continues to influence our diet. Not only may we prefer a peach to a serving of squash, we also add concentrated sugar to many prepared foods. In addition to sucrose, the main sugars we consume are fructose, glucose, maltose, and lactose (milk sugar, a disaccharide); the first three are the ones added to foods as sweeteners. The United States is the world's leading market for such sweeteners, with the average American consuming about 125 pounds a year.

People have known how to concentrate sucrose from sugarcane for at least 5000 years, but concentrated sugars did not become cheap and widely available until the industrialization of sugar refining in the nineteenth century. A worldwide sugar industry soon arose. By 1950, it was split about 60–40 between cane sugar from the tropics and beet sugar from temperate regions. Occupying only a small part of the sweetener market was corn syrup, which was made by breaking down cornstarch, a (nonsweet) polysaccharide, to its glucose monomers. Because glucose is only half as sweet as sucrose, corn syrup was not a serious rival to sucrose.

The balance among sweeteners was drastically upset in the 1980s, when corn syrup producers developed a commercial method for converting much of the glucose in corn syrup to fructose, an isomer of glucose even sweeter than sucrose. The resulting high-fructose corn syrup (HFCS), which is about half fructose, is inexpensive and has replaced sucrose in many prepared foods. Coca-Cola and PepsiCo, once the largest commercial users of sucrose in the world, have almost completely replaced sucrose with HFCS. The changeover shook up the sucrose industry and hurt the developing economies of tropical countries where sugarcane is a major crop. Corn syrup is produced mainly in developed countries.

While many people continue to eat lots of sugar, a growing number are concerned about the possible health effects. Scientists have shown that sugars are the major cause of tooth decay: Bacteria living in the mouth convert sugars to acids, which eat away tooth enamel. However, studies investigating whether sugar consumption causes such diseases as diabetes, cancer, and heart disease have not revealed any direct connections. Similar efforts have failed to show any link between sugar intake and behavioral problems in children.

Besides tooth decay, high sugar consumption seems to be hazardous only to the extent that sugars replace other, more varied foods in our diet. The description of sugars as "empty calories" is accurate in the sense that even the less-refined sweeteners such as brown sugar and honey contain only negligible amounts of nutrients other than carbohydrates. For good health, we also require protein, fats, vitamins, and minerals. Evidence is also mounting that we need to include substantial amounts of "complex carbohydrates"—that is, polysaccharides—in our diet.

> **?** How is the high-fructose corn syrup that is listed as an ingredient on a container of soda produced?
>
> A commercial process is used to convert much of the glucose in corn syrup to fructose, a much sweeter isomer.

Polysaccharides are long chains of sugar units

3C

Polysaccharides are polymers of a few hundred to a few thousand monosaccharides linked together by dehydration synthesis. Some polysaccharides are storage molecules, which cells break down as needed to obtain sugar. **Starch,** a storage polysaccharide in plant roots and other tissues, consists entirely of glucose monomers, as shown in the figure below. (The functional groups that extend from the glucose rings are omitted.) Starch molecules coil into a helical shape because of the angles of the bonds joining their glucose units. A starch helix may be unbranched (as shown below) or branched.

Plant cells often contain starch granules—actually masses of coiled starch molecules—that serve as sugar stockpiles. Plant cells need sugar for energy and as a raw material for building other molecules. They break starch down into glucose by hydrolyzing the bonds between the glucose monomers. Humans and most other animals are able to use plant starch as food by hydrolyzing it within their digestive systems. Potatoes and grains, such as wheat, corn, and rice, are the major sources of starch in the human diet.

Animals store excess sugar in the form of a polysaccharide called **glycogen.** Glycogen is identical to starch except that it is more extensively branched. Most of our glycogen is stored as granules in our liver and muscle cells, which hydrolyze the glycogen to release glucose when it is needed. Also, our digestive system can hydrolyze glycogen in the meat we eat.

Many polysaccharides serve as building material for structures that protect cells and support whole organisms. **Cellulose,** the most abundant organic compound on Earth, forms cable-like fibrils in the tough walls that enclose plant cells and is a major component of wood. Cellulose resembles starch and glycogen in being a polymer of glucose, but its glucose monomers are linked together in a different orientation; they form an unbranched rod, rather than a coil. Arranged parallel to each other, several thousand cellulose molecules are joined by hydrogen bonds, forming part of a fibril. In wood, layers of cellulose fibrils combine with other polymers, making a material strong enough to support trees hundreds of feet high.

Unlike the glucose linkages in starch and glycogen, those in cellulose cannot be hydrolyzed by most animals. The cellulose in plant foods that passes unchanged through our digestive tract is commonly known as "fiber." It may help keep our digestive system healthy, but it does not serve as a nutrient. Most animals that do derive nutrition from cellulose, such as cows and termites, have cellulose-hydrolyzing microorganisms inhabiting their digestive tracts.

Compare and contrast starch and cellulose, two plant polysaccharides.

Both molecules are polymers of glucose, but the glucose monomers are arranged differently. Starch functions mainly for sugar storage. Cellulose is a structural polysaccharide that is the main material of cell walls.

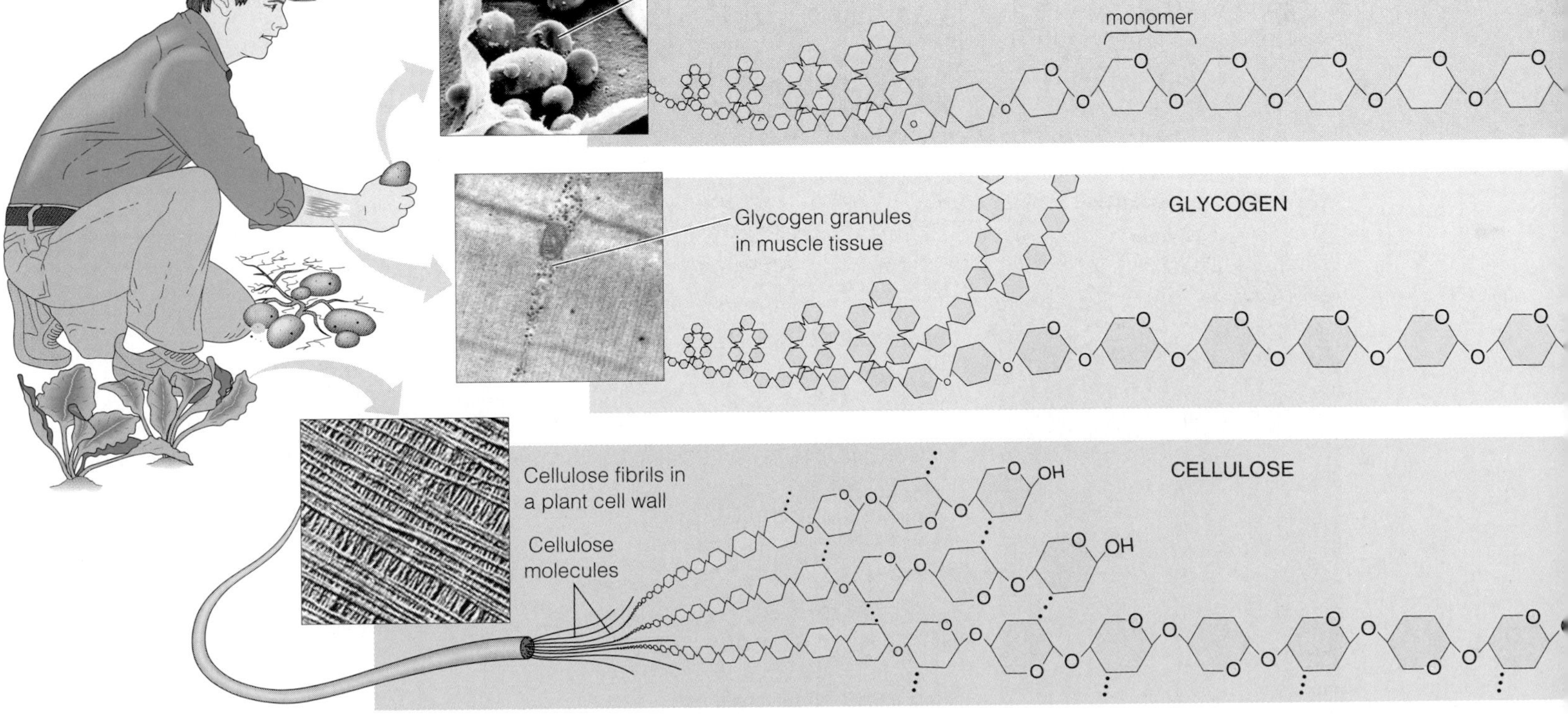

Figure 3.7
Polysaccharides

Starch granules in potato tuber cells

STARCH

Glucose monomer

Glycogen granules in muscle tissue

GLYCOGEN

Cellulose fibrils in a plant cell wall

Cellulose molecules

CELLULOSE

3.8 Lipids include fats, which are mostly energy-storage molecules

3D

Lipids are diverse compounds that consist mainly of carbon and hydrogen atoms linked by nonpolar covalent bonds. Being mostly nonpolar, lipid molecules are not attracted to water molecules, which are polar. You can see the effect of this chemical difference in a bottle of salad dressing: The oil (a type of lipid) separates from the vinegar (which is mostly water). Other oils make the feathers in Figure 3.8A repel water, which you see as beads on the surface. By keeping feathers from absorbing water, oils help waterfowl stay afloat. Because they do not mix with water, lipids are said to be **hydrophobic** (water-fearing).

Oils are a type of fat. A **fat** is a large lipid made from two kinds of smaller molecules: glycerol and fatty acids. Shown on the left in Figure 3.8B, glycerol is an alcohol with three carbons, each bearing a hydroxyl group. A fatty acid consists of a carboxyl group and a hydrocarbon chain with about 15 other carbon atoms. The carbons in the chain are linked to hydrogen atoms by nonpolar covalent bonds, making the hydrocarbon chain hydrophobic. The main function of fats is energy storage. A gram of fat stores more than twice as much energy as a gram of a polysaccharide such as starch.

Figure 3.8B shows how one fatty acid molecule can link to a glycerol molecule by dehydration synthesis. Figure 3.8C shows an actual fat molecule. It consists of three fatty acids hooked to one glycerol molecule as a result of dehydration synthesis occurring at all three hydroxyl sites on the glycerol. A synonym for "fat" is thus **triglyceride,** a term you may see on food labels or on medical tests for fat in the blood. The three fatty acids in a fat are often of different kinds, as in our example.

As shown by the third fatty acid in Figure 3.8C, some fatty acids contain double bonds, which cause kinks in the carbon chain. Double bonds prevent a carbon skeleton from bonding to the maximum number of hydrogen atoms. Fatty acids and fats with double bonds are said to be **unsaturated**—that is, having less than the maximum number of hydrogens. Fats with the maximum number of hydrogens are said to be **saturated.** The kinks in unsaturated fats prevent the molecules from packing tightly together and solidifying at room temperature. Corn oil, olive oil, and other vegetable oils are unsaturated fats. When you see "hydrogenated vegetable oils" on a margarine label, it means that unsaturated fats have been converted to saturated fats by adding hydrogen. This addition gives the lipids the consistency of margarine.

Most plant fats are unsaturated oils, whereas most animal fats are saturated. This is why butter and lard are solids at room temperature. Diets rich in saturated fats may contribute to cardiovascular disease by promoting a condition called atherosclerosis. In this condition, lipid-containing deposits called plaques build up on the inside surfaces of blood vessels, reducing blood flow.

On a food package, what is the meaning of the term "unsaturated fats"?

Unsaturated fats are fats with some double bonds in the hydrocarbons of their fatty acids. These fats have fewer hydrogen atoms than they would without the double bonds.

Figure 3.8A Water beading on the naturally oily coating of feathers

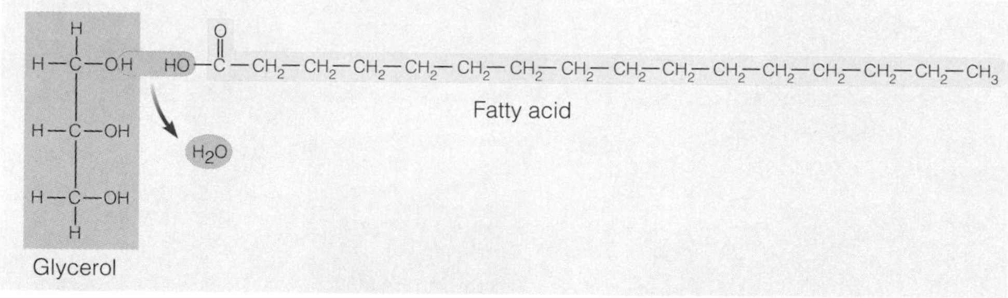

H–C–OH   HO–C(=O)–CH_2–CH_2–CH_2–CH_2–CH_2–CH_2–CH_2–CH_2–CH_2–CH_2–CH_2–CH_2–CH_2–CH_2–CH_3

H–C–OH H_2O

H–C–OH

Glycerol   Fatty acid

Figure 3.8B Dehydration synthesis linking a fatty acid to glycerol

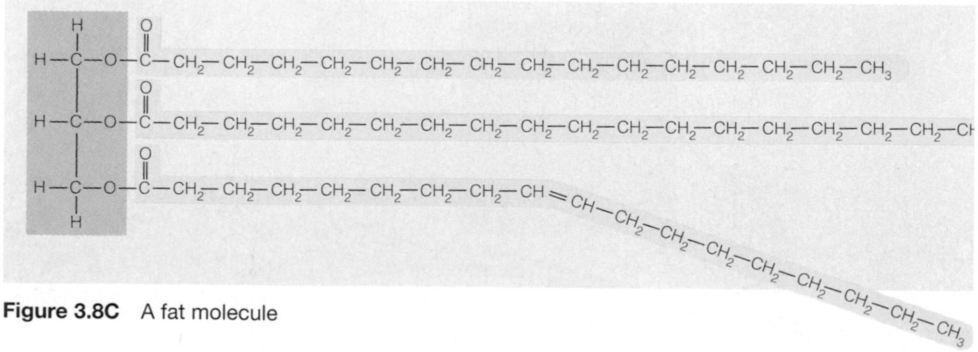

H–C–O–C(=O)–CH_2–CH_2–CH_2–CH_2–CH_2–CH_2–CH_2–CH_2–CH_2–CH_2–CH_2–CH_2–CH_2–CH_2–CH_3

H–C–O–C(=O)–CH_2–CH_2–CH_2–CH_2–CH_2–CH_2–CH_2–CH_2–CH_2–CH_2–CH_2–CH_2–CH_2–CH_2–CH_2–CH_2–CH

H–C–O–C(=O)–CH_2–CH_2–CH_2–CH_2–CH_2–CH_2–CH_2–CH=CH–CH_2–CH_2–CH_2–CH_2–CH_2–CH_2–CH_2–CH_3

Figure 3.8C A fat molecule

3.9 Phospholipids, waxes, and steroids are lipids with a variety of functions

3D

Fats are only one type of lipid important in living organisms. Other lipids are major constituents of cell membranes and perform such vital functions as protecting body surfaces and regulating cellular and body functions.

Phospholipids, a major component of cell membranes, are structurally similar to fats, but they contain the element phosphorus and have only two fatty acids instead of three. Phospholipids are very important biological molecules, as we will see when we discuss cell membranes in Chapter 5.

Figure 3.9 Cholesterol, a steroid

Waxes consist of one fatty acid linked to an alcohol. They are more hydrophobic than fats, and this characteristic makes waxes effective natural coatings for fruits such as apples and pears. Many animals, especially insects, also have waxy coats that help keep them from drying out.

Steroids are lipids whose carbon skeleton is bent to form four fused rings, as shown in the structural diagram of cholesterol in Figure 3.9. All steroids have the same ring pattern: three 6-sided rings and one 5-sided ring. (The diagram omits the carbons making up the rings and the hydrogens attached to these carbons.) Cholesterol is a common substance in animal cell membranes, and animal cells also use it as a starting material for making other steroids, including the female and male sex hormones. Too much cholesterol in the blood may contribute to atherosclerosis.

Human sex hormones belong to what family of lipids?

Steroids

3.10 Anabolic steroids make big bodies and big problems

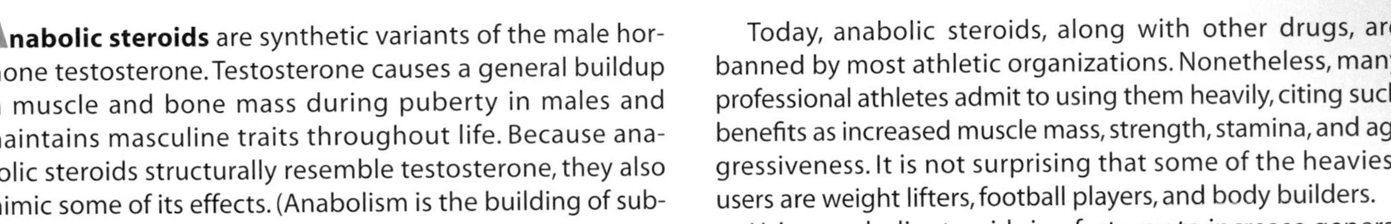

Anabolic steroids are synthetic variants of the male hormone testosterone. Testosterone causes a general buildup in muscle and bone mass during puberty in males and maintains masculine traits throughout life. Because anabolic steroids structurally resemble testosterone, they also mimic some of its effects. (Anabolism is the building of substances by the body.)

Pharmaceutical companies first produced and marketed anabolic steroids in the early 1950s as a treatment for general anemia and for certain diseases that destroy body muscle. About a decade later, some athletes began using anabolic steroids to build up their muscles quickly and enhance their performance.

Figure 3.10

Today, anabolic steroids, along with other drugs, are banned by most athletic organizations. Nonetheless, many professional athletes admit to using them heavily, citing such benefits as increased muscle mass, strength, stamina, and aggressiveness. It is not surprising that some of the heaviest users are weight lifters, football players, and body builders.

Using anabolic steroids is a fast way to increase general body size. They enable an athlete who is willing to cheat at a sport to increase body mass with less hard work. But at what cost? Medical research indicates that these substances can cause serious physical and mental problems. Overdosing can bloat the face and produce violent mood swings and deep depression. Internally, there may be liver damage leading to cancer. Anabolic steroids can also make blood cholesterol levels rise, perhaps increasing a user's chances of developing serious cardiovascular problems. Heavy users may also experience a reduced sex drive and become infertile, just the opposite of what one might expect from using a variant of a sex hormone. The reason is that anabolic steroids often make the body reduce its normal output of sex hormones. The health hazards of anabolic steroids strongly support the argument for banning their use in athletics.

What sort of carbon ring structure would you find in an anabolic steroid?

Four fused carbon rings: three 6-sided and one 5-sided, as in all steroids

3.11 Proteins are essential to the structures and activities of life

3E

The name protein, from the Greek word *proteios,* meaning "first place," suggests the importance of this class of macromolecules. A **protein** is a biological polymer constructed from amino acid monomers. Every one of us has tens of thousands of different kinds of proteins, each with a unique, three-dimensional structure that corresponds to a specific function. Proteins are important to the structures of cells and organisms and participate in everything they do.

There are seven major classes of proteins. One class, called structural proteins, includes the silk of spiders, the hair of mammals (including our own), and fibers that make up our tendons and ligaments. Working together with such structural elements is a second class, called contractile proteins; the proteins that provide muscular movement are one example.

A third class of proteins is storage proteins, such as ovalbumin, the main substance of egg white. Ovalbumin serves as a source of amino acids for developing embryos. A fourth class, the defensive proteins, includes the antibodies, which fight infections and are carried in the blood. Transport proteins, a fifth class, include hemoglobin, the iron-containing protein in blood that conveys oxygen from our lungs to other parts of the body. Certain hormones, which help coordinate body activities by serving as messages from one cell to another, are examples of signal proteins, a sixth class.

Perhaps the most important class of proteins is the enzymes. An **enzyme** is a protein that serves as a chemical catalyst, an agent that changes the rate of a chemical reaction without itself being changed into a different molecule in the process. Enzymes promote and regulate virtually all chemical reactions in cells. You will learn more about enzymes as we pursue the chemistry of life in later chapters.

Figure 3.11 Hair, made of structural proteins

Which of the following is *not* a protein: hemoglobin, cholesterol, an enzyme, an antibody?

Cholesterol

3.12 Proteins are made from just 20 kinds of amino acids

3F

Of all of life's molecules, proteins are the most diverse in structure and function. Protein diversity is based on differing arrangements of a universal set of amino acids. **Amino acids** all have an amino group and a carboxyl group, as you can see in the general structure shown in Figure 3.12A. Both these functional groups are covalently bonded to a central carbon atom called the alpha carbon (red). Also bonded to the alpha carbon is a hydrogen atom and a chemical group symbolized by the letter R. The R group is the variable part of an amino acid. In the simplest amino acid (glycine), the R group is just a hydrogen atom. In others, such as the three in Figure 3.12B, it is one or more carbon atoms with various functional groups attached. The structure of the R group determines the specific properties of each of the 20 amino acids in proteins (see Appendix 2).

The amino acids in Figure 3.12B represent two main types. Leucine (abbreviated Leu) is an example of a type of amino acid whose R group is nonpolar and hydrophobic. Serine (Ser), with a hydroxyl group in its R group, represents a type with polar, hydrophilic R groups. Cysteine (Cys) is one of two amino acids whose R groups include a sulfur atom (S); it is also hydrophilic. Hydrophilic amino acids help proteins dissolve in solutions within cells.

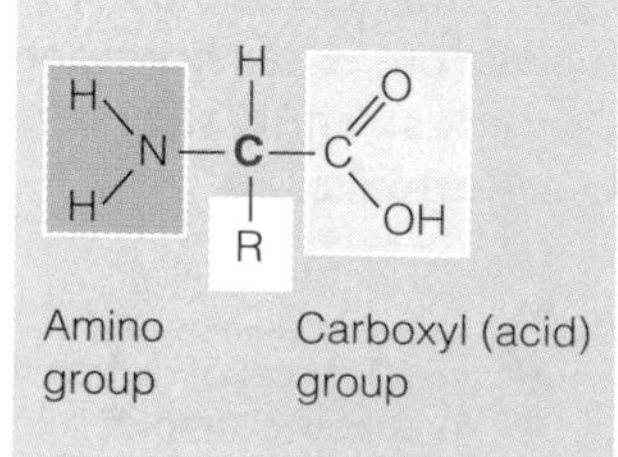

Figure 3.12A General structure of an amino acid

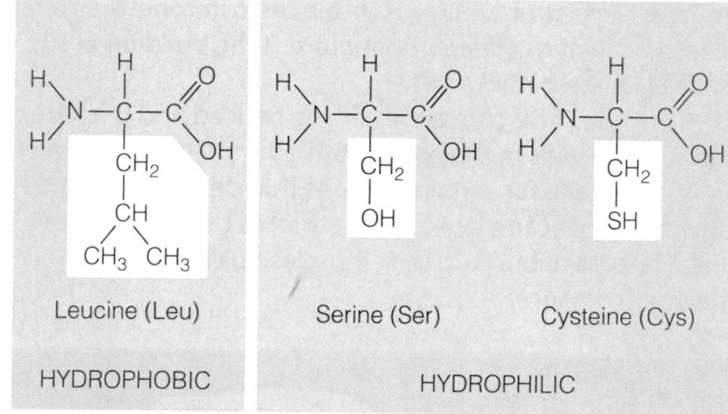

Figure 3.12B Examples of amino acids

What is the basis of the term "amino acid"?

The molecule has both an amino group and a carboxyl group, which is an acidic functional group.

Amino acids can be linked by peptide bonds 3.13

3F

Cells link amino acids together by dehydration synthesis. For this linkage to occur, the carboxyl group of one amino acid must be positioned next to the amino group of another, as shown on the left side of the figure below. A water molecule is then removed as the carboxyl-group carbon atom bonds to the amino-group nitrogen of its neighbor. The resulting covalent linkage is called a **peptide bond.** The product of the reaction in the figure is called a *di*peptide, because it was made from *two* amino acids. Additional amino acids can be added by the same process to form a chain of amino acids, a **polypeptide.** To release amino acids from the polypeptide by hydrolysis, a molecule of H_2O must be added back to each peptide bond.

Polypeptides range in length from a few monomers to a thousand or more. Each polypeptide has a unique sequence of amino acids and assumes a unique three-dimensional shape in a protein, as we see next. 

? In what way is the production of a dipeptide similar to the production of a disaccharide?

In both cases the monomers are joined by dehydration synthesis.

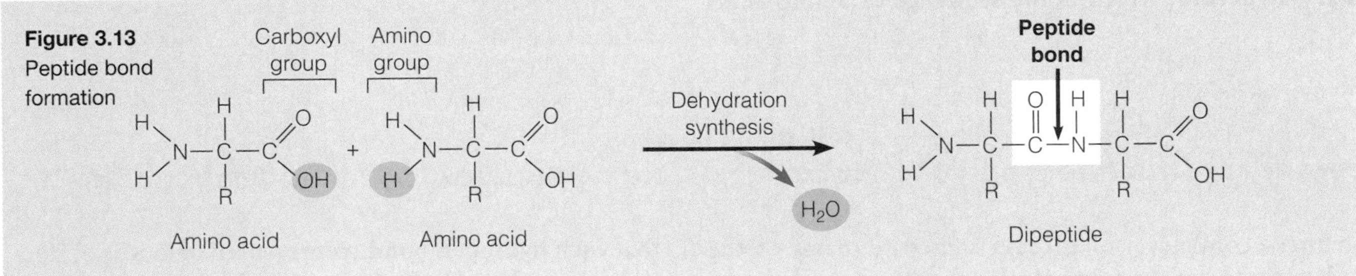

Figure 3.13 Peptide bond formation

Overview: A protein's specific shape determines its function 3.14

3F

A protein consists of one or more polypeptide chains folded into a unique shape. Figure 3.14A shows a ribbon model of the protein lysozyme, an enzyme found in our tears and white blood cells. Lysozyme consists of one long polypeptide, represented by the ribbon. Roughly spherical, lysozyme's general shape is called globular. This overall shape is more apparent in Figure 3.14B, a space-filling model of lysozyme. Most proteins are globular, although structural proteins are typically long and thin—fibrous.

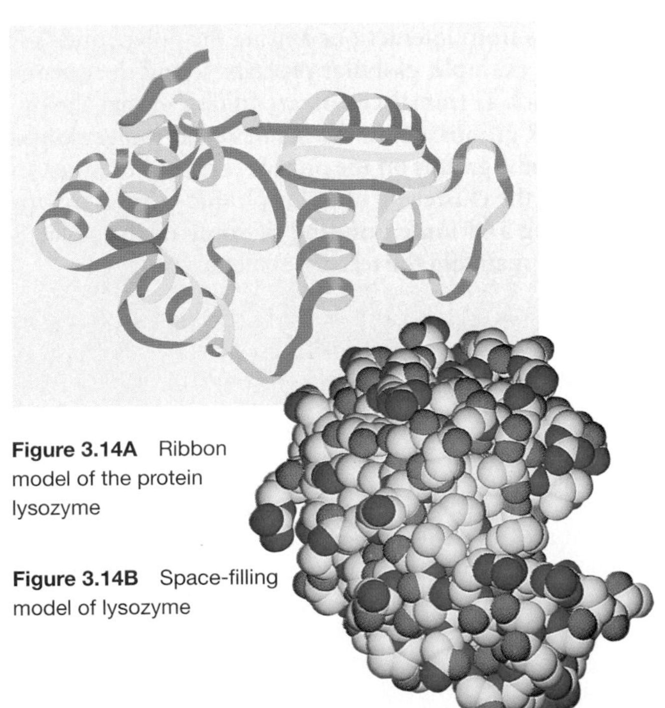

Figure 3.14A Ribbon model of the protein lysozyme

Figure 3.14B Space-filling model of lysozyme

General shape is one thing; specific shape is another. The coils and twists of lysozyme's polypeptide ribbon appear haphazard, but they represent the molecule's specific, three-dimensional shape, and this shape is what determines its specific function. Nearly all proteins must recognize and bind to some other molecule in order to function. Lysozyme, for example, can destroy bacterial cells, but first it must bind to specific molecules on the bacterial cell surface. Lysozyme's specific shape enables it to recognize and attach to its molecular target, which fits into the cleft you see on the right in Figure 3.14B.

The dependence of protein function on a protein's specific shape becomes clear when proteins are altered. In a process called **denaturation,** polypeptide chains unravel, losing their specific shape and, as a result, their function. For example, visualize what happens when you fry an egg. Heat quickly denatures the clear proteins surrounding the yolk, making them solid, white, and opaque. In this state, the proteins are insoluble in water and would be useless to a developing bird embryo. Changes in salt concentration and pH can also denature many proteins.

We examine the important subject of protein structure more closely in the modules that follow. 

? Why does a denatured protein no longer function normally?

The function of each protein is a consequence of its specific shape, which is lost when a protein denatures.

3.15 A protein's primary structure is its amino acid sequence

3F

The specific shape that determines a protein's function comprises four successive levels of structure, each determining the next one. We describe these levels in the four modules on this page. The figures on the facing page accompany these modules; Figures 3.15–3.18 illustrate the four levels of structure in a single protein called transthyretin. Found in blood, transthyretin is an important transport protein. It is a globular molecule whose specific shape enables it to transport two key chemicals throughout the body, a hormone from the thyroid gland and vitamin A.

Transthyretin, like every other protein, has a unique **primary structure,** which is the sequence of amino acids forming its polypeptide chains. Figure 3.15 on the facing page shows part of the primary structure of transthyretin. The three-letter abbreviations represent amino acids. A complete molecule of transthyretin has four polypeptide chains, each made up of 127 amino acids. In order for this or any other protein to perform its specific function, it must have the correct collection of amino acids arranged in a precise order. Even a slight change in a protein's primary structure may affect its overall shape and its ability to function. For instance, a single amino acid change in hemoglobin, the oxygen-carrying blood protein, causes sickle-cell disease, a serious blood disorder.

3.16 Secondary structure is polypeptide coiling or folding produced by hydrogen bonding

3F

In the second level of protein structure, parts of the polypeptide coil or fold into local patterns called **secondary structure.** As you can see in Figure 3.16, coiling of a polypeptide chain results in a secondary structure called an **alpha helix;** a certain kind of folding leads to a **pleated sheet.** Both of these patterns are maintained by regularly spaced hydrogen bonds between the —N—H groups and the —C=O groups along the polypeptide chain. Notice that each hydrogen bond (represented by a row of dots) links the —N—H of one amino acid with the —C=O of *another* amino acid, farther along the chain. Because the R groups of the amino acids and the H atoms attached to their alpha carbons are not important in these secondary structures, they are omitted from the diagrams. (The exception is the amino acid leucine, shown in full.)

3.17 Tertiary structure is the overall shape of a polypeptide

3F

The term **tertiary structure** refers to the overall, three-dimensional shape of a polypeptide. Most tertiary structures can be roughly described as either globular or fibrous. A transthyretin polypeptide has a generally globular shape, which results from the compact combination of an alpha helix and several pleated-sheet regions (Figure 3.17). The indentations and bulges arising from its particular arrangement of coils and folds give the polypeptide the specific shape appropriate to its function. Many proteins with globular tertiary structure have both helical and pleated-sheet regions. In contrast, many fibrous proteins, such as our hair proteins, are almost entirely helical. Tertiary structure generally results from interactions among the polypeptide's R groups. For example, globular proteins found in aqueous solution, such as transthyretin, are folded so that the hydrophobic R groups are on the inside of the molecule and the hydrophilic groups on the outside, exposed to water. In addition to the clustering of hydrophobic groups, hydrogen bonding and ionic bonding of some of the polar R groups help maintain the tertiary structure.

3.18 Quaternary structure is the relationship among multiple polypeptides of a protein

3F

Many proteins consist of two or more polypeptide chains, or subunits. Such proteins have a **quaternary structure,** resulting from bonding interactions among the subunits. Figure 3.18 shows a complete transthyretin molecule with its four identical subunits. Many other proteins have subunits that are different from one another. For example, the oxygen-transporting molecule hemoglobin has four subunits of two distinct types (see Figure 22.10B).

LEVELS OF PROTEIN STRUCTURE

Figure 3.15 Primary structure

Figure 3.16 Secondary structure

Figure 3.17 Tertiary structure

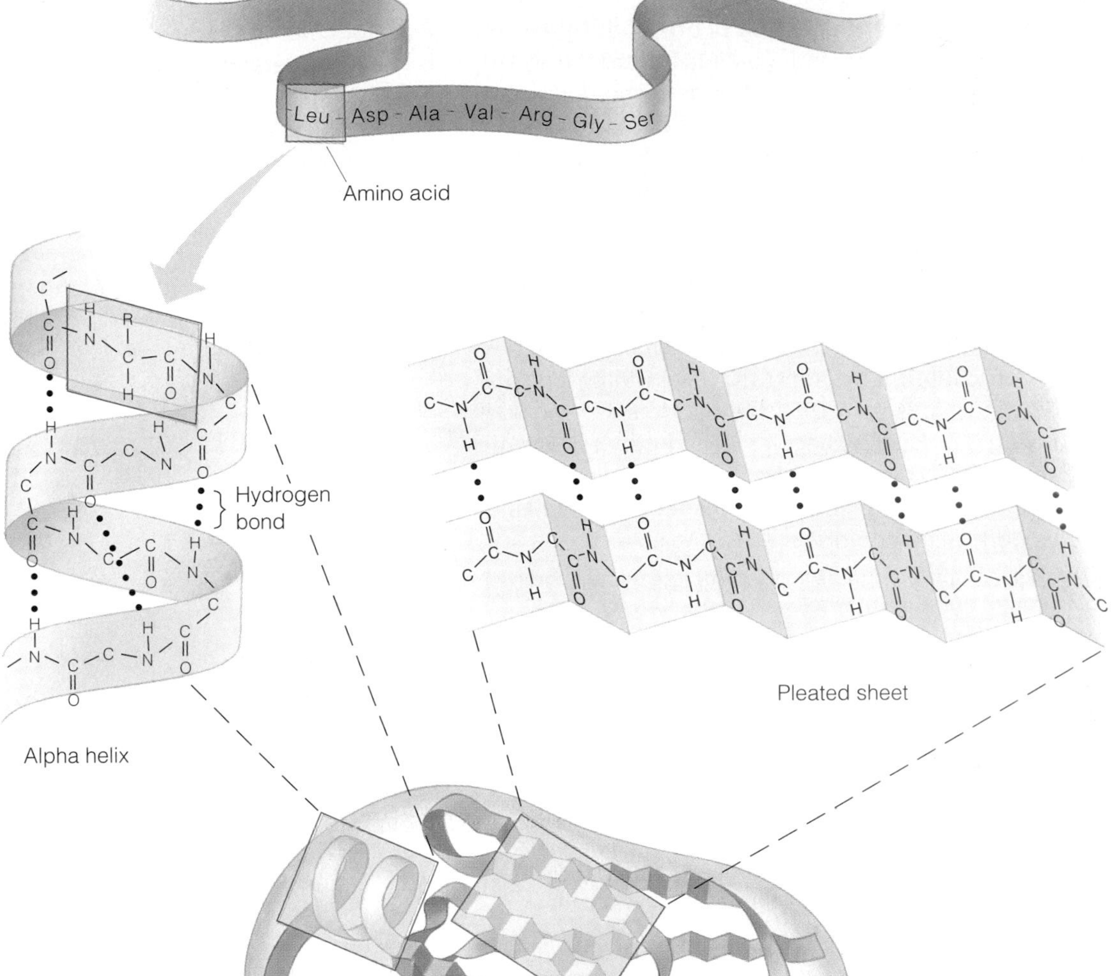

Figure 3.18 Quaternary structure

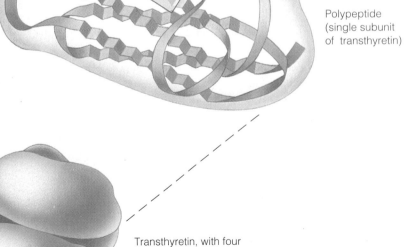

? A genetic mutation can change the primary structure of a protein. How can this destroy the protein's function?

Primary structure, the amino acid sequence, affects the secondary structure, which affects the tertiary structure, which affects the quaternary structure (if any). In short, the amino acid sequence affects the shape of the protein, and the function of a protein depends on its shape.

TALKING ABOUT SCIENCE

3.19 Linus Pauling contributed to our understanding of the chemistry of life

The year 1994 marked the passing of one of the giants of twentieth-century science, Linus Pauling. Toward the end of his 93 years, Pauling was most often associated with his controversial belief that large doses of vitamin C can help prevent the common cold, cancer, and other diseases. Earlier in his career, Pauling made extraordinary contributions to both science and human affairs. His connection with this chapter is his early work on protein structure and function.

Driven by the desire "to understand the world," Pauling started out in chemistry and physics. While a professor at the California Institute of Technology, he published a series of papers on chemical bonding that eventually led to a Nobel Prize in chemistry in 1954. By that time, Pauling was also studying biological molecules. In an interview several years ago, we asked him about the value of studying molecules for understanding life, and he responded by talking about his work on hemoglobin:

> Life is too complicated to permit a complete understanding through the study of whole organisms. Only by simplifying the problem—breaking it down into a multitude of individual problems—can you get the answers. In 1935, Dr. Charles Coryell and I made our discovery about how oxygen molecules are attached to the iron atoms of hemoglobin, not by getting a cow and putting it through our magnetic apparatus, but by getting some blood from the cow and studying this blood and the hemoglobin from it. . . .
>
> Of course, the study of different parts of an organism leads to the question, Do these parts interact? Can we learn more about the living organism by putting two parts together to see to what extent the properties of the combination are different from those of the two separated parts? This approach permits further progress in understanding the organism. And yet I, myself, have confidence that all of the properties of living organisms could ultimately be discovered by the process of attempting to reduce the organism . . . to a combination of the different parts: essentially, the molecules that make up the organism.

Besides finding out how hemoglobin carries oxygen, Pauling discovered how an abnormal hemoglobin molecule causes sickle-cell disease. And it was Pauling who first described the two fundamental secondary structures of proteins, the alpha helix and the pleated sheet. His recounting of that work reveals more about his scientific attitudes and efforts:

> I began trying to find the structure of proteins in 1937, and didn't succeed. So I began working with my collaborators to determine the three-dimensional structures of amino acids and simple peptides. In 1937, no one had yet determined such structures. . . . And then in 1948 I found the alpha-helix and pleated-sheet structures in proteins. . . . I'm surprised that nobody else had done this job in the 11 years that intervened—in a sense, surprised that I hadn't done it in '37, when my ideas were all the right ones. I just hadn't worked hard enough.

Pauling's efforts were not limited to science. He also became the scientific community's leading advocate for halting the testing of nuclear weapons, resulting in the accusation that he was a Communist and the revocation of his passport. Nevertheless, in 1963 he received the Nobel Peace Prize for helping produce a ban on nuclear testing. Pauling is the only person who has ever received two unshared Nobel Prizes.

The cancellation of Pauling's passport in 1952 kept him from working directly with scientists in Europe. This may have prevented Pauling from earning a third Nobel Prize, for at that time he was hot on the trail of the structure of DNA. But his triple-helix model was wrong, and it was James Watson and Francis Crick, working in England, who came up with the correct solution—a double helix, which we describe in the next module.

Figure 3.19 Linus Pauling with a model of the alpha helix in 1948

A decade after being awarded a Nobel Prize for chemistry, Pauling received the Nobel Peace Prize for ________________.

his efforts leading to a ban on nuclear testing

Nucleic Acids

Nucleic acids are information-rich polymers of nucleotides 3.20

3G

The **nucleic acids** are polymers that serve as the blueprints for proteins. There are two types: **deoxyribonucleic acid (DNA)** and **ribonucleic acid (RNA).** The genetic material that organisms inherit from their parents consists of DNA. Within the DNA are **genes,** specific stretches of the molecule that program the amino acid sequences (primary structure) of proteins. In determining primary structure, genes determine the specific three-dimensional structures and therefore the functions of proteins. Thus, through the actions of proteins, DNA controls the life of the cell and the organism.

DNA does not put its genetic information to work directly. It works through an intermediary—RNA. DNA's information is transcribed into RNA, which is then translated into the primary structure of proteins. We return to this chain of command and the functions of DNA and RNA later in the book. Here, we just want to introduce the structure of nucleic acids.

The monomers that make up nucleic acids are called **nucleotides.** As indicated in Figure 3.20A, each nucleotide has three parts. One part is a five-carbon sugar (blue); DNA has the sugar deoxyribose, whereas RNA has a closely related sugar called ribose. Linked to one end of the sugar in both types of nucleic acid is a functional group called a **phosphate group** (yellow). At the other end of the sugar is one of a number of chemical units called **nitrogenous bases** (like the one in Figure 3.20A, they all contain nitrogen). DNA has the nitrogenous bases adenine (A), thymine (T), cytosine (C), and guanine (G). RNA also has A, C, and G, but instead of thymine has uracil (U).

Like polysaccharides and polypeptides, a nucleic acid polymer—a polynucleotide—forms from its monomers by dehydration synthesis. In this process, the phosphate group of one nucleotide bonds to the sugar of the next monomer. The result is a repeating sugar-phosphate backbone in the polymer, as shown in Figure 3.20B.

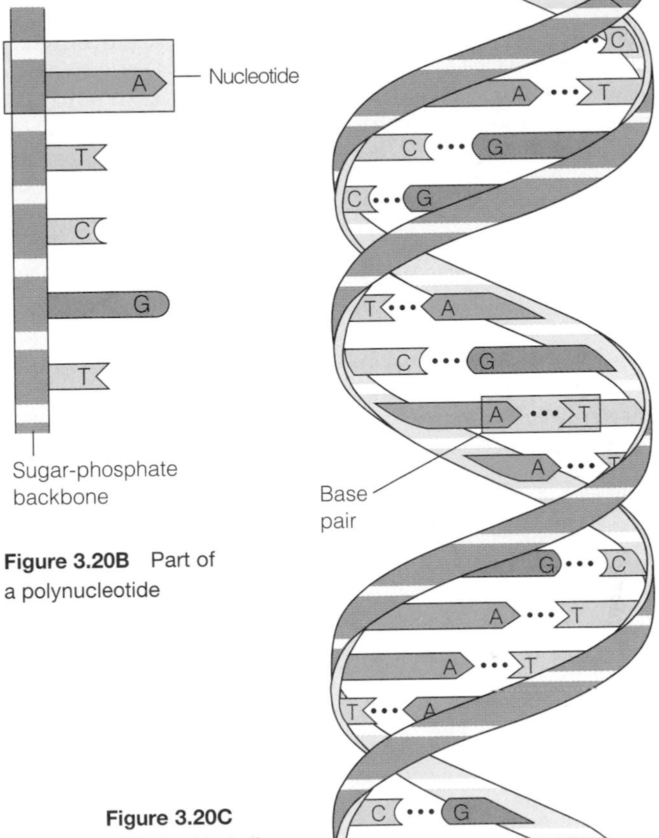

Figure 3.20B Part of a polynucleotide

Figure 3.20C DNA double helix

RNA usually consists of a single polynucleotide strand, but DNA is a **double helix,** in which two polynucleotides wrap around each other (Figure 3.20C). The nitrogenous bases protrude from the two sugar-phosphate backbones into the center of the helix. There they always pair up as shown: A pairs with T, and C pairs with G.

The two DNA chains are held in a double helix by hydrogen bonds (dotted lines) between their paired bases. Most DNA molecules are very long, with thousands or even millions of base pairs. One long DNA molecule contains many genes, each a specific series of hundreds or thousands of nucleotides along one of the polynucleotide strands. The specific sequence of nucleotides in a gene is the information that programs the primary structure of a protein.

This introduction to nucleic acids concludes our look at the four major classes of biological molecules. In the next chapter, we set the scene of molecular action—the cell.

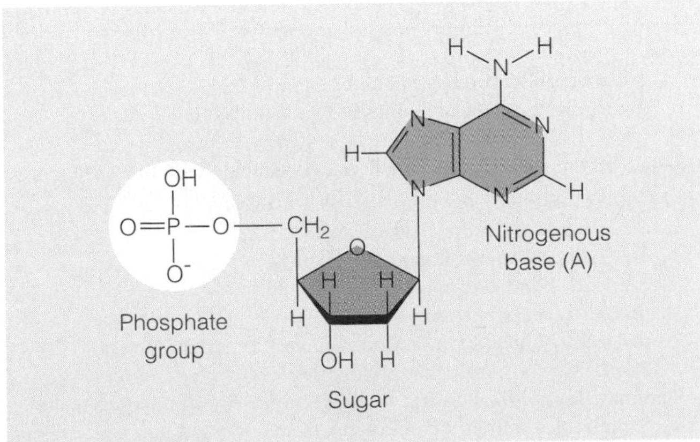

Figure 3.20A A nucleotide

In a double helix, a region along one DNA strand has this sequence of nitrogenous bases: TAGGCCT. What is the base sequence along the other strand of the double helix?

ATCCGGA

CHAPTER REVIEW

Chapter Summary

Introduction to Organic Compounds and Their Polymers (Introduction–3.3) Life's structural and functional diversity results from a great variety of molecules. A relatively small number of structural patterns underlies life's molecular diversity **(Introduction).** Virtually all of an organism's molecules are organic compounds, carbon-containing compounds made by living cells. A carbon atom forms four covalent bonds, so it can join with other carbon atoms to make chains or rings. Carbon chains vary in length, degree of branching, number and location of single and double bonds, and in the presence or absence of other elements **(3.1).** Functional groups are particular groupings of atoms that give organic molecules specific properties. For example, hydroxyl groups are characteristic of alcohols, and carboxyl groups are acidic **(3.2).** Three of the four kinds of large molecules in living things are polymeric macromolecules, long chains of smaller molecular units called monomers. A huge number of different polymers can be made from a small number of monomers. Cells link monomers to form polymers by dehydration synthesis. One monomer loses a hydroxyl group, the other loses a hydrogen atom, a covalent bond links the monomers, and a molecule of water is released. Macromolecules are broken down to their component monomers by the reverse process, hydrolysis **(3.3).**

Carbohydrates (3.4–3.7) Carbohydrates range from small sugars to large polysaccharides, which are macromolecules. A monosaccharide, or single sugar, typically has a formula that is a multiple of CH_2O and contains hydroxyl groups and a carbonyl group. The monosaccharides glucose and fructose, both with the formula $C_6H_{12}O_6$, are isomers; that is, they contain the same atoms but in different arrangements, and thus they have different properties. Monosaccharides are the fuels for cellular work **(3.4).** Monosaccharides can join to form disaccharides, such as sucrose (table sugar) **(3.5).** High sugar consumption can cause tooth decay and may be associated with poor diet **(3.6).** Polysaccharides are polymers of hundreds or thousands of monosaccharides linked by dehydration synthesis. Starch and glycogen are polysaccharides that store sugar for later use; cellulose is a polysaccharide that forms plant cell walls **(3.7).**

Lipids (3.8–3.10) Lipids, diverse compounds composed largely of carbon and hydrogen, are not true polymers. They are grouped together mainly because they do not mix with water. Fats, also called triglycerides, are lipids whose main function is energy storage. A triglyceride molecule consists of glycerol linked to three fatty acids. The fatty acids of unsaturated fats such as plant oils contain double bonds, which prevent them from solidifying at room temperature. Saturated fats, such as those in lard, lack double bonds and are solid at room temperature **(3.8).** Other lipids include phospholipids (found in cell membranes), waxes (which form waterproof coatings), and steroids (including sex hormones) **(3.9).** Use of synthetic variants of the male sex hormone, called anabolic steroids, can cause serious health problems **(3.10).**

Proteins (3.11–3.19) Proteins are involved in cellular structure, movement, defense, transport, and communication; and, as enzymes, they regulate chemical reactions **(3.11).** Proteins are the most structurally and functionally diverse of life's molecules. Their diversity is based on different arrangements of amino acids. Each amino acid contains an amino group, a carboxyl group, and an R group. The R groups in proteins distinguish 20 different amino acids, each with specific properties **(3.12).** Cells link amino acids together by dehydration synthesis; the bonds between amino acid monomers are called peptide bonds. Amino acid chains are called polypeptides **(3.13).** A protein consists of one or more polypeptide chains folded into a unique shape that determines the protein's function. A protein loses its specific functions when its polypeptides unravel **(3.14).** A protein's primary structure is the specific sequence of amino acids forming its polypeptide chains **(3.15).** A protein's secondary structure is the coiling or folding of its polypeptides. Secondary structure includes helical coiling and pleated-sheet folding, stabilized by hydrogen bonds between amino acids **(3.16).** Tertiary structure is the overall three-dimensional shape of a polypeptide, resulting from interactions between R groups **(3.17).** A protein's quaternary structure results from bonding interactions among its polypeptide subunits **(3.18).** Linus Pauling made important contributions to our understanding of protein structure and function **(3.19).**

Nucleic Acids (3.20) Nucleic acids—DNA and RNA—serve as the blueprints for proteins and thus ultimately control the life of a cell. The monomers of nucleic acids are nucleotides. Each nucleotide is composed of a sugar, a phosphate, and a nitrogenous base. DNA consists of two polynucleotides twisted around each other in a double helix. The sequence of the four kinds of nitrogenous bases in DNA carries genetic information. Stretches of a DNA molecule called genes program the amino acid sequences of proteins. DNA information is transcribed into RNA, a single-stranded nucleic acid, which is then translated into the primary structure of proteins.

Testing Your Knowledge

Multiple Choice

1. A glucose molecule is to starch as *(Explain your answer.)*
   a. a steroid is to a lipid
   b. a protein is to an amino acid
   c. a nucleic acid is to a polypeptide
   d. a nucleotide is to a nucleic acid
   e. an amino acid is to a nucleic acid

2. What makes a fatty acid an acid?
   a. It does not dissolve in water.
   b. It is capable of bonding with other molecules to form a fat.
   c. It has a carboxyl group that donates a hydrogen ion to a solution.
   d. It contains only two oxygen atoms.
   e. It is a polymer made of many smaller subunits.

3. Where in the tertiary structure of a water-soluble protein would you be most likely to find a hydrophobic amino acid R group?
   a. at both ends of the polypeptide chain
   b. on the outside, in the water
   c. covalently bonded to another R group
   d. on the inside, away from water
   e. covalently bonded to the amino group of the next amino acid

4. The enzyme called pancreatic amylase is a protein whose job is to attach to starch molecules in food and help break them down to disaccharides. Amylase cannot break down cellulose. Why not?

a. Cellulose is a kind of fat, not a carbohydrate like starch.
b. Cellulose molecules are much too large.
c. Starch is made of glucose; cellulose is made of other sugars.
d. The bonds between sugars in cellulose are much stronger.
e. The sugars in cellulose bond together differently than in starch, giving cellulose a different shape.

5. A shortage of phosphorus in the soil would make it especially difficult for a plant to manufacture

a. DNA
b. proteins
c. cellulose
d. fatty acids
e. sucrose

6. Lipids differ from other large biological molecules in that they

a. are much larger
b. are not truly polymers
c. do not have specific shapes
d. do not contain carbon
e. contain nitrogen atoms

7. Which functional group(s) act(s) as an acid?

a. carbonyl
b. amino
c. hydroxyl
d. carboxyl
e. all of the above

Describing, Comparing, and Explaining

1. List four different kinds of lipids and briefly describe their functions.

2. Explain why heating, changes in pH, and other environmental changes can interfere with the function of proteins.

3. How can a cell make many different kinds of protein out of only 20 amino acids? Of the myriad possibilities, how does the cell "know" which proteins to make?

4. Briefly describe the various functions performed by proteins in a cell.

5. When you eat a candy bar, the disaccharide sucrose in the candy is broken down in your intestine to two monosaccharide molecules (glucose and fructose), which are then absorbed into your blood. Starting with the sucrose molecule illustrated below, show how it would break down to produce glucose and fructose. What is the name of this reaction?

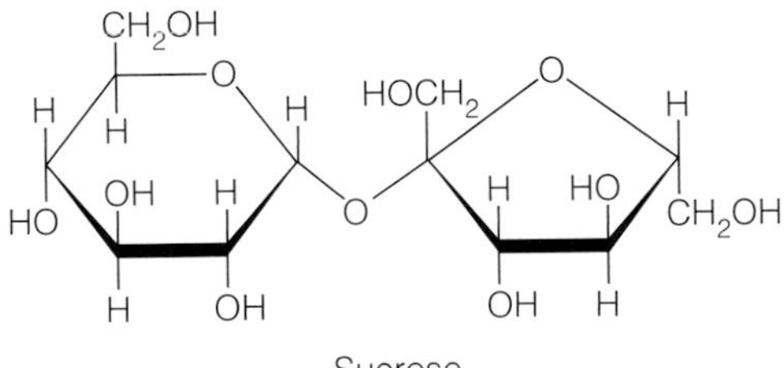

Sucrose

Thinking as a Scientist

1. Linus Pauling believed that large doses of vitamin C can help prevent the common cold, cancer, and other diseases. Imagine you have been given a research grant by the National Institutes of Health to evaluate Pauling's claims. How would you go about setting up an experimental study to determine whether vitamin C can prevent colds? How would you evaluate the results of your study?

2. Lysozyme is a small protein consisting of a single polypeptide chain of 129 amino acids. How would you calculate the number of possible different proteins 129 amino acids long that could be built using 20 amino acids? (*Hint:* How many different choices are there for the first amino acid? How many choices for the second? Then how many possible proteins could there be with only two amino acids? Can you extend this logic to 129 amino acids?)

3. A food manufacturer is advertising a new cake mix as fat-free. Scientists at the U.S. Food and Drug Administration (FDA) are testing the product to see if it truly lacks fat. Hydrolysis of the cake mix yields glucose, fructose, glycerol, a number of amino acids, and several kinds of molecules with long hydrocarbon chains. Further analysis shows that most of the hydrocarbon chains have a carboxyl group at one end. What would you tell the food manufacturer if you were the spokesperson for the FDA?

4. A short polypeptide is 10 amino acids long and contains only 3 kinds of amino acids: leucine (L), serine (S), and cysteine (C). The polypeptide is treated with a series of enzymes that break particular peptide bonds. The following molecular fragments are recovered from the reaction mixture:

Enzyme Cuts Peptide Bonds	Fragments Recovered
Between L and S	L, S, CS, and LCCC
Between L and C	CCC and LSLSCSL
Between C and S	SLCCC and LSLSC

What is the order of the 10 amino acids in the polypeptide?

a. LSCLSCLSCL
b. LSLSCSLCCC
c. CCCLSLSCSL
d. SLCCCLSLSC

Science, Technology, and Society

1. Each year, industrial chemists develop and test thousands of new organic compounds for use as pesticides, such as insecticides, fungicides, and weed killers. In what ways are these chemicals useful and important to us? In what ways can they be harmful? Is your opinion of pesticides positive or negative, in general? What influences have shaped your feelings about these chemicals?

2. Module 3.6 mentions that the economies of some developing tropical countries were hurt when food and beverage companies in the developed countries switched from using sucrose (derived from sugarcane) to high-fructose corn syrup (from corn grown in the developed countries). Many developing countries are dependent on single cash crops like sugarcane or coffee for most of their export income. Do the developed countries have a responsibility to protect the economies of developing countries from the impact of major product shifts like this one? Why or why not? Do consumers of the products have any say in the matter? Why or why not?

Media Resources

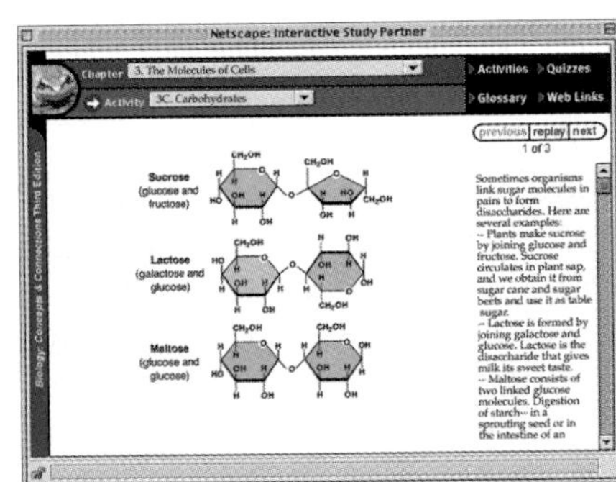

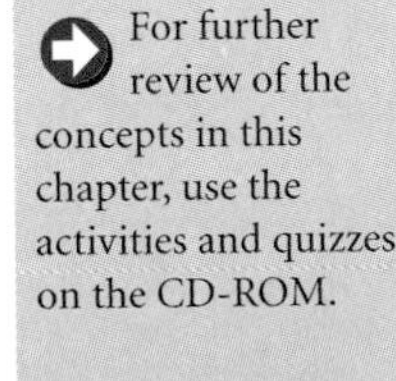
For further review of the concepts in this chapter, use the activities and quizzes on the CD-ROM.

Visit the Special Edition of The Biology Place web site for activities, quizzes, news articles, and web links.

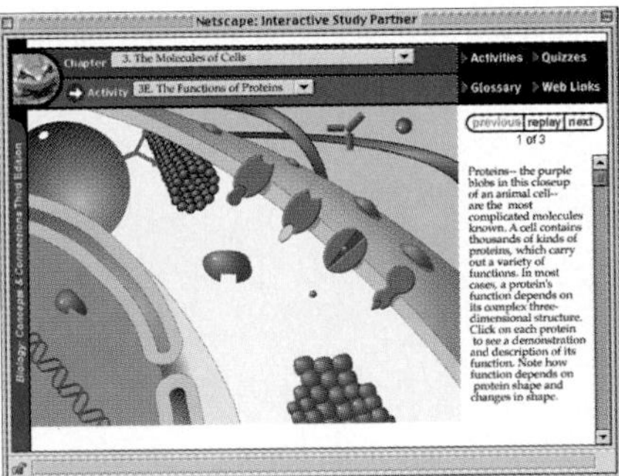

4 A Tour of the Cell

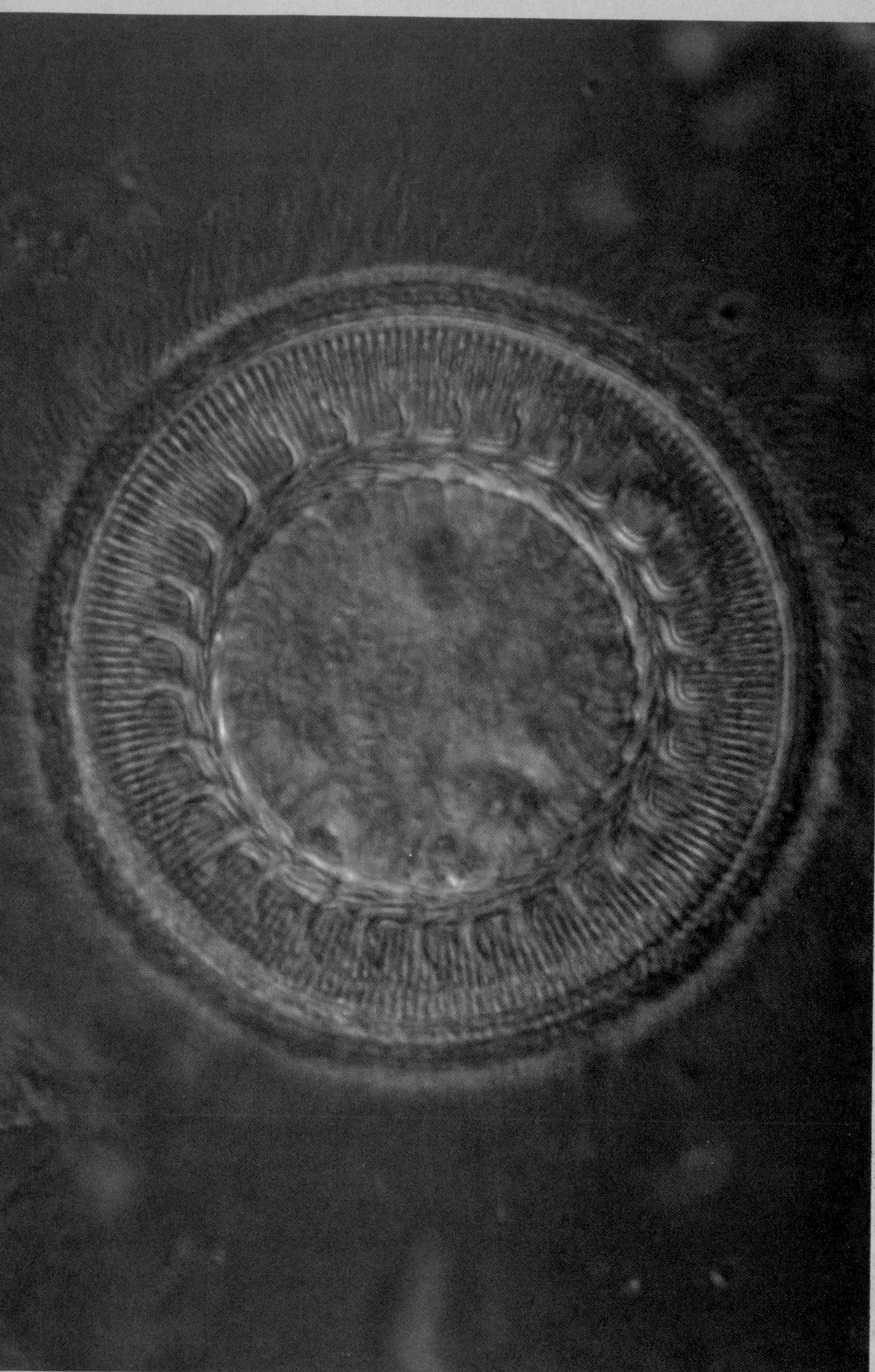

Outline

Spinning like a top in the water, the single-celled organism called *Trichodina* (Greek for whirling hairs) skims over the slippery surface of a fish and touches down. You see the underside of *Trichodina* in the photograph at the left—a complex circular clamp made of toothlike hooks and spines well suited for gripping fish skin. Fortunately for the fish, *Trichodina* eats mostly bacteria. Though the organism looks like a monster in this enlarged photograph, *Trichodina* is microscopic; about 100 of them would fit inside a small letter o on this page.

Trichodina is common in aquatic environments and often hitches a ride on fish. It is usually harmless, but sometimes its numbers build up to the point where it virtually covers a fish. When this happens, the host fish may die from tiny skin wounds inflicted by *Trichodina* or from suffocation. When *Trichodina* organisms coat a fish, they may cover the gills and prevent them from absorbing oxygen from the water.

With its specialized attachment device, *Trichodina* is one of the most complex of all single-celled organisms. But this protist also has a number of features in common with the cells that make up the bodies of multicelled organisms, including humans. In the drawing on this page, notice the rows of hairlike cilia ringing the organism's upper and lower surfaces. Beating back and forth, the cilia propel the cell through the water and also sweep bacteria into its cell mouth. Human cells lack a cell mouth, but many, such as those that line the windpipe, have cilia.

Another feature common to both *Trichodina* and our body cells is a nucleus containing DNA. The DNA specifies the proteins made in the cell and thus controls what the organism looks like and what it does. For instance, certain genes in *Trichodina's* DNA specify special hard proteins that become assembled into the hooks and spines of the organism's attachment device. Other genes specify proteins that function in the beating of the cilia.

Every cell, *Trichodina* included, also has a **plasma membrane,** which forms its outside border and sets the cell off from its fluid environment. More than just a boundary, the plasma membrane enables the cell to take up needed molecules from the surroundings, provides sites where important chemical reactions occur, and disposes of wastes. It thus helps maintain a life-sustaining collection of molecules inside the cell quite different from that of the outside environment.

Everything inside a cell between the plasma membrane and the nucleus is called the **cytoplasm.** The cytoplasm consists of a semifluid medium and various structures called organelles suspended in it. An **organelle** (Latin for small organ) is a cellular structure whose anatomy gives it a specific role to play in the life of the cell. The nucleus, specialized for controlling cellular activities, and the cilia, specialized for movement, are both organelles. Cilia are enclosed in extensions of the plasma membrane.

In this chapter, we examine the structure of living things at the cellular level. We begin by looking at the main kinds of microscopes that biologists use to explore structural details we can't see with the unaided eye. We then show the cellular structures that microscopes have revealed and describe their functions in the cell.

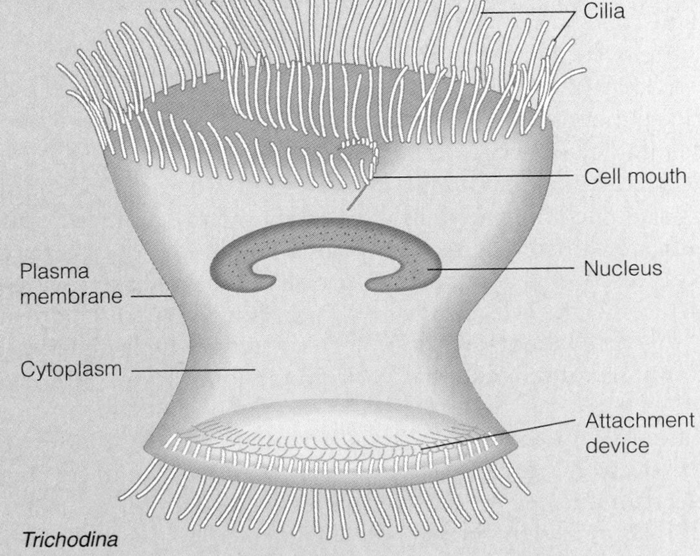

Trichodina

4.1 Microscopes provide windows to the world of the cell

Before microscopes were first used in the seventeenth century, no one knew for certain that living organisms were composed of cells. The first microscopes, like the ones you may have used in a biology laboratory, were light microscopes. A **light microscope (LM)** works by passing visible light through a specimen, such as a living *Trichodina* or a piece of animal or plant tissue. As Figure 4.1A shows, glass lenses in the microscope bend the light to magnify the image of the specimen and project the image into the viewer's eye or onto photographic film. A photograph taken through a microscope is called a **micrograph.** The notation "LM 109×" printed along the right edge of the micrograph of *Trichodina* in Figure 4.1A is in a form we use throughout this book. It tells you that the photograph was taken through a light microscope and that this image is about 109 times the actual size of the organism. *Trichodina* is actually about 1⁄20 of a millimeter in diameter. This image could be magnified many more times than shown here. Beyond a certain point, though, the image would begin to blur, and additional magnification would only cause more blurring. Light microscopes can magnify objects only about 1000 times without causing blurriness.

Magnification, the increase in the apparent size of an object, is only one important factor in microscopy (the use of a microscope). Also important is **resolving power,** a measure of the clarity of an image. Resolving power is the ability of an optical instrument to show two objects as separate. For example, what looks to your unaided eye like a single star in the sky may be resolved as two stars with the help of a telescope. Any optical device is limited by its resolving power. The light microscope cannot resolve detail finer than 0.2 micrometer (abbreviated µm; 1 µm = $\frac{1}{1000}$ mm), about the size of the smallest bacterium. Consequently, no matter how many times its image of such a bacterium is magnified, the light microscope cannot show the details of the cell's internal structure. (The µ in the abbreviation for micrometer is the Greek letter mu.)

From the year 1665, when English microscopist Robert Hooke discovered cells, until the middle of the twentieth century, biologists had only light microscopes for viewing cells. But they discovered a great deal, including the cells composing animal and plant tissues, microscopic organisms—for example, *Trichodina* was discovered about 200 years ago—and some of the structures within cells. By the mid-1800s, these discoveries led to the **cell theory,** which states that all living things are composed of cells and that all cells come from other cells.

Our knowledge of cell structure took a giant leap forward as biologists began using the electron microscope in the 1950s. Instead of light, the **electron microscope (EM)** uses a beam of electrons. The EM has a much higher resolving power than the light microscope. In fact, the most powerful modern EMs can distinguish objects as small as 0.2 nanometer (abbreviated nm; 1 nm = $\frac{1}{1,000,000}$ mm), a thousandfold improvement over the light microscope. The period at the end of this sentence is about a million times bigger than an object 0.2 nm in diameter, which is the size of a large atom. Only under special conditions can EMs detect individual atoms. However, cells, cellular organelles, and even molecules like DNA and protein are much larger than single atoms. The highest-power electron micrographs you will see in this book have magnifications of about 100,000 times.

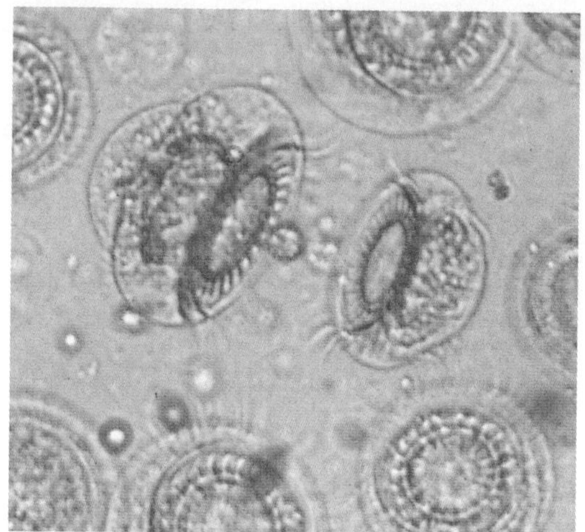

Light micrograph of *Trichodina*

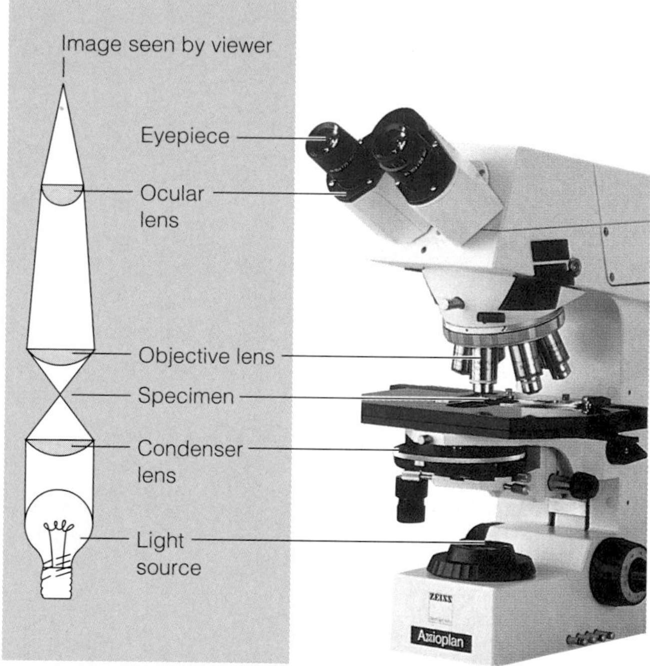

Figure 4.1A Light microscope (LM)

Figures 4.1B and 4.1C show two kinds of electron microscopes, along with images they have produced of cilia.

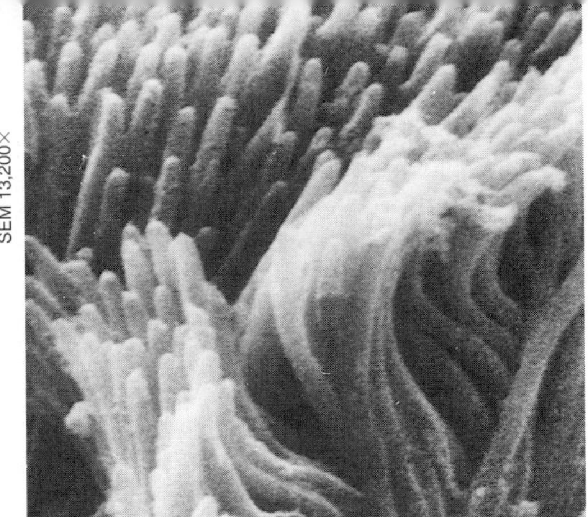

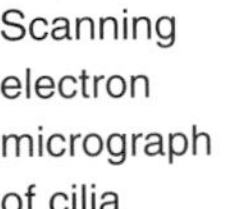

Scanning electron micrograph of cilia

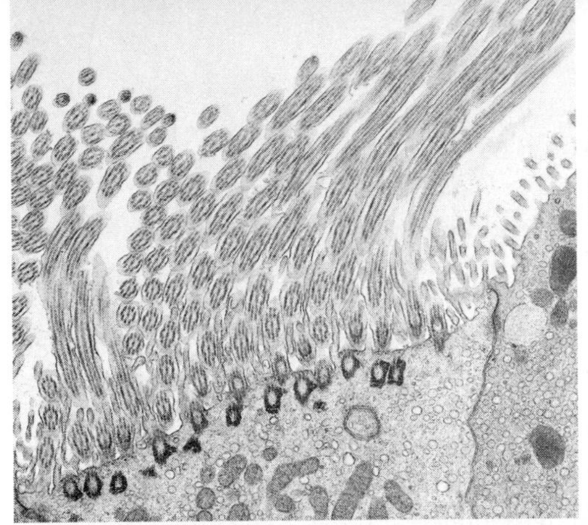

Transmission electron micrograph of cilia

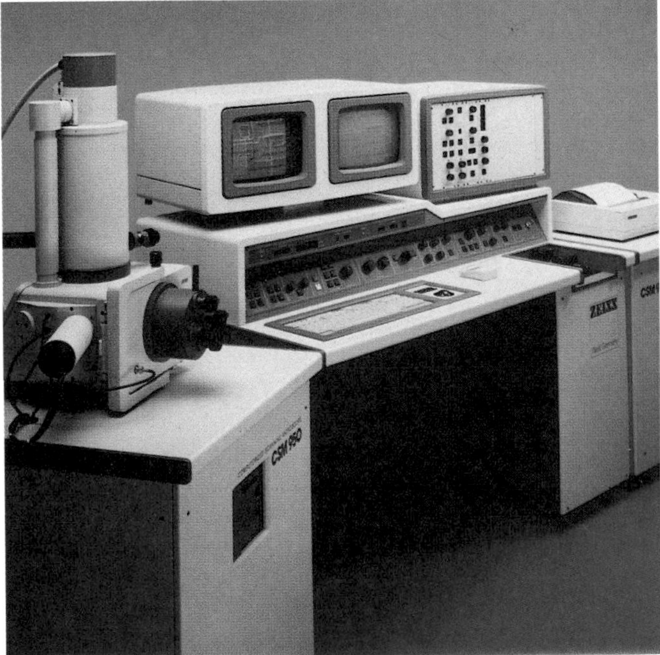

Figure 4.1B Scanning electron microscope (SEM)

Figure 4.1C Transmission electron microscope (TEM)

(The cilia in the photographs are from cells of a rabbit, but they are identical in structure to the cilia of *Trichodina* and other single-celled organisms.) Biologists use the **scanning electron microscope (SEM)** to study the detailed architecture of cell *surfaces.* The SEM uses an electron beam to scan the surface of a cell or group of cells that have been coated with metal. The metal stops the beam from going through the cells. When the metal is hit by the beam, it emits electrons. The electrons are focused to form an image of the outside of the cells. The scanning electron micrograph in Figure 4.1B shows the shapes and arrangement of cilia covering a cell. Many structural details of cell surfaces have been discovered using the SEM. As you can see, the SEM produces images that look three-dimensional.

The **transmission electron microscope (TEM)** is used to study the details of *internal* cell structure. Specimens are cut into extremely thin sections, and the TEM aims an electron beam through a section, just as a light microscope aims a beam of light through a specimen. However, instead of lenses made of glass, the TEM uses electromagnets as lenses, as do all electron microscopes. The electromagnets bend the electron beam to magnify and focus an image onto a viewing screen or photographic film. The micrograph in Figure 4.1C shows internal details of the cilia of a rabbit cell as seen with the TEM.

Electron microscopes have truly revolutionized the study of cells and cell organelles. Nonetheless, they have not replaced the light microscope. One problem with electron microscopes is that they cannot be used to study living specimens because the specimen must be held in a vacuum chamber; that is, all the air and liquid must be removed. For a biologist studying a living process, such as the whirling movement of *Trichodina,* a light microscope equipped with a video camera might be better than either an SEM or a TEM. Thus, the light microscope remains a useful tool, especially for studying living cells. The size of a cell often determines the type of microscope a biologist uses to study it. The next two modules discuss cell size.

Which type of microscope would you use to study: (a) the changes in shape of a living human white blood cell; (b) the finest details of surface texture of a human hair; (c) the detailed structure of an organelle in the cytoplasm of a human liver cell?

(a) light microscope; (b) scanning electron microscope; (c) transmission electron microscope

4.2 Cell sizes vary with their function

The figure at the right shows the size range of cells compared with objects both larger and smaller. The smallest cells are bacteria called mycoplasmas, with diameters between 0.1 μm and 1.0 μm. The bulkiest cells are bird eggs, and the longest cells are certain muscle and nerve cells. Most cells lie between these extremes, in the range indicated by the white area on the figure. *Trichodina,* for instance, is about 50 μm in diameter. The scale is logarithmic, with the length labels on the left ascending in powers of ten, to accommodate the range of sizes shown. Thus, except for the lengthy nerve and muscle cells and the egg cells of many animals, the biggest plant and animal cells, with diameters of about 100 μm, are ten times larger than the smallest, at about 10 μm.

Cell size and shape are related to cell function. Bird eggs are bulky because they contain a large amount of nutrient material for the developing young. Long muscle cells are efficient in pulling different body parts together. Lengthy nerve cells can transmit nerve signals rapidly between distant parts of an animal's body. On the other hand, small size also has many benefits. For example, human red blood cells are only about 8 μm in diameter and therefore can fit through our tiniest blood vessels.

Below is a list of the most common units of length biologists use. As you can see, they are metric, so interconversions are easy.

Using a light microscope to examine a thin section of a spherical cell, you find that it is 0.3 mm in diameter. The nucleus is about one-fourth as wide. What is the diameter of the nucleus in micrometers?

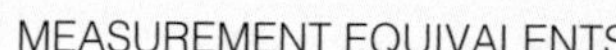

About 75

MEASUREMENT EQUIVALENTS

1 centimeter (cm) = 10^{-2} meter (m) = 1/100 m = about 0.4 inch

1 millimeter (mm) = 10^{-3} m = 1/1000 m = 1/10 cm

1 micrometer (μm) = 10^{-6} m = 1/1,000,000 m = 1/10,000 cm

1 nanometer (nm) = 10^{-9} m = 1/1,000,000,000 m = 1/10,000,000 cm

1 meter = 10^2 cm = 10^3 mm = 10^6 μm = 10^9 nm

Figure 4.2 The sizes of cells and related objects

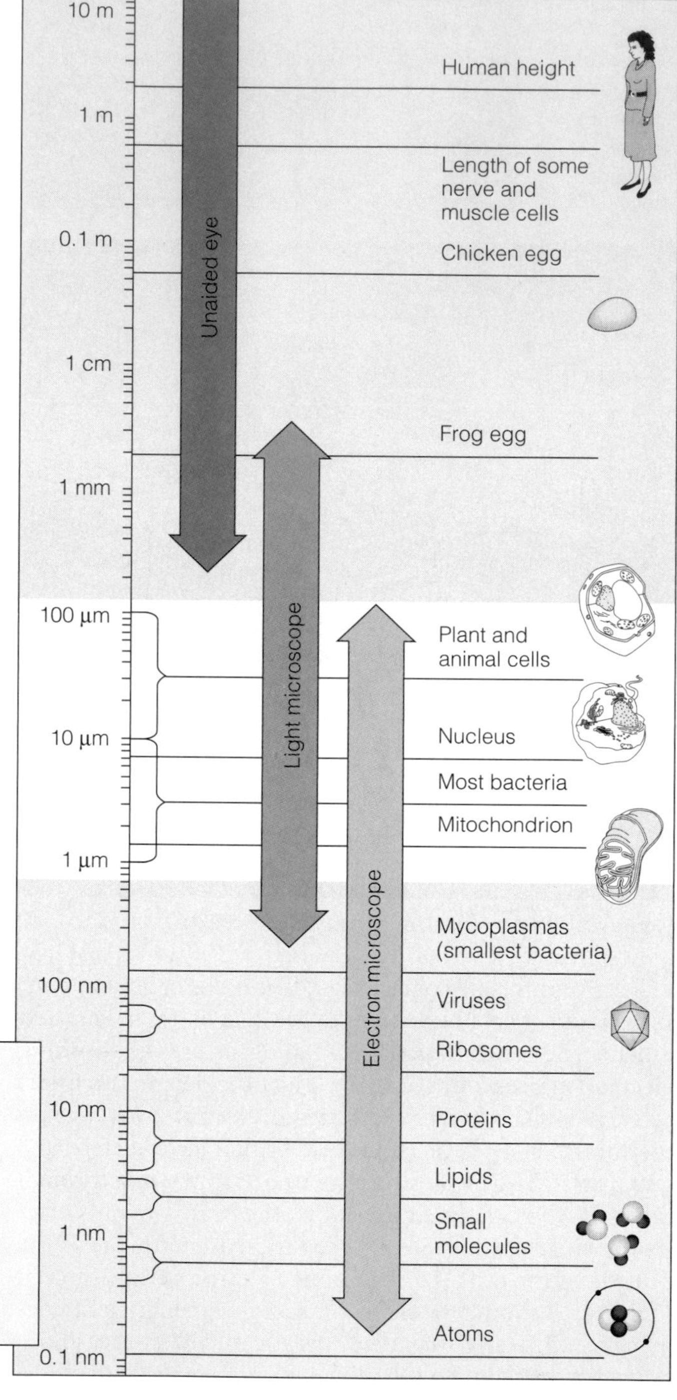

4.3 Natural laws limit cell size

There are lower and upper limits to cell size. At minimum, a cell must be able to house enough DNA, protein molecules, and internal structures to survive and reproduce. The maximum size of a cell is limited by its requirement for enough surface area to obtain adequate nutrients from the environment and dispose of wastes. Large cells have more surface area than small cells, but large cells have much less surface area *relative to their volume* than small cells of the same shape.

Figure 4.3 on the top of the facing page illustrates the surface-to-volume relationship using cube-shaped cells. The figure shows 1 large cubic cell and 27 small ones. (The purple spheres are cell nuclei.) In both cases, the total volume is the same:

$$\text{Volume} = 30\ \mu m \times 30\ \mu m \times 30\ \mu m = 27{,}000\ \mu m^3$$

In contrast to the total volume, the total surface areas are very different. Because a cube has six sides, its surface area is six times the area of one side. The surface areas of the cubes are as follows:

$$\text{Area of large cube} = 6 \times (30\ \mu m \times 30\ \mu m) = 5400\ \mu m^2$$

$$\text{Area of small cube} = 6 \times (10\ \mu m \times 10\ \mu m) = 600\ \mu m^2$$

For all 27 of the small cubes, the total surface area is $27 \times 600\ \mu m^2$, which equals $16{,}200\ \mu m^2$—three times the surface area of the large cube.

Thus, we see that a large cell has a much smaller surface area relative to its volume than smaller cells have. In fact, *it is the ratio of cell surface to cell volume that imposes upper limits on cell size.* If these were living cells, the plasma membranes of the small cells would service their small volumes of cytoplasm much more easily than the membrane of the large cell, with its single large volume. As living cells evolved, only those with sufficient surface area to serve their volume survived and reproduced. Of course, no cells are perfect cubes or spheres, and many cell shapes have evolved that affect the size restriction. Muscle and nerve cells can be very long because they are thin and therefore have more surface area per unit of volume than spherical cells.

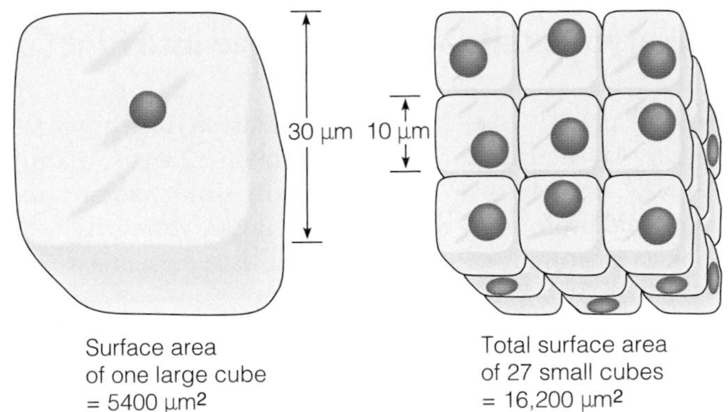

Figure 4.3 Effect of cell size on surface area

Red blood cells take up O_2 in the lungs and then give up the O_2 as they pass through the blood vessels of other organs. These cells are among the smallest of human cells. Explain one functional advantage of their small size.

A large number of small cells has a much greater total surface area than a small number of larger cells. The greater combined plasma-membrane area of the small cells is more readily crossed by O_2 molecules entering or leaving the cells.

Prokaryotic cells are small and structurally simple

4.4

4A

Two kinds of structurally different cells have evolved over time. Bacteria and archaea consist of **prokaryotic cells,** whereas all other forms of life are composed of **eukaryotic cells.** This module and the next give overviews of the structures of prokaryotic and eukaryotic cells.

It takes an electron microscope to clearly see the structural details of any cell, and this is especially true of prokaryotic cells because they are so small. Most prokaryotic cells range from 2 to 8 μm in length, averaging about one-tenth of the size of a typical eukaryotic cell. A prokaryotic cell (Greek *pro,* before, and *karyon,* kernel) lacks a nucleus; its DNA is coiled into a **nucleoid** (nucleuslike) **region,** as shown in Figure 4.4. Because no membrane surrounds the nucleoid region, the DNA is in direct contact with the rest of the cell contents. Notice the ribosomes (dark brown dots) inside the cell. Using instructions dispatched from the DNA, **ribosomes** assemble amino acids into polypeptides, the polymers that make up proteins. As mentioned in Chapter 3, DNA controls cells by using RNA intermediaries to specify what proteins are made.

A plasma membrane encloses the cytoplasm of the prokaryotic cell. Surrounding the plasma membrane of most bacteria is a fairly rigid, chemically complex **bacterial cell wall.** The wall protects the cell and helps maintain its shape. In some prokaryotes, another layer, a sticky outer coat called a **capsule,** surrounds the cell wall and further protects the cell surface. Capsules also help glue prokaryotes to surfaces, such as sticks and rocks in fast-flowing streams, or tissues within the human body. In addition to outer coats, some bacteria have surface projections. Short projections called **pili** (singular, *pilus*) help attach bacteria to surfaces. Longer projections called **prokaryotic flagella** (singular, *flagellum*) propel the prokaryotic cell through its liquid environment.

How is the nucleoid region of a prokaryotic cell unlike the nucleus of a eukaryotic cell?

There is no membrane enclosing the DNA of the nucleoid region.

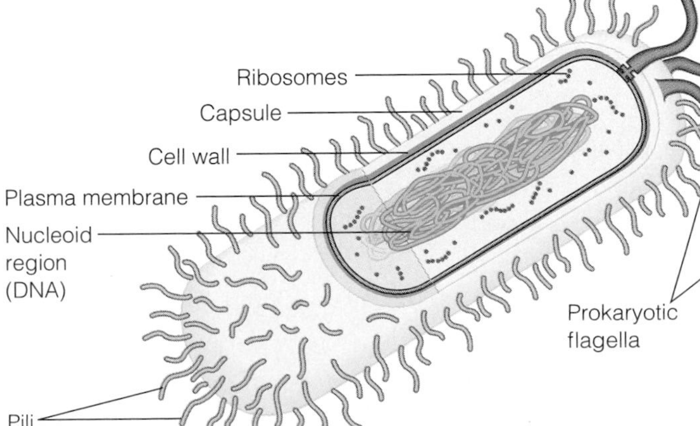

Figure 4.4 A prokaryotic cell (35,000x)

4.5

Eukaryotic cells are partitioned into functional compartments

4B

4C

See Activities 4B and 4C on the CD-ROM.

All eukaryotic cells (Greek *eu,* true, and *karyon,* kernel, referring to the nucleus)—whether from animals, plants, protists, or fungi—are fundamentally similar to one another and profoundly different from prokaryotic cells. Let's look at an animal cell and a plant cell as representatives of the eukaryotes.

Figure 4.5A illustrates an idealized animal cell, showing the details visible with the transmission electron microscope. A glance at the figure confirms that eukaryotic cells are much more complex than prokaryotic cells. The most obvious difference is the variety of structures in the cytoplasm. Notice that most of these structures are composed of membranes (recognizable in the drawing as white lines in cross section). In eukaryotes, membranes partition the cytoplasm into compartments, which biologists call *membranous* organelles (their names are underlined in the figures, as is the term *plasma membrane*). The membranous organelles shown in Figure 4.5A are the nucleus, endoplasmic reticulum, Golgi apparatus, mitochondria, lysosomes, and peroxisomes. Although most organelles are colorless, we color-code them here and in many places throughout the book for easier identification. For example, we consistently use purple for the nucleus, the control center of the eukaryotic cell, when it is featured in a figure.

Many of the chemical activities of cells—activities known collectively as **cellular metabolism**—occur in the fluid-filled spaces within membranous organelles. These spaces are important as sites where specific chemical conditions are maintained, conditions that vary from one organelle to another. Metabolic processes that require different conditions can take place simultaneously in a single cell because they occur within separate organelles. For example, while the endoplasmic reticulum is engaged in making a steroid hormone, neighboring peroxisomes may be making hydrogen peroxide (H_2O_2) as a poisonous by-product of certain metabolic reactions. Only because the H_2O_2 is confined within peroxisomes, where it is quickly converted to H_2O, are the steroids protected from destruction.

Another benefit of internal membranes is that they greatly increase a eukaryotic cell's total membrane area. A typical eukaryotic cell, with a diameter about ten times greater than that of a typical prokaryotic cell, has a thousand times the cytoplasmic volume but only a hundred times the plasma membrane area of the prokaryotic cell. In eukaryotic cells, internal (organelle) membranes provide the surfaces where many important metabolic processes occur; in fact, many enzymatic proteins essential for metabolic processes are components of organelle membranes.

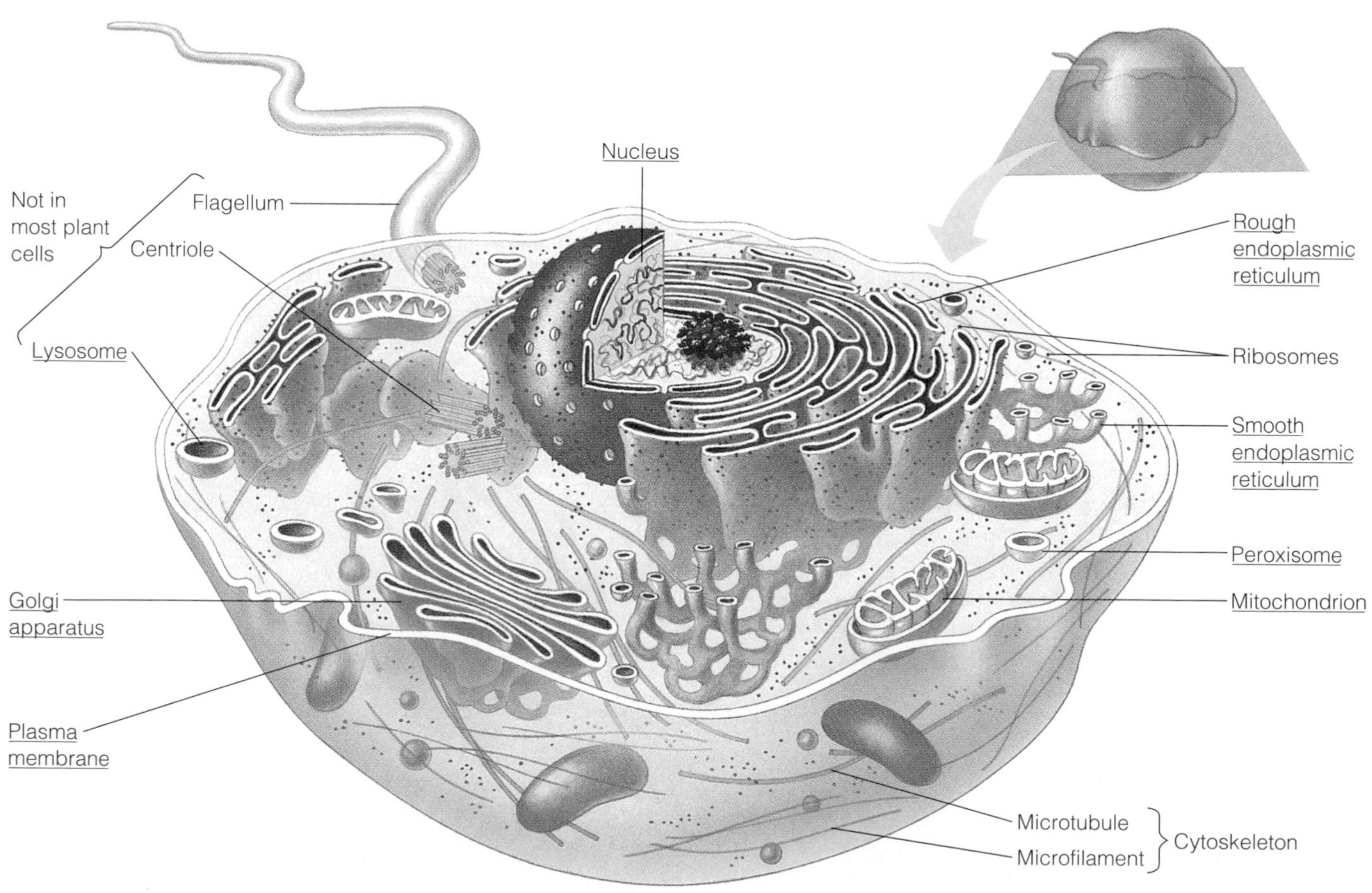

Figure 4.5A An animal cell (8,000x)

Without their internal membranes, eukaryotic cells probably would not have enough membrane surface area to meet their metabolic needs.

Almost all of the organelles mentioned so far are present in the cells of both animals and plants. As you can see by comparing Figures 4.5A and 4.5B, the exception is the lysosome, which is not found in plant cells. But there are also other differences between plant and animal cells. One difference is that an animal cell has a pair of centrioles, which plant cells lack. An animal cell may also have a flagellum, and many animal cells have more than one. In contrast, among the plants, only sperm cells in a few species have flagella. (The eukaryotic flagellum is different from the prokaryotic flagellum in both structure and operation.)

A plant cell has some structures that an animal cell lacks. For one, a plant cell has a rigid, rather thick cell wall (as do the cells of fungi and many protists). Cell walls protect cells and help maintain their shape. Chemically different from prokaryotic cell walls, plant cell walls contain the polysaccharide cellulose. Unlike typical animal cells, many mature plant cells have the polygonal shape shown in Figure 4.5B.

Another organelle found in plant cells but not in animal cells is the chloroplast, where photosynthesis occurs. (Chloroplasts are also found in some protists.) Unique to plant cells is a large central vacuole, a sac that stores water and a variety of other chemicals. Evident in most mature plant cells, the central vacuole contains enzymes that carry out cellular digestion, fulfilling the role of an animal cell's lysosome. Furthermore, by taking up additional water and expanding, the central vacuole can help the cell enlarge.

Although we have emphasized membranous organelles, eukaryotic cells contain nonmembranous structures as well (those with labels not underlined). Among them are the centriole, flagellum, and cytoskeleton, all of which contain protein tubes called microtubules. Also, you can see by the many dark brown dots in both figures that ribosomes, the sites of protein synthesis, occur throughout the cytoplasm, as they do in prokaryotic cells. In addition to ribosomes in the cytoplasmic fluid, eukaryotic cells have many ribosomes attached to parts of the endoplasmic reticulum (making it "rough") and to the outside of the nucleus. Ribosome structure is presented in Chapter 10. In the remaining modules of this chapter, we discuss in more detail the other organelles of eukaryotic cells, starting with the nucleus.

Which of the following organelles does not belong in the list: mitochondrion, chloroplast, ribosome, lysosome, peroxisome? Why?

Ribosome; because it is the only organelle in the list that is not bounded by a membrane

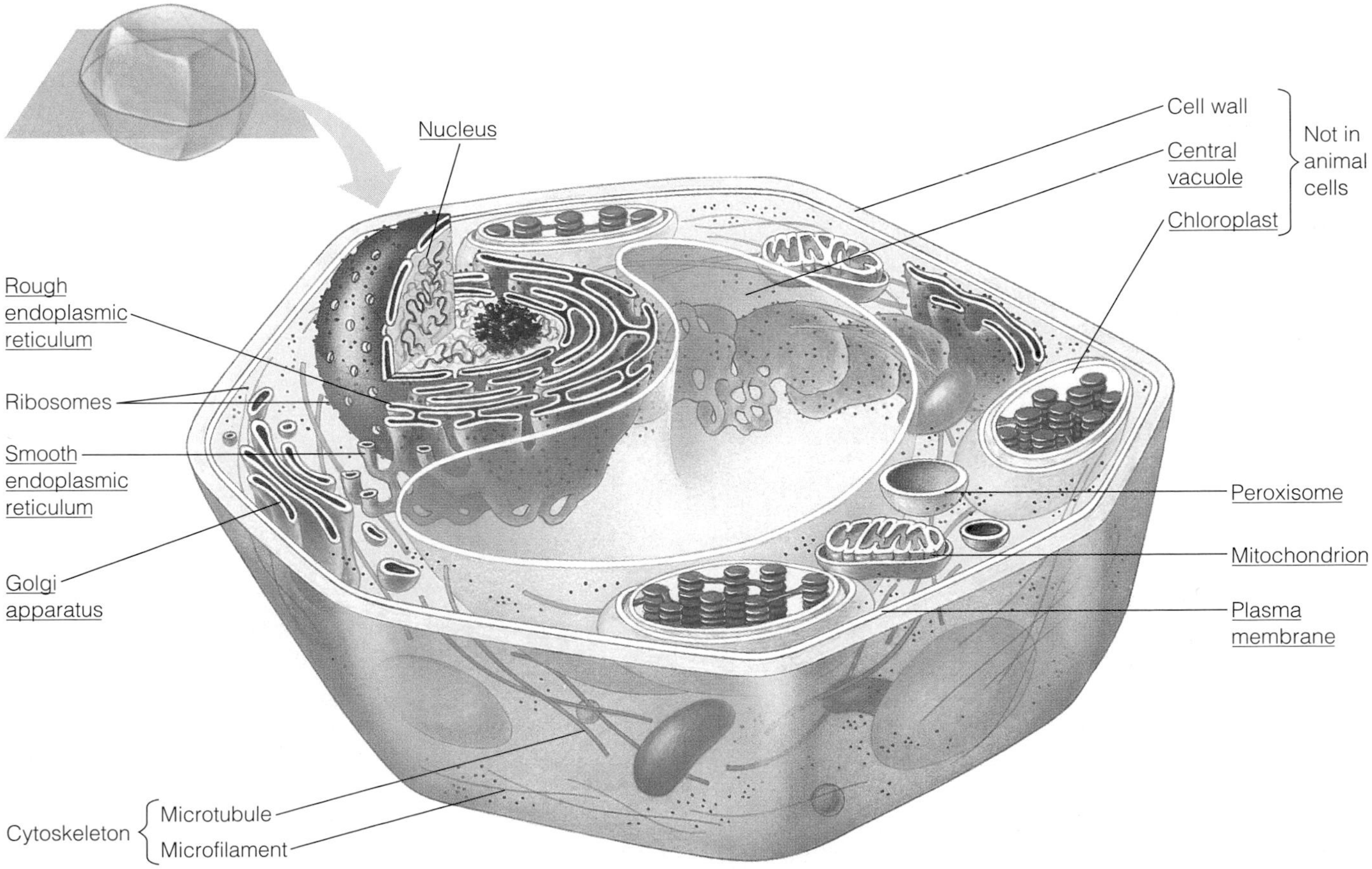

Figure 4.5B A plant cell (8,000x)

Organelles of the Endomembrane System

4.6 The nucleus is the cell's genetic control center

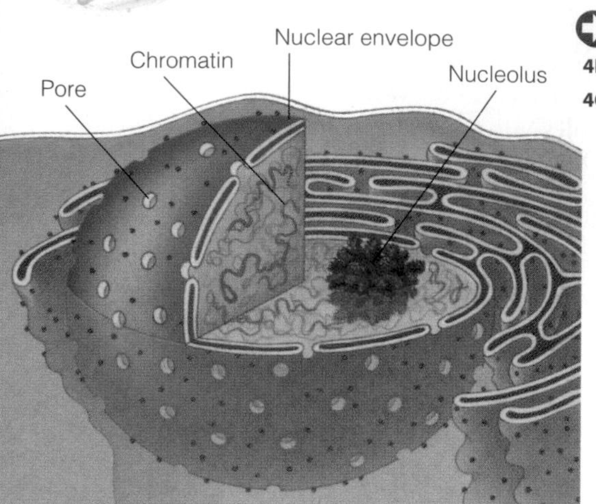

4B 4C The **nucleus** is the genetic control center of a eukaryotic cell. Its DNA, like a prokaryote's DNA, is the cell's hereditary blueprint. Nuclear DNA is attached to proteins, forming very long fibers called **chromatin** (the purple threads in Figure 4.6). Each fiber constitutes a **chromosome.** During cell reproduction, the chromatin coils up, forming thicker structures that are discernible with a light microscope as individual chromosomes.

Figure 4.6 The nucleus

Enclosing the nucleus is a **nuclear envelope,** a double membrane perforated with pores that control the flow of materials into and out of the nucleus. Adjoining the chromatin within the nucleus is a mass of fibers and granules called the **nucleolus.** The nucleolus, consisting of parts of the chromatin DNA combined with RNA and proteins, is where ribosomes are made.

What is the relationship of chromosomes to chromatin?

Chromosomes are made of chromatin, which is a combination of DNA and proteins. The chromatin fibers in a cell become more compact and recognizable as individual chromosomes during reproduction of the cell.

4.7 Overview: Many cell organelles are related through the endomembrane system

We now focus on eukaryotic organelles that are formed of interrelated membranes (Modules 4.7–4.14). Some of these membranes are physically connected and some are not, but collectively they constitute a cytoplasmic network that biologists call the **endomembrane system.** Many of the organelles of this system work together in the synthesis, storage, and export of important molecules.

One of the organelles, the **endoplasmic reticulum (ER),** is the prime example of the direct interrelatedness of parts of the endomembrane system. (The term *endoplasmic reticulum* comes from Greek words meaning "network within the cell.") As we will discuss, there are two kinds of ER: rough ER and smooth ER. These organelles differ in structure and function, but the membranes that form them are continuous. Membranes of the rough ER are also continuous with the outer membrane of the nuclear envelope. The space within the ER is separated from the cytoplasmic fluid by the ER membrane (see Figure 4.8A). Thus, the interconnected membranes of the ER and nuclear envelope partition the inside of the cell into two separate compartments. Dividing the cell into compartments is a major function of the endomembrane system.

Which structure includes all others in the list: rough ER, smooth ER, endomembrane system, nuclear envelope?

Endomembrane system

4.8 Rough endoplasmic reticulum makes membrane and proteins

4B 4C The "rough" in **rough endoplasmic reticulum,** or **rough ER,** refers to the appearance of this organelle in electron micrographs. As Figure 4.8A shows, the roughness results from ribosomes, which stud the membranes of the organelle. Rough ER is a network of interconnected flattened sacs, with two main functions. One is to make more membrane. Some of the proteins made by ER ribosomes are inserted into the ER membrane, as are phospholipids made by ER enzymes. As a result, the ER membrane enlarges, and some of it later ends up in other organelles.

The other major function of rough ER is to make proteins that are secreted by the cell. An example of such a **secretory protein** is an antibody, a defensive molecule made and secreted by white blood cells. Ribosomes of the rough

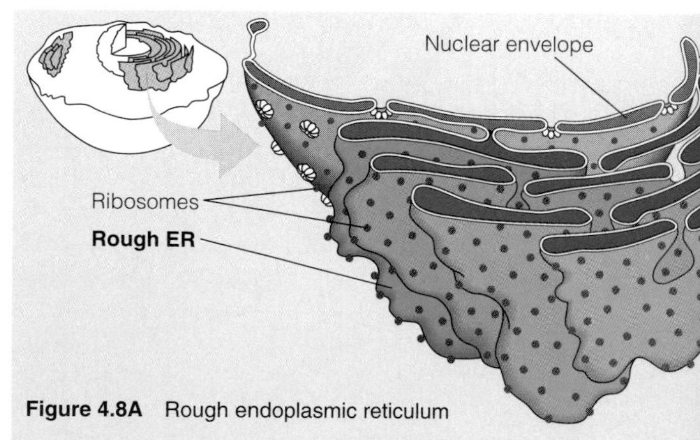

Figure 4.8A Rough endoplasmic reticulum

ER synthesize the antibody's polypeptides, which assemble into functional proteins inside the ER. Figure 4.8B shows the synthesis and packaging of a secretory protein made of a single polypeptide. ① As the polypeptide is synthesized, it passes into the ER. ② Short chains of sugars are then linked to the polypeptide, making the molecule a **glycoprotein** (*glyc-* means sugar). When the molecule is ready for export from the ER, ③ the ER packages it in a tiny sac called a **transport vesicle.** This vesicle ④ buds from the ER membrane.The secretory protein will now travel to the Golgi apparatus for further processing. From there, a transport vesicle containing the finished molecule will make its way to the plasma membrane and release its contents from the cell.

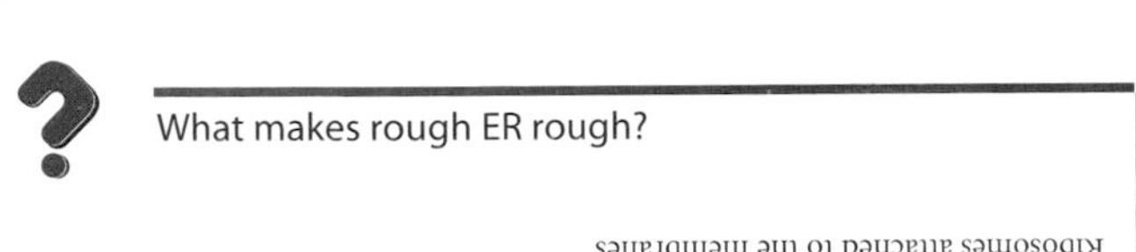

?
What makes rough ER rough?
Ribosomes attached to the membranes

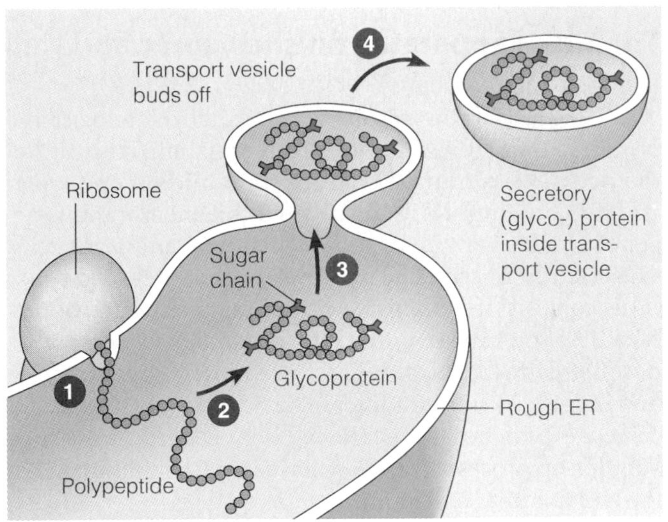

Figure 4.8B Synthesis and packaging of a secretory protein by the rough ER

4.9 Smooth endoplasmic reticulum has a variety of functions

4B
4C

As Figure 4.9 here indicates, **smooth endoplasmic reticulum, or smooth ER,** is continuous with rough endoplasmic reticulum. Smooth ER is a network of interconnected tubules that lack ribosomes. Much of its activity results from enzymes embedded in its membrane. One of the most important functions of smooth ER is the synthesis of lipids, including fatty acids, phospholipids, and steroids. Each of these products is made by particular kinds of cells. In mammals, for example, smooth ER in cells of the ovaries and testes synthesizes the steroid sex hormones.

Our liver cells also have large amounts of smooth ER, with additional kinds of functions. Certain enzymes in the smooth ER of liver help regulate the amount of sugar released from liver cells into the bloodstream. Other liver enzymes help break down drugs and other potentially harmful substances. The drugs detoxified by these enzymes include, among others, sedatives such as barbiturates, stimulants such as amphetamines, and certain antibiotics.

Undesirable complications may result when liver cells respond to drugs. As the cells are exposed to such chemicals, the amounts of smooth ER and its detoxifying enzymes increase, thereby increasing the body's tolerance to the drugs. This means that higher and higher doses of a drug are required to achieve a particular effect, such as sedation. Another complication is that detoxifying enzymes often cannot distinguish among related chemicals and therefore respond to many of them in virtually the same way, breaking down a wide variety of foreign substances in the blood. As a result, the growth of smooth ER in response to one drug can increase tolerance to other drugs, including important medicines. Barbiturate use, for example, may decrease the effectiveness of certain antibiotics.

Smooth ER has yet another function, the storage of calcium ions. In muscle tissue, these are necessary for contraction. When a nerve signal stimulates a muscle cell, calcium ions leak from the smooth ER into the cytoplasmic fluid, where they trigger contraction of the cell.

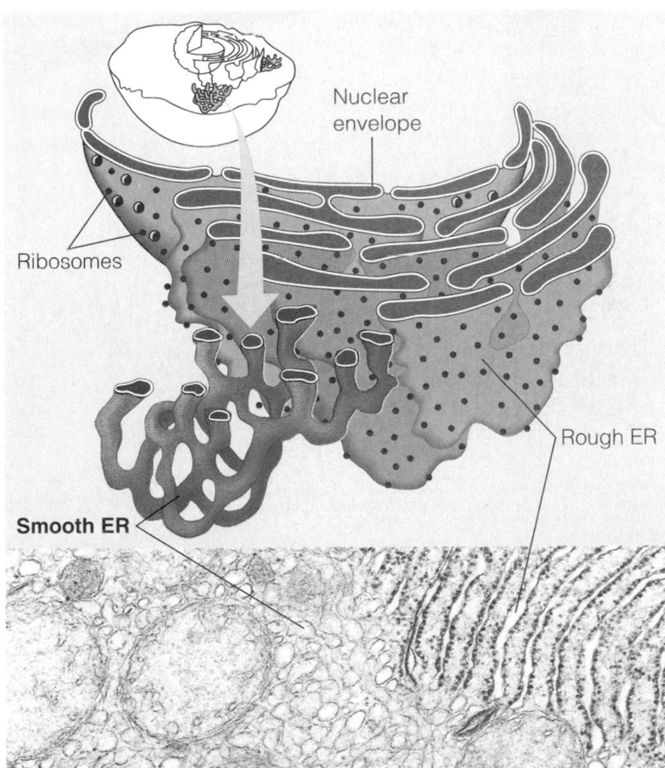

Figure 4.9 Smooth endoplasmic reticulum

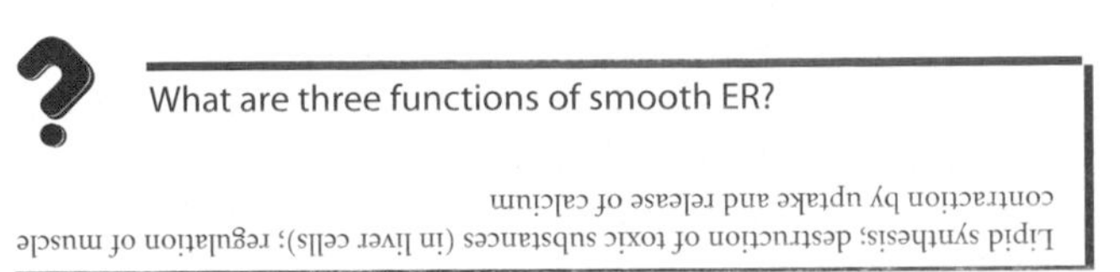

?
What are three functions of smooth ER?
Lipid synthesis; destruction of toxic substances (in liver cells); regulation of muscle contraction by uptake and release of calcium

4.10 The Golgi apparatus finishes, sorts, and ships cell products

4B
4C

The Golgi apparatus was named after Italian biologist and physician Camillo Golgi, whose career spanned the turn of the twentieth century. Using the light microscope, Golgi and his contemporaries discovered this membranous organelle and a number of others in animal and plant cells. The electron microscope has revealed that the Golgi apparatus consists of flattened sacs looking like a stack of pita bread. As you can see in the illustration below, the sacs are not interconnected like ER sacs. A cell may contain only a few Golgi stacks or hundreds. The number of Golgi stacks correlates with how active the cell is in secreting proteins—a multistep process that, as we have just seen, is initiated in the endoplasmic reticulum.

The **Golgi apparatus** performs several functions in close partnership with the ER. Serving as a molecular warehouse and finishing factory, a Golgi apparatus receives and modifies substances manufactured by the ER. One side of a Golgi stack serves as a receiving dock for transport vesicles produced by the ER. When a Golgi receives transport vesicles containing glycoprotein molecules, for instance, it takes in the materials and then modifies them chemically. One function of this chemical modification seems to be to mark and sort the molecules into different batches for different destinations. According to one popular hypothesis, molecules move from sac to sac in the Golgi within transport vesicles. The "shipping" side of the Golgi stack serves as a depot from which finished secretory products, also packaged in transport vesicles, move to the plasma membrane for export from the cell. Alternatively, finished products may become part of the plasma membrane itself or part of another organelle, such as a lysosome.

What is the relationship of the Golgi apparatus to the ER in a protein-secreting cell?

The Golgi receives transport vesicles that bud from the ER and that contain proteins synthesized in the ER. The Golgi finishes processing the proteins and then dispatches transport vesicles that secrete the proteins to the outside of the cell.

Figure 4.10 The Golgi apparatus

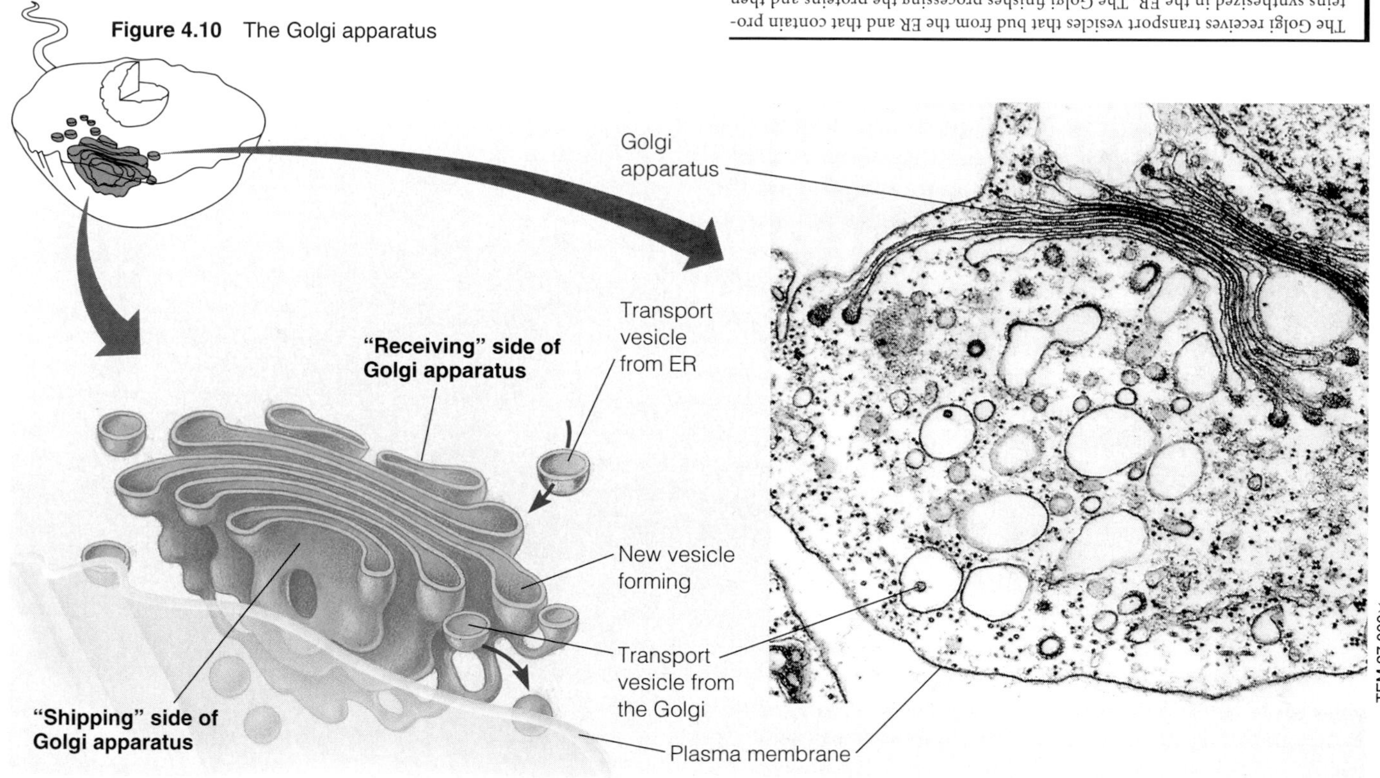

4.11 Lysosomes digest the cell's food and wastes

4B

A fourth component of the endomembrane system, the lysosome, is produced by the rough ER and the Golgi apparatus. The name **lysosome** is derived from two Greek words meaning "breakdown body," and lysosomes consist of digestive (hydrolytic) enzymes enclosed in a membranous sac (Figure 4.11A on the next page). The formation and functions of lysosomes are diagrammed in Figure 4.11B. The ER and Golgi cooperate to make lysosomes. First (bottom of Figure 4.11B), the rough ER puts the enzymes and membranes together; then the Golgi apparatus chemically refines the enzymes and releases mature lysosomes. Lysosomes illustrate the main theme of eukaryotic cell structure—compartmentalization. The lysosomal membrane encloses a compartment where digestive enzymes are stored and safely isolated from the rest of the cytoplasm. Without lysosomes, a cell could not contain active hydrolytic enzymes without digesting itself.

Lysosomes have several types of digestive functions, as Figure 4.11B shows. Many cells engulf nutrients into tiny cytoplasmic sacs called food vacuoles. Lysosomes fuse with the food vacuoles, exposing the nutrients to hydrolytic enzymes that digest them. Small molecular products of digestion, such as amino acids, leave the lysosome and are reused by the cell. Lysosomes also help destroy harmful bacteria. Our white blood cells ingest bacteria into vacuoles, and lysosomal enzymes emptied into these vacuoles rupture the bacterial cell walls. Moreover, lysosomes serve as recycling centers for damaged organelles. Without harming the cell, a lysosome can engulf and digest parts of another organelle, making its molecules available for the construction of new organelles. Lysosomes also play vital roles in embryonic development. For example, lysosomal enzymes destroy cells of the webbing that joins the fingers of early human embryos. 

? When lysosomes were first discovered, they were sometimes called "suicide capsules." In what way does that nickname fit?

If a cell's lysosomes break open, the hydrolytic enzymes released kill the cell.

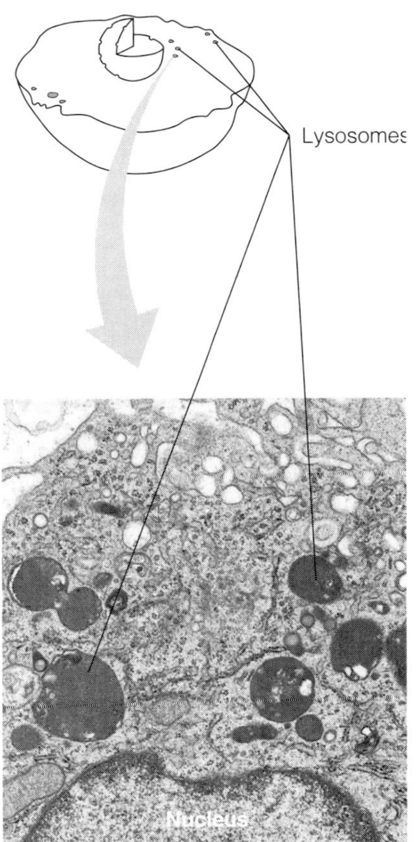

Figure 4.11A Appearance of lysosomes in a white blood cell

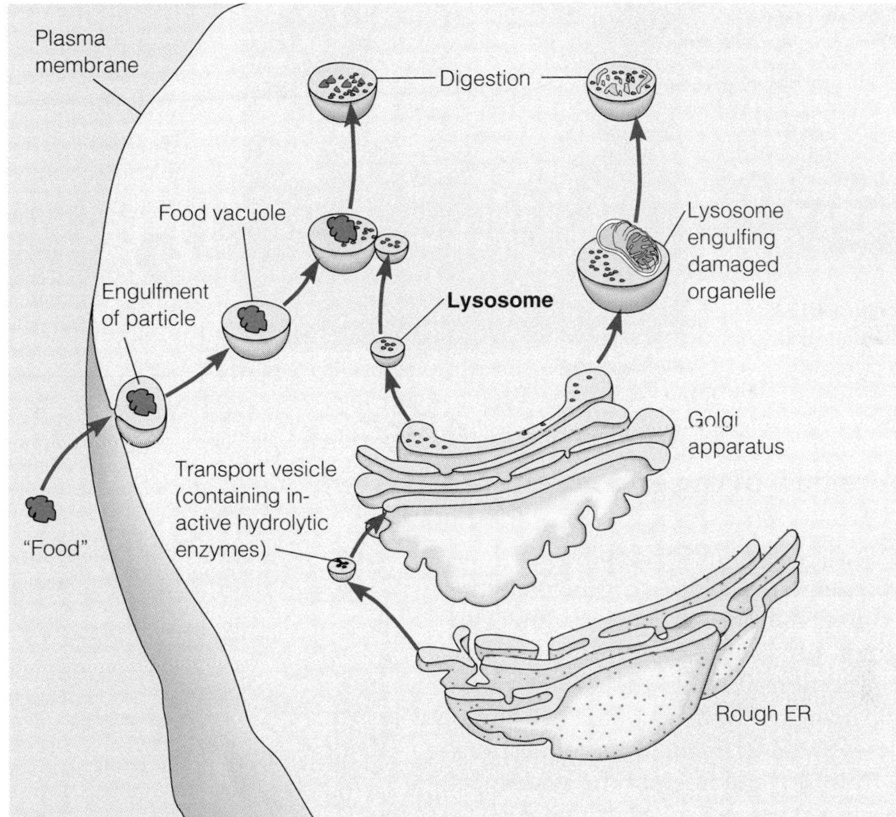

Figure 4.11B Lysosome formation and functions

4.12 Abnormal lysosomes can cause fatal diseases

The importance of lysosomes to cell function and human health is made strikingly clear by the serious hereditary disorders called **lysosomal storage diseases.** A person afflicted with a lysosomal storage disease is missing one of the hydrolytic enzymes of the lysosome. The abnormal lysosomes become engorged with indigestible substances, which eventually interfere with other cellular functions.

Most of these diseases are fatal in early childhood. In Pompe's disease, harmful amounts of the polysaccharide glycogen accumulate in liver cells because lysosomes lack a glycogen-digesting enzyme. Tay-Sachs disease ravages the nervous system. In this disorder, lysosomes lack a lipid-digesting enzyme, and nerve cells in the brain are damaged as they accumulate excess lipids. Fortunately, storage diseases are rare in the general population. For Tay-Sachs disease, carriers of the abnormal gene that causes it can be identified before they decide whether to have children.  

? How can defective lysosomes result in excess accumulation of a particular compound in a cell?

If the lysosomes lack an enzyme needed to hydrolyze the compound, the cell will accumulate an excess of the compound.

4.13 Vacuoles function in the general maintenance of the cell

4C

Like lysosomes, **vacuoles** are membranous sacs that belong to the endomembrane system. Vacuoles come in different shapes and sizes and have a variety of functions. In Module 4.11, we saw that the food vacuole functions in collaboration with a lysosome. Here, in Figure 4.13A, we see a plant cell's **central vacuole,** which can serve as a large lysosome. The central vacuole may also help the plant cell grow in size by absorbing water, and it can store vital chemicals or waste products of cell metabolism. Central vacuoles in flower petals may contain pigments that attract pollinating insects. Others contain poisons that protect against plant-eating animals.

TEM 3,600×

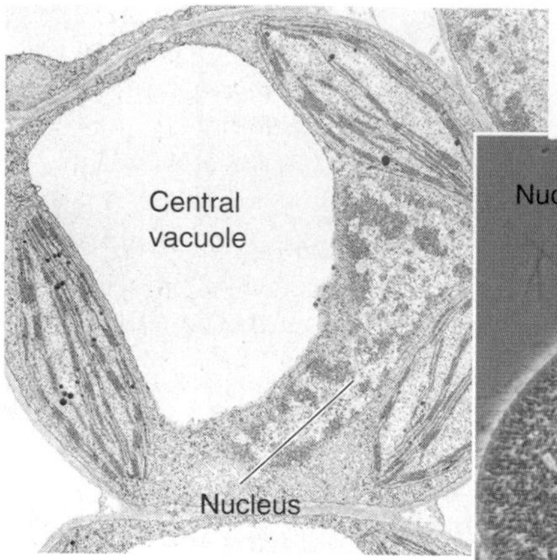

Central vacuole

Nucleus

Figure 4.13A Central vacuole in a plant cell

LM 880×

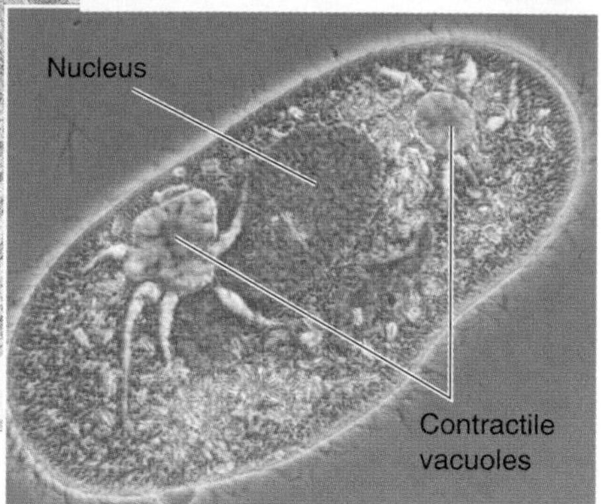

Figure 4.13B Contractile vacuoles in a protist

Figure 4.13B shows a very different kind of vacuole, in the protist *Paramecium.* (Like *Trichodina* in the chapter's opening essay, *Paramecium* eats bacteria and moves by means of cilia.) Notice the two contractile vacuoles, looking somewhat like wheel hubs with radiating spokes. The "spokes" collect excess water from the cell, and the hub expels it to the outside. This function is necessary for freshwater protists because they constantly take up water from their environment. Without a way to get rid of the excess water, the cell fluid would become too dilute to support life, and eventually the cell would swell and burst. Thus, the contractile vacuole is vital in maintaining the cell's internal environment.

The *Paramecium* in Figure 4.13B is about 0.08 mm long. Estimate the diameter in micrometers of the larger contractile vacuole. (Don't include the "spokes.")

10–15 μm

4.14 A review of the endomembrane system

Figure 4.14 summarizes the relationships among the major organelles of the endomembrane system (all shown in gray). You can see the direct *structural* connections between the nuclear envelope, rough ER, and smooth ER. The red arrows show the *functional* connections within the endomembrane system, as transport vesicles travel from the ER to the Golgi and from there to other destinations. Some vesicles develop into lysosomes and vacuoles.

Other transport vesicles fuse with the plasma membrane and contribute their membranes to it. The blue color inside the endomembrane organelles and outside the cell highlights the fact that an ER product can get out of the cell without ever actually crossing a membrane. A transport vesicle containing the product fuses with the plasma membrane and releases the product to the outside.

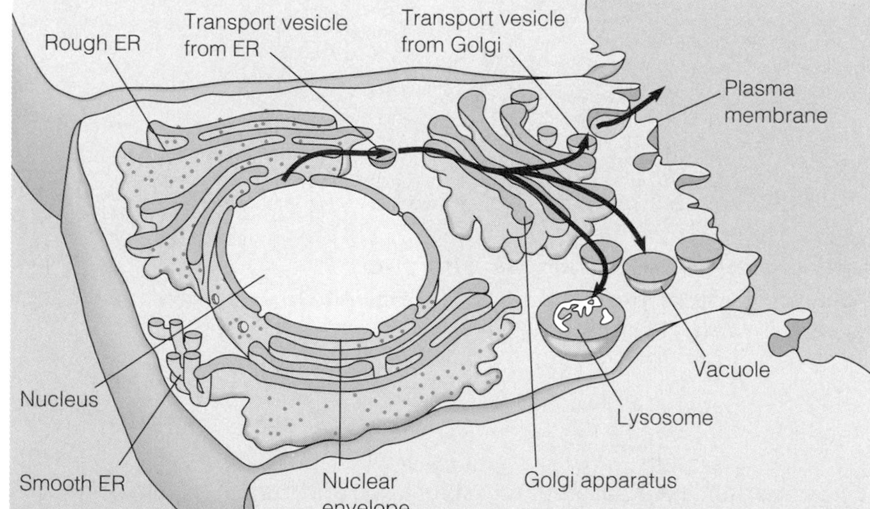

Figure 4.14 Connections among the organelles of the endomembrane system

How do transport vesicles integrate the endomembrane system?

Transport vesicles move membranes and substances they enclose between other components of the endomembrane system.

* * *

Next, we look at two membranous organelles that are not part of the endomembrane system—the chloroplast and the mitochondrion. Both contain some DNA and ribosomes and make some of their own proteins; their other proteins are made by free ribosomes in the cytoplasm. Chloroplasts and mitochondria are fuel processors: They convert energy to forms that living cells can use.

Energy-Converting Organelles

Chloroplasts convert solar energy to chemical energy 4.15

4C

Most of the living world runs on the energy provided by photosynthesis, the conversion of light energy from the sun to the chemical energy of sugar molecules. **Chloroplasts** are the photosynthesizing organelles of plants and protists. The chloroplast's solar power system is much more successful than anything yet produced by human ingenuity.

Befitting an organelle that carries out complex, multistep processes, internal membranes partition the chloroplast into three major compartments. The narrow intermembrane space, between the outer and inner membranes of the chloroplast, is one compartment. A second, the space enclosed by the inner membrane, contains a thick fluid called **stroma** and a network of tubules and hollow disks formed of membranes. The space inside the tubules and disks constitutes a third compartment. Notice that the disks occur in stacks, each called a **granum** (plural, *grana*). The grana are the chloroplast's solar power packs—the sites where chlorophyll actually traps solar energy. As we will discuss in Chapter 7, each part of the chloroplast plays a particular role in converting solar energy to chemical energy.

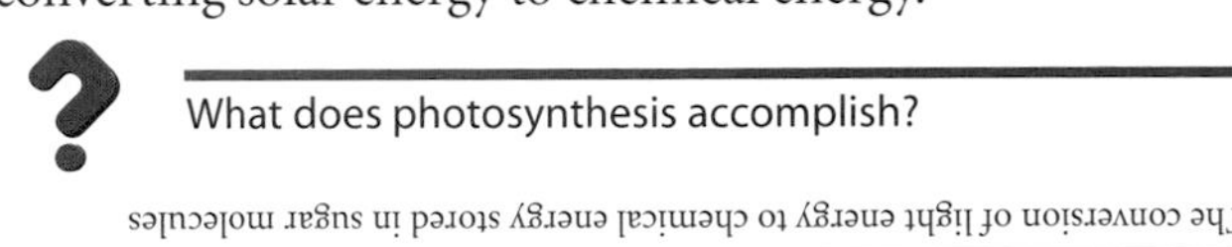

What does photosynthesis accomplish?
The conversion of light energy to chemical energy stored in sugar molecules

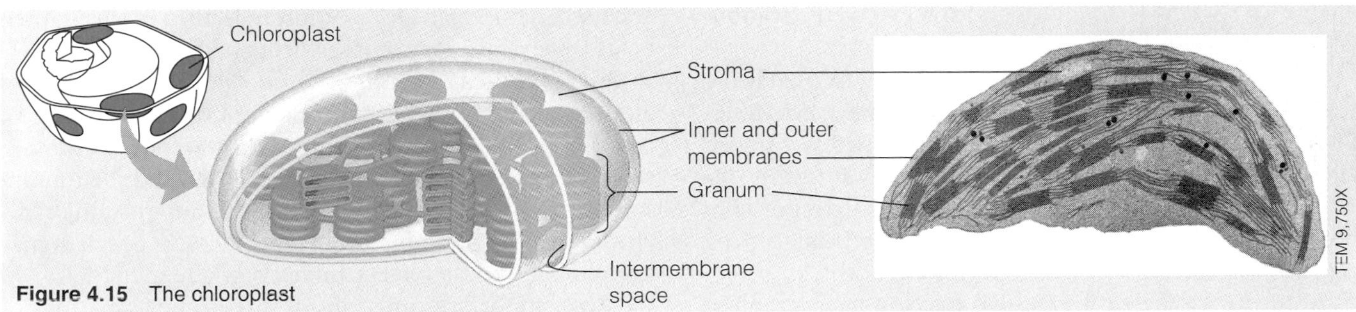

Figure 4.15 The chloroplast

Mitochondria harvest chemical energy from food 4.16

4B
4C

Mitochondria (singular, *mitochondrion*) are organelles that convert energy from one chemical form to another. They carry out cellular respiration, in which the chemical energy of foods such as sugars is converted to the chemical energy of a molecule called ATP (adenosine triphosphate). ATP is the main energy source for cellular work.

The structure of the mitochondrion suits its function. You can see in Figure 4.16 that the mitochondrion, like the chloroplast, is enclosed by two membranes. However, the mitochondrion has only two compartments. The **intermembrane space** forms one fluid-filled compartment. The inner membrane encloses the second compartment, containing a fluid called the **mitochondrial matrix.** Many of the chemical reactions of cellular respiration occur in the matrix. The inner membrane is highly folded, and enzyme molecules that make ATP are embedded in it. The folds, called **cristae** (singular, *crista*), increase the membrane's surface area, enhancing the mitochondrion's ability to produce ATP. We will discuss the role of mitochondria in cellular respiration in more detail in Chapter 6.

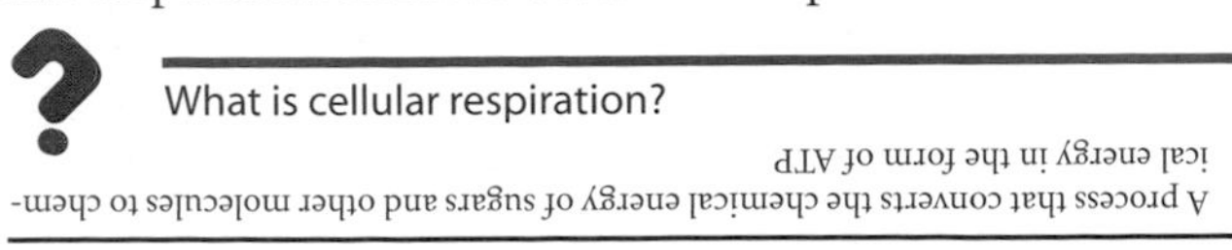

What is cellular respiration?
A process that converts the chemical energy of sugars and other molecules to chemical energy in the form of ATP

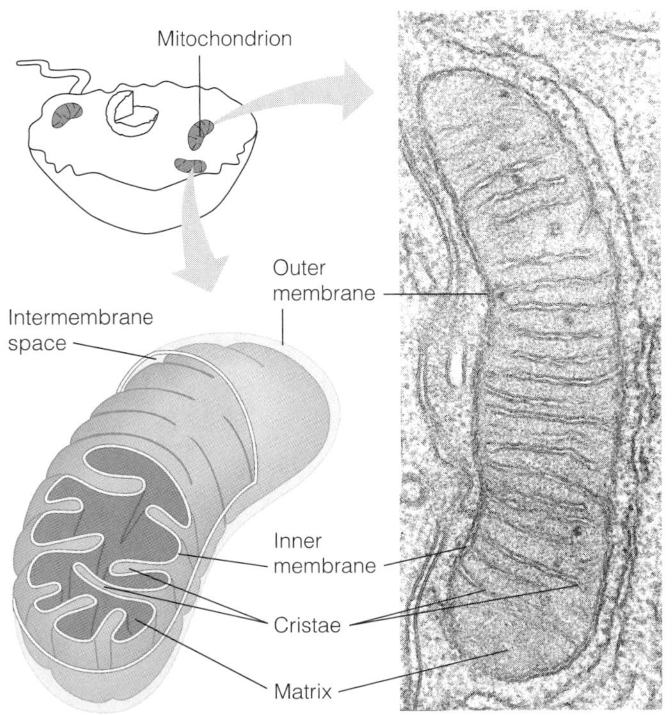

Figure 4.16 The mitochondrion

The Cytoskeleton and Related Structures

4.17 The cell's internal skeleton helps organize its structure and activities

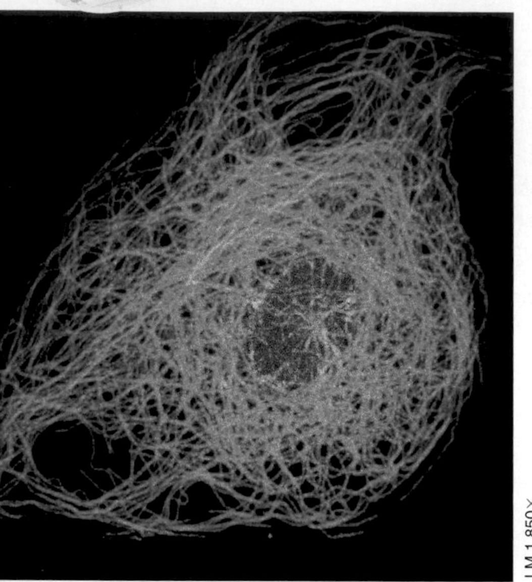

LM 1,850×

Figure 4.17A Cytoskeleton

Many of the organelles we have already described provide some structural support for cells. In addition, eukaryotic cells contain a supportive meshwork of fine fibers, collectively called the **cytoskeleton.** The micrograph in Figure 4.17A shows a cell stained to highlight its cytoskeleton in green. The sphere near the center, stained orange, is the nucleus. The fibers of the cytoskeleton extend throughout the cytoplasm.

In addition to providing structural support, cytoskeletal fibers are involved in cell movement. And recent research suggests that they help regulate cellular activities by mechanically transmitting signals from the cell's surface to its interior.

At the top of Figure 4.17B is an electron micrograph of part of a cytoskeleton. Three main kinds of fibers make up the cytoskeleton: microfilaments, the thinnest type of fiber; microtubules, the thickest; and intermediate filaments, in between in thickness.

Microfilaments are solid helical rods composed mainly of a globular protein called actin. Notice at the bottom left of Figure 4.17B that a microfilament is a twisted double chain of actin molecules. Actin microfilaments can help cells change shape and move by assembling (adding subunits) at one end while disassembling (losing subunits) at the other. The amoeboid (crawling) movement of the protist *Amoeba* and certain of our white blood cells involves this sort of process. In addition, actin microfilaments often interact with other kinds of protein filaments to make cells contract. This function of microfilaments is best known from studies of muscle cells, as we will see in Chapter 30.

Intermediate filaments are a varied group. They are made of fibrous proteins rather than globular ones and have a ropelike structure. Intermediate filaments serve mainly as reinforcing rods for bearing tension but also help anchor certain organelles. For instance, the nucleus is often held in place by a cage of intermediate filaments.

Microtubules are straight, hollow tubes composed of globular proteins called tubulins. Microtubules elongate by adding subunits consisting of tubulin pairs. They are readily disassembled in a reverse manner, and the tubulin subunits can then be reused in another microtubule. Microtubules that provide rigidity and shape in one area may disassemble and then reassemble elsewhere in the cell.

Other important functions of microtubules are to provide anchorage for organelles and to act as tracks for organelle movement within the cytoplasm. For example, a lysosome might move along a microtubule to reach a food vacuole. Microtubules also guide the movement of chromosomes when cells divide, and, as we see next, they are the basis of ciliary and flagellar movement. 

? Which component of the cytoskeleton is most important in (a) holding the nucleus in place within the cell; (b) guiding transport vesicles from the Golgi to the plasma membrane; (c) contracting muscle cells?

(a) intermediate filaments; (b) microtubules; (c) microfilaments

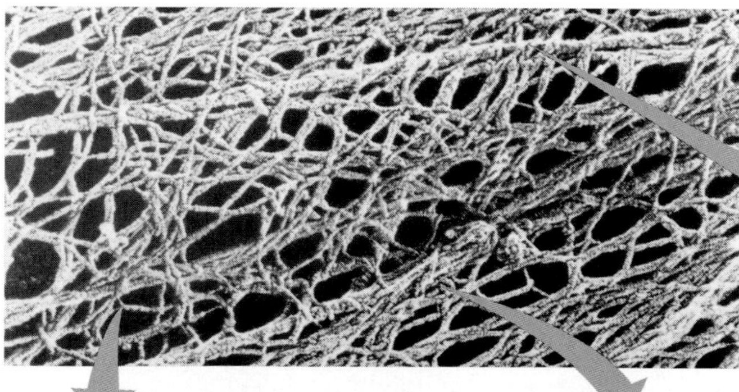

TEM 51,000×

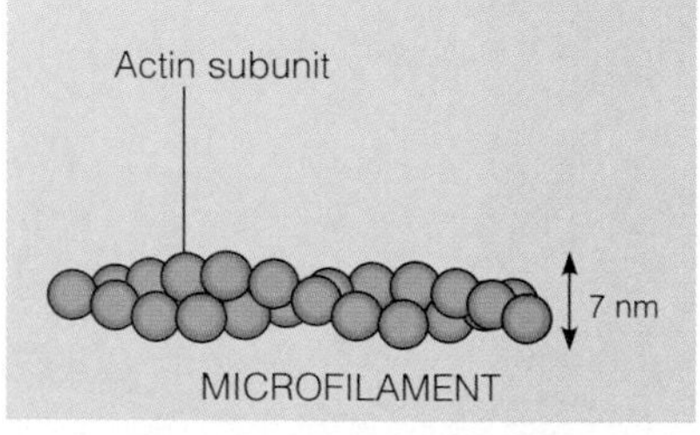

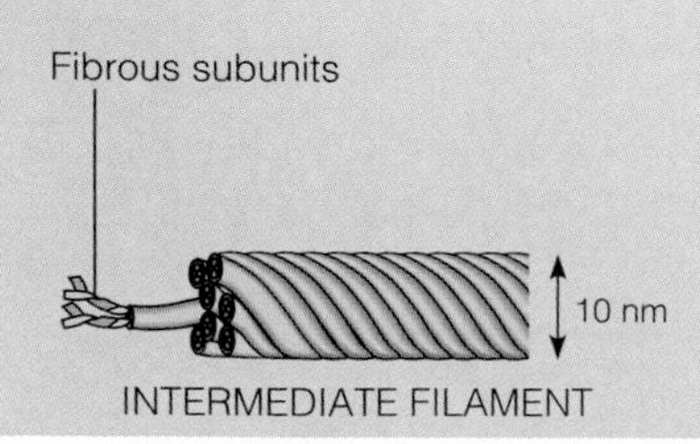

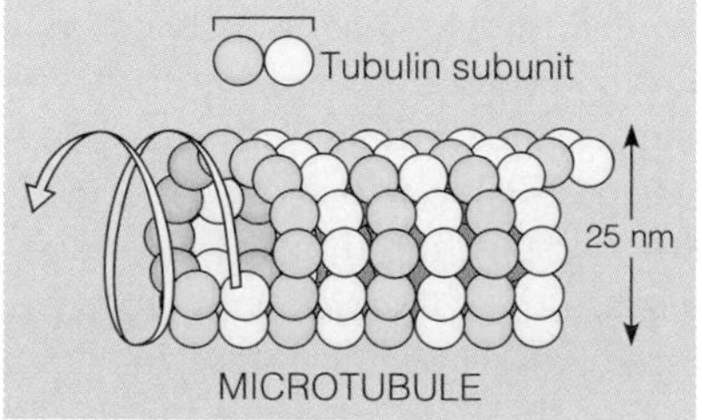

Figure 4.17B Fibers of the cytoskeleton

Cilia and flagella move when microtubules bend

The role of the cytoskeleton in movement is clearly seen in eukaryotic flagella and cilia, the locomotor appendages that protrude from certain cells. Eukaryotic flagella and cilia have a common structure and mechanism of movement. The short, numerous appendages that propel protists such as *Trichodina* through water are called **cilia** (singular, *cilium*). Longer, generally less numerous appendages on other protists are called **flagella.** Some cells of multicellular organisms also have cilia or flagella. For example, cilia on cells lining the human windpipe sweep mucus with trapped debris out of our breathing tubes. Most animals and some plants have flagellated sperm.

As you can see in Figure 4.18A, a cilium or flagellum is composed of a core of microtubules wrapped in an extension of the plasma membrane. A ring of nine microtubule doublets surrounds a central pair of microtubules. This arrangement, found in nearly all eukaryotic flagella and cilia, is called the 9 + 2 pattern. Notice that this pattern extends the length of the organelle but is different at the base. Here, the nine doublets extend into an anchoring structure called a **basal body,** which has a pattern of nine microtubule triplets. The central pair of microtubules terminates above the basal body. When a cilium or flagellum begins to grow, the basal body may act as a foundation for microtubule assembly from tubulin subunits. Basal bodies are identical in structure to **centrioles** (shown in Figure 4.5A). We will encounter centrioles again in Chapter 8, when we discuss cell division.

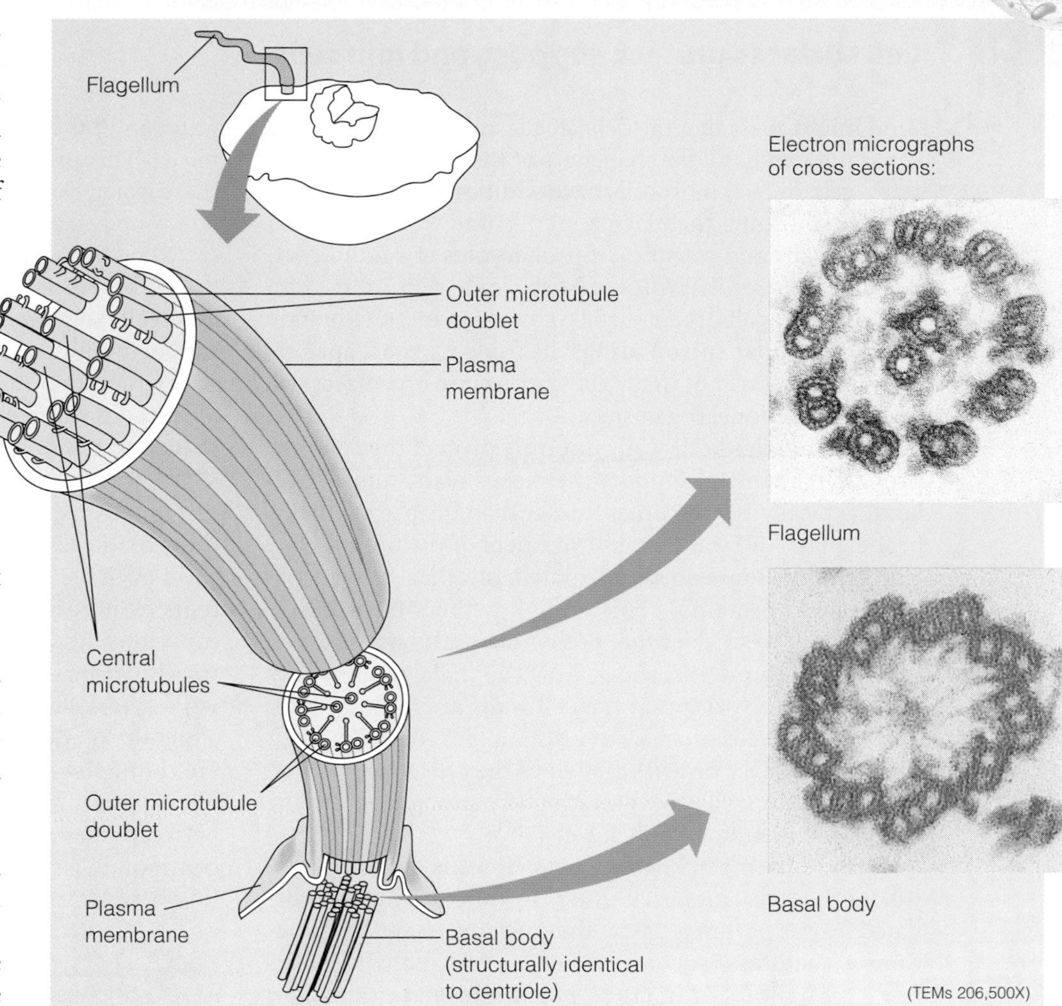

Figure 4.18A Structure of a eukaryotic flagellum or cilium

In cilia and flagella, microtubules provide both support and the locomotor mechanism underlying the whipping action of these organelles. Figure 4.18B shows the position of two microtubule doublets in a flagellum that is stationary (left) and in the process of bending (right). Bending involves protein knobs attached to each microtubule doublet—the **dynein arms** colored red in the drawings. Using energy from ATP, the dynein arms grab an adjacent doublet and exert a sliding force as they start to "walk" along it. The doublets are held together by cross-links (not illustrated); if they were not held in place, the walking action would make one doublet slide past the other. Instead, the microtubules (and consequently the flagellum or cilium) bend. 

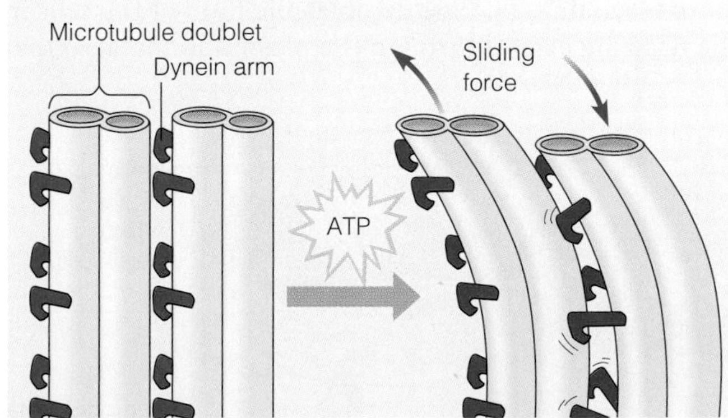

Figure 4.18B The mechanism of microtubule bending in cilia and flagella

? How do cilia and flagella bend?

Dynein arms, powered by ATP, move neighboring doublets of microtubules relative to one another. Because they are anchored within the organelle, the doublets bend instead of sliding past one another.

4.19 Cell surfaces protect, support, and join cells

4C

You might guess that the delicate plasma membrane alone could not handle all the challenges of the environment outside a cell. In fact, most cells have additional surface coverings surrounding the plasma membrane. We introduced the cell walls and capsules of prokaryotes in Module 4.4. Because most prokaryotes exist as single cells or as loose aggregates of cells, their surface coverings interact mainly with noncellular surroundings. In contrast, most eukaryotes are composed of many cells, which are organized into a single, functional organism.

In plants, rigid cell walls not only protect the cells but provide the skeletal support that keeps plants upright on land. Typically 10–100 times thicker than the plasma membrane, plant cell walls consist of fibers of the polysaccharide cellulose embedded in a matrix of other polysaccharides and proteins. This tough, fibers-in-a-matrix construction resembles that of fiberglass, also noted for its strength. Figure 4.19A shows how the walls of plant cells are arranged. Notice that the cell walls are multilayered. Between the walls of adjacent cells is a layer of sticky polysaccharides (dark brown) that glues the cells together. The walls of mature plant cells may be very strong; they are the main component of wood, for instance.

Despite their thickness, plant cell walls do not totally isolate the cells from each other. To function in a coordinated way as part of a tissue, the cells must have **cell junctions,** structures that connect them to one another. As Figure 4.19A shows, numerous **plasmodesmata** (singular, *plasmodesma*), channels between adjacent plant cells, form a circulatory and communication system connecting the cells in plant tissues. Notice that the plasma membrane (thin black line) and the cytoplasmic fluid of the cells in the drawing extend through the plasmodesmata, so that water and other small molecules can readily pass from cell to cell. Through plasmodesmata, the cells of a plant tissue share water, nourishment, and chemical messages.

Animal cells lack rigid cell walls, but most of them secrete and are embedded in a sticky layer of glycoproteins, the **extracellular matrix** (Figure 4.19B). This layer helps hold cells together in tissues and can have protective and supportive functions, too. Researchers have discovered that the extracellular matrix helps regulate cell behavior, probably by contacting proteins in the plasma membrane, which in turn contact fibers of the cytoskeleton (see Figure 5.12).

Adjacent cells in many animal tissues also connect by cell junctions; there are three general types. **Tight junctions** bind cells together, forming a leakproof sheet. Such a sheet of tissue lines the digestive tract, preventing the contents from leaking into surrounding tissues. **Anchoring junctions** attach adjacent cells to each other or to the extracellular matrix. Anchoring junctions rivet cells together with cytoskeletal fibers but still allow materials to pass along the spaces between cells. **Communicating junctions** are channels similar in function to the plasmodesmata of plants; they allow water and other small molecules to flow between neighboring cells. These junctions are especially common in animal embryos, where chemical communication between cells is essential for development.

How is a plant or animal tissue different from just an aggregate of similar cells?

Cell junctions integrate the cells of a tissue.

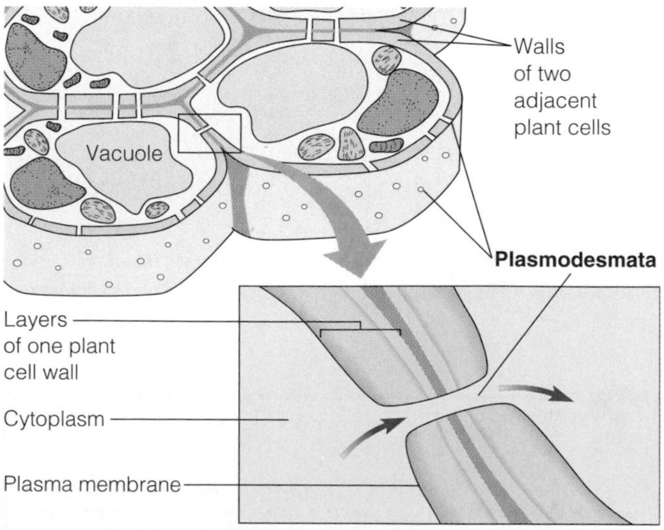

Walls of two adjacent plant cells
Vacuole
Plasmodesmata
Layers of one plant cell wall
Cytoplasm
Plasma membrane

Figure 4.19A Plant cell walls and cell junctions

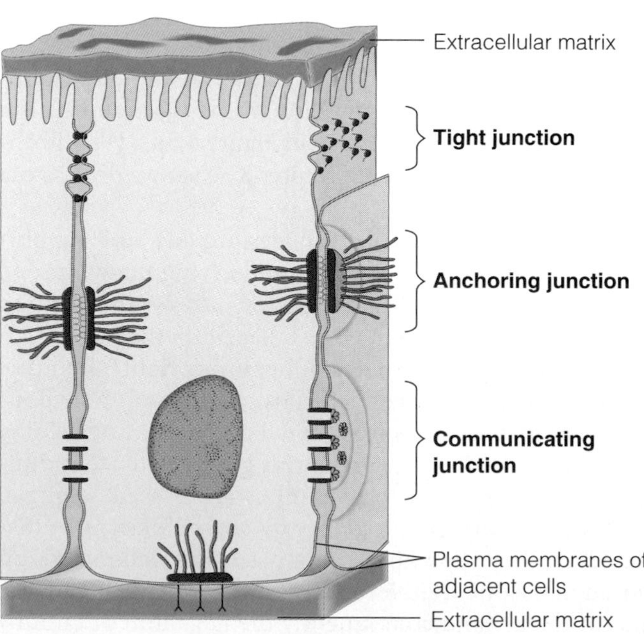

Figure 4.19B Animal cell surfaces and cell junctions

Eukaryotic organelles comprise four functional categories 4.20

We have introduced many important cell structures in this chapter. To provide a framework for this information and to reinforce the theme that structure is correlated with function, the table at the right groups the eukaryotic cell organelles into four categories by general function.

The first category is manufacture. Here we include not only the synthesis of molecules, but also their transport within the cell. The second category includes three organelles that break down and recycle materials that are harmful or no longer needed. (Vacuoles are included here, although, being multifunctional, they do not fit neatly into any one of our categories.) The third category contains the two energy-processing organelles. The fourth category is support, movement, and intercellular communication. These three functions are related because, for movement to occur, there must be some sort of rigid support against which force can be applied. And when a supporting structure forms the cell's outer boundary, it is necessarily involved in the cell's communication with its neighbors.

Within each of the four categories, a structural similarity underlies the general function of the organelles. In the first category, manufacture depends heavily on a network of metabolically active membranes. In the second category, all the organelles listed are composed of single membranous sacs, inside which materials can be broken down. In the third category, expanses of metabolically active membranes within the organelles make it possible for chloroplasts and mitochondria to perform complex energy conversions that power the cell. Even in the diverse fourth category, there is a common structural theme in the various fibers involved in the functioning of most of the organelles listed.

We can summarize further by emphasizing that these four categories of organelles form an integrated team, and that properties of life at the cellular level emerge from the coordinated functions of the team members. Cell movement is one such emergent property. In that case, mitochondria provide the energy that makes microtubules bend and cilia or flagella beat. As we will see in later chapters, the coordinated actions of cellular organelles underlie most of the emergent properties of life. 

EUKARYOTIC ORGANELLES AND THEIR FUNCTIONS

Organelle	Function
General Function: Manufacture	
Nucleus	DNA synthesis; RNA synthesis; assembly of ribosomal subunits (in nucleolus)
Ribosomes	Polypeptide (protein) synthesis
Rough ER	Synthesis of membrane proteins, secretory proteins, and hydrolytic enzymes; formation of transport vesicles
Smooth ER	Lipid synthesis; carbohydrate metabolism in liver cells; detoxification in liver cells; calcium ion storage
Golgi apparatus	Modification, temporary storage, and transport of macromolecules; formation of lysosomes and transport vesicles
General Function: Breakdown	
Lysosomes	Digestion of nutrients, bacteria, and damaged organelles; destruction of certain cells during embryonic development
Peroxisomes	Diverse metabolic processes, with breakdown of H_2O_2 by-product
Vacuoles	Digestion (like lysosomes); storage of chemicals; cell enlargement; water balance
General Function: Energy Processing	
Chloroplasts (in plants and some protists)	Conversion of light energy to chemical energy of sugars
Mitochondria	Conversion of chemical energy of food to chemical energy of ATP
General Functions: Support, Movement, and Communication Between Cells	
Cytoskeleton (including cilia, flagella, and centrioles in animal cells)	Maintenance of cell shape; anchorage for organelles; movement of organelles within cells; cell movement; mechanical transmission of signals from exterior of cell to interior
Cell walls (in plants, fungi, and some protists)	Maintenance of cell shape and skeletal support; surface protection; binding of cells in tissues
Extracellular matrix (in animals)	Binding of cells in tissues; surface protection; regulation of cellular activities
Cell junctions	Communication between cells; binding of cells in tissues

? How do mitochondria, smooth ER, and the cytoskeleton cooperate in contraction of a muscle cell?

Mitochondria supply energy in the form of ATP. The smooth ER helps regulate contraction by the uptake and release of calcium. Microfilaments function as the actual contractile apparatus.

4.21 All life forms share fundamental features

Cells are complicated, as the micrograph in Figure 4.21 demonstrates. Nevertheless, we could argue that cell structure is simpler than we might expect, considering the complex functions that cells perform and the diversity of life forms that exist on Earth.

Although it is almost certain that Earth is the only life-bearing planet in our solar system, it is conceivable that conditions on some of the moons of the outer planets or on planets elsewhere have allowed the evolution of life. Suppose for a moment that we have learned that alien organisms actually do exist. Such organisms might be very different from the ones we know. What could we predict about them from our knowledge of prokaryotes and eukaryotes on Earth? We could, at least, predict that extraterrestrial organisms would be highly structured and would demonstrate the structure-function theme common to Earth's organisms. We could also expect that they would consist of some type of fundamental units, like cells, that are set off from their environment by some sort of membrane. To enable the organisms to reproduce and adapt to their environment, they would have to have some type of genetic machinery in each "cell," perhaps housed in a control center like a nucleus. The molecular basis of heredity on another planet might not be DNA; the biological molecules there might even be based on an element other than carbon. But whatever the details of their chemistry, the alien forms of life would have to carry out metabolism.

Speculating about extraterrestrial life helps us summarize what we have learned in this chapter about cellular life as we do know it: All life forms on our planet share the fundamental features of (1) consisting of cells, each enclosed by a membrane that maintains internal conditions very different from the surroundings; (2) having DNA as the genetic material; and (3) carrying out metabolism, which involves the interconversion of different forms of energy and of chemical materials. We expand on the subjects of membranes and metabolism in Chapter 5. 

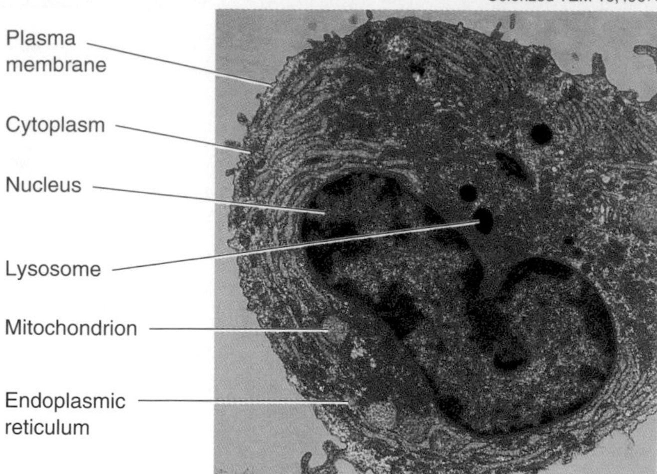

Figure 4.21 The complex interior structure of a human bone cell

? What three major characteristics are shared by prokaryotic and eukaryotic cells?

A plasma membrane that keeps the inside of the cell different from the outside; DNA as the genetic material; the ability to carry out metabolism

Chapter Review

Chapter Summary

Introduction to the World of the Cell (Introduction–4.5) All living things are made of cells (Introduction). The light microscope enables us to see the overall shape and structure of a cell. The greater resolving power of electron microscopes allows greater magnification and reveals cellular details (4.1). Cell size and shape relate to function (4.2). At minimum, a cell must be large enough to house the parts it needs to survive and reproduce. The maximum size of a cell is limited by the amount of surface needed to obtain nutrients from its environment and dispose of wastes. A small cell has a greater ratio of surface area to volume than a large cell of the same shape (4.3).

There are two kinds of cells: prokaryotic and eukaryotic. Prokaryotic cells—bacteria and archaea—are small, relatively simple cells that do not have a nucleus. A prokaryotic cell is enclosed by a plasma membrane and usually encased in a rigid cell wall. The cell wall may be covered with a sticky capsule. Inside the cell, its DNA and other parts are suspended in semifluid cytoplasm. Prokaryotic flagella may propel the organism (4.4). All other forms of life are composed of one or more larger and more complex eukaryotic cells, which are distinguished by the presence of a true nucleus. The plasma membrane controls the cell's interaction with its environment. The cytoplasm of a eukaryotic cell contains complex structures called organelles. Membranes form the boundaries of many of the organelles, compartmentalizing the interior of the cell and thereby enabling the cell to carry out a variety of metabolic activities simultaneously (4.5).

Organelles of the Endomembrane System (4.6–4.14) The largest organelle is the nucleus, which is separated from the cytoplasm by a porous nuclear envelope. The nucleus is the cellular control center, containing the DNA that carries the cell's hereditary blueprint and directs its activities (4.6). The endomembrane system is a collection of membranous organelles that manufacture and distribute cell products (4.7). The endoplasmic reticulum (ER) is a membranous network consisting of rough ER and smooth ER. Rough ER manufactures membranes, and ribosomes on its surface produce proteins that are secreted from the cell, as well as membrane proteins (4.8). Smooth ER synthesizes lipids. In liver cells, it also regulates carbohydrate metabolism and breaks down toxins and drugs. In muscle cells, its storage of calcium ions plays a role in contraction (4.9). The Golgi apparatus, stacks of membranous sacs, receives and modifies ER products and membranes, then sends them on to other organelles or to the cell surface (4.10). Lysosomes are sacs of digestive enzymes budded off the Golgi. Lysosomal enzymes digest food, destroy bacteria, recycle damaged organelles, and function in embryonic development (4.11–4.12). Plant cells contain a large central vacuole, which has both lysosomal and storage functions and also can help the cell enlarge. Protists have contractile vacuoles that pump out excess water (4.13). The various organelles of the endomembrane system are interconnected structurally and functionally (4.14).

Energy-Converting Organelles (4.15–4.16) Chloroplasts, in plants and some protists, convert solar energy to chemical energy in sugars (4.15). Mitochondria carry out cellular respiration, a process that uses the chemical energy in food to make ATP for cellular work (4.16).

The Cytoskeleton and Related Structures (4.17–4.18) A network of protein fibers makes up the cytoskeleton, the cell's structural framework. Microfilaments of actin enable cells to change shape and move, as in muscle contraction. Intermediate filaments reinforce the cell and anchor certain organelles. Microtubules are thick, hollow tubes that give the cell rigidity, provide anchorage for organelles, and act as tracks for organelle movement. In addition to its structural roles, the cytoskeleton may mechanically transmit signals through the cell (4.17). Eukaryotic cilia and flagella are locomotor appendages that protrude from certain cells, such as sperm. Clusters of microtubules drive the whipping action of these organelles (4.18).

Eukaryotic Cell Surfaces and Junctions (4.19) Cells interact with their environments and with each other via their surfaces. Plant cells are supported by rigid cell walls made largely of cellulose. Plant cells connect by plasmodesmata, channels that allow them to share water, food, and chemical messages. Animal cells are embedded in an extracellular matrix consisting mainly of glycoproteins. This matrix binds cells together in tissues and may help regulate cell behavior. Tight junctions can bind them together into leakproof sheets. Anchoring junctions link animal cells but leave spaces between the cells. Communicating junctions are channels that allow substances to flow from cell to cell.

Functional Categories of Organelles (4.20–4.21) Eukaryotic organelles fall into four functional groups; (1) manufacture; (2) breakdown; (3) energy processing; and (4) support, movement, and communication between cells (4.20). All living things consist of cells, have DNA as their genetic material, and carry out metabolism (4.21).

Testing Your Knowledge

Multiple Choice

1. Sara would like to film the movement of chromosomes during cell division. Her best choice for a microscope would be a
   a. light microscope, because of its resolving power.
   b. transmission electron microscope, because of its magnifying power.
   c. canning electron microscope, because the specimen is alive.
   d. transmission electron microscope, because of its great resolving power.
   e. light microscope, because the specimen is alive.

2. The cells of an ant and a horse are, on average, the same small size; a horse just has more of them. What is the main advantage of small cell size?
   a. Small cells are less likely to burst than large cells.
   b. Small cells are less likely to be infected by bacteria.
   c. Small cells can better take up what they need from their environment.
   d. It takes less energy to make an organism out of small cells.
   e. Small cells can change shape easily.

3. Which of the following clues would tell you whether a cell is prokaryotic or eukaryotic?
   a. the presence or absence of a rigid cell wall
   b. whether or not the cell is partitioned by internal membranes
   c. the presence or absence of ribosomes
   d. whether or not the cell carries out cellular metabolism
   e. whether or not the cell contains DNA

4. Which of the following structures is *not* directly involved in cell support or movement?
   a. microfilament
   b. flagellum
   c. microtubule
   d. lysosome
   e. cell wall

5. A type of cell called a lymphocyte makes proteins that are exported from the cell. You can track the path of these proteins within the cell by labeling them with radioactive isotopes. Which of the following might be the path of a protein from the site where its polypeptides are made to the lymphocyte's plasma membrane?
   a. chloroplast . . . Golgi . . . plasma membrane
   b. Golgi . . . rough ER . . . plasma membrane
   c. rough ER . . . Golgi . . . plasma membrane
   d. smooth ER . . . lysosome . . . plasma membrane
   e. nucleus . . . Golgi . . . rough ER . . . plasma membrane

Describing, Comparing, and Explaining

1. Briefly describe the three kinds of junctions between animal cells, and compare their functions.

2. What general function do the chloroplast and mitochondrion have in common? How are their functions different?

3. How does a eukaryotic cell benefit from its internal membranes?

Thinking as a Scientist

Imagine a spherical cell with a radius of 10 μm. What is the cell's surface area in μm^2? Its volume, in μm^3? What is the ratio of surface area to volume for this cell? Now do the same calculations for a second cell, this one with a radius of 20 μm. Compare the surface-to-volume ratios of the two cells. How is this comparison significant to the functioning of cells? (*Note:* For a sphere of radius r, surface area = $4\pi r^2$ and volume = $4/3\ \pi r^3$. The value of π is 3.14.)

Science, Technology, and Society

Doctors at UCLA removed John Moore's spleen, standard treatment for his type of leukemia, and the disease did not recur. Researchers kept the spleen cells alive in a nutrient medium. They found that some of the cells produced a blood protein that showed promise as a treatment for cancer and AIDS. The researchers patented the cells. Moore sued, claiming a share in profits from any products derived from his cells. The U.S. Supreme Court ruled against Moore, stating that his suit "threatens to destroy the economic incentive to conduct important medical research." Moore argued that the ruling left patients "vulnerable to exploitation at the hands of the state." Do you think Moore was treated fairly? Is there anything else you would like to know about this case that might help you decide?

Media Resources

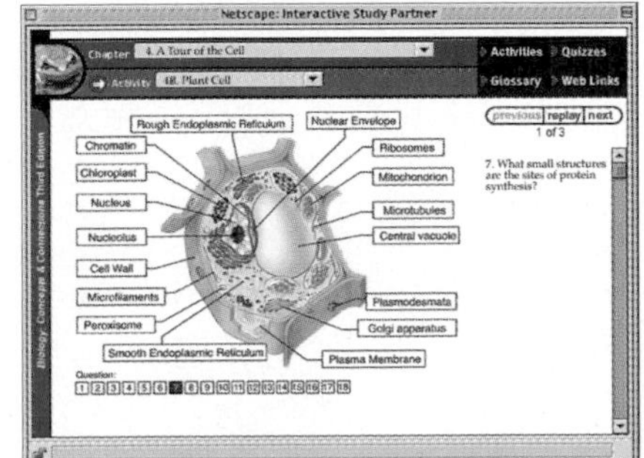

For further review of the concepts in this chapter, use the activities and quizzes on the CD-ROM.

Visit the Special Edition of The Biology Place web site for activities, quizzes, news articles, and web links.

5 The Working Cell

Outline

The points of light "decorating" the tree at the left are flashing insects commonly known as fireflies or lightning bugs. Our photograph was taken in southeast Asia, but fireflies live in many parts of the world. In the eastern United States, for example, the flashes of fireflies in a field are a common sight on a summer night. Males do most of the flashing.

When a female sees flashes of light from a male of her species, she reacts with flashes of her own. If the male sees her flashes, he automatically gives another display and flies in the female's direction. Members of both sexes are responding to particular patterns of light flashes characteristic of their species. Mating occurs when the female's display leads a male to her, and most females stop flashing after they mate. But in a few species, a mated female will continue to flash, using a pattern that attracts males of *other* firefly species. A veritable *femme fatale,* she waits until an alien male gets close, then grabs and eats him. In the photograph on the next page, the firefly clinging to the leaf is a female dining on a luckless male of another species.

Each of the 2000 or so species of firefly has its own way to signal a mate. Some flash more often than others or during different hours, while other species give fewer but longer flashes. Many species produce light of a characteristic color: yellow, bluish-green, or reddish. In areas where fireflies congregate—often on lawns, golf courses, or open meadows—you can usually see several different species signaling. Because the insects respond instinctively to specific flash patterns, you can even attract males to artificial light if you flash it a certain way.

From a biological standpoint, fireflies are poorly named. They are beetles, not flies, and their light is almost cold, not fiery. In fact, in emitting light, they give off only about one hundred-thousandth of the amount of heat that would be produced by a candle flame of equal brightness.

What happens in a firefly that makes this light? The light comes from a set of chemical reactions that occur in light-producing organs at the rear of the insect. Light-emitting cells in these organs contain an acidic substance called luciferin and an enzyme called luciferase (both named from the Latin for "light bearer"). In the presence of oxygen (O_2) and chemical energy from ATP molecules, luciferase catalyzes the conversion of luciferin to a molecule that emits light.

The luciferin-luciferase system is one example of how living cells put energy to work by means of enzyme-controlled chemical reactions. Light is a form of energy, and the firefly's ability to make light energy from chemical energy is an example of life's dependence on energy conversions.

Many of the enzymes that control a cell's chemical reactions, including the firefly's luciferase, are located in membranes. Membranes thus serve as sites where chemical reactions can occur in an orderly manner. Also, as we mentioned in Chapter 4, membranes control the passage of crucial substances into and out of cells, and they partition the eukaryotic cell into useful compartments. In fact, the firefly could not produce light, nor could any cell or organism survive, without membranes. Everything that happens in the firefly's light organs has some relation to this chapter's subject: how working cells use energy, enzymes, and membranes.

A female firefly devouring a male of another species

Energy and the Cell

5.1 Energy is the capacity to perform work

Energy can only be described and measured by how it affects matter. **Energy** is the capacity to perform work—that is, to move matter in a direction it would not move if left alone. Energy makes change possible, such as the movement of an animal from one place to another. And all organisms require energy to stay alive.

Figures 5.1A and 5.1B compare two forms of energy. **Kinetic energy** is energy that is actually doing work, such as pedaling a bicycle. A mass of matter that is moving performs work by transferring its motion to other matter, whether it is leg muscles pushing bicycle pedals or firefly wings moving air as they beat. **Heat,** the energy associated with the movement of molecules in a body of matter, is one kind of kinetic energy. Light is another kind of kinetic energy.

The second form of energy is stored energy, or **potential energy.** This is the capacity to perform work that matter possesses as a result of its location or arrangement. A cyclist motionless at the top of a hill and water behind a dam both have potential energy due to their altitude. Likewise, the negatively charged electrons of an atom have potential energy due to their positions in electron shells at a distance from the positively charged nucleus. The molecules in a living cell have potential energy due to the arrangement of their atoms; this potential energy is the chemical energy that does the work of the cell. **Chemical energy** is the potential energy of molecules, the most important type of energy for living organisms. 

? How can an object at rest have energy?

It can have potential energy due to its location.

Figure 5.1A Kinetic energy

Figure 5.1B Potential energy

5.2 Two laws govern energy conversion

Life depends on the fact that energy can be converted from one form to another. Chemical energy is tapped when a chemical reaction rearranges the atoms of molecules in such a way that potential energy is transformed into kinetic energy. This kind of transformation occurs in an automobile engine when gasoline (containing chemical energy) burns (reacts with oxygen), releasing kinetic energy that pushes the pistons. An organism taps chemical energy when chemical reactions in its cells rearrange atoms in organic molecules such as the sugar glucose, converting it to other molecules. In the process, some of the energy stored in the sugar is made available for cellular work. Whether energy transformations occur in a car engine or an organism, they are governed by the same two laws of the universe, known as the first and second laws of thermodynamics.

Thermodynamics is the study of energy transformations that occur in a collection of matter. In discussing energy transformations, we call the collection of matter under study the system and the rest of the universe the surroundings. A system could be an automobile engine, a single cell, or the entire planet. Like those examples, a living organism is an open system; that is, it exchanges both energy and matter with its surroundings. The firefly, for instance, takes in food and oxygen and releases heat, light, and chemical waste products, such as carbon dioxide.

According to the **first law of thermodynamics,** also known as the law of energy conservation, the total amount of energy in the universe is constant. Energy can be transferred and transformed, but it cannot be created or destroyed. An electrical company does not make energy; it merely converts it to a form that is convenient to use. The car in Figure 5.2A will transform some of the chemical (potential) energy in its gasoline fuel to kinetic energy of movement, and the cells in the woman's body will do the same with her sugar "fuel."

What actually happens to gasoline fuel as a car runs down the road? Some of the energy in the fuel remains in chemical form, in unburned hydrocarbons spewed into the air. The rest of the fuel energy is converted by the car's engine to kinetic energy. The second law of thermodynamics tells us what happens to this kinetic energy during the conversion.

The **second law of thermodynamics** states that energy conversions reduce the order of the universe. Put another way, energy changes, such as the conversion of chemical energy to kinetic energy, are accompanied by an increase in disorder, or randomness. The amount of disorder in a system is called **entropy.** Heat, which is random molecular motion, is one form of disorder. The more heat that is generated when one form of energy is converted to another, the more the entropy of the system increases.

In a broad sense, the second law of thermodynamics tells us that the entropy of the universe as a whole is increasing; every energy transfer or transformation increases the entropy of the universe. For instance, about 75% of the chemical energy in the fuel tank of the car in Figure 5.2B will become disordered by being converted to heat. Moreover, *all* of the chemical energy converted to kinetic energy will eventually be transformed into heat: The organized energy of the car's forward or backward movement becomes heat when friction between the brakes and the wheels, and between the tires and the road, stops the car. Thus, even energy that performs useful work is finally converted to heat. All this disordered energy is added to the surroundings.

The second law of thermodynamics implies that if a particular system becomes more ordered, its surroundings become more disordered, and this concept has direct application to cellular activities. A cell creates ordered structures from less organized starting materials. For example, amino acids are ordered into the specific sequences of polypeptide chains. However, this increase in order, which corresponds to a decrease in entropy, is accompanied by an increase in the entropy of the surroundings. In a thermodyamic sense, a cell or an organism is an island of low entropy in an increasingly random universe.

Because of the second law of thermodynamics, a cell cannot transfer or transform energy with 100% efficiency. When a chemical reaction occurs in the cell, chemical energy is transferred between molecules; when light energy is trapped for photosynthesis, it is converted to chemical energy; when a flagellum moves, chemical energy is converted to the kinetic energy of movement. As any such transfer or energy conversion occurs, some energy always escapes from the system as heat. Cells do not have the machinery necessary to put the disordered molecular movement of heat energy to work, and even if they did, some of the heat would still be lost to the surroundings. As we see in the modules that follow, the work of cells is powered by the potential energy contained in molecules. 

Figure 5.2A First law of thermodynamics: Energy can be transformed but not created or destroyed

Figure 5.2B Second law of thermodynamics: The entropy (disorder) of the universe is increasing

? Describe the energy transformations that occur when you climb to the top of a stairway.

You convert the chemical energy of food to the kinetic energy of your upward climb. At the top of the stairs, some of the energy has been stored as potential energy due to your higher elevation. The rest has been converted to heat.

5.3 Chemical reactions either store or release energy

5A
See Activity 5A on the CD-ROM.

Chemical reactions, including those that occur in cells, are of two types. One type, called **endergonic reactions,** requires a net input of energy (endergonic means "energy-in"). Endergonic reactions yield products that are rich in potential energy. As you can see in Figure 5.3A, an endergonic reaction starts out with reactant molecules that contain relatively little potential energy. Energy is absorbed from the surroundings as the reaction occurs, so that the products of an endergonic reaction store more energy than the reactants did. The energy is actually stored in the covalent bonds of the product molecules . And as the graph shows, the amount of additional energy stored in the products equals the difference in potential energy between the reactants and the products.

Photosynthesis, the process whereby plant cells make sugar, is one example of a strongly endergonic process. Photosynthesis starts with energy-poor reactants (carbon dioxide and water molecules) and, using energy absorbed from sunlight, produces energy-rich sugar molecules.

Some other chemical processes are exergonic. An **exergonic reaction** is a chemical reaction that releases energy (exergonic means "energy-out"). As indicated in Figure 5.3B, an exergonic reaction begins with reactants whose covalent bonds contain more energy than those in the products. The reaction releases to the surroundings an amount of energy equal to the difference in potential energy between the reactants and products.

As an example of an exergonic reaction, consider what happens when wood burns. One of the major components of wood is cellulose, a large carbohydrate composed of many glucose monomers. Each glucose monomer is rich in potential energy. When wood burns, the potential energy is released as heat and light. Carbon dioxide and water are the products of the reaction.

Burning is one way to release energy from chemicals. Cells release energy by means of a different exergonic process, called cellular respiration. **Cellular respiration** is the energy-releasing chemical breakdown of glucose molecules and the storage of the energy in a form that the cell can use to perform work. Burning and cellular respiration are alike in being exergonic. They differ in that burning is essentially a one-step process that releases all of a substance's energy at once. Cellular respiration, on the other hand, involves many steps, each a separate chemical reaction; you could think of it as a "slow burn." Some of the energy released from glucose by cellular respiration escapes as heat, but a substantial amount of released energy is converted to the chemical energy of molecules of ATP, which we discuss in the next module. Cells use ATP as an immediate source of fuel.

Every working cell in every organism carries out thousands of endergonic and exergonic reactions, the sum of which is known as **cellular metabolism.** In a firefly, for instance, the light display discussed earlier is exergonic. In this case, light energy is released when reactants are converted to product molecules with less energy than the reactants. In addition to generating light, fireflies, like all animals, must find, eat, and digest food; escape predators; repair damage to the body; grow; and reproduce. All these activities require energy, which is obtained from sugar and other food molecules by the exergonic reactions of cellular respiration. Cells then use that energy in endergonic reactions to make molecules that perform specific tasks. To digest food, for instance, an animal's cells use chemical energy to synthesize digestive enzymes. To repair damaged tissues, cells make other proteins that seal up wounds. In the next module, we see that the connection between exergonic and endergonic reactions in cellular metabolism is made by ATP molecules.

Cellular respiration is an exergonic (energy-releasing) process. Remembering that energy must be conserved, what becomes of the energy extracted from food during cellular respiration?

Some of it is stored in ATP molecules; the rest is released as heat.

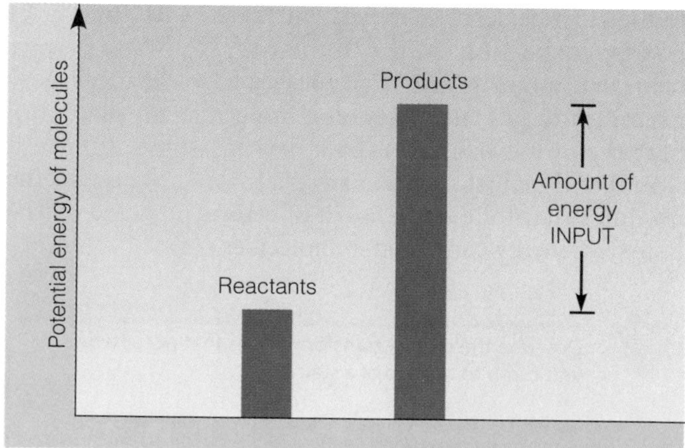

Figure 5.3A Endergonic reaction (energy required)

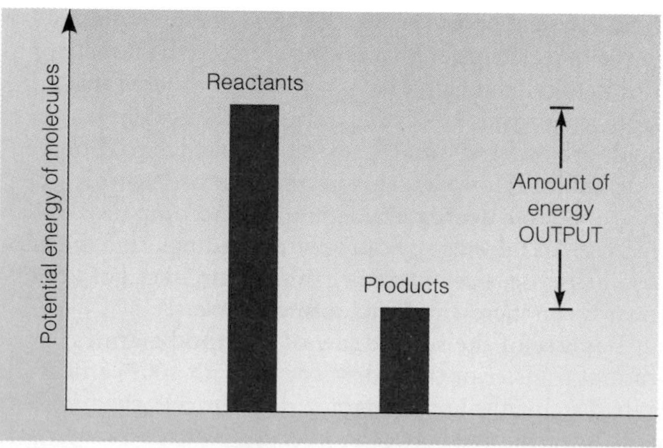

Figure 5.3B Exergonic reaction (energy released)

ATP shuttles chemical energy within the cell

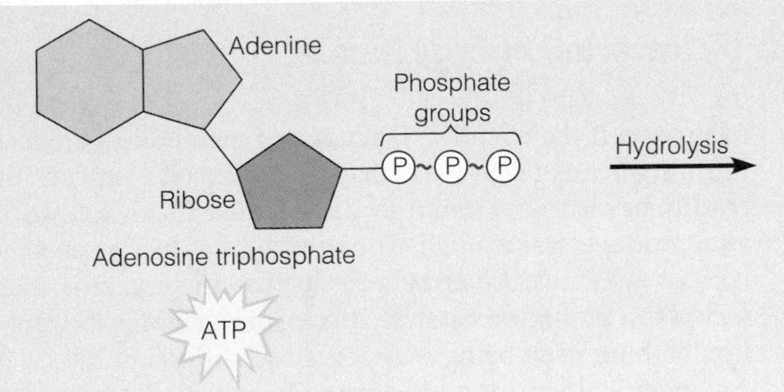

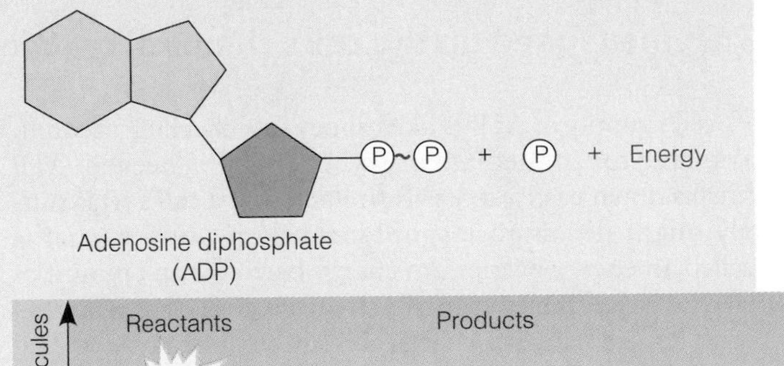

Figure 5.4A ATP structure and hydrolysis

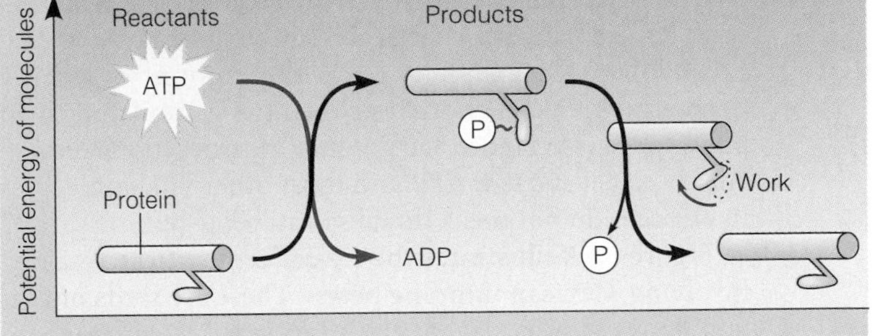

Figure 5.4B How ATP powers cellular work

ATP powers nearly all forms of cellular work, from the generation of light by fireflies to the movements of muscle cells that enable you to pedal a bicycle. Even as you read this page, nerve cells in your brain are using the chemical energy of ATP, energy your cells obtained earlier from food molecules.

When a cell uses chemical energy to perform work, it couples an exergonic reaction with an endergonic one. It first obtains chemical energy from an exergonic reaction and then uses the energy to drive an endergonic reaction. **Energy coupling**—using energy released from exergonic reactions to drive essential endergonic reactions—is a crucial ability of all cells. ATP molecules are the key to energy coupling. The usable energy released by most exergonic reactions, such as the breakdown of glucose molecules, is stored in ATP, and the energy used in most endergonic reactions comes from ATP. Let's see how the structure of ATP fits this function.

As shown in Figure 5.4A, ATP (adenosine triphosphate) has three parts, connected by covalent bonds: (1) adenine, a nitrogenous base; (2) ribose, a five-carbon sugar; and (3) a chain of three phosphate groups (symbolized by Ⓟ).

The covalent bonds connecting the second and third phosphate groups of ATP are unstable, as symbolized in the figure by a red wavy line: ~. These bonds can readily be broken by hydrolysis (see Module 3.3). Notice in Figure 5.4A that three things happen when the third bond breaks:

1. A phosphate is removed.
2. ATP becomes ADP (adenosine diphosphate).
3. Energy is released.

We call the hydrolysis of ATP an exergonic reaction because it releases energy. The cell can couple this reaction to an endergonic reaction. It does so by using the third phosphate group as an energy shuttle. In Figure 5.4B, the crossed arrows indicate two coupled reactions. In the figure, a protein molecule gains energy when it acquires a phosphate from ATP, changing shape in the process. The added energy enables the molecule to perform work, here changing back to its original shape. This kind of shape change is what helps a muscle cell contract. Notice that the energized molecule loses the phosphate in performing the work.

The transfer of a phosphate group to a molecule is called **phosphorylation**, and most cellular work depends on ATP energizing other molecules by phosphorylating them. Work can be sustained because ATP is a renewable resource that cells can regenerate. Figure 5.4C shows the ATP cycle. Energy released in exergonic reactions, such as glucose breakdown during cellular respiration, is used to regenerate ATP from ADP. In the process, a phosphate group is bonded to ADP, forming ATP by dehydration synthesis. This is an endergonic (energy-storing) reaction, just the reverse of a hydrolysis reaction. An organism or cell at work uses ATP continuously. In fact, a working cell consumes and regenerates its entire pool of ATP about once each minute.

Explain how ATP transfers energy from exergonic to endergonic processes in the cell.

By phosphorylation, the addition of phosphate groups: Exergonic processes phosphorylate ADP to form ATP, which transfers the energy to endergonic processes by phosphorylating other molecules.

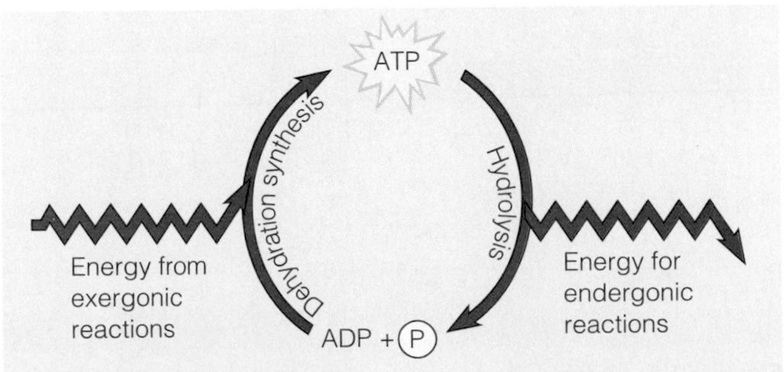

Figure 5.4C The ATP cycle

How Enzymes Work

5.5 Enzymes speed up the cell's chemical reactions by lowering energy barriers

5B

A cell's supply of ATP is like money in a checking account, available for immediate use whenever it is needed. ATP breaks down easily, so easily, in fact, that a cell's ATP supply might decompose spontaneously if not for what is called an energy barrier. An energy barrier is an amount of energy, called the **energy of activation (E_A)**, that reactants must absorb to start a chemical reaction. In the case of ATP breakdown, for instance, E_A is the amount of energy needed to break the bond between the second and third phosphates (see Figure 5.4A). Since most reactions require energy to get started, ATP and most other vital molecules in our cells do not break down spontaneously.

Figure 5.5A illustrates the E_A concept with an analogy involving Mexican jumping beans. These are seeds of certain desert shrubs, each containing an insect larva whose wriggling makes the bean jump. Jumping beans in the left-hand chamber of container 1 represent reactant molecules in a chemical reaction; those in the right-hand chamber represent product molecules. The partition symbolizes the E_A of the chemical reaction. Notice that the level of the left-hand chamber is higher than that of the right-hand chamber. Because they have a higher position, beans on the left have more potential energy than those on the right. But even though the beans that reach the lower chamber end up with less energy, they first had to have extra energy to make it over the partition.

Jumping beans vary in how much energy they have. Those with the most actively wriggling larvae jump the most often and the highest. At any particular instant, only a few beans may jump high enough to clear the partition between the two chambers and become "products." Thus, it may take a very long time for a significant number of reactant beans to reach the right-hand chamber.

Now we have a dilemma. Most of the essential reactions of metabolism must occur quickly and precisely for a cell to survive. If the chemical reactants of metabolism are like jumping beans faced with an energy barrier that cannot readily be cleared, a cell may die because it cannot make vital products fast enough. The solution lies in the capabilities of enzymes. An **enzyme** is a protein molecule that serves as a biological catalyst, increasing the rate of a reaction without itself being changed into a different molecule. An enzyme does not add energy to a cellular reaction; it speeds up a reaction by lowering the E_A barrier. Without enzymes, many metabolic reactions would occur too slowly to sustain life.

In Figure 5.5A, container 2 symbolizes the same chemical reaction as container 1, but catalyzed by an enzyme. The enzyme, in effect, pushes down the partition, making it possible for beans with less energy to clear it. As a result, in a given period of time, more beans will end up in the right-hand product chamber than would without the enzyme.

In Figure 5.5B, a graph shows the effect of an enzyme on the reaction it catalyzes. The black curve represents the course of the reaction without an enzyme; the E_A barrier is higher than in the reaction with an enzyme (red curve). Notice that the net change in energy from start to finish, analogous to the difference in heights of the two chambers in the jumping-bean model, is the same for both curves.

In catalyzing metabolic reactions in cells, enzymes are essential to life. One way enzymes lower the energy of activation is by holding reactant molecules in a particular position, as we see next. 

? Explain why an enzyme cannot change an endergonic reaction into an exergonic one.

Although an enzyme speeds a reaction by lowering the energy of activation, it has no effect on the relative energy content of products versus reactants.

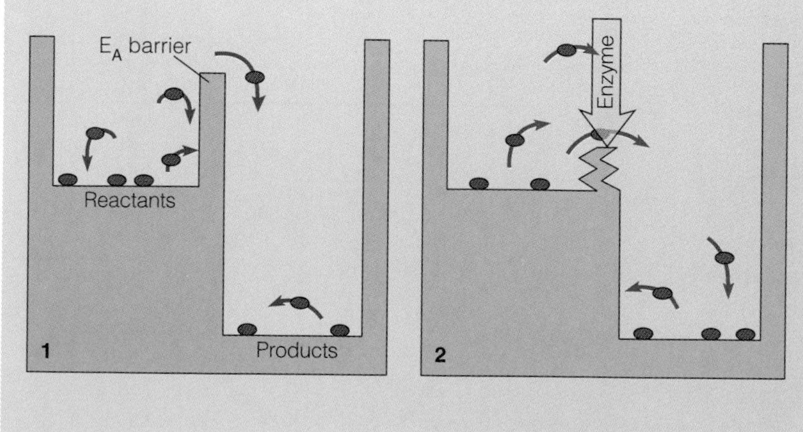

Figure 5.5A Jumping-bean analogy for energy of activation (E_A) and the role of enzymes

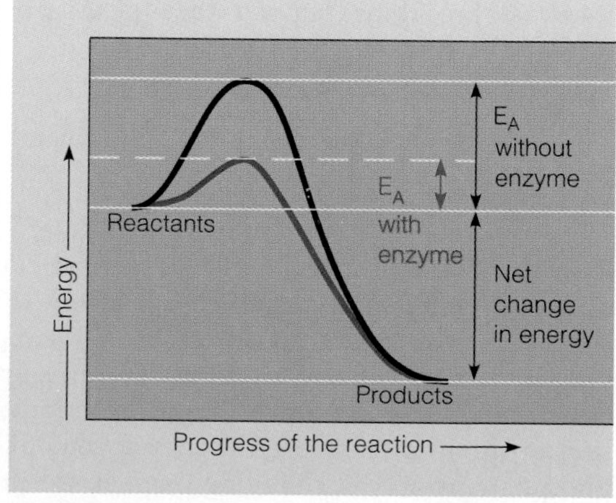

Figure 5.5B The effect of an enzyme on E_A

A specific enzyme catalyzes each cellular reaction 5.6

5B

Enzymes are very selective in the reactions they catalyze, and their selectivity determines which chemical processes occur in a cell at any particular time. As a protein, an enzyme has a unique three-dimensional shape, and that shape determines the enzyme's specificity. A substance that an enzyme acts on—a reactant in a chemical reaction—is called the enzyme's **substrate.** Each enzyme recognizes only the specific substrate or substrates of the reaction it catalyzes. It therefore takes many different kinds of enzymes to catalyze all the reactions that occur in a cell.

In catalyzing a reaction, an enzyme binds to its substrate. While the two are joined, the substrate changes into the product (or products) of the reaction. Only a small part of an enzyme molecule, called the **active site,** actually binds to substrate. The active site is typically a pocket or groove on the surface of the enzyme. The enzyme is specific because its active site fits only one kind of substrate molecule.

The figure here illustrates how an enzyme works. We use the enzyme sucrase as an example. Its specific substrate is table sugar (sucrose), and the reaction it catalyzes is the hydrolysis of sucrose to glucose and fructose. ① Sucrase starts with an empty active site. ② Sucrose enters the active site, attaching by weak bonds. The interaction with sucrose induces the enzyme to change shape slightly so that the active site fits even more snugly around the sucrose. This "induced fit" is like a clasping handshake. It holds the substrate in a position that facilitates the reaction, and ③ the substrate is converted to the products glucose and fructose. ④ The enzyme releases the products and emerges unchanged from the reaction. Its active site is now available for another substrate molecule, and another round of the cycle can begin. A single enzyme molecule may act on thousands or even millions of substrate molecules per second.

Figure 5.6 How an enzyme works

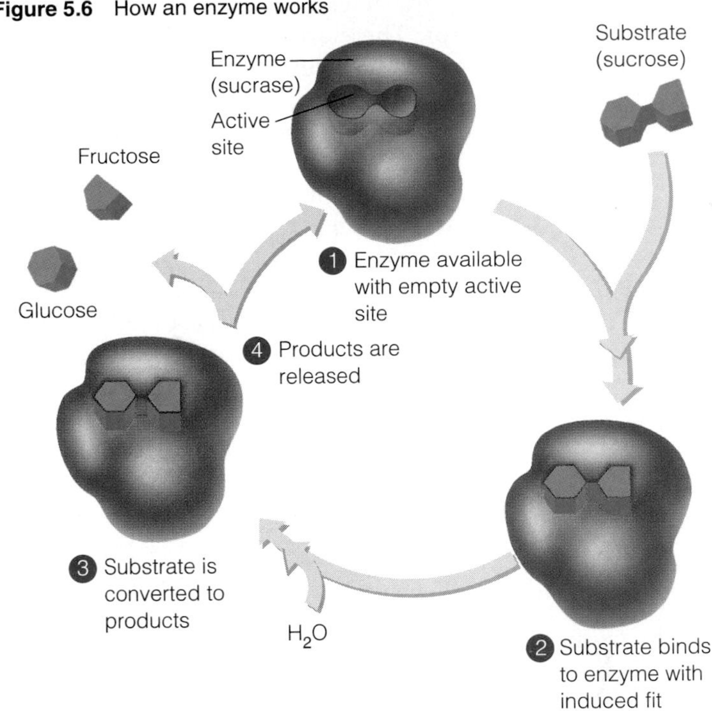

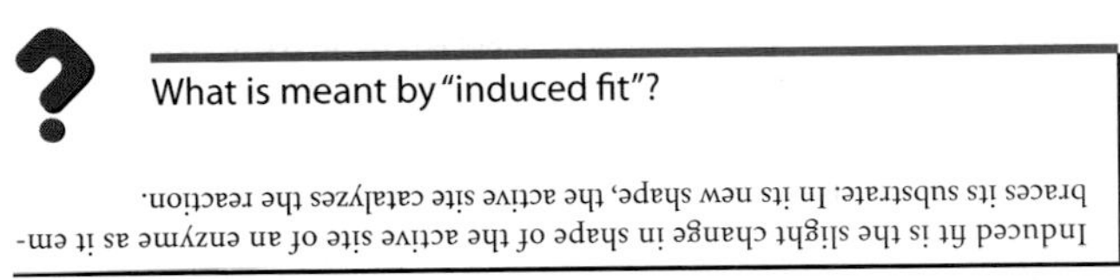

?
What is meant by "induced fit"?
Induced fit is the slight change in shape of the active site of an enzyme as it embraces its substrate. In its new shape, the active site catalyzes the reaction.

The cellular environment affects enzyme activity 5.7

The activity of an enzyme is affected by its environment, and for every enzyme there are conditions under which it is most effective. Temperature, for instance, affects molecular motion, and an enzyme's optimal temperature produces the highest rate of contact between reactant molecules and the enzyme's active site. Higher temperatures denature the enzyme, altering its specific three-dimensional shape and destroying its function. Most human enzymes have temperature optima of 35–40°C (close to our normal body temperature).

Salt concentration and pH also influence enzyme activity. Few enzymes can tolerate extremely salty solutions because the salt ions interfere with some of the chemical bonds that maintain protein structure. The same is true of the extra hydrogen ions present at very low pH. The optimal pH for most enzymes is near neutrality, in the range of 6–8. Outside this range, enzyme action, and thus the normal chemical functioning of cells, may be impaired. Entire lakes are threatened by acid precipitation, which can make the water so acidic that aquatic organisms experience enzyme failure (see Module 2.16).

Many enzymes will not work unless they are accompanied by nonprotein helpers called **cofactors.** Cofactors may be inorganic substances such as atoms of zinc, iron, or copper. If the cofactor is an organic molecule, it is called a **coenzyme.** Most coenzymes are made from vitamins or are vitamins themselves. For example, vitamin B_6 is a coenzyme required by enzymes involved in converting one amino acid to another.

Next we discuss chemicals that interfere with enzyme activity.

A few human enzymes have very low pH optima, about 2. Where in the body do you think these enzymes are located?

In the stomach

5.8 Enzyme inhibitors block enzyme action

A chemical that interferes with an enzyme's activity is called an inhibitor. There are two types of enzyme inhibitors. A **competitive inhibitor** resembles the enzyme's normal substrate and competes with the substrate for the active site on the enzyme. As shown in the figure, when a competitive inhibitor sits in the active site, it blocks the substrate from entering and thereby prevents the enzyme from acting.

A **noncompetitive inhibitor** does not enter the active site. Instead, it binds to the enzyme somewhere outside the active site. Its binding changes the shape of the enzyme so that the active site no longer fits the substrate.

The action of any inhibitor can be irreversible or reversible, depending on the kind of bonds formed with the enzyme. Inhibition is irreversible when covalent bonds form between inhibitor and enzyme. Inhibition is reversible when only weak bonds, such as hydrogen bonds, form. Weak bonds eventually break as neighboring molecules jostle together. When the concentration of substrate is higher than that of the inhibitor, it is more likely that a substrate molecule will be nearby when an active site becomes vacant, and the reaction will proceed. Conversely, when there is more inhibitor than substrate, the reaction will slow down.

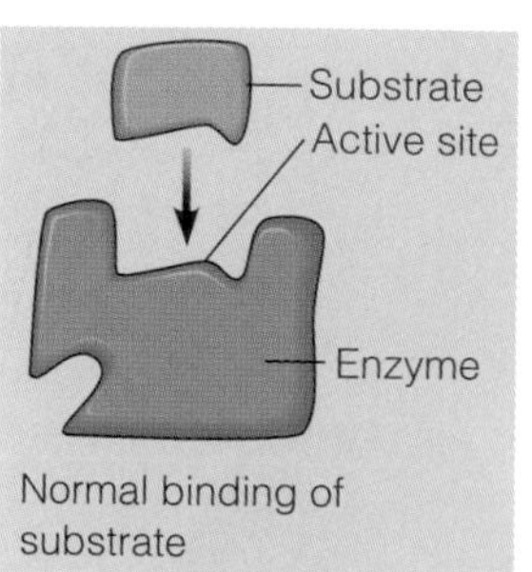

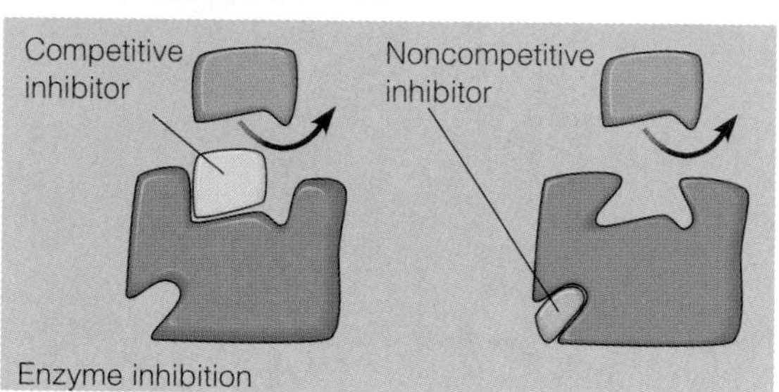

Figure 5.8 How inhibitors interfere with substrate binding

Enzyme inhibitors, especially reversible ones, are important regulators of cell metabolism. In some cases, the inhibitor of an enzyme is the very substance the reaction produces. For instance, when a cell's supply of ATP exceeds demand, ATP itself acts as a noncompetitive inhibitor, interfering with the enzymes that drive ATP synthesis. This sort of inhibition, whereby a metabolic reaction is blocked by its products, is called **negative feedback,** and it is one of the most important mechanisms that regulate metabolism. As we will see in Chapter 6, cells actually make most of their molecules by sequences of reactions, called metabolic pathways. Negative feedback controls many such pathways.

A competitive inhibitor of the enzyme sucrase (see Module 5.6) slows the production of glucose and fructose in a test-tube reaction. How could you overcome the effect of the inhibitor?

Add a lot more sucrose

5.9 Some pesticides and antibiotics inhibit enzymes

When an inhibitor, especially an irreversible one, prevents an enzyme from catalyzing a crucial metabolic reaction, an organism may be poisoned. This effect can be turned to human advantage. Certain pesticides are toxic to insects because they irreversibly inhibit key enzymes in the nervous system. The insecticide malathion, for instance, inhibits a nervous system enzyme called acetylcholinesterase. The inhibition prevents nerve cells from transmitting signals and kills the insect. For the same reason, malathion can also be toxic to other animals, including humans, though doses that kill insects are not generally harmful to people.

Many antibiotics also work by inhibiting enzymes—in this case, enzymes that are essential to the survival of disease-causing bacteria. Penicillin, for example, inhibits an enzyme that bacteria use in making cell walls. Since humans lack this enzyme, we are not harmed by the drug. The actions of pesticides and antibiotics help us appreciate how important enzymes are in living cells. Let's now turn to a discussion of membranes, which help organize the enzyme-catalyzed reactions that make up the cell's metabolism.

How does the antibiotic penicillin work?

It inhibits a bacterial enzyme that functions in cell wall production.

Membrane Structure and Function

5.10 Membranes organize the chemical activities of cells

So many metabolic reactions occur simultaneously in a cell that utter chaos would result if the cell were not highly organized and able to time its metabolic reactions precisely. Some of the organization comes from teams of enzymes that function like assembly lines. A product from one enzyme-catalyzed reaction becomes the substrate for a neighboring enzyme, and so on, until a final product is made. For a cell's assembly lines to operate, the right enzymes have to be present at the right time and in the right place.

Membranes provide the structural basis for metabolic order. In eukaryotes, membranes form most of the cell's organelles and partition the cell into compartments that contain enzymes in solution. For all cells, the plasma membrane is literally the edge of life, forming a fine boundary between the living cell and its surroundings and controlling the traffic of molecules into and out of the cell. Like all cell membranes, the plasma membrane exhibits **selective permeability;** that is, it allows some substances to cross more easily than others and blocks passage of some substances altogether. The plasma membrane takes up substances the cell needs and disposes of the cell's wastes.

For a structure that separates life from nonlife, the plasma membrane is amazingly thin, twenty times too thin to be resolved by the light microscope. It shows up quite clearly in electron micrographs, however. Figure 5.10A displays part of a red blood cell in cross section, magnified 200,000 times. At this magnification, you can see that the plasma membrane has three zones: an outermost dark band (upper arrow), a light zone just inside it, and a second dark band (lower arrow) bordering the outer edge of the cytoplasm.

Much of our knowledge about plasma membranes comes from studies of red blood cells. Cell biologists can prepare large amounts of pure plasma membrane by breaking open these cells and washing away the cytoplasm. The empty sacs remaining are called ghosts (Figure 5.10B). Chemical studies on these ghosts have revealed that the three layers seen in electron micrographs come from the way membrane lipid molecules are arranged, as we see next.

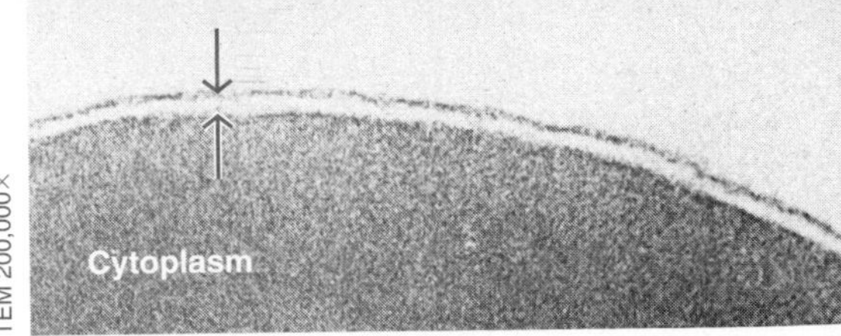

Figure 5.10A Plasma membrane in cross section

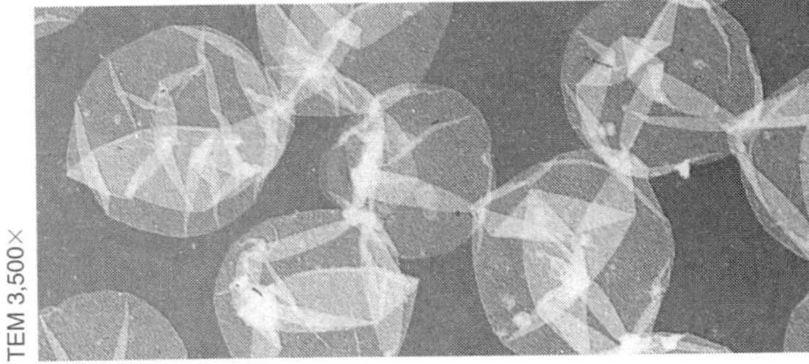

Figure 5.10B "Ghosts" of red blood cells

Use a metric ruler to measure the thickness of the membrane in Figure 5.10A (distance between the two arrow tips). Now, based on the magnification given and the conversion table in Module 4.2, estimate the plasma membrane's thickness in nanometers.

About 10 nm

Membrane phospholipids form a bilayer

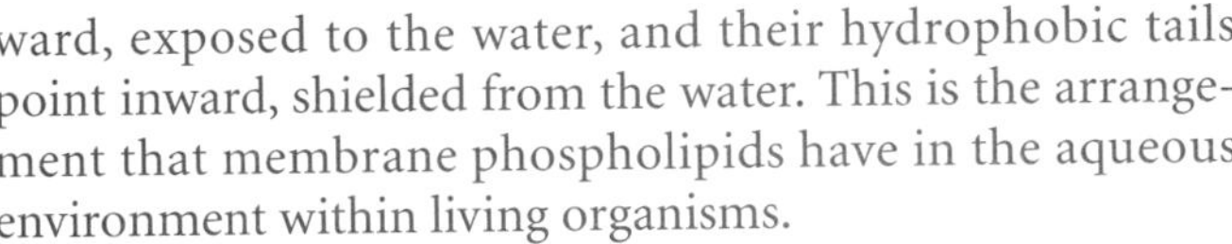

Head

Tails

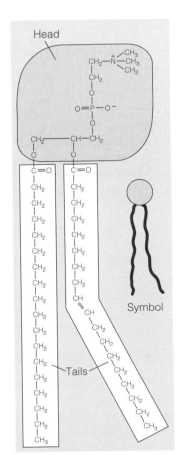

Symbol

Figure 5.11A Phospholipid molecule

Lipids, mainly phospholipids, are the main structural components of membranes. Figure 5.11A illustrates the structure of a phospholipid. The structural formula on the left highlights the two parts of the molecule, which have opposite interactions with water. The head (gray) is polar and therefore hydrophilic. The double tail (yellow) is nonpolar and hydrophobic. Next to it is the phospholipid symbol commonly used in depicting membranes.

The structure of phospholipid molecules is well suited to their role in membranes. In water, phospholipids spontaneously form a stable two-layer sheet called a phospholipid bilayer (Figure 5.11B). Their hydrophilic heads face outward, exposed to the water, and their hydrophobic tails point inward, shielded from the water. This is the arrangement that membrane phospholipids have in the aqueous environment within living organisms.

The hydrophobic interior of the bilayer is one reason membranes are selectively permeable. Nonpolar, hydrophobic molecules are soluble in lipids and can easily pass through membranes. In contrast, polar, hydrophilic molecules are not soluble in lipids. Whether polar molecules pass through the membrane depends on protein molecules in the phospholipid bilayer. In fact, much of a membrane's selective permeability depends on membrane proteins, which we examine in the next module.

Why do phospholipids tend to organize into a bilayer in an aqueous environment?

This structure shields the hydrophobic tails of the phospholipids from water, while exposing the hydrophilic heads to water.

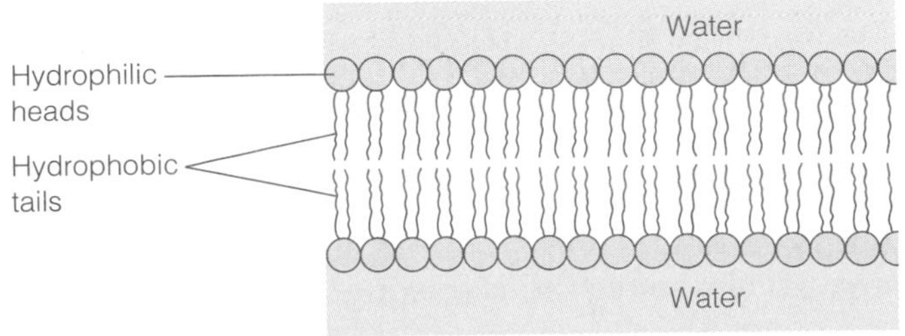

Figure 5.11B Phospholipid bilayer

5.12 The membrane is a fluid mosaic of phospholipids and proteins

5C

The drawing below shows the structure of the plasma membrane. Like other cellular membranes, it is commonly described as a **fluid mosaic.** The word "mosaic" denotes a surface made of small fragments, like pieces of colored tile cemented together in a mosaic floor or picture. A membrane is a *mosaic* in having diverse protein molecules embedded in a framework of phospholipids. The membrane mosaic is *fluid* in that most of the individual proteins and phospholipid molecules can drift laterally in the membrane. Helping to hold the delicate membrane in place, some of the proteins are linked both to the cytoskeleton and to fibers of the adjacent extracellular material—in the figure, the extracellular matrix of an animal cell.

Notice the kinked tails of many of the phospholipids. Resulting from double bonds in lipid tails, the kinks make the membrane more fluid by keeping adjacent phospholipids from packing tightly together. The steroid cholesterol (see Module 3.9), wedged into the bilayer, helps stabilize the phospholipids at body temperature (about 37°C for humans) but helps keep the membrane fluid at lower temperatures. In a living cell, the phospholipid bilayer remains about as fluid as salad oil at room temperature.

The two surfaces of a cellular membrane are different. For the plasma membrane, the outside surface has carbohydrates (chains of sugars, shown here in green) covalently bonded to proteins and lipids in the membrane. A protein with attached sugars is called a glycoprotein, whereas a lipid with sugars is called a glycolipid. The carbohydrate portions of glycoproteins and glycolipids vary from species to species, from one individual to another in the same species, and even from one cell type to another in a single individual. Many function as cell identification tags that are recognized by other cells. The ability to distinguish among different cells is crucial to life. It allows cells in an embryo to sort themselves into tissues and organs. It also enables cells of the immune system to recognize and reject foreign cells, such as infectious bacteria. The protein portions as well as the carbohydrates of glycoproteins play a role in making cell surfaces recognizable. Next, we look at some other functions of membrane proteins.

The carbohydrates attached to the proteins and lipids of the plasma membrane are added as the membrane is refined in the Golgi apparatus; the new membrane then forms transport vesicles that travel to the cell surface (see Figure 4.14). On which side of the vesicle membrane are the carbohydrates?

They are inside the transport vesicles.

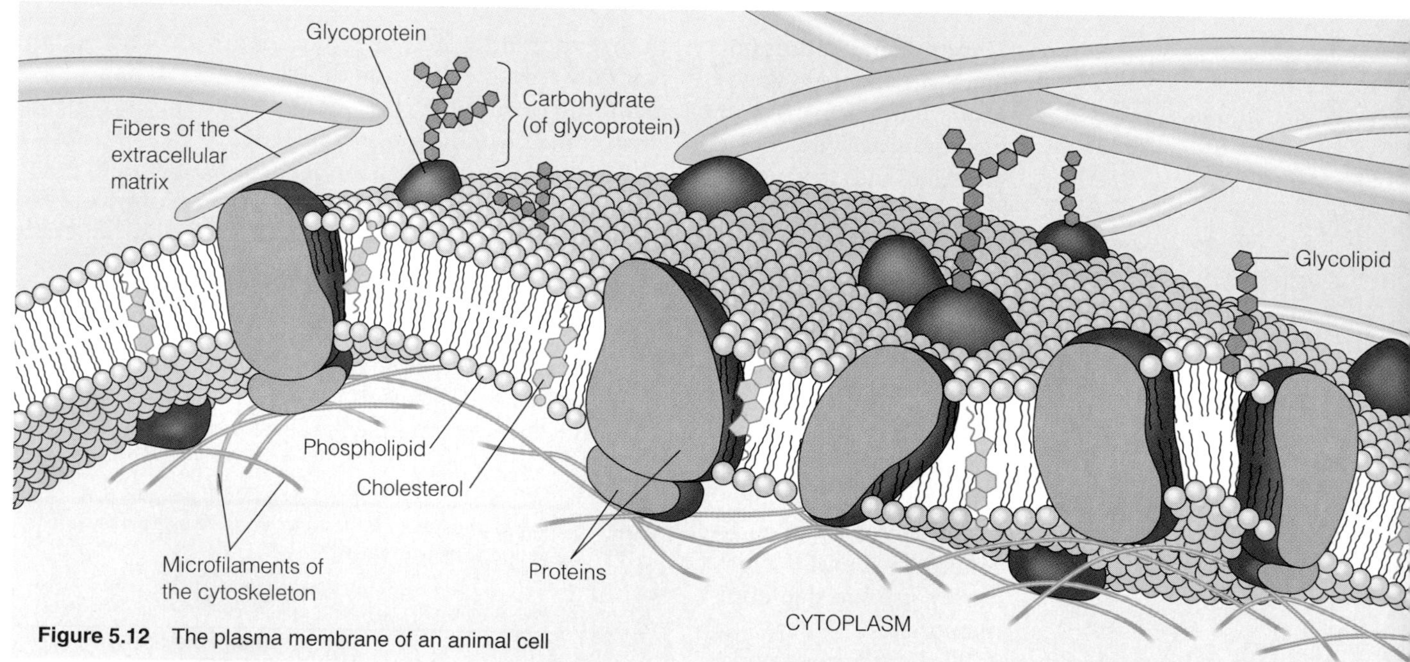

Figure 5.12 The plasma membrane of an animal cell

5.13 Proteins make the membrane a mosaic of function

5D

The word "mosaic" as applied to a membrane refers not only to the positioning of proteins in the phospholipid bilayer but also to the varied activities of these proteins. In fact, proteins perform most of the functions of the membrane. More than 50 different kinds of proteins have been found in the plasma membrane of human red blood cells, and there are probably many more that are just too scarce to have been detected so far.

We have already mentioned several functions of membrane proteins, including attaching the membrane to the cytoskeleton and external fibers, providing identification tags, and forming junctions between adjacent cells (see Module 4.19). What other roles do membrane proteins play? Many membrane proteins are enzymes, which may function in catalytic teams for molecular assembly lines (Figure 5.13A). Other proteins function as **receptors** for chemical messengers from other cells (Figure 5.13B). A receptor protein has a shape that fits the shape of a specific messenger, such as a hormone, just as an enzyme fits its substrate. Often the binding of the messenger to the receptor triggers a chain reaction involving other proteins, which relay the message to a molecule that performs a specific activity inside the cell. This is called **signal transduction.**

Finally, some membrane proteins help move substances across the membrane (Figure 5.13C). Although small molecules such as O_2 pass freely through the phospholipid bilayer, many essential molecules need assistance from proteins to enter or leave the cell. In the next seven modules, we look at molecular transport of all sorts.

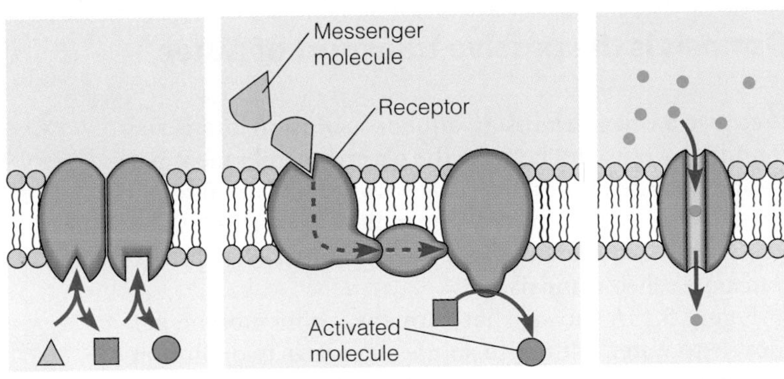

Figure 5.13A Enzyme activity

Figure 5.13B Signal transduction

Figure 5.13C Transport

The hormone epinephrine can cause a liver cell to hydrolyze its stored glycogen and release sugar without the hormone even entering the cell. Explain.

Epinephrine binds to a receptor on the liver cell surface, activating a signal transduction pathway inside the cell that leads to sugar release.

Passive transport is diffusion across a membrane 5.14

5E

The nature of phospholipids and the kinds of proteins in a membrane determine whether a particular substance can cross the membrane. Some molecules can cross without the cell doing any work, but even in these cases, transport proteins may be needed, and physical principles dictate when the molecules can cross and in which direction they will go.

Diffusion is the tendency for particles of any kind to spread out spontaneously to regions where they are less concentrated. Diffusion requires no work; it results from the random motion (kinetic energy) of atoms and molecules and is driven by the universal tendency of order to deteriorate into disorder (entropy). Because a cell does not perform work when molecules diffuse across its membrane, the diffusion of a substance across a biological membrane is called **passive transport.**

The figures here illustrate passive transport, showing how concentration affects the direction in which a substance crosses a membrane. In Figure 5.14A, molecules of a colored dye in solution tend to diffuse from the side of the membrane where they are more concentrated to the side where they are less concentrated, until the solutions on both sides have equal concentrations of dye. Put another way, the dye diffuses down its **concentration gradient** until equilibrium is reached. At equilibrium, molecules continue to move back and forth, but there is no *net* change in concentration on either side of the membrane. Figure 5.14B illustrates the important point that two or more substances diffuse independently of each other; that is, each diffuses down its own concentration gradient.

Passive transport is extremely important to all cells. In our lungs, for example, passive transport along concentration gradients is the sole means by which oxygen (O_2, essential for metabolism) enters red blood cells and carbon dioxide (CO_2, a metabolic waste) passes out of them. Water is another substance that crosses membranes by passive transport, and we focus on this topic next.

Explain how the second law of thermodynamics (see Module 5.2) helps explain diffusion of a substance across a membrane.

The second law is the trend toward randomness. Equal concentrations of a substance on both sides of a membrane is a more random distribution than unequal concentrations. Diffusion of a substance to a region where it is initially less concentrated increases entropy, as mandated by the second law.

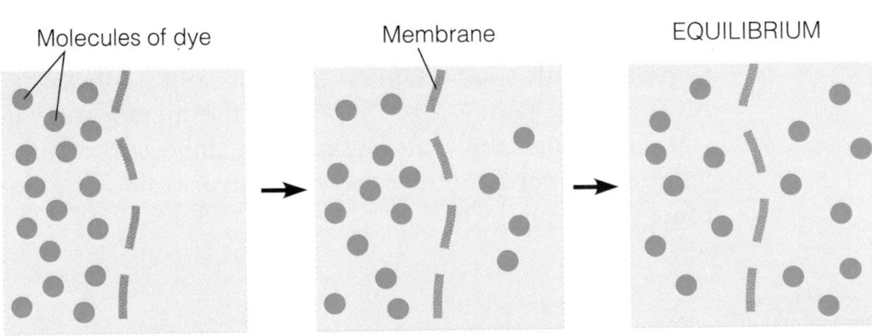

Figure 5.14A Passive transport of one type of molecule

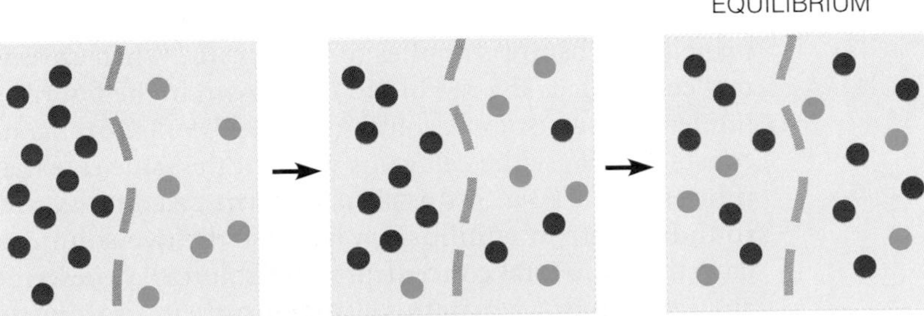

Figure 5.14B Passive transport of two types of molecules

5.15 Osmosis is the passive transport of water

5F

Because a cell contains an aqueous solution and is surrounded by one, and because the plasma membrane is permeable to water, water molecules can readily pass into and out of cells. Diffusion of water molecules across a selectively permeable membrane is a special case of passive transport called **osmosis.**

Figure 5.15A shows what happens if a membrane permeable to water but not to solutes separates two solutions with different concentrations of solute, here a molecule such as glucose. The solution with a higher concentration of solutes is said to be **hypertonic** (*hyper,* above; *tonos,* tension). The solution with the lower solute concentration is **hypotonic** (*hypo,* below). Water crosses the membrane until the solute concentrations (molecules per milliliter solution) are equal on both sides.

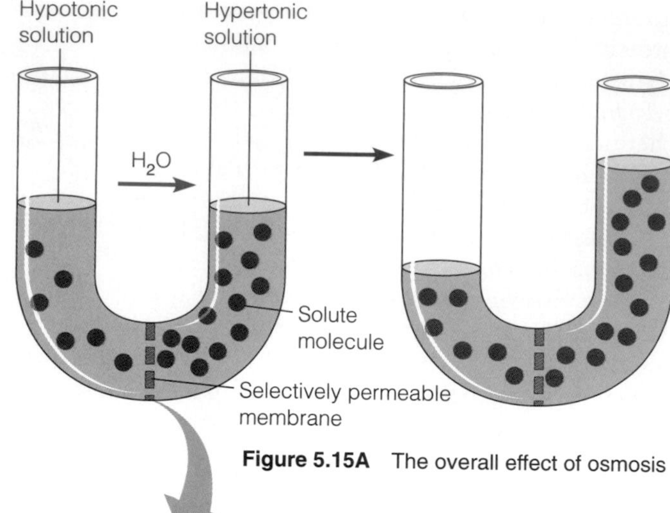

Figure 5.15A The overall effect of osmosis

In the close-up view in Figure 5.15B, you can see what happens at the molecular level. The polar solute molecules attract clusters of water molecules, so that fewer water molecules are free to diffuse across the membrane. With fewer solute molecules, the hypotonic solution has more free water molecules, and there is a net movement of water from the hypotonic solution to the hypertonic one. The result is the difference in water levels you see in Figure 5.15A.

Here we show only one type of solute molecule, but the same net movement of water would occur no matter how many kinds of solutes were present. The direction of osmosis is determined only by the difference in *total* solute concentration, not by the nature of the solutes. For example, seawater has a great variety of solutes, but it will lose water to a solution containing a high enough concentration of a single solute. Only if the total solute concentrations are the same on both sides of the membrane will water molecules move at the same rate in both directions. Solutions of equal solute concentration are said to be **isotonic** (*isos,* equal).

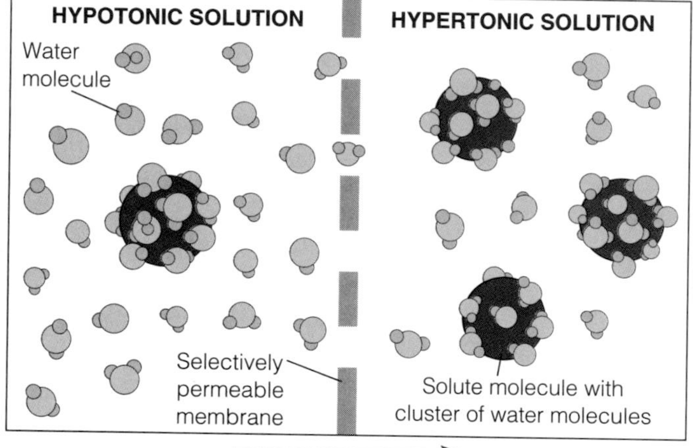

Figure 5.15B Osmosis at the molecular level

Explain why it is not enough just to say that a solution is "hypertonic."

Hypertonic and hypotonic are relative terms: A solution that is hypertonic to tap water could be hypotonic to seawater. You must say what the solution is compared to.

5.16 Water balance between cells and their surroundings is crucial to organisms

5F

Figure 5.16 at the top of the next page illustrates how the principle of osmosis applies to living cells. When an animal cell, such as the red blood cell shown in the figure, is immersed in an isotonic solution, the cell's volume remains constant because the cell gains water at the same rate that it loses it (part 1). In this situation, the cell and its surroundings are in equilibrium because the two solutions have the same total concentration of solutes. We describe an organism or a cell in this situation as being isotonic to the surrounding solution. Many marine animals, such as sea stars and crabs, are isotonic to seawater. Part 2 shows what can happen to an animal cell in a hypotonic solution, which has a lower solute concentration than the cell. The cell gains water, swells, and may pop (lyse) like an overfilled balloon. Part 3 shows the opposite case: An animal cell in a hypertonic solution shrivels and can die from water loss.

For an animal to survive if its cells are exposed to a hypertonic or hypotonic environment, the animal must have a way to prevent excessive uptake or excessive loss of water.

The control of water balance is called **osmoregulation.** For example, a freshwater fish, which lives in a hypotonic environment, has kidneys and gills that work constantly to prevent an excessive buildup of water in the body. (We discuss osmoregulation further in Chapter 25.)

Water balance problems are somewhat different for plant cells because of their rigid cell walls. A plant cell immersed in an isotonic solution (part 4) is flaccid, and a plant wilts in this situation. In contrast, a plant cell is turgid, and plants are healthiest, in a hypotonic environment, as in part 5. To become turgid, a plant cell needs a net inflow of water. Although the somewhat elastic cell wall expands a bit, the pressure it exerts prevents the cell from taking in too much water and bursting, as an animal cell would in this environment. Part 6 shows that in a hypertonic environment a plant cell is no better off than an animal cell. As a plant cell loses water, it shrivels, and its plasma membrane pulls away from the cell wall. This usually kills the cell. 

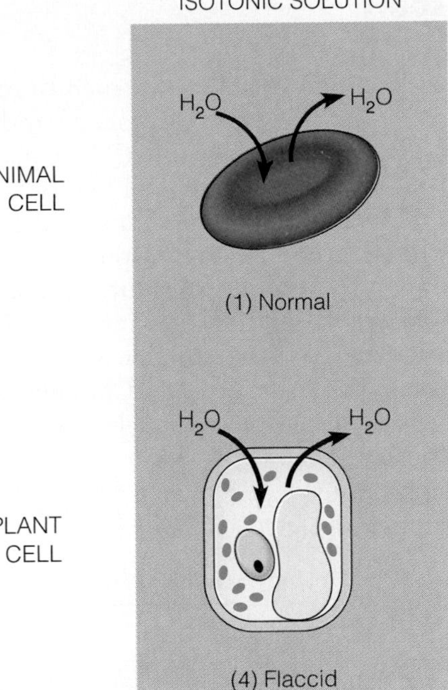

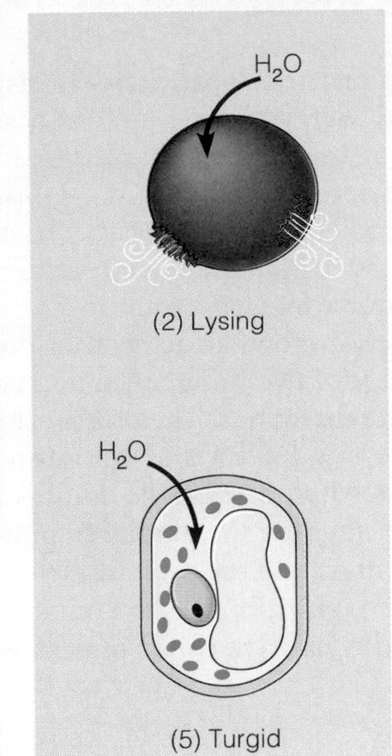

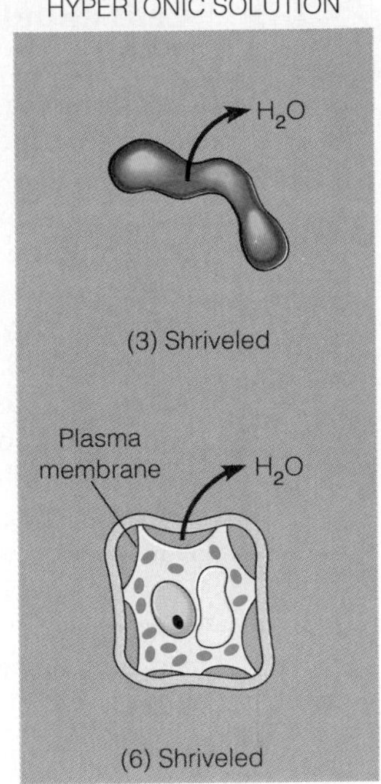

Figure 5.16 How cells behave in different solutions

? Explain the function of the contractile vacuoles in the *Paramecium* of Figure 4.13B in terms of what you have learned here about water balance in cells.

The pondwater in which the *Paramecium* lives is hypotonic to the cell, and thus there is a constant net osmosis of water into the cell. The contractile vacuoles expel this excess water, thus preventing the cell from bursting.

Specific proteins facilitate diffusion across membranes 5.17

5G

In discussing traffic across membranes, we have focused so far on small molecules, including water, that diffuse freely across biological membranes. Numerous substances that do not diffuse freely across membranes, because of their size or charge, can cross by special mechanisms that promote and regulate their movement.

Many molecules move across a membrane with the help of transport proteins in the membrane. When one of these proteins makes it possible for a substance to move down its concentration gradient, the process is called **facilitated diffusion.** Without the protein, the substance does not cross the membrane, or it diffuses across it too slowly to be useful to the cell. Like osmosis, facilitated diffusion is a type of passive transport because it does not expend energy. As in all passive transport, the driving force is the concentration gradient.

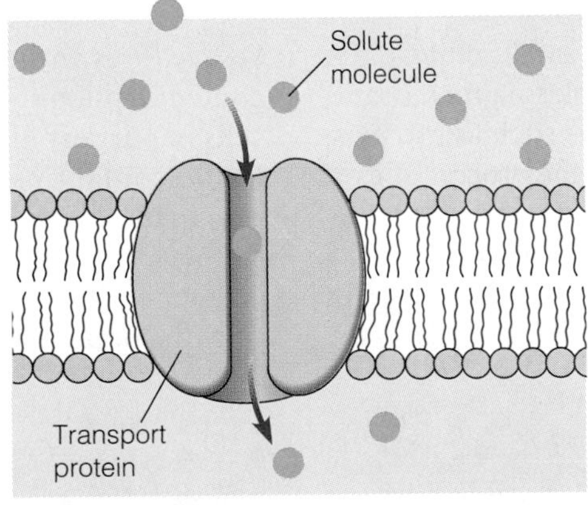

Figure 5.17 Transport protein providing a pore for solute passage

Figure 5.17 shows the most common way membrane proteins facilitate diffusion. The transport protein (violet) spans the membrane and provides a pore for the passage of a particular solute (orange). The rate of facilitated diffusion depends on how many transport-protein molecules are available in the membrane and how fast their specific solute travels through them. 

? How do transport proteins contribute to a membrane's selective permeablility?

Transport proteins are specific for the solutes they transport. Thus, the numbers and kinds of different transport proteins embedded in the membrane affect its permeability to various solutes.

5.18 Cells expend energy for active transport

5H

In contrast to passive transport, **active transport** requires that a cell expend energy to move molecules across a membrane. In this situation, a transport protein actively pumps a specific solute across a membrane *against* the solute's concentration gradient—that is, away from the side where it is less concentrated. Membrane proteins usually use ATP as their energy source for active transport.

Figure 5.18 below shows an active transport system involving the passage of two different solutes across a membrane, in opposite directions. The transport protein has a separate binding site for each of the solutes. ① Active transport begins when one of the solutes binds to the transport protein. ② After the binding, a phosphate group is transferred from ATP to the protein. ③ An energy surge from the transfer (phosphorylation) makes the protein change shape and release the solute molecule on the other side of the membrane. ④ As the protein releases the first solute, its shape and position allow the second solute to bind to it. ⑤ The phosphate group is released. This causes the protein to return to its original shape, ⑥ releasing the second solute on the opposite side of the membrane.

Active transport systems like this, coupling the passage of two solutes, are common in cell membranes, and they transport ions as well as uncharged molecules. A very important one is the sodium-potassium ($Na–K^+$) pump (discussed in Chapter 28), which helps nerve cells generate nerve signals.

In what way is active transport an endergonic process?

Active transport is energetically "uphill" in the sense that it requires a net input of energy (in the form of ATP).

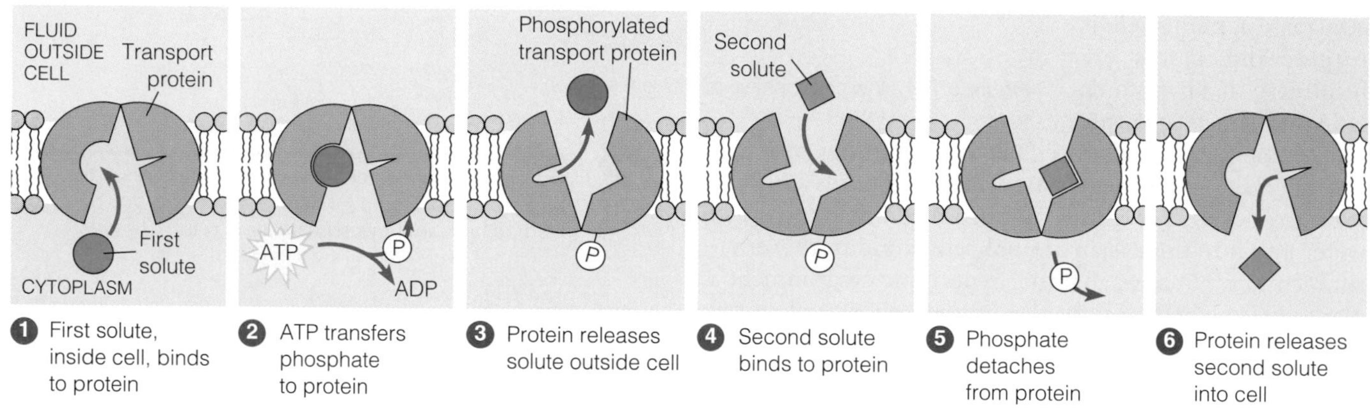
FLUID OUTSIDE CELL
Transport protein
First solute
CYTOPLASM
ATP
P
ADP
Phosphorylated transport protein
Second solute
1 First solute, inside cell, binds to protein
2 ATP transfers phosphate to protein
3 Protein releases solute outside cell
4 Second solute binds to protein
5 Phosphate detaches from protein
6 Protein releases second solute into cell

Figure 5.18 Active transport of two solutes across a membrane

5.19 Exocytosis and endocytosis transport large molecules

5I

So far we've focused on how water and small solutes enter and leave cells by moving through the plasma membrane. The story is different for large molecules such as proteins.

As shown in Figure 5.19A, a cell uses the process of **exocytosis** (from the Greek *exo*, outside, and *kytos*, cell) to export bulky materials. In the first step of this process, a membrane-enclosed vesicle filled with macromolecules (green) moves to the plasma membrane. Once there, the vesicle fuses with the plasma membrane, and the vesicle's contents spill out of the cell. When we weep, for instance, cells in our tear glands use exocytosis to export a salty solution containing proteins. In another example, certain cells in the pancreas manufacture the hormone insulin and secrete it into the bloodstream by exocytosis.

Figure 5.19B illustrates a transport process that is basically the opposite of exocytosis. In **endocytosis** (*endo*, in-

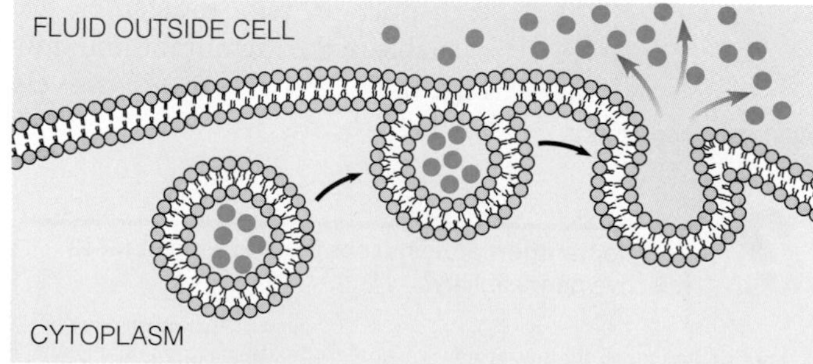
FLUID OUTSIDE CELL
CYTOPLASM

Figure 5.19A Exocytosis

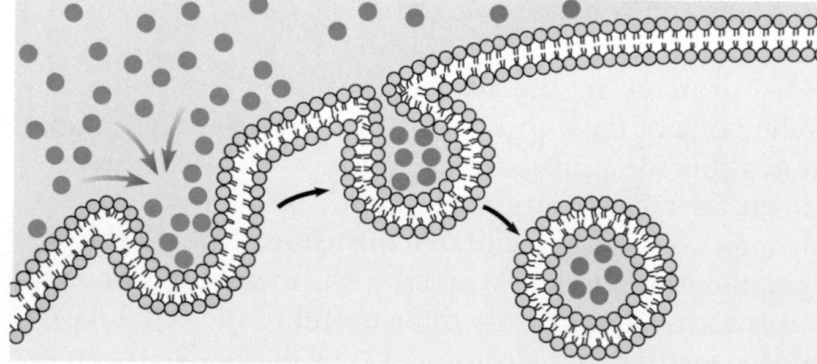

Figure 5.19B Endocytosis

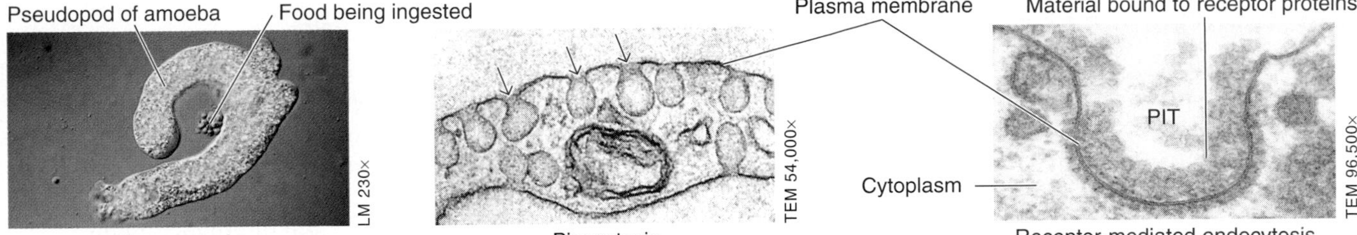

Figure 5.19C Three kinds of endocytosis

side), a cell takes in macromolecules or other particles by forming vesicles or vacuoles from its plasma membrane. Figure 5.19C shows three kinds of endocytosis. The color micrograph shows an amoeba taking in a food particle, an example of **phagocytosis,** or "cellular eating." The amoeba engulfs its prey by wrapping extensions, called pseudopodia, around it and packaging it within a vacuole. As Module 4.11 described, the vacuole then fuses with a lysosome, and the lysosome's hydrolytic enzymes digest the contents.

The center micrograph shows **pinocytosis,** or "cellular drinking." The cell is in the process of taking droplets of fluid from its surroundings into tiny vesicles (arrows). Pinocytosis is not specific, in that it takes in any and all solutes dissolved in the droplets.

In contrast to pinocytosis, a third type of endocytosis, called **receptor-mediated endocytosis,** is highly specific. The micrograph on the right shows one stage of the process. The plasma membrane has indented to form a pit. The pit is lined with receptor proteins that have picked up particular molecules from the surroundings. The pit will pinch closed to form a vesicle that will carry the molecules into the cytoplasm. In the next module, we examine a human hereditary condition in which receptor-mediated endocytosis malfunctions.

Explain how a protein-secreting cell can synthesize and secrete its product without the protein ever having to cross a membrane (see Modules 4.8 and 4.14).

From the time the protein is made by rough ER, it is topographically "outside" the cell, first in the ER interior, then within the Golgi and transport vesicles, and finally outside the plasma membrane as the vesicles release their contents by exocytosis.

5.20 Faulty membranes can overload the blood with cholesterol

Few molecules make the news as often as cholesterol does. Cholesterol is notorious as a possible factor in heart disease, yet it is essential for the normal functioning of all our cells. As mentioned earlier, cholesterol is a component of membranes as well as a starting material for making other steroids, such as sex hormones. Problems tend to arise only when we have too much cholesterol in our blood.

In most of us, cells in the liver remove excess cholesterol from the blood by receptor-mediated endocytosis. Consequently, we have only small amounts of the substance in our blood. Cholesterol circulates in the blood mainly in particles called low-density lipoproteins, or LDLs. As Figure 5.20 shows, an LDL is a globule of cholesterol (and other lipids) surrounded by a single layer of phospholipids in which proteins are embedded. One of the LDL proteins (pink) fits a specific type of receptor protein (purple) on cell membranes. Normally, liver cells take up LDLs from the blood by receptor-mediated endocytosis in which the receptors are these LDL receptor proteins.

Worldwide, about one in 500 human babies inherits a disease called **hypercholesterolemia,** which is characterized by an excessively high level of cholesterol in the blood. In severe cases (about one in a million people), there are no functional LDL receptors. People with a milder form of the disease have functional receptors but in low numbers. In either case, LDLs tend to accumulate in the blood. High levels of LDLs in the blood are life-threatening because the LDLs can deposit cholesterol on the lining of blood vessels, eventually obstructing them. The blockage of blood vessels that nourish the heart leads to heart disease. Individuals with severe hypercholesterolemia may die from heart disease in early childhood. Milder cases are also dangerous, but they can be treated with medications and a low-fat diet. We discuss the genetics of hypercholesterolemia in Module 9.12.

Explain how a person can have an excessive cholesterol level in the blood even without producing more cholesterol than other people.

The person has an impairment in the ability of cells to remove cholesterol from the blood because of the absence or shortage of LDL receptors in plasma membranes.

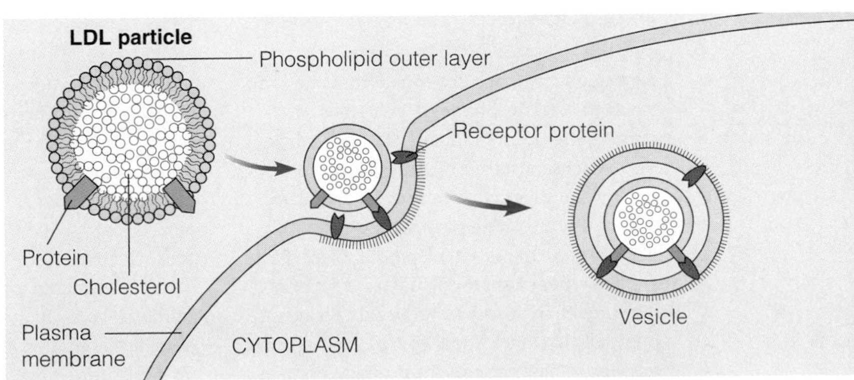

Figure 5.20 A cell using receptor-mediated endocytosis to take up an LDL

5.21 Chloroplasts and mitochondria make energy available for cellular work

This chapter's main topics—energy, enzymes, and membranes—lead us to see connections that are important to all organisms. Two organelles, the chloroplast and the mitochondrion, are both composed of membranes with enzyme assembly lines. Both play central roles in harvesting energy and making it available for cellular work. As shown in the diagram here, chloroplasts in photosynthetic organisms use solar energy to make glucose from carbon dioxide and water. In the process of photosynthesis, light energy is converted to chemical energy, and oxygen gas escapes as a by-product. Mitochondria, present in all eukaryotes, consume oxygen in cellular respiration, using the chemical energy stored in glucose to make ATP.

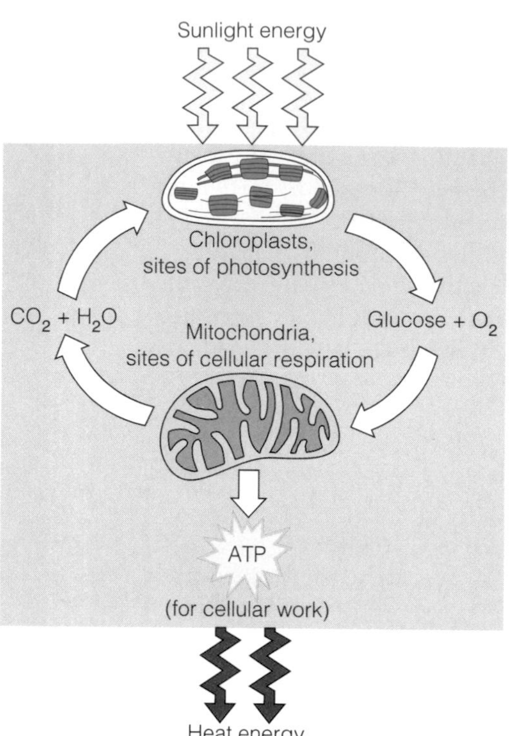

Figure 5.21 Energy flow and chemical recycling

Nearly all the chemical energy that organisms use comes ultimately from sunlight. Chloroplasts in plants and photosynthetic protists convert light energy to chemical energy. No animal can do this. Some animals obtain chemical energy by eating plants. Others, like the firefly, obtain it by eating other animals. In either case, the chemical energy was originally derived from sunlight through the process of photosynthesis.

The figure also makes the important point that chemicals recycle among living organisms and their environment. Even the oxygen that plants give off is not lost, but remains available in the air for mitochondrial use. Energy, however, does not recycle. In carrying out the activities of life, organisms dissipate some energy to the environment in the form of heat. This constitutes an energy loss, because heat cannot be used to make ATP. For this reason, organisms must be constantly supplied with energy. In the next chapter, we discuss how cells maintain their supplies of ATP. 

? In what way is the diagram in Figure 5.21 consistent with the second law of thermodynamics?

The living world takes in organized energy in the form of sunlight and replaces it with randomized energy in the form of heat.

Chapter Review

Chapter Summary

Energy and the Cell (Introduction–5.4) Compartmentalized by membranes, living cells process energy by means of enzyme-controlled chemical reactions **(Introduction)**. All organisms require energy, which is defined as the capacity to perform work. Kinetic energy is energy that is actually doing work. Potential energy is stored energy—energy that matter possesses because of its location or arrangement **(5.1)**. According to the laws of thermodynamics, energy can be changed from one form to another, but it cannot be created or destroyed. Energy changes are not 100% efficient; they increase disorder, or entropy, and some energy is always lost as heat **(5.2)**. Cells carry out thousands of chemical reactions, the sum of which constitutes cellular metabolism. There are two types of chemical reactions. An endergonic reaction absorbs energy and yields products rich in potential energy. An exergonic reaction releases energy; its products contain less potential energy than its reactants. Burning and cellular respiration are both exergonic **(5.3)**. In cellular respiration, some energy is stored in ATP molecules. When the bond joining a phosphate group to the rest of an ATP molecule is subsequently broken by hydrolysis, the reaction supplies energy for cellular work **(5.4)**.

How Enzymes Work (5.5–5.9) In order for a chemical reaction to begin, reactants must absorb some energy, called the energy of activation (E_A). This represents an energy barrier that helps prevent molecules from breaking down spontaneously. A protein catalyst called an enzyme can decrease the energy barrier **(5.5)**. Enzymes are selective, and this selectivity determines which chemical reactions occur in a cell. The reactant an enzyme acts on—its substrate—attaches to a specifically shaped active site on the enzyme. The substrate is held in a position that facilitates the reaction and is converted to a product, which is then released. The enzyme is unchanged and can rapidly repeat the process **(5.6)**. Temperature, salt concentration, and pH influence enzyme activity. Some enzymes require nonprotein cofactors. Cofactors that are organic molecules are called coenzymes **(5.7)**. Inhibitors interfere with enzymes. A competitive inhibitor takes the place of the substrate in the active site. A noncompetitive inhibitor alters an enzyme's function by changing its shape **(5.8)**. Some pesticides and antibiotics are enzyme inhibitors **(5.9)**.

Membrane Structure and Function (5.10–5.21) Membranes order metabolism. They are selectively permeable, controlling the flow of substances into and out of a cell and among compartments within a cell. They can also hold teams of enzymes that function in metabolism **(5.10)**. Phospholipids, each with a hydrophilic head and two hydrophobic tails, are the main structural components of membranes. In water, phospholipids form a stable bilayer with the heads facing outward and the tails facing inward **(5.11)**. The membrane is a fluid mosaic: Phospholipid molecules form a flexible bilayer, with cholesterol and protein molecules embedded in it. Attached carbohydrates act as cell identification tags **(5.12)**. Some of the proteins in membranes form cell junctions or transport substances across the membrane. Others are enzymes or receptors for chemical messages from other cells. The binding of a messenger to a receptor may trigger signal transduction, a chain reaction involving other proteins that relay the message to a molecule that performs a specific activity in the cell **(5.13)**.

In passive transport, substances simply diffuse through membranes, spreading from areas of higher concentration to areas of lower concentration **(5.14)**. Osmosis is the passive transport of water, from a solution of lower solute concentration to one of higher solute concentration **(5.15)**. Osmosis causes cells to shrink in a hypertonic solution and swell in a hypotonic solution. The control of water balance, called osmoregulation, is essential for organisms **(5.16)**.

Small and nonpolar molecules diffuse freely through the phospholipid bilayer. Other molecules may pass through selective protein pores by facilitated diffusion **(5.17)**. Transport proteins can move solutes against a concentration gradient; this active transport process requires ATP energy **(5.18)**. To move large molecules or particles through the membrane, a vesicle may fuse with the membrane and expel its contents (exocytosis), or the membrane may fold inward, trapping material from the outside (endocytosis). Receptor-mediated endocytosis traps specific substances **(5.19)**. Harmful levels of cholesterol can accumulate in the blood if membranes lack cholesterol receptors **(5.20)**.

Enzymes and membranes are central to the processes that make energy available to a cell. Chloroplasts carry out photosynthesis, using solar energy to produce glucose and oxygen from carbon dioxide and water. Mitochondria consume oxygen in cellular respiration, using the energy stored in glucose to make ATP **(5.21)**.

Testing Your Knowledge

Multiple Choice

1. Consider the following situations: chemical bonds in the gasoline in a car's gas tank, and the movement of the car along the highway; the tension of a stretched rubber band, and the pull that stretched it; a climber poised at the top of a hill, and the hike he took to get there. In each of these situations, the first part illustrates ____, and the second part illustrates ____.
   a. the first law of thermodynamics . . . the second law
   b. kinetic energy . . . potential energy
   c. an exergonic reaction . . . an endergonic reaction
   d. potential energy . . . kinetic energy
   e. the second law of thermodynamics . . . the first law

2. Which best describes the general structure of a cell membrane?
   a. proteins sandwiched between two layers of phospholipid
   b. proteins embedded in two layers of phospholipid
   c. a layer of protein coating a layer of phospholipid
   d. phospholipids sandwiched between two layers of protein
   e. phospholipids embedded in two layers of protein

3. The total solute concentration in a red blood cell is about 2%. Sucrose cannot pass through the membrane, but water and urea can. Osmosis would cause such a cell to shrink the most when immersed in which of the following?
   a. a hypertonic sucrose solution
   b. a hypotonic sucrose solution
   c. a hypertonic urea solution
   d. a hypotonic urea solution
   e. pure water

4. The calcium concentration in a cell is 0.3%; the concentration in the surrounding fluid is 0.1%. How could the cell obtain more calcium? *(Explain.)*
   a. passive transport
   b. diffusion
   c. active transport
   d. osmosis
   e. any of the above

Describing, Comparing, and Explaining

Food rots when microbes break down food molecules. Food preservation methods interfere with the enzyme activity of microbes and prevent them from surviving. Explain how each of the following would interfere with enzyme activity: canning (heating), freezing, pickling (soaking in acetic acid), salting.

Thinking as a Scientist

A biologist isolated a sample of an enzyme called lactase from the intestinal lining of a calf. The substrate of lactase is the disaccharide lactose. The enzyme breaks a lactose molecule in two, producing a molecule of glucose and a molecule of galactose. The biologist performed two series of experiments. First, she made up 10% lactose solutions with different concentrations of enzyme and measured the rate at which galactose was produced (in terms of grams of galactose produced per minute; a bigger number means a faster breakdown of lactose). Results of these experiments are shown in Table A below. In the second series of experiments, as shown in Table B, she prepared 2% enzyme solutions with different concentrations of lactose and again measured the rate of galactose production.

Table A: Rate and Enzyme Concentration

Lactose concentration	10%	10%	10%	10%	10%
Enzyme concentration	0%	1%	2%	4%	8%
Reaction rate	0	25	50	100	200

Table B: Rate and Substrate Concentration

Lactose concentration	0%	5%	10%	20%	30%
Enzyme concentration	2%	2%	2%	2%	2%
Reaction rate	0	25	50	65	65

a. Explain the relationship between the reaction rate and the enzyme concentration. b. Explain the relationship between the reaction rate and the substrate concentration. (*Hint:* Sketch graphs of rate versus concentration.) What might be happening between individual enzyme and substrate molecules? How and why did the results of the two experiments differ?

Science, Technology, and Society

Lead acts as an enzyme inhibitor, and it can interfere with the development of the nervous system. The Johnson Controls Company, a maker of lead-acid batteries, had a "fetal protection policy" that banned female employees of childbearing age from working in areas where they might be exposed to high levels of lead. Under this policy, women were involuntarily transferred to lower-paying jobs in lower-risk areas. A group of employees challenged the policy in court, claiming that it deprived women of job opportunities available to men. The U.S. Supreme Court ruled the policy illegal. Nonetheless, many people are uncomfortable about the "right" to work in an unsafe environment. What rights and responsibilities of employers, employees, and government agencies are in conflict in this situation? Whose responsibility should it be to determine what makes a safe environment and who should or should not work there? What criteria should be used to decide?

Media Resources

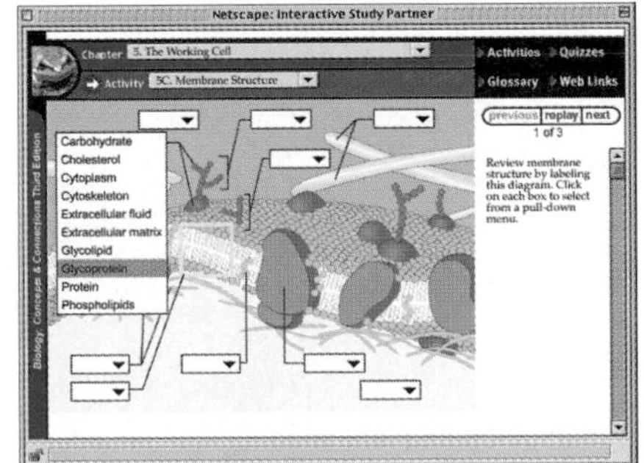

For further review of the concepts in this chapter, use the activities and quizzes on the CD-ROM.

Visit the Special Edition of The Biology Place web site for activities, quizzes, news articles, and web links.

6 How Cells Harvest Chemical Energy

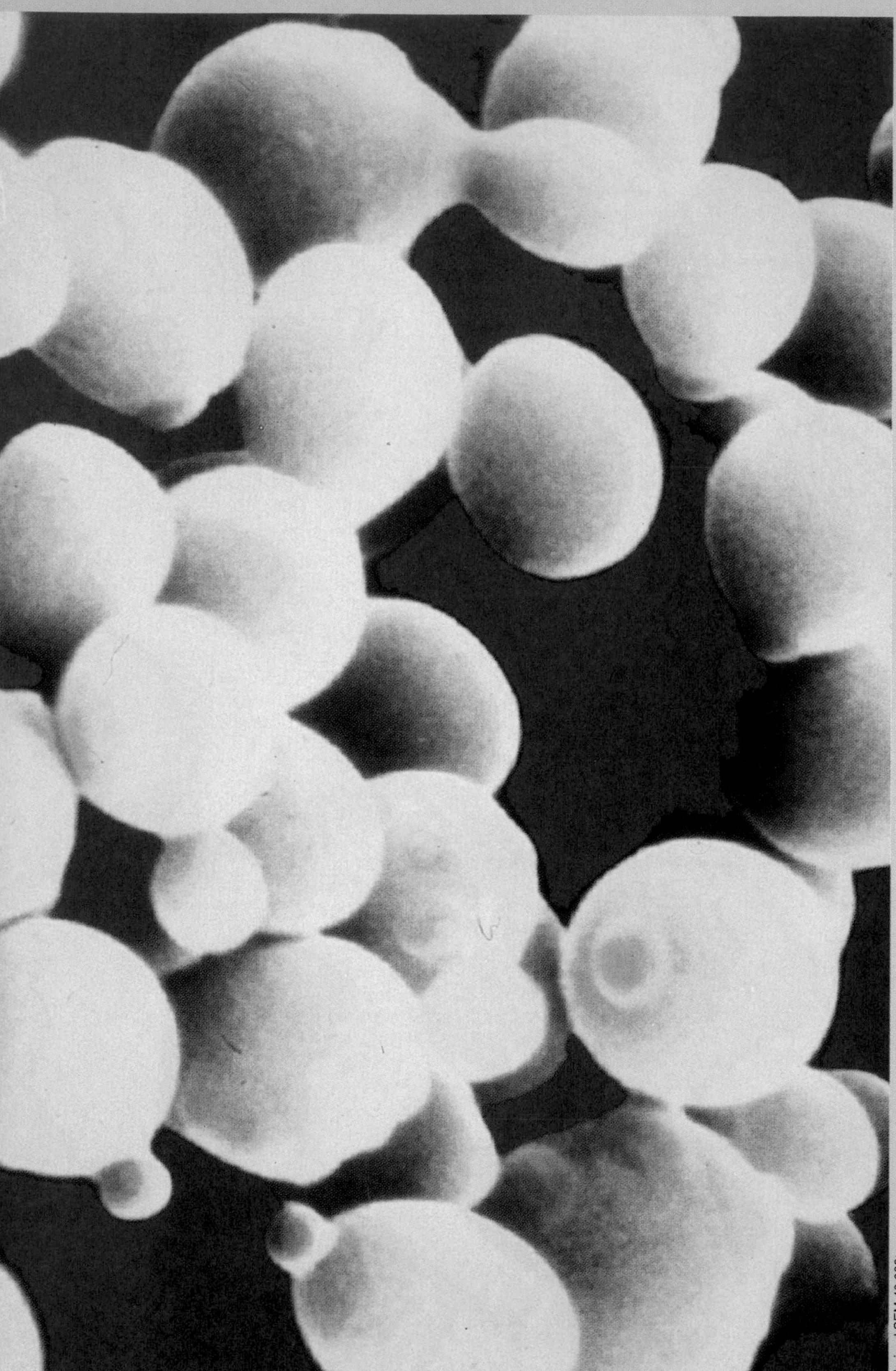

Colorized SEM 40,000×

Outline

Working on the revision of this chapter for the new edition, the authors debated.

"Not yeast *again*!" one of us said.

Should we again begin this chapter with a photograph of yeast cells?

"Let's face it: They're not exactly dynamic looking—just a bunch of blobs."

"But with color added to a scanning electron micrograph, they're attractive, and the buds are kind of neat."

"Well, what better organism to introduce this chapter?"

We finally decided—a well-known species of yeast would continue as our focal organism for this chapter. Even though the various yeasts are among the simplest eukaryotes, they perform all the activities this chapter covers. At the molecular level, yeasts are quite similar to other eukaryotes. This similarity has made them important research organisms. Discoveries about the molecular activities of yeasts have shed light on the biology of many other organisms, including humans.

Members of the kingdom Fungi, yeasts grow on moist surfaces and in liquids that are rich in nutrients. The cells in the photograph at the left were growing in a sugar solution. Yeasts exist as single cells, as cell clusters, and as short chains of cells. They usually reproduce by budding, a cell division process that produces a small new cell (the bud) from the larger parent cell. The yeast cells here belong to the species *Saccharomyces cerevisiae,* which has been used to make bread, beer, and wine throughout recorded history. You may recognize part of the name from our earlier discussion of carbohydrates; *Saccharomyces* means "sugar fungus" in Greek.

Saccharomyces cerevisiae and other yeasts normally live in **aerobic** (oxygen-containing) environments, but they can also live in **anaerobic** (oxygen-lacking) environments. When yeast is first added to bread dough or a wine vat, the cells are in an aerobic environment. As they grow, the cells consume oxygen (O_2) in the process of obtaining energy from sugars in the dough or grape juice. They give off carbon dioxide (CO_2) gas as a waste product. For the baker, the CO_2 bubbles make the dough rise; the yeast cells are then killed by the heat when the bread is baked.

In winemaking, the yeast's environment—a sealed vat—becomes anaerobic as the cells use up all the available O_2. The cells then obtain energy from sugars by another process, which yields CO_2 and ethyl alcohol as waste products. When the alcohol content reaches 12–16%, the cells die, poisoned by their own wastes.

In their ability to thrive in both aerobic and anaerobic environments, yeast cells are metabolically versatile. However, they can harvest much more energy from food molecules by using oxygen than they can without it.

Like yeast cells, our own muscle cells thrive when they have plenty of oxygen. But unlike yeasts, our muscle cells function efficiently for only a brief time without oxygen, even if sugars are available. When we exercise moderately—by jogging, for instance—our muscles use O_2 in harvesting energy from sugar molecules the same way yeasts do in an aerobic environment. The jogging is "aerobic exercise," meaning that it is vigorous but slow enough that the blood can deliver O_2 to the muscle cells at least as fast as they use it. In contrast, when we exercise very strenuously, our muscle cells soon use up their O_2 supply and become anaerobic. They continue to function for a while, but, like yeasts, they produce a toxic waste. In muscles, the waste is lactic acid, which makes the muscles ache and fatigue quickly.

We begin this chapter on how cells harvest energy from food with a look at how our own cells obtain O_2 and dispose of CO_2.

Aerobic exercise

Introduction to Cellular Respiration

6.1 Breathing supplies oxygen to our cells and removes carbon dioxide

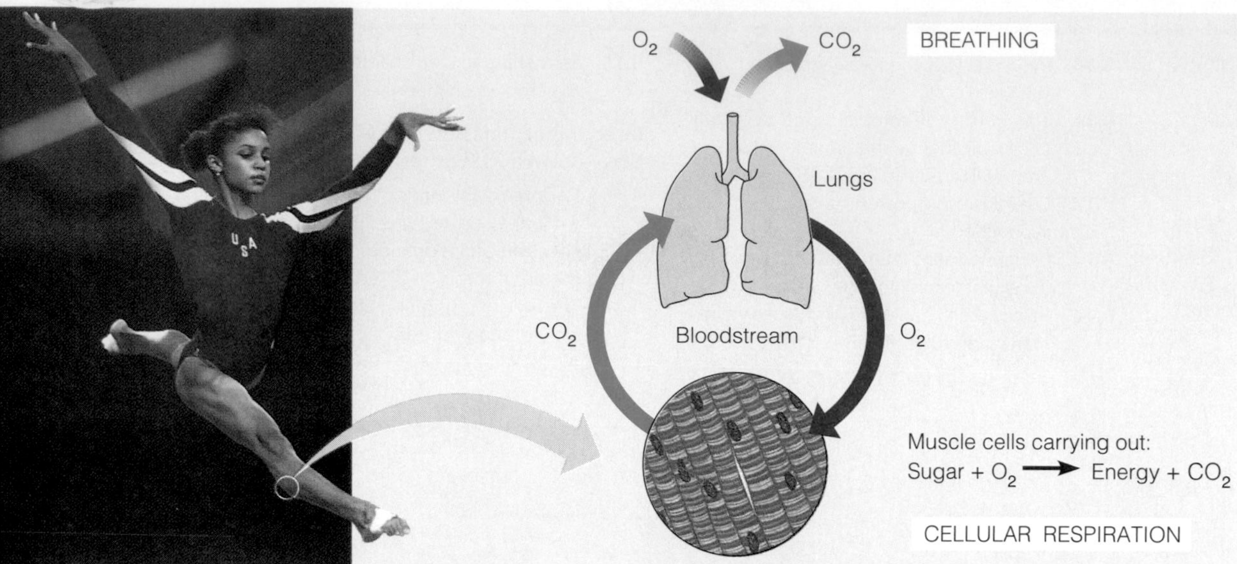

Figure 6.1 The connection between breathing and cellular respiration

We often use the word "respiration" as a synonym for breathing, the meaning of its Latin root. In this sense, respiration refers to an exchange of gases: An organism obtains O_2 from its environment and releases CO_2. Biologists also define respiration as the aerobic harvesting of energy from food molecules by cells. To distinguish it from breathing, this process is called **cellular respiration.**

Breathing and cellular respiration are closely related. As the gymnast above goes through her routine, her lungs take up O_2 from the air and pass it to her bloodstream. The bloodstream carries the O_2 to her muscle cells. Mitochondria in the muscle cells use the O_2 in cellular respiration, harvesting energy from sugar and other organic molecules the gymnast obtained from food several hours before. The muscle cells then use the energy to contract. Simultaneous contraction of many thousands of muscle cells, precisely controlled by the nervous system, make the gymnast's body move. Her bloodstream and lungs also perform the vital function of disposing of the CO_2 waste produced by cellular respiration in the muscle cells.

How is your breathing related to your cellular respiration?

In breathing, your lungs exchange CO_2 and O_2 between your body and the atmosphere. In cellular respiration, your cells consume the O_2 in extracting energy from food and release CO_2 as a waste product.

6.2 Cellular respiration banks energy in ATP molecules

As the gymnast example implies, oxygen usage is only a means to an end. Harvesting energy is the fundamental function of cellular respiration. The chemical equation in Figure 6.2A summarizes cellular respiration as carried out by muscle cells, yeasts, and all other cells that use O_2 in harvesting energy from the sugar glucose. Throughout this chapter, we use glucose as a convenient representative food molecule, although cells also "burn" many other organic molecules in cellular respiration. The summary equation indicates that the starting (reactant) molecules glucose and O_2 come apart, and their atoms regroup to form the products CO_2 and H_2O. In the process, glucose releases chemical-bond energy, which the cell stores (or "banks") in the chemical bonds of ATP.

Glucose contains a lot of chemical energy. For instance, 10 grams (roughly a tablespoon) contain about 40 kilocalories (kcal) of energy that can be made available for work. An adult human might use about this much energy by exercising vigorously for 15 minutes. A cell deals in tiny parcels of energy. Each ATP molecule it makes contains only about 1% of the amount of chemical energy present in a single glucose molecule.

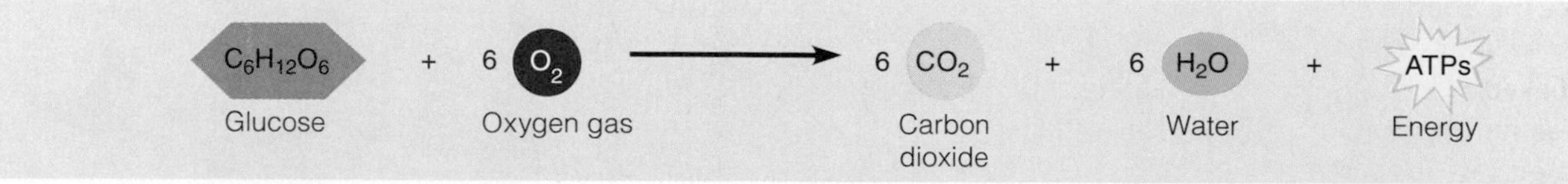

Figure 6.2A Summary equation for cellular respiration

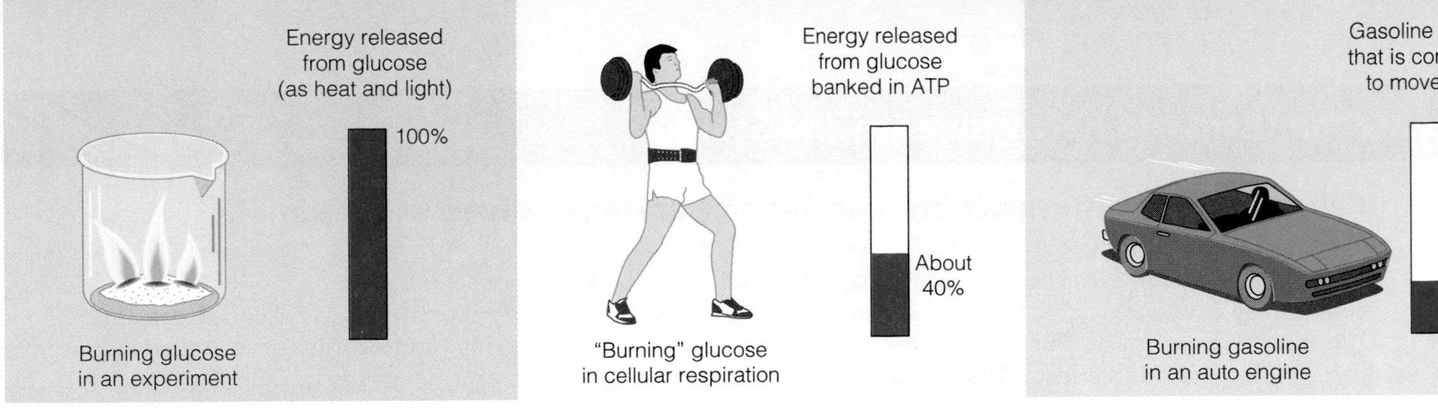

Figure 6.2B The efficiency of cellular respiration (and comparison with an auto engine)

Does this 1% figure mean that the cell makes 100 molecules of ATP from breaking down one molecule of glucose? Unfortunately not. Cellular respiration is not able to harvest all of the energy of glucose in usable form. When glucose is burned in a chemistry lab, 100% of its energy is released (Figure 6.2B, left). By contrast, a typical cell banks only about 40% of glucose's energy in ATP molecules (Figure 6.2B, center). Most of the rest is converted to heat. This may seem inefficient, but it compares very well with the efficiency of most energy-conversion systems. The average automobile engine, for instance, is able to convert only about 25% of the energy in gasoline into the kinetic energy of movement. Cellular respiration is also much more efficient than any process a cell can perform without O_2. For example, a yeast cell in an anaerobic environment harvests only about 2% of the energy in glucose. 

? Why are sweating and other body-cooling mechanisms necessary during vigorous exercise?

The demand for ATP is supported by an increased rate of cellular respiration, but about 60% of the energy from food produces body heat instead of ATP.

The human body uses energy from ATP for all its activities 6.3

Our body cells require a continuous supply of energy just to stay alive—to keep the heart pumping blood, to breathe, to maintain body temperature, and to digest food. These and other life-sustaining activities use as much as 75% of the energy a person takes in as food during a typical day. At any particular time, whether you are sleeping or active, most of your cells are busy with cellular respiration, producing ATP just to maintain the body.

Above and beyond the energy we need for body maintenance, the ATP made during cellular respiration provides energy for voluntary activities. The table on the right shows the amounts of energy it takes to perform some of these activities. The energy units are kilocalories (kcal), commonly referred to simply as "calories" in nontechnical sources. The values shown do not include the energy the body consumes for its life-sustaining activities. Thus, sleeping or lying quietly does not consume any energy except the energy used in maintenance.

The U.S. National Academy of Sciences estimates that the average adult human needs to take in food that provides about 2200 kcal of energy per day. This is a very general estimate of the total amount of energy a person of average weight expends, or "burns," in both maintenance and voluntary activity.

Walking at 3 mph, how far would you have to travel to "burn off" the equivalent of an extra slice of pizza, which has about 475 kcal? How long would that take?

About 9 miles; about 3 hours (Now you understand why it is said that the most effective exercise for losing weight is pushing away from the table!)

ENERGY CONSUMED BY VARIOUS ACTIVITIES (IN KCAL)

Activity	Kcal Consumed per Hour by a 67.5-kg (150-lb) Person*
Bicycling (racing)	514
Bicycling (slowly)	170
Dancing (slow)	202
Dancing (fast)	599
Driving a car	61
Eating	28
Gymnastics	186
Laboratory work	73
Piano playing	73
Running (7 min/mi)	865
Sitting (writing)	28
Sitting (playing chess)	30
Sleeping or lying still	0
Standing (relaxed)	32
Swimming (2 mph)	535
Walking (3 mph)	158
Walking (4 mph)	231

* Not including kcal necessary for body maintenance

6.4 Cells tap energy from electrons transferred from organic fuels to oxygen

Just how *do* our cells extract energy from organic fuel molecules? The energy available to a cell is contained in the specific arrangement of electrons in the chemical bonds that hold an organic molecule like glucose together. But the energy in even a single glucose molecule is far too great for use in a typical cellular job. Before we describe in detail how the process of cellular respiration solves this problem, let's examine the underlying mechanisms of energy release and harvest in the cell.

Simply stated, cellular respiration dismantles glucose in a series of steps and taps the energy carried by electrons that are rearranged whenever old bonds break and new ones form. Respiration works mainly by shuttling electrons through a series of energy-releasing reactions. At each step in the sequence, electrons start out in a molecule where they have more energy and end up in one where they have less energy. The reactions release energy in small amounts, and the cell stores some of the energy in ATP (see Module 5.4). Thus, a cell transfers energy from glucose to ATP by coupling energy-releasing (exergonic) chemical reactions to energy-storing (endergonic) ones. As we discussed in Chapter 5, this sort of energy transfer is a common theme in the life of the cell.

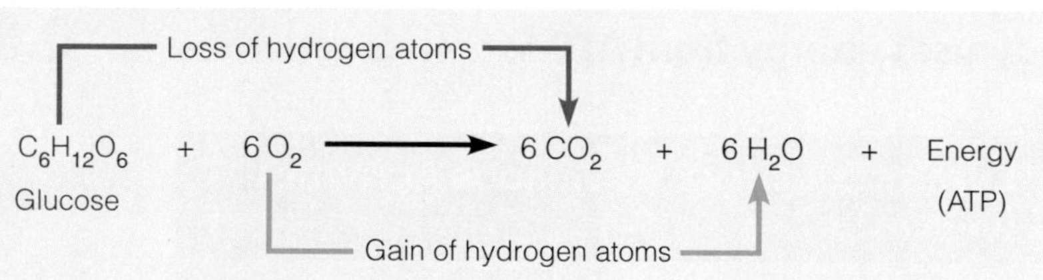

Figure 6.4 Rearrangement of H atoms (with their electrons) in cellular respiration

In the cellular respiration equation in Figure 6.4, you cannot see any electron transfers. What you do see are changes in hydrogen atom distribution. Glucose loses hydrogen atoms as it is converted to carbon dioxide; simultaneously, molecular oxygen (O_2) gains hydrogen atoms in being converted to water. These hydrogen movements represent electron transfers because, as we saw in Chapter 2, each hydrogen atom consists of an electron and a proton. As we will see later, the O_2 serves as the ultimate recipient of electrons in cellular respiration. O_2 can play this role because its atoms have a strong tendency to pull electrons away from other atoms. As a result, the electrons stripped from glucose finally end up in H_2O.

Figure 6.4 provides an overview of electron movement in cellular respiration, but it masks some important details. For instance, what happens when electrons move from molecule to molecule? We take a look at this process next.

What chemical characteristic of oxygen accounts for its function in cellular respiration?

Compared to other elements, oxygen is very electronegative (see Module 2.9), meaning that it is very powerful in removing electrons from other elements, including the atoms of organic fuels.

6.5 Hydrogen carriers such as NAD^+ shuttle electrons in redox reactions

The movement of electrons from one molecule to another is an oxidation-reduction reaction, or **redox reaction** for short. In a redox reaction, the loss of electrons from one substance is called **oxidation,** and the addition of electrons to another substance is called **reduction.** A molecule is said to be oxidized when it loses one or more electrons and reduced when it gains one or more electrons. Because an electron transfer requires both a donor and an acceptor, oxidation and reduction always go together. An electron leaves one molecule only when it contacts another molecule that attracts it more strongly. You can see the overall results of the redox reactions of cellular respiration in Figure 6.4 above: Glucose loses electrons (in H atoms), while O_2 gains electrons (in H atoms).

The top equation in Figure 6.5 (at right) depicts the oxidation of one of the organic molecules involved in cellular respiration. We show only its four carbon atoms (gray balls) and a few of its other atoms. Two key players in the process are an enzyme called a **dehydrogenase** and a coenzyme called NAD^+. **NAD^+** (nicotinamide adenine dinucleotide) is an organic molecule that cells make from the vitamin niacin and use to shuttle electrons in redox reactions.

With the help of a dehydrogenase, NAD^+ removes hydrogen atoms, with their electrons, from molecules such

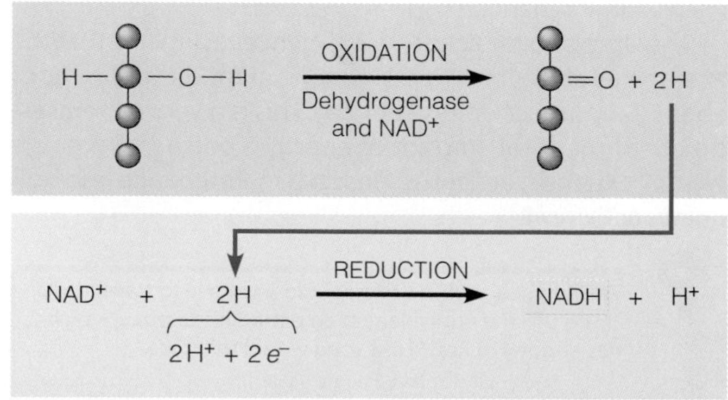

Figure 6.5 A pair of redox reactions, which occur simultaneously

as those of our example compound. Losing the electrons contained in two hydrogen atoms (2H), the compound is oxidized.

Meanwhile, as shown in the lower equation, NAD^+ is reduced (picks up electrons). When the dehydrogenase strips two hydrogen atoms from an organic molecule, it is actually removing two protons (written here as 2 H^+) and two electrons (2 e^-). NAD^+ picks up the two electrons and one H^+ and becomes NADH; it is thus a hydrogen carrier. The other H^+ goes into the surrounding solution in the cell.

The electrons added to NAD^+ in making NADH carry energy the cell has harvested and can eventually use. In the next module, we see how the cell processes the energy in NADH electrons in preparing to make ATP.

In the following redox reaction, which compound is oxidized and which is reduced?

$C_4H_6O_5 + NAD^+ \rightarrow C_4H_4O_5 + NADH + H^+$

$C_4H_6O_5$ is oxidized and NAD^+ is reduced.

6.6 Redox reactions release energy when electrons "fall" from a hydrogen carrier to oxygen

Glucose is like an electron bank; it is a rich source of electrons for the energy-yielding redox reactions in cells. NAD^+ and dehydrogenase enzymes work together to withdraw these electrons. The NADH molecules that result are loaded with energy, though each contains much less energy than glucose. NADH conveys electrons (and H^+) from glucose to other molecules in the cell.

Figure 6.6A shows NADH delivering its electron load to an **electron carrier** molecule, represented by a blue ball. The electron transfer is a redox reaction; NADH gives up electrons and becomes NAD^+ (which is recycled), while an electron carrier gains the electrons. This reaction starts an electron cascade, in which electrons "fall" down an energy "hill" consisting of a series of electron carriers. Each carrier is a different molecule; most are proteins. The first one is oxidized as the next one is reduced, and so on, down to the last molecule, which is O_2, at the bottom of the hill. What keeps the electrons moving is that each carrier molecule has a greater affinity for electrons than its uphill neighbor. In a sense, O_2, with the greatest affinity for electrons of all the carriers, pulls electrons down the energy hill. All along the way, the redox steps in the cascade release energy in amounts small enough to be used by the cell. In contrast, Figure 6.6B shows what happens if oxygen is reduced all in one step, by reacting directly with hydrogen. An explosion releases all the energy as heat and light—which a cell could not use.

Figure 6.6A is a highly simplified model of the actual series of electron carriers in the cell. Called **electron transport chains,** these ordered groups of molecules are embedded in membranes in the eukaryotic cell's mitochondria. (In prokaryotes, they are in the plasma membrane.) As electrons pass along the chains, they gradually lose energy, which the cell can use to make ATP. In cellular respiration, most of the energy used for ATP synthesis comes from the fall of electrons from glucose to NADH to an electron transport chain to O_2.

At the downhill end of the electron transport chain in Figure 6.6A, when electrons from NADH are finally passed to oxygen, what "waste" product of cellular respiration is produced? (*Hint:* Review Figure 6.4.)

Water (H_2O)

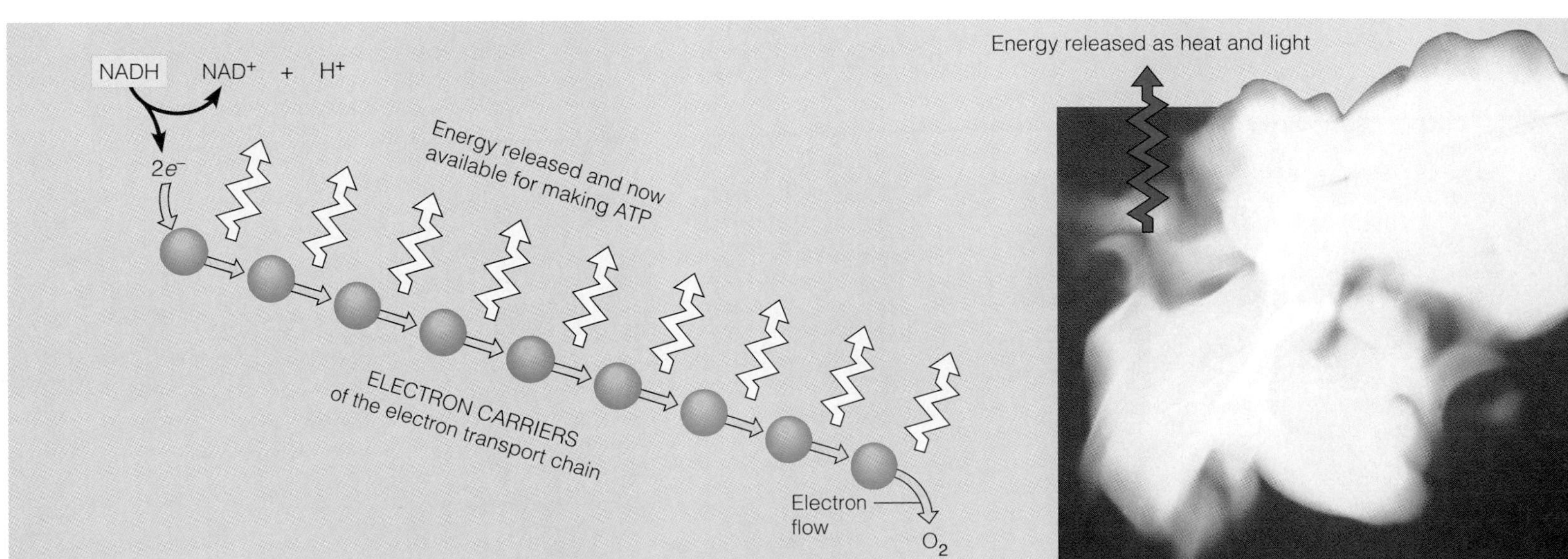

Figure 6.6A In cellular respiration, cascading electrons release energy in small increments and finally reduce O_2

Figure 6.6B In an explosion, O_2 is reduced in one step

6.7 Two mechanisms generate ATP

Virtually every cell in every organism relies on energy from ATP molecules. Every movement we make, every thought or memory we have, and every molecule our cells manufacture depend directly on ATP energy. Recall from Module 5.4 that cells generate ATP by phosphorylation, that is, by adding a phosphate group to ADP. A cell has two ways to do this: chemiosmotic phosphorylation, also called chemiosmosis, and substrate-level phosphorylation.

In 1978, British biochemist Peter Mitchell was awarded the Nobel Prize for developing the theory of **chemiosmosis.** Mitchell's theory describes how cells use the potential energy in concentration gradients to make ATP. A concentration gradient of a solute stores energy resulting from the tendency of the solute molecules to diffuse from where they are more concentrated to where they are less concentrated. The theory of chemiosmosis centers on membranes, and in particular on the activity of **ATP synthases,** protein complexes (clusters) that reside in membranes. ATP synthases synthesize ATP using the energy stored in concentration gradients of H^+ ions (that is, protons) across membranes. Cells generate most of their ATP in this way.

Let's look at an overview of the relationship between membrane structure and chemiosmotic ATP synthesis. As shown in Figure 6.7A, ATP synthase is built into the same membrane as the molecules of an electron transport chain. This structural connection allows the energy that NADH delivers to the electron transport chain to drive the production of ATP by the ATP synthase. Not shown in the figure are the details of the electron transport chain that make this energy transfer possible. Within the chain, redox reactions release energy from electrons cascading down the series of electron carriers mentioned in Module 6.6. As these exergonic reactions release energy, some of the proteins built into the chain use the energy to actively transport H^+ ions across the membrane. This flow results in a concentration gradient of H^+ ions across the membrane (notice the higher concentration of H^+ on the top side of the membrane). The ATP synthase then uses the potential energy in the concentration gradient to drive the endergonic (energy-storing) reaction that generates ATP from ADP and phosphate (Ⓟ). We discuss the details of this process in Module 6.12.

Substrate-level phosphorylation (Figure 6.7B) is much simpler than chemiosmosis and does not involve a membrane. In substrate-level phosphorylation, an enzyme transfers a phosphate group from an organic substrate molecule to ADP. The substrate is one of several substances produced as cellular respiration converts glucose to CO_2. The reaction occurs because the bond holding the phosphate group in the substrate molecule is less stable than the new bond holding it in ATP. The reaction products are a new organic molecule and a molecule of ATP. Substrate-level phosphorylation accounts for only a small percentage of the ATP that a cell generates. 

? What is the potential energy source that directly drives ATP production by ATP synthase?

A concentration gradient of hydrogen ions across a membrane

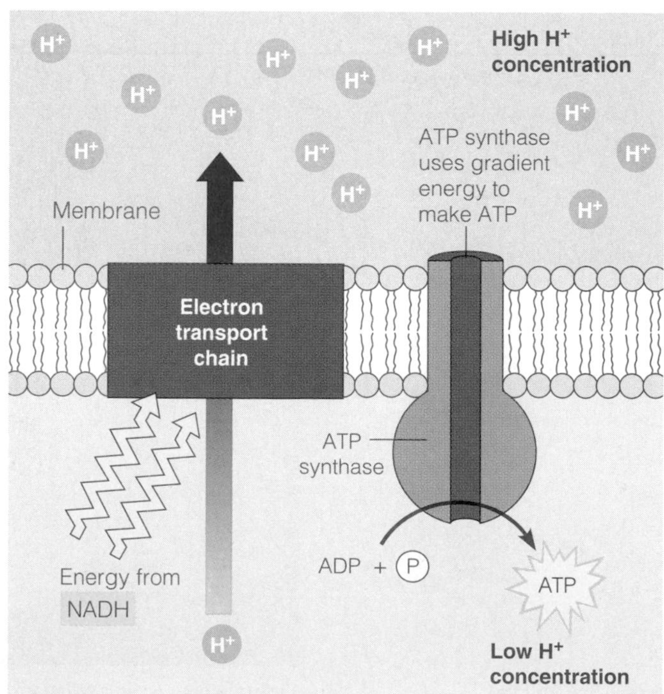

Figure 6.7A Chemiosmosis

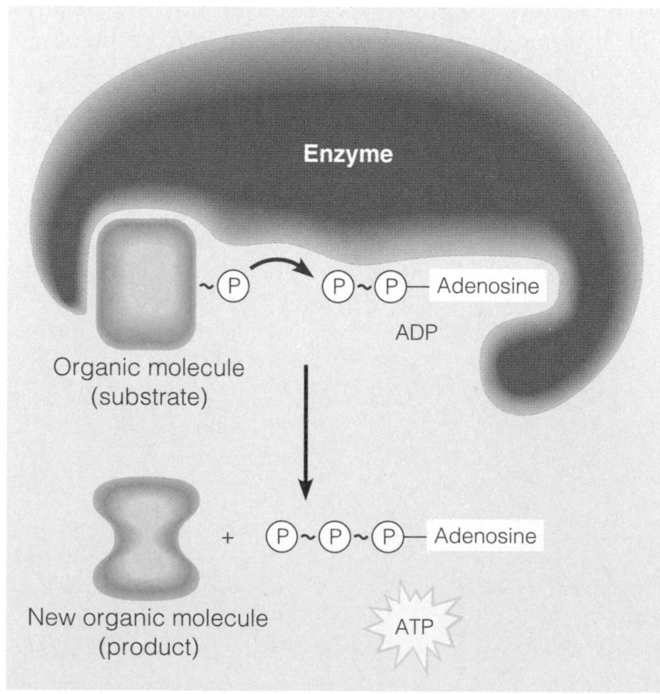

Figure 6.7B Substrate-level phosphorylation

Stages of Cellular Respiration and Fermentation

6.8 Overview: Respiration occurs in three main stages

6A
See Activity 6A on the CD-ROM.

Having examined the mechanisms that cells use to harvest energy, we are ready to look at the sequence of steps that make up cellular respiration. The process is a continuous one, but for study purposes we can divide it into three main stages. Figure 6.8 below gives an overview of these stages and shows where they occur in the cell.

The first two stages of cellular respiration, glycolysis and the Krebs cycle, are exergonic (energy-releasing) processes that break down glucose and other organic fuels. **Glycolysis** (shown with an aqua background throughout the chapter) occurs in the cytoplasmic fluid of the cell, that is, outside the organelles. Glycolysis begins respiration by breaking glucose into two molecules of a compound called pyruvic acid.

The **Krebs cycle** (light orange color), which takes place within the mitochondria, completes the breakdown of glucose by decomposing a derivative of pyruvic acid to carbon dioxide. As suggested by the smaller ATP symbols in the diagram, the cell makes a small amount of ATP (by substrate-level phosphorylation) during glycolysis and the Krebs cycle. The main function of glycolysis and the Krebs cycle, however, is to supply the third stage of respiration with electrons (gold arrows).

The third stage of cellular respiration is the electron transport chain (purple). As we have seen, the electron transport chain obtains electrons from the hydrogen carrier NADH, the reduced form of NAD^+. A related hydrogen carrier called FAD (flavin adenine dinucleotide) also shuttles some electrons from the Krebs cycle to the electron transport chain. The reduced form of FAD is $FADH_2$.

Thus, glycolysis and the Krebs cycle are energy-releasing stages that extract electrons from food molecules while breaking these molecules down to CO_2. NAD^+ and FAD temporarily capture the electrons and relay them to the top of the electron transport chain. The chain then uses the downhill flow of electrons from NADH and $FADH_2$ to O_2 to pump H^+ ions across a membrane. This process stores energy that ATP synthase uses to make most of the cell's ATP by chemiosmosis. We look more closely at the three stages of cellular respiration in Modules 6.9–6.12.

Of the three main stages of cellular respiration represented in Figure 6.8, which one uses oxygen directly to extract chemical energy from organic compounds?

The electron transport chain and chemiosmosis

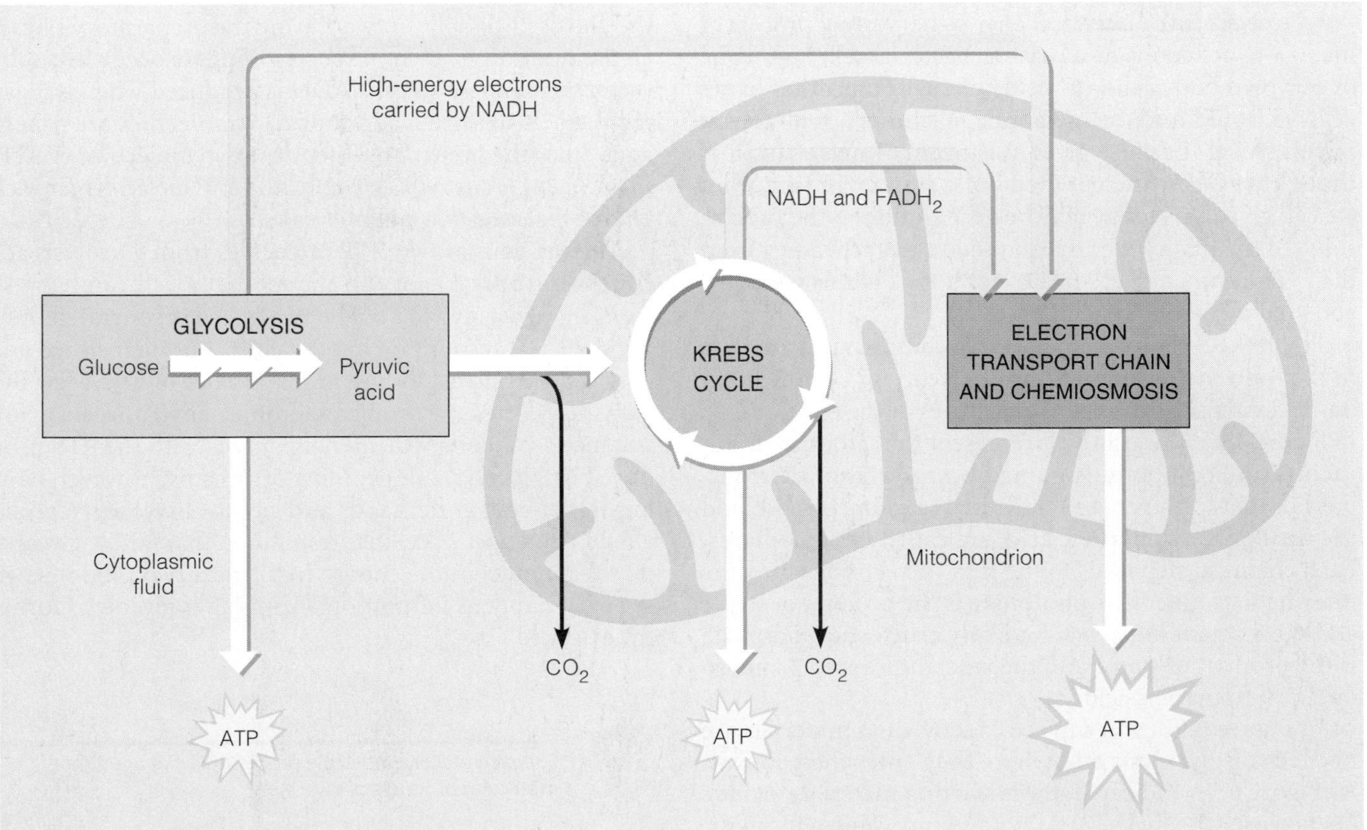

Figure 6.8 An overview of cellular respiration

6.9 Glycolysis harvests chemical energy by oxidizing glucose to pyruvic acid

6B

Now that we have introduced the major players and processes in cellular respiration, it's time to focus on the individual stages in the breakdown of a fuel molecule. The term for the first stage, glycolysis, means "splitting of sugar," and that's exactly what happens during this phase of respiration.

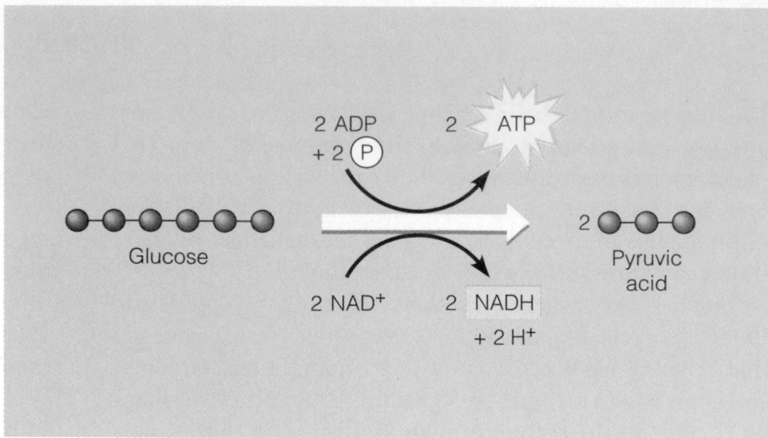

Figure 6.9A An overview of glycolysis

Figure 6.9A gives an overview of glycolysis in terms of input and output. Glycolysis begins with a single molecule of glucose and concludes with two molecules of another organic compound, pyruvic acid. The gray balls represent the carbon atoms in each molecule; glucose has six, and these same six end up in the two pyruvic acid molecules (three in each). The straight arrow from glucose to pyruvic acid represents nine chemical steps. During these steps, a number of organic compounds form, as enzymes catalyze the rearrangement of chemical bonds and the splitting of the carbon skeleton of glucose in half. As these reactions occur, the cell produces two molecules of ATP by substrate-level phosphorylation and reduces two molecules of NAD$^+$, forming two molecules of NADH. Thus, the energy extracted from glucose during glycolysis is banked in a combination of ATP and NADH. The cell can use the energy in ATP immediately, but for it to use the energy banked in NADH, electrons from NADH must pass down the electron transport chain.

Glycolysis is the universal energy-harvesting process of life. If we looked inside a yeast or bacterial cell, inside one of our own body cells, or inside virtually any other living cell, we would find the metabolic machinery of glycolysis in full swing. Because glycolysis occurs universally, it is thought to be an ancient metabolic system. In fact, what we call glycolysis today may be very similar to the process some of the first cells on Earth used to extract energy from their environment. Let's take a closer look at this venerable and vital metabolic system.

Figure 6.9B shows all the organic compounds that form in the nine chemical reactions of glycolysis. Commentary on the left highlights the main features of the reactions. As in Figure 6.9A, the gray balls represent the carbon atoms in each of the compounds named on the right. The compounds that form between the initial reactant, glucose, and the final product, pyruvic acid, are called **intermediates.** Each chemical step leads to the next one. For instance, the intermediate glucose 6-phosphate is the product of step 1 and the reactant for step 2. Similarly, fructose 6-phosphate is the product of step 2 and the reactant for step 3, and so on down to pyruvic acid.

In Figure 6.9B, you can see exactly what materials are needed for glycolysis and where they enter the pathway. As Figure 6.9A indicated, these starting materials include (1) glucose (the fuel), (2) ADP (and inorganic phosphate, or Ⓟ), and (3) the hydrogen-shuttle molecule NAD$^+$. Notice that ATP is also needed as a starting material; this tells us that the cell must expend some energy to get glycolysis started. Also essential are specific enzymes that catalyze each of the chemical steps; however, to keep the figures simple, we have not included enzymes in the diagrams.

As indicated by the two aqua background areas in Figure 6.9B, the individual steps of glycolysis can be grouped into two main phases. Steps ①–④, the first phase, are preparatory and *consume* energy. In this phase, ATP energy is used to split one glucose molecule into two small sugars that are primed to release some energy. Steps ⑤–⑨, the second phase, *yield an energy payoff* for the cell. Because the glucose was split in two during the preparatory phase, all the reactions shown in the payoff phase occur in duplicate. During these steps, NADH is produced when a sugar molecule is oxidized, and four ATP molecules are generated. Since the preparatory steps use two molecules of ATP, *the net gain to the cell is actually two ATP molecules for each glucose molecule that enters glycolysis.*

The net gain of two ATP molecules from glycolysis accounts for only 5% of the energy that a cell can harvest from a glucose molecule. The two NADHs generated during steps 5–9 account for another 16%, but their stored energy is not available for use in the absence of O_2. A few organisms—yeasts in an anaerobic environment, for instance—can satisfy their energy needs with the ATP produced by glycolysis alone. Most organisms, however, have far higher energy demands and cannot live on glycolysis alone. The stages of cellular respiration that follow glycolysis release much more energy. In the next two modules, we see what happens in most organisms after glycolysis forms pyruvic acid.

For each glucose molecule processed, what are the net molecular products of glycolysis?

Two molecules of pyruvic acid, two molecules of ATP, and two molecules of NADH

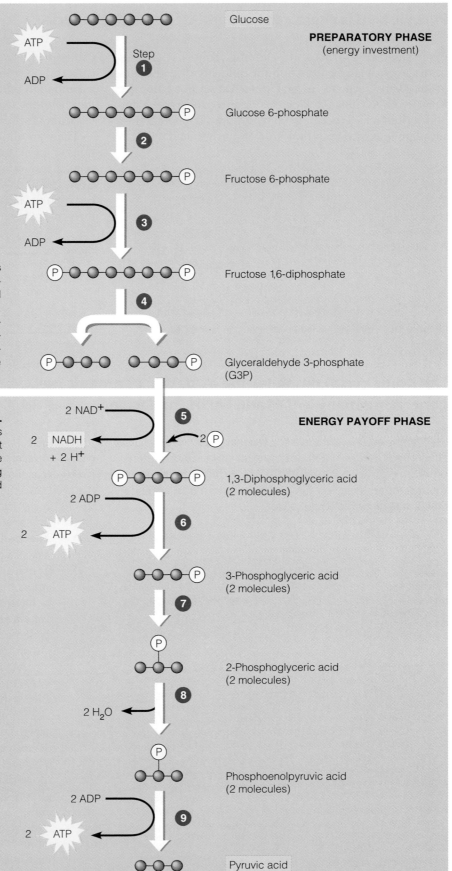

Steps 1 – 3 A fuel molecule is energized, using ATP. A sequence of three chemical reactions converts glucose to a molecule of the intermediate fructose 1,6-diphosphate. The coupled arrows indicate the transfer of a phosphate group with high energy content from ATP to another molecule. In these preparatory steps, the cell invests two ATP molecules, one at step 1 and one at step 3, to energize a fuel molecule. In becoming energized, the molecule becomes less stable and thus more reactive.

Step 4 A six-carbon intermediate splits into two three-carbon intermediates. Fructose 1,6-diphosphate is highly reactive and breaks into two three-carbon intermediates. Two molecules of glyceraldehyde 3-phosphate (G3P) emerge from each glucose molecule that enters glycolysis. The two G3P molecules enter step 5, so steps 5–9 occur twice per glucose molecule.

Step 5 A redox reaction generates NADH. The first step in the payoff phase of glycolysis is a redox reaction. The cell harvests its first parcel of energy. The coupled arrows indicate the transfer of hydrogen atoms (containing high-energy electrons) as G3P is oxidized and NAD^+ is reduced to NADH.

Steps 6 – 9 ATP and pyruvic acid are produced. This series of four chemical reactions completes glycolysis, producing two molecules of pyruvic acid for each initial molecule of glucose. During steps 6–9, specific enzymes make four molecules of ATP (per glucose molecule) by substrate-level phosphorylation, and water is produced (at step 8 as a by-product).

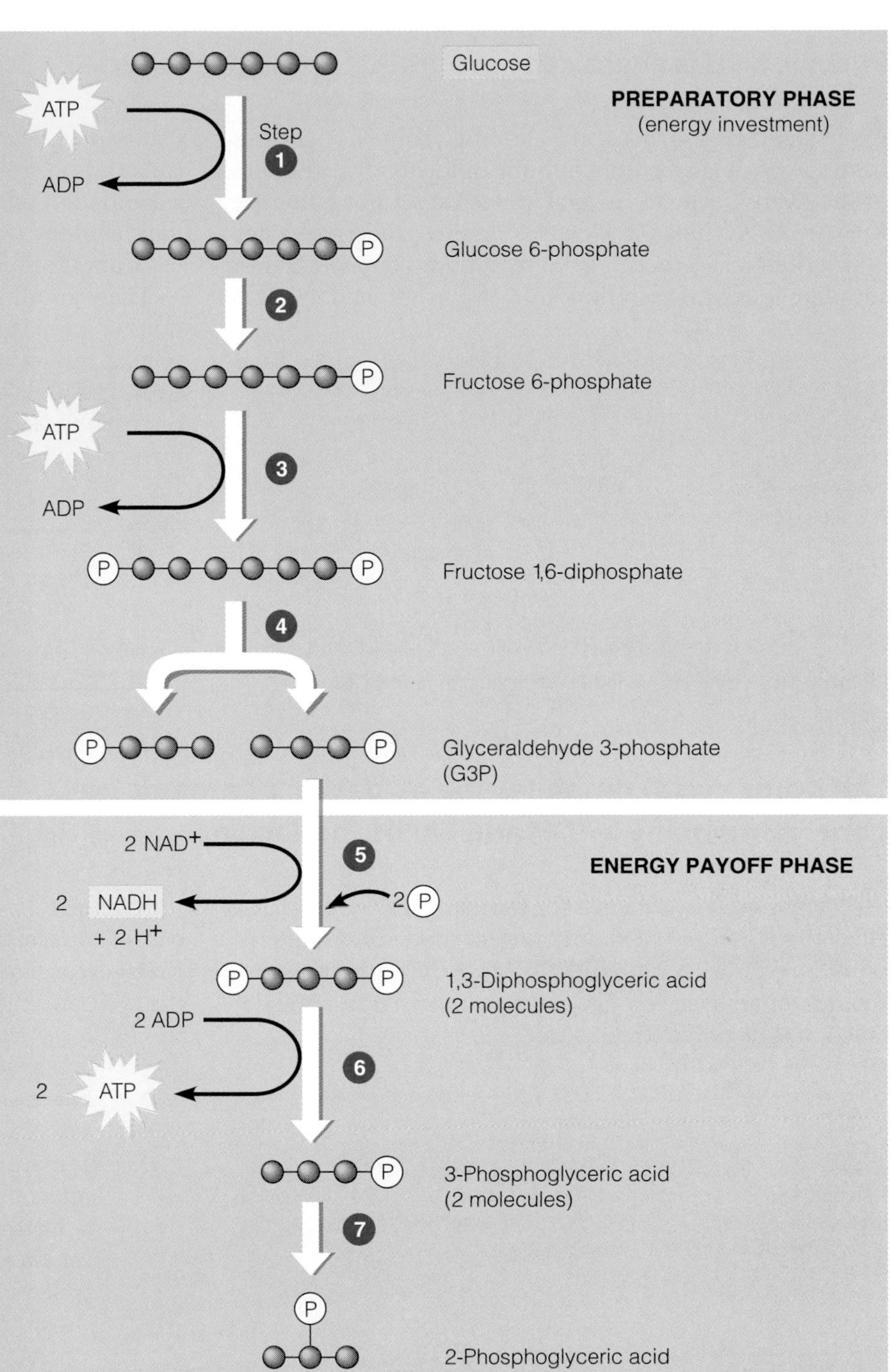

Figure 6.9B Details of glycolysis

6.10 Pyruvic acid is chemically groomed for the Krebs cycle

6C

As pyruvic acid forms at the end of glycolysis, it diffuses from the cytoplasmic fluid into the mitochondria, the sites of the Krebs cycle. Pyruvic acid itself does not enter the Krebs cycle. As shown in Figure 6.10, it first undergoes some major chemical "grooming." Several things happen almost simultaneously to each molecule of pyruvic acid: (1) It is oxidized while a molecule of NAD^+ is reduced to NADH; (2) a carbon atom is removed and released in CO_2; and (3) a compound called coenzyme A, derived from a B vitamin, joins with the two-carbon fragment remaining from pyruvic acid to form a molecule called acetyl coenzyme A.

These grooming steps—a chemical "haircut and conditioning" of pyruvic acid—set up the second major stage of cellular respiration. Acetyl coenzyme A, abbreviated **acetyl CoA,** is a high-energy fuel molecule for the Krebs cycle. For each molecule of glucose that entered glycolysis, two molecules of acetyl CoA enter the Krebs cycle.

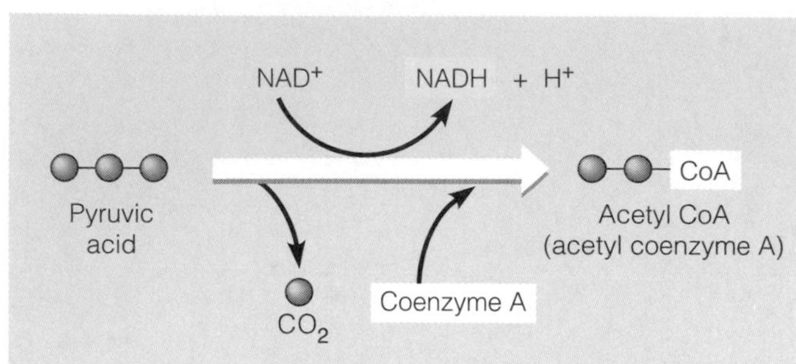

Figure 6.10 The conversion of pyruvic acid to acetyl CoA

The oxidation of pyruvic acid by NAD^+ extracts some chemical energy from the pyruvic acid. How can the cell harness that energy to make ATP?

By chemiosmosis: The NADH can supply electrons to the electron transport chain, which generates a hydrogen ion gradient that drives ATP synthesis (see Module 6.7).

6.11 The Krebs cycle completes the oxidation of organic fuel, generating many NADH and $FADH_2$ molecules

6C

The Krebs cycle was named for German-British researcher Hans Krebs, who worked out much of this cyclical phase of cellular respiration in the 1930s. As we did with glycolysis, we present an overview figure first, followed by a more detailed outline of the Krebs cycle.

As shown in Figure 6.11A, only the two-carbon acetyl part of the acetyl CoA molecule actually participates in the Krebs cycle. Coenzyme A helps the acetyl fragment enter the cycle and then splits off and is recycled. Not shown in this figure are the multiple steps that follow, each catalyzed by a specific enzyme in the mitochondrial matrix. The cycle completely disassembles acetyl CoA, stripping away its electrons and casting off two carbon atoms as CO_2 for every acetyl fragment that enters the Krebs cycle.

Compared with glycolysis, the Krebs cycle pays big energy dividends to the cell. Each turn of the cycle makes one ATP molecule by substrate-level phosphorylation (shown at the bottom in Figure 6.11A). It also produces four other energy-rich molecules: three NADH molecules and one molecule of the cell's other hydrogen carrier, $FADH_2$. Since the Krebs cycle processes two molecules of acetyl CoA for each initial molecule of glucose, the overall yield per molecule of glucose is: 2 ATP, 6 NADH, and 2 $FADH_2$. This yield is considerably more than the 2 ATP plus 2 NADH produced by glycolysis alone. In fact, the difference between the energy the Krebs cycle releases from organic fuel and the energy output of glycolysis is like the difference between the energy released by a blast furnace and a tiny campfire.

Overall, how many energy-rich molecules has the cell gained by processing one molecule of glucose through glycolysis and the Krebs cycle? Up to this point, the cell has gained a total of 4 ATP, 10

Figure 6.11A An overview of the Krebs cycle

NADH, and 2 $FADH_2$. Still, to be able to put to use the energy banked in these molecules, the cell has to transfer the energy now in NADH and $FADH_2$ to ATP, so that all the energy is stored in ATP. Before we look at how that comes about, you may want to examine the inner workings of the Krebs cycle "furnace" in Figure 6.11B, below.

What is the total number of NADH molecules generated during the complete breakdown of 1 glucose molecule to 6 carbon dioxide molecules? (*Hint:* Combine the outputs of Modules 6.9–6.11.)

10 NADH (Did you remember to double the output after the sugar-splitting step—step 4—of glycolysis?)

Acetyl CoA

CoA

2 carbons enter cycle

Oxaloacetic acid

Citric acid

CO_2 leaves cycle

NADH + H^+

NAD^+

KREBS CYCLE

Malic acid

NAD^+

NADH + H^+

ADP + (P)

ATP

Alpha-ketoglutaric acid

$FADH_2$

FAD

CO_2 leaves cycle

Succinic acid

NAD^+

NADH + H^+

Step 1

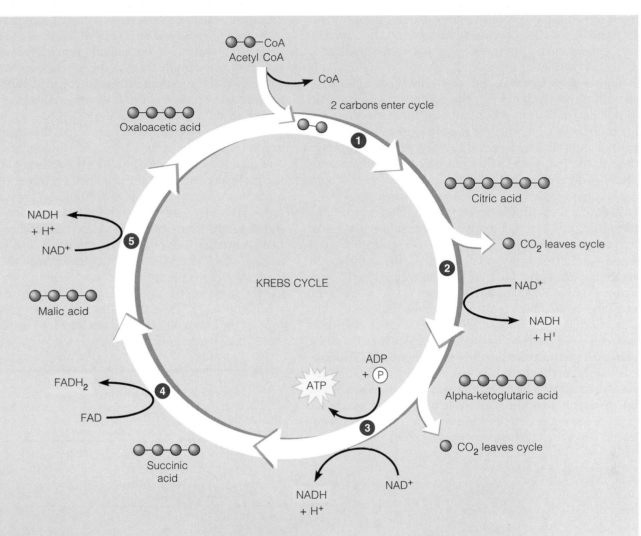

Acetyl CoA stokes the furnace.

A turn of the Krebs cycle begins (top center) as enzymes strip the CoA portion from acetyl CoA and combine the remaining two-carbon acetyl fragment with oxaloacetic acid (top left) already present in the mitochondrion. The product of this reaction is the six-carbon molecule citric acid, which contains most of the acetyl's fuel energy. Adding acetyl to the cycle is like shoveling coal into a furnace: The Krebs cycle "furnace" runs on the chemical energy that was stored in the acetyl group.

Steps 2 and 3

NADH, ATP, and CO_2 are generated during redox reactions.

Successive redox reactions harvest some of the energy of the acetyl group by stripping hydrogen atoms from organic acid intermediates (such as alpha-ketoglutaric acid) and producing energy-laden NADH molecules. The redox reactions dispose of two carbon atoms that came from oxaloacetic acid. The carbons are completely oxidized and released as two molecules of CO_2. Energy is also harvested by substrate-level phosphorylation of ADP to produce ATP. A four-carbon compound called succinic acid emerges at the end of step 3.

Steps 4 and 5

Redox reactions generate $FADH_2$ and NADH.

Enzymes rearrange chemical bonds, eventually completing the cycle by regenerating oxaloacetic acid. Concurrently, the hydrogen (electron) carriers FAD and NAD^+ are reduced to $FADH_2$ and NADH, respectively. One turn of the Krebs cycle is completed with the conversion of a molecule of malic acid to one of oxaloacetic acid. This compound is then ready to start the next turn of the cycle by accepting another acetyl group from acetyl CoA.

Figure 6.11B Details of the Krebs cycle

6.12 Chemiosmosis powers most ATP production

6D

The final stage of cellular respiration is the electron transport chain and the synthesis of ATP by chemiosmosis. As we discussed in Module 6.7, chemiosmosis is a clear illustration of structure fitting function: The spatial arrangement of membrane proteins makes it possible for the mitochondrion to use chemical energy to create an H^+ gradient and then use the energy stored in that gradient to drive ATP synthesis.

Figure 6.12 expands on our earlier discussion of chemiosmosis. It shows that the electron transport chain of cellular respiration is built into the inner membrane of the mitochondrion. The folds (cristae) of this membrane enlarge its surface area, providing space for many copies of the electron transport chain and many ATP synthase complexes. With all these ATP-making "machines," a mitochondrion can produce many ATP molecules simultaneously.

Let's first focus on the series of bright blue electron carrier molecules. Starting on the left, the gold arrow traces the path of electron flow from the shuttle molecule NADH down the electron transport chain to O_2, the final electron acceptor in the chain. Each of the oxygen atoms in O_2 combines with two electrons and with two H^+ ions (from the surrounding solution) to form H_2O, one of the final products of cellular respiration. Most of the carrier molecules reside in three protein complexes (purple spheres), which span the inner membrane of the mitochondrion. Two mobile carriers in the chain (the ones outside the protein complexes) transport electrons between the complexes. All of the carriers bind and release electrons in redox reactions.

As redox occurs, the protein complexes use energy released from the electrons to actively transport H^+ ions from one side of the membrane to the other. The pink vertical arrows indicate that the H^+ ions are transported from the matrix of the mitochondrion (its innermost compartment) into the mitochondrion's intermembrane space. Though the diagram shows only four H^+ ions in the intermembrane space, in fact many of them are stockpiled there. The resulting H^+ gradient—more H^+ ions on one side of the membrane than on the other—stores potential energy. With this potential energy, the mitochondrion is poised for the final act of cellular respiration: the production of a large amount of ATP.

If you follow the dashed arrows across the top of the figure, you will arrive at the site of ATP production. The H^+ ions tend to be pushed back across the membrane into the matrix by the energy of the gradient. However, the membrane is not very permeable to H^+, and H^+ ions can only cross back by passing through a special protein port. As shown on the far right, ATP synthase, the protein complex that synthesizes ATP, provides that port—a channel for the passage of H^+ ions. ATP synthase also contains the enzyme that catalyzes the phosphorylation of ADP to form ATP. As the H^+ ions move through the port, their flow drives the synthesis of ATP. Thus, by means of chemiosmosis, a cell couples the exergonic reactions of electron transport to the endergonic synthesis of ATP. 

? What effect would an absence of oxygen (O_2) have on the process illustrated in Figure 6.12?

There would be no chemiosmotic production of ATP. Without oxygen to "pull" electrons down the electron transport chain, the energy stored in NADH cannot be extracted and harnessed for ATP synthesis.

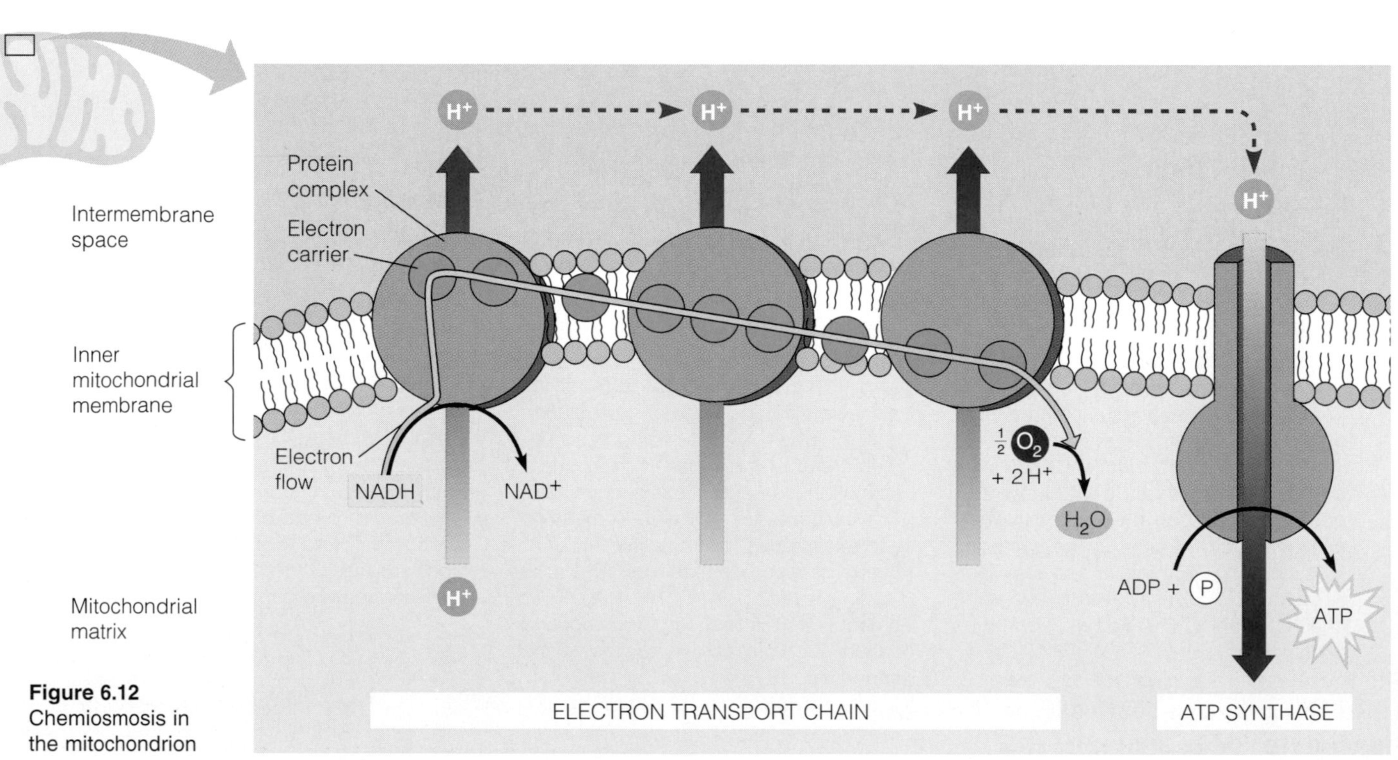

Figure 6.12 Chemiosmosis in the mitochondrion

Certain poisons interrupt critical events in cellular respiration 6.13

A number of poisons produce their deadly effects by interfering with some of the events we have just discussed. Figure 6.13 shows the places where three different categories of poisons obstruct cellular respiration.

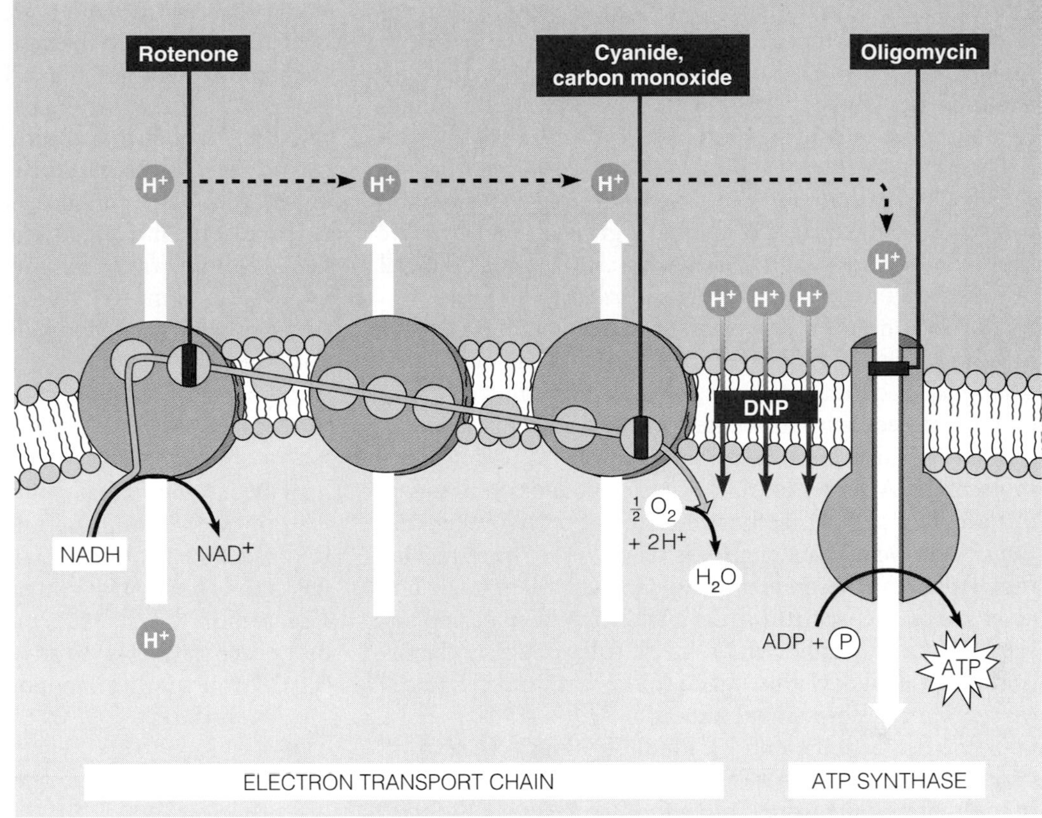

Figure 6.13 The effects of five poisons on the electron transport chain and chemiosmosis

Poisons in one category block the electron transport chain. A substance called rotenone, for instance, binds tightly with one of the electron carrier molecules in the first protein complex, preventing electrons from passing to the next carrier molecule. Rotenone is often used to kill pest insects and fish. By blocking the electron transport chain near its start and thus preventing ATP synthesis, rotenone literally starves an organism's cells of energy. Two other electron transport blockers, cyanide and carbon monoxide, bind with an electron carrier in the third protein complex. Here they block the passage of electrons to oxygen. This blockage is like turning off a faucet; electrons cease to flow throughout the "pipe." The result is similar to that of rotenone: No H^+ gradient is generated, and no ATP is made.

A second kind of respiratory poison inhibits ATP synthase. On the right side of the figure, the antibiotic oligomycin blocks the passage of H^+ ions through the channel in ATP synthase. Oligomycin is used on the skin to combat fungal infections. It kills fungal cells by preventing them from using the potential energy of the H^+ gradient to make ATP. (Because the drug cannot get into the living skin cells, they are protected from its effects.)

A third kind of poison, collectively called uncouplers, makes the membrane of the mitochondrion leaky to H^+ ions. Electron transport continues, but ATP cannot be made because leakage of H^+ ions through the membrane abolishes the H^+ gradient. Cells continue to consume oxygen, often at a higher than normal rate—but to no avail, for they cannot make any ATP.

One uncoupler, dinitrophenol (DNP), is highly toxic to humans. DNP poisoning produces an enormous increase in metabolic rate, profuse sweating as the body attempts to dissipate excess heat, collapse, then death. For a short time in the 1940s, some physicians prescribed DNP in low doses as weight-loss pills, but fatalities soon made it clear that there were far safer ways to lose weight. The pills caused weight loss by making the body's cells break down *all* fuel molecules in the diet, as well as stored molecules, including fat. When DNP is present, all steps of cellular respiration except chemiosmosis continue to run, consuming fuel molecules, even though almost all the energy is lost as heat.

Poisons do have some good points. Substances that are toxic to certain organisms may be medically beneficial as antibiotics, as is true for oligomycin. In addition, discovering exactly what these toxic substances do to the cell's respiratory machinery has, in many cases, helped biochemists understand how the machinery works. The effects of uncouplers, for example, made it clear that ATP synthesis is a complicated activity involving the distinct, but related, processes of electron transport and generation of a membrane H^+ gradient. 

? The poison DNP causes what one biochemist calls "mitochondrial wheel-spinning." Explain this metaphor.

Like an automobile wasting fuel by spinning its wheels and going nowhere, a mitochondrion poisoned with DNP consumes fuel and powers electron transport but makes no ATP. The DNP destroys the hydrogen ion gradient required for ATP synthesis. In the presence of DNP, hydrogen ions can cross the mitochondrial inner membrane without using the ATP synthase port. As a result, the H^+ gradient required for ATP synthesis is dissipated.

6.14 Review: Each molecule of glucose yields many molecules of ATP

6D

Now that we have looked at all stages of aerobic respiration, let's review what the cell accomplishes by oxidizing a molecule of glucose. Figure 6.14 puts all the stages together and indicates where they occur in the cell. At the bottom of the figure is a tally of ATP molecules, showing the potential energy payoff for a typical working cell. If you wish to refer back to earlier modules, this diagram summarizes glycolysis (Module 6.9), the chemical grooming of pyruvic acid (Module 6.10), and the Krebs cycle (Module 6.11), all of which occur *twice per glucose molecule.*

Starting on the left, glycolysis, occurring in the cytoplasmic fluid, and the Krebs cycle, occurring in the mitochondrial matrix, contribute a net total of 4 ATP molecules per glucose molecule by substrate-level phosphorylation. The cell harvests much more energy than this from the carrier molecules NADH and $FADH_2$ (gold rectangles), which are produced by glycolysis, the grooming of pyruvic acid, and the Krebs cycle. This energy is used to make numerous molecules of ATP (an estimated 34 in the diagram) when acted on by the electron transport chain and chemiosmosis. The actual number of ATP molecules made by chemiosmotic phosphorylation varies somewhat from cell to cell and with a cell's functional state.

Let's see where the numbers in the diagram come from. Our model assumes that each NADH that transfers a pair of high-energy electrons from a food molecule to the electron transport chain contributes enough to the mitochondrion's H^+ gradient to generate 3 ATP. Another assumption is that each $FADH_2$ molecule yields 2 ATP. These numbers are maximums; the actual amounts vary. Also, as indicated in the diagram, the cell may use energy equivalent to nearly 2 ATP to shuttle NADH electrons from glycolysis out of the cytoplasmic fluid into the mitochondrion. However, many cells use a shuttle mechanism that consumes no energy. Thus the total net yield of ATP molecules per glucose molecule has a theoretical maximum of about 38.

More important than the actual numbers of ATP molecules is the point that a cell can harvest a great deal of energy from glucose—up to about 40% of the molecule's potential energy—when oxygen is available for the electron transport chain and chemiosmosis can proceed.

Because most of the ATP generated by cellular respiration results from chemiosmosis, the ATP yield depends heavily on an adequate supply of oxygen (O_2) to the cell. Without oxygen to function as the final electron acceptor in the electron transport chain, chemiosmosis ceases, and death may ensue from energy starvation. However, as we discussed in the chapter's opening essay, yeasts and certain other cells have ways to survive when they run out of oxygen. We look at that situation next.

What would a cell's net ATP yield per glucose molecule be in the presence of the poison DNP? (See Module 6.13 to review DNP action.)

4 ATP, all from substrate-level phosphorylation

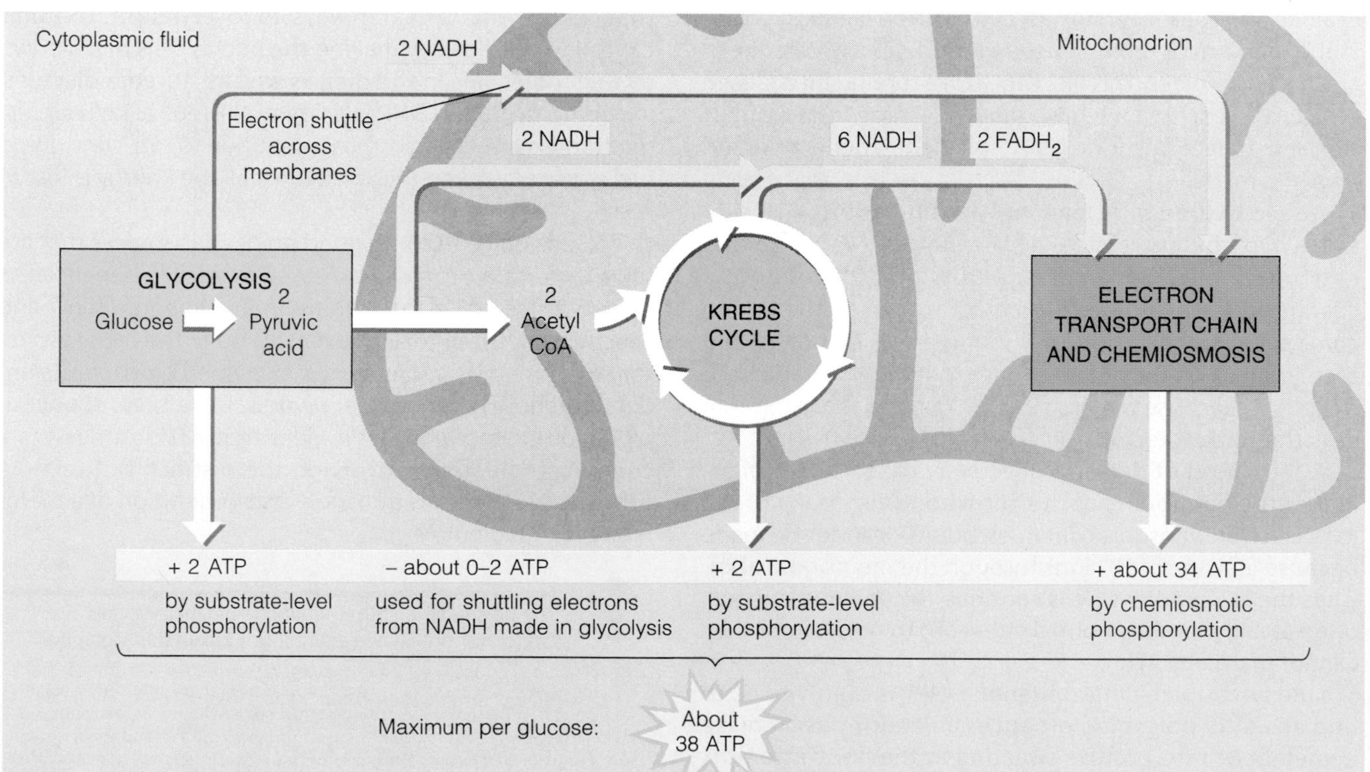

Figure 6.14 A tally of the ATP yield from cellular respiration

Fermentation is an anaerobic alternative to aerobic respiration 6.15

Yeasts normally use aerobic respiration to process their food, harvesting about 38 ATP per glucose molecule the way most organisms do. They are also able to survive without oxygen, on the two molecules of ATP per glucose molecule that come from glycolysis. Remember that glycolysis uses no oxygen; it simply generates a net gain of 2 ATP while converting glucose to two molecules of pyruvic acid and reducing NAD^+ to NADH. This is an inefficient way to use fuel—something like running an automobile on one poorly tuned cylinder—but yeasts can thrive on it in anaerobic environments where there is plenty of glucose to keep glycolysis operating.

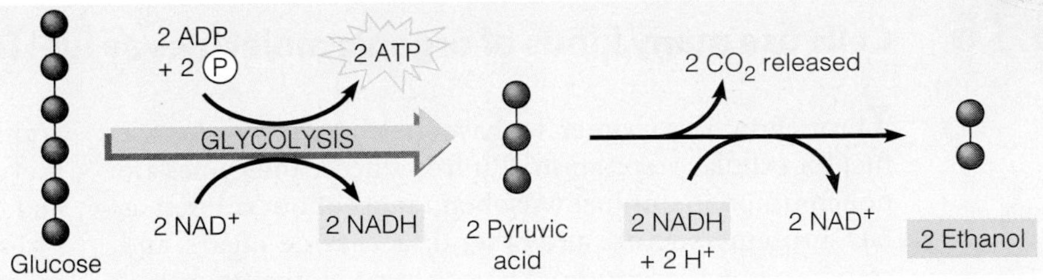

Figure 6.15A Alcoholic fermentation

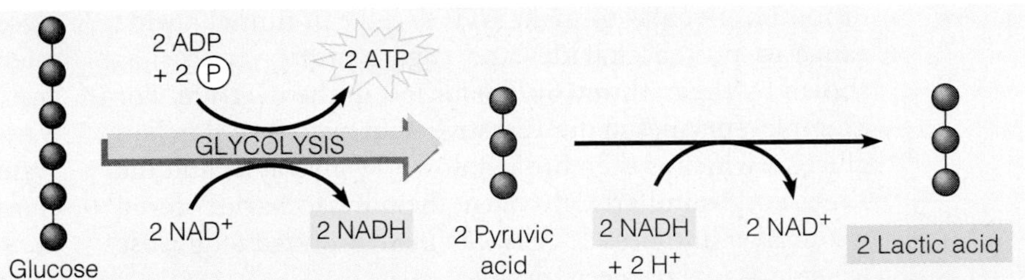

Figure 6.15B Lactic acid fermentation

There is one catch in using glycolysis as the sole means of producing ATP: A cell must have a way of replenishing its supply of NAD^+ as this molecule is reduced. Yeasts and certain bacteria do this by converting the pyruvic acid produced by glycolysis to CO_2 and ethanol (ethyl alcohol). The production of ethanol from glucose is called **alcoholic fermentation,** and it is catalyzed by specific microbial enzymes.

As indicated in Figure 6.15A, during alcoholic fermentation, CO_2 is removed from pyruvic acid and NADH is oxidized, thus recharging the cell with a supply of NAD^+ to keep glycolysis working. Notice that ethanol, formed when NADH is oxidized, is a reduced molecule. Thus, unlike the energy-poor H_2O and CO_2 molecules remaining after aerobic respiration, ethanol is energy-rich. However, it is toxic to the organisms that produce it. Yeasts release their alcohol wastes to their surroundings but die if the alcohol becomes too concentrated.

Figure 6.15B shows a different type of fermentation, used by many kinds of cells. It is called **lactic acid fermentation** because lactic acid, rather than alcohol, is produced when NADH from glycolysis is oxidized. The ATP yield is the same as in alcoholic fermentation, since glycolysis generates it, but no CO_2 is given off, and lactic acid retains all three carbons from pyruvic acid. Lactic acid fermentation is used in the dairy industry to make cheese and yogurt.

As mentioned in the chapter-opening essay, human muscle cells can also make ATP by lactic acid fermentation when oxygen is scarce, as it often is during strenuous exercise. The accumulation of lactic acid is the cause of muscle ache, but eventually the lactic acid is carried in the blood to the liver, where it is converted back to pyruvic acid.

Unlike muscle cells and yeasts, many bacteria that live in stagnant ponds and deep in the soil are **strict anaerobes,** meaning they require anaerobic conditions and are poisoned by oxygen. Yeasts and many other bacteria, including one called *E. coli* that thrives in the human intestine, are facultative anaerobes. A **facultative anaerobe** can make ATP either by fermentation or by chemiosmosis, depending on whether O_2 is available.

Figure 6.15C Fermentation vats for wine

For a facultative anaerobe, pyruvic acid is a fork in the metabolic road. If oxygen is available, the organism will always use the more efficient process—aerobic respiration. Thus, to make wine and beer, yeasts must be grown anaerobically so that they will ferment sugars and produce ethanol. For this reason, the large fermentation vats in Figure 6.15C are equipped with one-way gas valves that vent off CO_2 but keep air out. 

? A glucose-fed yeast cell is moved from an aerobic environment to an anaerobic one. If the cell continues to generate ATP at the same rate, how will its rate of glucose consumption compare to consumption in the aerobic environment?

The cell must consume glucose at a rate about 19× the consumption rate in the aerobic environment (2 ATP by fermentation vs. 38 ATP by cellular respiration).

Interconnections Between Molecular Breakdown and Synthesis

6.16 Cells use many kinds of organic molecules as fuel for cellular respiration

Throughout this chapter, we have spoken of glucose as the fuel for cellular respiration. But free glucose molecules are not common in our diet. We obtain most of our calories as fats, proteins, sucrose and other disaccharide sugars, and starch (a polysaccharide). You consume all these types of molecules when you eat a bag of peanuts, for instance.

Figure 6.16 illustrates how the cell uses three main kinds of food molecules to make ATP. A cell can funnel a wide range of polysaccharides and sugars into glycolysis, as shown by the pathway on the far left in the diagram. For example, enzymes in our digestive tract hydrolyze starch to glucose, which is then broken down by glycolysis and the Krebs cycle. Similarly, glycogen, the polysaccharide stored in our liver and muscle cells, can be hydrolyzed to glucose to serve as fuel between meals.

Proteins (far right) can be used for fuel, but first they must be digested to their constituent amino acids. Typically, a cell will use most of the amino acids to make its own proteins, but enzymes will convert excess amino acids to other organic compounds. During the conversion, the amino groups are either used in the synthesis of other compounds or stripped off and disposed of in urine. The other parts of amino acid molecules are usually converted to pyruvic acid, acetyl CoA, or one of the organic acids in the Krebs cycle, and their energy is then harvested by cellular respiration.

Fats make excellent cellular fuel because they contain many hydrogen atoms and thus many energy-rich electrons. As the diagram shows (center), the cell first hydrolyzes fats to glycerol and fatty acids. It then converts the glycerol to glyceraldehyde 3-phosphate (G3P), one of the intermediates in glycolysis. The fatty acids are changed into acetyl CoA and then enter the Krebs cycle. Processed this way, a gram of fat yields more than twice as much ATP as a gram of starch. This explains why it is so difficult for a dieter to lose excess fat. To get rid of fat, we have to expend the same large amount of energy we stored in the fat in the first place.

Animals store most of their energy reserves as fats, not as polysaccharides. What is the advantage of this mode of storage for an animal?

Because most animals are mobile, they benefit from a compact form of energy storage. A gram of fat stores about twice as much energy as a gram of carbohydrate.

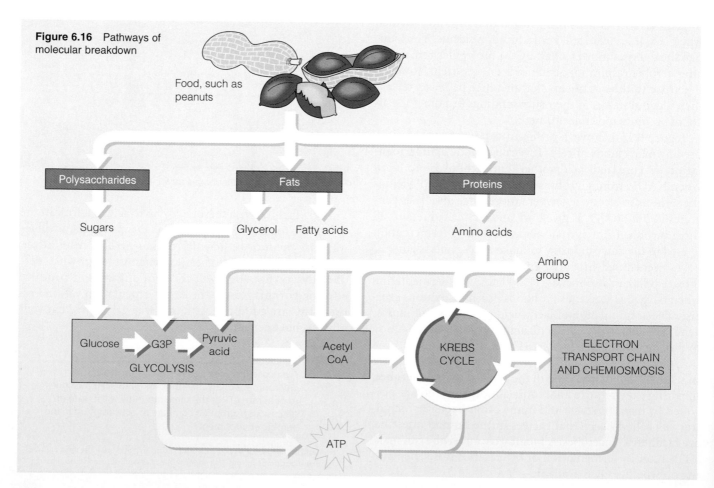

Figure 6.16 Pathways of molecular breakdown

Food molecules provide raw materials for biosynthesis

Not all food molecules are destined to be oxidized as fuel for making ATP. Food also provides the raw materials a cell uses for biosynthesis, to make its own molecules for repair and growth. Cells obtain some raw materials directly from food. Amino acids, for instance, can be incorporated without further change into an organism's own proteins. However, the body also needs certain molecules that are not present in food. Our cells make these molecules using some of the compounds of glycolysis and the Krebs cycle.

Figure 6.17 outlines biosynthetic pathways by which cells make three classes of macromolecules, starting with some of the small organic molecules produced in glycolysis and the Krebs cycle. These pathways consume ATP energy, rather than generate it. Notice that the pathways below appear to be the exact reverse of those in Figure 6.16. However, the details of the pathways can be different. In particular, glucose synthesis (pink box) is *not* the exact reverse of glycolysis (Figure 6.16), although some of the key intermediates—such as G3P—are the same in the two processes.

Thus, we see an important distinction as well as some clear connections between two aspects of metabolism that are central to life: the energy-harvesting process of cellular respiration and the biosynthetic pathways used to construct all parts of the cell.

? Explain how people can gain weight in the form of stored fat even though they are on a low-fat, carbohydrate-rich diet. (*Hint:* Look for G3P and acetyl CoA in *both* Figures 6.16 and 6.17.)

If caloric intake is excessive, whatever its form, body cells use a combination of metabolic pathways to convert the excess material to fat. Fats are made from glycerol and fatty acids, which are made from G3P and acetyl CoA, respectively (Figure 6.17). As Figure 6.16 shows, the acetyl CoA and G3P can come from the oxidation of carbohydrates (polysaccharides and sugars).

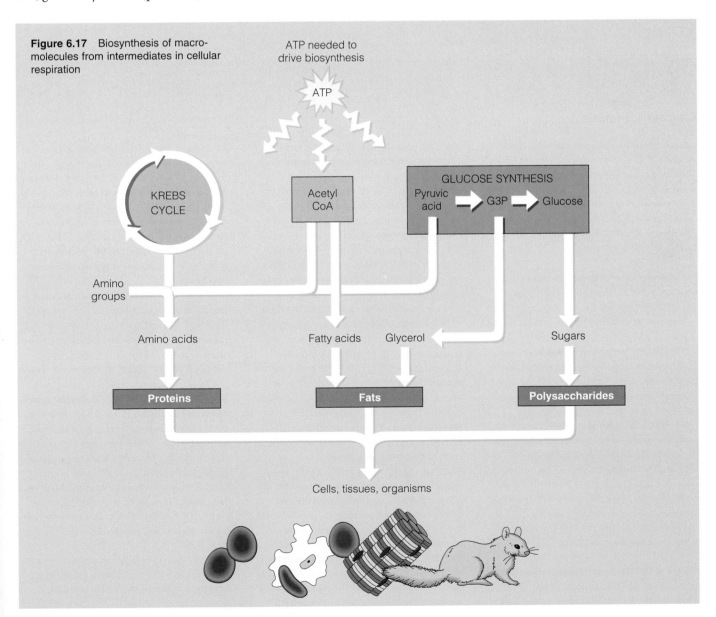

Figure 6.17 Biosynthesis of macromolecules from intermediates in cellular respiration

6.18 The fuel for respiration ultimately comes from photosynthesis

A giant panda eating its favorite food, a bamboo shoot, is an appropriate ending to this chapter and a good starting point for the next one. Almost entirely a vegetarian, a panda can get all the energy and nutrients it needs by eating a few armloads of bamboo leaves and shoots daily. The panda's digestive tract breaks the plant material down into fuel molecules, and its cells harvest energy from the molecules using cellular respiration.

The cells of all living organisms—those of pandas and bamboo plants included—have the ability to *harvest* energy from the breakdown of organic fuel molecules. When the breakdown process is cellular respiration, the atoms of the starting materials end up in CO_2 and H_2O. In contrast, the ability to *make* organic molecules from CO_2 and H_2O is not universal. Giant pandas—in fact, all animals—lack this ability, but plants have it. Animal cells can only convert the energy in the chemical bonds of organic molecules to other chemical forms, but plant cells can actually produce organic molecules from inorganic ones using the energy of sunlight. This process, photosynthesis, is the subject of Chapter 7.

Figure 6.18 Photosynthesis creates the plant material (bamboo) that nourishes a panda

? The chemical ingredients from which a plant makes food by photosynthesis are ______ and ______.

carbon dioxide . . . water

Chapter Review

Chapter Summary

Introduction to Cellular Respiration (Introduction–6.3) Cells harvest chemical energy from food molecules such as sugars **(Introduction)**. Cellular respiration breaks down glucose molecules and banks their energy in ATP, in the process using O_2 and releasing CO_2 **(6.1, 6.2)**. ATP powers almost all cell and body activities **(6.3)**.

Basic Mechanisms of Energy Release and Storage (6.4–6.7) Glucose gives up energy as it is oxidized: Enzymes remove electrons (as part of hydrogen atoms) from glucose molecules (an oxidation) and transfer them to a coenzyme, usually NAD^+ (a reduction) **(6.4, 6.5)**.

The NADH formed then delivers the electrons to a series of electron carriers in an electron transport chain. As electrons move "downhill" from carrier to carrier, their energy is released in small quantities **(6.6)**. Cells use the energy released by "falling" electrons to pump H^+ ions across a membrane, creating an H^+ gradient. The energy of the gradient is harnessed to make ATP from ADP and phosphate by the process of chemiosmosis. ATP can also be made by transferring phosphate groups from organic molecules to ADP. This process is substrate-level phosphorylation **(6.7)**.

Stages of Cellular Respiration and Fermentation (6.8–6.15) Cellular respiration oxidizes sugar and produces ATP in three main stages: glycolysis in the cytoplasm, and the Krebs cycle and the electron transport chain in mitochondria **(6.8)**. In the first phase of glycolysis, ATP energy is used to split a glucose molecule in two. In the second phase, these parts give up some of their energy, yielding some ATP and giving up hydrogens (electrons and H^+) to NAD^+. Glycolysis converts each original glucose into two molecules of pyruvic acid **(6.9)**. Each pyruvic acid molecule then breaks down to form CO_2 and a two-carbon acetyl group, which enters the Krebs cycle **(6.10)**. The Krebs cycle is a series of reactions in which enzymes strip away electrons and H^+ from each acetyl group. The carbons are completely oxidized and are released as CO_2. Two ATP molecules are produced directly, but most of the glucose energy is captured in NADH and $FADH_2$, which carry electrons and H^+ to the electron transport chain **(6.11)**. The electrons from NADH and $FADH_2$ travel down the electron transport chain to oxygen, which combines with the electrons and H^+ to form water. Energy released by the electrons' journey is used to pump H^+ into the space between the membranes of the mitochondrion. In chemiosmosis, the H^+ ions diffuse back across the inner membrane (down their concentration gradient) by passing through ATP synthase complexes, which capture their energy to make ATP **(6.12)**. Various poisons can block the movement of electrons, block the flow of H^+ through ATP synthase, or allow H^+ to leak through the membrane **(6.13)**. Chemiosmosis produces up to about 38 ATP molecules for every glucose molecule that enters cellular respiration **(6.14)**.

Under anaerobic (oxygen-lacking) conditions, many organisms, such as yeasts and certain bacteria, can use glycolysis alone to produce small amounts of ATP from each glucose molecule. The pyruvic acid produced may be converted to other substances, such as alcohol and CO_2 (alcoholic fermentation) or lactic acid (lactic acid fermentation), recycling the NAD^+ needed to keep glycolysis working. Human muscle cells can use lactic acid fermentation to make ATP for short periods when oxygen is in short supply **(6.15)**.

Interconnections Between Molecular Breakdown and Synthesis (6.16–6.18) Molecules other than glucose can fuel cellular respiration. Polysaccharides can be hydrolyzed to monosaccharides and then converted to glucose for glycolysis. Proteins can be digested to amino acids, their amino groups disposed of in urine, and their remains oxidized in the Krebs cycle. Fats, rich in hydrogen, electrons, and energy, are broken up and fed into glycolysis and the Krebs cycle **(6.16)**.

In addition to energy, cells need raw materials for growth and repair. Some are obtained directly from food. Others are made from intermediates in glycolysis and the Krebs cycle.

Biosynthesis consumes ATP **(6.17)**. All organisms have the ability to harvest energy from organic molecules. Plants (but not animals) can also make these molecules from inorganic sources, by the process of photosynthesis **(6.18)**.

Testing Your Knowledge

Multiple Choice

1. Which of the following processes produces the most ATP molecules per glucose molecule consumed?
   a. lactic acid fermentation
   b. the Krebs cycle
   c. electron transport and chemiosmosis
   d. alcoholic fermentation
   e. glycolysis

2. When the poison cyanide blocks the electron transport chain, glycolysis and the Krebs cycle soon grind to a halt as well. Why do you think they stop?
   a. They run out of ATP.
   b. Unused O_2 interferes with glycolysis and the Krebs cycle.
   c. They run out of NAD^+ and FAD.
   d. Electrons are no longer available from the electron transport chain.
   e. They run out of ADP.

3. A biochemist wanted to study how various substances were used and changed in cellular respiration. In one experiment, he allowed a mouse to breathe air containing O_2 "labeled" by a particular isotope of oxygen, (a procedure harmless to the mouse). In the mouse, the labeled oxygen atoms first showed up in
   a. ATP
   b. glucose, $C_6H_{12}O_6$
   c. NADH
   d. carbon dioxide, CO_2
   e. water, H_2O

4. In glycolysis, ____ is oxidized and ____ is reduced.
   a. NAD^+ . . . glucose
   b. glucose . . . oxygen
   c. ATP . . . ADP
   d. glucose . . . NAD^+
   e. ADP . . . ATP

5. Which of the following is the immediate source of energy used to make most of the ATP in your cells?
   a. the breakdown of ADP
   b. the transfer of phosphates from glucose breakdown products to ADP
   c. the movement of hydrogen ions through a membrane
   d. the splitting of glucose into two molecules of pyruvic acid
   e. the movement of electrons along the electron transport chain

6. Sports physiologists at an Olympic training center wanted to monitor athletes to determine at what point their muscles were functioning anaerobically. They could do this by checking for a buildup of
   a. ATP
   b. lactic acid
   c. carbon dioxide
   d. ADP
   e. oxygen

Describing, Comparing, and Explaining

1. What is the biggest disadvantage of making ATP by means of fermentation? What is the biggest advantage of making ATP this way?

2. Which of the three stages of cellular respiration is considered to be the most ancient? What is the reasoning behind this conclusion?

3. Explain in terms of cellular respiration why we need oxygen and why we exhale carbon dioxide.

Thinking as a Scientist

1. An average adult human requires 2200 kcal of energy per day. Suppose your diet provides an average of 2300 kcal per day. How many hours per week would you have to walk to burn off the extra calories? Swim? Run? (See the table in Module 6.3.)

2. Although your body can convert excess carbohydrates in the diet to fats, it cannot convert a carbohydrate or fat alone into protein; some input of protein from the diet must participate in this conversion. What does the dietary protein contribute?

3. Your body makes NAD^+ and FAD from two B vitamins, niacin and riboflavin. You need only tiny amounts of vitamins; the recommended dietary allowance for niacin is 20 mg daily and for riboflavin, 1.7 mg. These amounts are thousands of times less than the amount of glucose your body needs each day to fuel its energy needs. How many NAD^+ and FAD molecules are needed for the breakdown of each glucose molecule? Why do you think your daily requirement for these substances is so small?

4. In a detail of the Krebs cycle not shown in Figure 6.11B, succinic acid is converted to a compound called fumaric acid, with the release of H^+ ions. You are studying this reaction using a suspension of bean-cell mitochondria and a blue dye that loses its color as it takes up H^+ ions. You know from previous experiments that the higher the concentration of succinic acid, the more rapid the decolorization of the blue dye. You set up reaction mixtures with mitochondria, dye, and three different concentrations of succinic acid (0.1 mg/L, 0.2 mg/L, and 0.3 mg/L). Which of the following graphs represent the results you would expect, and why?

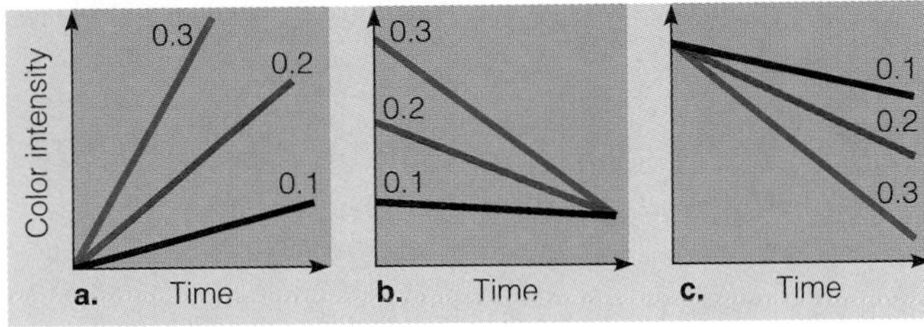

Science, Technology, and Society

The consumption of alcohol by a pregnant woman can cause a complex of birth defects called fetal alcohol syndrome (FAS). Symptoms of FAS include head and facial irregularities, heart defects, mental retardation, and behavioral problems such as hyperactivity. The U.S. Surgeon General's Office recommends that pregnant women abstain from drinking alcohol, and the government has mandated that a warning label be placed on liquor bottles: "Women should not drink alcoholic beverages during pregnancy because of the risk of birth defects." Imagine the following dilemma for a server in a restaurant. An obviously pregnant woman orders a strawberry daiquiri. The server shows a warning label to the customer and asks her, "Ma'am, are you sure you want this drink?" The customer becomes angry, demands her drink, and goes to the restaurant manager, who then fires the server. What would you have done, if you were in the server's place? Serve the customer, tell her of the possible harm her drinking alcohol might cause, or refuse to serve her and risk losing your job? Is a restaurant responsible for monitoring the health habits of its customers?

Media Resources

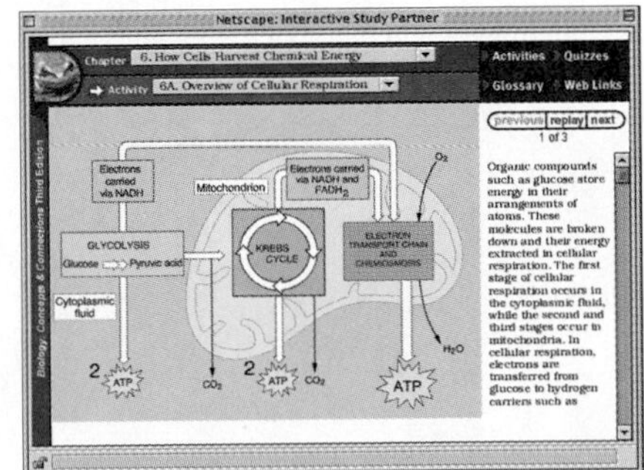

For further review of the concepts in this chapter, use the activities and quizzes on the CD-ROM.

Visit the Special Edition of The Biology Place web site for activities, quizzes, news articles, and web links.

7 Photosynthesis: Using Light to Make Food

Outline

An Overview of Photosynthesis

Life on Earth is almost entirely solar-powered, with nearly all organisms depending ultimately on food made by photosynthesis, which uses energy from sunlight. Knowledge about photosynthesis has accumulated rapidly in recent years, although the process was not well defined until the twentieth century. The ancient Greeks believed that the soil satisfies all of a plant's needs, and this idea was generally accepted as fact for nearly 2000 years.

The two types of plants in the photographs here played key roles in early research on photosynthesis. The tree at the left is a weeping willow, a native of Northern China that is now widely grown in Europe and the Western Hemisphere. The small plant on the facing page is a member of the mint family, a group of fragrant herbs that includes peppermint, spearmint, thyme, basil, oregano, and sage.

The origins of our understanding of photosynthesis extend back to a mid-seventeenth-century Belgian physician and part-time chemist named Jan Baptista van Helmont. Curious about whether plants really do get all their nourishment from the soil, van Helmont grew a small willow tree in a pot, adding nothing to the soil except water as the tree grew. Five years later, he found that the tree had gained nearly 75 kilograms while the soil had lost only about 60 grams. Van Helmont concluded correctly that a plant does not gain most of its substance from the soil. However, he concluded incorrectly that his willow tree gained most of its substance from the water he gave it.

A mint plant entered the picture in the early 1770s. English clergyman-chemist Joseph Priestley discovered that, although a candle burned out in a closed container, when he added a living sprig of mint to the container, the candle would continue to burn. At the time, Priestley did not know of O_2, but he correctly concluded that the mint sprig "restored" the air that the burning candle had depleted. Priestley also noted that his result was not always repeatable; in some of his experiments, the plant did not restore the air.

It was left to Dutch physician Jan Ingenhousz later in the decade to discover that plants require light and to explain Priestley's conflicting results. Ingenhousz demonstrated that plants can restore air only when their green parts are exposed to light. Although Ingenhousz did not point it out directly, he understood that Priestley had not provided adequate light for the plants he used in the failed experiments.

Van Helmont, Priestley, and Ingenhousz set the stage for understanding photosynthesis, but it was not until well into the twentieth century that the details of the process were sorted out. The equation below summarizes photosynthesis by showing the most important reactants and products. This equation reveals that green plants—willows, mints, and all others—need only energy from sunlight, carbon dioxide from the air, and water from the soil to make the food molecule glucose ($C_6H_{12}O_6$). The other main product of the photosynthesis reaction, O_2, is given off as a by-product.

Photosynthesis is thus the process by which plants use light energy to make food molecules from carbon dioxide and water. It is the most important chemical process on Earth, for it provides the food supply for virtually all organisms—plants, animals, protists, fungi, and bacteria alike. We examine this remarkable food-making process in this chapter.

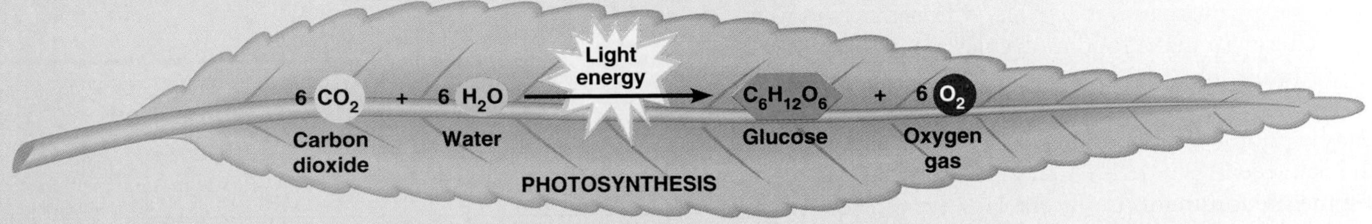

An Overview of Photosynthesis

7.1 Autotrophs are the producers of the biosphere

Figure 7.1A Oak tree

Figure 7.1B Cactus

Figure 7.1C Kelp, a large alga (a photosynthetic protist)

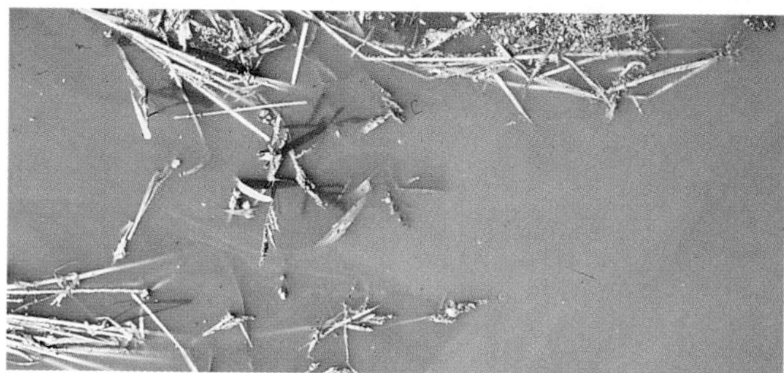

Figure 7.1D Photosynthetic bacteria in a pond

Plants are **autotrophs** (meaning "self-feeders" in Greek) in that they make their own food and thus sustain themselves without eating other organisms or even organic molecules. The chloroplasts of plant cells capture light energy that has traveled 150 million kilometers from the sun and convert it to chemical energy that is stored in glucose and other organic molecules made from carbon dioxide and water.

Because plants make organic food molecules from very simple raw materials, they are often referred to as the **producers** of the biosphere (the part of Earth occupied by living organisms). Plants produce the biosphere's food supply. Actually, plants are not the only producers in this sense; certain bacteria, archaea, and protists also make food molecules from inorganic materials. All organisms that use light energy to make food molecules are called photosynthetic autotrophs.

These photographs illustrate some of the diversity among photosynthetic autotrophs. On land, plants, such as the oak tree in Figure 7.1A and the cactus in Figure 7.1B, are the predominant producers. However, in aquatic environments, algae (photosynthetic protists) and photosynthetic bacteria are the main food producers. Figure 7.1C shows part of a large alga (a kelp) that forms extensive underwater and floating "forests" off the coast of California. Figure 7.1D shows a growth of photosynthetic bacteria in a pond. The color of the water results from pigment molecules that the bacteria use in photosynthesis.

Plants, algae, and photosynthetic bacteria all use light energy to drive the synthesis of organic molecules from carbon dioxide and water. In plants and algae, this process goes on in the cellular organelles called chloroplasts. The next module is an overview of the locations and structure of chloroplasts in plants.

Although they are "self-feeders," autotrophs are not totally self-sufficient. What do they require from the environment in order to synthesize sugar?

Light, carbon dioxide, and water (plants also require soil minerals, as you'll learn in Chapter 32)

Photosynthesis occurs in chloroplasts

All green parts of a plant have chloroplasts and can carry out photosynthesis, although in most plants, the leaves have the most chloroplasts and are the major sites of the process. The green color in plants is from chlorophyll pigments in the chloroplasts. Chlorophyll absorbs the light energy that the chloroplast puts to work in making food molecules.

Figure 7.2 below zooms in on a willow leaf to show the actual sites of photosynthesis. The top center drawing is a cross section (slice) of a leaf as it would look with a light microscope. Chloroplasts are concentrated in the cells of the **mesophyll,** the green tissue in the interior of the leaf. Carbon dioxide enters the leaf, and oxygen exits, by way of tiny pores called **stomata** (singular, *stoma*). As shown in the drawing and in the micrograph of a single mesophyll cell (on the left), each mesophyll cell has numerous chloroplasts.

The bottom micrograph and drawing show the structures in a chloroplast that constitute the machinery of photosynthesis. Membranes in the chloroplast form the structural framework where many of the reactions of photosynthesis occur, just as mitochondrial membranes do for the energy-harvesting machinery we discussed in Chapter 6. Like the mitochondrion, the chloroplast has an outer membrane and an inner membrane, with an intermembrane space between them. The chloroplast's inner membrane encloses a second compartment, which is filled with **stroma,** a thick fluid. The stroma is where sugars are made from carbon dioxide. In the stroma is suspended an elaborate system of disklike membranous sacs, called **thylakoids,** which contain the third chloroplast compartment. The thylakoids are concentrated in stacks called **grana** (singular, *granum*). Built into the thylakoid membranes are the chlorophyll molecules that capture light energy. The thylakoid membranes also house much of the molecular machinery that converts light energy to chemical energy.

Later in the chapter, we examine the function of these structures in more detail. But first, let's look more closely at the general equation for photosynthesis.

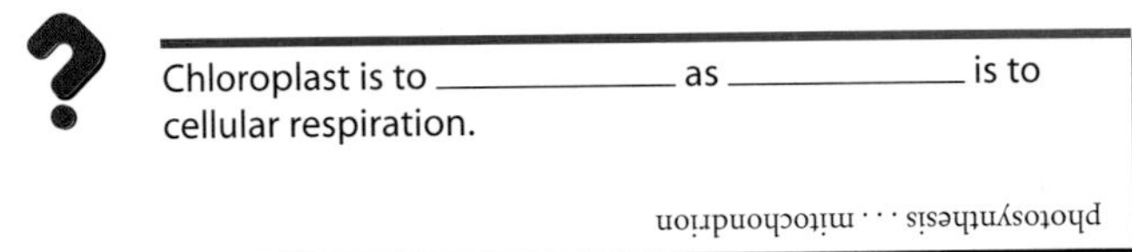

?
Chloroplast is to ________ as ________ is to cellular respiration.
photosynthesis . . . mitochondrion

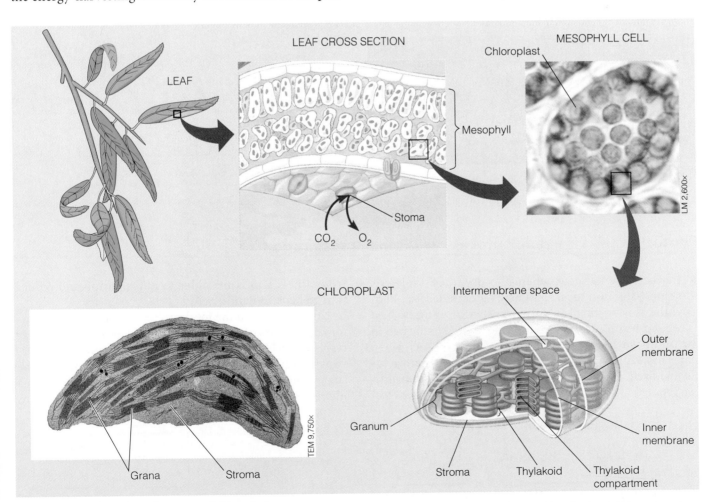

Figure 7.2 The location and structure of chloroplasts

7.3 Plants produce O_2 gas by splitting water

The leaves of plants that live in lakes and ponds are often covered with bubbles like the ones in Figure 7.3A. The bubbles are oxygen (O_2) gas produced during photosynthesis.

In the 1700s, Jan Ingenhousz suggested that plants produce O_2 by extracting it from CO_2. In the 1950s, scientists tested this hypothesis by using an isotope of oxygen, ^{18}O, to follow the fate of oxygen atoms during photosynthesis. (To review isotopes and their use as tracers, see Module 2.5.) In the photosynthesis equations in Figure 7.3B, the red type denotes ^{18}O. The photosynthesis equations here (and the photosynthesis and respiration equations in the next module) are written in a slightly more detailed form than you saw earlier. Notice that water is shown as both a reactant and a product.

In Experiment 1, a plant given carbon dioxide containing ^{18}O gave off no labeled (^{18}O-containing) oxygen gas. But in Experiment 2, a plant given water containing ^{18}O did produce labeled O_2 gas. These experiments showed that the O_2 produced during photosynthesis comes from water not, as Ingenhousz predicted, from CO_2. It takes two water (H_2O) molecules to make each molecule of O_2.

Knowing where the O_2 comes from gives us a hint of what else happens during photosynthesis. Additional experiments have revealed that the oxygen atoms in CO_2 and the hydrogens in the reactant H_2O molecules end up in the sugar molecule and in water that is formed anew. The carbon in CO_2 ends up in the sugar molecule. Figure 7.3C summarizes the fates of all the atoms that start out in the reactant molecules of photosynthesis.

? What step in cellular respiration essentially reverses the water-splitting step of photosynthesis? (*Hint:* Review Module 6.12.)

The formation of water when electrons and hydrogen are added to oxygen at the "bottom" of the electron transport chain

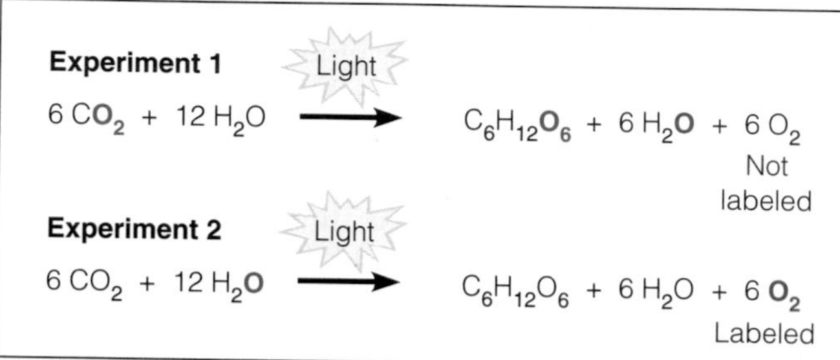

Figure 7.3B Experiments tracking the oxygen atoms in photosynthesis

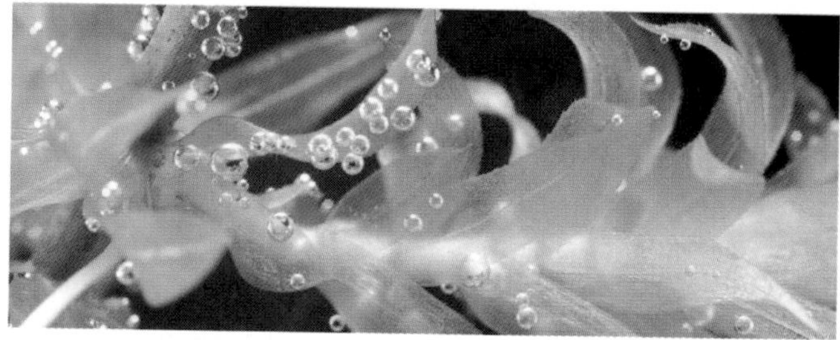

Figure 7.3A Oxygen bubbles on the leaves of an aquatic plant

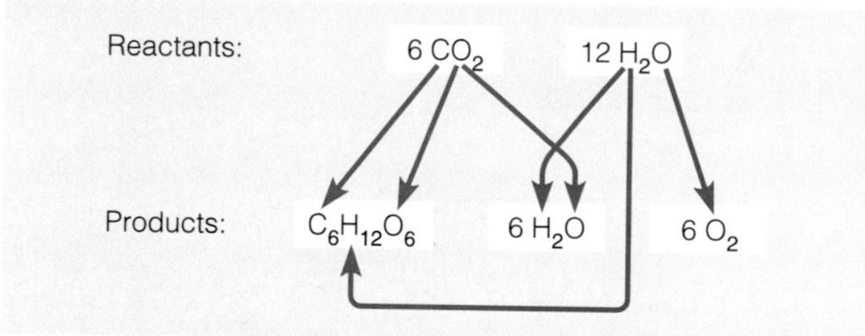

Figure 7.3C Fates of all the atoms in photosynthesis

7.4 Photosynthesis is a redox process, as is cellular respiration

What actually happens when photosynthesis converts CO_2 and water into sugars and new water molecules? Photosynthesis is a redox (oxidation-reduction) process, just as cellular respiration is (see Module 6.5). As indicated in the photosynthesis equation below (Figure 7.4A), when water molecules are split apart, yielding O_2, they are actually oxidized; that is, they lose electrons, along with hydrogen ions (H^+). Meanwhile, CO_2 is reduced to sugar as electrons and H^+ ions are added to it. Oxidation and reduction go hand in hand.

Now move from the food-producing equation for photosynthesis to the energy-releasing equation for cellular respiration (Figure 7.4B). Overall, cellular respiration harvests energy stored in a glucose molecule by oxidizing the sugar and reducing O_2 to H_2O. This process involves a number of energy-releasing redox reactions, with electrons

Reduction

$$6\,CO_2 + 12\,H_2O \longrightarrow C_6H_{12}O_6 + 6\,H_2O + 6\,O_2$$

Oxidation

Figure 7.4A Photosynthesis (uses light energy)

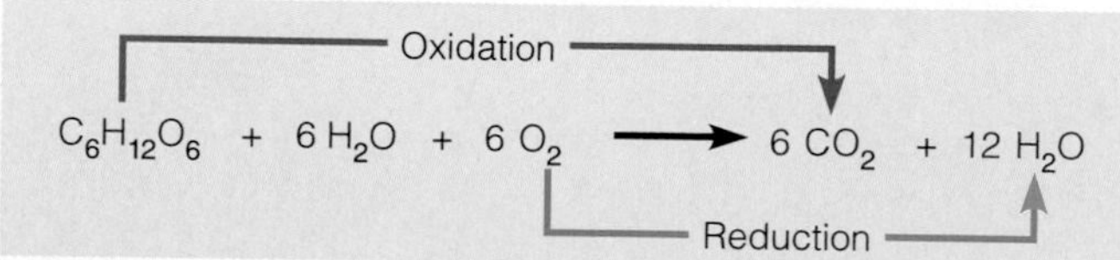

Figure 7.4B Cellular respiration (releases chemical energy)

losing potential energy as they travel down an energy hill from sugar to O_2. Along the way, the mitochondrion uses some of the energy to synthesize ATP, as we saw in Chapter 6.

In contrast, the food-producing redox reactions of photosynthesis involve an uphill climb. As water is oxidized and CO_2 is reduced during photosynthesis, electrons gain energy by being boosted up an energy hill. The light energy captured by chlorophyll molecules in the chloroplast provides the boost for the electrons. Photosynthesis converts the light energy to chemical energy and stores it in sugar molecules.

Which redox process, photosynthesis or cellular respiration, is endergonic? (*Hint:* See Module 5.3.)

Photosynthesis

Overview: Photosynthesis occurs in two stages linked by ATP and NADPH

7A
See Activity 7A on the CD-ROM.

The equation for photosynthesis is a simple summary of a very complex process. Actually, photosynthesis is not a single process, but has two stages, each with multiple steps. The steps of the first stage are known as the **light reactions;** these are the reactions that convert light energy to chemical energy and produce O_2 gas as a waste product. The steps of the second stage are known as the **Calvin cycle;** this is a cyclic series of reactions that assemble sugar molecules using CO_2 and the energy-containing products of the light reactions. The second stage of photosynthesis is named for American biochemist and Nobel laureate Melvin Calvin. In the 1940s, Calvin and his colleagues traced the path of carbon in the cycle, using the radioactive isotope ^{14}C to label the carbon from CO_2. The word "photosynthesis" capsulizes the two stages. *Photo-*, from the Greek word for light, refers to the light reactions; *synthesis,* meaning "putting together," refers to sugar construction by the Calvin cycle.

As indicated in Figure 7.5, the light reactions of photosynthesis occur in the thylakoid membranes of the chloroplast's grana. Light absorbed by chlorophyll in the thylakoid membranes furnishes the energy that eventually powers the food-making machinery of photosynthesis. Light energy is used to make ATP from ADP and phosphate. It is also used to drive a transfer of electrons from water to $NADP^+$, a hydrogen carrier similar to the NAD^+ that carries hydrogens in cellular respiration. Enzymes reduce $NADP^+$ to NADPH by adding a pair of light-excited electrons along with an H^+. This reaction temporarily stores the energized electrons. As $NADP^+$ is reduced to NADPH, water is split (oxidized), giving off O_2.

In summary, the light reactions of photosynthesis are the steps that absorb solar energy and convert it to chemical energy stored in ATP and NADPH. Notice that these reactions produce no sugar; sugar is not made until the Calvin cycle, the second stage of photosynthesis.

The Calvin cycle occurs in the stroma of the chloroplast. The incorporation of carbon from CO_2 into organic compounds, shown in the figure as CO_2 entering the Calvin cycle, is called **carbon fixation.** After carbon fixation, enzymes of the cycle make sugars by further reducing the fixed carbon—by adding high-energy electrons to it, along with H^+.

As the figure suggests, it is NADPH produced by the light reactions that provides the high-energy electrons for reduction in the Calvin cycle. And ATP from the light reactions provides chemical energy that powers several of the steps of the Calvin cycle. The Calvin cycle does not require light directly. However, in most plants, the Calvin cycle runs during daytime, when the light reactions power the cycle's sugar assembly line by supplying it with NADPH and ATP.

For chloroplasts to produce sugar from carbon dioxide in the dark, they would require an artificial supply of ________ and ________.

ATP . . . NADPH

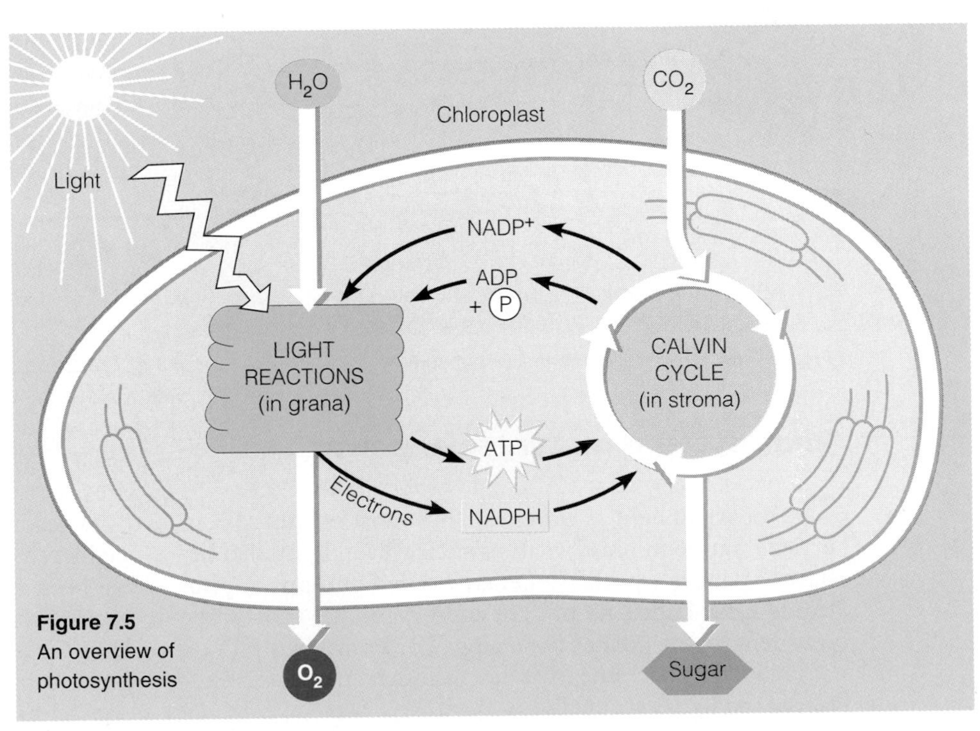

Figure 7.5
An overview of photosynthesis

The Light Reactions: Converting Solar Energy to Chemical Energy

7.6 Visible radiation drives the light reactions

What exactly do we mean when we say that photosynthesis is powered by light energy from the sun? Sunlight is a type of energy called radiation, or **electromagnetic energy.** Electromagnetic energy travels in space as rhythmic waves analogous to those made by a pebble dropped in a puddle of water. The distance between the crests of two adjacent waves is called a **wavelength.**

Figure 7.6A shows the full range of electromagnetic wavelengths, known as the electromagnetic spectrum. Visible light forms only a small fraction of the spectrum. It consists of different wavelengths (measured in nanometers, or nm) that our eyes see as different colors. As sunlight shines on a plant leaf, the light of some wavelengths is absorbed and put to use in photosynthesis, while the light of other wavelengths is reflected back from the leaf or transmitted through it. The light reactions of photosynthesis use only certain components (wavelengths, or colors) of visible light.

Figure 7.6B shows what happens to visible light in the chloroplast. Light-absorbing molecules called pigments in the membranes of a granum absorb mainly blue-violet and red-orange wavelengths. We do not see these absorbed wavelengths. What we see when we look at a leaf are the green wavelengths that the pigments transmit and reflect.

Different pigments absorb light of different wavelengths, and chloroplasts contain several kinds of pigments. One, chlorophyll *a*, absorbs mainly blue-violet and red light. Chlorophyll *a* participates directly in the light reactions. It looks grass-green because it reflects mainly green light. A very similar molecule, chlorophyll *b*, absorbs mainly blue and orange light and reflects (appears) yellow-green. Chlorophyll *b* does not participate directly in the light reactions, but it broadens the range of light that a plant can use by conveying absorbed energy to chlorophyll *a*, which then puts the energy to work in the light reactions.

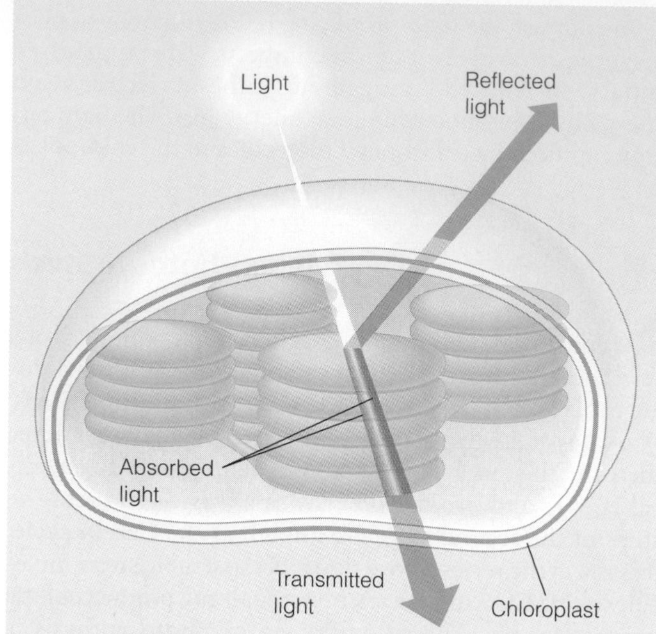

Figure 7.6B The interaction of light with a chloroplast

Chloroplasts also contain a family of yellow-orange pigments called carotenoids, which absorb mainly blue-green light. Some may pass energy to chlorophyll *a*, as chlorophyll *b* does. Other carotenoids have a protective function: They absorb and dissipate excessive light energy that would otherwise damage chlorophyll. (Similar carotenoids may help protect our eyes from very bright light.) 

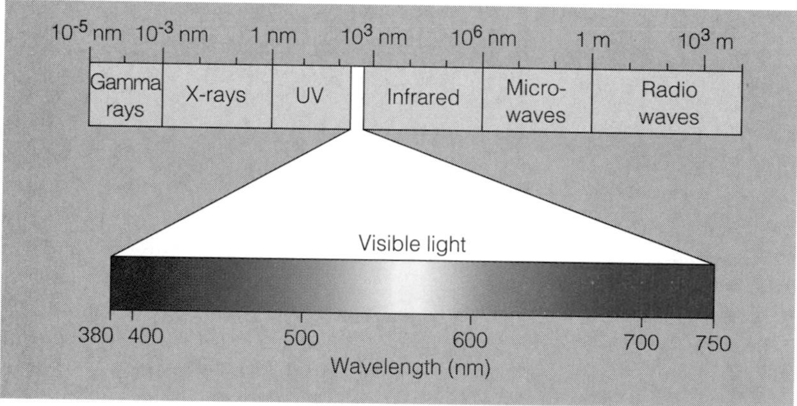

Figure 7.6A The electromagnetic spectrum

? What color of light is least effective in driving photosynthesis?

Green

7.7 Photosystems capture solar power

The theory of light as waves explains most of light's properties relative to photosynthesis. However, light also behaves as discrete packets of energy called photons. A **photon** is a fixed quantity of light energy, and the shorter the wavelength, the greater the energy. For example, a photon of violet light packs nearly twice as much energy as a photon of red light.

When a pigment molecule absorbs a photon, one of the pigment's electrons gains energy—we say that the electron has been raised from a ground state to an excited state. The excited state is very unstable, and generally the electron loses the excess energy and falls back to its ground state almost immediately. Several things may happen to the energy released in the process. For instance, the excess energy

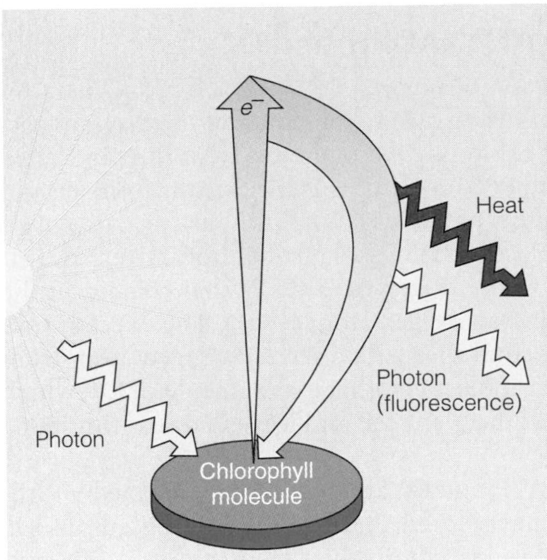

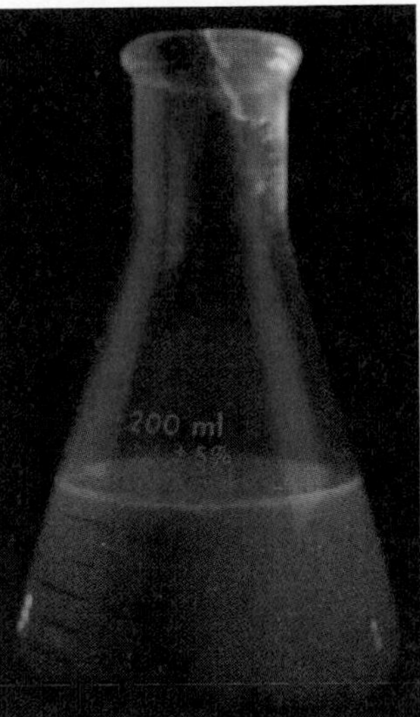

Figure 7.7A Fluorescence of isolated chlorophyll in solution

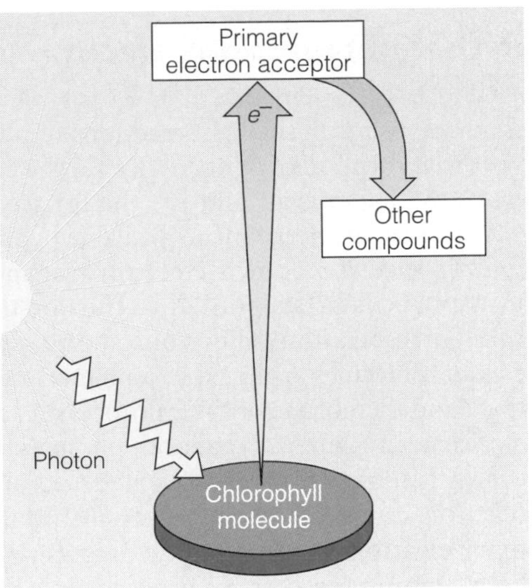

Figure 7.7B Excitation of chlorophyll in a chloroplast

may be released as heat. Some pigments emit light as well as heat after absorbing photons. In this case, the excited electron gives off a photon, in addition to heat, as it reverts to the ground state. We can demonstrate this phenomenon in the laboratory with a chlorophyll solution, as shown in Figure 7.7A (above). When illuminated, the chlorophyll emits heat and photons of light that produce a reddish afterglow, as electrons fall from an excited state to the ground state. This afterglow is called fluorescence.

In contrast to pure chlorophyll in solution, illuminated chlorophyll in an intact chloroplast passes its excited electron to a neighboring molecule (Figure 7.7B). The neighboring molecule, called the primary electron acceptor, is reduced as chlorophyll is oxidized.

The solar-powered electron transfer from chlorophyll to the primary electron acceptor is the first step in the light reactions and the first of many redox reactions in photosynthesis. The box labeled "other compounds" in Figure 7.7B represents the molecular machinery that uses the redox reactions to make ATP and NADPH. (We describe these reactions in Module 7.8.)

Chlorophyll *a*, chlorophyll *b*, and the carotenoid pigments are clustered in the thylakoid membrane of each chloroplast in assemblies of 200–300 pigment molecules. Evidence suggests that only one of the chlorophyll *a* molecules in each assembly actually donates excited electrons to the primary electron acceptor, thus triggering the light reactions. This chlorophyll *a* molecule and the primary electron acceptor make up what is called the **reaction center** of the pigment assembly (Figure 7.7C). The reaction center and the other pigment molecules function collectively as a light-gathering antenna that absorbs photons and passes the energy from molecule to molecule until it reaches the reaction center. The combination of the antenna molecules, the reaction center, and the primary electron acceptor is called a **photosystem.** This is the light-harvesting unit of the chloroplast's thylakoid membrane.

Two types of photosystems have been identified. They are referred to as photosystem I and photosystem II, in order of their discovery. In photosystem I, the chlorophyll *a* molecule of the reaction center is called P700 because the light it absorbs best is red light with a wavelength of 700 nm. The reaction-center chlorophyll of photosystem II is called P680 because the wavelength of light it absorbs best is 680 nm (a more orange shade of red). These two reaction-center pigments are actually identical chlorophyll *a* molecules, but their association with different proteins in the thylakoid membrane accounts for the slight difference in their light absorption. Let's now see how the two photosystems work together in the light reactions.

In comparison to a solution of isolated chlorophyll, why do intact chloroplasts release less heat and fluorescence when illuminated?

In the chloroplasts, the light-excited electrons are trapped by a primary electron acceptor rather than immediately giving up all their energy as heat and light.

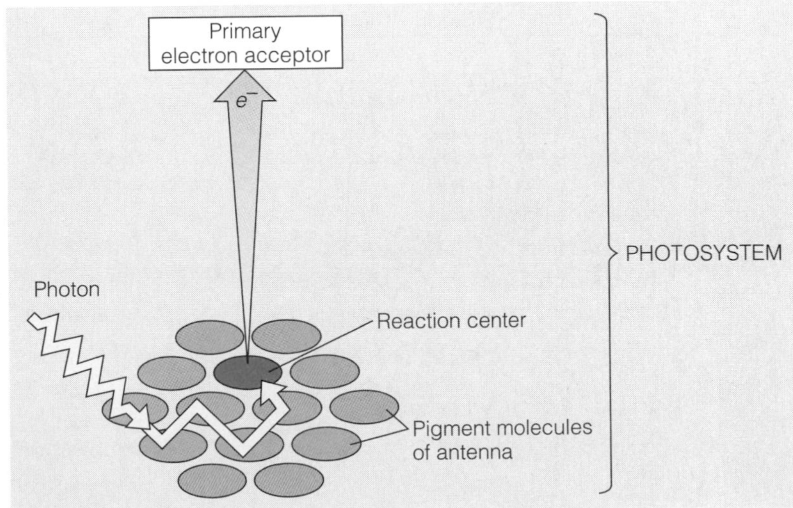

Figure 7.7C Components of a photosystem

7.8

7B

In the light reactions, electron transport chains generate ATP, NADPH, and O_2

The key events in the light reactions of photosynthesis are (1) the absorption of light energy, (2) the excitation of electrons by that energy, and (3) the formation of ATP and NADPH using energy made available by the cascade of energized electrons down electron transport chains. The transport chains are similar to the one that functions in cellular respiration. They consist of a series of electron-carrier molecules arranged in a membrane, in this case the thylakoid membrane of the chloroplast.

Notice in Figure 7.8 that both photosystems absorb light energy (yellow arrows), and excited electrons pass from the reaction-center chlorophylls (P680 and P700) to the primary electron acceptors (gold arrows represent electron flow). In turn, each primary acceptor is oxidized as it donates high-energy electrons to the first electron carrier of an electron transport chain. Additional redox reactions then shuttle the electrons from one electron-carrier molecule to the next down an energy cascade. At each step in the cascade, the electrons lose energy, some of which ends up temporarily stored in either ATP or NADPH molecules. On the far right in the diagram, you see NADPH being formed when $NADP^+$ is reduced. $NADP^+$ gains two high-energy electrons (the reduction) and an H^+.

Every molecule of NADPH formed in the light reactions requires two electrons from photosystem I. Where do all the electrons come from that keep the light reactions running? The answer for photosystem I is that electrons from the bottom of the cascade in the center of the diagram pass into its P700 chlorophyll. For photosystem II, as indicated on the far left, P680 regains its lost electrons from a water molecule. In this process, H_2O splits apart, and two of its electrons replace those missing from photosystem II. Left behind are two hydrogen ions (H^+) that remain in the chloroplast and an oxygen atom ($\frac{1}{2}O_2$). The oxygen atom immediately combines with a second oxygen atom from another water molecule to form a molecule of O_2, which diffuses out of the plant cell and leaves the leaf through a stoma.

NADPH, ATP, and O_2 are the products of the light reactions. We have noted that the high-energy molecule NADPH and the waste product O_2 both result directly from redox reactions. The synthesis of ATP is different; its synthesis is driven by chemiosmosis—the same mechanism that generates ATP in cellular respiration. We look at chemiosmosis in the chloroplast in the next module.

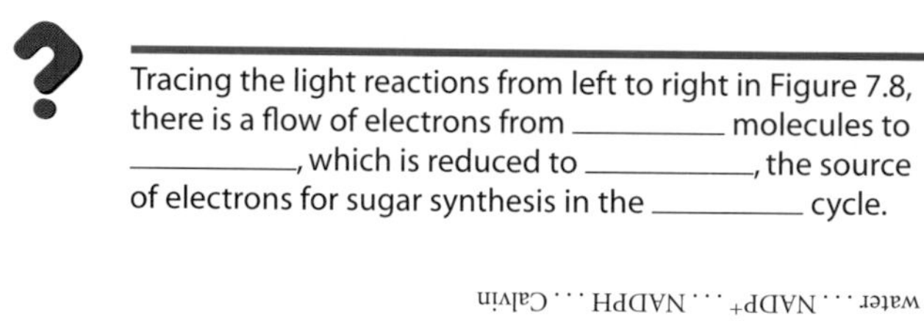

?

Tracing the light reactions from left to right in Figure 7.8, there is a flow of electrons from ________ molecules to ________, which is reduced to ________, the source of electrons for sugar synthesis in the ________ cycle.

water . . . $NADP^+$ . . . NADPH . . . Calvin

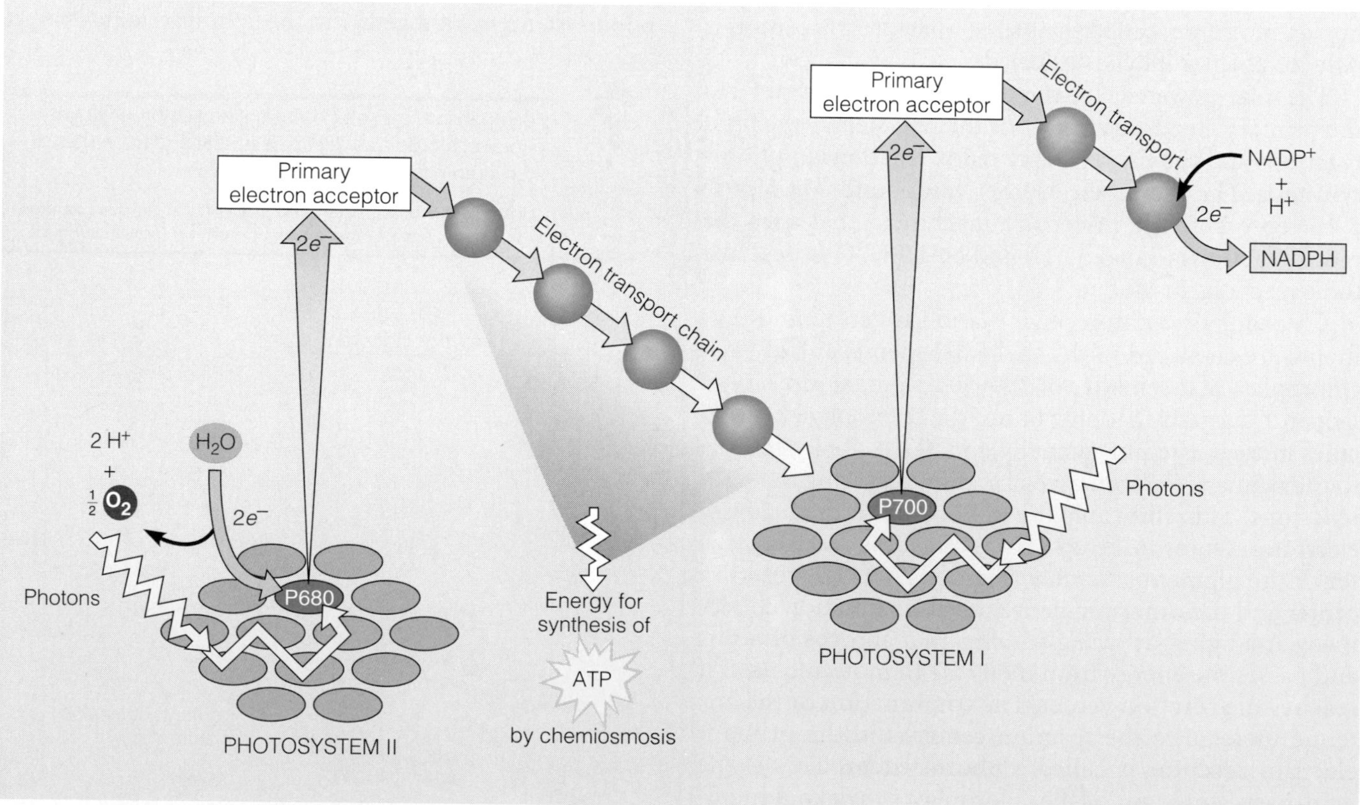

Figure 7.8 Photon and electron flow in the light reactions of photosynthesis

Chemiosmosis powers ATP synthesis in the light reactions

7.9

7B

The figure below illustrates, in a highly diagrammatic way, the relationship between chloroplast structure and function in the light reactions of photosynthesis. It shows the two photosystems and electron transport chains, all located within the thylakoid membrane of a chloroplast. The illustration is very schematic, but it makes the point that the photosystems are arranged in such a way that energy released during electron flow drives the transport of hydrogen ions (H^+) across the thylakoid membrane. The arrangement is very much like the one in our model for the electron transport chain of cellular respiration in the mitochondrion (see Figure 6.12). In both cases, excited electrons (gold line) pass along a series of electron carriers (bright blue circles) within a membrane, as redox reactions occur. The electrons give up energy on the way, and some of the energy is used to make ATP by chemiosmosis.

Although photosynthesis is a food-making process and cellular respiration is an energy-harvesting one, electron transport in the chloroplast drives chemiosmosis the same way it does in the mitochondrion. Specifically, some of the electron carriers use energy released from the electrons to actively transport H^+ ions from one side of a membrane to the other. In the chloroplast, the carriers move H^+ across the thylakoid membrane from the stroma into the thylakoid compartment. This generates a concentration gradient of H^+ across the membrane. (The higher H^+ concentration is indicated by the darker shade of gray in the diagram.) As in the mitochondrion, energy stored in this concentration gradient is used to drive ATP synthesis.

The flask-shaped structure on the right in the figure represents the protein complex ATP synthase, which is like the one we saw in the mitochondrion. ATP synthase provides a port through which H^+ can diffuse back into the stroma from the thylakoid compartment. The energy of the H^+ gradient drives H^+ back across, and energy is released in the process. ATP synthase uses some of this energy to phosphorylate ADP, making ATP. In photosynthesis, the chemiosmotic production of ATP is called **photophosphorylation** because the initial energy input is light energy.

Notice that the final electron acceptor is $NADP^+$, not O_2 as in cellular respiration. Rather than being consumed, O_2 is produced, as explained in Module 7.8, when water is split to provide replacement electrons for photosystem II.

We have now examined the major events of the light reactions. The ATP and NADPH produced during these reactions are used in the next stage of photosynthesis, the Calvin cycle. Module 7.10 describes how that cycle makes sugar.

? In terms of the spatial organization of photosynthesis within the chloroplast, what is the advantage of the light reactions producing NADPH and ATP on the stroma side of the thylakoid membrane? (*Hint:* Review Module 7.5.)

The Calvin cycle, which consumes the NADPH and ATP, occurs in the stroma.

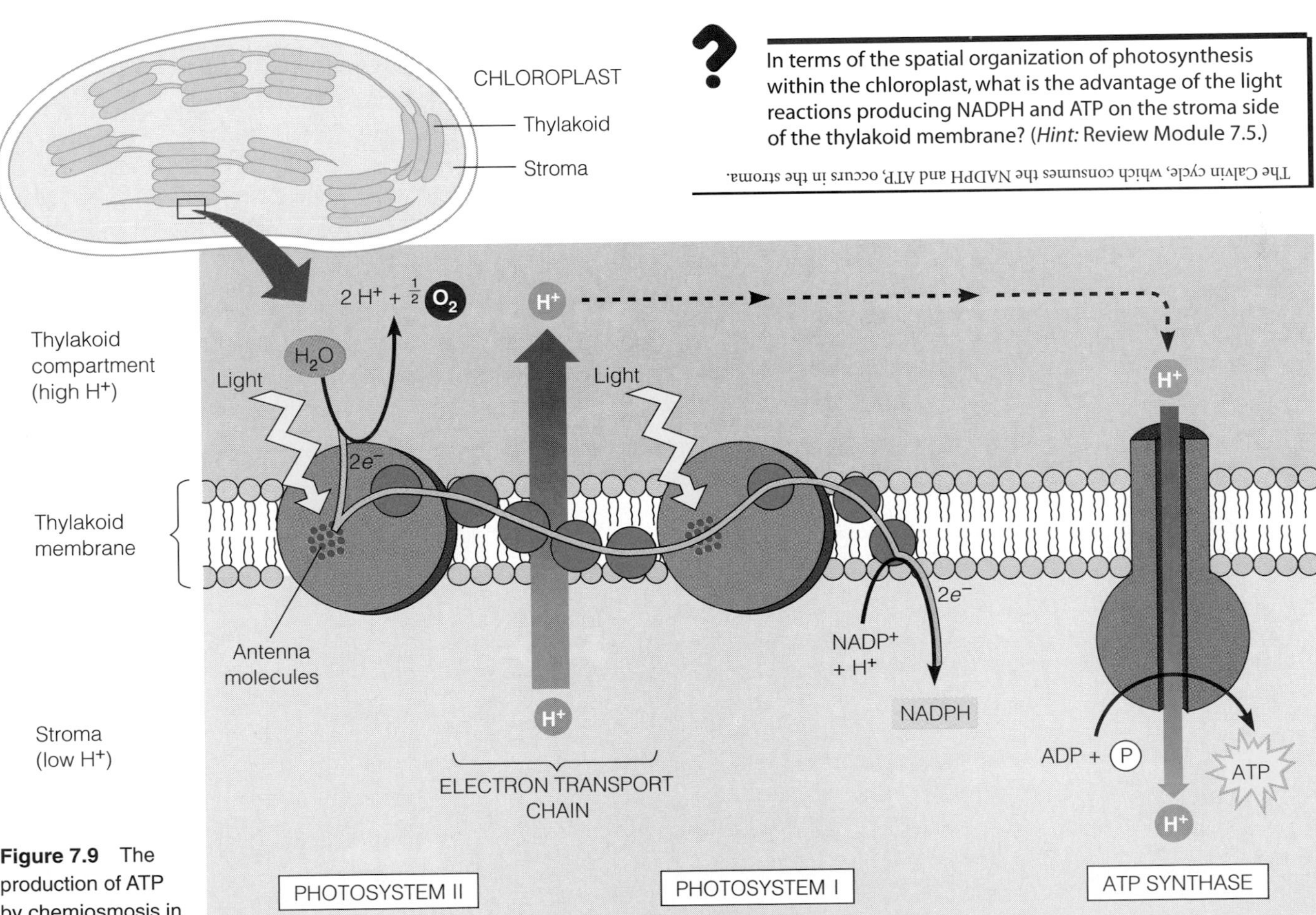

Figure 7.9 The production of ATP by chemiosmosis in photosynthesis

7.10 ATP and NADPH power sugar synthesis in the Calvin cycle

7C

The Calvin cycle functions like a sugar factory within a chloroplast. As Figure 7.10A shows, inputs to this all-important food-making process are CO_2 (from the air) and ATP and NADPH (both generated by the light reactions). Using carbon from CO_2, energy from ATP, and high-energy electrons from NADPH, the Calvin cycle constructs an energy-rich sugar molecule, glyceraldehyde 3-phosphate (G3P). The plant cell can use G3P to make glucose or other organic molecules as needed.

Figure 7.10B presents the details of the Calvin cycle. To make a molecule of G3P, the cycle must incorporate the carbon atoms (gray balls) from three molecules of CO_2. The cycle actually incorporates one carbon at a time, but we show it starting with three CO_2 molecules so that we end up with a complete G3P molecule.

INPUT { CO_2, ATP, NADPH

CALVIN CYCLE

OUTPUT: G3P

Figure 7.10A An overview of the Calvin cycle

The completion of the Calvin cycle, the final phase of photosynthesis, is an

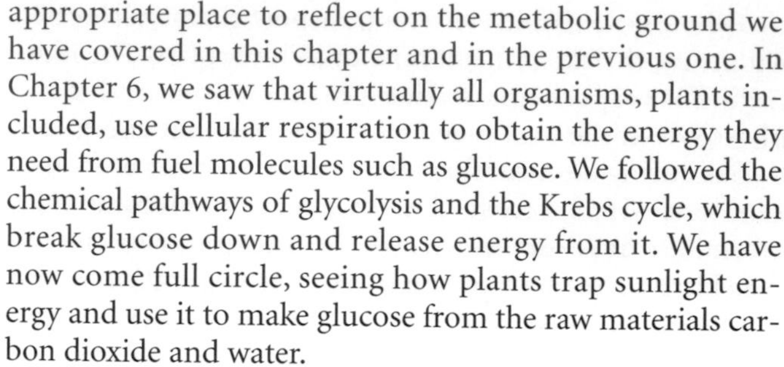

appropriate place to reflect on the metabolic ground we have covered in this chapter and in the previous one. In Chapter 6, we saw that virtually all organisms, plants included, use cellular respiration to obtain the energy they need from fuel molecules such as glucose. We followed the chemical pathways of glycolysis and the Krebs cycle, which break glucose down and release energy from it. We have now come full circle, seeing how plants trap sunlight energy and use it to make glucose from the raw materials carbon dioxide and water.

In tracing glucose synthesis and its breakdown, we have also seen that cells use several of the same mechanisms—electron transport, redox reactions, and chemiosmosis—in energy storage (photosynthesis) and energy harvest (cellular respiration).

To synthesize 1 glucose molecule, the Calvin cycle uses _____ molecules of CO_2, _____ molecules of ATP, and _____ molecules of NADPH.

6 . . . 18 . . . 12

Step 1 Carbon fixation. An enzyme called rubisco combines three molecules of CO_2 with three molecules of a five-carbon sugar called ribulose bisphosphate (abbreviated RuBP). Six molecules of the three-carbon organic acid 3-phosphoglyceric acid (3-PGA) result.

Step 2 Energy consumption and redox. Two chemical reactions (indicated by the two white arrows) consume energy from six molecules of ATP and oxidize six molecules of NADPH. Six molecules of 3-PGA are reduced, producing six molecules of the energy-rich three-carbon G3P.

Step 3 Release of one molecule of G3P. Five of the G3Ps from step 2 remain in the cycle. The single molecule of G3P you see leaving the cycle is the net product of photosynthesis. A plant cell uses two G3P molecules to make one molecule of glucose, which has six carbons. Since the Calvin cycle incorporates only one molecule of CO_2—and thus only one carbon—at a time, it takes six complete turns of the cycle to make the two molecules of G3P that go into one glucose molecule.

Step 4 Regeneration of RuBP. A series of chemical reactions uses energy from ATP to rearrange the atoms in the five G3P molecules, forming three RuBP molecules. These can start another turn of the cycle.

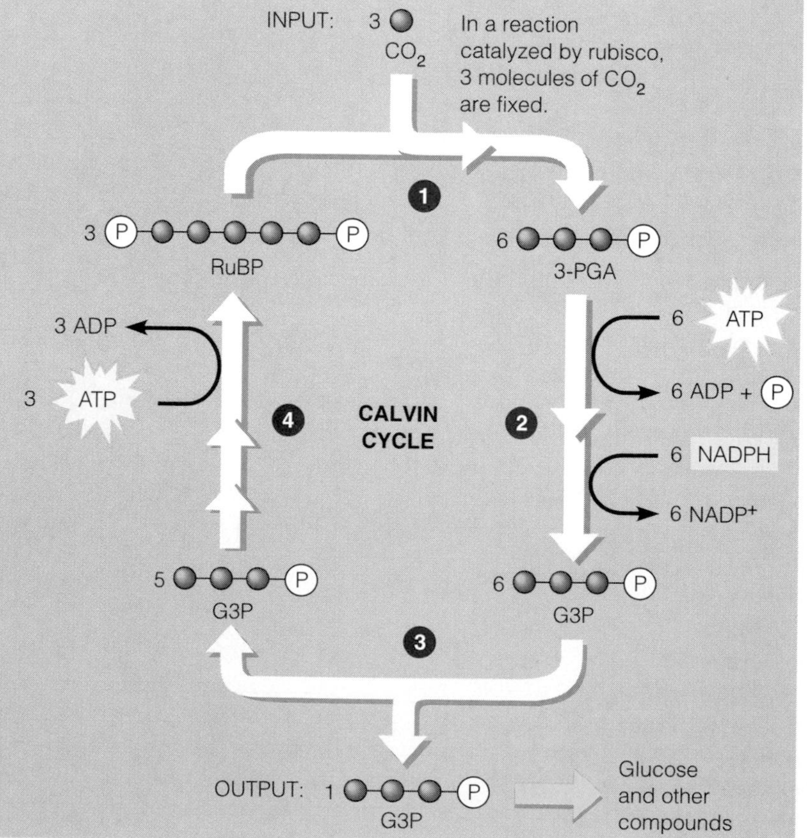

Figure 7.10B Details of the Calvin cycle

Review: Photosynthesis uses light energy to make food molecules 7.11

As we have discussed, most of the living world depends on the food-making machinery of photosynthesis. Figure 7.11 summarizes the two stages of this vital process and reviews where they occur in the chloroplast.

Starting on the left in the diagram, you see a summary of the light reactions, which occur in the thylakoid membranes (green). Two photosystems in the membranes capture solar energy, using it to energize electrons. Simultaneously, water is split, and O_2 is released. The photosystems transfer energized electrons to electron transport chains where energy is harvested and used to make the high-energy molecules NADPH and ATP.

The chloroplast's sugar factory is the Calvin cycle, the second stage of photosynthesis. In the stroma, enzymes of the cycle combine CO_2 with RuBP and produce G3P. Sugar molecules made from G3P are a plant's own food supply. Plants use sugars as fuel for cellular respiration and as starting material for making other organic molecules, such as the structural molecule cellulose. Most plants make considerably more sugar than they need. They stockpile the excess sugar as starch, storing some in chloroplasts and some in roots, tubers, and fruits. These stored sugars are a major source of food for many animals.

On a global scale, photosynthesis by plant chloroplasts creates billions of tons of organic matter each year, an output that no other chemical process on Earth can match. And no other process is more important to life than photosynthesis.

Plants (and other photosynthesizers) not only feed themselves, they are also the ultimate source of food for virtually all other organisms. Humans and other animals, for example, make none of their own food and are totally dependent on the organic matter made by photosynthesizers. Even the energy we acquire when we eat meat was originally captured by photosynthesis. The energy in a hamburger, for instance, came from sunlight that was originally converted to a chemical form in chloroplasts in the cells of the grasses eaten by cattle.

Explain why a poison that inhibits an enzyme of the Calvin cycle will also inhibit the light reactions.

The light reactions require ADP and $NADP^+$, which are not formed from ATP and NADPH when the Calvin cycle stops.

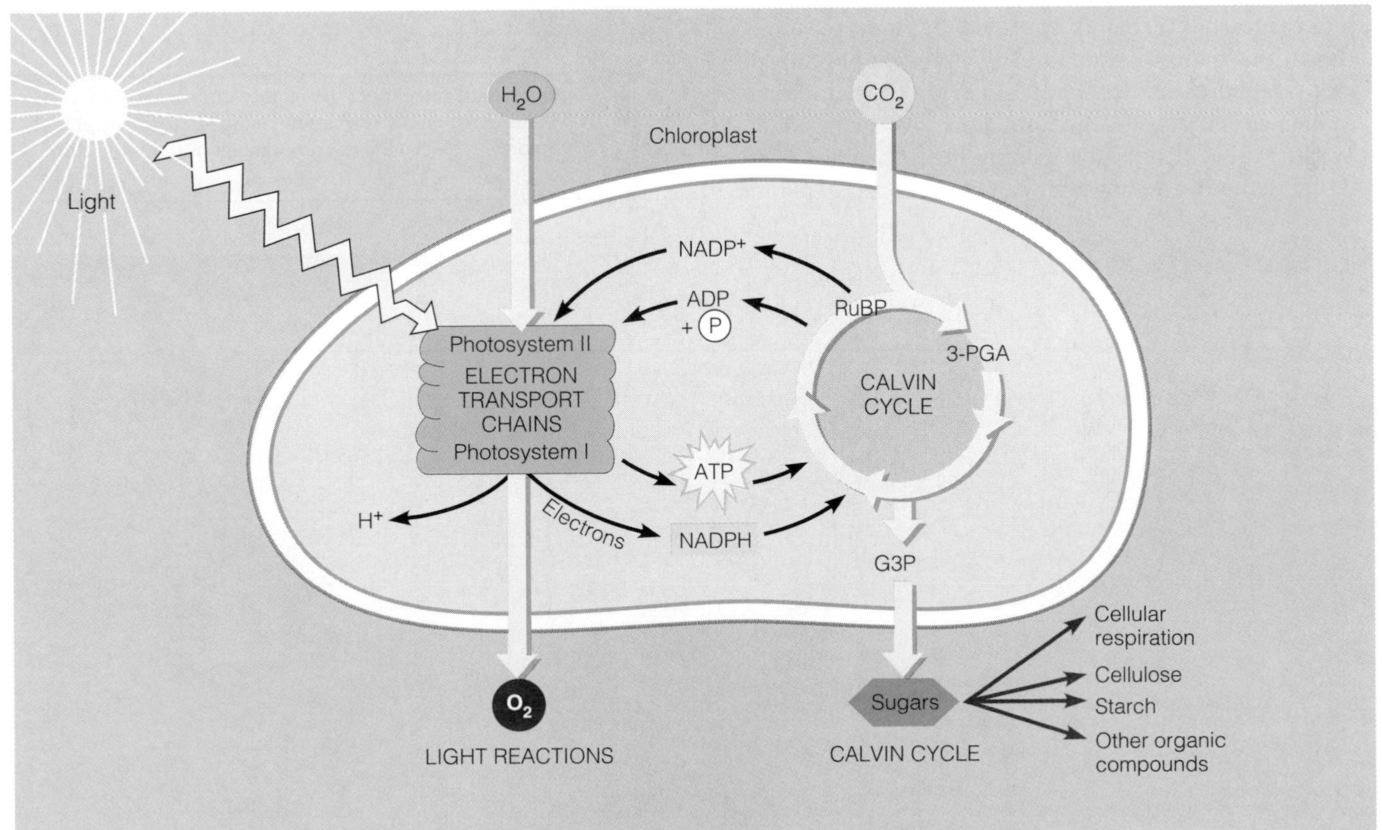

Figure 7.11 A summary of the chemical processes of photosynthesis

7.12 C_4 and CAM plants have special adaptations that save water

Plants are the main food-producing organisms on land, and they thrive in diverse environments, on all continents except Antarctica. One of the reasons for their great success has been the evolution of different ways of fixing CO_2 and saving water during photosynthesis.

Plants in which the Calvin cycle uses CO_2 directly from the air are called **C_3 plants** because the first organic compound produced is the three-carbon compound 3-PGA (step 1 in Figure 7.10B). C_3 plants are common and widely distributed, and some of them, such as soybeans, oats, wheat, and rice, are important in agriculture. One of the problems that farmers face in growing C_3 plants, however, is that dry weather can reduce the rate of photosynthesis and decrease crop productivity. On a hot, dry day, a C_3 plant closes its stomata, the pores in the leaf surface.

Closing stomata is an adaptation that reduces water loss, but it also prevents CO_2 from entering the leaf and O_2 from leaving. As a result, CO_2 levels can get very low in the leaf, while O_2 from the light reactions builds up. When this happens, the first enzyme of the Calvin cycle (called rubisco) incorporates O_2 instead of CO_2 (Figure 7.12A), and the Calvin cycle produces a two-carbon (2-C) compound instead of its usual three-carbon product. The plant cell then breaks the two-carbon compound down to CO_2 and H_2O. The entire process, starting with the fixation of O_2, is called **photorespiration.** Unlike photosynthesis, photorespiration yields no sugar molecules. Unlike cellular respiration, it produces no ATP.

In contrast to C_3 plants, so-called **C_4 plants** have special adaptations that save water and also prevent photorespiration. When the weather is hot and dry, a C_4 plant keeps its stomata closed most of the time, thus conserving water. At the same time, it continues making sugars by photosynthesis, using the route shown in Figure 7.12B. A C_4 plant has an enzyme that fixes carbon into a four-carbon (4-C) compound instead of into 3-PGA. This enzyme, unlike the carbon-fixing enzyme in C_3 plants, cannot switch over to fixing O_2. As a result, the C_4 enzyme can continue to fix carbon even when the CO_2 concentration in the leaf is much lower than the O_2 concentration. The four-carbon compound acts as a carbon shuttle; it donates the CO_2 to the Calvin cycle in a nearby cell, which therefore keeps on making sugars even though the plant's stomata are closed most of the time. Corn, sorghum, and sugarcane are examples of agriculturally important C_4 plants. All three evolved in the tropics, and their method of carbon fixation is advantageous in hot, dry climates.

A third mode of carbon fixation and water conservation has evolved in pineapples, many cacti, and most of the so-called succulent plants (those with very juicy tissues), such as ice plants and jade plants. Collectively called **CAM plants,** most such species are adapted to very dry climates. A CAM plant (Figure 7.12C) conserves water by opening its stomata and admitting CO_2 only at night. When CO_2 enters the leaves it is fixed into a four-carbon compound, as in C_4 plants. The four-carbon compound in a CAM plant banks CO_2 at night and releases it to the Calvin cycle during the day. This keeps photosynthesis operating during the day, even though the leaf admits no more CO_2. CAM stands for crassulacean acid metabolism, after the plant family Crassulaceae (jade plants and others), in which this important water-saving adaptation was first discovered.

How would you expect the relative abundance of C_3 versus C_4 and CAM species to change in a geographic region whose climate becomes much hotter and drier?

C_4 and CAM species would replace many of the C_3 species.

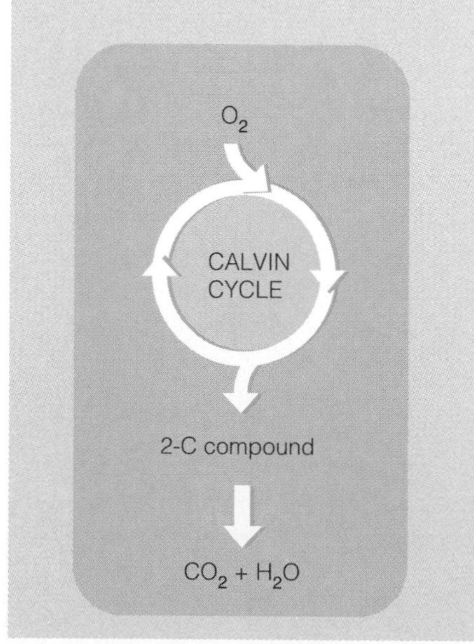

Figure 7.12A Photorespiration in a C_3 plant

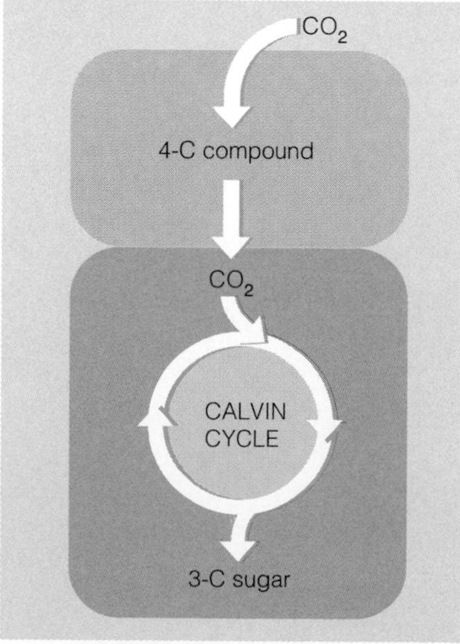

Figure 7.12B Carbon fixation in a C_4 plant

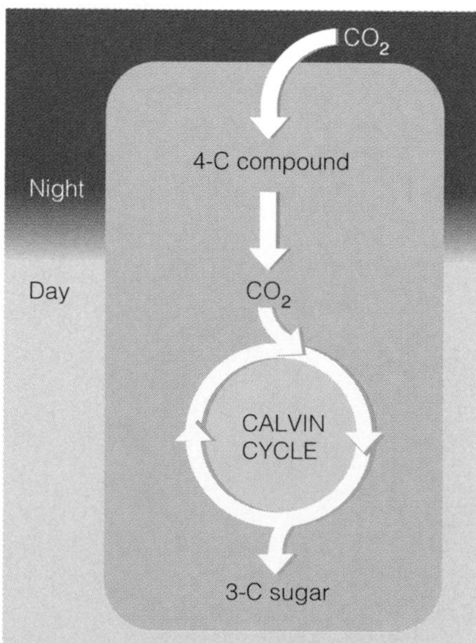

Figure 7.12C Carbon fixation in a CAM plant

Photosynthesis, Solar Radiation, and Earth's Atmosphere

7.13 Photosynthesis moderates the greenhouse effect; deforestation can intensify it

The trees in the photograph (Figure 7.13A) form part of an **old-growth forest** in the Pacific Northwest. Such a forest is an ancient one that has never been seriously disturbed by humans. Many of the trees—mainly Douglas fir here—are thousands of years old. The forest they dominate is one of a few remaining undisturbed areas that contain harvestable timber in the United States.

Old-growth forests are the focus of a long-standing controversy between the timber industry and conservationists. Trees like these contain a lot of marketable lumber, and economic arguments can be made for harvesting them. On the other side of the controversy, conservationists argue that old-growth forests are home to many species that can survive nowhere else, and that we should save these remnants of our ancient forests for future generations.

The photosynthesis carried out by old-growth forests has direct bearing on the controversy. At center stage is CO_2, the gas that plants use to make sugars in photosynthesis and that all organisms give off as waste from cellular respiration.

Carbon dioxide normally makes up about 0.03% of the air we breathe. This amount of CO_2 in the atmosphere provides plants with plenty of carbon. It also helps moderate world climates, because CO_2 retains heat from the sun that would otherwise radiate from Earth back into space. Warming induced by CO_2 is called the **greenhouse effect** because atmospheric CO_2 traps heat and warms the air just as clear glass does in a greenhouse (Figure 7.13B). The greenhouse effect keeps the average temperature on Earth some 10°C warmer than it would be otherwise.

Ironically, planet Earth may now be overheating from the greenhouse effect. The amount of atmospheric CO_2 has been on the rise in the past century, mainly because of worldwide industrialization and the increased use of oil, gas, coal, and wood as fuels. When these substances are burned, the carbon in them is released as CO_2. The steady increase in atmospheric CO_2 seems to be contributing to a rise in global temperatures. Photosynthesis consumes CO_2, tending to counteract the greenhouse effect, but the global rate of photosynthesis may decline as we clear huge tracts of forest for farming and urban expansion. Extensive deforestation continues, especially in the tropics, the northwestern and southeastern United States, Canada, and Siberia.

Figure 7.13A An old-growth forest in the Pacific Northwest

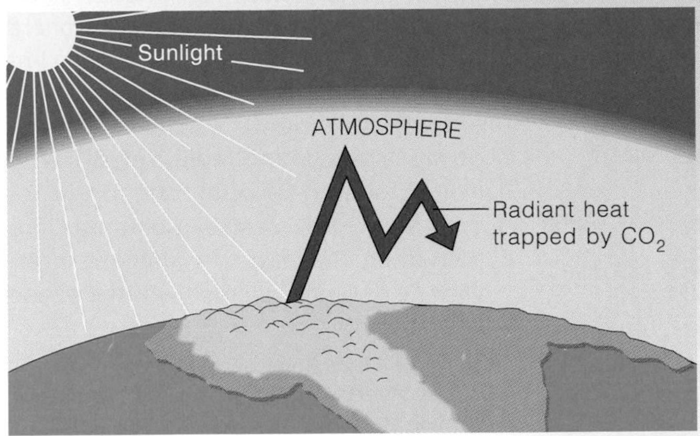

Figure 7.13B The greenhouse effect of CO_2 in the atmosphere

The greenhouse effect has come up on both sides of the old-growth forest controversy. Those who favor saving the old-growth trees argue that these large photosynthesizers remove a lot of potentially harmful CO_2 from the atmosphere. Those favoring harvesting argue that replacing the old trees with seedlings would actually increase photosynthesis and reduce atmospheric CO_2.

Relative to their size and weight, young, rapidly growing trees do, in fact, take up CO_2 at a faster rate than old ones. However, when old trees are harvested, much less than half their bulk becomes lumber. All the roots and many branches are left behind to decompose. Most of the wood itself is turned into paper, sawdust, or fuel—products that usually decompose or are burned within a few years. Decomposition and burning turn the carbon compounds that were in the tree into CO_2, and this puts much more CO_2 into the atmosphere than young trees can take up.

In the next module, we'll hear more about harmful effects of human activities on Earth's atmosphere.

Explain the greenhouse effect.

Atmospheric carbon dioxide allows sunlight through, which heats Earth's surface, but slows the escape of the heat back into space.

TALKING ABOUT SCIENCE

7.14 Mario Molina talks about Earth's protective ozone layer

As photosynthesis reduces the CO_2 in Earth's atmosphere, it also produces the O_2 on which plants, animals, and most other organisms depend for cellular respiration. This O_2 benefits organisms in another way: High in the atmosphere it is converted to ozone, a form of oxygen that plays an important protective role for life on Earth. Among the scientists who study the ozone layer is Mario Molina of MIT, who in 1995 shared a Nobel Prize for his research on how certain pollutants are damaging that layer. In a recent interview, Dr. Molina explained why the ozone layer is important.

Figure 7.14 Mario Molina

> The ozone layer shields the Earth's surface from powerful ultraviolet radiation that comes from the sun. This UV radiation is harmful to organisms, including humans. For example, UV radiation causes sunburn, and skin cancer can be a cumulative result of exposure. There is also evidence that UV can damage crops. Certain developing animals, such as the larvae of fish, seem to be particularly sensitive.

He described how the ozone forms and how it is destroyed.

> The ozone forms when high-energy solar radiation breaks apart O_2 molecules and frees oxygen atoms. These then react with unbroken O_2 molecules. The result is ozone, which has *three* oxygen atoms (O_3). So ozone is continuously forming by the action of sunlight on the atmosphere. This is balanced by continuous destruction of the ozone molecules when they react with other chemical compounds that are naturally present in the atmosphere. Humans have disrupted that balance by releasing certain industrial chemicals that hasten this destruction.

The research that won Dr. Molina the Nobel Prize dealt with the destruction of ozone by one particular class of chemicals, called chlorofluorocarbons, or CFCs. In 1974, when Dr. Molina and his colleagues first published their CFC-ozone depletion theory, CFCs were used in large amounts as refrigerants, as propellants in spray cans, as solvents, and in the process for making plastic foams.

> We predicted that the continuous release of CFCs would damage the protective ozone layer. CFCs are very stable compounds, and this stability allows them to make it up to the ozone layer, which is about 15 miles above the surface of the Earth. There solar radiation converts them to very reactive chemicals called free radicals, which then destroy the ozone.

How did scientists and the general public react to the alarming possibility that the ozone layer was thinning?

> At first, there was very little reaction because not many people were aware of the importance of invisible things like the ozone layer and UV radiation. Experts in our field quickly realized that the prediction of the CFC-ozone depletion theory was something to worry about, but many other scientists were skeptical, which is natural for scientists. Then as we and others began doing experiments to test the idea, the evidence became strong, and more and more people, including politicians, became concerned about ozone depletion. There was also growing awareness about the role of UV radiation in skin cancer and other biological damage. Then, in 1985, scientists documented a drastic depletion of the ozone layer over Antarctica—an ozone hole. This caused the United Nations to act in 1987.
>
> [Eventually, international agreement was reached] to require a complete phasing out of CFC production by developed countries by 1995. So that's already happened. These countries now use other refrigerants, which are destroyed in the atmosphere before they reach the ozone layer. There is a grace period for developing countries ... [but] CFCs will be phased out everywhere within the next decade. Because CFCs are so stable, however, we predict that the ozone layer won't recover until the middle of the twenty-first century.

We asked Dr. Molina what he has learned since 1974 about the interface between science and politics.

> One lesson is that science is not always something that politicians care very much about. Often, supporting science is considered a luxury, though actually it is a good investment. Another challenge is that many issues related to science and technology are long-term issues that require patience and long-term commitment. [In addition to the CFC problem,] an example is our excessive production of carbon dioxide and the related possibility of global warming.

Whether an environmental pollution problem involves CFCs or carbon dioxide emissions from the burning of fossil fuels, the solutions are often expensive and technologically sophisticated. Are strict international standards unfair to developing nations that are trying to improve their standard of living?

> That's an issue, but there is another important element that I've observed from my own experience. When I was going to the university in Mexico, no one cared about pollution. Dealing with pollution seemed like a luxury. Today, everybody in Mexico City cares about pollution, so solving this problem no longer seems like something that industrialized countries want to impose upon them. The costs of pollution are enormous in Mexico City, where I was born. If only there had been insight about what would happen to the city, pollution could have been reduced at a much lower cost. It's much more difficult to repair the environment now, but I think there is a strong incentive to do that in developing countries. It's not easy because of the complications of social and economic issues. Science and technology are very important components, but not the only ones, so scientists will have to be humble and realize their limits.

The connections between science, technology, and society, so clearly exemplified by the work of Mario Molina, are a major theme of this book. This theme will come up again in the next unit, on cellular reproduction and genetics. 

? Where does the ozone layer in Earth's atmosphere come from?

Photosynthesis puts oxygen into the atmosphere in the form of O_2. High in the atmosphere, radiation from the sun converts some of the O_2 to ozone, O_3.

Chapter Summary

An Overview of Photosynthesis (Introduction–7.5) Photosynthesis is the process by which autotrophic organisms use light energy to make sugar and oxygen gas from carbon dioxide and water. The following chemical equation shows the net input and output of photosynthesis:

$$6\,CO_2 + 6\,H_2O + \text{Light energy} \longrightarrow C_6H_{12}O_6 + 6\,O_2$$

Plants, some protists (algae), and some bacteria are photosynthetic autotrophs, the ultimate producers of food consumed by virtually all organisms **(7.1)**. In most plants, photosynthesis occurs primarily in the leaves, in chloroplasts. A chloroplast contains stroma, a fluid, and grana, stacks of thylakoids. The thylakoid membranes contain chlorophyll, the green pigment that captures light for photosynthesis **(7.2)**.

The O_2 liberated by photosynthesis is made from the oxygen atoms in H_2O **(7.3)**. Like cellular respiration, photosynthesis is a redox process. Water molecules are split apart, and electrons and H^+ ions are removed (oxidation), leaving O_2 gas. These electrons and H^+ are transferred to CO_2 (reduction), producing sugar. Unlike cellular respiration, which releases energy, photosynthesis stores energy **(7.4)**. The complete process of photosynthesis consists of two linked sets of reactions: the light reactions and the Calvin cycle. The light reactions convert light energy to chemical energy and produce O_2. The Calvin cycle assembles sugar molecules from CO_2 using the energy-carrying products of the light reactions **(7.5)**.

The Light Reactions: Converting Solar Energy to Chemical Energy (7.6–7.9) Certain wavelengths of visible light drive the light reactions of photosynthesis **(7.6)**. Each of the many light-harvesting units of the chloroplast's thylakoid membranes, called photosystems, consists of an "antenna" of chlorophyll and other pigment molecules that absorb light and a primary electron acceptor that receives excited electrons from the reaction-center chlorophyll **(7.7)**. Two connected photosystems (I and II) collect photons of light and transfer the energy to chlorophyll electrons. The excited electrons are passed from the primary electron acceptor to electron transport chains. Electrons shuttle from photosystem II to photosystem I, and their energy ends up in ATP and NADPH. Photosystem II regains electrons by splitting water, leaving O_2 gas as a by-product **(7.8)**.

The ATP is synthesized by chemiosmosis. The electron transport chains are arranged with the photosystems in the thylakoid membranes and pump H^+ through that membrane. The flow of H^+ back through the membrane is harnessed by ATP synthase to make ATP (photophosphorylation). In the stroma, the H^+ ions, electrons from the electron transport chain, and $NADP^+$ combine to form NADPH **(7.9)**.

The Calvin Cycle: Converting CO_2 to Sugars (7.10) The NADPH and ATP produced in the light reactions are used in the Calvin cycle, which occurs in the chloroplast's stroma. This is where carbon fixation takes place and sugar is manufactured. Using carbon from atmospheric CO_2, high-energy electrons and H^+ from NADPH, and energy from ATP, enzymes of the Calvin cycle construct G3P, an energy-rich sugar. In turn, G3P is used to build glucose and other organic molecules.

Photosynthesis Reviewed and Extended (7.11–7.12) Figure 7.11 reviews photosynthesis. Many plants make more sugar than they need; the excess is stored in roots, tubers, and fruits, and is a major food source for animals. Photosynthesis produces billions of tons of organic matter each year and thus sustains almost all life on Earth **(7.11)**. Most plants are C_3 plants, which take carbon directly from CO_2 in the air and use it in the Calvin cycle to build a three-carbon organic molecule. In such plants, stomata in the leaf surface usually close when the weather is hot and dry. Although this mechanism saves water, it can cause a drop in CO_2 and a buildup of O_2 in the leaf, diverting the Calvin cycle to an inefficient process called photorespiration. Some plants have special adaptations that enable them to save water and avoid photorespiration. Special cells in C_4 plants, such as corn and sugarcane, incorporate CO_2 into a four-carbon compound that can donate CO_2 to the Calvin cycle in another kind of cell, compensating for the shortage of CO_2 when the stomata are closed. Pineapples, many cacti, and most succulents—the CAM plants—employ a different mechanism. They open their stomata at night and make a four-carbon compound that is used as a CO_2 source by the same cell during the day **(7.12)**.

Photosynthesis, Solar Radiation, and Earth's Atmosphere (7.13–7.14) Because of the increased burning of fossil fuels, atmospheric CO_2 is increasing. CO_2 warms Earth's surface by trapping heat in the atmosphere; this is called the greenhouse effect. An excess of CO_2 may cause the planet to overheat. Because photosynthesis removes CO_2 from the atmosphere, it could moderate the greenhouse effect; unfortunately, deforestation may cause a decline in the global rate of photosynthesis **(7.13)**. The O_2 in the atmosphere results from photosynthesis. Solar radiation converts O_2 high in the atmosphere to ozone (O_3), which shields organisms on Earth's surface from the damaging effects of UV radiation. Industrial chemicals called CFCs have hastened ozone breakdown, causing dangerous thinning of the ozone layer, but international restrictions on these chemicals are allowing recovery **(7.14)**.

Testing Your Knowledge

Multiple Choice

1. The process of photosynthesis consumes ____ and produces ____.
   a. chlorophyll . . . H_2O
   b. H_2O . . . CO_2
   c. CO_2 . . . chlorophyll
   d. H_2O . . . O_2
   e. glucose . . . O_2

2. Which of the following are produced by reactions that take place in the thylakoids and consumed by reactions in the stroma?
   a. CO_2 and H_2O
   b. $NADP^+$ and ADP
   c. ATP and NADPH
   d. glucose and O_2
   e. CO_2 and ATP

3. In photosynthesis, ____ is oxidized and ____ is reduced.
   a. glucose . . . oxygen
   b. carbon dioxide . . . water
   c. water . . . carbon dioxide
   d. glucose . . . carbon dioxide
   e. water . . . oxygen

4. Why is it difficult for most plants to carry out photosynthesis in very hot, dry environments such as deserts?
   a. The light is too intense and overpowers pigment molecules.
   b. The closing of stomata keeps CO_2 from entering and O_2 from leaving the plant.
   c. They are forced to rely on photorespiration to make ATP.
   d. The greenhouse effect is intensified in a desert environment.
   e. CO_2 builds up in the leaves, blocking carbon fixation.

5. When light strikes chlorophyll molecules, they lose electrons, which are ultimately replaced by

a. splitting water.
b. breaking down ATP.
c. removing them from NADPH.
d. fixing carbon.
e. oxidizing glucose.

6. What is the role of $NADP^+$ in photosynthesis?

a. It assists chlorophyll in capturing light.
b. It acts as the primary electron acceptor for the photosystems.
c. As part of the electron transport chain, it manufactures ATP.
d. It assists photosystem II in the splitting of water.
e. It carries electrons to the Calvin cycle.

7. The reactions of the Calvin cycle are not directly dependent on light, but they usually do not occur at night. Why? *(Explain your answer.)*

a. It is often too cold at night for these reactions to take place.
b. Carbon dioxide concentrations decrease at night.
c. The Calvin cycle depends on products of the light reactions.
d. Plants usually open their stomata at night.
e. At night, plants cannot produce the water needed for the Calvin cycle.

Describing, Comparing, and Explaining

1. Create a diagram identifying the two major stages of photosynthesis. Show with arrows the major inputs and outputs of each process, including the products of one process that are consumed in the other.

2. Which take up CO_2 faster (per unit of weight), young, rapidly growing trees or old ones? What effect does harvesting old trees and replacing them with seedlings have on the greenhouse effect? Why?

3. Compare the electron transport chain in the thylakoid of a chloroplast with the electron transport chain in a mitochondrion: Where do the electrons come from? Where do the electrons get their energy? What picks up the electrons at the end of the chain? How is the energy given up by the electrons used?

4. What do plants do with the sugar they produce in photosynthesis?

Thinking as a Scientist

1. Do the oxygen atoms in the glucose produced by photosynthesis come from water or carbon dioxide? If you did not already know the answer, how could you use a radioactive isotope to find out? If the oxygen comes from water, how would your experiment turn out? How would it turn out if the oxygen comes from carbon dioxide?

2. A group of students are investigating the effect of light intensity on photosynthesis by the aquatic plant *Elodea*. The plant produces a stream of O_2 bubbles as it photosynthesizes. The students vary the light intensity by moving a sunlamp closer and closer to the aquarium in which the plant is growing. Leaving the room lights on and the window shades up, they measure the rate of O_2 gas production by the plant at five different lamp distances. They predict that the rate of gas production would increase as the sunlamp gets closer. The data recorded are shown below. (The numbers below the data points indicate the order in which the measurements are made.)

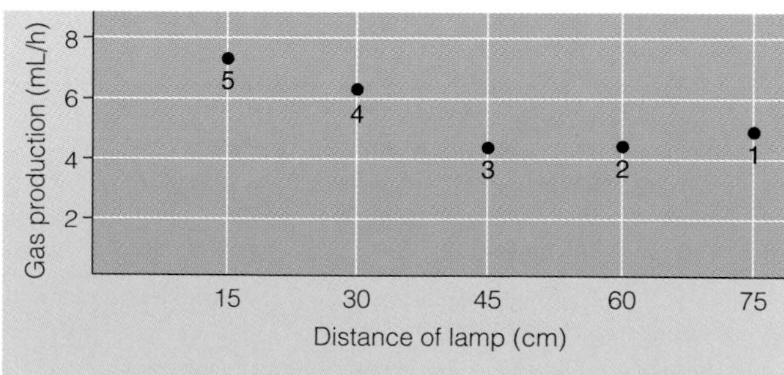

The students are puzzled because their data do not show an increase in photosynthesis as the lamp is moved from 75 cm to 45 cm from the aquarium. One student, who has just had a physics course, calculates that the light intensity increases 278% as the lamp is moved through that distance. Other students propose the following explanations for the results. Which is most convincing to you and why?

a. There is so little light at distances greater than 45 cm that the plant cannot photosynthesize at all.
b. *Elodea* photosynthesizes better in dim light and is inhibited by brighter light.
c. At close positions, the light from the lamp is so intense that the plant's photosynthetic machinery is "saturated"; that is, it reaches its maximum output.
d. The plant is using the room lights and light from the window to photosynthesize.

3. Tropical rain forests cover only about 3% of Earth's surface, but they are estimated to be responsible for more than 20% of global photosynthesis. For this reason, rain forests are often referred to as the "lungs" of the planet, providing O_2 for life all over the planet. However, most experts believe that rain forests make little or no *net* contribution to global oxygen production. From your knowledge of photosynthesis and cellular respiration, can you explain why they might think this? (What happens to the food produced by a rain forest tree when it is eaten by animals or when the tree dies?)

Science, Technology, and Society

Many scientists believe that Earth is getting warmer, and the cause could be an intensified greenhouse effect resulting from increased CO_2 from industry, vehicles, and the burning of forests. The year 1990 was the warmest in nearly two centuries. Seven of the ten warmest years in the last century have occurred since 1980. Global warming could influence agriculture and perhaps even melt polar ice and flood coastal regions. Most experts now agree that global warming is occurring and, in response to its potential threat, several European countries, the United States, Australia, and New Zealand have made a commitment to reduce carbon dioxide emissions significantly by the year 2000. However, some countries oppose taking strong actions at this time. Several reasons are cited: First, some experts think the apparent warming trend may be just a random fluctuation in temperature. Second, if the temperature increase is real, it has yet to be proven that it is caused by increased CO_2. Some people also believe it would be difficult to cut CO_2 emissions without sacrificing comfort, convenience, and economic growth. Do you think we should have more evidence that greenhouse warming is real before taking action? Or is it better to play it safe and act now to reduce CO_2 emissions? What are the possible costs and benefits of following each of these two strategies? Which do you favor, and why?

Media Resources

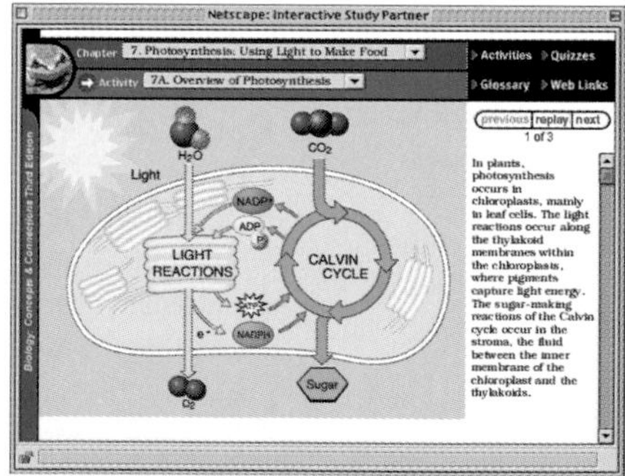

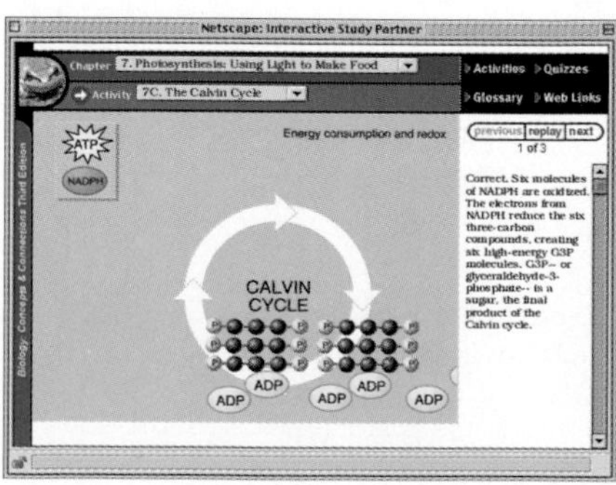

For further review of the concepts in this chapter, use the activities and quizzes on the CD-ROM.

Visit the Special Edition of The Biology Place web site for activities, quizzes, news articles, and web links.

UNIT II

Cellular Reproduction and Genetics

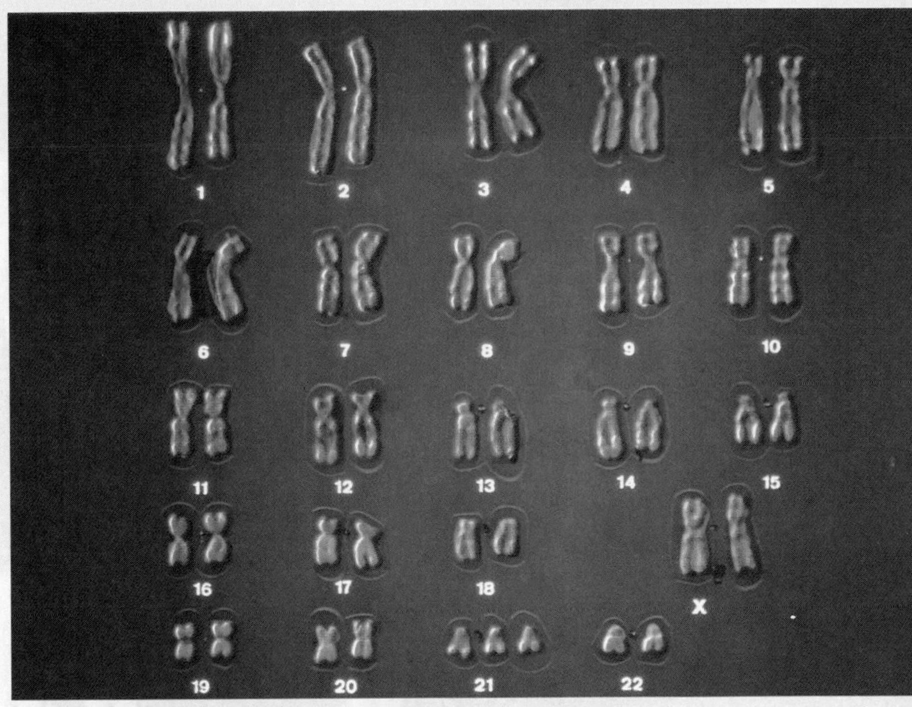

8 The Cellular Basis of Reproduction and Inheritance

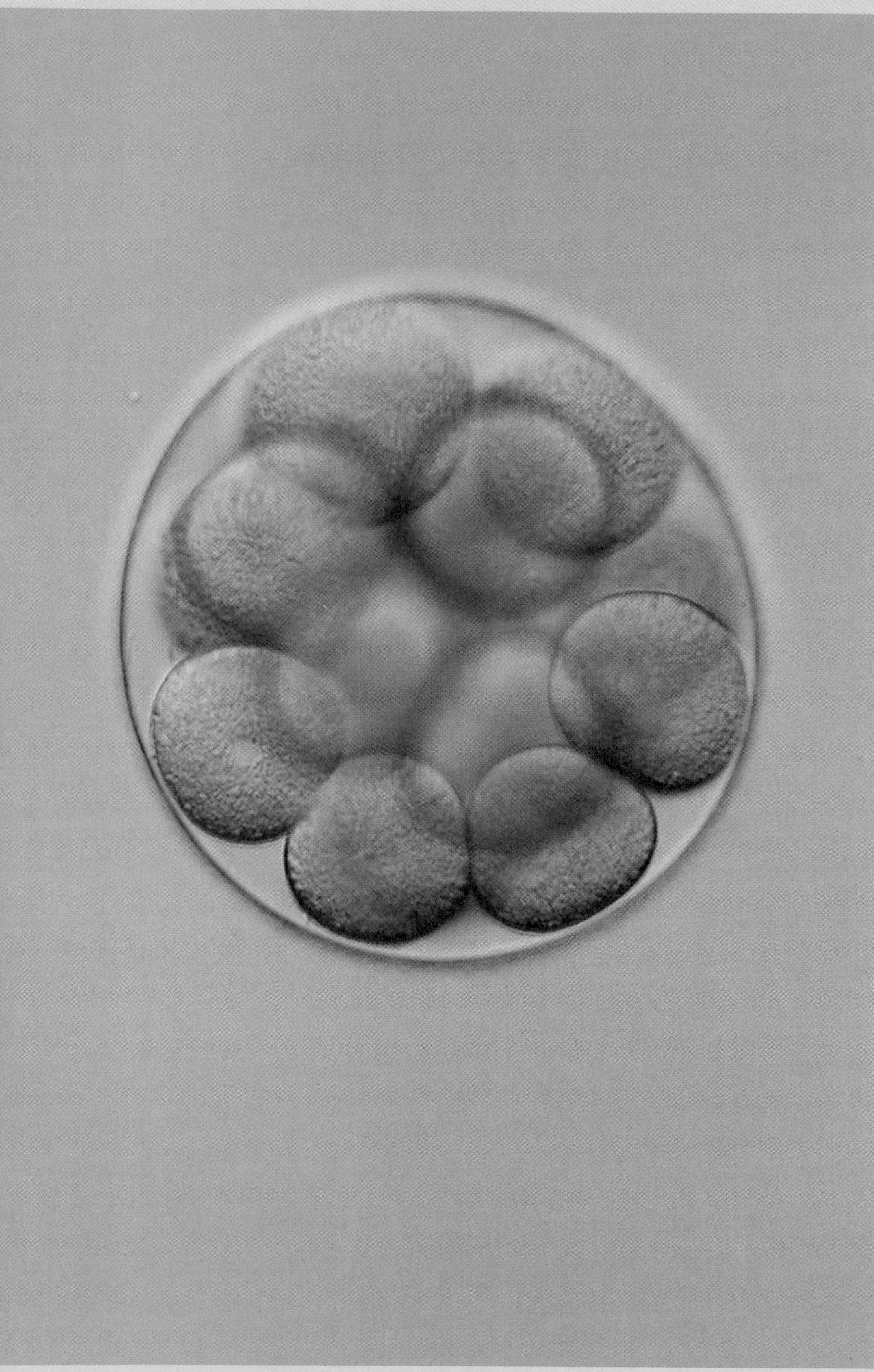

Outline

What would you say if you were asked to describe the object in the photograph at the left? Early microscopists called this a morula, from the Latin word for mulberry, and the name stuck. In fact, the photo has nothing to do with berries or even plants. What you are seeing is a sea-star embryo consisting of 32 cells. Sea stars belong to a group of marine animals called echinoderms. Most animals, humans included, have a morula stage early in their embryonic development.

A morula looks static in a photograph, but in real life it is a dynamic stage in animal development, part of a process that ensures the continuation of life from generation to generation. For a sea star, the process begins when an adult female casts thousands or millions of eggs into the ocean and they are joined by sperm discharged by a nearby male. The chance meeting of one sperm cell and one egg cell results in a fertilized egg. Within about an hour, the fertilized egg divides in two. The resulting cells stay together and divide again, their four descendants divide in turn, and so on. A morula is a cluster of 16–64 cells. Every one of its cells is about to divide again. As development continues, the ball of cells will transform into a juvenile form called a larva and later into an adult.

Development, from a fertilized egg to a new adult organism, is one phase of a multicellular organism's **life cycle,** the sequence of life stages leading from the adults of one generation to the adults of the next. The other phase of the life cycle is reproduction, the formation of new individuals from preexisting ones; for many animals, a new individual first appears as a fertilized egg. The reproductive phase of the life cycle entails the creation of offspring carrying genetic information, in the form of DNA, from their parents. The reproductive process that involves the union of a sperm and an egg is called **sexual reproduction.** Sperm and egg each carry one set of genetic information—one copy of the organism's **genome.** Thus, the offspring of sexual reproduction inherit traits from two parents.

Asexual reproduction is the production of offspring by a single parent, without the participation of sperm and egg. Offspring produced by asexual reproduction inherit DNA from only one parent. The life cycles of many organisms involve asexual reproduction, either instead of sexual reproduction or in addition to it. For example, some sea stars can reproduce asexually by dividing into two pieces that regrow into two whole new individuals. This remarkable process, called regeneration, is also used to regrow body parts. The sea star in the photograph here is regenerating an arm it lost, probably in an encounter with a predator.

A sea star's life cycle illustrates the central theme of this chapter: Cell division is at the heart of organismal reproduction. Regeneration of new sea stars from a fragmented parent results from repeated cell divisions, as does the more common development of a multicellular larva from a fertilized egg. And eggs and sperm themselves result from cell division of a special kind. In this chapter, we discuss the two main types of cell division and their functions in the organism.

A sea star regenerating an arm

Connections Between Cell Division and Reproduction

8.1 Like begets like, more or less

Only sea stars produce more sea stars, only people make more people, and only maple trees produce more maple trees. These simple facts of life have been recognized for thousands of years and are summarized by the age-old saying "Like begets like."

In a strict sense, "Like begets like" applies only to asexual reproduction, such as sea star regeneration. In this case, because offspring inherit all their DNA from a single parent, they are exact (or virtually exact) genetic replicas of that one parent and of each other.

Single-celled organisms, such as *Amoeba,* also reproduce asexually. The amoeba in Figure 8.1A is reproducing by dividing in half. The amoeba's **chromosomes,** the structures that contain most of the organism's DNA, have been duplicated, and identical chromosomes have been allocated to opposite sides of the parent cell. When the parent cell divides, the two daughter amoebas that result will be genetically identical to each other and to the original parent. (Biologists traditionally use the word "daughter" in this context only to indicate offspring, not to imply gender.)

Figure 8.1B The Hoffman family: parents Lisa and Dustin (back row) and their nonidentical offspring, produced by sexual reproduction

The photograph of the family in Figure 8.1B makes the point that, in a *sexually* reproducing species, like does not precisely beget like. Offspring produced by sexual reproduction generally resemble their parents more closely than they resemble unrelated individuals of the same species, but they are not identical to their parents or to each other. Each offspring inherits a unique combination of genes from its two parents, and this one-and-only set of genes programs a unique combination of traits. As a result, sexual reproduction can produce great variation among offspring. Notice in the photograph that, despite the family resemblances, each member of the Hoffman family has a unique appearance. You will find the same sort of similarities and differences in pictures of your relatives.

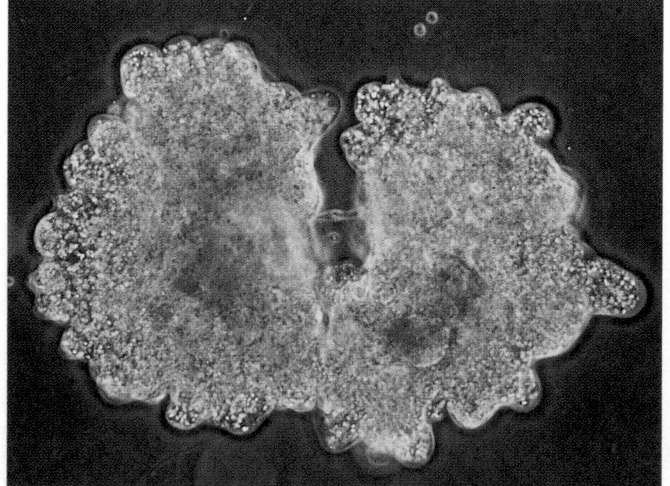

Figure 8.1A An amoeba producing genetically identical offspring

Long before anyone knew about genes, chromosomes, or the underlying principles of inheritance, people recognized that individuals of sexually reproducing species are highly varied. Furthermore, people learned to develop domestic breeds of plants and animals by controlling sexual reproduction. A domestic breed displays particular traits from among the great variety of traits found in the species as a whole. For example, though all domestic dogs belong to a single species, each breed of dogs exhibits much less variability than the species as a whole. In producing each kind of domestic dog, breeders selected from a varied population of dogs certain individuals with specific traits and allowed these individuals to mate and produce offspring. The ancestry of a dog breed such as the German shepherd can be traced back for many generations, in which breeders reduced variability in the breed by rigidly selecting only those dogs with specific (German shepherd) traits for mating. In a sense, selective breeding is an attempt to make like beget like more than it does in nature. 

? In terms of the "like begets like" adage, contrast asexual with sexual reproduction.

Asexual reproduction produces genetically identical offspring that inherit all their DNA from a single parent. The offspring of sexual reproduction show a family resemblance, but siblings vary because they inherit different combinations of genes from the two parents.

Cells arise only from preexisting cells 8.2

In 1858, German physician Rudolf Virchow stated an important biological principle: All cells come from cells. Virchow's principle is illustrated graphically by a dividing amoeba, a regenerating sea star, and any fertilized egg that has begun to divide.

Like many important ideas that we now take for granted, Virchow's principle is both simple and profound. It tells us that the perpetuation of life, including all aspects of reproduction and inheritance, is based on the reproduction of cells. We commonly refer to cellular reproduction as **cell division.**

Cell division plays two main roles in perpetuating the life cycle of animals and other multicellular organisms. First, as we mentioned earlier, cell division makes it possible for a fertilized egg to develop through various embryonic stages, and for an embryo to develop into an adult organism. Second, cell division ensures the continuity of life from generation to generation; it is the basis of both asexual reproduction and the formation of sperm and eggs in sexual reproduction.

So far, we have discussed only the division of eukaryotic cells, and we will emphasize them in this chapter. Prokaryotes, however, also illustrate Virchow's principle, and in the next module, we take a brief look at prokaryotic cell division.

? Starting with a fertilized egg cell, a series of five cell divisions would produce an early embryo with how many cells?

32 cells

Prokaryotes reproduce by binary fission 8.3

Prokaryotes (bacteria and archaea) reproduce by a type of cell division called **binary fission,** meaning "dividing in half." In prokaryotes, most genes are carried on a circular DNA molecule that, with associated proteins, constitutes the organism's single chromosome. Prokaryotic chromosomes are much less complicated than those of eukaryotes, though both types consist of DNA and proteins. Prokaryotic cells are generally smaller and simpler than eukaryotic cells. Nonetheless, replicating their DNA in an orderly fashion and distributing the copies equally to two daughter bacteria is a formidable task. Consider, for example, that, when stretched out, the chromosome of the bacterium *Escherichia coli* is some 500 times longer than the cell itself. Accurately duplicating this molecule when it is coiled and packed inside the cell is no small achievement.

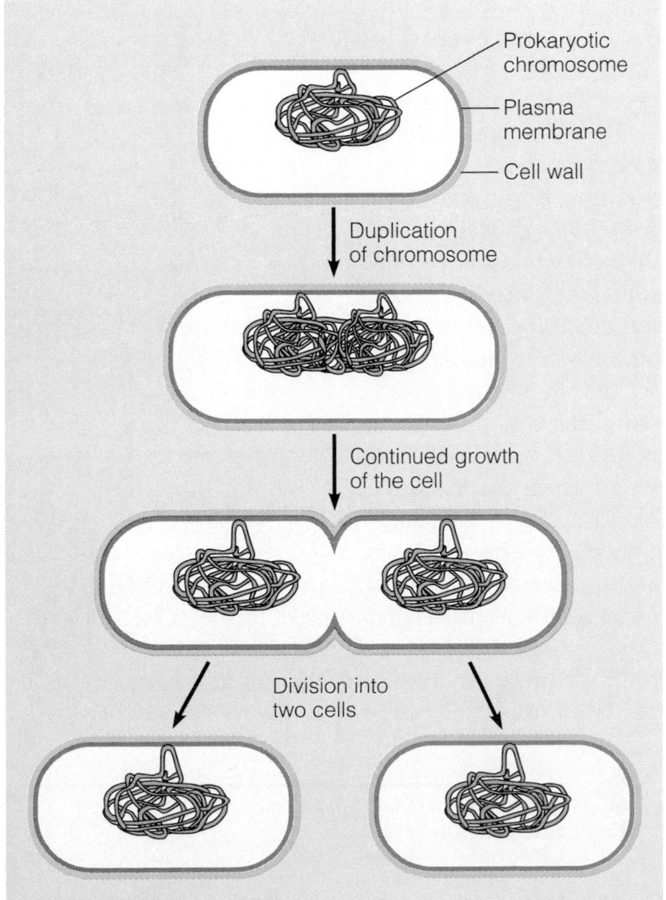

Figure 8.3A Binary fission of a prokaryotic cell

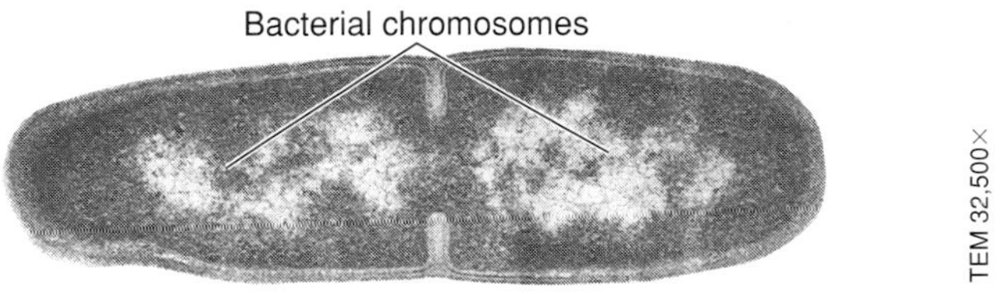

Bacterial chromosomes

TEM 32,500×

Figure 8.3B Electron micrograph of a dividing bacterium

Figure 8.3A illustrates how a prokaryotic cell allocates DNA to its daughter cells. When the DNA in the cell is replicating (top), it is attached to the plasma membrane. Continued growth of the cell gradually separates the daughter chromosomes, which remain attached to the membrane at separate points. Eventually, the plasma membrane and cell wall grow inward, dividing the cell in two. Figure 8.3B is an electron micrograph of a dividing bacterium, at a stage similar to the third step in Figure 8.3A.

? Why is binary fission classified as asexual reproduction?

Because the genetically identical offspring inherit their DNA from a single parent

The Eukaryotic Cell Cycle and Mitosis

8.4 The large, complex chromosomes of eukaryotes duplicate with each cell division

Eukaryotic cells are more complex and generally much larger than prokaryotic cells, and they have many more genes. Human cells, for example, carry about 100,000 genes, versus about 3000 for a typical bacterium. Almost all the genes in the cells of humans, and in all other eukaryotes, are found in the cell nucleus, grouped into multiple chromosomes. (The exceptions include genes on the small DNA molecules of mitochondria.) Chromosomes get their name (Greek *chroma,* colored, and *soma,* body) from their attraction for certain stains used in microscopy. In Figure 8.4A, which is a micrograph of a plant cell (of the African blood lily) that is about to divide, the chromosomes are stained dark purple. Each dark thread is an individual chromosome. Chromosomes are clearly visible under the light microscope as individual structures like these only when a cell is in the process of dividing. The rest of the time, the chromosomes exist as a diffuse mass of very long, very thin fibers. This material, called **chromatin,** is a combination of DNA and protein molecules. As a cell prepares to divide, its chromatin coils up, forming compact, distinct chromosomes.

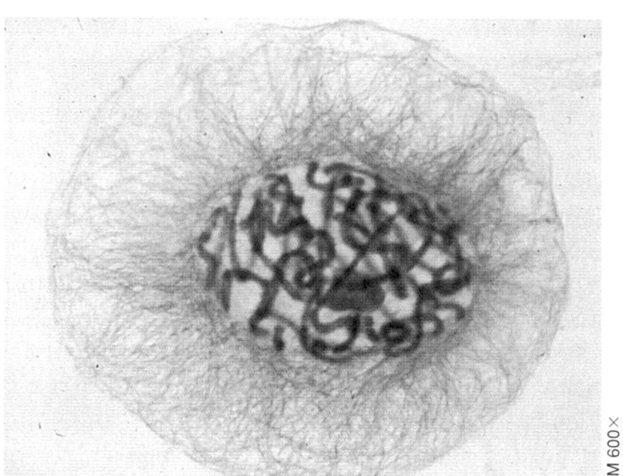

Figure 8.4A A plant cell just before division

Like a prokaryotic chromosome, each eukaryotic chromosome contains one long DNA molecule bearing thousands of genes. However, the eukaryotic chromosome has a much more complex structure than the prokaryotic chromosome. This structure includes many more protein molecules, which help organize the chromatin of the chromosome and control the activity of its genes. The number of chromosomes in a eukaryotic cell depends on the species. For example, human body cells generally have 46 chromosomes.

Well before a eukaryotic cell begins to divide, it duplicates all of its chromosomes. The DNA molecule of each chromosome is copied, and new protein molecules attach as needed. The result is that each chromosome now consists of two copies called **sister chromatids,** which contain identical genes. Figure 8.4B is an electron micrograph of a human chromosome that has duplicated. The two chromatids are joined together especially tightly at a region called the **centromere.** The fuzzy appearance of the chromosome comes from the intricate twists and folds of its chromatin fiber.

Figure 8.4C is a simple diagram making the point that, when the cell divides, the sister chromatids of a duplicated chromosome separate from each other. Once separated from its sister, each chromatid is called a chromosome, and it is identical to the chromosome we started with. One of the new chromosomes goes to one daughter cell, and the other goes to the other daughter cell. In this way, each daughter cell receives a complete and identical set of chromosomes. In humans, for example, a typical dividing cell has 46 duplicated chromosomes, and each of the two daughter cells that results from it has 46 single chromosomes. 

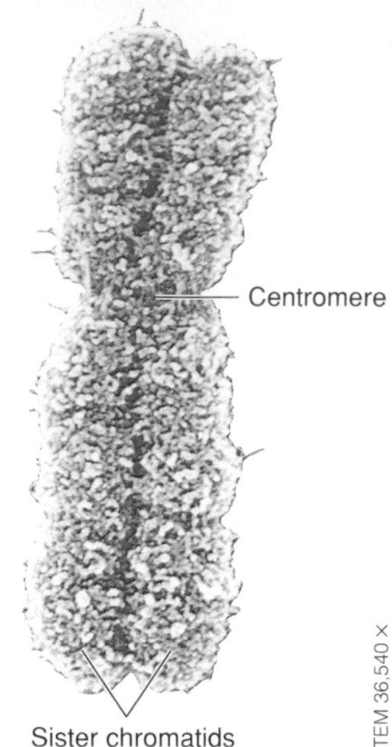

Figure 8.4B Electron micrograph of a duplicated chromosome

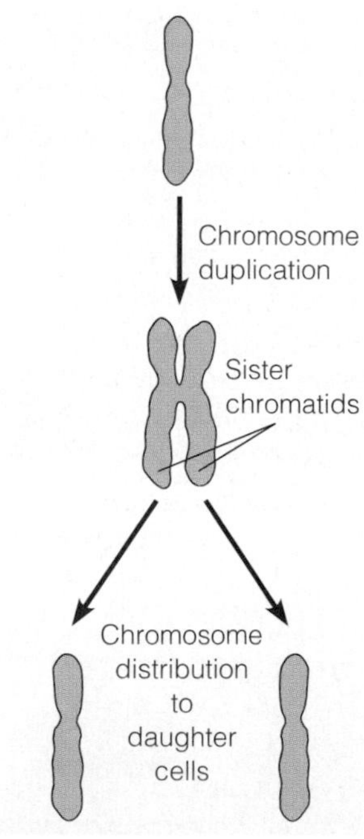

Figure 8.4C Chromosome duplication and distribution

? When would a chromosome consist of two identical chromatids?

When the cell is preparing to divide and has duplicated its chromosomes, but before the duplicates actually separate

The cell cycle multiplies cells

8A
See Activity 8A on the CD-ROM.

How do chromosome duplication and cell division fit into the life of a cell—and the life of an organism? As you already know, cell division is essential to life. Cell division is the basis of reproduction for every organism. It enables a multicellular organism to grow to adult size. It also replaces worn-out or damaged cells, keeping the total cell number in a mature individual relatively constant. In your own body, for example, millions of cells must divide every second to maintain the total number of about 60 trillion cells. Some cells divide once a day, others less often, and highly specialized cells, such as our mature muscle cells, not at all.

Eukaryotic cells that divide undergo a **cell cycle,** an orderly sequence of events that extends from the time a cell divides to form two daughter cells to the time those daughter cells divide again. Before cell division occurs, the cell roughly doubles everything in its cytoplasm and precisely duplicates its chromosomal DNA in preparation for division.

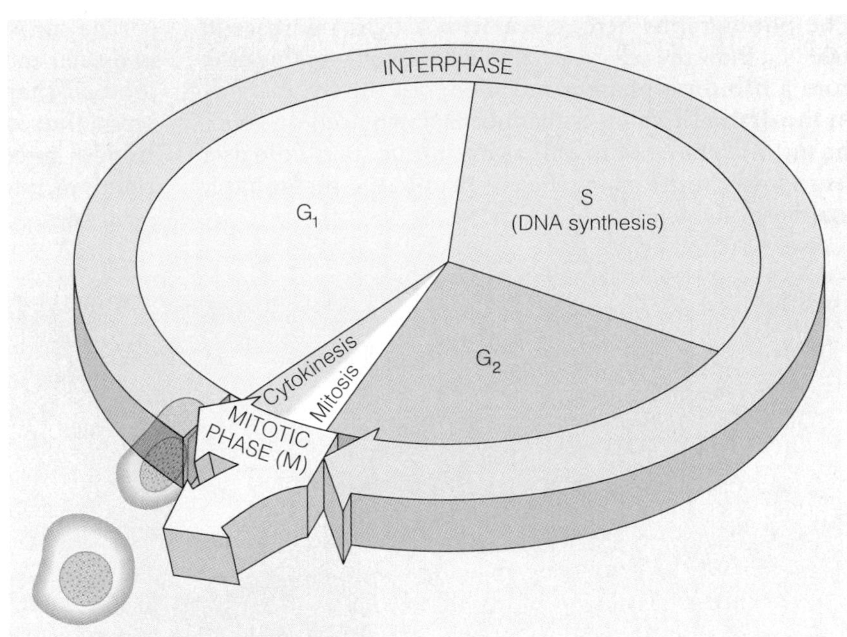

Figure 8.5 The eukaryotic cell cycle

As Figure 8.5 shows, most of the cell cycle is spent in **interphase.** This is a time when a cell's metabolic activity is very high, as it performs its various functions within the organism. Moreover, chromosomes duplicate during this period, many cell parts are made, and the cell does most of its growing. Typically, interphase lasts for at least 90% of the total time required for the cell cycle.

DNA synthesis, the main event in chromosome duplication, occurs in the middle of interphase and serves as the basis for dividing interphase into three subphases. These subphases were named by early microscopists before much was understood about the cell cycle, so their names are not especially revealing. The first subphase, designated G_1, is the period before DNA synthesis begins. G stands for *gap,* and G_1 refers to the gap between cell division and DNA synthesis. We now know that this subphase is anything but a gap in cell activity. G_1 is a time when the cell increases its supply of proteins, increases the numbers of many of its organelles (such as mitochondria and ribosomes), and grows in size.

Following G_1 is the subphase called the S phase, when DNA *synthesis* (replication) actually occurs. At the beginning of the S phase, each chromosome is single. At the end of this phase, after DNA replication, the chromosomes are double, each consisting of two sister chromatids.

The third subphase, called G_2, spans the time from the completion of DNA synthesis to the onset of cell division. Like G_1 and S, G_2 is a time of metabolic activity. Among the proteins synthesized during G_2 are some that are essential to cell division.

Eukaryotic cell division involves two processes. In a process called **mitosis** (light yellow area in the figure), the nucleus and its contents, including the duplicated chromosomes, divide and are evenly distributed to form two daughter nuclei. In a second process, called **cytokinesis,** the cytoplasm is divided in two. Cytokinesis usually begins before mitosis is completed. The combination of mitosis and cytokinesis produces two genetically identical daughter cells, each with a single nucleus, surrounding cytoplasm, and plasma membrane. Taken together, mitosis and cytokinesis make up the **mitotic phase** (the **M phase**) of the cell cycle.

Mitosis is unique to eukaryotes and may be an evolutionary solution to the problem of allocating identical copies of a large amount of genetic material, in a number of separate chromosomes, to two daughter cells. Mitosis is a remarkably accurate mechanism. Experiments with yeast, for example, indicate that an error in chromosome distribution occurs only once in about 100,000 cell divisions.

A living cell viewed through a light microscope undergoes dramatic changes in appearance during the mitotic phase. During interphase, the cell's individual chromosomes are not distinguishable because they are in the form of loosely packed chromatin fibers. With the onset of mitosis, however, striking changes are visible in the chromosomes and other structures, as we see in the next module. 

? A researcher treats cells with a chemical that prevents DNA synthesis from starting. This treatment would trap the cells in which part of the cell cycle?

G_1

8.6 Cell division is a continuum of dynamic changes

8B
8C

The photographs here, taken with a light microscope (600 ×), show the cell cycle for an animal cell—in this case, from a fish. Interphase is included, but the emphasis is on the dramatic changes that occur during cell division, the mitotic phase. Mitosis is a continuum, but biologists distinguish four main stages: **prophase, metaphase, anaphase,** and **telophase.**

The chromosomes are the stars of the mitotic drama, and their movements depend on the **mitotic spindle,** a football-shaped structure of microtubules that guide the separation of the two sets of daughter chromosomes. The spindle microtubules emerge from two **centrosomes,** clouds of cytoplasmic material that in animal cells contain centrioles. (Centrosomes are also known as *microtubule-*

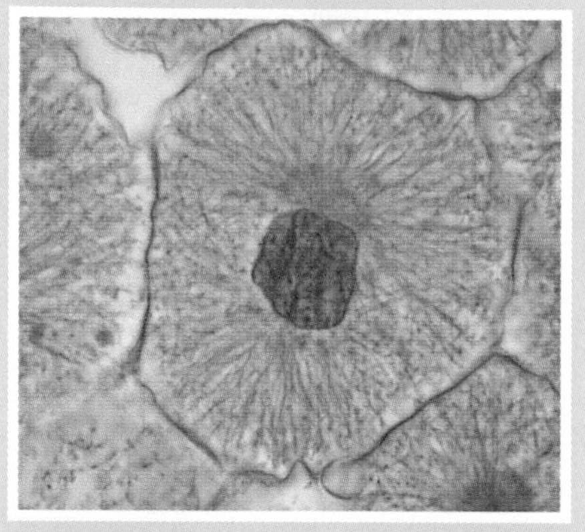

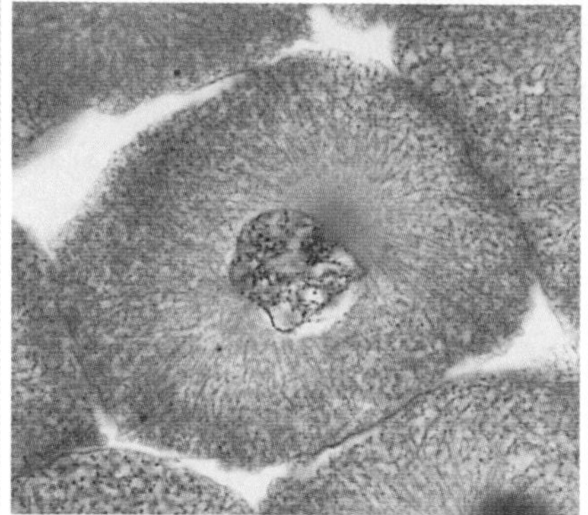

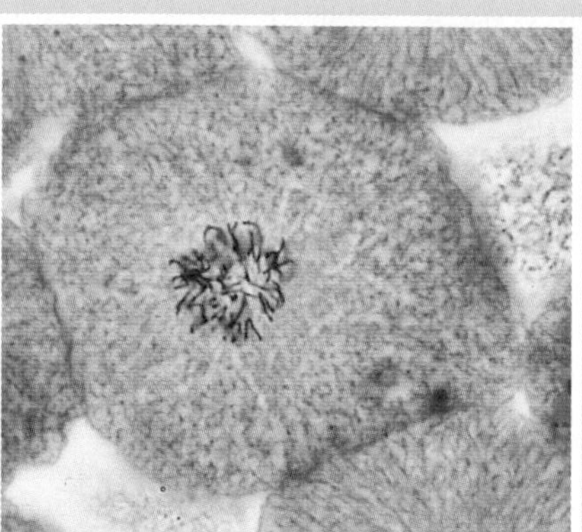

INTERPHASE

PROPHASE

Centrosomes (with centriole pairs)
Chromatin
Nucleolus
Nuclear envelope
Plasma membrane

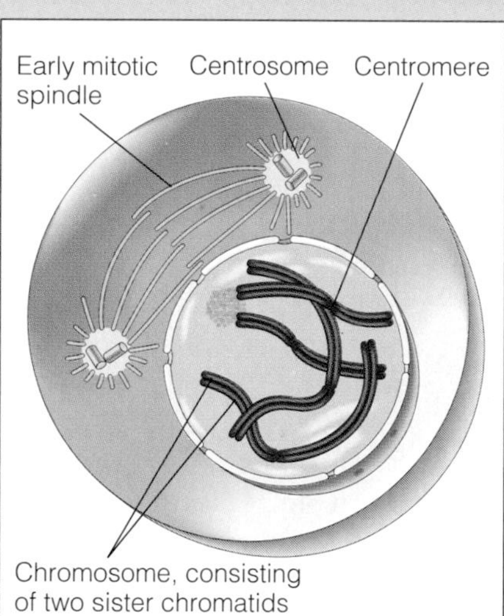

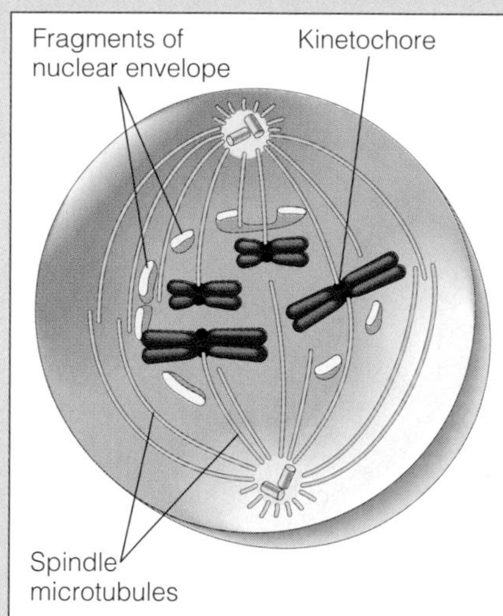

Interphase Interphase is the period of cell growth when the cell synthesizes new molecules and organelles. At the point shown here, late interphase (G_2), the cell looks much the same as it does throughout interphase. Nonetheless, by the G_2 stage the cell has doubled much of its earlier contents and the cytoplasm contains two centrosomes. Within the nucleus, the chromosomes are duplicated, but they cannot be distinguished individually because they are still in the form of loosely packed chromatin. The nucleus also contains one or more nucleoli, an indication that the cell is actively making proteins. Nucleoli are where the parts of ribosomes are assembled before export to the cytoplasm (see Chapter 10).

Prophase During prophase, changes occur in both the nucleus and the cytoplasm. Within the nucleus, the chromatin fibers become more tightly coiled and folded, forming discrete chromosomes that can be seen with the light microscope. The nucleoli disappear. Each duplicated chromosome appears as two identical sister chromatids joined together, with a narrow "waist" at the centromere. In the cytoplasm, the mitotic spindle begins to form as microtubules rapidly grow out from the centrosomes, which begin to move away from each other. Late in prophase, the nuclear envelope breaks into fragments.

With the nuclear envelope gone, microtubules emerging from the centrosomes at the poles (ends) of the spindle can reach the chromosomes, now highly condensed. At the centromere region, each sister chromatid has a protein structure called a kinetochore (shown as a black dot). Some of the spindle microtubules attach to the kinetochores, throwing the chromosomes into agitated motion. Other spindle microtubules make contact with microtubules coming from the opposite pole. Forces exerted by protein "motors" associated with spindle microtubules move the chromosomes toward the center of the cell.

organizing centers, a term describing their function.) The role of centrioles in cell division is a mystery; destroying them experimentally does not interfere with normal spindle formation, and plant cells lack them entirely.

The drawings show details not visible in the micrographs. For simplicity, only four chromosomes are drawn.

An organism called a plasmodial slime mold is one large cytoplasmic mass with many nuclei. Explain how such a "megacell" could form.

Mitosis occurs repeatedly without cytokinesis.

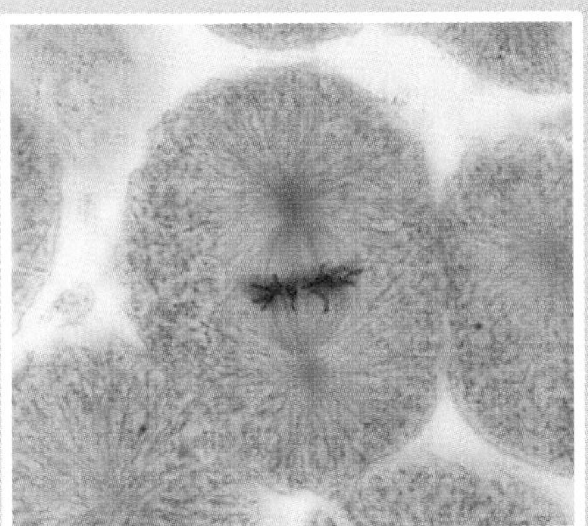

METAPHASE

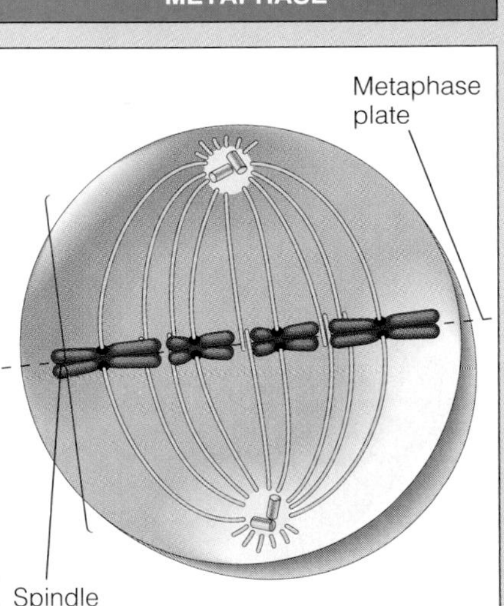

Metaphase At metaphase, the mitotic spindle is fully formed, with its poles at opposite ends of the cell. The chromosomes convene on the metaphase plate, an imaginary plane equidistant between the two poles of the spindle. (The metaphase plate is represented by the dashed black line.) The centromeres of all the chromosomes are lined up on the metaphase plate. For each chromosome, the kinetochores of the two sister chromatids face opposite poles of the spindle. The microtubules attached to a particular chromatid all come from one pole of the spindle, and those attached to its sister chromatid come from the opposite pole.

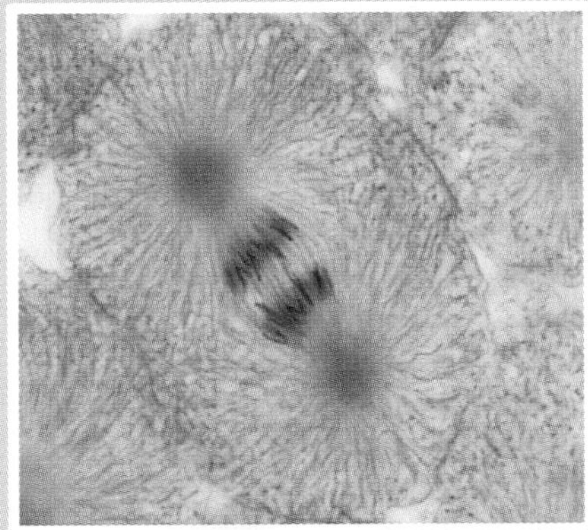

ANAPHASE

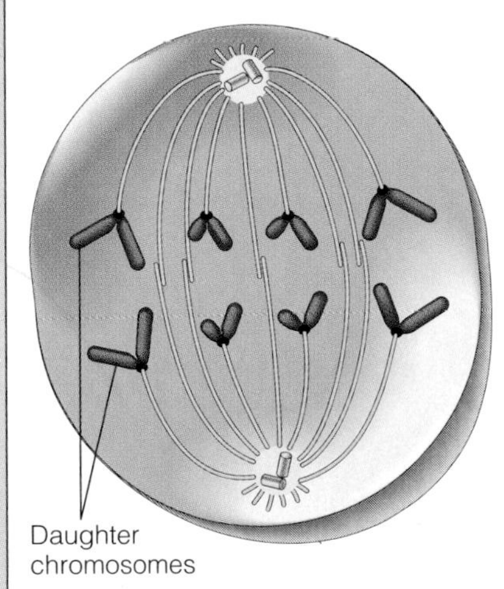

Anaphase Anaphase begins when the two centromeres of each chromosome come apart, separating the sister chromatids. Once separate, each sister chromatid is considered a full-fledged (daughter) chromosome. Motor proteins of the kinetochores, powered by ATP, "walk" the daughter chromosomes centromere-first along the microtubules toward opposite poles of the cell. As this happens, the spindle microtubules attached to the kinetochores shorten. However, the spindle microtubules not attached to chromosomes lengthen. The poles are moved farther apart, elongating the cell. Anaphase is over when equivalent—and complete—collections of chromosomes have reached the two poles of the cell.

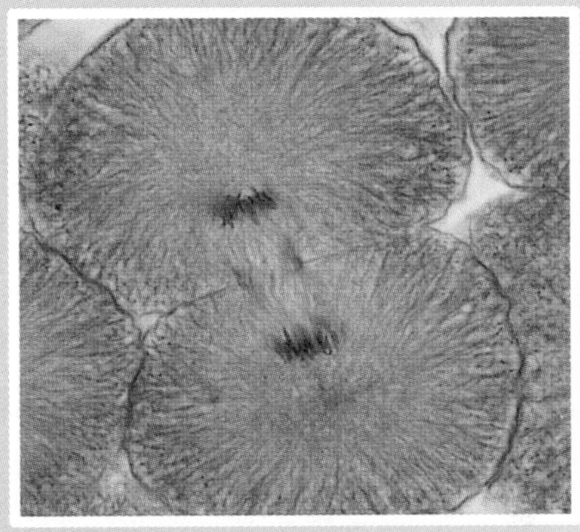

TELOPHASE AND CYTOKINESIS

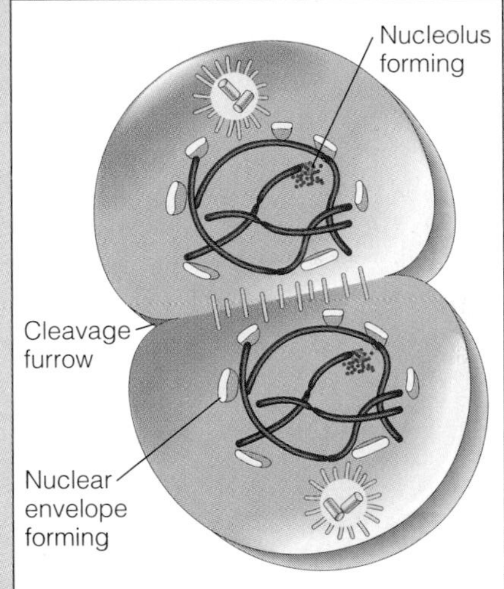

Telophase and Cytokinesis Telophase is roughly the reverse of prophase. The cell elongation that started in anaphase continues. Daughter nuclei appear at the two poles of the cell as nuclear envelopes form around the chromosomes. Meanwhile, the chromatin fiber of each chromosome uncoils, and nucleoli reappear. At the end of telophase, the mitotic spindle disappears. Mitosis, the equal division of one nucleus into two genetically identical daughter nuclei, is now finished.

Cytokinesis, the division of the cytoplasm, usually occurs along with telophase, with two daughter cells completely separating soon after the end of mitosis. In animal cells, cytokinesis involves a cleavage furrow, which pinches the cell in two.

8.7 Cytokinesis differs for plant and animal cells

8B

8C

Cytokinesis, or division of the cell into two, typically occurs with telophase, although it may actually begin in late anaphase. In animal cells, cytokinesis occurs by a process known as cleavage. As shown in Figure 8.7A, the first sign of cleavage is the appearance of a **cleavage furrow,** which begins as a shallow groove in the cell surface. At the site of the furrow, the cytoplasm has a ring of microfilaments made of actin, a protein that functions in several types of cellular contraction. The ring contracts much like the pulling of drawstrings, deepening the cleavage furrow and eventually pinching the parent cell in two.

Cytokinesis in a plant cell occurs differently, as Figure 8.7B shows. First, membrane-enclosed vesicles containing cell wall material collect at the middle of the parent cell. The vesicles then fuse, forming a membrane-enclosed disc called the **cell plate.** The cell plate grows outward, accumulating more cell wall materials as more vesicles fuse with it. Eventually, the membrane of the cell plate fuses with the plasma membrane, and the cell plate's contents join the parental cell wall. The result is two daughter cells, each bounded by its own continuous plasma membrane and cell wall. 

? Contrast cytokinesis in animals with cytokinesis in plants.

In animals, cytokinesis involves a cleavage furrow in which contracting microfilaments pinch the cell in two. In plants, it involves formation of a cell plate, a fusion of vesicles that forms new membrane and walls between the cells.

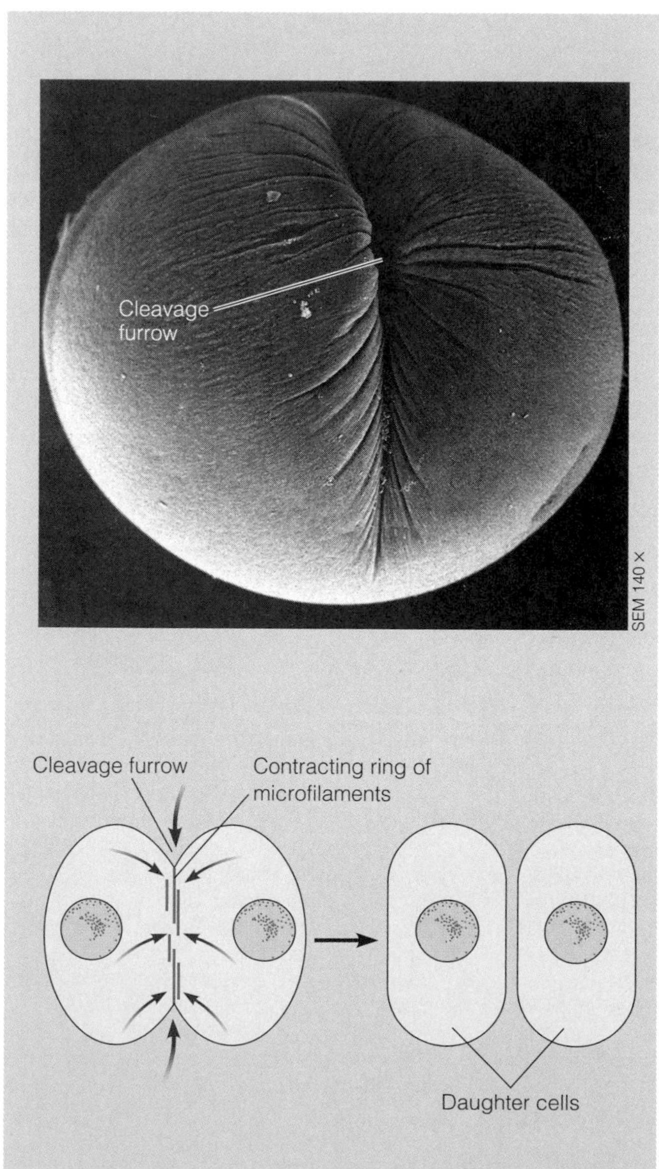

Figure 8.7A Cleavage of an animal cell

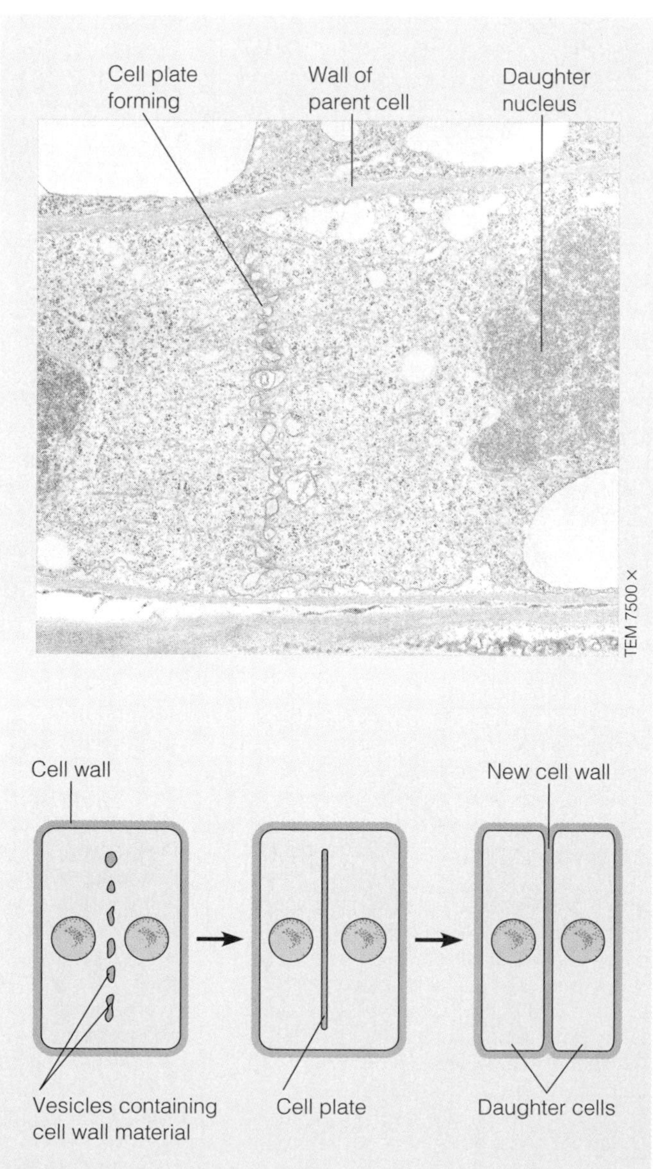

Figure 8.7B Cell plate formation in a plant cell

Anchorage, cell density, and chemical growth factors affect cell division 8.8

For a plant or an animal to grow and develop normally, and to maintain its tissues once full-grown, it must be able to control the timing of cell division in different parts of its body. For example, in the adult human, skin cells and the cells lining the digestive tract divide frequently throughout life, replacing cells that are constantly being abraded and sloughed off. In contrast, cells in the human liver usually do not divide unless the liver is damaged. Cell division in this case repairs wounds.

Many questions about how cell division is controlled are still unanswered, but biologists have learned a great deal by studying cells grown in laboratory cultures. They have learned, for instance, that most animal and plant cells do not divide unless they are in contact with a solid surface. If cells are suspended in a liquid medium, they rarely divide. However, when the same cells are poured onto a solid surface and allowed to attach, or "anchor" themselves, they start dividing immediately. In the body, this **anchorage dependence** of cell division may keep cells that become separated from their normal surroundings from dividing inappropriately. Most cells of the animal body are normally anchored to an extracellular matrix (see Module 4.19) or to other cells of the same tissue.

Scientists have also found that animal cells growing on the surface of a laboratory dish multiply to form a single layer and usually stop dividing when they touch one another (Figure 8.8A). This phenomenon is called **density-dependent inhibition,** implying that cell division slows down as the cell population becomes denser. But when some of the cells are scraped off a cell-covered surface, cells bordering the open space begin dividing again. They continue to grow and divide until the space is filled again.

Clearing a space in a cell culture is analogous to cutting your skin. When you cut yourself, skin cells all around the cut immediately begin dividing, and healing occurs as the cells fill in the gap. The cells stop dividing when they encounter other cells.

What actually causes density-dependent inhibition? Studies of cultured cells suggest that it is primarily an inadequate supply of certain proteins, called growth factors, rather than physical contact with other cells, that is responsible for inhibiting cell division. A **growth factor** is a protein secreted by certain body cells that stimulates cells in the vicinity to divide. Most cells require growth factors in order to begin dividing, and they stop dividing when they run out of these substances. For cells in culture, growth factors are provided in the nutrient medium. Apparently, density-dependent inhibition occurs when the cells become so crowded that each cell uses up the supply of growth factors in its immediate environment. As indicated in Figure 8.8B, flooding a layer of nondividing cells with a more concentrated solution of growth factors stimulates the cells to grow to a greater density than they would otherwise. The cells are still in a single layer but are smaller and more numerous.

Density-dependent inhibition mediated by the availability of growth factors is probably an important regulatory mechanism in the body's tissues, as well as in cell culture. It may help keep cell populations at optimal levels in the tissues. How exactly do growth factors work? We pursue answers to this question in the next module.

Cells anchor to dish surface and divide.

When cells have formed a complete single layer, they stop dividing (density-dependent inhibition).

If some cells are scraped away, the remaining cells divide to fill the dish with a single layer and then stop (density-dependent inhibition).

Figure 8.8A An experiment demonstrating density-dependent inhibition, using animal cells grown in culture

Compared to a control culture, the cells in an experimental culture are fewer but much larger when they cover the dish surface and stop growing. What is a reasonable hypothesis for this difference?

The experimental culture is deficient in one or more growth factors.

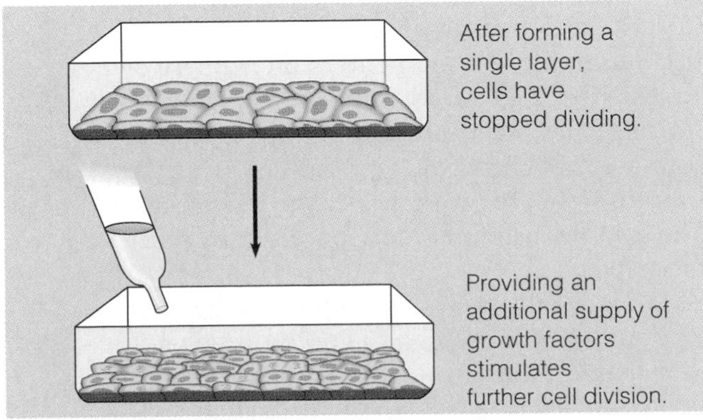

After forming a single layer, cells have stopped dividing.
Providing an additional supply of growth factors stimulates further cell division.

Figure 8.8B An experiment demonstrating the effect of growth factors on the division of cultured animal cells

8.9 Growth factors signal the cell-cycle control system

In a living animal, most cells are anchored in a fixed position and bathed in a solution of nutrients supplied by the blood, yet they usually do not divide unless they are signaled by other cells to do so. Growth factors are the main signals, and their role in promoting cell division leads us back to our earlier discussion of the cell cycle.

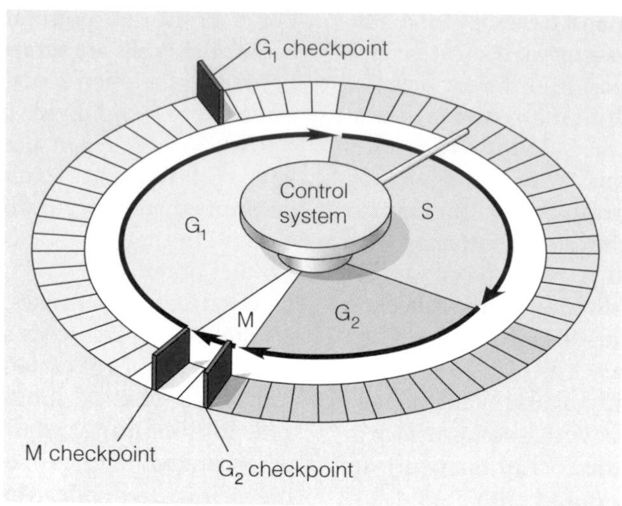

Figure 8.9A Mechanical model for the cell-cycle control system

The sequential events of the cell cycle, represented by the circle of flat blocks in Figure 8.9A, are directed by a distinct cell-cycle control system, represented by the knob in the center. Analogous to the control device of an automatic washing machine, the **cell-cycle control system** is a cyclically operating set of proteins in the cell that both triggers and coordinates major events in the cell cycle. The cell cycle is not like a row of falling dominoes, with each event causing the next one in line. Within the M phase, for example, metaphase does not automatically lead to anaphase. Instead, proteins of the cell-cycle control system must trigger the separation of sister chromatids that marks the start of anaphase.

The red barriers in Figure 8.9A represent three key checkpoints in the cell cycle. These are critical control points in the G_1 and G_2 subphases of interphase and in the M phase, when stop and go-ahead signals can be applied to the cycle. Most animal cells have built-in brakes that block the cell cycle at the checkpoints until overridden by go-ahead signals. Intracellular signals detected by the control system tell it whether key cellular processes have been completed, and thus whether or not to proceed past that point. The control system also receives messages from outside the cell, indicating both general environmental conditions and the presence of specific signal molecules from other cells. When the cell-cycle control system gets a go-ahead signal at the G_1 checkpoint, the cell soon enters the S phase of the cell cycle. A go-ahead signal at the G_2 checkpoint initiates the M phase. Finally, a go-ahead signal at the M checkpoint, indicating that all the chromosomes are properly attached to the mitotic spindle, prompts the separation of sister chromatids.

For many cells, the G_1 checkpoint is the most important. If a cell receives a go-ahead signal—for example, from a growth factor—at the G_1 checkpoint, it will usually go on through its cycle and divide. Alternatively, if it does not receive a go-ahead signal at G_1, it will switch into a nondividing state. Our nondividing nerve cells and muscle cells are permanently arrested at the G_1 checkpoint.

Figure 8.9B shows a simplified model for how a growth factor might affect the cell-cycle control system at the G_1 checkpoint. The process is one of *signal transduction* (see Figure 5.13B). A cell that responds to a growth factor has molecules of a specific receptor protein in its plasma membrane. Binding of the growth factor to the receptor triggers a signal-transduction pathway in the cell, a pathway that in this case leads to cell division. The "signals," represented by black arrows, are changes that a molecule induces in the next molecule in the pathway. Via a series of relay molecules, a signal finally reaches the cell-cycle control system and overrides the brakes that otherwise prevent progress of the cell cycle. In Figure 8.9B, the cell cycle is set off from the cell in a separate diagram because, unlike the components of the timer in a washing machine, the proteins making up the control system in the cell are not actually located together in one place.

Research on the control of the cell cycle is one of the hottest areas in biology today. This research is leading to a better understanding of cancer, which we discuss next.

A particular poison blocks the cell cycle at the M checkpoint. The cells will be arrested in which phase of mitosis?

Metaphase

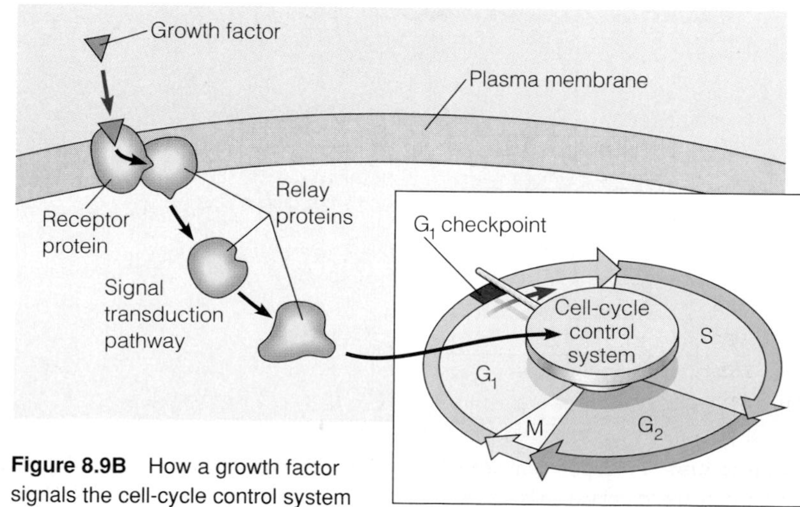

Figure 8.9B How a growth factor signals the cell-cycle control system

Growing out of control, cancer cells produce malignant tumors 8.10

Cancer, which currently claims the life of one out of every five people in the United States and other developed nations, is a disease of the cell cycle. Unlike normal cells of the body, **cancer cells** do not have a properly functioning cell-cycle control system; they divide excessively and can invade other tissues of the body. This excessive growth can result in an abnormal mass of cells called a **tumor.** Not all tumors are cancerous, however. A **benign tumor** is an abnormal mass of essentially normal cells. Benign tumors can cause problems if they grow in certain organs, such as the brain, but usually they can be completely removed by surgery. They always remain at their original site in the body.

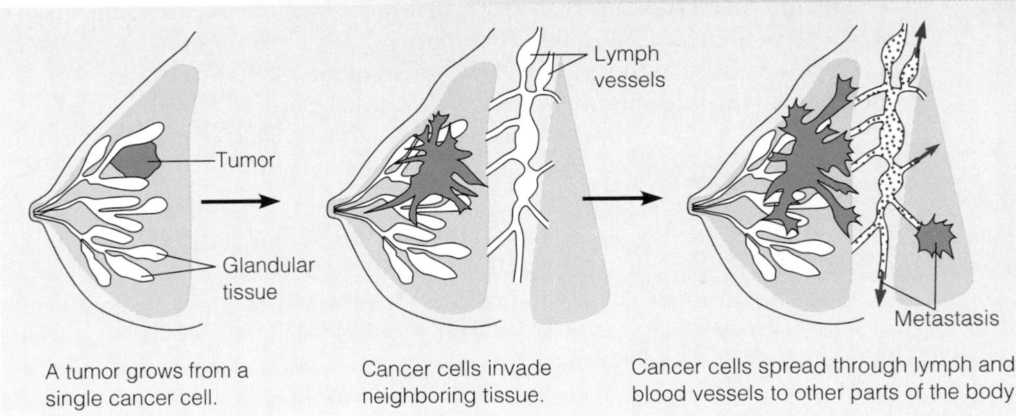
Lymph vessels
Tumor
Glandular tissue
Metastasis
A tumor grows from a single cancer cell.
Cancer cells invade neighboring tissue.
Cancer cells spread through lymph and blood vessels to other parts of the body.

Figure 8.10 Growth and metastasis of a malignant (cancerous) tumor of the breast

In contrast to a benign tumor, a **malignant tumor** is cancerous. It is a mass of cancer cells, which are capable of spreading into neighboring tissues and often to other parts of the body. Arising from a single cancer cell, a malignant tumor displaces normal tissue as it grows (Figure 8.10). If the tumor is not killed or removed, some of the cancer cells spread into surrounding tissues, enlarging the tumor. Cells may also split off from the tumor, invade the circulatory system (lymph vessels and blood vessels), and travel to new locations, where they can form new tumors. The spread of cancer cells beyond their original site is called **metastasis.**

Cancers are named according to the organ or tissue in which they originate. For simplicity, they are grouped into four categories. **Carcinomas** are cancers that originate in the external or internal coverings of the body, such as the skin or the lining of the intestine. **Sarcomas** arise in tissues that support the body, such as bone and muscle. Cancers of blood-forming tissues, such as bone marrow, spleen, and lymph nodes, are called **leukemias** and **lymphomas.**

From studying cancer cells in culture, researchers have learned that these cells are largely unrestrained by the system that normally controls cell division. Cancer cells are not usually affected by density-dependent inhibition; they continue to divide even at high densities, piling up on one another. Many cancer cells have cell-cycle control systems that proceed past checkpoints even in the absence of growth factors. Other cancer cells themselves synthesize growth factors that make them divide continuously, and the dividing cells do not respond to inhibiting signals. Cancer cells have a reduced need for anchorage and often grow without being attached to a surface.

Normal mammalian cells grow and divide in culture for only about 50 generations, but cancer cells can go on dividing indefinitely, as long as they have a supply of nutrients. If cancer cells do stop dividing, they seem to do so at random points in the cell cycle, rather than just at the normal cell-cycle checkpoints.

Two types of cancer treatment, chemotherapy and radiation therapy, attempt to halt the spread of cancer cells by stopping them from dividing. In radiation therapy, parts of the body that have cancerous tumors are exposed to high-energy radiation, which disrupts cell division. Because cancer cells divide more often than most normal cells, they are more likely to be dividing at any given time. So radiation can often destroy cancer cells without seriously injuring the normal cells of the body. However, there is sometimes enough damage to normal body cells to produce bad side effects. For example, damage to cells of the ovaries or testes can lead to sterility.

Chemotherapy generally uses the same strategy as radiation; in this case, drugs that disrupt cell division are administered to the patient. These drugs work in a variety of ways. Some, called antimitotic drugs, prevent cell division by interfering in one way or another with the mitotic spindle. One antimitotic drug, vinblastin, prevents the spindle from forming in the first place; another, taxol, "freezes" the spindle after it forms, keeping it from functioning.

Vinblastin was first obtained from the periwinkle, a flowering plant native to tropical rain forests in Madagascar. Taxol was discovered in the bark of the Pacific yew, a tree found mainly in the northwestern United States. Taxol has fewer side effects than many anticancer drugs and seems to be effective against some hard-to-treat cancers of the ovary and breast. Both vinblastin and taxol are now manufactured by pharmaceutical companies.

We will return to the topic of cancer in Chapter 11, after studying the structure and function of genes. We will see then that cancer results from changes in genes for proteins that control cell division. 

? What is metastasis?

Metastasis is the spread of cancer cells via the circulatory system from their original site of formation to another site in the body.

8.11 Review of the functions of mitosis: Growth, cell replacement, and asexual reproduction

8D

The pictures here summarize the roles that mitotic cell division plays in the lives of multicellular organisms. Figure 8.11A shows some of the cells in the tip of an onion plant root. Notice the large number of cells whose nuclei are in various stages of mitosis. This root was growing rapidly when it was harvested and prepared for microscopy. Cell division in the root tip produces new cells, which elongate to bring about growth of the root.

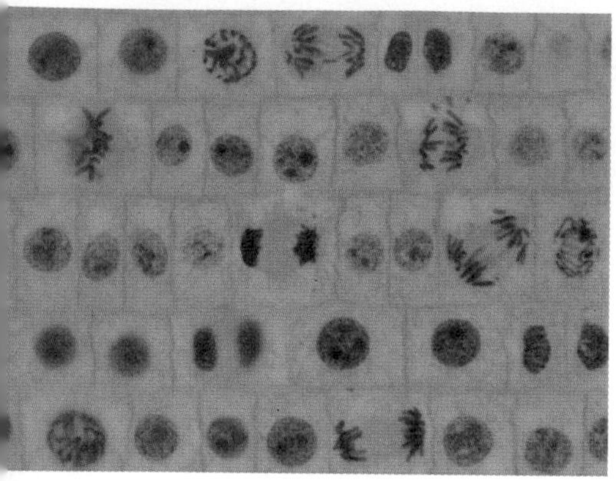

LM 600×

Figure 8.11A Growth (in an onion root)

Figure 8.11B shows a cross section of human skin. The layers of dead cells on the surface protect the body against injuries and infections and help prevent it from drying out. These surface layers are constantly abraded and sloughed off, but they are replaced by cells from the living layers underneath. This regeneration of the skin goes on throughout a person's life. New cells generated near the base of the epidermis are constantly moving outward toward the skin surface. As they do, they gradually flatten, become hardened, and die, regenerating the protective surface layers.

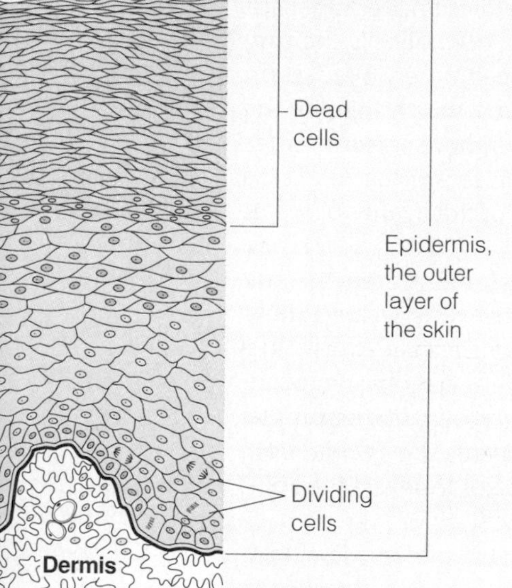

Figure 8.11B Cell replacement (in skin)

Figure 8.11C is a photograph of a hydra, a common inhabitant of freshwater lakes. A hydra is a tiny multicellular animal that reproduces by either sexual or asexual means. This individual is reproducing asexually by budding. A bud starts out as a mass of mitotically dividing cells growing on the side of the parent. The bud develops into a small hydra like the one in the photo here. Eventually, the offspring detaches from the parent and takes up life on its own. The offspring is literally a "chip off the old block," being genetically identical to its parent. (It is a clone of its parent.)

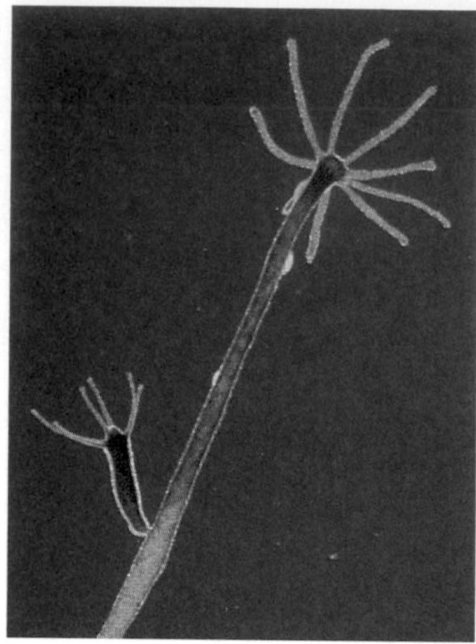

Figure 8.11C Asexual reproduction (of a hydra)

In all three of these cases—budding hydra, regenerating human skin, and growing onion—the new cells have exactly the same number and types of chromosomes as the parent cells because of the way duplicated chromosomes divide during mitosis. Mitosis makes it possible for organisms to grow, regenerate and repair tissues, and reproduce asexually by producing cells that carry the same genes as the parent cells.

If we examine the cells of any individual organism, we see that almost all of them contain the same number and types of chromosomes. Likewise, if we examine cells from different individuals of any one species, we see that they have the same number and types of chromosomes. In the next module, we take a closer look at the numbers and types of chromosomes in cells.

If a human skin cell with 46 chromosomes divides by mitosis, each daughter cell will have __ chromosomes.

46

Meiosis and Crossing Over

8.12 Chromosomes are matched in homologous pairs

In humans, a typical body cell, called a **somatic cell,** has 46 chromosomes. If we use a microscope to examine human chromosomes in metaphase of mitosis, we see that each duplicated chromosome has a twin that nearly always is identical in length and centromere position. Altogether, we see 23 such matched pairs of duplicated chromosomes. Other species have different numbers of chromosomes, but these too usually match in pairs. Furthermore, when treated with special dyes, the chromosomes of a pair display matching staining patterns. Figure 8.12 illustrates one pair of metaphase chromosomes, with the staining pattern represented by colored stripes. Notice that each chromosome consists of two sister chromatids joined at the centromere. The two chromosomes of such a pair are called

homologous chromosomes because they both carry genes controlling the same inherited characteristics. For example, if a gene for eye color is located at a particular place, or **locus** (plural, *loci*), on one chromosome—within the narrow orange band in our drawing, for instance—then the other chromosome of the homologous pair also has a gene for eye color at that locus. (However, the two homologues may have different versions of the eye-color gene, perhaps specifying different colors.)

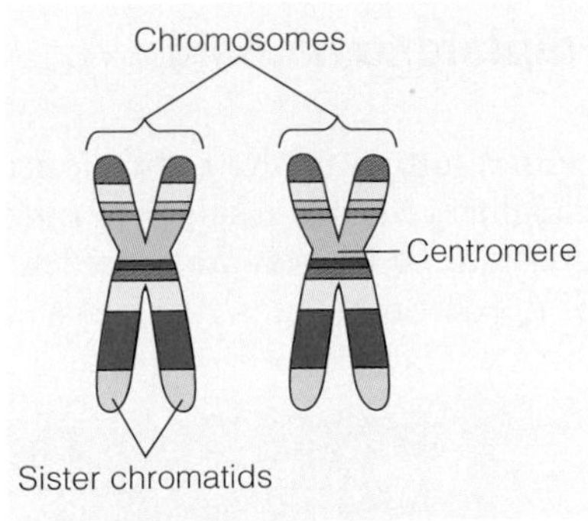

Figure 8.12 A homologous pair of chromosomes

Our 23 pairs of homologous chromosomes are of two general types. Twenty-two pairs consist of chromosomes called **autosomes,** found in both males and females. The other pair of chromosomes, the **sex chromosomes,** determines a person's gender. Like all mammals, human females have a pair of sex chromosomes called *X* chromosomes. By contrast, mammalian males have one *X* and one *Y*. These two chromosomes differ in size and shape. Only small parts are homologous; most of the genes carried on the *X* chromosome do not have counterparts on the *Y*. However, the homologous regions are large enough that the *X* and *Y* chromosomes do behave as a homologous pair (in meiosis; see Module 8.14). You will see an actual micrograph showing all the human chromosomes in Figure 8.19.

For both autosomes and sex chromosomes, we inherit one chromosome of each pair from our mother and the other from our father, as discussed below. 

? What is the explanation for our having two of each type of chromosome?

We inherit one of each kind from each parent.

Gametes have a single set of chromosomes

8.13

8D

Having two sets of chromosomes, one inherited from each parent, is a key factor in the human life cycle, outlined in Figure 8.13, and in the life cycles of all other species that reproduce sexually. (We examine other life cycles in Unit IV.)

Cells whose nuclei contain two homologous sets of chromosomes are called **diploid cells,** and the total number of chromosomes is called the diploid number (abbreviated $2n$). In humans, the diploid number is 46 (that is, $2n = 46$). Humans are said to be diploid organisms because almost all our cells are diploid. The exceptions are the egg and sperm cells, collectively known as **gametes.** Each gamete has a single set of chromosomes: 22 autosomes plus a single sex chromosome, either *X* or *Y*. A cell with a single chromosome set is called a **haploid cell.** For humans, the haploid number (abbreviated n) is 23.

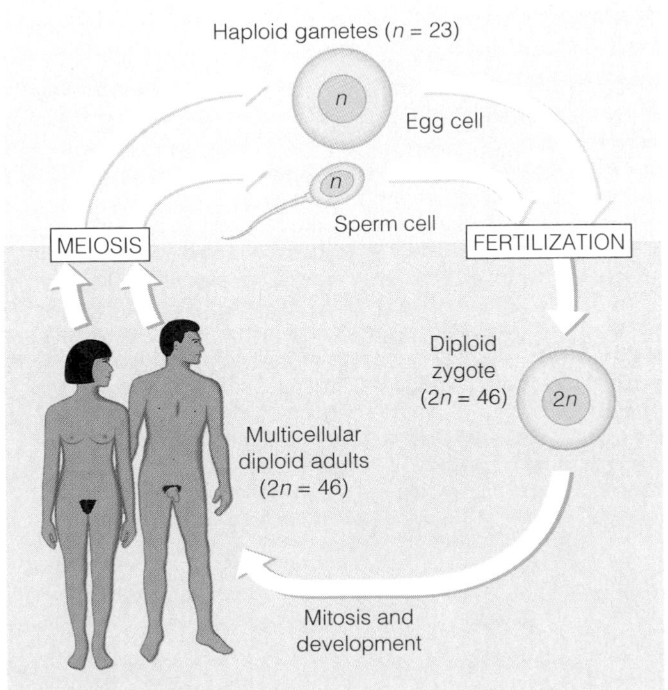

Figure 8.13 The human life cycle

In the human life cycle, sexual intercourse allows a haploid sperm cell from the father to reach and fuse with a haploid egg cell of the mother in the process of **fertilization.** The resulting fertilized egg, called a **zygote,** is diploid. It has two haploid sets of chromosomes: one set from the mother and a homologous set from the father. The life cycle is completed as a sexually mature adult develops from the zygote. Mitotic cell division ensures that all somatic cells of the human body receive copies of all of the zygote's 46 chromosomes.

All sexual life cycles, including our own, involve an alternation of diploid and haploid stages. Having haploid gametes keeps the chromosome number from doubling in each generation. Gametes are made by a special sort of cell division called **meiosis,** which occurs only in reproductive organs (ovaries and testes in animals). Whereas mitosis produces daughter cells with the same numbers of chromosomes as the parent cell, meiosis reduces the chromosome number by half. We turn to meiosis next.

? ________ is to somatic cells as haploid is to ________.

Diploid . . . gametes

8.14 Meiosis reduces the chromosome number from diploid to haploid

8E

Meiosis, the process that produces haploid gametes in diploid organisms, resembles mitosis, but there are important differences. In meiosis, a cell undergoes two consecutive divisions, called meiosis I and meiosis II. Four daughter cells result from these divisions, in contrast to the two daughter cells that result from mitotic cell division. The two divisions of meiosis are preceded by only one duplication of the chromosomes. As a result, *each of the four daughter*

MEIOSIS I: Homologous chromosomes separate

INTERPHASE

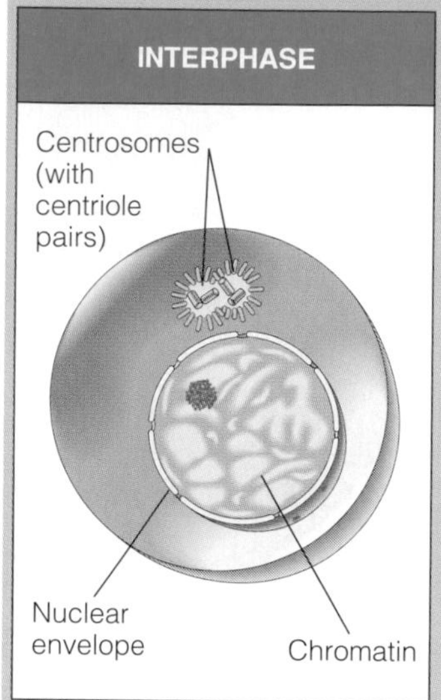

PROPHASE I

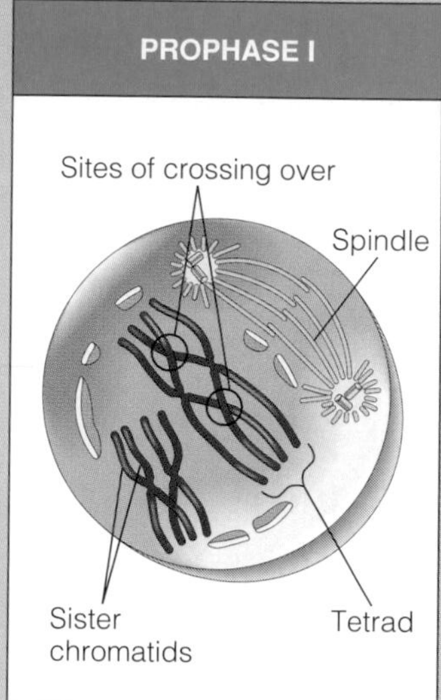

METAPHASE I

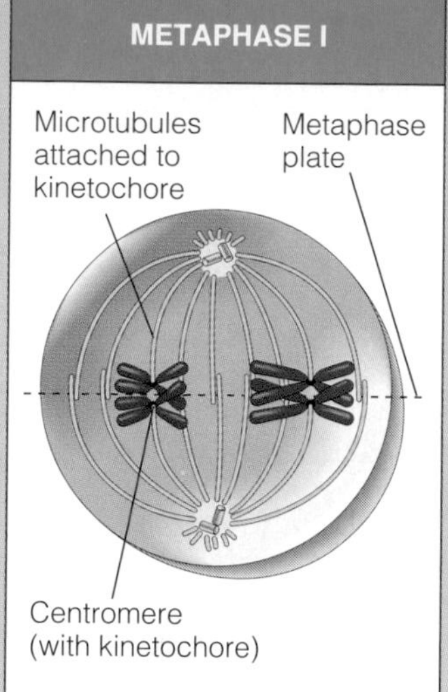

ANAPHASE I

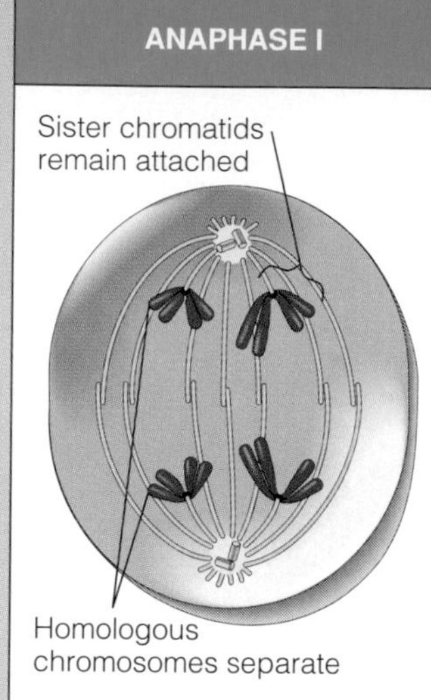

Interphase

Like mitosis, meiosis is preceded by an interphase, during which the chromosomes duplicate. At the end of this interphase, each chromosome consists of two genetically identical sister chromatids, attached together. But at this stage, the chromosomes are not yet visible under the microscope except as a mass of chromatin. The cell's centrosome has also duplicated by the end of this interphase.

Prophase I

Prophase I is the most complex phase of meiosis and typically occupies over 90% of the time required for meiotic cell division. Early in this phase, the chromatin coils up, so that individual chromosomes become visible with the microscope. In a process called synapsis, homologous chromosomes, each composed of two sister chromatids, come together as pairs. The resulting structure, consisting of four chromatids, is called a tetrad. During synapsis, chromatids of homologous chromosomes exchange segments, in a process called crossing over. Because the versions of the genes on a chromosome (or one of its chromatids) may be different from those on its homologue, crossing over rearranges genetic information. As we will discuss in Module 8.18, the genetic shuffling produced by crossing over can make an important contribution to the genetic variability resulting from sexual reproduction.

As prophase I continues, the chromosomes condense further, as the nucleoli disappear. Now the centrosomes move away from each other, and a spindle starts to form between them. The nuclear envelope breaks into fragments, and the chromosome tetrads, captured by spindle microtubules, are moved toward the center of the cell.

Metaphase I

At metaphase I, the chromosome tetrads are aligned on the metaphase plate, midway between the two poles of the spindle. Each chromosome is condensed and thick, with its sister chromatids still attached at their centromeres. Spindle microtubules are attached to kinetochores at the centromeres. In each tetrad, the homologous chromosomes are held together at sites of crossing over. Notice that, for each tetrad, the spindle microtubules attached to one of the homologous chromosomes come from one pole of the cell, and the microtubules attached to the other homologous chromosome come from the opposite pole. With this arrangement, the homologous chromosomes of each tetrad are poised to move toward opposite poles of the cell.

Anaphase I

Like anaphase of mitosis, anaphase I of meiosis is marked by the migration of chromosomes toward the two poles of the cell. In contrast to mitosis, however, the sister chromatids making up each doubled chromosome remain attached at their centromeres. Only the tetrads (pairs of homologous chromosomes) split up. Thus, in the drawing you see two still-doubled chromosomes (the haploid number) moving toward each spindle pole. If this were anaphase of mitosis, you would see four daughter chromosomes moving toward each pole.

cells resulting from meiosis has only half as many chromosomes as the starting cell—a single haploid set of chromosomes. The actual halving of chromosome number occurs during meiosis I. The drawings here show the two meiotic divisions for an animal cell with a diploid number of 4.

A cell has the haploid number of chromosomes, but each chromosome has two chromatids. The chromosomes are arranged at the center of the spindle. What is the meiotic stage?

Metaphase II

MEIOSIS II: Sister chromatids separate

TELOPHASE I AND CYTOKINESIS

Cleavage furrow

PROPHASE II

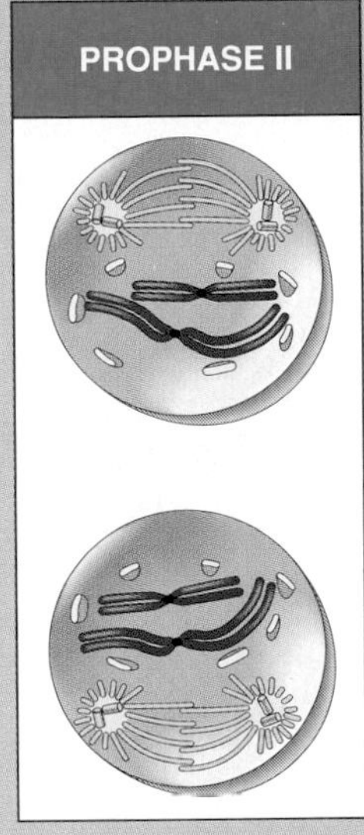

METAPHASE II

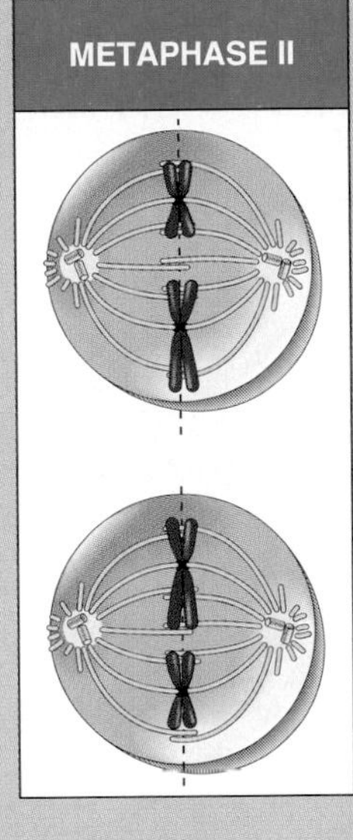

ANAPHASE II

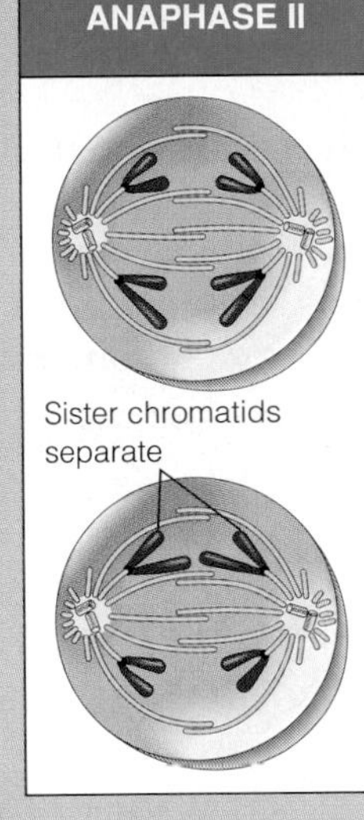

Sister chromatids separate

TELOPHASE II AND CYTOKINESIS

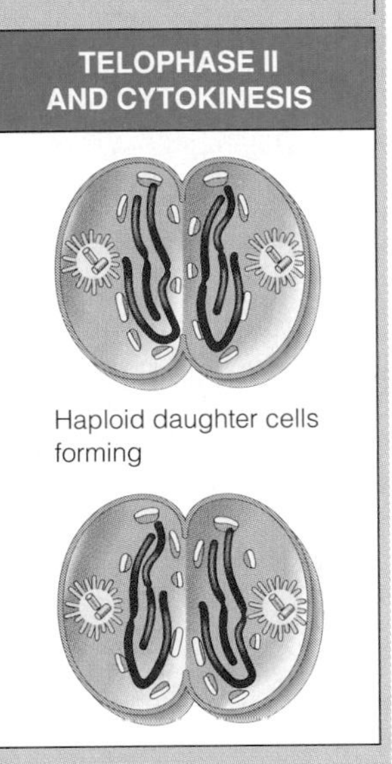

Haploid daughter cells forming

Telophase I and Cytokinesis

In telophase I, the chromosomes arrive at the poles of the cell. When the chromosomes finish their journey, each pole of the cell has a haploid chromosome set, although each chromosome is still in duplicate form at this point. In other words, each chromosome still consists of two sister chromatids. Usually, cytokinesis occurs along with telophase I, and two haploid daughter cells are formed.

Following telophase I in some organisms, the chromosomes uncoil and the nuclear envelope re-forms, and there is an interphase before meiosis II begins. In other species, daughter cells produced in the first meiotic division immediately begin preparation for the second meiotic division. In either case, no chromosome duplication occurs between telophase I and the onset of meiosis II.

Meiosis II

In organisms having an interphase after meiosis I, the chromosomes condense again and the nuclear envelope breaks down during prophase II. In any case, meiosis II is essentially the same as mitosis. The important difference is that meiosis II starts with a haploid cell.

During prophase II, a spindle forms and moves the chromosomes toward the middle of the cell. During metaphase II, the chromosomes are aligned on the metaphase plate as they are in mitosis, with the kinetochores of the sister chromatids of each chromosome pointing toward opposite poles. In anaphase II, the centromeres of sister chromatids finally separate, and the sister chromatids of each pair, now individual daughter chromosomes, move toward opposite poles of the cell. In telophase II, nuclei form at the cell poles, and cytokinesis occurs at the same time. There are now four daughter cells, each with the haploid number of (single) chromosomes.

8.15 Review: A comparison of mitosis and meiosis

We have now described the two ways that cells of eukaryotic organisms divide. Mitosis, which provides for growth, tissue repair, and asexual reproduction, produces daughter cells genetically identical to the parent cell. Meiosis, needed for sexual reproduction, yields haploid daughter cells—cells with one member of each homologous chromosome pair.

For both mitosis and meiosis, the chromosomes replicate only once, in the preceding interphase. Mitosis involves one division of the nucleus, and it is usually accompanied by cytokinesis, producing two diploid cells. Meiosis entails two nuclear and cytoplasmic divisions, yielding four haploid cells.

Figure 8.15 compares mitosis and meiosis, tracing these two processes for a diploid parent cell with four chromosomes. Homologous chromosomes are those matching in size.

All the events unique to meiosis occur during meiosis I. In prophase I, duplicated homologous chromosomes pair to form tetrads, and crossing over occurs between homologous (nonsister) chromatids. In metaphase I, tetrads (rather than individual chromosomes) are aligned at the metaphase plate. During anaphase I, sister chromatids of each chromosome stay together and go to the same pole of the cell, as homologous pairs of chromosomes separate. At the end of meiosis I, there are two haploid cells, but each chromosome still has two sister chromatids.

Meiosis II is virtually identical to mitosis and separates sister chromatids. But unlike mitosis, meiosis II yields daughter cells with a *haploid* set of chromosomes.

Explain how mitosis conserves chromosome number while meiosis reduces the number from diploid to haploid.

In mitosis, a single replication of chromosomes is followed by one division of the cell. In meiosis, a single chromosome replication is followed by two cell divisions.

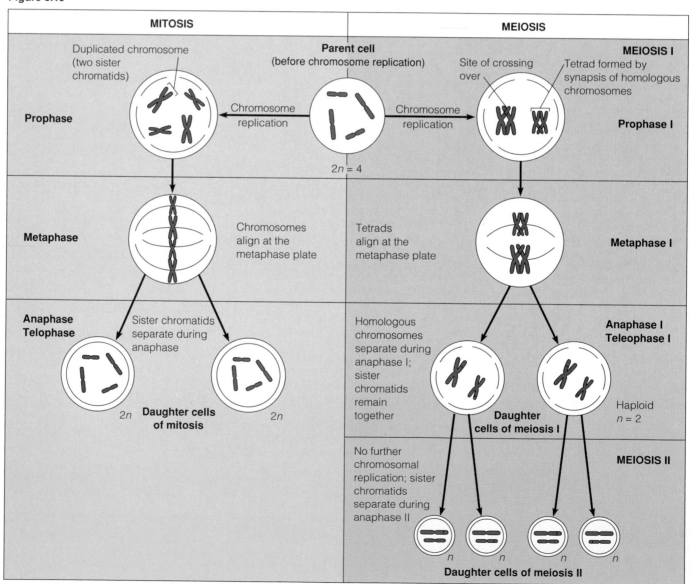

Figure 8.15

MITOSIS

MEIOSIS

Parent cell
(before chromosome replication)

$2n = 4$

MEIOSIS I

Prophase

Duplicated chromosome (two sister chromatids)

Chromosome replication

Chromosome replication

Site of crossing over

Tetrad formed by synapsis of homologous chromosomes

Prophase I

Metaphase

Chromosomes align at the metaphase plate

Tetrads align at the metaphase plate

Metaphase I

Anaphase Telophase

Sister chromatids separate during anaphase

$2n$ **Daughter cells of mitosis** $2n$

Homologous chromosomes separate during anaphase I; sister chromatids remain together

Anaphase I Telophase I

Daughter cells of meiosis I

Haploid $n = 2$

MEIOSIS II

No further chromosomal replication; sister chromatids separate during anaphase II

n n n n

Daughter cells of meiosis II

Independent orientation of chromosomes in meiosis and random fertilization lead to varied offspring

As we discussed in Module 8.1, offspring that result from sexual reproduction are highly varied; they are genetically different from their parents and from one another. When we discuss natural selection and evolution in Unit III, we will see that this genetic variety in offspring is the raw material for natural selection. For now, let's take another look at meiosis and fertilization to see how genetic variety arises.

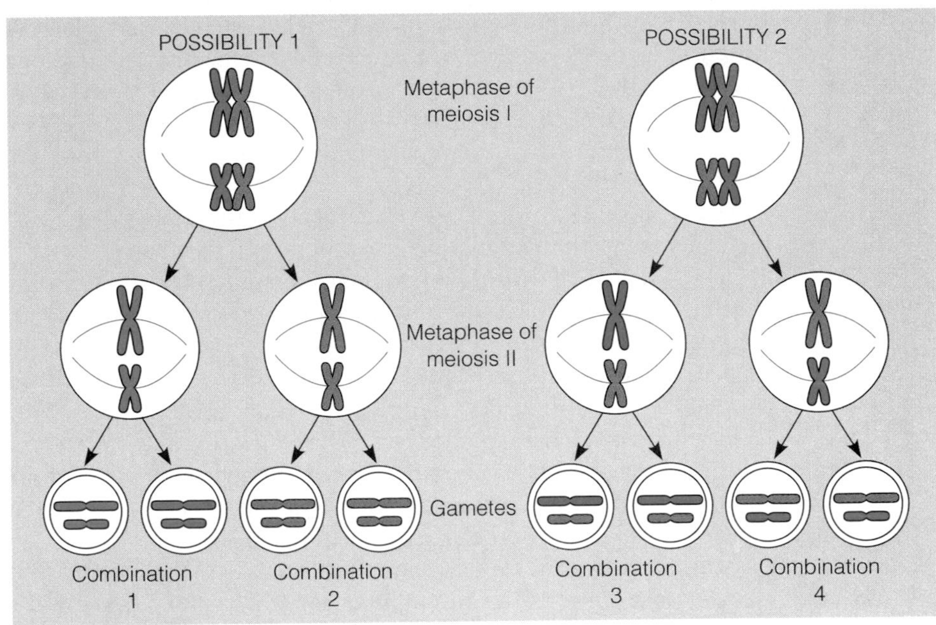

Figure 8.16 Results of the independent orientation of chromosomes at metaphase

Figure 8.16 illustrates one way in which the process of meiosis contributes to genetic differences in gametes. The figure shows how the arrangement of homologous chromosome pairs at metaphase of meiosis I affects the resulting gametes. Once again, our example is an organism with a diploid chromosome number of four ($2n = 4$). Here we use red and blue to distinguish chromosomes that originated from the organism's two parents. The colors highlight the important fact that each maternal chromosome differs genetically from its paternal homologue, although the two look alike under a microscope. (In fact, each chromosome you received from your mother carries many genes that are different versions from those on the homologous chromosome you received from your father.)

The orientation of the homologous pairs of chromosomes (tetrads) at metaphase I is a matter of chance, like the flip of a coin. In this example, there are two possible ways that the two tetrads can align during metaphase I. In possibility 1, the tetrads are oriented with both red chromosomes on the same side of the metaphase plate. In this case, one of the two daughter cells of meiosis I receives only red chromosomes, and the other cell receives only blue chromosomes (you can see these cells, at the metaphase II stage, in the second row of the figure). Therefore, the gametes produced in possibility 1 can each have only red *or* blue chromosomes (bottom row, combinations 1 and 2).

In possibility 2, the tetrads are oriented differently, with one red and one blue chromosome on each side of the metaphase I plate. This arrangement produces gametes that each have one red and one blue chromosome. Furthermore, half the gametes have a big blue chromosome and a small red one (combination 3), and half have a big red one and a small blue one (combination 4).

So we see that for this example a total of four chromosome combinations is possible in the gametes, and in fact the organism will produce gametes of all four types. This variety in gametes arises because each homologous pair of chromosomes orients itself on the metaphase I plate independently of the other pair. For a species with more than two pairs of chromosomes, such as the human, *all* the chromosome pairs orient independently at metaphase I. (Chromosomes *X* and *Y* behave as a homologous pair in meiosis.)

For any species, the total number of combinations of chromosomes that meiosis can package into gametes is 2^n, where n is the haploid number. For the organism in this figure, $n = 2$, so the number of chromosome combinations is 2^2, or 4. For a human ($n = 23$), there are 2^{23}, or about 8 million possible chromosome combinations! This means that every gamete a human produces contains one of about 8 million possible combinations of maternal and paternal chromosomes.

How many possibilities are there when a gamete from one individual unites with a gamete from another individual in fertilization? A human egg cell, representing one of about 8 million possibilities, will be fertilized at random by one sperm cell, representing one of about 8 million other possibilities. By multiplying 8 million times 8 million, we find that a man and a woman can produce a diploid zygote with any of 64 trillion combinations of chromosomes! Thus, the random nature of fertilization adds a huge amount of potential variability to the offspring of sexual reproduction.

These large numbers suggest that independent orientation of chromosomes at metaphase I and random fertilization could account for all the variety we see among people. Actually, these two events are only part of the picture, as we see in the next two modules.

? A particular species of worm has a diploid number of 10. How many chromosomal combinations are possible for gametes formed by meiosis?

32

8.17 Homologous chromosomes carry different versions of genes

So far, we have focused on genetic variability in gametes and zygotes at the whole-chromosome level. We have yet to discuss the actual genetic information—the genes—contained in the chromosomes of gametes and zygotes. The question we need to answer now is this: What is the significance of the independent orientation of metaphase chromosomes at the level of genes?

Let's take a simple example, the single tetrad in Figure 8.17A. The letters on the homologous chromosomes represent genes. Recall that homologous chromosomes have genes for the same characteristic at corresponding loci. Our example involves some imaginary genes controlling the appearance of mice (the precise effects of the actual genes are too complex for our purposes here). *C* and *c* are different versions of a gene for one characteristic, coat color; *E* and *e* are different versions of a gene for another characteristic, eye color. (As you'll learn in later chapters, different versions of a gene are called *alleles* and are slightly different nucleotide sequences in the chromosomal DNA.)

Let's say that letter *C* represents the gene for the brownish color of the two mice on the left in Figure 8.17B and that letter *c* represents the gene for white coat color (mouse on right). In the chromosome diagram, notice that *C* is at the same locus on the red homologue as *c* is on the blue one. Likewise, gene *E* (for black eyes) is at the same locus as *e* (pink eyes).

The fact that homologous chromosomes can bear two different kinds of genetic information for the same characteristic (for instance, coat color) is what really makes gametes—and therefore offspring—different from one another. In our example, a gamete carrying a red chromosome would have genes specifying brownish coat color and black eye color, while a gamete with the homologous blue chromosome would have genes for white coat and pink eyes. Thus, we see how a tetrad with genes shown for only two characteristics can yield two genetically different kinds of gametes. In the next module, we go a step further and see how this same tetrad can actually yield *four* different kinds of gametes.

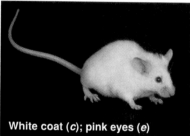

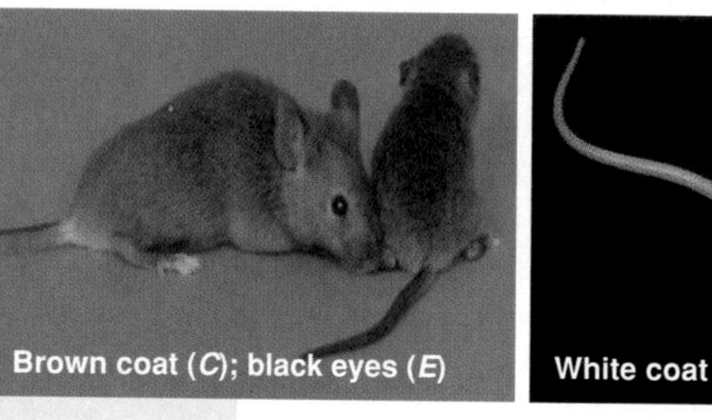

Figure 8.17B Coat-color and eye-color traits in mice

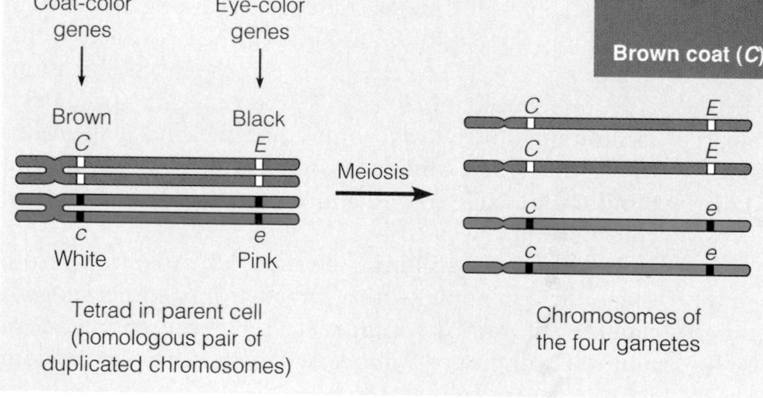

Figure 8.17A Differing genetic information on homologous chromosomes

In the tetrad of Figure 8.17A, use labels to distinguish the homologous pair of chromosomes from sister chromatids.

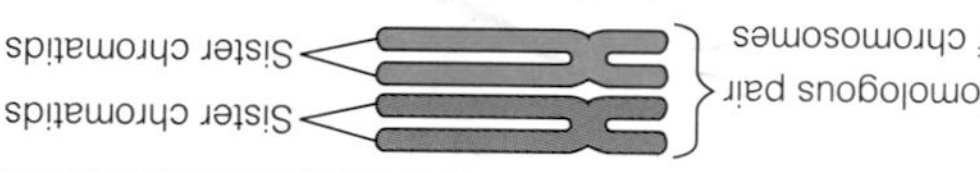

8.18 Crossing over further increases genetic variability

8F

Crossing over is the exchange of corresponding segments between two homologous chromosomes. The micrograph and drawing in Figure 8.18A show the results of crossing over between two homologous chromosomes during prophase I of meiosis. The chromosomes are a tetrad—four chromatids, with each pair of sister chromatids joined at their centromeres. The sites of crossing over appear as X-shaped regions; each is called a **chiasma** (Greek for cross). A chiasma (plural, *chiasmata*) is a place where two homologous (nonsister) chromatids are attached to each other.

Crossing over takes place during synapsis. At that time, homologous chromosomes are closely paired, with a precise, gene-by-gene alignment. The exchange of segments by homologous chromatids adds to the genetic variability that will result from the independent orientation of chromosomes at metaphase I.

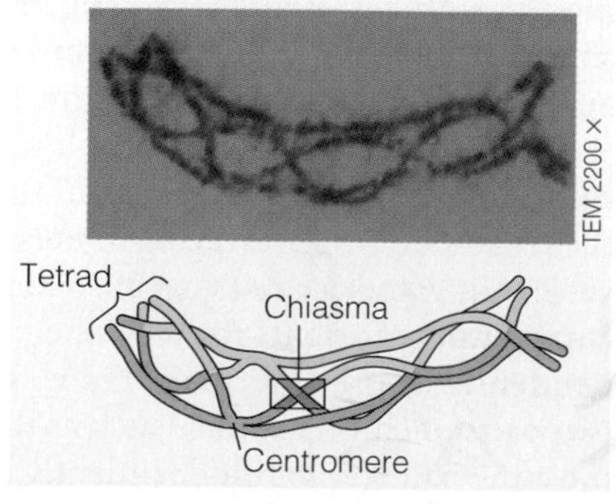

Figure 8.18A Chiasmata

Figure 8.18B illustrates how crossing over can produce new combinations of genes, using as examples the mouse genes mentioned in the previous module. The process begins during prophase I of meiosis. At the top of the figure is a tetrad with coat-color (*C*, *c*) and eye-color (*E*, *e*) genes labeled. In step ①, a chromatid from each homologous chromosome breaks in two; notice that the two chromatids break at corresponding points. ② Immediately, the two broken chromatids join together in a new way; the result is a chiasma. ③ When the homologous chromosomes separate in anaphase I, the joined homologous chromatids come completely apart. However, as the colors indicate, crossing over has changed the content of these two chromatids. A segment of one chromatid has changed place with the equivalent segment of its homologue. ④ Finally, in meiosis II, the sister chromatids separate, each going to a different gamete.

In this example, if there were no crossing over, meiosis could produce only two genetic types of gametes. These would be the ones ending up with the "parental" types of chromosomes, carrying either genes *C* and *E* or genes *c* and *e*. These are the same two kinds of gametes we saw in Figure 8.17A. With crossing over, two other types of gametes can result. One of these carries genes *C* and *e*, and the other carries genes *c* and *E*. The chromosomes carried by these gametes are called "recombinant" because they result from **genetic recombination,** the production of gene combinations different from those carried by the original chromosomes.

But our two-gene model oversimplifies the situation. The recombined segments of these chromatids carry many genes rather than just two; consequently, a single crossover event would affect many genes. When we consider that most chromosomes contain thousands of genes and that multiple crossovers can occur in each tetrad, it's no wonder that gametes and the offspring that result from them can be so varied. In fact, it's surprising that even siblings resemble one another as much as they do.

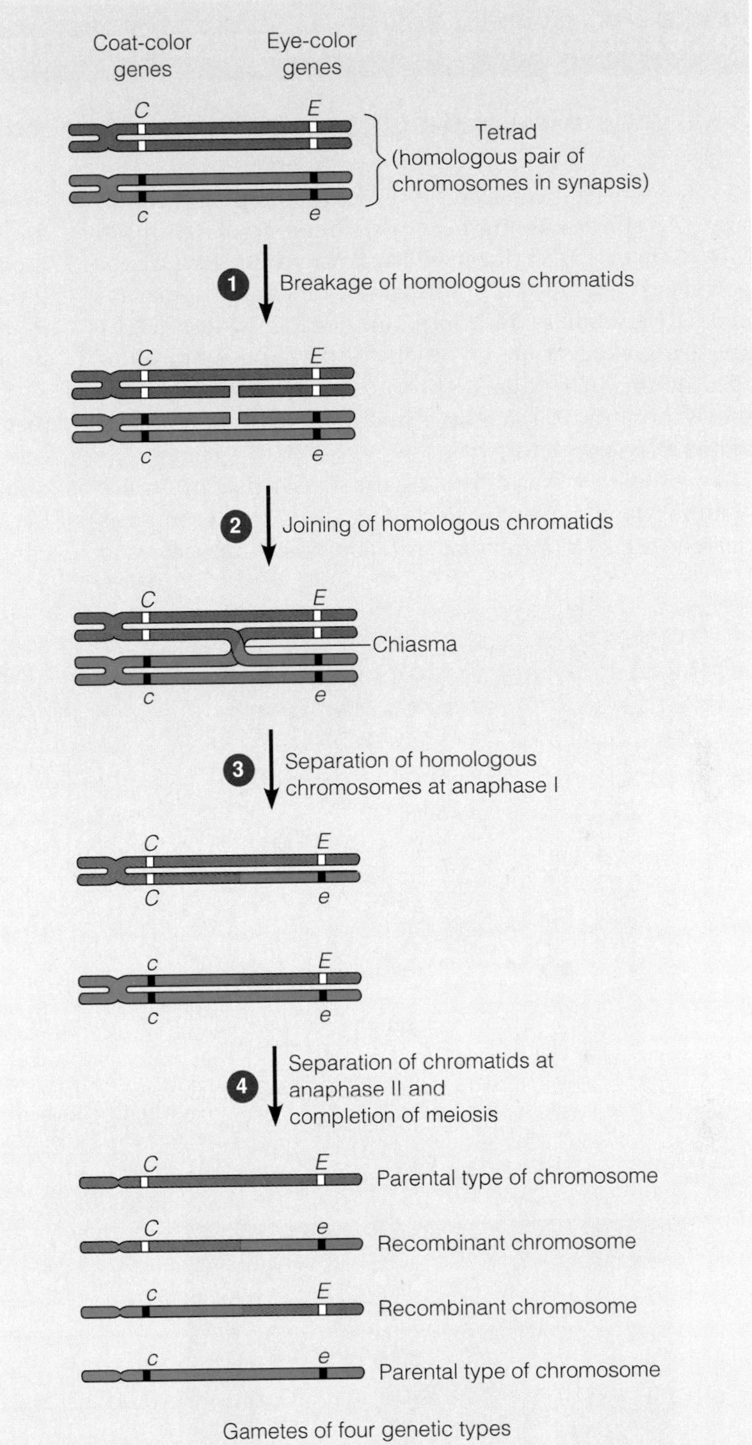

Figure 8.18B How crossing over leads to genetic recombination

We have now examined three sources of genetic variability in sexually reproducing organisms: crossing over during prophase I of meiosis, independent orientation of chromosomes at metaphase I, and random fertilization. When we take up molecular genetics in Chapter 10, we will see yet another source of variability—mutations, or rare changes in the DNA of genes. The different versions of genes that homologous chromosomes may have at each locus originally arise from mutations, so mutations are ultimately responsible for genetic diversity in living organisms.

Our discussion of meiosis to this point has focused on the process as it normally and "correctly" occurs. In the next, and last, major section of the chapter, we consider some of the consequences of errors in the process. 

? What two processes that you have learned about over the last several pages account for the genetic variation among gametes formed by meiosis?

(1) Crossing over creates recombinant chromosomes having a combination of genes that were originally on different, though homologous, chromosomes. (2) Homologous chromosome pairs are oriented randomly at metaphase of meiosis I (see Figure 8.16).

Alterations of Chromosome Number and Structure

8.19 A karyotype is a photographic inventory of an individual's chromosomes

Errors in meiosis can lead to gametes containing chromosomes in abnormal numbers or with major alterations in their structures. Fertilization involving these sorts of abnormal gametes results in offspring with chromosomal abnormalities. Such conditions can be readily detected by preparing a **karyotype,** an orderly display of magnified images of the individual's chromosomes. The karyotype shows the chromosomes thick and doubled, as they appear in metaphase of mitosis.

To prepare a karyotype, medical scientists often use lymphocytes, a type of white blood cell. A blood sample is treated with a chemical that stimulates the lymphocytes to divide. After growing in culture for several days, the cells are treated with another chemical to arrest mitosis at metaphase, when the chromosomes, each consisting of two joined sister chromatids, are most highly condensed. The figure below outlines the steps in one method for the preparation of a karyotype from a blood sample.

The photograph in step 5 shows the karyotype of a normal human male. The 46 chromosomes of a single, diploid cell are arranged in 23 homologous pairs: autosomes numbered from 1 to 22, and one pair of sex chromosomes. Because this is a male karyotype, it has one *X* and one *Y* chromosome. The chromosomes have been stained to re-

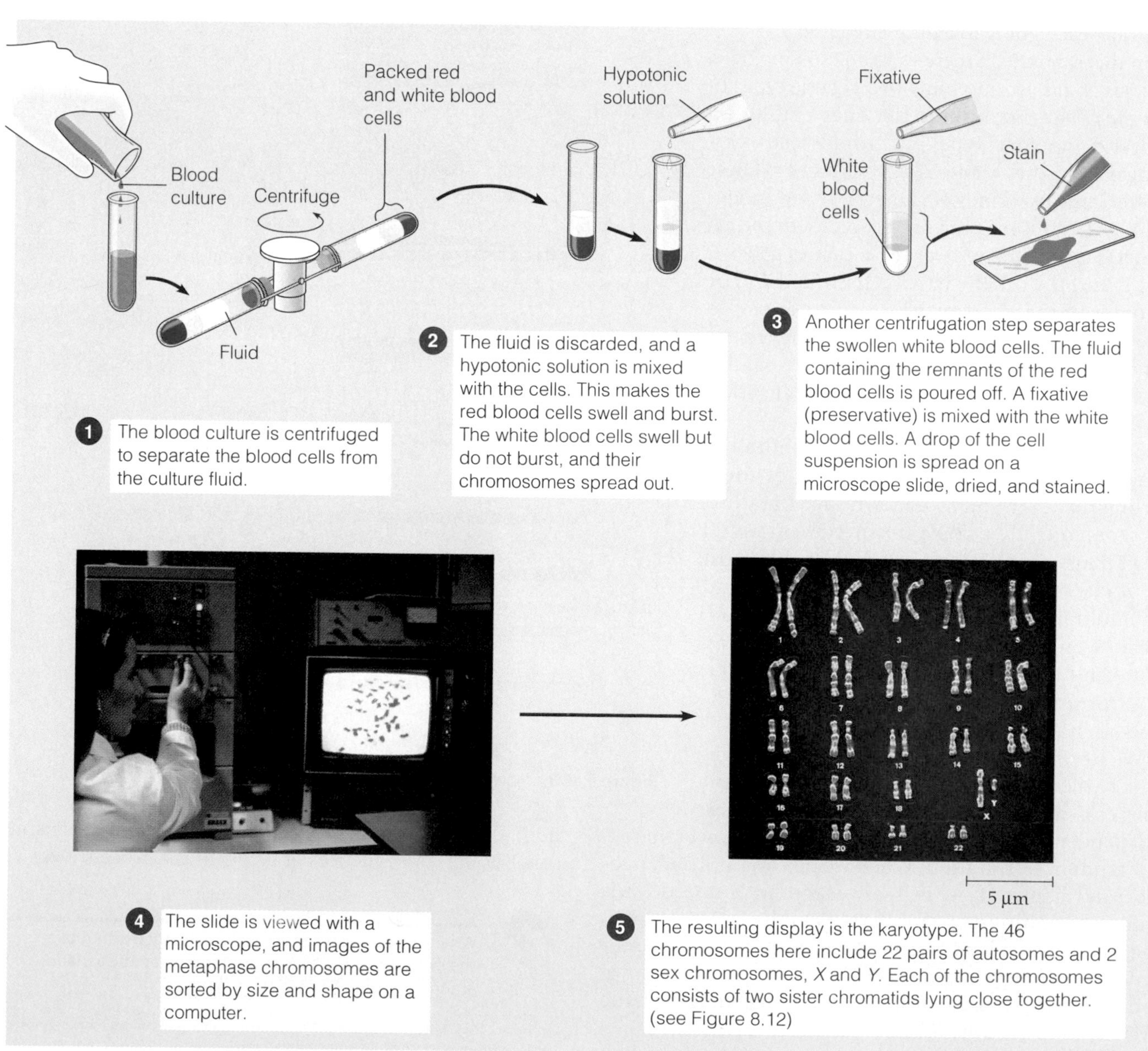

Blood culture
Centrifuge
Packed red and white blood cells
Fluid
Hypotonic solution
Fixative
White blood cells
Stain
1 The blood culture is centrifuged to separate the blood cells from the culture fluid.
2 The fluid is discarded, and a hypotonic solution is mixed with the cells. This makes the red blood cells swell and burst. The white blood cells swell but do not burst, and their chromosomes spread out.
3 Another centrifugation step separates the swollen white blood cells. The fluid containing the remnants of the red blood cells is poured off. A fixative (preservative) is mixed with the white blood cells. A drop of the cell suspension is spread on a microscope slide, dried, and stained.
4 The slide is viewed with a microscope, and images of the metaphase chromosomes are sorted by size and shape on a computer.
5 The resulting display is the karyotype. The 46 chromosomes here include 22 pairs of autosomes and 2 sex chromosomes, X and Y. Each of the chromosomes consists of two sister chromatids lying close together. (see Figure 8.12)
5 μm

Figure 8.19 Preparation of a karyotype from a blood sample

veal band patterns, which are helpful in differentiating the chromosomes and in detecting structural abnormalities. Among the alterations in chromosome number detected by karyotyping is trisomy 21, the basis of Down syndrome.

> **?** How would the karyotype of a human female differ from the male karyotype in Figure 8.19?
>
> Instead of an XY combination for the sex chromosomes, there would be a homologous pair of X chromosomes (XX).

An extra copy of chromosome 21 causes Down syndrome 8.20

The karyotype on the facing page shows the normal human complement of 23 pairs of chromosomes. The karyotype in Figure 8.20A is different; notice that there are three number-21 chromosomes. This condition is called **trisomy 21.**

In most cases, a human embryo with an abnormal number of chromosomes is spontaneously aborted (miscarried) long before birth. However, aberrations in chromosome number, including trisomy 21, appear to upset the genetic balance less drastically, so that individuals carrying them can survive. These people usually have a characteristic set of symptoms, called a syndrome. A person with an extra copy of chromosome 21, for instance, is said to have **Down syndrome** (named after John Langdon Down, who characterized it in 1866).

Trisomy 21 is the most common chromosome-number abnormality. Affecting about one out of every 700 children born, it is the most common serious birth defect in the United States. Chromosome 21 is one of our smallest chromosomes, but an extra copy produces a number of effects. Down syndrome (Figure 8.20B) includes characteristic facial features, notably a round face, flattened nose bridge, and small, irregular teeth, as well as short stature, heart defects, and susceptibility to respiratory infection, leukemia, and Alzheimer's disease.

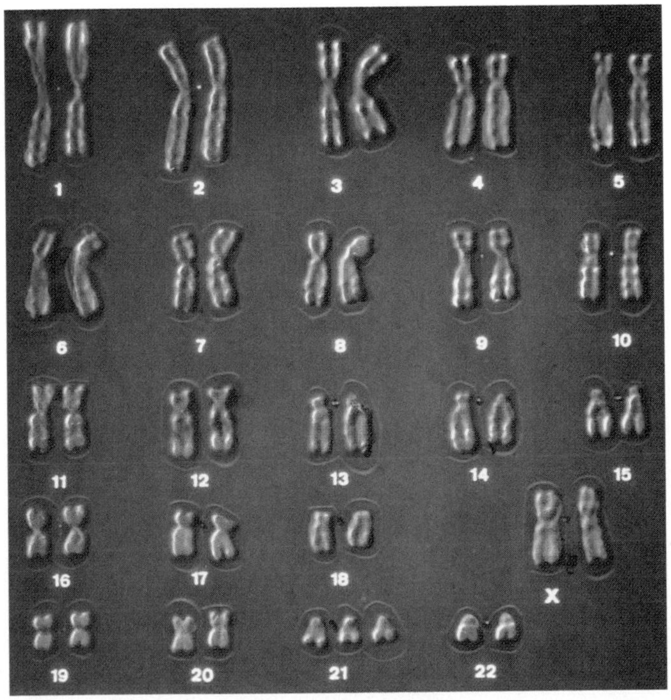

Figure 8.20A A karyotype for Down syndrome

People with Down syndrome usually have a life span much shorter than normal. They also exhibit varying degrees of mental retardation. However, some individuals with the syndrome live to middle age or beyond, and many are socially adept and able to hold jobs. A few women with Down syndrome have had children, though most people with the syndrome are sexually underdeveloped and sterile. Half the eggs produced by a woman with Down syndrome will have the extra chromosome 21, so there is a 50% chance that she will transmit the syndrome to her child.

As indicated in Figure 8.20C, the incidence of Down syndrome in the offspring of normal parents increases markedly with the age of the mother. Down syndrome strikes less than 0.05% of children (fewer than one in 2000) born to women under age 30. The risk climbs to 1% for mothers in their late 30s and is even higher for older mothers. Because of this relatively high risk, pregnant women over 35 are candidates for fetal testing for trisomy 21 and other chromosomal abnormalities (see Module 9.10).

What causes trisomy 21? We address that question in the next module.

Figure 8.20B A child with Down syndrome

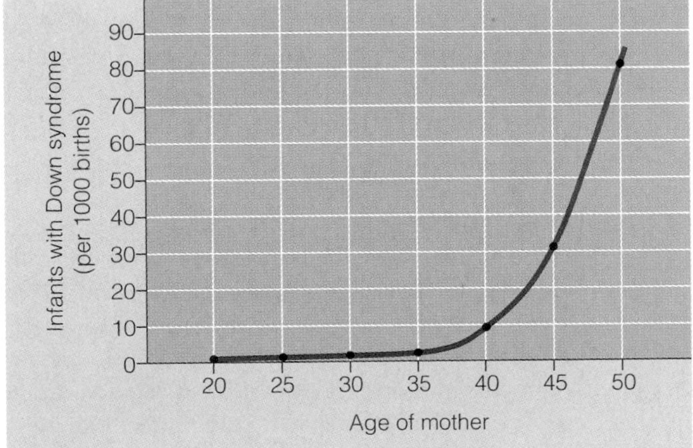

Figure 8.20C Maternal age and Down syndrome

> **?** For mothers of age 47, the risk of having a baby with Down syndrome is about __ per thousand births.
>
> 40

8.21 Accidents during meiosis can alter chromosome number

Meiosis occurs repeatedly in our lifetime, as our testes or ovaries produce gametes. Almost always, the meiotic spindle distributes chromosomes to daughter cells without error. But occasionally there is an accident, called a **nondisjunction,** in which the members of a chromosome pair fail to separate. Figures 8.21A and B illustrate two ways that nondisjunction can occur. For simplicity, we use a hypothetical organism whose diploid chromosome number is 4. In both figures, the cell at the top is diploid ($2n$), with two pairs of homologous chromosomes undergoing anaphase of meiosis I.

Sometimes, as in Figure 8.21A, a pair of homologous chromosomes does not separate during meiosis I. In this case, even though the rest of meiosis occurs normally, all the resulting gametes end up with abnormal numbers of chromosomes. Two of the gametes have three chromosomes, two of which are homologous; the other two gametes have only one chromosome each. In Figure 8.21B, meiosis I is normal, but a pair of sister chromatids fails to move apart in one of the cells during anaphase of meiosis II. In this case, two gametes have the normal complement of two chromosomes each, but the other two gametes are abnormal.

Figure 8.21C shows an example of what can happen when an abnormal gamete produced by nondisjunction unites with a normal gamete in fertilization. Here, an egg cell with two copies of one of its chromosomes (a total of $n + 1$ chromosomes) is fertilized by a normal sperm cell (n). The resulting zygote has an extra chromosome (a total of $2n + 1$ chromosomes). Mitosis will then transmit the abnormality to all embryonic cells. If this were a real organism and it survived, it would have an abnormal karyotype and probably a syndrome of disorders caused by the abnormal number of genes.

Nondisjunction can lead to an abnormal chromosome number in either sex of any sexually reproducing, diploid organism, including humans. If, for example, there is nondisjunction affecting human chromosome 21 during meiosis I, half the resulting gametes will carry an extra chromosome 21. Then, if one of these gametes unites with a normal gamete, trisomy 21 will result.

Nondisjunction explains how abnormal chromosome numbers come about, but what causes nondisjunction in the first place? We do not yet know the answer, nor do we fully understand why offspring with trisomy 21 are more likely to be born as a woman ages. We do know, however, that meiosis begins in a woman's ovaries before she is born but is not completed until years later, at the time of an ovulation. Because only one egg cell usually matures each month, a cell might remain arrested in the mid-meiosis state for decades. Perhaps damage to the cell during this time leads to meiotic errors. It seems that the longer the time lag, the greater the chance that there will be errors such as nondisjunction when meiosis is completed.

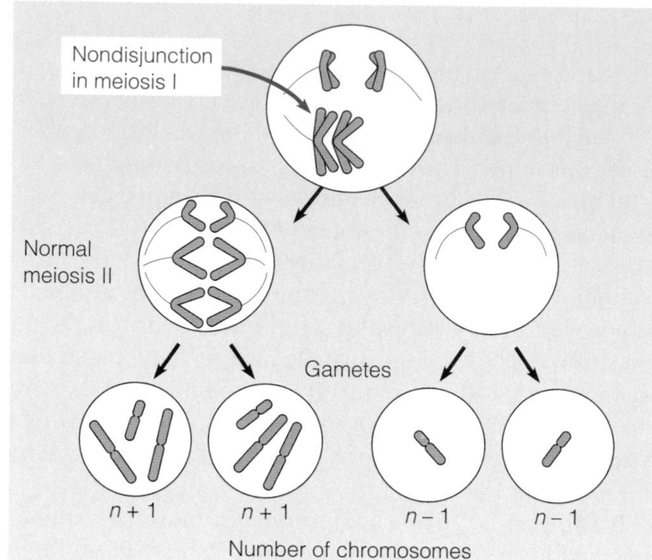

Figure 8.21A Nondisjunction in meiosis I

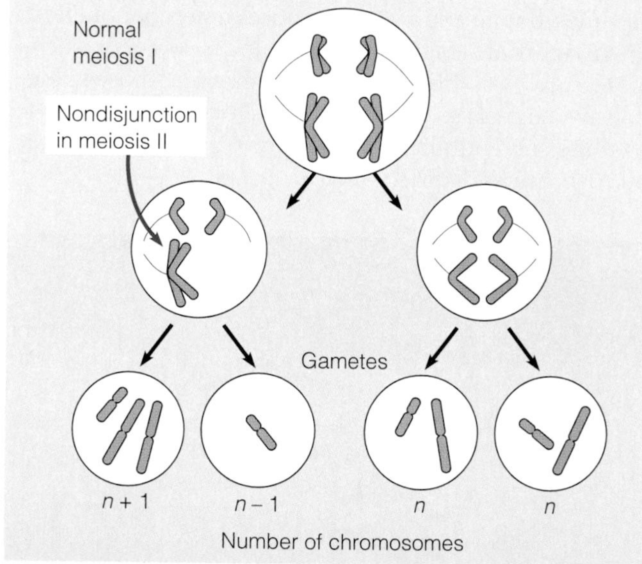

Figure 8.21B Nondisjunction in meiosis II

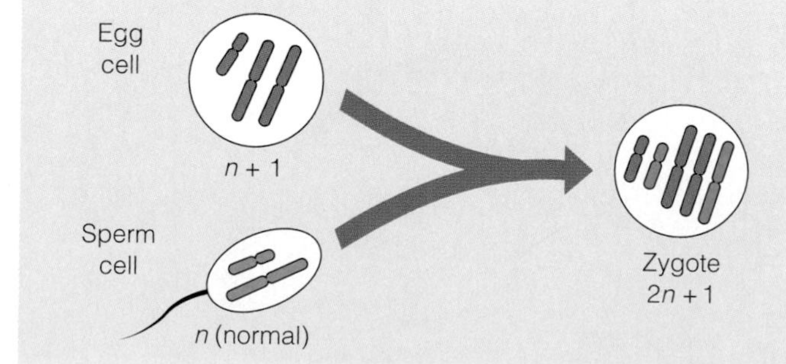

Figure 8.21C Fertilization after nondisjunction in the mother

? Explain how nondisjunction could result in a diploid gamete.

A diploid gamete would result if the nondisjunction affected all the chromosomes during one of the meiotic divisions.

Abnormal numbers of sex chromosomes do not usually affect survival

Nondisjunction in meiosis does not affect just autosomes, such as chromosome 21. It can also lead to abnormal numbers of sex chromosomes, *X* and *Y*. Unusual numbers of sex chromosomes seem to upset the genetic balance less than unusual numbers of autosomes. This may be because the *Y* chromosome is very small and carries fewer genes than other chromosomes; furthermore, most of the genes on the *Y* chromosome affect maleness but not functions that are essential to the person's survival. A peculiarity of *X* chromosomes in humans and other mammals also helps an individual tolerate unusual numbers of *X* chromosomes: In mammals, the cells usually operate with only one functioning *X* chromosome because extra copies of the chromosome become inactivated in each cell (see Module 11.5).

Abnormalities of Sex Chromosome Number in Humans

Sex Chromosomes	Syndrome	Origin of Nondisjunction	Frequency in Population
XXY	Klinefelter syndrome (male)	Meiosis in egg or sperm formation	$\frac{1}{2000}$
XYY	Normal male	Meiosis in sperm formation	$\frac{1}{2000}$
XXX	Metafemale	Meiosis in egg or sperm formation	$\frac{1}{1000}$
XO	Turner syndrome (female)	Meiosis in egg or sperm formation	$\frac{1}{5000}$

The table here lists the most common sex chromosome abnormalities. An extra *X* chromosome in a male, producing *XXY*, occurs approximately once in every 2000 live births (once in every 1000 male births). Men with this disorder, called *Klinefelter syndrome*, have male sex organs, but the testes are abnormally small and the individual is sterile. The syndrome often includes breast enlargement and other feminine body contours (Figure 8.22A). The person is usually of normal intelligence. Klinefelter syndrome is also found in individuals with more than one additional sex chromosome, such as *XXYY*, *XXXY*, or *XXXXY*. These abnormal numbers of sex chromosomes probably result from multiple nondisjunctions; such men are more likely to be mentally retarded than *XY* or *XXY* individuals.

Human males with a single extra *Y* chromosome *(XYY)* do not have any well-defined syndrome, although they tend to be taller than average. Females with an extra *X* chromosome *(XXX)* are called *metafemales;* they have limited fertility but are otherwise apparently normal.

Females who are lacking an *X* chromosome are designated *XO;* the *O* simply indicates the absence of a second sex chromosome. These women have *Turner syndrome*. They have a characteristic appearance, including short stature and often a web of skin extending between the neck and the shoulders (Figure 8.22B). Women with Turner syndrome are sterile, because their sex organs do not fully mature at adolescence, and they have poor development of breasts and other secondary sex characteristics. However, they are usually of normal intelligence. The *XO* condition occurs in about 1 in 5000 babies born (about 1 in 2500 female births).

The sex chromosome abnormalities described here illustrate the crucial role of the *Y* chromosome in determining a person's sex. In general, a single *Y* chromosome is enough to produce "maleness," even in combination with several *X* chromosomes. The absence of a *Y* chromosome results in "femaleness."

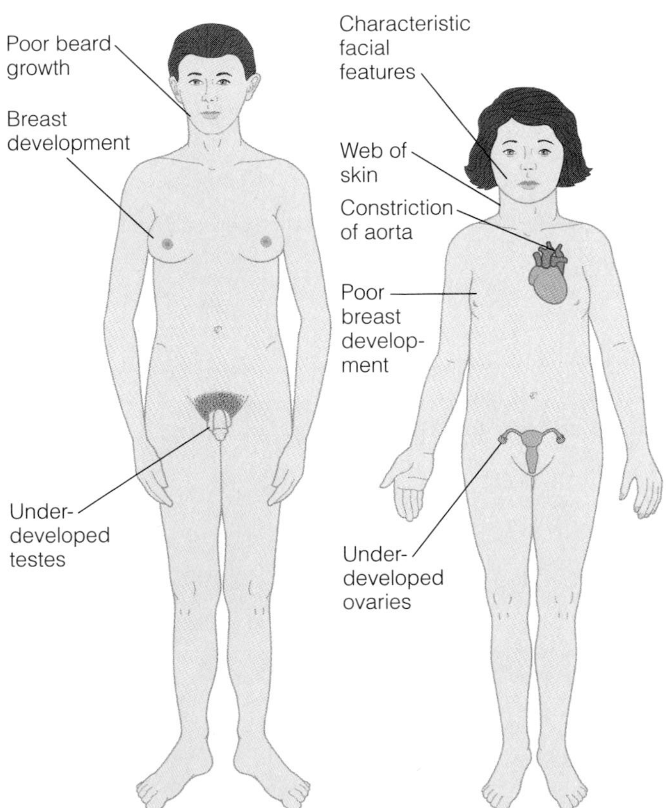

Figure 8.22A A man with Klinefelter syndrome *(XXY)*

Figure 8.22B A woman with Turner syndrome *(XO)*

What is the *total* number of chromosomes you would expect to find in the karyotype of a female with Turner syndrome?

45

8.23 Alterations of chromosome structure can cause birth defects and cancer

Even if all chromosomes are present in normal numbers, abnormalities in chromosome structure may cause disorders. Breakage of a chromosome can lead to a variety of rearrangements affecting the genes of that chromosome. Figure 8.23A shows three types of rearrangement. (The small red arrows indicate chromosome breaks.) If a fragment of a chromosome is lost, the remaining chromosome will then have a **deletion.** If a fragment from one chromosome joins to a homologous chromosome, it will produce a **duplication** there. If a fragment reattaches to the original chromosome but in the reverse direction, an **inversion** results.

Inversions are less likely than deletions or duplications to produce harmful effects, because in inversions all genes are still present in their normal number. Deletions, especially large ones, tend to have the most serious effects. One example is a specific deletion in chromosome 5 that causes the *cri du chat* ("cat-cry") syndrome. A child born with this syndrome is mentally retarded and has a small head and a cry like the mewing of a cat. Death usually occurs in infancy or early childhood.

Another type of chromosomal change is chromosomal **translocation,** the attachment of a chromosomal fragment to a nonhomologous chromosome. Figure 8.23B shows a translocation that is reciprocal; that is, two nonhomologous chromosomes exchange segments. Like inversions, translocations may or may not be harmful. Some people with Down syndrome have only part of a third chromosome 21; as the result of a translocation, it is attached to another (nonhomologous) chromosome.

Whereas chromosomal changes present in sperm or egg can cause congenital disorders, such changes in a somatic cell may contribute to the development of cancer. For example, a chromosomal translocation in somatic cells in the bone marrow is associated with *chronic myelogenous leukemia (CML).* CML is the most common of the leukemias, the cancers affecting cells that give rise to white blood cells (leukocytes). In the cancerous cells of most CML patients, a part of chromosome 22 has switched places with a small fragment from chromosome 9 (Figure 8.23C). This reciprocal translocation activates a gene that leads to leukemia. The chromosome ending up with the activated cancer-causing gene is called the "Philadelphia chromosome," after the city where it was discovered.

Because the chromosomal changes in cancer are usually confined to somatic cells, cancer is not usually inherited. We'll return to cancer in Chapter 11. In the next chapter, Chapter 9, we continue our study of genetic principles, looking first at the historical development of the science of genetics and then at the rules governing the way traits are passed from parents to offspring.

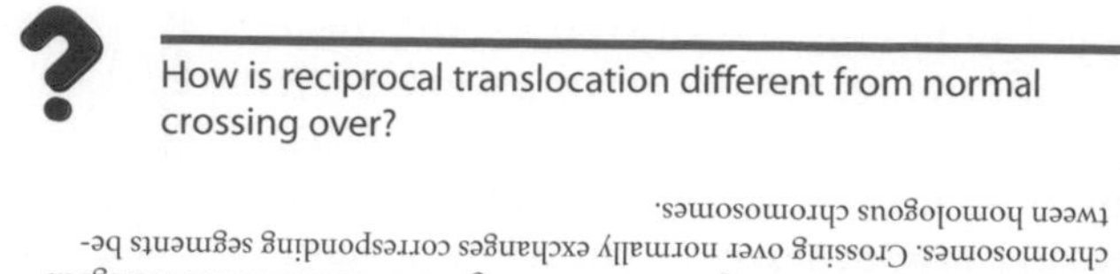

?
How is reciprocal translocation different from normal crossing over?
Reciprocal translocation swaps chromosome segments between nonhomologous chromosomes. Crossing over normally exchanges corresponding segments between homologous chromosomes.

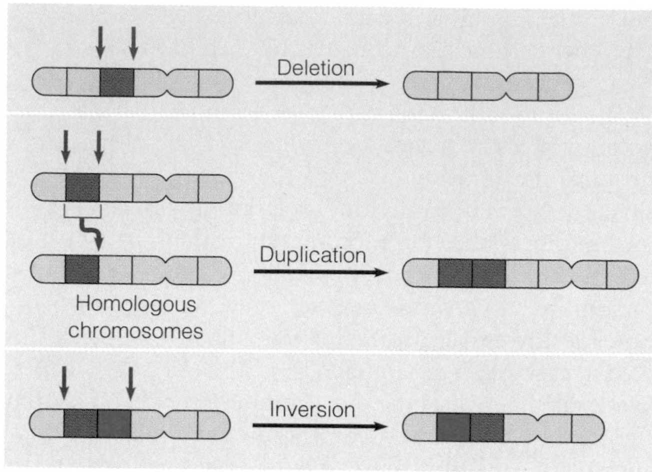

Figure 8.23A Alterations of chromosome structure involving one chromosome or a homologous pair

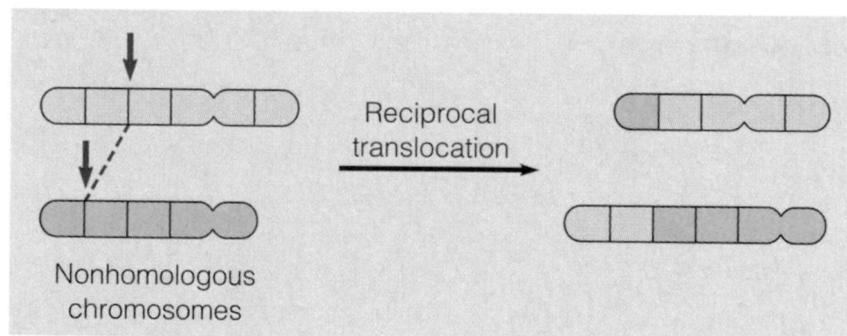

Figure 8.23B Chromosomal translocation between nonhomologous chromosomes

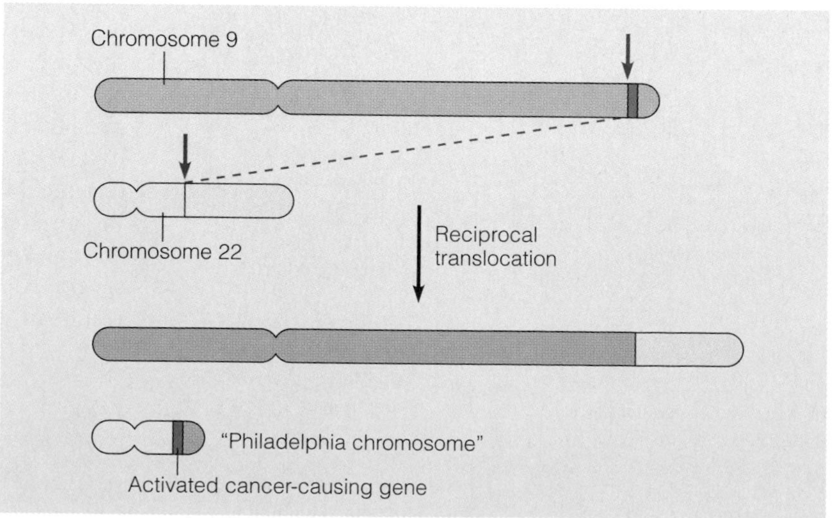

Figure 8.23C The translocation associated with chronic myelogenous leukemia

APTER REVIEW

Chapter Summary

Connections Between Cell Division and Reproduction (Introduction–8.3) Cell division is at the heart of the reproduction of cells and organisms, because cells come only from preexisting cells. Some organisms reproduce asexually, and their offspring are all genetic copies of the parent and of each other. Others reproduce sexually, producing a variety of offspring, each with a unique combination of traits **(Introduction–8.2).** Prokaryotic cells divide asexually by binary fission. A prokaryotic cell possesses a single chromosome—the structure containing its genes, which consist of DNA. The cell replicates the DNA, leaving both copies attached to the plasma membrane. Growth of the cell separates the chromosomes, and the membrane and cell wall divide the cell in two **(8.3).**

The Eukaryotic Cell Cycle and Mitosis (8.4–8.11) A eukaryotic cell has many more genes than a prokaryotic cell, and they are grouped into multiple chromosomes in the nucleus. Each chromosome contains a very long DNA molecule with thousands of genes. Individual chromosomes are visible only when the cell is in the process of dividing; otherwise, they are in the form of thin, loosely packed chromatin fibers. Before a cell starts dividing, the chromosomes replicate, producing sister chromatids (containing identical DNA) joined together at the centromere. Cell division involves the separation of sister chromatids, and results in two daughter cells, each containing a complete and identical set of chromosomes **(8.4).**

Cell division is only one phase, called the mitotic phase, of the eukaryotic cell cycle. Most of the cycle is spent in interphase, when metabolic activity is high, chromosomes duplicate, many cell parts are made, and the cell does most of its growing **(8.5).** Eukaryotic cell division consists of two processes: mitosis and cytokinesis. In mitosis, the duplicated chromosomes of one nucleus are distributed into two daughter nuclei. After the chromosomes coil up (becoming thick enough to be visible with a light microscope), a mitotic spindle made of microtubules moves them to the middle of the cell. The sister chromatids then separate and move to opposite poles of the cell, where two new nuclei form. The daughter cells receive identical sets of chromosomes. The process of cytokinesis, in which the cell divides in two, overlaps the end of mitosis. Mitosis and cytokinesis produce genetically identical cells **(8.6).** In animals, cytokinesis occurs by cleavage, which pinches the cell apart. In plants, a membranous cell plate splits the cell in two **(8.7).**

Most animal cells divide only when stimulated, and some not at all. In laboratory cultures, most normal cells divide only when attached to a surface. They continue dividing until they touch one another. Growth factors are proteins secreted by cells that stimulate other cells to divide **(8.8).** A set of proteins within the cell controls the cell cycle. Signals affecting critical checkpoints in the cell cycle determine whether a cell will go through the complete cycle and divide. The binding of growth factors to specific receptors on the plasma membrane is usually necessary for cell division **(8.9).**

Cancer cells have abnormal cell cycles. They divide excessively and can form abnormal masses called tumors. Malignant tumors can invade other tissues and may kill the organism. Radiation and chemotherapy are effective as cancer treatments because they interfere with cell division **(8.10).** When the cell cycle operates normally, mitotic cell division functions in growth, replacement of damaged and lost cells, and asexual reproduction **(8.11).**

Meiosis and Crossing Over (8.12–8.18) The somatic cells (non-gametes) of each species contain a specific number of chromosomes; for example, human cells have 46, making up 23 pairs (two sets) of homologous chromosomes. The chromosomes of a homologous pair carry genes for the same characteristics at the same place, or locus **(8.12).** Cells with two sets of homologous chromosomes are said to be diploid. Gametes—eggs and sperm—are haploid cells. Each gamete contains a single set of chromosomes. At fertilization, a sperm fuses with an egg, forming a diploid zygote. Repeated mitotic cell divisions lead to a multicellular adult made of diploid cells. The diploid adult produces haploid gametes by meiosis, a kind of cell division that reduces the chromosome number by half. Sexual life cycles involve the alternation of haploid and diploid stages **(8.13).**

Meiosis, like mitosis, is preceded by chromosome duplication, but in meiosis the cell divides twice to form four daughter cells. The first division, meiosis I, starts with synapsis, the pairing of homologous chromosomes. In crossing over, homologous chromosomes exchange corresponding segments. Meiosis I separates each homologous pair and produces two daughter cells, each with one set of chromosomes. Meiosis II is essentially the same as mitosis: In each of the cells, the sister chromatids of each chromosome separate; the result is a total of four haploid cells **(8.14).** For meiosis, as well as mitosis, chromosomes replicate only once, during the preceding interphase. Figure 8.15 reviews and compares mitosis and meiosis **(8.15).**

Each chromosome of a homologous pair comes from a different parent and hence differs at many points from the other member of the pair. The large number of possible arrangements of chromosome pairs at metaphase I of meiosis leads to many different combinations of chromosomes in eggs and sperm. This is one source of the variation in offspring that results from sexual reproduction. Random fertilization of eggs by sperm greatly increases the variation **(8.16).** The differences between homologous chromosomes are based on the fact that they can bear different versions of a gene at corresponding loci **(8.17).** Genetic recombination, which results from crossing over during prophase I of meiosis, increases variation still further **(8.18).**

Alterations of Chromosome Number and Structure (8.19–8.23) To study human chromosomes microscopically, researchers stain and display them as a karyotype, which usually shows 22 pairs of autosomes and one pair of sex chromosomes **(8.19).** Sometimes a person has an abnormal number of chromosomes, which causes problems. Down syndrome is caused by an extra copy of chromosome 21 **(8.20).** The abnormal chromosome count is a product of nondisjunction, the failure of a homologous pair of chromosomes to separate during meiosis I or of sister chromatids to separate during meiosis II **(8.21).** Nondisjunction can also produce gametes with extra or missing sex chromosomes, which lead to varying degrees of malfunction in humans but do not usually affect survival **(8.22).** Chromosome breakage can lead to rearrangements—deletions, duplications, inversions, and translocations—that can produce genetic disorders or, if the changes occur in somatic cells, cancer **(8.23).**

Testing Your Knowledge

Multiple Choice

1. If an intestinal cell in a grasshopper contains 24 chromosomes, a grasshopper sperm cell would contain ____ chromosomes.

a. 3
b. 6
c. 12
d. 24
e. 48

2. Which of the following phases of mitosis is essentially the opposite of prophase in terms of nuclear changes?

a. telophase
b. metaphase
c. S phase
d. interphase
e. anaphase

3. A biochemist measured the amount of DNA in cells growing in the laboratory and found that the quantity of DNA in a cell doubled

a. between prophase and anaphase of mitosis
b. between the G_1 and G_2 phases of the cell cycle
c. during the M phase of the cell cycle
d. between prophase I and prophase II of meiosis
e. between anaphase and telophase of mitosis

4. Which of the following is *not* a function of mitosis in humans?

a. repair of wounds
b. growth
c. production of gametes from diploid cells
d. replacement of lost or damaged cells
e. multiplication of somatic cells

5. A micrograph of a dividing cell from a mouse showed 19 chromosomes, each consisting of two sister chromatids. During which of the following stages of cell division could this picture have been taken? *(Explain your answer.)*

a. prophase of mitosis
b. telophase II of meiosis
c. prophase I of meiosis
d. anaphase of mitosis
e. prophase II of meiosis

6. Cytochalasin B is a chemical that disrupts microfilament formation. This chemical would interfere with

a. DNA replication
b. formation of the mitotic spindle
c. cleavage
d. formation of the cell plate
e. crossing over

7. It is difficult to observe individual chromosomes during interphase because

a. the DNA has not been replicated yet
b. they have uncoiled to form long, thin strands
c. they leave the nucleus and are dispersed to other parts of the cell
d. homologous chromosomes do not pair up until division starts
e. the spindle must move them to the metaphase plate before they become visible

8. A fruit fly somatic cell contains 8 chromosomes. This means that ____ different combinations of chromosomes are possible in its gametes.

a. 4
b. 8
c. 16
d. 32
e. 64

9. If a fragment of a chromosome breaks off and then reattaches to the original chromosome but in the reverse direction, the resulting chromosomal abnormality is called

a. a deletion
b. an inversion
c. a translocation
d. a nondisjunction
e. a reciprocal translocation

10. Why are individuals with an extra chromosome 21, which causes Down syndrome, more numerous than individuals with an extra chromosome 3 or chromosome 16?

a. There are probably more genes on chromosome 21 than on the others.
b. Chromosome 21 is a sex chromosome and 3 and 16 are not.
c. Down syndrome is not more common, just more serious.
d. Extra copies of the other chromosomes are probably fatal.
e. Nondisjunction of chromosome 21 probably occurs more frequently.

Describing, Comparing, and Explaining

1. Briefly describe how three different processes that occur during the sexual life cycle increase the genetic diversity of offspring.

2. What are the major differences between mitosis and meiosis?

3. In the light micrograph below of dividing cells near the tip of an onion root, identify a cell in interphase, prophase, metaphase, anaphase, and telophase. Describe the major events occurring at each stage.

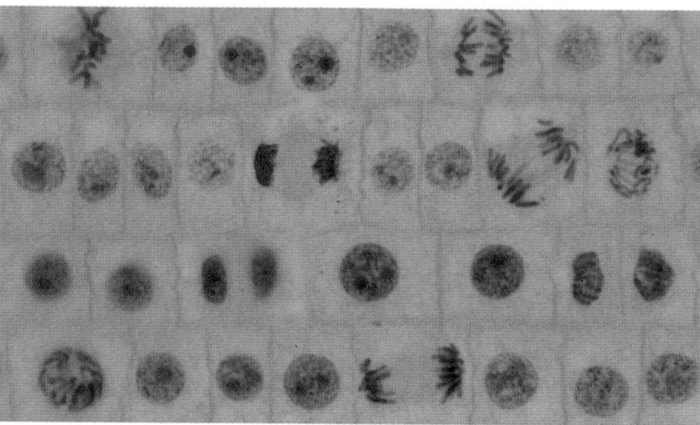

4. Discuss the factors that control the division of eukaryotic cells grown in the laboratory. Cancer cells are easier to grow in the lab than other cells. Why do you suppose this is the case?

5. Compare cytokinesis in plant and animal cells.

6. Sketch a cell with three pairs of chromosomes undergoing meiosis, and show how nondisjunction can result in the production of gametes with extra or missing chromosomes.

Thinking as a Scientist

1. Suppose you read in the newspaper that a genetic engineering laboratory has developed a procedure for fusing two gametes from the same person (two eggs or two sperm) to form a zygote. The article mentions that an early step in the procedure prevents crossing over from occurring during the formation of the gametes in the donor's body. The researchers are in the process of determining the genetic makeup of one of their new zygotes. Which of the following predictions do you think they would make? Justify your choice, and explain why you rejected each of the other choices.

a. The zygote would have 46 chromosomes, all of which came from the gamete donor (its one parent), so the zygote would be genetically identical to the gamete donor.
b. The zygote *could* be genetically identical to the gamete donor, but it is much more likely that it would have an unpredictable mixture of chromosomes from the gamete donor's parents.
c. The zygote would not be genetically identical to the gamete donor, but it would be genetically identical to one of the donor's parents.
d. The zygote would not be genetically identical to the gamete donor, but it would be genetically identical to one of the donor's grandparents.

2. Bacteria are able to divide on a much faster schedule than eukaryotic cells. Some bacteria can divide every 20 minutes, while the minimum time required by eukaryotic cells in a rapidly developing embryo is about once per hour, and most cells divide much less often than that. State several testable hypotheses explaining why bacteria can divide at a faster rate than eukaryotic cells.

3. Red blood cells, which carry oxygen to body tissues, live for only about 120 days. Replacement cells are produced by cell division in bone marrow. How many cell divisions must occur each second in your bone marrow just to replace red blood cells? Here is some information to use in calculating your answer. There are about 5 million red blood cells per mm^3 of blood. An average adult has about 5 L (5000 cm^3) of blood. (*Hints:* What is the total number of red blood cells in the body? What fraction of them must be replaced each day if all are replaced in 120 days?)

4. A mule is the offspring of a horse and a donkey. A donkey sperm contains 31 chromosomes and a horse egg 32 chromosomes, so the zygote contains a total of 63 chromosomes. The zygote develops normally. The combined set of chromosomes is not a problem in mitosis, and the mule combines some of the best characteristics of horses and donkeys. However, a mule is sterile; meiosis cannot occur normally in its testes (or ovaries). Explain why mitosis is normal in cells containing both horse and donkey chromosomes, but the mixed set of chromosomes interferes with meiosis.

5. The incidence of Down syndrome appears to increase with the increasing age of the mother. Could the increasing age of the father also be a factor? What kinds of data could you look at to determine the role of the father's age in Down syndrome?

Science, Technology, and Society

Every year about a million Americans are diagnosed as having cancer. This means that about 75 million Americans now living will eventually have cancer, and one in five will die of the disease. There are many kinds of cancers and many causes of the disease. For example, smoking causes most lung cancers. Overexposure to ultraviolet rays in sunlight causes most skin cancers. There is evidence that a high-fat, low-fiber diet is a factor in breast, colon, and prostate cancers. And agents in the workplace such as asbestos and vinyl chloride are also implicated as causes of cancer. Hundreds of millions of dollars are spent each year in the search for effective treatments for cancer; far less money is spent on preventing cancer. Why might this be the case? What kinds of lifestyle changes could we make to help prevent cancer? What kinds of prevention programs could be initiated or strengthened to encourage these changes? What factors might impede such changes and programs? Should we devote more of our resources to treating cancer or preventing it? Why?

Media Resources

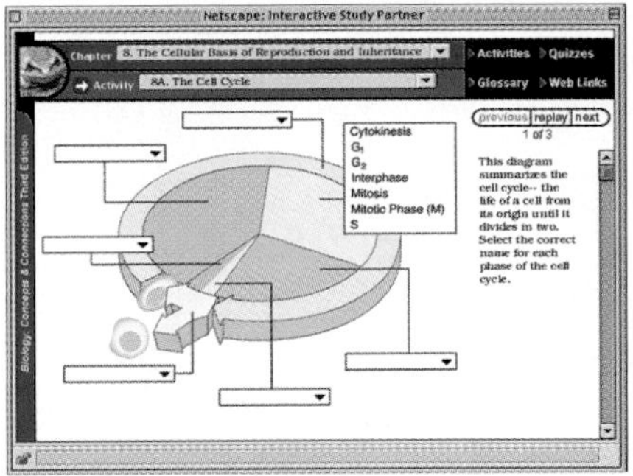

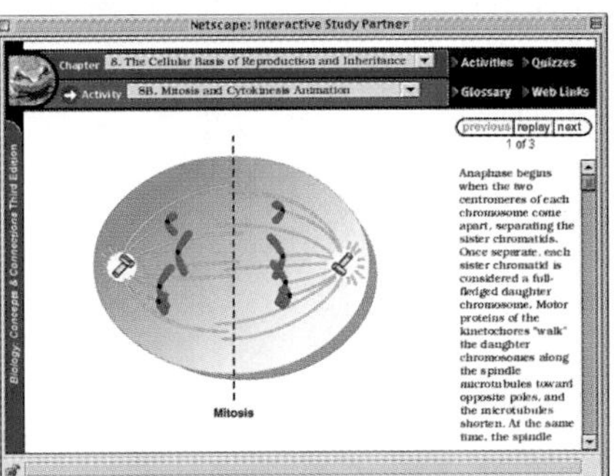

For further review of the concepts in this chapter, use the activities and quizzes on the CD-ROM.

Visit the Special Edition of The Biology Place web site for activities, quizzes, news articles, and web links.

9 Patterns of Inheritance

Outline

Budgies, also known as parakeets, are small, long-tailed parrots native to Australia and kept as pets throughout the world. Female budgies (short for budgerigars) have pinkish legs and a brownish cere, the featherless area above the beak. Male budgies have gray-blue legs and a bluish cere. Wild budgies typically have green underparts and yellow upperparts barred with black, like the bird in the top row (second from right) in the photograph at the left. Geneticists call such traits, the ones found most commonly in nature, the **wild type.** The white, blue, and yellow birds in the photo are uncommon in nature, but large numbers of these and other color variants are raised in captivity and are available in most pet stores. Blue budgies called sky-blues are especially popular.

Feather color in budgies is an inherited characteristic, and pet breeders can predict which color variants will result from particular parents. For instance, if two wild-type (green and yellow) budgies are mated, as shown in the top figure on the next page, it's likely that all their offspring will be green and yellow. The same is true when a wild-type bird and a sky-blue are mated (bottom figure). However, if two *offspring* of a wild-type budgie and a sky-blue bird are mated, their offspring, the "second generation" in the figure, will most likely include both green and yellow birds and sky-blues. Knowing how these feather-color traits are inherited—that is, their inheritance patterns—a breeder would predict that about one out of four second-generation birds would be sky-blue.

The heritable variation we see in budgies is minor compared to some of the dramatic differences breeders have developed in certain other species. For instance, of the hundreds of domestic dog breeds, some, such as Saint Bernards and Chihuahuas, are so different that they can hardly be recognized as members of the same species. In the plant kingdom, our domestic corn plant is so different from its original form that biologists have only recently identified its ancestor as a species of wild Mexican grass called teosinte.

Every species has a spectrum of heritable variation, and people have exploited this variation for centuries in developing domestic strains of plants and animals. The historical roots of **genetics,** the science of heredity, date back to the earliest attempts at selective breeding. In this chapter, we examine the rules that govern how inherited characteristics are passed from parents to offspring. We will look at several kinds of inheritance patterns, including the patterns shown here, and see how to predict the ratios of offspring with particular traits. Most importantly, we will uncover a basic biological concept—that the behavior of chromosomes during gamete formation and fertilization, discussed in Chapter 8, accounts for the patterns of inheritance that we observe.

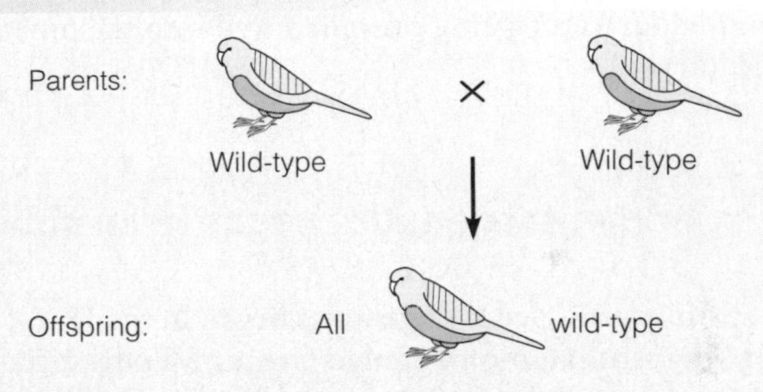

Offspring from the mating of two wild-type birds

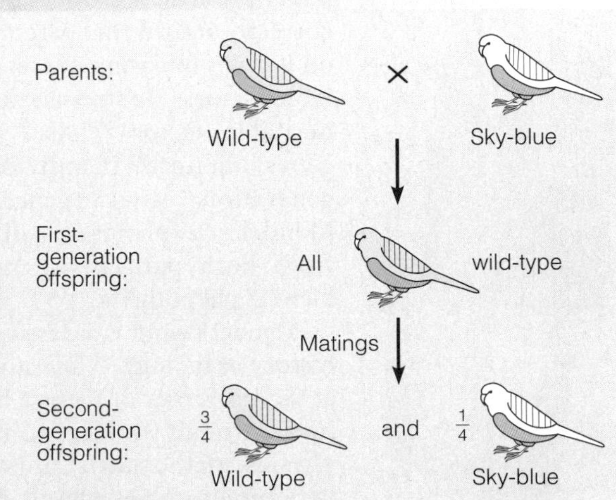

Two generations of offspring from the mating of wild-type with sky-blue birds

Mendel's Principles

9.1 The science of genetics has ancient roots

Attempts to explain inheritance date back at least to ancient Greece. Hippocrates, known as the father of medicine, suggested an explanation called pangenesis. According to this idea, particles called pangenes travel from each part of an organism's body to the eggs or sperm and are then passed to the next generation; moreover, changes that occur in various parts of the body during an organism's life are passed on in this way. The Greek philosopher Aristotle rejected this idea as simplistic, saying that what is inherited is the potential to produce body features, rather than particles of the features themselves.

Actually, pangenesis proves incorrect on several counts. The reproductive cells are not composed of particles from somatic (body) cells, and changes in somatic cells do not influence eggs and sperm. For instance, no matter how much you enlarge your biceps by lifting weights, muscle cells in your arms do not transmit genetic information to your gametes, and your offspring will not be changed by your weight-lifting efforts. This may seem like common sense today, but the pangenesis hypothesis and the idea that traits acquired during an individual's lifetime are passed on to offspring prevailed well into the nineteenth century.

By carefully tracking inheritance patterns in ornamental plants, biologists of the early nineteenth century established that offspring inherit traits from both parents. The favored explanation of inheritance then became the "blending" hypothesis, the idea that the hereditary materials contributed by the male and female parents mix in forming the offspring in the way that blue and yellow paints blend to make green. According to this hypothesis, once the genetic information for the colors of blue and yellow budgies blended in the hereditary material, they would be as inseparable as paint pigments. All offspring of the green budgies would be green, and eventually all budgies everywhere would be green. But as we saw in the introduction, green parakeets can produce blue offspring. The blending hypothesis was finally found to be untenable because it does not explain how traits that disappear in one generation can reappear in later ones. 

? Horse breeders sometimes speak of "mixing the bloodlines" of two pedigrees. How does this language hark back to the "blending" model of inheritance?

It implies that offspring are a blend of two parents as in a liquid mixture.

9.2 Experimental genetics began in an abbey garden

The modern science of genetics began in the 1860s, when an Augustinian monk named Gregor Mendel discovered the fundamental principles of genetics by breeding garden peas (Figure 9.2A). Mendel lived and worked in an abbey in Brunn, Austria (now Brno, in the Czech Republic). In a paper published in 1866, Mendel correctly argued that parents pass on to their offspring discrete heritable factors. He stressed that the heritable factors (today called genes) retain their individuality generation after generation. Mendel's explanation differed from both pangenesis and the blending hypothesis.

Figure 9.2A Gregor Mendel in his garden

Mendel's work is a classic in the history of biology. While studying at the University of Vienna, he had been strongly influenced by his physics, mathematics, and chemistry professors. As a result, his research was both experimentally and mathematically rigorous, and these qualities were largely responsible for his success.

Mendel probably chose to study garden peas because they were easy to grow and available in many readily distinguishable varieties. Also, with pea plants, Mendel was able to exercise strict control over plant matings. As Figure 9.2B shows, the petals of the pea flower almost completely enclose the female and male parts—carpel and stamens, respectively. (For better visibility, the drawing omits one of the petals.) Consequently, in nature, pea plants usually **self-fertilize,** when sperm-carrying pollen grains released from the stamens land on the tip of the egg-containing carpel of the same flower. Mendel could ensure self-fertilization by covering a flower with a small bag so that no pollen from another plant could reach the carpel. When he wanted **cross-**

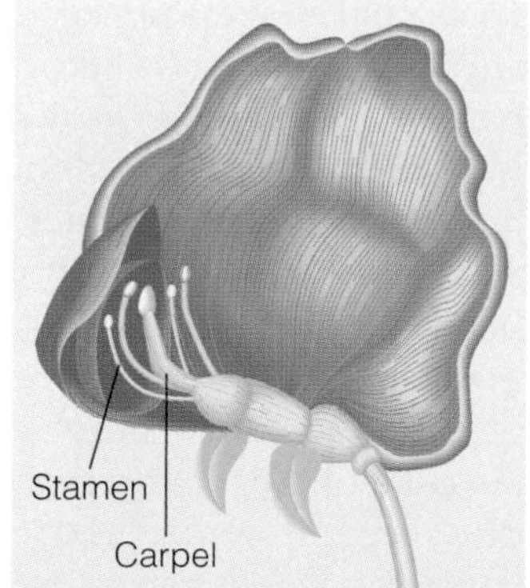

Figure 9.2B Anatomy of a pea flower

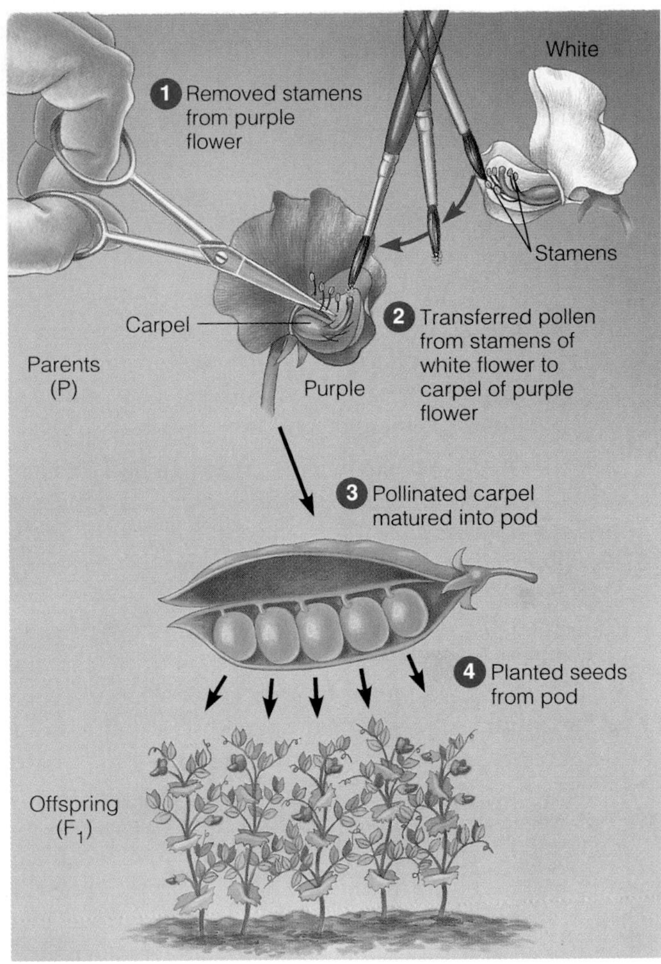

Figure 9.2C Mendel's technique for cross-fertilization of pea plants

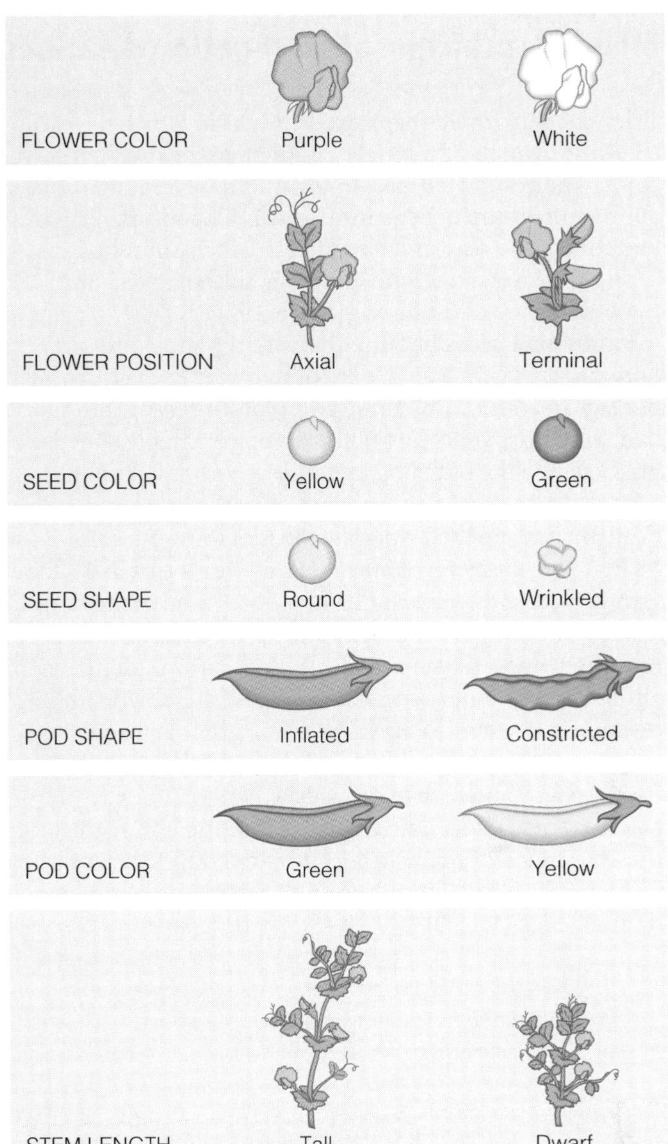

Figure 9.2D The seven pea characteristics studied by Mendel

fertilization (fertilization of one plant by pollen from a different plant), he used the method shown in Figure 9.2C. ① He first prevented self-fertilization by cutting off the stamens from an immature flower before they produced pollen. This stamenless plant would be the female parent in the experiment. ② To cross-fertilize this female, he dusted its carpel with pollen from another plant. After pollination, ③ the carpel developed into a pod, containing seeds (peas) that ④ he planted. The seeds grew into offspring plants.

Thus, Mendel could either let a pea plant self-fertilize or cross-fertilize it with a known source of pollen. In either case, he could always be sure of the parentage of new plants.

Mendel's success was due not only to his experimental approach and choice of organism, but also to his selection of characteristics to study. He chose to follow seven characteristics, each of which occurs in two distinct forms (Figure 9.2D). Mendel worked with his plants until he was sure he had *true-breeding* varieties—that is, varieties for which self-fertilization produced offspring all identical to the parent. For instance, he identified a purple-flowered variety that, when self-fertilized, produced offspring plants that all had purple flowers.

Now Mendel was ready to ask what would happen when he crossed his different varieties with each other. For example, what offspring would result if plants with purple flowers and plants with white flowers were cross-fertilized as shown in Figure 9.2C? In the language of plant and animal breeders and geneticists, the offspring of two different varieties, such as these, are called **hybrids,** and the cross-fertilization itself is referred to as a hybridization, or simply a **cross.** The parental plants are called the **P generation** (P for parental), and their hybrid offspring are the $\mathbf{F_1}$ **generation** (F for filial, from the Latin word for son). In the next module, we discuss the F_1 generation results of Mendel's flower-color crosses. We also see what happened when he allowed the F_1 offspring to self-fertilize or fertilize each other to produce the next generation of plants, the $\mathbf{F_2}$ **generation.** 

? What is a "true-breeding" variety of pea plant?

A variety that, when self-fertilized, produces offspring identical to one another and to the parent

9.3 Mendel's principle of segregation describes inheritance of a single characteristic

9A
See Activity 9A on the CD-ROM.

Mendel performed many experiments in which he tracked the inheritance of a single characteristic such as flower color. The results led him to formulate several hypotheses about inheritance. Let's look at some of his experiments and follow the reasoning that led to his hypotheses.

Figure 9.3A starts with a cross between a pea plant with purple flowers and one with white flowers. This is called a **monohybrid cross** because the parent plants differ in only one characteristic. Mendel discovered that F_1 plants (monohybrids) produced by these two true-breeding parents all had truly purple flowers. They were not a lighter purple, as predicted by the blending hypothesis. Was the heritable factor for white flowers now lost as a result of the hybridization? By mating the F_1 plants, Mendel found the answer to be no. Out of 929 F_2 plants, Mendel found that 705 (about $\frac{3}{4}$) had purple flowers and 224 (about $\frac{1}{4}$) had white flowers, a ratio of about three plants with purple flowers to every one with white flowers in the F_2 generation. Mendel concluded that the heritable factor for white flowers did not disappear in the F_1 plants, but that only the purple-flower factor was affecting F_1 flower color. He also deduced that the F_1 plants must have carried two factors for the flower-color characteristic, one for purple and one for white. From these results and many others, Mendel developed four hypotheses. Using modern terminology (including "gene" instead of "heritable factor"), his hypotheses are:

1. *There are alternative forms of genes, the units that determine heritable traits.* For example, the gene for flower color in pea plants exists in one form for purple and another for white. We now call alternative forms of genes **alleles.**
2. *For each inherited characteristic, an organism has two genes, one from each parent. These genes may both be the same allele, or they may be different alleles.*
3. *A sperm or egg carries only one allele for each inherited trait, because allele pairs separate (segregate) from each other during the production of gametes.* Moreover, when sperm and egg unite at fertilization, each contributes its allele, restoring the paired condition in the offspring.
4. *When the two genes of a pair are different alleles and one is fully expressed while the other has no noticeable effect on the organism's appearance, the alleles are called the* ***dominant allele*** *and the* ***recessive allele,*** *respectively.*

Figure 9.3B shows how Mendel explained the results given in Figure 9.3A. We use uppercase and lowercase letters to distinguish dominant from recessive alleles. *P* represents the dominant allele, for purple flowers, and *p* stands for the recessive allele, for white flowers. At the top in Figure 9.3B, you see the alleles carried by the parental plants. Both plants were true-breeding, and Mendel's first two hypotheses propose that one parental variety had two alleles for purple flowers *(PP)*, while the other variety had two alleles for white flowers *(pp)*. A true-breeding organism, which has a pair of identical alleles for a characteristic, is said to be **homozygous** for that characteristic.

Consistent with hypothesis 3, the gametes of Mendel's parental plants each carried one allele; thus, the parental

P GENERATION (true-breeding parents)
Purple flowers × White flowers
F_1 GENERATION
All plants have purple flowers
Fertilization among F_1 plants (F_1 X F_1)
F_2 GENERATION
$\frac{3}{4}$ of plants have purple flowers
$\frac{1}{4}$ of plants have white flowers

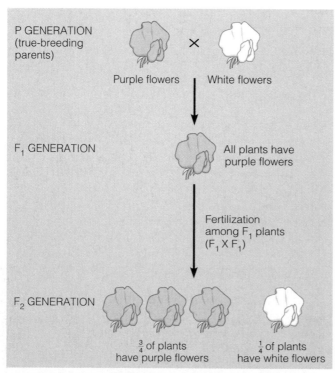

Figure 9.3A Crosses tracking one characteristic (flower color)

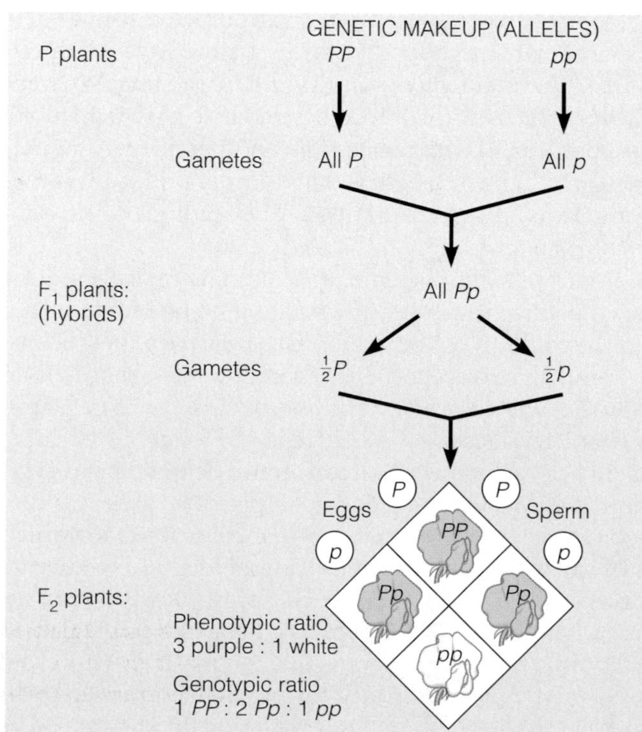

Figure 9.3B Explanation of the crosses in Figure 9.3A

gametes in Figure 9.3B are either *P* or *p*. As a result of fertilization, the F_1 hybrids each inherited one allele for purple flowers and one for white. Hypothesis 4 explains why all of the F_1 hybrids *(Pp)* had purple flowers: The dominant *P* allele is fully expressed, while the recessive *p* allele has no effect on flower color. An organism with two different alleles for a characteristic, such as a pea plant with alleles *P* and *p*, is said to be **heterozygous** for that characteristic.

Mendel's hypotheses also explain the 3:1 ($\frac{3}{4}$ purple flowers to $\frac{1}{4}$ white flowers) ratio in the F_2 generation. Because the F_1 hybrids are *Pp*, they make gametes *P* and *p* in equal numbers. The diamond in Figure 9.3B, called a **Punnett square,** shows the four possible combinations of gametes.

The Punnett square shows the proportions of F_2 plants predicted by Mendel's hypotheses. If a sperm carrying allele *P* fertilizes an egg carrying allele *P*, the offspring *(PP)* will produce purple flowers. Mendel's hypotheses predict that this combination will occur in $\frac{1}{4}$ of the offspring. As shown in the Punnett square, the hypotheses also predict that $\frac{2}{4}$ of the offspring will inherit one *P* allele and one *p* allele. These offspring *(Pp)* will all have purple flowers because *P* is dominant. The remaining $\frac{1}{4}$ of the F_2 plants will inherit two *p* alleles and will have white flowers.

Because an organism's appearance does not always reveal its genetic composition, geneticists distinguish between an organism's expressed, or physical, traits, called its **phenotype** (such as purple or white flowers), and its genetic makeup, its **genotype** (in this example, *PP*, *Pp*, or *pp*).

So now we see that Figure 9.3A shows the phenotypes and Figure 9.3B the genotypes in our sample crosses. For the offspring of our F_1 cross, the ratio of plants with purple flowers to those with white flowers (3:1) is called the phenotypic ratio. The genotypic ratio, as shown by the Punnett square, is 1*(PP)* : 2*(Pp)* : 1*(pp)*.

Mendel found that each of the seven characteristics he studied exhibited the same inheritance pattern. (In Figure 9.2D, on p. 157, the traits in the left column are all determined by dominant alleles, in the right column by recessive alleles.) One parental trait disappears in the F_1 generation of heterozygotes, only to reappear in one-fourth of the F_2 offspring. The mechanism underlying this inheritance pattern is stated by Mendel's **principle of segregation:** *Pairs of genes segregate (separate) during gamete formation; the fusion of gametes at fertilization pairs genes once again.* Research since Mendel's time has established that the principle of segregation applies to all sexually reproducing organisms, including humans. We return to Mendel and his experiments with pea plants in Module 9.5. Before that, let's see how some of the concepts we discussed in Chapter 8 fit with what we have said about genetics so far.

How can two plants that have different genotypes for a particular inherited characteristic be identical in phenotype?

One could be homozygous for the dominant allele, while the other is heterozygous.

Homologous chromosomes bear the two alleles for each characteristic 9.4

The diagrams in Figure 9.4 show a pair of homologous chromosomes. Recall from Chapter 8 that every diploid individual, whether pea plant or human, has two sets of homologous chromosomes. One set comes from the organism's female parent, the other from the male parent.

The labeled bands on the chromosomes in the figure represent a few gene loci, specific locations of genes along the chromosome. The matching colors of corresponding loci on the two homologues highlight the fact that homologous chromosomes have genes for the same characteristics located at the same positions along their lengths. However, as the uppercase and lowercase letters next to the loci indicate, the two chromosomes may bear either the same alleles or different ones. Thus, we see the connection between Mendel's principles and homologous chromosomes: *Alleles (alternative forms) of a gene reside at the same locus on homologous chromosomes.*

The diagram here also serves as a review of some of the genetic terms we have encountered to this point. We will return to the chromosomal basis of inheritance in more detail beginning with Module 9.17.

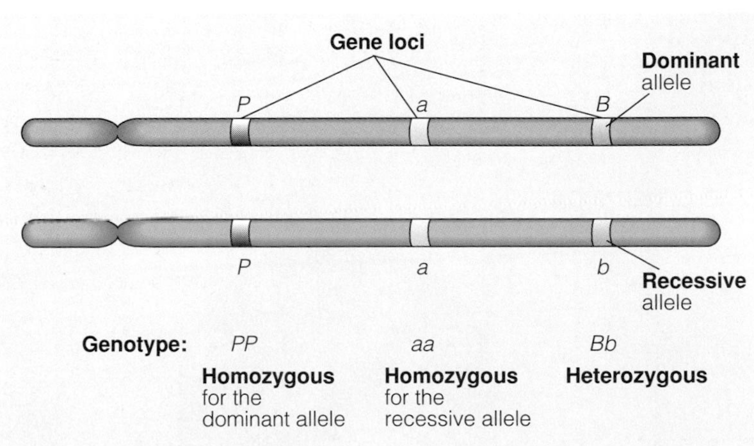

Figure 9.4 Homologous chromosomes

A particular organism is heterozygous, *Gg*, for a particular genetic locus. According to Mendel's principle of segregation, each gamete formed by this individual will have *either* the *G* allele *or* the *g* allele, but not both. Recalling what you learned about meiosis in Chapter 8, explain the physical basis for this segregation of alleles.

The *G* and *g* alleles are located on homologous chromosomes, which separate during meiosis and are packaged in separate gametes.

9.5 The principle of independent assortment is revealed by tracking two characteristics at once

9A

Two of the seven characteristics Mendel studied were seed shape and seed color. Mendel's seeds were either round or wrinkled in shape, and they were either yellow or green in color. From monohybrid crosses, Mendel knew that the allele for round shape (designated *R*) was dominant to the allele for wrinkled shape *(r)* and that the allele for yellow seed color *(Y)* was dominant to the allele for green seed color *(y)*. What would result from a mating of parental varieties differing in two characteristics—a **dihybrid cross**? Mendel crossed homozygous plants having round yellow seeds (genotype *RRYY*) with plants having wrinkled green seeds *(rryy)*. As shown in Figure 9.5A, the union of *RY* and *ry* gametes yielded hybrids heterozygous for both characteristics *(RrYy)*—that is, *dihybrids*. As we would expect, all of these offspring, the F_1 generation, had round yellow seeds. But were the two characteristics transmitted from parents to offspring as a package, or was each characteristic inherited independently of the other?

The question was answered when Mendel allowed fertilization to occur among the F_1 plants. If the genes for the two characteristics were inherited together, as shown on the left in the figure, then the F_1 hybrids would produce only the same two kinds (genotypes) of gametes that they received from their parents. In that case, the F_2 generation would show a 3:1 phenotypic ratio (three plants with round yellow seeds for every one with wrinkled green seeds), as in the left Punnett square. If, however, the two seed characteristics segregated independently, then the F_1 generation would produce four gamete genotypes—*RY, rY, Ry,* and *ry*—in equal quantities. The Punnett square on the right shows all possible combinations of alleles that can result in the F_2 generation from the union of four kinds of sperm with four kinds of eggs. If you study the Punnett square, you will see that there are nine different genotypes in the F_2. However, there are only four kinds of phenotypes, with a ratio of 9:3:3:1.

The right-hand Punnett square also reveals that a dihybrid cross is equivalent to two monohybrid crosses occurring simultaneously. From the 9:3:3:1 ratio, we can see that there are 12 plants with round seeds to 4 with wrinkled seeds, and 12 yellow-seeded plants to 4 green-seeded ones. These 12:4 ratios each reduce to 3:1, which is the F_2 ratio for a monohybrid cross. Mendel tried his seven pea characteristics in various dihybrid combinations and always observed a 9:3:3:1 ratio (or two simultaneous 3:1 ratios) of phenotypes in the F_2 generation. These results supported the hypothesis that *each pair of alleles segregates independently during gamete formation.* This is called Mendel's **principle of independent assortment.**

Figure 9.5B shows how this principle applies to the inheritance of feather color in budgies. Two genes are involved, each with two alleles *(B* and *b, C* and *c)*. Birds with the *B* allele have yellow pigment in their feathers. Those with the *C* allele have a pigment called melanin, which

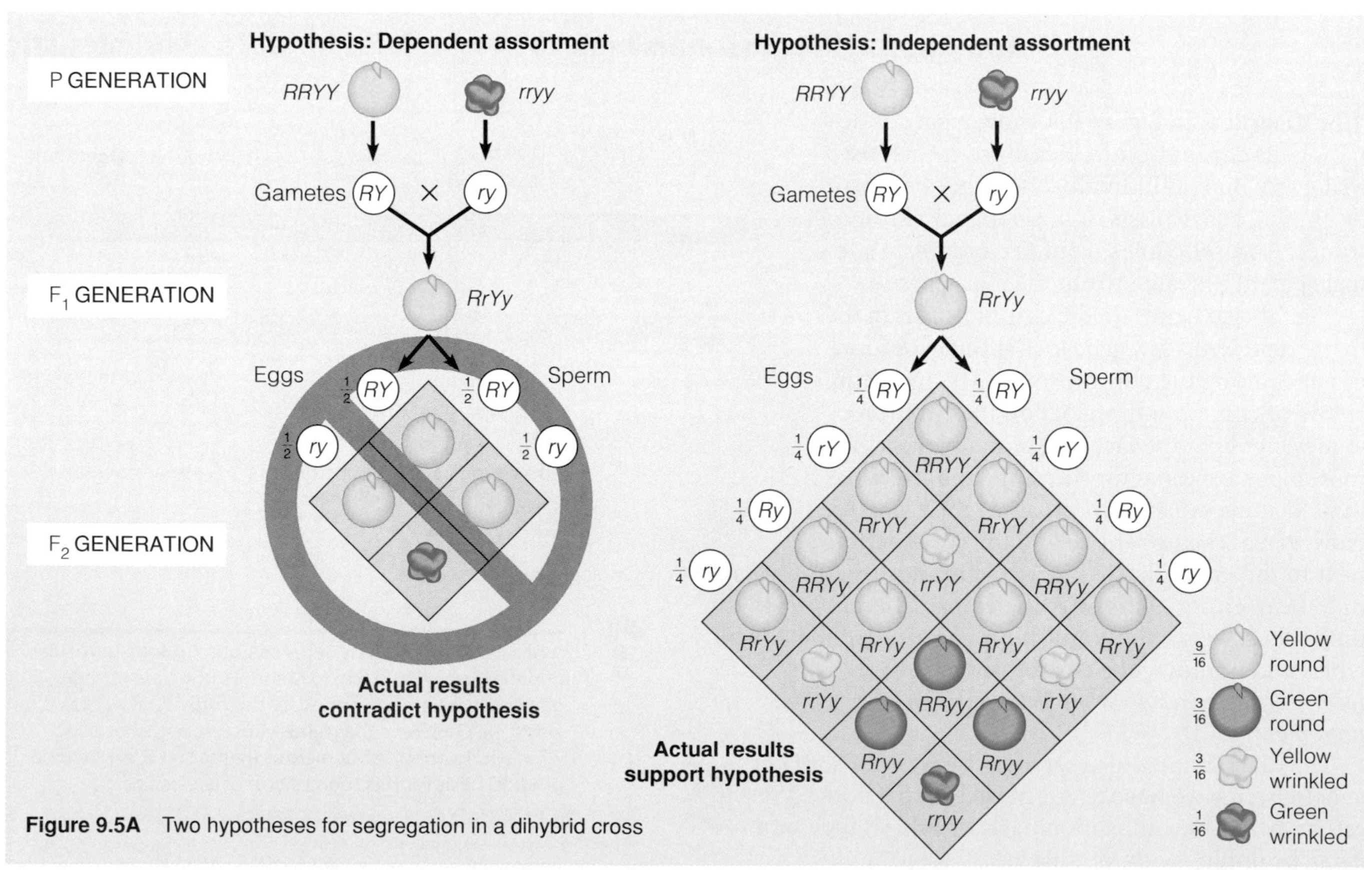

Figure 9.5A Two hypotheses for segregation in a dihybrid cross

Phenotypes	Green and yellow	Blue ("sky-blue")	Yellow ("black-eyed yellow")	White
Genotypes	$B\_C\_$	$bbC\_$	B_cc	$bbcc$

Mating of heterozygotes: (Green and yellow) *BbCc* × *BbCc* (Green and yellow)

Phenotypic ratio of offspring: 9 Green and yellow | 3 Blue | 3 Yellow | 1 White

Figure 9.5B Independent assortment in the budgie

gives feathers a blue or blackish color. The recessive alleles *c* and *b* produce no pigments in the feathers. Thus, the white budgie's genotype is *bbcc*. The other three budgies have color in their feathers because they have at least one *B* or one *C* allele. (The blanks indicate that the other alleles are unknown.) The all-yellow bird has the *B* allele but no *C*, thus no melanin pigment in its feathers. A sky-blue budgie has *C* but no *B*, thus no yellow. The green and yellow bird has both *B* and *C*; its green feathers result from the combination of yellow pigment and melanin. Its upper parts are colored by the separate expression of the *B* and *C* alleles—yellow color from *B* and black stripes from *C*.

The lower part of Figure 9.5B shows what happens when we mate two heterozygous budgies, both of genotype *BbCc*. The F_2 phenotypic ratio will be 9 green-and-yellows to 3 sky-blues to 3 pure yellows to 1 white. These results are analogous to the results in Figure 9.5A and demonstrate that the *B* and *C* genes are inherited independently.

Predict the phenotypes of offspring obtained by mating homozygous green-bodied budgies with white budgies.

All offspring would be green-bodied (*BBCC* × *bbcc* → *BbCc*.)

9.6 Geneticists use the testcross to determine unknown genotypes

Suppose you have a sky-blue budgie and you want to determine its genotype. You know that it carries two copies of the recessive allele *b* because it shows no evidence of yellow pigment. But is it homozygous (*CC*) or heterozygous (*Cc*) for the melanin allele? To answer this question, you need to perform what geneticists call a **testcross,** a mating between an individual of unknown genotype (your bird) and a homozygous recessive individual—in this case, a white budgie.

Figure 9.6 shows the offspring that could result from a mating between a white, *bbcc* budgie and a sky-blue one (*bbCC* or *bbCc*). If, as shown on the left, the sky-blue parent's genotype is *CC*, we would expect all the offspring to be sky-blue, because a cross between genotypes *CC* and *cc* can produce only *Cc* offspring. On the other hand, if the sky-blue parent is *Cc*, we would expect both sky-blue (*Cc*) and white (*cc*) offspring. Thus, the appearance of the offspring reveals the sky-blue bird's genotype. The figure also shows that you could expect the white and sky-blue offspring of a *Cc* × *cc* cross to exhibit a 1:1 (1 sky-blue to 1 white) phenotypic ratio.

Mendel used testcrosses to determine whether he had true-breeding varieties of plants. The testcross continues to be an important tool of geneticists for determining genotypes.

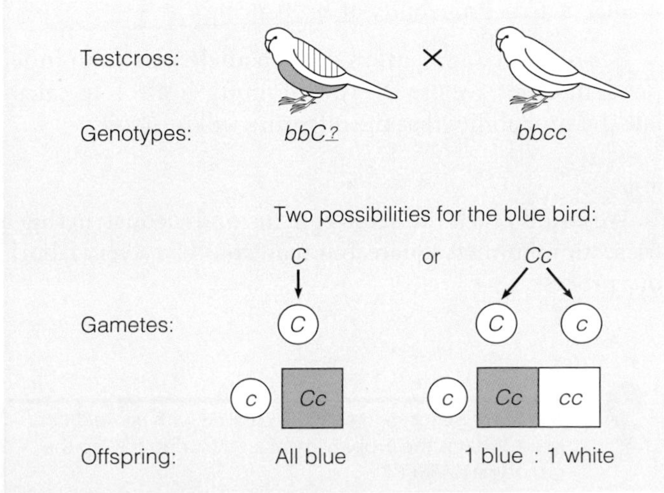

Testcross: (blue bird) × (white bird)

Genotypes: *bbC?* × *bbcc*

Two possibilities for the blue bird: *CC* or *Cc*

Gametes: *C* | *C*, *c*

Offspring: *c* × *C* → *Cc* — All blue | *c* × *C*, *c* → *Cc*, *cc* — 1 blue : 1 white

Figure 9.6 Using a testcross to determine genotype

You use a testcross to determine the genotype of a yellow budgie. Half of the offspring of the testcross are yellow and half are white. What is the genotype of the yellow parent?

Bbcc.

9.7 Mendel's principles reflect the rules of probability

Mendel's strong background in mathematics served him well in his studies of inheritance. He understood, for instance, that the segregation of allele pairs during gamete formation and the re-forming of pairs at fertilization obey the rules of probability—the same rules that apply to the tossing of coins, the rolling of dice, and the drawing of cards. Mendel also appreciated the statistical nature of inheritance. He knew that he needed to obtain large samples—count many offspring from his crosses—before he could begin to interpret inheritance patterns.

Let's see how the rules of probability apply to inheritance.The probability scale ranges from 0 to 1. An event that is certain to occur has a probability of 1, while an event that is certain not to occur has a probability of 0. The probabilities of all possible outcomes for an event must add up to 1. With a coin, the chance of tossing heads is $\frac{1}{2}$, and the chance of tossing tails is $\frac{1}{2}$. In a standard deck of 52 playing cards, the chance of drawing a jack of diamonds is $\frac{1}{52}$, and the chance of drawing any card other than the jack of diamonds is $\frac{51}{52}$.

An important lesson we can learn from coin tossing is that for each and every toss of the coin, the probability of heads is $\frac{1}{2}$. In other words, the outcome of any particular toss is unaffected by what has happened on previous attempts. Each toss is an *independent event.*

If two coins are tossed simultaneously, the outcome for each coin is an independent event, unaffected by the other coin. What is the chance that both coins will land heads up? The probability of such a *compound event* is the product of the separate probabilities of the independent events—for the coins, $\frac{1}{2} \times \frac{1}{2} = \frac{1}{4}$. This is called the **rule of multiplication,** and it holds true for genetics as well as coin tosses, as illustrated in Figure 9.7. In our dihybrid cross of budgies, the genotype of the F_1 birds for melanin pigment was *Cc*. What is the probability that a particular F_2 bird will have the *cc* genotype? To produce a *cc* offspring, both egg and sperm must carry the *c* allele. The probability that an egg will have the *c* allele is $\frac{1}{2}$, and the probability that a sperm will have the *c* allele is also $\frac{1}{2}$. By the rule of multiplication, the probability that two *c* alleles will come together at fertilization is $\frac{1}{2} \times \frac{1}{2} = \frac{1}{4}$. This is exactly the answer given by the Punnett square. If we know the genotypes of the parents, we can predict the probability for any genotype among the offspring.

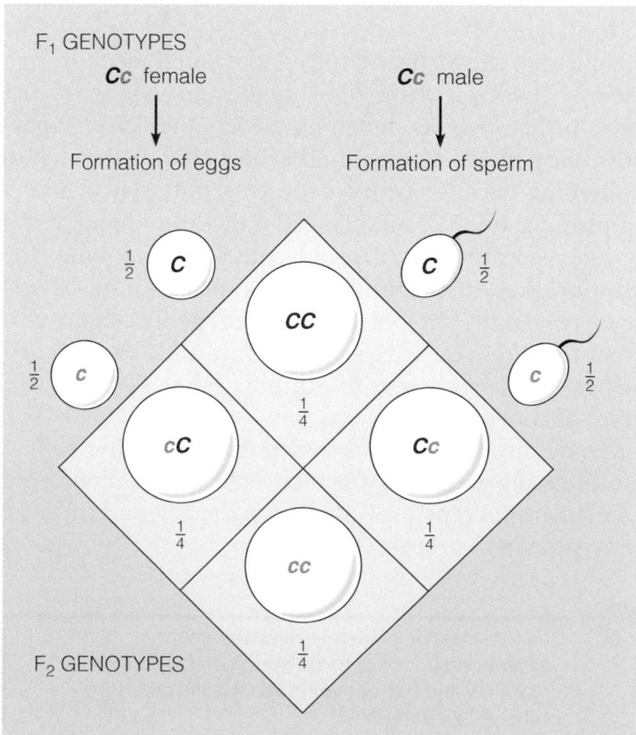

Figure 9.7 Segregation and fertilization as chance events

Now let's consider the probability that an F_2 budgie will be heterozygous for the melanin gene. As the figure shows, there are two ways in which F_1 gametes can combine to produce a heterozygous offspring. The dominant *(C)* allele can come from the egg and the recessive *(c)* allele from the sperm, or vice versa. The probability that an event can occur in two or more alternative ways is the *sum* of the separate probabilities of the different ways; this is known as the **rule of addition.** Using this rule, we can calculate the probability of an F_2 heterozygote as $\frac{1}{4} + \frac{1}{4} = \frac{1}{2}$.

By applying the rules of probability to segregation and independent assortment, we can solve some rather complex genetics problems. For instance, we can predict the results of trihybrid crosses, in which three different characteristics are involved. Consider a cross between two organisms that both have the genotype *AaBbCc*. What is the probability that an offspring from this cross will be a recessive homozygote for all three genes *(aabbcc)*? Since each allele pair assorts independently, we can treat this trihybrid cross as three separate monohybrid crosses:

$$Aa \times Aa\text{: Probability of } aa \text{ offspring} = \tfrac{1}{4}$$
$$Bb \times Bb\text{: Probability of } bb \text{ offspring} = \tfrac{1}{4}$$
$$Cc \times Cc\text{: Probability of } cc \text{ offspring} = \tfrac{1}{4}$$

Because the segregation of each allele pair is an independent event, we use the rule of multiplication to calculate the probability that the offspring will be *aabbcc:*

$$\tfrac{1}{4}\, aa \times \tfrac{1}{4}\, bb \times \tfrac{1}{4}\, cc = \tfrac{1}{64}$$

We could reach the same conclusion by constructing a 64-section Punnett square, but that would be a very laborious process. 

? A plant of genotype *AABbCC* is crossed with an *AaBbCc* plant. What is the probability of an offspring having the genotype *AABBCC*?

1/16 (that is, 1/2 × 1/4 × 1/2)

Genetic traits in humans can be tracked through family pedigrees 9.8

Mendel's principles apply to the inheritance of many human traits. Figure 9.8A illustrates alternative forms of three human characteristics that are thought to be determined by simple dominant-recessive inheritance at one gene locus. If we call the dominant allele of any such gene *A*, the dominant phenotype results from either the homozygous genotype *AA* or the heterozygous genotype *Aa*. Recessive phenotypes result from the homozygous genotype *aa*. In genetics, the word "dominant" does not imply that a phenotype is either normal or more common than a recessive phenotype; wild-type traits (those prevailing in nature) are not necessarily specified by dominant alleles. In genetics, dominance means that a heterozygote *(Aa)*, carrying only a single copy of a dominant allele, displays the dominant phenotype. By contrast, the phenotype of the corresponding recessive allele is seen only in a homozygote *(aa)*. Recessive traits are often more common in the population than dominant ones. For example, the absence of freckles is more common than their presence.

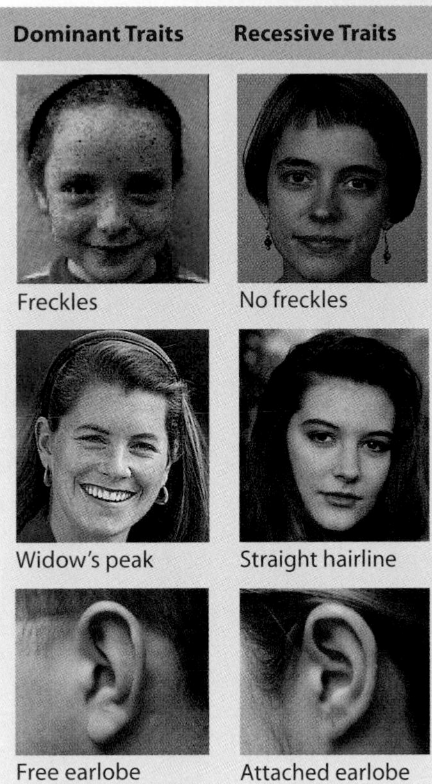

Figure 9.8A Examples of inherited traits in humans

How do we know how particular human traits are inherited? Unlike researchers working with pea plants or budgies, geneticists who study humans cannot control the mating of their subjects. Instead, they must analyze the results of matings that have already occurred. Suppose you wanted to study the inheritance of a type of deafness that is inherited as a recessive trait. First you would collect as much information as possible about a family's history for the trait. Then you would assemble this information into a family tree—the family pedigree. Finally, to analyze the pedigree, you would use Mendel's concept of dominant and recessive alleles, and his principle of segregation.

Let's apply this approach to the example in Figure 9.8B, which shows part of the pedigree from a family that lived on Martha's Vineyard, an island off the coast of Massachusetts, where a particular kind of inherited deafness was once prevalent. In the pedigree, squares represent males, and circles represent females; colored symbols here indicate deafness. The earliest generation studied is at the top of the pedigree. Notice that deafness did not appear in this generation and that it showed up in only two of the seven children in the generation at the bottom. By applying Mendel's principles, we can deduce that the deafness allele is recessive.

Mendel's principles also enable us to deduce the genotypes that are shown for most of the people in the pedigree. The letter *D* stands for the hearing allele, and *d* symbolizes the recessive allele for deafness. The first deaf individual to appear in this pedigree is Jonathan Lambert. Because only people who are homozygous for the recessive allele are deaf, his genotype must have been *dd*. Therefore, both his parents must have carried a *d* allele along with the *D* allele that gave them normal hearing. Two of Jonathan's children were deaf *(dd)*, so his wife, who had normal hearing, must also have carried a *d* allele. Likewise, all of the couple's normal children must have been heterozygous *(Dd)*. People who have one copy of the allele for a recessive disorder and do not exhibit symptoms are called **carriers** of the disorder.

What are the genotypes of Elizabeth Eddy's parents and Jonathan's sister Abigail? These three people had normal hearing, so they must have carried at least one *D* allele. And at least one of Elizabeth's parents must have had a *d* allele. But more than that we cannot say without additional information.

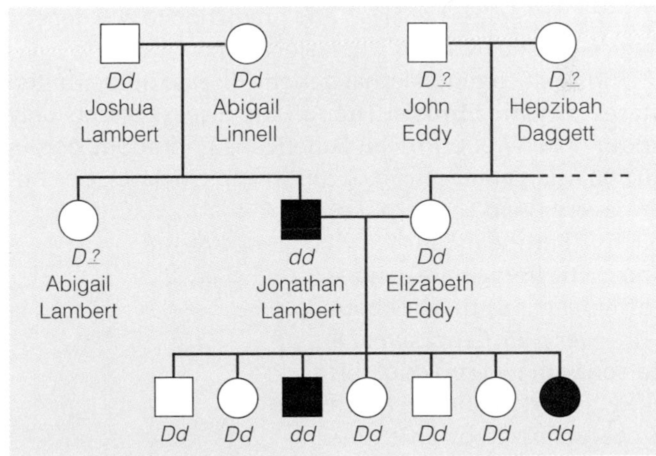

Figure 9.8B Pedigree showing inheritance of deafness in a family from Martha's Vineyard

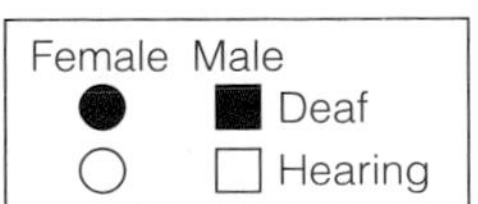

A man and a woman who are both carriers of the Martha's Vineyard deafness allele have had three children who are not deaf. If the couple has a fourth child, what is the probability that the child will be deaf?

1/4

9.9 Many inherited disorders in humans are controlled by a single gene

The hereditary deafness of Martha's Vineyard is just one of over 1000 human genetic disorders currently known to be inherited as dominant or recessive traits controlled by a single gene locus; ten more examples are listed in the table on the next page. These disorders show simple inheritance patterns like the traits Mendel studied in pea plants. The genes discussed in this module are all located on autosomes.

Recessive Disorders Most human genetic disorders are recessive. They range in severity from relatively harmless conditions, such as albinism (lack of pigmentation), to deadly diseases. The vast majority of people afflicted with recessive disorders are born to normal parents who are both heterozygotes, that is, who are carriers of the recessive allele for the disorder but are phenotypically normal. For example, in the last module, the hearing parents of Jonathan Lambert were both carriers.

Using Mendel's principles, we can predict the fraction of affected offspring likely to result from a marriage between two carriers. Suppose one of Jonathan Lambert's hearing sons *(Dd)* married a hearing cousin whose pedigree indicated that her genotype was also *Dd*. What is the probability that they would have deaf children? As the Punnett square in Figure 9.9A shows, each child of two carriers has a $\frac{1}{4}$ chance of inheriting two recessive alleles. Thus, we can say that about one-fourth of the children of this marriage are likely to be deaf. We can also say that a hearing ("normal") child from such a marriage has a $\frac{2}{3}$ chance of being a carrier (that is, 2 out of 3 of the offspring with the hearing phenotype are likely to be carriers). We can apply this same method of pedigree analysis and prediction to any genetic trait controlled by a single gene locus.

The most common lethal genetic disease in the United States is **cystic fibrosis.** Though the disease affects only about 1 in 17,000 African Americans and about one in 90,000 Asian Americans, it occurs in approximately one out of every 1800 Caucasian births (European ancestry). The cystic fibrosis allele is recessive and is carried by about 1 in every 25 Caucasians. A person with two copies of this allele has cystic fibrosis, which is characterized by an excessive secretion of very thick mucus from the lungs, pancreas, and other organs. This mucus can interfere with breathing, digestion, and liver function, and makes the person vulnerable to pneumonia and other infections. Untreated, most children with cystic fibrosis die by the time they are 5 years old. However, a special diet, antibiotics to prevent infection, frequent pounding of the chest and back to clear the lungs, and other treatments can prolong life to adulthood.

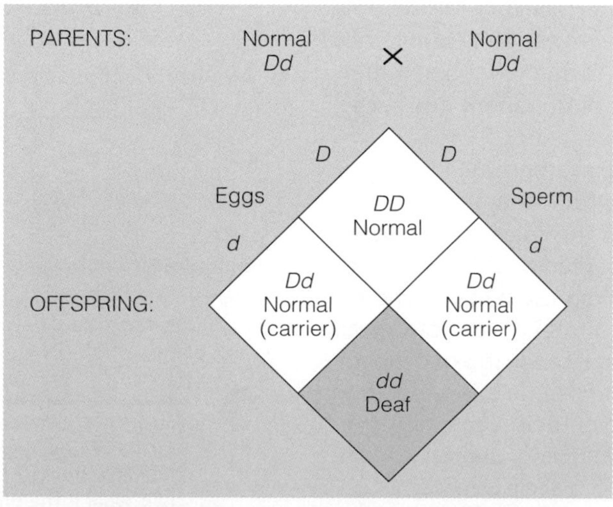

Figure 9.9A Offspring produced by parents who are both carriers for a recessive disorder

Like cystic fibrosis, most genetic disorders are not evenly distributed across all ethnic groups. Such uneven distribution is the result of prolonged geographic isolation of certain populations. For example, the isolated lives of the Martha's Vineyard inhabitants between 1700 and 1900 led to frequent marriage between close relatives. Consequently, the frequency of deafness remained high, and the deafness allele was not transmitted to outsiders.

With the increased mobility in most societies today, it is relatively unlikely that two carriers of a rare, harmful allele will meet and mate. However, the probability increases greatly if close relatives marry and have children. People with recent common ancestors are more likely to carry the same recessive alleles than are unrelated people. Therefore, a mating of close relatives is more likely to produce offspring homozygous for a harmful recessive trait.

Most societies have taboos and laws forbidding marriages between very close relatives. These rules may have arisen out of the observation that stillbirths and birth defects are more common when parents are closely related. Such effects can also be observed in many types of inbred domesticated and zoo animals. For example, dogs that have been inbred for appearance may have serious genetic disorders, such as weak hip joints and undesirable behaviors. The detrimental effects of inbreeding are also seen in some endangered species (see Module 13.16).

Although inbreeding is clearly dangerous, geneticists debate the extent to which human inbreeding increases the risk of inherited diseases. Many harmful mutations have such severe effects that a homozygous embryo spontaneously aborts long before birth—perhaps so early that the miscarriage goes undetected. Furthermore, as some geneticists argue, marriage between relatives is just as likely to concentrate favorable alleles as harmful ones. There are, in fact, some human populations, such as the Tamils of India, in which marriage between first cousins has long been common and has not produced ill effects.

Dominant Disorders Although most harmful alleles are recessive, a number of human disorders are due to dominant alleles. Some are nonlethal conditions, such as extra fingers and toes, or fingers and toes that are webbed.

One serious disorder caused by a dominant allele is

achondroplasia, a form of dwarfism. The head and torso of the body develop normally, but the arms and legs are short. (Figure 9.9B shows the late David Rappaport, an actor.) About 1 out of every 25,000 people has achondroplasia. Only heterozygotes, individuals with a single copy of the defective allele, have this disorder. The homozygous dominant genotype causes death of the embryo. Therefore, all those who do not have achondroplasia, more than 99.99% of the population, are homozygous for the normal, recessive allele. This example makes it clear that the term "dominant" does not imply that a dominant allele is somehow better than the corresponding recessive allele, or that a dominant allele will be more plentiful in a population.

Dominant alleles that are lethal are, in fact, much less common than lethal recessives. One reason for this difference is that the dominant lethal allele cannot be carried by heterozygotes without affecting them. Many lethal dominant alleles result from mutations in a sperm or egg that subsequently kill the embryo. And if the afflicted individual is born but does not survive long enough to reproduce, he or she will not pass on the lethal allele. This is in contrast to lethal recessive mutations, which are perpetuated from generation to generation by the reproduction of heterozygous carriers.

A lethal dominant allele can escape elimination, however, if it does not cause death until a relatively advanced age. By the time the symptoms become evident, the afflicted individual may have already transmitted the lethal gene to his or her children. **Huntington's disease,** a degeneration of the nervous system that usually does not begin until middle age, is one example. As the disease progresses, it causes uncontrollable movements in all parts of the body. Loss of brain cells leads to loss of memory and judgment and contributes to depression. Loss of motor skills eventually prevents swallowing and speaking. Death usually ensues 10–20 years after the onset of symptoms. (We discuss the genetic basis of Huntington's disease in more detail in Module 12.13.) 

Figure 9.9B Achondroplasia, a dominant trait

? Peter is a 30-year-old individual whose father died of Huntington's disease. Neither Peter's mother nor a much older sister, who is 48 years old, show any signs of Huntington's. What is the probability that Peter has inherited Huntington's disease?

1/2

Some Autosomal Disorders in Humans

Disorder	Major Symptoms	Incidence	Comments
Recessive disorders			
Albinism	Lack of pigment in skin, hair, and eyes	$\frac{1}{22,000}$	Very easily sunburned
Cystic fibrosis	Excess mucus in lungs, digestive tract, liver; increased susceptibility to infections; death in infancy unless treated	$\frac{1}{1800}$ Caucasians	See Modules 9.9 and 12.11
Galactosemia	Accumulation of galactose in tissues; mental retardation; eye and liver damage	$\frac{1}{100,000}$	Treated by eliminating galactose from diet
Phenylketonuria (PKU)	Accumulation of phenylalanine in blood; lack of normal skin pigment; mental retardation	$\frac{1}{10,000}$ in U.S. and Europe	See Module 9.10
Sickle-cell disease (homozygous)	Sickled red blood cells; damage to many tissues	$\frac{1}{500}$ African Americans	Alleles are codominant; see Modules 9.13–15
Tay-Sachs disease	Lipid accumulation in brain cells; mental deficiency; blindness; death in childhood	$\frac{1}{3500}$ Jews from central Europe	See Module 4.12
Dominant disorders			
Achondroplasia	Dwarfism	$\frac{1}{25,000}$	See Module 9.9
Alzheimer's disease (one type)	Mental deterioration; usually strikes late in life	Not known	
Huntington's disease	Mental deterioration and uncontrollable movements; strikes in middle age	$\frac{1}{25,000}$	See Modules 9.9 and 12.11
Hypercholesterolemia	Excess cholesterol in blood; heart disease	$\frac{1}{500}$ is heterozygous	Incomplete dominance; see Module 9.12

9.10 Fetal testing can spot many inherited disorders early in pregnancy

Many genetic disorders can be detected before birth. Tests done in conjunction with **amniocentesis** (Figure 9.10A) can determine, between the fourteenth and sixteenth weeks of pregnancy, whether the developing fetus has the condition. By this time, the fetus is about 15 cm (6 inches) long and is surrounded by a pool of liquid called amniotic fluid. To perform amniocentesis, a physician first determines the position of the fetus and then carefully inserts a needle through the mother's abdomen into her uterus, avoiding the fetus. The physician extracts a small sample—about 10 mL (2 teaspoonsful)—of the amniotic fluid. The sample is then centrifuged to separate the fluid from cells suspended in it. The cells come from the fetus, sloughed off from the skin and mouth cavity.

Once the sample is taken, some genetic disorders can be detected by immediate biochemical tests because of the presence of certain telltale chemicals in the amniotic fluid. Tests for other disorders are performed on the fetal cells. Before testing, these cells are usually cultured in the laboratory for several weeks. By then, enough dividing cells can be harvested so that karyotyping can be done and chromosomal abnormalities such as Down syndrome can be detected (see Modules 8.19 and 8.20). Biochemical tests can also be performed on the cultured cells, revealing conditions such as Tay-Sachs disease (see the table in Module 9.9).

In a newer technique called **chorionic villus sampling (CVS)** (Figure 9.10B), the physician inserts a narrow, flexible tube through the mother's vagina and cervix into the uterus and suctions off a small amount of fetal tissue (chorionic villi) from the placenta, the organ that transmits nourishment and wastes between the fetus and the mother. Because the cells of chorionic villi are proliferating rapidly, enough cells are undergoing mitosis to allow karyotyping to be carried out immediately. Results of the karyotyping, along with some biochemical tests, are available within a few hours. The speed of CVS is an advantage over amniocentesis. Another advantage is that CVS can be performed early, between the eighth and tenth weeks of pregnancy. It is appropriate for some biochemical tests, but not for those requiring amniotic fluid. Chorionic villus sampling is less

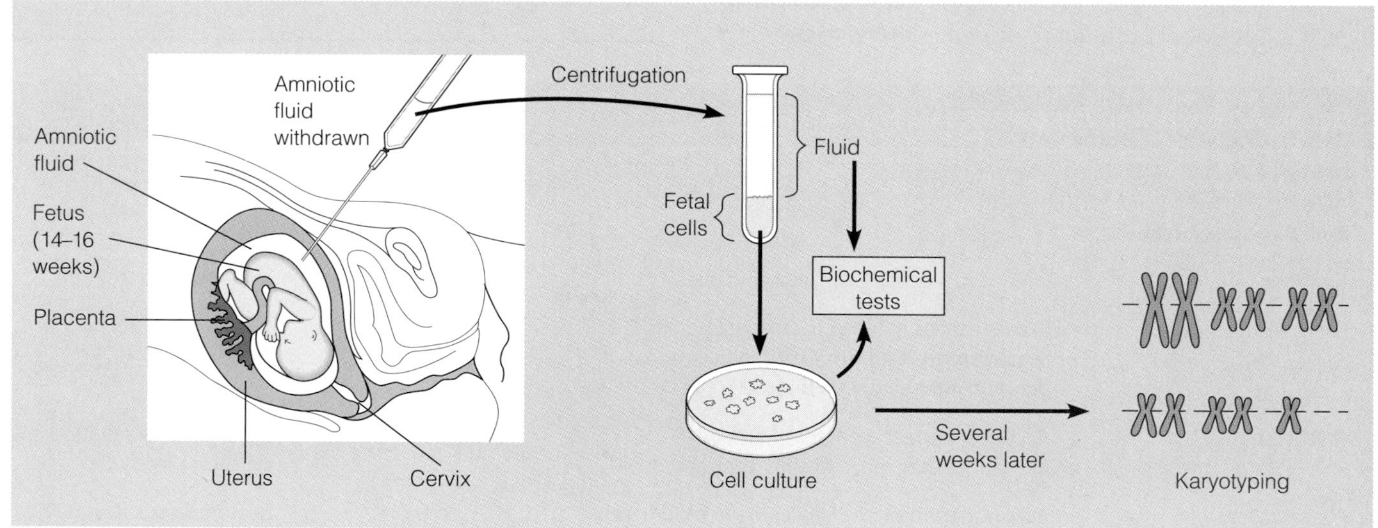

Figure 9.10A Amniocentesis

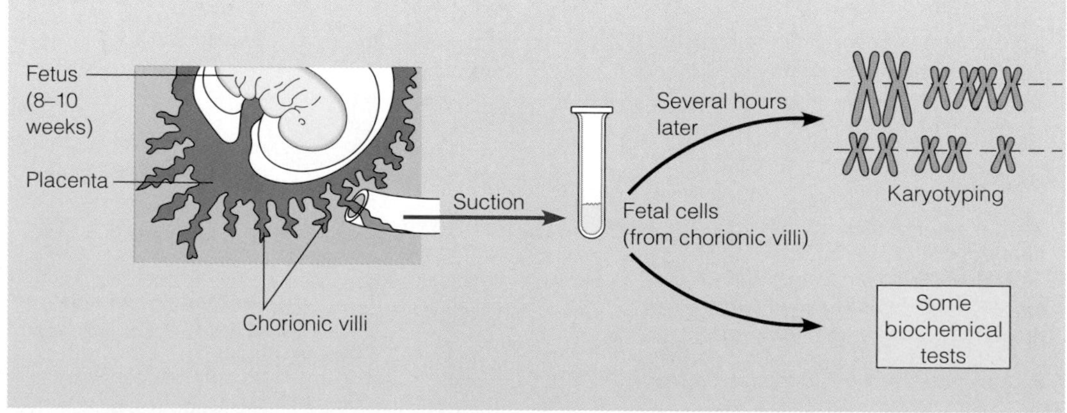

Figure 9.10B Chorionic villus sampling

widely available than amniocentesis. Recently medical scientists have developed methods for isolating fetal cells that have escaped into the mother's blood. Although very few in number, the cells can be cultured and tested.

Other techniques enable a physician to examine a fetus directly for anatomical deformities. One such technique is **ultrasound imaging,** which uses sound waves to produce an image of the fetus. Figure 9.10C shows an ultrasound scanner, which produces high-frequency sounds, beyond the range of hearing. When the sound waves bounce off the fetus, the echoes produce an image on the monitor. The color-enhanced image in Figure 9.10D shows a fetus at about eighteen weeks. Ultrasound imaging is noninvasive (no foreign objects are inserted into the mother's body) and has no known risk. This procedure is used during amniocentesis and CVS to determine the position of the fetus and of the needle or tube. With another technique, **fetoscopy,** a needle-thin tube containing a viewing scope is inserted into the uterus, giving the physician a direct view of the fetus.

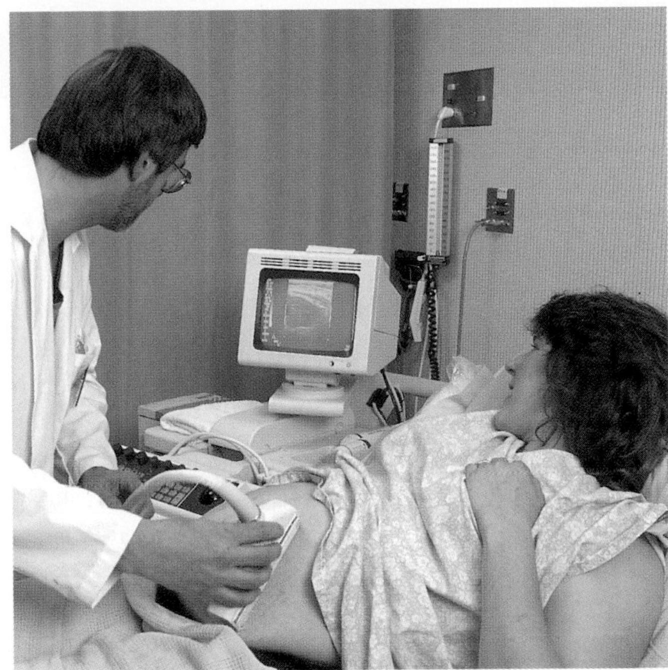

Figure 9.10C Ultrasound scanning of a fetus

Fetoscopy, amniocentesis, and chorionic villus sampling all involve some risk of complications, such as maternal bleeding, miscarriage, or premature birth. The complication rate for fetoscopy is the highest (up to about 10%). Complication rates for CVS and amniocentesis are about 2% and 1%, respectively. Because of the risks, all these techniques are usually reserved for situations in which the possibility of a genetic disorder or other type of birth defect is significant. For example, amniocentesis or chorionic villus sampling to test for Down syndrome is usually reserved for pregnant women age 35 or older.

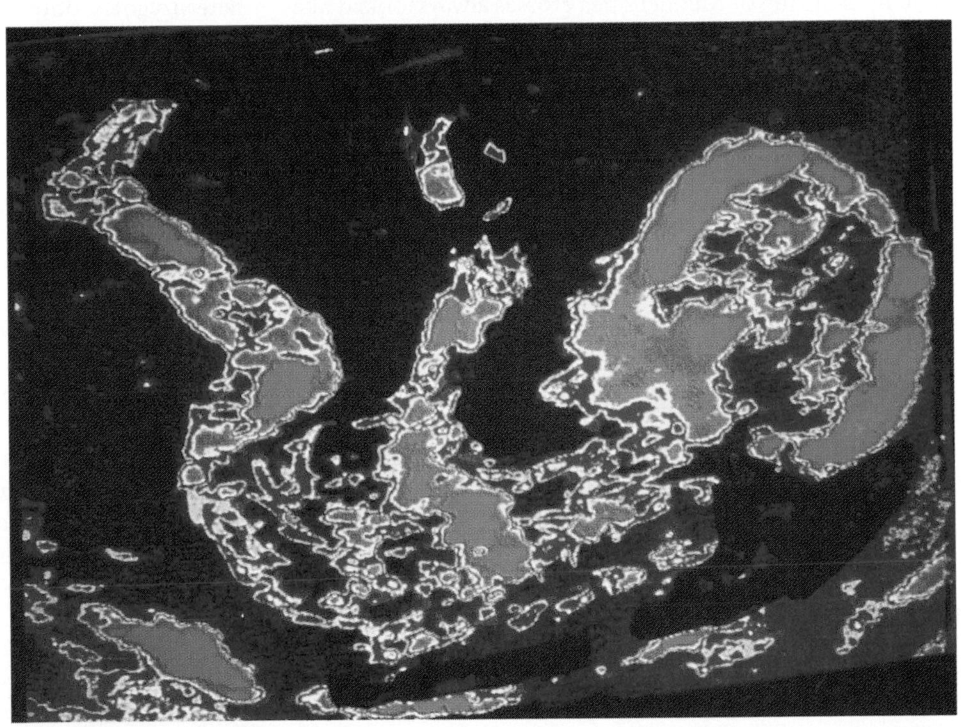

Figure 9.10D An ultrasound image

Family histories, blood tests, genetic counseling, and fetal testing offer couples a great deal of information about their unborn children. If fetal tests reveal a genetic disorder that cannot be helped by routine surgery or other therapy, the parents must choose between terminating the pregnancy and preparing themselves for a baby with a serious birth defect. CVS provides a chance to make a decision while the fetus is still quite young.

For some, no matter how undeveloped the fetus, an abortion for any reason is an unthinkable waste of human life. For others, it is a regrettable but permissible way to prevent suffering and to avoid a situation that could adversely affect their own lives, the life of the child, and the lives of other family members and close friends. Many genetic disorders, even treatable ones, require constant parental vigilance. Treatment for phenylketonuria (PKU), for instance, demands a strict diet during early childhood and reduced intake of the amino acid phenylalanine throughout life (see the table in Module 9.9). A disorder such as Down syndrome can lead to the agony of childhood leukemia or heart attack. 

?

Review the genetic basis of Down syndrome in Module 8.20. (a) In what circumstances would fetal testing be particularly advisable for detecting a fetus with Down syndrome? (b) How would the genotype for Down syndrome reveal itself in a karyotype of a fetal cell?

(a) If the mother is in her late 30s or older; (b) there would be 3 copies of chromosome 21, instead of 2.

9.11 The relationship of genotype to phenotype is rarely simple

Mendel's two principles explain inheritance in terms of discrete factors—genes—that are passed along from generation to generation, according to simple rules of probability. Mendel's principles are valid for all sexually reproducing organisms, including garden peas, birds, and human beings. But the patterns of inheritance we have described so far are simpler than most. Other inheritance patterns include cases in which one allele is not completely dominant to the other allele; cases in which there are more than two alternative alleles for a characteristic; and still other cases in which the genotype does not always dictate the phenotype in the way that Mendel's principles prescribe. Modules 9.12–9.16 consider some of these more complex inheritance patterns.

If a characteristic does not follow a Mendelian pattern, does that mean the characteristic is not inherited? Explain your answer.

No. Many inherited characteristics do not follow a simple Mendelian pattern in the relationship between genotype and phenotype.

9.12 Incomplete dominance results in intermediate phenotypes

The F_1 offspring of Mendel's pea crosses always looked like one of the two parental varieties because the dominant alleles had the same effect in one or two copies. But for some characteristics, the F_1 hybrids have an appearance *in between* the phenotypes of the two parental varieties, an effect called **incomplete dominance.** For instance, as Figure 9.12A illustrates, when red snapdragons are crossed with white snapdragons, all the F_1 hybrids have pink flowers. In heterozygotes, both the red allele and the white allele affect the flowers.

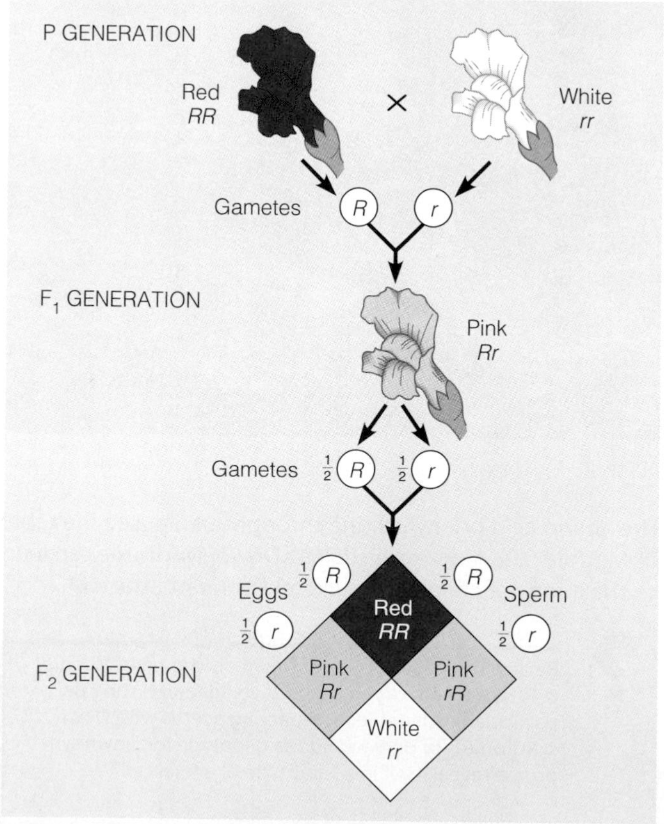

Figure 9.12A Incomplete dominance in snapdragon color

Incomplete dominance does *not* support the blending hypothesis, which would predict that the red and white traits could never be retrieved from the pink hybrids. As the Punnett square in Figure 9.12A shows, the F_2 offspring appear in a phenotypic ratio of 1 red to 2 pink to 1 white, because the red and white alleles segregate during gamete formation in the pink F_1 hybrids. In incomplete dominance, heterozygotes are phenotypically distinct from the two homozygous varieties, and the genotypic ratio and the phenotypic ratio are both 1:2:1.

We also see examples of incomplete dominance in humans. One case involves a recessive allele *(h)* responsible for *hypercholesterolemia,* dangerously high levels of cholesterol in the blood (see Module 5.20). Normal individuals are *HH.* Heterozygotes (*Hh;* about 1 in 500 people) have blood cholesterol levels about twice normal. They are unusually prone to atherosclerosis, the blockage of arteries by choles-

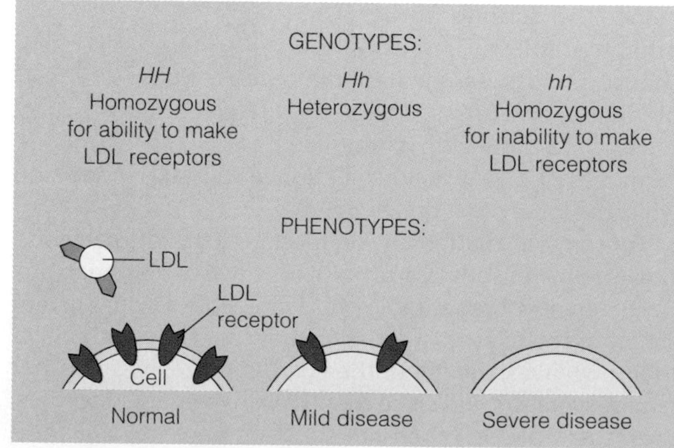

Figure 9.12B Incomplete dominance in human hypercholesterolemia

terol buildup in artery walls, and they may have heart attacks from blocked heart arteries by their mid-thirties. Hypercholesterolemia is even more serious in homozygous individuals (*hh;* about one in a million people). Homozygotes have about five times the normal amount of blood cholesterol and may have heart attacks as early as age two.

Figure 9.12B illustrates the molecular basis for hypercholesterolemia. The dominant allele, which normal individuals carry in duplicate *(HH)*, specifies a cell-surface protein called an LDL receptor. LDLs, or low-density lipoproteins, are cholesterol-containing particles in the blood. The LDL receptors pick up LDL particles from the blood and promote their uptake by cells that break down the cholesterol. This process helps prevent the accumulation of cholesterol in arteries. Without the receptors, lethal levels of LDL build up in the blood. Heterozygotes *(Hh)* have only half the normal number of LDL receptors, and homozygotes *(hh)* have none.

Why is a testcross unnecessary to determine whether a snapdragon with red flowers is homozygous or heterozygous?

Because the homozygotes and heterozygotes differ in phenotype: red flowers for the dominant homozygote and pink flowers for the heterozygote

9.13 Many genes have more than two alleles in the population

So far, we have discussed inheritance patterns involving only two alleles per gene. But many genes have *multiple* alleles. Although each individual carries, at most, two different alleles for a particular gene, in cases of multiple alleles, more than two possible alleles exist in the population. The **ABO blood groups** in humans are one example of multiple alleles. There are three alleles for this characteristic, which in various combinations produce four phenotypes: A person's blood type may be O, A, B, or AB. These letters refer to two carbohydrates, designated A and B, which are found on the surface of red blood cells. A person's red blood cells may be coated with one substance or the other (type A or B), with both (type AB), or with neither (type O). Matching compatible blood groups is critical for blood transfusions. If a donor's blood cells have a carbohydrate (A or B) that is foreign to the recipient, then the recipient produces blood proteins called antibodies that bind specifically to the foreign carbohydrates and cause the donor blood cells to clump together. This clumping can kill the recipient. Figure 9.13 shows which combinations of blood groups result in clumping. (You'll see blood-cell clumping if you type your blood in the laboratory.)

The four blood types result from various combinations of the three different alleles, symbolized as I^A (for the ability to make substance A), I^B (for B), and i (for neither A nor B). Each person inherits one of these alleles from each parent. Because there are three alleles, there are six possible genotypes, as listed in the figure. Both the I^A and I^B alleles are dominant to the i allele. Thus, I^AI^A and I^Ai people have type A blood, and I^BI^B and I^Bi people have type B. Recessive homozygotes, ii, have type O blood because neither the A nor the B substance is produced. The I^A and I^B alleles are said to exhibit **codominance,** meaning that both alleles are expressed in heterozygous individuals (I^AI^B), who have type AB blood.

Because a person's ABO blood group can be determined by a simple test, it is sometimes used as evidence in paternity suits. Such evidence can prove that a man is not the father of a certain baby, and it can suggest that he could be. For example, if a baby has type AB blood (genotype I^AI^B) and a man has type O (genotype ii), he cannot possibly be the father. If the man has type B (genotype I^BI^B or I^Bi), he *could* be the father—but his blood type does not prove that he is.

Maria has type O blood and her sister has type AB blood. The girls know that both of their maternal grandparents are type A. What are the genotypes of the girls' parents?

Their mother is I^Ai; their father is I^Bi.

Blood Group (Phenotype)	Genotypes	Antibodies Present in Blood	Reaction When Blood from Groups Below Is Mixed with Antibodies from Groups at Left			
			O	A	B	AB
O	ii	Anti-A Anti-B				
A	I^AI^A or I^Ai	Anti-B				
B	I^BI^B or I^Bi	Anti-A				
AB	I^AI^B	—				

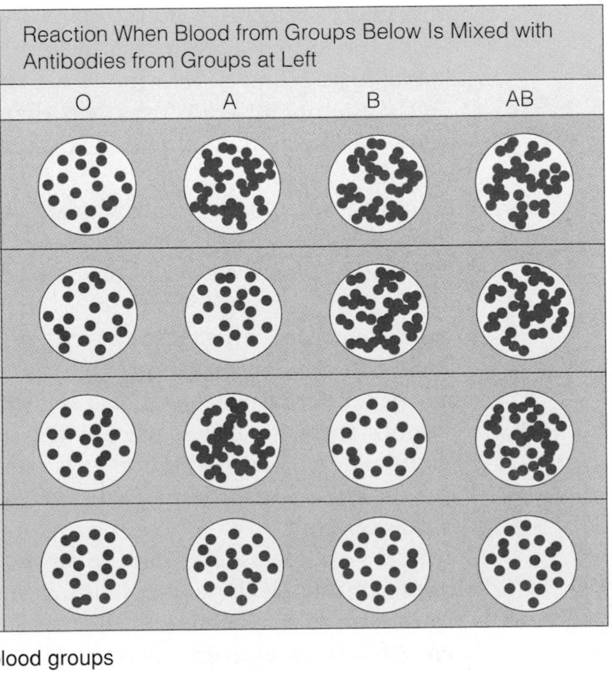

Figure 9.13 Multiple alleles for the ABO blood groups

9.14 A single gene may affect many phenotypic characteristics

All of our genetic examples to this point have been cases in which each gene specified one hereditary characteristic. But in many cases, one gene influences several characteristics. The impact of a single gene on more than one characteristic is called **pleiotropy** (from the Greek *pleion*, more).

An example of pleiotropy in humans is sickle-cell disease, a disorder characterized by the diverse symptoms shown in Figure 9.14. All of these possible phenotypic effects result from the action of a single kind of allele when it is present on both homologous chromosomes. The direct effect of the sickle-cell allele is to make red blood cells produce abnormal hemoglobin molecules. These molecules tend to link together and crystallize, especially when the oxygen content of the blood is lower than usual because of high altitude, overexertion, or respiratory ailments. As the hemoglobin crystallizes, the normally disk-shaped red blood cells deform to a sickle shape with jagged edges, as shown in the micrograph. Sickled cells are destroyed rapidly by the body, and the destruction of these cells may seriously lower the individual's red cell count, causing anemia and general weakening of the body. Also, because of their angular shape, sickled cells do not flow smoothly in the blood and tend to accumulate and clog tiny blood vessels. Blood flow to body parts is reduced, resulting in periodic fever, severe pain, and damage to various organs, including the heart, brain, and kidneys. Sickle cells accumulate in the spleen, damaging it. Blood transfusions and certain drugs may relieve some of the symptoms, but there is no cure, and sickle-cell disease kills about 100,000 people in the world annually.

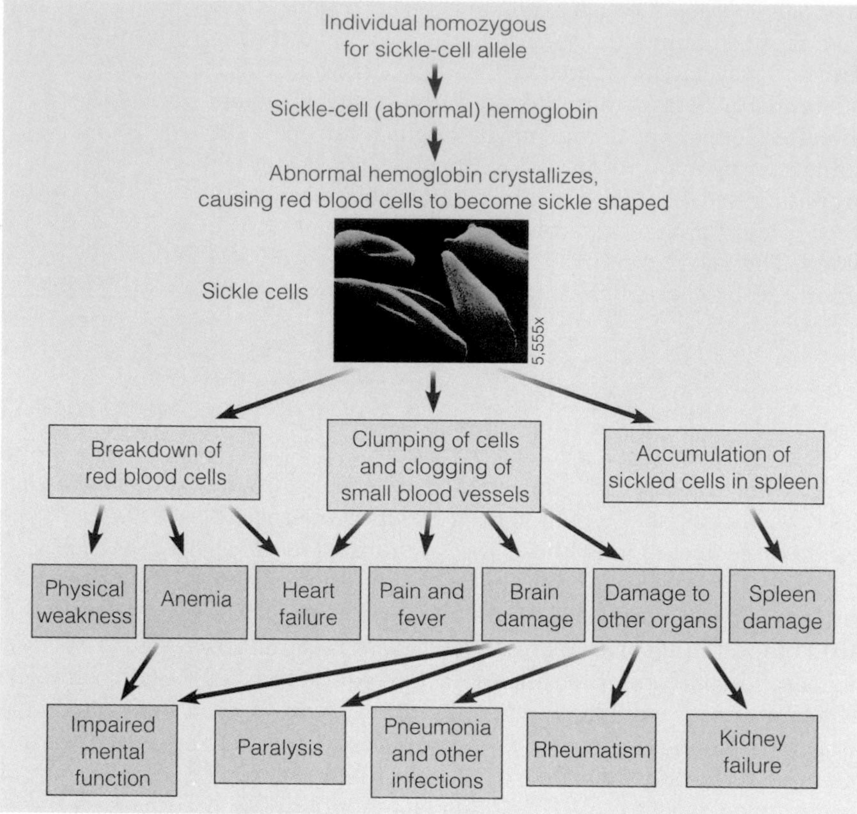

Figure 9.14 Sickle-cell disease, multiple effects of a single human gene

In most cases, only people who are homozygous for the sickle-cell allele suffer from the disease. Heterozygotes, who have one sickle-cell allele and one nonsickle allele, are usually healthy, although in rare cases they may experience some pleiotropic effects when oxygen in the blood is severely reduced, such as at very high altitudes. These effects may occur because the nonsickle and sickle-cell alleles are codominant: Both alleles are expressed in heterozygous individuals, and their red blood cells contain both normal and abnormal hemoglobin. Heterozygotes are said to have "sickle-cell trait."

Sickle-cell disease is by far the most common inherited illness among black people, striking one in 500 African-American children born in the United States. About one in ten African Americans is a carrier—a heterozygote. Among Americans of other ancestry, the sickle-cell allele is extremely rare.

One in ten is an unusually high frequency of carriers for an allele with such harmful effects in homozygotes. We might expect that the frequency of the sickle-cell allele in the population would be much lower because many homozygotes die before passing their genes to the next generation. The high frequency appears to be a vestige of the roots of African Americans. Sickle-cell disease is most common in tropical Africa, where the deadly disease malaria is also prevalent. The protistan parasite that causes malaria spends part of its life cycle inside red blood cells. When it enters those of a person with the sickle-cell allele, it triggers sickling. The body destroys most of the sickled cells, and the parasite does not grow well in those that remain. Consequently, sickle-cell carriers are resistant to malaria, and in many parts of Africa they live longer and have more offspring than noncarriers who are exposed to malaria. In this way, malaria has kept the frequency of the sickle-cell allele relatively high in much of the African continent. 

? How does sickle-cell disease exemplify the concept of pleiotropy?

Homozygosity for the sickle-cell allele causes abnormal hemoglobin, and the impact of the abnormal hemoglobin on the shape of red blood cells leads to a cascade of symptoms in multiple organs of the body.

TALKING ABOUT SCIENCE

9.15 Dr. David Satcher discusses sickle-cell disease and genetic screening

Dr. David Satcher, the Surgeon General of the United States, grew up on an Alabama farm, the son of parents who didn't finish elementary school. Even as a child, however, he aspired to become a physician. He attended Morehouse College and went on to earn a combined M.D.-Ph.D. at Case Western Reserve University. With a research background in medical genetics, Dr. Satcher then helped develop the King-Drew Sickle-Cell Research Center in Los Angeles and served as its director for 6 years. A gifted leader, he has since held a succession of important positions, including the presidency of Meharry Medical College. In 1993, Dr. Satcher was appointed director of the U.S. Centers for Disease Control and Prevention (CDC), one of the most influential health science positions in the world. In late 1997, he left the CDC to become Surgeon General, a position with greater visibility to the general public. In an interview, Dr. Satcher discussed some of the highlights of his career, beginning with the research he carried out as a graduate student:

Figure 9.15A David Satcher, Surgeon General of the United States

> My Ph.D. was in cytogenetics, which is the study of chromosomes located in cells. I studied the effects that X-radiation and radioactive iodine, I-131, have on chromosomes and was able to show a linear relationship between radiation dosage—and, to a certain extent, I-131 dosage—and chromosome damage. That was some of the earlier work looking directly at the impact of radiation on chromosomes.

The battle against sickle-cell disease (see Module 9.14) has been a major focus for Dr. Satcher:

> I directed one of the ten national sickle-cell research centers starting in 1972. The center I directed, the King-Drew Sickle-Cell Research Center, emphasized community education, early diagnosis, screening, and counseling programs, as well as treatment. I think we made major contributions in all those areas.

Genetic screening, testing people for alleles associated with a particular genetic disorder, can be useful to couples at risk for having an affected child. Because the sickle-cell allele is known to be relatively common in one segment of the American population (African Americans) and is easily detectable by a blood test, it is an obvious target for screening programs (Figure 9.15B). In our interview, Dr. Satcher emphasized the need to combine screening with community education and genetic counseling:

> The sickle-cell gene is recessive, which means that a person can be a carrier—have "sickle-cell trait"—but not be affected by the disease. Communities didn't understand that in the early days of screening programs; in fact, there were laws saying that people with sickle-cell trait couldn't go to public schools because it was thought to be contagious! This is one of the best examples of what's necessary if you're going to develop a mass screening program. We don't recommend mass screening, but if you are going to do it, you have to make sure you have the education first so that you are prepared to deal with the results.
>
> We have a similar problem right now with HIV and AIDS, although in this case the cause is not genetic. You might ask why only 15% of the people in this country have been tested for HIV. Part of the answer is that people are afraid to find out the results. Similarly, if you don't have the appropriate support mechanisms in place, genetic screening programs can do more harm than good. Along with education, you have to have trained people, in addition to physicians and geneticists, to do genetic counseling. That is happening now, all over the country.

At the CDC, Dr. Satcher helped the nation make strides against AIDS and teenage smoking, and he helped improve childhood immunization rates. As Surgeon General, Dr. Satcher hopes to contribute to further progress in these and other areas. In an interview published in the *New York Times,* he said:

> I want to be the Surgeon General who reaches our citizens with cutting-edge technology and plain, old-fashioned straight talk. Whether talking about smoking or poor diets, I want to send messages of good health to our cities and suburbs, our barrios and reservations, and even our prisons.

A dynamic leader with a passion for making the world a better place, Dr. Satcher may well achieve considerable success toward these ambitious goals. 

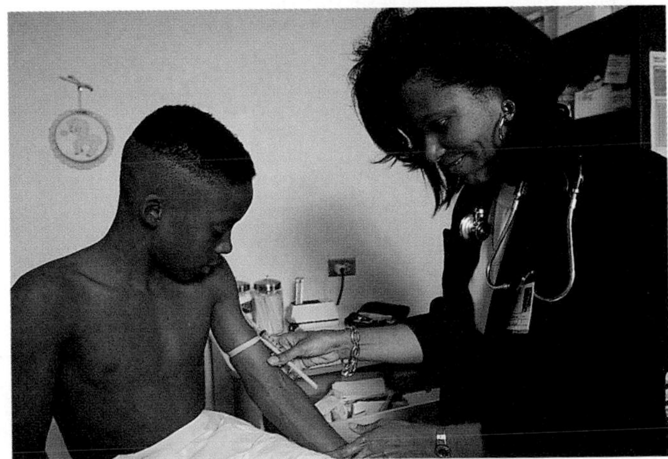

Figure 9.15B Collection of a blood sample for sickle-cell testing

? In what context does David Satcher believe genetic screening would do more harm than good?

When neither the individuals being screened nor the general public is educated about what particular results of the testing mean

9.16 A single characteristic may be influenced by many genes

Mendel studied genetic characteristics that could be classified on an either-or basis, such as purple or white flower color. However, many characteristics, such as human skin color and height, vary in a population along a continuum. Many such features result from **polygenic inheritance,** the additive effects of two or more genes on a single phenotypic characteristic. (This is the converse of pleiotropy, in which a single gene affects several characteristics.)

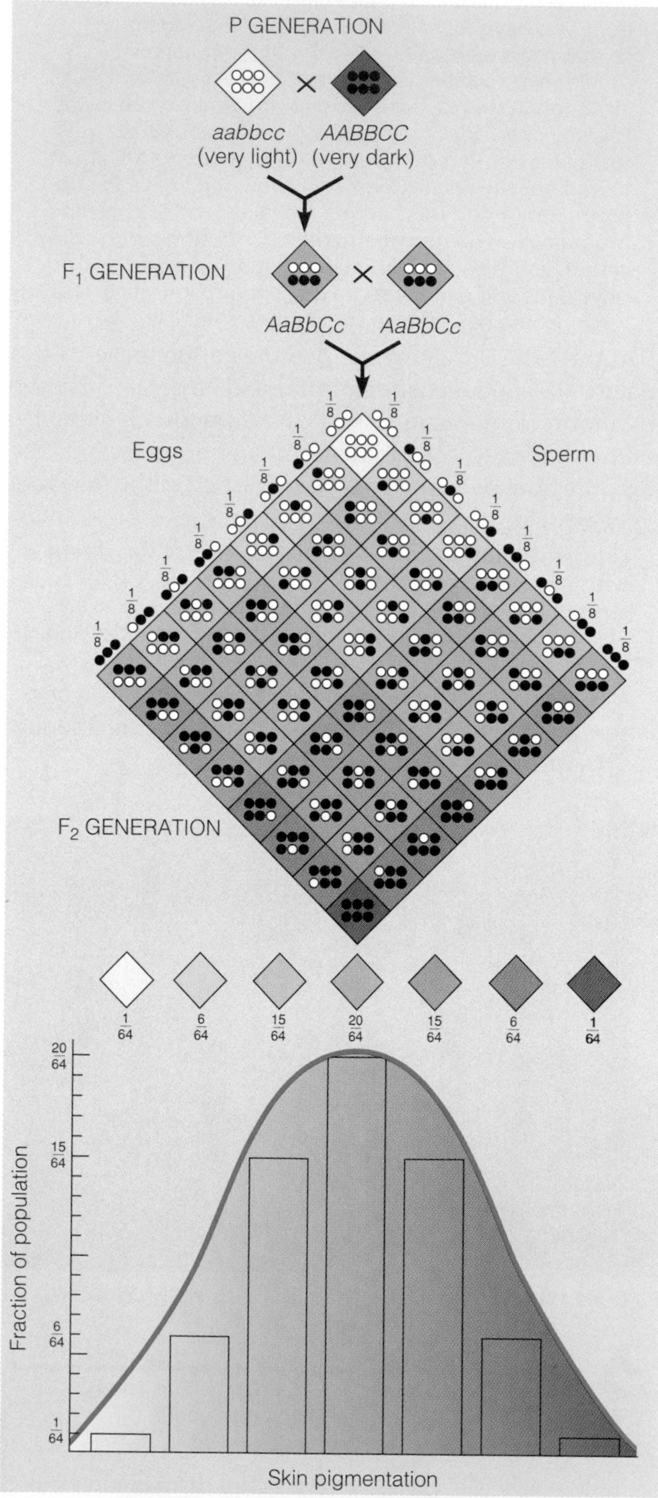

Figure 9.16 is a model that illustrates how three hypothetical genes could produce some of the continuous variation we observe in human skin pigmentation. The three genes are inherited separately, like Mendel's pea genes. The "dark-skin" allele for each gene *(A, B, C)* contributes one "unit" of darkness to the phenotype and is incompletely dominant to the other alleles *(a, b, c)*. A person who is *AABBCC* would be very dark, while an *aabbcc* individual would be very light. An *AaBbCc* person (resulting, for example, from a mating between an *AABBCC* person and an *aabbcc* person) would have skin of an intermediate shade. Because the alleles have an additive effect, the genotype *AaBbCc* would produce the same skin color as any other genotype with just three dark-skin alleles, such as *AABbcc*.

The Punnett square shows all possible genotypes of offspring from a mating of two triple heterozygotes (the F_1 generation here). The row of diamonds below the Punnett square shows the seven skin-pigmentation phenotypes that would theoretically result from this mating. The seven bars in the graph at the bottom of the figure depict the relative numbers of each of the phenotypes in the F_2.

If this were a real human population, the range of skin color would probably be even more of a continuum, perhaps similar to the entire spectrum of color under the bell-shaped curve in the graph. Most likely, there would be intermediate types between each of our seven color variants. Such intermediate colors would result from the effects of environmental factors, such as sun-tanning, on the seven variants. Of course, any variation resulting from environmental factors rather than genes would not be passed on to the next generation.

Figure 9.16 A model for polygenic inheritance of skin color

Based on the skin-tone model in Figure 9.16, an *AaBbcc* individual would be indistinguishable in phenotype from which of the following individuals?
AAbbcc, aaBBcc, AabbCc, Aabbcc, aaBbCc

All except *Aabbcc*

* * *

To this point in the chapter, we have presented four types of inheritance pattern that are extensions of Mendel's principles of inheritance: incomplete dominance, codominance, pleiotropy, and polygenic inheritance. It is important to realize that these patterns are extensions of Mendel's model, rather than exceptions to it. From Mendel's garden pea experiments came data supporting a particulate theory of inheritance, with the particles (genes) being transmitted according to the same rules of chance that govern the tossing of coins. The particulate theory holds true for all inheritance patterns.

9.17 Chromosome behavior accounts for Mendel's principles

Mendel published his results in 1866, but not until long after he died did biologists understand the significance of his work. Cell biologists worked out the processes of mitosis and meiosis in the late 1800s (see Chapter 8 to review these processes). Then, around 1900, researchers began to notice parallels between the behavior of chromosomes and the behavior of Mendel's heritable factors. One of biology's most important concepts—the **chromosome theory of inheritance**—was emerging. The chromosome theory states that genes are located on chromosomes and that the behavior of chromosomes during meiosis and fertilization accounts for inheritance patterns. Indeed, it is chromosomes that undergo segregation and independent assortment during meiosis and thus account for Mendel's principles.

We can see the chromosomal basis of Mendel's principles by following the fates of two genes during meiosis and fertilization in pea plants. In Figure 9.17, we picture the genes for seed shape (alleles *R* and *r*) and seed color (*Y* and *y*) as black bars on different chromosomes. We start with the F_1 generation, in which all individuals have the *RrYy* genotype. To simplify the diagram, we show only two of the seven pairs of pea chromosomes and three of the stages of meiosis: metaphase I, anaphase I, and metaphase II.

To see the chromosomal basis of the principle of segregation, let's follow just the pair of long chromosomes, the ones carrying *R* and *r*, taking either the left or the right branch from the F_1 cell. Whichever arrangement the chromosomes assume at metaphase I, the two alleles *segregate* as the homologous chromosomes separate in anaphase I. And at the end of meiosis II, a single long chromosome ends up in each of the gametes. Fertilization then recombines the two alleles at random, resulting in F_2 offspring that are $\frac{1}{4}$ *RR*, $\frac{1}{4}$ *rr*, and $\frac{1}{2}$ *Rr*. The ratio of round/wrinkled phenotypes is thus 3:1 (12 round to 4 wrinkled), the ratio Mendel observed (see Figure 9.3A).

To see the chromosomal basis of the principle of independent assortment, follow both the long and the short chromosomes through the figure below. Two alternative, equally likely arrangements of tetrads can occur at metaphase I. The nonhomologous chromosomes (and their genes) assort independently, leading to four gamete genotypes. Random fertilization leads to the 9:3:3:1 phenotypic ratio in the F_2 generation, as you saw in Figure 9.5A.

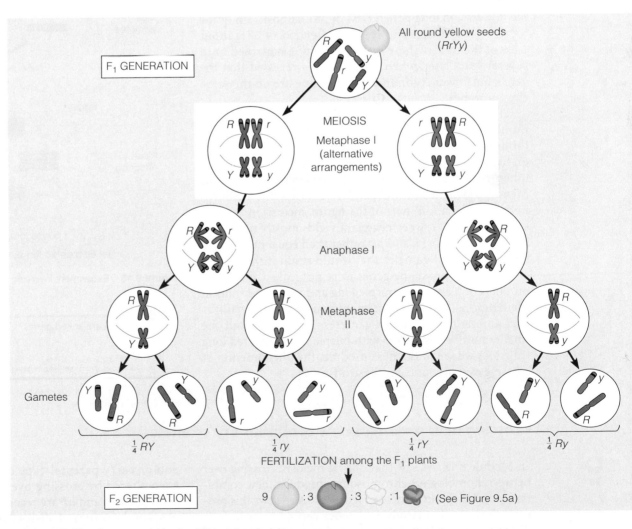

Which of Mendel's principles have their physical basis in the following phases of meiosis? (a) the orientation of homologous chromosome pairs in metaphase I; (b) the separation of homologues in anaphase I

(a) the principle of independent assortment; (b) the principle of segregation

F_1 GENERATION
All round yellow seeds (*RrYy*)
MEIOSIS
Metaphase I (alternative arrangements)
Anaphase I
Metaphase II
Gametes
$\frac{1}{4}$ *RY*
$\frac{1}{4}$ *ry*
$\frac{1}{4}$ *rY*
$\frac{1}{4}$ *Ry*
FERTILIZATION among the F_1 plants
F_2 GENERATION
9 :3 :3 :1 (See Figure 9.5a)

Figure 9.17 The chromosomal basis of Mendel's principles

9.18 Genes on the same chromosome tend to be inherited together

9B

In 1908, British biologists William Bateson and Reginald Punnett (originator of the Punnett square) discovered an inheritance pattern that seemed totally inconsistent with Mendelian principles. Bateson and Punnett were working with two characteristics in sweet peas, flower color and pollen shape. They crossed doubly heterozygous plants (*PpLl*) that exhibited the dominant traits: purple flowers (expression of the *P* allele) and long pollen grains (expression of the *L* allele). The corresponding recessive traits are red flowers (in *pp* plants) and round pollen (in *ll* plants).

The top part of Figure 9.18 illustrates Bateson and Punnett's experiment. When they looked at just one of the two characteristics (that is, either cross *Pp* × *Pp* or cross *Ll* × *Ll*), they found that the dominant and recessive alleles segregated, producing a phenotypic ratio of approximately 3:1 for the offspring, in agreement with Mendel's segregation principle. However, when the biologists combined their data for the two characteristics, they did not see the 9:3:3:1 ratio they predicted for a dihybrid cross. Instead, as shown in the table, they found a disproportionately large number of plants (339 of 381 observed offspring) with either purple flowers and long pollen (284 of 381, almost 75% of the total) or red flowers and round pollen (55 of 381, about 14% of the total). These results were not explained until several years later, when other studies revealed that the genes for flower color and pollen shape are on the same chromosome.

The number of genes in a cell is far greater than the number of chromosomes; in fact, each chromosome has thousands of genes. Genes that are located close together on the same chromosome, called **linked genes**, tend to be inherited together. As a result, they generally do not follow Mendel's principle of independent assortment. As shown in the "Explanation" part of the figure, meiosis in the heterozygous (*PpLl*) sweet-pea plant yields mostly two genotypes of gametes (*PL* and *pl*) rather than equal numbers of the four types of gametes that would result if the flower-color and pollen-shape genes were not linked. The large numbers of plants with purple long and red round traits in the Bateson-Punnett experiment resulted from fertilization among the *PL* and *pl* gametes. But what about the smaller numbers of plants with purple round and red long traits? As we see in the next module, the phenomenon of crossing over accounts for these offspring.

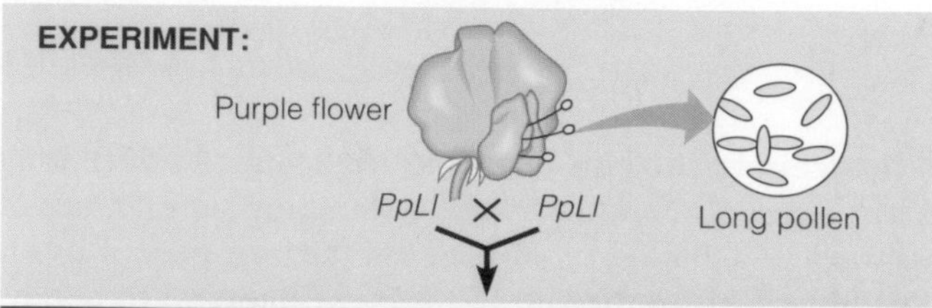

PHENOTYPES	OBSERVED OFFSPRING	PREDICTION (9:3:3:1)
Purple long	284	215
Purple round	21	71
Red long	21	71
Red round	55	24

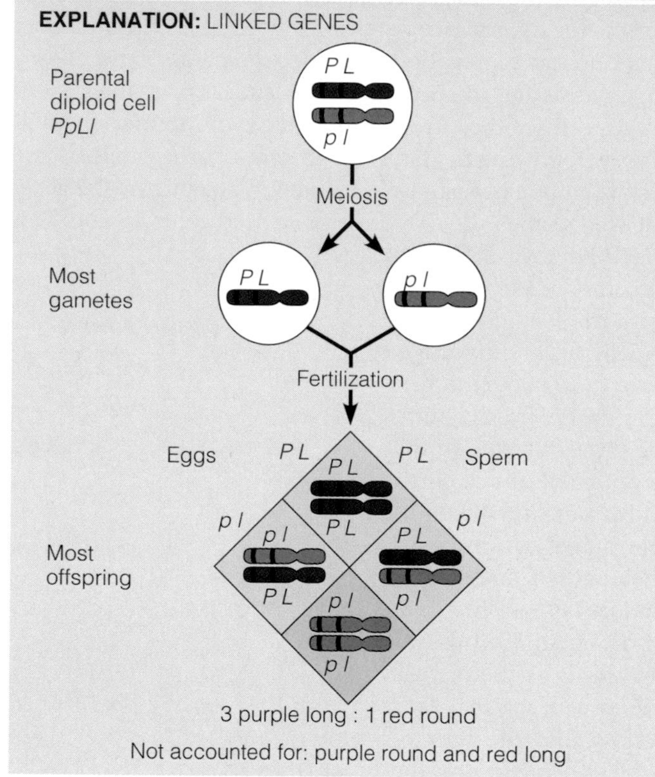

Figure 9.18 Experiment involving linked genes in the sweet pea

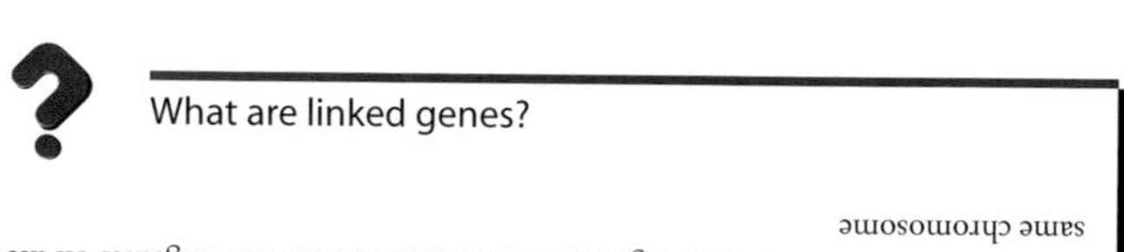

What are linked genes?

Genes that tend to be inherited together because their loci are close together on the same chromosome

9.19 Crossing over produces new combinations of alleles

9B

In Module 8.18, we saw that during meiosis, crossing over between homologous chromosomes produces new combinations of alleles in gametes. Figure 9.19A reviews this process, showing that two linked genes can give rise to four different gamete genotypes. Gametes with genotypes *AB* and *ab* carry parental-type chromosomes that have not been altered by crossing over. In contrast, gametes with genotypes *Ab* and *aB* are recombinant gametes. They carry new combinations of alleles that result from the exchange of chromosome segments in crossing over. Crossing over

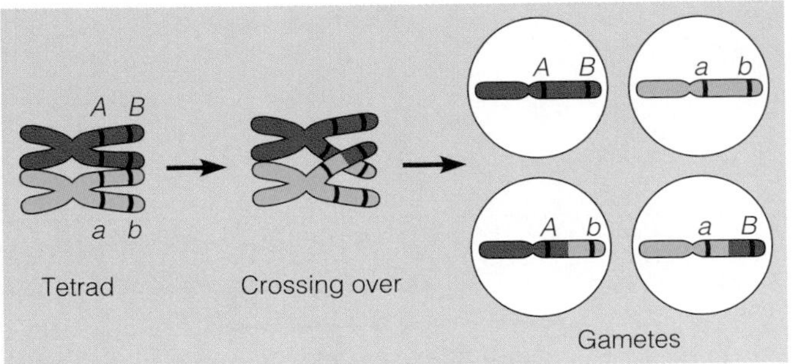

Figure 9.19A Review: Production of recombinant gametes

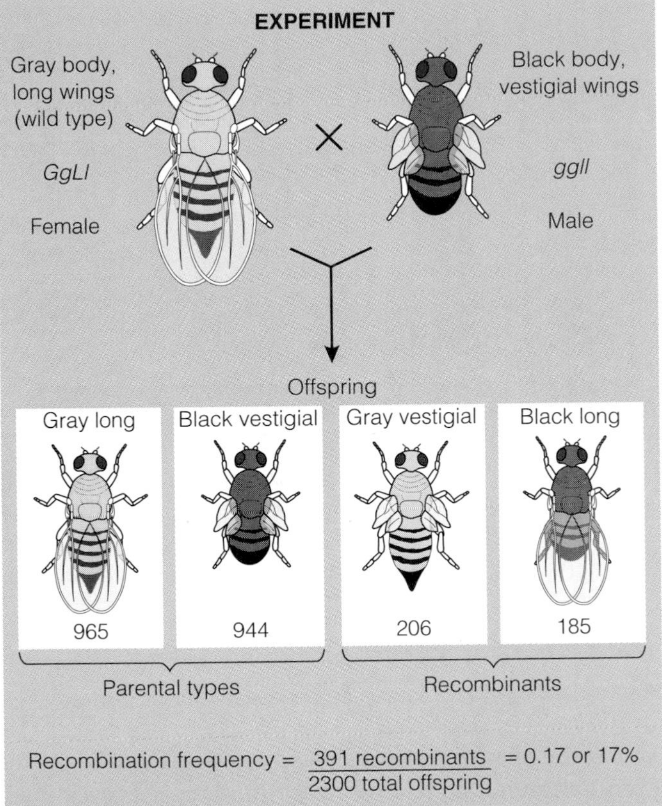

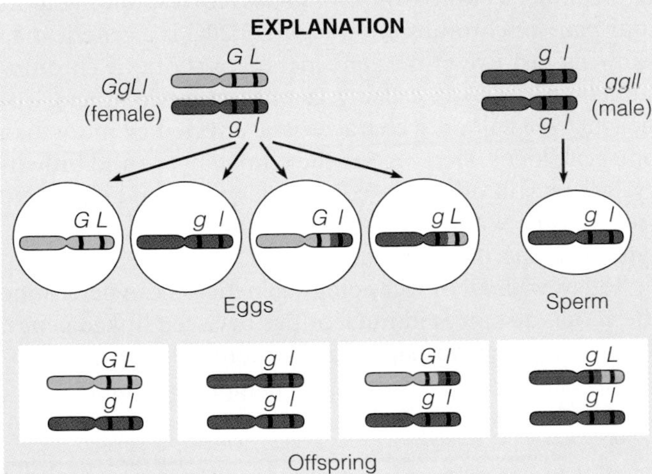

Figure 9.19B Fruit-fly experiment demonstrating the role of crossing over in inheritance

recombines linked genes into assortments of alleles not found in parents.

The discovery of how crossing over creates gamete diversity confirmed the relationship between chromosome behavior and inheritance. Some of the first experiments to demonstrate the effects of crossing over were performed in the laboratory of American embryologist Thomas Hunt Morgan in the early 1900s. Morgan and his colleagues used the fruit fly *Drosophila melanogaster* in many of their experiments. Often seen flying around overripe fruit, *Drosophila* is a good research animal for studies of inheritance. It can be grown in small containers on a mixture of cornmeal and molasses and will produce hundreds of offspring in a few weeks. Using fruit flies, geneticists can trace the inheritance of a trait through several generations in a matter of months.

Figure 9.19B shows one of Morgan's experiments, a cross between a wild-type fruit fly (gray body and long wings) and a fly with a black body and undeveloped, or vestigial, wings. Morgan knew the genotypes of these flies from previous studies. Here we use the following gene symbols:

G = gray body (dominant)
g = black body (recessive)
L = long wings (dominant)
l = vestigial wings (recessive)

In mating a gray fly with long wings (genotype *GgLl*) with a black fly with vestigial wings (genotype *ggll*), Morgan performed a testcross (see Module 9.6). If the genes had not been linked, then independent assortment would have produced offspring in a phenotypic ratio of 1:1:1:1 ($\frac{1}{4}$ gray body, long wings; $\frac{1}{4}$ black body, vestigial wings; $\frac{1}{4}$ gray body, vestigial wings; and $\frac{1}{4}$ black body, long wings). But because these genes were linked, Morgan obtained the results shown in Figure 9.19B: Most of the offspring had parental phenotypes, but 17% of the offspring flies were recombinants. The percentage of recombinants is called the **recombination frequency.**

When Morgan first obtained these results, he did not know about crossing over. To explain the ratio of offspring, he hypothesized that the genes were linked and that some mechanism occasionally broke the linkage. Tests of the hypothesis proved him correct, establishing that crossing over was the mechanism that "breaks linkages" between genes.

The lower part of Figure 9.19B explains Morgan's results in terms of crossing over. A crossover between chromatids of homologous chromosomes in parent *GgLl* broke linkages between the *G* and *L* alleles and between the *g* and *l* alleles, forming the recombinant chromosomes *Gl* and *gL*. Later steps in meiosis distributed the recombinant chromosomes to gametes, and random fertilization produced the four kinds of offspring Morgan observed.

Return to the data for sweet-pea plants in Figure 9.18. What is the recombination frequency for the flower-color and pollen-length genes?

11% (42/381)

9.20 Geneticists use crossover data to map genes

9B

Working mostly with *Drosophila,* T. H. Morgan and his students produced a virtual explosion in our understanding of genetics. In the photo in Figure 9.20A, Morgan (back row, far right), several students, and a skeleton are celebrating the return of Alfred H. Sturtevant (left foreground) from World War I military service. One of Sturtevant's major contributions to genetics was an approach for using crossover data to map gene loci. Sturtevant started by assuming that the chance of crossing over is approximately equal at all points on a chromosome. He then hypothesized that the farther apart two genes are on a chromosome, the higher the probability that a crossover would occur between them. His reasoning was elegantly simple: The greater the distance between two genes, the more points there are between them where crossing over can occur. With this principle in mind, Sturtevant began using recombination data from fruit-fly crosses to assign to genes relative positions on chromosomes—that is, to *map* genes.

Figure 9.20B represents a part of the chromosome that carries the linked genes for black body *(g)* and vestigial wings *(l)* that we described in Module 9.19. This same chromosome also carries a gene that has a recessive allele (we'll call it *c*) determining cinnabar eye color, a brighter red than the wild-type color. Figure 9.20B shows the actual crossover (recombination) frequencies between these alleles, taken two at a time: 17% between the *g* and *l* alleles, 9% between *g* and *c*, and 9.5% between *l* and *c*. Sturtevant reasoned that these values represent the relative distances between the genes. Because the crossover frequencies between *g* and *c* and between *l* and *c* are approximately half that between *g* and *l*, gene *c* must lie roughly midway between *g* and *l*. Thus, the sequence of these genes on one of the fruit-fly chromosomes must be *g-c-l* (or the equivalent *l-c-g*).

Years later it was learned that Sturtevant's assumption that crossovers are equally likely at all points on a chromosome was not exactly correct. Still, his method of mapping genes worked, and it proved extremely valuable in establishing the relative positions of many other fruit-fly genes. Eventually, enough data were accumulated to reveal that *Drosophila* has four groups of genes, corresponding to its four pairs of chromosomes. Figure 9.20C is a genetic map showing just five of the gene loci on part of one chromosome: the loci we've called *g*, *c*, and *l*, and two others. Notice that eye color is a characteristic affected by more than one gene locus. Here we see the cinnabar-eye and brown-eye genes; still other eye-color genes are found elsewhere (see Module 9.22). For all these genes, however, the wild-type allele specifies red eyes.

Figure 9.20A A party in Morgan's fly room

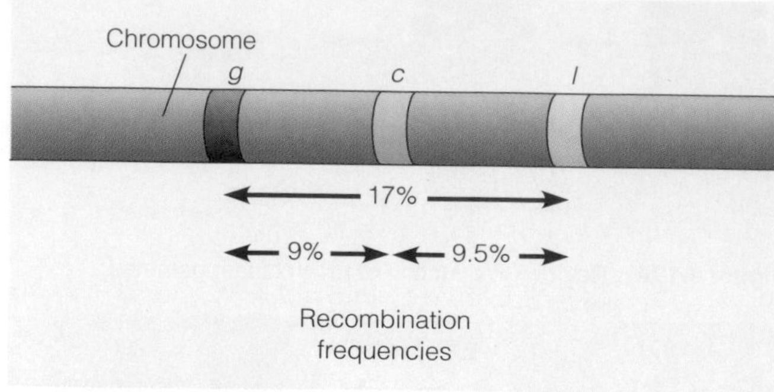

Figure 9.20B Mapping genes from crossover data

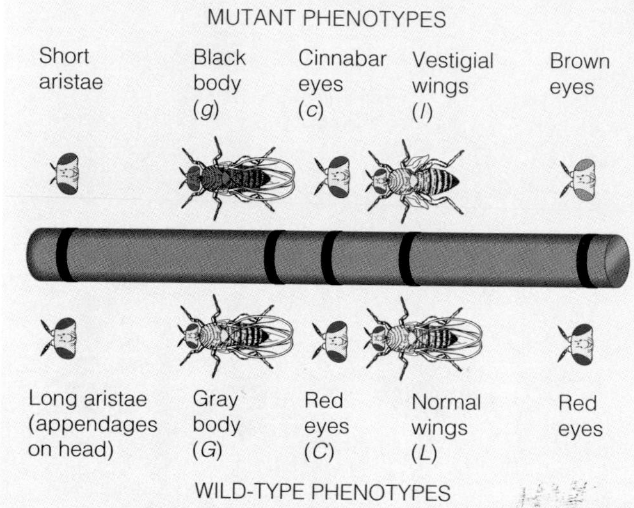

Figure 9.20C A partial genetic map of a fruit-fly chromosome

Today, with DNA technology, geneticists can determine the actual distances in nucleotides between linked genes. These newer genetic maps generally confirm the relative positions established by Sturtevant's mapping method.

? You design *Drosophila* crosses to provide recombination data for a gene not included in Figure 9.20C. The gene has recombination frequencies of 5% with the vestigial-wing *(l)* locus and 5% with the cinnabar-eye *(c)* locus. Where is it located on the chromosome?

About halfway between the vestigial and cinnabar loci

9.21 Chromosomes determine sex in many species

Many animals, including fruit flies and humans, have a pair of **sex chromosomes,** designated *X* and *Y*, that determine an individual's sex. Figure 9.21A reviews what you learned in Chapter 8 about sex determination in humans. Individuals with one *X* chromosome and one *Y* chromosome are males; *XX* individuals are females. Human males and females both have 44 autosomes (nonsex chromosomes). As a result of chromosome segregation during meiosis, each gamete contains one sex chromosome and a haploid set of autosomes (22 in humans). All eggs contain a single *X* chromosome. Of the sperm cells, half contain an *X* chromosome and half contain a *Y* chromosome. An offspring's sex depends on whether the sperm cell that fertilizes the egg bears an *X* or a *Y*.

The genetic basis of sex determination in humans is not yet completely understood, but one gene on the *Y* chromosome plays a crucial role. This gene, discovered by a British research team in 1990, is called *SRY* and triggers testis development. In the absence of a functioning version of *SRY*, an individual develops ovaries rather than testes. Other genes on the *Y* chromosome are also necessary for normal sperm production. The *X–Y* system in other mammals is similar to that in humans. In the fruit fly's *X–Y* system, however, some genetic details are different, although a *Y* chromosome is still essential for sperm formation.

The *X–Y* system is only one of several sex-determining systems. Grasshoppers, crickets, and roaches, for example, have an *X–O* system, in which *O* stands for the absence of a sex chromosome (Figure 9.21B). Females have two *X* chromosomes *(XX);* males have only one sex chromosome, giving them genotype *XO*. Males produce two classes of sperm (*X*-bearing and lacking any sex chromosome), and sperm cells determine the sex of the offspring at fertilization.

In contrast to the *X–Y* and *X–O* systems, *eggs* determine sex in certain fishes, butterflies, and birds (Figure 9.21C). The sex chromosomes in these animals are designated *Z* and *W*. Males have the genotype *ZZ;* females are *ZW*. Sex is determined by whether the egg carries a *Z* or a *W*.

Some organisms lack sex chromosomes altogether. In most ants and bees, sex is determined by chromosome *number,* rather than by sex chromosomes (Figure 9.21D). Females develop from fertilized eggs and thus are diploid. Males develop from unfertilized eggs—they are fatherless—and are haploid.

Most animals have two separate sexes; that is, individuals are either male or female. Many plants also have separate sexes, with male and female flowers borne on different individuals. Some plants with separate sexes, such as date palms, spinach, and marijuana, have the *X–Y* system of sex determination; others, such as the wild strawberry, have the *Z–W* system.

But not all organisms have separate sexes. Most plant species and some animal species have individuals that produce both sperm and eggs. Plants of this type—corn, for example—are said to be **monoecious** (from the Greek *monos,* one, and *oikos,* house). Animals of this type, such as earthworms and garden snails, are said to be **hermaphroditic** (from the names of the Greek god Hermes and goddess Aphrodite). In monoecious plants and hermaphroditic animals, all individuals of a species have the same complement of chromosomes.

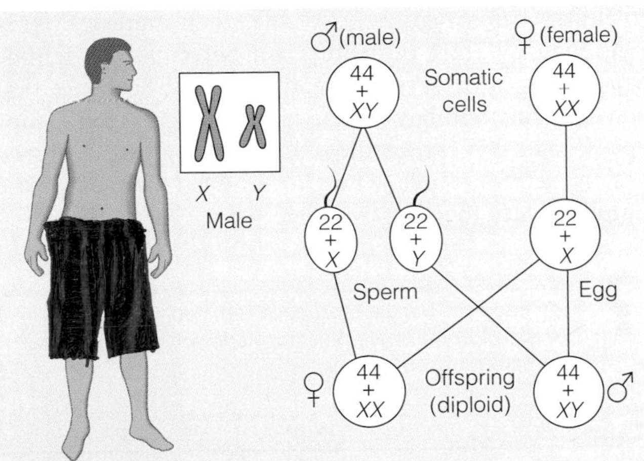

Figure 9.21A The *X–Y* system

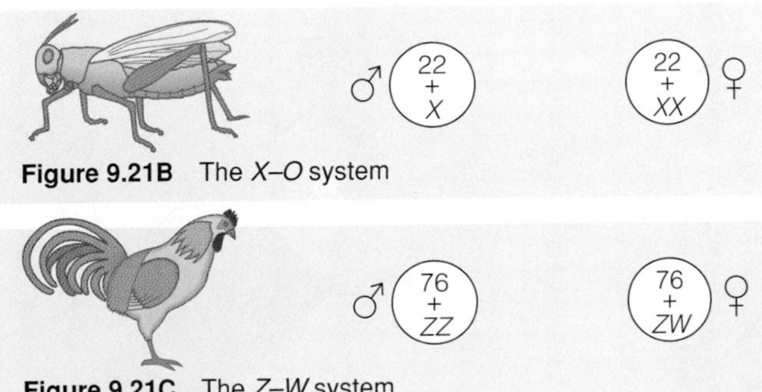

Figure 9.21B The *X–O* system

Figure 9.21C The *Z–W* system

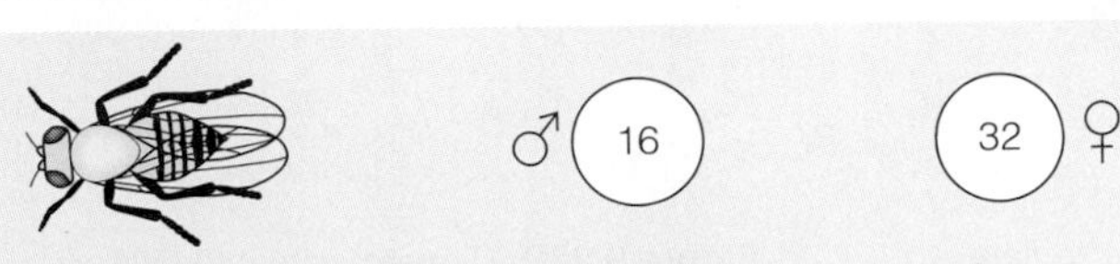

Figure 9.21D Sex determination by chromosome number

At the moment of conception in humans, what determines the sex of the offspring?

Whether the egg is fertilized by a sperm bearing an X chromosome (producing a female offspring) or by a sperm with a Y chromosome (producing a male)

9.22 Sex-linked genes exhibit a unique pattern of inheritance

9C

Besides bearing genes that determine sex, the so-called sex chromosomes of many species also contain genes for characteristics unrelated to femaleness or maleness. Any gene located on a sex chromosome is called a **sex-linked gene.** Sex-linked genes unrelated to sex determination are most often found on *X* chromosomes. However, in some animals, such as fruit flies and certain fishes, the *Y* chromosome does carry some genes unrelated to sex determination. Such genes give rise to traits appearing only in males and passed only from father to son. In humans, there is as yet no conclusive evidence that the *Y* chromosome carries genes unrelated to male sex determination.

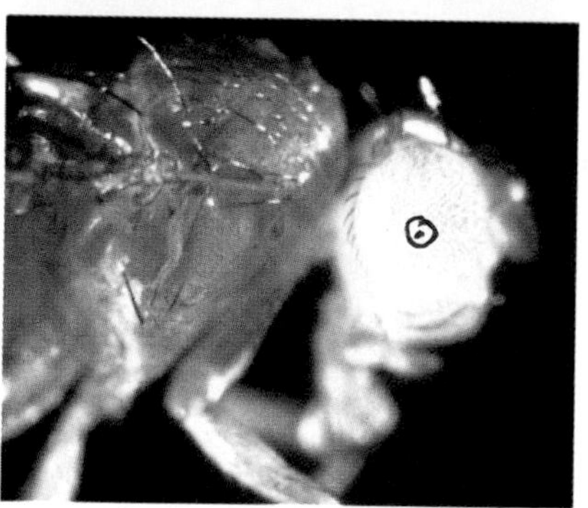

Figure 9.22A Fruit-fly eye color, a sex-linked characteristic

The figures here illustrate inheritance patterns for white eye color in the fruit fly, an *X*-linked recessive trait. Wild-type fruit flies have red eyes; white eyes are very rare (Figure 9.22A). We use uppercase letter *R* for the dominant, wild-type, red-eye allele and *r* for the recessive, white-eye allele. Because these alleles are carried on the *X* chromosome, we show them as superscripts to the letter *X*. Thus, red-eyed male fruit flies have the genotype X^RY; white-eyed males are X^rY. The *Y* chromosome does not have a gene locus for eye color; therefore, the male's phenotype results entirely from his single *X*-linked gene. In the female, X^RX^R and X^RX^r flies have red eyes, and X^rX^r flies have white eyes.

A white-eyed male (X^rY) will transmit his X^r to all of his female offspring, but to none of his male offspring. This is because, to be female, his daughters must inherit his *X* chromosome, but his sons must inherit his *Y* chromosome. As shown in Figure 9.22B, when the female parent is a dominant homozygote (X^RX^R) and the male parent is X^rY, all the offspring have red eyes, but the female offspring are all carriers of the allele for white eyes.

In Figure 9.22C, we see that when a heterozygous female (X^RX^r) mates with a red-eyed male (X^RY), half the male offspring are white-eyed and half are red-eyed. All the female offspring of this cross have red eyes because they inherit at least one dominant allele (from their father). Half of the female offspring are homozygous dominant because they inherit their mother's *R* allele. The other half of the females are heterozygotes (X^rX^R) like their mother.

As Figure 9.22D indicates, if a heterozygous female mates with a white-eyed male, there is a 50% chance that each offspring will have white eyes (resulting from genotype X^rX^r or X^rY), regardless of sex. Daughters with red eyes are heterozygotes, whereas red-eyed male offspring completely lack the recessive allele.

A white-eyed female *Drosophila* is mated with a red-eyed (wild-type) male. What result do you predict for the numerous offspring?

All female offspring will be red-eyed but heterozygous (X^RX^r); all male offspring will be white-eyed (X^rY).

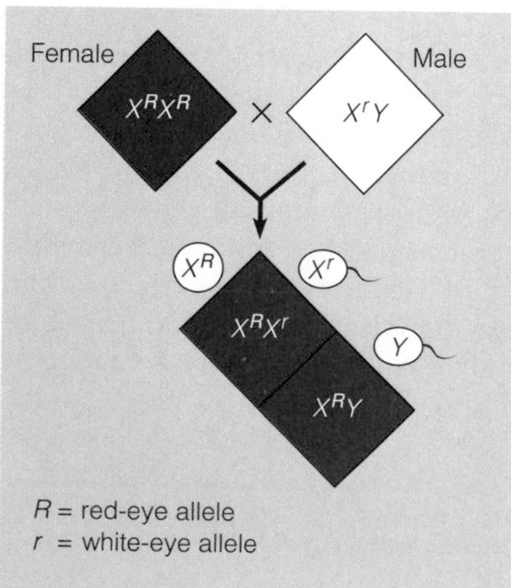

Figure 9.22B Homozygous, red-eyed female × white-eyed male

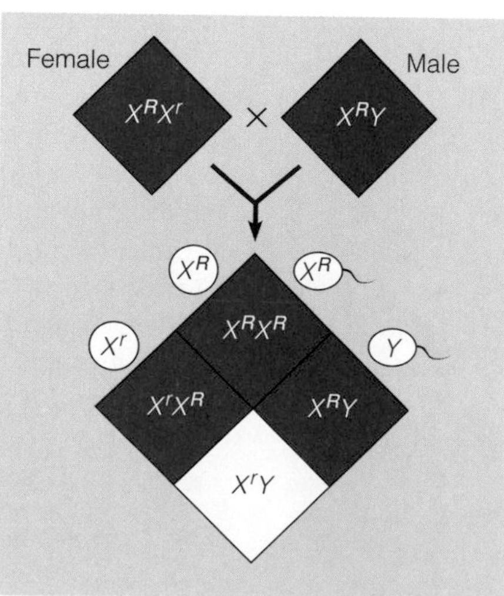

Figure 9.22C Heterozygous female × red-eyed male

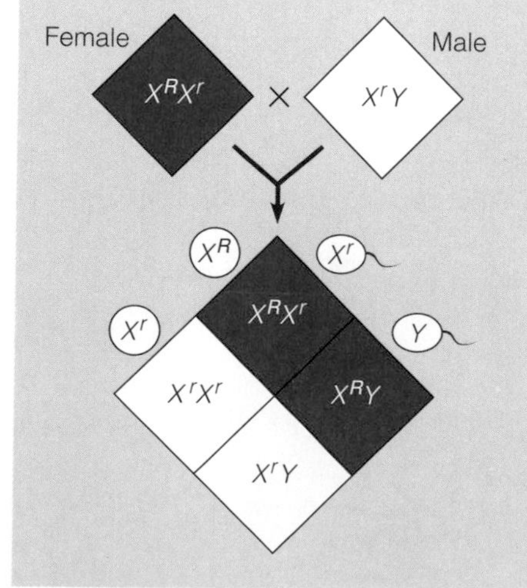

Figure 9.22D Heterozygous female × white-eyed male

Sex-linked disorders affect mostly males 9.23

Fruit-fly genetics has taught us much about human inheritance. A number of human conditions, including red-green color blindness, hemophilia, and a type of muscular dystrophy, result from sex-linked (*X*-linked) recessive alleles that are inherited in the same way as the white-eye trait in fruit flies. The fruit-fly model also shows us why recessive sex-linked traits are expressed much more frequently in men than in women. Like a male fruit fly, if a man inherits only one sex-linked recessive allele—from his mother—the allele will be expressed. In contrast, a woman has to inherit two such alleles—one from each parent—to exhibit the trait.

Red-green color blindness is a common sex-linked disorder characterized by a malfunction of light-sensitive cells in the eyes. It is actually a complex of disorders, involving several *X*-linked genes. A person with normal color vision can see more than 150 colors. In contrast, someone with red-green color blindness can see fewer than 25. For some affected people, red hues appear gray; others see gray instead of green; still others are green-weak or red-weak, tending to confuse shades of these colors. Mostly males are affected, but heterozygous females have some defects. (If you have red-green color blindness, you probably cannot see the numeral 7 in Figure 9.23A.)

Hemophilia is a sex-linked recessive trait with a long, well-documented history. Hemophiliacs bleed excessively when injured because they have inherited an abnormal allele for a factor involved in blood clotting. The most seriously affected individuals may bleed to death after relatively minor bruises or cuts.

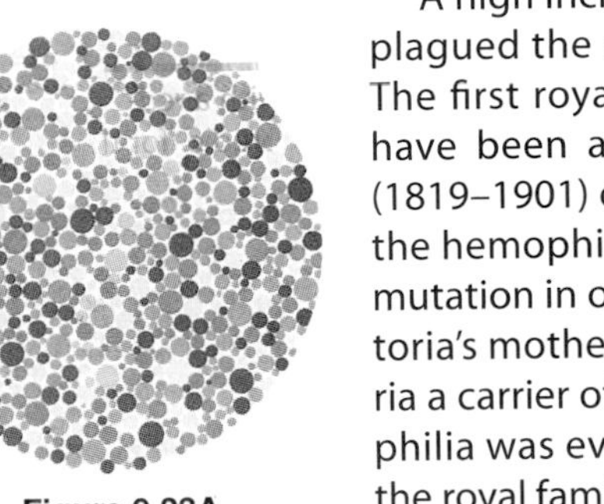

Figure 9.23A A test for red-green color blindness

A high incidence of hemophilia has plagued the royal families of Europe. The first royal hemophiliac seems to have been a son of Queen Victoria (1819–1901) of England. It is likely that the hemophilia allele arose through a mutation in one of the gametes of Victoria's mother or father, making Victoria a carrier of the deadly allele. Hemophilia was eventually introduced into the royal families of Prussia, Russia, and Spain through the marriages of two of Victoria's daughters who were carriers. Thus, the age-old practice of strengthening international alliances by marriage effectively spread hemophilia through the royal families of several nations. The photograph in Figure 9.23B shows Queen Victoria's granddaughter Alexandra, her husband Nicholas, who was the last czar of Russia, and their son Alexis. The pedigree uses half-colored symbols to represent heterozygous carriers of the hemophilia allele. As you can see in the pedigree, Alexandra, like her mother and grandmother, was a carrier, and Alexis had the disease.

Another sex-linked recessive disorder is **Duchenne muscular dystrophy**, a condition characterized by a progressive weakening and loss of muscle tissue. Almost all cases are males, and the first symptoms appear in early childhood, when the child begins to have difficulty standing up. He is inevitably wheelchair-bound by age 12. Eventually, he becomes severely wasted, and normal breathing becomes difficult. Death usually occurs by age 20.

Figure 9.23B Hemophilia in the royal family of Russia

For such a severe disease, Duchenne muscular dystrophy is relatively common. In the general U.S. population, about one in 3500 male babies is affected, and the disease is even more common in some inbred populations. In one Amish community in Indiana, for instance, one out of every 100 males is born with the disease.

With the help of DNA technology (discussed in Chapter 12), the gene whose defectiveness causes Duchenne muscular dystrophy has been mapped at a particular point on the *X* chromosome. The gene's wild-type allele codes for a protein called dystrophin, which is present in normal muscle but missing in Duchenne patients. 

? Neither Rudy nor Carla has Duchenne muscular dystrophy, but their first son does have it. If the couple has a second child, what is the probability that he or she will also have the disease?

1/4 (1/2 chance of a male child × 1/2 chance that a male will inherit the mutant *X*)

* * *

The discovery of sex-linked genes and their pattern of inheritance in fruit flies and humans was one of many breakthroughs in understanding how genes are passed from one generation to the next. During the first half of the twentieth century, Mendel's work was rediscovered, his principles were reinterpreted in light of chromosomal behavior during meiosis, and the chromosome theory of inheritance was firmly established. The chromosome theory set the stage for another explosion of experimental work in the second half of the twentieth century. This work was mostly in molecular genetics, an area we explore in the next three chapters.

CHAPTER REVIEW

Chapter Summary

Mendel's Principles (9.1–9.10) The historical roots of genetics, the science of heredity, date back to ancient attempts at selective breeding. Until the twentieth century, however, many biologists believed that characteristics acquired during an organism's lifetime could be passed on, and that characteristics of both parents blended irreversibly in their offspring (9.1). Modern genetics began with Gregor Mendel's quantitative experiments. Mendel crossed pea plants that differed in characteristics such as flower color and seed shape, and he traced traits from generation to generation. He hypothesized that there are alternative forms of genes (although he did not use that term), the units that determine heritable traits (9.2). From his experimental data, Mendel deduced that an organism has two genes (alleles) for each inherited characteristic, one from each parent. A sperm or egg carries only one allele of each pair, because allele pairs separate when gametes form—Mendel's principle of segregation (9.3). We now know that alleles (alternative forms) of a gene reside at the same locus, or position, on homologous chromosomes (9.4). Mendel noted that when two alleles of a gene are different, one (the dominant allele) may determine the inherited trait, while the other (the recessive allele) has no effect. By looking at two characteristics at once, Mendel found that the alleles of a pair segregate independently of other allele pairs during gamete formation—Mendel's principle of independent assortment (9.5). The offspring of a testcross, a mating between an individual of unknown genotype and a homozygous recessive individual, often reveal the unknown's genotype (9.6). Inheritance follows the rules of probability. The chance of inheriting a recessive allele *(a)* from a heterozygous *(Aa)* parent is $\frac{1}{2}$. The chance of inheriting it from both of two heterozygous parents is $\frac{1}{2} \times \frac{1}{2} = \frac{1}{4}$, illustrating the rule of multiplication for calculating the probability of two independent events. There are two ways a dominant and recessive allele from heterozygous parents can combine in offspring: *A* from the father and *a* from the mother, or *a* from the father and *A* from the mother. The probability of this occurring is $\frac{1}{4} + \frac{1}{4} = \frac{1}{2}$, illustrating the rule of addition for calculating the probability of an event that can occur in alternative ways (9.7). The inheritance of many human traits, from freckles to genetic diseases, follows Mendel's principles and the rules of probability. To study human genetics, we use family pedigrees to determine patterns of inheritance and individual genotypes (9.8). Many inherited disorders in humans are controlled by a single gene (two alleles). Most such disorders, such as cystic fibrosis and sickle-cell disease, are caused by autosomal recessive alleles. A few, such as achondroplasia and Huntington's disease, are caused by dominant alleles (9.9). Genetic and biochemical tests on fetal cells and molecules, and examination of the fetus with ultrasound or fetoscopy can help people make reproductive decisions (9.10).

Variations on Mendel's Principles (9.11–9.16) Mendel's principles are valid for all sexually reproducing species, but the genotype often does not dictate the phenotype in the simple way his principles describe (9.11). When an offspring's phenotype—flower color, for example—is in between the phenotypes of its parents, it exhibits incomplete dominance (9.12). In the population, there often exist multiple alleles for a characteristic, such as the three alleles for the ABO blood group. The alleles determining the A and B blood factors are codominant; that is, both are expressed in a heterozygote (9.13). A single gene, such as the allele for sickle-cell disease, may affect phenotype in multiple ways, a phenomenon called pleiotropy (9.14). Genetic screening for carriers of potentially harmful recessive alleles can be useful if supported by education and counseling (9.15). A single characteristic, such as skin color, may be affected by multiple genes, creating a continuum of phenotypes (9.16).

The Chromosomal Basis of Inheritance (9.17–9.20) Genes are located on chromosomes, whose behavior during meiosis and fertilization accounts for inheritance patterns (9.17). Certain genes are linked; they tend to be inherited together because they reside close together on the same chromosome. In fact, there are many genes on each chromosome (9.18). Crossing over can separate linked alleles, producing gametes with recombinant chromosomes (9.19). Because crossing over is more likely to occur between genes that are farther apart, recombination frequencies can be used to map the relative positions of genes on chromosomes (9.20).

Sex Chromosomes and Sex-Linked Genes (9.21–9.23) A pair of sex chromosomes determines sex in many species. In humans (and many other organisms, including fruit flies), a male has one *X* and one *Y* sex chromosome, and a female has two *X* chromosomes. Whether a sperm cell contains an *X* or a *Y* determines the sex of the offspring. The *Y* chromosome has genes for the development of testes, whereas an absence of the *Y* allows ovaries to develop. Other systems of sex determination exist in other animals and plants (9.21). All genes on the sex chromosomes are said to be sex-linked. However, in both fruit flies and humans, the *X* chromosome carries many genes unrelated to sex. Their inheritance pattern reflects the fact that females have two homologous *X* chromosomes, but males have only one (9.22). Most sex-linked human disorders, such as red-green color blindness and hemophilia, are due to recessive alleles and are seen mostly in males. A male receiving a single *X*-linked recessive allele from his mother will have the disorder; a female has to receive the allele from both parents to be affected (9.23).

Testing Your Knowledge

Multiple Choice

1. Edward was found to be heterozygous *(Ss)* for sickle-cell trait. The alleles represented by the letters *S* and *s* are
   a. on the *X* and *Y* chromosomes
   b. linked
   c. on homologous chromosomes
   d. both present in each of Edward's sperm cells
   e. on the same chromosome but far apart

2. Whether an allele is dominant or recessive depends on
   a. how common the allele is, relative to other alleles
   b. whether it is inherited from the mother or the father
   c. which chromosome it is on
   d. whether it or another allele determines the phenotype when both are present
   e. whether or not it is linked to other genes

3. Two fruit flies with eyes of the usual red color are crossed and their offspring are as follows: 77 red-eyed males, 71 ruby-eyed males, 152 red-eyed females. The allele for ruby eyes is
   a. autosomal (carried on an autosome) and dominant
   b. autosomal and recessive

c. sex-linked and dominant

d. sex-linked and recessive

e. impossible to determine without more information

4. All the offspring of a white hen and a black rooster are gray. The simplest explanation for this pattern of inheritance is

a. pleiotropy
b. sex linkage
c. linkage
d. independent assortment
e. incomplete dominance

5. In some of his experiments, Mendel studied the inheritance patterns of two characteristics at once—flower color and pod color, for example. He did this to find out

a. whether genes for the two characteristics are inherited together or separately
b. how many genes are responsible for determining a characteristic
c. whether genes are on chromosomes
d. the distance between genes on a chromosome
e. how many different genes a pea plant has

6. A man who has type B blood and a woman who has type A blood could have children of which of the following phenotypes?

a. A or B only
b. AB only
c. AB or O
d. A, B, or O
e. A, B, AB, or O

Additional Genetics Problems

1. Describe a genetic cross that illustrates the principle of independent assortment.

2. Why are there more men than women with color blindness?

3. In fruit flies, the genes for wing shape and body stripes are linked. In a fly whose genotype is *WwSs*, *W* is linked to *S*, and *w* is linked to *s*. Show how this fly can produce gametes containing four different combinations of alleles. Which are parental-type gametes? Which are recombinant gametes? What process produces recombinant gametes?

4. Adult height in humans is at least partially hereditary; tall parents tend to have tall children. But humans come in a range of sizes, not just tall and short. Explain an extension of Mendel's model that could produce the hereditary variation in human height.

5. A brown mouse is repeatedly mated with a white mouse, and all their offspring are brown. If two of these brown offspring are mated, what fraction of the F_2 mice will be brown?

6. How could you determine the genotype of one of the brown F_2 mice in problem 5? How would you know whether a brown mouse is homozygous? Heterozygous?

7. Tim and Jan both have freckles (see Module 9.8), but their son Michael does not. Show with a Punnett square how this is possible. If Tim and Jan have two more children, what is the probability that both of them will have freckles?

8. Both Tim and Jan (problem 7) have a widow's peak (Module 9.8), but Michael has a straight hairline. What are their genotypes? What is the probability that Tim and Jan's next child will have freckles and a straight hairline?

9. In rabbits, black hair depends on a dominant allele, *B*, and brown on a recessive allele, *b*. Short hair is due to a dominant allele, *S*, and long hair to a recessive allele, *s*. If a true-breeding black, short-haired male is mated with a brown, long-haired female, what will their offspring look like? What will be the genotypes of the offspring? If two of these F_1 rabbits are mated, what phenotypes would you expect among their offspring, in what proportions?

10. Incomplete dominance is seen in the inheritance of hypercholesterolemia. Mack and Toni are both heterozygous for this characteristic, and both have elevated levels of cholesterol. Their daughter Zoe has a cholesterol level six times normal; she is apparently homozygous, *hh*. What fraction of Mack and Toni's children are likely to have elevated but not extreme levels of cholesterol, like their parents? If Mack and Toni have one more child, what is the probability that the child will suffer from the more serious form of hypercholesterolemia seen in Zoe?

11. A fruit fly with a gray body and red eyes (genotype *BbPp*) is mated with a fly having a black body and purple eyes (genotype *bbpp*). What offspring, in what proportions, would you expect if the body color and eye color genes are on different chromosomes (unlinked)? When this mating is actually carried out, most of the offspring look like the parents, but 3% have gray body and purple eyes, and 3% have black body and red eyes. Are these genes linked or unlinked? What is the recombination frequency?

12. A series of matings shows that the recombination frequency between the black-body gene (problem 11) and the gene for dumpy (shortened) wings is 36%. The recombination frequency between purple eyes and dumpy wings is 41%. What is the sequence of these three genes on the chromosome?

13. A female fruit fly whose body is covered with forked bristles is mated with a male fly with normal bristles. Their offspring are 121 females with normal bristles and 138 males with forked bristles. Explain the inheritance pattern for this trait.

14. A couple are both phenotypically normal, but their son suffers from hemophilia, a sex-linked recessive disorder. What fraction of their children are likely to suffer from hemophilia? What fraction are likely to be carriers?

15. Heather was not able to see the numeral 7 in Figure 9.23A and was surprised to discover she suffered from red-green color blindness. She told her biology professor, who said, "Your father is color blind too, right?" How did her professor know this? Why did her professor not say the same thing to the color-blind males in the class?

Thinking as a Scientist

In 1981, a stray black cat with unusual rounded, curled-back ears was adopted by a family in Lakewood, California. Hundreds of descendants of this cat have since been born, and cat fanciers hope to develop the "curl" cat into a show breed. The curl allele is apparently dominant and autosomal (carried on an autosome). Suppose you owned the first curl cat and wanted to breed it in order to develop a true-breeding variety. Describe tests that would determine whether the curl gene is dominant or recessive and whether it is autosomal or sex-linked. Explain why you think your tests would be conclusive. Describe a test to determine that a cat is true-breeding.

Science, Technology, and Society

Gregor Mendel never saw a gene, yet he concluded that "heritable factors" were responsible for the patterns of inheritance he observed in peas. Similarly, Morgan and Sturtevant never actually observed the linkage of genes on chromosomes. Their maps of *Drosophila* chromosomes (and the very idea that genes are carried on chromosomes) were conceived by observing the patterns of inheritance of linked genes, not by observing the genes directly. Is it legitimate for biologists to claim the existence of objects and processes they cannot actually see? How do scientists know whether an explanation is correct?

Media Resources

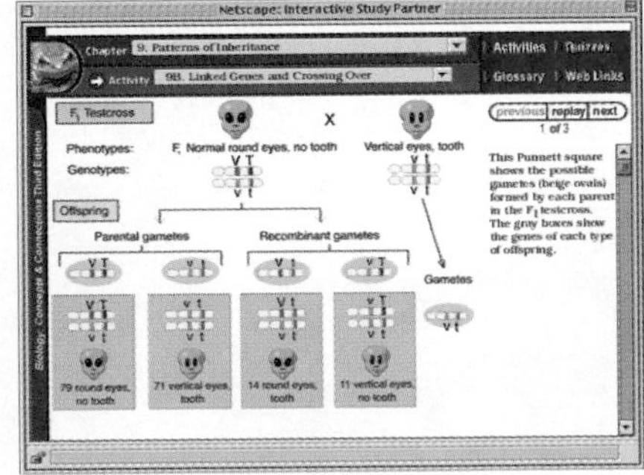

For further review of the concepts in this chapter, use the activities and quizzes on the CD-ROM.

Visit the Special Edition of The Biology Place web site for activities, quizzes, news articles, and web links.

10 Molecular Biology of the Gene

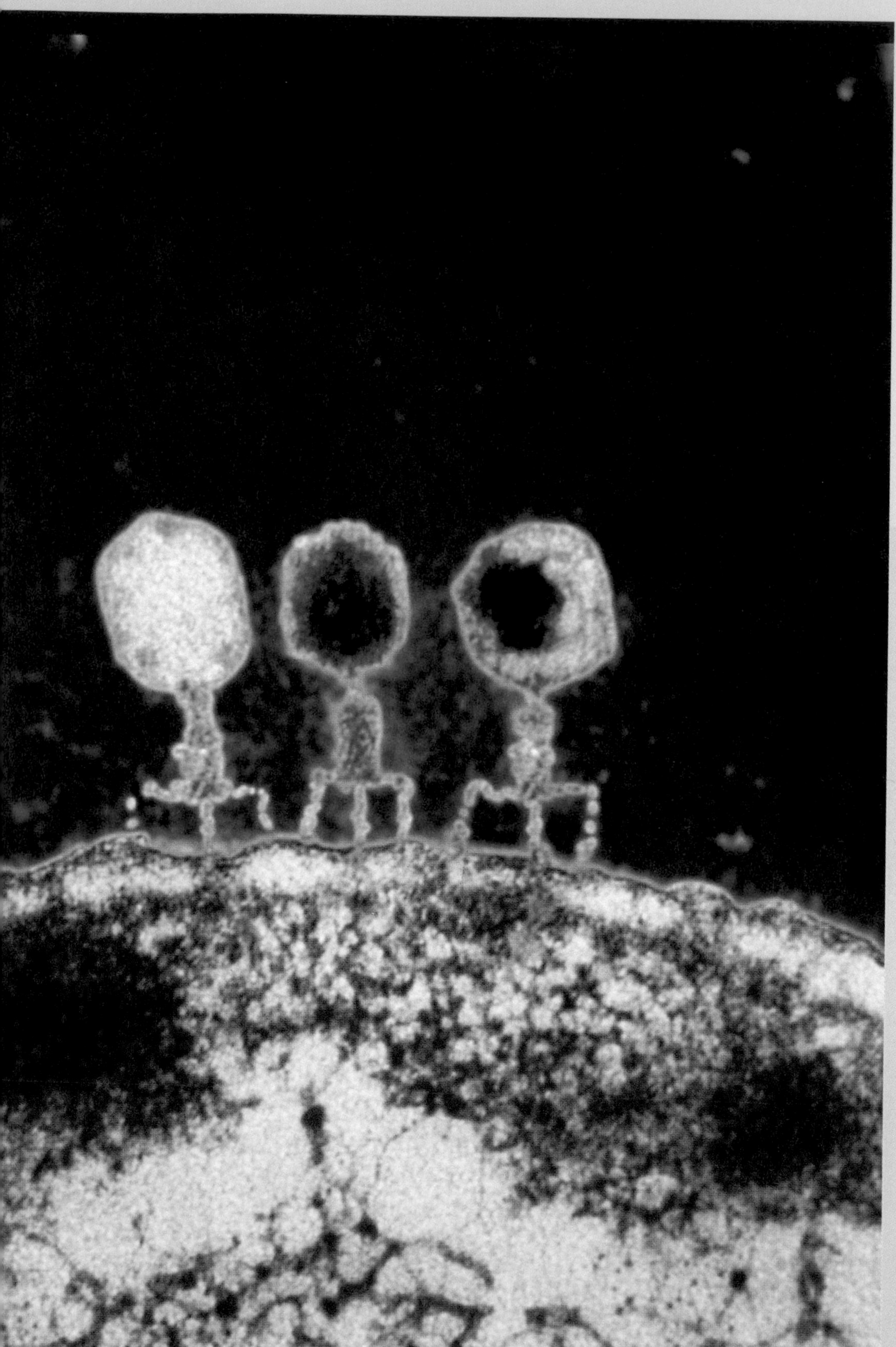

Outline

The three landing crafts drift slowly downward. The life forms—whatever they are—must be in the 20-sided cabins at the top. Maybe the tubelike underparts are entryways or exits. Thin legs touch the surface and anchor there.

Microscopists often say that viewing tiny objects and organisms magnified thousands of times gives them a sense of peering in on other worlds. In the picture at the left, we're actually looking at three viruses that have landed on the surface of a bacterial cell. The image has been magnified about 250,000 times with an electron microscope. The drawing on the facing page shows how this virus, called T4, infects a bacterium. The "legs" (called tail fibers) bend when they touch the cell surface. The tail is a hollow rod enclosed in a springlike sheath. As the legs bend, the spring compresses, and the bottom of the rod touches the cell surface, attaching to a protein landing pad (the purple "donut" in the cell membrane). The rod punctures the cell membrane, and the viral DNA passes from inside the head of the

The Flow of Genetic Information from DNA to RNA to Protein

Viruses: Genes in Packages

virus through the hollow rod and into the cell.

The scale here is very small. The virus is only about 200 nanometers (2 ten-thousandths of a millimeter) long. The bacterium, called *Escherichia coli (E. coli)* is only 1–2 μm (1–2 thousandths of a mm) across. Over a million *E. coli* would fit on the head of a pin.

On a smaller scale yet—at the molecular level—the T4 virus performs a deadly act. It takes control of the bacterial cell and then kills it. The DNA that it injects is the virus's genetic material. Once inside the bacterial cell, the viral DNA uses the cell's resources to produce hundreds of new T4s. The new viruses eventually break open (lyse) their host cell and infect any other *E. coli* cells they encounter. Viruses that attack bacteria are called **bacteriophages** (meaning "bacteria-eaters"), or **phages** for short.

T4 and other viruses sit on the fence, so to speak, between life and nonlife. A virus is lifelike in having DNA and a highly organized structure, but it differs from a living organism in not being cellular or able to reproduce on its own. A virus can survive only by infecting a living cell and using the cell's molecular machinery to make more viruses. T4 and other viruses are essentially genetic material enclosed in a protein coat. Because viruses are so simple, they are relatively easy to study on the molecular level, far easier than Mendel's peas or Morgan's fruit flies. For this reason, we owe our first glimpses of the functions of DNA, the molecule that carries hereditary traits, to phages and bacteria.

This chapter is about the DNA molecule and how it serves as the molecular basis of heredity—the subject called **molecular biology.** Here we explore the structure of DNA, how it replicates (the molecular basis of why offspring resemble their parents), how it controls the cell by directing protein synthesis, and how it can change. We also look at some viruses that infect animals and plants. We begin with the story of how we know that DNA is the genetic material, a story in which a close relative of T4 played a major role.

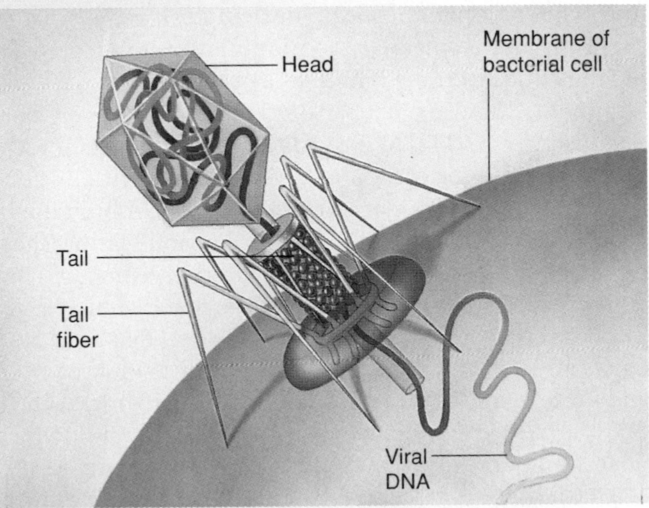

Bacteriophage T4 infecting a bacterial cell

The Structure of the Genetic Material

10.1 Experiments showed that DNA is the genetic material

Our knowledge of T4 and DNA is relatively new. Viruses were discovered as agents of disease a century ago but were not actually seen with a microscope until 1942. Similarly, DNA was known as a substance in cells 100 years ago, but Mendel, Morgan, and other early geneticists did all their work without any knowledge of DNA's role in heredity.

We can trace the discovery of the genetic role of DNA back to 1928. That year, English bacteriologist Frederick Griffith reported results of experiments on a species of bacterium that causes pneumonia. Griffith studied two varieties of the bacterium, a pathogenic (disease-causing) one and a variant that was harmless. He found that when he killed the pathogenic bacteria with heat and then mixed the cell remains with living bacteria of the harmless variety, some of the living cells were converted to the pathogenic form. Furthermore, this new trait of pathogenicity was inherited by all the descendants of the transformed bacteria. Clearly, some chemical component of the dead pathogenic cells—some "transforming factor"—caused this heritable change, but the identity of the factor was not known.

By the late 1930s, other experimental studies had convinced most biologists that a specific kind of molecule, rather than some complex chemical mixture, was at the root of inheritance. Attention focused on the chemical nature of chromosomes, which had been known since the turn of the century to carry hereditary information. By the 1940s, scientists knew that eukaryotic chromosomes consisted of two substances, DNA and protein. Most researchers thought the protein was the material of genes. Proteins were known to be made of 20 kinds of building blocks (amino acids) and to have elaborate and varied structures and functions. DNA, on the other hand, has only four types of building blocks (nucleotides), and its chemical properties seemed far too monotonous to account for the multitude of traits inherited by every organism. These arguments were so persuasive that even after researchers established in 1944 that Griffith's transforming factor was DNA, the scientific community remained skeptical.

Gradually, the evidence in support of DNA built up. In 1952, American biologists Alfred Hershey and Martha Chase performed one of the most convincing experiments. They demonstrated that DNA is the genetic material of a bacteriophage called T2 (a close relative of T4). At that time, biologists already knew that T2 was composed solely of DNA and protein, and that it could somehow reprogram its host cell to produce new phages. But they did not know which phage component—DNA or protein—was responsible.

Hershey and Chase found the answer by devising an experiment to determine what the phage transferred to *E. coli* during infection. Their experiment used only a few, relatively simple tools: chemicals containing radioactive isotopes (see Module 2.5); a device for detecting radioactivity; a kitchen blender; and a centrifuge, which is an extremely fast merry-go-round for test tubes that is used to separate particles of different weights. (These are still basic tools of molecular biology.)

Hershey and Chase used different radioactive isotopes to label the DNA and protein in T2. First, they grew T2 with *E. coli* in a growth medium containing radioactive sulfur (yellow in Figure 10.1A). Protein contains sulfur but DNA does not, so as the phages multiplied, the radioactive sulfur atoms were incorporated only into their proteins (the head exterior, tail, and tail fibers). Next, the researchers grew a separate batch of phages in medium containing radioactive phosphorus (green in Figure 10.1A). Because nearly all the phages' phosphorus is in their DNA, this procedure labeled the phage DNA.

Armed with radioactively labeled T2, Hershey and Chase were ready to perform the experiment outlined in Figure 10.1B on the facing page. ① They allowed the protein-labeled and DNA-labeled batches of T2 to infect separate samples of nonradioactive bacteria. ② Shortly after the onset of infection, they agitated the cultures in a blender to shake loose any parts of the phages that remained outside the bacterial cells. ③ They then spun the mixtures in a centrifuge. The bacterial cells, being heavier, were deposited as a pellet at the bottom of the centrifuge tubes, but the phages and phage parts remained suspended in the liquid. ④ The researchers then measured and compared the radioactivity in the pellet and the liquid.

Hershey and Chase found that when the bacteria had been infected with the T2 phages containing labeled proteins, most of the radioactivity was in the liquid, which contained phages but not bacteria. This result suggested that the phage protein did not enter the host cells. But when the bacteria had been infected with phages whose DNA was tagged with radioactive phosphorus, then most of the radioactivity was in the pellet, made up of bacteria. When these bacteria were returned to liquid growth medium, the bacterial cells lysed and released new phages containing radioactive phosphorus in their DNA but no radioactive sulfur in their proteins.

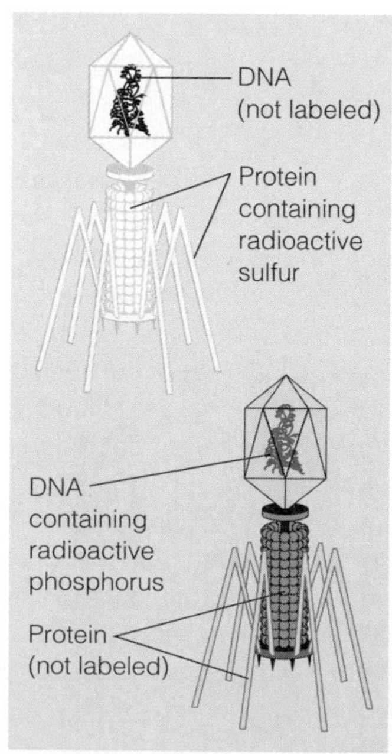

Figure 10.1A Labeled phages

Hershey and Chase concluded that the phage DNA was injected into the host cell but that most of the phage proteins remained outside (as indicated by the diagrams of bacteria and phages in Figure 10.1B). More important, they showed that it is the injected DNA molecules that cause the cells to produce additional phage DNA and proteins—indeed, new, complete phages. Figure 10.1C outlines the reproductive cycle for phages T2 and T4.

The Hershey-Chase results, added to earlier evidence, convinced the scientific world that DNA was the hereditary material. What happened next was one of the most celebrated quests in the history of science—the effort to figure out the structure of DNA and how this structure enables the molecule to store genetic information and transmit it from parents to offspring.

How did researchers use radioactively labeled virus molecules to show that DNA, not protein, is the genetic material?

They found that radioactively labeled viral DNA, but not labeled protein, entered the host cell during infection and directed the synthesis of new viruses.

Phage
Radioactive protein (yellow)
Bacterium
DNA
Empty protein shell
Radioactivity in liquid
Pour mixture into blender
Centrifuge
Test for radioactivity

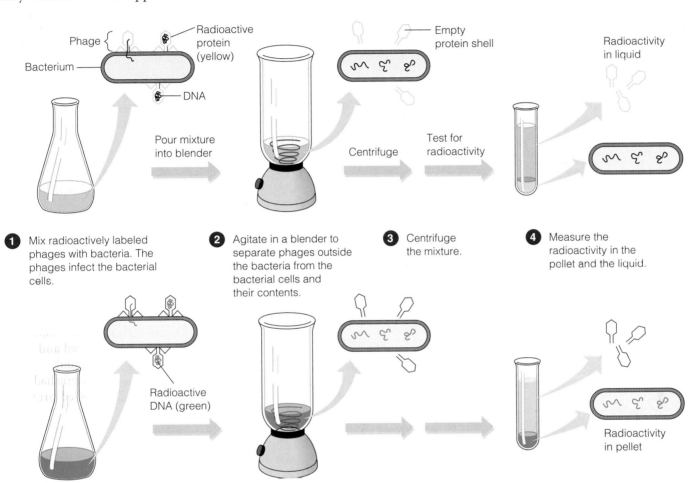

1 Mix radioactively labeled phages with bacteria. The phages infect the bacterial cells.

2 Agitate in a blender to separate phages outside the bacteria from the bacterial cells and their contents.

3 Centrifuge the mixture.

4 Measure the radioactivity in the pellet and the liquid.

Radioactive DNA (green)
Radioactivity in pellet

Figure 10.1B The Hershey-Chase experiment

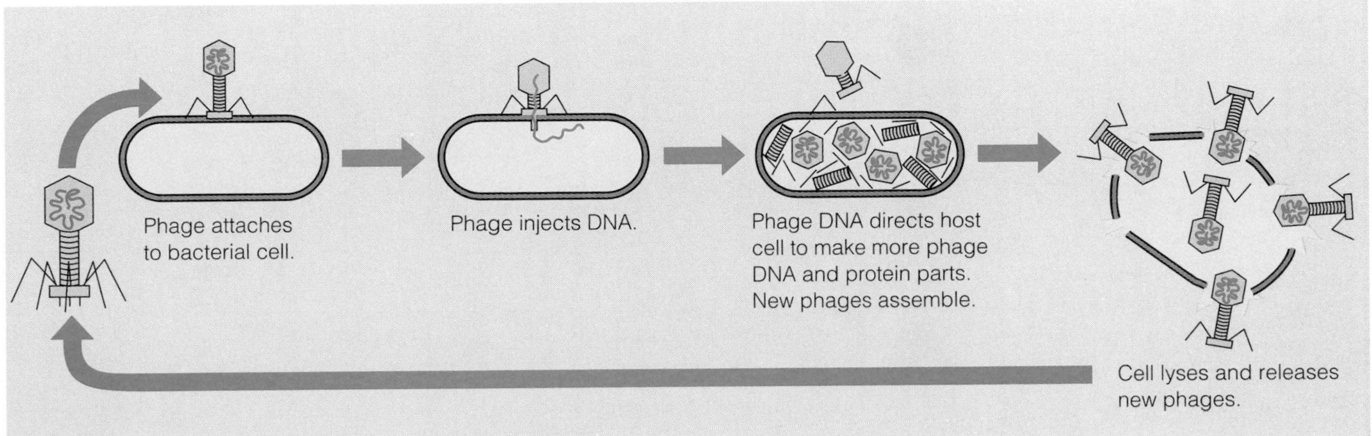

Phage injects DNA.

Cell lyses and releases new phages.

Figure 10.1C Phage reproductive cycle

10.2 DNA and RNA are polymers of nucleotides

10A
See Activity 10A on the CD-ROM.

By the time Hershey and Chase performed their experiments, a good deal was already known about DNA. Scientists had identified all its atoms and knew how they were covalently bonded to one another. What was not understood was the specific arrangement of parts that gave DNA its unique properties—the capacity to store genetic information, copy it, and pass it from generation to generation. However, it was only one year after Hershey and Chase published their results that scientists figured out DNA's three-dimensional structure and the basic strategy of how it works. We will examine that momentous discovery in Module 10.3. First, let's look at the underlying chemical structure of DNA and its chemical cousin RNA.

Recall from Module 3.20 that DNA and RNA are nucleic acids, which consist of long chains (polymers) of chemical units (monomers) called **nucleotides.** A very simple diagram of such a polymer, or **polynucleotide,** is shown on the left in Figure 10.2A. This sample polynucleotide chain shows only one possible arrangement of the four different types of nucleotides (abbreviated A, C, T, and G) that make up DNA. Because nucleotides can occur in a polynucleotide in any sequence and polynucleotides vary in length from long to very long, the number of possible polynucleotides is very great.

Zooming in on our polynucleotide, we see in the center of Figure 10.2A that each nucleotide consists of three components: a nitrogenous base (gray), a sugar (blue), and a phosphate group (yellow). The nucleotides are joined to one another by covalent bonds between the sugar of one nucleotide and the phosphate of the next. This results in a **sugar-phosphate backbone,** a repeating pattern of sugar-phosphate-sugar-phosphate. The nitrogenous bases are arranged as appendages all along this backbone.

Examining a single nucleotide even more closely (on the right in Figure 10.2A), we note the chemical structure of its three components. The phosphate group has a phosphorus atom (P) at its center and is the source of the *acid* in nucleic acid. The sugar has five carbon atoms (shown in red here for emphasis)—four in its ring and one extending above the ring. The ring also includes an oxygen atom. The sugar is called deoxyribose because, compared to the sugar ribose, it is missing an oxygen atom. (Notice that the C atom in the lower right corner of the ring is bonded to an H atom instead of to an —OH group, as it is in ribose; see Figure 10.2C.) The full name for DNA is deoxyribonucleic acid, with the "nucleic" part coming from DNA's location in the nuclei of eukaryotic cells. The nitrogenous base (thymine, in our example) has a ring of nitrogen and carbon atoms with various functional groups attached. In contrast to the acidic phosphate group, nitrogenous bases are basic, hence their name.

The four nucleotides found in DNA differ only in their nitrogenous bases. Figure 10.2B shows the structures of DNA's four nitrogenous bases. At this point, the structural details are not as important as the fact that the bases are of two types. **Thymine (T)** and **cytosine (C)** are single-ring structures called *pyrimidines.* **Adenine (A)** and **guanine (G)** are larger, double-ring structures called *purines.* Note

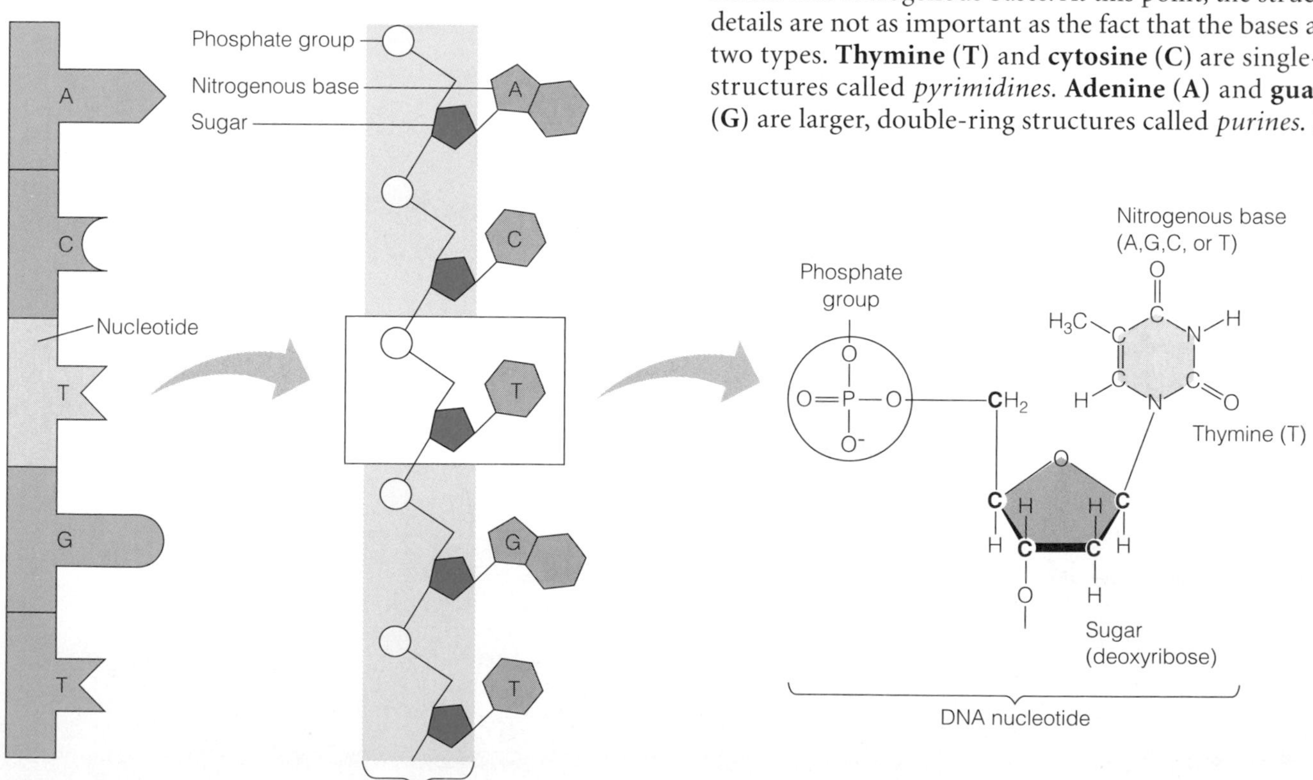

Figure 10.2A DNA polynucleotide

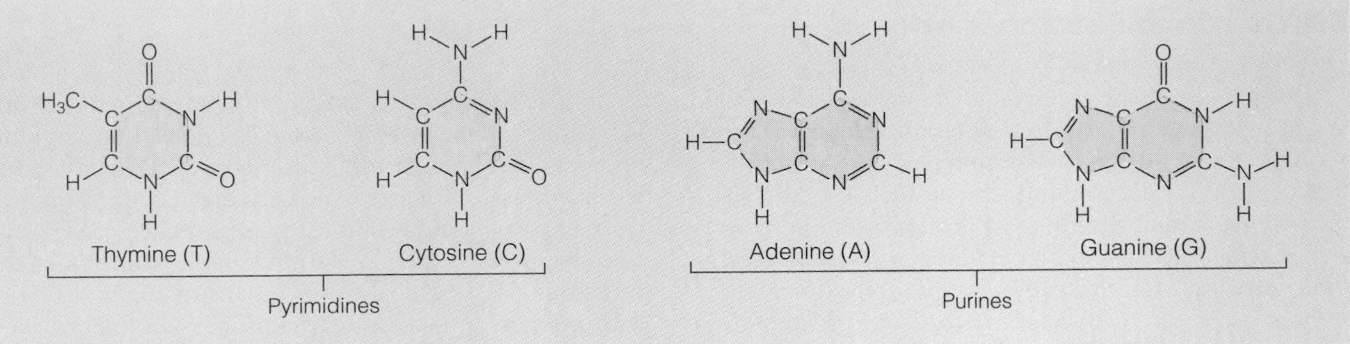

Figure 10.2B Nitrogenous bases of DNA

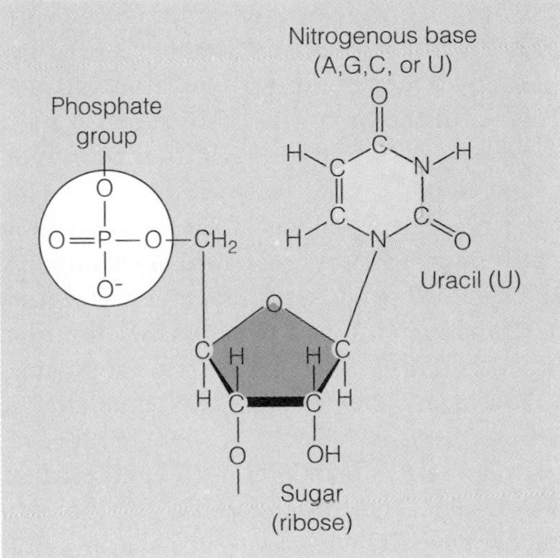

Figure 10.2C RNA nucleotide

that the one-letter abbreviations can be used for either the bases alone or for the nucleotides containing them.

What about RNA? As its name—ribonucleic acid—implies, its sugar is ribose rather than deoxyribose. Notice the ribose in the RNA nucleotide in Figure 10.2C; the sugar ring has an —OH group attached to the C atom at its lower right corner. Another difference between RNA and DNA is that, instead of thymine, RNA has a nitrogenous base called **uracil (U).** (You can see the structure of uracil in Figure 10.2C; it is very similar to thymine.) Except for the presence of ribose and uracil, an RNA polynucleotide chain is identical to a DNA polynucleotide chain. Figure 10.2D is a computer graphic of a piece of RNA polynucleotide six nucleotides long. The yellow color of the phosphorus atoms at the center of the phosphate groups makes it easy to spot the sugar-phosphate backbone.

? Compare and contrast DNA with RNA.

Both are polymers of nucleotides. A nucleotide consists of a sugar + a nitrogenous base + a phosphate group. In RNA, the sugar is ribose; in DNA it is deoxyribose. Both RNA and DNA have the bases A, G, and C; for a fourth base, DNA has T and RNA has U.

Figure 10.2D RNA polynucleotide

10.3 DNA is a double-stranded helix

10A

Once biologists were convinced that DNA was the genetic material, knowing the molecule's building blocks was not enough. A race was on to determine how the structure of this molecule could account for its role in heredity. The three-dimensional structures of proteins were starting to yield fascinating clues about the functions of those macromolecules, and biologists hoped the three-dimensional structure of DNA might do the same. Among the scientists working on the problem were several who had already made discoveries in deciphering protein structure: Linus Pauling in California, and Maurice Wilkins and Rosalind Franklin (Figure 10.3A) in London. First to the finish line with DNA, however, were two scientists who were relatively unknown at the time—American James D. Watson (left in Figure 10.3B) and Englishman Francis Crick. In 1962, Watson, Crick, and Wilkins received the Nobel Prize for their work. (Franklin probably would have received the prize as well, but for her death from cancer in 1958.)

Figure 10.3A Rosalind Franklin

The celebrated partnership that determined the structure of DNA began soon after the 23-year-old Watson journeyed to Cambridge University, where Crick was studying protein structure with a technique called X-ray crystallography. While visiting the laboratory of Maurice Wilkins at King's College in London, Watson saw an X-ray crystallographic photograph of DNA, produced by Wilkins's colleague Rosalind Franklin. The photograph clearly revealed the basic shape of DNA to be a helix. On the basis of Watson's later recollection of the photograph, he and Crick deduced that the helix had a uniform diameter of 2 nm, with its nitrogenous bases stacked about a third of a nanometer apart. (For comparison, recall that the plasma membrane of a cell is about 8 nm thick.) The diameter of the helix suggested that it was made up of two polynucleotide strands. The presence of two strands accounts for the now familiar term **double helix.**

Using wire models of the nucleotides, Watson and Crick began trying to construct a double helix that would conform both to Franklin's data and to what was then known about the chemistry of DNA. After failing to make a satisfactory model that placed the sugar-phosphate backbones inside the double helix, Watson tried putting the backbones on the outside and forcing the nitrogenous bases to swivel to the interior of the molecule. It occurred to him that the four kinds of bases might pair in a specific way. This idea of specific base pairing was a flash of inspiration that enabled Watson and Crick to solve the DNA puzzle.

At first, Watson imagined that the bases paired like with like—for example, A with A, and C with C. But that kind of pairing did not fit with the fact that the DNA molecule has a *uniform* diameter. An AA pair (made of double-ringed bases) would be almost twice as wide as a CC pair, causing bulges in the molecule. It soon became apparent that a double-ringed base (purine) must always be paired with a single-ringed base (pyrimidine) on the opposite strand. Moreover, Watson and Crick realized that the individual structures of the bases dictated the pairings even more specifically. Each base has chemical side groups that can best form hydrogen bonds with one appropriate partner (to review the hydrogen bond, see Module 2.10). Adenine can best form hydrogen bonds with thymine, and guanine with cytosine. In the biologist's shorthand, A pairs with T, and G pairs with C. A is also said to be "complementary" to T, and G to C.

Watson and Crick's pairing scheme not only fit what was known about the physical attributes and chemical bonding of DNA, it also explained some data obtained several years earlier by American biochemist Erwin Chargaff. Chargaff and his co-workers had discovered that the amount of adenine in the DNA of any one species was equal to the amount of thymine, and that the amount of guanine was equal to that of cytosine. Chargaff's rules, as they are called, are explained by the fact that A on one of DNA's polynucleotide chains always pairs with T on the other polynucleotide chain, and G on one chain pairs only with C on the other chain.

You can picture the model of the DNA double helix proposed by Watson and Crick as a rope ladder having rigid, wooden rungs, with the ladder twisted into a spiral (Figure 10.3C). The side ropes are the equivalent of the sugar-

Figure 10.3B Watson and Crick in 1953 with their model of the DNA double helix

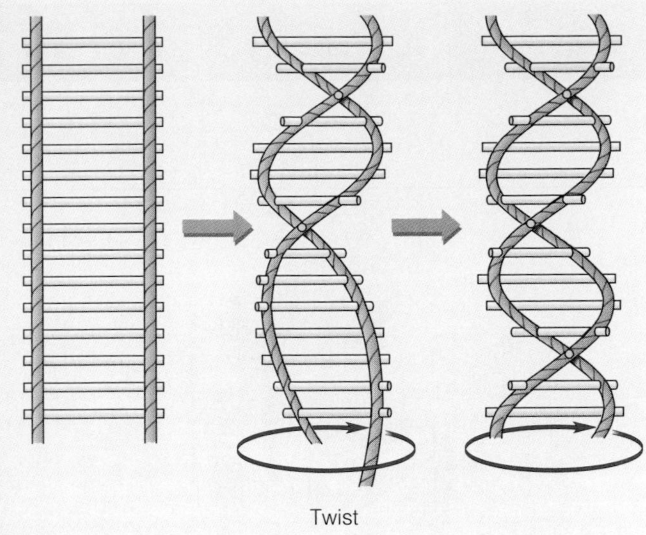

Figure 10.3C A rope-ladder model for the double helix

phosphate backbones, and the rungs represent pairs of nitrogenous bases joined by hydrogen bonds.

Figure 10.3D shows three representations of the double helix. The ribbonlike diagram on the left symbolizes the bases with shapes that emphasize their complementarity. In the center is a more chemical version showing four base pairs, with the helix untwisted and the hydrogen bonds specified by dotted lines; you can see that the two sugar-phosphate backbones of the double helix are oriented in opposite directions. (Notice that the sugars on the two strands are upside-down with respect to each other.) On the right is a computer graphic showing every atom of part of a double helix. Atoms composing the deoxyribose sugars are bright blue, phosphate groups are yellow, and nitrogenous bases are shades of green, light blue, and orange.

Although the Watson-Crick base-pairing rules dictate the side-by-side combinations of nitrogenous bases that form the rungs of the double helix, they place no restrictions on the *sequence* of nucleotides along the length of a DNA strand. In fact, the sequence of bases can vary in countless ways. Consequently, it is not surprising that the DNA of different species, which have different genes, have different proportions of the bases in their DNA.

In April 1953, Watson and Crick shook the scientific world with a succinct, two-page announcement of their molecular model for DNA in the journal *Nature*. Few milestones in the history of biology have had as broad an impact as their double helix, with its AT and CG base pairing.

The Watson-Crick model gave new meaning to the words genes and chromosomes—and to the chromosome theory of inheritance. With a complete picture of DNA, we can see that the genetic information in a chromosome must be encoded in the nucleotide sequence of the molecule. The structure of DNA also suggests a molecular explanation for life's unique properties of reproduction and inheritance, as we see next. 

? Along one strand of a double helix is the nucleotide sequence GGCATAGGT. What is the complementary sequence for the other DNA strand?

CCGTATCCA

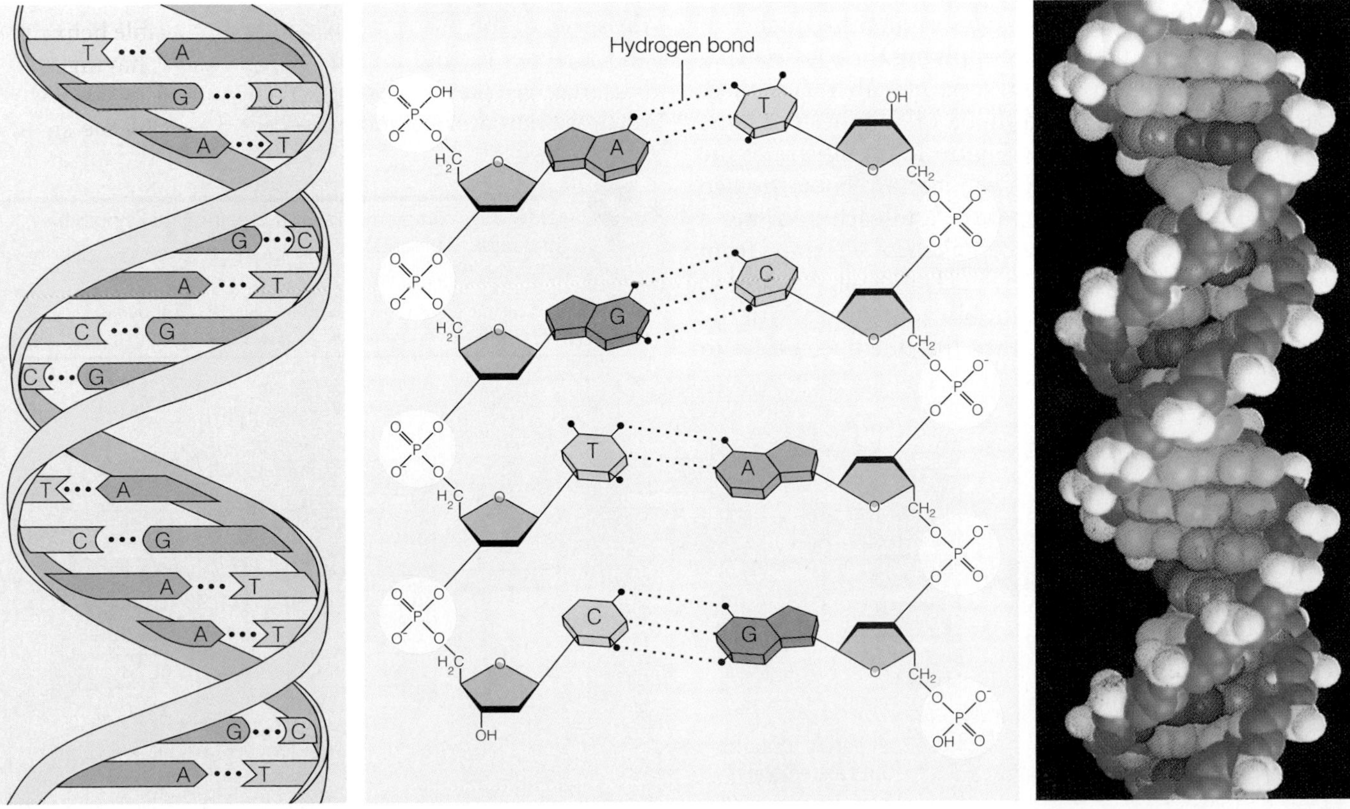

Figure 10.3D Three representations of DNA

10.4 DNA replication depends on specific base pairing

10B

An essential aspect of reproduction and inheritance is that a complete set of genetic instructions passes from one generation to the next. In order for this to occur, there must be a means of copying the instructions. Long before DNA was identified as the genetic material, some people argued that gene replication must be based on a concept called complementary surfaces. According to this idea, when a gene replicates, a "negative image" is created along the original ("positive") surface, just as clay or plaster forms a negative shape when it is packed around an object. The gene's negative image, like the plaster, could serve as a template (mold) for making copies of the original positive image. Photography provides another example of the template principle: A print can be used to make a negative, which can then be used to make copies of the original print. Until 1953, the template idea was discounted by many geneticists. However, Watson and Crick's model for DNA structure suggested a template mechanism for DNA replication. As they said in the conclusion of their first paper, "It has not escaped our notice that the specific pairing we have postulated immediately suggests a possible copying mechanism for the genetic material."

The logic behind the Watson-Crick proposal for how DNA is copied—by specific pairing of complementary bases—is quite simple. You can see this by covering one of the strands in the parental DNA molecule in Figure 10.4A with a piece of paper. You can determine the sequence of bases in the covered strand by applying the base-pairing rules to the unmasked strand: A pairs with T, G with C. Watson and Crick predicted that a cell applies the same rules when copying its genes. Figure 10.4A illustrates the template hypothesis for DNA replication. First, the two strands of parental DNA separate, and each becomes a template for the assembly of a complementary strand from a supply of free nucleotides. The nucleotides line up one at a time along the template strand in accordance with the base-pairing rules. Enzymes then link the nucleotides to form the new DNA strands. The completed new molecules, identical to the parental molecule, are known as daughter DNA. This hypothesis was confirmed by experiments performed in the 1950s.

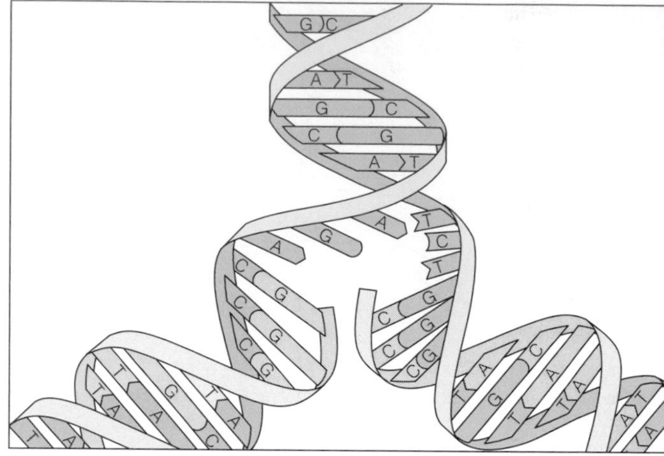

Figure 10.4B Untwisting and replication of DNA

Although the general mechanism of DNA replication is conceptually simple, the actual process involves complex biochemical gymnastics. Some of the complexity arises from the fact that the helical DNA molecule must untwist as it replicates and must copy its two strands roughly simultaneously (Figure 10.4B, above). Another challenge is the speed of the process. Nucleotides are added at a rate of about 50 per second in mammals and 500 per second in bacteria. We take a closer look at the mechanisms of DNA replication in the next module. 

? How does complementary base pairing make possible the replication of DNA?

When the two strands of the double helix separate, each serves as a "mold" upon which nucleotides can be arranged by specific base pairing into new complementary strands.

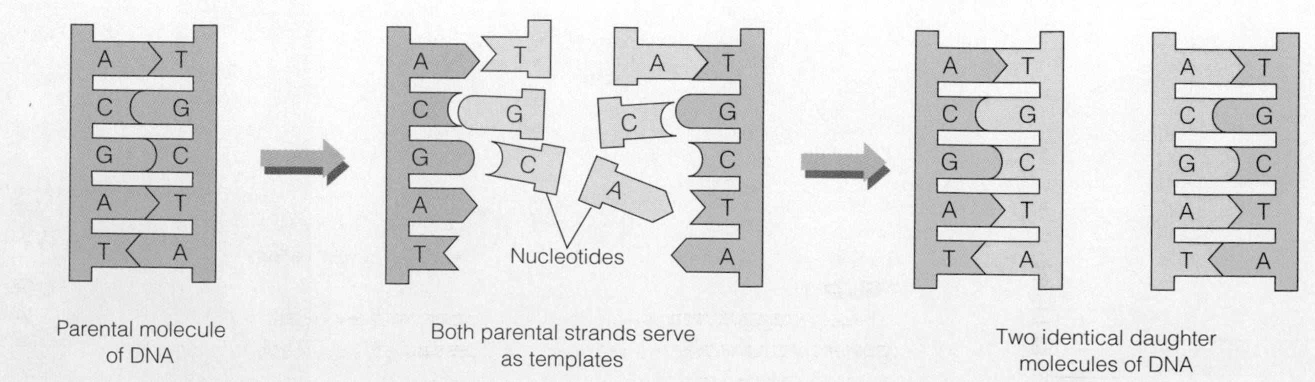

Figure 10.4A A template model for DNA replication

DNA replication: A closer look 10.5

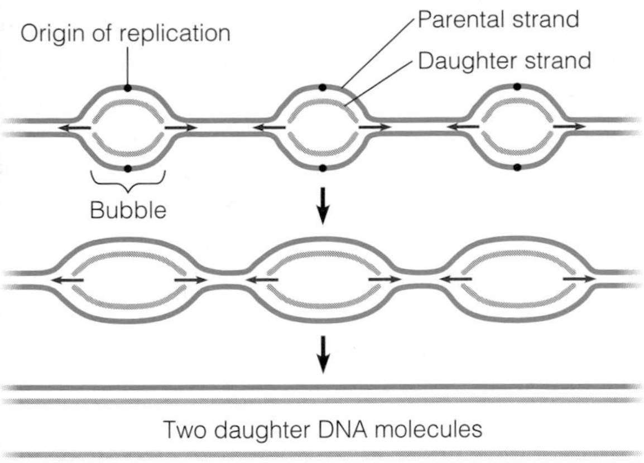

Figure 10.5A Multiple "bubbles" in replicating DNA

DNA replication begins at specific sites on the double helix, called origins of replication, where proteins that start the process attach to the DNA and separate the strands. As Figure 10.5A shows, replication then proceeds in both directions, creating what are called replication "bubbles." The parental DNA strands (blue) open up as daughter strands (gray) elongate on both sides of each bubble. The DNA molecule of a eukaryotic chromosome has many origins where replication can start simultaneously, shortening the total time needed for the process. Thus, thousands of bubbles can be present at once. Eventually, all the bubbles merge, yielding two completed daughter DNA molecules.

Figure 10.5B shows the molecular building blocks of a tiny segment of DNA, reminding us that the DNA's sugar-phosphate backbones run in opposite directions. Notice that each strand has a 3′ ("three-prime") end and a 5′ end. The primed numbers refer to the carbon atoms of the nucleotide sugars. At one end of each DNA strand, the sugar's 3′ carbon atom is attached to an —OH group; at the other end, the sugar's 5′ carbon has a phosphate group.

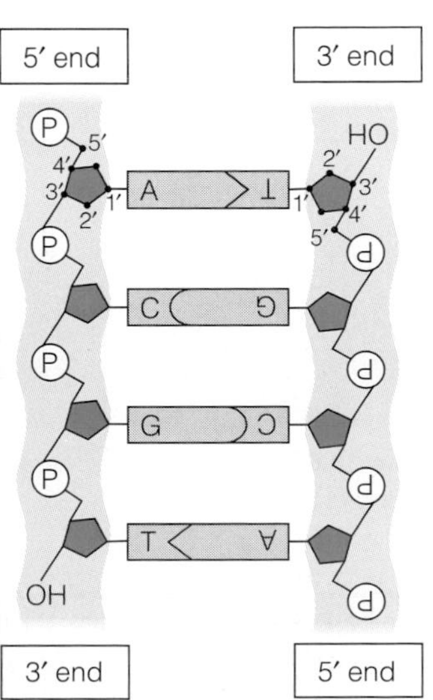

Figure 10.5B The opposite orientations of DNA strands

The opposite orientation of the strands is important in DNA replication. The enzymes that link DNA nucleotides to a growing daughter strand, called **DNA polymerases,** add nucleotides only to the 3′ end of the strand, never to the 5′ end. Thus, a daughter DNA strand can only grow in the 5′→3′ direction. You see the consequences of this enzyme specificity in Figure 10.5C. The forked structure represents one side of a replication bubble. One of the daughter strands (shown in light gray) can be synthesized in one continuous piece, by a DNA polymerase working toward the forking point of the parental DNA. However, to make the other daughter strand (dark gray), polymerase molecules must work outward from the forking point. This new strand is synthesized in short pieces as the fork opens up. Another enzyme, called **DNA ligase,** then ties (ligates) the pieces together into a single DNA strand.

Altogether, DNA replication requires the cooperation of more than a dozen enzymes and other proteins. The process is not only fast but amazingly accurate; typically, only about one in a billion nucleotides in DNA is incorrectly paired. In addition to their roles in linking nucleotides together, DNA polymerases also carry out a proofreading step that quickly removes nucleotides that have base-paired incorrectly during replication. DNA polymerases and DNA ligase are also involved in repairing DNA damaged by harmful radiation (such as ultraviolet light and X-rays) or toxic chemicals in the environment.

DNA replication ensures that all the somatic cells in a multicellular organism carry the same genetic information. It is also the means by which genetic instructions are copied for the next generation of the organism. In the next module, we begin to pursue the connection between DNA instructions and an organism's phenotypic traits.

? What is the function of DNA polymerase in DNA replication?

This enzyme covalently connects the nucleotides to form a new strand as the nucleotides line up along an existing strand according to the base-pairing rules.

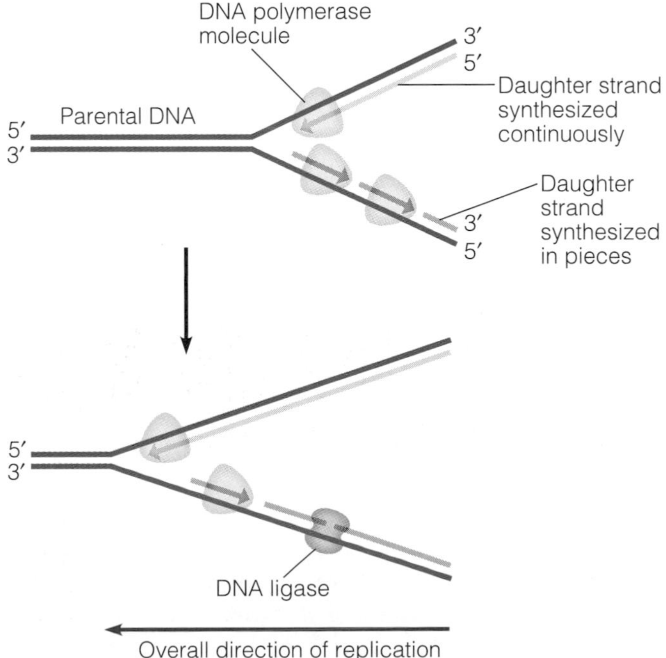

Figure 10.5C How daughter DNA strands are synthesized

The Flow of Genetic Information from DNA to RNA to Protein

10.6 The DNA genotype is expressed as proteins, which provide the molecular basis for phenotypic traits

10C With our knowledge of DNA, we can now define genotype and phenotype more precisely than we did in Chapter 9. An organism's genotype, its genetic makeup, is the heritable information contained in its DNA. The phenotype is the organism's specific traits. The molecular basis of the phenotype lies in proteins with a variety of functions. For example, structural proteins help make up the body of an organism, and enzymes catalyze its metabolic activities.

What is the connection between the genotype and the protein molecules that more directly determine the phenotype? The answer is that DNA specifies the synthesis of proteins. A gene does not build a protein directly, but rather dispatches instructions in the form of RNA, which in turn programs protein synthesis. This central concept in biology is summarized in Figure 10.6A. The chain of command is from DNA in the nucleus of the cell (purple area) to RNA to protein synthesis in the cytoplasm (tan area). The two main stages are **transcription,** the transfer of genetic information from DNA into an RNA molecule, and **translation,** the transfer of the information in the RNA into a protein. The next nine modules describe the steps in this flow of molecular information from gene to protein.

The relationship between genes and proteins was first proposed in 1909, when English physician Archibald Garrod suggested that genes dictate phenotypes through enzymes, the proteins that catalyze chemical processes in the cell. Garrod's idea came from his observations of inherited diseases. He hypothesized that an inherited disease reflects a person's inability to make a particular enzyme, and he referred to such diseases as "inborn errors of metabolism." He gave as one example the hereditary condition called alkaptonuria, in which the urine appears dark red because it contains a chemical called alkapton. Garrod reasoned that normal individuals have an enzyme that breaks down alkapton, whereas alkaptonuric individuals lack the enzyme. Garrod's hypothesis was ahead of its time, but research conducted decades later proved him right. In the intervening years, biochemists accumulated evidence that cells make and break down biologically important molecules via metabolic pathways, as in the synthesis of an amino acid or the breakdown of a sugar. As we described in Unit I, each step in a metabolic pathway is catalyzed by a specific enzyme. Therefore, individuals lacking one of the enzymes for a pathway are unable to complete the pathway.

DNA
TRANSCRIPTION
RNA
TRANSLATION
Protein

Figure 10.6A The flow of genetic information in a eukaryotic cell

The major breakthrough in demonstrating the relationship between genes and enzymes came in the 1940s, from the work of American geneticists George Beadle and Edward Tatum with the orange bread mold *Neurospora crassa* (Figure 10.6B). Beadle and Tatum studied strains of the mold that were unable to grow on the usual simple growth medium. Each of these so-called nutritional mutants turned out to lack an enzyme in a metabolic pathway that produced some molecule the mold needed, such as an amino acid. Beadle and Tatum also showed that each mutant was defective in a single gene. Accordingly, they formulated the one gene–one enzyme hypothesis, which stated that the function of an individual gene is to dictate the production of a specific enzyme.

The one gene–one enzyme hypothesis has been amply confirmed, but with some important modifications. First it was extended beyond enzymes to include *all* types of proteins. For example, alpha-keratin, the structural protein of your hair, and the structural proteins that make up the outside of phage T4 are as much the products of genes as enzymes are. So biologists soon began to think in terms of one gene–one *protein.* Then it was discovered that many proteins consist of two or more different polypeptide chains (see Module 3.18), and each polypeptide is specified by its own gene. Thus, Beadle and Tatum's hypothesis has come to be restated as one gene–one *polypeptide.* 

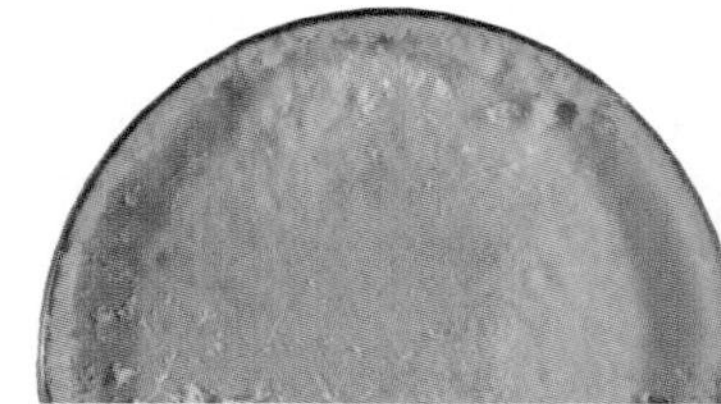

Figure 10.6B *Neurospora crassa* growing in a culture dish

? In the information flow from DNA to protein, what are the functions of transcription and translation?

Transcription is the transfer of genetic information from DNA to RNA. Translation is the use of the RNA as information for synthesis of a specific protein.

Genetic information is written as codons and translated into amino acid sequences 10.7

Stating that genetic information in DNA is transcribed into RNA and then translated into polypeptides does not tell us *how* these processes occur. Transcription and translation are linguistic terms, and it is useful to think of nucleic acids and polypeptides as having languages. To understand how genetic information passes from genotype to phenotype, we need to see how the chemical language of DNA is translated into the different chemical language of polypeptides.

What, exactly, is the language of nucleic acids? Both DNA and RNA are polymers made of monomers in specific sequences that carry information, much as specific sequences of letters carry information in English or Russian, for example. In DNA, the monomers are the four types of nucleotides, which differ in their nitrogenous bases (A, T, C, and G). The same is true for RNA, although it has the base U instead of the T in DNA.

Figure 10.7 focuses on a small region of one of the genes ("Gene 3," shown in dark blue) carried by a DNA molecule. DNA's language is written as a linear sequence of nucleotide bases on a polynucleotide, a sequence such as the one you see on the enlarged DNA strand in the figure. Specific sequences of bases, each with a beginning and an end, make up the genes on a DNA strand. A typical gene consists of hundreds or thousands of nucleotides, and a molecule of DNA may contain thousands of genes.

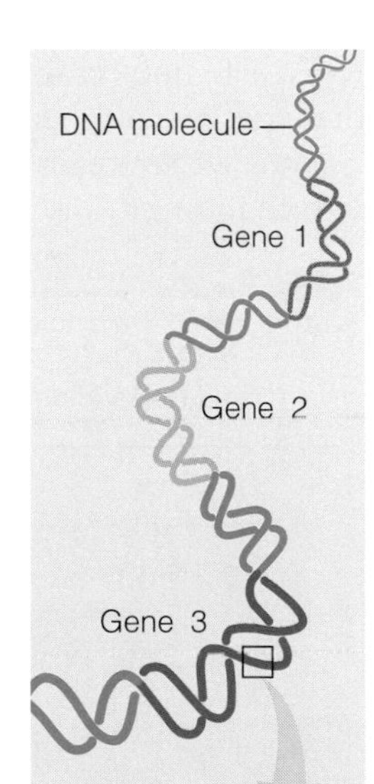

The pink strand underneath the enlarged DNA region represents the results of transcription: an RNA molecule. The process is called transcription because the nucleic acid language of DNA has simply been rewritten as a sequence of bases on RNA; the language is still that of nucleic acids.

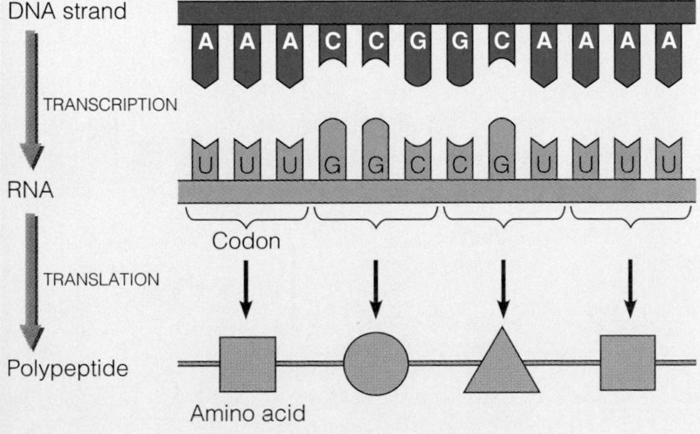

Figure 10.7 Transcription and translation of codons

Notice that the nucleotide bases on the RNA molecule are complementary to those on the DNA strand. As we will see in Module 10.9, this is because the RNA was synthesized using the DNA as a template.

The purple chain represents the results of translation, the conversion of the nucleic acid language into the polypeptide language. Like nucleic acids, polypeptides are polymers, but the monomers that make them up—the letters of the polypeptide alphabet—are the 20 amino acids common to all organisms. Again, the language is written in a linear sequence, and the sequence of nucleotides of the RNA molecule dictates the sequence of amino acids of the polypeptide. RNA is only a messenger; the genetic information that dictates the amino acid sequence is based in DNA.

The brackets below the RNA indicate how genetic information is coded in nucleic acids. Notice that each bracket encloses *three* nucleotides on RNA. Recall that there are only four different kinds of nucleotides in DNA (A, G, C, T) and RNA (A, G, C, U). In translation, these four must somehow specify 20 amino acids. If each nucleotide base specified one amino acid, only four of the 20 amino acids could be accounted for. What if the language consisted of two-letter code words? If we read the bases of a gene two at a time, AG, for example, could specify one amino acid, while AT could designate a different amino acid. However, when the 4 bases are taken in doublets, there are only 16 (that is, 4^2) possible arrangements—still not enough to specify all 20 amino acids.

Triplets of bases are the smallest "words" of uniform length that can specify all the amino acids. Suppose each code word in DNA consists of a triplet, with each arrangement of three consecutive bases specifying an amino acid. Then there can be 64 (that is, 4^3) possible code words—more than enough to specify the 20 amino acids. Indeed, there are enough triplets to allow more than one coding for each amino acid. For example, the base triplets AAT and AAC could both code for the same amino acid—and, in fact, they do.

Experiments have verified that the flow of information from gene to protein is based on a **triplet code:** The genetic instructions for the amino acid sequence of a polypeptide chain are written in DNA and RNA as a series of three-base words, called **codons.** Notice in the figure that three-base codons in the DNA are transcribed into complementary three-base codons in the RNA, and then the RNA codons are translated into amino acids that form a polypeptide. We turn to the codons themselves in the next module.

A particular protein is 100 amino acids long. In the gene for this protein, how many nucleotides are necessary to code for this protein?

300

10.8 The genetic code is the Rosetta stone of life

In 1799, a large stone tablet was found in Rosetta, Egypt, carrying the same lengthy inscription in three ancient languages: Egyptian written in hieroglyphics and two forms of Greek. This stone provided the key that enabled scholars to crack the previously indecipherable hieroglyphic code.

In cracking the genetic code, scientists wrote their own Rosetta stone. It was based on information gathered from a series of elegant experiments that disclosed the amino acid translations of each of the nucleotide-triplet code words. The first codon was deciphered in 1961 by American biochemist Marshall Nirenberg. He synthesized an artificial RNA molecule by linking together identical RNA nucleotides having uracil as their base. No matter where this message started or stopped, it could contain only one type of triplet codon: UUU. Nirenberg added this "poly-U" to a test-tube mixture containing ribosomes and the other ingredients required for polypeptide synthesis. This mixture translated the poly-U into a polypeptide containing a single kind of amino acid, phenylalanine. Thus, Nirenberg learned that the RNA codon UUU specifies the amino acid phenylalanine (Phe). By variations on this method, the amino acids specified by all the codons were determined.

As Figure 10.8A shows, 61 of the 64 triplets code for amino acids. The triplet AUG has a dual function: It not only codes for the amino acid methionine (Met) but also can provide a signal for the start of a polypeptide chain. Three of the other codons (in white boxes in the figure) do not designate amino acids. They are the stop codons that instruct the ribosomes to end the polypeptide.

Notice in Figure 10.8A that there is redundancy in the code but no ambiguity. For example, although codons UUU and UUC both specify phenylalanine (redundancy), neither of them ever represents any other amino acid (no ambiguity). The codons in the figure are the triplets found in RNA. They have a straightforward, complementary relationship to the codons in DNA. The nucleotides making up the codons occur in a linear order along the DNA and RNA, with no gaps or "punctuation" separating the codons.

As an exercise in translating the genetic code, consider the 12-nucleotide segment of DNA in Figure 10.8B. Let's read this as a series of triplets. Using the base-pairing rules (with U in RNA instead of T), we see that the RNA codon corresponding to the first transcribed DNA triplet, TAC, is AUG. AUG says, "Place Met as the first amino acid in the polypeptide." The second DNA triplet, TTC, dictates RNA codon AAG, which designates lysine (Lys) as the second amino acid. We continue until we reach the stop codon.

Almost all of the genetic code is shared by all organisms, from the simplest bacteria to the most complex plants and animals. In experiments, bacteria can translate human genetic messages, and human cells can translate bacterial RNA. The universality of the genetic vocabulary suggests that it was established very early in evolution.

Now we are ready to look more closely at how genetic information is transcribed and translated.

An mRNA molecule contains the nucleotide sequence CCAUUUACG. Translate this sequence into the corresponding amino acid sequence.

Pro-Phe-Thr

FIRST BASE	SECOND BASE: U		SECOND BASE: C		SECOND BASE: A		SECOND BASE: G		THIRD BASE
U	UUU	Phe	UCU	Ser	UAU	Tyr	UGU	Cys	U
U	UUC	Phe	UCC	Ser	UAC	Tyr	UGC	Cys	C
U	UUA	Leu	UCA	Ser	UAA	Stop	UGA	Stop	A
U	UUG	Leu	UCG	Ser	UAG	Stop	UGG	Trp	G
C	CUU	Leu	CCU	Pro	CAU	His	CGU	Arg	U
C	CUC	Leu	CCC	Pro	CAC	His	CGC	Arg	C
C	CUA	Leu	CCA	Pro	CAA	Gln	CGA	Arg	A
C	CUG	Leu	CCG	Pro	CAG	Gln	CGG	Arg	G
A	AUU	Ile	ACU	Thr	AAU	Asn	AGU	Ser	U
A	AUC	Ile	ACC	Thr	AAC	Asn	AGC	Ser	C
A	AUA	Ile	ACA	Thr	AAA	Lys	AGA	Arg	A
A	AUG	Met or start	ACG	Thr	AAG	Lys	AGG	Arg	G
G	GUU	Val	GCU	Ala	GAU	Asp	GGU	Gly	U
G	GUC	Val	GCC	Ala	GAC	Asp	GGC	Gly	C
G	GUA	Val	GCA	Ala	GAA	Glu	GGA	Gly	A
G	GUG	Val	GCG	Ala	GAG	Glu	GGG	Gly	G

Figure 10.8A Dictionary of the genetic code (RNA codons)

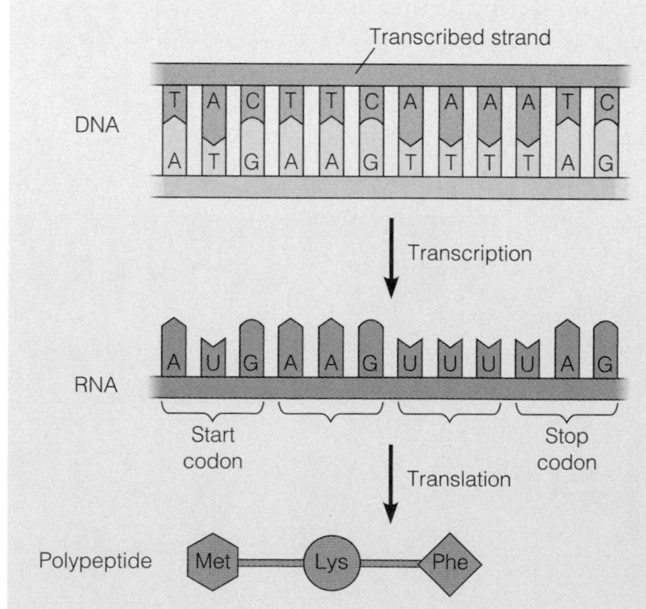

Figure 10.8B Deciphering the genetic information in DNA

Transcription produces genetic messages in the form of RNA 10.9

10D

Transcription, the transfer of genetic information from DNA to RNA, occurs in the cell nucleus, as we indicated in Figure 10.6A. An RNA molecule is transcribed from a DNA template by a process that resembles the synthesis of a DNA strand during DNA replication. Figure 10.9A is a close-up view of this process. As with replication, the two DNA strands must first separate at the place where the process will start. In transcription, however, only one of the DNA strands serves as a template for the newly forming molecule. The nucleotides that make up the new RNA molecule take their places one at a time along the DNA template strand by forming hydrogen bonds with the nucleotide bases there. Notice that the RNA nucleotides follow the same base-pairing rules that govern DNA replication, except that U, rather than T, pairs with A. The RNA nucleotides are linked by the transcription enzyme **RNA polymerase,** symbolized in the figure by the large gray shape in the background.

Figure 10.9B is an overview of the transcription of an entire gene. RNA polymerase must be instructed where to start and where to stop the transcribing process. The "start transcribing" signal is a nucleotide sequence called a **promoter,** which is located in the DNA next to the beginning of the gene. A promoter is a specific binding site for RNA polymerase. The first phase of transcription, called initiation, is the attachment of RNA polymerase to the promoter and the start of RNA synthesis. For any gene, the promoter dictates which of the two DNA strands is to be transcribed (the particular strand varying from gene to gene).

During a second phase of transcription, the RNA elongates. As RNA synthesis continues, the RNA strand peels away from its DNA template, allowing the two separated DNA strands to come back together in the region already transcribed. Finally, in the third phase, termination, the RNA polymerase reaches a special sequence of bases in the DNA template called a **terminator.** This sequence signals the end of the gene; at that point, the polymerase molecule detaches from the RNA molecule and the gene.

In addition to producing RNA that encodes amino acid sequences, transcription makes two other kinds of RNA that are involved in building polypeptides. We discuss these three kinds of RNA in the next three modules.

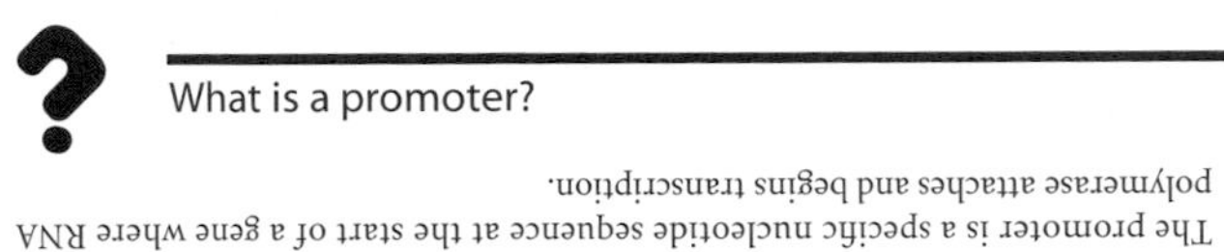

What is a promoter?
The promoter is a specific nucleotide sequence at the start of a gene where RNA polymerase attaches and begins transcription.

Figure 10.9A A close-up view of transcription

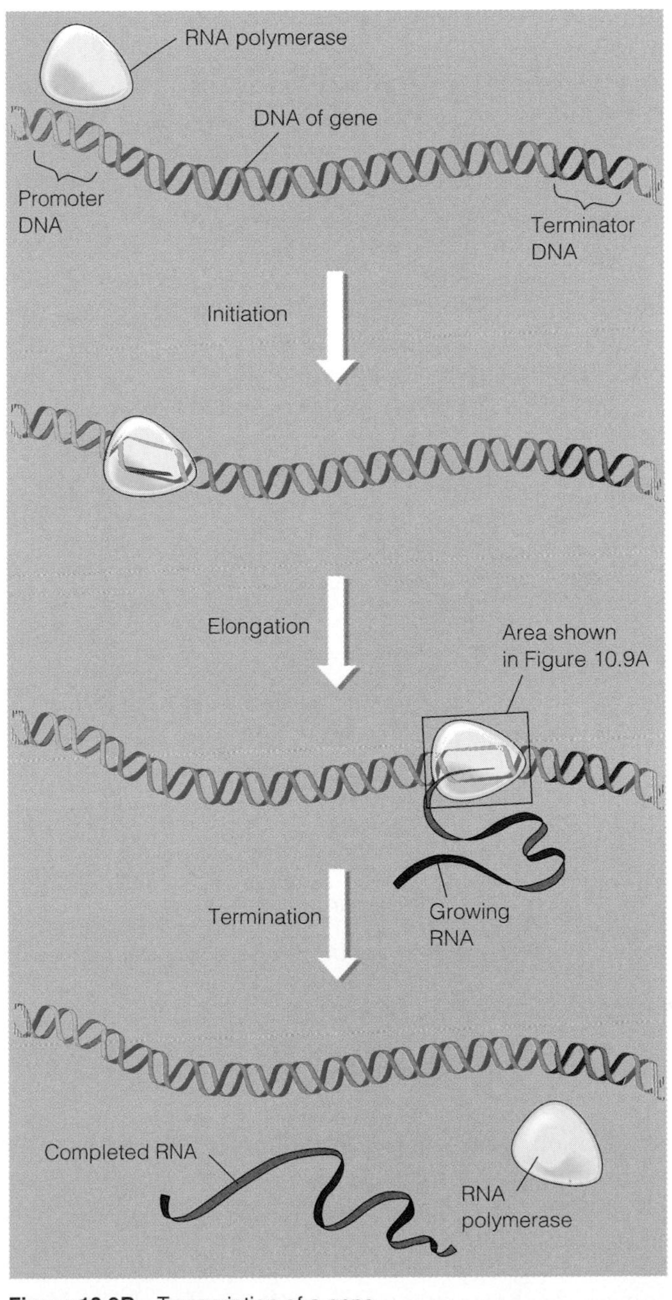

Figure 10.9B Transcription of a gene

10.10 Genetic messages are translated in the cytoplasm

The kind of RNA that encodes amino acid sequences is called **messenger RNA (mRNA)** because it conveys genetic information from DNA to the translation machinery of the cell. Messenger RNA is transcribed from DNA, and the genetic message the RNA carries is then translated into polypeptides. In prokaryotes, which lack a nucleus, transcription and translation both take place in the cytoplasm. As we indicated in Figure 10.6A, the situation is different in eukaryotic cells. Because DNA stays in the nucleus, eukaryotic mRNA molecules and the other RNA molecules required for translation must cross the nuclear envelope into the cytoplasm, where the equipment for polypeptide synthesis is located.

Figure 10.10 shows the results of an experiment that demonstrates the migration of RNA from the nucleus to the cytoplasm. Researchers first made radioactive RNA nucleotides and then grew eukaryotic cells in culture with the nucleotides. The cells used the radioactive nucleotides in synthesizing RNA. Radioactive emissions from the resulting RNA produced the black spots on the light micrographs shown here. Thus, the dots pinpoint the location of radioactive RNA in the cells.

The cell on the left was killed after 15 minutes of growing with the radioactive nucleotides, and the excess nucleotides were washed away. As you can see, the newly made RNA is heavily concentrated in the nucleus.

The cell on the right was also allowed to make radioactive RNA for 15 minutes. It was then washed free of excess radioactive nucleotides, but it was not killed. Instead, it was transferred for an additional 90 minutes to a growth liquid containing no radioactive nucleotides. As you can see, the radioactive RNA moved into the cytoplasm.

We are now ready to begin examining how the translation process works. Translation is a conversion between different languages—from nucleic acid language to protein language—and it involves more complex machinery than transcription. The machinery required for translating mRNA includes the following:

- Transfer RNA, another kind of RNA molecule
- Ribosomes, the organelles where polypeptide synthesis occurs
- Enzymes and a number of protein "factors"
- Sources of chemical energy, such as ATP

In the next two modules, we take a closer look at transfer RNA and ribosomes.

What hypothesis is supported by the experiment described in this module?

That RNA is synthesized in the nucleus and then migrates to the cytoplasm

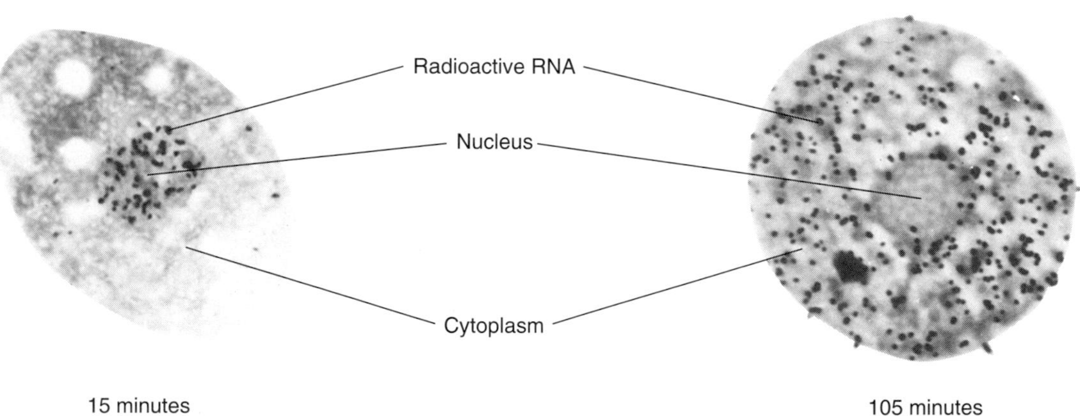

Figure 10.10 Movement of mRNA from the nucleus to the cytoplasm

10.11 Transfer RNA molecules serve as interpreters during translation

10E

Translation of any language into another language requires an interpreter, someone who can recognize the words of one language and convert them into the other. Translation of the genetic message carried in mRNA into the amino acid language of proteins also requires an interpreter. To convert the three-letter words (codons) of nucleic acids to the one-letter, amino acid words of proteins, a cell employs a molecular interpreter, a special type of RNA called **transfer RNA (tRNA).**

A cell that is ready to have some of its genetic information translated into polypeptides has a supply of amino acids in its cytoplasm. It has either synthesized them from other chemicals or obtained them from food. The amino acids themselves cannot recognize the codons arranged in sequence along messenger RNA. The amino acid tryptophan, for example, is no more attracted by codons for tryptophan than by any other codons. It is up to the cell's molecular interpreters, tRNA molecules, to match amino acids to the appropriate codons to form the new polypeptide. To perform this task, tRNA molecules must carry out two distinct functions: (1) picking up the appropriate amino acids, and (2) recognizing the appropriate codons in the mRNA. The unique structure of tRNA molecules enables them to perform both tasks.

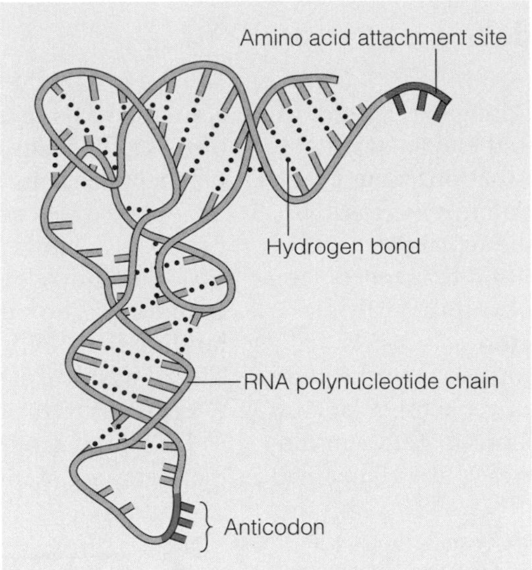

Figure 10.11A The structure of tRNA

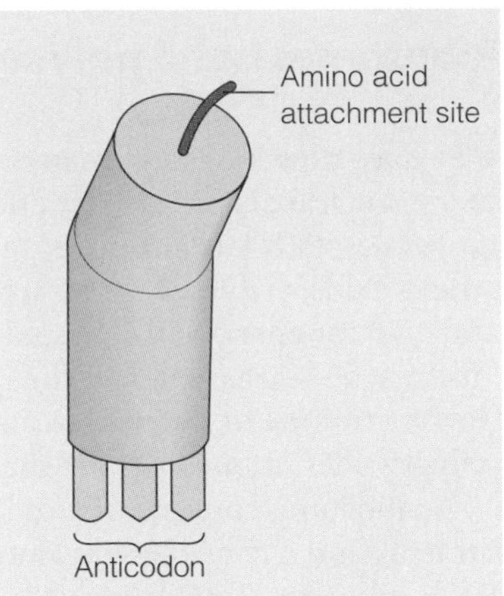

Figure 10.11B The symbol for tRNA used in this book

As shown in Figure 10.11A, a tRNA molecule is made of a single strand of RNA—one polynucleotide chain—consisting of only about 80 nucleotides. By twisting and folding upon itself, tRNA forms several double-stranded regions in which short stretches of RNA base-pair with other stretches. A single-stranded loop at one end of the folded molecule contains a special triplet of bases called an **anticodon.** The anticodon triplet is complementary to a codon triplet on mRNA. During translation, the anticodon on tRNA recognizes a particular codon on mRNA by using base-pairing rules. At the other end of the tRNA molecule is a site where an amino acid can attach.

In the modules that follow, in which we trace the process of translation, we represent tRNA with the simplified shape shown in Figure 10.11B (above right). This symbol emphasizes two parts of the molecule—the anticodon and the amino acid attachment site—that give tRNA its ability to match a particular nucleic acid word (codon) with its corresponding protein word (amino acid). Although all tRNA molecules are similar, there is a slightly different variety of tRNA for each amino acid.

Translating any language is a complex task. Likewise, the conversion of genetic information into the exact polypeptide it specifies is a complex process—too complex, in fact, for the cell's tRNA interpreters to carry out by themselves. By itself, a tRNA molecule cannot directly recognize an amino acid. What ensures that the appropriate amino acid attaches to a tRNA is an enzyme. There is a whole family of these enzymes, with at least one enzyme for each amino acid. Each enzyme specifically binds one type of amino acid to the appropriate tRNA molecule, using a molecule of ATP as energy to drive the reaction. The resulting amino acid–tRNA complex can then furnish its amino acid to a growing polypeptide chain, a process we describe in Module 10.12.

The computer graphic in Figure 10.11C shows a tRNA molecule (red and yellow) and an ATP molecule (green) bound to the enzyme molecule (blue). In this picture, you can see the proportional sizes of these three molecules. The amino acid that would attach to the tRNA is not shown; it would be less than half the size of the ATP.

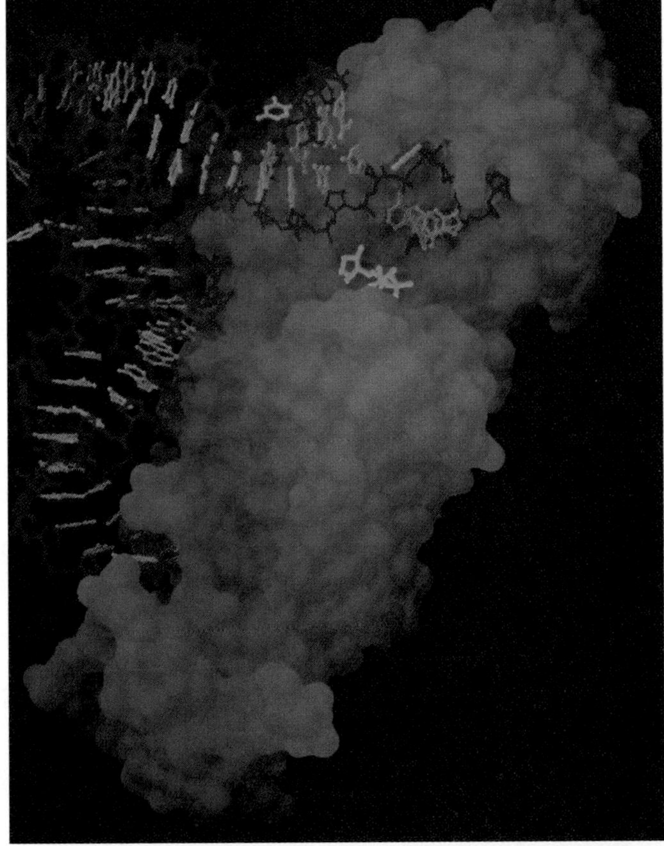
Figure 10.11C A molecule of tRNA binding to an enzyme molecule

What is an anticodon, and what is its function?

It is the base triplet of a tRNA molecule that couples the tRNA to a complementary codon in the mRNA. This is a key step in translating mRNA to polypeptide.

10.12 Ribosomes build polypeptides

10E

We have now looked at many of the things a cell needs to carry out translation: instructions in the form of mRNA molecules, tRNA to interpret the instructions, a supply of amino acids, enzymes for attaching amino acids to tRNA, and ATP for energy. Still needed are the actual polypeptide "factories"—organelles in the cytoplasm that coordinate the functioning of the mRNA and tRNA and actually make polypeptides. Ribosomes are these factories.

A ribosome consists of two subunits, each made up of proteins and a considerable amount of yet another kind of RNA, **ribosomal RNA (rRNA).** In Figure 10.12A, you can see the actual shapes and relative sizes of the ribosomal subunits. You can also see where mRNA, tRNA, and the growing polypeptide are located during translation.

The simplified drawings in Figures 10.12B and 10.12C indicate how tRNA anticodons and mRNA codons fit together on ribosomes. As Figure 10.12B shows, each ribosome has a binding site for mRNA on its small subunit. In addition, its large subunit has binding sites for tRNA. One of these, the P site, holds the tRNA carrying the growing polypeptide chain, while another, the A site, holds a tRNA carrying the next amino acid to be added to the chain. Figure 10.12C shows tRNA molecules occupying these two sites. The anticodon on each tRNA base-pairs with a codon on mRNA. The subunits of the ribosome act like a vise, holding the tRNA and mRNA molecules close together. The ribosome can then connect the amino acid from the A-site tRNA to the growing polypeptide.

Let's now examine translation in more detail, starting at the beginning of the process.

How does a ribosome function in protein synthesis?

A ribosome holds mRNA and tRNAs together and connects amino acids from the tRNAs to the growing polypeptide chain.

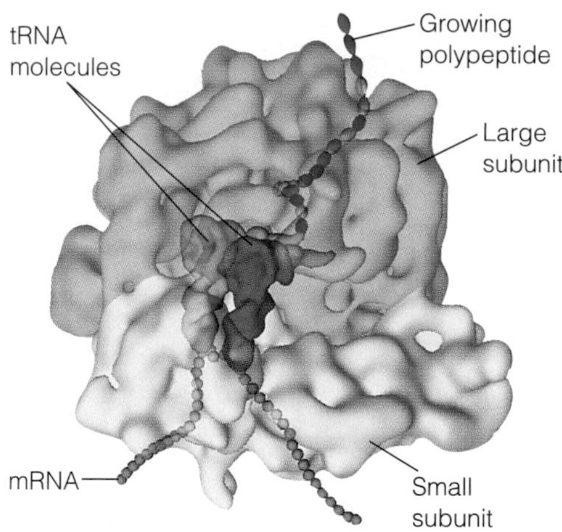

Figure 10.12A The true shape of a functioning ribosome

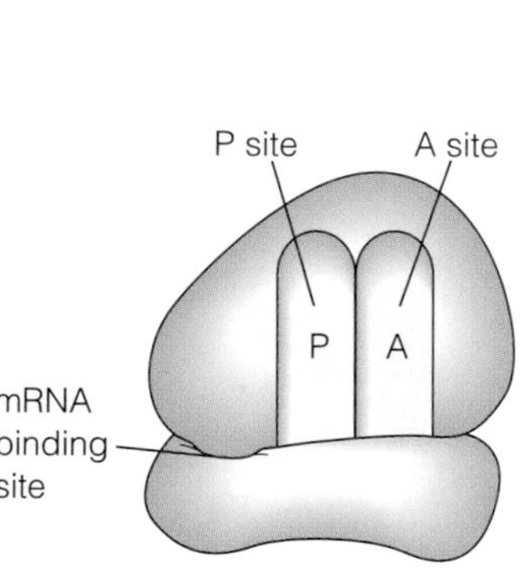

Figure 10.12B Binding sites of a ribosome

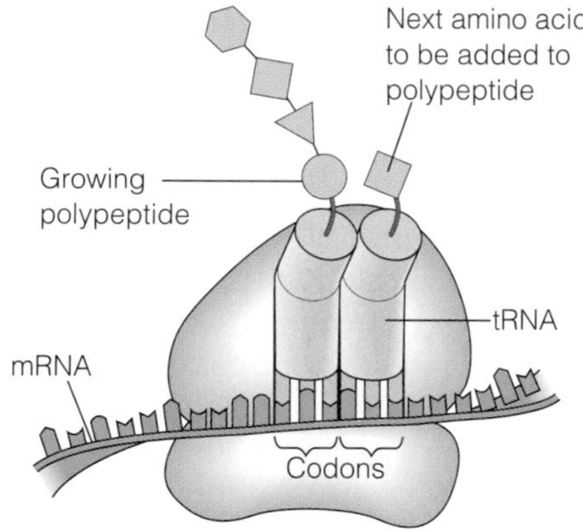

Figure 10.12C A ribosome with occupied binding sites

10.13 An initiation codon marks the start of an mRNA message

10E

Translation can be divided into the same three phases as transcription: initiation, elongation, and termination. The process of polypeptide initiation brings together the mRNA, the first amino acid with its attached tRNA, and the two subunits of a ribosome.

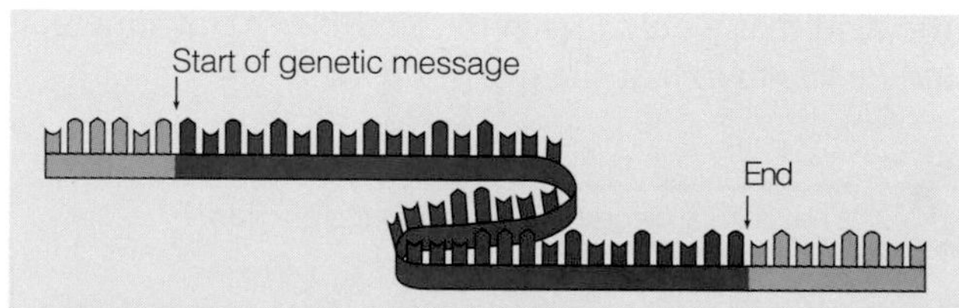

Figure 10.13A A molecule of mRNA

As indicated in Figure 10.13A, an mRNA molecule transcribed from DNA is longer than the genetic message it carries. A sequence of nucleotides (light pink) at either end of the molecule is not part of the message but helps the mRNA bind to the ribosome. The role of the initiation process is to determine exactly where translation will begin, so that the mRNA codons will be translated into the correct sequence of amino acids.

Initiation occurs in two steps, as shown in Figure 10.13B at the top of the next page:

Step ① An mRNA molecule binds to a small ribosomal subunit. A special initiator tRNA binds to the specific codon, called the **start codon,** where translation is to begin on the mRNA molecule. The initiator tRNA carries the

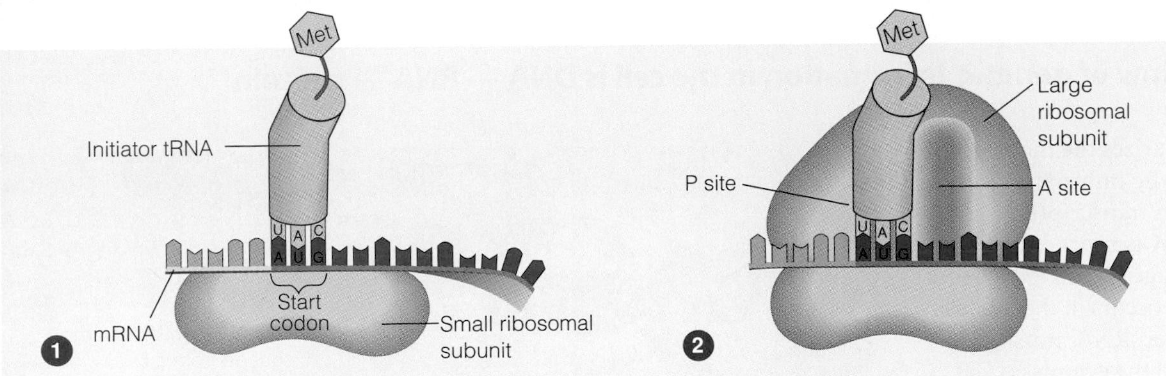

Figure 10.13B The initiation of translation

amino acid methionine (Met); its anticodon, UAC, binds to the start codon, AUG.

Step ② A large ribosomal subunit binds to the small one, creating a functional ribosome. The initiator tRNA fits into the P site on the ribosome.

> **?** What would happen if a genetic mutation changed a start codon to some other codon?
>
> Messenger RNA transcribed from the mutated gene, if any, would be nonfunctional because ribosomes could not initiate translation correctly.

10.14 Elongation adds amino acids to the polypeptide chain until a stop codon terminates translation

Once initiation is complete, amino acids are added one by one to the first amino acid. Each addition occurs in a three-step elongation process, shown in the figure here. (Red arrows indicate movement.)

Step ① Codon recognition. The anticodon of an incoming tRNA molecule, carrying its amino acid, pairs with the mRNA codon in the A site of the ribosome.

Step ② Peptide bond formation. The polypeptide separates from the tRNA to which it was bound (the one in the P site) and attaches by a peptide bond to the amino acid carried by the tRNA in the A site. The ribosome catalyzes formation of the bond. Thus, one more amino acid is added to the chain.

Step ③ Translocation. The P-site tRNA now leaves the ribosome, and the ribosome translocates (moves) the A-site tRNA, carrying the growing polypeptide, to the P site. The codon and anticodon remain bonded, and the mRNA and tRNA move as a unit. This movement brings into the A site the next mRNA codon to be translated, and the process can start again with step 1.

Elongation continues until a **stop codon** reaches the ribosome's A site. Stop codons—UAA, UAG, and UGA—do not code for amino acids but instead tell translation to stop. This is the termination stage of translation. The completed polypeptide, typically about one hundred amino acids long, is freed from the last tRNA and from the ribosome, which then splits into its subunits.

> **?** What happens as a tRNA passes through the A and P binding sites on the ribosome?
>
> In the A site, its amino acid receives the growing polypeptide from the tRNA that precedes it. In the P site, it gives up the polypeptide to the tRNA that follows it.

10E

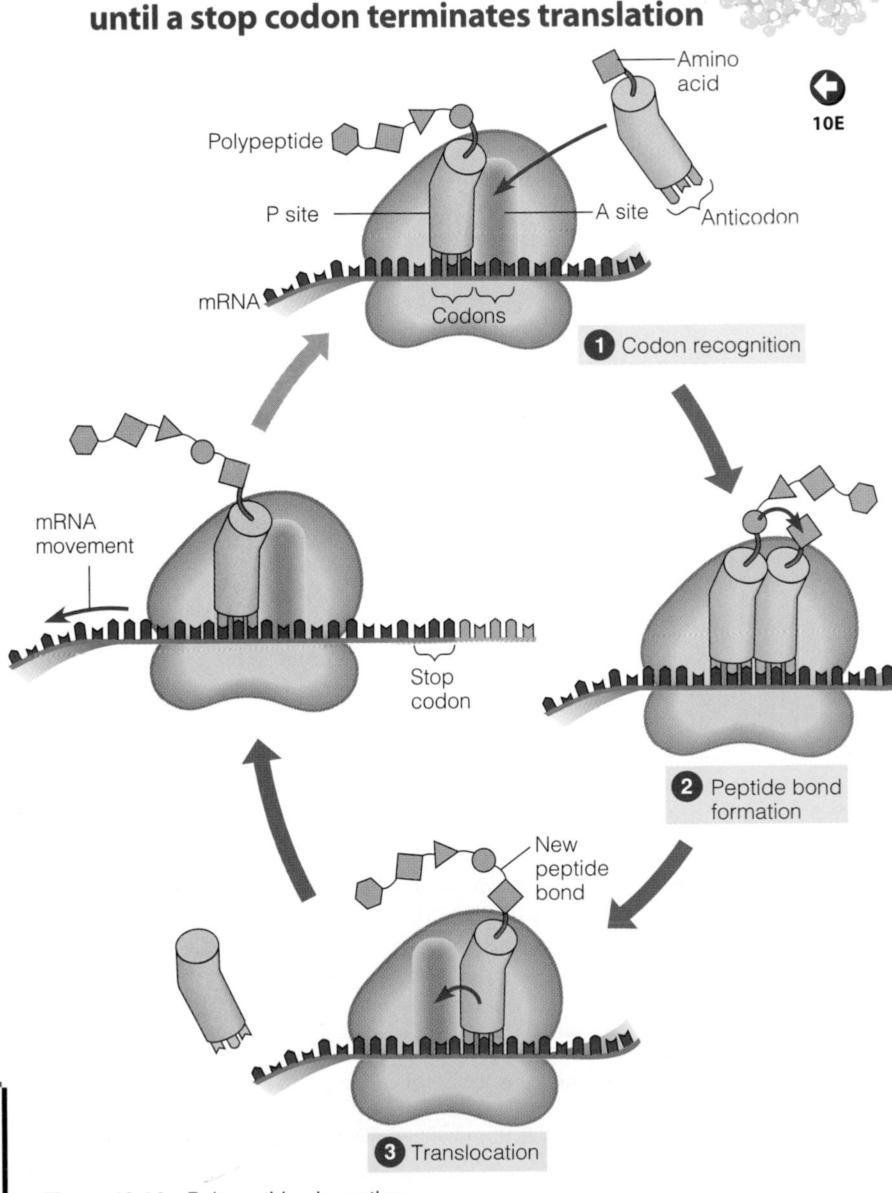

Figure 10.14 Polypeptide elongation

10.15 Review: The flow of genetic information in the cell is DNA → RNA → protein

10E

This figure summarizes the main stages in the flow of genetic information from DNA to protein. In transcription (DNA → RNA), the RNA is synthesized on a DNA template (stage ①). In eukaryotic cells, transcription occurs in the nucleus, and the messenger RNA must travel from the nucleus to the cytoplasm.

Translation (RNA → protein) can be divided into four stages (②–⑤), all of which occur in the cytoplasm. When the polypeptide is complete, the two ribosomal subunits come apart, and the tRNA and mRNA are released (not shown in this figure). Translation is rapid; a single ribosome can make an average-sized polypeptide in less than a minute. Typically, an mRNA molecule is translated simultaneously by a number of ribosomes. Once the start codon emerges from the first ribosome, a second ribosome can attach to it, and thus several ribosomes may trail along on the same mRNA molecule.

Each polypeptide coils and folds, assuming a three-dimensional shape, its tertiary structure. Several polypeptides may come together, forming a protein with quaternary structure (see Module 3.18).

What is the overall significance of transcription and translation? These are the processes whereby genes control the structures and activities of cells, or, more broadly, the way the genotype produces the phenotype. The chain of command originates with the information in a gene, a specific linear sequence of nucleotides in DNA. The gene serves as a template, dictating transcription of a complementary sequence of nucleotides in mRNA. In turn, mRNA dictates the linear sequence in which amino acids appear in a specific polypeptide. Finally, the proteins that form from the polypeptides determine the appearance and the capabilities of the cell and organism.

Which of the following molecules or structures does not participate directly in translation: ribosomes, transfer RNA, messenger RNA, DNA, ATP, enzymes?

DNA

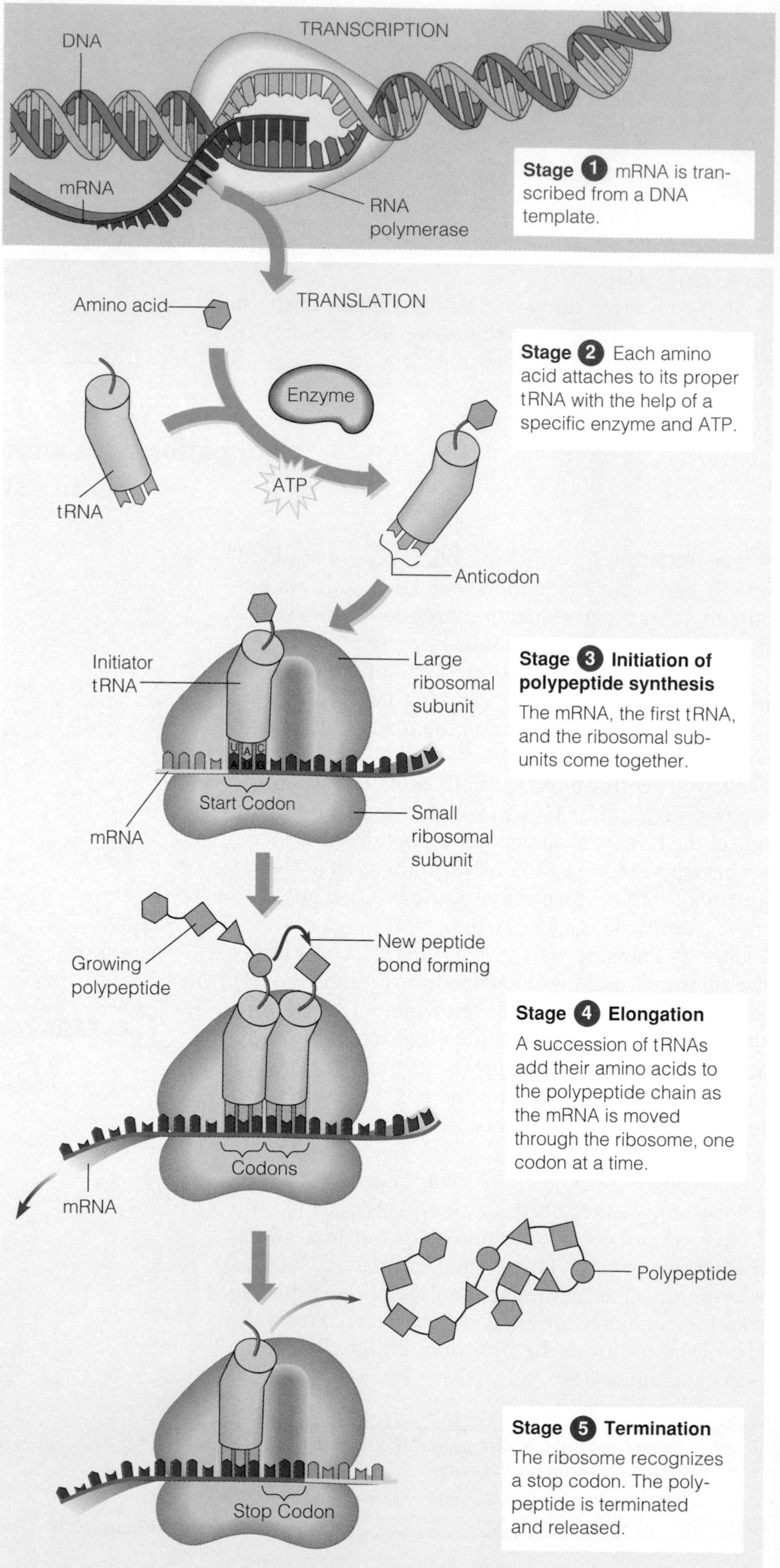

TRANSCRIPTION

DNA

mRNA

RNA polymerase

Stage ① mRNA is transcribed from a DNA template.

TRANSLATION

Amino acid

Enzyme

ATP

tRNA

Anticodon

Stage ② Each amino acid attaches to its proper tRNA with the help of a specific enzyme and ATP.

Initiator tRNA

Large ribosomal subunit

Start Codon

mRNA

Small ribosomal subunit

Stage ③ Initiation of polypeptide synthesis
The mRNA, the first tRNA, and the ribosomal subunits come together.

Growing polypeptide

New peptide bond forming

Codons

mRNA

Stage ④ Elongation
A succession of tRNAs add their amino acids to the polypeptide chain as the mRNA is moved through the ribosome, one codon at a time.

Polypeptide

Stop Codon

Stage ⑤ Termination
The ribosome recognizes a stop codon. The polypeptide is terminated and released.

Figure 10.15 Summary of transcription and translation

Mutations can change the meaning of genes 10.16

Since discovering how genes are translated into proteins, scientists have been able to describe many heritable differences in molecular terms. For instance, when a child is born with sickle-cell disease (see Module 9.14), the condition can be traced back through the difference in a protein to one tiny change in a gene. In one of the two kinds of polypeptides in the hemoglobin protein, the sickle-cell child has a single different amino acid, a Val instead of a Glu. This difference is caused by the change of a single nucleotide in the coding strand of DNA (Figure 10.16A). In the double helix, a base *pair* is changed.

The sickle-cell allele is not a unique case. We now know that the alternative alleles of many genes result from changes in single base pairs in DNA. Any change in the nucleotide sequence of DNA is called a **mutation.** Mutations can involve large regions of a chromosome or just a single nucleotide pair, as in sickle-cell disease. Here we consider how mutations involving only one or a few nucleotide pairs can affect gene translation.

Mutations within a gene can be divided into two general categories: base substitutions and base insertions or deletions (Figure 10.16B). A base substitution is the replacement of one nucleotide with another. In the second row in Figure 10.16B, A replaces G in the fourth codon of the mRNA molecule. Depending on how a base substitution is translated, it can result in no change in the protein, in an insignificant change, or in a change that might be crucial to the life of the organism. It is because of the redundancy of the genetic code that some substitution mutations have no effect. For example, if a mutation causes an mRNA codon to change from GAA to GAG, no change in the protein product would result, because GAA and GAG both code for the same amino acid (Glu). Other changes of a single nucleotide may alter an amino acid but have little effect on the function of the protein.

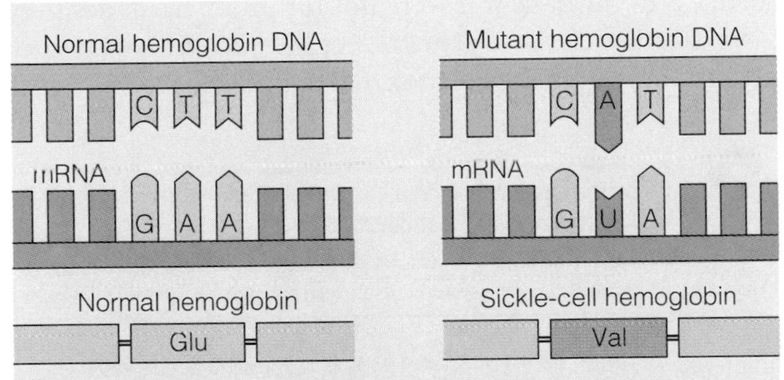

Figure 10.16A The molecular basis of sickle-cell disease

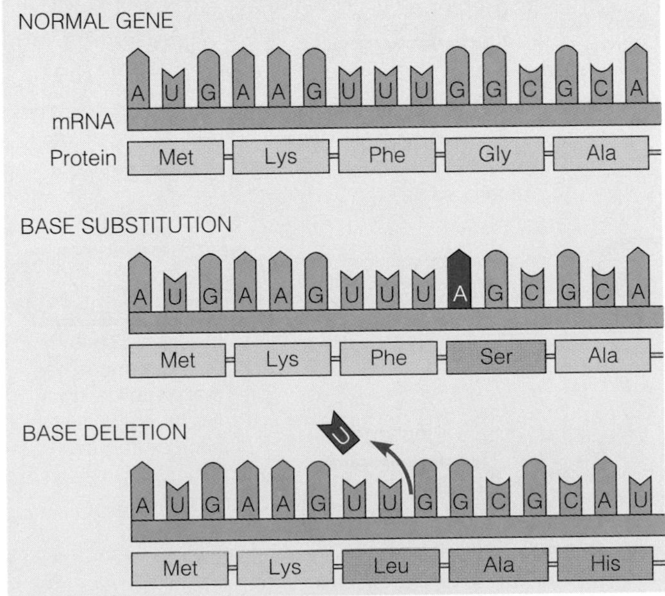

Figure 10.16B Types of mutations and their effects

Some base substitutions, as we saw in the sickle-cell case, cause changes in the protein that prevent it from performing normally. Occasionally, a base substitution leads to an improved protein or one with new capabilities that enhance the success of the mutant organism and its descendants. Much more often, though, mutations are harmful.

Mutations involving the insertion or deletion of one or more nucleotides in a gene often have disastrous effects. Because mRNA is read as a series of nucleotide triplets during translation, adding or subtracting nucleotides may alter the **reading frame** (triplet grouping) of the genetic message. All the nucleotides that are "downstream" of the insertion or deletion will be regrouped into different codons, as you can see for the deletion shown at the bottom in Figure 10.16B. The result will most likely be a nonfunctional polypeptide.

The creation of mutations, the process of **mutagenesis,** can occur in a number of ways. Mutations resulting from errors during DNA replication or recombination are called spontaneous mutations, as are other mutations of unknown cause. Another source of mutation is a physical or chemical agent, called a **mutagen.** The most common physical mutagen in nature is high-energy radiation, such as X-rays and ultraviolet light. Chemical mutagens are of various types. One type, for example, consists of chemicals that are similar to normal DNA bases but that pair incorrectly.

Although mutations are often harmful, they are also extremely useful, both in nature and in the laboratory. It is because of mutations that there is such a rich diversity of genes in the living world, a diversity that makes evolution by natural selection possible. Mutations are also essential tools for geneticists. Whether naturally occurring (as in Mendel's peas) or created in the laboratory (Morgan used X-rays to make most of his fruit-fly mutants), mutations create the different alleles needed for genetic research. 

? What is the molecular basis of sickle-cell disease?

A single base difference in a hemoglobin gene results in an amino acid substitution in the protein, altering the structure and behavior of the hemoglobin.

10.17 Viral DNA may become part of the host chromosome

10F
10G

As we discussed earlier, viruses provided some of the first glimpses into the molecular details of heredity. Now let's take a closer look at viruses, focusing on the relationship between viral structure and the processes of nucleic acid replication, transcription, and translation.

In a sense, viruses are nothing more than packaged genes. For a researcher in molecular genetics, they may seem almost ideal tools—some nucleic acid wrapped in a protein coat—all you need to study the flow of genetic information and nothing more. It's really not that simple, however, because viruses can reproduce only by infecting cells. In fact, the host cell provides most of the components necessary for replicating, transcribing, and translating the viral nucleic acid.

Earlier in the chapter, we described what happens after the DNA of phage T2 or T4 enters a host cell. The way T phages reproduce is called a **lytic cycle** because it always leads to the lysis (breaking open) of the host cell. The study of another phage of *E. coli,* called lambda, led to the discovery that some viruses can also reproduce by an alternative route called a lysogenic cycle. During a **lysogenic cycle,** viral DNA replication occurs without phage production or the death of the host cell.

Below you see the two kinds of cycles for phage lambda. Like T4, lambda has a head containing DNA and a tail, but no long tail fibers. Both cycles begin when the phage DNA enters the bacterium (top of the figure) and forms a circle (center). The DNA then embarks on one of the two pathways. In the lytic cycle (left), lambda's DNA immediately turns the cell into a virus-producing factory. In the lysogenic cycle, the DNA inserts by genetic recombination into the bacterial chromosome (bottom right). Once inserted, the phage DNA is referred to as a **prophage,** and most of its genes are inactive. Survival of the prophage depends on reproduction of the host cell. The host cell replicates the prophage DNA along with its cellular DNA (right side of figure) and then, upon dividing, passes on both the prophage and the cellular DNA to its two daughter cells. A single infected bacterial cell can quickly give rise to a large population of bacteria carrying prophages. The prophages may remain in the bacterial cells indefinitely. Occasionally, however, one leaves its host chromosome (top right); this event may be triggered by environmental conditions. Once separate, the lambda DNA usually switches to the lytic cycle.

Sometimes the few prophage genes active in a bacterial cell can cause medical problems. For example, the bacteria that cause diphtheria, botulism, and scarlet fever would be harmless to humans if it were not for prophage genes they carry. Certain of these genes direct the bacteria to produce toxins that are responsible for making people ill.

Describe one way a virus can perpetuate its genes without destroying its host cell.

Some viruses can insert their DNA into a chromosome of the host cell, which replicates the viral genes when it replicates its own DNA prior to cell division.

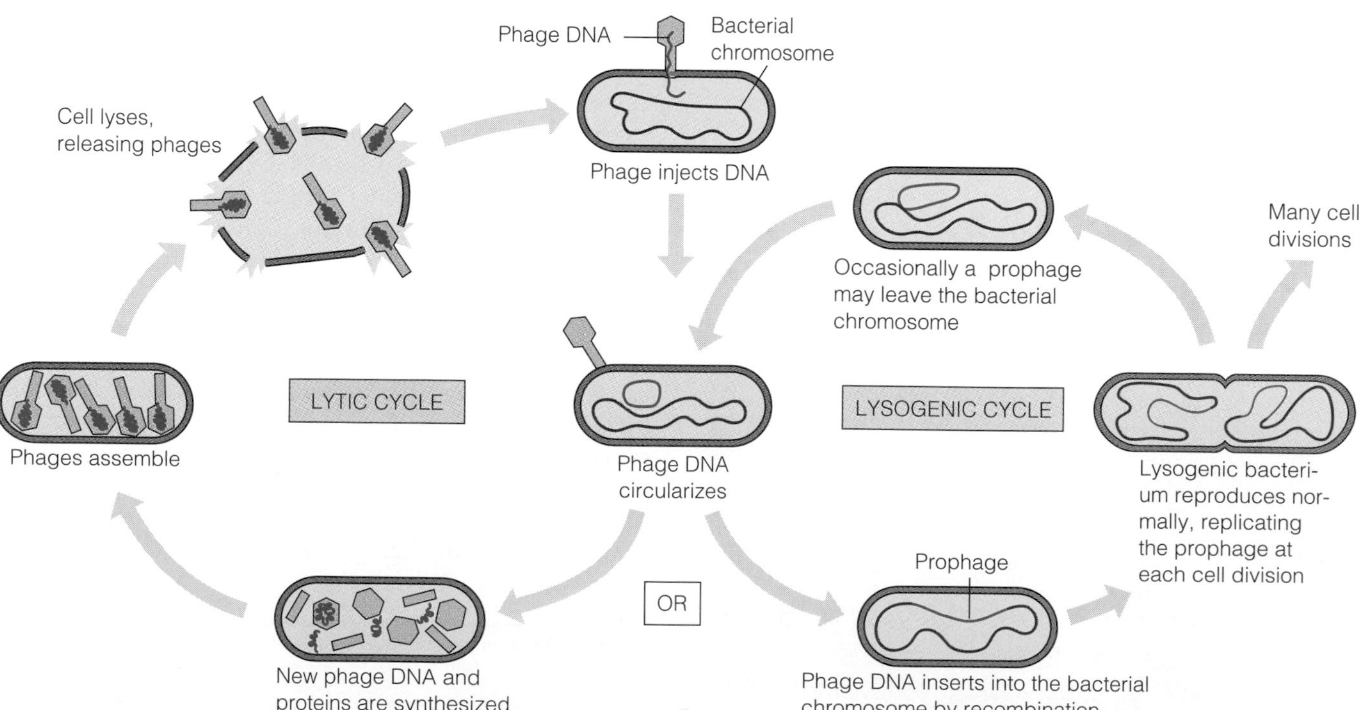

Figure 10.17 Two types of phage reproductive cycles

Many viruses cause disease in animals 10.18

Viruses that infect animal cells are common causes of disease. We have all suffered from virus infections. Figure 10.18A shows the structure of an influenza (flu) virus. Like many viruses infecting animal cells, this one has a membranous outer envelope and projecting spikes made of glycoprotein. The envelope helps the virus enter and leave the host cell. Like many viruses, flu viruses have RNA rather than DNA as their genetic material; the RNA of flu viruses is actually in eight pieces, each wrapped in protein. Other RNA viruses include those that cause the common cold, measles, and mumps, as well as ones that cause some more serious human diseases, such as AIDS and polio. Examples of diseases caused by DNA viruses are hepatitis, chicken pox, and herpes infections.

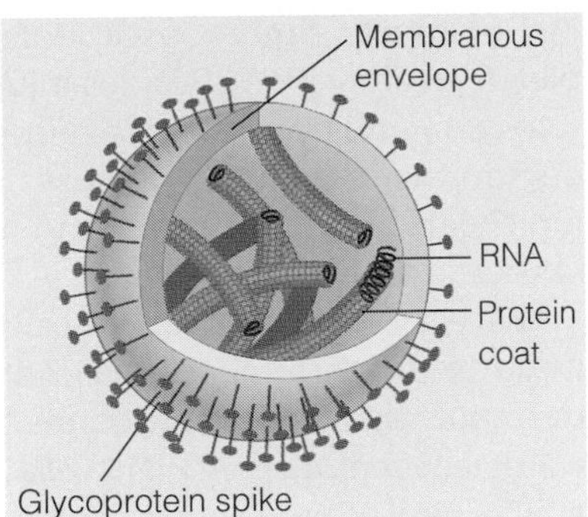

Figure 10.18A An influenza virus

Figure 10.18B shows the reproductive cycle of an enveloped RNA virus (the mumps virus). When the virus contacts a host cell, the glycoprotein spikes attach to receptor proteins on the cell's plasma membrane. The envelope fuses with the cell's membrane, allowing the protein-coated RNA to ① enter the cytoplasm. ② Enzymes then remove the protein coat. ③ An enzyme that entered the cell as part of the virus uses the virus's RNA genome as a template for making complementary strands of RNA (purple strand). The new strands have two functions: ④ They serve as mRNA for the synthesis of new viral proteins, and ⑤ they serve as templates for synthesizing new viral-genome RNA. ⑥ The new coat proteins assemble around the new viral RNA. ⑦ Finally, the viruses leave the cell by cloaking themselves in plasma membrane. Thus the virus obtains its envelope from the host cell, leaving the cell without necessarily lysing it.

Not all animal viruses reproduce in the cytoplasm. The herpesviruses that cause chicken pox, shingles, cold sores, and genital herpes, for example, are enveloped DNA viruses that reproduce in the host cell's nucleus, and they get their envelopes from the cell's nuclear membranes. While inside the nucleus, herpesvirus DNA may insert itself into the cell's DNA as a *provirus* (analogous to a prophage) and remain dormant within the body. From time to time, physical stress factors, such as a cold or sunburn, or emotional stress may cause the herpesvirus DNA to begin production of the virus, resulting in unpleasant symptoms. Once acquired, herpes infections may flare up repeatedly throughout a person's life.

The amount of damage a virus causes our body depends partly on how quickly our immune system responds to fight the infection and partly on the ability of the infected tissue to repair itself. We usually recover completely from colds because our respiratory tract tissue can efficiently replace damaged cells by mitosis. In contrast, the poliovirus attacks nerve cells, which do not divide. The damage to such cells by polio, unfortunately, is permanent. In such cases, we try to prevent the disease with vaccines (see Chapter 24). The antibiotic drugs that help us recover from bacterial infections are powerless against viruses. The development of antiviral drugs has been slow, because it is difficult to find ways to kill a virus without killing its host cell.

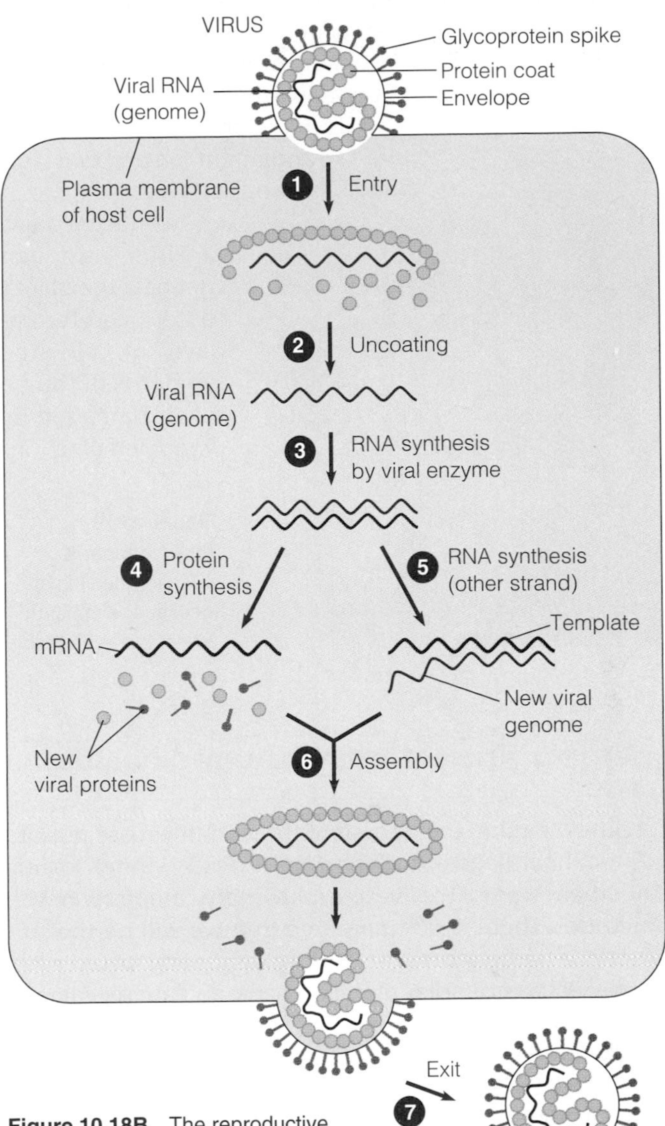

Figure 10.18B The reproductive cycle of an enveloped virus

Explain how some viruses replicate without ever having DNA.

The genetic material of these viruses is RNA, which is replicated inside the host cell by special enzymes encoded by the virus. The viral genome (or its complement) serves as mRNA for the synthesis of viral proteins.

10.19 Plant viruses are serious agricultural pests

Viruses that infect plant cells can stunt plant growth and diminish crop yields. Most plant viruses discovered to date are RNA viruses. Many of them, like the tobacco mosaic virus in Figure 10.19, are rod-shaped with a spiral arrangement of proteins surrounding the nucleic acid.

To infect a plant, a virus must first get past the plant's outer protective layer of cells (the epidermis). Thus, a plant damaged by wind, chilling, injury, or insects is more susceptible to infection than a completely healthy plant. Besides injuring plants, some insects also carry and transmit plant viruses. Farmers and gardeners, too, may spread plant viruses through the use of pruning shears and other tools. And infected plants may pass viruses to their offspring.

Once a virus enters a plant cell and begins reproducing, it can spread throughout the entire plant through plasmodesmata, the cytoplasmic connections that penetrate the walls between adjacent plant cells (see Figure 4.19A). As with animal viruses, there are no cures for most viral diseases of plants, and agricultural scientists focus on reducing the number of plants that become infected and on breeding genetic varieties of crop plants that resist viral infection.

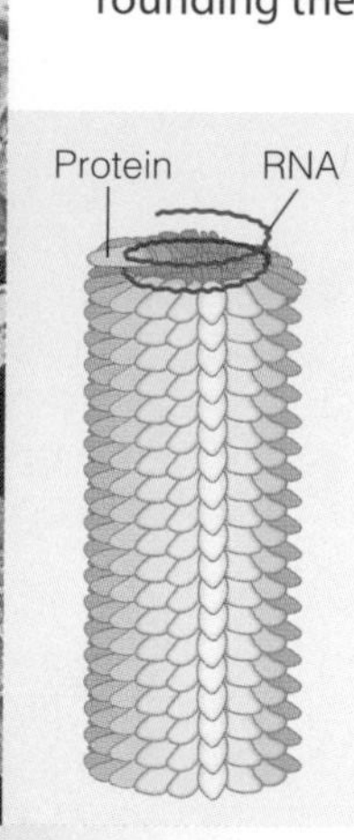

Figure 10.19 Tobacco mosaic disease (mottling of leaves) and the structure of the virus (right)

What are three ways viruses can enter plant cells?

Through lesions caused by injuries, through transfer by insects that feed on the plants, or through contaminated farming or gardening tools

10.20 Emerging viruses threaten human health

Acknowledging the persistent threat that viruses pose to human health, geneticist and Nobel Prize winner Joshua Lederberg warns, "We live in evolutionary competition with microbes. There is no guarantee that we will be the survivors." Lederberg cites the AIDS epidemic and recurrent flu epidemics as examples of the human population's vulnerability to viral attacks.

The AIDS virus (HIV), which seemed to arise abruptly in the early 1980s, and the new varieties of flu virus that frequently appear, are not the only examples of newly dangerous viruses. A deadly virus called the Ebola virus (Figure 10.20A) menaces central African nations periodically, and many biologists fear its emergence as a global threat. Another newly dangerous virus was responsible for killing dozens of people in the southwestern United States in 1993. It was first thought to be a new virus but turned out to have been known to western medicine for about 45 years, since the Korean War. Named for a river in Korea, it is called hantavirus (Figure 10.20B). Hantavirus is a potential problem worldwide, although the number of cases reported recently in the U.S. is small (about 20 in 1998).

Although the viruses mentioned here may not have originated as recently as once thought, they still raise intriguing questions: How do new viruses arise, and what causes certain viruses to emerge as major threats? Most biologists favor the hypothesis that the very first viruses arose from fragments of cellular nucleic acid that had some means of moving from one cell to another. Consistent with this idea, viral nucleic acid usually has more in common with its host cell's DNA than with the nucleic acid of viruses that infect other hosts. Indeed, some viral genes are virtually identical to genes of their host. The earliest viruses may have been naked bits of nucleic acid that traveled from one cell to another via injured cell surfaces. Genes coding for viral coat proteins may have evolved later.

Today, the mutation of existing viruses is a major source of new viral diseases. The RNA viruses tend to have an unusually high rate of mutation because the replication of their nucleic acid does not involve the proofreading steps of DNA replication. Some mutations may enable existing viruses to

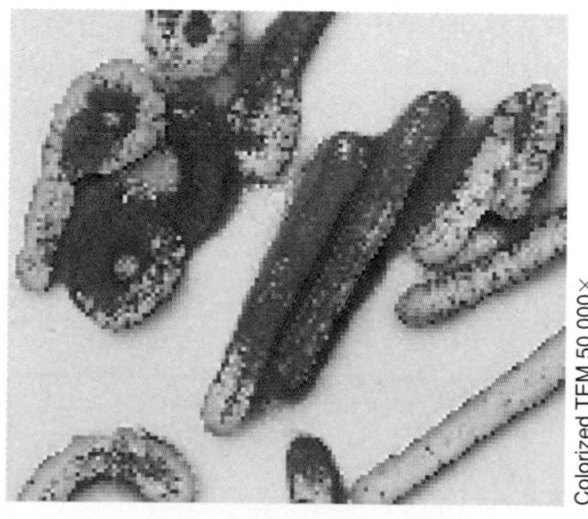

Colorized TEM 50,000×

Figure 10.20A Ebola virus, each an enveloped thread of protein-coated RNA

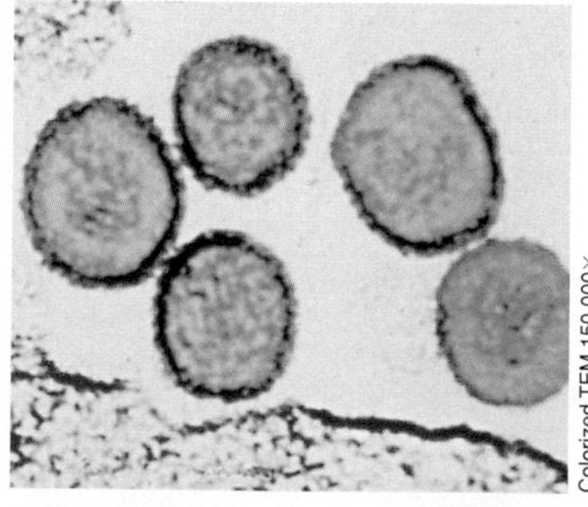

Colorized TEM 150,000×

Figure 10.20B Hantavirus, another enveloped RNA virus

evolve into new genetic varieties that can cause disease in individuals who had developed immunity to the ancestral virus. Flu epidemics are caused by viruses that are genetically different enough from earlier years' viruses that people have little immunity to them.

Another source of new viral diseases is the spread of existing viruses to a new host species. For example, hantavirus is common in rodents, especially deer mice. The population of deer mice in the southwestern U.S. exploded in 1993 after unusually wet weather increased the rodents' food supply. Humans acquired hantavirus when they inhaled dust containing traces of urine and feces from infected mice.

Finally, a viral disease may start out in a small, isolated population and then rather suddenly become widespread. AIDS, for example, went unnamed and virtually unnoticed for decades before starting to spread around the world. In this case, technological and social factors, including affordable international travel, blood transfusion technology, sexual promiscuity, and the abuse of intravenous drugs, allowed a previously rare human disease to become a global epidemic. Let's now take a closer look at the AIDS virus. 

? Why doesn't a bout of flu give us immunity to flu in subsequent years?

Influenza viruses evolve rapidly by frequent mutation; thus the strains that infect us later will most likely be different from the ones to which we've developed immunity.

The AIDS virus makes DNA on an RNA template 10.21

10H

The devastating disease AIDS is caused by a type of RNA virus with some special twists. In outward appearance, the AIDS virus (HIV) resembles the flu or mumps virus. As illustrated in Figure 10.21A, HIV has a membranous envelope and glycoprotein spikes. These components enable HIV to enter and leave a host cell much the way the mumps virus does. Notice, however, that HIV contains two copies of its RNA instead of one. HIV also has a different mode of reproduction. It is a **retrovirus,** an RNA virus that reproduces by means of a DNA molecule. Retroviruses are so named because they reverse the usual DNA → RNA flow of genetic information. They carry molecules of an enzyme called **reverse transcriptase,** which catalyzes reverse transcription, the synthesis of DNA on an RNA template.

Figure 10.21B illustrates what happens after HIV RNA is uncoated in the cytoplasm of a host cell. The reverse transcriptase (yellow) ① uses the RNA as a template to make a DNA strand and then ② adds a second, complementary DNA strand. ③ The resulting double-stranded DNA then enters the host cell nucleus and inserts itself into the chromosomal DNA, becoming a provirus. The provirus usually remains part of the cell's DNA. But occasionally the provirus is ④ transcribed into RNA and ⑤ translated into viral proteins. ⑥ New viruses assembled from these components eventually leave the cell. They can then infect other cells. This is the standard reproductive cycle for retroviruses.

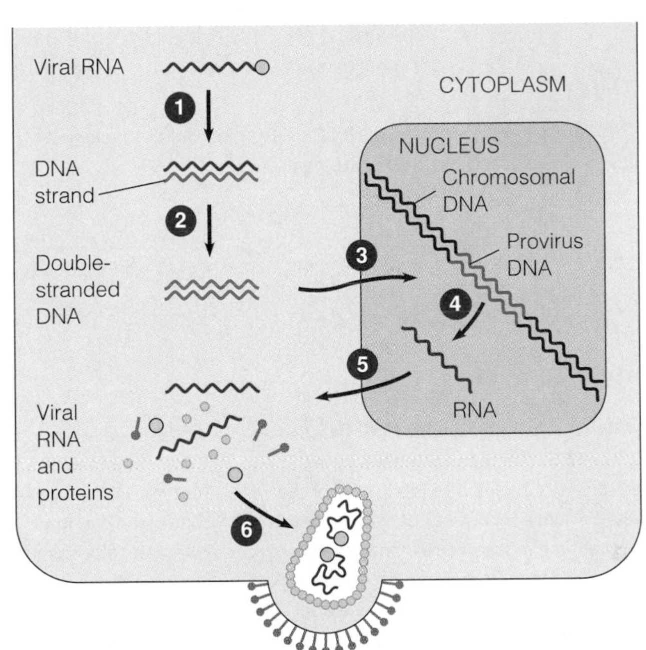

Figure 10.21B The behavior of HIV nucleic acid in a host cell

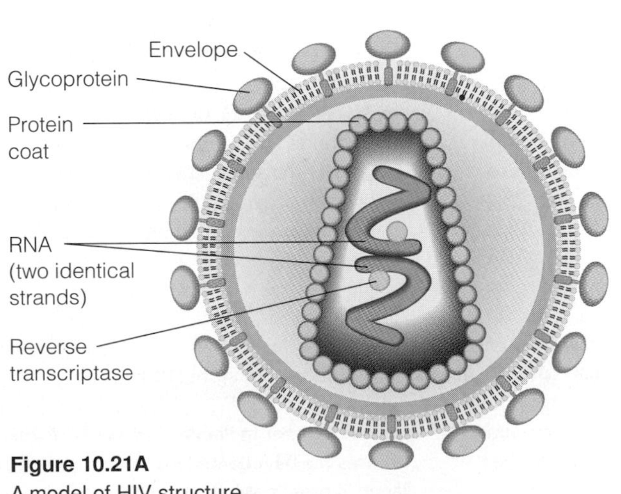

Figure 10.21A A model of HIV structure

AIDS stands for acquired immune deficiency syndrome, and **HIV** for human immunodeficiency virus; these terms describe the main effect of the virus on the body. HIV infects and eventually kills several kinds of white blood cells that are important in our immune system. We discuss AIDS and its deadly effects in more detail when we take up the immune system in Chapter 24. 

? Why is HIV classified as a retrovirus?

Because it uses the enzyme reverse transcriptase to synthesize DNA from its RNA genome. This is the reverse ("retro") of the usual DNA→RNA information flow.

10.22 Virus research and molecular genetics are intertwined

Molecular geneticists have a love-hate relationship with viruses. Studies of viruses have been of great benefit, essentially launching the science of molecular genetics some 45 years ago. The Hershey-Chase experiments, confirming that the DNA of phage T2 infects and usurps control of bacterial cells, focused attention on the central role of nucleic acids in inheritance and set the stage for Watson and Crick. After the structure of DNA was defined, studies of viral reproduction helped show how genes function and how genetic information flows from DNA to proteins.

Figure 10.22 HIV (blue dots) attacking a white blood cell

The micrograph in Figure 10.22 represents the other side of the virus–molecular genetics relationship: a battle instead of a love affair. The blue dots are AIDS viruses (HIV) attacking a human white blood cell. HIV invades and kills cells like this one, weakening the immune system of the infected person. AIDS is one of the deadliest diseases humankind has ever faced and also one of the most difficult to combat. Fighting viruses that cause disease is an important practical goal of molecular genetics. Effective vaccines have already been developed against some viral diseases, including mumps, measles, and polio. It is likely that when we find the means to control HIV and other deadly viruses (including some that cause certain types of cancer), research in molecular genetics will be responsible for their discovery.

We continue our study of molecular genetics in Chapter 11, where we explore what is known about how genes themselves are controlled.

In what sense is the relationship between molecular biology and virology a "two-way street"?

As simple genetic systems, viruses have been important research models for molecular biology. In turn, understanding molecular biology is helping researchers devise ways to combat pathogenic viruses.

Chapter Review

Chapter Summary

The Structure of the Genetic Material (Introduction–10.3) Bacteria and their viruses, called bacteriophages, provided some of the earliest evidence that genes of many viruses and all cells are made of DNA. One key experiment showed that DNA from pathogenic bacteria transforms harmless bacteria into pathogenic ones. Another experiment showed that certain phages reprogram host cells to produce more phages by injecting their DNA **(Introduction–10.1)**. DNA is a nucleic acid, made of long chains of nucleotide monomers. Each nucleotide consists of a sugar connected to a phosphate group and a nitrogenous base. Alternating sugars and phosphates form a backbone, with the bases as appendages. DNA has four kinds of bases, abbreviated A, T, C, and G. RNA is also a nucleic acid, with a slightly different sugar and U instead of T **(10.2)**. James Watson and Francis Crick worked out the three-dimensional structure of DNA, which is two polynucleotide strands wrapped around each other in a double helix. Hydrogen bonds between bases hold the strands together. Each base pairs with a complementary partner: A with T, and G with C **(10.3)**.

DNA Replication (10.4–10.5) In DNA replication, the DNA strands separate, and enzymes use each strand as a template to assemble new nucleotides into a complementary strand. Using the enzyme DNA polymerase, the cell synthesizes one of the daughter DNA strands as a continuous piece, the other as a series of short pieces, which are then connected by the enzyme DNA ligase. The two daughter DNA molecules are identical to the parent molecule. This is how genetic instructions are copied for the next generation.

The Flow of Genetic Information from DNA to RNA to Protein (10.6–10.16) The information constituting an organism's genotype is carried in the sequence of its DNA bases. Studies of inherited metabolic defects first suggested that phenotype is expressed through proteins. A particular gene—a linear sequence of many nucleotides—specifies a polypeptide. The DNA of the gene is transcribed into RNA, which is translated into the polypeptide **(10.6)**. The "words" of the DNA "language" are triplets of bases called codons. The codons in a gene specify the amino acid sequence of a polypeptide. Virtually all organisms share the same genetic code **(10.7–10.8)**.

In transcription, the DNA helix unzips, and RNA nucleotides line up and hydrogen-bond along one strand of the DNA, following the base-pairing rules. As the single-stranded messenger RNA (mRNA) peels away from the gene, the DNA strands rejoin **(10.9)**. Translation occurs in the cytoplasm **(10.10)**. A ribosome attaches to the mRNA and translates its message into a specific polypeptide, aided by transfer RNAs (tRNAs) that act as interpreters. Each tRNA is a folded molecule bearing a base triplet called an anticodon on one end; a specific amino acid is added to the other end **(10.11)**. The mRNA moves a codon at a time relative to the ribosome, and a tRNA with a complementary anticodon pairs with each codon, adding its amino acid to the peptide chain **(10.12–10.14)**. Thus, the sequence of codons in DNA, via the sequence of codons in mRNA, spells out the primary structure of a polypeptide. Polypeptides form proteins that determine the appearance and functions of the cell and organism. Figure 10.15 summarizes transcription and translation **(10.15)**.

Mutations are changes in the DNA base sequence, caused by errors in DNA replication or by mutagens. Substituting, inserting, or deleting nucleotides alters a gene, with varying effects on the organism **(10.16)**.

Viruses: Genes in Packages (10.17–10.22) Viruses can be regarded simply as genes packaged in protein. When phage DNA enters a lytic cycle inside a bacterium, it is replicated, transcribed, and translated; the new viral DNA and protein molecules then assemble into new phages, which burst from the host cell. In the lysogenic cycle, phage DNA inserts into the host chromosome and is passed on to generations of daughter cells. Much later, it may initiate phage production **(10.17)**. Many viruses cause disease when they invade animal or plant cells. Many, such as flu viruses, have RNA, rather than DNA, as their genetic material. Some animal viruses steal a bit of host-cell membrane as a protective envelope. Some insert their DNA into a host chromosome and remain latent for long periods **(10.18)**. Most plant viruses have RNA. They enter their hosts via wounds in the plant's outer layers **(10.19)**. Emerging viral diseases pose a threat to human health **(10.20)**. HIV, the AIDS virus, is a retrovirus; inside a cell it uses its RNA as a template for making DNA, which then inserts into a host chromosome **(10.21)**. Virus studies helped establish molecular genetics. Now molecular genetics helps us understand viruses **(10.22)**.

Testing Your Knowledge

Multiple Choice

1. Scientists have discovered how to put together a bacteriophage with the protein coat of phage T2 and the DNA of phage T4. If this composite phage were allowed to infect a bacterium, the phages produced in the host cell would have ______. *(Explain your answer.)*
   a. the protein of T2 and the DNA of T4
   b. the protein of T4 and the DNA of T2
   c. a mixture of the DNA and proteins of both phages
   d. the protein and DNA of T2
   e. the protein and DNA of T4

2. A geneticist found that a particular mutation had no effect on the polypeptide coded by a gene. This mutation probably involved
   a. deletion of one nucleotide
   b. alteration of the start codon
   c. insertion of one nucleotide
   d. deletion of the entire gene
   e. substitution of one nucleotide

3. Which of the following correctly ranks the structures in order of size, from largest to smallest?
   a. gene-chromosome-nucleotide-codon
   b. chromosome-gene-codon-nucleotide
   c. nucleotide-chromosome-gene-codon
   d. chromosome-nucleotide-gene-codon
   e. gene-chromosome-codon-nucleotide

4. The nucleotide sequence of a DNA codon is GTA. A messenger RNA molecule with a complementary codon is transcribed from the DNA. In the process of protein synthesis, a transfer RNA pairs with the mRNA codon. What is the nucleotide sequence of the tRNA anticodon?
   a. CAT
   b. CUT
   c. GUA
   d. CAU
   e. GT

Describing, Comparing, and Explaining

1. Describe the process of DNA replication: the "ingredients" needed, the steps in the process, and the final product.

2. Describe the process by which the information in a gene is transcribed and translated into a protein. Correctly use these words in your description: tRNA, amino acid, start codon, transcription, mRNA, gene, codon, RNA polymerase, ribosome, translation, anticodon, peptide bond, stop codon.

Thinking as a Scientist

1. A cell containing a single chromosome is placed in a medium containing radioactive phosphate, so that any new DNA strands formed by DNA replication will be radioactive. The cell replicates its DNA and divides. Then the daughter cells (still in the radioactive medium) replicate their DNA and divide, so that a total of four cells is present. Sketch the DNA molecules in all four cells, showing a normal (nonradioactive) DNA strand as a solid line and a radioactive DNA strand as a dashed line.

2. The base sequence of the gene coding for a short polypeptide is C T A C G C T A G G C G A T T G A C T. What would be the base sequence of the mRNA transcribed from this gene? Using the genetic code chart in Figure 10.8A, give the amino acid sequence of the polypeptide translated from this mRNA. (*Hint:* What is the start codon?)

3. In the early 1950s, scientists considered several different hypotheses about how DNA is replicated. One hypothesis, called "conservative" replication, was that the double-stranded parental DNA molecule is conserved intact but somehow directs the synthesis of a daughter DNA molecule consisting of two entirely new strands. Another hypothesis, "semiconservative" replication, was that the two strands of the parental DNA separate and each acts as a template for the construction of a new strand, yielding two daughter molecules that each have one old and one new strand (this is the hypothesis that turned out to be correct). Assume you grow bacteria in a medium containing radioactive phosphorus so all their DNA is labeled, and then you transfer them to a medium containing only nonradioactive phosphorus. After precisely one cell division in the nonradioactive medium, you test the bacterial DNA for radioactivity. The results that would support the conservative replication hypothesis would be that ______; the results that would support the semiconservative replication hypothesis would be that ______. Select one of the answers below, giving a short justification of why you chose it and why you rejected each of the other choices.
   a. none of the DNA would be radioactive . . . all the DNA would be radioactive
   b. all the DNA would be radioactive . . . half the DNA would be radioactive
   c. half the DNA would be radioactive . . . all the DNA would be radioactive
   d. None of these; this experiment would not be able to distinguish between these two hypotheses.

Science, Technology, and Society

Researchers on the Human Genome Project are determining the nucleotide sequences of human genes. In some cases, they are identifying the proteins encoded by the genes. Scientists at the National Institutes of Health (NIH) have worked out thousands of sequences, and similar analysis is being carried out at universities and at private companies. Knowledge of the nucleotide sequence of genes might be used to treat genetic defects or produce life-saving medicines. The NIH and some U.S. biotechnology companies have applied for patents on their discoveries. In Britain, the courts have ruled that a naturally occurring gene cannot be patented. Do you think individuals and companies should be able to patent genes and gene products? Before answering, consider the following: What are the purposes of a patent? How might the discoverer of a gene benefit from a patent? How might the public benefit? What negative impacts might result from patenting genes?

Media Resources

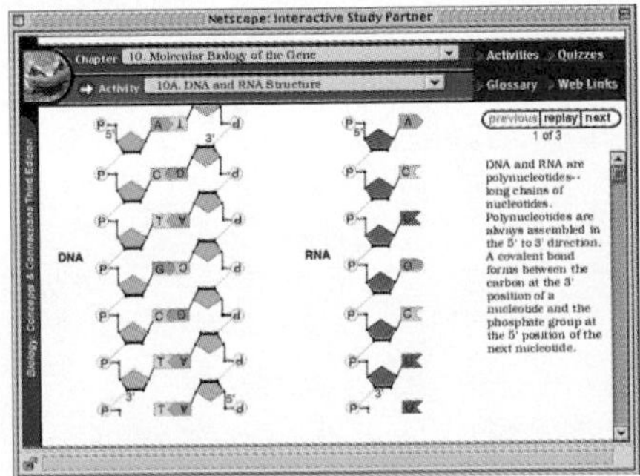

For further review of the concepts in this chapter, use the activities and quizzes on the CD-ROM.

Visit the Special Edition of The Biology Place web site for activities, quizzes, news articles, and web links.

11 The Control of Gene Expression

SEM 55×

Outline

Fruit flies of the species *Drosophila melanogaster* are only about 3 millimeters long, and they often go unnoticed as they hover around fruit in the kitchen or grocery store. Fruit growers consider them a nuisance but not a threat, because they eat the yeasts that grow on ripe fruit, not the fruit itself. (By contrast, the Mediterranean fruit fly, or medfly, a distant relative of *Drosophila,* is a serious threat to fruit crops.) *Drosophila* is a widely studied organism, immensely important in many areas of genetic research.

Something is wrong with the fruit fly in the photograph at the left. It is similar to the ones we saw in Chapter 9, but there's something different about it. One way you can tell that an insect is a fly—a fruit fly, house fly, gnat, mosquito, or any other kind of fly—is that it has two wings. Other insects, such as bees, beetles, butterflies, and moths have four wings. Unlike the normal fruit fly in the drawing on the facing page, the four-winged fly in the photo is a mutant. Its mutation has helped biologists uncover important secrets of how genes control the development of a complete, multicellular organism from a fertilized egg. (The scientific pioneers who first studied this and similar mutants of *Drosophila* were awarded a Nobel Prize in 1995.)

What determines that a fertilized egg, or zygote, will develop into a particular kind of organism? We know that DNA in the zygote nucleus contains all the genes that dictate what an organism will look like. But this is not the whole story. Every cell that develops from a zygote by mitotic division (that is, every cell except the organism's gametes) has DNA that is virtually identical to the DNA in the zygote. If all that DNA in every cell in a developing embryo were constantly active, every cell would look like every other cell. The only way that cells with the same genetic information can develop into cells with different structures and functions is if gene activity is regulated. In some way, control mechanisms must determine that certain genes will be active while others remain inactive in a particular cell. For instance, for cells in a fruit fly's eye to develop into a lens, genes that produce lens proteins must be turned on, and other genes must be turned off.

Discovering the connections between gene regulation and embryonic development is one of the major challenges of biology today. A combination of two powerful approaches is making rapid progress possible. One approach is biochemical: the use of DNA technology, which we discuss in Chapter 12. The other, older approach is genetic: the use of mutants that look or behave abnormally as clues to the normal. Researchers use mutants to find out what the normal (nonmutated) version of a gene does. A four-winged fruit fly is an example of a mutant that can teach us something about fruit-fly development. At some point in this fly's development from a zygote, its DNA directed the formation of four wings instead of two. Although wing formation involves many genes,

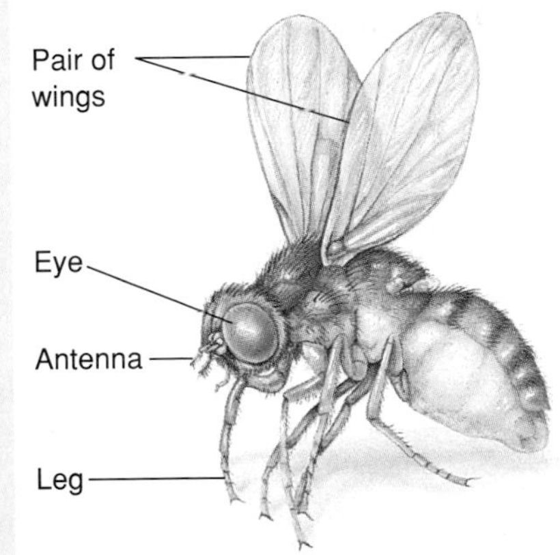

Normal anatomy of *Drosophila melanogaster*

the genetic basis for this abnormality turns out to be surprisingly simple—mutation of a single gene. Because that one gene controls the activity of many other genes in fruit-fly development, its mutation, and the resulting malfunctioning, can bring about a large-scale abnormality. Gene regulation—how cells control the expression of their genes—is the subject of this chapter.

11.1 Proteins interacting with DNA turn prokaryotic genes on or off in response to environmental changes

What do we actually mean when we say that genes are turned on or off? As we discussed in Chapter 10, genes determine the nucleotide sequences of specific RNA molecules; if this RNA is mRNA, it in turn determines the sequences of amino acids in protein molecules. Thus, a gene that is turned on is being transcribed into RNA, and that message is being translated into specific protein molecules. The overall process by which genetic information flows from genes to proteins—that is, from genotype to phenotype—is called **gene expression.** The turning on and off of transcription is the main way that gene expression is regulated. The control of gene expression makes it possible for cells to produce specific kinds of proteins when and where they are needed.

Our earliest understanding of gene control came not from multicellular organisms such as fruit flies but from the bacterium *Escherichia coli* (Figure 11.1A). As a single cell, this organism does not require the elaborate regulation of gene expression that leads to cell specialization in fruit flies and other multicellular creatures. Nevertheless, *E. coli* changes its activities from time to time in response to changes in its environment. Let's look at how the regulation of gene transcription helps *E. coli* accomplish these changes.

The *lac* Operon Picture an *E. coli* cell living in your intestine. If you eat only a sweet roll with butter for breakfast, the bacterium will be bathed in the sugars glucose and fructose and the digestion products of fats. Later on, if you have a fat-free milkshake and a salad for lunch, *E. coli*'s environment will change drastically.

Let's focus on your milkshake for a moment. One of the main nutrients in milk is the sugar lactose. When lactose is plentiful in the intestine, *E. coli* can make the enzymes necessary to absorb the sugar and use it as an energy source. Conversely, when lactose is not plentiful, *E. coli* does not waste its energy producing these enzymes.

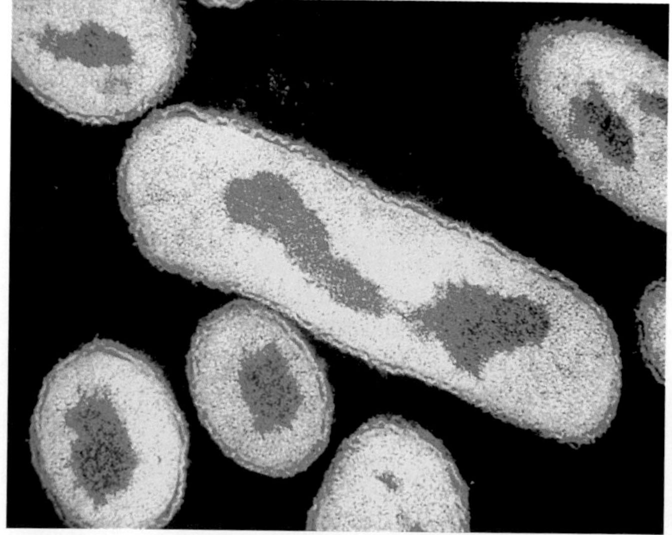

Figure 11.1A Cells of the bacterium *E. coli*

Remember that enzymes are proteins; their production is an example of gene expression. *E. coli* can make lactose-utilization enzymes because it has genes that code for these enzymes. In 1961, French biologists François Jacob and Jacques Monod proposed a hypothesis describing how an *E. coli* cell can adjust its enzyme production in response to changes in its environment. The Jacob and Monod model, shown in Figure11.1B on the facing page, explains how genes coding for lactose-utilization enzymes are turned off or on, depending on whether lactose is available.

E. coli uses three enzymes to take up and start metabolizing lactose, and the genes coding for these enzymes are regulated as a unit. In the DNA represented at the top of Figure 11.1B, which is a small segment of the bacterium's chromosome, notice that the three genes that code for the lactose-utilization enzymes (light blue) are next to each other in the DNA.

Adjacent to the group of lactose enzyme genes are two *control sequences,* short sections of DNA that help control them. One stretch of nucleotides is a **promoter** (green), a site where the transcription enzyme, RNA polymerase, attaches and initiates transcription—in this case, transcription of all three lactose enzyme genes (as depicted in the bottom panel of Figure 11.1B). Between the promoter and the enzyme genes, a DNA segment called an **operator** (yellow) acts as a switch. The operator determines whether RNA polymerase can attach to the promoter and start transcribing the genes.

Such a cluster of genes with related functions, along with a promoter and an operator, is called an **operon;** operons exist only in prokaryotes. The key advantage to the grouping of genes into operons is that the expression of these genes can be easily coordinated. The operon discussed here is called the *lac* operon, short for lactose operon. When an *E. coli* encounters lactose, all the enzymes needed for its use are made at once, because the operon's genes are all controlled by a single switch, the operator. But what determines whether the operator switch is on or off?

The top panel of Figure 11.1B shows the *lac* operon in "off" mode, its status when there is no lactose in the cell's environment. Transcription is turned off by a molecule called a **repressor** (red), a protein that functions by binding to the operator and blocking the attachment of RNA polymerase to the promoter. On the left side of the figure, you can see where the repressor comes from. A gene called a **regulatory gene** (dark blue), located outside the operon, codes for the repressor. The regulatory gene is expressed continually, so the cell always has a supply of repressor molecules. (However, because only a very small number of repressor molecules is involved, their production is much less wasteful for the cell than making unnecessary lactose enzymes.)

How can an operon be turned on if its repressor is always present? As the bottom panel of Figure11.1B indicates, lactose interferes with the attachment of the *lac* repressor to the promoter by binding to the repressor and changing its shape. With its new shape, the repressor cannot bind to the operator, and the operator switch remains on. RNA polymerase can now bind to the promoter and from there transcribe the genes of the operon. The resulting mRNA carries coding sequences for all three enzymes (purple) needed for lactose use. The cell can translate this message into separate polypeptides because the mRNA has codons signaling the start and stop of translation.

The *lac* mRNA and protein molecules remain intact for only a short while before cellular enzymes break them down. When their synthesis stops because lactose is no longer present, they quickly disappear.

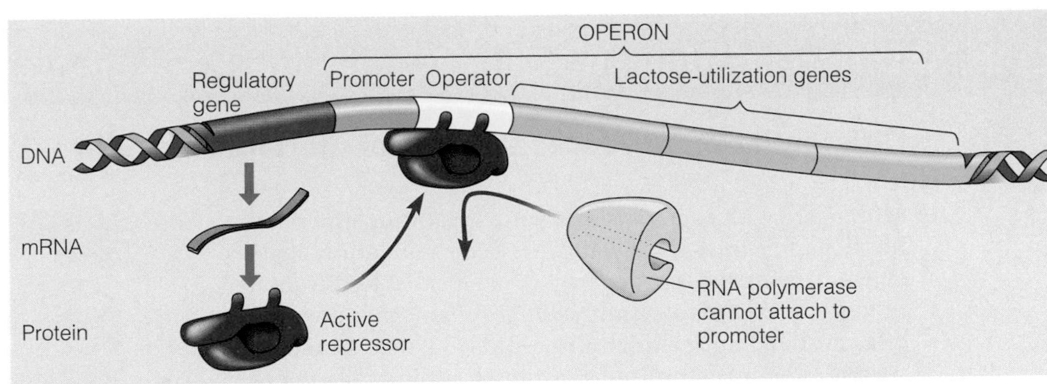

Operon turned off

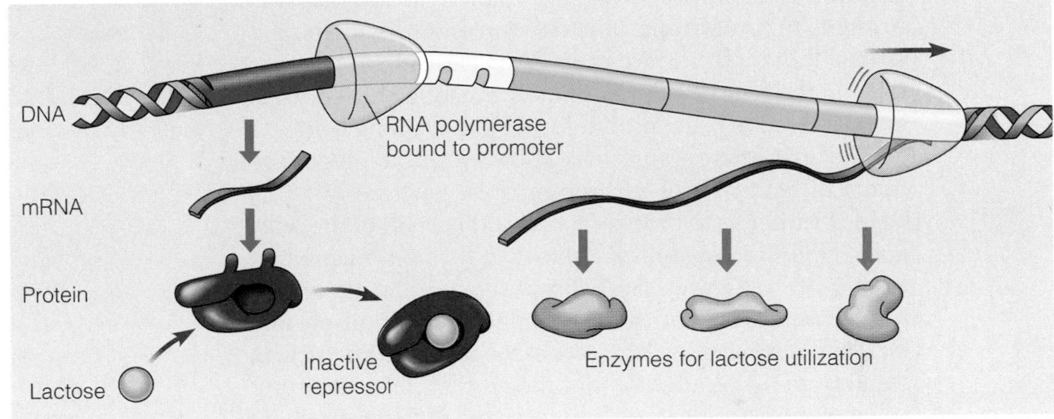

Operon turned on (repressor inactivated by lactose)

Figure 11.1B The *lac* operon

Other Kinds of Operons The *lac* operon is only one type of operon in bacteria. Other types also have a promoter, an operator, and several adjacent genes, but they differ in the way the operator switch is controlled. Figure 11.1C shows two types of repressor-controlled operons. As we discussed, the *lac* operon's repressor is active alone and inactive when bound to lactose. A second type of operon, represented here by the *trp* operon, is controlled by a repressor that is *inactive* alone. To be active, this type of repressor must combine with a specific small molecule. In our example, the small molecule is tryptophan, an amino acid essential for protein synthesis. *E. coli* can make tryptophan from scratch if it has to, using enzymes encoded in the *trp* operon. But it will stop making tryptophan and simply absorb it from its surroundings whenever possible. When *E. coli* is swimming in tryptophan, which occurs in large amounts in such foods as milk and poultry, tryptophan binds to the repressor of the *trp* operon. This activates the *trp* repressor, enabling it to switch off the operon. Thus, this type of operon allows bacteria to stop making certain essential molecules when the molecules are already present in the environment, saving materials and energy for the cells.

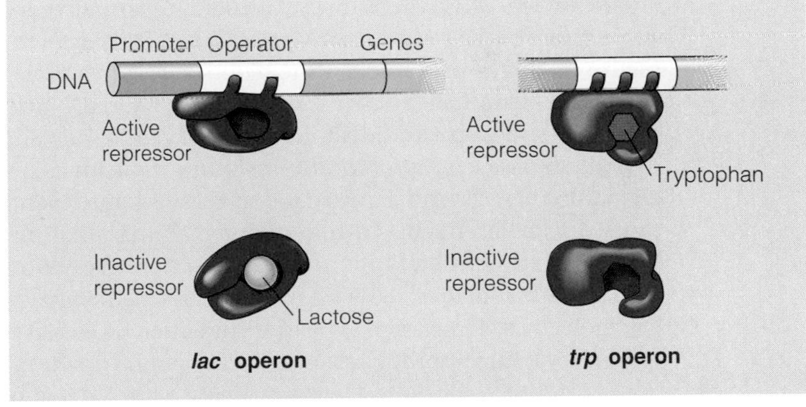

Figure 11.1C Two types of repressor-controlled operons

A third type of operon uses **activators,** proteins that turn operons *on* by binding to DNA. These proteins act by somehow making it easier for RNA polymerase to bind to the promoter, rather than by blocking RNA polymerase, as repressors do. Armed with a variety of operons, regulated by repressors and activators, *E. coli* and other prokaryotes can thrive in frequently changing environments. 

? A certain mutation in *E. coli* impairs the ability of the *lac* operator to bind to the repressor. How would this affect the cell?

The cell would wastefully produce the enzymes for lactose metabolism continuously, even in the absence of lactose.

Cellular Differentiation and Gene Regulation in Eukaryotes

11.2 Differentiation produces a variety of specialized cells in eukaryotes

Compared with bacteria, eukaryotic organisms, especially multicellular ones, face elaborate gene-regulation challenges. During the repeated cell divisions that lead from a unicellular zygote to a multicellular organism, individual cells must undergo **cellular differentiation,** that is, become specialized in structure and function. It is the regulation of genes that leads to this specialization. Active genes in the cells of a developing fly wing, for example, are expressed as proteins that make the cells flat and smooth, forming a strong, thin flight surface resembling clear plastic. In contrast, in the cells of a developing fly eye, other active genes are expressed as proteins that form light-focusing lenses.

The light micrographs here show four examples of the many different types of human cells. Each sample was stained with dyes to bring out structural details of the cells, and all four micrographs are shown at the same magnification (750×). Figure 11.2A shows short segments of two muscle cells, which are very long fibers with multiple nuclei (which appear as dark vertical rods here). The stripes result from the arrangement of the proteins that enable the cell to contract. The large orange shape in Figure 11.2B is part of a nerve cell, with its nucleus appearing as a lighter orange disk. The dark spot in the nucleus is a nucleolus. The cell has three long extensions, which carry nerve signals from one part of the body to another. Figure 11.2C shows sperm cells; their long flagella can propel them through a woman's reproductive tract to meet an egg cell. Figure 11.2D shows a single white blood cell (with a three-part, darkly stained nucleus) surrounded by red blood cells. The small size and disklike shape of red blood cells help them pass through blood vessels and transport oxygen molecules. Mammalian red blood cells lack nuclei and other organelles, having lost them during differentiation. Without organelles taking up space, a mature red blood cell can be literally packed with oxygen-carrying hemoglobin molecules. Thus, each type of differentiated cell has a unique structure appropriate for carrying out its function, and together the various types of cells enable a multicellular organism to function as a whole.

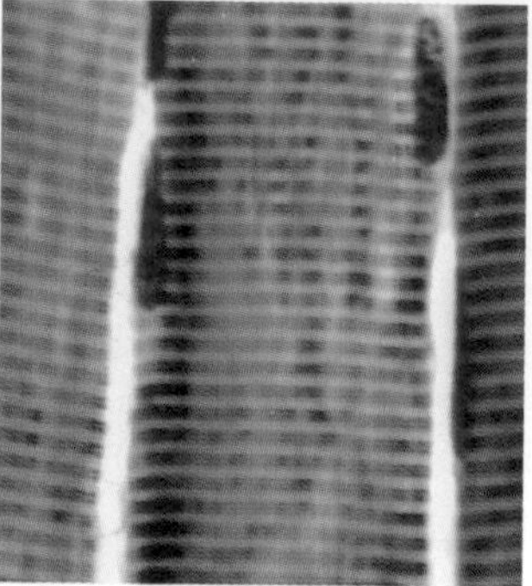

Figure 11.2A Muscle cells

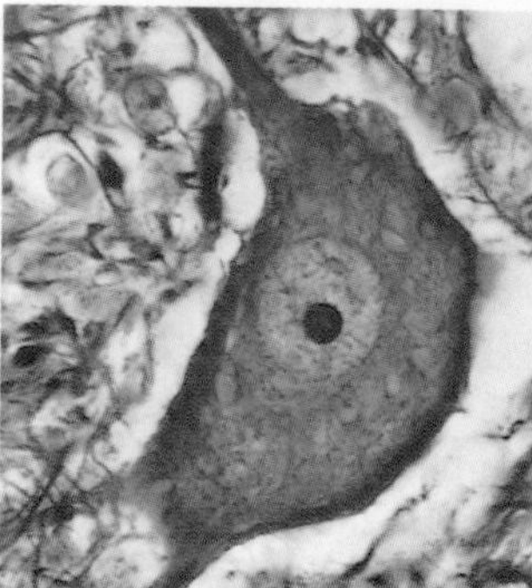

Figure 11.2B A nerve cell

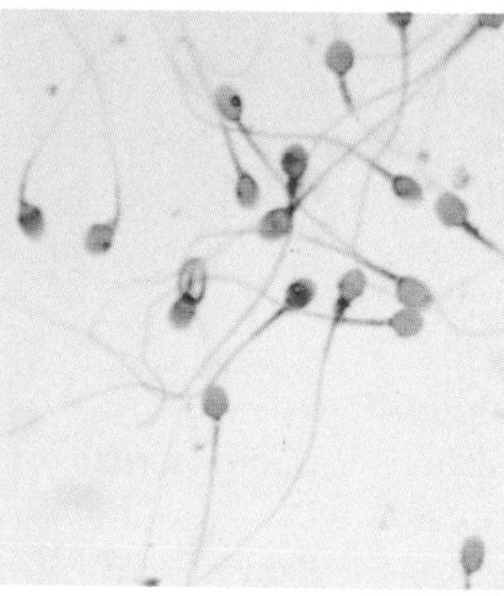

Figure 11.2C Sperm cells

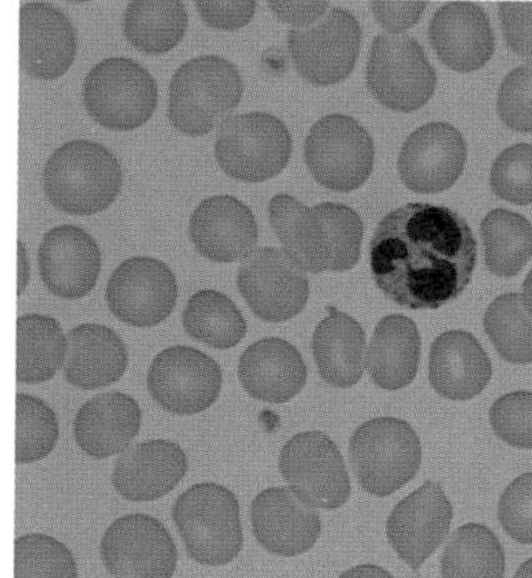

Figure 11.2D Blood cells

In the next module, we look at some of the evidence for the idea—our assumption up to now—that differentiated cells carry the entire genome.

What is cellular differentiation?

The structural and functional specialization of diverse cell types during development of a multicellular organism

11.3 Specialized cells may retain all of their genetic potential

The zygote has a complete set of genes for making all the types of specialized cells the organism will need. What happens to these genes as a cell differentiates? A red blood cell loses its nucleus and virtually all its DNA. But most differentiated cells retain a nucleus and a complete set of chromosomes. Are all the genes still present, and, if they are, do the differentiated cells retain the potential to express them?

One way to approach these questions is to replace the nucleus of an egg or zygote with the nucleus from a differentiated cell. If genes are lost or irreversibly turned off during differentiation, the transplanted nucleus will not allow the development of a normal embryo. The pioneering experiments in nuclear transplantation were performed by American embryologists Robert Briggs and Thomas King in the 1950s. You can follow one experiment in Figure 11.3A. The researchers destroyed the nuclei of frog egg cells with ultraviolet (UV) light and then transplanted nuclei from tadpole intestinal cells into the eggs. (The tadpole is the larva of the frog; its intestinal cells are already differentiated.) Many of the eggs containing transplanted nuclei started to develop; a few even developed into normal tadpoles. Thus researchers were able to **clone** frogs—produce

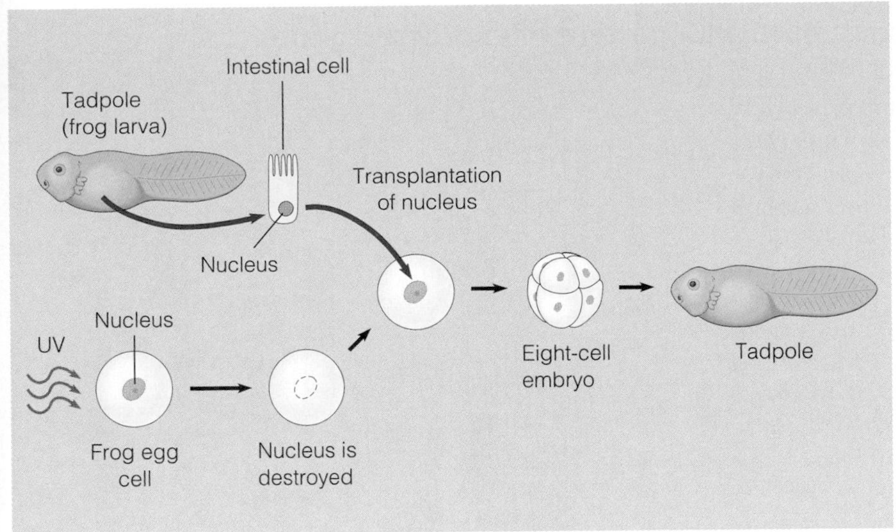

Figure 11.3A A nuclear transplantation experiment using the frog

Figure 11.3B Dolly, the result of mammalian cloning, with her surrogate mother

genetically identical copies—using nuclei of differentiated cells. Such studies showed that nuclei from differentiated cells can retain all of their genetic potential.

Additional evidence came in 1997 with the first cloning of a mammal using a differentiated nucleus. In this case, the researchers used an electric shock to fuse a specially treated sheep mammary cell with an egg from which they had removed the nucleus. The egg began to divide, was implanted in the uterus of another sheep, and developed into the celebrated "Dolly." As predicted, Dolly resembles her genetic parent, the mammary cell donor, not the egg donor or the surrogate mother (Figure 11.3B). Within two years of Dolly's birth, researchers in Hawaii cloned mice by transplanting ovary-cell nuclei into enucleated eggs, and a Japanese team cloned a cow.

Another indication that differentiation need not impair a cell's genetic potential is the natural process of regeneration, the replacement of lost body parts. When a salamander loses a leg, for example, certain cells in the leg stump dedifferentiate, divide, and then redifferentiate to give rise to a new leg. Many animals can regenerate lost parts, and in a few animals, isolated differentiated cells can be made to dedifferentiate and then develop into an organism.

In plants, the ability of a differentiated cell to develop into a whole new individual is common. Figure 11.3C shows this process using a carrot. A single cell removed from the root and placed in culture medium may begin dividing and eventually grow into an adult plant. This technique can be used to clone plants, reproducing hundreds or thousands of genetically identical organisms from the somatic cells of a single plant. In this way, it is possible to propagate large numbers of plants that have desirable traits such as high fruit yield or resistance to disease. The fact that a mature plant cell can dedifferentiate and then give rise to all the specialized cells of a new plant is further evidence that differentiation does not necessarily involve irreversible changes in the DNA. 

? If a nerve cell and a skin cell in your body have all the same genes, then how can they be so different?

Each cell type must be expressing certain genes that are present in, but not expressed in, the other cell type.

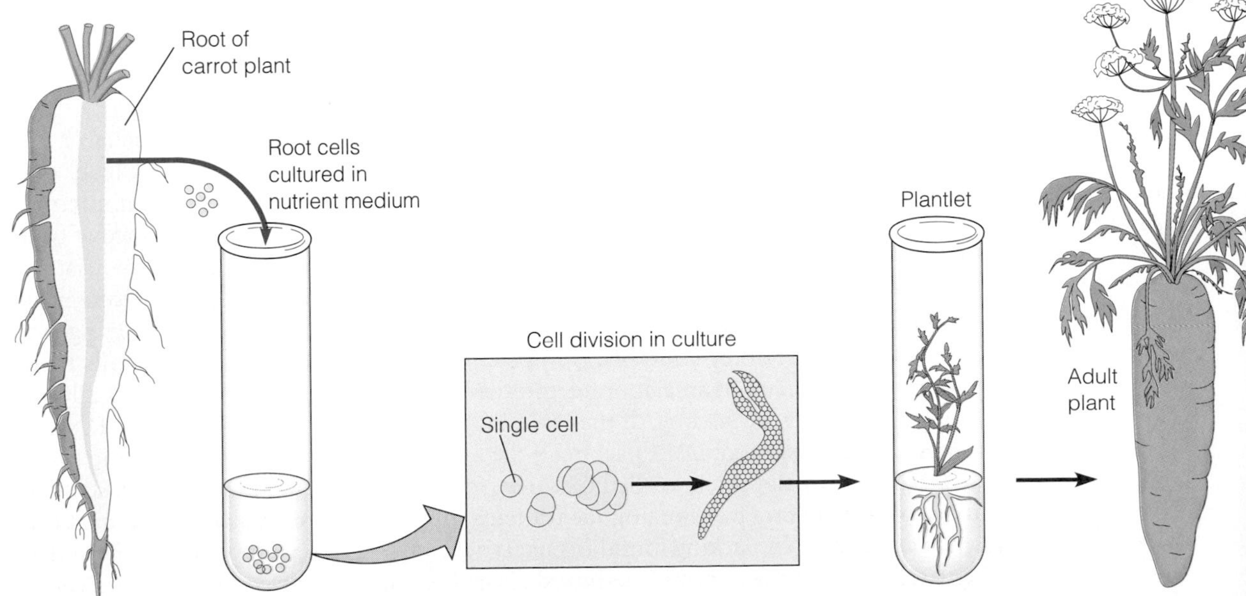

Figure 11.3C Growth of a carrot plant from a differentiated cell

11.4 Each type of differentiated cell has a particular pattern of expressed genes

If all the differentiated cells in an individual organism—say, a carrot plant or a fruit fly—contain the same genes, and all the genes have the potential of being expressed, how do cells become specialized? As we have already said, the great differences among cells in an organism result from the selective expression of genes. As a developing embryo undergoes successive cell divisions, specific genes must be activated in different cells at different times. Groups of cells follow diverging developmental pathways, and each group develops into a particular kind of tissue. Finally, in the mature organism, each cell type—nerve or pancreas, for instance—has a different pattern of turned-on genes.

Figure 11.4 illustrates patterns of gene expression for a few genes in three different specialized cells of a mammal. The genes for the enzymes of the metabolic pathway of glycolysis are active in all metabolizing cells, including the pancreas cell, embryonic lens cell, and nerve cell shown here. However, the genes for specialized proteins are expressed only by particular cells.

The specialized proteins we use as examples are the transparent protein crystallin, which forms the lenses of our eyes; the hormone insulin; and the oxygen-transport protein hemoglobin. Notice that the hemoglobin genes are not active in any of the cell types shown here. They are turned on only in cells that are developing into red blood cells. Insulin genes are activated only in the cells of the pancreas that produce that hormone. Nerve cells express genes for other specialized proteins not shown here. Mature lens cells, like mature red blood cells, achieve the ultimate in differentiation. After a frenzy of expressing crystallin genes and accumulating the protein products, these cells lose their nuclei, and thus all their genes.

So we see that eukaryotic cells become specialized because they express only certain genes. Thus, cellular differentiation in multicellular organisms results from selective gene expression, as does the ability of bacteria to produce different enzymes as needed. Now it's time to examine the control of gene expression in eukaryotes more closely. We begin by looking at how DNA is packaged in a eukaryotic cell, because this influences transcription. 

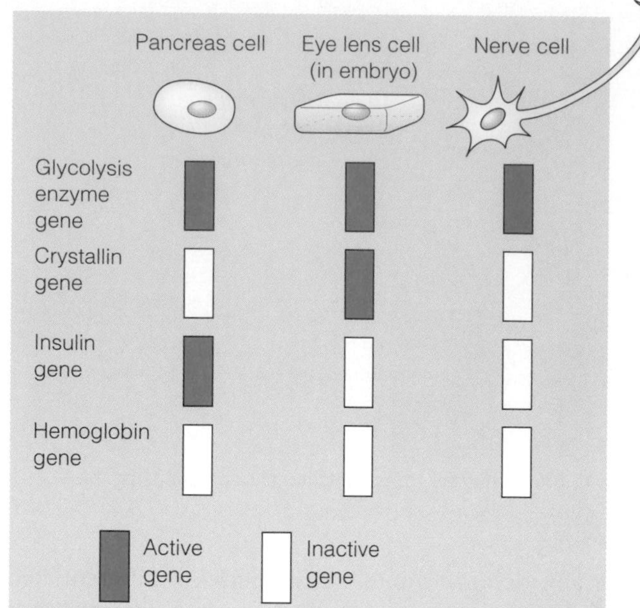

Figure 11.4 Gene expression patterns in three kinds of cells

? What evidence supports the view that differentiation is based on the control of gene expression rather than on irreversible changes in the genome?

The nuclear transplantation experiments in animals and the cloning of plants from differentiated cells described in Module 11.3

11.5 DNA packing in eukaryotic chromosomes helps regulate gene expression

11A *See Activity 11A on the CD-ROM.*

The nucleus in any one of your cells would be the envy of a computer manufacturer. Computer chips can store large amounts of information, but they are still a long way from the extraordinary storage capacity of a eukaryotic cell. Each eukaryotic chromosome contains a single DNA molecule that is thousands of times longer than the 5-µm diameter of a typical nucleus. The total DNA in a human cell's 46 chromosomes would stretch for 3 meters! All this DNA can fit into the nucleus because of an elaborate, multilevel system of coiling and folding, or *packing,* of the DNA in each chromosome. A crucial aspect of DNA packing is the association of the DNA with small proteins called **histones,** found only in eukaryotes. (Bacteria have analogous proteins, but they lack the degree of DNA packing found in eukaryotes.)

Figure 11.5A outlines a simplified model for the main levels of DNA packing. As shown at the top, the double helix has a diameter of 2 nm, which is not altered by packing. At the first level of packing, histones attach to the DNA. In electron micrographs, the DNA-histone complex has the appearance of beads on a string. Each "bead," called a **nucleosome,** consists of DNA wound around a protein core of eight histone molecules. At the next level of packing, the beaded string is wrapped into a tight helical fiber. Then this fiber coils further into a thick supercoil with a diameter of about 200 nm. Looping and folding can further compact the DNA, as you can see in the metaphase chromosome at the bottom of the figure. (DNA packing during mitosis is so specific and precise that individual genes always end up at the same positions in the metaphase chromosome.) Viewed as a whole, Figure 11.5A gives a sense of how successive levels of coiling and folding enable a huge amount of DNA to fit into a cell nucleus.

DNA packing tends to prevent gene expression, presumably by preventing RNA polymerase and other transcription proteins from contacting the DNA. At the nucleosome level, histones participate in the short-term switching on and off of genes: For a gene to be transcribed, the histones must loosen their grip on the DNA. Nonhistone chromosomal proteins (not shown in Figure 11.5A) control how tightly the histones bind to the DNA.

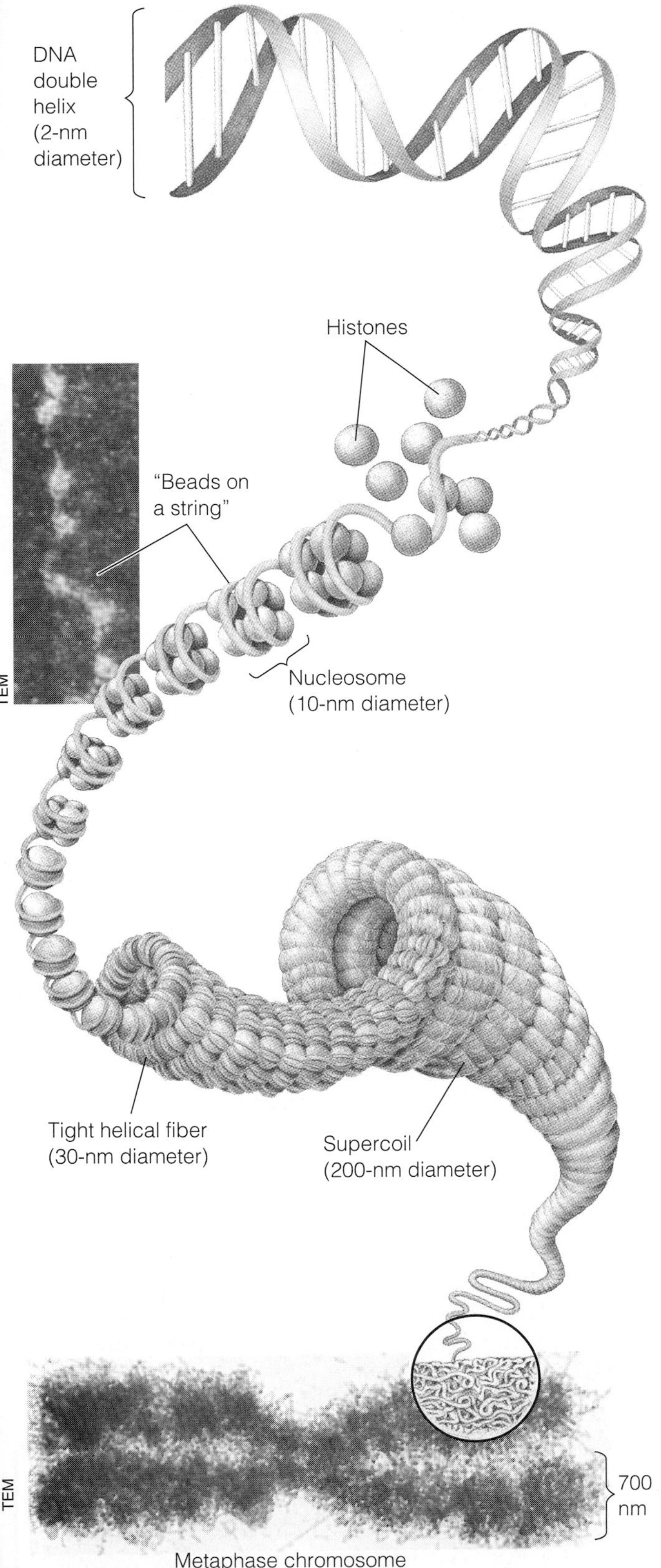

Figure 11.5A DNA packing in a eukaryotic chromosome

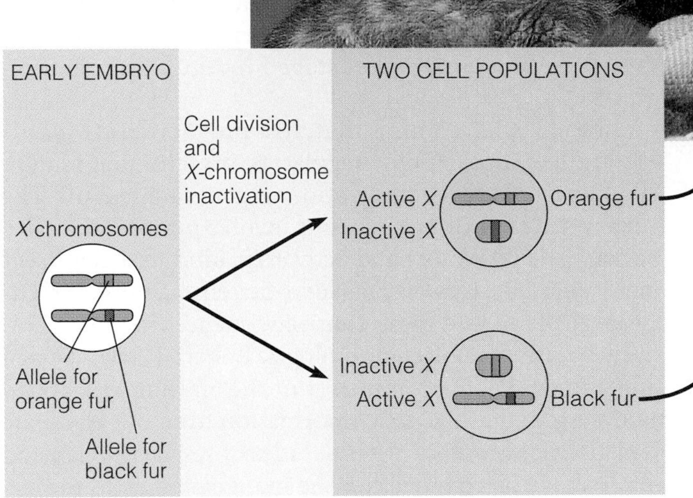

Figure 11.5B Tortoiseshell pattern on a cat, a result of *X*-chromosome inactivation

Cells may use higher levels of packing for long-term inactivation of genes. Highly compacted chromatin, which is found not only in metaphase chromosomes but also in varying regions of interphase chromosomes, is generally not expressed at all. One intriguing case is seen in female mammals, where one *X* chromosome in each somatic cell is highly compacted and almost entirely inactive, even during interphase. This ***X*-chromosome inactivation** is initiated early in embryonic development, when one of the two *X* chromosomes in each cell is inactivated at random. The inactivation is inherited by a cell's descendants. Consequently, a female heterozygous for genes on the *X* chromosome has populations of cells that express different *X*-linked alleles.

A striking example of the phenotypic effect of *X*-chromosome inactivation is the tortoiseshell cat, which has orange and black patches of fur (Figure 11.5B). The tortoiseshell gene is on the *X* chromosome, and the tortoiseshell phenotype requires the presence of two different alleles, one for orange fur and one for nonorange (black) fur. Normally, only females can have both alleles, because only they have two *X* chromosomes. If a female is heterozygous for the tortoiseshell gene, she is tortoiseshell. Orange patches are formed by populations of cells in which the *X* chromosome with the orange allele is active; black patches have cells in which the *X* chromosome with the nonorange allele is active. ("Calico" cats also have white areas, which are determined by another gene.) 

? How does dense packing of DNA in chromosomes prevent gene expression?

RNA polymerase and other proteins required for transcription do not have access to the DNA in tightly packed regions of a chromosome.

11.6 Complex assemblies of proteins control eukaryotic transcription

11A

The packing and unpacking of chromosomal DNA provide a coarse adjustment for eukaryotic gene expression by making a region of DNA either more or less available for transcription. The fine-tuning begins with the initiation of transcription. In both prokaryotes and eukaryotes, this is the most important stage for regulating gene expression.

Biologists have found that, like prokaryotes (Module 11.1), eukaryotes employ regulatory proteins that bind to DNA and turn the transcription of genes on and off. The eukaryotic control mechanisms involve proteins that, like prokaryotic repressors and activators, bind to specific segments of DNA. However, eukaryotic cells have more regulatory proteins and more control sequences in their DNA. In the words of molecular biologist Robert Tjian, "We had underestimated the complexity of the machinery that had evolved to retrieve genetic information from our elaborate [eukaryotic] genome." After decades of research, Tjian and others have pieced together the main elements of the eukaryotic control system that regulates the transcription of genes. Shown in Figure 11.6, their model features an intricate array of regulatory proteins that interact with DNA and with one another to turn genes on or off.

In contrast to the genes of bacterial operons, each eukaryotic gene usually has its own promoter and other control sequences. Moreover, activator proteins seem to be more important in eukaryotes than repressors are. In multicellular eukaryotes, the "default" state for most genes seems to be "off." A typical animal or plant cell needs to turn on (transcribe) only a small percentage of its genes, those required for the cell's specialized structure and function. Housekeeping genes, those continually active in virtually all cells for routine activities such as glycolysis, may be in an "on" state by default.

Transcription Factors Turning on a eukaryotic gene involves regulatory proteins called **transcription factors** in addition to RNA polymerase. Activators are one type of transcription factor. In the model depicted in Figure 11.6, the first step in initiating gene transcription is the binding of activators (green) to DNA sequences called **enhancers** (yellow). In contrast to the operators of prokaryotic operons, enhancers are usually far away from the gene they help regulate and may be on either side of the gene. How does the binding of an activator protein to an enhancer sequence in the DNA influence a distant gene? Apparently the DNA bends, and the bound activators interact with other transcription factors (purple), which then bind as a complex at the gene's promoter. This large assembly of proteins somehow facilitates the correct attachment of RNA polymerase to the promoter and the initiation of transcription. As shown in the figure, several enhancers and activators may be involved. Not shown are *repressor* proteins that may bind to DNA sequences called **silencers** and function analogously to *inhibit* the start of transcription.

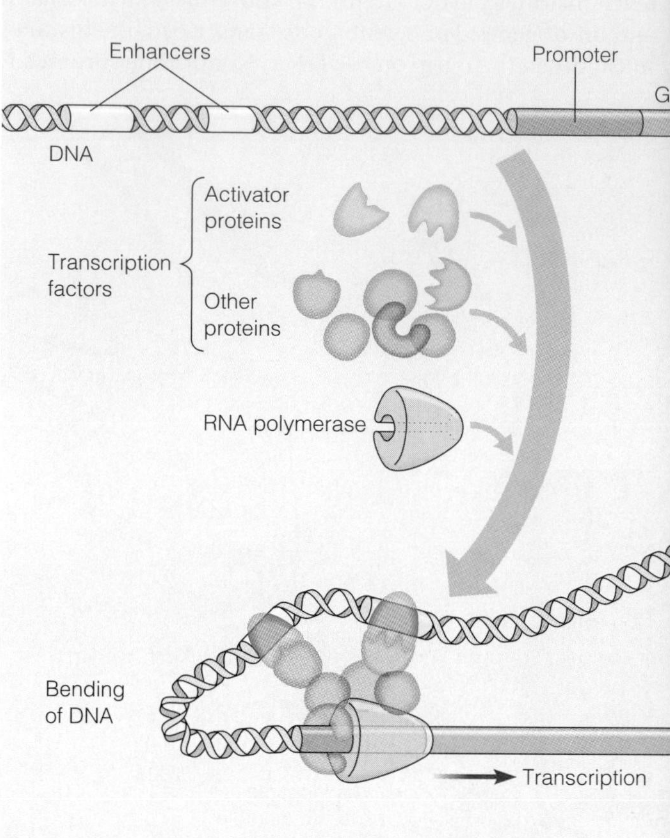

Figure 11.6 A model for the turning on of a eukaryotic gene

In summary, both eukaryotes and prokaryotes regulate transcription by using regulatory proteins that bind to DNA. However, many more regulatory proteins are involved in eukaryotes, and the interactions among these proteins are far more complex.

Coordinating Eukaryotic Gene Expression If eukaryotic genomes do not have operons, how does the eukaryotic cell deal with genes of related function that need to be turned on or off at the same time? Genes coding for the enzymes of a metabolic pathway, for example, are often scattered over different chromosomes. Coordinated gene expression in eukaryotes seems to depend on the association of a specific enhancer (or collection of enhancers) with every gene of a dispersed group. Copies of the transcription factors that recognize these DNA sequences bind to them, promoting simultaneous transcription of the genes. 

? In stimulating transcription of a specific eukaryotic gene, an enhancer does not act directly on the gene's promoter, but has its effect via DNA-binding proteins called ________ ________.

transcription factors

Eukaryotic RNA is processed: Cap and tail are added and noncoding segments removed 11.7

11A

Both prokaryotes and eukaryotes regulate gene expression at transcription, but the eukaryotic cell also has additional opportunities for control at later stages. The eukaryotic cell not only localizes transcription in the nucleus but also modifies, or *processes,* the RNA transcripts there, before they move to the cytoplasm for translation by the ribosomes.

One kind of RNA processing is the addition of extra nucleotides to the ends of the RNA transcript. These additions include a small *cap* (a single G nucleotide) at one end and a long *tail* (a chain of A's) at the other end (Figure 11.7A). The cap and tail protect the RNA from attack by cellular enzymes and will help ribosomes recognize the RNA as mRNA. (The cap and tail themselves are not translated.)

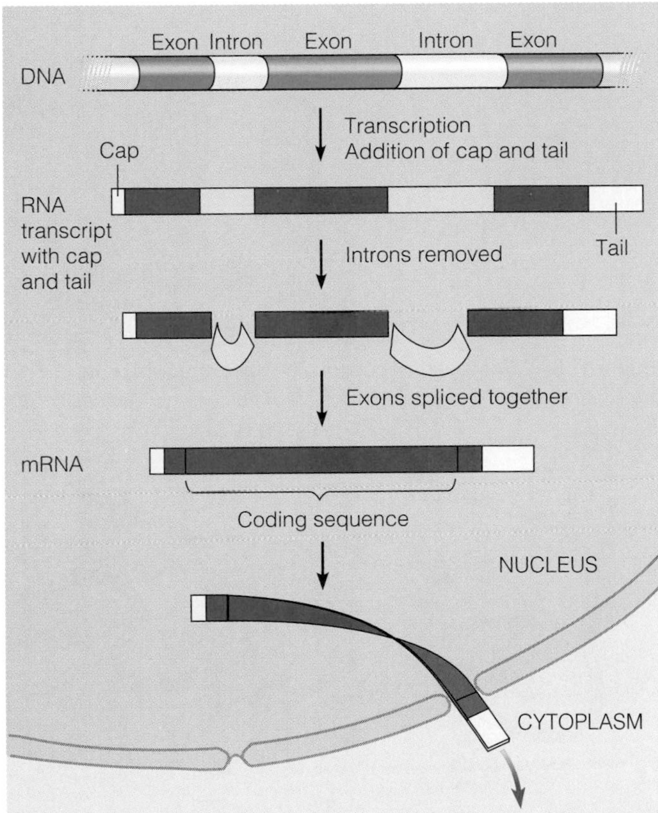

Figure 11.7A The production of eukaryotic mRNA

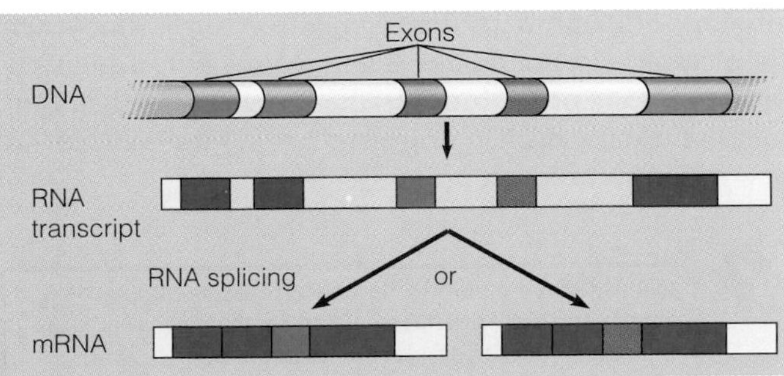

Figure 11.7B Production of two different mRNAs from the same gene

Another type of RNA processing is made necessary in eukaryotes by noncoding stretches of nucleotides that interrupt the nucleotides that actually code for amino acids. When these interruptions were first discovered, researchers were puzzled. It was as if unintelligible sequences of letters were randomly interspersed in an otherwise intelligible written document.

Most genes of plants and animals, it turns out, include such internal noncoding regions, which are called **introns** (for "intervening sequences"). The coding regions—the parts of a gene that are expressed—are called **exons.** As Figure 11.7A illustrates, both exons (darker color) and introns (lighter color) are transcribed from DNA into RNA. However, before the RNA leaves the nucleus, the introns are removed, and the exons are joined to produce an mRNA molecule with a continuous coding sequence. (The short noncoding regions at the ends, just inside the cap and tail, are considered parts of the first and last exons.) This process is called **RNA splicing.** It is also used by eukaryotic cells in producing transfer RNA and ribosomal RNA, although the final tRNA and rRNA molecules are never translated.

RNA splicing can occur by two mechanisms. When the product is mRNA, splicing is catalyzed by a complex of proteins and small RNA molecules. In certain other cases, splicing occurs completely without proteins or extra RNA molecules; the RNA transcript itself catalyzes the process. In other words, RNA acts as an enzyme!

Introns provide several possible ways for regulating gene expression. Some scientists think that the splicing process itself may help control the flow of mRNA from nucleus to cytoplasm, because until splicing is completed, the RNA is attached to the molecules of the splicing machinery and cannot pass through the nuclear pores. Moreover, in some cases, the cell can carry out splicing in more than one way, generating different mRNA molecules from the same RNA transcript. Notice in Figure 11.7B, for example, that one mRNA molecule ends up with the green exon and the other with the brown exon. With this sort of **alternative splicing,** an organism can get more than one type of polypeptide from a single gene. The fruit fly provides an interesting example of two-way mRNA splicing. In this animal, the differences between males and females are controlled by two sets of regulatory proteins, one set specific to the male and one to the female. The two sets of proteins, however, are coded by the same set of genes; the sex differences are largely due to different patterns of RNA splicing.

With capping, tailing, and splicing completed, the "final draft" of eukaryotic mRNA is ready for translation.

? Explain why many eukaryotic protein-coding genes are longer than the final forms of their RNA transcripts, the versions of mRNA that leave the nucleus.

Many eukaryotic genes have regions of noncoding DNA called introns. RNA splicing removes these regions from the RNA transcripts before the RNA leaves the nucleus.

11.8 Translation and later stages of gene expression are also subject to regulation

11A

After eukaryotic mRNA is transported to the cytoplasm, there are additional opportunities for regulation. These include mRNA breakdown, translation initiation, protein activation, and protein breakdown.

The lifetime of mRNA molecules is one important factor regulating the amounts of various proteins that are produced in the cell. Long-lived mRNAs are translated into many more protein molecules than are short-lived ones. Prokaryotic mRNAs have very short lifetimes; they are degraded by enzymes after only a few minutes. This is one reason bacteria can change their proteins relatively quickly in response to environmental changes. In contrast, the mRNA of eukaryotes can have lifetimes of hours or even weeks.

A striking example of long-lived mRNA is found in vertebrate red blood cells, which act like factories for manufacturing the protein hemoglobin. In most species of vertebrates, the mRNAs for hemoglobin are unusually stable. They probably last as long as the red blood cells that contain them—about a month in birds and perhaps longer in reptiles, amphibians, and fishes—and are translated repeatedly. Mammals are an exception. When their red blood cells mature, they lose their ribosomes (along with their other organelles) and thus cease to make new hemoglobin. However, mammalian hemoglobin itself lasts about as long as the red blood cells last, around four months.

The process of translation also offers opportunities for regulation. Among the molecules involved in translation are a great many proteins that have regulatory functions. Red blood cells, for instance, have an inhibitory protein that prevents translation of hemoglobin mRNA unless the cell has a supply of heme, the iron-containing chemical group essential for hemoglobin function. As indicated in Figure 11.8A, the binding of heme to the inhibitor inactivates the inhibitor, allowing hemoglobin genes to be translated. The polypeptides produced are then assembled with heme groups to form hemoglobin molecules.

The final opportunities for the control of gene expression are after translation. Post-translational control mechanisms in eukaryotes often involve cutting eukaryotic polypeptides into smaller, active final products. The hormone insulin, for example, is synthesized in pancreas cells as one long polypeptide, which is not active as a hormone. Then, as illustrated in Figure 11.8B, a large center portion is cut away, leaving two shorter chains, held together by chemical bonds that link up sulfur atoms (S in the figure stands for sulfur). This combination of two shorter polypeptides is the active form of insulin.

Another control mechanism operating after translation is the selective breakdown of proteins. Though mammalian hemoglobin may last as long as the red blood cell housing it, the lifetimes of many other proteins are closely regulated. Some of the proteins that trigger metabolic changes in cells are broken down within a few minutes or hours. This regulation allows a cell to adjust the kinds and amounts of its proteins in response to changes in its environment. It also enables the cell to maintain its proteins in prime working order. Indeed, when proteins are damaged, they are usually broken down right away and replaced by new ones that function properly. 

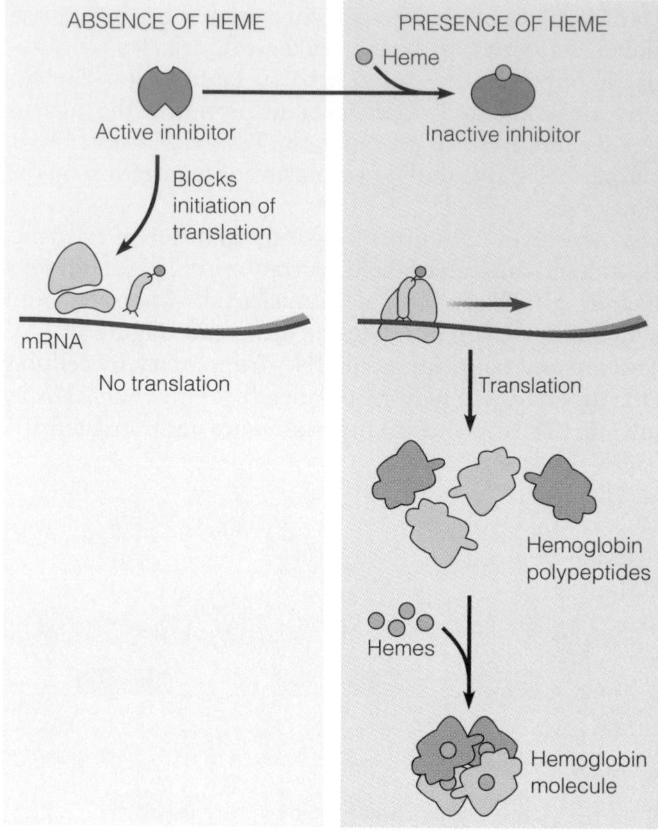

Figure 11.8A Translational control: The roles of an inhibitory protein and heme in regulating the start of hemoglobin synthesis

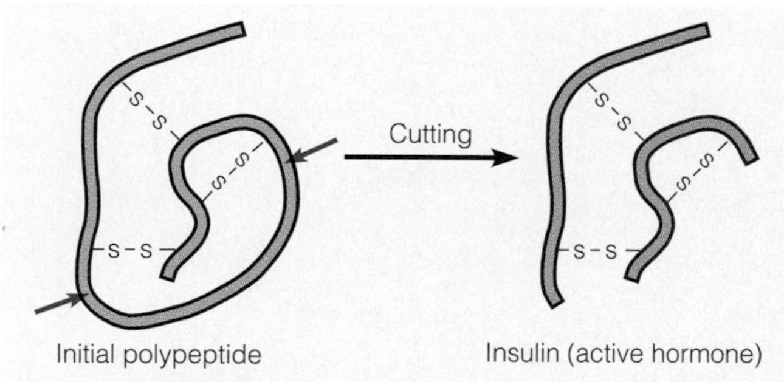

Figure 11.8B Post-translational control: Polypeptide cutting, the final step in forming the active insulin protein

? Once mRNA encoding a particular protein reaches the cytoplasm, what are four mechanisms that can regulate the amount of the active protein in the cell?

Breakdown of the mRNA; regulation of translation initiation; activation of the protein; and breakdown of the protein

Review: Multiple mechanisms regulate gene expression in eukaryotes 11.9

11A

The figure at the right provides a review of eukaryotic gene expression and highlights the multiple control points where the process can be turned on or off, speeded up, or slowed down. Picture the series of pipes that carry water from your local water supply, perhaps a reservoir, to a faucet in your home. At various points, valves control the flow of water. We use this model in the figure to illustrate the flow of genetic information from a chromosome—a reservoir of genetic information—to an active protein that has been synthesized in the cell's cytoplasm. The multiple mechanisms that control gene expression are analogous to the control valves in water pipes. In the figure, each gene-expression "valve" is indicated by a control knob. Note that these knobs represent *possible* control points; for most proteins, only a few control points are probably important. As we have seen, the most important control point, in both eukaryotes and prokaryotes, is usually the start of transcription. In the diagram, the yellow knob represents the mechanisms that regulate the start of transcription.

Of course, this diagram oversimplifies the control of gene expression. What it does not show is the web of control that connects different genes, often through their products. We have seen examples in both prokaryotes and eukaryotes of the actions of gene products (usually proteins) on other genes or on other gene products within the same cell. The genes of operons in *E. coli*, for instance, are controlled by repressor or activator proteins encoded by regulatory genes on the same DNA molecule. In eukaryotes, many genes are controlled by proteins encoded by regulatory genes on different chromosomes.

In a multicellular eukaryote, starting in the early embryo, the web of regulation crosses cell boundaries. A cell of the embryo can produce chemicals such as hormones that induce a neighboring or distant cell to develop along a certain pathway. In this web of interactions, some types of genes stand out as having a major impact on cell differentiation; they have been identified in situations where normal development has gone awry. In the next several modules, we look at a class of genes whose malfunctioning causes large-scale developmental abnormalities.

Of the nine regulatory "valves" in Figure 11.9, which five can also operate in a prokaryotic cell?

(1) Control of transcription; (2) control of mRNA breakdown; (3) control of translation; (4) control of protein activation; and (5) control of protein breakdown

Chromosome
DNA unpacking
Other changes to DNA
GENE
Gene
Transcription
Exon
Intron
RNA transcript
Addition of cap and tail
Splicing
Tail
Cap
mRNA in nucleus
NUCLEUS
Flow through nuclear envelope
mRNA in cytoplasm
CYTOPLASM
Breakdown of mRNA
Broken-down mRNA
Translation
Polypeptide
Cleavage / modification / activation
ACTIVE PROTEIN
Breakdown of protein
Broken-down protein

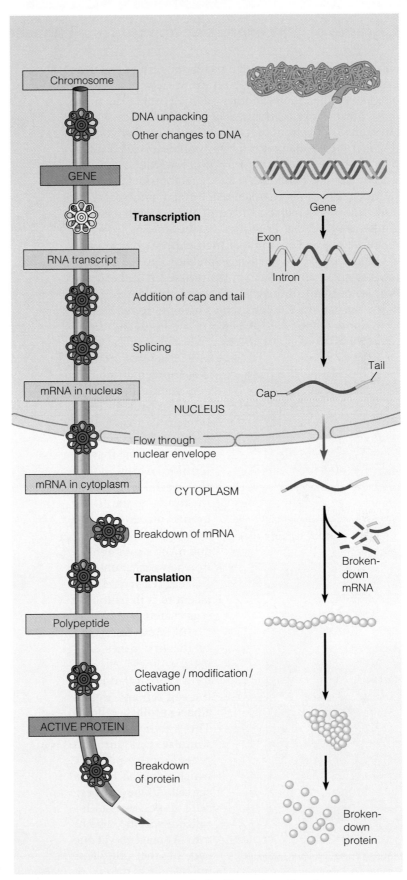

Figure 11.9 The gene-expression "pipeline" in a eukaryotic cell

11.10 Cascades of gene expression and cell-to-cell signaling direct the development of an animal

11B

We began this chapter with a picture of a fruit fly with an extra set of wings. Figure 11.10A, front views of the heads of two fruit flies, shows another oddity. The fly head on the top is normal; notice its antennae, the two knobby sensory structures on the front of its head. The fly head on the bottom is from a mutant; it exhibits a condition called antennapedia (antenna-feet): two legs on the front of its head where antennae should be. If these mutant fruit flies do not sound strange to you, imagine a human with analogous body features—a person with four arms or with legs dangling from the head! Fruit flies like the antennapedia mutant gave researchers some of their earliest glimpses into the relationship between gene expression and the development of an organism. Such mutants have led to the identification of many of the genes that program development in the normal fly. The researchers who pioneered this genetic approach, Edward B. Lewis in the 1940s and Christiane Nüsslein-Volhard and Eric Wieschaus in the 1970s and '80s, revolutionized developmental biology.

Among the earliest events in fruit-fly development is a cascade of gene expression and cell-to-cell signaling that determines which ends of the egg cell will become the head and tail. These events occur in the ovaries of the mother fly and involve communication between an unfertilized egg cell and cells adjacent to it in its follicle (egg chamber). As Figure 11.10B indicates, ① one of the first genes activated in the egg cell codes for a protein that leaves the egg cell and signals adjacent follicle cells. ② These follicle cells are stimulated to turn on genes for other proteins, which signal back to the egg cell. ③ One of the egg cell's responses is to localize a specific type of mRNA (magenta) at one end of the cell. This mRNA marks the end of the egg where the fly's

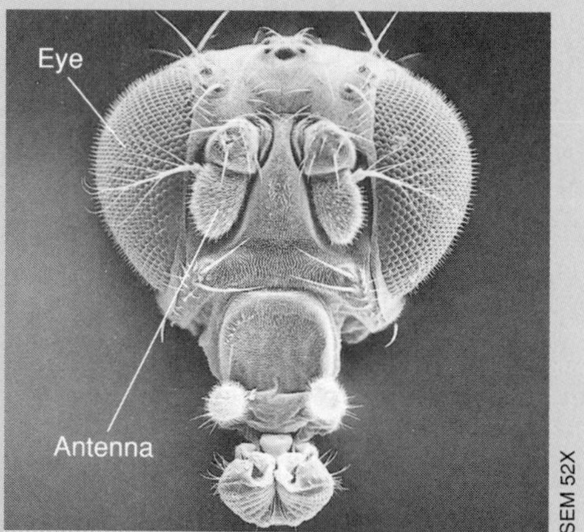

Head of a normal fruit fly

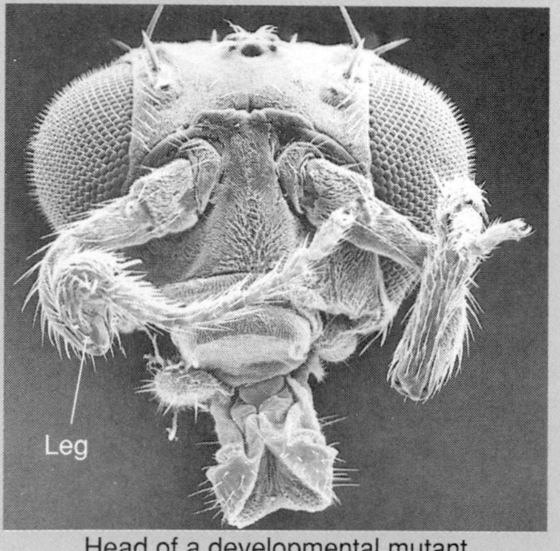

Head of a developmental mutant

Figure 11.10A Antennapedia in a mutant fruit fly, compared with a normal fruit fly

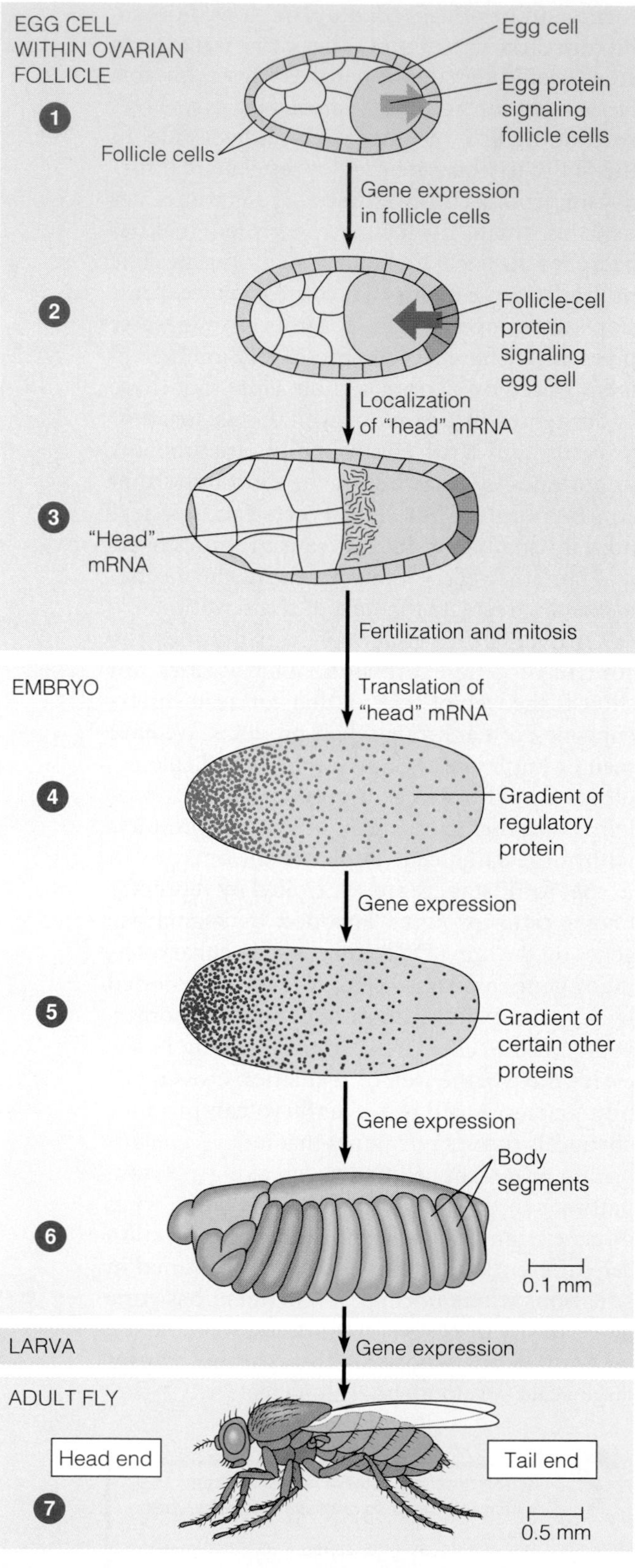

Figure 11.10B Key steps in the early development of head-tail polarity in a fruit fly

head will develop and thus defines the animal's head-to-tail axis. (One piece of evidence from mutant studies is that a defective gene for the "head" mRNA causes the embryo to develop tails at both ends!) Similar events establish the other axes (top to bottom and side to side) and thus lay out a plan for the positioning of body parts as they develop.

After the egg is fertilized and laid, repeated mitoses transform the zygote into an embryo. ④ Translation of the "head" mRNA in the early embryo produces a regulatory protein (aqua dots) that diffuses through the embryo but remains most concentrated at the head end. In turn, this protein gradient triggers a corresponding gradient of transcription in the embryo's nuclei. ⑤ The proteins (red dots) resulting from translation of this RNA initiate more rounds of gene expression. Cell signaling—now among the cells of the embryo—helps drive the process. ⑥ The result is the subdivision of the embryo's body into segments.

Now finer details of the fly can take shape. Protein products of some of the axis-specifying genes and segment-forming genes activate yet another set of genes. These genes, called homeotic genes, determine what body parts will develop from each segment. A **homeotic gene** is a master control gene that regulates batteries of other genes that actually create the anatomical identity of parts of the body. For example, one set of homeotic genes in fruit flies, the antennapedia complex, instructs cells in the segments of the head and thorax (midbody) to form antennae and legs, respectively. Elsewhere, these homeotic genes remain turned off, while others are turned on. The eventual outcome, shown in step ⑦ of the figure, is an adult fly. Notice that the adult's body segments correspond to those of the embryo in step 6.

Cascades of gene expression, with the protein products of one set of genes activating another set of genes, and so on, are a common theme in development. Many of the proteins, such as the one in step 1 of Figure 11.10B, do not act on genes directly. Instead, they act as molecular signals that indirectly trigger expression of a gene in a neighboring cell. Let's next take a closer look at how this happens.

? What determines which end of a developing fruit fly will become the head?

A specific kind of mRNA localizes at the end of the unfertilized egg that will become the head. After fertilization of the egg, this mRNA is translated into a regulatory protein that forms a gradient starting at the head end.

11.11 Signal-transduction pathways convert messages received at the cell surface into responses within the cell

11C

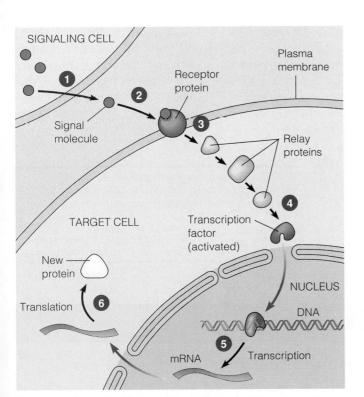

Figure 11.11 A signal-transduction pathway that turns on a gene

Cell-to-cell signaling, with proteins or other kinds of molecules carrying messages from signaling cells to receiving (target) cells, is a key mechanism in development, as well as in the coordination of cellular activities in the mature organism. In most cases, a signal molecule acts by binding to a receptor protein in the plasma membrane of the target cell and initiating a signal-transduction pathway in the cell. A **signal-transduction pathway** is a series of molecular changes that converts a signal on a target cell's surface into a specific response inside the cell.

Figure 11.11 shows the main elements of a signal-transduction pathway in which the target cell's response is the transcription (turning on) of a gene. ① The signaling cell first secretes the signal molecule. ② This molecule binds to a specific receptor protein embedded in the target cell's plasma membrane. ③ The binding activates the first in a series of relay proteins (green) within the target cell. Each relay molecule activates another. ④ The last relay molecule in the series activates a transcription factor that ⑤ triggers transcription of a specific gene. ⑥ Translation of the mRNA produces a protein.

Signal-transduction pathways are crucial to many cellular functions. We encountered them when we studied the cell-cycle control system earlier in this unit, and we'll revisit them when we discuss cancer shortly and when we study hormone function in Chapters 26 and 33.

? How can a signal molecule from one cell alter gene expression in a target cell without even entering the target cell?

By binding to a receptor protein in the membrane of the target cell and triggering within the cell a signal-transduction pathway that activates transcription factors

11.12 Key developmental genes are very ancient

Among the most exciting biological discoveries in recent years is that a class of similar genes—homeotic genes—help direct embryonic development in a wide variety of organisms. Researchers studying homeotic genes in fruit flies found a common structural feature: Every homeotic gene they looked at contained a common sequence of 180 nucleotides. Very similar sequences have since been found in virtually every eukaryotic organism examined so far, including yeasts, plants, earthworms, frogs, chickens, mice, and humans. These nucleotide sequences are called **homeoboxes,** and each is translated into a segment (60 amino acids long) of the protein product of the homeotic gene. The homeobox polypeptide segment binds to specific sequences in DNA, enabling homeotic proteins that contain it to turn groups of genes on or off during development.

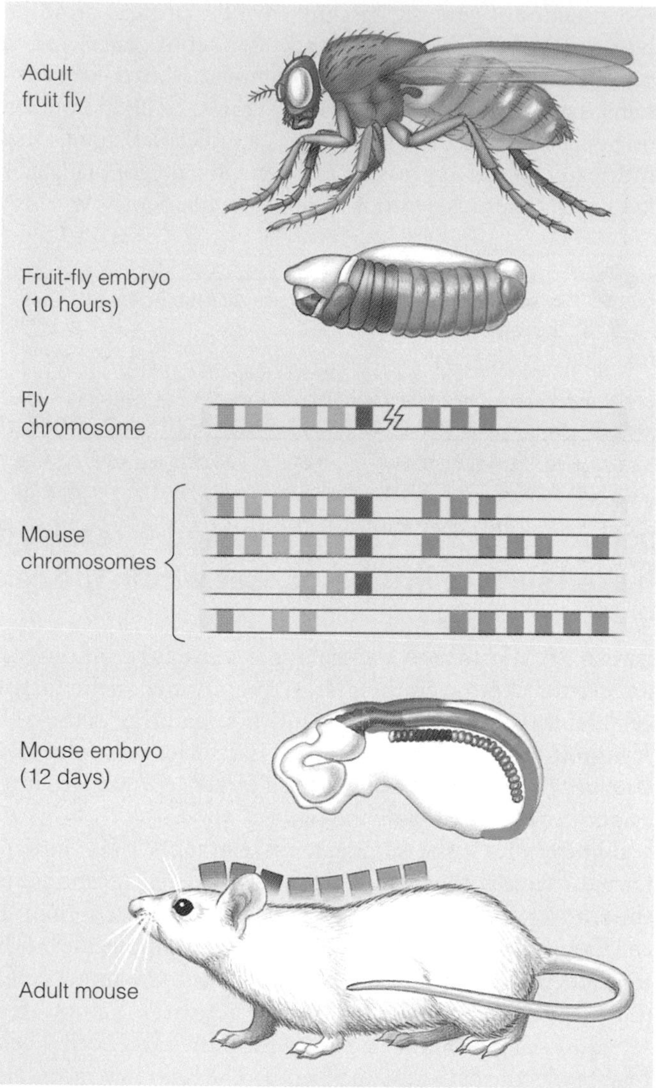

Figure 11.12 Comparison of fruit-fly and mouse homeotic genes

Figure 11.12 highlights some striking similarities in the chromosomal locations and developmental roles of homeobox-containing homeotic genes in two quite different animals. The figure shows portions of chromosomes that carry homeotic genes in the fruit fly and the mouse. The colored boxes represent homeotic genes that are very similar in flies and mice. Notice that the order of genes on the fly chromosome is the same as on the four mouse chromosomes and, further, that the gene order on the chromosomes corresponds to analogous body regions in both animals. These similarities suggest that the original version of these homeotic genes arose very early in the history of life and that the genes have remained remarkably unchanged for eons of animal evolution.

By their presence in such diverse creatures, homeotic genes illustrate one of the central themes of biology: unity in diversity (see Module 1.5). The fact that these key genes are *control* genes underscores the importance of regulation in the lives of organisms. In the next section of the chapter, we turn to regulatory genes of another type, genes that control cell growth and division. When mutations make the functioning of these genes go awry, the result can be cancer. 

? If the DNA sequences called homeoboxes, which help homeotic genes direct development, are common to flies and mice, then why aren't flies and mice more alike?

Homeotic genes have much DNA besides their homeoboxes, and so there can be much variation in homeotic genes. Moreover, the genes regulated by the protein products of the homeotic genes can be very different in different organisms.

The Genetic Basis of Cancer

11.13 Cancer results from mutations in genes that control cell division

In Chapter 8, we introduced cancer as a variety of diseases in which cells escape from the control mechanisms that normally limit their growth and division. In recent years, scientists have learned that this escape from normal controls is due to changes in some of the cells' genes or to changes in the way certain genes are expressed.

The abnormal behavior of cancer cells was observed years before anything was known about the cell cycle, its control, or the role genes play in making cells cancerous. One of the earliest clues to the cancer puzzle was the discovery, in 1911, of a virus that causes cancer in chickens. Recall that viruses are simply molecules of DNA or RNA coated with protein and in some cases a membranous envelope. (We referred to viruses as "packaged genes" in Chapter 10.) Viruses that cause cancer can become permanent residents in host cells by inserting their nucleic acid into the DNA of host chromosomes. It is now known that a number of viruses that can produce cancer carry specific cancer-causing genes in their nucleic acid. When inserted into a host cell, these genes can make the cell cancerous. A gene that causes cancer is called an **oncogene** (Greek *onkos*, tumor).

In 1976, American molecular biologists J. Michael Bishop, Harold Varmus, and their colleagues made a startling discovery. They found that a virus that causes cancer in chickens contains an oncogene that is an altered version of a gene found in normal chicken cells. Apparently, the virus picked up the gene from a former host cell. Subsequent research has shown that the chromosomes of many animals, including humans, contain genes that can be converted to oncogenes. A normal gene with the potential to become an oncogene is called a **proto-oncogene.** A cell can acquire an oncogene from a virus or from the conversion of one of its own genes. In 1989, Bishop and Varmus received the Nobel Prize for the pioneering work that led to our understanding that organisms have genes with the potential to cause cancer.

The work by Bishop and Varmus focused cancer research on proto-oncogenes. Searching for the normal role of these genes in the cell, researchers found that many proto-oncogenes code for growth factors—proteins that stimulate cell division—or for other proteins that somehow affect growth factor function or some other aspect of the cell cycle. As we discussed earlier, when all these proteins are functioning normally, in the right amounts at the right times, they help control cell division and cellular differentiation.

For a proto-oncogene to become an oncogene, a mutation must occur in the cell's DNA. Mutations that produce most types of cancer occur in somatic cells, those not involved in gamete formation. Therefore, cancer-causing mutations are generally not inherited.

Figure 11.13 illustrates three kinds of changes in somatic cell DNA that can produce active oncogenes. Let's assume that the starting proto-oncogene codes for a protein that stimulates cell division. On the left in the figure, a mutation (red) in the proto-oncogene itself creates an oncogene that codes for a hyperactive protein, one whose stimulating effect is stronger than normal. In the center, an error in DNA replication or recombination generates multiple copies of the gene, which are all transcribed and translated; the result is an excess of the normal stimulatory protein. On the right, the proto-oncogene has been moved from its normal location in the cell's DNA to another location. At its new site, the gene is under the control of a different promoter and other transcriptional control sequences, which cause it to be transcribed more often than normal; and the normal protein is again made in excess. So in all three cases, normal gene expression is changed, and the cell is stimulated to divide excessively.

Changes in genes whose products *inhibit* cell division are also involved in cancer. These genes are called **tumor-suppressor genes,** because the proteins they encode normally help prevent uncontrolled cell growth. Any mutation that keeps the normal tumor-repressor protein from being made or functioning may contribute to uncontrolled cell division.

Thus, we see several ways that genetic change may be connected to a malfunction of the cell cycle. We pursue this connection next.

In what respect is "proto-oncogene" a misnomer?

The term does not describe the normal function of the gene, which is generally the regulation of the cell cycle. The conversion of a proto-oncogene to an oncogene is an aberration, not a natural progression.

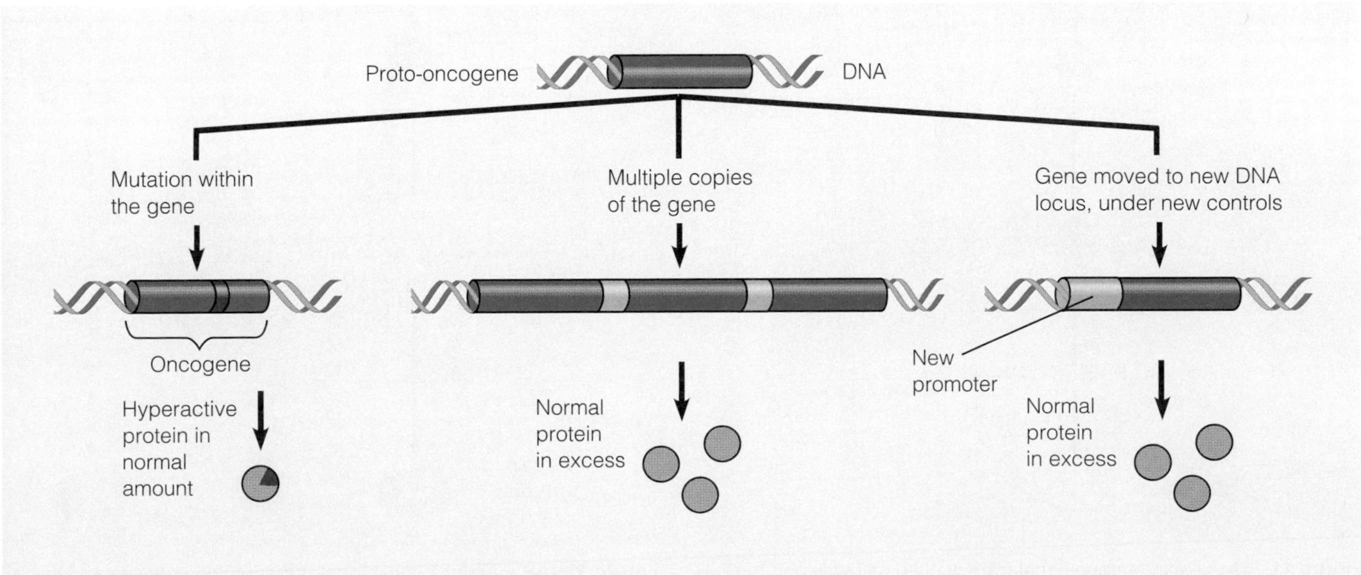

Figure 11.13 Alternative ways to make oncogenes from a proto-oncogene (all lead to excessive cell growth)

11.14 Oncogene proteins and faulty tumor-suppressor proteins can interfere with normal signal-transduction pathways

To understand how oncogenes and defective tumor-suppressor genes can contribute to uncontrolled cell growth, we need to look more closely at the normal functions of proto-oncogenes and tumor-suppressor genes. Genes in both categories often code for proteins involved in signal-transduction pathways leading to gene expression, pathways similar to the one described in Module 11.11.

The figures below (excluding, for the moment, the white boxes) illustrate two types of signal-transduction pathways leading to the synthesis of proteins that influence cell division. In Figure 11.14A, the pathway leads to the stimulation of cell division. The initial signal is a growth factor, and the target cell's ultimate response is the production of a protein that stimulates the cell to divide. By contrast, Figure 11.14B shows an inhibitory pathway, in which a growth-*inhibiting* factor causes the target cell to make a protein that inhibits cell division. In both cases, the newly made proteins function by interacting with components of the cell-cycle control system (see Module 8.9). The figures here do not show these interactions.

Now let's see what can happen when the target cell undergoes a cancer-causing mutation. The white box in Figure 11.14A shows the protein product of an oncogene resulting from mutation of a proto-oncogene. Normally, a stimulatory pathway like this will not operate unless the growth factor is available. However, an oncogene protein that is a hyperactive version of a protein in the pathway may trigger the pathway even in the absence of growth factor. In this ex-

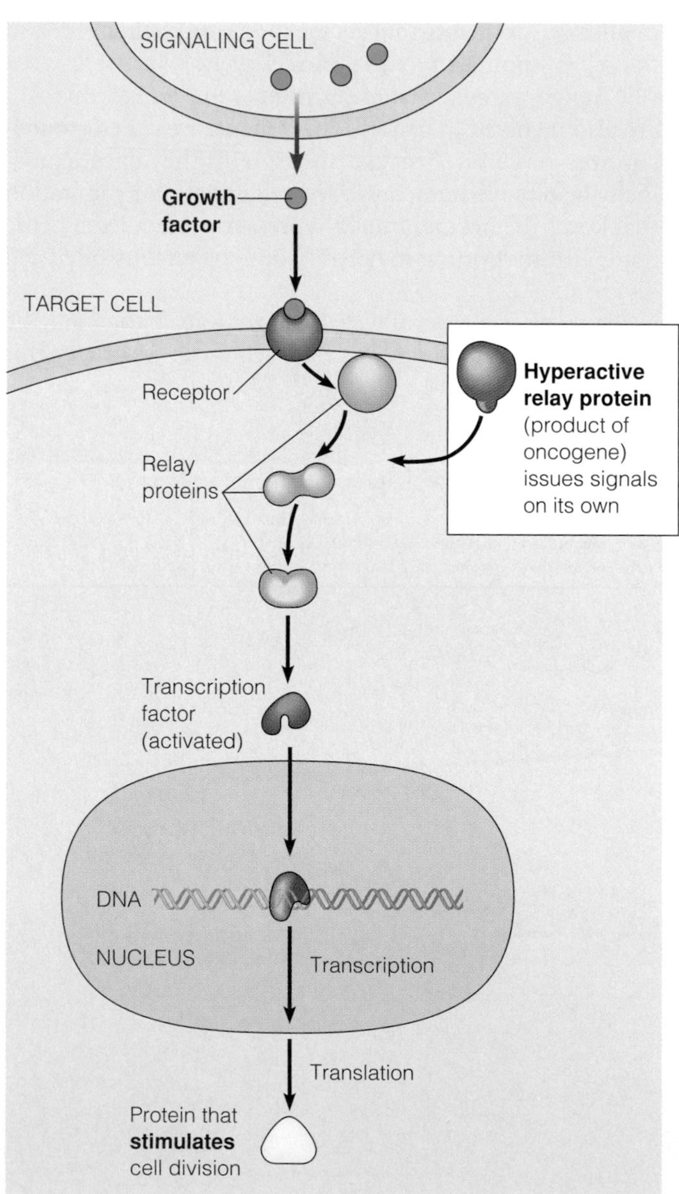

Figure 11.14A A stimulatory signal-transduction pathway and the effect of an oncogene protein

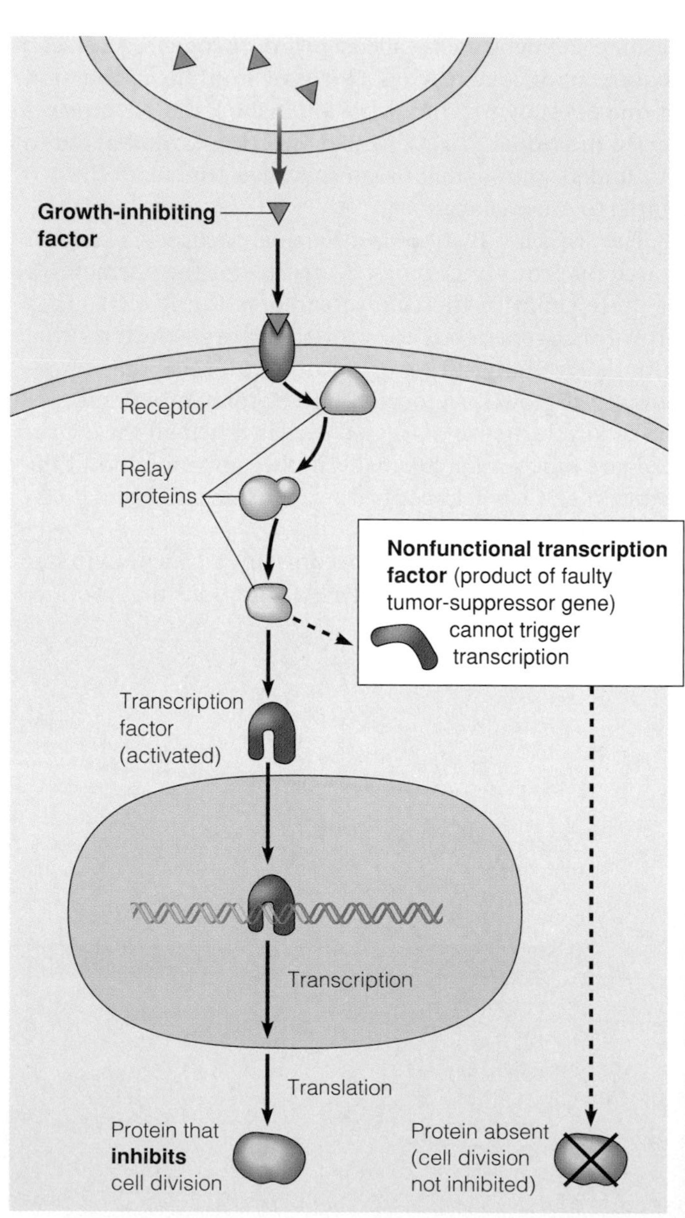

Figure 11.14B An inhibitory signal-transduction pathway and the effect of a faulty tumor-suppressor protein

ample, the oncogene protein is a hyperactive relay protein that issues signals on its own. However, abnormal versions or amounts of any of the pathway's components—from the growth factor itself to the transcription factor—could have the same final effect: overstimulation of cell division.

The white box in Figure 11.14B indicates how a mutant tumor-suppressor protein can affect cell division. In this case, the faulty tumor-suppressor protein is a nonfunctional transcription factor, one that the signal-transduction pathway cannot activate. As a result, the gene for the protein that inhibits cell division remains turned off, and excessive cell division may occur.

Examples of actual genes whose mutation gives rise to the kinds of abnormal proteins highlighted in our white boxes are a proto-oncogene named *ras* and a tumor-suppressor gene called *p53*. Mutations of these two genes have been implicated in many kinds of cancer. As we see next, most forms of cancer probably result from a series of such changes.

Contrast the action of an oncogene with that of a cancer-causing mutation in a tumor-suppressor gene.

An oncogene produces an abnormal protein that stimulates cell division via a signal-transduction pathway; a mutant tumor-suppressor gene produces a defective protein unable to function in a pathway that normally inhibits cell division.

Multiple genetic changes underlie the development of cancer 11.15

Including baseball player Daryl Strawberry and perhaps some of your own friends or relatives, over 130,000 Americans were stricken by cancer of the colon (large intestine) or rectum in 1998. One of the best understood types of human cancer, colon cancer illustrates an important principle about how cancer develops: *More than one somatic mutation is needed to produce a full-fledged cancer cell.* As in many cancers, the development of a colon cancer that metastasizes is gradual. (See Module 8.10 to review cancer terms.)

As shown in Figure 11.15A, ① the first sign of a colon cancer is the unusually frequent division of apparently normal cells in the colon lining. ② Later, a benign tumor (polyp) appears in the colon wall, and ③ eventually the malignant tumor (a carcinoma). These cellular changes parallel changes at the DNA level, including the activation of a cellular oncogene and the inactivation of two tumor-suppressor genes. These genetic changes (mutations) result in altered signal-transduction pathways like the ones outlined in Module 11.14. The requirement for several mutations—the actual number is usually four or more—explains why cancers can take a long time to develop.

Figure 11.15B indicates how mutations leading to cancer may accumulate in a lineage of somatic cells. Colors distinguish the normal cell (tan) from cells (shades of green) with one or more mutations leading to increased cell division and cancer. Once a cancer-promoting mutation occurs (green band on chromosome), it is passed to all the descendants of the cell carrying it. In our example, the first two mutations make the cells divide more rapidly; otherwise, the cells appear normal. The third mutation increases the rate of cell division more and also causes some changes in the cells' appearance. Finally, a cell accumulates a fourth cancer-promoting mutation and begins dividing uncontrollably. The structure of this cell and of its descendants is grossly altered.

Our understanding of the genetic basis of cancer has grown by leaps and bounds in recent years. Among the recent discoveries is a new category of tumor-suppressor genes. Rather than being components of pathways like the one in Figure 11.14B, the normal proteins encoded by these tumor-suppressor genes function in the repair of damaged DNA. When they are mutated, other cancer-causing mutations are more likely to accumulate.

One exposure to a particular carcinogenic (cancer-causing) chemical at a certain concentration for a certain length of time has a 10% chance of causing any one of the four mutations in Figure 11.15B. What is the chance (in %) that one exposure to the carcinogen will result in a malignant cell?

0.1 × 0.1 × 0.1 × 0.1 = 0.0001, or 0.01%

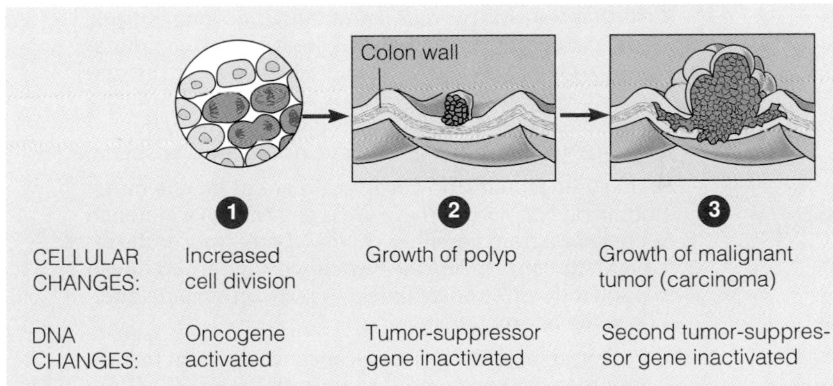

Figure 11.15A Stepwise development of a typical colon cancer

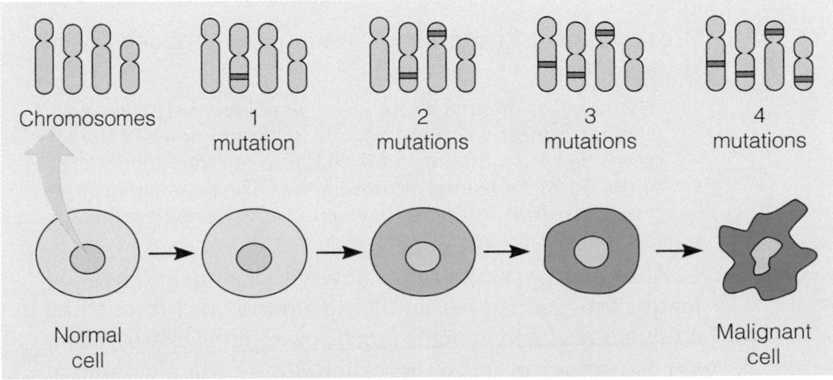

Figure 11.15B Accumulation of mutations in the development of a cancer cell

TALKING ABOUT SCIENCE

11.16 Mary-Claire King discusses mutations that cause breast cancer

Now a professor at the University of Washington, Mary-Claire King has spent over 25 years exploring the genetic basis of breast cancer, a disease that strikes one out of every ten American women. Dr. King was a math major at Carleton College and did not become interested in genetics until she arrived at UC Berkeley to study statistics in the mid-'60s. For her Ph. D. research, with the late Allan Wilson, she carried out a genetic comparison of humans and chimpanzees, finding extraordinary similarities (see Module 19.2). In 1973, she went to teach at the University of Chile in a Ford Foundation program, but her stay was cut short by the overthrow of the Allende government.

Figure 11.16 Mary-Claire King

> I came back to Berkeley thinking, "What shall I do next?" I knew that I needed to do work that was important to people in an everyday sort of way. The first thing that came up was the opportunity to set up a modern genetics lab in the cancer epidemiology group at UC San Francisco. I soon became captivated by the question of what causes cancer, in particular breast cancer.

Like all cancers, breast cancer is sometimes referred to as a "genetic disease." We asked Dr. King to explain.

> It means two things. Cancer is always genetic in the sense that cancer is always the consequence of changes in DNA. Cells that have cancer-causing mutations no longer divide and develop as they should. The great majority of mutations that lead to cancer arise in the tissue where the cancer starts—the colon or the breast, for example. These mutations are somatic mutations; they are in the body but not in the germ line, the cells that give rise to eggs or sperm.
>
> In some families, however, there are germ line mutations in one or more of these same genes. Such a mutation is passed on from parent to child and predisposes the recipient to cancer. We call this cancer "inherited," even though it doesn't appear unless the person acquires additional, somatic mutations.
>
> The vast majority of breast-cancer cases seem to have nothing to do with inherited mutations. But there are many accounts, going back to the ancient Greeks, of families in which breast cancer appears frequently, suggesting that an inherited trait might be involved.

Breast Cancer Research Dr. King was intrigued by the stories of familial breast cancer.

> It occurred to me that if mutations were involved in familial breast cancer, we might be able to identify some of the genes that are mutated and then figure out the functions of the normal versions of these genes. The results might give us insight into the more common, nonhereditary forms of the disease.

After almost two decades of work, Dr. King and her colleagues succeeded in identifying a gene on chromosome 17 that is mutated in many families with familial breast cancer. Mutations in this gene, called *BRCA1*, put a woman at high risk of breast cancer (and ovarian cancer, as well), a more than 80% risk of developing cancer during her lifetime. Recent research from other laboratories suggests that the protein encoded by the normal version of *BRCA1* acts as a tumor suppressor by helping repair damaged DNA in the cell. But however mutation of *BRCA1* contributes to cancer, it is likely that certain environmental influences, by tending to cause new somatic mutations, also play a role. Dr. King is attempting to sort out these environmental factors.

> We're now working with Jewish families in New York City and Israel, families with known mutations in *BRCA1* or *BRCA2*, a gene found soon after *BRCA1*. (Among breast cancer patients of Jewish ancestry, about 10% have mutations in one of these genes.) With the participants' permission, we test the DNA of breast-cancer patients from these families. For each patient who carries a mutation in *BRCA1* or *BRCA2*, we trace the history of the mutation in her family back as far as possible. Then we ask, "Did all of the women in the family who inherited the mutation develop breast cancer? At what age did cancer appear?" If some women developed breast cancer at 70 and some at 30, were there any differences in their environmental exposures? If some women with the mutation lived to a very old age and never developed breast cancer at all, can we explain that?

Genetic Testing Genetic testing of research participants has a clear goal: to learn more about the disease in the hope of helping prevent or cure the disease in future generations. But what of testing women in the general population for *BRCA* mutations? Dr. King has strong views.

> Does testing women for mutations in these genes make sense? To me there are three principles that are important to consider. One is that testing must be accompanied by patient education. Testing is useful only if it is presented in such a way that the person understands what the limitations of the test are and what the results mean. Genetic counselors are very helpful in conveying this information.
>
> The second critical principle is one of social justice: We need to respect absolutely the rights of the individual. A person's genetic background should have no bearing at all on his or her ability to obtain health insurance, for instance. For every one of us is predisposed to *something.*
>
> The third principle has to do with the pace of research in biology. For a woman today to know that she is predisposed to breast and ovarian cancer offers her a problem—but no solution except preventive surgical removal of her breasts and/or ovaries. I wish research moved so quickly that disease-predisposing genes could be identified and treatments developed within months of each other, but that is clearly not the case. The development of a useful intervention is enormously important but is also usually enormously difficult. So we need to work out a rational policy for genetic testing that recognizes the need for education and justice, and the realities of biological research.

(a) In what sense is breast cancer always "genetic"?
(b) Why is most breast cancer considered to be "nonhereditary"?

(a) DNA change is always involved. (b) Most breast cancers are associated with somatic mutations, not inherited mutations that are passed from parents to offspring via gametes.

Changes in lifestyle can reduce the risk of cancer 11.17

By the year 2000, it is likely that cancer will be the number-one cause of death in most developed countries, including the United States. Death rates due to certain forms of cancer (including stomach, cervical, and uterine cancers) have decreased, but the overall cancer death rate is still on the rise, currently increasing at about 1% per decade.

Cancer-causing agents, factors that alter DNA and make cells cancerous, are called **carcinogens.** Most mutagens are carcinogens, agents capable of bringing about cancer-causing DNA changes like the ones in Figure 11.15B. Two of the most potent carcinogens (and mutagens) are X-rays and ultraviolet radiation in sunlight. X-rays are a significant cause of leukemia and brain cancer. And exposure to UV radiation from the sun is known to cause skin cancer, including a deadly type called melanoma.

The largest group of carcinogens are mutagenic chemical compounds and substances containing them. Among the most important carcinogens, the one substance known to cause more cases and types of cancer than any other single agent is tobacco. In 1900, lung cancer was a rare disease. Since then, largely because of an increase in cigarette smoking, lung cancer has steadily increased, and today more people die of lung cancer than any other form of cancer. Most tobacco-related cancers come from actually smoking, but the passive inhalation of secondhand smoke is also a risk. As the table here indicates, tobacco use, sometimes in combination with alcohol consumption, causes a number of other types of cancer in addition to lung cancer. In nearly all cases, cigarettes are the main culprit, but smokeless tobacco products (snuff and chewing tobacco) are linked to cancer of the mouth and throat.

Cancer in the United States

Cancer	Examples of Known or Likely Carcinogens	Estimated Number of Cases in 1999
Prostate	Testosterone; possibly dietary fat	179,300
Breast	Estrogen; possibly dietary fat	176,300
Lung	Cigarette smoke	171,600
Colon and rectum	High dietary fat; low dietary fiber	129,400
Lymphomas	Viruses (for some types)	64,000
Bladder	Cigarette smoke	54,200
Melanoma of skin	Ultraviolet light	44,200
Uterus	Estrogen	37,400
Leukemias	X-rays, benzene; virus (for one type)	30,200
Kidney	Cigarette smoke	30,000
Mouth and throat	Tobacco in various forms; alcohol	29,800
Pancreas	Cigarette smoke	28,600
Ovary	(Large number of ovulation cycles)	25,200
Stomach	Table salt; cigarette smoke	21,900
Brain and nerve	Trauma; X-rays	16,800
Liver	Alcohol; hepatitis viruses	14,500
Cervix	Viruses; cigarette smoke	12,800
All others		155,600

How do carcinogens cause cancer? As we have seen, most cancers result from multiple genetic changes, including the activation of oncogenes and inactivation of tumor-suppressor genes. In many cases, these changes result from decades of exposure to the mutagenic effects of carcinogens. Carcinogens can also produce their effect by promoting cell division. Generally, the higher the rate of cell division, the greater the chance for mutations resulting from errors in DNA replication or recombination. Some carcinogens seem to have both effects. For instance, the hormones that cause breast and uterine cancers promote cell division and may also cause genetic changes that lead to cancer. In other cases, several different agents, such as viruses and one or more carcinogens, may together produce cancer.

We are still a long way from knowing all the factors that contribute to cancer. The effects of many environmental pollutants have yet to be evaluated, although some substances, such as asbestos, are definitely known to be carcinogens. Exposure to some of the most important known carcinogens is often a matter of individual choice. Tobacco use, consumption of animal fat and alcohol, and time spent in the sun, for example, are all behavioral factors that affect our cancer risk.

Avoiding carcinogens is not the whole story, for there is growing evidence that some food choices significantly reduce cancer risk. For instance, eating 20–30 grams of plant fiber daily (about twice the amount the average American consumes) and at the same time reducing animal-fat intake may help prevent colon cancer. There is also evidence that other substances in fruits and vegetables, including vitamins C and E and certain compounds related to vitamin A, may offer protection against a variety of cancers. Cabbage and its relatives, such as broccoli and cauliflower (see Figure 13.4A), are thought to be especially rich in substances that help prevent cancer, although the identities of these substances are not yet established. Determining how diet influences cancer has become a major research goal.

The battle against cancer is being waged on many fronts, and there is reason for optimism in the progress being made. It is especially encouraging that we can help reduce our risk of acquiring some of the most common forms of cancer by the choices we make in daily life. 

? Of all known environmental factors, which one causes the most cancer cases?

Tobacco

Chapter Review

Chapter Summary

Gene expression—the flow of information from genes to proteins—is subject to control, mainly by the turning on and off of genes. Cells become specialized in structure and function because only certain genes of the genome are expressed. Mutant organisms give researchers clues to the functions of normal genes—for example, genes that direct embryonic development **(Introduction).**

Gene Regulation in Prokaryotes (11.1) In prokaryotes, genes for related enzymes are often controlled together by being grouped into regulatory units called operons. Regulatory proteins bind to control sequences in the DNA and turn operons on or off, in response to environmental changes. The *lac* operon, for example, produces enzymes that break down lactose only when lactose is present.

Cellular Differentiation and Gene Regulation in Eukaryotes (11.2–11.9) In multicellular eukaryotes, cells become specialized, or differentiated, as a zygote develops into a mature organism. Different kinds of cells make different proteins **(11.2).** Most differentiated cells retain a complete set of genes, so a carrot plant, for example, can be made to grow from a single carrot cell. In general, the somatic cells of a multicellular organism all have the same genes **(11.3);** the cells are different because different genes are expressed **(11.4).** A chromosome contains a DNA double helix wound around clusters of histone proteins, forming a string of beadlike nucleosomes. This beaded fiber is further wound and folded, packing about 3 m of DNA into the nucleus of each human cell. DNA packing tends to block gene expression, presumably by preventing access of transcription proteins to the DNA. An extreme example is *X*-chromosome inactivation in the cells of female mammals **(11.5).** A variety of regulatory proteins interact with DNA and with each other to turn the transcription of eukaryotic genes on or off **(11.6).** There are also opportunities for the control of eukaryotic gene expression after transcription. Most eukaryotic genes include noncoding introns. After transcription, the introns are cut out of the RNA, leaving exons that code for a polypeptide **(11.7).** The lifetime of an mRNA molecule helps determine how much protein is made, as do protein factors involved in translation **(11.8).** Finally, the cell may activate and later break down the finished protein (Figure 11.9 reviews the multiple stages of eukaryotic gene expression.) **(11.9).**

The Genetic Control of Embryonic Development (11.10–11.12) Studies of mutant fruit flies show that a cascade of gene expression, involving genes for regulatory proteins, determines how an animal develops from a fertilized egg. Homeotic genes, for example, control batteries of genes that shape anatomical parts such as wings **(11.10).** Cell-cell signaling is key to development. Signal-transduction pathways convert molecular messages to cell responses **(11.11).** Homeotic genes contain nucleotide sequences, called homeoboxes, that are very similar in many kinds of organisms. This suggests that homeobox-containing genes arose very early in the history of life **(11.12).**

The Genetic Basis of Cancer (11.13–11.17) Cancer cells, which divide uncontrollably, can result from mutations in genes whose protein products regulate the cell cycle. A mutation can change a proto-oncogene (a normal gene that promotes cell division) into an oncogene, which causes cells to divide excessively. Mutations that inactivate tumor-suppressor genes have similar effects **(11.13).** Many proto-oncogenes and tumor-suppressor genes code for proteins active in signal-transduction pathways regulating cell division. Mutations of these genes cause malfunction of the pathway. Other cancer-causing mutations seem to impair the cell's ability to repair damaged DNA, allowing other mutations to accumulate. Cancers result from a *series* of genetic changes in a cell lineage **(11.14–11.15).** Researchers have gained insight into the genetic basis of breast cancer by studying families in which a disease-predisposing mutation is inherited **(11.16).** Reducing exposure to carcinogens (which induce cancer-causing mutations) and other lifestyle choices can help reduce cancer risk **(11.17).**

Testing Your Knowledge

Multiple Choice

1. The control of gene expression is more complex in multicellular eukaryotes than in prokaryotes because _____. *(Explain your answer.)*
   a. eukaryotic cells are much smaller
   b. in a multicellular eukaryote, different cells are specialized for different functions
   c. prokaryotes are restricted to stable environments
   d. eukaryotes have fewer genes, so each gene must do several jobs
   e. the genes of eukaryotes are information for making proteins

2. Your bone cells, muscle cells, and skin cells look different because
   a. different kinds of genes are present in each kind of cell
   b. they are present in different organs
   c. different genes are active in each kind of cell
   d. they contain different numbers of genes
   e. different mutations have occurred in each kind of cell

3. Which of the following methods of gene regulation do eukaryotes and prokaryotes appear to have in common?
   a. elaborate packing of DNA in chromosomes
   b. activator and repressor proteins, which attach to DNA
   c. the addition of a cap and tail to mRNA after transcription
   d. *lac* and *trp* operons
   e. the removal of noncoding portions of RNA

4. A eukaryotic gene was inserted into the DNA of a bacterium. The bacterium then transcribed this gene into mRNA and translated the mRNA into protein. The protein produced was useless; it contained many more amino acids than the protein made by the eukaryotic cell. Why?
   a. The mRNA was not spliced as it is in eukaryotes.
   b. Eukaryotes and prokaryotes use different genetic codes.
   c. Repressor proteins interfered with transcription and translation.
   d. The lifetime of the bacterial mRNA was too short.
   e. Ribosomes were not able to bind to tRNA.

5. A homeotic gene does which of the following?
   a. It serves as the ultimate control for prokaryotic operons.
   b. It regulates the expression of groups of other genes during development.
   c. It represses the histone proteins that package eukaryotic DNA.
   d. It helps splice mRNA after transcription.
   e. It inactivates one of the *X* chromosomes in a female mammal.

6. All your cells contain proto-oncogenes, which can change into cancer-causing genes. Why do cells possess such potential time bombs?
   a. Viruses infect cells with proto-oncogenes.
   b. Proto-oncogenes are genetic "junk" with no known function.

c. Proto-oncogenes are unavoidable environmental carcinogens.
d. Cells produce proto-oncogenes as a by-product of mitosis.
e. Proto-oncogenes are necessary for normal control of cell division.

Describing, Comparing, and Explaining

1. What kinds of evidence demonstrate that differentiated cells in a plant or animal retain the same full genetic potential?

2. A mutation in a single gene may cause a major change in the body of a fruit fly, such as an extra pair of legs or wings. Yet it probably takes the combined action of hundreds or thousands of genes to produce a wing or leg. How can a change in just one gene cause such a big change in the body?

Thinking as a Scientist

1. You are studying the changing pattern of gene expression during the embryonic development of a certain animal. To follow the proteins produced, you are using a technique called gel electrophoresis. In this technique, proteins migrate through a gel in an electrical field (left to right in the figure below), and are visible as bands spanning each horizontal "lane." Larger proteins migrate more slowly. The thick bars at the left edge of the gel represent the places where the proteins start their migration. You have run a "standard" lane containing a mixture of known proteins of molecular weight 10,000, 20,000, 30,000, 40,000, and 50,000 (10K–50K). The other lanes contain proteins from a certain area on the embryo taken at 8-hour intervals after fertilization. As different genes turn on and off, the mixture of proteins from this area changes. You are interested in the disappearance of two early-stage proteins, with molecular weights of 13K and 22K, and the appearance of two proteins characteristic of a later stage of development, with weights 25K and 36K. Look at the gel below, and decide which statement best interprets the results. Give a short justification of why you chose it and why you rejected each of the other choices.

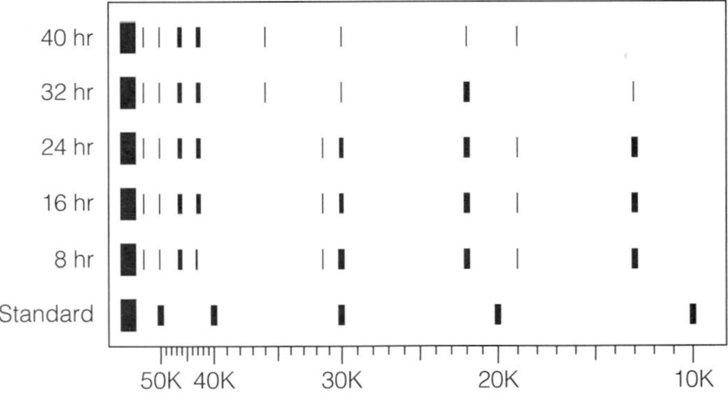

a. The 22K and 13K proteins continue to be made through 40 hours. The 36K and 25K proteins both start to be produced at 40 hours.
b. The 22K protein is not produced, and the 13K protein only appears at 40 hours. There is no sign of the "later" proteins (25K and 36K).
c. The 22K protein continues to be made up to 40 hours; the 13K protein fades out before that. The 36K protein starts to be produced at 32 hours. There is no sign of the 25K protein.
d. The 36K protein is made only from 8 to 24 hours, and the 25K protein first appears at 40 hours. The 22K and 13K proteins are produced from 24 hours on.

2. Study the illustrations of the *lac* operon in Module 11.1. Normally, the genes are turned off when lactose is not present. Lactose activates the genes, which code for enzymes that enable the cell to use lactose. Mutations can alter the function of this operon; in fact, it was the effects of various mutations that enabled Jacob and Monod to figure out how the operon works. Predict how the following mutations would affect the function of the operon in the presence and absence of lactose:

a. Mutation of regulatory gene; repressor will not bind to lactose.
b. Mutation of operator; repressor will not bind to operator.
c. Mutation of regulatory gene; repressor will not bind to operator.
d. Mutation of promoter; RNA polymerase will not attach to promoter.

3. Because a cat must have both orange and nonorange alleles to be tortoiseshell (see Module 11.5), we would expect only female cats, which have two X chromosomes, to be tortoiseshell. Normal male cats *(XY)* can carry only one of the two alleles. Male tortoiseshell cats are occasionally seen, although they are usually sterile. What might you guess their genotype to be? (*Hint:* Some of the abnormal human genotypes described in Module 8.22 can also be found in cats.)

Media Resources

For further review of the concepts in this chapter, use the activities and quizzes on the CD-ROM.

Visit the Special Edition of The Biology Place web site for activities, quizzes, news articles, and web links.

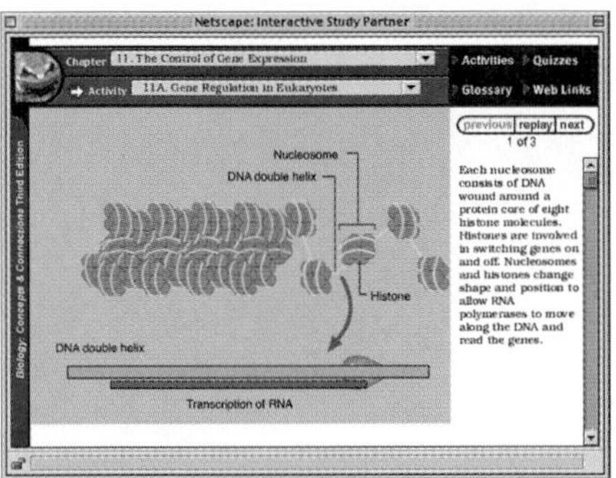

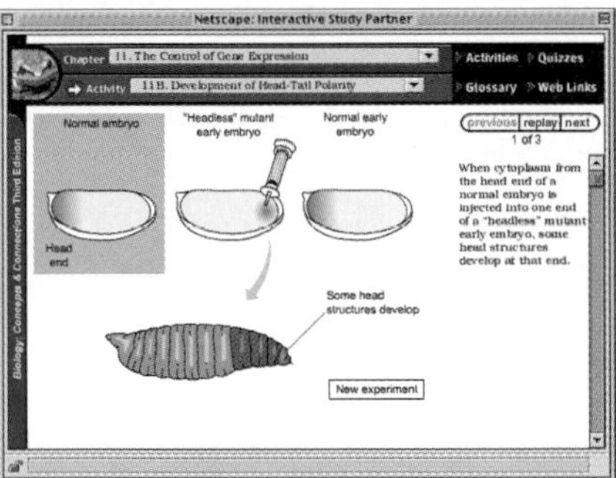

12 DNA Technology and the Human Genome

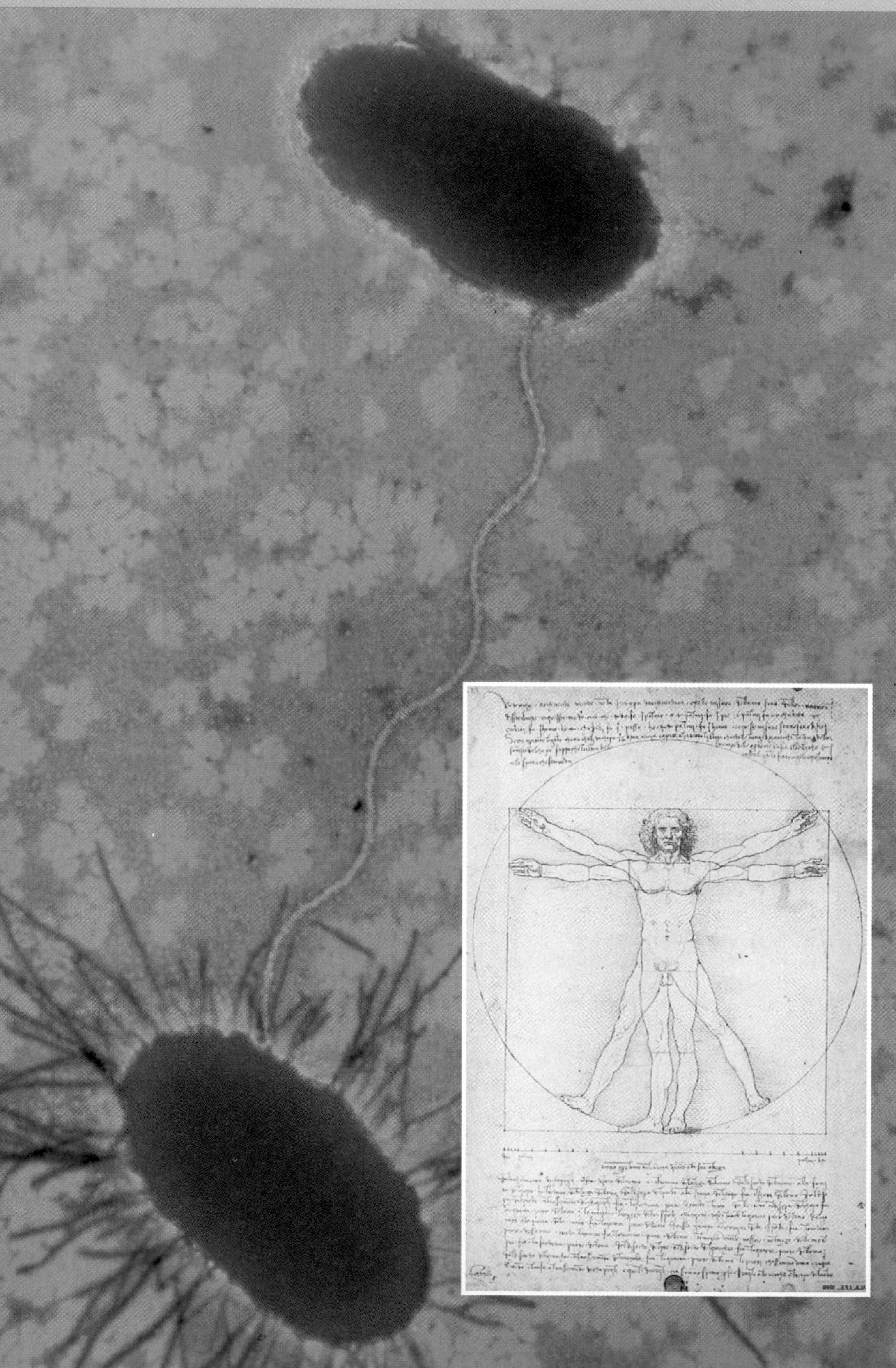

Colorized TEM 35,000×

Outline

Bacteria as Tools for Manipulating DNA

The scene in the background photograph at the left could be taking place right now in your large intestine. It is a pair of cells of the bacterium *Escherichia coli* in the process of mating. Bacteria have a sex life? Yes, *E. coli* bacteria can have an active one. The cells in this picture are joined by a long, threadlike sex pilus, an appendage of the "male" bacterium (the lower bacterium in the photo, which also has many shorter pili). The cells will soon join together, and the "male" will pass part of its DNA to the "female" through a cytoplasmic bridge.

Billions of *E. coli* inhabit the human large intestine (colon), where, along with countless other bacteria, they thrive on nutrients left over from our food. There are many different strains, or varieties, of *E. coli* that differ in their dietary requirements. Studies of such strains led to the discovery of bacterial sex and what it accomplishes. In 1946, American geneticists Joshua Lederberg and Edward Tatum concluded from experiments with *E. coli* that these bacteria have a sexual mechanism that results in the combining of genes from two different cells.

In their groundbreaking experiment, Lederberg and Tatum used two strains of *E. coli*. Each strain carried several mutations that prevented it from synthesizing some of the amino acids it needed. Because the two strains had mutations in different genes, they required the addition of different amino acids to their growth media. However, Lederberg and Tatum found that after samples of the two strains were mixed together, cells appeared that did not require the addition of *any* amino acids for growth. Apparently, cells of the two strains had somehow combined genes, creating a number of recombinant bacteria with functional versions of all the genes for amino acid synthesis.

Lederberg, Tatum, and their colleagues were pioneers of bacterial genetics, a field that within 20 years made *E. coli* the most thoroughly understood of all organisms at the molecular level. Eventually, in the 1970s, research on this bacterium led to the development of **recombinant DNA technology.** This is a set of techniques for combining, in a test tube, genes from different sources—even different species—and transferring the resulting recombinant DNA into cells, where it can be replicated and may be transcribed and translated into protein.

In the past two decades, **DNA technology**—including recombinant DNA technology and other methods for studying and manipulating DNA—has paid high dividends in basic research. For example, the noncoding sequences within eukaryotic genes were discovered using DNA technology. This technology has also helped uncover mutations that lead to cancer. Perhaps the most exciting use of DNA technology in basic research is the Human Genome Project, symbolized on the facing page by Leonardo da Vinci's drawing of a man. The Human Genome Project, the mapping of all the human DNA, is revealing the genetic basis of what it means to be human. On a more practical level, this ambitious endeavor is expected to help us understand and cure many diseases.

As we hear in the news, DNA technology is already widely used to engineer the genes of cultured cells for practical purposes. Scientists have genetically engineered bacteria to mass-produce many useful chemicals, from cancer drugs to pesticides; the photo on this page shows an industrial apparatus for growing engineered bacteria. Other kinds of engineered cells are

being used to clean up toxic wastes and improve crop plants.

This chapter examines the basic concepts of DNA technology, the Human Genome Project, other applications of DNA methods, and some social and ethical issues raised by these endeavors. We start with a look at some of the natural capabilities of bacteria, organisms that continue to play crucial roles in DNA technology. Among these capabilities are mechanisms for transferring DNA from one cell to another, which we discuss in the first two modules. We will then see how biologists use some of the molecules and processes of natural gene transfer to manipulate genes in the laboratory.

Bacteria as Tools for Manipulating DNA

12.1 In nature, bacteria can transfer DNA in three ways

The gene transfer process described in the chapter's opening essay involves the fusion of two bacterial cells and the movement of DNA from one cell to the other. After the DNA is transferred, the two cells separate. Unlike the sexual processes in plants and animals, bacterial mating is not reproductive, for the number of individuals does not increase. But although bacteria do not have meiosis, gametes, or fertilization, they do not lack for ways to produce new combinations of genes. In fact, in the bacterial world there are three mechanisms by which genes can move from one cell to another. These mechanisms of gene transfer are called transformation, transduction, and conjugation.

Most of a bacterium's DNA is found in a single bacterial chromosome, which is a closed loop of DNA with associated proteins. In the diagrams here, we show the chromosome much smaller than it actually is relative to the cell. A bacterial chromosome is hundreds of times longer than its cell; it fits inside the cell because it is highly folded.

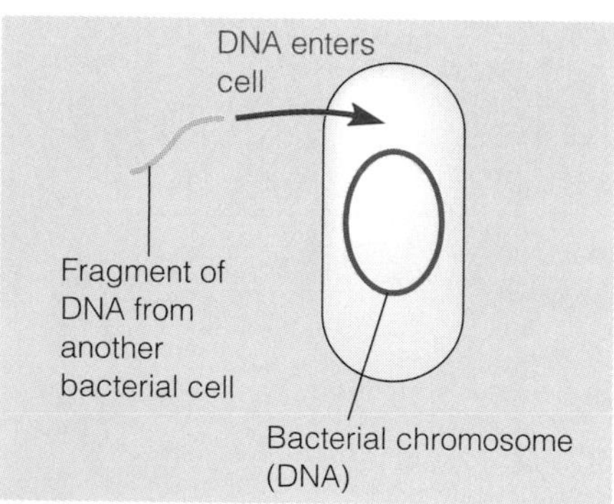

Figure 12.1A Transformation

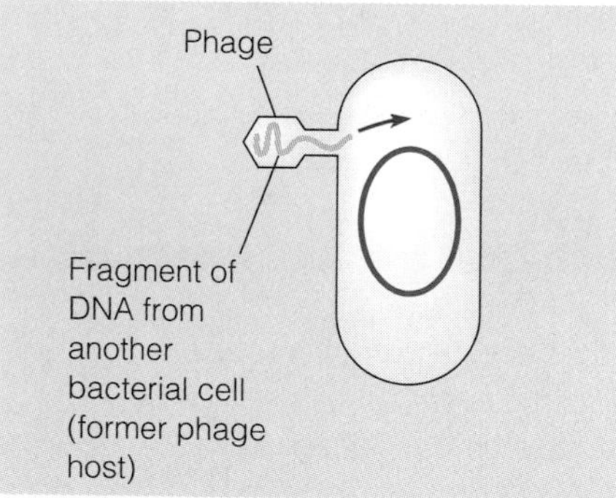

Figure 12.1B Transduction

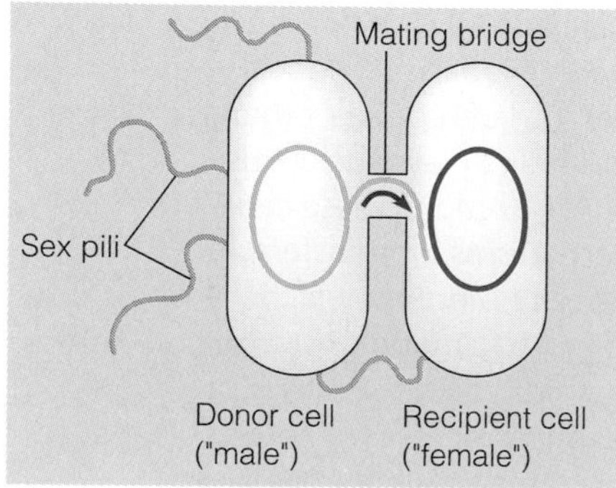

Figure 12.1C Conjugation

Figure 12.1A illustrates **transformation,** which is the taking up of DNA from the fluid surrounding a cell. This is what happened in the famous experiment with pneumonia-causing bacteria performed by Frederick Griffith in the 1920s (see Module 10.1). A harmless strain of bacteria took up pieces of DNA left from the dead cells of a disease-causing strain. The DNA from the pathogenic bacteria carried a gene that made the cells resistant to an animal's defenses, and the previously harmless bacteria could cause pneumonia in infected animals when they acquired this gene.

Bacteriophages provide the second means of bringing together genes of different bacteria. The transfer of bacterial genes by a phage is called **transduction.** In this process, a fragment of DNA belonging to a phage's previous host cell has been accidentally packaged within the phage's coat instead of phage DNA. When the phage infects a new bacterial cell, the DNA stowaway from the former host cell is injected into the new host (Figure 12.1B).

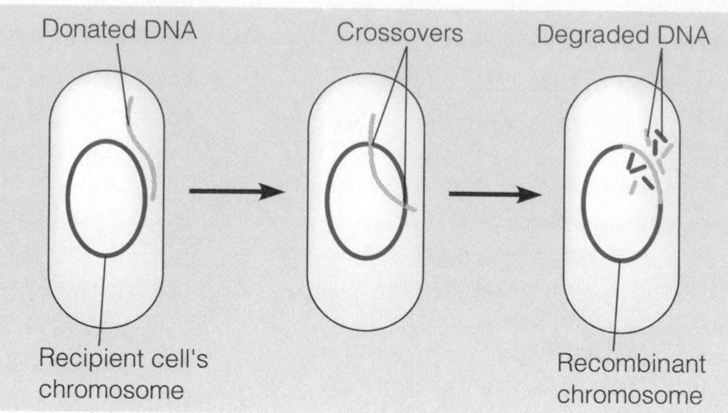

Figure 12.1D Integration of donated DNA into the recipient cell's chromosome

Figure 12.1C shows what happens at the DNA level when two bacterial cells mate, as we described in the opening essay. This union of cells and the DNA transfer between them is called **conjugation** (Latin *conjugatus*, united). Notice that the "male" donor cell has sex pili, one of which is attached to the "female" recipient cell. The outside layers of the cells have fused, and a cytoplasmic bridge has formed between them. Through this mating bridge, donor-cell DNA (bright blue in the figure) passes to the recipient cell. The donor cell replicates its DNA as it transfers it, so the cell doesn't end up lacking any genes. The DNA replication is a special type that allows one copy to peel off and transfer into the recipient cell.

Once new DNA gets into a bacterial cell, by whatever mechanism, part of it may then integrate into the recipient's chromosome. As Figure 12.1D indicates, integration occurs by crossing over between the two DNA molecules, a process similar to crossing over between eukaryotic chromosomes (see Module 8.18). Here we see that two crossovers result in a piece of the donated DNA replacing part of the recipient cell's original DNA. The leftover pieces of DNA are broken down, leaving the recipient bacterium with a recombinant chromosome. We look more closely at gene transfer by conjugation in the next module.

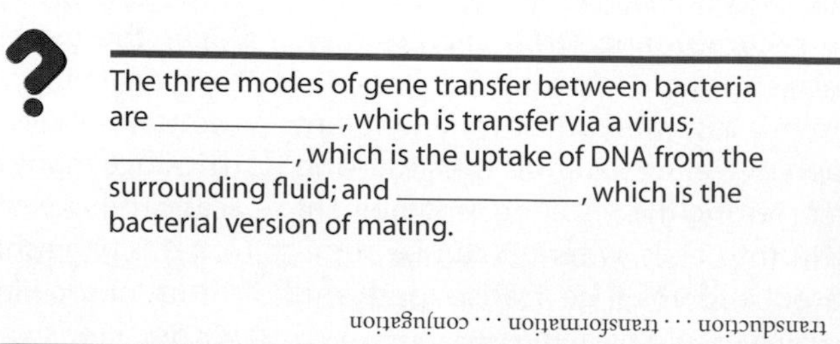

?

The three modes of gene transfer between bacteria are __________, which is transfer via a virus; __________, which is the uptake of DNA from the surrounding fluid; and __________, which is the bacterial version of mating.

transduction . . . transformation . . . conjugation

Bacterial plasmids can serve as carriers for gene transfer

The ability of a male *E. coli* cell to carry out conjugation is due to a specific piece of DNA, usually a piece of DNA called the **F factor** (F for fertility). The F factor carries genes for making sex pili and other things needed for conjugation; it also contains an origin of replication, where DNA replication can start.

Let's see how the F factor behaves during conjugation. In Figure 12.2A, the F factor (bright blue) is integrated into the male bacterium's chromosome. When this cell conjugates with a recipient cell, the male chromosome starts replicating at the F factor's origin of replication, indicated by the blue dot on the DNA. The growing copy of the DNA peels off the chromosome and heads into the recipient cell. Thus, part of the F factor serves as the leading end of the transferred DNA, but right behind it are genes from the donor's original chromosome. The rest of the F factor stays in the donor cell. Once inside the recipient cell, the transferred donor genes can recombine with the corresponding part of the recipient chromosome by crossing over. If crossing over occurs, the recipient cell may be genetically changed, but it usually remains female because the two cells break apart before the rest of the F factor transfers.

Alternatively, as Figure 12.2B shows, an F factor can exist as a **plasmid,** a small, circular DNA molecule separate from the much larger bacterial chromosome. Every plasmid has an origin of replication, required for its replication within the cell, and some plasmids, including the F-factor plasmid, can bring about conjugation and move to another cell. When the male cell in Figure 12.2B mates with a recipient cell, the F factor replicates and at the same time transfers one whole copy of itself, in linear rather than circular form, to the recipient cell. The transferred plasmid re-forms a circle in the recipient cell, and the cell becomes male. A transferable plasmid may carry genes other than those needed for replication and conjugation; when such a plasmid carries these extra genes to another cell, it is acting as a **vector.**

E. coli and other bacteria have many different kinds of plasmids. You can see several from one cell in Figure 12.2C, along with part of the bacterial chromosome, which extends in loops from the ruptured cell. Some plasmids carry genes that can affect the survival of the cell. Plasmids of one class, called **R plasmids,** pose serious problems for human medicine. Transferable R plasmids carry genes for enzymes that destroy antibiotics such as penicillin and tetracycline. Bacteria containing R plasmids are resistant (hence the designation R) to antibiotics that would otherwise kill them. The widespread use of antibiotics in medicine and agriculture has tended to kill off bacteria that lack R plasmids, while those with R plasmids have multiplied. As a result, an increasing number of bacteria that cause human diseases, such as food poisoning and gonorrhea, are becoming resistant to antibiotics. The transfer of R plasmids from resistant to nonresistant bacteria has contributed to the problem. However, R plasmids can be useful vectors for genetic engineering, in the role we describe next.

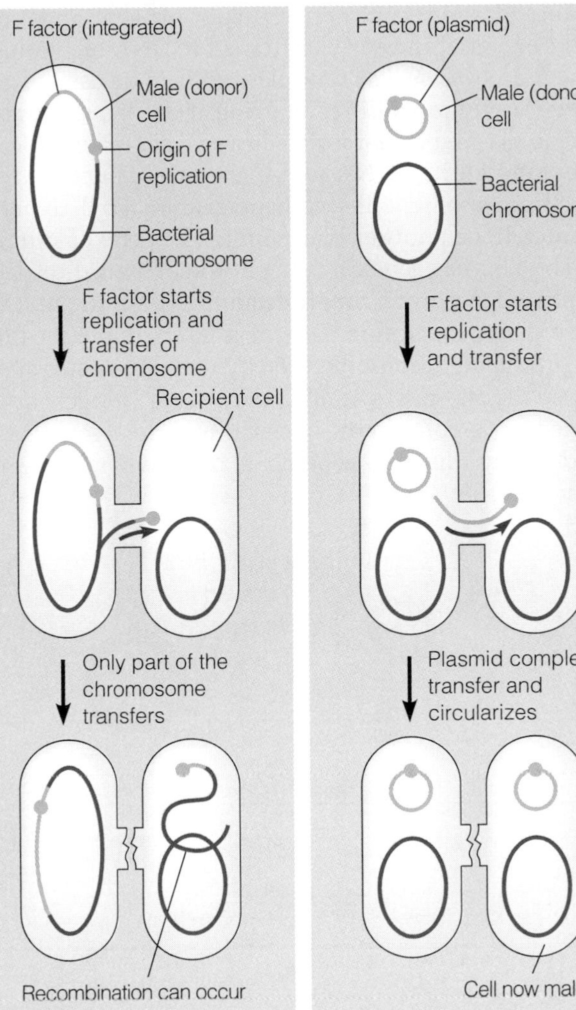

Figure 12.2A Transfer of chromosomal DNA by an integrated F factor

Figure 12.2B Transfer of an F-factor plasmid

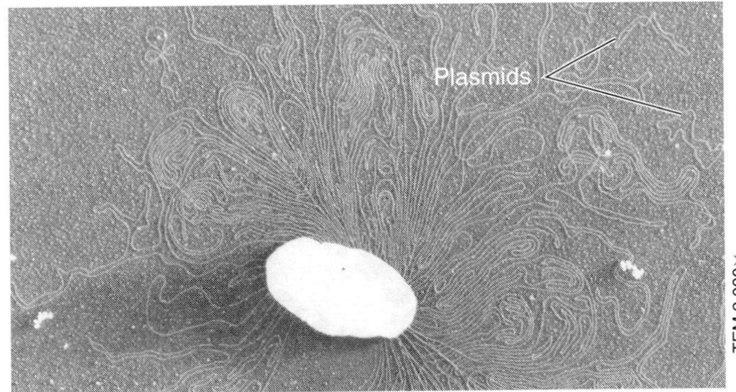

Figure 12.2C Plasmids and part of a bacterial chromosome released from a ruptured *E. coli* cell

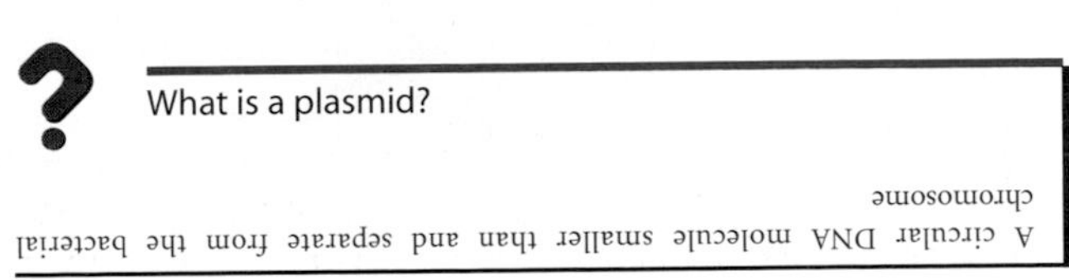

What is a plasmid?

A circular DNA molecule smaller than and separate from the bacterial chromosome

12.3 Plasmids are used to customize bacteria: An overview

12A
See Activity 12A on the CD-ROM.

Because plasmids can carry virtually any gene and replicate in bacteria, they are key tools for DNA technology. Below is an overview of how plasmids can be used to give bacteria useful capabilities. Starting at the top left, we see that ① a plasmid is first isolated from a bacterium. Meanwhile (top right), ② DNA carrying a gene of interest is obtained from another cell—perhaps an animal cell (as here), a plant cell, or another bacterium. The gene of interest could be a human gene encoding a protein of medical value or a plant gene conferring resistance to pest insects. ③ A piece of DNA containing the gene is inserted into the plasmid, producing recombinant DNA, and ④ a bacterial cell takes up the plasmid by transformation. ⑤ This genetically engineered, recombinant bacterium is then cloned (allowed to reproduce) to generate many copies of the gene.

At the bottom you can see a few applications of genetically engineered bacteria. In the examples on the left, copies of the gene itself are the desired product. In those on the right, the protein product of the gene is harvested from bacteria. Many pharmaceutical companies employ biologists to customize bacteria for making specific products. In fact, genetic engineering has launched a revolution in **biotechnology,** the use of organisms to perform practical tasks. In the next several modules, we examine in more detail some tools and procedures of this technology.

How does the rapid reproduction of bacteria make them a good choice for cloning a foreign gene?

A foreign gene located within plasmid DNA inside a bacterium is replicated each time the cell divides, resulting in rapid accumulation of many copies of the gene.

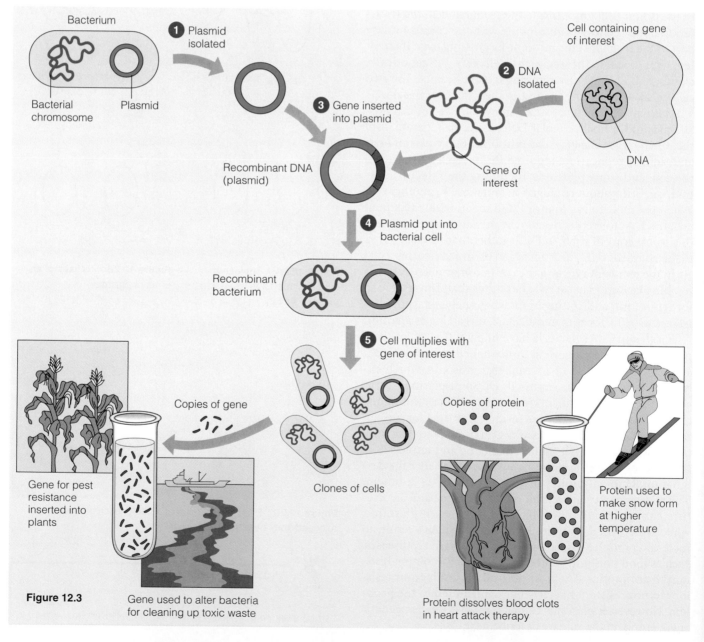

Figure 12.3

Enzymes are used to "cut and paste" DNA 12.4

12A

Extracting a gene from one DNA molecule and inserting it into another require precise "cutting and pasting." To carry out the procedure outlined in Module 12.3, a piece of DNA containing the gene of interest must be cut out of a chromosome and "pasted" into a bacterial plasmid.

The cutting tools for making recombinant DNA in a test tube are bacterial enzymes called **restriction enzymes,** which were first discovered in the late 1960s. In nature, these enzymes protect bacteria against intruding DNA from other organisms and phages. They work by chopping up the foreign DNA, a process called restriction because it *restricts* foreign DNA from surviving in the cell. Other enzymes chemically modify the cell's own DNA in a way that protects it from the restriction enzymes.

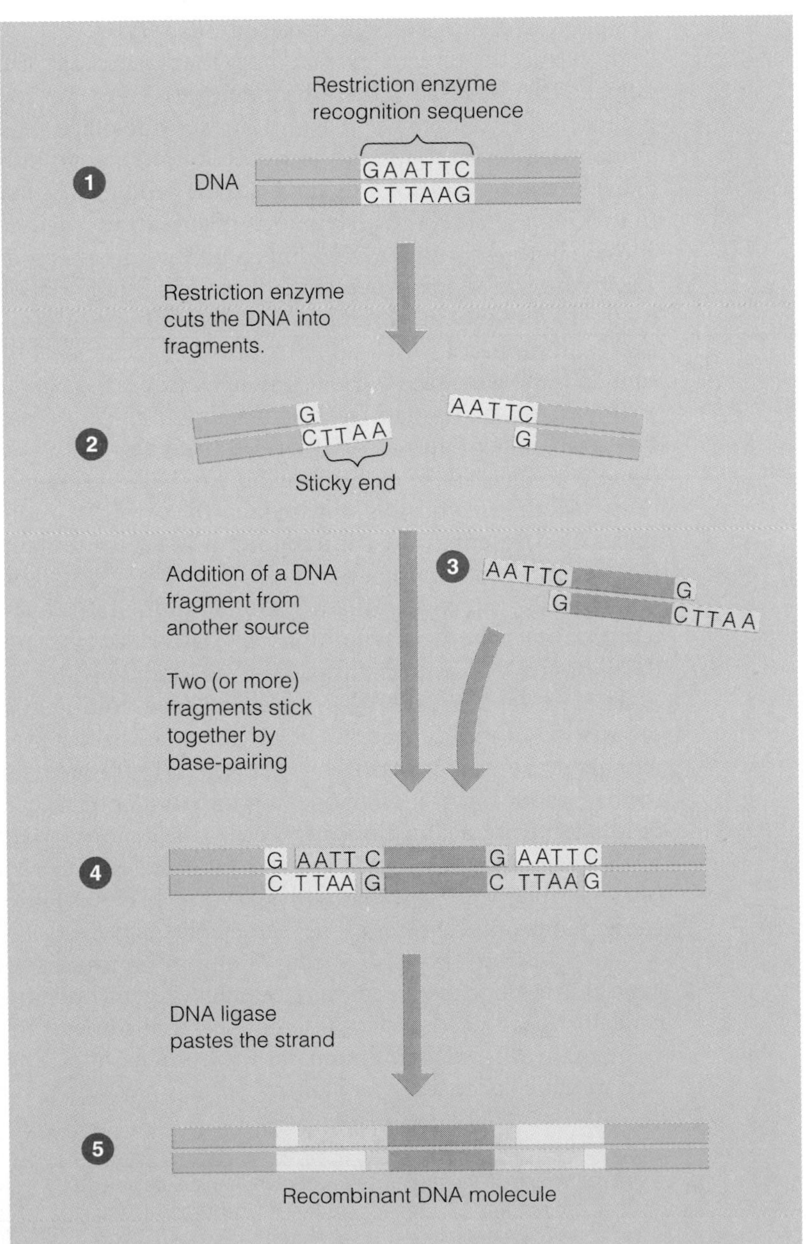

Figure 12.4 Creating recombinant DNA using a restriction enzyme and DNA ligase

Most restriction enzymes recognize short nucleotide sequences in DNA molecules and cut at specific points within these recognition sequences. Several hundred restriction enzymes and about a hundred different recognition sequences are known. In the figure here, step ① shows a piece of DNA containing one recognition sequence for a particular restriction enzyme. In this case, the restriction enzyme will cut the DNA strands between the bases A and G within the recognition sequence. (The places where DNA is cut are called restriction sites.) ② The staggered cuts yield two double-stranded DNA fragments with single-stranded ends, called "sticky ends." Sticky ends are the key to joining DNA restriction fragments originating from different sources. These short extensions will form hydrogen-bonded base pairs with complementary single-stranded stretches of DNA.

Step ③ shows a piece of DNA (gray) from another source. Notice that the gray DNA has single-stranded ends identical in base sequence to the sticky ends on the blue DNA. The gray, "foreign" DNA has ends with this particular base sequence because it was cut from a larger molecule by the same restriction enzyme used to cut the blue DNA. ④ The complementary ends on the blue and gray fragments allow them to stick together by base-pairing. (The hydrogen bonds that hold the base pairs together are not shown.)

The union between the blue and gray DNA fragments shown in step 4 is only temporary, because only a few hydrogen bonds hold the fragments together. The union can be made permanent, however, by the "pasting" enzyme **DNA ligase.** This enzyme, which the cell normally uses in DNA replication (see Module 10.5), catalyzes the formation of covalent bonds between adjacent nucleotides, sealing the breaks in the DNA strands. ⑤ The final outcome is **recombinant DNA,** a DNA molecule carrying a new combination of genes.

Our molecular toolbox now contains enzymes for cutting and pasting DNA, plasmids for carrying that DNA into cells and providing an origin of replication, and bacterial cells for replicating the DNA. We are ready to make multiple copies of a gene. 

? In making recombinant DNA, what is the benefit of using a restriction enzyme that cuts DNA in a staggered fashion?

Such an enzyme creates DNA fragments with "sticky ends," single-stranded regions whose unpaired bases can hydrogen-bond to the complementary sticky ends of other fragments created by the same enzyme.

12.5 Genes can be cloned in recombinant plasmids: A closer look

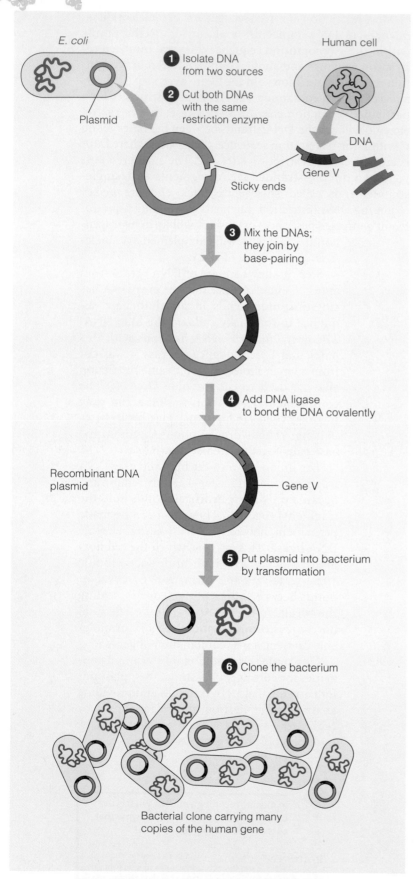

Figure 12.5 Cloning a gene in a bacterial plasmid

Making recombinant DNA in large enough amounts to be useful requires several steps. Consider a typical genetic engineering challenge: A molecular biologist at a pharmaceutical company has identified a human gene that codes for a valuable product—a hypothetical substance called protein V that kills certain human viruses. The biologist wants to set up a system for making large amounts of the gene, so that the protein can be manufactured on a large scale. Figure 12.5 illustrates a way to make many copies of the gene using the techniques we have been discussing.

In step ①, the biologist isolates two kinds of DNA: the bacterial plasmid that will serve as the vector, and human DNA containing the protein-V gene. In this example, the DNA containing the gene of interest comes from human tissue cells that have been growing in laboratory culture. The plasmid comes from the bacterium *E. coli.*

In step ②, the researcher treats both the plasmid and the human DNA with the same restriction enzyme. An enzyme is chosen that cleaves the plasmid in only one place. The human DNA, with thousands of restriction sites, is cut into many fragments, one of which carries the protein-V gene. In making the cuts, the restriction enzyme creates sticky ends on both the human DNA fragments and the plasmid. The figure shows the processing of just one human DNA fragment and one plasmid, but actually millions of plasmids and human DNA fragments (most of which do not contain the V gene) are treated simultaneously.

In step ③, the human DNA is mixed with the cut plasmid. The sticky ends of the plasmid base-pair with the complementary sticky ends of the human DNA fragment. In step ④, the enzyme DNA ligase joins the two DNA molecules by covalent bonds, and the result is a recombinant DNA plasmid containing gene V.

In step ⑤, the recombinant plasmid is added to a bacterium. Under the right conditions, the bacterium takes up the plasmid DNA by transformation (see Module 12.1).

The last step here, step ⑥, is the actual **gene cloning,** the production of multiple copies of the gene. The bacterium, with its recombinant plasmid, is allowed to reproduce. As the bacterium forms a cell clone (a group of identical cells descended from a single ancestral cell), any genes carried by the recombinant plasmid are also "cloned" (copied). In our example, the biologist will grow a cell clone large enough to produce protein V in marketable quantities.

The procedure here is called a "shotgun" approach to gene cloning because it doesn't actually target a specific gene. Instead it yields many different bacterial clones, carrying many different fragments of the human DNA. The next module discusses these clones. 

? Which of the three modes of gene transfer described in Module 12.1 is exploited by scientists to introduce a foreign gene into a bacterial cell?

1 Isolate DNA from two sources

4 Add DNA ligase to bond the DNA covalently

Transformation

Cloned genes can be stored in genomic libraries 12.6

The bacterial cells making up each clone produced in the shotgun procedure of Module 12.5 have identical recombinant plasmids, each carrying the same fragment of human DNA. The entire collection of cloned DNA fragments from such a shotgun experiment, in which the starting material is bulk DNA from whole cells, is called a **genomic library.** A typical cloned DNA fragment is big enough to carry one or a few genes, and together the fragments include the entire genome of the organism from which the DNA was derived—in our example, a human.

Bacterial plasmids are one type of vector that can be used in the shotgun cloning of genes, but not the only type. Phages can also serve as vectors. When a phage is used, the DNA fragments are inserted into phage DNA molecules. Each recombinant DNA molecule is then put into a bacterial cell. There it replicates and produces many new phages (a clone) carrying the same foreign DNA "passenger."

Figure 12.6 illustrates the two types of genomic libraries. On the left, the red, yellow, and green DNA segments (three of the thousands of different library "books") are shelved in plasmids in bacterial cells. On the right, the same books are shelved in phages. In Module 12.8, you will learn how to find a book in this library. But first, in the next module, we look at another source of DNA for cloning.

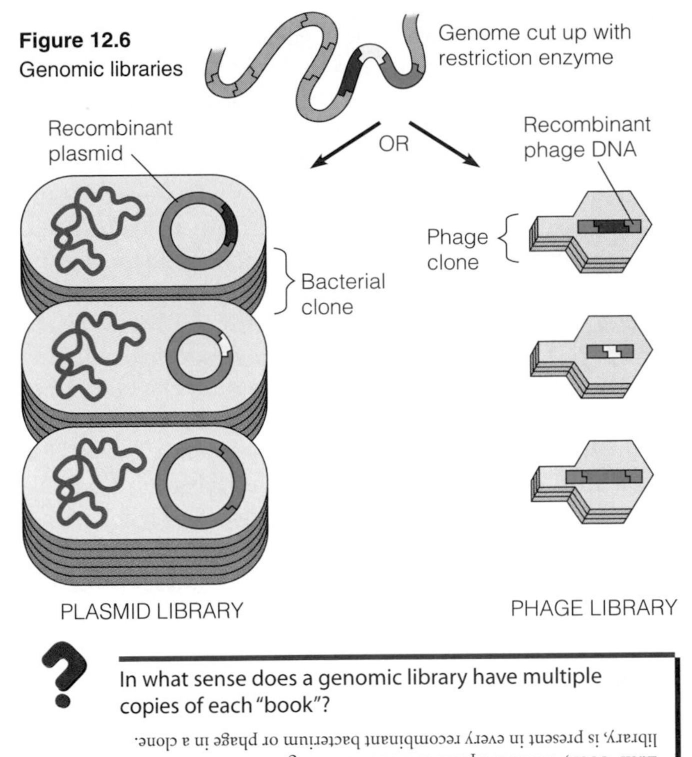

Figure 12.6
Genomic libraries

? In what sense does a genomic library have multiple copies of each "book"?

Each "book," which is a piece of DNA from the genome that was the source of the library, is present in every recombinant bacterium or phage in a clone.

Other Tools of DNA Technology

Reverse transcriptase helps make genes for cloning 12.7

Not all DNA that is cloned comes directly from cells. Rather than starting with an entire eukaryotic genome, researchers can focus in on the genes *expressed* in a particular kind of cell (for example, muscle cells) by using mRNA as the starting material. As reviewed in steps ① and ② of Figure 12.7, the cells transcribe the genes and process the transcripts to produce mRNA. The researcher then isolates the mRNA and, with appropriate enzymes, uses it as template for synthesizing DNA (steps ③ – ⑤). The key enzyme is reverse transcriptase, obtained from retroviruses (see Module 10.21), because making DNA on an RNA template is the reverse of transcription.

The DNA that results from this procedure represents pure genes—and only the genes actually transcribed in the starting cells. Furthermore, because these artificial genes lack introns, they can be correctly transcribed and translated by bacterial cells, which do not have RNA-splicing machinery. Thus, after the genes are cloned, the "host" bacteria can make the eukaryotic protein product.

Why is an artificial gene made using reverse transcriptase often shorter than the natural form of the gene?

Because it does not contain introns

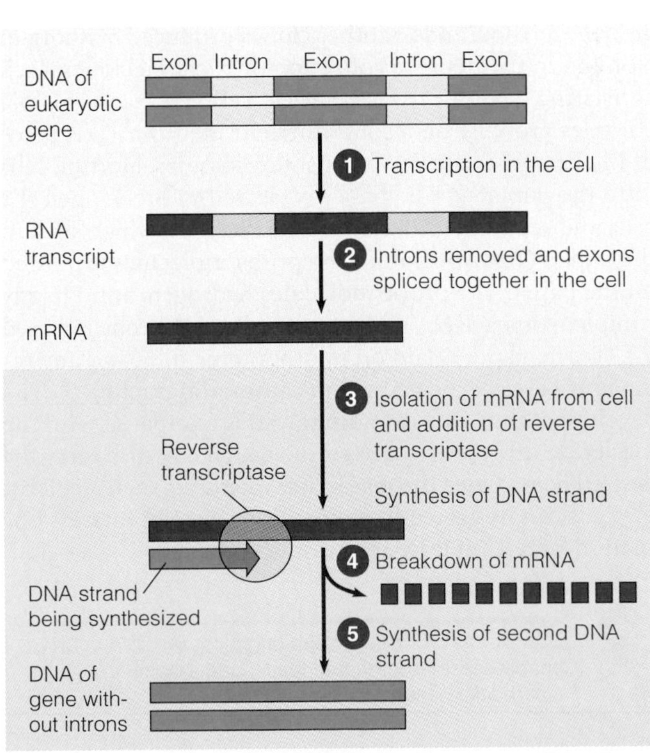

Figure 12.7 Making an intron-lacking gene from eukaryotic mRNA

12.8 Nucleic acid probes identify clones carrying specific genes

Often the most difficult task in gene cloning is finding the right shelf in a genomic library—that is, identifying the bacterial or phage clone containing a desired gene. If bacterial clones containing a specific gene actually translate the gene into protein, they can be identified by testing for the protein. This is not always the case, however. Fortunately, researchers can also test directly for the gene itself.

Methods for detecting genes all depend on base-pairing between the gene and a complementary sequence on another nucleic acid molecule, either DNA or RNA. When at least part of the nucleotide sequence of a gene is already known or can be guessed, this information can be used to advantage. Taking a simplified example, if we know that our hypothetical gene V contains the sequence TAGGCT, a biochemist can use nucleotides labeled with a radioactive isotope to synthesize a short single strand of DNA with a complementary sequence (ATCCGA). This sort of labeled nucleic acid molecule is called a **probe** because it is used to find a specific gene or other nucleotide sequence within a mass of DNA. (In practice, a probe molecule would be considerably longer than six nucleotides.)

Figure 12.8A shows how a probe works. The DNA sample to be tested is treated with heat or alkali to separate the DNA strands. When the radioactive DNA probe is added to these strands, it tags the correct molecule—finds the needle in the haystack—by hydrogen-bonding to the complementary sequence in the gene of interest.

Nucleic acid probes are powerful tools that do not require the targeted DNA to be a pure preparation. Figure 12.8B shows how a researcher might actually use such a probe to find a bacterial clone carrying a gene of interest among the thousands of other clones produced by shotgun cloning. In the figure, a collection of bacterial clones, each consisting of millions of identical cells, appear as visible colonies growing on a solid nutrient medium. ① A piece of filter paper is pressed against the colonies, blotting cells onto the paper. ② The paper is treated to break open the cells and separate the strands of their DNA, which stick to the paper. ③ A solution of the probe molecules is poured on the paper. The probe molecules hydrogen-bond to any complementary DNA sequences, and excess probe is rinsed off. ④ The paper is laid on photographic film, and any radioactive areas expose the film (autoradiography). ⑤ The developed film, an autoradiograph, is compared with the master culture plate to determine which colonies carry the desired gene. Once the researcher identifies such a colony, the cells can be grown further and the gene of interest isolated in large amounts.

How does a probe consisting of radioactive DNA or RNA enable a researcher to find the bacterial clones carrying a particular gene?

The probe molecules bind to and label DNA only in the specific bacterial clone containing the gene of interest.

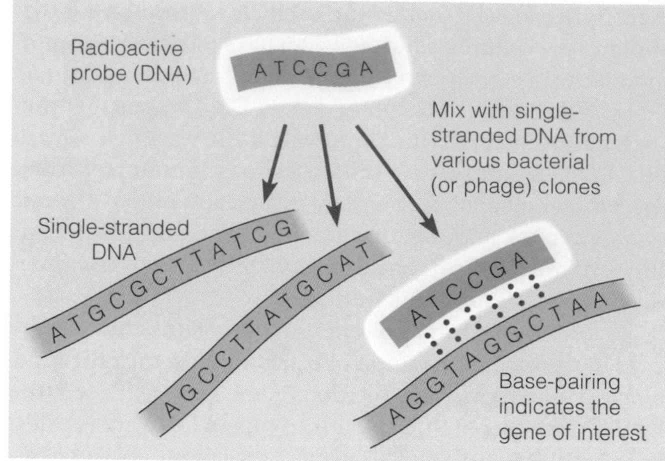

Figure 12.8A How a DNA probe tags a gene

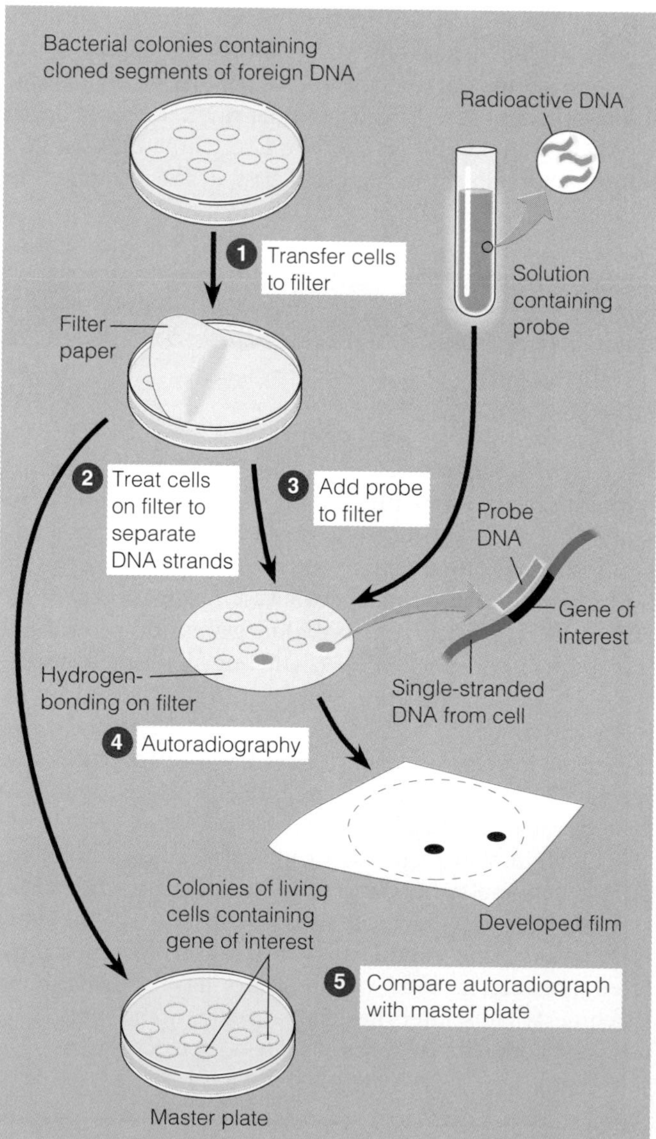

Figure 12.8B Using a DNA probe to identify a bacterial clone (colony) carrying a specific gene

Automation makes rapid DNA synthesis and DNA sequencing possible 12.9

As mentioned in the last module, nucleic acid probes can be chemically synthesized in the laboratory. In some cases, scientists have made entire genes in this way, without using mRNA or any other nucleic acid template. If the order of amino acids in a protein is known, it is easy to use the genetic code (see Figure 10.8) to determine a nucleotide sequence that would code for it. More difficult is actually synthesizing gene-length DNA. Among the first genes synthesized artificially were two that encoded the two polypeptide chains making up human insulin. (In nature, the insulin polypeptides are encoded in one long gene; see Module 11.8.) The task of making these first artificial genes was arduous, feasible only because the two genes together had only about 150 nucleotides. Today, DNA-synthesizing machines are available that can rapidly produce DNA hundreds of nucleotides long.

Automation has also revolutionized DNA sequencing, the process of determining the exact sequence of nucleotides in a stretch of DNA. Sequencing even a short gene used to be very difficult, but thanks to methods developed in the 1970s, the sequences of even large genes can now often be determined in a day or so. DNA sequencing is made possible by restriction enzymes, which are used to cut the very long DNA of cells into discrete fragments that can be analyzed by machine.

Thousands of DNA sequences are now being collected in computer data banks, and they are proving of great value for understanding genes and gene control, as well as for biotechnology. With the help of computers, long sequences can be scanned to find particular genes or control sequences, such as promoters. Computers also help researchers compare DNA of different species.

Both DNA sequencing and virtually every kind of DNA analysis depends on the technique we describe next.

Your assignment is to synthesize a gene for a specific human protein and then clone the gene in bacteria. What do you need to know about the protein to make the synthetic DNA?

Its amino acid sequence (primary structure)

Gel electrophoresis sorts DNA molecules by size 12.10

An essential tool of DNA technology, **gel electrophoresis** is a method for physically sorting macromolecules—proteins or nucleic acids—primarily on the basis of their electric charge and size. The figure below shows how we would use gel electrophoresis to separate the various DNA molecules in three different mixtures. A sample of each mixture is placed in a well at one end of a flat, rectangular gel, a thin slab of jellylike material. The gel is supported by glass plates. A negatively charged electrode is then attached to the DNA-containing end of the gel and a positive electrode to the other end. Because the DNA molecules have negative charge due to their phosphate groups, they move through the gel toward the positive pole. However, the longer DNA molecules are held back by the molecules of the gel, so they move more slowly, and therefore not as far in a given time period, as the shorter DNA molecules. When the current is turned off, the result is a series of bands in each "lane" of the gel. Each band consists of DNA molecules of one size.

(a) What causes DNA molecules to move toward the positive pole during electrophoresis? (b) Why do large molecules move more slowly than smaller ones?

(a) The negatively charged phosphate groups of the DNA are attracted to the positive pole. (b) The gel resists their movement.

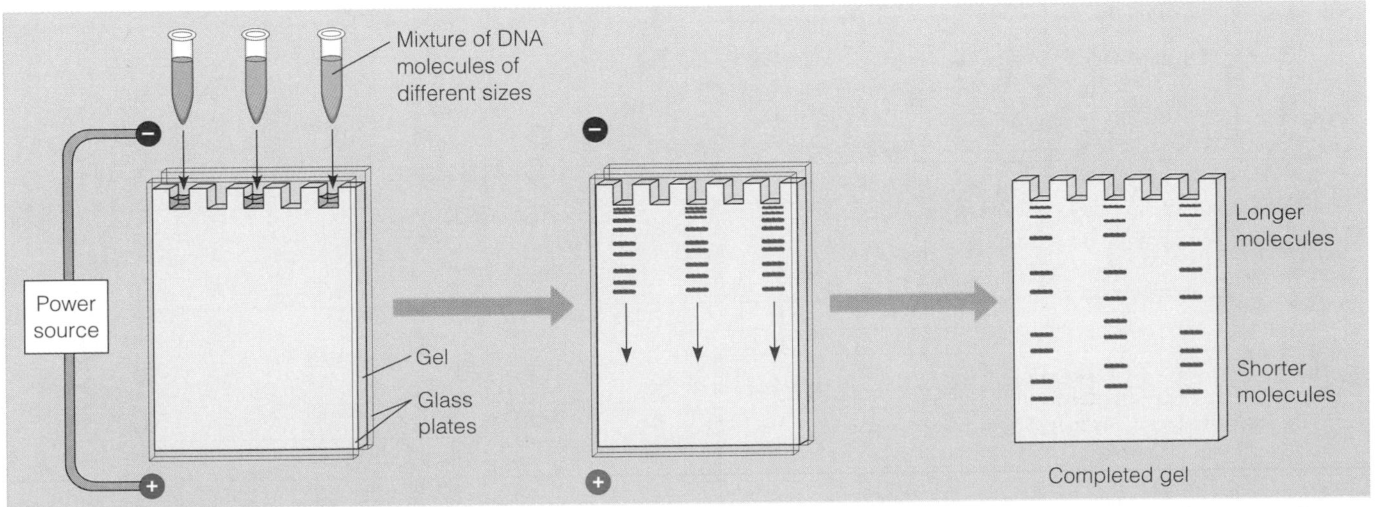

Figure 12.10 Gel electrophoresis of DNA

12.11 Restriction fragment analysis is a powerful method that detects differences in DNA sequences

Unless you have an identical twin, your DNA is different from everyone else's; its total nucleotide sequence is unique. Some of your DNA consists of your particular sets of alleles for all the human genes, and even more of it is composed of noncoding regions. Whether a segment of DNA codes for amino acids or not, it is inherited just like any other part of a chromosome. For this reason, geneticists can use *any* DNA segment that varies from person to person as a **genetic marker,** a chromosomal landmark whose inheritance can be studied. And just like a gene, a noncoding segment of DNA is more likely to be an exact match to the comparable segment in a relative than to the segment in an unrelated individual.

With the help of restriction enzymes and gel electrophoresis, a molecular geneticist could demonstrate the uniqueness of your DNA and also identify particular features of your DNA sequence that are shared by members of your family. To do this, he would extract DNA from some of your cells and treat it with a restriction enzyme, creating a mixture of DNA pieces called **restriction fragments.** *The number of restriction fragments and their sizes reflect the specific sequence of nucleotides in your DNA.* To understand this statement, we need to examine the figures below, which illustrate the principles underlying restriction fragment analysis.

Preparing Restriction Fragments In Figure 12.11A, we see corresponding segments of DNA from two chromosomes; let's imagine that they are two different alleles, 1 and 2, of the same gene. Allele 1 might be from one of your chromosomes; allele 2 might be from your homologous chromosome or from another person. Notice that the alleles differ by a single base pair (highlighted in yellow). In this case, the restriction enzyme cuts the DNA between two cytosine (C) bases in the sequence CCGG or in its complement, GGCC. Because allele 1 has two recognition sequences for the restriction enzyme, it is cleaved in two places, yielding three restriction fragments (labeled *w*, *x*, and *y*). Allele 2, however, has only one recognition sequence and yields only two restriction fragments (*z* and *y*). Notice that fragments *w* and *x* from allele 1 differ in length from both of the allele 2 fragments. In other words, the lengths of restriction fragments, as well as their numbers, differ depending on the exact sequence of bases in the DNA.

Analyzing Restriction Fragments by Electrophoresis

To detect the differences between the collections of restriction fragments that come from alleles 1 and 2, we need to separate the restriction fragments in the two mixtures and compare their lengths. We can accomplish these things with the method described in Module 12.10, gel electrophoresis. As shown in Figure 12.11B, the restriction fragments from allele 1 separate into three bands in the gel, while those from allele 2 separate into only two bands. Notice that the smallest fragment from allele 1 (*y*) produces a band at the same location as the identical small fragment from allele 2. Thus, electrophoresis allows us to see similarities as well as differences between mixtures of restriction fragments—and thus similarities as well as differences between the base sequences in DNA from two individuals.

The differences in DNA sequences on comparable (homologous) chromosomes that result in restriction fragments of different lengths are called restriction fragment length polymorphisms, or **RFLPs** (pronounced "rif-lips"). The DNA segments being compared may be alleles of a gene, as in our example, or noncoding DNA that carries inherited differences in nucleotide sequence. (*Polymorphism* refers to the *many forms* of such sequences found in the population.) Scientists usually use the term *RFLP analysis* for the comparison of the restriction fragment patterns produced by DNA from different individuals.

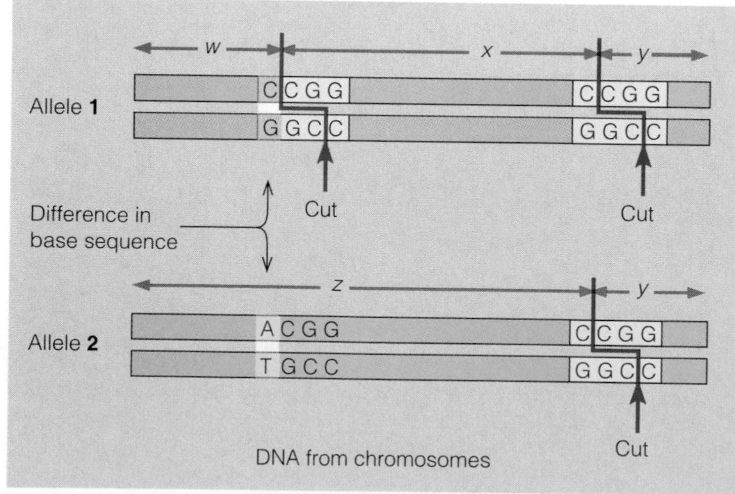

Figure 12.11A Restriction-site differences between two alleles

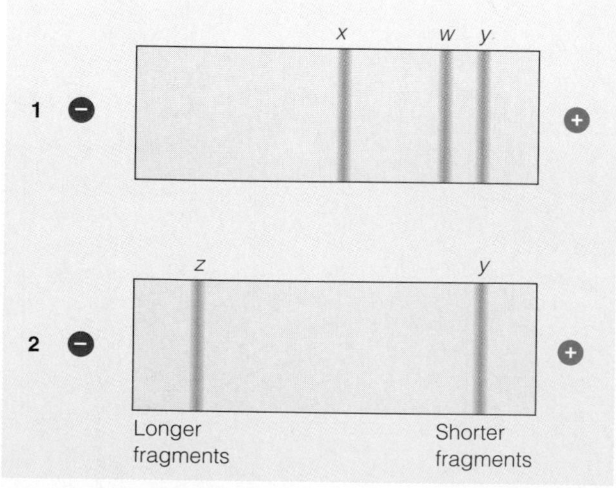

Figure 12.11B Electrophoresis of restriction fragments

Using RFLP Analysis to Detect Harmful Alleles An important application of RFLP analysis is the detection of potentially harmful alleles in heterozygous individuals who are free of symptoms. The heterozygotes may be carriers of a harmful recessive allele, such as for cystic fibrosis, or a dominant allele that is not expressed until later in life, such as for Huntington's disease (see Module 9.9). A key to detecting the harmful allele is that, within a particular family, the allele for a disease is generally identical in all family members who carry it. That allele often contains one or more restriction sites that are different from the ones in the normal allele.

Once the restriction fragment patterns for the normal and harmful alleles are known, RFLP analysis can be used to test family members who are suspected carriers of the harmful allele. Figure 12.11C shows the use of RFLP analysis on blood samples from three relatives: a known carrier and two suspected carriers. Notice that the five steps all involve techniques that you have seen before. The entire procedure is often called **Southern blotting,** after E. M. Southern, who developed it. Note the crucial blotting step (step 3, which is similar to the first step in Figure 12.8B). The paper to which the DNA is transferred (blotted) holds the DNA stationary while it is being exposed to the probe.

RFLP analysis not only allows the testing of possible carriers for a defective gene, it can also help researchers pinpoint the chromosomal location of such a gene. The method has enabled researchers to map a number of genes whose defective alleles cause fatal diseases, including cystic fibrosis and Huntington's disease.

Restriction fragment analysis requires only about 1 μg of DNA—the amount in a small drop of blood. Still, even this tiny amount of DNA may not be available. Fortunately, another powerful method, the topic of the next module, can help out in those situations.

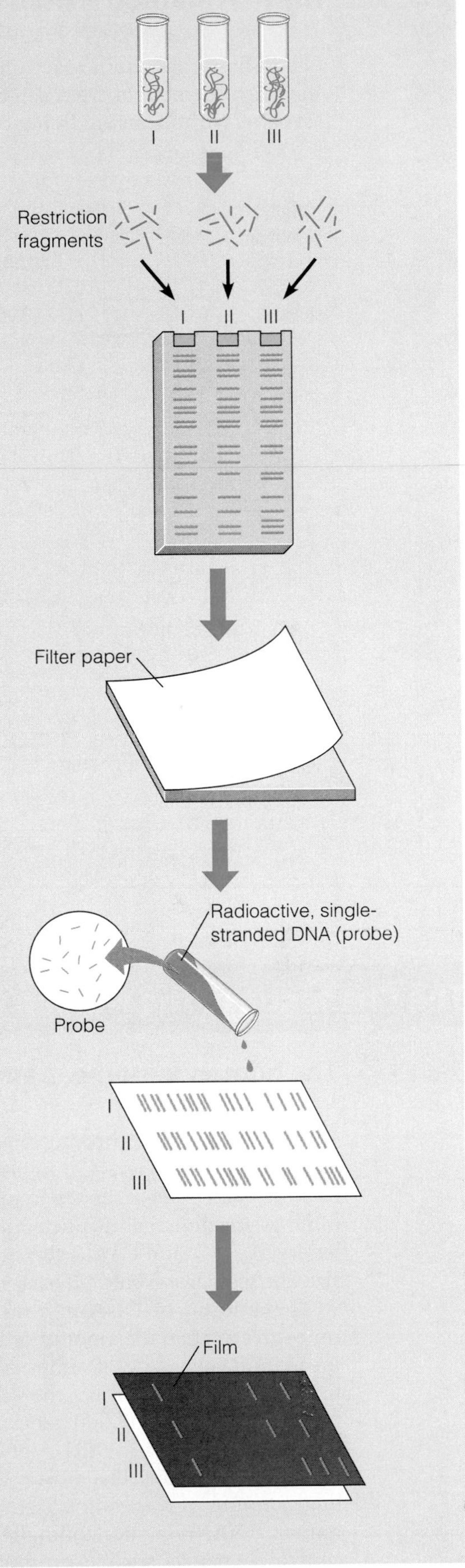

1 **Restriction fragment preparation**. DNA is extracted from white blood cells taken from individuals I, II, and III. Individual I is known to carry a disease allele. A restriction enzyme is added to the three samples of DNA to produce restriction fragments.
I
II
III
Restriction fragments
2 **Gel electrophoresis**. The mixture of restriction fragments from each sample is separated by electrophoresis. Each sample forms a characteristic pattern of bands. (Since all the DNA in the blood cells was used, there would be many more bands than shown here.)
I
II
III
3 **Blotting**. The DNA bands are treated to separate the strands of the molecules, and the single strands are transferred by simple blotting onto special filter paper.
Filter paper
4 **Radioactive probe**. The paper blot is immersed in a solution of a radioactive probe, a single-stranded DNA molecule that is complementary to the DNA sequence of the genetic marker of interest. The probe attaches by base-pairing to restriction fragments arising from the marker DNA. (Thus it may stick to several bands in a lane.)
Radioactive, single-stranded DNA (probe)
Probe
I
II
III
5 **Detection of radioactivity (autoradiography)**. The unattached probe is rinsed off, and a sheet of photographic film is laid over the paper. The radioactivity in the probe exposes the film to form an image corresponding to specific bands—the bands containing DNA that base-pairs with the probe. In this example, the band pattern is the same for individuals I and II but different for individual III. Since we know that individual I carries the disease allele, the test shows that individual II is also a carrier but that individual III is not.
Film
I
II
III

Figure 12.11C A procedure for restriction fragment analysis (RFLP analysis)

? You use a restriction enzyme to cut a DNA molecule. The base sequence of this DNA is known, and the molecule has a total of three restriction sites, clustered close together near one end. When you separate the restriction fragments by electrophoresis, how do you expect the bands to be distributed along the electrophoresis lane?

Three bands near the positive pole (small fragments) and one band near the negative pole (large fragment)

12.12 The PCR method is used to amplify DNA sequences

The **polymerase chain reaction (PCR)** is a technique by which any segment of DNA can be amplified (cloned) in a test tube, without using living cells. In principle, PCR is surprisingly simple. The DNA is mixed with a special heat-resistant version of the DNA-replication enzyme DNA polymerase, nucleotide monomers, and a few other crucial ingredients. In this mixture, the DNA replicates, producing two daughter DNAs; the daughter molecules, in turn, replicate; and the process continues as long as nucleotides remain. Each time replication occurs, the amount of DNA doubles, as indicated in Figure 12.12. (Colors distinguish new and old DNA strands for each replication.) Starting with a single DNA molecule, PCR can generate 100 billion similar molecules in a few hours—enough for RFLP analysis or certain other uses. This is a much shorter time than the days it takes to clone a piece of DNA by attaching it to a plasmid or viral genome. Moreover, PCR can copy a specific segment from within a mass of DNA.

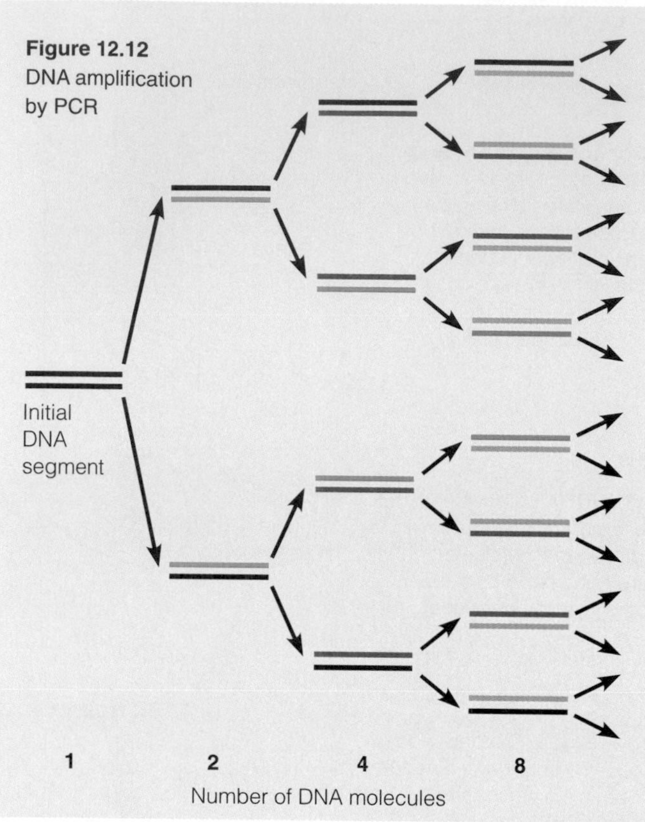

Figure 12.12 DNA amplification by PCR

PCR is being applied in a great number of ways. In evolution research, the technique has been used to amplify DNA pieces recovered from an ancient mummified human, from a 40,000-year-old woolly mammoth frozen in a glacier, and from a 30-million-year-old plant fossil. In medicine, PCR has been used to amplify DNA from single embryonic cells for rapid prenatal diagnosis, and it has made possible the detection of viral genes in cells infected with HIV. We'll discuss other applications of PCR later.

If it takes 3 minutes for each replication cycle of a PCR procedure, how many copies of DNA can be produced in 18 minutes from a single starting DNA molecule?

64

The Challenge of the Human Genome

12.13 The human genome is enormous, and most of the DNA does not consist of genes

Within the set of 23 chromosomes making up the haploid human genome, there are approximately 3 billion nucleotide pairs of DNA. To try to get a sense of this quantity of DNA, imagine that its nucleotide sequence is printed in letters (A, T, C, and G) like the letters in this book. At this size, the sequence would fill over 500 volumes!

The amount of DNA in a human cell is about 1000 times greater than the amount of DNA in *E. coli*. Does this mean humans have 1000 times as many genes as the 2000 in *E. coli?* The answer is no; the human genome carries between 50,000 and 100,000 genes, which code for various proteins, as well as for tRNA and rRNA. In addition to these genes, humans, like most complex eukaryotes, have a huge amount of noncoding DNA, about 97% of the total human DNA. Some noncoding DNA is made up of gene-control sequences such as promoters and enhancers. The remaining DNA has been called "junk DNA," but it is more accurate to say that we do not yet understand its functions. This DNA includes introns (whose total length may be ten times greater than the exons of a gene) and noncoding DNA located between genes.

Much of the DNA between genes consists of **repetitive DNA,** nucleotide sequences present in many copies in the genome. There are two main types of repetitive DNA. In one type, a unit of just a few nucleotide pairs is repeated many times in a row. Stretches of DNA with thousands of such repetitions are prominent at the centromeres and ends of chromosomes, suggesting that this DNA plays roles in chromosome structure. It may help keep the rest of the DNA properly organized during DNA replication and mitosis. Recent research supports the idea that the repetitive DNA at chromosome ends—called **telomeres**—also has a protective function; a significant loss of telomeric DNA quickly leads to cell death. Furthermore, abnormal lengthening of this DNA may help "immortal" cancer cells evade normal cell aging.

A number of genetic disorders affecting the nervous system are caused by abnormally long stretches of repeated nucleotide triplets. One is Huntington's disease (see Module 9.9), in which a long string of CAG triplets is actually located within a coding region of the gene. The protein produced has a long string of the amino acid glutamine.

In the second main type of repetitive DNA, each repeated unit is hundreds of nucleotide pairs long, and the copies are scattered around the genome. Scientists know little about the functions of this DNA, but they do have an idea how it came to be both abundant and dispersed in genome. Most of these sequences seem to be associated with transposons, the "jumping genes" discussed next. 

? About how many nucleotide pairs are present in the average human chromosome?

1.3×10^8, or 130 million

TALKING ABOUT SCIENCE

Barbara McClintock discovered "jumping genes" 12.14

In the 1940s, studying inheritance in corn plants, American geneticist Barbara McClintock (Figure 12.14A) made a startling discovery. She found that certain genetic elements (now known to be DNA segments) can move from one location to another in a chromosome and even from one chromosome to another. McClintock discovered that these "jumping genes," now called **transposons,** can land in the middle of other genes and disrupt them. In Indian corn, for instance, transposons can disrupt pigment genes in some of the cells, leading to spotted kernels like the ones in Figure 12.14B.

McClintock worked largely alone, and few other geneticists appreciated the significance of her discoveries until the 1970s. By that time, transposons had been found in *E. coli* and the era of recombinant DNA technology was beginning. Finally, in 1983, McClintock received a Nobel Prize for her pioneering work.

Current evidence suggests that all organisms, prokaryotic and eukaryotic alike, have transposons. Figure 12.14C shows the simplest kind of transposon and how its movement can interfere with another gene. As part ① indicates, this transposon includes one gene (purple), encoding an enzyme that catalyzes movement of the transposon. Identical noncoding nucleotide sequences (yellow) mark the ends of the transposon. In part ②, the enzyme attaches to the ends of the transposon and simultaneously to another site on the DNA. It cuts the DNA and catalyzes insertion of the transposon at the new site. (The DNA of the transposon is never free in the cell.) ③ Insertion of the transposon at its new site disrupts the nucleotide sequence of the gene shown in green.

In addition to the cut-and-paste type of transposon shown here, molecular biologists have discovered copy-and-paste transposons, which leave the original copy behind when they move. It is this type of transposon that is apparently responsible for the proliferation of dispersed repetitive DNA in the human genome.

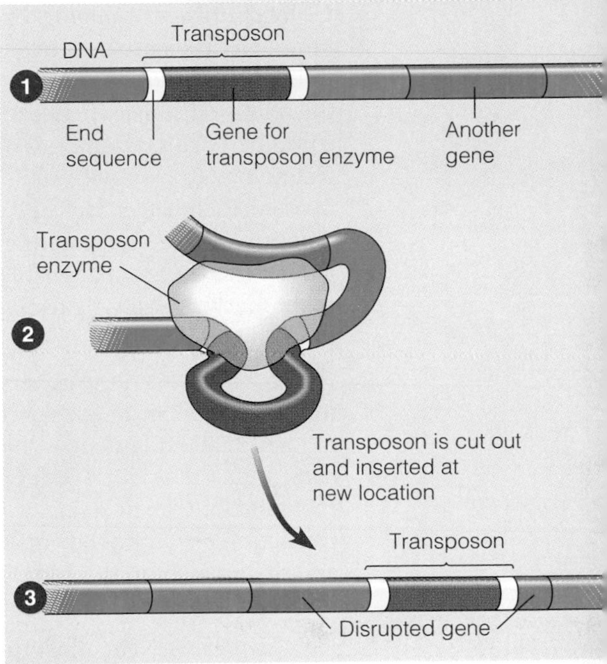

DNA
Transposon
1
End sequence
Gene for transposon enzyme
Another gene
Transposon enzyme
2
Transposon is cut out and inserted at new location
Transposon
3
Disrupted gene

Figure 12.14C One type of transposon movement

Realizing that transposons act as natural mutagens, McClintock suggested that they may help generate genetic diversity and could thus be a significant factor in evolution. In recent years, transposons have been implicated in some cases of cancer (see Module 11.13).

One of the great biologists of the twentieth century, Barbara McClintock continued working until her death at age 90, in 1992. Like Mendel, she was one of those rare scientists who have profound insights years ahead of their time. 

Figure 12.14A Barbara McClintock in 1947

Figure 12.14B Corn kernels with spots caused by transposons

? How might the movement of a transposon contribute to the development of cancer? (*Hint:* See Module 11.13.)

By interrupting and thereby inactivating a tumor-suppressor gene, or by bringing a proto-oncogene close to a control element that excessively activates transcription

12.15 The Human Genome Project is a scientific adventure

Only 20 years ago, the task of completely deciphering the DNA blueprint for human life was almost inconceivable. But by 1990, advances in DNA technology had led an international consortium of scientists to launch the **Human Genome Project,** with the ultimate goal of determining the nucleotide sequence of all the DNA in the human genome. Researchers now expect the project to be completed well before its target date of 2005, and the results to date have already yielded exciting insights into the genes of humans and other organisms.

The mapping of the human genome involves three major stages:

1. *Genetic (linkage) mapping.* In Module 9.17, you learned how geneticists use data from genetic crosses to determine the order of linked genes on a chromosome and the relative distances between the genes. Scientists have combined pedigree analysis of large families with DNA technology to map over 5000 genetic markers. These markers include genes, RFLP sites, and stretches of repetitive DNA. The resulting low-resolution map provides a framework that enables researchers to map other markers easily, by testing for genetic linkage to the known markers.
2. *Physical mapping.* Researchers use restriction enzymes to break the DNA of each chromosome into a number of identifiable fragments, which they clone. They then determine the original order of the fragments in the chromosome. Their strategy: Using several different restriction enzymes, they make fragments that overlap and then match up the ends. They use probes to relate the fragments to the markers mapped in stage 1.
3. *DNA sequencing.* The most time-consuming part of the project is determining the nucleotide sequences of a set of DNA fragments covering the entire genome, ideally the fragments already mapped in stage 2. Advances in automatic DNA sequencing have been crucial to this endeavor. Figure 12.15 shows the computer screen of a DNA sequencer. Sequencers use gel electrophoresis; notice the band patterns. At the bottom of the screen is a partial sequence.

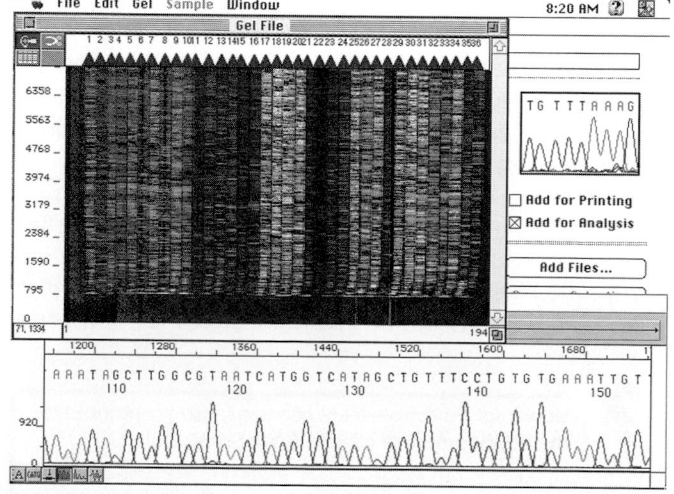

Figure 12.15 Computer screen from a DNA sequencer

While some scientists are mapping the human genome, others are pursuing another important goal of the project, mapping the genomes of certain other important research organisms: *E. coli* and other prokaryotes, yeast, nematode, fruit fly, and mouse. Work on these genomes has been useful for developing the strategies and techniques needed to map the much larger human genome. Moreover, comparative analysis of the genes of other species helps scientists interpret the human data. For example, when scientists find a nucleotide sequence in the human genome similar to a yeast gene whose function is known, they have a valuable clue to the function of the human sequence. Many genes of disparate organisms are turning out to be astonishingly similar, to the point that one researcher has joked that he now views fruit flies as "little people with wings."

As of early 1999, the genomes of almost two dozen organisms have been fully sequenced. So far, most are prokaryotes, including *E. coli,* a number of other bacteria (some of medical importance), and several archaea. Yeast was the first eukaryote to have its sequence completed and the nematode *Caenorhabditis elegans,* a simple worm, the first multicellular organism. Meanwhile, work on the human sequence is ahead of schedule.

The potential benefits of having a complete map of the human genome are great. For basic science, the information is already providing insight into such fundamental mysteries as embryonic development and evolution. For human health, the identification of genes will aid in the diagnosis, treatment, and possibly prevention of many of our more common ailments, including heart disease, allergies, diabetes, schizophrenia, alcoholism, Alzheimer's disease, and cancer. Roughly 200 disease-associated genes have already been identified as a result of the project.

The DNA sequences are deposited in a database that is available to researchers all over the world via the Internet. The scientists use software to scan and analyze the sequences for genes, control elements, and other features. Then comes the most exciting challenge: figuring out the functions of the genes and other sequences and how they work together to direct the structure and function of a living organism. This challenge and the applications of the new knowledge should keep scientists busy well into the twenty-first century. 

? How is the Internet facilitating the Human Genome Project?

By making the database of DNA sequences easily accessible to researchers all over the world (It also helps researchers communicate directly with each other.)

DNA technology is used in courts of law 12.16

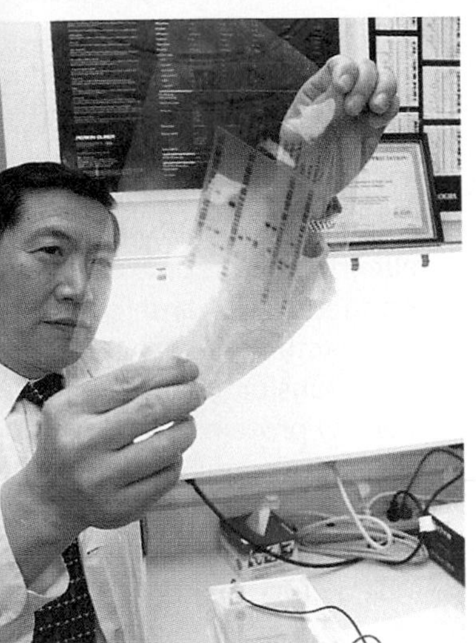

Figure 12.16A DNA data for forensic use

"DNA FINGERPRINTING SOUGHT IN MURDER PROSECUTION" blare the headlines. In the last few years, the news media have heralded the use of DNA technology as a new tool for forensic (legal) science. From small-town attorneys' offices to the FBI, lawyers, judges, and juries are having to learn about the new molecular techniques. Figure 12.16A shows the head of the Connecticut State Forensic Laboratory examining DNA data that will be stored in a state database.

In violent crimes, blood or small fragments of other tissue may be left at the scene of the crime or on the clothes or other possessions of the victim or assailant. If rape is involved, small amounts of semen may be recovered from the victim's body. If enough tissue or semen is available, forensic scientists can determine its blood type or tissue type using older methods that test for proteins. However, such tests require a relatively large amount of fairly fresh tissue. Also, because there are many people in the population with the same blood type or tissue type, this approach can only *exclude* a suspect; it cannot prove guilt.

DNA testing, on the other hand, can theoretically identify the guilty individual with certainty because the DNA sequence of every person is unique (except for identical twins). RFLP analysis is a powerful method for the forensic detection of similarities and differences in DNA samples and requires only tiny amounts of blood or other tissue (about 1000 cells). For a murder case, for example, RFLP analysis can be used to compare DNA samples from the defendant, the victim, and bloodstains on the defendant's clothes (Figure 12.16B). Radioactive probes mark the electrophoresis bands that contain certain RFLP markers. Usually about five markers are tested, only a few selected portions of the DNA. However, even such a small set of markers from an individual can provide a **DNA fingerprint,** or specific pattern of bands, that is of forensic use, because the probability that two people would have exactly the same set of RFLP markers is very small. The autoradiograph in Figure 12.16B resembles the type of evidence presented (with explanation) to juries in murder trials. As you can see, DNA from blood on the defendant's clothes matches the DNA of the victim but differs from that of the defendant, providing strong evidence of guilt.

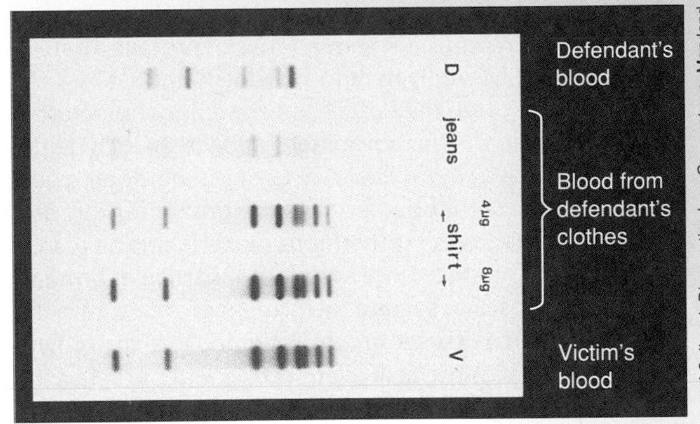

Courtesy of Cellmark Diagnostics Inc., Germantown, Maryland.

Figure 12.16B DNA fingerprints from a murder case

The forensic use of DNA fingerprinting extends beyond violent crimes. For instance, comparing the DNA of a mother, her child, and the purported father can conclusively settle a question of paternity. Sometimes paternity is of historical interest: Recently, DNA fingerprinting provided strong evidence that Thomas Jefferson fathered at least one of the children of his slave Sally Hemings.

Today, the markers most often used in DNA fingerprinting are inherited variations in the lengths of repetitive DNA (see Module 12.13). For example, one person may have the nucleotides AC repeated 65 times at one genome locus, 118 times at a second locus, and so on, whereas another person is likely to have different numbers of repeats at these loci. PCR is often used to amplify particular repetitive-DNA loci before electrophoresis. With PCR, a quantity of DNA sufficient for analysis can be generated from a tissue sample as small as 20 cells!

Just how reliable is DNA fingerprinting? In most legal cases, the probability of two people having identical DNA fingerprints is between one chance in 100,000 and one in a billion. The exact figure depends on how many markers are compared and on how common those markers are in the population. Growing collections of data on the frequencies of various markers—both in the population as a whole and in particular ethnic groups—are enabling forensic scientists to make extremely accurate statistical calculations. Thus, despite problems that can arise from insufficient statistical data, human error, or flawed evidence, DNA fingerprints are now accepted as compelling evidence by legal experts and scientists alike. Many argue that DNA evidence is more reliable than eyewitnesses in placing a suspect at the scene of a crime. 

? Why is only the slightest trace of DNA at a crime scene sufficient for forensic analysis?

Because PCR can produce enough copies for analysis

12.17 Bacteria, yeasts, and mammalian cells are used to mass-produce gene products

DNA technology has been used to create many useful pharmaceutical products. Most of these products are proteins. By transferring the gene for a desired protein product into a bacterium, yeast, or other kind of cell that is easy to grow, one can produce large quantities of proteins that are present naturally in only minute amounts.

Bacteria are often the best organisms for manufacturing a protein product. Major advantages of bacteria include the plasmids and phages available for use as gene cloning vectors and the fact that bacteria can be grown rapidly and cheaply in large tanks. Furthermore, bacteria can be readily engineered to produce large amounts of particular proteins and in some cases to secrete the protein products into the medium in which they are grown. Secretion into the growth medium simplifies the task of collecting and purifying the products. As the table below shows, a number of proteins of importance in human medicine and agriculture are being produced in the bacterium *E. coli*.

Despite the advantages of bacteria, it is sometimes desirable or necessary to use eukaryotic cells to produce a protein product. Often the first-choice eukaryotic organism for protein production is the yeast *Saccharomyces cerevisiae*, the single-celled fungus featured in Chapter 6's opening essay. As bakers and brewers have recognized for centuries, yeast cells are easy to grow. And like *E. coli*, yeast cells can take up foreign DNA and integrate it into their genomes. Yeast also have plasmids that can be used as gene vectors, and yeast are often better than bacteria at synthesizing and secreting eukaryotic proteins. *S. cerevisiae* is currently used to produce a number of proteins. In some cases, the same product (for example, interferons used in cancer research) can be made in either yeast or bacteria. In other cases, such as the hepatitis B vaccine, yeast alone is used.

The cells of choice for making some gene products come from mammals (see the table). Genes for these products are often cloned in bacteria as a preliminary step. For example, the genes for colony-stimulating factor (CSF) and for two proteins that affect blood clotting, Factor VIII and t-PA, are cloned in a bacterial plasmid before transfer to mammalian cells for large-scale production. Many proteins that mammalian cells normally secrete are glycoproteins, meaning that they have chains of sugars attached to the protein. Because only mammalian cells can attach the sugars correctly, mammalian cells must be used for making these products.

Why can't glycoproteins be mass-produced by engineered bacteria or yeast cells?

Because bacteria and yeast cells cannot correctly attach the sugar groups to the protein of glycoproteins

Some Protein Products of Recombinant DNA Technology

Product	Made in	Use
Human insulin	*E. coli*	Treatment for diabetes
Human growth hormone (GH)	*E. coli*	Treatment for growth defects
Epidermal growth factor (EGF)	*E. coli*	Treatment for burns, ulcers
Tumor necrosis factor	*E. coli*	Killing of certain tumor cells
Interleukin-2 (IL-2)	*E. coli*	Possible treatment for cancer
Prourokinase	*E. coli*	Treatment for heart attacks
Porcine growth hormone (PGH)	*E. coli*	Improving weight gain in hogs
Bovine growth hormone (BGH)	*E. coli*	Improving weight gain in cattle
Cellulase	*E. coli*	Breaking down cellulose for animal feeds
Taxol	*E. coli*	Treatment for ovarian cancer
Interferons (alpha and gamma)	*S. cerevisiae; E. coli*	Possible treatment for cancer and virus infections
Hepatitis B vaccine	*S. cerevisiae*	Prevention of viral hepatitis
Colony-stimulating factor (CSF)	Mammalian cells	Treatment for leukemia; boosts resistance to AIDS and other infectious diseases
Erythropoietin (EPO)	Mammalian cells	Treatment for anemia
Factor VIII	Mammalian cells	Treatment for hemophilia
Tissue plasminogen activator (t-PA)	Mammalian cells	Treatment for heart attacks

DNA technology is revolutionizing the pharmaceutical industry and human medicine 12.18

Commercial applications of DNA technology have not proceeded as far or fast as some biologists once predicted (to the dismay of certain Wall Street investors). Nevertheless, the new technology has already had a major impact on the pharmaceutical industry and on human medicine.

Consider the first two products in the table on the previous page, human insulin and human growth hormone (GH). These were the first pharmaceutical products made using recombinant DNA technology. In the U.S. alone, about two million people with diabetes depend on insulin treatment. Before 1982, the main sources of this hormone were pig and cattle tissues obtained from slaughterhouses. Insulin extracted from these animals is chemically similar, but not identical, to human insulin, and it causes harmful side effects in some people. Genetic engineering has largely solved this problem by developing bacteria that actually synthesize and secrete human insulin.

GH was harder to produce than insulin because the GH molecule is about twice as big. Since growth hormones from other animals are not effective growth stimulators in humans, however, GH was urgently needed. The gene that codes for GH consists of nearly 600 nucleotides. In 1985, molecular biologists produced GH in *E. coli* after combining a DNA fragment taken from human cells with a chemically synthesized piece of DNA. Before this genetically engineered version became available in 1985, children with a GH deficiency had to rely on scarce supplies from human cadavers or else face dwarfism.

DNA technology is also helping medical researchers develop vaccines. A **vaccine** is a harmless variant or derivative of a pathogen (usually a bacterium or virus) that is used to prevent an infectious disease. When a person—a potential host for the pathogen—is inoculated, the vaccine stimulates the immune system to develop lasting defenses against the pathogen. Especially for the many viral diseases for which there is no effective drug treatment, prevention by vaccination is virtually the only medical way to fight the disease.

Figure 12.18 Equipment used in the production of a vaccine against hepatitis B

Genetic engineering can be used in several ways to make vaccines. One approach is to use genetically engineered cells to produce large amounts of a protein molecule that is found on the pathogen's outside surface. This method has been used to make the vaccine against hepatitis B virus. Hepatitis is a disabling and sometimes fatal liver disease, and the hepatitis B virus can cause liver cancer. The photograph on this page shows a tank for growing yeast cells that have been engineered to carry hepatitis B genes.

Another way to use DNA technology in vaccine development is to make a harmless artificial mutant of the pathogen by altering one or more of its genes. When a harmless mutant is used as a vaccine, it multiplies in the body and may trigger a stronger immune response than the protein-molecule type of vaccine. Artificial-mutant vaccines may cause fewer side effects than those that have traditionally been made from natural mutants.

Yet another scheme for making vaccines employs a virus related to the one that causes smallpox. Smallpox was once a dreaded human disease, but it was eradicated worldwide in the 1970s by widespread vaccination with a harmless variant (natural mutant) of the smallpox virus. Using this harmless virus, genetic engineers can replace some of the genes encoding proteins that induce immunity to smallpox with genes that induce immunity to other diseases. In fact, the virus could be engineered to carry the genes needed to vaccinate against several diseases simultaneously. In the future, one inoculation may prevent a dozen diseases.

Among the potential new treatments for viral diseases and some cancers are short, synthetic pieces of single-stranded nucleic acid. Readily taken up by cells, these molecules are designed to base-pair with specific sequences on the DNA or mRNA in cancer cells or cells infected with viruses. One type, called **antisense nucleic acid,** is engineered to bind to, and prevent translation of, mRNA molecules coding for disease-causing proteins. Trials of antisense nucleic acid are promising, and along with new vaccines, they will probably be part of the future of human medicine.

In the next two modules, we move from the genetic engineering of microbes and single cells to the use of DNA technology to alter the genotypes and phenotypes of plants and animals.

Human GH produced by DNA technology is used in the treatment of ________________.

dwarfism

12.19 For agriculture, DNA technology helps make growth-enhancing products and plants and animals with desired traits

Scientists concerned with feeding the growing human population are using DNA technology to improve the productivity of plants and animals important to agriculture.

Plant Crops Agricultural scientists have already provided a number of crop plants with genes for desirable traits such as delayed ripening and resistance to spoilage and disease. In one striking way, plants are easier to engineer than most animals. For many plant species, a single tissue cell grown in culture can give rise to an adult plant (see Figure 11.3C). Thus, genetic manipulations can be performed on a single cell and the cell then used to generate an organism with new traits.

The vector used to introduce new genes into plant cells is most often a plasmid from the soil bacterium *Agrobacterium tumefaciens*. This is the **Ti plasmid,** so called because in nature it induces tumors in plants infected by the bacterium. For using Ti as a vector, researchers have eliminated its tumor-causing properties while keeping its ability to transfer DNA into plant cells.

Figure 12.19 shows how a plant with a new trait can be created using the Ti plasmid. ① With the help of a restriction enzyme and DNA ligase, the gene for the desired trait (red) is inserted into a segment of the plasmid called T DNA (black). ② Then the recombinant plasmid is put into a plant cell, where the T DNA carrying the new gene integrates into a plant chromosome. ③ Finally, the recombinant cell is cultured and grows into a whole plant. If the newly acquired gene is from another species, the recombinant organism is called a **transgenic organism.**

The Ti vector does not work with many grain-producing species. However, researchers can make transgenic varieties of these plants by using a "gene gun" to fire pieces of foreign DNA directly into cultured cells.

Genetic engineering is quickly replacing traditional plant breeding programs, especially in cases where useful traits are determined by one or only a few genes. Of the transgenic plants now in field trials, over 40% have received genes for herbicide resistance. For example, several companies have developed varieties of cotton carrying a bacterial gene that makes the plants resistant to the herbicides used by many farmers to control weeds. This gene can make it easier to grow crops while still ensuring that weeds are destroyed. In addition, a number of crop plants are being engineered to resist infectious microbes and pest insects. Growing insect-resistant plants will reduce the need to apply chemical insecticides to crops.

One of the most exciting potential uses of genetic engineering in agriculture involves nitrogen fixation. Nitrogen fixation is the conversion of atmospheric nitrogen gas, which plants cannot use, into nitrogen compounds that plants can convert to essential nitrogen-containing molecules such as amino acids. In nature, nitrogen fixation is performed by certain bacteria that live in the soil or in plant roots. Even so, the level of nitrogen compounds in the soil is often so low that fertilizers must be applied in order for crops to grow. Nitrogen-providing fertilizers are costly and contribute to water pollution. DNA technology offers ways to increase bacterial nitrogen fixation and eventually, perhaps, to engineer crop plants to fix nitrogen themselves.

Animal Husbandry DNA technology is now routinely used to make vaccines and growth hormones for treating farm animals and, on an experimental basis, to make transgenic animals. To produce such animals, scientists first remove egg cells from a female and fertilize them in vitro. Meanwhile, they clone the desired gene from another organism. They then inject the DNA directly into the nuclei of the eggs. Some of the cells integrate the foreign DNA into their genomes and are able to express the foreign gene. The engineered eggs are then surgically implanted in a surrogate mother. If an embryo develops successfully, the result is a transgenic animal, containing a gene from a third "parent" of another species.

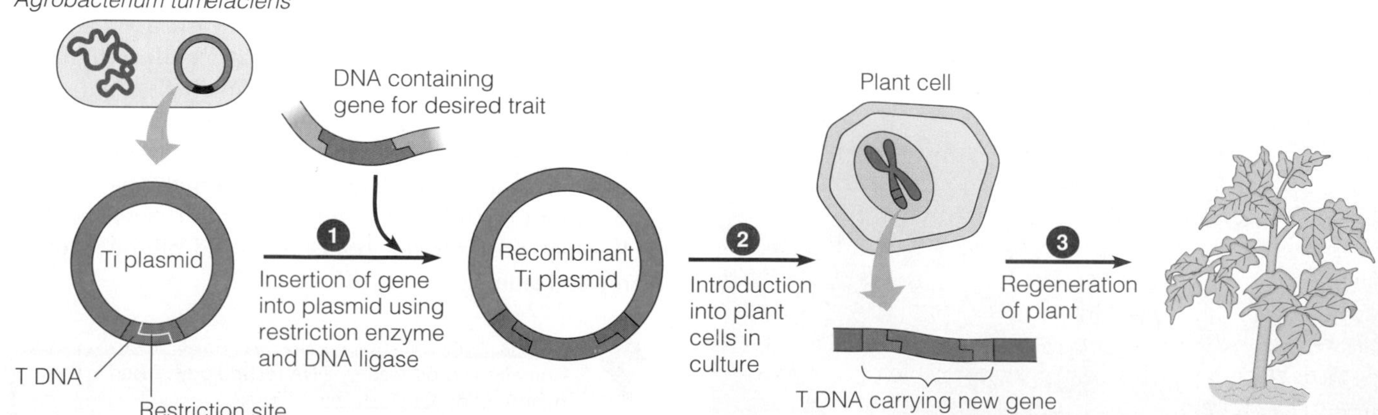

Figure 12.19 Using the Ti plasmid as a vector for genetically engineering plants

The development of transgenic farm animals is proceeding slowly, partly because of unanticipated problems in the animals, such as low fertility or increased susceptibility to disease. Other issues include the safety of the final product (meat or milk, for instance) for the human consumer.

What is the function of the Ti plasmid in the creation of transgenic plants?

It is used as the vector for introducing foreign genes into a plant cell.

Gene therapy may someday help treat genetic disorders and other kinds of disease 12.20

Techniques for manipulating DNA hold great potential for treating a variety of diseases by gene therapy—alteration of an afflicted individual's genes. In people with disorders traceable to a single defective gene, it should theoretically be possible to replace or supplement the defective gene with a normal allele. Unfortunately, contrary to some reports in the news media, such therapy is not yet a reality.

Ideally, the normal allele would be put into cells that multiply throughout a person's life. Bone marrow cells, which include the "stem cells" that give rise to all the cells of the blood and immune system, are prime candidates. Figure 12.20 outlines one procedure for a situation in which bone marrow cells are failing to produce a vital protein product because of a defective gene. ① The normal gene is cloned by recombinant DNA techniques. It is then inserted into the nucleic acid of a retrovirus vector that has been rendered harmless. ② Bone marrow cells are taken from the patient and infected with the virus. ③ The virus inserts its nucleic acid, including the human gene, into the cells' DNA (see Module 10.21). ④ The engineered cells are then injected back into the patient, where they colonize the bone marrow. If the procedure succeeds, the cells will multiply throughout the patient's life and express the normal gene. The engineered cells will supply the missing protein, and the patient will be cured.

A number of gene therapy trials are under way in humans. Most experiments to date have been preliminary, designed to test the safety and effectiveness of a procedure rather than to attempt a cure. One highly publicized clinical trial involves an immune deficiency disease that results from the lack of an enzyme called ADA. Patients in the trial are given injections of their own white blood cells that have been engineered to carry the normal ADA allele. The procedure is similar to that in Figure 12.20, but because the mature white blood cells do not multiply in the body, repeated treatments are given. Furthermore, the patients also receive injections of the enzyme. Thus their continued health does not actually demonstrate the success of the gene therapy. As of early 1999, we still await the first scientifically sound evidence of benefit to a patient from gene therapy.

Currently, the most promising gene therapy trials are not aimed at correcting genetic defects. Instead, researchers are engineering cells from bone marrow in attempts to enhance the abilities of immune cells to fight off cancer or resist infection by HIV. This approach may lead to effective treatments for many nonhereditary diseases.

Human gene therapy raises both technical and ethical issues. One important technical question is: How can researchers build in gene control mechanisms to ensure that cells with the transferred gene make appropriate amounts of the gene product at the right time and in the right parts of the body? And how can they be sure that the gene's insertion does not harm some other necessary cell function?

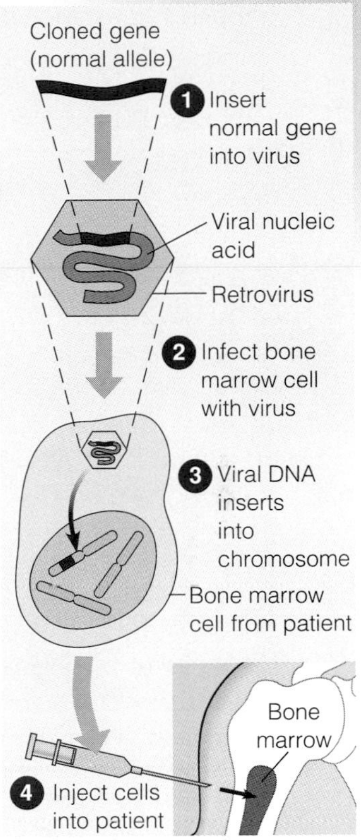

Among the ethical questions posed by gene therapy is: Who will have access to it? The procedures now being tested are expensive and require expertise and equipment found only in major medical centers. A related question is: Should gene therapy be reserved for treating serious diseases? And what about its potential use for enhancing athletic ability, physical appearance, or even intelligence?

Technically easier than modifying genes in somatic cells of children or adults is the genetic engineering of germ cells or zygotes—already accomplished in lab animals. But this possibility raises the most difficult ethical question of all: whether we should try to eliminate genetic defects in our children and their descendants. Should we interfere with evolution in this way? From a biological perspective, the elimination of unwanted alleles from the gene pool could backfire. Genetic variety is a necessary ingredient for the survival of a species as environmental conditions change with time. Genes that are damaging under some conditions may be advantageous under others (one example is the sickle-cell allele; see Module 9.14). Are we willing to risk making genetic changes that could be detrimental to our species in the future? We may have to face this question soon.

Cloned gene (normal allele)
1 Insert normal gene into virus
Viral nucleic acid
Retrovirus
2 Infect bone marrow cell with virus
3 Viral DNA inserts into chromosome
Bone marrow cell from patient
Bone marrow
4 Inject cells into patient

Figure 12.20 One type of gene therapy procedure

So far, how many inherited diseases in humans can definitely be treated by gene therapy?

Zero

Risks and Ethical Questions

12.21 Genetic engineering involves some risks

As soon as scientists realized the power of DNA technology, they began to worry about potential dangers. Early concerns focused on the possibility that genetic engineering might create hazardous new pathogens. What might happen, for instance, if cancer cell genes were transferred into bacteria or viruses? Scientists developed a set of guidelines that in the United States and some other countries have become formal government regulations.

One type of safety measure is a set of strict laboratory procedures designed to protect researchers from infection by engineered microbes and to prevent the microbes from accidentally leaving the laboratory. In addition, strains of microorganisms to be used in recombinant DNA experiments are genetically crippled to ensure that they cannot survive outside the laboratory. Finally, certain obviously dangerous experiments have been banned.

Figure 12.21 Pollen might transfer genes from genetically engineered crop plants to wild relatives nearby

The debate about the engineering of crop plants by DNA technology centers on the ecological dangers of introducing new organisms into the environment. A major concern is that transgenic plants might pass their new genes to close relatives in nearby wild areas. We know that lawn and crop grasses, for example, commonly exchange genes with wild relatives via pollen transfer (Figure 12.21). If domestic plants carrying genes for resistance to herbicides, diseases, or insect pests pollinated wild ones, the offspring might become "superweeds" very difficult to control. However, researchers may be able to prevent the escape of such plant genes by engineering plants so that they cannot hybridize.

Today, governments and regulatory agencies throughout the world are grappling with how to facilitate the use of biotechnology in agriculture, industry, and medicine while ensuring that new products and procedures are safe. In the U.S., all projects are closely scrutinized and evaluated for potential risks by regulatory agencies, including the Food and Drug Administration, the Environmental Protection Agency, the National Institutes of Health, and the Department of Agriculture. These agencies also consider some of the ethical questions we discuss next.

What is one of the concerns about engineering crop plants by adding genes for herbicide resistance?

The possibility that the genes could escape, via cross-pollination, to weeds that are closely related to the crop species

12.22 DNA technology raises important ethical questions

DNA technology has opened the door to a new world of novel organisms and gene products created by scientists instead of by the mechanisms of evolution. And the results of the Human Genome Project are revolutionizing medicine and are likely to have many other benefits. It is reassuring that the potential risks are being carefully weighed. But some important ethical questions remain.

The child in Figure 12.22A is growing at a normal rate, thanks to regular injections of human growth hormone (GH) that was made by genetically engineered *E. coli*. Like any new drug, this GH was subjected to exhaustive laboratory tests before it was released for human use. However, because it is a powerful hormone that affects the body in a number of ways, this drug may be more likely than most ordinary drugs to produce unanticipated side effects in years to come. The increased availability of GH has also raised ethical questions by leading some parents of short but hormonally normal children to seek GH treatment.

A much broader ethical question is: How do we really feel about wielding one of nature's singular powers—the ability to make new microorganisms, plants, and even animals? Some might ask: Do we have any right to alter an organism's genes? Or to add our new creations to an already beleaguered environment?

Such questions must be weighed against the apparent benefits to humans and the environment that can be brought

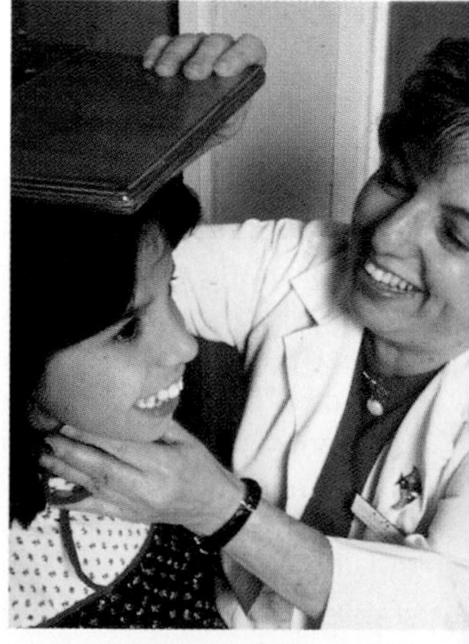

Figure 12.22A A child treated with human growth hormone made by bacteria

about by using DNA technology. For example, bacteria are being engineered to clean up mining wastes and a number of industrial and domestic pollutants that threaten our soil, water, and air. These organisms may be the only feasible solutions to some of our most pressing environmental problems.

Ethical issues relating to the human genome are perhaps the most difficult for us. As discussed in Module 12.20, human gene therapy is rife with ethical dilemmas, whether it involves changing the DNA in germ-line cells or simply in somatic cells. And what of the information being obtained in the Human Genome Project? The potential benefits to human health provide strong ethical support for the project. But there is a danger that information about disease-associated genes—especially data about *individual* genomes that are collected outside the official project—could be abused. In the words of James D. Watson of DNA fame:

> We must work to ensure that society learns to use [genetic] information only in beneficial ways and ... pass laws ... to prevent invasion of privacy of an individual's genetic background by either employers, insurers, or government agencies and to prevent discrimination on genetic grounds.... We have only to look at how the Nazis used leading members of the German human genetics and psychiatry communities to justify their genocide programs, first against the mentally ill and then the Jews and the Gypsies. We need no more vivid reminders that science in the wrong hands can do incalculable harm.*

Largely because of the events in Nazi Germany, our society today rejects the notion of **eugenics**—the effort to control the genetic makeup of human populations. The possibility of gene therapy on germ cells raises the greatest fears in this regard. However, many people argue that providing genetically engineered somatic cells to individuals with life-threatening diseases is basically not different from other medical interventions that save lives. Some compare the transplantation of genes in this case to the transplantation of organs.

With the establishment of databases holding information about individual genomes, the potential for genetic discrimination is already becoming a thorny issue. Insurance companies and employers are increasingly interested in gene testing. Breast-cancer researcher Mary-Claire King mentions this issue in Module 11.16. Nancy Wexler, a leading Huntington's disease researcher, has put it bluntly:

> A very big question is the problem of possible discrimination and stigmatization. An insurance company could refuse coverage or an employer could refuse a job if a person was found to carry the gene for a particular disease. People might also be coerced into taking a test that they wouldn't want in order to be considered for a job or an insurance policy.

To what extent should we allow genetic information to be used this way? If we allow its use at all, how do we prevent the information from being used in a discriminatory manner? These are complex issues. As Nancy Wexler has said:

> The question could be asked, Do you want people who have genetic susceptibility in positions in which they could have a major impact on other individuals? For example, do you want an airline pilot with a genetic predisposition toward heart attack? My own feeling is that all of us have genes for something [harmful] that may be quiescent for a major part of our lives. If you start kicking out everybody who has some genetic susceptibility, then you're going to be in tough shape because there won't be anyone left. It's better to provide excellent medical care and preventive measures or early treatment for problems as they arise.

Some people have suggested that the dangers of abusing genetic information are so great that we should cease research in certain areas. But most scientists would probably tend to agree with the following statement by molecular biologist Leroy Hood: "What science does is give society opportunities. What we have to do is look at these opportunities and then set up the constraints and the rules that will allow society to benefit in appropriate ways."

As citizens in the twenty-first century, we must all participate in making the decisions that Watson, King, Wexler, and Hood call for. 

Figure 12.22B James D. Watson

Figure 12.22C Nancy Wexler

Figure 12.22D Leroy Hood

? The quotes from James Watson and Leroy Hood in this module both highlight the distinction between __________, the basic pursuit of knowledge about the natural world, and __________, the application of scientific knowledge. (*Hint:* Review Module 1.8.)

science . . . technology

*Excerpted with permission from James Watson, *Science*, April 6, 1990. Copyright 1990 American Association for the Advancement of Science.

Chapter Review

Chapter Summary

Bacteria as Tools for Manipulating DNA (Introduction–12.6) Recombinant DNA technology is a set of techniques for combining genes from different sources and transferring this new DNA into cells where the genes may be expressed. These and other DNA techniques are making possible the Human Genome Project and other important research, the manufacture of useful protein products, new ways to improve agriculture, and new treatments for disease **(Introduction)**. Bacterial genes can transfer from cell to cell by one of three processes: transformation, transduction, or conjugation **(12.1)**. Plasmids, small circular DNA molecules separate from the bacterial chromosome, can serve as carriers for the transfer of genes **(12.2)**. Researchers use restriction enzymes (which cut DNA at specific points) and DNA ligase (which "pastes" DNA fragments together) to insert genes into plasmids, creating recombinant DNA **(12.3–12.4)**. Bacteria take up the recombinant plasmids from their surroundings by transformation and reproduce, thereby cloning the plasmids and the genes they carry. Copies of a gene or quantities of a gene's protein product may then be harvested **(12.5)**. Recombinant DNA techniques allow the construction of genomic libraries, sets of DNA fragments containing all of an organism's genes. Multiple copies of each fragment are stored in a cloned bacterial plasmid or phage **(12.6)**.

Other Tools of DNA Technology (12.7–12.12) Reverse transcriptase can be used to make smaller DNA libraries, containing only the genes that are transcribed by a particular type of cell **(12.7)**. A nucleic acid probe can tag a desired gene in a library. The probe is a short, single-stranded molecule of radioactively labeled DNA or RNA whose nucleotide sequence is complementary to part of the gene (or other DNA of interest) **(12.8)**. Automation speeds up the sequencing and synthesis of genes in the laboratory **(12.9)**. Gel electrophoresis is a method used to sort DNA molecules (such as restriction fragments) by size **(12.10)**. Scientists can compare the DNA sequences of different individuals by exploiting the presence of restriction fragment length polymorphisms (RFLPs). A radioactive probe reveals the bands of interest on an electrophoretic gel **(12.11)**. When a DNA sample is very small, the polymerase chain reaction (PCR) can be used to clone it quickly in a test tube **(12.12)**.

The Challenge of the Human Genome (12.13–12.15) The 23 chromosomes in the haploid human genome contain about 3 billion nucleotide pairs. This DNA includes 50,000–100,000 genes and a huge amount of noncoding DNA, much of which consists of repetitive nucleotide sequences **(12.13)**. Barbara McClintock discovered that segments of DNA called transposons can move about within a cell's genome, changing the organism's phenotype in the process **(12.14)**. The Human Genome Project involves genetic and physical mapping of chromosomes, DNA sequencing, and comparison of human genes to those of other species. The data will give insight into development and evolution and will aid in the diagnosis and treatment of disease **(12.15)**.

Other Applications of DNA Technology (12.16–12.20) DNA fingerprinting can help solve crimes **(12.16)**. Recombinant DNA technology is used to mass-produce gene products such as human insulin and other hormones, cancer-fighting drugs, and new vaccines **(12.17–12.18)**. Recombinant DNA technology is also being used to produce new genetic varieties of plants and animals **(12.19)**. Human gene therapy, though still not a reality, raises ethical questions. The biggest question is whether we should treat reproductive cells, such as gametes, and thus alter the genotypes of future generations **(12.20)**.

Risks and Ethical Questions (12.21–12.22) Genetic engineering involves some risks, such as a harmful recombinant organism escaping into the environment **(12.21)**. Our new genetic knowledge will affect our lives in many ways. There are reasons to press forward, but we must use the information wisely **(12.22)**.

Testing Your Knowledge

Multiple Choice

1. Which of the following would be considered a transgenic organism?
   a. a bacterium that has received genes via conjugation
   b. a human given a corrected human blood-clotting gene
   c. a fern grown in cell culture from a single fern root cell
   d. a rat with rabbit hemoglobin genes
   e. a human treated with insulin produced by *E. coli* bacteria

2. A microbiologist found that some bacteria infected by phages had developed the ability to make a particular amino acid that they could not make before. This new ability was probably a result of
   a. transformation
   b. natural selection
   c. conjugation
   d. mutation
   e. transduction

3. When a typical restriction enzyme cuts a DNA molecule, the cuts are uneven, so that the DNA fragments have single-stranded ends. These ends are useful in recombinant DNA work because
   a. they enable a cell to recognize fragments produced by the enzyme
   b. they serve as starting points for DNA replication
   c. the fragments will bond to other fragments with complementary ends
   d. they enable researchers to use the fragments as molecular probes
   e. only single-stranded DNA segments can code for proteins

4. DNA fingerprints used as evidence in a murder trial look something like supermarket bar codes. The pattern of bars in a DNA fingerprint shows
   a. the order of bases in a particular gene
   b. the presence of various-sized fragments from chopped-up DNA
   c. the presence of dominant or recessive alleles for particular traits
   d. the order of genes along particular chromosomes
   e. the exact location of a specific gene in a genomic library

5. A biologist isolated a gene from a human cell, attached it to a plasmid, and inserted the plasmid into a bacterium. The bacterium made a new protein, but it was nothing like the protein normally produced in a human cell. Why? *(Explain your answer.)*
   a. The bacterium had undergone transformation.
   b. The gene did not have sticky ends.
   c. The gene contained introns.
   d. The gene did not come from a genomic library.
   e. The biologist should have cloned the gene first.

6. A paleontologist has recovered a bit of organic material from the 400-year-old preserved skin of an extinct dodo. She would like to compare DNA from the sample with DNA from living birds. Which of the following would be most useful for increasing the amount of DNA available for testing?
   a. restriction fragment analysis
   b. polymerase chain reaction
   c. molecular probe analysis
   d. electrophoresis
   e. Ti plasmid technology

7. How many genes are there in a human sperm cell?
   a. 23
   b. 46
   c. 5,000–10,000
   d. 50,000–100,000
   e. about 3 billion

Describing, Comparing, and Explaining

1. Explain how you might engineer *E. coli* to produce human growth hormone (GH), using the following: *E. coli* containing plasmids, DNA carrying a gene for GH, DNA ligase, a restriction enzyme, equipment for manipulating and growing bacteria, a method for extracting and purifying the hormone.

2. Recombinant DNA techniques are used to custom-build bacteria for two main purposes: to obtain multiple copies of certain genes, and to obtain useful proteins produced by certain genes. Give an example of the use of each of these in medicine and in agriculture.

Thinking as a Scientist

1. A biochemist hopes to find a gene in human liver cells that codes for an important blood-clotting protein. She knows that the nucleotide sequence of a small part of the gene is CTGGACTGACA. Briefly explain how to obtain the desired gene.

2. You are investigating a genetic disorder characterized by paralysis and death by age 20. There are two forms of the disease: type 1 and type 2, differentiated by the additional presence of heart degeneration in type 2. You have found a large family afflicted with the disease and have recruited 6 normal members, 9 members with type 1 disease, and 3 members with type 2 disease. You are using restriction fragment length polymorphisms to search for a RFLP marker that would reliably identify those with the disease even before symptoms become apparent. One of your electrophoretic gels is shown below.

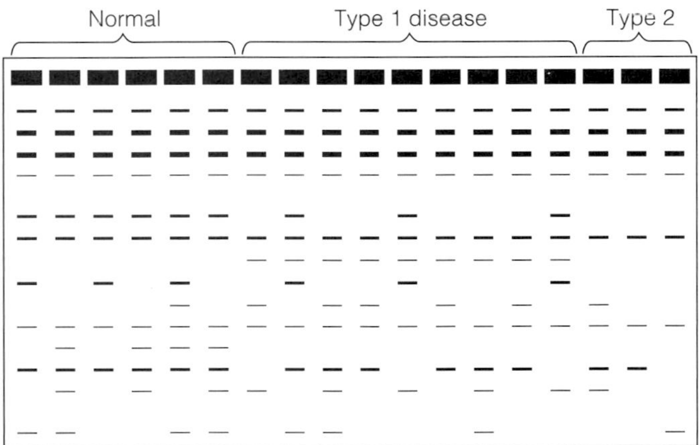

Which of the following choices best states your preliminary conclusions from the data? Explain your choice and your rejection of the other choices.
   a. Success! You have found a marker whose presence indicates that either type 1 or type 2 disease will develop.
   b. Partial success: There is no marker for type 1 disease, but you have found a marker for type 2 disease. Type 1 and type 2 may actually be entirely different diseases.
   c. Partial success: There is no marker for type 2 disease, but you have found a marker for type 1 disease. Type 1 and type 2 may actually be entirely different diseases.
   d. There is no evidence for a marker for either type of disease.

3. Will there be anything left for genetic researchers to do once the Human Genome Project has determined the nucleotide sequences of all of the human chromosomes? Explain.

Science, Technology, and Society

1. Today, it is fairly easy to make transgenic plants and animals. What are some important safety and ethical issues raised by this use of recombinant DNA technology? What are some of the possible dangers of introducing genetically engineered organisms into the environment? What are some reasons for and against leaving decisions in these areas to scientists? To business owners and executives? What are some reasons for and against more public involvement? How might these decisions affect you? How do you think these decisions should be made?

2. In the not-too-distant future, gene therapy may be an option for the treatment and cure of many inherited disorders. What do you think are the most serious ethical issues that must be dealt with before human gene therapy is used on a large scale? Why do you think these issues are important?

3. The possibility of extensive genetic testing raises questions about how personal genetic information should be used. For example, should employers or potential employers have access to such information? Why or why not? Should the information be available to insurance companies? Why or why not? Is there any reason for the government to keep genetic files? Is there any obligation to warn relatives who might share a bad gene? Might some people avoid being tested, for fear of being labeled genetic outcasts? Or might they be compelled to be tested against their wishes? Can you think of other reasons to proceed with caution?

Media Resources

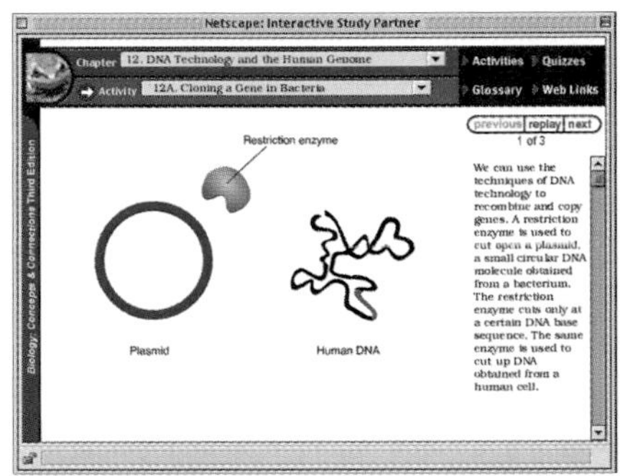

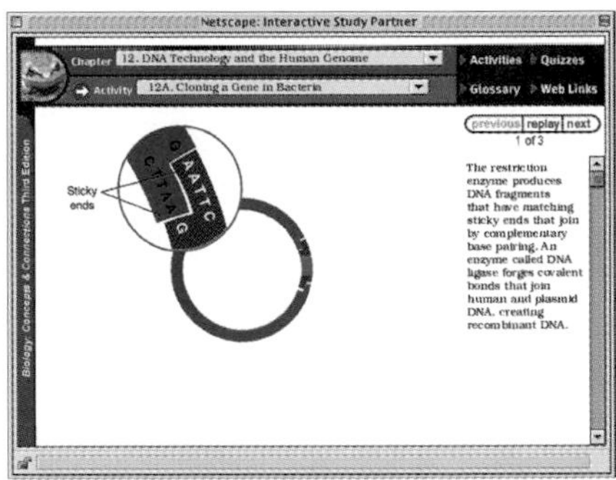

For further review of the concepts in this chapter, use the activities and quizzes on the CD-ROM.

Visit the Special Edition of The Biology Place web site for activities, quizzes, news articles, and web links.

UNIT III
Concepts of Evolution

13 How Populations Evolve

Outline

In the 1800s, whaling ships prowling the tropical Pacific Ocean often stopped by the Galápagos Islands, where they stocked up on fresh meat and water. The crew of one whaler included 22-year-old Herman Melville, the author of *Moby Dick*. Unimpressed with the islands and its inhabitants, Melville wrote, "Little but reptile life is here found, the chief sound of life is a hiss." Melville, however, underestimated the biological significance of Galápagos life.

Located some 950 km (570 mi) west of Ecuador, the Galápagos are home to many organisms, such as the meter-long marine iguana shown at the left, that are found nowhere else on Earth. Created by volcanoes only about five million years ago, the islands were probably first colonized by a few plants and animals that chanced to reach them from the South American mainland. The unique species that arose on the islands from the early "colonists" have played a large role in the development of evolutionary theory.

Just before Melville's visit, British naturalist Charles Darwin spent part of his 26th year on the Galápagos. Darwin saw that there was much to be learned from the islands' unique forms of life.

Variation and Natural Selection

He was fascinated by the marine iguana's body features and behavior. Watching the iguanas dive into the ocean to feed on algae, Darwin wrote, "This lizard swims with perfect ease and quickness, by serpentine movement of its body and flattened tail." Darwin contrasted the marine iguana's tail with the rounded tail of typical terrestrial iguanas. He also noted that the strong claws and partial webbing on the marine iguana's feet probably help steer and pull it along the seafloor. Darwin found he could not frighten the iguanas into the sea. When he chased them to the water's edge, they would stop, and he could easily pick them up by the tail. He wrote, "They do not seem to have any notion of biting; but when much frightened they squirt a drop of fluid from each nostril." Darwin tossed one of the lizards into the water repeatedly; each time, the animal quickly returned to its place on the rocky shore. Darwin supposed that the lizard was safe from predators on land and, except when searching for food, remained on the lava rocks to avoid being eaten by sharks.

Many biologists since Darwin have studied the marine iguana, and we now have answers to some of the questions he raised. For instance, the iguana's tendency to remain on dark rocks instead of diving into the sea when approached seems to be related more to keeping warm than to the danger of shark attack. The ocean around the Galápagos is quite cold for marine iguanas. Although they can withstand the water temperature long enough to feed, they must sunbathe to maintain a suitable metabolic rate. When an iguana is out on the warm rocks, blood vessels near its body surface expand, speeding up heat absorption by the blood. Its heart rate also increases, distributing the heat to the rest of its body. When the lizard dives into the ocean, its circulatory system adjusts to cooling. Its surface vessels are constricted and its heart rate slows, keeping much of its blood and heat in deeper tissues.

Biologists have also learned that the fluid Darwin saw squirting from the iguana's nostrils is a highly concentrated salt solution. The marine iguana has salt-excreting glands inside its head just above the eyes. The salt solution drips into its nostrils and passes out in a cloudlike mist as the lizard exhales. This feature enables the marine iguana to dispose of excess salt it obtains from eating marine algae.

Salt-excreting glands, heat-absorbing mechanisms, and behavior that helps conserve body heat are examples of **evolutionary adaptations,** inherited characteristics that enhance an organism's ability to survive and reproduce in a particular environment. In this chapter, we examine some of the processes by which the inhabitants of the Galápagos, and species elsewhere, develop evolutionary adaptations to their environment. We thus begin our study of **evolution,** inherited change in organisms over time, the process that has transformed life on Earth from its earliest forms to the vast diversity that we see today.

Concentrated salt solution squirted into the air from the nostrils of marine iguanas

13.1 A sea voyage helped Darwin frame his theory of evolution

13A
See Activity 13A on the CD-ROM.

Today, when we think of an animal, plant, or other organism, we tend to envision it in its natural surroundings. Nature films and guides don't just show us interesting organisms; they also tell us how special structures or behaviors help the organisms survive and reproduce in a particular environment. If you visited the Galápagos Islands today, you would see many of the same sights that fascinated Darwin over a century ago. Marine iguanas still bask on the shoreline. Inland, you would probably see the land iguana (Figure 13.1A), whose snout is more tapered and tail more rounded than the marine iguana's. You would also find long-necked giant tortoises (whose Spanish name is *galápagos*) feeding on thorny cactuses. Being aware of each organism's evolutionary adaptations and how they fit the particular conditions of the environment would help you appreciate everything you saw and heard. An appreciation for the close ties between organisms and their environment is part of Charles Darwin's legacy.

Like many concepts in science, the main ideas Darwin advanced—that species change over time, and that living species have arisen from earlier life forms—can be traced back to the ancient Greeks. About 2500 years ago, the Greek philosopher Anaximander promoted the idea that life arose in water and that simpler forms of life preceded more complex ones. The road from Anaximander to Darwin was long and tortuous. The Greek philosopher Aristotle, whose views had an enormous impact on Western culture, generally held that species are fixed, or permanent, and do not evolve. Judeo-Christian culture fortified this idea with the common interpretation of the biblical book of Genesis, that all species were individually designed by a divine creator. The ideas that Earth and all living species are static and perfect, and that planet Earth is, at most, about 6000 years old, dominated the intellectual climate of the Western world for centuries.

In the century prior to Darwin, only a few scientists questioned the biblical story of creation or the belief that species are fixed (unchanging) and perfect. In the mid-1700s, the study of **fossils** (the imprints or remnants of organisms that lived in the past) led French naturalist Georges Buffon to suggest that Earth might be much older than 6000 years. In addition, scientists observed some similarities between fossils and living animals. In 1766, Buffon raised the possibility that different species had arisen from common ancestors, although he went on to argue against the idea. In the early 1800s, French naturalist Jean Baptiste Lamarck suggested that the best explanation for fossils and for the current diversity of life is that organisms evolve. Today, we remember Lamarck mainly for his erroneous view of *how* species evolve. He proposed that by using or not using its body parts, an individual tends to develop certain characteristics, which it passes on to its offspring. Lamarck's proposal is known as the inheritance of acquired characteristics. As an example, he suggested that the giraffe acquired its long neck because its ancestors stretched higher and higher into the trees to reach leaves, and that the animal's increasingly lengthened neck was passed on to its offspring. This mistaken idea obscures the important fact that Lamarck helped set the stage for Darwin by strongly advocating evolution and by proposing that species evolve as a result of interactions with their environment.

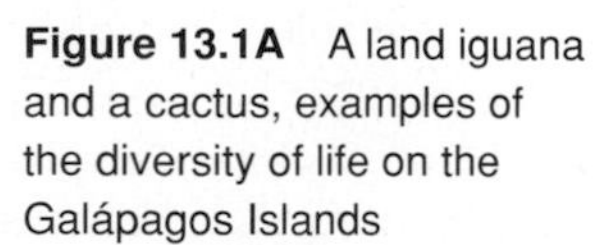

Figure 13.1A A land iguana and a cactus, examples of the diversity of life on the Galápagos Islands

Charles Darwin (Figure 13.1B) was born in 1809, the same year Lamarck published some of his ideas on evolution. In December 1831, at the age of 22, Darwin began a round-the-world sea voyage (Figure 13.1C) that profoundly influenced his thinking and eventually the thinking of the entire world. He joined the crew of HMS *Beagle*, a surveying ship, on a mission to chart poorly known stretches of the South American coastline. Darwin actually spent most of his time on shore, collecting thousands of specimens of fossils and living plants and animals and noting the unique adaptations of the South American organisms. Despite growing up in the generally antievolutionist climate of the Victorian era, Darwin had a questioning mind. He asked himself why fossils of the South American continent were more similar to modern South American species than to fossils of other continents. Could this mean that species in the Americas were more closely related to each other than to species elsewhere? Could contemporary South American species have arisen from South American ancestors? Other questions arose from Darwin's visit to the Galápagos Islands, where the *Beagle* also stopped. Referring to the islands and their unique inhabitants, he later wrote, "Both in space and time, we seem to be brought somewhat near to that great fact—that mystery of mysteries—the first appearance of new beings on the earth."

While on the voyage, Darwin read and was strongly influenced by the recently published *Principles of Geology*, by Scottish geologist Charles Lyell. Lyell's work led Darwin to realize that natural forces gradually change Earth's surface and that the forces of the past are still operating in modern

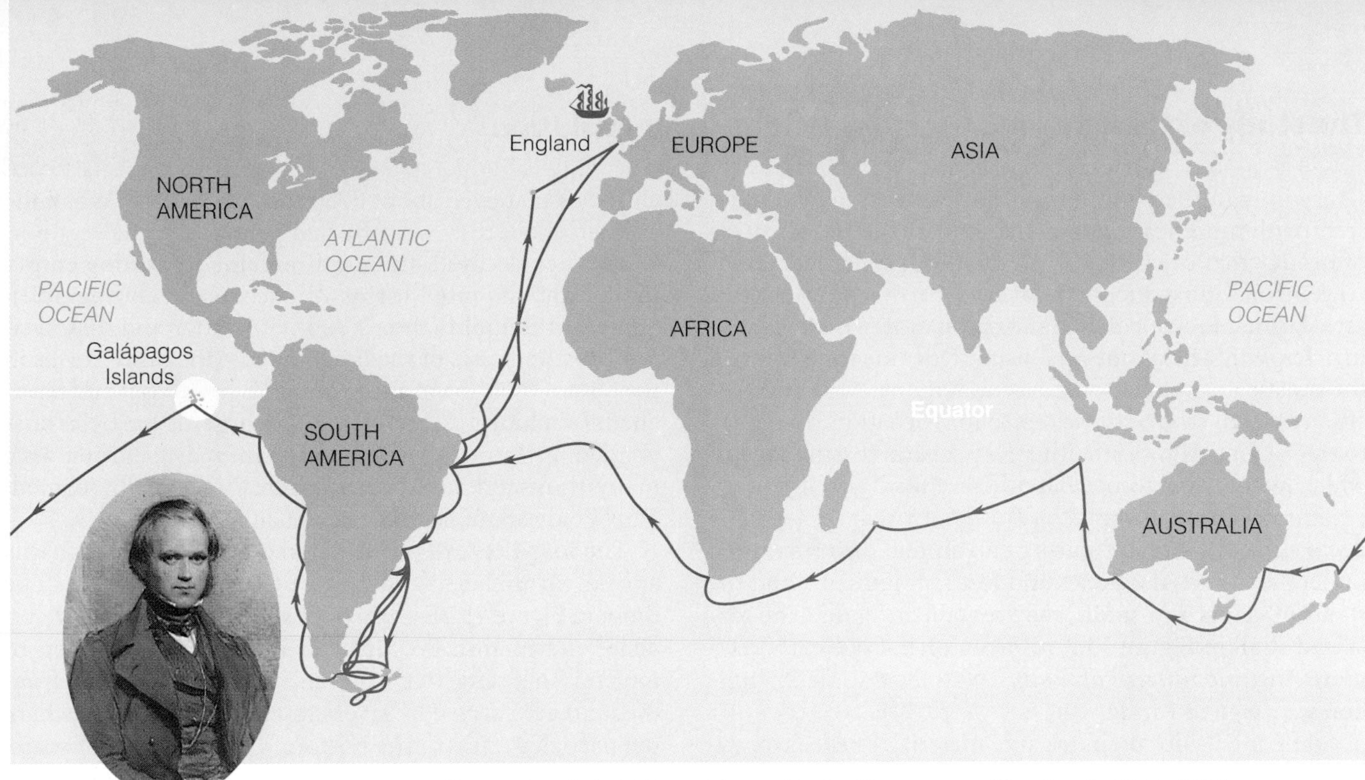

Figure 13.1B Charles Darwin in 1840

Figure 13.1C The voyage of the *Beagle* (1831–1836)

times. These ideas ran counter to the prevailing nineteenth-century view that the most important geological events in Earth's history were certain rare catastrophes and sudden changes. Darwin had collected fossils of marine snails in the Andes Mountains. Having read and known Lyell, he came to believe that slow, natural processes such as the growth of mountains as a result of earthquakes could account for the snails' being there.

By the time Darwin returned to Great Britain five years after the *Beagle* first set sail, his experiences and reading had led him to doubt seriously that Earth and living organisms were unchangeable and had been specially created only a few thousand years earlier. By then he believed that Earth was very old and constantly changing, views that were heretical at that time. In questioning these views, Darwin had taken an important step toward recognizing that life on Earth had evolved.

By the early 1840s, Darwin had composed a long essay describing the major features of his theory of evolution. He realized that his theory would cause a social furor, however, and he delayed publishing it. Then, in the mid-1850s, Alfred Wallace, a British naturalist doing fieldwork in Indonesia, conceived a theory identical to Darwin's. When Wallace sent Darwin a manuscript he had written about his theory, Darwin thought "all my originality will be smashed." However, in 1858, two of Darwin's colleagues presented Wallace's paper and excerpts of Darwin's earlier essay together to the scientific community. With the publication in 1859 of his complete text, *On the Origin of Species by Means of Natural Selection,* Darwin presented the world with an avalanche of evidence and a strong, logical argument for evolution. He also developed his theory of natural selection, an explanation of how evolution occurs. Darwin's book was a monumental contribution to science and to human thought in general. Evolution is the great unifying theme of biology, and *The Origin of Species* fueled an explosion in biological research and knowledge that continues today.

In the first edition of his book, Darwin did not actually use the word evolution, referring instead to "descent with modification." This phrase summarized Darwin's view of life. He perceived a unity among species, with all organisms related through descent from some unknown species that lived in the remote past. He postulated that as the descendants of the earliest organism spread into various habitats over millions of years, they accumulated diverse modifications, or adaptations, that accommodated them to diverse ways of life. The history of life seemed to resemble a tree, with multiple branching and rebranching from a common trunk to the tips of the twigs. At each fork of the evolutionary tree is an ancestor common to all lines of evolution branching from that fork. Species that are closely related, such as the domestic dog and the wolf, share many characteristics because their lineage of common descent extends to the smallest branches of the tree of life.

As we will see in this chapter and the next two, evolutionary theory has been greatly expanded beyond Darwin's basic ideas. Nonetheless, few contributions in all of science have explained as much, withstood as much repeated testing over the years, and stimulated as much other research as those of Darwin.

Darwin's phrase for evolution, ________ with ________, captured the idea that an ancestral species could diversify into many descendant species by the accumulation of different ________ to various environments.

descent . . . modification . . . adaptations

13.2 The study of fossils provides strong evidence for evolution

Darwin compiled enough support for his theory of descent with modification to convince many of the scientists of his day that organisms evolve without supernatural intervention. Subsequent discoveries further support this great principle—one that connects an otherwise bewildering chaos of facts about organisms. Documenting some of the drastic changes that life has undergone over time, fossils are a rich source of evidence for evolution. The photographs in this module illustrate several fossils, each of which formed in a somewhat different way.

The organic substances of a dead organism usually decay rapidly, but hard parts of an animal that are rich in minerals, such as the bones and teeth of dinosaurs and the shells of clams and snails, may remain as fossils. The fossilized skull in Figure 13.2A is from one of our early relatives, *Australopithecus africanus,* who lived some 2.5 million years ago in Africa.

Some fossils do retain organic material. The leaf on the left in Figure 13.2B is about 40 million years old. It is a thin film pressed in rock, still greenish with remnants of its chlorophyll and well-enough preserved that biologists can analyze its molecular and cellular structure. In rare instances, an entire organism, including its soft parts, is fossilized. This can happen only if the individual is buried in a medium that prevents bacteria and fungi from decomposing the corpse. The insect on the right in Figure 13.2B got stuck in the resin of a tree about 40 million years ago. The resin hardened into amber (fossilized resin), preserving the insect.

Sometimes the remains of dead organisms are actually turned into stone by a process called petrification. Petrification occurs when minerals dissolved in groundwater seep into the tissues of dead organisms and replace organic matter. The petrified trees in Figure 13.2C stood about 190 million years ago in what is now a desert in eastern Arizona.

The fossils resembling wrinkled butterfly wings in Figure 13.2D are imprints of the shells of animals called brachiopods that lived about 375 million years ago. When the animals died, they were buried in mud on the seafloor. When they decayed, the mud remained, forming empty molds (the indented forms you see in the photograph). Some of the molds then filled with water and dissolved minerals, and casts of the brachiopods (the raised forms in the photo) took shape as the minerals settled out and hardened. Brachiopods, which feed on organic particles suspended in seawater, were a dominant marine group with many thousands of species when these fossils formed. Today, only about 300 species remain.

The **fossil record**—the ordered array in which fossils appear within layers, or strata, of sedimentary rocks like those in Figure 13.2E—provides some of the strongest evidence of evolution. Sedimentary rocks form from layers of minerals that settle out of water. Sand and silt eroded from the land are carried by rivers to seas and swamps, where the particles settle to the bottom. Over millions of years, deposits pile up and compress the older sediments below into rock. When aquatic organisms die, they settle along with the sediments and may leave imprints in the rocks. Many organisms living on land may be swept into swamps and seas. Land organisms that remain in place when they die may first be covered by windblown silt and then buried in waterborne sediments when sea levels rise over them.

Rock strata form when the rate of sedimentation and the types of particles that settle vary over time. Each of the strata you see in Figure 13.2E may bear a unique set of fossils representing a local sampling of the organisms that lived when the sediment was deposited. Younger strata are on top of older ones; thus, the position of fossils in the strata reveals their relative age.

The fossil record shows that organisms have appeared in a historical sequence. The oldest known fossils, dating from about 3.5 billion years ago, are prokaryotes; from molecular and cellular evidence, prokaryotes are thought to have been the ancestors of all life. Fossils in younger layers of rock reveal the evolution of various groups of eu-

Figure 13.2A A fossilized hominid skull

Figure 13.2B Fossilized organic material: a leaf (left) and an insect fossilized in amber (right)

Figure 13.2C Petrified trees

Figure 13.2D Fossilized imprints: molds (right) and casts (left)

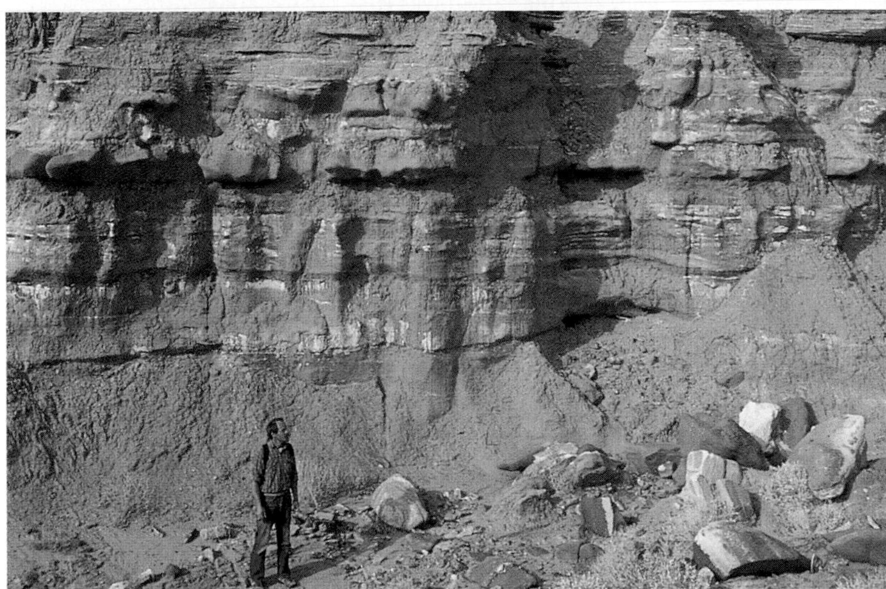

Figure 13.2E Strata of sedimentary rock

karyotic organisms. One example is the successive appearance of the different classes of vertebrates (animals with backbones) in the fossil record. Fishlike fossils are the oldest vertebrates in the fossil record; amphibians are next, followed by reptiles, then mammals and birds.

The evolutionary view of life predicts that we would find signs in the fossil record of linkages between extinct organisms and species living today. Indeed, paleontologists (scientists who study fossils) have discovered many fossils that link past and present. For example, a series of fossils documents the changes in skull shape and size that occurred as mammals evolved from reptiles. Another series of fossils indicates that whales evolved from four-legged land mammals that lived some 55 million years ago. Whales living today have forelegs in the form of flippers but lack hind legs, although they do have small, functionless hind-leg and foot bones that do not extend from the body. Just a few years ago, paleontologists digging in Egypt and Pakistan found fossils of extinct whales that actually had hind legs. In one type, shown in Figure 13.2F, the legs were about half a meter long and included bones similar to those of land mammals. Researchers hypothesize that these hind limbs functioned in copulation and represent a transitional stage in the evolution of whales from their land-dwelling ancestors.

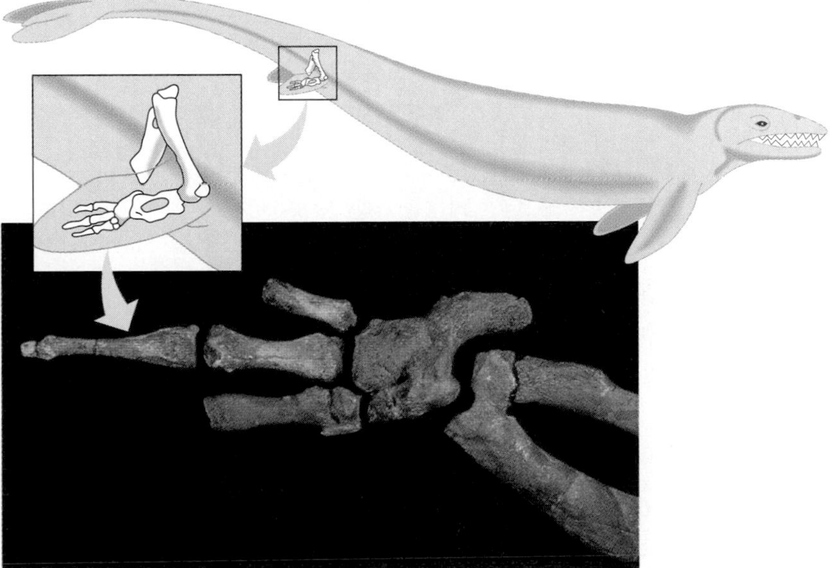

Figure 13.2F *Basilosaurus,* an extinct whale whose hind legs link living whales with their land-dwelling ancestors

In what chronological order do the different vertebrate groups appear in the fossil record?

First fishes, then amphibians, followed by reptiles, and then mammals and birds

13.3 A mass of evidence validates the evolutionary view of life

Every aspect of life shows signs of evolutionary change, and, as we discussed in Chapter 1, evolution is biology's core theme. Let's now take a look at some of the evidence that reinforces the fossil evidence for evolution.

Biogeography It was the geographical distribution of species, known as **biogeography,** that first suggested to Darwin that organisms evolve from common ancestors. Darwin noted that, although the environment of the Galápagos was more like that of certain tropical islands in distant parts of the world than like the environment of the nearby South American mainland, Galápagos animals nonetheless resembled species of the mainland more than they resembled animals on similar but distant islands. The logical explanation was that the Galápagos species evolved from South American immigrants. Today, it is clearer than ever that biogeography makes sense only in the historical context of evolution.

Comparative anatomy Also providing support for evolution, and cited extensively by Darwin, is the science of **comparative anatomy,** the comparison of body structures in different species. Anatomical similarities among many species give signs of common descent. As Figure 13.3A shows, the same skeletal elements make up the forelimbs of humans, cats, whales, and bats, all of which are mammals. The functions of these forelimbs differ. A whale's flipper does not do the same job as a bat's wing, so if these structures had been uniquely engineered, we would expect that their basic designs would be very different. However, their structural similarity would be unsurprising if all mammals descended from a common ancestor with the same basic limb elements. The logical explanation is, in fact, that the arms, forelegs, flippers, and wings of different mammals are variations on a common anatomical theme that has become adapted to different functions. Biologists call such features on different organisms **homologous structures**—features that often have different functions but are structurally similar because of common ancestry. Comparative anatomy testifies that evolution is a remodeling process in which ancestral structures that originally functioned in one capacity become modified as they take on new functions—the kind of process that Darwin called descent with modification. We see many signs that evolution remodels structures, rather than creating them anew. For example, the human spine and knee joint were derived from ancestral structures that supported four-legged mammals. Almost none of us will reach old age without experiencing knee or back problems. If these structures had been designed specifically to support our bipedal posture, we would expect them to be less subject to sprains, spasms, and other common injuries.

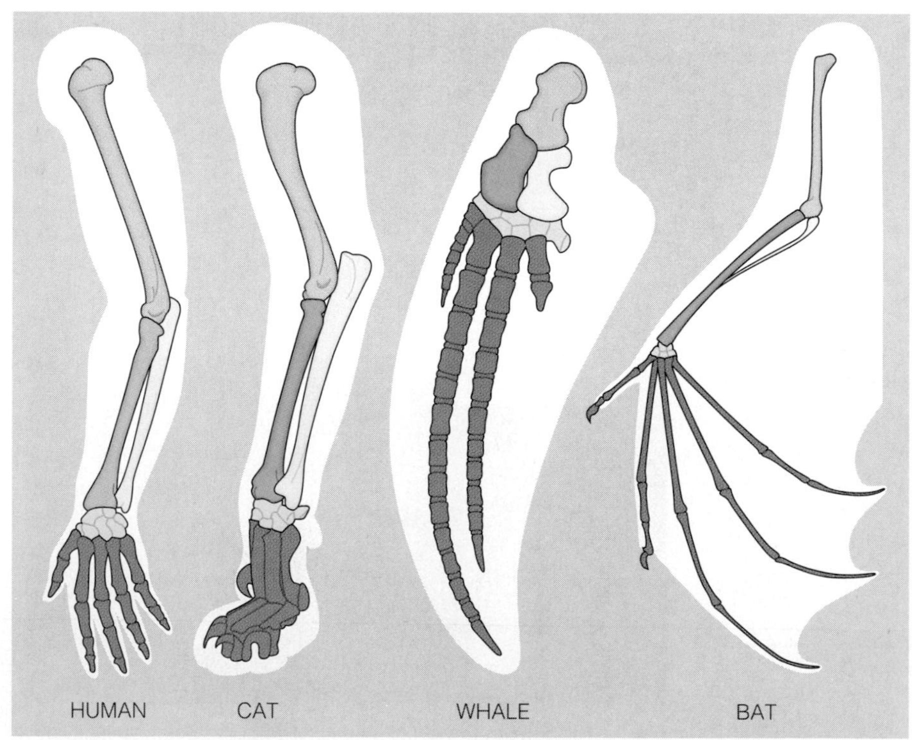

Figure 13.3A Homologous structures: vertebrate forelimbs

Comparative embryology The study of structures that appear during the development of different organisms, called **comparative embryology,** is another major source of evidence for the common descent of organisms. Closely related organisms often have similar stages in their embryonic development. (An embryo is any of the early developmental stages of a multicellular organism.) One sign that vertebrates evolved from a common ancestor is that all of them have an embryonic stage in which structures called gill pouches appear on the sides of the throat. At this stage, the embryos of fishes, frogs, snakes, birds, apes—indeed, all vertebrates—look more alike than different. They take on more and more distinctive features as development progresses. In fishes, for example, most of the gill-pouch structures develop into gills. In land vertebrates, however, these embryonic features develop into other kinds of structures, such as bones of the skull, bones supporting the tongue, and the voice box of mammals.

Molecular biology Recently, much support for evolution has come from **molecular biology,** the study of the molecular basis of genes and gene expression. The universality of the genetic code is strong evidence that all life is related. As we saw in Chapter 10, the hereditary background of an organism is documented in its DNA and in the proteins encoded in the DNA. Molecular biologists have shown that, consistent with the idea of common descent, related individuals have greater similarity in their DNA and proteins than do unrelated individuals of the same species. And two species judged to be closely related by other criteria have a greater proportion of their DNA and proteins in common than more distantly related species.

Research in molecular biology, especially studies of the amino acid sequences of similar proteins in different species, has been a rich source of data about evolutionary relationships. Figure 13.3B illustrates the results of studies on one of the polypeptide chains of hemoglobin, the protein that carries oxygen in the blood of humans and other vertebrates. Researchers have isolated this polypeptide from a number of different animal species and have found that its amino acid sequence differs somewhat from species to species. The graph shows how the polypeptide chain of humans compares with that of five other vertebrate animals. By matching the length of the vertical bar for each animal with the vertical scale on the left, you can see how many amino acids are different from those in the human polypeptide. For example, the rhesus monkey polypeptide differs from the human polypeptide by only 8 amino acids, whereas the lamprey version differs from the human version by 125 amino acids. (Since the human polypeptide is 146 amino acids long, these numbers represent differences of 5% and 86%, respectively.)

These amino acid comparisons lead to a hypothesis about evolutionary relationships: Rhesus monkeys are much more closely related to humans than are lampreys; mice, chickens, and frogs fall in between. It turns out that this hypothesis agrees with earlier conclusions from comparative anatomy and embryology. Furthermore, the hypothesis agrees with fossil evidence. As indicated in Figure 13.3B, the lineage that led to humans diverged from the one leading to monkeys only about 26 million years ago, whereas lampreys (some of the most primitive living vertebrates) branched off the trunk of vertebrate evolution some 450 million years ago.

Another molecular indication of the relatedness of organisms comes from master control genes, those that regulate the activity of groups of other genes during embryonic development. Research on the genetics of development has turned up very similar nucleotide sequences in the genes of organisms as different as fruit flies and mammals, as we saw when we examined the homeotic genes in Module 11.12. Biologists predict that most, if not all, eukaryotic organisms will be found to have similar genes regulating their early development. The logical conclusion is that these genes first arose in an ancestor common to all eukaryotes.

Having examined some of the evidence that validates evolution, our next task is to see how evolution operates. We begin in the next module.

? (a) What is homology? (b) How does the concept of homology relate to Figure 13.3B?

(a) Homology is similarity between different species due to their evolution from a common ancestor. (b) Proteins are gene products, and thus similarities in their amino acid sequences reflect the hereditary connection that is the basis of homology.

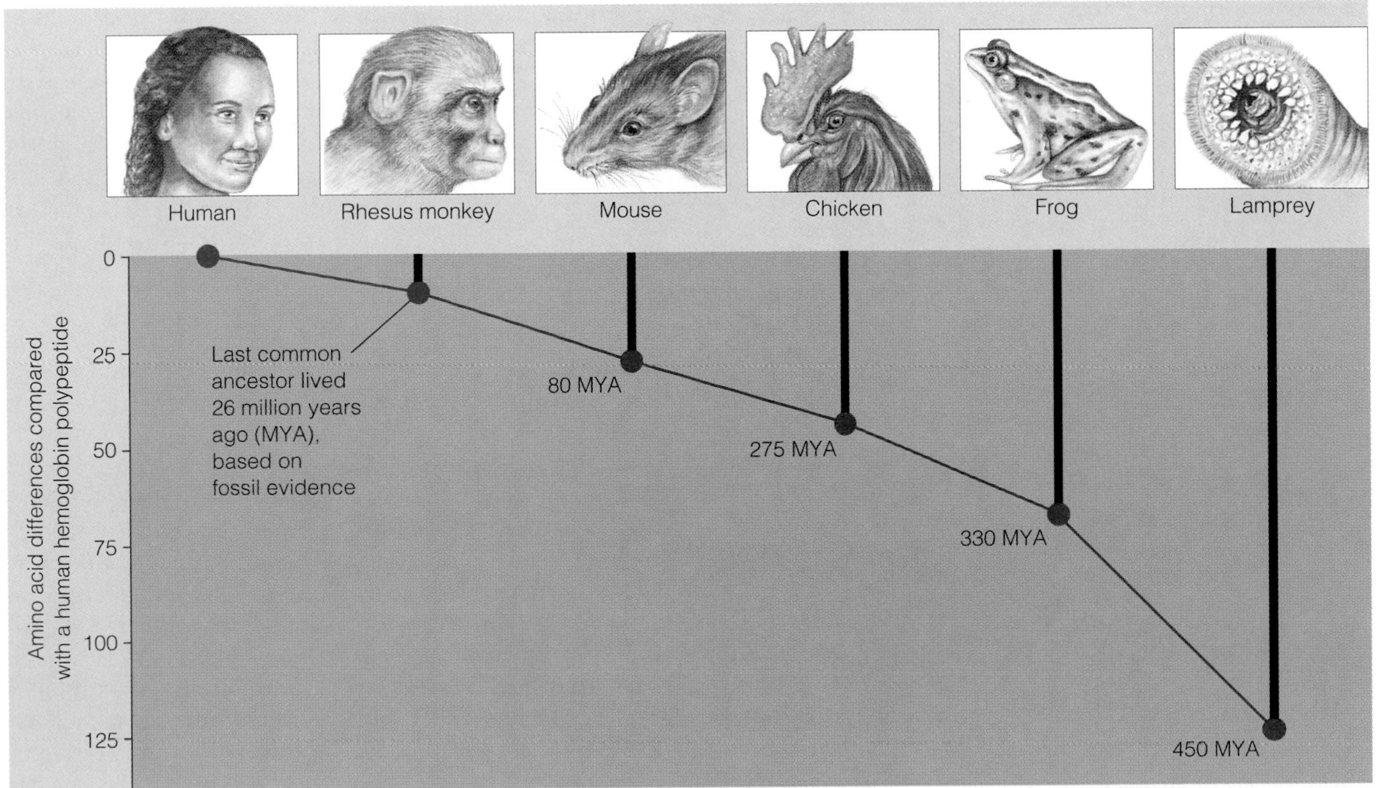

Figure 13.3B Evolutionary relationships between humans and five other vertebrates, based on hemoglobin comparisons

Darwin's Theory and the Modern Synthesis

13.4 Darwin proposed natural selection as the mechanism of evolution

Darwin devoted much of *The Origin of Species* to the ways that organisms become adapted to their environment. His theory of how this happens arose from several key observations. First of all, Darwin recognized that all species tend to produce excessive numbers of offspring. He had read an influential essay on human population written in 1798 by British economist Thomas Malthus. Malthus contended that much of human suffering—disease, famine, homelessness, and war—was the inescapable consequence of the human population's potential to grow much faster than the rate at which supplies of food and other resources could be produced. It was apparent to Darwin that Malthus's concepts applied to all species. Darwin deduced that, because natural resources are limited, the production of more individuals than the environment can support leads to a struggle for existence among the individuals of a population, with only a percentage of offspring surviving in each generation. Many eggs are laid, young born, and seeds spread, but only a tiny fraction complete their development and leave offspring of their own. The rest are starved, eaten, frozen, diseased, unmated, or unable to reproduce for other reasons.

In addition to the overproduction of offspring, two other important observations by Darwin were that individuals of a population vary extensively in their characteristics, and that many of the varying traits are passed from one generation to the next.

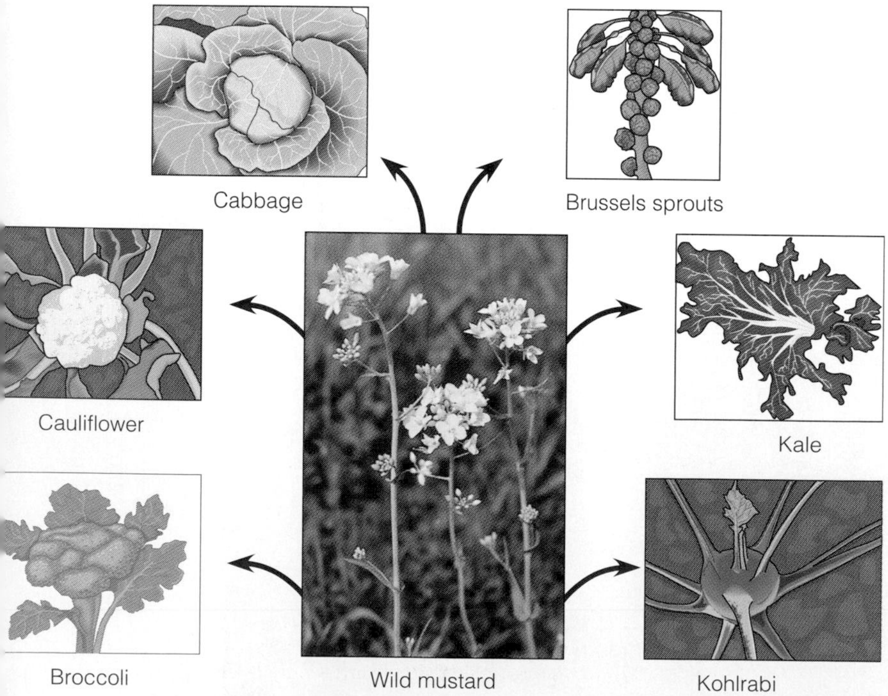

Figure 13.4A Artificial selection in plants: six vegetables derived from wild mustard

What do overproduction of offspring, limited natural resources, and heritable variations have to do with organisms becoming adapted to their environment? Darwin saw that every environment has only a limited supply of resources, and that survival in a limited environment depends in part on the features the organisms inherit from their parents. He concluded that, within a varied population, individuals whose inherited characteristics adapt them best to their environment are most likely to survive and reproduce; these individuals thus tend to leave more offspring than less fit individuals do. Reproduction is central to what Darwin saw as the basic mechanism of evolution, the process he called **natural selection.** Darwin perceived that the essence of natural selection is differential, or unequal, success in reproduction.

Darwin's insight was both simple and profound: Reproduction is unequal, with the individuals that best meet specific environmental demands having the greatest reproductive success. Put another way, differential reproductive success (natural selection) is the means by which the environment screens variations, favoring some over others. Individuals better adapted to a particular environment are more likely to survive and leave offspring that in turn will reproduce than are those less well adapted. As Darwin reasoned, natural selection results in favored traits being represented more and more and unfavored ones less and less in ensuing generations. Thus, the unequal ability of individuals to survive and reproduce leads to a gradual change in the characteristics of a population of organisms, with favored characteristics accumulating over the generations.

Darwin found convincing evidence for his ideas in the results of **artificial selection,** the selective breeding of domesticated plants and animals. He saw that by selecting individuals with the desired traits as breeding stock, humans were playing the role of the environment and bringing about differential reproduction. They were, in fact, modifying species. We see evidence of what Darwin was talking about in the vegetables we eat. For example, broccoli, cauliflower, cabbages, Brussels sprouts, kale, and kohlrabi (Figure 13.4A) are all varieties of a single species of wild mustard, and all were produced by artificial selection. These and many other domesticated plants and animals bear little resemblance to the wild species from which they were derived. Figure 13.4B shows some of the enormous diversity that breeders have produced in a few thousand years within a single species, *Canis familiaris,* the domestic dog.

Darwin reasoned that, if so much change could be achieved in a relatively short period of time by artificial selection, then over hundreds or thousands of generations, natural selection should be able to modify species considerably. With natural selection operating over vast spans of

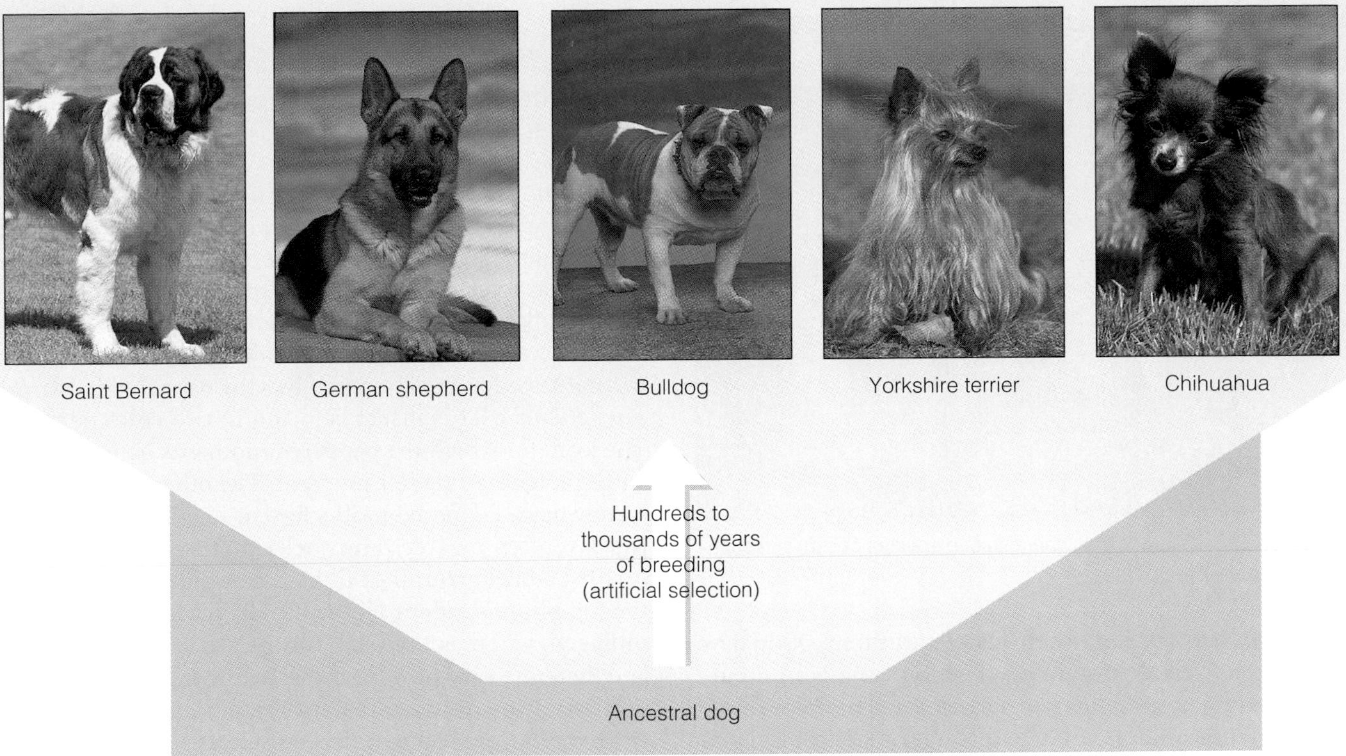

Figure 13.4B Five breeds of dogs, the results of hundreds to thousands of years of artificial selection

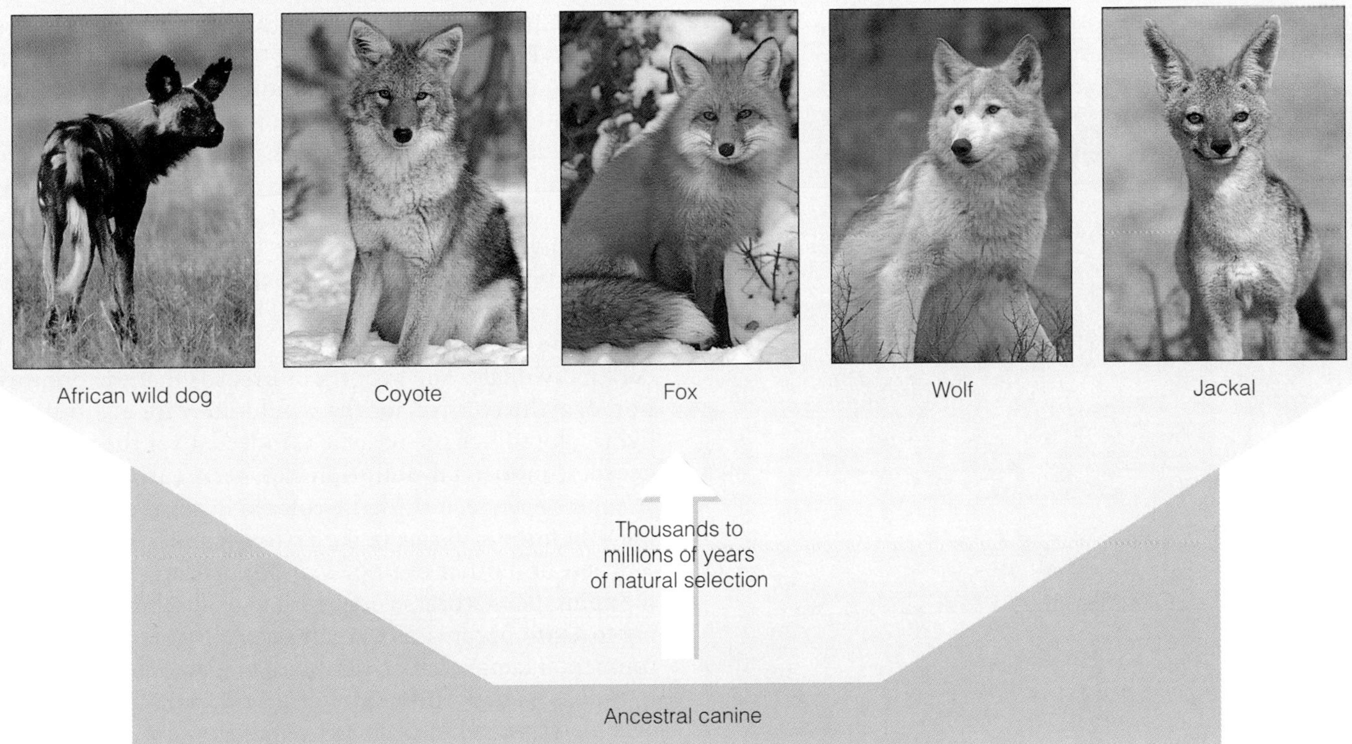

Figure 13.4C Five species of canines, the results of thousands to millions of years of natural selection

time, heritable changes would gradually accumulate. Such changes would account for the evolution of new species—for example, the five species of canines in Figure 13.4C—from preexisting ancestors. In the next module, we examine some documented cases of natural selection. 

? Differential reproductive success among a population's varying individuals in their natural environment is called ______ ______.

natural selection

13.5 Natural selection is a prominent force in nature

Figure 13.5A Different shell colors in one species of snails, resulting from natural selection

Well over 100 cases of the effects of natural selection have been extensively documented. The land snails in Figure 13.5A belong to a single species, *Cepaea nemoralis,* found in woodlands and open fields throughout Europe. Their shells differ in color and in the presence or absence of stripes. The snails in the photograph are not usually found together. Snails with light-colored, striped shells are found mainly in well-lighted areas, whereas dark-colored individuals lacking stripes are found in shady places. The shell color pattern camouflages the snails in their respective habitats, hiding them from birds that prey on snails. If a snail appears in an area where its color pattern does not conceal it, birds are more likely to find that snail than a camouflaged snail. Thus, snails whose shells match their environment are likely to survive and reproduce, passing along characteristics to their offspring, while those with uncamouflaged shells will likely be eaten.

One of the most extensively documented examples of natural selection in action involves the peppered moth, *Biston betularia.* This insect is found in two color varieties. The form for which the peppered moth was named is light, with splotches of darker pigment. The other variety is uniformly dark. Peppered moths feed at night and rest during the day, often on trees and rocks encrusted with lichens. Against a background of light-colored lichens, shown in the top photograph of Figure 13.5B, light moths are camouflaged, whereas the dark moths are very conspicuous and therefore easy prey for birds. Before the Industrial Revolution in the nineteenth century, dark peppered moths were very rare in certain parts of Great Britain. Probably no single factor was responsible for the scarcity, but it is likely that many of the dark moths became bird food before they could reproduce and pass the genes for darkness on to the next generation. As industrial pollution in the late 1800s killed large numbers of lichens, uncovering dark tree bark, dark moths became more abundant. As shown in the bottom photograph of Figure 13.5B, the dark moth is camouflaged on a tree trunk that lacks lichens, but the light moth stands out. By the early 1900s, the population of *Biston* near some British industrial centers consisted almost entirely of dark moths.

Experiments have supported the hypothesis that increased predation by birds on light-colored moths contributed to this population's shift to a higher proportion of dark individuals. The hypothesis predicts that the proportion of light-colored moths would increase again if the light-colored lichens returned. Indeed, over the past two decades, industrial pollution has been curbed, and researchers have seen the light-colored lichens reappear and the light form of *Biston* make a strong comeback.

These and other examples of population change make the point that natural selection is regional and timely, tending to adapt organisms to their local environments. Environmental factors vary from place to place and from time to time. An adaptation in one situation may be useless or even detrimental in different circumstances. Changes in the peppered moth populations also show that significant evolutionary change can occur in a short time.

Figure 13.5B Natural selection over a short time span in the peppered moth

? In what sense is natural selection more of an "editing" process than a "creating" process?

Natural selection cannot create beneficial traits on demand, but "edits" variation in a population by selecting for those traits that work best in the current environment.

Populations are the units of evolution 13.6

A population is a group of individuals of the same species living in the same place at the same time. It is the smallest unit that can evolve. We can, in fact, measure evolution as a change in the prevalence of certain heritable traits in a population over a succession of generations. The increasing proportion of the dark-pigment trait in peppered moths in industrial areas is one example. When natural selection favored the dark trait, dark moths left more offspring, and the population changed, or evolved, from being mostly light to mostly dark. Note that the individual moths did not evolve. In evolution, the individual interacts with the environment and is screened by natural selection. An individual may contribute to population change by its success or failure at reproduction.

Figure 13.6 Human population centers in the United States

Darwin understood that it is populations that evolve. He saw that natural selection, in favoring some heritable traits over others, changes populations over successive generations. What eluded him was the genetic basis of population change. Without this basis, he could not explain the cause of variation among the individuals making up a population, nor could he account for the perpetuation of parents' traits in their offspring. Today, we know that heritable traits are carried by genes on chromosomes and that mutations may produce new traits. Also, we understand how Mendel's principles of segregation and independent assortment of alleles operate during meiosis (see Module 9.17). In Darwin's time, however, little was known about chromosomes, and nothing about gene mutations or the details of meiosis. The significance of Mendel's work did not come to light until 20 years after Darwin's death, some 40 years after the publication of *The Origin of Species.*

An important turning point for evolutionary theory was the birth in the 1920s of the science of genetic change in populations—**population genetics.** A theory of evolution that includes genetics was developed in the early 1940s. Known as the **modern synthesis,** this theory focuses on populations as the units of evolution and includes most of Darwin's ideas. Most importantly, it melds population genetics with the theory of natural selection.

Central to the modern synthesis is the relationship between populations and species. We will have more to say about this in Chapter 14, but for now let's limit our discussion to organisms that reproduce sexually. In this context, we may define a **sexual species** (also called "biological" species) as a group of populations whose individuals have the potential to interbreed and produce fertile offspring. Each species has a geographical range, where individuals usually concentrate in localized populations. One population may be isolated from others of the same species. If individuals in the isolated population interbreed only rarely with those in other populations, there will be little exchange of genes. Such isolation is common for populations confined to widely separated islands, unconnected lakes, or mountain ranges separated by lowlands. However, populations are not always isolated, nor do they necessarily have sharp boundaries. One population center may blur into another in an intermediate region where members of both populations occur but are less numerous.

Figure 13.6 illustrates the tendency for humans to concentrate locally. This is a nighttime satellite photograph showing the lights of population centers in the United States. We know that these populations are not really isolated; people move around, and there are suburban and rural communities between cities. Nevertheless, people are most likely to choose mates locally. As a result, for humans and other species, individuals in one population center are, on average, more closely related to one another than to members of other populations.

Understanding what populations are and how they behave sets the stage for understanding how populations and species evolve. To analyze populations in an evolutionary context, we must consider the genes in the population. Our next step, then, is to focus on population genetics. Before we begin, you might wish to review Mendelian genetics in Chapter 9 (especially Modules 9.2 and 9.4).

? The modern evolutionary synthesis of the 1940s brought together Darwin's theory of natural selection and ________ theory of inheritance.

Mendel's

13.7 Microevolution is change in a population's gene pool over time

In studying evolution at the population level, geneticists focus on what is called the **gene pool,** the total collection of genes in a population at any one time. The gene pool is the reservoir from which members of the next generation of that population derive their genes; it consists of all alleles (alternative forms of genes) in all the individuals making up a population.

For most gene loci, there are two or more alleles in the gene pool, and individuals may be either homozygous (having two identical alleles) or heterozygous (having two different alleles) for each locus. For example, in peppered moths there are two alleles for body color: One allele determines light body color, and the other determines dark body color. The relative frequencies of these and other alleles in a gene pool may change over time. In a peppered moth population in an unpolluted region, the allele for light color tends to have a much higher frequency in the population than the allele for dark color. But in industrially polluted regions, the allele for dark color tends to increase in frequency at the expense of the allele for light color.

When the relative frequencies of alleles in a population change like this over a number of generations, evolution is occurring on its smallest scale. Such a change in a gene pool is called **microevolution.** To understand how microevolution works, it helps to begin by examining the genetics of a simple hypothetical population whose gene pool is *not* changing.

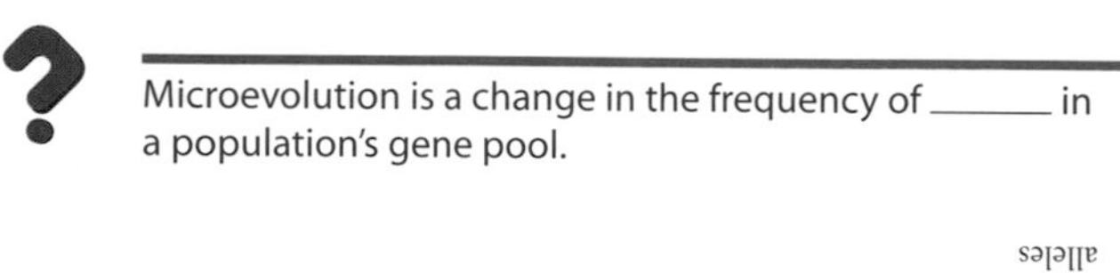
?
Microevolution is a change in the frequency of ______ in a population's gene pool.
alleles

13.8 The gene pool of an idealized, nonevolving population remains constant over the generations

Let's consider an imaginary, nonevolving population of iguanas with two varieties that differ in foot webbing (Figure 13.8A). Let's also imagine that foot webbing is controlled by a single gene. We'll assume that the allele for nonwebbed feet *(W)* is completely dominant to the allele for webbed feet *(w)*. The term "dominant" (see Module 9.3) may seem to suggest that, over many generations of sexual reproduction, the *W* allele will somehow come to "dominate" the population, becoming more and more common at the expense of the recessive *w* allele. In fact, this is not what happens. In an idealized case, the shuffling of genes that accompanies sexual reproduction does not alter the genetic makeup of the population. In other words, sexual reproduction alone does not lead to microevolution. No matter how many times alleles are segregated into different gametes by meiosis and united in different combinations by fertilization, the frequency of each allele in the gene pool will remain constant unless acted on by other agents. This principle is **Hardy-Weinberg equilibrium,** named for two scientists who derived it in 1908.

To test Hardy-Weinberg equilibrium, let's look at two generations of our iguana population. Figure 13.8B shows the genetic situation in the first generation. We have a total population of 500 iguanas; of these, 320 animals have the genotype *WW* (nonwebbed feet), 20 have the genotype *ww* (webbed feet), and 160 have the heterozygous genotype, *Ww* (nonwebbed feet, since *W* is dominant). The proportions of the three possible genotypes (the genotype frequencies) are 0.64 for *WW* ($\frac{320}{500} = 0.64$), 0.04 for *ww* ($\frac{20}{500} = 0.04$), and 0.32 for *Ww* ($\frac{160}{500} = 0.32$).

From the genotype frequencies, we can calculate the frequency of each allele in the first-generation population. Each iguana carries two genes for foot type, so the population has 1000 genes for this characteristic. To find the number of *W* alleles, we add the number carried by the *WW*

Webbing — No webbing

Figure 13.8A Imaginary iguanas, with and without foot webbing

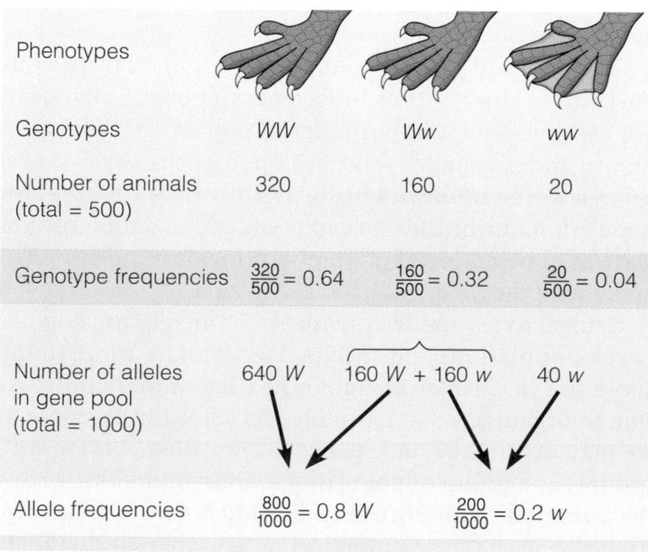

Phenotypes			
Genotypes	*WW*	*Ww*	*ww*
Number of animals (total = 500)	320	160	20
Genotype frequencies	$\frac{320}{500} = 0.64$	$\frac{160}{500} = 0.32$	$\frac{20}{500} = 0.04$
Number of alleles in gene pool (total = 1000)	640 *W*	160 *W* + 160 *w*	40 *w*
Allele frequencies	$\frac{800}{1000} = 0.8\ W$		$\frac{200}{1000} = 0.2\ w$

Figure 13.8B A gene pool of first-generation iguanas

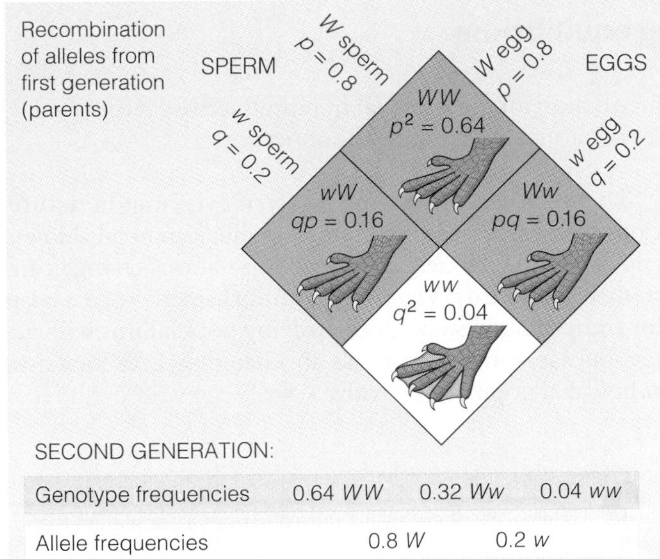

Figure 13.8C The gene pool of second-generation iguanas

iguanas, $2 \times 320 = 640$, to the number carried by the *Ww* iguanas, 160. The total number of *W* alleles is thus 800. The frequency of the *W* allele, which we will call *p*, is $\frac{800}{1000}$, or 0.8. We can calculate the frequency of the *w* allele in a similar way; this frequency, called *q*, is 0.2. (The letters *p* and *q* are often used to represent allele frequencies.)

What happens when the first-generation iguanas form gametes? At the end of meiosis, each gamete has one foot-type allele, either *w* or *W*. The frequency of the two alleles in the gametes is the same as it is in the parental population, 0.8 for *W* and 0.2 for *w*.

Figure 13.8C uses these gamete allele frequencies and the rule of multiplication (see Module 9.7) to calculate the frequencies of the three possible genotypes in the second generation. The probability of producing a *WW* individual (by combining two *W* alleles from the pool of gametes) is $p \times p = p^2$, or $0.8 \times 0.8 = 0.64$. Thus, the frequency of *WW* iguanas in the next generation would be 64%. Likewise, the frequency of *ww* individuals would be $q^2 = 0.04$, or 4%.

For heterozygous individuals, *Ww*, the genotype can form in two ways, depending on whether the sperm or egg supplies the dominant allele. In other words, the frequency of *Ww* would be $2pq = 2 \times 0.8 \times 0.2 = 0.32$, or 32%. Thus, the three possible genotypes have the same frequency in the second generation as they did in the first generation.

Finally, what about the frequencies of the alleles in the second generation? Since the genotype frequencies are the same as in the first generation, the allele frequencies, *p* and *q*, are the same, too. In fact, we could follow the frequencies through many generations, and the results would continue to be the same. Thus, the gene pool of this population is in a state of equilibrium, Hardy-Weinberg equilibrium.

Now, let's write a general formula for calculating the frequencies of alleles in a gene pool from the frequencies of genotypes, and vice versa. In our imaginary iguana population, the frequency of the *W* allele (p) is 0.8, and the frequency of the *w* allele (q) is 0.2. Note that $p + q = 1$; this is always true since the combined frequencies of all alleles must be 100% of the alleles for that gene in the population. If there are only two alleles and we know the frequency of one, we can calculate the frequency of the other one:

$$p = 1 - q \quad \text{and} \quad q = 1 - p$$

Notice in Figures 13.8B and 13.8C that the frequencies of all possible genotypes in the populations also add up to 1 (that is, $0.64 + 0.32 + 0.04 = 1$). We can represent this symbolically with the Hardy-Weinberg equation:

$$\underset{\text{Frequency of } WW}{p^2} + \underset{\text{Frequency of } Ww}{2pq} + \underset{\text{Frequency of } ww}{q^2} = 1$$

An imaginary iguana population of the same species shown in Figure 13.8A is in Hardy-Weinberg equilibrium. The frequency of individuals with webbed feet is 0.01. What is the frequency of individuals that are homozygous for nonwebbed feet?

0.81

The Hardy-Weinberg equation is useful in public health science 13.9

The Hardy-Weinberg equation has broad application. For instance, public health scientists use it to estimate how many people carry alleles for certain inherited diseases. Consider the case of phenylketonuria (PKU), which is an inherited inability to break down a certain amino acid. PKU occurs in about one out of 10,000 babies born in the United States and, if untreated, results in severe mental retardation.

PKU is due to a recessive allele, and thus the frequency of individuals in the U.S. population born with PKU corresponds to the q^2 term in the Hardy-Weinberg equation. Given one PKU occurrence per 10,000 births, $q^2 = 0.0001$. Therefore, the frequency of the recessive allele for PKU in the population, *q*, equals the square root of 0.0001, or 0.01. And the frequency of the dominant allele, *p*, equals $1 - q$, or 0.99. The frequency of carriers, heterozygous people who are normal but may pass the PKU allele on to offspring, is $2pq$, which equals $2 \times 0.99 \times 0.01$, or 0.0198. Thus, the equation tells us that about 2% (actually 1.98%) of the U.S. population carries the PKU allele. Estimating the frequency of a harmful allele is useful for any public health program dealing with genetic diseases.

Which term in the Hardy-Weinberg equation — p^2, $2pq$, or q^2—corresponds to the frequency of individuals who have no alleles for PKU?

p^2

13.10 Five conditions are required for Hardy-Weinberg equilibrium

Hardy-Weinberg equilibrium tells us that something other than the reshuffling processes of sexual reproduction is required to alter a gene pool—that is, to change allele frequencies in a population from one generation to the next. One way to find out what factors *can* change a gene pool is to identify the conditions that must be met if genetic equilibrium is to be maintained. Hardy-Weinberg equilibrium will be maintained in nature only if all five of the following conditions are met:

1. The population is very large.
2. The population is isolated; that is, there is no migration of individuals, and the alleles they carry, into or out of the population.
3. Mutations (changes in genes) do not alter the gene pool.
4. Mating is random.
5. All individuals are equal in reproductive success; that is, natural selection does not occur.

These five conditions are rarely, if ever, met in nature. Consequently, Hardy-Weinberg equilibrium breaks down, and allele frequencies in natural populations change constantly. But Hardy-Weinberg equilibrium gives us a basis for comparing idealized, nonevolving populations with actual ones in which gene pools are changing. Let's look now at how real populations evolve.

Hardy-Weinberg equilibrium describes a population that is not ________.

evolving

13.11 There are five potential causes of microevolution

13B

Five agents, each a deviation from one of the five conditions for Hardy-Weinberg equilibrium, can cause changes in gene pools, or microevolution. These agents are genetic drift, gene flow, mutation, nonrandom mating, and natural selection.

Genetic drift is a change in the gene pool of a small population due to chance. As we have seen, in an idealized population that is not evolving, the gene pool will remain the same from one generation to the next. However, if a population of organisms is small, a chance event can have a disproportionately large effect, making the gene pool different in the next generation. For example, consider what could happen in an imaginary iguana population consisting of only ten individuals. Assume that three of the lizards have the genotype *WW* (nonwebbed feet), two are *Ww* (also nonwebbed), and five are *ww* (webbed). Now imagine that three of the iguanas are killed in an earthquake. If, by chance, these three lizards are *WW* individuals, the result will be a marked decrease in the frequency of the *W* allele, with a corresponding increase in the *w* allele frequency. This change in relative allele frequencies would be a case of microevolution caused solely by chance. Natural selection is not involved.

A population must be infinitely large for genetic drift to be ruled out completely as an agent of microevolution; however, the populations in which genetic drift is most likely to play a major role typically have 100 or fewer individuals.

One situation that involves a population small enough for genetic drift is known as a bottleneck. The **bottleneck effect** is genetic drift resulting from an event that drastically reduces population size. Events such as earthquakes, floods, or fires may kill large numbers of individuals unselectively, producing a small surviving population that is unlikely to have the same genetic makeup as the original population. Figure 13.11A shows how a bottlenecking event works, representing alleles by balls of different colors. Certain alleles (red balls) will be present at higher frequency in the surviving population than in the original population, others (white balls) will be present at lower frequency, and some (green balls) may not be present at all. In a real example, human hunters in the 1890s reduced the population of northern elephant seals in California to about 20 individuals. Since then, this mammal (pictured in Figure 13.11B) has become a protected species, and the population has grown back to over 30,000 members. However, in examining 24 gene loci in a representative sample

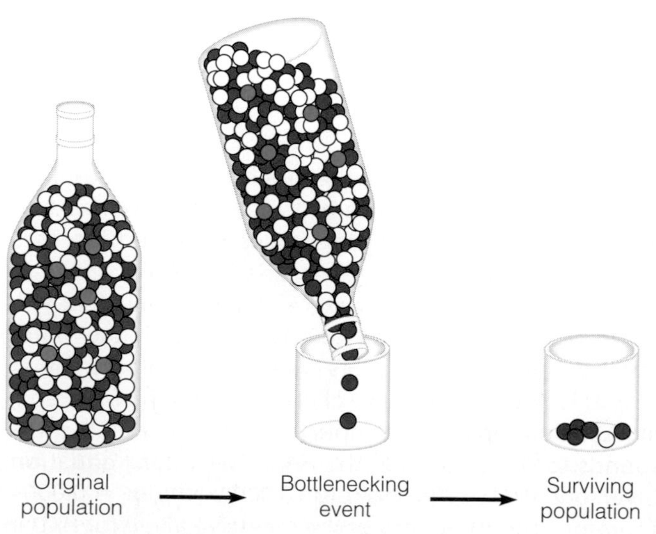

Figure 13.11A The bottleneck effect

of the seals, researchers found *no* variation. For each of the 24 genes they found only one allele, probably because of bottlenecking. In contrast, genetic variation abounds in populations of a closely related species, the southern elephant seal, which was not bottlenecked.

Another situation that can produce a population small enough for genetic drift is the colonization of a new location by a small number of individuals. Random change in the gene pool that occurs in a small colony is called the **founder effect.** In the most extreme case, a single pregnant animal or a single plant seed might found a new population. The founder effect was important in the evolution of many species in the Galápagos Islands. For instance, the ancestors of the marine iguana were probably a few stray land iguanas that arrived there from the South American mainland. The gene pool of the marine iguana would have been derived from those few individuals.

Figure 13.11B Elephant seals descended from bottleneck survivors

The second agent of microevolution, **gene flow,** is the gain or loss of alleles from a population by the movement of individuals or gametes. Gene flow occurs when fertile individuals move into or out of a population, or when gametes (such as the sperm of plant pollen) are transferred from one population to another. Gene flow tends to reduce genetic differences between populations. Over the history of our own species, for example, local groups have sometimes become reproductively isolated. This isolation has reduced gene flow, resulting in genetic distinctions among groups of people living in different parts of the world. Reflecting the genetic differences are phenotypic variation in characteristics such as skin color and other racial and ethnic differences within our species. In a case that we examined in Modules 9.8 and 9.9, isolation resulted in an unusually high frequency of hereditary deafness in a small group of island dwellers. Working against reproductive isolation has been the influence of migrations and wars, which tends to increase interbreeding among groups. Today, air travel makes it possible for people all over the globe to interact, and there is more gene flow among geographically isolated populations than ever before.

A third agent of microevolution is **mutation.** A mutation is a random change in an organism's DNA that creates a new allele (see Module 10.16). Mutations of a given gene are rare events, typically occurring only about once per gene locus per 10^5 or 10^6 gametes. As a result, in a large population, mutation alone does not have much effect in a single generation. Generally, if some new allele rapidly increases its frequency in a population, it is not because mutation is generating the allele in abundance, but because individuals carrying the mutant allele are producing a disproportionate number of offspring as a result of genetic drift or natural selection. Over the long term, however, mutation is vital to evolution because it is the only force that actually generates new alleles. Thus, mutation is the ultimate source of the genetic variation that serves as raw material for evolution.

A fourth agent that can change gene pools is the selection of mates other than by chance, or **nonrandom mating.** In order for Hardy-Weinberg equilibrium to be maintained, every male in the population must have an equal chance of mating with every female in the population. But the fact is, nonrandom mating is the rule in most populations. For example, in some species, males and females with similar phenotypic traits tend to mate. (In the human species, short women often tend to marry short men and tall women, tall men.) Nonrandom mating is also the rule for organisms that stay in one place, such as plants, or that move about but only locally, such as land iguanas living on islands. In such cases, individuals generally mate with their neighbors rather than with more distant members of the population. This tendency promotes inbreeding, mating between closely related partners. The most extreme case of inbreeding is self-fertilization ("selfing"), which is common in plants.

Natural selection, or differential success in reproduction, is the fifth agent of microevolution. It is the factor that is most likely to result in adaptive changes in a gene pool. 

? Random change in the frequencies of alleles in a population is called ________.

genetic drift

13.12 Adaptive change results when natural selection upsets genetic equilibrium

One condition for Hardy-Weinberg equilibrium—that all individuals in a population be equal in their ability to reproduce—is probably never met in nature. Populations of sexually reproducing organisms consist of varied individuals, and some variants leave more offspring than others. In our imaginary iguana population, lizards with webbed feet (genotype *ww*) might produce more offspring because they are more efficient at finding food or mates than lizards without webbed feet (genotype *Ww* or *WW*). This would disturb genetic equilibrium, causing the frequency of the *W* allele to decline in the gene pool. In this way, natural selection results in the accumulation and maintenance of traits that adapt a population to its environment. If the environment should change, natural selection would favor traits adapted to the new conditions. The degree of adaptation that can occur is limited by the amount and kind of genetic variation in the population.

A population that is in the process of evolutionary adaptation due to ________ cannot be in ________ equilibrium because some ________ are increasing in frequency compared to others in the population's gene ________.

natural selection . . . Hardy-Weinberg . . . alleles . . . pool

Variation and Natural Selection

13.13 Variation is extensive in most populations

We are very conscious of human diversity, but individuality in populations of other animals and plants may escape our notice. Nonetheless, some degree of individual variation occurs in all populations. In addition to anatomical differences, most populations have a great deal of variation that can be detected only by biochemical means. Researchers often use the technique of electrophoresis (see Module 12.10) to study molecular variation in populations.

Not all variation in a population is heritable. The phenotype results from a combination of the genotype, which is inherited, and many environmental influences. For instance, a strength-training program can build up your muscle mass beyond what would naturally occur from your genetic makeup, but you would not pass this environmentally induced phenotypic change on to your offspring. Only the genetic component of variation can lead to evolution as a result of natural selection.

Many of the variable traits of the individuals in a population result from the combined effect of several genes. As we saw in Module 9.16, polygenic inheritance produces traits that vary more or less continuously—in human height, for instance, from very short individuals to very tall ones. By contrast, other features, such as red versus white flowers in some plants, are determined by a single gene locus with different alleles that produce only distinct phenotypes (red or white flower color, for example); there are no in-between types. In such cases, when a population includes two or more forms of a phenotypic characteristic, the contrasting forms are called **morphs.** A population is said to be **polymorphic** for a characteristic if two or more morphs are present in readily noticeable numbers (that is, if neither morph is extremely rare). Figure 13.13 illustrates a striking example of polymorphism in a population of California king snakes. The two morphs differ markedly in their color patterns. Polymorphism is also extensive in human populations, both in physical characteristics, such as the presence or absence of freckles, and in biochemical features, such as the ABO blood groups (see Module 9.13).

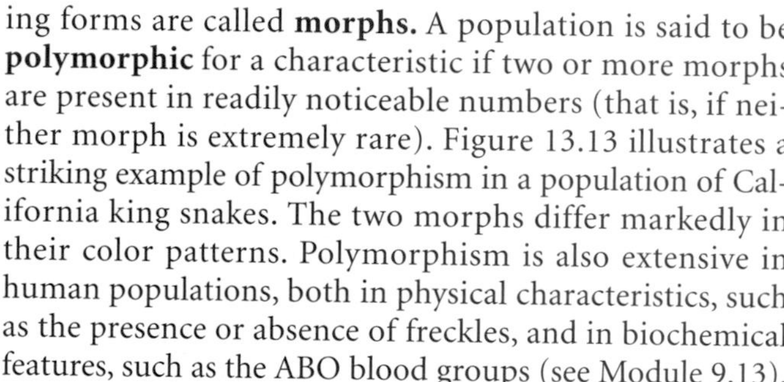

Most species exhibit geographic variation among populations. For example, a wild plant population with two morphs—red flower color (genotypes *RR* and *Rr*) and white color (genotype *rr*)—could have a higher frequency of recessive alleles than another population of the same plant because of a local prevalence of pollinating insects that favor white flowers.

Sometimes geographic variation occurs in what is called a **cline,** a graded change in an inherited characteristic along a geographical continuum. Clines often occur where there is a gradation in some environmental variable. For example, the body size of many birds and mammals, such as white-tailed deer, tends to increase gradually with increasing latitude and colder climate in North America. Large size is adaptive in northern latitudes because it reduces the ratio of body surface area to volume and helps conserve body heat. Smaller body size helps dissipate heat in warmer regions. How do clines and other inherited variations arise? We address this question in the next module.

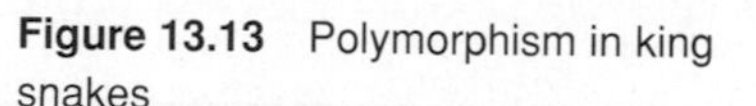

Figure 13.13 Polymorphism in king snakes

Which of the following variations in a human population is the best example of polymorphism: height; blood group (A, B, AB, or O); number of fingers; math proficiency?

Blood group

Two random processes generate variation 13.14

13C

Mutations and sexual recombination, which are both random processes, produce genetic variation. As we saw in Module 10.16, mutations can actually create new alleles. For example, in Figure 13.14A, a point mutation in a gene substitutes one nucleotide (T) for another (G). This type of change will be harmless if it does not affect the function of the protein the DNA encodes. However, if it does affect the protein's function, the mutation will probably be harmful. An organism is a refined product of thousands of generations of past selection, and a random change in its DNA is not likely to improve its genome any more than shooting a bullet through the hood of a car is likely to improve engine performance.

On rare occasions, however, a mutant allele may actually improve the adaptation of its bearer to the environment and enhance reproductive success. This kind of effect is more likely when the environment is changing in such a way that mutations that were once disadvantageous are favorable under the new conditions. For instance, mutations that endow house flies with resistance to the pesticide DDT also reduce their growth rate. Before DDT was introduced, such mutations were harmful. But once DDT was part of the environment, the mutant alleles were favored, and natural selection increased their frequency in fly populations.

On a larger scale are so-called chromosomal mutations, which are changes involving stretches of DNA long enough to be detected microscopically. A single chromosomal mutation affects many genes and is almost certain to be harmful. Very rarely a translocation (such as the one shown in Figure 8.23B) might bring benefits. For instance, the movement of a piece of one chromosome to another chromosome might link genes that affect the organism in some positive way when they are inherited together as a package.

Organisms with very short generation spans, such as prokaryotes, can evolve rapidly by means of mutations alone. Prokaryotes multiply so rapidly that a beneficial mutation can increase its frequency in descendant populations in a matter of hours or days. Furthermore, prokaryotes have only a single allele for each inherited trait. Thus, a newly created allele can have an effect immediately, because it cannot be obscured by the expression of another allele for the same trait.

For most animals and plants, however, their long generation times and generally diploid condition prevent most mutations from significantly affecting genetic variation from one generation to the next, at least in large populations. Consequently, animals and plants depend mainly on sexual recombination for the genetic variation that makes adaptation possible. As we saw in Modules 8.14 and 8.15, fresh assortments of existing alleles arise every generation from three random components of sexual recombination: independent assortment of homologous chromosomes during meiosis, crossing over during meiosis, and random fertilization of an egg by a sperm. In Figure 13.14B, we see the results of a mating of parents with the genotypes A^1A^1 and A^2A^3, where A^1, A^2, and A^3 are three different alleles for a gene. Even without considering independent assortment or crossing over, the offspring have different combinations of alleles than their parents do.

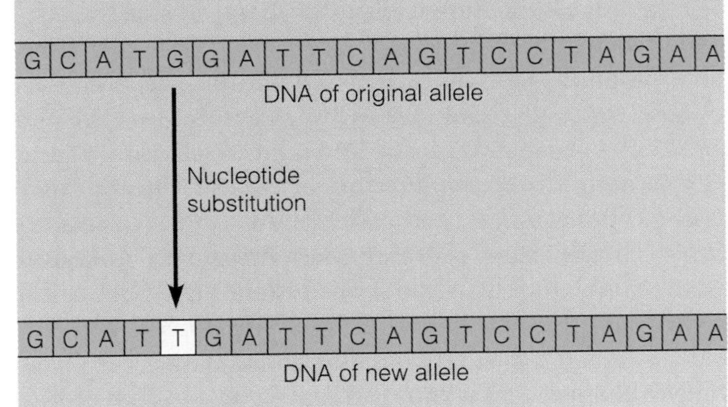

Figure 13.14A Creating a new allele by mutation

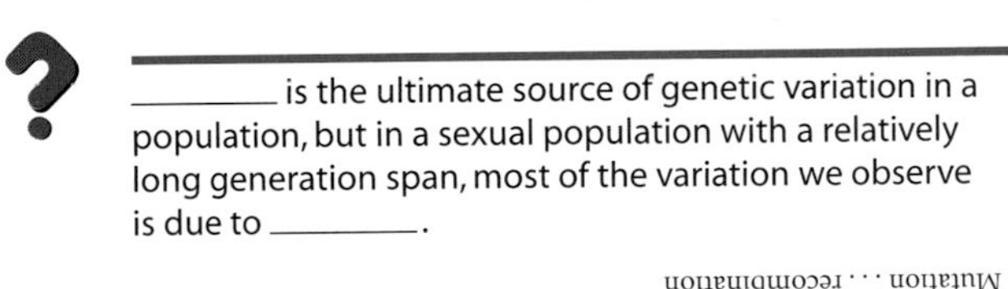

_______ is the ultimate source of genetic variation in a population, but in a sexual population with a relatively long generation span, most of the variation we observe is due to _______.

Mutation . . . recombination

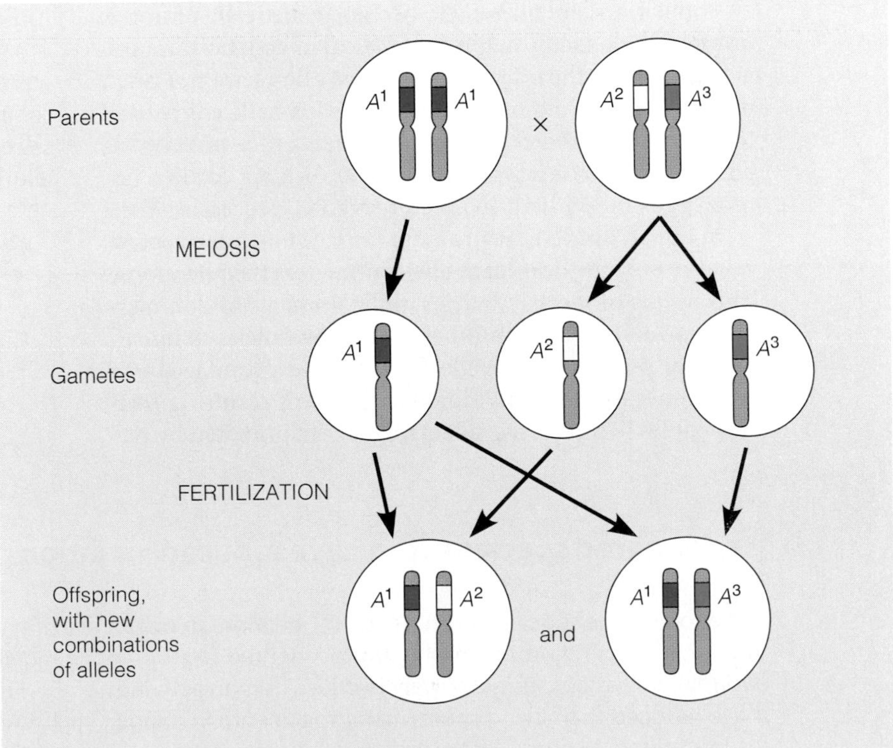

Figure 13.14B Shuffling alleles by sexual recombination

13.15 Overview: How natural selection affects variation

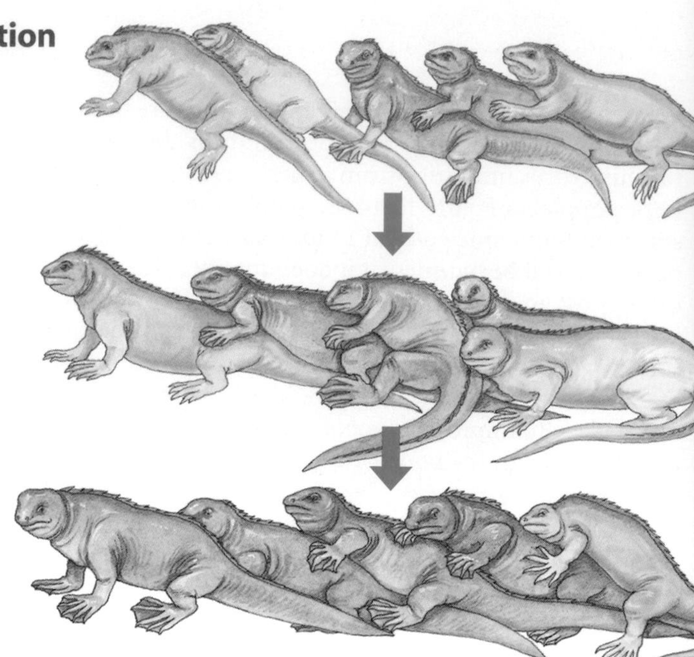

Figure 13.15 Natural selection in successive populations of iguanas

The subject of variation leads us back to Darwin, who devoted two whole chapters of *The Origin of Species* to heritable variations before he discussed natural selection. Following Darwin's lead, we now start to look at the connection between variation and natural selection. Figure 13.15 is an idealized model summarizing the effect of natural selection on inherited variation in successive generations. The top row represents an imaginary ancestral iguana population living on a small island. The population is varied, with most individuals having nonwebbed feet and a rounded tail, and two (color-coded dark green) having webbed feet and a flattened tail. Food is scarce on the island itself but abundant in the surrounding ocean. Natural selection favors webbed feet and flattened tails, traits that help the animals swim and exploit the rich food supply at sea. Over the generations, individuals with webbed feet and flattened tails leave proportionately more offspring, and we see the results in the population at the bottom. Natural selection has eliminated most of the individuals with nonwebbed feet and rounded tails. The bottom population is clearly better adapted to life on the island than was the ancestral population. This imaginary model indicates how natural selection adapts populations to the environment.

Our imaginary iguana model also makes the point that natural selection tends to reduce variability in populations. Notice that the population at the bottom of the figure is less varied than the ancestral population at the top. In fact, most populations tend to lose genetic variability as a result of the culling effect of natural selection.

Populations are unlikely to become genetically uniform, however. One factor helping to maintain genetic variation is the fact that the effects of recessive alleles are not often displayed in diploid organisms. A recessive allele is subject to natural selection only when it influences the phenotype, and this occurs only when two copies of it appear in a homozygous individual. In a heterozygote, a recessive allele is, in effect, hidden, or protected, from natural selection because only the dominant allele influences the phenotype. This hiding of recessive alleles in the presence of dominant ones can allow a large number of recessive alleles to remain in a gene pool, and these alleles may prove advantageous in later generations. Individuals with a trait resulting from two copies of a recessive allele might be eliminated by natural selection in one environment, but if the environment changes, such individuals might have greater reproductive success.

Genetic variability in diploid organisms can also be preserved by natural selection, the very force that generally reduces it. For some genes, natural selection tends to maintain the two or more alleles that appear in the heterozygote and therefore preserve genetic variability. In such cases, there is a **heterozygote advantage**—that is, heterozygous individuals have greater reproductive success than homozygotes. One example is the resistance to malaria conferred by the recessive sickle-cell allele (see Module 9.14). In areas where malaria is a major cause of death, natural selection favors heterozygotes (carriers of the sickle-cell allele) because they are resistant to malaria. 

? Some organisms, such as mosses and many algae, are haploid as adult organisms. Why would natural selection tend to narrow genetic variation more in populations of such haploid organisms than it would for populations of diploid organisms?

All alleles in a haploid organism are phenotypically expressed and are hence screened by natural selection.

13.16 Endangered species often have reduced variation

The topic of genetic variation has direct bearing on one of our most pressing problems, the unprecedented rate of the worldwide loss of species. Largely because human activities have reduced their living space, many species are in danger of becoming extinct. Without significant efforts to curb habitat destruction, the rate will climb even more. Competition with exotic (nonnative) species and unregulated hunting also threaten many species.

Endangered species typically have low genetic variability. As their populations are severely reduced, the diversity of their gene pools also declines. One such species is the cheetah *(Acinonyx jubatus)*, pictured in Figure 13.16. The fastest of all running animals, the cheetah can run down the swiftest antelope. These magnificent cats were formerly widespread in Africa and Asia. Like many African mammals, their numbers fell drastically during the last ice age some 10,000 years ago. At that time, the species may have suffered a severe bottleneck, possibly as a result of disease, human hunting, and periodic droughts. Today, only three cheetah populations exist in the wild, one in East Africa, another in South Africa, and a third in northern Iran. All three populations are small and endangered, primarily because their habitat is still being reduced.

Figure 13.16 The cheetah, a species with low genetic variability

Studies of the African cheetah populations show very low genetic variation. Today, only about 1.4% of the East African population's gene loci are heterozygous, a very low percentage compared with other species of the cat family. Also, only about 0.04% of the South African population's gene loci are heterozygous. This represents extreme genetic uniformity—far greater than average for mammals, and even greater than for some highly inbred varieties of laboratory mice. Some researchers think that the South African cheetah population suffered a second bottleneck during the nineteenth century when South African farmers hunted the animals to near extinction. Some researchers also posit that this species has always had low genetic variability.

However we interpret the cheetah's gene pool, its current genetic uniformity, coupled with an increasing loss of habitat, makes its future seem precarious. The cheetahs remaining in Africa are being crowded into nature preserves and parks as human demands on the land increase. Along with crowding comes greater opportunity for predation by other large cats, such as lions, and increased potential for the spread of disease. With so little variability, the cheetah may have a reduced capacity to adapt to such environmental challenges. Captive breeding programs are already under way and may be required for the cheetah's long-term survival. We look at the problem of endangered species in more detail in Chapter 38. 

? Why might new strains of pathogens pose a greater threat to cheetah populations than to mammalian populations having more genetic variation?

Because cheetah populations have so little genetic variation, there is potential for some new disease against which no individuals are resistant.

Not all genetic variation may be subject to natural selection 13.17

Some genetic variations in populations seem to have a trivial impact on reproductive success and therefore may not be subject to natural selection. The diversity of human fingerprints, for example, seems to be an example of **neutral variation**—variation in a heritable characteristic that provides no apparent selective advantage for some individuals over others. The neutral variation hypothesis proposes that species typically have numerous alleles that confer no selective advantage or disadvantage. Some of these supposedly neutral alleles will increase their frequency in the gene pool and others will decrease by the chance effects of genetic drift, but natural selection will not affect them.

There is little consensus among evolutionary biologists about how much genetic variation is neutral. Some researchers do not believe that any variation is truly neutral. They point out that variations appearing to be neutral may influence reproductive success in ways that are difficult to measure. We may be able to show that a particular allele is harmful or beneficial, but we cannot demonstrate that an allele has no effect at all on an organism. Also, a variation may be neutral in one environment but not in another. We can never know the degree to which genetic variation is neutral. But we can be certain that even if only a fraction of the extensive variation in a gene pool significantly affects the organism, that is still an enormous resource of raw material for natural selection to act on. 

? Why isn't it possible to determine with certainty that a particular hereditary variation is neutral?

A phenotype may confer some advantage or disadvantage that we do not yet understand.

Figure 13.17 Human fingerprints, probably an example of neutral variation

13.18 The perpetuation of genes defines evolutionary fitness

In discussing natural selection in later editions of *The Origin of Species,* Darwin wrote: "This preservation of favourable individual differences and variations, and the destruction of those which are injurious, I have called Natural Selection, or the Survival of the Fittest." The phrase "survival of the fittest" caught on and is often quoted when evolution is discussed. As long as we keep in mind what the words "survival" and "fittest" mean in an evolutionary sense, the phrase is still a useful summary of the process of natural selection. The emphasis must be on the survival of genes, not individual organisms. Today, we define evolutionary **fitness** as the contribution an individual makes to the gene pool of the next generation relative to the contribution of other individuals. Thus, the fittest individuals in the context of evolution are those that pass on the greatest number of genes to the next generation.

We examine the connections among fitness, genes, and individuals in more depth in the next module. 

? What number equals the Darwinian fitness of a sterile eagle?

Zero

13.19 Natural selection acts on whole organisms and affects genotypes as a result

Evolutionary fitness has to do with genes, but it is the phenotype of an organism—its physical traits, metabolism, and behavior—not the genotype, that is directly exposed to the environment. Natural selection therefore targets the whole organism, which is an integrated composite of many phenotypic traits. Individuals with a high degree of fitness are those whose phenotypic traits enable them to reproduce and contribute genes to more offspring than other individuals. For example, marine iguanas with the flattest tails and most foot webbing may produce more offspring because they swim faster and therefore can exploit marine food resources more efficiently.

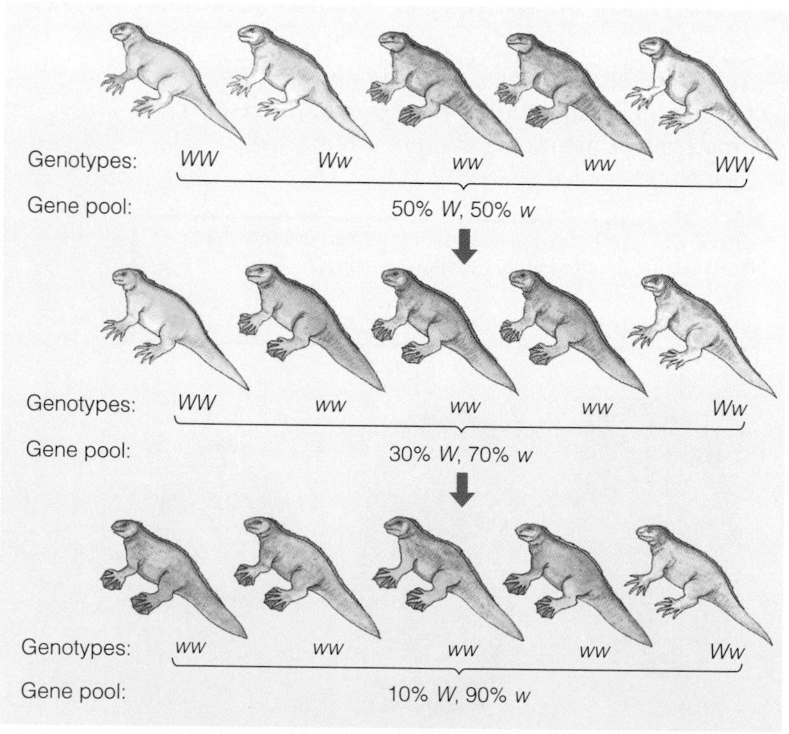

Genotypes: *WW* *Ww* *ww* *ww* *WW*
Gene pool: 50% *W*, 50% *w*

Genotypes: *WW* *ww* *ww* *ww* *Ww*
Gene pool: 30% *W*, 70% *w*

Genotypes: *ww* *ww* *ww* *ww* *Ww*
Gene pool: 10% *W*, 90% *w*

Figure 13.19 Changes in a gene pool as natural selection culls iguanas with nonwebbed feet

Figure 13.19 illustrates the connections between natural selection, genotypes, and changes in gene pool, using our imaginary iguana populations from Figure 13.15 and focusing on the hypothetical foot-webbing gene. Natural selection does not target genes or genotypes directly, but, in favoring certain phenotypes over others, it has the effect of favoring certain genotypes. The favored genotypes are those whose positive phenotypic effects outweigh any harmful effects they may have on the reproductive success of the organism. Put another way, in culling less fit individuals, natural selection also culls unfavorable genotypes. In this example, you see that the eliminated genotypes are those that produce nonwebbed feet. The result in the population at the bottom is a gene pool in which 90% of the gene loci for foot webbing have the *w* (webbing) allele. In this idealized model, the change in the gene pool is reflected in the relative abundance of the webbed-foot phenotype.

There is one other very important point about genotypes and natural selection: The fitness of a genotype at any one gene locus depends on many other genes. For instance, in the marine iguana, alleles that enhance the ability of an individual to swim and search for food at sea may be useless or even harmful in the absence of alleles at other loci that enhance the animal's ability to get rid of excess salt and survive cold temperatures. Overall, the entire genome of an organism is a unit of functionally interrelated genes, in the same sense that the whole organism is an integrated composite of many phenotypic traits. Thus, natural selection culls or favors whole genomes as it targets and acts on whole phenotypes.

? Natural selection screens genotypes indirectly by favoring some ________ over others.

phenotypes

Keeping in mind that evolution acts on populations and that natural selection targets whole organisms, let's see what the culling effects of natural selection can actually do to the frequencies of phenotypic variants in a population. Here we use an imaginary population of snails, with individuals varying in the intensity of their shell color. The bell-shaped curve in the top graph of Figure 13.20 depicts an idealized population that could result from a polygenic inheritance pattern for variation in shell color. In this case, which we use as our starting population, shell color varies along a continuum from very light (only a few individuals) through various intermediate shades (many individuals) to very dark (few individuals). The other three graphs show three different ways in which natural selection could alter the phenotypic variation in the idealized population.

Stabilizing selection favors intermediate variants. It typically occurs in relatively stable environments where conditions tend to reduce phenotypic variation. In the snail population depicted in the graph on the bottom left, stabilizing selection has eliminated the extremely light and dark individuals, and the population has a greater number of intermediate phenotypes, which are best suited to a stable environment. Stabilizing selection probably prevails most of the time in most populations. For example, this type of selection keeps the majority of human birth weights in the range of 3–4 kg (6.5–9 lb). For babies much smaller or larger than this size, infant mortality is greater.

Directional selection shifts the overall makeup of the population by acting against individuals at *one* of the phenotypic extremes. For the snail population in the bottom center graph, the trend is toward darker shell color, as might occur if the landscape were blackened by volcanic ash. Directional selection is most common during periods of environmental change, or when members of a species migrate to some new habitat with different environmental conditions. Changes we described in populations of the peppered moth are an example of directional selection.

Diversifying selection typically occurs when environmental conditions are varied in a way that favors individuals at *both* extremes of a phenotypic range over intermediate individuals. For the snails in the graph on the bottom right, individuals with light and dark shells have increased their numbers relative to intermediate variants. We would expect this to occur if the snails had recently colonized a patchy habitat where a background of white sand was studded with dark lava rocks.

The effects of natural selection on populations have direct bearing on some important environmental problems, as we see next. 

? Of the three modes of natural selection, which is most common?

Stabilizing selection

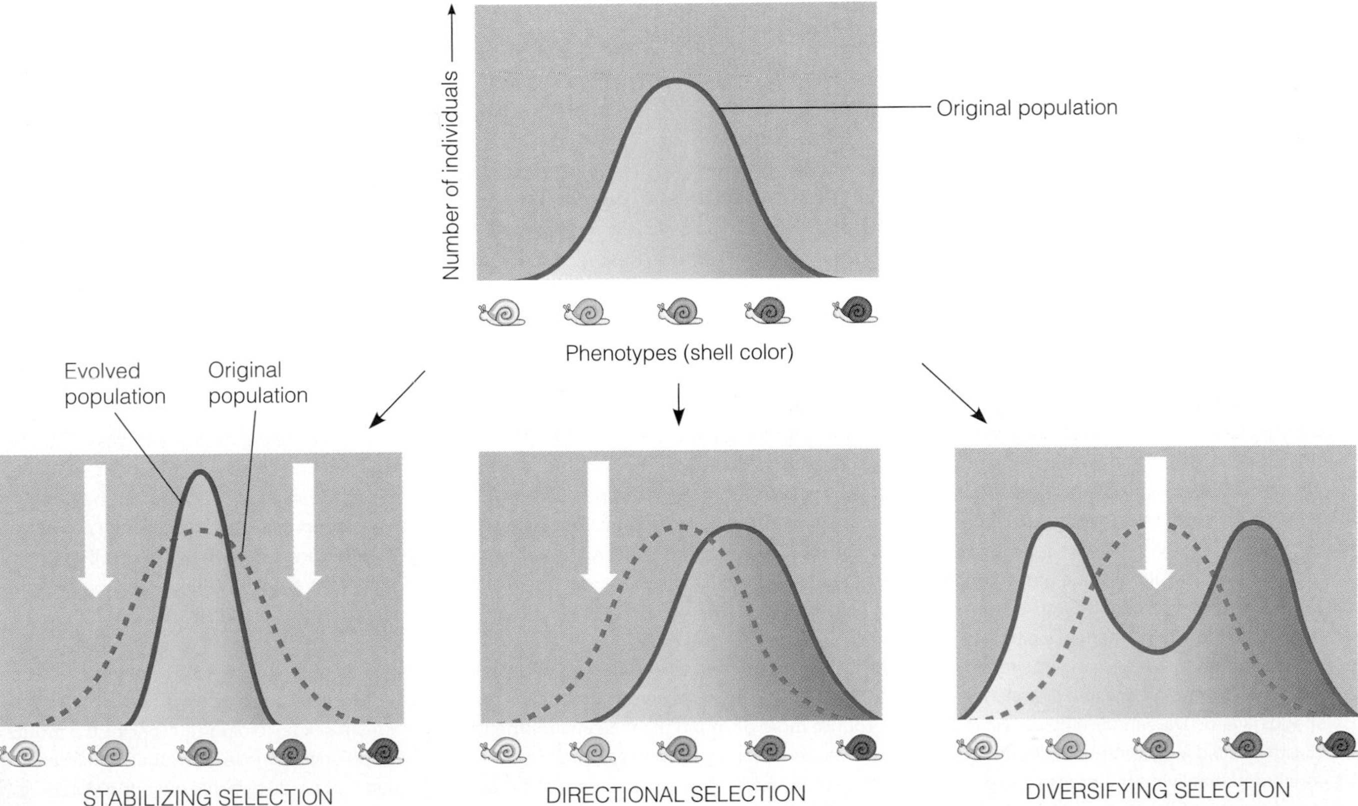

Figure 13.20 Three modes of natural selection

13.21 Directional selection has produced resistant populations of pests and parasites

Over the past several decades, the world has witnessed some very unsettling examples of natural selection: the development of pesticide resistance in insects, and antibiotic resistance in bacteria that cause human disease. Some of the protists that cause diseases such as malaria have also evolved resistance to drugs used to control them. Pesticide-resistant mosquitoes carrying malarial or bacterial pathogens that may themselves be drug-resistant pose a serious public health problem with no immediate solutions.

Additional applications of the same insecticide will not be effective, and the population of resistant insects will grow.

Figure 13.21 Formation of an insecticide-resistant population

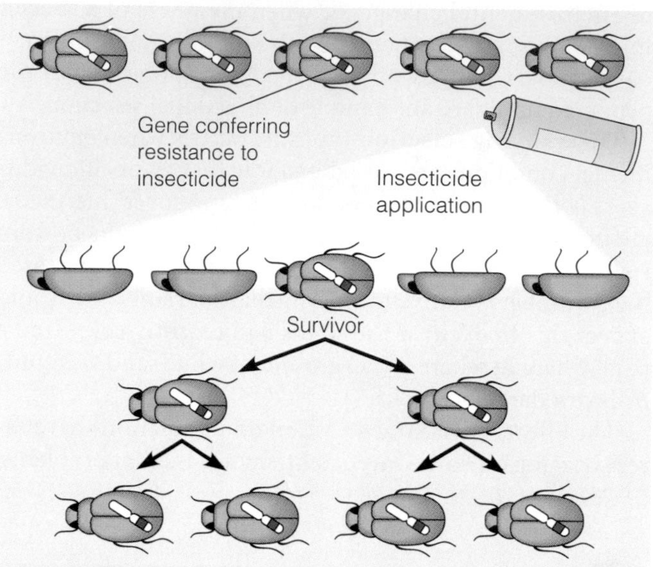

Figure 13.21 illustrates how directional selection has produced resistant populations. When a new insecticide, antibiotic, or drug is first used, it is usually effective in small amounts, killing all but a few individuals in a population. The survivors live and reproduce because by chance they have genes that protect them from the poison. Eventually, most of the population consists of resistant individuals, and the poison is ineffective.

When challenged with new problems, any population adapts to the changes, moves, or becomes extinct. Insects and microorganisms can usually adapt because they are numerous and varied and can multiply rapidly. As the fossil record shows, however, many more species have become extinct than currently exist. Those populations that do survive crises often change enough to become new species, as we will see in the next chapter.

? Explain why the following statement is incorrect: "Pesticides have created resistant insects."

Pesticides have not created resistant insects, but have increased the frequency of alleles for resistance that were already naturally present among the varying members of insect populations.

Chapter Review

Chapter Summary

Evidence of Evolution (Introduction–13.3) Studies of historical change in many kinds of organisms have led to the theory of evolution. All organisms have evolutionary adaptations, inherited characteristics that enhance their ability to survive and reproduce in a particular environment **(Introduction).** Countering the idea that species evolve, Aristotle and the Judeo-Christian culture held that species are fixed. Fossils suggested that life forms change, a concept embraced by J. B. Lamarck in the early 1800s. Lamarck proposed the erroneous idea that phenotypic characteristics acquired during an individual's lifetime are passed on to offspring. As a crew member on the voyage of HMS *Beagle* in the 1830s, Charles Darwin observed similarities between living and fossil organisms of South America and the nearby Galápagos Islands. Darwin became convinced that Earth was old and continually changing. He concluded that living things also could change, or evolve, over generations and that living species are descended from earlier life forms; this concept he called descent with modification **(13.1).** Fossils (the remnants of extinct organisms) and the fossil record (the order in which fossils appear in the strata of sedimentary rocks) strongly support the theory of evolution. The fossil record shows that organisms have appeared in a historical sequence, and many fossils link early extinct species with those living today **(13.2).** Other evidence for evolution comes from biogeography, comparative anatomy, comparative embryology, and molecular biology. Conclusions based on comparisons of DNA and proteins generally agree with those based on fossils and anatomy **(13.3).**

Darwin's Theory and the Modern Synthesis (13.4–13.12) Darwin observed that organisms produce more offspring than the environment can support, that organisms vary, and that their variations can be inherited. He concluded that the individuals best suited for a particular environment are more likely to survive and reproduce than those less well adapted. Darwin saw natural selection, the essence of which is differential (unequal) reproductive success, as the basic mechanism of evolution. As a result of natural selection, the proportion of individuals with favorable characteristics increases, and populations gradually change in response to the environment. Darwin also saw that when humans choose organisms with desired characteristics as breeding stock, they are taking the role of the environment, bringing about differential reproduction and modifying species **(13.4).** Natural selection has been observed in populations of snails, insects, microorganisms, and many other organisms **(13.5).**

The modern synthesis connects Darwin's theory with the science of population genetics, which focuses on collections of genes in populations. A sexual (biological) species is a group of populations whose individuals can interbreed and produce fertile offspring **(13.6).** A gene pool is the total collection of genes in a population at any one time. Microevolution is a change in the relative frequencies of alleles in a gene pool **(13.7)**. As prescribed by Hardy-Weinberg equilibrium, the shuffling of genes during sexual reproduction does not alter the proportions of different genes in a gene pool **(13.8)**. Public health scientists use the Hardy-Weinberg equation to estimate frequencies of disease-causing alleles in human populations **(13.9)**. A population's gene pool will remain the same from generation to generation only if the population is large and isolated from other populations, mutations do not occur, mating is random, and all genotypes have equal reproductive success **(13.10)**. Allele frequencies in populations change (microevolution occurs) because of one or more of the following: genetic drift (change in a gene pool due to chance), gene flow (movement of genes carried by individuals or gametes from one population to another), mutations, nonrandom mating, and natural selection (differential success in reproduction) **(13.11)**. Natural selection probably occurs in all populations. With some individuals in a varied population leaving more offspring than others, natural selection upsets genetic equilibrium and results in the accumulation and maintenance of traits that adapt a population to its environment **(13.12)**.

Variation and Natural Selection

(13.13–13.21) Phenotypic variation may be environmental or genetic in origin, but only genetic changes result in evolutionary adaptation. Many populations exhibit polymorphism—different forms of phenotypic characteristics—and geographic variation **(13.13)**. Genetic variation is generated by mutation and by sexual recombination **(13.14)**. Natural selection tends to reduce variability in populations, while the diploid condition preserves it by "hiding" recessive alleles **(13.15)**. Low genetic variability may reduce the capacity of endangered species to survive as humans continue to alter the environment **(13.16)**. Some variations may be neutral, providing no apparent advantage or disadvantage **(13.17)**. An individual's evolutionary fitness is the contribution it makes to the gene pool of the next generation relative to the contribution made by other individuals **(13.18)**. Natural selection culls less fit individuals, thereby also culling unfavorable genotypes **(13.19)**. In a stable environment, stabilizing selection tends to favor intermediate forms and cull extreme forms. During periods of environmental change, directional selection acts against one extreme, while diversifying selection favors extreme types over intermediate forms **(13.20)**. Directional selection has produced pesticide-resistant insects and antibiotic-resistant bacteria **(13.21)**.

Testing Your Knowledge

Multiple Choice

1. The processes of ____ and ____ generate variation, and ____ produces adaptation to the environment.
   a. sexual recombination . . . natural selection . . . mutation
   b. mutation . . . sexual recombination . . . genetic drift
   c. genetic drift . . . mutation . . . sexual recombination
   d. mutation . . . natural selection . . . sexual recombination
   e. mutation . . . sexual recombination . . . natural selection

2. Natural selection is sometimes described as "survival of the fittest." Which of the following most accurately measures an organism's fitness?
   a. how strong it is when pitted against others of its species
   b. its mutation rate
   c. how many fertile offspring it produces
   d. its ability to withstand environmental extremes
   e. how much food it is able to make or obtain

3. A geneticist studied a grass population growing in an area of erratic rainfall. She found that plants with alleles for curled leaves reproduced better in dry years, and plants with alleles for flat leaves reproduced better in wet years. This situation would tend to ____. *(Explain your answer.)*
   a. cause genetic drift in the grass population
   b. preserve the variability in the grass population
   c. lead to directional selection in the grass population
   d. lead to uniformity in the grass population
   e. cause gene flow in the grass population

4. Birds with average-sized wings survived a severe storm more successfully than other birds in the same population with longer or shorter wings. This illustrates
   a. the founder effect
   b. stabilizing selection
   c. artificial selection
   d. gene flow
   e. diversifying selection

5. Which of the following is a true statement about Charles Darwin?
   a. He was the first to discover that living things can change, or evolve.
   b. He based his theory on the inheritance of acquired characteristics.
   c. He worked out the principles of population genetics.
   d. He proposed natural selection as the mechanism of evolution.
   e. He was the first to realize that Earth is billions of years old.

Describing, Comparing, and Explaining

1. Write a paragraph briefly describing the kinds of evidence for evolution.

2. Sickle-cell disease is caused by a recessive allele. Roughly one out of every 500 African Americans (0.2%) is afflicted with sickle-cell disease. Use the Hardy-Weinberg equation to calculate the percentage of African Americans who are carriers of the sickle-cell allele. (*Hint:* $0.002 = q^2$.)

Thinking as a Scientist

A population of snails is preyed upon by birds that break the snails open on rocks, eat the soft bodies, and leave the shells. The snails occur in both striped and unstriped forms. In one area, researchers counted both live snails and broken shells. Their data are summarized below:

	Striped	Unstriped	Total	Percent Striped
Living	264	296	560	47.1
Broken	486	377	863	56.3

Which snail form seems to be better adapted to this environment? Why? Predict how the frequencies of striped and unstriped individuals might change in the future.

Science, Technology, and Society

School districts in several states have been criticized by groups demanding that science classes give "equal time" to alternative, usually fundamentalist Christian, interpretations of the origin and history of life. They argue that it is only fair to let students evaluate both evolution and the idea that all species were created by God as the Bible relates. Do you think religious views about the origin of species should receive the same emphasis as evolution in science courses? Why or why not?

Media Resources

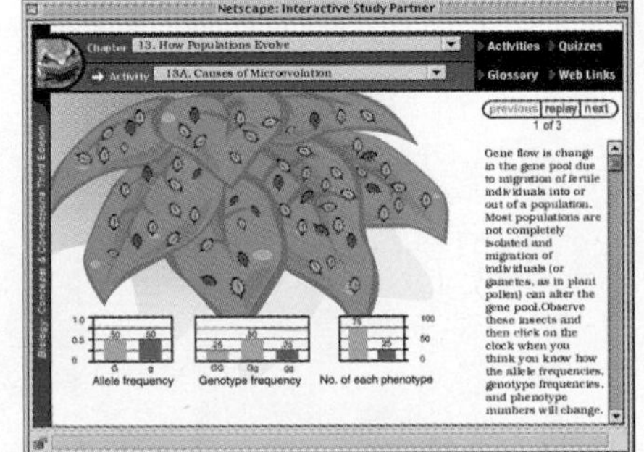

For further review of the concepts in this chapter, use the activities and quizzes on the CD-ROM.

Visit the Special Edition of The Biology Place web site for activities, quizzes, news articles, and web links.

14 The Origin of Species

Outline

North America has ten species of skunks. The one most people have seen—or at least smelled—is the abundant and widespread striped skunk *Mephitis mephitis* (from the Latin for "bad odor"). The skunk in the photograph at the left is a spotted skunk, rarely seen but especially interesting because it illustrates some important concepts about biological species. This particular skunk belongs to a species called the western spotted skunk (*Spilogale gracilis,* from the Latin for "slender spotted cat"). The adult is only about the size of a house cat, but it has a potent chemical arsenal that makes up for its small size. Before spraying her potent musk, a female guarding her young usually warns an intruder by raising her tail, stamping her forefeet, raking the ground with her claws, or even doing a handstand, as shown on the facing page. When all else fails, she can spray her penetrating odor for 3 meters with considerable accuracy.

The western spotted skunk inhabits a variety of environments in the United States, from the Pacific coast to the western Great Plains. It is closely related

Mechanisms of Speciation

to the eastern spotted skunk, *Spilogale putorius* ("foul-smelling cat"), which occurs throughout the southeastern and midwestern United States. The ranges of these two species overlap, and the two species look so much alike that even experts can have a difficult time telling them apart. Both are black with broken white stripes and spots. Individuals of the western species are, on average, slightly smaller, and some have a white tip on the tail, but these and other minor differences in body form are not always present.

For many years, biologists debated whether all spotted skunks belong to one species. But in the 1960s, studies of sexual reproduction in these animals showed that they are indeed two species. Reproduction in the eastern spotted skunk is a straightforward affair. Mating occurs in late winter, and young are born between April and July. In marked contrast, the western spotted skunk includes what is called delayed development in its reproductive cycle. Mating takes place in the late summer and early fall, and zygotes begin to develop in the uterus of the female. Further development, however, is temporarily arrested at an early point called the blastocyst stage. Blastocysts remain dormant in the female's uterus throughout the winter months and resume growth in the spring, with the young (usually 5–7) being born in May or June. Because mating occurs at different times of the year for the two species, there is no opportunity for gene flow between populations of eastern and western spotted skunks. Thus, they are separate species, despite the pronounced similarities in their body form and coloration.

Spotted skunks show us that looks can be deceiving. Without knowledge of the mating cycles, we could interpret the minor differences between the two species as insignificant phenotypic variation and conclude that there is only one species of spotted skunk in North America. Many other species are also difficult to distinguish and define. So before we pursue this chapter's main subject—how different species evolve—we need to examine more closely what we mean by the term "species."

A spotted skunk *(Spilogale)* warning an intruder

Concepts of Species

14.1 What is a species?

The word species is from the Latin for "kind" or "appearance," and indeed we learn to distinguish between kinds of plants or animals—between dogs and cats, for instance—from differences in their appearance. **Taxonomy** is the branch of biology concerned with naming and classifying the diverse forms of life. In the eighteenth century, Swedish physician and botanist Carolus Linnaeus developed the two-part, or binomial, system of naming organisms, which we still use. For our own species, the binomial is *Homo sapiens* (Latin for "wise human being"). Linnaeus defined each of the species he named by its physical appearance.

As we saw with spotted skunks, there are potential pitfalls to defining species by their appearance alone. Like the spotted skunks, the two birds in Figure 14.1A look much the same, even though they represent two species; the one on the left is an eastern meadowlark *(Sturnella magna)*, and the one on the right is a western meadowlark *(Sturnella neglecta)*. Though their body shapes and colorations are very similar, the songs of the two species are different, a feature that helps prevent interbreeding between them.

Figure 14.1A Similarity between two species of songbirds, the eastern meadowlark, *Sturnella magna,* (left) and the western meadowlark, *Sturnella neglecta*

Whereas the individuals of a species of spotted skunk or meadowlark seem to exhibit fairly limited variation in physical appearance, certain other species—our own, for example—seem extremely varied. If we did not know that humans all belong to one species, *Homo sapiens,* the physical diversity within our species (partly illustrated in Figure 14.1B) might lead us to guess that there are several human species. Taxonomists usually have to rely on a limited number of phenotypic features (such as physical appearance and songs) to tell one species from another. Whichever features are used, it is important to realize that phenotype reflects both genotype and environmental influences. When all we know about a population are some of the phenotypic traits of its members, we may not be able to tell whether the population should be considered a species.

Figure 14.1B Diversity within one species

What actually determines the gap between two species? One view of species, called the **biological species concept,** centers on sexual reproduction. It defines a species as a population or group of populations whose members have the potential to interbreed and produce fertile offspring (offspring who themselves can reproduce). Members of one species may mate with members of other species, but the offspring (or those of later generations) will not be fertile. In effect, a failure of sexual reproduction (an inability to produce successive generations of fertile offspring) prevents gene flow and is the gap between species. Western and eastern spotted skunks, for example, remain distinct species even where they share territory, because they are reproductively isolated; individuals of these species do not interbreed and produce fertile hybrids. Next we examine some situations that are less clear-cut and that demonstrate the limitations of the biological species concept.

Although horses and donkeys can mate and produce viable offspring, they are considered separate species according to the ________ species concept because the offspring of these matings, mules, are ________.

biological . . . infertile (or sterile)

The biological species concept does not work in all situations 14.2

When it can be applied, the biological species concept provides a useful definition of a species. However, it does not work in all cases. For example, the classification of extinct forms of life must rely on the appearance and chemical analysis of fossils. Also, the criterion of interbreeding is useless for organisms that are completely asexual in their reproduction, as are prokaryotes and most single-celled protists. As one of these organisms (a single cell) divides, it produces a lineage of genetically identical cells. Some asexual organisms can exchange genes in processes resembling sex (for instance, conjugation among prokaryotes, described in Chapter 12), but otherwise there is no gene flow among the various lineages resulting from reproduction. Consequently, asexual organisms can be assigned to species only by the grouping of cell lineages that have the same appearance and biochemical features.

Even when populations reproduce sexually, are not extinct, and are not geographically isolated, the biological species concept often cannot be applied unambiguously. For example, consider the deer mouse *(Peromyscus maniculatus)*, which is common throughout most of North America. The four names on the map in Figure 14.2 indicate that there are four populations of this species in and near the Rocky Mountains. Each population is phenotypically distinct; taxonomists call each one a subspecies and give each a third name. The deer mouse populations overlap in certain places. Some interbreeding occurs in these zones; therefore, according to the biological species concept, we would consider the interbreeding populations to belong to a single species.

There is a complication, however. Two of the subspecies—*Peromyscus maniculatus nebrascensis* and *Peromyscus maniculatus artemisiae*—overlap but do not interbreed (red symbol). Yet their gene pools are not completely isolated, because each of these populations interbreeds with its other neighboring populations *(sonoriensis* and *borealis)*, and some genes could flow indirectly from the *artemisiae* population to the *nebrascensis* population. Actually, the extent of gene flow by this roundabout route is probably slight, and if the two populations *(sonoriensis* and *borealis)* connecting them became extinct, *artemisiae* and *nebrascensis* could be named separate species without reservation. This is a situation where two phenotypically distinct populations seem to be on the brink of becoming unable to interbreed. Cases like this may represent **speciation,** the evolution of new species, in progress.

Population biologists are discovering more and more cases where the distinction between populations with limited gene flow and species with fully separated gene pools blurs. Whenever there is a trickle of genes between two different populations that are in contact, the biological species concept is difficult to apply. As with the deer mouse, it is as though we are seeing populations at different stages in their evolutionary descent from common ancestors. This is to be expected if new species arise by the gradual divergence of populations.

Even more common than cases like the deer mouse, where reproductive isolation is not complete, are those about which we have little or no information. Thus a species definition that does not rely on reproductive isolation may have broader application than the biological species concept.

One alternate approach for defining species that evolutionary biologists are evaluating focuses on evolutionary history instead of on reproductive isolation. This **evolutionary species concept** defines a species as a cluster of organisms that represents a genealogy, or a lineage of descent. The evolutionary species concept depends on finding unique, diagnostic genotypic or phenotypic features, such as DNA sequences or body structures, for the clusters and their lineages. Species defined by evolutionary criteria may be sexual or asexual and may exchange some genes with other clusters. The main criterion for species status is that a cluster and its unique lineage be recognizably distinct from others.

In practice, for most organisms—sexual, asexual, and fossils alike—*classification* is based mainly on observable and measurable phenotypic traits and, increasingly, on DNA analysis of genotype as well. However, when reproductive isolation exists and can be documented, it provides clear definition of species. Furthermore, reproductive isolation is an essential factor in the evolution of many species, and we need to take a closer look at it.

? By defining a species by its reproductive ________ from other populations, the biological species concept can only be applied to organisms that reproduce ________.

isolation . . . sexually

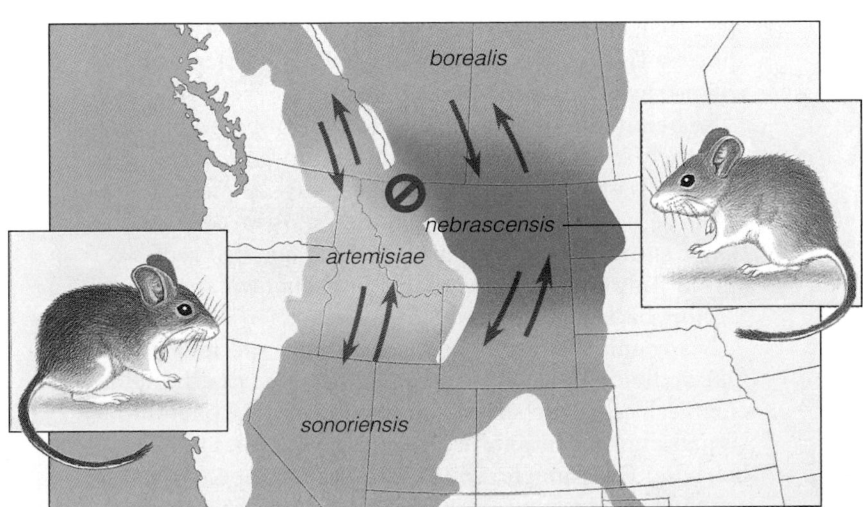

Figure 14.2 Overlapping subspecies of deer mice

14.3 Reproductive barriers keep species separate

Two different kinds of barriers can prevent closely related species from interbreeding. A geographical barrier isolates two species physically, so they cannot interbreed even though they may have the potential to do so. In contrast, a **reproductive barrier** is a biological feature of the organisms themselves. Reproductive barriers prevent populations belonging to closely related species from interbreeding even when their ranges overlap. As shown in the table at the right, the various types of reproductive barriers that isolate the gene pools of species can be categorized as either prezygotic or postzygotic, depending on whether they function before or after zygotes (fertilized eggs) form.

Reproductive Barriers Between Species

PREZYGOTIC BARRIERS:	Prevent mating or fertilization
Temporal isolation:	Mating or flowering occurs at different times for different species.
Habitat isolation:	Species breed in different habitats.
Behavioral isolation:	There is little or no sexual attraction between individuals of different species.
Mechanical isolation:	Structural differences between species (in animal sex organs or plant flowers, for example) prevent copulation or pollen transfer.
Gametic isolation:	Male and/or female gametes die before uniting with gametes of other species, or the gametes fail to unite.
POSTZYGOTIC BARRIERS:	Prevent the development of fertile adults
Hybrid inviability:	Hybrid zygotes fail to develop, or the hybrids fail to reach sexual maturity.
Hybrid sterility:	Hybrids fail to produce functional gametes.
Hybrid breakdown:	Offspring of hybrids are weak or infertile.

Prezygotic Barriers Prezygotic barriers actually prevent mating or fertilization between species. There are five main types of prezygotic barriers. One type, called **temporal isolation,** occurs when two species breed at different times—during different seasons, at different times of the day, or even in different years. Spotted skunks illustrate seasonal isolation. As mentioned in the opening essay, the eastern and western species overlap in the Great Plains, but the western spotted skunk breeds in the fall and the eastern species in late winter. Many plants also exhibit seasonal differences in breeding time. For example, two species of pine trees, the Monterey pine *(Pinus radiata)* and Bishop's pine *(P. muricata)* inhabit some of the same areas of central California. The two species are reproductively isolated, however: The Monterey pine releases pollen in February, while the Bishop's pine does so in April. Some plants are temporally isolated because their flowers open at different times of the day, so pollen cannot be transferred from one to another.

In a second type of prezygotic barrier, called **habitat isolation,** two species live in the same general area but not in the same kinds of places. For example, two closely related species of garter snake found in western North America are generally isolated by habitat: One lives mainly in water and the other on land. Habitat isolation also affects parasites that are confined to certain plant or animal host species. Two species of parasites living in different hosts will not have a chance to interbreed.

In **behavioral isolation,** a third type of prezygotic barrier, there is little or no sexual attraction between females and males of different species. For many species, specific visual signals identify potential mates. As we discussed in the opening essay for Chapter 5, most female fireflies respond only to flashes of males of their own species, by flashing back and attracting the males.

Many animals recognize mates of their species by odor. One individual, usually a female, will emit a perfumelike chemical, called a pheromone, into the air. The pheromone signals another individual of the same species to alter its behavior. For example, females of the insect species *Porthetria dispar* (the gypsy moth) attract males by emitting a pheromone. Male gypsy moths are tuned in to the specific

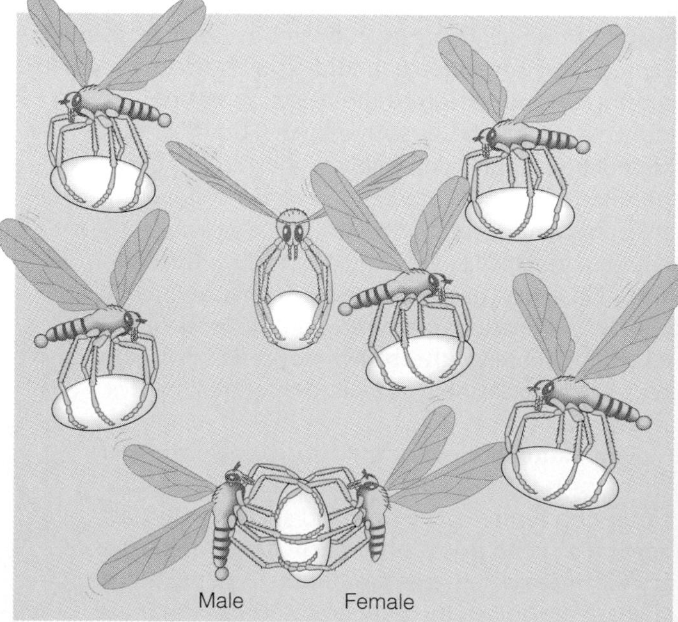

Figure 14.3A Behavioral isolation by courtship ritual in dance flies

odor of the pheromone produced by females of their species. They do not confuse it with sex attractants given off by females of closely related species.

Figure 14.3A shows another form of behavioral isolation, called a courtship ritual. Many species will not copulate until the male and female have performed an elaborate ritual that is unlike that of any other species. The insects in this figure are called dance flies because the males fly up and down in the air in swarms that attract females. The males of this particular species also make balloons from silk secreted by glands in their feet and carry the balloons as they swarm. When a female selects a male from the swarm, the male offers his balloon, and the female accepts it. Only then will the couple leave the "dance floor" and copulate. In other species of dance flies, males offer the females bits of prey or pieces of flower petals, sometimes wrapped in silk. Dance flies are common around streams and lakeshores throughout North America.

A fourth type of prezygotic barrier, called **mechanical isolation,** occurs when female and male sex organs are not compatible; for instance, the male copulatory organs of many insect species have a unique and complex structure that fits the female parts of only one species. In the plant kingdom, many species have flower structures that prevent certain animals from pollinating them. Mechanical isolation is often important for plants that depend on animal pollinators. Figure 14.3B shows a hummingbird obtaining nectar from a flowering plant. As it does so, its head is dusted with pollen, which the bird then transfers to the next flower it visits. In some cases, the beak of a particular species of hummingbird is just the right length for the flower tube of the one plant species it pollinates. Consequently, the bird transfers pollen only among plants of that species.

Gametic isolation is a fifth type of prezygotic barrier. A male and a female from two different species may copulate, but the gametes do not unite to form a zygote. In many mammals, for example, the sperm cannot survive in the female of a different species. Gametic isolation can operate even in species that do not copulate. Male and female sea urchins, of many species, release eggs and sperm into the sea, but fertilization occurs only if species-specific molecules on the surfaces of egg and sperm attach to each other.

Figure 14.3B Mechanical isolation of a flowering plant

Figure 14.3C Hybrid sterility: a horse and a donkey, representing two species that remain separate because their hybrid offspring, the mule, does not interbreed with them

Postzygotic Barriers In contrast to prezygotic barriers, postzygotic barriers operate after hybrid zygotes are formed. (Hybrid zygotes are fertilized eggs resulting from the union between gametes of two different species.) In some cases, there is **hybrid inviability;** that is, genes of the two parent species are not compatible, and the hybrids do not survive. Hybrid inviability may occur, for example, in certain frogs of the genus *Rana* that live in the same regions and habitats. Occasional hybrids are produced, but they do not complete development or are extremely frail. Another type of postzygotic barrier is **hybrid sterility,** in which hybrids of two species reach maturity and are vigorous but sterile and therefore unable to bring about gene flow between the parent species. A mule, for example, is the offspring of a female horse and a male donkey (Figure 14.3C). Smaller than a mule is the hinny, a hybrid of a male horse and a female donkey. The horse and donkey remain separate species because a mule or a hinny virtually never interbreeds with a horse or a donkey. Therefore, the gene pools of the horse and donkey remain isolated. In a third type of postzygotic barrier, called **hybrid breakdown,** the first-generation hybrids are viable and fertile, but when these hybrids mate with one another or with either parent species, the offspring are feeble or sterile. For example, different species of cotton plants can produce fertile hybrids, but the offspring of the hybrids do not survive.

In summary, reproductive barriers form the boundaries around many closely related species. The process of speciation sometimes depends on the formation of these barriers that prevent the spreading of genes. Next, we examine situations that make reproductive isolation, and the speciation that results from it, possible.

? Two closely related tropical bird species live in the same forest, but one feeds and mates in the forest canopy and the other on the forest floor. This is an example of ________ isolation, which is a ________zygotic reproductive barrier.

habitat . . . pre

14.4 Geographical isolation can lead to speciation

A key event in the origin of many species is the separation of a population—with its gene pool—from other populations of the same species. With its gene pool isolated, the splinter population can follow its own evolutionary course. Changes in its allele frequencies caused by natural selection, genetic drift, and mutations are unaffected by gene flow from other populations. In the formation of many species, the initial block to gene flow seems to have been a geographical barrier that isolated a population. This mode of speciation is called **allopatric speciation** (from the Greek *allos,* other, and *patra,* fatherland). Populations separated by a geographical barrier are known as allopatric populations.

Many factors can isolate a population geographically. For example, a mountain range may emerge and gradually split a population of organisms that can inhabit only lowland lakes; certain fish populations might become isolated in this way. Similarly, a creeping glacier may gradually divide a population, or a land bridge such as the Isthmus of Panama may form and separate the marine life on either side.

How formidable must a geographical barrier be to keep allopatric populations apart?The answer depends on the ability of the organisms to move about. Birds, mountain lions, and coyotes can easily cross mountain ranges, rivers, and canyons. The wind-blown pollen of pine trees is also not hindered by such barriers, and the seeds of many plants may be carried back and forth on animals. In contrast, small rodents may find a deep canyon or a wide river a formidable barrier. For example, the Grand Canyon (pictured below) separates the range of the white-tailed antelope squirrel *(Ammospermophilus leucurus)* from that of the closely related Harris's antelope squirrel *(Ammospermophilus harrisi).* Slightly smaller, with a shorter tail that is white underneath, *A. leucurus* inhabits deserts north of the canyon and west of the Colorado River in southern California. Harris's antelope squirrel has a more limited range, south of the Grand Canyon.

Geographical isolation creates opportunities for speciation, but it does not necessarily lead to new species. Indeed, even when gene pool changes result in the adaptation of an isolated population to a local environment, speciation may or may not occur. Speciation occurs only when the gene pool undergoes changes that establish *reproductive* barriers between the isolated population and its parent population.

The likelihood of allopatric speciation increases when a population is both small and isolated. A small, isolated population is more likely than a large population to have its gene pool changed substantially by factors such as genetic drift or natural selection. For example, in less than 2 million years, small populations of stray animals and plants from the South American mainland that managed to colonize the Galápagos Islands gave rise to all the species that now inhabit the islands. Let's examine some hypotheses explaining how this probably happened.

The more mobile an animal, the more formidable a new geographic barrier must be for ____________ speciation to lead to different species on opposite sides of the barrier.

allopatric

Figure 14.4 Geographically isolated species of antelope squirrels

When oceanic islands are far enough apart to permit populations to evolve in isolation, but close enough to allow occasional dispersions to occur, they are effectively outdoor laboratories of evolution. Located west of Ecuador, the Galápagos Archipelago is one of the world's great showcases of evolution. Each island was born naked from underwater volcanoes and was gradually clothed by plants, animals, and microorganisms derived from strays that rode the ocean currents and winds from other islands and continents. Organisms can also be carried to islands by other organisms, such as sea birds that travel long distances with seeds clinging to their feathers (Figure 14.5A).

The species on the Galápagos Islands today, most of which occur nowhere else, descended from stragglers that floated, flew, or were blown over the sea from the South American mainland. For instance, the Galápagos island chain has a total of 13 species of closely related birds called Galápagos finches (also known as Darwin's finches because Darwin wrote about them). These birds have many similarities but differ in their feeding habits and their beak type, which is correlated with what they eat. The three examples shown in Figure 14.5B illustrate the range of beak shape and size. (Also see Figure 15.9.)

Evidence accumulated since Darwin's time indicates that all 13 finch species evolved from a single small population of ancestral birds that colonized one of the islands. Figure 14.5C illustrates how this might have happened. Completely isolated on the island after ① migrating from the mainland, the founder population (species A in Figure 14.5C) may have undergone significant changes in its gene pool and become a new species, which we'll call species B. ② Later, a few individuals of species B may have been blown by storms to a neighboring island. Isolated on this second island, the second founder population could have evolved into a second new species, species C. ③ Species C could later recolonize the island from which its founding population emigrated, and it might coexist there with its ancestral species if reproductive barriers kept the species distinct. In the case of the Galápagos finches, the two species came to rely on different foods, besides being unable to interbreed. In this simplified model, after species C evolved on the second island, ④ it colonized a third island, where it evolved into species D. ⑤ Species D then dispersed to the two islands of its ancestors. Actually, the Galápagos finches colonized and speciated repeatedly on many separate islands of the Galápagos.

Today, each of the Galápagos Islands has multiple species of finches, with as many as ten on some islands. In contrast, the island of Cocos, about 700 km (420 mi) north of the Galápagos, has only one finch species, found nowhere else. This single species was apparently derived from an ancestral founder population in the same way as the Galápagos birds, but Cocos Island is so isolated that there apparently has been no opportunity for its finch species to colonize other islands, or for other finches to become established on Cocos.

The emergence of numerous species from a common ancestor introduced to new and diverse environments is called **adaptive radiation.** The effects of adaptive radiation of Darwin's finches are evident in the many types of beaks, specialized for different foods.

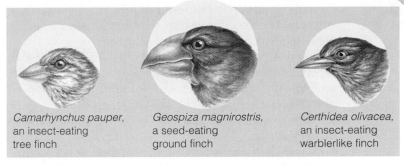

Camarhynchus pauper, an insect-eating tree finch

Geospiza magnirostris, a seed-eating ground finch

Certhidea olivacea, an insect-eating warblerlike finch

Figure 14.5B Examples of differences in beak shape and size in Darwin's finches

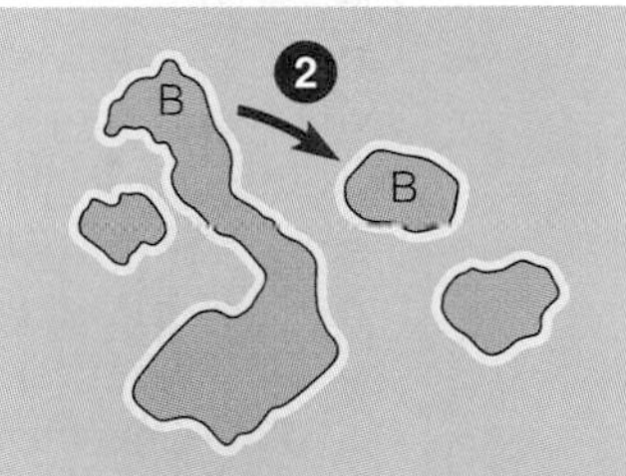

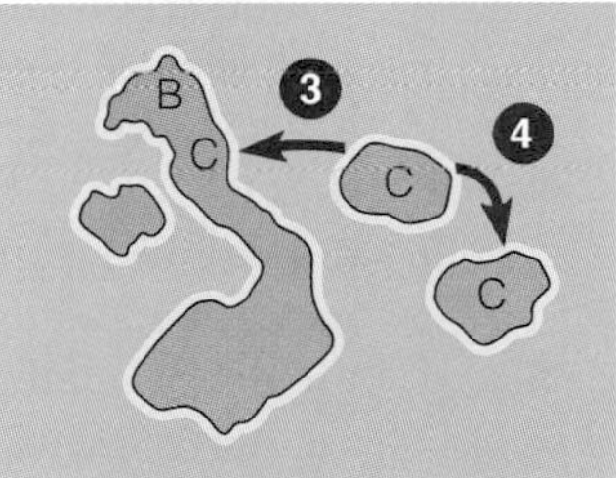

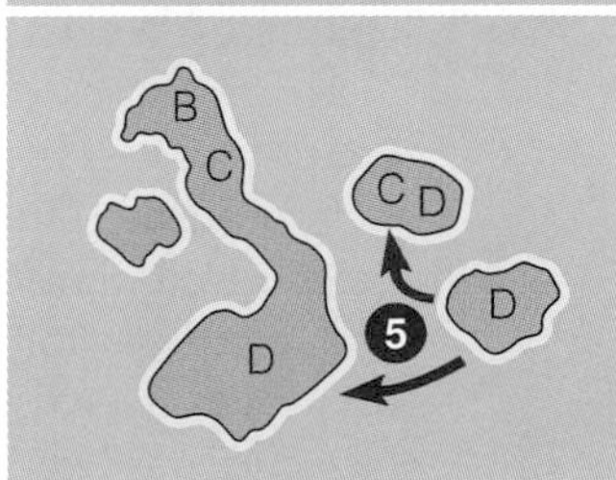

Figure 14.5C Adaptive radiation on an island chain

Figure 14.5A Long-distance dispersal of plant seeds (black spots) on the feathers of a tern

Why would allopatric speciation be less common on an island close to a mainland than on a more isolated island of the same size?

Continued gene flow between mainland populations and those on nearby islands reduces the chance of enough genetic divergence for speciation.

14.6 New species can also arise within the same geographic area as the parent species

14A
See Activity 14A on the CD-ROM.

Not all species arise as a result of geographical isolation. In **sympatric speciation** (Greek *syn*, together, and *patra*, fatherland), reproductive isolation develops and new species arise without geographical isolation. A new species can arise in a single generation if a genetic change produces a reproductive barrier between mutants and the parent population. Sympatric speciation does not seem to be widespread among animals, but it has been important in plant evolution.

Many plant species have originated from accidents during cell division that result in extra sets of chromosomes. In this type of sympatric speciation, the new species has **polyploid cells,** meaning that each cell has more than two complete sets of chromosomes. Figure 14.6A shows how a polyploid zygote can result from a single parent species that is diploid. The key abnormality is that meiosis fails to occur properly during gamete formation; instead, the cells divide by mitosis. Consequently, the chromosome number is not reduced, and diploid, rather than haploid, gametes are produced. If self-fertilization occurs (as it commonly does in plants), the resulting zygote is tetraploid; that is, it has four of each type of chromosome. This zygote may develop into a mature plant that can reproduce by self-fertilization.

These new tetraploid plants will also be able to breed with diploid plants of the parental type, but the resulting hybrids will be triploid ($3n$). The triploid zygote comes from the fusion of a diploid ($2n$) gamete from the tetraploid parent and a haploid (n) gamete from the diploid parent. Triploid individuals are sterile; they cannot produce normal gametes because the odd number of chromosomes cannot form homologous pairs and separate normally during meiosis. Thus, the creation of the tetraploid plant has been an instantaneous speciation event: A new species, reproductively isolated from its parent species, has been produced in just one generation.

Sympatric speciation by polyploidy was first discovered by Dutch botanist Hugo de Vries. Working in the early 1900s, de Vries studied genetic diversity in evening primroses (Figure 14.6B). In one study he started with *Oenothera lamarckiana* (upper left), a diploid species of primrose with 14 chromosomes. From this plant he developed, by self-fertilization, a polyploid—a tetraploid with 28 chromosomes. De Vries named this new primrose, which could not interbreed with its parent species, *Oenothera gigas*, for its large size (lower left).

Polyploids do not always come from a single parent species. In fact, most polyploid species arise from the hybridization of two parent species. As we see in the next module, the creation of a polyploid species in this way requires the coupling of two accidental events: the hybridization of the two parent species, and a failure of meiosis in the resultant hybrid.

Return to the table of reproductive barriers in Module 14.3 and choose the one that reproductively isolates a viable polyploid plant from its parental species.

Hybrid sterility

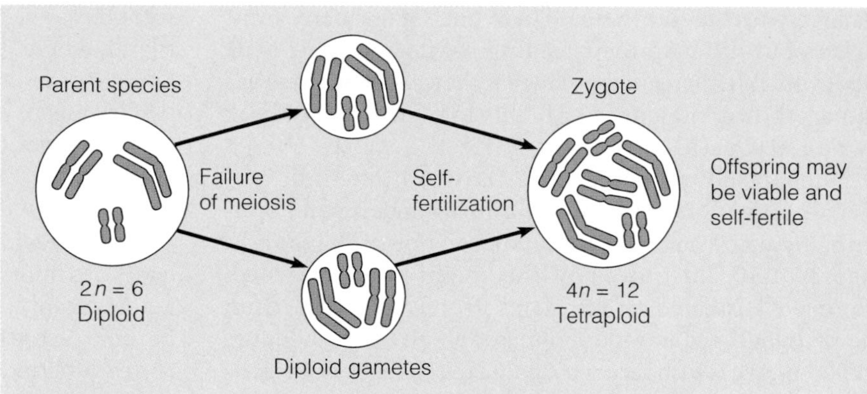

Figure 14.6A Sympatric speciation by polyploid formation

O. lamarckiana

O. gigas

Figure 14.6B Botanist Hugo de Vries with two species of evening primrose

14A

Plant biologists estimate that 25–50% of all plant species are polyploids. Hybridization between two species accounts for most of this polyploidy, perhaps because the unusually diverse assortment of genes a hybrid inherits from parents of different species can be advantageous.

Many of the plants we grow for food are polyploids, including oats, potatoes, bananas, peanuts, barley, plums, apples, sugarcane, coffee, and wheat. Cotton, also a polyploid, remains the source of one of the world's most popular clothing fibers, despite the great numbers of synthetic materials that have been on the market for decades. Cotton cloth is made from the long white plumes that extend from the seeds of the plant.

Wheat, the most widely cultivated plant in the world, occurs as 20 different species of *Triticum*. We know that humans domesticated diploid species of wheat at least 11,000 years ago because wheat grains of the diploid species *Triticum monococcum* (with $2n = 14$) have been found in the remains of Middle Eastern farming villages about this old. These diploids have small seed heads and are not highly productive, but some grow wild and others are still cultivated in the Middle East.

Our most important wheat species today is bread wheat (*Triticum aestivum,* Figure 14.7A), a polyploid with 42 chromosomes. Figure 14.7B illustrates how this species evolved; the uppercase letters represent not genes but *sets of chromosomes* that have been traced through the lineage. The process began with hybridization between two diploid wheats, one the domesticated species *T. monococcum* (AA), the other one of several wild species that probably grew as weeds at the edges of cultivated fields (BB). Chromosome sets A and B of the two species would not have been able to pair at meiosis, making the AB hybrid sterile. However, a failure of meiosis in this sterile hybrid and self-fertilization among the resulting gametes produced a new species (AABB) with 28 chromosomes. Today, we know this species as emmer wheat *(T. turgidum)*. Varieties of emmer wheat are grown widely in Eurasia and western North America. It is used mainly for making macaroni and other noodle products, because its proteins have a tendency to hold their shape better than bread-wheat proteins.

The final step in the evolution of bread wheat is believed to have occurred in early farming villages on the shores of European lakes over 8000 years ago. At that time, the cultivated emmer wheat, with its 28 chromosomes, hybridized spontaneously with the closely related wild species *T. tauschii* (DD), which has 14 chromosomes. The hybrid (ABD, with 21 chromosomes) was sterile, but a failure of meiosis in this hybrid and self-fertilization doubled the chromosome number to 42. The result was bread wheat, with two each of the three ancestral sets of chromosomes (AABBDD). 

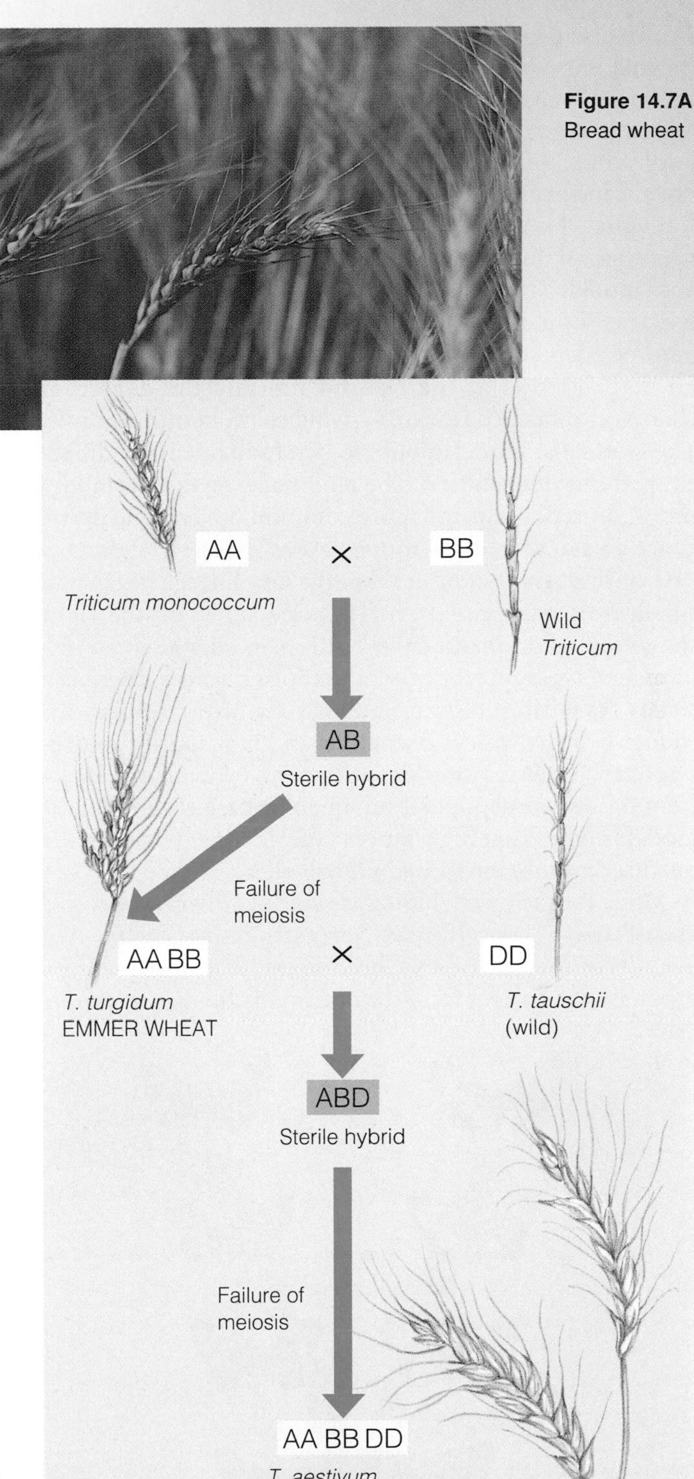

Figure 14.7A Bread wheat

Figure 14.7B The evolution of wheat

?

Each speciation episode in the evolution of bread wheat is an example of __________ speciation, which is the origin of a new species without geographic isolation from the parent species.

sympatric

14.8 The tempo of speciation can appear steady or jumpy

As we discussed in Chapter 13, evidence of evolution over the millions of years of geological time comes mainly from the fossil record, the chronicle of extinct organisms engraved in layers of rock. Let's take a look at two models that have proved useful in interpreting the patterns in evolutionary history suggested by the fossil record.

Figure 14.8A illustrates the evolution of two lineages of hypothetical butterflies by what has been called the **gradualist model.** This model fits Darwin's view of the origin of species: Populations evolve differences gradually as they become adapted to their local environments. The horizontal black arrows at the bottom indicate the degree of change in inherited features—wing color, in our example. The gradualist model applies to two fundamentally different patterns of evolution. The blue arrow on the left in Figure 14.8A represents the more common pattern: the divergence of a new lineage from a parent lineage; the narrow base of this arrow indicates that the new lineage began as a small, isolated population. The blue arrow on the right shows the second pattern: evolutionary change in an unbranched lineage. Notice that, in both lineages, there is a steady transition in inherited features, with populations changing little by little over time, and new species (represented by the two butterflies at the top) evolving gradually from the ancestral population. In either case, the gradualist model implies that big changes (speciations) occur by the steady accumulation of many small ones.

Many evolutionary biologists since Darwin's time, and even Darwin himself, have been struck by how few sequences of fossils have ever been found that clearly show a gradual, steady accumulation of small changes in evolutionary lineages. Instead, most fossil species appear suddenly, without transitional forms, in a layer of rock and persist essentially unchanged until disappearing from the record of the rocks as suddenly as they appeared.

Within the past 50 years, some evolutionary biologists have addressed the nongradual appearance of the fossil record by developing a model now known as **punctuated equilibrium.** Illustrated in Figure 14.8B, this model suggests that the evolution of our hypothetical butterflies actually occurs in spurts. The narrow bases of the blue branches represent small, isolated populations that diverge from the ancestral lineage in the middle. The horizontal portions of the branches indicate that evolutionary changes occurred abruptly, with the new species diverging from the ancestral lineage in a relatively brief time. Notice that, in contrast to the gradualist model, Figure 14.8B shows no transitional stages in the lineages. The butterflies look the same at the bottom and top of each lineage; the vertical portions of the blue arrows indicate that the species change little, if at all, once they appear. Moreover, there is no speciation within an unbranched lineage (middle blue arrow). The punctuated equilibrium model holds that evolution's tempo has been mainly jumpy, with abrupt episodes of speciation punctuating long periods of little change, or equilibrium.

Is it likely that most species evolve abruptly and then remain essentially unchanged for most of their existence? Rapid speciation certainly occurs in some cases. As we saw earlier, abrupt speciation can occur by polyploidy. Also, mutation of just a few of the genes that regulate embryonic development may produce radically new body features,

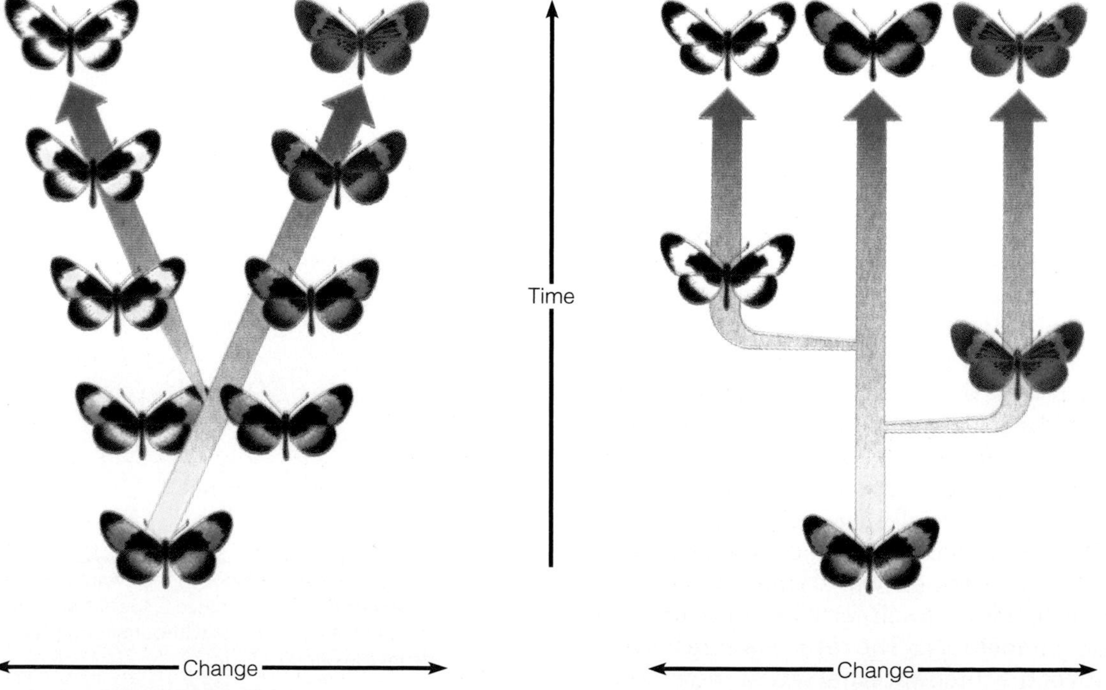

Figure 14.8A The gradualist model

Figure 14.8B The punctuated equilibrium model

and recent evidence suggests that such changes can generate new species. We might also predict that genetic drift and natural selection could significantly alter the gene pool of a small population isolated in a challenging new environment in a few hundred to a few thousand generations—a relatively short period in geological time.

The photograph at the right in Figure 14.8C shows the results of a change in climate that led to geographical isolation and rather abrupt speciation. About 50,000 years ago, what is now the Death Valley region of California and Nevada had a wet climate and an extensive system of interconnected lakes and rivers. A drying trend began about 10,000 years ago, and by 4000 years ago, the region had become a desert. Today all that is left of the network of lakes and rivers are isolated springs scattered in the desert, mostly in deep clefts between rocky walls. The springs vary greatly in water temperature and salinity. Some of them are home to small animals called pupfishes (Figure 14.8C). In certain cases, a single spring is home to a species of pupfish found nowhere else in the world. Apparently, these desert pool fishes evolved from one ancestral species whose range was broken up when the region became arid, isolating a number of small populations. By either genetic drift or natural selection, and in just a few thousand years, the isolated populations evolved into separate species. The lower right photo in Figure 14.8C shows ecologists counting pupfish in a Death Valley spring.

Can we really consider speciation that takes several thousand years "abrupt"? Compared to abrupt speciation by polyploidy, we certainly cannot. However, our perception may change if we consider the overall life of a species. The fossil record suggests that successful species last for a few million years, on average. Let's say that a particular species survived for 5 million years, but that the fossil record indicates that most of the changes in its body features (the changes associated with speciation) occurred during the first 50,000 years of its existence. In this case, speciation took up only 1% of the overall history of the species. Fossils of the species might appear abruptly in rocks of a certain age and then continue to appear, apparently unchanged, in rocks spanning millions of years, before disappearing. Thus, the overall history of the lineage as depicted in the fossil record may seem to fit the punctuated equilibrium model.

During the 1970s and 1980s, evolutionary biologists actively debated the relative importance of the punctuated equilibrium model and the gradualist model in interpreting evolutionary history. The debate is muted today, and many biologists find both models useful in interpreting the fossil record. For example, the history of a lineage that generally seems to fit the punctuated model may also include periods of gradual change, with speciation actually occurring during those periods. In many species, the evolution of some features seems to fit the punctuated model, while other features seem to evolve more gradually.

In summary, evolution's tempo may be jumpy, especially when viewed over very long time spans, but still be relatively steady for substantial periods. Controversy over how to interpret the fossil record will always be part of evolutionary biology, and it is important to realize that the controversy in no way implies disagreement about the reality of evolution. Biologists agree that evolution occurs, and their debate about its patterns and mechanisms serves a useful purpose, catalyzing research and ultimately leading to a better understanding of the process. Arguing about the details of evolution is like arguing about different theories of gravity: Objects keep right on falling, even as the controversy continues. We hear from one of evolutionary biology's most controversial thinkers next. 

An isolated spring in the Death Valley desert

A pupfish found in an isolated spring

Researchers studying pupfish

Figure 14.8C Geographical isolation and speciation in Death Valley

? How does the punctuated equilibrium model account for the relative rarity of transitional fossils linking newer species to older ones?

According to this model, speciation generally takes place in a relatively short time compared to the overall time the species exists. Thus, on the vast geological time scale of the fossil record, the transition of one species to another seems abrupt.

TALKING ABOUT SCIENCE

14.9 Evolutionary biologist Richard Dawkins expounds on the power of natural selection

Best-selling, award-winning books by Oxford University professor Richard Dawkins have made significant contributions to the public's understanding of natural selection. They have also ignited controversies among evolutionary biologists. In his 1976 book, *The Selfish Gene,* Dawkins argued that genes, not whole organisms, are the units of natural selection. In this view of Darwinism, organisms are "survival machines" for the "replicators," molecules (genes) that "selfishly" use organisms to make more copies of themselves. Many biologists disagree, countering that a strictly gene-centered concept of natural selection is too simplistic. In a 1998 interview, Dawkins commented:

> In *The Selfish Gene* I emphasized the importance of genes as the central units of natural selection. Genes are the only things that pass through generations. But what's special about genes is that they are replicated; they function as what I called "replicators" in *The Selfish Gene.* So, it's replicators that matter, not specifically genes. Anything that is self-replicating anywhere in the universe is fair game for natural selection. There probably is life on other planets, and if there is, then I absolutely bet my shirt that it's based upon natural selection. That would require a replicator. It doesn't have to be DNA, but it would have the fundamental property of self-replication.

In another of Dawkins's books, *The Blind Watchmaker*, he elaborates on the cumulative power of natural selection in the adaptation of organisms, an idea that is not controversial at all in biology, but still stirs some public debate.

> The "blind watchmaker" is natural selection. Natural selection doesn't look ahead. Natural selection is totally blind to the future. It's always just picking the best of available alternatives in the present. It can never say, "Well, if we tolerate a bit of unpleasantness for the next 2 million years, then that will set us up wonderfully to improve in 2 million years' time." Natural selection can't do that. It's always optimizing in the short term, though it's blind to the future.

Dawkins elaborated more on the theme of how natural selection can lead to elegant structure by optimizing for the short term in yet another best-seller, *Climbing Mount Improbable.*

Figure 14.9 Richard Dawkins

> "Mount Improbable" is a metaphor for adaptation occurring gradually, in increments. The metaphor is that of a mountain which has an absolutely sheer cliff face. If we relate this cliff to adaptation, to the most complicated piece of biological machinery you can think of, which for many people is an eye, then you say to yourself that it's impossible to leap from the bottom of this mountain to the top, which indeed it is. Leaping from the bottom of the cliff to the top would correspond to having the sheer luck to get that eye coming into place in one fell swoop. Many people wrongly think that Darwinism is a theory of chance, that it means that eyes and other complex organs come about by sheer luck. So no wonder these people don't believe natural selection. Of course an eye couldn't possibly come about like that. But on the other side of the mountain, you've got a slow, gradual slope, and it is very easy to get to the top of the mountain if you go around the other side and just walk up the slope. Relating this to adaptation, you have gradual, incremental improvement. You begin with hardly any eye at all and then each step of the way up the mountain is a gradual improvement. It may not be much, but it's enough to be better than your predecessors, who didn't have even that improvement. *Climbing Mount Improbable* emphasizes that evolution of complex adaptations has got to be gradual.

Professor Dawkins has demonstrated repeatedly that he has a gift for stimulating thoughtful debate and research in a scientific field, as well as engaging and challenging non-scientists. Summarizing his thoughts on natural selection, Dawkins had this to say:

> Natural selection is a bewilderingly simple idea. And yet what it explains is the whole of life, the diversity of life, the complexity of life, the apparent design of life. It all flows from this one remarkably simple idea. 

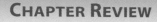

? Based on Dawkins's metaphor, how is the evolution of a complex structure such as the vertebrate eye like climbing a gentle slope rather than scaling a sheer cliff to reach a mountain top?

A structure such as the eye does not arise in one evolutionary leap, but by incremental refinement as natural selection chooses variations that work slightly better than others.

CHAPTER REVIEW

Chapter Summary

Concepts of Species (Introduction–14.3) We generally distinguish between kinds, or species, of organisms by their appearance. Linnaeus also used physical appearance to identify species when he established the binomial system for naming plants and animals. The similarities between different species and variation within a species make defining species difficult. The biological species concept is one way of defining a species. A biological species is a population or group of populations whose members can interbreed and produce fertile offspring. **(Introduction–14.1).** The concept of biological species is not applicable to fossils or asexual organisms. The occurrence of limited interbreeding between populations also makes defining species difficult, and such situations may illustrate speciation, the formation of new species. The evolutionary species concept defines a species as a cluster of organisms representing a specific evolutionary lineage **(14.2).**

Prezygotic and postzygotic reproductive barriers prevent individuals of different species from interbreeding. Prezygotic barriers are habits or mechanisms that prevent mating or fertilization. Postzygotic barriers operate after fertilization; they result in inviability or sterility of hybrids or their offspring **(14.3)**.

Mechanisms of Speciation (14.4–14.9) When a population is cut off from its parent stock, it may veer off on its own evolutionary course. Species evolution that occurs in this manner is called allopatric speciation. Often the initial block to gene flow is a geographical barrier such as a canyon or mountain range. Cut off from other populations, a small splinter population may become genetically unique as its gene pool is changed by natural selection, genetic drift, or mutation. However, gene pool changes, including those leading to the adaptation of an isolated population to a local environment, do not necessarily result in speciation **(14.4)**.

Small populations sometimes colonize islands and evolve in isolation, making islands natural laboratories of evolution. For example, the Galápagos Islands have become populated by plants and animals from South America. On these islands, repeated isolation and adaptation have resulted in adaptive radiation of 13 species of Darwin's finches from a common ancestral population **(14.5)**. In sympatric speciation, relatively common in plants, new species may arise without geographical isolation. A failure of meiosis can produce diploid gametes. Self-fertilization can then produce a tetraploid zygote, which may develop into a plant that can reproduce by self-fertilization. Because it has a polyploid set of chromosomes, this plant is an instant new species, isolated from its parent **(14.6)**. Many plants, including many food plants, are polyploid. They are often the products of hybridization, followed by meiotic failure and self-fertilization. Two such episodes appear to have occurred in the evolution of modern bread wheat **(14.7)**.

According to the gradualist model of the origin of species, new species evolve by the gradual accumulation of changes brought about by natural selection. Gradual change may occur in a lineage that has branched off from a parent lineage or in an unbranched lineage. However, few gradual transitions are found in the fossil record; new species seem to appear suddenly and persist unchanged for most of their history. The punctuated equilibrium model suggests that speciation occurs in spurts: Rapid change occurs when an isolated population diverges from the ancestral stock, and then virtually no change occurs for the rest of the species' existence. Both the gradualist model and the punctuated equilibrium model are useful, and the history of most species may include periods of gradual change and periods of little or no change **(14.8)**. Complex structures, such as vertebrate eyes, arise by incremental refinements selected because they work slightly better than others, not as a result of one evolutionary leap **(14.9)**.

Testing Your Knowledge

Multiple Choice

1. Biologists have found more than 500 species of fruit flies on the various Hawaiian Islands, all apparently descended from a single ancestor species. This example illustrates
   a. polyploidy
   b. temporal isolation
   c. adaptive radiation
   d. hybrid breakdown
   e. meiotic failure

2. Bird guides once listed the myrtle warbler and Audubon's warbler as distinct species that lived side by side in parts of their ranges. However, recent books show them as eastern and western forms of a single species, the yellow-rumped warbler. Apparently, it has been found that the two kinds of warblers
   a. live in the same areas
   b. successfully interbreed
   c. are almost identical in appearance
   d. are merging to form a single species
   e. live in different places

3. Which of the following is an example of a postzygotic reproductive barrier?
   a. One *Ceanothus* shrub lives on acid soil, another on basic soil.
   b. Mallard and pintail ducks mate at different times of year.
   c. Two species of leopard frogs have different mating calls.
   d. Hybrid offspring of two species of jimsonweeds always die before reproducing.
   e. Pollen of one kind of tobacco cannot fertilize another kind.

4. A small, isolated population is more likely to undergo speciation than a large one, because a small population
   a. contains a greater amount of genetic diversity
   b. is more susceptible to gene flow
   c. is more affected by genetic drift
   d. is more subject to errors during meiosis
   e. is more likely to survive in a new environment

Describing, Comparing, and Explaining

1. Explain how each of the following makes it difficult to clearly define a species: variation within a species, geographically isolated populations, asexual reproduction, fossil organisms.

2. The mating of a horse and a donkey produces a mule. Does this mean that horses and donkeys are the same species? Why or why not?

Thinking as a Scientist

Cultivated American cotton plants have a total of 52 chromosomes ($2n = 52$). In each cell, there are 13 pairs of large chromosomes and 13 pairs of smaller chromosomes. Old World cotton plants have 26 chromosomes ($2n = 26$), all large. Wild American cotton plants have 26 chromosomes, all small. Propose a testable hypothesis to explain how cultivated American cotton probably originated.

Science, Technology, and Society

The red wolf, *Canis rufus*, formerly widespread in the southeastern and south-central U.S., nearly became extinct in the late 1970s. Saved by a captive breeding program under the authority of the Endangered Species Act (ESA), it has been reintroduced in areas such as the Great Smoky Mountains National Park. Recent genetic evidence indicates that the red wolf may not be a separate species, but a hybrid of the coyote, *Canis latrans*, and the gray wolf, *Canis lupus*. Though the original intent of the ESA was to protect all endangered groups—whether species, subspecies, or hybrids—the costs may be prohibitive. What criteria should be applied if we must decide which organisms to protect? Are there reasons to preserve hybrids, subspecies, or local populations of species when the species as a whole is not at risk?

Media Resources

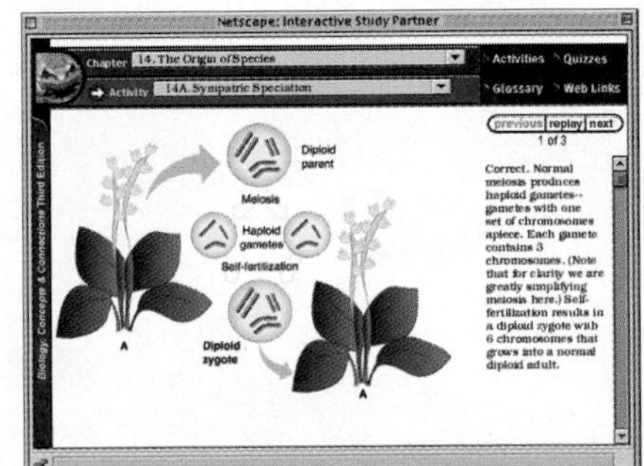

For further review of the concepts in this chapter, use the activities and quizzes on the CD-ROM.

Visit the Special Edition of The Biology Place web site for activities, quizzes, news articles, and web links.

15 Tracing Evolutionary History

Outline

Some 80 million years ago, when this duck-billed dinosaur laid her eggs, the North American continent was split in half by a shallow inland sea, now called the Western Interior Seaway. West of the seaway, land that is now part of eastern Montana and the Dakotas rose gently toward the newly formed Rocky Mountains, creating a coastal plain some 400 kilometers wide. Today, this region is arid and sparsely vegetated, with hot summers and cold winters. Eighty million years ago, it was densely vegetated and subtropical—a dinosaur's paradise, with warm rivers meandering down from the Rockies, carrying silt to seaside deltas.

In the late 1970s, paleontologist Jack Horner and his colleagues at the Museum of the Rockies in Bozeman, Montana, made some startling discoveries on the dry plains east of the Rocky Mountains in Montana and Alberta, Canada. Near the small town of Choteau, Montana, they began unearthing entire dinosaur nesting grounds, with many large mud nests and fossil skeletons of duck-billed dinosaurs, also called hadrosaurs (Greek for "stout lizards"). Apparently, a large number of hadrosaurs had nested near a river, and a flash flood had devastated the area, covering everything with silt. The researchers named the species of hadrosaur pictured here *Maiasaura peeblesorum* (Greek *maia*, good mother, and *sauros*, lizard; "of Peebles," the owners of the land where the fossils were found). The "good mother lizard" was about 7 meters long, and its nests were about 2 m wide and 1 m deep.

Maiasaura peeblesorum had a whole new story to tell about dinosaurs. Until its nesting grounds were unearthed, dinosaur eggs and nests were almost unheard of. A few had been found in the 1920s in the Gobi Desert in East Asia, but North America seemed to lack them altogether. Earlier prevailing opinion held that all dinosaurs, like many reptiles of today, laid their eggs, covered them with sand, and then promptly abandoned them. But some of the *Maiasaura* nests contained fossilized skeletons of well-developed young up to three months old. *Maiasaura* apparently remained with its nest and cared for its young for at least this much time. Some of the nests also contained small bits of fossilized eggshells, another indication of parental care. Today, the nests of some birds that nurture their young contain similar shell fragments; the developing young pulverize the shell fragments as they move around in the nest. In contrast, we see much larger shell fragments in the nests of birds, such as ducks, whose hatchlings leave the nest almost immediately.

Paleontologist Jack Horner at a fossilized dinosaur nesting site in Montana

Fossil finds like the *Maiasaura* nests have generated a lively debate about dinosaurs. Why were some groups, such as the hadrosaurs, so numerous and successful for millions of years? Was it because they lived in large groups and cared for their young? Hadrosaurs were prey to many carnivorous dinosaurs. Neither the young nor the adults were armored, and the structure of their skeleton indicates that they were not fast runners. Adults and active young could seek safety in the water when attacked, and their island nesting sites undoubtedly afforded some protection. But, perhaps more often, a herd or a large nesting group of adults discouraged predators by swinging their heavy tails and stamping their feet. This group behavior, coupled with parental feeding and care of the young, may have been key to their success.

Fossils have much to tell us about extinct organisms and how they lived. They can also tell us about the major milestones in the evolutionary history of life, the main subject of this chapter.

Earth History and Macroevoution

15.1 The fossil record chronicles macroevolution

As we discussed in Chapter 13, the fossil record, the sequence in which fossils appear in rock strata, is a catalog of evolutionary history. By studying fossils in rock strata in one area, we gain a local glimpse of long-term evolutionary change. On a broader scale, studying the order of fossils in strata from many locations enables us to trace **macroevolution,** the main events in the evolutionary history of life on Earth. By compiling a list of the organisms present in a

Figure 15.1

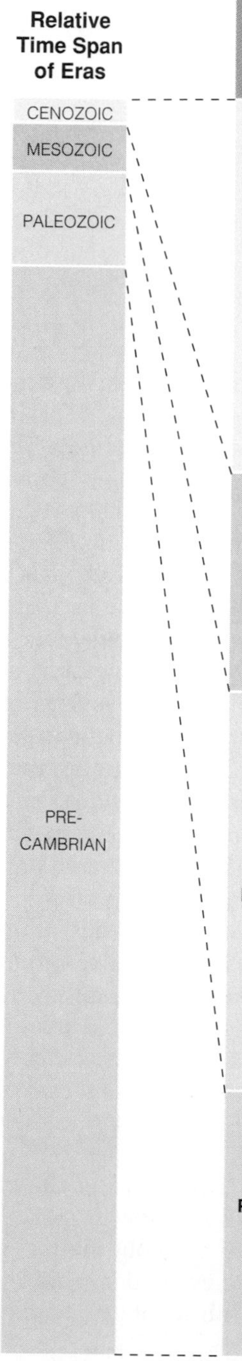

Relative Time Span of Eras

CENOZOIC
MESOZOIC
PALEOZOIC
PRE-CAMBRIAN

The Geological Time Scale

Era	Period	Epoch	Age (Millions of Years Ago)	Some Important Events in the History of Life
CENOZOIC	Quaternary	Recent	0.01	Historic time
		Pleistocene	1.8	Ice ages; humans appear
	Tertiary	Pliocene	5	Apelike ancestors of humans appear
		Miocene	23	Continued radiation of mammals and angiosperms
		Oligocene	35	Origins of many primate groups, including apes
		Eocene	57	Angiosperm dominance increases; origins of most modern mammalian orders
		Paleocene	65	Major radiation of mammals, birds, and pollinating insects
MESOZOIC	Cretaceous		145	Flowering plants (angiosperms) appear; many groups of organisms, including most dinosaur lineages, become extinct at end of period
	Jurassic		208	Gymnosperms continue as dominant plants; dinosaurs dominant
	Triassic		245	Cone-bearing plants (gymnosperms) dominate landscape; first dinosaurs, mammals, and birds
PALEOZOIC	Permian		290	Extinction of many marine and terrestrial organisms; radiation of reptiles; origins of mammal-like reptiles and most modern orders of insects
	Carboniferous		363	Extensive forests of vascular plants; first seed plants; origin of reptiles; amphibians dominant
	Devonian		409	Diversification of bony fishes; first amphibians and insects
	Silurian		439	Diversity of jawless fishes; first jawed fishes; colonization of land by vascular plants and arthropods
	Ordovician		510	Origin of plants; marine algae abundant
	Cambrian		570	Origin of most modern animal phyla
PRECAMBRIAN			610	Diverse invertebrate animals; first vertebrates; diverse algae
			700	Oldest animal fossils
			1700	Oldest eukaryotic fossils
			2500	Oxygen begins accumulating in atmosphere
			3500	Oldest fossils known (prokaryotes)
			4600	Approximate time of origin of Earth

particular stratum at widely separated sites, a paleontologist obtains a broad picture of the kinds of organisms that existed when the stratum formed. Additional information comes from fossils present in different kinds of strata that were deposited at the same time in geological history.

Paleontologists trace the main events in the history of life, using evidence from the sequence of fossils at many different sites. Their results are summarized in the geological time scale in Figure 15.1. The table provides an overview of macroevolution, measured in millions of years. Notice that the time scale is divided into major groupings called eras and that most eras are subdivided into periods. The diagram to the left of the table shows the relative lengths of the eras. The boundaries between the eras mark major transitions in the forms of life fossilized in the rocks; this is also true to some extent for periods.

Rocks from the Precambrian era (at the bottom of the time scale) have undergone extensive change over time, and much of their fossil contents is no longer visible. Nonetheless, paleontologists have found some fossil-rich Precambrian strata and have pieced together the most ancient events in life's history. The earliest fossils discovered, dating from 3.5 billion years ago, represent microorganisms only. However, strata from the late Precambrian (some 610 million years ago) bear highly diverse fossils, including ones of jellyfish, corals, and worms that resemble certain modern species, as well as other animals that bear little or no resemblance to species alive on Earth today.

Dating from about 570 million years ago, rocks of the Paleozoic ("ancient animal") era contain fossils of lineages that gave rise to modern organisms, as well as many lineages that have become extinct. During the early Paleozoic, virtually all life was aquatic, but by about 400 million years ago, plants and animals were well established on land.

Following the Paleozoic era was the Mesozoic ("middle animal") era, also known as the age of reptiles. More accurately, it should be called the age of dinosaurs and cone-bearing plants (gymnosperms), because both of these groups of organisms dominated the Mesozoic landscape. The Mesozoic also saw the beginnings of mammals, birds, and flowering plants (angiosperms). By the end of the Mesozoic, all the dinosaur lineages except one, which apparently gave rise to birds, had become extinct, and an explosive period of evolution of mammals, birds, and angiosperms began.

The Cenozoic ("recent animal") era began about 65 million years ago and continues today. Because much more is known about the Cenozoic than about earlier eras, our table subdivides the two Cenozoic periods into finer intervals called epochs. Our own species, *Homo sapiens*, appeared on the scene only about 100,000–200,000 years ago, during the Pleistocene (Ice Age) epoch. Thus, our tenure on the planet is only a tiny portion of the immense saga of geological time.

Use the table in Figure 15.1 to estimate how many hundreds of millions of years prokaryotes inhabited Earth before eukaryotes evolved.

About 1800 million years, or 1.8 billion years

The actual ages of rocks and fossils mark geological time 15.2

The record of the rocks chronicles the *relative* ages of fossils, the order in which groups of species present in a sequence of strata evolved. However, the sequence alone does not tell the *actual* ages in years of the embedded fossils.

Geologists and paleontologists use several techniques to determine the actual ages of rocks and the fossils they contain. The method most often used, called **radiometric dating,** is based on the fact that living organisms contain certain atomic isotopes in certain ratios. For instance, living organisms have the same constant ratio of carbon-14 (^{14}C), a radioactive isotope, to carbon-12 (^{12}C), a stable isotope, as does Earth's atmosphere. However, when an organism dies, its ratio of ^{14}C to ^{12}C starts to drop, because ^{14}C decays to other chemical elements, and the organism no longer obtains any ^{14}C from the atmosphere. Each radioactive isotope has a fixed rate of decay, known as its half-life. For example, ^{14}C has a half-life of 5600 years, meaning that half the ^{14}C in a specimen decays in about 5600 years, half the remaining ^{14}C decays in the next 5600 years, and so on, until all the ^{14}C is gone. Knowing both the half-life of a radioactive isotope and the ratio of radioactive to stable isotope in a fossil enables us to tell how old the fossil is. For instance, if a fossil has a ^{14}C-to-^{12}C ratio half that of the atmosphere, it is about 5600 years old; a fossil with one-fourth the atmosphere's ratio is about 11,200 years old.

The ^{14}C-to-^{12}C ratio is reliable for dating fossils less than about 50,000 years old. To date older fossils, paleontologists use radioactive isotopes with longer half-lives. For instance, potassium-40, a radioactive isotope with a half-life of 1.3 billion years, can be used to date rocks and fossils hundreds of millions of years old. Radiometric dating has an error factor of plus or minus about 10%.

The dates you see on the geological time scale in Figure 15.1 were established by dating rocks and fossils. Notice that the geological periods span unequal intervals of time. The boundaries between periods mark distinct changes in the species composition of the sedimentary rocks. In the next module, we examine some of the geological processes that can produce these distinct changes.

Your measurements indicate that a fossilized skull you unearthed has a ^{14}C-to-^{12}C ratio about one-sixteenth that of the atmosphere. What is the approximate age of the skull?

22,400 years (four half-life reductions)

15.3 Continental drift has played a major role in macroevolution

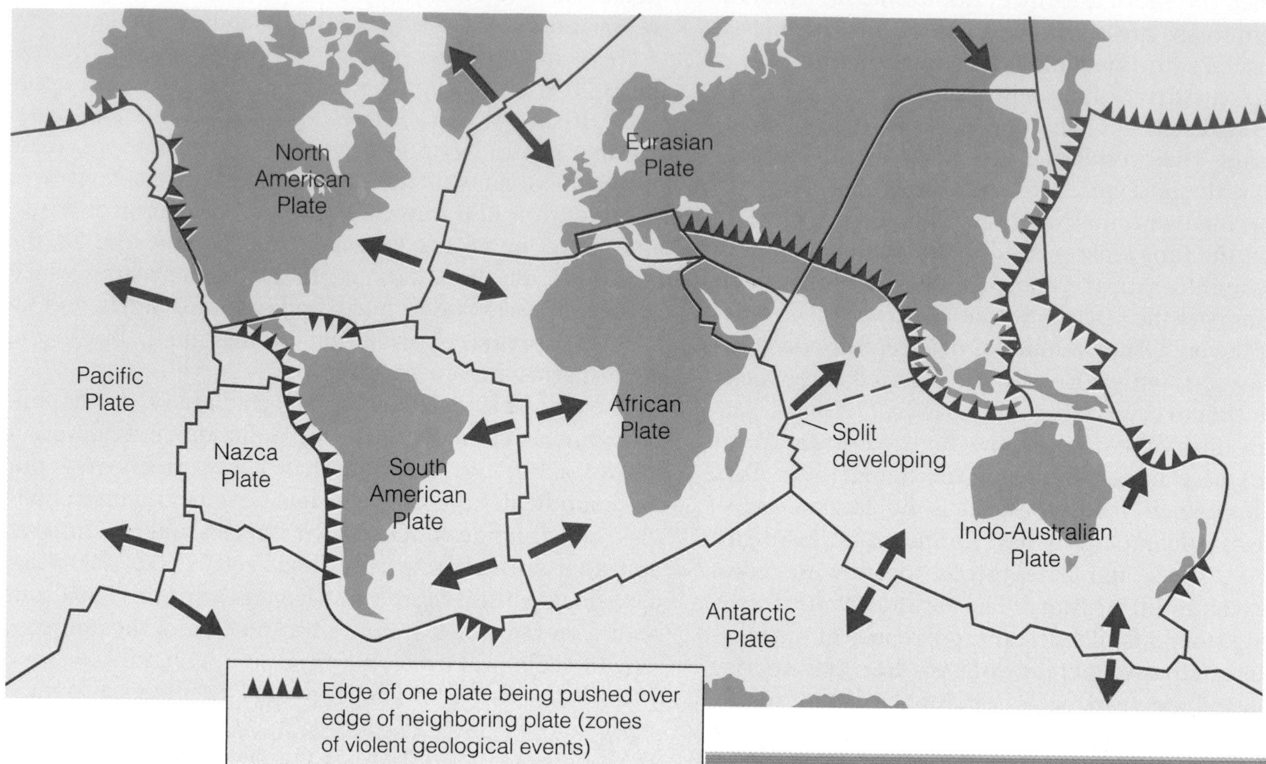

Figure 15.3A Earth's crustal plates

In 1912, German meteorologist Alfred Wegener proposed the hypothesis of **continental drift.** Wegener postulated that all land on Earth was once one great mass, which broke up into continents that drifted like rafts to their present positions. Wegener believed that the shapes of our modern continents, like pieces of a jigsaw puzzle, reflect their former positions in the original supercontinent. Like many ideas generated before their time, Wegener's were not taken seriously for decades. Until the 1960s, the general belief was that the continents have always been fixed in their present positions.

In recent decades, geologists, paleontologists, and biologists have accumulated overwhelming support for the concept of continental drift. We now know that the continents and seafloors form a thin outer layer of planet Earth, called the crust, and that under the crust is a mass of hot material called the mantle. Furthermore, the crust is divided into giant, irregularly shaped plates (outlined in red in Figure 15.3A). Because the mantle circulates constantly, the crustal plates move about slowly but incessantly—they literally float—on the underlying mantle. As a result of plate movements, world geography changes constantly, for unless landmasses are embedded in the same crustal plate (China and Japan, for instance), their positions relative to each other do not remain the same. North America and

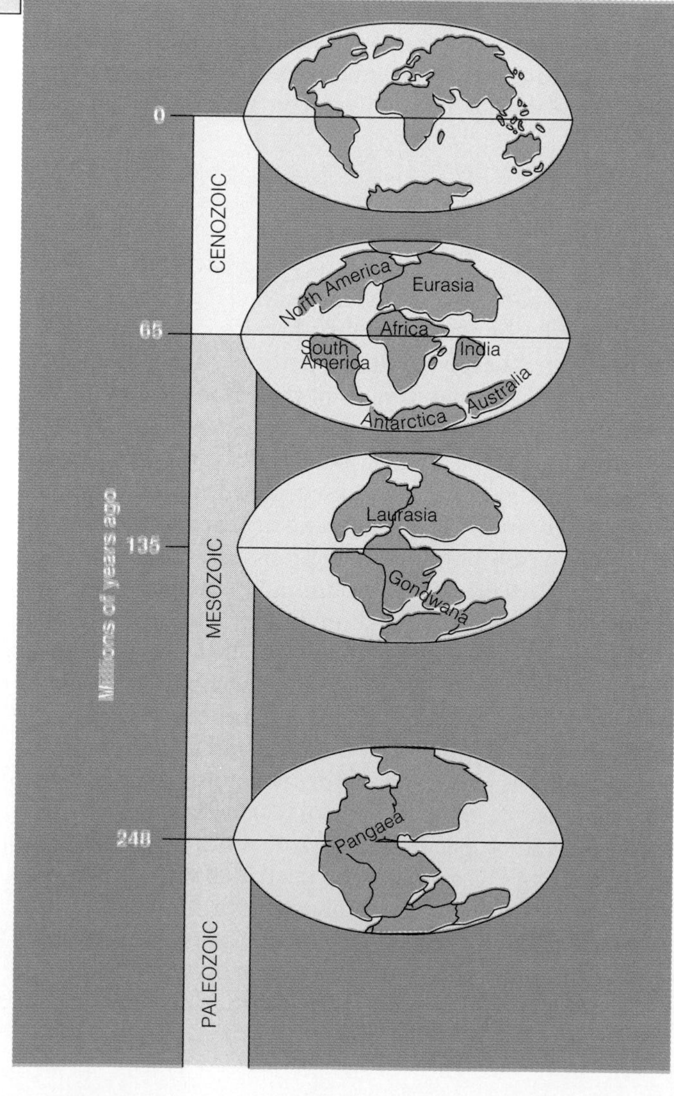

Figure 15.3B Continental drift

Eurasia, for example, are presently drifting apart at a rate of about 2 cm per year. Throughout geological time, continental movements have greatly influenced the distribution of organisms around the world, and continental drift explains much of the history of life.

Two chapters in the continuing saga of continental drift seem to have been especially significant in their influence on life. The first occurred about 250 million years ago, near the end of the Paleozoic era, when plate movements brought all the landmasses back together into the supercontinent Wegener had originally proposed. Shown at the bottom in Figure 15.3B, this supercontinent is called **Pangaea,** meaning "all land." Imagine some of the possible effects on life as massive continents joined. Species that had been evolving in isolation came together and competed. When the landmasses fused, the total amount of shoreline and shallow coastal areas was reduced. Then, as now, most marine species inhabited shallow waters, and many of these organisms probably died out as their habitats shrank. The formation of Pangaea also would have altered terrestrial environments, because ocean currents, which affect climates on land, would have changed drastically. Moreover, with Pangaea's reduced coastline, less land area would have been exposed to the ocean's moderating effect on air temperatures. As a result, much of Pangaea may have had a drier and more varied climate. Overall, the fossil record indicates that the formation of Pangaea reshaped biological diversity, causing great numbers of extinctions. These in turn provided new opportunities for organisms that survived the crisis.

The second dramatic chapter in continental drift began about 180 million years ago, during the early Mesozoic era. Pangaea started to break apart again, causing colossal changes for life forms of the time. As the continents drifted apart, each became a separate evolutionary arena—a huge island on which organisms evolved in isolation from their previous neighbors. At first, Pangaea split into northern and southern landmasses, which we call **Laurasia** and **Gondwana,** respectively. This split was more or less complete by about 135 million years ago, as shown in Figure 15.3B. By the end of the Mesozoic era (and the Cretaceous period), some 65 million years ago, the modern continents were beginning to take shape. Then, just 10 million years ago, India collided with Eurasia, and the slow, steady crunching of the Indo-Australian and Eurasian plates formed the Himalayas, the tallest and youngest of Earth's major mountain ranges. (Regions where two plates crunch together are marked by red arrowheads in Figure 15.3A.) Today, Australia seems to be moving independently of India, apparently because the ongoing collision between plates in the Himalayan region is creating forces that are splitting the Indo-Australian plate in two (dashed line in Figure 15.3A).

The pattern of continental separation solves many puzzles. One example is the distribution of a group of ancient vertebrates called lungfishes (Figure 15.3C). Today, there are six species of lungfishes in the world, four in Africa, and one each in Australia and South America (orange areas in Figure 15.3D). What is the evolutionary history of these fishes? As the triangles in Figure 15.3D indicate, fossil lungfishes have been found on all continents except Antarctica. This widespread fossil record indicates that lungfishes evolved when Pangaea was intact.

In what way is North America's location relative to Europe gradually changing? Why?

North America and Europe are moving farther apart as the North American plate drifts northwestward and the European plate drifts southeastward.

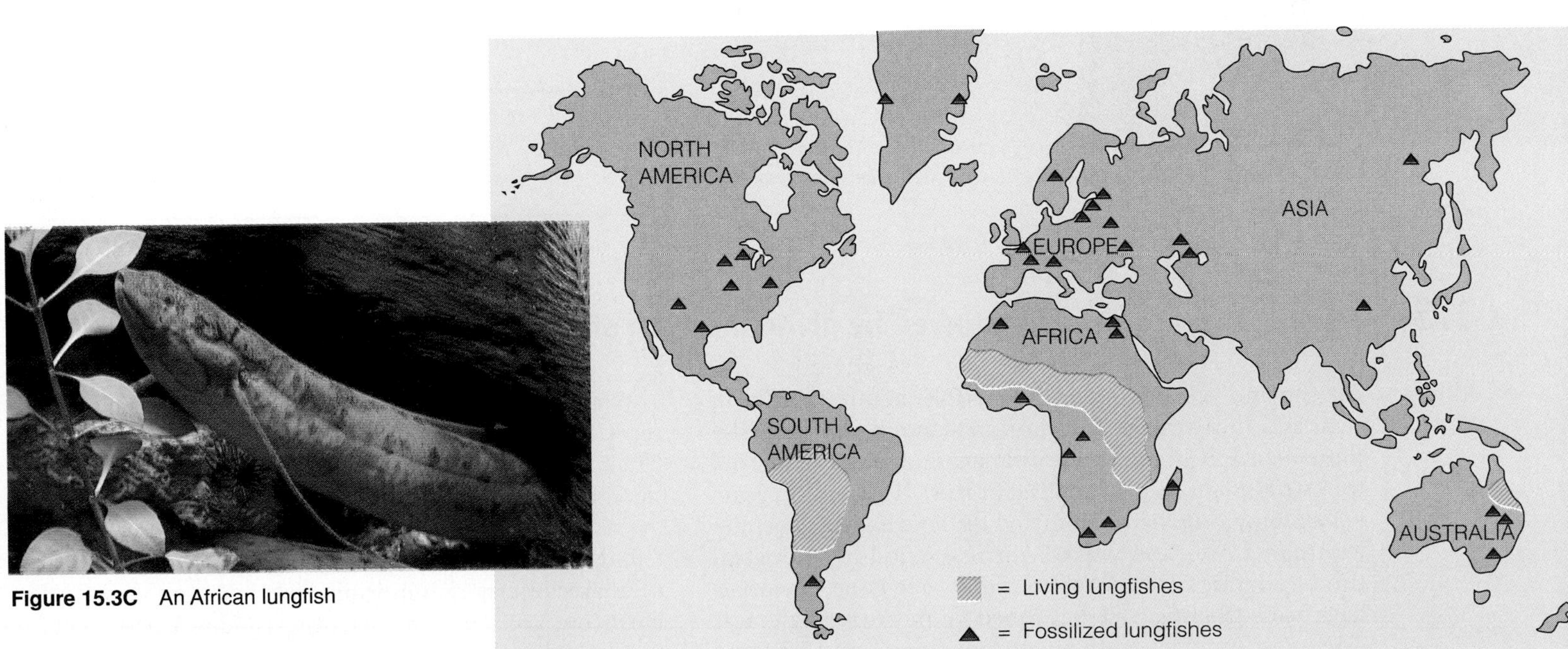

Figure 15.3C An African lungfish

Figure 15.3D Lungfish distribution, a result of continental drift

15.4 Tectonic trauma imperils local life

Geologists call the forces within Earth that cause movements of the crust **plate tectonics.** These forces affect life in many ways. Not only do they move continents and cause mountain ranges to rise, they also produce volcanoes and earthquakes. The boundaries of crustal plates are hotspots of geological activity. California's frequent earthquakes result from movement along the infamous San Andreas fault, which is part of the border where the Pacific plate and the American plate grind together and gradually slide past each other (Figure 15.4A). Earthquakes can also occur in the interiors of crustal plates, but these movements ultimately result from forces exerted at the edges of plates.

Erupting volcanoes, emitting hot, molten rock from beneath Earth's crust, can cause tremendous devastation. Volcanoes can also create opportunities for living organisms. For instance, volcanoes at sea can produce islands, such as the Galápagos, but the same volcanic activity that creates an oceanic island may destroy life that evolved there. In 1883, fiery pumice from a volcano covered the small volcanic island of Krakatau, near the boundary between the Indo-Australian plate and the Eurasian plate (Figure 15.4B). Before the eruption, Krakatau was covered with a dense tropical rain forest. Afterward, it was virtually devoid of life. Despite the devastation, life from neighboring islands began recolonizing Krakatau soon after its surface cooled. Within 50 years after the eruption, tropical forest with a great diversity of plants and animals again covered the island.

Figure 15.4B The volcanic eruption that destroyed the island of Krakatau

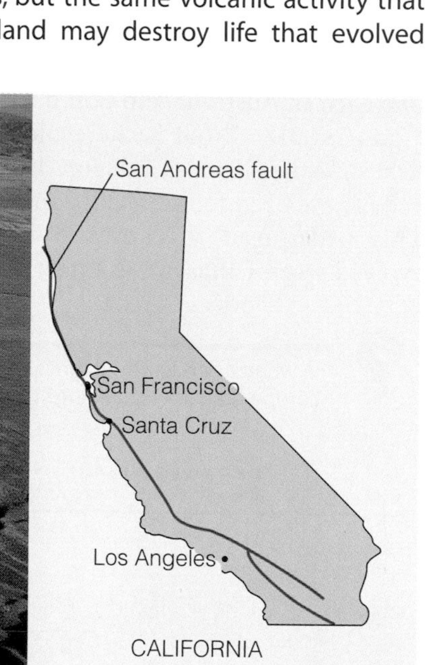

Figure 15.4A The San Andreas fault, a boundary between two crustal plates

Plate tectonics has played a role in many of the major changes that characterize evolutionary history. As we see in the next module, extraterrestrial objects may also be very important. 

? The Andes Mountains are associated with tectonic activity near which plate boundary?

Boundary between the Nazca and South American plates

15.5 Mass extinctions were followed by diversification of life forms

At the end of the Cretaceous period, about 65 million years ago, the world lost an enormous number of species—more than half of its marine animals, and many lineages of terrestrial plants and animals. For some 150 million years before, dinosaurs had dominated the land and air. Then, in less than 10 million years—a brief period in geological time—all dinosaur lineages but one were gone. Scientists have been debating what happened for more than a decade. The question mark in the panorama in Figure 15.5 symbolizes the debate.

What do we know about this unusual time? The fossil record shows us that the climate cooled late in the Cretaceous and that shallow seas were receding from continental lowlands. We also know that many plants required by plant-eating dinosaurs died out first. Perhaps most telling of all, the sediments deposited at the time contain a thin layer of clay rich in iridium, an element very rare on Earth but common in meteorites. Many paleontologists conclude that the iridium layer is the result of fallout from a huge cloud of dust that billowed into the atmosphere when a large meteorite or asteroid hit Earth. The cloud would have blocked light and disturbed climate severely for months, perhaps killing off many plant species.

The asteroid hypothesis has many supporters, and a large asteroid crater that dates from the late Cretaceous has been found under the Caribbean Sea. Many scientists believe the impact that produced that crater could have caused global climate change and mass extinctions. Other researchers propose that climatic changes due to continental drift could have caused the extinctions, whether an asteroid collided with Earth or not. Still others point to evidence in the fossil record in India indicating that, during the late Cretaceous, massive volcanic activity released particles into the atmosphere, blocking sunlight and thereby contributing to climatic cooling. The various hypotheses are not mutually exclusive, and researchers continue to debate the extent to which each contributed to the extinctions.

Extinction is inevitable in a changing world. A species may become extinct because its habitat has been destroyed or because of unfavorable climatic changes. Extinctions occur all the time, but extinction rates have not been steady. There have been six distinct periods of mass extinction over the last 600 million years. During these times, losses escalated to nearly six times the average rate.

Of all the mass extinctions, the ones marking the ends of the Cretaceous and the Permian periods have been the most intensively studied. The Permian extinctions, at about the time the continents merged to form Pangaea, claimed over 90% of the species of marine animals and took a tremendous toll on terrestrial life as well. Whatever their causes, mass extinctions affect biological diversity profoundly.

But there is a creative side to the destruction. Each massive dip in species diversity has been followed by an explosive *increase* in diversity. Extinctions seem to have provided the surviving organisms with new environmental opportunities. For example, mammals existed for at least 75 million years before undergoing an explosive increase in diversity just after the Cretaceous. Their rise to prominence was undoubtedly associated with the void left by the extinction of nearly all dinosaurs. The world might be a very different place today if many dinosaur lineages had escaped the Cretaceous extinctions, or if none of the mammals that lived in the Cretaceous had survived.

The Permian extinctions and the Cretaceous extinctions mark the ends of the ____________ and ____________ eras, respectively. (*Hint:* Refer back to Figure 15.1.)

Paleozoic . . . Mesozoic

Figure 15.5 Animals and plants before and after the Cretaceous extinctions

15.6 Key adaptations may enable species to survive and proliferate after mass extinctions

Geologists and paleontologists have found evidence that each of the six periods of mass extinction over the past 600 million years was followed by a virtual explosion of organisms that had previously been much less prevalent. What enabled some species to survive and proliferate when great numbers of others died out? There is no simple answer. Chance undoubtedly played a major role in determining whether organisms survived or became extinct, and survival does not imply that one species is somehow better than another. It is likely that certain features—key adaptations—enabled surviving species to withstand the great changes of the mass extinction periods and to multiply and evolve after many other species died out. For example, hair and the ability to nurse young on milk are unique mammalian features that evolved long before the Cretaceous extinctions. So are feathers in birds. These key adaptations probably helped mammals and birds adapt to an environment in which most of the dinosaurs could not survive.

How do key adaptations arise? One way is by the gradual refinement of existing structures for new functions. The term **exaptation** refers to a structure that evolved in one context and later was adapted for another function. This term suggests that a structure can become adapted to alternative functions; it does not mean that a structure somehow evolves in anticipation of future use. Indeed, natural selection can only result in the improvement of a structure in the context of its current function.

The plants in Figure 15.6 illustrate the concept of exaptation. They are members of a large group (about 2000 species) called bromeliads, common throughout tropical and subtropical America. A few bromeliads thrive in arid soils, the kind of environment that ancestral bromeliads inhabited. Most bromeliads, however, are epiphytes, growing on other plant surfaces, often on tree limbs high off the ground, as we discussed in Module 1.3.

Being arid-adapted, the ancestors of bromeliads had a number of exaptations that helped them spread into habitats lacking soil. We see evidence of these exaptations in modern bromeliads. Consider, for example, the pineapple (*Ananas*, Figure 15.6, top photo), which grows in arid soil. The bases of its leaves form a catch-basin that holds rainwater, and the pineapple's roots absorb water from the catch-basin. Furthermore, pineapple leaves have hairlike projections, called trichomes, that reduce water loss from the plant. Among the epiphytic bromeliads, the vase plant (*Aechmea*, Figure 15.6, bottom photo) also has a catch-basin. Trichomes on its leaves absorb water and dissolved nutrients from the catch-basin. The spread of bromeliads into aerial environments seems to have hinged largely on the presence of catch-basins and trichomes in the ancestral plants and on the adaptation of these features for living on other plants without being grounded in the soil. With its roots in the ground, the pineapple may be reminiscent of an ancestral bromeliad.

A pineapple plant, a bromeliad adapted to arid soils

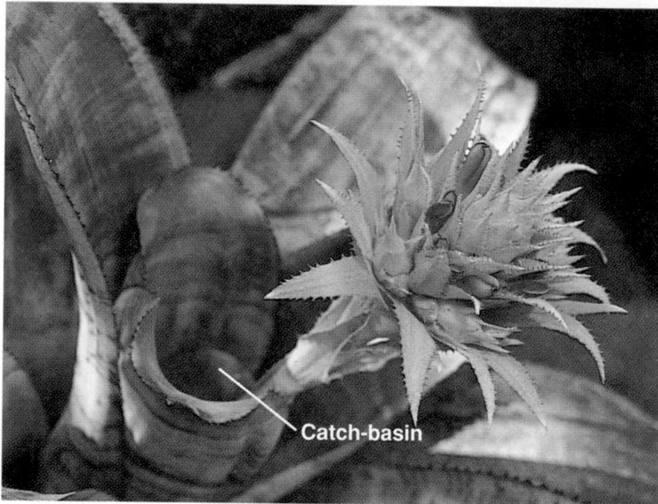

A vase plant, a bromeliad that lives on tree limbs in tropical forests

Figure 15.6 Two closely related species of plants with a catch-basin, an adaptation to dry environments

Another way that key adaptations can arise is by changes that affect an organism's development. Such changes may result from the mutation of only a few genes. As we saw in Module 11.10, genes that are expressed very early in development regulate the expression of numerous other genes acting later. Thus, a change in just a single early-acting gene may have a significant effect on an organism's development. Many species that differ significantly in development and body form, such as the human and the chimpanzee, are genetically quite similar. In cases like this, the two species may have diverged rather rapidly as changes in a few genes produced major changes in their ancestors' development and adult structure. 

? Explain why the concept of exaptation does not imply that a structure evolves *in anticipation of* some future environmental change.

Although an exaptation is co-opted for new or additional functions in a new environment, it existed because it worked as an adaptation to the old environment.

Delayed maturity was a key novelty in human evolution

A subtle change in a species' developmental program can have profound effects. For instance, a slowing of the development of some organs relative to others may produce a very different kind of organism. Figure 15.7A is a photograph of an axolotl, a salamander that illustrates a phenomenon called **paedomorphosis** (from the Greek *paedos,* child, and *morphosis,* shaping), the retention of juvenile body features in the adult. The axolotl grows to full size and reproduces without losing its external gills, a juvenile feature in most species of salamanders.

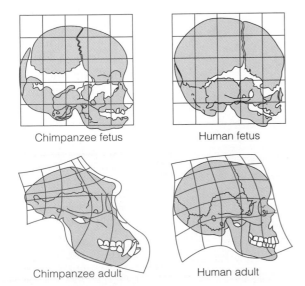

Figure 15.7A An axolotl, a paedomorphic salamander

Paedomorphosis has also been important in human evolution. Humans and chimpanzees are much more alike as fetuses than they are as adults, as the skulls in Figure 15.7B show. In the fetuses of both species, the skulls are rounded and the jaws are small, making the face rather flat and rounded. As development proceeds, uneven bone growth makes the chimpanzee skull sharply angular, with heavy brow ridges and massive jaws. The adult chimpanzee has much greater jaw strength than we have, and its teeth are proportionately larger. In contrast, the adult human has a skull with decidedly rounded, more fetuslike contours. The jaw is more pronounced than in the fetus, but otherwise the contours of the skull do not change significantly. Put another way, our skull is paedomorphic; it retains fetal features even after we are sexually mature.

Our large, paedomorphic skull is one of our most distinctive features. Our large, complex brain is another. Our brain develops to its unparalleled size and complexity during childhood, a period of development unique to humans. All mammals have a period of infant dependency, when the young are fed milk and require parental protection. After weaning, most mammals mature rapidly to adulthood. Apes, including chimpanzees, have a longer period of infancy than most mammals, but only humans have a true *childhood,* a prolonged period between infancy and adolescence when we remain dependent on parental care. The main function of childhood may be in providing more time to learn from adults.

Evolutionary biologist Stephen Jay Gould contends that there is a connection between our juvenile physical traits and our unusually long period of juvenile dependency. Gould suggests that our juvenile physical traits—rounded head with large forehead, bulging cheeks, and small chin—may be visual clues that make adults feel affectionate, and thus caring and protective during our childhood. Gould uses the cartoon character Mickey Mouse to illustrate his points. Early Mickey Mouse renditions were not nearly as popular as later ones (Figure 15.7C). Gould contends that the present-day Mickey Mouse is successful because he elicits affectionate, parental responses. Compared to earlier versions, he *is* more youthful: His head and eyes are larger relative to his body and his legs are shorter.

Whether or not the Mickey Mouse analogy is appropriate, the long retention of juvenile features, coupled with a prolonged period of dependency, are hallmarks of our species. Genetic changes that led to paedomorphosis were probably key events in the divergence of our ancestors from the lineage we share with the great apes.

Chimpanzee fetus

Human fetus

Chimpanzee adult

Human adult

Figure 15.7B Chimpanzee and human skulls compared

How are our large, roundish skulls analogous to the gills of axolotls?

Both structures are paedomorphic; juvenile anatomies retained into adulthood.

© Disney Enterprises, Inc.

Figure 15.7C The "evolution" of paedomorphosis in Mickey Mouse

15.8 What accounts for evolutionary trends?

The fossil record seems to reveal trends in the evolution of many species. For instance, two trends in the human lineage were toward a more complex brain and toward a more paedomorphic life cycle. Figure 15.8 summarizes fossil evidence for trends in the evolution of horses from ancestors about the size of a large dog that lived some 40 million years ago. Named *Hyracotherium,* these ancestors had four toes on each foot and teeth adapted to browsing on shrubs and trees. In contrast, modern horses *(Equus)* have only one toe per foot and teeth modified for grazing on grasses.

As the rectangles in the figure indicate, the fossil record includes many species that descended from *Hyracotherium.* The yellow highlighter tracks a sequence of fossil horses that were intermediate in form between *Hyracotherium* and *Equus.* If these were the only fossils known, they could create the illusion of a single trend in an unbranched lineage, progressing toward larger size, reduced number of toes, and teeth modified for grazing. However, this would ignore other fossil horses in lineages that died out (the terminal arrows in the figure); an example is the lineage represented by browsers of the genus *Megahippus.* Actually, *Equus* represents the only surviving twig of a multibranched evolutionary bush with several divergent trends.

What accounts for the continuance of some evolutionary trends, such as the accumulation of traits characteristic of modern horses, and the extinction of others? In one view, species are analogous to individuals. Speciation is their birth and extinction is their death. According to this model, unequal generation of new species and unequal survival of species play roles in macroevolution similar to the role of differential reproduction in microevolution (see Module 13.7). In other words, the species that endure the longest and generate the greatest number of new species determine the direction of major evolutionary trends. Evolutionary biologists continue to debate the relative importance of unequal survival and unequal speciation in the macroevolutionary trends we see in the fossil record.

It is important to recognize that the existence of an evolutionary trend does not imply that the trend is preordained or unchangeable. Evolution is a response to interactions between organisms and their current environment. If conditions change, an evolutionary trend may cease or even reverse itself.

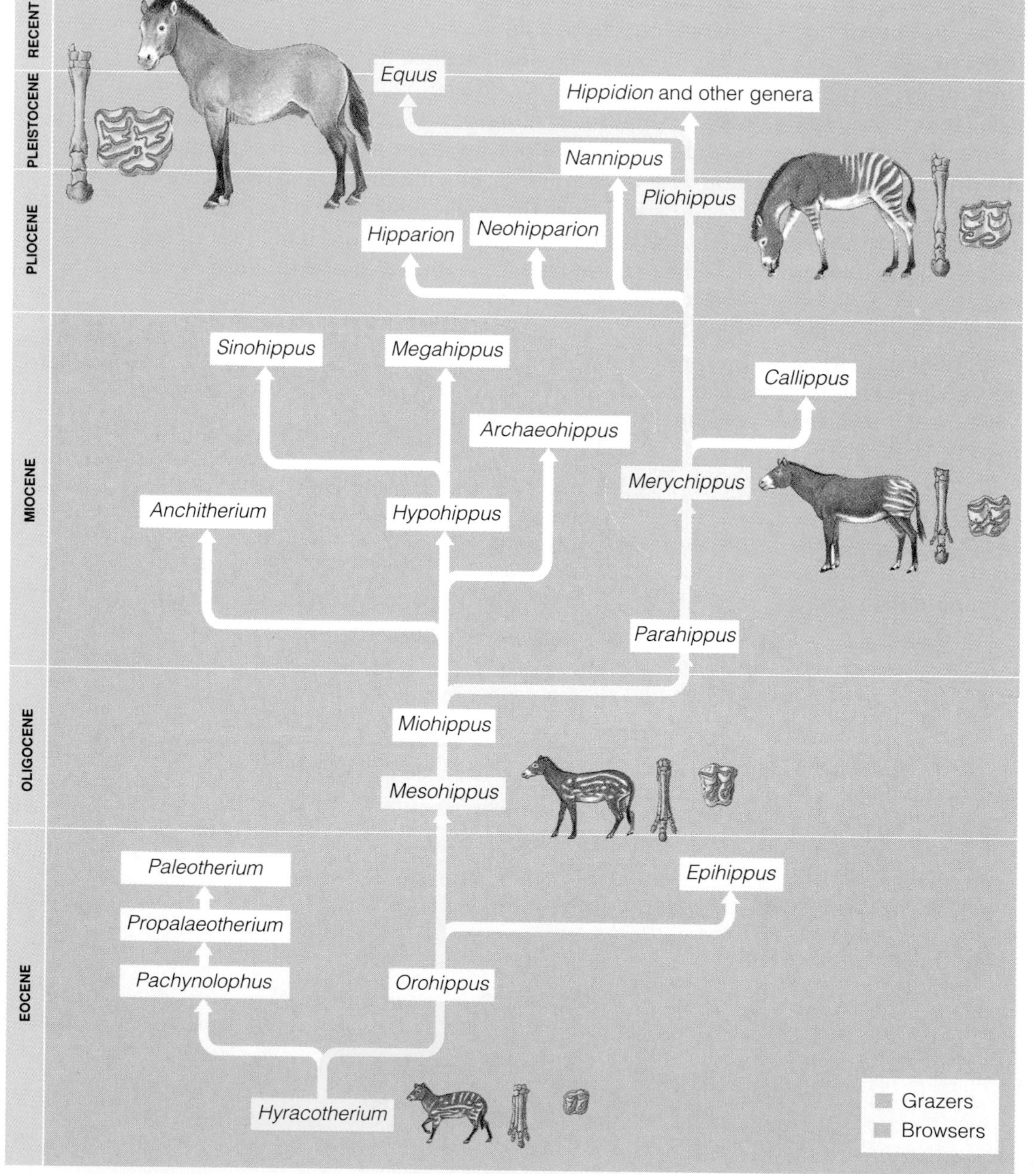

Figure 15.8 Trends in the evolution of horses

? A general trend in the evolution of mammals was larger brain size relative to body size. How could such a trend occur?

This would occur if, on average, those species with larger brains persisted longer before extinction and gave rise to more "daughter" species than did species with smaller brains.

Phylogenetic trees symbolize evolutionary history

The evolutionary history of a group of organisms is called **phylogeny** (from the Greek *phylon*, tribe, and *genesis*, origin). Biologists traditionally represent the genealogies of organisms as **phylogenetic trees,** diagrams that trace evolutionary relationships as best they can be determined. The more we know about an organism and its relatives, the more accurately we can portray its phylogeny. We can be much less certain about the evolutionary history of an extinct group represented only by fossils, than we can about a group with living representatives, which can be thoroughly studied. However, even the best phylogenetic tree only represents the most likely evolutionary relationships based on the available evidence.

Below is a phylogenetic tree for 13 species of finches living on the Galápagos Islands (see Module 14.5). This tree is based on detailed studies of body structures, especially beak shape and size, and extensive field studies of reproductive isolation and feeding behavior. As indicated at the bottom of the tree, all Galápagos finches probably arose from one ancestor.

Each branch point on a phylogenetic tree has meaning. For example, the lowest branch on the right indicates that the warbler finch is fundamentally distinct from the other 12 species, and that its lineage diverged from theirs early in the history of the Galápagos finches. The next branch point upward shows a more recent divergence of two large groups, the ground finches and the tree finches. Up one more level (more recent yet), notice that one of three ground finch lineages is represented by a single species, the sharp-beaked ground finch. Among the tree finches, five species are insect-eaters, more closely related to one another than to the vegetarian tree finch, which eats plant buds. Among the insect-eaters, the mangrove finch's closest relative is the woodpecker finch. Both of these species have the unusual habit of probing insects out of tree bark using a small twig or cactus spine.

We will return to phylogenetic trees, but before we do, we need to examine the connection between phylogeny and the classification system used in biology.

Is the medium ground finch more closely related to the small ground finch or to the large ground finch? How is that indicated by this phylogenetic tree?

The medium ground finch is more closely related to the large ground finch. That is indicated in the tree by a more recent branch point (common ancestor) than the branch point shared with the small ground finch.

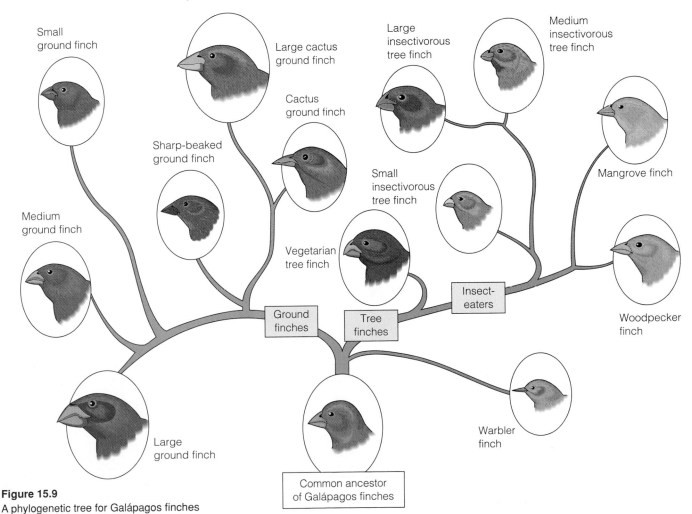

Figure 15.9
A phylogenetic tree for Galápagos finches

15.10 Systematists classify organisms by phylogeny

Classification of the Domestic Cat

Category	Domestic Cat
Domain	Eukarya
Kingdom	Animalia (animals)
Phylum	Chordata (chordates)
(Subphylum)	Vertebrata (vertebrates)
Class	Mammalia (mammals)
Order	Carnivora (carnivorans)
Family	Felidae (cats)
Genus	*Felis* } (domestic cat species)
Specific name	*catus* } (domestic cat species)

Reconstructing phylogeny is part of the science of **systematics,** the study of biological diversity and its classification. Systematists employ taxonomy, the identification and classification of species, and they attempt to arrange organisms in categories that reflect phylogeny.

Assigning scientific names to species is an essential part of systematics. Common names, such as monkey, fruit fly, crayfish, and garden pea, may work well in everyday communication, but they can be ambiguous because there are many species of each of these kinds of organisms. The system of scientific names developed by Linnaeus in the eighteenth century remains in use today. In fact, we still use many of the more than 11,000 names that Linnaeus originally proposed.

As we have seen, Linnaeus's system assigns to each species a two-part latinized name, or **binomial.** The first part of a binomial is the **genus** (plural, *genera*) to which the species belongs. The Galápagos ground finches illustrated in Figure 15.9, for instance, are all members of the genus *Geospiza* (a latinized name derived from the Greek words *ge,* land, and *spiza,* finch). The second part of a binomial refers to one **species** within the genus. For example, the scientific name for the medium ground finch species is *Geospiza fortis* (*fortis* is Latin for strong). Notice that the first letter of the name of the genus is capitalized.

In addition to defining and naming species, a major objective of systematics is to group species into broader taxonomic categories, using a system also devised by Linnaeus. Beyond the grouping of species within genera, the Linnaean system extends to progressively broader categories of classification. It places similar genera in the same **family,** groups families into **orders,** orders into **classes,** classes into **phyla** (singular, *phylum*), and phyla into **kingdoms.** Many taxonomists, and this book, also group kingdoms into a higher taxonomic category called the **domain.**

The table here illustrates the taxonomy of the domestic cat. The genus *Felis,* which includes the domestic cat (*Felis catus*) and several closely related wild cats, is grouped with the genus *Panthera* in the cat family, Felidae (*Panthera* includes the tiger, leopard, jaguar, and African lion). Family Felidae belongs to the order Carnivora, which also includes the family Canidae (e.g., the wolf and domestic dog) and several other families. Order Carnivora is grouped with many other orders in the class Mammalia, the mammals. Class Mammalia is one of several classes belonging to the phylum Chordata in the kingdom Animalia, which is one of several kingdoms in the domain Eukarya. Each taxonomic level is more comprehensive than the one below it in the hierarchy. For instance, not all mammals are cats, but all members of the family Felidae belong to the order Carnivora and the class Mammalia.

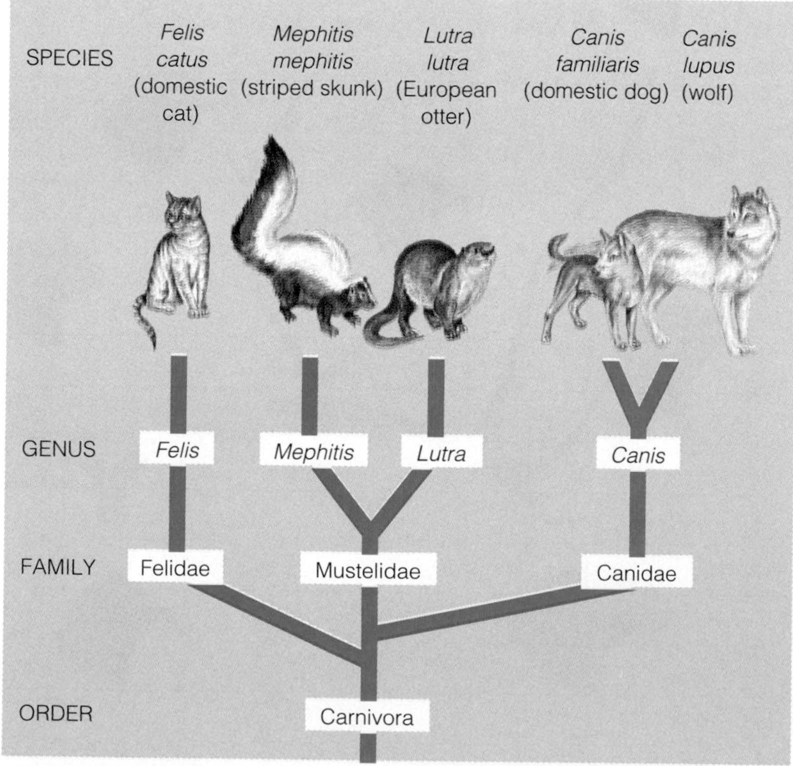

Figure 15.10 The relationship of classification and phylogeny for some carnivorans

Each proper name in the taxonomic hierarchy—family Felidae, order Carnivora, or class Mammalia, for instance—is called a **taxon** (plural, *taxa*). Defining taxa usually requires judgment calls; it is often difficult to determine the dividing lines between taxa even at the species level. As we saw in Module 14.1, some species are clearly separated from all others because their individuals cannot produce fertile offspring with individuals of other species, but it is often difficult to tell if a group of organisms constitutes a species. Many species do not reproduce sexually, and for those that do, we may not know enough to be certain about their identity.

Identifying species often requires judgment calls, but classifying species into higher taxa *always* does. Systematists must weigh all the available evidence about which taxon fits where in the classification system, and decisions about classification often involve heated debate. One systematist may value fine distinctions and favor a relatively large number of taxa for each category above species, whereas another may stress similarities and propose a minimal number of taxa. Even after a classification system is agreed upon, it is subject to updating when new information about the organisms appears.

Ever since Darwin, systematics has had a goal beyond simple organization: to have classification reflect the evolutionary connections among species. As a systematist classifies species in groups subordinate to other groups in the taxonomic hierarchy, the final product takes on the branching pattern of a phylogenetic tree. Figure 15.10 on the facing page shows the pattern for some of the taxa in the order Carnivora. Following conventional practice, the "highest," or most inclusive, taxa are at the bottom. Systematists judge from current evidence that species classified in each genus shown here evolved from a common ancestor that was also a member of that genus. Likewise, they judge that all the genera in any one family arose from a member of that family, and that all the families in the order Carnivora arose from an ancestral carnivoran. Systematists use many kinds of evidence, such as structural and developmental features, molecular data, and behavioral traits of organisms, to make decisions about classification and its relationship with phylogeny. We look at two major lines of evidence in the next two modules.

How much of the classification in the table on the facing page do we share with the domestic cat?

We are classified the same down to and including the class level: Both the cat and human are mammals. We do not belong to the same order.

Homology indicates common ancestry, but analogy does not 15.11

One of the best sources of information about phylogenetic relationships are homologous structures. As we discussed in Module 13.3, homologous structures may look different and function very differently in different species, but they exhibit fundamental similarities because they evolved from the same structure in a common ancestor. Among the vertebrates, for instance, the whale limb is adapted for steering in the water; the bat wing is adapted for flight. Nonetheless, there is a basic similarity in the bones supporting these two structures (see Figure 13.3A).

A systematist always searches for homologies between species, as they are often keys for determining phylogenetic relationships. Generally, the greater the number of homologous structures between two species, the more closely the species are related.

Figure 15.11 Analogous structures, resulting from convergent evolution

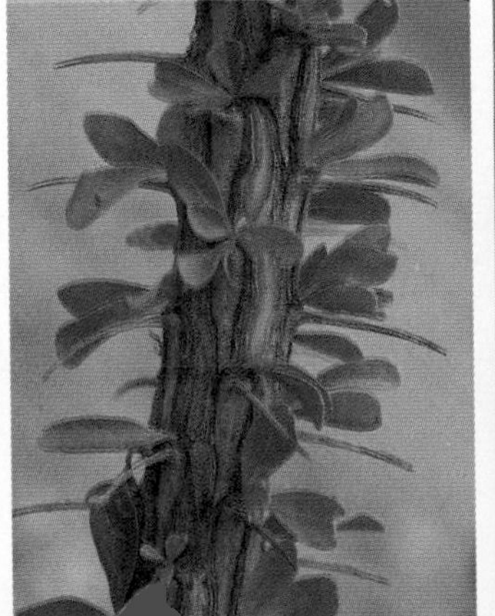

The search for homologies is not without pitfalls, however, for not all likenesses are inherited from a common ancestor. Evolution often produces *non*homologous similarities among organisms. In a process called **convergent evolution,** species from different evolutionary branches may come to resemble one another if they live in very similar environments. In such cases, natural selection may result in body structures and even whole organisms that look very similar but that are *analogous,* rather than homologous. For example, the two plants in Figure 15.11 look remarkably similar, although they are not closely related. On the left is the ocotillo, which is common on the coastal desert of Baja California. On the right is the allauidia, which grows in desert areas in Madagascar. These two plants are on widely separated lineages and have evolved in isolation for millions of years. Nonetheless, as the plants became adapted to similar environments, analogous equipment evolved, including short, water-retentive leaves and cactuslike thorns. The task for the systematist is to distinguish homologies, which indicate common ancestry, from analogies, which do not.

Our forearms and a bat's wings (13.3A) are derived from the same ancestral prototype, thus they are ______. A bat's wings and a bee's wings are derived from unrelated structures, thus they are ______.

homologous . . . analogous

15.12 Molecular biology is a powerful tool in systematics

15A
See Activity 15A on the CD-ROM.

Systematists use all available evidence when classifying organisms and developing phylogenetic hypotheses. Anatomical comparisons continue to be a mainstay, but molecular biology now provides systematists with powerful new ways to study evolutionary relationships. As we saw in Module 10.7, the characteristics of an organism are ultimately determined by the sequences of nucleotides in its DNA. The reason is that the nucleotide sequences code for sequences of amino acids in proteins, which give an organism its unique characteristics. A comparison of the DNA and proteins from different organisms highlights the organisms' basic differences or similarities. The phylogenetic tree for the family Ursidae (bears) and the family Procyonidae (raccoons) shown in Figure 15.12A was constructed from comparisons of DNA and blood proteins.

Bears and raccoons are closely related mammals, but systematists can also use DNA and protein analyses to assess relationships between groups of organisms that are so phylogenetically distant that structural similarities are absent—human beings and yeasts, for instance. Today, both the amino acid sequences for many proteins and the nucleotide sequences for a rapidly increasing number of genomes are in electronic databases that are available via the Internet for comparative studies.

Protein comparisons Among the methods researchers can use to compare proteins, **amino acid sequencing**—determining the sequence of amino acids in a polypeptide—is the most precise. A close match in the amino acid sequences of comparable proteins from different species indicates that the genes that program those proteins evolved from a common gene, inherited from a shared ancestor (see Figure 13.3B). Accordingly, the degree of similarity in amino acid sequences indicates the degree of phylogenetic relationship between different species. For example, the amino acid sequence of cytochrome *c*, an electron transport protein common to all aerobic organisms, has been determined for a wide variety of species ranging from bacteria to plants and animals. The sequences for humans and chimpanzees match perfectly for all 104 amino acid positions along the cytochrome *c* polypeptide chain, and this version of cytochrome *c* differs by just one

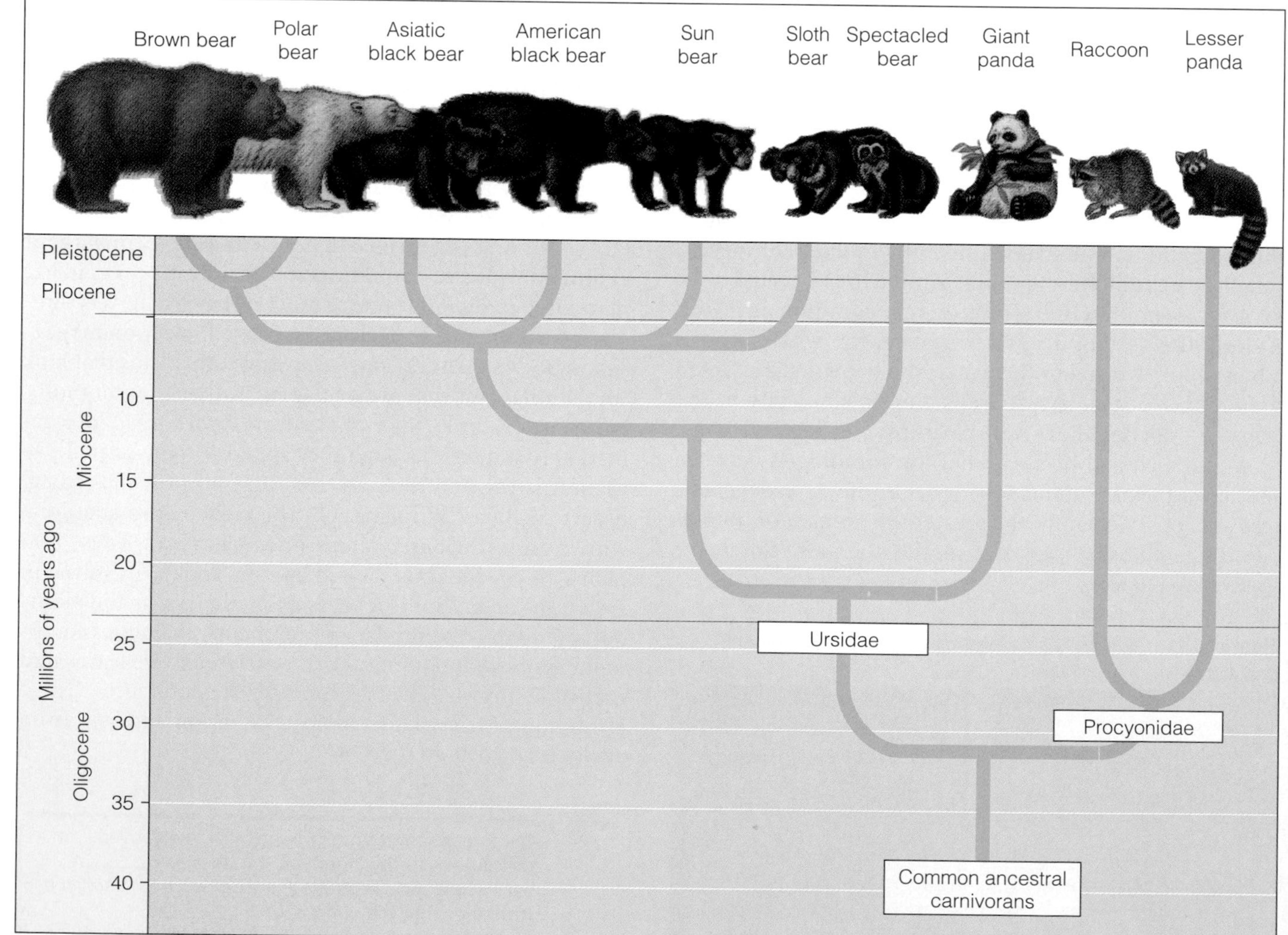

Figure 15.12A A phylogenetic tree based on molecular data

amino acid from the cytochrome *c* in the rhesus monkey. Humans, chimpanzees, and rhesus monkeys all belong to the same mammalian order, Primates. Differences between the cytochromes of these species and more distantly related vertebrates are greater. For example, human cytochrome *c* differs from that of the dog by 13 amino acids, from rattlesnake cytochrome *c* by 20 amino acids, and from tuna cytochrome *c* by 31 amino acids. Phylogenetic trees based on cytochrome *c* are generally consistent with those based on comparative anatomy and the fossil record.

Different proteins evolve at different rates, but the rate of evolution for any one protein seems to be nearly constant over time. Compared to many proteins, cytochrome *c* seems to change very slowly and consequently shows relatively few amino acid differences among closely related organisms. Proteins that change more rapidly—for instance, certain albumins in the blood—vary more from one species to the next, and they are more useful in assessing relationships among closely related organisms, such as those in Figure 15.12A.

DNA and RNA comparisons The most direct way to determine how closely two species are related is by comparing their nucleic acids. Several methods are available for comparing DNA and RNA from different organisms.

One method, called **DNA-DNA hybridization,** measures the extent of hydrogen bonding between single strands of DNA obtained from different species. When single strands from two species are mixed, the more alike the DNA molecules are, the more hydrogen bonds will form between the strands in a hybrid molecule, and the harder it is to separate them. Heat is used to test for hydrogen bonding between strands. The higher the temperature (the greater the heat energy) required to separate paired strands, the better the match. More to the point, the higher the temperature needed to separate hybrid DNA strands, the closer the evolutionary connection between species. The phylogenetic tree in Figure 15.12A reflects evidence from DNA-DNA hybridization that the giant panda is more closely related to bears than to raccoons and that the lesser panda is a member of the raccoon family.

The most direct and precise method for comparing nucleic acids is **DNA sequence analysis.** A researcher first uses recombinant DNA techniques to prepare comparable DNA segments from two species and then determines the actual nucleotide sequences of the segments. Comparison of the sequences can reveal the most fundamental similarities or differences between the two species—exactly which bases are alike or different within equivalent sequences. The application of polymerase chain reaction (PCR; see Module 12.12) to clone traces of DNA, coupled with automated sequencing, has made the collection of DNA sequence data relatively simple and fast. Systematists increasingly use nucleotide sequence data to classify organisms and deduce phylogenetic relationships.

The DNA segments compared using DNA sequencing and other methods involving recombinant DNA technology are usually only a few thousand nucleotides long. Yet the DNA derived from cell nuclei is much longer than this, and it can be unwieldy. Consequently, many researchers use DNA extracted from mitochondria. Mitochondrial DNA is relatively small and more manageable. Moreover, mitochondrial DNA seems to mutate about ten times faster than nuclear DNA, making it possible to sort out genetic relationships among populations of the same species.

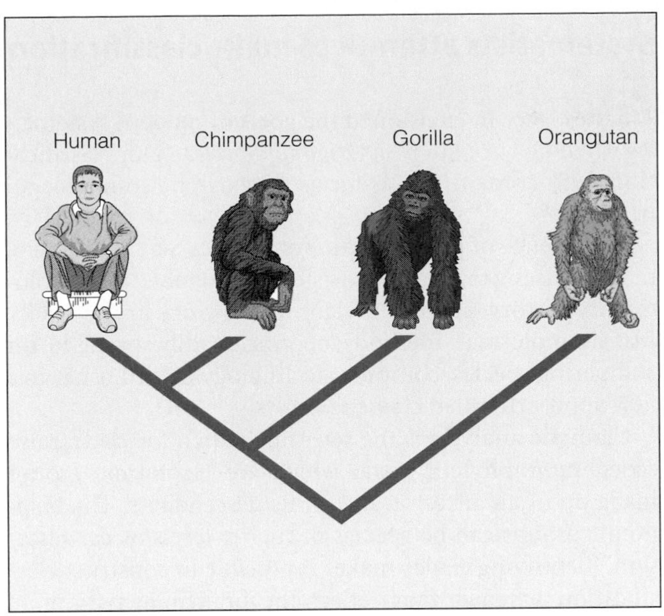

Figure 15.12B Phylogenetic tree illustrating relationships between humans and apes, based on molecular studies

Related to DNA sequence analysis, **ribosomal RNA (rRNA) sequence analysis** is another way to use macromolecules to assess evolutionary relationships and classify organisms. The DNA that codes for rRNA changes slowly relative to most other DNA, thus differences in rRNA sequences can be used to trace some of the earliest branching in the tree of life.

Systematists are using molecular comparisons more and more to help resolve difficult questions in systematics. The phylogenetic tree in Figure 15.12B illustrates a hypothesis about the relationships between humans and our closest living relatives, three great apes. Based on numerous comparisons of proteins, nuclear DNA, and mitochondrial DNA, the tree indicates that we are more closely related to chimpanzees than to gorillas. The strongest support for any such hypothesis is agreement between molecular data and other means of comparing species. In this case, one detailed anatomical comparison supports the tree in the figure, but numerous other structural studies favor the view that chimpanzees and gorillas are more closely related to each other than either is to humans. Debate continues, but the tree in the figure is favored widely. 

? If proteins are not genes, then how can comparisons between the proteins of two species yield data about their evolutionary relationship?

Although proteins are not genes, they are gene products—their amino acid sequences are determined by nucleotide sequences in the DNA.

15.13 Systematists attempt to make classification consistent with phylogeny

Charles Darwin envisioned the goals of modern systematics when he wrote in *The Origin of Species,* "Our classifications will come to be, as far as they can be so made, genealogies."

The science of phylogenetic systematics, with the goal of making classification as consistent as possible with evolutionary history, entered a vigorous new era in the 1960s. Just as molecular methods became readily available for comparing species, computer technology helped usher in a new approach called cladistic analysis.

Cladistic analysis is the scientific search for clades, also called **monophyletic taxa,** which are taxonomic groups made up of an ancestor and all its descendants. The taxonomic groups can be species or higher levels of classification. Identifying clades makes it possible to construct classification schemes that reflect the branching pattern of evolution. The simplified example in Figure 15.13A illustrates the basic features of cladistic analysis. The figure compares four taxa (all vertebrate animal species) according to the presence or absence in these taxa of a set of four homologous features (characters). (Cladistic analysis centers on homologous features, those that are shared by a group of species and their common ancestor, see Module 15.11.) The color coding in Figure 15.13A highlights how these four characters are shared among the four vertebrates.

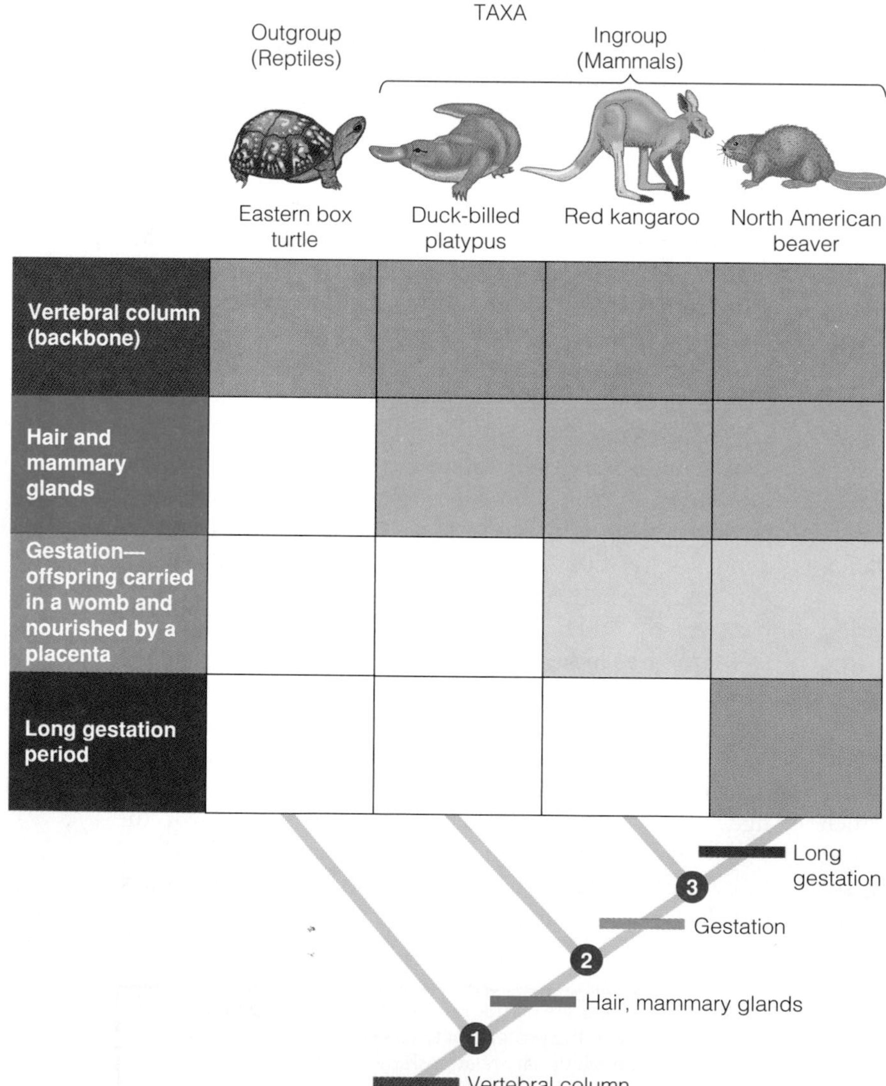

Figure 15.13A Constructing a phylogenetic tree using cladistic analysis

An important part of cladistic analysis is a comparison between a so-called ingroup and an outgroup. The **ingroup** (the three mammals in this simplified example) is the group of taxa that is actually being analyzed. The **outgroup** has a known relationship to the ingroup, but is not a member of it. In our example, the turtle (representing reptiles, the outgroup) and the mammals (collectively the ingroup) are all related in that they are vertebrates; the presence in all of them of a vertebral column (backbone) is a structural feature that represents this relationship. However, the turtle lacks hair and mammary glands, two structural features that unite all the mammals.

Every species (or other taxon) has two types of homologous features, called primitive characters and derived characters. For any group of taxa, **primitive characters** are those that existed in the ancestor common to all. The outgroup serves as a reference, helping us recognize characters that are primitive for the ingroup as a whole. In our example, a vertebral column (backbone) is a shared primitive character. It is present in all the taxa, including the outgroup.

Now let's see how having the outgroup for reference helps us formulate a hypothesis about the relationships among the taxa making up the ingroup. Consider the third character in the figure—gestation, the carrying of offspring in a womb within the female parent. The outgroup does not exhibit gestation. Instead, turtles and most other reptiles lay eggs with a shell (not indicated in the figure). One of the mammals, the duck-billed platypus, also lays eggs with a shell; and we might infer from this that egg-laying is a primitive feature of the vertebrates we are studying and that the duck-billed platypus is a primitive mammal. In fact, this hypothesis is strongly supported by structural and molecular evidence, and the duck-billed platypus is more like the ancestors of mammals (which were reptilelike) than the other mammals in the figure.

In contrast to primitive characters, **derived characters** are homologous features that have changed from a primitive, or ancestral, condition. Again, let's consider gestation. Mammals that have gestation do not lay eggs. (The embryos develop in the womb instead of inside an eggshell.) Our outgroup comparison tells us that egg-laying is primitive, so we can deduce that gestation is a derived character. As indicated in the table, gestation is shared by the kangaroo and beaver but not the platypus. Thus, we can deduce that this character was derived from the primitive egg-laying condition in the lineage that gave rise to the kangaroo and the beaver. Shared derived characters are unique to a lineage, and cladistic analysis involves developing hypotheses about which derived characters are shared by which taxa.

Phylogenetic trees are drawings that represent hypotheses about evolutionary history and the relationships among organisms. The tree at the bottom of Figure 15.13A is based on comparisons of the four homologous characters. Each node (branch point) on the tree represents an ancestor common to all taxa above that node. For instance, the tree posits that the ancestor at node 2 gave rise to all the mammals. The tree also indicates the historical sequence of changes from primitive characters to the derived condition. For instance, the green bar indicates that hair and mammary glands evolved in the lineage that led to all the mammals; gestation and long gestation were derived later in the course of mammalian evolution. In a phylogenetic tree based on cladistic analysis, the ancestor at each node and all of the taxa above each node represent a monophyletic taxon. For instance, the tree posits that the ancestor at node 3, along with the mammals represented by the kangaroo and the beaver, are a monophyletic taxon.

Another key aspect of cladistic analysis is called **parsimony,** the quest for the simplest (and thus probably the most likely) explanation for observed phenomena. Useful in many areas of science, parsimony in systematics means that the simplest hypotheses that are consistent with the comparative data are likely to be the correct ones. For instance, parsimony leads to the hypothesis that a beaver is more closely related to a kangaroo than to a platypus because the beaver and the kangaroo both have gestation. It is possible that gestation evolved twice, once in the kangaroo lineage and independently in the beaver lineage, but this explanation is more complicated, and less likely. When we revisit these hypotheses in Chapter 18, we will see that a more detailed set of evidence strongly supports the parsimonious hypotheses we have made here. Typical cladistic analyses involve much more complex data sets and are usually handled by computer programs designed to construct parsimonious trees.

Cladistic analysis has become the most widely used method in systematics, and its strict application underscores some taxonomic dilemmas. For instance, the familiar taxa class Aves (birds) and class Reptilia (reptiles), as we conventionally know them, are inconsistent with strict cladistic analysis. Studies of shared derived characters and the fossil record indicate that birds represent a lineage of dinosaurs more closely related to one group of reptiles, the crocodiles, than are lizards and snakes. Thus, a classification scheme that reflects phylogenetic branchings would recognize that birds and crocodiles make up one clade and lizards and snakes make up another. The phylogenetic tree shown in Figure 15.13B reflects this cladistic view.

In contrast to cladistic analysis, classical systematics takes into account the apparent degree of divergence between lineages. This approach underlies combining lizards, snakes, and crocodiles in the class Reptilia and assigning birds to their own class (Aves). In this view (Figure 15.13C), flight ability in birds and all the adaptations that make flight possible were major evolutionary changes that allowed birds to diverge so far from any reptiles that they should be classified in their own, nonreptilian class. While debate about the classification of birds continues, most scientific and popular literature (and this text) continue to use the traditional, and convenient, Class Aves.

Classical, noncladistic approaches to systematics often require subjective judgments about the importance of the degree of divergence between taxa. By contrast, cladistic analysis tends to be more objective and focused on testable hypotheses. Most systematists subscribe to the cladistic approach as a more certain way to make classification schemes reflect evolutionary history. However, debate about the strict use of cladistics versus the more subjective approach of classical systematics continues to invigorate the science of classification and phylogeny.

To distinguish a particular clade of mammals within the larger clade that corresponds to class Mammalia, why is hair not a useful characteristic?

Hair is a shared primitive character common to all mammals and cannot be helpful in distinguishing different mammalian subgroups.

Lizards Snakes Crocodiles Birds

Common ancestor

Figure 15.13B Phylogenetic tree of four groups of vertebrates according to cladistic analysis

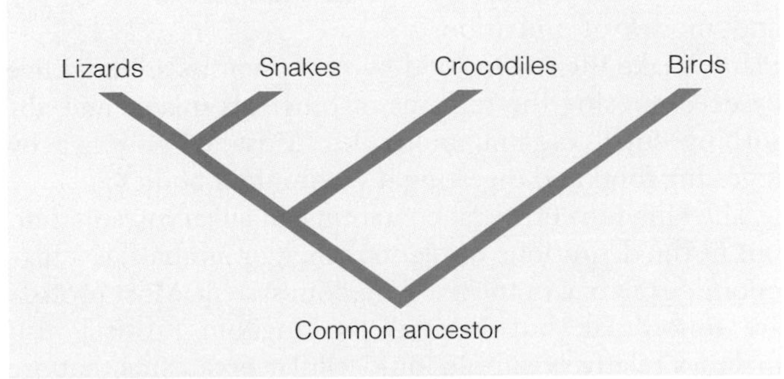

Figure 15.13C Phylogenetic tree of four groups of vertebrates according to classical systematics

15.14 Arranging life into kingdoms is a work in progress

15B

In the next unit, we examine the enormous diversity of organisms that have populated Earth since life first arose about 3.5 billion years ago. As we do so, systematics will help us trace the evolutionary relationships among the organisms.

Over the years, many schemes have been proposed for classifying organisms into kingdoms. A two-kingdom system that divided all organisms into plants and animals was first proposed by Linnaeus and was popular for over 200 years. But it was beset with problems. Where do prokaryotes fit in such a system? Can they be considered members of the plant kingdom? And what about the fungi?

In 1969, American ecologist Robert H. Whittaker argued effectively for a five-kingdom system that places prokaryotes in the kingdom Monera (Figure 15.14A). Organisms of the other four kingdoms all consist of eukaryotic cells. Kingdoms Plantae, Fungi, and Animalia consist of multicellular eukaryotes that generally differ in structure, development, and modes of nutrition. Plants make their own food by photosynthesis. Fungi live by decomposing the remains of other organisms and absorbing small organic molecules. Most animals live by ingesting food and digesting it within their bodies.

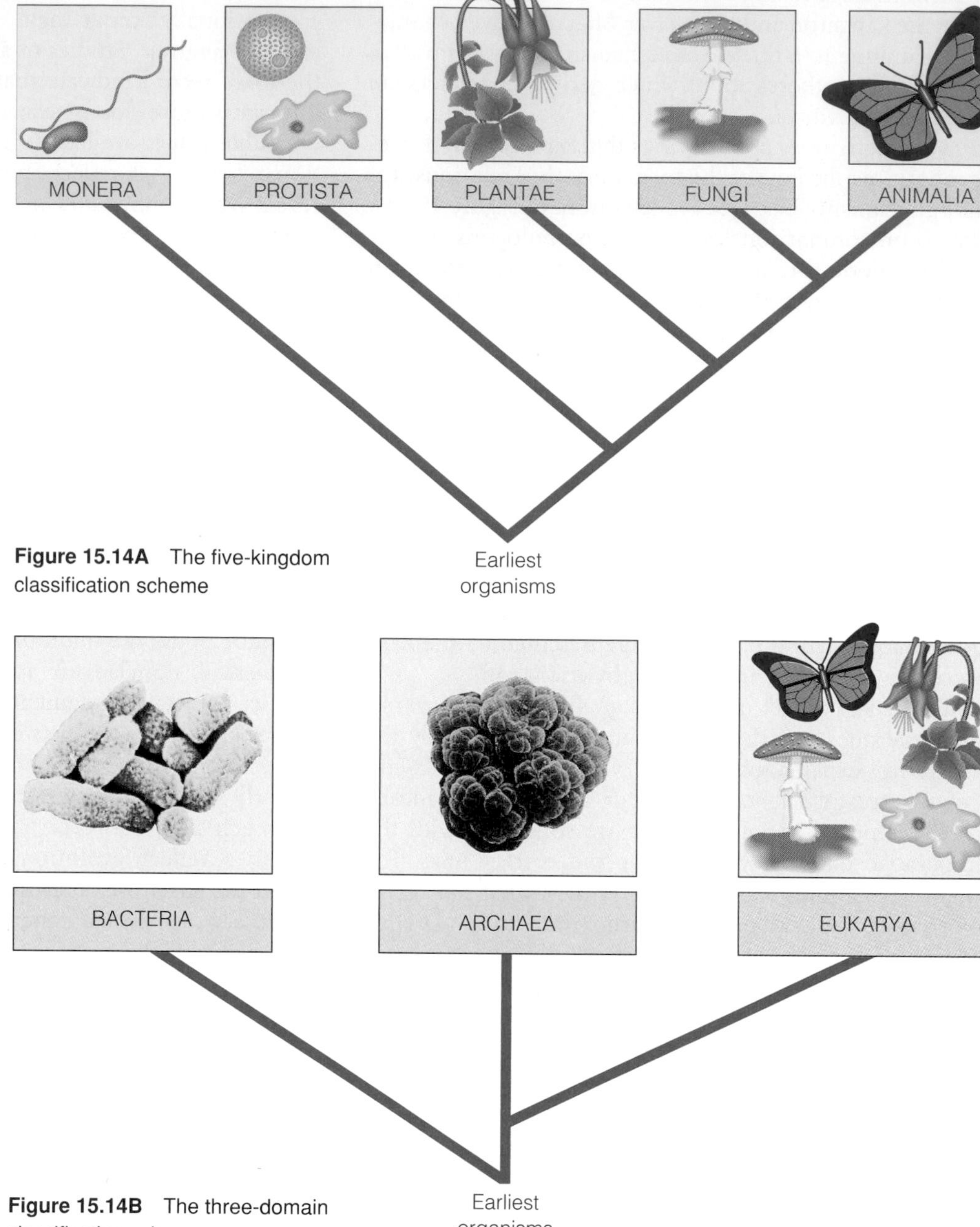

Figure 15.14A The five-kingdom classification scheme

Figure 15.14B The three-domain classification scheme

The kingdom Protista, containing all eukaryotes that do not fit the definitions of plants, fungi, or animals, is a taxonomic grab bag in the five-kingdom system. Most protists are unicellular, but Whittaker's kingdom Protista also includes relatively simple multicellular organisms that are believed to be direct descendants of unicellular protists.

It is important to keep in mind that classification schemes, including the five-kingdom system, are not facts of nature, but human constructions. The five-kingdom system was one attempt to classify the diversity of life into a scheme that is useful and reflective of evolutionary history. In the last decade, molecular studies have highlighted serious flaws in the five-kingdom system, and most systematists now favor replacing it with classification schemes that are more in line with these new data.

Figure 15.14B shows an alternative to the five-kingdom system that has gained support from many biologists in recent years. Whereas the five-kingdom system reflects the concept that there are two fundamentally different groups of living organisms—prokaryotes and eukaryotes—this newer scheme recognizes three basic groups: two domains of prokaryotes, the Bacteria (also called Eubacteria) and Archaea (or Archaebacteria), and one domain of eukaryotes, called Eukarya. The Bacteria and the Archaea differ in a number of important structural, biochemical, and functional features, which we will discuss in Chapter 16.

Molecular and cellular evidence supports the phylogenetic hypothesis outlined in Figure 15.14B—that two lineages of prokaryotes (the bacteria lineage and the archaea lineage) diverged very early in the evolutionary history of life. Molecular evidence also suggests that the Archaea are more closely related to eukaryotes than to the Bacteria. Thus, Figure 15.14B indicates that the lineage of the domain Eukarya diverged from the Archaea after the archaeans split off from the Bacteria.

The color coding in Figures 15.14A and 15.14B highlights the relationship between the five-kingdom system and the three-domain system. Notice that all the eukaryotic kingdoms are included in the domain Eukarya. As we discuss further in the next chapter, the number of kingdoms and their evolutionary relationships are subjects of ongoing research and debate. For example, nucleic acid evidence indicates that Kingdom Protista, as defined in the five-kingdom system, contains several groups of eukaryotes that are so phylogenetically distinct that they should be classified in separate kingdoms.

A strong consensus favors replacing the five-kingdom system, and we follow the domain scheme outlined in Figure 15.14B in this textbook. However, what is most important to understand is that defining the higher categories of classification (the kingdoms and domains of life) will always be a work in progress. The most important function of a classification system is to serve as a focal point for discussion about how to make classification fit phylogeny. 

? In comparing the five-kingdom system with the three-domain system, how many of the kingdoms fall into domain Eukarya?

Four

Chapter Review

Chapter Summary

Earth History and Macroevolution (Introduction–15.9) The fossil record documents macroevolution, the main events in the history of life. Major transitions in life forms mark separate geological eras; smaller changes divide eras into periods. Life began in the Precambrian era, at the end of which animals first appeared. During the Paleozoic era, plants and animals diversified and moved onto land. During the Mesozoic era, reptiles and cone-bearing plants dominated the land. Extinction of the dinosaurs marked the beginning of the Cenozoic era, during which mammals and flowering plants proliferated **(15.1)**. The sequence of fossils in rock strata indicates the relative ages of extinct species. Radiometric dating, which measures the decay of radioactive isotopes, can gauge the actual ages of fossils **(15.2)**.

Continental drift, the slow, incessant movement of Earth's crustal plates on the hot mantle, has influenced the distribution of organisms and greatly affected the history of life. Continental drift brought the continents together near the end of the Permian period, forming the supercontinent Pangaea. This continental merger altered coastlines and climate and triggered extinctions. Early in the Mesozoic era, separation of the continents caused the isolation and diversification of organisms **(15.3)**. Plate tectonics, the movements of Earth's crustal plates, are also associated with volcanoes and earthquakes **(15.4)**.

At the end of the Cretaceous period, many life forms disappeared, including nearly all lineages of the dinosaurs. This mass extinction may have resulted mainly from the impact of an asteroid or from an increase in volcanic activity. Every mass extinction reduced the diversity of life, but each was followed by a rebound in diversity. Mammals filled the void left by the dinosaurs **(15.5)**.

Distinctive features, such as the hair and mammary glands of mammals, may enable particular organisms to prosper after mass extinctions. Adaptations (also called exaptations) that have evolved in one environmental context may be able to perform new functions when conditions change. Major adaptations may arise rapidly if mutations occur in certain genes that control early events in an organism's development **(15.6)**. Mutations in developmental genes may result in paedomorphosis, the retention of juvenile characteristics in the adult. Paedomorphosis seems to have played a key role in human evolution **(15.7)**. The fossil record seems to reveal trends in the evolution of species and lineages, such as a gradual size increase in the evolution of dinosaurs. Trends or apparent trends are evident in extinct as well as surviving lineages. An evolutionary trend may reflect unequal speciation or unequal survival of species on a branching evolutionary tree **(15.8)**. Phylogeny, which can be depicted by phylogenetic trees, is the evolutionary history of a group of organisms **(15.9)**.

Systematics and Phylogenetic Biology (15.10–15.13) Reconstructing phylogeny is part of systematics, the study of biological diversity and classification. Taxonomists assign a two-part name, or binomial, to each species. The first name, the genus, covers a group of related species. The second name refers to a species within the genus. Genera are grouped into progressively larger categories: family, order, class, phylum, kingdom, and domain. Taxonomists often debate the particular placement of organisms in categories, as they strive to make their categories reflect evolutionary relationships **(15.10)**. Homologous structures are evidence that organisms have evolved from a common ancestor. In contrast, nonhomologous (analogous) similarities are evidence that organisms from different evolutionary lineages have undergone convergent evolution. Their resemblances have resulted from living in similar environments, not from common ancestry **(15.11)**. Systematists increasingly use molecular techniques (especially DNA and RNA sequence analyses) to classify organisms and develop phylogenetic hypotheses **(15.12)**. The goal of systematics is to make classification reflect evolutionary history (phylogeny). Using homologous features to compare organisms, cladistic analysis attempts to define monophyletic taxa, groups consisting of an ancestor and all its descendants. Shared primitive characters are homologous features that existed in an ancestor common to all members of a monophyletic taxon. Unique to an evolutionary lineage, shared derived characters are those that have changed from a primitive, or ancestral, condition. Cladistic analysis often is a search for the simplest (most parsimonious) hypotheses about phylogeny. Debate about cladistic hypotheses and those developed by classical, noncladistic approaches to systematics continues to invigorate research **(15.13).**

The Domains of Life (15.14) For several decades, systematists have classified life into five kingdoms, with all prokaryotes making up one kingdom, the Monera. The multicellular eukaryotes (plants, animals, and fungi) differ in structure, development, and modes of nutrition and thus are grouped into separate kingdoms. The kingdom Protista, a grab bag of organisms, consists of all the remaining eukaryotes, mostly unicellular forms. A newer system recognizes two basically distinctive groups of prokaryotes, the domain Bacteria and the domain Archaea. A third domain, the Eukarya, includes all the kingdoms of eukaryotes.

Testing Your Knowledge

Multiple Choice

1. Many species of plants and animals adapted to desert conditions probably did not arise there. Their success in living in deserts could be due to

   a. paedomorphosis
   b. convergent evolution
   c. exaptation
   d. phylogeny
   e. mass extinction

2. Mass extinctions that occurred in the past

   a. cut the number of species to the few survivors left today
   b. resulted mainly from the separation of the continents
   c. occurred regularly, about every million years
   d. were followed by diversification of the survivors
   e. wiped out land animals, but had little effect on marine life

3. The animals and plants of India are almost completely different from the species in nearby Southeast Asia. Why might this be true?

   a. They have become separated by convergent evolution.
   b. The climates of the two regions are completely different.
   c. India is in the process of separating from the rest of Asia.
   d. Life in India was wiped out by ancient volcanic eruptions.
   e. India was a separate continent until relatively recently.

4. Two worms in the same class must also be grouped in the same

   a. order
   b. phylum
   c. genus
   d. family
   e. species

5. Suppose you checked the Galápagos finch phylogenetic tree in Module 15.9 using DNA-DNA hybridization. Which of the following hybrid DNAs would you expect to come apart at the lowest temperature? *(Explain your answer.)*

   a. medium ground/small ground
   b. large ground/mangrove
   c. large cactus ground/sharp-beaked ground
   d. woodpecker/vegetarian tree
   e. mangrove/woodpecker

Describing, Comparing, and Explaining

1. Explain how the amino acid sequences of proteins can be used to trace evolutionary relationships among organisms.

2. Discuss various hypotheses that have been suggested to explain the Cretaceous mass extinction that ended the reign of the dinosaurs. What kind of evidence supports each hypothesis?

Thinking as a Scientist

1. Scientists have calculated that a particular piece of rock contained about 12 g of radioactive potassium-40 when it was formed. It now contains 3 g of potassium-40. The half-life of potassium-40 is 1.3 billion years. About how old is the rock?

2. Imagine yourself writing a news article about dinosaurs. All you have are the notes that another reporter jotted down during an interview with a researcher. The reporter did not really understand the research, but his notes say: "This guy has figured out how to isolate DNA from dinosaur fossils; gave me the data shown below; the numbers = temperatures obtained in DNA-DNA hybridization tests of evolutionary relationships between duck-billed dinosaur *Maiasaura (M)* and four close relatives *(A, B, C, D)*. To read the table, the melting point of hybrid DNA between species *M* and *A* (85°C) is under the *A* at the top of table and to the right of the *M* on the left side of table, etc."

	M	*A*	*B*	*C*	*D*
M	86	85	81	82	85
A		87	82	81	84
B			86	85	80
C				87	81
D					85

The reporter also made three little sketches, shown below. Explain why each of the following fits or does not fit the data in the table:

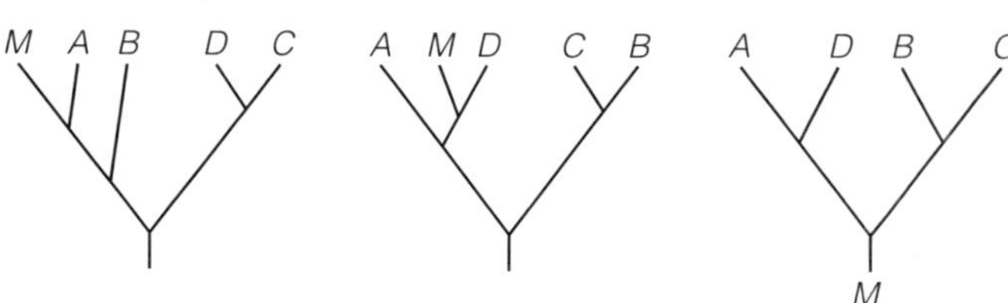

Science, Technology, and Society

Experts estimate that human activities cause the extinction of hundreds of species every year. The natural "background" rate of extinction is thought to be a few species per year. As we continue to alter the environment, especially by destroying tropical rain forests, the resulting mass extinction will probably rival the end of the Cretaceous period, when perhaps half the species on Earth disappeared, including the dinosaurs. Many scientists and environmentalists are alarmed at this prospect, which they see as a catastrophe of unprecedented proportions. Other people are not as worried, because life has endured numerous mass extinctions in the past and has always bounced back. Some environmentalists respond that the situation is different this time. Is the current mass extinction different? Why or why not? What might be the eventual consequences for most species? For the survivors? For humans?

Media Resources

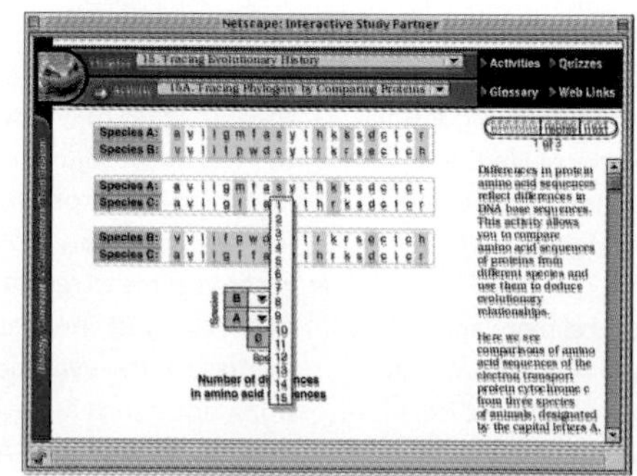

For further review of the concepts in this chapter, use the activities and quizzes on the CD-ROM.

Visit the Special Edition of The Biology Place web site for activities, quizzes, news articles, and web links.

UNIT IV

The Evolution of Biological Diversity

16 The Origin and Evolution of Microbial Life: Prokaryotes and Protists

Outline

Cruising in shallow water off western Australia, an undersea research vessel glides over an array of what look like pockmarked boulders. Are they lava rocks from an ancient volcano? Mounds of coral? Massive sponges? Their surfaces are greenish and sticky. They seem rigid, and an edge of one shows evidence of layering. Back in the laboratory, a look at a sample through a microscope shows that prokaryotic organisms coat the mounds. Most of the prokaryotes turn out to be cyanobacteria, photosynthetic bacteria that commonly grow on the surface of layered mats.

The photograph on the facing page shows part of a cyanobacterial mat that was growing in a warm lagoon in Baja California. The reddish-brown layers are bands of sediment that the bacteria accumulated. Coating the top of the mat where they are exposed to sunlight, sticky cyanobacteria concentrate sand grains and other fine particles from the seawater. While a sediment layer builds up on them, the bacteria keep migrating to the surface and growing over it. A layered mat builds up as this process is repeated over and over. The large Australian mounds formed in this way over thousands of years, hardening as their older sediment layers solidified.

The cyanobacteria that produce mats and mounds are descended from some of the oldest organisms known—photosynthetic prokaryotes that dominated Earth some 3 billion years ago. The fossil record indicates that at that time, greenish lawns of prokaryotes covered virtually every wet, sunlit surface on the planet. Animals that might have grazed the prokaryotic lawns had not evolved yet, and the unchecked growth of ancient prokaryotes built up huge expanses of thick, layered mats. Many of the mats eventually became fossilized, and the layers of sediment became sedimentary rocks. Geologists call fossilized prokaryotic mats **stromatolites** (from the Greek *stroma,* bed, and *lithos,* rock).

Stromatolites occur throughout the fossil record. Their prominence in rocks around 2.5 billion years old marks the time when ancient photosynthetic prokaryotes were so numerous that the gaseous oxygen (O_2) they produced made Earth's atmosphere aerobic. This atmospheric change set the stage for the evolution of all aerobic life.

Photosynthetic prokaryotes dominated Earth from nearly 3 billion years ago to about 1 billion years ago and then declined significantly. Their direct descendants, the cyanobacteria and other photosynthetic bacteria of today, remain abundant in freshwater lakes and ponds and in shallow oceans, but they only form thick mats or mounds where there is little animal life to eat them. Thus, great expanses of stromatolites are a thing of the past. Nevertheless, the grand, global legacy of ancient photosynthetic bacteria—the aerobic atmosphere — remains with us today. And the aerobic atmosphere is but one illustration of the profound effects that living organisms can have on the environment.

Tracing the roles played by various organisms in the history of life on Earth is one of two main objectives in this unit of chapters. Our other goal is to introduce you to the diversity of life. We start in this chapter with the prokaryotes and the protists. As mentioned in Modules 1.4 and 15.14, the prokaryotes are the members of the domains Bacteria and Archaea; the protists are the unicellular eukaryotes and their direct multicellular descendants.

A cross section of a mat of layered sediment produced by cyanobacteria

Early Earth and the Origin of Life

16.1 Life began on a young Earth

Figure 16.1A A common scene on Earth when mats of prokaryotes were the main form of life

Earth some 3 billion years ago—it bristles with volcanoes spewing dust and gases into the atmosphere and molten rock onto the surface. Yellowish-green "stepping stones," actually thick mats of prokaryotes, dominate a shoreline. Oxygen released from photosynthetic bacteria in the mats will change the atmosphere forever. The painting above illuminates the inseparable histories of planet Earth and its living organisms.

Biological and geological history have been closely intertwined since life began. As we saw in Chapter 15, the formation and subsequent breakup of the supercontinent Pangaea had a tremendous effect on the diversity and geographical distribution of life. Conversely, life has changed the planet it inhabits, sometimes profoundly. The photosynthetic prokaryotes that first released oxygen to the air completely altered Earth's atmosphere. Much more recently, the emergence of *Homo sapiens* has changed the land, water, and air on a scale and at a rate unprecedented for a single species.

Earth is one of nine planets orbiting the sun, which is one of billions of stars in the Milky Way. The Milky Way, in turn, is one of billions of galaxies in the universe. Gazing at stars gives us a look back in time. The star closest to our sun is 4 light years—40 trillion kilometers—away; we see the light it emitted 4 years ago. Some stars are so far away that even if they burned out millions of years ago, we would still see them in the sky tonight. And there are new stars that are invisible because their light has not yet reached Earth.

The universe has not always been so spread out. Apparently, before the universe existed in its present form, all of its matter was concentrated in one mass. The mass seems to have blown apart with a "big bang" sometime between 10 and 20 billion years ago and to have been expanding ever since. Our solar system probably arose from a swirling cloud of dust. Most of the dust condensed in the center to form the sun, but some matter was left orbiting around the infant sun in several concentric rings. Within each orbit, kernels of matter had enough gravity to draw neighboring dust and ice particles together, forming the planets.

Earth probably began about 4.5 billion years ago as a cold world. Later on, heat generated by the impact of meteorites, radioactive decay, and tight compaction by gravity thawed Earth and eventually turned it into a molten mass. The mass then sorted into layers of varying densities, with most of the nickel and iron sinking to the center and forming a core. Less dense material became concentrated in a mantle surrounding the core, and the least dense material settled on the surface, solidifying into a thin crust.

The first atmosphere was probably composed mostly of hot hydrogen gas (H_2). Because the gravity of Earth was not strong enough to hold such small molecules, the H_2 soon escaped into space. Volcanoes and other vents through the crust belched gases that formed a new atmosphere. Analyses of gases vented by modern volcanoes have led scientists to speculate that this second early atmosphere consisted of carbon monoxide (CO), carbon dioxide (CO_2), nitrogen (N_2), and water vapor (H_2O), with possibly some methane (CH_4) and ammonia (NH_3). The first seas were created by torrential rains that began when the planet had cooled enough for water in the atmosphere to condense. Not only was the atmosphere of the young Earth very different from the one we know today, but lightning, volcanic activity, and ultraviolet radiation were much more intense. It was in such an environment that life began.

Fossil evidence indicates that nonliving matter evolved into primitive cellular organisms within a few hundred million years after Earth's crust solidified. We would guess from the relatively simple structure of prokaryotic cells (compared with that of eukaryotic cells) that early organisms were primitive prokaryotes, and the fossil record supports this idea. Fossils that appear to be prokaryotes about the size of some bacteria living today have been discovered in southern Africa in a rock formation called the Fig Tree Chert, which is 3.4 billion years old. Evidence of even more ancient prokaryotic life occurs in stromatolites. Notice how similar the fossilized mat in Figure 16.1B is to the still-growing one on p. 317. The stromatolite in Figure 16.1B is from western Australia. Dating from the Precambrian period about 3.5 billion years ago, it is among the oldest signs of life found thus far. The timeline in Figure 16.1C puts these most ancient of fossils in historical perspective.

Figure 16.1D shows a fossilized prokaryote from a Precambrian stromatolite. It is a filament composed of a chain of prokaryotic cells. Fossils like this are common in stromatolites and probably represent some of the vast numbers of photosynthetic prokaryotes that produced the atmosphere's O_2.

Photosynthetic prokaryotes are among the simplest kinds of organisms we know that are autotrophic (produce their own food molecules). However, photosynthesis is not a simple metabolic process, and it is unlikely that photosynthetic bacteria were the first forms of life on Earth. In fact, the evidence that photosynthetic prokaryotes thrived 3.5 billion years ago is strong support for the hypothesis that life in a simpler form, unable to make its own food molecules, arose much earlier, perhaps as early as 4 billion years ago.

Figure 16.1B A cross section of a stromatolite

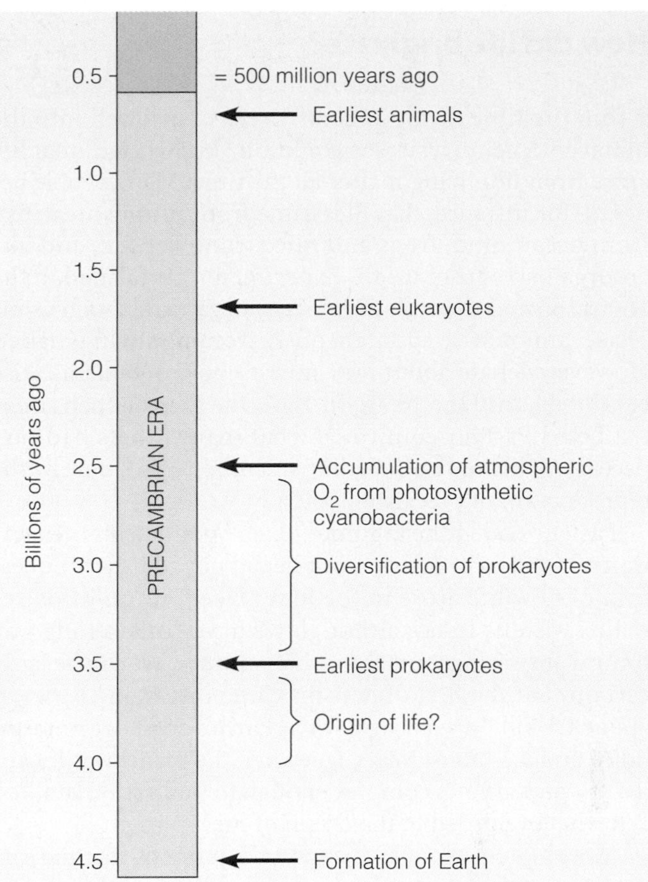

Figure 16.1C Major episodes in the early history of life

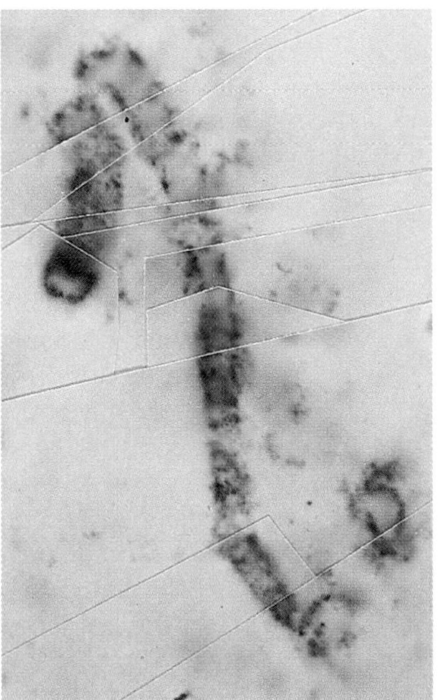

Figure 16.1D A fossilized prokaryote

What is the relationship between stromatolites, such as the one in Figure 16.1B, and bacterial mats, such as the one on page 317?

Stromatolites are fossilized bacterial mats.

16.2 How did life originate?

From the time of the ancient Greeks until well into the nineteenth century, it was common "knowledge" that life arose from nonliving matter all the time. Many people believed, for instance, that flies came from rotting meat, fish from ocean mud, frogs and mice from wet soil, and microorganisms from broth. Experiments performed in the 1600s showed that relatively large organisms, such as insects, cannot arise spontaneously from nonliving matter. However, debate about how microscopic organisms arise continued until the 1860s. In 1862, the great French scientist Louis Pasteur confirmed what many others had suspected: All life today, including microbes, arises only by the reproduction of preexisting life.

Pasteur ended the argument over spontaneous generation of modern organisms, but he did not address the question of how life arose in the first place. The question remains with us today, although we have come a long way toward answering it in the last half century. We can be fairly certain that the first organisms came into being between about 4.1 billion years ago, when Earth's crust began to solidify, and 3.5 billion years ago, when the planet was inhabited by prokaryotes complex enough to build stromatolites. But what events led to the origin of life?

Most biologists subscribe to the hypothesis that the earliest form of life was much simpler than anything alive today, and that life did first develop from nonliving materials. The earliest lifelike entities may have been aggregates of molecules with a particular arrangement that made simple metabolism and self-replication possible. Because living organisms consist of polymers formed from small organic molecules (monomers), the origin and accumulation of small organic molecules must have been the earliest chemical steps preceding the origin of life. Some scientists propose that meteorites and comets seeded the planet with small organic molecules, but the prevailing opinion is that most of the first organic molecules arose from inorganic materials on the early Earth. The formation of polymers, including proteins and nucleic acids, from organic monomers would have been a second stage preceding the origin of life. Once this occurred, the polymers and monomers may have formed aggregates having chemical characteristics different from their surroundings. Heredity may also have originated during this stage. In the next four modules, we examine several stages in chemical evolution, the process by which the molecules of life may have originated on the early Earth.

According to the model introduced in this module, the first cells were preceded by a chemical evolution that first produced small ________ molecules, which subsequently joined to form ________.

organic . . . polymers

TALKING ABOUT SCIENCE

16.3 Stanley Miller's experiments showed that organic molecules could have arisen on a lifeless Earth

In 1953, when Stanley Miller was a 23-year-old graduate student in the laboratory of Harold Urey at the University of Chicago, he performed some experiments that would attract global attention. Miller was the first to show that amino acids and other organic molecules could have been generated on a lifeless Earth.

Miller's experiments were a test of a hypothesis about the origin of life developed in the 1920s by Russian biochemist A. I. Oparin and British geneticist J. B. S. Haldane. Oparin and Haldane proposed that the conditions on the early Earth could have generated a collection of organic molecules that in turn could have given rise to the first living organisms. They reasoned that present-day conditions on Earth do not allow the spontaneous synthesis of organic compounds simply because the atmosphere is rich in O_2. O_2 is corrosive: As a strong oxidizing agent, it tends to disrupt chemical bonds by extracting electrons from them. However, before the early prokaryotes added O_2 to the air, Earth probably had a reducing atmosphere instead of an oxidizing one. A reducing environment tends to *add* elec-

Figure 16.3A Stanley Miller in his laboratory

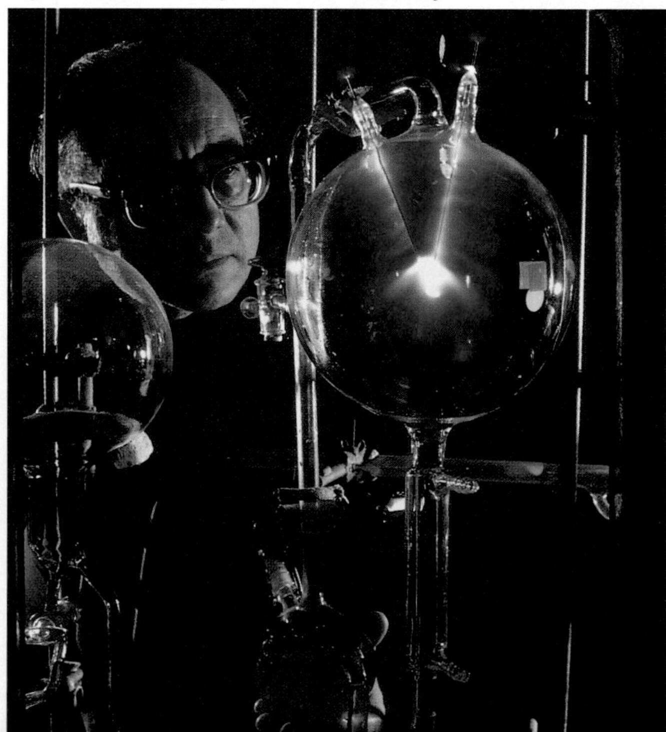

trons to molecules. Thus, an early reducing atmosphere could have caused simple molecules to combine to form more complex ones. As Miller, now a professor at the University of California, San Diego, told us several years ago,

> Oparin proposed that the primitive atmosphere contained the gases methane, ammonia, hydrogen, and water, and that chemical reactions in that primitive atmosphere produced the first organic molecules. That hypothesis had a good deal of appeal, but without the experiments, it was talked about but not very well accepted.

The construction of complex molecules from simple ones also requires energy, and Miller and Urey reasoned that there were abundant energy sources in the environment of the early Earth. Besides lightning discharges, ultraviolet radiation probably reached Earth's surface with much greater intensity than it does today. Evidence indicates that young suns emit more ultraviolet radiation than older suns. Also, in our modern atmosphere, a layer of ozone (O_3) screens out most UV radiation. Ozone, which forms from ordinary oxygen (O_2), would not have been present before photosynthesis arose.

Miller and Urey predicted that organic molecules would form from inorganic ones under conditions like those on the early Earth. The apparatus shown with Dr. Miller in Figure 16.3A on the preceding page is similar to the one he used to test the prediction. Figure 16.3B indicates how the apparatus simulated conditions on the early Earth. A flask of warmed water represented the primeval sea. The "atmosphere" consisted of a mixture of water vapor, H_2, CH_4, and NH_3—the gases that scientists in the 1950s believed prevailed in the ancient world. Electrodes discharged sparks into the gas mixture to mimic lightning. Below the spark

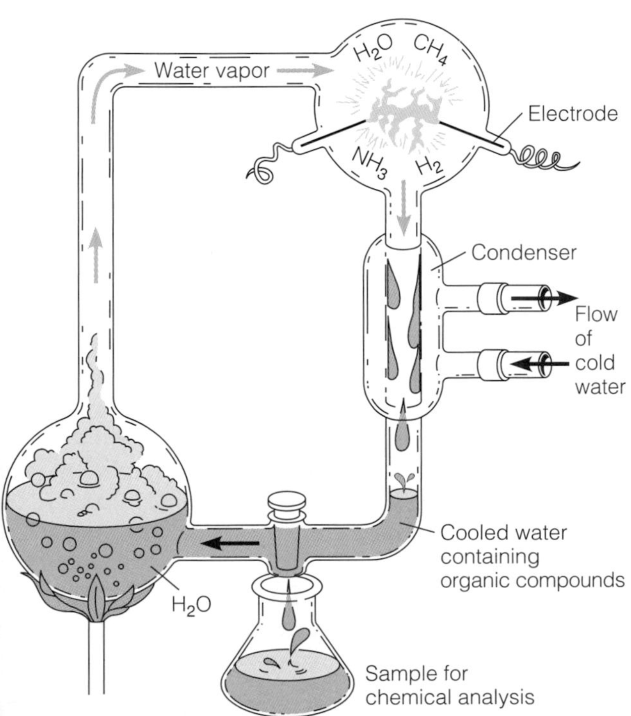

Figure 16.3B The synthesis of organic molecules in Miller's apparatus

chamber, a glass jacket called a condenser surrounded the apparatus. Filled with cold water, the condenser cooled and condensed the water vapor in the gas mixture, causing "rain," along with any dissolved compounds, to fall back into the miniature sea. As material circulated through the apparatus, the solution in the flask slowly changed color. As Dr. Miller described it,

> The first time I did the experiment, it turned red. Very dramatic! And then, after it turned red, it got more yellow and then brown as the sparking went on.

After the experiment proceeded for a week, Miller found a variety of organic compounds in the solution, including some of the amino acids that make up the proteins of organisms.

> The surprise was that we ... got mainly organic compounds of biological significance. And the amino acids were formed, not in trace quantities, but abundantly! The experiment went beyond our wildest hopes.

Miller's early experiments stimulated a great deal of interest and research on the prebiotic (before-life) origin of organic compounds. Since the 1950s, Miller and other researchers using modifications of Miller's apparatus have made most of the 20 amino acids commonly found in organisms, as well as sugars, lipids, the nitrogenous bases present in nucleotides of DNA and RNA, and even ATP. These laboratory studies support the idea that many of the organic molecules that make up living organisms could have formed before life itself arose on the early Earth.

Scientists now believe that the composition of the atmosphere of the early Earth was somewhat different from what Miller assumed in his historic first experiment. Modern volcanoes emit CO, CO_2, N_2, and water vapor, and it is likely that these gases were abundant in the atmosphere when life first arose. H_2, CH_4, and NH_3 were probably not major components. Furthermore, many scientists today doubt that the early atmosphere played a direct role in early chemical reactions. Instead, submerged volcanoes and deep-sea vents—gaps in the Earth's crust where hot water and minerals gush into deep oceans—may have provided the initial chemical resources.

We asked Dr. Miller in what sort of place he thought life actually began on Earth.

> The usual assumption is that it began in the ocean. But you can legitimately propose that some of the processes occurred in different areas. For example, some of the polymerization reactions that made larger organic molecules probably occurred on beaches that had dried out and heated up, and some may have occurred in hot springs. But the oceans form the bulk of the area where organic reactions could take place, and I think most of the chemistry took place there. 

? What was the hypothesis Stanley Miller was testing with his experiments?

The hypothesis that conditions on the early Earth favored synthesis of organic molecules from inorganic ingredients

16.4 The first polymers may have formed on hot rocks or clay

After small organic molecules formed on the prebiotic Earth, the second major chemical step before life arose must have been polymerization, the formation of organic polymers, such as nucleic acids and proteins, from their monomers.

The polymers of life are synthesized by dehydration reactions that release a water molecule for each monomer added to the chain (see Module 3.3). In the living cell, specific enzymes catalyze these reactions. Polymerization also occurs in laboratory situations without enzymes, for example, when dilute solutions of organic monomers are dripped onto hot sand, clay, or rock. The heat vaporizes the water in the solutions and concentrates the monomers on the underlying substance. Some of the monomers then spontaneously bond together in chains, forming polymers.

Using this method, biochemist Sidney Fox, at the University of Miami, has succeeded in making polypeptides. On the early Earth, in a similar fashion, raindrops or waves may have splashed dilute solutions of organic monomers onto fresh lava or other hot rocks, and then rinsed polypeptides and other polymers back into the sea.

Clay surfaces may have been especially important as early polymerization sites. Even cool clay concentrates amino acids and other organic monomers from dilute solutions, because the monomers bind to electrically charged sites on the clay particles. Such binding sites could have brought monomers close together. Clay also contains metal atoms, such as iron and zinc, that can function as catalysts. On the primitive Earth, metal atoms might have catalyzed the first dehydration reactions that linked monomers together to form polymers.

Before enzymes existed, clays with metal atoms might have catalyzed the __________ reactions that joined organic monomers to form __________.

dehydration . . . polymers

16.5 The first genetic material and enzymes may both have been RNA

The formation of polymers on the early Earth set the stage for the origin of early forms of life. But which polymers were most important? Stanley Miller gives us a hint: "The essential difference between life and nonlife is replication. There are other differences, but this is the essential one."

Miller's message is that nucleic acids, the only biological polymers that can replicate and store genetic information, were the essential polymers. Today's cells store their genetic information as DNA, transcribe the information into RNA, and then translate RNA messages into specific enzymes and other proteins. As we have seen in earlier chapters, this DNA → RNA → protein assembly system is extremely intricate. Most likely, it emerged gradually through a series of refinements to much simpler processes. What were the first genes like?

One popular hypothesis is that the first genes were short strands of RNA that replicated themselves without the assistance of proteins, perhaps on clay surfaces. Laboratory experiments support this idea. Short RNA molecules can assemble spontaneously from nucleotide monomers without the presence of cells or enzymes. Furthermore, when RNA is added to a solution containing a supply of RNA monomers, new RNA molecules complementary to parts of the starting RNA sometimes assemble. So we can imagine a scenario on the early Earth like the one in Figure 16.5: ① RNA monomers—nucleotides—spontaneously join to form the first, small genes. ② Then, an RNA chain complementary to one of these genes assembles. If the new chain, in turn, serves as a template for another round of RNA assembly, the result is a replica of the original gene.

This RNA replication process might have been aided by RNA molecules that acted as catalysts. Scientists have discovered that some RNAs, which they call **ribozymes,** can act as enzymes. Modern ribozymes can catalyze RNA splicing (see Module 11.7) and even some limited RNA polymerization.

Scientists use the term **RNA world** for the hypothetical period in the evolution of life when RNA served as both rudimentary genes and the sole catalytic molecules.

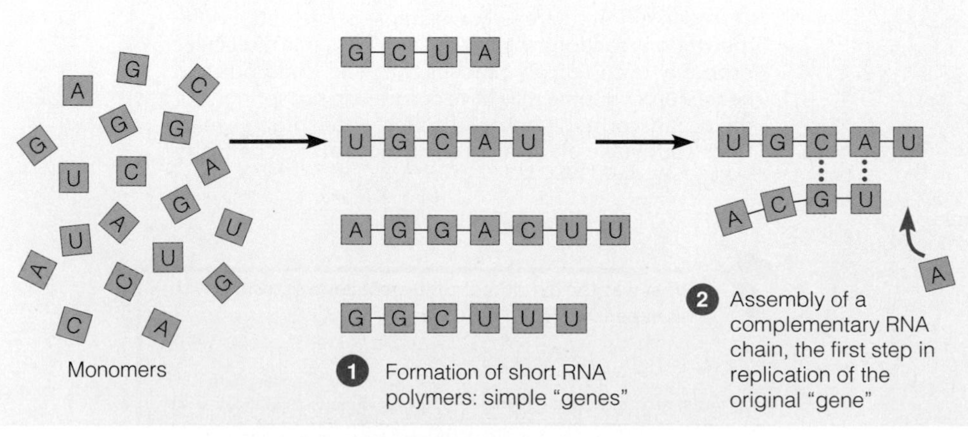

Monomers
G-C-U-A
U-G-C-A-U
A-G-G-A-C-U-U
G-G-C-U-U-U
1 Formation of short RNA polymers: simple "genes"
2 Assembly of a complementary RNA chain, the first step in replication of the original "gene"

Figure 16.5 A hypothesis for the origin of the first genes

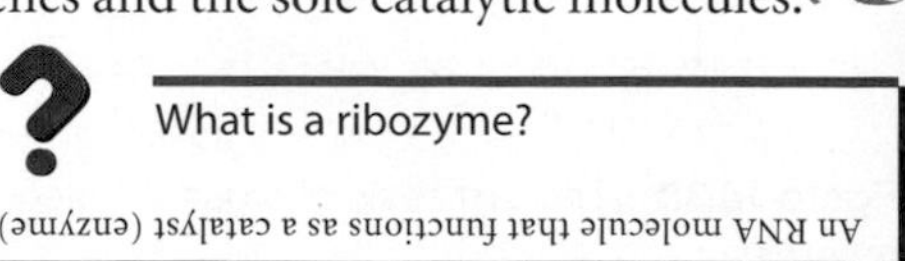

What is a ribozyme?

An RNA molecule that functions as a catalyst (enzyme)

Molecular cooperatives enclosed by membranes probably preceded the first real cells 16.6

If self-replicating RNA molecules did arise as we have described, they were still a far cry from a living cell. Life as we know it requires a great number of complex organic molecules, and the molecules must interact and cooperate in precise ways. Put another way, life depends on intricate metabolic machinery that derives from the cooperation of many complex organic molecules. It is likely that some amount of molecular cooperation preceded life's origin.

The earliest form of molecular cooperation may have involved a primitive form of translation of simple RNA genes into polypeptides, translation that did not use ribosomes or tRNA. Suppose a strand of RNA acted as a rough template for polypeptide synthesis by weakly binding to amino acids and holding a few of them together long enough for them to be linked. Zinc or some other metal in the vicinity may have acted as a catalyst for the linkage. If the sequence of nucleotides in the RNA influenced the sequence of amino acids, the resulting polypeptide would be a rough translation of the RNA gene. If, in turn, the polypeptide behaved as an enzyme for helping RNA molecules replicate (Figure 16.6A), then reciprocal molecular cooperation between nucleic acids and polypeptides would have begun.

Let's suppose that cooperative associations between RNA and polypeptides became common on the prebiotic Earth. We might imagine a world where molecular "co-ops" often formed in aquatic environments, especially in small puddles or films of water on clay surfaces. These sites may have harbored large numbers of organic molecules, including RNA, polypeptides, and others.

Experiments show that in an aqueous environment, polypeptides can self-assemble into microscopic spheres filled with fluid (Figure 16.6B). These spheres are not alive, but they display some of the properties of living cells. They have a selectively permeable membranelike surface, can grow by absorbing polypeptide molecules from their surroundings, divide when they reach a certain size, and swell or shrink osmotically when placed in solutions of different salt concentrations. Furthermore, certain kinds of lipid molecules, such as phospholipids, can form spheres with similar properties. On the early Earth, spheres with enclosing membranes of either sort may have formed. Eventually, some may have had a membrane made of both types of molecules, roughly like the plasma membrane of a true cell.

It is easy to imagine that certain of these cell-like entities on the early Earth might have contained some of the preexisting RNA-polypeptide co-ops (Figure 16.6C). The enclosure of such a co-op by a membrane would have been a major milestone in the evolution of the first cells. Isolated from other co-ops in its surroundings, the enclosed co-op would be the sole beneficiary of its own molecular products. A membrane-enclosed co-op that grew and replicated more efficiently than others would have been favored by natural selection. In other words, the molecular co-ops could have begun evolving in a Darwinian sense.

We have now plotted a scenario whereby local concentrations of organic chemicals on the prebiotic Earth may have given rise to cooperative associations of molecules enclosed by membranes. These molecular co-ops would have developed the ability to replicate and to carry out essential chemical reactions—a primitive metabolism. But millions of years probably passed before the first living cells appeared on Earth. During this long period, the molecular co-ops would have evolved into complex metabolic machines containing DNA and capable of efficiently using a variety of raw materials from the environment. Even the simplest cell would have been vastly more complex than any molecular co-op we have described. Whatever the evolutionary steps that led to the first cellular life, we can be fairly certain that the earliest cells were prokaryotes, as we discuss next. 

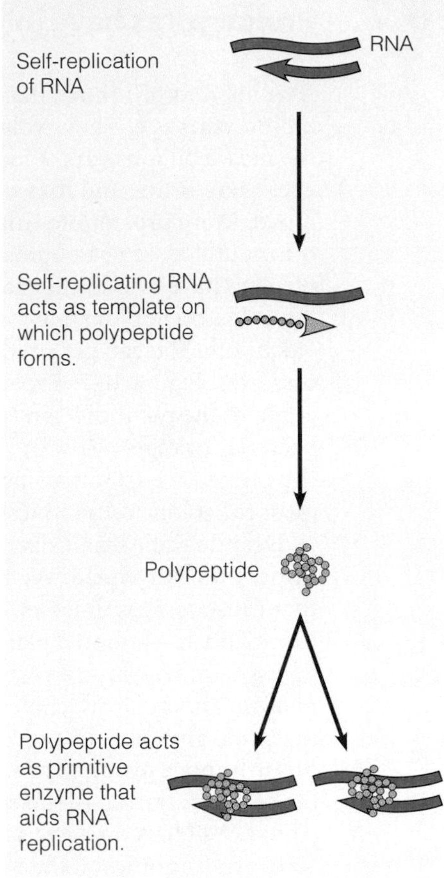

Figure 16.6A The beginnings of cooperation among macromolecules, in the absence of membranes

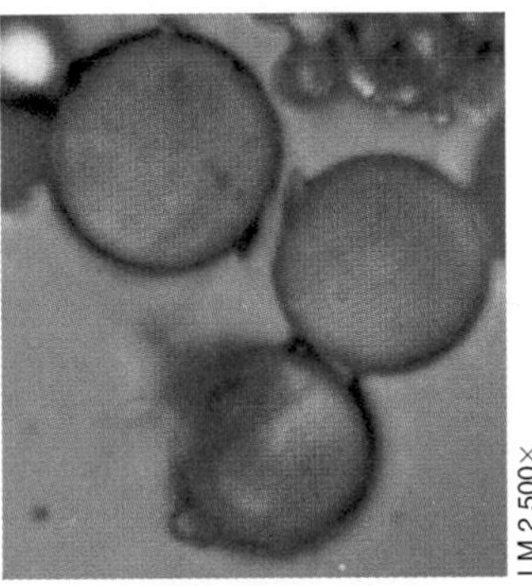

Figure 16.6B Microscopic spheres formed by polypeptides

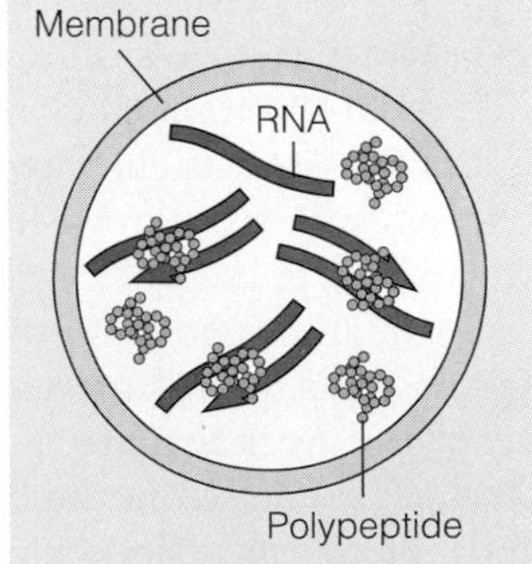

Figure 16.6C Macromolecular cooperation within a membrane-enclosed compartment

? According to the hypothesis presented in this module, why was the enclosure of protein–nucleic acid cooperatives by membranes a key step in the onset of Darwinian evolution (natural selection)?

Segregation of molecular systems within membrane-enclosed compartments allowed competition (selection) for the most successful self-replicating collections of molecules.

16.7 Prokaryotes have inhabited Earth for billions of years

The fossil record shows that prokaryotes were abundant 3.5 billion years ago. They evolved all alone on Earth for about the next 2 billion years. Today, prokaryotes are found wherever there is life, and they outnumber all eukaryotes combined. More prokaryotes inhabit a handful of fertile soil or the mouth or skin of a human than the total number of people who have ever lived. Prokaryotes also thrive in habitats too cold, too hot, too salty, too acidic, or too alkaline for any eukaryote. You can get an idea of the size of most prokaryotes from Figure 16.7, a colorized scanning electron micrograph of the point of a pin (purple), covered with numerous bacteria (orange). Most prokaryotic cells have diameters in the range of 1–10 micrometers (μm), much smaller than most eukaryotic cells (typically 10–100 μm).

Despite their small size, prokaryotes have an immense impact on our world. We hear most about a few species that cause serious illnesses. During the fourteenth century, Black Death—bubonic plague, a bacterial disease—spread across Europe, killing an estimated 25% of the human population. Tuberculosis, cholera, many sexually transmissible diseases, and certain types of food poisoning are some other human problems caused by bacteria. Bacteria also cause many kinds of diseases in other animals and in plants. We focus on bacterial diseases in Module 16.15.

Far more common than harmful bacteria are those that are benign or beneficial. We have bacteria in our intestines that provide us with important vitamins, and others living in our mouth help prevent harmful fungi from growing there. Essential to all life on Earth are prokaryotes that decompose dead organisms. Found in soil and at the bottom of lakes, rivers, and oceans, these prokaryotes—bacteria and archaea—return vital chemical elements to the environment in the form of inorganic compounds that can be used by plants, which in turn feed animals. If prokaryotic decomposers were to disappear, the chemical cycles that sustain life would halt, and all forms of eukaryotic life would also be doomed. In contrast, prokaryotic life would undoubtedly persist in the absence of eukaryotes, as it once did for billions of years. 

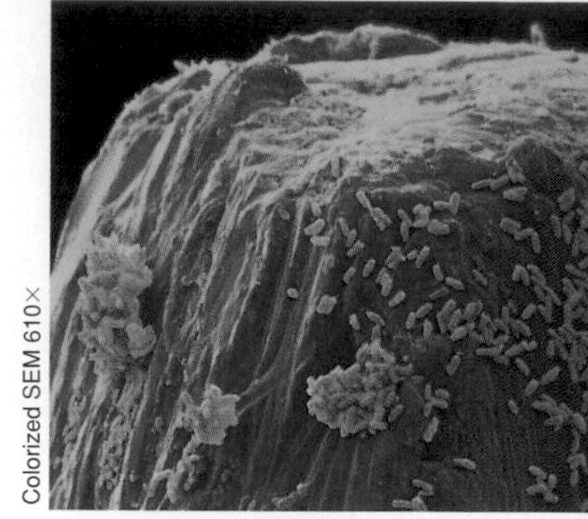

Figure 16.7 Bacteria on the point of a pin

? What are decomposers?

Organisms that recycle the chemical ingredients of life by breaking down dead organisms and organic wastes to inorganic compounds that can be reassimilated by plants and other photosynthetic organisms

16.8 Archaea and bacteria are the two main branches of prokaryotic evolution

Prokaryotes have a cellular organization fundamentally different from that of eukaryotes, as we saw in Modules 4.4 and 4.5. Whereas eukaryotic cells have a membrane-enclosed nucleus and numerous other membrane-enclosed organelles, prokaryotic cells lack these structural features.

As we discussed in Module 15.14, however, two very different kinds of prokaryotes, classified in the domains **Bacteria** and **Archaea,** are found on Earth today. The name Archaea comes from the Greek *archaios* ("ancient"), and most biologists believe that these prokaryotes and bacteria diverged from each other in very ancient times.

The most fundamental differences between the organisms of these two domains are in their nucleic acids. Researchers first focused on one type of ribosomal RNA (rRNA), a type found in all prokaryotes and eukaryotes. Comparing the nucleotide sequences, they made some interesting discoveries. For instance, near nucleotide number 910 (of 1500) in this rRNA, they have found the following difference in sequence:

Bacteria	AAACUCAAA
Archaea	AAACUUAAAG

Researchers have identified about a dozen such short rRNA sequences that distinguish bacteria from archaea. Intriguingly, in a number of cases, the archaeal sequence is identical to that of eukaryotes.

More recently, researchers have focused on DNA and have completely sequenced a number of bacterial and archaeal genomes (see Module 12.15). When compared with each other and with the genomes of eukaryotes such as yeast, these genome sequences strongly support the three-domain view of life. Some genes of archaea are similar to bacterial genes and others to eukaryotic genes; still others seem unique to archaea.

The table on the next page summarizes some of the main differences between bacteria and archaea. In addition to the rRNA sequences, several other differences involve the cellular machinery for gene expression. These include differences in RNA polymerases (enzymes catalyzing the synthesis of RNA), in the presence of introns within genes, and in sensitivity to certain antibiotics that inhibit protein synthesis. Subtle differences between archaeal and bacterial ribosomes—in both rRNA and pro-

Differences Between Bacteria and Archaea		
Main Features	**Bacteria**	**Archaea**
rRNA sequences	Many unique to bacteria	Many match eukaryotic ones
RNA polymerase	Relatively small and simple	Complex; similar to eukaryotic
Introns (noncoding parts of genes)	Absent	Present in some genes
Antibiotic sensitivity (to streptomycin, chloramphenicol)	Inhibited	Not inhibited
Peptidoglycan in cell wall	Present	Absent
Membrane lipids	Carbon chains unbranched	Carbon chains branched

teins—undoubtedly account for the insensitivity of archaea to these antibiotics.

Other differences between bacteria and archaea show up in their cell walls and membranes. Nearly all prokaryotes have a cell wall outside their plasma membrane. As in plants, the wall maintains cell shape and provides physical protection. Bacterial cell walls contain a unique material called **peptidoglycan,** a polymer of sugars cross-linked by short polypeptides. No archaea have true peptidoglycan. Furthermore, the lipids forming the backbone of plasma membranes differ between the two domains.

Notice that in most of the features in the table, *archaea are more like eukaryotes than like bacteria.* This may mean that the archaea are more closely related to eukaryotes than to bacteria, the other prokaryotes. As you saw in Figure 15.14B, a current hypothesis is that modern archaea and eukaryotes evolved from a common ancestor. But the situation is complicated by evidence of gene swapping among the three domains.

Probably more important than the outcome of the taxonomic debate on this issue is to realize that there are two very different kinds of prokaryotic organisms. We will discuss the diversity within these two groups after a look at some more general features of prokaryotes.

? As different as archaea and bacteria are, both groups are characterized by ________ cells that lack nuclei and other membrane-bounded organelles.

prokaryotic

16.9 Prokaryotes come in a variety of shapes

16A See Activity 16A on the CD-ROM.

Determining cell shape by microscopic examination is an important step in identifying prokaryotes. The micrographs below show three of their most common cell shapes. Spherical prokaryotic cells are called **cocci** (from the Greek word for berries). Cocci (singular, *coccus*) that occur in clusters, like the ones in Figure 16.9A, are called staphylococci (from the Greek *staphyle,* cluster of grapes). Other cocci occur in chains; they are called streptococci (from the Greek *streptos,* twisted). The bacterium that causes strep throat in humans is a streptococcus.

Figure 16.9B shows rod-shaped prokaryotes, which are called **bacilli** (singular, *bacillus*). Most bacilli occur singly, but the cells of some species occur in pairs (diplobacilli) and in chains (streptobacilli). The species shown here, which is common in fertile soil, exists as solitary cells.

A third prokaryotic cell shape is curved or spiral. Some bacteria in this category resemble commas and are called *vibrios.* Other bacteria and archaea have a helical shape, like a corkscrew. Helical prokaryotes that are relatively short and rigid are called *spirilla;* those with longer, more flexible cells are called *spirochetes* (Figure 16.9C). The bacterium that causes syphilis, for example, is a spirochete. Spirochetes include some giants by prokaryotic standards—cells 0.5 mm long (though very thin).

? How could a microscope help you distinguish the cocci that cause "staph" infections from those that cause "strep" throat?

It would show clusters of cells for staphylococcus and chains of cells for streptococcus.

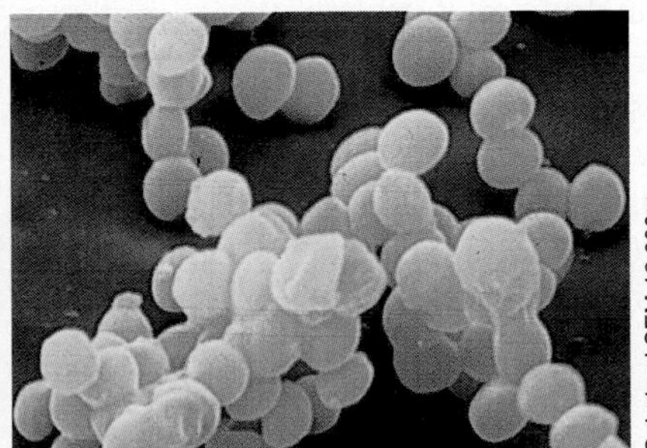
Colorized SEM 12,000×

Figure 16.9A Cocci

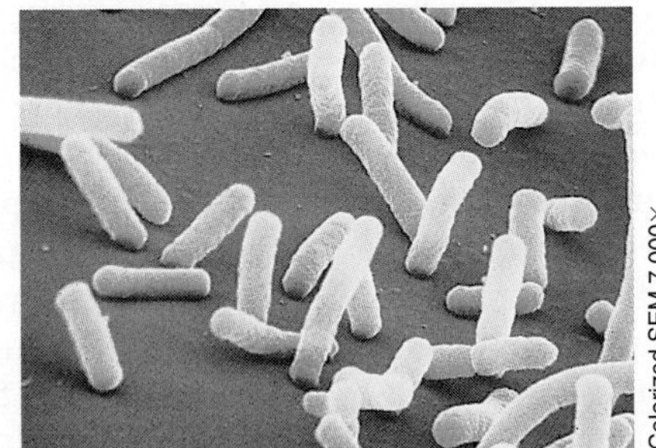
Colorized SEM 7,000×

Figure 16.9B Bacilli

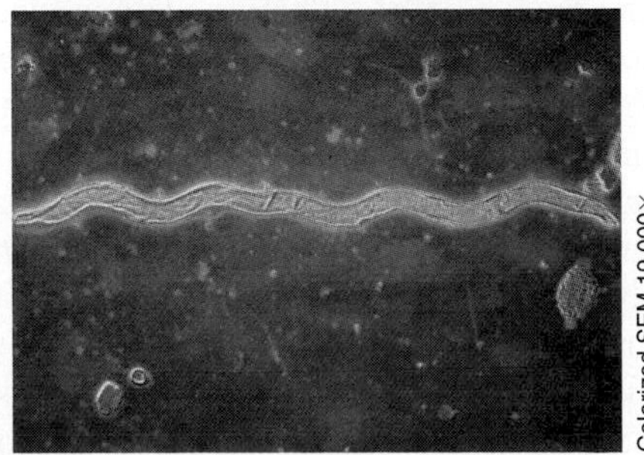
Colorized SEM 19,000×

Figure 16.9C Spirochete

16.10 Prokaryotes obtain nourishment in a variety of ways

16B

When classifying diverse organisms, biologists often use the phrase "mode of nutrition" to describe how an organism obtains two main resources: carbon (for synthesizing organic compounds) and energy. As a group, prokaryotes exhibit much more nutritional diversity than eukaryotes.

Many prokaryotes are **autotrophs** ("self-feeders"), making their own organic compounds from inorganic sources. As shown in the top half of the table here, autotrophs obtain their carbon atoms from carbon dioxide (CO_2). They get their energy from sunlight or from inorganic chemicals, such as hydrogen sulfide (H_2S), elemental sulfur (S), or compounds containing iron (Fe). Autotrophs that harness sunlight for energy and use CO_2 for carbon, such as the cyanobacteria, do so by photosynthesis; they are called **photoautotrophs.** (Cyanobacteria use H_2O as a source of electrons for photosynthesis and produce O_2 as a waste product, just like plants.) Autotrophic organisms that obtain energy from inorganic chemicals instead of sunlight are called **chemoautotrophs.**

Most prokaryotes are **heterotrophs** ("other-feeders"), meaning they obtain their carbon atoms from organic compounds. Some heterotrophs, called **photoheterotrophs,** can obtain energy from sunlight. By far the largest group of prokaryotes, however, are nutritionally similar to animals in that they obtain both energy and carbon from organic molecules. Called **chemoheterotrophs,** these bacteria are so diverse that almost any organic molecule can serve as food for some species. Many species, such as *Escherichia coli,* a resident of the human intestine, can thrive on a variety of organic nutrients. The photograph in Figure 16.10 shows a culture of *E. coli* grown with only glucose as an organic nutrient. Each round spot in the culture dish is a colony, a clone of millions of bacterial cells.

When nutrients are available, *E. coli* and other prokaryotes multiply exponentially: one cell divides to form 2, 2 cells form 4, 4 form 8, and so on. With generation times as short as a few hours or less, prokaryotes have enormous growth potential. Their actual growth is limited by environmental factors (such as nutrient availability) and by the buildup of toxic metabolic wastes from the microbes themselves.

NUTRITIONAL CLASSIFICATION OF ORGANISMS

Nutritional Type	Energy Source	Carbon Source
Photoautotroph (photosynthesizers)	Sunlight	CO_2
Chemoautotroph	Inorganic chemicals	CO_2
Photoheterotroph	Sunlight	Organic compounds
Chemoheterotroph	Organic compounds	Organic compounds

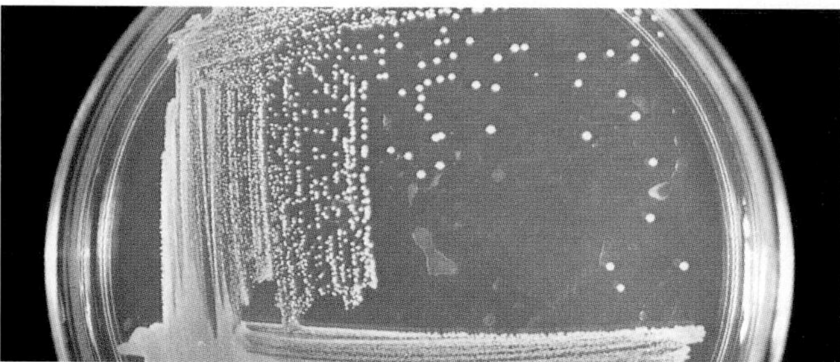

Figure 16.10 *E. coli* colonies grown on glucose

Chemoheterotrophs are the dominant prokaryotes today—and may have been since the dawn of life. However, there are other possibilities, as we see next.

? A bacterium requires only the amino acid methionine as an organic nutrient and lives deep in the soil where no light penetrates. Based on its mode of nutrition, this bacterium would be classified as a ______________.

chemoheterotroph

16.11 The first cells probably used chemicals for both carbon and energy

With nutrition in mind, let's return to our discussion of the origin of life. Earlier, we proposed that the first life forms, which were prokaryotes, evolved from membrane-enclosed molecular co-ops. The first prokaryote would have had a very simple metabolism requiring only a few enzymes. Its environment contained almost no O_2, so its metabolism would have been anaerobic. It is unlikely that the earliest organisms were able to use sunlight as an energy source because doing so requires a complex set of enzymes. More likely, the early life forms would have simply obtained their carbon and energy from the rich soup of molecules and ions in which they evolved.

Figure 16.11 illustrates two hypotheses about the earliest kinds of energy metabolism. Both hypotheses center on the fact that all modern forms of life use ATP as their main energy currency; thus, it is likely that prokaryotes started using this molecule very early. Hypothesis A proposes that a substantial supply of ATP was present in the early "soup" of organic molecules. In this scenario, the earliest life form was a chemoheterotroph that met all its carbon and energy needs by absorbing organic molecules, including ATP, from its surroundings. Such a life form would have had enzymes for converting the absorbed organic molecules to other compounds and for breaking down ATP and using the released energy. Later, enzymes may have evolved that could regenerate ATP from ADP using energy released from the breakdown of other organic nutrients.

A more likely hypothesis (B) assumes that the early environment was less rich in organic molecules and that the earliest life form was a chemoautotroph that synthesized its own ATP. The primitive cell may have obtained its energy from chemical reactions involving inorganic sulfur and iron compounds, which were abundant. A membrane enzyme may have used energy and H_2 released from the reaction of these inorganic materials to generate H^+ ions just outside the cell. ATP might then have been made by a primitive form of chemiosmosis (see Module 6.12). CO_2 or perhaps abiotically formed organic molecules may have served as the carbon source.

Scientists got the idea for hypothesis B from the metabolic activities of certain archaea living on Earth today, which we discuss next.

What was the main function of ATP in early cells?

Probably the same as today: the transfer of energy for cellular work

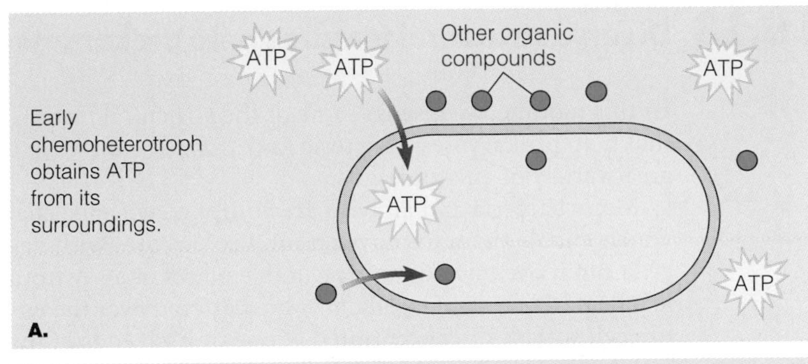

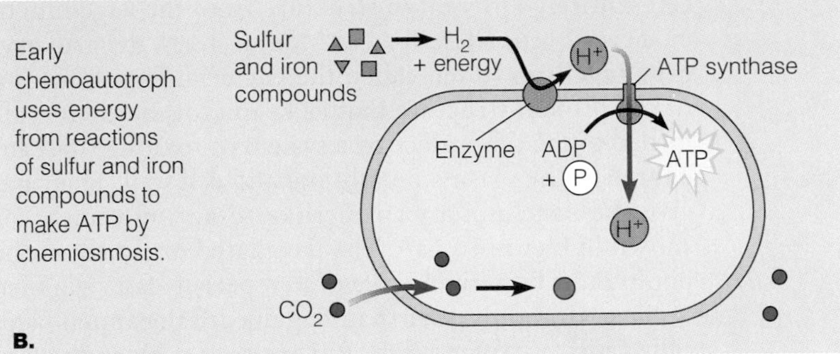

Figure 16.11 Two hypotheses for the earliest kinds of energy metabolism

Archaea thrive in extreme environments 16.12

Archaea are abundant in many habitats, including places where few other organisms can survive. The archaeal inhabitants of extreme environments have unusual proteins and other molecular adaptations that enable them to metabolize and reproduce effectively. Scientists are only beginning to learn about these adaptations.

A group of archaea called the extreme halophiles ("salt lovers") thrive in very salty places, such as the Great Salt Lake in Utah and seawater-evaporating ponds used to produce salt. Figure 16.12 shows some ponds of this sort next to San Francisco Bay. The purplish color is due to a huge population of an archaeon called *Halobacterium halobium*. A unique photosynthesizer, *H. halobium* lacks chlorophyll; instead, it has a purple molecule called bacteriorhodopsin that traps solar energy.

Another group of archaea, the hyperthermophiles ("heat lovers"), thrive in very hot water; some even live near deep-ocean vents where temperatures are above 100°C, the boiling point of water at sea level! Other hyperthermophiles thrive in acid. Many hot, acidic pools in Yellowstone National Park harbor such archaea, which give the pools a vivid greenish color. One of these organisms, *Sulfolobus*, can obtain energy by oxidizing sulfur or a compound of sulfur and iron; the mechanisms may be ancient.

A third group of archaea, the methanogens, live in anaerobic environments and give off methane as a waste product. Many thrive in anaerobic mud at the bottom of lakes and swamps. You may have seen methane, also called marsh gas, bubbling up from a swamp. Great numbers of methanogens also inhabit the digestive tracts of animals. In humans, intestinal gas is largely the result of their metabolism. More importantly, methanogens aid digestion in 16A
cattle, deer, and other animals that depend heavily on cellulose for their nutrition. Normally, bloating does not occur, because these animals regularly belch out large volumes of gas produced by methanogens and other microorganisms that enable them to utilize cellulose.

Some archaea are referred to as "extremophiles." Why?

Because they can thrive in extreme environments too hot, too salty, or too acidic for other organisms

Figure 16.12 Seawater-evaporating ponds colored purple by photosynthetic archaea

16.13 Diverse structural features help prokaryotes thrive almost everywhere

In this module, we discuss some of the structural features that help prokaryotes—bacteria and archaea—thrive in a great variety of environments.

Many bacteria and archaea are equipped with flagella, which enable them to move about. Prokaryotes with flagella can move toward more favorable places or away from less favorable ones. Flagella may be scattered over the entire cell surface or concentrated at one or both ends of the cell. Entirely different in structure from the flagellum of eukaryotic cells (described in Module 4.18), the **prokaryotic flagellum** (often called the bacterial flagellum) is a naked protein structure that lacks microtubules. It is attached to the cell surface by a system of rotating rings anchored in the plasma membrane and cell wall. The rings give the flagellum a propellerlike rotary movement, as shown in Figure 16.13A. The flagellated organism in the photo is the bacterium *Proteus*, an especially fast swimmer.

Shorter and thinner than flagella are the appendages called **pili** (singular, *pilus*). Pili are not visible in the Figure 16.13A micrograph but show up clearly in the more highly magnified micrograph of another bacterium in Figure 16.13B. Pili help bacteria stick to each other and to surfaces, such as rocks in flowing streams or the lining of human intestines. As we mentioned in Chapter 12, special pili called *sex pili* are required for initiating bacterial "mating" (conjugation). Sex pili are fewer and longer than ordinary pili.

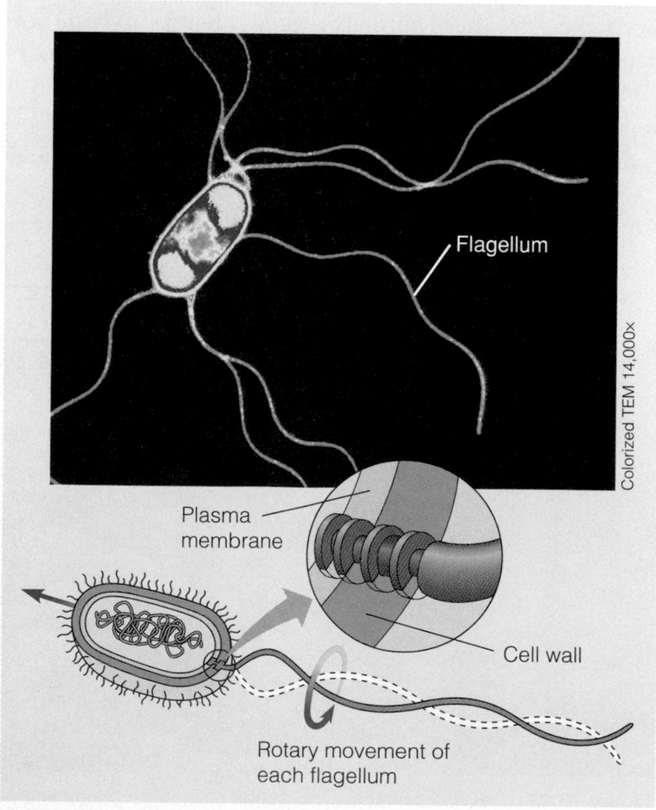

Flagellum
Colorized TEM 14,000×
Plasma membrane
Cell wall
Rotary movement of each flagellum

Figure 16.13A Bacterial flagella

Although few bacteria can thrive in the extreme environments favored by many archaea, bacteria of several genera can survive extended periods of very harsh conditions by forming specialized "resting" cells. Figure 16.13C (top of the facing page) shows an example of such an organism, *Bacillus anthracis*, the bacterium that produces the deadly disease called anthrax in cattle, sheep, and humans. There are actually two cells here, one inside the other. The outer cell produced the specialized inner cell, called an **endospore.** The endospore has a thick, protective coat, its cytoplasm is dehydrated, and it does not metabolize. Under harsh conditions, the outer cell may disintegrate, but the endospore survives all sorts of trauma, including lack of water and nutrients, extreme heat or cold, and most poisons. When the environment becomes more hospitable, the endospore absorbs water and resumes growth.

Some endospores can remain dormant for centuries. Not even boiling water kills most of these resistant cells. To sterilize laboratory equipment, microbiologists use an appliance called an autoclave, a pressure cooker that kills endospores by heating to a temperature of 121°C (250°F) with high-pressure steam. The food-canning industry uses similar methods to kill endospores of dangerous bacteria such as *Clostridium botulinum*, which produces the potentially fatal disease botulism.

The mass of branching cell chains (filaments) in Figure 16.13D is a structural feature unique to the bacterial group called **actinomycetes.** These bacteria are very common in soil, where they break down organic substances. The filaments enable the organism to bridge dry gaps between soil particles. Actinomycetes were once mistaken for fungi, which also grow in branching filaments; this similarity explains their name, which is from the Greek for "ray fungus." The actinomycete in Figure 16.13D is of the genus *Streptomyces*, a common soil organism. *Streptomyces* secretes the antibiotic streptomycin and a number of other antibiotics, which inhibit the growth of competing bacteria. Pharmaceutical companies use various species of actinomycetes to produce many antibiotics.

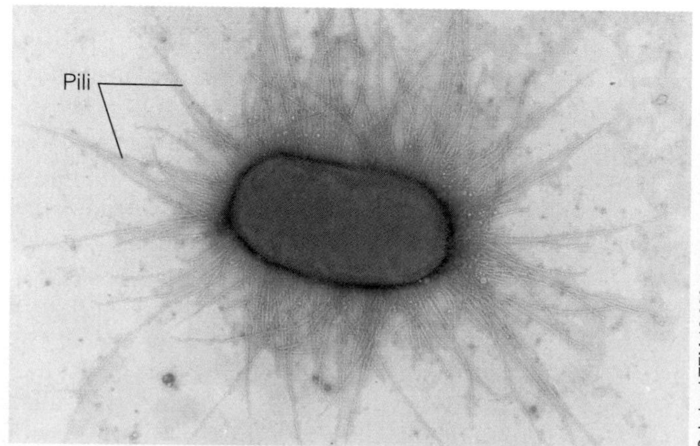

Figure 16.13B Pili

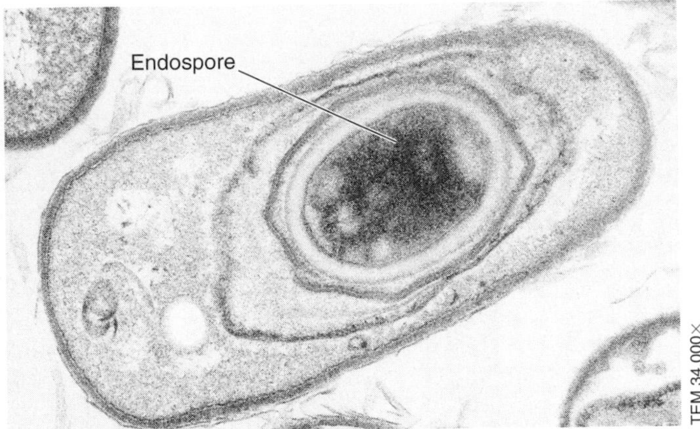

Figure 16.13C An endospore within a bacterial cell

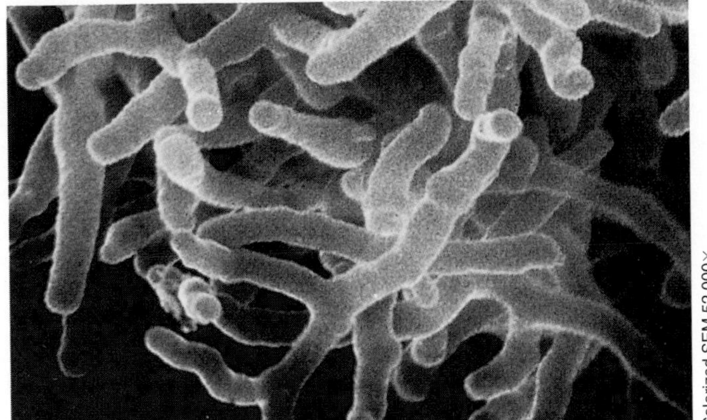

Figure 16.13D Filaments of an actinomycete

Branching chains of cells are unusual, but many other prokaryotes, both bacteria and archaea, grow in unbranched chains. Next, we look at a cyanobacterium that grows in unbranched filaments, and at the effect it can have on a polluted lake.

Why do microbiologists autoclave lab instruments and glassware, rather than simply boiling them?

To kill bacterial endospores, which can survive boiling water

Cyanobacteria sometimes "bloom" in aquatic environments 16.14

16A

The larger lake pictured in Figure 16.14A below is undergoing a population explosion—often called a "bloom"—of **cyanobacteria.** The blue-green color results from the presence of trillions of cyanobacterial cells. Figure 16.14B is a micrograph of *Anabaena,* the predominant cyanobacterium in the lake.

Cyanobacteria make up one group of photosynthetic bacteria. They are common in lakes, ponds, and tropical oceans. Extensive blooms of cyanobacteria in a lake usually indicate polluted water conditions. In the case shown here, the water was loaded with organic wastes from agricultural runoff. Phosphates and nitrates from the waste material acted as fertilizers, stimulating explosive multiplication of the cyanobacteria.

This lake's condition may be reminiscent of the age of cyanobacteria, a time when these prokaryotes dominated Earth. At that time, from 3.0 to 1.5 billion years ago, ancient cyanobacteria gave Earth its first greenish coat, generated the stromatolites we discussed in the chapter's opening essay, and made the atmosphere aerobic. Some of the molecular machinery for photosynthesis housed in cyanobacteria today may be much like that which first added O_2 to Earth's atmosphere.

How did early cyanobacteria change Earth's atmosphere?

Their water-splitting mechanism of photosynthesis liberated O_2, which eventually accumulated in the atmosphere.

Figure 16.14A A bloom of cyanobacteria in a lake

Figure 16.14B *Anabaena*

16.15 Some bacteria cause disease

All organisms, humans included, are almost constantly exposed to bacteria, some of which are potentially harmful. In fact, most of us are well most of the time only because our body defenses check the growth of bacterial **pathogens,** disease-causing agents. Occasionally, the balance shifts in favor of a pathogen, and we become ill. Even some of the bacteria that are normal residents of the human body can make us ill when our defenses have been weakened by poor nutrition or by a viral infection.

Pathogenic bacteria cause about half of all human diseases. Most cause disease by producing poisons, which are of two types: exotoxins and endotoxins. **Exotoxins,** toxic proteins secreted by bacterial cells, include some of the most potent poisons known. A single gram of the exotoxin that causes botulism, for instance, could kill a million people.

The culture dish in Figure16.15A shows yellow colonies of *Staphylococcus aureus,* another exotoxin producer (Latin *aureus,* golden). *S. aureus* is a common, usually harmless resident of our skin surface. If it enters the body through a cut or other wound, however, or is swallowed in contaminated food, it can cause serious diseases. One type of *S. aureus* produces exotoxins that cause layers of skin to slough off; another can cause vomiting and severe diarrhea; yet another can produce the potentially deadly toxic shock syndrome.

Bacterial species that are generally harmless can also develop strains that cause illness. In the past decade there have been a number of outbreaks of severe illness and several thousand deaths caused by a pathogenic, exotoxin-producing strain of *E. coli* designated O157:H7. Commonly found in cattle, these bacteria do not hurt the cattle, but in infected humans, the exotoxin selectively enters the cells that line blood vessels and kills them. Bloody diarrhea ensues and, in extreme cases, kidney failure. Vaccines for both people and cattle are being developed, but the best preventive measure at present is to avoid eating undercooked meat.

In contrast to exotoxins, **endotoxins** are not cell secretions, but components of the cell walls of certain bacteria. Endotoxins are glycolipids, large molecular complexes of polysaccharides and lipids. All endotoxins induce the same general symptoms: fever, aches, and sometimes a dangerous drop in blood pressure (shock). The severity of symptoms varies with the host's condition and with the bacterium. Different species of *Salmonella,* for example, produce endotoxins that cause food poisoning and typhoid fever.

During the last 100 years, following the nineteenth-century discovery that "germs" cause disease, the incidence of bacterial diseases has declined, particularly in developed nations. Sanitation is generally the most effective way to prevent bacterial diseases, and the installation of water-treatment and sewage systems continues to be a public health priority throughout the world. Antibiotics have been discovered that can cure most bacterial diseases, but many pathogenic bacteria have evolved resistance to widely used antibiotics and now pose a serious challenge to human health (see Module 13.21).

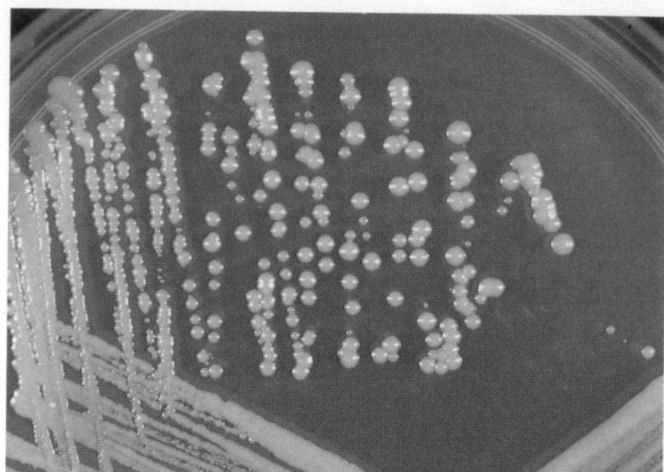

Figure 16.15A An exotoxin producer, *Staphylococcus aureus*

Tick that carries the Lyme disease bacterium

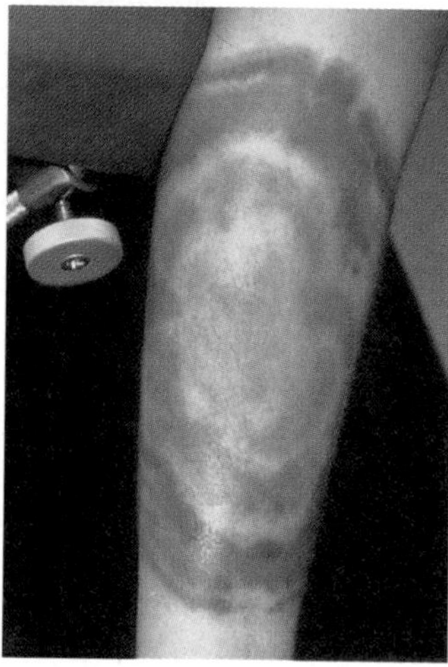

"Bull's-eye" rash

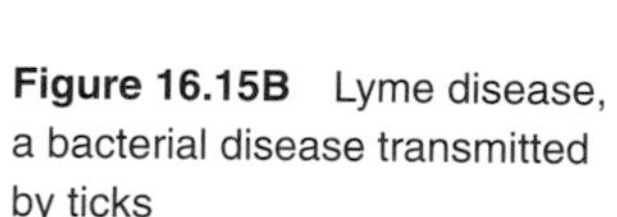

Figure 16.15B Lyme disease, a bacterial disease transmitted by ticks

In addition to sanitation and antibiotics, a third defense against bacterial diseases is education. A case in point is **Lyme disease,** currently the most widespread pest-carried disease in the U.S. The disease is caused by *Borrelia burgdorferi,* a bacterium carried by ticks that live on deer and field mice. As shown in Figure 16.15B, the disease usually starts as a red rash shaped like a bull's-eye around a tick bite. Antibiotics can cure the disease if administered within about a month after exposure. If untreated, Lyme disease can cause debilitating arthritis, heart disease, and nervous disorders. A vaccine is now available, but it does not give full protection. The best defense tactic is still public education about avoiding tick bites and the importance of seeking treatment if a rash develops. Using insect repellent and wearing light-colored clothes reduce contact with ticks.

Contrast exotoxins with endotoxins.

Exotoxins are poisons secreted by pathogenic bacteria; endotoxins are components of the cell walls of pathogenic bacteria.

Koch's postulates are used to identify disease-causing bacteria 16.16

Discovering what kind of bacterium causes a disease is a key step in developing ways to prevent or cure it. The diagnosis of bacterial diseases has been on firm scientific footing for over 100 years, thanks to the work of German physician Robert Koch (Figure 16.16A). In 1876, Koch discovered rod-shaped bacteria, which we now call *Bacillus anthracis,* in the blood of cattle suffering from anthrax. Koch used a set of diagnostic criteria—now called **Koch's postulates**—to prove that the bacteria were the cause of the disease.

Koch's postulates are as follows:

1. The same, specific pathogen must be identified in each animal (host) that has the disease.
2. The pathogen must be isolated from a host and grown in a pure culture, one in which no other kinds of cells are present.
3. The original disease must be produced in experimental hosts that are inoculated with the pathogen from the pure culture.
4. The same pathogen must be isolated from the experimental hosts after the disease develops in them.

Figure 16.16B outlines the procedure that microbiologists use to apply Koch's postulates. Like all scientific work, it hinges on the *repeatability of results.* It is the repeated production of a specific disease by a bacterium and the cultivation of the same bacterium from a number of experimentally infected hosts that confirm the identity of the pathogen.

Koch's postulates have been used to identify most bacterial pathogens, but there are some exceptions. For example, no one has yet been able to grow the bacterium *Treponema pallidum* in pure culture, but strong circumstantial evidence leaves no doubt that this microorganism causes syphilis.

Microbiologists have used Koch's criteria to identify pathogens outside the bacterial world—viruses, fungi, protists, and parasitic worms—but frequently they must modify the procedures. Viruses, for instance, reproduce only inside host cells, and protists may require other living cells for food. Koch's postulates are nevertheless the basis for determining the causative agents of most infectious diseases.

Why would medical ethics preclude strict application of Koch's postulates to a pathogen that can survive only in humans?

Postulate #3 would require experimental inoculation of the pathogen into humans, which is unacceptable.

Figure 16.16A Robert Koch studying bacterial diseases in Africa in 1906

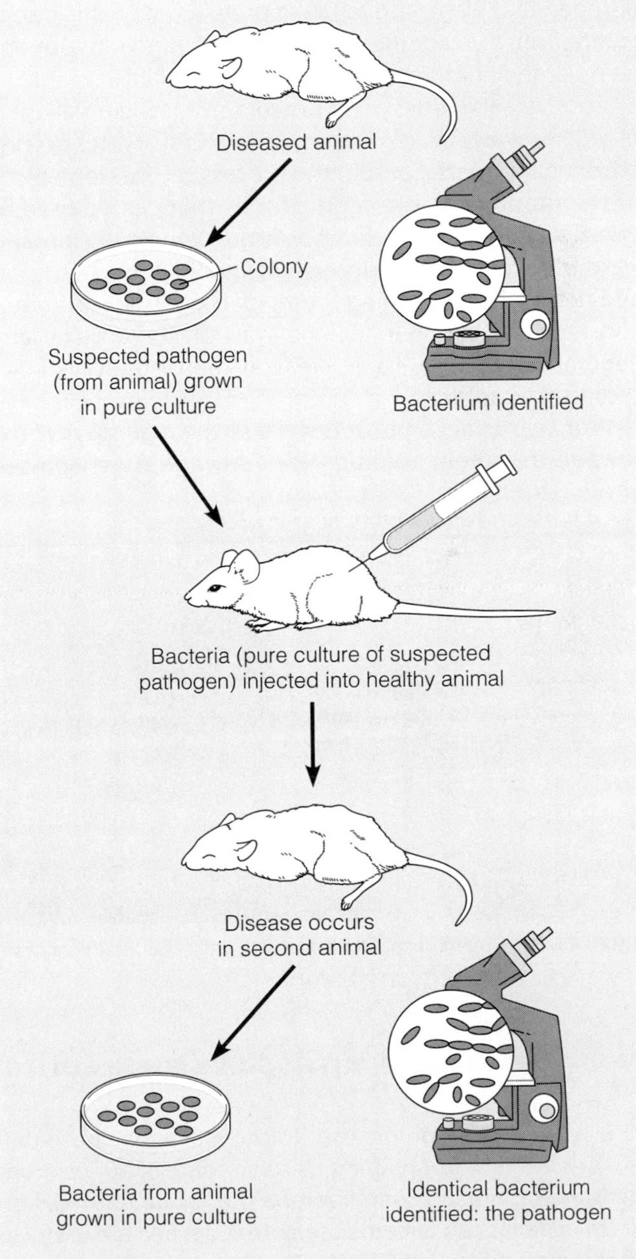

Figure 16.16B The usual procedure for demonstrating Koch's postulates

16.17 Chemical cycles in our environment depend on prokaryotes

Despite the attention they demand, prokaryotes that cause disease are in the minority. Far more common are species that are vital to our well-being, either directly or indirectly. All life depends on the cycling of chemical elements between organisms and the nonliving components of our environment. Prokaryotes are indispensable components of chemical cycles. For example, cyanobacteria not only restore oxygen to the atmosphere, some of them also convert nitrogen gas (N_2) in the atmosphere to nitrogen compounds (nitrates and nitrites) in soil and water. Other prokaryotes, including bacteria living in nodules on the roots of bean plants and other legumes, also contribute large amounts of nitrogen compounds to soil. In fact, all the nitrogen that plants use to make proteins and nucleic acids comes from prokaryotic metabolism in the soil. In turn, animals get their nitrogen compounds from plants.

Another vital function of prokaryotes is the breakdown of organic wastes and dead organisms. Prokaryotes decompose organic matter and, in the process, return elements to the environment in inorganic forms that can be used by other organisms. If it were not for decomposers, organic wastes and dead organisms would literally pile up on Earth, and the atoms composing them, such as carbon, nitrogen, and oxygen, would not become available for reuse by later generations of organisms. We'll talk more about chemical cycling in Chapter 36.

Prokaryotic decomposers are also the mainstays of our sewage-treatment facilities. Raw sewage is first passed through a series of screens and shredders, and solid matter is allowed to settle out from the liquid waste. This solid matter, called sludge, is then gradually added to a culture of anaerobic prokaryotes, including both bacteria and archaea. The microbes decompose the organic matter in the sludge, converting it to material that can be used as landfill or fertilizer, after chemical sterilization.

Liquid wastes are treated separately from the sludge. In Figure 16.17, you can see a trickling filter system, one type of mechanism for treating liquid wastes. The long horizontal pipes rotate slowly, spraying liquid wastes through the air onto a thick bed of rocks, the filter. Aerobic bacteria and fungi growing on the rocks remove much of the organic material dissolved in the waste. Outflow from the rock bed is sterilized and then released, usually into a river or ocean. In some sewage systems, such as one at Lake Tahoe, between California and Nevada, dissolved nitrates and phosphates are also removed from the liquid waste. Systems like this one greatly reduce environmental contamination, but their high cost prevents many cities from building them. 

? How do bacteria help restore the atmospheric CO_2 required by plants for photosynthesis?

By decomposing the organic molecules of dead organisms and organic refuse such as leaf litter, the metabolism of bacteria releases carbon from the organic matter in the form of CO_2.

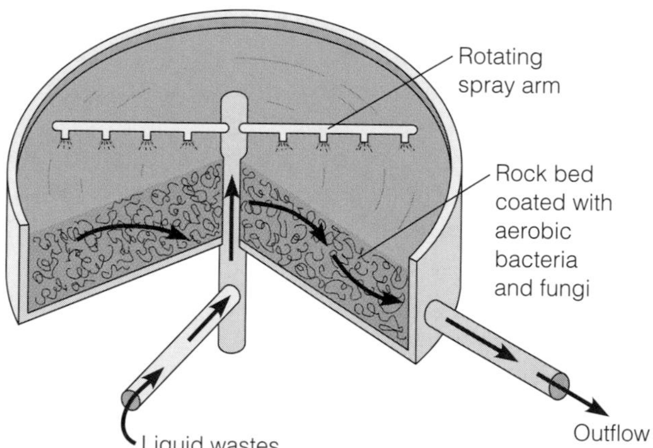

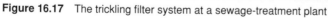

Figure 16.17 The trickling filter system at a sewage-treatment plant

16.18 Prokaryotes may help us solve some environmental problems

We are just beginning to explore the great potential prokaryotes have for helping us solve some of our environmental problems. Certain bacteria that occur naturally on ocean beaches can decompose petroleum and are useful in cleaning up oil spills. In Figure 16.18A on the facing page, workers are spraying nitrogen and phosphorus compounds on an oil-soaked beach in Alaska. The chemicals act as fertilizers, stimulating the reproduction of "oil-eating" bacteria that occur naturally in the soil. The bacteria obtain their energy and carbon from the oil molecules, breaking them down into smaller, less toxic compounds. Although there is some danger of water pollution from the fertilizers, this technique is the most rapid and inexpensive way yet devised to restore oil-fouled beaches. Researchers are also at-

Figure 16.18A Treatment of an oil spill in Alaska

Figure 16.18B Water containing toxic chemicals in an old copper mine in Montana

tempting to genetically engineer bacteria to degrade oil more rapidly than the naturally occurring oil-eaters.

Bacteria may also help us clean up old mining sites. The water that drains from mines is highly acidic and is also laced with poisons—often compounds of arsenic, copper, zinc, and the heavy metals lead, mercury, and cadmium. Contamination of our soils and groundwater by these toxic substances poses a widespread threat, and cleaning up the mess is extremely expensive.

The photograph in Figure 16.18B shows the remains of a half-mile-deep open-pit mine known as the Berkeley Pit, in Butte, Montana. The mine has not operated since 1983, when its high-quality copper ore was used up. Water in the pit is highly acidic and loaded with heavy metal ions. It threatens groundwater, soils, adjacent rivers, and communities for hundreds of miles downstream. The Berkeley Pit is one of the most troublesome cleanup sites in the United States, and the federal government has spent millions to initiate waste removal at this site.

Although there are no simple solutions to environmental problems like this one, prokaryotes may be able to help. Bacteria called *Thiobacillus* thrive in the acidic waters that drain from mines, and some mining companies use these microbes to extract copper and other valuable metals from low-grade ores. While obtaining energy by oxidizing sulfur or sulfur-containing compounds, the bacteria also accumulate metals from the mine waters. Unfortunately, their use in cleaning up mine wastes is limited, because their metabolism also adds sulfuric acid to the water. If this problem is solved, perhaps through genetic engineering, *Thiobacillus* and other prokaryotes may help us overcome some environmental dilemmas that seem intractable today.

What is the pitfall in believing that genetically-engineered bacteria will cure all our problems with chemical pollution?

They won't, and falsely believing in such an elixir might increase our public tolerance for pollution.

Summary: Prokaryotes are at the foundation of life on Earth 16.19

Summarizing the vital roles of prokaryotes on Earth, we could simply say that all life depends on them. This is true today and has always been so. Another truism is that environmental change and biological evolution are closely connected. Both have been episodic, marked by what were, in essence, revolutions that opened up new ways of life. Prokaryotes have been at the center of these revolutions and have had a greater impact on the environment and on evolution than all other forms of life combined. Only prokaryotes have produced an environmental change as radical and significant to the evolution of life as making the atmosphere aerobic. Ancient prokaryotes not only created an oxygen revolution, they were the first organisms to tolerate the corrosive effects of atmospheric oxygen and the first to use oxygen to metabolize organic molecules. If the oxygen revolution had not occurred, it is likely that Earth would still be populated only by anaerobic prokaryotes.

Cyanobacteria continue to play a major role in maintaining the atmospheric reservoir of oxygen and in removing carbon dioxide from the atmosphere. The role of cyanobacteria and other prokaryotes in converting atmospheric nitrogen to chemical forms that plants can use is essential to all life. Indeed, the cycling of nitrogen between the atmosphere and living organisms would stop altogether without these prokaryotes.

Prokaryotes are at the foundation of life in both an environmental and an evolutionary sense. In the next module, we plot a scenario illustrating the essential role of prokaryotes in yet another monumental event—the formation of the first eukaryotic cells.

How do we know that it is possible for prokaryotic life to thrive in the absence of eukaryotes?

We know from the fossil record that, for almost two billion years, Earth was populated only by prokaryotes.

16.20 The eukaryotic cell probably originated as a community of prokaryotes

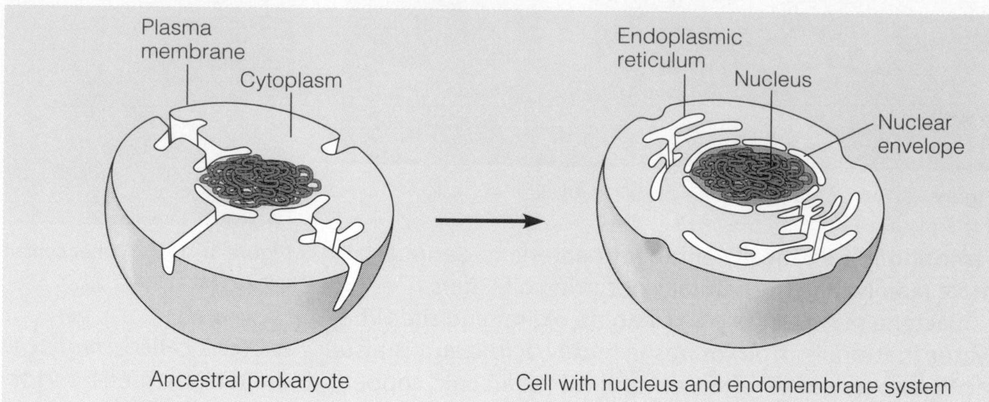

Figure 16.20A Infolding of a prokaryotic plasma membrane, giving rise to endoplasmic reticulum and a nuclear envelope

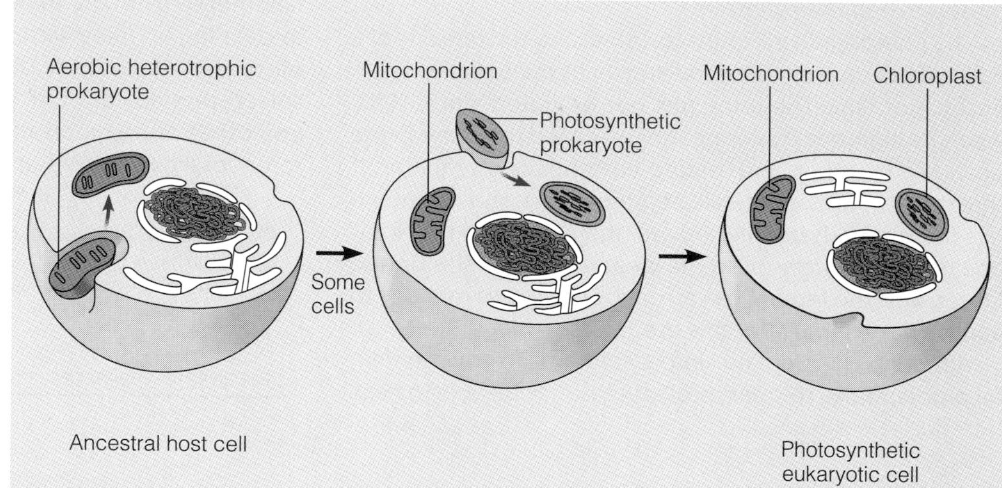

Figure 16.20B Endosymbiotic bacteria giving rise to mitochondria and chloroplasts

The fossil record indicates that eukaryotes evolved from prokaryotes more than 1.7 billion years ago. One of biology's most basic questions is how this happened—in particular, how the membrane-enclosed organelles of eukaryotic cells arose. A widely accepted theory is that eukaryotic cells arose through a combination of two processes. In one process, **membrane infolding,** the eukaryotic cell's endomembrane system—all the membrane-enclosed organelles except mitochondria and chloroplasts (see Chapter 4)—evolved from inward folds of the plasma membrane of a prokaryotic cell. Figure 16.20A suggests how the nuclear envelope and endoplasmic reticulum (ER) may have developed by infolding. The Golgi apparatus and other parts of the endomembrane system may then have evolved from the ER.

A second, very different process, called **endosymbiosis,** is believed to have generated the first mitochondria and chloroplasts. **Symbiosis** is a close association between organisms of two or more species. (The word is from the Greek for "living together," and *endo*symbiosis refers to one species living *within* another, called the host.) Chloroplasts and mitochondria seem to have evolved from small prokaryotes that established residence within other, larger prokaryotes (Figure 16.20B). The ancestors of mitochondria may have been small heterotrophic prokaryotes that were able to use O_2 to release large amounts of energy from organic molecules by cellular respiration. At some point, such a prokaryote might have been an internal parasite of a larger heterotroph, or an ancestral host cell may have ingested some of these aerobic cells for food. If some of the smaller cells were indigestible, they might have remained alive and continued to perform respiration in the host cell. In a similar way, small photosynthetic prokaryotes ancestral to chloroplasts may have come to live inside a larger host cell. Because almost all eukaryotes have mitochondria but only some have chloroplasts, scientists think that mitochondria evolved first.

It's not hard to imagine how a symbiosis between engulfed aerobic cells and a larger host cell might have become mutually beneficial. In both cases, the engulfed cells may have grown increasingly dependent on the host cell for molecules and inorganic ions needed to carry out their biochemical activities. Likewise, the host cell may have derived increasing proportions of its ATP and organic molecules from the engulfed cells. As the cells in these prokaryotic communes became more interdependent, they may have become truly a single organism, its parts inseparable.

Developed most extensively by Lynn Margulis of the University of Massachusetts, the endosymbiosis model is supported by strong evidence. For instance, present-day mitochondria and chloroplasts are similar to prokaryotic cells in a number of ways. Both types of organelles contain small amounts of DNA, RNA, and ribosomes, all of which resemble their counterparts in prokaryotes more than those in eukaryotes. These components enable chloroplasts and mitochondria to exhibit some autonomy in their activities. These organelles transcribe and translate their DNA into polypeptides contributing to some of their own enzymes. They also replicate their own DNA and reproduce within the cell by a process resembling the binary fission of prokaryotes.

Endosymbiosis can also explain how chloroplasts and mitochondria came to be enclosed by two membranes. As Figure 16.20B indicates, the inner membranes of these organelles could have been derived from the plasma membranes of the engulfed prokaryotes, and their outer membranes could have come from the infolded plasma membranes of the original host cells. In fact, the inner membranes of mitochondria and chloroplasts have several enzymes and electron transport molecules that resemble those found in the plasma membranes of modern prokaryotes, presumably a result of their endosymbiotic origin.

Let's now look at the most direct descendants of the first eukaryotes—the protists living on Earth today.

Which organelles of eukaryotic cells probably descended from endosymbiotic bacteria?

Mitochondria and chloroplasts

16.21 Protists—unicellular eukaryotes and their close multicellular relatives—probably represent multiple kingdoms

The photograph in Figure 16.21—a drop of pond water viewed with the light microscope—illustrates a variety of **protists,** a diverse group of mostly unicellular eukaryotes. Traditionally, biologists have classified all protists in a kingdom called Protista, but many now believe that these organisms constitute several kingdoms within Domain Eukarya. However, while our knowledge of evolutionary relationships among the unicellular eukaryotes remains incomplete, "protist" remains a useful word.

Some protists synthesize their own food by photosynthesis; these are called algae (another term no longer regarded as taxonomically meaningful). Others, called protozoans, are heterotrophic, eating bacteria, other protists, or organic matter suspended or dissolved in the water. Also regarded as protists are a number of colonial and multicellular eukaryotes whose immediate ancestors were unicellular.

Almost any aquatic environment is home to great numbers of protists. Most species, including those shown here, are aerobic. Some protists, however, are anaerobic, living in mud at the bottom of lakes and stagnant ponds or thriving in the digestive tracts of animals, including humans.

As eukaryotes, protists are more complicated than any prokaryotes. Their cells have a membrane-enclosed nucleus (containing multiple chromosomes) and other organelles characteristic of eukaryotic cells. Flagella and cilia on protistan cells have a 9 + 2 pattern of microtubules, another typical eukaryotic trait (see Module 4.18).

Protists occupy a pivotal position in the history of life: The first ones arose from prokaryotes, and their descendants gave rise to all plants, fungi, and animals, as well as to all modern protists.

Because most protists are unicellular, they are justifiably considered to be the simplest eukaryotes. However, the cells of many protists are among the most elaborate in the world. This level of cellular complexity is not really surprising, for each unicellular protist is a complete eukaryotic organism analogous to an entire animal or plant.

During the past decade, molecular and cellular studies have shaken the foundations of protistan taxonomy as much as they have that of the prokaryotes. Comparisons of nucleic acid sequences suggest that different groups of protists evolved from different prokaryotic ancestors. Altogether, the protists seem to represent several ancient "experiments" in the evolution of eukaryotic cells. Several coherent kingdoms are emerging, although researchers have not yet reached a consensus for a complete classification scheme. For the purposes of surveying protistan diversity in this chapter, we'll look at four major types of protists, grouped more by lifestyle than by taxon: protozoans, slime molds, unicellular algae, and multicellular algae (seaweeds, which are direct descendants of unicellular algae).

Why are protists especially important to biologists investigating the evolution of eukaryotic life?

Because the first eukaryotes were protists, and protists were ancestral to all other eukaryotes, including plants, fungi, animals, and modern protists

Figure 16.21 Protists in pond water

16.22 Protozoans are protists that ingest their food

Protists that live primarily by ingesting food, a heterotrophic mode of nutrition that is animal-like, are called **protozoans** (from the Greek *protos*, first, and *zoion*, animal). Protozoans thrive in all types of aquatic environments, including the tiny water droplets in wet moss and soil, and the watery environment inside animals. Most species eat bacteria or other protozoans, and some can absorb nutrients dissolved in the water. Protozoans that live as parasites on animals, though in the minority, cause some of the world's most harmful human diseases. The organisms featured in this module represent the four most common groups of protozoans: flagellates, amoebas, apicomplexans, and ciliates.

Flagellates are protozoans that move by means of one or more flagella. Most species are free-living (nonparasitic), but we picture two parasitic examples here because of their importance in evolution and human disease. The protozoan on the left in Figure 16.22A, called *Giardia*, is a flagellate that lives in the human intestine and can cause abdominal cramps and severe diarrhea. People become infected mainly by drinking water contaminated with feces from infected animals. *Giardia* belongs to a group of protists that is receiving a lot of attention from evolutionary biologists. Analysis of the ribosomal RNA of these organisms suggests that they are the eukaryotes most closely related to prokaryotes. The two eyelike spots visible in the micrograph are haploid nuclei. *Giardia* lacks mitochondria, but vestiges of mitochondrial genes in its genome suggest that it once had those organelles.

On the right in Figure 16.22A, the purple squiggles are cells of another protozoan that is flagellated, although not closely related to *Giardia*. This flagellate is *Trypanosoma*, which lives in the bloodstream of vertebrate animals. The disk-shaped cells in the micrograph are human red blood cells. Trypanosomes obtain their nutrients from the host blood. The ones shown here cause sleeping sickness, a debilitating disease common in parts of Africa. Living in the bloodstream, trypanosomes are under constant attack by the host's immune system. Trypanosomes escape being killed by their host's defenses by being quick-change artists. They alter the molecular structure of their coats frequently, thus preventing immunity from developing in the host.

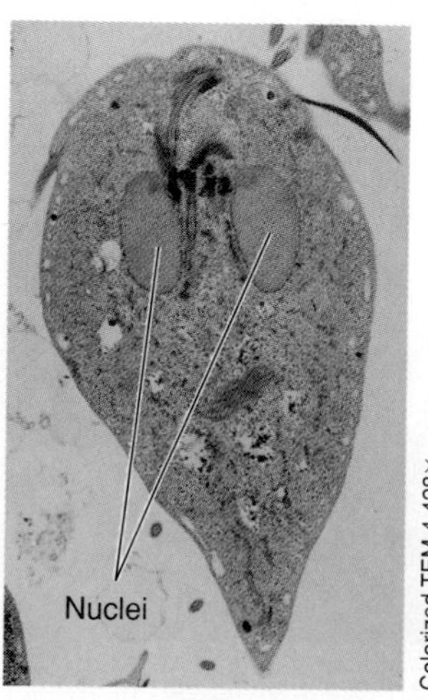

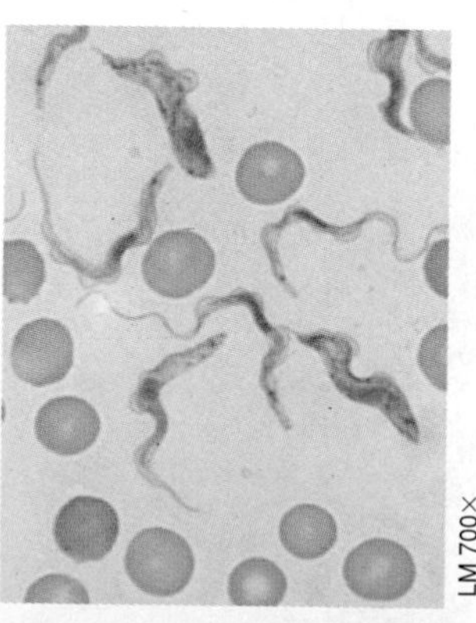

Figure 16.22A Flagellates: *Giardia* (left) and *Trypanosoma* (right, with blood cells)

Amoebas are characterized by great flexibility and the absence of permanent locomotor organelles. Most species move and feed by means of **pseudopodia** (singular, *pseudopodium*), which are temporary extensions of the cell. The individual in Figure 16.22B is ingesting another, much smaller protozoan for food. The amoeba's pseudopodia arch around the prey, engulfing it into a food vacuole (see Module 4.11). Amoebas can assume virtually any shape, as they creep over rocks, sticks, or mud at the bottom of a pond or ocean. Researchers expect that the amoebas will eventually sort into several distinct taxonomic groups.

Scientists do know that **apicomplexans,** another large group of protozoans, are closely related to each other. Apicomplexans are all parasitic, and some cause serious human diseases. As seen with the electron microscope, one end (the *apex*) of the infectious cell of these parasites contains a *complex* of organelles specialized for penetrating host cells and tissues—thus the name apicomplexan. The micrograph in Figure 16.22C shows *Plasmodium*, the apicomplexan that causes malaria. This parasite actually enters red blood cells, feeding on them from within and eventually destroying them. Spread by mosquitoes, malaria is one of the most debilitating and widespread human diseases. Each year in the tropics, more than 200 million people become infected, and at least a million die in Africa alone.

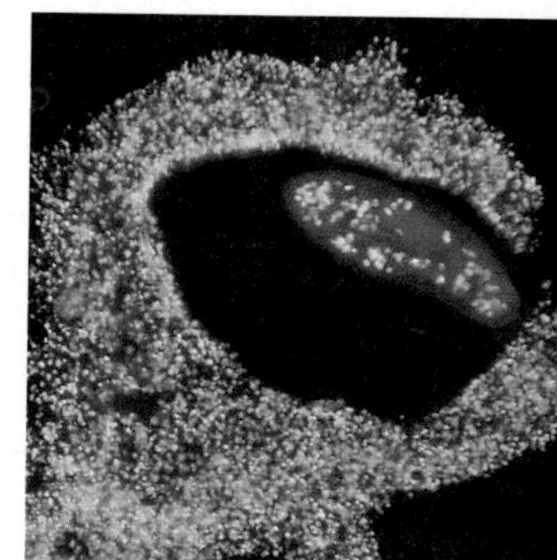

Figure 16.22B An amoeba ingesting a smaller protozoan

A fourth group of protozoans are the **ciliates.** These use cilia to move and feed and are common in all types of aquatic environments. Nearly all ciliates are free-living, including the very common freshwater protist *Paramecium* (Figure 16.22D, left). Ciliates are unique in having two types of nuclei: a single, large polyploid macronucleus, which controls everyday activities, and from one to as many as 80 tiny micronuclei, which are diploid and function in sexual reproduction.

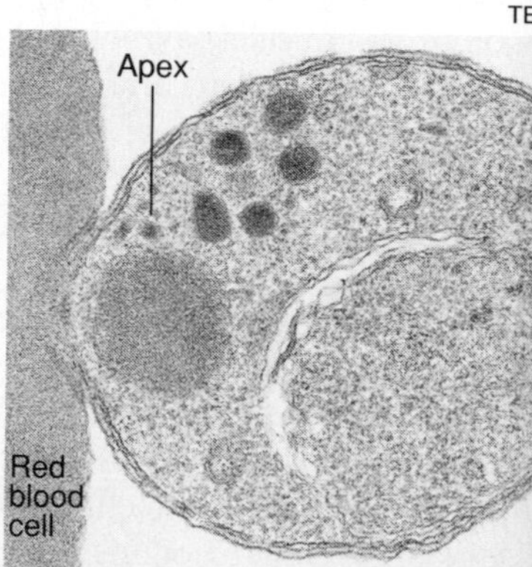

Figure 16.22C An apicomplexan: *Plasmodium*

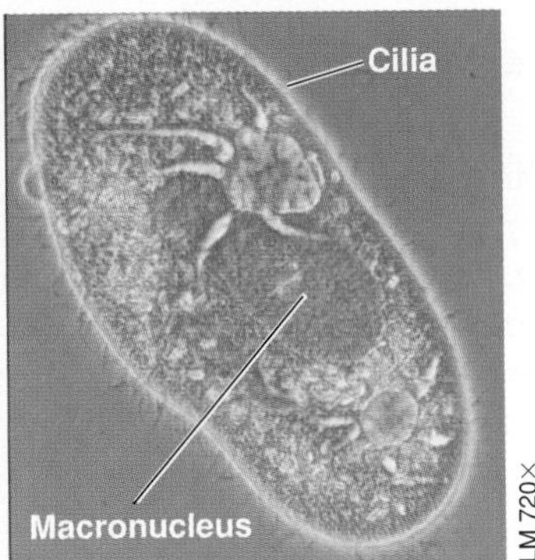

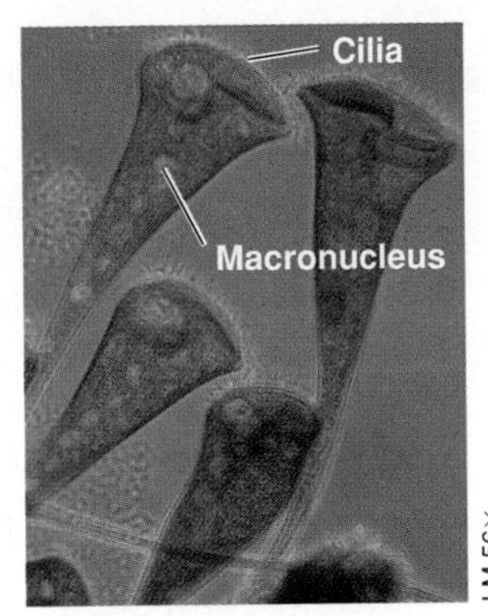

Figure 16.22D Ciliates: *Paramecium* (left) and *Stentor* (right)

You can see the macronucleus both in *Paramecium* and in the horn-shaped ciliate *Stentor* (Figure 16.22D, right). The macronucleus of *Stentor* resembles a string of beads. Like *Paramecium, Stentor* is common in freshwater ponds.

Protozoans are highly diverse protists. Indeed, they show us how structurally varied single-celled life can be. 

? What three modes of locomotion occur among protozoans?

Movement using flagella, cilia, and pseudopodia

Cellular slime molds have both unicellular and multicellular stages 16.23

In contrast to protozoans, protists called **cellular slime molds** lead a dual existence; they have both unicellular and multicellular life stages. Cellular slime molds are common on rotting logs and other decaying organic matter.

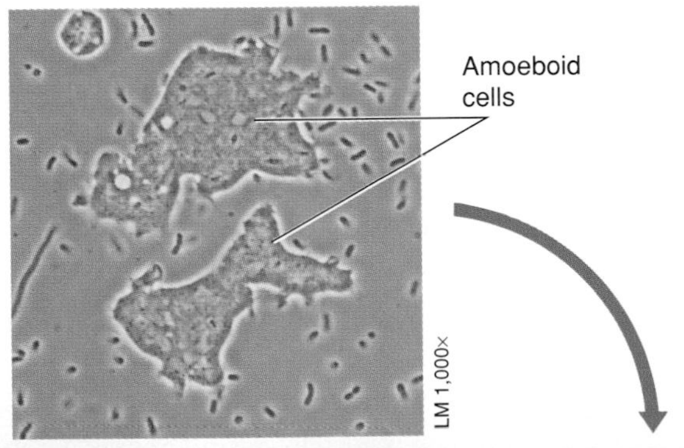

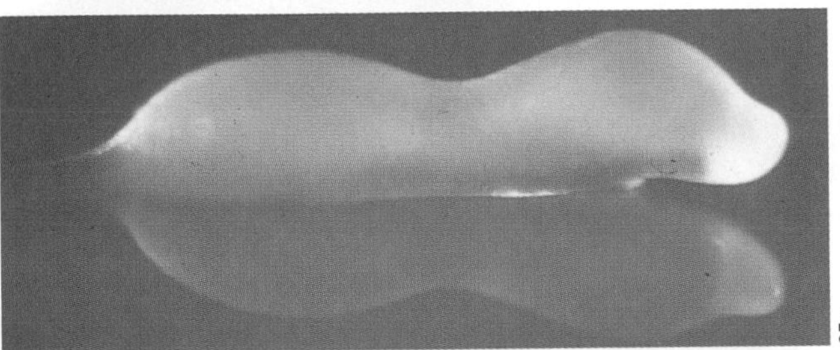

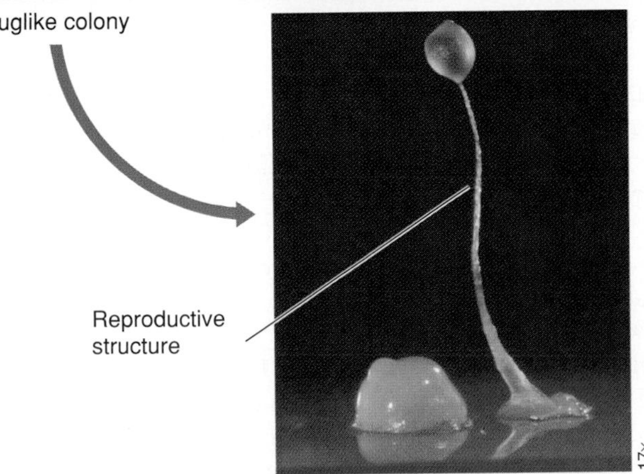

Figure 16.23 Stages in the life cycle of *Dictyostelium*

The micrographs in Figure 16.23 show three stages in the life cycle of a typical cellular slime mold, *Dictyostelium*. Most of the time, this organism exists as solitary amoeboid cells, creeping about by pseudopodial movement and engulfing bacteria as they go. The top picture shows two amoeboid cells. The small, dark rods are bacteria; the bacteria inside the slime-mold cells are being digested within food vacuoles.

When bacteria are plentiful, amoeboid slime-mold cells multiply by mitotic cell division but remain solitary. However, when bacteria are in short supply, the amoeboid cells swarm together, forming a colony. The center picture shows a sluglike colony of *Dictyostelium*. Cells in the colony secrete a cellulose covering coated with a slimy sheath. After wandering around for a short time, the colony develops into a multicellular reproductive structure, shown in the bottom picture.

Dictyostelium is easily cultured in the laboratory, and its relatively simple structure, compared to that of most multicellular eukaryotes, makes it an attractive research organism. Because the amoeboid cells in the sluglike colony develop into specialized cells when they form the reproductive structure, *Dictyostelium* is a useful model for researchers studying the genetic mechanisms and chemical changes underlying cellular differentiation.

Some cellular slime molds superficially resemble molds, but these organisms have virtually nothing in common with true molds or any other fungi. Since all their life stages are composed of amoeboid cells or their derivatives, it is likely that the closest living relatives of cellular slime molds are the single-celled amoebas. 

? Which protozoans are most similar in their movement to the unicellular stage of slime mold?

Amoebas

16.24 Plasmodial slime molds have brightly colored stages with many nuclei

The yellow, branching growth on the dead log in Figure 16.24A is a protist called a **plasmodial slime mold.** Plasmodial slime molds are common almost everywhere there is moist, decaying organic matter. They exist as solitary single cells and also as organisms that grow to several centimeters in diameter, like the one shown here.

Large and branching as it is, the organism in Figure 16.24A is not multicellular; it is a **plasmodium,** a single mass of cytoplasm, undivided by membranes and containing many nuclei. This is an amoeboid life stage that extends pseudopodia for feeding on bacteria, yeasts, and bits of dead organic matter. Its weblike form is an adaptation that enlarges the organism's surface area, increasing its contact with food, water, and oxygen. Within the fine channels of the plasmodium, cytoplasm streams first one way and then the other, in pulsing flows that are beautiful to watch with a microscope. The cytoplasmic streaming probably helps distribute nutrients and oxygen. (This "plasmodium" is not to be confused with the protozoan genus *Plasmodium,* which causes malaria.)

As long as a plasmodial slime mold has ample food and water, it usually stays in the plasmodial stage. When food and water are in short supply, however, the organism stops growing and differentiates into reproductive structures, such as the yellow ones in Figure 16.24B. When conditions again become favorable, the life cycle continues with amoeboid cells or flagellated cells emerging from the reproductive structures.

Like the cellular slime molds, plasmodial slime molds superficially resemble certain fungi. They are clearly not fungi, however. Plasmodial slime molds are probably not very closely related to any other contemporary organisms. 

? Contrast the plasmodium of a plasmodial slime mold with the sluglike stage of a cellular slime mold.

A plasmodium is not multicellular, but is one cytoplasmic mass with many nuclei; the sluglike stage of a cellular slime mold consists of many cells.

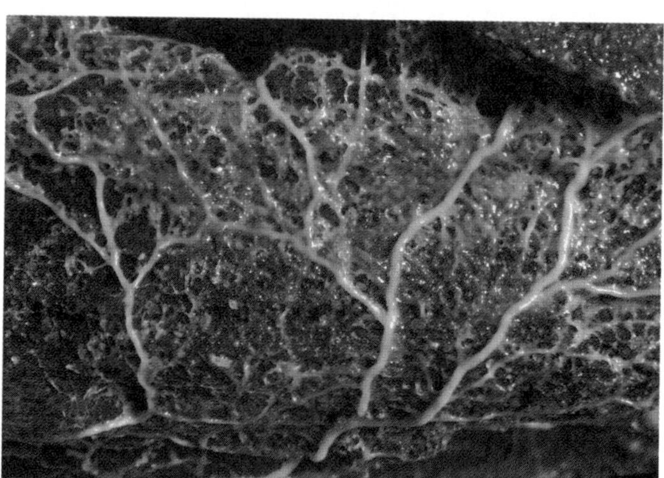

Figure 16.24A Plasmodial stage of a plasmodial slime mold, *Physarum*

Figure 16.24B Reproductive stage of *Physarum*

16.25 Photosynthetic protists are called algae

Whereas slime molds and most protozoans obtain their organic food molecules from other organisms, most **algae** are plantlike in that they synthesize their own food molecules from carbon dioxide and water.

Photosynthetic algae—the vast majority of algae—have chloroplasts containing the pigment chlorophyll *a,* the same type of chlorophyll plants have. In addition, certain heterotrophic protists are regarded as algae because many of their features suggest a close evolutionary relationship with photosynthetic algae. Many algae are unicellular, others live in colonies, and, as we will see in the next module, still others are multicellular. In this module, we focus on three groups of the unicellular and colonial types: dinoflagellates, diatoms, and green algae. The first two groups are noteworthy because they are numerous and widespread, the third because it was of pivotal importance in the history of life.

Dinoflagellates (Figure 16.25A on the facing page) are unicellular algae that are very common in marine and freshwater environments. Some are photosynthesizers (photoautotrophs); others are chemoautotrophs or chemoheterotrophs (see Module 16.10). Each dinoflagellate species has a characteristic shape reinforced by plates made of cellulose. The beating of two flagella in perpendicular grooves in the cellulose plates produces the spinning movement for which these organisms are named

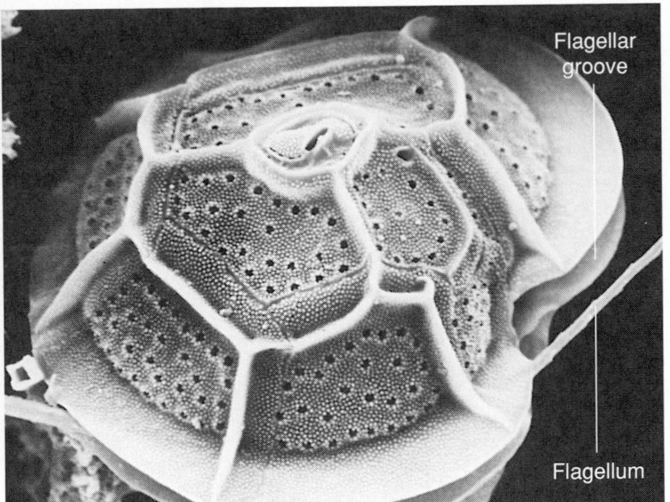

Figure 16.25A A dinoflagellate, a unicellular alga

(Greek *dinos,* whirling). Dinoflagellate blooms—population explosions—sometimes cause warm coastal waters to turn pinkish-orange, a phenomenon known as red tide. Toxins produced by some red-tide organisms have produced massive fish kills, especially in the tropics, and are deadly to humans as well.

Diatoms (Figure 16.25B) are unicellular, photosynthetic algae with a unique, glassy cell wall containing silica, the mineral actually used to make glass. The cell wall consists of two halves that fit together like the bottom and lid of a shoe box. Both freshwater and marine environments are rich in diatoms, and the organic molecules these microscopic algae produce are a key source of food in all aquatic environments. Indeed, diatoms are as important a food source for marine animals as plants are for land animals. Diatoms store their food reserves in the form of an oil, which also provides buoyancy, keeping diatoms floating near the surface, in the sunlight. Massive accumulations of fossilized diatoms make up thick sediments known as diatomaceous earth, which is mined for use as both a filtering medium and an abrasive.

The micrographs in Figure 16.25C show two types of **green algae.** *Chlamydomonas* is a unicellular alga common in freshwater lakes and ponds. Surrounded by a cellulose wall, it is propelled through the water by two flagella. (Cells with two flagella are said to be biflagellated.) *Volvox* is a colonial green alga, also common in fresh water. Each *Volvox* colony is a hollow ball composed of hundreds or thousands of biflagellated cells. The large colonies shown here will eventually release the small green and red daughter colonies within them. Each of the cells in a *Volvox* colony closely resembles the biflagellated cell of *Chlamydomonas.* It is likely that complex colonial protists such as *Volvox* evolved from unicellular green algae that were structurally similar to *Chlamydomonas.* The cells of *Volvox* and *Chlamydomonas* also resemble the biflagellated gametes of many multicellular algae and certain plants. This feature and others common to green algae and plants—cellulose cell walls, starch as a food-storage compound, and chloroplasts that are virtually the same—strongly suggest that ancient green algae gave rise to the first plants. In fact, many biologists believe the green algae should be classified in the plant kingdom. In the next module, we look at the most complicated of the modern algae, the multicellular forms.

What metabolic process mainly distinguishes algae from protozoans?

Photosynthesis

LM 1,782×

Figure 16.25B Diatoms, unicellular algae

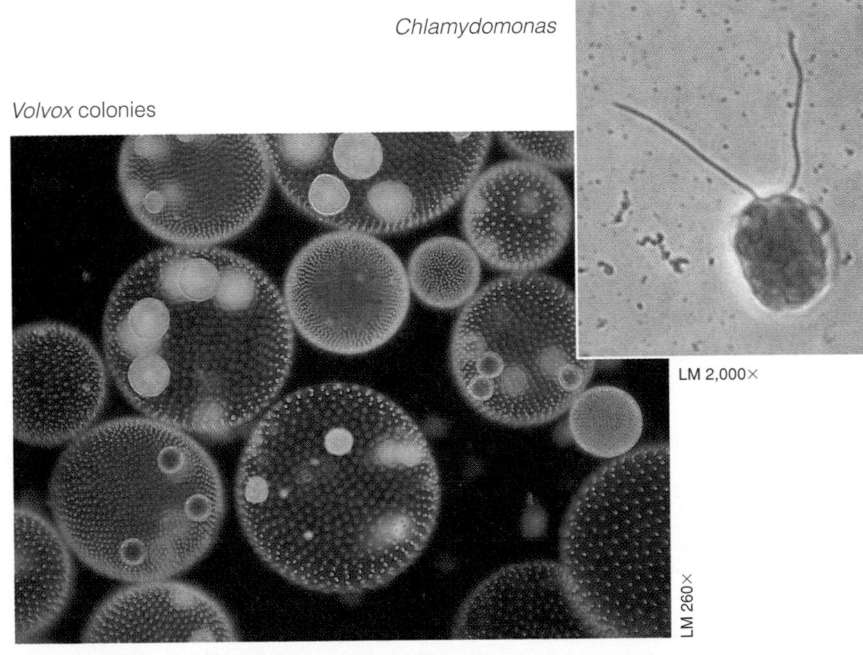

Figure 16.25C Green algae, unicellular (top) and colonial (bottom)

16.26 Seaweeds are multicellular marine algae

Figure 16.26A A brown alga: a kelp "forest"

Figure 16.26B A red alga: an encrusted type, on a coral reef

We use the word "seaweeds" here to refer to marine algae that have large multicellular bodies. Some of these organisms can be as large and complex as many plants. However, even the most complex seaweeds lack true stems, leaves, roots, and the internal tubes (vascular systems) that transport nutrients and water in most plants. Partly because seaweeds lack these structures and partly because they are direct descendants of unicellular protists, some biologists consider them all to be protists. However, recent molecular and cellular studies suggest that the three groups of seaweeds—brown algae, red algae, and multicellular green algae—may be members of three separate kingdoms.

Seaweeds grow on rocky shores and just offshore beyond the zone of pounding surf. The biggest and most complex seaweeds belong to the **brown algae,** named for the brown or olive color of many of them. Based on the molecules they use in photosynthesis, brown algae are closely related to diatoms (see Figure 16.25B). These two groups and some related organisms may belong to a distinct kingdom.

Figure 16.26A is a photograph of an underwater forest of brown algae called **kelp,** off the coast of California. Anchored to the seafloor by rootlike structures called holdfasts, kelp may grow to heights of 100 m. Fish, sea lions, sea otters, and gray whales regularly use these kelp "forests" as their feeding grounds.

The warm coastal waters of the tropics are home to the majority of species of a second group of seaweeds, the **red algae.** Red algae are typically soft-bodied, but some have cell walls encrusted with hard, chalky deposits. Encrusted species, such as the one in Figure 16.26B, are common on coral reefs, and their hard parts are important in reef-building. Taxonomic debate seems to be leaning toward classifying the red algae as a separate kingdom.

The third group of seaweeds, the multicellular **green algae,** is represented here by *Ulva,* the sea lettuce (Figure 16.26C). *Ulva*'s life cycle, the sequence of life stages leading from the adults of one generation to those of the next generation, follows a pattern called **alternation of generations.** This is a pattern in which a multicellular diploid ($2n$) alternates with a multicellular haploid (n), and it occurs in a number of multicellular algae and in all plants.

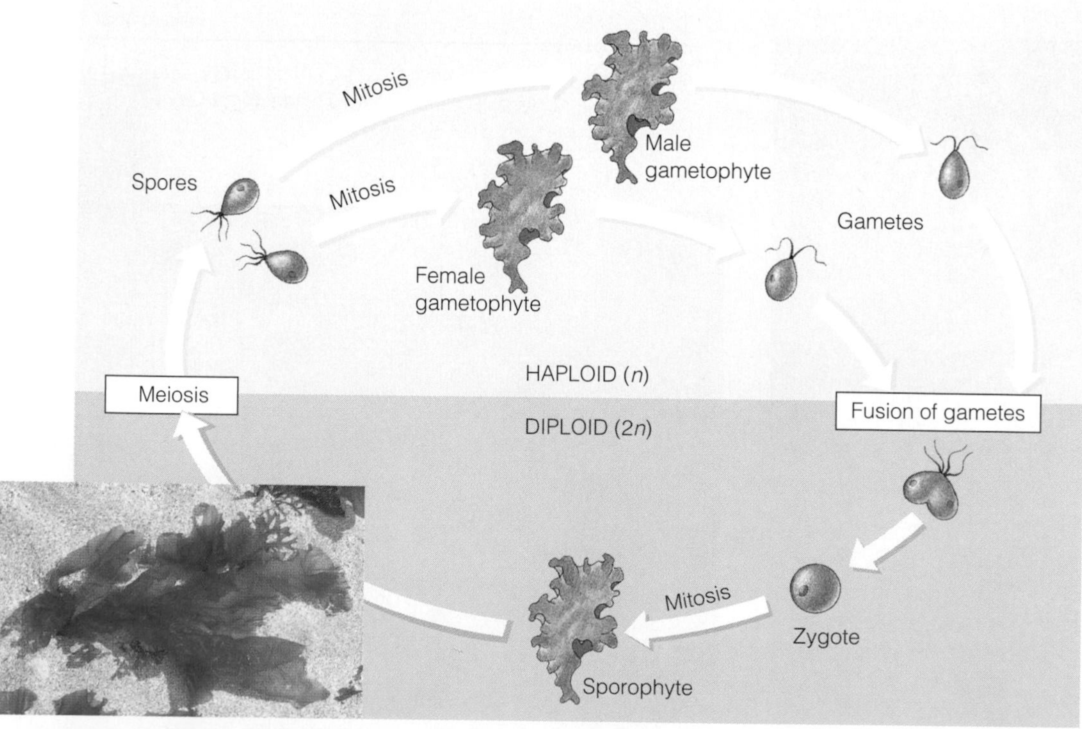

Figure 16.26C A multicellular green alga: *Ulva* (sea lettuce) and its life cycle

Notice in the figure that the multicellular haploids are called **gametophytes.** The gametophyte generation alternates with a diploid generation that features a multicellular diploid called a **sporophyte.** In *Ulva,* the gametophyte and sporophyte organisms are identical in appearance; both look like the one in the photograph, although they differ in chromosome number. The haploid gametophyte produces gametes by mitosis, and fusion of the gametes begins the sporophyte generation. In turn, cells in the sporophyte undergo meiosis and produce haploid, flagellated spores. The life cycle is completed when a spore settles to the bottom of the ocean and develops into a gametophyte.

A growing number of biologists believe that the multicellular green algae, along with the unicellular and colonial green algae, belong in the plant kingdom. Indeed, seaweeds such as *Ulva,* with their alternation of generations, may resemble a stage in the evolution of plants from green algae.

Gametophyte is to ________ and production of reproductive cells called gametes as ________ is to diploid and production of reproductive cells called ________.

haploid . . . sporophyte . . . spores

Multicellular life may have evolved from colonial protists 16.27

Multicellular organisms—seaweeds, plants, animals, and most fungi—are fundamentally different from unicellular ones. In a unicellular organism, all of life's activities occur within a single cell. In contrast, a multicellular organism has various specialized cells, which perform different functions and are dependent on each other. For example, some cells give the organism its shape, while others make or procure food, transport materials, or provide movement.

Multicellularity probably evolved on many separate occasions among the ancient protists, with today's multicellular organisms arising from several kinds of unicellular protists. Three or more different protists may have given rise to the seaweeds, with plants being an offshoot of the green-algae lineage. Molecular comparisons suggest that animals and fungi, which we study in later chapters, came from a common protistan ancestor. (If true, you are more closely related to a mushroom than to a plant.)

The most widely held view is that the organisms linking multicellular organisms to their unicellular ancestors were loose colonial aggregates of interdependent cells. The figure below suggests how a unicellular protist with flagellated cells may have formed colonies that eventually gave rise to multicellular organisms. ① An ancestral colony may have formed, as colonial protists do today, when a cell divided and its offspring remained attached to one another. ② Next, the cells in the colony may have become somewhat specialized and interdependent, with different cell types becoming more and more efficient at performing specific, limited tasks. Cells that retained a flagellum may have become specialized for locomotion, while others that lost their flagellum could have assumed functions such as ingesting or synthesizing food. ③ Later on, additional specialization among the cells in the colony may have led to distinctions between sex cells (gametes) and nonreproductive cells (somatic cells).

We see specialization and cooperation among cells today in several colonial protists, such as the green alga *Volvox* in Figure 16.25C. *Volvox* produces gametes, which depend on somatic cells while developing. Cells in truly multicellular organisms, as we know them today, are specialized for many more nonreproductive functions, including feeding, waste disposal, gas exchange, and protection, to name a few. Evolution of the division of labor to this extent involved many additional steps in somatic cell specialization.

What characteristic distinguishes true multicellularity from colonies of cells?

A greater extent of cellular specialization in multicellular organisms

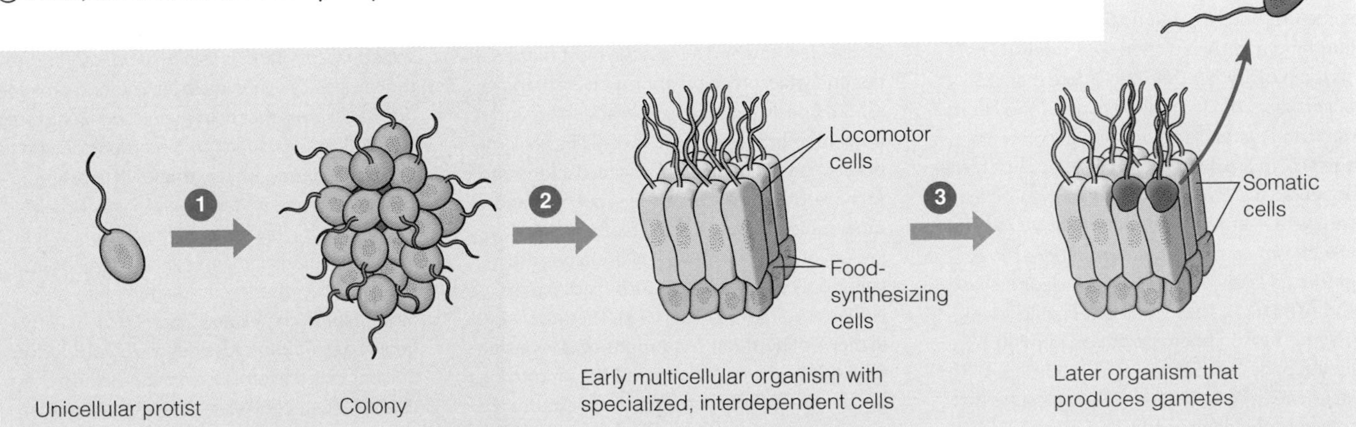

Figure 16.27 A model for the evolution of a multicellular organism from a unicellular protist

16.28 Multicellular life has diversified over hundreds of millions of years

Based on the fossil record, the timeline here summarizes some key stages in the evolution of multicellular algae, plants, animals, and fungi. Notice that the earliest multicellular organisms on the timeline, dating from about 700 million years ago, during the Precambrian era, were red algae and animals (mostly organisms resembling corals, jellyfishes, and worms). Although not yet found in the fossil record, other kinds of multicellular algae probably also existed at that time, and the first multicellular organisms may have appeared over a billion years ago. A period of mass extinction separated the Precambrian from the Paleozoic era, but multicellular life again flourished soon thereafter. By about 500 million years ago, diverse animals, fungi, and multicellular algae populated aquatic environments. All life was aquatic.

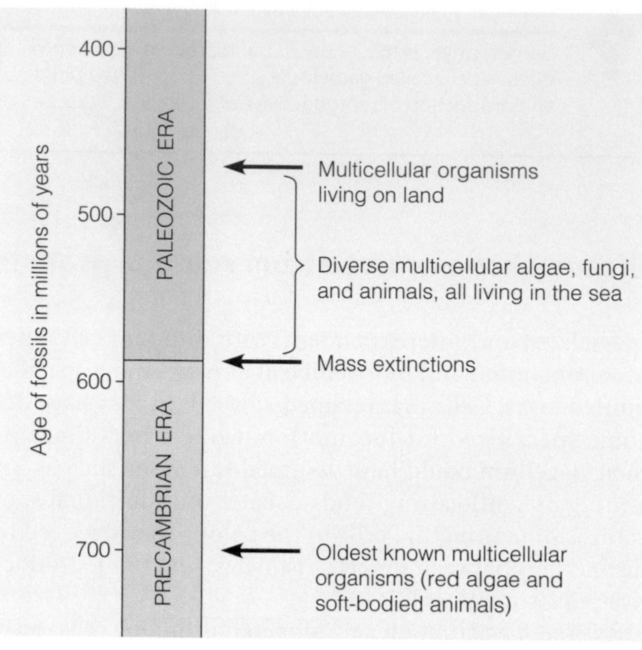

Figure 16.28 A timeline of early multicellular life

Between 500 and 475 million years ago, the move onto land began, probably as certain green algae living in the company of fungi along the edges of lakes gave rise to primitive plants. In the next chapter, we trace the long evolutionary movement of plants onto land and their diversification there. After that, we pick up the threads of animal evolution in Chapter 18. 

? About how long did life evolve in aquatic habitats before the algal ancestors of plants first began to colonize land? (*Hint:* See Module 16.1.)

About 3 billion years

Chapter Review

Chapter Summary

Early Earth and the Origin of Life (Introduction–16.6) Biological and geological history are closely intertwined. Fossilized mats of prokaryotes 2.5 billion years old mark the time when photosynthetic bacteria produced the O_2 that made the atmosphere aerobic **(Introduction).** Planet Earth formed some 4.5 billion years ago. The early atmosphere probably contained H_2O, CO, CO_2, and N_2, and possibly some CH_4 and NH_3, but no O_2. Volcanic activity, lightning, and UV radiation were intense. Fossilized prokaryotes date back more than 3.5 billion years, but life may have developed from nonliving materials as long as 4 billion years ago **(16.1).** Small organic molecules must have appeared first, probably when inorganic chemicals were energized by lightning or UV radiation **(16.2).** Simulations of such conditions have produced amino acids, sugars, nucleotide bases, and ATP **(16.3).** These molecules could have polymerized on hot rocks or clay, forming polypeptides and short nucleic acids **(16.4).** The first genes may have been RNA molecules that catalyzed their own replication in a prebiotic RNA world **(16.5).** These molecules might have acted as rough templates for the formation of polypeptides, which in turn assisted in RNA replication. Surrounding membranes may have protected some of these molecular co-ops as they evolved a rudimentary metabolism. Natural selection would have favored the most efficient co-ops, which may have eventually evolved into the first prokaryotic cells **(16.6).**

Prokaryotes (16.7–16.19) Prokaryotes are the oldest life forms and remain the most numerous and widespread organisms on Earth today **(16.7).** Consisting of cells that lack nuclei and other membrane-enclosed organelles, prokaryotes are classified in two domains, Bacteria and Archaea, based on nucleotide sequences and other molecular and cellular features **(16.8).** Prokaryotes come in a variety of shapes, most commonly spheres (cocci), rods (bacilli), and curves or spirals (spirilla) **(16.9).** They also vary in mode of nutrition. Autotrophs obtain carbon from CO_2 and are of two types: Photoautotrophs obtain energy from sunlight; chemoautotrophs obtain energy from inorganic compounds. Heterotrophs obtain carbon from organic compounds, with photoheterotrophs getting energy from sunlight and chemoheterotrophs getting energy, as well as carbon, from organic compounds **(16.10).** The first cells were most likely chemoautotrophs, which may have obtained energy from sulfur and iron compounds **(16.11).** Archaea are common in extreme environments, such as anaerobic swamps, salt lakes, acidic hot springs, deep-ocean vents, and animal digestive systems **(16.12).** Among the structural adaptations that help prokaryotes to thrive virtually everywhere are rotating prokaryotic flagella, pili that help them cling to surfaces, and endospores, which allow certain bacteria to survive environmental extremes in a resting stage. Many prokaryotes grow in linear chains (filaments); the bacteria called actinomycetes form branching filaments **(16.13).** Cyanobacteria photosynthesize in a plantlike way and often "bloom" in polluted water **(16.14).** Pathogenic bacteria can cause disease by producing poisons called exotoxins and endotoxins **(16.15).** Koch's postulates are a set of criteria that can prove that bacteria are the cause of a disease **(16.16).** Many prokaryotes, both bacteria and archaea, are environmentally important in decomposition and chemical cycles. We exploit bacterial and archaeal decomposers in sewage treatment **(16.17).** Prokaryotes hold great potential for solving environmental problems, such as oil spills and toxic mine wastes **(16.18).** Prokaryotes are the foundations of all life, in both an environmental and an evolutionary sense **(16.19).**

Protists (16.20–16.28) Fossils indicate that eukaryotic cells evolved from prokaryotic cells more than a billion years ago. The nucleus and endomembrane system of eukaryotes probably evolved from infoldings of the plasma membrane of ancestral prokaryotes. Mitochondria and chloroplasts probably evolved from symbiotic prokaryotes that took up residence inside larger prokaryotic cells **(16.20).**

Protists are single-celled eukaryotes and their close multicellular relatives. Early protists were the ancestors of plants, animals, and fungi. The taxonomy of the protists is in a state of flux, although it is likely that they comprise multiple kingdoms **(16.21).** Protozoans are unicellular protists that mostly ingest their food. They include flagellates, amoebas, apicomplexans, and ciliates. Most protozoans live freely in water or moist soil, but some live in humans and other animals and cause disease. The parasitic flagellate *Giardia* may resemble the earliest eukaryotes **(16.22).** Cellular slime molds have both unicellular and multicellular life stages **(16.23).** Plasmodial slime molds have unicellular stages and stages when they exist as plasmodia, multinuclear masses of cytoplasm undivided by membranes **(16.24).** Algae, such as the unicellular dinoflagellates, diatoms, and green algae, are protists that are mostly photosynthetic **(16.25).** Seaweeds—brown algae, red algae, and some green algae—are multicellular photosynthetic organisms that lack the structural specializations of plants. Brown algae seem closely related to diatoms, and these groups may eventually be classified in a separate kingdom. Many biologists favor classifying the red algae in their own kingdom and all the green algae in the plant kingdom. The life cycles of many seaweeds involve the alternation of haploid gametophyte and diploid sporophyte generations **(16.26).** Multicellular organisms probably evolved by specialization of the cells of colonial protists **(16.27).** Multicellular life may have originated over a billion years ago. All life was aquatic until almost 500 million years ago **(16.28).**

Testing Your Knowledge

Multiple Choice

1. Ancient cyanobacteria, found in fossil stromatolites, were very important in the history of life because they
   a. were probably the first living things to exist on Earth
   b. produced the oxygen in the atmosphere
   c. are the oldest known archaea
   d. were the first multicellular organisms
   e. extracted heat from the atmosphere, cooling Earth

2. You set your time machine for 3 billion years ago and push the start button. When the dust clears, you look out the window. Which of the following describes what you would probably see?
   a. plants and animals very different from those alive today
   b. a cloud of gas and dust in space
   c. green scum in the water
   d. land and water sterile and devoid of life
   e. an endless expanse of red-hot molten rock

3. In terms of nutrition, autotrophs are to heterotrophs as
   a. algae are to slime molds
   b. archaea are to bacteria
   c. slime molds are to algae
   d. kelp are to diatoms
   e. pathogenic bacteria are to harmless bacteria

4. The bacteria that cause tetanus can be killed only by prolonged heating at temperatures considerably above boiling. This suggests that tetanus bacteria
   a. have cell walls containing peptidoglycan
   b. protect themselves by secreting antibiotics
   c. secrete endotoxins
   d. are autotrophic
   e. produce endospores

5. Glycolysis is the only metabolic pathway common to nearly all organisms. To scientists, this suggests that it
   a. evolved many times during the history of life
   b. was first seen in early eukaryotes
   c. first appeared early in the history of life
   d. must be very complex
   e. appeared rather recently in the evolution of life

Describing, Comparing, and Explaining

1. How do most biologists think that the mitochondria and chloroplasts of eukaryotic cells originated? What is the evidence for this idea?

2. *Chlamydomonas* is a unicellular green alga. How does it differ from a photosynthetic bacterium, which is also single-celled? How does it differ from a protozoan, such as an amoeba? How does it differ from larger green algae, such as sea lettuce *(Ulva)?*

Thinking as a Scientist

1. Imagine you are on a team designing a moon base that will be self-contained and self-sustaining. Once supplied with building materials, equipment, and organisms from Earth, the base will be expected to function indefinitely. One of the members of your team has suggested that everything sent to the base be chemically treated or irradiated so that no bacteria of any kind are present. Do you think this is a good idea? Predict some of the consequences of eliminating all bacteria from an environment.

2. Billions of years ago, life apparently arose from inorganic chemicals, but all life today apparently arises only by the reproduction of preexisting life. If life could come from nonlife on the ancient Earth, why do you think this does not continue to happen today?

Media Resources

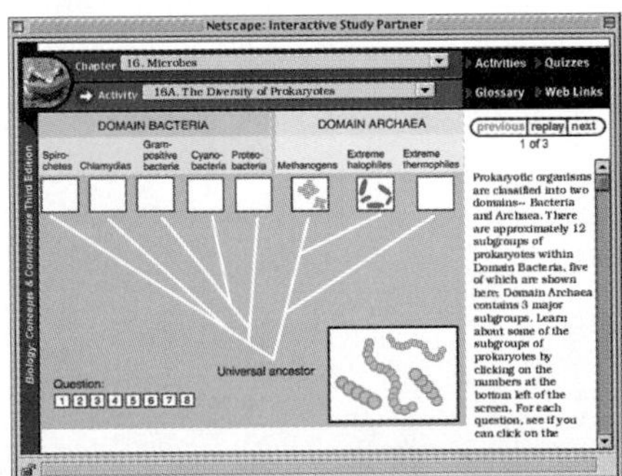

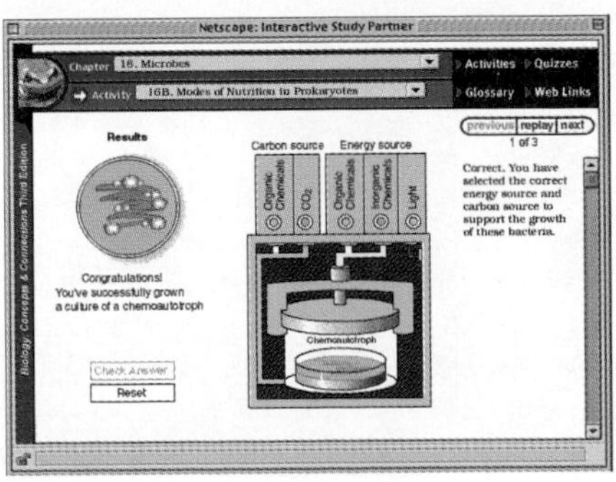

For further review of the concepts in this chapter, use the activities and quizzes on the CD-ROM.

Visit the Special Edition of The Biology Place web site for activities, quizzes, news articles, and web links.

17 Plants, Fungi, and the Colonization of Land

Outline

We tend to take our orange juice for granted, but it is no small feat for citrus growers to produce it at a reasonable cost. Orange groves are found in Florida, Texas, and California. An enormous investment, trees like the one pictured at the left take 3–7 years to start producing fruit and require a rich supply of fertilizer. They are also vulnerable to freezing and to a long list of pathogenic bacteria, insects, and, especially, fungi.

Fungi are not always harmful to plants. There is another kind of association between fungi and plants in which the fungus plays the role of vital benefactor. You can see an example of this relationship in the micrograph on the next page, which shows a fungus growing inside a piece of a small onion root. Everything that is darkly stained in the micrograph is part of the fungus: a dense network of thin strands and a number of round vesicles, which are used for lipid storage. (In order to make the fungus more visible, the sample was chemically treated to destroy the root cells.) This fungus is called a mycorrhizal fungus; together, the root and the fungus form an intimate, mutually beneficial association, called a **mycor-**

rhiza (pronounced MY-koh-RY-za; the word means "fungus root"). For their part, mycorrhizal fungi absorb phosphorus and other essential minerals from the soil, and these nutrients are then available to the plant. The sugars produced by the plant nourish the fungi.

Citrus trees can also have mycorrhizae, and the fungi may offer a way to cut the high economic and environmental costs of producing citrus fruits. Mycorrhizae can make a tree more resistant to disease, thereby reducing the need for pesticides that kill disease-causing organisms. Also, by enhancing a tree's uptake of nutrients, mycorrhizae can reduce, or even eliminate, the need for fertilizers. Unfortunately, the conditions in a typical citrus grove undermine the growth of mycorrhizae. Citrus growers use fungus-killing chemicals (fungicides) to control fungi that cause disease, and the fungicides poison the mycorrhizal fungi as well. As a result, the grower loses the benefits of mycorrhizae and must apply expensive fertilizers. The environment also suffers, because fungicides harm many kinds of organisms, and excess fertilizers can pollute streams, lakes, and groundwater.

There is no immediate solution to this dilemma. The citrus industry has relied on fungicides and fertilizers for many years, and few growers are willing to risk replacing the use of chemicals with the cultivation of mycorrhizae. One solution would be for researchers to find a way to control disease without hurting beneficial fungi.

Cultivated citrus groves are unnatural in lacking mycorrhizae. In nature, nearly all plants have mycorrhizae. In fact, mycorrhizae appear in fossils of the oldest known plants, suggesting that these beneficial relationships with fungi may have been essential for plant evolution. Some 475 million years ago, when plants first appeared, soil would have been poor in nutrients, and early plants or the algae that gave rise to them probably would have had difficulty obtaining enough phosphorus or other nutrients without the help of fungi. Thus, as the first plants evolved from green algae and adapted to land, symbiotic fungi probably played an important role.

LM 473×

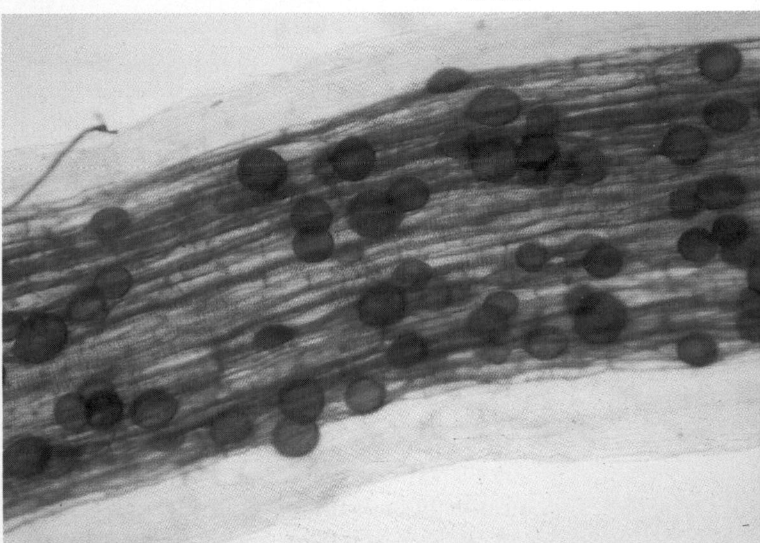

A mycorrhizal fungus in an onion root

The colonization of land by plants was a major event in the history of life. Plants transformed the landscape, creating new environmental opportunities for protists and prokaryotes, and making it possible for herbivorous animals and their predators to evolve on land. This chapter continues our account of the evolution of life's diversity, focusing on plants and fungi.

17.1 What is a plant?

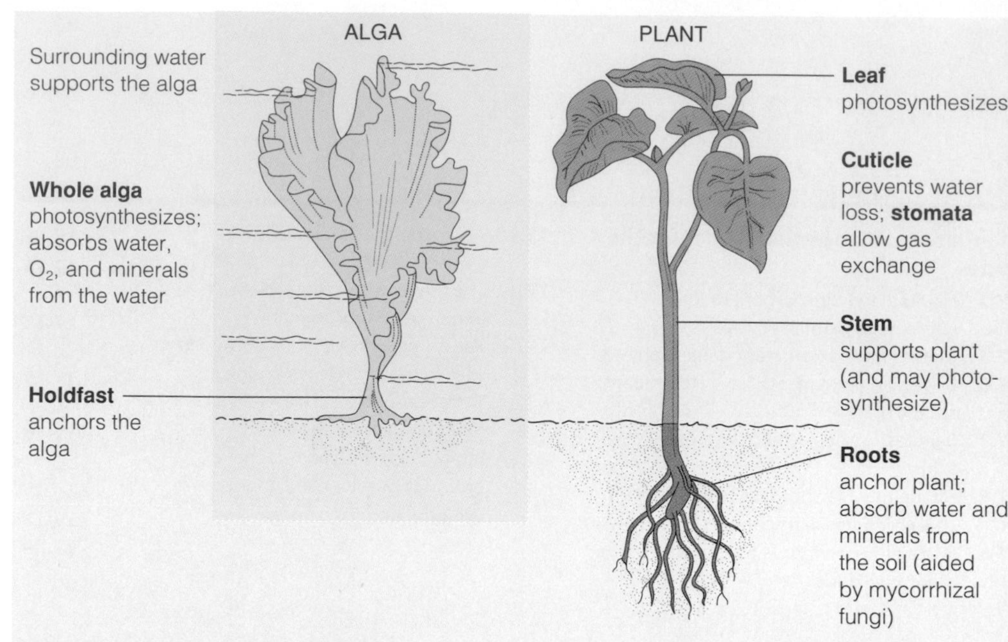

Figure 17.1A A multicellular green alga and a plant compared

Plants are multicellular eukaryotes that make organic molecules by photosynthesis. Trees, grasses, and many other green organisms we see around us fit this definition and are clearly members of the kingdom **Plantae.** But, as we saw in Modules 16.25 and 16.26, multicellular algae also fit the definition. The multicellular green algae, such as the seaweed *Ulva,* are especially similar to plants.

What are the differences between a plant and a multicellular green alga? Most arise from the fact that the alga is adapted for aquatic life and the plant for terrestrial life.

As indicated in Figure 17.1A, an alga is supported by the surrounding water. Many multicellular algae are anchored by a holdfast, but generally they have no rigid, supporting tissues. The entire algal body has direct access to water and obtains CO_2 and minerals from the water. Because almost none of the organism is underground, almost all of it receives light and can perform photosynthesis.

The aquatic ancestors of plants changed drastically as they became adapted to the challenges of living on land. One such challenge is loss of water to the air. Helping plants retain water is a waxy **cuticle** that covers their aerial parts (stems and leaves). Gas exchange cannot occur directly through the cuticle, but CO_2 and O_2 diffuse across the leaf surfaces through tiny pores called **stomata** (singular, *stoma*).

Posing other challenges, the body of a land plant is partly below ground, in soil, and partly above ground, in air. A land plant must be able to hold itself upright, because air provides no support. Moreover, a plant must be able to obtain chemicals from both soil and air, two very different media. The discrete organs of a plant—its roots, stems, and leaves—help the plant meet these challenges. (The organs of a plant are only analogous, not homologous, to the holdfast and other parts of a multicellular alga.) Plant roots provide anchorage and absorb water and mineral nutrients from the soil, and in most plants, as we have noted, mycorrhizae greatly enhance this absorption. The stems and leaves contain rigid support elements. Their cells have chloroplasts, which obtain CO_2 from the air and light from the sun.

While obtaining light and chemicals from the environment, a plant must also be able to conduct water and minerals upward from its roots to its leaves and to distribute sugars produced in the leaves throughout its body. Performing these functions, **vascular tissue,** a network of cells joined into narrow tubes, extends throughout the plant body. The photograph of part of an aspen leaf in Figure 17.1B shows the leaf's network of veins, which are fine branches of the vascular tissues. There are two types of vascular tissue. One, called **xylem,** is made of dead cells forming microscopic pipes that convey water and minerals up from the roots. The other, called **phloem,** consists of living cells and distributes sugars throughout the plant. Algae have no comparable transport tissues.

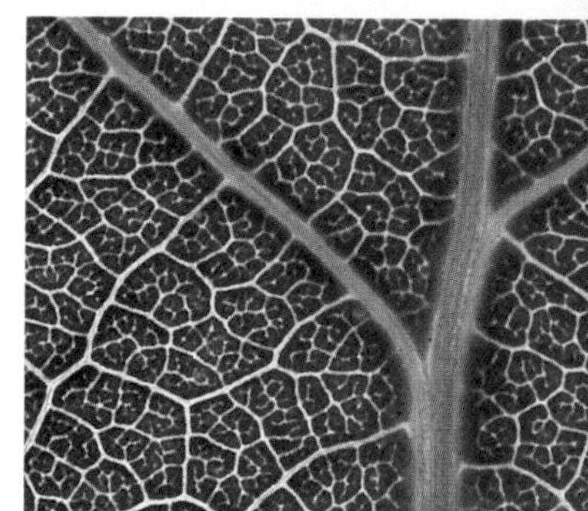

Figure 17.1B The network of veins in a leaf

Still other differences between algae and plants are seen during reproduction and development. For an alga, the surrounding water ensures that gametes and offspring stay moist, while providing the means for their dispersal. Plants, however, must keep their gametes and developing embryos from drying out in the air. Both plants and many algae produce gametes in **gametangia** (singular, *gametangium*), structures that consist of protective jackets of cells surrounding the gamete-producing cells. The egg remains in the female gametangium and is fertilized there. Either the sperm swim to the egg through water, or sperm-producing cells contained in pollen are conveyed, by wind or animals, close to the egg, where the sperm are then produced. In plants, but not in algae, the embryo develops inside the female gametangium, where it is kept moist. Most plants rely on wind or animals (such as fruit-eating birds or mammals) to disperse their offspring, which are in the form of embryos contained in seeds.

In summary, we can define a plant as a multicellular photosynthetic eukaryote with adaptations to life on land, including roots, stems, and leaves, a cuticle, stomata, transport tissues, and gametangia in which embryos develop. Focusing on these adaptations, we see that plants *are* quite different from algae, and this is why the five-kingdom system of classification places the algae in the kingdom Protista. But are plants and all the algae really different enough to be assigned to separate kingdoms? Focusing on the evolutionary ancestry of land plants, many biologists prefer to unite plants with the green algae, including both multicellular and unicellular species, in the plant kingdom.

What are three adaptations of plants for living on land?

Any three of the following: cuticle; stomata; vascular tissue; gametangia that protect gametes; and differentiation of the body into a subterranean root system and stems and leaves above ground

Plant Evolution and Diversity

Plants probably evolved from green algae called charophytes 17.2

Despite their differences, plants and green algae have a number of homologous features—for instance, chloroplasts and a particular combination of photosynthetic pigments. Also, most green algae and plants have cell walls made of cellulose, both store carbohydrates in the form of starch, and during cell division in both, vesicles derived from the Golgi apparatus form a cell plate that divides the cytoplasm. Because of these and other homologies, biologists generally agree that plants evolved from green algae.

The algal ancestors of plants may have carpeted the moist fringes of lakes or coastal salt marshes as early as 500 million years ago. At that time, the continents were relatively flat and may have been flooded periodically. Natural selection would have favored algae that could survive periodic droughts. Eventually, some species might have accumulated adaptations that enabled them to live permanently above the water line. The modern green alga *Coleochaete,* shown in Figure 17.2A, may resemble an early plant ancestor. It grows at the edges of lakes, and its disklike, multicellular colonies resemble plants in having jacketed zygotes. *Coleochaete* belongs to a group of green algae called the **charophytes**. Analyses of nucleic acid sequences, cell structure, reproduction, and cell metabolism indicate that charophytes are close relatives of plants. Plants and modern charophytes probably evolved from an ancestor that would be classified as a charophyte.

Figure 17.2A A charophyte named *Coleochaete,* a modern alga that may resemble a plant ancestor

Whatever organisms first colonized land, early plant life would have thrived in the new environment. Bright sunlight was virtually limitless on land, and at first there may not have been any plant-eating animals. Among the earliest terrestrial organisms known were simple plants called *Cooksonia,* which have been found only as 415-million-year-old fossils. Growing along the shores of lakes, *Cooksonia* (Figure 17.2B) had a branched, upright stem containing primitive vascular tissues, though it lacked leaves. The tips of some of its branches bore a bulbous structure called a **sporangium** (plural, *sporangia*) that produced reproductive cells called spores. A plant **spore** is a haploid cell that can develop into a haploid multicellular adult without fusing with another cell.

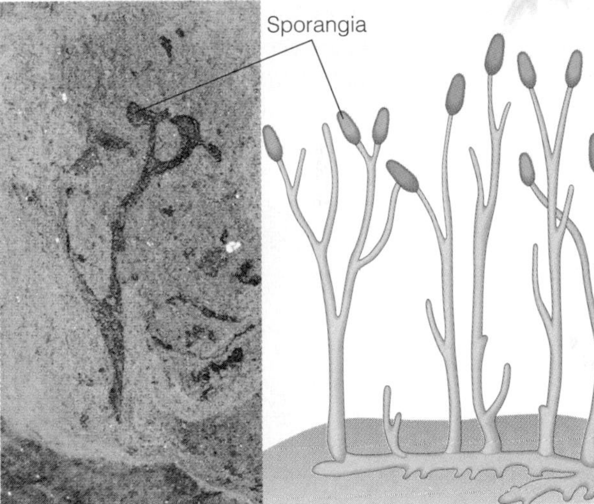

Figure 17.2B *Cooksonia,* one of the earliest land plants (fossil at left)

Cooksonia was a truly primitive plant. Slightly more complex plants flourished about 400 million years ago. They still lacked leaves but branched more profusely, had more complex sporangia, and were more firmly anchored in the soil than *Cooksonia.* By 375 million years ago, plants with well-developed leaves and roots were numerous and diverse.

What are some key homologies between plants and green algae?

Chloroplasts with homologous photosynthetic pigments; cell walls of cellulose; carbohydrate storage in the form of starch; and a similar mechanism of cell plate formation during cell division

17.3 Plant diversity provides clues to the evolutionary history of the plant kingdom

17A
See Activity 17A on the CD-ROM.

The phylogenetic tree in Figure 17.3A highlights some of the major events in the history of the plant kingdom. The diversity of modern plants, represented by the four major groups of plants shown along the top of the tree, gives us a broad picture of plant evolution.

An initial period of plant evolution saw the origin of plants from aquatic ancestors, most likely green algae called charophytes. As the geological timeline on the left of Figure 17.3A indicates, two distinct lineages arose from ancestral plants about 400 million years ago. One of these lineages gave rise to modern plants called **bryophytes,** a group that includes the mosses. Bryophytes resemble other plants in having a cuticle and embryos that develop within gametangia. Unlike other plants, however, many bryophytes lack vascular tissue, although some have water-conducting tubes. Bryophytes also generally lack internal support, which the rigid cell walls of vascular tissue provide. Thus, a mat of moss, like the one shown in Figure 17.3B, actually consists of many plants growing in a tight pack, holding one another up. The mat is spongy and can retain water. Each moss plant has many small stems and leaves and grips the underlying ground with elongated cells or rootlike extensions from cells. Mosses and other bryophytes also have flagellated sperm, which closely resemble those of some green algae. The sperm must swim to the eggs to fertilize them, so fertilization requires the plant to be covered with a film of water.

Back on our phylogenetic tree, the other ancient lineage of plants, the **vascular plants,** have xylem and phloem as well as embryonic development within gametangia. Their vascular tissues provide strong support, enabling stems to stand upright and grow tall on land. Ancestral vascular plants may have resembled *Cooksonia,* which we discussed in Module 17.2. As indicated on the phylogenetic tree, the vascular plant lineage split in two about 360 million years ago. Plants in one of these lineages have **seeds,** which consist of an embryo packaged with a food supply within a protective covering. Seeds did not evolve in the other lineage.

Ferns belong to the seedless vascular plant lineage. A fern has well-developed roots and rigid stems. In many

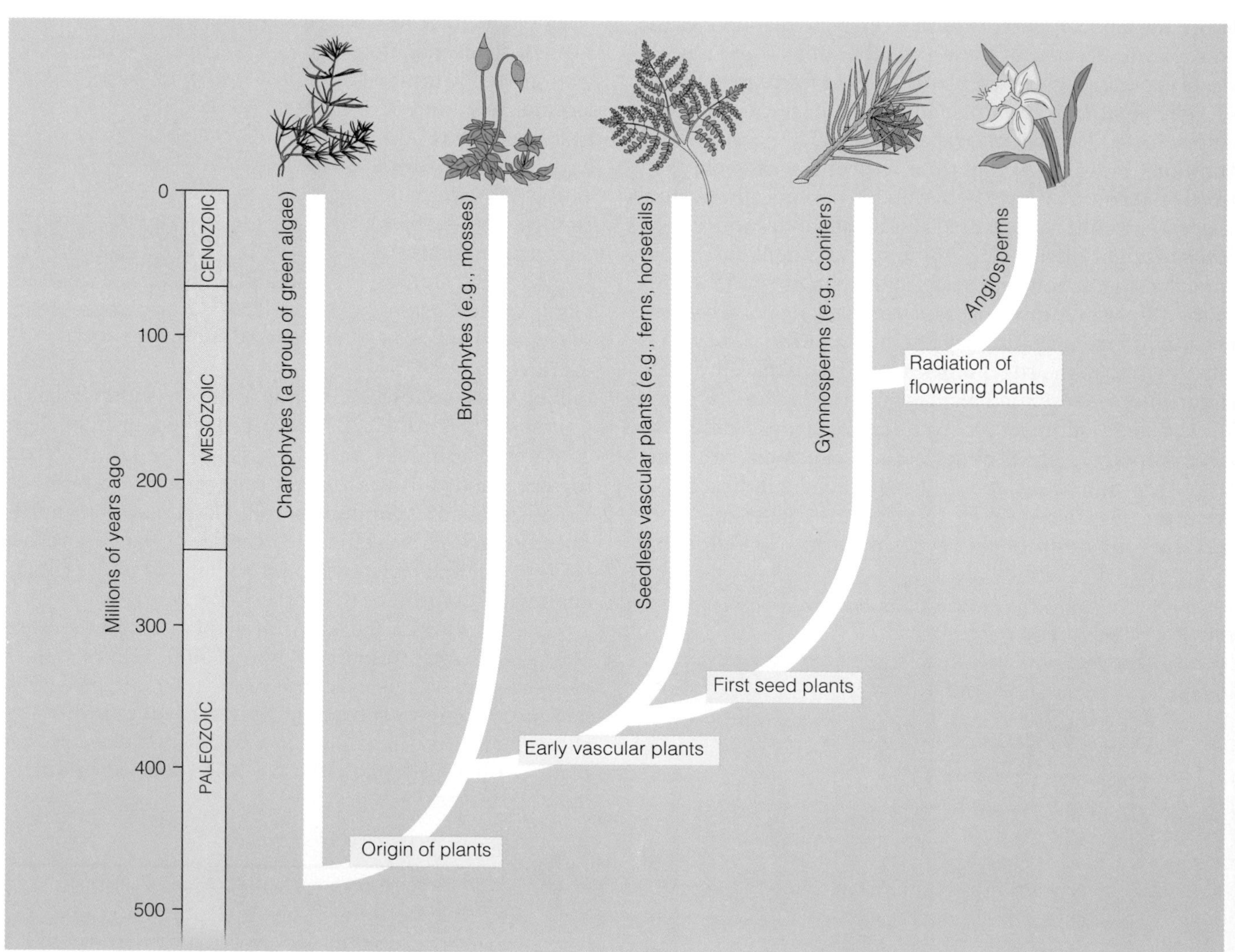

Figure 17.3A Some highlights of plant evolution

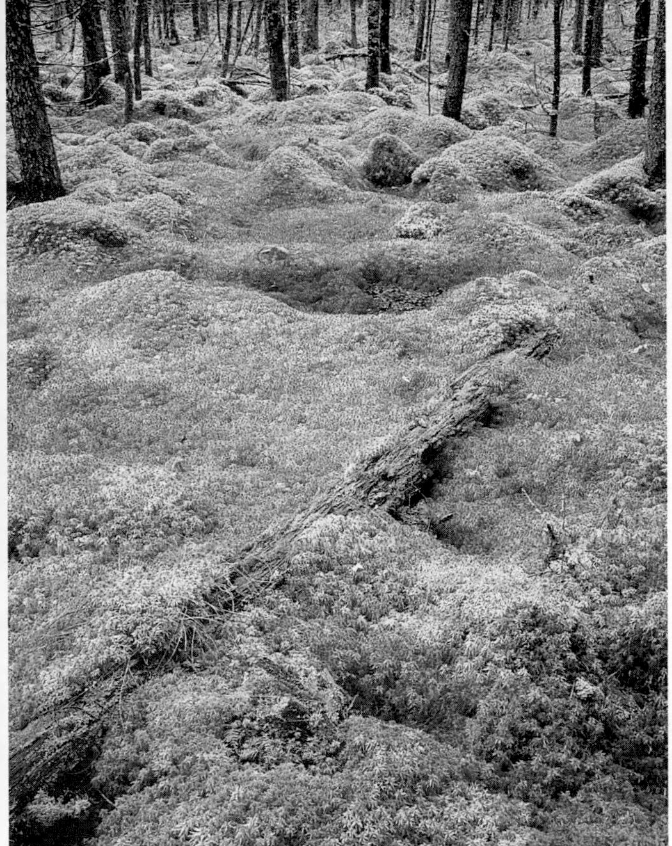

Figure 17.3B Mosses (bryophytes) blanketing the ground in a forest

species, the leaves, commonly called fronds, sprout from stems that grow along the ground (Figure 17.3C; the "fiddleheads" in the inset are young fronds ready to unfurl). Ferns are common in shady areas in temperate forests, but they are most diverse in the tropics. In some of the tropical species, called tree ferns, upright stems grow several meters tall. In common with mosses, ferns have flagellated sperm that require a layer of water to reach the eggs.

The other branch of the phylogenetic tree that arose from the same ancestor as the ferns led to a great variety of plants that produce seeds. Today, the seed plant lineage accounts for nearly 90% of the approximately 265,000 species of living plants.

Several key adaptations underlie the enormous success of seed plants. First, they make seeds, which are survival packets for life on land. Second, seed plants do not require a water layer for fertilization. Instead of producing sperm that swim to the eggs, they produce pollen, a vehicle that transfers nonflagellated sperm-forming cells to the female parts of the plant. Pollen is carried passively by wind or animals, and the arrival of pollen at the female is called **pollination.** As we will see, fertilization occurs some time after pollination.

Among the earliest seed plants to appear were the **gymnosperms** (Greek *gymnos,* naked, and *sperma,* seed). The seed of a gymnosperm is said to be naked because it is not contained in a fruit. Actually, a gymnosperm seed has a thin protective covering that is part of the seed itself. Gymnosperms coexisted with ferns and other seedless plants in great forests that dominated the landscape for more than 200 million years. Today, the conifers—pine, spruce, fir, and many other kinds of trees with seed-bearing cones and needlelike leaves—are the largest group of gymnosperms.

The most recent major branching point in plant evolution was a split in the seed plant lineage. About 130 million years ago, the flowering plants, or **angiosperms** (Greek *angeion,* vessel, and *sperma,* seed) diverged from the gymnosperm lineage. Flowers are complex reproductive structures that develop seeds within protective chambers. The great majority of modern plants—some 235,000 species—are angiosperms.

In summary, four key adaptations for life on land mark the main lineages of the plant kingdom. (1) Gametangia, which protect gametes, zygotes, and embryos from drying out, are present in all plants. The earliest species of plants probably nurtured their developing embryos inside gametangia. (2) Vascular tissues (phloem and xylem) mark a lineage that gave rise to most modern plants. Vascular tissues transport water and nutrients throughout the plant body and also provide support for upright stems. (3) Seeds appeared in a lineage that dominates the plant kingdom today. (4) Flowers mark the angiosperm lineage, which is the dominant group of seed plants. As we will see in the next several modules, the life cycles of modern plants reveal additional details about plant evolution.

Which of the following structures is common to all four major plant groups: flowers, seeds, cuticle, pollen?

Cuticle

Figure 17.3C Ferns (seedless vascular plants)

Alternation of Generations and Plant Life Cycles

17.4 Haploid and diploid generations alternate in plant life cycles

Plants have life cycles very different from ours. Each of us is a diploid individual; the only haploid stages in the human life cycle, as for nearly all animals, are sperm and eggs. By contrast, plants have alternating generations: Diploid ($2n$) individuals called sporophytes and haploid (n) individuals called gametophytes generate each other in the life cycle. Starting at the top of Figure 17.4, you can see that haploid male and female gametophytes produce gametes by mitosis. Fertilization results in a diploid zygote. The zygote undergoes mitosis and develops into the diploid sporophyte. The sporophyte produces haploid spores by meiosis. Completing one life cycle, a spore develops by mitosis into a multicellular haploid gametophyte.

Many algae undergo alternation of generations (for example, the seaweed *Ulva* in Figure 16.26C), and plants probably inherited it from their algal ancestors. However, plants exhibit changes associated with life on land. In most cases, the sporophyte is physically joined with the gametophyte and protects and supports it. Land plants also have other adaptations that protect the delicate reproductive stages. None have flagellated eggs or spores, and mosses and ferns are unusual in having flagellated sperm. The next module highlights another unusual feature of mosses: Unlike most plants, including ferns, mosses have a gametophyte that is larger than the sporophyte and upon which the sporophyte develops.

Figure 17.4

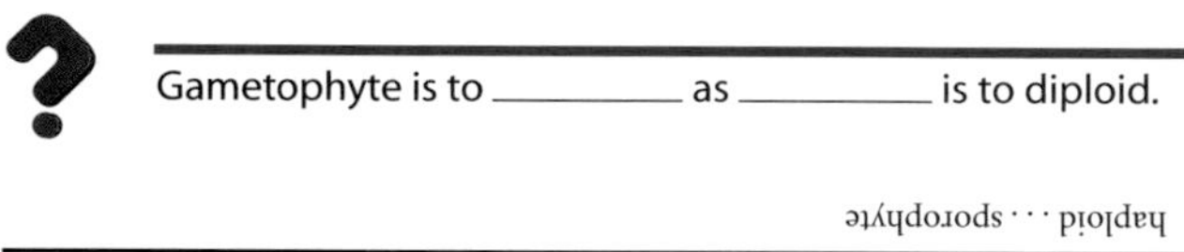

?
Gametophyte is to ________ as ________ is to diploid.
haploid . . . sporophyte

17.5 Mosses have a dominant gametophyte

(See text at top of facing page.)

17B

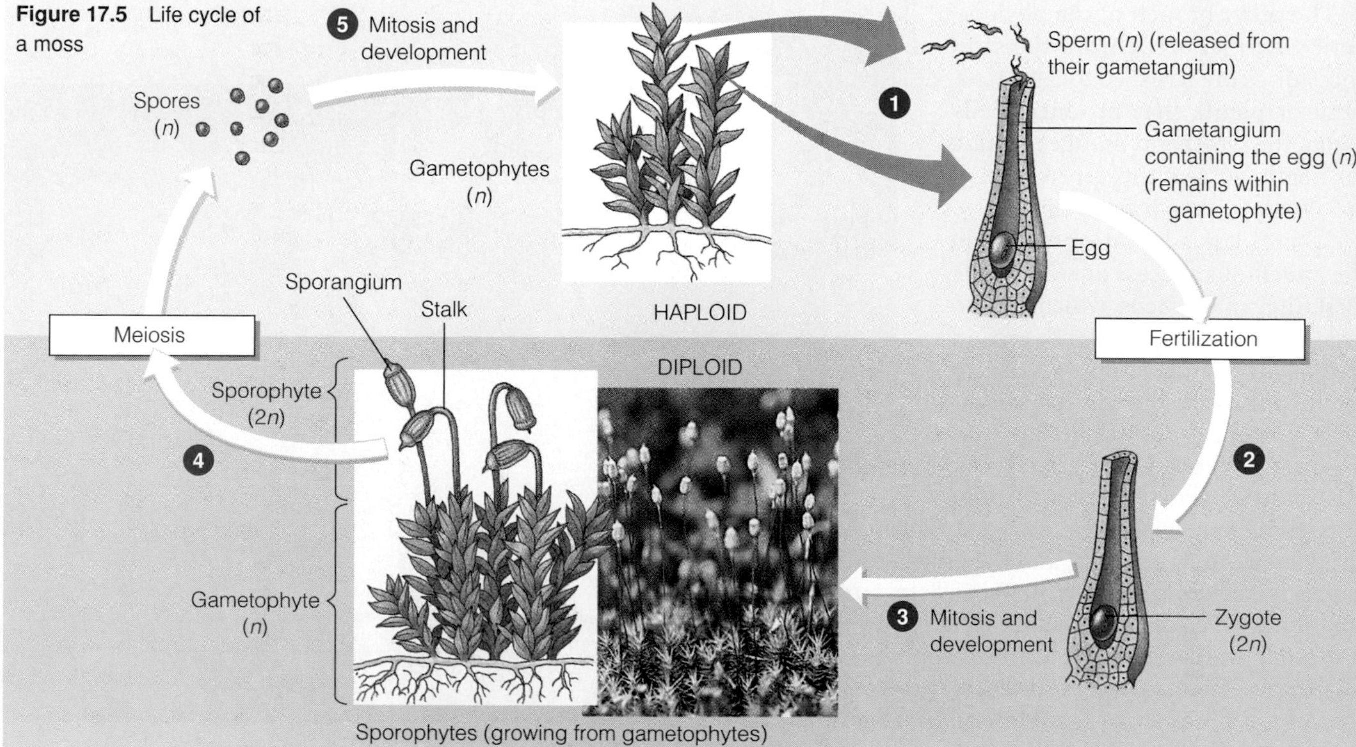

Figure 17.5 Life cycle of a moss

In a moss, most of the green, cushiony growth we see consists of gametophytes. ① Gametes develop in gametangia on the gametophytes. The flagellated sperm require a film of water in which to swim to the egg, which stays in its gametangium on the gametophyte. After fertilization, ② the zygote remains in the gametangium. ③ There it divides by mitosis and develops into a sporophyte. Each sporophyte remains attached to a gametophyte, as you can see in the photograph, and ④ meiosis occurs in the sporangia at the tips of the sporophyte stalks. Haploid spores resulting from meiosis are released. Later, ⑤ they undergo mitosis and develop into gametophytes, completing one life cycle. 

? How do moss sperm travel from male gametangia to female gametangia, where fertilization of eggs occurs?

The flagellated sperm swim through a film of water.

17.6 Ferns, like most plants, have a dominant sporophyte

17C

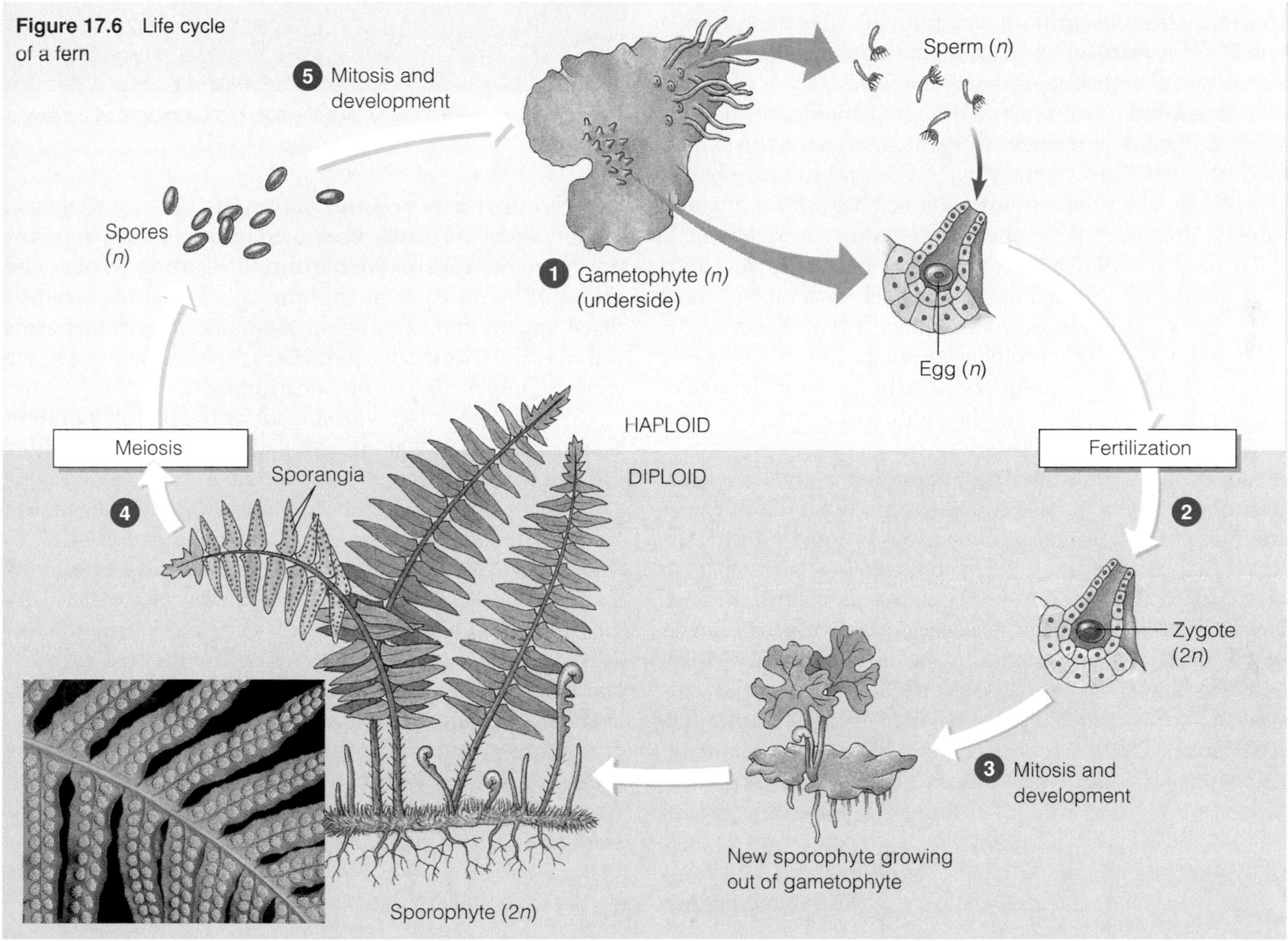

Figure 17.6 Life cycle of a fern

5 Mitosis and development
Sperm (*n*)
Spores (*n*)
1 Gametophyte (*n*) (underside)
Egg (*n*)
HAPLOID
DIPLOID
Meiosis
Fertilization
Sporangia
4
2
Zygote (2*n*)
3 Mitosis and development
New sporophyte growing out of gametophyte
Sporophyte (2*n*)

The life cycle of a fern illustrates a dominant sporophyte generation. In fact, all we usually see of a fern is the sporophyte. But let's start the fern life cycle with the gametophyte. ① Fern gametophytes often have a distinctive heart-like shape (top of Figure 17.6), but they are quite small (about 0.5 cm across) and inconspicuous. Like mosses, ferns have flagellated sperm that require moisture to reach the egg. ② The zygote remains on the gametophyte, where ③ it develops into the sporophyte. ④ Cells in sporangia undergo meiosis, producing haploid spores. ⑤ The spores are released and develop into gametophytes by mitosis.

Today, about 95% of all plants, including all seed plants, have a dominant sporophyte in their life cycle. As seed plants evolved, their sporophyte became adapted to house the gametophyte and all reproductive stages (including eggs, sperm, spores, zygotes, and embryos). Before we resume this story, let's glance back to a time in plant history, before seed plants rose to dominance, when ferns and other seedless plants covered much of the land surface. 

? How is it possible for the fern gametophyte to produce haploid gametes without meiosis?

All the gametophyte's cells are haploid, so there is no need to reduce chromosome number by meiosis to produce haploid gametes; they are produced by mitosis.

17.7 Seedless plants formed vast "coal forests"

Ferns and other seedless plants have a long history. During the Carboniferous period (about 285–360 million years ago), vast forests of them grew in swampy areas that covered much of what is now Eurasia and North America. At that time, these continents were close to the equator and had tropical climates. The painting in Figure 17.7, based on fossil evidence, reconstructs one of those great seedless forests. Most of the large trees with straight trunks are seedless plants called lycopods. On the far right, the feathery branches belong to a horsetail, another seedless plant. Tree ferns were also prominent, although they are not featured in this picture. Animals, including the giant dragonfly you see, also thrived in the swamp forests.

The tropical swamp forests of the Carboniferous period generated great quantities of organic matter. As the plants died, they fell into stagnant wetlands and did not decay completely. Their remains formed thick deposits of organic rubble called peat. Later, the swamps were covered by seawater, marine sediments covered the peat, and pressure and heat gradually converted the peat to coal. Coal is black sedimentary rock made up of fossilized plant material. It formed during several geological periods, but the most extensive coal beds formed from Carboniferous deposits. (Carboniferous comes from the Latin, *carbo,* coal, and *fer-*, bearing.)

Coal, oil, and natural gas are **fossil fuels**—fuels formed from the remains of extinct organisms. Fossil fuels generate much of our electricity. As we deplete our oil and gas reserves, the use of coal is likely to increase.

"Coal forests" dominated the North American and Eurasian landscapes until near the end of the Carboniferous period. At that time, the world climate turned drier and colder, and the vast swamps and forests began to disappear. The climatic change provided an opportunity for seed plants, which can complete their life cycles on dry land and withstand long, harsh winters. The evolution of pollen, produced by the dominant sporophyte, was a key step in the adaptation of seed plants to dry land. By providing transport for sperm-producing cells, pollen makes it possible for sperm to reach and fertilize eggs without being immersed in water.

Of the earliest seed plants, the most successful were the gymnosperms, and several kinds grew along with the seedless plants in the Carboniferous swamps. You can see one gymnosperm in Figure 17.7, the low-growing shrub with large fernlike leaves near the bottom left. Called a seed fern, although only its leaves resembled ferns, this type of gymnosperm did not survive beyond the Carboniferous. The group of gymnosperms that prevailed after the swamps dried up and that is still dominant among gymnosperms today is the **conifers**—naked-seed plants that produce cones. (Conifer means cone-bearing.)

Figure 17.7 An artist's view of an extinct forest dominated by seedless plants

Conifers have been widespread for the past 250 million years, especially in regions with dry, cool climates and short growing seasons. The pine tree on the facing page is a typical conifer. Adapted to harsh winter storms, its branches readily shed snow; when wet, heavy snow does stick to them, they usually bend rather than break. The tree's needlelike leaves resist drying because they have little surface area for evaporation; a thick cuticle covering the leaf surface also helps retain water. Because the tree keeps its leaves throughout the year, it can start photosynthesizing as soon as the short growing season begins.

Let's now return to our evolutionary story and see how the pine tree life cycle fits with the major trends in plant history: dominance of the sporophyte generation and protection of the delicate reproductive stages.

Why are coal, oil, and natural gas called "fossil" fuels?

Because they are derived from ancient organisms that did not decay completely after dying

17.8 A pine tree is a sporophyte with tiny gametophytes in its cones

17D

Pines and other conifers illustrate how drastically the relative roles of the haploid and diploid generations changed as plants evolved on land. A pine tree itself is a sporophyte; the gametophyte generation consists of microscopic stages that grow inside the tree's cones.

Cones are a significant adaptation to land, for they harbor all of a conifer tree's reproductive structures. These are the same structures we described earlier for the mosses and ferns: diploid sporangia, which produce haploid spores by meiosis; haploid female and male gametophytes; gametes, which are produced by the gametophytes; and zygotes, resulting from fertilization.

A pine tree bears two types of cones. The hard, woody ones we usually notice are female cones (step ① in Figure 17.8). The female cone has many hard, radiating scales, each bearing a pair of **ovules.** An ovule starts out as a sporangium and a covering, or integument. ② Male cones are generally much smaller than female cones; they are also soft and short-lived. Each scale on a male cone produces many sporangia, each of which makes numerous haploid spores by meiosis. Male gametophytes, or **pollen grains,** develop from the spores. When male cones are mature, the scales open and release a cloud of pollen (millions of microscopic grains). You may have seen yellowish conifer pollen covering car tops and windshields or floating on ponds in the spring.

Carried by the wind and independent of water, pollen grains house the cells that will develop into sperm. ③ Pollination occurs when a pollen grain lands on and enters an ovule. After pollination, meiosis occurs in the ovule, and ④ a haploid spore cell begins developing into the female gametophyte. Not until months later do eggs appear within the female gametophyte. It also takes months for sperm to develop in the pollen grain. ⑤ A tiny tube grows out of the pollen grain and eventually releases a sperm into the egg. Fertilization does not occur until more than a year after pollination.

⑥ Following fertilization, the zygote develops into a sporophyte embryo, and the whole ovule transforms into the seed. The seed contains the embryo's food supply (the remains of the female gametophyte) and has a tough seed coat (the ovule's integument). In a typical pine, seeds are shed from the cones about two years after pollination. ⑦ The seed falls to the ground, or is dispersed by wind or animals, and when conditions are favorable, it germinates (the embryo starts growing). Eventually, the embryo grows into a tree.

In summary, all the reproductive stages of conifers are housed in cones borne on sporophytes. The ovule is a key adaptation—a protective device for all the female stages in the life cycle, as well as the site of pollination, fertilization, and embryonic development. The ovule becomes the seed, a major factor in the success of the conifers and flowering plants on land.

Contrast the mode of sperm delivery in ferns with sperm delivery in conifers.

The flagellated sperm of ferns must swim through water to reach eggs. In contrast, the airborne pollens of conifers bring male and female gametophytes together without need of an aqueous route, and a pollen grain releases sperm near the egg after pollination occurs.

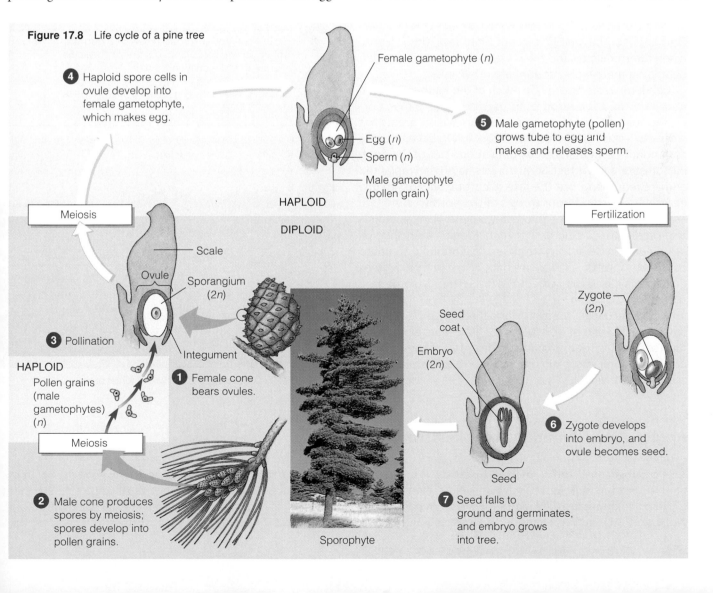

Embryo (2n)

1 Female cone bears ovules.

Figure 17.8 Life cycle of a pine tree

17.9 Sustaining coniferous forests poses many challenges

Figure 17.9A A forest of conifers

Attesting to the success of the conifers, a broad band of coniferous forests covers much of northern Eurasia and North America and extends southward in mountainous regions. Figure 17.9A shows a coniferous forest in Oregon.

Today, about 190 million acres in the United States, mostly in the western states and Alaska, are designated National Forest lands. Some of these areas are set aside as wilderness; most are managed by the U.S. Forest Service for lumbering, grazing, mining, public recreation, and wildlife habitat. Balancing these diverse, often conflicting, uses while trying to sustain coniferous forests for future generations is a formidable challenge. As Forest Service chief Michael Dombeck emphasized in a recent interview, "This is a time when forested lands and the variety of things they provide people are more highly sought after than ever."

Coniferous forests provide much of our lumber. They are also harvested for wood pulp used to make paper. Currently, our demand for wood and paper is so great that clear-cut areas like the one in Figure 17.9C have become commonplace. In many areas, only about 10% of the original forest remains intact. Some forests have been replanted and are regrowing, but the rate of cutting and loss to diseases in many coniferous forests of the northwestern U.S. and Canada exceeds the reforestation rate. Many scientists also predict that an increase in global temperatures, which now seems to be occurring, poses an additional threat to coniferous forests. An increase of only one or two degrees Celsius may adversely affect reproduction by some tree species.

The loss of coniferous forests also threatens other species. The original forests of North America were more biologically diverse—that is, they contained many more different kinds of trees, animals, and other organisms—than the forests that are now regrowing.

Figure 17.9B Michael Dombeck

Sustaining coniferous forests and their diversity is a challenge for private citizens as well as public institutions. Speaking for his agency, Michael Dombeck put it this way: "These forested lands contain some of the best and last habitats for rare and endangered species. The Forest Service and other federal agencies are mandated by law to manage their lands to restore and sustain biological integrity."

The average U.S. citizen now consumes about 50 times more paper than the average person in less developed nations. Reducing paper waste, increasing recycling, and greater use of electronic media can help reduce that trend. In Dombeck's words, "We have to learn to live within the limits of the land. An old proverb goes something like: 'We haven't inherited the land from our forefathers; we have borrowed it from our children.'" 

?

(a) In what way are forests "renewable" resources?
(b) Under what conditions does reality belie the "renewable" designation?

(a) Forests are renewable in the sense that new trees can grow where old growth has been cut. (b) Currently, wood is harvested at a much faster rate than new growth can replace, and thus "renewable" is an overly optimistic term for forest resources.

Figure 17.9C A clear-cut in a coniferous forest area

17.10 Angiosperms dominate most landscapes today

The photograph of the coniferous forest above could give us a somewhat distorted view of today's plant world. Conifers do dominate the land in the northern parts of the globe, but it is the angiosperms, or flowering plants, that dominate most other land areas. Today, worldwide, there are about 400 times more species of angiosperms than gymnosperms, and nearly 80% of all plants are angiosperms. Whereas gymnosperms supply most of our lumber and paper, angiosperms supply nearly all our food and much of our fiber for textiles. Cereal grains, including wheat, corn, oats, and barley, are flowering plants, as are citrus and other fruit trees, garden vegetables, cotton, and flax. Fine hardwoods from flowering plants, such as oak, cherry, and walnut trees, supplement the lumber we get from conifers.

Several unique adaptations account for the success of angiosperms. The leaves of most species are broad and flat, a shape that makes them very effective collectors of solar energy. Angiosperms also have vascular tissues with thicker and therefore stronger cell walls than gymnosperms. Above all, it is the flower that accounts for the unparalleled success of the angiosperms.

_________ are to conifers as flowers are to _________.

Cones . . . angiosperms

The flower is the centerpiece of angiosperm reproduction 17.11

17E

No organisms make a showier display of their sex life than angiosperms. From roses to dandelions, flowers expose a plant's male and female parts and are the sites of pollination and fertilization. They also generate fruits, which contain the angiosperm's seeds. Figure 17.11A shows a tulip flower.

Figure 17.11A A close-up view of a tulip

Figure 17.11B shows the anatomy of a flower. Although the generalized flower drawn here looks different from the tulip, these and all other kinds of flowers have a common basic anatomy. A **flower** is actually a short stem with four kinds of modified leaves called sepals, petals, stamens, and carpels. At the bottom of the flower are the **sepals,** which are usually green. They enclose the flower before it opens (think of a rosebud). Above the sepals are the **petals,** which are usually the most striking part of the flower and are often important in attracting animal pollinators. The actual reproductive structures are multiple stamens and one or more carpels. Each **stamen** consists of a stalk bearing a sac called an **anther,** the male organ in which pollen grains develop. The **carpel** consists of a stalk with an ovary at the base and a sticky tip known as the **stigma,** which traps pollen. The **ovary** is a protective chamber containing one or more ovules, in which the eggs develop. As we see in the next module, a seed develops from each ovule, and the fruit develops from the ovary.

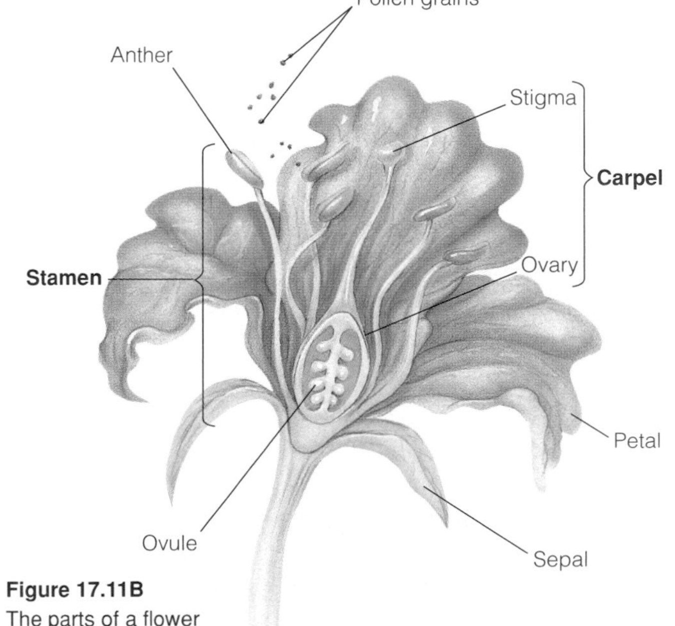

Figure 17.11B The parts of a flower

What are the four main organs of a flower?

Sepals, petals, stamens, and carpels

17.12 The angiosperm plant is a sporophyte with gametophytes in its flowers

17E

In broad outline, the angiosperm life cycle resembles that of a gymnosperm. The plant we see is a sporophyte, and the tiny gametophyte generation lives on it. In contrast to a gymnosperm, whose gametophytes grow in its cones and whose seeds are naked, an angiosperm has its gametophytes in its flowers and its seeds packaged inside fruits.

Figure 17.12, illustrating the life cycle of a flowering plant, highlights features that have been especially important in angiosperm evolution. (We will discuss these, and the unique angiosperm adaptation called double fertilization, in more detail in Modules 31.10–31.14.) Starting at the "Meiosis" box on the left side of Figure 17.12, ① meiosis occurring in the anthers of the flower leads to the haploid spores that undergo mitosis and form the male gametophytes, or pollen grains. ② Meiosis in the ovules leads to haploid spores that undergo mitosis and form the female gametophytes, each of which produces an egg. ③ Pollination occurs when a pollen grain, carried by the wind or an animal, lands on the stigma. As in gymnosperms, a tube grows from the pollen grain to an egg, and a sperm fertilizes the egg, creating ④ a zygote. Also as in gymnosperms, ⑤ a seed develops from each ovule. Each seed consists of an embryo (a new sporophyte) surrounded by a store of food and a seed coat. While the seeds develop, ⑥ the ovary's wall thickens, forming the fruit that encloses the seeds. Fruits help disperse the seeds. When conditions are

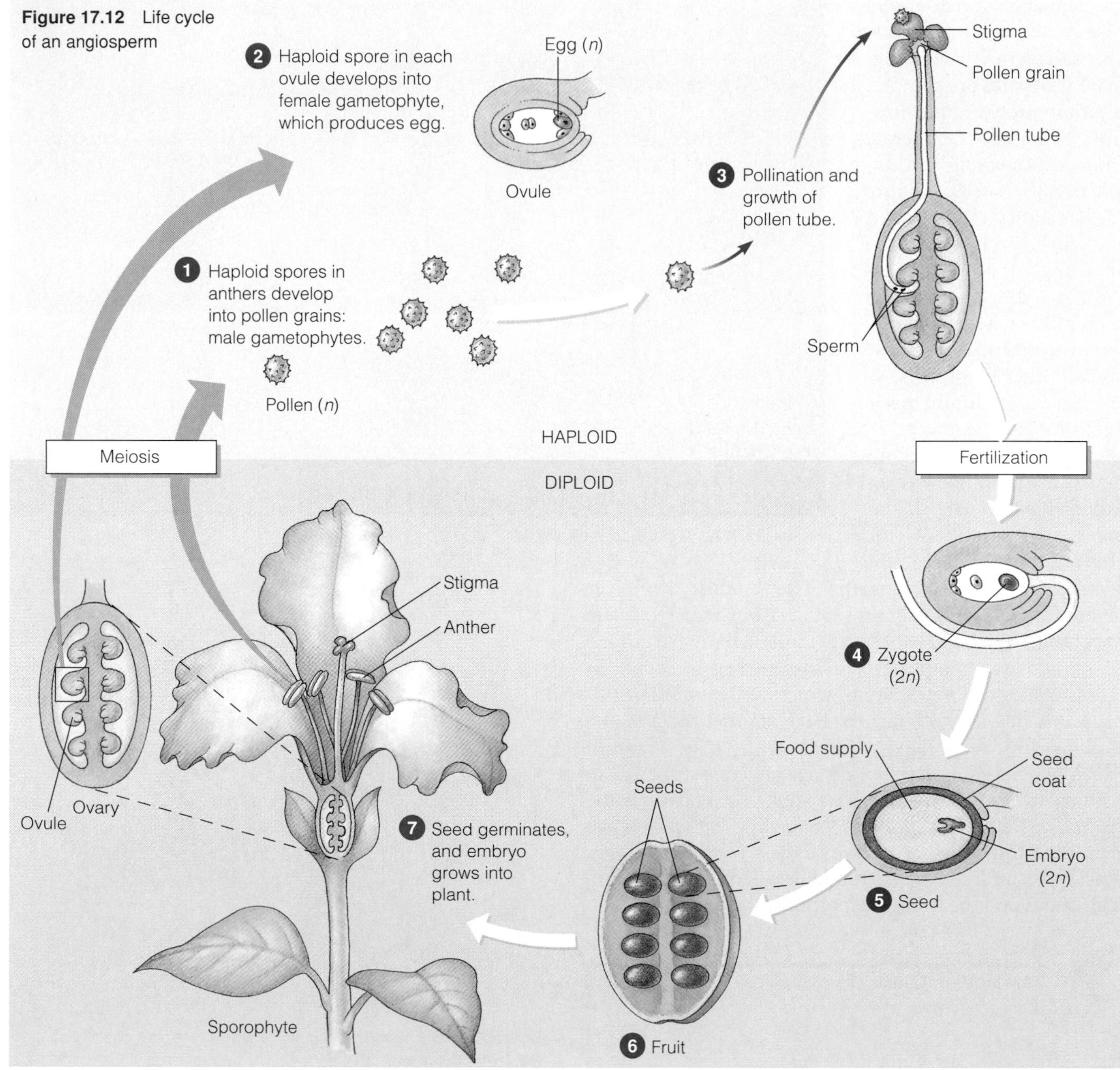

Figure 17.12 Life cycle of an angiosperm

favorable, ⑦ the seed germinates and the embryo grows into a mature sporophyte, completing the life cycle.

Besides the packaging of angiosperm seeds inside fruit, several other features of angiosperm life have enhanced the success of these plants. One is the evolution of mutually dependent relationships with animals, which carry pollen more reliably than the wind. Another is the ability to reproduce rapidly. Fertilization in angiosperms usually occurs about 12 hours after pollination, making it possible for the plant to produce seeds in only a few days or weeks. As we mentioned in Module 17.8, a typical gymnosperm usually takes well over a year to produce seeds. Rapid seed production is advantageous, particularly in environments such as deserts, where growing seasons are extremely short.

What is the difference between pollination and fertilization?

Pollination is the transfer of pollen by wind or animals from stamens to the tips of carpels. Fertilization is the union of egg and sperm, which is released from the pollen tube after pollination.

17.13 The structure of a fruit reflects its function in seed dispersal

Figure 17.13A Dandelion fruit dispersed by the wind

A **fruit,** the ripened ovary of a flower, is a special adaptation that helps disperse seeds. Some angiosperms depend on wind for dispersal. For example, the dandelion fruit (Figure 17.13A) acts like a kite, carrying a seed away from the parent plant on wind currents. Some other angiosperms produce fruits that hitch a free ride on animals. The cockleburs attached to the fur of the dog in Figure 17.13B are fruits that may be carried miles before they open and release their seeds.

Many angiosperms produce fleshy, edible fruits that are attractive to animals as food. When the mouse in Figure 17.13C eats a berry, it digests the fleshy part of the fruit, but most of the tough seeds pass unharmed through its digestive tract. The mouse may then deposit the seeds, along with a supply of natural fertilizer, some distance from where it ate the fruit. Many types of garden produce, including tomatoes, squash, and melons, as well as strawberries, apples, cherries, and oranges, are edible fruits.

The fruit of flowering plants usually develops and ripens quickly, so the seeds can be produced and dispersed in a single growing season. The dispersal of seeds in fruits is one of the main reasons angiosperms are so numerous and widespread.

What is a fruit?

A ripened ovary of a flower, which contains, protects, and aids in the dispersal of seeds.

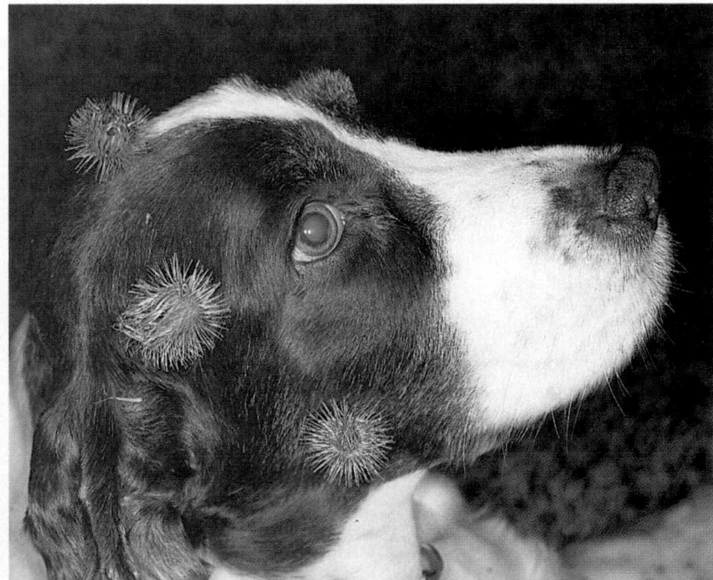

Figure 17.13B Cockleburs (fruit) carried by animal fur

Figure 17.13C A mouse eating fruit containing seeds that will be dispersed later with the animal's feces

17.14 Interactions with animals have profoundly influenced angiosperm evolution

Flowering plants and land animals have had mutually beneficial relationships throughout their evolutionary history. Most angiosperms depend on insects, birds, or mammals for pollination and seed dispersal. And most land animals depend on angiosperms for food. These mutual dependencies tend to improve the reproductive success of both the plants and the animals and, thus, are favored by natural selection. Let's examine some cases of pollination.

Many angiosperms produce flowers that attract pollinators that rely entirely on the flowers' nectar and pollen for food. Nectar is a high-energy fluid that is of use to the plant only for attracting pollinators. The color and fragrance of a flower are usually keyed to a pollinator's sense of sight and smell. Many flowers also have markings that attract pollinators, leading them past pollen-bearing organs on the way to nectar. For example, flowers that are pollinated by bees often have markings that reflect ultraviolet light. Such markings are invisible to us, but vivid to bees. The bee in Figure 17.14A is harvesting nectar and pollen from a scotch broom flower. The flower parts arched over the insect are the pollen-bearing stamens. Some of the pollen the bee picks up here will rub off onto the female parts of the next flower it visits.

Many flowers pollinated by birds are red or pink, colors to which bird eyes are especially sensitive. The shape of the flower may also be important. Flowers that depend largely on hummingbirds, for example, typically have their nectar located deep in a floral tube, where only the long, thin beak and tongue of the bird are likely to reach. As a hummingbird (Figure 17.14B) flies among flowers in search of nectar, its feathers and beak pick up pollen from the anthers of the flowers. It will deposit the pollen in other flowers of the same shape, and so probably of the same species, as it continues to feed.

Insects and birds are active mainly during the day. Some flowering plants, however, depend on nocturnal pollinators, such as bats. These plants typically have large, light-colored, highly scented flowers that can easily be found at night. Taken at night, the photograph in Figure 17.14C shows a bat approaching the large, white flower of a baobab tree in tropical Africa. While the bat eats part of the flower, the yellow anthers dust its hair with pollen, which it passes on as it visits other flowers.

To a large extent, flowering plants are as diverse and successful as they are today because they have close connections with other organisms. As we discussed in the chapter's introduction, these connections include ones with fungi.

How are an angiosperm and its pollinators mutually rewarded by their relationship?

The pollinators obtain food and the plant has its pollen targeted much more efficiently than if it were carried by the wind.

Figure 17.14B A rufous hummingbird obtaining nectar from a red flower that fits the shape of its long, thin beak

Figure 17.14A A bee picking up pollen as it feeds on nectar

Figure 17.14C A bat, a nighttime pollinator

Fungi and plants moved onto land together 17.15

If plants had not evolved mutually beneficial relationships with fungi, they might never have been able to colonize land. Mycorrhizae, the associations of plant roots with fungi that we discussed in the opening essay, helped make the colonization of land possible. Conversely, a connection with plants probably also helped fungi make the move onto land. Fungi are heterotrophic; like animals, they cannot make their own food molecules and must obtain them from other organisms. Before plants colonized land and began stocking soil with organic molecules that heterotrophs could use for food, fungi may have thrived only in aquatic environments. The first fungi on dry land may have been the mycorrhizal partners of the first land plants.

Today, fungi are found virtually everywhere. They abound in the soil as well as in all types of aquatic environments. Not all fungi have beneficial associations with plants. Some are parasites, obtaining their nutrients at the expense of plants or other organisms. The rust-colored spots on the wheat leaf in Figure 17.15A are growths of black stem rust, a fungus that infects a number of agriculturally important grains and is highly destructive.

Figure 17.15A A fungal parasite, black stem rust, growing on a leaf of a wheat plant

The micrograph in Figure 17.15B illustrates another kind of fungus, one that catches and eats small animals. The thick entity in the picture is an animal called a roundworm, which lives in soil. It has been snared by a predatory fungus (the thin structures in the picture). The fungus constricts around the worm in a fraction of a second. The fungus then penetrates its prey and digests it.

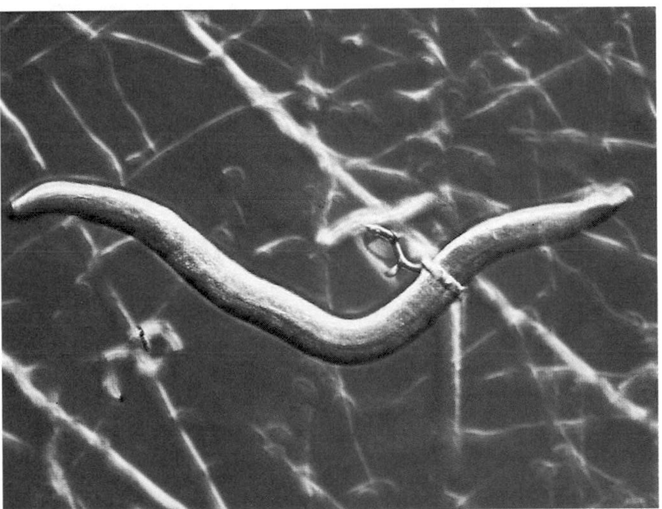

Figure 17.15B A fungal predator trapping a roundworm

Many fungi decompose (decay) dead organisms. Fungi that decompose organic matter are essential to all forms of life because they restock the environment with inorganic nutrients essential for plant growth. The white growths on the rotting log in Figure 17.15C are a fungus that digests wood. If this and other fungi and bacteria in a forest suddenly stopped decomposing for just a few years, leaves, logs, feces, and dead animals would pile up on the forest floor. Plants and the animals they feed would starve because elements taken from the soil would not be returned. Eventually, the forest and all its inhabitants would die.

Figure 17.15C A fungal decomposer breaking down a dead log

Fortunately, fungi are among the most adaptable of all living organisms. Many fungi that decompose wood, for example, also decompose many toxic pollutants, including the pesticide DDT and certain chemicals that cause cancer. Fungal decomposers can usually thrive in any place where there is organic matter.

What are mycorrhizae?

Root–fungus associations that enhance the uptake of water and minerals by the plant and provide organic nutrients to the fungus

17.16 Fungi absorb food after digesting it outside their bodies

Fungi are classified in their own kingdom, the **kingdom Fungi.** These organisms were formerly thought to be plants that lack chlorophyll, but about the only trait fungi share with plants is that most of them grow in the ground. Just what are fungi, and what makes them so distinctive that we put them into their own kingdom?

In brief, **fungi** are heterotrophic eukaryotes that digest their food externally and absorb the small nutrient molecules that result. Fungi have body structures and modes of reproduction unlike those of any other organism. We concentrate on fungal structure in this module and on fungal reproduction in Module 17.17.

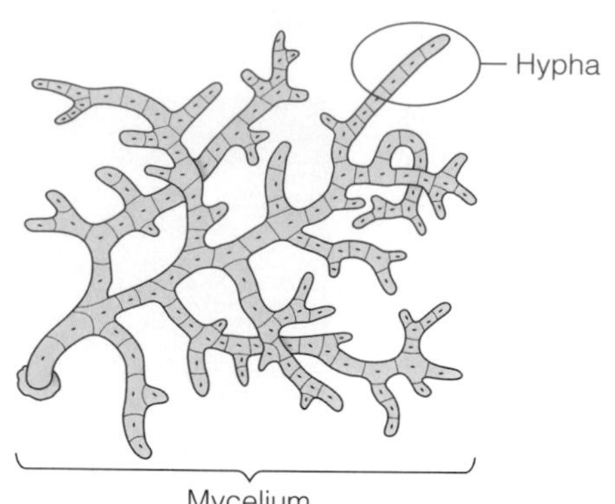

Figure 17.16A A mycelium, made of numerous hyphae (in this case, consisting of chains of cells)

Figure 17.16C Mushrooms, reproductive structures made of packed hyphae

Figure 17.16B The mycelium of a mold growing on decaying leaves

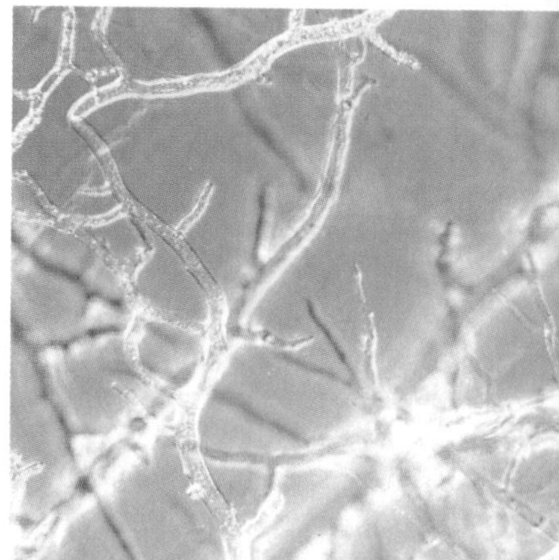

Figure 17.16D Hyphae of a bread mold (continuous cytoplasm with many nuclei)

Most fungi, including the molds and mushrooms, are multicellular. Yeasts are unicellular fungi, but their simple structure probably evolved from multicellular ancestors. A typical fungus consists of a netlike mass of filaments called **hyphae** (singular, *hypha*). Hyphae start as single filaments and then branch repeatedly, forming a network known as a **mycelium** (plural, *mycelia*), illustrated in Figure 17.16A. Figure 17.16B is a photograph of the mycelium of a mold growing on decaying leaves. Hyphae actually grow into the cells of the leaf and digest the cytoplasm. The mushrooms in Figure 17.16C, solid as they seem, are actually made of tightly packed hyphae. A mushroom is just an aboveground reproductive structure attached to a much more extensive underground mycelium.

Figure 17.16D is a micrograph of hyphae in a culture of a bread mold. In this mold, the hyphae have many nuclei within a single mass of cytoplasm (the nuclei are not visible here). In many other fungi, the hyphae consist of chains of cells. In either case, the hyphae are surrounded by a plasma membrane covered by a cell wall. Unlike plants, which have cellulose cell walls, most fungi have cell walls made of chitin, a strong, flexible polymer of a nitrogen-containing sugar.

Most fungi are nonmotile; they do not move about in search of food or mates. And, unlike animals, most protists, and many plants, most fungi do not have flagellated or amoeboid cells at any stage in their life cycles. But the mycelium makes up for the lack of mobility by being able to grow at a phenomenal rate. A fungus concentrates its resources on extending its hyphae around and through food sources. Because its hyphae grow longer without getting thicker, the fungus develops a huge surface from which it can secrete digestive enzymes and through which it can absorb food. A mycelium can add as much as a kilometer of hyphae each day. A mushroom can grow to its full size in one night. The tips of the hyphae are just the right size and shape to invade plant cells or grow between them. Whether a fungus is a decomposer infiltrating the remains of a dead plant, a parasite infecting a living plant, or a mutualist that participates in mycorrhizae, the hyphae secrete enzymes that digest plant cell walls and then grow into them.

Contrast the heterotrophic nutrition of a fungus with your own heterotrophic nutrition.

A fungus digests its food externally by secreting digestive juices onto the food and then absorbing the small nutrients that result from digestion. In contrast, humans and most other animals "eat" relatively large pieces of food and digest the food within their bodies.

Many fungi have three distinct phases in their life cycle 17.17

17F

Fungal life cycles range from simple to complex. For example, many yeasts reproduce only by mitotic cell division. By contrast, mushrooms and many other kinds of fungi have three distinct phases in their life cycles. As illustrated below, they have diploid and haploid phases, and they also have a unique third phase, called the **dikaryotic phase,** in which cells contain two haploid nuclei.

Let's follow the life cycle of a mushroom, starting at the top left of Figure 17.17. ① The mushroom itself is called a **fruiting body.** ② Numerous zygotes develop in specialized cells on the underside of the fruiting body's cap. The zygotes are the only diploid stage in the life cycle. Each is simply a diploid nucleus resulting from fertilization, which in this case is the fusion of the two haploid nuclei present in a cap cell (there are no sperm or eggs). Then, without going through any mitotic divisions, each zygote undergoes meiosis, and haploid spores are formed. ③ The fruiting body releases enormous numbers of these spores. Carried by wind, water, or animals, the spores may happen to land on moist matter that can serve as food. ④ There they germinate and grow into haploid mycelia.

Much of a mushroom's sex life occurs underground. The haploid mycelia are of discrete kinds, called **mating types.** The different mating types contain genetically distinct nuclei, and only certain types are sexually compatible. ⑤ The dikaryotic stage begins when hyphae of two compatible mycelia (indicated here by red and yellow nuclei) grow together. The hyphae fuse, but the nuclei do not. You see the result, a dikaryotic mycelium, in ⑥. Note that each of its cells contains two *genetically different* nuclei.

The fruiting body is an extension of the dikaryotic mycelium, and cells making up the fruiting body are dikaryotic. The dikaryotic phase ends and a new diploid stage begins when haploid nuclei in the cap fuse to form the zygotes.

Much of the success of fungi is due to their reproductive capacity. Dikaryotic fruiting bodies lead to the formation of genetically different kinds of mycelia. In this way, the fungus retains genetic variability and its capacity to adapt to the changing conditions of its environment. A single round of the three-phase life cycle of a mushroom also increases the number of organisms enormously. A massive boost in numbers occurs when the haploid spores form. As mycelia develop from the spores (step 4 in the life cycle) and from the dikaryotic hyphae (step 6), the fungus spreads extensively through new soil areas. Dikaryotic mycelia often last for years, branching widely as they decompose organic matter in soil, rotting logs, or leaf litter. The dikaryotic mycelia of some mushrooms are among the world's oldest and largest organisms. One growing in forest soil in northern Michigan extends over about 30 acres, and scientists estimate that it began growing from a single spore about 1500 years ago.

What is the dikaryotic phase of a fungus?

The phase in which each cell has two nuclei from two different parents, with the nuclei not yet fused

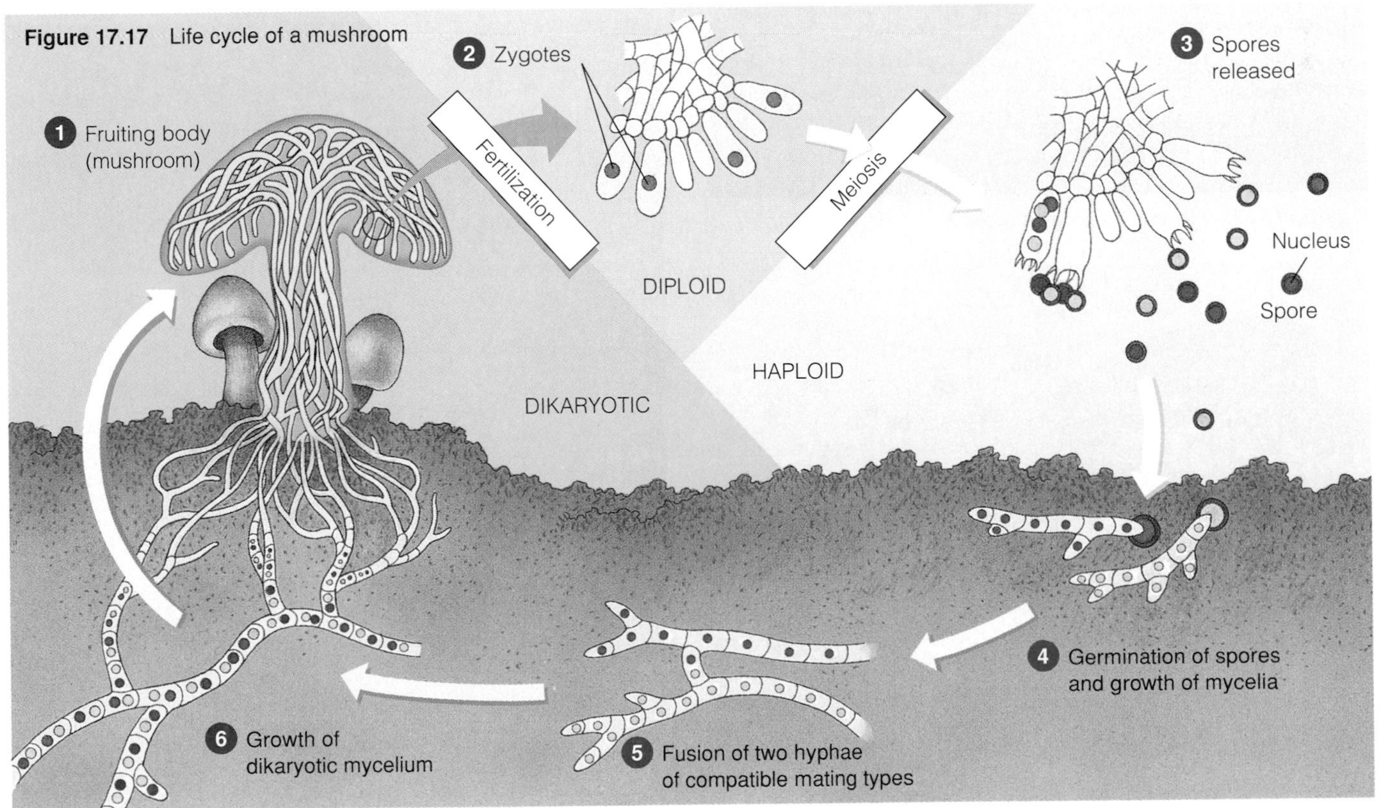

Figure 17.17 Life cycle of a mushroom

17.18 Lichens consist of fungi living mutualistically with photosynthetic organisms

Mutually beneficial connections are a recurring theme in both the plant and the fungal kingdoms. The bright yellow and red forms that stripe the surface of the rock outcropping in Figure 17.18A are another example of a mutual relationship involving fungi. These lichens superficially resemble mosses or other simple plants, but they are not plants at all—nor are they even individual organisms. **Lichens** are associations of millions of green algae or cyanobacteria held in a tangled network of fungal hyphae (Figure 17.18B). Not all lichens are as vividly colored as those in Figure 17.18A. Bright colors in lichens usually result from pigments in the hyphal walls of the fungi.

The lichen association is still not completely understood. The fungus is known to receive food from its photosynthetic partners, and the algae or cyanobacteria receive housing, water, and certain minerals from the fungus. But much remains to be learned about exactly what molecules are exchanged and how the exchanges occur. A few of the algae that live in lichens can grow independently, but lichen fungi cannot survive on their own.

Lichens reproduce asexually by fragmenting or, as shown in Figure 17.18B, by dispersing tiny airborne units containing both the fungus and some photosynthetic cells. The fungus and alga or bacterium can also reproduce independently, either sexually or asexually, although they must reassociate for the fungus to survive in nature.

Whatever the precise nature of the lichen mutualism, it gives the two organisms the ability to survive in habitats that are inhospitable to either organism alone. Lichens are rugged and able to live where there is little or no soil. As a result, they are important pioneers on new land. Lichens grow into tiny rock crevices, adding to the forces that erode hard surfaces and paving the way for future plant growth. Some lichens can tolerate severe cold, and carpets of them cover the arctic tundra. The caribou in Figure 17.18C eats a lot of lichens and is especially dependent on them in its winter feeding grounds in Alaska.

Lichens can also withstand severe drought. They are opportunists, growing in spurts when conditions are favorable. When it rains, a lichen quickly absorbs water and photosynthesizes at a rapid rate. In dry air, it dehydrates and photosynthesis may stop, but the lichen remains alive more or less indefinitely. Some lichens are thousands of years old, rivaling the oldest plants and fungi as the oldest organisms on Earth.

Figure 17.18A Lichens growing on rock

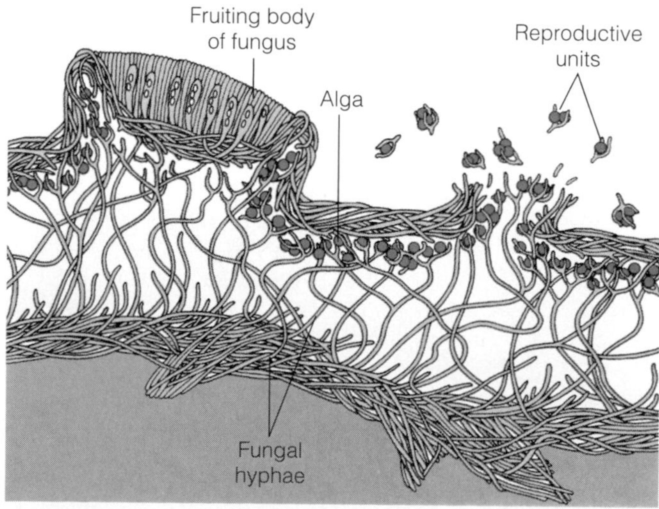

Figure 17.18B Asexual reproduction of a lichen

Figure 17.18C A caribou eating lichens in the arctic tundra

As tough as lichens are, many do not withstand air pollution very well. Because they get most of their minerals from the air, in the form of dust or compounds dissolved in raindrops, lichens are very sensitive to airborne pollutants such as sulfur dioxide. The death of lichens may be a sign that air quality in an area is deteriorating.

? Why is the health of lichens an indicator of air quality?

Lichens are very sensitive to air pollution because they absorb mineral compounds, both nutrients and pollutants, from the air.

Parasitic fungi harm plants and animals 17.19

About 100,000 species of fungi are known to science. Approximately one-third of these are mutualists, in either lichens or mycorrhizae. About another third are decomposers living in soil and rotting organic material. The rest—some 30,000 species—are parasitic, mostly in or on plants. In some cases, fungi that infect plants have literally changed landscapes. The dead trees in Figure 17.19A are American elms killed by the parasitic fungus that causes Dutch elm disease. The fungus evolved with European species of elm trees, and it is relatively harmless to them. But it is deadly to American elms. Accidentally introduced into the United States on logs sent from Europe just after World War I, the fungus was carried from tree to tree by bark beetles. Since then, it has destroyed elm trees all across North America. The spread of the Dutch elm fungus continues despite decades of research and attempts to control it.

Fungi are a serious problem as agricultural pests. Species called smuts and rusts (see Figure 17.15A) are common on grain crops and cause tremendous economic losses each year. The ear of corn shown in Figure 17.19B is infected with a widespread fungal pathogen called corn smut. The grayish growths are called galls. Analogous to the fruiting body of a mushroom, a gall is made up of dikaryotic hyphae that grow into a developing corn kernel and eventually displace it. When a gall matures, it breaks open and releases thousands of blackish spores. In parts of Central America, the smutted ears are cooked and eaten as a delicacy, but generally corn smut is regarded as a scourge. Fortunately, certain genetic strains of corn are resistant to it.

Figure 17.19A Elm trees killed by Dutch elm disease

Figure 17.19B Corn smut

The seed heads of many other kinds of grain, including rye, wheat, and oats, are sometimes infected with fungal growths called ergots. Consumption of flour made from ergot-infested grain can cause gangrene, nervous spasms, burning sensations, hallucinations, temporary insanity, and death. One epidemic in Europe in the year 944 killed more than 40,000 people. Several kinds of toxins have been isolated from ergots. One called lysergic acid is the raw material from which the hallucinogenic drug LSD is made. Certain others are medicinal in small doses. An ergot compound is useful in treating high blood pressure, for example.

Animals are much less susceptible to parasitic fungi than plants are. Only about 50 species of fungi are known to be parasitic in humans and other animals. However, their effects are significant enough to make us take them seriously. Among the diseases that fungi cause in humans are yeast infections of the lungs, some of which can be fatal, and of the vagina. Other fungal parasites produce a skin disease called ringworm, so named because it appears as circular red areas on the skin. The ringworm fungi can infect virtually any skin surface. Most commonly, they attack the feet and cause intense itching and sometimes blisters (Figure 17.19C). This condition, known as athlete's foot, is highly contagious but can be treated with various fungicidal preparations.

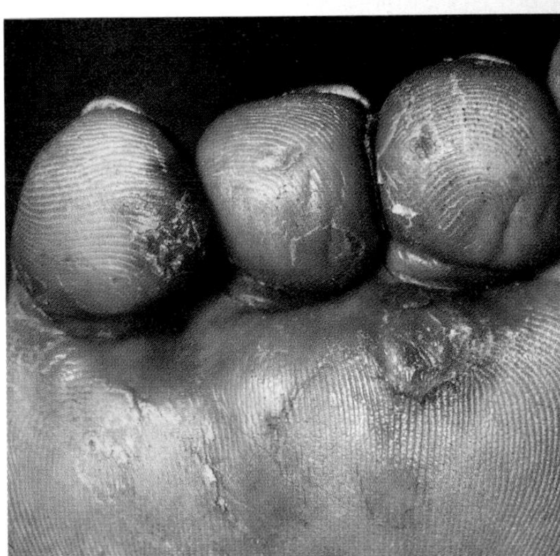

Figure 17.19C Lesions on a human foot due to a fungal infection; also called athlete's foot or ringworm of the foot

What is "athlete's foot"?

Infection of the foot's skin with ringworm fungus

17.20 Fungi have enormous ecological and practical impacts

Figure 17.20A Blue cheese

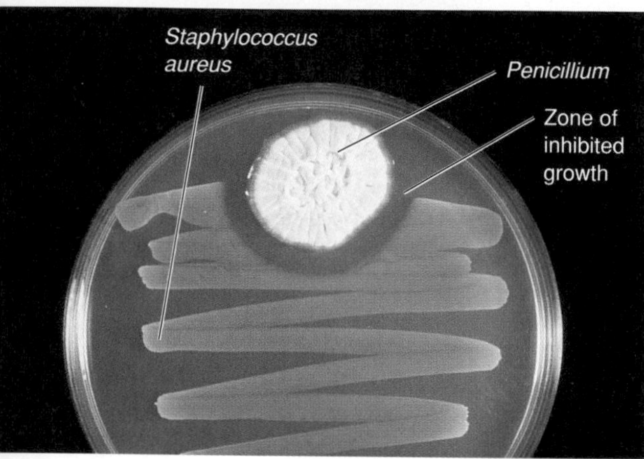

Figure 17.20B A culture of *Penicillium* and bacteria

It would not be fair to fungi to end our discussion with an account of diseases. Far more important are the benefits we derive from these interesting eukaryotes. On a global scale, we depend on them as decomposers and recyclers of organic matter.

Fungi also have a number of practical uses for humans. Most of us have eaten mushrooms, although we may not have realized that we were ingesting the fruiting bodies of subterranean fungi. And mushrooms are not the only fungi we eat. The distinctive flavors of certain kinds of cheeses, including Roquefort and blue cheese (Figure 17.20A), come from the fungi used to ripen them. Highly prized by gourmets are truffles, the fruiting bodies of certain mycorrhizal fungi associated with tree roots.

More important in food production are unicellular fungi, the yeasts. As discussed in Chapter 6, yeasts are used in baking, brewing, and winemaking. And fungi are medically valuable as well. Like the bacteria called actinomycetes (see Module 16.13), some fungi produce antibiotics that are used to treat bacterial diseases. In fact, the first antibiotic discovered was penicillin, which is made by the common mold called *Penicillium*. In Figure 17.20B, the clear area between the mold and the bacterial colony is where the antibiotic produced by *Penicillium* inhibits the growth of the bacteria *(Staphylococcus aureus)*.

As producers of antibiotics and food, as decomposers, and as mutualistic partners in mycorrhizae and lichens, fungi are vital contributors to the living world.

Fungi are the third group of eukaryotes we have surveyed so far. Strong evidence suggests that they evolved from protistan ancestors that also gave rise to the fourth and most diverse group of eukaryotes, the animals, which we study next. 

? What do you think is the function of the antibiotics that fungi produce in their natural environments?

The antibiotics probably block the growth of microorganisms, especially bacteria, that compete with the fungi for nutrients and other resources.

Chapter Review

Chapter Summary

Introduction to Plants (Introduction –17.1) Mutually beneficial associations of plant roots and fungi, called mycorrhizae, are common and may have enabled ancestral plants to colonize land **(Introduction)**. Plants are multicellular photosynthetic eukaryotes. They share many characteristics with green algae, but plants evolved unique features as they colonized land. Most plants have roots, stems, and leaves—organs that function in absorption, support, and photosynthesis. Stems and leaves are covered by a protective cuticle, and leaves have stomata, which allow gas exchange. Vascular tissues transport water and nutrients throughout the plant body. Most plants have a vascular system with xylem tissue conducting water and minerals and phloem tissue distributing sugars. Vascular tissue also provides internal support. Furthermore, in plants, the female gametangia not only produce gametes (eggs) but also protect the developing embryos after the eggs are fertilized **(17.1)**.

Plant Evolution and Diversity (17.2–17.3) Plants probably evolved from green algae. Molecular studies indicate that green algae called charophytes are the closest relatives of plants **(17.2)**. Two main lineages arose early from ancestral plants. One lineage gave rise to bryophytes, plants that lack the vascular tissues xylem and phloem. These tissues characterize the other main plant lineage, the vascular plants. Bryophytes include the mosses, which, unsupported by vascular tissues, grow in a low, spongy mat. Derived from early vascular plants, ferns and seed plants have xylem and phloem, well-developed roots, and rigid stems. Ferns are seedless plants whose flagellated sperm require moisture to reach the egg. A major step in plant evolution was the appearance of seed plants, gymnosperms and angiosperms. These vascular plants have pollen grains for producing and transporting sperm and protect their embryos in seeds. Gymnosperms, such as the pines, are called naked-seed plants because their seeds do not develop inside a fruit. The seeds of angiosperms (flowering plants) develop in protective chambers within fruits **(17.3)**.

Alternation of Generations and Plant Life Cycles (17.4–17.14) The life cycles of all plants include alternation of haploid and diploid generations. The haploid gametophyte produces eggs and sperm by mitosis. The eggs and sperm unite, and the zygote develops into the diploid sporophyte. Meiosis in the sporophyte produces haploid spores, which grow into gametophytes **(17.4)**. Most of a mat of moss consists of gametophytes, which produce eggs and swimming sperm. The zygote stays on the gametophyte and develops into the less conspicuous sporophyte **(17.5)**. The sporophyte dominates the fern life cycle, as it does that of all seed plants. Ferns, like mosses, have swimming sperm. The fern zygote remains on the small, inconspicuous gametophyte, where it develops into the sporophyte **(17.6)**. Ferns and other seedless plants once dominated ancient forests; their remains formed coal. Gymnosperms that produce cones, called conifers, largely replaced the ancient forests of seedless plants and remain the dominant gymnosperms today **(17.7)**. The pine tree, for example, is a sporophyte; tiny gametophytes grow in its cones. Sporangia in male cones make spores that develop into male gametophytes, the pollen grains. Sporangia called ovules in female cones produce female gametophytes. A sperm from a pollen grain fertilizes an egg in the female gametophyte. The zygote becomes a sporophyte embryo, and the ovule becomes a seed, with stored food and a protective coat. The conifer life cycle takes at least two years **(17.8)**. Coniferous forests are important sources of timber and paper **(17.9)**. Most plants are angiosperms, whose hallmarks are flowers **(17.10)**. An angiosperm is a sporophyte with gametophytes in its flowers **(17.11)**. The angiosperm life cycle is similar to that of conifers, except that it is much more rapid, and angiosperm seeds are protected and dispersed in fruits, which develop from ovaries **(17.12)**. Fruits are adaptations that help disperse seeds **(17.13)**. Interactions with animals influenced the evolution of flowers and fruits. Angiosperms are a major food source for animals, while animals aid plants in pollination and seed dispersal **(17.14)**.

Fungi (17.15–17.20) Plants probably moved onto land along with mycorrhizal fungi, which help plants absorb water and nutrients. Mycorrhizal fungi are mutualistic organisms; other fungi are parasites, predators, or decomposers of dead organisms **(17.15)**. Fungi are heterotrophic eukaryotes that digest their food externally and absorb the nutrients. A fungus usually consists of a mass of threadlike hyphae, forming a network called a mycelium. Most fungi cannot move, but they can grow around and through their food very rapidly **(17.16)**. Fungal spores germinate to form haploid hyphae. In some fungi, such as mushrooms, the fusion of hyphae results in the unique dikaryotic phase of the life cycle, in which each cell contains two unfused nuclei. The dikaryotic mycelium forms a fruiting body—the mushroom—where fertilization occurs. The diploid zygotes then undergo meiosis, producing a new generation of spores **(17.17)**. Lichens are associations of algae or cyanobacteria with a network of fungal hyphae. The fungus receives food in exchange for housing, water, and minerals. Lichens survive in hostile environments. They cover rocks and frozen tundra soil, and they are pioneers on new land **(17.18)**. Parasitic fungi cause diseases, such as Dutch elm disease, corn smut, and athlete's foot **(17.19)**. Numerous fungi are beneficial, with many being essential in the decomposition of organic matter and nutrient recycling. Fungi are also commercially important as food, in baking and beer and wine production, and in the manufacture of antibiotics **(17.20)**.

Testing Your Knowledge

Multiple Choice

1. Angiosperms are different from all other plants because only they have
   a. a vascular system
   b. flowers
   c. a life cycle that involves alternation of generations
   d. seeds
   e. a sporophyte phase

2. Which of the following produce eggs and sperm? *(Explain your answer.)*
   a. the fruiting bodies of a fungus
   b. fern sporophytes
   c. moss gametophytes
   d. the anthers of a flower
   e. moss sporangia

3. The eggs of seed plants are fertilized within ovules, and the ovules then develop into
   a. seeds
   b. spores
   c. gametophytes
   d. fruit
   e. sporophytes

4. The diploid sporophyte stage is dominant in the life cycles of all of the following except
   a. a pine tree
   b. a dandelion
   c. a rose bush
   d. a fern
   e. a moss

5. Under a microscope, a piece of a mushroom would look most like
   a. jelly
   b. a tangle of string
   c. grains of sugar or salt
   d. a piece of glass
   e. foam

Describing, Comparing, and Explaining

1. Compare a seed plant with an alga in terms of adaptations for life on land versus life in the water.

2. How do animals help flowering plants reproduce? What do the animals get in return?

3. Why are fungi and plants classified in different kingdoms?

Thinking as a Scientist

1. Many fungi produce antibiotics, such as penicillin, which are valuable in medicine. But of what value might the antibiotics be to the fungi? Similarly, fungi often produce compounds with unpleasant tastes and odors as they digest their food. What might be the value of these chemicals to the fungi? How might the production of antibiotics and odors have evolved?

2. In April 1986, an accident at a nuclear power plant in Chernobyl, Ukraine, scattered radioactive fallout for hundreds of miles. In assessing the biological effects of the radiation, researchers found mosses to be especially valuable as organisms for monitoring the damage. As mentioned in Module 10.16, radiation damages organisms by causing mutations. Explain why it is faster to observe the genetic effects of radiation on mosses than on plants from other groups. Imagine that you are conducting tests shortly after a nuclear accident. Using potted moss plants as your experimental organisms, design an experiment to test the hypothesis that the frequency of mutations decreases with the organism's distance from the source of radiation.

Science, Technology, and Society

Much of the conifer forest in the U.S. Pacific Northwest has been clear-cut; less than 10% of the original ancient forest, dominated by giant firs and hemlocks, remains. There is no law protecting endangered habitats, so to protect the northern spotted owl, which lives only in old-growth conifers, conservationists have sued to stop logging under the Endangered Species Act. The lawsuits have halted logging in many National Forest areas. Lumber companies buy trees from National Forests, loggers work there, and the economies of many small communities depend on logging. The reduction in timber supply has driven up the cost of lumber. Imagine you have been named by the U.S. president to deal with this situation. What are the opposing issues? What would you suggest to resolve this conflict, and how would you defend your policy?

Media Resources

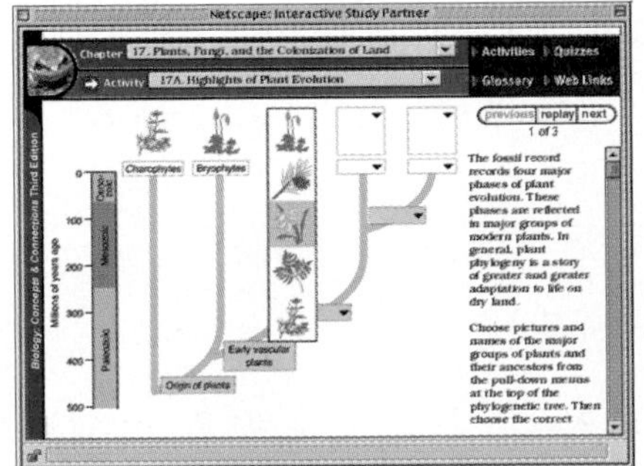

For further review of the concepts in this chapter, use the activities and quizzes on the CD-ROM.

Visit the Special Edition of The Biology Place web site for activities, quizzes, news articles, and web links.

18 The Evolution of Animal Diversity

Outline

Of some 1.5 million species of organisms known to science, over two-thirds are animals. As a group, animals thrive in nearly all environments on Earth. Most species of animals live on land, but the greatest number of animal *phyla* are found in the sea, where animals undoubtedly arose. A diver on a tropical coral reef like the one at the left could see more different phyla in half an hour than most people see in a lifetime.

The photograph at the left was taken on a reef near the Fiji Islands. To experience even a fraction of the teeming life in these waters, a diver has to stop swimming and focus on small areas one at a time. Schools of fishes explode out of caves. A startled moray eel wriggles back into a crevice. Other brightly colored fishes nibble on the coral surfaces. Sea stars, multicolored shrimps, and ornate worms crawl slowly around clusters of sponges.

Of all the different organisms living on a coral reef, often just two kinds actually build the reef, multicellular algae that are encrusted with limestone (see Module 16.26) and coral animals. Many of the colorful branching forms in the photo at the left are skeletons of coral

Vertebrates

Phylogeny of the Animal Kingdom

animals. The reef builders occupy only the outer surface of a reef, which consists of dense layers of limestone built up over many generations. The algae add limestone to the reef surface when they die. The coral animals build skeletal support for themselves by secreting a layer of limestone. Gradually, over many years, as one generation of algae and coral animals is replaced by another, layer upon layer of limestone gradually build up, and a reef forms.

The photograph at the right gives you an impression of what coral animals look like up close. They live in interconnected colonies, with the individuals, each called a polyp, housed in pockets formed of their limestone skeletons. When feeding, the polyps (each only about 3 mm in diameter) extend the whitish tentacles you see here. Mucus on the tentacles traps fine organic food particles suspended in the seawater. When disturbed, the polyps pull back into the coral skeleton.

Coral animals also obtain some of their food from dinoflagellates that live in the polyp tissues. (Dinoflagellates are photosynthetic protists; see Module 16.25.) In turn, the dinoflagellates make use of some of the coral animals' waste products, such as carbon dioxide and nitrogen compounds. Coral animals can survive without their dinoflagellates, but they cannot form reefs without them. Apparently, the metabolic exchange between the animal and the protist makes it possible for the animal to produce enough limestone to build the reef. In a coral reef, we see many such mutually beneficial connections among organisms.

Mutually beneficial relationships between animals and other kinds of organisms are common in the living world. We encountered another example of such a relationship in Chapter 17, where we saw that interactions between flowering plants and pollinating animals account for much of plant diversity. The association between coral animals and dinoflagellates also contributes to diversity, helping create reef habitats for a host of animals. A diver on a reef might actually see animals representing as many as 27 of some 35 phyla in the kingdom **Animalia.**

In this chapter, we look at nine major animal phyla, those that contain the greatest number of species and that are the most abundant and widespread. Along the way, we will give special attention to the major milestones in animal evolution.

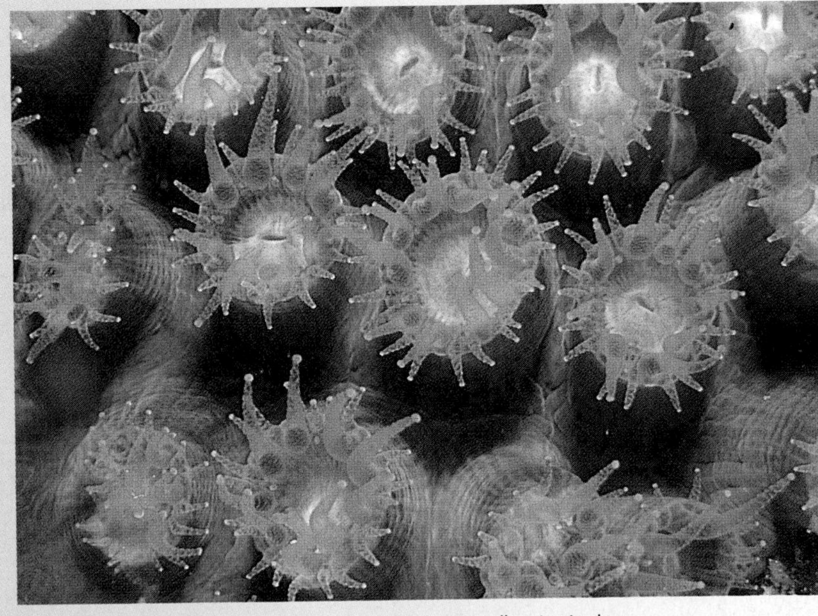

Coral polyps (each about 3 mm in diameter) extending tentacles

18.1 What is an animal?

We know when we see one—or do we? What about the coral animals on the previous page, or the sea star in the figure here? Can we put into words why we call these organisms animals? We can distinguish animals from organisms in the other four kingdoms very generally by saying that animals are eukaryotic, multicellular heterotrophs that lack cell walls. Let's start by examining that statement more closely.

First, being *eukaryotic* unites animals with protists, fungi, and plants, and separates them from all prokaryotes. Second, being *multicellular* separates animals, fungi, and plants from most protists, which are unicellular. Third, being *heterotrophic* separates animals and fungi from plants and plantlike protists (the algae), which are photosynthetic. Finally, *lacking cell walls* distinguishes animals from plants, algae, and fungi. So we have four features that, taken together, distinguish animals from other organisms.

Animals also have other distinctive features. Most animals take food molecules into their bodies and digest them there. And most animals have a digestive tract, an internal cavity or tube where hydrolytic enzymes are secreted and digestion occurs. At the cellular level, animals have unique types of junctions connecting the cells in their tissues (see Module 4.20).

Other important features show up in the life cycles of animals. Figure 18.1 shows the life cycle of a sea star. Except for the haploid eggs and sperm, all stages are diploid, as is true for most animals. ① Male and female adult animals make haploid gametes by meiosis, and ② an egg and a sperm fuse to produce a zygote. The zygote then undergoes mitosis and starts developing, going through a series of stages that are characteristic of the animal kingdom. ③ Early mitotic divisions lead to an embryonic stage called a **blastula,** common to all animals. Typically, the blastula consists of a single layer of cells surrounding a hollow cavity. ④ Later, in the sea star and many other animals, one side of the blastula folds inward, forming an embryonic stage called a **gastrula.** In its early stages, the gastrula looks like an indented blastula. ⑤ The gastrula develops into a saclike embryo with an opening at one end. At this point, the gastrula has an outer cell layer and an inner cell layer that results from infolding. Eventually, the outer layer develops into the animal's epidermis and nervous system, and the inner layer forms the digestive tract. Still later in development, in most animals, a third layer forms between the other two and develops into most of the other internal organs (not shown in the figure).

Following the gastrula, many animals develop directly into adults. But others, including the sea star, ⑥ develop into one or more larval stages first. A **larva** is an immature individual that looks very different from an adult. The larva ⑦ undergoes a major change of body form, called **metamorphosis,** in becoming an adult—a mature animal capable of reproducing sexually.

Thus, we have a number of characteristics that answer the question, What is an animal? To our earlier description, we can add that animals have unique intercellular junctions; generally ingest their food and digest it in an internal digestive tract; are diploid; and have unique embryonic stages, including a blastula, a gastrula with distinct layers of cells, and, in many species, a larva. How might organisms with these distinctive, animal-like features have evolved? We start to address this question in the next module.

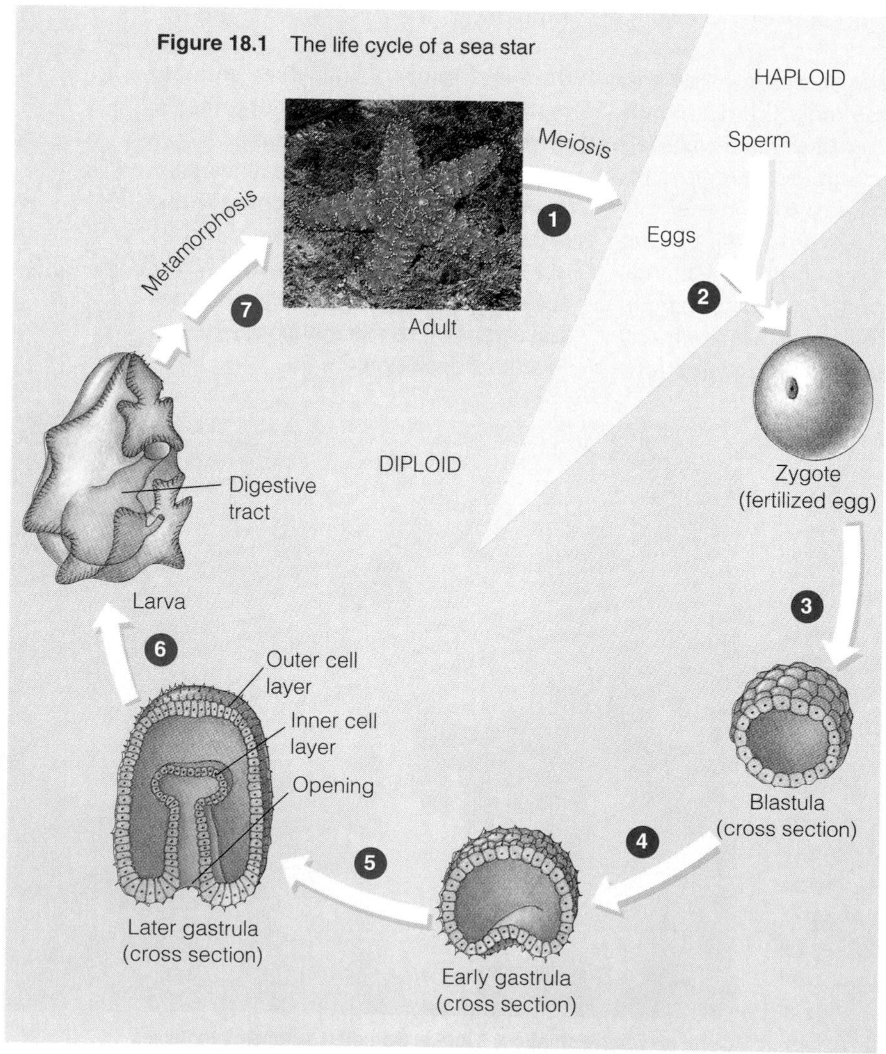

Figure 18.1 The life cycle of a sea star

? What embryonic stage is common to the development of all animals?

Blastula

The animal kingdom probably originated from colonial protists

Biologists have speculated about the origins of animals ever since Darwin's time. Fossils of the oldest known animals have been found in rocks from the late Precambrian era, about 700 million years ago. However, most of the modern animal phyla—the basic types of animals we see today—probably evolved during the Cambrian period, which began about 600 million years ago. Neither the Cambrian fossils nor the Precambrian ones tell us much about the earliest animals: All fossils found so far are the remains of animals whose body structure was too complex to have been the first animals. So let's speculate a bit.

Animals, like fungi and plants, are multicellular organisms. They probably evolved from protists that lived as colonies of cells. Figure 18.2 below shows a sequence of stages that may have led to the first animal. ① The earliest colonial aggregates may have been only a few cells, all of which were flagellated and virtually identical. Colonies form when cells divide but do not separate, and early colonies in the animal lineage may have grown larger by simply adding cells. At some point, colonies probably appeared that consisted of hundreds or thousands of cells. ② Such colonies may have been hollow spheres—floating aggregates of heterotrophic cells—that ingested organic nutrients from the water. ③ Eventually, cells in the colony may have undergone a division of labor, with some cells adapted for reproduction and others for somatic (nonreproductive) functions, such as locomotion and feeding. This simple way of dividing up the tasks of staying alive involves a degree of specialization—into two types of cells, reproductive and somatic.

A colony of cells, even one with some division of labor, is still a long way from a multicellular organism. The cells of truly multicellular organisms are highly specialized and interdependent. In addition, multiple layers of cells are a fundamental feature of animals, as we saw in Module 18.1. If and when a hollow colony became animal-like, it first had to develop internal layers of cells. Stage ④ in the figure shows how a simple multicellular organism with cell layers might have evolved from a hollow colony of heterotrophic cells. Layers may have developed as cells on one side of the colony folded inward, the way they do in the gastrula we discussed in Module 18.1. Originally, the infolding may have provided a temporary digestive cavity, a region where cells were specialized for feeding and digestion.

Eventually, as shown in stage ⑤, the infolding may have almost completely eliminated the original hollow space, producing an organism with two cell layers. The main advantage of the layers is to enable further division of labor among the cells. The outer flagellated cells would have provided locomotion and some protection, while the inner cells could have specialized in reproduction or feeding. Junctions holding the cells together and enabling the transport of nutrients among them may also have evolved by this time. It is probably appropriate to call organisms with this level of organization primitive animals, or at least "protoanimals" (organisms that gave rise to animals). Our protoanimal was probably a crawler rather than a floater. Even today, most animals live and feed on some type of surface, and the bottom of an ocean seems a likely candidate for a protoanimal's domain. With its specialized cells and a simple digestive tract formed by infolding, the protoanimal shown here could have fed on organic matter on the seafloor.

What protoanimals and their forerunners really looked like remains a mystery. The two main stages we have proposed—the hollow colony and the "infolded" stage—are modeled after the blastula and gastrula stages in animal development. Indeed, these embryonic stages may resemble organisms that were early ancestors of animals.

We now begin looking at some real animals. We turn first to those with relatively simple body organization and then proceed to more complicated ones. We'll focus on the major landmarks in animal evolution, as reflected in body features of animals living today. As we will see, most animals are **invertebrates,** so called because they lack a vertebral column (backbone).

What is the main difference between a colonial organism and an organism that is truly multicellular?

The cells of multicellular organisms are more extensively specialized and interdependent than are the cells of colonial organisms.

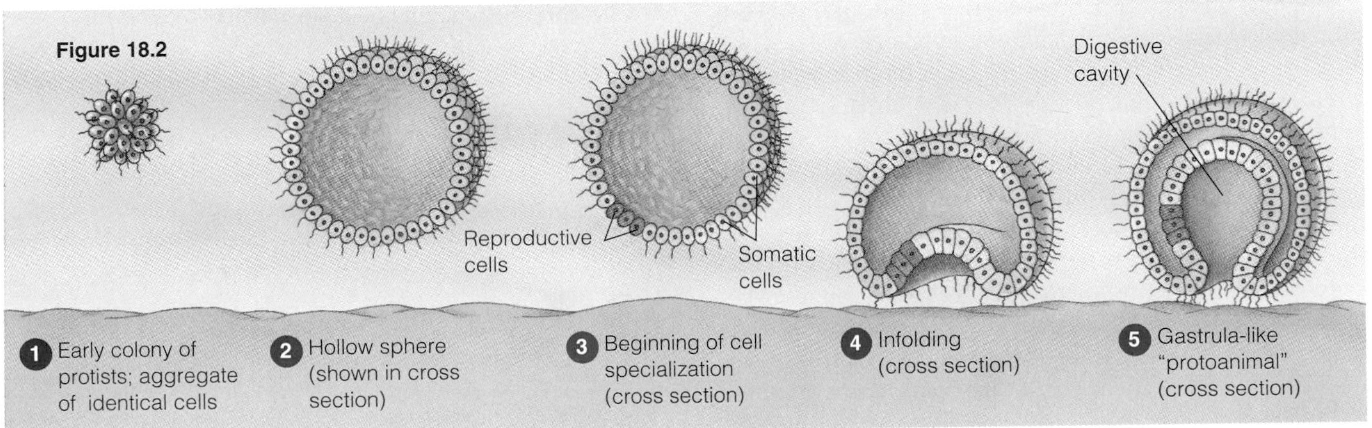

18.3 Sponges have relatively simple, porous bodies

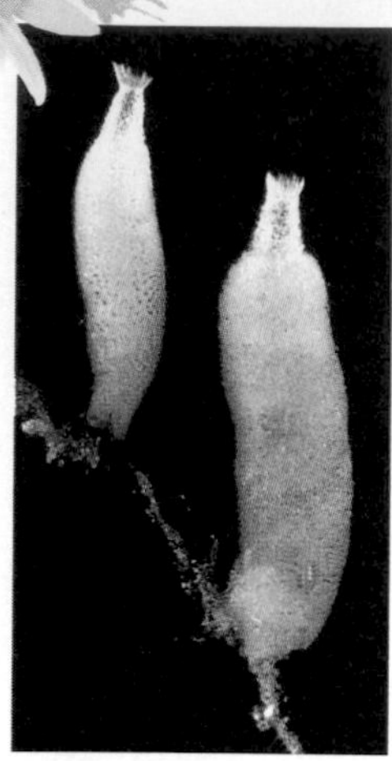

Figure 18.3A *Scypha* (each about 2 cm high)

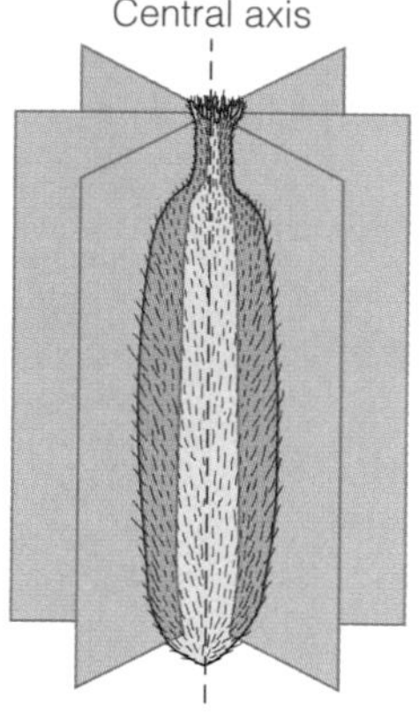

Figure 18.3B Radial symmetry

There are about 5000 species of **sponges** (phylum **Porifera**). Most are marine, but about 150 species live in fresh water. Sponges live singly or in clusters formed by budding. Most species have great powers of regeneration and often reproduce by this means. If a sponge body is fragmented into several pieces, each piece can often develop into a whole new sponge.

Figure 18.3A shows two individuals of the genus *Scypha,* a small marine sponge measuring only about 1–3 cm high. Certain other sponges reach heights of 2 m or spread irregularly over large areas on seafloors and lake bottoms. Vase-shaped or cylindrical sponges like *Scypha* have **radial symmetry.** This means that the body parts are arranged like pieces of a pie around an imaginary central axis. As Figure 18.3B shows, any imaginary slice passing longitudinally through the central axis of *Scypha* will divide it into mirror images.

Sponges are among the simplest animals. *Scypha,* for example, resembles a simple tube open at one end and perforated with holes. (Porifera means pore-bearers in Latin.) The entire body of *Scypha* and most other sponges consists of three loosely associated layers of cells, which you can see in Figure 18.3C. An outside layer of flattened cells, marked by numerous pores (pink), provides protection. A middle body layer has motile cells called **amoebocytes** (blue) and skeletal components (yellow) composed of either a flexible protein called spongin or mineral-containing particles. A third cell layer, consisting of flagellated cells called choanocytes (purple), lines a large central cavity. A **choanocyte,** also called a collar cell, has a collarlike ring surrounding the base of its flagellum. Aided by the flagella on choanocytes, water flows through the pores into the central cavity and then out of the sponge through its one large opening. More complex sponges have folded body walls and branching water canals.

Most sponges feed by collecting bacteria from water that streams through their porous bodies. A choanocyte traps bacteria in mucus on its collar and then engulfs the food by phagocytosis (see Module 5.19). The cells called amoebocytes are constantly on the move within the sponge body. They pick up food packaged in food vacuoles from the choanocytes, digest it, and carry the nutrients to other cells. Amoebocytes are the "do-all" cells of sponges. Moving about by means of pseudopodia, they digest and distribute food, transport oxygen, dispose of wastes, and manufacture skeletal elements. They can even change into other cell types.

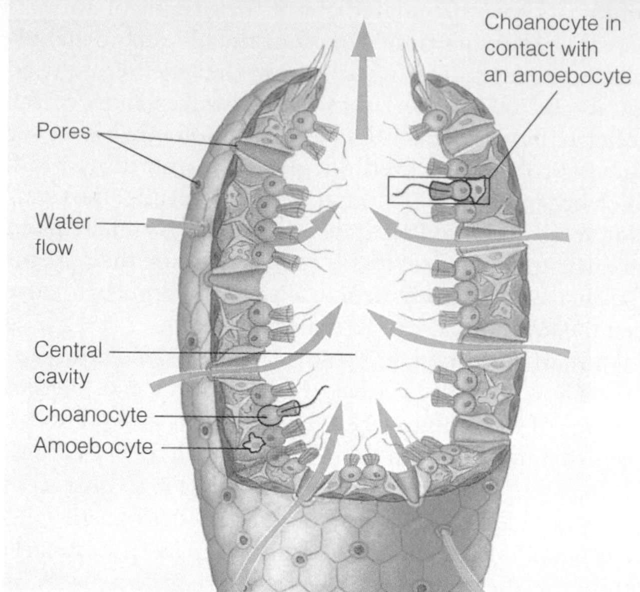

Figure 18.3C Structure and feeding of a simple sponge

Biologists generally believe the sponge line arose very early from the multicellular organisms that gave rise to the animal kingdom. Although sponges are clearly animals, they have several protist-like features. For example, sponges lack a digestive tract and digest their food inside their cells. Moreover, they lack a gastrula stage of development, their three cell layers are not homologous to the body layers of other animals, and they lack nerves and muscles.

Figure 18.3D is a drawing of a colonial protist called a **choanoflagellate.** Organisms like this one live at the bottom of ponds and shallow seas. Each colony consists of a small cluster of flagellated collar cells attached to a stalk, which is often anchored to a rock or submerged stick. Sponge choanocytes are remarkably similar to choanoflagellate cells, and it is likely that sponges arose from choanoflagellates that lived in late Precambrian seas. In fact, molecular studies point to these ancient choanoflagellates as the most likely ancestors of all animals. 

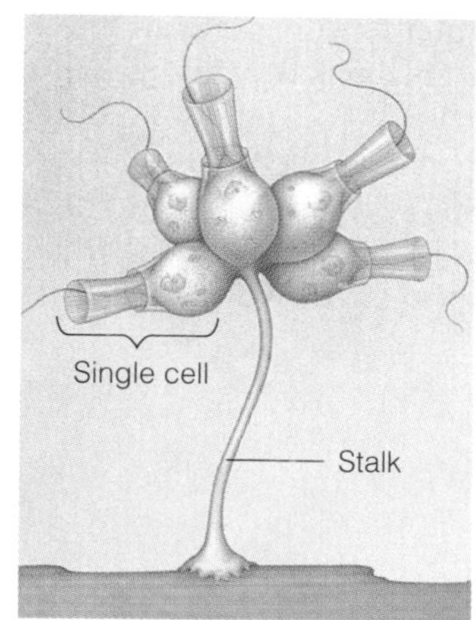

Figure 18.3D A choanoflagellate colony (about 0.02 mm high)

? Your circulatory system distributes nutrients, wastes, and oxygen between different parts of your body. How do sponges accomplish these transport functions?

Transport is carried out mainly by the mobile cells called amoebocytes.

Cnidarians are radial animals with stinging threads

18.4

18A
See Activity 18A on the CD-ROM.

Radial symmetry, common in sponges, is a hallmark of the phylum **Cnidaria:** the hydras, jellies (also called "jellyfish"), sea anemones, and corals. Cnidarians can have two kinds of radially symmetrical body forms. Hydras, common in freshwater ponds and lakes, have a cylindrical body with arms, called tentacles, projecting from one end (Figure 18.4A). This body form is a **polyp,** like the coral animals we saw earlier.

Figure 18.4A Polyp body form: a hydra (about 2–25 mm high)

The other type of cnidarian body is the **medusa,** exemplified by the marine jelly in Figure 18.4B. While polyps are mostly stationary, medusas move freely about in the water. They are shaped like umbrellas with fringes of tentacles around the lower edge. This type of radial body is unlike the hydra's cylindrical form, but any vertical cut through the center of the umbrella will divide it into mirror images. Some cnidarian life cycles include both a medusa and a polyp. Others, such as the sea nettle in Figure 18.4B, exist only as medusas; still others, such as hydras and sea anemones (Figure 18.4C), exist only as polyps.

Figure 18.4B Medusa body form: a marine jelly called a sea nettle (about 5 cm in diameter)

Figure 18.4C Sea anemones, such as this *Anthopleura* (about 6 cm in diameter), are all marine

Cnidarians are carnivores that use their tentacles to capture small animals and protists and to push the prey into their mouths. In a polyp, the mouth is on the top of the body, at the hub of the radiating tentacles (see Figure 21.3A). In a medusa, the mouth is in the center of the undersurface of the umbrella. In both polyp and medusa, the mouth leads into a digestive compartment called the **gastrovascular cavity** (Greek *gaster,* belly, and Latin *vas,* vessel). Undigested food and other wastes exit through the mouth; there is no anus, and therefore the digestive system is said to be incomplete. The cavity also circulates fluid that services internal cells (hence the "vascular" in gastrovascular; see Figure 23.2A). Fluid in the cavity provides body support and helps give cnidarians shape, much like water in a balloon.

If we were to list two traits that, taken together, mark an animal as a cnidarian, one would be radial symmetry. The other would be the specialized cells, called cnidocytes, for which the phylum is named. **Cnidocytes** (meaning "stinger cells"), found on the surface of the tentacles of polyps and medusas, function in defense and prey capture. Each cnidocyte contains a fine thread coiled within a capsule (Figure 18.4D). When it is discharged, the thread can sting or entangle prey. A hydra uses these threads to capture prey that swim close to its tentacles; certain large marine cnidarians often use their stinging threads to catch fish.

Cnidarians have several features that are absent in sponges but present in nearly all other animals. One is a digestive cavity; another is a gastrula stage in development; and a third is the presence of tissues. A sea anemone, for example, has well-developed nervous and muscle tissue, allowing it to creep, swim, roll, burrow, or somersault toward or away from external stimuli.

As a group, cnidarians are what we might call "tissue animals." Although sea anemones and certain jellies have organs, most vital functions in cnidarians are performed by individual tissues. In contrast, the rest of the animals we will discuss have most of their tissues organized into organs, and many of their organs organized into functional units called organ systems.

What are three functions of a cnidarian's gastrovascular cavity?

(1) Digestion; (2) circulation; and (3) physical support

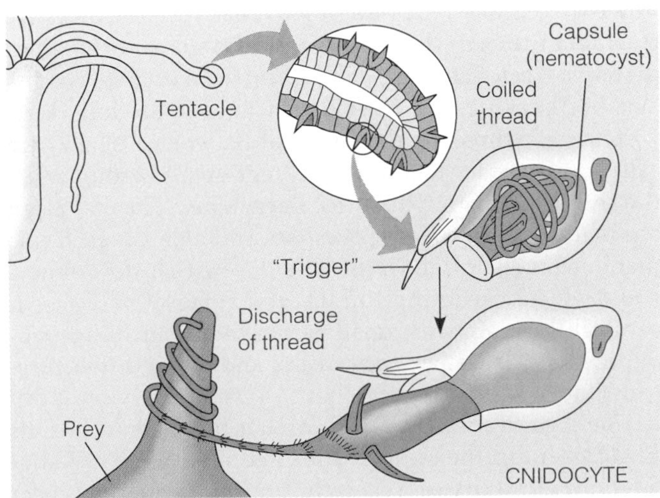

Figure 18.4D Cnidocyte action

18.5 Most animals are bilaterally symmetrical

18A

In contrast to radially symmetrical animals such as cnidarians, most animals are bilaterally symmetrical. **Bilateral symmetry** (Latin *bi-*, double, and *latus,* side) means that an animal can be divided equally by a single cut, as shown in Figure 18.5, and has mirror-image right and left sides. A bilaterally symmetrical animal, such as this crayfish, has a distinct head, or **anterior,** end, and tail, or **posterior,** end. It also has a back, or **dorsal,** surface; a bottom, or **ventral,** surface; and two side, or **lateral,** surfaces. The head is a prominent part of a bilaterally symmetrical animal. It houses its main sensory structures (such as eyes), its brain, and usually its mouth. The brain and sensory structures are organs that, along with nerves that branch throughout the body, form an organ system—the nervous system.

Bilateral animals are fundamentally different from radial ones. A radial animal lacks a head or any forward orientation of its body and typically spends most of its time sitting on the seafloor or drifting about in water currents. In contrast, most bilaterally symmetrical animals are quite active and travel head-first through the environment. Their eyes and other sense organs are up front on the head, where they contact the environment first and help the animal respond appropriately. Humans are bilateral, but because we walk on two feet instead of four, we move ventral-surface–first instead of head-first.

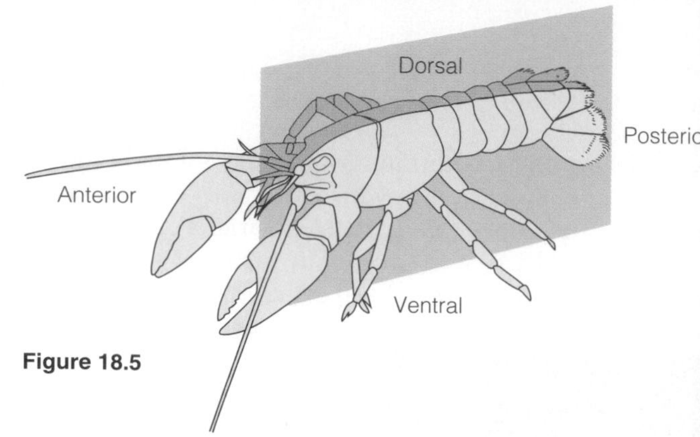

Figure 18.5

Bilateral symmetry and a head end are prerequisites for the forward movement typical of animals. Most animals crawl, walk, run, burrow, swim, or fly in a head-first direction. The evolution of bilateral symmetry, a head end, and forward movement were important milestones in the history of the animal kingdom.

? Head end is to __________ as __________ end is to posterior.

anterior . . . tail (or rear)

18.6 Flatworms are the simplest bilateral animals

18A

Flatworms, of the phylum **Platyhelminthes** (Greek *platys,* flat, and *helmis,* worm), are leaflike or ribbonlike animals, ranging in length from about 1 mm to 20 m. They are bilaterally symmetrical, but with bodies that are unusually simple for bilateral animals. In common with cnidarians, most flatworms have an incomplete digestive tract (no anus). Most bilateral animals, by contrast, have a complete digestive tract with both a mouth and an anus. Also in common with cnidarians, the digestive cavity is the only space inside the flatworm body. Typical bilateral animals have another space, called the body cavity, between the digestive tract and the body wall. Except for the space *within* their digestive tract, flatworms lack an internal cavity. (We discuss body cavities and their significance in Module 18.7.)

There are three major groups of flatworms. The worm called a planarian (Figure 18.6A) represents a group called the **free-living** (nonparasitic) **flatworms.** The planarian has a head with two large eyespots and a flap at each side that detects chemicals in the water. Dense clusters of nervous tissue form a simple brain, and a pair of nerve cords connect the brain with small nerves that branch throughout the body. Together, the sensory and nervous structures constitute a nervous system.

The digestive tract of a planarian is highly branched. Its single opening, the mouth, is located not on the head but on the ventral surface. When the animal feeds, a muscular tube projects through the mouth (as shown in the figure) and pulls food in. Planarians live on the undersurfaces of rocks in freshwater ponds and streams. Using cilia on their ventral surface, they crawl about in search of food. They also have muscles that enable them to twist and turn.

A second group of flatworms, the **flukes,** are parasites. The photograph in Figure 18.6B shows a male and female blood fluke *(Schistosoma)*. The female spends much of her time in a groove running the length of the body of the

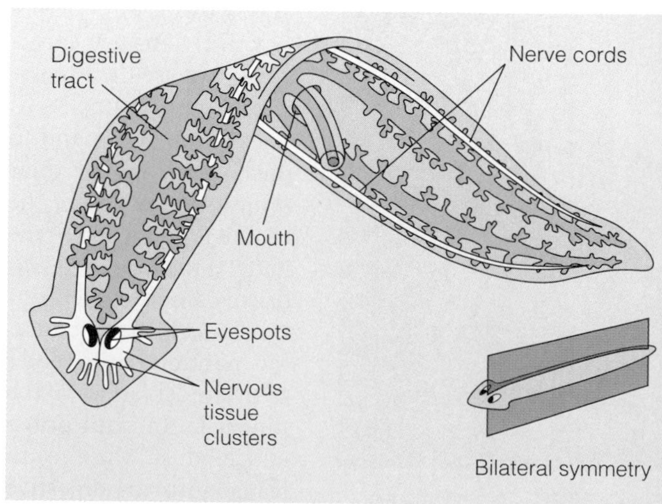

Figure 18.6A A free-living flatworm, the planarian (most are about 5–10 mm long)

larger male. In this position, the worms copulate frequently, and a single pair can produce over a thousand eggs a day. Both the female and the male have suckers that attach to the inside of the blood vessels near the host's intestines. Blood flukes infect humans and cause a severe, long-lasting disease called schistosomiasis (blood fluke disease). Blood fluke disease is widespread in Africa, Southeast Asia, and South America. It afflicts some 250 million people in 70 countries, causing severe abdominal pain, anemia, and dysentery.

Most flukes, including *Schistosoma,* have a complex life cycle that includes reproduction in more than one host. As outlined in Figure 18.6B, ① blood flukes living in a human host reproduce sexually, and fertilized eggs pass out in the host's feces. If an egg lands in a pond or stream, ② a ciliated larva hatches and ③ can enter a snail, the next host. ④ Asexual reproduction in the snail eventually produces ⑤ other larvae that can infect humans. ⑥ A person becomes infected when these larvae penetrate the skin.

Tapeworms, which are also parasitic, make up the third group of flatworms. Adult tapeworms inhabit the digestive

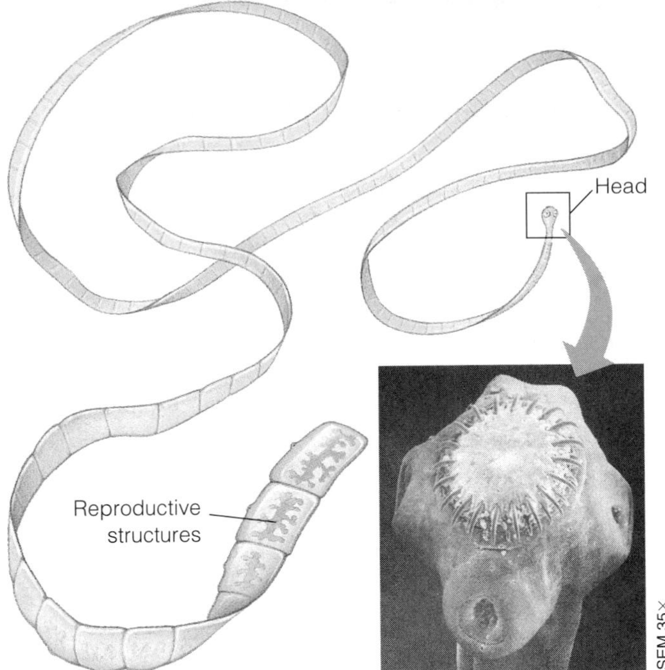

Figure 18.6C A tapeworm, a parasitic flatworm

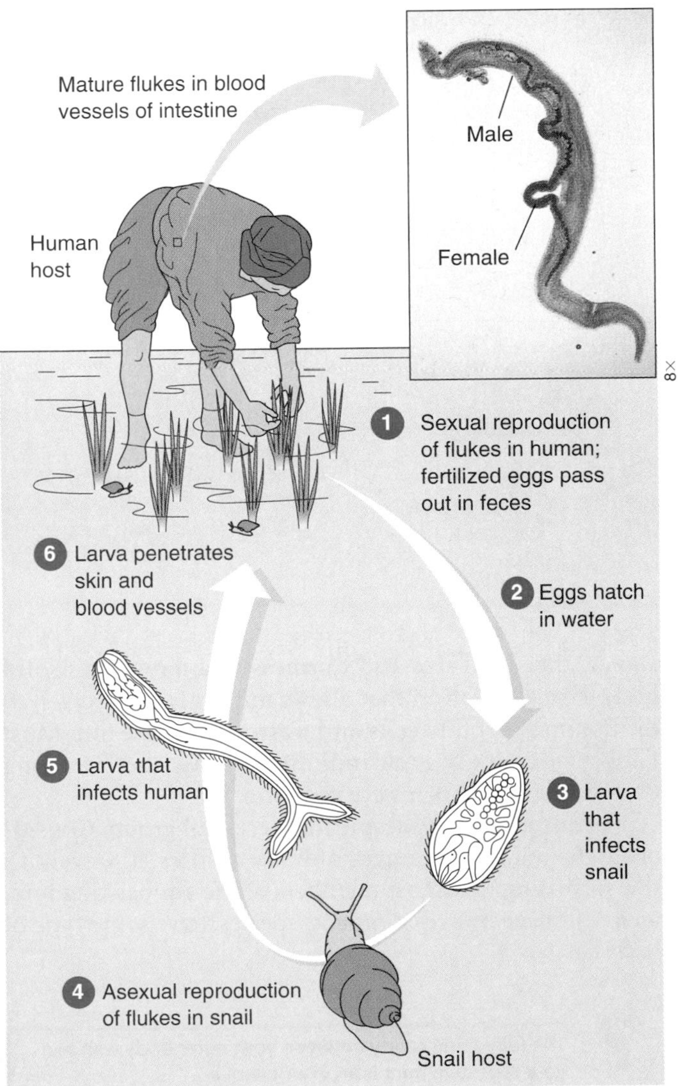

Figure 18.6B A fluke *(Schistosoma)* and its life cycle (adults are about 1 cm long)

tracts of vertebrate animals, including reptiles, birds, and mammals. In contrast to planarians and flukes, most tapeworms have a very long, ribbonlike body with repeated parts. They also differ from other flatworms in not having any digestive tract at all. Living in partially digested food in the intestines of their hosts, they simply absorb nutrients across their body surface. As the drawing in Figure 18.6C above shows, the head is the smallest part of the tapeworm body and is armed with suckers and teeth that grasp the host. Behind the worm's head, a short neck generates the repeated parts of the worm. The youngest part is just behind the neck, and the oldest is at the posterior end. The repeated parts are filled with both male and female reproductive structures. Full of ripe eggs, those at the posterior end break off and pass out of the host's body in feces.

Like parasitic flukes, tapeworms have a complex life cycle, usually involving more than one host. Most species benefit from the predator-prey relationships of their hosts. A prey species—a sheep or a rabbit, for example—may become infected by eating grass contaminated with tapeworm eggs. Larval tapeworms develop in these hosts, and a predator—a coyote or a dog, for instance—becomes infected when it eats an infected prey animal. The adult tapeworms develop in the predator's intestine.

Several kinds of tapeworms infect humans. We can be infected, for example, by a large tapeworm called *Taeniarhynchus* by eating rare beef infected with the worm's larvae. The larvae are microscopic, but the adults can reach lengths of 20 m in the human intestine. An orally administered drug called niclosamide kills the adult worms.

? Flatworms and cnidarians differ in symmetry, with flatworms being __________ and cnidarians being __________, but the animals of both phyla have __________ cavities.

bilateral . . . radial . . . gastrovascular

18.7 Most animals have a body cavity

18A

The evolution of a **body cavity,** a fluid-filled space between the digestive tract and the body wall, was significant in animal history. Sponges, cnidarians, and flatworms lack a body cavity, but nearly all other animals have one.

The figures at the right compare the internal structure of three animals. In all three cross sections, the colors indicate the same tissue layers: the outside covering (blue), a middle region (pink), and the lining of the digestive tract (yellow). These tissue layers form from the cell layers of the gastrula stage during embryonic development (see Figure 18.1).

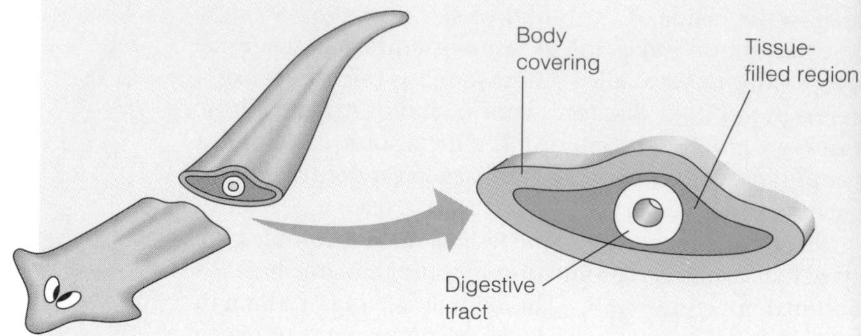

Figure 18.7A A flatworm

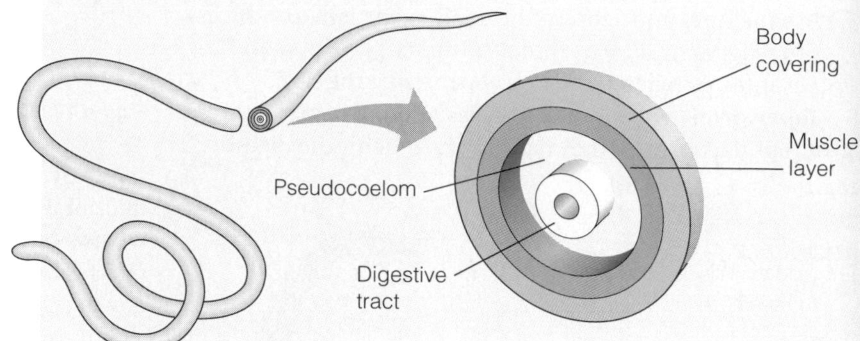

Figure 18.7B A roundworm

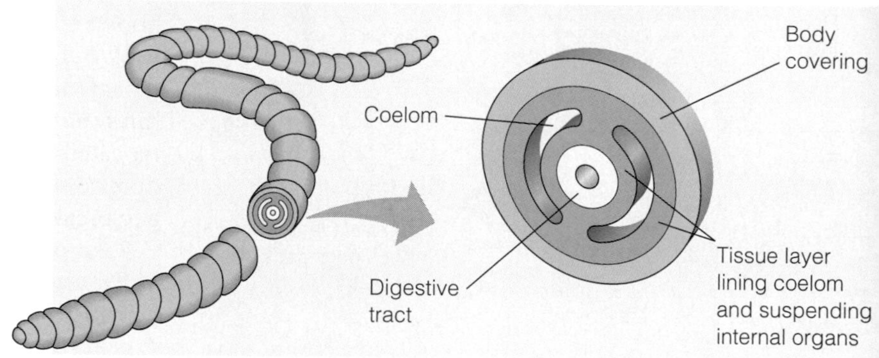

Figure 18.7C An earthworm

The cross section through a flatworm (Figure 18.7A) reveals a body that is solid—filled with cells of the three tissue layers—except for the cavity of the digestive tract. The other two animals have a body cavity. The roundworm (Figure 18.7B) has a body cavity called a pseudocoelom. A **pseudocoelom** (Greek *pseudes,* false, and *koilos,* hollow) is an internal space in direct contact with the wall of the digestive tract. The outer edge of the pseudocoelom contacts a muscle layer that is part of the body wall.

The third animal, an earthworm (Figure 18.7C), has a more complex body than either the flatworm or the roundworm. Its body cavity, called a **coelom,** is completely lined by the middle tissue layer (shown in pink). This tissue layer extends from the body wall and wraps around the digestive tract. In effect, this middle tissue layer suspends the digestive tract and other internal organs from the body wall.

There are many advantages to having a body cavity. Having an internal cavity, instead of being solid, makes an animal more flexible and thus better able to crawl and burrow. It also allows the internal organs to grow and move independently of the outer body wall. The fluid in the body cavity cushions the internal organs, helping prevent internal injury when the animal receives a sharp blow or is pinched. In an animal with a hard skeleton—especially one with internal bones like ours—even mild exercise could harm internal organs if it were not for the fluid in the coelom. In soft-bodied animals such as earthworms, fluid in the body cavity functions as a skeleton against which muscles can exert force to move the body. The fluid may also help circulate nutrients and O_2 throughout the body and assist in waste disposal. Amoeboid cells in the fluid help with these functions.

There is a connection between an animal's shape and size and the presence or absence of a body cavity. For instance, the small size and thinness of a flatworm's solid body is an adaptation that allows nutrients and oxygen to diffuse into all body cells and wastes to diffuse out. Most flatworms are only a few millimeters long, and even long ones such as tapeworms are very thin.

Flatworms are a widespread, successful group. Considering the many advantages of body cavities, however, it's not surprising that most members of the animal kingdom, an assemblage of over a million species, have some type of body cavity.

The fully-lined cavity between your outer body wall and your digestive tract is an example of a ________.

coelom

Roundworms have a pseudocoelom and a complete digestive tract 18.8

18A

Roundworms, also called nematodes, make up the phylum **Nematoda.** These are cylindrical worms with a finely tapered tail; the head is more blunt, as you can see at the right end of the worm in Figure 18.8A. The nematode body is covered by a tough, nonliving skin, or **cuticle,** that resists drying and crushing. The roundworm here has a transparent cuticle, and you can see some of its internal organs.

Nematodes have a pseudocoelom. They also have a complete digestive tract, extending as a straight tube from a mouth at the tip of the head to an anus near the tip of the tail. Food travels only one way through the system, and regions are specialized for certain functions. In animals with a complete digestive tract, the anterior region of the tract churns and mixes food with enzymes, while the posterior region absorbs nutrients and disposes of wastes. This division of labor allows each part of the digestive tract to be highly efficient at its particular function.

Nematodes are among the most numerous of all animals in both number of species and number of individuals. Nematodes live virtually every place there is rotting organic matter, and these worms are important decomposers in soil and on the bottom of lakes and oceans. Other nematodes thrive as parasites in the moist tissues of plants and in the body fluids and tissues of animals. About 90,000 species are known, and it is likely that at least ten times that number actually exist.

Little is known about most free-living nematodes. A notable exception is the soil-dwelling species *Caenorhabditis elegans,* an important research organism and one of the best-understood animals. An adult *C. elegans* consists of only about 1000 cells—in contrast to the human body, which consists of some 60 trillion cells. Because of the simplicity of this worm, researchers have been able to trace the lineage of individual cells in the adult back to parts of the zygote. These studies are contributing to our understanding of how genes control animal development.

Though free-living nematodes are the most abundant, many species of roundworms are serious agricultural pests that attack the roots of plants or parasitize animals. Humans are host to at least 50 species of roundworms, including a number of disease-causing organisms that are responsible for major health problems. Among these are hookworms, nematodes that attach to the intestinal wall and suck blood. Humans, dogs, cats, and many other mammals are susceptible to hookworms. Nematodes called heartworms are deadly to dogs. Spread by mosquitoes and also infectious to humans, heartworms seem to be on the increase in the United States.

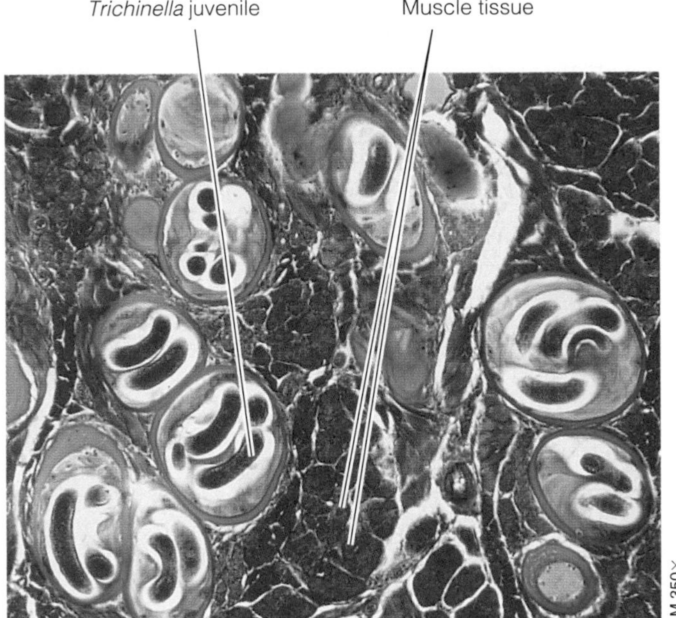

Figure 18.8B A parasitic roundworm *(Trichinella spiralis);* juvenile worms (sausagelike in this microscopic section) encysted in muscle

One of the most notorious roundworms is *Trichinella spiralis,* which causes a disease called trichinosis in a wide variety of mammals, including humans. People usually acquire the worms by eating undercooked pork containing the juvenile worms. You can see some of these juveniles in the section of muscle shown in Figure 18.8B. Trichinosis causes severe nausea and sometimes death when large numbers of the worms penetrate heart muscle. Cooking meat until it is no longer pink kills the worms.

We might expect that as numerous and widespread an animal group as the nematodes would include a great diversity of body form. In fact, the opposite is true. Most species of nematodes look very much alike: cylindrical, with a blunt head and a tapered tail. In sharp contrast, animals in the phylum Mollusca, which we examine next, exhibit enormous diversity in body form.

Figure 18.8A A free-living roundworm

Why is it a mistake for someone who eats pork chops to order them "rare" in a restaurant?

Incomplete cooking doesn't kill nematodes and other parasites that might be present in the meat.

18.9 Diverse mollusks are variations on a common body plan

18A

Snails, slugs, oysters, clams, octopuses, and squids are just a few of the great variety of animals known as **mollusks.** The phylum **Mollusca** comprises more than 150,000 species. Most mollusks have soft bodies protected by a hard shell; their phylum name comes from the Latin *molluscus,* meaning soft.

It may seem that animals as different as squids and clams could not belong in the same phylum, but these and other mollusks have inherited several common features from their ancestors. Figure 18.9A illustrates the basic body plan of a mollusk. Two hallmarks of the phylum are the muscular **foot** (gray in the drawing), which functions in locomotion, and the **mantle** (purple), an outgrowth of the body surface that drapes over the animal. A key molluskan feature, the mantle produces the shell in mollusks such as clams and snails. The mantle also functions in respiration, waste disposal, and sensory reception. In addition, the mantle creates another distinctive feature, the mantle cavity. In many mollusks, the mantle cavity houses a gill (left side in Figure 18.9A). The gill extracts oxygen dissolved in the water and may also dispose of fluid wastes.

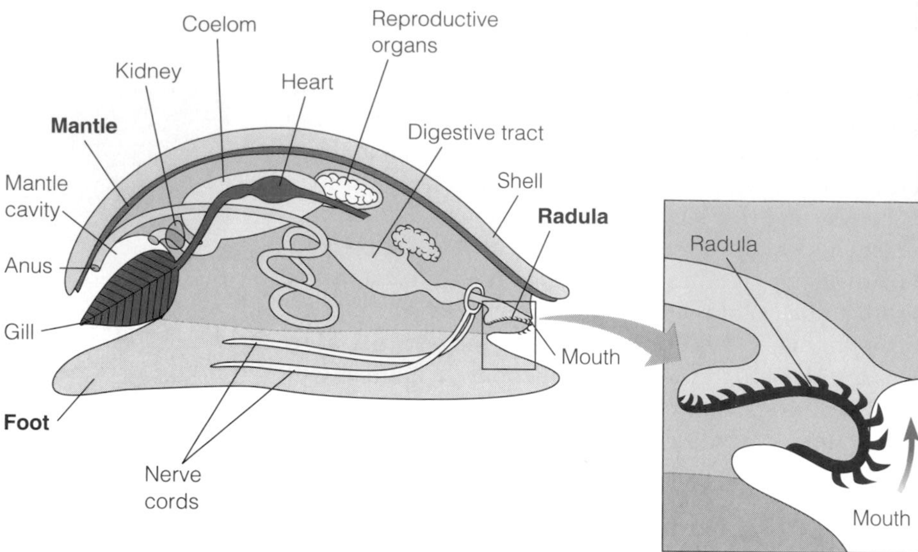

Figure 18.9A The general body plan of a mollusk

Figure 18.9A shows yet another body feature found in many mollusks—a unique rasping organ called a **radula,** which is used to scrape up food. In a snail, for example, the radula extends from the mouth and slides back and forth like a backhoe, scraping and scooping algae off rocks.

In contrast to roundworms and flatworms, mollusks have a coelom (introduced in Module 18.7). Their coelom (light blue in Figure 18.9A) consists of three small cavities: one each around the heart and reproductive organs, and one that forms part of the kidney. Also unlike roundworms and flatworms, mollusks have a circulatory system—an organ system that distributes materials such as nutrients and water throughout the body. Only the heart and main vessels are shown in the diagram, but the circulatory system branches throughout the body and has taken over the functions of a large coelom.

The basic body features have evolved in markedly different ways in different groups of mollusks. The three most diverse groups (classes) are the gastropods (including snails and slugs), bivalves (such as clams, scallops, and oysters), and cephalopods (including squids and octopuses).

Gastropods (Greek *gaster,* belly, and *pous,* foot), the largest group of mollusks, with over 75,000 species, are found in fresh water, salt water, and terrestrial environments; in fact, they are the only mollusks that live on land. Terrestrial slugs and snails like the one in Figure 18.9B lack the gills typical of aquatic mollusks; instead, their mantle

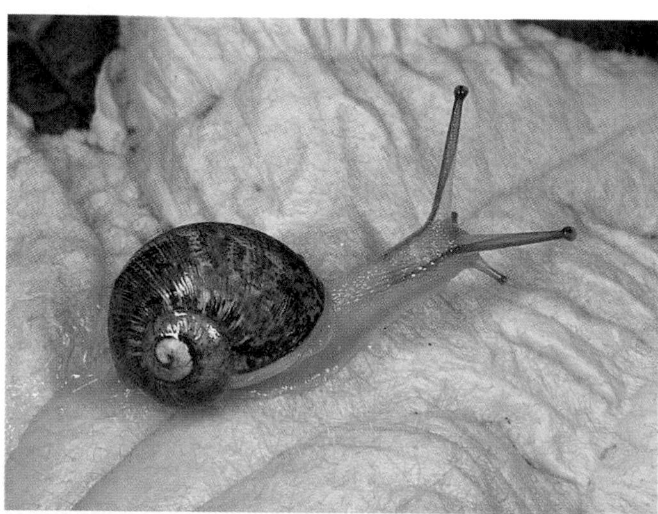

Figure 18.9B A terrestrial gastropod: a land snail

Figure 18.9C A marine gastropod: a sea slug

cavity has evolved into a lung that breathes air. Most gastropods are marine, and the group includes some of the most colorful animals in the sea. The gastropod shown in Figure 18.9C (about 5 cm long) is called a sea slug. Sea slugs are unusual mollusks in that they lack a mantle, mantle cavity, and shell. The long projections on this species serve as gills, obtaining oxygen and disposing of wastes.

Bivalves (Latin *bi-*, double, and *valva*, "leaf of a folding door"), including numerous species of clams, oysters, mussels, and scallops, have shells divided into two halves that are hinged together. Most bivalves are sedentary, living in sand or mud. They use their muscular foot for digging and anchoring, and most have mucus-coated gills that trap fine food particles suspended in the water. The scallop in Figure 18.9D (about 10 cm in diameter) is an unusual bivalve in that it sits on the ocean floor rather than digging into it. Notice the many bluish eyes peering out between the two halves of the hinged shell. The eyes are set into the fringed edges of the animal's mantle. When the eyes detect movement nearby, or if a predator touches one of the long projections of the mantle shown in the photograph, the scallop can jet a short distance away by clapping its valves together and squirting water out of its mantle cavity.

Cephalopods (Greek *kephale*, head, and *pous*, foot) generally differ from gastropods and sedentary bivalves in being built for speed and agility. A few cephalopods have large, heavy shells, but in most the shell is small and internal (as in squids) or missing altogether (as in octopuses). Cephalopods are marine predators, using beaklike jaws and a radula to crush or rip prey apart. Their mouth is at the base of their foot, which is drawn out into several long tentacles for catching and holding prey.

The squid in Figure 18.9E (about 20 cm long) ranks with fishes as a fast, streamlined predator. It darts about by drawing water into its mantle cavity and then firing a jet of water back out. Squids have a large, complex brain, rivaling that of fishes, and their eyes (one on each side of the body) are among the most complex sense organs in the animal kingdom. Each squid eye has a lens that focuses light and a retina on which clear images form.

All cephalopods have large brains and sophisticated sense organs, and these contribute to their being successful, mobile predators. The octopus in Figure 18.9F (about 30 cm long) lives on the seafloor, where it scurries about in search of crabs and other food. Its brain is larger and more complex, proportionate to body size, than that of any other animals except the vertebrates. Octopuses are highly intelligent and have shown remarkable learning abilities in laboratory experiments. 

?

As representatives of classes of mollusks, a garden snail is an example of a __________; a clam is an example of a __________; and a squid is an example of a __________.

gastropod . . . bivalve . . . cephalopod

Figure 18.9D A bivalve: a scallop

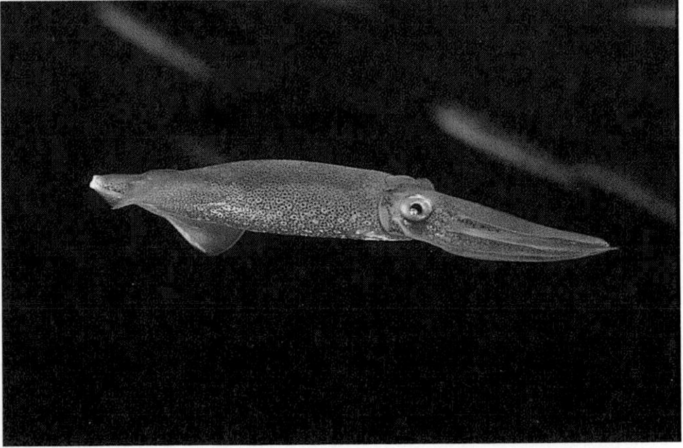

Figure 18.9E A cephalopod with an internal shell: a squid

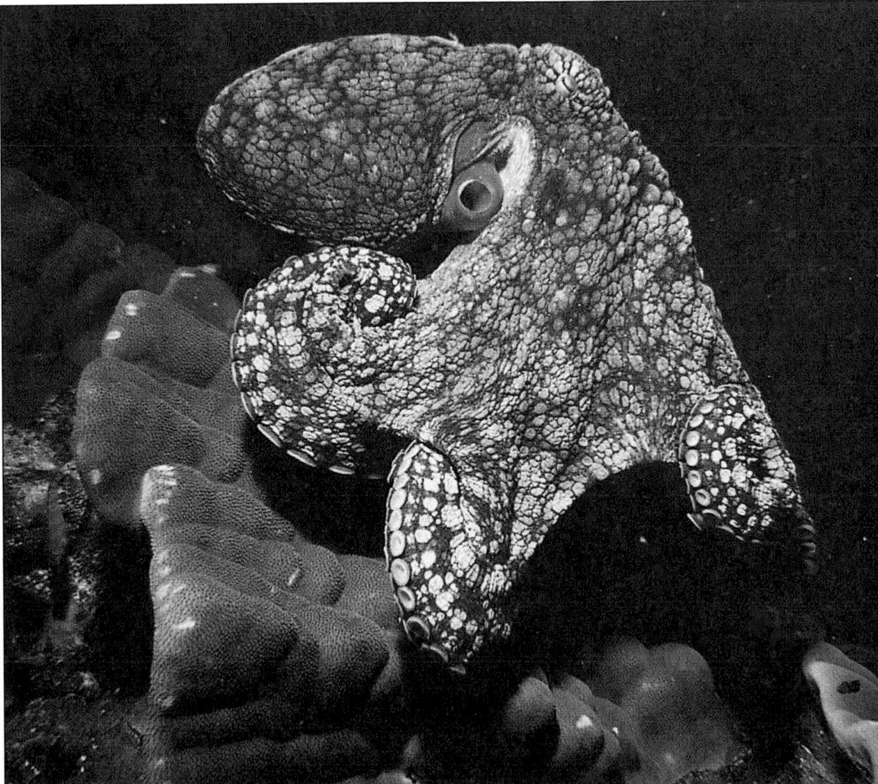

Figure 18.9F A cephalopod without a shell: an octopus

18.10 Most animals have segmented bodies

18A

We now begin to discuss another major body feature that has had profound influence on animal history—body **segmentation,** the subdivision of the body along its length into a series of repeated parts (segments). This feature played a central role in the evolution of many complex animals.

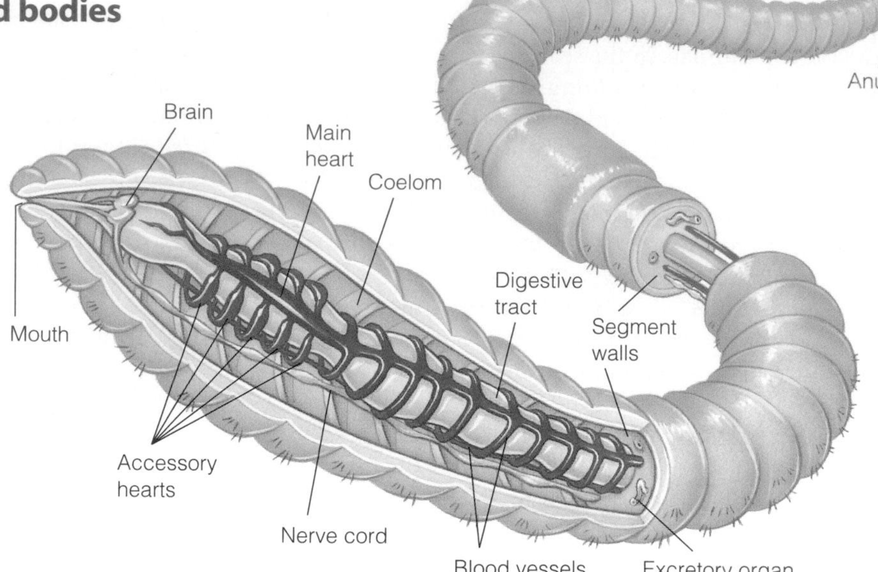

Figure 18.10A Segmentation in an earthworm

Segmentation is an obvious feature of an animal like an earthworm (Figure 18.10A), in which the segments are marked off externally by grooved rings. Internally, the coelom is partitioned by walls (only two are fully shown here). The nervous system (yellow) includes a ventral nerve cord with a cluster of nerve cells in each segment. Excretory organs (green), which dispose of fluid wastes, are also repeated in each segment. The digestive tract, however, is not segmented; it passes through the segment walls from the mouth to the anus. The main channels of the circulatory system—a dorsal blood vessel and a ventral blood vessel—are also unsegmented. But they are connected by segmental vessels, including five pairs of accessory hearts near the anterior end. The main heart is simply the enlarged anterior region of the dorsal blood vessel.

The dragonfly (Figure 18.10B) is also segmented, though less uniformly than the earthworm. Its segments are most pronounced in its abdomen; its head and midregion (thorax) are each formed from several fused segments. Each pair of its six walking legs and each pair of its four wings emerge from a body segment in the thorax.

Segmentation also occurs in the human body (Figure 18.10C), although our body segments are not homologous with those of earthworms or insects. We have a backbone formed of a repeated series of bones called vertebrae, and muscles associated with our vertebrae are segmented. We also have segmented abdominal muscles, clearly visible in body builders as "stomach ripples."

A segmented body is advantageous in many ways. It allows great body flexibility and mobility, and it probably evolved as an adaptation for movement. The earthworm uses its flexible, segmented body to crawl over wet surfaces and burrow rapidly into the soil. In the dragonfly, segmentation occurs in both its tough external skeleton and the muscles that move its head, thorax, and abdomen. The segments provide flexibility for flying, perching, mating, and laying eggs.

What feature of your skeleton is an example of segmentation?

The vertebrae of your backbone

Figure 18.10B Dragonfly segmentation

Figure 18.10C Indications of segmentation in the human body

Earthworms and other annelids are segmented worms 18.11

18A

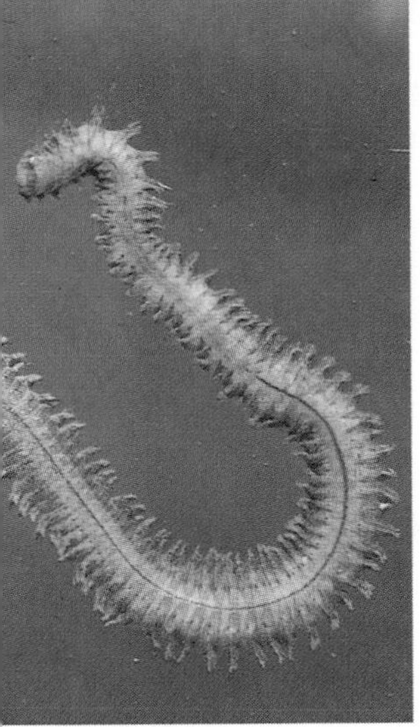
Figure 18.11A A sandworm (about 15 cm long), a polychaete that lives on the seafloor

Figure 18.11B A Christmas tree worm, a tube-building polychaete (5–10 cm long)

Phylum **Annelida** (Latin *anellus,* ring), comprising earthworms and other segmented worms, is named for its characteristic ringlike body parts. Except for a distinct head and tail, a typical **annelid** consists of body segments that are all very similar.

There are about 15,000 annelid species, ranging in length from less than 1 mm to a 3-m-long giant earthworm that lives in Australia. Annelids live in the sea, in most freshwater habitats, and in damp soil. Some aquatic annelids swim in pursuit of food, but most are bottom-dwelling scavengers that burrow in sand and mud.

Earthworms are one of three large groups of annelids. An earthworm eats its way through the soil, extracting nutrients as soil passes through its digestive tube. Undigested material passes out the anus at the posterior end of the worm. Farmers value earthworms because the animals actually till the soil and because earthworm feces improve the soil's texture. Darwin estimated that a single acre of British farmland had about 50,000 earthworms, producing 18 tons of feces per year.

The largest group of annelids are **polychaetes** (Greek *polys,* many, and *chaeta,* hair). Figure 18.11A shows a polychaete called a sandworm (about 15 cm long), which lives on the seafloor. Segmental appendages and hard bristles that project from them help the worm wriggle about in search of small invertebrates to eat. The appendages also increase the animal's surface area for taking up oxygen and disposing of wastes. Most polychaetes are marine, although a few inhabit estuaries and fresh water. Many marine species live in tubes and extend feathery appendages that trap suspended food particles. Tube-dwellers usually build their tubes out of sticky proteins secreted by glands near their mouth. Some, such as the Christmas tree worm (Figure 18.11B), bore their tubes in the limestone of coral reefs. The brightly colored spire you see in the photograph is made up of feeding appendages extending from the worm's head. The funnel-like projection on the right is a stopper that seals off the tube when the worm withdraws.

Leeches make up a third large group of annelids (Figure 18.11C). Leeches are notorious for their blood-sucking habits. However, most species are free-living carnivores that eat small invertebrates such as snails and insects. The majority of leeches inhabit fresh water, but a few terrestrial species inhabit moist vegetation in the tropics.

Until this century, blood-sucking leeches were frequently used by physicians for bloodletting, removing what was considered "bad blood" from sick patients. Some leeches have razorlike jaws that cut through the skin, and they secrete saliva containing a strong anesthetic and an anticoagulant into the wound. The anesthetic makes the bite virtually painless, and the anticoagulant keeps the blood from clotting. Leech anticoagulant is now being produced commercially by genetic engineering and may find wide use in human medicine. Tests show that it prevents blood clots that can cause heart attacks.

Leeches are still occasionally used to remove blood from bruised tissues and to help relieve swelling in fingers or toes that have been sewn back on after accidents. Blood tends to accumulate and cause swelling in a reattached finger or toe until small veins have a chance to grow back into it. Leeches are applied to remove the excess blood.

What is the main difference between the digestive tract of an earthworm and the gastrovascular cavity of a cnidarian, such as a jelly?

A complete digestive tract with separate mouth and anus for the earthworm; a digestive sac with a single opening for the jelly

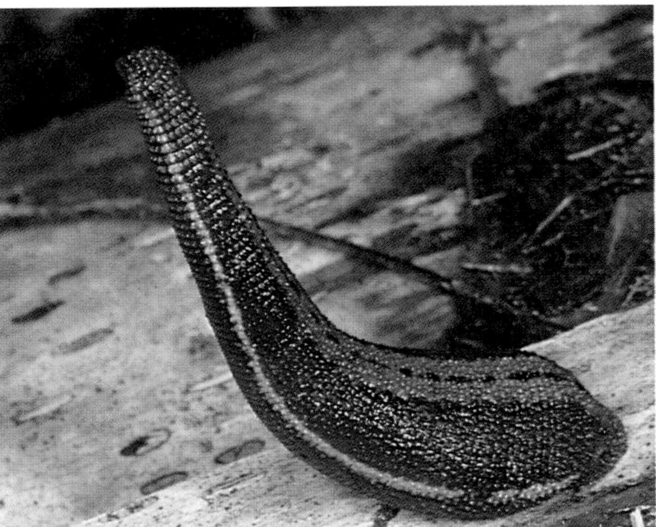
Figure 18.11C A terrestrial leech (about 3 cm long), common in tropical forests

18.12 Arthropods are the most numerous and widespread of all animals

18A

Nearly a million types of segmented animals—including crayfish, lobsters, crabs, barnacles, spiders, ticks, and insects—are members of the phylum **Arthropoda.** It is estimated that the arthropod population of the world numbers about a billion billion (10^{18}) individuals! In terms of species diversity, geographical distribution, and sheer numbers, Arthropoda is the most successful phylum of animals that has ever existed.

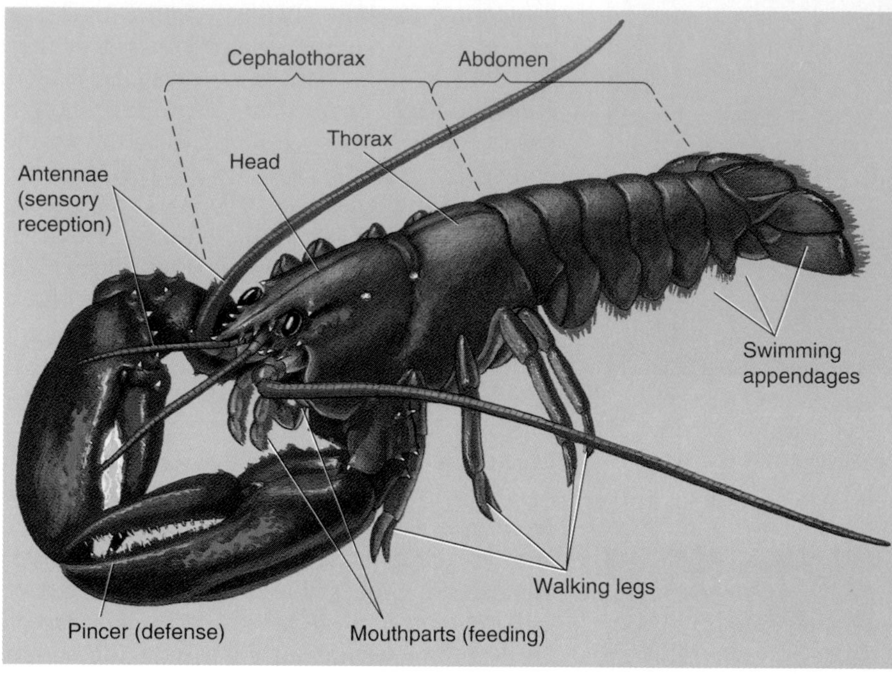

Figure 18.12A The structure of an arthropod, a lobster

According to one hypothesis, **arthropods** evolved from annelids or from a segmented ancestor of annelids. A number of fossils from the Cambrian period (some 550 million years ago) seem intermediate between annelids and arthropods. However, molecular evidence indicates otherwise—that arthropods and annelids evolved as separate branches from earlier bilateral animals, and that segmentation first arose in those animals.

Arthropods are equipped with jointed appendages, for which the phylum is named (Greek *arthron,* joint, and *pous,* foot). As indicated in the drawing of a lobster in Figure 18.12A, the appendages are variously adapted for walking, swimming, feeding, sensory reception, and defense. The arthropod body, including the appendages, is covered by a hard external skeleton, the **exoskeleton.** The exoskeleton consists of layers of chitin, a polysaccharide, mixed with proteins. It protects the animal and provides points of attachment for the muscles that move the appendages. The exoskeleton is thick around the head, where its main function is to house and protect the brain. It is paper-thin and flexible in many other locations, such as the joints of the legs. To grow, an arthropod must periodically shed its old exoskeleton and secrete a larger one, a complex process called **molting.**

In contrast to annelids, which have similar segments throughout the body, the body of most arthropods is formed of several distinct groups of segments. The lobster, for example, has three groups: head, thorax, and abdomen. (Actually, as shown, the exoskeleton of the head and thorax are partly fused, forming what is called the cephalothorax.) Each of the segment groups is specialized for a different function. The head bears sensory antennae, eyes, and jointed mouthparts underneath. The thorax bears a pair of defensive appendages (the pincers) and four pairs of legs for walking. The abdomen has swimming appendages.

Figures 18.12B–18.12E illustrate representatives of four major groups of arthropods. Figure 18.12B shows a number of horseshoe crabs. The **horseshoe crab** is a "living fossil"; that is, it has survived with little change for hundreds of millions of years. It is the only surviving member of a group of spiderlike arthropods that were abundant in the sea some 300 million years ago. Horseshoe crabs are common on the Atlantic and Gulf coasts of the United States.

The closest living relatives of horseshoe crabs are the scorpions, spiders, ticks, and mites, collectively called **arachnids,** a second major arthropod group. Most arachnids live on land. Scorpions (Figure 18.12C, left) are nocturnal hunters. Their ancestors were among the first terrestrial carnivores, preying on herbivorous arthropods that fed, in turn, on the early land plants. Scorpions have a large pair of pincers (analogous to those of lobsters) for defense and the capture of prey. The tip of the tail bears a poiso-

Figure 18.12B Horseshoe crabs (up to about 30 cm wide)

A scorpion (about 8 cm long)

A garden spider (about 1 cm wide)

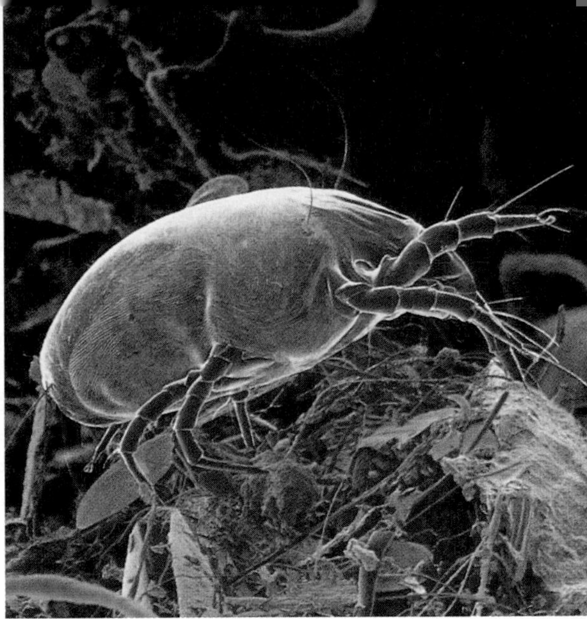

A dust mite SEM 150×

Figure 18.12C Arachnids

nous stinger. Scorpions eat mainly insects and spiders and will attack people only when prodded or stepped on. Only a few species are dangerous to humans.

Spiders, a diverse group of arachnids, are usually active during the daytime, hunting insects or trapping them in webs. The garden spider (Figure 18.12C, center) spins an orb web (described in the introduction to Chapter 3).

Mites make up another large group of arachnids. On the right in Figure 18.12C is a micrograph of a house-dust mite, a ubiquitous scavenger in our homes. Thousands of these microscopic animals can thrive in a few square centimeters of carpet or in one of the dust balls that form under a bed. Dust mites do not carry infectious diseases, but many people are allergic to them.

A third major group of arthropods, the **crustaceans,** are nearly all aquatic. Lobsters and crayfish are in this group, along with numerous crabs, shrimps, and barnacles. Barnacles (Figure 18.12D) are marine crustaceans that live in a limestone shell. Their jointed appendages project from the shell and capture small invertebrates and organic particles suspended in seawater. Each of the appendages and the main body of a barnacle are covered with a chitinous exoskeleton that is separate from the surrounding shell.

Millipedes and centipedes make up a fourth group of arthropods. They have similar segments over most of the body and superficially resemble annelids; however, their jointed legs identify them as arthropods. **Millipedes** (Figure 18.12E) are wormlike landlubbers that eat decaying plant matter. They have two pairs of short legs per body segment. **Centipedes** are terrestrial carnivores, with a pair of poison claws used in defense and to paralyze prey, such as cockroaches and flies. Each of their body segments bears a single pair of long legs.

The four groups of arthropods illustrated in this module account for about 170,000 living species. We turn next to a fifth major group of arthropods, the insects, whose numbers dwarf all other groups combined.

The phylum Arthropoda is named for its members' ___________ ___________.

jointed appendages

Figure 18.12D Crustaceans: barnacles (about 5 cm high)

Figure 18.12E A millipede (about 7 cm long)

18.13 Insects are the most diverse group of organisms

18A

The total number of insect species is greater than the total of all other species combined. About a million insect species are known today; probably at least twice this many exist (mostly in tropical forests) but have not yet been discovered. Insects have been prominent on land for the last 400 million years. They have been much less successful in aquatic environments; there are only about 20,000 species in freshwater habitats and far fewer in the sea.

Insects have a number of common features. Like the grasshopper in Figure 18.13A, most have a three-part body, consisting of a head, a thorax, and an abdomen. The head usually bears a pair of sensory antennae and a pair of eyes. Several pairs of mouthparts are adapted for particular kinds of eating—for example, for biting and chewing plant material in grasshoppers; for lapping up fluids in houseflies; and for piercing skin and sucking blood in mosquitoes and other biting flies. Most adult insects have three pairs of legs and one or two pairs of wings, all borne on the thorax. Insects are the only animals other than bats and birds that have wings, and the ability to fly has been a major factor in their evolutionary success.

Many insects undergo metamorphosis in their development (see Module 18.1). For instance, the three groups on this page undergo **incomplete metamorphosis;** the young resemble adults but are smaller with different body proportions. By contrast, the groups on the facing page undergo **complete metamorphosis;** their larval stages (such as caterpillars, which are the larvae of moths and butterflies, and maggots, which are fly larvae) are specialized for eating and growing and look very different from the adults, which are specialized for dispersal and reproduction.

Systematists classify insects into about 26 orders, mostly on the basis of wing and mouthpart structure. The drawings in Figures 18.13A–18.13G illustrate representative insects in seven of the most common orders.

? Most of Earth's species of living organisms are ____________.

insects

A. Order Orthoptera. The grasshopper represents this group, which contains about 30,000 species. Other orthopterans are the crickets, katydids, locusts, cockroaches, walking sticks, and praying mantises. These insects have biting and chewing mouthparts, and most species are herbivorous. Among the carnivorous species are the praying mantises, which use their forelegs to grip prey. Some orthopterans lack wings, but most have two pairs: a pair of forewings and a pair of hindwings. The forewings are often thickened.

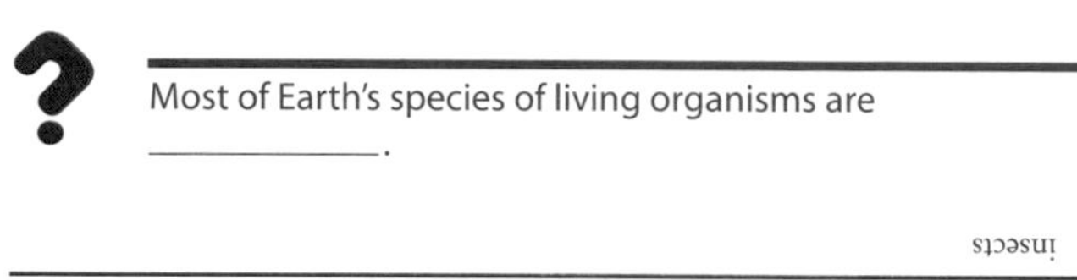

Figure 18.13A Insect anatomy, as seen in a grasshopper

B. Order Odonata. This order includes about 5000 species of dragonflies and damselflies. These insects have two pairs of similar wings. They have biting mouthparts and are carnivorous, often catching and eating other insects on the wing. The larvae of larger species sometimes eat tadpoles and small fishes.

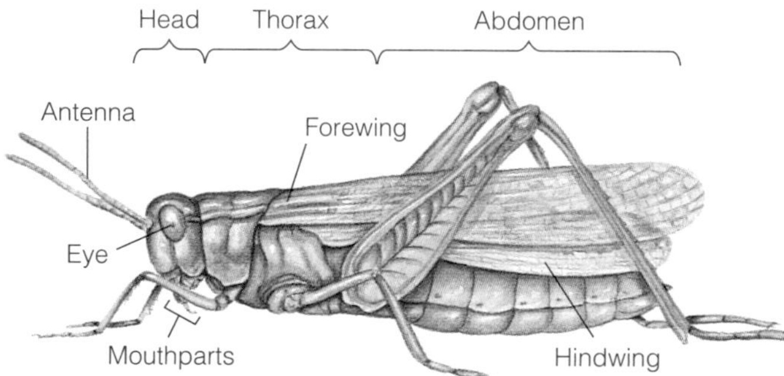

Figure 18.13B A damselfly

C. Order Hemiptera. Often called the true bugs, hemipterans (about 55,000 species) include bedbugs, plant bugs, stinkbugs, and water striders. They have piercing, sucking mouthparts, and most species feed on plant sap. A few, such as bedbugs, feed on blood. The true bugs have two pairs of wings, and the front half of each forewing is thickened and leathery. Water striders walk on water by taking advantage of surface tension (see Module 2.11).

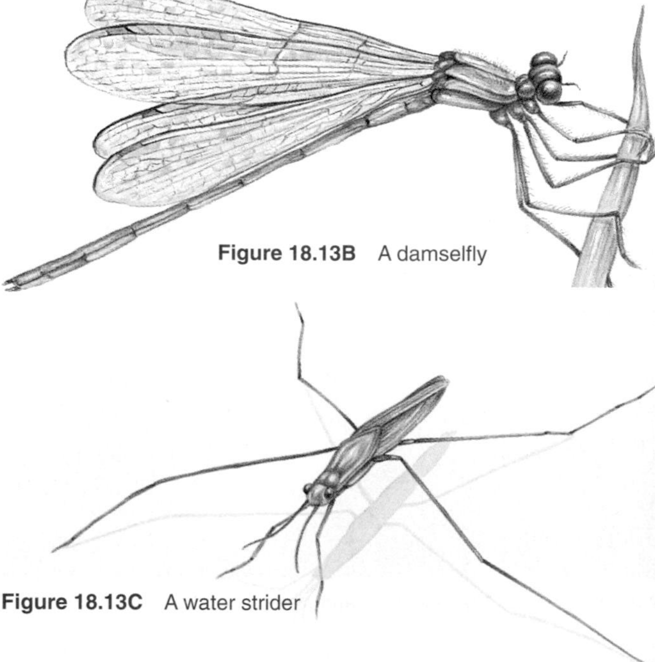

Figure 18.13C A water strider

D. Order Coleoptera. Beetles make up the largest order in the animal kingdom. There are about 500,000 species known worldwide and some 30,000 species in the United States. Beetles occur almost everywhere, from high mountains to the seashore, and their habitats are extremely varied. Forests, streams, ponds, soil, dung, carrion, and plant material all support their quota of species. Beetles have biting and chewing mouthparts, and they include carnivores, herbivores, and omnivores. Beetles vary in length from less than 1 mm to about 12 cm. They have two pairs of wings, but only the hindwings function in flight. The forewings are hardened and thickened as protective covers for the hindwings.

Figure 18.13D A ground beetle

E. Order Lepidoptera. These are the moths and butterflies (about 140,000 species). They have two pairs of wings, with the hind pair smaller. Typically, the wings and body are covered by scales—the dust you find on your fingers after holding a butterfly. The mouthparts form a long drinking tube that is adapted for drinking nectar from flowers. The tube is coiled under the head when not in use. When the insect drinks nectar, the tube is uncoiled and extended deep into a flower.

Figure 18.13E A hawk moth

F. Order Diptera. Dipterans (about 120,000 species) are the flies, including fruit flies, houseflies, gnats, and mosquitoes. Flies have a single pair of wings; instead of hindwings, they have small, club-shaped organs called halteres, which function in maintaining balance during flight. Mosquitoes have piercing, sucking mouthparts, and the females suck blood. Most other dipterans have lapping mouthparts and feed on nectar or other liquids. Mosquitoes and other bloodsuckers may transmit disease, such as malaria or African sleeping sickness, to their animal hosts.

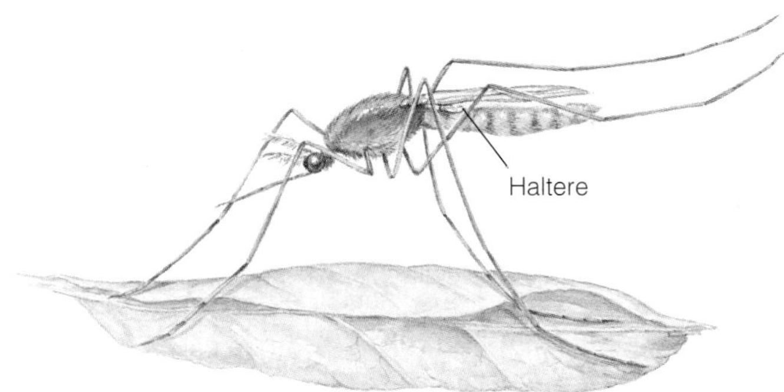

Figure 18.13F A mosquito

G. Order Hymenoptera. The hymenopterans (about 110,000 species) are the ants, bees, and wasps. They have two pairs of wings, both used in flight. The hindwings are smaller and are hooked to the rear of the forewings. This arrangement improves flight efficiency by making the wings behave as single functional units. Generally, the thorax and abdomen are separated by a narrow waist, which, along with the four translucent wings, makes it easy to identify an insect as a member of this order. With chewing and sucking mouthparts, some hymenopterans are herbivorous; others are carnivorous, eating mainly other insects. Many hymenopterans display complex behavior, including social organization (see Modules 37.1 and 37.19).

Figure 18.13G A paper wasp

18.14 Echinoderms have spiny skin, an endoskeleton, and a water vascular system for movement

18A

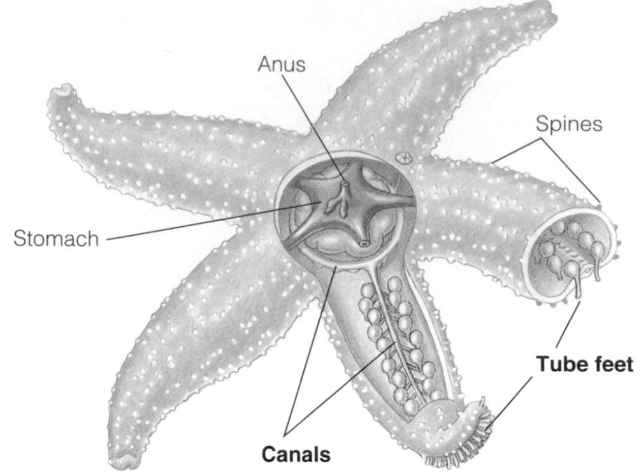

Figure 18.14A The water vascular system of a sea star (canals and tube feet)

Annelids and arthropods represent one large evolutionary branch of the animal kingdom. **Echinoderms,** such as sea stars and sea urchins, represent a very different evolutionary branch.

Echinoderms are all marine. They lack body segments, and most are radially symmetrical as adults. Both the external and the internal parts of a sea star, for instance, radiate from the center like spokes of a wheel. In contrast to the adult, the larval stage of echinoderms is bilaterally symmetrical. This tells us that echinoderms are not closely related to cnidarians or other animals that never show bilateral symmetry.

The phylum name **Echinodermata** is derived from the Greek words meaning "like a sea urchin's skin," and most echinoderms do have a rough or spiny surface. The spininess of a sea star or sea urchin comes from hard spines or plates embedded under the skin. The spines and plates are actually components of a hard internal skeleton, the **endoskeleton.**

Unique to echinoderms is the **water vascular system,** a network of water-filled canals that branch into extensions called tube feet (Figure 18.14A). A sea star pulls itself slowly over the seafloor using its suction-cup–like tube feet. Its mouth is centrally located on its undersurface. When a sea star encounters an oyster or clam, its favorite food, it grips the mollusk's shell with its tube feet and positions its mouth next to the narrow opening between the two valves of the shell. The sea star then pushes its stomach out through its mouth. The stomach enters the mollusk through its shell opening and proceeds to digest the soft parts of the prey (Figure 18.14B).

Sea stars and other echinoderms have strong powers of regeneration. Tube feet and whole arms that are damaged or lost are readily regrown. One group of echinoderms, the sea cucumbers, can even regenerate some of their internal organs, such as parts of the digestive system.

In contrast to sea stars, sea urchins are spherical and have no arms (Figure 18.14C). They do have five rows of tube feet that project through tiny holes in the animal's globelike case. If you look carefully, you can see the long, threadlike tube feet projecting among the spines of this purple sea urchin (both spines and tube feet are purple). Sea urchins move by pulling with their tube feet. They also have muscles that pivot their spines, and some species can walk on their spines. In contrast to sea stars, which are carnivorous, most sea urchins eat algae.

Though echinoderms have many unique features, we see evidence of their relation to other animals in their embryonic development. In sea stars, for instance, the digestive tract and coelom develop from the same parts of the gastrula as they do in animals called lancelets, which are invertebrate members of the phylum Chordata. We examine lancelets and other invertebrate chordates next, followed by a survey of the vertebrate chordates. We return to the subject of echinoderm-chordate relationships when we summarize animal phylogeny in Module 18.23. 

? Contrast the skeleton of an echinoderm with that of an arthropod.

An echinoderm has an endoskeleton; an arthropod has an exoskeleton.

Figure 18.14B A sea star feeding on a clam

Figure 18.14C A sea urchin (about 12 cm in diameter)

Our own phylum, Chordata, is distinguished by four features 18.15

Four distinctive features appear in the embryos, and sometimes in the adults, of animals in the phylum **Chordata.** These four chordate hallmarks are: (1) a **dorsal, hollow nerve cord;** (2) a **notochord,** a flexible, longitudinal rod located between the digestive tract and the nerve cord; (3) **gill structures** (gill slits and structures supporting them) in the pharynx, the region of the digestive tube just behind the mouth; and (4) a **post-anal tail** (a tail posterior to the anus). Together, these features identify an animal as a **chordate.** (We discuss these features in human embryos in Module 27.17.)

The most diverse chordates are the **vertebrates,** animals with a segmented backbone. All other animals, including two groups of chordates, lack a backbone and are called **invertebrates.** Let's look first at the invertebrate chordates, the tunicates and the lancelets.

Adult **tunicates** are stationary and look more like small sacs than anything we usually think of as chordates (Figure 18.15A). Tunicates often adhere to rocks and boats, and they are common on coral reefs. The adults have no trace of a notochord, nerve cord, or tail, but they do have a large gill apparatus. Each of the short, white, vertical lines inside the adult animal in the photo surrounds a gill slit. Very different from the adult, the tunicate larva (drawing) is a swimming, tadpolelike organism that exhibits all four chordate trademarks.

Tunicates eat fine organic particles suspended in seawater. Seawater enters the adult animal through an opening at the top (see the photo in Figure 18.15A). Food particles are trapped in mucus around the gills and then transported to the intestine, where they are digested. Water passes out of the gill slits into a large cavity in the animal and back into the ocean via an opening on the side of the body (upper right in photo).

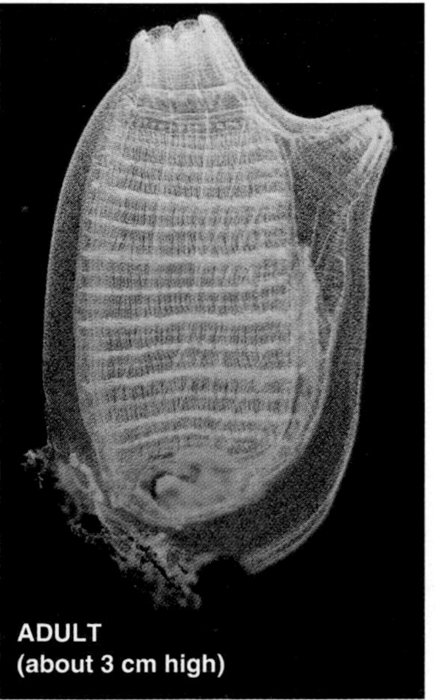

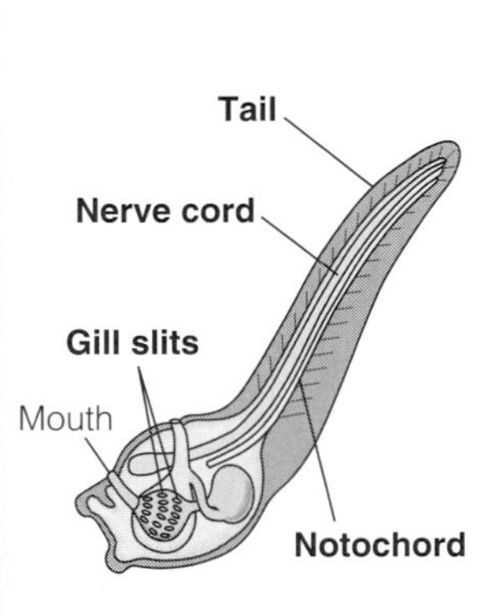

Figure 18.15A Tunicates

18B

Lancelets, another group of marine chordates, also feed on suspended particles (Figure 18.15B). Resembling tunicate larvae, lancelets are small (5–15 cm), bladelike chordates that live in marine sands. They illustrate the four chordate features more clearly than any other members of the phylum. Lancelets also have segmental muscles, which flex the body from side to side, producing slow swimming movements. When feeding, a lancelet sits vertically in sand with its head sticking out. As in tunicates, the gill apparatus traps food particles. Water flowing through the gill slits exits via an opening in front of the anus.

What is the relationship between the invertebrate chordates and the vertebrates? Molecular evidence indicates that the lancelets are the closest living relatives of vertebrates, and that the two groups probably had a common ancestor. The ancestor probably had all four of the chordate hallmarks, and many biologists think it was a stationary gill feeder looking something like an adult tunicate. If the swimming larva of such an ancestor developed the ability to reproduce, natural selection may have eliminated the stationary stage, creating a life cycle similar to that of lancelets. As we saw in Chapter 11, mutations in only one or a few of the genes controlling development can produce significant changes. Perhaps some such mutations led to an extreme form of paedomorphosis (see Module 15.7) in which the adult, reproducing animal was similar to the larva. Paedomorphosis may have been key to the evolution of both lancelets and vertebrates.

The invertebrate chordates have helped us highlight the four chordate hallmarks. Let's turn next to the vertebrates and the features that make them unique. 

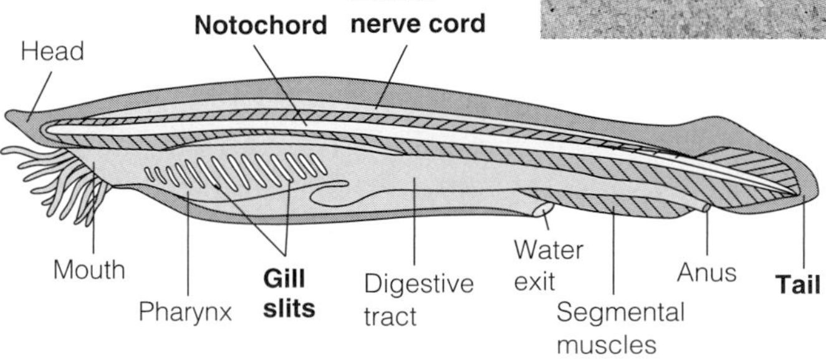

Figure 18.15B Lancelets (5–15 cm long)

? What four features do we share with invertebrate chordates such as lancelets?

(1) Dorsal, hollow nerve chord; (2) notochord; (3) gill structures at some time during development; (4) post-anal tail at some time during development

Vertebrates

18.16 A skull and a backbone are hallmarks of vertebrates

18B

Vertebrates (Latin *vertebra,* joint) comprise most of the phylum Chordata. The cat skeleton in Figure 18.16 illustrates two distinguishing features of vertebrates: a **skull** and a **backbone** composed of a series of segmented units called **vertebrae** (singular, *vertebra*). These skeletal elements enclose the main parts of the nervous system. The skull forms a case for the brain, and the vertebrae enclose the nerve cord. In addition to the vertebrae and skull, most vertebrates have skeletal parts supporting their body appendages (legs or fins).

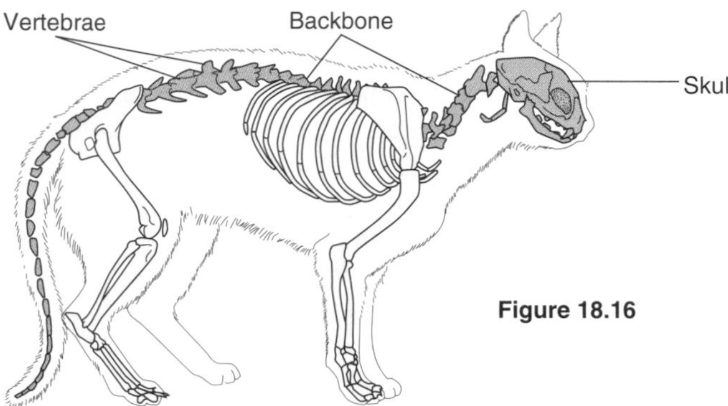

Figure 18.16

The vertebrate skeleton, an endoskeleton, is made of either flexible cartilage or a combination of hard bone and cartilage, as in the cat. Bone and cartilage are mostly nonliving material, but they contain living cells that secrete the nonliving material. Because of its living cells, the endoskeleton can grow with the animal; it is unlike the arthropod's nonliving exoskeleton, which must be shed periodically. We illustrate the major groups of vertebrates (agnathans, fishes, amphibians, reptiles, birds, and mammals) in the next six modules.

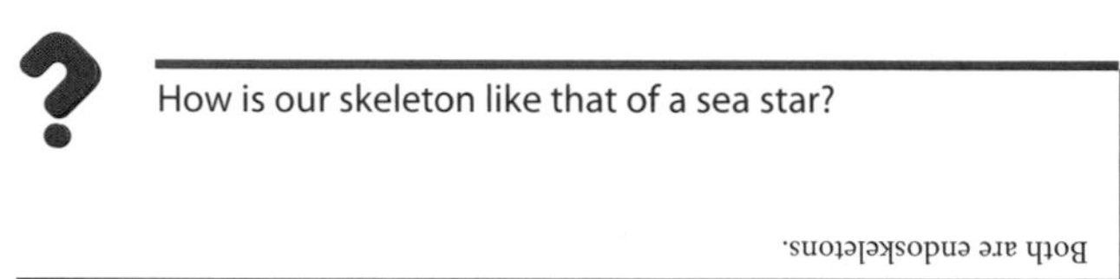

How is our skeleton like that of a sea star?

Both are endoskeletons.

18.17 Most vertebrates have hinged jaws

18B

Just seeing the mouth of a sea lamprey (Figure 18.17A, below) almost tells us what it can do. It bores a hole in the side of a fish and sucks its victim's blood. Introduced into the Great Lakes in the early 1900s, these voracious vertebrates multiplied rapidly, decimating fish populations as they spread. By mid-century, sea lampreys, along with an increase in commercial fishing and water pollution, had destroyed virtually all the large fish in the lakes. Since the 1960s, streams that flow into the lakes have been treated with a chemical that reduces lamprey numbers, and fish populations have been recovering.

Lampreys, about 60 species worldwide, range from about 75 mm to 1 m long. They belong to a group of primitive vertebrates called the class **Agnatha** (Greek *a-*, without, and *gnathos,* jaw). Agnathans are distinctly different from all other vertebrates, including the fishes. Though they are superficially fishlike, agnathans lack paired fins, and their mouth lacks jaws. Fishes, amphibians, reptiles, birds, and mammals—the vast majority of living vertebrates—have jaws supported by two skeletal parts held together by a hinge. In contrast, the lamprey's toothed, sucking disk is an unhinged, circular outgrowth of the mouth. Lampreys are more closely related to extinct jawless vertebrates than to any other living vertebrate. Larval lampreys, perhaps like those ancestors, trap suspended particles with their gills.

Figure 18.17A The mouth of a lamprey (no jaws)

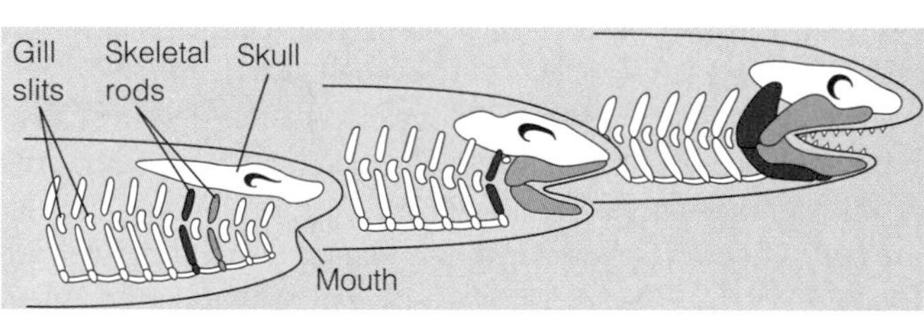

Figure 18.17B The origin of vertebrate jaws

Where did the hinged jaws of vertebrates come from? They evolved by modification of skeletal supports of the gill slits. The first part of Figure 18.17B shows the skeletal rods supporting the gills in a hypothetical ancestor. The main function of these gills was trapping suspended food particles. The other two parts show changes that probably occurred as jaws evolved. By following the red and green structures, you can see that the jaws and their supports evolved from two pairs of skeletal rods located between gill slits that were near the mouth. Similar events occur during embryonic development in fishes today.

The oldest fossils of jawed vertebrates, an extinct group of fishes, appear in rocks about 450 million years old. Jaws were of paramount importance in vertebrate evolution, enabling vertebrates to catch and eat a wide variety of prey, instead of feeding on suspended particles.

Vertebrate jaws evolved from the skeletal rods supporting ________.

gills

Fishes are jawed vertebrates with gills and paired fins

Fishes, the first jawed vertebrates, have been numerous and diverse for over 400 million years. Today, they make up about 30,000 of the 45,000 species of vertebrates. Besides having hinged jaws, a fish has gills, which extract oxygen from water, and paired forefins and hindfins, which help maneuver the body when swimming. Nearly all fishes are carnivores, thriving virtually everywhere in the watery two-thirds of Earth's surface. There are two major groups of living fishes: the class **Chondrichthyes,** the cartilaginous fishes (sharks, rays, and skates) and the class **Osteichthyes,** the bony fishes, a group that includes the familiar trout and goldfish.

Cartilaginous fishes have a flexible skeleton made of cartilage. Most sharks, the blacktip reef shark in Figure 18.18A, for example, are adept predators—fast swimmers with a streamlined body, acute senses, and powerful jaws. A shark does not have keen eyesight, but its sense of smell is very acute, and on its head it has special electrosensors, organs that can detect minute electrical fields produced by muscle contractions in nearby animals. Sharks also have a **lateral line system,** a row of sensory organs running along each side of the body. Sensitive to changes in water pressure, the lateral line system enables a shark to detect minor vibrations caused by animals swimming in its neighborhood. There are fewer than 1000 living species of cartilaginous fishes, nearly all of them marine.

Bony fishes have a stiff skeleton reinforced by hard calcium salts. They also have a lateral line system, a keen sense of smell, and excellent eyesight. Figure 18.18B highlights the diagnostic features of a bony fish (the labels in bold type). On each side of the head, a protective flap called the **operculum** (plural, *operculi*) covers a chamber housing the gills. Movement of the operculum allows the fish to breathe without swimming. (By contrast, sharks lack operculi and must swim to pass water over their gills.) Bony fishes also have a specialized organ that helps keep them buoyant—the **swim bladder,** a gas-filled sac. Some bony fishes have a connection between the swim bladder and the digestive tract that enables them to gulp air and extract oxygen from it when the dissolved oxygen level in the water gets too low.

Figure 18.18A A cartilaginous fish: a shark

The largest class of vertebrates (about 29,000 species), bony fishes are common in the seas and in freshwater habitats. Most bony fishes, including trout, bass, perch, and the bluefin tuna in Figure 18.18C, are **ray-finned fishes.** Their fins are supported by thin, flexible skeletal rays. In contrast to the rayfins, **lobe-finned fishes** have muscular fins supported by stout bones. Lobefins are all extinct except for one species, the coelacanth (Figure 18.18C). A deep-sea dweller, the coelacanth may use its fins to waddle along the seafloor. As we see next, freshwater lobe-finned fishes played a key role in the colonization of land by vertebrates.

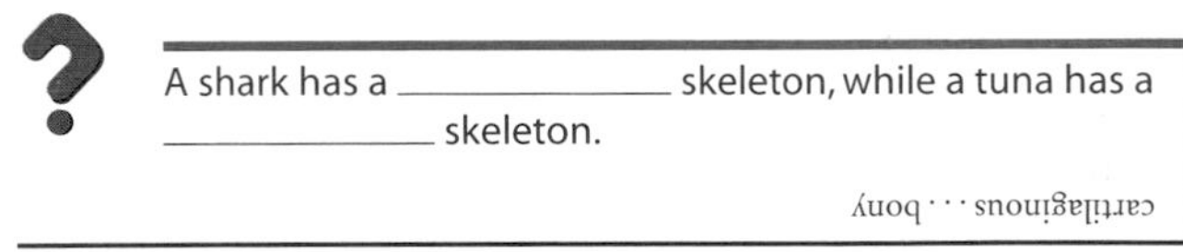

? A shark has a ________ skeleton, while a tuna has a ________ skeleton.

cartilaginous . . . bony

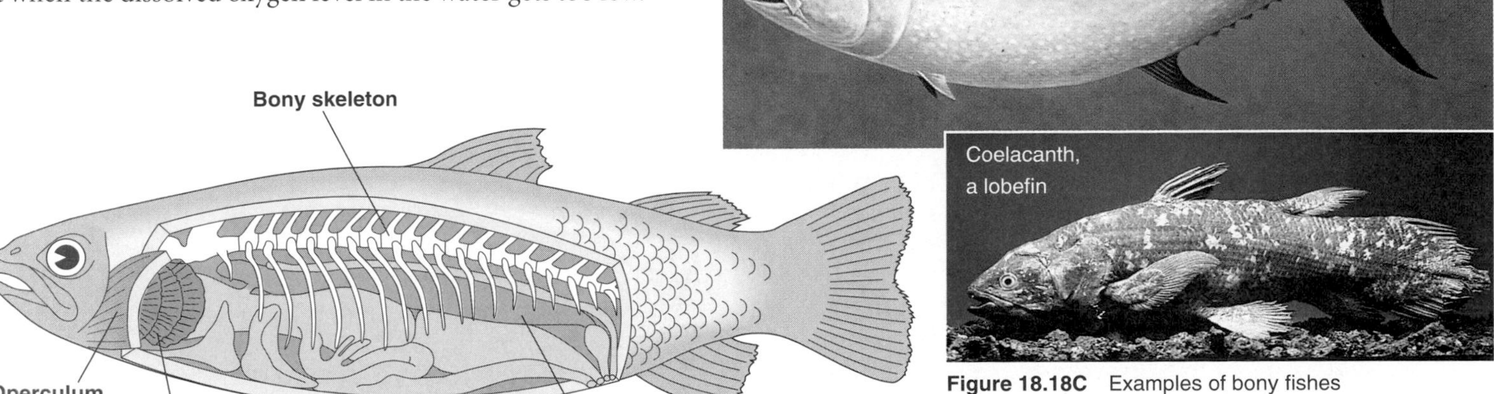

Figure 18.18C Examples of bony fishes

Figure 18.18B Diagnostic features of a bony fish

18.19 Amphibians were the first land vertebrates

18B

Figure 18.19A An adult frog

Figure 18.19B A tadpole, a frog larva

In Greek, the word *amphibios* means "living a double life," and most members of the class **Amphibia** exhibit a mixture of aquatic and terrestrial adaptations. Most species are tied to water because their eggs dry out quickly in the air. The frog in Figure 18.19A spends much of its time on land, but it lays its eggs in water. An egg develops into a larva called a tadpole (Figure 18.19B), a legless, aquatic algae-eater with gills that strain food from the water, a lateral line system resembling that of fishes, and a long finned tail. In changing into a frog, the tadpole undergoes a radical metamorphosis. When a young frog crawls onto shore and begins life as a terrestrial insect-eater, it has four legs, air-breathing lungs instead of gills, a pair of external eardrums, and no lateral line system. Because of metamorphosis, amphibians truly live a double life.

The amphibians of today (frogs, toads, and salamanders) account for only about 8% of all existing vertebrate species. Historically, amphibians are of major importance because they were the first terrestrial vertebrates. Figure 18.19C illustrates some structural changes that may have taken place as vertebrates began to move onto land. The drawing on the left shows a lobe-finned fish that lived in freshwater lakes and rivers some 400 million years ago. With their strong fins, these bony fishes may have wriggled over mud flats on drying lakeshores and riverbeds. One hypothesis is that extinct lobefins gave rise to the first land vertebrates. Lobefins also had saclike lungs (as well as gills), so they could have breathed air when they were out of the water. An alternate hypothesis, supported by molecular evidence, is that a group closely related to the lobefins, called lungfishes, were ancestral to the first land vertebrates. Lungfishes also had saclike lungs. Most lungfishes are extinct; one of six living species is shown in Figure 15.3C.

The drawing on the right in Figure 18.19C shows an early amphibian that may have evolved from lobe-finned fishes or lungfishes. This creature, which lived some 350 million years ago, had limbs much like those of modern amphibians. With legs instead of fins, it could have walked on land rather than just wriggling around in mud the way its lobe-finned ancestors did.

The early amphibians encountered favorable habitats; they probably feasted on insects and and other invertebrates in the lush forests of the Carboniferous period (see Module 17.7). As a result, amphibians became so widespread and diverse that the Carboniferous period is sometimes called the age of amphibians. With the decline of the coal forests about 300 million years ago, the early amphibians also waned. Many lineages became extinct, some gave rise to modern amphibians, and others gave rise to the reptiles.

What is metamorphosis?

The transformation from a larval form, such as a tadpole, to an adult form, such as the frog in Figure 18.19A.

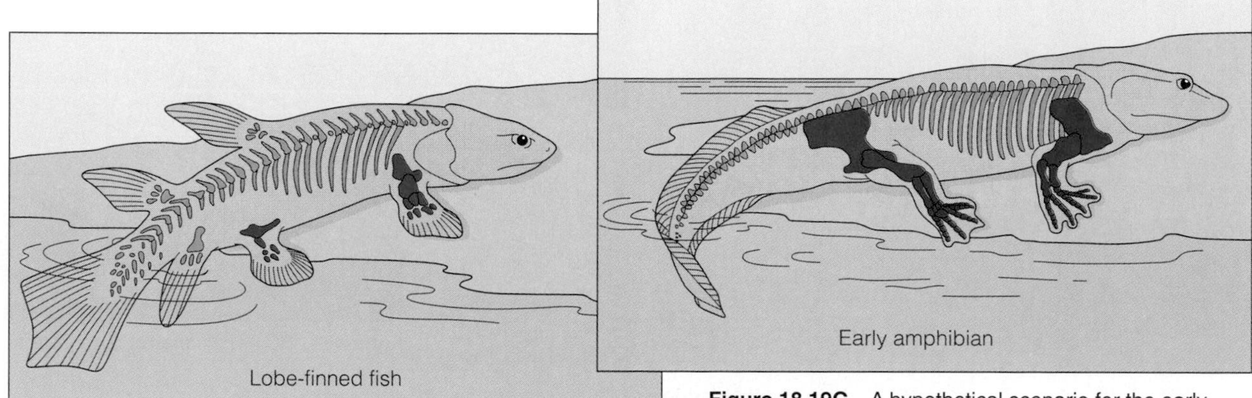

Figure 18.19C A hypothetical scenario for the early evolution of land vertebrates

Reptiles have more terrestrial adaptations than amphibians 18.20

18B

The reptiles, Class **Reptilia,** include snakes, lizards, turtles, crocodiles, and alligators. Reptiles have several adaptations for terrestrial living not generally found in amphibians. Reptilian skin, covered with scales waterproofed with the tough protein keratin, keeps the body from drying out. Most reptiles also have eggs covered with parchmentlike shells that retain water. Unlike most amphibians, which have to lay eggs in a pool of water, a reptile such as the bull snake in Figure 18.20A can lay them in a rotten log or in sandy soil under leaves. Inside each egg, the embryo develops within a protective, fluid-filled sac called the amnion (see the introduction to Chapter 27). The developing embryo is nourished by yolk until it hatches as a juvenile, ready to move about freely and feed itself. The evolution of this self-contained, drought-resistant **amniotic egg** enabled reptiles to complete their life cycles on land and sever ties with their aquatic origins. As we saw in Module 17.3, the seed played a similar role in the evolution of plants.

Reptiles in existence today are sometimes called cold-blooded animals because they do not use their metabolism to control body temperature. Nonetheless, reptiles do have ways to raise or lower their body temperature. The horned lizard of our southwestern deserts (Figure 18.20B) commonly warms up in the morning by sitting on warm rocks. Its skin keeps it from drying out as it basks in the sun. If the lizard starts getting too hot, it seeks shade. Because reptiles warm up by absorbing external heat rather than by generating much of their own, they are said to be **ectothermic** (Greek *ektos,* outside, and *therme,* heat), a term more appropriate than cold-blooded.

Like the amphibians from which they evolved, reptiles were once much more prominent than they are today. Following the decline of amphibians, reptilian lineages expanded rapidly, creating a dynasty that lasted until about 65 million years ago. During that age of reptiles, dinosaurs, the most diverse group, included the largest animals ever to inhabit land. Some were "gentle giants" like the large dinosaur in Figure 18.20C, lumbering about while browsing vegetation. Others, like the 3-m-long *Deinonychus* (Greek for "terrible claw"), were voracious carnivores that ran on two legs. Probably a pack hunter, *Deinonychus* may have used its sickle-shaped claws to slash at large prey. Unlike modern reptiles, *Deinonychus* and many other dinosaurs may have been **endothermic,** using heat generated by metabolism to maintain a warm, constant body temperature.

The age of reptiles began to wane about 70 million years ago. As we discussed in Module 15.5, periods of mass extinction killed off all but one of the dinosaur lineages by about 65 million years ago. That lineage survives today as the birds.

What is an amniotic egg?

An egg with a shell, housing an embryo contained in a fluid-filled sac, the amnion

Figure 18.20A A bull snake laying eggs

Figure 18.20B A horned lizard

Figure 18.20C A pack of *Deinonychus* attacking a larger dinosaur

18.21 Birds share many features with their reptilian ancestors

18B

Birds, Class **Aves,** may seem very unlike reptiles, but there are more similarities than differences between these two groups of vertebrates. Strong fossil evidence indicates that birds evolved from a lineage of two-legged dinosaurs about 150–200 million years ago. Birds evidently inherited amniotic eggs, scales on their legs, toenails containing keratin, and their general body form from their reptilian ancestors. Even feathers, a hallmark of birds, were derived from reptilian scales, and recent discoveries indicate that some dinosaurs had downlike feathers.

Figure 18.21A is an artist's reconstruction based on a 150-million-year-old fossil of an extinct bird. Called *Archaeopteryx* (Greek *archaios,* ancient, and *pteryx,* wing), this animal lived near tropical lagoons in central Europe. Like modern birds, it had flight feathers, but otherwise it was more like some small bipedal dinosaurs of its era; for instance, like those dinosaurs, *Archaeopteryx* had teeth, wing claws, and a tail with many vertebrae.

Flight feathers shape bird wings into airfoils, providing lift and maneuverability in the air (see Figure 30.1E). *Archaeopteryx*'s flight feathers could have enabled it to fly by flapping its wings. However, it had a fairly heavy body and probably relied mainly on gliding from trees.

Despite its feathers, *Archaeopteryx* is not considered an ancestor of modern birds. Instead, it probably represents an extinct side-branch of the bird lineage. Many questions remain about the origin of birds. Recent fossil discoveries indicate that many groups of birds became extinct about 65 million years ago, when most dinosaurs died out. Modern birds seem to have evolved from a few groups that survived the bottleneck effect of the mass extinctions (see Module 13.11).

Nearly every part of the body of most birds is adapted to enhance flight. Many features help reduce weight for flight: Modern birds lack teeth, their tail is supported by only a few small vertebrae, their wings lack claws, their feathers have hollow shafts, and some of their bones are hollow and contain air sacs. Providing power for flight are large breast (flight) muscles, which are anchored to a keel-like breastbone. Most of what we call white meat on a turkey or chicken are the flight muscles.

Figure 18.21B A turkey vulture

Flying requires a great amount of energy, and modern birds have a high rate of metabolism. Birds also are endothermic, with their feathers helping maintain body temperature by providing insulation. Supporting their high metabolic rate, birds have a highly efficient circulatory system, and their lungs are even more efficient at extracting oxygen from the air than the lungs of mammals.

With wings driven by powerful flight muscles, modern birds are masterly flyers. Some species, such as the turkey vulture in Figure 18.21B, have wings adapted to soaring on air currents, and they flap their wings only occasionally. Others, such as hummingbirds, excel at maneuvering but must flap almost continuously to stay aloft. Flight ability is typical of birds, but there are a few flightless species, including the ostrich and the emu of Australia (Figure 18.21C). Modern flightless birds evolved from flying ancestors.

Teeth (like reptile)
Wing claw (like reptile)
Feathers
Long tail with many vertebrae (like reptile)

Figure 18.21A *Archaeopteryx,* an extinct bird

Birds and reptiles differ in their main source of body heat, with birds being ______________ and reptiles being ______________.

endothermic . . . ectothermic

Figure 18.21C An emu, a flightless bird

Mammals also evolved from reptiles 18.22

18B

Vertebrates of the class **Mammalia** evolved from reptiles about 225 million years ago, long before there were any dinosaurs. During the peak of the age of reptiles, there were a number of mouse-sized, nocturnal mammals, which lived on a diet of insects. Mammals became much more diverse after the downfall of the dinosaurs. Most mammals are terrestrial, but there are nearly 1000 species of winged mammals, the bats, and about 80 species of totally aquatic forms: the dolphins, porpoises, and whales. The blue whale, an endangered species, grows to lengths of nearly 30 m and is the largest animal that has ever existed.

Like birds, mammals are endothermic, with a high rate of metabolism. Two features—hair and mammary glands that produce milk to nourish young—are mammalian hallmarks. The main function of hair is to insulate the body and help maintain a warm, constant body temperature. There are three major groups of mammals: the monotremes, the marsupials, and the eutherians.

The duck-billed platypus (Figure 18.22A) is one of only three existing species of **monotremes,** the egg-laying mammals. The platypus lives along rivers in eastern Australia and on the nearby island of Tasmania. It eats mainly small shrimps and aquatic insects. The female usually lays two eggs and incubates them in a leaf nest. After hatching, the young nurse by licking up milk secreted onto the mother's fur.

Most mammals are born rather than hatched. During gestation in marsupials and eutherians, the embryos are nurtured inside the mother by an organ called the placenta. Consisting of both embryonic and maternal tissues, the **placenta** joins the embryo to the mother within the mother's uterus. The embryo is nurtured by maternal blood that flows close to the embryonic blood system in the placenta.

Marsupials, the so-called pouched mammals, have a brief gestation and give birth to tiny, embryonic offspring that complete development while attached to nipples on the abdomen of the mother. The nursing young are usually housed in an external pouch, called the **marsupium,** on the mother's abdomen (Figure 18.22B). Nearly all marsupials live in Australia, New Zealand, and Central and South America. Australia has been a marsupial sanctuary for much of the past 60 million years, and marsupials have diversified extensively there, filling terrestrial habitats that on other continents are occupied by eutherian mammals.

Eutherians are commonly called **placentals** because their placentas provide more intimate and long-lasting association between the mother and her developing young than do marsupial placentas. Eutherians make up almost 95% of the 4500 species of living mammals. The placenta is the reddish portion of the afterbirth still clinging to the newborn zebra in Figure 18.22C. The large silvery membrane is the amniotic sac, which held the developing embryo in a bath of protective amniotic fluid. The amniotic tissues are homologous with those of reptiles.

Humans are eutherians that belong to Order Primates, along with monkeys and apes. We examine human evolution in more detail in Chapter 19.

What are two hallmarks of mammals?

Hair and mammary glands

Figure 18.22A Monotremes: a duck-billed platypus with her young

Figure 18.22B A marsupial: a gray kangaroo with young, called a joey, in her pouch

Figure 18.22C Eutherians: a zebra and newborn

18.23 A phylogenetic tree gives animal diversity an evolutionary perspective

18A
18C

We have looked at examples of an immense diversity of animals in this chapter—representatives of about one-third of the known animal phyla. The table at the right recaps the phyla we discussed and key animals within those phyla. The phylogenetic tree in Figure 18.23 summarizes the evolutionary threads that tie all this diversity together.

At the bottom of the tree are the colonial protists, probably choanoflagellates, that were the ancestors of the earliest animals. The position of Phylum Porifera (the sponges) on the lowest branch of the tree reflects the relatively simple body structure of sponges and their marked similarity to choanoflagellates. Sponges consist of aggregrates of only a few types of cells, and they lack tissues. Also, they have larvae that are unlike those of any other group of animals.

Above the sponge lineage, the main trunk of the animal kingdom represents ancestral animals whose bodies consisted of specialized cells organized into tissues—a more complex level of organization than sponges have. These "tissue-level" animals gave rise to two distinct lineages, represented by the second branch point on the phylogenetic tree. In one lineage, radial symmetry evolved as the adult body form. Radial animals of today include the hydras, jellies, sea anemones, and corals of Phylum Cnidaria. The other lineage, the mainstream of the animal kingdom, gave rise to the bilateral animals—those that swim, crawl, run, or fly in a head-first direction.

The bilateral animals with the simplest body construction are the flatworms (Phylum Platyhelminthes). Judging from their simplicity, they diverged from the main trunk of bilateral animals soon after bilateral symmetry evolved. Recall that a typical flatworm, in common with cnidarians, lacks a body cavity and has an incomplete digestive tract.

As we discussed in Module 18.7, body cavities played a significant role in animal history. Above the flatworm side-branch on the tree, bilateral animal phyla divide on the basis of the type of body cavity they have. A body cavity called a pseudocoelom makes the roundworms (Phylum Nematoda) fundamentally different from other phyla, and they are placed on another side-branch of the tree.

After the nematode branch, the main trunk of the phylogenetic tree continues with animals that have a coelom, a body cavity lined with a sheet of tissue that wraps around the digestive tract and other internal organs. The coelom was a key adaptation, for it provided a way to stabilize body organs while still allowing them to move within the body.

An animal's embryonic development may provide some clues to its evolutionary history (phylogeny). For example, animals with a coelom sort out into two large groups, the protostomes and the deuterostomes, according to the way their body cavity develops. On the **protostome** limb of the tree, the phyla Arthropoda, Annelida, and Mollusca have a coelom that develops from solid masses of cells that form between the digestive tube and the embryonic body wall. On the **deuterostome** limb, the phyla Echinodermata and Chordata have a coelom that forms from hollow outgrowths of the digestive tube of the early embryo. Other embryological evidence supports the distinction between deuterostomes and protostomes.

We now come to body segmentation, which we discussed in Module 18.10 as a major body feature in several animal phyla. According to one hypothesis, segmentation arose in the deuterostome lineage and in the protostome lineage as completely independent evolutionary events. Let's consider the deuterostomes first. Embryological evidence places the echinoderms in the same lineage as the chordates, but echinoderms are not segmented and chordates are. This may indicate that segmentation as we see it in the body muscles and skeleton of vertebrates did not evolve until after the echinoderms and chordates diverged. An alternative hypothesis is that segmentation evolved in an ancestor common to chordates and echinoderms but was lost by the echinoderms as they evolved radial symmetry. In either view, the echinoderms are placed on a branch of the phylogenetic tree separate from the chordates. Similarly, in the protostome part of the tree, there are two phyla with segmentation (annelids and arthropods). Segmentation in annelids and arthropods may have evolved after mollusks diverged as a phylum.

Some Major Animal Phyla: A Review

Phylum	Representative Animals
Porifera	Sponges
Cnidaria	Hydras, jellies, sea anemones, corals
Platyhelminthes	Flatworms (planarians, flukes, tapeworms)
Nematoda	Roundworms
Mollusca	Snails, slugs, bivalves (such as clams), squids, octopuses
Annelida	Segmented worms (earthworms, polychaetes, leeches)
Arthropoda	Horseshoe crabs, arachnids (such as spiders), crustaceans (such as lobsters), insects, centipedes, millipedes
Echinodermata	Sea stars, sea urchins
Chordata	Lancelets, tunicates, lampreys, fishes, amphibians, reptiles, birds, mammals

Segmentation has recently become a controversial topic in animal biology. For instance, molecular evidence indicates that this body feature first arose very early in animal evolution—perhaps in the ancestor of all bilateral animals. If so, the appearance or absence of segmentation in various phyla may result from differential expression of genes that regulate early development.

Our phylogenetic tree linking the major phyla of animals is based on what appear to be fundamental structural features in adult animals and on patterns of embryological development. Built on this circumstantial evidence, the tree represents widely held hypotheses for the key events in evolutionary history that led to nine of the animal phyla now living on Earth. Like other phylogenetic trees, this one serves to stimulate research and discussion, and it is subject to revision as new information is acquired. Whether or not some of its branch points are revised, however, the tree's overall message remains the same: The animal kingdom's great diversity arose through the process of evolution, and all animals exhibit features reminiscent of their evolutionary history.

In key features of embryonic development, humans and other chordates are most like animals of the phylum ________________.

Echinodermata

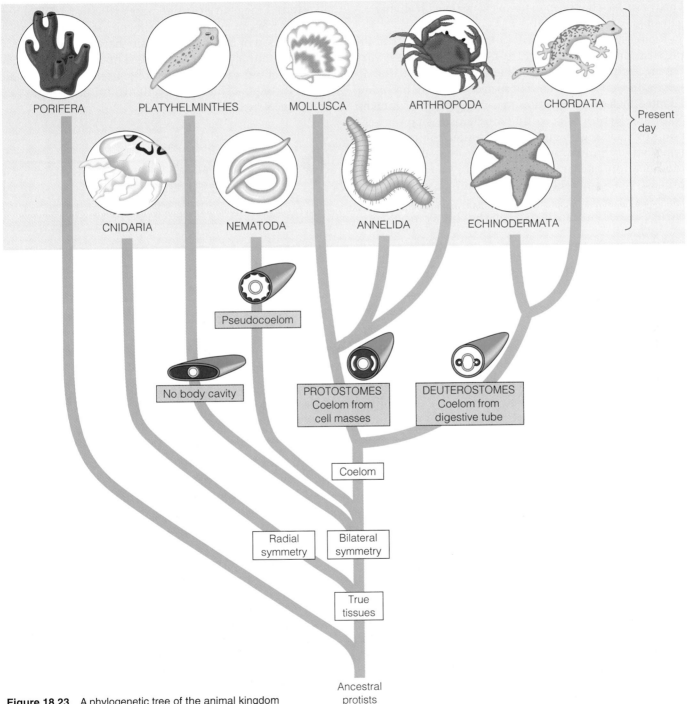

Figure 18.23 A phylogenetic tree of the animal kingdom

18.24 Change is a fundamental feature of the environment and of life

In this photograph, we see an aerial view of a portion of the Great Barrier Reef, which parallels the northeastern Australian coast for nearly 2000 km. This reef is a major feature of our planet's surface, visible from satellites orbiting hundreds of kilometers above Earth. Built mainly by tiny coral animals and algae, the reef is a showplace of life's diversity. It is also a dramatic measure of life's capacity to alter Earth's surface. For all of the billions of years it has existed, life has had a profound impact on the globe. From the time of the earliest life forms to the bacteria, archaea, protists, fungi, plants, and animals of today, every part of Earth's surface has been extensively changed by living organisms.

Biological change is closely connected to change in the physical environment, and to a large extent the word "change" capsulizes this book's units on evolution and diversity. In Unit III, we saw that species change in becoming adapted by natural selection to a changing environment. Natural selection is an editing mechanism, shaping gene pools by selecting phenotypes best suited to the environment while eliminating others. Evolution—descent with modification, as Darwin put it—is thus a restructuring process, changing species and fashioning new ones from preexisting ones. The connection between biological change and environmental change has been crucially important in the history of life and of planet Earth, and in this unit, we have documented the role of change in shaping the vast diversity of life.

We have also chronicled the role of chance. Chance has affected the evolutionary process in the generation of genetic diversity through mutation. Chance has also played a role at every major milestone in the history of life. Before life began, over 3.5 billion years ago, the chance union of certain small organic molecules may have ignited a chain of events that led to the first genes. Much later—about 65 million years ago—a chance collision between Earth and an asteroid may have caused mass extinctions. What are the odds for and against asteroid collisions? How would the evolution of biological diversity have been different if most dinosaur lineages had not died out? One of the great wonders of our existence and of life itself is that it has all arisen through a combination of evolutionary processes and chance events.

Here's a review question: How does the process of natural selection change organisms?

Natural selection shapes gene pools by selecting phenotypes best suited to the environment while eliminating other, less-suited phenotypes.

Figure 18.24 Part of the Great Barrier Reef, along the northeastern coast of Australia

PTER REVIEW

Chapter Summary

Animal Evolution and Diversity (Introduction–18.2) Animals are eukaryotic, multicellular heterotrophs that lack cell walls, have unique intercellular junctions, and digest their food internally. Most animals are diploid except for haploid eggs and sperm. The zygote often grows into a hollow blastula, which folds inward to form a gastrula. The animal may go through a larval stage, which metamorphoses into an adult **(18.1)**. Animals probably evolved from colonial protists, whose cells gradually became more specialized and layered **(18.2)**.

Invertebrates (18.3–18.15) The approximately 35 phyla in the kingdom Animalia all contain invertebrates, animals without a backbone. Sponges (Phylum Porifera) are among the simplest animals. Many sponges are radially symmetrical; their parts are arranged around a central axis. Choanocytes (flagellated collar cells) collect food from the water passing through the porous sponge body. The sponge lineage arose very early from multicellular protists, probably from choanoflagellates, the same group that most likely gave rise to the animal kingdom **(18.3)**. Cnidarians (Phylum Cnidaria) are the simplest animals with tissues. Cnidarians exist in two radially symmetrical body forms: cylindrical polyps (such as coral animals) and umbrella-shaped medusas (jellies). Cnidocytes on their tentacles sting prey, and the tentacles, controlled by nerves, push the food through the mouth into a gastrovascular cavity, where it is digested and distributed **(18.4)**.

Most animals are bilaterally symmetrical, having mirror-image right and left sides. In contrast to sponges and cnidarians, a bilaterally symmetrical animal has a head with sensory structures, and it moves head-first through its environment **(18.5)**. Flatworms (Phylum Platyhelminthes), such as the free-living planarian, are the simplest bilaterally symmetrical animals. A planarian has a simple nervous system consisting of a brain and sense organs in the head, nerve cords, and branching nerves. As in cnidarians, the mouth of a flatworm is the only opening for its gastrovascular cavity. Flukes and tapeworms are parasitic flatworms with complex life cycles **(18.6)**.

Most animals also have a body cavity, a fluid-filled space between the digestive tract and body wall. The body cavity aids in movement, cushions internal organs, and may stiffen the body and help in circulation **(18.7)**. Roundworms (Phylum Nematoda) have a pseudocoelom, a body cavity in contact with the digestive tract. Like most animals, they also possess a complete digestive tract, a tube with both a mouth and an anus. Many nematodes are free-living; others are plant or animal parasites **(18.8)**.

Mollusks (Phylum Mollusca) are a large and diverse phylum that includes gastropods (such as snails and slugs), bivalves (such as clams and scallops), and cephalopods (such as octopuses and squids). All have a muscular foot and a mantle, which secretes the shell. Mollusks have a coelom, a body cavity lined by a sheet of tissue that encloses the digestive tract. They also have a true circulatory system. Many mollusks feed with a rasping radula **(18.9)**.

Most animals exhibit segmentation, subdivision of some or most of the body into a series of repeated parts (segments). Segmentation enhances flexibility and mobility **(18.10)**. The segmented bodies of annelids (Phylum Annelida), such as earthworms, give them added mobility for burrowing and swimming. An earthworm eats its way through soil. Polychaetes, the largest group of annelids, search for prey on the seafloor, or live in tubes and filter food particles from the water. Their appendages aid in locomotion and oxygen uptake. Most leeches, the third group of annelids, are free-living carnivores, but some suck blood **(18.11)**. Arthropods (Phylum Arthropoda) are segmented animals with exoskeletons and jointed appendages. In terms of numbers, distribution, and diversity, they are the most successful phylum of animals. There are several groups of arthropods. Horseshoe crabs are ancient marine arthropods. Their closest relatives are the arachnids: scorpions, spiders, ticks, and mites. Most arachnids live on land, and most are carnivorous. Crustaceans, such as lobsters, crabs, and barnacles, are nearly all aquatic. Another group of arthropods is the millipedes and centipedes **(18.12)**. Insects are the most numerous and successful arthropods. They have a three-part body (consisting of head, thorax, and abdomen) and three pairs of legs, and most have wings. Insects are grouped into about 26 orders, primarily on the basis of their wings and mouthparts. The four largest orders are the beetles, the moths and butterflies, the ants and their relatives, and the flies. The development of many insects includes metamorphosis. Insects, such as grasshoppers, with incomplete metamorphosis, have young that resemble adults but are smaller with different body proportions. Insects, such as bees and butterflies, with complete metamorphosis, have larvae specialized for eating and growing and that look very different from the adults, which are specialized for dispersal and reproduction **(18.13)**. Sea stars and sea urchins are echinoderms (Phylum Echinodermata), radially symmetrical marine creatures with spiny skins and endoskeletons. Their unique water vascular system moves an array of suction-cup–like tube feet. Sea stars are carnivores, and sea urchins eat algae **(18.14)**. Chordates (Phylum Chordata) are segmented animals with a dorsal hollow nerve cord, a stiff notochord, gill structures behind the mouth, and a post-anal tail. The simplest chordates are lancelets and tunicates, marine invertebrates that feed on suspended particles **(18.15)**.

Vertebrates (18.16–18.22) Most chordates are vertebrates, with endoskeletons that include a skull and a backbone composed of vertebrae. There are seven classes of vertebrates **(18.16)**. Lampreys and hagfishes (vertebrates lacking hinged jaws and paired fins or legs) are classified in the class Agnatha **(18.17)**. Two classes of fishes are the Chondrichthyes (cartilaginous fishes, such as sharks) and the Osteichthyes (bony fishes, such as tuna and trout). Bony fishes are more diverse and have more mobile fins, operculi that move water over the gills, and a buoyant swim bladder. Among the bony fishes, ray-finned fishes have thin skeletal rods supporting their fins, whereas lobe-finned fishes have muscular fins supported by stout bones **(18.18)**. Amphibians (Class Amphibia), represented today by frogs, toads, and salamanders, were the first terrestrial vertebrates. Their limbs allow them to move on land, but amphibian embryos and larvae still must develop in water. Extinct lobe-finned fishes or lungfishes probably gave rise to the first amphibians **(18.19)**. Waterproof scales and a shelled amniotic egg enable members of Class Reptilia (snakes, lizards, turtles, crocodiles, and extinct dinosaurs) to live on dry land. Modern reptiles are ectothermic, warming their bodies by absorbing heat from their environment. Some dinosaurs may have been endothermic, using metabolic heat to maintain a warm, constant temperature **(18.20)**.

Birds (Class Aves) share many characteristics with their reptilian ancestors, including amniotic eggs, scales, and overall body form. Birds have wings, feathers, an endothermic metabolism, and many other adaptations (such as hollow bones and a highly efficient circulatory system) related to flight **(18.21)**. Mammals (Class Mammalia) also descended from reptiles and are endothermic. Two unique characteristics of mammals are hair, which insulates their bodies, and mammary glands, which produce milk that nourishes their young. A few mammals, the monotremes, lay eggs, but most give birth to young after a period of embryonic development inside the body of the mother. During gestation, the embryos of marsupials and eutherian mammals are nurtured by an organ called the placenta. Marsupials, such as kangaroos, have a brief gestation; the tiny offspring complete their development attached to their mothers' nipples, usually in a pouch. Most mammals are eutherians (also called placentals). Eutherians have a relatively long gestation and complete embryonic development within the mother **(18.22).**

Phylogeny of the Animal Kingdom (18.23–18.24) A phylogenetic tree showing hypotheses about the evolutionary history of animals is based on patterns of embryonic development and fundamental structural features of adult animals, such as tissues, bilateral symmetry, and body cavities **(18.23).** Evolutionary change and chance events, such as asteroid collisions, have shaped planet Earth and its organisms **(18.24).**

Testing Your Knowledge

Multiple Choice

1. Which of the following most clearly demonstrates the evolutionary relationship between a lobster and a millipede?
   a. a complete digestive tract
   b. an exoskeleton
   c. body segmentation
   d. bilateral symmetry
   e. radial symmetry

2. Jon found an organism in a pond, and he thinks it's a freshwater sponge. His friend Liz thinks it looks more like an aquatic fungus. How can they decide whether it is an animal or a fungus?
   a. see if it can swim
   b. figure out whether it is autotrophic or heterotrophic
   c. see if it is a eukaryote or prokaryote
   d. look for cell walls under a microscope
   e. determine whether it is unicellular or multicellular

3. Reptiles are much more extensively adapted to life on land than amphibians in that reptiles
   a. have a complete digestive tract
   b. lay shelled eggs
   c. are endothermic
   d. have legs
   e. go through a larval stage

4. A small group of worms classified in the phylum Gnathostomulida (not discussed in this chapter) live between sand grains on ocean beaches. These animals are unsegmented and lack a body cavity, but some have a complete digestive tract. These characteristics suggest that gnathostomulid worms might branch from the main trunk of the animal phylogenetic tree between
   a. flatworms and nematodes
   b. annelids and arthropods
   c. mollusks and annelids
   d. cnidarians and flatworms
   e. nematodes and annelids

5. A lamprey, a shark, a lizard, and a rabbit share all the following characteristics except
   a. gill structures in the embryo or adult
   b. vertebrae
   c. hinged jaws
   d. a dorsal, hollow nerve cord
   e. a post-anal tail

6. Which of the following categories includes the largest number of species? *(Explain your answer.)*
   a. invertebrates
   b. chordates
   c. arthropods
   d. insects
   e. vertebrates

7. Which of the following pairs of animals undergoes metamorphosis that includes a larval stage?
   a. human, hydra
   b. grasshopper, frog
   c. sea star, tunicate
   d. roundworm, lancelet
   e. human, sea anemone

Matching

1. Include the vertebrates	a. annelids
2. Medusa and polyp body forms	b. nematodes
3. The simplest bilateral animals	c. sponges
4. May have a separate origin from other animals	d. arthropods
5. Earthworms, polychaetes, and leeches	e. flatworms
6. Largest phylum of all	f. cnidarians
7. Closest relatives of chordates	g. mollusks
8. Body cavity is a pseudocoelom	h. echinoderms
9. Have a muscular foot and a mantle	i. chordates

Describing, Comparing, and Explaining

1. Birds are a lot like their reptile ancestors, but their bodies are highly modified for flight. What characteristics do birds share with reptiles? What are their adaptations for flight?

2. Compare the structure of a planarian (a flatworm) and an earthworm with regard to the following: digestive tract, body cavity, and segmentation.

3. Name two phyla of animals that are radially symmetrical and two that are bilaterally symmetrical. How do the overall lifestyles of radial and bilateral animals differ?

4. One of the key characteristics of arthropods is their jointed appendages. Describe four functions of these appendages, in four different arthropods.

Thinking as a Scientist

1. Construction of a dam and irrigation canals in an African country has enabled farmers to increase the amount of food they can grow. In the past, crops were planted only after spring floods; the fields were too dry the rest of the year. Now fields can be watered year-round. Improvement in crop yield has had an unexpected cost—a tremendous increase in the incidence of schistosomiasis, or blood fluke disease. Look at the blood fluke life cycle in Figure 18.6B and imagine that your Peace Corps assignment is to help local health officials control the disease. Why do you think the irrigation project increased the incidence of schistosomiasis? It is difficult and expensive to control the disease with drugs. Suggest three other methods that could be tried to prevent people from becoming infected.

2. A marine biologist has dredged up an unknown animal from the seafloor. Describe some of the characteristics she should look at to determine the animal phylum to which the creature should be assigned.

3. On the basis of relationships described in the text, sketch a phylogenetic tree like the one in Figure 18.23 for the seven major groups of vertebrates: mammals, reptiles, birds, amphibians, bony fishes, cartilaginous fishes, and agnathans. Use the ancestral group for the trunk of the tree. Show which groups branched from the ancestral group and which groups branched from them. (*Hint:* Bony and cartilaginous fishes are separate branches that probably arose at about the same time.)

Science, Technology, and Society

Coral reefs harbor a greater diversity of animals than any other environment in the sea. Australia's Great Barrier Reef has been protected as a marine reserve and is a mecca for scientists and nature enthusiasts. Elsewhere, such as Indonesia and the Philippines, coral reefs are in danger. Many reefs have been depleted of fish, and runoff from the shore has covered coral heads with sediment. Nearly all the changes in the reefs can be traced back to human activities. What kinds of activities do you think might be contributing to the decline of the reefs? What are some reasons to be concerned about this decline? Do you think the situation is likely to improve or worsen in the future? Why? What might the local people do to halt the decline? Should the developed countries help? Why or why not?

Media Resources

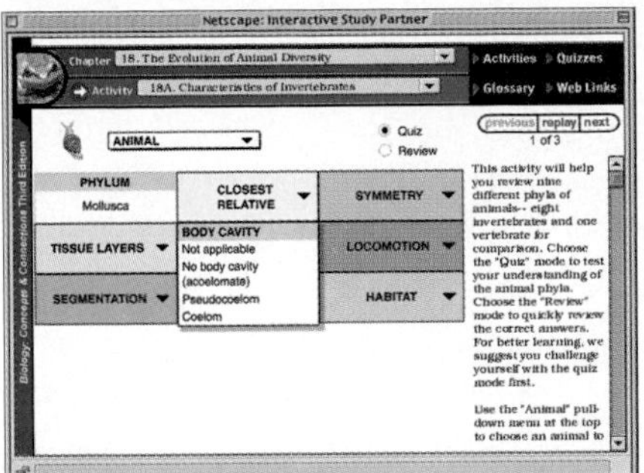

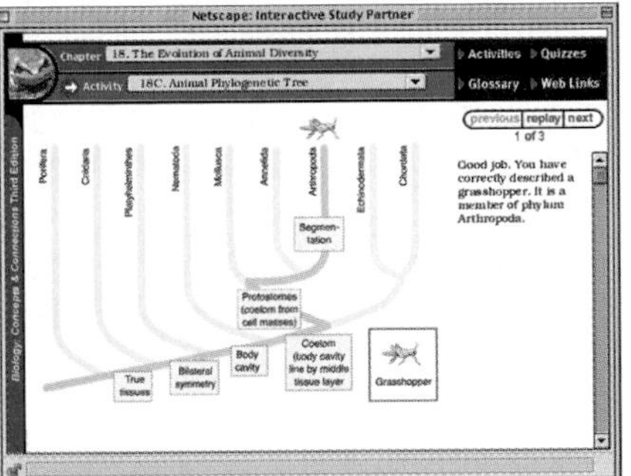

For further review of the concepts in this chapter, use the activities and quizzes on the CD-ROM.

Visit the Special Edition of The Biology Place web site for activities, quizzes, news articles, and web links.

19 Human Evolution

Outline

In early autumn of 1991, two hikers working their way along the edge of a melting glacier in the high Alps of northern Italy found what seemed to be the weathered remains of an unlucky mountain climber. It was a man clothed in hand-sewn leather, frozen in glacial ice. Next to him were a bow and several arrows, a wooden backpack, and a metal ax. A closer look turned up a leather pouch and other tools, shown in the photograph on the facing page. The "Ice Man" turned out to be a leftover from the Stone Age, a young hunter who may have died from exhaustion and exposure some 5000 years ago.

The question of the Ice Man's antiquity was settled soon after he was found. It took longer to mute debate about where he came from. Was he an early European or had someone with a strange sense of humor hauled an Egyptian or South American mummy to the site and heaved it into the glacier? Finally, in 1994, researchers reported that mitochondrial DNA recovered from the Ice Man closely matched that of central and northern Europeans, not Native Americans. His geographic ties confirmed, the Ice Man now re-

Hominid Evolution

19.3 The human branch of the primate tree is only a few million years old

19.4 Upright posture evolved well before our enlarged brain

19.5 *Homo erectus* gave rise to *Homo sapiens*

19.6 When and where did modern humans arise?

Our Cultural History and Its Consequences

19.7 Culture gives us enormous power to change our environment

19.8 Scavenging-gathering-hunting was the first major stage of culture

19.9 Agriculture was a second major stage of culture

19.10 The machine age is the third major stage of culture

poses in a frozen state in the anatomy department of the University of Innsbruck. Continuing analysis of his DNA may provide more clues about his place in human evolution.

The analysis of DNA from fossils provides an opportunity to better understand extinct life and to trace the ancestry of genes found in modern organisms, including humans. Scientists have reported finding fragments of ancient DNA from many fossils, including a 40-million-year-old insect preserved in amber (fossilized plant resin), a 65-million-year-old dinosaur fossil, and a 30,000-year-old fossil arm bone of an extinct member of the human family tree (one of a group found in northern Europe, called the Neanderthals).

Science advances by the ebb and flow of ideas. Hypotheses are proposed, predictions made, and test results evaluated. Too often, however, the advance is highlighted by the popular press as though it involves only a forward march, and the importance of disproving hypotheses is lost. Some of the research on fossilized DNA illustrates this key aspect of science. Because it is difficult to obtain uncontaminated samples from fossils, ancient DNA research is very challenging. Every report of success in isolating ancient DNA has been met with skepticism and further analyses to make sure the DNA traces were not contaminated with DNA from bacteria, fungi, or other organisms. DNA is unlikely to remain intact, except when organisms fossilize in extremely dry or cold places where organic material tends to be preserved. Contamination is a potential pitfall even in ideal conditions. In the case of the Ice Man, the chance of contamination was great because many people had touched the body before samples were taken. DNA may have been left behind with skin cells that everyone sheds constantly.

Exhaustive tests have shown that many of the early reports of ancient DNA, including that from dinosaurs, were in fact the results of contamination. However, DNA samples from most of the specimens in amber, from the Ice Man, and from the Neanderthal arm bone have withstood repeated testing.

In this chapter, we examine one of the most popular and controversial subjects in biology, the evolutionary history of our species, *Homo sapiens* (Latin for "wise man"). Molecular biology, including the search for DNA from human fossils, is part of the front line of research on human ancestry. But more traditional studies, such as comparisons of the structural features of fossils, are equally important in the quest to understand our heritage. We look first at some of the structural features that indicate our kinship with other members of the animal kingdom.

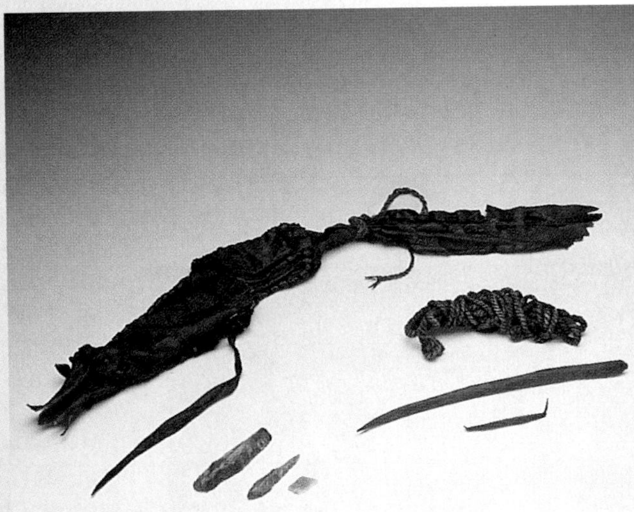

Pouch, string, antler scraper, bone awl, and flint materials found on the body of the Ice Man

Primate Diversity

19.1 The human story begins with our primate heritage

19A
See Activity 19A on the CD-ROM.

Humans, apes, monkeys, and lemurs all belong to the mammalian order Primates. The earliest primates were probably small arboreal (tree-dwelling) mammals that arose when dinosaurs still dominated the planet. Most living primates are arboreal, and the primate body has a number of features that were shaped, through natural selection, by the demands of living in the trees. Although humans probably never lived in trees, the human body retains many of the traits that evolved in our arboreal ancestors, and we can learn a great deal about ourselves and our relationship with the environment by studying other primates.

The squirrel-sized slender loris in Figure 19.1A illustrates a number of the basic primate features. It has limber shoulder and hip joints, enabling it to climb and move gracefully in trees. Also, the five digits of its hands and feet are highly mobile; its thumbs and big toes are opposable to the other digits, giving it the ability to hang onto branches and manipulate food. The great sensitivity of the hands and feet to touch also aids in manipulation. Moreover, lorises have a short snout and eyes set close together on the front of the face. The position of the eyes makes their fields of vision overlap, enhancing depth perception, an important trait for arboreal maneuvering. We humans share all these basic primate traits with the slender loris except for opposable big toes.

The slender loris belongs to a group of primates called the **prosimians** (Greek *pro,* before, and Latin *simia,* ape). Fossil prosimians, the oldest known primates, date back 65 million years, to the period of mass extinctions that saw the end of the dinosaurs (see Module 15.5). The prosimians are one of the mammalian lineages that expanded extensively after the dinosaurs became extinct.

Figure 19.1A The slender loris, a prosimian

Today, the prosimians are represented by about 35 species, including lorises, bushbabies, and tarsiers of Africa, India, and Southeast Asia and the lemurs of Madagascar. Ranging from the size of a mouse to the 10-kg (22-lb) weight of the colorful indri of Madagascar, shown in Figure 19.1B, lemurs are the most diverse prosimians. Most are agile climbers and leapers that spend nearly all their time in trees. All prosimians live in tropical forests, and nearly all are threatened by habitat destruction. Lemurs, for instance, occur in nature only in Madagascar, a Texas-sized island in the Indian Ocean, about 420 km off the eastern coast of Africa. Severed by plate tectonics from the African continent, the island has been an isolated hotbed of tropical speciation for well over 100 million years. Of about 40 species of lemurs originally present, 18 have become extinct since humans first colonized Madagascar about 2000 years ago.

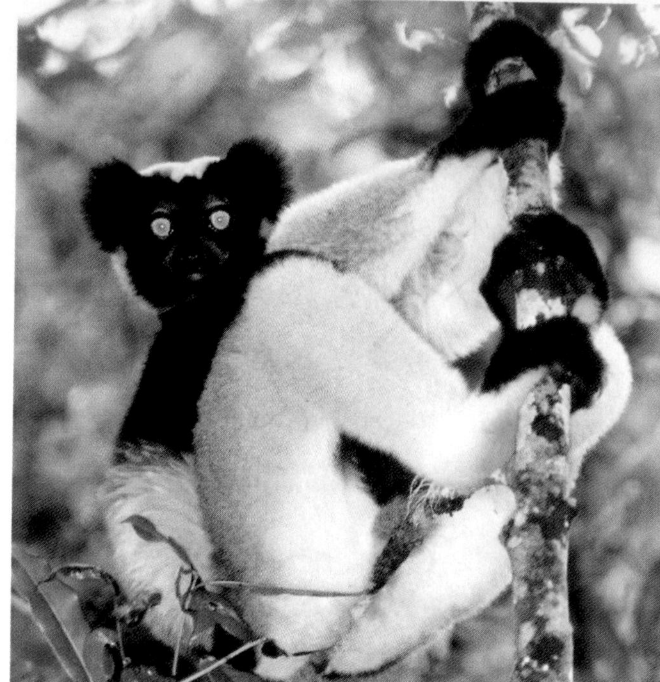
Figure 19.1B An indri, one species of lemur

The other group of primates, the **anthropoids** (Greek *anthropos,* man, and *eidos,* form), includes monkeys, apes, and humans. Anthropoids generally have a larger brain relative to body size and rely more on eyesight and less on sense of smell than prosimians. The earliest anthropoids probably were monkeylike primates that arose from prosimian ancestors some 40 million years ago in Africa and Asia. Monkeys differ from humans and from most apes in having forelimbs that are about equal in length to their hind limbs. Monkeys originated in Africa, spread throughout that continent and much of Asia, and finally reached the Americas. They may have migrated to the Americas on floating logs or via a bridge of land that stretched from Asia to what is now Alaska. The so-called Old World (Africa and Eurasia) and the New World (the Americas) have different types of monkeys, which have been evolving along separate pathways for millions of years.

Old World monkeys include some ground-dwelling species such as the baboons of the African savanna (Figure 19.1C, left), as well as many arboreal species, such as the

Baboons

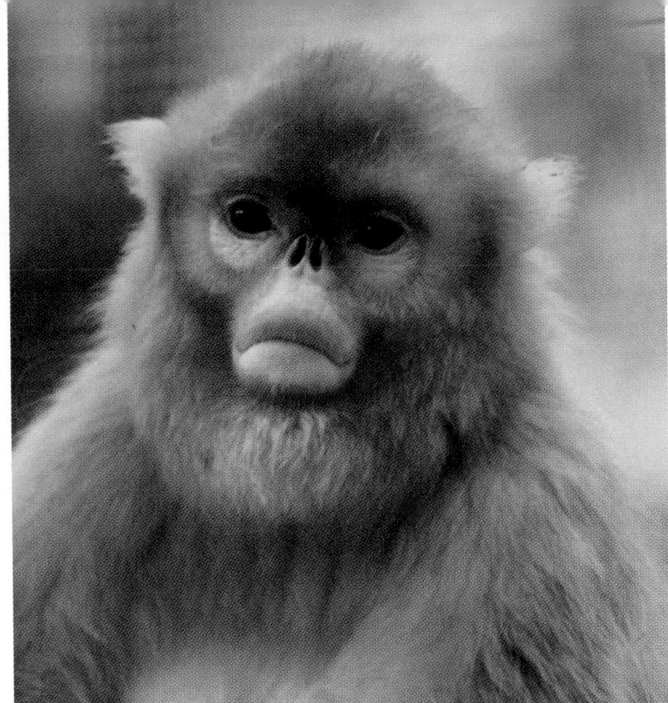

Hanuman langur

Figure 19.1C Old World monkeys

Hanuman langur of the southern Himalayas (Figure 19.1C, right). Defining features of Old World monkeys are nostrils that are narrow and close together. Many (baboons, for example) also have a tough seat pad.

New World monkeys, found in Central and South America, are all arboreal. Their nostrils are wide open and far apart, none has a seat pad, and many, such as Geoffrey's spider monkey, an inhabitant of rain forests in Central America (Figure 19.1D, left), have a long tail that is prehensile—specialized for grasping tree limbs. (Many Old World monkeys have a tail, but it is not prehensile.) A truly striking New World monkey, the squirrel-sized golden lion tamarin (Figure 19.1D, right), inhabits lowland rain forests of eastern Brazil. With most of its habitat destroyed by housing developments, this species has been reduced to only about 100 individuals in the wild and is the subject of an intense international conservation effort to save them from destruction. 

?

(a) To which mammalian order do we belong?
(b) What are the two main subgroups of this order?

(a) primates; (b) prosimians and anthropoids

Figure 19.1D New World monkeys

Geoffrey's spider monkey

Golden lion tamarin

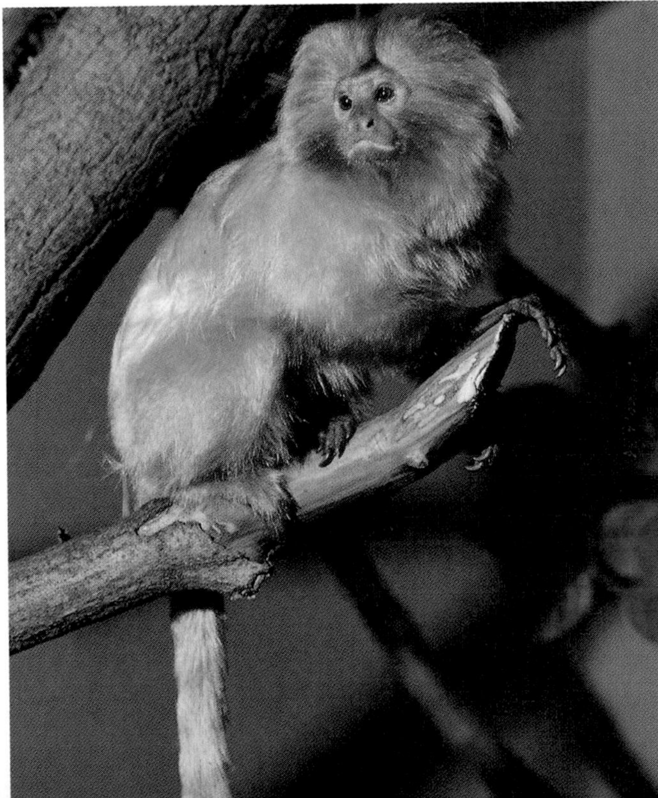

19.2 Apes are our closest relatives

19A

Apes, including the gibbons, orangutan, gorilla, and chimpanzee, are closely related to humans. In contrast to the monkeys, the apes are all confined to tropical areas (mainly rain forests) in Southeast Asia and Africa. Apes lack a tail and have forelimbs that are longer than their hind limbs. They are chiefly vegetarians, although chimpanzees also eat insects and some vertebrates, such as young antelope, pigs, and monkeys.

Nine species of **gibbons** (Figure 19.2A), all found in Southeast Asia, are the only entirely arboreal apes. Gibbons are the smallest, lightest, and most agile of the apes. They are also the only apes that are monogamous, with mated pairs remaining together for life.

Figure 19.2A A gibbon

Figure 19.2B An orangutan

The **orangutan** is a shy, solitary species that lives in the rain forests of Sumatra and Borneo. The largest living arboreal mammal, it moves rather slowly through the trees, supporting its stocky body with all four limbs (Figure 19.2B). In contrast to the gibbons, the orangutan may often venture onto the forest floor.

The **gorilla** (Figure 19.2C) is the largest of all primate species, with some males weighing over 200 kg (440 lb). Living in African rain forests, the gorilla can climb trees, but with its massive body, it spends nearly all of its time on the ground. The gorilla is often called a knuckle walker because when it is on all fours, the knuckles of its hands contact the ground. Gorillas also can stand upright on their hind legs.

Figure 19.2C A gorilla

Figure 19.2D A chimpanzee

Like the gorilla, the **chimpanzee** and a very similar species called the **bonobo** are knuckle walkers. These apes spend as much as a quarter of their time on the ground. Both species inhabit rain forests in central Africa, and the chimpanzee is also found in African savannas. Chimpanzees have been studied extensively, and many aspects of their behavior resemble human behavior. For example, chimpanzees make and use simple tools. The individual in Figure 19.2D is using a blade of grass to "fish" for termites. Chimpanzees also raid other social groups of their own species, exhibiting behavior formerly believed to be uniquely human. Researchers have demonstrated repeatedly that chimpanzees can learn human sign language. However, we do not yet know what role symbolic communication plays in the behavior of wild chimpanzees.

One of our most entrenched beliefs is that humans are the only thinking, self-aware beings. The behavior of chimpanzees in front of mirrors, however, challenges this belief. When first introduced to a mirror, a chimpanzee responds the way most other animals do—as if it were seeing another individual of its species. After several days, though, a chimp will begin using a mirror in ways that indicate it has a concept of self. It will inspect its face and other parts of its body that it cannot see without the mirror. It will also make faces at the mirror, using expressions different from those used in communicating with others.

Recent biochemical evidence indicates that the chimpanzee and the gorilla are more closely related to humans than they are to other apes. Humans and chimpanzees are especially closely related; human DNA differs from chimpanzee DNA by less than 3%. Primate researchers are acutely aware of the special significance of the great apes to us. In the words of chimpanzee authority Jane Goodall, "The most important spin-off of the chimp research is probably the humbling effect it has on us who do the research. We are not, after all, the only aware, reasoning beings on this planet." 

? The apes most closely related to humans are the ______________.

chimpanzees

19.3 The human branch of the primate tree is only a few million years old

19B

A popular misconception is that humans descended from apes that were just like chimpanzees or gorillas. Instead, apes and **hominids** (species on the human branch of the evolutionary tree; green in Figure 19.3A) diverged from a common, ape*like* ancestor, probably between 6 and 8 million years ago.

Figure 19.3B is a rough timeline of human evolution. The evolutionary connections are still being debated, but we humans clearly have several different species in our lineage, including two species of the genus *Homo* in addition to *Homo sapiens*. The vertical bars indicate the approximate time when each species existed (see the time scale on the left), as judged from the fossil record. The overlaps in the bars represent the coexistence of two or more hominid species.

Much debate centers on the evolutionary relationships among the extinct hominids. One very old hominid, a species called *Ardipithecus ramidus*, has been identified from fossilized bone fragments about 4.4 million years old. Other early hominids were a diverse group called **australopithecines** (genus *Australopithecus*, meaning "southern ape"). As the figure indicates, there were probably several australopithecine species; their fossils are distinguished by skull and tooth anatomy. Some of the australopithecines were probably contemporaries of *A. ramidus*. Were these species closely related? The australopithecines also overlapped in time with species of *Homo*. Were the australopithecines all evolutionary sidelines or were some of them on a direct line of descent to modern humans? Our own species, *Homo sapiens*, is the sole remaining member of the hominid family tree. We trace the threads of human evolution in more detail in the next three modules.

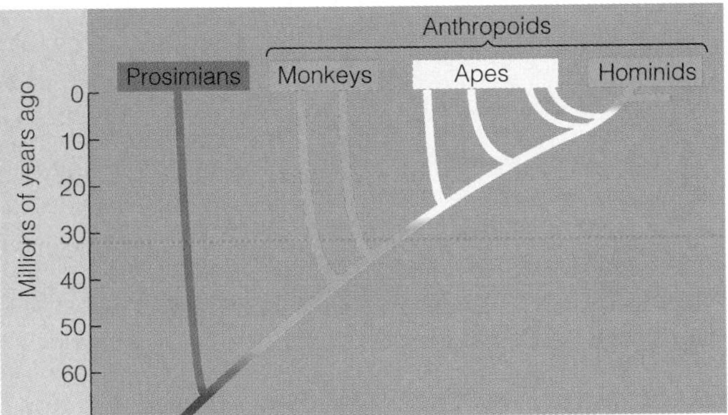

Figure 19.3A Primate evolution

Based on the fossil evidence in Figure 19.3B, how many hominid species existed 1.7 million years ago?

Four

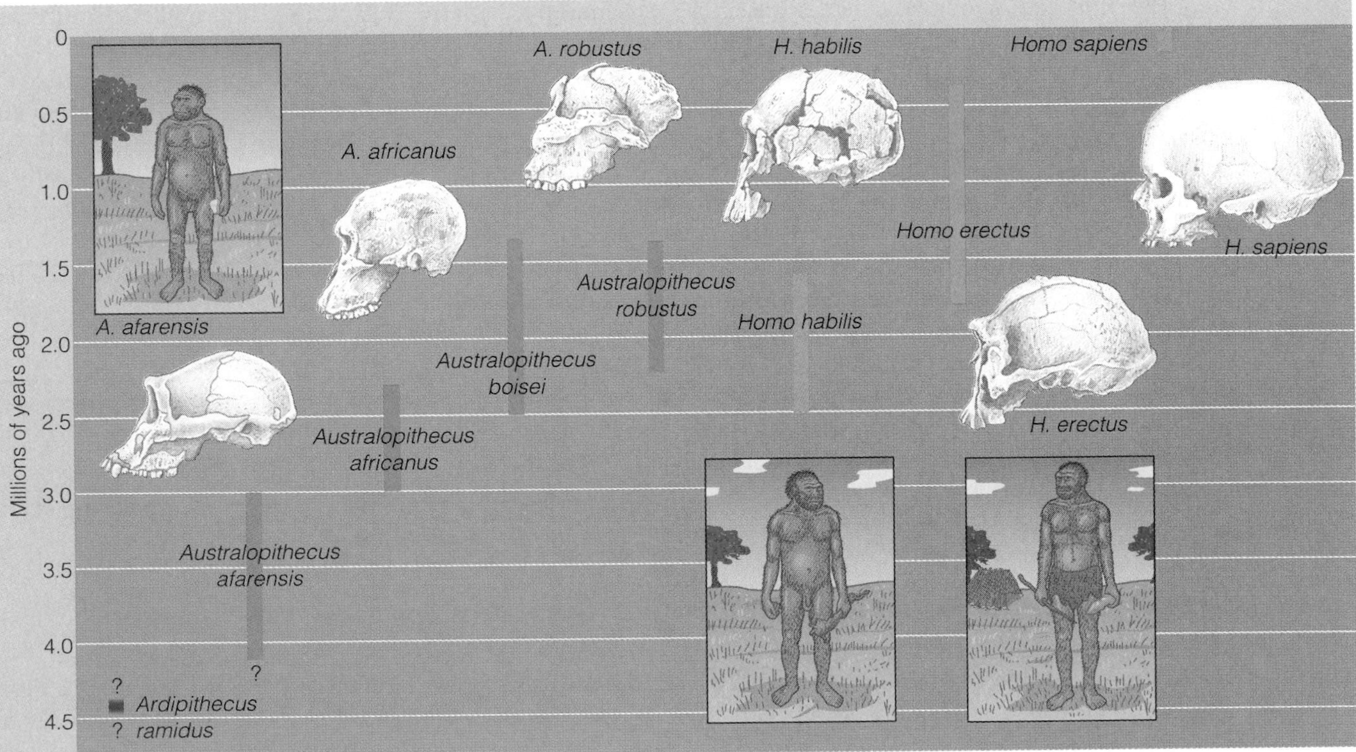

Figure 19.3B Human evolution timeline

19.4 Upright posture evolved well before our enlarged brain

19B

Some 3.7 million years ago, several bipedal (upright-walking) human animals of the species *Australopithecus afarensis* left footprints in damp volcanic ash in what is now Tanzania in East Africa (Figure 19.4A). The prints fossilized and were discovered by British anthropologist Mary Leakey in 1978. The footprints are part of the strong evidence that upright walking (bipedalism) is a very old human trait. Doubt remains whether the 4.4-million-year-old *Ardipithecus* was bipedal, but australopithecines dating from about 4 million years ago clearly were. *Ardipithecus* and some of the oldest australopithecines were forest dwellers, and some anthropologists now question the long-standing hypothesis that the first two-footed hominids lived on savannas (grasslands interspersed with trees). Bipedalism may have arisen in forest-dwelling hominids and allowed them to expand their range into savannas. One of the most complete fossil skeletons of an australopithecine, an *afarensis* type, dates to about 3 million years ago in East Africa. Nicknamed Lucy by her discoverers, the individual was a female, only about 3 feet tall, with a head about the size of a softball. Lucy and her kind lived in savanna areas (Figure 19.4B) and may have subsisted on nuts and seeds, bird eggs, and whatever animals they could catch or scavenge from kills made by more efficient predators such as large cats and dogs. They probably spent most of their time foraging on the ground, but they may have taken to the trees to escape predators, pick fruits, and steal animal carcasses stored there by leopards and other large cats.

All australopithecine species became extinct by about 1.4 million years ago. Whether they were on the lineage that gave rise to *Homo sapiens* or not, these early hominids show us that the fundamental human trait of bipedalism evolved millions of years before the other major human trait—an enlarged brain—became evident. As evolutionary biologist Stephen Jay Gould puts it: "Mankind stood up first and got smart later."

Figure 19.4A Ancient footprints

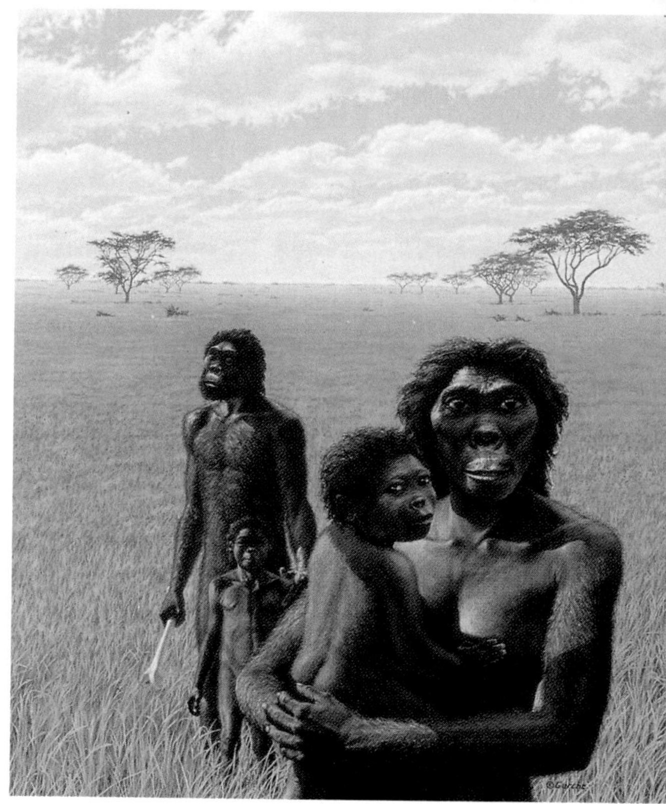

Figure 19.4B An artist's depiction of an australopithecine family

Enlargement of the human brain is first evident in fossils from East Africa dating to the latter part of australopithecine times, about 2.5 million years ago. Skulls have been found with brain capacities intermediate in size between those of the australopithecines and those of *Homo sapiens*. Simple stone tools are sometimes found with the larger-brained fossils, which have been dubbed *Homo habilis* ("handy man"). *Homo habilis* existed on the African savanna with the australopithecines for nearly a million years, probably scavenging, gathering, and hunting in much the same way as its smaller-brained relatives. *H. habilis* may have given rise to another ancestor of ours, called *Homo erectus*.

According to the evidence we now have, the first hominid genus that was definitely bipedal was ________.

Australopithecus

19.5 *Homo erectus* gave rise to *Homo sapiens*

19B

Fossils of an extinct hominid called *Homo erectus* ("upright man") range in age from about 1.8 million years to 250,000 years. *H. erectus* was taller than *H. habilis*, had a larger brain, and had a more advanced culture. According to most accounts, *H. erectus* was the first hominid to spread out of Africa to Eurasia. To survive in the colder climates of the north, *H. erectus* lived in huts or caves, built fires, wore clothes made of animal skins, and designed more elaborate stone tools than those of *H. habilis.* Spread over much of the Old World, populations of *H. erectus* became regionally diverse. One or more of the *H. erectus* populations gave rise to *Homo sapiens,* and we track some of the ways that might have occurred in the next module.

? Based on fossil evidence, what was the approximate duration of *Homo erectus*'s existence as a species?

About 1.5 million years

When and where did modern humans arise? 19.6

The oldest known fossils of *Homo sapiens,* dating back over 300,000 years, are found in Africa. Humans of that time, apparently the earliest descendants of *H. erectus,* were generally heavier boned, with a markedly thicker skull and more pronounced brow ridges, than people living today. Experts often call these early forms of our species "archaic *Homo sapiens.*" By about 100,000 years ago, archaic *H. sapiens* were found throughout the Old World. The archaic group we know most about were the Neanderthals, who lived in Europe, the Middle East, and parts of Asia from about 130,000–35,000 years ago. The Neanderthals were relatively short and stocky, with a heavily muscled body. They were skilled toolmakers, and they participated in rituals that required abstract thought. Figure 19.6 is a museum reconstruction of a neanderthal family group.

The oldest known fossils of fully modern *Homo sapiens*—humans that if dressed in modern clothes would not stand out in a crowd—date to about 100,000 years ago in Africa. Similar fossils nearly as old have been found in caves in Israel. Apparently, these modern humans coexisted with the archaic types until about 35,000 years ago, when archaic humans disappeared. There is heated debate over what happened to the archaic groups and over the related questions of when and where modern *Homo sapiens* arose.

Some anthropologists argue from fossil evidence that populations of modern *Homo sapiens* evolved from archaic populations in Africa, Europe, Asia, and Australia. Proponents of this "multiregional hypothesis" contend that the modern races of humankind stem from the regional diversity that evolved over a million years ago in *H. erectus* populations. Despite racial differences, however, all living humans are very similar genetically. Multiregionalists believe that interbreeding among neighboring populations accounts for our genetic likeness.

In sharp contrast to the multiregional hypothesis is a proposal that modern *Homo sapiens* arose from a single archaic group in Africa. According to this "monogenesis hypothesis," the Neanderthals and archaic peoples outside Africa were evolutionary dead ends. Proponents of this hypothesis argue that modern humans spread out of Africa about 100,000 years ago and completely replaced the archaic peoples living elsewhere. Most of the support for this idea stems from molecular data—specifically, analyses of DNA found in the mitochondria of human cells (mtDNA). The mtDNA in today's global human population seems to be very uniform, and supporters of monogenesis argue that such uniformity could only stem from a single ancestral stock, not from diverse, geographically isolated populations. Researchers are also finding that nuclear DNA is more variable in modern humans in Africa south of the Sahara than in human populations elsewhere in the world. Monogenesis predicts this result: If modern humans arose only in Africa, the population there would have had the longest time to develop genetic diversity.

Figure 19.6 Neanderthals, about 50,000 years ago. (Chewing animal skins was part of a leather-making process.)

Proponents of monogenesis argue that the African-derived population of *H. sapiens* not only replaced the archaic peoples elsewhere but did not interbreed with them at all. One way to test this idea is to isolate DNA from human fossils found in various regions and compare it with DNA from living humans. Monogenesis would be supported if the comparisons indicated that the DNA from archaic peoples outside Africa was markedly different from that of the modern population. As we mentioned in the introduction to this chapter, molecular biologists have successfully isolated mitochondrial DNA from a Neanderthal fossil found in Europe. Their results showed some distinct differences between the Neanderthal DNA and that of modern humans and seemed to support the monogenesis hypothesis. However, as the scientists themselves were quick to point out, these results are far from conclusive because the DNA was obtained from only one fossil. We can expect to hear considerably more debate before the question of when and where modern humans arose is settled.

? In spite of their debate, proponents of the "multiregional" and "monogenesis" hypotheses at least agree that *Homo sapiens* originated on the continent of ________.

Africa

Our Cultural History and Its Consequences

19.7 Culture gives us enormous power to change our environment

Three major milestones highlight the evolution of *Homo sapiens:* (1) the evolution of our erect stance, which required major remodeling of the entire skeleton; (2) enlargement of the brain, which paralleled a prolonged period of growth of the skull and its contents after birth (see Module 15.7); and (3) the evolution of a prolonged period of parental care. Humans care for their offspring far longer than any other species, giving young children the chance to learn from the experiences of earlier generations. This is the basis of **culture**—the accumulated knowledge, customs, beliefs, arts, and other human products that are socially transmitted over the generations. More than any other factor in human history, culture has made *Homo sapiens* a unique force in the history of life on Earth—a species that can defy its physical limitations and alter nature at a rate far exceeding that of biological evolution. Our culture enables us to change the environment to meet our needs. As we will see in Chapter 35, it has also allowed the human population to explode and, in effect, threaten the existence of many other species and challenge Earth's capacity to sustain us. In the next three modules, we examine three stages of culture, each of which gave humans increasing power to alter the environment. 

? What role does a long period of parental care play in culture?

Extends the opportunity for parents to transmit the lessons of the past to offspring

19.8 Scavenging-gathering-hunting was the first major stage of culture

The first stage of culture, called scavenging-gathering-hunting for its three central activities, began with the earliest hominids in Africa. It continued to be the way of life for the australopithecines and species of *Homo,* including modern humans, during most of the last 100,000 years. Actually, for most of human existence, people have probably relied more on scavenging, especially stealing fresh kills from other predators, and gathering wild fruits, seeds, and vegetables, than on hunting. Only in the last 50,000 years or so did toolmaking become sophisticated enough to allow hunting to become a major food-producing activity. Gathering and hunting continue successfully to this day in various societies. Figure 19.8 shows a !Kung tribesman, a skilled bow hunter. The !Kung, who live in southwestern Africa, are one of several hunter-gatherer peoples of modern times. (The exclamation point represents a clicking sound used in their language.)

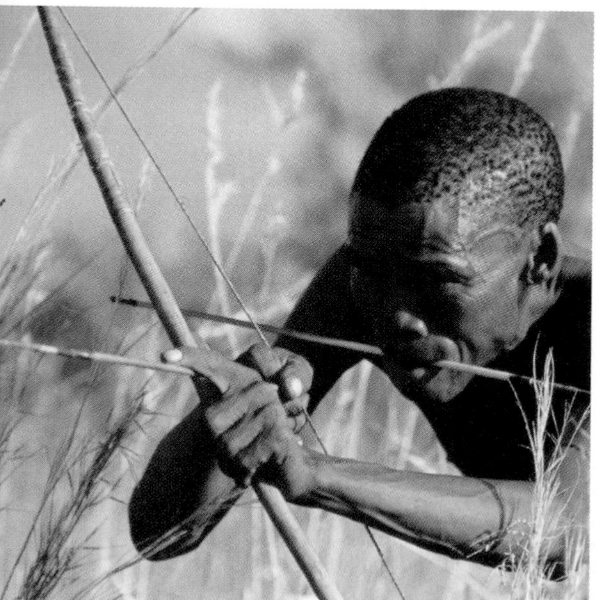

Figure 19.8 A !Kung hunter with bow and arrows

When did human culture begin to have a major influence on the environment? It has been suggested that even scavenging by bands of australopithecines or *Homo habilis* could have created serious problems for other species. Several lion-sized predators, saber-toothed cats, became extinct in Africa some 1.5 million years ago. Could early hominids have become so adept at stealing the kills of these carnivores that they seriously reduced the cats' food supply? Some anthropologists think it is likely.

With the development of tools that could be thrown and cooperative hunting techniques, humans began to have profound effects on certain other species. Fossil records indicate that *Homo sapiens* decimated populations of woolly rhinoceroses and giant deer in Europe. About 50,000 years ago, modern humans reached Australia by boat and may have killed off that continent's giant kangaroos. Nomadic hunters migrated from Asia to North America via the Bering land bridge about 30,000 years ago. These early migrants may also have pushed to extinction some large mammals that they hunted.

Hunter-gatherers not only made tools, they also organized communal activities and divided labor. Many groups also developed semipermanent residences near rich hunting grounds and along seacoasts and lakes. Some grew crops and traded with other groups. These activities ushered in a new stage of culture—the rise of agriculture. 

? Humans became better hunters as they developed ________.

tools

Agriculture was a second major stage of culture 19.9

The 5000-year-old Ice Man discussed in this chapter's introduction probably lived in a small village whose inhabitants grew some food and supplemented it by hunting and gathering. Farming was already an old practice, with roots going back to about 10,000–15,000 years ago in Eurasia and the Americas.

Major agricultural centers sprang up in the Middle East, in an area called the Fertile Crescent. Extending northeast from the Nile River in Africa and including the modern nations of Israel, Syria, Iraq, and Kuwait, the Fertile Crescent was once covered with forests and grasses and had rich soils. Farmers cleared the forests and exploited the soil for years, growing mostly wheat, without serious loss in fertility. About 5000 years ago, farmers in the area began using primitive plows, like the one shown here of an Egyptian farmer. The plow broke up the thick sod of grasslands and opened vast new areas for farming. With more intensive agriculture, local populations increased, placing an ever-increasing burden on the soil to produce food. Herds of domestic animals overgrazed the land, leaving it exposed to eroding winds and rains, while the crops eventually depleted the soil of nutrients. Today, much of the Fertile Crescent is a desert.

In adopting farming as a way of life, *Homo sapiens* took the first big step toward becoming the dominant species on Earth. Along with agriculture came permanent settlements and the first cities. Fewer people were needed to grow a group's food, and many people could specialize in other activities. Agriculture thus paved the way for the development of industry and technology—the next great cultural wave, which continues today. 

Figure 19.9 A primitive plow

? What is the connection between agriculture and cities?

Agriculture made food sources more localized so that larger settlements were possible, and the increasing efficiency of agriculture provided time for people to engage in activities other than food gathering.

The machine age is the third major stage of culture 19.10

The Industrial Revolution, which began in England in the eighteenth century, brought a switch from small-scale, hand production of tools and other goods to large-scale machine production. With the machine age came increasing demands for energy sources to fuel the machines—mostly timber and coal at first, then oil, and recently also hydroelectric power and nuclear power. The invention of the tractor and other farm machinery reduced the need for farm laborers, and many people migrated to cities in search of work. As more food was produced and medical advances reduced the number of deaths from diseases, the human population began to grow faster throughout the world. Both the human population and technology are now escalating rapidly.

Through the cultural changes from scavenging-gathering-hunting to high-tech societies, humans have not changed biologically in any significant way. We are probably no more intelligent than our forebears who lived in caves. The same toolmaker who chipped away at stones now fashions microchips for computers. The know-how to build computers and spaceships is stored not in our genes but in what is passed along by parents, teachers, and books. We are presently the most numerous and widespread of all large animals, and everywhere we go, we bring change. As we saw when we discussed biological evolution in Unit III, there is nothing new about environmental change. What is new is the *speed* of change, for cultural changes outpace biological evolution by orders of magnitude. As we discuss in the final chapter of this book, we are changing the world so quickly that many species cannot adapt—in fact, we may be jeopardizing our own existence as well. 

Figure 19.10 From eighteenth-century looms to twentieth-century computer chips

? How has technology contributed to growth of the human population?

By making more food available and by reducing disease

CHAPTER REVIEW

Chapter Summary

Primate Diversity (Introduction–19.2) Humans are members of an order of mammals, the primates, that first appeared about 65 million years ago. The first primates lived in trees, and we have inherited some of their characteristics: limber joints, sensitive grasping hands, a short snout, and forward-pointing eyes that enhance depth perception. There are two groups of living primates: the prosimians (such as lorises and lemurs) and the anthropoids (monkeys, apes, and humans) **(19.1)**. Humans are most closely related to the apes, primates that lack tails and have forelimbs longer than their hind limbs. The apes include gibbons, orangutans, gorillas, chimpanzees, and bonobos. We share more than 97% of our genes with chimpanzees, our closest living relatives; our behavior also has some similarities **(19.2)**.

Hominid Evolution (19.3–19.6) Apes and hominids, species on the human branch of the evolutionary tree, probably diverged from a common apelike ancestor between 6 and 8 million years ago. The earliest hominids were a diverse group. The oldest so far discovered, a forest dweller called *Ardipithecus ramidus,* dates from about 4.4 million years ago. The oldest member of another group, the australopithecines (genus *Australopithecus*), dates from about 4 million years ago. Our own species *Homo sapiens* is the only hominid that has not become extinct **(19.3)**. The australopithecines, including the earliest species, walked upright on two legs and lived in Africa. Their small brains, relative to body size, indicate that the evolution of the basic human trait of bipedalism preceded the evolution of the enlarged brain. Anthropologists debate whether the australopithecines were ancestors of the genus *Homo* or were on a separate evolutionary line. *Homo habilis,* an early African hominid, coexisted with some of the australopithecines, had a larger brain, and made simple tools. *Homo habilis* may have given rise to the more advanced *Homo erectus* **(19.4)**. *Homo erectus* had a larger brain than *H. habilis* and a more complex culture. *H. erectus* spread out of Africa over most of the Old World and eventually gave rise to *Homo sapiens* **(19.5)**. The oldest fossils of modern humans, *Homo sapiens,* are found in Africa and date back about 300,000 years. By about 100,000 years ago, archaic groups of *Homo sapiens,* with heavier bones and thicker skulls than modern humans, had spread throughout the Old World. One archaic group, the stocky and muscular Neanderthals, appeared about 130,000 years ago and died out about 35,000 years ago. Fully modern *Homo sapiens* date back about 100,000 years. Some anthropologists think that modern humans arose from geographically separate populations of archaic *Homo sapiens* in Africa, Europe, Asia, and Australia, and that Neanderthals and other archaic peoples outside Africa were all transitional stages in human evolution. Others believe that modern *Homo sapiens* arose only from an archaic group in Africa, and the African-derived modern group replaced archaic peoples elsewhere **(19.6)**.

Our Cultural History and Its Consequences (19.7–19.10) Major milestones in the evolution of *Homo sapiens* are the evolution of an erect stance, a large brain, and a prolonged period of parental care. We have not changed much biologically since modern humans first appeared. But our culture—the accumulated knowledge, customs, beliefs, arts, and other products—has evolved enormously. There have been three main stages in cultural change **(19.7)**. The first humans survived by scavenging, gathering, and hunting—the three main components of the first stage of human culture. Early scavengers and hunters may have depleted populations of some of their prey and some competing carnivores **(19.8)**. The second major stage in cultural change was the rise of agriculture about 10,000 to 15,000 years ago, when people settled down, gradually began growing some food, and began domesticating animals. Early farmers in the Fertile Crescent allowed their land to be overgrazed and depleted the soil, leaving much of the Middle East a desert **(19.9)**. The third stage in cultural change, the Industrial Revolution, began in the 1700s. Industrialization brought a change from hand production to energy-intensive, large-scale machine production. Mechanized farming and improved medicine then accelerated the growth of the human population and our impact on the environment **(19.10)**.

Testing Your Knowledge

Multiple Choice

1. The two major groups of primates are
   a. monkeys and anthropoids
   b. prosimians and apes
   c. monkeys and apes
   d. prosimians and anthropoids
   e. Old World monkeys and New World monkeys

2. Which of the following correctly lists probable ancestors of modern humans from the oldest to the most recent?
   a. *Homo erectus, Australopithecus, Homo habilis*
   b. *Australopithecus, Homo habilis, Homo erectus*
   c. *Australopithecus, Homo erectus, Homo habilis*
   d. *Homo erectus, Homo habilis, Australopithecus*
   e. *Homo habilis, Homo erectus, Australopithecus*

3. Fossils suggest that the first major trait distinguishing human primates from other primates was
   a. forward-facing eyes with depth perception
   b. a larger brain
   c. erect posture
   d. grasping hands
   e. toolmaking

4. Some anthropologists believe that the modern races of *Homo sapiens* evolved from separate populations of archaic *Homo sapiens* in different geographic areas. How, then, do proponents of this multiregional hypothesis explain the great degree of genetic similarity among all modern humans?
   a. The same mutations occurred in populations in different locations.
   b. There probably was interbreeding among neighboring populations.
   c. All *Homo sapiens* populations were shaped by similar environments.
   d. The ancestral *Homo erectus* originally came from Africa.
   e. Modern races of humans are not at all genetically similar.

5. Which species seems to have survived the longest?
   a. *Homo erectus*
   b. *Australopithecus africanus*
   c. *Homo habilis*
   d. *Homo sapiens*
   e. *Australopithecus robustus*

6. The earliest hominids
   a. had a very large brain compared to all other primates
   b. probably hunted dinosaurs
   c. were prosimians
   d. lived about 1.5 million years ago
   e. were bipedal

7. In which way are we significantly different from our 100,000-year-old ancestors?

a. We use tools.
b. We are much more intelligent.
c. We have accumulated more information.
d. We converse.
e. We are more emotional.

Describing, Comparing, and Explaining

1. What adaptations inherited from our primate ancestors enable humans to make and use tools?

2. In what ways is chimpanzee behavior similar to human behavior?

3. Explain why humans have been able to expand our numbers and distribution to a greater extent than any other animal.

4. Describe the lifestyle of an early hominid.

Thinking as a Scientist

1. Anthropologists are interested in locating areas in Africa where fossils 4–10 million years old might be found. Why?

2. Some researchers think that drying and cooling of the climate caused expansion of the African savanna, and that this environment favored upright-walking early humans. Why might bipedalism be advantageous in the savanna? How might an erect posture relate to the evolution of a larger brain?

Science, Technology, and Society

The human body has not changed much in the last 100,000 years, but human culture has changed a great deal. As a result of our culture, we change the environment at a rate far greater than many species, including our own, can evolve. What evidence of rapid environmental change do you see regularly? What aspect(s) of human culture are responsible for these changes? Do you see any evidence of a decrease in the rate of human-caused environmental changes?

Media Resources

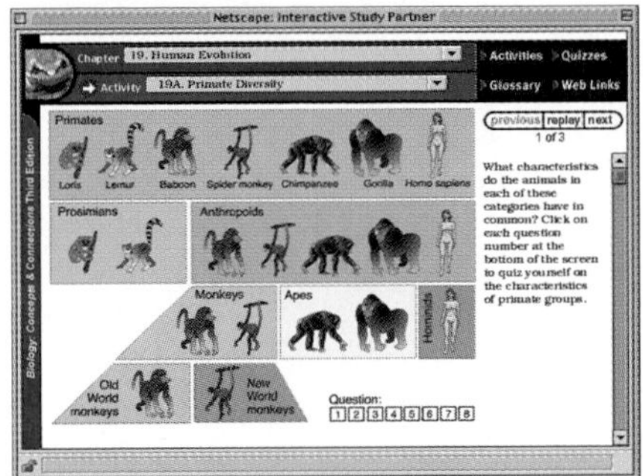

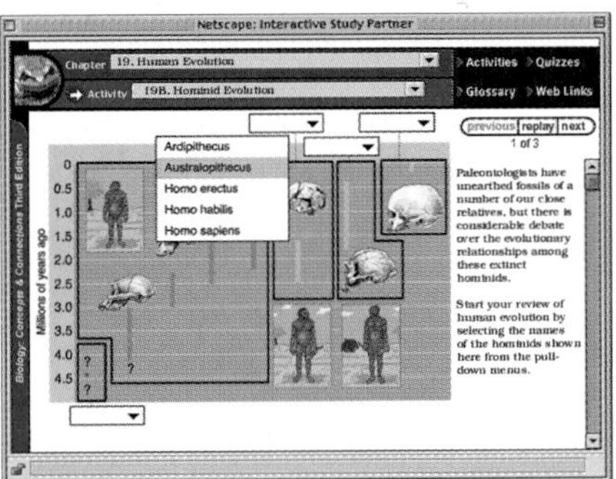

For further review of the concepts in this chapter, use the activities and quizzes on the CD-ROM.

Visit the Special Edition of The Biology Place web site for activities, quizzes, news articles, and web links.

UNIT V

Animals: Form and Function

20 Unifying Concepts of Animal Structure and Function

Outline

Flying over the frigid waters of the South Atlantic, the two birds at the left are living up to their common name: wandering albatross. Their scientific name, *Diomedea exulans,* means "exiled warrior" in Greek and, indeed, an albatross spends most of its time far from land, often alone on the open ocean.

Only a few birds can live at sea, but the albatross is a model of fitness for this environment. It even drinks seawater, which is much saltier than its body fluids. Few animals can tolerate such salty liquid, because the salt draws water out of their tissues, and they become severely dehydrated. An albatross can thrive on salt water because it has special salt-excreting glands in its nostrils. The glands dispose of excess salts, allowing the bird to eat salty fish and squid and drink all it needs without ever visiting land.

Along with its remarkable salt tolerance, the wandering albatross has extraordinary flight abilities specially suited to its seafaring life. Its wing span is greater than that of any other bird—about 3.6 meters (nearly 12 feet)! Most birds have to flap their wings to stay up in the air very long, but the albatross's long wings provide so much lift that the bird can stay aloft for hours by gliding up and down on wind currents. Albatrosses frequent areas where steady winds blow over the ocean. The wandering albatross spends most of its time in an area of high winds and rough seas, between about 40° and 50° latitude in the Southern Hemisphere. The winds carry it from west to east, and in a year's time, an albatross may circle the globe, seeing little land except New Zealand and the tip of South America.

The drawing below illustrates how an albatross uses the energy of ocean winds to stay aloft. Because of the friction between air and water, wind speed is lowest where the air meets the ocean. On the open sea, where there are no obstructions, the wind speed regularly increases from the surface up to an altitude of about 15 meters. (The thickness of the blue arrows in the diagram is proportional to wind speed.) An albatross takes full advantage of these conditions. Almost effortlessly, it glides downward in the same direction as the wind, picking up momentum as it descends. Near the ocean surface, it turns sharply into the wind and uses its downward momentum to soar upward, the way a bicyclist uses downhill momentum to coast partway up the next hill. As the bird rises, the increasing wind helps lift it back up to its starting altitude. An albatross can glide up and down like this over and over—with just a few turns of its body—as long as the wind holds.

The albatross's special flight ability and salt tolerance are functions that result from special structural adaptations of its body. The connection between structure and function is a basic concept of biology, and this chapter is an introduction to how the concept applies to animals.

The Hierarchy of Structural Organization in an Animal

20.1 Structure fits function in the animal body

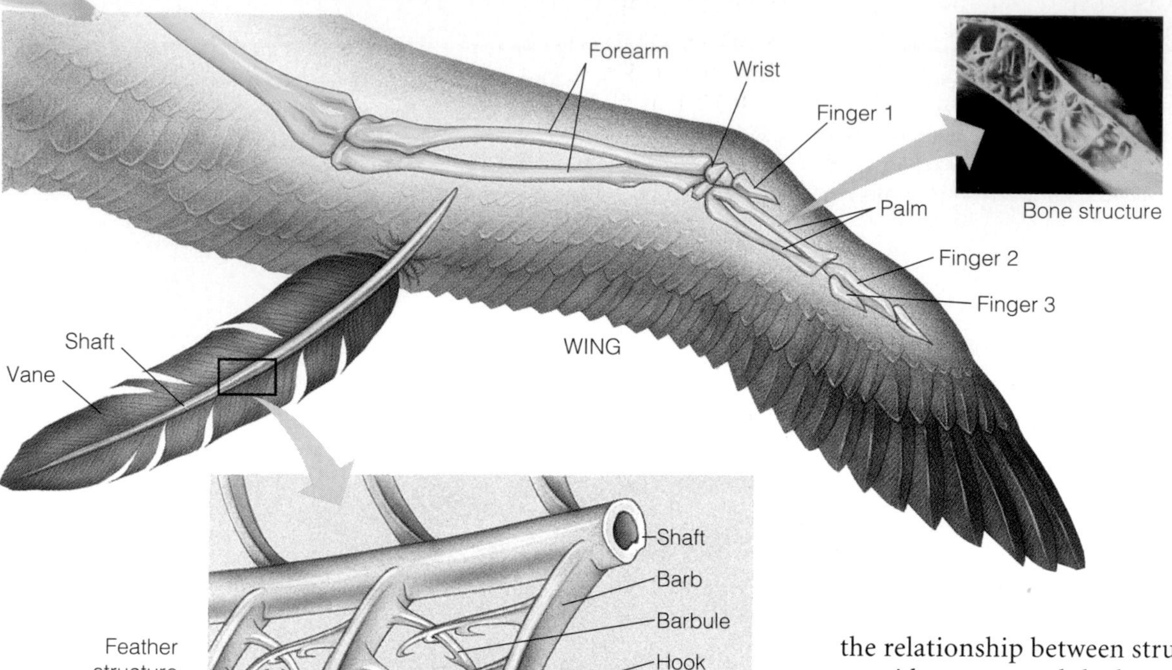

Figure 20.1 The structure of a bird wing

We can approach the study of an animal in several ways. An anatomist studying the albatross, for instance, would focus on structure, perhaps on the shape and size of the wing or the anatomy of its muscles, bones, feathers, and skin. A physiologist (a biologist who studies the function of body parts) might study the mechanics of flight or the action of muscles in holding the wings out for hours on end. Despite their different approaches, the anatomist and the physiologist are both working toward a better understanding of the connection between structure and function—that is, how structural adaptations give the albatross its remarkable ability to exploit the energy of ocean winds.

Let's look at how some of the parts of an albatross illustrate the correlation between structure and function. Take the feathers of the wing, for example. Feathers give the wing its shape and insulate it against the chill of the strong sea wind, without adding much weight to the body. They remain dry because they are lightly coated with oil, and they also trap air, which helps the bird float on the water.

The functions of feathers result from their unique structure. Produced by special pits in the bird's skin, feathers consist entirely of nonliving material, mainly the protein keratin. A flight feather, like the one enlarged in the figure above, has a hollow keratin shaft that provides trunklike support with minimum weight. A row of small flat rods called barbs extend from both sides of the shaft, forming the vane of the feather. Still finer rods called barbules extend from the sides of the barbs. Each barbule has tiny hooks that interlock with adjacent barbules. With all of its barbules interlocked, a feather has a particular shape and rigidity that support flight. When barbules become detached, these features are lost, and the bird's flight ability is impaired. When a bird preens, it draws its feathers through its beak, hooking the barbules back together in a zipperlike manner.

The muscles and bones in a bird wing also show the relationship between structure and function. Muscles provide power, and the bones provide support for flight. The flight muscles are situated on the breast and around the base of the wings, keeping most of the weight off the wings and helping the bird maintain balance in flight.

The bones in a bird wing are homologous with those in the human arm, but the number of bones in the wing diminished as birds evolved. The bird wing has only three fingers (numbered in the figure), and only the middle one (finger 2) has a complete set of bones. The bird's wrist and palm also have fewer bones than ours do. This adaptation helps make the wing lighter but less mobile than the human arm. The reduced mobility stabilizes the wing and makes it better able to function as a unit in flight. The photograph of the cut-open bone in the upper right corner of Figure 20.1 illustrates another significant adaptation in the bird skeleton. Many bones are hollow but reinforced internally with trusses similar to those used in airplane wings. This structure provides maximum strength with minimum weight, the ideal combination for flight.

The ability to fly thus results from special structural features of bones, muscles, feathers, and skin—arranged in a particular way to form the bird wing. It is one illustration of the universal theme of a specific function correlated with, and emerging from, specific structures. 

? You learned in this module that the skeleton of a bird's wing is homologous to the skeleton of a human arm. What is the evolutionary implication of this point? (*Hint:* Review Module 13.3.)

Both the bird wing and the human arm evolved from a vertebrate limb that was present in an ancestor common to birds and mammals.

Animal structure has a hierarchy 20.2

Structure in the living world is organized in a series of hierarchical levels. We saw one example in Module 1.1 and another in Module 20.1, on the facing page. At one level, a feather consists of molecules of keratin. At another level, resulting from different arrangements of the keratin molecules, a feather consists of a shaft, barbs, and barbules. At still another level, the way the barbs and barbules fit together forms the feather's vane.

The same hierarchy seen in an albatross exists in all animal bodies. The figure here illustrates a structural hierarchy in a zebra. Part 1 shows a single muscle cell in the zebra's heart. This cell's main function is to contract and relax in a coordinated way with the other heart muscle cells. Each heart muscle cell is branched, so that it can connect with several other similar cells. The multiple connections allow the cells to cooperate, coordinating their contractions, so the heart can pump blood effectively. Many thousands of these cells make up a tissue (part 2), the second level of structure and function. Heart muscle tissue is the main component of the heart walls.

Part 3, the heart itself, illustrates the organ level of the hierarchy. As an organ, the heart is made up of several types of tissue, including muscle tissue, nervous tissue, and connective tissue. Part 4 shows the cardiovascular system, the organ system of which the heart is a part. Organ systems have multiple parts. The other parts of the cardiovascular system are the blood vessels: arteries, veins, and capillaries. All the parts combine to carry out the function of this organ system, transporting blood throughout the body.

In part 5, the zebra itself forms the final level of this hierarchy: the organism—in this case, a large mammal. The whole animal consists of a number of organ systems, each specialized for certain functions and all functioning together as an integrated, cooperative unit. In Modules 20.3–20.7, we focus on the tissue level of the biological hierarchy.

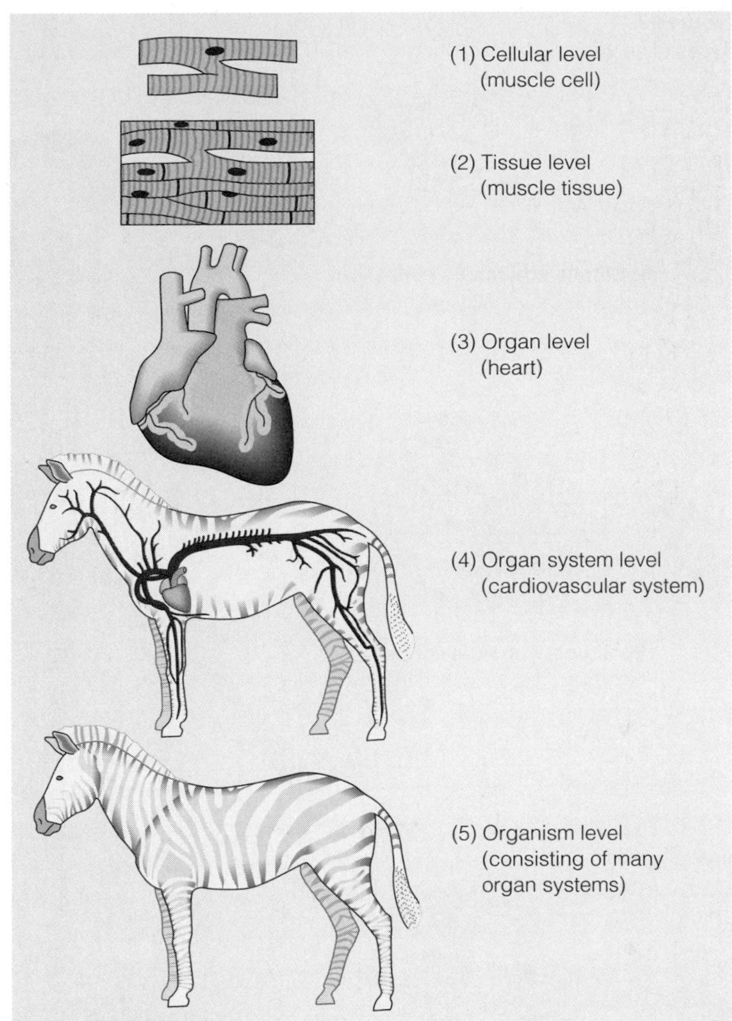

Figure 20.2 A structural hierarchy in a zebra

What level of animal structure does the human brain represent?

Organ

Tissues are groups of cells with a common structure and function 20.3

A tissue is a cooperative unit of many very similar cells that perform a specific function. Most of the cells in multicellular organisms, including animals, are organized into tissues. The cells composing a tissue are specialized; they have a particular structure that enables them to perform a specific task. As we saw in the last module, for example, each of the heart's muscle cells has branches that connect to several other muscle cells. The branches help the cells contract in a coordinated manner.

The term *tissue* is from a Latin word meaning "weave," and some tissues resemble woven cloth in that they consist of a meshwork of nonliving fibers surrounding living cells. Other tissues are held together by a sticky glue that coats the cells or by special junctions between adjacent plasma membranes (see Module 4.19). An animal has four major categories of tissue: epithelial tissue, connective tissue, muscle tissue, and nervous tissue. We examine each of these separately in the next four modules.

20A
See Activity 20A on the CD-ROM.

An organ is usually constructed from two or more different ________.

tissues

20.4 Epithelial tissue covers and lines the body and its parts

20B

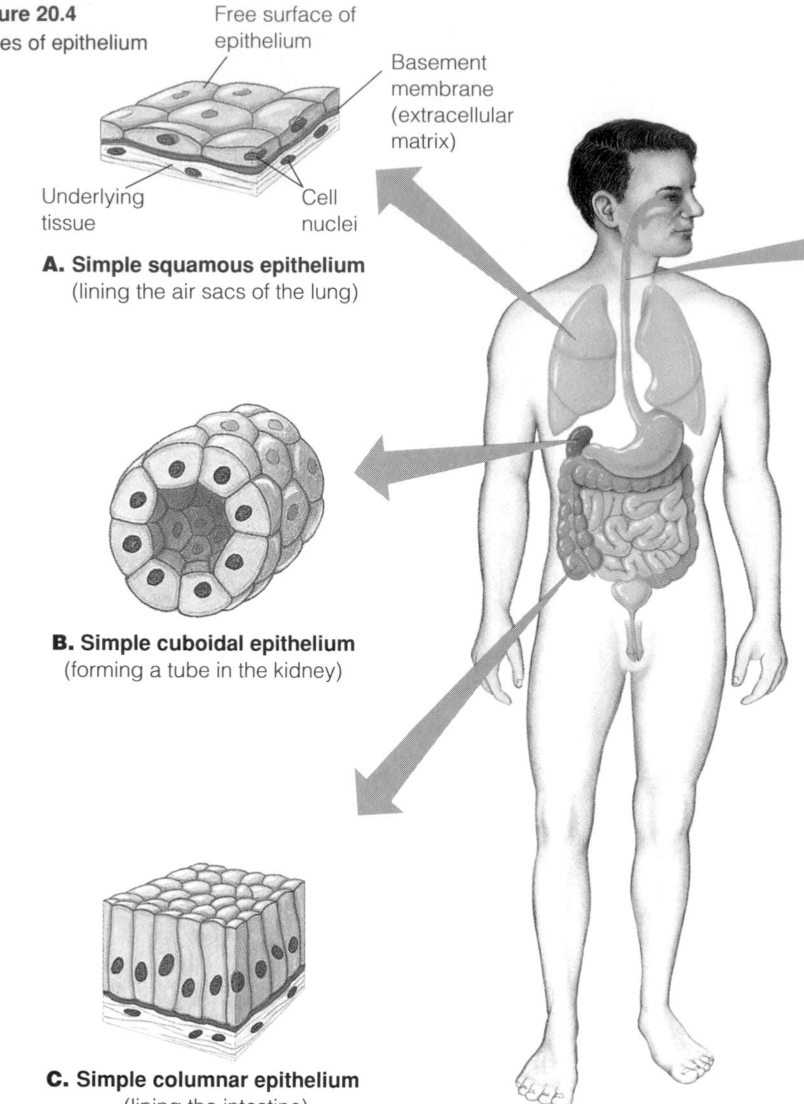

Figure 20.4
Types of epithelium

Epithelial tissue, also called **epithelium,** occurs as sheets of closely packed cells that cover body surfaces and line internal organs. The epidermis, the outer portion of our skin, is one example. Other epithelial tissues line the passageways and air sacs of our lungs, the tiny tubes where urine forms in our kidneys, and the organs of our digestive tract, including the esophagus, stomach, and intestines. One side of an epithelium—the "free" surface—forms the actual lining of the passageway. The other side is anchored to underlying tissues by a **basement membrane,** extracellular matrix that is a dense mat of fibrous proteins and sticky polysaccharides. (This "membrane" is *not* a phospholipid bilayer.) Together, the tightly knit cells and basement membrane form a barrier and, in some cases, an exchange surface between underlying tissues and the air or fluid in a passageway or hollow organ.

Epithelial tissues are named according to the number of cell layers they have and according to the shape of most of their cells. A simple epithelium has a single layer of cells, whereas a stratified epithelium has multiple layers. The shape of the cells may be squamous (like floor tiles), cuboidal (like dice), or columnar (like bricks or cones on end). Parts A, B, and C of the figure here show examples of simple epithelia with the three cell shapes; part D shows stratified squamous epithelium. In each case, the pink color identifies the cells of the epithelium itself.

The structure of each type of epithelium fits its function. Stratified squamous epithelium regenerates rapidly by division of the cells at its attached surface. New cells move toward the free surface as older cells slough off. Stratified squamous epithelium is well suited for covering and lining surfaces subject to abrasion. The esophagus, for instance, can be abraded by rough food. Our epidermis is also stratified squamous epithelium, with a thick layer of dead cells at the free surface. In contrast, simple squamous epithelium is thin and leaky, suitable for exchanging materials by diffusion. We find it in such places as our blood vessels and the air sacs of our lungs.

Cuboidal epithelium and columnar epithelium both have cells with a relatively large amount of cytoplasm, where secretory products are made, and a large surface area, where substances are secreted or absorbed. Lining our digestive tract and the air tubes of our lungs, such cells form a smooth, moist epithelium called a **mucous membrane.** The mucous membrane of our air tubes helps keep our lungs clean by trapping dust, pollen, and other particles in its secretions. The beating of cilia on this mucous membrane then sweeps the mucus-trapped materials upward and out of the breathing passageways.

? Epithelial tissues are classified according to the ________ of their cells and the number of cell ___________.

shapes . . . layers

Connective tissue binds other tissues and provides support matrices 20.5

20C

Unlike epithelium, **connective tissue** consists of a sparse population of cells scattered through an extracellular matrix. The cells synthesize the matrix, which is usually a web of fibers embedded in a liquid, jelly, or solid.

There are six major types of connective tissue, and the figure below illustrates one example of each. The most common type in the human body is called **loose connective tissue** (part A), because its matrix is a loose weave of fibers. Many of the fibers consist of the strong, ropelike protein collagen. Loose connective tissue serves mainly as a binding and packing material, holding other tissues and organs in place. In the figure, we show the loose connective tissue that lies directly under the skin, where it helps bind the skin to underlying muscles.

Adipose tissue (B) contains fat, which pads and insulates the body and stores energy. Each adipose cell contains a large fat droplet that swells when fat is stored and shrinks when fat is used as fuel. There is relatively little matrix.

Blood (C) is a connective tissue with a fluid rather than a solid matrix. The blood matrix, called plasma, consists of water, salts, and dissolved proteins. Red and white blood cells are suspended in the plasma. Blood functions mainly in transporting substances from one part of the body to another, and in immunity.

The other three types of connective tissue have dense matrices. **Fibrous connective tissue** (D) has a matrix of densely packed parallel bundles of collagen fibers. It forms tendons, which attach muscles to bone, and ligaments, which join bones together. The matrix of **cartilage** (E), a connective tissue that forms a strong but flexible skeletal material, consists of an abundance of collagen fibers embedded in a rubbery substance. Cartilage commonly surrounds the ends of bones, where it forms a smooth, flexible surface. It also supports the nose and the ears, and it forms the cushioning discs between our vertebrae. **Bone** (F), a rigid connective tissue, has a matrix of collagen fibers embedded in calcium salts. This combination makes bone hard without being brittle. As shown here, bones may contain repeating circular units of matrix, each with a hollow central canal. Blood vessels and nerves enter the bone through the canals and keep the bone cells alive. Because bone contains living cells, it can grow with the animal.

Why does blood qualify as a type of connective tissue?

Because it consists of a relatively sparse population of cells surrounded by a noncellular matrix, which is plasma in the case of blood

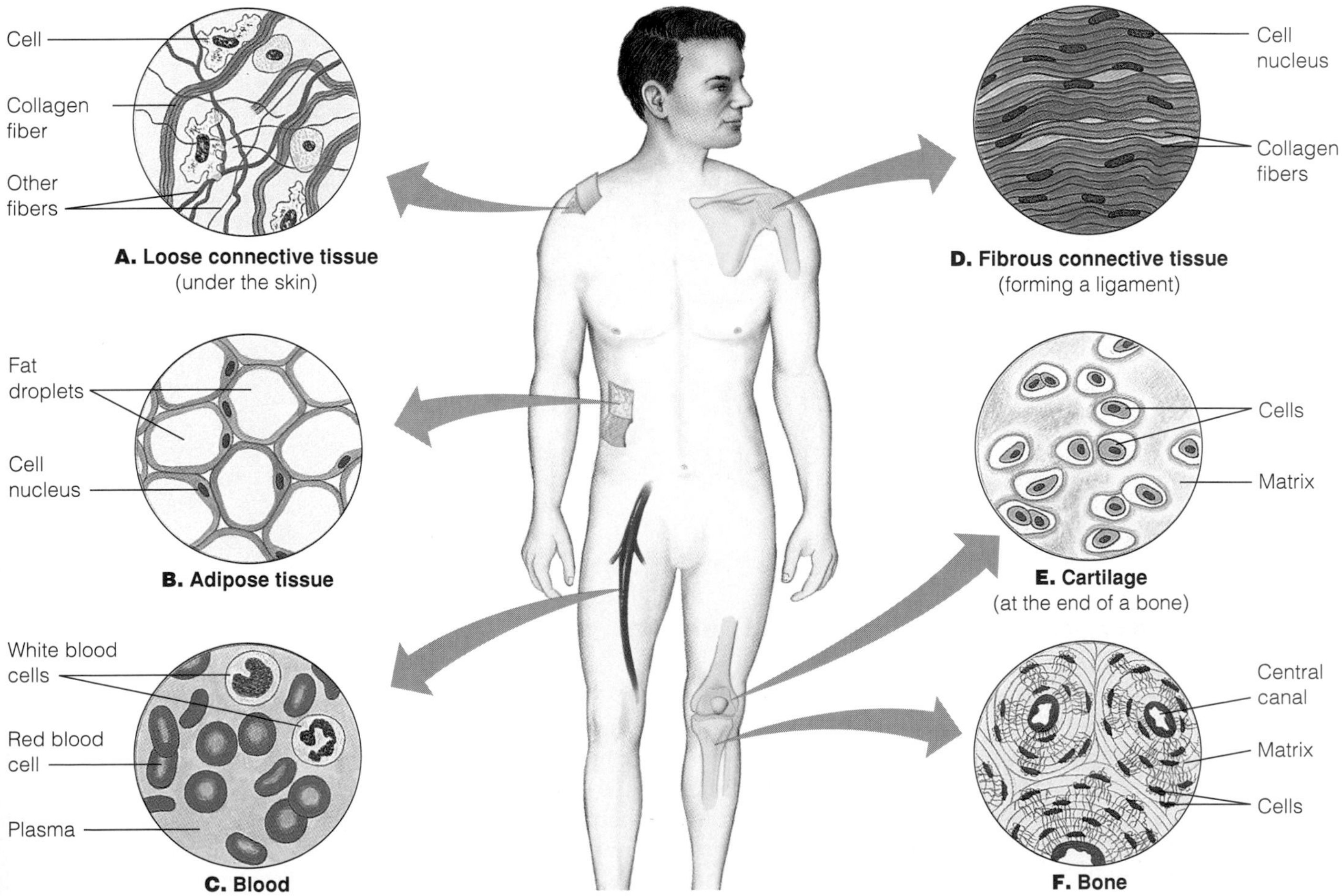

Cell
Collagen fiber
Other fibers

A. Loose connective tissue (under the skin)

Fat droplets
Cell nucleus

B. Adipose tissue

White blood cells
Red blood cell
Plasma

C. Blood

Cell nucleus
Collagen fibers

D. Fibrous connective tissue (forming a ligament)

Cells
Matrix

E. Cartilage (at the end of a bone)

Central canal
Matrix
Cells

F. Bone

Figure 20.5 Types of connective tissue

20.6 Muscle tissue functions in movement

20D

Muscle tissue, which consists of bundles of long cells called muscle fibers, is the most abundant tissue in a typical animal. Albatrosses, zebras, humans, and all other vertebrates have three types of muscle tissue: skeletal muscle, cardiac muscle, and smooth muscle.

Skeletal muscle is attached to bones by tendons, cords of fibrous tissue. Skeletal muscles are called "voluntary" muscles because an animal can generally contract them at will. As you can see in part A of Figure 20.6 below, a skeletal muscle fiber is packed with strands that have alternating light and dark bands. Because of these bands, skeletal muscle is said to be striated (striped). The bands occur in repeated groups, each of which is a structural and functional unit of muscle contraction. Adults have a fixed number of skeletal muscle cells. Exercise does not increase the number of our muscle cells; it simply enlarges those already present.

Cardiac muscle (part B) forms the contractile tissue of the heart. It is striated like skeletal muscle, but its cells are branched. Also, the ends of the cells mesh tightly together, forming relay structures that carry the signals to contract from cell to cell during the heartbeat.

Smooth muscle (part C) gets its name from its lack of striations. This type of muscle is found in the walls of the digestive tract, urinary bladder, arteries, and other internal organs. The cells (fibers) are shaped like spindles. They contract more slowly than skeletal muscles, but they can sustain contractions for a longer period of time.

Smooth muscles and cardiac muscles are mostly involuntary; in contrast to skeletal muscles, they are not generally subject to conscious control. We can decide to use our skeletal muscles to step forward or raise a hand, but our smooth muscles churn our stomach and our cardiac muscles pump blood without our conscious command.

The muscles responsible for the flapping of a bird's wings are __________ muscles.

skeletal

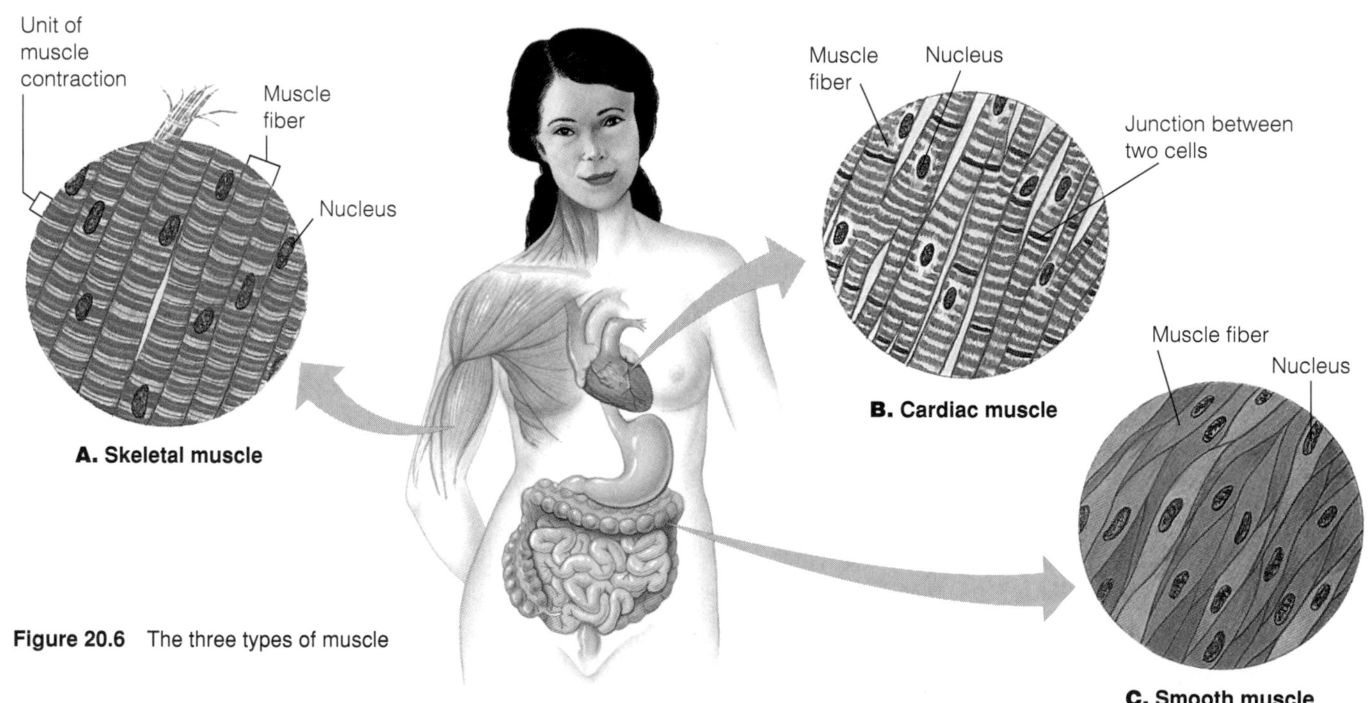

Figure 20.6 The three types of muscle

20.7 Nervous tissue forms a communication network

20E

An animal's survival depends on its ability to respond appropriately to stimuli from its environment. In most cases, responding to stimuli requires the animal to relay information from one part of its body to another. **Nervous tissue** forms a communication system that rapidly accomplishes this, transmitting information as nerve signals. The nervous system senses stimuli, determines and directs responses, and enables the body to function as a coordinated whole.

The structural and functional unit of nervous tissue is the nerve cell, or **neuron,** which is uniquely specialized to conduct nerve signals. The micrograph in Figure 20.7 at the top of the next page shows three neurons. Each neuron consists of a cell body (containing the cell's nucleus) and a number of slender extensions. One type of extension, called a dendrite, generally conveys signals toward the cell body; another type, the axon, usually transmits signals

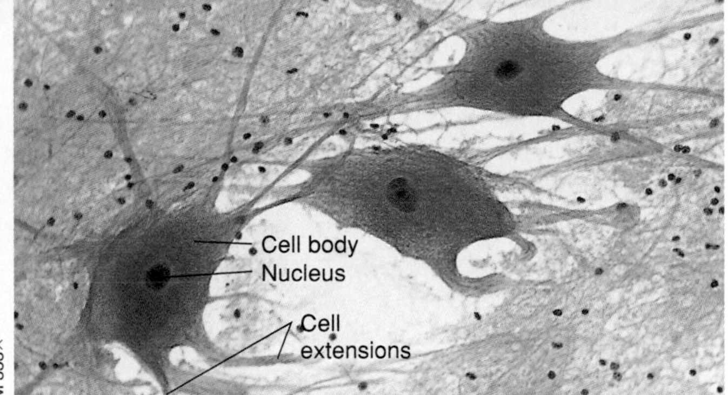

Figure 20.7 Neurons in the spinal cord

away from the cell body, often to another neuron. Some axons, such as those extending from our brain into our lower abdomen, are half a meter or more in length.

Nerve signals generally travel from one part of the body to another via a chain of several neurons. Transmission of a signal from neuron to neuron is usually brought about by chemicals that diffuse from one cell to the next.

Nervous tissue is not made up entirely of neurons. It also contains many cells that support the neurons. Some of these cells help nourish the neurons. Others surround and insulate axons and dendrites, promoting faster transmission of signals.

Having discussed tissues, we now turn to organs, the next level in the structural hierarchy of an animal.

The two types of extensions of neurons are ________, which usually transmit signals toward the cell body, and ________, which generally transmit away from the cell body.

dendrites . . . axons

Several tissues arranged in a particular way form an organ 20.8

Virtually all animals except sponges, which have a very simple body construction, have organs. An **organ** consists of several tissues adapted to perform specific functions as a group. The heart, for example, while mostly muscle, also has epithelial, connective, and nervous tissues. Epithelial tissue lining the heart chambers prevents leakage and provides a smooth surface over which blood can flow with little friction. Connective tissue makes the heart elastic and strengthens its walls and valves. Neurons direct the rhythmic contractions of cardiac muscles.

Another organ, the stomach, consists mainly of three types of tissue. As you can see in Figure 20.8, the tissues are arranged in multiple layers. The lumen, or space, within the stomach is lined by thick, columnar epithelium that secretes mucus and digestive juices. (Notice that the epithelium bends to form fingerlike projections, increasing its surface area.) Surrounding this layer is a zone of connective tissue that contains nerves and blood vessels (shown in red). Responsible for churning food and moving it through the digestive tract, three layers of smooth muscle surround the connective tissue, and they, in turn, are surrounded by a layer of connective tissue.

An organ represents a higher level of structure than the tissues composing it, and it performs functions that none of its component tissues can carry out alone. These organ-level functions emerge from the cooperative interaction of tissues. Cooperative interaction is a basic feature at all levels in an animal's structural hierarchy.

Explain why a disease that damages connective tissue can impair most of the body's organs.

Connective tissue is a component of most organs.

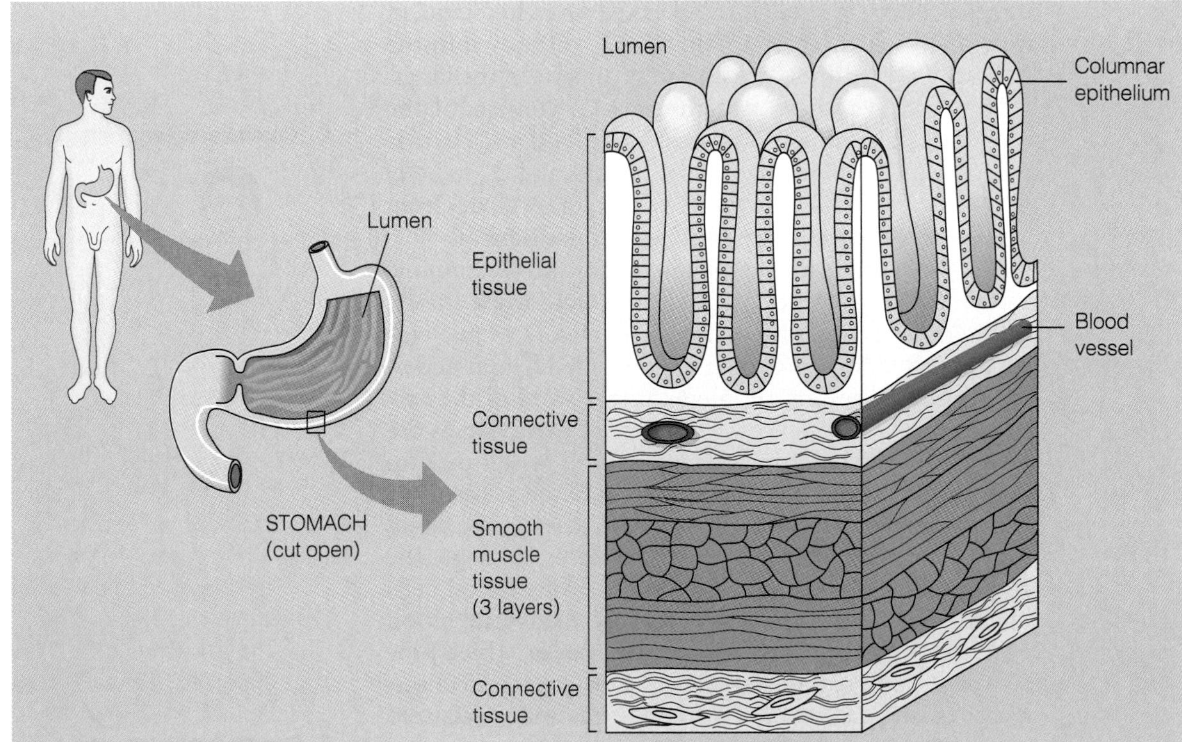

Figure 20.8 Tissue layers of the stomach wall

20.9 The body is a cooperative of organ systems

A level of organization still higher than an organ is an **organ system,** a group of several organs that work together to perform a vital body function. There are twelve major organ systems in vertebrate animals. The figure on these two pages introduces the main elements of these systems, using the human as an example. We examine each system in detail in the chapters of this unit.

Part A of the figure shows the main components of the **digestive system,** which ingests food and breaks it down into smaller chemical units. Food enters the mouth and travels via the esophagus to the stomach. Digestion occurs mainly in the stomach and small intestine. Nutrient molecules and some water are absorbed into the bloodstream through the walls of the small intestine. The large intestine absorbs additional water and compacts indigestible material into feces, which leave the body through the anus.

The liver is the largest organ in the body and has multiple functions. As part of the digestive system, it discharges bile, which aids in fat digestion, into the small intestine. Among its other functions, the liver carries out metabolic reactions involving carbohydrates and lipids, produces a number of important blood proteins, and removes toxins and worn-out cells from the blood.

The **respiratory system** (part B) is the body's organ system for exchanging gases with the environment. It supplies the blood with oxygen (O_2) and disposes of carbon dioxide (CO_2), a waste product of cellular metabolism. Air enters and exits the system through the nose and mouth, passing through the larynx (voicebox) into the trachea (windpipe). The trachea is a single, large air tube that branches into two smaller tubes called bronchi (singular, bronchus), which enter the lungs. O_2 diffuses into the blood and CO_2 diffuses out through air sacs in the lungs.

The **cardiovascular system** (part C) consists of the heart, which pumps blood, and the blood vessels that transport it. The blood supplies nutrients and O_2 to body cells. It also carries CO_2 to the lungs and other wastes from body cells to other disposal sites, such as the kidneys.

Part D illustrates the lymphatic system and the immune system, which share several structures and work closely together. The **lymphatic system** is a network of fine vessels connected to many small organs called lymph nodes. The lymphatic system supplements the work of the cardiovascular system, and together these two systems are known as the **circulatory system.** Lymph vessels pick up fluid, called lymph, that leaks through blood vessels into tissue spaces and return it to the blood, keeping the blood volume constant. Lymph also percolates through the lymph nodes, which are packed with white blood cells called lymphocytes and macrophages. The white blood cells are components of the **immune system,** which protects the body by attacking foreign substances, infectious microbes, and cancer cells. Lymphocytes and specialized proteins they secrete, known as antibodies, are transported throughout the body in the blood and lymph. The thy-

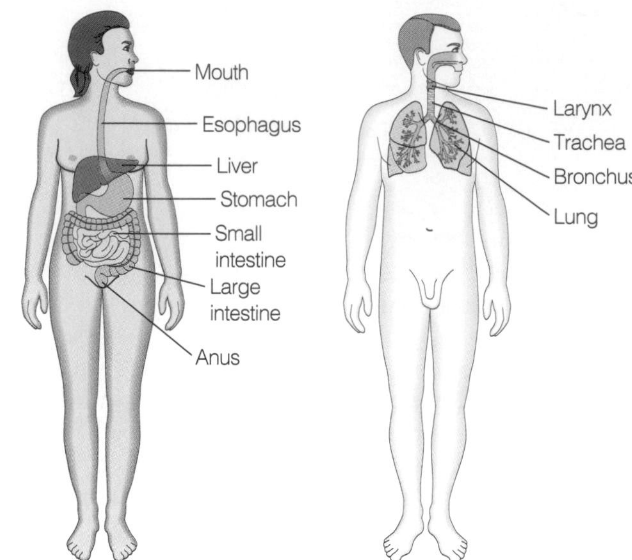

A. Digestive system

B. Respiratory system

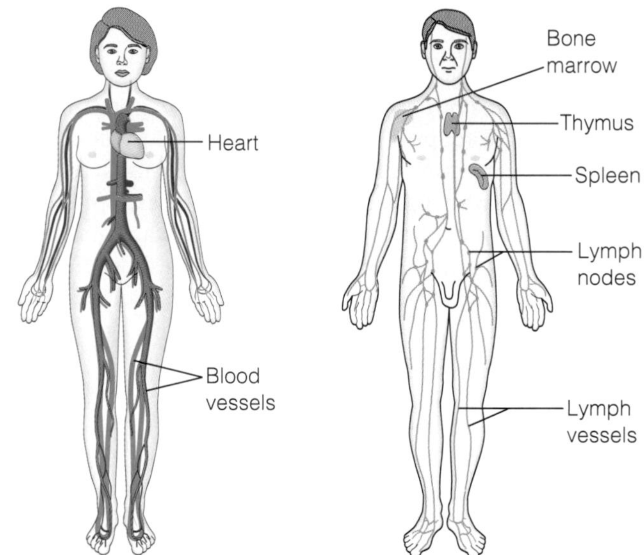

C. Cardiovascular system

D. Lymphatic and immune systems

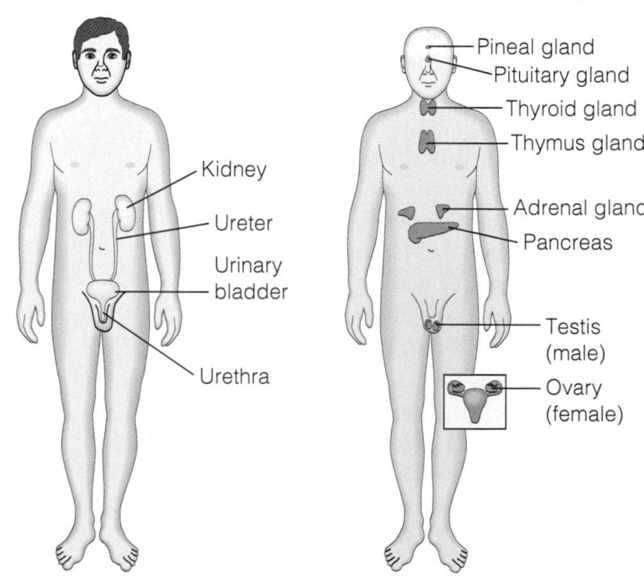

E. Excretory system

F. Endocrine system

Figure 20.9 Main components of vertebrate organ systems

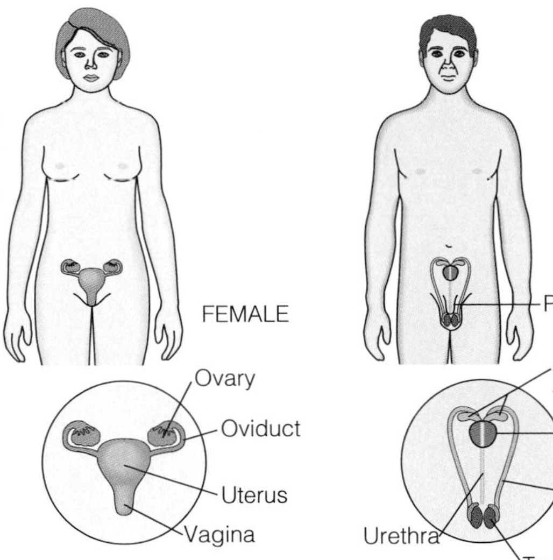

G. Reproductive systems

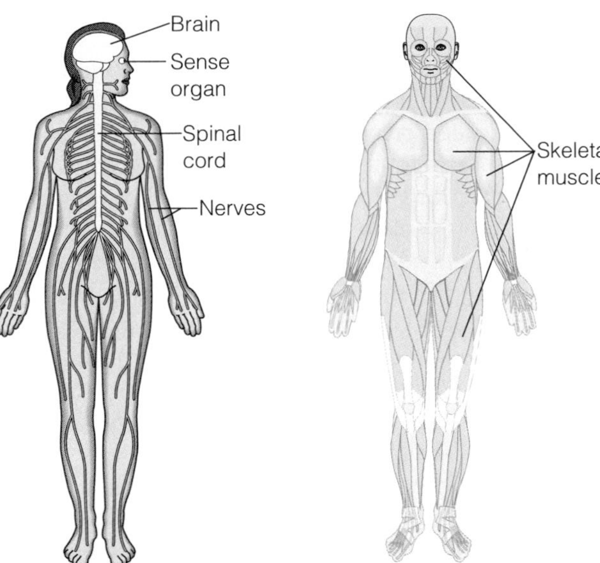

H. Nervous system

I. Muscular system

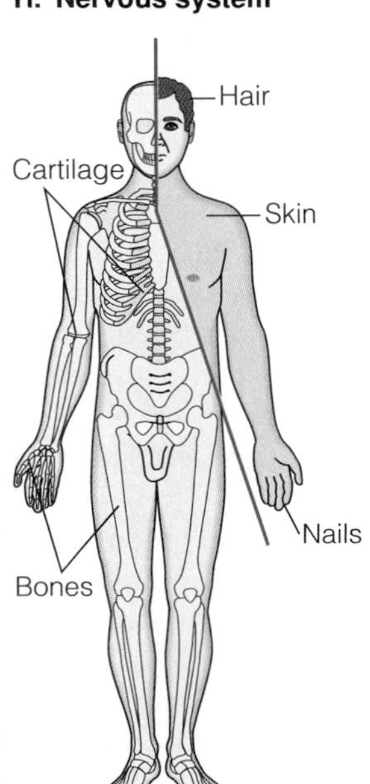

J. Skeletal and integumentary systems

mus, the bone marrow, and the spleen also play roles in the immune system.

The body's main waste-disposal system is the **excretory system** (part E). The kidneys remove the nitrogen-containing waste products of cellular metabolism from the blood. In urine, these wastes pass through the ureters to the bladder for temporary storage; they finally leave the body via the urethra. The kidneys also have the vital function of regulating the osmotic balance of the blood.

A number of organs produce chemicals, called hormones, that regulate the activity of organ systems. Hormone-producing organs are called endocrine glands; as a group, they constitute the **endocrine system** (part F). The endocrine glands secrete hormones into the blood, and the blood transports the hormones throughout the body. Hormones affect the behavior of specific cells, thereby regulating such activities as digestion, metabolism, growth, reproduction, heart rate, and water balance. Some of the organs that are part of the endocrine system perform double duty. Endocrine tissue of the pancreas, for instance, produces hormones that regulate the amount of sugar in the blood, while neighboring nonendocrine tissue produces juices that aid in digestion. Likewise, the ovaries and testes, which produce sex hormones, also produce gametes.

The gamete-producing portions of the ovaries and testes are part of the female and male **reproductive systems** (part G). Whereas all other organ systems are essential to the survival of the individual organism, an animal can live without its reproductive system. These systems help the species, rather than the individual, survive. In the female, the ovaries produce egg cells and release them into the oviducts (fallopian tubes), where they may be fertilized. A fertilized egg develops into an embryo in the uterus. The vagina accepts the male's penis during sexual intercourse and acts as a birth canal. In the male, the testes produce sperm, and the other organs shown in the figure help keep the sperm viable and convey it into the female's body.

The **nervous system** (part H) works together with the endocrine system to coordinate body activities. The brain receives information from the sense organs, such as the eyes. In response, the brain sends signals to muscles or glands via the spinal cord and nerves. The nervous system also responds to internal information from the body itself.

The **muscular system** (part I) consists of all the skeletal muscles in the body. Skeletal muscles can move parts of the body because they are attached to rigid bones or cartilage structures. The muscular system enables us to move about, to manipulate our environment, and to change our facial expressions. (Cardiac muscle and smooth muscle are components of other organ systems; see Module 20.6.)

Part J illustrates the main parts of the skeletal system and the integumentary system (the body covering). The main function of the **skeletal system** is to provide body support, but it also has a protective function. The skull houses and protects the brain, and the rib cage protects the lungs and heart. The **integumentary system** consists of the skin and its derivatives, the hair and nails. Its major function is to protect the internal body parts from mechanical injury, infection, excessive heat or cold, and drying out.

? The two organ systems most directly involved in regulating all other systems are the ____________ and ____________ systems.

nervous . . . endocrine

20.10 New imaging technology reveals the inner body

Among the most exciting recent developments in medical technology are techniques that allow physicians to "see" the organs and organ systems we have just surveyed, without resorting to surgery. We mentioned one of these techniques—ultrasound—in Module 9.10. Some of the others are new versions of X-ray technology.

X-rays, discovered in 1895, were the first means of producing a photographic image of internal organs and the only imaging method available until the 1950s. X-rays are a type of high-energy radiation (see Module 7.6). They pass readily through soft tissues, such as skin, nerves, and muscle. The photographic film on which an X-ray image is recorded is placed behind the body, and what show up most distinctly are the shadows of hard structures that block the rays—bones and dense tumors, for instance.

Conventional X-rays are used routinely to check for broken bones and tooth cavities. However, there are three major problems with routine X-rays. One obvious shortcoming is their failure to make soft tissues clearly visible. In addition, the standard X-ray technique produces only a flat, two-dimensional image, with anatomical structures often confusingly overlapped. Finally, the X-rays themselves, in large enough doses, can cause cancer.

Modern X-ray techniques have overcome some of these disadvantages. Today, X-rays use much lower doses of radiation than in the past and also reveal more details of soft tissues. An exciting extension of the standard technique has conquered the overlapping-image problem. This newer X-ray method is called **computed tomography (CT),** a computer-assisted technique that produces images of a series of thin cross sections through the body. ("Tomography" comes from the Greek words *tomos,* meaning slice, and *graphe,* drawing.) The patient is slowly moved through a doughnut-shaped CT machine (Figure 20.10A), as the X-ray source circles around the body, illuminating successive sections from many angles. The CT scanner's computer then produces high-resolution video images of the cross sections, which can be studied individually or combined into various three-dimensional views.

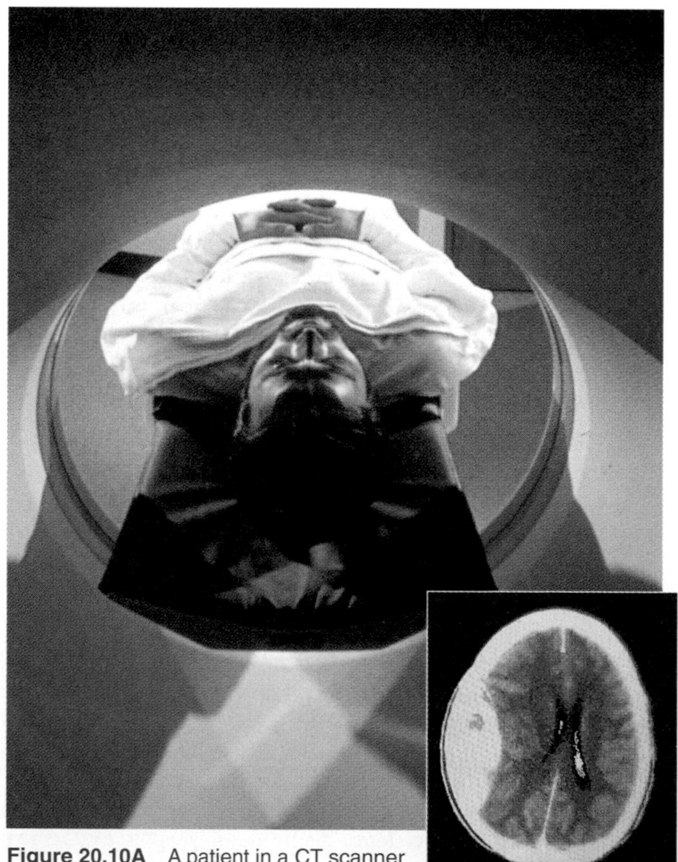

Figure 20.10A A patient in a CT scanner and a two-dimensional CT image of a brain

Figure 20.10B A three-dimensional CT image of a pelvis

CT scans can detect small differences between normal and abnormal tissues in many organs, but they are especially useful for evaluating brain problems—an area where conventional X-ray procedures are of little help. In the CT image shown in Figure 20.10A, you can see a brain hemorrhage (pale oval at far left), the result of a ruptured blood vessel. This CT image is two-dimensional; it shows only a single, thin slice of the brain.

A promising new use for the three-dimensional capability of CT is in the field of reconstructive surgery. Surgeons can study lifelike images of a patient's bones, for example, before operating. Figure 20.10B shows a three-dimensional CT image of the pelvis of a young woman who was born with a dislocated right hip (left side of photo). In this case, the surgeon used the information in the CT scan to plan an artificial hip implant that fit perfectly.

A completely different technique, **magnetic resonance imaging (MRI),** uses no X-rays or any other high-energy radiation. Instead, MRI takes advantage of the behavior of the hydrogen atoms in water molecules. The nuclei of hydrogen atoms are usually oriented in random directions, but in a magnetic field they align in the same direction. MRI uses powerful magnets to align the hydrogen nuclei, then knocks the nuclei out of alignment with a brief pulse of radio waves. Still under the magnets' influence, the hydro-

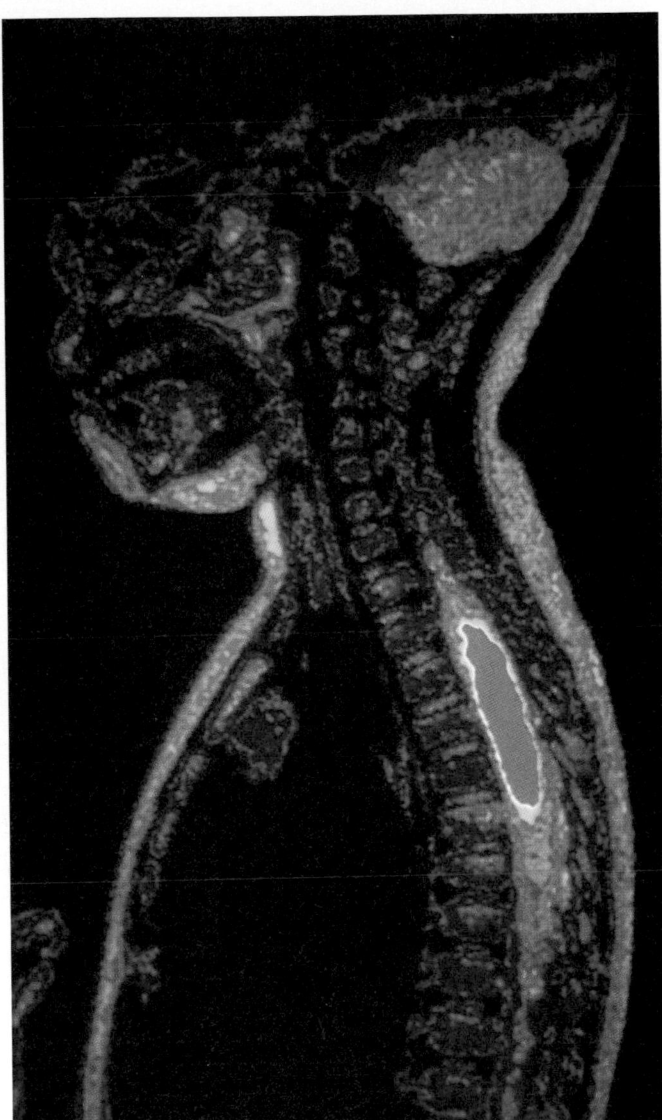

Figure 20.10C An MRI scan showing a spinal tumor

gen atoms immediately spring back into alignment, giving out faint radio signals of their own. These signals are picked up by the MRI scanner and translated by computer into an image.

Since water is a major component of all of our soft tissues, MRI visualizes them well. At the same time, dense structures such as bone, which contains little water, are nearly invisible to MRI. These qualities make MRI particularly good for detecting problems in nervous tissue that is surrounded by bone. For example, Figure 20.10C shows a tumor (artificially colored red) in the spinal cord of a 5-year-old girl who had lost the use of her legs. With this information, her physician was able to remove the tumor safely, restoring the girl's ability to walk.

Positron-emission tomography (PET) is an imaging technology that differs from both CT and MRI in its ability to yield information about metabolic processes at specific locations in the body. In preparation for a PET scan, the patient is injected with a biological molecule—glucose, for example—labeled with a radioactive isotope (see Module 2.4). Used only in small quantities, the isotope is not dangerous. Metabolically active cells take up more of the labeled glucose than less active cells. The isotope emits positively charged subatomic particles called positrons. When the positrons collide with electrons inside the cells, enough high-energy radiation (gamma rays) is released to be detected by an instrument called a PET scanner. Thus, PET pinpoints metabolic hot spots by highlighting the sites of most intense radiation.

PET is proving most valuable for measuring the metabolic activity of various parts of the brain. This technique is providing insights into brain activity in people affected by illnesses such as schizophrenia and Alzheimer's disease, and in stroke patients. (A stroke is a loss of brain function caused by the blockage or rupture of a blood vessel.) Equally exciting is the use of PET to learn about the healthy brain. Figure 20.10D, for example, shows PET scans of a person's brain during four different kinds of mental activity involving language. Metabolic hot spots in the brain appear white, orange, and yellow in these scans, clearly revealing the regions of the brain that are most active during each activity.

Because of the high cost of the necessary equipment, PET is still primarily a research tool. However, some less expensive versions are in clinical use, and we will probably see wider application of PET in the future.

Why are all three imaging techniques described in this module referred to as "noninvasive," in contrast to such invasive diagnostic methods as exploratory surgery or biopsy? (*Hint:* It's fair to use a dictionary.)

The imaging techniques require no penetration of the body with instruments such as needles or scalpels.

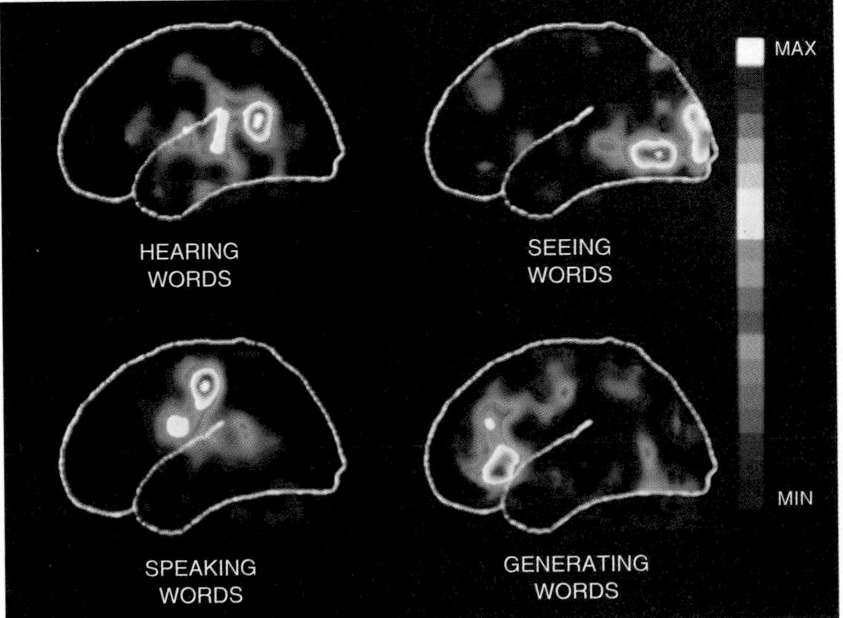

Figure 20.10D PET scans of a brain engaged in different mental activities

20.11 Structural adaptations enhance chemical exchange between animals and their environment

Although animals are covered with protective skin and are chemically distinct from their environment, they are not closed systems. They cannot survive unless they can exchange materials with their environment. This holds true down to the cellular level, where exchanges occur across the plasma membrane. O_2 and nutrients must be able to enter the cell, and CO_2 and other metabolic wastes must be able to exit. Because a living cell must be bathed in aqueous fluid for its plasma membrane to remain intact, only molecules dissolved in water can diffuse across the membrane.

The freshwater invertebrate animal hydra has a body wall only two cell layers thick (Figure 20.11A). The outside layer is in direct contact with the environment. The inner layer contacts fluid in the animal's saclike gastrovascular cavity. The gastrovascular cavity opens directly to the outside via the mouth, and water flushes in and out of the cavity, bathing the inner layer of cells. As indicated by the arrows in the figure, materials diffuse back and forth between the cells, the hydra's surroundings, and the gastrovascular cavity. With this arrangement, almost every cell in the animal's body has some of its plasma membrane exposed to an aqueous environment with which it can directly exchange materials. In a hydra, each cell has enough exposed surface area to service its entire volume of cytoplasm by direct diffusion and active transport.

The saclike body of a hydra, or a paper-thin one like the flatworms we discussed in Module 18.6, works well for animals with a simple body structure. However, most animals have an outer surface that is relatively small compared with the animal's overall volume. As an extreme example, the surface-to-volume ratio (see Module 4.3) of a whale is millions of times smaller than that of a hydra. Still, every cell in the whale's body must be bathed in fluid, have access to oxygen and nutrients, and be able to dispose of its wastes. How is all this accomplished?

Instead of relying on the general body surface, most animals have specialized surfaces for exchanging materials with the environment. Figure 20.11B is a schematic model illustrating four of the organ systems of an animal with a structurally complex body and a large internal surface area that services it. Each system has a specialized internal exchange surface. We have placed the circulatory system in the middle because of its central role in transporting substances among the other three systems. The blue arrows indicate exchange of materials between the circulatory system and the other systems.

Actually, direct exchange does not occur between the blood and the cells making up tissues and organs. At each exchange surface (in Figure 20.11B, wherever blue arrows enter or exit the circulatory system), the body cells are bathed in a solution called **interstitial fluid** (see the circular enlargement). Materials are exchanged between the blood and the interstitial fluid and between the interstitial fluid and the body cells. In other words, to get from the blood to body cells or vice versa, materials must pass through interstitial fluid.

The digestive system, especially the intestine, of this model animal has an expanded surface area resulting from folds and projections of its cells and tissues. Finely

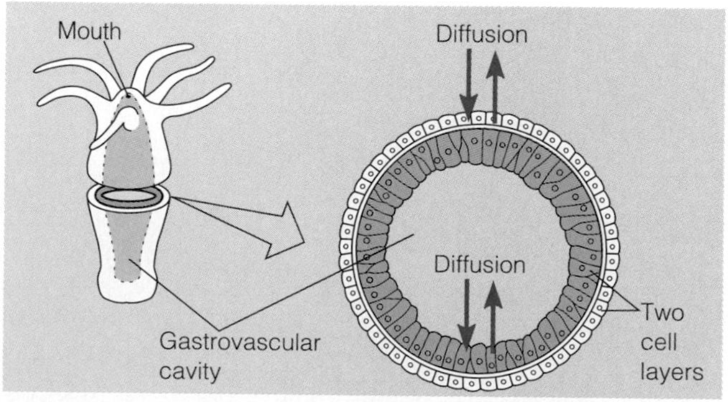

Mouth
Diffusion
Diffusion
Gastrovascular cavity
Two cell layers

Figure 20.11A Direct exchange between the environment and the cells of a structurally simple animal (a hydra)

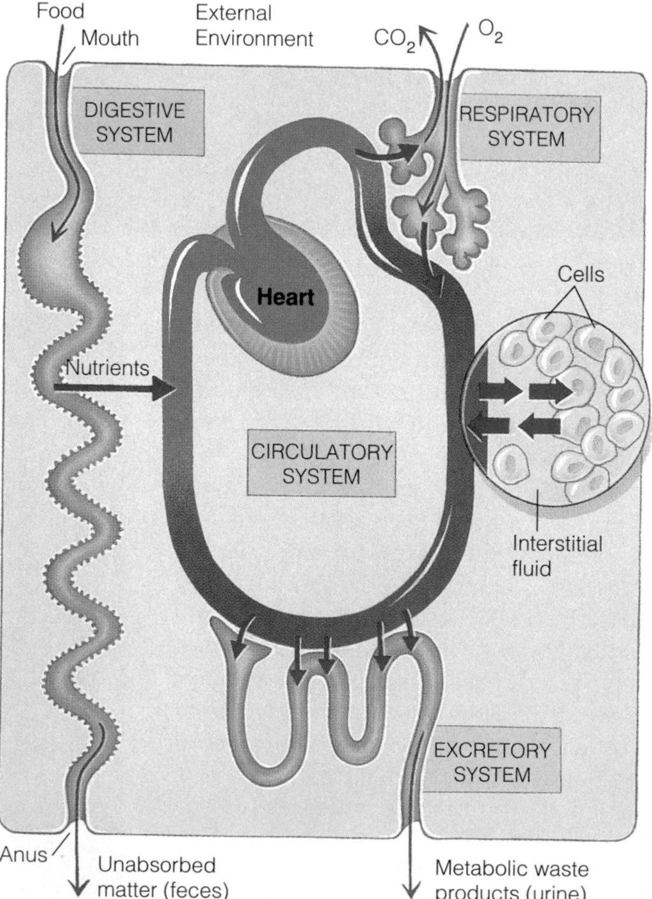

Figure 20.11B Indirect exchange between the environment and the cells of a complex animal

branched vessels of the circulatory system (not shown here) form an exchange network with the digestive surfaces. This system of exchange from the cells of the intestine to the surrounding interstitial fluid to the blood is so effective that enough nutrients diffuse from it into the circulatory system to support the rest of the cells in the body.

The folded tubes of the excretory system are equally effective. Enmeshed in tiny blood vessels, they extract metabolic wastes that the blood brings from cells throughout the body. The wastes diffuse out of the blood into the excretory tubes and pass out of the body in the urine.

The respiratory system also has an enormous internal surface area with a vast number of fine blood vessels. Figure 20.11C shows a model of the interior of the human lungs. The white branches represent tiny air tubes, and the red branches represent the fine blood vessels that convey blood from the heart through the lungs.

Figure 20.11C highlights a basic concept in animal biology: Any animal with a complex body—one with most of its cells not in direct contact with the outside environment—must have internal structures that provide enough surface area to service those cells.

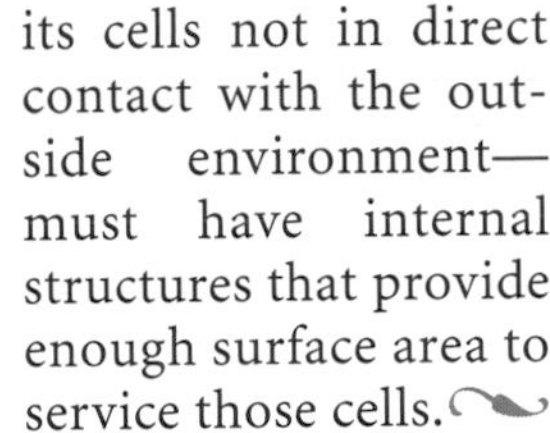

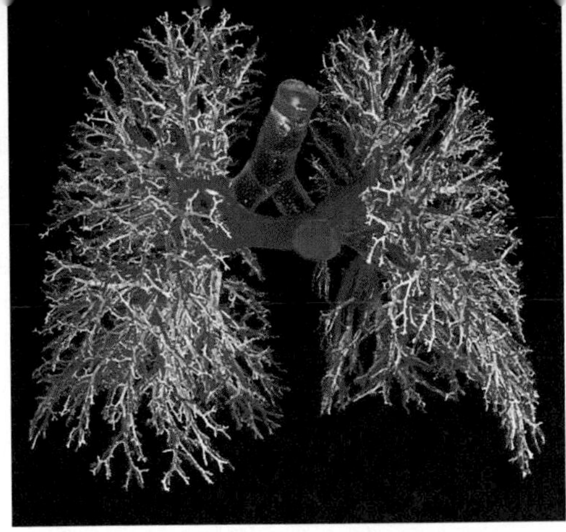

Figure 20.11C The interior surface of the human lungs

As organs specialized for the uptake and release of certain chemicals, the lungs, small intestine, and kidneys all have an extensive __________ __________ of epithelium.

surface area

Animals regulate their internal environment 20.12

Figure 20.12A An albatross in snow

Over a century ago, French physiologist Claude Bernard recognized that *two* environments are important to an animal: the external environment surrounding the animal, and the internal environment where its cells actually live. The internal environment of a vertebrate is the interstitial fluid that fills the spaces around the cells. Many animals can maintain relatively constant conditions in their internal environment. Our own bodies maintain the salt and water balance of our internal fluids and also keep the fluids at about 37°C (98.6°F). A bird like the albatross also maintains salt and water balance and temperature (about 40°C, or 104°F), even in winter when the temperature can drop to −50°C (−58°F). The bird uses energy from its food to generate body heat, and it has a thick, insulating coat of down feathers (Figure 20.12A). A much simpler animal such as a hydra is powerless to affect the temperature of the fluid that soaks its cells, but it does regulate the salt and water content of its cells.

Today, Bernard's concept of the constant internal environment is incorporated into the broader principle of **homeostasis,** which means "a steady state." Figure 20.12B illustrates this principle in very general terms, using a pink box to represent an animal. Outside the box, the large double-headed arrow indicates that conditions such as air temperature may fluctuate widely in the animal's external environment. The small double-headed arrow stands for the smaller fluctuations in the animal's internal environment, which are regulated by the animal's control systems (purple box). For example, birds and mammals have a control system that keeps body temperature within a narrow range, despite wide fluctuations in the external environment. Control systems also regulate factors, such as salt concentration, that may not fluctuate much outside the body but that must be maintained at a different level inside.

The internal environment of an animal always fluctuates slightly. Homeostasis is a dynamic state, an interplay between outside forces that tend to change the internal environment and internal control mechanisms that oppose such changes. An animal's homeostatic control systems maintain internal conditions within a range where life's metabolic processes can occur. We focus on homeostatic control systems next.

The ability of your body to regulate its internal temperature is an example of __________.

homeostasis

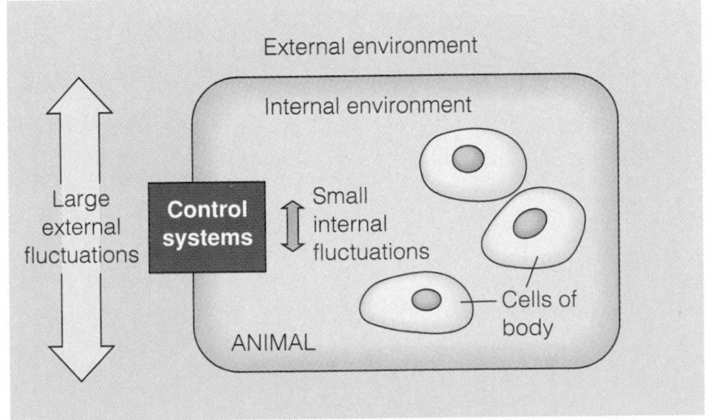

Figure 20.12B A model of homeostasis

20.13 Homeostasis depends on negative feedback

Most of the known control mechanisms of homeostasis are based on **negative feedback,** in which the results of a process inhibit the process. We discussed negative feedback in the control of cellular metabolism in Module 5.8. Figure 20.13A shows a mechanical example of negative feedback. A thermostat is the control center for regulating the temperature of a room. When the room temperature falls below a set point, such as 20°C (68°F), a sensor (thermometer) turns on the thermostat's switch, which sends a signal to turn on the heater. Heat is produced, warming the room. When the sensor detects that the temperature is above the set point, the thermostat switches the heater off. Physiologists would say that the sensor is triggered by a *stimulus* (room temperature, in this case) and would call the heater an *effector.* An effector produces a *response* (here, heat).

This mechanism is called *negative* feedback because a change in one condition (in this case, temperature) triggers the control mechanism to *counteract* further change in the same direction. Negative feedback prevents small changes from becoming too large.

The topic of homeostatic control brings us back to the connection between structure and function. Like all biological functions, homeostatic control results from the interactions of specialized structures. For example, consider the control of temperature in a healthy human body, which actually fluctuates between about 36.1°C and 37.8°C (97–100°F). As Figure 20.13B indicates, our brain has a thermostatic control center that is sensitive to slight changes in the temperature of our blood. Starting at the arrow that extends to the right and upward from the brain, we see that when the temperature is "too hot" (above the set point of 37°C), the control center sends signals to the skin that increase the activity of its sweat glands and cause its dense network of blood vessels to dilate. Evaporative cooling occurs as sweating increases, and heat radiates from the blood vessels as they fill with warm blood. The skin loses heat until the blood cools back to the set point. When this happens, the control center turns off its signals to the skin.

The same structures are involved when blood temperature drops below the set point (bottom half of figure). Signals from the brain's control center shut off the sweat glands and constrict the skin's blood vessels. Blood is then shunted to deeper tissues, reducing heat loss from the skin until the blood temperature returns to the set point and again turns off the control center. If the body does not warm up, shivering—involuntary contractions of skeletal muscles—may occur. These muscle contractions generate a lot of heat, and body temperature increases as a result. We will see other examples of homeostatic control and negative feedback as we examine each of the body's organ systems in detail in the chapters of this unit.

When the level of a sex hormone in your body reaches a set point, it turns off its own production by acting on a brain control center, the hypothalamus, which in turn regulates sex hormone production by your testes or ovaries. This control circuit is an example of ________________.

negative feedback

Figure 20.13A Control of room temperature

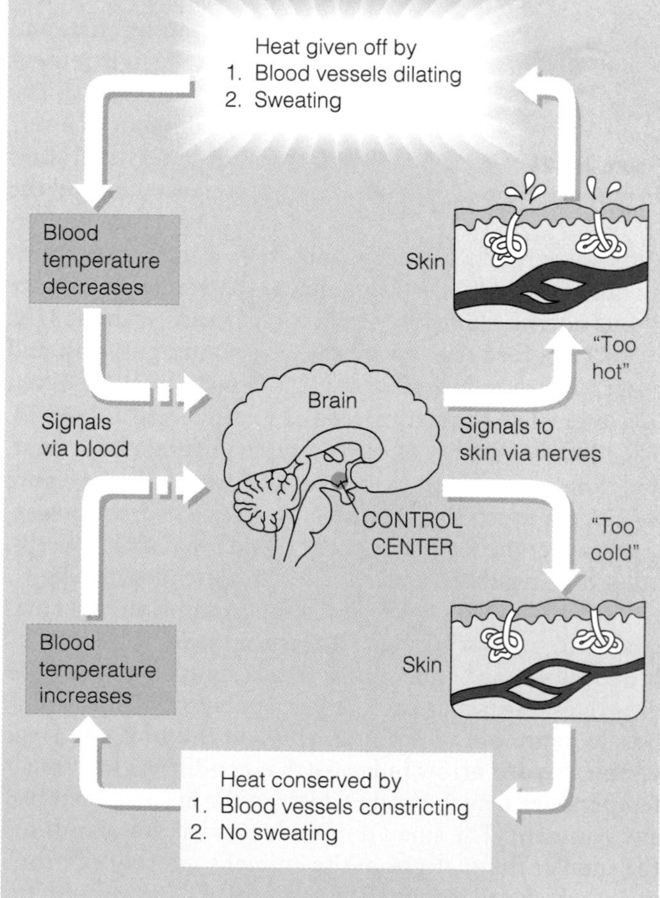

Figure 20.13B Control of body temperature

Chapter Summary

The Hierarchy of Structural Organization in an Animal (Introduction–20.10) The function of any part of an animal results from the unique structure of that part. The correlation between structure and function is one of biology's most fundamental concepts **(Introduction–20.1)**. Structure and function are correlated at each level in the structural hierarchy of an animal's body: cell, tissue, organ, organ system, and whole animal **(20.2)**. A tissue is a cooperative of many similar cells that perform a specific function. An animal has four main kinds of tissues: epithelial, connective, muscle, and nervous **(20.3)**. Epithelial tissue occurs as sheets of closely packed cells that cover surfaces and line the cavities and tubes of internal organs **(20.4)**. Connective tissue binds other tissues and provides support matrices **(20.5)**.

Skeletal muscle tissue is responsible for voluntary body movements; cardiac muscle pumps blood; and smooth muscle moves the walls of internal organs such as the stomach **(20.6)**. The branching neurons of nervous tissue transmit nerve signals that help control body activities **(20.7)**.

Each organ is made of several tissues that cooperate to perform specific functions **(20.8)**. Each organ system has one or more functions. The integumentary system covers and protects the body. Skeletal and muscular systems support and move it. The digestive and respiratory systems gather food and oxygen, and the cardiovascular system, aided by the lymphatic system, transports them. The excretory system disposes of certain wastes, while the immune system protects the body from infection and cancer. The nervous and endocrine systems control and coordinate body functions. The reproductive system perpetuates the species **(20.9)**. New technologies, such as computed tomography (CT), magnetic resonance imaging (MRI), and positron-emission tomography (PET), enable us to see body organs and systems without surgery **(20.10)**.

Exchanges with the External Environment (20.11–20.13) An animal must exchange materials and heat with its environment. Small animals with simple body construction have enough surface to meet their cells' needs, but larger, more complex animals have specialized structures that increase surface area **(20.11)**. In response to changes in external conditions, animals regulate their internal environment to achieve homeostasis, an internal steady state **(20.12)**. Control systems sense change, and negative feedback mechanisms keep fluctuations in internal conditions, such as temperature and salt balance, within the narrow range compatible with life **(20.13)**.

Testing Your Knowledge

Multiple Choice

1. Which of the following pairs of body systems primarily regulate the activities of the other systems?
   a. cardiovascular and muscular systems
   b. nervous and endocrine systems
   c. lymphatic and integumentary systems
   d. endocrine and lymphatic systems
   e. integumentary and nervous systems

2. Every living cell in the human body is in contact with an internal environment consisting of
   a. blood
   b. connective tissue
   c. interstitial fluid or blood
   d. matrix
   e. mucous membranes

3. Which of the following best illustrates homeostasis? *(Explain your answer.)*
   a. Most adult human beings are between 5 and 6 feet tall.
   b. The lungs and intestines have large surface areas for exchange.
   c. When blood salt concentration goes up, the kidney expels more salt.
   d. All the cells of the body are about the same size.
   e. When oxygen in the blood decreases, you may feel light-headed.

Matching *(Terms in the right-hand column may be used more than once.)*

1. Closely packed cells covering a surface
2. Neurons
3. Adipose tissue, blood, and cartilage
4. May be simple or stratified
5. Scattered cells embedded in matrix
6. Senses stimuli and transmits signals
7. Long cells called fibers
8. Cells may be squamous, cuboidal, or columnar
9. Skeletal, cardiac, or smooth

a. connective tissue
b. muscle tissue
c. nervous tissue
d. epithelial tissue

Describing, Comparing, and Explaining

1. Briefly explain how the structure of each of the following tissues is well suited to its function: stratified squamous epithelium in the skin, neurons in the brain, simple squamous epithelium lining the lung, bone in the skull.

2. Describe ways in which the bodies of large, complex animals are structured for exchanging materials with the environment. Why can some smaller creatures get along without such structural features?

Thinking as a Scientist

You are studying the body temperature of a large, active lizard. Your hypothesis is that the lizard has homeostatic mechanisms that maintain its temperature within a narrow range. You decide to test your hypothesis by comparing the lizard's body temperature in a range of surrounding air temperatures with the body temperatures of a turtle and a rat in the same air temperatures. (The body temperatures of rats and most other mammals remain fairly constant, whereas the body temperatures of turtles generally fluctuate with the temperature of their surroundings.) You obtain the following results in your first test:

	Temperature (°C)			
Time	**Air**	**Turtle**	**Lizard**	**Rat**
6 A.M.	22	21	27	36
Noon	31	29	30	37
6 P.M.	27	26	29	37
Midnight	23	22	27	37
6 A.M.	21	21	26	36

Do these data support your hypothesis? How does the lizard's temperature-regulating ability compare with the turtle's? With the rat's?

Media Resources

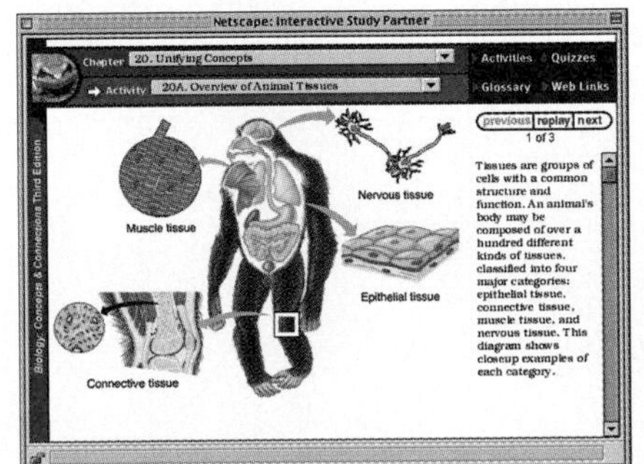

For further review of the concepts in this chapter, use the activities and quizzes on the CD-ROM.

Visit the Special Edition of The Biology Place web site for activities, quizzes, news articles, and web links.

21 Nutrition and Digestion

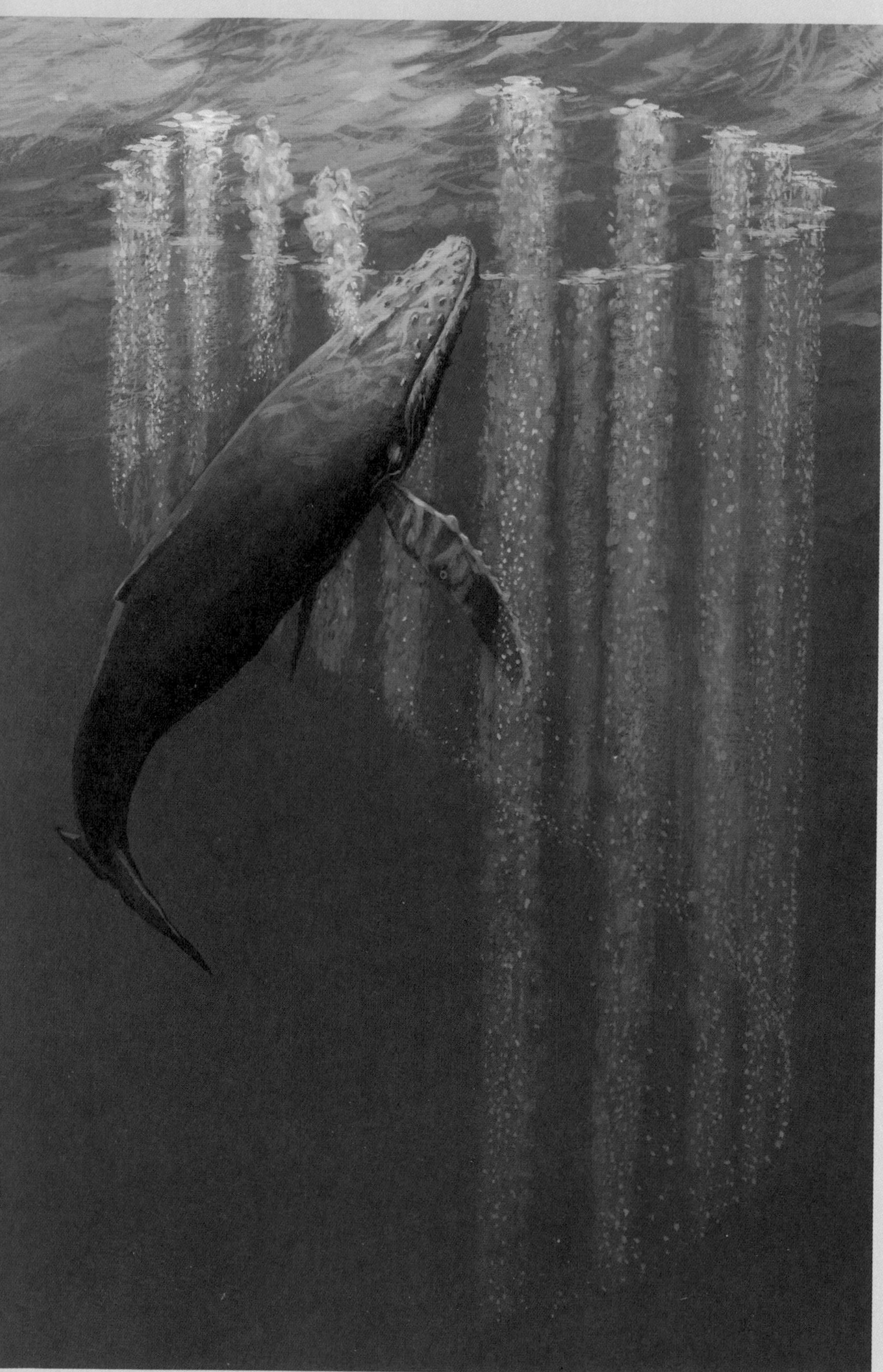

Outline

Obtaining and Processing Food

Whales are the largest animals in the world. Few other species, living or extinct, even approach their great size. The humpback whale, shown in the pictures here, is a medium-sized member of the whale clan. It can be 16 meters (53 feet) long and weigh up to 65,000 kg (72 tons), about as much as 70 midsize cars.

It takes an enormous amount of food to support a 72-ton animal. Humpback whales eat small fishes and crustaceans called krill. The painting at the left shows a remarkable technique they often use to corral food organisms before gulping them in. Beginning about 20 meters below the ocean surface, a humpback swims slowly in an upward spiral, blowing air bubbles as it goes. The rising bubbles form a cylindrical screen, or "bubble net." Krill and fish inside the bubble net swim away from the bubbles and become concentrated in the center of the cylinder. The whale then surges up through the center of the net with its mouth open, harvesting the catch in one giant gulp.

Humpback whales strain their food from seawater. Instead of teeth, these giants have an array of brushlike plates called baleen on each side of their upper jaw. You can see the white, comblike baleen in the open mouth of the whale in the photograph at the right. The baleen is used to sift food from the ocean. To start feeding, a humpback whale opens its mouth, expands its throat, and takes a huge gulp of seawater. When its mouth closes, the water squeezes out through spaces in the baleen, and a mass of food is trapped in the mouth. The food is then swallowed whole, passing into the stomach, where digestion begins. The humpback's stomach can hold about half a ton of food at a time, and in a typical day, the animal's digestive system will process as much as 2 tons of krill and fish.

The humpback and most other large whales are endangered species, having been hunted almost to extinction for meat and whale oil by the 1960s. Today, most nations honor an international ban on whaling, and some species are showing signs of recovery. Humpbacks still roam the Atlantic and Pacific oceans. They feed in polar regions during summer months and migrate to warmer oceans to breed when temperatures begin to fall. The photograph on this page was taken during summer in the Pacific Northwest. Food is so abundant there that humpbacks harvest much more energy than they burn each day. Much of the excess is stored as a thick layer of fat, or blubber, just under their skin. After a summer of feasting, humpback whales leave Glacier Bay and head south to breeding and calving grounds off the Hawaiian Islands, some 6000 km (3600 mi) away. Living off body fat, they eat little, if at all, until they return to Alaskan waters eight months later.

In about four months, a humpback whale eats, digests, and stores as fat enough food to keep its 72-ton body active for an entire year—a remarkable feat, and a fitting introduction to this chapter on animal nutrition and digestion. We will return to the whale as we examine the diverse ways that animals obtain and process nutrients.

Obtaining and Processing Food

21.1 Animals ingest their food in a variety of ways

Animal diets vary enormously, and so do methods of feeding. Certain parasites—tapeworms, for instance—are absorptive feeders; lacking a mouth or digestive tract, they absorb nutrients through their body surface. In contrast, the majority of animals, including the great whales, are ingestive feeders; they eat (ingest) living or dead organisms, either plants or animals or both, through a mouth. We will concentrate on ingestive feeders for the rest of this chapter.

Animals that ingest *both* plants and animals are called **omnivores** (from the Latin *omnis,* all, and *-vorus,* devouring). We humans are omnivores, as are crows, cockroaches, and raccoons. In contrast, plant-eaters, such as cattle, deer, gorillas, and a vast array of aquatic species that graze on algae—sea urchins, for instance—are called **herbivores** (Latin *herba,* green crop). **Carnivores** (Latin *carne,* flesh), such as lions, sharks, hawks, spiders, and snakes, eat other animals. Figure 21.1A illustrates both an herbivore, the impala, and a carnivore, the oxpecker clinging to the side of the impala's head. The oxpecker plucks blood-engorged ticks and other parasites from the impala's skin and fur.

Ingestive feeders use several different mechanisms to obtain their food. **Suspension feeders** extract food particles suspended in the surrounding water. For example, the humpback whale actively sifts its food from the water, using its baleen. Clams, oysters, and scallops (Figure 21.1B) are also suspension feeders. A film of mucus on their gills traps tiny morsels suspended in the water, and beating cilia on the gills sweep the food along to the mouth.

Substrate feeders live in or on their food source and eat their way through the food. Figure 21.1C shows a caterpillar eating its way through the soft green tissue inside an oak leaf. The dark spots are a trail of feces that the caterpillar leaves in its wake. Earthworms are also substrate feeders. They eat their way through the soil, digesting partially decayed organic material as they go. In doing so, they help aerate the soil, making it more suitable for plants.

Fluid feeders obtain food by sucking nutrient-rich fluids from a living host, either a plant or an animal. Aphids, for example, tap into the sugary sap in plants. Blood suckers, such as mosquitoes and ticks, pierce animals with needlelike mouthparts. The female mosquito in Figure 21.1D has just filled her abdomen with a meal of human blood. (Only female mosquitoes suck blood; males live on plant nectar.)

Rather than filtering food from water, eating their way through a substrate, or sucking fluids, most animals are **bulk feeders,** meaning they ingest relatively large pieces of

Figure 21.1A An herbivore (impala) and a carnivore (oxpecker)

Figure 21.1B A suspension feeder (scallop)

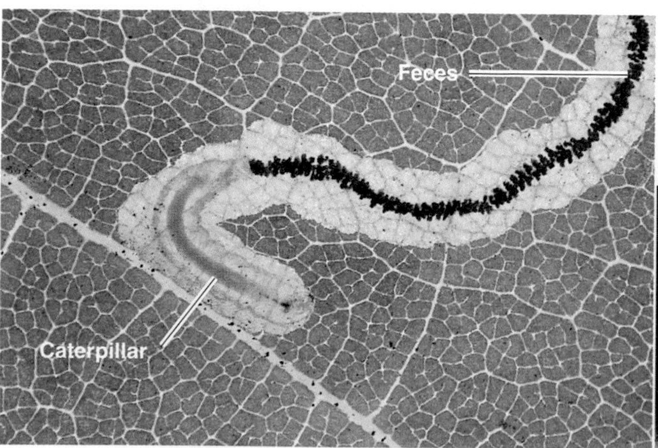

Figure 21.1C A substrate feeder (caterpillar)

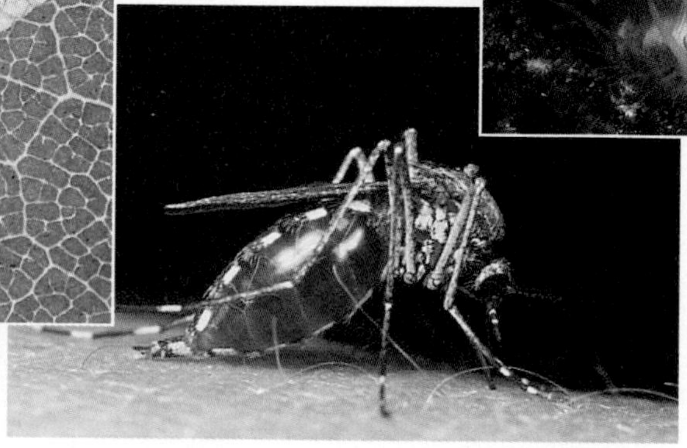

Figure 21.1D A fluid feeder (mosquito)

food. A bulk feeder uses equipment such as tentacles, pincers, claws, poisonous fangs, or jaws and teeth to kill their prey, to tear off pieces of meat or vegetation, or to take mouthfuls of animal or plant products. Figure 21.1E shows that humans clearly fit into this category.

Figure 21.1E Bulk feeding, ingestion of mouthfuls of food

? The largest animals ever to live are __________ feeders.

suspension

21.2 Overview: Food processing occurs in four stages

So far we have discussed what animals eat and how they feed. As shown in Figure 21.2 below, ① **ingestion,** the act of eating, is only the first of four main stages of food processing. ② **Digestion,** the second stage, is the breaking down of food into molecules small enough for the body to absorb. Most of the organic matter in food consists of proteins, fats, and carbohydrates—all large polymers (multi-unit molecules made up of small monomers). Animals cannot use these materials directly for two reasons. First, as macromolecules, these polymers are too large to pass through plasma membranes and enter the cells. Second, an animal needs monomers to make the polymers of its own body. Most of the polymers in food (for instance, the proteins in beans) are different from those making up the animal's body. Therefore, an animal has to break the food polymers into monomers and then use the monomers to make its own brand of polymers (see Modules 6.16 and 6.17).

All organisms use the same monomers. For instance, humpback whales, humans, and bean plants all make their proteins from the same 20 kinds of amino acids. Digestion in an animal breaks the macromolecules in food into their component monomers. Proteins are split into amino acids, polysaccharides and disaccharides are split into simple sugars, fats are split into glycerol and fatty acids, and nucleic acids are split into nucleotides.

As Figure 21.2 shows, digestion typically occurs in two phases. First, food is mechanically broken into smaller pieces. In a baleen whale, this occurs in the stomach. In animals with teeth, the process of chewing breaks large chunks of food into smaller ones. The second phase of digestion is the chemical breakdown process called hydrolysis. Catalyzed by specific enzymes, hydrolysis breaks chemical bonds in the food polymers by adding water to them (see Module 3.3). In the process, polymers are broken down into monomers, which are small molecules.

The last two stages of food processing occur after digestion. ③ In the third stage, **absorption,** the cells lining the digestive tract take up (absorb) small nutrient molecules. From here the molecules travel in the blood to other body cells, where they are incorporated into the cells or broken down further to provide energy. In a humpback whale, as in almost any animal that eats much more than its body immediately uses, many of the nutrient molecules are converted into fat for storage. ④ In the fourth and last stage of food processing, **elimination,** undigested material passes out of the digestive tract.

? What are the two main digestive processes?

Mechanical digestion and chemical digestion, or enzymatic hydrolysis

Figure 21.2 The four main stages of food processing

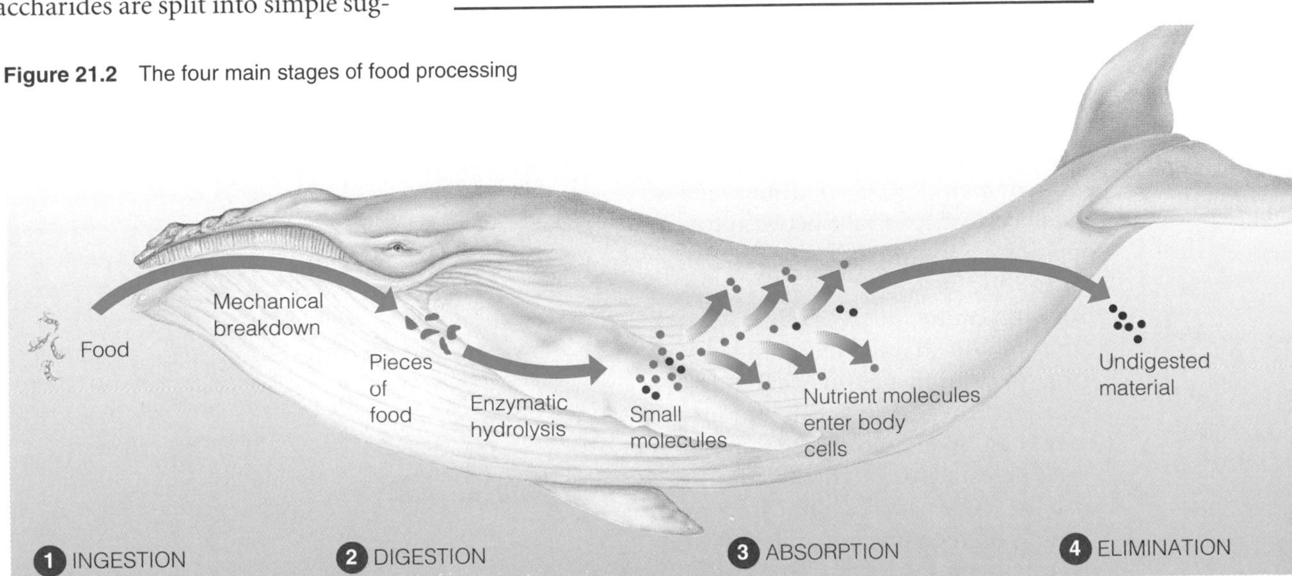

21.3 Digestion occurs in specialized compartments

To process food, an animal's body must provide an environment that favors the action of digestive enzymes. In addition, that environment must be contained in some type of compartment where the enzymes will not attack the organism's own macromolecules.

Even single-celled organisms have digestive compartments. An amoeba, for instance, has food vacuoles in which the cell can digest food without the hydrolytic enzymes mixing with its own cytoplasm. Among the animals, most sponges are like amoebas in carrying out all of their digestion within their cells (see Module 18.3). In contrast, most other animals have a specific compartment within the body, but outside of cells, in which at least some digestion occurs.

As we saw in Chapter 18, relatively simple animals such as hydras have a **gastrovascular cavity,** a digestive compartment with a single opening, the **mouth.** The gastrovascular cavity functions in both digestion and the distribution of nutrients throughout the body. A hydra's gastrovascular cavity enables it to ingest prey much larger than any of its cells could take in directly.

Figure 21.3A illustrates the main food processing events in a hydra. A hydra is a carnivore that stings its prey (here, a small crustacean called *Daphnia*). It then uses its tentacles to stuff the food into its mouth, which expands to accommodate it. Once the prey is in the gastrovascular cavity, cells lining the cavity secrete digestive enzymes (represented by green dots in the figure). Flagella on the cells keep the food mixed with the enzymes, and hydrolysis breaks down the soft tissues of the prey into tiny particles. Once the pieces are small enough, the cells lining the gastrovascular cavity engulf them into food vacuoles, where additional enzymes complete the digestion of the food into simple nutrient molecules. After the hydra has digested its meal, undigested materials remaining in the gastrovascular cavity are eliminated through the mouth.

In contrast to the gastrovascular cavity, a second type of digestive compartment, called an **alimentary canal,** consists of a tube between two openings, a mouth and an anus. Because food moves in one direction from mouth to anus, the alimentary canal can be adapted along the way into specialized regions that carry out digestion and absorption of nutrients in sequence. For example, food ingested through the mouth usually passes into a **pharynx,** or throat, and then into a channel called the **esophagus.**

Depending on the species, the esophagus may channel food to a crop, a gizzard, or a stomach; many animals have all three of these organs. A **crop** is a pouchlike organ in which food is usually softened and stored temporarily. ("Crop" comes from the Greek word for "bulge.")

Stomachs and **gizzards** are also pouchlike and may store food temporarily, but they are more muscular than crops and actively churn and grind the food. Gizzards often contain teeth or grit to assist in grinding. Chemical digestion and nutrient absorption occur mainly in the **intestine,** the region of the digestive tract between the stomach or gizzard and the anus. The **anus** is the opening through which undigested materials are expelled.

The regions of the alimentary canal vary according to the type of food they process. Figure 21.3B on the facing page illustrates the alimentary canals of animals with three different types of diet. The earthworm is an omnivorous substrate feeder. As it burrows through the ground, its muscular pharynx sucks food—actually, soil—into its mouth. The food passes through the esophagus and is stored and moistened in the crop. The muscular gizzard retains small bits of sand and gravel, which pulverize the food. Organic matter in the food is chemically digested and nutrients are absorbed in the intestine. As the enlargement to the right of the worm shows, the intestine is not a simple cylinder. The intestinal wall has a large inward fold, which increases the intestine's inside surface area. As a result, a greater number of cells are exposed to the food passing through. Without this increased surface area, the intes-

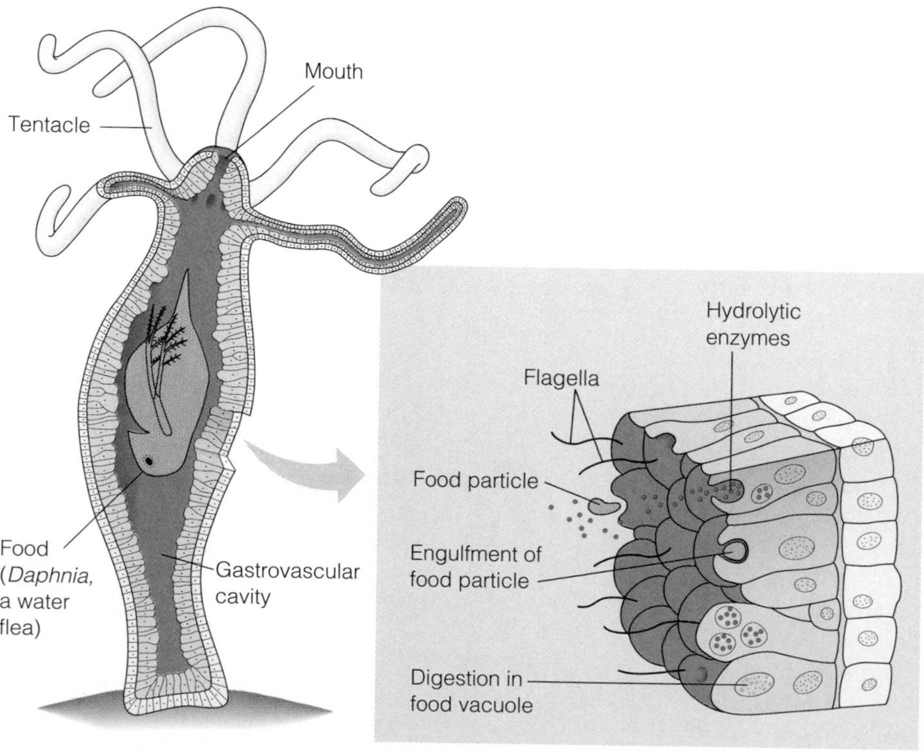

Figure 21.3A Digestion in the gastrovascular cavity of a hydra

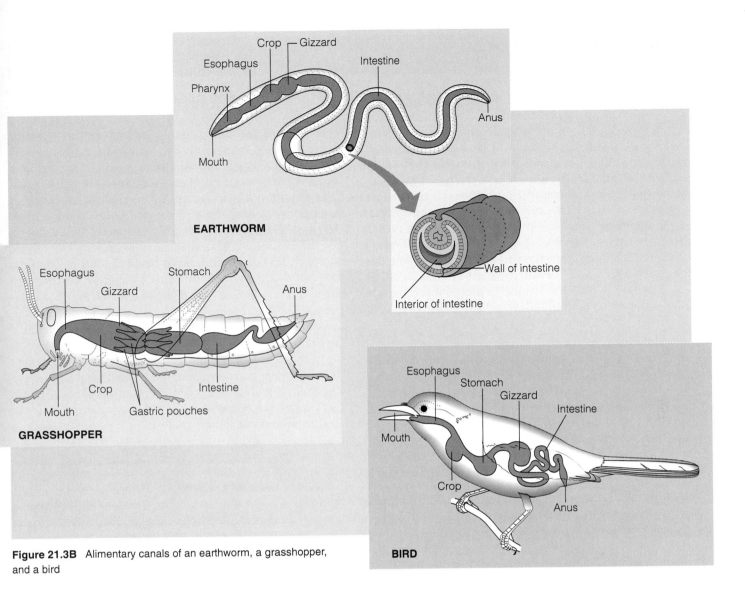

Figure 21.3B Alimentary canals of an earthworm, a grasshopper, and a bird

tine would be much less effective in digesting food and absorbing nutrients and water. The indigestible material in the soil that an earthworm eats is eliminated through the anus. This is mostly sand, fine gravel, and other inorganic matter.

A grasshopper is herbivorous. It has jawlike mouthparts that cut and chew plant leaves into small pieces. Like the earthworm, it also has a crop where food is stored and moistened, and a gizzard where hard teeth reduce the food to a pulverized mass. From the gizzard, food passes into the stomach, where most chemical digestion occurs. Nutrients are absorbed in the stomach and also in a cluster of gastric pouches, which extend from the stomach. The short intestine functions mainly to absorb water and to compact undigested solids, which are expelled through the anus.

The alimentary tract of a typical bird consists of the same major organs as those in the earthworm and grasshopper. Birds eat a variety of foods. Hawks and owls, for instance, are carnivores, catching and eating mice, small birds, snakes, and insects. Many other birds, such as robins and chickens, are omnivores, eating worms, insects, nuts, and fruits. Most birds have a crop, which enables them to eat a large amount of food quickly. Lacking teeth, birds swallow food whole into the crop. The crop stores and softens the food, then passes it into the stomach. Mechanical and chemical digestion begin in the stomach and continue in the gizzard. Many birds eat gravel, as you may see them doing along roadsides. The gravel collects in the gizzard and helps pulverize tough plant fibers and hard parts of insects. (If you've ever kept a parakeet or other seed-eating bird, you know they must have fine gravel in their diet.) After food passes through the gizzard, chemical digestion is completed in the bird's intestine. Nutrients and water are absorbed through the intestinal wall, and undigested materials (often including gravel) pass out through the anus.

Like earthworms, robins, and chickens, humans are omnivorous. We turn to the main features of the human digestive system next.

The main organ of nutrient absorption in most animals is the ________.

intestine

21.4 The human digestive system consists of an alimentary canal and accessory glands

As an introduction to our own digestive system, Figure 21.4 below provides an overview of the human alimentary canal and the digestive glands associated with it. The main parts of the canal are the mouth, oral cavity, tongue, pharynx, esophagus, stomach, small intestine, large intestine, rectum, and anus. The digestive glands—the salivary glands, pancreas, and liver—are labeled in blue on the figure. They secrete digestive juices that enter the alimentary canal through ducts. Secretions from the liver are stored in the gallbladder before they are released into the intestine.

Once food is swallowed, muscles propel it through the alimentary canal by **peristalsis,** rhythmic waves of contraction of smooth muscles in the walls of the digestive tract. In only 5–10 seconds, food passes from the pharynx down the esophagus and into the stomach. Constriction at the base of the esophagus keeps food in the stomach.

A muscular ring, called the **pyloric sphincter,** regulates the passage of food out of the stomach and into the small intestine. The sphincter works like a drawstring, closing off the tube and keeping food in the stomach long enough for stomach acids and enzymes to begin digestion. The final steps of digestion, and nutrient absorption, occur in the small intestine over a period of 5–6 hours. Undigested material passes through the large intestine, where water is taken into the body from the remains of the food and digestive juices, and feces are formed.

In the next several modules, we follow a snack through the alimentary canal, to see in more detail what happens to the food in each of the processing stations along the way. Let's start with you walking into a snack bar for a slice of cheese pizza and a soft drink.

? What is peristalsis and what is its function in our digestive system?

Peristalsis is the wavelike contraction of smooth muscles; the waves move food along our alimentary canal.

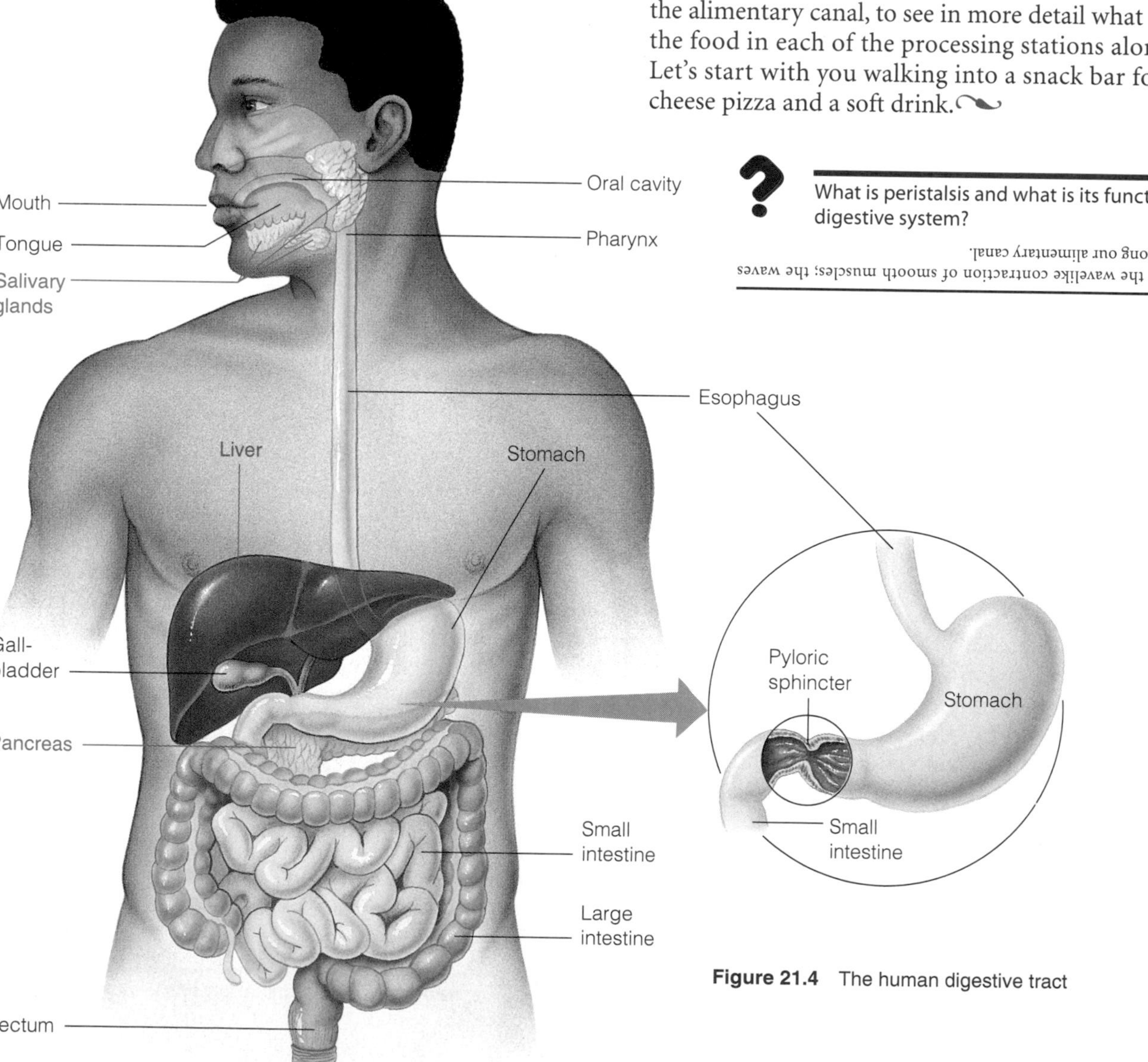

Figure 21.4 The human digestive tract

Digestion begins in the oral cavity 21.5

In the snack bar, your salivary glands may start delivering saliva through ducts to the oral cavity even before you place your order. This is a response to the sight or smell of food, or to your usual eating time. In a typical day, our salivary glands secrete over a liter of saliva.

Saliva contains several substances important in food processing. A slippery glycoprotein protects the soft lining of the mouth and lubricates the solid food for easier swallowing. Buffers neutralize food acids, such as those in soft drinks and tomato sauce, helping prevent tooth decay. Antibacterial agents kill many potentially harmful bacteria that may enter the mouth with food. Saliva also contains a digestive enzyme called salivary amylase that begins hydrolyzing starch, a major ingredient of your pizza crust.

Mechanical and chemical digestion begin in the oral cavity, as we chew our food. Chewing cuts, smashes, and grinds solid food, making it easier to swallow and exposing more food surface to digestive enzymes. As Figure 21.5 shows, we have four kinds of teeth. Starting at the front and proceeding backward on each side of the upper or lower jaw, there are two bladelike incisors. These you use for biting. Behind the incisors, a single pointed canine tooth helps you tear loose a bite of your pizza. Next come two premolars and three molars, which grind and crush the morsel. (The third molar, a "wisdom" tooth, does not appear in some people.)

Also prominent in the oral cavity is the tongue, a muscular organ covered with taste buds. Besides enabling you to taste your meal, the tongue manipulates food and helps shape it into a ball called a bolus. In swallowing, the tongue pushes the bolus to the back of the oral cavity and into the pharynx.

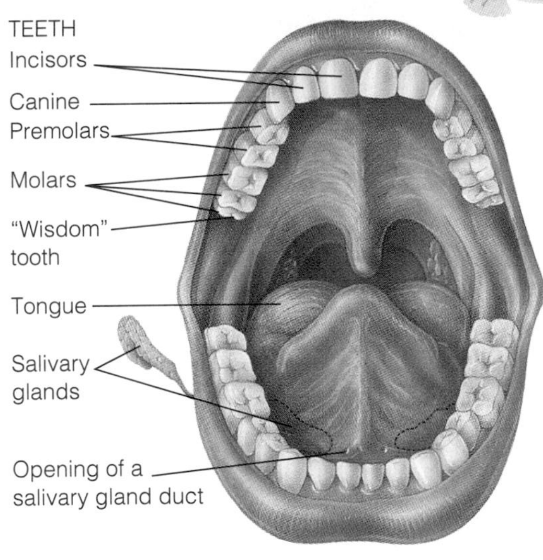

Figure 21.5 The human oral cavity

Chewing functions in ____________ digestion, and salivary amylase initiates the chemical digestion of ____________.

mechanical . . . starch

The food and breathing passages both open into the pharynx 21.6

Openings into both our esophagus and our **trachea** (windpipe) are in the pharynx. Most of the time, as shown on the left drawing in Figure 21.6, the esophageal opening is closed off by a sphincter (blue arrows), and the trachea is open for breathing. This situation changes when you start to swallow some of the pizza you've just finished chewing. As you do, a bolus of the food enters the pharynx, triggering the swallowing reflex; as shown in the center drawing, the esophageal sphincter relaxes and allows the bolus to enter the esophagus. At the same time, the larynx (voicebox) moves upward and tips the epiglottis (a flap of cartilage and fibrous connective tissue) over the tracheal opening. In this position, the epiglottis prevents food from passing into the windpipe. You can see this motion in the bobbing of your larynx (also called your Adam's apple) during swallowing. After the bolus has entered the esophagus, the larynx moves downward, and the breathing passage reopens (right drawing). The esophageal sphincter contracts above the bolus.

When we start coughing because food or drink "went down the wrong pipe," the material has entered the ____________ instead of the ____________.

trachea (windpipe) . . . esophagus

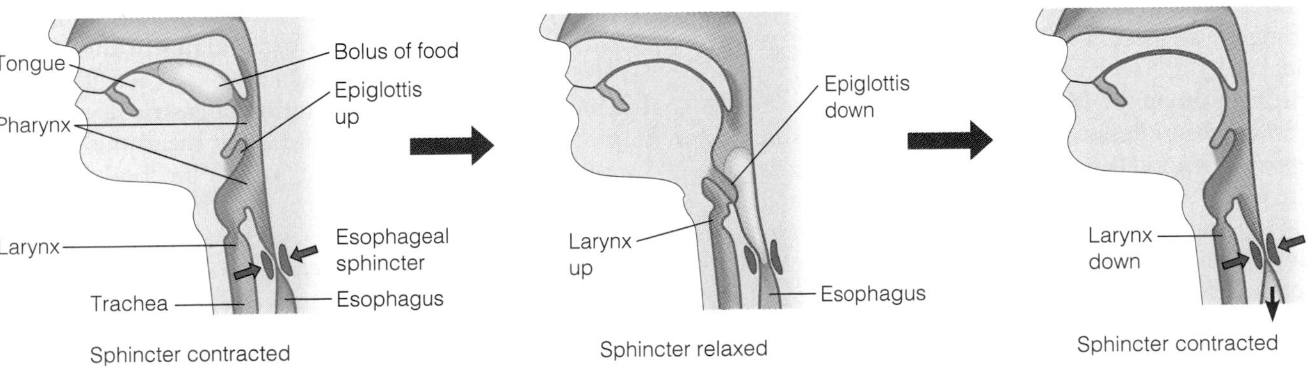

Figure 21.6 The swallowing reflex

21.7 The esophagus squeezes food along to the stomach

The **esophagus** is a muscular tube that conveys food boluses from the pharynx to the stomach. The esophageal muscles are arranged in two layers. A circular muscle layer runs around the esophagus (blue); a longitudinal layer runs the length of the esophagus (yellow). Contraction of the circular layer constricts the esophagus. Contraction of the longitudinal layer shortens the esophagus. Both layers are smooth muscle, which contracts involuntarily. Wherever one muscle layer contracts, the other is relaxed.

Figure 21.7 shows how wavelike contractions—peristalsis—of the circular and longitudinal muscles squeeze a bolus toward the stomach. As food is swallowed, circular muscles above the bolus contract (blue arrows), pushing the bolus downward (longitudinal muscles remain relaxed there). At the same time, longitudinal muscles below the bolus contract (yellow arrows), shortening the passageway ahead of the bolus (the circular muscles there are relaxed, allowing the passageway to be open). These contractions continue in waves until the bolus enters the stomach.

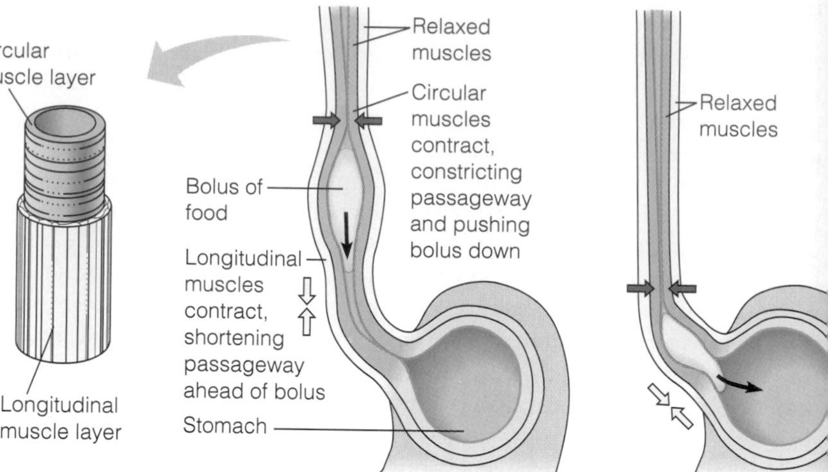

Figure 21.7 Muscle layers of the esophagus and their roles in peristalsis

How could an astronaut get food from the pharynx to the stomach in the weightless environment of a space shuttle?

By peristalsis

21.8 The stomach stores food and breaks it down with acid and enzymes

21A
See Activity 21A on the CD-ROM.

Having a stomach is the main reason we do not need to eat constantly. Our stomach is highly elastic and can stretch to accommodate about 2 liters of food and drink, usually enough to satisfy our body's needs for many hours.

Some chemical digestion occurs in the stomach. The stomach secretes gastric juice, which is made up of mucus, enzymes, and strong acid. The interior surface of the stomach wall is highly folded, and, as Figure 21.8 shows, it is dotted with pits leading down into tubular **gastric glands.** The gastric glands have three types of cells that secrete different components of the gastric juice. Mucous cells (dark pink) secrete mucus, which lubricates and protects the cells lining the stomach; parietal cells (gold color) secrete hydrochloric acid (HCl); and chief cells (rust color) secrete pepsinogen, an inactive form of the digestive enzyme pepsin.

The diagram on the far right indicates how pepsinogen, HCl, and pepsin interact during digestion in the stomach. ① Pepsinogen and HCl are secreted into the interior of the gastric gland. ② Next, the HCl converts pepsinogen to pepsin. ③ Pepsin itself then activates more pepsinogen, starting a chain reaction. Pepsin begins the chemical digestion of proteins—those in the cheese on your pizza, for instance. It splits the polypeptide chains of the proteins into smaller polypeptides. This action primes the proteins for further digestion, which will occur in the small intestine.

What prevents gastric juice from digesting away the stomach lining? Secreting pepsin in the inactive form of pepsinogen helps protect the cells of the gastric glands, and mucus helps protect the stomach lining from both pepsin and acid. Still, the epithelium is constantly eroded. Mitosis must generate enough new cells to replace the stomach lining completely about every three days.

Cells in our gastric glands do not secrete gastric juice constantly. Their activity is regulated by a combination of nerve signals and hormones. When you see, smell, or taste food, a signal from your brain to your stomach stimulates your gastric glands to secrete gastric juice. Once you have food in your stomach, substances in the food stimulate cells in the stomach wall to release the hormone **gastrin** into the circulatory system. Gastrin circulates in the bloodstream, returning to the stomach wall. When it arrives there, it stimulates further secretion of gastric juice. Thus, an initial burst of gastric secretion at mealtime triggers more secretion, adding gastric juice to the food for some time. A negative-feedback mechanism like the one we described in Module 20.13 inhibits the secretion of gastric juice when the stomach contents become too acidic. The acid inhibits the release of gastrin, and, with less gastrin in the blood, the gastric glands secrete less gastric juice.

Contraction of muscles in the stomach wall aids chemical digestion. The active stomach churns the food with the gastric juice, forming a mixture called **acid chyme.** Most of the time, the stomach is closed off at both ends. The opening between the esophagus and the stomach is closed except when a bolus driven downward by peristalsis arrives there. Acid chyme is thus kept from flowing backward into the esophagus. Occasional backflow of acid chyme into the lower end of the esophagus causes the feeling we call heartburn. During vomiting, peristalsis reverses direction and drives the stomach contents upward into the oral cavity. Between the stomach and the small intestine, the pyloric sphincter helps regulate the passage of acid chyme from the stomach into the small intestine. With the acid chyme leaving the stomach only a squirt at a time, the stomach takes about 2–6 hours to empty after a meal.

?

If you added pepsinogen to a test tube containing protein dissolved in distilled water, not much protein would be digested. What inorganic substance could you add to the tube to accelerate protein digestion?

Hydrochloric acid or some other acid that will convert inactive pepsinogen to active pepsin

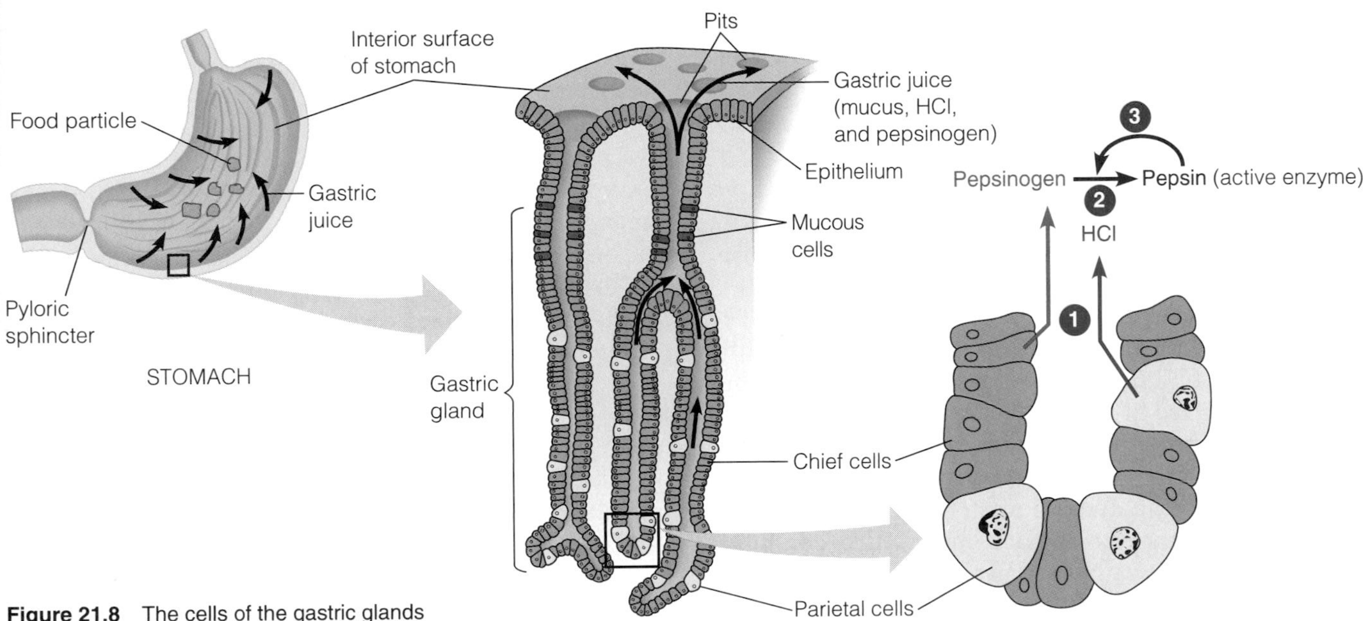

Figure 21.8 The cells of the gastric glands

Bacterial infections can cause ulcers 21.9

A stomachful of digestive juice laced with strong acid lets us digest a diverse array of foods. At the same time, these chemicals, acidic enough to dissolve steel, can be harmful. A coat of mucus normally protects the stomach wall from the corrosive effect of digestive juice, but this is not foolproof protection. When it fails, open sores called **gastric ulcers** can develop in the stomach wall. The symptoms are usually a gnawing pain in the upper abdomen, which may come and go for several hours after eating.

Gastric ulcers were formerly thought to result from the production of too much pepsin and/or acid, or too little mucus, by the stomach lining. However, strong evidence now points to a spiral-shaped prokaryote called *Helicobacter pylori* as the primary culprit. The low pH of the stomach kills most microbes, but not *H. pylori,* which attaches to the stomach lining and seems to surround itself with chemicals that neutralize acid. Growth of *H. pylori* seems to result in a localized loss of protective mucus and damage to the stomach lining. Numerous white blood cells move into the stomach wall to fight the infection, and their presence is associated with mild inflammation of the stomach, called gastritis. Researchers estimate that 50% of the world's population is infected with *H. pylori.* In about 10% of those infected, the gastritis worsens, and gastric ulcers develop when pepsin and hydrochloric acid destroy cells faster than they can regenerate. Eventually, the stomach wall may erode to the point that it actually has a hole in it. This hole can lead to infection within the abdomen or life-threatening internal bleeding. Evidence also links *H. pylori* to the development of certain kinds of stomach cancer.

Gastric ulcers usually respond to a combination of antibiotics and bismuth (the active ingredient in Pepto Bismol®), which eliminate the bacteria. Drugs that reduce stomach acidity may also help, and researchers are making progress toward developing a vaccine to prevent *H. pylori* infection.

When digesting food leaves the stomach, it is accompanied by gastric juices, and so the first section of the small intestine—the duodenum—is also susceptible to ulcers.

?

In contrast to most microbes, the species that causes ulcers thrives in an environment with a very low ______.

pH

21.10 The small intestine is the major organ of chemical digestion and nutrient absorption

21A

Returning to our journey through the digestive tract, what is the status of your meal as it passes out of the stomach into the small intestine? The food has been mechanically reduced to smaller pieces and mixed with liquid; it now resembles a thick soup. Chemically, starch digestion began in the mouth, and protein breakdown began in the stomach. Aside from this, virtually all chemical digestion of the original macromolecules in the pizza and soft drink occurs in the **small intestine.** Nutrients are also absorbed into the blood from the small intestine. With a length of over 6 m, the small intestine is the longest organ of the alimentary canal. (Its name is based not on its length but on its diameter, which is only about 2.5 cm; the large intestine is much shorter but has twice the diameter.)

Two large glandular organs, the pancreas and the liver, contribute to digestion in the small intestine (Figure 21.10A). The **pancreas** produces digestive enzymes and an alkaline solution rich in bicarbonate. The alkaline solution neutralizes acid chyme as it enters the small intestine. The **liver** performs a wide variety of functions, including the production of bile. **Bile** contains no digestive enzymes, but bile salts dissolved in it make fats more susceptible to enzyme attack. The **gallbladder** stores bile until it is needed in the small intestine. The first 25 cm or so of the small intestine is called the **duodenum.** This is where the acid chyme squirted from the stomach mixes with bile from the gallbladder and digestive enzymes from the pancreas and from the wall of the intestine itself.

The table below summarizes the processes of enzymatic digestion that occur in the duodenum. All four types of macromolecules (carbohydrates, proteins, nucleic acids, and fats) are digested. As we discuss the digestion of each, the table will help you keep track of the enzymes involved.

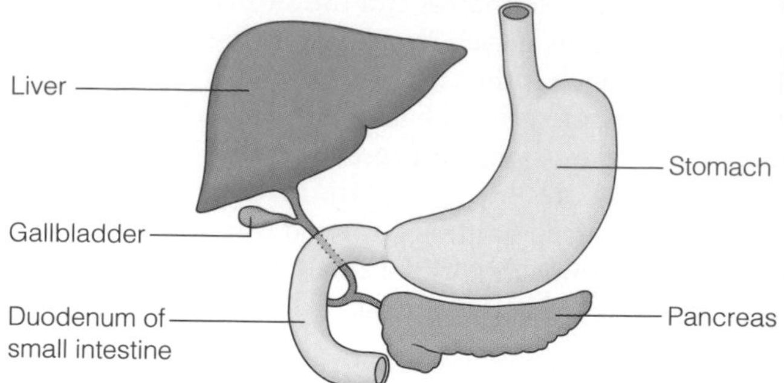

Figure 21.10A The relation of the small intestine to other organs involved in digestion

The digestion of carbohydrates begun in the oral cavity is completed in the small intestine. An enzyme called pancreatic amylase hydrolyzes starch (a polysaccharide) into the disaccharide maltose. The enzyme maltase then splits maltose into the monosaccharide glucose. Maltase is one of a family of enzymes, each specific for the hydrolysis of a different disaccharide. Another enzyme, sucrase, hydrolyzes table sugar (sucrose), and lactase digests milk sugar (lactose, common in milk and cheese). Children generally have much more lactase than adults. Some adults lack lactase altogether, and ingesting milk products can give them cramps and diarrhea because they cannot digest the lactose.

The small intestine also completes the digestion of proteins begun in the stomach. The pancreas and the duodenum secrete hydrolytic enzymes that completely dismantle polypeptides into amino acids. The enzymes trypsin and chymotrypsin break polypeptides into shorter chains than those resulting from pepsin digestion. Two other enzymes, aminopeptidase and carboxypeptidase, split off one amino

Enzymatic Digestion in the Small Intestine

Substrate	Enzyme	Product	Substrate	Enzyme	Product
CARBOHYDRATES					
Starch (a polysaccharide)	Pancreatic amylase →	Maltose (a disaccharide)	Disaccharides	Maltase, Sucrase, Lactase (etc.) →	Monosaccharides
PROTEINS					
Polypeptides	Trypsin, Chymotrypsin →	Smaller polypeptides	Small polypeptides and dipeptides	Aminopeptidase, Carboxypeptidase, Dipeptidase →	Amino acids
NUCLEIC ACIDS					
DNA and RNA	Nucleases →	Nucleotides	Nucleotides	Other enzymes →	Nitrogenous bases, sugars, and phosphates
FATS					
Fat globules	Bile salts →	Fat droplets (emulsified)	Fat droplets	Lipase →	Fatty acids and glycerol

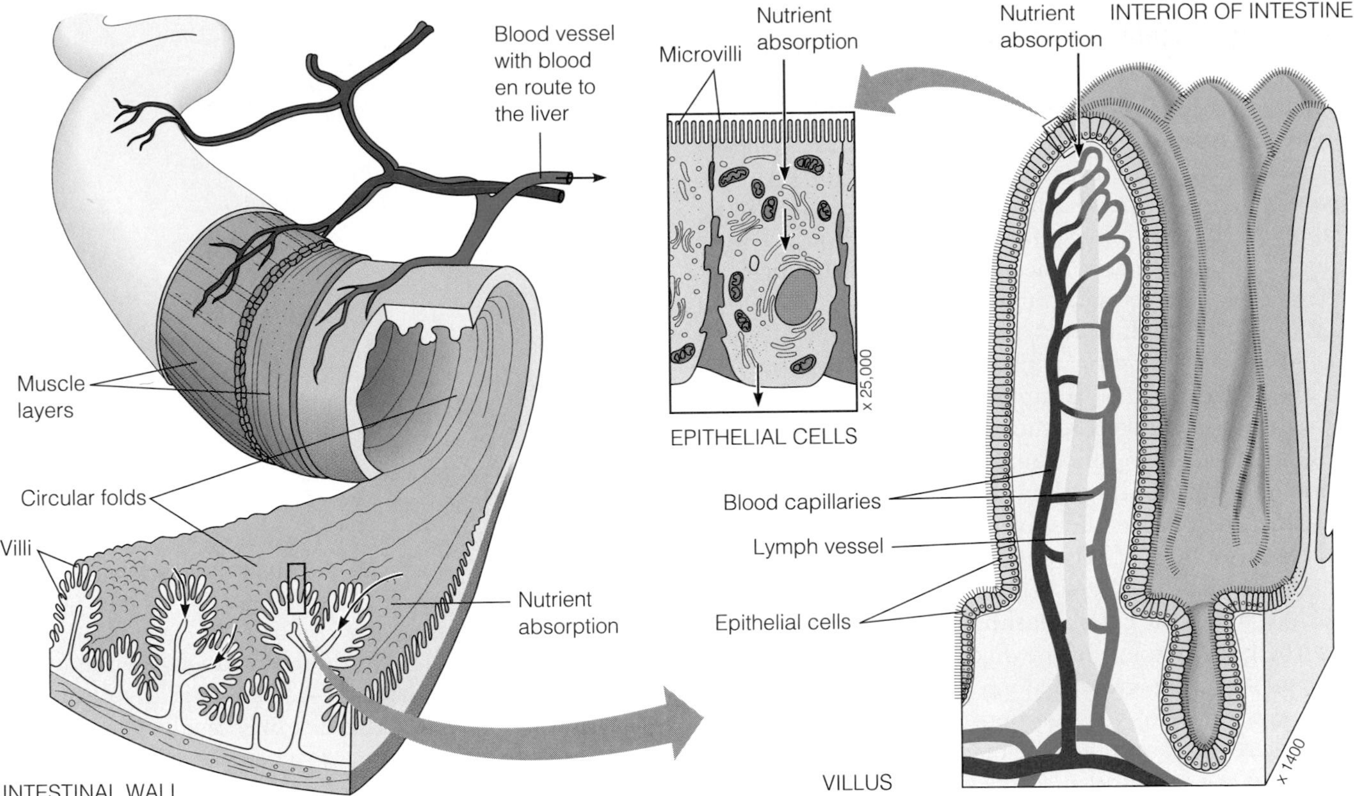

Figure 21.10B Structure of the small intestine

acid at a time, working from the ends of the polypeptides. A fourth type of enzyme, dipeptidase, hydrolyzes fragments only two or three amino acids long. Working together, this enzyme team digests proteins much faster than any single enzyme could.

Another team of enzymes, the nucleases, hydrolyzes the nucleic acids in food. Nucleases from the pancreas split DNA and RNA (which would be present in any meat or vegetables on your pizza) into their component nucleotides. The nucleotides are then broken down into nitrogenous bases, sugars, and phosphates by other enzymes produced by the duodenal cells.

In contrast to starch and proteins, nearly all the fat in your pizza remains completely undigested until it reaches the duodenum. Hydrolysis of fats is a special problem because fats are insoluble in water. First, bile salts from the gallbladder coat tiny fat droplets and keep them separate from one another, a process called emulsification. When there are many small droplets, a large surface area of fat is exposed to lipase, an enzyme that breaks fat molecules down into fatty acids and glycerol.

By the time peristalsis has moved the chyme mixture through the duodenum, chemical digestion of our meal is just about complete. The remaining regions of the small intestine are adapted for the absorption of nutrients.

Structurally, the small intestine is well suited for the task of absorbing nutrients. Its lining has a huge surface area—roughly 300 m^2, about the size of a tennis court. As Figure 21.10B indicates, the extensive surface area results from several kinds of folds and projections. Around the wall of the small intestine are large circular folds with numerous small, fingerlike projections called **villi** (singular, *villus*). If we look at the epithelial cells of a villus with an electron microscope, we see many tiny surface projections, called **microvilli.** The microvilli extend into the interior of the intestine. Notice that a small lymph vessel and a network of capillaries (microscopic blood vessels, shown in red and blue) penetrate the core of each villus. Nutrients pass first across the intestinal epithelium and then through the thin walls of the capillaries or lymph vessel. Some nutrients simply diffuse from the digested food into the epithelial cells and then into the blood or lymph. Others are pumped against concentration gradients by the membranes of epithelial cells.

The capillaries that drain nutrients away from the villi converge into larger blood vessels and eventually a main vessel (not shown) that leads directly to the liver. The liver thus gets first access to nutrients absorbed from a meal. The liver converts many of the nutrients into new substances that the body needs. One of its main functions is to remove excess glucose from the blood and convert it to glycogen (a polysaccharide), which is stored in liver cells. From the liver, blood travels to the heart, which pumps the blood and the nutrients it contains to all parts of the body. The nutrients from your pizza and soft drink are now on their way to being incorporated into your body.

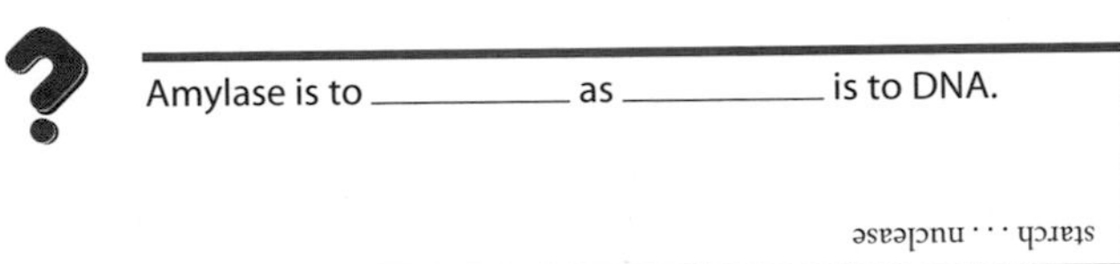

Amylase is to ________ as ________ is to DNA.
starch . . . nuclease

21.11 The large intestine reclaims water

The **large intestine,** or **colon,** is about 1.5 m long and 5 cm in diameter. As the enlargement in Figure 21.11 shows, it joins the small intestine at a T-shaped junction, where a sphincter controls the passage of unabsorbed food material out of the small intestine. One arm of the T is a blind pouch called the **cecum.** The **appendix,** a small, fingerlike extension of the cecum, contains a mass of white blood cells that make a minor contribution to immunity. Despite this role, the appendix itself is prone to infection (appendicitis). If this occurs, the appendix can be surgically removed without weakening the immune system.

The colon's main function is to absorb water from the alimentary canal. Altogether, about 7 liters of fluid enter the canal each day as the solvent of the various digestive juices. About 90% of this water is absorbed back into the blood and tissue fluids, with the small intestine reclaiming much of it and the colon finishing the job. As the water is absorbed, the remains of the digested food become more solid as they are conveyed along the colon by peristalsis. These waste products of digestion, the **feces,** consist mainly of indigestible plant fibers (cellulose from the peppers on your pizza, for instance) and prokaryotes that normally live in the colon. Some of our colon bacteria, such as *E. coli,* produce important vitamins, including biotin, folic acid, several B vitamins, and vitamin K. These vitamins are absorbed into the bloodstream through the colon.

The terminal portion of the colon is the **rectum,** where the feces are stored until they can be eliminated. Strong contractions of the colon create the urge to defecate. Two rectal sphincters, one voluntary and the other involuntary, regulate the opening of the anus.

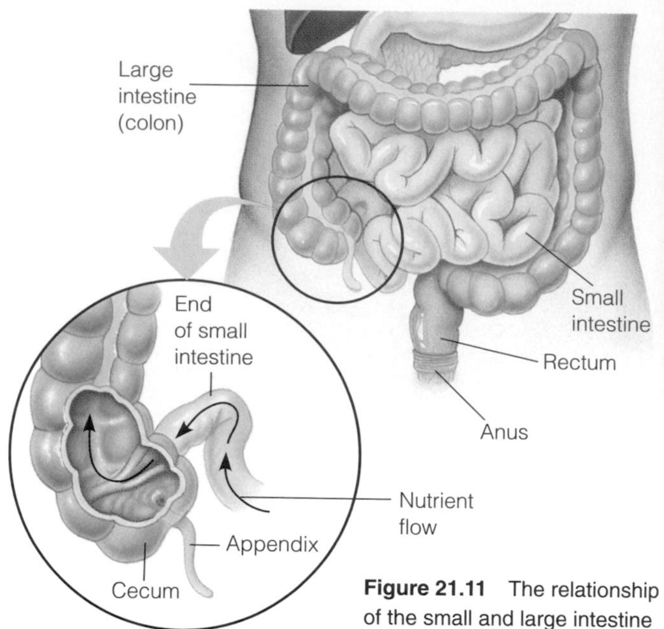

Figure 21.11 The relationship of the small and large intestine

If the lining of the colon is irritated—by a viral or bacterial infection, for instance—the colon is less effective in reclaiming water, and diarrhea may result. The opposite problem, constipation, occurs when peristalsis moves the feces along too slowly; the colon reabsorbs too much water, and the feces become too compacted. Constipation often results from a diet that does not include enough plant fiber, or from a lack of exercise. 

? Explain why treatment of a chronic infection with antibiotics for an extended period of time may cause a vitamin K deficiency.

By killing symbiotic bacteria that synthesize vitamin K in the colon

Diets and Digestive Adaptations

21.12 Adaptations of vertebrate digestive systems reflect diet

In following a meal through our own alimentary canal, we have seen the basic plan of the vertebrate digestive system. As a group, the vertebrates exhibit many variations on this basic plan. In every case, the structure and function of the digestive system are keyed to the kind of food the animal eats.

The length of an animal's digestive tract tells us something about its diet. In general, herbivores and omnivores have longer alimentary canals, relative to their body size, than carnivores. A longer canal provides the extra time it takes to extract nutrients from vegetation, which is more difficult to digest than meat because of the cell walls in plant material. A longer canal also provides more surface area for absorbing nutrients, which are usually less concentrated in vegetation than in meat. A model case is the frog, which is carnivorous as an adult but mainly herbivorous as a tadpole. A tadpole's intestine is long relative to its body size. When a tadpole transforms into an adult, the rest of its body grows more than its intestine, leaving the adult frog with an intestine that is shorter relative to its overall size.

Herbivorous mammals typically have very long alimentary canals. Most of them also have special chambers in the canal that house great numbers of microbes—prokaryotes (bacteria and archaea) and protists. The mammals themselves cannot digest cellulose in the plants they eat. The microbes convert the cellulose to simple sugars and other nutrients, which the mammals then absorb directly or obtain by digesting the microbes.

Many herbivorous mammals—horses and elephants, for example—house cellulose-digesting microbes in the colon and in a large cecum, the pouch where the small and large intestines connect. Some of the nutrients produced by the microbes are absorbed in the cecum and colon. Most of the

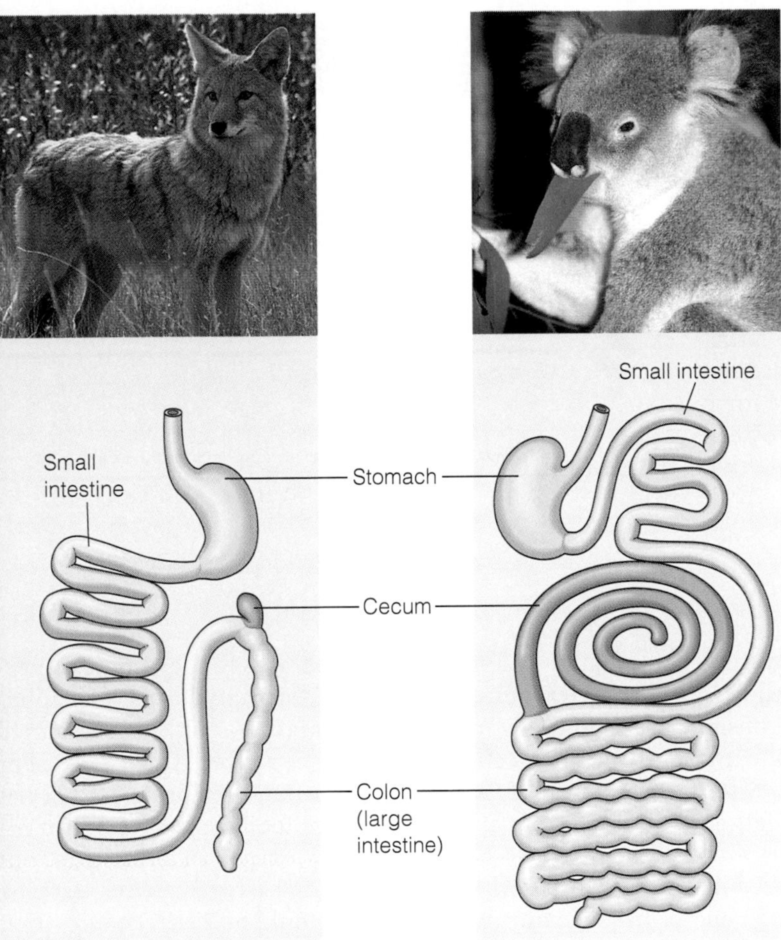

Figure 21.12A The alimentary canal in a carnivore (coyote) and an herbivore (koala)

nutrients, however, are lost in the feces, because they do not go through the small intestine, the main site of nutrient absorption. Rabbits and some rodents obtain these nutrients by eating some of their feces, thus passing the food through the alimentary canal a second time. The feces in the second round are more compact and are not reingested. Many desert rodents conserve water by eating their first round of feces.

Figure 21.12A compares the digestive tract of a carnivore, the coyote, with that of an herbivore, the koala. The koala is an Australian marsupial (see Module 18.22). These two mammals are about the same size, but the koala's intestine is much longer, and includes the longest cecum (about 2 m) of any animal of this size. Prokaryotes in the cecum digest plant material, making it possible for the koala to get almost all its food and water from the leaves of eucalyptus trees.

Ruminant mammals, such as cattle, sheep, and deer, have a more elaborate system for cellulose digestion. The stomach of a ruminant has four chambers, which we show in pink in Figure 21.12B. The arrows in this figure indicate the pathway of food. When the cow first chews and swallows a mouthful of grass, the food enters the rumen ① and then the reticulum (green arrows) ②. Prokaryotes and protists in the rumen and reticulum immediately go to work on the cellulose-rich meal, and the cow helps by periodically regurgitating and rechewing her food (red arrows). This rumination, or "chewing the cud," softens and helps break down plant fibers, making them more accessible to digestion by the microbes.

As the blue arrows indicate, the cow swallows her cud into the omasum ③, where water is absorbed. The cud finally passes to the abomasum ④, where the cow's own enzymes complete digestion. Here, the cow obtains many of her nutrients by digesting the microbes along with nutrients they produce. The microbes reproduce so rapidly that their numbers remain stable despite this constant loss. With its microbes and multistage food-processing system, a ruminant harvests more energy and nutrients from the cellulose in hay or grass than a nonruminant herbivore like a horse or an elephant can.

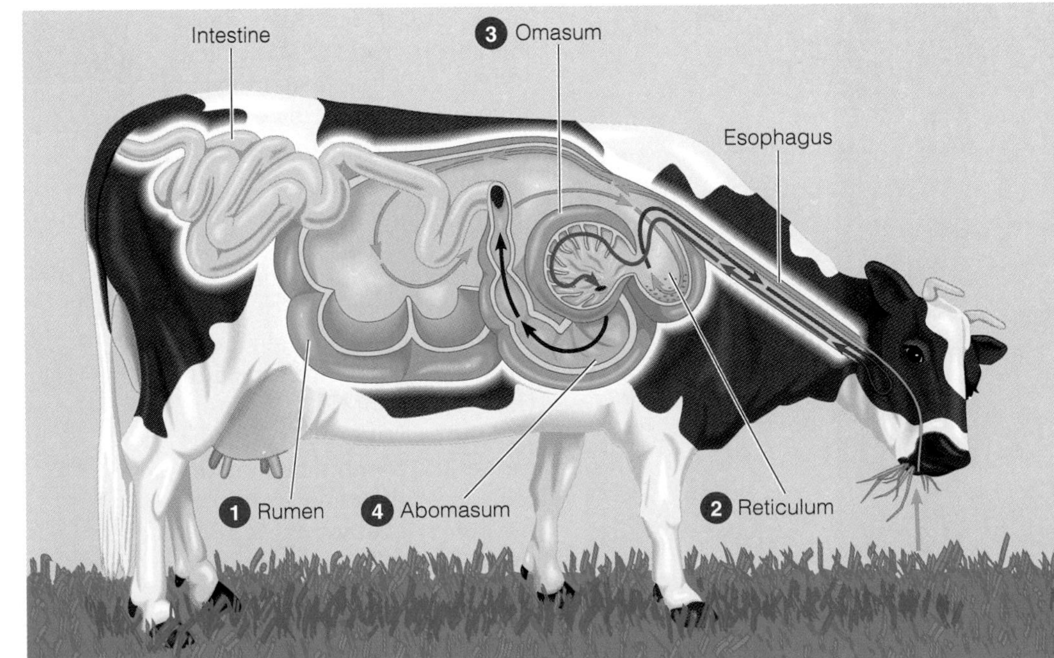

Figure 21.12B The digestive system of a ruminant mammal

When a tadpole becomes an adult frog during metamorphosis, there is little growth of the intestine relative to the rest of the body. Thus, compared to the adult frog, a tadpole has a longer intestine relative to its body size. What does this suggest about the diets of these two stages in the frog's life history?

The tadpole is mainly herbivorous (eats mostly algae), while the adult frog is carnivorous (eats insects, for example).

Nutrition

21.13 Overview: A healthful diet satisfies three needs

All animals—whether herbivores like cows, carnivores like cats, or omnivores like humans—have the same basic nutritional needs. All animals must obtain (1) fuel to power all body activities, (2) organic raw materials needed to make the animal's own molecules, and (3) essential nutrients, or substances the animal cannot make for itself from any raw material but must obtain in prefabricated form from food. The different digestive systems we have been examining represent diverse evolutionary adaptations that meet these three basic needs. Starting with the need for fuel and paying particular attention to humans, we discuss basic nutritional needs for the rest of this chapter.

What is an "essential nutrient"?

A substance an organism requires but cannot make by its own metabolism

21.14 Chemical energy powers the body

Reading a book, walking to class, eating and digesting food, and every other activity your body performs requires fuel in the form of chemical energy. Cellular metabolism produces the body's energy currency, ATP, by oxidizing small molecules (such as glucose) digested from food (see Module 6.14). Usually cells use carbohydrates and fats as fuel sources, but when these are in short supply, they will use proteins. The energy content of food is measured in **kilocalories** (1 **kcal** = 1000 calories). ("Calories" listed on food labels are kilocalories.)

Cellular metabolism must continuously drive several processes in order for an animal to remain alive. These include breathing, the beating of the heart, and, in birds and mammals, the maintenance of body temperature. The number of kilocalories a resting animal requires to fuel these essential processes for a given time is called the **basal metabolic rate (BMR).** The BMR for adult humans averages 1300–1800 kcal per day. This is about equivalent to the daily energy consumption of a 100-watt light bulb. But this is only a basal (base) rate—the amount of energy we "burn" lying motionless. Any activity, even working quietly at your desk, consumes kilocalories in addition to the BMR. The more strenuous the activity, the greater the energy demand. The table here gives you an idea of the amount of activity it takes for a 150-lb (68-kg) person to use up the kilocalories contained in several common foods. The numbers in the table are above and beyond the BMR.

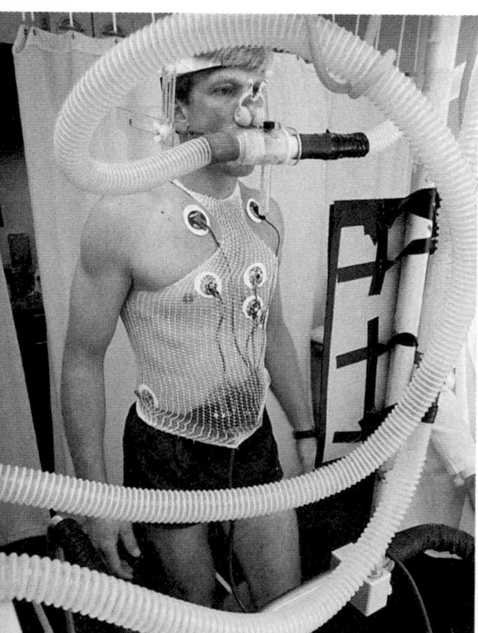

Figure 21.14 Determining metabolic rate by measuring O_2 consumption

Figure 21.14 shows an apparatus that estimates a person's metabolic rate at various activity levels. The apparatus actually measures the amount of oxygen consumed as the body cells oxidize food in a given time period. For every liter of O_2 consumed, cellular respiration liberates about 4.83 kcal of energy from food molecules. If, for instance, your body used 16 L of oxygen in an hour (about the rate at which you consume oxygen while washing dishes), your metabolic rate would be 77.28 kcal/hr (that is, 16 L/hr × 4.83 kcal/L).

What happens when we take in more kilocalories than we consume in meeting our energy requirements? Rather than discarding the extra energy, our cells store it in various forms. Our liver and muscles store energy in the form of glycogen, a polymer of glucose molecules. Most of us

FOOD–EXERCISE ENERGY EQUIVALENTS

	Jogging	Swimming	Walking
Speed	9 min/mi	30 min/mi	20 min/mi
Kcal/kg/min	0.173	0.132	0.039
Big Mac® 560 kcal	47 min	1 hr, 3 min	3 hr, 31 min
Cheese pizza (1 slice) 450 kcal	38 min	50 min	2 hr, 50 min
Coca-Cola Classic® (10 oz) 144 kcal	12 min	16 min	54 min
Whole wheat bread (1 slice) 86 kcal	7 min	10 min	32 min

These data are for a person weighing 68 kg (150 lb).

can store enough glycogen to supply about a day's worth of basal metabolism. Our cells also store excess energy as fat. This happens even if our diet contains little fat, because the liver converts excess carbohydrates and proteins into fat. We discuss fat storage and its consequences next.

What is the metabolic rate in kcal/hr of a person consuming 1 L of O_2 kcal per minute while walking briskly on a treadmill?

289.8 kcal/hr (4.83 kcal/min × 60 minutes)

Body fat and fad diets 21.15

In our culture, excess body fat is considered unattractive. In contrast, some other cultures tend to equate a well-rounded or plump body with beauty and prosperity. Whatever our cultural perspective, fats and related lipids are essential components of the human body. For instance, body fat helps insulate us against cold, and, in moderate amounts, it seems to correlate with a healthy immune system. Extremely thin people tend to have lower levels of vitamin A and beta-carotene in their blood, which may make them more susceptible to certain forms of cancer. Healthy women may have as much as 20–25% of their body weight in fat; for healthy men, the amount is typically 15–19%. When we are too fat, our body fat may be 20% or more above these amounts. Being too fat, or obese, can increase our chances of developing certain diseases (such as heart disease) and decrease our life span.

Our cells make fats and other lipids by combining fatty acids with other molecules, such as glycerol (see Module 3.8). We can make most of the required fatty acids, in turn, from simpler molecules; those we cannot make, called **essential fatty acids,** we must obtain in our diet. One essential fatty acid, linoleic acid, is especially important because it is needed to make some of the phospholipids of cell membranes. Most diets furnish ample amounts of essential fatty acids, and deficiencies are rare.

The human body tends to hoard fat—to immediately store any excess fat molecules obtained from food instead of using them for fuel. By contrast, when we eat an excess of carbohydrates, the body tends to increase its rate of carbohydrate use. Thus, the amount of fat in the diet may have a more direct effect on weight gain than the amount of carbohydrate. Although the tendency to hoard fat is a potential problem today, it may once have helped our ancestors survive periods of famine.

Despite its tendency to store fat, the human body seems to impose limits on weight gain and loss. Some people hold a more-or-less constant weight no matter how much they eat. Even obese people usually attain a fairly stable weight. Recent studies suggest that complex feedback mechanisms, involving the brain and chemical signals in the blood, regulate fat storage and use. Some of the signals and their antagonists are under development as potential medications for obesity.

Fad diets like the first three in the table below are designed to take pounds off fast, but they rarely succeed. Whatever weight is lost in fad dieting is usually quickly regained once the diet ends and we return to our usual eating habits. Fad diets are not only ineffective in the long run, but, as the table indicates, they can also be harmful.

The best approach to weight control is a combination of exercise and a restricted but balanced diet. As indicated in the table, a balanced diet provides at least 1200 kcal per day and adequate amounts of all essential nutrients. Such diets meet the **Recommended Dietary Allowances (RDAs),** minimal standards established by nutritionists for preventing nutrient deficiencies. A restricted, balanced diet, along with regular aerobic exercise, can trim the body gradually and keep extra fat off without harmful side effects.

State a hypothesis that could explain why the human body tends to hoard fat.

Among our ancestors, individuals with genes promoting the storage of high-energy molecules during feasts may have been those that survived famines and consequently passed more genes to offspring.

Diet Type	Health Effects and Potential Problems
Extremely low-carbohydrate diets Less than 100 g of carbohydrates per day.	Initial loss of weight is primarily water. Problems may include fatigue and headaches. Long-term use of diet may be associated with muscle loss.
Extremely low-fat diets Less than 20% of kilocalories from fat. Elimination of most or all animal protein sources and all fats, nuts, and seeds.	May be inadequate in essential fatty acids, protein, and certain minerals. May decrease absorption of fat-soluble vitamins. May result in irregular menstrual periods in women.
Formula diets Based on formulated or packaged products. Many are very low in kilocalories.	If very low in kilocalories (less than 800 kcal per day), may result in loss of body protein and may cause dry skin, thinning hair, constipation, and salt imbalance. Usually weight is regained when diet is ended.
Balanced diet of 1200 kcal or more	If carefully chosen, such a diet can meet all nutrient needs. Weight loss is usually 1–2 lb per week; dieter may become discouraged.

21.16 Nine amino acids are essential nutrients

Speaking of diets, is it possible for a person to be a true vegetarian—that is, obtain all necessary nutrients by eating only plant material, without animal products of any kind, including eggs, milk, or cheese? The answer is yes, and many people do so, but they have to know how to get all the essential nutrients.

Given sources of organic carbon (such as sugar) and nitrogen (such as amino acids from the digestion of protein), the human body can make a great variety of organic molecules for its own use. For instance, it can make a number of different amino acids using the nitrogen from a single type of amino acid in food. Nonetheless, there are some substances that the body needs but cannot make on its own, and we must obtain these substances from food.

Adult humans cannot make nine of the 20 kinds of amino acids needed to synthesize proteins. These nine, known as the **essential amino acids,** must be obtained from the diet. Because the body cannot store excess amino acids, a deficiency of a single essential amino acid limits the use of other amino acids, impairs protein synthesis, and can lead to protein deficiency, a serious type of malnutrition. Malnutrition due *solely* to protein deficiency is very rare, but protein deficiency accompanied by an inadequate intake of other energy-yielding foods is fairly widespread in developing countries. The victims are usually children, who, if they survive infancy, are likely to be retarded mentally and underdeveloped physically.

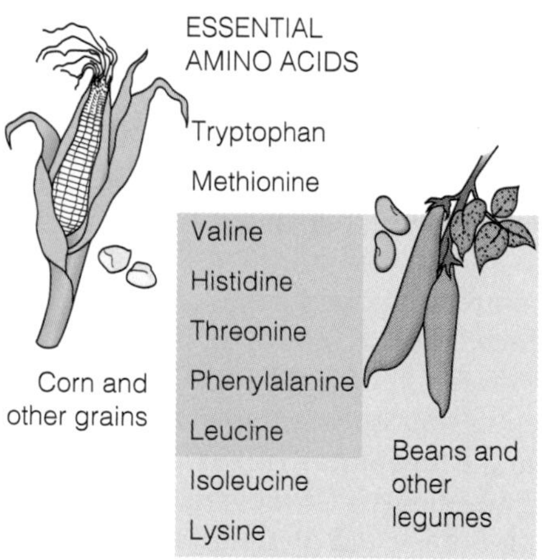

ESSENTIAL AMINO ACIDS
Tryptophan
Methionine
Valine
Histidine
Threonine
Phenylalanine
Leucine
Isoleucine
Lysine
Corn and other grains
Beans and other legumes

Figure 21.16

The simplest way to get all the essential amino acids is to eat meat and animal by-products such as eggs, milk, and cheese. The proteins in these products are said to be complete, meaning they provide all the essential amino acids in the proportions needed by the human body. In contrast, most plant proteins are incomplete, or deficient in one or more essential amino acids. People may become vegetarians by choice or, more commonly, because they simply cannot afford animal protein. Animal protein is more expensive to produce, and usually to buy, than plant protein, and most of the human population is primarily vegetarian. Nutritional problems can result when people have to rely on a single type of plant food—just corn, rice, or wheat, for instance. When they do, they stand a good chance of becoming protein-deficient.

The key to being a healthy vegetarian is to eat a variety of plant foods that together supply sufficient quantities of all the essential amino acids. Simply by eating a combination of beans and corn, for example, vegetarians can get all nine essential amino acids (Figure 21.16). Most societies have, by trial and error, developed balanced diets that prevent protein deficiency. The Mexican staple diet of corn tortillas and beans is one example.

? A diet consisting strictly of various legumes and water would probably result in a deficiency of the essential amino acids ________ and ________.

tryptophan . . . methionine

21.17 A healthful diet includes 13 vitamins

How much of each vitamin do we need? Do we need vitamin supplements? Some nutritionists argue that vitamin supplements are a waste of money for healthy people who eat a balanced diet. But do most of us eat a balanced diet? Are there dangers in taking too much of a supplement? What do we really know about vitamins?

A **vitamin** is an organic nutrient that is essential but required in much smaller quantities than the essential amino acids. Most vitamins serve as coenzymes or parts of coenzymes; they have catalytic functions and are used over and over in metabolic reactions (see Module 5.7). The table on the facing page lists 13 vitamins that are essential in the human diet. Even though needed in only tiny amounts, these substances are absolutely necessary; deficiencies can cause serious problems. On the other hand, vitamins in excess can be dangerous.

Vitamins are either water-soluble or fat-soluble. Water-soluble vitamins include the B complex (the first eight vitamins in the table), which are important in cellular metabolism. Vitamins B-1, B-6, B-12, folic acid, and biotin are coenzymes used in the metabolism of sugars, amino acids, nucleic acids, and fats. Vitamin B-2, niacin, and pantothenic acid form parts of the coenzymes central to cellular respiration: NAD^+, $NADP^+$, FAD, and coenzyme A. Vitamin C, also water-soluble, seems to play many roles, including that of an antioxidant, preventing cell damage caused by certain toxic ions produced by cellular respiration. In general, consuming more water-soluble vitamins than we need is not harmful because the excess passes out in urine and feces.

Fat-soluble vitamins play a variety of important roles. Vitamin A is part of the visual pigment of the eye and helps maintain epithelial tissues. Vitamin D aids in calcium and phosphorus absorption and bone growth. The functions of vitamin E are not yet well understood, but it seems to protect the phospholipids in cell membranes from oxidation. Vitamin K is required for blood clotting. Unlike the water-soluble vitamins, excessive amounts of the fat-soluble vitamins do not readily leave the body but instead are deposited in body fat. As a result, overdoses of at least some may accumulate and cause toxic effects. Excess vitamin A, for instance, can cause liver and bone damage.

Many nutritionists believe that the vitamin RDAs are sufficient. Health-food enthusiasts argue that the RDAs are set too low for at least some vitamins. Debate is especially intense over optimal amounts of vitamins C and E. We can say with some certainty that people who eat a balanced diet are unlikely to develop the symptoms of vitamin deficiency listed in the table. We can say with more certainty that we still have much to learn about the biological roles of vitamins.

Why are vitamins required in such small doses compared to other essential organic nutrients such as essential amino acids?

Because vitamins generally have catalytic functions as coenzymes, and thus each vitamin molecule can repeat its function many times

Vitamin	Major Dietary Sources	Functions in the Body	Symptoms of Deficiency or Extreme Excess*
Water-Soluble Vitamins			
Vitamin B-1 (thiamine)	Pork, legumes, peanuts, whole grains	Coenzyme used in removing CO_2 from organic compounds	Beriberi (nerve disorders, emaciation, anemia)
Vitamin B-2 (riboflavin)	Dairy products, meats, enriched grains, vegetables	Component of coenzyme FAD	Skin lesions such as cracks at corners of mouth
Niacin	Nuts, meats, grains	Component of coenzymes NAD^+ and $NADP^+$	Skin and gastrointestinal lesions, nervous disorders Flushing of face and hands, liver damage
Vitamin B-6 (pyridoxine)	Meats, vegetables, whole grains	Coenzyme used in amino acid metabolism	Irritability, convulsions, muscular twitching, anemia Unstable gait, numb feet, poor coordination
Pantothenic acid	Most foods: meats, dairy products, whole grains, etc.	Component of coenzyme A	Fatigue, numbness, tingling of hands and feet
Folic acid (folacin)	Green vegetables, oranges, nuts, legumes, whole grains	Coenzyme in nucleic acid and amino acid metabolism; essential for normal neural tube development in the embryo	Anemia, gastrointestinal problems Masks deficiency of vitamin B-12
Vitamin B-12	Meats, eggs, dairy products	Coenzyme in nucleic acid metabolism; needed for maturation of red blood cells	Anemia; nervous system disorders
Biotin	Legumes, other vegetables, meats	Coenzyme in synthesis of fat, glycogen, and amino acids	Scaly skin inflammation; neuromuscular disorders
Vitamin C (ascorbic acid)	Fruits and vegetables, especially citrus fruits, broccoli, cabbage, tomatoes, green peppers	Used in collagen synthesis (e.g., for bone, cartilage, gums); antioxidant; aids in detoxification; improves iron absorption	Scurvy (degeneration of skin, teeth, blood vessels), weakness, delayed wound healing, impaired immunity Gastrointestinal upset
Fat-Soluble Vitamins			
Vitamin A	Dark green and orange vegetables and fruits, dairy products	Component of visual pigments; needed for maintenance of epithelial tissues	Vision problems; dry, scaly skin Headache, irritability, vomiting, hair loss, blurred vision, liver and bone damage
Vitamin D	Dairy products, egg yolk (also made in human skin in presence of sunlight)	Aids in absorption and use of calcium and phosphorus; promotes bone growth	Rickets (bone deformities) in children; bone softening in adults Brain, cardiovascular, and kidney damage
Vitamin E (tocopherol)	Vegetable oils, nuts, seeds	Antioxidant; prevents damage to lipids of cell membranes and probably other molecules	None well documented
Vitamin K	Green vegetables, tea	Important in blood clotting	Defective blood clotting Liver damage and anemia

*Symptoms of extreme excess are given in red type.

21.18 Essential minerals are required for many body functions

In addition to amino acids and vitamins, which are organic nutrients, a number of minerals are essential for body functioning. In the area of nutrition, **minerals** are chemical elements *other than* carbon, hydrogen, oxygen, and nitrogen, the four staples of organic compounds. We must acquire the essential minerals listed in the table below from our diet, and some of the major dietary sources are listed.

Along with other vertebrates, humans require relatively large amounts of calcium and phosphorus to construct and maintain our skeleton. Calcium is also necessary for the normal functioning of nerves and muscles, and phosphorus is an ingredient of ATP and nucleic acids. The third mineral in the table, sulfur, is a necessary component of several amino acids. Iron is a component of several electron-carrier molecules that function in cellular respiration. It is also a component of hemoglobin, the oxygen-carrying protein of red blood cells. Small amounts of fluorine help maintain bones and teeth. Magnesium, manganese, zinc, copper, cobalt, se-

Mineral	Dietary Sources	Functions in the Body	Symptoms of Deficiency*
Calcium (Ca)	Dairy products, dark green vegetables, legumes	Bone and tooth formation, blood clotting, nerve and muscle function	Stunted growth, possibly loss of bone mass
Phosphorus (P)	Dairy products, meats, grains	Bone and tooth formation, acid-base balance, nucleotide synthesis	Weakness, loss of minerals from bone, calcium loss
Sulfur (S)	Proteins from many sources	Component of certain amino acids	Symptoms of protein deficiency
Potassium (K)	Meats, dairy products, many fruits and vegetables, grains	Acid-base balance, water balance, nerve function	Muscular weakness, paralysis
Chlorine (Cl)	Table salt	Acid-base balance, gastric juice	Muscle cramps, reduced appetite
Sodium (Na)	Table salt	Acid-base balance, water balance, nerve function	Muscle cramps, reduced appetite
Magnesium (Mg)	Whole grains, green leafy vegetables	Component of certain enzymes	Nervous system disturbances
Iron (Fe)	Meats, eggs, legumes, whole grains, green leafy vegetables	Component of hemoglobin and of electron carriers in energy metabolism	Iron-deficiency anemia, weakness, impaired immunity
Fluorine (F)	Drinking water, tea, seafood	Maintenance of tooth (and probably bone) structure	Higher frequency of tooth decay
Zinc (Zn)	Meats, seafood, grains	Component of certain digestive enzymes and other proteins	Growth failure, scaly skin inflammation, reproductive failure, impaired immunity
Copper (Cu)	Seafood, nuts, legumes, organ meats	Component of enzymes in iron metabolism	Anemia, bone and cardiovascular changes
Manganese (Mn)	Nuts, grains, vegetables, fruits, tea	Component of certain enzymes	Abnormal bone and cartilage
Iodine (I)	Seafood, dairy products, iodized salt	Component of thyroid hormones	Goiter (enlarged thyroid)
Cobalt (Co)	Meats and dairy products	Component of vitamin B-12	None, except as B-12 deficiency
Selenium (Se)	Seafood, meats, whole grains	Component of enzyme; functions in close association with vitamin E	Muscle pain, maybe heart muscle deterioration
Chromium (Cr)	Brewer's yeast, liver, seafood, meats, some vegetables	Involved in glucose and energy metabolism	Impaired glucose metabolism
Molybdenum (Mo)	Legumes, grains, some vegetables	Component of certain enzymes	Disorder in excretion of nitrogen-containing compounds

*All of these minerals can be harmful when consumed in extremely excessive amounts.

lenium, and molybdenum are components of various enzymes. Vertebrates need iodine to make a hormone called thyroxin, which regulates metabolic rate.

Sodium, potassium, and chlorine are important in nerve function and help maintain the osmotic balance of cells. A common source of sodium and chlorine is sodium chloride, which we know as table salt. Because salt is often in limited supply in natural environments, many herbivores such as deer and elk do not get enough salt from plants. To satisfy their needs, they regularly visit salt licks, places where the soil or rocks have a high salt content. Sometimes you even see deer and rabbits licking the salt left over from winter de-icing efforts on the gravel along the sides of highways. In contrast to many other animals, humans often ingest far more salt than they need. In the U.S., the average person eats enough salt to provide about twenty times the required amount of sodium. Too much sodium may cause high blood pressure, and low-sodium foods are becoming increasingly popular. 

? What is an essential mineral?

A chemical element—other than carbon, oxygen, hydrogen, and nitrogen—that is required by the body

What do food labels tell us? 21.19

Have you ever found yourself sitting at the breakfast table reading the label on a box of cereal or loaf of bread? It can be interesting to compare the way a product is advertised on the front of the package with the actual nutritional information on the side or back. For instance, so-called fiber-enriched white bread may have less fiber than an unenriched whole-grain product. Similarly, products like cheese and chips are often labeled "low fat" or "lite." This merely means that they have less fat than the same products without these labels. In fact, their actual fat content may be surprisingly high, and by eating just a few more "lite" chips than regular ones, you may consume just as much fat.

Public pressure has led to regulations requiring more informative food labels. Various types of information are given on packaged-food labels, as shown in Figure 21.19 here. One is simply the ingredients, listed in order from the greatest amount (by weight) to the least. More informative are several kinds of "nutrition facts" found on food labels. First, a serving size of the food is defined according to standards set by the U.S. Food and Drug Administration. Next listed is the amount per serving of the food's energy content in "calories" (that is, kilocalories). Selected nutrients are then listed as amounts per serving and as percentages of a daily value. The daily values are based on a diet containing 2000 kcal per day. For example, the 1 gram of fat in a slice of this bread provides 2% of the daily fat allowance for a person needing 2000 kcal per day.

Food labels emphasize nutrients believed to be associated with disease risks (fats, cholesterol, and sodium) and with a healthy diet (such as dietary fiber, protein, certain vitamins, and minerals). From the data shown, you can tell that each serving of this bread contains 1 g of unsaturated fat (1 g of total fat minus 0 g of saturated fat). Dietary fiber consists of indigestible complex carbohydrates, mainly cellulose. Subtracting 2 g of dietary fiber and 3 g of sugars (simple carbohydrates) from the 17 g of total carbohydrate tells you that each serving of this bread contains 12 g of digestible complex carbohydrate. This is chiefly starch.

To help consumers compare nutrient amounts in a particular food with their total daily needs, food labels also provide some general nutritional information. For example, the lower part of the label prescribes less than 20 g of saturated fat and at least 25 g of dietary fiber for those following a 2000-kcal daily diet. Reducing the amount of saturated fat we eat may lower our risk of developing certain diseases, including cancer, as we discuss in the next module. 

? What percent of the daily requirements for the fat-soluble vitamins is provided by a slice of the bread in Figure 21.19? (*Hint:* Review the table in Module 21.17.)

0%

Ingredients: water, wheat, enriched flour [flour, malted barley flour, niacin, reduced iron, thiamine mononitrate (vitamin B-1), riboflavin (vitamin B-2)], honey, raisin juice concentrate, raisins, butter, salt, wheat bran, partially hydrogenated soybean oil, wheat gluten, yeast, wheat germ, vinegar, mono- and diglycerides.

NUTRITION FACTS

Serving size	1 slice (approx. 34g)
Servings per container	20
Calories	90
Calories from Fat	10
Amount/Serving	**%Daily Value***
Total Fat 1g	2%
Saturated Fat 0g	0%
Cholesterol 0mg	0%
Sodium 170mg	7%
Total Carbohydrate 17g	6%
Dietary Fiber 2g	7%
Sugars 3g	
Protein 3g	
Vitamin A	0%
Vitamin C	0%
Thiamine	6%
Riboflavin	2%
Niacin	6%
Calcium	0%
Iron	4%

* Percent Daily Values are based on a 2000 calorie diet. Your daily values may be higher or lower depending on your calorie needs:

	Calories:	2000	2500
Total Fat	Less than	65g	80g
Sat. Fat	Less than	20g	25g
Cholest.	Less than	300mg	300mg
Sodium	Less than	2400mg	2400mg
Total Carb.		300g	375g
Fiber		25g	30g

Figure 21.19

21.20 Diet can influence cardiovascular disease and cancer

Figure 21.20 at the right shows some of the risk factors associated with cardiovascular disease. Though certain factors are unavoidable, we can influence others through our behavior. Diet is an example of a behavioral factor that may affect cardiovascular health. For instance, a diet rich in saturated fats is linked to high blood cholesterol levels, which in turn may be linked to cardiovascular disease.

Cholesterol travels through the body in blood lipoproteins, which are particles made up of thousands of molecules of cholesterol and other lipids, and one or more protein molecules (see Figure 5.20). High blood levels of a family of lipoproteins called **low-density lipoproteins (LDLs)** generally correlate with a tendency to develop blocked blood vessels, high blood pressure, and consequent heart attacks. In contrast, cholesterol carriers called **high-density lipoproteins (HDLs)** may decrease the risk of vessel blockage, perhaps because some HDLs convey blood cholesterol to the liver, where it is broken down. Many researchers believe that reducing LDLs while maintaining or increasing HDLs lowers the risk of cardiovascular disease. Exercise tends to increase HDL levels, while smoking lowers them.

A diet high in saturated fats tends to increase LDL levels. Saturated fats are found in eggs, butter, lard, and most other animal products, and also the artificially saturated ("hydrogenated") vegetable oils in margarine and shortening. By contrast, eating mainly unsaturated fats, such as fish oil and most liquid vegetable oils (including corn, soybean, and olive oils), tends to lower LDL levels and raise HDL levels. These oils are also important sources of vitamin E, whose antioxidant effect may help prevent blood vessel blockage.

Dietary Guidelines for Reducing Cancer Risk

- Limit fat kilocalories to no more than 30% of daily total.
- Eat plenty of high-fiber foods, such as whole-grain breads and cereals, fruits, and vegetables.
- Eat foods rich in vitamins A and C daily. Include vegetables from the mustard family, such as broccoli and cabbage.
- To avoid possible carcinogens, minimize consumption of cured and smoked foods, such as hot dogs, salami, and bacon. Avoid moldy foods. Avoid charred foods.
- If you drink alcoholic beverages, limit yourself to a maximum of one ounce of alcohol (two typical drinks) a day.

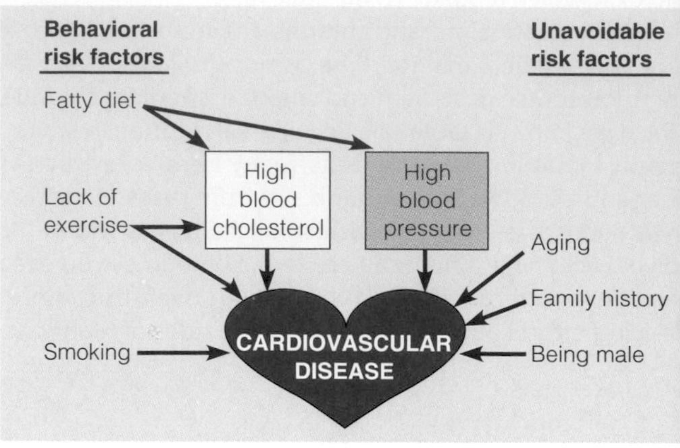

Figure 21.20 Risk factors associated with cardiovascular disease

As discussed in Module 11.17, diet also seems to be involved in some forms of cancer. Some research suggests a link between "high-energy" diets—those heavy in fats or carbohydrates—and the incidence of breast cancer. The incidence of colon cancer and prostate cancer may be linked to a diet rich in saturated fat or red meat.

The relationship between foods and health is complex, and much remains to be learned. The table at the left lists some dietary guidelines that may lower cancer risk. Such lists are controversial, but researchers generally agree that every diet should emphasize vegetables and fruits.

If you are trying to minimize the damaging effects of blood cholesterol on your cardiovascular system, your goal is to increase/decrease *(choose one)* your ratio of HDLs to LDLs.

Increase

* * *

In this chapter, we have seen that a sound diet supplies enough raw materials to make all the macromolecules we need, the proper amounts of prefabricated essential nutrients, and enough kilocalories to satisfy our energy needs. Energy remains a theme in Chapter 22, as we look at how the body obtains the oxygen it needs to harvest energy from food molecules.

Chapter Review

Chapter Summary

Obtaining and Processing Food (Introduction–21.3) Animal diets are highly varied, as are the methods animals use to obtain food. Some animals are herbivores (plant-eaters), some are carnivores (meat-eaters), and some are omnivores (eaters of both plants and other animals). Most animals ingest chunks of food, but some are suspension feeders (consuming particles from water), some are substrate feeders (living in or on their food source), and some are fluid feeders (sucking liquids) **(21.1)**. In the four stages of nutrient processing, food is ingested, digested, and absorbed into the body, and undigested materials are eliminated **(21.2)**. Food is digested in compartments housing hydrolytic enzymes. Most animals have a specialized digestive tract. In relatively simple animals like the hydra, the digestive compartment is a gastrovascular cavity, a sac with a single opening. In most animals, the digestive compartment is an alimentary canal, a tube running from mouth to anus, divided into specialized regions that process food sequentially **(21.3)**.

Human Digestive System (21.4–21.11) The human digestive tract consists of an alimentary canal and accessory glands. Once food is swallowed, the rhythmic muscle contractions of peristalsis squeeze food toward the stomach and through the rest of the alimentary canal. Ringlike sphincter muscles regulate its passage **(21.4)**. Digestion begins in the oral cavity. The teeth break up food, saliva moistens it, and salivary enzymes begin the hydrolysis of starch. The tongue pushes the chewed food into the pharynx **(21.5)**. The swallowing reflex moves food from the pharynx into the esophagus, while keeping it out of the trachea **(21.6)**. Peristalsis in the esophagus moves food into the stomach **(21.7)**. The stomach stores food and mixes it with acidic gastric juice. Pepsin in gastric juice begins the hydrolysis of protein **(21.8)**. Bacterial infections in the stomach and duodenum are associated with ulcers **(21.9)**. Most digestion and nutrient absorption occur in the small intestine. Alkaline pancreatic juice neutralizes stomach acid, and its enzymes digest polysaccharides, proteins, nucleic acids, and fats. Bile, made in the liver and stored in the gallbladder, emulsifies fat droplets for attack by pancreatic enzymes. Enzymes from the walls of the small intestine complete the digestion of many nutrients. The lining of the small intestine is folded and covered with tiny, fingerlike villi that increase its absorptive surface. Nutrients pass through the epithelium of the villi and into the blood, which flows to the liver. The liver can store nutrients and convert them to other substances the body can use **(21.10)**. Undigested material passes to the large intestine, or colon, where water is absorbed and feces are produced **(21.11)**.

Diets and Digestive Adaptations (21.12) Herbivores and omnivores, which eat plant material that is more difficult to digest and has less-concentrated nutrients than meat, generally have longer alimentary canals. Some mammals house cellulose-digesting microbes in the colon or cecum, and some re-ingest their feces to recover nutrients. Ruminants such as cows process food in a four-chambered stomach **(21.12)**.

Nutrition (21.13–21.20) An animal's diet provides fuel for its activities, raw materials for making the body's own molecules, and essential nutrients that the body cannot make itself **(21.13)**. Once nutrients are inside cells, they can be oxidized by cellular metabolism to generate energy in the form of ATP. The energy a resting animal requires each day to stay alive is its basal metabolic rate (BMR). More energy is required for an active life. Excess energy is stored as glycogen or fat **(21.14)**. The human body tends to store excess fat molecules instead of using them for fuel. A balanced diet includes adequate amounts of all nutrients. Fad diets are often ineffective and can be harmful **(21.15)**. The nine essential amino acids are easily obtained from animal protein and are also available from the proper combination of plant food **(21.16)**. Thirteen vitamins are essential in the human diet. Most function as coenzymes **(21.17)**. Essential minerals are elements other than carbon, hydrogen, oxygen, and nitrogen that play a variety of roles in the body **(21.18)**. Food labels provide important nutritional information about packaged foods **(21.19)**. Choice of diet may reduce the risk of cardiovascular disease and cancer **(21.20)**.

Testing Your Knowledge

Multiple Choice

1. Which of the following sequences correctly traces the passage of food through the human digestive tract?
   a. pharynx, esophagus, stomach, small intestine, large intestine
   b. esophagus, stomach, small intestine, large intestine, pharynx
   c. esophagus, pharynx, stomach, large intestine, small intestine
   d. pharynx, stomach, esophagus, small intestine, large intestine
   e. pharynx, esophagus, stomach, large intestine, small intestine

2. Which of the following is considered an essential nutrient in the human diet?
   a. pepsin
   b. glucose
   c. starch
   d. fat
   e. vitamin A

3. A human requires hundreds of grams of carbohydrates every day. The daily requirement for most vitamins is in the milligram range. Why are vitamins needed in such small quantities? *(Explain your answer.)*
   a. Vitamins are not very important in metabolism.
   b. The energy content of vitamins is so great that you don't need much.
   c. The body can store large quantities of most vitamins.
   d. Vitamins are reusable.
   e. Every cell needs carbohydrates, but only a few need vitamins.

Describing, Comparing, and Explaining

A peanut butter and jelly sandwich contains carbohydrates, proteins, and fats. Describe what happens to the sandwich when you eat it. Discuss ingestion, digestion, absorption, and elimination.

Thinking as a Scientist

1. Most large, complex animals, from worms to mammals, have alimentary canals with a mouth and an anus. State a hypothesis to explain why digestion and absorption in an alimentary canal might be a more efficient way for a large animal to process food than processing it in a gastrovascular cavity.

2. Some essential mineral nutrients, such as selenium and molybdenum, are required in the human diet in miniscule amounts, thousandths of a milligram per day. What kinds of experiments would you have to carry out to show that one of these minerals is required by humans? Why might this be difficult?

3. Using the information in the table and text of Module 21.14, calculate how many liters of oxygen you would consume in liberating the energy from one slice of cheese pizza.

Science, Technology, and Society

According to recent surveys, 75% of Americans aged 18–35 think they are fat. At any given time, 30 million women and 18 million men are on diets. Only 25% are actually medically overweight. Other data show that more than twice as many high school girls as boys see themselves as overweight, and 45% of underweight women see themselves as fat. Why do you think there is a difference between men and women regarding the perception of their weight? To the extent that unwarranted concern over weight is a problem, what can be done about it?

Media Resources

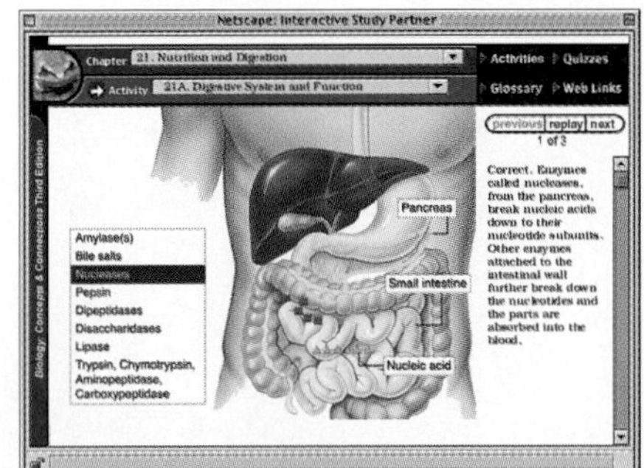

For further review of the concepts in this chapter, use the activities and quizzes on the CD-ROM.

Visit the Special Edition of The Biology Place web site for activities, quizzes, news articles, and web links.

22 Respiration: The Exchange of Gases

Outline

Mechanisms of Gas Exchange

Imagine yourself in a jetliner. The pilot announces that you have reached an altitude of 30,000 feet. You glance out the window and see a flock of geese just below the plane. Your cabin is warm and pressurized. Outside, the temperature is far below zero, and the air is so thin you would pass out almost instantly from lack of oxygen (O_2). But the geese are doing quite well out there.

If you were ever to fly over the Himalaya Mountains between India and China, you might actually see geese this high up. Twice a year, the bar-headed goose migrates over the Himalayas, sometimes at nearly 30,000 feet, en route between its winter quarters in India and its summer breeding grounds in northern Russia.

Closer to home, it is not unusual to see migrating ducks and geese, such as the ones in the photograph at the left, flying as high as 20,000 feet. This altitude is still far out of the comfort range for most humans. Warm clothes insulate our bodies almost as effectively as a bird's feathers, but most of us start feeling light-headed and lethargic at about 10,000 feet and pass out from a lack of oxygen at about 17,000 feet.

How do geese and ducks manage to fly long distances and stay alive at such great heights? One factor is the efficiency of bird lungs, which can extract considerably more oxygen from a given quantity of air than our lungs can. (We'll compare bird and human lungs later in this chapter.) Birds such as the mallard duck and the bar-headed goose also have hemoglobin with a very high affinity for oxygen, picking it up in the lungs and carrying it to body tissues. They also have a very large number of capillaries (tiny blood vessels) that carry oxygen-rich blood to their flight muscles. The flight muscles themselves contain a protein called myoglobin that picks up oxygen from the blood and keeps a ready supply of it in the muscle. All these adaptations combine to allow high-flying birds to obtain enough oxygen even when the air is very thin.

Throughout the world, most people live well below 10,000 feet and are helpless at very high elevations without an oxygen mask. But there are permanent villages as high as 19,000 feet in the Himalayas and nearly that high in the Andes. Like the Nepalese Sherpa shown here, most people born at such heights are small of stature and have unusually large lungs and a large heart. In addition, their blood carries an unusually large amount of oxygen, because they have more hemoglobin and more red blood cells than most of us.

Even if you live at sea level, chances are good that you can condition yourself for fairly high altitudes. If you drive from sea level up into the mountains, your body starts adjusting immediately: Your heart pumps faster, and your capillaries may increase in diameter. If you stay in the mountains more than a few days, your heart slows down, as other changes begin to compensate for the thin air. In a matter of weeks, the rate and depth of your breathing increase, bringing more air into the lungs. At the same time, your body may actually develop more capillaries, and your hemoglobin level and red blood cell count may go up, enabling your blood to carry more oxygen. After rigorous, long-term training, a few mountain climbers have been able to survive for a short time without oxygen masks on the top of the world's highest peak, Mount Everest, at 29,028 feet.

Our study of cellular respiration in Chapter 6 showed why animals require O_2. Without a continuous supply of this gas, most animals cannot obtain enough energy from their food to survive. Without O_2, the metabolic machinery that releases energy from food molecules shuts down or runs far too slowly to keep a goose or human awake, much less flying or running. Energy metabolism also generates waste—notably, the gas carbon dioxide (CO_2). Thus, cells must dispose of CO_2 as well as obtain O_2. The process of **gas exchange,** or **respiration,** the interchange of O_2 and CO_2 between an animal and its environment, is the subject of this chapter.

Mechanisms of Gas Exchange

22.1 Overview: Gas exchange involves breathing, the transport of gases, and the servicing of tissue cells

Gas exchange makes it possible for animals to put to work the food molecules the digestive system provides. Figure 22.1 presents an overview of three phases of gas exchange, all of which we examine in detail in this chapter. ① Breathing is one phase of the gas-exchange process. When a bird or mammal inhales air, O_2 passes into cells lining the inside of its lungs. As it exhales, CO_2 passes from lung cells into the environment. (Not shown in the figure are important air sacs that connect to the bird's lungs; we will discuss the air sacs and their function in Module 22.8.)

② A second phase of gas exchange is the transport of gases by the circulatory system. It begins with the diffusion of O_2 from the lung cells into blood vessels. The O_2 attaches to hemoglobin and is carried in the blood from the lungs to the body's other tissues. At the same time, the blood also conveys CO_2 from the tissues to the lungs.

③ A third phase of gas exchange occurs in the body tissues, where the cells take in O_2 from the blood and give up CO_2 to the blood. Gaining O_2 and giving up CO_2 makes it possible for the body cells to obtain energy from food molecules by cellular respiration. The O_2 helps a cell "burn" fuel—that is, release energy from the food molecules the body has digested and absorbed. As we saw in Module 6.4, O_2 is a key ingredient in the cell's energy-harvesting operation because it tends to pull electrons away from organic molecules. The CO_2 the cell gives up to the blood is the main waste product that results when food molecules are split apart as the cell harvests energy-rich electrons from them.

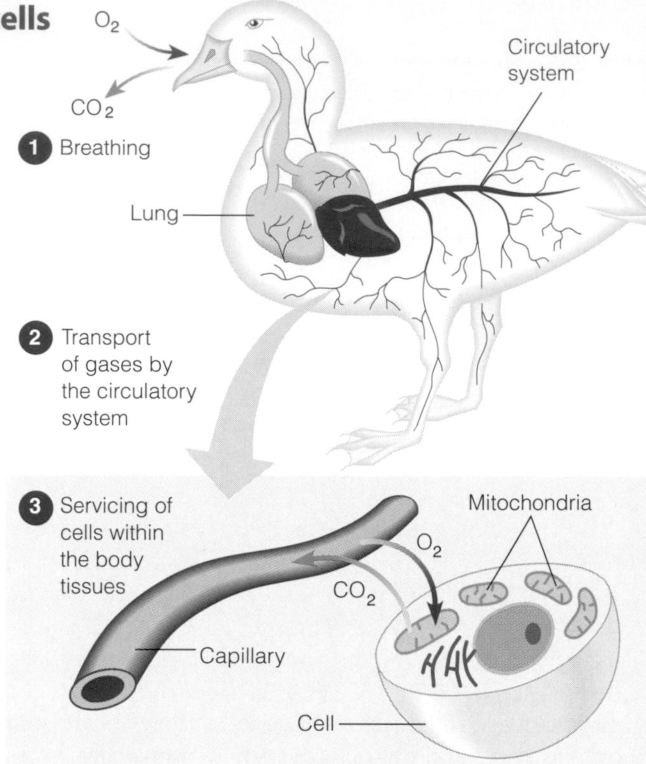

Figure 22.1 Three phases of gas exchange

What role does the circulatory system play in gas exchange?

It transports O_2 from the respiratory organs (lungs in birds for example) to all other parts of the body, and it carries CO_2 from all parts of the body to the respiratory organs.

22.2 Animals exchange O_2 and CO_2 through moist body surfaces

The part of an animal where O_2 diffuses into the animal and where CO_2 diffuses out to the surrounding environment is called the **respiratory surface.** (In this context, the word "respiratory" refers to the process of breathing, not to cellular respiration.) Living cells cover respiratory surfaces, and O_2 and CO_2 diffuse through the cells' plasma membranes, which must be wet to function properly. Thus, the gases must be dissolved in water before they can diffuse in or out. Moreover, the respiratory surface must be extensive enough to take up sufficient O_2 for every cell in the body and to dispose of all waste CO_2. Usually a single layer of moist cells covers or lines the entire respiratory surface. Being thin as well as moist, the layer allows O_2 to diffuse rapidly into the circulatory system or directly into body tissues and allows CO_2 to diffuse out.

The four figures on the facing page illustrate, in simplified form, four types of animal respiratory organs. In each case, the circle represents a cross section of the animal's body through the respiratory surface. The yellow areas represent the respiratory surfaces; the green circles represent body surfaces with little or no role in respiration. The boxed enlargements show a portion of the respiratory surface in the process of exchanging O_2 and CO_2.

Some animals use their entire outer skin as a gas-exchange organ. The earthworm in Figure 22.2A is an example. Notice in the cross-sectional diagram that its whole body surface is yellow; there are no specialized gas-exchange surfaces. Oxygen diffuses into a dense net of thin-walled capillaries lying just beneath the skin. Earthworms and other "skin-breathers" must live in damp places or in water because their whole body surface has to stay moist. Animals that breathe only through their skin and lack gas-exchange organs are generally small, and many are long and thin or flattened—the flatworms we saw in Module

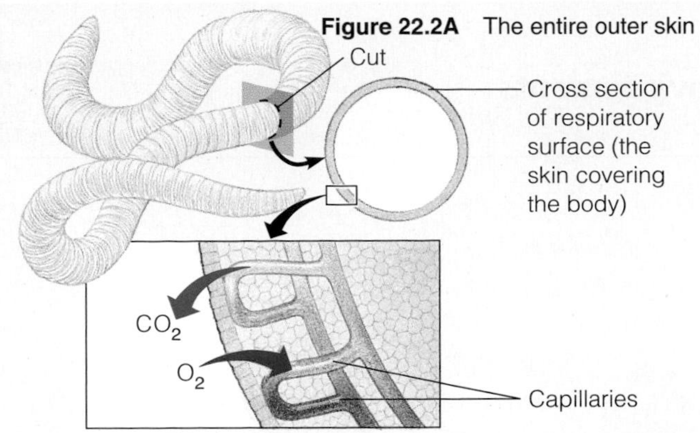

Figure 22.2A The entire outer skin

18.6, for example. Small size or flatness provides a high ratio of respiratory surface to body volume, ensuring that all body cells will receive enough O_2 and be able to dispose of their waste CO_2.

In most animals, either the skin surface is not extensive enough to exchange gases for the whole body, or much of it is covered with dead cells or another impermeable material. Consequently, certain parts of the body have become adapted as respiratory surfaces. Gills have evolved in most aquatic animals. Lungs or an internal system of gas-exchange tubes have evolved in most terrestrial animals. Gills, lungs, and internal tubes all have extensive surfaces for gas exchange, as shown in Figures 22.2B–22.2D.

Gills are featherlike extensions of the body surface specialized for gas exchange. A fish (Figure 22.2B) has a set of gills on each side of its head. As indicated in the enlargement, O_2 diffuses across the gill surfaces into capillaries, and CO_2 diffuses in the opposite direction, out of the capillaries into the external environment. Since the respiratory surfaces of aquatic animals extend into the surrounding water, keeping the surface moist and its cells alive is not a problem. Gills are generally unsuitable for an animal living on land because the wet surfaces tend to stick together and collapse when exposed to air. Nevertheless, a number of invertebrates with gills live on land at least part of the time. In such animals (certain crabs, for example), the gills are housed in special chambers that keep the respiratory surfaces moist and prevent the gills from collapsing.

In most terrestrial animals, the respiratory surfaces are folded into the body rather than projecting from it. The infolded surfaces open to the air only through narrow tubes, an arrangement that helps retain the moisture that keeps the cells of the respiratory surfaces alive.

Insects have an extensive system of internal tubes called **tracheae** (singular, *trachea*) (Figure 22.2C). As we will see in Module 22.5, tracheae branch throughout the body, exchanging gases directly with body cells and requiring no assistance from the circulatory system.

Most terrestrial vertebrates have **lungs** (Figure 22.2D), which are internal sacs lined with moist epithelium. As the diagram indicates, the inner surfaces of the lungs branch extensively, forming a large respiratory surface. Gases are carried between the lungs and the body cells by the circulatory system.

We examine gills, tracheae, and lungs more closely in the next several modules.

Figure 22.2B Gills

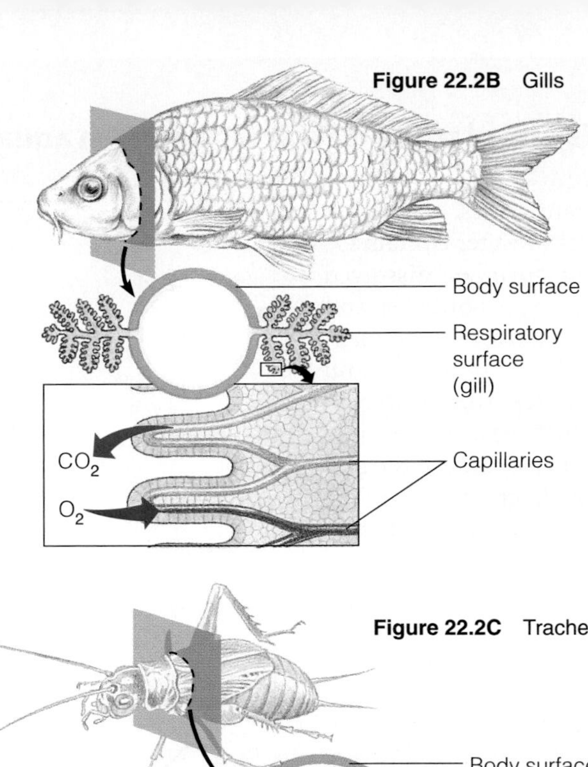

Figure 22.2C Tracheae

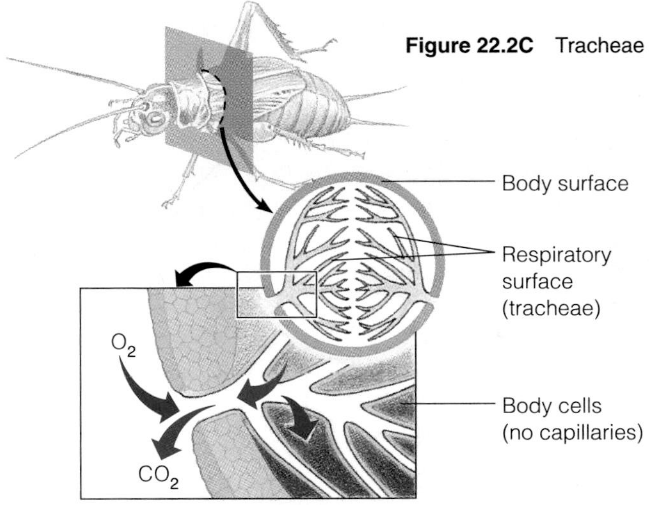

Figure 22.2D Lungs

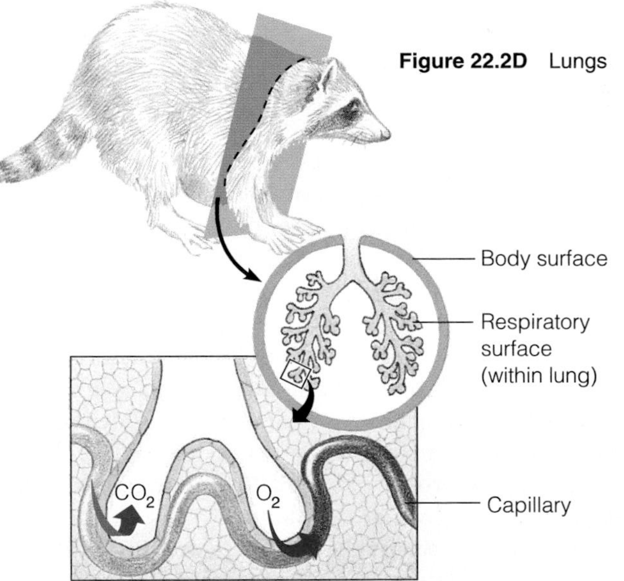

What is the main difference between gills and lungs as respiratory organs in terms of their spatial relationships to the rest of animals' bodies?

The extensive respiratory surface of gills extends outward from the body into the surrounding environment (water); in contrast, lungs are internal sacs with respiratory surfaces.

22.3 Gills are adapted for gas exchange in aquatic environments

Oceans, lakes, and other bodies of water contain O_2 in the form of dissolved gas. The gills of fishes and many invertebrate animals, including lobsters and clams, tap this source of O_2. In fishes, the respiratory surface area of the gills is much greater than the whole remaining body surface. The gills are full of tiny blood vessels covered by only one or a few layers of cells; the vessels are so narrow that red blood cells must pass through them in single file. As a result, every red blood cell comes in close contact with oxygen dissolved in the surrounding water.

The chief advantage of exchanging gases in water is that the animal does not have to spend energy to keep its respiratory surface wet. On the other hand, the amount of available oxygen (dissolved O_2) in water is only about 3–5% of what it is in the air, and the warmer and saltier the water, the less dissolved O_2 it holds. Thus, gills—especially those of large, active animals in warm oceans—must be very efficient to obtain enough oxygen from water.

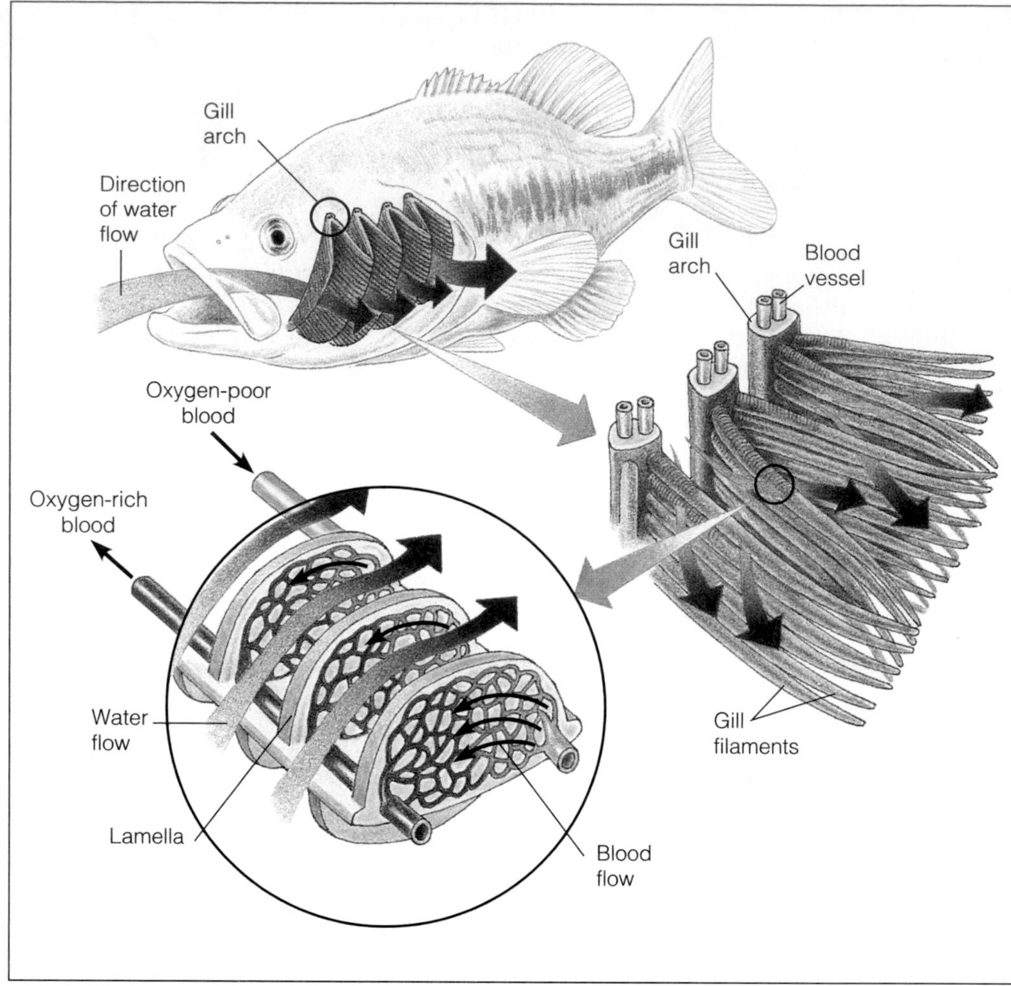

Figure 22.3 The structure of fish gills

The drawings in Figure 22.3 show the architecture of fish gills, which are among the most efficient gas-exchange organs in the aquatic world. There are four gill arches on each side of the body. Two rows of gill filaments project from each gill arch. Each filament bears many platelike structures called lamellae (singular, lamella), which are the actual respiratory surfaces. The blue arrows in the drawings represent the route of water flowing over these surfaces.

What you can't see in the drawings are the breathing movements (inhalation and exhalation) that ventilate the gills of the fish. We use the term **ventilation** to refer to any mechanism that increases contact between the surrounding water or air and the respiratory surface (gills or lungs). A fish inhales water by opening its mouth. As it does this, the gill coverings (opercula) on the sides of its body close tightly over the gills. The animal exhales water by closing its mouth and pumping water from its mouth cavity over its gills and out the sides of its body. The gill coverings open during exhalation, allowing the water to escape. These ventilatory movements enhance the gills' gas-exchange efficiency by keeping the water around the gills from stagnating and becoming depleted of oxygen.

In the circular enlargement at the bottom of Figure 22.3, notice that the direction of water flow over the fish gills (blue arrows) is opposite that of the blood flow (black arrows) in the gill lamellae. Because of the way the blood vessels are arranged, blood that is low in oxygen (oxygen-poor blood from the body, shown in blue) arrives at one side of a lamella. Oxygen-rich blood (blood laden with oxygen; red) then leaves the gills from the other side of the lamella. The opposite flows of blood and water here constitute countercurrent flow, which, as we see in the next module, further increases the efficiency of the gills. 

? What is the function of ventilation?

Ventilation replaces O_2-depleted/CO_2-enriched water (for gills) or air (for lungs) with a fresh supply of the environmental medium (water or air) with which the animal exchanges gases.

Countercurrent flow in the gills enhances O_2 transfer **22.4**

Countercurrent exchange is the transfer of something from a fluid moving in one direction to another fluid moving in the opposite direction. The name comes from the fact that the two fluids are moving *counter* to each other. As the two fluids pass each other, their opposite flows maintain a diffusion gradient that enhances transfer of the substance. Let's see how this principle works in a fish gill.

Figure 22.3 showed us that water flows over the surface of a lamella in one direction, while blood is flowing in the opposite direction within the lamella. In Figure 22.4, the changing intensities of the large arrows and the numbers on them indicate the changing amount of O_2 dissolved in each fluid: the darker the color, the more O_2. The small red arrows represent O_2 diffusing from the water into the blood. Notice that as the blood flows through a lamella and picks up more and more O_2, the blood comes in contact with water that has more and more O_2 available. As a result, a diffusion gradient is maintained that favors the transfer of O_2 from the water to the blood along the entire surface of the lamella. Because of this, O_2 diffuses from the water into the blood all along the lamella. So efficient is this countercurrent exchange mechanism that fish gills can remove more than 80% of the oxygen dissolved in the water flowing through them.

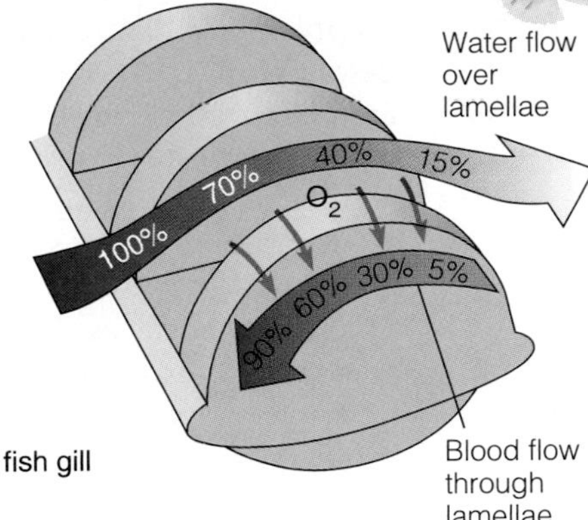

Figure 22.4 Countercurrent flow in a fish gill

(This is a tough one!) What would be the maximum percentage of the water's oxygen a gill could extract if its blood flowed in the same direction as the water instead of counter to it? Explain your answer.

50%—Once the blood reached the O_2 concentration of the water, there would be no concentration gradient making diffusion possible.

The tracheal system of insects provides direct exchange between the air and body cells **22.5**

Unlike water-dwellers, land animals exchange gases by breathing air. There are two big advantages to doing so. For one thing, a given volume of air contains much more O_2 than an equal volume of water. Also, air is much lighter and easier to move than water. Thus, a terrestrial animal expends much less energy than an aquatic animal ventilating its respiratory surface.

The tracheal system of insects minimizes energy expense in gas exchange. Insect tracheae are different from the vertebrate trachea (windpipe). Whereas the vertebrate windpipe is a single straight tube, insect tracheae are tubes that branch throughout the body (Figure 22.5A). The tubes convey air directly to the body cells, and no energy is spent in moving gases via the circulatory system.

Figure 22.5A also shows that an insect's tracheal system includes a number of enlargements called air sacs. The air sacs work like bellows. When muscles around them contract, air is forced out of the system through openings on the body surface; when the muscles relax, air is drawn in.

The main problem facing any air-breathing animal is the drying effect of the atmosphere. Figure 22.5B indicates how the tracheal system minimizes water loss. Air passes inward through smaller and smaller tubes. The narrowest tubes, called **tracheoles,** extend to nearly every cell in the insect's body. The tiny tips of the tracheoles are closed and contain fluid (dark blue). The O_2 diffuses into the body cells from the tracheoles, and CO_2 diffuses from the cells into the tracheoles. When the animal is active, and using more O_2, most of the fluid in the tracheoles is drawn back into the body, increasing the inward diffusion of O_2. With the respiratory surfaces in tiny tubes deep in the body, very little water can evaporate from the respiratory system.

When Washington's Mt. Saint Helens erupted in 1980, the fine ash in the air devastated insect populations. What do you suppose was the cause of death?

Suffocation due to the ash plugging up the tracheal systems of the insects

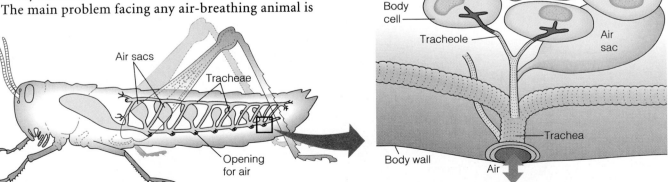

Figure 22.5A The tracheal system of an insect

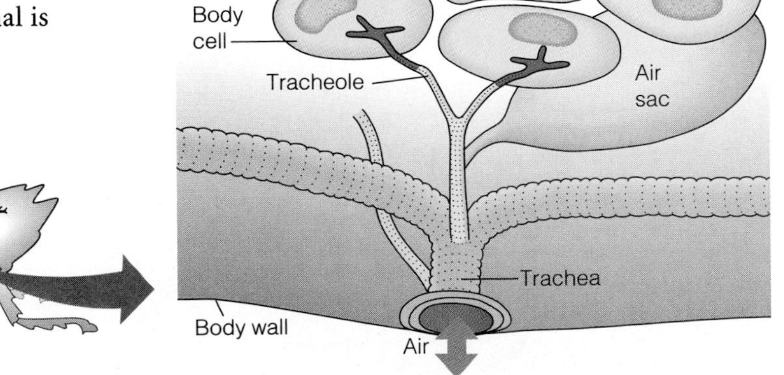

Figure 22.5B Tracheal connections to body cells

22.6 Terrestrial vertebrates have lungs

22A
See Activity 22A on the CD-ROM.

Reptiles, birds, mammals, and many amphibians exchange gases in lungs. In contrast to the tracheae of insects, lungs are restricted to one location in the body. Therefore, the circulatory system must transport gases between the lungs and the rest of the body.

Amphibians have small lungs (some lack lungs altogether) and rely heavily on the diffusion of gases across other body surfaces. The skin of frogs, for example, supplements gas exchange in the lungs. Compared to the lungs of amphibians, those of reptiles, birds, and mammals are very large and contain a large respiratory surface relative to the animal's body size. The total respiratory surface of human lungs, for instance, is about equal to the surface area of a racquetball court. Most reptiles and all birds and mammals rely entirely on their lungs for gas exchange. Turtles are an exception; their rigid shell restricts breathing movements, and they supplement lung breathing with gas exchange across moist surfaces in their mouth and anus.

Figure 22.6A shows the organs of the human respiratory system (along with the esophagus and heart, for orientation). Our lungs are in the chest cavity, which is bounded at the bottom by a sheet of muscle called the **diaphragm.** Air passes to our lungs via a system of branching tubes. The tubes are narrow, reducing loss of water by evaporation from the lung's moist respiratory surfaces.

Air usually enters our respiratory system through the nostrils. It is filtered by hairs, and warmed, humidified, and sampled for odors as it flows through a maze of spaces in the nasal cavity. Air can also be drawn in through the mouth, but mouth breathing does not allow the air to be processed by the nasal cavity. From the nasal cavity or mouth, air passes to the **pharynx,** where the paths for air and food cross. As we saw in Module 21.6, the air passage in the pharynx is open for breathing except when we swallow.

From the pharynx, air is inhaled into the **larynx** (voicebox). When we exhale, the outgoing air rushes by a pair of **vocal cords** in the larynx, and we can produce sounds by voluntarily tensing muscles in the voicebox, stretching the cords and making them vibrate. We produce high-pitched sounds when our vocal cords are tense and therefore vibrating very fast. When the cords are less tense, they vibrate slowly and produce low-pitched sounds.

From the larynx, inhaled air passes toward the lungs through the **trachea,** or windpipe. Rings of cartilage maintain the shape of the trachea, much as metal rings keep the hose of a vacuum cleaner from collapsing. The trachea forks into two **bronchi** (singular, *bronchus*), one leading to each lung. Within the lung, the bronchus branches repeatedly into finer and finer tubes called **bronchioles.**

Though gas exchange occurs only in the lungs, our respiratory system is lined by a moist epithelium from the nasal cavity and mouth all the way to the lungs. The epithelium lining the trachea and all but the smallest bronchioles are covered by cilia and a thin film of mucus. The cilia and mucus are the system's cleaning elements. The mucus traps dust, pollen, and other contaminants, and the beating cilia move the mucus upward to the pharynx, where it is usually swallowed.

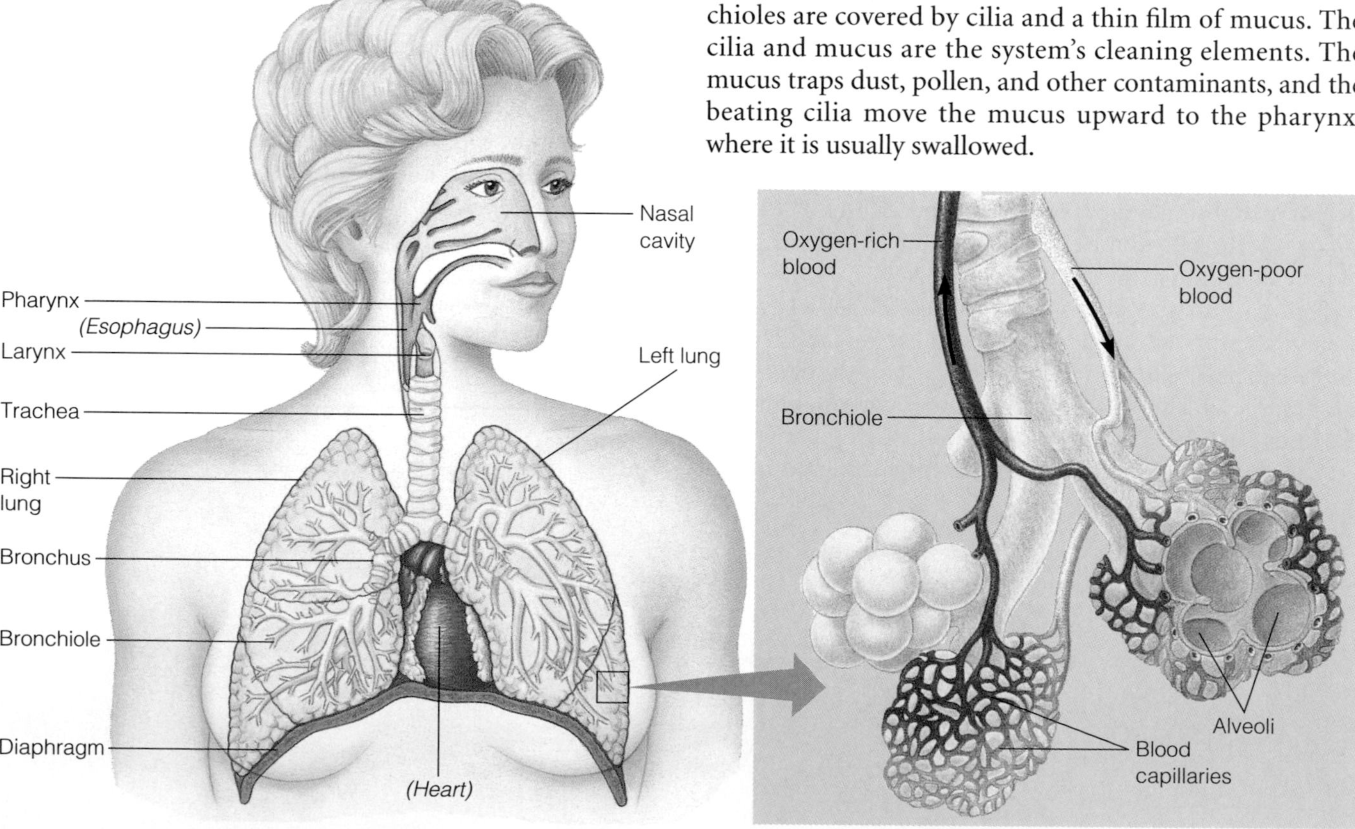

Figure 22.6A The human respiratory system

Figure 22.6B The structure of alveoli

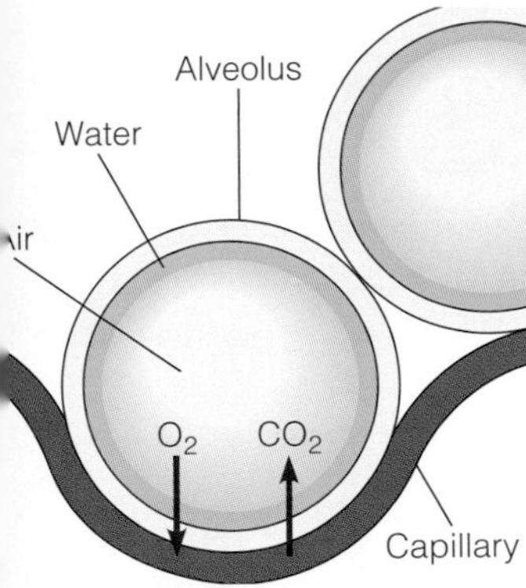

Figure 22.6C Gas exchange in an alveolus

As Figure 22.6B shows, the bronchioles dead-end in grapelike clusters of air sacs called **alveoli** (singular, *alveolus*). Each of our lungs contains millions of these tiny sacs. Figure 22.6C indicates the gas-exchange process in an alveolus. The inner surface of each alveolus is lined with a thin layer of epithelial cells that form the respiratory surface, where gases are actually exchanged. The O_2 in inhaled air dissolves in a film of moisture on the epithelial cells. It then diffuses across the epithelium and into a web of blood capillaries that surrounds each alveolus (shown in Figure 22.6C). The CO_2 diffuses the opposite way—from the capillaries, across the epithelium of the alveolus, into the air space of the alveolus, and finally out in the exhaled air. 

> **?** Arrange the following parts of the respiratory system into the order encountered by an inhaled breath of air: trachea, pharynx, bronchus, alveolus, bronchiole, larynx, nasal cavity.
>
> Nasal cavity→pharynx→larynx→trachea→bronchus→bronchiole→alveolus

22.7 Smoking is one of the deadliest assaults on our respiratory system

The epithelial tissue lining our respiratory system is extremely delicate. Its main protection is the mucus covering the cells and the beating cilia that sweep dirt particles and microorganisms off their surfaces. Also helping to cleanse the inhaled air are defensive amoeboid cells called macrophages that reside in the tissues lining the respiratory system. Macrophages engulf fine particles and microorganisms.

Virtually everywhere today, the air we breathe exposes the cells in our respiratory system to chemicals that they are not adapted to tolerate. Air pollutants such as sulfur dioxide, carbon monoxide, and ozone are associated with serious respiratory diseases. One of the worst sources of toxic air pollutants is tobacco smoke. A single drag on a cigarette exposes a person to over 4000 different chemicals, many of which are known to be harmful and potentially deadly. The visible smoke from a cigarette, cigar, or pipe is mainly microscopic particles of carbon. Sticking to the carbon particles are many different kinds of toxic chemicals. Tobacco smoke irritates the cells lining the bronchi, inhibiting or destroying their cilia and macrophages. This interferes with the normal cleansing mechanism of the respiratory system and allows more toxin-laden smoke particles in the inhaled air to reach the lung's delicate alveoli. Frequent coughing—common in heavy smokers—then becomes the system's attempt to clean itself.

Every year in the United States, smoking kills about 350,000 people, more than all the deaths caused by traffic accidents or illegal drug abuse. Some of the toxins in tobacco smoke cause lung cancer, which nearly always kills its victims. The photograph in Figure 22.7A shows a cutaway view of a pair of healthy human lungs (and heart). Figure 22.7B shows the lungs of a smoker with cancer. The lungs are black from the long-term buildup of smoke particles, except where pale cancerous tumors appear.

Smokers also have a markedly greater risk than nonsmokers of developing cancers of the bladder, pancreas, mouth, throat, and several other organs. Another disease called **emphysema** can develop as cigarette smoke makes the thin-walled alveoli brittle. Many of the alveoli eventually rupture, reducing the lungs' capacity for gas exchange. Breathlessness and constant fatigue result, as the body is forced to spend more and more energy just breathing. The heart is also forced to work harder, and emphysema often leads to heart disease. Probably every cigarette a person smokes takes about five minutes away from his or her life expectancy. Moreover, some studies indicate that nonsmokers exposed to secondary cigarette smoke are also at risk.

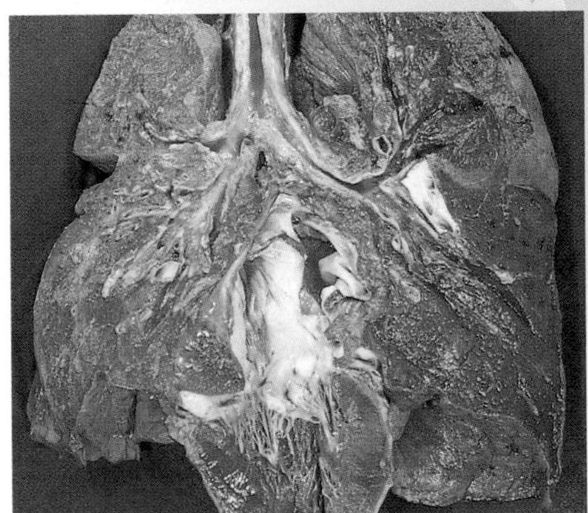

Figure 22.7A Healthy lungs

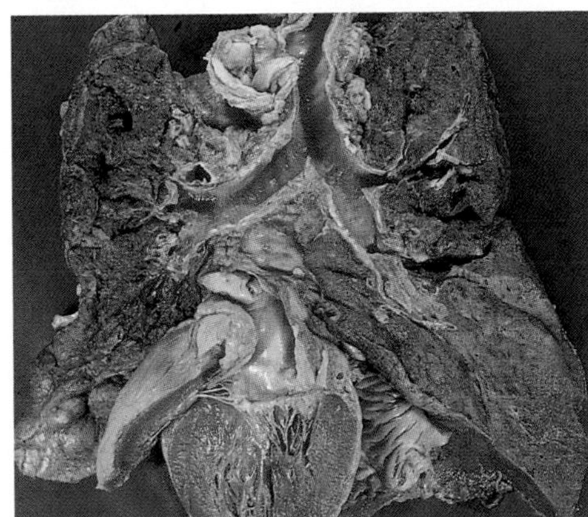

Figure 22.7B Cancerous lungs

Clearly, efforts to reduce smoking and secondary exposure to smoke can have positive effects on personal and public health. For former smokers, it is encouraging that after about 10 years of not smoking, the chances of developing lung cancer and heart disease seem to drop to levels similar to those of people who have never smoked. 

> **?** What effect does emphysema have on the surface area of the lungs?
>
> Surface area for gas exchange decreases due to rupturing of alveoli.

22.8 Breathing ventilates the lungs

Breathing is the alternation of inhalation and exhalation. It supplies our lungs with O_2-rich air and expels the CO_2-rich gas that remains after O_2 and CO_2 are exchanged in the alveoli. Like all mammals, we ventilate our lungs by sucking air into the alveoli and then pushing it back out.

Figure 22.8A shows the changes that occur in our rib cage, chest cavity, and lungs during breathing. During inhalation (left diagram), both the rib cage and chest cavity expand, and the lungs follow suit. You can feel what happens if you touch your rib cage and inhale deeply. The ribs move upward and the rib cage expands as muscles between the ribs contract. Meanwhile, the diaphragm contracts, moving downward and expanding the chest cavity as it goes.

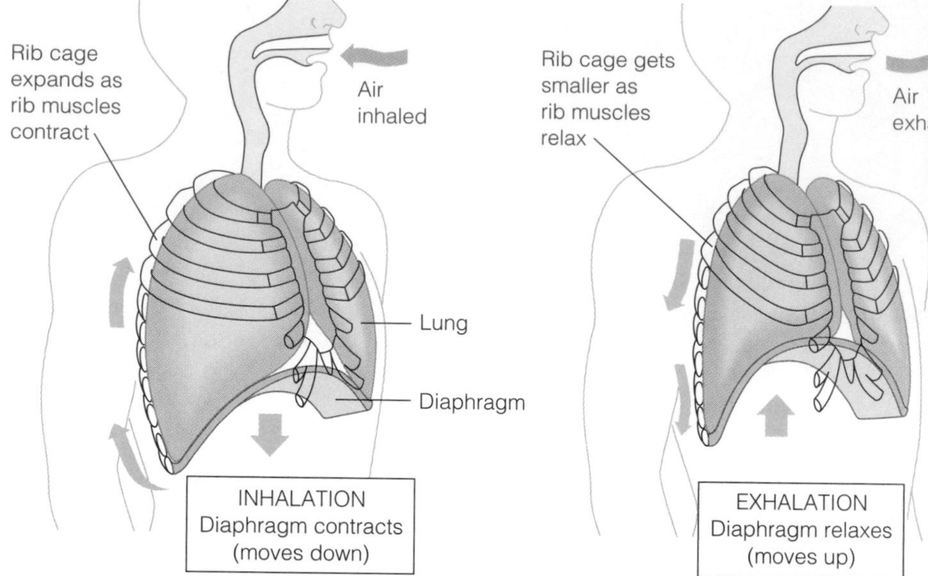

Figure 22.8A How a human breathes

The lungs expand and fill with air during inhalation because the enlargement of the rib cage and chest cavity reduces the air pressure in the alveoli to less than that of the atmosphere. Flowing from a region of higher pressure to one of lower pressure, air rushes through the nostrils and down the breathing tubes to the alveoli.

The diagram on the right in Figure 22.8A shows exhalation. The rib muscles and diaphragm both relax, decreasing the volume of the rib cage and chest cavity and forcing air out of the system. Notice that the diaphragm is higher in the chest cavity when relaxed.

Each year, a human adult may take between 4 million and 10 million breaths. The volume of air in each breath is about 500 mL when we breathe quietly. The maximum volume of air that we can inhale and exhale—as in strenuous exercise—is called **vital capacity.** It averages about 3500 mL and 4800 mL for college-age females and males, respectively. (Women tend to have smaller rib cages and lungs.) Vital capacity depends heavily on the resilience (springiness) of the lungs. The lungs actually hold more air than the vital capacity. Because the alveoli do not completely collapse, a residual volume of "dead" air remains in the lungs even after we blow out as much air as we can. As lungs lose resilience with age or as the result of disease, such as emphysema, our residual volume increases at the expense of vital capacity.

As we mentioned in the chapter's introduction, the gas-exchange system of birds is different from ours. Let's return now to the bird to see some of the adaptations that make its respiratory system a more efficient one.

In contrast to the in-and-out flow of air in the human alveoli, birds have a one-way flow of air through the lungs. Birds have several large air sacs in addition to their lungs. As the simplified diagrams in Figure 22.8B indicate, the air sacs' function is to ventilate the respiratory surface in the lungs. During inhalation, both sets of air sacs expand. The posterior sacs fill with fresh air (blue) from the outside, while the anterior sacs fill with stale air (gray) from the lungs. During exhalation, both sets of air sacs deflate, forcing air from the posterior sacs into the lungs, and air from the anterior sacs out of the system via the trachea.

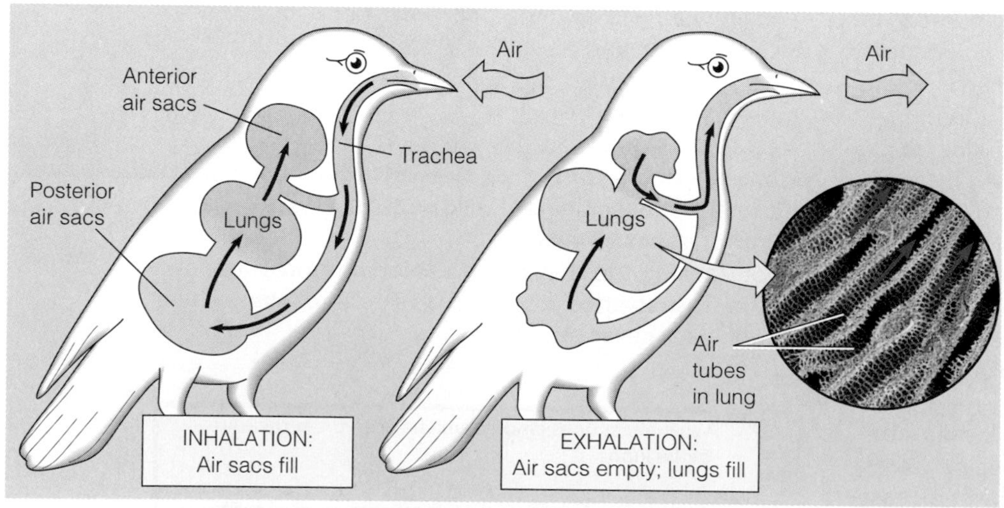

Figure 22.8B How a bird breathes

The electron micrograph in the circular inset shows another key feature in the bird. Instead of alveoli, bird lungs contain tiny, parallel tubes (each about 0.5 mm across). Air passes one-way through the tubes (red arrows), blood flows in the opposite direction (not shown here), and gases are exchanged efficiently across the walls of the tubes (see countercurrent flow in Module 22.4). Also, because of the one-way flow of air, there is no dead air (residual volume) in the bird lung. Birds can extract about 5% more oxygen from a volume of inhaled air than we can.

Contrast ventilation in mammals and birds in terms of the pathway of air flow.

In mammals, air that was inhaled reverses its pathway through the respiratory system during exhalation; in birds, air flows unidirectionally through the system, except for through the trachea.

Breathing is automatically controlled 22.9

What controls our breathing? We obviously have some conscious control over it because we can voluntarily hold our breath or breathe faster and deeper. Most of the time, however, automatic control centers in our brain regulate our breathing movements. Automatic control is essential, for it ensures coordination between the respiratory system and other body systems, especially the circulatory system.

Our **breathing control centers** (represented by the yellow circles in Figure 22.9) are located in parts of the brain called the pons and the medulla oblongata (medulla, for short). Nerves from the medulla's control center (solid yellow arrows) signal the diaphragm and rib muscles to contract, making us inhale. These nerves send out signals that result in about 10–14 inhalations per minute when we are at rest. Between inhalations, the muscles relax, and we exhale. The control center in the pons smooths out the basic rhythm of breathing set by the medulla.

The medulla's control center also regulates how much CO_2 our alveoli dispose of. It does so by monitoring the pH of the blood and of the fluid bathing the brain (cerebrospinal fluid). The pH starts to drop when the amount of CO_2 increases in the blood (red arrow). When we exercise vigorously, for instance, our metabolism speeds up and our body cells generate more CO_2 as a waste product. The CO_2 goes into the blood, where it reacts with water to form carbonic acid. The acid lowers the pH of the blood and cerebrospinal fluid slightly. When this happens, the medulla senses the pH drop, and its breathing control centers increase the breathing rate and depth. As a result, more CO_2 is eliminated in the exhaled air, and the pH returns to normal.

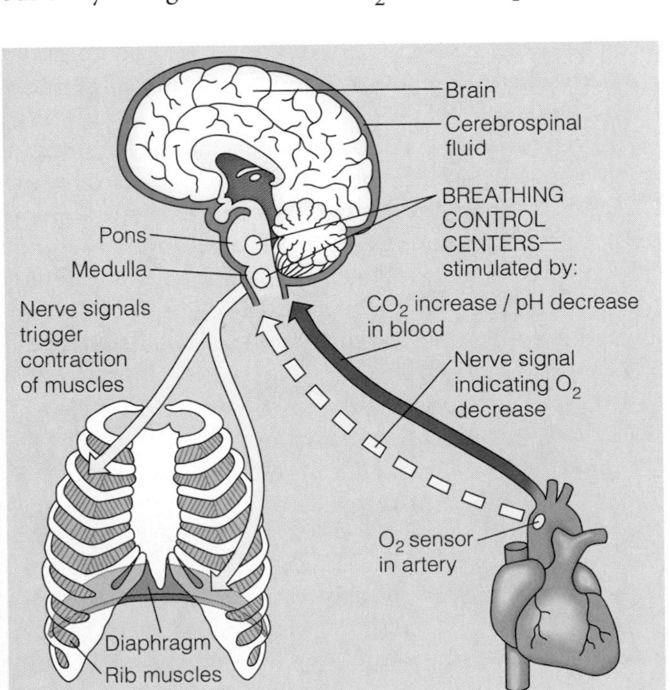

Figure 22.9 Control centers that regulate breathing

When you were a kid, did you ever make yourself dizzy by **hyperventilating,** rapidly taking several deep breaths? Hyperventilating demonstrates the action of your breathing control centers, but it's hard on your body. Deep, rapid breathing can purge the blood of so much CO_2 that the control centers temporarily cease to send signals to the rib muscles and diaphragm. Breathing stops until the CO_2 level increases enough to switch the breathing centers back on.

Our breathing control centers respond directly to CO_2 levels, but they usually do not respond directly to oxygen levels. So how is the level of blood *oxygen* controlled? Since the same process that consumes O_2, cellular respiration, also produces CO_2, a drop in pH (rise in CO_2) in the blood and cerebrospinal fluid is generally a good indication of a drop in blood oxygen. Thus, by responding to lowered pH, the breathing control centers also control blood oxygen level. Also, some of our large arteries have O_2 sensors, and, when the O_2 level in the blood is severely depressed, these sensors signal the control centers via nerves (dashed yellow arrow in the diagram) to increase the rate and depth of breathing. This response may occur, for example, at high altitudes, where the air is so thin that we cannot get enough O_2 by breathing normally.

By responding quickly to the signals they receive from the blood and the arterial sensors, our breathing control centers keep the breathing rate and depth in tune with the changing demands of the body. The breathing centers are only effective, however, if their regulatory activity is coordinated with the activity of the circulatory system. During exercise, the rate at which our heart beats and the amount of blood it pumps with each beat must be increased to match the increased breathing rate. We examine the role of the circulatory system in gas exchange more closely in the next module.

Explain how hyperventilation disrupts the control of breathing.

By purging the blood of CO_2 (and hence carbonic acid), which indirectly stimulates inhalation via its action on the breathing center, hyperventilation temporarily suspends breathing.

Transport of Gases in the Body

22.10 Blood transports the respiratory gases, with hemoglobin carrying the oxygen

We have yet to focus on how O_2 gets from our lungs to all the other tissues in our body, and on how CO_2 travels from the tissues to the alveoli. To do so, we must jump ahead a bit, to the subject of Chapter 23, and look at the basic organization of our cardiovascular system.

Figure 22.10A is a schematic diagram showing the main components of the human cardiovascular system and their roles in gas exchange. Let's start with the heart, in the middle of the diagram. One side of the heart handles oxygen-poor blood (colored blue). The other side handles oxygen-rich blood (red). As indicated in the lower portion of the diagram, oxygen-poor blood returns to the heart from capillaries in body tissues. The heart pumps this blood to the alveolar capillaries in the lungs. At the top of the diagram, gases are being exchanged between the air in the alveolar spaces and the blood in the alveolar capillaries. Off to the right, the diagram shows blood leaving the alveolar capillaries, having lost CO_2 and gained O_2. This oxygen-rich blood returns to the heart, and the heart pumps it out to the body tissues.

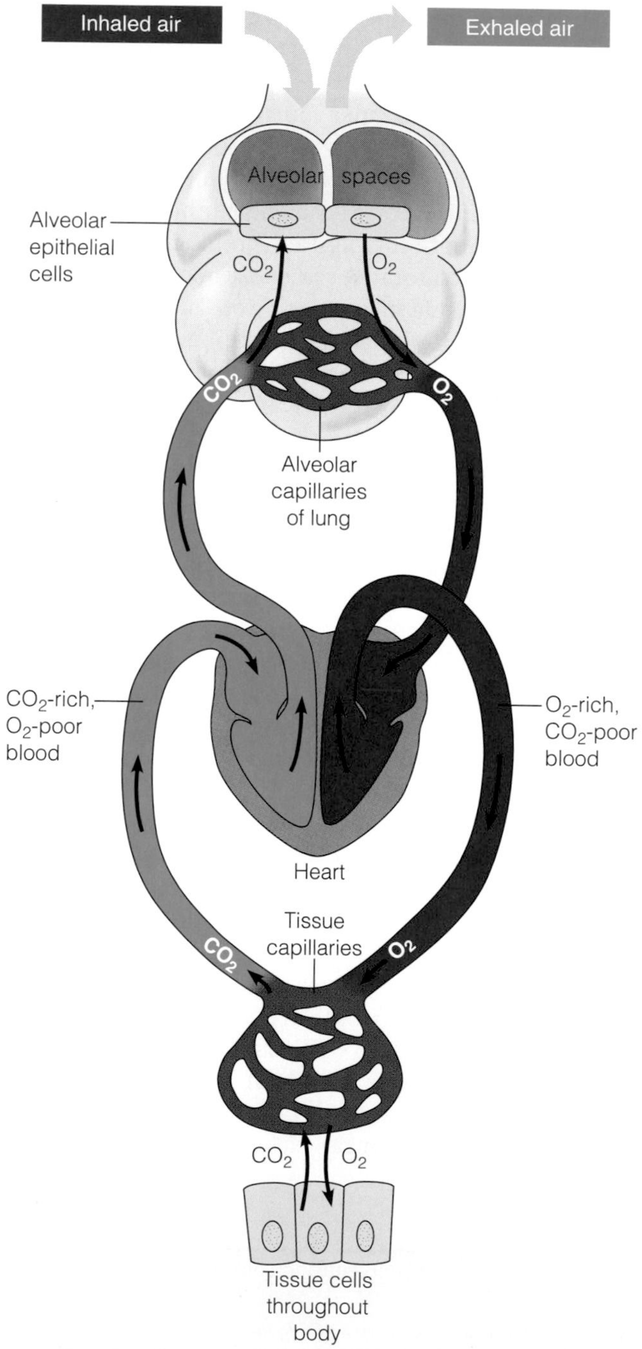

The exchange of gases between capillaries and the cells around them occurs by diffusion. Gases actually diffuse along gradients of pressure. A mixture of gases, such as air, exerts a pressure; the more gas molecules present in a particular volume of gas and the higher the temperature, the greater the pressure. (You see evidence of gas pressure whenever you open a can of pop, releasing the pressure of the CO_2 it contains.) Each kind of gas in a mixture accounts for a portion, called the **partial pressure,** of the mixture's total pressure. Molecules of each kind of gas will diffuse down a gradient of its own partial pressure. At the bottom of Figure 22.10A, for instance, O_2 moves out of the oxygen-rich blood into the tissue cells because it diffuses from a region of higher partial pressure to a region of lower partial pressure. The tissue cells maintain the partial-pressure gradient by consuming O_2 in cellular respiration. Meanwhile, as O_2 enters the tissue cells, CO_2, produced as a waste product of cellular respiration, diffuses down its own partial-pressure gradient out of the cells and into the capillaries. Diffusion also accounts for gas exchange in the alveoli.

Oxygen is not very soluble in water, and the blood actually transports very little of it as dissolved O_2. Most of the O_2 in blood is carried by **hemoglobin** in the red blood cells. As indicated in Figure 22.10B, a hemoglobin molecule consists of four polypeptide chains (of two different types, distinguished by the two shades of purple). Attached to each polypeptide is a chemical group called a heme

Inhaled air
Exhaled air
Alveolar spaces
Alveolar epithelial cells
CO_2
O_2
Alveolar capillaries of lung
CO_2-rich, O_2-poor blood
O_2-rich, CO_2-poor blood
Heart
Tissue capillaries
Tissue cells throughout body

Figure 22.10A Gas exchange in the body

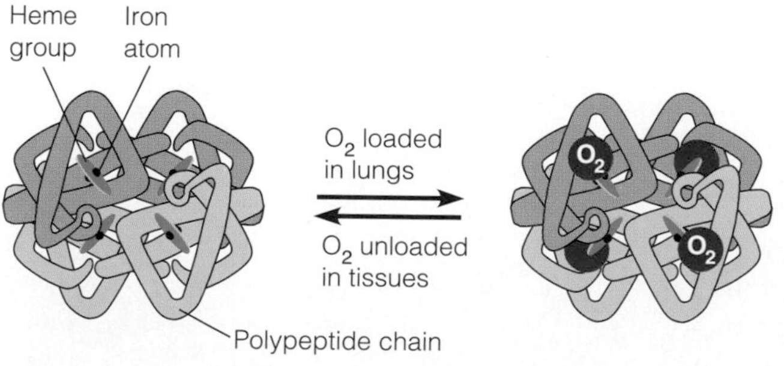

Figure 22.10B Hemoglobin loading and unloading of O_2

(green), at the center of which is an iron atom (black). Each iron atom can carry an O_2 molecule. Thus, a hemoglobin molecule can carry up to four molecules of oxygen. Hemoglobin loads up with oxygen in the lungs, retains it while traveling in the bloodstream, and then unloads it in the body tissues.

What is the physical process that causes respiratory gases to move down their partial-pressure gradients in the body?

Diffusion

Hemoglobin helps transport CO_2 and buffer the blood 22.11

22B

Hemoglobin is a multipurpose molecule. It transports oxygen, loading and unloading it when diffusion allows, and it also helps the blood transport carbon dioxide. A third important function of hemoglobin is helping to buffer the blood—that is, prevent changes in pH that a cell could not survive.

The diagrams here illustrate how CO_2 is carried in the blood. As shown in Figure 22.11A, when CO_2 leaves a tissue cell, it first diffuses into the interstitial fluid (light blue) and then across the wall of a capillary into the blood fluid (plasma, shown in yellow). A small amount of the CO_2 remains dissolved in the plasma, but most of it enters the red blood cells, where some of it combines with hemoglobin. The rest reacts with water molecules, forming carbonic acid (H_2CO_3). The H_2CO_3 forms more readily inside the red blood cells than in the plasma because the cells contain an enzyme that hastens the reaction.

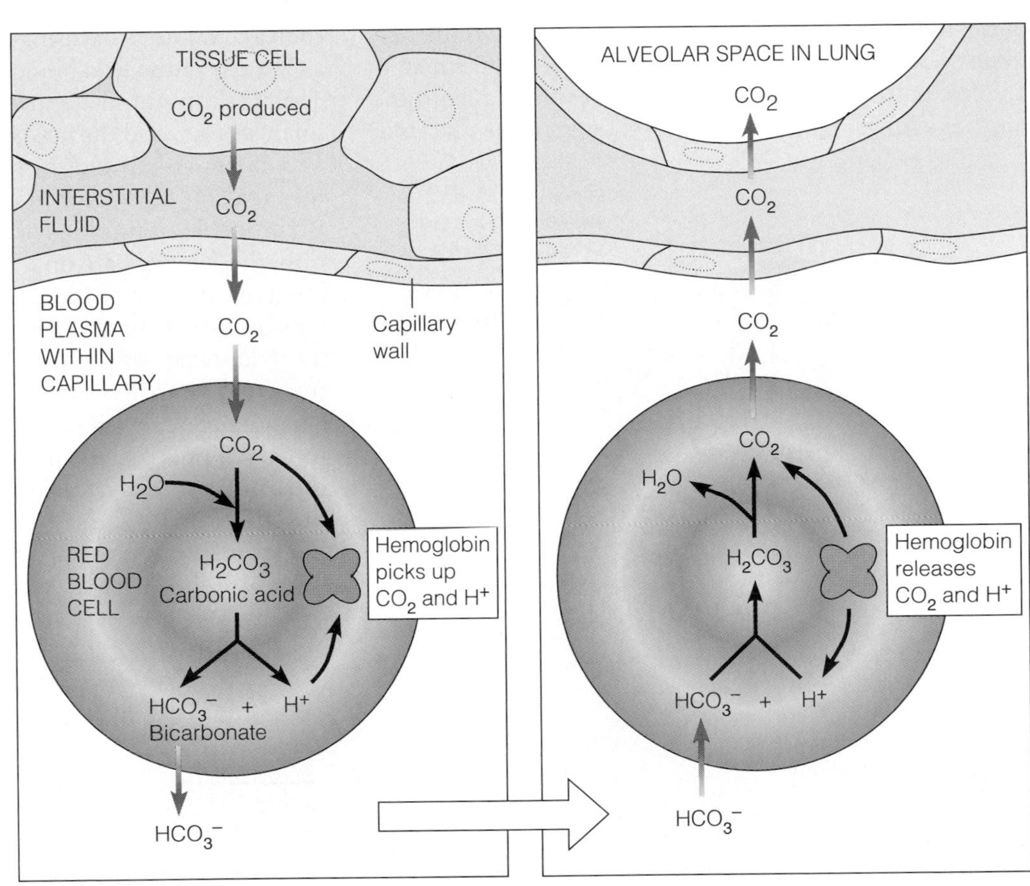

Figure 22.11A CO_2 transport in the body tissues

Figure 22.11B CO_2 transport in the lungs

As shown in the center of the red blood cell in Figure 22.11A, H_2CO_3 does not remain intact. It breaks apart into a hydrogen ion (H^+) and a bicarbonate ion (HCO_3^-). Hemoglobin acts as a buffer by picking up most of the H^+ ions, thus preventing them from acidifying the blood. Most of the bicarbonate ions diffuse out of the red blood cell into the blood plasma and are carried in this form to the lungs.

Figure 22.11B shows what happens to the bicarbonate ions and CO_2 as the blood flows through the lungs. The processes in Figure 22.11A are reversed. Carbonic acid forms when bicarbonate combines with H^+ ions given up by hemoglobin. The carbonic acid is then converted back into CO_2 and water. Finally, the CO_2 diffuses from the blood into the alveoli, and out of the body in exhaled air.

Carbon dioxide is a waste gas, but as we see here, the body puts CO_2 to good use before disposing of it. The CO_2 built into the bicarbonate ions in the blood is an important part of the blood-buffering system. If the pH of the plasma drops (that is, if the concentration of H^+ increases), the bicarbonate ions remove H^+ ions by combining with them to form carbonic acid. When the pH rises, the carbonic acid releases H^+ back into the blood.

?

Most oxygen in the blood is transported bound to the protein __________ within __________ __________ cells, while carbon dioxide is mainly transported as dissolved __________ ions within the plasma.

hemoglobin . . . red blood . . . bicarbonate

22.12 The human fetus exchanges gases with the mother's bloodstream

Figure 22.12 shows a human fetus inside the mother's uterus. The fetus literally swims in a protective watery bath, the amniotic fluid. The lungs are full of fluid and are nonfunctional. How does the fetus exchange gases with the outside world? The answer lies in the function of the placenta, a composite organ that includes tissues from both the fetus and the mother. A large net of capillaries fans out into the placenta from blood vessels in the umbilical cord of the fetus. These capillaries exchange gases with the maternal blood that circulates in the placenta, and the maternal circulatory system carries the gases to and from the mother's lungs. Enhancing gas exchange in the placenta, the maternal and fetal blood flows in opposite directions, providing countercurrent exchange, as we saw in fish gills in Module 22.4. Also aiding O_2 uptake by the fetus is the fetal hemoglobin, a special type that attracts O_2 more strongly than does the mother's hemoglobin.

What happens when a baby is born? Suddenly placental gas exchange ceases, and the baby's lungs must begin to work. Carbon dioxide in the fetal blood acts as a signal. As soon as CO_2 stops diffusing from the fetus into the placenta, a CO_2 rise in the fetal blood causes a drop in the blood pH. The pH drop stimulates the breathing control centers in the infant's brain, and the newborn takes its first breath.

A human birth and the radical changes in gas-exchange mechanisms that accompany it are extraordinary events. Resulting from millions of years of evolutionary adaptation, these events are on a par with the remarkable flight ability of the bar-headed goose we discussed in the chapter's introduction. For a goose to breathe the thin air and fly great distances high above Earth, or for a human baby to switch almost instantly from living in water and exchanging gases with maternal blood to breathing air directly, requires truly remarkable adaptations in the organism's respiratory system. Also required are adaptations of the circulatory system, which, as we have seen, supports the respiratory system in its gas-exchange function. We turn to circulatory systems themselves in Chapter 23.

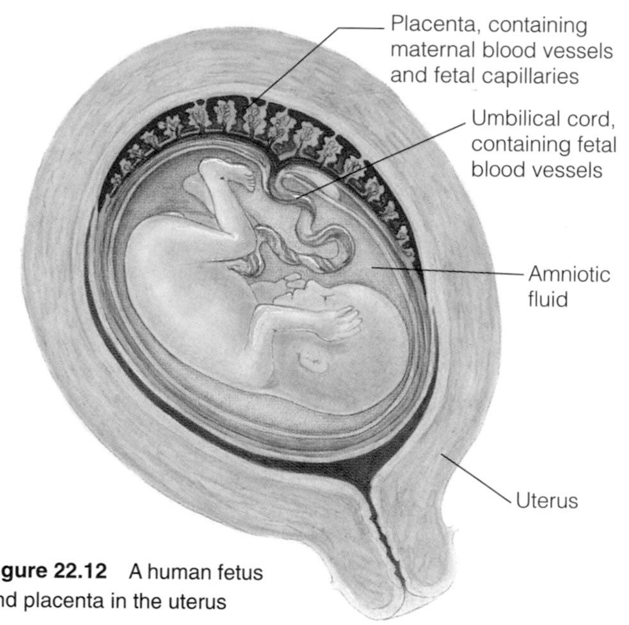

Figure 22.12 A human fetus and placenta in the uterus

How does fetal hemoglobin enhance oxygen transfer from mother to fetus across the placenta?

Fetal hemoglobin has a greater affinity for O_2 than does adult hemoglobin, which helps "pull" the O_2 across the placenta.

Chapter Review

Chapter Summary

Mechanisms of Gas Exchange (Introduction–22.9) Gas exchange is the interchange of O_2 and CO_2 between an organism and its environment **(Introduction).** An animal must exchange these gases with the environment because energy metabolism requires O_2 and produces CO_2. The three main phases of gas exchange are the diffusion of O_2 into the animal, and CO_2 out, during breathing; the transport of dissolved gases within the body; and the uptake of O_2 and the disposal of CO_2 by body cells **(22.1).** O_2 enters an animal and CO_2 leaves by diffusion through a respiratory surface consisting of living cells. Some animals, like the earthworm, use their entire skin as a gas-exchange organ. In most animals, specialized body parts—such as gills in fish, tracheal tubes in insects, or lungs in birds and mammals—carry out gas exchange **(22.2).**

Gills are extensions of the body that absorb O_2 dissolved in water. In a fish, gill filaments bear numerous platelike lamellae, which are packed with blood vessels **(22.3).** Blood flows through the lamellae in a direction opposite to water flow; this countercurrent maintains a diffusion gradient that maximizes the uptake of O_2 **(22.4).**

Land animals exchange gases by breathing air. Air contains more O_2 and is easier to move than water, but water loss from the respiratory surfaces can be a problem. In insects, a network of tracheal tubes carries out gas exchange. The movements of air sacs pump air in and out, and O_2 diffuses from the finely branched tubes directly into cells **(22.5).**

Most terrestrial vertebrates have lungs. In humans and other mammals, air enters through the nasal cavity and passes through the pharynx and larynx (the site of the vocal cords) into the trachea (windpipe). The trachea forks to form two bronchi, each of which branches into numerous bronchioles. The bronchioles end in clusters of tiny sacs called alveoli, which form the respiratory surface of the lungs. Oxygen diffuses through the thin walls of the alveoli into the blood **(22.6).** Mucus and cilia in the respiratory passages protect the lungs, but pollutants can destroy these protections. Smoking kills many people every year, mostly by causing lung cancer and contributing to heart disease. In addition, smoking can make alveoli brittle, leading to their rupture, a disease called emphysema **(22.7)**. Breathing, the alternation of inhalation and exhalation, supplies the lungs with O_2-rich air and expels CO_2-rich air. During inhalation, muscles expand the rib cage, and the diaphragm contracts and moves downward. The chest cavity expands, reducing air pressure in the alveoli. Air rushes through the respiratory passages into the lungs. During exhalation, the rib muscles and diaphragm relax, forcing air out. Vital capacity is the maximum volume of air we can inhale and exhale, but our lungs hold more air than this amount **(22.8)**.

Breathing control centers in the pons and medulla of the brain keep breathing in tune with body needs. During exercise, the CO_2 level in the blood rises, lowering blood pH. The medulla senses this drop and increases the rate and depth of breathing, eliminating CO_2 and returning blood pH to normal. Since CO_2 is produced by the same process that uses O_2, pH changes reflect O_2 levels; the control centers do not normally respond directly to a lack of oxygen **(22.9)**.

Transport of Gases in the Body (22.10–22.12) The heart pumps oxygen-poor blood to the lungs, where it picks up O_2 and drops off CO_2. Then it pumps the oxygen-rich blood to body cells, where it drops off O_2 and picks up CO_2. Gases diffuse down pressure gradients. A mixture of gases, such as air, exerts a particular pressure; the more gas molecules present in a particular volume and the higher the temperature, the greater the pressure. Each kind of gas in a mixture exerts a portion, called its partial pressure, of the mixture's total pressure. At the tissues, O_2 diffuses down its partial-pressure gradient to the cells, and CO_2 diffuses down its partial-pressure gradient into the blood. In the lungs, the gradients are reversed; CO_2 leaves the blood, and O_2 enters from the air. The protein hemoglobin in red blood cells carries most of the oxygen in the blood. Hemoglobin loads up with oxygen in the lungs and unloads it in the tissues **(22.10)**. Hemoglobin also helps buffer the pH of blood and carries some CO_2. Most CO_2 in the blood combines with water to form carbonic acid. In turn, carbonic acid breaks down to form bicarbonate ions and H^+ ions, which help buffer blood pH. Most CO_2 is transported to the lungs in the form of bicarbonate ions **(22.11)**. A human fetus depends on the placenta for gas exchange. A network of capillaries exchanges O_2 and CO_2 with maternal blood that carries gases to and from the mother's lungs. At birth, increasing CO_2 in the fetal blood stimulates the fetus's breathing control centers to initiate breathing **(22.12)**.

Testing Your Knowledge

Multiple Choice

1. When you exhale, air passes through the respiratory structures in which of the following sequences?
   a. alveolus, trachea, bronchus, bronchiole, larynx, pharynx, nasal cavity
   b. alveolus, bronchiole, bronchus, trachea, pharynx, larynx, nasal cavity
   c. alveolus, bronchiole, bronchus, trachea, larynx, pharynx, nasal cavity
   d. alveolus, bronchus, bronchiole, trachea, larynx, pharynx, nasal cavity
   e. alveolus, bronchiole, bronchus, larynx, trachea, pharynx, nasal cavity

2. Countercurrent exchange in the gills of a fish
   a. speeds up the flow of water through the gills
   b. maintains a gradient that enhances diffusion
   c. enables the fish to obtain oxygen while swimming backward
   d. means that blood and water flow in the same direction
   e. interferes with the efficient absorption of oxygen

3. When you inhale, the diaphragm
   a. relaxes and moves upward
   b. relaxes and moves downward
   c. contracts and moves upward
   d. contracts and moves downward
   e. is not involved in the breathing movements

4. In ______, oxygen diffuses directly from the air through a respiratory surface to cells, without being carried by the blood.
   a. an ant
   b. a whale
   c. an earthworm
   d. a sparrow
   e. a mouse

Describing, Comparing, and Explaining

1. What are two advantages of breathing air, compared to obtaining dissolved oxygen from water? What is a comparative disadvantage of breathing air?

2. Trace the path of an oxygen molecule from the air to a muscle cell in your arm, naming all the structures involved along the way.

Thinking as a Scientist

As we discussed in Module 22.10, partial pressure is a measure of the relative amount of gas in a mixture. A biologist has the precision equipment needed to determine the partial pressure of oxygen (abbreviated PO_2) in an insect's body fluids. She is attempting to compare the oxygen consumption of an active insect with that of a resting insect of the same species. One of her new technicians has recently determined the body-fluid PO_2 of an insect that had just been vigorously exercised and of the same insect when fully rested. He has also measured the PO_2 of the surrounding air. The technician obtained the PO_2 values 60, 159, and 40 (in mm of mercury), but unfortunately he forgot to record them in the appropriate columns on the results sheet. The PO_2 of the air was maintained at the same level throughout the experiment, but the technician does not know the PO_2 of the air. Before instructing the technician to repeat the experiment, the biologist asks him which columns he *thinks* the numbers belong in. How would you respond to this question? Explain your answer.

Science, Technology, and Society

In 1990, the tobacco industry briefly succeeded in having a respected scientist, Dr. David Burns, removed from a U.S. Environmental Protection Agency advisory panel reviewing a report on the effects of secondhand smoke. Scientists agreed that Burns was a leading authority in the field, having worked on several studies and reports on the effects of smoking on health. The Tobacco Institute, an industry group, protested his selection, stating, "Frankly, we are mystified how an individual with Dr. Burns's long and intense involvement with the antismoking movement can be expected to contribute to a reasonable, objective, examination" of the EPA report. Do you agree that this is a reason for dismissal from the panel of experts? Six of the members of the same panel came under fire because they had conducted research financed by the tobacco industry. Would they be expected to be reasonable and objective? How should experts be selected?

Media Resources

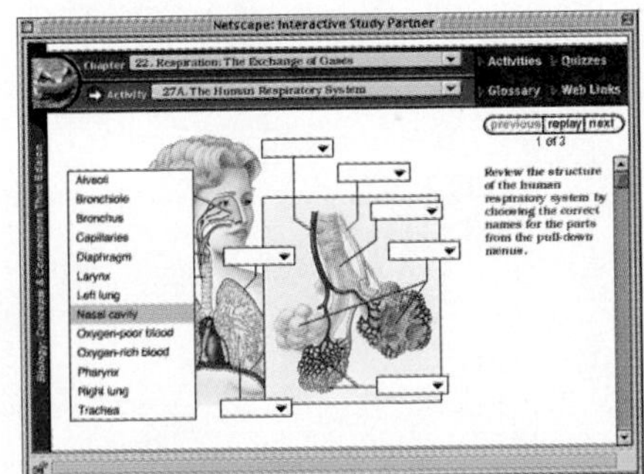

For further review of the concepts in this chapter, use the activities and quizzes on the CD-ROM.

Visit the Special Edition of The Biology Place web site for activities, quizzes, news articles, and web links.

23 Circulation

Outline

Few animals seem less alike than those on these two pages. At the left is a corn snake, found throughout much of the U.S. This nonpoisonous predator eats mostly rats and mice, but it can also climb trees and dine on bird eggs. The giraffe at the right lives in central Africa. It is an herbivore, using its long neck to browse trees.

Despite their differences, snakes and giraffes have many features in common. As land vertebrates, they both have a backbone, lungs, and a circulatory system that carries oxygen, carbon dioxide, and other materials. They also have something in common with *all* animals that live on land—every part of their body is subject to the persistent, unwavering force of gravity.

Gravity does not greatly affect aquatic animals because their bodies are supported by water, but it has profound effects on terrestrial species. Our own bodies show signs of constant, long-term exposure to gravity, such as the tendency of the skin on our face to sag with age. As is generally true for vertically oriented animals, our circulatory system is strongly affected by gravity. Gravity tends to pull blood downward into the lower parts of the body. The pull of gravity is a problem for us because we stand upright on two legs; the giraffe is affected because of its long neck and legs; the corn snake faces the problem when it climbs a tree after bird eggs. If gravity prevailed, blood would accumulate in our legs and in the giraffe's, and when a snake climbed upward, its long, thin body would bulge like a balloon at the tail.

What are the solutions to these problems? Mammals, including humans and giraffes, have very strong hearts that keep blood circulating despite gravity's pull. The blood flow slows as it travels uphill through the veins back to the heart, but muscle contractions help it on its way. As we walk or run, our leg muscles squeeze the veins and force the blood upward toward the heart. Our veins also have valves that allow the blood to flow in only one direction, preventing it from flowing back down the legs. Giraffes and humans are also endowed with tight skin and abundant connective tissue in the legs. These adaptations help keep blood vessels there from enlarging. In the same way, the tight suits worn by fighter pilots counteract the g-forces pulling on their blood during high-speed maneuvers.

Despite the safeguards that help prevent blood from pooling in our legs, gravity sometimes wins. If we don't exercise adequately or if we stand too long in one position, our circulation may slow down, and blood can pool in our leg veins. This is often a problem for people who work standing up. It can also be a problem for a giraffe standing for long periods of time in a zoo. Over time, reduced circulation weakens the leg veins, causing them to stretch and enlarge. The valves in the veins also weaken and become less effective at preventing the backflow of blood. As a result, some people develop varicose veins in their legs—veins just under the skin that become visibly swollen because they have too little support from other tissues.

The corn snake does not have valves in its veins, but its blood still circulates, even when it is climbing. Because its heart is located close to its head, the snake's brain receives enough blood even when it is vertical. Also, its thin, tight body helps prevent blood pooling, and blood vessels in its tail constrict when it climbs. After a climb, a corn snake wriggles vigorously. This motion contracts muscles all over its body, squeezing veins and increasing circulation.

The ability to keep the blood circulating despite the pull of gravity is a matter of life or death for most animals. All organisms must exchange materials with their environment and distribute materials within their bodies. Most animals have a system of internal transport—a **circulatory system**—that transports oxygen and carbon dioxide, distributes nutrients to body cells, and conveys the waste products of metabolism to specific sites for disposal. This chapter focuses on the structure, function, and evolution of circulatory systems.

23.1 The circulatory system associates intimately with all body tissues

A circulatory system is necessary in any animal whose body is too large or too complex for vital chemicals to reach all its parts by diffusion alone. Diffusion is inadequate for transporting chemicals over distances greater than a few cell widths—far less than the distance oxygen must travel between our lungs and brain, or the distance nutrients must go between our small intestine and the muscles in our arms or legs. Without our circulatory system, nearly every cell in the body would quickly die from lack of oxygen.

To be effective, the circulatory system must have an intimate connection with the tissues. In the human body, for instance, the heart pumps blood that has just been oxygenated in the lungs through a system of blood vessels into microscopic vessels called **capillaries.** The capillaries form an intricate network among the tissue cells, such that no substance has to diffuse far to enter or leave a cell. The micrograph in Figure 23.1A illustrates the relationship between capillaries and our body tissues. This particular capillary supplies freshly oxygenated, nutrient-rich blood to smooth muscle cells. (The oval structures outside the capillary, each a little bigger than a red blood cell, are nuclei of the muscle cells; see Module 20.6.) Notice that red blood cells pass single-file through the capillary. Each red blood cell comes close enough to the surrounding tissue that O_2 and nutrients can diffuse out of it into the muscle cells.

In Figure 23.1B, the downward arrows show the route that molecules take in diffusing from the blood into tissue cells. As we discussed in Module 20.11, materials are not exchanged directly between the blood and the body cells. Each body cell is immersed in a watery interstitial fluid. Molecules such as O_2 (red dots in the figure) diffuse first out of a capillary into the interstitial fluid, and then from the interstitial fluid into a tissue cell.

The circulatory system has several other major functions, besides transporting O_2 and nutrients. It conveys metabolic wastes to waste-disposal organs: CO_2 to the lungs, and a variety of other metabolic wastes to the kidneys. The upward arrows in Figure 23.1B represent the diffusion of waste molecules out of a tissue cell into the capillary.

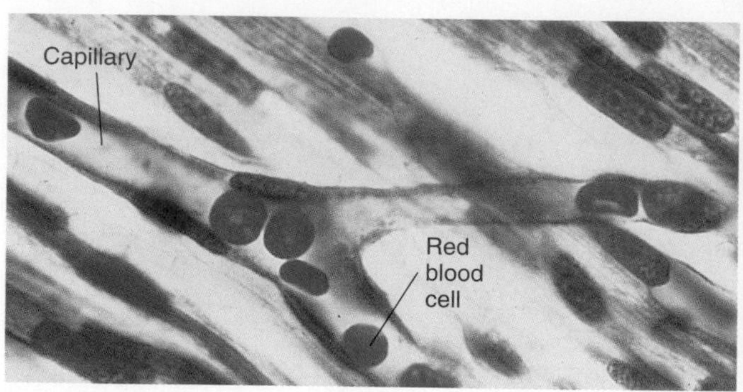

Figure 23.1A A capillary in muscle tissue

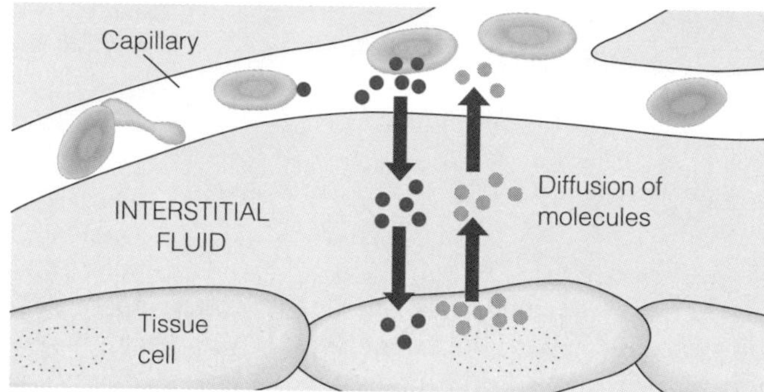

Figure 23.1B Diffusion between blood and tissue cells

The circulatory system also plays two key roles in maintaining a constant internal environment (homeostasis). First, by exchanging molecules with the interstitial fluid, the circulatory system helps control the makeup of the environment in which the tissue cells live. Second, the circulatory system helps control the makeup of the blood by continuously moving it through organs, such as the liver and kidneys, that regulate the blood's contents.

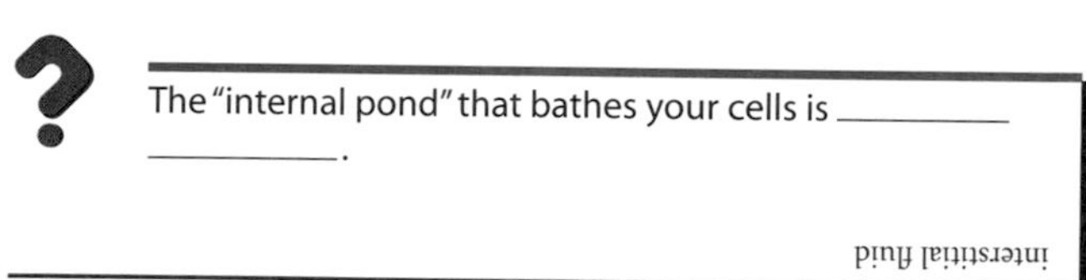

?
The "internal pond" that bathes your cells is ___________ ___________.
interstitial fluid

Mechanisms of Internal Transport

23.2 Several types of internal transport have evolved in animals

Not all animals have a circulatory system like ours. Cnidarians such as hydras and jellies (also known as "jellyfish"), for example, have no true circulatory system—that is, no organ system specifically for internal transport. As we saw in Module 21.3, the body wall of a hydra is only two or three cells thick, so all the cells can exchange materials directly with the water surrounding the animal or with the water in the gastrovascular cavity. Water is drawn in through the mouth, circulates in the gastrovascular cavity, and then passes back out through the mouth. Hydras have no blood.

The jelly in Figure 23.2A (facing page) has an elaborate gastrovascular cavity (gold), with branches radiating to and from a circular canal. Cells lining all these branches bear flagella, whose beating helps circulate the gastrovascular fluid. Digestion also occurs in the gastrovascular cavity and in the cells lining it. Only these cells have direct access to nutrients, but the nutrients have only a short distance to diffuse to other body cells.

A gastrovascular cavity provides adequate internal transport for animals whose body cells are not arranged in numerous layers, but it is not adequate for animals with thick, multiple layers of cells. Such animals have a true circulatory *system* containing a specialized circulatory fluid, **blood.**

Two basic types of circulatory systems have evolved in animals. Many invertebrates, including most mollusks and all arthropods, have what is called an **open circulatory system.** The system is termed "open" because blood is pumped through open-ended vessels and flows out among the cells; there is no separate interstitial fluid. In an insect, such as the grasshopper (Figure 23.2B), pumping of the tubular heart drives the blood into the head and the rest of the body (red arrows). Nutrients diffuse from the blood directly into the body cells, as contractions of body muscles move the blood toward the tail. When the heart relaxes, blood returns to it (blue arrows) through several pores. Each pore has a valve that closes when the heart contracts, preventing backflow of the blood. As we saw in Module 22.5, respiratory gases are conveyed to and from the insect's body cells by the tracheal system (not shown here).

Vertebrates, including humans, have a **closed circulatory system,** often called a **cardiovascular system** because it consists of a heart and a network of tubelike vessels (Greek *kardia,* heart, and Latin *vas,* vessel). The blood is confined to the vessels, which keep it distinct from the interstitial fluid. There are three kinds of vessels in a closed circulatory system. **Arteries** carry blood away from the heart to organs throughout the body, **veins** return blood to the heart, and **capillaries** convey blood between arteries and veins within each organ. Arteries and veins are distinguished by the *direction* in which they carry blood, not by the quality of the blood they contain. Although most arteries convey O_2-rich blood and most veins transport blood depleted of oxygen (O_2-poor blood), there are important exceptions. For example, we have two arteries, called pulmonary arteries, that carry O_2-poor blood from our heart to our lungs; and we have four pulmonary veins that carry freshly oxygenated blood from the lungs to the heart.

The cardiovascular system of a fish illustrates the main features of a closed circulatory system. Figure 23.2C shows a simplified version. (A bony fish actually has four gills on each side of its head, instead of just two as shown, and there are hundreds of thousands of gill capillaries.) The heart has two main chambers. The **atrium** (plural, *atria*) receives blood from the veins, and the **ventricle** pumps blood to the gills via large arteries. As in all the closed circulatory system figures in this chapter, red represents O_2-rich blood and blue represents O_2-poor blood. After passing through the gill capillaries, the O_2-rich blood flows into other large arteries that carry it to all other parts of the body. The large arteries branch into **arterioles,** small vessels that give rise to capillaries. Networks of capillaries called **capillary beds** infiltrate every organ and tissue in the body. The thin walls of the capillaries allow chemical exchange between the blood and the interstitial fluid. The capillaries converge into **venules,** which in turn converge into veins that return blood to the heart.

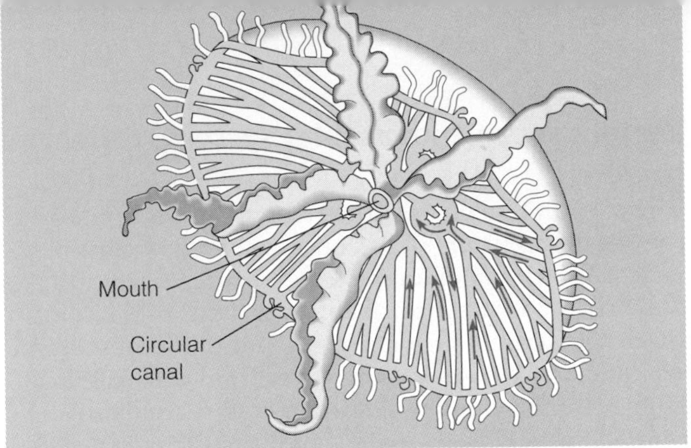

Figure 23.2A The gastrovascular cavity (gold color) in a jellyfish

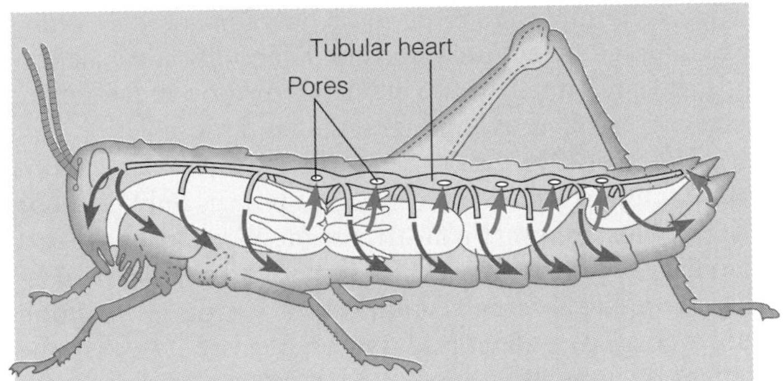

Figure 23.2B The open circulatory system (vessels in gold) in a grasshopper

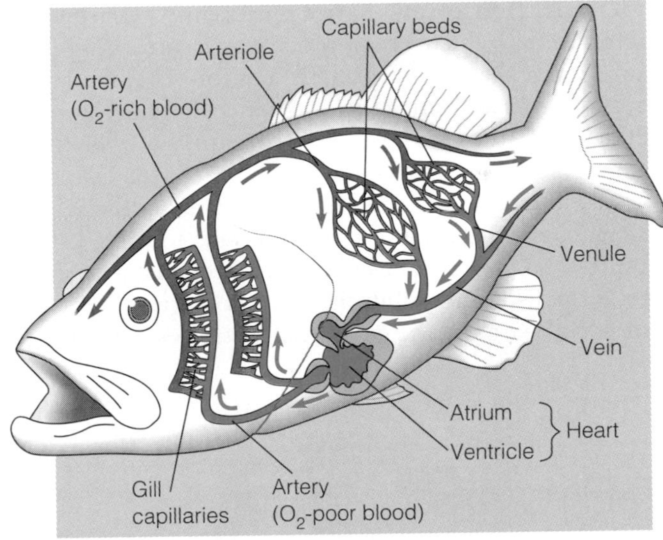

Figure 23.2C The closed circulatory system in a fish

? A (An) ________ is a blood vessel that carries blood toward the heart, to a heart chamber called a (an) ________. A (An) ________ is a blood vessel that carries blood away from the heart, from a heart chamber called a (an) ________.

vein . . . atrium . . . artery . . . ventricle

23.3 Vertebrate cardiovascular systems reflect evolution

The colonization of land by vertebrates was a momentous episode in the history of life, opening vast new opportunities for this group of animals. As aquatic vertebrates became adapted for terrestrial life, nearly all of their organ systems underwent major changes. One of the most drastic evolutionary changes was the switch from gill breathing to lung breathing, and this switch was accompanied by equally drastic changes in the cardiovascular system.

These diagrams compare the cardiovascular systems of a fish and a mammal. Other terrestrial vertebrates—amphibians, reptiles, and birds—also have lungs and a cardiovascular system adapted for life on land. By comparing a fish and a mammal, we see the major cardiovascular differences between vertebrates with gills and those with lungs.

As indicated in Figure 23.3A, a fish has a single circuit of blood flow. Its heart receives and pumps only O_2-poor blood. After leaving the heart, the blood passes through a capillary bed in the gills, where it picks up oxygen. In passing through the many capillaries in the gills, the blood slows down considerably, but it is helped on its way to the other organs by the animal's swimming movements. In the other organs, it passes through other beds of capillaries, known as systemic capillaries, before returning to the heart.

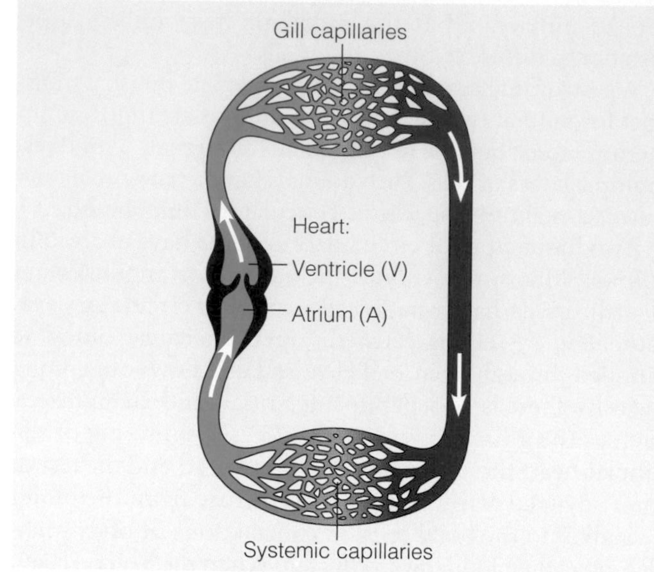

Figure 23.3A Diagram of the cardiovascular system of a fish

Terrestrial vertebrates have a more complex cardiovascular system, which accommodates air breathing. The mammalian heart (Figure 23.3B) has four chambers: two atria (A) and two ventricles (V). Notice that the right side of the heart (from the animal's viewpoint) is on the left in the diagram, and the left side of the heart is on the right. (It is customary to draw the system this way, as though the heart is in a body facing you from the page.) Terrestrial vertebrates also have two blood circuits instead of the single one in the fish. The **pulmonary circuit** carries blood between the heart and the gas-exchange tissues in the lungs, and the **systemic circuit** carries blood between the heart and the rest of the body. With this double circulation, only the right side of the heart handles blood that has been depleted of oxygen. The right atrium receives this blood from body tissues, and the right ventricle pumps it to capillary beds in the lungs.

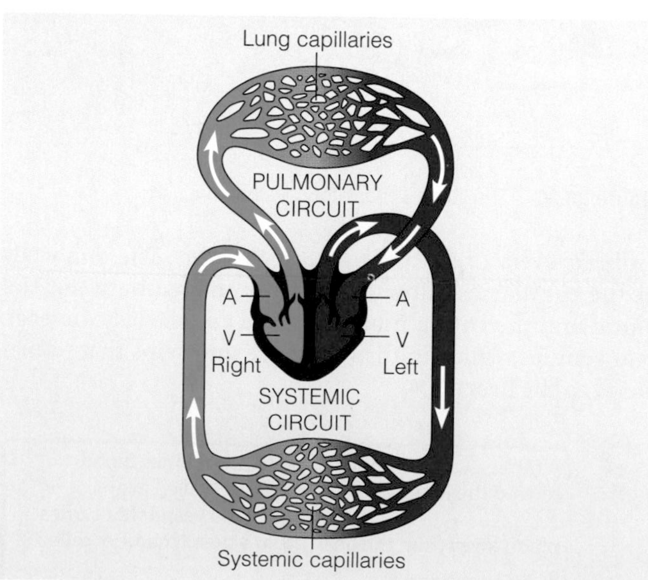

Figure 23.3B Diagram of the cardiovascular system of a mammal

Completely separate from the right side of the mammalian heart, the left side acts as a second pump, handling only O_2-rich blood. As indicated in Figure 23.3B, the left atrium receives blood from the lungs, and the left ventricle pumps it out to body organs via the systemic circuit. Thus, the mammalian heart is actually two pumps in one; the system accommodates lung breathing, while keeping O_2-rich blood flowing rapidly to body organs. Blood returns to the mammalian heart just after being oxygenated in the lungs, and the left ventricle gives this blood a strong boost that propels it rapidly out to the systemic capillaries. Rapid delivery of O_2-rich blood to the body tissue supports the high metabolic rate characteristic of many land vertebrates, especially birds and mammals, which are endothermic.

The differences between the fish and mammalian cardiovascular systems represent the kinds of changes that occurred as terrestrial vertebrates evolved from their fishlike aquatic ancestors. Along with the change from gills to lungs was the switch from a single-circuit cardiovascular system into a dual system with a double-pump heart. Let's now take a look at the mammalian system in more detail.

Some babies are born with a small hole in the septum between the left and right ventricles. How would this affect the oxygen content of the blood pumped out of the heart into the systemic circuit?

This reduces the oxygen content by mixing O_2-depleted blood returned to the right ventricle from the systemic circuit with the O_2-rich blood of the left ventricle.

A trip through the mammalian cardiovascular system 23.4

23A
23B
23C

The human heart and cardiovascular system typify those of mammals. The human heart (Figure 23.4A) is about the size of a clenched fist. It is enclosed in a sac just under the breastbone. The heart is formed mostly of cardiac muscle tissue. Its thin-walled atria collect blood returning to the heart and pump it only the short distance into the ventricles. The thicker-walled ventricles pump blood to all other body organs. The valves in the heart regulate the direction of blood flow, as we will see in Module 23.6.

Let's follow the blood through the entire circulatory system, shown diagrammatically in Figure 23.4B. Beginning with the pulmonary (lung) circuit, ① the right ventricle pumps blood to the lungs via ② two **pulmonary arteries.** As the blood flows through ③ capillaries in the lungs, it unloads CO_2 and loads up on O_2. The O_2-rich blood then flows back to the left atrium via the **pulmonary veins** (actually two from each lung, though only one per lung is shown here). Next, the O_2-rich blood flows from ④ the left atrium into ⑤ the left ventricle (as the ventricle opens and the atrium contracts).

In Figure 23.4A, you can see that the walls of the left ventricle are thicker than those of the right ventricle. The powerful muscles in the left ventricle pump blood to all body organs (including the lungs) through the systemic circuit. As Figure 23.4B shows, oxygen-rich blood leaves the left ventricle through ⑥ the **aorta.** The aorta is our largest blood vessel, with a diameter of roughly 2.5 cm, about the same as that of a quarter. Several large arteries branch from the aorta (only one is shown in Figure 23.4B) and lead to ⑦ the upper body. The aorta then curves down, behind the heart, and more arteries branching from it supply blood to ⑧ abdominal organs and the lower body. For simplicity, Figure 23.4B does not show the individual organs, but within each one, arteries carry the blood into arterioles, which in turn convey it to capillaries. The capillaries rejoin as venules, which carry the blood back into veins. Oxygen-poor blood from the upper body and head is channeled into a large vein called ⑨ the **superior vena cava.** Another large vein, the ⑩ **inferior vena cava,** drains blood from the lower body. The two venae cavae empty their blood into the ⑪ right atrium. As the blood flows from the right atrium into the right ventricle, we complete our journey.

Now that we have surveyed the cardiovascular system as a whole, let's take a closer look at the structure and function of its parts, first the vessels and then the heart.

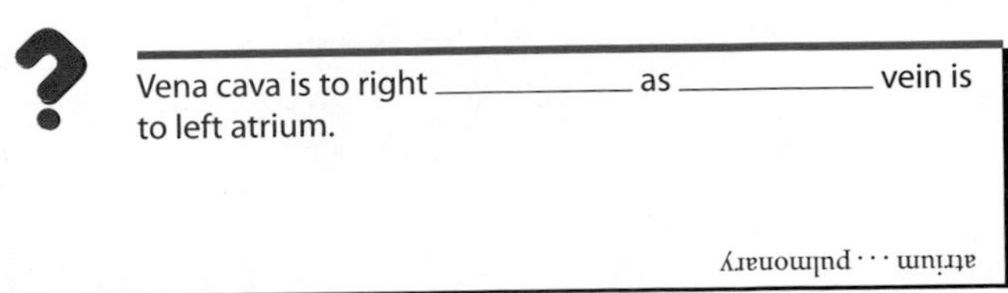

Vena cava is to right ________ as ________ vein is to left atrium.

atrium ··· pulmonary

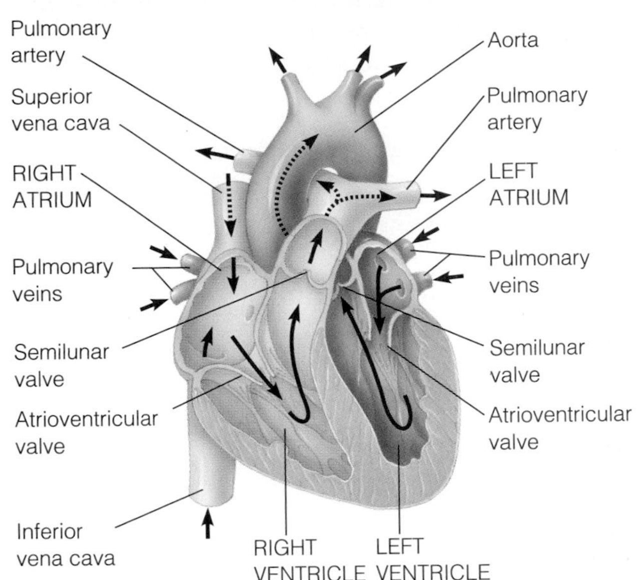

Figure 23.4A The path of blood flow through the human heart

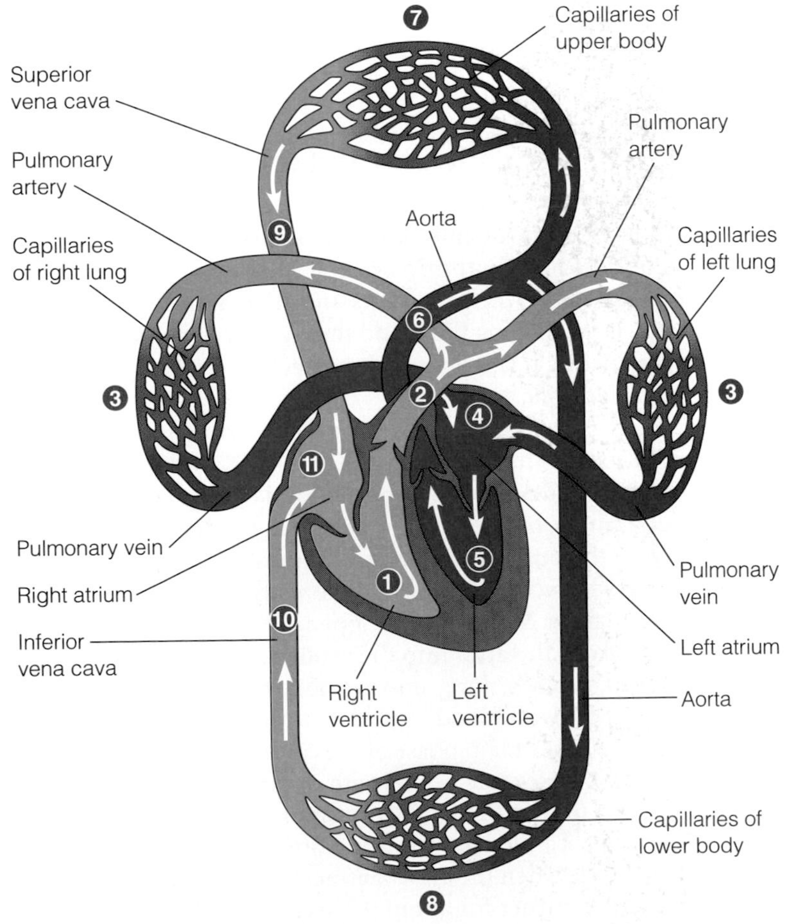

Figure 23.4B The path of blood flow through the cardiovascular system

23.5 The structure of blood vessels matches their functions

Figure 23.5 illustrates the structures of the different kinds of blood vessels and how the vessels are connected. Look first at the capillaries (center), which form fine branching networks where materials are exchanged between the blood and tissue cells. Appropriate to this function, capillaries have very thin walls formed of a single layer of epithelial cells, which is wrapped in a thin basement membrane. The inner surface of the capillary is smooth and keeps the blood cells from being abraded as they tumble through the system.

Arteries, arterioles, veins, and venules have walls that are thicker than those of capillaries. The walls have the same epithelium as capillaries, but they are reinforced by two other tissue layers, both of which are thicker and sturdier in the arteries than in the veins. The middle layer, mainly smooth muscle, allows arteries and some veins to regulate blood flow by constricting. The thick muscle layer in the large arteries near the heart also allows these vessels to withstand surges of blood carrying the full force of the heartbeat. An outer layer of connective tissue is elastic and enables the vessels to stretch. Many of the veins have valves, flaps of tissue projecting toward the heart. The valves prevent backflow, permitting blood to flow only *toward* the heart.

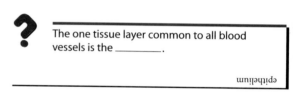

Figure 23.5 Structural relationships of blood vessels

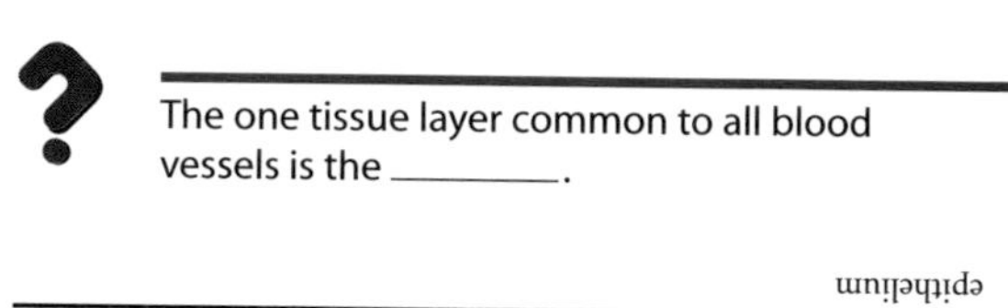

?
The one tissue layer common to all blood vessels is the ________.
epithelium

23.6 The heart contracts and relaxes rhythmically

The heart is the hub of the circulatory system. In a continuous, rhythmic cycle, it passively fills with blood from the large veins and then actively contracts, propelling the blood throughout the body. Its alternating relaxations and contractions make up the **cardiac cycle.**

As shown in Figure 23.6, when the entire heart is relaxed, in the phase called ① **diastole,** blood flows into all four of its chambers. Blood enters the right atrium from the venae cavae and the left atrium from the pulmonary veins. Also during diastole, the valves between the atria and the ventricles (atrioventricular, or AV, valves) are open, allowing blood to flow from the atria into the ventricles. Diastole lasts about 0.4 sec, long enough for the ventricles to nearly fill with blood.

The other main phase of the cardiac cycle is called **systole.** ② Systole begins with a very brief (0.1-sec) contraction of the atria that completely fills the ventricles with blood. This is the only time in the cardiac cycle when the atria contract. Finally, ③ the ventricles contract for about 0.3 sec. The force of their contraction closes the AV valves and opens the semilunar valves located at the exit from each ventricle.

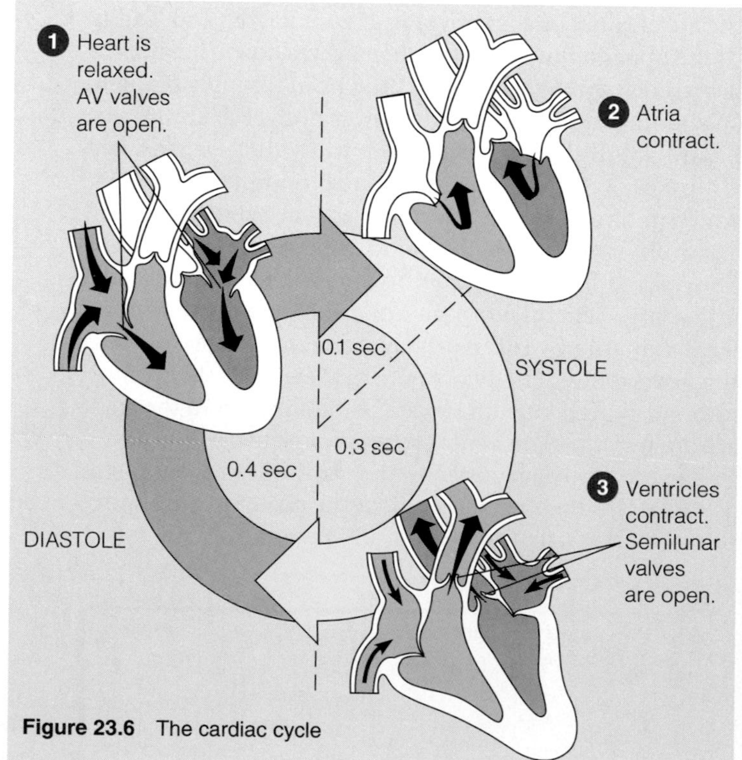

Figure 23.6 The cardiac cycle

With its semilunar valves open, the heart pumps O_2-poor blood to the lungs via the pulmonary artery and O_2-rich blood to the rest of the body through the aorta. Blood also flows into the atria during the second part of systole, as the small arrows in part 3 indicate.

The volume of blood per minute the left ventricle pumps into the aorta is called **cardiac output.** This volume is equal to the amount of blood pumped by the left ventricle each time it contracts (about 75 mL per beat for the average person) times the heart rate. An average person at rest might have a heart rate of about 70 beats per minute. At this rate, the cardiac output would be $70 \times 75 = 5.25$ L/min. Your heart rate and cardiac output will vary, depending on your level of activity and other factors. Both will increase, for instance, when you consume stimulants such as caffeine.

At different times in its cycle, the heart makes distinctive sounds, which you can hear if you press your ear against the breastbone of a friend. The sound pattern is "lub-dupp, lub-dupp." The "lub" sound comes from the contraction of the ventricles and the discharge of the blood against the AV valves. The "dupp" comes from the discharge of the blood against the semilunar valves.

A trained ear can also detect the sound of a heart murmur, indicating a defect in one or more of the heart valves. A murmur sounds like a hiss, and it occurs when a stream of blood squirts backward through a valve. Some people are born with murmurs, while others have their valves damaged by infection (from rheumatic fever, for instance). Most cases of heart murmur are not serious, and those that are can be corrected by replacing the damaged valves with artificial ones or with valves taken from an organ donor.

During a cardiac cycle of 0.8 sec, the atria are generally relaxed for _____ sec.

0.7

23.7 The pacemaker sets the tempo of the heartbeat

23C

What we call our heartbeat results from the contraction of cardiac muscles in the walls of the atria and ventricles. A specialized region of cardiac muscle called the **pacemaker,** or **SA (sinoatrial) node,** maintains the heart's pumping rhythm by setting the rate at which the heart contracts.

The pacemaker is situated in the wall of the right atrium (Figure 23.7). ① The pacemaker generates electrical signals (orange arrows) much like those produced by nerve cells. ② The signals spread (yellow color) quickly through both atria, making them contract in unison. The signals also pass to a relay point called the **AV (atrioventricular) node,** in the wall between the right atrium and right ventricle. Here the signals are delayed about 0.1 sec. The delay ensures that the atria will contract first and empty completely before the ventricles contract. ③ Specialized muscle fibers (gold) then relay the signals to the tips of the ventricles and up through their walls, triggering the strong contractions that drive the blood out of the heart to the lungs (from the right ventricle) and into the aorta (from the left ventricle). The electrical signals in the heart generate electrical changes in the skin, which can be detected by electrodes and recorded as an electrocardiogram, or ECG. The yellow color in the graphs under the hearts indicates the components of an ECG that match the electrical events in the heart.

In certain kinds of heart disease, the heart's self-pacing system fails to maintain a normal heart rhythm. The remedy is an **artificial pacemaker,** a tiny electronic device surgically implanted near the AV node. Artificial pacemakers emit electrical signals that trigger normal heartbeats.

Actually, there is more to the heart rate story than the pacemaker/AV node system. The brain also exerts some influence on heart rate. When we exercise or become excited, for example, control centers in our brain send nerve signals to both the pacemaker and the AV node, making them increase the heart rate. In contrast, when we are asleep or depressed, the brain's control centers decrease the activity of the pacemaker and AV node and the tempo of the heartbeat. Thus, the brain's control centers provide a way for the heart to respond to stimuli from our surroundings—something the pacemaker and AV node cannot do.

A slight decrease in blood pH causes the pacemaker to speed up. What is the function of this control mechanism? (*Hint:* See Module 22.9.)

The increased heart rate in response to lower pH enhances delivery of CO_2-rich blood to the lungs for removal of the CO_2.

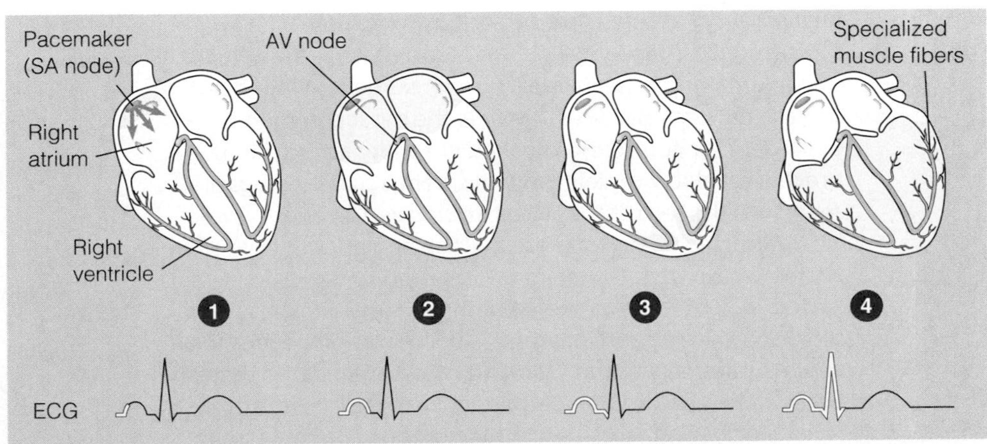

Figure 23.7 Control of the heart's rhythm (top) and electrocardiogram (bottom)

23.8 What is a heart attack?

Diseases of the heart and blood vessels—cardiovascular diseases—are the leading cause of death in the United States today. Every year, several hundred thousand Americans die from heart attacks. A **heart attack** is the death of cardiac muscle cells and the resulting failure of the heart to deliver enough blood to the rest of the body. Cardiac muscle cells are nourished and supplied with oxygen by blood vessels called **coronary arteries.** Shown in red in Figure 23.8A, the coronary arteries branch from the aorta just as it emerges from the heart. If these arteries become clogged, the heart muscle dies from lack of oxygen. The blockage shown in the figure, for example, would cut off the blood supply to the part of the heart muscle shown in blue.

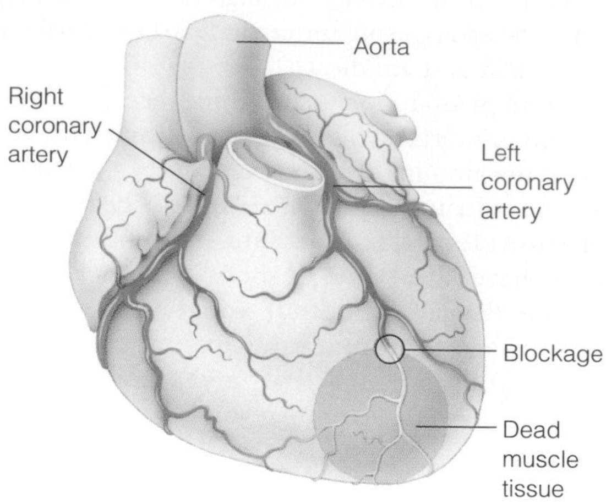

Figure 23.8A Blockage of a coronary artery, resulting in a heart attack

What causes blockage of the coronary arteries? Sometimes, a gradual narrowing of the channel within these arteries results from atherosclerosis, the buildup of lipids on the inside of the vessels. In such cases, a person may feel occasional chest pains, a condition known as angina pectoris (Latin *angere,* to choke, and *pector,* breast). Angina is most likely to occur when the heart is laboring hard because of physical or emotional stress. Angina is a signal that part of the heart is not receiving a sufficient supply of oxygen, and that a heart attack could occur in the future.

Many heart attacks occur without warning. A blood clot may completely block a coronary artery, or atherosclerosis may reach a critical level, causing massive damage to the heart muscle. All of a sudden, the person feels a heavy squeezing ache or discomfort in the center of the chest. The pain may radiate to the shoulder, arm, neck, or jaw. Other symptoms may include sweating, nausea, shortness of breath, and dizziness or fainting.

When heart muscles die, they are not replaced, because cardiac muscle cells do not divide. When a person survives a heart attack, scar tissue (a type of connective tissue) grows into the areas where the heart muscle has died. This is the body's way of repairing the damaged areas. However, the scar tissue cannot contract, as cardiac muscle does. As a result, the damaged heart is permanently weakened, and its ability to pump blood throughout the body, including to its own coronary arteries, may be seriously impaired.

Corrective surgeries, such as coronary bypasses, heart transplants, and artificial hearts, offer some relief for heart disease patients. Bypass surgery is the most common type of surgery for correcting blocked coronary arteries. In this procedure, blood vessels removed from a patient's legs are sewn into the heart to replace the clogged arteries. Newer, less radical techniques include balloon angioplasty, in which a tube with a deflated balloon attached to one end is inserted in a blocked artery. When the tube runs into the blockage, the surgeon inflates the balloon, which opens the vessel by squeezing the blocking tissue against the vessel wall, where it sticks. Laser beams are also used to destroy arterial blocks. Unfortunately, surgery of any kind only treats the disease symptoms, not the underlying causes, and heart disease often develops again, even after these expensive procedures.

On the positive side, over the past three decades, the death rate from cardiovascular disease in the U.S. has declined by more than 25%. Health education and early diagnosis seem to have been the main reasons for this decrease. In recent years, many people have become more conscious of their health; as a group, Americans are smoking less, exercising more, watching their diet, and having regular medical checkups. Early diagnosis and treatment of problems such as high blood pressure may prevent a deadly heart attack. Figure 23.8B shows another important factor in preventing many deaths from heart attacks: Paramedics, skilled emergency-aid specialists, help save thousands of lives every year.

What is angina?

Pain, which may signal the possibility of a future heart attack, due to oxygen deprivation of heart muscle

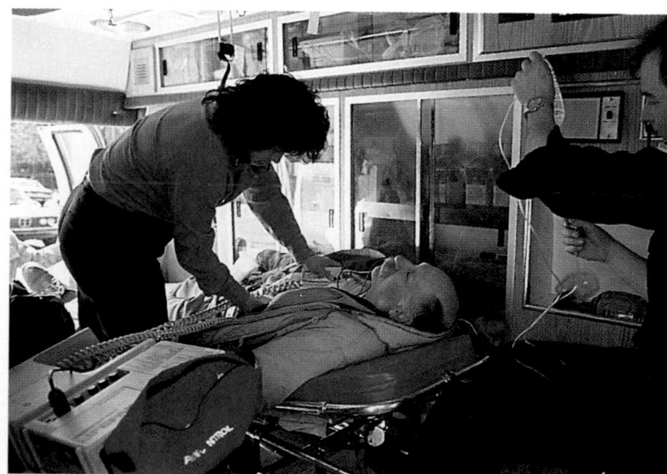

Figure 23.8B Paramedics aiding a heart attack victim

Blood exerts pressure on vessel walls 23.9

23C

Blood pressure is the force that blood exerts against the walls of our blood vessels. Created by the beating of the heart, blood pressure is the main force driving the blood from the heart through the arteries and arterioles to the capillary beds. When the ventricles contract, blood is forced into the arteries faster than it can flow into the arterioles. This stretches the elastic walls of the arteries. You can feel this effect of blood pressure when you measure your heart rate by taking your pulse. The **pulse** is the rhythmic stretching of the arteries caused by the pressure of blood forced through the arteries by the powerful contractions of the ventricles during systole. What you actually feel is an artery bulging with each heartbeat. Between heartbeats, the arteries snap back because the pressure is reduced when the heart relaxes.

Blood pressure depends partly on cardiac output (the volume of blood per minute the left ventricle pumps into the aorta) and partly on the resistance to blood flow imposed by the blood vessels. As Figure 23.9A indicates, blood pressure (expressed in millimeters of mercury, mm Hg) and the blood's velocity (rate of flow, expressed in centimeters per second, cm/sec) are highest nearest the heart, in the aorta, and change little as the blood passes through the other arteries. However, blood pressure and velocity both decline abruptly as the blood enters the arterioles. The pressure drop results mainly from a resistance to blood flow caused by friction between the blood and the inner walls of the arterioles. The resistance is high because the blood contacts a large amount of wall surface in the numerous, tiny arterioles. By lowering the blood pressure, the high resistance eliminates pressure peaks in the arterioles. This is why you can feel your pulse only in your arteries.

Friction also reduces the velocity of the blood in the arterioles. But velocity declines mainly because of the structural arrangement indicated at the top of Figure 23.9A. The total, combined width of all the openings into a set of arterioles is much greater than the width of the artery that feeds blood into them. If there were only one arteriole per artery, the blood would flow faster through the arteriole, the way water does when you pinch a garden hose. However, there are many arterioles per artery, so the effect is like letting go of a pinched hose: As you increase the width of the opening, the flow rate goes down. In the circulatory system, the combined width of the openings of the arterioles is analogous to the larger opening in the hose.

The overall result of the decline in velocity and pressure in the arterioles is a steady, leisurely flow of blood in the capillaries. The gentle flow allows the exchange of substances between the blood and the interstitial fluid.

By the time the blood reaches the veins, its pressure has dropped to near zero. The blood has encountered so much resistance as it passes through the millions of tiny arterioles and capillaries that the force from the pumping heart no longer propels it. How, then, does blood return to the heart, especially when it must travel up from the legs,

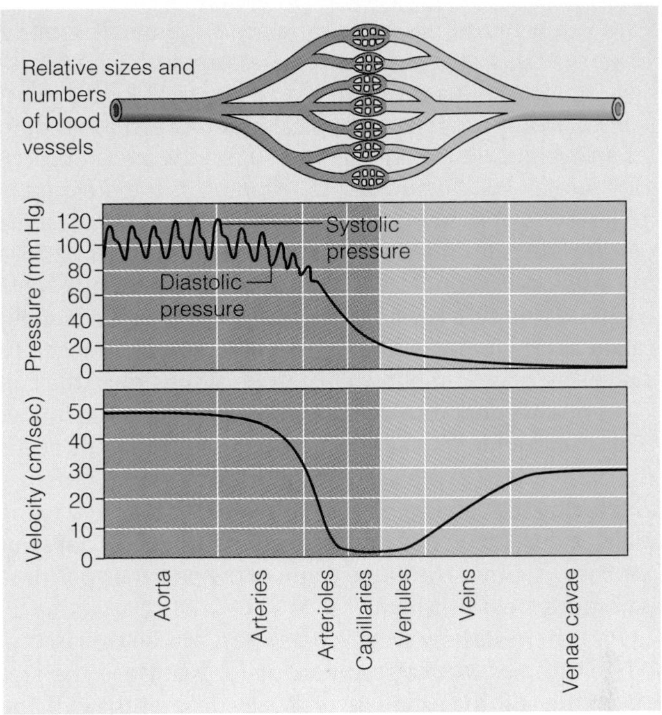

Figure 23.9A Blood pressure and velocity in the blood vessels

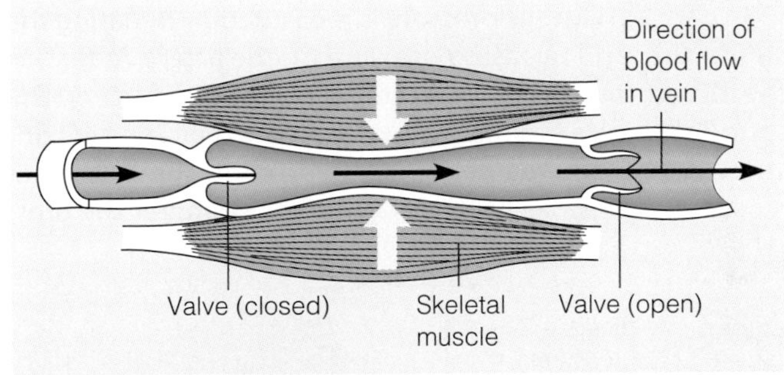

Figure 23.9B Blood flow in a vein

against gravity? As we discussed in the chapter's introduction, the veins of mammals such as humans or giraffes (and corn snakes, as well) are sandwiched between skeletal muscles (Figure 23.9B). Consequently, whenever the body moves, the muscles pinch the veins and squeeze blood along toward the heart. The large veins of mammals have valves that allow the blood to flow only toward the heart. Breathing also helps return blood to the heart. When we inhale, the change in pressure within our chest cavity causes the large veins near our heart to expand and fill. 

? (a) In which type of vessel is blood pressure and velocity lowest? (b) How does this enhance chemical exchange between the blood and interstitial fluid?

(a) capillaries; (b) provides more time for passage of substances across the thin capillary walls

23.10 Measuring blood pressure can reveal cardiovascular problems

Although blood pressure varies throughout the body, when we talk about a person's blood pressure, we are usually referring to the pressure exerted by the blood against the walls of the arteries. A typical blood pressure reading for an average young adult is 120/70, as indicated in part ① of Figure 23.10. The number 120 represents the pressure in millimeters of mercury during systole, and 70 represents the pressure during diastole. A blood pressure of 120 is a force that can support a column of mercury 120 mm high.

Blood pressure is usually measured with an instrument called a sphygmomanometer, an inflatable cuff attached to a pressure gauge. As shown in part ② of the figure, the cuff is wrapped around the upper arm. By squeezing the rubber bulb, the examiner inflates the cuff until its pressure closes the main artery in the arm. When this occurs, the pressure exerted by the cuff is greater than the blood pressure in the main artery (120 in this case, though the examiner does not yet know it). Since the blood cannot flow past the cuff, there is no pulse below the cuff.

Part ③ shows how systolic pressure is actually measured. While listening with a stethoscope for a pulse in the arm below the cuff, the examiner gradually deflates the cuff. The systolic pressure (120) is recorded when the examiner hears the first sounds of blood spurting through the constricted artery—a mild tapping.

In part ④, diastolic pressure is measured by deflating the cuff further and releasing pressure on the artery. At first, as the blood moves more rapidly through the artery, the sounds become louder. When the artery is no longer constricted, the sounds cease. At this point, the examiner records the diastolic pressure (70, in our example), the pressure remaining in the artery when the heart is relaxed between beats.

Blood pressure in a typical healthy adult ranges between 110 and 140 mm Hg systolic and between 70 and 90 mm Hg diastolic. Although individual differences occur, pressures significantly below or above these levels may indicate cardiovascular problems. A person with a systolic blood pressure persistently below 100 mm Hg is considered to have **low blood pressure.** This may be no cause for concern and may even suggest that the person will live a long, healthy life. However, some cases of low blood pressure are associated with poor nutrition or disorders of the thyroid or adrenal glands.

High blood pressure, or hypertension, is a type of cardiovascular disorder and an important cause of heart disease. **Hypertension** is defined as a persistent blood pressure of 140/90 or higher; the higher the values, the greater the risk of serious cardiovascular disease. Elevated pressure caused by reduced flexibility or partial blockage of the arteries makes the heart work harder, because it has to pump against greater resistance in the arteries. As a result, the heart muscle may tend to wear out prematurely. Hypertension also may contribute to arterial blockage by causing small tears in the epithelium that lines the blood vessels

Listening with a stethoscope below a sphygmomanometer cuff on your friend's arm, you hear sounds that begin at 135 mm Hg and cease at 80 mm Hg. What are your friend's systolic and diastolic blood pressures?

Systolic = 135; diastolic = 80 (135/80)

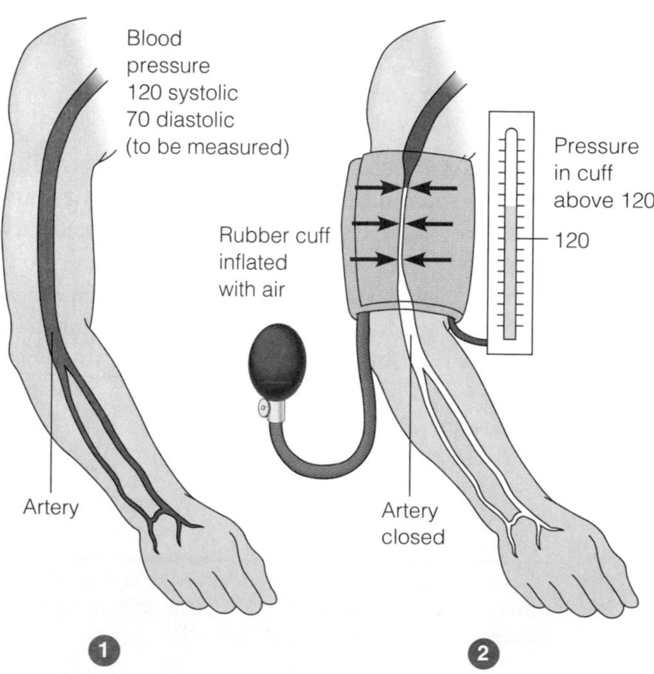

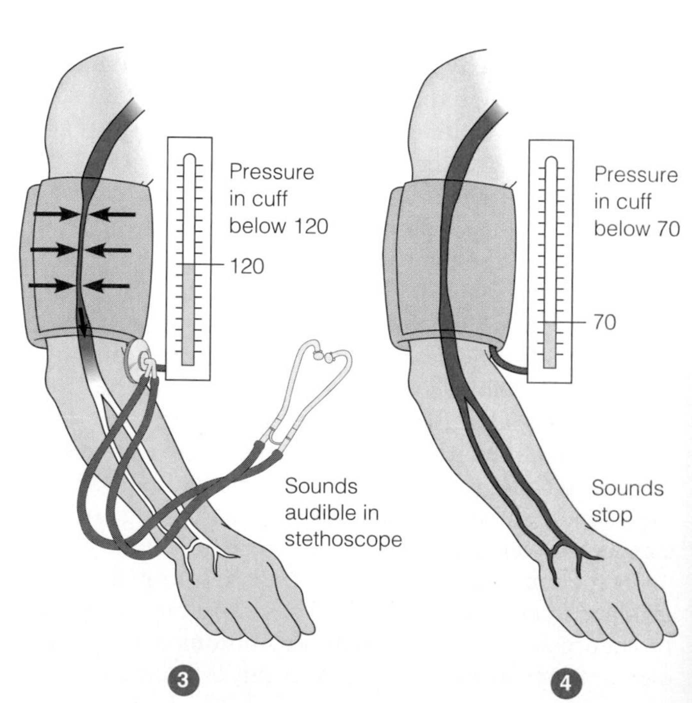

Figure 23.10 Measuring blood pressure

and increasing the tendency of atherosclerotic deposits to form.

Hypertension is sometimes called the "silent killer" because many people with the disease experience no symptoms until tragedy strikes. Even those who know they have high blood pressure may tend to ignore their condition and not seek medical treatment. Fortunately, hypertension can usually be controlled by diet, exercise, and medication.

Smooth muscle controls the distribution of blood 23.11

Because the heart beats constantly and each tissue has many capillaries, every part of the body is supplied with blood at all times. Nonetheless, at any given time, only about 5–10% of the body's capillaries have blood flowing through them. Capillaries in a few organs, such as the brain and heart, usually carry a full load of blood, but in many other sites, the blood supply varies from time to time.

Two mechanisms control the distribution of blood to capillaries of the various organs; both depend on smooth muscle tissue (see Module 20.6). Figure 23.11A shows one mechanism, which involves contraction and relaxation of the smooth muscle layer of the arteriole wall. ① When the muscle cells in this layer relax, the arteriole opens (dilates), allowing blood to enter the capillaries. ② Contraction of the muscle layer, under the influence of nerves and hormones, constricts the arteriole, decreasing blood flow through it to a capillary bed.

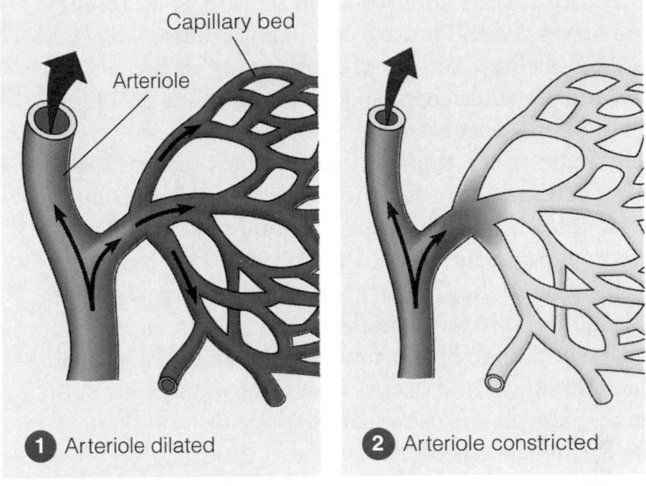

Figure 23.11A The control of capillary blood flow by arteriole constriction

The other controlling mechanism is illustrated in Figure 23.11B. Notice that in both parts of this figure there is a capillary called a thoroughfare channel, through which blood streams directly from arteriole to venule. This channel is always open. Capillaries branching off from thoroughfare channels form the bulk of the capillary bed. Passage of blood into these branching capillaries is regulated by rings of smooth muscle called precapillary sphincters. As you can see in Figure 23.11B, ① blood flows through a capillary bed when its precapillary sphincters are relaxed. ② It bypasses the bed when the sphincters are contracted. After a meal, for instance, precapillary sphincters in the wall of the digestive tract let a larger quantity of blood pass through the capillary beds there than when food is not being digested. During strenuous exercise, many of the capillaries in the digestive tract are closed off, and blood is supplied more generously to skeletal muscles.

Going for a vigorous swim right after eating a heavy meal is probably more likely to cause indigestion than cramping of the muscles. Explain.

The vigorous exercise routes more blood to the capillary beds of muscles, resulting in less blood flow to the capillaries of the digestive tract.

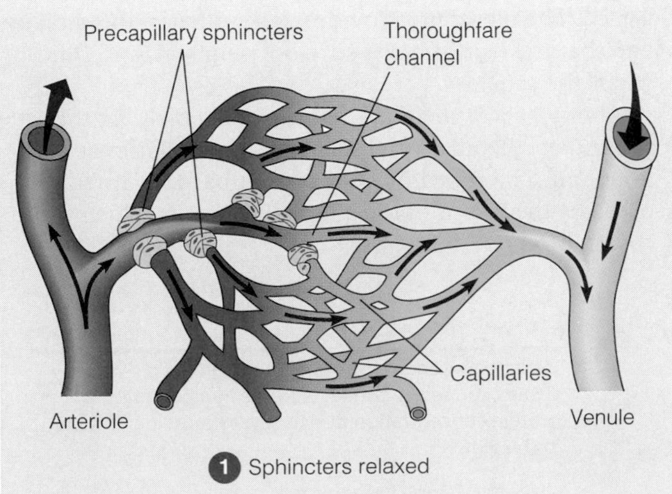

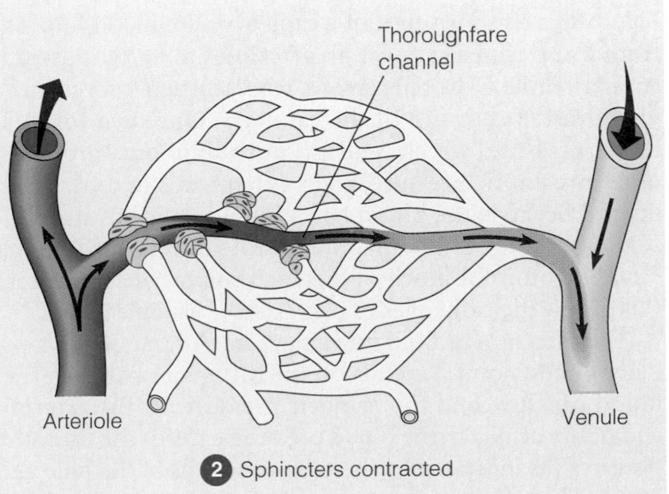

Figure 23.11B The control of capillary blood flow by precapillary sphincters

23.12 Capillaries allow the transfer of substances through their walls

Capillaries are the only blood vessels with walls thin enough for substances to cross between the blood and the interstitial fluid that bathes the body cells. The transfer of materials between the blood and interstitial fluid is the most important function of the circulatory system, so let's take a closer look at it.

Figure 23.12A shows a cross section of a capillary that serves skeletal muscle cells. The capillary wall (tan in the drawing) consists of overlapping epithelial cells that enclose a lumen, or space, that contains the blood. The nucleus you see here belongs to one of the two cells making up this portion of the capillary. (The other cell's nucleus does not appear in this particular cross section.) The blue area around the capillary is a space containing interstitial fluid. (The basement membrane of the capillary wall is not clearly visible in the micrograph; see Module 23.5.)

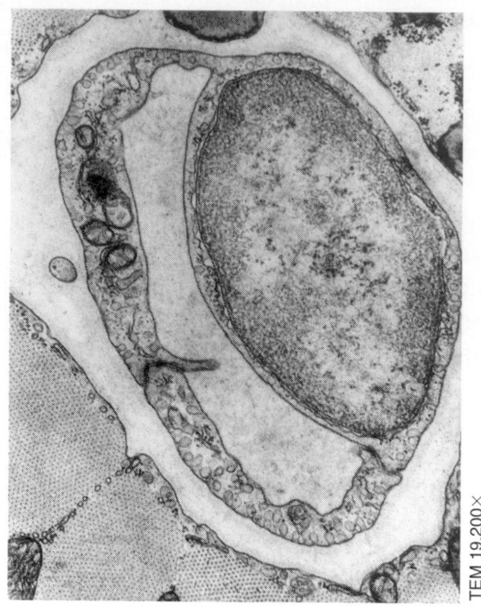

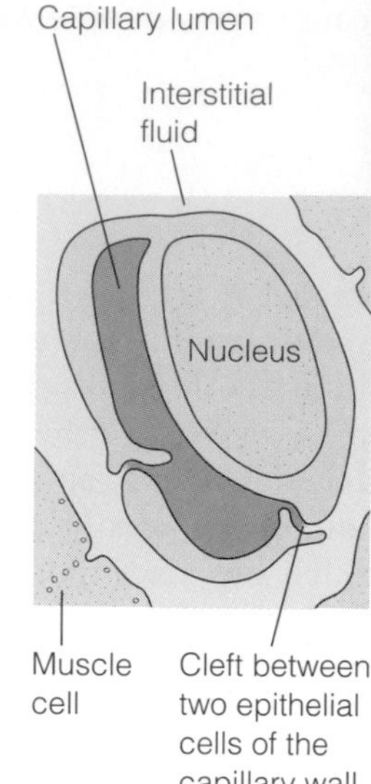

Figure 23.12A A capillary in cross section

The exchange of substances between the blood and the interstitial fluid occurs in several ways. Some substances simply diffuse through the epithelial cells of the capillary wall or are carried in endocytotic vesicles across these cells (see Module 5.19). In addition, the capillary wall is leaky; there are narrow clefts between the epithelial cells making up the wall. Through these clefts, water and small solutes, such as sugars, salts, and oxygen, pass freely. Blood cells and dissolved proteins remain inside the capillary because they are too large to pass.

Figure 23.12B The movement of fluid into and out of a capillary

In addition to the passive diffusion of substances out of the capillary, active forces also push fluid through the leaky capillary wall. The diagram in Figure 23.12B shows part of a capillary with blood flowing from its arterial end (near an arteriole) to its venous end (near a venule). The blue arrows represent active forces driving fluid into or out of the capillary. One such force is blood pressure, which tends to push fluid outward. Another is osmotic pressure, a force that tends to draw fluid inward because the blood has a higher concentration of solutes than the interstitial fluid. Proteins dissolved in the blood account for much of this high solute concentration. (To review the principles of osmosis, see Module 5.15.)

The direction of fluid movement into or out of the capillary at any point depends on the difference between the blood pressure and the osmotic pressure. At the arterial end of the capillary, the blood pressure exceeds the osmotic pressure (as indicated by the relative widths of the blue arrows). Thus, there is a net pressure forcing fluid outward, and more fluid moves out of the capillary than in.

At the venous end of the capillary, the situation is reversed. The blood pressure drops so much in the capillary bed that the osmotic pressure outweighs it, and fluid reenters the capillary.

About 99% of the fluid that leaves the blood at the arterial end of a capillary bed reenters the capillaries at the venous end. The remaining 1% is eventually returned to the blood by the vessels of the lymphatic system, which we discuss in Module 24.3.

Explain how edema, the accumulation of fluid in body tissues, can result from a decrease in blood plasma protein concentration due to severe protein deficiency in the diet.

Decreased protein concentration reduces capillary uptake of fluid from interstitial fluid by reducing the osmotic gradient across the capillary wall.

Blood consists of cells suspended in plasma 23.13

Now that we have examined the structure and function of the heart and blood vessels, let's focus on the composition of blood itself. In an average adult human, the circulatory system contains 4–6 liters of blood. Blood consists of several types of cellular elements suspended in a liquid called **plasma.** When a blood sample is taken, as shown below, the cellular elements, which make up about 45% of the blood, can be separated from the plasma by spinning the sample in a centrifuge. Red and white blood cells and small cell pieces called platelets settle to the bottom of the centrifuge tube, underneath the transparent, straw-colored plasma. **Platelets** are bits of cytoplasm pinched off from large cells in the bone marrow. As we will see in Module 23.16, platelets are important in blood clotting.

Plasma, making up just over half the volume of the blood, is about 90% water. The other 10% is made up of dissolved salts, proteins, and various other substances being transported by the blood. Salts are dissolved in the plasma as inorganic ions. The dissolved ions have several functions, such as maintaining the osmotic balance between the blood and the interstitial fluid and keeping the pH of the blood at about 7.4. Inorganic ions also help regulate the permeability of cell membranes.

Plasma proteins work together with the salts in maintaining osmotic balance and pH. The protein fibrinogen functions with the platelets in blood clotting, as we will see. Another group of plasma proteins, the immunoglobulins, are important in body defense (immunity), which we discuss in Chapter 24.

What about the blood cells shown below? We discuss their functions in the next two modules.

For every white blood cell in normal human blood, there are about ____ to ____ red blood cells.

500 . . . 1000

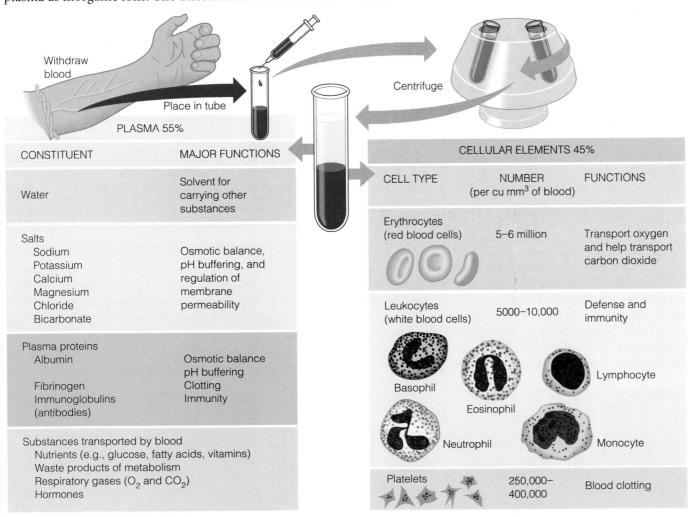

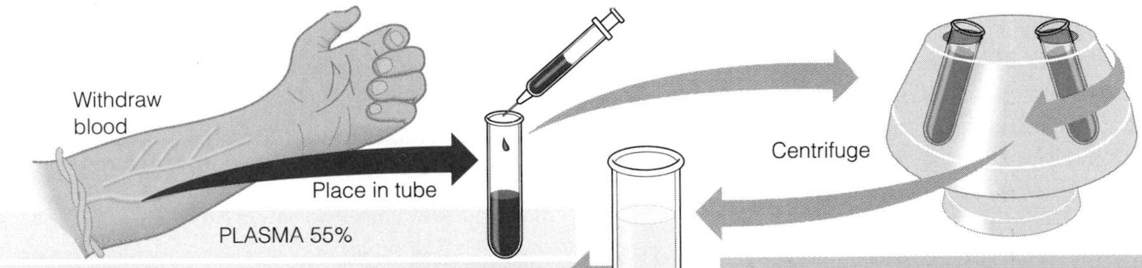

Withdraw blood

Place in tube

Centrifuge

PLASMA 55%

CONSTITUENT	MAJOR FUNCTIONS
Water	Solvent for carrying other substances
Salts Sodium Potassium Calcium Magnesium Chloride Bicarbonate	Osmotic balance, pH buffering, and regulation of membrane permeability
Plasma proteins Albumin Fibrinogen Immunoglobulins (antibodies)	 Osmotic balance pH buffering Clotting Immunity
Substances transported by blood Nutrients (e.g., glucose, fatty acids, vitamins) Waste products of metabolism Respiratory gases (O_2 and CO_2) Hormones	

CELLULAR ELEMENTS 45%

CELL TYPE	NUMBER (per cu mm^3 of blood)	FUNCTIONS
Erythrocytes (red blood cells)	5–6 million	Transport oxygen and help transport carbon dioxide
Leukocytes (white blood cells) Basophil, Eosinophil, Lymphocyte, Neutrophil, Monocyte	5000–10,000	Defense and immunity
Platelets	250,000–400,000	Blood clotting

Figure 23.13 The composition of blood

23.14 Red blood cells transport oxygen

Red blood cells, also called **erythrocytes,** are by far the most numerous blood cells. There are about 25 trillion of these tiny cells in the average person's bloodstream. Our red blood cells and those of other mammals lack nuclei and mitochondria, an unusual characteristic for living cells. These organelles are lost as the cells develop.

The structure of a red blood cell suits its main function, which is to carry oxygen. The micrograph in Figure 23.14 shows that human red blood cells are biconcave disks, thinner in the center than at the sides. A biconcave disk has more surface area for gas exchange than a flat disk or a sphere has. Red blood cells are small relative to other cells—white blood cells, for instance—and their small size also gives them greater total surface area for gas exchange.

As small as a red blood cell is, it contains about 250 million molecules of hemoglobin. As red blood cells pass through the capillary beds of lungs or gills, oxygen diffuses into the red blood cells, where it binds to the iron in hemoglobin. This process is reversed in the capillaries of the systemic circuit, where the hemoglobin unloads its cargo of oxygen to the tissues (see Module 22.10).

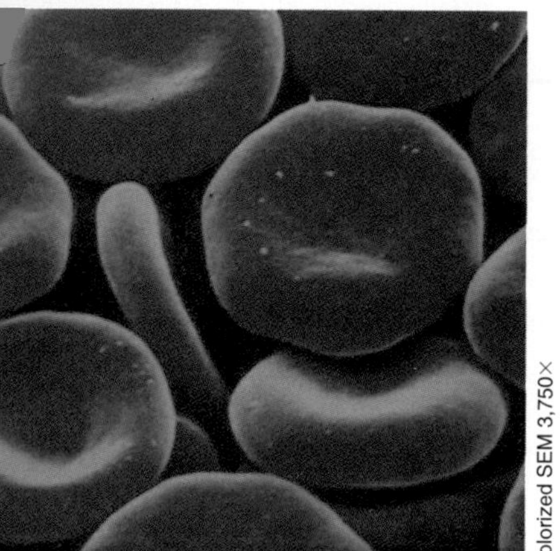

Figure 23.14 Mammalian red blood cells

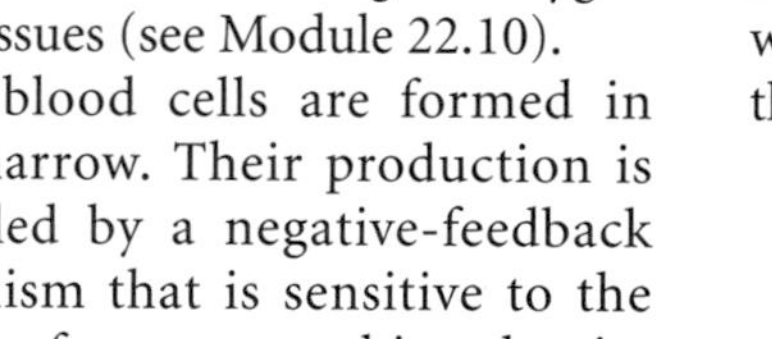

Red blood cells are formed in bone marrow. Their production is controlled by a negative-feedback mechanism that is sensitive to the amount of oxygen reaching the tissues via the blood. If the tissues are not receiving enough oxygen, the kidneys secrete a hormone called erythropoietin that stimulates the bone marrow to produce more red blood cells. On the other hand, if the tissues are receiving more oxygen than they can use, the kidneys stop secreting the hormone, and erythrocyte production slows.

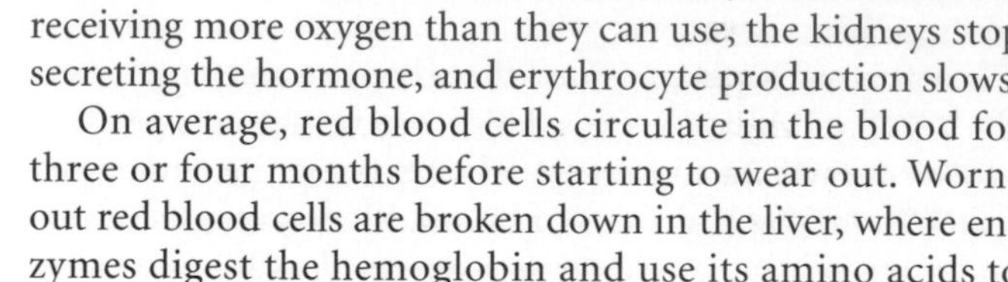

On average, red blood cells circulate in the blood for three or four months before starting to wear out. Worn-out red blood cells are broken down in the liver, where enzymes digest the hemoglobin and use its amino acids to make other proteins. Much of the iron returns to the bone marrow, where it is used in making more red blood cells.

Maintaining adequate amounts of iron, hemoglobin, and red blood cells is essential to the normal functioning of the body. An abnormally low amount of hemoglobin or a low number of red blood cells is a condition called **anemia.** An anemic person feels constantly tired and run down and is often susceptible to infections, because the body cells do not get enough oxygen. Anemia can result from a variety of factors, including excessive blood loss, vitamin or mineral deficiencies, and bone marrow cancer. Iron deficiency is the most common cause. A person may not get enough iron in the diet, or the digestive tract may not absorb enough of it. Women are more likely to develop iron deficiency than men because of blood loss during menstruation. Pregnant women generally benefit from iron supplements to support the developing fetus and placenta.

? (This is a tough one!) About how many red blood cells does the bone marrow of a human produce per day, assuming an average longevity of 4 months for the cells?

About 208 billion (208,000,000,000, or 2.08×10^{11}), calculated by dividing 25 trillion cells (25,000,000,000,000, or 25×10^{12}) by 120 days.

23.15 White blood cells help defend the body

There are five kinds of **white blood cells,** or **leukocytes.** As you can see in the micrographs of stained cells shown in Figure 23.15, the different kinds of leukocytes are distinguished by their staining properties and the shape of their nuclei. As a group, leukocytes fight infections and prevent cancer cells from growing.

Basophils help fight infection by releasing chemicals—for example, histamine. Histamine dilates blood vessels and allows other white blood cells to move out of capillaries into surrounding tissues. Two of the most common white blood cells that move into body tissues are neutrophils and monocytes. These are **phagocytes;** they "eat" bacteria, foreign proteins that enter the body through wounds, and the debris from other body cells that have died. The blood carries large numbers of neutrophils and monocytes to sites of injury, where the cells combat bacte-

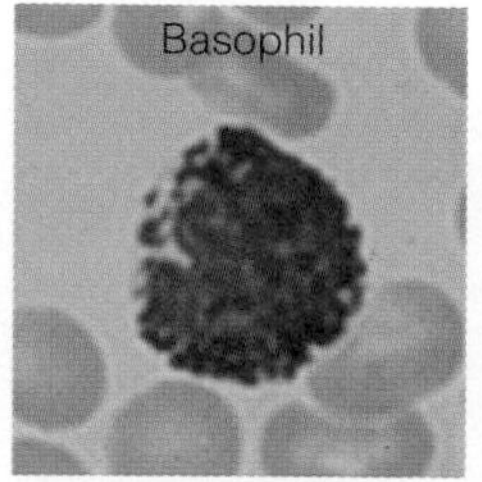

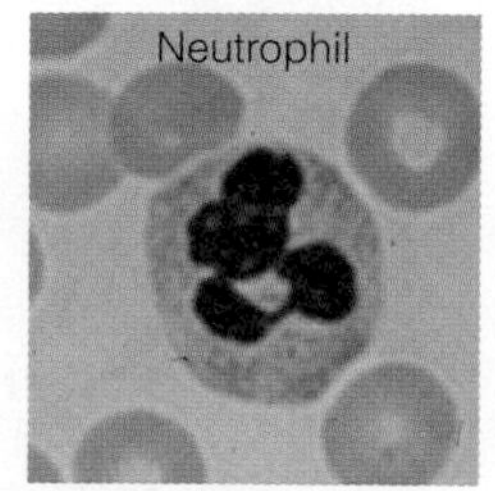

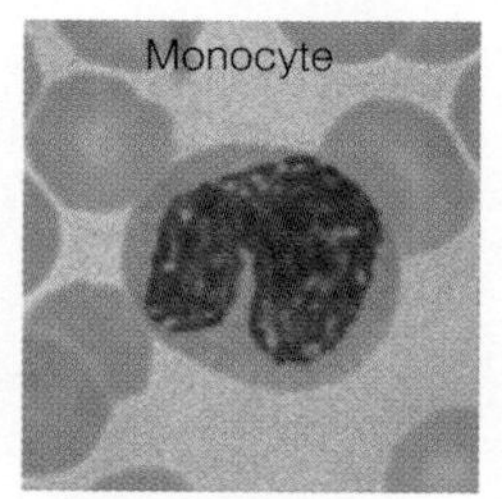

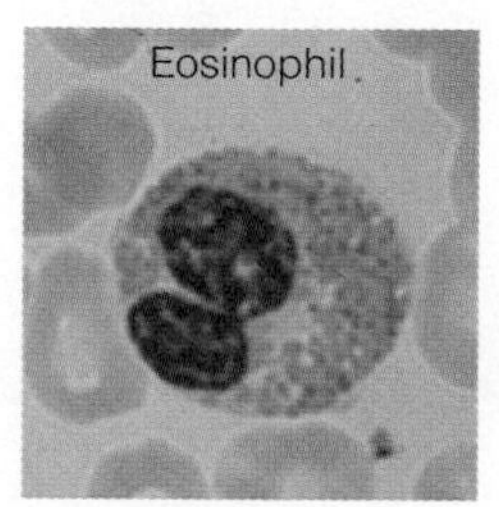

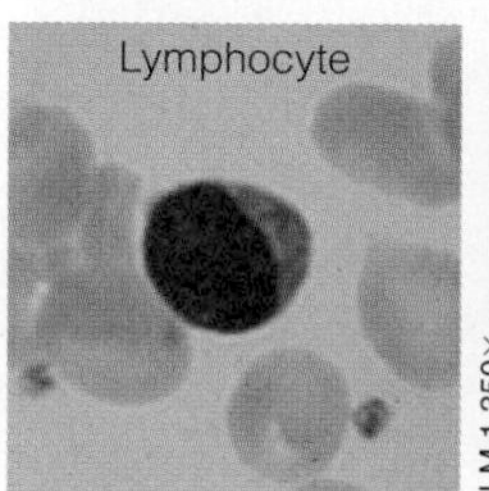

Figure 23.15 White blood cells (leukocytes)

rial infections and help the tissue heal by removing cellular debris.

Less is known about a fourth type of leukocyte, the eosinophil. Eosinophils are phagocytic and seem to be important in fighting infections by parasitic protozoans and worms. They may also help reduce allergy attacks.

The fifth type of leukocyte, the lymphocyte, is the key cell in immunity, the defense against specific invaders. Some lymphocytes produce antibodies, proteins that react against foreign substances. Other lymphocytes fight viruses and cancer cells.

As a group, white blood cells actually spend most of their time outside the circulatory system, moving through interstitial fluid, where most of the battles against infection are waged. There are also great numbers of white cells in the lymphatic system. Like red blood cells, leukocytes arise in bone marrow. Their numbers increase whenever the body is fighting an infection. 

? Among white blood cells are phagocytes, basophils, and ___________, and cells that produce antibodies, the ___________.

neutrophils . . . lymphocytes

Blood clots plug leaks when blood vessels are injured 23.16

We all have cuts and scrapes from time to time, yet we don't bleed to death because our blood contains self-sealing materials that plug leaks in our vessels. The sealants—platelets and the plasma protein **fibrinogen**—are always present in our blood. They are activated to form a clot when a blood vessel is injured.

Figure 23.16A shows the stages of the clotting process. ① It begins when the epithelium (tan) lining a blood vessel is damaged and connective tissue in the wall of the vessel is exposed to blood. Platelets (light purple) respond immediately. Those in the vicinity adhere to the exposed connective tissue and release a substance that makes nearby platelets sticky. ② Soon a cluster of sticky platelets forms a plug that seals minor breaks in the blood vessel.

A platelet plug provides fast protection against blood loss; a tiny wound, such as from a pinprick, might require nothing more. However, when damage is more severe—an open cut, for instance—a chain of reactions is set off that culminates in the formation of a more complex plug called a fibrin clot. First, as shown in the yellow box in Figure 23.16A, clotting factors released by platelets and damaged cells mix with other factors in the plasma. The mixture activates a protein called prothrombin, converting it into the enzyme thrombin. The thrombin then converts fibrinogen into a threadlike protein called **fibrin.** As shown in part ③ of the figure, threads of fibrin (white) trap blood cells. In this way, the injured vessel is sealed until connective tissue forms a permanent patch in it. Figure 23.16B is a micrograph of a fibrin clot.

The clotting mechanism is so important that any defect in it can be life-threatening. In the inherited disease hemophilia, excessive, sometimes fatal, bleeding occurs from even minor cuts and bruises. Another type of defect can lead to a blood clot in the *absence* of injury; such a clot is called a thrombus. If a thrombus is carried in the blood to a coronary artery and is large enough to block it, a heart attack occurs. 

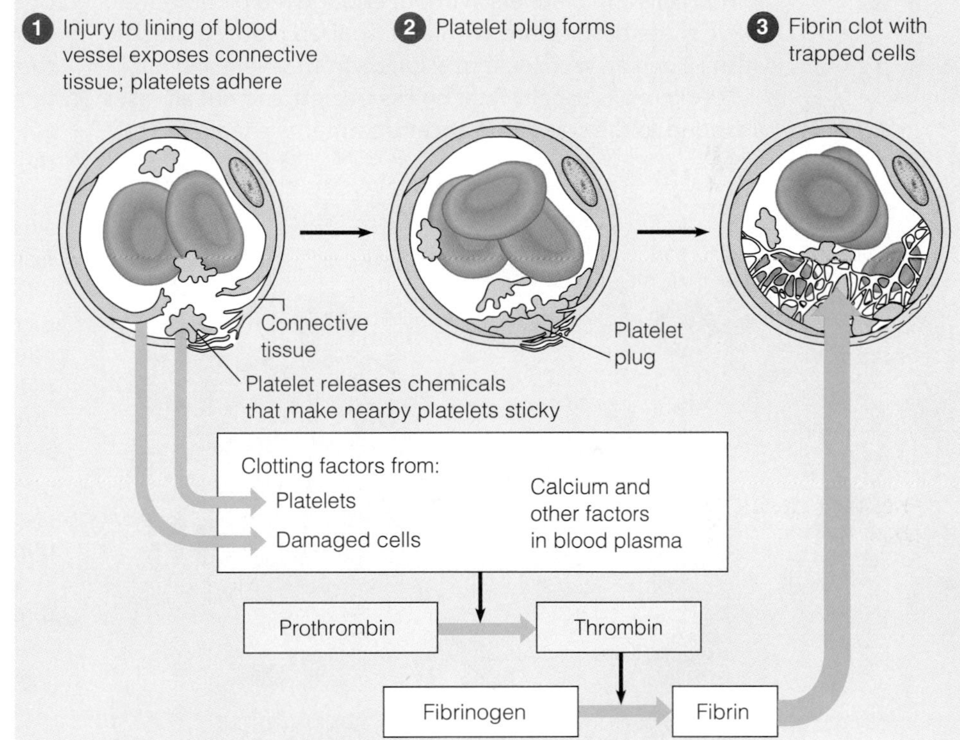

Figure 23.16A The blood-clotting process

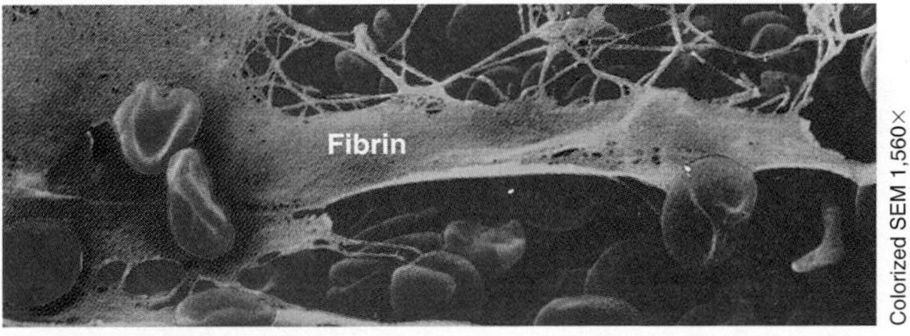

Figure 23.16B A fibrin clot

? What is the advantage of having a stock of fibrinogen in the blood versus what would happen if the body had to synthesize fibrin protein when an injury occurs?

Fibrinogen can be activated to fibrin much faster than fibrin can be synthesized.

23.17 Stem cells offer a potential cure for leukemia and other blood cell diseases

Our white blood cells form a main line of defense against infections and against cancer cells that arise in the body. This cellular defense system is effective but not foolproof. As we saw in Module 10.22, for example, the AIDS virus can overwhelm the body's defenses by infecting and killing white blood cells. In addition, the very tissues that produce defensive cells may become cancerous themselves. White blood cells, as well as red blood cells and platelets, arise from cells in the bone marrow called **stem cells** (see Figure 23.17). When cells that produce white blood cells (leukocytes) become malignant, the resulting cancer is called **leukemia.** A person with leukemia has an unusually high number of leukocytes, many of which are abnormal. Also, the cancerous cells may crowd out cells that make red blood cells and platelets. Without enough red blood cells to oxygenate tissues and with an impaired clotting mechanism, a person with leukemia suffers from severe anemia.

Leukemia is usually fatal unless treated, and not all cases respond to the standard cancer treatments—radiation and chemotherapy. Another approach to leukemia treatment is transplanting healthy bone marrow tissue from a suitable donor into a patient whose own, cancerous marrow has been purposely destroyed. However, such a patient requires lifelong treatment with drugs that suppress the tendency of some of the transplanted cells to "reject" the cells of the recipient. These drugs have the side effect of suppressing immunity to infections. To avoid the rejection problem, patients may be treated with their own bone marrow: Marrow from the patient is removed, processed to remove as many of the cancerous cells as possible, and then injected back into the patient.

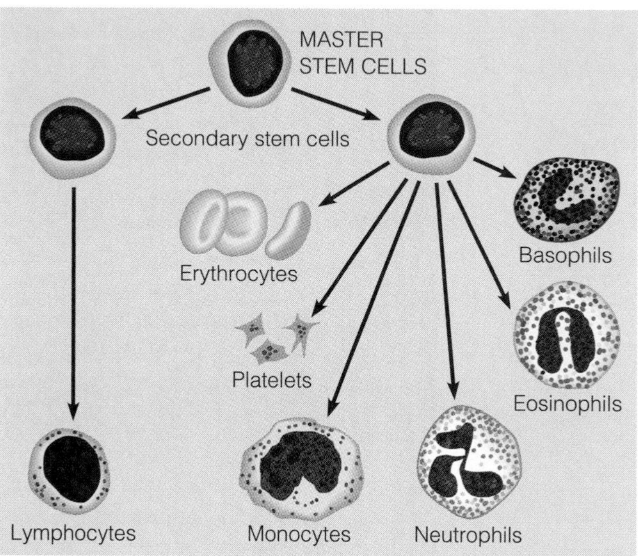

Figure 23.17 The development of blood cells in the bone marrow

A sophisticated variation of this approach is now being used in experimental trials to treat leukemia and certain other diseases of the immune system. Several research groups have developed ways to concentrate and/or isolate the master blood-forming stem cells, the ones that give rise to all of our blood cells and platelets. These stem cells are very rare, only one in several thousand marrow cells, but they are very powerful. Injection of as few as 30 of these cells can completely repopulate an animal's blood and immune system. Because a complete new immune system arises from the injected stem cells, there is no danger of rejection.

Stem cells might also be used to treat genetic blood disorders, such as sickle-cell disease, if some of the stem cells could be genetically engineered before reinjection. Moreover, AIDS, which is caused by a virus (HIV) that affects lymphocytes, may conceivably be fought by stem cells. Researchers are experimenting with inserting HIV-resistance genes into stem cells. The lymphocytes and other white blood cells arising from these genetically altered stem cells may be resistant to HIV.

We pursue the diverse roles of white blood cells in more detail when we study the immune system in Chapter 24.

Why is anemia often a complication of leukemia?

Because the excess of leukocyte-producing stem cells in the marrow is at the expense of erythrocyte-producing stem cells

Chapter Review

Chapter Summary

Most animals have a circulatory system that transports O_2 and nutrients to cells, takes away CO_2 and other wastes, and regulates the makeup of the interstitial fluid in which the cells live **(Introduction).** In many animals, microscopic blood vessels called capillaries form an intricate network among the tissue cells **(23.1).**

Mechanisms of Internal Transport (23.2–23.3) Several means of internal transport have evolved. In cnidarians, the gastrovascular cavity functions in both digestion and internal transport. Most animals have a separate circulatory system, either open or closed. In open systems, a heart pumps blood through open-ended vessels into spaces between the cells. In closed systems, a heart pumps blood through arteries to capillary beds, and the blood returns to the heart via veins. The fish heart pumps blood through arteries to capillaries in the gills; the blood then goes via other arteries to capillaries in other organs and returns to the heart via large veins **(23.2).** In contrast, cardiovascular systems of land vertebrates have a pulmonary circuit, which conveys blood between the heart and gas-exchange tissues, and a systemic circuit, which carries blood between the heart and the rest of the body (23.3).

The Mammalian Cardiovascular System (23.4–23.12) The mammalian heart has two thin-walled atria that pump blood into the ventricles and two thick-walled ventricles that pump blood to all other body organs (23.4). A single layer of epithelial cells and a thin basement membrane form the walls of capillaries. Arteries and veins are thicker and muscular. Valves in large veins prevent backflow (23.5).

The heart's activity follows a rhythmic cycle: During diastole (relaxation), blood flows from the veins into all four heart chambers; during systole, contractions of the atria push blood into the ventricles, and then stronger contractions of the ventricles propel blood into the pulmonary artery and aorta. Cardiac output is the volume of blood pumped into the aorta by the left ventricle each minute. Heart valves prevent the backflow of blood (23.6). The pacemaker (SA node), consisting of specialized muscles in the wall of the right atrium, generates electrical signals that trigger contraction of the atria. The AV node then relays these signals to the ventricles. An electrocardiogram (ECG) is a recording of electrical changes in the skin resulting from the electrical signals in the heart. Control centers in the brain adjust heart rate to body needs (23.7). A heart attack is damage to cardiac muscle that occurs when a coronary artery feeding the heart is blocked, usually by lipid deposits or a blood clot (23.8).

Blood pressure, the force blood exerts on vessel walls, depends on cardiac output and the resistance of vessels. Pressure is highest in the arteries, and it drops as blood flows into arterioles. Blood pressure drops nearly to zero by the time the blood reaches the veins, but muscle contractions, breathing, and one-way valves keep blood moving back to the heart (23.9). Blood pressure measured in the walls of the arteries can reveal cardiovascular problems (23.10).

Smooth muscle regulates blood flow. Muscular constriction of arterioles and precapillary sphincters controls the flow through capillary beds (23.11).

The transfer of materials between the blood and interstitial fluid can occur by leakage through clefts in the capillary walls or by diffusion through the wall. In addition, blood pressure forces fluid out through the walls of a capillary at the arterial end, and osmotic pressure draws fluid in at the venous end (23.12).

Structure and Function of Blood (23.13–23.17) Blood consists of cellular elements in a fluid plasma, which is an aqueous solution of various salts, proteins, nutrients, wastes, gases, and hormones (23.13). Red blood cells (erythrocytes) contain hemoglobin, which enables them to transport O_2 (23.14). White blood cells (leukocytes) function both inside and outside the circulatory system to fight infections and cancer (23.15). When a blood vessel is damaged, platelets (fragments of bone marrow cells) help trigger the conversion of soluble fibrinogen to an insoluble fibrin clot that plugs the leak (23.16). All blood cells develop from stem cells in bone marrow; such cells may prove valuable for treating certain blood disorders (23.17).

Testing Your Knowledge

Multiple Choice

1. Blood pressure is highest in ____, and blood moves most slowly in ____.
   a. veins . . . capillaries
   b. arteries . . . capillaries
   c. veins . . . arteries
   d. capillaries . . . arteries
   e. arteries . . . veins

2. When the doctor listened to Janet's heart, he heard "lub-hisss, lub-hiss" instead of the normal "lub-dupp" sounds. The hiss is most likely due to ____. *(Explain your answer.)*
   a. a clogged coronary artery
   b. a defective atrioventricular (AV) valve
   c. a damaged pacemaker
   d. a defective semilunar valve
   e. high blood pressure

3. Which of the following is the biggest difference between your cardiovascular system and the cardiovascular system of a fish?
   a. In a fish, blood is oxygenated by passing through a capillary bed.
   b. Your heart has two chambers; a fish heart has four.
   c. Your circulation has two circuits; fish circulation has one circuit.
   d. Your heart chambers are called atria and ventricles.
   e. Yours is a closed system; the fish's is an open system.

4. Paul's blood pressure is 125/80. The 125 indicates ______, and the 80 indicates ______.
   a. pressure in the left ventricle . . . pressure in the right ventricle
   b. arterial pressure . . . heart rate
   c. pressure during ventricular contraction . . . pressure during heart relaxation
   d. systemic circuit pressure . . . pulmonary circuit pressure
   e. pressure in the arteries . . . pressure in the veins

5. Which of the following *initiates* the process of blood clotting?
   a. damage to the lining of a blood vessel
   b. exposure of blood to the air
   c. conversion of fibrinogen to fibrin
   d. attraction of leukocytes to a site of infection
   e. conversion of fibrin to fibrinogen

Describing, Comparing, and Explaining

1. Trace the path of blood starting in a pulmonary vein, through the heart and around the body, returning to the pulmonary vein. Name, in order, the heart chambers and types of vessels through which the blood passes.

2. Explain how the structure of capillaries relates to their function of exchanging substances with the surrounding interstitial fluid.

Thinking as a Scientist

Juan has a disease in which damaged kidneys allow some of his normal plasma proteins to be removed from the blood. How might this condition affect the osmotic pressure of blood in capillaries, compared to the surrounding interstitial fluid? One of the symptoms of this kidney malfunction is an accumulation of excess interstitial fluid, which causes Juan's arms and legs to swell. Can you explain why this occurs?

Science, Technology, and Society

Recently, a 19-year-old woman received a bone marrow transplant from her 1-year-old sister. The woman was suffering from a deadly form of leukemia and was almost certain to die without a transplant. Their parents had decided to have another child in a final attempt to provide their daughter with a matching donor. Although the ethics of the parents' decision were criticized, doctors reported that this situation is not uncommon. In your opinion, is it acceptable to have a child in order to provide an organ or tissue donation? Why or why not?

Media Resources

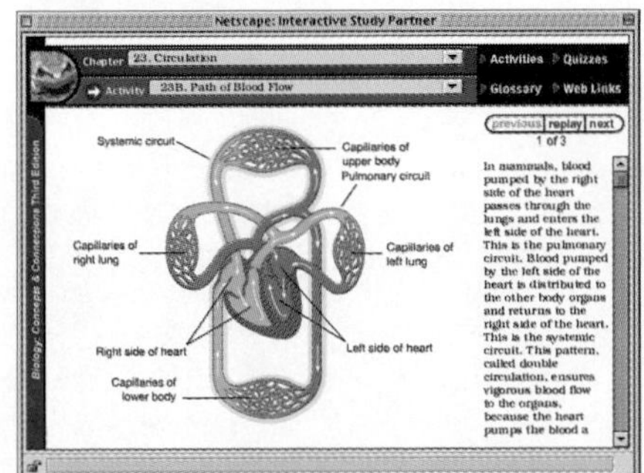

For further review of the concepts in this chapter, use the activities and quizzes on the CD-ROM.

Visit the Special Edition of The Biology Place web site for activities, quizzes, news articles, and web links.

24 The Immune System

Outline

Nonspecific Defenses Against Infection

First identified in the United States in 1981, AIDS (acquired immune deficiency syndrome) is now epidemic throughout much of the world. Globally, about 16,000 people are infected with the AIDS virus each day, and in the U.S. today, over a million people have tested positive for HIV. One of those is Earvin (Magic) Johnson, basketball superstar for twelve years with the Los Angeles Lakers. About a month after he was diagnosed, Johnson told of his initial shock in a *Sports Illustrated* interview (November 18, 1991):

> [On] October 25 [1991], as I walked into the offices of [my physician] Dr. Michael Mellman, I was more curious than worried. I had been rejected for a life-insurance policy. . . .
>
> "Earvin, sit down. I have your test results," Dr. Mellman said. "You're HIV-positive. You have the AIDS virus."
>
> Suddenly I felt sick. I was numb. In shock. And, yes, I was scared. Dr. Mellman quickly told me that I didn't have AIDS, that I was only infected with the virus that could someday lead to the disease. But I didn't really hear him. Like almost everyone else who has not paid attention to the growing AIDS epidemic in the U.S. and the rest of the

world, I didn't know the difference between the virus and the disease. While my ears heard HIV-positive, my mind heard AIDS.

As we discussed in Module 10.21, HIV, or human immunodeficiency virus, attacks the immune system, the body's main defense against infectious diseases and cancer. HIV enters the cells of the immune system and may lurk there for years, safe from the body's attempts to fight back. As Magic Johnson points out, testing positive for HIV is different from having AIDS, although, tragically, it appears that nearly all people with HIV will eventually develop AIDS.

Once HIV starts to multiply, it kills its host cells, infects other cells, and eventually destroys the body's ability to fight even the mildest infections. It may take ten years or more for AIDS to develop after the initial HIV infection. In someone with AIDS, the immune system may be totally undermined, and diseases that it ordinarily fights off become deadly. Most AIDS patients die from other infectious diseases or from certain types of cancer, usually within three years of an AIDS diagnosis.

The AIDS virus is transmitted mainly in blood and semen. Most often, it enters the body through imperceptible wounds during sexual contact or via needles contaminated with infected blood. Routine screening of donated blood has greatly reduced the risk from blood transfusions, but the sharing of needles to inject drugs remains a major source of infection. Unfortunately, many people in developed countries still think of AIDS as a disease confined to drug addicts and homosexuals. However, the rate of new infections continues to rise among heterosexual people who do not inject drugs, including teenagers and college students. Worldwide, about 5.8 million people were newly infected with HIV in 1998—44,000 in North America—and anyone who has sex with a person who has had any chance of exposure to HIV in the past 20 years risks exposure to the virus.

Soon after being diagnosed as HIV-positive, Magic Johnson retired from the Lakers. However, he made brief comebacks in 1992 and 1996 and starred in the 1992 Olympics. Johnson has become a strong advocate for safer sex practices. He believes he acquired HIV and may have infected many others during the years before he was married, when he was sexually promiscuous and did not always use condoms. Condoms do, in fact, minimize the AIDS risk, but they do not eliminate it. Not having sex is the best defense against the AIDS virus; for a sexually active person, using condoms and having only one, monogamous partner can be equally effective.

The AIDS story demonstrates how much we depend on our body's built-in defense systems. As we will see in this chapter, our immune system is a very specific defense system, in that it recognizes an invader, such as a virus, and then produces large numbers of cells that combat that particular agent. In a healthy individual, this specific defense backs up several mechanisms of nonspecific resistance, which provide nondiscriminating protection against a variety of invaders. Following a brief look at our nonspecific defenses in the first three modules, we concentrate on the mechanisms of immune defense. At the end of the chapter, we return to AIDS and examine how the insidious HIV can defeat the body's defenses.

Nonspecific Defenses Against Infections

24.1 Nonspecific defenses against infection include the skin and mucous membranes, phagocytic cells, and antimicrobial proteins

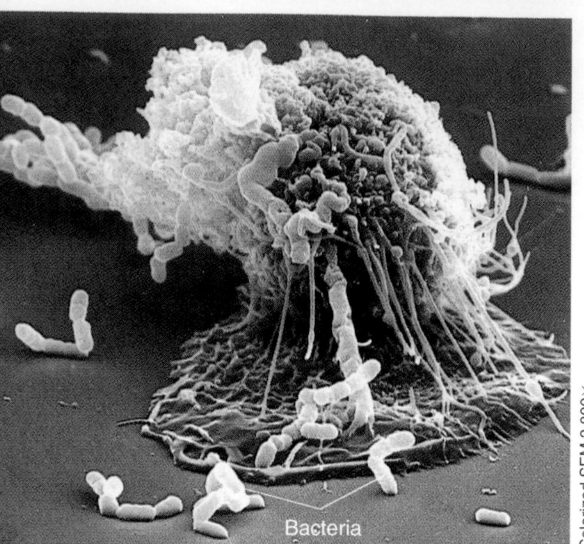

Figure 24.1A Phagocytosis of bacteria by a macrophage

The body's first lines of defense against infection are nonspecifc, meaning that they do not distinguish one invader from another. The skin plays an important role in nonspecific resistance to infection. The outer layer of intact skin is a tough barrier of dead cells that most bacteria and viruses cannot penetrate. Moreover, acids in sweat and oils secreted by glands in the skin inhibit the growth of many microorganisms. Sweat, saliva, and tears also contain lysozyme, an enzyme that attacks the cell walls of many bacteria.

Two organ systems that open to the external environment—the digestive and respiratory systems—are guarded by the mucous membranes that line them (see Module 20.4) and have other nonspecific defenses as well. Stomach acid kills most bacteria swallowed with food. Guarding the respiratory route, the hairs in our nostrils filter incoming air, and mucus in our respiratory tubes traps most microbes and dirt that get past the nasal filter. Cilia on cells lining the tubes sweep the mucus upward and out of the system.

Microbes that do penetrate the skin or enter the tissues of the digestive or respiratory system are soon confronted by nonspecific defensive cells. These are all classified as white blood cells (see Module 23.15), although they are found in interstitial fluid as well as blood vessels. **Neutrophils** and **monocytes** are phagocytic white blood cells; they engulf bacteria and viruses in infected tissues. **Macrophages** ("big eaters") are large phagocytic cells that develop from monocytes. Macrophages wander actively in the interstitial fluid, eating any bacteria and virus-infected cells they encounter. In Figure 24.1A, a single large macrophage is using multiple extensions (pseudopodia) to snare small bacteria. Other white blood cells, called **natural killer cells,** attack cancer cells and infected body cells, especially those harboring viruses.

Other nonspecific defenses include antimicrobial proteins that either attack microoganisms directly or impede their reproduction. Especially important are interferons and the complement proteins.

Interferons are proteins produced by virus-infected cells that help other cells resist viruses. Figure 24.1B shows how the interferon mechanism seems to work. ① The virus infects a cell and ② turns on interferon genes in the cell's nucleus, ③ causing the cell to make interferon. The infected cell then dies, but ④ its interferon molecules may diffuse to neighboring healthy cells, ⑤ stimulating them to produce other proteins that inhibit viral reproduction. The defense is not virus-specific; interferon made in response to one virus confers resistance to unrelated viruses. The resistance is short-term and seems to be most effective against viruses that cause influenza and the common cold.

Interferons have been in the news fairly regularly since they were discovered in the late 1950s. The body makes these protective proteins in very small quantities, but recombinant DNA technology has made it possible to produce amounts large enough for use in treating viral infections. Also, in some experimental studies, interferons have been useful in treating certain cancers.

Antimicrobial proteins of another sort, called **complement proteins,** circulate in inactive form in the blood plasma. Named for their cooperation (complementation) with other defense mechanisms, complement proteins are activated by the immune system or by microbes. Some of these proteins coat the surfaces of microbes, making them

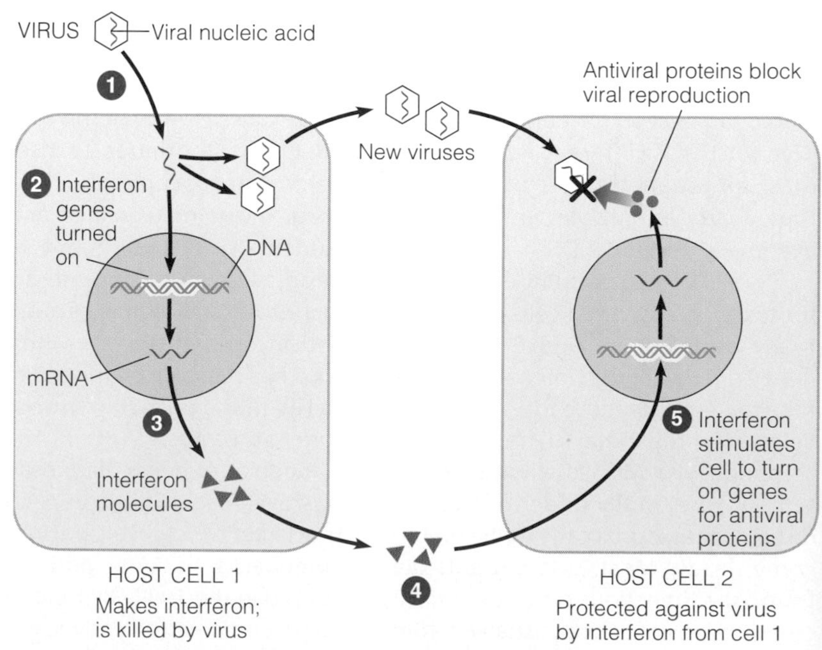

Figure 24.1B The interferon mechanism against viruses

easier for macrophages to engulf. Other complement proteins cut lethal holes in microbial membranes. Also, complement proteins amplify another nonspecific defense, the inflammatory response.

? What do interferons interfere with?

The replication of viruses that have infected cells of the body

The inflammatory response mobilizes nonspecific defense forces 24.2

The **inflammatory response** is a major component of our nonspecific defense system. Any damage to tissue, whether caused by microorganisms or by physical injury, even just a scratch or insect bite, triggers this response. You can see signs of the inflammatory response if you watch your skin react to a mosquito bite. The bite area becomes red, swollen, and warmer than the surrounding area. This reaction is inflammation, which literally means "setting on fire."

The figure below shows the chain of events that make up the inflammatory response, in a case where a pin has broken the skin and infected it with bacteria. ① The damaged cells soon release chemical alarm signals, such as **histamine.** ② The chemicals spark the mobilization of various defenses. Histamine, for instance, induces neighboring blood vessels to dilate and become leakier. Blood flow to the damaged area increases, and blood plasma passes out of the leaky vessels into the interstitial fluid of the affected tissues. Other chemicals attract phagocytes and other white blood cells to the area. Squeezing between the cells of the blood vessel wall, these white cells (yellow in the figure) migrate out of the blood into the tissue spaces. The local increase in blood flow, fluid, and cells produces the redness, heat, and swelling characteristic of inflammation.

The major results of the inflammatory response are to disinfect and clean injured tissues. ③ The white blood cells mustered into the area engulf bacteria and the remains of any body cells killed by them or by the physical injury. Many of the white cells die in the process, and their remains are also engulfed and digested. The pus that often accumulates at the site of an injury or infection consists mainly of dead white cells and fluid that has leaked from the capillaries during the inflammatory response.

The inflammatory response also helps prevent the spread of infection to surrounding tissues. Clotting proteins (see Module 23.16) present in blood plasma pass into the interstitial fluid during inflammation. Along with platelets, these substances form local clots that help seal off the infected region and allow repair of the damaged tissue to begin. Healing is now under way.

The inflammatory response may be localized, as we have just described, or widespread (systemic). Sometimes microorganisms such as bacteria or protozoans get into the blood or release toxins that are carried throughout the body in the bloodstream. The body may react with one or several inflammatory weapons. For instance, the number of white blood cells circulating in the blood may increase. Another response is fever, an abnormally high body temperature. Toxins themselves may trigger the fever, or certain white blood cells may release compounds that set the body's thermostat at a higher temperature. A very high fever is dangerous, but a moderate one may contribute to defense by stimulating phagocytosis and inhibiting the growth of many kinds of microorganisms.

Why is the inflammatory response considered to be *nonspecific?*

Because the response is the same regardless of the cause of tissue damage

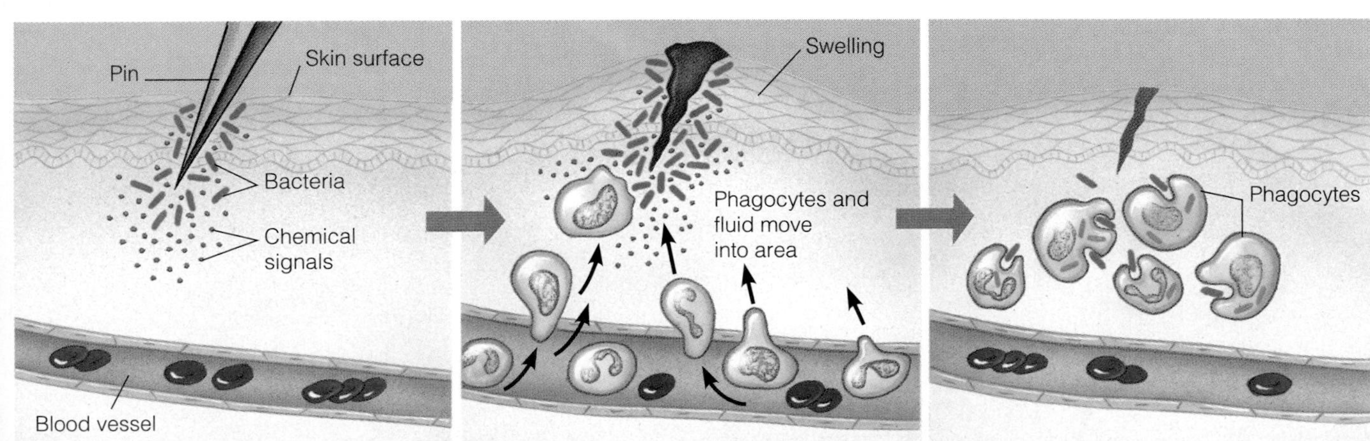

❶ Tissue injury; release of chemical signals such as histamine

❷ Dilation and increased leakiness of local blood vessels; migration of phagocytes to the area

❸ Phagocytes (macrophages and neutrophils) consume bacteria and cell debris; tissue heals

Figure 24.2 The inflammatory response

24.3 The lymphatic system becomes a crucial battleground during infection

Involved in both nonspecific and specific resistance to infection, the **lymphatic system** (Figure 24.3A) consists of a branching network of vessels, numerous lymph nodes (saclike organs packed with the white blood cells called lymphocytes), the thymus, the tonsils (including the adenoids), the appendix, the spleen, and the bone marrow. The lymphatic vessels carry a fluid called **lymph,** which is similar to interstitial fluid but contains less oxygen and fewer nutrients. The lymphatic system has two main functions: to return tissue fluid to the circulatory system, and to fight infection.

As we noted in Module 23.12, about 1% of the fluid that enters the tissue spaces from the blood in a capillary bed does not reenter the blood capillaries. Instead, this small amount of fluid is returned to the blood via lymphatic vessels. The enlargement in Figure 24.3B shows a branched lymphatic vessel in the process of taking up fluid from tissue spaces in the skin. As shown here, fluid enters the lymphatic system by diffusing into tiny, dead-end lymphatic capillaries that are intermingled among the blood capillaries.

Lymph drains from the lymphatic capillaries into larger and larger lymphatic vessels. It reenters the circulatory system via two large lymphatic vessels, which fuse with veins in the shoulders. These large lymphatic vessels are the thoracic duct and the right lymphatic duct, which you can see in Figure 24.3A. As Figure 24.3B indicates, the lymphatic vessels resemble veins in having valves that prevent the backflow of fluid toward the capillaries. Also, like veins, lymphatic vessels depend mainly on the movement of skeletal muscles to squeeze their fluid along. The black arrows in Figures 24.3B and 24.3C indicate the flow of lymph.

The infection-fighting activities of the lymphatic system occur in the lymph nodes and other lymphatic organs shown in Figure 24.3A. As shown in Figure 24.3D for a lymph node, these organs are densely packed with lymphocytes (smaller cells) and macrophages (larger cells). Lymph circulates through the lymphatic organs, carrying microbes from infection sites in the body, and sometimes cancer cells as well. Within the lymphatic organs, macrophages may engulf the invaders in a nonspecific fashion, and lymphocytes may be activated to mount a specific immune response against them.

Thus, when your body is fighting an infection, the lymphatic system becomes a major battleground. You can see evidence of the cellular maneuvers on this battleground

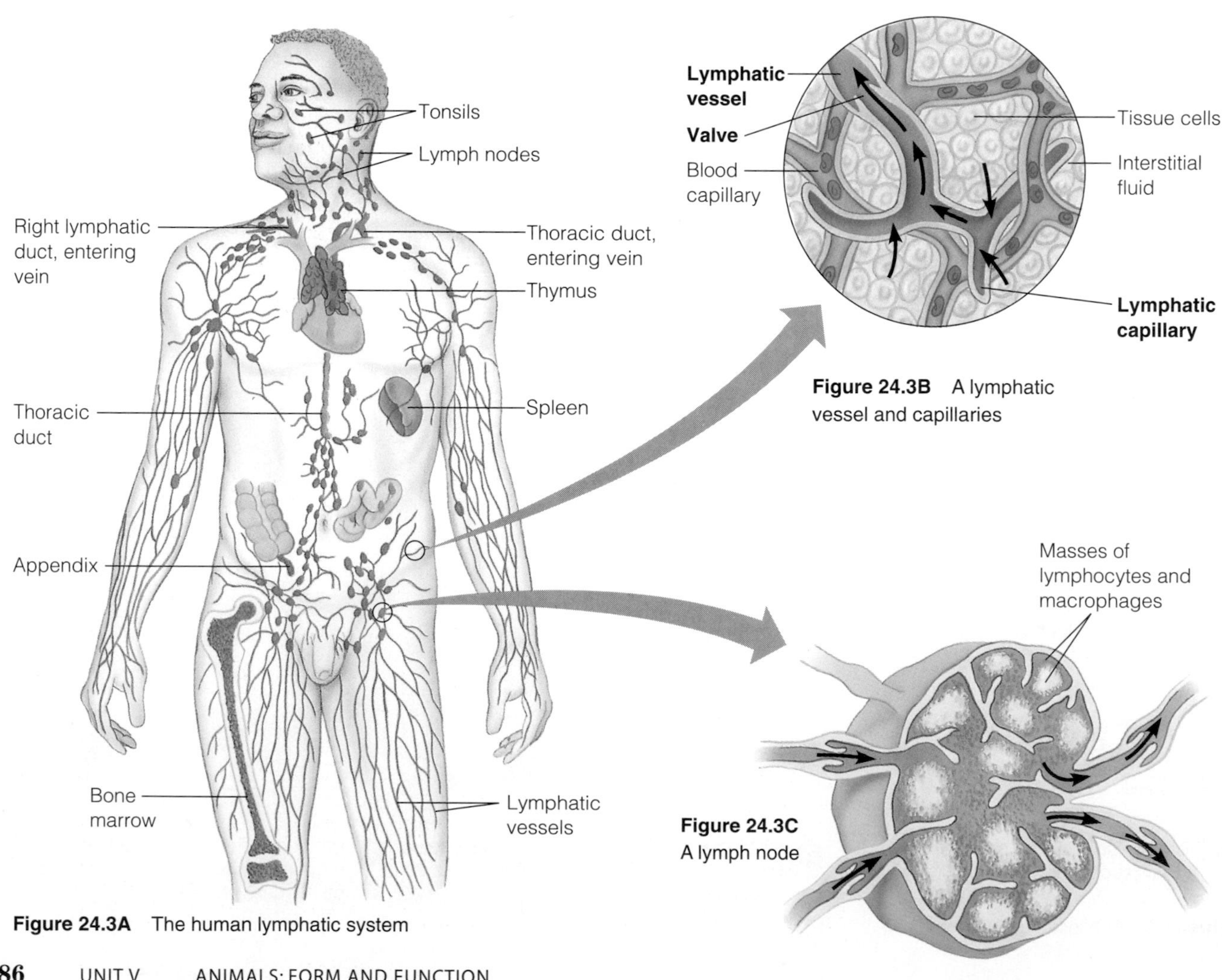

Figure 24.3A The human lymphatic system

Figure 24.3B A lymphatic vessel and capillaries

Figure 24.3C A lymph node

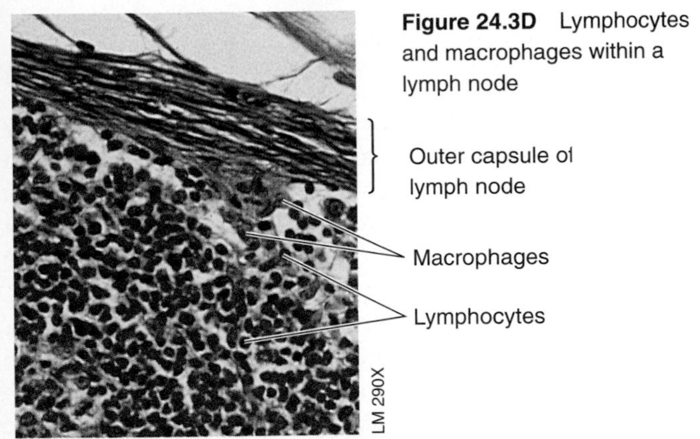

Figure 24.3D Lymphocytes and macrophages within a lymph node

when your lymph nodes become swollen and tender, mainly as a result of the multiplication of defensive lymphocytes. Swollen lymph nodes, often referred to as "swollen glands," are often most obvious in the neck. To learn how lymphocytes are activated to combat an infection, we next examine the specific resistance to infection provided by the immune system. 

? Why do your lymph nodes swell when you are fighting certain kinds of infections?

Because of the proliferation of lymphocytes in the nodes

Specific Immunity

24.4 The immune response counters specific invaders

When our nonspecific defenses fail to ward off an infectious agent, the **immune system** provides another line of defense. The immune system recognizes and defends against invading microbes and against cancer cells (which the body usually identifies as foreign). Our immune system often acts more effectively than nonspecific resistance. Moreover, it can amplify certain nonspecific responses, such as inflammation and the complement reactions.

Whereas our nonspecific defenses are always ready to fight a variety of infections, the immune response must be primed by the presence of a foreign substance, called an antigen. When the immune system detects an antigen, it responds with an increase in the number of cells that either attack the invader directly or produce defensive proteins called antibodies. The defensive cells and antibodies produced against that antigen are usually ineffective against any other foreign substance.

An **antigen** is defined as a molecule that elicits an immune response. (The word "antigen" is a contraction of "*anti*body-*gen*erating," a reference to the fact that the foreign agent provokes the immune response.) Antigens include certain molecules on the surfaces of viruses, bacteria, mold spores, cancer cells, pollen, and house dust, as well as molecules on the cell surfaces of transplanted organs. An **antibody** is a protein found in blood plasma that attaches to one particular kind of antigen and helps counter its effects.

The immune system is extremely specific, and it has a remarkable "memory." It can "remember" antigens it has encountered before and react against them more promptly and vigorously on second and subsequent exposures. For example, if a person gets rubella (German measles), the immune system remembers certain molecules on the virus that causes this disease. The person is then immune to reinfection because the body will recognize and destroy the rubella virus before it can produce symptoms of illness. Thus, the immune response, unlike nonspecific defenses, is adaptive; exposure to a particular foreign agent enhances future response to that same agent.

The term **immunity,** in the context of our immune system, means resistance to *specific* invaders. Immunity is usually acquired by natural infection, but it can also be achieved by the procedure known as **vaccination.** In this procedure, the immune system is confronted with a **vaccine** composed of a harmless variant of a disease-causing microbe. The vaccine stimulates the immune system to mount defenses against this variant, defenses that will also be effective against the actual pathogen because it has similar antigens. Once we have been successfully vaccinated, our immune system will respond quickly if it is exposed to the microbe. Vaccination has been particularly effective in combating viral diseases, including smallpox, polio, mumps, and measles. Researchers are trying to develop a vaccine for AIDS, but so far results are not encouraging.

Whether antigens enter the body naturally (if you catch the flu) or artificially (if you get a flu shot), the resulting immunity is called **active immunity,** because the body is stimulated to produce antibodies in its own defense. It is also possible to acquire **passive immunity.** For example, a fetus obtains antibodies by transfer from its mother's bloodstream to its own, and travelers sometimes get a shot containing antibodies to pathogens they are likely to encounter (rather than a vaccine supplying antigens). Both the fetus and the travelers have acquired antibodies passively. This immunity is temporary because the person's immune system is not stimulated by antigens; the antibodies remain effective for only a few weeks or months. 

? Why is protection resulting from a vaccination considered to be *active* immunity rather than *passive* immunity?

Because the body itself produces the immunity by mounting an immune response and generating antibodies, even though the stimulus consists of artificially introduced antigens

24.5 Lymphocytes mount a dual defense

Lymphocytes, white blood cells that spend most of their time in the tissues and organs of the lymphatic system, produce the immune response. Like all blood cells, lymphocytes originate from stem cells in the bone marrow (see Module 23.17). Some immature lymphocytes (left side of the figure below) continue developing in the bone marrow; these become specialized as B lymphocytes, or **B cells.** Other immature lymphocytes (right side) are carried by the blood from the bone marrow to the thymus, a gland in the upper chest region. There the lymphocytes become specialized as T lymphocytes, or **T cells.** Both B cells and T cells eventually make their way via the blood to the lymph nodes and other lymphatic organs.

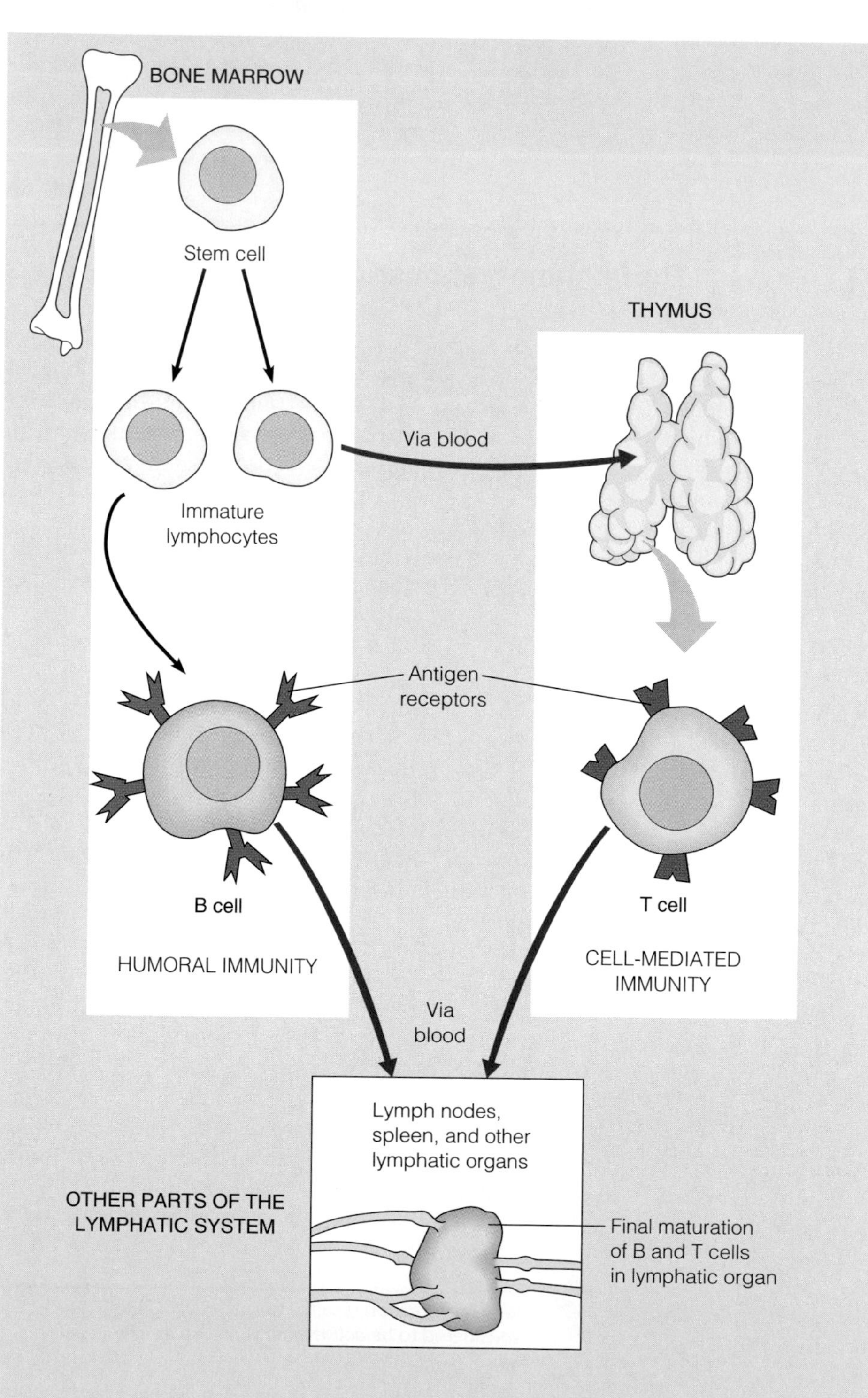

Figure 24.5 The development of B cells and T cells

The B cells and T cells of our immune system mount a dual defense. The B cells secrete antibodies, and because antibodies become dissolved in the blood, immunity conferred by B cells is called **humoral immunity.** (The body fluids—blood, lymph, and interstitial fluid—were formerly called "humors.") The humoral system defends primarily against bacteria and viruses present in body fluids. Antibodies are carried in the lymph and blood to sites of infection wherever they occur in the body. Humoral immunity can be passively transferred by injecting blood plasma (containing antibodies) from an immune individual into a nonimmune individual.

The second type of immunity, produced by the T cells, is called **cell-mediated immunity.** In contrast to humoral immunity, cell-mediated immunity cannot be transferred passively with plasma. Cell-mediated immunity can be passively transferred only by giving actual T cells from an immune individual to a nonimmune one. T cells circulate in the blood and lymph, attacking body cells that have been infected with bacteria or viruses. T cells also work against infections caused by fungi and protozoans and are thought to be important in protecting the body from its own cells if they become cancerous. In addition, T cells function indirectly by promoting phagocytosis by other white blood cells and by stimulating B cells to produce antibodies. Thus T cells are involved in both cell-mediated and humoral immunity.

When a T cell develops in the thymus or a B cell develops in bone marrow, certain genes in the cell are turned on, and the cell synthesizes molecules of a specific protein and builds them into its plasma membrane. As indicated in Figure 24.5, these protein molecules (shown in dark purple) stick out from the cell's surface. The molecules are **antigen receptors,** capable of binding one specific type of antigen. In the case of a B cell, the receptors are actually molecules of the particular antibody that the B cell will secrete. Once a B cell or T cell has its surface proteins in place, it can recognize a specific antigen and mount an immune response against it. One cell may recognize an antigen on the mumps virus, for instance, while another detects a particular antigen on a tetanus-causing bacterium.

We see in the figure that after the B cells and T cells have developed their antigen receptors, these lymphocytes leave the bone marrow and thymus, and move via the bloodstream to the lymph nodes, spleen, and other parts of the lymphatic system. In these organs, many B and T cells take up residence and encounter infectious agents that have penetrated the body's outer defenses. Because lymphatic capillaries extend into virtually all the tissues of the body (see Figure 24.3A), bacteria or viruses infecting nearly any part of the body eventually enter the lymph and are carried to the lymphatic organs. As we will describe in Module 24.7, when a mature B or T cell within a lymphatic organ first confronts the specific antigen it is programmed to recognize, it differentiates further and becomes a fully mature component of the immune system.

An enormous diversity of B cells and T cells develops in each individual. Researchers estimate that each of us has between 100 million and 100 billion different kinds—enough to recognize and bind virtually any kind of antigen we would ever encounter. A small population of each kind of lymphocyte lies in wait in our body, genetically programmed to recognize and respond to a specific antigen. A key feature of our immune system is its preparedness for an almost unlimited variety of potentially harmful antigens. 

? Contrast the targets of humoral immunity with those of cell-mediated immunity.

Humoral immunity works against pathogens in the body fluids; cell-mediated immunity attacks infected cells or abnormal cells (cancer cells, for example).

24.6 Antigens have specific regions where antibodies bind to them

As molecules that elicit the immune response, antigens usually do not belong to the host animal. Most antigens are proteins or large polysaccharides on the surfaces of viruses or foreign cells. Common examples are protein-coat molecules of viruses, parts of the capsules and cell walls of bacteria, and macromolecules on the surface cells of other kinds of organisms, such as protozoans and parasitic worms. Sometimes a particular microbe is called an antigen, but this usage is misleading because the microbe will almost always have several kinds of antigenic molecules. Blood cells or tissue cells from other individuals (of the same species or a different species) can also provide antigenic molecules. Foreign macromolecules dissolved in body fluids sometimes act as antigens.

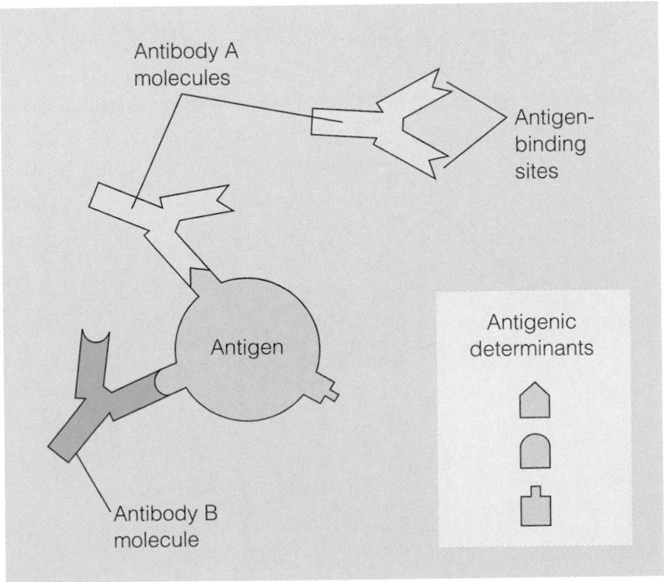

Our immune system can recognize millions, perhaps billions, of different antigens, as noted in Module 24.5. In fact, we have so many different kinds of B and T cells that many of them never encounter the antigen they would recognize and simply remain idle throughout our lives. As shown in Figure 24.6, antibodies usually identify localized regions, called **antigenic determinants,** on the surface of an antigen molecule. An antigen-binding site, a specific region on the antibody molecule, recognizes an antigenic determinant by the fact that the binding site and antigenic determinant have complementary shapes, like an enzyme and substrate, or a lock and key. An antigen usually has several different determinants (there are three in the diagram here), so different antibodies (two, in this case) can bind to the same antigen. A single antigen molecule may stimulate the immune system to make several distinct antibodies against it. Notice that each antibody molecule has two identical antigen-binding sites. We'll return to antibody structure in Module 24.10. 

Antibody A molecules
Antigen-binding sites
Antigen
Antigenic determinants
Antibody B molecule

Figure 24.6 The binding of antibodies to antigenic determinants

? Why is it inaccurate to refer to a pathogen such as a virus as an antigen?

Because antigens are not whole pathogens; they are molecules, which may be chemical components of a pathogen's surface

24.7 Clonal selection musters defensive forces against specific antigens

The immune system's ability to defend against an almost infinite variety of antigens depends on a process called **clonal selection.** At first, an antigen introduced into the body activates only a tiny number of lymphocytes. These "selected" cells then proliferate, forming a clone of cells (a population of genetically identical B cells or T cells) that are specific for the stimulating antigen.

The figure below indicates how clonal selection works. The row of cells at the top represents a diverse population of B cells in a lymph node. Each B cell has its own specific type of antigen receptor embedded in its surface. The cells have their receptors in place before they ever encounter an antigen. The surface of each cell has multiple copies of its particular receptor (actually about 100,000, though only a few are shown here). Because the cells shown in this figure are B cells, the receptors are antibody molecules.

Once an antigen enters the body and is swept into the lymph node, it binds with the receptors that fit it. Binding of the antigen activates the lymphocyte. The figure shows only one cell (with purple receptors) being activated. In real life, a larger number of cells would respond—but still only a small fraction of the cells in the body's lymph nodes. Other lymphocytes, without the appropriate binding sites, would not be affected.

The rest of the figure shows what happens after antigen molecules bind with the specific receptors on one of the B cells. Primed by the interaction with the antigen, the selected cell grows, divides, and differentiates further. The result is a clone of effector cells specialized for defending against the very antigen that triggered the response. The effector cells shown here secrete antibody molecules, all of the same specific type. (These effector cells have large amounts of endoplasmic reticulum, a characteristic of cells actively synthesizing and secreting proteins.) The same kind of clonal selection mechanism operates on T cells to produce the mature ones that carry out cell-mediated immunity and help in humoral immunity.

Thus, we see that the versatility of the immune system—its ability to defend against a virtually unlimited variety of invaders—depends on a great diversity of preexisting lymphocytes with different antigen receptors.

How does an antigen trigger its own destruction?

By binding to receptors on specific B cells and stimulating cloning of those cells, which develop into plasma cells that secrete antibodies against the antigen

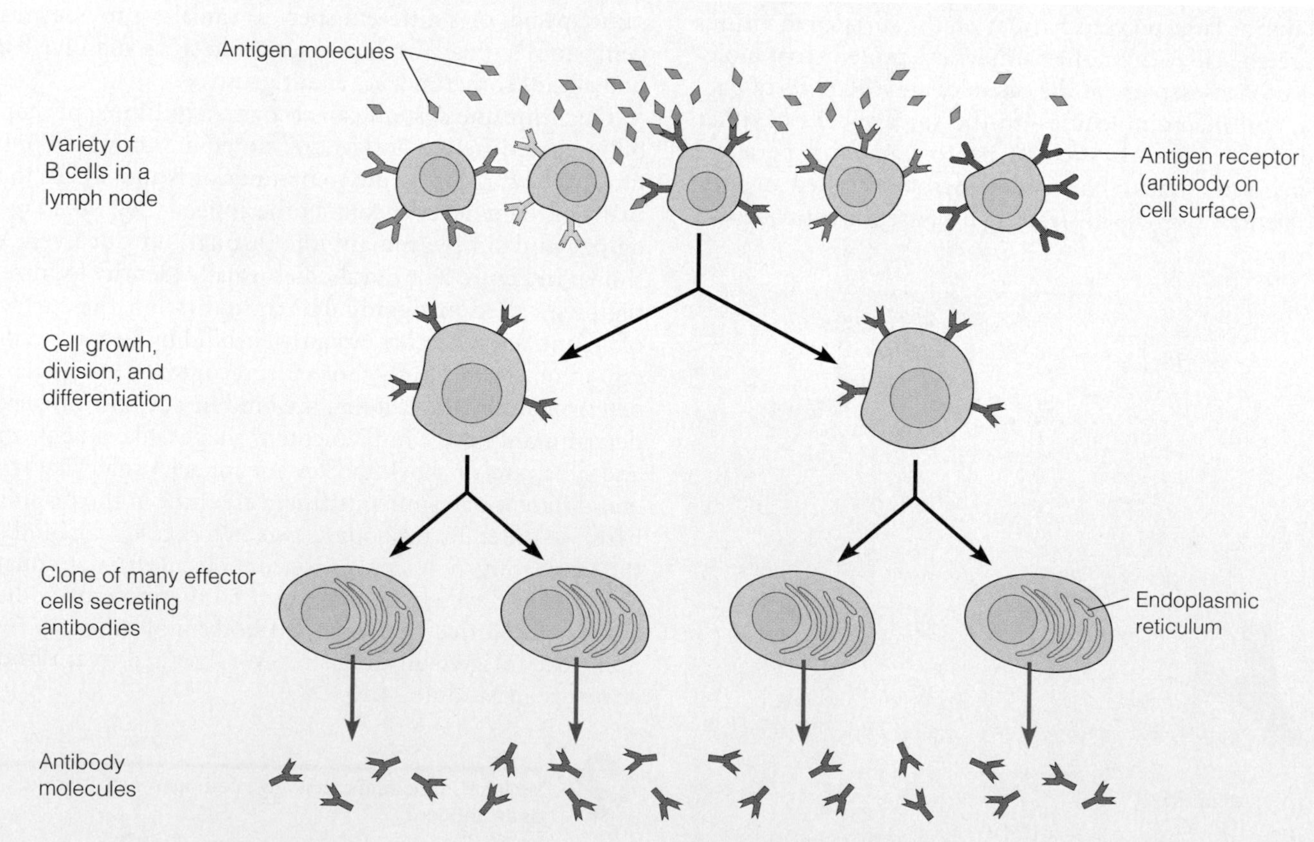

Figure 24.7 Clonal selection of B cells

The initial immune response results in a type of "memory" 24.8

Now that we have seen how clonal selection works, we can see how it fits into the rest of the immune system's response to an antigen. The immune system can mount an effective response the first time it encounters an antigen, but it takes two exposures to elicit the strongest response. As indicated by the blue curve in Figure 24.8A, the two exposures trigger two distinct phases of immune response. The initial phase, called the **primary immune response,** occurs when lymphocytes are first exposed to an antigen and form a clone of effector cells. On the far left of the graph, you can see that the primary response does not start right away; it usually takes several days for the lymphocytes to become activated by an antigen (X) and form clones of effector cells. When the effector cell clone forms, antibodies start showing up in the blood, as the graph shows. The antibody level is highest about two weeks after the initial exposure.

After the primary immune response, a second exposure to the *same* antigen elicits a faster and stronger response, called the **secondary immune response.** In the case of humoral immunity (the antibody response shown here), the secondary response produces very high levels of antibodies that are often more effective against the antigen than those produced during the primary response. The secondary response also lasts much longer than the primary response, as you can see in the graph.

The red curve in Figure 24.8A illustrates the specificity of the immune response. If the body is exposed to a different antigen (Y), even after it has already responded to antigen X, it responds with another primary response, this one directed against antigen Y. The response to Y is not enhanced by the response to X.

Figure 24.8B outlines the cellular events that produce the primary and secondary immune responses. Each exposure to the antigen triggers a clonal selection, resulting in a clone of lymphocytes. The cells of each clone, while all identical genetically, fall into two sets. One set consists of effector cells; the other set consists of cells called **memory cells,** which differ from effector cells in both appearance and function. During the primary response, the effector cells of the clone combat the antigen by producing antibodies if they are B cells. These effector cells usually survive only a few days. In contrast, the memory cells of the clone may last for decades. They remain in the lymph nodes, ready to be activated by a second exposure to the antigen. When this happens, the memory cells initiate the secondary immune response: They multiply quickly, producing a large new clone of lymphocytes that mount the secondary response. Like the first clone, the second clone includes effector cells that actually produce the antibodies of the secondary response and memory cells capable of responding to future exposures to the antigen. In some cases, memory cells seem to confer lifetime immunity, as they may in such childhood diseases as mumps and measles.

Although we have focused on humoral immunity (B cells) in this module, clonal selection, effector cells, and memory cells are features of cell-mediated immunity (T cells) as well. In the next four modules, we discuss humoral immunity further. After that, we focus on how the cell-mediated arm of the immune system helps defend the body against pathogens.

? What is the immunological basis for referring to certain diseases, such as mumps, as *childhood* diseases?

Because one bout with the pathogen, which most often occurs during childhood, is usually enough to confer immunity for the rest of that individual's life

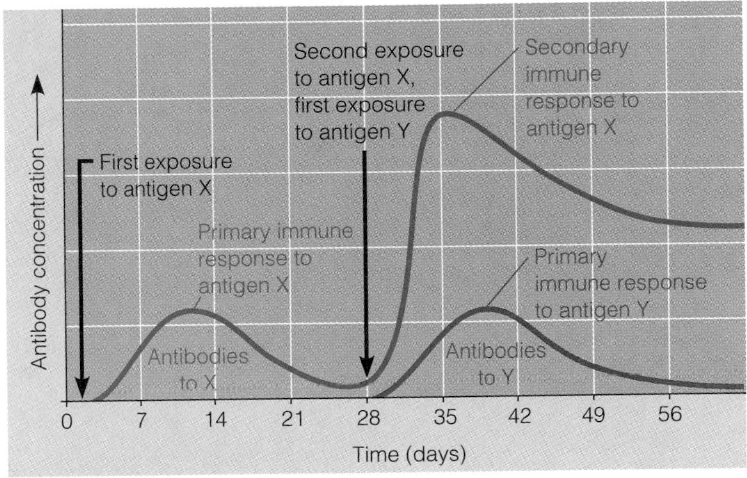

Figure 24.8A Immunological memory

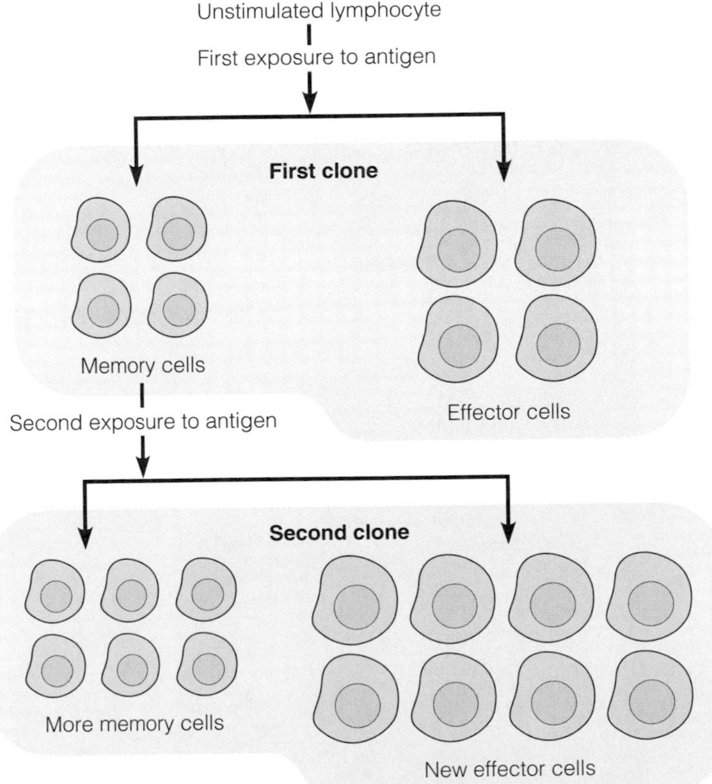

Figure 24.8B The cellular basis of immunological memory

24.9 Overview: B cells are the main warriors of humoral immunity

24A
See Activity 24A on the CD-ROM.

By connecting the concepts of clonal selection and immunological memory, we can obtain an overview of the "defensive machine" we call humoral immunity. Continuing our military analogy, B cells are the main "warriors" of this machine.

Starting with the first encounter between an antigen and a collection of B cells, the figure below summarizes the primary and secondary responses in humoral immunity. Recall that a B cell is first "selected" by the antigen. In the first step of this process, surface receptors (specific antibodies) on the B cell bind with antigen molecules (actually antigenic determinants on the antigen's surface). This binding triggers the growth, division, and further differentiation of the selected cell. The resulting clone contains many effector B cells, called **plasma cells,** and a smaller number of memory B cells; only a few of the two types of cells are shown in the figure. The plasma cells secrete antibody molecules, completing the primary response. In producing antibodies that can actually combat antigen molecules, plasma cells are the "front-line warriors" of humoral defense. They may secrete as many as 2000 antibody molecules per second for their brief lifetime of four or five days. These antibodies circulate in the blood and lymphatic fluid, binding to antigens and contributing to their destruction or blocking their harmful effects. The primary response subsides as the plasma cells die out.

In a military sense, we could consider memory B cells the "reservists" of humoral immunity. Rather than actively combating antigens, they await future exposure to antigens. The secondary response in humoral immunity can occur only if these cells contact the same antigen that triggered their production, perhaps not until years later. If this contact does occur, the events in the bottom part of the figure are set in motion. The memory cells bind antigens and are stimulated to quickly produce

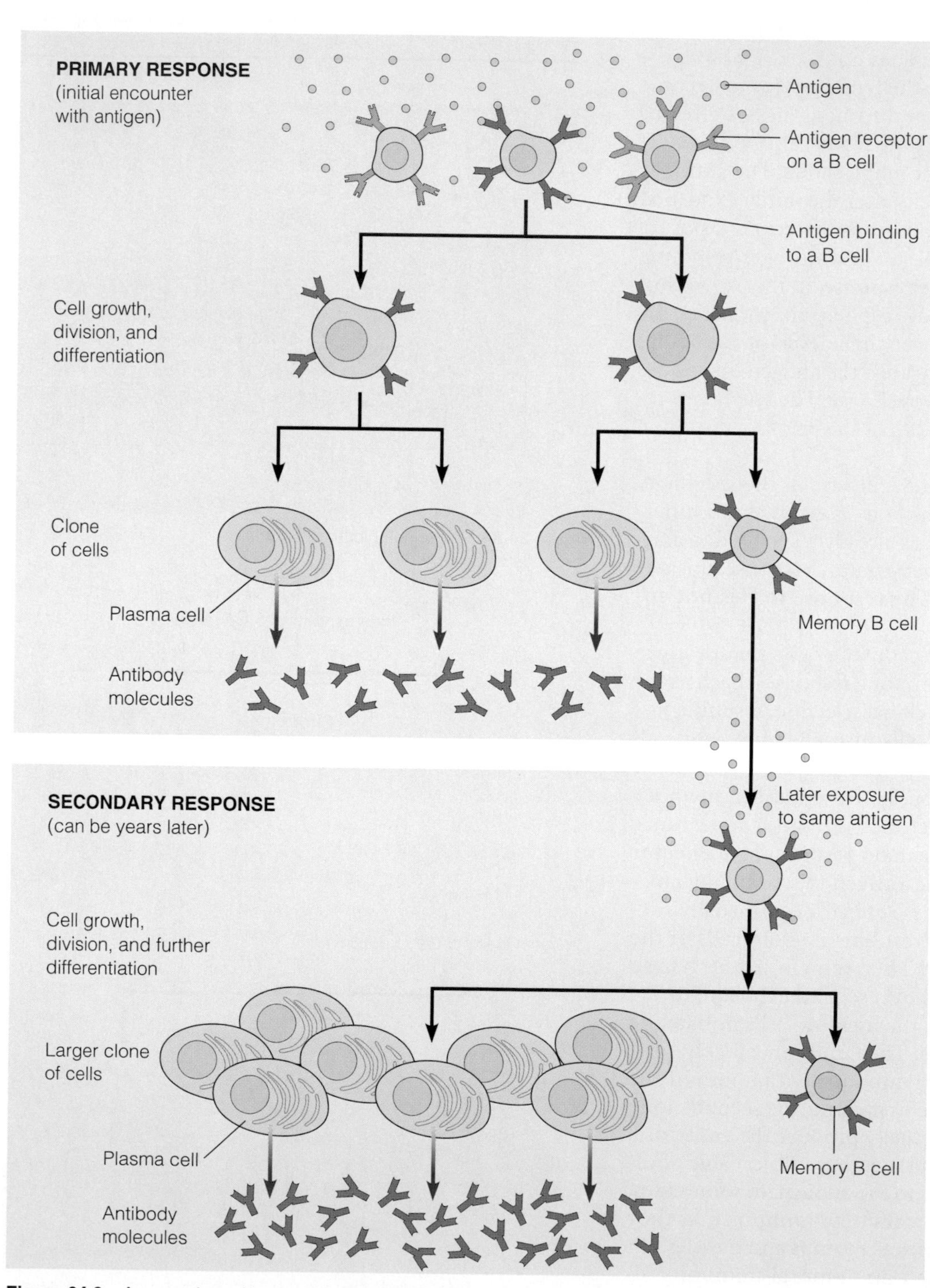

Figure 24.9 An overview of humoral immunity

large new clones of cells. The cloning process is the same as in the primary response, but it occurs more rapidly and yields more plasma cells. As a result, antibody levels in the blood and lymph are much higher than during the primary response. Also, as the next module discusses, the secondary response may involve antibodies differing in structure. 

? (a) What are the effector cells of the humoral immune response? (b) How do these effector cells combat antigens in body fluids?

(a) plasma cells; (b) by producing specific antibodies

Antibodies are the weapons of humoral immunity 24.10

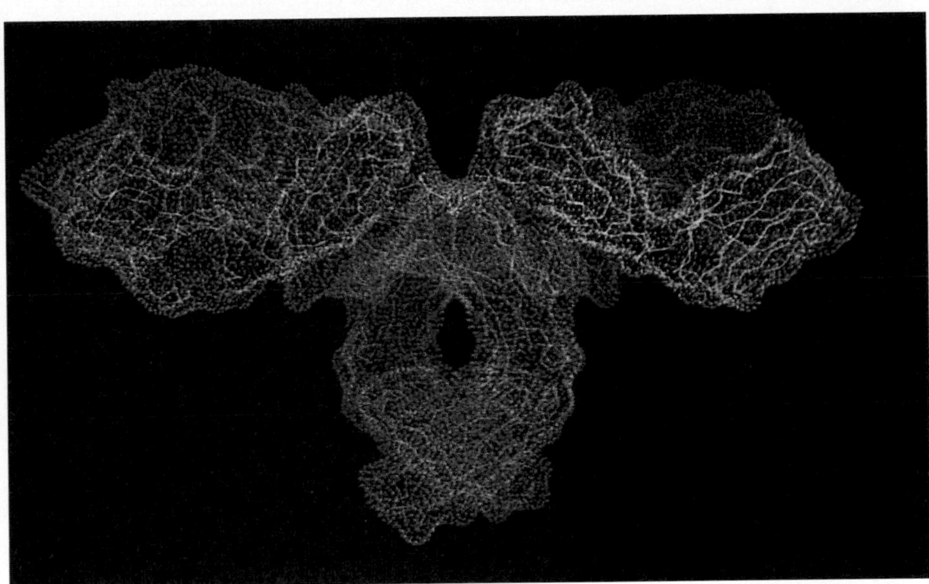

Figure 24.10A A computer graphic of an antibody molecule

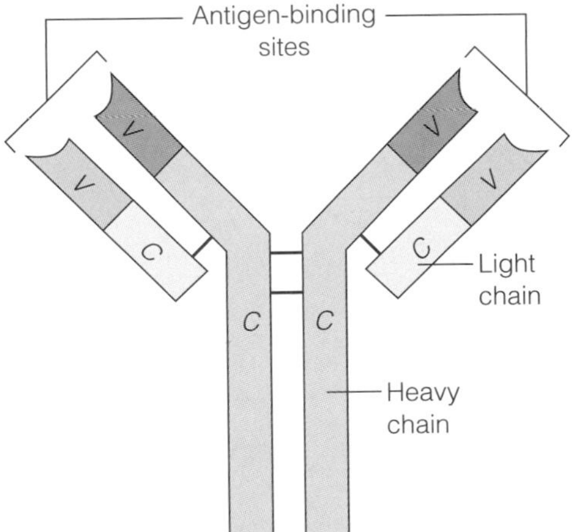

Figure 24.10B Antibody structure

Antibodies are proteins that serve as molecular weapons of defense. We have been using Y-shaped symbols to represent these molecules, and the Y actually does resemble their shape. Figure 24.10A is a computer graphic of a single antibody molecule. Figure 24.10B is a simplified diagram explaining its structure. Each antibody molecule is made up of four polypeptide chains, two "heavy" chains and two "light" chains. In both figures, the parts colored in shades of pink represent the fairly long, heavy chains of amino acids that give the molecule its Y shape. Bonds (black lines in Figure 24.10B) at the fork of the Y hold these chains together. The two green regions in each figure are shorter chains of amino acids, the light chains. Each of the light chains is bonded to one of the heavy chains. As the computer graphic indicates, the bonded chains actually intertwine.

An antibody molecule has two related functions in humoral immunity: to recognize and bind to a certain antigen, and, in so doing, to assist in neutralizing the antigen it recognizes. The structure of an antibody allows it to perform these functions. Notice in Figure 24.10B that each of the four chains of the molecule has a *C* (constant) region and a *V* (variable) region. At the tip of each arm of the Y, a pair of *V* regions forms an **antigen-binding site,** a region of the molecule responsible for the antibody's recognition-and-binding function. A huge variety in the three-dimensional shapes of the binding sites of different antibody molecules arises from a similarly large variety in the amino acid sequences in the *V* regions, hence the term "variable." This structural variety gives the humoral immune system the ability to react to virtually any kind of antigen.

The tail of the antibody molecule, formed by the constant regions of the heavy chains, helps mediate the disposal of the bound antigen. Antibodies with different kinds of *C* regions are grouped into different classes. Humans and other mammals have five major classes of antibodies, and each of the five classes has a particular role in humoral immunity. For instance, one type functions mainly in the primary immune response; another comes into play during the secondary response. Next, we look at how antibody-antigen binding leads to the destruction of invading microbes. 

? How is the specificity of an antibody for an antigen analogous to an enzyme's specificity for its substrate?

Both antibodies and enzymes are proteins with binding sites of specific shape that recognize and bind to other molecules (antigens for antibodies; substrates for enzymes).

24.11 Antibodies mark antigens for elimination

24A

The main role of antibodies in eliminating invading microbes or molecules is to mark the invaders. An antibody marks an antigen by combining with it to form an antigen-antibody complex. Weak chemical bonds between antigen molecules and the antigen-binding sites on antibody molecules hold the complex together.

As the figure below illustrates, it is the binding of antibodies to antigens that actually triggers mechanisms to neutralize or destroy an invader. Such a mechanism is called an effector mechanism. Several effector mechanisms are depicted in the figure. In neutralization, the binding of antibodies physically blocks harmful antigens, making them harmless. For example, antibodies may bind to the surface molecules a virus uses to attach to a host cell. Antibodies may also bind to toxin molecules on bacterial cells. Phagocytes such as macrophages then dispose of the complexes.

Another effector mechanism is the agglutination (clumping together) of viruses, bacteria, or foreign eukaryotic cells. Because each antibody molecule has at least two binding sites, antibodies can hold a clump of invading cells together. Agglutination makes the cells easy for phagocytes to capture.

A third effector mechanism, precipitation, is similar to agglutination, except that the antibody molecules link *dissolved* antigen molecules together. This makes the antigen molecules precipitate out of solution as solids. The precipitated antigens, like clumps of agglutinated cells, are easily engulfed by phagocytes.

One of the most important effector mechanisms in humoral immunity is the activation of complement proteins (see Module 24.1) by antigen-antibody complexes. Activated complement proteins (green in the diagram) can attach to a foreign cell and open holes in its plasma membrane, causing cell lysis (rupture).

Taken as a whole, this figure illustrates a fundamental concept of immunity: All effector mechanisms involve a *specific* recognition and attack phase followed by a *nonspecific* destruction phase. Thus, the antibodies of humoral immunity, which identify and bind to foreign invaders, work with nonspecific defenses, such as phagocytes and complement, to form a complete defense system.

How does specific humoral immunity interface with the body's nonspecific defense system?

Antibodies mark specific antigens for destruction, but it is usually complement, phagocytes, or other components of nonspecific defense that destroy the antigens.

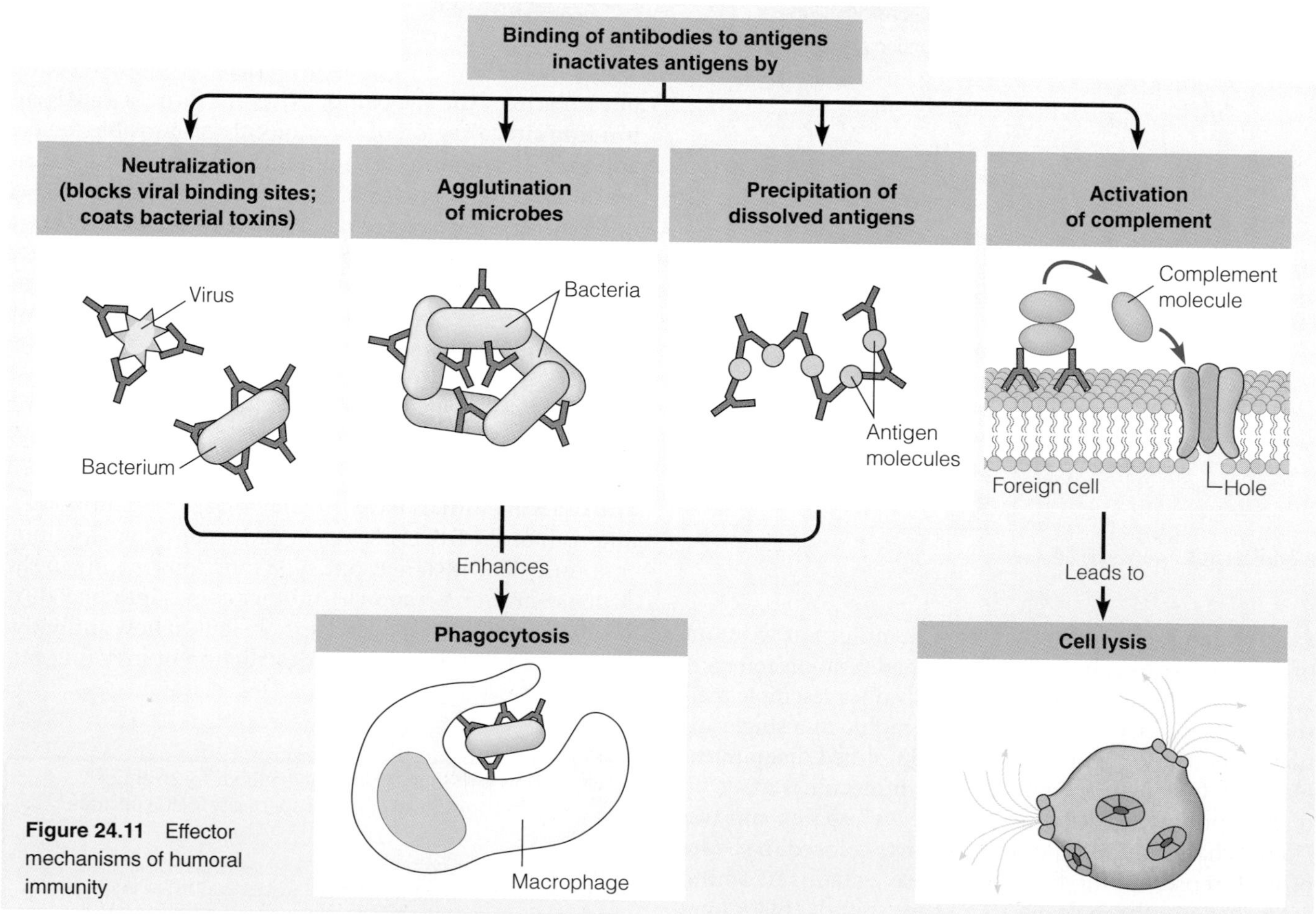

Binding of antibodies to antigens inactivates antigens by

Neutralization (blocks viral binding sites; coats bacterial toxins)

Agglutination of microbes

Precipitation of dissolved antigens

Activation of complement

Virus

Bacterium

Bacteria

Antigen molecules

Complement molecule

Foreign cell

Hole

Enhances

Leads to

Phagocytosis

Cell lysis

Macrophage

Figure 24.11 Effector mechanisms of humoral immunity

Monoclonal antibodies are powerful tools in the lab and clinic 24.12

Because of their ability to tag specific molecules or cells, antibodies are widely used in biological research and clinical testing. In the original procedure for preparing antibodies, a small sample of antigen is injected into a rabbit or mouse. In response to the antigen, the animal produces antibodies, which can be collected directly from its blood. However, because the antigen usually has many different antigenic determinants, the result is a mixture of different antibodies rather than a pure preparation of one. And to make a large amount of antibody in this way, more than one animal has to be used.

Such problems limited the use of antibodies until the late 1970s, when a technique for making **monoclonal antibodies** was developed. The term "monoclonal" means that all the cells producing the antibodies are descendants of a single cell; thus, they all produce identical antibody molecules. Monoclonal antibodies are harvested from cell cultures rather than from animals.

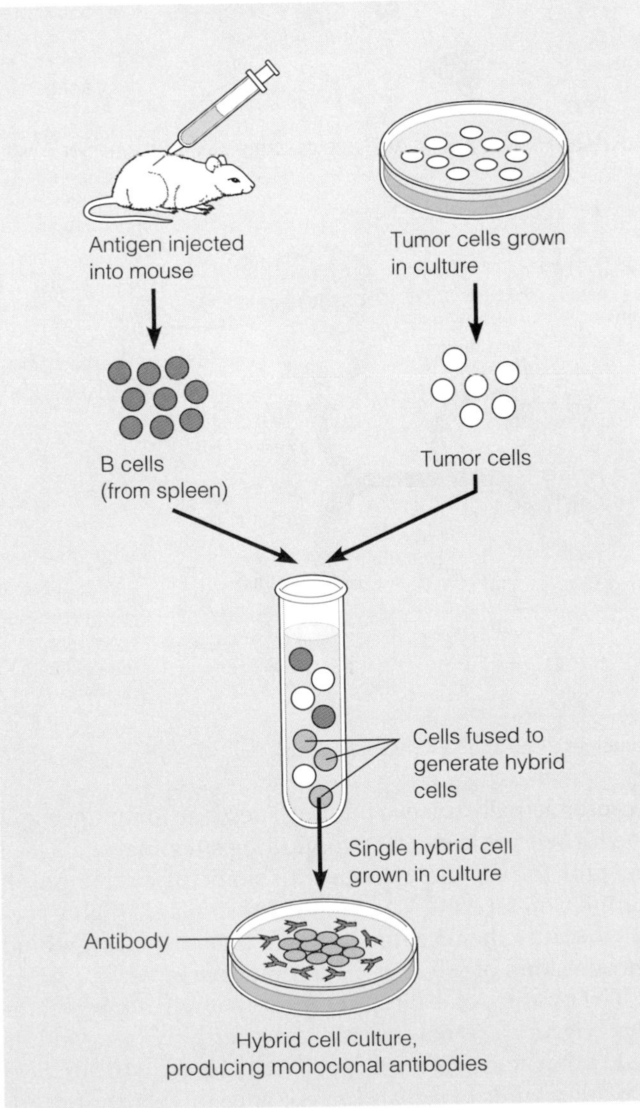

Figure 24.12A The procedure for making monoclonal antibodies

As shown in Figure 24.12A, the trick to making monoclonal antibodies is the fusion of two cells to form a hybrid cell with a combination of desirable properties. First, an animal is injected with an antigen that will stimulate B cells that will make the desired antibody. At the same time, cancerous tumor cells, which can multiply indefinitely, are grown in a culture. The scientist then fuses a tumor cell with a normal antibody-producing B cell from the animal. The hybrid cell makes antibody molecules specific for a single antigenic determinant and is able to multiply indefinitely in a laboratory dish. Thus, large amounts of identical antibody molecules are easily prepared.

The ability to make monoclonal antibodies has spawned a new industry. One area of application is medical diagnosis. For example, monoclonal antibodies may bind with bacteria that cause a sexually transmissible disease, or with a hormone that indicates pregnancy. If the antibodies have been labeled (by a dye, for instance) for easy detection, they will reveal the presence of the bacteria or hormone. This procedure is the basis for a sensitive home pregnancy test. The photograph in Figure 24.12B shows a woman testing a sample of her urine for a hormone produced during early pregnancy. If the hormone is present, the monoclonal antibodies will bind with it, producing a color change in the sample.

Monoclonal antibodies also have potential use in the treatment of certain diseases, including cancer. In late 1998, Herceptin, a genetically engineered monoclonal antibody, became available for treating a common form of aggressive breast cancer. The Herceptin antibody molecules act by binding to growth-factor receptors that are present in excess on the cancer cells. They thus prevent the receptors from transmitting "grow" signals to the cells. Another way to combat cancer might be to combine a cancer-specific monoclonal antibody with a cell-killing drug. An effective antibody-drug combination would act like a molecular guided missile, killing only the targeted cancer cells. 

Figure 24.12B A home pregnancy test

? How do home pregnancy tests based on monoclonal antibodies work?

The monoclonal antibodies produce a color change in the test solution when they react specifically with a hormone present in the female's urine during early pregnancy.

24.13 T cells mount the cell-mediated defense and aid humoral immunity

24A

The antibody-producing B cells of humoral immunity make up one army of the body's defense network. The humoral defense system identifies and helps destroy invaders that are in our blood, lymph, or interstitial fluid—in other words, outside our body cells. But many invaders, including all viruses, enter cells and reproduce there. It is the cell-mediated immunity produced by T cells that battles pathogens that have already entered body cells.

Whereas B cells respond to free antigens present in body fluids, T cells respond only to antigens present on the surfaces of the body's own cells. There are two main kinds of T cells. **Cytotoxic T cells** attack body cells that are infected with pathogens; we'll discuss these T cells shortly. **Helper T cells** have roles in many aspects of immunity. They help activate cytotoxic T cells and macrophages, and even help stimulate B cells to produce antibodies.

Helper T cells interact with other white blood cells that function as **antigen-presenting cells (APCs).** All of cell-mediated immunity and much of humoral immunity depend on the precise interaction of APCs and helper T cells. This interaction activates the helper T cells, which can then go on to activate other cells of the immune system.

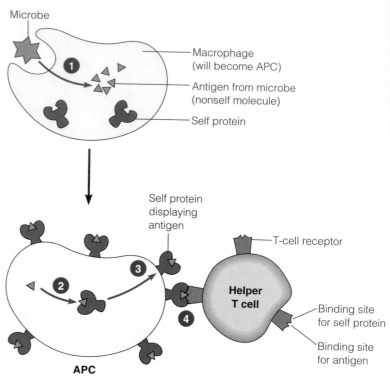

Figure 24.13A Development of an APC and its interaction with a helper T cell

As its name implies, an APC *presents* a foreign antigen to a helper T cell. A typical APC is a macrophage. As shown in Figure 24.13A, ① the macrophage ingests a microbe (or other foreign particle) and breaks it into fragments—foreign antigens. Then, molecules of a special protein belonging to the macrophage, which we will call **self protein** (because it belongs to the body itself), ② bind the foreign antigens—**nonself molecules**—and ③ display them on the cell's surface. Each of us has a unique set of self proteins, which serve as identity markers for our body cells. ④ Helper T cells recognize and bind to the *combination* of a self protein and a foreign antigen displayed on an APC. The double-recognition system used by helper T cells is like the system banks use for safe-deposit boxes: Opening your box requires the banker's key along with your specific key.

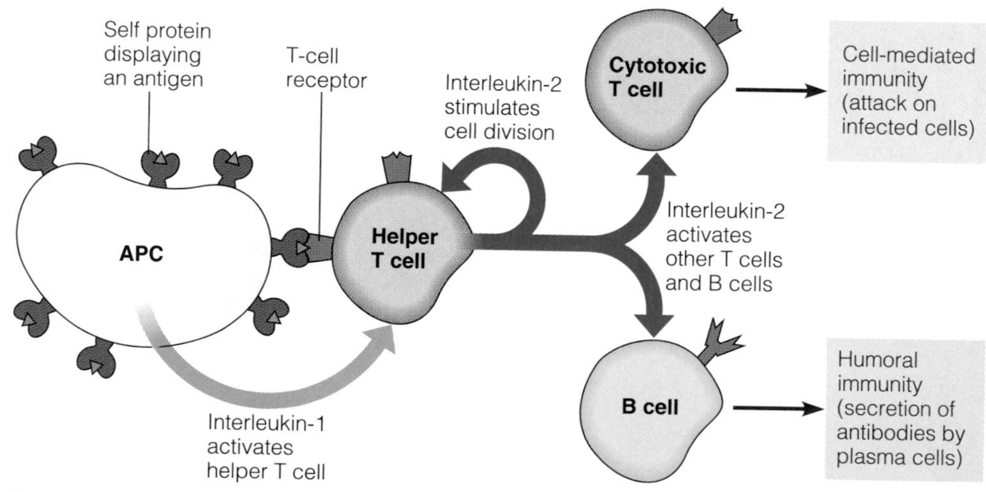

Figure 24.13B The activation of a helper T cell and its roles in immunity

The ability of a helper T cell to recognize a unique self-nonself complex on an APC depends on the T-cell receptors (purple in the figures here) embedded in the T cell's plasma membrane. As indicated in Figure 24.13A, a T-cell receptor actually has two binding sites: one for antigen and one for self protein. The two binding sites enable a T-cell receptor to recognize the overall shape of a self-nonself complex on an APC. The immune response is highly specific because the receptors on each helper T cell can bind only one kind of self-nonself complex on an APC.

The binding of a T-cell receptor to a self-nonself complex triggers a signal transduction pathway (see Module 11.11) that activates the helper T cell (Figure 24.13B). Several other kinds of signals can enhance this activation. For example, certain proteins secreted by the APC, such as interleukin-1 (green arrow), diffuse to the helper T cell and stimulate it.

What do activated helper T cells do? These cells promote the immune response in several ways, with a major mechanism being the secretion of additional stimulatory proteins. One such protein, interleukin-2 (blue arrows), has three major effects. First, it makes the helper T cell itself grow and divide, producing both memory cells and additional active helper T cells. This positive feedback loop amplifies the cell-mediated defenses against the antigen at hand. Second, interleukin-2 also stimulates the activity of cytotoxic T cells. And third, it helps activate B cells, thus stimulating humoral immunity as well.

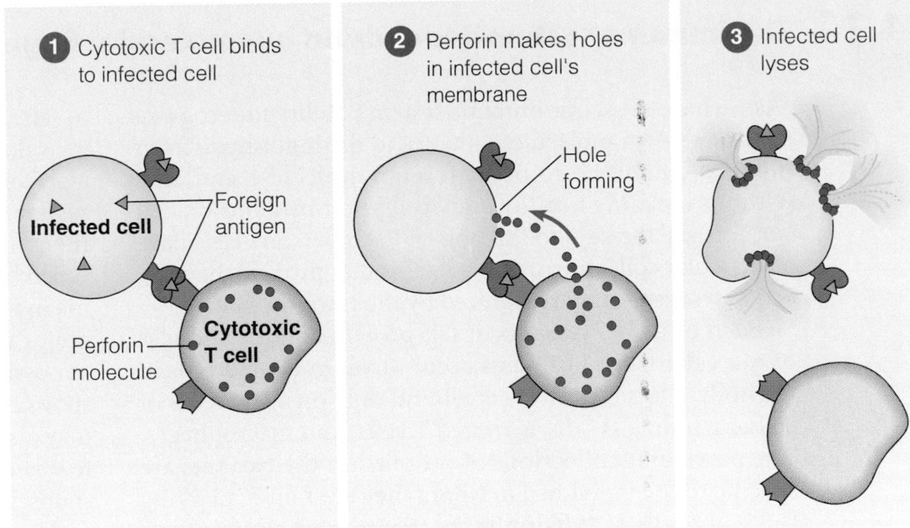

Figure 24.13C How a cytotoxic T cell kills an infected cell

Cytotoxic T cells are the only T cells that actually kill other cells. Infected body cells are important targets. Once activated, cytotoxic T cells identify the infected cells in the same way that helper T cells identify APCs. An infected cell has foreign antigens—molecules belonging to the viruses or bacteria infecting it—attached to self proteins on its surface (Figure 24.13C). Like a helper T cell, a cytotoxic T cell carries receptors (purple) that can bind with a self-nonself complex on the infected cell.

The self-nonself complex on an infected body cell is like a red flag to cytotoxic T cells that have matching receptors. As shown in Figure 24.13C, ① the cytotoxic T cell binds to the infected cell. The binding initiates signal-transduction pathways that activate the T cell. The activated cytotoxic T cell synthesizes a protein called **perforin.** ② Perforin is discharged and attaches to the infected cell's membrane, making holes in it. ③ The infected cell lyses and dies. 

? Compare and contrast the T-cell receptor with the antigen receptor on the surface of a B cell.

Both receptors bind to a specific antigen, but the T-cell receptor only recognizes that antigen when it is presented along with a "self" marker on the surface of one of the body's own cells.

Cytotoxic T cells may help prevent cancer 24.14

People with immune deficiencies are often unusually susceptible to cancer, a fact suggesting that the immune system plays a watchdog role against at least some forms of cancer. Researchers are therefore actively engaged in studying the interactions between the immune system and cancer cells.

As we discussed in Module 8.10 and Chapter 11, the genetic changes that lead to cancer produce changes in normal body cells, and some of these changes take place on the outer membrane surfaces of the cells. Surface molecules may be altered in such a way that the cell-mediated branch of the immune system identifies the cancer cells as foreign intruders and musters cytotoxic T cells to destroy them.

Figure 24.14 Cytotoxic T cells attacking a cancer cell

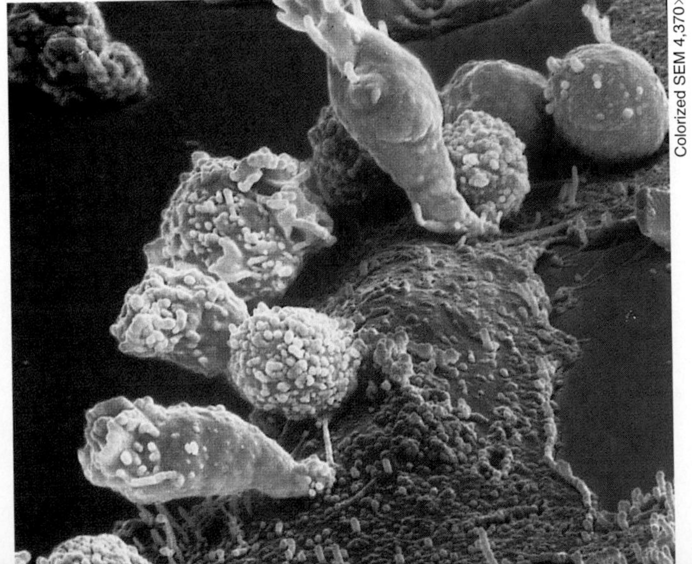

The micrograph here shows a troop of cytotoxic T cells (light blue) attacking a tumor cell (gold) in a laboratory culture. This process may also occur in the body; cytotoxic T cells may attack and kill cancer cells whenever they appear. Why this built-in surveillance system sometimes fails, allowing tumors to develop, is still largely a mystery. Some scientists suggest that tumors develop when cancer cells shed surface molecules that mark them as foreign, or when cancer cells evade the immune system. Studying how the immune system normally helps prevent cancer is an important part of the cancer research effort. 

? Stress causes the adrenal glands to secrete steroid hormones that, among many other functions, suppress the immune response. How might this link prolonged and severe stress to a greater probability of developing cancer?

Suppressing the immune response might make it possible for new cancer cells to avoid destruction by cytotoxic T cells.

24.15 The immune system depends on our molecular fingerprints

As we have seen, our immune system's ability to recognize the body's own molecules—that is, to distinguish *self* from *nonself*—enables it to battle foreign molecules and cells without harming healthy body cells. Self proteins on cell surfaces are the keys to this ability. Each person's cells have a particular collection of self proteins that provide the molecular "fingerprints" recognized by the immune system.

Each of us has two sets of self proteins on the surfaces of our cells. Class I proteins occur on all nucleated cells in the body. Class II proteins are found only on a few types of cells, including B cells, activated T cells, and macrophages. The particular collections of proteins in the two sets are specific to the individual in whom they are found, marking the body cells as "off-limits" to the immune system; our lymphocytes do not attack these molecules.

The immune system not only distinguishes body cells from microbes, it also can tell your cells from those of other people. Genes at six chromosomal loci determine the structure of the main self proteins—thus, in diploid cells, 12 genes in all. Because there are hundreds of alleles in the human population for each of the gene loci, it is virtually impossible for any two people (except identical twins) to have completely matching sets of self proteins.

The immune system's ability to recognize foreign antigens does not always work in our favor. For example, when a person receives an organ transplant, the person's immune system recognizes the donor's cells as foreign and attacks them. (For this reason, the group of self-protein genes is called the *major histocompatibility complex*, or *MHC; histo-* means tissue.) To minimize rejection, doctors look for a donor with self proteins matching the recipient's as closely as possible, and they use drugs to suppress the immune response against the transplant. Unfortunately, these drugs may also reduce the ability to fight infections. However, a few, such as cyclosporine, can suppress cell-mediated responses without crippling humoral immunity.

There are two promising new approaches to preventing transplant rejection. One uses monoclonal antibodies as "guided missiles" to target and destroy the T cells that attack the transplant. The other uses stem cells to establish a new immune system that recognizes the transplant as self.

In what sense is a cell's set of MHC surface markers analogous to a fingerprint?

The set of MHC ("self") markers is unique to each individual.

Disorders of the Immune System

24.16 Malfunction or failure of the immune system causes disease

Overall, our immune system is highly effective, protecting our bodies against a vast array of potentially harmful invaders. When the immune system malfunctions or fails, a number of serious diseases can result.

A number of **autoimmune diseases**—for example, systemic lupus erythematosus (lupus), insulin-dependent diabetes (see Module 26.9), rheumatoid arthritis, and multiple sclerosis (MS)—result when the immune system goes awry and turns against the body's own molecules. In lupus, for example, B cells make antibodies against many sorts of molecules, such as histones and DNA released by the normal breakdown of body cells. Rheumatoid arthritis is another antibody-mediated autoimmune disease; it leads to damage and painful inflammation of the cartilage and bone of joints. In insulin-dependent diabetes, the insulin-producing cells of the pancreas are the targets of autoimmune cell-mediated responses. In MS, the most common chronic neurological disease in developed countries, T cells react against myelin, a protein that insulates the axons of neurons (see Figure 28.2), and destroy the function of neurons in the spinal cord and brain. Most medicines currently available for treating autoimmune diseases either suppress immunity in general or are limited to the alleviation of specific symptoms. However, as research scientists learn more about these diseases and about the normal operation of the immune system, they hope to develop more effective therapies.

In contrast to autoimmune diseases are a variety of defects called **immunodeficiency diseases.** Immunodeficient people lack one or more of the components of the immune system and, as a result, are susceptible to infections that would ordinarily not cause a problem. In the rare congenital disease called severe combined immunodeficiency (SCID), both T cells and B cells are absent or inactive. People with SCID are extremely sensitive to even minor infections. Until recently, their only hope for survival was to live behind protective barriers or to receive a successful bone marrow transplant that would continue to supply functional lymphocytes. Now there are hopes of simply supplementing the patient's immune system with injections of purified, normal stem cells. Cases of SCID involving a single faulty gene may some day be treated with gene therapy (see Module 12.20).

Immunodeficiency is not always an inborn condition. For instance, Hodgkin's disease, a type of cancer that affects the lymphocytes, can depress the immune system. Radiation therapy and the drug treatments used against many cancers can have the same effect. Another well-known acquired immune deficiency is AIDS, as we'll discuss in Module 24.18.

In addition to autoimmune and immunodeficiency diseases, physical and emotional stress may also weaken both the immune system and the nonspecific defenses. In one study, students were examined just after a vacation and then again during exams. Their natural killer cells were less effective, and they produced less interferon, during exams.

How might stress affect our immune system? It turns out that nerve fibers penetrate the organs that produce lymphocytes and that lymphocytes have receptors for chemical signals secreted by nerve cells. Might signals from our nerve cells enhance our immunity when we are relaxed and happy? Research on the role of stress in immunity may yield some answers in the near future. 

? What is a probable side effect of autoimmune disease treatments that suppress the immune system?

Lowered resistance to infections

Allergies are overreactions to certain environmental antigens 24.17

Allergies are abnormal sensitivities to antigens in our surroundings. Antigens that cause allergies are called **allergens.** Protein molecules on pollen grains, on the surface of tiny mites that live in house dust, and in animal dander are common allergens. Many people who are allergic to cats and dogs are actually allergic to proteins in the animal's saliva. They become sensitized to salivary proteins deposited on the fur when the animal licks itself. Allergic reactions typically occur very rapidly and in response to tiny amounts of an allergen. A person allergic to cat or dog saliva, for instance, may react to a few molecules of the allergen in a matter of minutes. Allergic reactions can occur in many parts of the body, including the nasal passages, bronchi, digestive tract, and skin. Symptoms may include sneezing, coughing, wheezing, upset stomach, and itching.

The symptoms of an allergy result from a two-stage reaction sequence outlined in the figure below. The first stage, called sensitization, occurs when a person is first exposed to an allergen—pollen, for example. B cells make a special class of antibodies (one of five classes in mammals) in response to allergens. Some of these antibodies attach to receptor proteins on the surface of **mast cells,** normal body cells that produce histamine and other chemicals that trigger the inflammatory response (see Module 24.2). In our example, the affected mast cells are in the nose.

The second stage of an allergic response begins when the person is exposed to the same allergen later. The allergen binds to the antibodies attached to mast cells, causing the cells to release histamine, which triggers the allergic symptoms. As in inflammation, it causes blood vessels to dilate and leak fluid. Histamine also elicits other symptoms, such as nasal irritation, itchy skin, and tears. **Antihistamines** are drugs that interfere with histamine's action and give temporary relief from an allergy.

Allergies range from seasonal nuisances to severe, life-threatening responses. **Anaphylactic shock** is an especially dangerous type of allergic reaction. Some people are extremely sensitive to certain allergens, such as the venom from a bee sting. Any contact with these allergens makes their mast cells release inflammatory chemicals very suddenly. As a result, their blood vessels dilate abruptly, causing a precipitous drop in blood pressure (shock), which is potentially fatal. Fortunately, anaphylactic shock can be counteracted with injections of the hormone epinephrine. 

? The binding of ________ to ________ on the surface of a ________ cell causes that cell to release molecules called ________, which stimulate fluid loss from blood vessels.

allergens . . . antibodies . . . mast . . . histamines

Figure 24.17 The two stages of an allergic reaction

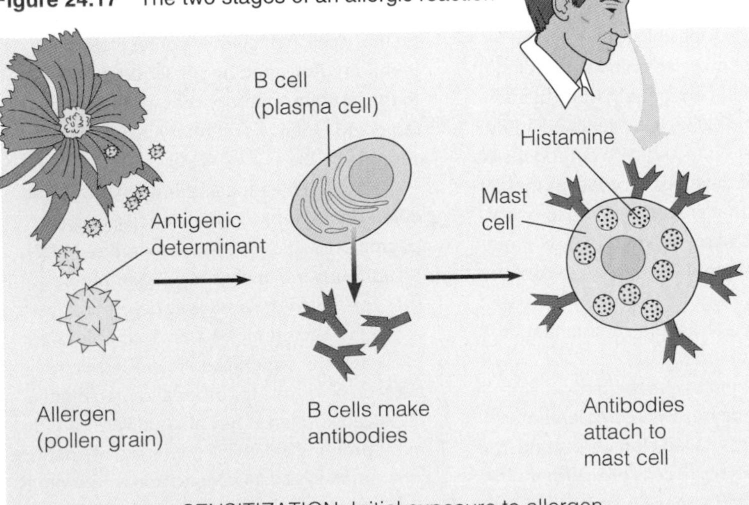

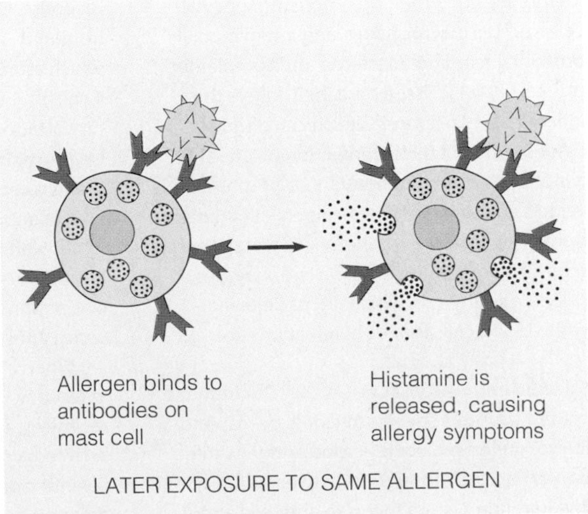

24.18 AIDS leaves the body defenseless

24B

In the past two decades, AIDS has killed more than 11 million people worldwide, and more than 33 million people are now infected with the AIDS virus, HIV. At present rates of infection, this number will double in the next few years.

HIV is deadly because it destroys the immune system, leaving the body defenseless against most invaders. HIV can infect a variety of cells, but it has a preference for helper T cells—the cells that activate other T cells and B cells as well. When HIV depletes the body of helper T cells, the immune system cannot carry out either the cell-mediated or the humoral response. Death usually results, not from AIDS itself but from another infectious agent or from cancer. We described the course of cell infection by HIV in Module 10.21. If you turn back to Figure 10.22, you will see a micrograph showing HIV attacking a helper T cell.

At present, AIDS is incurable, and the AIDS virus continues to evade a worldwide effort to control its transmission and deadly effects. In the past few years, new combinations of drugs have effectively slowed the progress of AIDS and markedly decreased the number of deaths from the disease in the U.S. and some other developed countries. These drug combinations combat the virus in different ways, thus making it much more difficult for drug-resistant varieties of the virus to arise—a common problem with single-drug treatments. However, the multidrug regimens are complicated for patients to follow and the drugs are very expensive, limiting their use even in the richer countries. Furthermore, some patients cannot tolerate the side effects of the drugs.

While some medical scientists are working to improve drug therapies, others are exploring different approaches, such as injecting patients with HIV-resistant stem cells that could give rise to HIV-resistant T cells. Clearly, an effective vaccine against HIV is of the highest priority, and a variety are being tested in animal and human studies. Some of these vaccines are of the traditional kind, consisting of inactivated whole virus. Others consist of HIV surface proteins, prepared in large quantities by recombinant DNA techniques. The main challenge is that HIV is continually mutating to new forms. Even within a single infected individual, there are usually multiple genetic types of the virus!

Time is a big problem in the fight against AIDS. Drugs and vaccines must be thoroughly tested for efficacy and for toxic side effects before they can be released for general use. While hearing the outcry of AIDS patients and activists, some experts predict that it may take decades to develop effective vaccines and affordable drugs. At present, education remains our best weapon against AIDS. The World Health Organization concurs with Magic Johnson's campaign for practicing safer sex. Reducing promiscuity and increasing the use of condoms could save millions of lives. To learn more about AIDS and its mode of transmission, anyone may call the National AIDS Hotline, 1-800-342-AIDS.

Why is it unusually difficult to develop an AIDS vaccine?

Because the HIV virus evolves so rapidly

Chapter Review

Chapter Summary

Nonspecific Defenses Against Infection (24.1–24.3) The body's nonspecific obstacles to infection include the skin and mucous membranes and their secretions, phagocytic cells such as macrophages, and antimicrobial proteins, including interferon and complement proteins (24.1). Tissue damage triggers the inflammatory response, which can disinfect tissues and limit further infection (24.2). The lymphatic system is a network of lymphatic vessels and organs. The vessels collect fluid from body tissues and return it as lymph to the blood. The organs include the lymph nodes, spleen, and thymus, which are packed with white blood cells that fight infections (24.3).

Specific Immunity (24.4–24.15) Our immune system counters specific invaders by responding to foreign molecules called antigens. Infection or vaccination triggers active immunity. The immune system reacts to antigens and "remembers" an invader. We can also temporarily acquire passive immunity (24.4). Two kinds of lymphocytes carry out the immune response. B cells, responsible for humoral immunity, secrete proteins called antibodies, which attack foreign antigens in body fluids. T cells are responsible for cell-mediated immunity, attacking cells infected with pathogens. As each lymphocyte develops, it produces membrane receptors that are able to bind to one kind of antigen. Millions of kinds of B cells and T cells wait in the lymphatic system to confront invaders (24.5). Antigen molecules have specific regions, called antigenic determinants, to which antibodies bind (24.6).

When an antigen enters the body, it activates only lymphocytes with complementary receptors, a process called clonal selection. The selected lymphocytes (B cells or T cells) multiply into clones of effector cells specialized for defending against the antigen that triggered the response (24.7). In the primary immune response, clonal selection produces not only effector cells, which are short-lived, but also memory cells, which may confer lifelong immunity. Activated by subsequent exposure to the antigen, memory cells mount a more rapid and massive secondary immune response (24.8).

Triggered by a specific antigen, a B cell differentiates into an effector cell called a plasma cell, which secretes antibodies (24.9). An antibody molecule has antigen-binding sites specific to the antigenic determinants that elicited its secretion (24.10). Antibodies may block harmful antigens on microbes, clump bacteria or viruses together, and precipitate dissolved antigens. They also activate complement proteins, which kill bacteria and enhance inflammation (24.11). Monoclonal antibodies, produced by fusing B cells specific for a single antigenic determinant with easy-to-grow tumor cells, are useful in research, diagnosis, and treatment of certain cancers **(24.12).**

Helper T cells and cytotoxic T cells are the main effectors of cell-mediated immunity, and helper T cells also stimulate the humoral responses. In cell-mediated immunity, an antigen-presenting cell (APC), such as a macrophage, first displays a foreign antigen (a nonself molecule) and one of the body's own self proteins to a helper T cell. The helper T cell's receptors recognize the self-nonself complexes on the APC, and the interaction activates the helper T cell. In turn, the helper T cell can activate cytotoxic T cells with the same receptors (and can stimulate B-cell activation, as well). Cytotoxic T cells bind to infected body cells displaying both self proteins and foreign antigens, and destroy them **(24.13).** Cytotoxic T cells may also attack cancer cells, whose surface molecules are altered by the disease **(24.14).** The immune system normally reacts only against nonself substances (foreign molecules and cells), not against self (the body's own macromolecules). It generally rejects transplanted organs, whose cells lack the recipient's unique "fingerprint" of self proteins **(24.15).**

Disorders of the Immune System (24.16–24.18) The immune system sometimes malfunctions or fails. In autoimmune diseases, the system turns against the body's own molecules. In immunodeficiency diseases, immune components are lacking, and infections recur. Physical and emotional stress may weaken the immune system **(24.16).** Allergies are abnormal sensitivities to antigens (allergens) in the surroundings **(24.17).** The AIDS virus attacks helper T cells, crippling both cell-mediated and humoral immunity. So far, AIDS is incurable, but drugs and vaccines offer hope for the future. Practicing safer sex could save many lives **(24.18).**

Testing Your Knowledge

Multiple Choice

1. Foreign molecules that evoke an immune response are called
   a. pathogens
   b. antibodies
   c. lymphocytes
   d. histamines
   e. antigens

2. Which of the following is *not* part of the body's nonspecific defense system?
   a. natural killer cells
   b. antibodies
   c. interferons
   d. complement
   e. inflammation

3. Which of the following best describes the difference in the way B cells and cytotoxic T cells deal with invaders?
   a. B cells confer active immunity; T cells confer passive immunity.
   b. B cells send out antibodies to attack; T cells themselves do the attacking.
   c. T cells handle the primary immune response; B cells handle the secondary response.
   d. B cells are responsible for cell-mediated immunity; T cells are responsible for humoral immunity.
   e. B cells attack the first time the invader is present; T cells attack subsequent times.

4. The antigen-binding sites of an antibody molecule are formed from the molecule's variable regions. Why are these regions called variable?
   a. They can change their shapes on command to fit different antigens.
   b. They change their shapes when they bind to an antigen.
   c. Their specific shapes are unimportant.
   d. They can be different shapes on different antibody molecules.
   e. Their sizes vary considerably from one antibody to another.

5. Researchers suspect that cytotoxic T cells are usually able to find and attack cancer cells because
   a. cancer changes the surfaces of cancerous cells
   b. B cells help them
   c. cancer is a bacterial infection
   d. cancer cells release antibodies into the blood
   e. cancer is an autoimmune disease

Matching

1. Attacks infected body cells	a. lymphocyte
2. Carries out humoral immunity	b. cytotoxic T cell
3. Triggers allergy symptoms	c. helper T cell
4. Nonspecific white blood cell	d. mast cell
5. General name for a B or T cell	e. neutrophil
6. Carries out the secondary immune response	f. B cell
7. Cell most commonly attacked by HIV	g. memory cell

Describing, Comparing, and Explaining

1. Describe (a) how AIDS is transmitted and (b) how immune system cells in an infected person are affected by HIV. Why is AIDS particularly deadly compared to other viral diseases?

2. What is inflammation? How does it protect the body?

Thinking as a Scientist

1. Influenza (flu) viruses undergo frequent genetic mutations and rearrangements that produce viruses with new combinations of surface antigens. How would this affect the ability of the immune system to defend against the flu?

2. Most biologists believe that the immune system's defense of the body against infections largely rests on its ability to distinguish self molecules from nonself molecules. This concept seems central to our understanding of immune function. However, like all scientific ideas, it is not beyond question. Several immunologists have developed an alternative hypothesis: that the immune system's effectiveness rests mostly on its ability to recognize damage to body tissues caused by the invaders, not on the ability to recognize nonself. If you were going to test the "damage" hypothesis, what might you look for? Which type of cell would you expect to be directly affected by damaged tissues? Why? Some proponents argue that the "damage" hypothesis makes more sense from an evolutionary perspective, since it is more advantageous for an organism's defense system to respond to tissue damage than to the mere presence of a foreign microbe. Do you agree? Why or why not?

Science, Technology, and Society

Concern is increasing over the rising rate of HIV infection among teenagers. Schools in some large cities have instituted programs to make condoms available to students, along with counseling about safer sex. These plans have divided school boards and communities. Some citizens and church groups are opposed to giving condoms to students, because it might appear to encourage sexual activity. By contrast, many school and public health officials view the situation as a health issue rather than a moral issue. The heart of the controversy seems to be whether the schools should take such a direct role in this part of student life. What are the reasons for and against distribution of condoms? What do you think the school's role should be?

Media Resources

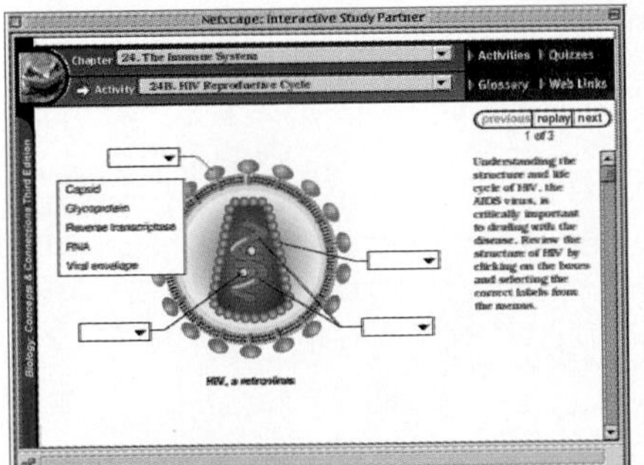

For further review of the concepts in this chapter, use the activities and quizzes on the CD-ROM.

Visit the Special Edition of The Biology Place web site for activities, quizzes, news articles, and web links.

25 Control of the Internal Environment

Outline

Thermoregulation

One of the most confusing notions about animals is that they are either "cold-blooded" or "warm-blooded." These terms are deeply entrenched, and when we think about the body temperature of an invertebrate, a fish, an amphibian, or a reptile, we typically think "cold-blooded." When we think about warm-blooded animals, we think only of mammals or birds. But many tropical fishes and invertebrates have body temperatures as warm as those of any mammal. How do we classify them? And what about the fish shown at the left, the great white shark? Is it cold-blooded? It cruises through cool and warm oceans alike, maintaining a warm body temperature wherever it goes. The great white's body temperature stays fairly constant. But so does the body of many "cold-blooded" fishes and invertebrates that live in tropical or polar seas where water temperatures stay about the same year-round.

Instead of calling animals cold-blooded or warm-blooded, biologists prefer the terms ectothermic and endothermic, referring to the main source of body heat instead of the animal's body temperature. An **ectotherm** is an animal that warms itself *mainly* by absorbing heat from its surroundings. An **endotherm** derives *most* of its body heat from its own metabolism. In general, endothermic animals maintain a fairly constant body temperature, whereas an ectotherm's body temperature usually fluctuates with the environmental temperature. The distinction between ectotherms and endotherms often blurs, however, and, as we will see in this chapter, many animals are not strictly ectotherms or endotherms.

Most fishes are ectothermic, but there are exceptions. The great white shark is an endotherm; it has adaptations that conserve metabolic heat produced by its muscles, keeping it warm—often warmer than the surrounding water. We examine these adaptations in Module 25.2.

The great white shark is a large predator, reaching 6–7 meters in length. One of the most dangerous of all sharks, it is the species usually depicted in films and popular novels about shark attacks. Despite its reputation as a "man-eater," however, most of its attacks on humans seem to be in defense of its territory or because it mistakes a diver for its preferred prey, a seal or sea lion. Its warm muscles give the great white shark the extra power and speed it often needs to catch fast-swimming marine mammals.

The great white shark's ability to maintain its body temperature is extraordinary for a fish, but it is a poor second to most mammals, especially the one pictured below. The polar bear is an endotherm that maintains a consistently warm body temperature despite spending most of its time in subzero weather and often not eating or drinking for months. Its dense fur and thick body fat provide such superb insulation that the bear can plunge into the Arctic Ocean among floating icebergs and swim long distances after seals, its favorite prey. Out of the water, the animal's hairs not only minimize heat loss, they also help its skin absorb heat. At a distance, the hairs appear white, but they are actually transparent and beam sunlight to the skin. The bear's skin itself is black, maximizing heat absorption.

Animals survive fluctuations in the external environment because they have homeostatic ("steady-state") control mechanisms, which allow only a narrow range of fluctuation in the fluid environment that bathes their cells. This chapter examines three elements of homeostasis: **thermoregulation,** the maintenance of internal temperature within a range that allows cells to function efficiently; **osmoregulation,** control of the gain and loss of water and dissolved solutes; and **excretion,** the disposal of nitrogen-containing wastes. Let's look at thermoregulation first.

The polar bear, a master of thermoregulation

Thermoregulation

25.1 Heat is gained or lost in four ways

In regulating its internal temperature, an animal must adjust to heat gained from or lost to the environment by four physical processes. An animal in actual contact with an object, such as a warm rock, can exchange heat with the object by direct transfer, a process called *conduction.* Heat is always conducted from a body of higher temperature to one of lower temperature. The lizard in Figure 25.1 is elevating its body temperature using heat conducted from the rock, a common practice among reptiles.

Convection
Radiation
Evaporation
Conduction

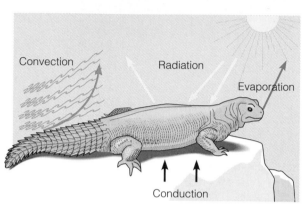

Figure 25.1 Mechanisms of heat exchange

In contrast to conduction, *convection* is the transfer of heat by the movement of air or liquid past a body surface. Here a breeze removes heat (orange arrow) from the lizard's tail.

Radiation, another means of heat exchange, is the emission of electromagnetic energy. Radiation can transfer heat between objects that are not in direct contact, as when an animal absorbs heat from the sun. The yellow arrow emerging from the lizard indicates that the animal radiates some of its own heat into the external environment.

The fourth type of heat transfer, *evaporative cooling,* is the loss of heat from the surface of a liquid that is transforming into a gas. A lizard has dry skin but may lose some heat as moisture evaporates from its nostrils (blue arrow). Evaporative cooling of the human body is increased greatly by sweating. The comfort a fan brings us on a hot day is mainly due to evaporative cooling.

A lecture hall that lacks air conditioning is warmer when full of students than when it is empty mainly because of __________ of heat from the students.

radiation

25.2 Thermoregulation depends on both heat production and heat gain or loss

Different animals are adapted to different external temperature ranges, and each species has an optimal range. Within this range, endotherms and many ectotherms maintain a fairly constant internal temperature as the external temperature fluctuates. They may do so in two ways: (1) Endotherms and even some ectotherms may alter their rate of heat production, and (2) both endotherms and ectotherms may change their rate of heat gain or loss by conduction, convection, radiation, or evaporative cooling.

In cold weather, hormonal changes tend to boost the metabolic rate of birds and mammals, increasing their heat production. Simply moving around more or shivering increases heat produced by the skeletal muscles as a metabolic by-product. Many ectotherms use a similar mechanism when temperatures become extreme. Honeybees, for instance, survive cold winters by clustering together and shivering in their hive. Each bee in the cluster in Figure 25.2A shivers almost constantly when it is cold. The metabolic activity of all the bees together generates enough heat to keep the cluster alive, although some bees on the outside usually die from the cold. In using metabolic heat to warm up, honeybees show that the distinction between ectotherms and endotherms often blurs.

Figure 25.2A A cluster of shivering honeybees generating heat

Several kinds of adaptations adjust the rate of an animal's heat loss or gain. Heat loss in mammals is often regulated by hairs—a thin coat of fur in summer and a thick, insulating coat in winter. Mammals also have muscles in the skin that raise their hairs in the cold. The raised hairs have more insulating power because they trap a layer of air next to the warm skin. (Although humans have little hair, we do have muscles that raise our hair in the cold, causing goose bumps, a vestige from our furry ancestors.)

Heat loss can also be altered by the amount of blood flowing to the skin. In a bird or mammal, nerves signal surface blood vessels to constrict or dilate, depending on the external temperature. When the vessels are constricted, less blood flows from the warm body core to the body surface, reducing the rate of heat loss. Conversely, dilation of surface blood vessels increases the rate of heat loss. Evaporative cooling, as in panting or sweating, also increases heat loss (see Module 20.13).

The circulatory system plays other roles in thermoregulation. Figure 25.2B indicates how the circulatory system conserves heat in the great white shark. Like other fishes, this shark loses some heat to the surrounding water when its blood passes through its gills. In most fishes, the dorsal aorta carries most of the cool blood from the gills directly

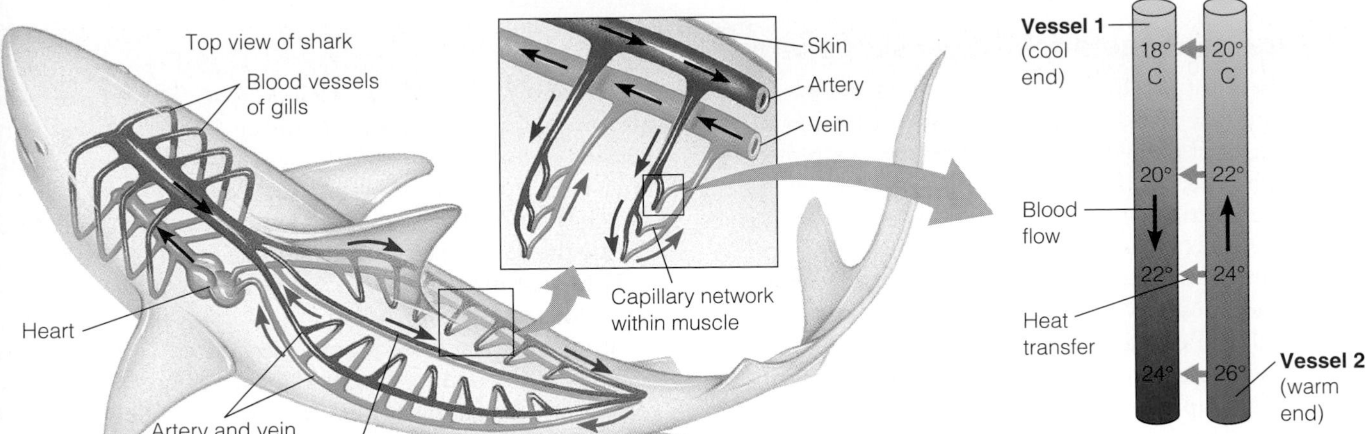

Figure 25.2B Circulatory mechanisms that conserve heat in the great white shark

Figure 25.2C Countercurrent heat exchange

inward along the core of the body. As a result, the typical fish's body temperature is about the same as its environment. In the great white shark, however, most of the blood from the gills flows into two large arteries that lie just under the skin. Figure 25.2B shows a segment of one of the large arteries and of a large vein that parallels it. A network of capillaries (small red vessels) carries cool blood inward from the artery. Other capillaries (blue) carry warm blood from the inner body to the vein. The two sets of capillaries parallel each other, forming what is called a **countercurrent heat exchanger.** Figure 25.2C shows how this heat exchanger prevents the loss of much of the metabolic heat generated by the shark's muscles. Warm and cold blood flow in opposite (countercurrent) directions in two adjacent blood vessels. Vessel 1 carries cool blood inward from the surface of the shark, and vessel 2 carries warmed blood outward. Because of the countercurrent flow, heat passes from the warmer blood in vessel 2 to the cooler blood in vessel 1 along the whole length of the vessels. In the shark, the warm, outflowing blood loses heat to the cooler, inflowing blood in the capillaries. As a result, much of the metabolic heat from the muscles is retained in the body instead of being lost, as it would if warmer blood reached the surface. The shark is endothermic because its muscles generate a large amount of heat and because its countercurrent heat exchanger conserves much of that heat.

Countercurrent heat exchange is important in controlling heat loss in many animals. Many birds, for instance, have heat-retaining countercurrent systems in their legs. Seals, sea lions, and whales have them in their flippers. Heat loss is minimal, even when the animal is swimming in frigid water or standing on ice, because heat carried by blood flowing outward from the animal's core is picked up by blood flowing inward. 

? Compare countercurrent heat exchange with the countercurrent exchange in fish gills. (*Hint:* See Module 22.4.)

In both cases, countercurrent enhances transfer all along the length of a blood vessel—transfer of heat from one vessel to another in the case of a heat exchanger and transfer of oxygen between water and vessels in the case of gills.

25.3 Behavior often affects body temperature

Behavior is often important in thermoregulation. Many animals increase or decrease body heat by simply relocating. Some species migrate to a more suitable climate. Others bask in the sun or huddle together when it is cold, and find cool, damp areas or burrow when it is hot. Bathing brings immediate relief from the heat by convection and continues to cool the surface for some time by evaporation. Elephants often seek relief from tropical heat by bathing and by flapping their ears to cool their blood. Packed with blood vessels, their ears can radiate a lot of heat. Dressing for the weather is a thermoregulatory behavior unique to humans.

By seeking warmer or cooler locations, many ectotherms, especially reptiles, keep their body temperature quite stable. If a sunny spot is too warm, for instance, a lizard will alternate between the sun and the shade, or it will turn in a direction that exposes less surface area to the sun. 

Figure 25.3 Behavior that affects thermoregulation

? How has the Florida economy benefited from a behavioral adaptation for thermoregulation?

The winter migration of northerners to the warmer Florida climate brings a lot of money to the state.

25.4 Reducing the metabolic rate saves energy

The gray tree frog, a North American ectotherm, can spend much of the winter frozen. Having an extremely low metabolic rate, a frozen frog burns almost no energy all winter. A solution of freeze-protectant proteins and glycerol keeps ice crystals from rupturing its cells.

In contrast to ectotherms, endotherms can generally remain active in severe weather. However, endothermy consumes a lot of energy. Endotherms spend much more of their food energy on heat production than ectotherms do. When food supplies are low and environmental temperatures are extreme, many birds and mammals temporarily abandon endothermy, thereby expending less energy on thermoregulation.

Torpor is a state of reduced activity in which body temperature and metabolic rate decrease and the heart and respiratory system slow down. Many endotherms have short periods of torpor, usually at times when they cannot feed. Most bats, for instance, go into torpor when they are inactive, most often during daylight hours. Hummingbirds feed during the day and often undergo torpor on cold nights.

Figure 25.4 The gray tree frog: active (left) and frozen (right)

Hibernation (Latin *hibernus,* winter) is a type of long-term torpor by some animals in cold weather. Many ground squirrels, for example, hibernate in a grass-lined burrow. With a reduced metabolic rate, they live on energy stored in body fat when food was abundant. Some ground squirrels also have a summer torpor, called **estivation** (Latin *aestas,* summer). Estivation allows an animal to survive long periods of high environmental temperatures and reduced food and water supplies.

Many animals can tolerate wide fluctuations in their body temperature, but no animal can withstand much change in the relative amounts of dissolved solutes and water in its internal fluids. We now turn to homeostasis in body fluids.

? Why doesn't your sleep qualify as a short hibernation?

Because there is no large decrease in body temperature or basal metabolic rate

Osmoregulation and Excretion

25.5 Osmoregulation: All animals balance the gain and loss of water and dissolved solutes

The metabolic reactions on which life depends require a precise balance of water and dissolved solutes. Among the solutes whose concentration must be regulated are a variety of amino acids, proteins, and dissolved ions such as sodium (Na^+), chloride (Cl^-), potassium (K^+), calcium (Ca^{2+}), and bicarbonate (HCO_3^-). Osmosis occurs whenever two solutions separated by a membrane differ in total solute concentration (see Module 5.15). There is a net movement of water from the hypotonic solution to the hypertonic one until the solute concentrations are equal on both sides of the membrane. Whether an animal inhabits land, fresh water, or salt water, its cells cannot survive a *net* water gain or loss that significantly alters the concentration of its dissolved solutes. Animal cells swell and burst if there is a net uptake of water; they shrivel and die if there is a net loss of water.

As terrestrial animals, we obtain most of our water from food and drink. Like other animals, we lose water by urinating and defecating. We also lose it by evaporation as we breathe and perspire. Evaporation is unimportant for aquatic animals. Water continuously enters and leaves their bodies by osmosis.

What prevents an aquatic animal from having a net gain or loss of water? Some aquatic animals that live in the sea have body fluids with a solute concentration equal to that of seawater. Called **osmoconformers,** such animals do not undergo a net gain or loss of water because equal amounts of water move back and forth between two solutions with equal solute concentrations. The sea, where animals first evolved, is the only environment that supports osmoconformers. The total solute concentration in jellies, scallops, lobsters, and most other marine invertebrates conforms to that of seawater, and thus these animals do not expend energy regulating their water content. However, the concentration of certain ions in their body fluids, especially the fluid within their cells, is different from that of seawater. For example, in order for their cell membranes to function properly, the concentration of potassium ions (K^+) must be higher within their cells than in either their interstitial fluid or seawater. Because it takes energy to actively transport ions into cells, even an osmoconformer expends some energy to maintain its ion concentrations.

All freshwater animals, all land animals, and most marine vertebrates—whales, seals, sea birds, and most fishes—

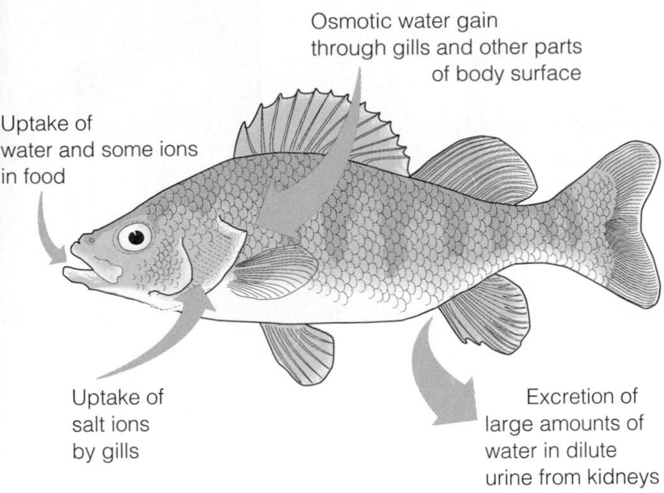

Figure 25.5A Osmoregulation in a freshwater fish, a perch

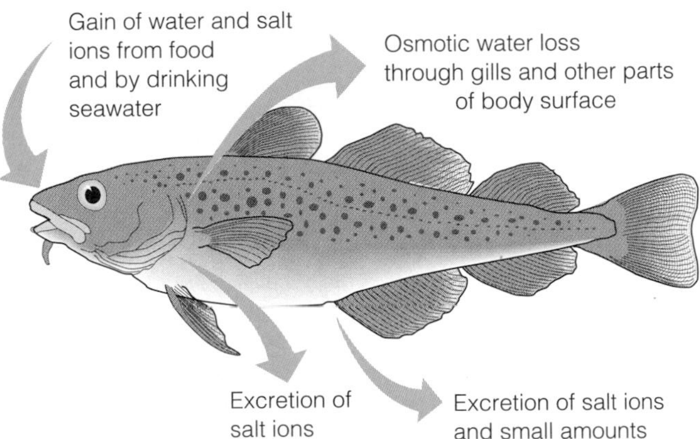

Figure 25.5B Osmoregulation in a saltwater fish, a cod

have body fluids whose solute concentration is different from that of their environment. Therefore, they must use energy in controlling water loss or gain. Such animals are called **osmoregulators.** The freshwater fish in Figure 25.5A has a very different internal solute concentration than its environment. The concentration of solutes in its internal fluids is much higher than that of fresh water, creating an osmotic problem for the animal. Because water flows by osmosis from a solution with a lower solute concentration to one with a greater solute concentration, the fish constantly takes in water from its surroundings. A freshwater fish gains water through its body surface, especially through its gills, and also in its food. It does not drink water except with its food. The freshwater fish also loses some solutes in its **urine,** the waste material produced by its excretory system.

It takes the work of three organ systems to achieve the proper water and solute balance in a freshwater fish. The animal's digestive system takes up ions from the food. Its gills (respiratory system) also take up salt ions, especially Na^+ and Cl^-. The fish's kidneys (excretory system) work constantly to produce large amounts of dilute urine—that is, urine with a much lower solute concentration than that of the animal's internal fluids. By excreting dilute urine, the fish disposes of excess water and conserves solutes.

The saltwater fish in Figure 25.5B has osmoregulatory problems opposite those of its freshwater relatives. Its internal fluids must be lower in total solutes than seawater. A saltwater fish loses water by osmosis through its body surfaces but compensates for the loss by drinking seawater and pumping excess salt ions, such as Na^+ and Cl^-, out through its gills. It also saves water and disposes of some salts by producing small amounts of urine with the same solute concentration as its body fluids.

Most fishes have little tolerance for changes in the solute concentration of their surroundings. The perch and the cod shown here, for instance, are restricted to fresh water and the ocean, respectively. In contrast, a few fishes, such as salmon, can migrate between seawater and fresh water. Salmon have remarkable osmoregulatory adaptations. While in the ocean, they drink seawater and excrete excess salts from their gills, osmoregulating like a cod. When salmon move into fresh water to spawn, their osmoregulatory mechanism switches to the freshwater mode of the perch; they cease drinking, and their gills take up salts from the dilute environment.

What about land animals? They are osmoregulators, but they are not surrounded by water; therefore, they cannot directly exchange water with the environment by osmosis. A terrestrial animal gains water by drinking and by eating moist foods, and it constantly loses water as it breathes and disposes of wastes. Its osmoregulatory situation resembles that of a marine fish like the cod; its paramount problem is losing water and becoming dehydrated. In fact, dehydration is such a severe problem on land that it may largely explain why only two groups of animals, arthropods and vertebrates, have colonized land with great success.

What is it about arthropods and vertebrates that gives them an edge against the dehydration threat? Insects, the prevalent arthropods on land, have tough exoskeletons impregnated with waterproof wax. Most terrestrial vertebrates, including humans, have an outer skin formed of multiple layers of dead, water-resistant cells. Also key to survival on land are adaptations that protect fertilized eggs and developing embryos from drying out. Many insects lay their eggs in humid or moist areas, and the eggs of many species are surrounded by a tough, watertight shell that keeps the developing embryo moist. Likewise, the embryos of terrestrial vertebrates—reptiles, birds, and mammals—develop in a water-filled amniotic sac surrounded by protective membranes. Behavioral adaptations can also save water. For instance, many desert mammals and arthropods spend much of their time in moist burrows and venture out only at night. And, as we will see in Module 25.11, the kidney also plays a major role in conserving water.

? Contrast osmoconforming and osmoregulating as mechanisms of homeostasis.

Osmoconforming prevents large gains or losses of water due to the animal being isotonic to the surrounding water; osmoregulating expends energy to balance water uptake and loss for an animal that is not isotonic to its surroundings.

25.6 Sweating can produce serious water loss

Despite cold weather, a cross-country skier will sweat heavily. Sweating helps the body thermoregulate, but it can also cause osmoregulatory problems. Knowing that replacing lost water is essential to preventing dehydration, the skier in Figure 25.6 stops in the middle of a race to drink.

Water loss is the main problem with sweating, although serious salt losses can also occur in extreme cases. Most of the salt we lose in sweat is NaCl (dissolved as the ions Na^+ and Cl^-). Sweat itself is much less concentrated than our body fluids; a volume of sweat contains only about one-third as many Na^+ and Cl^- ions as an equal volume of body fluid. As a result, sweating causes serious water loss well before ion losses become a problem.

The way to prevent dehydration from sweating is simply to drink water. Plain water is absorbed faster than beverages that contain sugar or dissolved ions (electrolytes). It is usually not necessary to take salt tablets or drink fluids that contain dissolved ions. In fact, if a person who has been sweating heavily takes salt tablets before water, the salts in the digestive tract tend to extract water from the body tissues by osmosis and dehydrate the body even more. This is why patients who are severely dehydrated are first given hypotonic fluids (solutions less concentrated than body fluids). Salt supplements are given later, if necessary. 

Figure 25.6 Rehydration

> ? What is the best way to prevent dehydration from heavy sweating in most cases?
>
> Drink water

25.7 Many freshwater animals face seasonal dehydration

When any animal loses more water than it takes in over a period of time, it becomes dehydrated. Even some aquatic animals—for instance, those that inhabit ponds or small lakes—are at risk, because their habitats tend to dry up seasonally. At even greater risk are many small invertebrates that live in moist soil or on wet plant surfaces. Droplets of water on the leaves of moss are seasonal homes to a great number of tiny animals. During wet months, the leaf surfaces come alive with tiny insects and other aquatic invertebrates. When the moss dries up, most of the animals lay eggs with tough waterproof coverings and die. But there are exceptions.

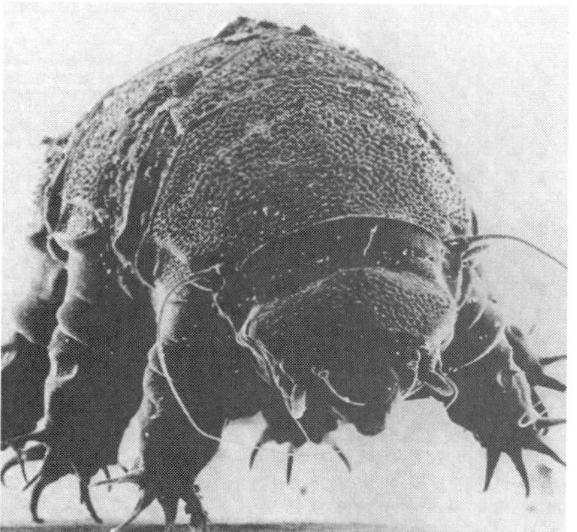

Figure 25.7 A tardigrade, active (left) and in a dehydrated, dormant state (right)

The micrographs in Figure 25.7 show a tiny (less than 1 mm long) invertebrate called a tardigrade, or water bear. Tardigrades thrive in droplets of water on mosses and other moist plants. When active, a tardigrade lumbers around on its stumpy legs, feeding on plant juices. When its habitat dries up, the tardigrade has the remarkable ability to lose more than 95% of its body water and survive in a dried-out, dormant state for decades (right micrograph). When a dried-out water bear is rehydrated, it can begin walking around again within minutes.

Biologists are just beginning to learn how this type of dehydration works. The problems for the animal are that fragile proteins tend to break down when they dry out, and cell membranes that are fluid when intact tend to collapse and stick together. Like a frozen frog (see Module 25.4), a dehydrated animal must have adaptations that keep proteins and membranes intact. Sugars are part of the solution for frozen animals, and studies of nematodes that can withstand dehydration indicate that a sugar called trehalose largely replaces the water that normally protects proteins and cell membranes from the effects of dehydration. Pharmaceutical companies have begun experimenting with sucrose and other sugars as preservatives for drugs composed of proteins. A mixture of a drug and sucrose is freeze-dried to a glasslike consistency, in some cases making the drug proteins stable over a wide range of temperatures. 

> ? The dormant water bear on the right in Figure 25.7 would be ______tonic to a puddle of rainwater.
>
> hyper

Animals must dispose of nitrogenous wastes

Waste disposal is as important to homeostasis as water and solute balance. Metabolism produces a number of toxic by-products, particularly the nitrogenous (nitrogen-containing) wastes that result from the breakdown of proteins and nucleic acids. An animal must dispose of (excrete) these metabolic wastes or be poisoned by them.

The form of an animal's nitrogenous wastes depends on the animal's evolutionary history and its habitat. As Figure 25.8 indicates, most aquatic animals dispose of their nitrogenous wastes as ammonia, which simply diffuses into the surrounding water. Small, soft-bodied invertebrates, such as planarian worms, excrete ammonia across their whole body surface. Fish excrete it mainly from the gills. Among the most toxic of all metabolic by-products, ammonia (NH_3) forms from the amino groups ($-NH_2$) removed from proteins and nucleic acids. Ammonia is highly toxic because it tends to raise the pH of body fluids and interfere with membrane transport functions. It is, in fact, too toxic to be stored in the body, but it is highly soluble in water and diffuses rapidly across cell membranes. If an animal is surrounded by water, ammonia readily diffuses out of cells and out of the body.

Ammonia excretion works only for aquatic animals. A terrestrial animal cannot dispose of ammonia fast enough to avoid poisoning because ammonia does not diffuse rapidly into the air. As indicated in the figure, land animals convert the amino groups from nitrogen-containing compounds into less toxic compounds such as urea or uric acid. These substances can be safely transported and stored in the body and then released periodically from the excretory system. The disadvantage of excreting urea or uric acid is that the animal must use energy to produce these compounds from amino groups. In contrast, no energy is expended when ammonia forms from an amino group.

Two groups of terrestrial vertebrates, mammals and amphibians, and some fishes excrete the organic molecule urea. Urea is highly soluble in water. It is also some 100,000 times less toxic than ammonia, so it can be held in a concentrated solution in the body and disposed of with relatively little water loss. Many animals that excrete urea also dispose of some nitrogenous wastes as ammonia. Moreover, some animals can switch between these waste products, depending on environmental conditions. Certain toads, for example, excrete ammonia (thus conserving energy) when in water, but they excrete mainly urea when on land.

Urea can be stored in a concentrated solution, but it still takes water to dispose of it. By contrast, land animals that excrete uric acid (birds, insects, many reptiles, snails, and a few amphibians living in deserts) avoid the water-loss problem almost completely. As you can see in the figure, uric acid is a considerably more complex molecule than either urea or ammonia. Uric acid is also thousands of times less soluble in water than either ammonia or urea. Uric acid can be excreted as small, solid crystals. In most cases, it is excreted as a paste or dry powder, combined with the feces. (The white material in bird droppings is mostly uric acid.) An animal must expend more energy to excrete uric acid than urea, but the higher energy cost is offset by the great savings in body water.

Why is ammonia safe as the main nitrogenous waste for most aquatic animals?

Because they can dispose of the very toxic ammonia continuously by secreting it across epithelia into the surrounding water

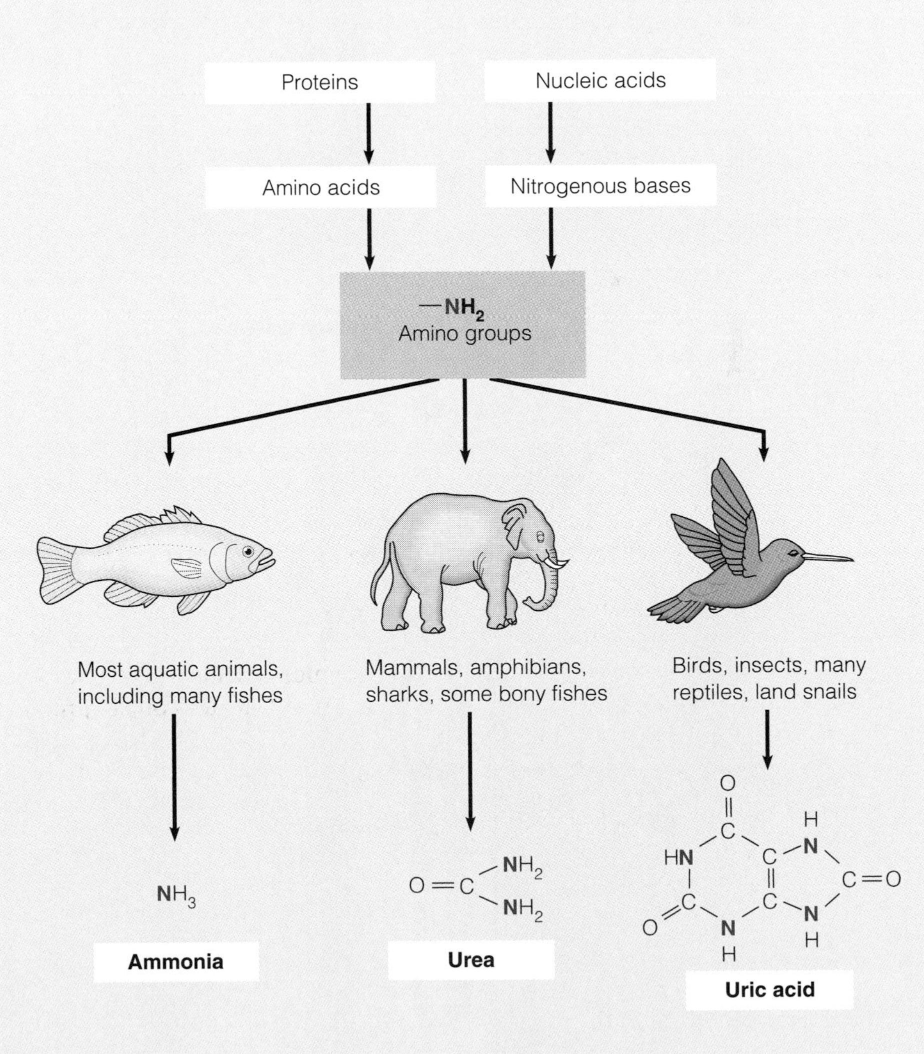

Figure 25.8 Nitrogen-containing metabolic waste products

25.9 The excretory system plays several major roles in homeostasis

25A
See Activity 25A on the CD-ROM.

Survival in any environment requires a precise balance between waste disposal and the animal's needs for water and salt. The excretory system plays a central role in homeostasis, forming and excreting urine while regulating the amount of water and salts in the body fluids.

Our two kidneys are the main processing centers of the excretory system. Each is a compact organ, about the size of your fist, nearly filled with about 80 km of fine tubes (tubules) and an intricate network of blood capillaries. The human body contains only about 5 liters of blood, but since this blood circulates repeatedly, about 1100–2000 L pass through the capillaries in our kidneys every day. As the blood circulates, our kidneys extract daily about 180 L (45 gal) of fluid, called **filtrate,** consisting of water, urea, and a number of valuable solutes, including Na^+, K^+, Cl^-, HCO_3^- (bicarbonate), glucose, and amino acids. If we excreted all the filtrate as urine, we would lose vital nutrients and dehydrate rapidly. But our kidneys refine the filtrate, concentrating the urea and returning most of the water and solutes to the blood. In a typical day, we excrete only about 1.5 L of urine.

Figure 25.9 below illustrates the "plumbing" plan and the blood supply of the human excretory system. Starting with the whole system in part A, blood to be filtered enters each kidney via a renal artery, shown in red; blood that has been filtered leaves the kidney in the renal vein, shown in blue. Urine leaves each kidney through a duct called a **ureter** and passes into the **urinary bladder.** The bladder empties periodically during urination. Urine actually leaves the body through a tube called the **urethra,**

Figure 25.9 Anatomy of the human excretory system

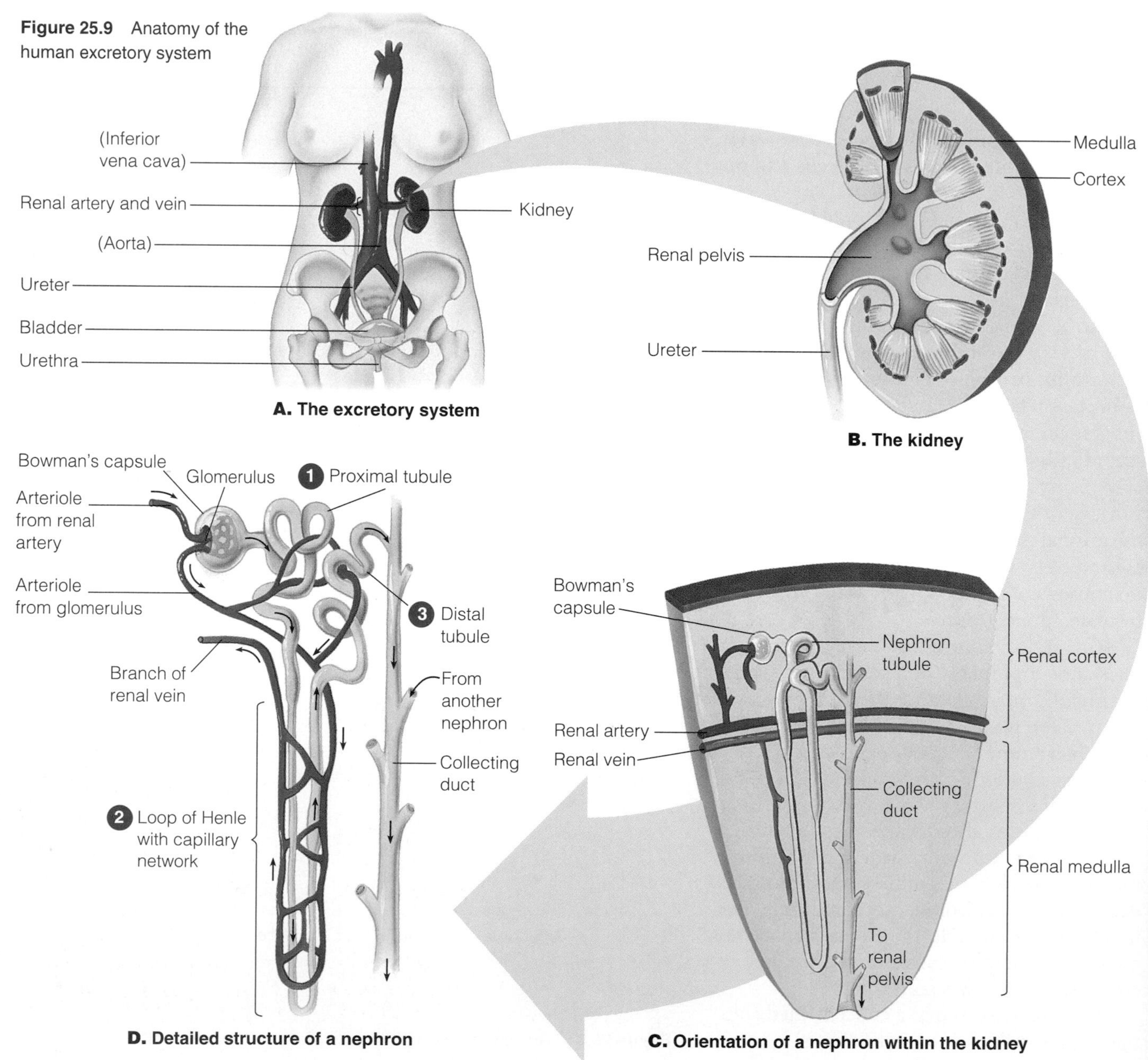

which empties near the female vagina or through the male penis.

As indicated in part B, the kidney has two main regions, the **renal cortex** (outer layer) and the **renal medulla** (inner region). From the medulla, the urine flows into a chamber called the renal pelvis, and from there into the ureter.

Each of our kidneys contains about a million tiny functional units called **nephrons,** one of which is shown in part C. A nephron consists of a tubule and its associated blood vessels. Performing the kidney's functions in miniature, the nephron extracts a tiny amount of filtrate from the blood and then refines the filtrate into a much smaller quantity of urine. Each nephron starts and ends in the kidney's cortex; some extend into the medulla, as in part C. The receiving end of the nephron is a cup-shaped swelling, called **Bowman's capsule.** At the nephron's other end is the **collecting duct,** which carries urine to the renal pelvis.

Part D in Figure 25.9 shows a nephron in more detail, along with its blood vessels. Bowman's capsule envelops a ball of capillaries called the **glomerulus** (plural, *glomeruli*). The glomerulus and Bowman's capsule make up the blood-filtering unit of the nephron. Here, blood pressure forces water and solutes from the blood in the glomerular capillaries across the wall of Bowman's capsule and into the nephron tubule. This process creates the filtrate, leaving blood cells and macromolecules behind in the capillaries.

The rest of the nephron refines the filtrate. There are three sections of the tubule: ① the **proximal tubule** (in the cortex); ② the **loop of Henle,** a hairpin loop carrying filtrate toward—in some cases, into—the medulla and then back toward the cortex; and ③ the **distal tubule** (called distal because it is the most distant from Bowman's capsule). The distal tubule empties its filtrate into a collecting duct, which receives filtrate from many nephrons. Passing through a collecting duct, the filtrate actually becomes urine. From the kidney's many collecting ducts, urine passes into the renal pelvis, and then into the ureter.

The intricate association between blood vessels and tubules is the key to nephron function. As shown in part D, the nephron has two distinct networks of capillaries. One network is the glomerulus, a finely divided portion of an arteriole that branches from the renal artery. Leaving the glomerulus, the arteriole re-forms and carries blood to the second capillary network, which surrounds the rest of the nephron tubule. This second network functions with its tubule in refining the filtrate. Notice how some of the vessels in this network are arranged around the loop of Henle. Blood is conveyed via these capillaries into the medulla along one side of the loop, then back along the other side of the loop. Leaving the nephron, the capillaries converge to form a small branch of the renal vein.

With the structure of a nephron in mind, we focus next on what actually happens as our excretory system filters blood, refines the filtrate, and excretes urine.

Which one of the following structures includes all others in the list? proximal tubule, Bowman's capsule, nephron, distal tubule, loop of Henle

Nephron

Overview: The key functions of the excretory system are filtration, reabsorption, secretion, and excretion 25.10

25B

Our excretory system produces and disposes of urine in four major processes, shown in Figure 25.10. First, during **filtration,** water and virtually all other molecules small enough to be forced through the capillary wall enter the nephron tubule from the glomerulus.

After filtration, two processes refine the filtrate. In **reabsorption,** water and valuable solutes, including glucose, salts, and amino acids, are reclaimed from the filtrate and returned to the blood. In **secretion,** certain substances are removed from the blood and added to the filtrate. When there is an excess of K^+ or H^+ in the blood, for example, these ions are transported into the cells of the nephron tubule, which secrete them into the filtrate. In removing excess H^+ from the blood, secretion keeps the blood from becoming acidic. Secretion also eliminates certain drugs (if they are present) and toxic substances from the blood.

Finally, in **excretion,** urine, the product of filtration, reabsorption, and secretion, passes from the kidneys to the outside via the ureters, urinary bladder, and urethra.

Urine differs in composition from the fluid that enters a nephron tubule by filtration because of the actions of ________ and ________.

reabsorption . . . secretion

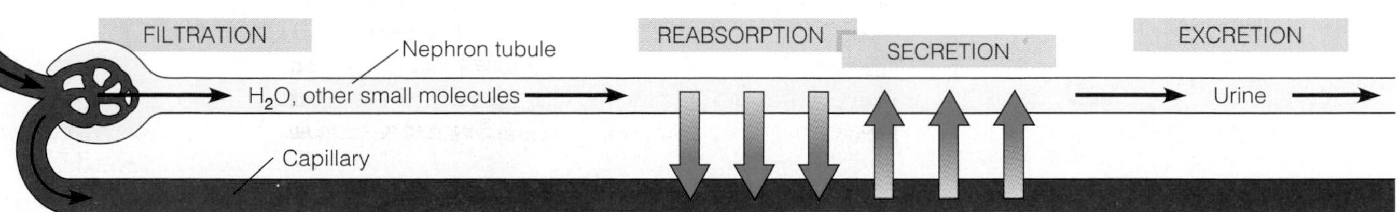

Figure 25.10 Major functions of the excretory system

25.11 From blood filtrate to urine: A closer look

25B
25C

Let's take a closer look at how a single nephron and collecting duct in the kidney produce urine from a blood filtrate that initially consists of a large amount of water and a number of valuable solutes, along with waste molecules.

The colored arrows in Figure 25.11 indicate where reabsorption and secretion occur along the nephron tubule. For simplicity, we have not illustrated the capillary network that surrounds the tubule of the nephron. The blue area around the nephron and collecting duct represents interstitial fluid. Small molecules and ions travel between the tubule and the capillaries by passing through the interstitial fluid during reabsorption and secretion.

The red and pink arrows pointing out of the tubule represent reabsorption. The blue arrows pointing into the tubule represent secretion. The red and blue arrows both indicate active transport; the pink arrows indicate passive transport (osmosis). The intensity of the blue color surrounding the tubule corresponds to the concentration of solutes in the interstitial fluid: The concentration is lowest in the cortex of the kidney and is progressively higher toward the inner medulla. We will see that it is by maintaining this solute gradient that the kidney can extract and save most of the water from the filtrate. All along the tubule, wherever you see water passing out of the filtrate into the interstitial fluid, it moves by osmosis. It does so because the solute concentration of the interstitial fluid exceeds that of the filtrate.

For our discussion, it is convenient to separate the activities of the proximal and distal tubules from the functions of the loop of Henle and collecting duct. As the figure indicates, the cells of the proximal and distal tubules reabsorb NaCl. They secrete some poisons, such as ammonia, and are the sites where controlled secretion of excess H^+ ions into the filtrate and reabsorption of HCO_3^- ions help regulate the blood's pH. The proximal tubule cells also reabsorb water and nutrients such as glucose and amino acids from the filtrate. And the distal tubule may also secrete drugs, such as penicillin, that circulate in the blood.

In contrast to the varied functions of the proximal and distal tubules, the loop of Henle and the collecting duct have one major function: water reabsorption. Although NaCl and some urea are also reabsorbed in these parts, their removal from the filtrate serves mainly to increase water reabsorption. Reabsorption of NaCl and urea maintains the concentration gradient of solutes in the interstitial fluid.

Notice that the long loop of Henle in the figure carries the filtrate deep into the medulla and then back to the cortex. As the filtrate passes into the medulla, water leaves the tubule because the interstitial fluid there has a higher solute concentration than the filtrate. As soon as the water passes into the interstitial fluid, it is carried away by nearby blood capillaries. This is essential, because the water would otherwise dilute the interstitial fluid surrounding the loop, destroying the concentration gradient.

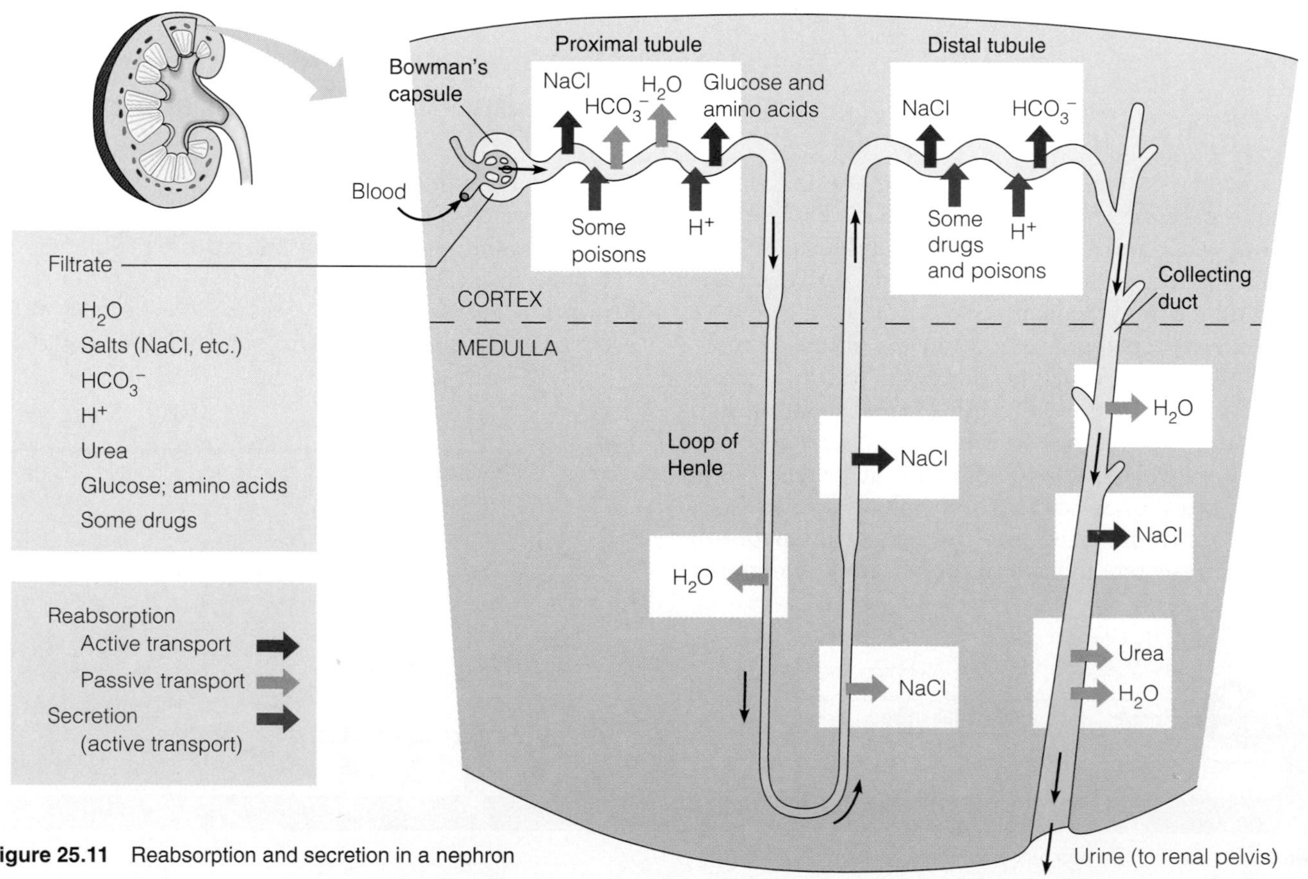

Figure 25.11 Reabsorption and secretion in a nephron

Just after the filtrate rounds the hairpin turn in the loop of Henle, water reabsorption stops, because the tubule there is impermeable to water. As the filtrate moves through the medulla back toward the cortex, NaCl is reabsorbed from it. At first, this reabsorption occurs passively, because the concentration of NaCl is higher in the filtrate than in the interstitial fluid. (The high solute concentration around the turn of the loop of Henle is largely due to urea, not to NaCl.) Active transport then takes over, and even more NaCl is reabsorbed as the filtrate nears the distal tubule. Because of this active transport, the NaCl concentration of the interstitial fluid surrounding the distal tubule and the first part of the collecting duct remains high. More water leaves the filtrate in the upper collecting duct as a result.

Final refining of the filtrate occurs in the collecting duct in the medulla, and urea is the main substance involved. Until the collecting duct enters the medulla, urea tends to remain in the filtrate because the nephron is relatively impermeable to it. As a result, the filtrate's concentration of urea has become quite high—in fact, higher than that of the interstitial fluid. In the medulla, however, the collecting duct is permeable to urea, and some urea leaks out of it into the interstitial fluid. Deep in the medulla, the urea makes the interstitial fluid so concentrated that even more water is reabsorbed before the urine passes into the renal pelvis.

So we see that the nephron can reabsorb and thus save much of the water that filters into it from the blood. Water conservation is a major part of body fluid homeostasis for a land animal, but our nephrons do much more than just save water. Under hormonal control, they maintain a precise balance between water and solutes in our body fluids. When the solute concentration of the body fluids rises above a set point, a control center in the brain increases the level of a hormone called ADH (antidiuretic hormone) in the blood. ADH signals the nephrons to reabsorb more water from the filtrate. Conversely, when the body fluid solutes are diluted below the set point, as when we drink too much water, blood levels of ADH drop, making the nephrons reabsorb less water. ADH is one of three hormones that provide an elaborate system of checks and balances on the kidney's effects on salt and water balance. We return to hormones and their roles in homeostasis in Chapter 26.

Some of the drugs classified as diuretics make the epithelium of the collecting duct less permeable to water. How would this affect kidney function?

The kidney medulla would reabsorb less water, and thus the diuretic would increase water loss in the urine.

Kidney dialysis can be a lifesaver 25.12

Knowing how the nephron works helps us understand how some of its functions can be performed artificially when the kidneys are damaged. Although kidney failure is not common, it means certain death from the buildup of toxic wastes in the blood if left untreated.

Figure 25.12 illustrates a type of artificial kidney, called a dialysis machine. **Dialysis** means "separation" in Greek, and, like the nephrons of the kidney, the machine sorts small molecules of the blood, keeping some and discarding others. The patient's blood is pumped from an artery through a series of tubes that operate like nephrons. The tubes are made of a selectively permeable membrane, which allows only water and certain solutes to pass through it, and they are immersed in a dialyzing solution much like the interstitial fluid that bathes the nephrons. As the blood circulates through the dialysis tubing, urea and excess salts diffuse out of it (instead of leaving by filtration and secretion, as in the nephron). Needed substances, such as bicarbonate ions, diffuse from the dialyzing solution into the blood (reabsorption). The machine continually discards the used dialyzing solution as wastes build up in it (excretion).

Dialysis may be used when the kidneys are damaged physically, become infected, or are chemically poisoned, but can be repaired or replaced by transplant. Unfortunately, dialysis machines are not always available when needed, and not everyone can afford them when they are.

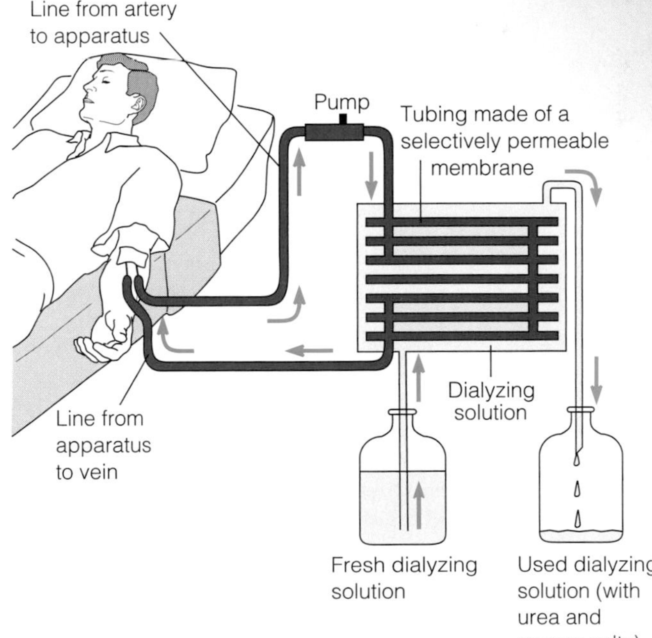

Figure 25.12 The principles of dialysis

Why isn't dialysis available to everyone who needs it?

It's expensive.

Homeostatic Functions of the Liver

25.13 The liver is vital in homeostasis

A discussion of homeostasis would not be complete without mention of the liver, which performs more functions than any other organ in the body. In addition to its roles in digestion (see Module 21.10), the liver supports the activities of the kidney. Specifically, the liver prepares nitrogenous wastes for disposal by synthesizing urea from ammonia. It also helps the kidney get rid of other toxins such as alcohol and other drugs. The liver converts these into inactive products that the kidney can filter from the blood and excrete in the urine.

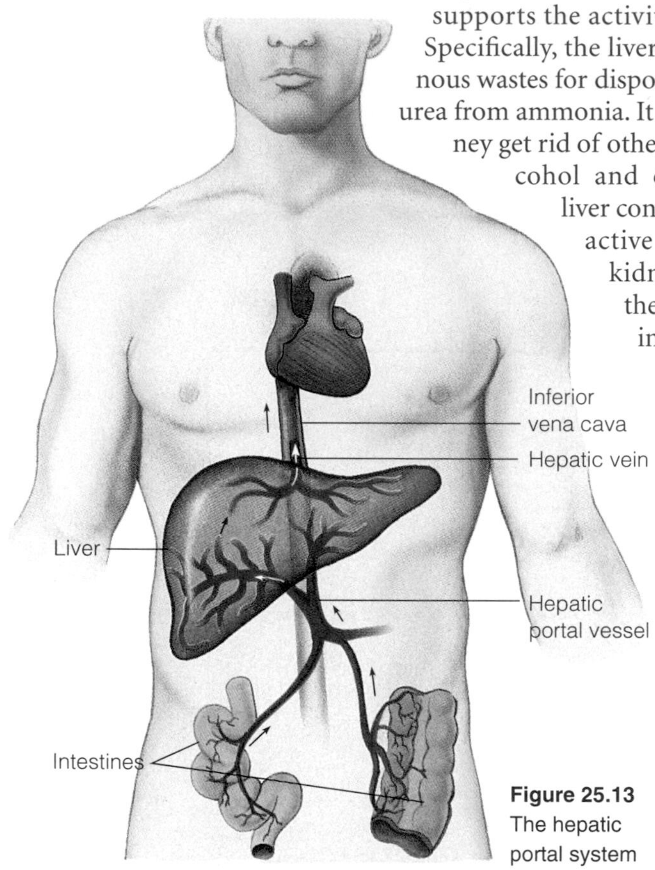

One reason the liver can do so much is that liver cells have the metabolic machinery to chemically alter a great number of substances. Another consideration is the liver's strategic location in the body—between the intestines and the heart. As indicated in Figure 25.13, the intestinal capillaries converge to form a single blood vessel, the **hepatic portal vessel,** which conveys blood directly to the liver. Nutrients absorbed by the intestine, as well as any harmful substances picked up by the intestinal capillaries, pass to the liver. (Alcohol is absorbed mainly by the stomach, and capillaries in the stomach wall pass it into the hepatic portal vessel.) Thus, the liver has a chance to modify and detoxify substances absorbed by the digestive tract before the blood carries these materials to the heart for distribution to the rest of the body. Blood travels from the liver to the heart via the hepatic vein and the inferior vena cava.

Figure 25.13 The hepatic portal system

The liver also contributes in a major way to homeostasis by regulating the amount of glucose in the blood. One of the liver's most important functions is to convert glucose into glycogen and store the glycogen for use at a later time. Blood exiting the liver usually has a very low glucose concentration (close to 0.1%). In balancing the amount of glycogen it stores with the amount of glucose in the blood, the liver plays a key role in regulating body metabolism.

The liver's diverse roles in homeostasis and its interaction with other organs make the point that homeostasis requires the concerted action of several body systems. The coordination of all the body's regulatory systems by hormones is the subject of Chapter 26.

What role does the liver play in the body's processing of nitrogenous waste?

The liver is the site of urea synthesis.

Chapter Review

Chapter Summary

Homeostasis (Introduction) Internal homeostatic mechanisms compensate for fluctuations in the external environment. Osmoregulation is the control of the gain and loss of water and dissolved solutes; excretion is the disposal of metabolic wastes; and thermoregulation is the maintenance of body temperature within a tolerable range. Endothermic animals derive most of their body heat from metabolism; ectothermic animals warm themselves mainly by absorbing heat from their surroundings.

Thermoregulation (25.1–25.4) Body temperature regulation requires adjustment to heat gained from or lost to an animal's environment by conduction, convection, radiation, and evaporative cooling **(25.1)**. Thermoregulation depends on heat production by the animal, as well as on heat gain and loss. Endotherms and some ectotherms can change their rate of heat production. Both endotherms and ectotherms may change their rate of heat gain or loss by conduction, convection, radiation, or evaporation. In cold weather, hormonal changes may increase heat production by raising the metabolic rate. Shivering also increases metabolic heat production. Fur and feathers help the body retain heat. Blood flow to the skin affects heat loss. In a countercurrent heat exchanger, blood from the core of the body warms cooler blood returning from the skin or limbs, conserving body heat **(25.2)**. Basking in the sun, sitting in the shade, bathing, burrowing, huddling, or migrating can also affect body temperature **(25.3)**. When exposed to extreme temperatures, many animals conserve energy by reducing their metabolic rate. Torpor, which includes hibernation in cold weather and estivation in warm weather, is a state of reduced activity that saves energy **(25.4)**.

Osmoregulation and Excretion (25.5–25.12) Many marine animals are osmoconformers; their body fluids have the same concentration of solutes as seawater. Osmoregulators control water and solute concentrations. Freshwater fishes gain water by osmosis and tend to lose solutes; they excrete excess water and conserve solutes. Many marine fishes lose water by osmosis, drink seawater, and excrete excess salts. Land animals gain water by drinking and eating but lose water and solutes by evaporation and waste disposal; their kidneys, behavior, and waterproof skin conserve water **(25.5)**. Water lost in thermoregulation can cause osmoregulatory problems **(25.6)**. Many small invertebrates can dehydrate and become dormant when their environment dries up **(25.7)**.

Nitrogen-containing wastes are toxic byproducts of protein and nucleic acid breakdown. Ammonia is poisonous but soluble, and it is easily disposed of by aquatic animals. Urea is less toxic and easier for many land animals to store and excrete. Some land animals save water by excreting a virtually dry waste, which is mainly uric acid **(25.8)**.

The excretory system expels wastes and regulates water and salt balance. The two human kidneys each contain about a million functional units called nephrons. Each nephron consists of a folded tubule and associated blood vessels. The nephrons extract a filtrate from the blood and refine the filtrate into a much smaller amount of urine. Urine leaves the kidneys via the ureters, is stored in the urinary bladder, and is expelled through the urethra **(25.9)**. The excretory system has four primary functions: filtration, reabsorption, secretion, and excretion. In filtration, blood pressure forces water and many small solute molecules from the blood into the nephron. In reabsorption, the nephron tubule reclaims valuable solutes, such as glucose and salts, from the filtrate. In secretion, the nephron removes substances, such as excess H^+ and other ions, and adds them to the filtrate. The product of these processes, urine, is then excreted **(25.10)**. Reabsorption of salts and urea by active and passive transport makes the interstitial fluid surrounding the tubules highly concentrated, promoting the osmotic reabsorption of water. Controlled secretion of hydrogen ions and reabsorption of bicarbonate ions help regulate the pH of the blood. Secretion also includes the active transport of drugs and poisons, such as ammonia. Antidiuretic hormone (ADH) and other hormones regulate the amount of salt and water the kidneys excrete **(25.11)**. Compensating for kidney failure, a dialysis machine performs the functions of the nephrons by removing wastes from the blood and maintaining its solute concentration **(25.12)**.

Homeostatic Functions of the Liver (25.13)
The liver performs many homeostatic functions. It assists the kidneys by making urea from ammonia and breaking down toxic chemicals. Blood from the intestines flows through the liver before distribution to the rest of the body, allowing the liver to adjust the blood's chemical content.

Testing Your Knowledge

Multiple Choice

1. The main difference between endotherms and ectotherms is
   a. how they conserve water
   b. whether they are warm or cold
   c. where they get most of their body heat
   d. whether they live in a warm or cold environment
   e. whether they live on land or in the water

2. In each nephron of the kidney, the glomerulus and Bowman's capsule
   a. filter the blood and capture the filtrate
   b. reabsorb water into the blood
   c. break down harmful toxins and poisons
   d. reabsorb salts and nutrients
   e. refine and concentrate the urine for excretion

3. As filtrate passes through the long loop of Henle, salt is removed and concentrated in the interstitial fluid of the kidney medulla. Because of this high salt concentration, the nephron is able to
   a. excrete the maximum amount of salt
   b. neutralize toxins that might accumulate in the kidney
   c. control the pH of the interstitial fluid
   d. excrete a large amount of water
   e. reabsorb water most efficiently

4. Birds and insects excrete uric acid, while mammals and amphibians excrete mainly urea. What is the chief advantage of uric acid over urea as a waste product?
   a. Uric acid is more soluble in water.
   b. Uric acid is a much simpler molecule.
   c. It takes less energy to make uric acid.
   d. Less water is lost excreting uric acid.
   e. More solutes are lost excreting uric acid.

Matching

Match each of the following components of blood with what happens to it as the blood is processed by the kidney.

1. Water
2. Glucose
3. Plasma protein
4. Hydrogen ion (H^+)
5. Red blood cell
6. Urea

a. passes into filtrate; partially reabsorbed; excreted in urine
b. remains in blood
c. passes into filtrate; mostly reabsorbed
d. passes into filtrate; also secreted; excreted in urine

Describing, Comparing, and Explaining

1. Compare the problems of water and salt regulation a salmon faces when it is swimming in the ocean and when it migrates into fresh water to spawn.

2. Compare the countercurrent heat exchanger with the countercurrent mechanism that enables a fish to maximize oxygen absorption from water. How do the opposing flows increase the efficiency of heat or gas exchange?

Thinking as a Scientist

Riding by a lake in midwinter, you notice a small flock of geese standing on the ice. Imagine what it would be like for you to stand there with no boots or warm pants. Propose a hypothesis to explain why the birds' legs do not freeze. You may assume you have equipment for measuring temperatures in the birds' legs. What results would you expect if your hypothesis is correct?

Science, Technology, and Society

The kidneys remove many drugs from the blood, and these substances show up in the urine. Some employers require a urine drug test at the time of hiring and/or at intervals during the term of employment. Why do some employers feel that drug testing is necessary? Do you think that passing a drug test is a valid criterion for employment? If so, for what types of jobs? Would you take a drug test to get or keep a job? Why or why not?

Media Resources

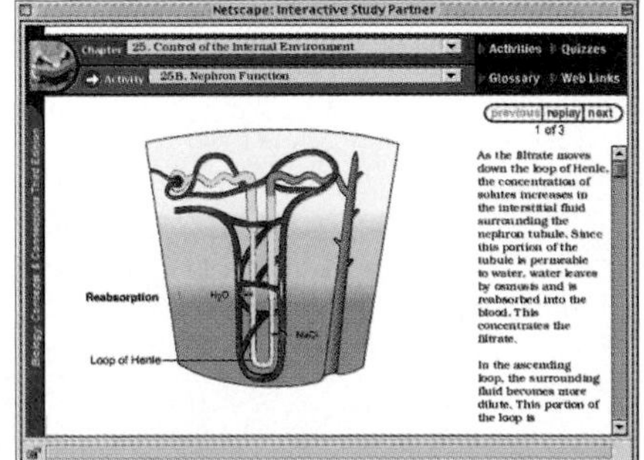

For further review of the concepts in this chapter, use the activities and quizzes on the CD-ROM.

Visit the Special Edition of The Biology Place web site for activities, quizzes, news articles, and web links.

26 Chemical Regulation

Outline

The Nature of Chemical Regulation

Frogs have three eyes: the two big ones that bulge from the head and a tiny one you can't see, between the other two. The third eye, called the frontal organ, is embedded in the skin. What can an eye covered with skin do? Scientists do not yet have a complete answer, but the frontal organ and a structure next to it, an outgrowth of the brain called the pineal gland, are known to detect light. The pineal gland also secretes a regulatory chemical called melatonin into the blood. The more light the pineal gland and frontal organ detect, the less melatonin the pineal puts out.

Most vertebrates lack a frontal organ, but nearly all have a pineal gland that secretes melatonin. The amount of melatonin in the blood generally cycles over a 24-hour period, varying with the amount of sunlight and thus the time of day. The constantly changing level of melatonin in the blood may help regulate daily rhythms in body activities. In humans, for example, a melatonin peak often occurs at about 2:00 A.M., and high blood levels of melatonin seem to promote sleep. Conversely, low melatonin levels occur while we are awake during the day. The sleepiness commonly experienced by jet travelers and workers on night shifts may be due to melatonin telling the body to sleep when it needs to be awake. A promising new therapy for jet lag and other problems of sleepiness and insomnia is timely exposure to bright artificial light. The light probably works by resetting the body's melatonin cycle.

For many vertebrates, melatonin also helps regulate reproductive events. Consider, for instance, the behavior of the bullfrog during its breeding season, in the spring (photograph on facing page). The male is silent for most of the day, but on a still night you can hear his distinctive, deep-voiced "jug o' rum" or "br' wum" from nearly a mile away. His calls repel other males and attract females. Changes in the frog's melatonin level at nightfall and daybreak probably control the timing of this behavior.

Many animals also have seasonal melatonin cycles. In humans, melatonin levels are high during the short days of late fall and early winter and lower in the spring and early summer. The role of these seasonal changes is not known. They may be an evolutionary holdover from an ancestral species whose reproductive activities were limited to a certain season, as they are in most mammals. More is known about some of the problems that may result when our melatonin system goes awry. Probably related to abnormal melatonin cycles are seasonal moodiness, such as the winter "blues," and a severe condition called seasonal affective disorder (SAD). For people with SAD, short winter days bring severe depression, extreme irritability, and anxiety, often with a craving for sweets, a tendency to eat ravenously between meals, and a resulting weight gain. SAD sufferers may go to bed early and still feel exhausted when they arise nine or ten hours later. Their condition is likely to worsen if they move north in the winter, but if they move south, they improve. SAD symptoms usually disappear in the spring.

SAD seems to be a relatively rare disorder, and it is not yet clear whether melatonin imbalances really cause it. But SAD sufferers do have unusual daily melatonin cycles, with peaks often occurring at midnight or dawn, instead of around 2:00 A.M. Also, people with SAD become more depressed if they take oral doses of melatonin. In contrast, their symptoms are relieved and the melatonin in their blood decreases if they are exposed to high-intensity light (about ten times normal room lighting) for an hour or more each morning. Similar light therapy can also relieve the irritability and drowsiness that many people experience during the short days of winter.

Melatonin is one of many chemical signals that make an animal's organ systems function cooperatively. Recent research suggests that melatonin acts on a part of the brain that functions as a biological clock. The hormone seems to decrease the activity of nerve cells there, and this effect may be related to its role in maintaining body rhythms. But key questions about how this chemical works remain unanswered. As we will see in this chapter, research on most other chemical signals is further along.

This chapter is specifically about chemical signals, but our overriding theme is homeostasis, with a focus on how chemical signals maintain an animal body's dynamic steady state. We begin with a look at the main kinds of chemicals that regulate body functions and the cells that secrete them.

The Nature of Chemical Regulation

26.1 Chemical signals coordinate body functions

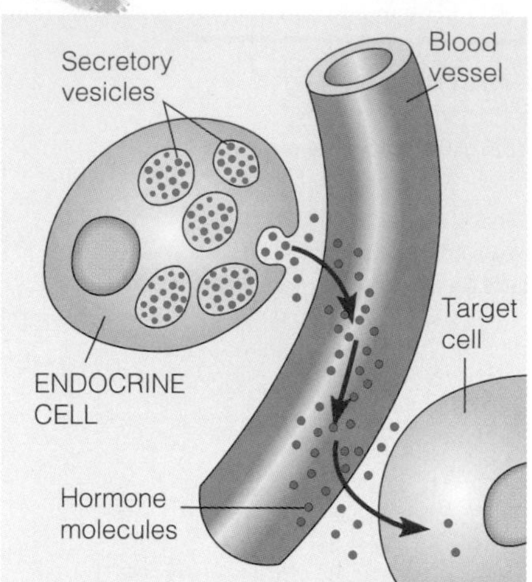

Figure 26.1A Hormone from an endocrine cell

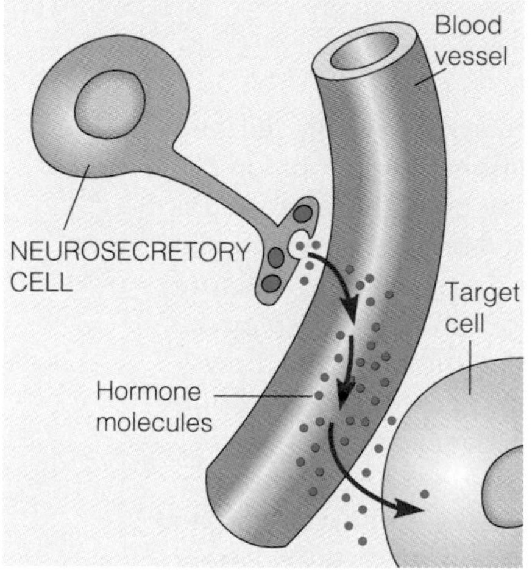

Figure 26.1B Hormone from a neurosecretory cell

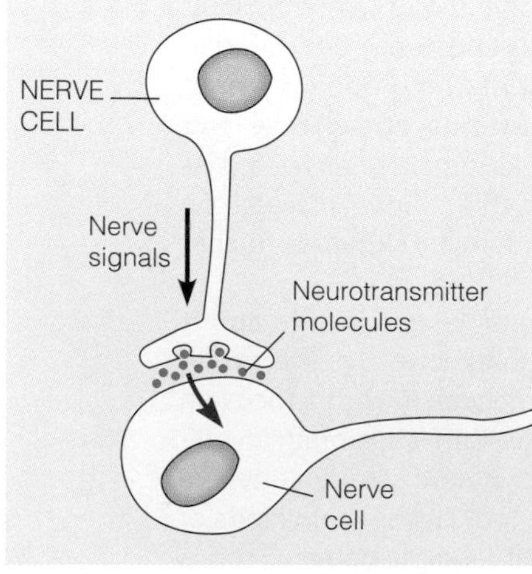

Figure 26.1C Neurotransmitter

Animals rely on many kinds of chemical signals to regulate their body activities. Melatonin is an example of one kind, a hormone (from the Greek *hormon*, to excite). A **hormone** is a regulatory chemical that travels in the blood from its production site and affects other sites in the body, often at some distance. Hormones are made and secreted mainly by organs called **endocrine glands.** Figure 26.1A sketches the activity of a hormone-secreting cell in an endocrine gland. Secretory vesicles in the endocrine cell are full of molecules of the hormone (blue). The endocrine cell secretes the molecules directly into the circulatory system. The molecules travel in the blood to **target cells,** cells that respond to this hormone.

Figure 26.1B shows a second type of hormone-secreting cell. This **neurosecretory cell** is a specialized nerve cell that, in addition to conducting nerve signals, makes and secretes hormones. Like endocrine cells, neurosecretory cells release hormones into the blood for transport to target cells.

Collectively, all hormone-secreting cells constitute the **endocrine system,** the body's main chemical-regulating system. Because hormones are carried in the blood, they reach all parts of the body, and the endocrine system is especially important in controlling whole-body activities. For example, hormones govern our metabolic rate, growth, maturation, and reproduction. In many cases, a single hormone molecule can dramatically alter a target cell's metabolism by turning on the production of a number of enzymes. A tiny amount of a hormone can govern the activities of enormous numbers of target cells in a variety of organs.

The endocrine system often collaborates with the body's other major coordinating system, the nervous system. The nervous system transmits electrical signals via nerve cells. When a nerve signal reaches the end of a nerve cell, it triggers the secretion of molecules called neurotransmitters (Figure 26.1C). **Neurotransmitters** are chemicals that carry information from one nerve cell to another, or from a nerve cell to another kind of cell that will react, such as a muscle cell or an endocrine cell. Thus, the nervous system, like the endocrine system, relies on chemical messengers. Unlike hormones, however, most neurotransmitters do not travel in the bloodstream.

Why does an animal need two kinds of regulatory systems? Timing is part of the answer. In many cases, the endocrine system takes minutes, hours, or even days to act, partly because of the time it takes for hormones to be made and carried in the blood to all their target organs. And some hormones act by stimulating their target cells to synthesize new proteins, adding even more time to the process. In contrast, the nervous system provides split-second control. The flick of a frog's tongue catching a fly and the jerk of your hand away from a flame result from high-speed nerve signals, not from hormonal control. In the case of your hand, a single nerve cell carries a signal that conveys the sensation of burning to your spinal cord. A second nerve cell transmits a signal into the arm muscle that actually pulls away. A chemical signal is involved: A neurotransmitter carries the signal from the first nerve cell to the second one. But the distance it travels and the time it takes are slight.

A neurotransmitter is a type of chemical signal called a local regulator. A **local regulator** is secreted into the interstitial fluid and affects cells very near the point of secretion—in contrast to a hormone, whose target cells typically are distant from the cells that secrete it. Interleukins, for example, are local regulators made by immune-system cells (see Module 24.13). Local regulators called **prostaglandins** are made by nearly all cells and have a variety of functions. For instance, prostaglandins secreted by the placenta cause the nearby muscles of the uterus to contract, helping induce labor during childbirth. We discuss neurotransmitters further in Chapter 28. For the rest of this chapter, we'll explore the endocrine system and its hormones. 

? How does a hormone travel between an endocrine gland and the hormone's target cells?

Via the bloodstream

Hormones affect target cells by two main signaling mechanisms

26A
26B
26C

Vertebrates make over 50 different hormones. Released into the blood, a hormone may come into contact with all the tissues in the body, but only cells with specific receptors for that hormone are affected by it. Despite the diversity of hormones, there are only two general mechanisms by which hormones trigger changes in target cells.

Our current understanding of the most common mode of hormone action stems from the pioneering work of American physiologist Earl W. Sutherland in the 1950s. Sutherland studied the effects of epinephrine on liver cells and muscle cells. Epinephrine (also known as adrenaline) is called the "fight-or-flight" hormone because it prepares the body for sudden action. Among other effects, it stimulates the breakdown of the storage polysaccharide glycogen in liver cells. Glycogen breakdown yields glucose, which provides body cells with a ready supply of energy.

Sutherland found that epinephrine brings about glycogen breakdown in a liver cell without ever entering the cell. Instead, as illustrated in Figure 26.2A, ① it binds to a specific receptor protein in the plasma membrane of the cell. The binding activates the receptor, which ② initiates a multistep signal-transduction pathway in the cell. A series of relay molecules are activated in sequence. ③ The final relay molecule activates a protein that carries out the cell's response. A liver cell's response to signaling from epinephrine is the breakdown of glycogen by enzymes in the cytoplasm. Other hormones bring about other kinds of cellular responses; one important category is the turning on of specific genes in the nucleus, as you saw in Module 11.11.

The hormones that bind to plasma-membrane receptors are all synthesized from amino acids, and there are three main classes of these substances. The *amine hormones* are modified versions of single amino acids. The *peptide hormones* are short chains of amino acids (as few as three). The *protein hormones* are made of polypeptides having as many as 200 amino acids. (For examples of amine, peptide, and protein hormones, see the table in Module 26.3.)

The most important class of hormones *not* derived from amino acids are the **steroid hormones,** which are lipids made from cholesterol (see Module 3.9). The steroid hormones, such as our sex hormones estrogen and testosterone, use the mechanism depicted in Figure 26.2B. As lipids, ① steroid hormones can diffuse through the phospholipid plasma membrane of cells. Once inside a target cell, a steroid hormone ② binds to a specific receptor protein in the cytoplasm. Rather than triggering a signal-transduction pathway, the receptor itself carries out the transduction of the hormonal signal: When activated by the hormone, it becomes a transcription factor—a gene activator (see Module 11.6). ③ The hormone-receptor complex enters the nucleus and attaches to specific sites on the cell's DNA (enhancers). ④ The binding to DNA activates transcription of certain genes into RNA, which is translated into new proteins. In contrast to nonsteroid hormones, steroid hormones *always* affect gene expression.

A hormone can affect a variety of receptors and other signal-transduction molecules in various kinds of target cells. Thus, different kinds of cells can respond differently to the same hormone. The main effect of epinephrine on heart muscle cells, for example, is not glycogen breakdown but cellular contraction, which speeds up the heartbeat.

What are two major differences between the mechanisms of action of steroid and nonsteroid hormones?

(1) Steroid hormones bind to receptors inside the cell; most other hormones bind to plasma-membrane receptors. (2) Steroid hormones always activate gene expression; other hormones have this or other effects.

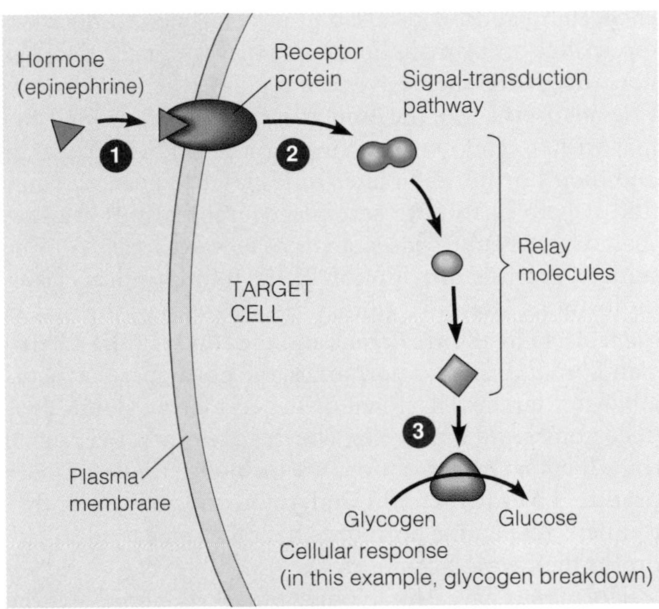

Figure 26.2A A hormone that binds a plasma-membrane receptor

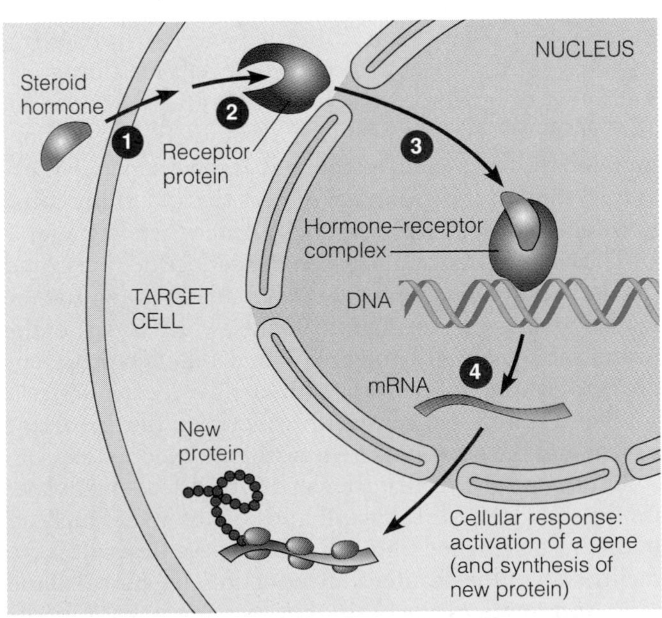

Figure 26.2B A hormone that binds an intracellular receptor

The Vertebrate Endocrine System

26.3 Overview: The vertebrate endocrine system

26D

The vertebrate endocrine system consists of more than a dozen glands. Some of these, such as the thyroid and the pituitary gland, are endocrine specialists; their sole or main function is secreting hormones into the blood. Several other glands have both endocrine and nonendocrine functions. The pancreas, for example, contains endocrine cells that secrete three hormones that influence the level of glucose in the blood. The pancreas also has nonendocrine cells that secrete digestive enzymes into the intestine via ducts (see Module 21.10).

Figure 26.3 shows the locations of most of the human endocrine glands. The table on the facing page summarizes the actions of the main hormones they produce, as well as how the glands themselves are regulated. The table provides an overview of the human endocrine system, and you may wish to refer to it as we focus on the individual glands and their hormones in later modules. For now, you may find it helpful to note several general features that are shared by the endocrine systems of all vertebrates. For one thing, notice the distribution of the four chemical classes of hormone (steroids, amines, peptides, and proteins) in the table. Only the sex organs and the cortex of the adrenal gland produce steroid hormones, the hormones that actually enter target cells. Most of the endocrine glands produce nonsteroid hormones, which generally act via signal transduction. Two exceptions are the nonsteroid hormones produced by the thyroid gland, thyroxine and triiodothyronine. These amine hormones have the same mode of action as the steroid hormones.

Hormones have a wide range of targets. Some, like the sex hormones, which promote male and female characteristics, affect most of the tissues of the body. Other hormones, such as glucagon from the pancreas, have only a few kinds of target cells (liver and fat cells for glucagon). The hormone gastrin, secreted by cells of the stomach wall, also has its target cells there (see Module 21.8). Some hormones have other endocrine glands as their targets. For example, the pituitary gland produces thyroid-stimulating hormone, which stimulates activity of the thyroid gland.

The close association between the endocrine system and the nervous system is apparent in both the drawing and the table. For example, the hypothalamus, which is part of the brain, secretes many hormones that regulate other endocrine glands, especially the pituitary. We'll explore structural and functional connections between the endocrine system and the nervous system further in Module 26.4.

Endocrine glands that we will not be discussing in later modules are the pineal gland and the thymus. Much remains to be learned about both of these organs. As we mentioned in the chapter's introduction, the **pineal gland** is an outgrowth of the brain that secretes melatonin, the hormone that links environmental light conditions with daily or seasonal rhythms. We know the most about melatonin's function in mammals that breed during certain seasons. For example, in sheep and deer that breed in the fall, when days are short, high levels of melatonin in the blood stimulate reproductive activity. In contrast, in mammals that breed in the spring, longer days and less melatonin in the blood promote reproductive activity. We do not yet know what specific effects melatonin has on the body cells that produce these rhythms.

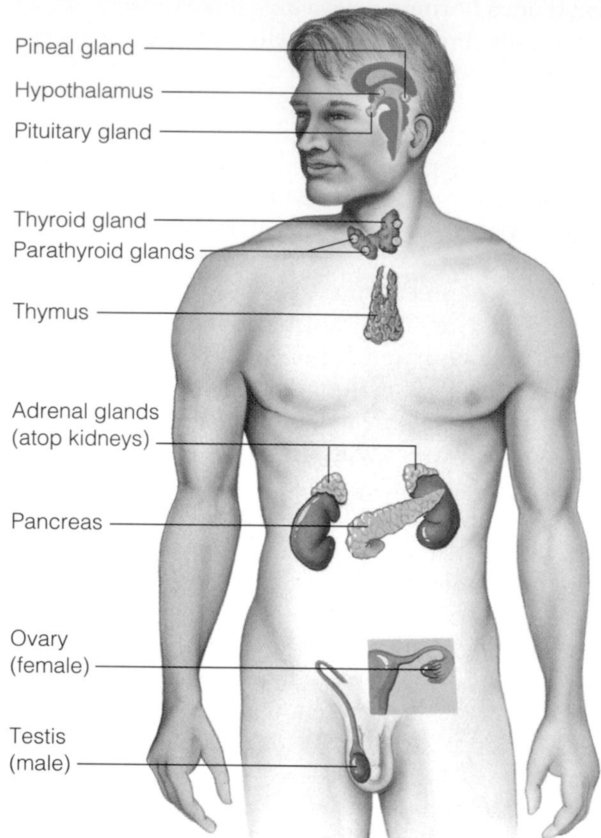

Figure 26.3 The major endocrine glands in humans

The **thymus gland** lies under the breastbone in humans and is quite large during childhood. Not until the 1960s was the important role of the thymus in the immune system discovered. Thymus cells secrete several important hormones, including a peptide that stimulates the development of T cells (see Module 24.5). Beginning at puberty, when the immune system is well established, the thymus shrinks drastically. However, it continues to secrete its T-cell–stimulating hormones throughout life. 

? The pituitary is sometimes called the "master gland" because its hormones control other endocrine glands. Which four endocrine glands are directly regulated by pituitary hormones? (*Hint:* Consult the table at the right.)

Ovaries, testes, adrenal cortex, and thyroid

Major Vertebrate Endocrine Glands and Some of Their Hormones

Gland	Hormone	Chemical Class	Representative Actions	Regulated by
Hypothalamus	Hormones released by the posterior pituitary and hormones that regulate the anterior pituitary (see below)			
Pituitary gland Posterior lobe (releases hormones made by hypothalamus)	Oxytocin	Peptide	Stimulates contraction of uterus and mammary gland cells	Nervous system
	Antidiuretic hormone (ADH)	Peptide	Promotes retention of water by kidneys	Water/salt balance
Anterior lobe	Growth hormone (GH)	Protein	Stimulates growth (especially bones) and metabolic functions	Hypothalamic hormones
	Prolactin (PRL)	Protein	Stimulates milk production	Hypothalamic hormones
	Follicle-stimulating hormone (FSH)	Protein	Stimulates production of ova and sperm	Hypothalamic hormones
	Luteinizing hormone (LH)	Protein	Stimulates ovaries and testes	Hypothalamic hormones
	Thyroid-stimulating hormone (TSH)	Protein	Stimulates thyroid gland	Thyroxine in blood; hypothalamic hormones
	Adrenocorticotropic hormone (ACTH)	Protein	Stimulates adrenal cortex to secrete glucocorticoids	Glucocorticoids; hypothalamic hormones
Pineal gland	Melatonin	Amine	Involved in rhythmic activities (daily and seasonal)	Light/dark cycles
Thyroid gland	Thyroxine (T_4) and triiodothyronine (T_3)	Amine	Stimulate and maintain metabolic processes	TSH
	Calcitonin	Peptide	Lowers blood calcium level	Calcium in blood
Parathyroid glands	Parathyroid hormone (PTH)	Peptide	Raises blood calcium level	Calcium in blood
Thymus	Thymosin	Peptide	Stimulates T-cell development	Not known
Adrenal glands Adrenal medulla	Epinephrine and norepinephrine	Amines	Increase blood glucose; increase metabolic activities; constrict certain blood vessels	Nervous system
Adrenal cortex	Glucocorticoids	Steroids	Increase blood glucose	ACTH
	Mineralocorticoids	Steroids	Promote reabsorption of Na^+ and excretion of K^+ in kidneys	K^+ (potassium) in blood
Pancreas	Insulin	Protein	Lowers blood glucose	Glucose in blood
	Glucagon	Protein	Raises blood glucose	Glucose in blood
Testes	Androgens	Steroids	Support sperm formation; development and maintenance of male secondary sex characteristics	FSH and LH
Ovaries	Estrogens	Steroids	Stimulate uterine lining growth; development and maintenance of female secondary sex characteristics	FSH and LH
	Progesterone	Steroid	Promotes uterine lining growth	FSH and LH

26.4 The hypothalamus, closely tied to the pituitary, connects the nervous and endocrine systems

26D

The distinction between the endocrine system and the nervous system often blurs, especially when we consider the diverse roles of the hypothalamus and its intricate association with the pituitary gland. As part of the brain (Figure 26.4A), the hypothalamus receives information from nerves about the internal condition of the body and about the external environment. It then responds to these conditions by sending out appropriate nervous or endocrine signals. The **hypothalamus** is the master control center of the endocrine system. Its endocrine signals directly control the pituitary gland, which in turn secretes hormones that influence numerous body functions (see Module 26.3).

As Figure 26.4A shows, the pituitary gland consists of two distinct parts: a posterior lobe and an anterior lobe, both situated in a pocket of skull bone just under the hypothalamus. The posterior lobe, or **posterior pituitary,** is composed of nervous tissue and is actually an extension of the hypothalamus. It stores and secretes hormones made in the hypothalamus.

The anterior lobe, or **anterior pituitary,** is composed of non-nervous, glandular tissue. Unlike the posterior pituitary, it synthesizes its own hormones, several of which control the activity of other endocrine glands. The hypothalamus exerts control over the anterior pituitary by secreting two kinds of hormones into the blood. **Releasing hormones** make the anterior pituitary secrete hormones, and **inhibiting hormones** make the anterior pituitary stop secreting hormones.

The left half of Figure 26.4B shows how the hypothalamus operates through the anterior pituitary to direct the activity of another endocrine organ, the thyroid gland. The hypothalamus secretes a releasing hormone known as TRH (TSH-releasing hormone). In turn, TRH makes the anterior pituitary secrete TSH (thyroid-stimulating hormone). Under the influence of TSH, the thyroid secretes the hormone thyroxine into the blood. Thyroxine increases the metabolic rate of most body cells, warming the body as a result.

Precise regulation of the TRH-TSH-thyroxine system keeps the hormones at levels that maintain homeostasis. The hypothalamus takes some cues from the environment; for instance, cold temperatures generally tend to increase its secretion of TRH. In addition, as the red arrows in Figure 26.4B indicate, negative-feedback mechanisms control TRH secretion and consequently the secretion of TSH and thyroxine. When TSH and thyroxine increase in the blood, they inhibit TRH secretion. Negative feedback is important throughout the endocrine system, as we'll see in this chapter.

What is the source of releasing hormones, and what is their function?

The hypothalamus is the source of releasing hormones, which trigger the release of certain other hormones from the anterior pituitary.

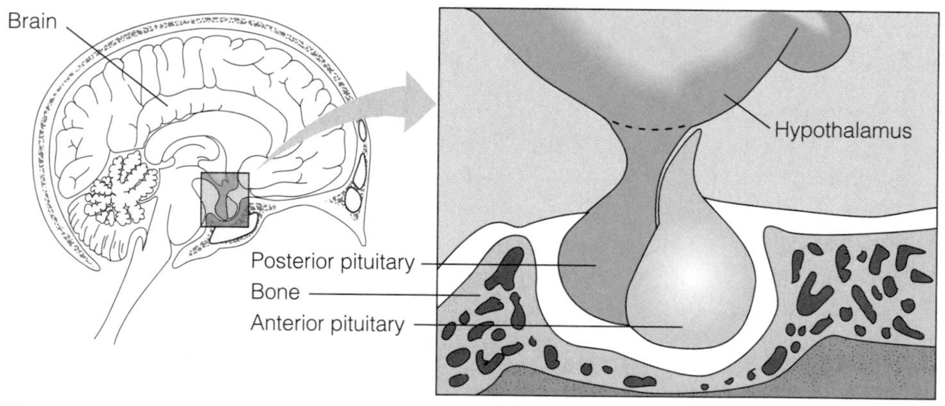

Figure 26.4A Location of the hypothalamus and pituitary

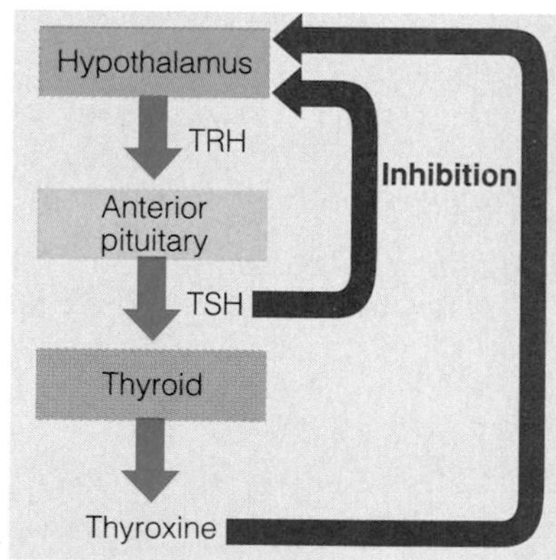

Figure 26.4B Control of thyroxine secretion

26.5 The hypothalamus and pituitary have multiple endocrine functions

26D

The TRH-TSH-thyroxine example illustrates a regulatory hierarchy characteristic of the vertebrate endocrine system. The hypothalamus exerts master control over the system and serves as a regulatory center for feedback control. It uses the pituitary gland to relay directives to other glands.

The two figures on the facing page emphasize the structural and functional connections between the hypothalamus and the pituitary. As Figure 26.5A indicates, a set of neurosecretory cells extend from the hypothalamus into the posterior pituitary. These cells synthesize the hormones **oxytocin** and **antidiuretic hormone (ADH).** These

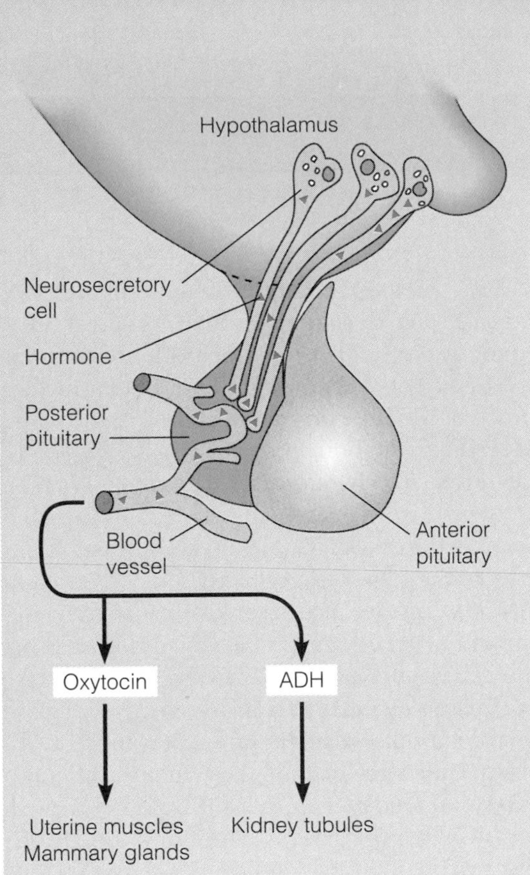

Figure 26.5A Hormones of the posterior pituitary

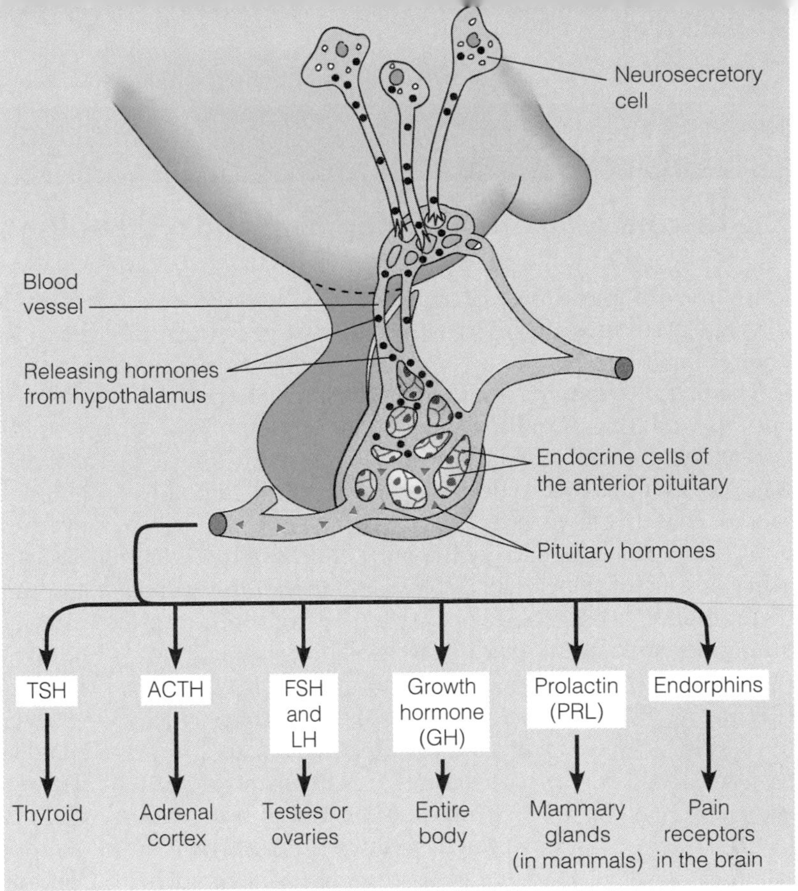

Figure 26.5B Hormones of the anterior pituitary

hormones (blue triangles in Figure 26.5A) are channeled along the neurosecretory cells into the posterior pituitary. When released into the blood from the posterior pituitary, oxytocin causes uterine muscles to contract during childbirth and mammary gland muscle to pump out milk during nursing. ADH helps cells of the kidney tubules reabsorb water, thus decreasing urine volume when the body needs to retain water (see Module 25.11). When the body has too much water, the hypothalamus responds to negative feedback, slowing the release of ADH from the posterior pituitary.

Figure 26.5B shows a second set of neurosecretory cells in the hypothalamus. These cells secrete releasing and inhibiting hormones (red dots) that control the anterior pituitary. A system of small blood vessels carries these hormones from the hypothalamus to the anterior pituitary. In response to hypothalamic-releasing hormones, the anterior pituitary synthesizes and releases many different peptide and protein hormones (blue triangles), which influence a broad range of body activities. **Thyroid-stimulating hormone (TSH), adrenocorticotropic hormone (ACTH), follicle-stimulating hormone (FSH),** and **luteinizing hormone (LH)** all activate other endocrine glands. Feedback mechanisms control the secretion of these hormones by the anterior pituitary.

Of all the pituitary secretions, none has a broader effect than the protein called **growth hormone (GH).** GH promotes protein synthesis and the use of body fat for energy metabolism in a wide variety of target cells. In young mammals, GH promotes the development and enlargement of all parts of the body. When the pituitary produces too much GH in a young person, giantism can result, while too little GH during development can lead to dwarfism. One of the most dramatic achievements of genetic engineering has been the artificial production of human growth hormone, which can be used to treat pituitary dwarfism and may be used to reverse some of the symptoms of aging.

Another anterior pituitary hormone, **prolactin (PRL),** produces very different effects in different species. In mammals, it stimulates mammary glands to produce milk; in birds, it controls fat metabolism and reproduction; in amphibians, it regulates larval development; and in freshwater fishes, it regulates salt and water balance. These diverse effects suggest that prolactin is an ancient hormone whose functions diversified during vertebrate evolution.

The **endorphins,** another kind of anterior pituitary hormone, are sometimes called the body's natural painkillers, or "natural opiates." These chemical signals have a pain-inhibiting effect on the nervous system similar to that of the drug morphine. They are produced by the brain and by the anterior pituitary. The so-called "runner's high" may result partly from the release of endorphins when stress and pain in the body reach critical levels. It has also been suggested that endorphins may be released during deep meditation and by acupuncture treatments. 

? Alcohol inhibits secretion of ADH by the anterior pituitary. Predict how this action of alcohol would affect urination.

Alcohol increases the volume of urine produced.

26.6 The thyroid regulates development and metabolism

26D

Our **thyroid gland** is located just under the voicebox. Thyroid hormones affect virtually all the tissues of vertebrate animals.

The thyroid produces two very similar amine hormones, both of which contain the element iodine. One of these, **thyroxine,** is often called $\mathbf{T_4}$ because it contains four iodine atoms; the other, **triiodothyronine,** is called $\mathbf{T_3}$ because it contains three iodine atoms.

T_3 and T_4 have essentially the same effects on their target cells. One of their crucial roles is in development and maturation. In a bullfrog, for example, they trigger the profound reorganization of body tissues that occurs as a tadpole—a strictly aquatic organism—transforms into an adult frog, which spends much of its time on land. Thyroid hormones are equally important in mammals, especially in bone and nerve cell development. In humans, a congenital thyroid deficiency known as cretinism results in retarded skeletal growth and poor mental development.

The thyroid hormones continue to play vital roles during adulthood. For example, T_3 and T_4 help maintain normal blood pressure, heart rate, muscle tone, digestion, and reproductive functions. Throughout the body, these hormones tend to increase the rate of oxygen consumption and cellular metabolism. Too much or too little thyroid hormone in the blood can result in serious metabolic disorders. An excess of T_3 and T_4 in the blood (*hyper*thyroidism) can make a person overheat, sweat profusely, become irritable, develop high blood pressure, and lose weight. Conversely, insufficient amounts of T_3 and T_4 (*hypo*thyroidism) can cause weight gain, lethargy, and intolerance to cold. Fortunately, hypothyroidism and hyperthyroidism are uncommon disorders, and both can be successfully treated.

Figure 26.6A Goiter in a Malaysian woman

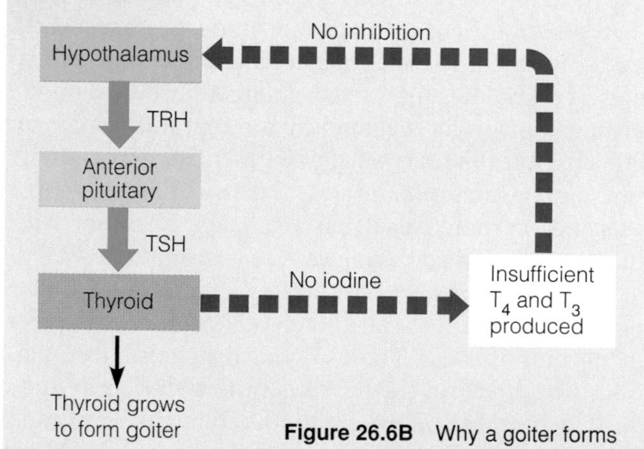

Figure 26.6B Why a goiter forms

Hypothyroidism can result from a defective thyroid gland or from dietary disorders. Figure 26.6A illustrates a severe condition called **goiter,** an enlargement of the thyroid that can result if dietary hypothyroidism is not treated. Goiter can occur when there is not enough iodine in the diet. In such cases, the thyroid gland cannot synthesize adequate amounts of its T_3 and T_4 hormones. The lack of T_3 and T_4 interrupts one of the feedback loops that control thyroid activity (Figure 26.6B). The blood never carries enough of the T_3 and T_4 hormones to shut off the secretion of TRH (TSH-releasing hormone) by the hypothalamus. The thyroid enlarges because TSH continues to stimulate it. Severe iodine deficiency during childhood can cause cretinism.

Goiter (and dietary cretinism) can be prevented simply by including iodine in the diet. Seawater is a rich source of iodine, and goiter rarely occurs in people living near the seacoast, where the soil is iodine-rich and a lot of seafood is consumed. Goiter has also been reduced in many nations by the incorporation of iodine into table salt. Unfortunately, goiter still affects many thousands of people in developing nations.

How does thyroxine switch off its own production?

By negative feedback: It inhibits the secretion of TRH from the hypothalamus, and hence TSH from the pituitary

26.7 Hormones from the thyroid and parathyroids maintain calcium homeostasis

26D

An appropriate level of calcium in the blood and interstitial fluid is essential for many body functions. Without calcium, nerve signals cannot be transmitted from cell to cell, muscles cannot function properly, blood cannot clot, and cells cannot transport molecules across their membranes. The thyroid and parathyroid glands function in homeostasis of calcium ions (Ca^{2+}), keeping the concentration of the ions within a narrow range (about 9–11 mg per 100 mL of blood).

There are four **parathyroid glands,** all embedded in the surface of the thyroid. Two peptide hormones, **calcitonin** from the thyroid gland and **parathyroid hormone (PTH),** secreted by the parathyroids, regulate the blood calcium level. Calcitonin and PTH are said to be **antagonistic hormones** because they have opposite effects. Calcitonin lowers the calcium level in the blood, whereas PTH raises it. As the figure below indicates, these two antagonistic hormones operate by means of feedback systems that keep the calcium level near a homeostatic set point of about 10 mg of Ca^{2+} per 100 mL of blood. To read the diagram, start with the yellow box on the left and follow the arrows to the top part of the figure. You can see that a rise in the blood Ca^{2+} level above 10 mg/100 mL induces the thyroid gland to secrete calcitonin. Calcitonin, in turn, has three effects: (1) It causes more Ca^{2+} to be deposited in the bones, (2) it makes the intestines absorb less Ca^{2+} from the diet, and (3) it makes the kidneys reabsorb less Ca^{2+} as they form urine. The result is a lower Ca^{2+} level in the blood.

Starting from the yellow box on the right, now follow the bottom part of the diagram to see how PTH from the parathyroid glands reverses all three of calcitonin's effects. When the blood Ca^{2+} level drops below 10mg/100 mL of blood, the parathyroids release PTH into the blood. PTH stimulates the release of calcium ions from bones and increases Ca^{2+} uptake by the kidneys and intestines. The diagram also indicates the important role that vitamin D plays in calcium homeostasis. Synthesized in an inactive form by skin exposed to sunlight, vitamin D is carried in the blood to the liver and kidneys, where it is converted to an active form. The active form enables PTH to increase calcium uptake by the intestines.

In summary, a sensitive balancing system maintains calcium homeostasis. The system depends on feedback control involving two antagonistic hormones. Failure of the system can have far-reaching effects in the body. For example, a shortage of PTH causes the blood level of calcium to drop dramatically, leading to convulsive contractions of the skeletal muscles. If unchecked, this condition, known as tetany, is fatal.

In the control of calcium ion levels in the blood by calcitonin and PTH, what are the three main target organs of the hormones?

Bones, intestines, and kidneys

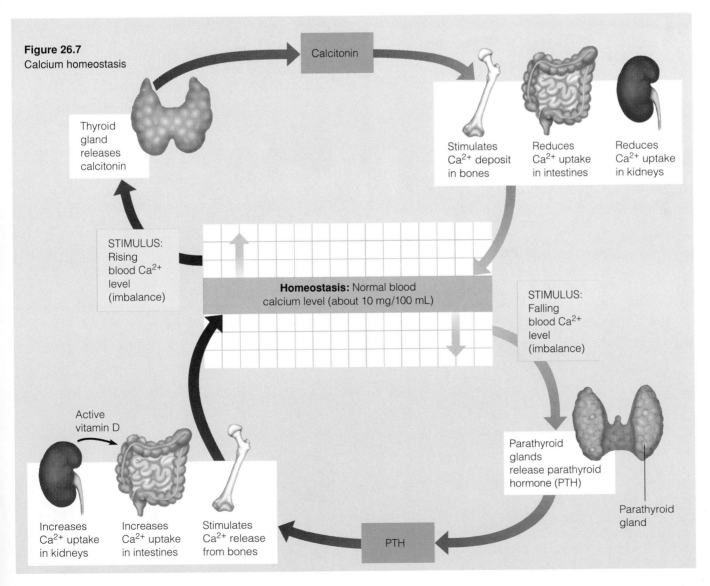

Figure 26.7
Calcium homeostasis

Calcitonin

Thyroid gland releases calcitonin

Stimulates Ca^{2+} deposit in bones

Reduces Ca^{2+} uptake in intestines

Reduces Ca^{2+} uptake in kidneys

STIMULUS: Rising blood Ca^{2+} level (imbalance)

Homeostasis: Normal blood calcium level (about 10 mg/100 mL)

STIMULUS: Falling blood Ca^{2+} level (imbalance)

Parathyroid glands release parathyroid hormone (PTH)

Parathyroid gland

Active vitamin D

Increases Ca^{2+} uptake in kidneys

Increases Ca^{2+} uptake in intestines

Stimulates Ca^{2+} release from bones

PTH

26.8 Pancreatic hormones manage cellular fuel

26D

The **pancreas** produces two hormones that play a large role in managing the body's energy supplies. One of the hormones is insulin, probably the most familiar of all chemical regulators. **Insulin** is a protein hormone produced by clusters of specialized pancreatic cells called **islet cells.** There are several distinct types of islet cells, but only those known as beta cells synthesize and secrete insulin. Islet cells of another type, called alpha cells, secrete a different hormone, a peptide called **glucagon.** Insulin and glucagon control the maintenance of a precise homeostatic balance between the amount of the cellular fuel glucose available in the blood and the amount of glucose stored in the polymer glycogen in body cells.

As shown in the figure below, insulin and glucagon are antagonists, countering each other in a feedback circuit that precisely manages both fuel storage and fuel use by body cells. By negative feedback, the concentration of glucose in the blood determines the relative amounts of insulin and glucagon secreted by the islet cells. In the top half of the diagram, you see what happens when the glucose concentration of the blood rises above the set point of about 90 mg/100 mL, as it does shortly after we eat a carbohydrate-rich meal. The rising blood glucose level (yellow box on the left) stimulates the beta cells in the pancreas to secrete more insulin. The insulin makes the body cells take up more glucose from the blood, thereby decreasing the blood glucose level. Liver cells (and skeletal muscle cells) take up much of the glucose and use it to form glycogen, which they store. Insulin also stimulates cells to metabolize the glucose for immediate energy use, for the storage of energy in fats, or for the synthesis of proteins. When the blood glucose level falls to the set point, the beta cells lose their stimulus to secrete insulin.

Following the bottom half of the diagram, you see what

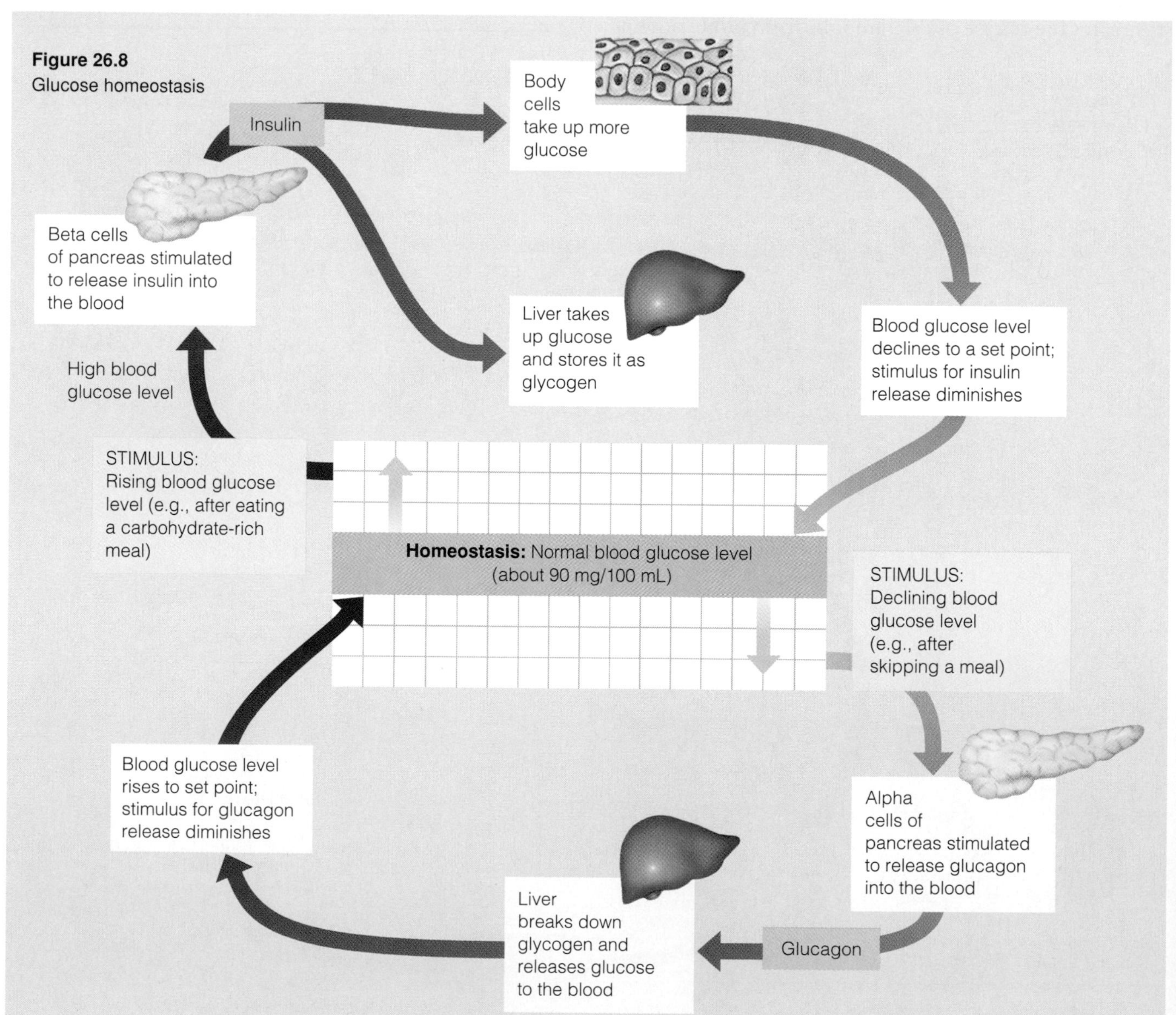
Insulin
Body cells take up more glucose
Beta cells of pancreas stimulated to release insulin into the blood
Liver takes up glucose and stores it as glycogen
Blood glucose level declines to a set point; stimulus for insulin release diminishes
High blood glucose level
STIMULUS: Rising blood glucose level (e.g., after eating a carbohydrate-rich meal)
Homeostasis: Normal blood glucose level (about 90 mg/100 mL)
STIMULUS: Declining blood glucose level (e.g., after skipping a meal)
Blood glucose level rises to set point; stimulus for glucagon release diminishes
Alpha cells of pancreas stimulated to release glucagon into the blood
Liver breaks down glycogen and releases glucose to the blood
Glucagon

Figure 26.8
Glucose homeostasis

happens when the blood glucose level starts to dip below the set point (yellow box on the right), as it may between meals or during strenuous exercise. The pancreatic alpha cells respond by secreting more glucagon. Glucagon is a fuel mobilizer, making liver cells break glycogen down into glucose and release the glucose into the blood. (It also makes liver cells convert fatty acids and amino acids to glucose.) Then, when the blood glucose level returns to the set point, the alpha cells slow their secretion of glucagon.

In the next module, we see what can happen when this delicately balanced system breaks down. 

? How is the insulin-glucagon relationship similar to the calcitonin-PTH relationship?

In both cases, the two hormones are antagonists that help maintain homeostasis by counteracting one another's effects. Their actions keep the blood concentration of a key chemical (glucose for insulin-glucagon; calcium ions for calcitonin-PTH) near the set point.

26.9 Diabetes is a common endocrine disorder

26D

Diabetes mellitus, a serious hormonal disease in which the body cells are unable to absorb glucose from the blood, affects as many as five out of every 100 people in the United States. This disease occurs when there is not enough insulin in the blood or when the body cells do not respond normally to blood insulin. In either case, the cells cannot obtain enough glucose from the blood, and thus, starved for fuel, they are forced to burn the body's supply of fats and proteins. Meanwhile, since the digestive system can continue to absorb glucose from the diet, the glucose concentration in the blood can become extremely high—so high, in fact, that glucose is excreted and can be detected in the urine. (Normally, the kidney leaves no glucose in the urine.) There are treatments for diabetes mellitus—insulin supplements and/or special diets—but no cure. Every year some 350,000 Americans die from the disease or from its complications, which include severe dehydration, cardiovascular and kidney disease, nerve damage, and gangrene.

There are two types of diabetes mellitus. Type I (insulin-dependent) diabetes is an autoimmune disease, in which white blood cells (T cells) of the body's own immune system attack and destroy the pancreatic beta cells. As a result, the pancreas does not produce enough insulin, and glucose builds up in the blood. Type I diabetes often develops before the age of 15. Patients require regular supplements of insulin, and most take the hormone by direct injection. The insulin in use now is made commercially by genetically engineered bacteria.

A second type of diabetes mellitus (called non–insulin-dependent) develops even though the pancreatic beta cells are functioning normally and there is plenty of insulin in the blood. Type II diabetes occurs because the body cells fail to respond adequately to insulin. The disease seems to be inherited and may result from genes that code for malfunctional insulin receptors on the cells. Type II diabetes accounts for about 90% of the cases in the United States. It is almost always associated with obesity and often does not show up until a person is over 40. The disease is managed by controlling sugar intake and by exercising and dieting to reduce weight; recommended diets are high in soluble fiber and low in fat and sodium. Oral drugs that reduce the blood glucose level are also available.

How is diabetes detected? The early signs of either type of diabetes are a lack of energy, a craving for sweets, frequent urination, and persistent thirst. A combination of these symptoms and a family history of diabetes indicate that a person should be tested for the disease. The diagnostic test for diabetes is a glucose-tolerance test: The person swallows a sugar solution and then has blood drawn at prescribed time intervals. Each blood sample is tested for glucose. In the graph below, you can compare the glucose tolerance of a diabetic with that of a normal individual. The normal person can keep the concentration of glucose in the blood more constant.

Diabetes is not the only disease that can result from problems with insulin. Some people have hyperactive beta cells that put too much insulin into the blood when sugar is eaten. As a result, their blood glucose level can drop well below normal. This condition, called **hypoglycemia,** usually occurs 2–4 hours after a meal and may be accompanied by hunger, weakness, sweating, and nervousness. In severe cases, when the brain receives inadequate amounts of glucose, a person may develop convulsions, become unconscious, and even die. Hypoglycemia is not common, and most forms of it can be controlled by reducing sugar intake and eating more frequently, in smaller amounts. 

? Three hours after glucose ingestion, the diabetic whose test is shown below has a blood glucose concentration about ___ times that of the normal individual.

3

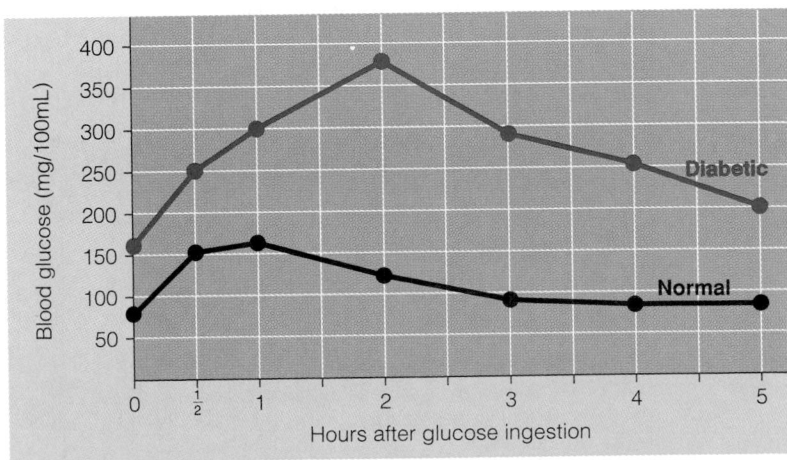

Figure 26.9 Results of glucose-tolerance tests

26.10 The adrenal glands mobilize responses to stress

26D

The human body has two **adrenal glands** sitting atop the kidneys. As you can see in the inset at the far left of Figure 26.10, each of these organs is actually two glands in one: a central portion called the **adrenal medulla** and an outer portion called the **adrenal cortex.** Though the cells they contain and the hormones they produce are different, both the medulla and the cortex secrete hormones that enable the body to respond to stress.

The adrenal medulla produces the "fight-or-flight" hormones, which ensure a rapid, short-term response to stress. You've probably felt your heart beat faster and your skin develop goose bumps when sensing danger or approaching a stressful situation, like speaking in public. Positive emotions—extreme pleasure, for instance—can produce the same effects. These reactions are triggered by two amine hormones secreted by the adrenal medulla, **epinephrine** (adrenaline) and **norepinephrine** (noradrenaline).

Stressful stimuli, whether negative or positive, activate certain nerve cells in the hypothalamus. As indicated on the left side of the diagram, these cells send signals to nerve cells in the spinal cord. The spinal cord cells extend to the adrenal medulla and stimulate it to secrete epinephrine and norepinephrine into the blood. Norepinephrine and epinephrine have somewhat different effects on tissues, but both contribute to the short-term stress response. Both hormones stimulate liver cells to release glucose, thus making more fuel available for cellular work. They also prepare the body for action by raising the blood pressure, the breathing rate, and the metabolic rate. In addition, epinephrine and norepinephrine change blood-flow patterns, making some organs more active and others less so. For example, epinephrine dilates blood vessels in the brain and skeletal muscles, thus increasing alertness and the muscles' ability to react to stress. At the same time, epinephrine and norepinephrine constrict blood vessels elsewhere, thereby reducing activities that are not immediately involved in the stress response, such as digestion. The short-term stress response occurs and subsides rapidly.

In contrast to hormones from the adrenal medulla, those secreted by the adrenal cortex can provide a slower, longer-lasting response to stress. The adrenal cortex responds to endocrine signals—chemical signals in the blood—rather than to nerve cell signals. As the right side of the diagram indicates, the hypothalamus secretes a releasing hormone that stimulates target cells in the anterior pituitary to secrete the hormone ACTH. In turn, ACTH stimulates cells

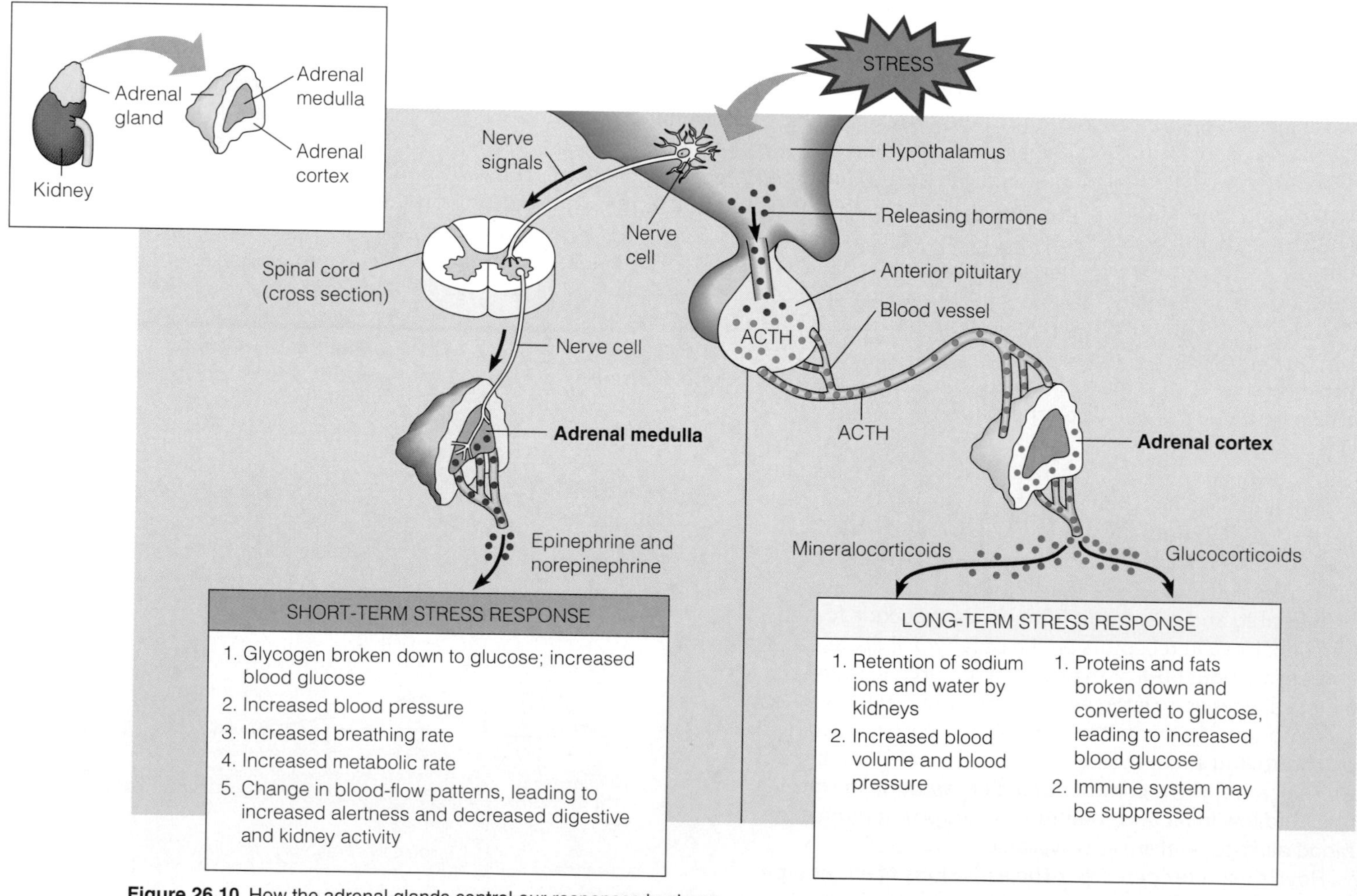

Figure 26.10 How the adrenal glands control our responses to stress

of the adrenal cortex to synthesize and secrete a family of steroid hormones called **corticosteroids.** The two main types in humans are the mineralocorticoids and the glucocorticoids. Both are essential to homeostasis, helping the body function normally whether or not it is stressed.

Mineralocorticoids have their main effects on salt and water balance. One of these hormones makes the kidney reabsorb sodium ions and water, with the overall effect of increasing the volume of the blood and raising blood pressure as a response to prolonged stress.

Glucocorticoids function mainly in mobilizing cellular fuel, thus reinforcing the effects of glucagon. Glucocorticoids promote the synthesis of glucose from noncarbohydrates such as proteins and fats. When the body cells consume more glucose than the liver can provide from glycogen stores, glucocorticoids stimulate the breakdown of muscle proteins, making amino acids available for conversion to glucose by the liver. This makes more glucose available in the blood as cellular fuel in response to stress.

Very high levels of glucocorticoids in the blood can suppress the body's defense system, including the inflammatory response that occurs at infection sites. For this reason, physicians may use glucocorticoids to treat diseases in which excessive inflammation is a problem. The glucocorticoid cortisone, for example, was once regarded as a miracle drug for treating serious inflammatory conditions such as arthritis. Cortisone and other glucocorticoids can relieve swelling and pain from inflammation, but by suppressing immunity, they can also make a person highly susceptible to infection. We discuss some other dangers of glucocorticoids next.

How would a deficiency of receptors in the hypothalamus for adrenal steroids affect levels of those hormones in the blood? (*Hint:* Apply to adrenal steroids what you learned in Module 26.4 about the regulation of thyroxine.)

This receptor deficiency would cause abnormally high levels of adrenal steroids.

Glucocorticoids offer relief from pain, but not without serious risks 26.11

Pain is often part of a professional athlete's life, and few are better acquainted with it than Bill Walton, former basketball superstar for UCLA, the Portland Trail Blazers, and the Boston Celtics. Walton was born with high arches and a malformed left foot. Running or jumping usually hurt, but for years he accepted the pain as part of his heavy workouts.

In 1977, Walton, a 6'11", 225-lb center, had been with the Blazers three years and led them to the 1977 National Basketball Association championship. But Walton's stardom with the Blazers was all too brief. Midway through the 1978 season, he was sidelined with painful injuries. Following the team physician's advice, he started taking glucocorticoids and other painkillers so he could stay active. During the '78 playoffs, primed with oral doses of dexamethasone, a glucocorticoid, and several other drugs, he played in two games. Though limping badly, Walton scored 27 points and got 22 rebounds. The morning after the second game, X-rays showed he had been playing with a fractured bone in the arch of his left foot. Walton's superstar days were over. Amid a storm of media attention, he made several comebacks in the 1980s, but he was never his former self on the basketball court. Nevertheless, in recognition of his early triumphs, Walton was elected to the Basketball Hall of Fame in 1993. He is now a TV sportscaster.

Figure 26.11 Bill Walton (right), who used a glucocorticoid for foot pain

Physicians often prescribe glucocorticoids to relieve pain from athletic injuries, and the oral use of these drugs is not uncommon. Potentially, glucocorticoids are very dangerous; taking them by mouth for more than five days can depress the activity of the adrenal glands and may cause psychological changes. It is safer, but still potentially dangerous, to inject a glucocorticoid at the site of an injury. With this treatment, the pain usually subsides, but its underlying cause remains. Masking the pain covers up the pain's message—that tissue is damaged and may get worse if not allowed to heal. If an athlete exercises an injured site before the tissue has recovered, the added stress can cause more serious damage.

Bill Walton's case was complicated. It never was firmly established that glucocorticoids worsened his condition, because he had been playing with foot pain and may have seriously injured his foot before he started using painkillers. One physician contends that Walton fractured the same bone in his left foot four times during his basketball career. An important outcome of Walton's plight was the widespread attention he drew, making more people aware of the potential dangers of painkilling drugs.

Some patients who are treated with high doses of glucocorticoids to reduce inflammation of joints have difficulty fighting infections. Explain this side effect. (*Hint:* Review Module 26.10.)

The very same action that reduces inflammation also suppresses the body's defense against disease.

26.12 The gonads secrete sex hormones

26D

The glucocorticoids and mineralocorticoids secreted by the adrenal cortex are steroid hormones. Some other steroids, the sex hormones, affect growth and development and also regulate reproductive cycles and sexual behavior. The **gonads,** or sex glands (ovaries in the female and testes in the male), secrete sex hormones, in addition to producing gametes.

There are three major categories of sex hormones: androgens, estrogens, and progestins. Both females and males have all three types, but in different proportions. Females have a high ratio of estrogens to androgens. **Estrogens** maintain the female reproductive system and promote the development of such female features as the generally smaller body size, higher-pitched voice, breasts, and wider hips of women. **Progestins,** such as progesterone, are primarily involved in preparing the uterus to support the embryo, at least in mammals.

Figure 26.12 Male elephant seals in combat

In general, **androgens** stimulate the development and maintenance of the male reproductive system. Males have a high ratio of androgens to estrogens, their main androgen being testosterone. Androgens produced by male embryos early in development stimulate the embryo to develop into a male rather than a female. High concentrations of androgens trigger the development of male characteristics: in humans, for instance, a lower-pitched voice, facial hair, and large skeletal muscles. (In Module 3.10, we discussed the powerful and potentially harmful effects of anabolic steroids, chemical relatives of testosterone, which athletes sometimes use to build large muscles.) Androgens have somewhat different effects in different animals. In bullfrogs, the effects include both large body size and the tendency to call loudly in the spring. In elephant seals (see photo), male androgens produce bodies weighing 2 tons and more, an inflatable enlargement of the nasal cavity, a thick hide that can withstand bloody conflicts, and aggressive behavior toward other males. The two males in the photo are fighting. One will establish dominance over the other and the right to mate with many females, the much smaller individuals in the background.

As with hormone production by the thyroid gland and the adrenal cortex, the synthesis of sex hormones by the gonads is regulated by the hypothalamus and anterior pituitary. In response to a releasing factor from the hypothalamus, the anterior pituitary secretes follicle-stimulating hormone (FSH) and luteinizing hormone (LH). These stimulate the ovaries or testes to synthesize and secrete the sex hormones, among other effects. We examine the complex effects of these hormones when we focus on human reproduction in the next chapter.

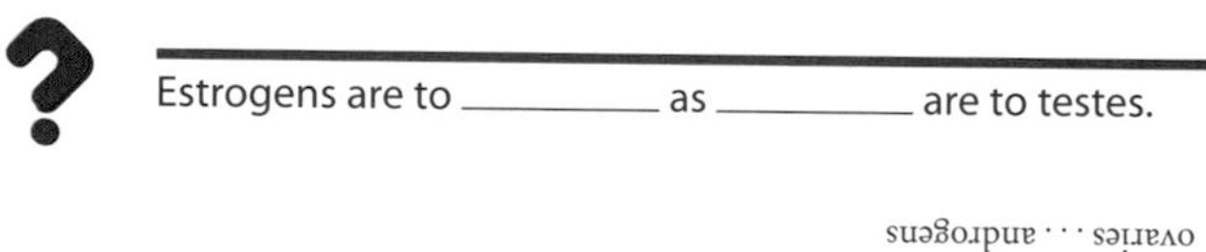

Estrogens are to ________ as ________ are to testes.
ovaries . . . androgens

Chapter Review

Chapter Summary

The Nature of Chemical Regulation (Introduction–26.2) Chemical signals make the body's organ systems function cooperatively and thus are essential to homeostasis. One of many chemical regulators in vertebrates, melatonin is an important regulator of daily and seasonal body activities **(Introduction).** Endocrine glands and neurosecretory cells secrete hormones, chemical signals that are carried by the blood and cause specific changes in target cells. The hormone-secreting cells make up the endocrine system, which works with the nervous system in regulating body activities. Local regulators, such as neurotransmitters and prostaglandins, are secreted into interstitial fluid and produce changes in cells close to the point of secretion **(26.1).**

Hormones trigger changes in target cells by two general mechanisms. Most hormones derived from amino acids bind to receptor proteins in target-cell plasma membranes and initiate signal-transduction pathways that ultimately cause changes inside the target cell. Steroid hormones bind to intracellular receptors; the steroid-receptor complex then binds to DNA, turning on specific genes. The effects of a hormone on different cells depend on the receptors and other molecules of the cells **(26.2).**

The Vertebrate Endocrine System (26.3–26.5) The vertebrate endocrine system consists of more than a dozen glands, secreting more than 50 hormones. Some glands are specialized for hormone secretion only; some also do other jobs. Only the sex glands and adrenal cortex secrete steroids; the remaining glands make nonsteroid hormones. Some hormones have a very narrow range of targets and effects; others have numerous effects on many kinds of target cells **(26.3).**

The hypothalamus is the master control center of the vertebrate endocrine system. It regulates the pituitary gland, which consists of two parts. The posterior pituitary, composed of nervous tissue, stores and secretes hormones made in the hypothalamus. The anterior pituitary, composed of glandular tissue, is controlled by releasing and inhibiting hormones carried by the blood from the hypothalamus. Hormones in the blood exert feedback control over the secretion of many hormones **(26.4)**. Neurosecretory cells extending from the hypothalamus into the posterior pituitary make the hormones oxytocin and ADH (antidiuretic hormone) and transmit nerve signals that trigger their release from the posterior pituitary. Releasing and inhibiting hormones secreted by the hypothalamus control the secretion of TSH (thyroid-stimulating hormone), ACTH (adrenocorticotropic hormone), GH (growth hormone), and other hormones from the anterior pituitary. The brain and anterior pituitary also produce endorphins, the body's natural painkillers **(26.5)**.

Hormones and Homeostasis (26.6–26.12) Two amine hormones from the thyroid gland, T_4 and T_3, regulate an animal's development and metabolism. Negative feedback maintains homeostatic levels of T_4 and T_3 in the blood. Thyroid imbalance can cause cretinism, metabolic disorders, and goiter **(26.6)**. Blood calcium level is regulated by a tightly balanced antagonism between calcitonin from the thyroid and parathyroid hormone from the parathyroid glands **(26.7)**. The pancreas secretes two hormones, insulin and glucagon, that control the blood glucose level. Insulin signals cells to use and store glucose. Glucagon causes cells to release stored glucose into the blood **(26.8)**. Diabetes mellitus results from a lack of insulin or a failure of cells to respond to it **(26.9)**. Hormones from the adrenal glands help maintain homeostasis when the body is stressed. Nerve signals from the hypothalamus stimulate the adrenal medulla to secrete epinephrine and norepinephrine, which quickly trigger the fight-or-flight response. ACTH from the pituitary causes the adrenal cortex to secrete glucocorticoids and mineralocorticoids, which boost blood pressure and energy in response to long-term stress **(26.10)**. Glucocorticoids such as cortisone relieve inflammation and pain, but they can mask injury and suppress immunity **(26.11)**. Estrogens, progestins, and androgens are steroid sex hormones produced by the ovaries in females and the testes in males. Estrogens and progestins stimulate the development of female characteristics and maintain the female reproductive system. Androgens, such as testosterone, trigger the development of male characteristics. The secretion of sex hormones is controlled by the hypothalamus and pituitary **(26.12)**.

Testing Your Knowledge

Multiple Choice

1. Which of the following controls the activity of all the others?
   a. thyroid gland
   b. pituitary gland
   c. adrenal cortex
   d. hypothalamus
   e. ovaries

2. The pancreas increases its output of insulin in response to
   a. an increase in body temperature
   b. changing cycles of light and dark
   c. a decrease in blood glucose
   d. a hormone secreted by the anterior pituitary
   e. an increase in blood glucose

3. Which of the following hormones have antagonistic (opposing) effects?
   a. parathyroid hormone and calcitonin
   b. glucagon and thyroxine
   c. growth hormone and epinephrine
   d. ACTH and cortisone
   e. epinephrine and norepinephrine

4. The body is able to maintain a relatively constant level of thyroxine in the blood because ______. *(Explain your answer.)*
   a. thyroxine stimulates the pituitary to secrete thyroid-stimulating hormone (TSH)
   b. thyroxine inhibits the secretion of TSH-releasing hormone (TRH) from the hypothalamus
   c. TRH inhibits the secretion of thyroxine by the thyroid gland
   d. thyroxine stimulates the hypothalamus to secrete TRH
   e. thyroxine stimulates the pituitary to secrete TRH

5. Which of the following hormones has the broadest range of targets?
   a. ADH
   b. oxytocin
   c. TSH
   d. epinephrine
   e. ACTH

Matching

Match each hormone (left column) with its effect on target cells (center column) and the gland where it is produced (right column).

1. thyroxine	a. lowers blood glucose	p. pineal gland
2. insulin	b. stimulates ovaries	q. testes
3. PTH	c. triggers "fight or flight"	r. parathyroid gland
4. epinephrine	d. promotes male traits	s. adrenal medulla
5. melatonin	e. regulates metabolism	t. hypothalamus
6. ADH	f. related to daily rhythm	u. pancreas
7. androgen	g. raises blood calcium level	v. anterior pituitary
8. FSH	h. boosts water retention	w. thyroid gland

Describing, Comparing, and Explaining

1. Explain how the hypothalamus controls body functions through its action on the pituitary gland. How do control of the anterior and posterior pituitary differ?

2. Explain how the same hormone might have different effects on two different target cells and no effect on a third type of cell.

Thinking as a Scientist

A strain of transgenic mice remains healthy as long as you feed them regularly and do not let them exercise. After they eat, their blood glucose level rises slightly and then declines to a homeostatic level. However, if these mice fast or exercise at all, their blood glucose drops dangerously. Which hypothesis best explains their problem? *(Explain your choice.)*

a. The mice have insulin-dependent diabetes.
b. The mice lack insulin receptors on their cells.
c. The mice lack glucagon receptors on their cells.
d. The mice cannot synthesize glycogen from glucose.

Science, Technology, and Society

A low rate of secretion of growth hormone (GH) causes pituitary dwarfism. Growth hormone made using recombinant DNA technology enables children who suffer from pituitary dwarfism to grow normally and reach a stature within the normal range. So far, no long-term side effects from GH use are known. With GH readily available and relatively inexpensive, some parents who are afraid their normal children are not growing fast enough want to use GH to make them grow faster and taller. Are there reasons to hesitate treating a child with growth hormone, or are the potential benefits worth the risk?

There is some evidence that GH injected into older adults may delay or even reverse some of the effects of aging. Should GH be freely available for any adult who wants to use it for that purpose?

Media Resources

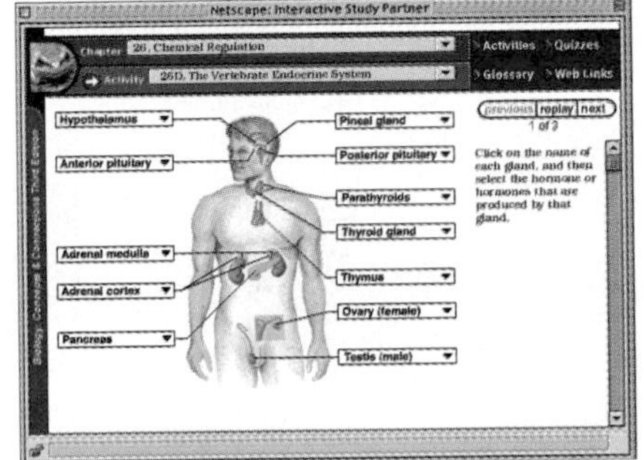

For further review of the concepts in this chapter, use the activities and quizzes on the CD-ROM.

Visit the Special Edition of The Biology Place web site for activities, quizzes, news articles, and web links.

27 Reproduction and Embryonic Development

Outline

These lizards seem to be mating, and in a sense they are, although both of them are females. There are no males in this species, called the desert-grassland whiptail. The two individuals here are involved in a complex ritual that primes the one on the bottom to lay her eggs. There is no copulation, but the female on top behaves much like a male in other species of whiptail lizards. She grabs her mate by the neck, mounts her, and wraps her tail around her abdomen. If these lizards find mates again a few weeks later, their roles will reverse. The one on top here will be on the bottom, and vice versa.

Desert-grassland whiptails inhabit dry prairies and deserts of the southwestern United States and northern Mexico. About 25 cm (10 in) long, they are active predators, darting quickly about in search of insects. Research indicates that this unusual species arose from a single female. DNA-sequencing studies have found that the ancestral female was a hybrid of two still-existing species having both male and female individuals. Researchers think that the ancestral female produced diploid egg cells that developed into lizards without having been fertilized, and that she passed this trait on to her offspring. The mating behavior of today's desert-grassland females seems to be an evolutionary leftover—a ritual derived from one or both of the ancestral species. Despite the lack of copulation and fertilization, the ritual has an important result: After mating, a female produces about three times as many eggs as she would if she did not mate. Apparently, desert-grassland whiptails still require the sexual stimuli their ancestors did to ensure maximum reproductive success.

Desert-grassland whiptails are truly unusual in the way they reproduce. However, they are not unusual at all in the way their embryo develops. The diagram on this page shows a whiptail embryo developing within an egg. Nourished by a large supply of yolk, it undergoes mitotic cell division. The embryo's body tissues take form as its cells differentiate—that is, take on the specific characteristics of various types of tissue. Body structures such as the whiptail's legs, head, and tail take form under the direction of master control genes like the ones we discussed in Module 11.12. These same basic processes occur during the development of all animal species. Also, the four membranes associated with the whiptail embryo—the chorion, amnion, allantois, and yolk sac—occur in all reptiles, birds, and mammals. Supporting the embryo and keeping it (and its food, the yolk) from drying out, these membranes are hallmarks of all terrestrial vertebrates except amphibians.

This chapter surveys animal reproduction and development. Following a brief introduction to the diverse ways that animals reproduce, we focus on the reproductive system of our own species. We examine how human eggs and sperm form, and then we return briefly to the subject of hormones, to see how they affect our reproductive activities. In the second half of the chapter, we discuss the processes of fertilization and embryonic development in vertebrates, concluding with several modules on human embryonic development and birth.

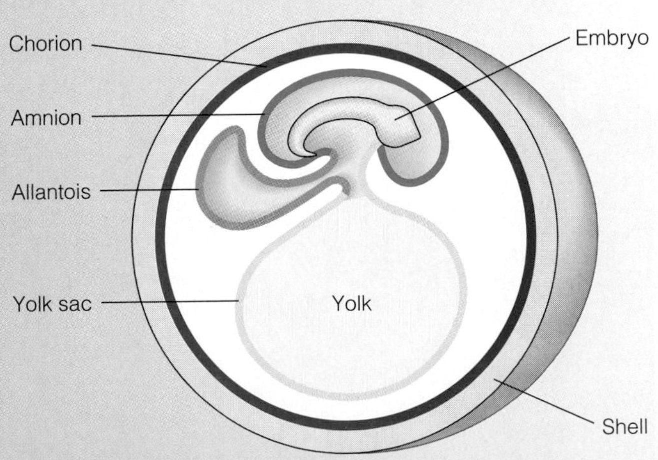

An embryo developing within an egg

27.1 Sexual and asexual reproduction are both common among animals

An individual animal has a finite life span. In contrast, a species transcends the finite life spans of its individual members by **reproduction,** the creation of new individuals from existing ones. Animals reproduce in a great variety of ways.

Asexual reproduction is the creation of offspring whose genes all come from one parent without the fusion of egg and sperm. The desert-grassland whiptail lizard, for example, reproduces only asexually; a female produces eggs that, without being fertilized, develop into a clone of female offspring.

Many invertebrates reproduce asexually by **budding,** splitting off new individuals from existing ones (see Figure 8.11C). The sea anemone in the center of Figure 27.1A, below, is undergoing **fission,** another means of asexual reproduction. In fission, one individual separates into two or more individuals of about equal size. The offspring of budding and of fission are genetic copies of the parent.

Asexual reproduction can also occur by **fragmentation,** the breaking of the parent body into several pieces. For an animal to reproduce this way, fragmentation must be accompanied by **regeneration,** the regrowth of body parts from pieces of an animal. Reproduction—an increase in the number of individuals—occurs if two or more pieces of a parent body regenerate into complete adults. In some animals, the entire parent body fragments, and the pieces develop into a clone of new individuals. In certain others, only parts of the parent body break off.Sea stars have remarkable powers of regeneration. If a sea star loses one of its arms, for example, it will regenerate a new one in a matter of weeks. In the sea star *Linckia,* a whole new individual can develop from a broken-off arm. Thus, a single animal with five arms, if broken apart, could asexually give rise to five offspring.

Asexual reproduction has a number of advantages. For one thing, it allows animals that do not move from place to place or that live in isolation to produce offspring without finding mates. Indeed, the desert-grassland whiptail survived as a species because its single female ancestor was capable of reproducing asexually. Another advantage of asexual reproduction is that it enables an animal to produce many offspring quickly; no time or energy is lost in gamete production or fertilization. Asexual reproduction perpetuates a particular genotype precisely and rapidly. Therefore, it can be an effective way for animals that are genetically well suited to a particular environment to quickly expand their populations and exploit available resources.

A potential disadvantage of asexual reproduction is that it produces genetically uniform populations. Genetically similar individuals may thrive in one particular environment, but if the environment changes and becomes less favorable to survival, all individuals may be affected equally, and the entire population may die out.

In contrast to asexual reproduction, **sexual reproduction** is the creation of offspring by the fusion of two haploid *(n)* sex cells, or **gametes,** to form a diploid *(2n)* **zygote.** The male gamete, the **sperm,** is generally a small cell that moves by means of a flagellum. The female gamete, the **ovum** (unfertilized egg), is usually a relatively large cell that is not self-propelled. The zygote and the new individual it develops into contain a unique combination of genes carried from the parents via the egg and sperm.

Figure 27.1A Asexual reproduction of a sea anemone by fission

Unlike asexual reproduction, sexual reproduction increases genetic variability among offspring. As we discussed in Modules 8.16 and 8.18, meiosis and random fertilization can generate enormous genetic variation. The variability produced by the reshuffling of genes in sexual reproduction may provide greater adaptability to changing environments. In theory, when an environment changes suddenly or drastically, there is a better chance that some of the variant offspring will survive and reproduce than if all offspring are genetically very similar.

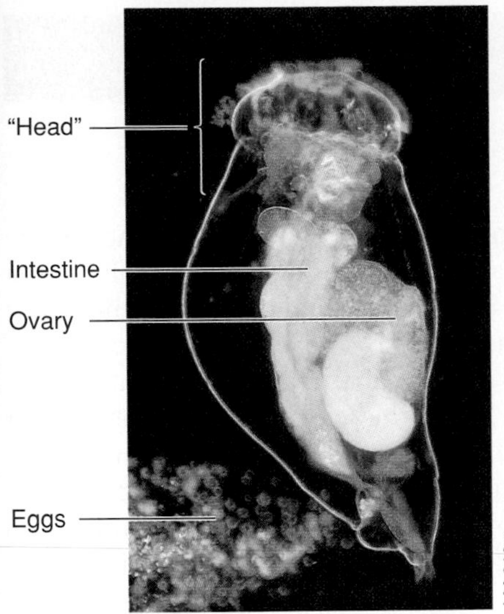

Figure 27.1B A rotifer laying eggs (side view)

Figure 27.1D Frogs in an embrace that triggers the release of eggs and sperm

Figure 27.1C Earthworms mating

Many animals can reproduce both sexually and asexually, benefiting from both modes. The microscopic animal in Figure 27.1B is a rotifer. Rotifers abound in freshwater ponds and lakes, where their diet consists mainly of algae, bacteria, and protozoans. Most rotifers reproduce asexually when there is ample food and when water temperatures are favorable for rapid growth and development. The female in the photograph is laying unfertilized eggs produced by mitosis. They will hatch almost immediately into a clone of new females. Asexual reproduction usually continues until cold temperatures signal the approach of winter, or until the food supply dwindles or a pond starts to dry up. The rotifers then reproduce sexually, producing a generation of genetically varied individuals. The fertilized eggs of the sexual generation have a thick shell and can withstand harsh conditions such as freezing and drying.

Although sexual reproduction has advantages, it presents a problem for nonmobile animals and for those that live solitary lives: How to find a mate? One solution that has evolved is **hermaphroditism.** Each hermaphroditic individual has both female and male reproductive systems. (The term comes from the Greek myth in which Hermaphroditus, son of the gods Hermes and Aphrodite, fused with a woman, forming a single, bisexual individual.)

In some species, such as the majority of tapeworms, hermaphrodites can fertilize their own eggs. However, mating must occur in many other hermaphroditic animals. When hermaphrodites mate (for example, the two earthworms seen in Figure 27.1C), each animal serves as both male and female, donating and receiving sperm. For hermaphrodites, every individual encountered is a potential mate, and many more offspring can be produced from a mating than if only one individual's eggs were fertilized.

The mechanics of fertilization play an important part in sexual reproduction. Many aquatic invertebrates, and most fishes and amphibians, have **external fertilization.** The parents discharge their gametes into the water where fertilization occurs, often without the male and female even making physical contact. Timing is crucial because the eggs must be ripe for fertilization when sperm contact them. For many species—certain clams that live in freshwater rivers and lakes, for instance—environmental cues such as temperature and day length cause both males and females to release gametes all at once. Males or females may also emit a chemical signal as they release their gametes. The signal triggers gamete release in members of the opposite sex. Most fishes and amphibians with external fertilization have specific courtship rituals that trigger simultaneous gamete release in the same vicinity by the female and male. An example of such a ritual is the clasping of a female frog by a male (Figure 27.1D; the sperm are too small to see in the photograph).

In contrast to external fertilization, **internal fertilization** occurs when sperm are deposited in or close to the female reproductive tract, and gametes unite within the female's body. Nearly all terrestrial animals have internal fertilization, which is an adaptation that protects developing eggs from excessive heat and drying. Internal fertilization usually requires **copulation,** or sexual intercourse. It also requires complex reproductive systems, including copulatory organs and receptacles for storing sperm and transporting them to the eggs. For examples of these complex structures, we turn next to the human female and male. 

?

What is the most important difference between the outcome of sexual reproduction and that of asexual reproduction?

The offspring of sexual reproduction are genetically diverse.

27.2 Reproductive anatomy of the human female

27A
See Activity 27A on the CD-ROM.

The drawings in this and the next module illustrate the structures of the human female and male reproductive systems. Both sexes have a pair of gonads (ovaries or testes) where the gametes are produced, a system of ducts that house and conduct the gametes, and structures that facilitate copulation.

As illustrated in Figure 27.2A, below, a woman's **ovaries** are each about an inch long, with a bumpy surface. The bumps are **follicles,** each consisting of a single developing egg cell surrounded by one or more layers of follicle cells that nourish and protect the developing egg cell. In addition to producing egg cells, the ovaries produce hormones, as we saw in Chapter 26. Specifically, the follicle cells produce the female sex hormone estrogen.

A woman is born with between 40,000 and 400,000 follicles, but only several hundred will release egg cells during her reproductive years. Starting at puberty and continuing until menopause, one follicle (or rarely two or more) matures and releases its egg cell about every 28 days. An egg cell is ejected from the follicle in the process called **ovulation,** shown in the photograph in Figure 27.2B, at the right. The orangish mass below the ejected egg cell is part of the ovary.

After ovulation, the remaining follicular tissue grows within the ovary to form a solid mass called a **corpus luteum** (Latin for "yellow body"); you can see one on the right side of Figure 27.2A. The corpus luteum secretes progesterone, the hormone that helps maintain the uterine lining during pregnancy, and additional estrogen. If the egg is not fertilized, the corpus luteum degenerates, and a new follicle matures during the next cycle. We discuss ovulation and female hormonal cycles further in later modules.

Notice in Figure 27.2A that each ovary lies next to the opening of an **oviduct,** also called a fallopian tube. The oviduct opening resembles a funnel fringed with fingerlike projections. The projections touch the surface of the ovary, but the ovary is actually separated from the opening of the

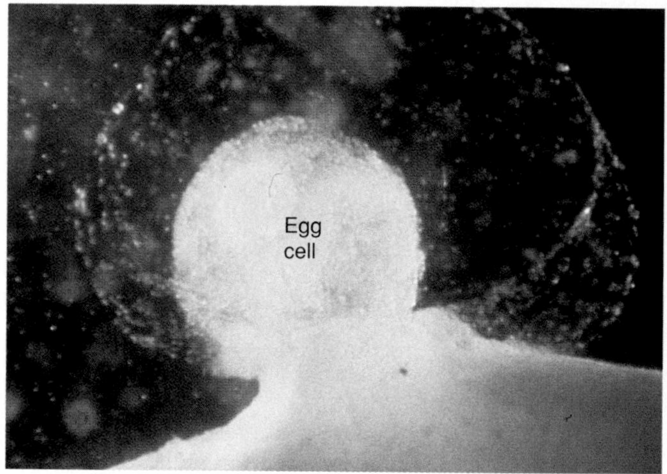

Figure 27.2B Ovulation

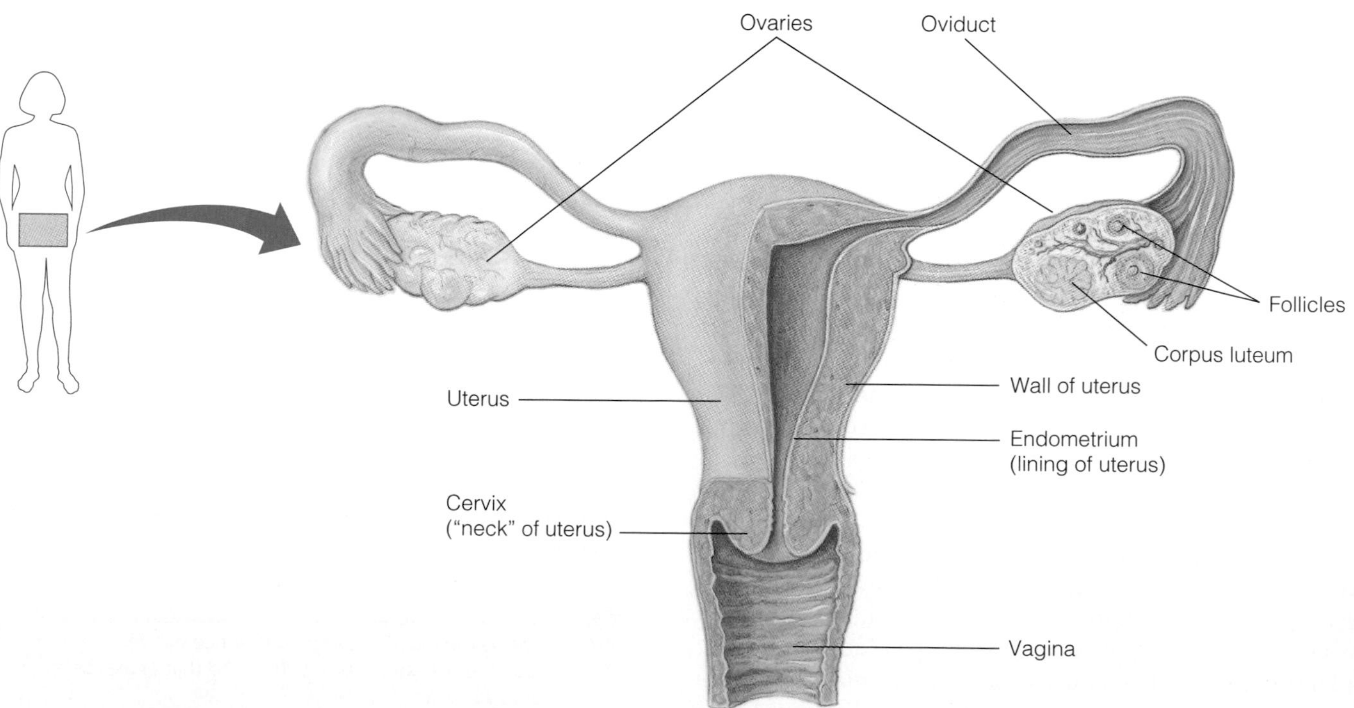

Figure 27.2A Front view of female reproductive anatomy (upper portion)

oviduct by a tiny space. When ovulation occurs, the egg cell passes across the space and into the oviduct, where cilia sweep it toward the uterus. Fertilization usually occurs in the upper third of the oviduct. The resulting zygote starts to divide, thus becoming an embryo, as it moves along within the oviduct.

The **uterus,** also known as the womb, is the actual site of pregnancy. The uterus is only about 3 inches long in a woman who has never been pregnant, but during pregnancy it can expand to accommodate a baby weighing 4 kg (8.8 lb) or more. The uterus has a thick muscular wall, and its inner lining, the **endometrium,** is richly supplied with blood vessels. The embryo implants (digests a place for itself) in the endometrium, and development is completed there. The term **embryo** is used for the stage in development from the first division of the zygote until body structures begin to appear, about the ninth week in humans. From the ninth week until birth, a developing human is called a **fetus.**

The uterus is the *normal* site of pregnancy. However, in about one out of 100 pregnancies, the embryo implants somewhere else, resulting in an **ectopic pregnancy** (Greek *ektopos,* out of place). Most ectopic pregnancies occur in the oviduct and are called tubal pregnancies. Ectopic pregnancies require surgical removal; otherwise, they can rupture surrounding tissues, causing severe bleeding and even death.

The narrow neck of the uterus is the **cervix,** which opens into the vagina. The **vagina** is a thin-walled, but strong, muscular chamber that serves as the birth canal through which the baby is expelled. The vagina also accommodates the male's penis and is a repository for sperm during copulation.

You can see more features of female reproductive anatomy in Figure 27.2C, a side view. Notice that the vagina opens to the outside just behind the opening of the urethra, the tube through which urine is excreted. A pair of skin folds, the **labia minora,** border the openings, and a pair of thick, fatty ridges, the **labia majora,** protect the entire genital region. Until sexual intercourse or vigorous physical activity ruptures it, a thin membrane called the **hymen** partly covers the vaginal opening; the hymen has no known function. **Bartholin's glands,** near the vaginal opening, secrete lubricating fluid during sexual arousal, as does the vaginal lining.

Several female reproductive structures are important in sexual arousal, and stimulation of them can produce highly pleasurable sensations. The vagina, labia minora, and a structure called the **clitoris** all engorge with blood and enlarge during sexual activity. The sole function of the clitoris is sexual arousal. It consists of a short shaft supporting a rounded **glans,** or head, covered by a small hood of skin called the **prepuce.** In Figure 27.2C, blue highlights the spongy tissue within the clitoris that fills with blood during arousal. The clitoris, especially the glans, has an enormous number of nerve endings and is very sensitive to touch. Accompanied by other arousing stimuli, gentle stimulation of the glans can often trigger orgasm. We discuss the human sexual response in more detail in Module 27.6.

In which organ of the human female does the fetus develop?

The uterus

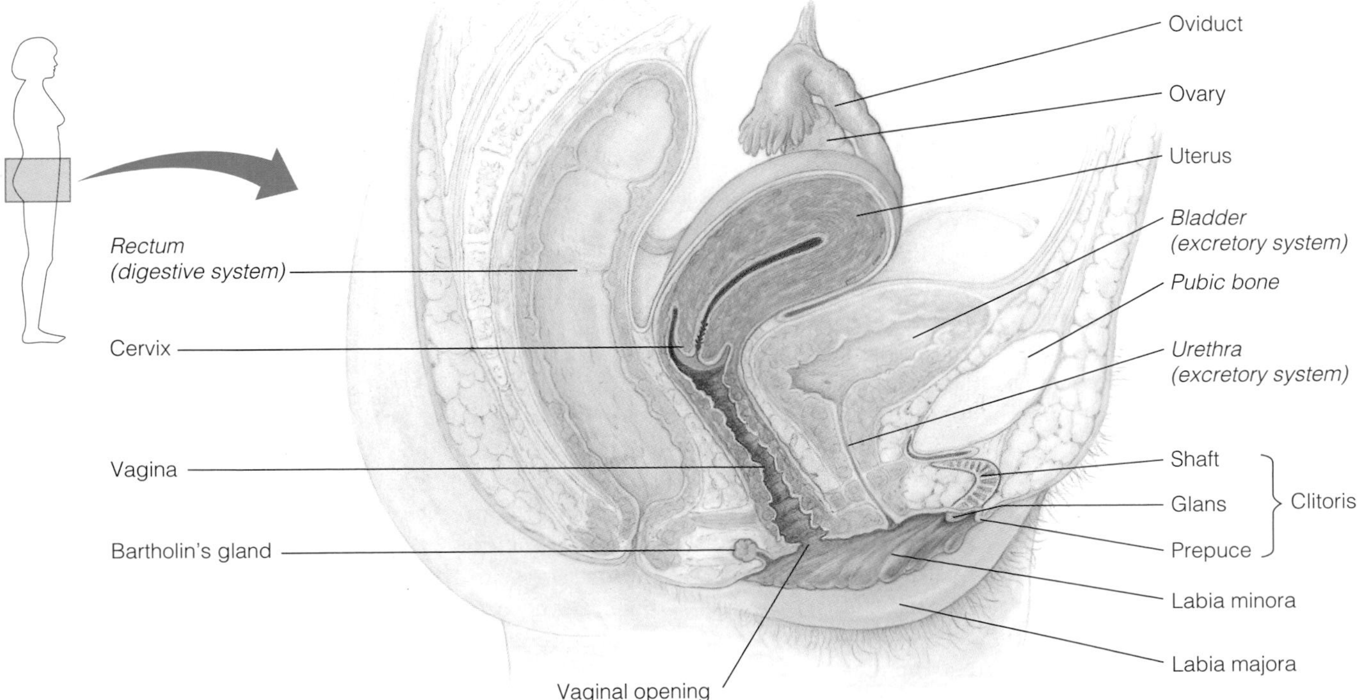

Figure 27.2C Side view of female reproductive anatomy

27.3 Reproductive anatomy of the human male

27B

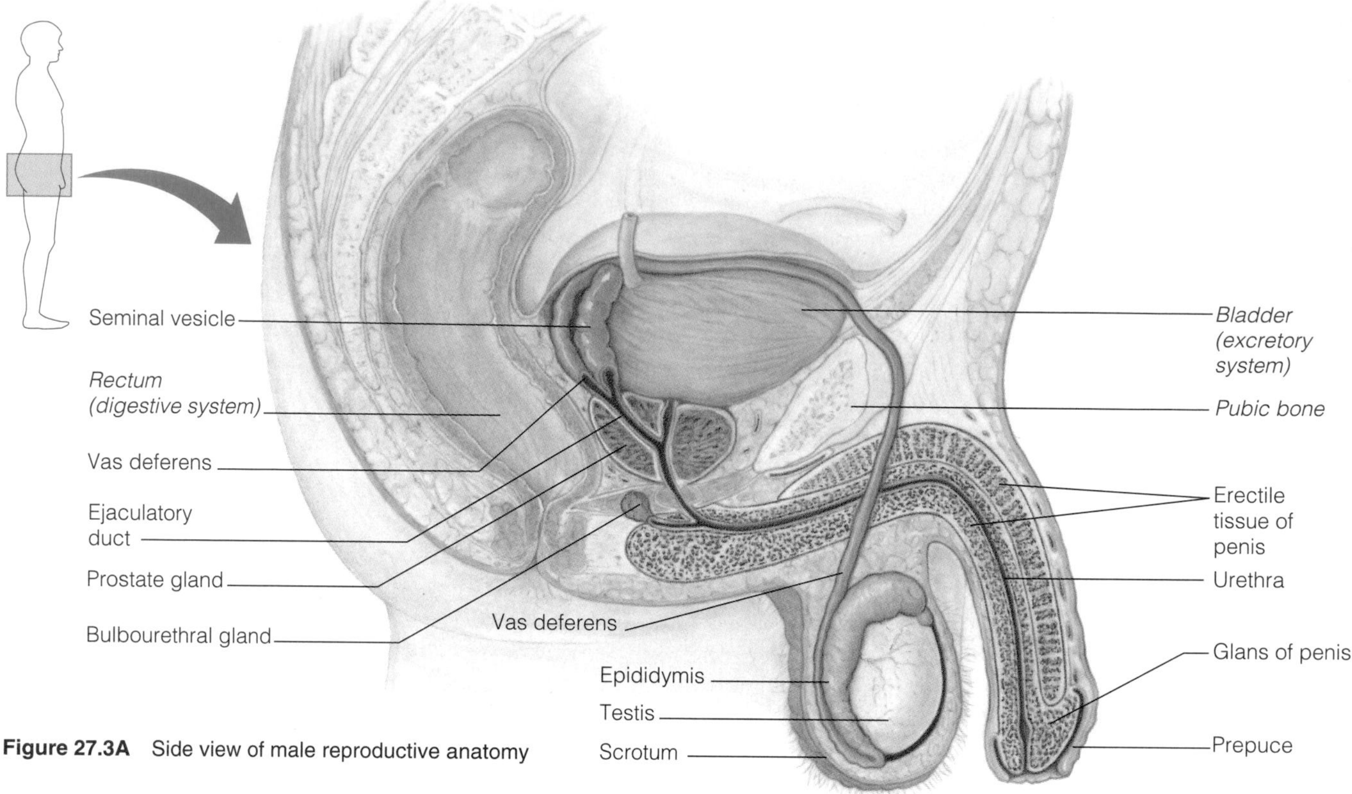

Figure 27.3A Side view of male reproductive anatomy

Figures 27.3A and 27.3B present two views of the male reproductive system. The male gonads, or **testes** (singular, *testis*), are both housed outside the abdominal cavity in a sac called the **scrotum.** Sperm cannot develop at human body temperature, but the scrotum keeps the sperm-forming cells cool enough to function normally.

Let's now track the path of sperm from one of the testes to the outside of the male's body. From each testis, sperm pass into a coiled tube called the **epididymis,** which stores sperm while they develop motility and fertilizing ability. Sperm leave the epididymis during **ejaculation,** the expulsion of sperm-containing fluid from the penis. At that time, muscular contractions propel the sperm from the epididymis through another duct called the **vas deferens.** The vas deferens passes upward into the abdomen and loops around the urinary bladder. Adjacent to the bladder, the vas deferens joins a short duct from a gland, the seminal vesicle (see Figure 27.3A). Union of these two ducts forms a short **ejaculatory duct,** which joins its counterpart conveying sperm from the other testis. The union of the two ejaculatory ducts forms the urethra, which conveys both urine and sperm to the outside through the penis, although not at the same time. Thus, unlike the female, the male has a connection between the reproductive and excretory systems.

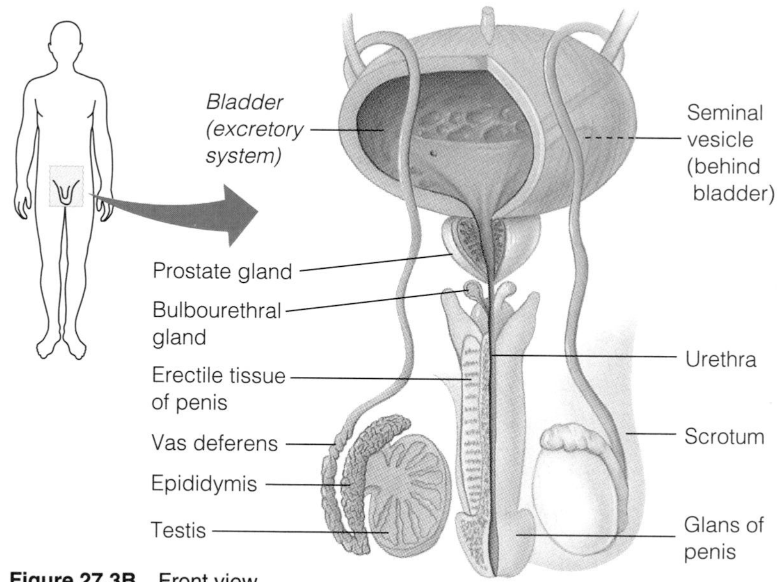

Figure 27.3B Front view

In addition to the testes and ducts, the male reproductive system contains three sets of glands: the seminal vesicles, the prostate gland, and the bulbourethral glands. The two **seminal vesicles** secrete a thick, clear fluid that lubricates and nourishes the sperm. The **prostate gland** secretes a milky, alkaline fluid that balances the acidity of any traces of urine in the urethra and helps protect the sperm from the natural acidity of the vagina. The two **bulbourethral glands** secrete only a few drops of fluid into the urethra, during sexual arousal. The fluid may help lubricate the urethra, helping sperm move through it.

Together, the sperm and the secretions of the glands make up **semen,** the fluid that is discharged (ejaculated) from the penis during orgasm. About 5 milliliters (1 tea-

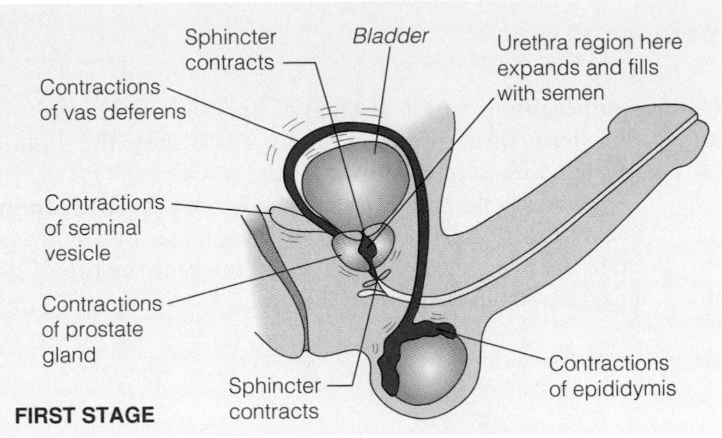

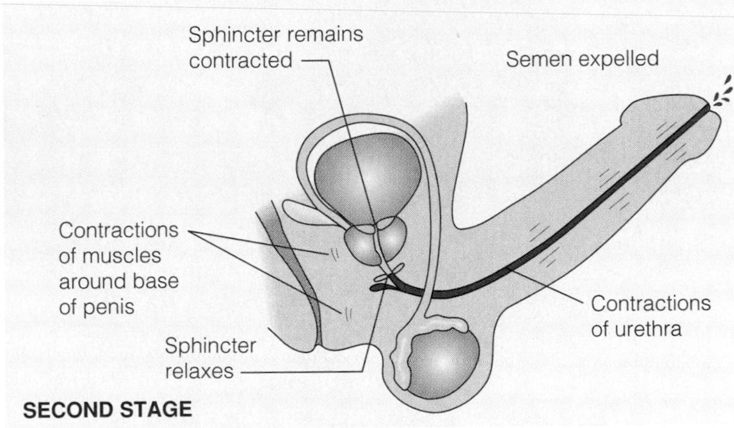

Figure 27.3C The two stages of ejaculation

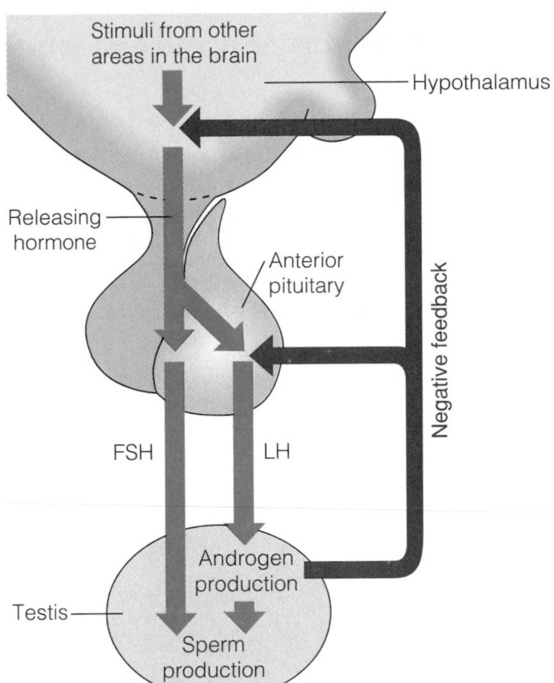

Figure 27.3D Hormonal control of the testis

spoonful) of semen are discharged during a typical ejaculation. About 95% of the fluid consists of glandular secretions. The other 5% is made up of 200–500 million sperm, only one of which may fertilize an egg cell. The other sperm seem to play an accessory role, probably causing chemical changes in the environment of the female reproductive tract that promote fertilization.

The human **penis** consists mainly of tissue that can fill with blood to cause an erection during sexual arousal. The erectile tissue is shown in blue in Figures 27.3A and 27.3B. Erection is essential for insertion of the penis into the vagina. (A signal molecule that plays a key role in erection is discussed in Module 28.8.) Like the clitoris, the penis consists of a shaft that supports the glans, or head. The glans is richly supplied with nerve endings and is highly sensitive to stimulation. As in the female, a fold of skin called the prepuce, or foreskin, covers the glans. Circumcision, the surgical removal of the prepuce, is commonly performed for religious or health reasons. However, scientific studies have not proved that circumcision has any effect on a man's health or on that of his sex partner.

Figure 27.3C illustrates the process of ejaculation and summarizes what we have said about the production of semen and its expulsion. Ejaculation occurs in two stages. At the peak of sexual arousal, muscles in the epididymis, seminal vesicles, prostate gland, and vas deferens contract (upper drawing). These contractions force secretions from the glands into the vas deferens and propel sperm from the epididymis. At the same time, a sphincter muscle at the base of the bladder contracts, preventing urine from leaking into the urethra from the bladder. Another sphincter also contracts, closing off the entrance of the urethra into the penis. The section of the urethra between the two sphincters fills with semen and expands. In the second stage of ejaculation, the expulsion stage (lower drawing), the sphincter at the base of the penis relaxes, admitting semen into the penis. Simultaneously, a series of strong muscle contractions around the base of the penis and along the urethra expels the semen from the body.

Figure 27.3D shows how hormones control sperm production by the testes. Influenced by signals from other parts of the brain, the hypothalamus secretes a releasing hormone that regulates release of follicle-stimulating hormone (FSH) and luteinizing hormone (LH) by the anterior pituitary (see Module 26.5). FSH increases sperm production by the testes, while LH promotes the secretion of androgens, mainly testosterone. Androgens stimulate sperm production. In addition, androgens carried in the blood help maintain homeostasis by a negative-feedback mechanism (red arrows), inhibiting secretion of both the releasing hormone and LH. Under the control of this chemical regulating system, the testes produce hundreds of millions of sperm every day, from puberty well into old age. Next we'll see how sperm and eggs are made.

? Arrange the following organs in the correct sequence for the travel of sperm: epididymis, testis, urethra, vas deferens.

Testis, epididymis, vas deferens, urethra

27.4 The formation of sperm and ova requires meiosis

8E

Both sperm and ova are haploid cells that develop by meiosis from diploid cells in the gonads. Before we turn to the formation of gametes, you may want to review Modules 8.12–8.14 as background for our discussion.

Spermatogenesis, the formation of sperm cells, takes about 65–75 days in the human male. Figure 27.4A outlines spermatogenesis. Though the diploid chromosome number in humans is 46 (that is, $2n = 46$), for the sake of simplicity, this figure and Figure 27.4B show the diploid cell with only four chromosomes ($2n = 4$).

Sperm develop in the testes in coiled tubes called **seminiferous tubules.** Diploid cells that begin the process are located near the outer wall of the tubules (at the top of the enlarged wedge of tissue in Figure 27.4A). These cells mul-

Figure 27.4A Spermatogenesis

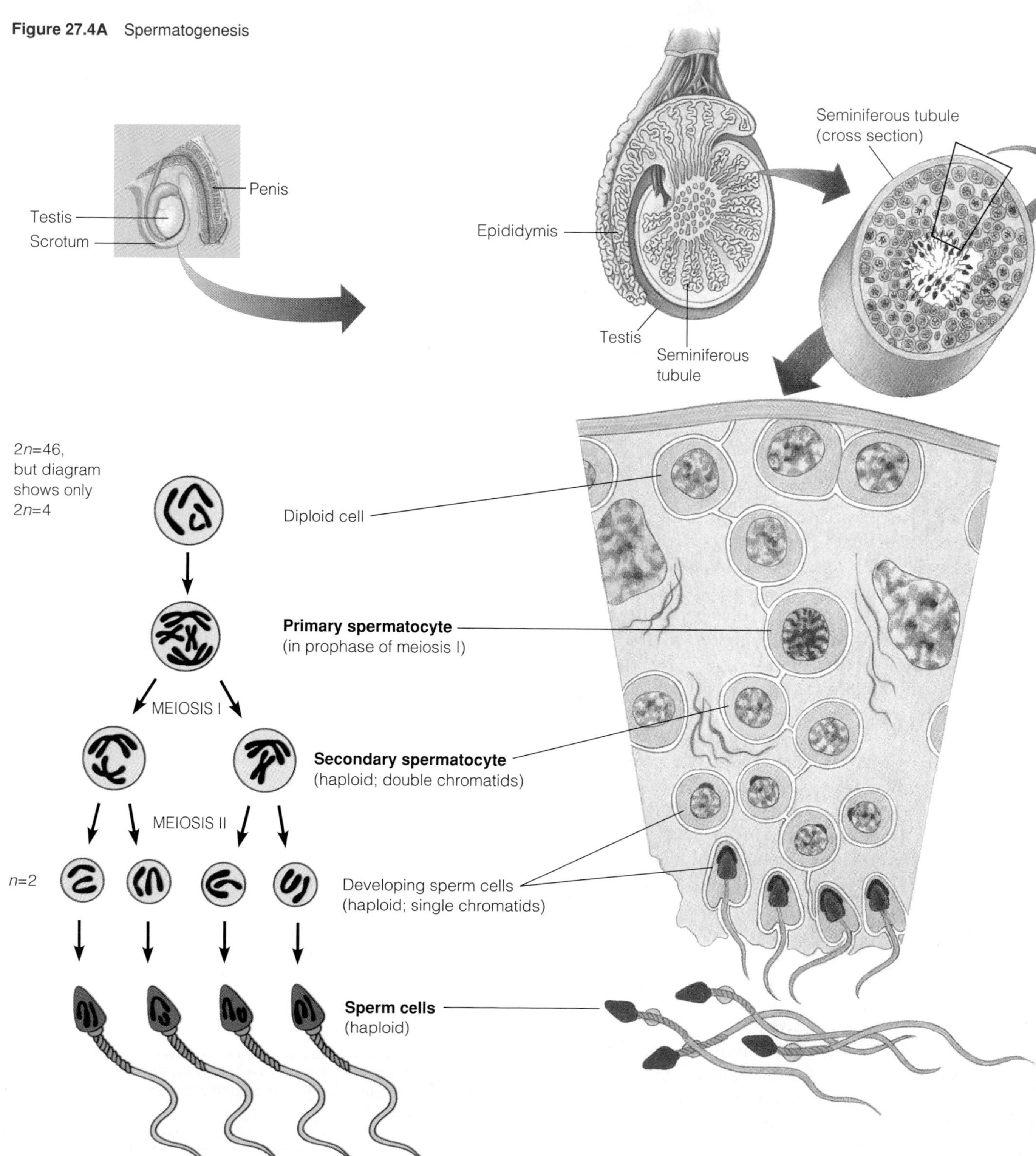

tiply constantly by mitosis, and each day about 3 million of them develop into **primary spermatocytes,** the cells that undergo meiosis. Meiosis I of a primary spermatocyte produces two **secondary spermatocytes,** each with the haploid number of chromosomes *(n)*. The chromosomes are still in their duplicated state, each consisting of two identical chromatids. Meiosis II then forms four cells, each with the haploid number of single-chromatid chromosomes. A sperm cell develops from each of these haploid cells and is gradually pushed toward the center of the seminiferous tubule. From there it passes into the epididymis, where it matures and becomes motile. Sperm are stored in the epididymis until ejaculation.

Figures 27.4B and 27.4C show **oogenesis,** the development of an ovum, which occurs in the ovary. As we indicated in Module 27.2, a female's ovary at birth contains all the follicles (structures in which ova form) she will ever have. Oogenesis actually begins prior to birth, when a diploid cell in each developing follicle begins meiosis. At birth, each follicle contains a dormant **primary oocyte,** a diploid cell that is resting in prophase of meiosis I. A primary oocyte can be hormonally triggered to develop further. After puberty, about every 28 days, FSH (follicle-stimulating hormone) from the pituitary stimulates one of the dormant follicles to develop. The follicle enlarges, and the primary oocyte completes meiosis I. In the female, the division of the cytoplasm in meiosis I is unequal, with a single **secondary oocyte** receiving almost all of it. The smaller of the two daughter cells, called the first polar body, receives almost no cytoplasm.

The secondary oocyte is the stage released by the ovary during ovulation. It enters the oviduct, and if a sperm cell penetrates it, the secondary oocyte undergoes meiosis II. Meiosis II yields a second polar body and the actual ovum. The haploid nucleus of the ovum can then fuse with the haploid nucleus of the sperm cell, producing a zygote.

As Figure 27.4B shows, the first polar body also undergoes meiosis II, forming two cells. These and the second polar body receive virtually no cytoplasm and degenerate soon after the ovum forms. Polar body formation enables the ovum to acquire nearly all the cytoplasm and thus the bulk of the nutrients contained in the original diploid cell.

Figure 27.4C is a cutaway view of an ovary. The series of follicles here represents the changes one follicle undergoes over time; the arrows indicate the sequence. An actual ovary would have thousands of dormant follicles, each containing a primary oocyte.Usually, only one follicle has a dividing oocyte at any one time, and as it develops, that follicle stays in one place in the ovary. Meiosis I occurs as the follicle matures. About the time the secondary oocyte forms, the pituitary hormone LH (luteinizing hormone) triggers ovulation, the expulsion of the secondary oocyte from the follicle. The ruptured follicle then develops into a corpus luteum. Unless fertilization occurs, the corpus luteum degenerates before another follicle starts to develop.

Oogenesis and spermatogenesis are alike in that they produce haploid gametes. However, these two processes differ in three important ways. First, only one ovum results from each diploid parent cell that undergoes oogenesis, whereas four sperm cells result from each parent cell that undergoes spermatogenesis. Second, an ovary at birth contains all the primary oocytes it will ever have, whereas the testes produce new primary spermatocytes throughout the male's reproductive years. Third, oogenesis is not completed without stimulation from a sperm cell, whereas spermatogenesis produces mature sperm in an uninterrupted sequence.

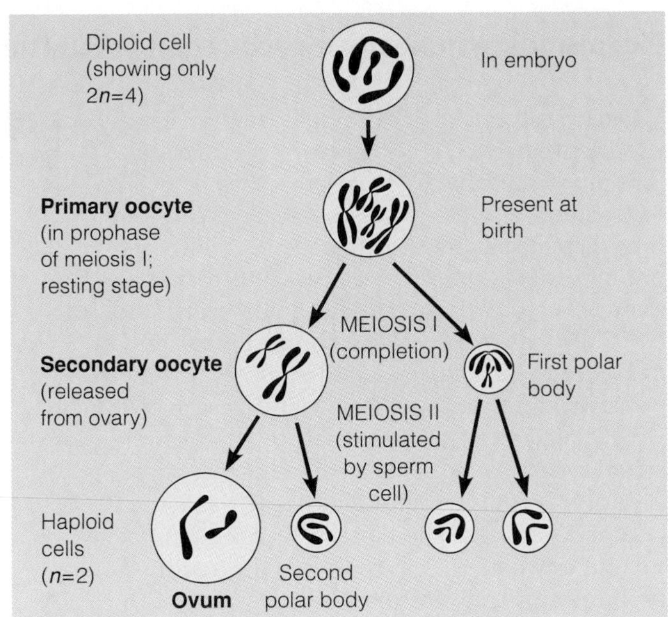

Figure 27.4B Meiosis in oogenesis

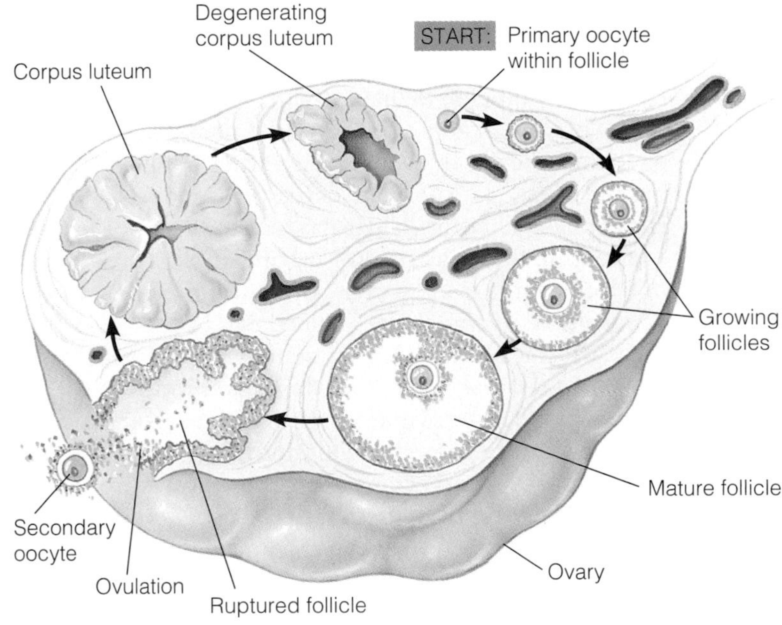

Figure 27.4C The development of an ovarian follicle

Which process in the development of sperm and ova is responsible for the genetic variation among gametes? (*Hint:* Review Module 8.17.)

Meiosis, specifically meiosis I

27.5 Hormones synchronize cyclic changes in the ovary and uterus

27A

Oogenesis is one part of a female mammal's reproductive cycle, a recurring sequence of events that produces gametes, makes them available for fertilization, and, as we will see, prepares the body for pregnancy. In discussing oogenesis in the last module, we described the cyclic events that occur about every 28 days in the human ovary; this **ovarian cycle** is represented in part 3 of Figure 27.5 on the facing page. As the figure indicates, hormonal messages synchronize the ovarian cycle with related events in the uterus called the **menstrual cycle** (part 5). The hormone story (parts 1, 2, and 4) is complex and involves intricate feedback mechanisms. Therefore, we'll need to move up and down the figure to follow the actions of the hormones. Notice that the time scale at the bottom of part 5 also applies to parts 2–4.

Hormones of the Ovarian and Menstrual Cycles

Hormone	Secreted by	Major Roles
Releasing hormone	Hypothalamus	Regulates secretion of LH and FSH by pituitary
FSH	Pituitary	Stimulates growth of ovarian follicle
LH	Pituitary	Stimulates growth of ovarian follicle and production of secondary oocyte; promotes ovulation; promotes development of corpus luteum and secretion of hormones
Estrogen	Ovarian follicle	Low levels inhibit pituitary; high levels stimulate hypothalamus; promotes endometrium
Estrogen and progesterone	Corpus luteum	Maintain endometrium; high levels inhibit hypothalamus and pituitary; sharp drops promote menstruation

An Overview of the Ovarian and Menstrual Cycles Let's begin with the straightforward, structural events of the ovarian and menstrual cycles. For simplicity, we have divided the ovarian cycle (part 3 of the figure) into two phases separated by ovulation: the pre-ovulatory phase, when a follicle is growing and a secondary oocyte is developing, and the post-ovulatory phase, after the follicle has become a corpus luteum.

Events in the menstrual (or uterine) cycle (part 5) occur in tune with the ovarian cycle. By convention, the first day of a woman's "period" is designated day 1 of the menstrual cycle. Uterine bleeding, called **menstruation,** usually persists for 3–5 days. Notice that this corresponds to the beginning of the pre-ovulatory phase of the ovarian cycle. During menstruation, the endometrium (inner lining of the uterus) breaks down and leaves the body through the vagina. The menstrual discharge consists of blood, small clusters of endometrial cells, and mucus. After menstruation, the endometrium regrows. It continues to thicken through the time of ovulation, reaching a maximum at about 20–25 days. If an embryo has not implanted in the uterine lining by this time, menstruation begins again, marking the start of the next ovarian and menstrual cycles.

Now let's consider the hormones that regulate the ovarian and menstrual cycles. The ebb and flow of the five hormones listed in the table on this page synchronize events in the ovarian cycle (the growth of the follicle and ovulation) with events in the menstrual cycle (preparation of the uterine lining for possible implantation of an embryo). As in the male, a releasing hormone from the hypothalamus in the brain regulates secretion of the two pituitary hormones FSH and LH. The blood levels of FSH, LH, and two other hormones—estrogen and progesterone—coincide with specific events in the ovarian and menstrual cycles.

Hormonal Events Before Ovulation Focusing on part 1 of Figure 27.5 at the right, we see that the releasing hormone from the hypothalamus stimulates the anterior pituitary to increase its output of FSH and LH. True to its name, FSH (black arrows) stimulates the growth of an ovarian follicle, in effect starting the ovarian cycle. In turn, the follicle secretes estrogen. Early in the pre-ovulatory phase, the follicle is small (part 3) and secretes relatively little estrogen (part 4). As the follicle grows, it secretes more and more estrogen, and the rising but still relatively low level of estrogen exerts negative feedback on the pituitary. This keeps the blood levels of FSH and LH low for most of the pre-ovulatory phase (part 2). As the time of ovulation approaches, hormone levels change drastically, with estrogen reaching a critical peak (part 4) just before ovulation. This high level of estrogen exerts positive feedback on the hypothalamus (green arrow in part 1), which then makes the pituitary secrete bursts of FSH and LH. By comparing parts 2 and 4 of the figure, you can see that the peaks in FSH and LH occur just after the estrogen peak. It may help to place a piece of paper over the figure and slide it slowly to the right. As you uncover the figure, you will see the follicle getting bigger and the estrogen level rising to its peak, followed almost immediately by the LH and FSH peaks. Then, just to the right of the peaks, comes the dashed line representing ovulation.

Hormonal Events at Ovulation and After The role of FSH after a follicle matures is unknown, but the LH peak has pronounced effects. It stimulates the completion of meiosis, transforming the primary oocyte in the follicle into a secondary oocyte. It also signals enzymes to rupture the follicle, allowing ovulation to occur, and triggers the development of the corpus luteum from the ruptured follicle (hence its name, luteinizing hormone). LH also promotes the secretion of progesterone and estrogen by the corpus luteum. In part 4 of the figure, you can see the progesterone peak and the second (lower and wider) estrogen peak after ovulation.

High levels of estrogen and progesterone in the blood following ovulation have a strong influence on both ovary and uterus. The combination of the two hormones exerts negative feedback on the hypothalamus and pituitary, producing the falling FSH and LH levels shown on the right side of the graph in part 2. The drops in FSH and LH prevent follicles from developing and ovulation from occurring during the post-ovulatory phase. Also, the LH drop is followed by the gradual degeneration of the corpus luteum. Near the end of the post-ovulatory phase, unless an embryo has implanted in the uterus, the corpus luteum stops secreting estrogen and progesterone. As the blood levels of these hormones decline, the hypothalamus once again can stimulate the pituitary to secrete more FSH and LH, and a new cycle begins.

Control of the Menstrual Cycle Hormonal regulation of the menstrual cycle is simpler than that of the ovarian cycle. The menstrual cycle (part 5) is directly controlled by estrogen and progesterone alone. You can see the effects of these hormones by comparing parts 4 and 5 of the figure. Starting around day 5 of the cycle, the endometrium thickens in response to the rising levels of estrogen and, later, progesterone. When the levels of these hormones drop, the endometrium begins to slough off. Menstrual bleeding begins soon thereafter, on day 1 of a new cycle.

We have now described what happens in the human ovary and uterus in the absence of fertilization. As we'll see later, the ovarian and menstrual cycles are put on hold if fertilization and pregnancy occur. Early in pregnancy, the developing embryo, implanted in the endometrium, releases a hormone (human chorionic gonadotropin, or HCG) that acts like LH. The hormone maintains the corpus luteum, which continues to secrete progesterone and estrogen, keeping the endometrium intact. We'll return to the events of pregnancy in Modules 27.16 and 27.17.

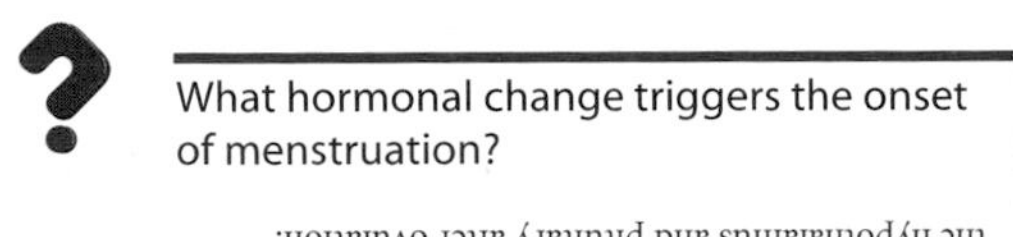

What hormonal change triggers the onset of menstruation?

The drop in the levels of estrogen and progesterone. These changes are caused by negative feedback of these hormones on the hypothalamus and pituitary after ovulation.

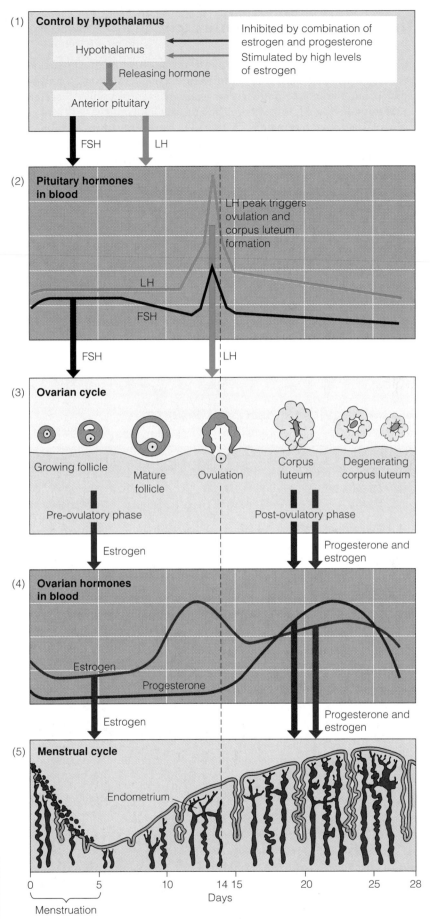

Figure 27.5 The reproductive cycle of the human female

27.6 The human sexual response occurs in four phases

Most female mammals are receptive to males only on certain days—in many species, for only a brief period once or a few times a year. A female deer or bear, for example, will mate only during a few weeks in the autumn. During specific mating times, a female is said to be in estrus, meaning she is at her peak of sexual readiness. This is the only time she ovulates and the only time her uterus is primed for implantation.

Humans and several other primates are unusual in having no distinct mating periods; females are potentially receptive to males throughout the year. In humans, the sexual behavior called "making love" may have evolved as a way to strengthen the bond between mates, as well as to promote the union of egg and sperm. Our sexuality is emotional as well as physical, and we have highly varied sexual expression.

The physical events of the human sexual response occur in a sequence of four phases. During the **excitement phase,** sexual passion builds, the penis and clitoris become erect, and the testes, labia, and nipples may swell. The vagina secretes lubricating fluid, and muscles tighten in the arms and legs. These responses continue during the **plateau phase,** which is marked by increases in breathing and heart rates. **Orgasm** follows, characterized by rhythmic contractions of the reproductive structures, extreme pleasure for both partners, and ejaculation by the male. The **resolution phase** reverses the previous responses; the structures return to normal size, muscles relax, and passion subsides.

How does the timing of mating in humans contrast with that of most other mammals?

Human females are potentially receptive to mating throughout the year, in contrast to the seasonal mating of most other mammals.

27.7 Sexual activity can transmit disease

STDs Common in the United States

Disease	Microbial Agent	Major Symptoms and Effects	Treatment
Chlamydial infections	*Chlamydia trachomatis* (bacterium)	Genital discharge, itching, and/or painful urination; often no symptoms in women; PID	Antibiotics
Gonorrhea	*Neisseria gonorrhoeae* (bacterium)	Genital discharge; painful urination; sometimes no symptoms in women; PID	Antibiotics
Syphilis	*Treponema pallidum* (bacterium)	Ulcer (chancre) on genitals in early stages; spreads throughout body and can be fatal if not treated	Antibiotics can cure in early stages
Genital herpes (see Module 10.18)	Herpes simplex virus type 2, occasionally type 1	Recurring symptoms: small blisters on genitals, painful urination, skin inflammation; linked to cervical cancer, miscarriage, birth defects	Acyclovir can shorten duration of symptoms
Genital warts	Papilloma-viruses	Painless growths on genitals; some of the viruses linked to cancer	Removal by freezing
AIDS and HIV infection	HIV (virus)	See Module 24.18	Combination of drugs
Trichomoniasis	*Trichomonas vaginalis* (protozoan)	Vaginal irritation, itching, and discharge; usually no symptoms in men	Antiprotozoan drugs
Candidiasis (yeast infections)	*Candida albicans* (fungus)	Similar to symptoms of trichomoniasis; frequently acquired nonsexually	Antifungal drugs

Speaking to the widespread fear of AIDS, one advertisement for condoms reads, "I enjoy sex, but I'm not willing to die for it." We discussed the importance of safer sex as a deterrent to AIDS in Chapter 24's introduction. AIDS is only one of many **sexually transmissible diseases (STDs),** contagious diseases spread by sexual contact. Latex condoms can usually prevent STD spread.

AIDS and most other viral STDs are not curable, but the bacterial, protozoan, and fungal STDs (listed in separate sections of the table) generally are. Usually both partners must be treated to prevent reinfection.

STDs are epidemic throughout the world. Many of these diseases cause long-term problems if they are not treated. A condition called pelvic inflammatory disease (PID) is a common secondary result of STDs caused by bacterial infections in women. In PID, bacteria spread from the vagina into the uterus, oviducts, and ovaries, and may cause acute pain, scarring, and sterility.

The table here includes most of the STDs common in the United States.

Besides abstinence from sexual contact, what can prevent the spread of STDs?

Latex condoms

Contraception is the deliberate prevention of pregnancy. Only complete abstinence (no intercourse) is totally effective, but other forms of contraception have varying degrees of effectiveness. Contraception works in one of three ways: (1) preventing the release of gametes from the gonads, (2) preventing fertilization, or (3) preventing the embryo from implanting in the uterus.

Methods that prevent the release of gametes are highly effective, as the table indicates. Birth control pills, now used by over 60 million women worldwide, have been available since the 1960s. The most widely used pills are combinations of a synthetic estrogen and a synthetic progesterone-like hormone called progestin. "The pill" prevents ovulation and keeps follicles from developing. What about long-term side effects? No solid evidence links the pill to cancers, but cardiovascular problems are a concern, especially in women who smoke or have a history of such problems.

A second type of birth control pill, called the minipill, contains only progestin. Slightly less effective than the combination pill, the minipill prevents fertilization by altering a woman's cervical mucus so that it blocks sperm from entering the uterus. Norplant®, a time-release capsule that is implanted under a woman's skin, releases progestin into the blood and is effective for five years.

Sterilization prevents conception permanently. In **vasectomy,** in men, a doctor cuts a section out of each vas deferens to prevent sperm from entering the urethra. In **tubal ligation,** in women, a doctor cuts a short section out of each oviduct (and may tie, or ligate, the remaining ends) to prevent eggs from reaching the uterus. Both forms of sterilization are considered relatively safe and free from side effects. They are difficult to reverse, however, and should be considered permanent.

The effectiveness of other methods that prevent fertilization depends on how they are used. Temporary abstinence, called the **rhythm method** or **natural family planning,** depends on refraining from intercourse during the few days around ovulation, when conception is most likely. This method is generally not reliable because it is difficult to predict or detect the time of ovulation. Equally unreliable is **withdrawal** of the penis from the vagina before ejaculation. Even before ejaculation, sperm may be present in the penis and may be deposited in the vagina.

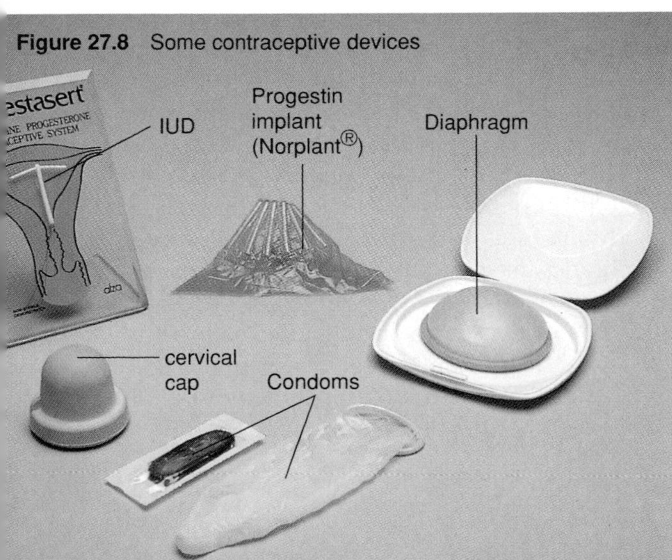

Figure 27.8 Some contraceptive devices

Contraceptive Methods

Method	Pregnancies/100 Women/Year* Used Correctly	Typically
Prevents Release of Gametes		
Birth control pill (combination)	0.1	3
Prevents Fertilization		
Vasectomy	0.1	0.15
Tubal ligation	0.2	0.4
Progestin minipill or implant	0.5	3
Rhythm	1–9	20
Withdrawal	4	18
Condom (male)	2	12
Diaphragm and spermicide	6	18
Cervical cap and spermicide	6	18
Spermicide alone	3	21
Prevents Implantation		
Intrauterine device (IUD)	1–2	3

*Without contraception, about 85 pregnancies would occur.

If used correctly, some **barrier methods** can be more effective than either rhythm or withdrawal in preventing sperm and egg from meeting. These methods physically stop sperm from moving into the uterus and oviducts. Condoms, now made for both men and women, are thin sheaths, usually of latex, that fit over the penis or within the vagina. The diaphragm is a dome-shaped rubber cap that covers the cervix; the cervical cap is similar, but it is smaller and fits more closely over the cervix. To be effective, diaphragms and cervical caps must be used with **spermicides,** sperm-killing chemicals in the form of cream, jelly, or foam. Spermicides alone are typically less effective.

Designed to prevent implantation, **intrauterine devices (IUDs)** are small, usually plastic objects that a doctor inserts into the uterine cavity. IUDs cause bleeding, infections, or uterine perforation in some women, discouraging wider use. Also preventing implantation, when taken within 3 days after intercourse, are **morning-after pills (MAPs),** which contain various combinations of estrogen and progesterone. High doses of certain ordinary birth control pills can also function as MAPs, with an effectiveness of about 75%.

Let's now return to our main subject—reproduction—and follow what happens in humans and other animals when an egg and sperm actually meet.

? ________ is to males as tubal ligation is to ________.

Vasectomy . . . females

Principles of Embryonic Development

27.9 Fertilization results in a zygote and triggers embryonic development

27C

Embryonic development begins with **fertilization,** the union of a sperm and an egg to form a diploid zygote. Fertilization introduces the sperm's haploid set of chromosomes into the egg and also activates the egg by triggering metabolic changes that start embryonic development.

Figure 27.9A is a micrograph of an unfertilized human egg almost covered by sperm. Of all these sperm, only a single one will enter and fertilize the egg. All the other sperm—the ones shown here and millions more that were ejaculated with them—will die. The sperm that penetrates the egg gains a chance to have its unique set of genes combine with those of the egg and contribute to the next generation.

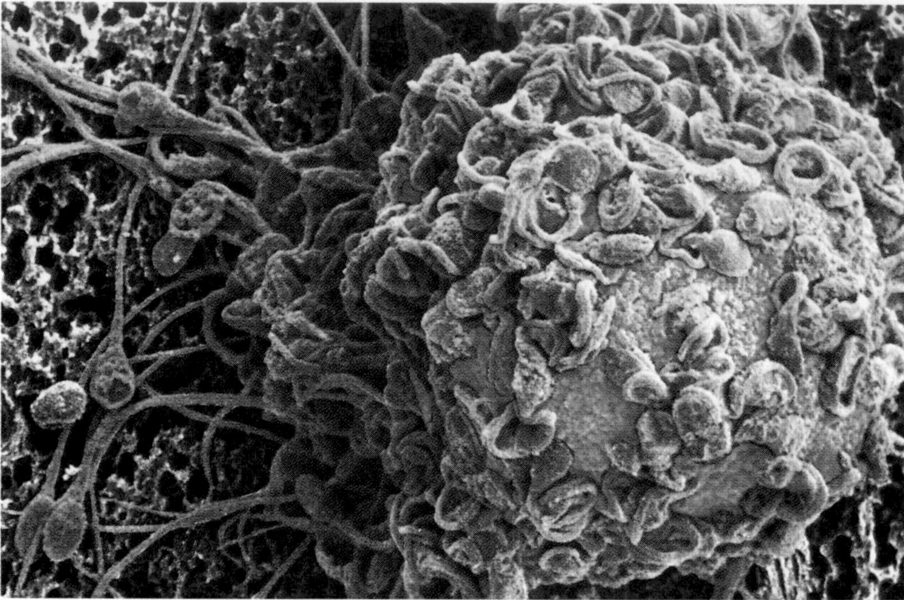

Figure 27.9A A human egg cell surrounded by sperm

Figure 27.9B illustrates the structure of a mature human sperm. Here is another case of form fitting function. The sperm's streamlined shape is an adaptation for swimming through fluids in the vagina, uterus, and oviduct of the female. The sperm cell's thick head contains a haploid nucleus and is tipped with a membrane-enclosed sac, the **acrosome,** which lies just inside the plasma membrane. The acrosome contains enzymes that help the sperm penetrate the egg. The neck and middle piece of the sperm contain a long, spiral mitochondrion. The sperm absorbs high-energy nutrients, especially the sugar fructose, from the semen. Thus fueled, its mitochondrion provides ATP for movement of the tail, which is actually a flagellum. By the time a sperm has reached the egg, it has consumed much of the energy available to it. But a successful sperm will have enough energy left to enter the egg and deposit its nucleus in the egg's cytoplasm.

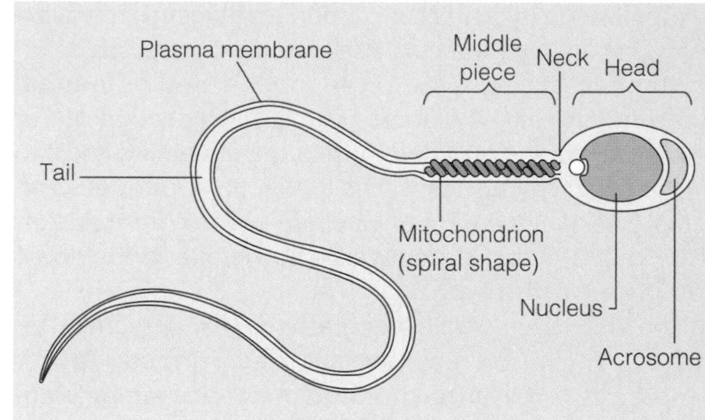

Figure 27.9B The structure of a human sperm cell

Figure 27.9C illustrates the sequence of events in fertilization. This diagram is based on fertilization in sea urchins, on which a great deal of research has been done. Similar processes occur in other animals, including humans. The diagram traces one sperm through the successive activities of fertilization. Notice that, to reach the egg nucleus, the sperm nucleus must pass through three barriers: the egg's jelly coat (yellow), a middle region of glycoproteins called the vitelline layer (pink), and the egg cell's plasma membrane (black line).

Let's follow the steps shown in the figure. As a sperm ① approaches and then ② contacts the jelly coat of the egg, the acrosome in the sperm head releases a cloud of enzyme molecules that digest a cavity in the jelly. When the sperm head reaches the vitelline layer, ③ species-specific protein molecules on its surface bind with specific receptor proteins on the vitelline layer. The specific binding between the proteins of the sperm and egg ensures that sperm of other species cannot fertilize the egg. This specificity is especially important when fertilization is external, because the sperm of other species may be present in the water. After the specific binding occurs, the sperm proceeds through the vitelline layer, and ④ the sperm's plasma membrane fuses with that of the egg. Fusion of the two membranes makes it possible for ⑤ the sperm nucleus to enter the egg.

Fusion of the sperm and egg plasma membranes triggers a number of important changes in the egg. Two such changes prevent other sperm from entering the egg. Less than a second after the membranes fuse, the entire egg plasma membrane becomes impenetrable to other sperm cells. Shortly thereafter, ⑥ the vitelline layer hardens and separates from the plasma membrane. The space quickly fills with water, and the vitelline layer becomes the so-called **fertilization membrane,** another barrier impenetrable to sperm. If these events did not occur and an egg were fertilized by more than one sperm, the resulting zygote nucleus would contain too many chromosomes, and the zygote could not develop normally.

Membrane fusion also triggers a burst of metabolic activity in the egg. In preparation for the enormous growth and development that will follow fertilization, the egg's metabolic machinery suddenly gears up from virtual dormancy. At the same time, ⑦ the egg and sperm nuclei fuse, producing the diploid nucleus of the zygote. In the next module, we begin to trace the development of the zygote into a whole new animal.

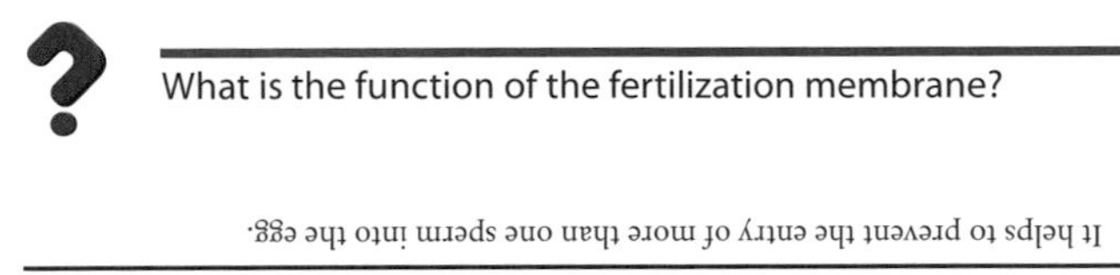

What is the function of the fertilization membrane?

It helps to prevent the entry of more than one sperm into the egg.

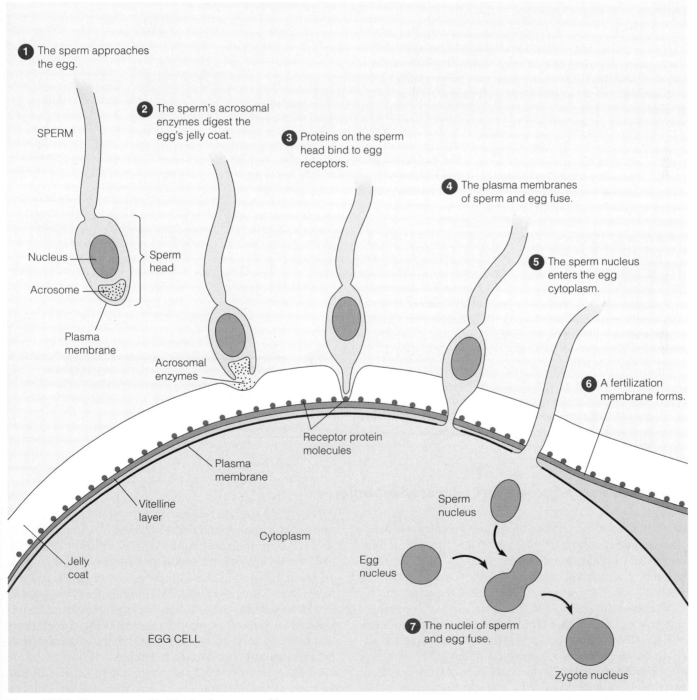

Figure 27.9C The process of fertilization for a sea urchin

27.10 Cleavage produces a ball of cells from the zygote

27C

An animal is made up of many thousands, millions, even trillions of cells organized into complex tissues and organs. The transformation to this multicellular state from a zygote is truly phenomenal. Order and precision are required at every step, and they are clearly displayed in the first two major phases of embryonic development: cleavage and gastrulation. We focus on cleavage in this module and gastrulation in the next.

Cleavage is a rapid succession of cell divisions that produces a ball of cells—a multicellular embryo—from the zygote. DNA replication, mitosis, and cytokinesis occur rapidly, but gene transcription is virtually shut down, and few new proteins are synthesized. As a result, the embryo of most animals does not grow larger during cleavage. Nutrients stored in the egg nourish the dividing cells, and the cell divisions partition the zygote into many smaller cells.

Figure 27.10 illustrates cleavage in a sea urchin. As the first three steps show, the number of cells doubles with each cleavage division. In a sea urchin, a doubling occurs about every 20 minutes, and the whole cleavage process takes about 3 hours to produce a solid ball of cells. Notice that each cell in the ball is much smaller than the zygote. As cleavage continues, a fluid-filled cavity called the **blastocoel** forms in the center of the embryo. At the completion of cleavage, there is a large cavity surrounded by one or more layers of cells. This hollow ball of cells is called the **blastula.**

Cleavage makes two very important contributions to early development. It creates a multicellular embryo, the blastula, from a single-celled zygote. Cleavage is also an organizing process, partitioning the multicellular embryo into developmental regions. As we discussed in Module 11.10, the cytoplasm of the zygote contains a variety of chemicals that control gene expression during early development. During cleavage, regulatory chemicals become localized in particular groups of cells, where they activate the genes that direct the formation of specific parts of the animal. Gastrulation, the next phase of development, further refines the embryo's cellular organization.

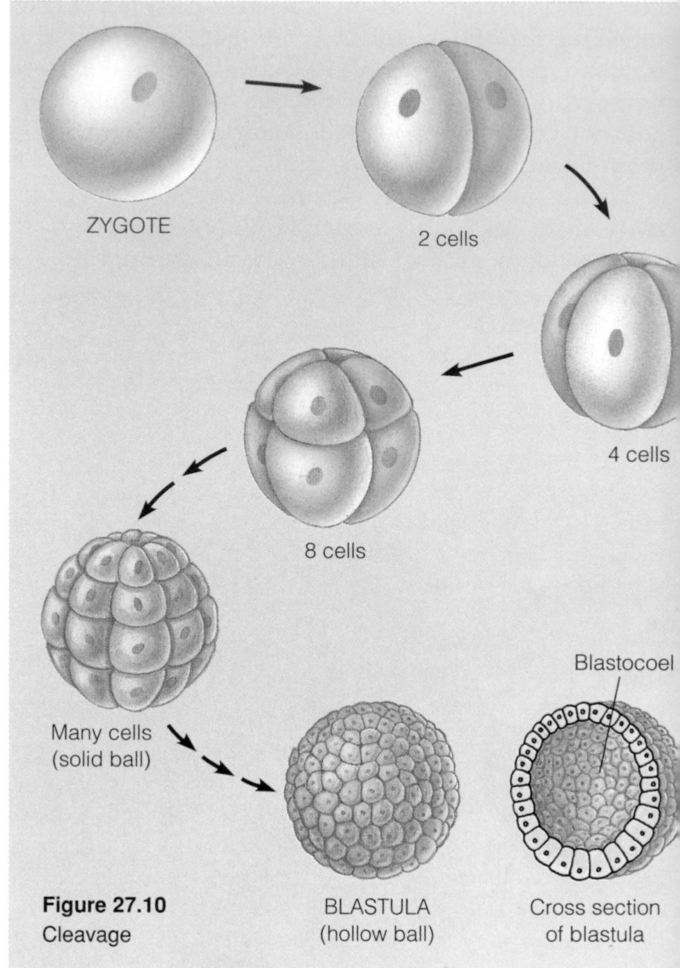

Figure 27.10
Cleavage

How does the reduction of cell size during cleavage increase oxygen supply to the cells' mitochondria? (*Hint:* Review Module 4.3.)

Smaller cells have a greater plasma membrane surface relative to cellular volume, and this facilitates diffusion of oxygen from the environment to the cells' cytoplasm.

27.11 Gastrulation produces a three-layered embryo

27C

Gastrulation, the second major phase of embryonic development, adds more cells to the embryo; more importantly, it sorts all the cells into distinct cell layers. In the process, the embryo is transformed from a hollow ball of cells—the blastula—into a three-layered stage called the **gastrula.**

The three layers produced in gastrulation are embryonic tissues called **ectoderm, endoderm,** and **mesoderm.** The ectoderm forms the outer layer (skin) of the gastrula. The endoderm forms an embryonic digestive tract. And the mesoderm partly fills the space between the ectoderm and the endoderm. Eventually, these three cell layers develop into all the parts of the adult animal. For instance, our nervous system and the outer layer (epidermis) of our skin come from ectoderm; the innermost lining of our digestive tract arises from endoderm; and most other organs and tissues, such as the kidney, heart, muscles, and the inner layer of our skin (dermis), develop from mesoderm.

The mechanics of gastrulation vary somewhat depending on the species. We have chosen the frog, a vertebrate that has long been a favorite of researchers, to demonstrate how gastrulation produces the three cell layers. Figure 27.11 takes us from a blastula at the top to a three-layered gastrula at the bottom. The diagrams in the left column show an external view of gastrulation; each drawing represents a multicellular embryo, and the arrows indicate movements of cell layers. The cutaway drawings on the right reveal the internal structures that develop as gastrulation occurs. The timing of these events varies with the species and the temperature of the lake or pond in which the frog develops. In many frogs, cleavage and gastrulation together take about 15–20 hours.

? Gastrulation forms a new cavity, the ________, which is lined by ________ and which develops into the animal's ________ tract.

archenteron . . . endoderm . . . digestive

1 The blastula. Formed by cleavage, the frog blastula is a partially hollow ball of unequally sized cells. As the cross section shows, the cells toward one end, called the animal pole, are smaller than those near the opposite end, the vegetal pole. The cells near the vegetal pole are larger because they contain yolk granules, which make them divide at a slower rate than those of the animal pole. The three colors on the blastula indicate regions of cells that will give rise to the primary cell layers: ectoderm (blue), endoderm (yellow), and mesoderm (pink). (Notice that each layer may be more than one cell thick.) In real embryos, these regions have been identified by dyeing the cells with harmless stains and observing where the cells go as development proceeds. A glance ahead at parts 2–4 will show you that, as gastrulation occurs, the cells that form mesoderm and endoderm move from the surface to the inside of the embryo.

2 Blastopore formation. Gastrulation begins when a small groove, called the **blastopore,** appears on one side of the blastula. The blastopore is the place where cells that will form endoderm and mesoderm move inward from the surface (the dashed part of the arrow indicates inward movement). Meanwhile, other cells, which will form ectoderm, spread over the surface of the embryo.

3 Cell migration to form layers. The beginnings of the three layers can now be seen in the cross section. Migrating endodermal cells (yellow) have produced a simple digestive cavity called the **archenteron.** The advancing endoderm and the archenteron have filled some of the space formerly occupied by the blastocoel. Cells that will form the mesoderm (pink) are located between the endoderm and the ectoderm (blue).

4 Completion of gastrulation. Gastrulation is completed when the embryo is three-layered. Ectoderm covers the surface except for a cluster of endodermal cells called the **yolk plug.** The yolk plug marks the site of the blastopore and of the future anus. At this stage, the endoderm and its archenteron have replaced the blastocoel. Mesoderm forms a layer between the ectoderm and the endoderm.

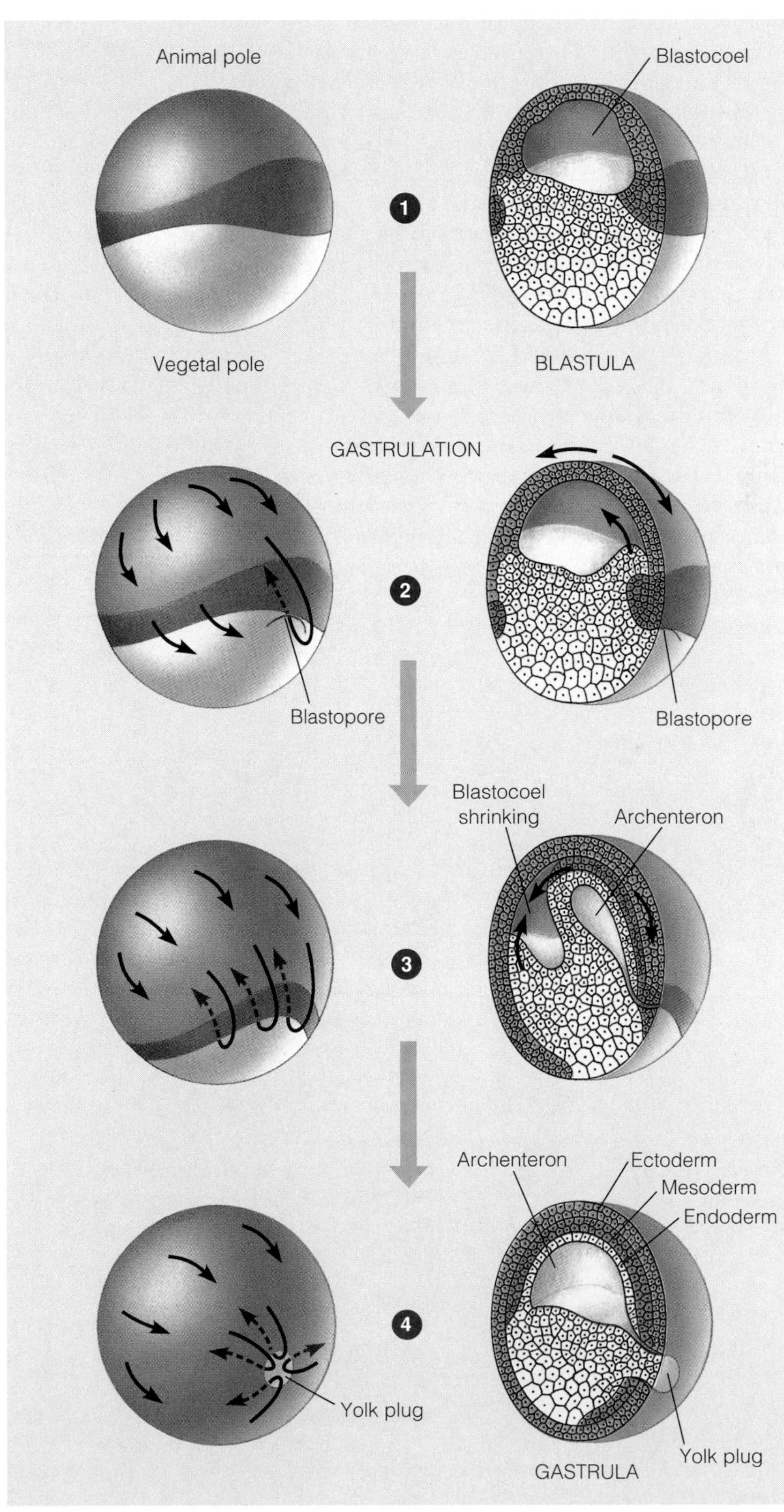

Figure 27.11C Development of the frog gastrula

27.12 Organs start to form after gastrulation

27D

In organizing the embryo into three layers, gastrulation sets the stage for the shaping of an animal. Once the ectoderm, endoderm, and mesoderm form, cells in each layer begin to differentiate into tissues and embryonic organs. The cutaway drawing in Figure 27.12A shows the developmental structures that appear in a frog embryo a few hours after the completion of gastrulation. The orientation drawing at the upper left of the figure indicates a corresponding cut through an adult frog.

We see two structures in the embryo in Figure 27.12A that were not present at the gastrula stage described in the last module. An organ called the notochord has developed in the mesoderm, and a structure that will become the hollow nerve cord is beginning to form in the ectoderm. Recall that the notochord and dorsal, hollow nerve cord are hallmarks of the chordates (see Module 18.15).

The notochord is visible in cross section in the drawing in Figure 27.12A. It forms from mesoderm just above the archenteron. Made of a cartilagelike substance, the **notochord** extends for most of the embryo's length and provides support for other developing tissues. Later in development, the notochord will function as a core around which mesodermal cells gather and form the frog's backbone.

You can also see in Figure 27.12A the beginnings of the frog's hollow nerve cord, formed from a portion of the ectoderm. The area shown in green in the cutaway drawing is a thickened region of ectoderm called the neural plate. From it arises a pair of pronounced ectodermal ridges, called neural folds, visible in both the drawing and the micrograph below it. If you now look at the series of diagrams in Figure 27.12B, you will see what happens as the neural folds and neural plate develop further. The neural plate rolls up and forms the neural tube, which then sinks beneath the surface of the embryo and is covered by an outer layer of ectoderm. The **neural tube** is destined to become the brain and spinal cord.

Figure 27.12C shows a later frog embryo (about 12 hours older than the one in Figure 27.12A), in which the neural tube has formed. Notice in the drawing that the neural tube lies directly above the notochord. The relative positions of the neural tube, notochord, and archenteron give us a preview of the basic body plan of a frog. The spinal cord will lie within extensions of the dorsal surface of the backbone (which will replace the notochord), and the digestive tract will be ventral to the backbone. We see this same arrangement of organs in all vertebrates.

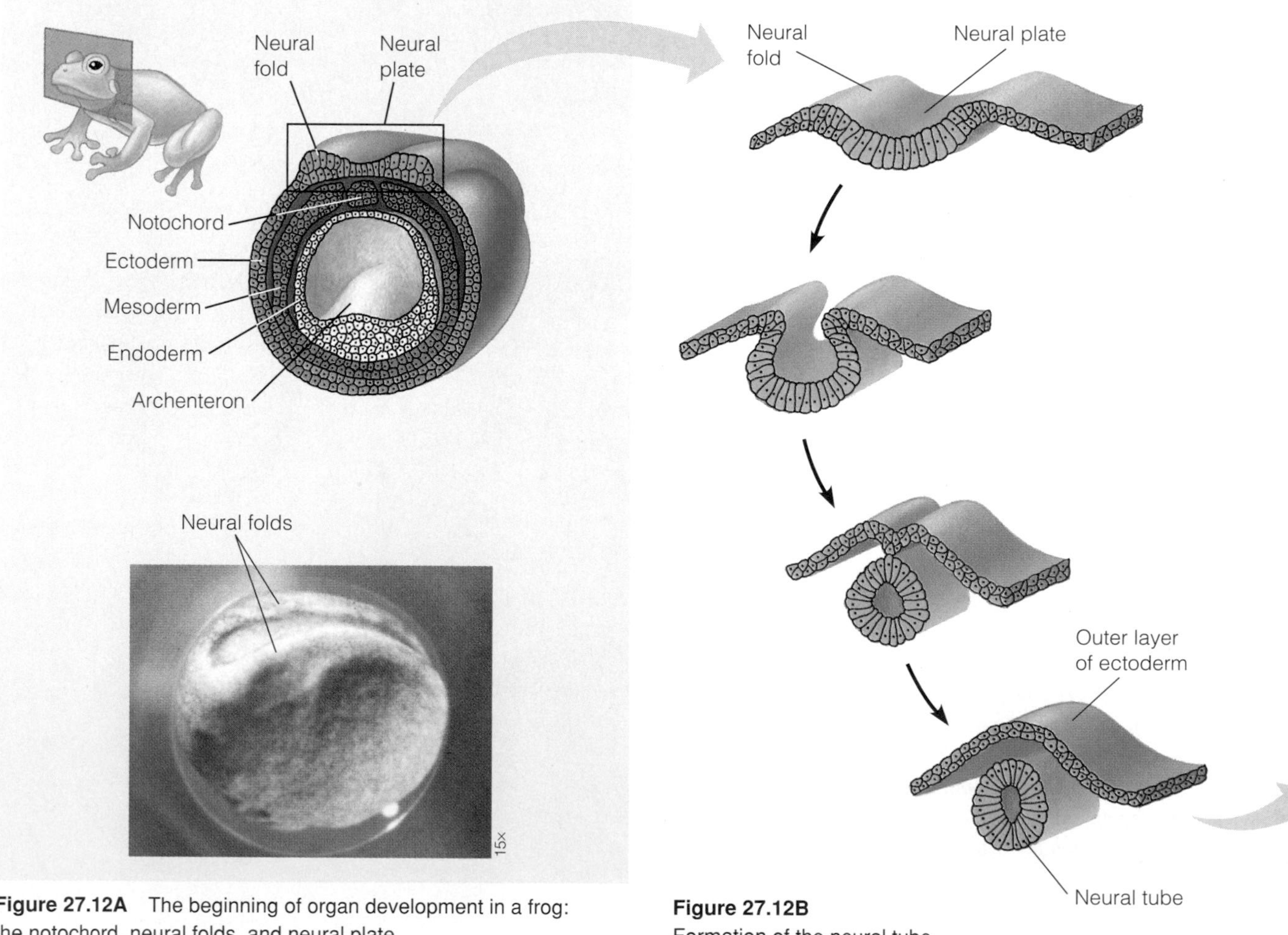

Figure 27.12A The beginning of organ development in a frog: the notochord, neural folds, and neural plate

Figure 27.12B Formation of the neural tube

Figure 27.12C shows several other fundamental changes. In the micrograph, which is a side view, you can see that the embryo is more elongated than the one in Figure 27.12A. You can also see the beginnings of an eye and a tail (called the tail bud). Part of the ectoderm has been removed to reveal a series of internal ridges called somites. The **somites** are blocks of mesoderm that will give rise to segmental structures, such as the vertebrae and associated muscles of the backbone. In the cross-sectional drawing, notice that the mesoderm next to the somites is developing a hollow space—the body cavity, or **coelom.** Body parts that are segmented (constructed of repeating units) and a coelom are basic features of chordates (see Module 18.15).

In this and the previous module, we have observed the sequence of changes that occur as an animal begins to take shape. To summarize, the key phases in embryonic development are cleavage (which creates a multicellular animal from a zygote), gastrulation (which organizes the embryo into three discrete layers), and organ formation (which generates embryonic organs from the three embryonic tissue layers). These same three phases occur in nearly all animals.

If we followed a frog's development beyond the stage represented in Figure 27.12C, within a few hours we would be able to monitor muscular responses and a heartbeat and see a set of gills with blood circulating in them. A long tail fin would grow from the tail bud. The timing of the

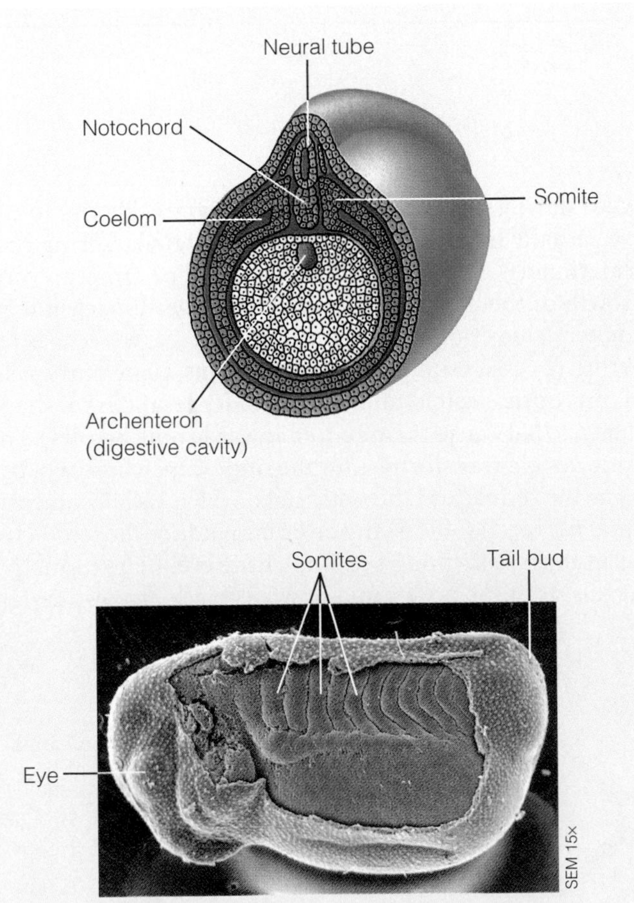

Figure 27.12C An embryo with completed neural tube, somites, and coelom

Derivatives of the Three Embryonic Tissue Layers

Embryonic Layer	Organs and Tissues in the Adult
Ectoderm	Epidermis of skin and its derivatives; epithelial lining of mouth and rectum; sense receptors in epidermis; cornea and lens of eye; nervous system; adrenal medulla; tooth enamel
Endoderm	Epithelial lining of digestive tract (except mouth and rectum); epithelial lining of respiratory system; liver; pancreas; thyroid; parathyroids; thymus; lining of urethra, urinary bladder, and reproductive system
Mesoderm	Notochord (in animals retaining it as adults); skeletal system; muscular system; circulatory system; excretory system; reproductive system (except germ cells, which differentiate during cleavage); dermis of skin; lining of body cavity; adrenal cortex

later stages in frog development varies enormously, but in many species, by 5–8 days after development begins, we would see all the body tissues and organs of a tadpole emerge from cells of the ectoderm, mesoderm, and endoderm. Eventually, the structures of the tadpole (Figure 27.12D) would transform into the tissues and organs of an adult frog. The table above lists the major organs and tissues that arise in frogs, and other vertebrates, from each of the three main embryonic tissue layers.

Figure 27.12D A tadpole

Watching embryos develop helps us appreciate the enormous changes that occur as one tiny cell, the zygote, gives rise to a highly structured, many-celled animal. Your own body, for instance, is a complex organization of some 60 trillion cells, all of which arose from a zygote smaller than the period at the end of this sentence. Discovering how this incredibly intricate arrangement is achieved is one of biology's greatest challenges. Through research that combines the experimental manipulation of embryos with molecular biology and genetics, developmental biologists have begun to work out the mechanisms that underlie development. We examine some of these mechanisms in the next three modules. 

? What is the embryonic basis for the dorsal, hollow nerve cord that is common to all members of our phylum?

The nerve cord, which becomes the brain and spinal cord, develops from a dorsal ectodermal plate that folds to form an interior tube.

27.13 Changes in cell shape, cell migration, and programmed cell death give form to the developing animal

Many events in development depend on a combination of several kinds of cellular processes. Figure 27.13A shows how two changes in cell shape bring about the formation of the neural tube (see Figure 27.12B). Cells of the ectoderm fold inward by first elongating and then becoming wedge-shaped. The result is a tube of ectoderm—the start of the brain and spinal cord.

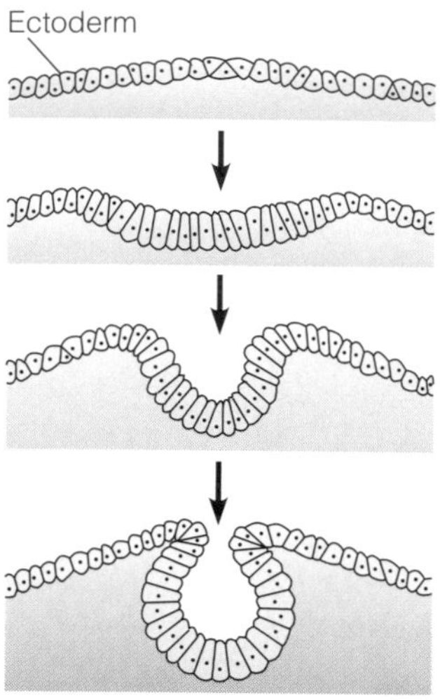

Ectoderm

Figure 27.13A Changes in cell shape in neural tube formation

Cell migration is also essential in development. For example, during gastrulation, ectodermal cells use fingerlike extensions (pseudopodia) to "crawl" to the embryo's surface. Migrating cells may follow chemical trails secreted by cells near their specific destination. Once a migrating cell reaches its destination, surface proteins enable it to recognize similar cells. The cells join together and secrete glycoproteins that glue them in place. Finally, they differentiate, taking on the characteristics of a particular tissue.

Another key developmental process is **programmed cell death,** the timely and tidy suicide of cells. Some, perhaps all, animals have suicide genes coding for proteins that kill the cell that produces them. In humans, the timely death of specific cells in developing arms and legs creates the spaces between fingers and toes. Cell death is also essential for the normal development of our nervous and immune systems. In Figure 27.13B, the cell on the left shrinks and dies because a suicide gene has been turned on. Meanwhile, signals from the dying cell make an adjacent cell phagocytic. This cell engulfs and digests the dead cell, keeping the embryo free of harmful enzymes and other debris.

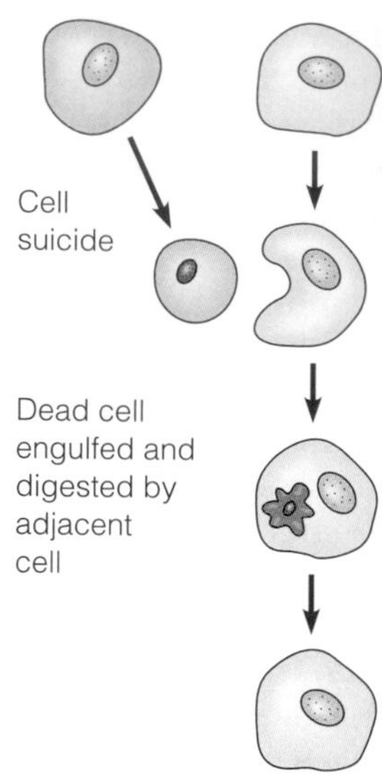

Figure 27.13B Programmed cell death

Which usually comes first in the developmental history of a cell, its migration within the embryo or its differentiation into a specialized cell?

Migration

27.14 Embryonic induction initiates organ formation

All developmental processes depend on signals passed between neighboring cells and cell layers, telling embryonic cells precisely what to do when. The mechanism by which one group of cells influences the development of an adjacent group of cells is called **induction.** Induction plays a major role in the early development of tissues and organs from ectoderm, endoderm, and mesoderm. Its effect is to switch on a set of genes whose expression makes the receiving cells differentiate into a specific tissue. A sequence of inductive signals leads to increasingly greater specialization of cells, as organs begin to take shape.

Figure 27.14 illustrates the differentiation of cells that form the vertebrate eye and two of the inductions that occur during this process. Cells destined to give rise to the eye actually begin receiving inductive signals during gastrulation. ① The eye begins to take shape from an outgrowth of the developing brain (the optic vesicle) and an adjacent cluster of cells on the body surface (the lens ectoderm). ② As a result of earlier inductions, some of the cells of the optic vesicle and lens ectoderm undergo shape changes that cause them to fold inward (pink arrows). The optic vesicle transforms into the optic cup, which will become the retina, and the optic stalk, which will become the optic nerve. ③ Cells of the optic cup induce the lens ectoderm to start forming the lens of the eye. (The inductive signals from the optic cup, shown as black arrows, are late

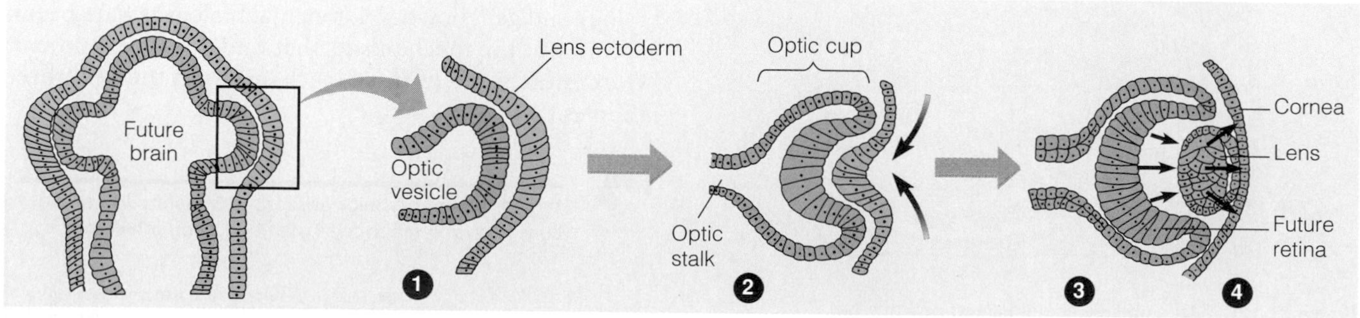

Figure 27.14 Induction during eye development

ones in a sequence begun during gastrulation.) Finally, ④ cells of the developing lens induce development of the cornea, the eye's transparent outer covering.

Induction plays a role in the early development of virtually all organs and tissues. Researchers are now focusing on the mechanisms of induction, including both the signal molecules and the signal-transduction pathways that transform inductive signals into cellular responses (see Module 11.11).

How do signal-transduction pathways function in induction?

They mediate between the chemical signal received by the cell and the resulting changes in gene expression and other responses by the cell.

Pattern formation organizes the animal body 27.15

Forming the parts of an eye is one thing, but what about the development of an entire region of the body? An arm and a leg, for instance, have the same kinds of tissues—muscle, connective tissue, cartilage, and skin—but these tissues are arranged somewhat differently in the two limbs. What directs the formation of major body parts?

The shaping of an animal's major parts involves **pattern formation,** the emergence of a body form with specialized organs and tissues all in the right places. Research indicates that master control genes respond to chemical signals that tell a cell where it is relative to other cells in the embryo (see Module 11.10). These positional signals determine which master control genes will be expressed and, consequently, which body parts will form.

Figure 27.15A indicates how positional signals affect the development of the limbs of vertebrates. Vertebrate limbs develop from embryonic structures called limb buds. Bird wings, for example, develop from the two anterior limb buds. In order for a wing to form properly, each embryonic wing cell must receive signals specifying its position in three dimensions: How close is it to the embryo's main axis, to the anterior or posterior edge of the developing wing, and to the dorsal or ventral surface of the embryo? Only with this information will the cell's genes direct the synthesis of the proteins needed for normal differentiation in that cell's specific location.

Experiments have revealed that vertebrate limbs have zones of cells that provide positional information to other cells via chemical signals. Researchers have located one such pattern-forming zone on the posterior surface of the wing-forming limb buds of birds. Cells nearest the zone—presumably those exposed to the highest concentration of chemical signals from it—develop into posterior wing structures; cells farthest from the zone form anterior structures. As indicated in Figure 27.15B, if a block of cells from this zone is removed from one bird embryo (the donor) and grafted onto the anterior part of the limb bud of another embryo (the host), the host will develop additional wing structures—almost a double wing.

A major goal of research on pattern formation and of other developmental studies is learning how the one-dimensional information encoded in the nucleotide sequence of a zygote's DNA directs the development of the three-dimensional form of an animal. In the next two modules, we'll see the results of this process as we watch an individual of our own species take shape.

How is pattern formation already apparent at the gastrula stage? (*Hint:* Review Figure 27.11C.)

The major axes of the animal—anterior/posterior, dorsal/ventral, and left/right—are already set at the gastrula stage.

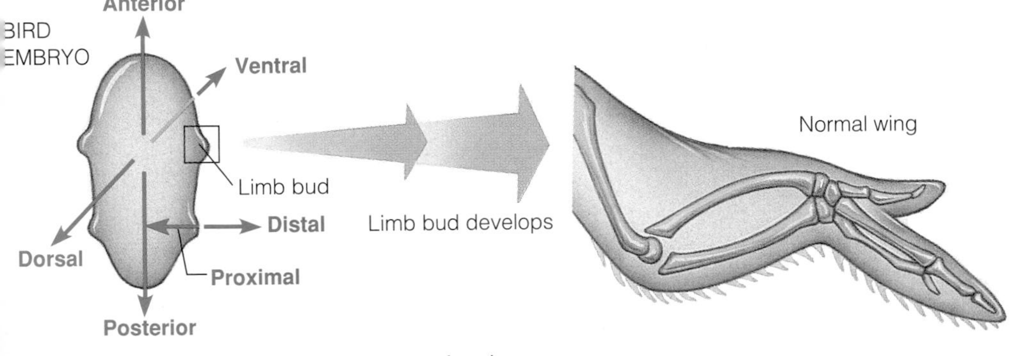

Figure 27.15A The normal development of a wing

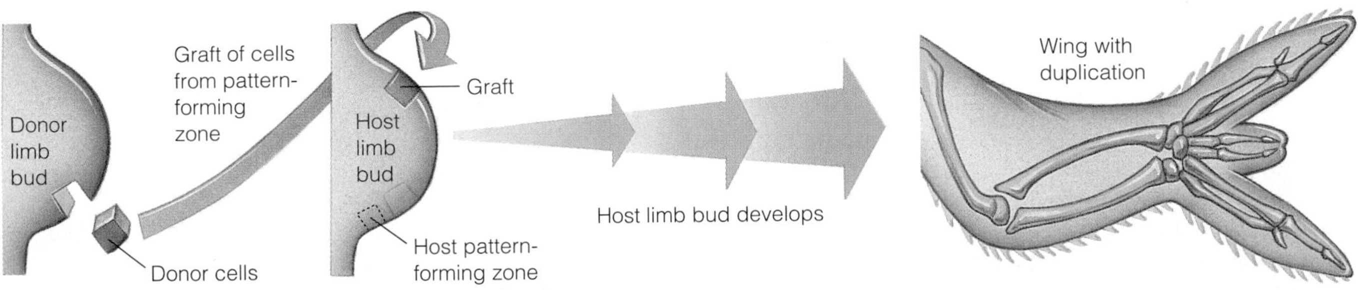

Figure 27.15B Experimental evidence for a pattern-forming zone

27.16 The embryo and placenta take shape during the first month of pregnancy

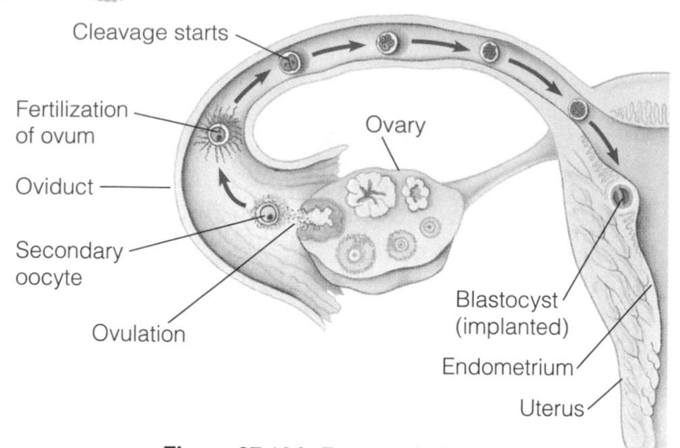

Figure 27.16A From ovulation to implantation

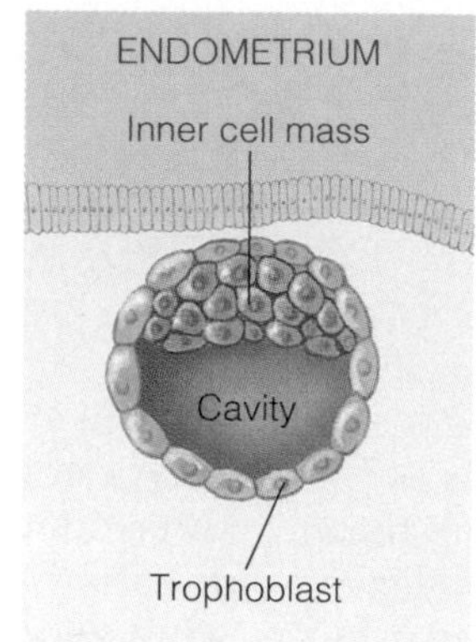

Figure 27.16B Blastocyst (6 days after conception)

Pregnancy, or **gestation,** is the carrying of developing young within the female reproductive tract. It begins at conception, the fertilization of the egg by a sperm, and continues until the birth of the baby. In humans, gestation lasts about 266 days (38 weeks). This is a long time compared to some other mammals; gestation in mice, for instance, lasts only about one month. At the other extreme, elephants have a 22-month gestation period.

An Overview of Developmental Events The figures on this and the next page illustrate, in cross section, the changes that occur during the first month of human development. The insets at the lower left of Figures 27.16C–27.16F show the actual size of the embryo at each stage.

Figure 27.16A takes us back to ovulation and fertilization in the oviduct. Cleavage starts about 24 hours after fertilization and continues as the embryo moves down the oviduct toward the uterus. By the sixth or seventh day after fertilization, the embryo has reached the uterus, and cleavage has produced about 100 cells. The embryo is now a hollow ball of cells called a **blastocyst** (the mammalian equivalent of the frog blastula we saw in Module 27.10).

The human blastocyst (Figure 27.16B) has a fluid-filled cavity, an inner cell mass that will actually form the baby, and an outer layer of cells called the **trophoblast.** The trophoblast secretes enzymes that enable the blastocyst to implant in the endometrium, the uterine lining (gray in all the figures).

The blastocyst starts to implant in the uterus about a week after conception. In Figure 27.16C, you can see extensions of the trophoblast spreading into the endometrium; these extensions consist of multiplying cells. The trophoblast cells eventually form part of the **placenta,** the organ that provides nourishment and oxygen to the embryo and helps dispose of its metabolic wastes. As we'll see, the placenta consists of both embryonic and maternal tissues.

In Figure 27.16C, the cells colored purple and yellow are derived from the inner cell mass. Most of the purple cells will give rise to the embryo. The yellow cells, some purple cells, and some trophoblast cells will give rise to four structures called the **extraembryonic membranes,** which develop as attachments to the embryo and help support it. You can see three of these membranes—the amnion (from purple cells), the yolk sac (from yellow cells), and the chorion (partly from trophoblast)—starting to take shape in Figure 27.16D. A later stage (Figure 27.16E) shows the fourth extraembryonic membrane, the allantois, developing as an extension of the yolk sac. In having the four extraembryonic membranes, the human embryo is like the embryos of all other mammals, birds, and reptiles (such as the lizard in this chapter's introduction).

Gastrulation is underway by 9 days after conception, the stage shown in Figure 27.16D. There is already evidence of the three embryonic layers—ectoderm (blue), endoderm (yellow), and mesoderm (pink). The embryo itself (not including the membranes) develops from the three inner cell layers shown in Figure 27.16E. The ectoderm layer will form the outer part of the embryo's skin. As indicated in the drawing, the ectoderm layer is continuous with the amnion. Similarly, the embryo's digestive tract will develop from the endoderm layer, which is continuous with the yolk sac. The bulk of most other organs will develop from the central layer of mesoderm.

Roles of the Extraembryonic Membranes Figure 27.16F shows the embryo about a month after conception, along with its life-support system, made up largely of the four extraembryonic membranes. By this time, the **amnion** has grown to enclose the embryo. The amniotic cavity is filled with fluid, which protects the embryo. The amnion

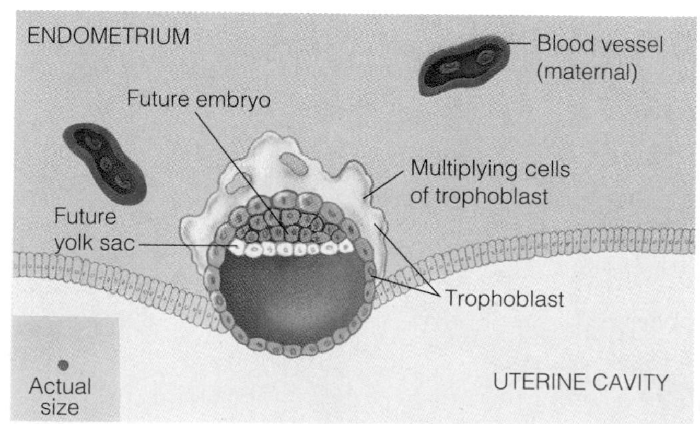

Figure 27.16C Implantation underway (about 7 days)

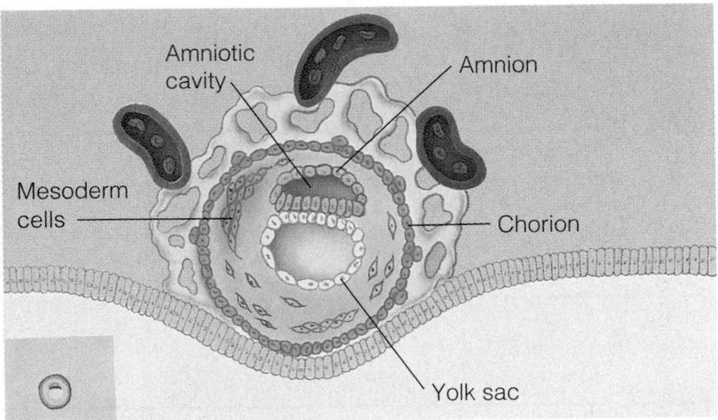

Figure 27.16D Embryonic layers and extraembryonic membranes starting to form (9 days)

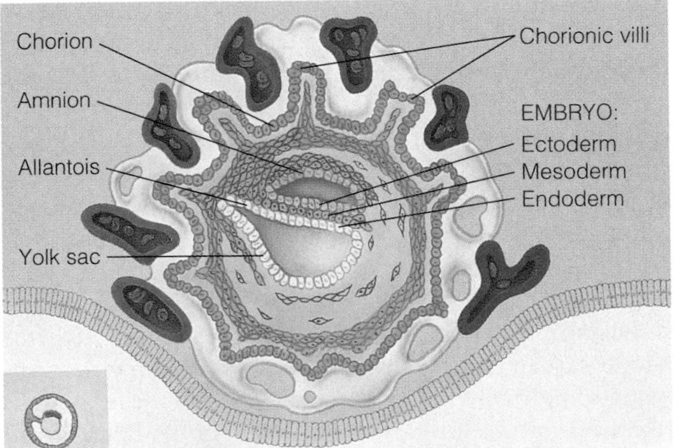

Figure 27.16E Three-layered embryo and four extraembryonic membranes (16 days)

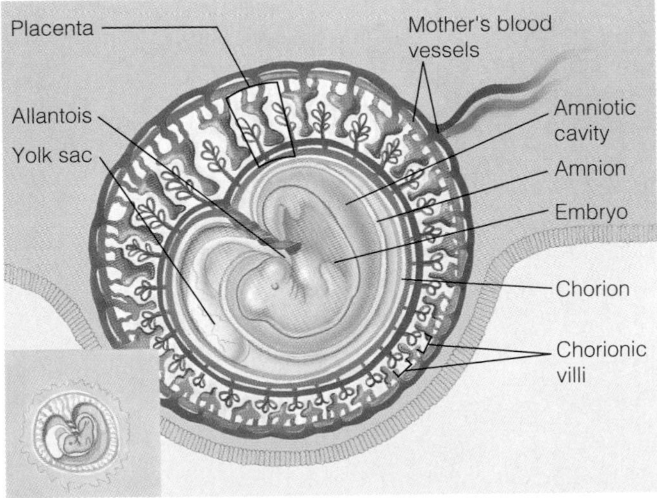

Figure 27.16F Placenta formed (31 days)

usually breaks just before childbirth, and the amniotic fluid ("water") leaves the mother's body through her vagina.

In humans and most other mammals, the **yolk sac** contains no yolk, but it is given the same name as the homologous structure in other vertebrates. In a bird or reptile egg, the yolk sac contains a large mass of yolk. Isolated within a shelled egg outside the mother's body, a developing bird or reptile obtains nourishment from the yolk rather than from a placenta. In mammals, the yolk sac, which remains small, has other important functions. It produces the embryo's first blood cells and its first germ cells, the cells that will give rise to the gamete-forming cells in the gonads.

The **allantois** also remains small in mammals. It forms part of the umbilical cord—the lifeline between the embryo and the placenta. It also forms part of the embryo's urinary bladder. In birds and reptiles, the allantois expands around the embryo and functions in waste disposal.

The outermost extraembryonic membrane, the **chorion,** develops from the trophoblast and from mesoderm cells derived from the yolk sac (see Figure 27.16D). The chorion becomes the embryo's part of the placenta. Cells in the chorion secrete a hormone called **human chorionic gonadotropin (HCG),** which maintains the corpus luteum of the ovary during the first three months of pregnancy. In turn, the corpus luteum continues to secrete estrogen and progesterone into the mother's blood. Without these hormones, menstruation would occur, and the embryo would abort spontaneously.

The Placenta Notice in Figure 27.16D the knobby outgrowths on the outside of the chorion. In Figure 27.16E, these outgrowths, now called **chorionic villi,** are larger and contain mesoderm. In Figure 27.16F, the chorionic villi contain embryonic blood vessels formed from the mesoderm. By this stage, the placenta is fully developed. Starting with the chorion and extending outward, the placenta is a composite organ consisting of chorionic villi closely associated with the blood vessels of the mother's endometrium. The villi are actually bathed in tiny pools of maternal blood (purple in the figure). The mother's blood and the embryo's blood are not in direct contact. However, the chorionic villi absorb nutrients and oxygen from the mother's blood and pass these substances to the embryo via the chorionic blood vessels colored red. The blue chorionic vessels carry wastes away from the embryo. The wastes diffuse into the mother's bloodstream and are excreted by her kidneys.

The placenta takes care of the embryo's every need. It even allows protective antibodies to pass from the mother to the fetus. Depending on what is circulating in the mother's blood, however, the placenta can also be a source of trouble. A number of viruses—the German measles virus and HIV are examples—can cross the placenta. German measles can cause serious birth defects; HIV-infected babies usually die of AIDS within a few years. Most drugs, both prescription and not, also cross the placenta, and many can harm the developing embryo. Alcohol and the chemicals in tobacco smoke, for instance, raise the chances of miscarriage and birth defects. Alcohol can cause a set of birth defects called fetal alcohol syndrome, which includes mental retardation. 

? Why does testing for HCG in a woman's urine or blood work as an early test of pregnancy?

Because this hormone is secreted by the chorion of an embryo

27.17 Human development from conception to birth is divided into three trimesters

27E

In this module, we use photographs to illustrate the rest of human development in the uterus. For convenience in studying human development, we divide the period from conception to birth into three **trimesters** of about three months each. The first trimester is the time of most radical change.

Certain drugs cause their most serious damage to an embryo very early in pregnancy, often before the mother realizes she is pregnant. Why?

Because organ systems begin to develop early in the first trimester

First Trimester

The photograph in Figure 27.17A shows a human embryo about one month after fertilization. In that brief time, this highly organized multicellular embryo has developed from a single cell. Not shown here are the extraembryonic membranes that surround the embryo or most of the umbilical cord that attaches it to the placenta. A month-old human embryo is about 7 mm (0.28 in) long and has a number of features in common with the somite stage of a frog embryo (see Figure 27.12C). The embryo has a notochord and a coelom, both formed from mesoderm. Its brain and spinal cord have begun to take shape from a tube of ectoderm, as in the frog. The human embryo also has four stumpy limb buds, a short tail, and elements of gill pouches. The gill pouches appear during embryonic development in all chordates; in land vertebrates, they eventually develop into parts of the throat and middle ear. Overall, a month-old human embryo is similar to other vertebrates at the somite stage of development.

Figure 27.17B shows a developing human, now called a fetus, about 9 weeks after fertilization. The large pinkish structure on the left is the placenta, attached to the fetus by the umbilical cord. The clear sac around the fetus is the amnion. By this time, the fetus is decidedly human, rather than generally vertebrate. It is about 5.5 cm (2.2 in) long and has all of its organs and major body parts, including a disproportionately large head. The somites have developed into the segmental muscles and the bones of the back and ribs. The limb buds have become tiny arms and legs with fingers and toes.

Beginning at about 9 weeks, the fetus can move its arms and legs, turn its head, frown, and make sucking motions with its lips. By the end of the first trimester, the fetus looks like a miniature human being, although its head is still oversized for the rest of the body. The sex of the fetus is usually evident at this time.

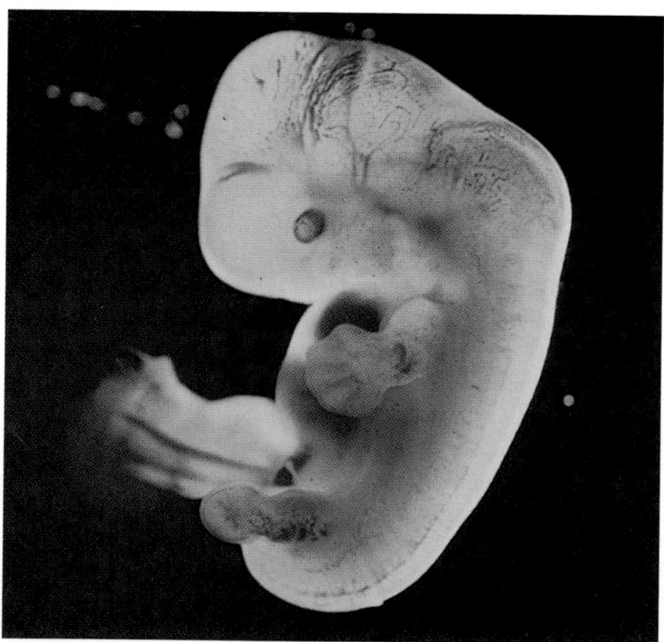

Figure 27.17A 5 weeks

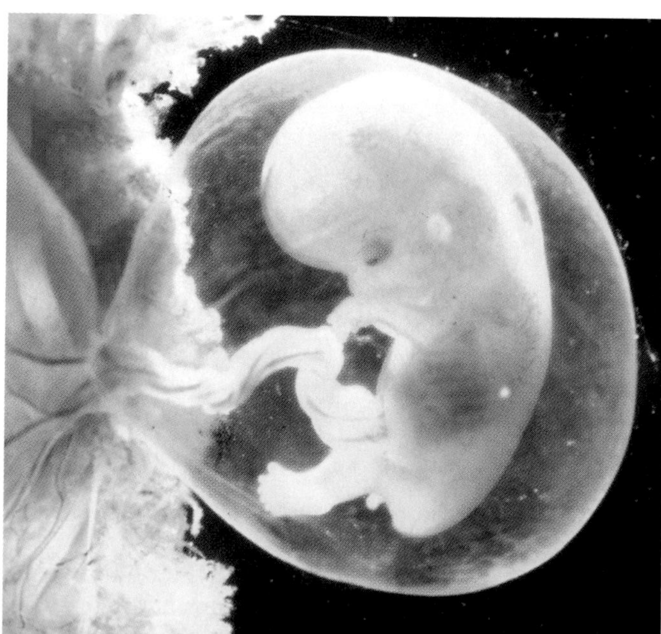

Figure 27.17B 9 weeks

Second Trimester

The main developmental changes during the second and third trimesters involve an increase in size and general refinement of the human features—nothing as dramatic as the changes of the first trimester. The photograph in Figure 27.17C shows a fetus at 14 weeks, 2 weeks into the second trimester. The fetus is now about 6 cm (2.4 in) long. During the second trimester, the placenta takes over the task of maintaining itself by secreting progesterone, rather than receiving it from the corpus luteum. At the same time, the placenta stops secreting HCG, and the corpus luteum, no longer needed to maintain pregnancy, degenerates.

At 20 weeks (Figure 27.17D), well into the second trimester, the fetus is about 19 cm (7.6 in) long, weighs about half a kilogram (1 lb), and has the face of an infant, complete with eyebrows and eyelashes. Its arms, legs, fingers, and toes have lengthened. It also has fingernails and toenails, and is covered with fine hair. By this time, the fetal heartbeat is readily detected, and the fetus usually is quite active. The mother's abdomen has become markedly enlarged, and she may often feel her baby move. Because of the limited space in the uterus, the fetus flexes forward into the so-called fetal position. By the end of the second trimester (about 30 weeks), the fetus's eyes are open and its teeth are forming.

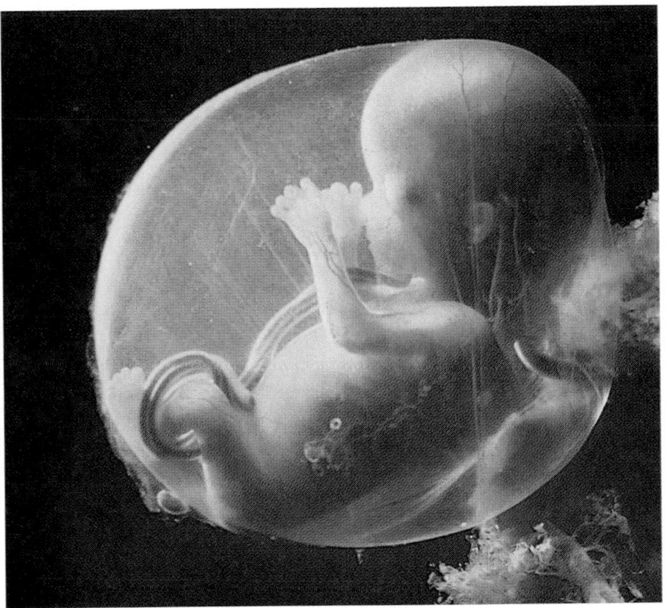

Figure 27.17C 14 weeks

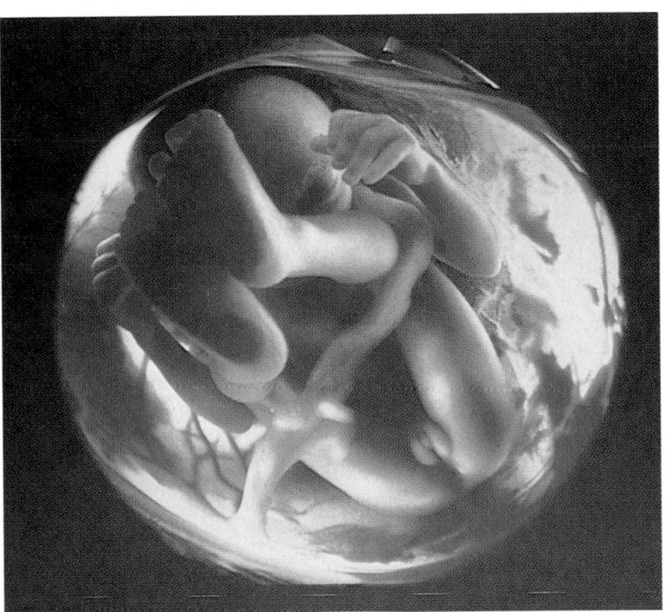

Figure 27.17D 20 weeks

Third Trimester

The third trimester (24 weeks to birth, Figure 27.17E) is a time of rapid growth, as the fetus gains the strength it will need to survive outside the protective environment of the uterus. Babies born prematurely—as early as the 24th week—may survive, but they require special medical care after birth. During the third trimester, the fetus's circulatory system and respiratory system undergo changes that will allow the switch to air breathing (see Module 22.12). The fetus also gains the ability to maintain its own temperature, its bones begin to harden, and its muscles thicken. It also loses much of its fine body hair, except on its head. The head itself changes its proportions. The fetus becomes less active as it fills the space in the uterus. At birth, a typical baby is about 50 cm (20 in) long and weighs 2.7–4.5 kg (6–10 lb).

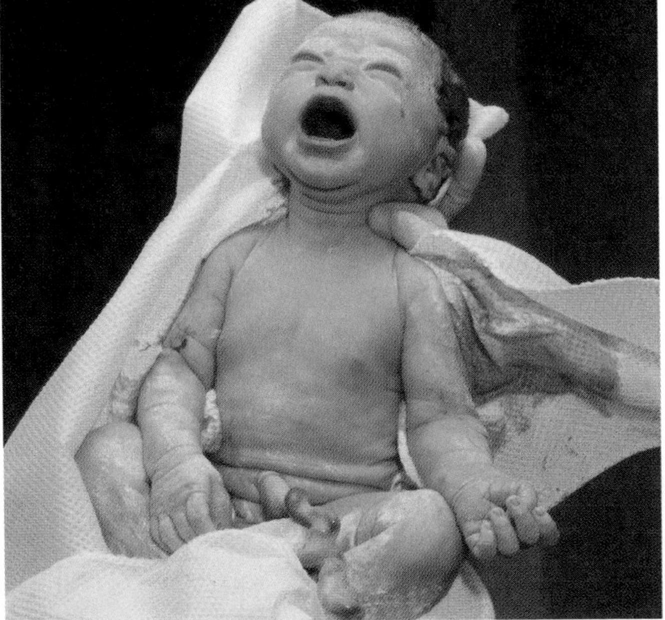

Figure 27.17E At birth

27.18 Childbirth is hormonally induced and occurs in three stages

The birth of a child is brought about by a series of strong, rhythmic contractions of the uterus, commonly called **labor.** As illustrated in Figure 27.18A, hormones play a role in inducing labor. One of the hormones, estrogen, reaches its highest level in the mother's blood during the last weeks of pregnancy. A key effect of all this estrogen is to trigger the formation of numerous oxytocin receptors on the uterus. Cells of the fetus produce the hormone oxytocin, and later in pregnancy, the mother's pituitary gland produces it in increasing amounts. Oxytocin is a powerful stimulant for the smooth muscles in the wall of the uterus, causing them to contract. It also stimulates the placenta to make prostaglandins, local tissue regulators that also stimulate uterine muscle cells, making the muscles contract even more.

The hormonal induction of labor involves positive-feedback control. In this case, oxytocin and prostaglandins cause uterine contractions that in turn stimulate the release of more and more oxytocin and prostaglandins. The result is climactic—the intense muscle contractions that propel a baby from the womb.

Figure 27.18B shows the three stages of labor. As the process begins, the cervix (neck of the uterus) gradually opens, or dilates. ① The first stage, dilation, is the time from the onset of labor until the cervix reaches its full dilation of about 10 cm. Dilation is the longest stage of labor, lasting 6–12 hours or even considerably longer.

② The period from full dilation of the cervix to delivery of the infant is called the expulsion stage. Strong uterine contractions, lasting about 1 minute each, occur every 2–3 minutes, and the mother feels an increasing urge to push or bear down with her abdominal muscles. Within a period of 20 minutes to an hour or so, the infant is forced down and out of the uterus and vagina. An attending physician or midwife clamps and cuts the umbilical cord after the baby is expelled.

③ The final stage of labor is the delivery of the placenta, which is usually accomplished within 15 minutes after the birth of the baby.

Hormones continue to play important roles after the baby and placenta are delivered. Decreasing levels of pro-

ESTROGEN
from ovaries
Induces oxytocin receptors on uterus
OXYTOCIN
from fetus and pituitary
Stimulates uterus to contract
Stimulates placenta to make
PROSTAGLANDINS
Stimulate more contractions of uterus
Positive feedback

Figure 27.18A The hormonal induction of labor

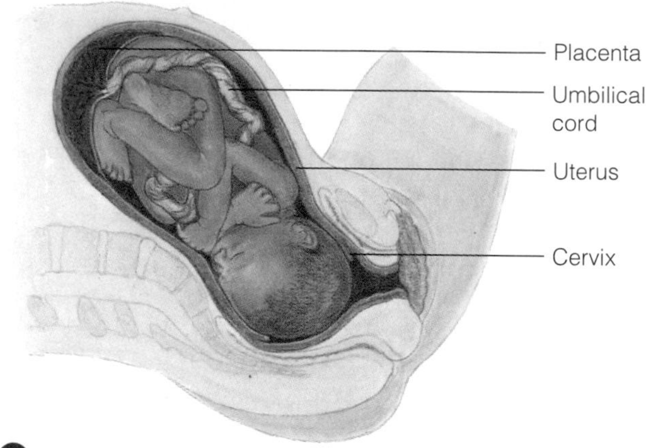

❶ Dilation of the cervix

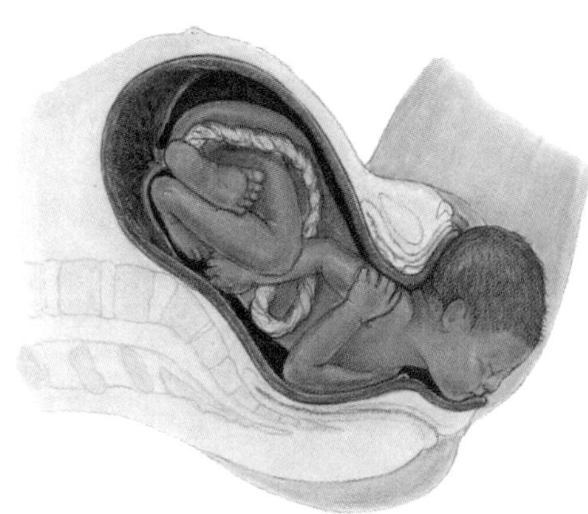

❷ Expulsion: delivery of the infant

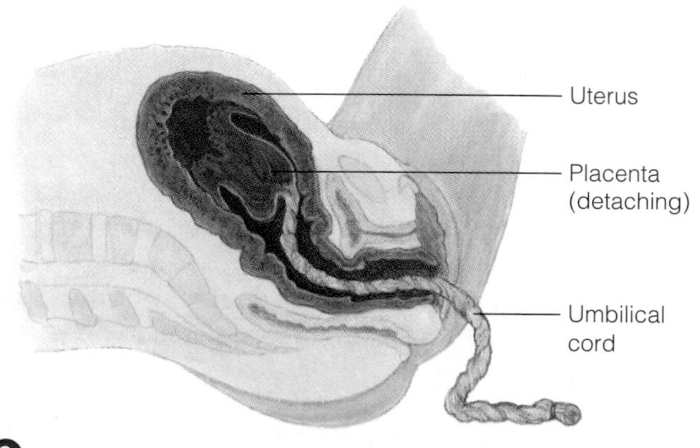

❸ Delivery of the placenta

Figure 27.18B The three stages of labor

gesterone and estrogen allow the uterus to start returning to its prepregnancy state. Less progesterone in the maternal blood also allows the pituitary hormone prolactin to promote milk production by the mammary glands. About 2–3 days after birth, the mother begins to secrete milk under the direct influence of both oxytocin and prolactin.

?

Why is "labor" an appropriate term?

It's hard, painful work for the mother!

Reproductive technology increases our reproductive options 27.19

Today, most couples have a choice about whether or not, and when, to have children. The choice is not available, however, to millions of couples who are unable to conceive children because of one or more physical problems. More often than not, the man is infertile; his testes may not produce enough sperm, or his sperm may not be vigorous enough to swim through the uterus and vagina and fertilize an egg. Female infertility can result from a failure to ovulate or from a blockage in the oviducts, preventing egg and sperm from meeting. Sexually transmissible diseases often produce scar tissue that blocks the oviducts. Some women also have antibodies that immobilize sperm in the uterus.

Reproductive technology can solve a number of infertility problems. Hormone therapy will sometimes increase sperm or egg production. Surgery can correct certain disorders, such as blocked oviducts. For some couples, it is simply a matter of timing sexual activity to maximize the chance for fertilization to occur. But for as many as half the couples who seek medical help, treatment is not effective or the cause of infertility remains unknown.

A breakthrough in reproductive technology has been **in vitro fertilization (IVF).** In the first step of this procedure, a woman whose oviducts are blocked has ova surgically removed from her ovaries. The ova are then mixed with sperm in culture dishes (in vitro means "in glass" in Latin). The photograph in Figure 27.19 shows an early human embryo that is undergoing cleavage in vitro. About 2 days after fertilization, when it has divided to form about 8 cells, it is carefully inserted into the woman's uterus and allowed to implant. Embryos can also be frozen, to be used later if the first attempt is aborted.

In vitro fertilization is now performed in major medical centers throughout the world. It costs thousands of dollars per attempt, but hundreds of children have been conceived this way. In no case has there been any evidence of abnormalities.

A spin-off of in vitro fertilization is surrogate motherhood. A woman may be able to conceive but unable to carry a fetus. The blastocyst may fail to implant in her uterus, or she may have repeated miscarriages. In such cases, a couple may produce an embryo by IVF and then enter into a contract with another woman, who agrees to become a surrogate mother. The surrogate mother has the embryo implanted in her uterus and carries it through to birth. This method has worked in a number of cases, but serious ethical and legal problems can arise if a surrogate mother

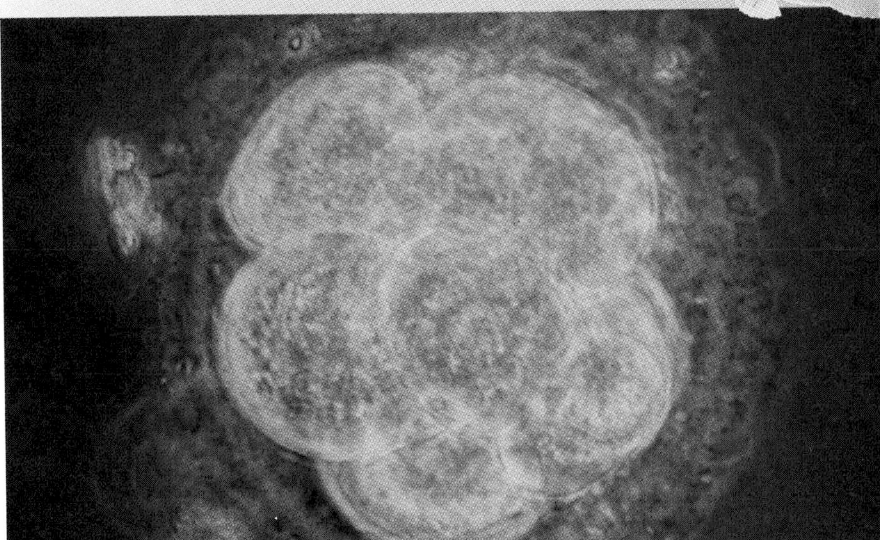

Figure 27.19 A human embryo in culture

changes her mind and wants to keep the baby she has carried for nine months. There is also the question of who is at fault if the child is born with genetic defects. In the United States, about one-third of the states have laws restricting surrogate motherhood.

Surrogate motherhood is one of many important social issues that center on human reproduction. In coming years, we can expect debate over reproductive issues to intensify. Where do you stand on these issues?

Why is the term "test-tube baby" an inaccurate reference to the product of in vitro fertilization?

Because the embryo is transferred very early to a human container, the uterus of a woman, where the rest of its development occurs

* * *

In this chapter, we have considered the structural and functional bases of animal reproduction and some of the mechanics of embryonic development. We have watched a single-celled product of sexual reproduction, the zygote, become transformed into a new organism, complete with all organ systems. One of the first of those organ systems to develop is the nervous system. In the next chapter, we see how the nervous system functions together with the endocrine system to regulate virtually all body activities.

Chapter Review

Chapter Summary

Asexual and Sexual Reproduction (Introduction–27.1) Animal reproductive mechanisms are highly varied, but embryonic development involves a common set of processes, including mitotic cell division, cellular differentiation, and the formation of the body and its structures **(Introduction).** Sexual reproduction involves the fusion of gametes from two parents. In asexual reproduction, one parent produces offspring by budding, fragmentation, or the development of unfertilized eggs. Asexual reproduction enables a single individual to produce many offspring rapidly. Sexual reproduction increases the variation among offspring, which may enhance reproductive success in changing environments **(27.1).**

Human Reproduction (27.2–27.8) The human reproductive system consists of a pair of ovaries (in females) or testes (in males), ducts that carry gametes, and structures for copulation. A woman's ovaries contain follicles that nurture eggs and produce sex hormones. Oviducts convey eggs to the uterus, where a fertilized egg develops. The uterus opens into the vagina, which receives the penis during intercourse and forms the birth canal **(27.2).** A man's testes produce sperm, which are expelled through ducts during ejaculation. Several glands contribute to the formation of fluid that carries, nourishes, and protects sperm. This fluid and the sperm constitute semen **(27.3).**

Spermatogenesis and oogenesis produce sperm and ova, respectively. Primary spermatocytes are made continuously in the testes; these diploid cells undergo meiosis to form four haploid sperm. A woman's ovaries contain her lifetime supply of primary oocytes at birth. Each month, one matures to form a secondary oocyte, which, if fertilized, completes meiosis and becomes a haploid ovum **(27.4).**

Hormones synchronize cyclic changes in the ovaries and uterus. Approximately every 28 days, the hypothalamus signals the anterior pituitary to secrete FSH and LH, which trigger the growth of a follicle and ovulation, the release of an egg. The follicle secretes estrogen; after ovulation, the follicle becomes the corpus luteum, which secretes both estrogen and progesterone. These two hormones stimulate the endometrium (the uterine lining) to thicken, preparing the uterus for implantation. They also inhibit the hypothalamus, reducing FSH and LH secretion. If the egg is not fertilized, the drop in LH shuts down the corpus luteum and its hormones. This triggers menstruation, the breakdown of the endometrium. The hypothalamus and pituitary then stimulate another follicle, starting a new cycle. If fertilization occurs, a hormone from the embryo maintains the uterine lining and prevents menstruation **(27.5).**

Human sexual behavior makes possible the union of egg and sperm and may promote bonding between mates. The human sexual response occurs in four phases: excitement, plateau, orgasm, and resolution **(27.6).** Sexual intercourse may carry risks of sexually transmissible diseases, such as AIDS, gonorrhea, and genital herpes **(27.7).** Contraception prevents pregnancy by blocking the release of gametes, preventing fertilization, or preventing implantation **(27.8).**

Principles of Embryonic Development (27.9–27.15) In fertilization, a sperm releases enzymes that pierce the egg's coat. Sperm surface proteins bind to egg receptor proteins, sperm and egg plasma membranes fuse, and the two nuclei unite. Changes in the egg membrane prevent entry of additional sperm, and the fertilized egg, called a zygote, is stimulated to develop into an embryo **(27.9).**

Cleavage is a rapid series of cell divisions that turns a zygote into a ball of cells called a blastula **(27.10).** In gastrulation, cells migrate inward and form a rudimentary digestive cavity. The resulting gastrula has three layers of cells: ectoderm, endoderm, and mesoderm **(27.11).** After gastrulation, the three embryonic tissue layers give rise to specific organ systems. In chordates, the notochord develops above the digestive cavity. The cell layer above the notochord rolls up to become the neural tube, which in turn becomes the brain and spinal cord **(27.12).**

Tissues and organs take shape in a developing embryo as a result of cell shape changes, cell migration, and programmed cell death **(27.13).** In a process called induction, adjacent cells and cell layers influence each other's differentiation via chemical signals **(27.14).** Pattern formation, the emergence of the parts of a structure in their correct relative positions, involves the response of genes to spatial variations of chemicals in the embryo **(27.15).**

Human Development (27.16–27.19) Human development begins with fertilization in the oviduct. Cleavage produces a blastocyst, whose inner cell mass becomes the embryo. The blastocyst's outer layer, the trophoblast, implants in the uterine wall. Gastrulation occurs, and organs develop from the three embryonic layers. Meanwhile, the four extraembryonic membranes develop: the amnion, the chorion, the yolk sac, and the allantois. The embryo floats in the fluid-filled amniotic cavity, while the chorion and embryonic mesoderm form the embryo's part of the placenta. The placenta's chorionic villi absorb food and oxygen from the mother's blood. Viruses and drugs may cross the placenta and harm the embryo **(27.16).**

Human embryonic development is divided into three trimesters of about three months each. The most rapid changes occur during the first trimester. By 9 weeks, all organs are formed, and the embryo is called a fetus. The second and third trimesters are times of growth and preparation for birth **(27.17).**

Hormonal changes induce birth. Estrogen makes the uterus more sensitive to oxytocin, which acts with prostaglandins to initiate labor. The cervix dilates, the baby is expelled by strong muscular contractions, and the placenta follows. Oxytocin and prolactin then stimulate milk secretion **(27.18).** In vitro fertilization and surrogate motherhood give hope to infertile couples, but these new technologies have raised ethical and legal questions **(27.19).**

Testing Your Knowledge

Multiple Choice

1. After a sperm penetrates an egg, the fertilization membrane
   a. secretes important hormones
   b. enables the fertilized egg to implant in the wall of the uterus
   c. prevents more than one sperm from entering the egg
   d. attracts additional sperm to the egg
   e. activates the egg for embryonic development

2. In an experiment, a researcher colored a bit of tissue on the outside of a frog gastrula with a fluorescent dye. The embryo developed normally. When the tadpole was placed under an ultraviolet light, which of the following glowed bright orange? *(Explain your answer.)*
   a. the heart
   b. the pancreas
   c. the brain
   d. the stomach
   e. the liver

3. How does a zygote differ from an ovum?
   a. A zygote has more chromosomes.
   b. A zygote is smaller.
   c. A zygote consists of more than one cell.
   d. A zygote is much larger.
   e. A zygote divides by meiosis.

4. Which of the following traces the path of sperm out of the body of a human male?
   a. epididymis, seminiferous tubule, vas deferens, urethra
   b. seminiferous tubule, vas deferens, epididymis, urethra
   c. epididymis, seminiferous tubule, urethra, vas deferens
   d. seminiferous tubule, epididymis, vas deferens, urethra
   e. seminiferous tubule, epididymis, urethra, vas deferens

5. A woman had several miscarriages. Her doctor suspected that a hormonal insufficiency was causing the lining of the uterus to break down, as it does during menstruation, terminating her pregnancies. Treatment with which of the following might help her remain pregnant?
   a. oxytocin
   b. follicle-stimulating hormone
   c. testosterone
   d. luteinizing hormone
   e. prolactin

6. Which of the following most reduces the chances of both conception and the spread of sexually transmissible diseases?
   a. condom
   b. birth control pill
   c. diaphragm
   d. intrauterine device
   e. withdrawal

Matching

1. Turns into the corpus luteum	a. vas deferens
2. Female gonad	b. prostate gland
3. Site of spermatogenesis	c. endometrium
4. Site of fertilization in humans	d. testis
5. Human gestation occurs here	e. follicle
6. Sperm duct	f. uterus
7. Secretes seminal fluid	g. ovary
8. Lining of uterus	h. oviduct

Describing, Comparing, and Explaining

1. The graph below plots the rise and fall of pituitary and ovarian hormones during the human ovarian cycle. Identify each hormone (A–D) and the reproductive events with which each one is associated (P–S). For A–D, choose from estrogen, LH, FSH, and progesterone. For P–S, choose from ovulation, growth of follicle, menstruation, and development of corpus luteum.

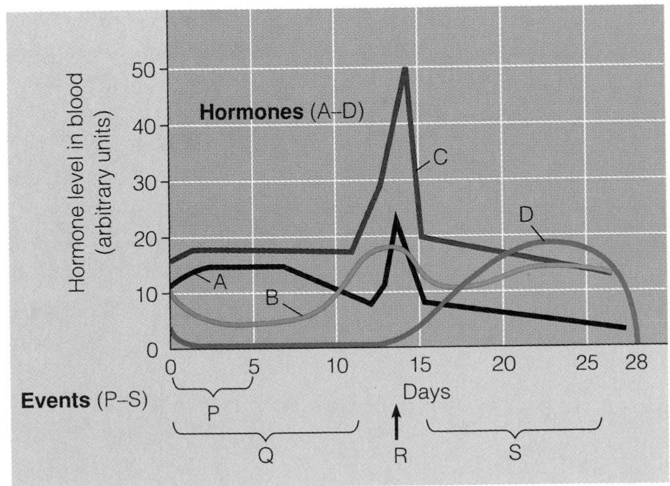

2. Some animals, such as rotifers and aphids, are able to alternate between sexual and asexual reproduction. Under what conditions might it be advantageous to reproduce sexually? Asexually?

3. Compare sperm formation with egg formation. In what ways are the processes similar? In what ways are they different?

4. The embryos of reptiles, birds, and mammals have systems of extraembryonic membranes. What are the functions of these membranes, and how do fish and frog embryos survive without them?

5. In an embryo, nerve cells grow out from the spinal cord and form connections with the muscles they will eventually control. What mechanisms described in this chapter might explain how these cells "know" where to go and which cells to connect with?

Thinking as a Scientist

1. As a frog embryo develops, the neural tube forms from ectoderm along what will be the frog's back, directly above the notochord. To study this process, a researcher carefully extracted a bit of notochord tissue and inserted it underneath the ectoderm where the belly of the frog would normally develop. What can the researcher hope to learn from this experiment? Predict the possible outcomes. What experimental control would you suggest?

2. Variation in the length of the human menstrual cycle results mainly from individual differences in the pre-ovulatory phase. The relatively uniform life span of the corpus luteum keeps the length of the post-ovulatory phase close to 14 days. A certain woman usually has 34-day cycles. If she last began menstruating on January 29 and it is now February 14, what is her reproductive status? Choose the best answer, and briefly explain why you chose it.
   a. Her endometrium has reached its peak development.
   b. An ovarian follicle has just started to grow.
   c. Unprotected intercourse now has a good chance of resulting in pregnancy.
   d. The LH concentration in her blood has surged upward, stopping menstruation.

Science, Technology, and Society

New technology has made it possible for doctors to save a small percentage of babies born 16 weeks prematurely. A baby born this early weighs just over a pound and faces months of care in an intensive-care nursery. The cost for care may be hundreds of thousands of dollars per infant. Some people wonder whether such a huge technological and personnel investment should be devoted to such a small number of babies. They feel that limited resources might be better directed at providing prenatal care that could prevent many premature births. What do you think? Why?

Media Resources

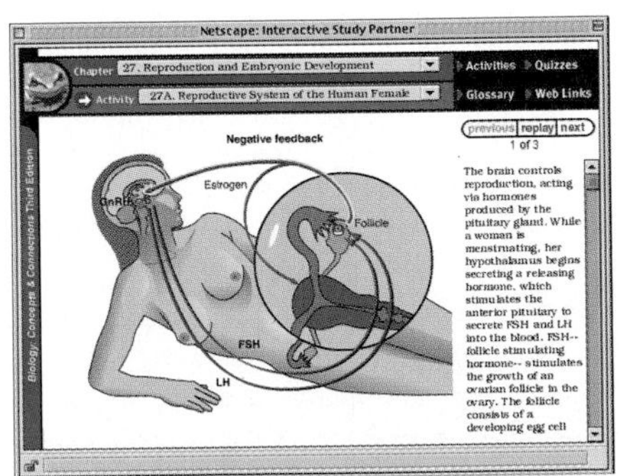

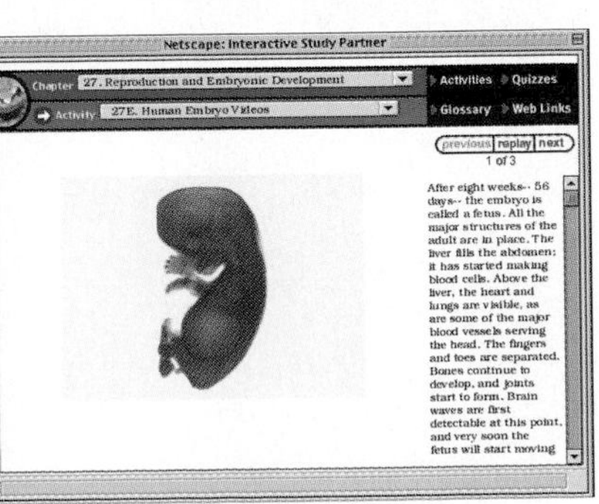

For further review of the concepts in this chapter, use the activities and quizzes on the CD-ROM.

Visit the Special Edition of The Biology Place web site for activities, quizzes, news articles, and web links.

28 Nervous Systems

Outline

Squids are superbly equipped predators of the open sea. They drift and glide—watching, waiting—then suddenly dart after a fish with knifelike precision. In a chase, a squid can reach almost 40 km/hr (about 25 mi/hr) in a matter of seconds, faster than any other invertebrate and most fish. Suction cups on a squid's tentacles and arms grasp prey, and beaklike jaws at the base of the tentacles quickly chop the food into bite-sized pieces.

The squid in this photograph *(Loligo pealei)* is only about 20 cm (8 in) long. Often pursued by sharks or other large fish, it can twist and quickly turn out of the way. It can also shoot a cloud of black ink into the water to divert attackers, and it can change color almost instantly to blend with the background. The squid changes color by changing its spots, which are due to pigment-containing cells in the skin. When tiny muscles attached to the cells contract, the cells stretch, spreading the pigment out so that the spots enlarge and darken the squid's skin. When the muscles relax, the pigment cells shrink, and the skin pales as the pigmented areas become smaller.

A squid's abilities to attack or escape depend on a sophisticated nervous system (colored gold in the drawing on this page). A squid has a large brain for its body size. Nerves carry sensory signals into the brain from the eyes and other sense organs. The brain quickly sorts the information and sends out commands, via other nerves, to the muscles. As indicated in the drawing, large nerves run forward from the brain into the tentacles and arms, and even larger nerves course backward into the main body. Some of the nerves carrying commands to the muscles contain unusually large nerve cells with long, thick extensions called giant fibers. The giant fibers can conduct nerve signals at great speed, from the brain to the tip of the tail in about a hundredth of a second. Command signals sent through these big nerve cells give the squid the quick reflexes it needs to catch fast-swimming prey and escape predators.

The discovery of the squid's giant fibers, over 50 years ago, set the stage for a revolution in research on animal nervous systems. The fibers are easy to remove and keep alive for study in the laboratory. The photograph to the right was taken with a light microscope. It shows a small piece of a giant fiber with a glass electrode inserted into it. A researcher can use the electrode to make direct measurements of tiny electrical changes that occur in the fiber when it is stimulated. From such research has come much of our knowledge about how nerve cells, including our own, conduct signals.

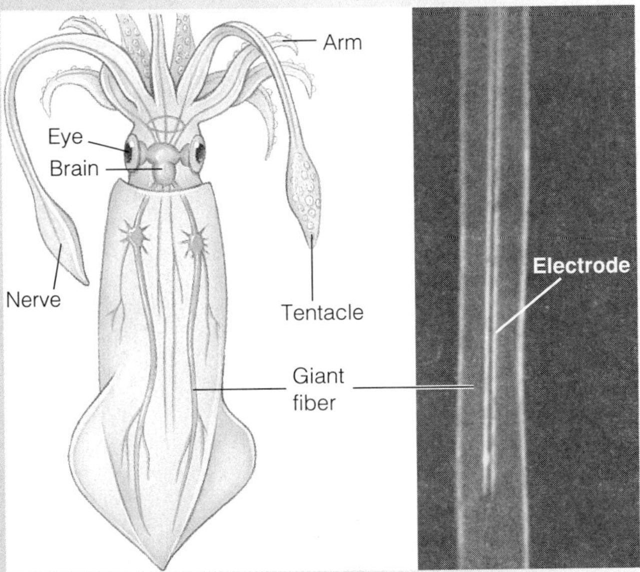

The nervous system of a squid

Electrode in a giant fiber

To survive and reproduce, a squid or any other animal must respond appropriately to environmental stimuli. Most animals have two coordinating systems, the endocrine (hormone) system, which we discussed in Chapter 26, and the nervous system. The two systems often cooperate in triggering responses to stimuli. In a squid, for instance, hormones sometimes help the nervous system alter skin color. In our own bodies, an interplay of hormones and nerve signals produces our fight-or-flight response to threats and stress. The nervous system alone, however, provides the pinpoint control underlying delicate body movements, as when a squid suddenly changes direction in the water, or when your eyes move back and forth between words and pictures as you read this chapter.

28.1 Nervous systems receive sensory input, interpret it, and send out appropriate commands

Nervous systems are the most intricately organized data-processing systems on Earth. A cubic centimeter of your brain, for instance, contains several million **neurons** (nerve cells), which are specialized for carrying signals from one location in the body to another. Each neuron may communicate with thousands of other nerve cells, forming networks that enable us to learn, remember, perceive our surroundings, and move.

A nervous system has three interconnected functions (Figure 28.1A): **Sensory input** is the conduction of signals from sensory receptors, such as light-detecting cells of the eye, to integration centers. **Integration** is the interpretation of the sensory signals and the formulation of responses. **Motor output** is the conduction of signals from the integration centers to **effectors,** such as muscle cells, which perform the body's responses.

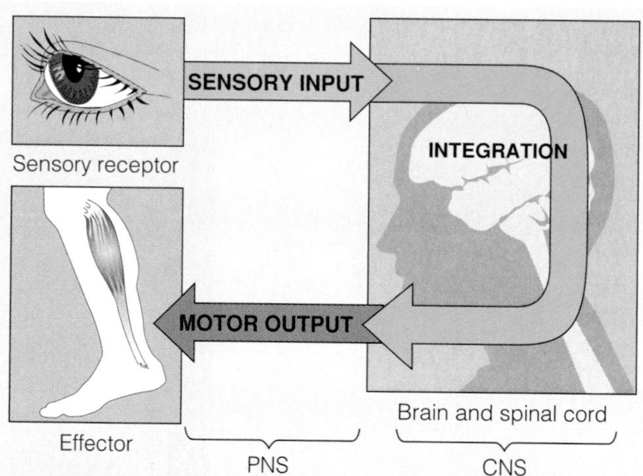

Figure 28.1A Organization of a nervous system

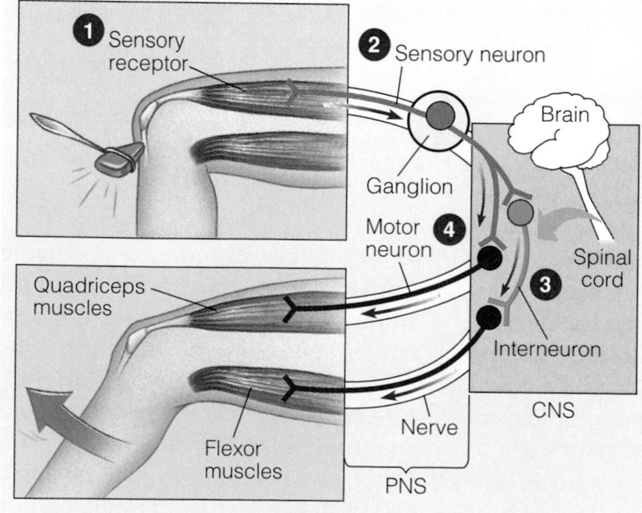

Figure 28.1B The knee-jerk reflex

With few exceptions, nervous systems have two main divisions. The division called the **central nervous system (CNS),** where most integration occurs, consists of the brain and, in vertebrates, the spinal cord. The other division, the **peripheral nervous system (PNS),** is made up mostly of communication lines called nerves that carry signals into and out of the CNS. A **nerve** is a cablelike bundle of neuron extensions tightly wrapped in connective tissue. (A neuron consists of a cell body and long, thin extensions called neuron fibers that convey signals.) The PNS also has **ganglia** (singular, *ganglion*), which are mainly clusters of neuron cell bodies in the nerves.

Figure 28.1B, a diagram of the human knee-jerk reflex, highlights the relationship between neurons and nervous system structure and function. The small colored balls represent neuron cell bodies; the thin lines represent neuron fibers. Three functional types of neurons (sensory neurons, interneurons, and motor neurons) correspond to a nervous system's three main functions. **Sensory neurons** convey signals, or information, (red arrows) from sensory receptors into the CNS. **Interneurons** are entirely within the CNS. They integrate data and then relay appropriate signals to other interneurons or to motor neurons. **Motor neurons** function in motor output, conveying signals from the CNS to effectors.

When the knee is tapped, ① a sensory receptor detects a stretch in muscle, and ② a sensory neuron conveys this information into the CNS (spinal cord). In the CNS, the information goes to ③ an interneuron and to ④ a motor neuron. One set of muscles (quadriceps) responds to motor signals conveyed by a motor neuron by contracting, jerking the lower leg forward. At the same time, another motor neuron, responding to signals from the interneuron, inhibits the flexor muscles in the lower leg, making them relax. (When the flexors contract, they pull the lower leg backward and resist the action of the quadriceps.)

One of our simplest body actions, the knee-jerk illustrates the cellular basis of nervous system functions. For simplicity, this figure shows only one neuron of each functional type, but virtually any body activity (including a simple knee-jerk) involves many sensory neurons, interneurons, and motor neurons. Let's take a closer look at these highly specialized cells.

(a) Arrange the following neurons into the correct sequence for information flow during the knee-jerk reflex: interneuron, sensory neuron, motor neuron.
(b) Which of the neuron types is located entirely within the central nervous system?

(a) sensory neuron → interneuron → motor neuron; (b) interneuron

28.2 Neurons are the functional units of nervous systems

28A
See Activity 28A on the CD-ROM.

Neurons vary widely in shape, but most of them share some common features. Figure 28.2 depicts a motor neuron, like those that carry command signals from your spinal cord to your skeletal muscles. A motor neuron has a large **cell body** housing the nucleus and most of its cytoplasmic organelles. Two types of fibers project from the cell body. Fibers of one type, called **dendrites** (Greek *dendron,* tree), are often short, numerous, and highly branched. Dendrites convey signals from their tips inward, toward the rest of the neuron. The signals may come from a sensory cell or from an interneuron. The second type of neuron fiber is the **axon** (Greek *axon,* axle), which on many neurons is a single fiber. The axon conducts signals toward another neuron or toward an effector. Many axons are long. Certain ones in your leg, for instance, stretch from the lower part of your spinal cord all the way to muscles in your toes. The giant fibers of a squid are also axons.

Neurons actually make up only part of a nervous system. Outnumbering neurons by as many as 50 to 1 are **supporting cells** that protect, insulate, and reinforce the neurons. This figure shows one kind of supporting cell, called a Schwann cell. In many animals, axons that convey signals very rapidly are enclosed along most of their length by a thick insulating material. In vertebrates, the insulating material, called the **myelin sheath,** resembles a chain of oblong beads. Each bead is actually a Schwann cell, and the myelin sheath is essentially a chain of Schwann cells, each wrapped many times around the axon. The spaces between Schwann cells are called **nodes of Ranvier,** and they are the only points on the axon where signals can be transmitted. Everywhere else, the myelin sheath insulates the axon, preventing signals from passing along it. When a signal travels along a myelinated axon, it jumps from node to node, as indicated in the figure. By jumping along the axon, the signal travels much faster than it could if it had to take the long route along the whole length of the axon. In the human nervous system, signals can travel along a myelinated axon about 150 m/sec (over 330 mi/hr), which means that a command from your brain can make your fingers move in just a few milliseconds (thousandths of a second). Without myelin sheaths, the signals could go only about 5 m/sec.

The debilitating disease called multiple sclerosis (MS) demonstrates the importance of myelin sheaths. MS leads to a gradual destruction of myelin sheaths by the individual's own immune system. The result is a progressive loss of signal conduction, muscle control, and brain function. MS is not yet curable, but drugs that suppress the immune system can relieve symptoms and slow its progress. Returning to the figure below, notice that the axon ends in a cluster of branches. A typical axon has hundreds or thousands of these branches, each with a bulblike synaptic knob at the very end. As we will see later, the **synaptic knobs** relay signals to another neuron or to an effector such as a muscle cell. With the basic structure of a neuron in mind, let's take a closer look at the signals that neurons convey.

What is the function of the myelin sheath?

It speeds up conduction of signals along some dendrites and axons.

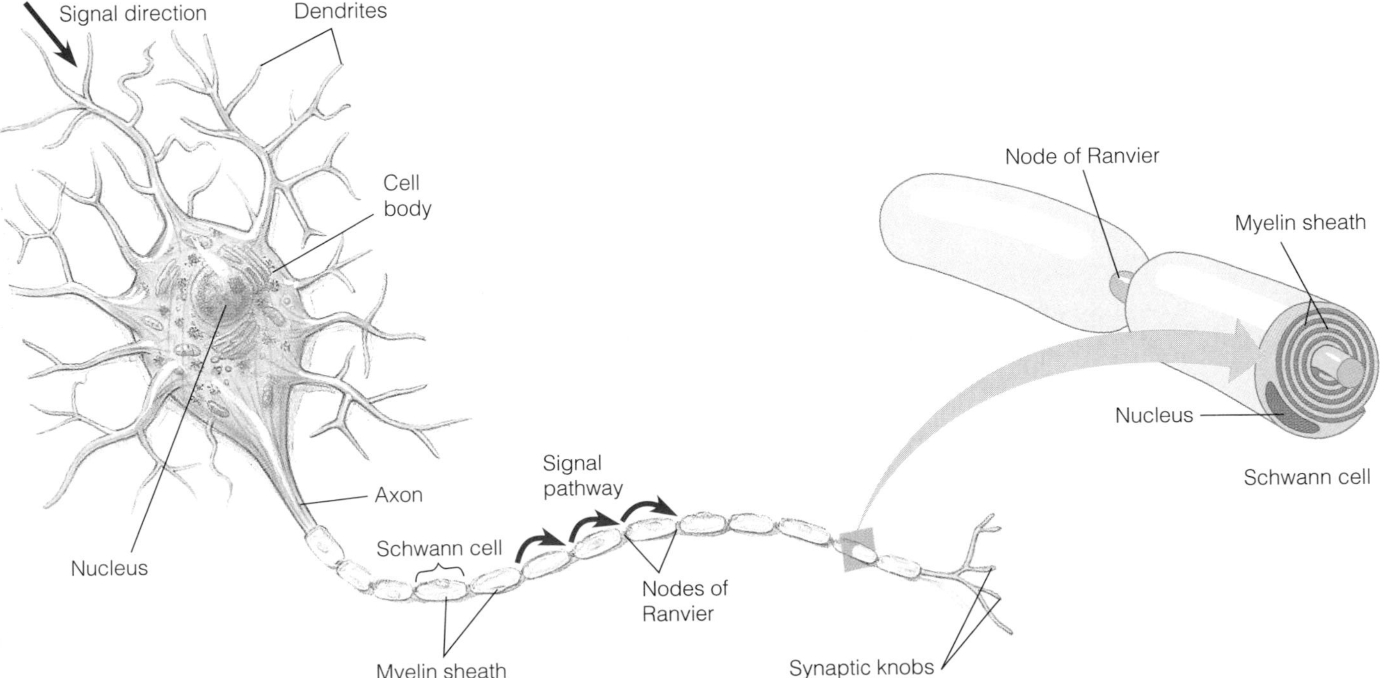

Figure 28.2 Structure of a motor neuron

Nerve Signals and Their Transmission

28.3 A neuron maintains a resting potential across its membrane

28B

To understand nerve signals, we must first study a resting neuron, one that is not transmitting a signal. A resting neuron contains potential energy, energy that can be put to work. The neuron's potential energy is used to send signals from one part of the body to another. This energy resides in an electrical charge difference across the neuron's plasma membrane. The cytoplasm just inside the membrane is negative in charge, and the fluid just outside the membrane is positive. Since opposite charges tend to move toward each other, a membrane stores energy by holding opposite charges apart. The strength (voltage) of a neuron's stored energy can be measured with microelectrodes connected to a voltmeter (Figure 28.3A). The voltage across the plasma membrane of a resting neuron is called the **resting potential.** A neuron's resting potential is about −70 millivolts (mV). (The resting potential is negative because the inside of the cell is negative relative to the outside.)

What causes the resting potential? The answer lies with the membrane itself. The membrane keeps dissolved proteins and other large organic molecules inside the cell. Most of these molecules are negatively charged. Also, the membrane has channels and pumps, made of proteins, that regulate the passage of inorganic ions. For instance, a resting membrane allows much more potassium (K^+) than sodium (Na^+) to diffuse across. Notice in Figure 28.3B that Na^+ (blue) is more concentrated outside the cell than inside; the Na^+ channels allow very little to diffuse in. But K^+ (green), which is more concentrated inside, can freely flow out. As the positively charged K^+ ions diffuse out, they leave behind an excess of negative charge. Together, the negative molecules trapped inside the cell and the diffusion of K^+ out of the cell create most of the resting potential. (For simplicity, Figure 28.3B omits Cl^- and other negative inorganic ions on both sides of the membrane.)

Also helping maintain the resting potential are membrane proteins called **sodium-potassium (Na^+-K^+) pumps.** These pumps actively transport Na^+ out of the cell and K^+ in, thereby helping keep the concentration of Na^+ low in the cell and K^+ high. The pumps also contribute slightly to the loss of positive charge from the cell by moving more Na^+ out than K^+ in.

As we see next, stimuli that trigger nerve signals act by triggering a release of the resting potential energy.

If a neuron's membrane suddenly becomes more permeable to sodium ions, there is a rapid net transport of the Na^+ into the cell. What are the two forces that drive the ions inward?

The greater concentration of Na^+ outside the cell than inside and the membrane potential (negatively charged inside vs. outside) favor the inward diffusion of Na^+.

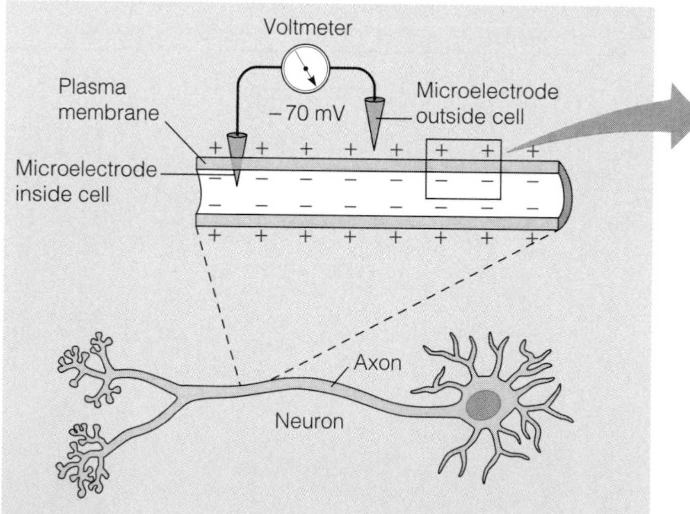

Figure 28.3A Measuring a neuron's resting potential

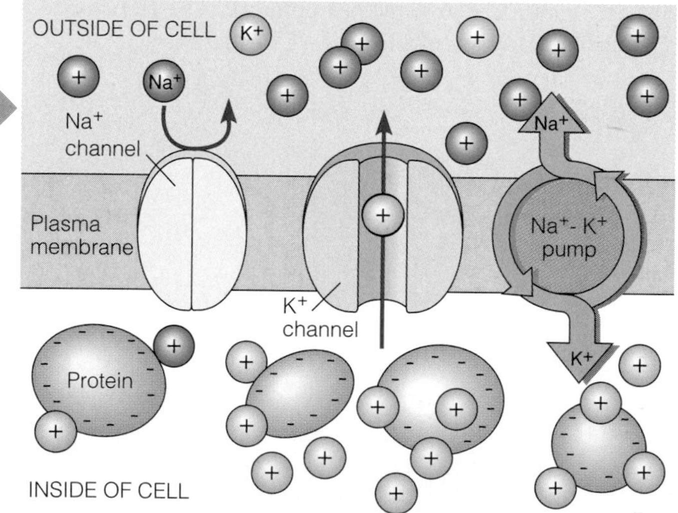

Figure 28.3B How the resting potential is generated

28.4 A nerve signal begins as a change in the membrane potential

28B

Turning a flashlight on uses the energy stored in a battery to create light. In a similar way, stimulating a neuron's plasma membrane can trigger the release and use of the membrane's potential energy to generate a nerve signal. A **stimulus** is any factor that causes a nerve signal to be generated. Examples of stimuli include the light from a flashbulb, an audible sound, a tap on the knee, an electrical shock, or a sudden temperature change.

The discovery of squid giant fibers (axons) gave researchers their first chance to study how stimuli trigger signals in a living neuron. From microelectrode studies with squid neurons, British biologists A. L. Hodgkin and A. F. Huxley worked out the details of nerve signal transmission in the 1940s. Their findings, summarized in Figure 28.4 below, apply to neurons in all animals.

Graphing electrical changes in neuron membranes was the first step in discovering how nerve signals are generated. The multicolored line on the graph below traces the electrical changes that make up the **action potential,** the technical name for a nerve signal. All the changes indicated by the graph occur at the place on the membrane where a stimulus is applied. The graph records electrical events over time (in milliseconds) at that particular place. ① The graph starts out at the membrane's resting potential (−70 mv). ② The stimulus is applied, at time 0, and in 2–3 msec, the voltage rises to what is called the **threshold potential** (−50 mV, in this case). The difference between the threshold potential and the resting potential is the minimum change in the membrane's voltage that must occur to generate the action potential. If the threshold potential is reached, the action potential is triggered. ③ The voltage then drops back down, ④ undershoots the resting potential, and finally returns to it.

What actually causes the electrical changes of the action potential? Huxley and Hodgkin found that specific ion movements coincide with the electrical changes. Later studies showed that the movements take place at Na^+ and K^+ channels (called *voltage gated channels* because they have special gates that open and close depending on changes in voltage across the membrane). The diagrams surrounding the graph show the ion movements. In ①, at the lower left, the resting membrane is positively charged on the outside, and the cytoplasm just inside the membrane is negatively charged. ② The stimulus applied at time 0 triggers the opening of Na^+ channels in the membrane. At first, only a few channels open, and a tiny amount of Na^+ enters the axon. Just this tiny change, however, makes the inside surface of the membrane slightly less negative than before, and as a result, the membrane's voltage begins to change. If the stimulus is strong enough, a sufficient number of Na^+ channels open to change the voltage to the threshold level. Once the threshold is reached, the increasing positive charge inside the membrane triggers the opening of more and more Na^+ channels. As more Na^+ moves in, the voltage soars to its peak. ③ The peak voltage triggers closing and inactivation of the Na^+ channels. Na^+ stops moving in, while the K^+ channels open, allowing K^+ to diffuse rapidly out. These changes produce the downswing on the graph. ④ A very brief undershoot of the resting potential results because the K^+ channels take longer to close than the Na^+ channels. Closing of the K^+ channels results in the membrane returning to its resting potential.

In what way is an action potential a self-boosted response—a chain reaction?

The opening of Na^+ gates caused by stimulation of the neuron changes the membrane potential, and this change causes more of the voltage-gated Na^+ channels to open.

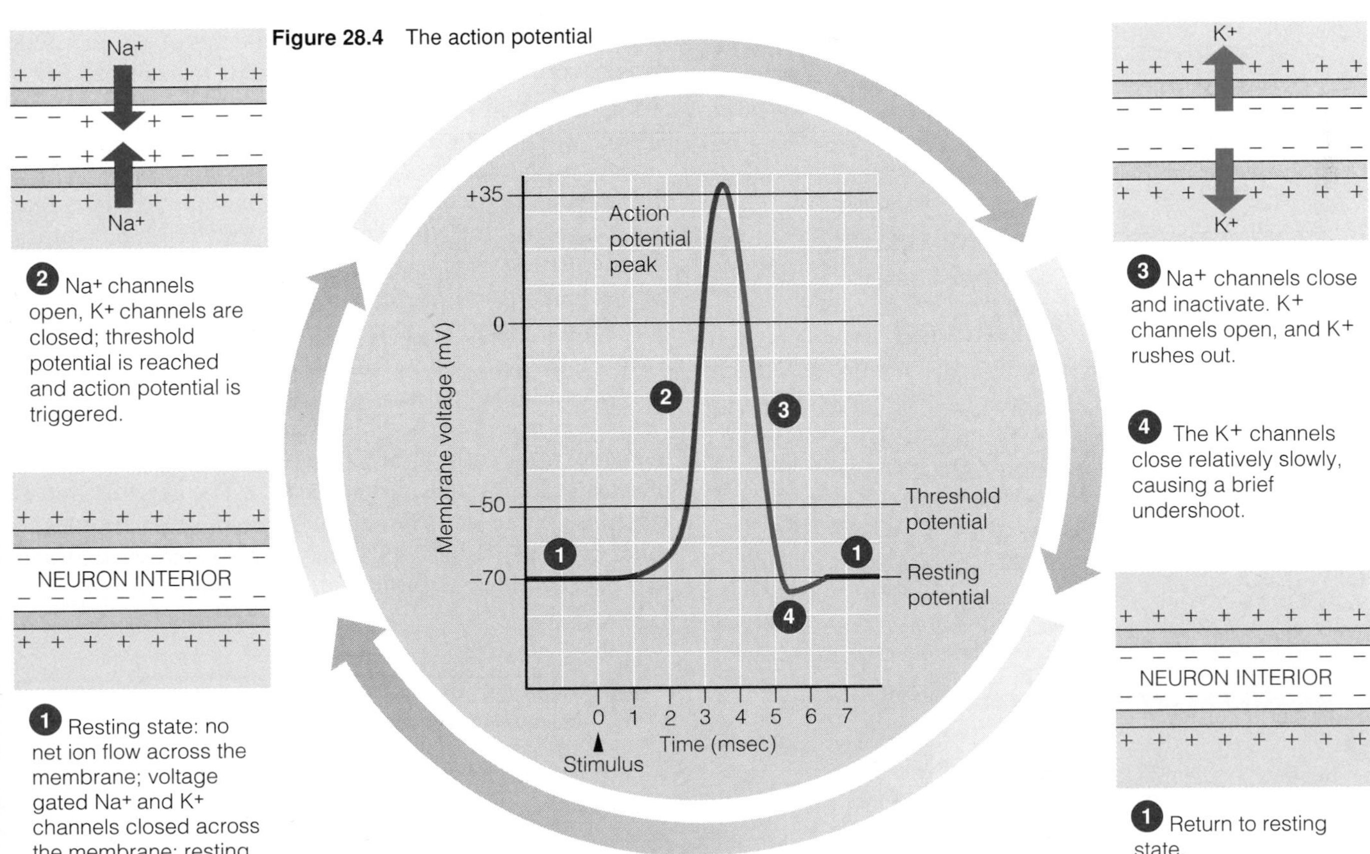

Na+
Na+
2 Na+ channels open, K+ channels are closed; threshold potential is reached and action potential is triggered.
NEURON INTERIOR
1 Resting state: no net ion flow across the membrane; voltage gated Na+ and K+ channels closed across the membrane; resting potential is maintained.
+35
0
−50
−70
Membrane voltage (mV)
Action potential peak
Threshold potential
Resting potential
0 1 2 3 4 5 6 7
Time (msec)
Stimulus
K+
K+
3 Na+ channels close and inactivate. K+ channels open, and K+ rushes out.
4 The K+ channels close relatively slowly, causing a brief undershoot.
NEURON INTERIOR
1 Return to resting state.

Figure 28.4 The action potential

28.5 The action potential regenerates itself along the neuron

28B

An action potential is a localized electrical event—a change from a neuron's resting potential at a specific point. To function as a signal, this local event must travel along the neuron. You can follow this movement in Figure 28.5. A nerve signal starts out as one action potential, generated on the axon near the cell body of the neuron. The effect of this action potential is like tipping the first of a row of standing dominoes: The first domino does not travel along the row, but its fall is relayed to the end of the row, one domino at a time.

The three parts of Figure 28.5 show the changes that occur in part of an axon at three successive times, as a nerve signal passes from left to right. As we saw in Module 28.4, all the ion movements associated with a particular action potential occur at one place on the axon. Let's first focus on the axon region on the far left. ① When this region of the axon (blue) has its Na^+ channels open, Na^+ rushes inward (blue arrows), and an action potential is generated. This corresponds to the upswing of the curve on the graph in Figure 28.4. ② When that same region has its K^+ channels open, K^+ diffuses out of the axon (green arrows); at this time, its Na^+ channels are closed and inactivated, and we would see the downswing of the action potential initiated in part 1. ③ A short time later, we would see no signs of an action potential at this (far-left) spot, because the axon membrane here has returned to its resting potential.

Now let's see how these events lead to the "domino effect" of a nerve signal. In part 1 of the figure, the blue arrows pointing sideways within the axon indicate local spreading of the electrical changes caused by the inflowing Na^+ ions associated with the first action potential. These changes trigger the opening of Na^+ channels in the membrane just to the right of the action potential. As a result, a second action potential is generated, as indicated by the blue region in part 2. In the same way, a third action potential is generated in part 3, and each action potential generates another all the way down the axon.

So why are action potentials propagated in only one direction along the axon (left to right in the figure). As the blue arrows indicate, local electrical changes do spread in both directions in the axon. However, these changes cannot open Na^+ channels and generate an action potential when the Na^+ channels are inactivated. Thus, an action potential cannot be generated in the regions where K^+ is leaving the axon (green in the figure).

So we see that a nerve signal, also known as an action potential, propagates itself in one direction by the electrical changes it produces in the neuron membrane. An action potential is a bit of coded information that can travel from one end of a neuron to another. What does this tell us about how a nervous system actually works? If you rap your finger on a desk, for instance, the contact is a stimulus that triggers action potentials in the tips of sensory neurons in your skin. The action potentials regenerate, carrying the information (that your finger has hit a hard object) into your central nervous system.

Action potentials are *all-or-none* events; that is, they are the same no matter how strong or weak the stimulus that triggers them. How, then, do action potentials relay different intensities of information to your central nervous system? If you rap your finger hard against the desk, your CNS receives many more action potentials per millisecond than after a soft tap. It is the *frequency* of action potentials that changes with the intensity of stimuli.

Once your central nervous system receives information in the form of action potentials, it can process the information and formulate a response to it. The nervous system depends on the sensory neurons' passing their signals to other neurons in the CNS. Our next step is to see how signals pass from one neuron to another. 

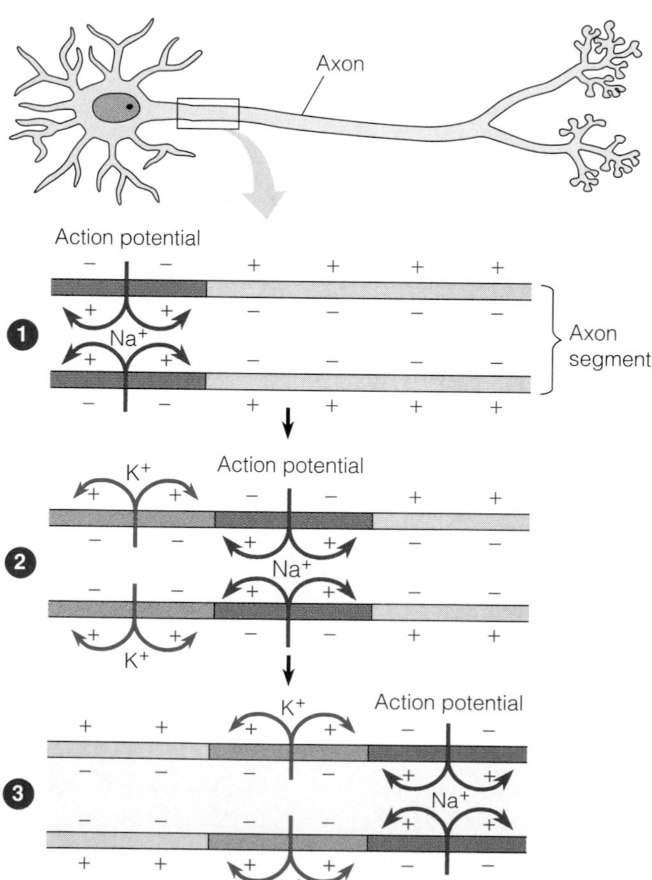

Figure 28.5 Propagation of the action potential along an axon

? During an action potential, ions move across the neuron membrane in a direction perpendicular to the direction of the impulse along the neuron. What is it that actually travels along the neuron as the signal?

The signal is the wavelike change in membrane potential, the self-perpetuated action potential that regenerates sequentially at points farther and farther away from the site of stimulation.

28C

The **synapse**—the junction, or relay point, between two neurons or between a neuron and an effector cell—is a key element of nervous systems. When action potentials arrive at the end of one neuron's axon, the information they represent passes to a receiving cell across the synapse.

Synapses are either electrical or chemical. In an electrical synapse, action potentials themselves pass from one neuron to the next. The receiving neuron is stimulated quickly and always at the same level (same frequency of action potentials) as the sending neuron. Lobsters, crayfish, and many fishes can flip their tails with lightning speed because the neurons that carry signals for these movements communicate by electrical synapses. In the human body, electrical synapses are common in the heart and digestive tract, where nerve signals maintain steady, rhythmic muscle contractions. In contrast, chemical synapses are prevalent in most other organs, including skeletal muscles, and in the central nervous system, where signaling among neurons is complex and varied.

Figure 28.6 Neuron communication

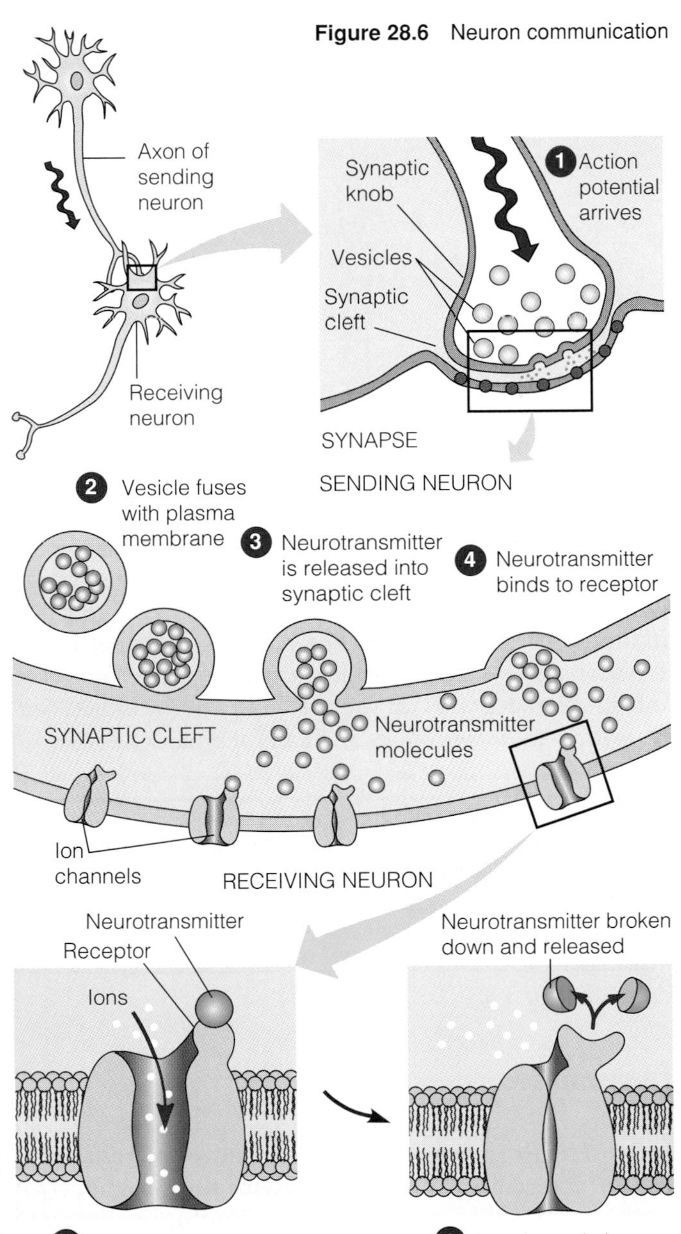

Unlike electrical synapses, chemical synapses have a narrow gap, called the **synaptic cleft,** separating a synaptic knob of the sending neuron from the receiving neuron. The cleft prevents the action potential in the sending neuron from spreading directly to the receiving neuron. Instead, the action potential (an electrical signal) is first converted to a chemical signal. The chemical signal, consisting of molecules of **neurotransmitter** (see Module 26.1), then may generate an action potential in the receiving cell.

Now let's follow the events that occur at a chemical synapse in Figure 28.6. The neurotransmitter is contained in vesicles in the synaptic knob of the sending neuron. Following the numbered sequence, ① an action potential (red arrow) arrives at the synaptic knob. ② The action potential triggers chemical changes that make neurotransmitter vesicles fuse with the plasma membrane of the sending cell. ③ The fused vesicles release their neurotransmitter molecules (green) into the synaptic cleft. ④ The released neurotransmitter molecules diffuse across the cleft and bind to receptor molecules on the receiving cell's plasma membrane. ⑤ The binding of neurotransmitter to receptor opens chemical-sensitive ion channels in the receiving cell's membrane. With the channels open, ions can diffuse into the receiving cell and trigger new action potentials. ⑥ The neurotransmitter is broken down by an enzyme, and the ion channels close. Step 6 ensures that the neurotransmitter's effect on the receiving cell is brief and precise.

You can review what we have covered so far by thinking about what is happening right now in your own nervous system. Action potentials carrying information about the words on this page are streaming along sensory neurons from your eyes to your brain. Arriving at synapses with receiving cells (interneurons in the brain), the action potentials are triggering the release of neurotransmitters at the ends of the sensory neurons. The neurotransmitters are diffusing across synaptic clefts and triggering changes in some of your interneurons—changes that lead to integration of the signals and ultimately to what the signals actually mean (in this case, the meaning of words and sentences). Next, motor neurons in your brain will send out action potentials to muscle cells in your fingers, telling them to contract in just the right way to turn the page.

? How does a synapse ensure that signals pass *only* in one direction, from a sending neuron to a receiving cell?

The signal can go only one way at any one synapse because only the sending neuron releases neurotransmitters, and only the receiving cell has receptors for the neurotransmitters.

28.7 Chemical synapses make complex information processing possible

The drawing and micrograph here indicate the many inputs one neuron can receive from other neurons. The synaptic knobs (red and green) are the tips of axons of other neurons, delivering information via neurotransmitters to the dendrites, axon, and cell body of the receiving cell. A receiving neuron may obtain input from hundreds of other neurons via thousands of synaptic knobs—many more than you see here. The inputs can be highly varied, because each sending neuron may secrete a different quantity or kind of neurotransmitter. The membrane of a neuron resembles a tiny circuit board, receiving and processing multiple bits of information in the form of neurotransmitter molecules. These living circuit boards account for the nervous system's ability to process data and formulate appropriate responses to stimuli.

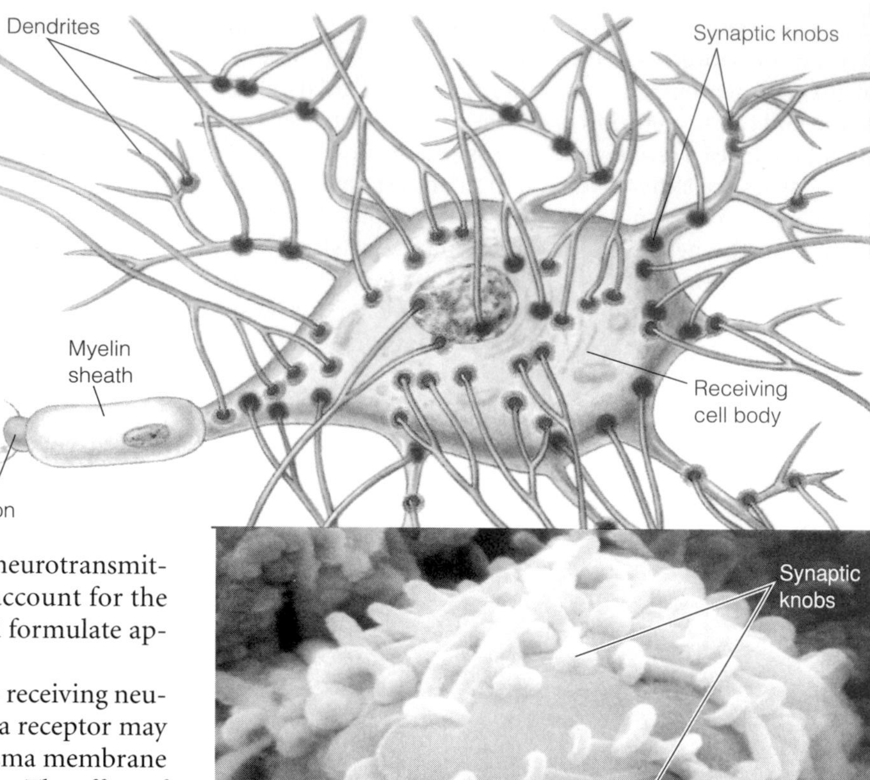

Figure 28.7 The multiple synaptic inputs that a neuron may receive

What do neurotransmitters actually do to receiving neurons? The binding of a neurotransmitter to a receptor may open ion channels in the receiving cell's plasma membrane or trigger a signaling mechanism that does so. The effect of the neurotransmitter depends on the kind of membrane channels they open. Neurotransmitters that open Na^+ channels, for instance, may trigger action potentials in the receiving cell. Such neurotransmitters, and the synapses where they are released, are referred to as excitatory (green in the drawing). In contrast, many neurotransmitters open membrane channels for ions (specifically Cl^-) that *decrease* the receiving cell's tendency to develop action potentials. These neurotransmitters and their synapses are inhibitory (red). The effects of both excitatory and inhibitory neurotransmitters can vary in magnitude. In general, the more neurotransmitter molecules that bind to receptors on the receiving cell and the closer the synapse is to the base of the receiving cell's axon, the stronger the effect.

Depending on the information other cells send it, a receiving neuron's membrane may or may not generate action potentials. At any time, the inhibitory signals may cancel out the excitatory signals. If the excitatory signals are collectively strong enough to overpower the inhibitory ones, the receiving cell may then send out action potentials to other cells. The neuron passes signals to the other cells at a rate that represents a **summation** of all the information it has received. (Signal frequency is key because action potentials are all-or-none events.) Each new receiving cell, in turn, processes this information along with other inputs it obtains. The capacity of the nervous system to process information depends on the variability in the effects that neurotransmitters produce at chemical synapses. 

? Contrast excitatory and inhibitory synapses in how they change a receiving cell's membrane potential relative to triggering an action potential.

Release of neurotransmitters from an excitatory synapse moves the receiving cell's membrane potential closer to threshold; neurotransmitters from an inhibitory synapse moves the receiving cell's membrane potential farther from threshold.

28.8 A variety of small molecules function as neurotransmitters

Neurotransmitters are essential in homeostasis. The precise signaling among neurons they provide enables the nervous system as a whole to coordinate the activities of all parts of the body. Dozens of neurotransmitters are known, and researchers expect to find many more.

Most neurotransmitters are small, nitrogen-containing organic molecules. One, called **acetylcholine,** is important in the brain and at synapses between motor neurons and muscle cells. Depending on the kind of receptors on receiving cells, acetylcholine may be excitatory or inhibitory. For instance, acetylcholine makes our skeletal muscles contract but also slows our heart rate.

A group of nitrogen-containing neurotransmitters called the **biogenic amines** are derived from amino acids. The biogenic amines include epinephrine, norepinephrine, serotonin, and dopamine, all of which also function as hormones (see Chapter 26). Biogenic amines are important neurotransmitters in the central nervous system. Epinephrine and norepinephrine increase the heart rate during stress. Serotonin and dopamine affect sleep, mood, attention, and learning. Certain psychoactive drugs, including LSD and mescaline, apparently produce their hallucinatory effects by binding to serotonin and dopamine receptors in the brain. Imbalances of biogenic amines are associated with various kinds of mental illness. For example, the degenerative disorder Parkinson's disease is associated with a lack of dopamine in the brain, and an excess of dopamine is linked to schizophrenia. Reduced levels of norepinephrine and serotonin seem to be linked with depression. The widely prescribed antidepressant drug fluoxetine (Prozac®) blocks the removal of serotonin from the synaptic cleft, thus increasing the amount of serotonin available to receiving cells at synapses.

Four other neurotransmitters—aspartate, glutamate, glycine, and GABA (gamma aminobutyric acid)—are actually amino acids. All are known to be important in the central nervous system. Aspartate and glutamate are excitatory; glycine and GABA are inhibitors. The brain has hundreds of times more GABA than any other neurotransmitter.

Several peptides, relatively short chains of amino acids, also serve as neurotransmitters. One, called substance P, is an excitatory neurotransmitter that underlies our perception of pain. The endorphins are peptides that function as both neurotransmitters and hormones, decreasing our perception of pain.

Neurons also use some dissolved gases, notably nitric oxide (NO), as chemical signals. NO may be a factor in memory storage and learning. It also relaxes smooth muscle; for example, this effect leads to the penis engorging with blood and becoming erect during sexual arousal. (The male impotence drug Viagra pharmaceutical appears to work by promoting this effect of NO.) Neurons produce NO molecules on demand, rather than storing them in synaptic vesicles. The dissolved gas diffuses into neighboring cells, produces a change, and is broken down, all within a few seconds.

What determines whether a neuron is affected by a specific neurotransmitter?

To be affected by a particular transmitter, a neuron must have specific receptors for that transmitter built into its membrane.

Many stimulants and depressants act on chemical synapses 28.9

Knowing about neurotransmitters helps us understand some of the effects on our nervous system of chemicals we call stimulants and depressants. Stimulants increase the activity of the central nervous system, often by altering the effect of neurotransmitters at synapses. Cocaine and amphetamines increase the normal stimulatory effect of the neurotransmitter norepinephrine. The most commonly used stimulant is caffeine, found in coffee, tea, chocolate, many soft drinks, and a number of over-the-counter drugs. (The related compounds theophylline and theobromine are also found in many caffeine food sources.) Caffeine, used in nearly all cultures, generally raises the heart rate and breathing rate and keeps us awake by countering the effects of inhibitory signals on synapses in our central nervous system. As a result, many of us cannot drink more than one or two cups of coffee at a sitting without becoming restless. Many people consume several hundred milligrams of caffeine a day, mainly from the sources listed in the table to the right. Several studies indicate that consuming more than 600 mg of caffeine a day can cause stomach problems, depression, and sleeplessness.

In contrast to stimulants, depressants often inhibit signal transmission at synapses. Among the depressant drugs, barbiturates and the tranquilizers Valium® and Librium® intensify the effect of the neurotransmitter GABA. Alcohol is a strong depressant. Its precise effect on the nervous system is not yet known, but it seems to increase GABA's inhibitory effect. As people become dependent on alcohol, their nervous system produces less and less GABA.

More commonly used than any other depressant, alcohol can be a very dangerous drug. Habitual heavy drinking (three or more drinks a day at least three times a week) can cause liver and brain damage and may be associated with cardiovascular disease and several forms of cancer. When a woman drinks during pregnancy, alcohol enters the fetus's bloodstream and can cause serious birth defects, as we mentioned in Module 27.16.

Caffeine Content of Foods*

Food	Caffeine
Brewed coffee (6 oz)	47–211 mg
Instant coffee (6 oz)	35–140 mg
Black tea (6 oz)	34–60 mg
Cola drink (12 oz)	30–58 mg
Green tea (6 oz)	11–43 mg
Cocoa (6 oz)	2–8 mg
Milk chocolate candy (1 oz)	1–15 mg
Decaffeinated coffee (6 oz)	1–10 mg

*The ranges in caffeine content are due to variations in starting materials and preparation methods.

When people say that "alcohol lowers a person's inhibitions," it is usually meant as a behavioral description. As a description of what happens at the neurological level, it would probably be more accurate to say that "alcohol raises inhibitions." Why?

Alcohol probably depresses the brain by enhancing the inhibitory effects of GABA, a neurotransmitter.

Nervous Systems

28.10 Nervous system organization usually correlates with body symmetry

To this point, we have concentrated on the cellular mechanisms that are fundamental to the nervous system. There is remarkable uniformity throughout the animal kingdom in the way nerve cells function. However, as we will now see, there is great variety in how nervous systems as a whole are organized. Some animals even lack a nervous system altogether; sponges, for instance, have no cells that are specialized for generating and transmitting nervous signals. Hydras have one of the simplest types of nervous systems (Figure 28.10A). A hydra has what is called a **nerve net,** a weblike system of neurons extending throughout the body. A hydra has no head or brain, and its nerve net has no central or peripheral divisions. It can glide slowly or somersault from place to place, but much of the time it is stationary, attached to submerged plant stems or rocks. The nerve net is quite adequate for a hydra's headless, radially symmetrical body and limited activity.

In contrast to a hydra's net of neurons, seastars and many other echinoderms (Figure 28.10B) have nerves, each of which is composed of many neurons. Radial nerves extend through each arm from a central nerve ring. Branches of the radial nerves form an interconnected network similar to the cnidarian nerve net.

Hydras and other cnidarians and many echinoderms are radially symmetrical, and so are their nervous systems. Radially symmetrical nervous systems tend to be uncentralized, but this does not imply that they are structurally or functionally simple, or that radially symmetrical animals are handicapped by their nervous system. The movements of hundreds of tube feet that allow a sea star to move in one direction, for example, require complex coordination.

Most animals are bilaterally symmetrical, with a head and tail and a tendency to move head-first through the environment. The head—most often the first part of the animal to encounter new stimuli—is usually equipped with sense organs and a brain. Flatworms are the simplest animals that show two evolutionary hallmarks of bilateral symmetry: **cephalization,** or concentration of the nervous system at the head end, and **centralization,** the presence of a central nervous system (CNS) distinct from a peripheral nervous system (PNS). The planarian worm in Figure 28.10C has a small brain composed of ganglia (masses of nerve cell bodies) and two parallel **nerve cords** (bundles of axons and dendrites). These elements constitute the worm's CNS. Other, smaller nerves make up the PNS.

Many bilaterally symmetrical invertebrates show greater cephalization and centralization of the nervous system than flatworms. They exhibit the basic pattern of a CNS composed of a brain and one or more nerve cords, but the CNS tends to be more complex and more distinctly set off from peripheral nerves. For instance, the brain and nerve cord of a leech (Figure 28.10D) contains a much greater concentration of neurons than in the flatworm CNS. The insect shown in Figure 28.10E has a brain composed of several fused ganglia and a ventral nerve cord with a ganglion in each body segment. Each of these ganglia directs the activity of muscles in its segment of the body. As we noted in this chapter's introduction, a squid's nervous system is highly sophisticated (see page 563). In fact, the high degree of centralization and cephalization of the squid's central nervous system parallels that of the vertebrates.

Why is it advantageous for the brain of most bilateral animals to be located at the head end?

Cephalization places the brain near major sense organs that are concentrated on the end of the animal that leads the way as the animal moves through its environment.

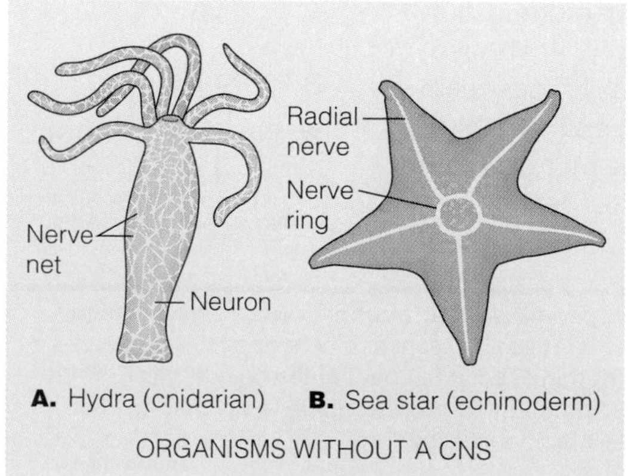

Figure 28.10 Invertebrate nervous systems

Vertebrate nervous systems are highly centralized and cephalized 28.11

Vertebrate nervous systems are diverse in both structure and level of sophistication. For instance, the brains of dolphins and humans are much more complex structurally and much more powerful integrators than the brains of frogs or fishes. However, all vertebrate nervous systems have some fundamental similarities. All have distinct central and peripheral elements and are highly centralized and cephalized. In all vertebrates, the brain and spinal cord make up the CNS (Figure 28.11A). The **spinal cord**, which lies inside the vertebral column, or spine, receives sensory information from the skin and muscles and sends out motor commands for movement. The master control center, the **brain,** includes homeostatic centers that keep the body functioning smoothly; sensory centers that integrate data from the sense organs; and (in humans, at least) centers of emotions and intellect. The brain also exerts control over the spinal cord and sends out its own motor commands to muscles.

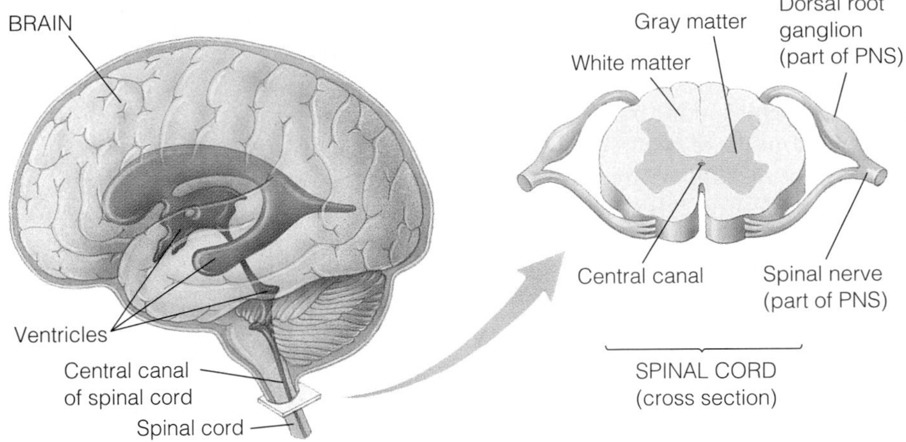

Figure 28.11B Fluid-filled spaces of the vertebrate CNS

A vast network of blood vessels services the CNS. Capillaries in the brain are more selective in what they allow to pass into the brain than those elsewhere in the body. Essential nutrients and oxygen pass freely into the brain, but many other chemicals, such as metabolic wastes from other parts of the body, are kept out. This selective mechanism, called the **blood-brain barrier,** maintains a stable chemical environment for the brain.

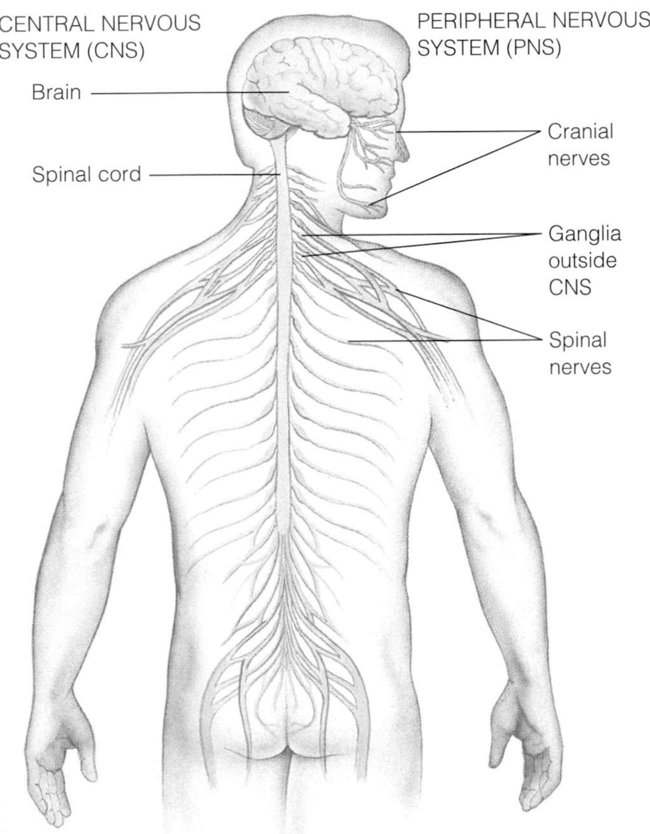

Figure 28.11A A vertebrate nervous system (back view)

Both the brain and spinal cord are hollow (Figure 28.11B). Fluid-filled spaces called **ventricles** in the brain are continuous with the **central canal** of the spinal cord. Each ventricle has a cluster of capillaries that secretes liquid, called **cerebrospinal fluid,** into the spaces. Circulating through the central canal and ventricles (and then draining back into the veins), the cerebrospinal fluid cushions the CNS and helps supply it with nutrients, hormones, and white blood cells. Also protecting the brain and spinal cord are layers of connective tissue, called **meninges.** In mammals, cerebrospinal fluid circulates between two of these layers, providing an additional protective cushion.

As shown in the spinal cord diagram in Figure 28.11B, the CNS has two distinct areas. **White matter** is mainly axons and dendrites; **gray matter** is mainly nerve cell bodies. In the mammalian brain, most of the gray matter is **cerebral cortex,** an outer layer that is the center for higher brain functions such as problem solving.

The ganglia and nerves of the vertebrate PNS are a vast communication network. **Cranial nerves** carry signals to or from the brain. Your eyes, nose, and ears, for instance, are serviced by branches of cranial nerves. **Spinal nerves** carry signals to or from the spinal cord. The muscles and skin of your arms and legs contain branches of your spinal nerves. All spinal nerves and most cranial nerves contain sensory neurons as well as motor neurons. Thousands of incoming and outgoing signals pass each other within the same nerves all the time. Let's take a closer look at the vertebrate PNS.

? A vertebrate's central nervous system consists of the ________ and the ________.

brain . . . spinal cord

28.12 The peripheral nervous system of vertebrates is a functional hierarchy

Figure 28.12A indicates the hierarchical nature of the vertebrate peripheral nervous system. Functionally, the **sensory division** (blue) of the PNS has two sets of neurons. One set brings in information about the outside environment (from the eyes, ears, and other external sense organs). The other set supplies the CNS with information about the body itself—for example, data about the acidity of the blood from sensors in large arteries. Both sets of neurons of the sensory division also provide the brain with sensations of pain, which is the body's warning that it has been or may be suffering tissue damage. Neurons carrying pain signals from the internal organs and those carrying pain signals from the skin and muscles synapse with the same neurons in the CNS. When the brain receives these signals along the same neural pathways, it usually interprets them as coming from the body surface. As a result, we may feel pain from an internal organ on the body surface, a phenomenon called **referred pain** (Figure 28.12B). Referred pain is important in the diagnosis of many disorders. For instance, a heart attack often produces the perception of pain in the left chest, shoulder, and arm.

The body's motor neurons make up the second major functional division of the PNS, the **motor division** (purple). Within the motor division, neurons of the so-called **somatic nervous system** carry signals to skeletal muscles, mainly in response to external stimuli. When you touch a hot stove, for instance, these neurons carry commands to your arm to pull away. The somatic nervous system is said to be voluntary because many of its actions (which we take up in Chapter 30) are under conscious control. In contrast, the motor neurons of the **autonomic** (self-governing) **nervous system** are generally involuntary. We discuss the varied functions of the autonomic nervous system next.

When you write the answer to this question, the muscles in your hand will be controlled by neurons of the ____________ nervous system, which is functional part of the ____________ division of the nervous system.

somatic . . . motor

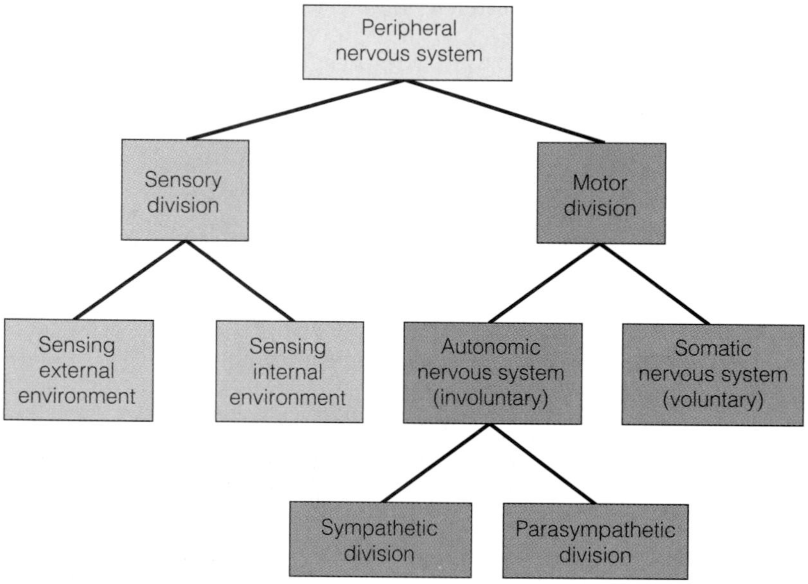

- Peripheral nervous system
  - Sensory division
    - Sensing external environment
    - Sensing internal environment
  - Motor division
    - Autonomic nervous system (involuntary)
      - Sympathetic division
      - Parasympathetic division
    - Somatic nervous system (voluntary)

Figure 28.12A Functional divisions of the vertebrate PNS

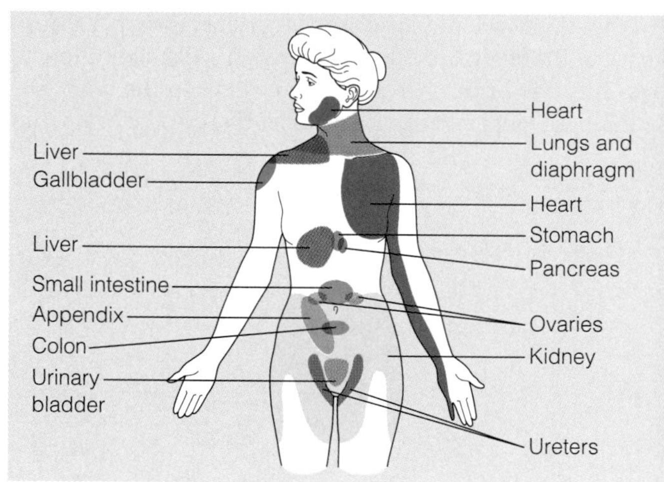

Figure 28.12B Referred pain

28.13 Opposing actions of sympathetic and parasympathetic neurons regulate the internal environment

Our autonomic nervous system consists of two sets of neurons with opposing effects on most body organs. One set, called the **parasympathetic division,** primes the body for digesting food and resting—activities that gain and conserve energy for the body. A sample of the effects of parasympathetic signals appears on the left in Figure 28.13. These include stimulating the digestive organs, such as the salivary glands, stomach, and pancreas; decreasing the heart rate; and narrowing the bronchi, which correlates with a decreased breathing rate.

The other set of neurons, the **sympathetic division,** tends to have the opposite effect, preparing the body for intense, energy-consuming activities, such as fighting, fleeing, or competing in a strenuous game. You see some of the effects of sympathetic signals on the right side of the figure. The digestive organs are inhibited, the bronchi are relaxed so that more air can pass through them, the heart rate is increased, the liver releases the high-energy compound glucose into the blood, and the adrenal glands secrete the fight-or-flight hormones epinephrine and norepinephrine.

Fight-or-flight and relaxation are opposite extremes. Our bodies usually operate at intermediate levels, with most of our organs receiving both sympathetic and parasympathetic signals (the liver and adrenal medulla are exceptions). The opposing signals adjust an organ's activity to a suitable level. Right now, for instance, your salivary glands are probably secreting just enough saliva to keep your mouth moist. The glands are receiving parasympathetic signals that tend to increase their activity, but these signals are counterbalanced by sympathetic signals with the opposite effect. The situation changes when you think about eating your favorite snack. Your thoughts then trigger an increase in parasympathetic signals without changing sympathetic signals. You will notice more saliva in your mouth as the parasympathetic signals overpower the sympathetic ones.

Sympathetic and parasympathetic neurons emerge from different regions of the CNS, and they use different neurotransmitters. As the green dots in the figure indicate, neurons of the parasympathetic system emerge from the brain and the lower part of the spinal cord. Most parasympathetic neurons produce their effects by releasing the neurotransmitter acetylcholine at synapses with target organs. In contrast, neurons of the sympathetic system emerge from the middle regions of the spinal cord (red dots). Most sympathetic neurons release the neurotransmitter norepinephrine at target organs.

As carriers of command signals, the motor neurons of the sympathetic and parasympathetic systems constitute lower levels of the nervous system's hierarchy. In the next several modules, we take a closer look at the highest level of the hierarchy, the brain.

How would a drug that inhibits the parasympathetic nervous system affect a person's pulse?

The pulse, or heart rate, would probably increase.

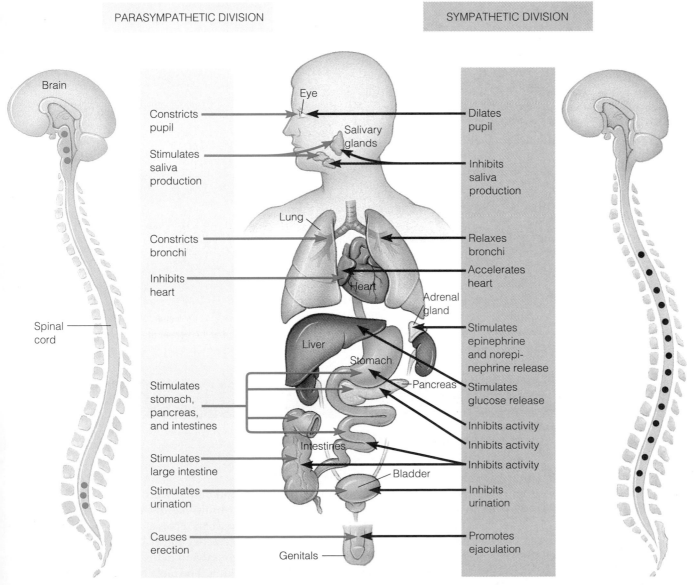

Figure 28.13 The autonomic nervous system

28.14 The vertebrate brain develops from three anterior bulges of the spinal cord

The brain and spinal cord of vertebrates are derived from the dorsal hollow nerve cord, a hallmark of the chordates. The brain evolved from a set of three hollow bulges at the anterior end of the spinal cord. These three ancestral regions, called the **forebrain, midbrain,** and **hindbrain,** still appear during early embryonic development in all vertebrates (Figure 28.14, left). As the vertebrate brain evolved, three major trends altered the three ancestral brain regions. First, the relative size of the brain increased in certain evolutionary lineages. Birds and mammals have bigger brains relative to body size than fishes, amphibians, and reptiles.

A second trend in brain evolution was the subdivision of the forebrain and hindbrain. We see evidence of this today when we watch the brain of a vertebrate developing from its embryonic form into its adult form. The forebrain and hindbrain gradually become subdivided into regions that assume specific responsibilities. The major subdivisions of the human brain are evident after only 3 months of development (Figure 28.14, right).

The third trend in brain evolution was the increasing integrative power of the forebrain. Evolution of the most complex vertebrate behavior paralleled evolution of the **cerebrum,** the dominant part of the forebrain. The cerebrum of birds and mammals is much larger relative to the other parts of the brain than the cerebrum of other vertebrates. And the sophisticated behavior of birds and mammals is directly correlated with their large cerebrum.

Among the mammals, porpoises and primates have a much larger and more complex cerebral cortex than any other vertebrates. Porpoises communicate using a large repertory of sounds. They also have the ability to locate objects, such as prey, using sound echoes. Much of a porpoise's cerebral cortex may be devoted to processing information about its sound-oriented world. In contrast, the brain of humans (one species of the primates) is strongly oriented toward visual perceptions. Humans have the largest brain surface area, relative to body size, of all animals. We take a look at the main components of the human brain next.

What three trends are evident in the evolution of vertebrate brains?

(1) Increase in brain size relative to the animal's size; (2) subdivision of the forebrain and hindbrain into different functional regions; (3) increase in size of cerebrum relative to the rest of the brain

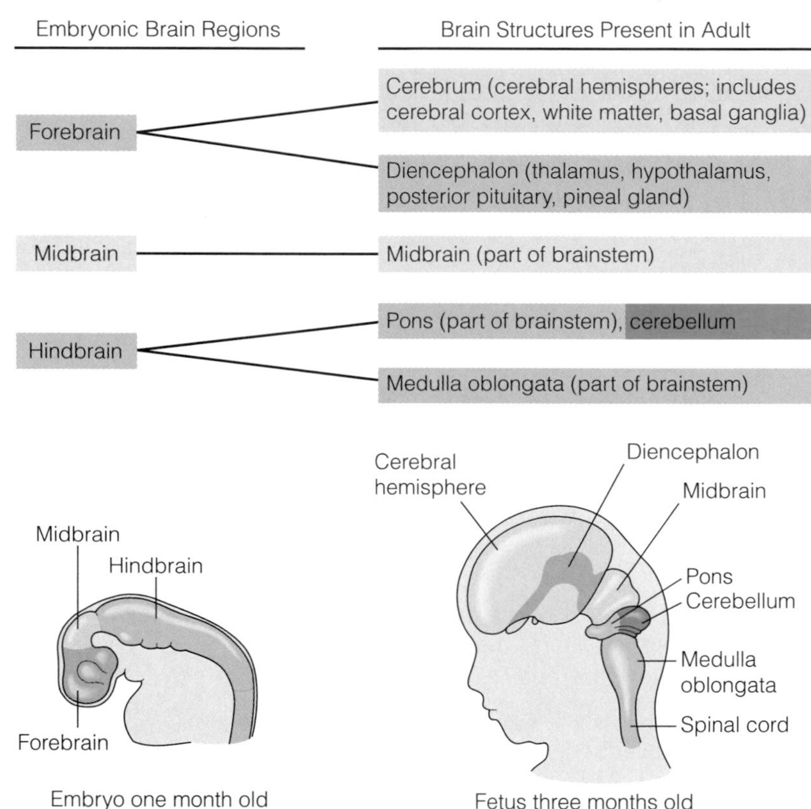

Figure 28.14 Embryonic development of the vertebrate brain

The Human Brain

28.15 The structure of a living supercomputer: The human brain

Composed of about 100 billion intricately organized neurons, with a much larger number of supporting cells, the human brain is more powerful than the most sophisticated computer. In Figure 28.15A and in the table on the facing page, we see that the three ancestral brain regions have evolved considerably from their original state. Looking first at the lower brain centers, two sections of the hindbrain (blue), the **medulla oblongata** and **pons,** and the midbrain (purple) make up a functional unit called the **brainstem.** All of the sensory and motor neurons carrying information to and from higher brain regions pass through the brainstem. Thus, the brainstem serves as a sensory filter, selecting which information reaches higher brain centers. It also regulates sleep and arousal and helps coordinate body movements, such as walking. The table lists some of the individual functions of the medulla oblongata, pons, and midbrain.

Brain Structure	Major Function
Brainstem	Conducts data to and from other brain centers; homeostatic control; coordinates body movement
Medulla oblongata	Controls breathing, circulation, swallowing, digestion
Pons	Controls breathing
Midbrain	Receives and integrates auditory data; major visual center in nonmammalian vertebrates; coordinates visual reflexes in mammals; codes sensory data and sends it to higher brain centers
Cerebellum	Coordinates body movement; learns and remembers motor responses
Thalamus	Input center for sensory data going to the cerebrum; output center for motor responses leaving the cerebrum; data sorting
Hypothalamus	Homeostatic control center; controls pituitary gland; biological clock
Cerebrum	Sophisticated integration; memory, learning, speech; emotions; formulates complex behavioral responses

Another part of the hindbrain, the **cerebellum** (blue), is a planning center for body movements. It also plays a role in learning and remembering motor responses. The cerebellum receives information from sensory receptors in muscles. At the same time, it receives signals from the cerebrum outlining changes that are needed in body movements. The cerebellum evaluates all these inputs and, in a few milliseconds, sends a plan for coordinated movements back to the cerebrum. When you step off a curb, for instance, your cerebellum evaluates your body position and relays to your cerebrum a plan for smoothly getting the rest of your body to follow. The cerebrum then fine-tunes the plan and sends appropriate commands to the muscles.

The most sophisticated integrating centers are those derived from the forebrain (gold)—the thalamus, the hypothalamus, and the cerebrum. The **thalamus** contains most of the cell bodies of neurons that relay information to the cerebral cortex. The thalamus first sorts data into categories (all the touch signals from a hand, for instance). It also suppresses some signals and enhances others. Thus, the thalamus exerts some control over what information gets from sense organs and other parts of the brain to the cerebrum.

In Module 26.4, we saw that the hypothalamus controls the pituitary gland and the secretion of many hormones. The hypothalamus also regulates body temperature, blood pressure, hunger, thirst, the sex drive, and fight-or-flight responses, and it helps us experience emotions such as rage

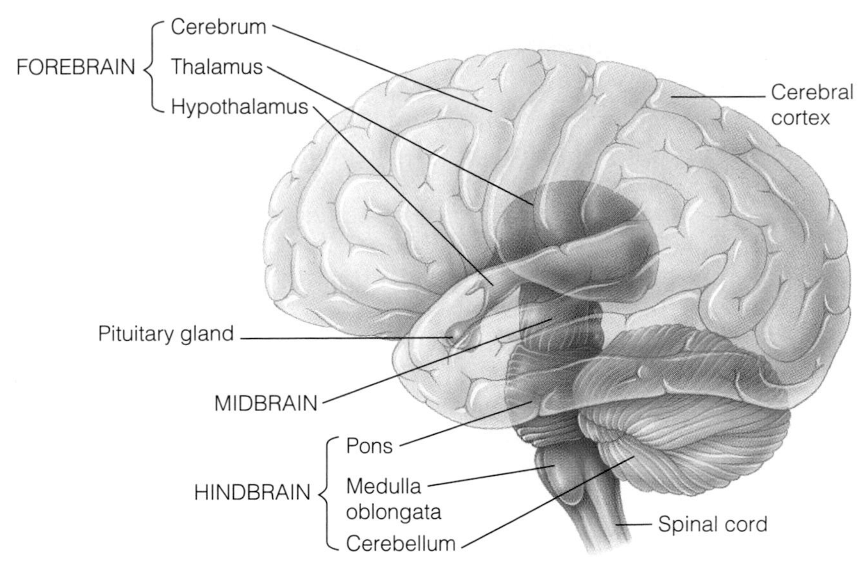

Figure 28.15A The main parts of the human brain

and pleasure. A "pleasure center" in the hypothalamus could also be called an addiction center, for it is strongly affected by certain addictive drugs. Cocaine, for example, stimulates this pleasure center, producing a short-term (5–20 min) high, often followed by depression. Cocaine addiction may involve chemical changes in the pleasure center and elsewhere in the hypothalamus. Another part of the hypothalamus functions as a timing mechanism, our **biological clock.** Receiving visual input from the eyes, the clock maintains our daily biorhythms, such as cycles of sleepiness and hunger.

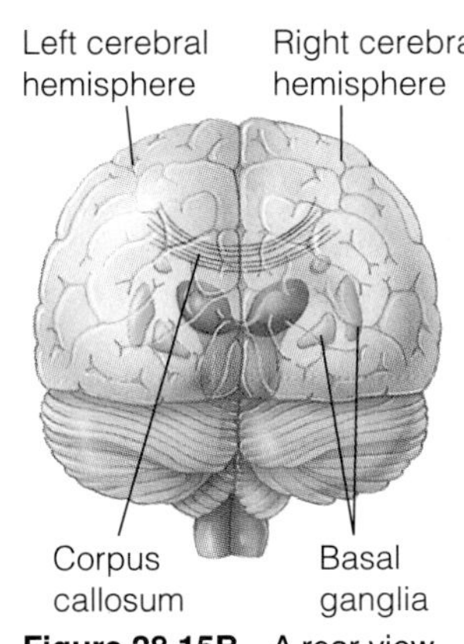

Figure 28.15B A rear view of the brain

The **cerebrum,** the largest and most sophisticated part of our brain, consists of right and left **cerebral hemispheres** (Figure 28.15B). A thick band of nerve fibers called the **corpus callosum** connects the cerebral hemispheres, enabling them to process information together. Under the corpus callosum, small clusters of neuron cell bodies, the **basal ganglia** (also called basal nuclei), are important in motor coordination. If they are damaged, a person may be immobilized. Degeneration of the basal ganglia occurs in Parkinson's disease.The most extensive portion of our cerebrum, the cerebral cortex, is the subject of the next module. 

? Choosing from the structures in the table here, identify the brain part most important in (a) monitoring body temperature; (b) managing your body movements for walking; (c) solving an algebra problem.

(a) hypothalamus; (b) cerebellum; (c) cerebrum

28.16 The cerebral cortex is a mosaic of specialized, interactive regions

Although less than 5 mm thick, the highly folded human **cerebral cortex** occupies over 80% of the total brain mass. It contains some 10 billion neurons and hundreds of billions of synapses. Its intricate neural circuitry produces our most distinctive human traits: reasoning and mathematical abilities, language skills, imagination, artistic talent, and personality traits. Assembling information it receives from our eyes, ears, nose, taste buds, and touch sensors, the cortex also creates our sensory perceptions—what we are actually aware of when we see, hear, smell, taste, or touch.

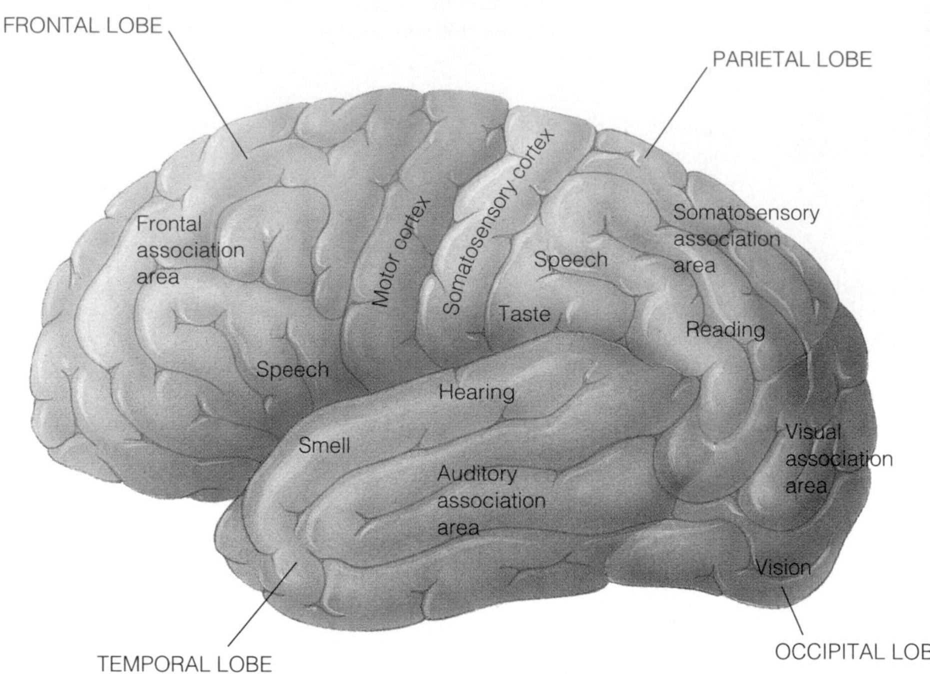

Figure 28.16 Functional areas of the cerebrum's left hemisphere

Discovering how the cerebral cortex works is a monumental task. Some of what we know comes from studies of the effects of tumors, strokes (death of brain cells due to blocked blood vessels), and accidental damage to parts of the cortex. The effects of electrical stimulation of the cortex performed during surgery have also advanced our knowledge. The cortex lacks cells that detect pain; thus, after anesthetizing the scalp, a neurosurgeon can operate on the cerebrum with the patient awake. Parts of the cortex can then be stimulated with a harmless electrical current, and a researcher can obtain information about the effects simply by questioning the patient.

At the forefront of brain research are studies using PET scans and some of the other imaging techniques we discussed in Module 20.10. These techniques reveal which parts of the cortex are most active and how the different parts interact during specific mental activities, such as recognizing words or visualizing pictures.

The surface of the cerebral cortex, for both the right and the left hemispheres, has four discrete lobes (represented by different colors in Figure 28.16). Researchers have identified a number of functional areas within each lobe. Figure 28.16 identifies the main functional areas in the brain's left cerebral hemisphere. Two of these areas are located where the frontal and parietal lobes meet. One, called the motor cortex, functions mainly in sending commands to skeletal muscles, signaling appropriate responses to sensory stimuli. Next to the motor cortex, the somatosensory cortex receives and partially integrates signals from touch, pain, pressure, and temperature receptors throughout the body. The cerebral cortex also has centers that receive and begin processing sensory information concerned with vision, hearing, taste, and smell. Each of these centers, as well as the somatosensory cortex, cooperates with an adjacent area, called an association area. Imaging techniques are beginning to show how a complicated interchange of signals among receiving centers and association areas produces our sensory perceptions.

Making up most of our cerebral cortex, the association areas are the sites of higher mental activities—roughly, what we call thinking. A large association area in the frontal lobe uses varied inputs from many other areas of the brain to evaluate consequences, make considered judgments, and plan for the future. Language results from some extremely complex interactions among several association areas. For instance, the parietal lobe of the cortex has association areas used for reading and speech. These areas obtain visual information (the appearance of words on a page) from the vision centers. Then, if the words are to be spoken aloud, they arrange the information into speech patterns and tell another speech center, in the frontal lobe, how to make the motor cortex move the tongue, lips, and other muscles to form words. When we hear words, the parietal areas perform similar functions using information from auditory centers of the cortex. You can see the locations of the various language centers in the PET scans in Figure 20.10.

What we have said about association areas so far applies to the left hemisphere of the cerebral cortex. As we see in the next module, the association areas in the right hemisphere have somewhat different functions.

? A stroke that causes loss of taste and numbness of limbs has probably damaged brain tissue in the ________ lobe.

parietal

Roger Sperry discovered that the right and left cerebral hemispheres function differently 28.17

Our two cerebral hemispheres look alike superficially. Their primary motor and sensory areas also function in the same way. However, the left and right association areas function so differently that we could almost say we have two brains in one. In most people, the left hemisphere is mainly responsible for language, logic, and mathematical abilities. In contrast, the right hemisphere is mainly responsible for our imagination, spatial perceptions, artistic and musical abilities, and emotions. (In about 10% of us, these roles of the left and right hemispheres are reversed or the roles are more alike.)

Much of what we know about the differing functions of our right and left cerebral hemispheres stems from the work of Nobel Prize winner Roger Sperry (above). Starting in the 1960s, Sperry and his colleagues performed extensive studies of "split-brain" patients, people who had a severed corpus callosum, the mass of neuron fibers that pass information back and forth between the right and left hemispheres (see Figure 28.15B). Most of Sperry's subjects were patients whose corpus callosum had been surgically cut in a now-outmoded treatment for severe epileptic seizures. The technique was drastic, but the patients improved rapidly and did not seem to suffer from personality changes or loss of intelligence. However, as Roger Sperry was to show, cutting masses of neurons does cause problems.

Sperry's first subject was a veteran whose war wounds had left him with uncontrollable seizures. In 1975, Sperry wrote:

> Upon recovery from the [corpus callosum] surgery, he was free of seizures and seemed quite normal in his everyday behavior.... But his surface normality seemed to overlie some startling changes in his inner mental makeup—a suspicion we were to confirm in studies of additional subjects who had had the same operation.

Figure 28.17 illustrates what Sperry was referring to. As shown in panel 1, a split-brain patient holding a key in the left hand, with both eyes open, can readily name it as a "key." If blindfolded, the subject can recognize the key by touch and use it to open a lock, but will be unable to say its name (panel 2). The problem results from the fact that the right side of the brain receives most of its information from sense organs on the left side of the body, and the left side of the brain from the right side of the body. Thus, signals indicating the sensation of a key in the left hand go to the right hemisphere (panel 3). But the speech center is in the left hemisphere. Because the patient lacks an intact corpus callosum, the information about the key cannot travel from the right hemisphere to the left, and he cannot name the key.

(1) Can name key viewed with both eyes

(2) Can use but not name key in left hand when blindfolded

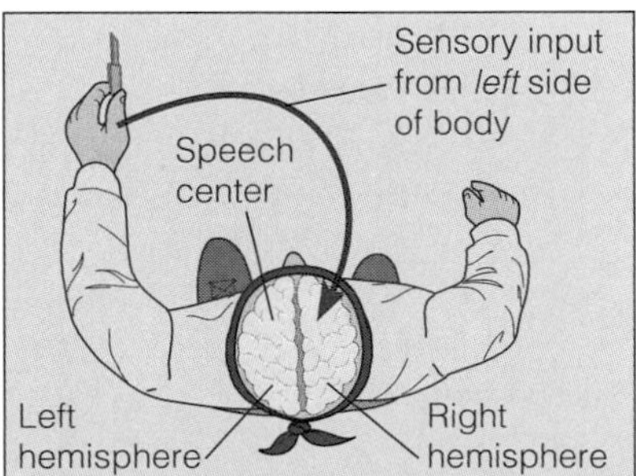

(3) Explanation

Figure 28.17 A split-brain patient demonstrating the different functions of the left and right hemispheres

Summarizing this and other findings, Sperry wrote:

> It is most impressive and compelling to watch a [split-brain] subject solve a given problem like two different people ... using two quite different strategies—depending on whether he is using his left or his right hemisphere. Both the left and right hemispheres of the brain have been found to have their own specialized forms of intellect. The left is highly verbal and mathematical, performing with analytic, symbolic, computerlike, sequential logic. The right, by contrast, is spatial and mute, performing with a ... kind of information processing that cannot yet be simulated by computers.

The realization that our right and left hemispheres contain different association areas has led to some interesting speculations. For instance, is each of us either right-brain or left-brain dominant? A related idea is that artists, for example, are right-brain dominant, whereas mathematicians and engineers are left-brain dominant. These hypotheses remain largely unsubstantiated, and they leave open the question of what might be the evolutionary advantages of having different functional areas on the right and left sides of our brain.

In the remaining modules of this chapter, we examine several other topics of active brain research: arousal and sleep, emotions, memory, and learning.

If the blindfolded individual represented by the cartoon in Figure 28.17 (3) must respond to the question, "What are you holding in your left hand?" he might guess, "a rock." How does the person realize that his guess is incorrect?

The right side of the brain, which knows it is holding a key, hears the left brain's incorrect response.

28.18 Several parts of the brain regulate sleep and arousal

As anyone who has sat through a lecture on a warm day knows, attentiveness and mental alertness vary from moment to moment. *Arousal* is a state of awareness of the outside world. Its counterpart is *sleep,* a state in which we continue to receive stimuli but are not conscious of them.

The hypothalamus, which regulates our sleep-wake cycles, is one of several parts of the brain involved in sleep and arousal.The pons and medulla oblongata contain centers that produce sleep when stimulated, and the midbrain has a center that causes arousal. Serotonin may be the neurotransmitter of the sleep-producing centers. Drinking milk before bedtime may induce sleep because milk contains large amounts of tryptophan, the amino acid cells use to synthesize serotonin.

Also important in sleep and arousal is a functional system of neurons that extends from the medulla oblongata through the brainstem to the thalamus. Called the **reticular formation** (Figure 28.18A), this functional system receives data from sensory receptors (blue arrows in the figure). It filters out some familiar and repetitive information that constantly enters the nervous system and sends useful data to the cerebral cortex (green arrows). Generally, the more input the cortex receives from the reticular formation, the more alert and aware we are.

Researchers study the patterns of electrical activity in the brain during arousal and sleep by a technique called electroencephalography. Electrical contacts placed on the head (Figure 28.18B) detect signals from action potentials in brain neurons. The wires lead to a recording device that converts streams of such electrical signals, called brain waves, to a graph called an **electroencephalogram,** or **EEG** (Figure 28.18C). In general, the less mental activity taking place, the more regular the graph appears. The top part of Figure 28.18C shows the fairly regular, slow brain waves (known as alpha waves) recorded when a person is awake but quiet, with closed eyes. The middle recording, made when a person was solving a complex mental problem, consists of faster, more irregular waves (beta waves).

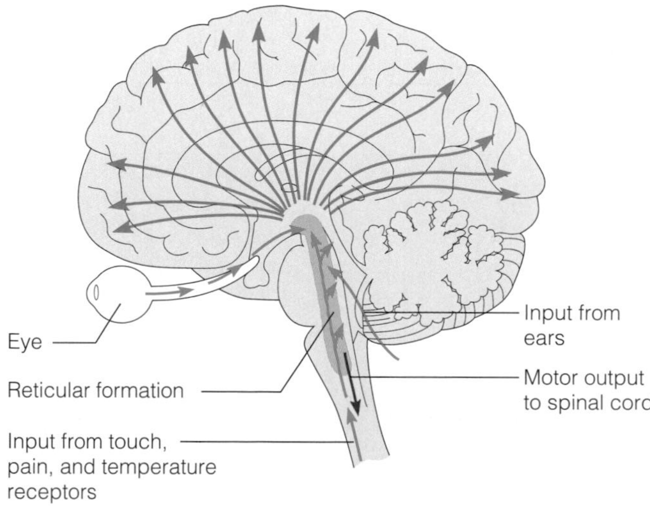

Figure 28.18A The reticular formation

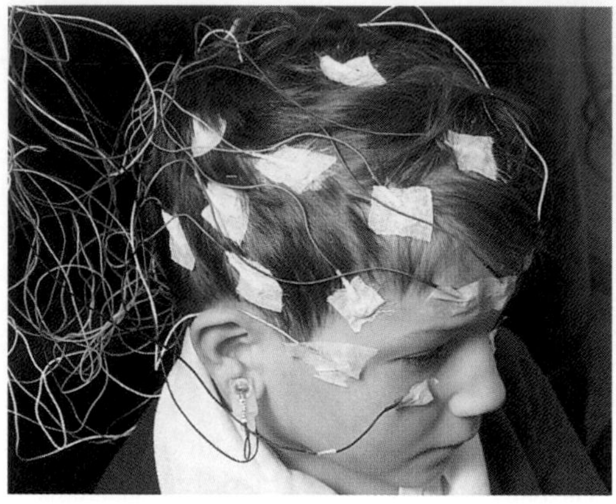
Figure 28.18B Electrodes placed on scalp

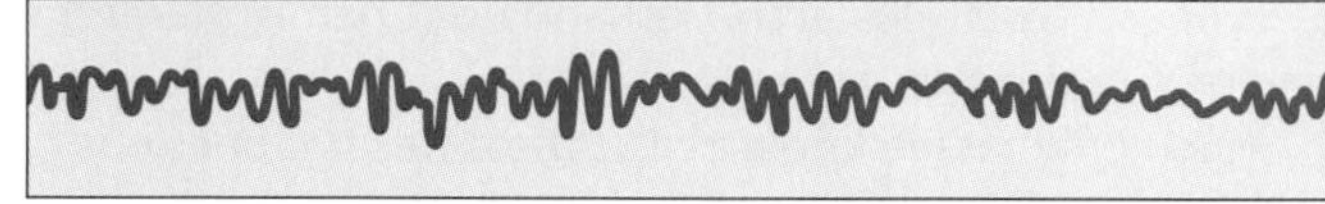
Awake but quiet (alpha waves)

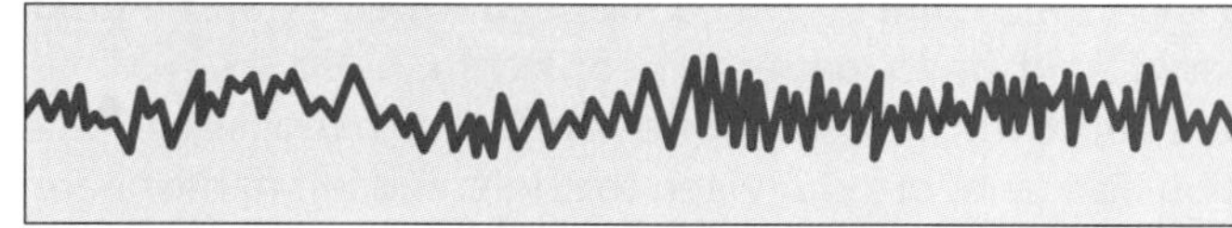
Awake during intense mental activity (beta waves)

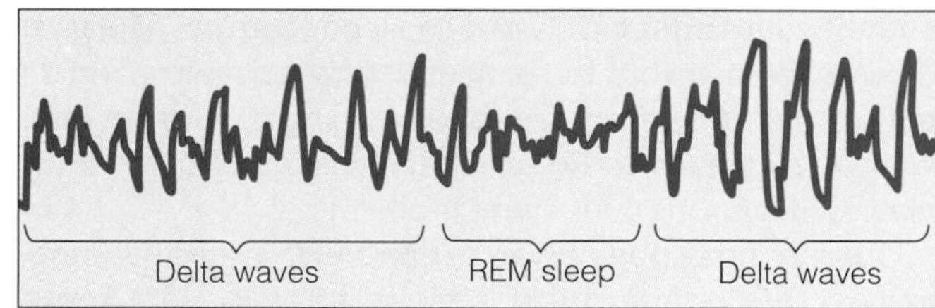

Asleep

Figure 28.18C Some EEG printouts

The bottom part, showing a portion of a sleep cycle, reveals two alternating types of deep sleep. One type, **slow-wave (SW) sleep,** is characterized by delta waves, which are fairly regular, with strong bursts of brain activity. The other type of deep sleep is **REM sleep,** in which the brain waves are rapid and less regular, more like those of the awake state. During REM (rapid-eye-movement) sleep, the eyes move rapidly under the closed lids. The brain itself is highly active and may consume more oxygen than it does when awake. We have most of our dreams during REM sleep, which typically occurs about six times a night for periods of 5–50 minutes each.

There are more questions than answers about what sleeping and dreaming actually do for us. Both REM and SW sleep seem to be important in learning and memory. Recent evidence suggests that the strong bursts of electrical activity during SW sleep may represent neurons in the brain "replaying" recent experiences very rapidly, thereby helping brain cells store the experiences as memories. REM sleep appears to strengthen our ability to learn and remember repetitive skills such as typing or bicycling. We examine learning and memory in more detail next.

> **?** If you are completely unaroused, you must be ________.
>
> asleep

The limbic system is involved in emotions, memory, and learning 28.19

Mapping the parts of the brain involved in human emotions, learning, and memory, and studying the interactions of these parts, are among the great challenges in biology today. Much of human emotion, learning, and memory depends on our **limbic system,** a functional unit of several integrating centers and interconnecting neuron tracts in our forebrain. The limbic system (Figure 28.19) includes parts of the thalamus and hypothalamus and two partial rings around them formed by portions of the cerebral cortex. Two cerebral structures, the amygdala and the hippocampus, play key roles in memory, learning, and emotions. Sensory data converge in the **amygdala,** which seems to act as a memory filter, somehow labeling information to be remembered by tying it to an event or emotion of the moment. The **hippocampus** interacts closely with the amygdala, hypothalamus, brainstem, and another part of the cerebral cortex, the prefrontal cortex, which is involved in complex learning, reasoning, and personality.

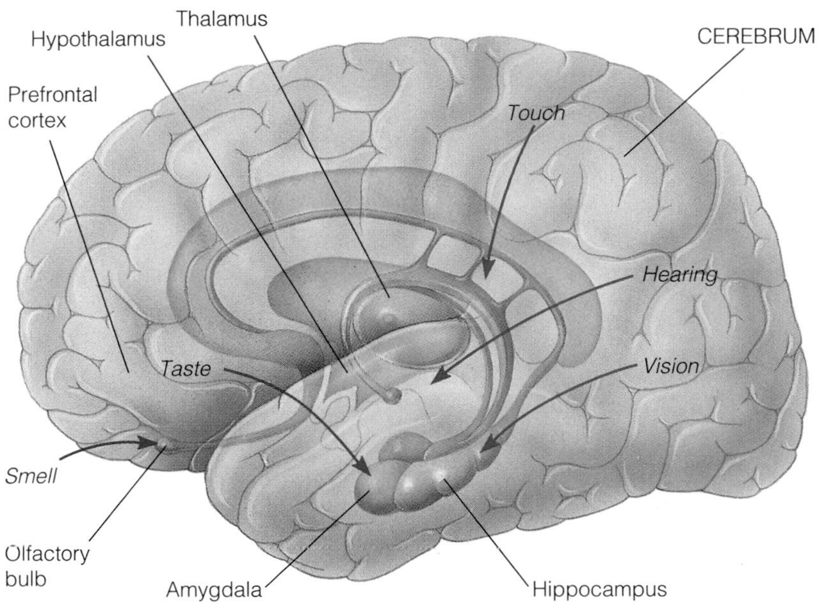

Figure 28.19 The limbic system (shown in shades of gold)

We sense our limbic system's role in both emotion and memory when certain odors bring back "scent memories." Have you ever had a particular smell suddenly make you nostalgic for something that happened when you were a child? As indicated in Figure 28.19, signals from your nose enter your brain through the olfactory bulb. Thus, a specific scent can immediately trigger activity in the memory and emotion centers of the limbic system.

Memory, which is essential for learning, is the ability to store and retrieve information related to previous experiences. We have more than one kind of memory. **Short-term memory,** as the name implies, lasts only a short time—usually only a few minutes. It is short-term memory that allows you to dial a phone number just after looking it up. If you call the number frequently, happen to associate it with some intense emotion, or try to memorize it, the number may become stored in **long-term memory.** You will be able to recall it weeks after you originally looked it up, or even longer. Using long-term memory requires the ability to retrieve stored information and apply it to current situations. The prefrontal cortex seems to play a major role in memory retrieval. Somehow it accesses long-term memory banks and works with other brain centers in using memories to modify behavior.

Information processing by the brain generally seems to involve a complex interplay of several integrating centers. By experimenting with animals, studying amnesia (memory loss) in humans, and using brain imaging techniques, scientists have begun to map some of the major brain pathways involved in memory. Let's consider the example of a telephone number you have just looked up. Sensory signals from your eyes go to the vision centers in your brain's occipital lobe, where a visual perception is formed. The sensory signals also pass through the brain's sensory filter, the reticular formation, to determine whether the information deserves any attention. Signals also go to parts of the limbic system, such as the amygdala, to determine whether emotion should be involved, and to the prefrontal cortex for higher level integration. The memory storage is completed when signals return to the vision centers where the perception first occurred. Storing memories of other kinds of facts probably involves similar pathways. Now let's take a brief look at some of the cellular changes that may underlie memory storage and learning.

> Which of the following brain structures includes all others in the list: amygdala, limbic system, forebrain, thalamus, cerebrum?
>
> Forebrain

28.20 The cellular changes underlying memory and learning probably occur at synapses

What actually happens in the brain when we learn and remember? A look at some recent attempts to answer this question helps us tie what we have discussed about brain structures to our earlier discussions of neurons and synapses. For example, studies of the hippocampus indicate that functional changes at chemical synapses are associated with memory storage and learning. One kind of change, called **long-term depression (LTD),** is decreased responsiveness to an action potential by a receiving neuron. Researchers have been able to induce LTD by repeated, weak stimulation of receiving neurons.

A second kind of synaptic change, called **long-term potentiation (LTP),** is an enhanced responsiveness to an action potential by a receiving neuron. As indicated in Figure 28.20, ① LTP can result when several sending neurons bombard a receiving neuron with repeated action potentials. As we saw in Module 28.6, ② action potentials trigger the release of neurotransmitters by the sending neurons. In LTP, sending neurons release excitatory neurotransmitters. When these chemical signals bind to receptors on the receiving neuron's plasma membrane, ③ calcium (Ca^{2+}) channels in the membrane open, and Ca^{2+} rushes into the cell. In turn, the increased concentration of Ca^{2+} in the receiving neuron triggers ④ enzymatic changes that increase the neuron's responsiveness to a single action potential. Lasting for hours, days, or weeks, depending on the number and frequency of repeated action potentials, LTP may be what happens when a memory is being stored or when learning occurs.

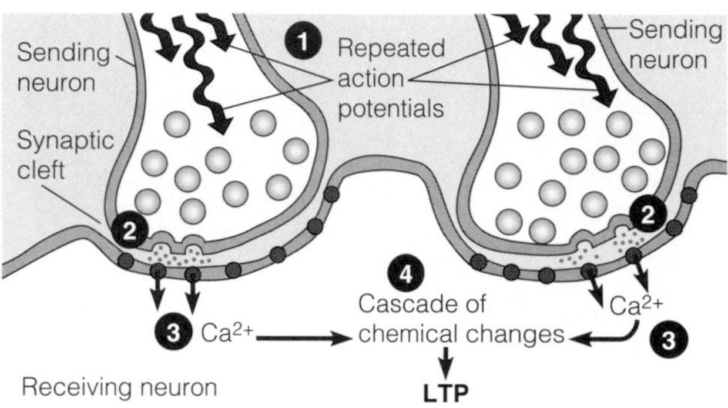

Figure 28.20 A possible cellular mechanism of memory storage

Unraveling how networks of neurons in the brain store, retrieve, and use memories, control the body's internal environment, and construct our thoughts and feelings is one of the most challenging and engaging aspects of modern biology. In the next chapter, we examine another aspect of nervous systems—how sense organs gather information about the environment.

How does the model in Figure 28.20 relate to the adage, "practice makes perfect"?

According to this model, repeated use of a neural pathway enhances subsequent transmission along that pathway by reinforcing the responsiveness of the receiving neuron.

Chapter Review

Chapter Summary

Nervous System Structure and Function (Introduction–28.2) An animal's nervous system enables it to respond quickly and precisely to environmental changes **(Introduction).** The nervous system obtains and processes sensory information and sends commands to effectors (such as muscles) that carry out appropriate responses. Three kinds of neurons illustrate the functional organization of the nervous system. Sensory neurons conduct signals from sensory receptors to the central nervous system (CNS), which consists of the brain and, in vertebrates, the spinal cord. Interneurons in the CNS integrate information and send it to motor neurons. Motor neurons, in turn, convey signals to effectors. Located outside the CNS, the peripheral nervous system (PNS) consists of nerves (bundles of axons and dendrites of sensory and motor neurons) and ganglia (clusters of cell bodies of the neurons in nerves) **(28.1).** The functional units of the nervous system are neurons, cells specialized for carrying signals. A neuron consists of a cell body and two types of extensions (fibers) that conduct signals, dendrites and axons. Many axons are enclosed by cellular insulation called the myelin sheath, which speeds up signal transmission **(28.2).**

Nerve Signals and Their Transmission (28.3–28.9) At rest, a neuron's plasma membrane has an electrical voltage called the resting potential. The resting potential is due to the membrane's ability to maintain a positive charge on its outer surface opposing a negative charge on its inner (cytoplasmic) surface **(28.3).** A stimulus alters the permeability of a portion of the membrane, allowing ions to pass through and changing the membrane's voltage. A nerve signal, called an action potential, is a change in the membrane voltage, from the resting potential to a maximum level and back to the resting potential **(28.4).** Action potentials are self-propagated in a one-way chain reaction along a neuron. An action potential is an all-or-none event; its size is not affected by differences in stimulus strength. The *frequency* of action potentials does change with the strength of the stimulus **(28.5).** The transmission of signals between neurons or between neurons and effector cells occurs at junctions called synapses. Action potentials pass between cells at electrical synapses. At chemical synapses, the sending cell secretes a chemical signal, a neurotransmitter, which crosses the synaptic cleft (a gap between the cells) and binds to a specific receptor on the surface of the receiving cell **(28.6).** Some neurotransmitters excite the receiving cell; others inhibit, by decreasing the receiving cell's ability to develop action potentials. A cell may receive differing signals from many neurons; the summation of excitation and inhibition determines whether or not it will transmit a nerve signal **(28.7).** Many small, nitrogen-containing molecules serve as neurotransmitters. Neurons also use some dissolved gases, such as NO, as signals **(28.8).** Many hallucinogens, stimulants, and depressants act at synapses **(28.9).**

Nervous Systems (28.10–28.14) Radially symmetrical animals, such as hydras, have a nervous system arranged in a weblike system of neurons called a nerve net. Most bilaterally symmetrical animals exhibit cephalization, the concentration of the nervous system in the head end, and centralization, the presence of a central nervous system **(28.10)**. Vertebrate nervous systems are highly centralized and cephalized. The brain and spinal cord contain fluid-filled spaces. Cranial and spinal nerves make up the peripheral nervous system **(28.11)**. Functionally, the vertebrate PNS has a sensory division and a motor division. In the motor division, the somatic nervous system exerts mostly voluntary control over the skeletal muscles, and the autonomic nervous system exerts mostly involuntary control over the internal organs **(28.12)**. The autonomic nervous system consists of two sets of neurons, the parasympathetic division and the sympathetic division, which have opposing effects on most body organs **(28.13)**. The vertebrate brain evolved by the enlargement and subdivision of three anterior bulges of the spinal cord: the hindbrain, midbrain, and forebrain. Three trends in brain evolution were an increase in the size of the brain relative to body size; subdivision of the forebrain and hindbrain; and an increase in the integrative power of the forebrain **(28.14)**.

The Human Brain (28.15–28.20) The midbrain and subdivisions of the hindbrain, and the thalamus and hypothalamus of the forebrain, function mainly in conducting information to and from higher brain centers, regulating homeostatic functions, keeping track of body position, and sorting sensory information. The forebrain's cerebrum is the largest and most complex part of the brain. Most of the cerebrum's integrative power resides in the cerebral cortex of the two cerebral hemispheres **(28.15)**. Specialized integrative regions of the cerebral cortex include the somatosensory cortex and centers for vision, hearing, taste, and smell. The motor cortex directs responses. Association areas, concerned with higher mental activities such as reasoning and language, make up most of the cerebrum **(28.16)**. The right and left cerebral hemispheres specialize in different mental tasks **(28.17)**. Sleep and arousal involve activity by the hypothalamus, medulla oblongata, pons, and midbrain **(28.18)**. The limbic system, a functional group of integrating centers in the cerebral cortex, thalamus, and hypothalamus, is involved in emotions, memory, and learning **(28.19)**. Research suggests that learning and memory involve structural and chemical changes at synapses **(28.20)**.

Testing Your Knowledge

Multiple Choice

1. Joe accidentally touched a hot pan. His arm jerked back, and an instant later, he felt a burning pain. How would you explain that his arm moved before he felt the pain?
   a. His limbic system blocked the pain momentarily, but the important pain signals eventually got through.
   b. His response was a spinal cord reflex that occurred before the pain signals got to the brain.
   c. It took a while for his brain to search long-term memory and figure out what was going on.
   d. The autonomic nervous system responded to the danger because the brain was too busy to react quickly.
   e. This scenario is not actually possible. The brain must register pain before a person can react.

2. Suppose a researcher carefully removes the myelin sheath from the axon of a neuron. The most likely result would be that the altered neuron would
   a. no longer be able to produce neurotransmitters
   b. be much more sensitive to stimuli
   c. be much less sensitive to stimuli
   d. transmit nerve signals more rapidly
   e. transmit nerve signals more slowly

3. The folds of the cerebrum increase its surface area. This is important in the brain's performance because it
   a. maximizes the number of cell bodies that process information
   b. prevents a "short circuit" between adjacent areas
   c. allows the cerebrum to absorb more oxygen
   d. increases myelin, which speeds up nerve signals
   e. more effectively protects the cerebrum from damage

4. Anesthetics block pain by blocking the transmission of nerve signals. Which of these three chemicals might work as anesthetics? *(Explain your answer.)*
   I. a chemical that prevents the opening of sodium channels in membranes
   II. a chemical that inhibits the enzymes that degrade neurotransmitters
   III. a chemical that blocks neurotransmitter receptors
   a. I b. II c. III d. II and III e. I and III

Describing, Comparing, and Explaining

As you hold this book, nerve signals are generated in nerve endings in your fingertips and sent to your brain. Once a touch has caused an action potential at one end of a neuron, what causes the nerve signal to move from that point along the length of the neuron to the other end? What is the nerve signal, exactly? Why can't the signal go backward? How is the nerve signal transmitted from one neuron to the next across a synapse? Write a short paragraph that answers these questions.

Thinking as a Scientist

Using microelectrodes, a researcher recorded nerve signals in four neurons in the brain of a snail. The neurons are called A, B, C, and D in the table below. A, B, and C all can transmit signals to D. In three experiments, the animal was stimulated in different ways. The numbers of nerve signals transmitted per second by each of the cells is recorded in the table. Write a short paragraph explaining the different results of the three experiments.

	Signals/sec			
	A	B	C	D
Experiment #1	50	0	40	30
Experiment #2	50	0	60	45
Experiment #3	50	30	60	0

Science, Technology, and Society

Alcohol's depressant effects on the nervous system cloud judgment and slow reflexes. Alcohol consumption is a factor in most fatal traffic accidents in the United States. What are some other impacts of alcohol abuse on society? What are some of the responses of people and society to alcohol abuse? Do you think this is primarily an individual or societal problem? Do you think our responses to alcohol abuse are appropriate and proportional to the seriousness of the problem?

Media Resources

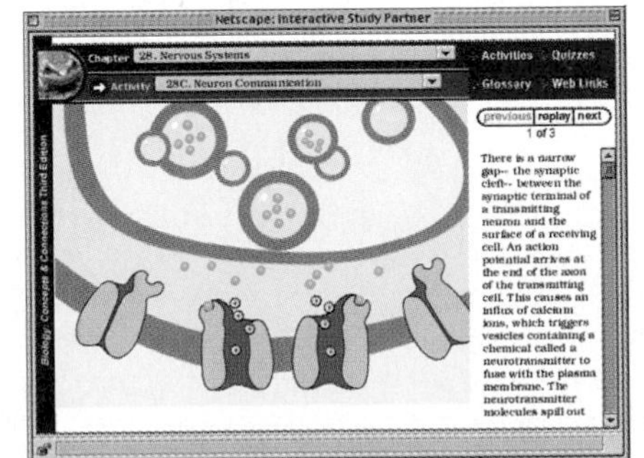

For further review of the concepts in this chapter, use the activities and quizzes on the CD-ROM.

Visit the Special Edition of The Biology Place web site for activities, quizzes, news articles, and web links.

29 The Senses

Outline

A fatal encounter between a grizzly bear and a migrating salmon in a glacial stream in Alaska: Salmon are fast and agile, but easy prey for a bear in shallow water. Despite the grizzly's rather poor eyesight and great size (up to about 450 kg, or nearly 1000 pounds), it has lightning reflexes and can snatch up a large fish with little effort.

What brought these animals together at this particular moment? The grizzly learned as a cub to head for the salmon streams in late autumn. It followed its mother there and learned how to fish by watching her. The bear's acute sense of smell helped it find the stream. For the salmon, becoming a meal for the bear suddenly ends a remarkable odyssey. After nearly eight years at sea, it was returning to reproduce in the stream where its life began. The salmon has a keen sense of smell, and odors led it to this particular stream among thousands of possibilities.

In the fall, a female salmon lays her eggs in a shallow depression on the bottom of a small, fast-flowing stream.

A male covers the eggs with sperm, and both parents die soon thereafter. The following spring, larval salmon hatch and begin drifting downstream toward the ocean, starting a journey that will take them far out to sea. While at sea, they feed in large schools with salmon from other river systems. When they are 4–6 years old and sexually mature, the salmon segregate into groups of common geographic origin and start migrating back toward the river from which they emerged as juveniles.

Research on how salmon locate their home stream has been going on since the early 1950s. The fish seem to navigate to the mouth of their home river system by using the angle of the sun for reference. Once there, however, their sense of smell takes over.

The water that flows from each stream into a river seems to carry a unique scent, a mixture of chemicals from the plants and soils in the area. The scent of its home stream apparently becomes fixed in the memory of a young salmon before it migrates to the sea. When a mature salmon arrives in the vicinity of its home river system, it swims along the coast until it detects the faint odors matching the scent memory in its brain. In response to perhaps only a few molecules of its "home chemicals," it enters that river and begins its upstream journey.

A salmon has a nostril on each side of its head. In the photograph below, showing a spawning male salmon, you can see that each nostril has two openings. As the fish swims, water flows in one opening and out the other, and highly sensitive cells in the nostril detect chemicals dissolved in the water. When researchers blocked the nostrils of migrating salmon with cotton, the fish lost their homing ability. They still migrated, but could not make the correct choices at forks of streams.

As a salmon swims upstream, its nostrils follow a scent trail in the water. The closer it gets to its home stream, the stronger the scent becomes. More and more of the scent molecules stimulate sensory cells in its nostrils, and the cells send more and more signals to the fish's brain, telling it that it's on the right track.

Sensory information gathered by sensory receptors and processed by the brain guide salmon and grizzly bears to specific stream sites. This chapter focuses on sensory structures and how they gather information. To begin, we examine the distinction between information gathering and information processing—the difference between sensing chemicals in the air or water, for instance, and recognizing them as odors.

29.1 Sensory inputs become sensations and perceptions in the brain

Sensory receptor cells are tuned to the condition of the external world and the internal organs. They detect stimuli, such as chemicals, light, tension in a muscle, sounds, electricity, cold, heat, and touch, and send information to the central nervous system. The sensory cells in a bear's nose, for instance, detect chemicals in the air and send reports about the chemicals to the brain. The reports take the form of action potentials, the same signals used throughout the nervous system (see Module 28.4). The sensory receptor's job is completed when it triggers action potentials that go to the central nervous system.

What happens to sensory information in the central nervous system—our brain, for example? When action potentials reach the brain, we may experience a **sensation,** an awareness of sensory stimuli. Sensations result when the brain integrates new information. If you detect the scent of a carnation, for instance, and you have never had any experience that would connect the scent with a flower (or with a substance that smells like one), you simply have a sensation; your brain integrates sensory signals from your nose and makes you aware of a pleasant odor.

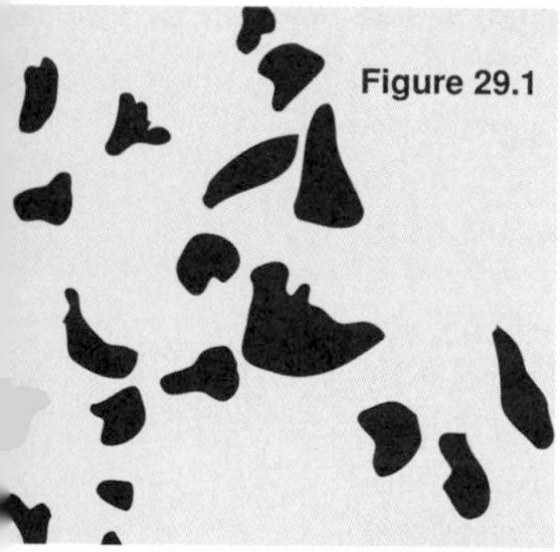

Figure 29.1

In most cases, the brain does much more than form sensations. It integrates the new information with other information and forms a **perception,** a meaningful interpretation or conscious understanding of sensory data. In the case of a carnation, for example, if you connect the scent with the flower, either because you see the flower and smell it or because you have seen one before, your brain has formed a perception.

Figure 29.1 may further demonstrate the difference between sensation and perception for you. What do you see when you first look at the figure? If you see only some black splotches and blue space, you have developed a sensation—in this case, a more or less meaningless interpretation of some sensory information. What if we say that the figure shows a person riding a horse? With this clue, the brain forms a perception; it converts the sensation into a meaningful image. The figure may not have worked this way for you. If you saw the horse and rider right away, you experienced both a sensation and a perception. Your brain integrated the new information with some of its stored data; perhaps you've had a lot of experience with horses or have seen this figure before.

What does the brain actually do with sensory information in creating a perception? As we discussed in Chapter 28, researchers using brain imaging techniques are beginning to find out. The perception of a sweet-smelling flower, for instance, results from communication among neurons arranged in an extremely complex circuitry within the vision and odor centers of our cerebral cortex. The vision centers analyze and connect information on the flower's color, form, and any detectable motion. Neuronal communication involving the visual centers, odor centers, and perhaps the memory banks creates the overall perception.

Perceptions are the product of a continuum of information processing, beginning with the detection of stimuli by sensory receptors. For the rest of this chapter, we concentrate on sensory receptors and how they function.

In comparing sensations and perceptions to a specific environmental stimulus, humans probably vary more in their __________ than in their __________.

perceptions . . . sensations

Sensory Reception

29.2 Sensory receptor cells convert stimuli into electrical energy

Sensory organs such as your eyes or the taste buds on your tongue contain specialized receptor cells that detect stimuli. The receptor cells in your eyes, for instance, detect light energy; those in your taste buds detect chemicals dissolved in saliva, such as salt or sugar.

What exactly do we mean when we say that a receptor cell detects a stimulus—a molecule of sugar or a photon of light, for instance? Stimulus detection means that the cell converts one type of signal (the stimulus) into an electrical signal. This conversion, called **sensory transduction,** occurs in the plasma membrane of the receptor cell.

Figure 29.2A shows sensory transduction occurring when receptor cells in a taste bud detect sugar molecules. ① The molecules first enter the taste bud, where ② they bind to specific protein molecules in a receptor cell membrane. The binding changes the membrane permeability by causing ion channels in the membrane to open. ③ Positively charged ions then flow into the cell from the surrounding fluid and alter the membrane voltage (the potential energy stored by the membrane; see Module 28.3), raising it to a higher level called the receptor potential. The **receptor potential** is the electrical signal produced by sensory transduction. In contrast to action potentials, which are all-or-none phenomena, receptor potentials vary; the stronger the stimulus, the larger the receptor potential.

Once a receptor cell converts a stimulus to a receptor potential, the receptor potential usually results in signals entering the central nervous system. In our taste bud example,

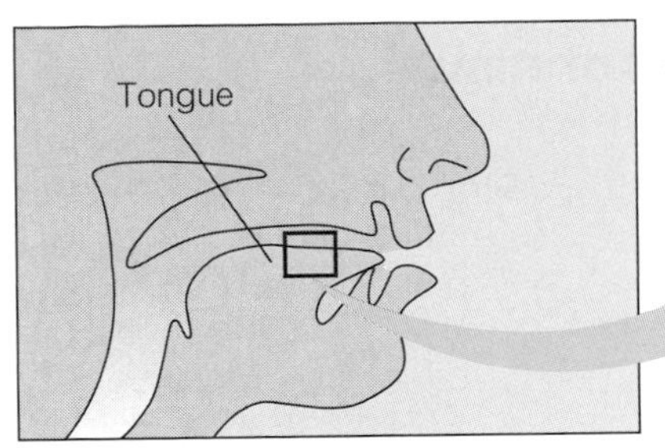

④ each receptor cell forms a synapse with a sensory neuron. This is a chemical synapse just like the one between neurons described in Module 28.6. Notice that the neurotransmitter molecules diffuse across the synapse from the receptor cell to the sensory neuron. The receptor cell constantly secretes neurotransmitter at a set rate, which triggers a steady stream of action potentials in the sensory neuron. ⑤ The graph on the left represents the rate at which the sensory neuron triggers action potentials when the taste bud is not detecting any sugar. The graph on the right shows what happens when there are enough sugar molecules to trigger a strong receptor potential. This receptor potential makes the receptor cell release more neurotransmitter than usual, enough to increase the rate of action potential generation in the sensory neuron. This change in the rate of action potentials signals the brain that the receptor cell detects the stimulus.

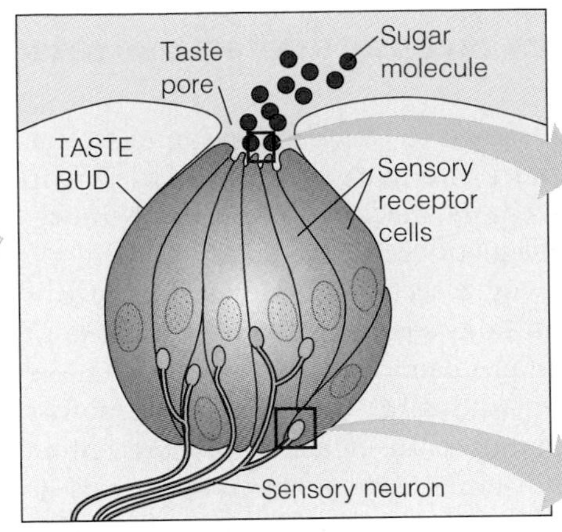

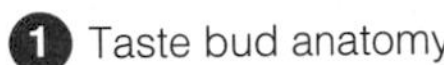

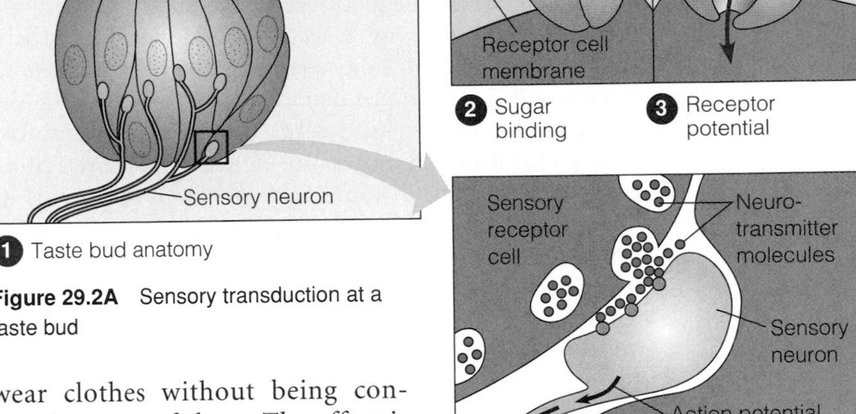

Figure 29.2A Sensory transduction at a taste bud

Thus we see that sensory receptors transduce (convert) stimuli into electrical signals and can trigger action potentials that go into the central nervous system for processing. Since action potentials are the same no matter where or how they are produced, how do they communicate different information, like a sweet taste instead of a salty one? In Figure 29.2B, the taste bud on the left has receptor molecules only for sugar, the one on the right only for salt. The sensory neurons from the salt-detecting taste bud synapse with interneurons in the brain that are different from those contacted by neurons from the sugar-detecting taste bud. The brain distinguishes stimulus types (in this case, salt from sugar) by which interneurons are stimulated.

The graphs in Figure 29.2B indicate how action potentials communicate information about the *intensity* of stimuli (for example, very sweet or less sweet). In each case, the graph on the left represents the rate at which the sensory neurons in the taste bud transmit action potentials when the receptor cells are not stimulated. The graphs on the right show that the rate of transmission depends on the intensity of the stimulus. The stronger the stimulus, the more frequently the receptor transmits action potentials to the brain. The brain interprets the intensity of the stimulus from the rate at which it receives action potentials. It gains additional information about stimulus intensity by keeping track of how many sensory neurons it receives signals from.

There is an important qualification to what we have just said about stimulus intensity. Have you ever noticed how an odor that is strong at first seems to fade with time, even when you know the substance is still there? The same effect helps you adjust to a hot or cold shower and enables you to wear clothes without being constantly aware of them. The effect is called **sensory adaptation,** the tendency of sensory receptor cells to become less sensitive when they are stimulated repeatedly. When receptors become less sensitive, they trigger fewer action potentials, and the brain may lose its awareness of stimuli as a result. Sensory adaptation keeps the body from reacting to normal background stimuli. Without it, our nervous system would become overloaded with useless information.

This overview of sensory transduction, transmission, and adaptation explains how sensory receptors work in general. Now let's look at the receptors themselves.

? (a) What enables a receptor cell in a taste bud to key on just a certain type of chemical in food? (b) What determines the "taste" that is perceived in response to that particular chemical?

(a) The specific receptors on the membrane of the taste receptor cell; (b) The part of the brain stimulated by sensory neurons from the taste bud

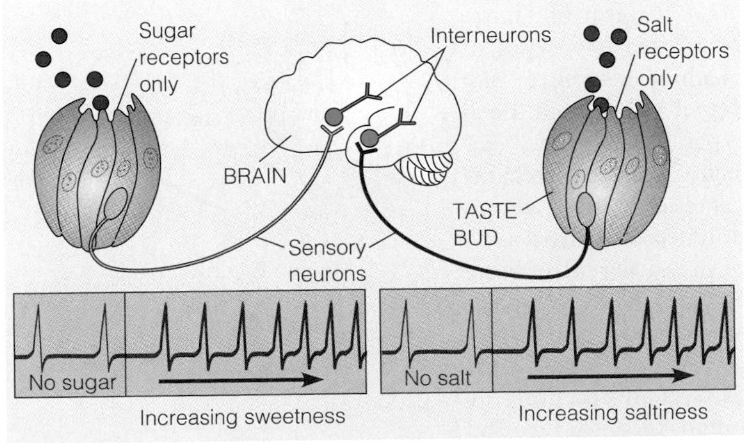

Figure 29.2B How action potentials represent different taste sensations

29.3 Specialized sensory receptors detect five categories of stimuli

Based on the type of signals to which they respond, we can group sensory receptors into five general categories: pain receptors, thermoreceptors, mechanoreceptors, chemoreceptors, and electromagnetic receptors.

Figure 29.3A, showing a section of human skin, reveals why the surface of our body is sensitive to such a variety of stimuli. Our skin contains pain receptors, thermoreceptors (sensors for both heat and cold), and mechanoreceptors (sensors for touch and pressure). Each of these receptors has dendrites of one or more neurons that transduce stimuli and also send action potentials to the central nervous system. In other words, each neuron in the receptor serves as both a receptor cell and a sensory neuron. Most of the dendrites in the dermis (the underlying region of the skin) are wrapped in one or more layers of connective tissue (gold areas in the figure); however, the pain and touch receptors in the epidermis (outer skin layer) and the touch receptors around the base of hairs are naked dendrites.

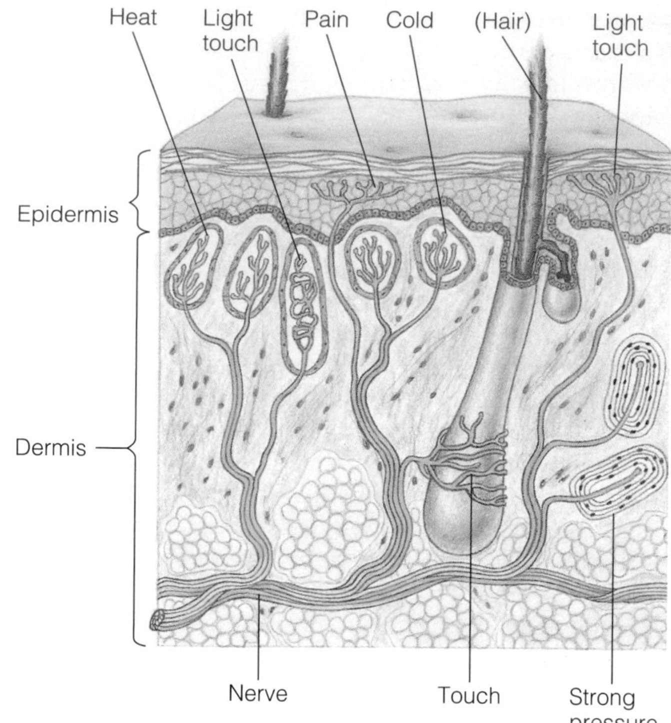

Figure 29.3A Sensory receptors in the human skin

Probably all animals have **pain receptors,** although we cannot say what nonhuman perceptions of pain are like. Pain is important because it often indicates danger and usually makes an animal withdraw to safety. Pain can also make us aware of injury or disease. All parts of the human body except the brain have pain receptors.

Thermoreceptors in the skin detect either heat or cold. Other temperature sensors located deep in the body monitor the temperature of the blood. The hypothalamus in the brain is the body's major thermostat. Receiving action potentials from both surface and deep sensors, the hypothalamus keeps a mammal's or bird's body temperature within a narrow range (see Module 20.13).

Mechanoreceptors are highly diverse. Different types are stimulated by various forms of mechanical energy, such as touch and pressure, stretching, motion, and sound. All these forces produce their effects by bending or stretching the plasma membrane of a receptor cell. When the membrane changes shape, it becomes more permeable to positive ions, and if the stimulus is strong enough, the mechanical energy of the stimulus is transduced into a receptor potential.

At the top of Figure 29.3A are two types of mechanoreceptors that detect light touch. Both types transduce very slight inputs of mechanical energy into action potentials. A third type of pressure sensor, lying deeper in the skin, is stimulated by strong pressure. A fourth type of mechanoreceptor, the touch receptor around the base of the hair, detects hair movements. Touch receptors at the base of the stout whiskers on a cat or a bear are extremely sensitive and enable the animal to detect close objects by touch in the dark. Two other types of mechanoreceptors (not shown) are blood pressure sensors, found in some of our blood vessels, and stretch receptors, found in our skeletal muscles. Sensitive to changes in muscle length, **stretch receptors** monitor the position of body parts (see Figure 28.1B).

A variety of mechanoreceptors collectively called **hair cells** detect sound waves and other forms of movement in air or water. The "hairs" on these sensors are either specialized types of cilia or cellular projections called microvilli. The sensory hairs project from the surface of a receptor cell into either the external environment, such as the water sur-

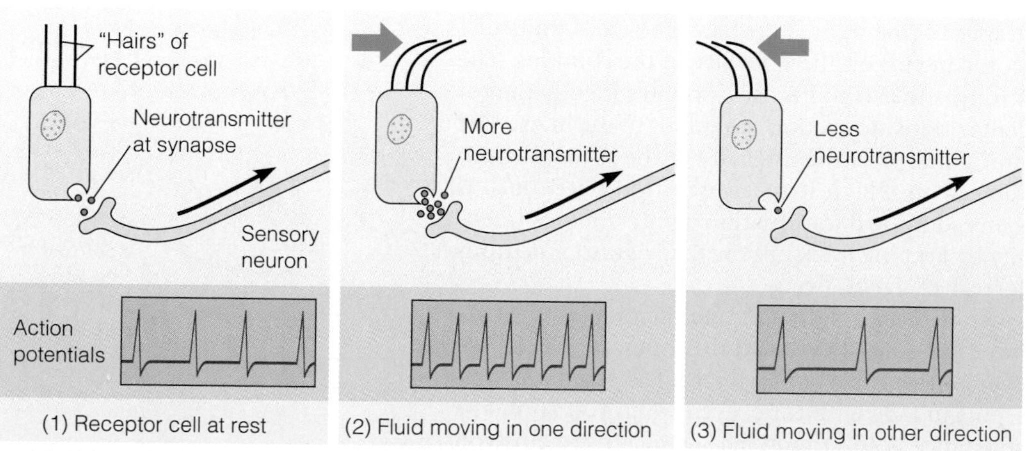

Figure 29.3B Mechanoreception by a hair cell

rounding a fish, or an internal fluid-filled compartment, such as our inner ear. Figure 29.3B indicates how hair cells work. Part 1 shows a receptor cell at rest. When fluid movement bends the hairs in one direction, as shown in part 2, the hairs stretch the cell membrane, increasing its permeability to positively charged ions. This makes the hair cell secrete more neurotransmitter molecules and increases the rate of action potential production by a sensory neuron. When the hairs bend in the opposite direction (part 3), ion permeability decreases, the hair cell releases fewer neurotransmitter molecules, and the rate of action potential generation decreases. We'll see later that hair cells are important in both hearing and balance.

Chemoreceptors include the sensory cells in our nose and taste buds, which are attuned to chemicals in the external environment, as well as some internal receptors that detect chemicals in the body's internal environment. Internal chemoreceptors include sensors in some of our arteries that can detect changes in the amount of CO_2 in the blood. In all types of chemoreceptors, a receptor cell develops receptor potentials in response to chemicals dissolved in fluid (for instance, blood, saliva, the fluid coating the inside surface of the nose, or the water surrounding a fish).

One of the most sensitive chemoreceptors is found on the antennae of the male silkworm moth *Bombyx mori* (Figure 29.3C). The antennae are covered with thousands of tiny bristles (visible in the micrograph below). Most of the bristles are chemoreceptors that detect a sex attractant called bombykol, produced by the female moth. A male begins to respond to a female when as few as 50 of his bristles detect one bombykol molecule per second.

Electromagnetic receptors are sensitive to energy of various wavelengths, which takes such forms as electricity, magnetism, and light. Certain fishes locate prey by sensing electric fields produced by the prey's muscle activity. The so-called electric fishes of Africa and Australia discharge electricity into the water, and electroreceptors in their skin can detect nearby obstacles and animals.

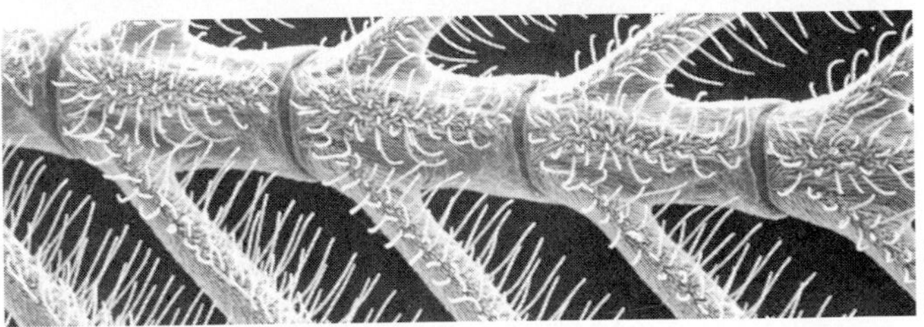

Figure 29.3C Chemoreceptors on insect antennae

Figure 29.3D Electromagnetic receptors in a snake

Evidence suggests that many species can detect Earth's magnetic fields and use this information to navigate from place to place. The abdomen of some insects and the head of certain fishes, birds, and mammals (including humans) contain the iron-containing mineral magnetite. A recent experimental study demonstrated that rainbow trout have magnetic sensitivity, and researchers are attempting to work out how the fish uses this ability to navigate. Once used by sailors as a primitive compass, magnetite may be part of an orienting mechanism in many organisms.

Photoreceptors, including eyes, are probably the most common type of electromagnetic receptor. Photoreceptors detect the electromagnetic energy we call light, which may be in the visible, infrared, or ultraviolet part of the electromagnetic spectrum (see Module 7.6).The rattlesnake in Figure 29.3D has two kinds of photoreceptors. Its prominent eyes detect visible light. Below its eyes are two receptors extremely sensitive to infrared radiation, a form of heat. These receptors detect the body heat of its preferred prey, small mammals and birds. The receptors can detect the infrared radiation emitted by a mouse a meter away. The snake moves its head from side to side until the radiation is detected equally by the two receptors, indicating that the mouse is straight ahead.

As we will see in the next module, eyes are diverse. However, molecular evidence indicates that the basic mechanism of photoreception is at least as old as the ancestor of all bilateral animals. If so, the different eyes we are about to study represent relatively recent evolutionary novelties superimposed on an ancient, homologous mechanism.

For each of the following senses in humans, identify the type of receptor: seeing, tasting, hearing, smelling.

Photoreceptor; chemoreceptor; mechanoreceptor; chemoreceptor

29.4 Three different types of eyes have evolved among invertebrates

Three main types of photoreceptors have evolved in the animal kingdom, and we find examples of all three among the invertebrates. The simplest type of photoreceptor is the **eye cup,** such as that of a planarian worm. As shown in Figure 29.4A, the eye cup is made up of a cuplike cluster of dark-colored cells that partially shield adjacent photoreceptor cells. The photoreceptor cells contain molecules of a substance called a visual pigment. These molecules absorb light and cause the light energy to be transduced. Action potentials are then sent through the photoreceptor cells to the brain. The photoreceptors do not provide data that the brain can use to form images. Rather, depending on the position of the animal and which part of the photoreceptors are shielded from the light, the cells simply detect the intensity and direction of light. Stimulated by action potentials from the photoreceptor cells, the animal's brain can determine which way the worm should crawl to escape bright light and find a dark hiding place.

Two other types of eyes have lenses that focus light and form images. A large number of invertebrates, including crayfish, crabs, and nearly all insects, have compound eyes. A **compound eye** consists of many tiny light-detecting units, called **ommatidia.** You can see some of the thousands of ommatidia (the tiny whitish dots) making up the two compound eyes of a fly in Figure 29.4B. Each ommatidium has its own cornea (a transparent covering) and a lens, which focuses light onto several photoreceptor cells. Every ommatidium picks up light from a tiny portion of the field of view. The animal's brain then forms a visual image by assembling the data from all the ommatidia.

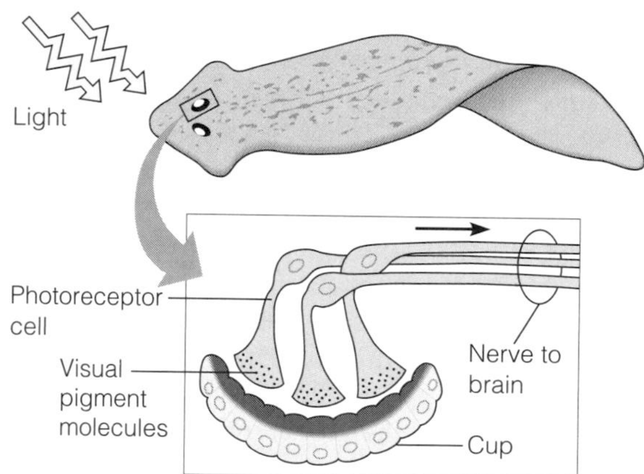

Figure 29.4A The eye cup of a planarian worm

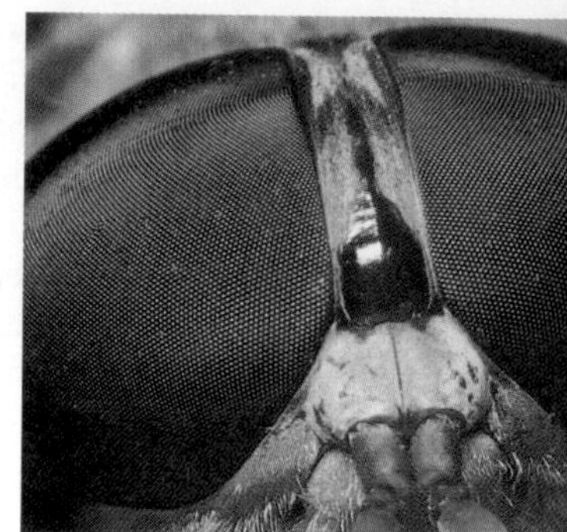

Figure 29.4B The two compound eyes of a fly

Compound eyes are extremely acute motion detectors, an important advantage for insects and other small animals that are often threatened by predators. The compound eyes of most insects also provide excellent color vision. Some species, such as honeybees, can see ultraviolet light (invisible to humans), which helps them locate certain nectar-bearing flowers.

Figure 29.4C The single-lens eye of a squid

The third type of eye, the **single-lens eye,** works on a principle similar to that of a camera. For example, the eye of a squid (Figure 29.4C) has a small opening, the pupil, through which light enters. Analogous to a camera's shutter, an adjustable iris changes the diameter of the pupil. Behind the pupil, a single lens focuses light onto the retina, which consists of many photoreceptor cells. A camera uses photosensitive film in place of a retina. We look at the single-lens eyes of vertebrates next. 

? What key optical feature is found in the eyes of both insects and squids but is not present in planarians?

Lenses, which focus light on the retina

29.5 Vertebrates have single-lens eyes

29A
See Activity 29A on the CD-ROM.

The vertebrate eye, including that of humans, is like the eye of a squid in that it has a single lens and is cameralike, but it evolved independently and differs in several details from any invertebrate photoreceptor. Our eyes are remarkable sense organs, able to detect a multitude of colors and respond to minute amounts of light energy.

The outer surface of the human eyeball is a tough, whitish layer of connective tissue called the **sclera** (Figure 29.5). The sclera surrounds a pigmented layer called the **choroid.** The transparent **cornea,** which lets light into the front of the eye, is part of the sclera. The cornea also helps focus light. Not shown in the figure, a thin mucous membrane called the **conjunctiva** helps keep the eye moist (see Module 20.4). The conjunctiva lines the inner surface of the eyelids and covers the front of the eye.

At the front of the eye, the choroid forms the **iris,** which gives the eye its color. Muscles in the iris regulate the size of the **pupil,** the opening in the center of the iris that lets light into the interior of the eye. After going through the pupil, light passes through the disklike **lens,** which is held in position by ligaments. As in the squid, the lens focuses images onto the **retina,** a layer just inside the choroid. Photoreceptor cells of the retina transduce light energy, and action potentials pass via sensory neurons in the optic nerve to the visual centers of the brain. Photoreceptor cells are highly concentrated at the retina's center of focus, called the **fovea.** Because there are no photoreceptor cells in the optic nerve, the place where the optic nerve passes through the back of the eye is a **blind spot.** We cannot detect light that is focused on the blind spot, but having two eyes with overlapping fields of view enables us to perceive uninterrupted images.

Two chambers make up the bulk of the eye. The large chamber behind the lens is filled with jellylike **vitreous humor.** The much smaller chamber in front of the lens is full of a liquid similar to blood plasma, called **aqueous humor.** The humors help maintain the shape of the eyeball. In addition, the aqueous humor circulates through its chamber. Secreted by capillaries, this fluid supplies nutrients and oxygen to the lens, iris, and cornea and carries off wastes. It reenters the blood through tiny ducts near the iris. Blockage of these ducts can cause glaucoma, increased pressure inside the eye that may lead to blindness by compressing the retina and optic nerve. Medication that increases the circulation of aqueous humor can help relieve glaucoma, if the condition is diagnosed early.

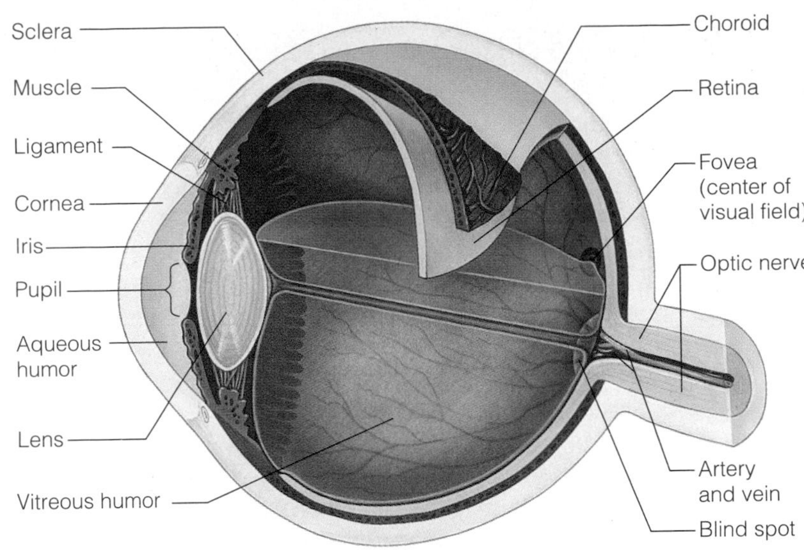

Figure 29.5 The single-lens eye of a vertebrate

? Arrange the following eye parts into the correct sequence encountered by photons of light traveling into the eye: pupil, retina, cornea, lens, vitreous humor, aqueous humor.

Cornea → aqueous humor → pupil → lens → vitreous humor → retina

29.6 To focus, a lens changes position or shape

29A

In almost every type of eye, the lens focuses light onto a retina by bending light rays. Focusing can occur in two ways. The lens may be rigid, as in many fishes, and focused by muscles that move it back or forth, as you might focus on an object using a magnifying glass. Or, as in the mammalian eye, the lens may be elastic and focused by muscles that change its shape. The thicker the lens, the more sharply it bends light.

The shape of the mammalian lens is controlled by the muscles attached to the choroid. When the eye focuses on a nearby object, these muscles contract. This makes the ligaments that suspend the lens slacken. With this reduced tension, the elastic lens becomes thicker and rounder, as shown in the left diagram below; this change is called **accommodation.** When the eye focuses on a distant object, the muscles controlling the lens relax, putting tension on the ligaments and flattening the lens, as shown in the right diagram.

? As you read this text, your lenses are relatively thick/thin (choose one).

Thick

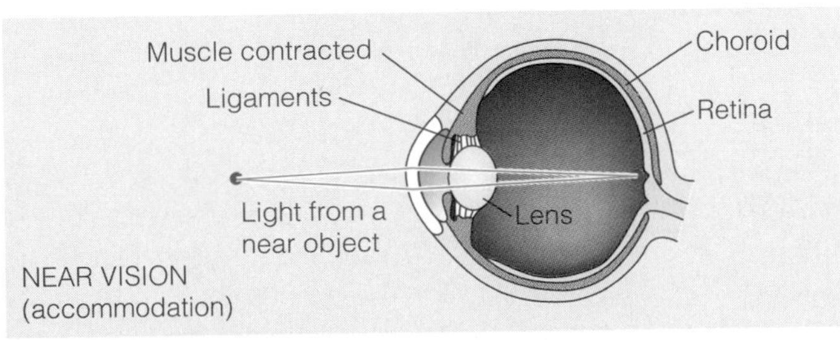

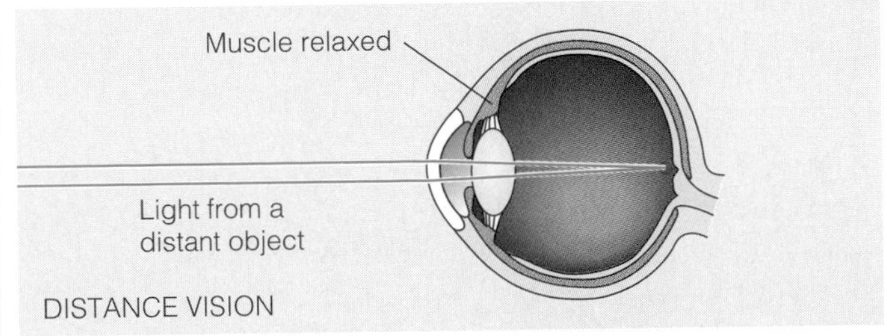

Figure 29.6 How lenses focus light

29.7 How do corrective lenses work?

When you have your vision tested, you are asked to read letters on a special chart. The chart measures your **visual acuity,** the ability of your eyes to distinguish fine detail. The examiner asks you to read a line of letters sized for legibility at a distance of 20 feet, using one eye at a time. If you can do this, you have so-called normal (20/20) acuity in each eye. This means that from a distance of 20 feet, each of your eyes can read the chart's line of letters designated for 20 feet.

Suppose you find out that your visual acuity is 20/10. This is actually better than normal; it means that you can read letters from a distance of 20 feet that a person with 20/20 vision can only read at 10 feet. On the other hand, someone with 20/50 acuity has worse than normal vision. He or she must stand at a distance of 20 feet to read what a person with normal acuity can read at 50 feet. Visual acuity tests can show that there is a vision problem, but they cannot determine the cause.

Three of the most common visual problems are nearsightedness, farsightedness, and astigmatism. All three are focusing problems, easily corrected with artificial lenses. Nearsighted people cannot focus well on distant objects, although they can see well at short distances. A nearsighted eyeball (Figure 29.7A) is longer than normal. The lens cannot flatten enough to compensate, and it focuses distant objects in front of the retina, instead of on it. As shown by the drawing on the right in Figure 29.7A, **nearsightedness** (also known as myopia) is corrected by glasses or contact lenses that are thinner in the middle than at the outside edge. The corrective lenses make the light rays from distant objects diverge slightly as they enter the eye. The focal point formed by the lens in the eye then falls directly on the retina.

Farsightedness (also known as hyperopia) is the opposite of nearsightedness. It occurs when the eyeball is shorter than normal, and the focal point of the lens is behind the retina (left drawing in Figure 29.7B). Farsighted people see distant objects normally, but they can't focus at short distances. Corrective lenses that are thicker in the middle than at the outside edge compensate for farsightedness by making light rays from nearby objects converge slightly before they enter the eye. Another type of farsightedness, called presbyopia (Greek for "old eye"), develops with age. Beginning around the mid-forties, the lens of the eye becomes less elastic. As a result, the lens gradually loses its ability to focus on nearby objects, and reading without glasses becomes difficult.

Astigmatism is blurred vision caused by a misshapen lens or cornea. Any such distortion makes light rays converge unevenly and not focus at any one point on the retina. Lenses that correct astigmatism are asymmetrical in a way that compensates for the asymmetry in the eye. 

? A person with 20/100 vision in both eyes must stand at ______ feet to read what someone with normal vision can read at ______ feet.

20 . . . 100

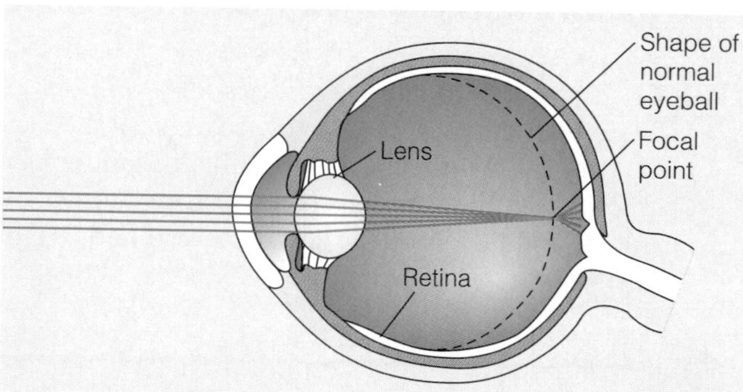

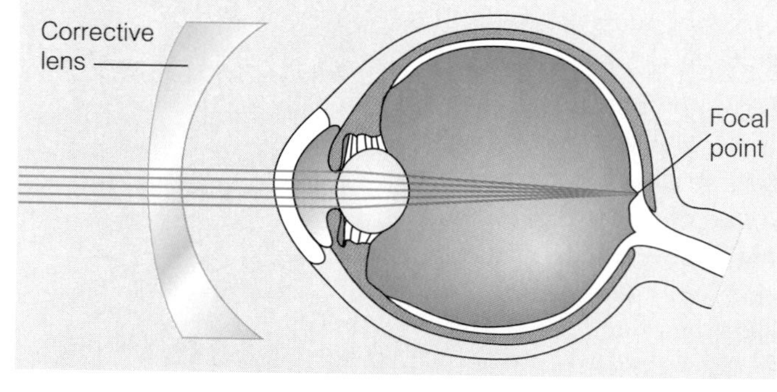

Figure 29.7A A nearsighted eye (eyeball too long)

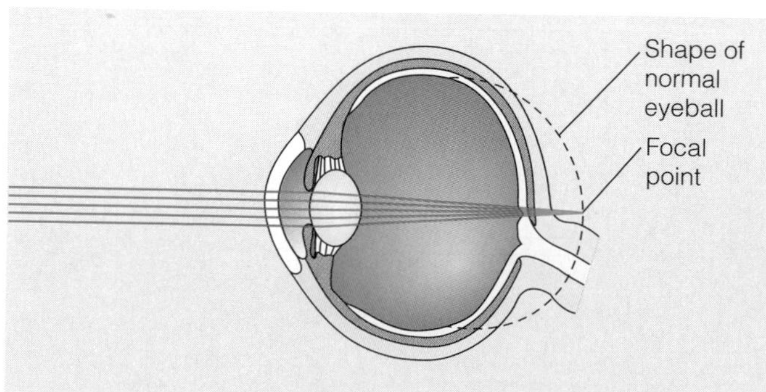

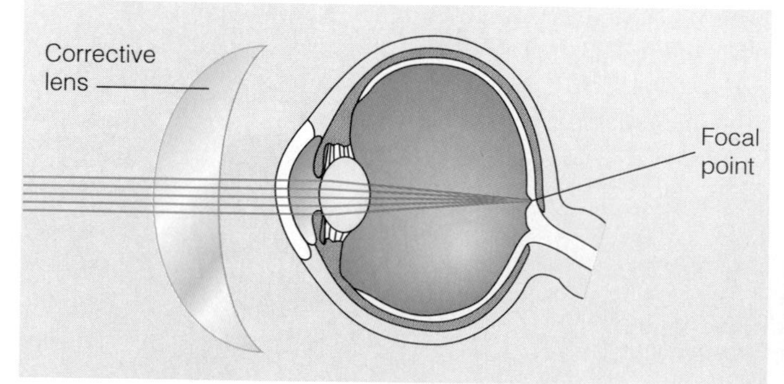

Figure 29.7B A farsighted eye (eyeball too short)

29.8 Our photoreceptor cells are rods and cones

Built into the human retina are about 130 million photoreceptor cells called rods and cones because of their shapes (Figure 29.8A). **Cones** are stimulated by bright light and can distinguish color, but they do not function in night vision. **Rods** are extremely sensitive to light and enable us to see in dim light at night, though only in shades of gray.

In humans, rods are found in greatest density at the outer edges of the retina, and are completely absent from the fovea, the retina's center of focus (Figure 29.8B). If you face directly toward a dim star in the night sky, the star is harder to see than if you look at it at an angle. Viewing it at an angle makes your lens focus the starlight onto the parts of the retina with the most rods. By contrast, you achieve your sharpest day vision by looking straight at the object of interest. This is because cones are densest (about 150,000 per mm^2) in the fovea. Some birds, such as hawks, have over ten times more cones in their foveas than we do, which enables them to spot small prey from high in the air.

How do rods and cones detect light? As Figure 29.8A shows, each rod and cone contains an array of membranous discs containing light-absorbing visual pigments. Rods contain a visual pigment called **rhodopsin,** which functions by absorbing dim light. Cones contain visual pigments called **photopsins,** which absorb bright, colored light. We have three types of cones, each containing a different type of photopsin. These cells are called blue cones, green cones, and yellow cones, referring to the colors absorbed best by their photopsin. All three types of cones actually absorb a wide range of colors, and together they can detect virtually any color in the visible spectrum (between ultraviolet and infrared). We can perceive a great number of colors because the light from each particular color triggers a unique pattern of stimulation in the millions of cones in our retina. Color blindness results from a deficiency in one or more types of cones.

Like all receptor cells, rods and cones are stimulus transducers. When rhodopsin and photopsin absorb light, they change chemically, and the change alters the permeability of the cell's membrane. The resulting receptor potentials trigger a complex integration process that actually begins in the retina. Notice in Figure 29.8B that the rods and cones have their tips embedded in the back of the retina (pink cells). Light must pass through several layers of neurons in the retina before reaching the pigments in the rods and cones. Visual information transduced by the rods and cones passes in the opposite direction (black arrows), from the photoreceptor cells through the network of neurons. Notice the numerous synapses between the photoreceptor cells and the neurons and among the neurons themselves. Integration in this maze of synapses helps sharpen images and increases the contrast between their light and dark areas. Action potentials carry the partly integrated information into the brain via the optic nerve. Three-dimensional perceptions of surroundings (what we actually see) result from further integration in several processing centers of the cerebral cortex.

Explain why our night vision is mostly in black-and-white rather than color.

Rods are more sensitive than cones to light, and thus the low light intensity at night stimulates far more rods than cones.

Figure 29.8A Photoreceptor cells

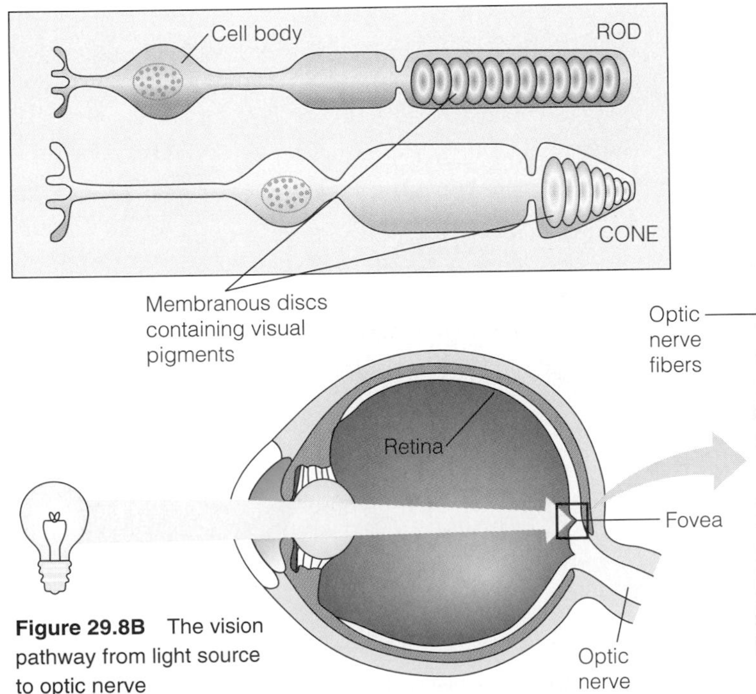

Figure 29.8B The vision pathway from light source to optic nerve

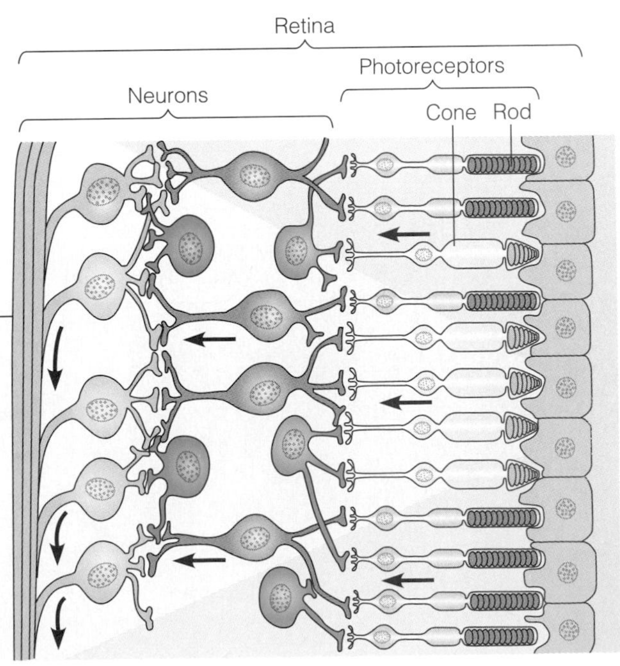

29.9 The ear converts air pressure waves into action potentials that are perceived as sound

The human ear is really two separate organs, one for hearing and the other for maintaining balance. We look at the structure and function of our hearing organ in this module and then turn to our sense of balance in Module 29.10. Both organs operate on the same basic principle, the stimulation of cilialike projections on hair cells in fluid-filled canals.

The ear is complex, and it helps to learn its basic structure before studying how it functions. The ear is composed of three regions: the outer ear, the middle ear, and the inner ear (Figure 29.9A). The **outer ear** consists of the flaplike **pinna**—the structure we commonly refer to as our "ear"—and the **auditory canal.** The pinna and the auditory canal collect sound waves and channel them to the **eardrum,** a sheet of tissue that separates the outer ear from the **middle ear** (Figure 29.9B). When sound waves strike the eardrum, it vibrates and passes the sound waves to three small bones: the hammer, anvil, and stirrup. From the stirrup, the vibrations pass into the **inner ear** through the **oval window,** a membrane-covered hole in the skull bone. The **Eustachian tube** conducts air between the middle ear and the back of the throat, ensuring that air pressure is kept equal on either side of the eardrum. The tube is what enables you to move air in or out to equalize pressure ("pop" your ears) when changing altitude rapidly in an airplane or car.

The inner ear consists of fluid-filled channels in the bones of the skull. Sound waves or movements of the head set the fluid in motion. One of the channels, the **cochlea** (Latin for snail), is a long, coiled tube that contains what is actually the hearing organ. The cross-sectional view of the cochlea in Figure 29.9C shows that inside it are three fluid-filled canals. Our hearing organ, the **organ of Corti,** is a long, thin spiral within the middle canal. The organ of Corti consists of an array of hair cells embedded in a **basilar membrane** (the floor of the middle canal). As you can see in the enlargement, a shelf of tissue projects over the hair cells from the wall of the middle canal. The hair cells are the receptor cells of the ear. Notice that they project into the fluid in the middle canal, and that the tips of most are in contact with the overlying shelf. Sensory neurons at the base of the hair cells carry action potentials from the organ of Corti into the brain via the auditory nerve.

Now let's see how the parts of the ear function in hearing. As indicated in Figure 29.9D, a vibrating object, such as a plucked guitar string, creates pressure waves in the surrounding air, represented by the up-and-down waves in the figure. Collected by the pinna and auditory canal of the outer ear, these waves make your eardrum vibrate with the same frequency as the sound. The frequency, measured in hertz (Hz), is the number of vibrations per second (1 Hz is equal to one vibration per second).

From the eardrum, the vibrations pass through the hammer, anvil, and stirrup in the middle ear. These small bones amplify the vibrations (heighten the sound waves) and transmit them to the oval window between the middle ear and inner ear. Vibrations of the oval window then produce pressure waves in the fluid within the cochlea. The vibrations first pass from the oval window into the fluid in the upper canal of the cochlea. Pressure waves travel through the upper canal to the tip of the cochlea, at the coil's center. The pressure waves then enter the lower canal and gradually fade away.

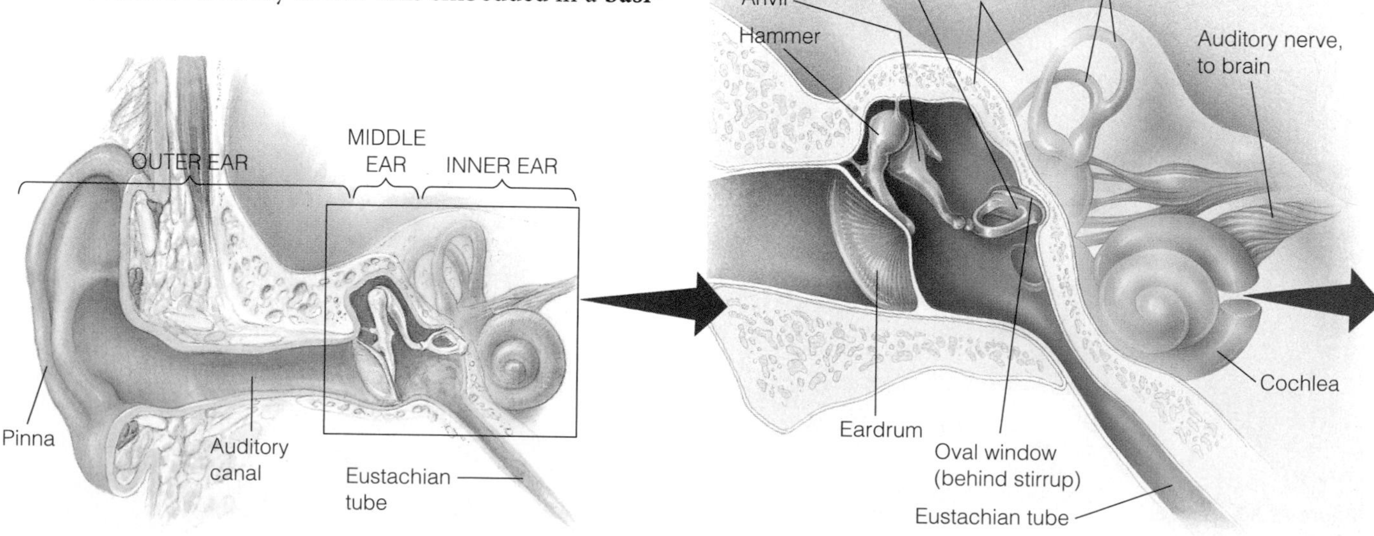

Figure 29.9A An overview of the human ear

Figure 29.9B The middle ear and the inner ear

As a pressure wave passes through the upper canal of the cochlea, it pushes downward on the middle canal, making the basilar membrane vibrate. Vibration of the basilar membrane makes the hairlike projections on the hair cells alternately brush against and draw away from the overlying shelf. When a hair cell's projections are bent, ion channels in its plasma membrane open, and positive ions (K^+ in this case) enter the cell. As a result, the hair cell develops a receptor potential and releases more neurotransmitter molecules at its synapse with a sensory neuron. In turn, the sensory neuron sends more action potentials to the brain.

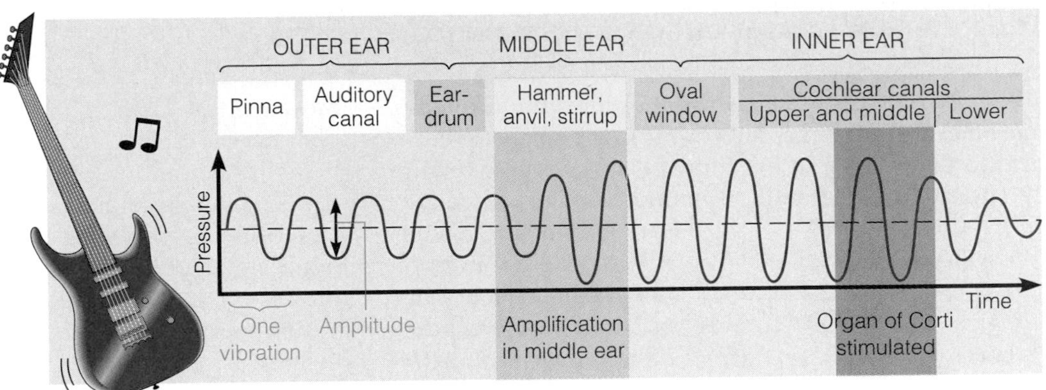

Figure 29.9D The route of sound waves through the ear

The brain perceives a sound as an increase in the frequency of action potentials it obtains from the auditory nerve. But how is the quality (volume and pitch) of the sound determined? The higher the volume (loudness) of sound, the higher the amplitude of the pressure wave it generates. In the ear, the higher the amplitude, the more vigorous the vibrations of fluid in the cochlea, the more pronounced the bending of the hair cells, and the more action potentials generated in the sensory neurons. The loudness of sound is measured in decibels (dB). The decibel scale for human hearing ranges from 0 to 120 dB, the loudest we can hear without intolerable pain.

The pitch of a sound depends on the frequency of the sound waves. High-pitched sounds, such as high notes sung by a soprano, generate high-frequency waves. Low-pitched sounds, like low notes sung by a bass, generate low-frequency waves. How does the cochlea distinguish sounds of different pitch? The key is that the basilar membrane is not uniform along its spiraling length. The end near the oval window is relatively narrow and stiff, while the other end, near the tip of the cochlea, is wider and more flexible. Each region of the basilar membrane is most sensitive to a particular frequency of vibration, and the region vibrating most vigorously at any instant sends the most action potentials to auditory centers in the brain. The brain interprets the information and gives us a sensation of pitch. Young people with healthy ears can hear pitches in the range of 20–20,000 Hz. Dogs can hear sounds as high as 40,000 Hz, and bats can emit and hear clicking sounds as high-pitched as 75,000 Hz. (A bat detects moths and other flying insects by sending out high-pitched clicks and plotting the position of its prey from the echoes.)

Few parts of our anatomy are more delicate than the organ of Corti. Quiet sounds make its hairs wave gently against the overlying shelf. But loud sounds can damage or destroy hair cells, causing hearing loss. Amplified rock music often exceeds 90 dB, an intensity that can cause hearing loss. 

? How does the ear convert sound waves in the air into pressure waves of the fluid in the cochlea?

Sound waves in air cause the eardrum to vibrate. The small bones attached to the inside of the eardrum amplify the movement and transmit it to the oval window on the wall of the inner ear. Vibrations of the oval window set in motion the fluid in the inner ear, which includes the cochlea.

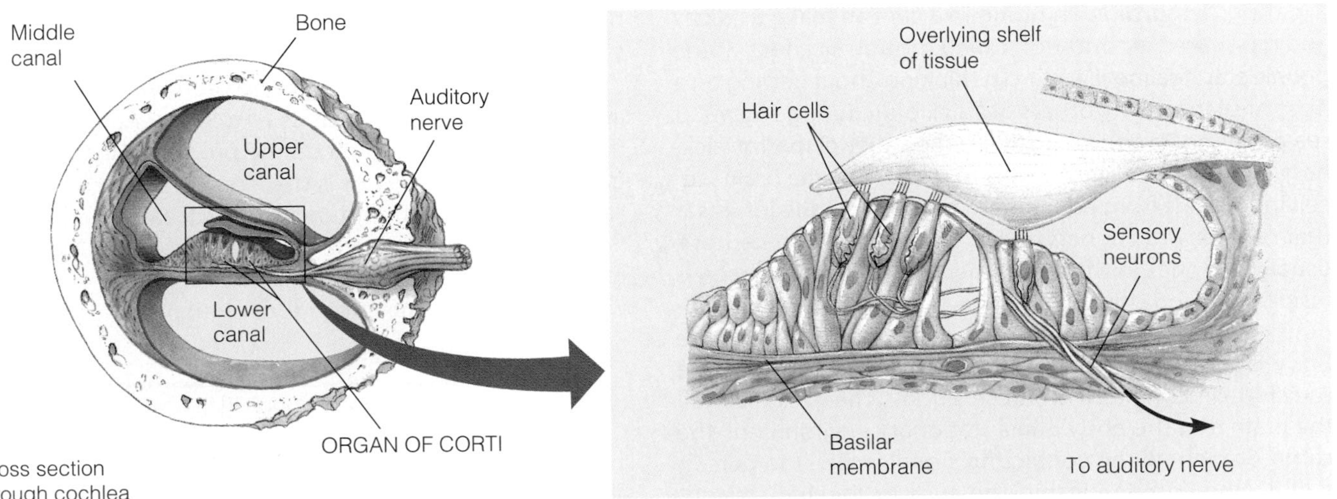

Figure 29.9C The organ of Corti, within the cochlea

29.10 The inner ear houses our organ of balance

We have two sets of balance, or equilibrium, receptors, one on each side of the skull, in the inner ear. Each set lies next to the cochlea in five fluid-filled structures (Figure 29.10): three semicircular canals and two chambers, the utricle and the saccule. All the equilibrium structures operate on the same principle, by the bending of hairs on hair cells.

The three **semicircular canals** detect changes in the head's rate of rotation or angular movement. As shown at the left, the canals are arranged in three perpendicular planes and can therefore detect movement in all directions. A swelling at the base of each semicircular canal contains a cluster of hair cells with their hairs projecting into a gelatinous mass called a cupula (shown in the enlargements). When you rotate your head in any direction, the thick, sticky fluid in the canals moves more slowly than your head. Consequently, the fluid presses against the cupula, bending the hairs. The faster you rotate your head, the greater the pressure and the higher the frequency of action potentials sent to the brain. If you rotate your head at a constant slow speed, the fluid in the canals can move with the head, and the pressure on the cupula is reduced. But if you rotate your head fast and then stop suddenly, the fluid continues to move after the head stops; as a result, you may feel dizzy.

Clusters of hair cells in the **utricle** and **saccule** detect the position of the head with respect to gravity. When you move your head, tiny limestone crystals in a gelatinous layer coating the hairs are pulled by gravity, and their movement bends the hair cells in a specific direction. This alters the rate at which action potentials are sent to the brain. The brain determines the new position of the head by interpreting the altered flow of action potentials.

The equilibrium receptors provide data the brain needs to determine the position and movement of the head. Using this information, the brain develops and sends out commands that make the skeletal muscles balance the body.

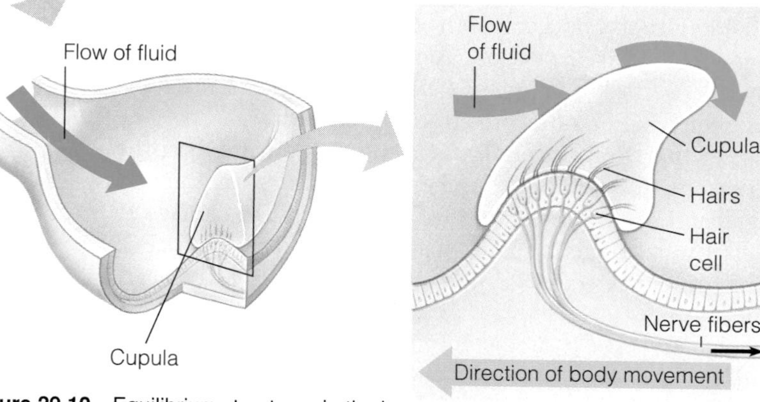

Figure 29.10 Equilibrium structures in the inner ear

What type of receptor cell is common to our senses of hearing and equilibrium?

The hair cells, or mechanoreceptors, of the inner ear

29.11 What causes motion sickness?

Boating, flying, or even riding in a car can make us dizzy and nauseated, a condition called motion sickness. Some people start feeling ill just from thinking about getting on a boat or plane. Many others get sick only during storms at sea or during "rough air" in flight. The cause of motion sickness is not known, but it seems to result from the brain's receiving signals from equilibrium receptors in the inner ear that conflict with signals from a different set of receptors, usually the ones in the eyes. When a susceptible person is inside a moving ship, for instance, signals from the equilibrium receptors in the inner ear indicate, correctly, that the body is moving (in relation to the environment outside the ship). In conflict with these signals, the eyes may tell the brain that the body is in a stationary environment, the cabin. Somehow the conflicting signals make the person feel ill. Sometimes, closing the eyes or looking straight ahead relieves symptoms. Many sufferers of motion sickness take a sedative such as Dramamine® to relieve their symptoms.

Motion sickness can be a severe problem for astronauts, and the National Aeronautics and Space Administration (NASA) conducts research on the problem. One of NASA's most interesting findings is that some people can learn to consciously control body functions, such as the vomiting reflex, that are ordinarily under involuntary control. Astronauts receive intensive training in how to exert "mind over body" when zero gravity starts to induce motion sickness.

Using the hypothesis put forth in this module, explain how someone could suffer motion sickness when watching a film shot from the front of a roller coaster.

There would be conflicting information between vision ("I'm moving") and the equilibrium sense ("I'm sitting still in my theater seat").

Taste and Smell

Odor and taste receptors detect categories of chemicals 29.12

Our senses of smell and taste depend on receptor cells that detect chemicals in the environment. Chemoreceptor cells in our nose detect airborne molecules; those in our taste buds detect molecules in food. In both cases, a cell responds to a group of chemically related molecules, not just to one kind of molecule. In the nose, for example, each type of receptor cell may detect one of about fifty general types of odor (such as spicy, musky, or putrid). Research indicates that a particular odor triggers a specific level of stimulation in the receptor cells. The brain perceives the odor of cinnamon, for instance, when it receives a specific pattern of action potentials from the spice receptors; it perceives the odor of cloves when it receives another pattern from the same receptor cells.

Figure 29.12A illustrates the mechanics of smell in a human. The olfactory (smell) receptor cells are sensory neurons that extend into the upper portion of our nasal cavity. Notice the cilia extending from the tips of the receptor cells into the mucus that coats the nasal cavity. When you smell an odor, molecules (shown as blue dots in Figure 29.12A) have entered your nose, dissolved in the mucus, and bound to receptor molecules on the cilia. The binding triggers receptor potentials, which alter the rate of action potentials passing into the brain. Integration of the signals in the brain results in an odor perception.

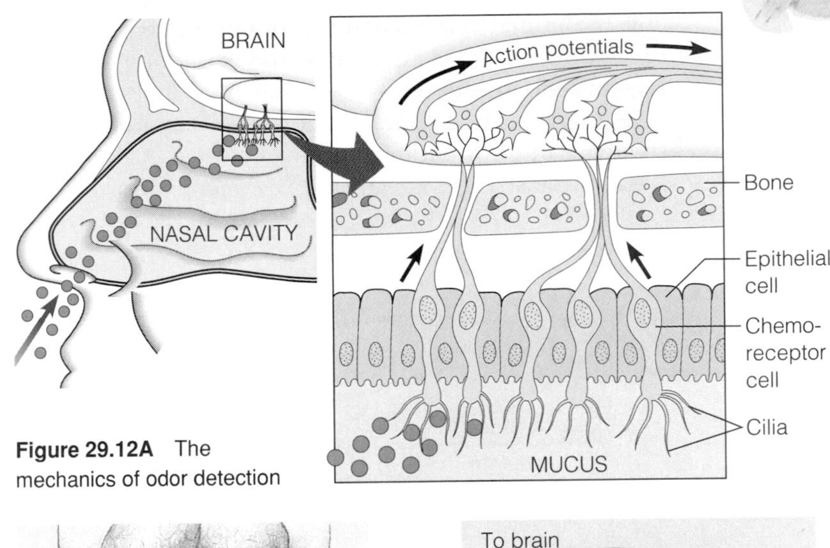

Figure 29.12A The mechanics of odor detection

Many animals, such as the salmon and grizzly bear in the chapter's opening essay, rely heavily on their sense of smell for survival. Odors often provide more information than visual images about food, the presence of mates, or danger. In contrast, humans often pay more attention to sights and sounds than to smells. The sense of smell was more important to our ancestors, whose survival may have depended on the ability to smell prey, edible plants, and danger such as wildfires.

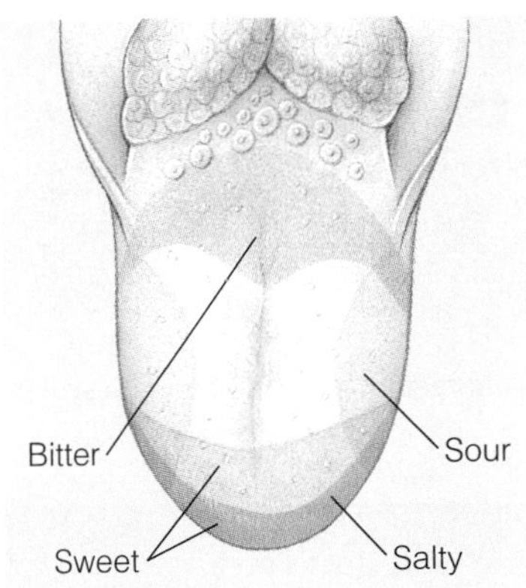

Figure 29.12B Taste buds on the human tongue

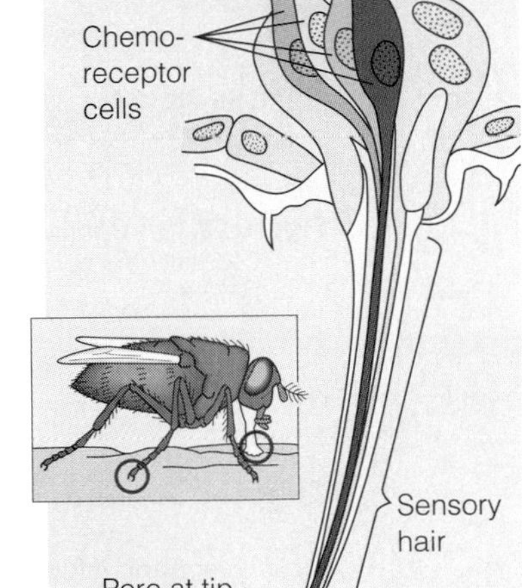

Figure 29.12C Taste in a fly

Our sense of taste depends on taste buds on the tongue (see Module 29.2). We have four types of taste buds, detecting the categories sweet, sour, salty, and bitter, and they are arranged on the tongue as in Figure 29.12B.

Each type of taste receptor cell in a taste bud can be stimulated by a broad range of chemicals in its category. Sweetness receptors, for example, detect several kinds of sugars, as well as some amino acids, peptides, and proteins. The artificial sweetener aspartame (sold under the trade names Equal® and NutraSweet®) is a dipeptide. When you taste something, your brain receives a variety of taste inputs, and the flavors you perceive usually result from a combination of the four taste categories in varying proportions.

Imagine tasting with your hands and feet, instead of with your tongue. Insects do just that. They have chemoreceptors in sensory hairs on their feet and can taste food simply by stepping in it. Some also have liplike mouthparts covered with sensory hairs. As shown in Figure 29.12C, a fly's sensory hairs each contain four chemoreceptor cells that extend to a pore. Like the cells in our taste buds, each of the fly's taste cells detects a category of chemicals and responds to a broad range of them. The fly's brain probably receives signals from two or more types of receptor cells for any food the insect touches.

Probably all animals can detect certain chemicals, and evidence indicates that in most—if not all—cases, receptor cells respond to a broad range of chemicals. 

? Both odor and taste depend on sensory receptors classified as __________ receptors. Binding of specific molecules to the membrane of one these cells can cause a receptor __________. This results in information traveling along __________ neurons to the brain in the form of __________ potentials.

chemo · · · potential · · · sensory · · · action

29.13 Review: The central nervous system couples stimulus with response

In this chapter and the previous one, we have focused on information gathering and processing. Sensory receptors provide an animal's nervous system with vital data that enable the animal to avoid danger, communicate with others of its kind, find food and mates, and maintain homeostasis—in short, to survive.

A grizzly bear catching a salmon helps us summarize the sequence of information flow in an animal. A bear sees a flash in the stream. Within milliseconds, photoreceptor cells in the bear's retinas transduce the light energy focused on them by the lens, and action potentials representing a glimpse of the salmon enter the brain. As the salmon swims close, movement of the water stimulates touch receptors at the base of the bear's whiskers. Action potentials triggered by these receptors reach the brain at the same time as those from the eyes. Before the salmon can swim out of reach, a vast network of neurons in the grizzly's brain, with thousands of synapses, integrates all the information and sends out command signals, again in the form of action potentials. The commands go out via motor neurons to muscles in the bear's paws, neck, and jaws. The bear lunges and grabs its meal.

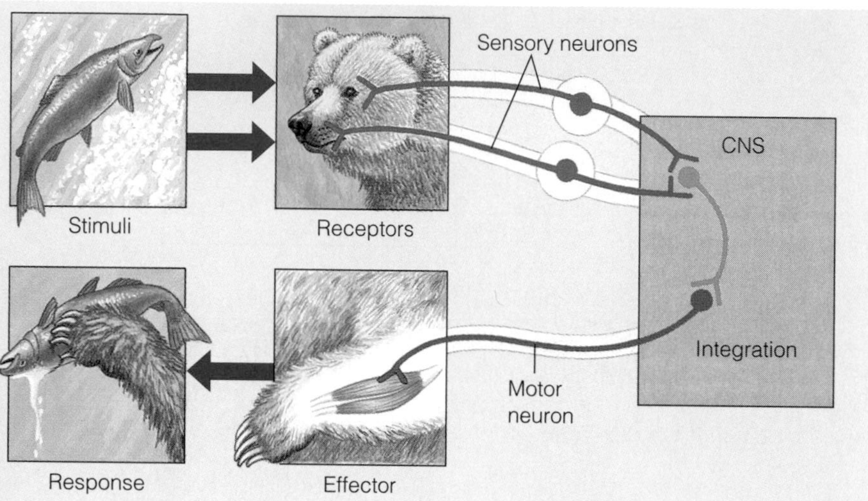

Figure 29.13 Coupling of stimuli to response by the nervous system

The nervous system links stimulus reception with response. It takes in information coded as action potentials, integrates it, plans a response, and sends out action potentials commanding an appropriate action. In doing so, the nervous system couples the various forms of stimulus signals to body response—in the case of the bear, to split-second muscle contractions. In the next chapter, we see how muscles carry out the commands they receive.

What type of neuron is symbolized by the green cell in Figure 29.13? (*Hint:* Review Module 28.2.)

Interneuron

CHAPTER REVIEW

Chapter Summary

Sensory receptors inform an animal about conditions inside and outside its body **(Introduction)**. The awareness of sensory stimuli is called sensation. Perception is the brain's full integration of sensory data **(29.1)**.

Sensory Reception (29.2–29.3) The process of sensory transduction is the conversion by sensory receptors of stimuli into electrical signals called receptor potentials. Action potentials representing the stimuli are transmitted via sensory neurons to the central nervous system for processing. The brain distinguishes different types of stimuli. The strength of the stimulus alters the rate of action potential transmission. Sensory neurons tend to become less sensitive when stimulated repeatedly, a phenomenon known as sensory adaptation **(29.2)**. There are five categories of sensory receptors. Pain receptors sense dangerous stimuli. Thermoreceptors detect heat or cold. Mechanoreceptors respond to mechanical energy (such as touch, pressure, and sound), and chemoreceptors to chemicals in the external environment or body fluids. Electromagnetic receptors respond to electricity, magnetism, and light. Photoreceptors, which sense light, are the most common type of electromagnetic receptor **(29.3)**.

Vision (29.4–29.8) The photoreceptors of flatworms are simple eye cups that sense the intensity and direction of light. The compound eyes of insects consist of many lenses that together produce a visual image. Vertebrates and some invertebrates, such as squids, have single-lens eyes **(29.4)**. In the human eye, the cornea, lens, and fluid in the eyeball focus light on photoreceptor cells in the retina. Having two eyes compensates for our blind spot, the place where the optic nerve passes through the retina **(29.5)**. The human lens changes shape to bring objects at different distances into sharp focus **(29.6)**. Nearsightedness results when the eyeball is longer than normal, and the lens cannot flatten enough to compensate. Farsightedness occurs when the eyeball is shorter than normal, and the focal point is behind the retina **(29.7)**. Photoreceptor cells called rods contain the visual pigment rhodopsin and function in dim light. Cones are photoreceptor cells that contain photopsin, which enables us to see color in full light **(29.8)**.

Hearing and Balance (29.9–29.11) The human ear functions in hearing and balance. The outer ear channels sound waves to the eardrum, which passes the vibrations to a chain of bones in the middle ear. The bones transmit the vibrations to fluid in the cochlea, a coiled tube in the inner ear. The cochlea houses the hearing organ, called the organ of Corti. The waves generated in the cochlear fluid move hair cells (mechanoreceptors) of the organ of Corti against an overlying shelf of tissue. Movement of the hair cells triggers nerve signals to the brain. Louder sounds cause greater movement and more action potentials; sounds of different pitches stimulate hair cells in different parts of the cochlea (29.9). The organ of balance consists of three semicircular canals and two chambers, the utricle and saccule, near the cochlea in the inner ear. The utricle and saccule contain clusters of hair cells that are bent by gravity. Fluid in the semicircular canals also bends hair cells when the head moves. Changes in neural output from these hair cells enable the brain to sense body position and movement (29.10). Conflicting signals from the inner ear and eyes may cause motion sickness (29.11).

Taste and Smell (29.12) The senses of smell and taste depend on chemoreceptors, which send nerve signals to the brain when specific molecules bind to them. In humans, olfactory (smell) receptors line the upper part of the nasal cavity. Taste receptors are in taste buds on the tongue. Olfactory receptors respond to chemicals that produce many different smell sensations, but there are only four basic types of taste receptors, responding to many combinations of chemicals that taste sweet, sour, salty, and bitter.

Review: The Connection Between Sensory Reception and Behavioral Response (29.13) The nervous system obtains information from the environment, integrates it, plans a response, and commands appropriate body actions.

Testing Your Knowledge

Multiple Choice

1. Mr. Johnson was becoming slightly deaf. To test his hearing, his doctor held a vibrating tuning fork tightly against the back of Mr. Johnson's skull. This sent vibrations through the bones of the skull, setting the fluid in the cochlea in motion. Mr. Johnson could hear the tuning fork this way, but not when it was held away from the skull a few inches from his ear. The problem was probably in the ______. *(Explain your answer.)*
   a. auditory center in Mr. Johnson's brain
   b. auditory nerve leading to the brain
   c. hair cells in the cochlea
   d. bones of the middle ear
   e. fluid of the cochlea

2. Which of the following correctly traces the path of light into your eye?
   a. lens, cornea, pupil, retina
   b. cornea, pupil, lens, retina
   c. cornea, lens, pupil, retina
   d. lens, pupil, cornea, retina
   e. pupil, cornea, lens, retina

3. If you look away from this book and focus your eyes on a distant object, the eye muscles ______ and the lenses ______ to focus images on the retinas.
   a. relax . . . flatten
   b. relax . . . become more rounded
   c. contract . . . flatten
   d. contract . . . become more rounded
   e. contract . . . relax

4. Which of the following receptors are *not* present in human skin?
   a. thermoreceptors
   b. chemoreceptors
   c. touch receptors
   d. pressure receptors
   e. pain receptors

5. Jim had his eyes tested and found that he has 20/40 vision. This means
   a. the muscles in his iris accommodate too slowly
   b. he is farsighted
   c. the vision in his left eye is normal, but his right eye is defective
   d. he can see at 40 feet what a person with normal vision can see at 20 feet
   e. he can see at 20 feet what a person with normal vision can see at 40 feet

Describing, Comparing, and Explaining

1. Listen for a moment to the sounds around you. How are the sound waves converted into action potentials in your ears? How does your brain determine the volume and pitch of the sounds?

2. As you read these words, the lenses of your eyes project patterns of light representing the letters onto your retinas. There the photoreceptors respond to the patterns of light and dark, and transmit nerve signals to the brain. The brain then interprets the words. In this example, what is the difference between sensation and perception?

Thinking as a Scientist

1. Sensory organs tend to come in pairs. We have two eyes and two ears. Similarly, a planarian worm has two eye cups, a rattlesnake has two infrared receptors, and a butterfly has two antennae. Propose a testable hypothesis that could explain the advantage of having two ears or eyes instead of one.

2. Sea turtles bury their eggs on the beach above the high-tide line. When the baby turtles hatch, they dig their way to the surface of the sand and quickly head straight for the water. How do you think the turtles know which way to go? Outline an experiment to test your hypothesis.

3. People with a rare condition called synesthesia interpret one kind of stimulus as another. For example, they might feel the shapes of certain objects when tasting certain kinds of food. The cause of synesthesia is unknown. Suppose a person saw a series of bright lights in the air when a telephone rang and an undulating chain of light when a kitten purred. Which of the following hypotheses would you choose to test as the most likely explanation for this person's synesthesia? Explain your answer, and briefly tell why you rejected the other choices.
   a. The person has rods and cones connected to the auditory centers in the brain.
   b. The person has defective organs of Corti.
   c. The person's auditory neurons cannot produce action potentials.
   d. The person has some sensory neurons connected to the wrong brain centers.
   e. Receptor potentials instead of action potentials are entering the person's central nervous system.

Science, Technology, and Society

Have you ever felt your ears ringing after listening to loud music from a stereo or at a concert? Can this music be loud enough to permanently impair your hearing? Do you think people are aware of the possible danger of prolonged exposure to loud music? Should anything be done to warn or protect them? If you think so, what action would you suggest? What effect might warnings have?

Media Resources

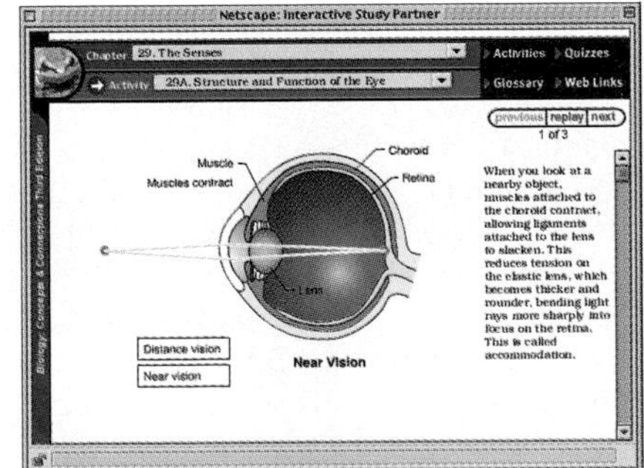

For further review of the concepts in this chapter, use the activities and quizzes on the CD-ROM.

Visit the Special Edition of The Biology Place web site for activities, quizzes, news articles, and web links.

30 How Animals Move

Outline

Ants rival humans in living just about any place on land except where there is permanent snow cover. Literally trillions of these insects walk the planet, and they are almost always on the move, engaged in some activity. Strenuous labor and almost constant body movement are two of their hallmarks.

The ants in this photograph are leaf-cutters that live in huge underground nests in tropical rain forests. A colony of these insects, often well over a million individuals, can strip a large tree of its foliage in a single night. The ants shown here are part of a "leaf parade" of workers—all females—returning home after a successful nocturnal raid high in the trees. Up on a high branch, the ant at the bottom of the photograph used her razor-sharp mouthparts to slice up a leaf. She then picked up a piece and joined the parade headed back to the nest. The other two smaller ants riding on the leaf fragment are workers that guard the leaf-carriers. They, too, are active, gnashing their mouthparts at parasitic flies that try to lay eggs on the load-bearers.

Adult leaf-cutter ants eat mostly plant sap, but their young eat fungi that grow on the leaves the adults cut. A leaf-cutter nest, often 4–5 meters deep and 7–8 meters in diameter at the soil surface, is a subterranean farm—a maze of tunnels and chambers where the ants cultivate a particular species of fungus for food. Returning from a leaf-cutting trip, the loaded workers descend into a tunnel and drop their cuttings off in a brood chamber. Other workers chew up the leaves and start new fungal growth by placing bits of live fungus on them. Still other members of the colony "weed" undesirable fungi out of the fungus gardens, cover the fungal growths with chemicals that inhibit competing fungi, harvest the food, feed it to the young, and carry refuse out of the nest. The work is nonstop.

Over a period of 4–5 years, a colony of leaf-cutters may excavate nearly 50 tons of forest soil in constructing and renovating a nest. It will also cut, haul, and process many tons of forest leaves and pile up large refuse heaps in keeping the nest clean. Constant activity in service of the colony is the life of every worker ant, and the ant body is a model of strength and mobility. Typical of insects, an ant's skeleton is a stiff outer coat called an exoskeleton. The insect exoskeleton is made of chitin, an unusually strong, durable polysaccharide. Chitin threads are embedded in a matrix of protein, forming a material analogous to fiberglass.

As shown below, each of the six legs of an ant consists of several exoskeletal pieces held together at flexible joints. The joints give the ant its mobility. Each exoskeletal leg piece is a tube that is virtually unbreakable by any force the insect confronts in hauling a bit of soil or a piece of leaf. The muscles in an ant's legs are tiny but powerful, and because the load of an ant's body is distributed over six legs, each leg uses only a tiny fraction of its muscle power in supporting the ant's own weight. The rest is used to power movements and perform useful work.

Movement is one of the most distinctive features of animals. Whether an animal walks or runs on two legs or six, swims, crawls, flies, or sits in one place and only moves its mouthparts, the interplay of three organ systems provides its movement. The nervous system plays a key role in issuing commands to the muscular system. The muscular system exerts the force that actually makes an animal or its parts move. The skeletal system is essential because the force exerted by muscles produces movement only when the force is applied against a firm structure, the skeleton. In this chapter, we focus on skeletons, muscles, and the movement their interactions produce.

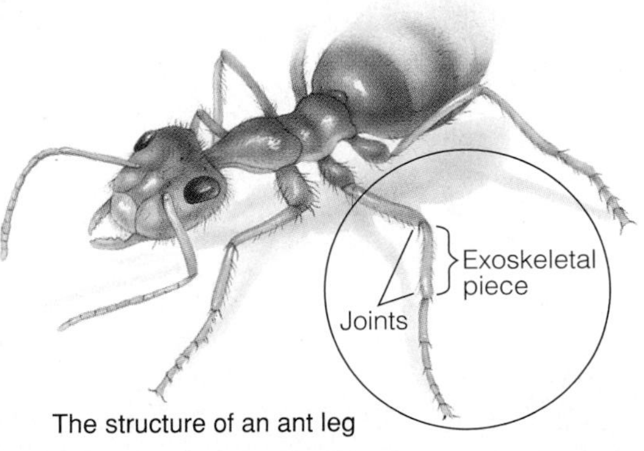

The structure of an ant leg

30.1 Diverse means of animal locomotion have evolved

Animal movement is extremely diverse. Many animals stay in one place and move only certain parts of their bodies. For example, an adult sponge's only movements are the opening and closing of cellular pores on its surface and the beating of flagella, drawing suspended food particles in through the pores. Most animals exhibit more movement than sponges, and animals that move about in search of food may spend much of their time and energy doing so. Active travel from place to place, also called **locomotion,** is our focus in this chapter. Locomotion in all its forms requires that an animal expend energy to overcome two forces that tend to keep it stationary: friction and gravity. The relative importance of these two forces varies, depending on the environment.

Swimming Gravity is not much of a problem for a swimming animal, because water supports much or all of the animal's weight. On the other hand, overcoming friction is more difficult for a swimmer, because water is dense and offers considerable resistance to a body moving through it.

Many different modes of swimming have evolved. Many insects, for example, swim the way we do, using their legs as oars to push against the water. Squids and some jellies are jet-propelled, taking in water and squirting it out in bursts. Fishes swim by moving their body and tail from side to side (Figure 30.1A). Whales and other aquatic mammals move their body and tail from top to bottom. A sleek, streamlined shape, like that of many fishes, is an adaptation that aids rapid swimming.

Locomotion on Land: Hopping, Walking, Running, and Crawling The problems of locomotion on land are more or less the opposite of those in the water. Air offers very little resistance to an animal moving through it. However, air provides little support for an animal's body, and a land animal must be able to support itself and overcome the force of gravity. When a land animal walks, runs, or hops, its leg muscles expend energy both to propel it and to keep it from falling down. To move on land, powerful muscles and strong skeletal support are more important than a streamlined shape.

The kangaroo travels mainly by hopping (Figure 30.1B). Large muscles in its hind legs generate a lot of power. Muscles and tendons (which connect muscle to bone) in the legs also momentarily store energy when the kangaroo lands—somewhat like the spring on a pogo stick. The higher the jump, the tighter the spring coils when a pogo stick lands, and the greater the tension in the muscles and tendons when a kangaroo lands. In both cases, the stored energy is available for the next jump. For the kangaroo, the tension in its legs is a cost-free energy boost that reduces the total amount of energy the animal expends to travel. The pogo stick analogy applies to hundreds of thousands of animal species. The legs of an ant or a dog, for instance, retain some spring during walking or running, although less than those of a hopping kangaroo.

The kangaroo illustrates a solution to another problem of terrestrial life: maintaining balance. The animal can sit upright with relatively little energy cost because it has three parts of its body touching the ground simultaneously, its two hind legs and the base of its tail. Similar to a camera tripod, this arrangement stabilizes the upright body. The same principle applies to the dog in Figure 30.1C. When walking, a dog keeps three feet on the ground at all times (in this case, both hind feet and the right front foot). Bipedal (two-footed) animals such as birds and humans are less stable on land, but keep part of at least one foot on the ground when walking. When an animal is running, all of its feet may be off the ground momentarily. At running

Figure 30.1A A fish swimming

Figure 30.1B Kangaroos, hopping and at rest

Figure 30.1C A dog walking

speeds, momentum, more than foot contact, stabilizes the body's position, just as a moving bicycle stays upright.

A crawling animal such as a snake or an earthworm faces very different problems. Because much of the animal's body is in contact with the ground, friction offers considerable resistance to movement. Many snakes crawl rapidly by undulating the entire body from side to side. In the process, the snake's body pushes against the ground, and this drives it forward. Snakes also have large, movable scales on their underside that assist in locomotion. Boa constrictors and pythons, for instance, creep forward in a straight line, driven by the leglike action of their belly scales. Muscles lift the scales away from the ground, tilt them forward, and then push them backward against the ground. The backward push drives the snake forward.

Earthworms crawl by peristalsis, a type of movement produced by rhythmic waves of muscle contractions passing from head to tail. (In Module 21.7, we saw how peristalsis squeezes food through our digestive tract.) To move by peristalsis, an animal needs a set of muscles that elongates the body, another set that shortens it, and a way to anchor itself to the ground. As illustrated in Figure 30.1D, the contraction of circular muscles constricts and elongates certain regions (groups of body segments) of a crawling earthworm, while longitudinal muscles shorten and thicken other regions. In position ①, segments at the head of the worm are short and thick (longitudinal muscles contracted) and anchored to the ground by bristles. Just behind the head, a group of segments is thin and elongated (circular muscles contracted), with bristles held away from the ground. In position ②, the head has moved forward because circular muscles in the head segments have contracted. Segments just behind the head and near the tail are now thick and anchored, thus preventing the head from slipping backward. In position ③, the head segments are thick again and anchored to the ground in their new position, well ahead of their starting point. The rear segments of the worm now release their hold on the ground and are pulled forward.

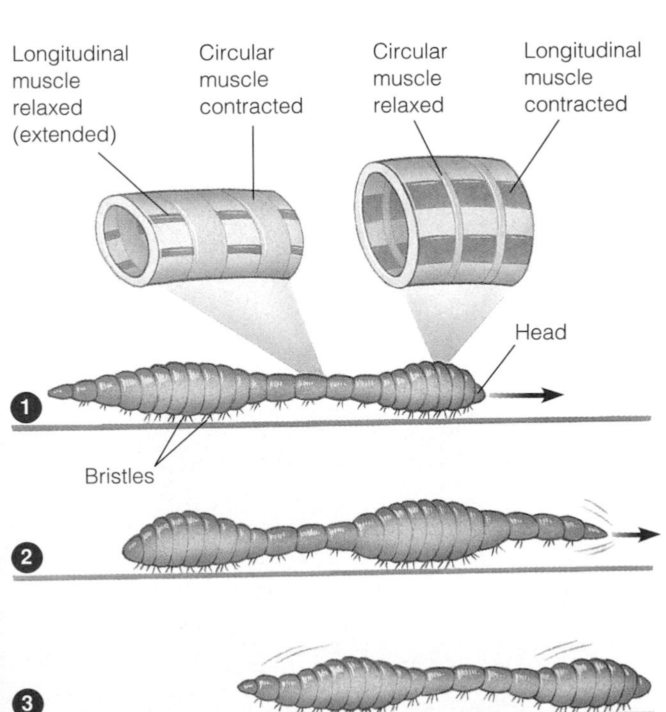

Figure 30.1D An earthworm crawling, by peristalsis

Figure 30.1E A bald eagle flying

Flying Many phyla of animals include species that crawl, walk, or run, and almost all phyla include swimmers. But flying has evolved in only a few animal groups: insects, reptiles, birds, and, among the mammals, bats. A large group of flying reptiles died out millions of years ago, leaving birds and bats as the only flying vertebrates.

For an animal to become airborne, its wings must develop enough lift to completely overcome the downward pull of gravity. The key to flight is in the shape of wings. All types of wings, including those of airplanes, are airfoils—structures whose shape alters air currents in a way that creates lift. As Figure 30.1E shows, an airfoil has a leading edge that is thicker than the trailing edge. It also has an upper surface that is somewhat convex and a lower surface that is flattened or concave. This shape makes the air passing over the wing travel farther than the air passing under the wing. As a result, air molecules are spaced farther apart above the wing than under it. The air pressure underneath is therefore greater, and this greater pressure lifts the wing.

All types of animal movement, including the diverse forms of locomotion, have certain underlying similarities. At the cellular level, every form of movement is based on one of two basic contractile systems, microtubules and microfilaments, both of which require energy-consuming cellular work to move protein strands against one another. The movements of cilia and flagella result from the bending of microtubules, as we discussed in Module 4.18. Microfilaments play a major role in amoeboid movement and are the contractile elements of muscle cells. Later in this chapter, we will look at the contraction of muscles and how it translates into movement when the muscles work against a firm skeleton. First, let's look at skeletons. 

? Contrast swimming with walking in terms of the forces an animal must overcome to move.

Friction resists an animal moving through water, but gravity has little effect because of the animal's buoyancy; air poses little resistance to an animal walking on land, but the animal must support itself against the force of gravity.

30.2 Skeletons function in support, movement, and protection

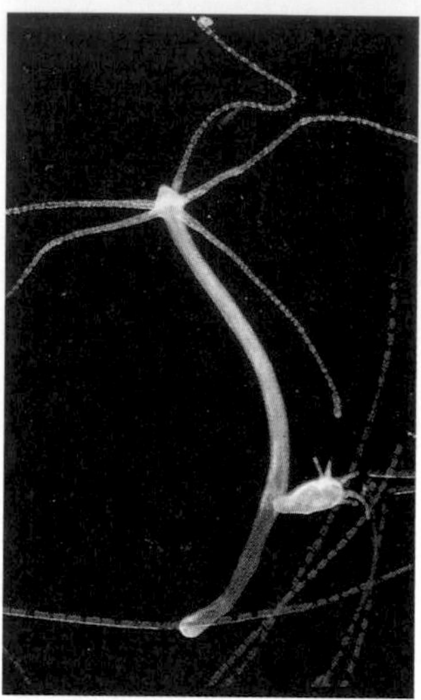

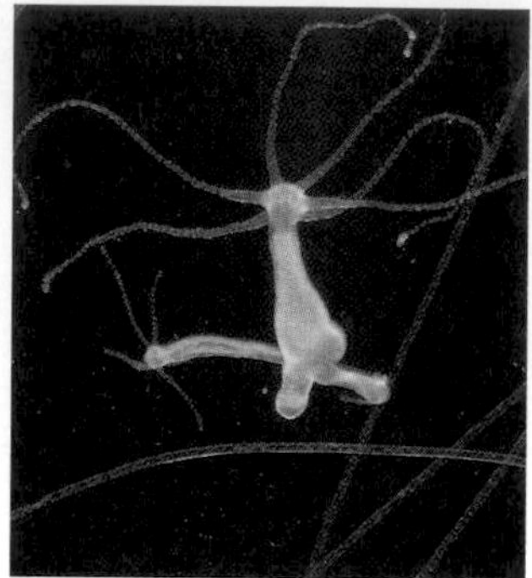

Figure 30.2A The hydrostatic skeleton of a hydra in two states

A skeleton has many functions. An animal could not move without its skeleton, and most land animals would sag from their own weight if they had no skeleton to support them. Even an animal in water would be a formless mass with no skeletal framework to maintain its shape. Skeletons also may protect an animal's soft parts. For example, the vertebrate skull protects the brain, and the ribs form a cage around the heart and lungs. There are three main types of skeletons: hydrostatic skeletons, exoskeletons, and endoskeletons. All three types have multiple functions.

A **hydrostatic skeleton** consists of fluid held under pressure in a closed body compartment. This is very different from the more familiar skeletons made of hard materials. Nonetheless, a hydrostatic skeleton helps protect other body parts, cushioning them from shocks. It also gives the body shape and provides support for muscle action, as does a hard skeleton.

Earthworms have a fluid-filled internal cavity—their body cavity, or coelom (see Module 18.7). As a segmented animal, the earthworm has its coelom divided into separate compartments. The fluid in this compartmented coelom functions as a hydrostatic skeleton, and the action of circular and longitudinal muscles working against the hydrostatic skeleton produces the peristaltic movement described in Module 30.1.

Cnidarians, such as hydras and jellies, also have a hydrostatic skeleton. A hydra, for example, holds fluid in its gastrovascular cavity and can alter its body shape drastically by contracting muscles in its body wall. When a hydra closes its mouth and the muscles encircling its body wall constrict, it elongates and its tentacles extend, as shown on the left in Figure 30.2A. Because the muscles cannot compress the water in the gastrovascular cavity, constricting the cavity forces the water to elongate the animal, somewhat like squeezing a water-filled balloon. A hydra often sits in this position for hours, waiting for prey to swim by. If it is disturbed, its mouth opens, allowing water to flow out, and longitudinal muscles in its body wall contract, shortening the body (Figure 30.2A, right photograph).

Most animals with hydrostatic skeletons are soft and flexible. A hydra can extend its body and spread out its tentacles, as well as expand its body around ingested prey. An earthworm can burrow through soil because it is flexible and has a hydrostatic skeleton. Similarly, having an expandable body and a hydrostatic skeleton enables many tube-dwelling animals to extend out of their tubes for feeding and gas exchange, and then quickly squeeze back into their tubes when threatened (see Figure 18.11B). Hydrostatic skeletons work well for many aquatic animals and for terrestrial animals that crawl or burrow by peristalsis. However, a hydrostatic skeleton cannot support the forms of terrestrial locomotion in which an animal's body is held off the ground, such as walking.

A great variety of aquatic and terrestrial animals have a rigid external skeleton, or **exoskeleton.** In insects and other arthropods (see Module 18.12), muscles attached to knobs and plates on the inner surfaces of the exoskeleton move the jointed body parts. At the joints of legs, the skeleton is thin and flexible, making possible a wide variety of body movements.

The armorlike protection, support, and flexibility of the exoskeleton have helped ensure the evolutionary success of the arthropods. The arthropod exoskeleton is secreted by living cells, but it is nonliving material that does not grow with the animal. It must be shed (molted) and replaced by a larger exoskeleton at intervals to allow for the animal's growth. Depending on the species, most insects molt from four to eight times before reaching adult size. A few insect species and certain other arthropods, such as lobsters and crabs, molt at intervals throughout life.

Figure 30.2B The exoskeleton of an arthropod: a crab molting

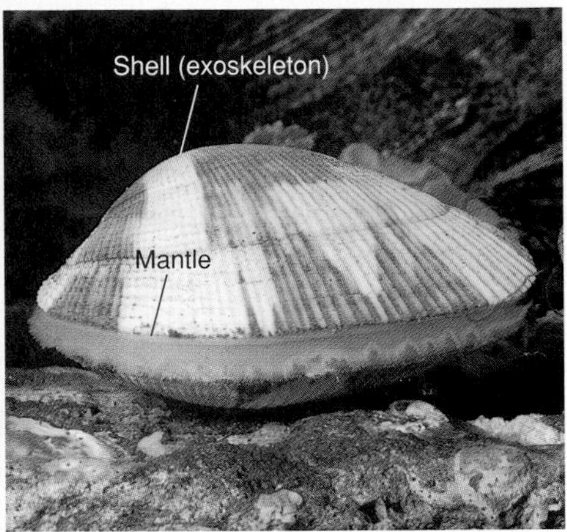

Figure 30.2C The exoskeleton of a mollusk: a clam

Figure 30.2D A sea urchin (above) and its endoskeleton (right)

In Figure 30.2B, you see a crab in the process of molting. Its old shell (the dark one on the right) split open when the crab outgrew it. An arthropod is never without an exoskeleton of some sort. A newly molted crab, for instance, has a new, soft, elastic exoskeleton, which formed under the old one. Soon after molting, the crab expands its body by gulping air or water. Its new exoskeleton then hardens in the expanded position, and the animal has room for further growth. Until its exoskeleton hardens, an arthropod is very susceptible to predation; besides being weakly armored, it is usually less mobile, because the soft exoskeleton cannot support the full action of its muscles.

The shells of mollusks such as clams (Figure 30.2C) are also exoskeletons, but unlike the chitinous arthropod exoskeleton, mollusk shells are made of a mineral, calcium carbonate. The mantle, a sheetlike extension of the animal's body wall, secretes the shell (see Module 18.9). As a mollusk grows, it does not molt; rather, it enlarges the diameter of its shell by adding to its outer edge.

An **endoskeleton** consists of hard or leathery supporting elements situated among the soft tissues of an animal. Sponges, for example, are reinforced by a framework of tough protein fibers or by hard structures called spicules. Usually microscopic and sharp-pointed, a spicule consists of inorganic material such as calcium salts or silica. Sea stars, sea urchins, and most other echinoderms have an endoskeleton of hard plates beneath their skin. In living sea urchins about all you see are movable spines, which are attached to the endoskeleton by muscles (Figure 30.2D, left). A dead urchin with its spines removed reveals the plates that form a rigid case (right photo).

Vertebrates have endoskeletons consisting of cartilage or a combination of cartilage and bone. One major lineage of vertebrates, the sharks, have cartilaginous skeletons. Figure 30.2E shows the more common condition. Bone makes up most of a frog's skeleton, as it does in most fishes and in land vertebrates. Cartilage (blue in the figure) remains in the frog skeleton, and in the skeletons of most other vertebrates, mainly in areas where flexibility is needed. Our own skeleton is an example, as we see next. 

? Compared to an endoskeleton, an exoskeleton offers greater ________ of soft tissues.

protection

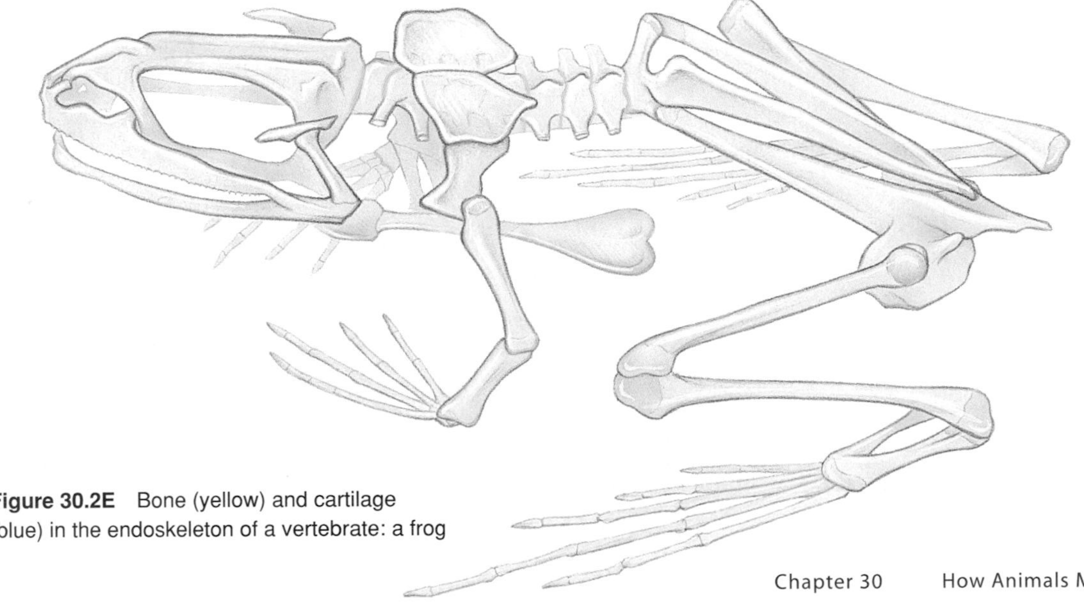

Figure 30.2E Bone (yellow) and cartilage (blue) in the endoskeleton of a vertebrate: a frog

30.3 The human skeleton is a unique variation on an ancient theme

30A *See Activity 30A on the CD-ROM.*

In contrast to the frog skeleton, which supports an animal that sits on all fours and hops on its hind legs, the human skeleton supports an upright body that sits on its hindquarters and walks or runs on two legs. Despite the differences, the skeletons of these and other vertebrates have a number of similarities. For instance, all vertebrates have an axial skeleton (green in Figure 30.3A) supporting the axis, or trunk, of the body. The **axial skeleton** consists of the skull, enclosing and protecting the brain; the backbone (vertebral column), enclosing the spinal cord; and, in most vertebrates, a rib cage around the lungs and heart.

Figure 30.3A The human skeleton

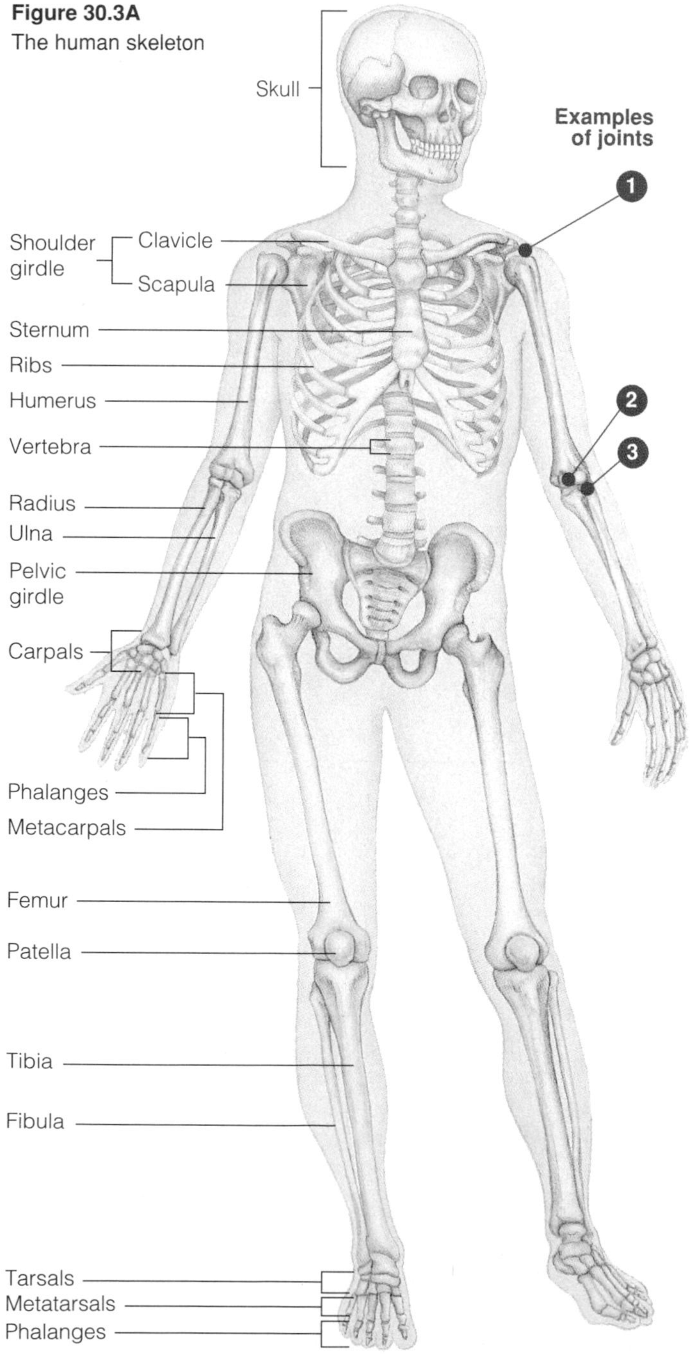

Most vertebrates also have appendages (arms, legs, wings, fins) and an appendicular skeleton (gold in Figure 30.3A) supporting the appendages. In a land vertebrate, the **appendicular skeleton** is made up of the bones of the forelimbs and hind limbs, the shoulder girdle, and the pelvic girdle. Bones of the girdles provide a base of support for the arm and leg bones. In the human, three bones support each arm: the humerus in the upper arm, and the radius and ulna in the forearm. Corresponding bones in the leg are the femur in the thigh, and the tibia and fibula in the lower leg. These same limb bones are found in most land vertebrates, although in the frog and certain others, the bones in the lower limbs are fused.

The skeletal features the human body has in common with other vertebrates stem from an ancient pattern, a basic skeletal model that probably originated in a group of fishes ancestral to land vertebrates. Hundreds of millions of years of adaptation have reshaped the model in a great variety of ways, but the main components remain. The skeleton of each species of land vertebrate, including the human, is a unique variation on the ancestral theme.

What are some of the distinctive features of the human skeleton? Our distant ancestors were quadrupedal (four-footed), and virtually every part of the skeleton changed drastically as upright posture and bipedalism evolved in the human lineage. Figure 30.3B contrasts the human skeleton (side view) and that of a modern quadrupedal primate, a baboon. Housing our large brain, our skull is large and flat-faced; its rounded part is the largest brain case relative to body size in the animal kingdom. In humans, the skull is balanced atop the backbone (red), whereas in quadrupeds, it is attached to the leading edge of the backbone. Our backbone is S-shaped, which helps balance the body in the vertical plane. In contrast, the baboon's is arched horizontally. Our pelvic girdle is shorter and rounder and oriented more vertically than the quadruped's. The bones of our hands and feet are also different from the

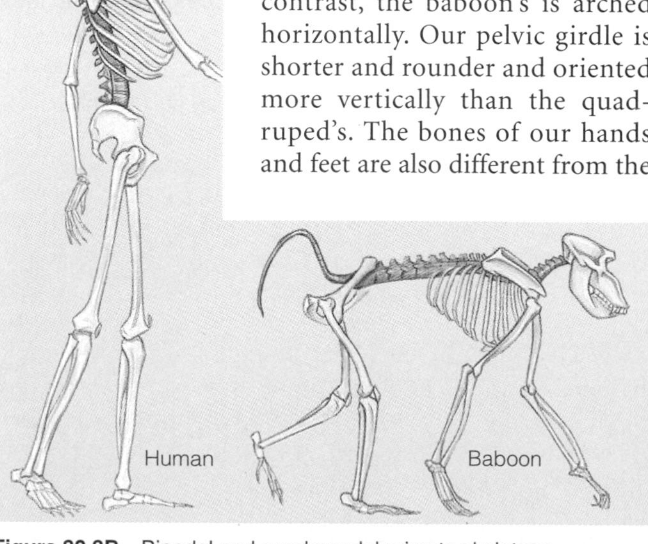

Figure 30.3B Bipedal and quadrupedal primate skeletons compared

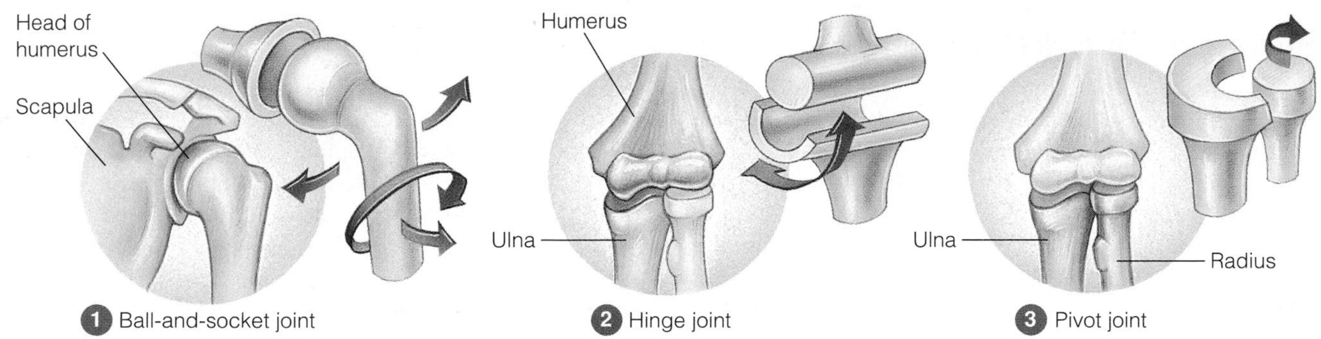

Figure 30.3C Three kinds of joints

baboon's. Free of locomotor functions, the human hand is adapted for strong gripping and precise manipulation. Our two feet are specialized for supporting the entire body and for bipedal walking and running.

Much of the versatility of the vertebrate skeleton comes from its movable joints (Figure 30.3C, and the numbered locations in Figure 30.3A). We have ① **ball-and-socket joints** where the humerus joins to the shoulder girdle and in the hip where the femur joins to the pelvic girdle. These joints enable us to rotate our arms and legs and move them in several planes. Two other kinds of joints provide flexibility at the elbow and knee. Shown here in the arm, ② a **hinge joint** between the humerus and the head of the ulna permits movement in a single plane. ③ A **pivot joint** enables us to rotate the forearm at the elbow. Hinge and pivot joints between the bones in our wrists and hands enable us to make precise manipulations.

Our skeleton is a showpiece of functional form. Many animals outperform us in specific activities—flying, swimming, running, digging. But the human body and the skeleton that supports it are unique in their versatility. We can swim, walk, run, crawl, jump, burrow, and manipulate objects with our hands with unparalleled precision. Indeed, our success on the planet is due partly to the diverse movements our skeleton makes possible. 

? Of the following, which is not part of the axial skeleton: skull, ribs, pelvic girdle, vertebrae?

Pelvic girdle

Skeletal disorders afflict millions 30.4

Nothing produced by evolution's restructuring process is perfect, and the human skeleton, despite its many adaptive advantages, is no exception. Our distant ancestors were quadrupedal and had an arched backbone similar to that of the baboon in Figure 30.3B. An arched backbone easily carries the weight of the ribs and organs suspended more or less evenly under it, much as a suspension bridge bears weight. In contrast, our vertical backbone, restructured from the arched condition, bears weight unevenly. Our internal organs hang down along our backbone, and our lower back (the lower part of the S curve) bears much of the load. This strains the lower back, especially when we bend over or lift heavy objects. Our tendency to have lower back problems is not surprising

Another common skeletal disorder, **arthritis**—inflammation of joints—affects one out of every seven people in the U.S. One form of arthritis seems to occur as part of aging. The joints become stiff and sore and often swell, as the cartilage between the bones wears down. Sometimes the bones thicken at the joints, producing crunching noises when they rub together and restricting movement. This form of arthritis is irreversible but not crippling in most cases, and moderate exercise, rest, and over-the-counter pain medications usually relieve it.

A much more serious form of arthritis, rheumatoid arthritis, is an autoimmune disease. The joints become highly inflamed, and their tissues may be destroyed by the body's immune system (see Module 24.16). Rheumatoid arthritis usually begins between ages 30 and 40 and affects more women than men. It may be triggered by a microbial infection, stress, or genetic factors. Anti-inflammatory drugs help relieve symptoms, but there is no cure. In some cases, patients are fitted with artificial joints.

Osteoporosis, another serious bone disorder, is usually related to hormonal changes that accompany aging. Affecting about 20 million Americans, it is most common in women after menopause, when estrogen levels drop. Estrogen contributes to normal bone maintenance, and with lowered production of the hormone, bones may become thinner, more porous, and easily broken. Insufficient exercise, an inadequate intake of protein and calcium (for maintaining bone mass), and diabetes mellitus may also contribute to the disease. Calcium and sometimes estrogen are prescribed as treatment, but many researchers feel the best approach is prevention. They encourage young women to maximize their calcium intake and exercise regularly in order to build bone mass. 

? What two things can a young woman do to help prevent osteoporosis in her future?

Ingest plenty of calcium and exercise regularly

30.5 Bones are complex living organs

Familiar expressions can seldom be taken literally, and "dry as a bone" is a good example. Bones are actually complex organs consisting of several kinds of moist, living tissues, amply supplied with blood. You can get a sense of some of a bone's complexity from these drawings of a human humerus (the upper arm bone). A sheet of fibrous connective tissue, shown in pink (most visible in the enlargement on the lower right), covers most of the outside surface. This tissue is able to form new bone in the event of a fracture. Replacing the connective tissue at either end of the bone, a thin sheet of cartilage (blue), also living tissue, forms a cushionlike surface for joints. The bone itself contains living cells that secrete a surrounding material, or matrix. Bone matrix consists of flexible fibers of the protein collagen embedded in hard calcium salts (see Figure 20.5). Analogous to the steel rods in reinforced concrete, the collagen fibers resist cracking; analogous to the mineral substance of concrete, the calcium salts resist compression.

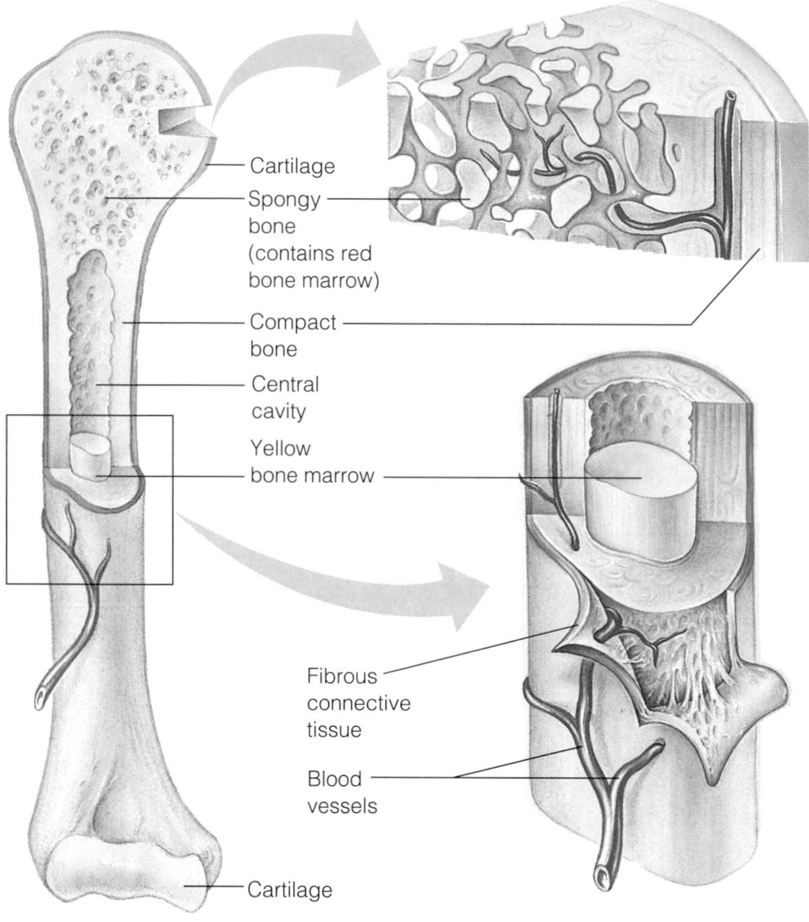

Figure 30.5 The structure of an arm bone

The shaft of this long bone is made of compact bone, so named because it has a dense matrix. Notice that the compact bone surrounds a central cavity. The central cavity contains **yellow bone marrow,** which is mostly stored fat brought into the bone by the blood. The ends, or heads, of the bone have an outer layer of compact bone and an inner layer of spongy bone, so named because it is honeycombed with small cavities. The cavities contain **red bone marrow** (not shown in the figure), a specialized tissue that produces our blood cells.

Like all living tissues, tissues in a bone require servicing. Blood vessels course through channels in the bone, transporting nutrients and regulatory hormones to its cells. Nerves (not shown) paralleling the blood vessels help regulate the traffic of materials between the bone and the blood.

As we see in the next module, the diverse living tissues in bones are key to bone growth.

What accounts for the color of yellow bone marrow? Of red bone marrow?

Stored fat; developing red blood cells

30.6 Bone growth is a major feature of human development

Our living endoskeleton is with us from about a month after conception. It starts out as fibrous connective tissue and cartilage, with actual bone starting to form when the embryo is about 6 weeks old. The X-ray photograph in Figure 30.6A shows the developing skeleton in a 12-week-old human fetus. The bones that form most of the skull develop from sheets of fibrous connective tissue. In contrast, long bones, such as those in the legs and arms, develop from shafts of cartilage, each surrounded by a sheet of connective tissue.

Figure 30.6B illustrates how a long bone develops. Part ① shows an early embryonic stage. The bone originates as a shaft of cartilage surrounded by a sheet of connective tis-

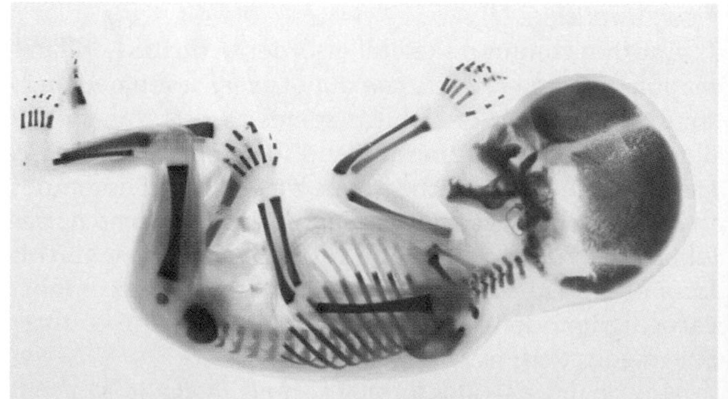
Figure 30.6A The skeleton of a 12-week-old fetus

sue (this sheet is not shown). A ring of bone then forms around the shaft. Bone also starts to replace cartilage at the center of the shaft. ② In the fetus, the bone grows in length and thickness as blood vessels penetrate the center of the shaft, and the yellow marrow cavity begins to form. Other blood vessels penetrate the cartilage near the ends of the shaft, and new bone formation also begins there.

Part ③ shows the growing bone of a child. The bone grows in overall width as bone cells produce new bone along the outer edge (green arrows); at the same time, bone on the inner surface (purple arrows) is broken down, and the marrow cavity increases in volume. The bone grows in length as long as cartilage continues to grow in the areas indicated in darker blue. Simultaneously, new bone replaces the cartilage in the areas colored dark gold.

Skeletal growth stops at about age 18 in women and about age 21 in men. Until those times, a complex interaction of hormones from the pituitary gland, thyroid gland, and gonads maintains cartilage growth and keeps the bones growing in proper proportion. The body reaches its adult size, and bones stop growing, when cells involved in bone growth begin to respond differently to the hormones.

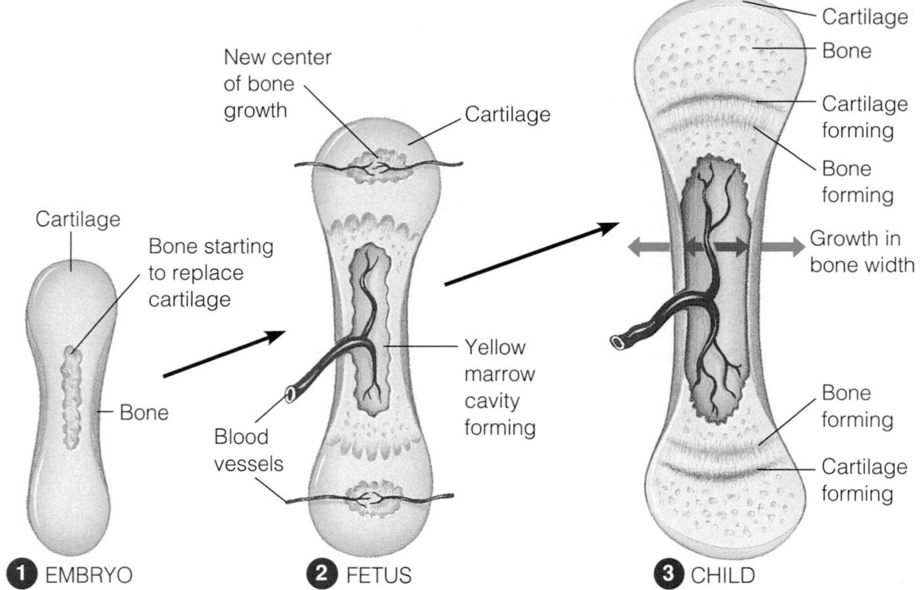

Figure 30.6B Development of a bone

Why are the bones of children more flexible than those of adults?

A child's bones have a greater proportion of cartilage compared to hard bone tissue.

Muscle Contraction and Movement

30.7 The skeleton and muscles interact in movement

We now focus on how an animal's skeleton interacts with its muscles in producing movement. Figure 30.7 illustrates how two muscles interact with parts of the skeleton and with each other to raise and lower the human forearm. As in this example, muscles are connected to bones by **tendons** (see Module 20.5). For instance, one end of the biceps muscle is attached by a tendon to bones of the shoulder. The other end is attached to one of the bones in the forearm.

Under its own power, a muscle can *only* contract, or shorten. Once a muscle is contracted, its relaxation to an extended position is a passive process. Often it is pulled by the action of another muscle. If we had only one muscle in our arm, for example, we could not move it back and forth or rotate it. The ability to move an arm in opposite directions requires that muscles be attached to the arm bones in antagonistic pairs—that is, two muscles working against each other. As shown here, contraction of the biceps muscle raises the forearm. The triceps muscle is the biceps' antagonist. The upper end of the triceps attaches to the shoulder, while its lower end attaches to the elbow. At the right, you see that contraction of the triceps lowers the forearm, extending the biceps in the process.

All animals—even very small ones like the ants in the chapter's introduction—have antagonistic pairs of muscles that apply opposite forces against parts of their skeleton. Next, let's see how the structure of a muscle explains its ability to contract.

In exercising your arms or legs to strengthen muscles, why is it important to impose resistance during both flexing and extending of the limbs?

This exercises both muscles of antagonistic pairs, which only do work when they are contracting.

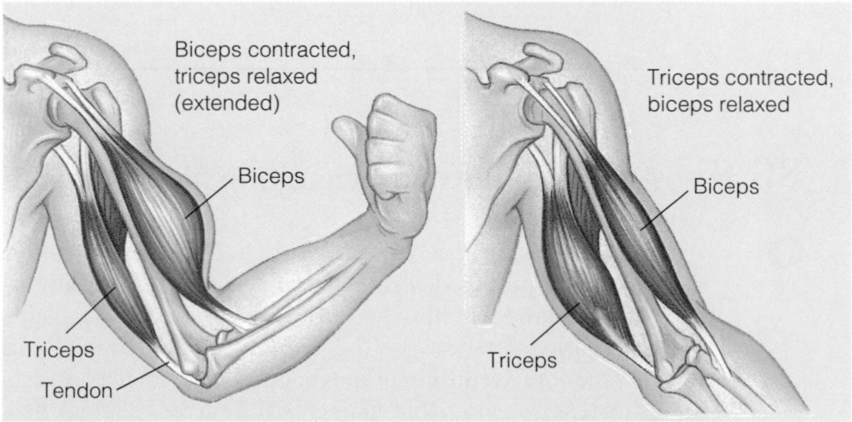

Figure 30.7 Antagonistic action of muscles in the human arm

30.8 Each muscle cell has its own contractile apparatus

30B

We looked briefly at the various types of muscle tissue in Module 20.6. **Skeletal muscle,** which is attached to the skeleton and produces body movements, is made up of a hierarchy of smaller and smaller parallel strands. As indicated at the top in Figure 30.8, a muscle consists of bundles of parallel muscle fibers. Each muscle fiber is a single cell with many nuclei.

Farther down in the drawing, notice that each muscle fiber is itself a bundle of smaller **myofibrils.** Skeletal muscle is also called striated (striped) muscle because the myofibrils exhibit alternating light and dark bands when viewed with a light microscope. A myofibril consists of repeating units called **sarcomeres.** Structurally, a sarcomere is the region between two dark, narrow lines, called Z lines, in the myofibril. Functionally, the sarcomere is the contractile apparatus in a myofibril—the muscle fiber's fundamental unit of action.

The micrograph and the diagram below it reveal the structure of a sarcomere in more detail. They show that a myofibril is composed of regular arrangements of two kinds of filaments: thin filaments, colored blue in the diagram, and thick filaments, colored pink. Biochemical analyses have shown that a **thin filament** is a double strand of the protein actin and one strand of a regulatory protein, coiled around each other. Each **thick filament** is a number of parallel strands of the protein myosin. The broad, dark band taking up most of the sarcomere is where the thick filaments are (bottom diagram); they are interspersed with thin filaments extending almost to the center of the sarcomere. In contrast, the light bands have only thin filaments. Within the light bands, the Z lines consist of proteins that connect adjacent thin filaments.

This specific arrangement of repeating units of thin and thick filaments is directly related to the mechanics of muscle contraction. In the next module, we see how this structure contributes to the functioning of a muscle cell, and hence how an entire muscle contracts.

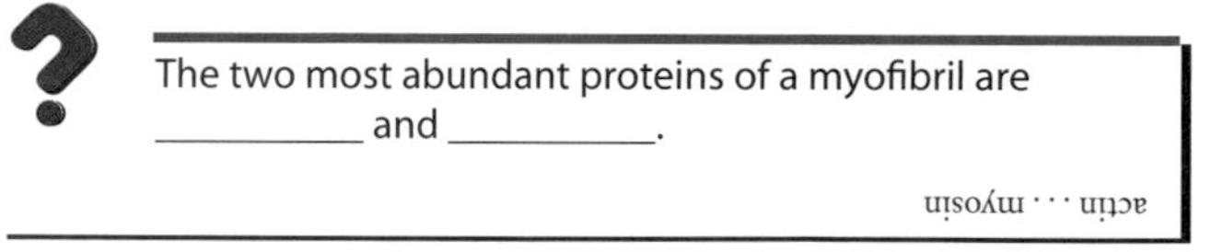

? The two most abundant proteins of a myofibril are __________ and __________.

actin . . . myosin

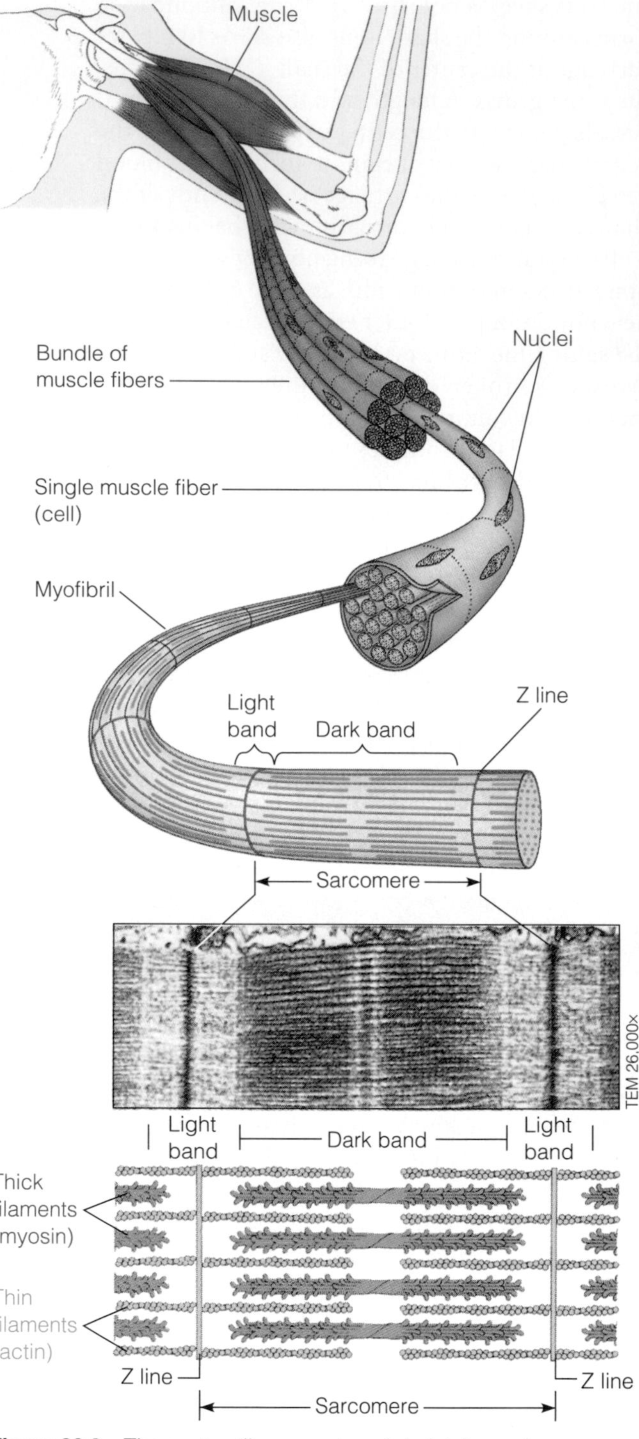

Figure 30.8 The contractile apparatus of skeletal muscle

30.9 A muscle contracts when thin filaments slide across thick filaments

30C

In the 1950s, British Nobel laureate A. F. Huxley and other researchers proposed what has become known as the **sliding-filament model** of muscle contraction. This model explains the relationship between the structure of a sarcomere and its function. According to the model, a sarcomere contracts (shortens) when its thin filaments slide across its thick filaments. Figure 30.9A, a simplified diagram of Huxley's model, shows a sarcomere in a relaxed muscle, in a contracting one, and in a fully contracted one. Notice in the contracting sarcomere that the Z lines and the thin filaments (blue) have moved toward the middle of the sarcomere. And when the muscle is fully contracted, the thin filaments overlap in the middle of the sarcomere. Contraction only shortens the sarcomere; it does not change the lengths of the thick

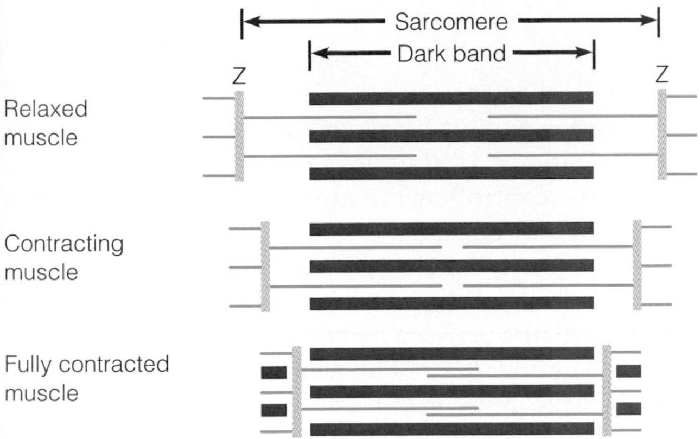

Figure 30.9A The sliding-filament model of muscle contraction

and thin filaments. A whole muscle can contract to about half of its resting length when all its sarcomeres shorten.

What makes the thin filaments slide when a sarcomere contracts? Following Huxley's lead, researchers have sought answers at the molecular level. The key events are energy-consuming interactions between the myosin molecules of the thick filaments and the actin of the thin filaments. Electron micrographs show that parts of myosin molecules, called heads, bind with specific sites on the thin filaments. Energy for sliding comes from ATP. Also essential for muscle contraction are calcium ions (Ca^{2+}), which trigger the initial events in sliding.

Figure 30.9B indicates how sliding seems to work. ① ATP binds to a myosin head, causing the head to detach from a binding site (dark blue) on actin. (Only two of many binding sites are indicated.) ② Next, energy is made available for contraction when the ATP is broken down (hydrolyzed) to ADP and Ⓟ (phosphate), which remain bound to the myosin head. The head gains some of the energy, and its position changes as a result. In its new position, the myosin head is cocked like a pistol ready to fire (actually, ready to bind with another site on the actin molecule). ③ Then Ca^{2+} enters the picture. Acting like a molecular trigger, it opens up a binding site on the actin molecule, making it possible for the myosin head to bind to the actin. ④ The molecular event that actually causes sliding is called the power stroke. The myosin head bends when ADP and Ⓟ are released from it. The bending pulls the thin filament toward the center of the sarcomere, in the direction of the blue arrow. After the power stroke, more ATP binds with the myosin head, and the whole process repeats. On the next power stroke, the myosin head attaches to another binding site ahead of the previous one on the thin filament (closer to the Z line). This process—detach, straighten out (cock), attach, bend—occurs again and again in a contracting muscle. Though we show only one myosin head in the figure, a typical thick filament has about 350 heads, each of which can bind and unbind to a thin filament about five times per second. Preventing the filaments from backsliding during contraction, some myosin heads hold the thin filaments in position, while others are reaching for new binding sites. The process continues until the muscle is fully contracted, or until it stops contracting.

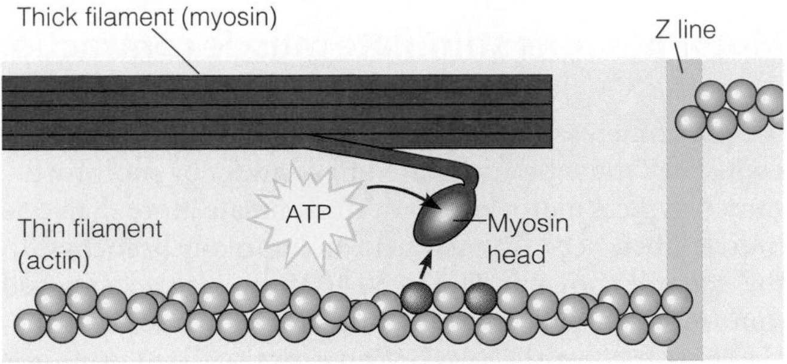

1 ATP binds to a myosin head, which is released from an actin filament.

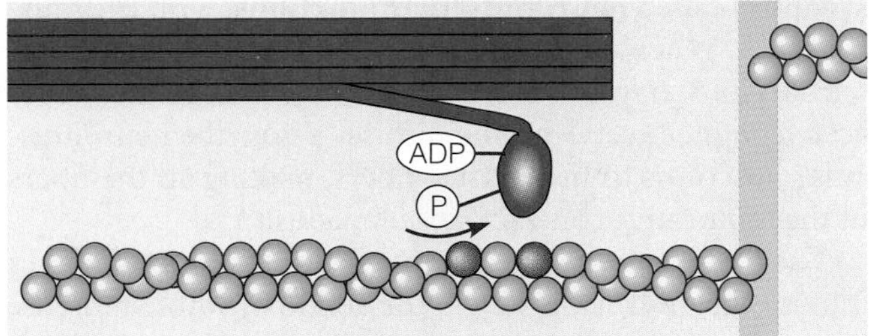

2 Hydrolysis of ATP cocks the myosin head.

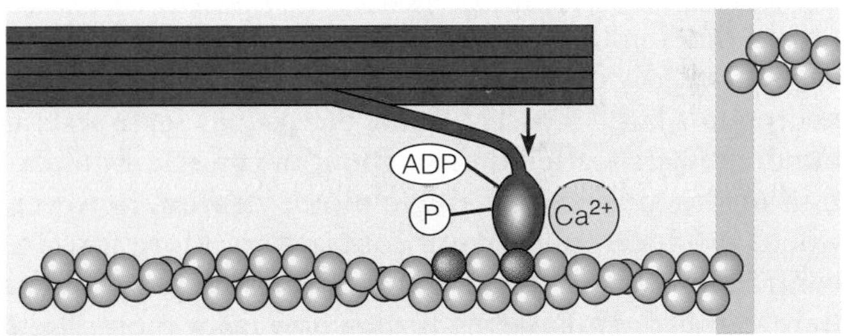

3 The myosin head attaches to an actin binding site, with the help of calcium.

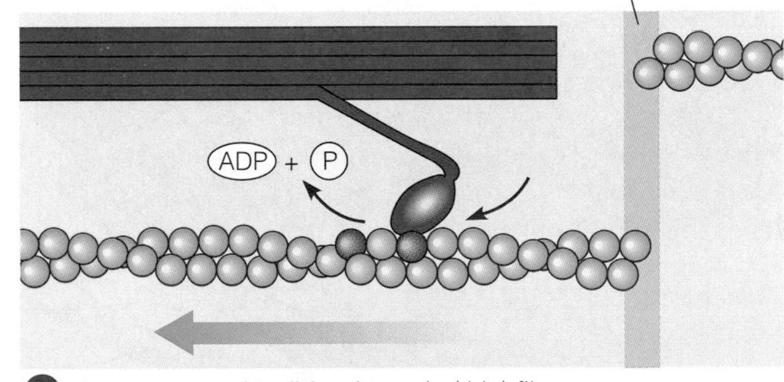

4 The power stroke slides the actin (thin) filament.

Figure 30.9B The mechanism of filament sliding

The sliding-filament model explains how the sarcomeres of skeletal muscles contract. Next, let's return to an earlier topic and see how the nervous system controls muscle contraction. 

?

Which part of a sarcomere becomes shorter during contraction of a muscle? (a) the dark band; (b) the light bands; (c) both the dark and light bands become shorter; (d) neither the dark nor light bands become shorter

(q)

30.10 Motor neurons stimulate muscle contraction

The sarcomeres of a muscle fiber do not contract on their own. They must be stimulated to contract by motor neurons. A typical motor neuron can stimulate more than one muscle fiber, because each neuron has many branches. In the example shown in Figure 30.10A, you see two so-called **motor units,** each consisting of a neuron and all the muscle fibers it controls (two or three, in this case). A motor neuron has its dendrites and cell bodies in the central nervous system (here, the spinal cord). Its axon forms synapses, called **neuromuscular junctions,** with the muscle fibers. When a motor neuron sends out an action potential, its synaptic knobs release the neurotransmitter acetylcholine. Acetylcholine diffuses across the neuromuscular junctions to the muscle fibers, making all the fibers of the motor unit contract simultaneously.

The organization of individual neurons and muscle cells into motor units is the key to the action of whole muscles. We know that we can vary the amount of force our muscles develop—an arm wrestler, for example, may change the amount of force developed by the biceps and triceps several times in the course of a match. The ability to do so depends mainly on the nature of motor units. Each motor neuron in a large muscle like the biceps may serve several hundred fibers scattered throughout the muscle. Stimulation of the muscle by a single motor neuron, however, would produce only a weak contraction. More forceful contractions would result when additional motor units were activated. Thus, depending on how many motor units your brain commands to contract, you can apply a small amount of force to lift a fork, or considerably more to lift, say, this textbook. In muscles requiring precise control, such as those controlling eye movements, a motor neuron may control only one fiber.

How does a motor neuron make a muscle fiber contract? The initial events of stimulation are the same as those that occur at a synapse between two neurons in the nervous system (see Module 28.6): The acetylcholine that diffuses across the neuromuscular junction changes the permeability of the muscle fiber's plasma membrane. The change triggers action potentials that sweep across the muscle cell membrane. As indicated in Figure 30.10B, the action potentials (black arrows) then pass deep into the muscle cell along membranous tubules that fold inward from the plasma membrane. Inside the cell, the action potentials make the endoplasmic reticulum (ER) release Ca^{2+} (green dots) into the cytoplasm. The Ca^{2+} then triggers the binding of myosin to actin, initiating filament sliding, as we saw in the previous module. When a muscle relaxes, the process reverses: Motor neurons stop sending action potentials to the muscle fibers, the ER pumps Ca^{2+} back out of the cytoplasm, and the sarcomeres stop contracting.

How does the endoplasmic reticulum help regulate muscle contraction?

By reversibly storing and releasing Ca^{2+}, the ER regulates the cytoplasmic concentration of this ion, which is required in the cytoplasm for the binding of myosin to actin.

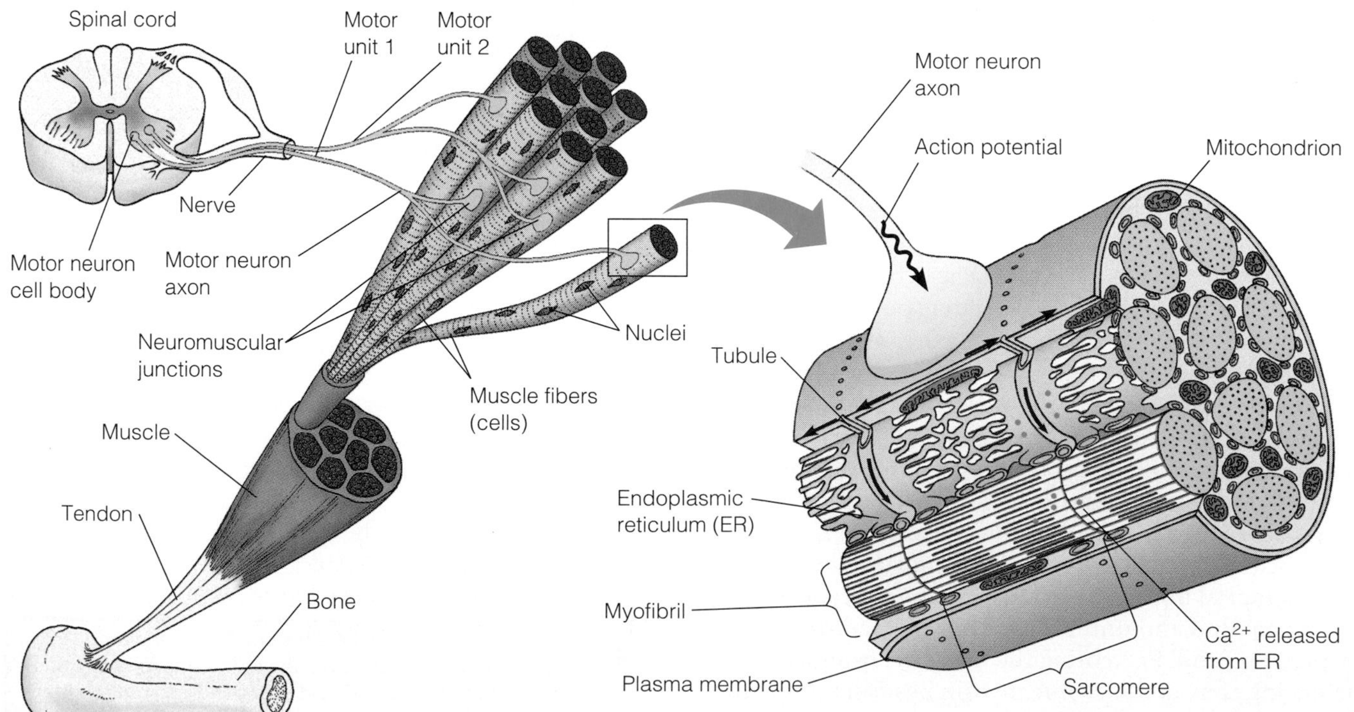

Figure 30.10A The relation between motor neurons and muscle fibers

Figure 30.10B Part of a muscle fiber (cell) at a neuromuscular junction

Probably no one appreciates the precise control that nervous systems exert over contracting muscles more than people who build robots. Robotic arms, hands, and legs are common on assembly lines, performing routine tasks, such as welding and riveting auto and airplane parts together. We see robotic dinosaurs in movies, and mobile robotic vehicles make it possible for us to explore the surface of other planets.

A robot capable of independent action must have a complex computer for a brain and a locomotor system that mimics that of a complex animal. So far, no computer comes close to the integrating capacity of the human brain, nor has robotic science devised any mechanical system of movement that closely mimics all the complex activities our muscles can perform. The robot in Figure 30.11A is a model built by artificial intelligence researcher Stanley Rosenschein of Palo Alto, California. Rosenschein used the task of buttering a piece of toast as a research model to construct a robot that can perform complex jobs requiring precise actions. Buttering toast is no great challenge for the human nervous system or for the manipulative skills of our hands. Our brain is usually so preoccupied with more complex thinking and problem solving that we pay little attention to what it is telling the muscles in our hands to do when we pick up a piece of toast, put it on a plate, decide on a certain amount of butter, turn a knife in just the right way, judge the hardness of the butter, apply a certain pressure to spread it on the toast, and make any other adjustments to perform all these tasks efficiently. A robot that can do all this smoothly and independently has to be programmed to do precisely this job.

Figure 30.11A A robot that butters toast

The difficulty of building robots with humanlike intelligence, movement, and manipulative abilities led some robotic scientists to simplify their goals. A number of years ago, Rodney Brooks, at the Massachusetts Institute of Technology, became intrigued with the relative simplicity of insect brains and muscle systems and with what insects can do with what they have. As we saw in the chapter's introduction, ants perform a variety of complex tasks with their tiny brains and small bodies. The challenge Brooks set for his research team was to build small robots with some of the abilities of insects—abilities that could be adapted to perform useful tasks.

Figure 30.11B Genghis on the move

The insectlike robot in Figure 30.11B, known as Genghis, was the brainchild of Rodney Brooks's robotics laboratory. Genghis has six legs with wires for muscles, and it walks like a clumsy ant. For sensory receptors, it has six infrared-sensing eyes and two mechanical whiskers. These sensors feed information into a brain consisting of only 57 electronic circuits. The brain does little integrating but manages to steer Genghis around objects in its path. Other tiny circuit boards between each pair of legs resemble the ganglia in an insect's ventral nerve cord (see Figure 28.10E). The artificial ganglia run tiny motors that propel the legs.

Most robots have circuitry modeled after animals with relatively simple nervous systems. Their electronic brains receive data from sensors and produce predictable, automatic actions in response. This eliminates the need for the highly complex circuitry it takes to process sensory data. It also makes it possible to build relatively cheap miniature robots that can do many useful things. Worldwide, robots make up a workforce of over 650,000. Most are programmed to perform specific tasks that are repetitive and, in some cases, dangerous or impossible for humans. Some tiny robots with very simple circuitry are designed to move through tubes that are only a few millimeters in diameter. Such micromachines will have increasing application in industry and medicine—for instance, in constructing microscopic computer parts or crawling through a patient's colon to remove a cancerous tumor. As the robotic workforce grows, it is likely that most designers will continue to focus on machines that perform relatively simple, specific jobs.

What limits insectlike robots to behavior that is predictable and automatic?

Its circuitry can only link specific sensory inputs to certain motor responses.

30.12 The structure-function theme underlies all the parts and activities of an animal

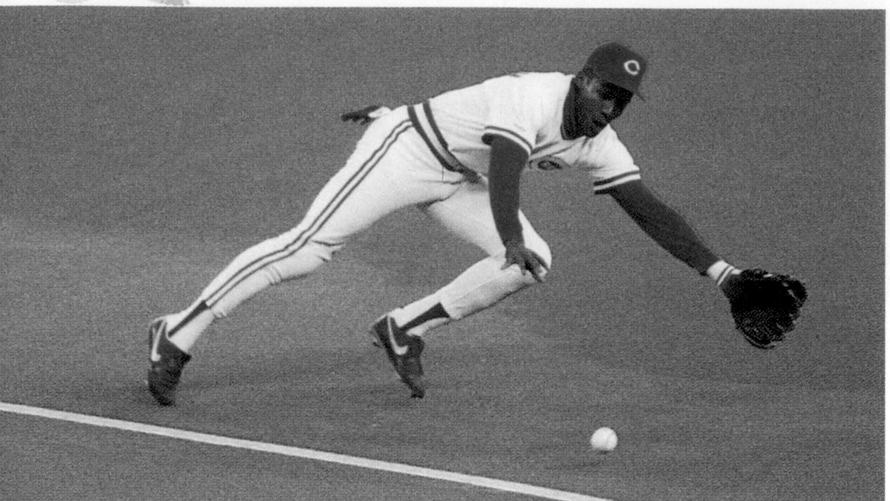

Figure 30.12A A baseball player in action

The batter drives the ball hard toward center field. Will it get out of the infield? The shortstop dives and robs the batter of a base hit.

Though far removed from the arena of natural selection, a baseball game demonstrates some of the remarkable evolutionary adaptations of the human body. A hit and a catch are both spectacular displays of our nervous system's ability to almost instantly link environmental stimuli with appropriate muscular activity. The batter hits a ball that was traveling over 90 miles per hour. The shortstop's eyes detect the ball as soon as it leaves the bat. In both cases, a supercomputer we call the human brain integrates information about the angle and speed of the ball and then signals multiple muscles to perform a very specific action—to swing a bat at precisely the right moment, or dive and thrust out a hand to meet the ball, again at just the right moment. These remarkable abilities result from adaptations of the human body, derived by natural selection and key to our survival as a species.

No less remarkable are the adaptations that underlie the cooperative work of a leaf-cutter ant colony. Movement is a common denominator of everything the million or so ants in a colony do. Each ant's tiny brain keeps it performing a specific set of movements that make it a contributing member of the colony as a whole. The chores of colony maintenance and fungus gardening involve an assembly line of cooperative legwork and delicate manipulations of mouthparts. Colony defense often takes the form of slashing or crushing invaders by the powerful mouthparts of detachments of soldier ants. Out in a rain forest at night (Figure 30.12B), the ants follow a precise route, a scent trail laid down earlier by scouts from their nest. Their senses are so acute that just a milligram of scent chemical would be enough to lead an entire colony of leaf-cutters three times around the world.

In a sentence, describe how Figure 30.12A illustrates the popular sports term "hand-eye coordination."

One possible answer: Ball meets glove because the brain coordinates visual input about the path and speed of the ball with motor output that places the hand at the right place at the right time.

* * *

We have now completed our unit on the organ systems of animals. Our overriding theme has been that structure underlies function—that the structural adaptations of a cell, tissue, organ, or organ system determine the job it can perform. We have seen numerous examples of this basic principle, from our early studies of the digestive system to our discussion of the cells of the immune system, and also in our last three chapters on nervous control, sensory receptors, and movement systems. Watching the precise movements of a baseball player, the work of an ant colony, or the outward activity of any whole animal gives us an overview of the structure-function relationship and a reminder of the evolutionary process that generated it. The visible activities of any animal result from a nervous system (composed of cells with a specific structure) directing muscles (capable of contracting because of their cellular structure) to respond to information that sensory receptors (specifically structured cellular detectors) have gathered about the environment. We'll see the structure-function theme emerge again in the context of adaptations to the environment when we take up the study of plants in the next unit.

Figure 30.12B Leaf-cutter ants in motion

Chapter Review

Chapter Summary

Movement and Locomotion (Introduction–30.1) Movement is one of the most distinctive features of animals. In all forms of movement, the nervous system issues commands to the muscular system, and the muscular system exerts propulsive force against the skeleton **(Introduction)**. Some aquatic animals, such as sponges, do not move about in search of food, but move body parts to generate currents that bring food to them. Most animals perform locomotion, active travel from place to place. Locomotion requires that an animal use energy to overcome friction and gravity, forces that tend to keep it stationary. Animals that swim use their limbs as oars, expel water for jet propulsion, move their tails from side to side, or move their bodies up and down. Aquatic animals are supported by water, so gravity has little effect on them, but they are slowed by friction. Many are smooth and streamlined. Animals that walk, hop, or run on land are not affected as much by friction, but they must be able to support themselves against the force of gravity. Legs stabilize the body and propel the animal forward. Muscles generate the power for movement, and springy legs store energy for each step or jump. Animals that crawl or burrow must overcome friction. They move by side-to-side undulation or by peristalsis (head-to-tail waves of muscle contraction). The wings of birds, bats, and flying insects are airfoils, which overcome the pull of gravity by generating lift **(30.1)**.

Skeletal Support (30.2–30.6) All animal movement is based on contractile systems working against a firm skeleton. Skeletons function in support, movement, and the protection of internal organs. Worms and cnidarians have hydrostatic skeletons, which consist of fluid held in closed compartments. These animals can change shape by contracting muscles in their body walls. Exoskeletons are hard or leathery external cases, such as the calcium carbonate shells of clams and the chitinous, armorlike skeletons of insects. Sponges, echinoderms, and vertebrates have endoskeletons. The vertebrate endoskeleton is composed of cartilage or cartilage and bone **(30.2)**. All vertebrates have an axial skeleton (skull, vertebrae, and in most species, ribs). Most vertebrates also have an appendicular skeleton supporting their paired appendages. In humans, the appendicular skeleton consists of bones of the shoulder girdle, upper limbs, pelvic girdle, and lower limbs. Movable joints provide flexibility in vertebrate skeletons. Derived from the skeleton of four-footed ancestors, the human skeleton changed dramatically as upright posture and bipedalism evolved **(30.3)**.

The human skeleton is versatile, but it is also subject to problems such as lower back pain, arthritis, and osteoporosis **(30.4)**. A bone is a living organ containing several kinds of tissues. It is covered with a connective tissue membrane. Cartilage at the ends of the bone cushions the joints. The bone is served by blood vessels and nerves. Bone cells live in a matrix of flexible protein fibers and hard calcium salts. Long bones have a central cavity that stores fat, and spongy bone at their ends that contains red marrow, where blood cells form **(30.5)**. Most of the skull bones develop from sheets of connective tissue. Long bones start out as shafts of cartilage, which are gradually replaced by bone. A long bone keeps growing until cartilage growth ceases **(30.6)**.

Muscle Contraction and Movement (30.7–30.12) Muscles pull on bones, which act as levers that produce movements. Antagonistic pairs of muscles produce opposite movements. Muscles perform work only when contracting. A muscle extends when pulled by another muscle or returned to an extended position by parts of the skeleton **(30.7)**. Each muscle cell, or fiber, consists of bundles of myofibrils, which, in striated muscle, exhibit alternating light and dark bands. Each myofibril contains bundles of overlapping thick (myosin) and thin (actin) protein filaments. Repeating units of thick and thin filaments, called sarcomeres, are the muscle fiber's contractile units **(30.8)**. Calcium ions trigger myosin heads on the thick filaments to pull themselves along the actin molecules of the thin filaments. ATP provides the energy for this movement. The sliding of thick and thin filaments across one another increases their degree of overlap, shortening the muscle fiber. This shortening of muscle fibers produces the pull, or contraction, of a muscle **(30.9)**. Motor neurons carry action potentials that initiate muscle contraction. A neuron releases the neurotransmitter acetylcholine at a neuromuscular junction, signaling a muscle fiber to contract. A neuron can branch to a number of muscle fibers; the neuron and the muscle fibers it controls constitute a motor unit. The strength of a muscle contraction depends on the number of motor units activated. Motor units vary in size, and the control of small motor units creates precise movements **(30.10)**. Robots can imitate, in a simple way, some of the senses and responses of animals **(30.11)**.

A hallmark of animals, body movement is a visible reminder that function emerges from structure. An animal's nervous system connects sensations derived from environmental stimuli to responses carried out by its muscles **(30.12)**.

Testing Your Knowledge

Multiple Choice

1. A human's internal organs are protected mainly by the
   a. hydrostatic skeleton
   b. motor unit
   c. axial skeleton
   d. exoskeleton
   e. appendicular skeleton

2. When your biceps muscle contracts, your arm bends at the elbow. Contraction of the muscle results from the ____ of thick and thin filaments inside its muscle fibers.
   a. collapse
   b. rotation
   c. shortening
   d. folding
   e. sliding

3. Gravity would have the least effect on the movement of which of the following? *(Explain your answer.)*
   a. a salmon
   b. a human
   c. a snake
   d. a sparrow
   e. a grasshopper

4. Arm and leg muscles are arranged in antagonistic pairs. How does this affect their functioning?
   a. It provides a backup if one of the muscles is injured.
   b. One muscle of the pair pushes while the other pulls.
   c. A single neuron controls both of them.
   d. It allows the muscles to produce opposing movements.
   e. It doubles the strength of contraction.

continued on next page

5. Which of the following bones in the human arm would correspond to the femur in the leg?

a. radius
b. tibia
c. humerus
d. metacarpal
e. ulna

6. Which of the following animals is correctly matched with its type of skeleton?

a. fly—endoskeleton
b. earthworm—exoskeleton
c. dog—exoskeleton
d. lobster—exoskeleton
e. bee—hydrostatic skeleton

7. When a horse is running fast, its body position is stabilized by

a. side-to-side undulation
b. energy stored in muscles and tendons
c. the lift generated by its movement through the air
d. foot contact with the ground
e. its momentum

8. What is the role of acetylcholine in muscle contraction?

a. It makes up part of the thin filaments inside a muscle fiber.
b. It provides energy for contraction.
c. It blocks contraction when the muscle relaxes.
d. It signals a muscle fiber to contract.
e. It forms the heads of the myosin molecules in the thick filaments inside a muscle fiber.

9. Muscle A and muscle B are the same size, but muscle A is capable of much finer control than muscle B. Which of the following is likely to be true of muscle A? *(Explain your answer.)*

a. It is controlled by more neurons than muscle B.
b. It contains fewer motor units than muscle B.
c. It is controlled by fewer neurons than muscle B.
d. It has larger sarcomeres than muscle B.
e. Each of its motor units consists of more cells than the motor units of muscle B.

10. Which of the following statements about skeletons is true?

a. Hydrostatic skeletons are soft and do not protect body parts.
b. Chitin is a major component of vertebrate skeletons.
c. Evolution of bipedalism involved little change in the axial skeleton.
d. Most cnidarians must shed their skeleton periodically in order to grow.
e. Vertebrate bones contain living cells.

Describing, Comparing, and Explaining

1. A hawk swoops down, seizes a mouse in its talons, and flies back to its perch. In a few sentences, explain how its wings enable it to overcome the downward pull of gravity as it flies upward.

2. In terms of both numbers of species and numbers of individuals, insects are the most successful land animals. Write a paragraph explaining how their exoskeletons help them live on land. Are there any disadvantages to having an exoskeleton?

3. Describe how you bend your arm, starting with action potentials and ending with the contraction of a muscle. How does a strong contraction differ from a weak contraction?

4. In what ways did the human skeleton change as an upright posture and bipedalism evolved? Describe the changes by comparing the human skeleton and the skeleton of a quadruped such as a baboon.

Thinking as a Scientist

1. Drugs are often used to relax muscles during surgery. Which of the following two chemicals do you think would make the best muscle relaxant, and why? Chemical A: Blocks acetylcholine receptors on muscle cells. Chemical B: Floods the cytoplasm of muscle cells with calcium.

2. An earthworm's body consists of a number of fluid-filled compartments, each with its own set of longitudinal and circular muscles. In a different kind of worm, the roundworm, a single fluid-filled cavity occupies the body, and there are only longitudinal muscles that run its entire length. Predict how the movement of a roundworm would differ from the movement of an earthworm.

3. When a person dies, muscles often become rigid and fixed in position—a condition known as rigor mortis, which often figures importantly in mystery novels. Rigor mortis occurs because muscle cells use up their supply of ATP as they die. The rigor disappears several hours after death because the biological molecules break down. Explain, in terms of the mechanism of contraction described in Module 30.9, why the lack of ATP would cause muscles to become rigid, rather than limp, after death.

Science, Technology, and Society

1. A goal of the Americans with Disabilities Act is to allow people with physical limitations to fully participate in and contribute to society. Perhaps you have a disability or know someone with a disability. Imagine that a neuromuscular disease or injury makes it impossible for you to walk. Think about your activities during the last 24 hours. How would your life be different if you had to get around in a wheelchair? What kinds of barriers or obstacles would you encounter? What kinds of changes would have to be made in your activities and surroundings to accommodate your change in mobility?

2. Athletes sometimes take anabolic steroids illegally to increase the buildup of muscle that occurs with training. In young people, whose bones are still growing, steroids also act to speed up the conversion of cartilage to bone. With regard to bone growth, what might be some negative consequences of steroid use by young athletes? Does steroid use seem to you to be worth the risks? Why or why not? As a competing athlete, would you use anabolic steroids if you knew you could get away with it? Why or why not?

3. You may know an elderly person who has broken a bone (often a hip) as a consequence of osteoporosis. To prevent osteoporosis, researchers recommend exercise and maximizing calcium intake beginning in the teens and twenties. Do you think young people think of themselves as future senior citizens? How would you recommend that they be encouraged to develop health habits that might not pay off for 40 or 50 years?

Media Resources

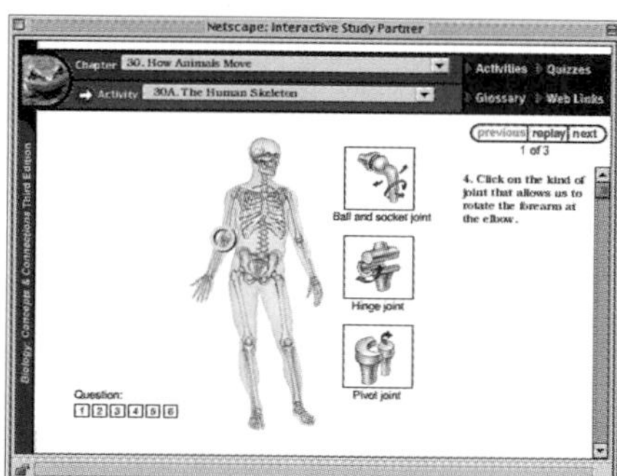

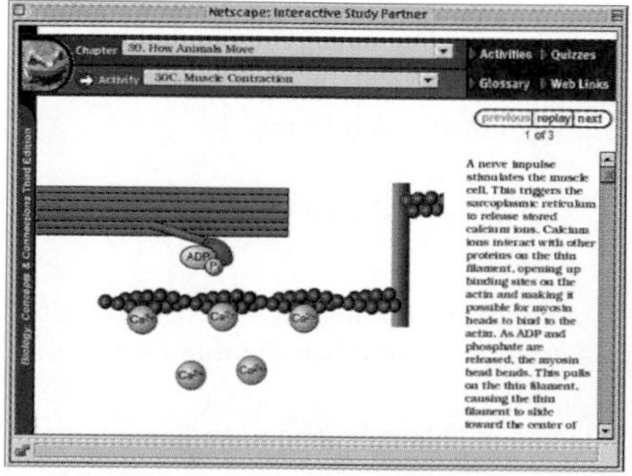

For further review of the concepts in this chapter, use the activities and quizzes on the CD-ROM.

Visit the Special Edition of The Biology Place web site for activities, quizzes, news articles, and web links.

UNIT VI

Plants: Form and Function

31 Plant Structure, Reproduction, and Development

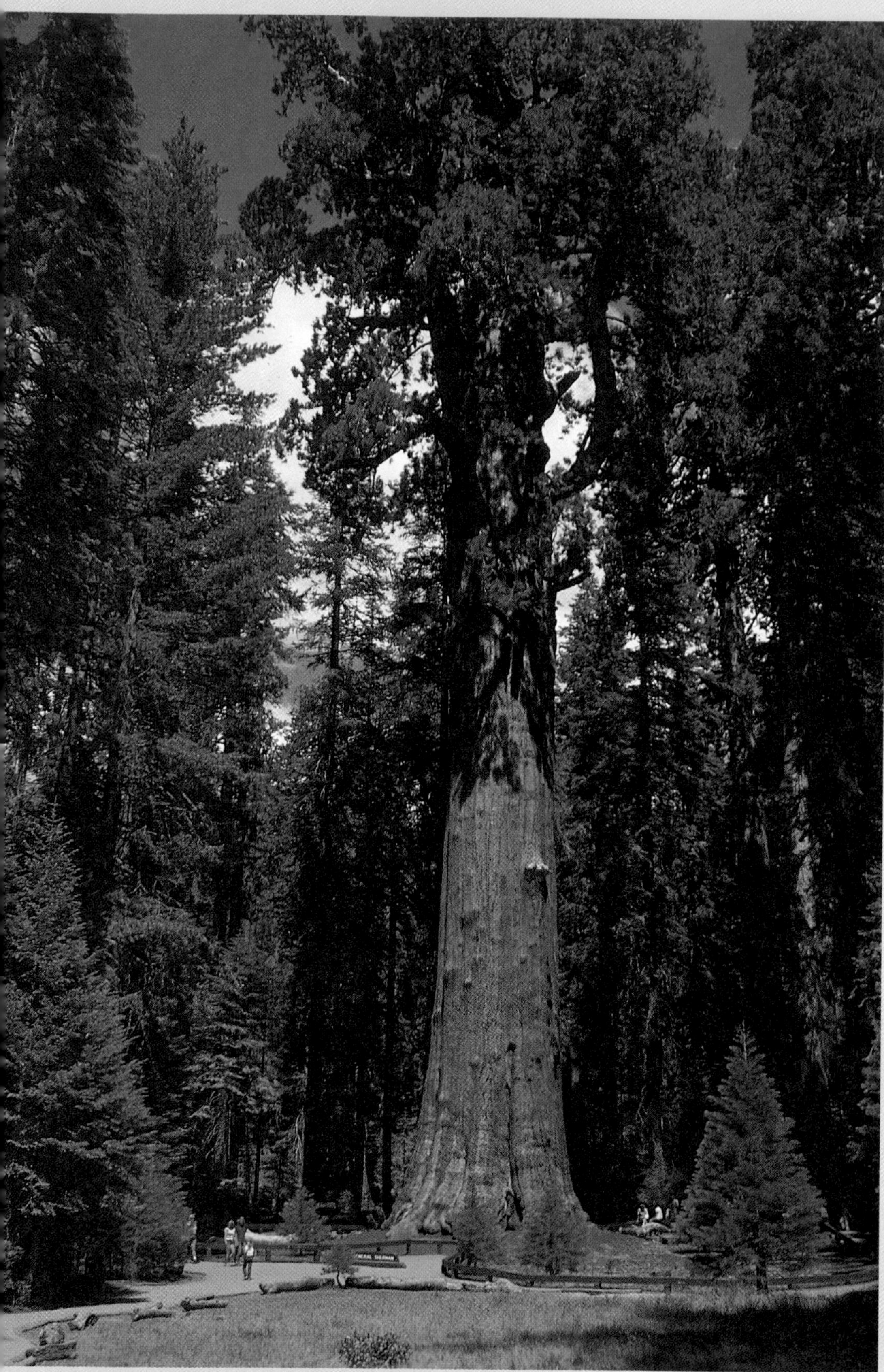

Outline

In 1951, author and conservationist Freeman Tilden wrote, "Not a single *Sequoiadendron gigantea* [giant sequoia] should be cut. They represent a unique survival; they belong to our remote and future histories." The giant sequoia at the left, named the General Sherman for the controversial Union commander in the Civil War, is a case in point; it has been growing for about 2500 years, and plant biologists believe it could live for many more.

A unique survivor and a truly awesome representative of the plant kingdom, the General Sherman resides in the Giant Grove in Sequoia–Kings Canyon National Park, in the Sierra Nevada Mountains of central California. At 83 m (272 ft) tall, it is over 100 feet taller than Niagara Falls. Its massive trunk, weighing nearly 1400 tons, is about 10 m in diameter at the base; its lowest limb, about 40 m above the ground, is over 2 m in diameter. This is the largest plant on Earth.

The photograph on the facing page shows the ancient trunk of another giant sequoia. The rings you see in this cross section are called growth rings because each one marks a year in the tree's life. The rings vary in thickness, depending on weather conditions during the growing seasons. About 1700 years old when it was felled in 1917, this tree was a seedling at the time of the Roman Empire. It was over 200 years old when the Saxons invaded England in A.D. 449. By 1492, when Columbus landed in the New World, it had laid down 80% of its growth rings. Having survived earthquakes, numerous fires, droughts, and storms, most of the giant sequoias were cut down between 1850 and 1900, a time of virtually uncontrolled lumbering. A tree like this one or the General Sherman would yield enough prime boards for nearly 50 four- or five-room houses, and many giant sequoias were cut down for just that purpose. Today, only a few groves remain, nearly all in national and state parks, where they are viewed as living treasures.

As we become more aware of the importance of environmental conservation, our population continues to grow and our demands for plant products increase; lumber, fabric, paper, most of our food, and numerous industrial chemicals all come from plants. Beyond our immediate needs, plants are vital to the planet's well-being. Virtually all land animals depend on them for food, either eating plants directly or eating other animals that eat plants. Above and below the ground, plants provide food, shelter, and breeding areas for animals, fungi, and microorganisms. Plant roots also prevent soil erosion, and photosynthesis in plant leaves helps reduce carbon dioxide levels in the atmosphere and adds oxygen to the air.

A cross section of the trunk of a giant sequoia

The giant sequoia is one of about 550 living species of conifers, or cone-bearing plants. As we saw in Chapter 17, conifers are gymnosperms, one of two groups of seed plants. Gymnosperms bear seeds in cones, whereas plants of the other group, the angiosperms (flowering plants), produce seeds enclosed in fruits. Because angiosperms make up nearly 90% of the plant kingdom, we concentrate on them in this unit. In this chapter, we continue our structure-function theme, studying plant structure in relation to how plants reproduce and grow. We begin with a look at one of plant biology's most distinguished historical figures.

TALKING ABOUT SCIENCE

31.1 Plant scientist Katherine Esau, preeminent student of plant structure and function

Figure 31.1A Katherine Esau with sugar beets, about 1924

Today, with researchers using modern molecular and analytical methods to answer questions about how plants work and how they function in ecosystems, plant biology is in a truly golden age. In this spirit, it seems fitting to honor a pioneer whose energy and focus helped lead the way. Dr. Katherine Esau was one of the twentieth century's most prolific plant scientists. During a career spanning more than 60 years, she published over 150 articles in research journals and six books about plant biology. Two of her texts, *Plant Anatomy* and *Anatomy of Seed Plants,* are classics that have been translated into many languages and used throughout the world.

Born in Ukraine in 1898, Katherine Esau was fascinated by plants at an early age. She entered the Women's Agricultural College in Moscow when she was 17. The Esau family moved to Germany in 1918, and Katherine graduated from the Agricultural College of Berlin in 1922 with a degree in plant breeding. That same year, she emigrated with her parents to the United States.

The Esau family settled in California, and soon thereafter, Katherine was hired as a plant breeder on a sugar-beet seed farm. The photograph in Figure 31.1A was taken in a field near Oxnard, California.

Sugar beets played a major role in Esau's career. Her early research led to an understanding of how viruses infect the plant and destroy its tissues. Her later studies of sugar beets and other angiosperms made Dr. Esau an international expert on phloem (the food-conducting tissue in plants). By the time the electron microscope became available to biologists, Dr. Esau was already in her sixties. Nevertheless, she became skilled with the instrument (Figure 31.1B) and continued to make important discoveries of the detailed structure of plants well into her nineties.

For much of her career, Katherine Esau was a professor of botany at the Davis and Santa Barbara campuses of the University of California. Soft-spoken, meticulous, witty, and a strong leader by example, she excelled in both teaching and research. Dr. Esau studied plant cell structure with the goal of understanding how plant parts work, and until her death in 1993, she remained one of plant biology's strongest advocates of the integrated study of biological structure and function. Among her many accomplishments, Dr. Esau's microscopy studies led to the discovery of plasmodesmata, the open connections between plant cells (see Figure 4.19A). Her research on plasmodesmata suggested how sugars are transported throughout the plant body. She also discovered that plant viruses are transmitted through plant tissues via plasmodesmata. Her dedication to discovering how a plant's structure fits its activities leads us into this chapter.

By studying the structure of a plant's cells and tissues, botanists gain insight into how those structures __________.

function

Figure 31.1B Dr. Esau with an electron microscope

There are two major groups of angiosperms: monocots and dicots 31.2

Angiosperms have dominated the land for over 100 million years, and there are about 235,000 species of flowering plants living today. Most of our foods come from a few hundred domesticated species of flowering plants. Among these foods are roots, such as beets and carrots; the fruits of trees and vines, such as apples, nuts, berries, and squashes; the fruits and seeds of legumes, such as peas and beans; and grains, the fruits of grasses such as rice, wheat, and corn.

Plant biologists classify angiosperms in two groups, called monocots and dicots, on the basis of several structural features, illustrated below. The names "monocot" and "dicot" refer to the first leaves that appear on the plant embryo. These embryonic leaves are called seed leaves, or **cotyledons.** A **monocot** embryo has one seed leaf; a **dicot** embryo has two seed leaves.

Monocots (about 65,000 species) include the orchids, bamboos, palms, and lilies, as well as the grains and other grasses. You can see the single cotyledon inside the seed on the top left in Figure 31.2. The leaves, stems, flowers, and roots of monocots are also distinctive. Most monocots have leaves with parallel veins. Monocot stems have vascular tissues (tissues that transport water and nutrients) arranged in a complex array of bundles. The flowers of most monocots have their petals and other parts, which we will describe later, typically in multiples of three. The roots of monocots form a fibrous system—a mat of threads—that spreads out below the soil surface. With most of their roots in the top few centimeters of soil, monocots, especially grasses, make excellent ground cover that reduces erosion. The roots of a rye-grass plant, for instance, have more than 20 times the surface area of its stems and leaves.

Most angiosperms (about 170,000 species) are dicots. This group includes most shrubs and trees (except for the conifers), as well as the majority of our ornamental plants and many of our food crops. You can see the two cotyledons of a typical dicot in the seed on the lower left. Dicot leaves have a multibranched network of veins, and dicot stems have vascular bundles arranged in a ring. The dicot flower usually has petals and other parts in multiples of four or five. The large, vertical root of a dicot, known as a taproot, goes deep into the soil, as you know if you've ever tried to pull up a dandelion.

As we saw in the preceding unit on animals, a close look at a structure often reveals its function. Conversely, function provides insight into the logic of a structure. In the modules that follow, we'll begin a detailed look at the correlation between plant structure and function.

The "cot" in monocot and dicot refers to the number of _________ in the seed.

cotyledons (or seed leaves)

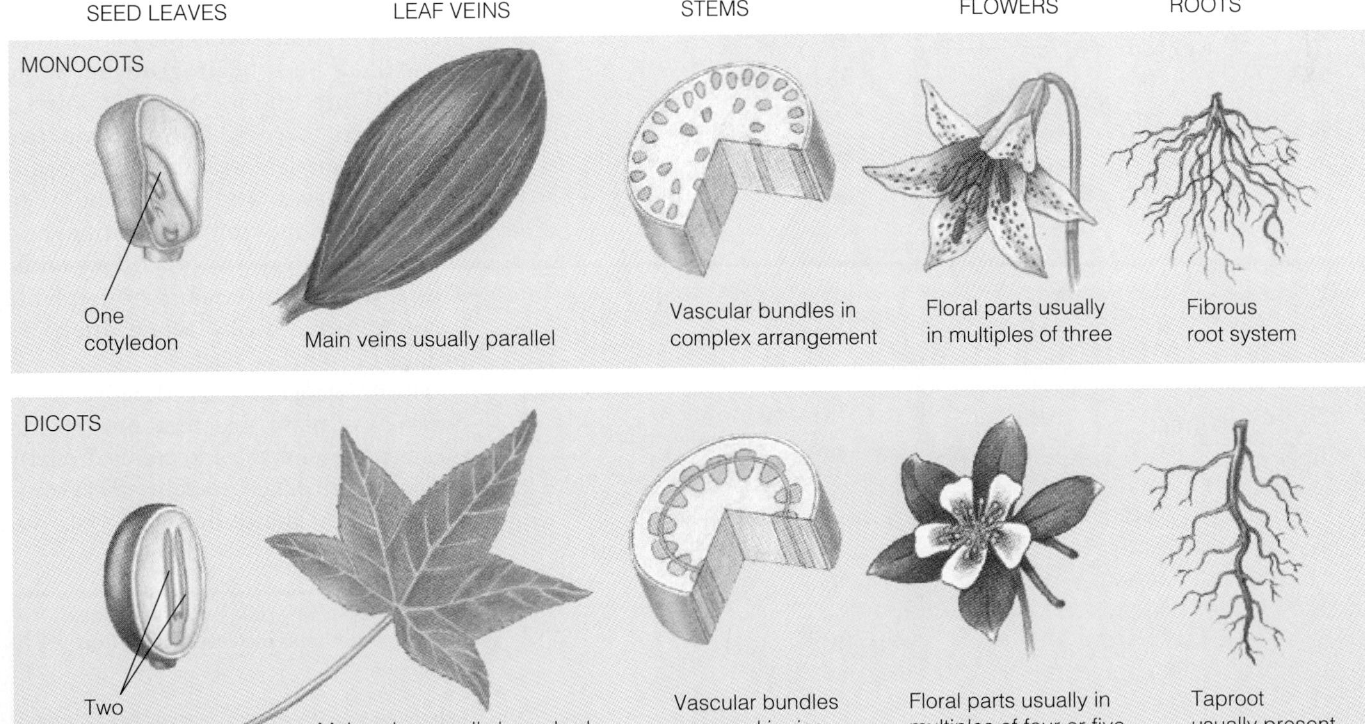

Figure 31.2 A comparison of monocots and dicots

31.3 The plant body consists of roots and shoots

Among the evolutionary adaptations that made it possible for plants to move onto land were the abilities to take up water and minerals from the soil, to absorb light and take in carbon dioxide from the air for photosynthesis, and to survive dry conditions. The subterranean roots and aerial shoots of a land plant, such as the generalized flowering plant shown in Figure 31.3, perform all these vital functions. Neither the root nor the shoot can survive without the other. Lacking chloroplasts and living in the dark, the roots would starve without sugar and other organic nutrients transported from the photosynthetic leaves of the shoot system. Conversely, stems and leaves depend on the water and minerals absorbed by roots.

A plant's **root system** anchors it in the soil, absorbs and transports minerals and water, and stores food. The fibrous root system of a monocot provides broad exposure to soil water and minerals, as well as firm anchorage. The root system of a dicot, with many small secondary roots growing out from one large taproot, is different but also effective at absorbing water and minerals. Near the root tips in both dicots and monocots, tiny projections called root hairs increase the surface area of the root. Each **root hair** is an outgrowth of an epidermal cell (a cell in the outer layer of the root). Root hairs absorb water and minerals.

The **shoot system** of a plant is made up of stems, leaves, and adaptations for reproduction—flowers, in angiosperms. (We'll return to the angiosperm flower in Module 31.9.) As indicated in the drawing, the **stems** are the parts of the plant that are generally above the ground and that support the leaves and flowers. In the case of a tree, the stems are the trunk and all the branches, including the smallest twigs. A stem has **nodes,** the points at which leaves are attached, and **internodes,** the portions of the stem between nodes. The **leaves** are the main site of photosynthesis in most plants, although some species have green, photosynthetic stems. A leaf consists of a flattened blade and a stalk, or petiole, which joins the leaf to the stem.

The two types of buds you see in the figure are undeveloped shoots. When a plant stem is growing in length, the **terminal bud,** at the apex (tip) of the stem, has developing leaves and a compact series of nodes and internodes. The **axillary buds,** one in each of the angles formed by a leaf and the stem, are usually dormant. In many plants, the terminal bud produces hormones that inhibit growth of the axillary buds, a phenomenon called **apical dominance.** By concentrating resources on growing taller, apical dominance is an evolutionary adaptation that increases the plant's exposure to light. This is especially important where vegetation is dense. However, branching is also important for increasing the exposure of the shoot system to the environment, and under certain conditions, the axillary buds begin growing. Some develop into shoots bearing flowers, and others become nonreproductive branches complete with their own terminal buds, leaves, and axillary buds. In some cases, removing the terminal bud stimulates the growth of axillary buds. This is why fruit trees and houseplants become more bushy when they are pruned or "pinched back."

The drawing in Figure 31.3 gives us an overview of plant structure, but it by no means represents the enormous diversity of angiosperms. Let's look briefly at some variations on the basic themes of root and stem structure.

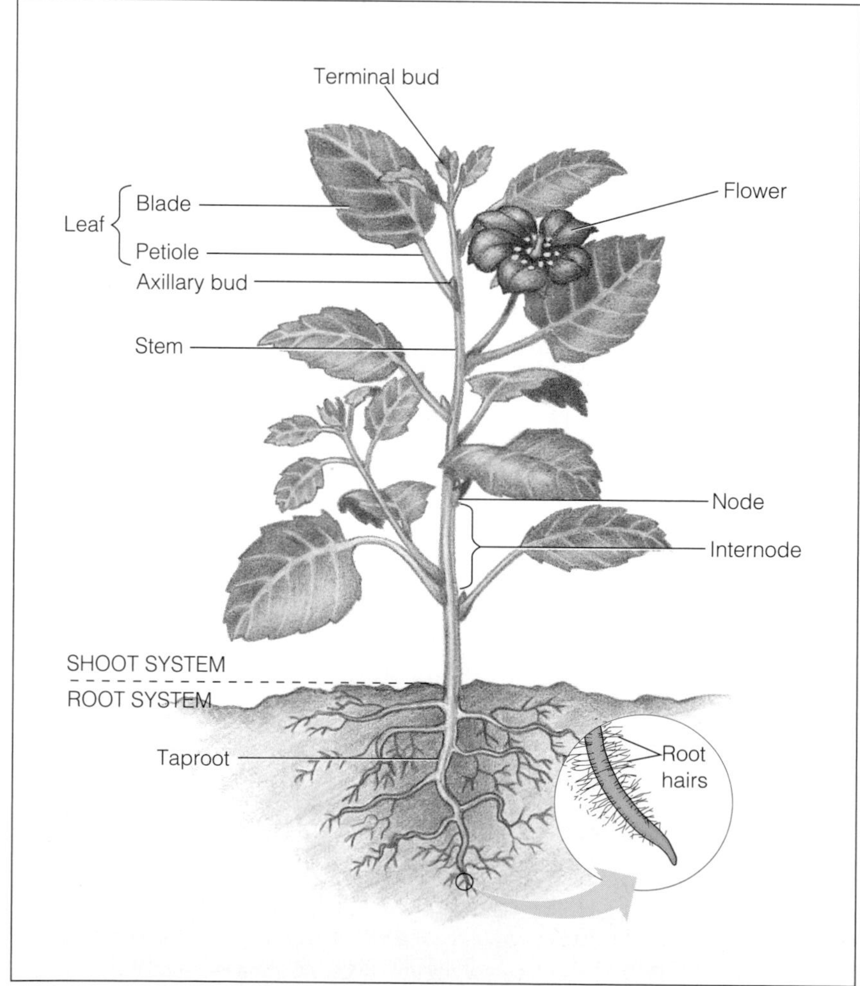

Figure 31.3 The body plan of a flowering plant (a dicot)

Explain why pruning certain types of fruit trees increases future fruit harvest.

Removal of terminal buds from major branches results in more branching by reducing inhibition of axillary buds. More branches produce more flowers, hence more fruit.

Many plants have modified roots and shoots

Roots, stems, and leaves come in a variety of sizes and shapes, and they are adapted for a variety of functions. Many are adapted for storing food. The vegetables we call carrots, turnips, sugar beets, and sweet potatoes, for instance, are unusually large taproots that store food in the form of carbohydrates such as starch. The plants use the stored sugars during periods of active growth and when they are producing flowers and fruits. Plant breeders have improved the yields of root crops by selecting varieties, such as the sugar beet plant in Figure 31.4A, with very large taproots.

Figure 31.4A The modified root of a sugar beet plant

Figure 31.4B shows some examples of modified stems. The strawberry plant has a horizontal stem, or runner, that grows along the ground surface. A runner is a means of asexual reproduction; as shown here, a new plant can emerge from its tip. You've seen a different stem modification if you have ever dug up an iris plant; the large, brownish, rootlike structures near the soil surface are actually horizontal stems called **rhizomes.** The rhizomes store food, and having buds, they can also spread and form new plants. The white potato plant has rhizomes that end in enlarged structures called **tubers** (the potatoes we eat), where food is stored in the form of starch.

Plant leaves, too, are highly varied. Grasses and most other monocots, for instance, have long leaves without petioles. Some dicots, such as celery, have enormous petioles—the stalks we eat—which contain a lot of water and stored food. The upper photograph in Figure 31.4C shows a modified leaf called a tendril, with its tip coiled around a wire. Tendrils help plants such as this cucumber climb. The spines of the prickly pear cactus (lower photo) are parts of modified leaves that may protect the plant from plant-eating animals. The main part of the cactus is the green stem, which is adapted for photosynthesis and water storage.

So far we have examined plants as we see them with the unaided eye. Next, we begin to dissect the plant and explore its microscopic organization.

The "eyes" of a white potato mark nodes with buds. If those buds break dormancy and the potato "sprouts," are the resulting appendages (choose one) *root* branches or *shoot* branches? Explain your answer.

Shoot branches; the potato tuber is a modified stem, part of the shoot system.

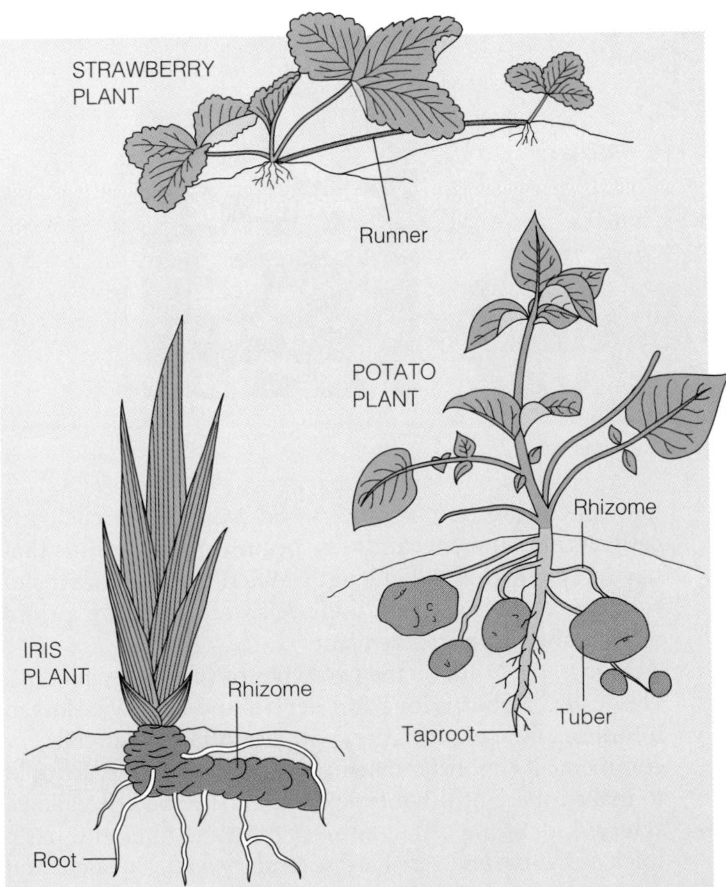

Figure 31.4B Three kinds of modified stems

Figure 31.4C Modified leaves: a cucumber tendril (top) and cactus spines (bottom)

31.5 Plant cells and tissues are diverse in structure and function

Plant cells are unique in several ways. Many have chloroplasts, which contain the photosynthetic pigments chlorophyll *a* and *b*. Mature plant cells often have a large central vacuole (Figure 31.5A). Fluid in the vacuole helps maintain the cell's firmness, or turgor, which we discuss in Chapter 32.

The single most distinctive feature of plant cells is their cell wall surrounding the plasma membrane. Plant cell walls are made mainly of the structural carbohydrate cellulose. Many plant cells, especially those that provide structural support, have a two-part cell wall; a primary cell wall is laid down first, and then a more rigid secondary cell wall is secreted between the plasma membrane and the primary wall. The enlargement on the right in Figure 31.5A shows the adjoining cell walls of two cells. The primary walls of adjacent cells in plant tissues are held together by a sticky layer called the middle lamella. Pits, where the cell wall is relatively thin, allow the contents of adjacent cells to lie close together. Plasmodesmata are channels of communication and circulation between adjacent plant cells.

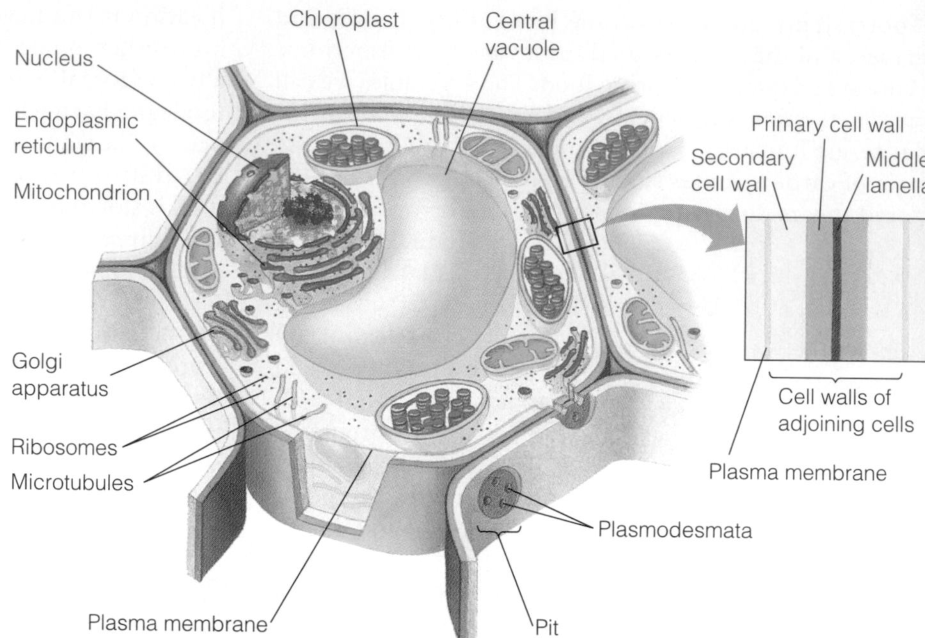

Figure 31.5A The structure of a plant cell

The structure of the cell and the nature of its wall often correlate with the cell's main functions. This is evident in the five major types of plant cells shown in Figures 31.5B–31.5F.

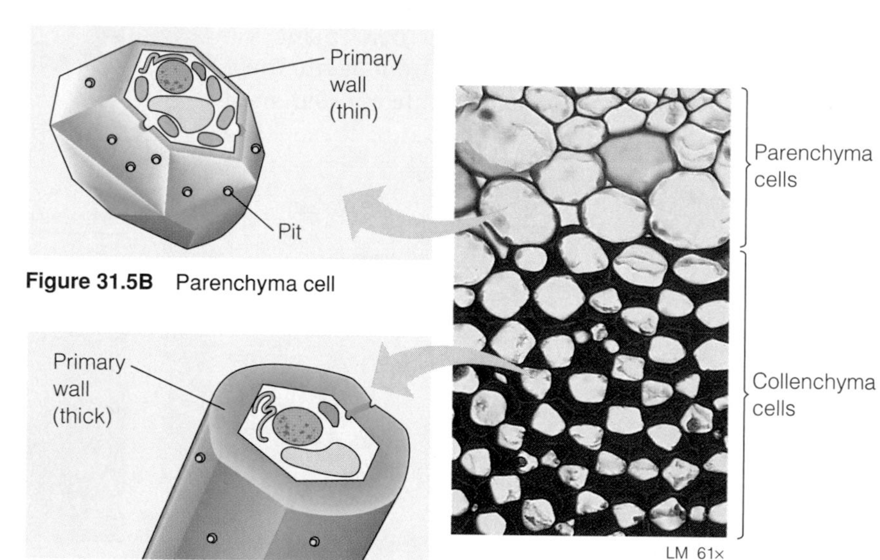

Figure 31.5B Parenchyma cell

Figure 31.5C Collenchyma cell

Parenchyma cells (Figure 31.5B) are the most abundant type of cell in most plants. They remain alive when mature and have only primary (often thin) walls. Parenchyma cells perform a variety of functions, such as food storage, photosynthesis, and aerobic respiration. Parenchyma cells come in a variety of shapes, but they are often multisided, like the one in Figure 31.5A. Most parenchyma cells can divide and differentiate into other types of plant cells, which they may do as part of the process of repairing an injury.

Collenchyma cells (Figure 31.5C) resemble parenchyma cells in lacking secondary walls, but they have unevenly thickened primary walls. Their main function is to provide support in parts of the plant that are still growing; young stems often have collenchyma cells just below their surface. These living cells elongate with the growing stem.

Sclerenchyma cells have rigid secondary cell walls (yellow in Figure 31.5D) hardened with lignin, which is the main chemical component of wood. Mature sclerenchyma cells cannot elongate, and they occur only in regions that have stopped growing in length. When mature, most sclerenchyma cells are dead, their cell walls forming a rigid scaffold that supports the plant.

Figure 31.5D shows the two types of sclerenchyma cells. One, called a **fiber,** is long and slender and usually occurs in bundles. Some plant tissues with abundant fiber cells are commercially important; hemp fibers, for example, are used to make rope. The other type of sclerenchyma cell, called a **sclereid,** or stone cell, is shorter than the fiber and has a thick, irregular, and very hard secondary wall. Nutshells and seed coats owe their hardness to sclereids, and sclereids scattered in the soft tissue of a pear feel gritty when eaten.

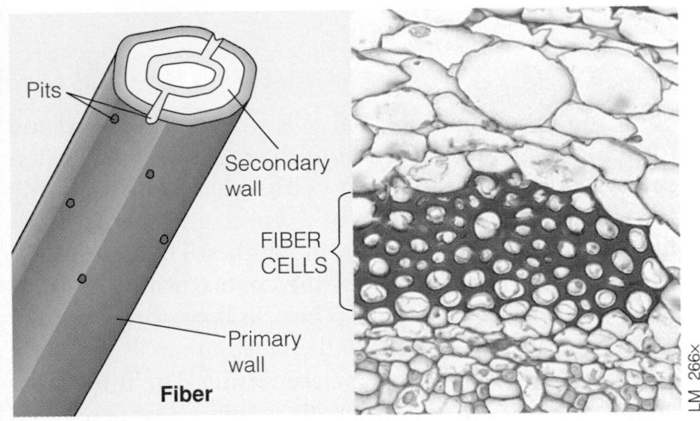

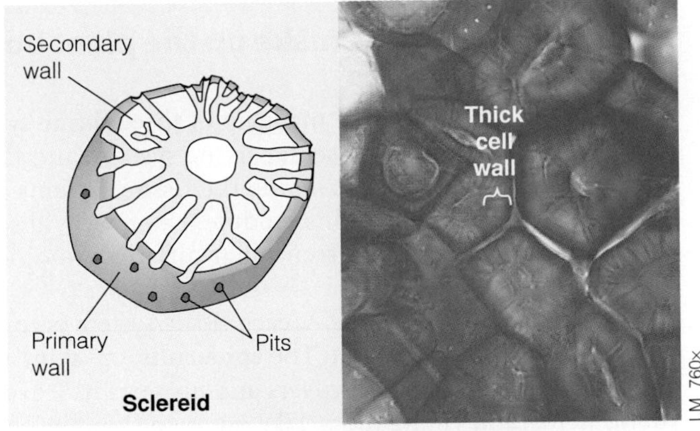

Figure 31.5D Sclerenchyma cells

Water-conducting cells are of two types, both having rigid, lignin-containing secondary cell walls. As Figure 31.5E shows, **tracheids** are long cells with tapered ends. **Vessel elements** are wider, shorter, and less tapered. Chains of tracheids or vessel elements with overlapping ends form a system of tubes that convey water from the roots to the stems and leaves. The tubes are hollow because, when mature, both tracheids and vessel elements are dead, and only their cell walls remain. Water passes through pits in the walls of tracheids and vessel elements, and through openings in the end walls of vessel elements. Because of their thick, rigid walls, these cells also function in support.

Food-conducting cells, also known as **sieve-tube members,** are also arranged end-to-end, forming tubes (Figure 31.5F). Unlike water-conducting cells, however, sieve-tube members have thin primary walls and no secondary walls, and they remain alive at maturity. Their end walls, which are perforated large plasmodesmata, form **sieve plates,** through which sugars, other compounds, and some mineral ions move between adjacent food-conducting cells. Each sieve-tube member is flanked by at least one **companion cell,** which is connected to the sieve-tube member by numerous plasmodesmata. The nucleus and ribosomes of the companion cells may make certain proteins for the sieve-tube member, which loses its nucleus, ribosomes, and some other organelles during development.

As in animals, the cells of plants are grouped into tissues with characteristic functions, such as photosynthesis. Simple tissues consist of a single cell type, such as parenchyma. Thus, we can speak of a parenchyma cell, but we can also refer to parenchyma tissue, an association of many parenchyma cells.

A plant tissue composed of more than one type of cell is called a complex tissue. The vascular tissues of plants—tissues that conduct water and food—are complex tissues. Vascular tissue called **xylem** contains water-conducting cells that convey water and dissolved minerals upward from the roots. Vascular tissue called **phloem** contains sieve-tube members that transport sugars from leaves or storage tissues to other parts of the plant. In addition to having cells that conduct food or water, phloem and xylem also contain sclerenchyma cells that provide support and parenchyma cells that store various materials.

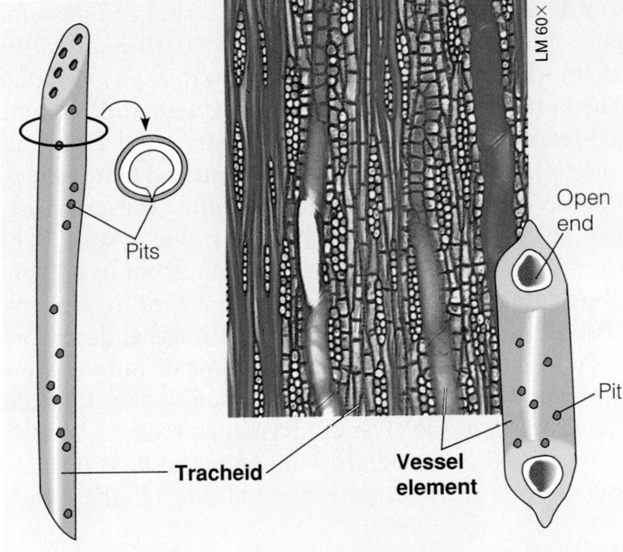

Figure 31.5E Water-conducting cells

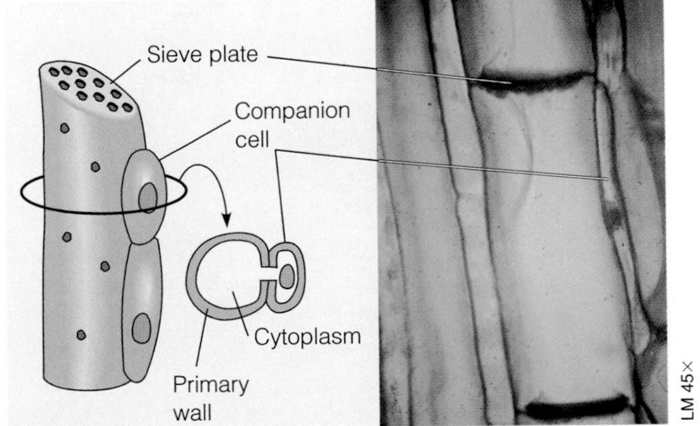

Figure 31.5F Food-conducting cells (sieve-tube members)

Plants, like animals, have several levels of structural organization. We examine the level above plant tissues, called tissue systems, next.

Which of the following cell types has the potential to give rise to all others in the list: collenchyma, sclereid, parenchyma, vessel element, companion cell?

Parenchyma

31.6 Three tissue systems make up the plant body

31A
See Activity 31A on the CD-ROM.

Roots, stems, and leaves are made up of three **tissue systems:** the epidermis, the vascular tissue system, and the ground tissue system. We examine the tissue systems of young roots and shoots in this module. Later, we will see that the tissue systems are somewhat different in older roots and stems.

As indicated in Figure 31.6A, each tissue system is continuous throughout the plant. The **epidermis,** or "skin," of a plant (blue in the figure), covers and protects its leaves, young stems, and young roots. Like our own skin, the epidermis is a plant's first line of defense against physical damage and infectious organisms. On the leaves and on some stems, epidermal cells secrete a waxy coating called the **cuticle,** which helps the plant retain water. The **vascular tissue system** (purple), made up of xylem and phloem, provides support and transports water and nutrients throughout the plant. The **ground tissue system** (yellow) makes up the bulk of a young plant, filling the spaces between the epidermis and vascular tissue system. The ground tissue system, mainly parenchyma, but usually including some collenchyma and sclerenchyma, has diverse functions, including photosynthesis, storage, and support.

Figure 31.6B, a cross section of a young dicot root (from a buttercup), shows what the three tissue systems look like under a microscope. The epidermis is a single layer of tightly packed cells covering the entire root. Water and minerals enter the plant from the soil through these cells. Some of the young epidermal cells will grow outward and form root hairs. In the center of the root, the vascular tissue system forms a cylinder, with xylem cells radiating from the center like spokes of a wheel, and phloem cells filling in the wedges between the spokes. The ground tissue system of the root forms the **cortex,** which consists mostly of parenchyma tissue. The cells here store food (the purple granules are starch) and take up minerals that have entered the root through the epidermis. The innermost layer of cortex is the **endodermis,** a thin cylinder one cell thick. The endodermis is a selective barrier, determining which substances pass between the rest of the cortex and the vascular tissue. (We discuss how this barrier works in Chapter 32.)

As Figure 31.6C indicates, the young stem of a dicot (sunflower) looks quite different from that of a monocot (corn). Both stems have their vascular tissue system arranged in numerous **vascular bundles.** However, the monocot stem has vascular bundles throughout its ground tissue system, whereas the dicot stem has a distinct ring of vascular bundles and a two-part ground tissue system. The dicot cortex fills the space between the vascular ring and the epidermis. The other part of the dicot ground tissue system, called the **pith,** fills the center of the stem and is often important in food storage.

Figure 31.6D illustrates the arrangement of the three tissue systems found in many dicot leaves. Leaves (and some stems) have pores, called **stomata** (singular, *stoma*), in their epidermis. You can see the stomata clearly in the micrograph on the right. Each is a gap between two specialized epidermal cells, called guard cells. **Guard cells** regulate the size of the stomata, allowing gas exchange between the surrounding air and the photosynthetic cells inside the leaf. The stomata are also the major avenues of water loss from the plant, as we discuss in Chapter 32.

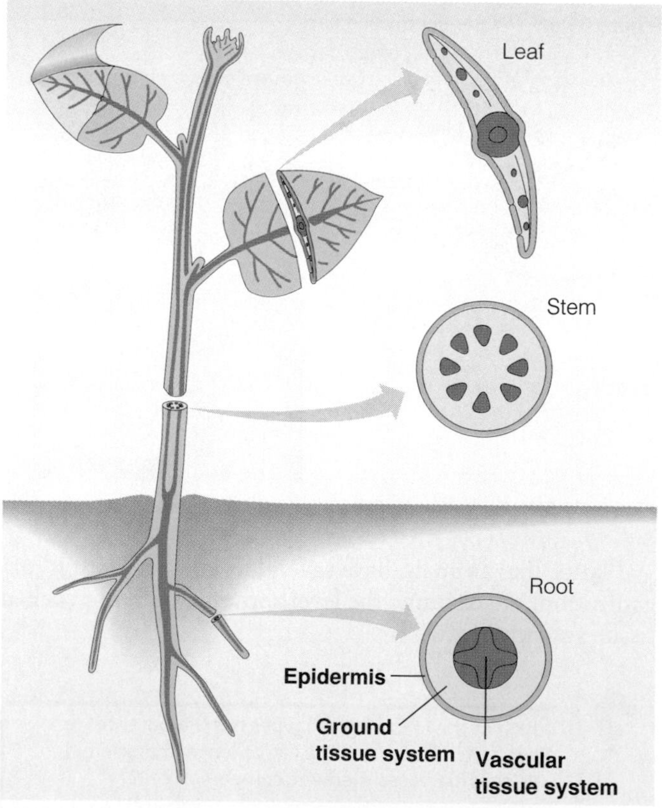

Leaf
Stem
Root
Epidermis
Ground tissue system
Vascular tissue system

Figure 31.6A The three tissue systems

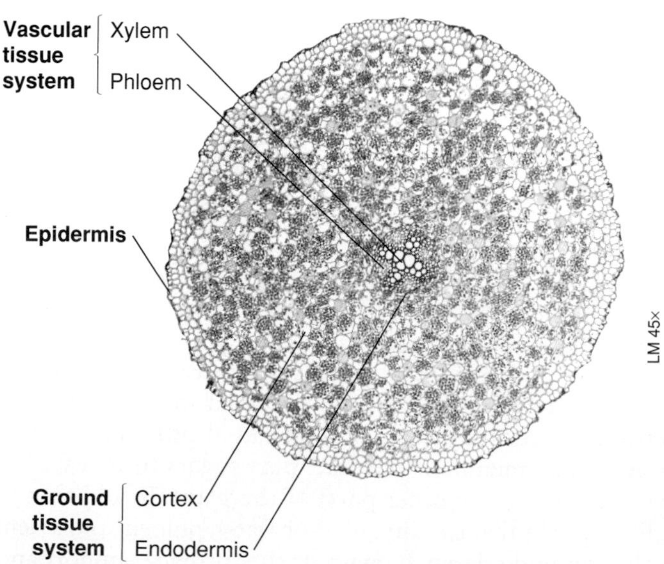

Figure 31.6B Tissue systems in a young root (a dicot)

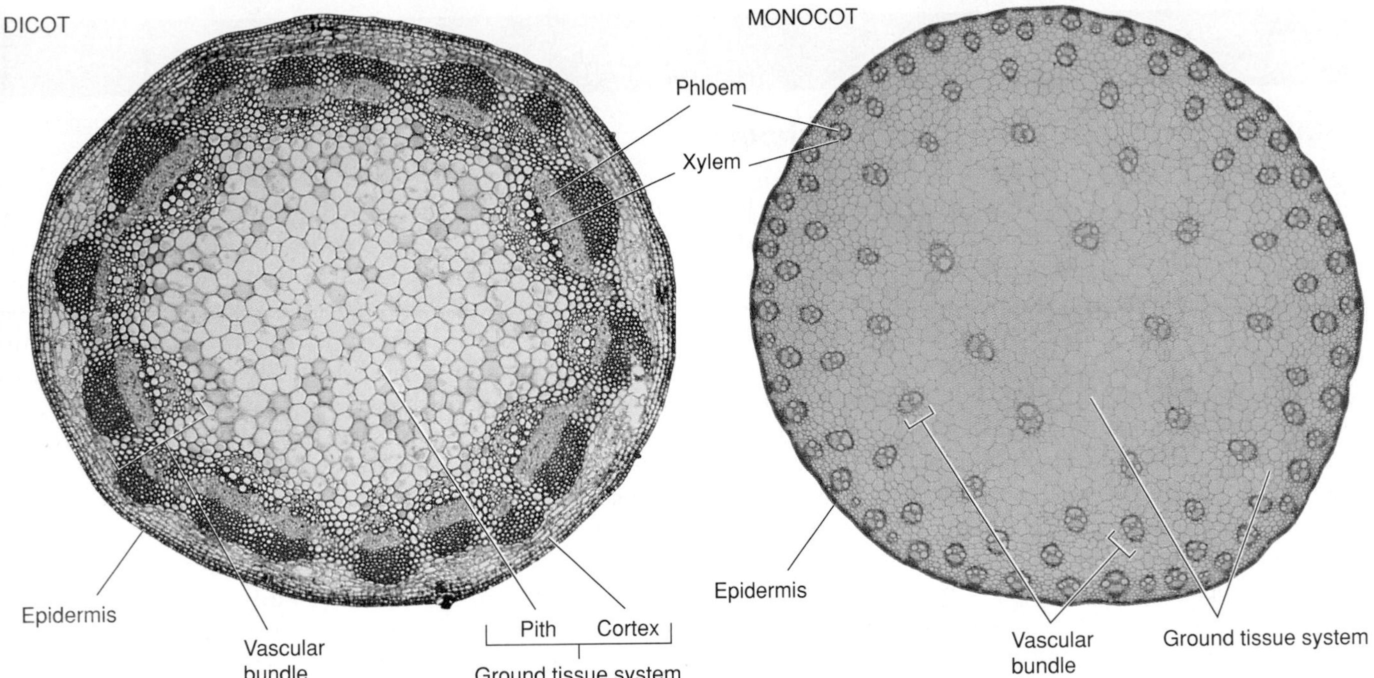

Figure 31.6C Tissue systems in young stems

The ground tissue system of a leaf is called **mesophyll.** Mesophyll consists mainly of parenchyma cells equipped with chloroplasts (green in Figure 31.6D) and specialized for photosynthesis. In this dicot leaf, notice that cells in the lower mesophyll are loosely arranged, with many air spaces. This is the chief site of gas exchange. Air enters through stomata and circulates freely in the spaces among the cells. The stomata are more numerous in the leaf's lower epidermis. This adaptation minimizes water loss, which occurs more rapidly from the sunlit upper surface of the leaf.

The leaf's vascular tissue system is made up of a network of veins. As you can see in the figure, each **vein** is a vascular bundle composed of xylem and phloem surrounded by a sheath of parenchyma cells. The veins' xylem and phloem, continuous with the vascular bundles of the stem, are in close contact with the leaf's photosynthetic tissues. This ensures that those tissues are supplied with water and mineral nutrients from the soil and that sugars made in the leaves are transported throughout the plant. In some plants (the C_4 plants; see Module 7.12), the sheath cells are the sites of the Calvin cycle in photosynthesis.

This completes our survey of basic plant anatomy. Next, we examine how plants grow in length and thickness. 

? Biologists generally define an "animal tissue" as a group of cells with common structure and function. How does this definition of a tissue contrast with what biologists call a "tissue system" in plants?

A plant tissue system may consist of several types of specialized cells, such as the different cell types of the vascular tissue system.

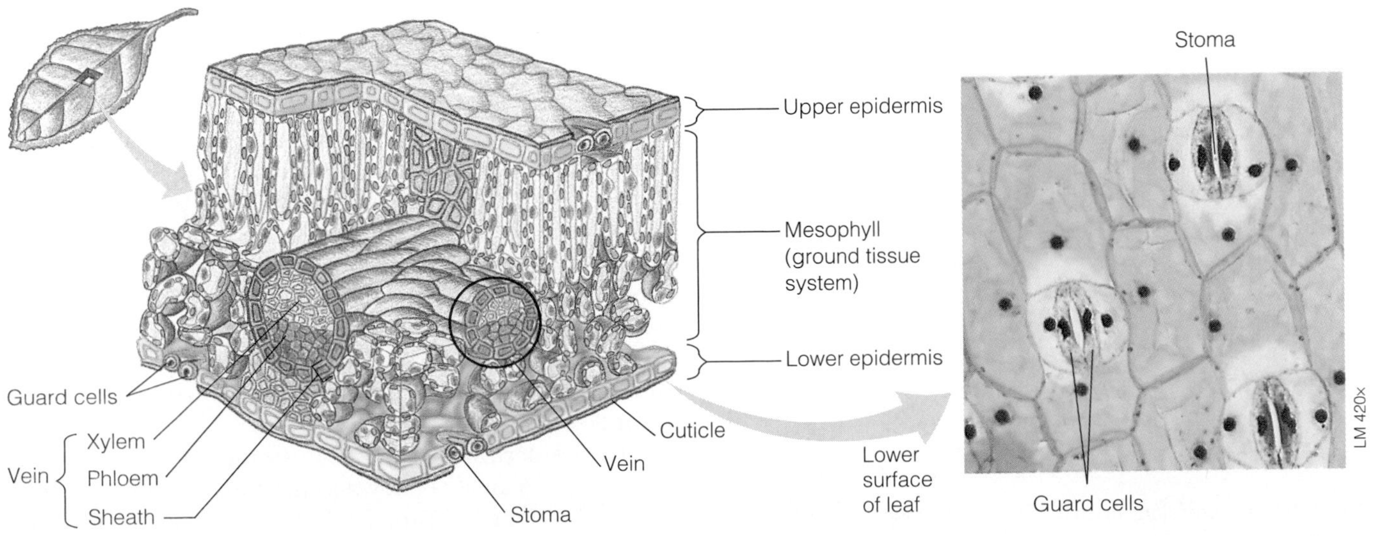

Figure 31.6D Tissue systems in a dicot leaf

31.7 Primary growth lengthens roots and shoots

Most species of plants continue to grow as long as they live, a condition known as **indeterminate growth.** Most animals, in contrast, are characterized by **determinate growth;** that is, they cease growing after reaching a certain size. These differences underlie a broader distinction between plants and animals. Most animals *move* through their environment. Plants, in contrast, *grow* through their environment. This indeterminate growth enables plants to increase their exposure to sunlight, air, and soil throughout life.

Indeterminate growth does not mean that plants are immortal. In fact, most plants have a finite life span. Species called **annuals,** for instance, complete their life cycle (the time from when they begin to grow through flowering and seed production to death) in a single year or growing season. Our most important food crops (wheat, corn, and rice, for example) are annuals, as are a great number of wildflowers. **Biennials** complete their life cycle in two years; flowering usually occurs during the second year. Beets and carrots are biennials, but we usually harvest them in their first year and miss seeing their flowers. Plants that live and reproduce for many years, including trees, shrubs, and some grasses, are known as **perennials.**

Some perennial plants are among the oldest organisms alive. The giant sequoias we discussed in the chapter's introduction are ancient (the oldest living one began growing about 3000 years ago), and another conifer, the bristlecone pine, can outlive the sequoias by thousands of years. Older still are clones of certain shrubs and grasses (see Figure 31.14C). It is likely that even the oldest plants could live for many more thousands of years if they could escape disease, fires, floods, climatic change, and human encroachment.

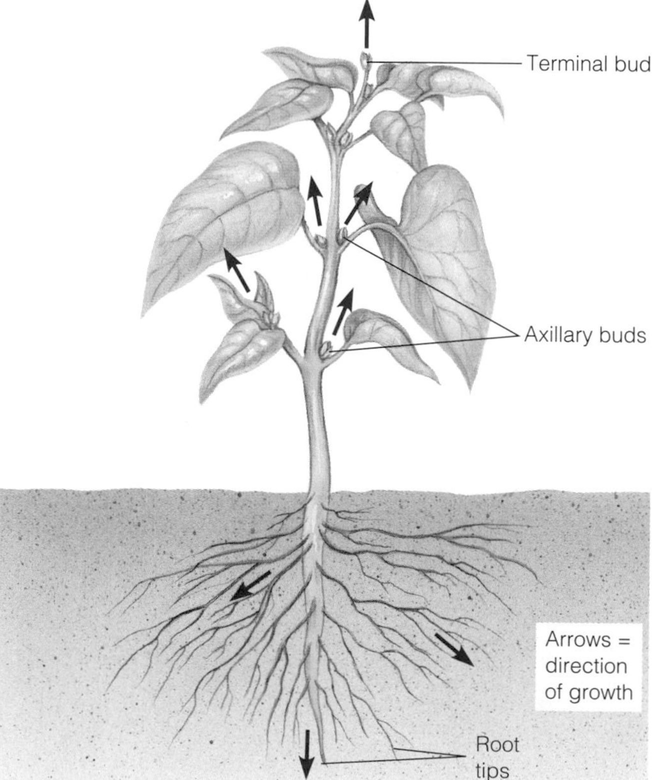

Figure 31.7A Locations of apical meristems, responsible for primary growth

Growth in all plants is made possible by tissues called meristems. A **meristem** (from the Greek *meristos,* divided) consists of localized, unspecialized cells that divide and generate new cells and tissues. Meristems are present throughout a plant's life. Those at the tips of roots and in the terminal (top) and axillary buds of shoots are called **apical meristems** (Figure 31.7A). Cell division in the apical meristems produces the new cells that enable a plant to grow in length.

The lengthwise growth produced by apical meristems is called **primary growth.** Figure 31.7B is a longitudinal section through a growing onion root. Primary growth actually helps push the root through the soil. The very tip of the root is the **root cap,** a thimblelike cone of cells that protect the delicate, actively dividing cells of the apical meristem. The root's apical meristem (outlined by the blue oval) has two roles: It replaces cells of the root cap that are scraped away by the soil (downward arrow), and it produces the cells for primary growth (upward arrow). Cells produced by primary growth form three concentric cylinders of developing ("embryonic") tissue. The outermost cylinder will differentiate into the epidermis of the root. The intermediate cylinder—the bulk of the root tip—will develop into the root's cortex. The innermost cylinder will become the vascular tissue.

The apical meristem sustains growth of the root by continuously adding cells to the three developing tissue cylinders. However, cell division alone does not make the root elongate. An equally important factor is the lengthening of cells. Notice in Figure 31.7B that the meristem grades upward into a zone of elongation. Cells in this zone can undergo a tenfold increase in length. They lengthen, rather than expand equally in all directions, mainly because of the circular arrangement of cellulose fibers in parallel bands in their cell walls. The enlargement diagrams at the left of the root indicate how this works. The cells elongate by taking up water, and as they do, the cellulose fibers (shown in red) separate, somewhat like an expanding accordion. The cells cannot expand greatly in width because the cellulose fibers do not stretch much. Elongation is what actually forces the root down through the soil.

The epidermis, cortex, and vascular cylinder actually take shape in an area known as the zone of differentiation, above the elongation zone. (One zone actually grades into the next.) Cells of the vascular cylinder differentiate into

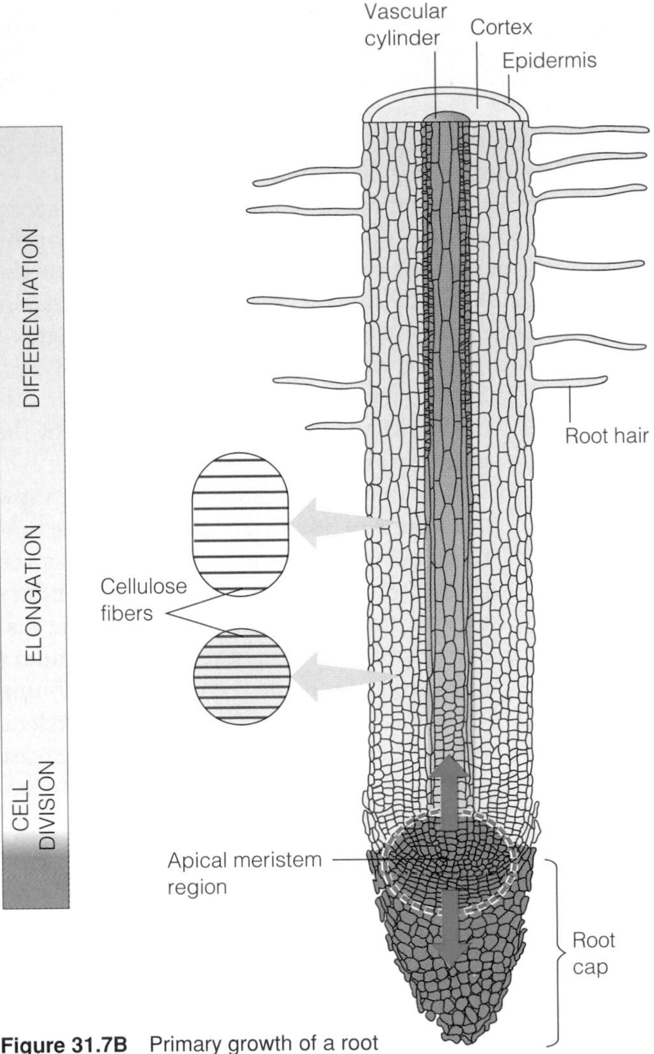

Figure 31.7B Primary growth of a root

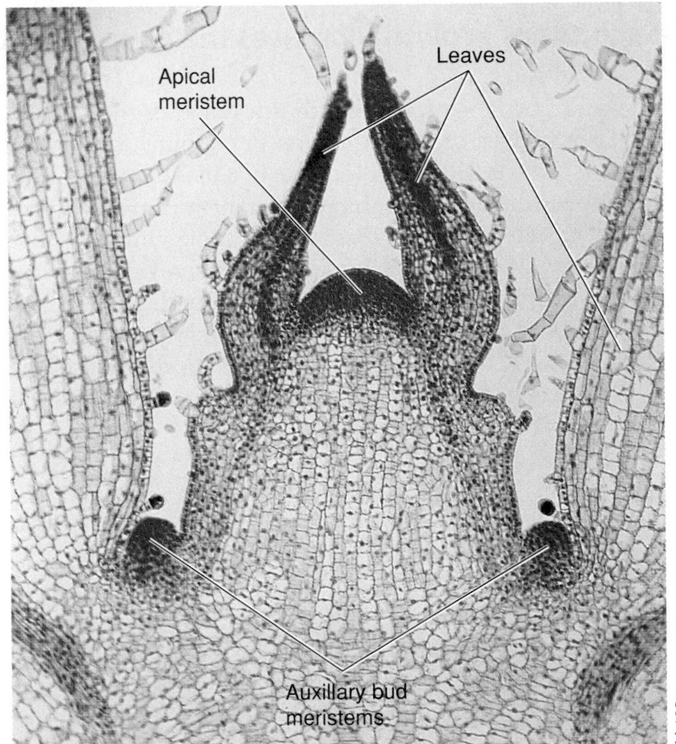

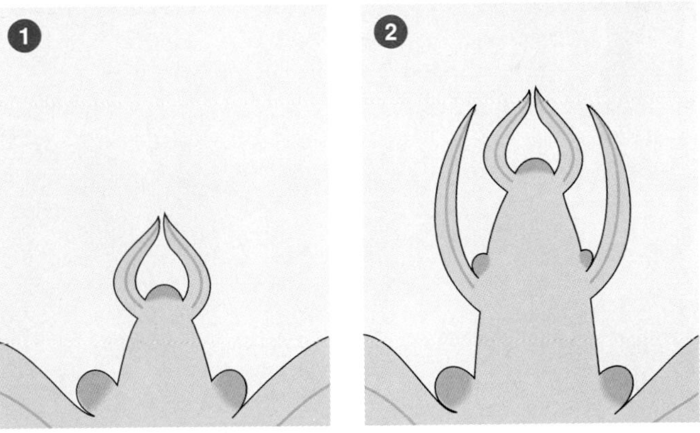

Figure 31.7C Primary growth of a shoot

vascular tissues called **primary xylem** and **primary phloem.** Differentiation of cells (specialization of their structure and function) results from differential gene expression (see Module 11.4). Cells in the vascular cylinder, for instance, develop into primary xylem or phloem cells because a certain set of genes is turned on and is therefore expressed as specific proteins, while other genes in these cells are turned off. Genetic control of differentiation is one of the most active areas of plant research. Recently, developmental biologists have discovered that plants, like animals, have homeotic genes (see Module 11.12), master control genes that determine major features of the body, such as the placement of sepals and petals on flowers.

The micrograph in Figure 31.7C shows a longitudinal section through the end of a growing shoot that was cut lengthwise from its tip to just below its uppermost pair of axillary buds. You can see the apical meristem, which is a dome-shaped mass of dividing cells at the very tip of the terminal bud. Elongation occurs just below this meristem, and the elongating cells push the apical meristem upward, instead of downward as in the root. As the apical meristem advances upward, some of its cells remain behind, and these become new axillary bud meristems at the base of the leaves. As in the root, the apical meristem forms three developing tissue cylinders. Below the zone of elongation in the stem, the cells of the cylinders differentiate into the plant's three tissue systems.

The sketches in Figure 31.7C show two stages in the growth of a shoot. Stage ① is just like the micrograph. At the later stage shown in sketch ②, the apical meristem has been pushed upward by elongating cells underneath.

Primary growth accounts for a plant's lengthwise growth. The stems and roots of many plants increase in thickness too, and in the next module, we see how this usually happens. 

? You have cells in the lower layers of your skin that continue dividing, replacing dead cells that slough from your surface. Why is it inaccurate to compare such regions of active cell division in your body to a plant meristem?

Your dividing cells normally are limited in the types of cells they can form. In contrast, the products of cell division in plant meristem differentiate into all the diverse cell types of a plant.

31.8 Secondary growth increases the girth of woody plants

31B

Stems and roots often begin to thicken after their apical meristems have produced the three developing tissue cylinders we just described. An increase in a plant's girth results from **secondary growth.** Increase in girth is most evident in the woody plants—trees, shrubs, and vines—whose stems last from year to year and consist mainly of thick layers of mature, dead xylem tissue, called wood.

Secondary growth involves cell division in two meristems we have not yet discussed: the vascular cambium and the cork cambium. The **vascular cambium,** shown in green in the dicot stem in Figure 31.8A, first appears as a cylinder of actively dividing cells (called a lateral meristem) between the primary xylem and primary phloem, as you can see in the pie section at the top of the figure. This region of the stem is just beginning secondary growth. Except for the vascular cambium, the stem at this stage of growth is virtually the same as a young stem growing only in length—that is, showing only primary growth (see Figure 31.6C). Secondary growth adds cells on either side of the vascular cambium, as indicated by the green arrows.

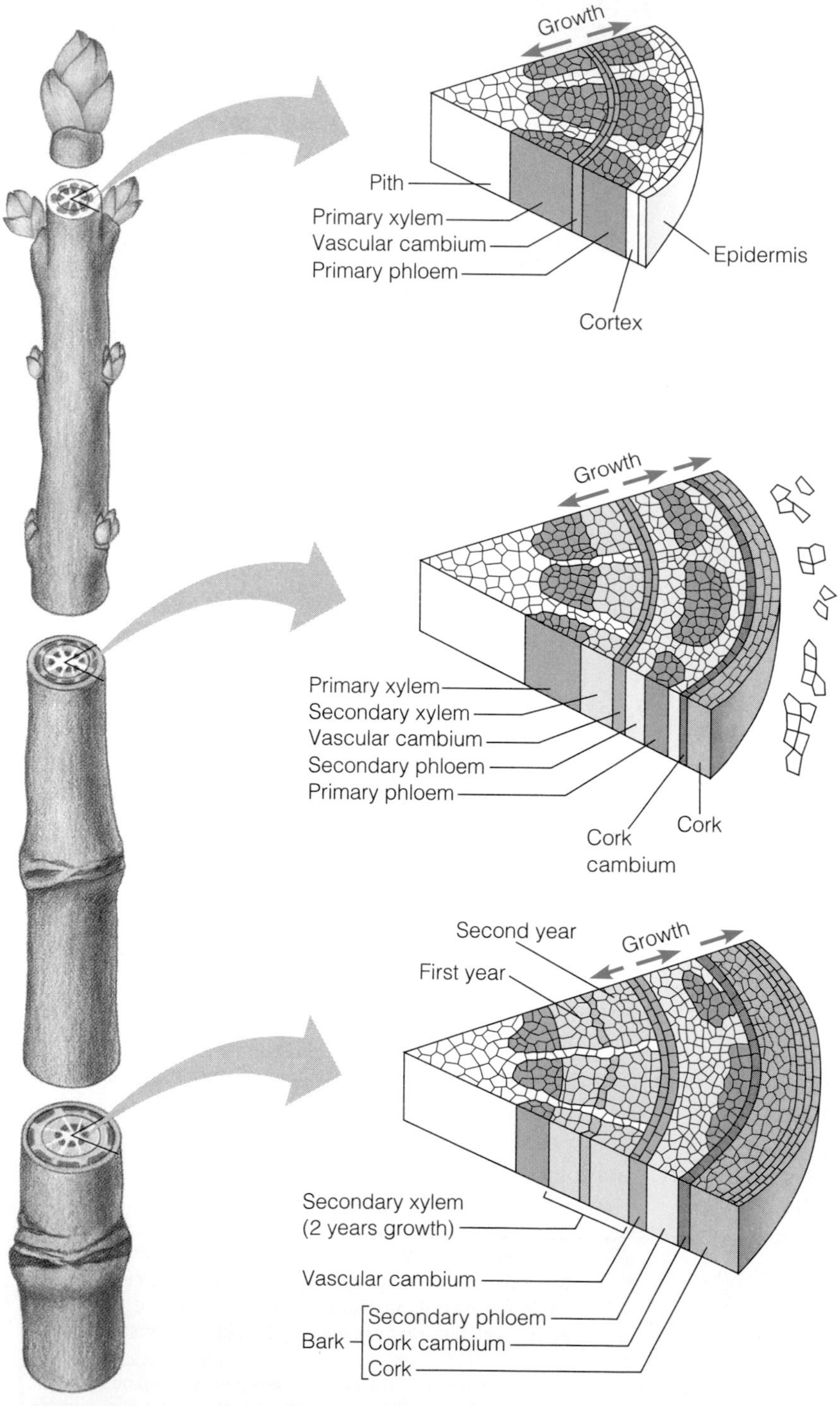

The center and bottom drawings show the results of secondary growth. In the center drawing, the vascular cambium has given rise to two new tissues. One is the **secondary xylem,** next to the inner surface of the vascular cambium. The other is the **secondary phloem,** just outside the vascular cambium.

By focusing first on the secondary xylem, you can see what causes most of the increase in a stem's thickness. Notice there are *two* layers of secondary xylem in the bottom drawing. The inner layer is the one produced by the vascular cambium during the first year of secondary growth. The outer layer was produced in the second year of secondary growth. This yearly production of a new layer of secondary xylem accounts for most of the growth in thickness of a perennial plant.

Consisting of xylem cells and fibers that have thick walls rich in lignin, the secondary xylem makes up the **wood** of a tree, shrub, or vine. Over the years, a woody stem gets thicker and thicker as its vascular cambium produces layer upon layer of secondary xylem. The annual growth rings, such as those in the trunk of a giant sequoia or the locust tree in Figure 31.8B, result from the layering of secondary xylem. The layers are visible as rings because of uneven activity of the vascular cambium during each year. In woody plants that live in temperate regions, such as most of the United States, the vascular cambium becomes dormant each year during winter, and secondary growth is interrupted. When secondary growth resumes in the spring, a cylinder of spring wood forms. Made up of the first new xylem cells to develop, spring wood cells are usually larger and thinner-walled than those produced later in summer (the summer wood). Each tree ring consists of a cylinder of spring wood surrounded by a cylinder of summer wood.

Figure 31.8A Secondary growth of a woody stem

Now let's return to Figure 31.8A and see what happens to the parts of the stem that are *external* to the vascular cambium. Unlike xylem, the external tissues do not accumulate over the years. Instead, they are sloughed off at about the same rate they are produced.

Notice at the top of the figure that the epidermis and cortex, both the result of primary growth, make up the young stem's external covering. When secondary growth begins, the epidermis and cortex start sloughing off. In the center drawing, you see that a new outer layer called **cork** (gray) has replaced the cortex and epidermis, which have sloughed off. Mature cork cells are dead and have thick, waxy walls, which protect the underlying tissues of the stem. Cork is produced by meristematic tissue called the **cork cambium** (dark gray), which first forms from parenchyma cells in the cortex. As the stem thickens and the secondary xylem expands, the original cork and cork cambium are pushed outward and sloughed off. You see their remains as the three outermost gray layers in the bottom diagram. By this time, a new cork cambium has developed from parenchyma cells in the secondary phloem.

As indicated in the bottom diagram of Figure 31.8A, everything external to the vascular cambium is called **bark:** all of the secondary phloem, the cork cambium, and the cork. The youngest secondary phloem (next to the vascular cambium) functions in sugar transport. The older secondary phloem dies, as does the cork cambium you see here. Pushed outward, these tissues and cork produced by the cork cambium help protect the stem until they, too, are sloughed off as part of the bark. Keeping pace with secondary growth, cork cambium keeps regenerating from the young secondary phloem, and keeps producing a steady supply of cork.

The log on the left in Figure 31.8B, from a locust tree, shows the results of several decades of secondary growth. The bulk of a trunk like this is dead tissue. The living tissues in it are the vascular cambium, the youngest secondary phloem, the cork cambium, and cells in the wood rays, which you can see radiating from the center of the log in the drawing (to the right). The **wood rays** consist of parenchyma cells that transport water to the outer living tissues in the trunk. The **heartwood,** in the center of the trunk, consists of older layers of secondary xylem. These cells no longer transport water; they are clogged with resins and other metabolic by-products that make the heartwood resistant to rotting. The lighter-colored **sapwood** consists of younger secondary xylem that actually conducts water.

Thousands of useful products are made from wood—from construction lumber to fine furniture, musical instruments, paper, insulation, and a long list of chemicals, including turpentine, alcohols, artificial vanilla flavoring, and preservatives. Among the qualities that make wood so useful are a unique combination of strength, hardness, lightness, high insulating properties, durability, and workability. In many cases, there is simply no good substitute for wood. A wooden oboe, for instance, produces far richer sounds than a plastic one. Fence posts made of locust tree wood last much longer in the ground than metal ones. Ball bearings are sometimes made of a very hard wood called lignum vitae. Unlike metal bearings, they require no lubrication, because a natural oil completely penetrates the wood.

In a sense, wood is analogous to the hard endoskeletons of many land animals. It is an evolutionary adaptation that enables a shrub or tree to remain upright and keep growing year after year on land—sometimes to attain enormous masses and heights, as we saw in the chapter's introduction. In the next few modules, we examine some of the adaptations that enable plants to reproduce on land.

(a) What type of plant tissue makes up wood?
(b) What is bark?

(a) secondary xylem; (b) all tissues exterior to the vascular cambium—secondary phloem, cork cambium, and cork

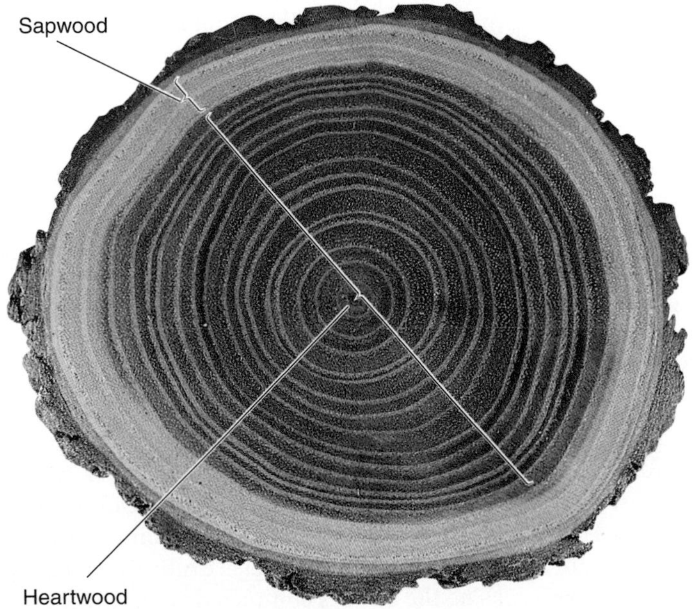

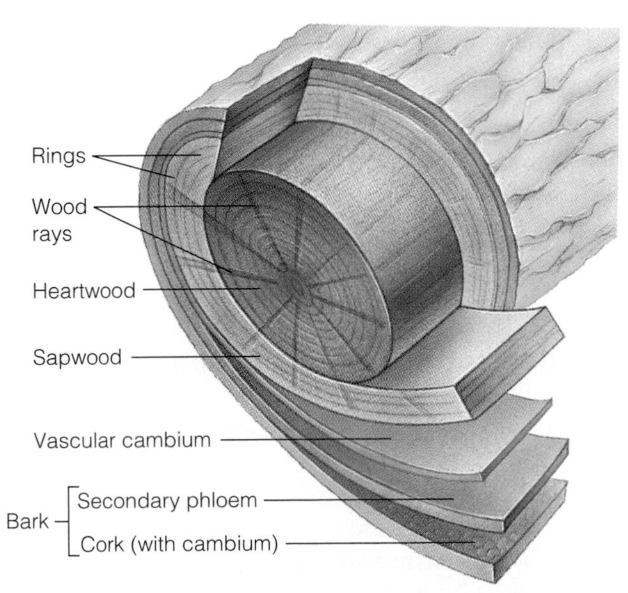

Figure 31.8B Anatomy of a log

31.9 Overview: The sexual life cycle of a flowering plant

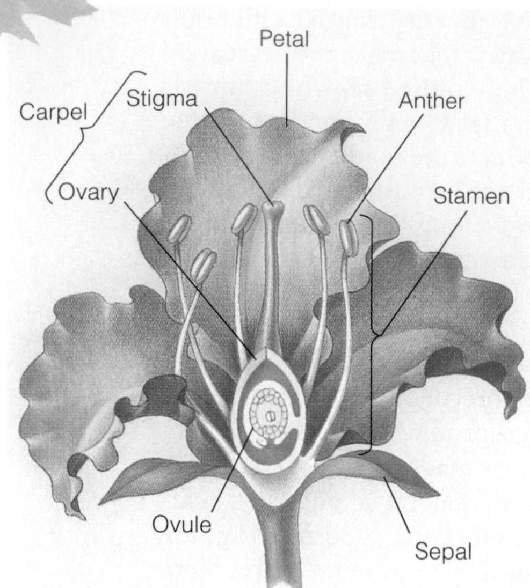

Figure 31.9A
The structure of a flower

Evolutionary fitness for any organism is measured only by its ability to replace itself with fertile offspring. Thus, from an evolutionary viewpoint, all the structures and functions of a plant can be interpreted as mechanisms contributing to reproduction.

In flowering plants, the structure specific to reproduction is the flower (Figure 31.9A). A flower is a compressed shoot, and its main parts, the sepals, petals, stamens, and carpels, are modified leaves. The **sepals** are usually green and look more like leaves than the other flower parts. Before the flower opens, the sepals enclose and protect the flower bud. The **petals** are often bright and colorful and advertise the flower to insects and other pollinators.

The flower's reproductive organs are the stamens and carpel. The **stamens** are the male organs. At the tip of each stamen is an **anther,** a sac in which meiosis occurs and in which pollen grains develop. Pollen grains house the cells that develop into sperm.

The **carpel** (more than one in some plants) is the female organ of the flower. The tip of the carpel, the **stigma,** is the receiving surface for pollen grains brought from other flowers, or from the same flower, by wind or animals. The base of the carpel is the **ovary,** which houses reproductive structures called the ovules. The **ovule** contains the developing egg and cells that support it.

Figure 31.9B shows the life cycle of a generalized angiosperm. Fertilization occurs in the ovule, which then matures into a seed containing the embryo. Meanwhile, the ovary develops into a fruit, which protects the seed and aids in dispersing it. Completing the life cycle, the seed **germinates** (begins to grow) in a suitable habitat, the embryo develops into a seedling, and the seedling grows into a mature plant.

In the next four modules, we examine key stages in the angiosperm sexual life cycle in more detail. We will see that there are a number of variations on its basic themes.

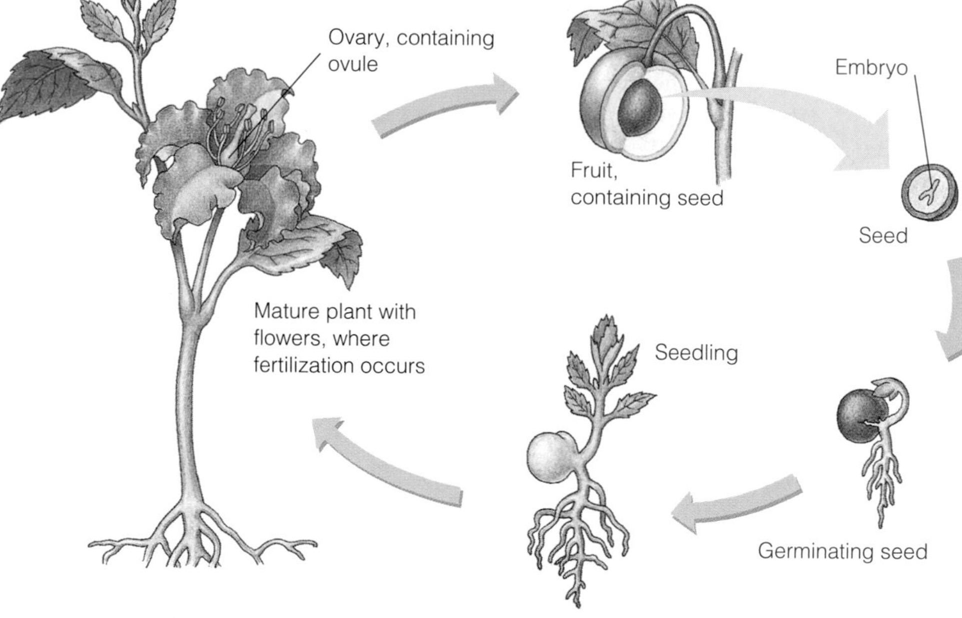

Figure 31.9B Life cycle of a generalized angiosperm

? Pollen develops within the __________ of __________. Ovules develop within the __________ of __________.

anthers · · · stamens · · · ovaries · · · carpels

31.10 The development of pollen and ovules culminates in fertilization

31C

Recall from Chapter 17 that the life cycles of all plants include alternation of haploid (n) and diploid ($2n$) generations. The roots, stems, leaves, and most of the reproductive structures of a rose plant, an oak tree, or a grass—in fact, all angiosperms and gymnosperms—are diploid. The diploid plant body is called the **sporophyte.** A sporophyte produces special structures, the anthers and ovules in angiosperms, in which cells undergo meiosis. Haploid spores result. Each of these then divides mitotically and becomes a multicellular **gametophyte,** the plant's haploid generation. The gametophyte produces gametes by mitosis. At fertilization, the male and female gametes unite, producing a diploid zygote. The life cycle is completed when the zygote divides by mitosis and develops into a new sporophyte.

Without a microscope, all we can see of the angiosperm and gymnosperm life cycles are the sporophytes and occasionally the pollen (male gametophytes) produced by them. In this module, we take a "microscopic look" at the gametophytes of a flowering plant.

We begin with the development of the male gametophyte, the pollen grain. The cells that develop into pollen grains are found in chambers within a flower's anthers (top left in Figure 31.10). Each cell first undergoes meiosis, forming four haploid cells called spores. Each spore then divides mitotically, forming two haploid cells, called the tube cell and the generative cell. A thick wall forms around these cells, and the resulting pollen grain is ready for release from the anther.

Moving to the top right of the figure, we can follow the development of the flower parts that form the female gametophyte and eventually the egg. In most species, the ovary of a flower contains several ovules, but only one is shown here. An ovule contains a central cell (pink) surrounded by a protective covering of smaller cells (yellow). The central cell enlarges and undergoes meiosis, producing four haploid spores. Three of the spores usually degenerate, but the surviving one enlarges and divides mitotically, producing a multicellular structure called the **embryo sac** (pink area). Housed in several layers of protective cells (yellow) produced by the sporophyte plant, the embryo sac is the female gametophyte. The sac contains a large central cell with two haploid nuclei. One of its other cells (dark pink in the drawing) is the haploid egg, ready to be fertilized.

The first step leading to fertilization is **pollination** (at the center of the figure), the delivery of pollen to the stigma of a carpel. Most angiosperms are dependent on animals to transfer their pollen. But the pollen of some plants, such as grasses, is windborne, as anyone with allergies knows.

After pollination, the pollen grain germinates on the stigma. Its tube cell gives rise to the pollen tube, which grows downward into the ovary. Meanwhile, the generative cell divides mitotically, forming two sperm. When the pollen tube reaches the base of the ovule, it enters the embryo sac through a pore and discharges both its sperm. One sperm fertilizes the egg, forming the zygote (purple). The other sperm contributes its haploid nucleus to the large central cell of the embryo sac. This cell, now with a triploid ($3n$) nucleus, will give rise to tissue that nourishes the embryo that develops from the zygote.

The formation of both a zygote and a cell with a triploid nucleus is called **double fertilization.** This occurs only in plants, mainly in angiosperms. Next, let's see what happens after double fertilization.

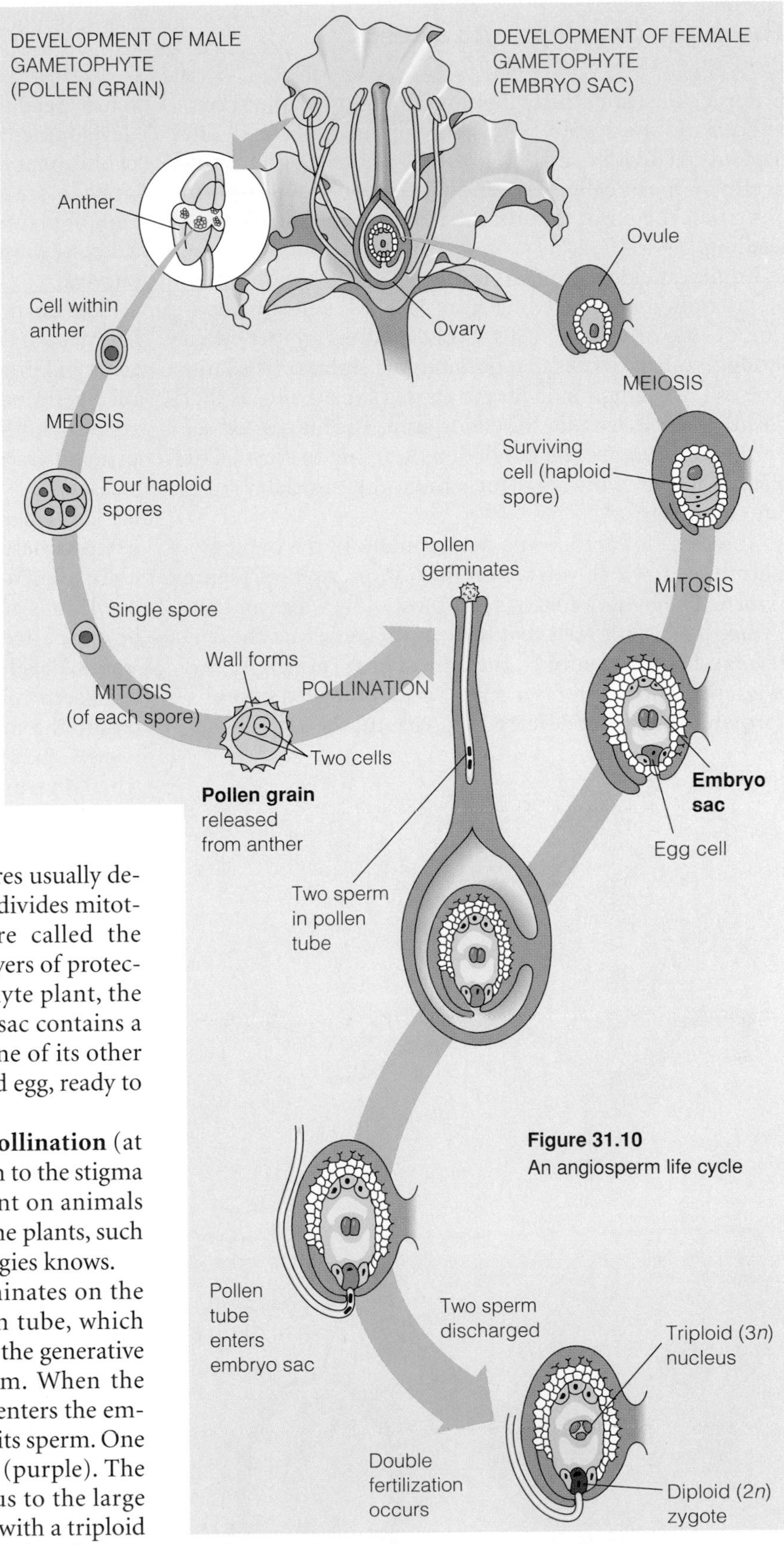

Figure 31.10
An angiosperm life cycle

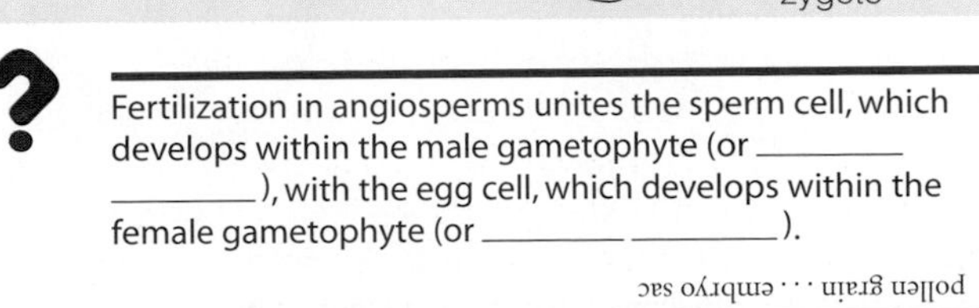

?
Fertilization in angiosperms unites the sperm cell, which develops within the male gametophyte (or ________ ________), with the egg cell, which develops within the female gametophyte (or ________ ________).
pollen grain . . . embryo sac

31.11 The ovule develops into a seed

After fertilization, the ovule, containing the triploid central cell and the zygote, begins developing into a seed. The triploid cell divides and develops into a nutrient-rich, multicellular mass called the **endosperm.** The endosperm nourishes the embryo until it becomes a self-supporting seedling.

Embryonic development begins when the zygote (purple) divides into two cells (Figure 31.11A). Repeated division of one of the two cells (still called the zygote) then produces a ball of cells that becomes the embryo. The large blue cell in the figure divides to form a thread of cells that pushes the embryo into the endosperm. The bulges you see on the embryo are the cotyledons starting to form. The plant in this drawing is a dicot; a monocot would have only one cotyledon.

The result of embryonic development in the ovule is a mature seed, which you see on the bottom right of Figure 31.11A. The ovule's coat has lost most of its water and has formed a resistant **seed coat** (brown) that encloses the embryo and its food supply, the endosperm (pink). At this point, the embryo stops developing, and the seed becomes dormant; it will not develop further until the seed germinates. **Seed dormancy,** a condition in which growth and development are suspended temporarily, is an important evolutionary adaptation. It allows time for a plant to disperse its seeds and increases the chance that a new generation of plants will begin growing only when environmental conditions, such as temperature and moisture, favor survival.

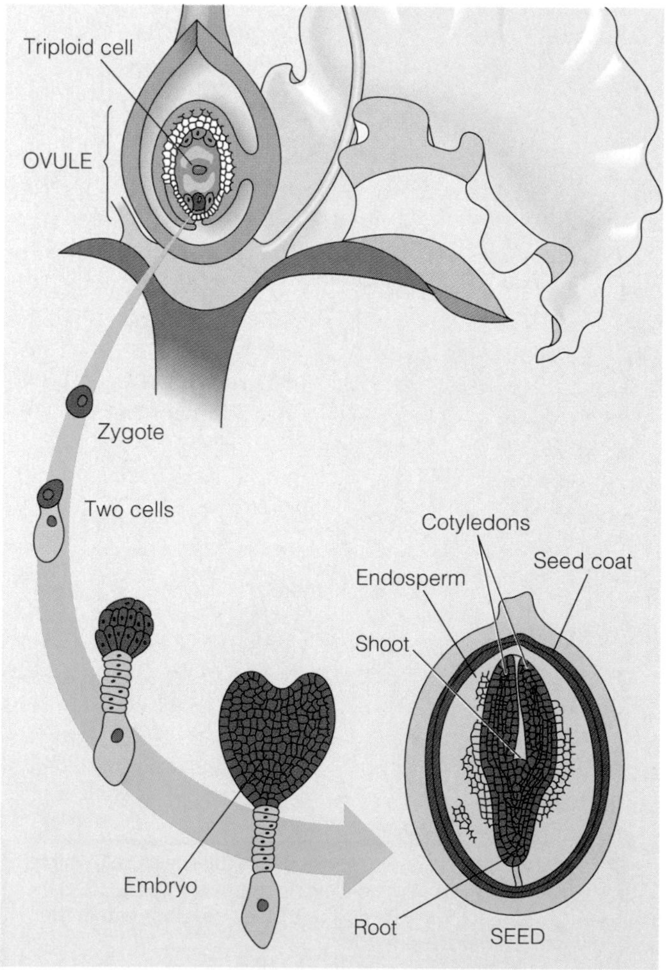

Figure 31.11A Development of a dicot plant embryo

The dormant embryo consists of a miniature root and shoot, each equipped with an apical meristem. The meristems will produce cells that elongate the embryo when the seed germinates. Also present in the embryo are the three tissue cylinders that will form the epidermis, cortex, and primary vascular tissues.

Figure 31.11B shows what two seeds look like when they are split open. In the bean, a dicot, the embryo is an elongated structure with two fleshy cotyledons (yellow). The embryonic root develops just below the point at which the cotyledons are attached to the rest of the embryo. The embryonic shoot, tipped by a pair of miniature embryonic leaves, develops just above the point of attachment. The bean seed contains no endosperm, because its cotyledons absorb the endosperm nutrients as the seed forms. The nutrients start passing from the cotyledons to the embryo when it germinates.

The kernel of corn, an example of a monocot, is actually a fruit containing one seed. Everything you see in the drawing is the seed, except the kernel's outermost covering. The covering is the dried tissue of the fruit, tightly bonded to the seed coat. Different from the bean, the corn seed contains a large endosperm and a single cotyledon. The cotyledon absorbs the endosperm's nutrients during germination. Also unlike the bean, the corn's embryonic root and shoot each have a protective sheath.

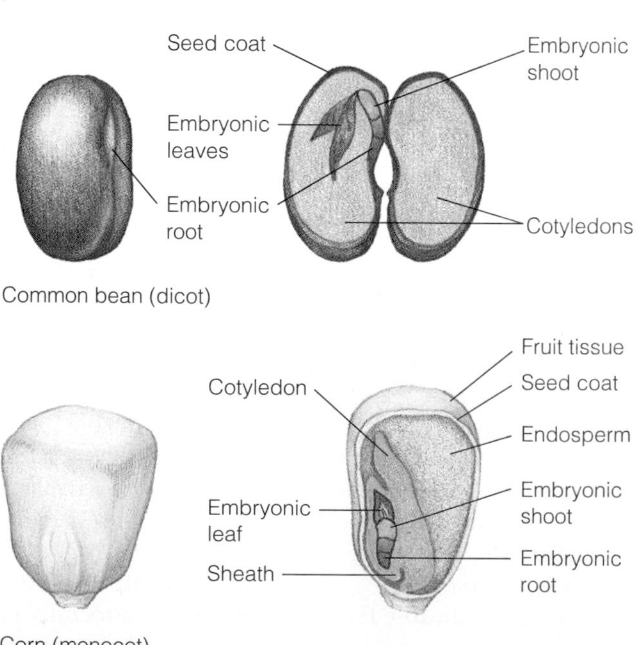

Figure 31.11B Seed structure

We have now followed the angiosperm life cycle from the flower on the sporophyte plant through the transformation of an ovule into a seed. Fruits, one of the most distinctive features of angiosperms, develop at the same time seeds do, but from the outer parts of the ovary. We study them next.

The two products of double fertilization are the ________, which gives rise to the embryo, and the ________, which stores nutrients.

zygote . . . endosperm

The ovary develops into a fruit 31.12

Figure 31.12A Development of a simple fruit, a pea pod

A **fruit** is a matured ovary, specialized as a vessel that houses and protects seeds and helps disperse them from the parent plant. A corn kernel is a fruit, as is a peach, orange, tomato, cherry, or pea pod.

The photographs in Figure 31.12A illustrate the changes in a pea plant leading to pod formation. ① Soon after pollination, ② the flower drops its petals, and hormonal changes make the ovary start growing. The ovary expands tremendously, and its wall thickens, ③ forming the pod, or fruit.

Figure 31.12B matches the parts of a pea flower with what they become in the pod. The wall of the ovary becomes the pod. The ovules, within the ovary, develop into the seeds. The small, threadlike structure at the end of the pod is what remains of the upper part of the flower's carpel. The sepals of the flower often stay attached to the base of the green pod. Peas are usually harvested at this stage of fruit development. If the pods are allowed to develop further, they become dry and brownish and will split open, releasing the seeds.

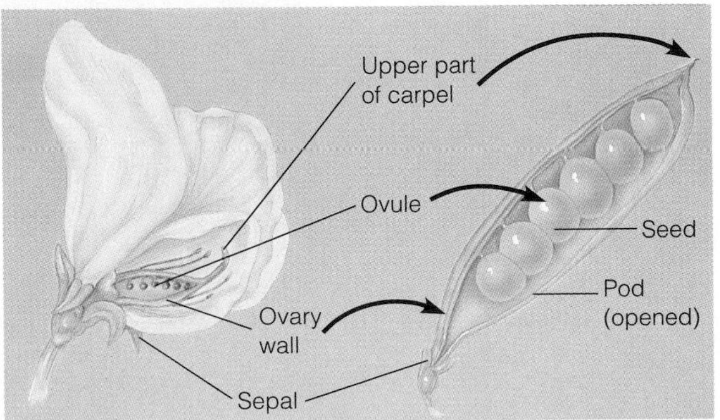

Figure 31.12B The correspondence between flower and fruit in the pea plant

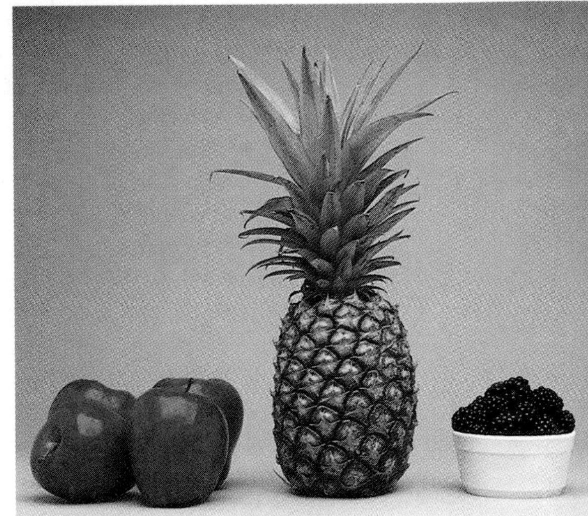

Figure 31.12C A variety of fruit

Fruits are highly varied, as Figure 31.12C illustrates. In some cases, what we commonly call fruits are actually more than developed ovaries. In an apple, for instance, the part we discard, the core, is the thickened ovary, and therefore the true fruit. The soft, fleshy part we eat develops from parts of the flower base that fuse with the ovary. In contrast to the pea pod, which dehydrates when it is fully developed, a fleshy fruit like an apple becomes softer as enzymes weaken the cell walls. It also may change color from green to red, orange, or yellow, and become sweeter as organic acids or starch molecules are converted to sugar. An unripe apple or pear is sour to the taste because of high acid concentrations. Fleshy, edible fruits such as apples serve as enticements to animals that help spread seeds.

Apples, pea pods, and cherries are examples of **simple fruits**—those that develop from a flower with a single carpel and ovary. In contrast, a blackberry is an **aggregate fruit,** because it develops from a flower with many carpels. Each of the small parts of the berry develops from a single ovary. Different yet is a **multiple fruit,** which develops from a group of separate flowers tightly clustered together. When the walls of the many ovaries start to thicken, they fuse together and become incorporated into one fruit. The pineapple is an example of a multiple fruit. Each of its many parts develops from a separate flower.

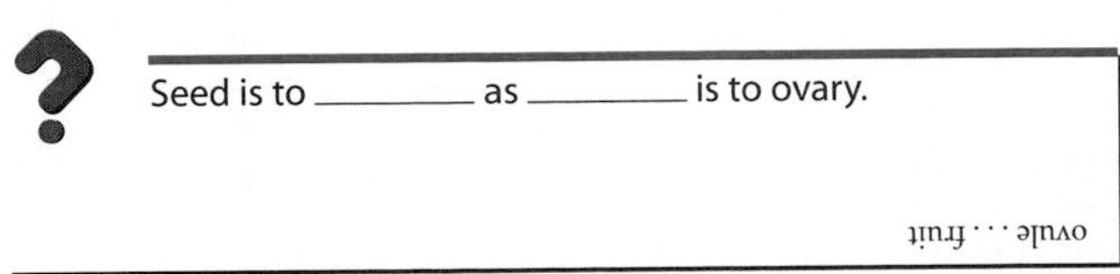

Seed is to ________ as ________ is to ovary.

ovule . . . fruit

31.13 Seed germination continues the life cycle

The germination of a seed is often used to symbolize the beginning of life, but as we have seen, the seed already contains a miniature plant, complete with embryonic root and shoot. Thus, at germination, the plant does not begin life but rather resumes the growth and development that was temporarily suspended during seed dormancy.

Germination usually begins when the seed takes up water. The hydrated seed expands, rupturing its coat and triggering metabolic changes in the embryo that make it start growing again. Enzymes in the endosperm or cotyledons begin digesting stored nutrients, and the nutrients are transported to the growing regions of the embryo.

The figures below trace germination in a dicot, the garden pea, and a monocot, corn. In Figure 31.13A, notice that the embryonic root of a pea emerges first and grows downward from the germinating seed. Next, the embryonic shoot emerges, and a hook forms near its tip. The hook protects the delicate shoot tip by holding it downward, rather than pushing it up through the abrasive soil. As the shoot breaks through the soil surface, its tip is lifted gently out of the soil as exposure to light stimulates the hook to straighten. The first foliage leaves then expand from the shoot tip and begin making food by photosynthesis. In the pea, the cotyledons, their food reserves used by the germinating embryo, remain behind in the soil and decompose.

In corn (Figure 31.13B), the embryonic root begins to grow just before the embryonic shoot. A hook does not develop on the shoot. Instead, a protective sheath surrounding the shoot pushes straight upward and breaks through the soil. Then, safe from soil abrasion, the shoot tip grows up through the tunnel provided by the sheath. As in the pea, the corn cotyledon remains in the soil and decomposes.

A germinating seed is fragile. In the wild, only a small fraction of seedlings endure long enough to reproduce. The great numbers of seeds produced by most plants compensate for the odds against their seedlings. Many plants also reproduce asexually, as we see next.

Which meristems provide additional cells for early growth of a seedling after germination?

The apical meristems of the shoot and root

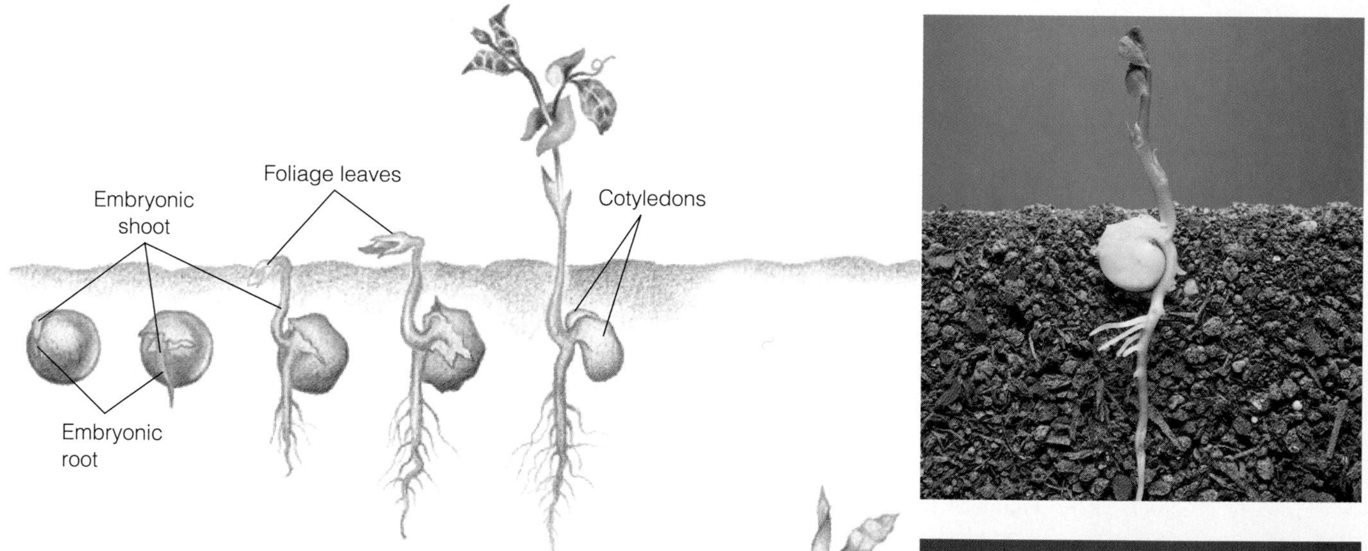

Embryonic shoot
Foliage leaves
Cotyledons
Embryonic root

Figure 31.13A Pea germination (a dicot)

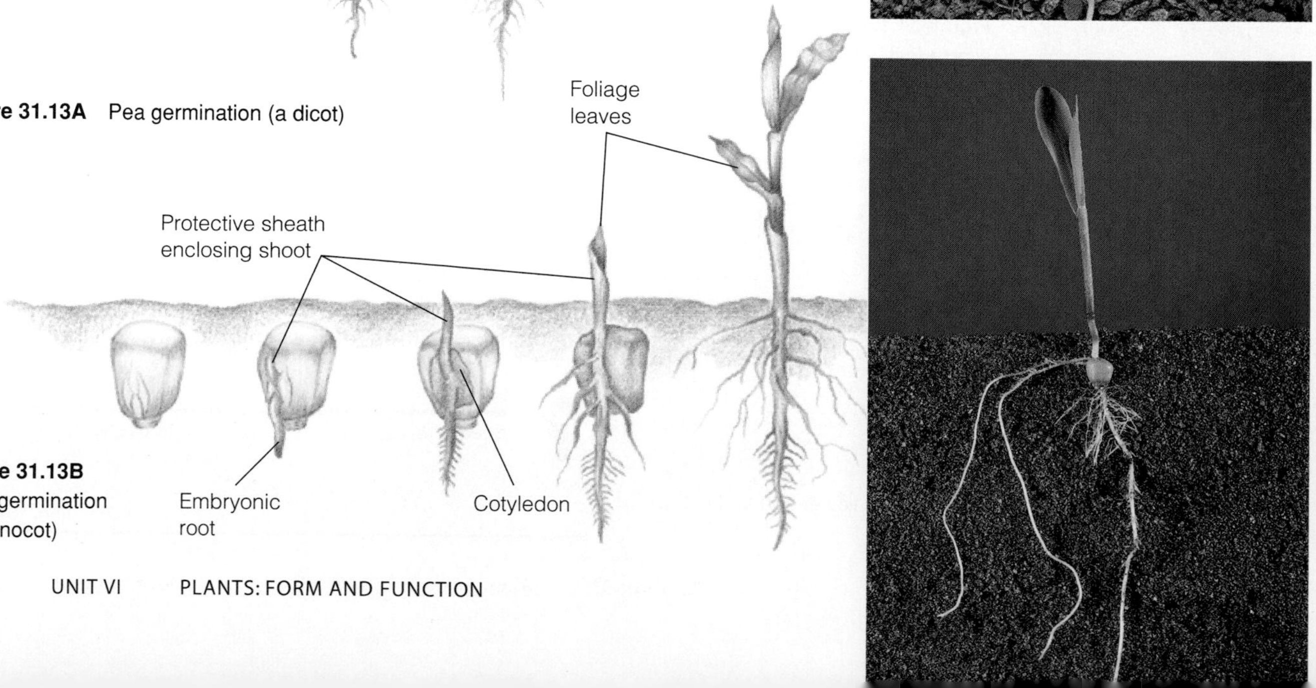

Foliage leaves
Protective sheath enclosing shoot
Embryonic root
Cotyledon

Figure 31.13B Corn germination (a monocot)

Asexual reproduction produces plant clones 31.14

Imagine some of your fingers separating from your body, taking up life on their own, and eventually developing into copies of yourself. This would be asexual reproduction, the creation of offspring derived from a single parent without fertilization. The result would be a clone, a population of asexually produced, genetically identical organisms.

As these photographs suggest, asexual reproduction, often called **vegetative reproduction,** is common among plants. Vegetative reproduction involves **fragmentation,** the separation of parts from the parent plant and regeneration of the parts into whole plants. The garlic bulb (Figure 31.14A) is actually an underground stem that functions in storage. A single large bulb fragments into several parts, called cloves. Each clove can give rise to a separate plant, as indicated by the green shoots emerging from some of them. The white sheaths are leaves attached to the stem.

Each of the small trees you see in Figure 31.14B is a sprout from the roots of a coast redwood tree, a close relative of the giant sequoia we discussed in the chapter's introduction. Eventually, one or more of these root sprouts may take the place of its parent in the forest.

The ring of plants in Figure 31.14C is a clone of creosote bushes growing in the Mojave Desert in southern California. All these bushes came from generations of vegetative reproduction by roots. Making the oldest sequoias seem youthful, this clone apparently began with a single plant that germinated from a seed about 12,000 years ago. The original plant probably occupied the center of the ring.

Figure 31.14D shows a patch of dune grass in Cape Cod, Massachusetts. This species and most other grasses can propagate asexually by sprouting shoots and roots from runners. A small patch of grass can spread in this way until it covers an acre or more of surface.

Vegetative reproduction is an extension of the capacity of plants to continue growing throughout their lives. Their meristematic tissues can sustain or renew growth indefinitely. In addition, the parenchyma cells throughout a plant can divide and differentiate into the various types of specialized cells.

Asexual reproduction has several potential advantages. For one thing, a parent plant well suited to its environment can clone many copies of itself. Also, early life for vegetative offspring, which are mature fragments of the parent, may be less hazardous than for seedlings. Both asexual reproduction and sexual reproduction have played important roles in the evolutionary adaptation of plant populations to their environments.

Many plant species can reproduce both sexually and asexually. Which mode of reproduction would generally be more advantageous in a location where the composition of the soil is changing for some reason? Why?

Sexual, because it generates genetic variation among offspring, which enhances the potential for adaptation to a changing environment

Figure 31.14A Cloves of a garlic bulb

Figure 31.14B Sprouts from the roots of coast redwood trees

Figure 31.14C A ring of creosote bushes

Figure 31.14D Dune grass

31.15 Vegetative reproduction is a mainstay of modern agriculture

The ability of plants to reproduce vegetatively provides many opportunities for producing large numbers of plants with minimal effort and expense. For example, most of our fruit trees, ornamental trees and shrubs, and houseplants are asexually propagated from stem or leaf cuttings. Several other plants are propagated from root sprouts (raspberries, for example) or pieces of underground stems (such as potatoes).

Plants can also be propagated by test-tube methods. The laboratory tube in Figure 31.15A contains an apple plantlet that was grown from a few meristem cells cut from a large tree and cultured on a chemical medium. Using this method, a single plant can be cloned into thousands of copies that will continue to grow when transferred to soil. Orchids and certain pine trees used for mass plantings are commonly propagated this way. Furthermore, test-tube methods enable researchers to grow genetically engineered plant cells into plants (see Module 12.19). Foreign genes are incorporated into a single parenchyma cell, and the cell is then cultured until it multiplies and develops into a new plantlet.

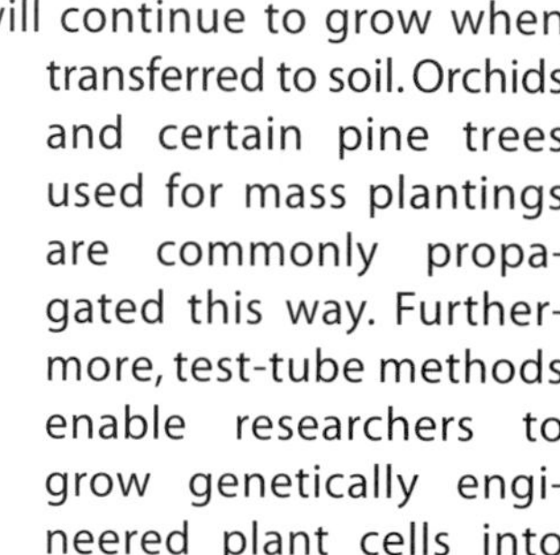

31.15A Test-tube cloning

31.15B An artificial seed germinating

Figure 31.15B shows another new development, a so-called artificial seed. It is a tiny alfalfa plant embryo cloned from a few cells of a parent plant. In the lab, the embryo is packaged with some nutrients in a polysaccharide capsule. Using artificial seeds, a farmer can grow a truly uniform crop, one in which all the plants have the same set of desirable traits and will be ready for harvest at the same time. The main disadvantage of artificial seeds is that they are more expensive than the natural ones produced by plants.

Despite the promising new developments, modern agriculture faces some potentially serious problems. Nearly all of today's crop plants have very little genetic variability, and the use of gene-cloning techniques may compound the problem. Furthermore, we grow most crops in **monocultures,** large areas of land with a single plant variety. Given these conditions, plant scientists fear that a small number of diseases could devastate large crop areas. In response, plant breeders are working to maintain "gene banks," storage sites for seeds of many different plant varieties that can be used to breed new hybrids. We'll have more to say about plants and agriculture when we take up the subject of plant nutrition in Chapter 32.

What is the most serious threat to an agriculture based on monoculture?

New or newly arrived pathogens to which the clone lacks resistance

Chapter Review

Chapter Summary

Plants are important to the global environment, and humans depend on them for such essentials as food, lumber, fabric, and paper. The most familiar and most diverse plants are the angiosperms, or flowering plants **(Introduction)**. The correlation of structure and function is a central theme in plant biology **(31.1)**.

Plant Structure and Function (31.2–31.6) There are two main types of angiosperms. The monocots include the grains and other grasses, orchids, palms, and lilies. The dicots are a larger group, including most trees, shrubs, ornamental plants, and many food crops. Monocots and dicots differ in the number of seed leaves and the structure of roots, stems, leaves, and flowers **(31.2)**. The body of a plant consists of a root system and a shoot system. Roots anchor the plant, absorb and transport minerals and water, and store food. Tiny root hairs increase the absorptive surface of roots. The shoot system of an angiosperm consists of stems, leaves, and flowers. Leaves are the plant's main photosynthetic organs. At the tip of a stem is a terminal bud, the growth point of the stem. Axillary buds can give rise to branches **(31.3)**. Many roots and stems function in storing food, asexual reproduction, and protection **(31.4)**. There are five major types of plant cells: parenchyma, collenchyma, sclerenchyma (including fiber and sclereid cells), water-conducting cells (including tracheids and vessel elements), and food-conducting cells (sieve-tube members). Two kinds of vascular tissue are xylem, which contains water-conducting cells that convey water and dissolved minerals, and phloem, which contains sieve-tube members that transport sugars **(31.5)**. Roots, stems, and leaves are made up of three tissue systems: the epidermis, the vascular tissue system, and the ground tissue system. The epidermis covers and protects the plant. The vascular tissue system contains xylem and phloem. The ground tissue system consists of parenchyma cells and supportive collenchyma and sclerenchyma cells. The ground tissues of stems and roots function mainly in storage. Mesophyll, the ground tissue of leaves, is where most photosynthesis occurs **(31.6)**.

Plant Growth (31.7–31.8) Most plants grow as long as they are alive. Growth originates in meristems, areas of unspecialized, dividing cells. Apical meristems at the tips of roots and in the terminal buds and axillary buds of shoots initiate lengthwise growth by producing new cells. A root or shoot lengthens further as the cells elongate and differentiate. This cell division and elongation is called primary growth (31.7). An increase in a plant's girth, called secondary growth, arises from cell division in a cylindrical meristem called the vascular cambium. The vascular cambium thickens a stem by adding layers of secondary xylem, or wood, next to its inner surface. Outside the vascular cambium, the bark consists of secondary phloem (also produced by the vascular cambium), a meristem called the cork cambium, and protective cork cells produced by the cork cambium. The outer layers of bark are sloughed off as the plant thickens. Secondary phloem does not thicken the stem because bark is sloughed off at about the same rate as new phloem is produced (31.8).

Plant Reproduction (31.9–31.15) The angiosperm flower is a reproductive shoot consisting of sepals, petals, stamens, and carpels. Pollen grains develop in anthers, at the tips of stamens. The tip of the carpel, the stigma, receives pollen grains. The ovary, at the base of the carpel, houses the female reproductive structure, the ovule (31.9). The plant life cycle alternates between diploid (2*n*) and haploid (*n*) generations. Most of the body of an angiosperm or gymnosperm is a diploid sporophyte. Haploid spores are formed within ovules and anthers. The spores in the anthers give rise to male gametophytes—pollen grains—which produce sperm. A spore in an ovule produces the female gametophyte, called an embryo sac. Each embryo sac contains an egg cell. Pollination is the arrival of pollen grains, carried by wind or animals, onto a stigma. A pollen tube grows into the ovule, and sperm pass through it and fertilize both the egg and a second cell. The second cell has two nuclei, and after fertilization it is triploid. This process, called double fertilization, is unique to plants (31.10). After fertilization, the ovule becomes a seed, and the fertilized egg within it divides to become an embryo. The other fertilized cell develops into the endosperm, which stores the food for the embryo. A tough seed coat protects the embryo and endosperm (31.11). While the ovule becomes a seed, the ovary develops into a fruit, which helps protect and disperse the seeds (31.12). A seed starts to germinate when it takes up water, expands, and bursts its seed coat. Metabolic changes cause the embryo to resume growth and absorb nutrients from the endosperm. An embryonic root emerges, and a shoot pushes upward and expands its leaves (31.13). Many plants can reproduce asexually, via bulbs, sprouts, or runners (31.14). Propagating plants vegetatively from cuttings or bits of tissue can increase agricultural productivity but can also reduce genetic diversity (31.15).

Testing Your Knowledge

Multiple Choice

1. Which of the following is closest to the center of a woody stem? *(Explain your answer.)*
   a. vascular cambium
   b. primary phloem
   c. secondary phloem
   d. primary xylem
   e. secondary xylem

2. A pea pod is formed from ____. A pea inside the pod is formed from ____.
   a. an ovule . . . a carpel
   b. an ovary . . . an ovule
   c. an ovary . . . a pollen grain
   d. an anther . . . an ovule
   e. endosperm . . . an ovary

3. While walking in the woods, you encounter an unfamiliar nonwoody flowering plant. If you want to know whether it is a monocot or dicot, it would *not* help to look at the
   a. number of seed leaves, or cotyledons, present in its seeds
   b. shape of its root system
   c. number of petals in its flowers
   d. arrangement of vascular bundles in its stem
   e. size of the plant

4. In angiosperms, each pollen grain produces two sperm. What do these sperm do?
   a. Each one fertilizes a separate egg cell.
   b. One fertilizes an egg, and the other fertilizes the fruit.
   c. One fertilizes an egg, and the other is kept in reserve.
   d. Both fertilize a single egg cell.
   e. One fertilizes an egg, and the other fertilizes a cell that develops into stored food.

Matching

1. Attracts pollinator	a. pollen grain
2. Develops into seed	b. ovule
3. Protects flower before it opens	c. anther
4. Produces sperm	d. ovary
5. Produces pollen	e. sepal
6. Houses ovules	f. petal

Describing, Comparing, and Explaining

1. The scent of apple blossoms and the buzzing of bees fill an orchard on a warm spring day. Describe the processes by which the pollen carried from flower to flower by the bees results in the apple you might pick in the fall.

2. Name three kinds of vegetative (asexual) reproduction. Explain two advantages vegetative reproduction has over sexual reproduction.

3. What part of a plant are you eating when you consume each of the following? Tomato, celery stalk, peanut, strawberry, lettuce, artichoke, beet.

Thinking as a Scientist

Plant scientists are searching Peru, Mexico, and the Middle East for the wild ancestors of potatoes, corn, and wheat. Why is this search important?

Science, Technology, and Society

Tropical forests contain a wealth of plants that are potential new sources of food, as well as sources of medicine and other useful products. The developing nations of the tropics cannot develop these resources themselves. Under pressure from growing populations and debt, they are cutting their forests for lumber and farmland, and many species are disappearing. Developed countries are pressuring the tropical countries to protect the forests before even more species are lost. Many people in the developing nations see little incentive to preserve the forests, only to have corporations from industrialized countries profit from new products obtained from the forests. Is there a way to preserve the tropical forests so that both the developed and developing nations will benefit from their abundance?

Media Resources

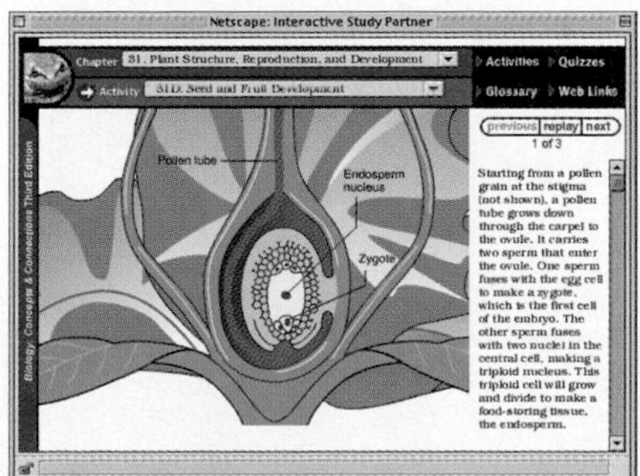

For further review of the concepts in this chapter, use the activities and quizzes on the CD-ROM.

Visit the Special Edition of The Biology Place web site for activities, quizzes, news articles, and web links.

32 Plant Nutrition and Transport

Outline

Peanuts, peanut butter, and peanut oil are part of our culture. Yet how many consumers know where peanuts come from? Do they grow on trees, shrubs, or nonwoody plants? The peanut plant, which is shown in the large photograph at the left, is short and nonwoody. But unless you live in the southeastern United States, where large peanut crops are grown, you may not know some of the truly unusual features of this important crop plant.

The peanut is one species in a large group called the legumes (about 10,000 species). Other legumes include peas, beans, lentils, clovers, alfalfa, and locust trees. Most species, including the peanut, bear seeds in pods.

Outside the U.S., peanuts are often called groundnuts, and this is actually a more appropriate name. The plant's scientific name, *Arachis hypogaea* (Greek *arakos*, legume plant, and *hypogaia*, "under earth") is also fitting. Unlike nearly all other legumes, which produce their pods on aerial stems, the peanut bears its fruit—the tough fibrous pods we crack open—about 5 cm *below* the ground. When the plant matures, small yellow flowers appear above the ground at the base of its leaf stalks. After fertilization, each flower's ovary grows a long stalk that points downward and pushes the developing pod into the ground. Notice the stalks attached to the ripe pods in the photograph. The seeds—peanuts—develop in the pods.

The history of the peanut industry in the U.S. is closely tied to the career of George Washington Carver (1864–1943). Carver's life was filled with accomplishments in education, literature, art, and science. Born a slave in strife-torn southwestern Missouri during the Civil War, he rose to become a professor at the Tuskegee Institute in Alabama and one of the world's most respected agricultural researchers.

One of Carver's major contributions to agriculture was to promote the practice of rotating crops. He recognized that by alternating a cotton or corn crop with a legume like peanuts, all the plants grew better, and the soil remained more fertile. Carver also championed the nutritional and economic potential of the peanut plant. In the late 1800s, the economic survival of the South depended on finding crops to replace or rotate with cotton, which had seriously diminished the soil's fertility. Carver's work helped open new markets for peanuts and peanut products. In his laboratory at Tuskegee, he separated peanut meal into its chemical constituents and showed that the meal could be used to make a host of marketable foods, among them a nutritious milk substitute, a rich oil, flour, dye, and cheese.

George Washington Carver (tall man in the center) with students at Tuskegee Institute

One of the main reasons peanuts and other legumes are grown so widely is that they are rich in protein; as Carver discovered, as much as 30% of a peanut's dry weight is protein. Small bumps, or nodules, on the roots of legumes play a key role in protein production. Root nodules house bacteria that supply nitrogen, needed for protein synthesis, to the plant. (The inset on the facing page shows the bacteria in root nodules of an alfalfa plant.) With their root nodule bacteria, legumes often thrive in nitrogen-poor soils and actually improve soil fertility by adding nitrogen compounds to it.

Nitrogen is one of several inorganic nutrients that plants must have in order to grow. In this chapter, we see how plants obtain nitrogen and other essential nutrients and how they transport them throughout their roots, stems, and leaves.

The Uptake and Transport of Plant Nutrients

32.1 Plants acquire their nutrients from soil and air

Watch a plant grow from a tiny seed, and you can't help wondering where all the mass comes from. Aristotle thought soil provided all the substance for plant growth, and, as we discussed in Chapter 7, seventeenth-century physician Jan Baptista van Helmont performed an experiment to find out if plants grew by absorbing material from soil. He planted a willow seedling in a pot containing 90.0 kg of soil. After five years, the willow had grown into a tree weighing 76.8 kg, but only 0.06 kg of soil had disappeared from the pot. Van Helmont concluded that the willow had grown mainly from the water he had added regularly. A century later, though, Stephen Hales, an English botanist, postulated that plants were nourished mostly by air.

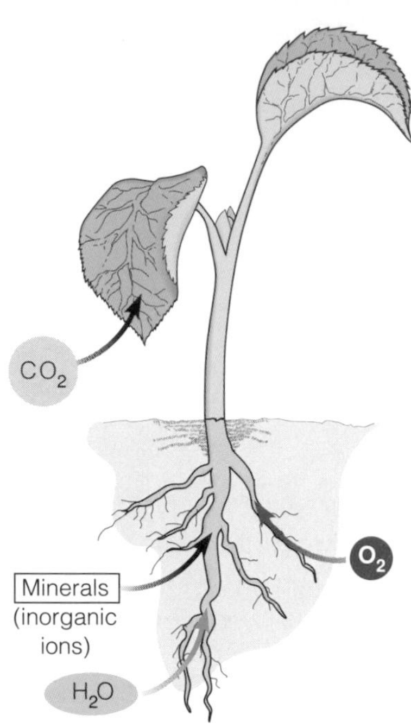

Figure 32.1A The uptake of nutrients by a plant

As it turns out, there is some truth in all these early ideas about plant nutrition. As indicated in Figure 32.1A, a plant's leaves absorb CO_2 from the air; in fact, about 95% of a plant's dry weight is organic (carbon-containing) material built mainly from CO_2. The figure also points out that a plant gets water, inorganic ions (minerals), and some oxygen (O_2) from the soil.

What happens to the materials a plant takes up from the air and soil? The sugars a plant makes by photosynthesis are composed of the elements carbon, oxygen, and hydrogen. In Chapter 7, we saw that the carbon and oxygen used in photosynthesis come from atmospheric CO_2 and that the hydrogen comes from water molecules. Plant cells use the sugars made by photosynthesis in constructing all the other organic materials they need. The giant trunks of the redwood trees in Figure 32.1B, for instance, consist mainly of sugar derivatives.

Plants use cellular respiration to break down some of the sugars they make, obtaining energy from them in a process that consumes O_2. A plant's leaves take up some O_2 from the air, but we do not show this in Figure 32.1A because plants are actually net producers of O_2, giving off more of this gas than they use. When water is split during photosynthesis, O_2 gas is produced and released through the leaves. The O_2 being taken up from the soil by the plant's roots in Figure 32.1A is actually atmospheric O_2 that has diffused into the soil; it is used in cellular respiration in the roots themselves.

What does a plant do with the minerals it absorbs from the soil? A look at three elements (nitrogen, phosphorus, and magnesium) that plant roots take up as inorganic ions provides a partial answer. Nitrogen is a component of many plant hormones and coenzymes and of ATP, all nucleic acids, and all proteins. Nitrogen and magnesium are both components of chlorophyll, the plant's key light-absorbing molecule. Phosphorus is a major component of nucleic acids, phospholipids, and ATP.

A plant's ability to move water from its roots to its leaves and to deliver sugars to specific areas of its body are staggering feats of evolutionary engineering. Figure 32.1B highlights the distance between the bottom of a tree and its leaves; the roots of a redwood can be over 100 m (330 ft) below the topmost leaves! In the next four modules, we follow the movements of water, dissolved mineral nutrients, and sugar throughout the plant body.

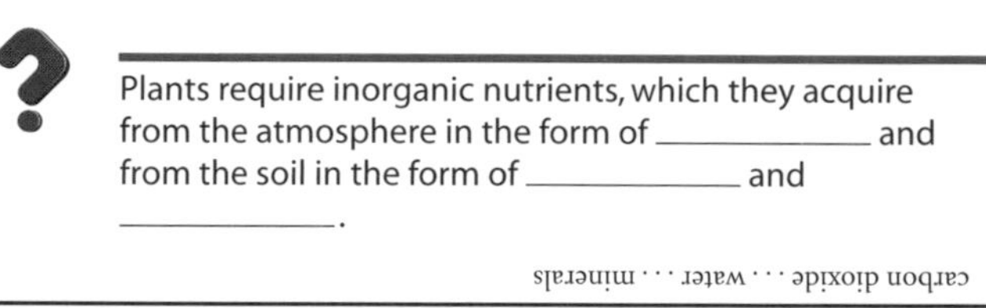

? Plants require inorganic nutrients, which they acquire from the atmosphere in the form of ________ and from the soil in the form of ________ and ________.

carbon dioxide . . . water . . . minerals

Figure 32.1B Redwood trees, giant products of photosynthesis

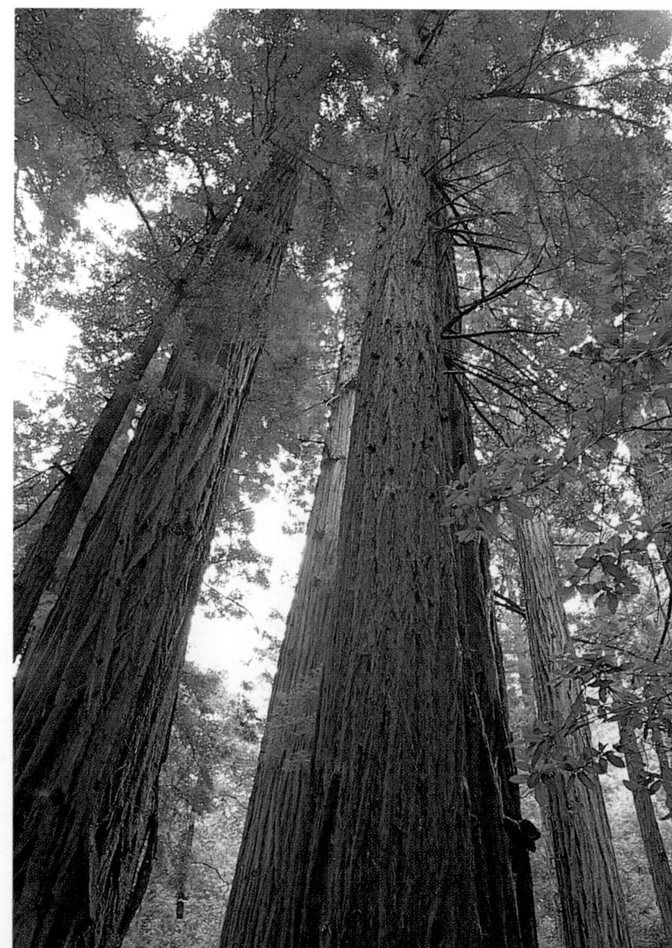

The plasma membranes of root cells control solute uptake 32.2

32A
See Activity 32A on the CD-ROM.

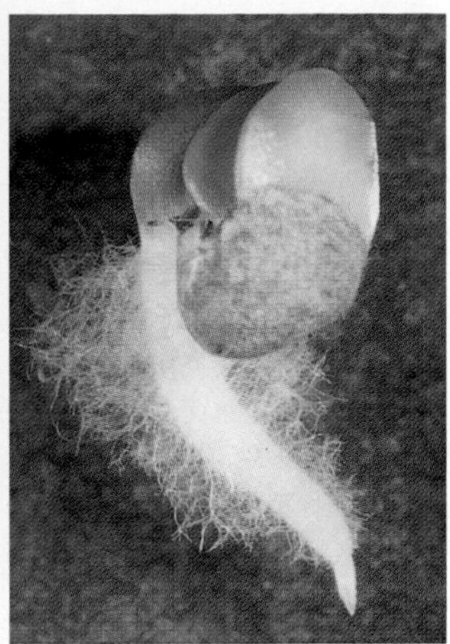

Figure 32.2A Root hairs of a radish seedling

With its surface area enormously expanded by thousands of root hairs (Figure 32.2A), a plant root has a remarkable ability to extract materials from soil. Recall from Chapter 31 that root hairs are extensions of epidermal cells that cover the root. Because of its large root surface area, a plant can absorb enough water and inorganic ions to survive and grow.

All substances that enter a plant root are in solution. In order for water and solutes to be transported from the soil throughout the plant, they must move through the epidermis and cortex of the root, and then into the water-conducting xylem tissue in the central cylinder of the root. (In the root cross section in Figure 32.2B, epidermis is gray, cortex is peach, and xylem is white.) Any route the water and solutes take from the soil to the xylem requires that they pass through some of the plasma membranes of the root cells. Because plasma membranes are selectively permeable, only certain solutes can enter the xylem.

You can see two possible routes to the xylem in the bottom part of Figure 32.2B. The blue arrows indicate an *intracellular* route. Water and selected solutes cross the cell wall and plasma membrane of an epidermal cell (usually at a root hair). The cells within the root are all interconnected by plasmodesmata (channels through the walls of adjacent cells); there is a continuum of living cytoplasm among the root cells. Therefore, once inside the epidermal cell, the solution can move inward from cell to cell without crossing any other plasma membranes, diffusing through the interconnected cytoplasm all the way into the root's endodermis (yellow). An endodermal cell then discharges the solution into the xylem (white).

The red arrows indicate an alternative route. This route is *extracellular;* the solution moves inward within the porous walls of the root cells but does not enter the cytoplasm of the epidermis or cortex cells. The solution crosses no plasma membranes, and there is no selection of solutes until they reach the endodermis. Here, a barrier called the **Casparian strip** stops water and solutes from entering the xylem via cell walls. Shown in black in Figure 32.2B, the Casparian strip is a waxy belt that extends through the walls of the endodermal cells and is continuous from cell to cell. Because of the Casparian strip, water and ions that travel the extracellular (red) route can enter the xylem only by crossing a plasma membrane into an endodermal cell. Ion selection occurs at this membrane instead of in the epidermis, and once the selected solutes and water are in the endodermal cell, they can be discharged into the xylem.

Actually, water and solutes rarely follow just the two kinds of routes in Figure 32.2B. In a real plant, they may take any combination of these routes, and they may pass through numerous plasma membranes and cell walls enroute to the xylem. Because of the Casparian strip, however, there are no nonselective routes; the water and solutes must cross a plasma membrane at some point. Next, we see how water and minerals move upward within the xylem from the roots to the shoots. 

? What is the function of the Casparian strip?

It regulates the passage of minerals (inorganic ions) into the xylem by blocking access via cell walls and requiring all minerals to cross a selectively permeable plasma membrane.

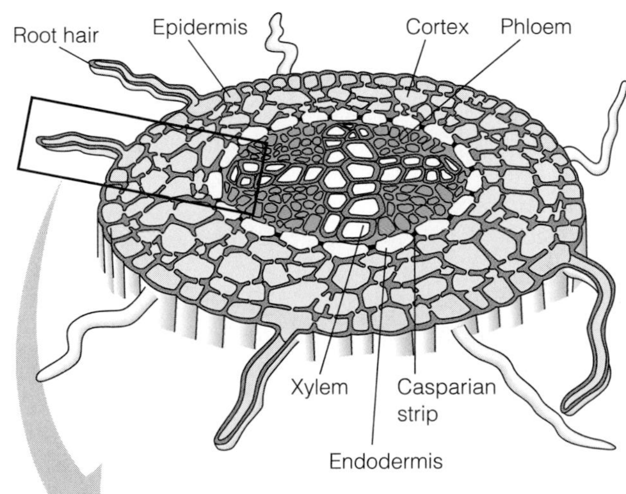

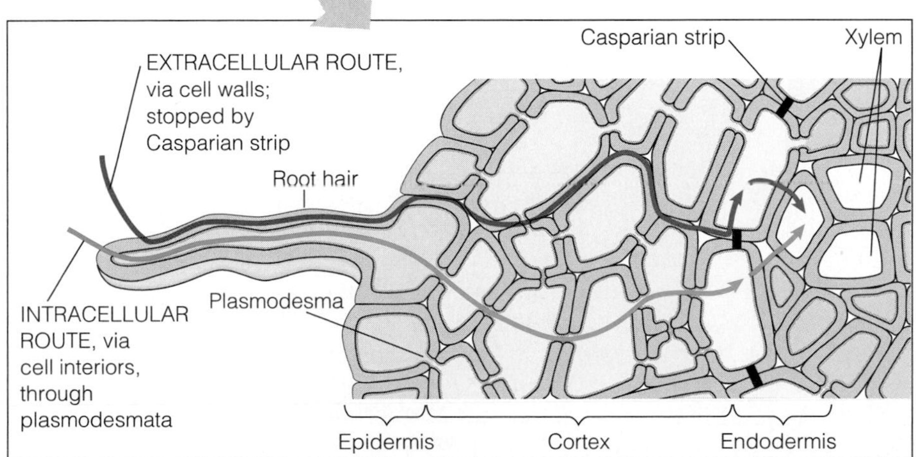

Figure 32.2B Routes of water and solutes from soil to root xylem

32.3 Transpiration pulls water up xylem vessels

32A

The ability to transport water and dissolved ions upward from the soil is a significant adaptation for a land plant. It enables the plant to supply nutrients to its stems and leaves while growing upward and exposing its leaves to sunlight. But how does a plant transport water and nutrients dissolved in it from its roots to its leaves?

We saw in Figure 31.5E that xylem tissue consists of two types of cells: tracheids and vessel elements. When mature, both types of cells are dead, consisting only of cell walls, and both are in the form of very thin, vertical tubes, which are arranged end to end. Because the cells have openings in their ends, a solution of inorganic nutrients, called **xylem sap,** can flow through these tubes all the way up from the plant's roots to the tips of the leaves. What comes to mind when you think about a fluid moving upward in tubes? In humans, for example, blood flows vertically through vessels into the head because the heart pumps it there. Do plants also have some kind of pump that pushes their xylem sap upward? Plant biologists have found that the roots of some plants do exert a slight upward push on xylem sap. The root cells actively pump inorganic ions into the xylem, and the root's endodermis holds the ions there. As ions accumulate in the xylem, water tends to enter by osmosis, pushing xylem sap upward ahead of it. This force, called **root pressure,** can push xylem sap up a few meters. But the push exerted by root pressure does not account for most of the sap's ascent, and some of the tallest trees, including redwoods and giant sequoias, generate no root pressure at all.

It turns out that xylem sap is mainly *pulled,* rather than pushed, from the roots of a plant to the leaves. Plant biologists have determined that the pulling force is **transpiration,** which is the loss of water from the leaves and other aerial parts of a plant.

Figure 32.3 illustrates transpiration and its effect on water movement in a tree. At the top right of the figure, water molecules (blue dots) are shown leaving the leaf through a stoma. This occurs as long as the stoma is open. The water diffuses out of the leaves because the concentration of water molecules is higher in the spaces between cells inside the leaf than in the surrounding air. Transpiration can pull xylem sap up the tree because of two special properties of water: cohesion and adhesion. **Cohesion** is the sticking together of molecules of the same kind. In the case of water, hydrogen bonds make the H_2O molecules stick to one another, as the circular enlargement in the figure shows (see also Module 2.10). The cohering water molecules in the xylem tubes form continuous strings, extending all the way from the leaves down to the roots. In contrast to cohesion, **adhesion** is the sticking together of molecules of different kinds. Water molecules tend to adhere to cellulose molecules in the walls of xylem cells.

What effect does transpiration have on a vertical string of water molecules that tend to adhere to the walls of xylem tubes? Before a water molecule can leave the leaf, it must break off from the top of the string. In effect, it is pulled off by a steep diffusion gradient between the moist interior of the leaf and the dryer surrounding air. Cohesion resists the pulling force of the diffusion gradient, but it is not strong enough to overcome it. The molecule breaks off, and the opposing forces of cohesion and transpiration put tension on the remainder of the string of water molecules. As long as transpiration continues, the string is kept tense and is pulled upward as one molecule exits the leaf and one right behind it is tugged up into its place. Adhesion of the water molecules to the walls of the xylem cells assists the upward

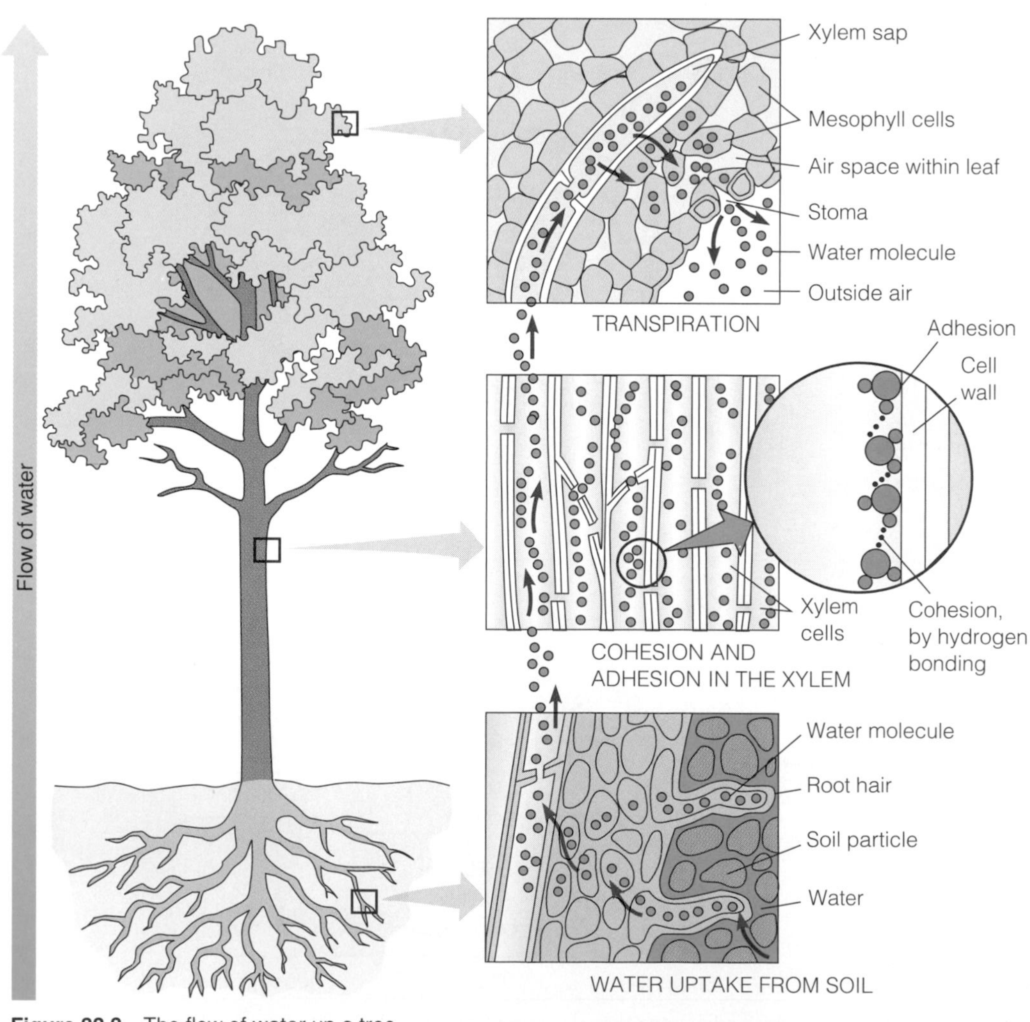

Figure 32.3 The flow of water up a tree

movement of the xylem sap by counteracting the downward pull of gravity. Adhesion also helps hold water in the xylem when transpiration is not occurring.

Plant biologists call this explanation for the ascent of xylem sap the **transpiration-cohesion-tension mechanism.** We can summarize it as follows: Transpiration exerts a pull that is relayed downward along a string of water molecules held together by cohesion and helped upward by adhesion. It is especially important to the plant that none of its own energy is required to transport xylem sap. A plant's xylem tissue is adapted to use outside forces—cohesion, adhesion, and the evaporating effect of the sun—to move water and dissolved minerals from its roots to its shoots.

? (a) Contrast cohesion and adhesion and (b) describe the role of each in the ascent of xylem sap.

(a) Cohesion is the sticking together of identical molecules—water molecules in the case of xylem sap. Adhesion is the sticking together of different kinds of molecules, as in the adhesion of water to the cellulose of xylem walls. (b) Cohesion enables transpiration to pull xylem sap up without the water in vessels separating; adhesion helps to support xylem sap against the downward pull of gravity.

Guard cells control transpiration 32.4

32A

Transpiration actually works both for and against plants. In using the pull of transpiration to move its xylem sap, a plant can lose an astonishing amount of water. Transpiration is usually greatest on days that are sunny, warm, dry, and windy, because these climatic factors increase evaporation. An average-sized maple tree (about 20 m high), for instance, can lose more than 200 L of water an hour during a summer day. As long as water moves up from the soil fast enough to replace the water that is lost, even this amount of transpiration presents no problem. But if the soil dries out and transpiration exceeds the delivery of water to the leaves, the leaves will wilt. Unless the soil and leaves are rehydrated, the plant will eventually die.

The leaf stomata, which can open and close, are adaptations that help plants regulate their water content and adjust to changing environmental conditions. As shown in the diagrams below, a pair of guard cells flanks each stoma. The guard cells control the opening of a stoma by changing shape. Guard cells usually keep the stomata open during the day and closed at night. During the day, CO_2 can enter the leaf from the atmosphere and thus keep photosynthesis going when sunlight is available. At night, when there is no light for photosynthesis and therefore no need to take up CO_2, the closed stomata save water.

What actually causes guard cells to change shape and thereby open or close stomata? Figure 32.4 illustrates the principle. A stoma opens (left) when its guard cells gain K^+ ions (red dots) and water (blue arrows) from surrounding cells (shown in light green). The cells actively take up K^+, and water then enters by osmosis. (For a review of osmosis, see Module 5.15.) When the vacuoles in the guard cells gain water, the cells swell and become turgid. The cell wall of a guard cell is not uniformly thick, as you can see in the drawing, and the cellulose molecules are oriented in such a way that the cell buckles away from its companion guard cell when it is turgid. The result is an increase in the size of the gap (stoma) between the two cells. Conversely, when the guard cells lose K^+, they also lose water by osmosis, become flaccid, and sag together, closing the space between them (right).

Several factors influence the opening and closing of stomata. One is sunlight, which stimulates guard cells to take up K^+ and water, opening the stomata in the morning. A low level of CO_2 in the leaf can have the same effect. A third factor is an internal timing mechanism, a biological clock, found in the guard cells. This clock triggers ion uptake (stomatal opening) in the morning and ion release (stomatal closing) at night. (We'll return to biological clocks in plants in Module 33.10.) The guard cells also close the stomata during the day if the plant loses water too fast. This response reduces further water loss and may prevent wilting, but it also slows down CO_2 uptake and photosynthesis—one reason that droughts reduce crop yields. Overall, the mechanisms that regulate stomatal opening enable a plant to strike a balance between the need to save water and the need to make sugars.

? Some leaf molds, fungi that parasitize plants, secrete a chemical that causes guard cells to accumulate K^+ ions. How does this adaptation help the mold infect the plant?

Accumulation of K^+ by guard cells results in osmotic water uptake, and the turgid condition of the cells keeps the stomata open. The mold can then grow into the leaf interior via the stomata.

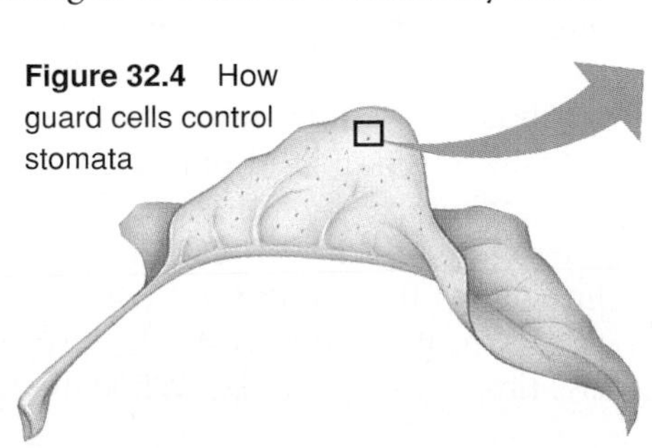

Figure 32.4 How guard cells control stomata

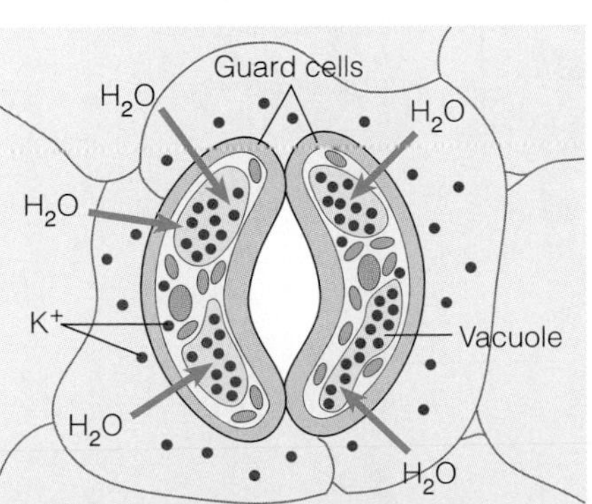

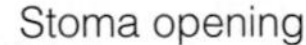

Stoma opening

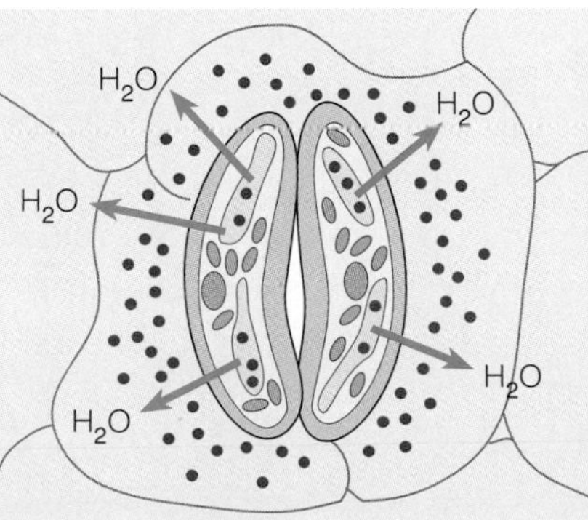

Stoma closing

32.5 Phloem transports sugars

32B

A plant has two separate transport systems: xylem (the topic of Module 32.3) and phloem. The main function of phloem is to transport food molecules, the sugars the plant makes by photosynthesis. Phloem contains food-conducting cells called sieve-tube members arranged end-to-end as tubes (see Figure 31.5F). The micrograph in Figure 32.5A shows two sieve-tube members and the sieve plate between them. Through the perforations in sieve plates, the sugary solution, called **phloem sap,** moves freely from one cell to the next (the cytoplasms of these living cells are continuous). Phloem sap may contain inorganic ions, amino acids, and hormones in transit from one part of the plant to another, but its main solute is usually the disaccharide sucrose.

In contrast to xylem sap, which only flows upward from the roots, phloem sap moves throughout the plant in various directions. A location in a plant where sugar is being produced, either by photosynthesis or by the breakdown of stored starch, is called a **sugar source.** Phloem moves sugar from a sugar source, such as a leaf or green stem, to other parts of the plant. A recipient location in the plant, where the sugar is stored or consumed, is called a **sugar sink.** Growing roots, shoot tips, and fruits are sugar sinks, as are nonphotosynthetic stems and the living cells in tree trunks. Storage structures such as the taproot of a beet, the tubers of a potato plant, and the bulb of a lily are sugar sinks during the summer, when the plant is stockpiling sugars. In early spring, when the plant renews its growth and consumes its stored sugars, beet roots, tubers, bulbs, and other storage structures become sugar sources, and phloem transports sugar away from them to growing organs. Thus, each food-conducting tube in phloem tissue has a source end and a sink end, but these may change with the season or the developmental stage of the plant.

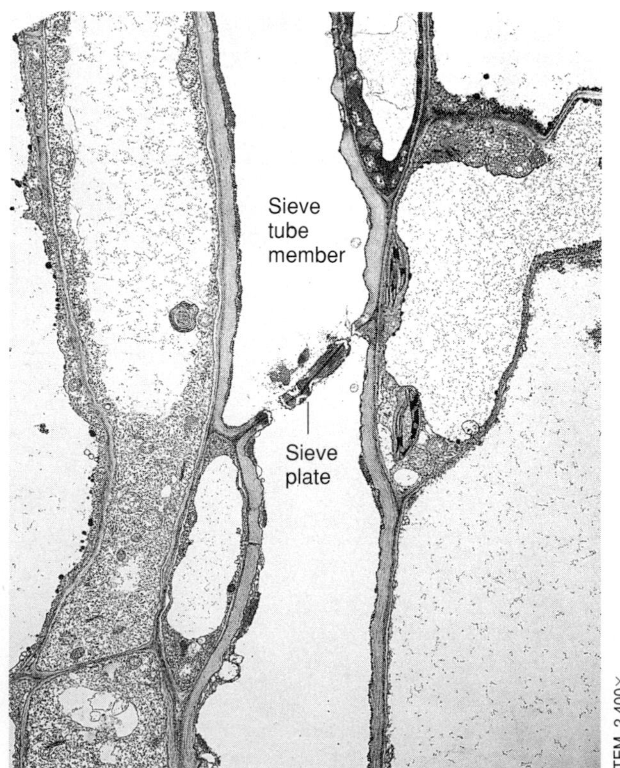

Figure 32.5A Food-conducting cells of phloem

What causes phloem sap to flow from a sugar source to a sugar sink? Flow rates may be as high as 1 m/hr, which is much too fast to be accounted for by diffusion. (It would take phloem sap 8 years to travel a meter if it moved by diffusion alone.) Plant biologists have tested a number of hypotheses for phloem sap movement. A model called the **pressure-flow mechanism** is now widely accepted. Figure 32.5B on the facing page illustrates how this works, using a beet plant as an example. The pink dots in the phloem tube represent sugar molecules; notice their concentration gradient from top to bottom. The blue color represents a parallel gradient of water (hydrostatic) pressure in the phloem sap.

At the sugar source (leaves, in this example), ① sugar is loaded into a phloem tube by active transport. Sugar loading at the source end raises the solute concentration inside the phloem tube. ② The high solute concentration draws water into the tube by osmosis. The inward flow of water raises the water pressure at the source end of the tube.

At the sugar sink (the beet root, in this case), both sugar and water leave the phloem tube. ③ As sugar departs from the phloem, ④ water follows by osmosis. The exit of sugar lowers the sugar concentration in the sink end; the exit of water lowers the hydrostatic pressure in the tube. The building of water pressure at the source end and the reduction of that pressure at the sink end cause water to flow from source to sink—down a gradient of hydrostatic pressure. Since the sugar is dissolved in the water and the sieve plates allow free movement of solutes as well as water, the sugar is carried along from source to sink at the same rate as the water. As indicated on the right side of Figure 32.5B, xylem tubes transport the water back from sink to source.

The pressure-flow mechanism explains why phloem sap always flows from a sugar source to a sugar sink, regardless of their locations in the plant. However, the mechanism is somewhat difficult to test because most experimental procedures disrupt the structure and function of the phloem tubes. Some of the most interesting studies have taken advantage of natural phloem probes: insects called aphids, which feed on phloem sap.

The three photographs in Figure 32.5C show how plant biologists have used aphids to study phloem sap. On the left, an aphid feeds by inserting its needlelike mouthpart, called a stylet, into the phloem of a tree branch. The aphid is releasing from its anus a drop of so-called honeydew—actually, a tiny amount of phloem sap lacking some solutes that the insect's digestive tract has removed for food. The micrograph in the center shows an aphid's stylet inserted into one of the plant's food-conducting cells. The pressure within the phloem force-feeds the aphid, swelling it to several times its original size. While the aphid is feeding, it can be anesthetized and severed from its stylet. The stylet then serves the researcher as a miniature tap that drips phloem sap for hours. The photograph on the right shows a droplet of phloem sap on the cut end of a stylet. Studies using this technique support the pressure-flow model: The closer the stylet to a sugar source, the faster the sap flows out, and the greater its sugar concentration. This is what we would expect if pressure is generated at the source end of the phloem tube by the active pumping of sugar into the tube.

We now have a broad picture of how a plant transports materials from one part of its body to another. Water and inorganic ions enter from the soil and are distributed by xylem. The xylem sap is pulled upward by transpiration. Carbon dioxide enters the plant through leaf stomata and is converted into sugars in the leaves. A second transport system, phloem, distributes the sugars. Pressure flow drives the phloem sap from leaves and storage sites to other parts of the plant, where the sugars are used or stored.

In Chapter 7, we discussed how plants convert raw materials into organic molecules by photosynthesis. We have yet to say much about the kinds of inorganic nutrients a plant needs and what it does with them. This is the subject of plant nutrition, which we discuss in the next section.

Contrast the forces that move phloem sap with the forces that move xylem sap.

Pressure is generated at the source end of a sieve tube by the loading of sugar and the resulting osmotic flow of water into the phloem. This pressure pushes phloem sap from the source end to the sink end of the tube. In contrast, transpiration generates a pulling force that drives the ascent of xylem sap.

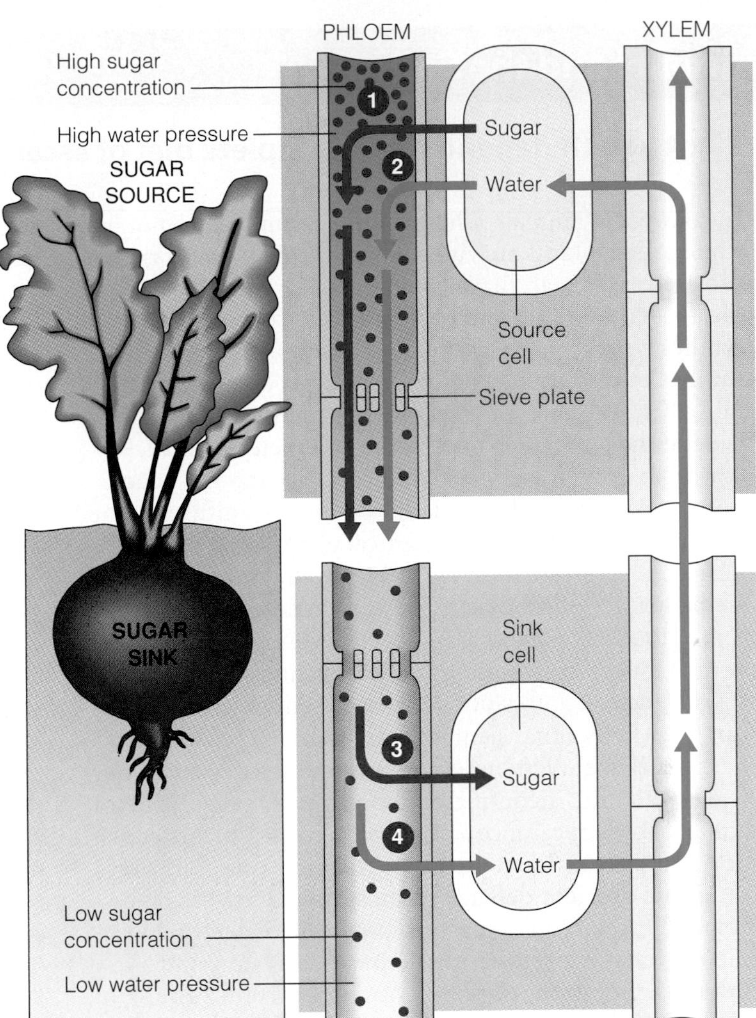

Figure 32.5B Pressure flow in plant phloem from a sugar source to a sugar sink (and the return of water to the source via xylem)

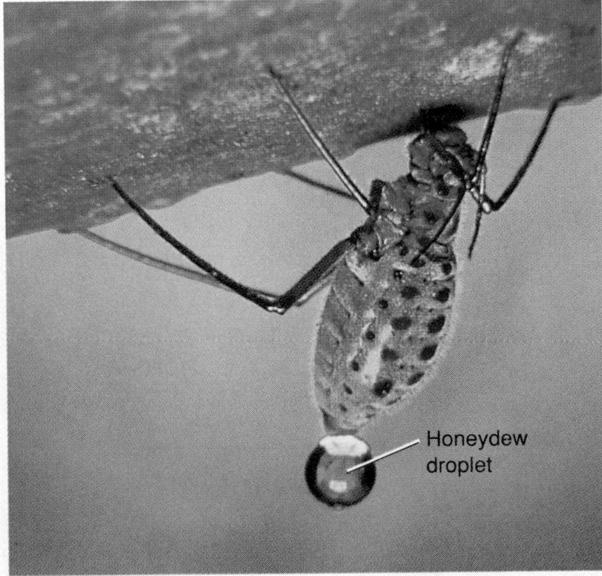

Aphid feeding on a small branch

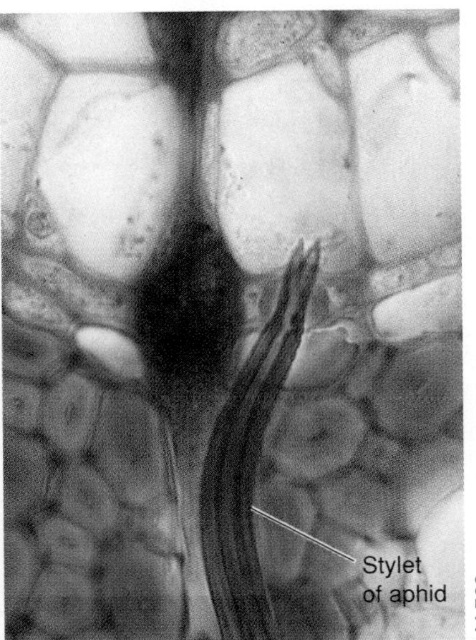

Aphid's stylet inserted into a phloem cell

Severed stylet dripping phloem sap

Figure 32.5C Tapping phloem sap with the help of an aphid

Plant Nutrients and the Soil

32.6 Plant health depends on a complete diet of essential inorganic nutrients

In contrast to animals, which require a complex diet of organic foods, plants survive and grow solely on inorganic substances. The ability of plants to assimilate CO_2 from the air, extract water and inorganic ions from the soil, and synthesize organic compounds is essential not only to the survival of plants but also to humans and other animals.

A chemical element is considered an essential plant nutrient if the plant must obtain it to complete its life cycle—that is, to grow from a seed and produce another generation of seeds. A method called hydroponic culture can be used to determine which chemical elements are essential nutrients. As shown in Figure 32.6A, the roots of a plant are bathed in solutions of various minerals in known concentrations. Air is bubbled into the water to give the roots oxygen for cellular respiration. By omitting a particular element, such as potassium, from the medium, a researcher can test whether that element is essential to the plant.

If the element left out of the solution is an essential nutrient, then the incomplete medium will make the plant abnormal in appearance compared to control plants grown on a complete nutrient medium. The most common symptoms of a nutrient deficiency are stunted growth and discolored leaves. Studies like this have helped identify 17 elements that are essential nutrients in all plants, and a few other elements that are essential to certain groups of plants. Most research has involved crop plants and houseplants; little is known about the nutritional needs of uncultivated plants.

Of the 17 essential elements, nine are called **macronutrients** because plants require relatively large amounts of them. Six of the nine macronutrients—carbon, oxygen, hydrogen, nitrogen, sulfur, and phosphorus—are the major ingredients of organic compounds. These six elements make up almost 98% of a plant's dry weight. The other three macronutrients—calcium, potassium, and magnesium—make up another 1.5%.

How does a plant use calcium, potassium, and magnesium? Calcium has several functions. For example, it is important in the formation of cell walls, and it combines with certain proteins to form a glue that holds plant cells together in tissues. Calcium also helps maintain the structure of cell membranes and helps regulate their selective permeability. Potassium is crucial as a cofactor required for the activity of several enzymes. (Recall from Module 5.7 that a cofactor is an atom or molecule that cooperates with an enzyme in catalyzing a reaction.) Potassium is also the main solute for osmotic regulation in plants; we saw in Module 32.4 how potassium ion movements regulate the opening and closing of stomata. Magnesium is a component of chlorophyll and thus essential for photosynthesis. Magnesium is also a cofactor for several enzymes.

Elements that plants need in very small amounts are called **micronutrients.** The eight known micronutrients are iron, chlorine, copper, manganese, zinc, molybdenum, boron, and nickel. These elements function in the plant mainly as components or cofactors of enzymes. Iron, for example, is a metallic component of cytochromes, proteins that function in the electron transport chains of chloroplasts and mitochondria. Because micronutrients function mainly in catalysis (and are therefore used over and over), plants need only minute quantities of these elements. The requirement for molybdenum, for example, is so modest

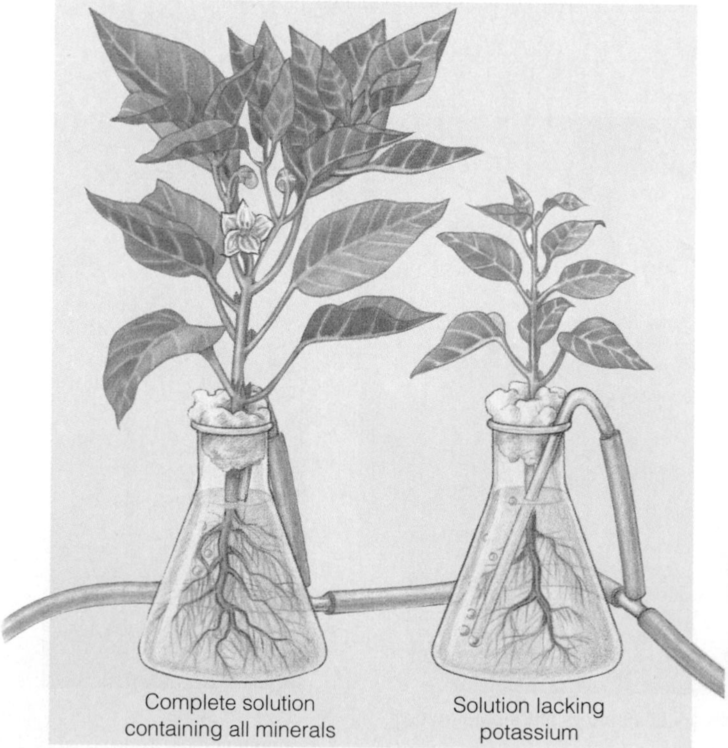

Complete solution containing all minerals

Solution lacking potassium

Figure 32.6A A hydroponic culture experiment

Figure 32.6B The effect of nitrogen availability on corn growth: Corn grown in nitrogen-rich soil (left) and nitrogen-poor soil (right)

that there is only one atom of this rare element for every 16 million atoms of hydrogen in dried plant material. Yet a deficiency of molybdenum or any other micronutrient can kill a plant.

The quality of soil—that is, the nutrients available to plants growing in it—determines the quality of our own nutrition. Figure 32.6B (preceding page) shows two corn crops. The plants on the left are growing in soil rich in nitrogen-containing compounds that the plants can use to build proteins. The small, lighter-colored plants on the right are growing in soil deficient in nitrogen. Even if the nitrogen-deficient plants produce grain, the crop will have a lower food value, and its nutrient deficiencies will then be passed on to livestock or human consumers. Maximizing the nutritional value of crops such as corn is one of the goals of research in plant nutrition.

You conduct an experiment like that in Figure 32.6A to test whether a certain plant species requires a particular chemical element as a micronutrient. Why is it important that the glassware be completely clean?

Because micronutrients are required in only minuscule amounts, even the smallest amount of dirt on the flask on the right may contain enough of the element you are testing to allow normal growth and invalidate your results.

32.7 You can diagnose some nutrient deficiencies in your house and garden plants

The symptoms of nutrient deficiency in plants are usually obvious, and although a number of deficiencies produce similar outward signs, it is possible to diagnose some problems quite readily. Many growers make visual diagnoses of their own ailing house or garden plants and then check their conclusions by having samples of soils and plants chemically analyzed at a state or local soils laboratory.

Deficiencies of the macronutrients nitrogen, phosphorus, and potassium are seen most often. The photographs here compare a healthy tomato plant (Figure 32.7A) with genetically identical plants suffering from these macronutrient deficiencies. Nitrogen shortage is the single most common nutritional problem for plants. Soils are usually not deficient in nitrogen, but they are often deficient in the nitrogen compounds that plants can use, dissolved nitrate ions (NO_3^-) and ammonium ions (NH_4^+). Stunted growth and yellow-green leaves (Figure 32.7B) are signs of nitrogen deficiency. The older leaves usually show the signs first.

Phosphorus deficiency is the second most common nutritional ailment in plants. As is the case with nitrogen, soils usually contain plenty of phosphorus, but it may not be in the ionic, water-soluble forms, $H_2PO_4^-$ or HPO_4^{2-}, that plants can use. A phosphorus-deficient plant may have green leaves, but its growth rate is markedly reduced, and its new growth is often spindly and brittle. Also, in some plants, such as the one in Figure 32.7C, phosphorus deficiency produces a purplish color on the undersides of the leaves.

Figure 32.7D illustrates potassium deficiency. Plants take up potassium as K^+ ions dissolved in the soil water. Once again, most potassium compounds in the soil are only slightly soluble in water and therefore not available to plants. The outward signs of a potassium shortage are generally more localized than those of nitrogen and phosphorus deficiencies. The older leaves usually show the most pronounced signs; they often turn yellow and develop dead, brownish tissue at the edges or in scattered spots. Stems and roots are also weakened, leading to stunting.

Once a diagnosis of a nutrient deficiency is made, treating the problem is usually simple. You can choose from a number of fertilizer products for enriching the soil. Many of these consist of the inorganic compounds plants can use directly, such as nitrates and phosphates; others contain organic materials that are broken down to the usable inorganic compounds by microbes in the soil.

What is the most common nutrient deficiency in plants?

Nitrogen deficiency

Figure 32.7A A healthy plant

Figure 32.7B Nitrogen deficiency

Figure 32.7C Phosphorus deficiency

Figure 32.7D Potassium deficiency

32.8 Soil in the life of plants: A closer look

Figure 32.8A Three soil horizons in a cotton field

32C

Soil characteristics determine how plants grow in a particular location. Fertile soil is soil that supports abundant plant growth by providing adequate water and dissolved nutrients. It also provides the particular conditions that enable plant roots to absorb the substances the plant needs.

Figure 32.8A shows a researcher photographing a cross section of the soil in a Tennessee cotton field. You can see three distinct layers, called **soil horizons,** in the cut. The A horizon, the top 20 cm (8 inches) in this case, is the **topsoil.** The topsoil is subject to extensive weathering (freezing, drying, and erosion, for example). Fertile topsoil contains rock particles of various sizes, including sand and even smaller particles called clays. These particles provide an extensive surface area that retains water and inorganic nutrients, while allowing oxygen to diffuse into plant roots. Topsoil also contains decomposing organic material, called **humus,** and living organisms. Humus is an important source of plant nutrients; it also tends to retain water while keeping the topsoil porous enough for good aeration of the plant roots. Fertile topsoils usually support teeming numbers of bacteria, protozoans, fungi, and small animals such as earthworms, roundworms, and burrowing insects. Along with plant roots, these organisms loosen and aerate the soil and contribute organic matter to the soil as they live and die. Nearly all plants depend on bacteria and fungi in the soil to break down organic matter into inorganic molecules that roots can absorb. Plant roots branch out in the A horizon and usually extend into the next layer, the B horizon.

The soil's B horizon contains many fewer organisms and much less organic matter than the topsoil and is less subject to weathering. Fine clay particles and nutrients dissolved in soil water drain down from the topsoil and often accumulate in the B horizon. Below the B horizon, the C horizon is composed mainly of partially broken-down rock.

Figure 32.8B illustrates the intimate association between a plant's root hairs, soil water, and the tiny particles of topsoil. The root hairs are in direct contact with soil water held in tiny spaces among the particles. The soil water is actually a solution containing dissolved oxygen (O_2) and inorganic ions, many of them plant nutrients. Oxygen diffuses into the water from small air spaces in the soil. The root hairs take up dissolved oxygen, ions, and water from the film of soil water that surrounds them.

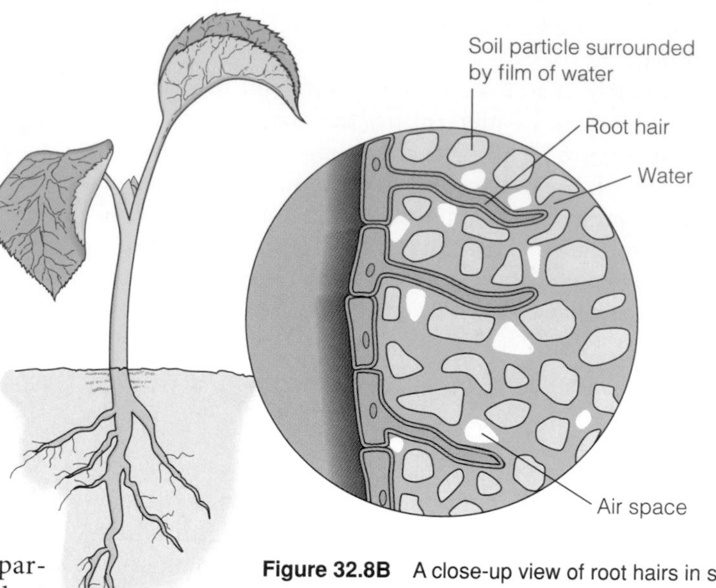

Figure 32.8B A close-up view of root hairs in soil

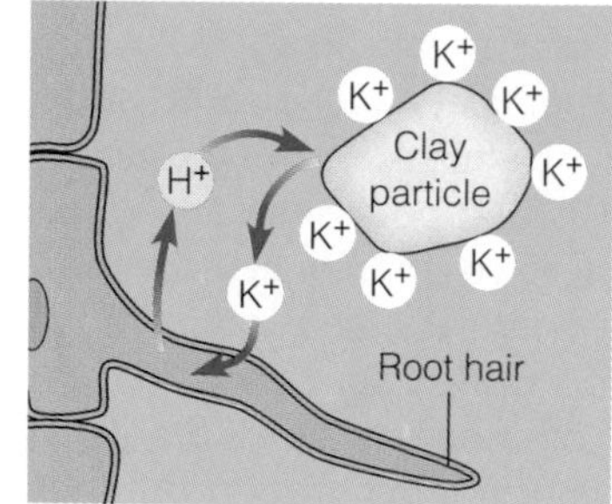

Figure 32.8C Cation exchange

Cation exchange is a mechanism by which root hairs take up certain positively charged ions (cations). Inorganic cations, such as calcium (Ca^{2+}), magnesium (Mg^{2+}), and potassium (K^+), adhere by electrical attraction to the negatively charged surfaces of clay particles. This adherence helps prevent these positively charged nutrients from draining away during heavy rain or irrigation. In cation exchange (Figure 32.8C), the root hairs release hydrogen ions (H^+) into the soil solution. The H^+ ions displace nutrient ions on the clay particle surfaces, and the root hairs can then absorb the free cations.

In contrast to cations, negative ions (anions), such as nitrate (NO_3^-), are usually not bound tightly by soil particles. Unbound ions are readily available to plants, but they tend to drain out of the soil quickly. This is often how soils become deficient in nitrogen.

It may take centuries for a soil to become fertile. The loss of soil fertility is one of our most pressing environmental problems, as we discuss next. 

? How do roots actively increase the availability of mineral nutrients that are cations?

By secreting H^+ ions (acid), which displace cations from soil particles by cation exchange

32.9 Soil conservation is essential to human life

Our survival as a species depends on soil, yet the loss of soil by erosion and chemical pollution threaten this vital resource throughout the world. As the human population continues to grow, and more and more land is cultivated, prudent farming practices that conserve soil fertility will become essential to our survival. Three critical aspects of soil conservation are irrigating properly, preventing erosion, and fertilizing.

Irrigation can turn a desert into a garden, but farming in dry regions is a huge drain on water resources. Irrigation also tends to make the soil salty. The whitish deposits on the soil in the photograph in Figure 32.9A are salts that were dissolved in irrigation water flooded onto a field. Left behind when the excess water evaporated, the deposits will eventually make the soil too salty for crop plants to tolerate. Instead of flooding fields, modern irrigation often employs perforated pipes that drip water slowly into the soil close to plant roots. This drip irrigation uses less water, allows the plants to absorb most of the water, and reduces water loss from evaporation and drainage.

Preventing erosion—the blowing or washing away of soil—is one of the most important challenges of modern agriculture. Thousands of acres of farmland are lost to water and wind erosion each year in the United States alone. Plowed soil is especially vulnerable. Plowing aerates the soil and buries weeds and crop stubble, which return nutrients to the soil when they decay, but it also exposes the soil to eroding winds and rains. A method known as minimal tillage farming can reduce erosion. Farmers using this method do not plow every year, and they usually rely on herbicides to kill weeds. Unfortunately, the herbicides contribute to chemical pollution of the soil. Other ways of reducing soil losses to erosion include planting trees along field edges to prevent wind erosion, and taking special care in hilly terrain. The crops in Figure 32.9B are planted in rows that go around, rather than up and down, the hill in the field. This contour tillage helps slow the runoff of water and topsoil after heavy rains.

Fertilizers have probably been applied to crops since prehistoric farmers noticed that grass grew faster and greener where animals had defecated. Today, in developed nations, most farmers use commercially produced fertilizers containing minerals that are either mined or prepared by industrial processes. These fertilizers contain inorganic compounds of nitrogen, phosphorus, and potassium, the three elements that are most commonly deficient in farm soils.

Figure 32.9A Flood irrigation

Figure 32.9B Planting to prevent soil erosion in a hilly area

Manure, fishmeal, and compost (decaying plant matter) are referred to as "organic" fertilizers because they are of biological origin. Before the nutrients in these substances can be used by plants, the organic material must be broken down by bacteria and fungi to inorganic nutrients that roots can absorb. In the end, the inorganic ions a plant extracts from the soil are in the same form whether they came from organic fertilizer or from a chemical factory. The difference is that organic fertilizers release nutrients gradually, whereas inorganic fertilizers make them available immediately. Problems arise when fields are overfertilized with inorganic products and excess nutrients are not taken up by plants. The nutrients are not usually retained in the soil, so they can enter and pollute the groundwater, as well as streams and lakes. In contrast, organic fertilizers are retained in the soil because their organic components are relatively insoluble in water.

Inorganic fertilizers have greatly increased agricultural productivity, and they are used so extensively today that if farmers suddenly stopped using them, widespread famine could result. Researchers are trying to find ways to reduce the amounts of inorganic fertilizers applied to croplands, while maintaining crop yields. In the next module, we discuss one farmer's use of exclusively organic fertilizers.

Irrigating fields by repeatedly flooding them often causes the soil to become very salty. Why?

Salts present in the irrigation water accumulate in the soil as water evaporates from the fields.

32.10 Organic farmer Stephen Moore uses no commercial chemicals

Stephen Moore has a Ph.D. in environmental engineering, was on the faculty at MIT, and spent five years as an environmental consultant. But in 1981, Moore changed direction and took over his parents' 60-acre southern California farm, which had been in the family since 1875. As he puts it, "I have a love for getting my hands right in it." With a work force of eight, he grows lemons, limes, avocados, persimmons, and a variety of vegetables, without any pesticides or commercial fertilizers.

Prior to the 1950s, soil fertility on Moore's land was maintained by regular applications of organic matter, mainly animal manure. As a result, the humus content and fertility of the topsoil remained high. But between the 1950s and 1980s, the Moores followed the nationwide trend of using more and more inorganic chemicals and less organic fertilizer. Steve Moore found that, after more than 30 years of treatment with commercial chemicals, the soil he inherited was virtually infertile. He decided to revert to "the old ways," rebuilding soil fertility with compost and animal manure, and by what he calls "growing our own fertilizer." Among his orchard trees and on recently harvested cropland, he raises legumes and grasses. While growing and when plowed under, these crops help rebuild soil fertility. With healthier soil, there is little trouble with pests. In 1985, Moore stopped using commercial chemicals altogether, and today he takes great pride in running a profitable, completely organic farm.

Figure 32.10 Stephen Moore

Organically grown foods are becoming increasingly popular. Many growers can market their produce locally, and there is an established system for national and international distribution. Moore sells most of his vegetables locally, and his avocados and citrus fruits sell regionally on the West Coast. The operating expenses of an organic farm are higher than those of conventional farms, and organic foods usually cost more than those produced conventionally. But a growing number of consumers are willing to pay extra for what they believe to be safer, perhaps more nutritious, and often tastier food. Steve Moore thinks of it this way:

> When you buy organically grown products, you benefit from eating nutritious food. You also make an investment in the future, because you are paying the real costs up front. It is cheaper to grow food the conventional way, but there are hidden costs. Pesticides and chemical fertilizers are costly to the environment. They contaminate groundwater, and we pass the bill for that on to future generations.

Moore believes, as do a growing number of agricultural experts, that the future of farming depends on an international effort in sustainable agriculture—growing crops and maintaining healthy, fertile soil at the same time. 

? Why do organic fertilizers generally contaminate water resources less than inorganic fertilizers?

The organic fertilizers release mineral nutrients gradually as they decompose, and thus there is less likelihood of the minerals leaching into the groundwater or running off into streams and lakes before they can be absorbed by plant roots.

32.11 Fungi help most plants absorb nutrients from the soil

Reliance on the soil for nutrients that may be in short supply makes it imperative that the roots of plants have a large absorptive surface area. As we have seen, root hairs add a great deal of surface to plant roots. Most plants gain even more absorptive surface by teaming up with fungi. The micrograph here shows a small root of a eucalyptus tree. The root is covered with a twisted mat of fungal filaments. Together, the roots of this plant and the fungus comprise a mutually beneficial association called a **mycorrhiza** (Greek *mykes*, fungus, and *rhiza*, root).

Mycorrhizae are particularly beneficial adaptations for plants growing in poor soils, but almost all plants are capable of forming this symbiotic relationship if their roots are exposed to appropriate species of fungi. The fungal filaments around the root provide an enormous surface, which absorbs water and inorganic ions, especially phosphate, more rapidly than the root alone can. Some of the water and ions taken up by the fungus are transferred to the plant. The fungus may also secrete acid that increases the solubility of some minerals in the soil and may convert them to forms that are more readily used by the plant. In turn, the plant's photosynthetic products nourish the fungus. The fungi may also help protect the plant against certain pathogenic microorganisms common in soil.

As more is learned about mycorrhizal associations, important agricultural applications are likely. As we men-

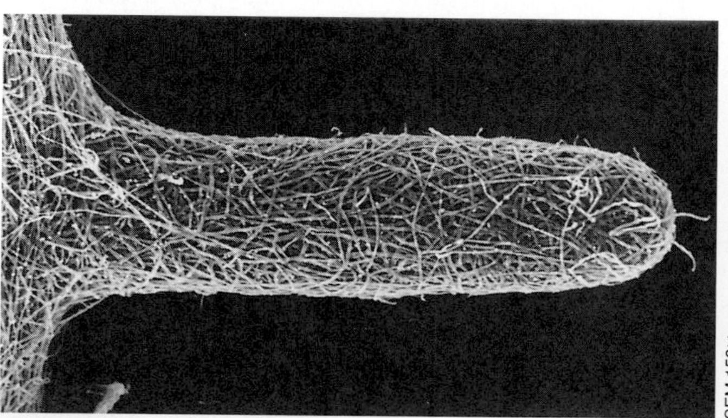

Figure 32.11 A mycorrhiza on a eucalyptus root

tioned in the introduction to Chapter 17, citrus trees require less fertilizer when grown with mycorrhizae. Techniques for supplying root fungi to other crops may eventually enable farmers to reduce fertilizer use to a greater extent.

The widespread presence of mycorrhizae is a reminder of the theme of connections among living organisms. Despite their ability to make their own food molecules, plants are not independent of other organisms. The fossil record shows that mycorrhizae have been common since plants first evolved. Indeed, the mycorrhizal connection probably altered the entire course of evolution by helping make possible the colonization of land.

What are mycorrhizae?

Symbiotic associations of roots and fungi

The plant kingdom includes parasites and carnivores 32.12

Mycorrhizae are not the only kind of symbiosis that can increase a plant's nutrient supply and chances of survival. Some plants have evolved ways of obtaining food from other plants or animals. Figure 32.12A shows a parasitic plant called dodder (the yellow-orange threads wound around the green plant). Dodder cannot photosynthesize; it obtains organic molecules from other plant species, using specialized roots that tap into the host's vascular tissue.

Figure 32.12B shows part of an oak tree parasitized by mistletoe, the plant we may tack above doorways during the Christmas season. There are about 1000 species of mistletoe, and one or more usually occur in areas where there are deciduous trees. All the leaves you see here are mistletoe; the oak has lost its leaves for winter. Mistletoe is photosynthetic, but it supplements its diet by siphoning sap from the vascular tissue of the host tree. Both dodder and mistletoe may kill their hosts by blocking light or taking too much food from them.

Figure 32.12A Dodder growing on a pickleweed

Figure 32.12B Mistletoe growing on an oak

Certain plants are carnivorous, obtaining some of their nutrients, especially nitrogen, from animal tissues. Carnivorous plants grow in bogs where the soil is highly acidic. Organic matter decays so slowly in acidic soils that there is little inorganic nitrogen available for plant roots to take up. Though they are photosynthetic, the sundew and Venus flytrap (Figures 32.12C and 32.12D) thrive by obtaining their nitrogen from insects.

Few species illustrate the correlation of structure and function better than carnivorous plants. The sundew plant (Figure 32.12C) has modified leaves, each bearing many club-shaped hairs. A sticky, sugary secretion at the tips of the hairs attracts insects and traps them. The presence of an insect triggers the hairs to bend and the leaf to cup around its prey. The hairs then secrete digestive enzymes, and the plant absorbs nutrients released as the insect is digested.

The Venus flytrap (Figure 32.12D) has hinged, V-shaped leaves that close around small insects. As insects enter the open V, they touch sensory hairs that trigger closure of the trap. The leaf then secretes digestive enzymes and absorbs nutrients from the prey.

Using insects as a source of nitrogen is a nutritional adaptation that enables carnivorous plants to thrive in soils where most other plants cannot. Only a few plants are carnivorous, however. Most plants obtain nitrogen with the help of soil bacteria, as we discuss next.

Figure 32.12C A sundew plant trapping a fly

Figure 32.12D A Venus flytrap digesting a katydid

Carnivorous plants are most common in locales where the soil is deficient in ________.

nitrogen

32.13 Most plants depend on bacteria to supply nitrogen

It is ironic that plants often have nitrogen deficiencies, for the atmosphere is nearly 80% nitrogen. Atmospheric nitrogen, however, is gaseous N_2, and plants cannot use nitrogen in that form. In fact, nearly all plants depend to some extent on nitrogen supplies in the soil.

For plants to absorb nitrogen from the soil, the nitrogen must first be converted to ammonium ions (NH_4^+) or nitrate ions (NO_3^-). Ammonium and nitrate in the soil are produced from atmospheric N_2 or from organic matter by bacteria. As shown in Figure 32.13, certain soil bacteria, called nitrogen-fixing bacteria, convert atmospheric N_2 to ammonium. This process, called **nitrogen fixation,** is vital to plants. A second group of bacteria, called ammonifying bacteria, adds to the soil's supply of ammonium by decomposing organic matter (humus).

The dashed red arrow in the diagram indicates that plant roots absorb only a small amount of nitrogen as ammonium. The availability of ammonium to the plant is limited, because it tends to remain bound to clay particles in the soil. Fortunately for plants, a third type of soil bacteria, nitrifying bacteria, converts soil ammonium to nitrate. Plants take up most of their nitrogen in this form. They then convert the nitrate back into ammonium, which they incorporate into amino acids. Amino acids are used to make proteins and other nitrogen-containing organic molecules.

Why would a pollutant that kills soil bacteria result in nitrogen deficiency in plants?

Because it is certain soil bacteria that make nitrogen available to plants in forms they can use—nitrate and ammonium ions

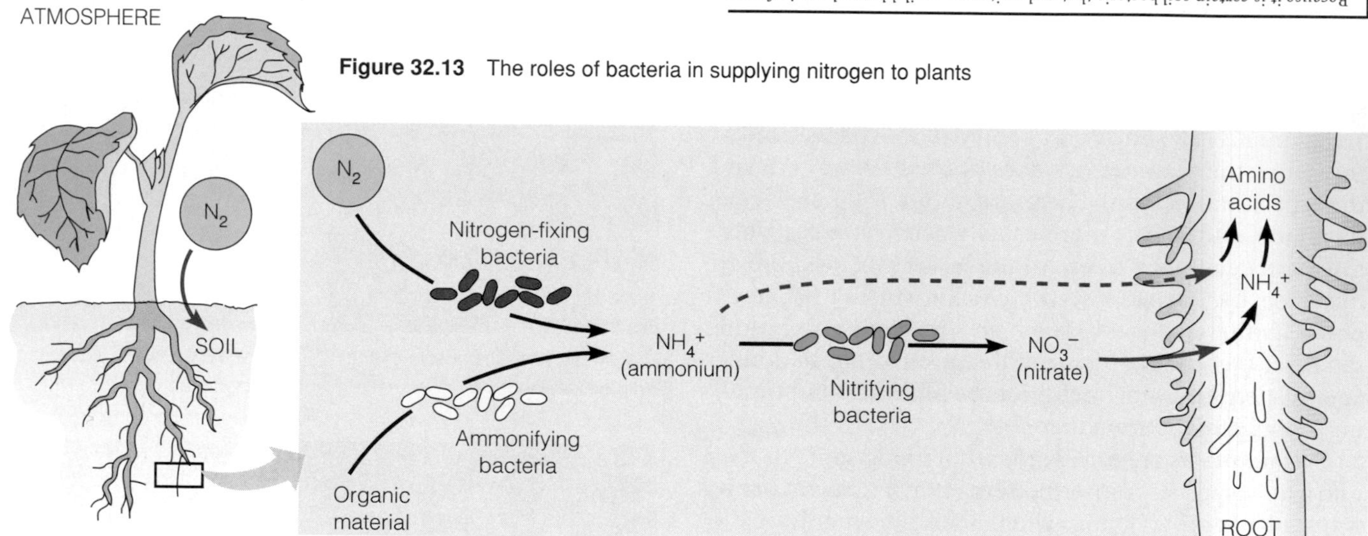

Figure 32.13 The roles of bacteria in supplying nitrogen to plants

32.14 Legumes and certain other kinds of plants have built-in nitrogen-fixing bacteria

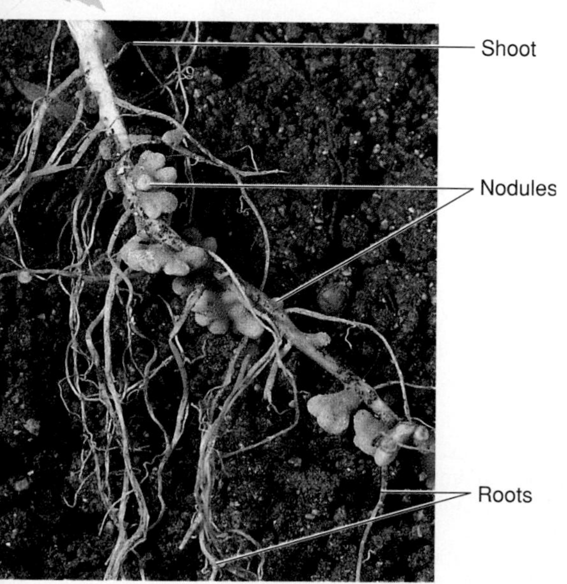

Figure 32.14A Root nodules on a pea plant

As we discussed in the chapter's introduction, the plants called legumes have their own source of fixed nitrogen—bacteria housed in nodules on their roots (Figure 32.14A). The legumes include peanuts, peas, beans, and many other plants that produce their seeds in pods. The **root nodules** are swellings consisting of plant cells that contain nitrogen-fixing bacteria. The micrograph at the top of the facing page (Figure 32.14B) shows a cross section of one such cell; notice the vesicles full of bacteria. Most of the nitrogen-fixing bacteria in legume nodules belong to the genus *Rhizobium* (from the Greek *rhiza,* root, and *bios,* life). Other nitrogen-fixing bacteria (actinomycetes) are found in root nodules of some nonlegume plants, such as alders.

The relationship between a plant and its nitrogen-fixing bacteria is mutually beneficial. The plant provides the bacteria with carbohydrates and other organic compounds. The bacteria have enzymes that catalyze the conversion of atmospheric N_2 to ammonium ions (NH_4^+). When conditions are favorable, root nodule bacteria actually fix so much nitrogen that the nodules secrete excess NH_4^+, which increases the fertility of the soil. This is one reason farmers rotate crops, one year planting a nonlegume such as corn, and the next year planting a legume to raise the concentration of usable nitrogen compounds in the soil. The legume crop is often soybeans or alfalfa. When the crop is harvested, the roots remain in the ground and add their fixed nitrogen and other nutrients to it. Instead of being harvested, the legume crop may be plowed under so that it will decompose and add even more fixed nitrogen to the soil.

How do the nitrogen-fixing bacteria of root nodules benefit from their symbiotic relationship with plants?

The bacteria, which do not photosynthesize, depend on the host plant for a supply of certain organic compounds produced by the photosynthesizing plant.

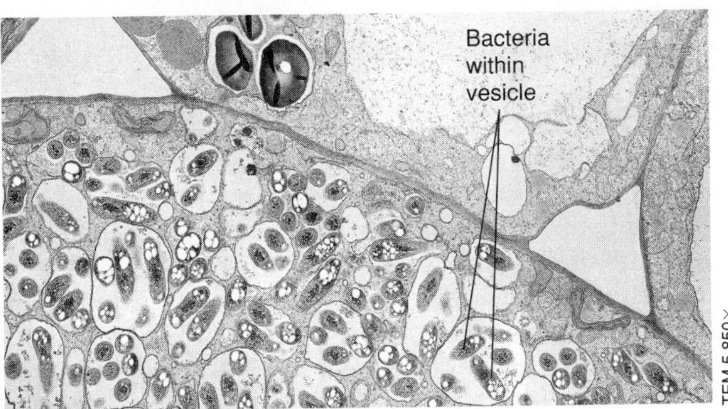

Figure 32.14B Bacteria within a root nodule cell

Plant Nutrients and Agriculture

A major goal of agricultural research is to improve the protein content of crops 32.15

The ability of plants to incorporate the element nitrogen into proteins and other organic substances has a major impact on human welfare. Either by choice or by economic necessity, the majority of people in the world have a predominantly vegetarian diet. Thus, particularly in the developing countries, people depend mainly on plants as immediate sources of protein. Unfortunately, many plants have a low protein content, and the proteins that are present in them may be deficient in one or more of the amino acids that humans need from their diet. Protein deficiency is a common form of malnutrition in humans.

Improving the quality and quantity of proteins in crops is a major goal of agricultural researchers. Among some of the most important advances to date are new varieties of corn, wheat, and rice with improved protein content (Figure 32.15A). However, many of these "super" varieties have an extraordinary demand for nitrogen, which is usually supplied to them in the form of commercial, inorganic fertilizer. Besides being hazardous to the environment, these fertilizers are expensive to produce. Most of the countries that have the greatest need for high-protein crops are the ones least able to afford to grow them.

Figure 32.15A Plant researchers with "super" rice

One of the most promising lines of agricultural research is directed toward improving the output of the *Rhizobium* bacteria that inhabit the root nodules of legumes. A negative-feedback mechanism normally regulates the rate at which these bacteria convert N_2 to nitrogen compounds. When the quantity of fixed nitrogen in a root nodule reaches a certain level (the green rectangle in Figure 32.15B), it switches off the bacterial genes that code for the enzymes needed for nitrogen fixation. Researchers have isolated certain *Rhizobium* mutants that continue to make these enzymes even after nitrogen compounds accumulate. Farmers may someday grow plants colonized by these mutant bacteria, a breakthrough that would increase the protein content of the legume crop and also add more fixed nitrogen to the soil. Genetic engineering also promises additional improvements in crops, as we discuss next.

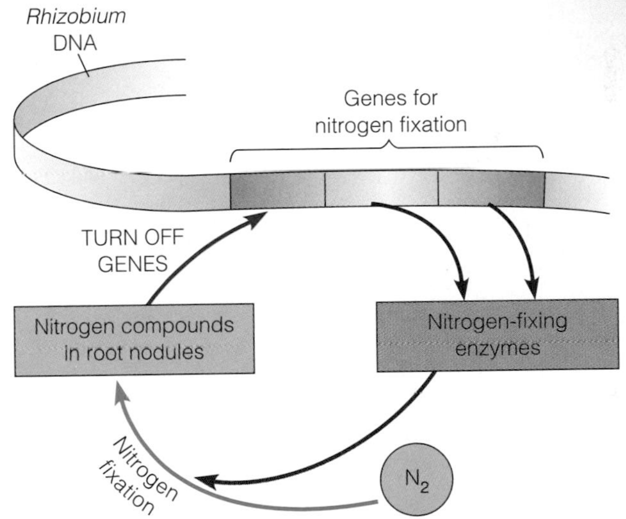

Figure 32.15B Regulation of nitrogen fixation in *Rhizobium* bacteria

Why is research on nitrogen metabolism in crop plants so important to human health worldwide?

Because the most common form of malnutrition is protein deficiency, and most people in the world get most of their protein from plants

32.16 Genetic engineering may greatly increase crop yields

A technician readies a .22-caliber gun for shooting foreign genes into plant cells. The gun fires plastic bullets containing tiny metal pellets coated with DNA. As Figure 32.16 shows, the bullet stays in the gun, but the particles are driven into the plant cells. There the foreign DNA integrates into the cells' DNA. A cell engineered in this way can be grown into a whole new plant that produces proteins encoded by the foreign DNA, along with its usual proteins.

Many new varieties of crop plants have already been produced with the gene gun and by an older technique that uses bacterial plasmids for gene transfer (see Module 12.19). Cotton and tobacco plants have been engineered that are resistant to viral attack. Potato plants have been engineered to synthesize their own insecticide, making them resistant to attack by beetles that can destroy whole crops. Tomato plants have been engineered to produce fruit that is slow to spoil. In the future, genetic engineering may also produce crop plants that can synthesize pharmaceutical products, industrial oils, and other useful chemicals.

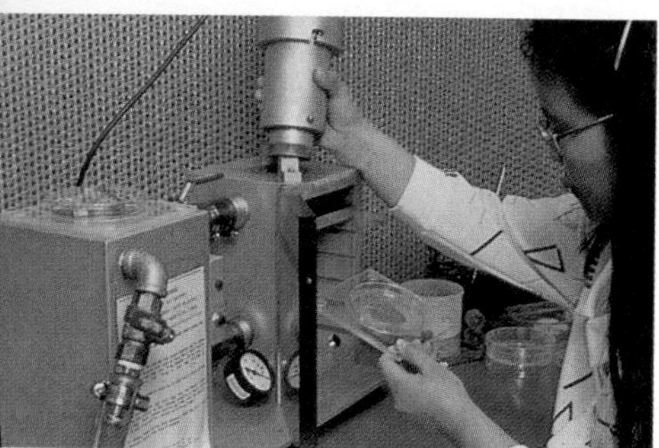

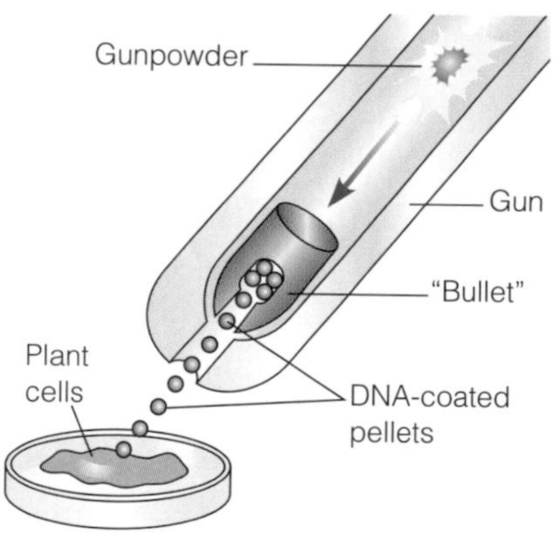

Figure 32.16
Using a gene gun

A major goal of genetic engineering is to create crop plants that provide more nutritious food—for instance, corn, wheat, and other grains that have a full complement of the amino acids humans need to make proteins. Another goal is to engineer varieties of nitrogen-fixing bacteria that are more efficient than naturally occurring varieties at making NH_4^+. Eventually, it may be possible to transplant genes for nitrogen fixation directly into the DNA of nonlegume crop plants.

Genetic engineering holds great potential for increasing agricultural production. There are potential problems, however. Crop plants containing genes that help the plant resist diseases, might, for example, escape into the wild and overgrow native species. Such plants might also hybridize (interbreed) with their wild relatives, creating weeds that grow out of control. Another concern is that new proteins in foods produced by gene-spliced plants could be toxic or cause serious allergies in some people. Governments throughout the world are grappling with how to proceed—whether to promote the agricultural revolution offered by gene splicing, or slow its progress until more information is available about the potential hazards. We will touch on the subject of agricultural productivity again in Chapter 33, which discusses plant hormones.

How does the ability of closely related plant species to hybridize (interbreed) pose a potential hazard of technology that introduces foreign genes into crop species?

The foreign genes could be transferred to weeds and other wild relatives of the crop species by interbreeding between the crop and wild relatives.

Chapter Review

Chapter Summary

The Uptake and Transport of Plant Nutrients (Introduction–32.5) The discovery that the roots of legume plants house bacteria that supply the plant with nitrogen usable for protein production was a milestone in the advancement of modern agriculture **(Introduction).** As a plant grows, its roots absorb water, minerals (inorganic ions), and some O_2 from the soil. Its leaves take carbon dioxide from the air **(32.1).** Root hairs greatly increase a root's absorptive surface. Water and solutes can move through the root's epidermis and cortex by going either through cells or between them. However, all water and solutes must pass through the plasma membranes of cells of the endodermis to enter the xylem (water-conducting tissue) for transport upward. The endodermal cell membranes admit only certain solutes **(32.2).**

In some plants, solute transport may raise water pressure in the xylem. This root pressure can push water a short way up the stem. Most of the force that moves water and solutes upward in the xylem comes from transpiration, the evaporation of water from the leaves. Cohesion causes water molecules to stick together, relaying the pull of transpiration along a string of water molecules all the way to the roots. The adhesion of water molecules to xylem cell walls helps counter gravity. Transpiration can move xylem sap, consisting of water and dissolved inorganic nutrients, to the top of the tallest tree **(32.3).** Guard cells surrounding stomata in the leaves control transpiration **(32.4).** Phloem transports food molecules made by photosynthesis by a pressure-flow mechanism. At a sugar source, such as a leaf, sugar is loaded into a phloem tube. The sugar raises the solute concentration in the tube, and water follows by osmosis, raising the pressure in the tube. As sugar is removed and stored or used in a sugar sink, such as the root, water follows. The increase in pressure at the sugar source and decrease at the sugar sink causes phloem sap to flow from source to sink **(32.5).**

Plant Nutrients and the Soil (32.6–32.14) A plant must obtain from its surroundings usable sources of the chemical elements—"nutrients"—it requires. Macronutrients, such as carbon, oxygen, nitrogen, and phosphorus, are needed in large amounts, mostly to build organic molecules. Micronutrients, including iron, copper, and zinc, act mainly as cofactors of enzymes. Growing plants in solutions of known composition enables researchers to determine nutrient requirements (32.6).

Stunting, wilting, and color changes indicate nutrient deficiencies (32.7). Soil characteristics determine whether a plant will be able to obtain the nutrients it needs to grow. Fertile soil contains a mixture of small rock and clay particles that hold water and ions and also allow O_2 to diffuse into plant roots. Humus (decaying organic material) provides nutrients, holds water and air, and supports the growth of organisms that enhance soil fertility. Anions (negatively charged ions), such as nitrate (NO_3^-), are readily available to plants because they are not bound to soil particles. However, anions tend to drain out of soil rapidly. Cations (positively charged ions), such as Ca^{2+} and Mg^{2+}, adhere to soil particles. In cation exchange, root hairs release H^+ ions, which displace cations from soil particles; the root hairs then absorb the free cations (32.8). Water-conserving irrigation, erosion control, and the prudent use of herbicides and fertilizers are aspects of good soil management (32.9). Organic farmers avoid the use of inorganic fertilizers, which can damage the environment (32.10). Relationships with other organisms help plants obtain nutrients. Many plants form mycorrhizae, mutually beneficial associations with fungi. A network of fungal threads increases a plant's absorption of nutrients and water, and the fungus receives some nutrients from the plant (32.11). Parasitic plants such as mistletoe siphon sap from host plants. Carnivorous plants can obtain nitrogen by digesting insects (32.12). Bacteria in the soil convert N_2 from the air and nitrogen compounds from decomposing organic matter to forms that plants can take up and use, chiefly nitrate ions (NO_3^-) and ammonium ions (NH_4^+). The conversion of N_2 to NH_4^+ is called nitrogen fixation (32.13). Legumes and certain other plants have nodules in their roots that house nitrogen-fixing bacteria (32.14).

Plant Nutrients and Agriculture
(32.15–32.16) Improving the protein content of crops is an important research goal because plants are the main nutritional source for most people in the world (32.15). Genetic engineering is already being used to develop new varieties of crop plants (32.16).

Testing Your Knowledge

Multiple Choice

1. Plants require the smallest amount of which of the following nutrients?
   a. oxygen
   b. phosphorus
   c. carbon
   d. iron
   e. hydrogen

2. Which of the following activities of soil bacteria does *not* contribute to creating usable nitrogen supplies for plant use?
   a. the fixation of atmospheric nitrogen
   b. the conversion of ammonium ions to nitrate ions
   c. the decomposition of dead animals
   d. the assembly of amino acids into proteins
   e. the generation of ammonium from proteins in dead leaves

3. By trapping insects, carnivorous plants obtain ____, which they need ____. *(Pick the best answer.)*
   a. water . . . because they live in dry soil
   b. nitrogen . . . to make sugar
   c. phosphorus . . . to make protein
   d. sugars . . . because they can't make enough in photosynthesis
   e. nitrogen . . . to make protein

4. A major long-term problem resulting from flood irrigation is the
   a. drowning of crop plants
   b. accumulation of salts in the soil
   c. erosion of fine soil particles
   d. encroachment of water-consuming weeds
   e. excessive cooling of the soil

Describing, Comparing, and Explaining

Explain how guard cells limit water loss from a plant on a hot, dry day. How can this be harmful to the plant?

Thinking as a Scientist

1. Acid rain contains an excess of hydrogen ions (H^+). One effect of acid rain is to deplete the soil of plant nutrients such as calcium (Ca^{2+}), potassium (K^+), and magnesium (Mg^{2+}). Offer a hypothesis to explain why acid rain washes these nutrients from the soil. How might you test your hypothesis?

2. In some situations, the application of nitrogen fertilizer to crops has to be increased each year because the fertilizer decreases the rate of nitrogen fixation in the soil. Propose a hypothesis to explain this phenomenon. Describe a test for your hypothesis. What results would you expect from your test?

3. Transpiration is fastest when humidity is low and temperature is high, but in some plants it seems to increase in response to light as well. During one 12-hour period when cloud cover and light intensity varied frequently, a scientist studying a certain crop plant recorded the data in the table below. (The transpiration rates are grams of water per square meter of leaf area per hour.)

Time (hr)	Temperature (°C)	Humidity (%)	Light (% of full sun)	Transpiration Rate ($g/m^2/hr$)
8	14	88	22	57
9	14	82	27	72
10	21	86	58	83
11	26	78	35	125
12	27	78	88	161
1	33	65	75	199
2	31	61	50	186
3	30	70	24	107
4	29	69	50	137
5	22	75	45	87
6	18	80	24	78
7	13	91	8	45

Do these data support the hypothesis that the plants transpire more when the light is more intense? If so, is the effect independent of temperature and humidity? Explain your answer. (*Hint:* Look for overall trends in each column, and then compare pairs of data within each column and between columns.)

Science, Technology, and Society

Agriculture is by far the biggest user of water in arid western states, including Colorado, Arizona, and California. The populations of these states are growing, and there is an ongoing conflict between cities and farm regions over water. To ensure water supplies for urban growth, cities are purchasing water rights from farmers. This is often the least expensive way for a city to obtain more water, and some farmers can make more money selling water than growing crops. Discuss the possible consequences of this trend. Is this the best way to allocate water for all concerned? Why or why not?

Media Resources

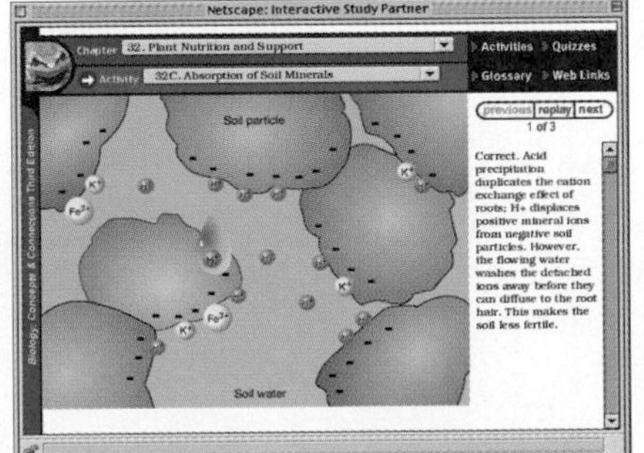

For further review of the concepts in this chapter, use the activities and quizzes on the CD-ROM.

Visit the Special Edition of The Biology Place web site for activities, quizzes, news articles, and web links.

33 Control Systems in Plants

Outline

Plant Hormones

The plant in the photograph at the left, a popular houseplant, is a tropical species originally from South America. It is called the sensitive plant because, as you can see, its leaves fold up when touched. Few people who know about it can resist the urge to touch a leaf and watch it react. As much for its fascinating behavior as for its beauty, the sensitive plant has been grown indoors for hundreds of years. In the eighteenth century, Swedish botanist Carolus Linnaeus gave the sensitive plant its scientific name. Linnaeus called it *Mimosa pudica,* which means "modest mimic" in Latin.

The smaller photographs on the facing page provide a closer view of the difference between an undisturbed and a disturbed *Mimosa*. The undisturbed plant on the left has its leaves extended. *Mimosa* has a main leaf stalk, which branches into four smaller stalks, each of which, in turn, bears numerous small leaflets. When touched, the leaflets fold upward, and the four small stalks move closer together, making the plant look wilted, as shown on the right. If you pinch just a single leaflet, you can watch the effect of the stimulus travel throughout the plant. Within a few seconds, all the plant's leaves appear wilted. This response results from a rapid loss of water from specialized cells in cushionlike thickenings, called pulvini, located at the base of each leaflet, small stalk, and main leaf stalk. The signals that trigger the response travel as action potentials, similar to, but slower than, those in the nervous systems of animals (see Module 28.5). The action potentials trigger cells on one side of the pulvinus to lose ions, and thus water, and become flaccid. After a sensitive plant reacts, it takes about 10 minutes for the cells to regain water and become turgid, restoring the plant's original form.

What is the function of this behavior? Scientists have studied *M. pudica* for centuries, but we still can only speculate. In the wild, the plant inhabits dry, windy areas in the tropics. Perhaps its ability to fold its leaves is an adaptation that reduces surface area and saves water. The folding response may also make the plant less attractive to herbivores.

The sensitive plant not only responds to touch, it also carries out what plant biologists call sleep movements. At dusk, the plant folds its leaves in the same way it does when disturbed. You may have noticed that dandelions, morning glories, and many other plants fold their flowers or leaves in the evening and unfold them in the morning. Again, we can only speculate about the significance of this behavior. Charles Darwin thought that the sleep movements of plants might serve to slow the loss of heat from the leaves after dark. Whatever their function, these movements demonstrate a daily regularity in a plant's activities. In fact, plants as well as animals have internal biological clocks that keep them on a regular daily cycle.

In this chapter, we look at the ability of plants to sense and respond to their environment and at the control mechanisms that make plant behavior possible. The first half of the chapter discusses the chemical messengers—hormones—of plants. Hormones play some key roles in plant movement and in controlling activities such as growth, flowering, and fruit production.

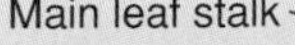

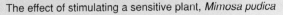

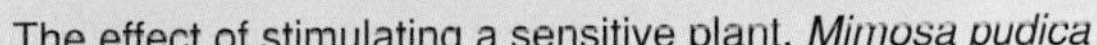
The effect of stimulating a sensitive plant, *Mimosa pudica*

33.1 Experiments on how plants turn toward light led to the discovery of a plant hormone

Figure 33.1A A houseplant growing toward light

A houseplant on a windowsill grows toward light (Figure 33.1A). If you rotate the plant, it will soon reorient its growth until its leaves again face the window. The growth of a shoot toward light is called **phototropism** (from the Greek *photos*, light, and *tropos*, turn). Phototropism is an adaptive response, directing growing seedlings and the shoots of mature plants toward the sunlight they need for photosynthesis.

Microscopic observations of plants growing toward light indicate the cellular mechanism that underlies phototropism. Figure 33.1B shows a grass seedling curving toward light coming from one side. As the enlargement shows, cells on the darker side of the seedling are larger—actually, they have elongated faster—than those on the brighter side. The different cellular growth rates made the shoot bend toward the light. Other experiments (not illustrated) have shown that the cells of a seedling all elongate at a similar rate if the plant is illuminated uniformly from all sides or if it is kept in the dark. In these situations, the seedling grows straight upward.

What causes plant cells in a shoot to grow at different rates? We found in Chapter 26 that hormones help coordinate internal activities, such as growth rates, in animals. We might predict, therefore, that plants also have hormones that regulate growth. Actually, the idea that plants have hormones emerged from a series of classic experiments on how shoots respond to light.

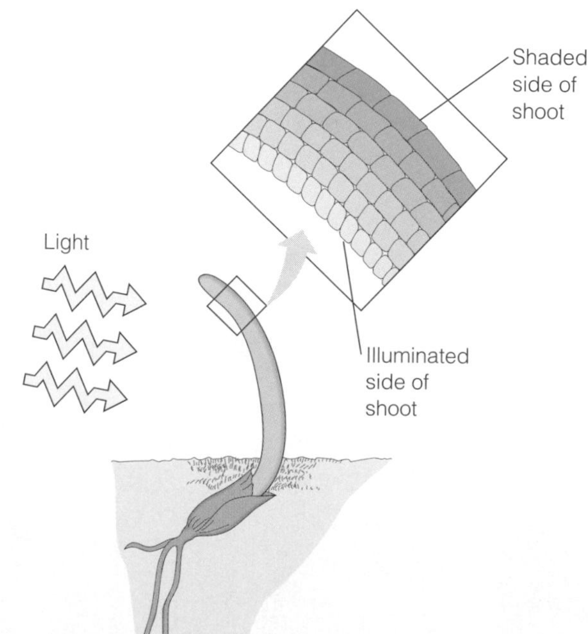

Figure 33.1B Phototropism in a grass seedling

Showing That Light Is Detected by the Shoot Tip

In 1880, Charles Darwin and his son, Francis, conducted some of the earliest experiments on phototropism and set the stage for the discovery of an important plant hormone. They found that grass seedlings would bend toward light only if the tips of their shoots were present. The first five grass plants in Figure 33.1C summarize the Darwins' findings. When they removed the tip of a grass shoot, the shoot grew straight up, rather than curving toward the light. The shoot also remained straight when the Darwins placed an opaque cap on its tip. However, the shoot curved normally when they placed a transparent cap on its tip or an opaque shield around its base. The Darwins concluded that the tip of the shoot was responsible for sensing light. They also recognized that the growth response that makes the seedling curve toward light actually occurs below the tip. Therefore, they speculated that some signal was transmitted downward from the tip to the growth region of the shoot.

In 1913, Danish botanist Peter Boysen-Jensen further tested the chemical signal idea of the Darwins; the last two plants in Figure 33.1C summarize his findings. In one group of grass seedlings, Boysen-Jensen inserted a block of gelatin between the tip and the lower part of the shoot. The gelatin block prevented cellular contact between the tip and the rest of the shoot but allowed chemicals to diffuse through. The seedlings with gelatin blocks behaved normally, bending toward the light. In a second set of seedlings, Boysen-Jensen inserted a thin piece of the mineral mica under the shoot tip. Mica is an impermeable barrier, and the seedlings with mica had no phototropic response. These experiments supported the hypothesis that the signal for phototropism is a mobile chemical.

Isolating the Chemical Signal

In 1926, Dutch botanist Fritz Went modified Boysen-Jensen's techniques and discovered the chemical messenger for phototropism in grasses. As shown in Figure 33.1D, Went first removed the tips of grass seedlings and placed them on blocks of agar, a gelatinlike material. He reasoned that the chemical messenger (pink in the figure) from the shoot tips should diffuse into the agar, and that the blocks should then be able to substitute for the shoot tips. Went tested the effects of the agar blocks on tipless seedlings, which he kept in the dark to eliminate the effect of light. First, he centered the treated agar blocks on the cut tips of a batch of seedlings. These plants grew straight upward, and significantly faster

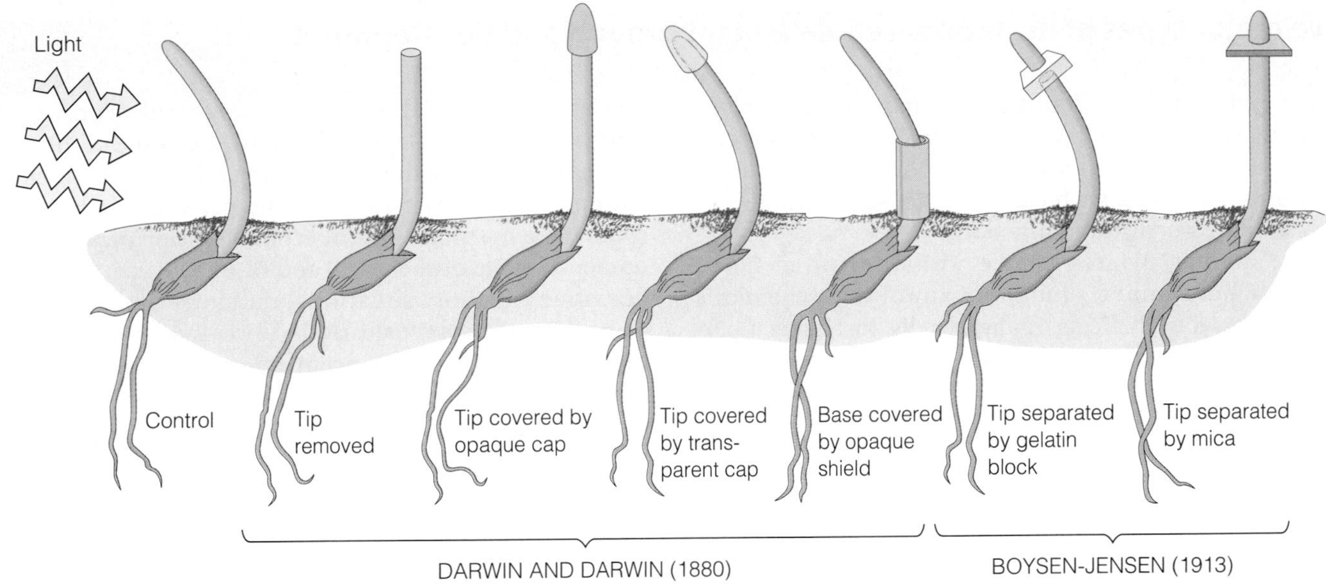

Figure 33.1C Early experiments on phototropism: Detection of light by shoot tips and evidence for a chemical signal

than the tipless control seedlings, which hardly grew at all. Went concluded that the agar had absorbed the chemical messenger produced in the shoot tip, and that the chemical had passed into the shoot and stimulated it to grow. He then placed agar blocks off-center on another batch of tipless seedlings. These plants bent away from the side with the chemical-laden agar block, as though growing toward light. Control seedlings with *blank* agar blocks (whether offset or not) grew no more than the first control. Went concluded that when shoots curve toward light, they do so because there is a higher concentration of the growth-promoting chemical on the darker side of the shoot. For this chemical messenger, or hormone, Went chose the name auxin, from the Greek *auxein*, "to increase."

Went's pioneering work sparked an explosion of research on plant growth regulators. Follow-up experiments showed that the tips of growing shoots secrete auxin in equal amounts in the dark or light. When a shoot is exposed to light from one direction, auxin travels from the lighted side to the shaded side of the shoot and stimulates cell growth there. At the same time, the growth rate on the lighted side may decrease because of a reduced auxin level there. In the 1930s, biochemists determined the chemical identity of Went's auxin, finding it to be a small organic molecule that plant cells make from the amino acid tryptophan. 

? How do the experiments illustrated in Figures 33.1C and 33.1D combine to provide evidence that phototropism depends on a chemical signal—that is, a hormone?

Light is detected by the shoot tip, but the bending response occurs some distance down from the tip, indicating transmission of a signal. The fact that the signal can pass from tip to responding region through a barrier that prevents cell contact but allows chemicals to pass suggests that the signal is a chemical.

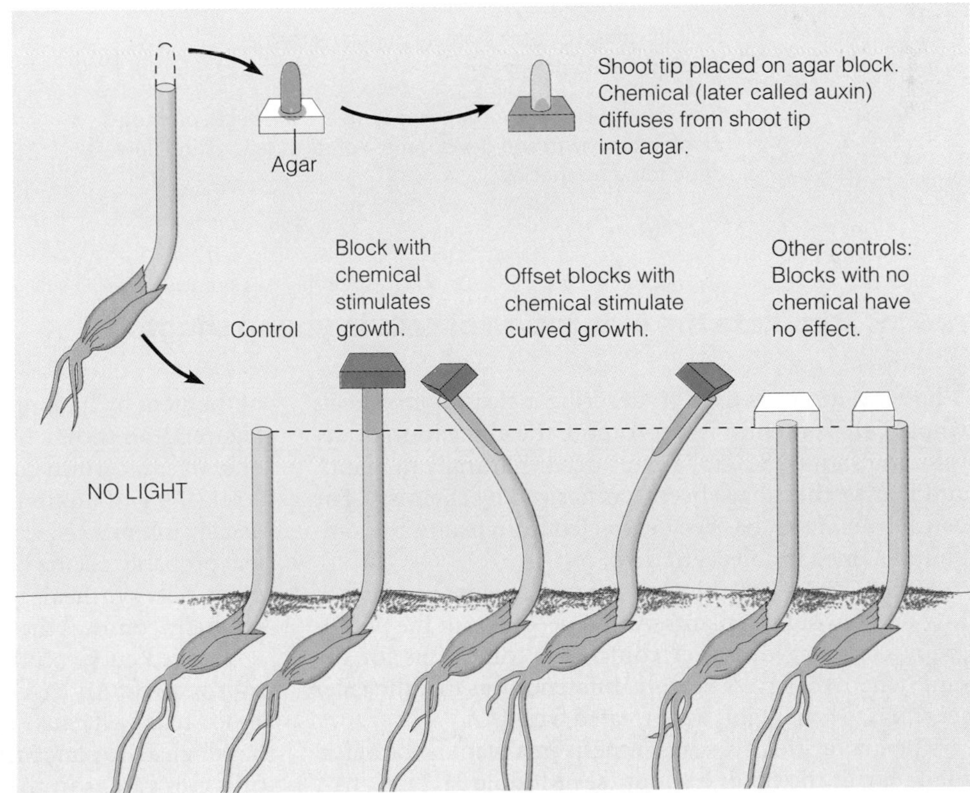

Figure 33.1D Went's experiments: Isolation of the chemical signal

33.2 Five major types of hormones regulate plant growth and development

33A
See Activity 33A on the CD-ROM.

Plant biologists have identified five major types of plant hormones, which are listed in the table below. Three of these—the auxins, cytokinins, and gibberellins—are actually hormone *classes*. Each class includes several chemicals with similar structure and function.

Like animals, plants produce hormones in very small amounts, but a minute amount of any of these chemicals can have profound effects on target cells. Just a few molecules of a hormone can redirect the metabolism and development of a plant's cells. Hormones do this by triggering signal-transduction pathways in target cells (see Module 11.11). In plants, as in animals, these pathways lead to cellular responses such as the turning on or off of genes, the inhibition or activation of enzymes, or changes in membranes.

As the table indicates, each type of hormone can produce a variety of effects. Notice that all five types of hormones influence growth, and four of them affect development (cell differentiation). Hormones stimulate growth by signaling target cells to divide or elongate; some of the hormones inhibit growth by depressing cell division or elongation. The effects of a hormone depend on the plant species, the hormone's site of action in the plant, the developmental stage of the plant, and the concentration of the hormone. In most situations, no single hormone acts alone. Instead, it is usually the balance of several plant hormones—their relative concentrations—that controls the growth and development of a plant. We will see several examples of hormones working together as we focus on each of the plant hormones in the next five modules.

The cells that respond to a particular hormone are called that hormone's ________ cells.

target

Hormone	Major Functions	Where Produced or Found in Plant
Auxins	Stimulate stem elongation; affect root growth, differentiation, and branching; development of fruit; apical dominance; phototropism and gravitropism (response to gravity)	Meristems of apical buds; young leaves; embryos within seeds
Cytokinins	Affect root growth and differentiation; stimulate cell division and growth in general; stimulate germination; delay aging	Made in roots, embryos, and fruits; move from roots to other organs
Gibberellins	Promote seed germination, bud development, stem elongation, and leaf growth; stimulate flowering and fruit development; affect root growth and differentiation	Meristems of apical buds and roots; young leaves; embryos
Abscisic acid (ABA)	Inhibits growth; closes stomata during water loss; helps maintain dormancy	Leaves, stems, roots, green fruits
Ethylene	Promotes fruit ripening; opposes some auxin effects; promotes or inhibits growth and development of roots, leaves, and flowers, depending on species	Ripening fruits, nodes of stems, dying leaves

33.3 Auxins stimulate the elongation of cells in young shoots

33A

The term **auxin** is used to describe a class of chemicals whose chief function is to promote the elongation of developing shoots. Several auxins occur naturally in plants, and many others have been synthesized by chemists. The natural auxin that has been extracted from plants is a compound named **indoleacetic acid,** or **IAA.**

Figure 33.3A (on the facing page) shows the effect of IAA on growing pea plants. All the seedlings in the photograph were grown under controlled conditions for the same length of time. The only difference was that the taller seedlings, on the right, were treated with IAA.

The major site of auxin synthesis in a plant is the apical meristem at the tip of a shoot (see Module 31.7). As IAA produced in the tip moves downward, it stimulates growth of the stem by making cells elongate. As the black curve in Figure 33.3B shows, IAA promotes cell elongation in stems only over a certain concentration range. Above a certain level (0.9 g of auxin per liter of solution, in this case), it usually inhibits cell elongation in stems. This inhibitory effect probably occurs because a high level of IAA makes the plant cells synthesize another hormone, ethylene, which generally counters the effects of IAA.

The red curve on the graph shows the effect of IAA on root growth. An IAA concentration too low to stimulate shoot cells will cause root cells to elongate. On the other hand, an IAA concentration high enough to make stem cells elongate is in the concentration range that inhibits root cell elongation. These effects of IAA on cell elonga-

Figure 33.3A The effect of auxin (IAA) on pea plants

Figure 33.3B The effect of auxin concentration on cell elongation

tion reinforce two points: (1) the same chemical messenger may have different effects at different concentrations in one target cell, and (2) a given concentration of the hormone may have different effects on different target cells.

How do auxins make plant cells elongate? One hypothesis is that auxins initiate elongation by weakening cell walls. As shown in Figure 33.3C, auxins may stimulate certain proteins in a plant cell's plasma membrane to pump hydrogen ions into the cell wall. The H^+ ions activate enzymes that break some of the hydrogen bonds cross-linking cellulose molecules in the wall (see Figure 3.7). The cell then swells with water and elongates because its weakened wall no longer resists the cell's tendency to take up water osmotically. After this initial elongation, the cell sustains the growth by synthesizing more wall material and cytoplasm.

Auxins produce a number of other effects, in addition to stimulating cell elongation and causing stems and roots to lengthen. These hormones can also trigger the development of vascular tissues and induce cell division in the vascular cambium, thus promoting growth in stem diameter (see Module 31.8). Furthermore, auxin (IAA) is produced by developing seeds, and, in many plants, IAA in seeds promotes the growth of fruit. Some plants will even develop fruits without being fertilized if they are sprayed with auxin. Farmers sometimes produce tomatoes, cucumbers, and eggplants, for example, by spraying the plants with synthetic auxins. The fruits that result are seedless.

Suppose you had a tiny pH electrode that could measure the pH of a plant cell's wall. How could you use it to test the hypothesis presented in Figure 33.3C for how auxins stimulate cell elongation?

The hypothesis predicts that addition of auxins to the cell should stimulate H^+ pumps and lower the pH of the wall (make it more acidic). You could test this prediction with your pH electrode by measuring the wall pH in the presence of (experimental group) or absence of (control group) auxins.

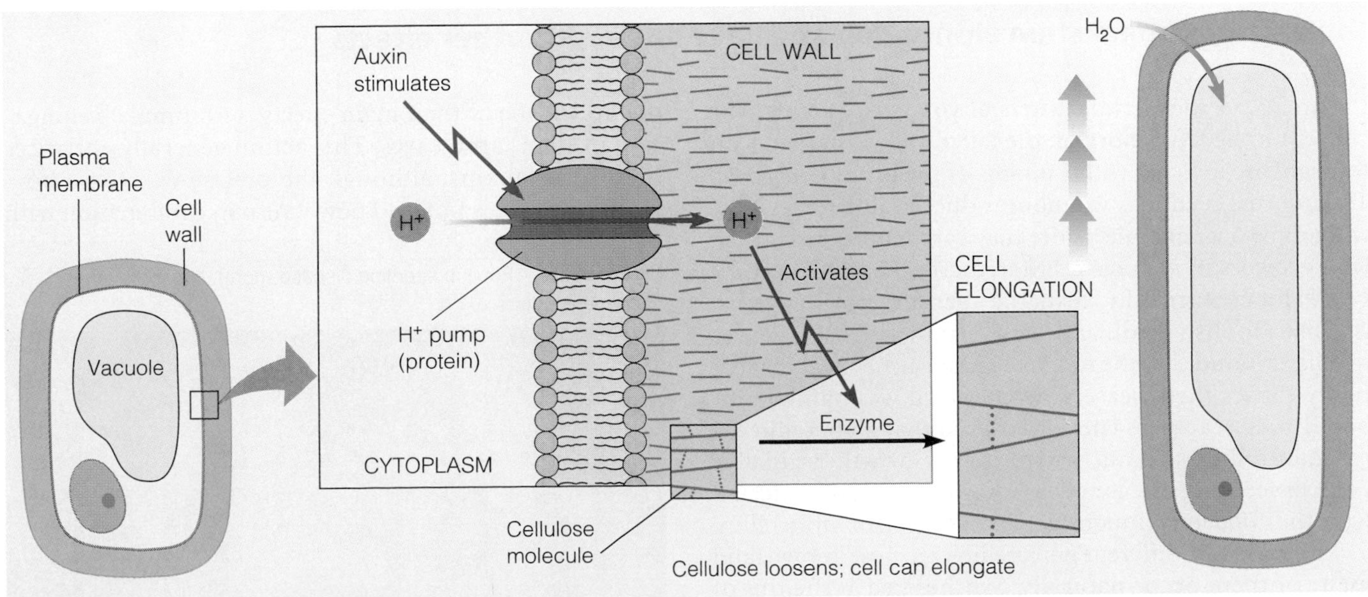

Figure 33.3C A hypothesis to explain how auxins stimulate cell elongation

33.4 Cytokinins stimulate cell division

33A

Cytokinins are growth regulators that promote cell division, or cytokinesis. A number of cytokinins have been extracted from plants, and several synthetic ones have been made. Natural cytokinins are produced in actively growing tissues, particularly in roots, embryos, and fruits. Cytokinins made in the roots reach target tissues in stems by moving upward in xylem sap.

Plant biologists have found that cytokinins enhance the division, growth, and development of plant cells grown in culture. Cytokinins can also retard the aging of flowers and fruits, and cytokinin sprays are used to keep cut flowers fresh.

In whole plants, the effects of cytokinins are often influenced by the concentration of auxins present. The photographs in Figure 33.4 show the results of a simple experiment that partially separates the different effects of auxins and cytokinins. Both basil plants pictured are the same age. The one on the left has an intact terminal bud; the one on the right had its terminal bud removed several weeks earlier. In the plant on the left, auxin transported down the stem from the terminal bud promoted lengthwise growth while inhibiting the growth of the axillary buds (the buds that produce side branches). As a result, the shoot grew in height but did not branch out to the sides very much. In the plant on the right, the lack of a terminal bud eliminated the inhibitory effect of auxin on the axillary buds. This allowed cytokinins transported up from the roots to activate the axillary buds, making the plant grow more branches and become bushy. Growers of Christmas trees sometimes use cytokinins to produce attractive branching in their trees.

Terminal bud

No terminal bud

Figure 33.4 The effects of naturally occurring auxins and cytokinins on plant growth

Most plants have complex growth patterns, with some lateral growth occurring even when terminal buds are intact. These patterns probably result from the interaction of auxins and cytokinins, with the ratio of the two hormones playing a critical role. Cytokinins entering the shoot system from the roots counter the effects of auxins coming down from the terminal buds. The lower axillary buds on a shoot usually begin to grow before those closer to the terminal bud, reflecting the higher ratio of cytokinins to auxins in the lower parts of the plant.

The antagonistic interaction of auxins and cytokinins may also be one way the plant coordinates the growth of its root and shoot systems. As roots become more extensive and produce more and more cytokinins, the increased level of the cytokinins would signal the shoot system to form more branches. Antagonistic interactions are common among plant hormones, as we will see in the next several modules.

The status of axillary buds—dormant or growing—depends on the relative concentrations of ________ moving down from the shoot tip and ________ moving up from the roots.

auxins . . . cytokinins

33.5 Gibberellins affect stem elongation and have numerous other effects

33A

Figure 33.5A shows two clusters of young rice plants. The cluster on the left is normal; the taller, yellowish plants on the right are infected with a fungus of the genus *Gibberella*. The infected seedlings will not produce grain; in fact, they will topple over and die before they can mature and flower. Rice growers in Asia have suffered crop losses from *Gibberella* for centuries. In Japan, the aberrant growth pattern is called "foolish seedling disease." In the 1920s, Japanese scientists found that the fungus releases a chemical that actually causes the disease. The chemical was named **gibberellin.** Researchers later discovered that gibberellin exists naturally in plants, where it is a growth regulator. Foolish seedling disease occurs when rice plants infected with the *Gibberella* fungus get an overdose of gibberellin.

More than 70 different gibberellins are now known, and many of them occur naturally. Synthesized at the tips of both stems and roots, gibberellins produce a wide variety of effects. One of their main effects is to stimulate elongation in stems and leaves. This action generally enhances that of the auxins, although the precise way these hormones interact is not yet known. Also in combination with

Figure 33.5A Foolish seedling disease (right), caused by the *Gibberella* fungus

Figure 33.5B An effect of gibberellin treatment on grapes (right)

auxins, gibberellins can influence fruit development, and gibberellin-auxin sprays can make apples, currants, and eggplants develop without fertilization. One of the most widespread uses of gibberellins is in the production of the Thompson variety of seedless grapes. Gibberellins make the grapes grow larger and farther apart in a cluster. The left cluster of grapes in Figure 33.5B is untreated; the right one shows the effect of gibberellin treatment.

Gibberellins are also important in seed germination in many plants. Many seeds that require special environmental conditions to germinate, such as cold temperatures, will germinate when sprayed with gibberellins. In nature, gibberellins in seeds are probably the link between environmental cues and the metabolic processes that renew growth of the embryo. For example, when water becomes available to a grass seed, it causes the embryo in the seed to release gibberellins, which promote germination by mobilizing nutrients stored in the seed. In some plants, gibberellins seem to interact antagonistically with another hormone, abscisic acid, which maintains seed dormancy.

? Researchers working with pea plants have connected a mutation that blocks gibberellin synthesis to one of the recessive traits Gregor Mendel studied in his famous experiments. Return to Figure 9.2D and identify the gibberellin-deficient mutant among Mendel's pea varieties.

Dwarf variety

33.6 Abscisic acid inhibits many plant processes

33A

There are times, such as the onset of winter or severe drought, when it is adaptive for a plant to become dormant, temporarily ceasing to grow. At such times, the hormone **abscisic acid (ABA),** produced in buds, inhibits cell division in buds and in the vascular cambium, thus suspending both primary and secondary growth. ABA also signals the buds to form scales that will protect them from harsh seasonal conditions. Abscisic acid was named when it was believed that this hormone caused abscission (breaking off) of leaves from deciduous trees in autumn. The name has prevailed even though research has never shown that ABA plays a role in leaf drop.

Another time in the life of a plant when it is advantageous to suspend growth is the onset of seed dormancy. Seed dormancy is especially important to annual plants in deserts and semiarid regions because germination without sufficient water for growth would quickly kill the plants. A number of environmental factors influence seed dormancy, but in many plants, ABA seems to act as the growth inhibitor. The seeds of such annuals remain dormant in parched soil until a downpour washes ABA out of the seeds, allowing them to germinate. For example, the dune primroses (white) and desert sunflowers (yellow) in Figure 33.6, photographed in the Anza-Borrego Desert in California, grew from seeds that germinated just after a hard rain.

Figure 33.6 The Anza-Borrego Desert blooming after a rain

As we saw in the previous module, gibberellins also participate in seed germination. For many plants, the ratio of ABA to gibberellins, which promote germination, determines whether the seed will remain dormant or germinate. Similarly, the dormancy of buds is controlled more by a balance of these growth regulators than by their absolute concentrations. In apple trees, for example, the ABA concentration is actually higher in growing buds than in dormant buds, but a high level of gibberellins in the growing buds overpowers the inhibitory effect of ABA.

In addition to its role in dormancy, ABA acts as a "stress hormone" in growing plants, helping them cope with adverse conditions. For instance, when a plant is dehydrated, ABA accumulates in the leaves and causes stomata to close. This reduces transpiration and prevents further water loss. However, it also reduces the plant's rate of sugar production by photosynthesis.

? How does abscisic acid help maintain winter dormancy in trees living in temperate climates?

By inhibiting cell division in buds and in the vascular cambium

33.7 Ethylene triggers fruit ripening and other aging processes

33A

Early in this century, oranges and grapefruits were ripened for market in sheds equipped with kerosene stoves. Fruit growers thought it was the heat that ripened the fruit, but when they tried newer, cleaner-burning stoves, the fruit did not ripen fast enough. Plant biologists learned later that ripening in the sheds was actually due to **ethylene,** a gaseous by-product of kerosene combustion. We now know that plants produce their own ethylene, which functions as a hormone that triggers a variety of aging responses, including fruit ripening.

Fruit Ripening Fruit ripening involves the breakdown of cell walls, changes in color (often from green to yellow or red), and sometimes drying. These are aging processes triggered by ethylene, which can be produced by all the cells of a plant. As fruits age, they produce more and more ethylene, thus hastening the ripening process. As a gas, ethylene diffuses through the fruit in the air spaces between cells. It also diffuses through the air from fruit to fruit, and, as a result, if one apple in a bin starts to spoil, it really can spoil the whole lot. You can make some fruits ripen faster if you store them in a plastic bag so that ethylene gas can accumulate. Figure 33.7A shows the results of a demonstration with bananas. Three unripe bananas were stored for the same time period in plastic bags: (1) with an ethylene-releasing orange, (2) with a beaker of an ethylene-releasing chemical, and (3) alone (the control). As you can see, the more ethylene present, the riper (darker) the banana. On a commercial scale, many kinds of fruit—tomatoes, for instance—are often picked green and then partially ripened in huge storage bins into which ethylene gas is piped—a modern variation on the old storage shed.

In other cases, growers take measures to *retard* the ripening action of natural ethylene. Stored apples are often flushed with CO_2, which inhibits the action of ethylene. Also, gas is circulated around the apples to prevent ethylene from accumulating. In this way, apples picked in autumn can be stored for sale the following summer.

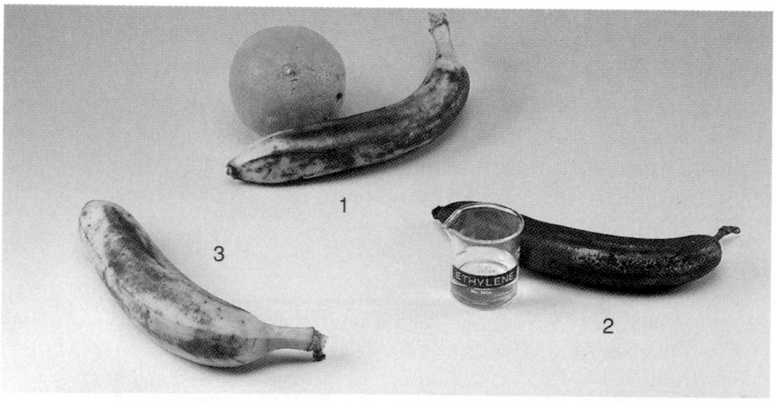

1

3

ETHYLENE

2

Figure 33.7A The effect of ethylene on the ripening of bananas

The Falling of Leaves Like fruit ripening, the changes that occur in deciduous trees each autumn—color changes, drying, and the loss of leaves—are also aging processes. Fall colors result from a combination of new pigments made in autumn and pigments that were already present in the leaf but masked by the dark-green chlorophyll. Leaves lose their green color in autumn because they stop making chlorophyll. Ethylene probably plays a role in fall color changes and in leaf drying, but little is known about its effects.

Much more is known about how ethylene promotes the loss (abscission) of leaves. When an autumn leaf falls, the base of the leaf stalk separates from the stem. The separation region is called the abscission layer. As indicated in Figure 33.7B, the abscission layer consists of a narrow band of small parenchyma cells with thin, weak walls. The leaf drops off when its weight, often helped by wind, splits the abscission layer apart. Notice the layer of protective cells adjacent to the abscission layer. Even before the leaf falls, these cells form a leaf scar on the stem. Dead cells covering the scar help protect the plant from infectious organisms.

Leaf drop is triggered by environmental stimuli, including the shortening days of autumn and, to a lesser extent, cooler temperatures. These stimuli apparently cause a change in

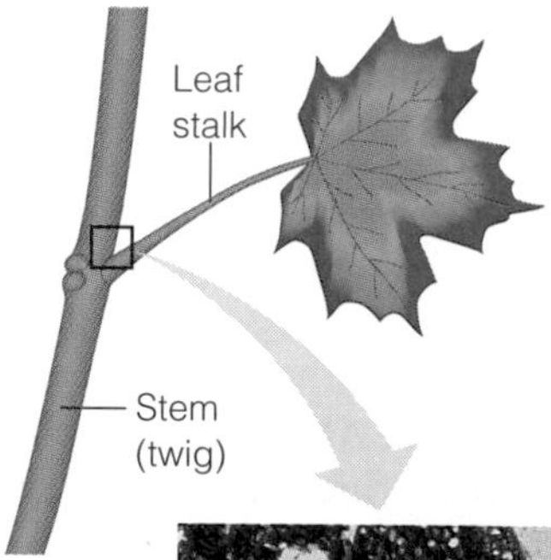

Leaf stalk

Stem (twig)

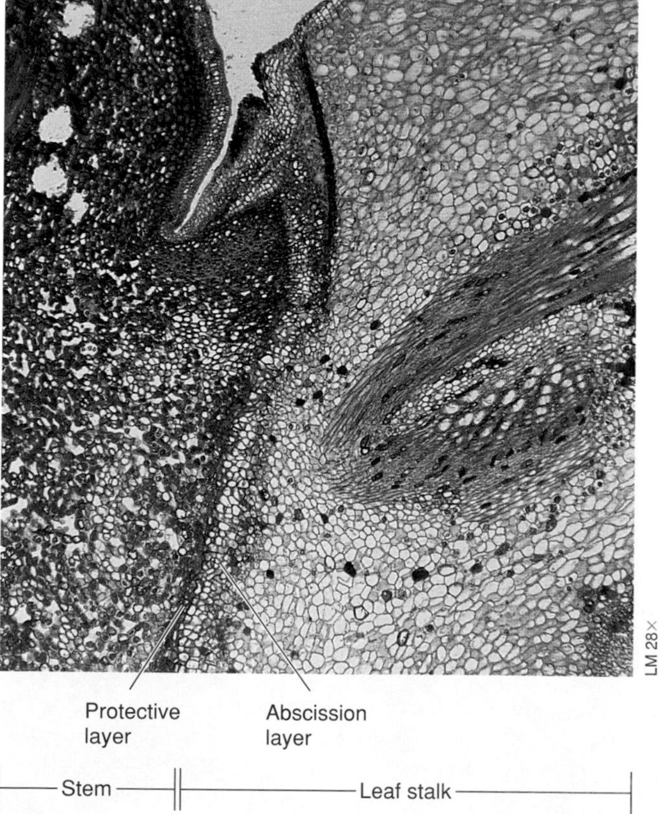

Figure 33.7B Abscission layer at the base of a leaf

the balance of ethylene and auxin—another case of antagonism between plant hormones. The auxin prevents abscission and helps maintain the leaf's metabolism, but as a leaf ages, it produces less and less auxin. Meanwhile, cells begin producing ethylene, which stimulates formation of the abscission layer. The ethylene then primes the abscission layer to split by promoting the synthesis of enzymes that digest cell walls in the layer. Autumn leaf drop is an adaptation that helps keep the tree from drying out in winter. Without its leaves, a tree loses less water by evaporation when its roots cannot take up water from the frozen ground.

We have now completed our survey of the five major types of plant hormones. Before moving on to the topic of plant behavior, let's look at some agricultural uses of these chemical regulators. 

? Botanists sometimes refer to ethylene as the "senescence hormone." What is the basis for this term?

Many of ethylene's functions, including fruit ripening and leaf abscission, are associated with aging-like changes in cells.

Plant hormones have many agricultural uses 33.8

Much of what we know about plant hormones—and there is much more to be learned—has a direct application to agriculture. As already mentioned, the control of fruit ripening and the production of seedless fruits are two of several major uses of these chemicals. Plant hormones also enable farmers to control when plants will drop their fruit. For instance, auxins are often used to prevent orange and grapefruit trees from dropping their fruit before they can be picked. Figure 33.8 shows a citrus grower in Florida spraying an orange grove with auxins. The quantity of auxin must be carefully monitored because too much of the hormone may stimulate the plant to release more ethylene, making the fruit ripen and drop off sooner.

Figure 33.8 Using auxins to prevent early fruit drop

Large doses of auxins are often used intentionally to promote premature fruit drop. For example, auxins may be sprayed on apple and olive trees to thin the developing fruits; the remaining fruits will grow larger. Ethylene is used to thin peaches and prunes, and it is sometimes sprayed on berries, grapes, and cherries to loosen the fruit so it can be picked by machines.

In combination with auxins, gibberellins are used to produce seedless fruits, as mentioned in Module 33.5. Sprayed on other kinds of plants, at an earlier stage, gibberellins can have the opposite effect: the *promotion* of seed production. A large dose of gibberellins will induce many biennial plants, such as carrots, beets, and cabbage, to flower and produce seeds during their first year of growth. Ordinarily, biennials do not produce seeds until their second year of growth.

Research on plant hormones has had other spin-offs. One of the most widely used herbicides, or weed killers, is 2,4-D, a synthetic auxin that disrupts the normal balance of hormones that regulate plant growth. Because dicots are more sensitive than monocots to this herbicide, 2,4-D can be used to selectively remove dandelions and other broadleaf dicot weeds from a lawn or grainfield. By applying herbicides to cropland, a farmer can reduce the amount of tillage required to control weeds, thus reducing soil erosion, fuel consumption, and labor costs.

Modern agriculture relies heavily on the use of synthetic chemicals. Without chemically synthesized herbicides to control weeds and synthetic plant hormones to help grow and preserve fruits, less food would be produced, and food prices could increase considerably. At the same time, there is growing concern that the heavy use of artificial chemicals in food production may pose environmental and health hazards. A chemical called dioxin, for example, is a by-product of 2,4-D synthesis. Though 2,4-D itself does not appear to be toxic to mammals, dioxin causes birth defects, liver disease, and leukemia in laboratory animals. Therefore, dioxin is a serious hazard when it leaks into the environment. Also, many consumers are concerned that foods produced with artificial help may not be as tasty or nutritious as those raised more naturally. At present, however, organic foods are relatively expensive to produce. As we discussed in Module 32.10, these issues involve both economics and ethics: Should we continue to produce cheap, plentiful food using artificial chemicals and tolerate the potential problems, or should we put more of our agricultural effort into farming without these potentially harmful substances, recognizing that foods may be less plentiful and more expensive as a result? 

? (a) What is the main commercial incentive for agribusiness to treat its products with plant hormones? (b) What behavior of consumers helps drive this use of hormone sprays in agriculture?

(a) the need to compete in the market by increasing production and lowering cost; (b) shopping for the lowest price on produce

33.9 Tropisms orient plant growth toward or away from environmental stimuli

Figure 33.9A Gravitropism

Having surveyed the hormones that carry signals within a plant, we now shift our focus to the responses of plants to physical stimuli from the environment. In the chapter's introduction, we discussed the rapid leaf movements and sleep activities of the sensitive plant, *Mimosa.* These dramatic movements result from reversible changes in turgor in specialized cells. Less obvious because they occur very slowly are plant movements called tropisms. **Tropisms** are directed growth responses that cause parts of a plant to grow toward or away from a stimulus. Phototropism, the growth of a plant shoot toward light, is one type of tropism. Two other types are gravitropism, a response to gravity, and thigmotropism, a response to touch.

Response to Light As we saw in Module 33.1, the mechanism for phototropism is a greater rate of cell elongation on the darker side of a stem. In grass seedlings, the signal linking the light stimulus to the cell-elongation response is auxin. Researchers have shown that illuminating a grass shoot from one side causes auxin to migrate across the tip from the bright side to the dark side. The shoot tips contain a protein pigment that detects the light and somehow passes the "message" to molecules that affect auxin transport. (We discuss the light-sensitive protein in Module 33.12.)

Response to Gravity A plant's growth response to gravity, **gravitropism,** is illustrated by the corn seedlings in Figure 33.9A. These seedlings were both germinated in the dark. The one on the left was left untouched; notice that its shoot grew straight up and its root straight down. The seedling on the right was germinated in the same way, but two days later it was turned on its side so that the shoot and root were horizontal. By the time the photo was taken, the shoot had turned back upward, exhibiting a negative response to gravity, and the root had turned down, exhibiting positive gravitropism.

Plant biologists have yet to work out how plants tell up from down. One hypothesis is that gravity pulls special organelles containing dense starch grains to the low points of cells. The uneven distribution of organelles may in turn signal the cells to redistribute auxin. At least in roots, the differential auxin concentrations may then control the direction of growth. Little is known about the role of auxin in shoot gravitropism.

Whatever the underlying mechanisms, gravitropism is an important adaptation, making the shoot of a germinating plant grow upward into the light and the root grow into the soil, no matter how the seed lands or is planted in the soil.

Response to Touch Growth movement in response to touch, **thigmotropism** (Greek *thigma,* touch), is illustrated in Figure 33.9B by the tendril of a pea plant coiling around a wire fence. The tendril (actually, a modified leaf) grew straight until it touched the support. Contact then stimulated the cells to grow at different rates on opposite sides of the tendril (slower in the contact area), making the tendril coil around the wire. Most climbing plants have tendrils that respond by coiling and grasping when they touch rigid objects. Thigmotropism enables these plants to use other objects for support while growing toward sunlight.

Tropisms all have one function in common: They help plant growth stay in tune with the environment. In the next module, we see that plants also have a way of keeping time with their environment. 

? Why are tropisms called "growth responses"?

Because the movement of a plant organ toward or away from an environmental stimulus takes place by growing. An organ bends when cells on one side grow faster than cells on the other side, and it extends in one direction when cells grow evenly.

Figure 33.9B Thigmotropism

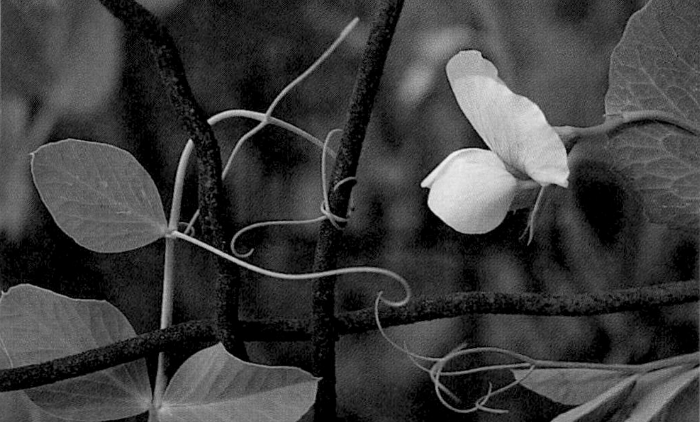

Your pulse rate, blood pressure, body temperature, rate of cell division, blood cell count, alertness, urine composition, metabolic rate, sex drive, and responsiveness to drugs all fluctuate rhythmically with the time of day. Plants also display rhythmic behavior; examples include the opening and closing of stomata (see Module 32.4) and the sleep movements of *Mimosa pudica* (see p. 659) and many other species.

Innate Biological Rhythms and Their Fine-Tuning by Environmental Cues An innate biological cycle of about 24 hours is called a **circadian rhythm** (from the Latin *circa,* about, and *dies,* day). A circadian rhythm persists even when an organism is sheltered from environmental cues. A *Mimosa* plant, for example, exhibits sleep movements at about the same intervals even if kept in constant light or darkness. Thus, circadian rhythms occur with or without external stimuli such as sunrise and sunset. Research on many organisms indicates that circadian rhythms are controlled by internal timekeepers called **biological clocks.**

Although a biological clock continues to mark time in the absence of environmental cues, to remain tuned to a period of *exactly* 24 hours, it requires daily signals from the environment. This is because innate circadian rhythms generally differ somewhat from a 24-hour period. Consider bean plants, for instance, which have sleep movements similar to those of the sensitive plant. As shown in Figure 33.10, the leaves of a bean plant are normally horizontal at noon and folded downward at midnight. When the plant is held in darkness, however, its sleep movements change to a cycle of about 26 hours.

The light-dark cycle of day and night provides the cues that most commonly keep biological clocks precisely synchronized with the outside world. But a biological clock cannot immediately adjust to a sudden major change in the light-dark cycle. We observe this problem ourselves when we cross several time zones in an airplane: When we reach our destination, we have "jet lag"—our internal clock is not synchronized with the clocks on the wall. Moving a plant across several time zones produces a similar lag. In the case of either the plant or the human traveler, resetting the clock usually takes several days.

Noon

Midnight

Figure 33.10 Sleep movements of a bean plant

The Nature of Biological Clocks Just what is a biological clock? Researchers are actively investigating this question. In humans and other mammals, the location of the clock is now known: It is a cluster of nerve cells in the hypothalamus of the brain (see Module 28.15). But for most other organisms, including plants, we know little about where the clocks are located or what kinds of cells are involved. However, by studying genes that are altered in mutants with abnormal circadian rhythms, scientists have discovered a molecular mechanism that may control all clocks, whatever their specific structure. Researchers have identified some intriguing genes that control clocks in rodents, fruit flies, fungi, bacteria, and the plant *Arabidopsis.* Each of these genes encodes a transcription factor that accumulates and at sufficient concentration turns off its own gene. When the concentration of the protein falls, transcription restarts. The result seems to be a cycling of the protein's concentration over a roughly 24-hour period—a clock!

Unlike most metabolic processes, biological clocks and the circadian rhythms they control are affected little by temperature. Somehow, a biological clock compensates for temperature shifts. This adjustment is essential, for a clock that speeds up or slows down with the rise and fall of outside temperature would be an unreliable timepiece.

In attempting to answer questions about biological clocks, it is essential to distinguish between the clock and the processes it controls. You could think of the sleep movements of leaves as the "hands" of a biological clock, but they are not the essence of the clockwork itself. You can restrain the leaves of a bean plant for several hours so that they cannot move. But on release, they will rush to the position appropriate for the time of day. Thus, we can interfere with an organism's rhythmic activity, but its biological clock goes right on ticking off the time. 

? If a bean plant is kept in constant darkness, its sleep-movement cycle lasts about 26 hours. How many days will it take for its leaves to be in the "noon position" when the time is actually midnight? Explain your answer.

About 6 days. The 26-hour cycle in a constant environment is 2 hours longer than a real day. Thus, it would take 6 days for the plant's leaf position to be 12 hours out of step with the actual time of day.

33.11 Plants mark the seasons by measuring photoperiod

33B

A biological clock not only times a plant's everyday activities, it may also influence seasonal events that are important in a plant's life cycle. Flowering, seed germination, and the onset and ending of dormancy are examples of stages in plant development that usually occur at specific times of the year. The environmental stimulus plants most often use to detect the time of year is called **photoperiod,** the relative lengths of day and night. In this module, we see how photoperiod affects flowering in certain plants. In the next module, we will focus on a mechanism that may set the biological clock and help the plant keep track of photoperiod.

Plants whose flowering is triggered by photoperiod fall into two groups. One group, the **short-day plants,** generally flower in late summer, fall, or winter, when light periods shorten. Chrysanthemums and poinsettias are examples of short-day plants. In contrast, **long-day plants,** such as spinach, lettuce, iris, and many cereal grains, usually flower in late spring or early summer, when light periods lengthen. Spinach, for instance, flowers only when daylight lasts at least 14 hours.

In the 1940s, researchers discovered that flowering and other responses to photoperiod are actually controlled by *night* length, not day length. In fact, the so-called short-day plants are actually long-night plants, and the so-called long-day plants are actually short-night plants. Unfortunately, the day-length terms are embedded firmly in the literature of plant biology.

Figure 33.11 illustrates the evidence for the night-length effect and also shows the difference between the flowering response of a short-day plant and a long-day plant. The left side of the figure represents short-day plants. The first two bars and corresponding plant drawings show that a short-day plant will not flower until it is exposed to a continuous dark period exceeding a critical length (about 10 hours, in this case). The continuity of darkness is important. The short-day plant will not blossom if the nighttime part of the photoperiod is interrupted by even a flash of light (third bar). (There is no effect if the daytime portion of the photoperiod is broken by a brief exposure to darkness.)

Florists apply this information about short-day plants to bring us flowers out of season. Chrysanthemums, for instance, are short-day plants that normally bloom in the autumn, but their blooming can be stalled until spring by punctuating each long night with a flash of light, thus turning one long night into two short nights.

The right side of the figure demonstrates the effect of night length on a long-day plant. In this case, flowering occurs when the night length is *shorter* than a critical length (less than 10 hours, in this example). As the third bar shows, flowering can be induced in a long-day plant by a flash of light during the night.

Notice that we distinguish long-day from short-day plants not by an *absolute* night length but by whether the the plants flower in response to a night that is shorter than a critical period (long-day plants) or longer than a critical period (short-day plants). The actual length of the critical period varies from species to species.

A particular short-day plant won't flower in the spring. We try to induce flowering by using a short dark interruption to split the long-light period of spring into two short-light periods. What result do you predict?

The plants still won't flower, because it is actually night length, not day length, that counts in the photoperiodic control of flowering.

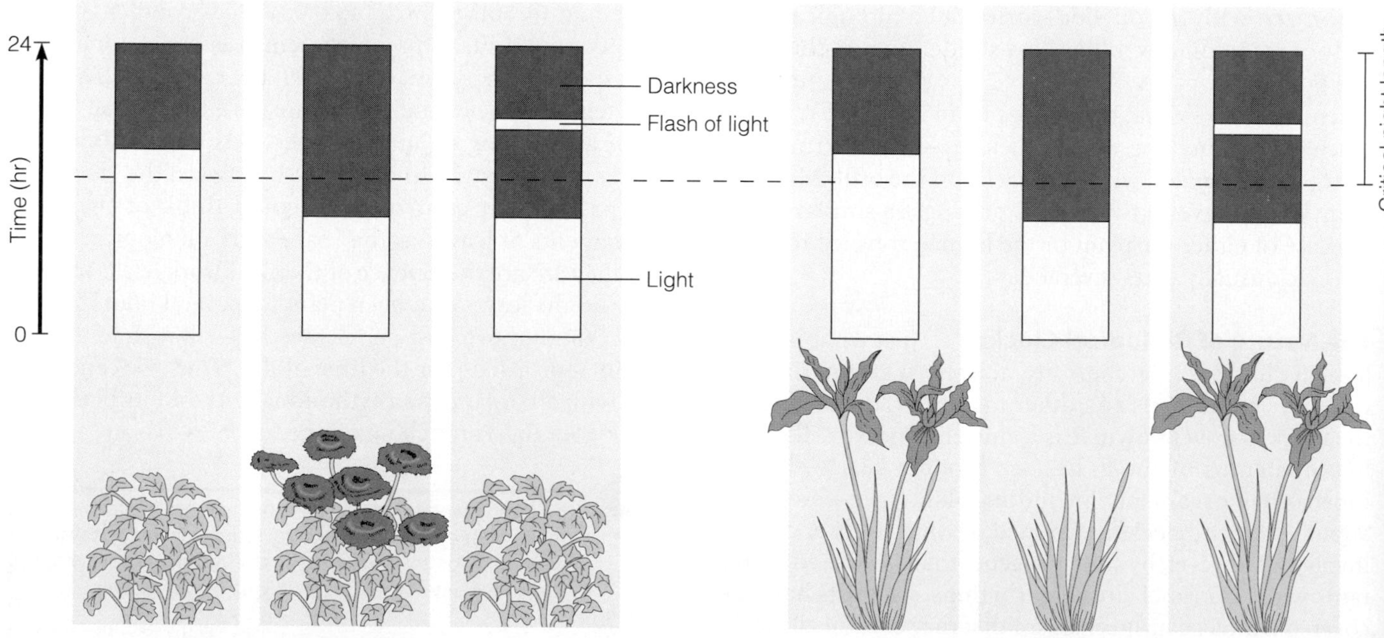

Figure 33.11 Photoperiodic control of flowering

Phytochrome is a light detector that may help set the biological clock 33.12

33B

The discovery that photoperiod (specifically night length) determines seasonal responses of plants poses another question: How does a plant actually measure photoperiod? Much remains to be learned about this, but pigments called phytochromes are part of the answer. **Phytochromes** are proteins with a light-absorbing component. Because the light absorbed is at the red end of the spectrum, the molecules appear blue or bluish green (see Module 7.6). Like the pigments in our eyes, phytochromes are pigments that function as photoreceptors—light detectors.

Phytochromes were discovered as a result of studies on how different wavelengths of light affect flowering in short-day and long-day plants. Bar 1 in Figure 33.12A repeats the effect we saw in the previous module for both short-day and long-day plants that receive a flash of light during their critical dark period. The letter R on the light flash stands for red light with a wavelength of 660 nanometers (nm). Researchers learned that this type of light (which is one component of white daylight) is the most effective wavelength for interrupting night length.

The other three bars in Figure 33.12A show how flashes of light of longer wavelength, around 730 nm, affect flowering. This type of light is called far-red (FR). As bar 2 shows, the effect of a flash of red light (R) that interrupts a period of darkness can be reversed by a subsequent flash of FR light: Both types of plants behave as though there is no interruption in the night length. Bars 3 and 4 indicate that

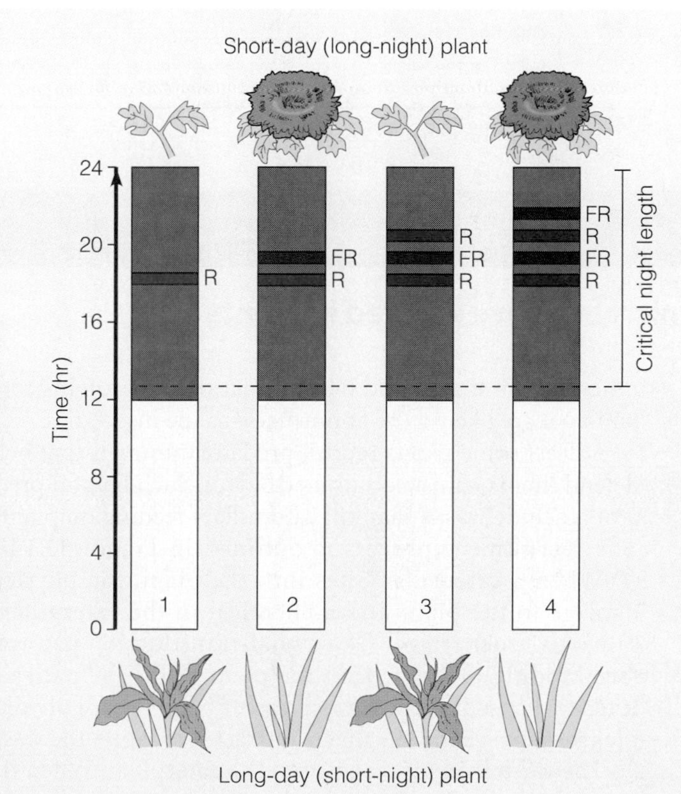

Figure 33.12A The reversible effects of red and far-red light

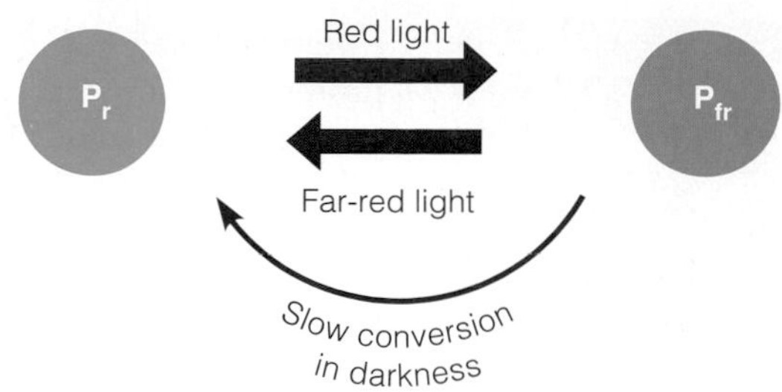

Figure 33.12B Interconversion of the two forms of phytochrome

no matter how many flashes of light a plant receives, only the wavelength of the last flash affects the plant's measurement of night length. Thus, the sequence R-FR-R produces the same results as in bar 1, and the sequence R-FR-R-FR yields the same effect as in bar 2.

Now we're ready to consider the role of phytochromes. A phytochrome, researchers have learned, alternates between two forms that differ only slightly in structure. One form absorbs red light and the other absorbs far-red light. The two forms of a phytochrome are designated P_r (red-absorbing) and P_{fr} (far-red–absorbing). As diagrammed in Figure 33.12B, when the P_r form absorbs red light (660 nm), it is quickly converted to P_{fr}, and when P_{fr} absorbs far-red light (730 nm), it is converted back to P_r. Also, P_{fr} in a plant reverts to P_r in the dark. These interconversions help account for the experimental results shown in Figure 33.12A.

In nature, of course, the night is not punctuated by flashes of red and far-red light. Instead, each day, the conversion of P_{fr} to P_r occurs in the continuous darkness that follows sunset, without the participation of FR light. At sunrise, much of the phytochrome is rapidly converted from the P_r form to P_{fr} because sunlight is richer in red than in far-red light. Apparently, these phytochrome conversions are cues that set a plant's biological clock. The clock may measure the time between the beginning of the P_{fr} to P_r change (sunset) and the P_r to P_{fr} change (sunrise). In doing so, the clock monitors photoperiod and becomes synchronized to the actual length of a day. In addition to flowering, physiological responses such as seed germination and the opening and closing of stomata also seem to be triggered by phytochrome conversions.

Found in all species of plants and algae examined to date, phytochromes may play a universal role in helping photosynthetic organisms stay synchronized with seasonal changes in light conditions. 

? How do phytochrome molecules help the plant distinguish daytime from nighttime?

Phytochrome molecules are mainly in the P_{fr} state during the day and mainly in the P_r state during the night.

TALKING ABOUT SCIENCE

33.13 Gloria Coruzzi studies the genetic basis of control systems in the model plant *Arabidopsis*

Figure 33.13 Gloria Coruzzi with *Arabidopsis* plants

A lifelong New Yorker and a professor at New York University, Gloria Coruzzi is a leader of a new wave of plant researchers who make use of DNA technology (see Chapter 12). In a recent interview, we asked her how she first began to study plants.

> I began working on plant molecular biology as a postdoctoral researcher. At the time, scientists were saying that plant genes couldn't be cloned—and then we cloned some and started to demonstrate how they are regulated by light. At the time, I was a "gene jock," focused on isolating DNA and sequencing it. Plants were just organisms to grind up to get out their DNA. But I've come to appreciate plant biology, and rather than just studying a gene, my lab now studies how the gene works in the context of the whole plant. The organism we study is *Arabidopsis thaliana,* which is easy to grow even in New York City!

Arabidopsis, a tiny weed in the mustard family, is now used as a model organism by many plant researchers, much as microbiologists use *E. coli* or animal biologists *Drosophila.* How did this modest plant become so important?

> One advantage of *Arabidopsis* is that it has a relatively short generation span, only 4–6 weeks, which makes genetic research easier. The plant also has a relatively small genome, and researchers have already sequenced more than a quarter of its DNA. The DNA sequence is a treasure trove of information for biologists. Genes for enzymes that we could never purify are popping out, and some of these genes are similar to genes in *E. coli* or humans. It's really a renaissance time in biology to be able to make these kinds of connections between organisms.

Dr. Coruzzi's lab group is using mutants and gene manipulation to figure out the roles of genes in regulating the growth of *Arabidopsis.* They hope that this basic research will eventually have practical benefits for agriculture.

> We're studying how plants take inorganic nitrogen and convert it to organic form, such as amino acids [see Module 32.13]. I thought it would be possible to engineer plants to increase their assimilation of nitrogen into organic compounds. These engineered plants could turn out to be more nutritious than wild-type plants.
>
> When we started, there was little known about the genes involved in making amino acids in plants. One possibility was that plants make amino acids continuously, with no regulation. But we discovered that amino acid production in plants is regulated by light, which affects the transcription of genes coding for the enzymes of amino acid synthesis. Also, we found that when no more amino acids are needed, the amino acids themselves turn off transcription of synthesis genes. This is reminiscent of regulation in bacteria and yeast.

Thus Dr. Coruzzi's research relates to the mechanisms of gene regulation you learned about in Chapter 11, as well as to the importance of light in regulating the lives of plants, which you read about earlier in this chapter. 

? What are some advantages of *Arabidopsis* as a model organism for basic research?

Small size, short generation time, and small genome. Also, the copious information about the plant already amassed by researchers facilitates further study.

Plant Defenses

33.14 Defenses against herbivores and infectious microbes have evolved in plants

In their natural environments, plants are continually threatened by other organisms, including munching animals and a variety of infectious microbes. These environmental "stresses" have led to the evolution of plant defense mechanisms that researchers are now investigating.

Defenses Against Herbivores Plants counter **herbivores**—plant-eating animals—with both physical defenses, such as thorns, and chemical defenses, such as distasteful or toxic compounds. For example, some plants produce an unusual amino acid called canavanine. Canavanine resembles arginine, one of the 20 amino acids normally used to make proteins. If an insect eats a plant containing canavanine, the molecule is incorporated into the insect's proteins in place of arginine. But canavanine is different enough from arginine that the shapes and hence functions of proteins containing it are likely to be abnormal—so the insect dies.

At least some plants recruit predatory animals that help defend them against certain herbivores. Such helpful predators include wasps that kill caterpillars feeding on plants. The recruitment process is outlined in Figure 33.14A. ① When a caterpillar bites into the plant, the physical damage to the plant and a chemical in the caterpillar's saliva together trigger ② a signal-transduction pathway (see Module 11.11) within the plant cells. The pathway leads to a specific cellular response: ③ the release of volatile (gaseous) chemicals that ④ attract ("recruit") the wasp. ⑤ The wasp injects its eggs into the caterpillar. When the eggs hatch, the wasp larvae eat their way out of the caterpillar, killing it.

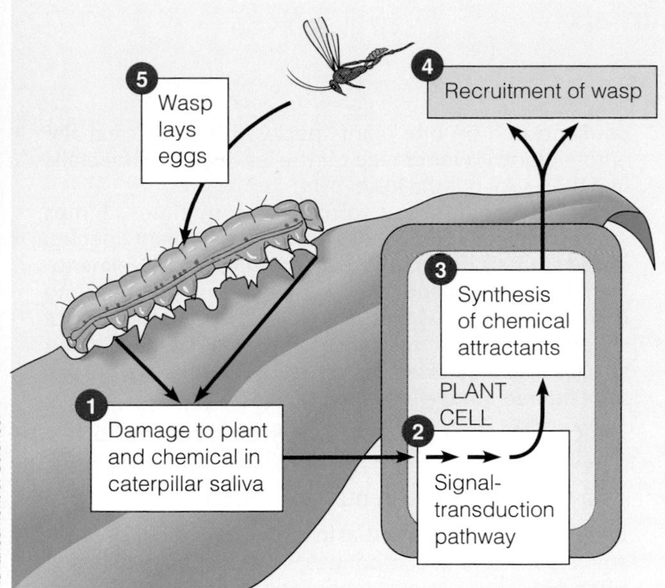

Figure 33.14A Recruitment of a wasp in response to a herbivore

Defenses Against Pathogens A plant, like an animal, is subject to infection by pathogenic microbes: viruses, bacteria, and fungi. And like an animal, a plant has defense systems that hinder infection and counter pathogens that do manage to infect the plant.

The first line of defense is the physical barrier of the plant's "skin," its epidermis (see Chapter 31). However, microbes can cross this barrier through wounds or through natural openings such as stomata. Once infected, the plant uses chemicals as a second line of defense. Plant cells damaged by the infection release microbe-killing molecules and, in addition, chemicals that signal nearby cells to mount a similar chemical defense. Infection also stimulates chemical changes in the plant cell walls, which toughen the walls and thus slow the spread of the microbe within the plant.

The plant's chemical defense system is enhanced by the plant's inherited ability to recognize certain pathogens. A kind of compromise has evolved between plants and most of their pathogens: The pathogen gains enough access to its host to perpetuate itself, but the plant survives. The plant is said to be resistant to that pathogen.

This resistance to destruction by a specific pathogen is based on the ability of the plant and the microbe to make a complementary pair of molecules. A plant has many *R* genes (for *resistance*), and each pathogen has a set of *Avr* genes (for *avirulence*, the state of being relatively harmless to the plant). Researchers think that an *R* gene encodes a receptor protein on the plant's cells and that the complementary *Avr* gene enables the pathogen to make a signal molecule that binds specifically to that receptor. When the products of an *R* gene and an *Avr* gene provide this sort of matchup between plant and pathogen, the plant is resistant.

Figure 33.14B shows this interaction and subsequent events in the plant. ① The binding of the pathogen's signal molecule (magenta) to the plant's receptor triggers ② a signal-transduction pathway, which leads to ③ a defense response in the infected plant tissue that is much stronger than would occur without the *R-Avr* matchup. The cells at the site of infection mount a vigorous chemical defense, tightly seal off the area, and then kill themselves. The spots on one of the leaves in the photo in Figure 33.14B result from this sort of local response. As sick as such a leaf appears, it will survive.

The defense response at the site of infection helps protect the rest of the plant in yet another way. Among the signal molecules produced there are ④ hormones that sound an alarm throughout the plant. ⑤ At destinations distant from the original site, these hormones trigger signal-transduction pathways leading to ⑥ the production of additional defensive chemicals. This defense response, called **systemic acquired resistance,** is actually nonspecific, providing protection against a diversity of pathogens for days.

Researchers suspect that one of the alarm hormones is salicylic acid, a compound whose pain-relieving effects led early cultures to use the salicylic-acid–rich bark of willows (*Salix*) as a medicine. Aspirin is a chemical derivative of this compound. With the discovery of systemic acquired resistance, biologists may have learned one function of salicylic acid in plants.

What is released at a site of infection that triggers the development of general resistance to pathogens elsewhere in the plant?

A hormone

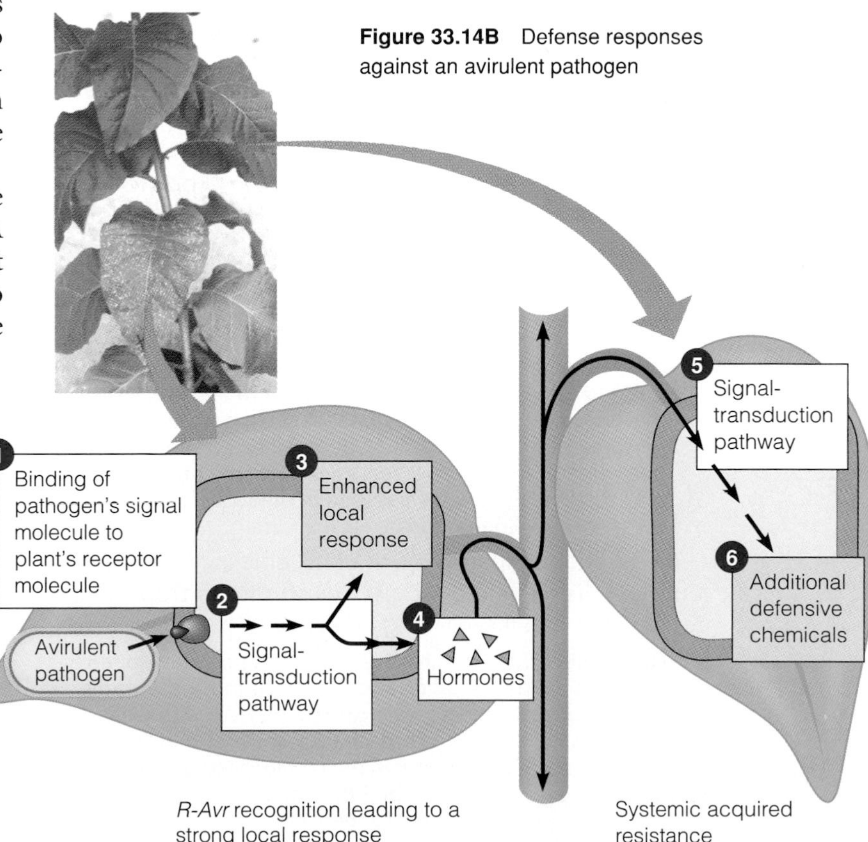

Figure 33.14B Defense responses against an avirulent pathogen

TALKING ABOUT SCIENCE

33.15 Plant biochemist Eloy Rodriguez studies how animals use defensive chemicals made by plants

Figure 33.15 Eloy Rodriguez in his research laboratory

Dr. Eloy Rodriguez, James A. Perkins Professor of Environmental Biology at Cornell University, is one of the world's leading experts on defensive chemicals produced by plants. In a 1995 interview, he described the significance of these chemicals this way:

> A plant's ability to survive is really due to chemistry. A large array of organisms eat plants, and a plant can't just get up and run. Natural selection favors those plants with the right kinds of chemical compounds that ward off fungi, bacteria, viruses, insects, and large herbivores.

Some of the defensive chemicals that plants produce are familiar to us all. For instance, as mentioned on the previous page, a compound very similar to aspirin probably helps ward off infections in many plants. In describing his research, Dr. Rodriguez often begins by talking about everyday chemicals derived from plants.

> Aspirin was originally derived from willow trees before it was synthetically made. Caffeine is not only a stimulant for humans; it also has physiological effects on fungi. These substances are small molecular compounds that function mainly in defense. Chiles are another intriguing case. Chiles are South American in origin and became an important food for the Aztecs and indigenous groups of Mexico, and now Asia is the biggest consumer of chiles. Studies have shown that chiles are very good antibiotics.

As Dr. Rodriguez points out, different parts of a plant have different chemicals. From your own knowledge of plant biology, you may be able to predict which plant organs have defensive chemicals and which do not.

> Defensive compounds aren't just randomly produced and packaged in the plant. Flower chemistry is very different from leaf chemistry, which is very different from root chemistry. In flowers you wouldn't expect to find a lot of poisonous chemicals; you would expect to find chemicals that attract pollinators. And that's what flowers have—nice perfumes and pigments. They generally aren't very toxic. But when you get into a leaf [which is essential to the plant's survival], that's where you begin to see a large diversity of toxic compounds.

Dr. Rodriguez spends much of his time in his laboratory, but he is also a field biologist. His studies in tropical rain forests have had far-reaching influence.

> One project that I got involved with has now developed into a discipline. It's called zoopharmacognosy, the study of how animals possibly medicate themselves with plants. I got involved with Richard Wrangham, a primatologist who was studying primates in the Kibale Forest of Uganda. In one observation, researchers followed a particular chimp for several days, and for two whole days this animal concentrated on one plant species. It wouldn't eat the whole plant. It would take off the leaves, crack the stalk, and then suck out the juice.
>
> Richard pointed out to me that sometimes chimps seem to select young leaves from certain plant species. These animals get up in the morning, make a beeline toward these plants, and take a certain amount of the young leaves. We calculated that they were more or less getting a set dosage of the drug or drugs in the leaf. Of the 15 plant species that we looked at while studying these chimps, we have figured out the chemistry of five of them.

Some of the compounds that Rodriguez and other researchers have discovered by following apes and other animals have potential use in human medicine:

> Clearly I see some potential in modern medicine. For example, some of the compounds we've identified kill parasitic worms, and some may be useful against tumors. There is no question that the templates for most drugs are in the natural world. . . . This new field is giving us clues about where we might find medically useful chemicals. A lot of plant collection is done randomly—you go out and collect a bunch of bark, for example. It is a very tedious way to go about getting drugs. It's nice when you have animals basically telling you, "Here, try these leaves."

Actually, as Dr. Rodriguez relates, modern scientists did not originate the idea of finding drugs by watching animals:

> I think the use of medicinal plants started when the first humans got sick. When you are sick, you try to figure out how to take care of yourself. Besides trial and error, there must have been some experimentation taking place and some very keen observation. Now when I go out to the tropical rain forest, I always ask the people, "Do you observe animals?" and they will tell me, "Oh yeah, sure we observe animals." . . . Indigenous people have been extremely successful. Anthropologists have documented that they extract plant materials, grind them up, filter them, and treat them with burned leaves (which act as a base)—almost the same process that I use. They get out materials that are relatively pure for use as medicine. A lot of the drug companies started out thanks to those folks.

Though his work focuses on the medicinal potential of rain forest chemicals, Rodriguez is quick to emphasize that the importance of rain forests goes beyond human concerns.

> The more we search the rain forest, the more in awe of it we are. We see how important it is. In the tropical rain forest, we are talking about the ultimate diversity in plants, animals, fungi—life! The arguments that have been made for preserving the forests are excellent. We are talking about the health of the planet.

Rodriguez's words lead us to our next and final unit, Ecology. In the chapters that follow, we will explore connections among organisms and between organisms and their environments. 

? What do researchers hope to learn by observing the feeding behavior of sick chimpanzees?

The selective feeding of the ill animals on certain parts of specific plants might lead researchers to medicines that could help humans too.

Chapter Review

Chapter Summary

Plant Hormones (Introduction–33.8) Plants can sense and respond to environmental stimuli. The search for mechanisms that control plant behavior has focused on hormones (Introduction). Hormones coordinate the activities of plant cells and tissues. The study of plant hormones began with observations of plants bending toward light, a phenomenon called phototropism. The bending results from faster cell growth on the shaded side of the shoot than on the lighted side. Experiments carried out by Darwin and others showed that the tip of a grass seedling detects light and transmits a signal down to the growing region of the shoot. This signal is a hormone named auxin; it accumulates on the dark side of the shoot and stimulates growth below the tip (33.1). Auxins are now known to be a class of chemicals, one of five main types of plant hormones. The hormones, produced in very small amounts, regulate plant growth, development, and metabolism (33.2). Much of the natural auxin, IAA, is produced in the apical meristems at the tips of shoots. At different concentrations, IAA stimulates or inhibits the elongation of shoots and roots. It may act by weakening cell walls, allowing them to stretch when cells take up water. Auxins also stimulate the development of vascular tissues and cell division in vascular cambium, promoting growth in stem diameter (33.3). Cytokinins, produced by growing roots, embryos, and fruits, are hormones that promote cell division. Cytokinins from roots may balance the effects of auxins from apical meristems, causing lower buds to develop into branches. Thus, the ratio of auxins to cytokinins may coordinate the growth of roots and shoots (33.4). Gibberellins stimulate the elongation of stems and leaves and the development of fruit. Gibberellins can induce some plants to flower and produce seeds early, or to develop seedless fruits without fertilization. Gibberellins released from embryos function in some of the early events of seed germination (33.5). Abscisic acid (ABA) inhibits the germination of seeds. The ratio of ABA to gibberellins often determines whether a seed will remain dormant or germinate. Seeds of many desert plants remain dormant until rain washes away their ABA. During winter or drought, ABA from buds inhibits primary and secondary growth in buds and vascular cambium. ABA also acts as a "stress hormone," causing stomata to close when a plant is dehydrated (33.6). As fruit cells age, they give off ethylene gas, which hastens ripening. A changing ratio of auxin to ethylene, triggered mainly by shorter days, probably causes autumn color changes and the loss of leaves from deciduous trees. Fruit growers use ethylene to control ripening (33.7). Plant hormones have a variety of agricultural uses. Farmers use auxins to delay or promote fruit drop. Auxins and gibberellins are used to produce seedless fruits. Large doses of gibberellins are used to force biennial plants to make seeds in their first year of growth. A synthetic auxin called 2,4-D is used to kill weeds. There are some questions about the safety of using such chemicals (33.8).

Growth Responses and Biological Rhythms in Plants (33.9–33.13) Plants sense and respond to environmental changes in a variety of ways. Tropisms are growth responses that change the shape of a plant or make it grow toward or away from a stimulus. Phototropism, bending toward light, seems to result from auxin moving from the light side to the dark side of a stem. A response to gravity, or gravitropism, may be caused by a buildup of special organelles on the low sides of shoots and roots, which may trigger a change in the distribution of hormones. Thigmotropism, a response to touch, seems to be responsible for the coiling of tendrils and vines around objects (33.9). An internal biological clock controls sleep movements and other daily cycles in plants. These cycles, called circadian rhythms, persist with periods of about 24 hours even in the absence of environmental cues, but such cues are needed to keep them synchronized with day and night. The exact mechanisms and locations of biological clocks in plants are not known (33.10). Plants mark the seasons by measuring photoperiod, the relative lengths of night and day. The timing of flowering is one of the seasonal responses to photoperiod. Short-day plants flower when nights exceed a certain critical length; long-day plants flower when nights are shorter than a critical length (33.11). A light-absorbing protein called phytochrome enables plants to monitor photoperiod (33.12). A small, wild mustard called *Arabidopsis* is a popular model organism for plant molecular biologists (33.13).

Plant Defenses (33.14–33.15) Plants use chemicals to defend themselves against both herbivores and pathogens. So-called avirulent plant pathogens interact with host plants in a specific way that stimulates both local and systemic defenses (33.14). Some animals may medicate themselves by eating plants containing certain defensive chemicals (33.15).

Testing Your Knowledge

Multiple Choice

1. During winter or periods of drought, which of the following plant hormones inhibits growth and seed germination?
   a. ethylene
   b. abscisic acid
   c. gibberellin
   d. auxin
   e. cytokinin

2. A certain short-day plant flowers only when days are less than 12 hours long. Which of the following would cause it to flower? *(Explain your answer.)*
   a. a 9-hr night and 15-hr day with 1 min of darkness after 7 hr
   b. an 8-hr day and 16-hr night with a flash of white light after 8 hr
   c. a 13-hr night and 11-hr day with 1 min of darkness after 6 hr
   d. a 12-hr day and 12-hr night with a flash of red light after 6 hr
   e. alternating 4-hour periods of light and darkness

3. Auxins cause a shoot to bend toward light by
   a. causing cells to shrink on the dark side of the shoot
   b. stimulating growth on the dark side of the shoot
   c. causing cells to shrink on the lighted side of the shoot
   d. stimulating growth on the lighted side of the shoot
   e. inhibiting growth on the dark side of the shoot

4. In the autumn, the amount of ____ increases and ____ decreases in fruit and leaf stalks, causing a plant to drop fruit and leaves.
   a. ethylene . . . auxin
   b. gibberellin . . . abscisic acid
   c. cytokinin . . . abscisic acid
   d. auxin . . . ethylene
   e. gibberellin . . . auxin

5. Plant hormones act by affecting the activities of
   a. genes
   b. membranes
   c. enzymes
   d. genes, membranes, and enzymes
   e. genes and enzymes

6. Buds and sprouts often form on tree stumps. Which of the following hormones would you expect to stimulate their formation?

a. auxin
b. cytokinins
c. abscisic acid
d. ethylene
e. gibberellins

7. A plant's defense response at the site of initial infection by a pathogen will be especially strong if

a. the pathogen is virulent
b. the plant makes a receptor protein that recognizes a signal molecule from the microbe
c. the pathogen is a fungus
d. the plant has an *Avr* gene that is the right match for one of the microbe's *R* genes
e. the right combination of hormones travel from the infection site to other parts of the plant

Matching

1. Bending of a shoot toward light	a. phytochrome
2. Growth response to touch	b. photoperiod
3. A cycle with a period of about 24 hours	c. sleep movement
4. Pigment that helps control flowering	d. circadian rhythm
5. Relative lengths of night and day	e. thigmotropism
6. Growth response to gravity	f. phototropism
7. Folding of plant leaves at night	g. gravitropism

Describing, Comparing, and Explaining

1. If apples are to be stored for long periods, it is best to keep them in a place with good air circulation. Explain why.

2. Write a short paragraph explaining why a houseplant becomes more bushy if you pinch off its terminal buds.

Thinking as a Scientist

1. Jon just started a new job as night watchman at a plant nursery. His boss told him to stay out of a room where chrysanthemums (which are short-day plants) were about to flower. Around midnight, looking for the restroom, Jon accidentally opened the door to the chrysanthemum room and turned on the lights for a moment. How might this affect the chrysanthemums? What could Jon do to correct his mistake?

2. A plant biologist observed a peculiar pattern when a tropical shrub was attacked by caterpillars. He noticed that after a caterpillar ate a leaf, it would skip over nearby leaves and attack a leaf some distance away. The researcher found that when a leaf was eaten, nearby leaves started making a chemical that deterred the caterpillars. Simply removing a leaf did not trigger the same change nearby. The biologist suspected that a damaged leaf sent out a chemical that signaled other leaves. How could he test his hypothesis?

3. In the 1950s, scientists discovered that many plants emit a gaseous compound called isoprene into the atmosphere. Isoprene is a precursor of the cytokinin hormones, but why plants emit it remains an open question. Researchers have found that plants emit isoprene when they are photosynthesizing, and the amount emitted increases as the temperature of a plant's leaves increases. Plants also synthesize more isoprene when they cannot obtain enough water. Darkness, substances that inhibit photosynthesis, and pure nitrogen gas (N_2) slow or stop isoprene emission. One hypothesis is that isoprene emissions help prevent damage to plants caused by high temperatures. Chlorophyll fluorescence has been used as a measure of irreversible leaf damage; the more fluorescence, the greater the damage. The graph below displays the results of some experimental tests of this hypothesis.

The red line on the graph shows the effect of increasing temperature on leaves in air with no isoprene. The green line shows the effect on leaves from the same species of plant in air containing isoprene. Do these results support the hypothesis? Explain your answer. The leaves were exposed to N_2 while the recordings were made. Why do you suppose the experimenters did this?

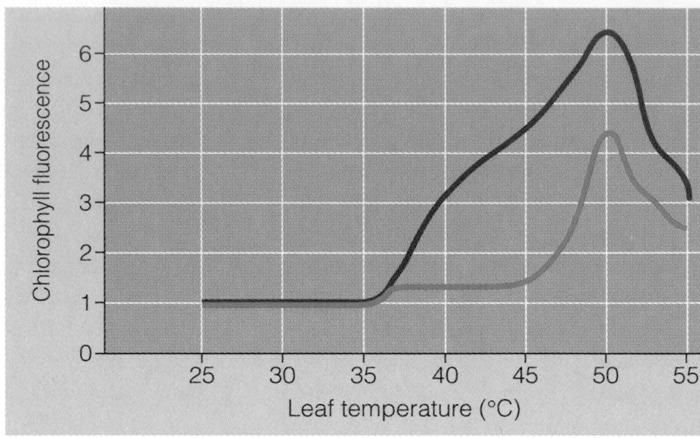

Science, Technology, and Society

Imagine the following scenario: A plant scientist has developed a synthetic chemical that mimics the effects of a plant hormone. The chemical can be sprayed on apples before harvest to prevent flaking of the natural wax that is formed on the skin. This makes the apples shinier and gives them a deeper red color. What kinds of questions do you think should be answered before farmers start using this chemical on apples? How might the scientist go about finding answers to these questions?

Media Resources

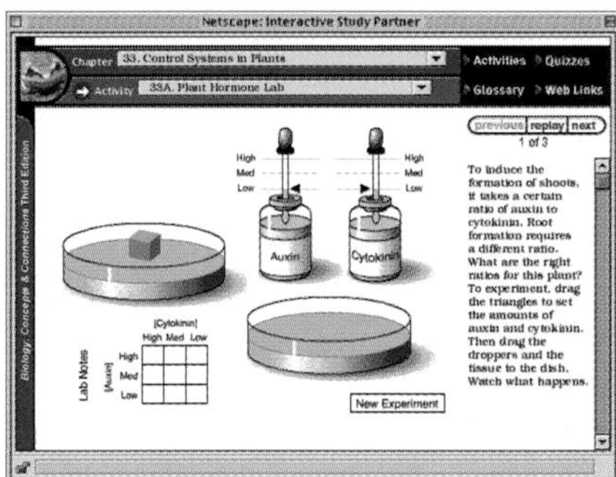

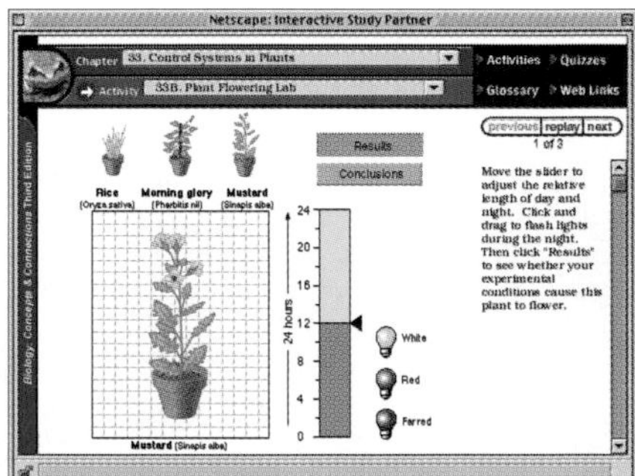

For further review of the concepts in this chapter, use the activities and quizzes on the CD-ROM.

Visit the Special Edition of The Biology Place web site for activities, quizzes, news articles, and web links.

UNIT VII
Ecology

34 The Biosphere: An Introduction to Earth's Diverse Environments

Outline

Imagine finding yourself in a strange, dark environment full of species never before seen by humans—species whose ultimate energy source is not sunlight, but energy that comes from the interior of their planet. Sound like science fiction? It actually happened to scientists in the battery-powered deep-sea research vessel *Alvin*, pictured at the left.

Alvin can withstand water pressures that would flatten an ordinary submarine. Accommodating a pilot and two other people, and equipped with spotlights and sampling arms, it has carried scientists down some 2500 m, over a mile deeper than sunlight penetrates. In the late 1970s, diving off the southern tip of Baja California in Mexico, scientists in *Alvin* discovered seafloor life that does not depend on photosynthesis. They found new organisms inhabiting the unique world of hydrothermal vents, sites near the adjoining edges of giant plates of Earth's crust where molten rock and hot gases surge upward from Earth's interior. Figure A shows one such site, where a chimney-like vent, perhaps as much as 30 m high, emits scalding water and hot gases such as hydrogen sulfide. The water temperature is 350°C, but boiling is prevented by immense water pressure. Just a few meters from the vents, the water temperature is only about 2°C. This is truly an environment of extremes.

A variety of animals, including sea anemones, giant clams (30 cm long), shrimps, crabs, a few fishes, and tube worms, thrive near hydrothermal vents. In Figure B, a crab crawls over a dense growth of 3-m-long tube worms. The worms were unknown to science until hydrothermal vents were explored. The bright red head of each worm projects from a rigid tube anchored in the seafloor. These and other vent animals live on energy extracted from chemicals by bacteria, rather than on light energy trapped by plants, algae, or photosynthetic bacteria, as other animals do.

The bacteria in hydrothermal vents also have an unusual existence. Living in crevices among the hot rocks, most of them are chemoautotrophs that obtain energy by oxidizing hydrogen sulfide (H_2S) to sulfates (SO_4^{2}) or elemental sulfur (S). These so-called sulfur bacteria use the energy to convert CO_2 from seawater into organic food molecules. Some of the vents give off a dull, yellowish glow.

Many of the animals in vent communities obtain nutrients by eating bacteria. Others, such as the giant tube worms, harbor sulfur bacteria within their bodies. The worms absorb sulfur compounds from the water; the bacteria use the compounds as an energy source and make organic food molecules.

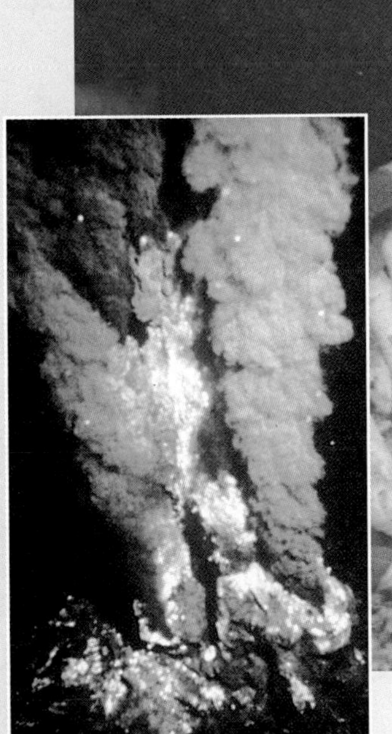
A. A hydrothermal vent

B. Tube worms

The unusual organisms living at hydrothermal vents indicate how diverse life is on our planet. Until the *Alvin* expeditions, no one knew that entire communities of organisms could live on energy from Earth itself. Scientists have since discovered hydrothermal vents at several locations in the Pacific and Atlantic Oceans.

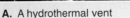

The scientific study of the interactions of organisms with their environments—from hydrothermal vents to the more typical solar-powered systems we depend on—is called **ecology** (from the Greek *oikos*, home). As we will see in this chapter and those that follow, ecology is a critically important field of biology, with implications for all forms of life on Earth.

34.1 Ecologists study how organisms interact with their environments at several levels

Figure 34.1 Research on giant clams near an ocean vent

The definition of ecology just given sounds straightforward, but it covers an enormously complex area of biology. The interactions between organisms and their environment are two-way. Organisms are affected by their environment, but, by their very presence and activities, they also change the environment, often dramatically. In transforming energy, bacteria change the ocean vent environment and make it habitable for animals. Likewise, plants drastically alter their environments, extracting CO_2 from the atmosphere and adding O_2 to it, in the process providing themselves and animals with food.

Ecologists study environmental interactions at several levels. At the **organism** level, they may examine how one kind of organism meets the challenges of its environment. An ecologist working at this level might study, for instance, the adaptations of clams to the extreme temperatures around hydrothermal vents, as shown in Figure 34.1.

Another level of study in ecology is the **population,** an interbreeding group of individuals belonging to the same species and living in a particular geographic area. The clams of one species living near a particular ocean vent would constitute a population. One ecologist might study a clam population's rate of growth relative to the temperature of the surrounding water. Someone else, interested in clam evolution, might look at genetic differences among populations from different vents.

A third level, the **community,** consists of all the organisms—that is, all the populations of different species—that inhabit a particular area. For example, all the organisms supported by a particular hydrothermal vent would constitute a community. An ecologist working at this level might focus on interactions among organisms, such as the effect of predation by crabs on tube worms or clams.

The fourth level of ecological study, the **ecosystem,** includes all the life forms existing in a certain area and all the nonliving factors as well. The nonliving factors, or **abiotic** factors, include temperature, forms of energy, gases, water, nutrients, and other chemicals. The organisms making up the community of species in the area are called **biotic** factors. Some critical questions at the ecosystem level concern how chemicals cycle, and how energy flows between organisms and their surroundings. For a vent community, one ecosystem-level question would be: How much of the energy available to them do giant clams and other animals actually use?

Ecological research at any level uses the elements of the scientific process: observations, questions, hypothesis, procedure for testing the hypothesis, predicted results, and testing. However, in contrast to the laboratory, where conditions can be simplified and experiments can be precisely controlled, environments are very complex. As we discussed in Module 1.3, testing ecological hypotheses usually requires adapting the scientific process to the multiple variables of the environment.

A (an) ________ consists of a biological ________, or all the biotic factors in the area, along with the nonliving environmental factors, or ________ factors.

ecosystem . . . community . . . abiotic

The Biosphere

34.2 The biosphere is the total of all of Earth's ecosystems

Recalling a view of Earth from space (as in Figure 34.2A), Apollo astronaut Rusty Schweickart once remarked: "On that small blue-and-white planet below is everything that means anything to you. National boundaries and human artifacts no longer seem real. Only the biosphere, whole and home of life."

The **biosphere** is the global ecosystem—that portion of Earth that is alive, or all of life and where it lives. The most complex level in ecology, the biosphere includes the atmosphere to an altitude of several kilometers, the land down to water-bearing rocks about 1500 meters deep, lakes and streams, caves, and the oceans to a depth of several kilometers. Isolated in space, the biosphere is self-contained, or

Figure 34.2A Earth as seen from the moon

closed, except that its photosynthesizers derive energy from sunlight, and it loses heat to space.

Another feature of the biosphere is its patchiness, and we can see this on several levels. On a global scale, we see it in the distribution of continents and oceans. On a regional scale, patchiness occurs in the distribution of deserts, grasslands, forests, lakes, and streams, for example. The aerial view of a wilderness area in Figure 34.2B shows patchiness on a local scale. Here we see a mixture of forest, small lakes, a meandering river, and open meadows. If we moved even closer, into any one of these different environments, we would find patchiness on yet a smaller scale. For example, we would find that each lake has several different **habitats** (environmental situations in which organisms live). Each habitat has a characteristic community of organisms. Abiotic factors, especially water depth, temperature, and dissolved O_2, largely determine the kinds of organisms that live in the different lake habitats.

Standing in a wilderness can be misleading; the lakes and streams appear untouched, and the forest seems almost boundless. Views from space are more sobering, for they show planet Earth as only a small sphere in the vastness of the universe. Unfortunately, we humans tend to treat the biosphere as an unlimited resource for our own consumption.

Figure 34.2B Patchiness of the environment in the Alaskan wilderness

? Why is it more accurate to define the biosphere as the global ecosystem rather than the global *community*?

Because the biosphere includes both abiotic and biotic components

34.3 Environmental problems reveal the limits of the biosphere

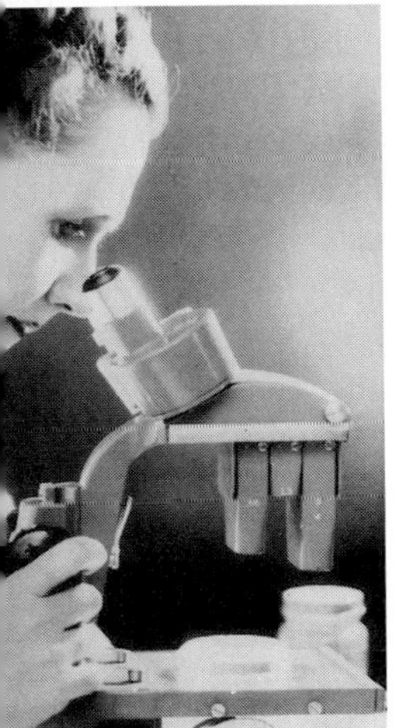

"The 'control of nature' is a phrase conceived in arrogance, born of the Neanderthal age of biology and philosophy, when it was supposed that nature exists for the convenience of man."

—Rachel Carson, *Silent Spring*

Our current awareness of the biosphere's limits stems mainly from the 1960s, a time of growing disillusionment with environmental practices of the past. In the 1950s, technology seemed poised to free humankind from several age-old bonds. New chemical fertilizers and pesticides, for example, showed great promise for increasing agricultural productivity and eliminating insect-borne diseases. Fertilizers were applied extensively, and pests were attacked by the use of massive aerial sprays. The immediate results were astonishing: Increases in farm productivity enabled developed nations such as the United States to grow surplus food and market it overseas, and the worldwide incidence of malaria and several other insect-borne diseases was markedly reduced. DDT and other chemicals were hailed as miracle weapons with potential use anywhere insects caused problems.

Our enthusiasm for chemical fertilizers and pesticides began to wane as some of the side effects of DDT and other widely used poisons began appearing in the late 1950s. One of the first to perceive the global dangers of pesticide abuse was the late Rachel Carson. Much of our current environmental awareness stems from her book *Silent Spring,* published in 1962. Her warnings were underscored when, shortly thereafter, scientists reported that DDT was threatening the survival of predatory birds and was showing up in human milk. Another serious problem to arise was genetic resistance to pesticides, evolving in an increasing number of pest populations. At the same time, some of the ill effects of other technological developments began to be widely publicized. By the early 1970s, disillusionment with chemicals and a realization that our finite biosphere could not tolerate unlimited exploitation had developed into widespread concern about environmental problems.

Today, it's clear that no part of the biosphere is untouched by human activities, and many people are concerned about the abusive impact of human populations and technology. The misuse of natural resources, localized famine aggravated by land misuse and expanding population, the growing list of species extinguished or endangered by loss of habitat, and the poisoning of soil and streams with toxic wastes—these are just a few of the problems that we have created and must solve. We will examine some of our environmental problems in Chapter 38. But first, let's explore the biosphere and the major concepts of ecology in more detail.

? As a call for environmental consciousness, Rachel Carson's *Silent Spring* focused on the destructive consequences of toxic pollutants, especially ________.

the pesticide DDT

34.4 Physical and chemical factors influence life in the biosphere

The biosphere is finite, but it is also extremely diverse. Understanding its structure and dynamics can help us understand our environmental dilemma. A variety of physical and chemical factors affect the organisms living in particular ecosystems and in the biosphere as a whole. Solar energy, water, temperature, and wind are among the most important of these abiotic factors.

Solar energy powers nearly all surface terrestrial and shallow water ecosystems. Other ecosystems, such as the hydrothermal vents and ecosystems in dark caves and in groundwater deep beneath Earth's surface, are powered by energy that bacteria extract from inorganic chemicals. In aquatic environments where sunlight reaches, the availability of light has a significant effect on the growth and distribution of plants and algae. Because water itself and the microorganisms in it absorb light and keep it from penetrating very far, most photosynthesis occurs near the surface of a body of water. In terrestrial environments, light is often not the most important factor limiting plant growth. In many forests, however, shading by trees creates intense competition for light at ground level.

Water, a second abiotic factor, is essential to all life. Aquatic organisms have a seemingly unlimited supply of water, but they face problems of water balance if their own solute concentration does not match that of their surroundings. As we saw in Module 25.5, aquatic organisms confront very different solute concentrations in the sea than in freshwater lakes and streams. For a terrestrial organism, the main water problem is the threat of drying out. Therefore, many land species have water-tight coverings that reduce water loss. Many also have kidneys that save water by excreting very concentrated urine.

Figure 34.4 Recovery after fire in Yellowstone National Park

Temperature is an important abiotic factor because of its effect on metabolism. Few organisms can maintain a sufficiently active metabolism at temperatures close to 0°C, and temperatures above 50°C destroy the enzymes of most organisms. Extraordinary adaptations enable some species to live outside this temperature range. For example, some of the frogs and turtles living in the northern U.S. and Canada can freeze during winter months and still survive. Bacteria living in hydrothermal vents and hot springs have enzymes that function optimally at extremely high temperatures. Mammals and birds can remain considerably warmer than their surroundings and can be active in a fairly wide range of temperatures, but even these animals function best at certain temperatures.

Wind is an important abiotic factor for several reasons. Some organisms—for example, bacteria, protists, and many insects that live on snow-covered mountain peaks—depend on nutrients blown to them by winds. Local wind damage often creates openings in forests, contributing to patchiness in ecosystems. Wind also increases an organism's rate of water loss by evaporation. The consequent increase in evaporative cooling can be advantageous on a hot summer day, but it can cause dangerous wind chill in the winter.

These abiotic factors may combine with others to produce the physical and chemical components of an organism's environment. For example, most soils are complex combinations of inorganic nutrients, organic materials in various stages of decomposition, water, and air. Such variables as soil structure, pH, and nutrient content often play a large role in determining the distribution of organisms.

Other combinations of abiotic factors include storms, such as hurricanes (wind and water), and fire (temperature and combustible materials). Storms and fires are infrequent and unpredictable in many ecosystems. However, fire recurs frequently in such ecosystems as grasslands and many dry forests. The photograph at the left was taken in 1989, only a short time after extensive fires swept through Yellowstone National Park. Small plants were already colonizing the area, taking advantage of nutrients released from the trees that burned.

In the next module, we examine the interaction between one animal species and the abiotic factors of its environment. We also look at some of the biotic factors that act with the abiotic ones in the animal's ecosystem.

________ energy is such an important abiotic factor because ________ provides the organic fuel and building material for the organisms of most ecosystems.

Solar . . . photosynthesis

Organisms are adapted to abiotic and biotic factors by natural selection 34.5

An organism's ability to survive and reproduce in a particular environment is a result of natural selection, as we discussed in Chapter 13. By eliminating the least fit individuals in populations, environmental forces help shape species to the mix of abiotic and biotic factors that they encounter.

The presence of a species in a particular place can come about in two ways: The species may evolve in that location, or it may disperse to that location and be able to survive once it is there. The pronghorn "antelope," pictured in Figure 34.5, evolved on the open plains and shrub deserts of North America. It is not found elsewhere and is only distantly related to the numerous species of antelopes in Africa. Taking the pronghorn as an example, let's see how some of its unique adaptations fit the environmental conditions in which it evolved.

Figure 34.5 Pronghorns

First, what about the major abiotic factors? The pronghorn's habitat is arid, windswept, and subject to extreme temperature fluctuations both daily and seasonally. The pronghorn is superbly adapted to these conditions. If you drive through Wyoming or parts of Colorado in the winter, you will see herds of these animals foraging in the open when temperatures are well below 0°C. The pronghorn has a thick coat made of hollow hairs that trap air and use it as insulation. Water is rarely a problem for a pronghorn because it can live on the moisture it obtains from vegetation.

What about the pronghorn's adaptations to the biotic components of its habitat? The pronghorn's main foods are short, coarse grasses and woody shrubs, and its teeth are adapted for biting and chewing these plants. Also, like a cow, it has a stomach containing cellulose-digesting bacteria. As the pronghorn eats plants, the bacteria digest cellulose, and the animal obtains most of its nutrients from the bacteria. As the pronghorn evolved, it became adapted to predation by wolves, coyotes, and cougars. The pronghorn's main adaptations to escape predators are great speed and endurance. Capable of sprinting about 95 km/hr on flat ground, it is one of the fastest mammals. An adult pronghorn can also keep up a pace of about 65 km/hr for at least 30 minutes—a definite advantage when being chased by long-distance runners such as wolves. Other adaptations that help the pronghorn foil predators include its tan-and-white coat, which often camouflages the animal on the open plains, and its keen eyes, which can detect movement at great distances. The pronghorn also derives protection from living in herds. When one pronghorn starts to run, its white rump patch seems to alert other herd members to danger.

Organisms vary a great deal in their ability to tolerate fluctuations and long-term changes in their environments. The pronghorn has survived significant environmental changes. During the pioneering days of the American West, its numbers were seriously reduced by human hunting. Unlike the bison, elk, wolf, and cougar (all of which were extinguished from the plains), wild populations of the pronghorn survived. Today, the species is common in several western states, where it competes mainly with domestic cattle and sheep for food and is hunted mainly by humans and coyotes.

The pronghorn is a highly successful, herbivorous, running mammal of open country. In a very different environment, such as a wooded area where predators would be more easily hidden by vegetation and could stalk them at close range, the pronghorn's adaptations for escaping predators might not be as effective as they are on the open plains. This suggests that an organism can usually tolerate environmental fluctuations only within the set of conditions to which it is adapted. Outside that set, the organism may not survive long enough to reproduce. Thus, in adapting populations to local environmental conditions, natural selection may limit the distribution of organisms. The absence of the pronghorn outside North America, however, does not necessarily imply that the species could not survive elsewhere; it may only mean they were never able to disperse beyond this region. 

? Why does the set of evolutionary adaptations characterizing a species tend to limit the geographic distribution of that species?

To the extent that the species is adapted to particular environmental conditions, it is not so well-equipped to survive and reproduce where the environment poses different problems to which the species is not adapted.

34.6 Regional climate influences the distribution of biological communities

When we ask what determines whether a particular organism or community of organisms lives in a certain area, the climate of the region—especially temperature and rainfall—is often a large part of the answer. A number of interacting factors determine climate on both global and regional scales.

Figure 34.6A shows that, because of its curvature, Earth receives an uneven distribution of solar energy. The sun's rays strike equatorial areas most directly (perpendicularly). Away from the equator, the rays strike Earth's surface at oblique angles. As a result, the same amount of solar energy is spread over a larger area. Thus, any particular area of land or ocean near the equator absorbs more heat than comparable areas in the more northern or southern latitudes. The uneven heating of Earth's surface is a major factor driving air movements and water currents.

The seasons of the year result from the permanent tilt of the planet on its axis as it orbits the sun. As Figure 34.6B shows, the globe's position relative to the sun changes through the year. The Northern Hemisphere, for instance, is tipped most toward the sun in June, creating the longest days of the year and summer in that hemisphere; at the same time, days are short and it's winter in the Southern Hemisphere. Conversely, the Southern Hemisphere is tipped farthest toward the sun in December, creating summer there and winter in the Northern Hemisphere.

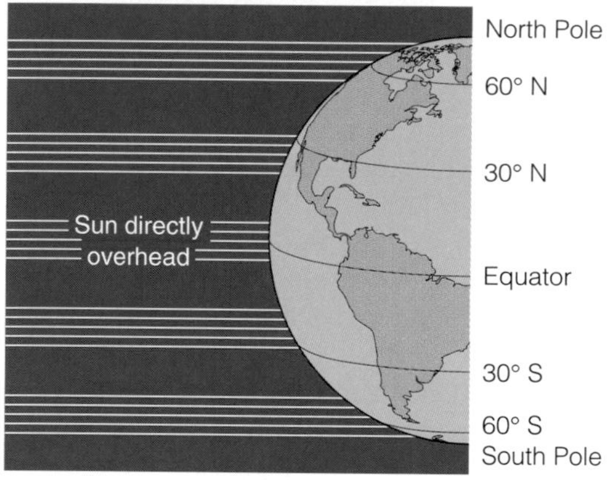

Figure 34.6A Uneven heating of Earth's surface

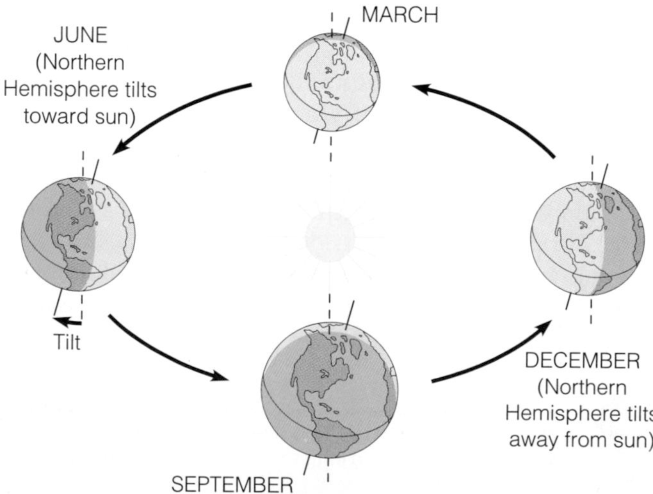

Figure 34.6B How Earth's tilt causes the seasons in the Northern Hemisphere

When air is warmed, it rises and tends to absorb moisture; when air is cooled, it falls and tends to lose moisture. Figure 34.6C shows some of the effects of uneven warming on global wind patterns and rainfall. Red arrows indicate air movements. Heated by the direct rays of the sun, air at the equator rises, creating an area of calm or of very light winds known as the **doldrums.** As warm equatorial air rises, it cools, forms clouds, and drops rain. Warm temperatures throughout the year and heavy rainfall largely explain why rain forests are concentrated near the equator.

After losing their moisture over equatorial zones, high-altitude air masses spread away from the equator until they cool and descend again at latitudes of about 30° north and south. Many of the world's great deserts—the Sahara in North Africa and the Arabian on the Arabian peninsula, for example—are centered at these latitudes because of the dry air they receive from the equator. As the dry air descends, some of it spreads back toward the equator. This movement creates the **trade winds,** which dominate the **tropics** (latitudes between 23.5° north and south). As the air moves back toward the equator, it warms and picks up moisture until it is uplifted again.

Latitudes between the tropics and the Arctic Circle in the north and the Antarctic Circle in the south are called **temperate zones.** Generally, these regions have milder climates than the tropics or the polar regions. Notice in Figure 34.6C that some of the descending dry air heads into the latitudes above 30°. At first these air masses pick up moisture, but they tend to drop it as they cool at higher latitudes. This is why the north and south temperate zones, especially latitudes between about 40° and 60°, tend to be moist. Deserts and arid regions in higher latitudes usually

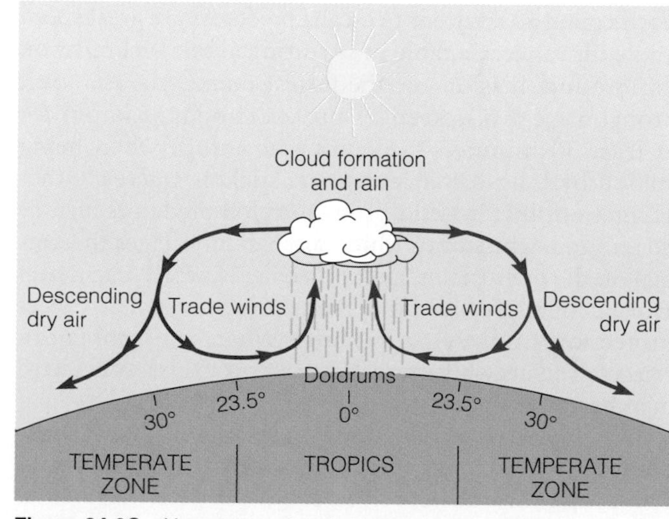

Figure 34.6C How uneven heating causes rain and winds

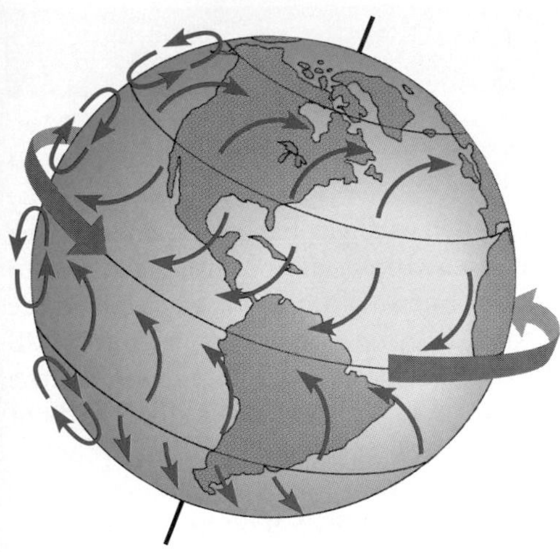

Figure 34.6D Prevailing wind patterns

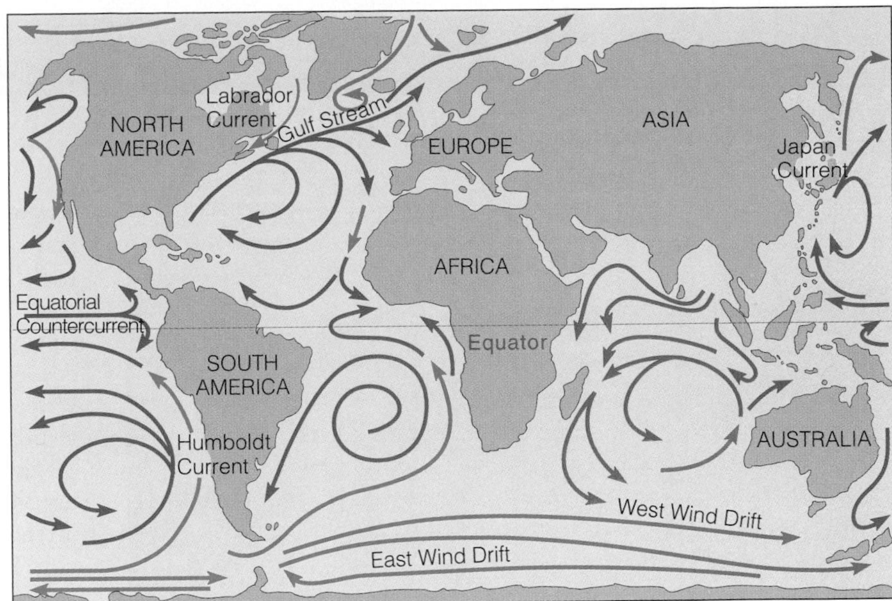

Figure 34.6E Ocean currents

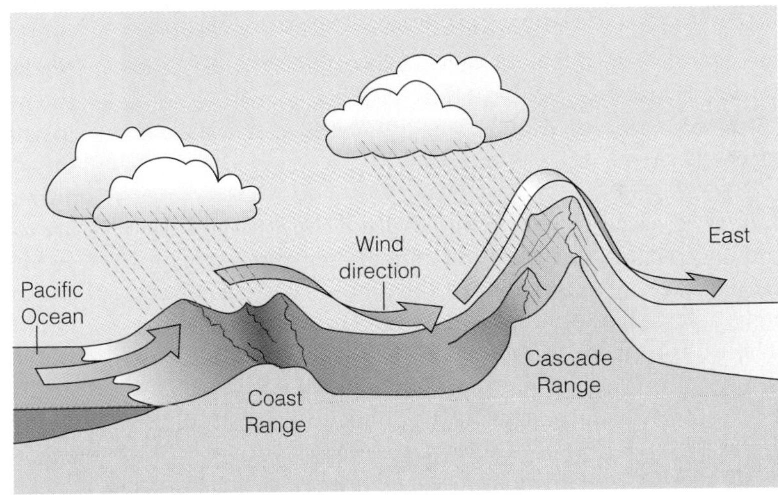

Figure 34.6F How mountains affect rainfall (Washington state)

result from mountain ranges that cut off the flow of moist air, as we'll see.

Figure 34.6D shows the major global air movements, called the prevailing winds. **Prevailing winds** result from the combined effects of the rising and falling of air masses (red arrows on the left side of the globe) and Earth's rotation (gray arrows). Because Earth is spherical, its surface moves faster at the equator than at other latitudes. In the tropics, Earth's rapidly moving surface deflects vertically circulating air, making the trade winds blow from the northeast in the Northern Hemisphere and from the southeast in the Southern Hemisphere. In temperate zones, the slower-moving surface produces the **westerlies,** winds that blow from west to east.

A combination of the prevailing winds, the planet's rotation, unequal heating of surface waters, and the locations and shapes of the continents creates **ocean currents,** riverlike flow patterns in the oceans. Figure 34.6E depicts the major ocean currents. Red arrows indicate warm currents; blue arrows, cold currents. Ocean currents have a profound effect on regional climates. For instance, the Gulf Stream circulates warm water northward from the Gulf of Mexico and makes the climate on the west coast of Great Britain warmer than many areas farther south. In contrast, the northward flowing Humboldt Current cools the west coast of South America.

Landforms can also affect local climate. Figure 34.6F presents one example, the effect of mountains on rainfall. This drawing represents major landforms across the state of Washington, but mountain ranges cause similar effects elsewhere. Washington is a temperate area in which the prevailing winds are westerlies. As moist air moves in off the Pacific Ocean and encounters the westernmost mountains (the Coast Range), it flows upward, cools at higher altitudes, and drops a large amount of water. The biological community in this wet region is a temperate rain forest. Some of the world's tallest trees, the Douglas firs, thrive here (see introduction to Chapter 1). Farther inland, precipitation increases again as the air moves up and over higher mountains (the Cascade Range). On the eastern side of the Cascades, there is little precipitation; as a result of this rain shadow, much of central Washington is virtually a desert.

Rain forests and deserts are among the world's major biological communities. Just as rain forests appear where there is abundant precipitation, and deserts where dry air descends over land, the appearance of other types of biological communities can be explained by regional climates. We'll see this clearly when we survey the biosphere's major terrestrial ecosystems. First, let's take a brief look at aquatic ecosystems. 

? What causes summer in the Northern Hemisphere?

Because of the fixed angle of Earth's polar axis relative to the orbital plane around the sun, the Northern Hemisphere is tilted toward the sun during the portion of the annual orbit that corresponds to the summer months.

34.7 Oceans occupy most of Earth's surface

Figure 34.7A An estuary in Maryland

Life originated in the sea and evolved there for almost 3 billion years before plants and animals began moving onto land. Covering about 75% of the planet's surface, oceans have always had an enormous impact on the biosphere. Their evaporation provides most of Earth's rainfall, and ocean temperatures have a major effect on climate and wind patterns. Photosynthesis by marine algae supplies a substantial portion of the biosphere's oxygen.

Figure 34.7A shows an **estuary,** an area where fresh water merges with seawater. This particular estuary is part of the Chesapeake Bay in Maryland. Several freshwater rivers empty into this bay, which merges with the Atlantic Ocean near Norfolk, Virginia. The saltiness of estuaries ranges from nearly that of fresh water to that of the ocean. Estuaries are among the most productive environments on Earth. Oysters, crabs, and many fishes live in estuaries or reproduce in them.

The shallow zone at the edge of an estuary or ocean where estuarine water or seawater meets land is called the **intertidal zone.** This area is often flooded by high tides and then left dry during low tides, about every 12 hours. Intertidal zones vary from salt marshes (the green area next to the mudflats in Figure 34.7A) to wave-splashed sandy or pebbly beaches. All of the intertidal zone is a type of **wetland,** an ecosystem that is intermediate between an aquatic ecosystem and a terrestrial one. Most wetlands have soil that is saturated with water, either permanently or periodically. The flow of water through a wetland, as well as the frequency, duration, depth, and season of flooding, determine the types of organisms that live there. The highly productive wetlands of the intertidal zone are home to many sedentary organisms such as algae, barnacles, mussels, sea stars, and sea anemones. Intertidal organisms attach to rocks or vegetation, or burrow into mud or sand, and are thus prevented from being washed away. Figure 34.7B shows some of the diverse organisms in a tidepool on the coast of central California. A tidepool is a small body of water that remains in a rock or sand depression in the intertidal zone during low tide.

The intertidal zone is one of several oceanic zones (Figure 34.7C, at the top of the facing page). Abiotic conditions often dictate the kinds of communities these zones support. The ocean water itself, called the **pelagic zone** (Greek *pelagos,* sea), supports communities dominated by highly motile animals such as fishes, squids, and marine mammals, including whales and dolphins. Diverse algae and cyanobacteria, collectively called **phytoplankton** (Greek *phyton,* plant, and *plankton,* wandering), drift passively in the pelagic zone. Phytoplankton are the ocean's main photosynthesizers, making most of the organic food molecules on which other ocean-dwellers depend. **Zooplankton** are animals that drift in the pelagic zone either because they are too small to resist ocean currents or because they don't swim. Zooplankton eat phytoplankton and, in turn, are consumed by other animals, including fishes. The seafloor is called the **benthic zone** (Greek *benthos,* "depth of the sea"). Depending on depth and light penetration, the benthic community consists of attached algae, fungi, bacteria, sponges, burrowing worms, sea anemones, clams, crabs, and fishes.

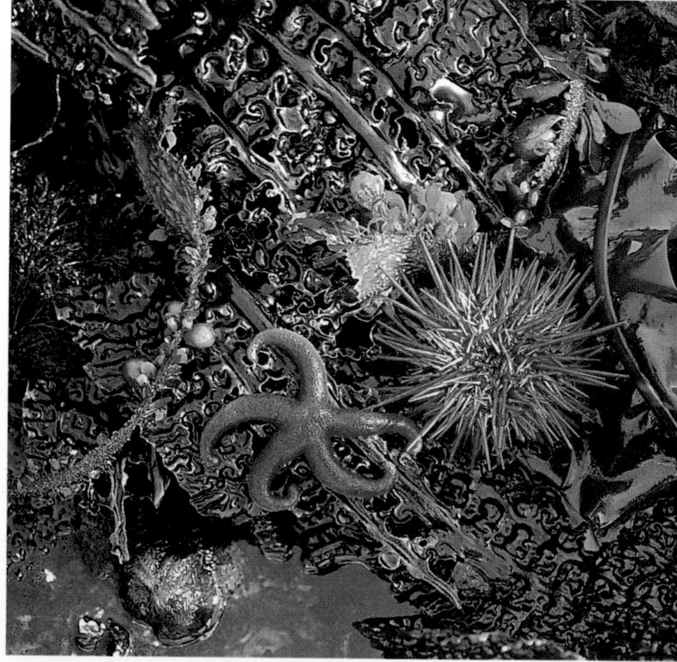

Figure 34.7B Intertidal zone organisms

The availability of light and inorganic nutrients has a huge impact on the distribution of organisms in the ocean. Water and particles suspended in it absorb light, and sunlight does not penetrate very far into the sea. On the submerged parts of continents called continental shelves, the pelagic and benthic communities usually receive some light, and nutrients from the seafloor circulate in the shallow water. Estuaries and coral reefs (described in the introduction to Chapter 18), harboring a great diversity of species, are found on continental shelves. Coral reefs are also common around oceanic islands. And some islands, called atolls, consist of a coral reef surrounding a lagoon (Figure 34.7D).

As Figure 34.7C indicates, marine biologists often group the illuminated regions of the benthic and pelagic communities together, calling them the photic zone. The **photic zone** is a relatively small portion of ocean water and bottom into which light penetrates and in which photosynthesis occurs. Underlying the photic zone is a vast, dark region called the **aphotic zone.** This is the most extensive part of the biosphere. Without light, there are no photosynthetic organisms, but life is still diverse in the aphotic zone. Many kinds of invertebrates, such as sea urchins and polychaete worms, and some fishes scavenge organic matter that sinks from the lighted waters above. The hydrothermal vent communities, powered by chemical energy rather than sunlight, are densely populated, as we discussed in the chapter's introduction.

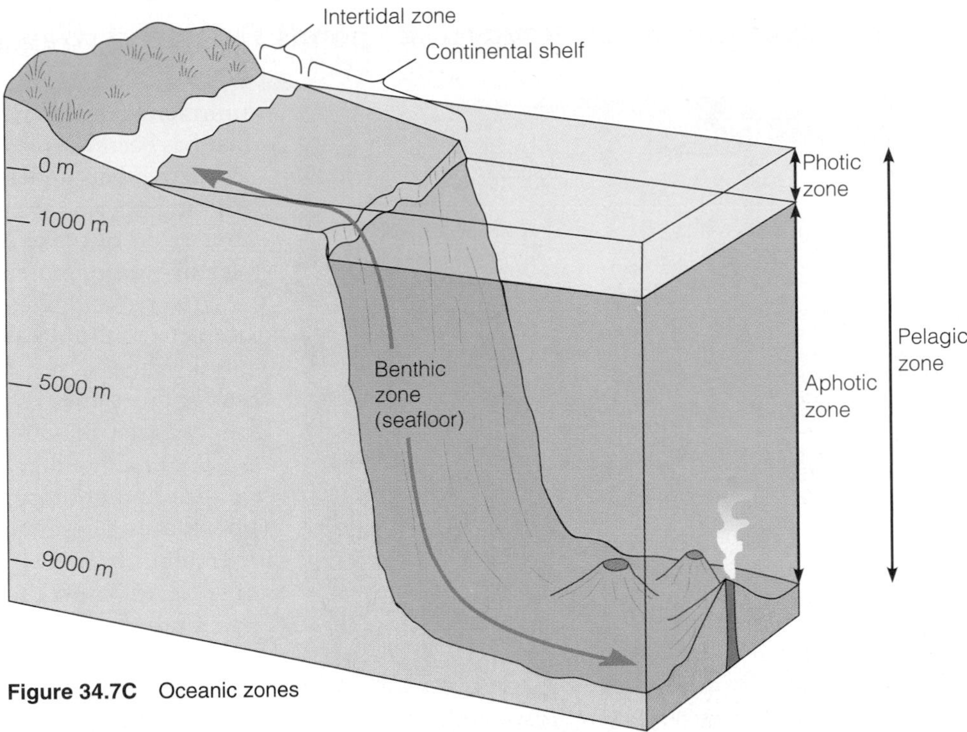

Figure 34.7C Oceanic zones

Figure 34.7D A coral atoll in the South Pacific

Until fairly recently, many people viewed the ocean as a bountiful, virtually limitless resource, and we have harvested the ocean heavily and used it as a dumping ground for wastes. Estuaries and intertidal wetlands have been especially abused, with few undisturbed areas remaining, and many totally replaced by commercial and residential developments on landfill. We are now seeing the effects of our disregard for marine communities, as seafood is becoming less plentiful, the result of overharvesting and pollution; whales are in danger of extinction, mainly from overhunting; and oil and other pollutants foul coastal areas. Laws in many countries, including the U.S., now prohibit whaling and the disposal of sewage and other wastes at sea. Many countries are also taking steps to restore and conserve estuaries and other wetlands. 

? The ________, small photosynthetic organisms inhabiting the ________ zone of the pelagic zone, provide most of the food for oceanic life.

phytoplankton . . . photic

34.8 Freshwater communities occupy lakes, ponds, rivers, and streams

Figure 34.8A A freshwater wetland in Pennsylvania

Figure 34.8B A stream in Georgia

The major difference between seawater and fresh water is salinity. As discussed in Module 25.5, the scarcity of dissolved ions in fresh water has a profound effect on organisms living there. Light also has a significant impact on the communities in freshwater ecosystems, as it does on ocean communities.

In all but the smallest lake or pond, there is usually a distinct photic (lighted) zone and an aphotic zone. Phytoplankton grow in the photic zone, and rooted plants often inhabit shallow waters, as in the marshy wetland in Figure 34.8A. Large populations of microorganisms in the benthic (bottom-dwelling) community decompose dead organisms that sink to the bottom. Respiration by microbes also removes oxygen from water near the bottom, and in some lakes, benthic areas are unsuitable for any organisms except anaerobes.

Temperature also has a profound effect on freshwater communities, especially in temperate areas. During the summer, lakes often have a distinct upper layer of water that has been warmed by the sun. The warm-water layer is less dense than underlying, cooler water and does not mix with it. Fishes often spend much of their time in the deep, cool waters of a lake, unless oxygen levels there become depleted by decomposers.

Nitrogen and phosphorus are the nutrients that usually limit the amount of phytoplankton growth in a lake or pond. When there are temperature layers in a lake, for instance, nutrients released by decomposers can become trapped near the bottom, out of reach of the phytoplankton. During the summer months, this may limit the growth of algae and photosynthesis in the photic zone. As winter approaches, the surface water becomes denser as it cools; it then tends to mix with the deeper water, allowing nutrients to return to the surface where phytoplankton can again use them. Seasonal mixing also restores oxygen to the depths.

Today, many lakes and ponds are affected by large inputs of nitrogen and phosphorus from sewage and runoff from fertilized lawns and agricultural fields. These nutrients often produce blooms, or population explosions of algae. Heavy algal growth reduces light penetration into the water, and, when the algae die and decompose, a pond or lake can suffer serious oxygen depletion.

Rivers and streams generally support quite different communities of organisms than lakes and ponds. A river or a stream changes greatly between its source (perhaps a spring or snowmelt) and the point at which it empties into a lake or the ocean. Near a source, the water is usually cold, low in nutrients, and clear (Figure 34.8B). The channel is often narrow, with a swift current that does not allow much silt to accumulate on the bottom. The current also inhibits the growth of phytoplankton; most of the organisms found here are supported by the photosynthesis of algae attached to rocks or organic material (such as leaves) carried into the stream from the surrounding land. The most abundant benthic animals are usually insects that eat algae, leaves, or one another. Trout are often the predominant fishes, locating their food, including insects, mainly by sight in the clear water.

Downstream, a river or stream generally widens and slows. Marshes, ponds, and other wetlands are common in downstream areas. There, the water is usually warmer and may be murkier because of sediments and phytoplankton suspended in it. Worms and insects that burrow into mud are often abundant, as are waterfowl, frogs, and catfish and other fishes that find food more by scent and taste than by sight. 

? Why does sewage cause algal blooms in lakes?

The sewage adds nutrients, such as nitrates and phosphates that stimulate growth of the algae.

34.9 The major terrestrial ecosystems are called biomes

34A
See Activity 34A on the CD-ROM.

The map below introduces the nine major types of terrestrial ecosystems, called **biomes,** which cover the land surface of Earth. Many of the biomes are named for their predominant vegetation, but each is also characterized by animals adapted to that particular environment. A grassland, for instance, is more likely than a mountain biome to be populated by grazing animals such as the pronghorn.

The distribution of the biomes largely depends on climate, with temperature and rainfall often the key factors determining the kind of biome that exists in a particular region. If the climate in two geographically separate areas is similar, the same type of biome may occur in them; notice on the map that each biome type occurs on at least two continents. Each biome is characterized by a *type* of biological community, not a specific assemblage of species. For example, the groups of species living in the Sahara Desert of Africa and in the Gobi Desert of eastern Asia are different, but the species in both are adapted to desert conditions. Widely separated biomes may look alike because of convergence, the evolution of similar traits in independently evolved species living in similar environments (see Module 15.11).

Biomes tend to grade into each other, and within each biome, there is local variation, giving the vegetation a patchy, rather than a uniform, appearance. For example, in northern coniferous forests, snowfall may break branches and small trees, causing openings where deciduous trees such as aspen and birch can grow. Local storms and fires also create openings in many biomes.

Fire has a very important effect in some biomes. Without periodic burning, many grasslands would be replaced by forests. Most grasses survive burning, because the growing points of their shoots are below ground and are not killed. Some species of pines that dominate coniferous forests produce cones that release their seeds only when subjected to extreme heat. Germinating soon after a forest fire has blackened an area, they have access to open space and a rich supply of nutrients from burned plants.

Today, in many areas, the natural biomes are broken up by human activity. In fact, as we discuss in Chapter 38, a high rate of biome alteration by humans is correlated with an unusually high rate of species loss throughout the globe.

We now begin a more detailed survey of the major biomes. To help you locate the biomes, we include with each module an orientation map that is color-coded to match the map here.

Here's a chance to apply your knowledge of world geography: Which biome is most closely associated with a "Mediterranean climate"?

Chaparral

Figure 34.9

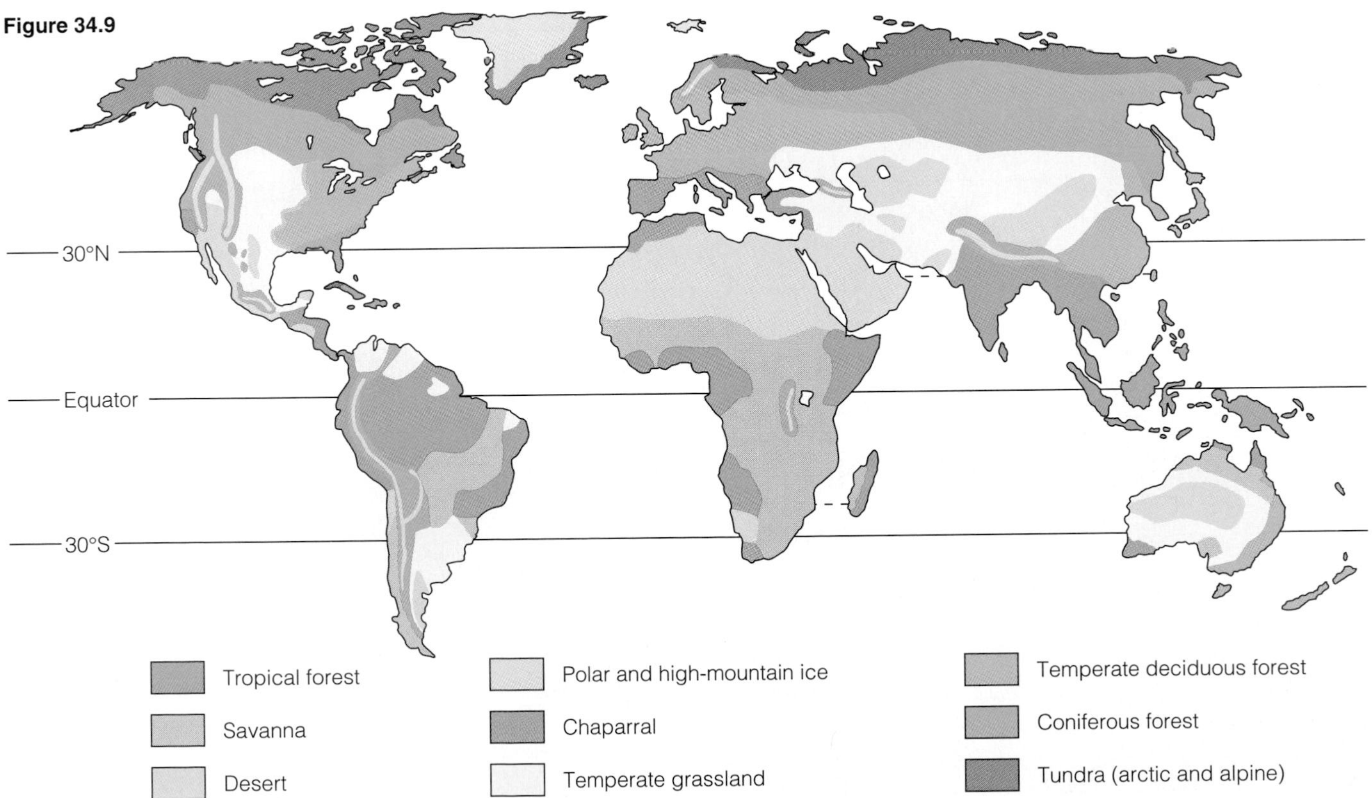

34.10 Tropical forests cluster near the equator

34A

Tropical forests occur in equatorial areas where the temperature is warm and days are 11–12 hours long year-round. Rainfall in these areas is quite variable, and this, rather than temperature or photoperiod, generally determines the kind of vegetation that grows in a particular kind of tropical forest. There are three major types of tropical forests. **Tropical thorn forests** are common in equatorial lowlands, such as parts of eastern Africa and northwestern India, where rainfall is scarce. These areas have prolonged dry seasons, and the plants found there are a mixture of thorny shrubs and trees, and nonwoody plants that retain water for long periods. In areas with distinct wet and dry seasons, such as central West Africa and much of India and Southeast Asia, **tropical deciduous forests** dominate. Tropical deciduous trees and shrubs drop their leaves during the long dry season and releaf only during the following heavy rains or monsoons. **Tropical rain forests** are found in very humid equatorial areas, such as Indonesia and the Amazon River basin in South America, where rainfall is abundant (greater than 250 cm per year), and a season of reduced rainfall lasts no more than a few months.

The tropical rain forest, such as the luxuriant area in Costa Rica shown below, is among the most complex of all biomes, harboring enormous numbers of different species. Up to 300 species of trees, many of them evergreen angiosperms 50–60 m tall, can be found in a single hectare (2.5 acres). Because of the density of large trees, the tropical rain forest often has a closed canopy, with little light reaching the forest floor. Where an opening does occur, perhaps because of a fallen tree, other trees and large, woody vines known as lianas grow rapidly. Many of the animals that live in tropical rain forests are tree-dwellers; monkeys, birds, insects, snakes, bats, and frogs find food and shelter many meters above the ground. The soils of tropical rain forests are typically poor, because high temperatures and rainfall lead to rapid decomposition and recycling rather than to a buildup of organic material. At any given time, almost all of the nutrients are incorporated in living organisms.

Human impact on the world's tropical rain forests is currently a source of great concern. It is a common practice to clear the forest for lumber or simply burn it, farm the land for a few years, and then abandon it. Mining has also devastated large tracts of rain forest. Once stripped, the tropical rain forest recovers very slowly because the soil is so nutrient-poor. The destruction of these forests is now proceeding at an alarming rate. We discuss the potential consequences of this destruction in Modules 38.1, 38.2, and 38.5. The devastation of tropical rain forests may also cause large-scale changes in world climate (Module 38.4). With international financial support, a number of countries, including Costa Rica, Belize, Mexico, Venezuela, and Brazil, have begun preserving some of their rain forests. In the next module, we hear from a scientist who is in the thick of the rain forest conflict.

Why are the soils in most tropical rain forests so poor in nutrients that they can only support farming for a few years after the forest is cleared?

The tropical conditions favor rapid decomposition of organic litter in the forest soil, and most of the ecosystem's nutrients are tied up in the vegetation that is cleared away for farming.

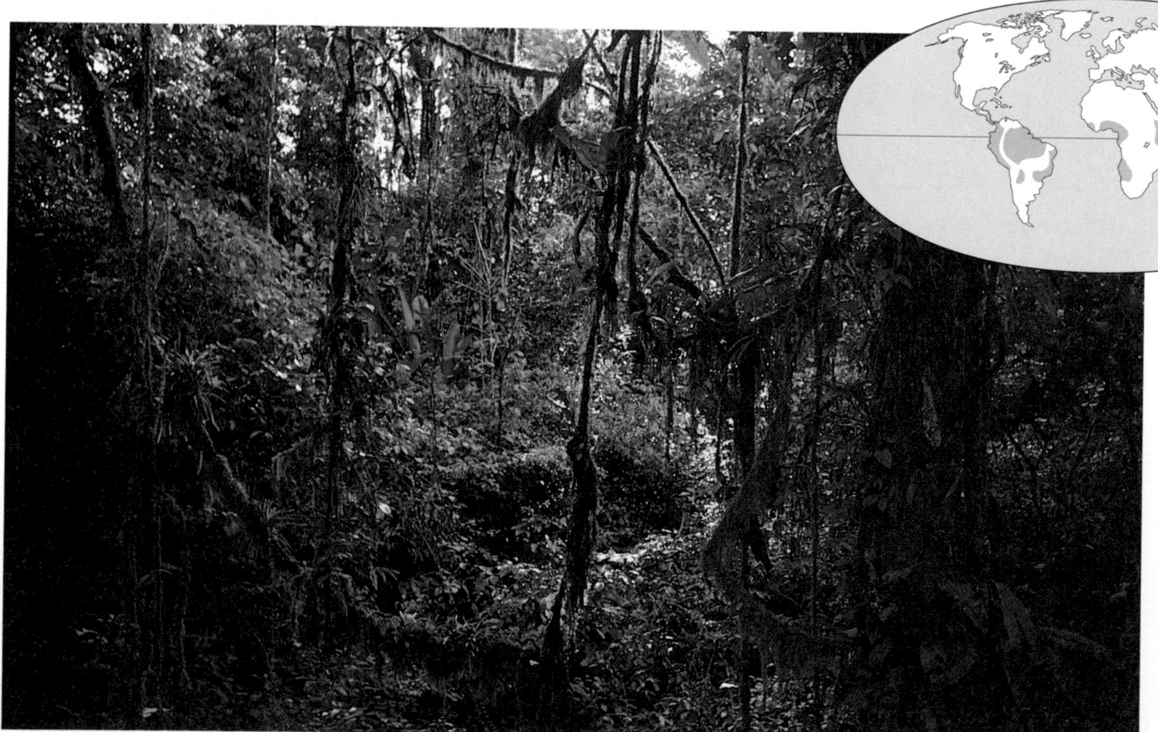

Figure 34.10
Tropical rain forest

34.11 U.S. Forest Service scientist Ariel Lugo shares his views on tropical forest issues

Figure 34.11 Ariel Lugo in the Caribbean National Forest

Dr. Ariel Lugo is a forest ecologist with a firm grasp of the problems and controversies of tropical forests. After an undergraduate education and a master's degree at the University of Puerto Rico, he earned his Ph.D. at the University of North Carolina and served as a biology professor at the University of Florida. Dr. Lugo is a research scientist and director of the U.S. Forest Service's Institute of Tropical Forestry in his native Puerto Rico. Dr. Lugo understands the global community's concern about the current rate of destruction of tropical forests and the loss of species in them. He tempers these concerns with those of local people who inhabit and use tropical forests. We asked Dr. Lugo to summarize the situation as he sees it.

> Each sector of society sees the forest in a different way. First of all, millions of people live in tropical forests, and they clearly have a plan for the forest that provides them with their survival. But a conservationist who is flying in an airplane over the forest sees all that green and says, "Ah, here's where I'm going to protect the biodiversity of the world. We're not going to touch this forest, because this is going to balance what we've been doing to the rest of the world." So for them, you want to leave the tropical forest alone. But if in the plane you have a timber baron, that person sees in this piece of forest one of the last of the Earth's untouched stores of timber. And you could say, if you were a timber baron, "Ah, I want to make money by cutting those trees and selling them." But there might be other people with different plans—the government, for example. The government sees the vast expanses of tropical forest as the future of the country. The United States made its future by cutting its forests. For the government, it's an area to be developed; the forest provides space for new cities for new economic development.

What do you see as the main reasons for tropical forest destruction?

> Lumber is really a minor factor in deforestation. The major [one] is people have to cut the forest down and plant their crops. The main reason for deforestation is to satisfy human needs for food. We have to admit that sometimes you need to deforest—we cannot be such purists as to say that we can't cut any forest.

Often, when tropical forests are cut, the land is farmed for only a short time before it becomes infertile and is abandoned. Is there an alternative to this destructive practice?

> To me, it is critical that when we convert land from one use to another, we make sure that the use we're after is sustainable. Sustainable productivity means that you benefit from what the land can produce without being so greedy as to wipe out this productivity in the process. How can we manage ecosystems within our own generation, get the benefits of the production, and then pass these ecosystems on with the same capacity so that the next generation is not handicapped by a reduction of productivity? The level to which we're going to achieve this depends on culture. I think different cultures have different expectations, different needs, and so some cultures push nature more to the limit than others.

Are tropical rain forests affected much by natural disturbances?

> Hurricanes are huge disturbances that can take out all the leaves in large areas of tropical forests. But when we studied hurricane damage in the Luquillo Experimental Forest in Puerto Rico, we found that hurricanes are the main organizing force of that forest. The forest goes through a cycle that averages sixty years, starting with great impact by winds and rain of a hurricane and then about sixty years of regrowth. In those sixty years, we see the species change, the growth rates change, the size of the trees change, and the density of the wood change. This is teaching us how the forest can incorporate the energy of the hurricane. In other words, the hurricane might appear destructive but it's actually constructive; it makes the forest more productive; it rejuvenates the forest.
>
> Now you ask me, how does this impact of natural disturbance compare with human impact on the rain forest? Well, there's no comparison! Life can adapt to these regular disturbances. For example, animal populations have actually exploded after a hurricane, which means they have behavioral adaptations, they have population dynamics that allow them to respond in a positive way. But human activity is unpredictable because humans act on impulse, in the short term, or without reason. So it's much harder for nature to adapt to human activity.

What are the consequences of pushing tropical rain forests to the limit?

> When deforestation occurs in an unplanned, uncontrolled way, then you lose species and productivity and degrade your soil and water resources. These regional effects fragment the landscape. And when you fragment the landscape, you disconnect nature—you lose the value of ecosystems working in synchrony. When you add up all the deforestation, you get into the global impact. Tropical forests are so huge that they help regulate climate, they help regulate the cycles of nutrients and water and gases, and so on. The main thing I want to tell you about the consequences of deforestation is just the waste. If we just wantonly destroy, then we are wasting a resource that could be used to improve the well-being of people.

What needs to be done?

> The challenge is to harmonize or integrate all the legitimate claims on the forest. Now, who's going to put it all together? Who's going to organize development such that you have indigenous people, and you have conservation of the resource, and you maintain climatic balances, and you develop some areas, and you have agriculture, and you get fuel wood? That requires management in my view, and it requires a lot of compromise, and a lot of hard thinking, and a lot of good will and political will. 

? Why can widespread destruction of tropical forests affect ecosystems far from the forests?

Because tropical forests help regulate climate and chemical cycling on a global scale

34.12 Savannas are grasslands with scattered trees

This photograph, taken in Kenya, shows a typical **savanna,** a biome dominated by grasses and scattered trees. Extensive savannas (from the Spanish *sabana,* meadow) cover wide areas of the tropics in central South America, central and South Africa, and parts of Australia. Scattered savannas also occur in temperate North America (roughly in a band from Minnesota to eastern Texas), where the temperate forests of the eastern states merge with the grasslands of the west.

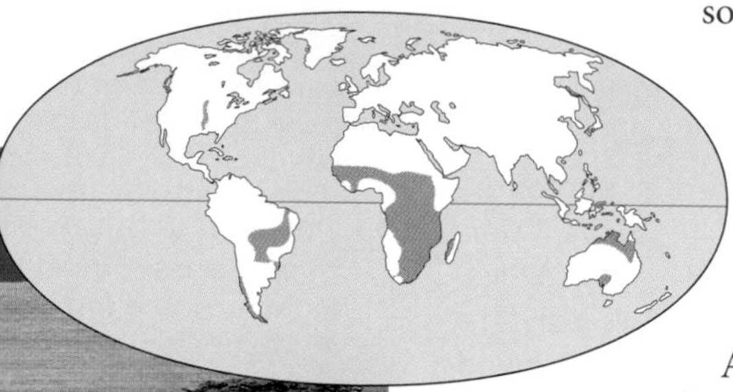

Savannas are simple in structure compared to tropical forests. Frequent fires, caused by lightning or human activity, and grazing animals inhibit further invasion by trees. Fires and grazers also maintain the small-growth form of grasses and nonwoody, broadleaf dicot plants that grow with them. Grasses are wind-pollinated, but the dicots produce showy flowers that attract insect pollinators, often in great numbers, in the summer.

Grasses grow rapidly, providing a good food source for many species of animals. The savanna is home to many of the world's large herbivores and their predators. African savannas are home to giraffes, zebras, and many species of antelope, as well as to baboons, lions, and cheetahs. Several species of kangaroos are the dominant herbivores of Australian savannas. Throughout much of our North American savannas, farms replace areas once inhabited by bison, deer, black bear, coyotes, and wolves.

Also common on savannas are many burrowing animals, including mice, moles, gophers, snakes, ground squirrels, worms, and numerous arthropods.

Figure 34.12 Savanna

How do fires help maintain savannas as grassland ecosystems?

By repeatedly preventing the spread of trees and other woody plants

34.13 Deserts are defined by their dryness

34A

Daytime temperatures in **deserts** sometimes get as high as 54°C, and nights often drop to freezing. However, it is sparse rainfall (less than 30 cm per year) and (typically) rapid evaporation, rather than extreme temperatures, that make an area a desert. The driest deserts are in central Australia and the central Sahara in Africa, where the average annual rainfall is less than 2 cm, there is no rain at all during some years, and evaporation greatly exceeds precipitation. But not all desert air is dry. The Namib Desert along the southwestern coast of Africa is often shrouded in fog, although there is little rainfall, and the ground remains extremely dry.

As we discussed in Module 34.6, large tracts of desert occur in two regions of descending dry air, centered around the 30° north and 30° south latitudes. In the Southern Hemisphere, these include the Kalahari in Africa and much of central Australia; in the Northern Hemisphere, they include the Sahara, the Arabian Desert, large areas of Mexico, and much of the southwestern United States. The photograph in Figure 34.13 (facing page) was taken in the Sonoran Desert in southern Arizona. Large deserts also occur at higher latitudes, in the rain shadows of mountains; these include much of central Asia east of the Caucasus Mountains, southern Argentina east of the Andes, and much of California and Nevada east of the Sierra Nevada.

Unlike other biomes, many deserts are actually growing in size. The process of **desertification,** the conversion of other biomes—especially savannas—to deserts, is a significant environmental problem. In central Africa, for example, a burgeoning human population, overgrazing, and dry land farming are converting large areas of savanna to desert.

The cycles of growth and reproduction in the desert are keyed to rainfall. The driest deserts have no perennial vegetation at all, but in less arid regions the dominant plants are scattered shrubs, often interspersed with cacti, as in the Sonoran Desert. Desert plants typically produce great numbers of seeds, which may remain dormant until a heavy rain triggers germination. Periods of rainfall (often in late winter) may produce spectacular blooms of annual plants.

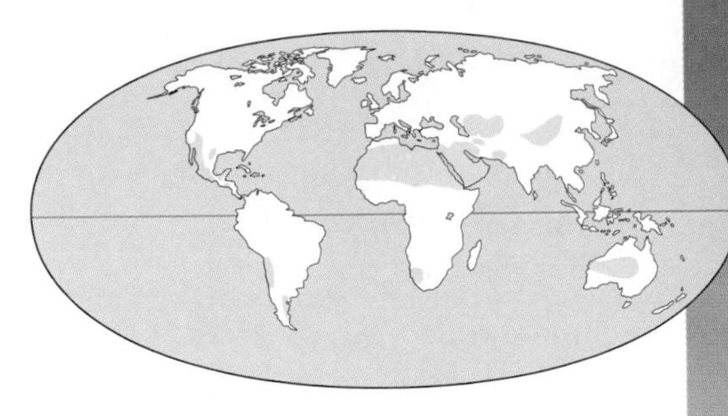

Figure 34.13
Desert

Like desert plants, desert animals are adapted to drought and extreme temperatures. Many live in burrows and are active only during the cooler nights, and most have special adaptations that conserve water. Seed-eaters such as ants, many birds, and rodents are common in deserts. Lizards, snakes, and hawks eat the seed-eaters.

Why isn't "cold desert" an oxymoron?

Because deserts are defined by relatively little precipitation and dry soil, not mainly by temperature

Spiny shrubs dominate the chaparral 34.14

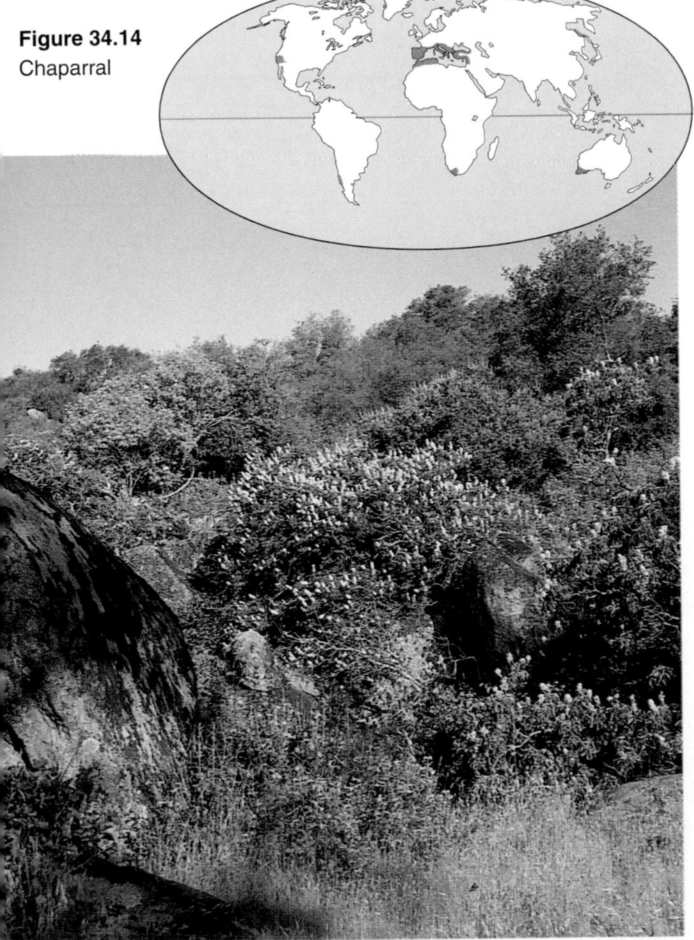

Figure 34.14
Chaparral

Chaparral (the Spanish word for "place of evergreen scrub oaks") is a region of dense, spiny shrubs with tough, evergreen leaves. The climate of chaparral areas results mainly from cool ocean currents circulating offshore, which usually produce mild, rainy winters and long, hot, dry summers.

First described in the Mediterranean region, chaparral vegetation is also found in coastal areas of Chile, southwestern Africa, southwestern Australia, and California. This photograph was taken in chaparral just south of San Francisco. In addition to the perennial shrubs that dominate chaparral, annual plants are also commonly seen, especially during the wet winter and spring months.

Chaparral vegetation is adapted to periodic fires, most often caused by lightning; in fact, the vegetation requires occasional fires for long-term maintenance. The shoot tips of many of the shrubs are adapted to fire and do not burn, and the plants regenerate quickly, using nutrients released by the fires. In addition, many chaparral plant species produce seeds that will germinate only after a hot fire. Others reproduce asexually without reliance on seeds.

Animals characteristic of the chaparral are browsers such as deer, fruit-eating birds, and seed-eating rodents, as well as lizards and snakes.

What is the main reason that homeowners' insurance is relatively expensive for people who chose to build in the chaparral?

High fire risk

34.15 Temperate grasslands include the North American prairie

34A

Temperate grasslands have some of the characteristics of tropical savannas, but they are mostly treeless, except along rivers or streams, and are found in regions of relatively cold winter temperatures. Temperate grasslands include the areas known as pampas in Argentina and Uruguay, steppes in Asia, and prairies in central North America. The keys to the persistence of most grasslands are seasonal drought, fires, and grazing by large mammals, all of which inhibit growth of woody plants but do not harm the belowground grass shoots.

Grasslands expanded in range following the retreat of the glaciers after the last ice age. Coupled with this expansion was the proliferation of large grazing mammals. The bison and pronghorn of North America, the gazelles and zebras of the African veldt, and the wild horses and sheep of the Asian steppes are some examples. Enriched by glacial deposits, the soil of grasslands supports a great diversity of microorganisms and animals, especially annelids, arthropods, and burrowing mammals.

The amount of annual rainfall influences the height of grassland vegetation. The photograph at the left was taken on the relatively dry, short-grass prairie of western South Dakota. Tall-grass prairie occurs in wetter areas, such as eastern Kansas. Whereas the vegetation of short-grass prairie is usually shorter than 1 m, tall-grass vegetation often attains heights of over 2 m. In some of the more arid grassland areas of the American West and Eurasia, a mix of short grasses and sagebrush predominates. Little remains of North American prairies today. Most of the region is intensively farmed, and it is one of the most productive agricultural regions in the world.

Figure 34.15 Temperate grassland (short-grass prairie)

? How do humans now use most of the North American land that was once temperate grasslands?

For farming

34.16 Deciduous trees dominate temperate forests

34A

Temperate deciduous forests grow in latitudes between about 35° and 50°, regions where there is sufficient moisture to support the growth of large trees. This includes most of the eastern United States, most of central Europe, and parts of eastern Asia and Australia. Broadleaf, deciduous trees characterize temperate deciduous forests. Some of the dominant trees include species of oak, hickory, beech, and maple. The mix of tree species varies widely, depending on such factors as climate in different latitudes and local soil conditions. This photograph was taken in the autumn in Great Smoky Mountains National Park in North Carolina.

Figure 34.16 Temperate deciduous forest in North Carolina

Temperatures in temperate deciduous forests range from very cold in the winter to hot in the summer (−30°C in higher latitudes to +30°C in the lower latitudes). Precipitation is relatively high and usually evenly distributed throughout the year. These forests usually have a growing season of 5–6 months and a distinct annual rhythm, in which the trees drop leaves and become dormant in late autumn, then produce new leaves each spring. The loss of leaves in winter prevents evaporation of water from the leaves at a time when freezing reduces the available water.

Temperate deciduous forests are more open than tropical rain forests and are not as tall or as diverse. However, their soils are rich in inorganic and organic nutrients, and these forests support a great variety of species. Many arthropods, for example, live in the soil or leaf litter or feed on the leaves and branches of small trees and shrubs. Temperate deciduous forest is also home to many species of birds, small mammals, and, where not eliminated by humans, bobcats, foxes, black bears, and mountain lions.

Virtually all the original deciduous forests in North America were destroyed by logging and by clearing land for agriculture and urban development. In contrast to drier ecosystems, these forests tend to recover after disturbance, and today we see deciduous trees dominating undeveloped areas over much of their former range. 

? How does the soil of a temperate deciduous forest differ from that of a tropical rain forest?

The soil in temperate deciduous forests is rich in inorganic and organic nutrients.

Coniferous forests are often dominated by a few species of trees 34.17

Coniferous forests are biomes in which the principal trees are cone-bearers, such as spruce, pine, fir, and hemlock. Many coniferous forests are dominated by just a few species of trees. Before agricultural and urban development, much of the southeastern U.S. resembled the scene at the right in Figure 34.17. The tall trees are longleaf pines, all one species. Fire is essential to the longleaf coniferous forest and to many others. Without periodic burning, deciduous trees will frequently replace the conifers.

By far the most extensive coniferous forest is the **taiga,** the northern (boreal) coniferous forest. It stretches in a broad band across North America and Eurasia, reaching the southern border of the arctic tundra. Taiga (from the Russian word for mountain) is also found on mountainsides in more temperate latitudes.

The taiga is characterized by harsh winters and short summers that are sometimes warm. The soil is thin and acidic. It forms slowly because of the low temperatures and the waxy covering of conifer needles, which decompose slowly. There may be considerable precipitation, mostly in the form of snow. The snow usually falls before the coldest temperatures occur, and it insulates the soil, keeping it from freezing to such depths that it would never thaw out during the short taiga summers. In addition, heavy snow breaks limbs and fells some trees, creating openings in the taiga. Clusters of deciduous trees, such as birch, willow, aspen, and alder, often grow in these openings. During summer, fires caused by lightning may burn extensive areas of taiga. Conifers with fire-resistant cones and seeds frequently colonize recently burned areas.

Coniferous forests of coastal North America (Northern California to Alaska) are actually temperate rain forests. Warm, moist air from the Pacific Ocean supports this unique biome, which like most coniferous forests is dominated by a few tree species. In some extremely wet areas, the dominant species include broadleaf trees as well as conifers. We highlighted temperate rain forests in the introduction to Chapter 1. 

Figure 34.17 Coniferous forest

34A

? Why would you find less diversity of insects that pollinate tree flowers in taiga than you would find in deciduous forests?

Because taiga is dominated by cone-bearing trees (conifers)—not flowering trees (angiosperms), which characterize deciduous forests.

34.18 Long, bitter-cold winters characterize the tundra

34A

At the northernmost limits of plant growth, and at high altitudes just below areas covered permanently with ice and snow, is the **tundra** (from the Russian word for "marshy plain"). Plant forms in the tundra include dwarf woody shrubs, grasses, mosses, and lichens. The arctic tundra encircles the North Pole, extending southward to the coniferous forests. Alpine tundras are found above the treeline on high mountains, even in the tropics. For example, tundra is found in the Andes Mountains in Ecuador, where the elevation is high enough to produce a very cold climate.

This photograph shows the arctic tundra in central Alaska in the autumn. The climate here is often extremely cold, with little light for long periods of time. During the brief, warm summers, when there is nearly constant daylight, plants grow quickly and flower in a rapid burst. The arctic tundra is characterized by **permafrost,** continuously frozen ground. Permafrost underlies about 80% of Alaska and almost half of Canada, Scandinavia, and Russia. The depth of the permafrost ranges from a few meters to nearly 1500 m (5000 ft) in northern Siberia. Only the upper part of tundra soil, from a few centimeters to several meters deep, thaws in the summer, and then only for a brief time. The permafrost prevents the roots of plants from penetrating very far into the soil. Extremely cold winter air temperatures and permafrost explain the absence of trees. The arctic tundra may receive as little precipitation as some deserts. But poor drainage, due to the permafrost, and slow evaporation, because of the low temperatures, keep the soil continually saturated.

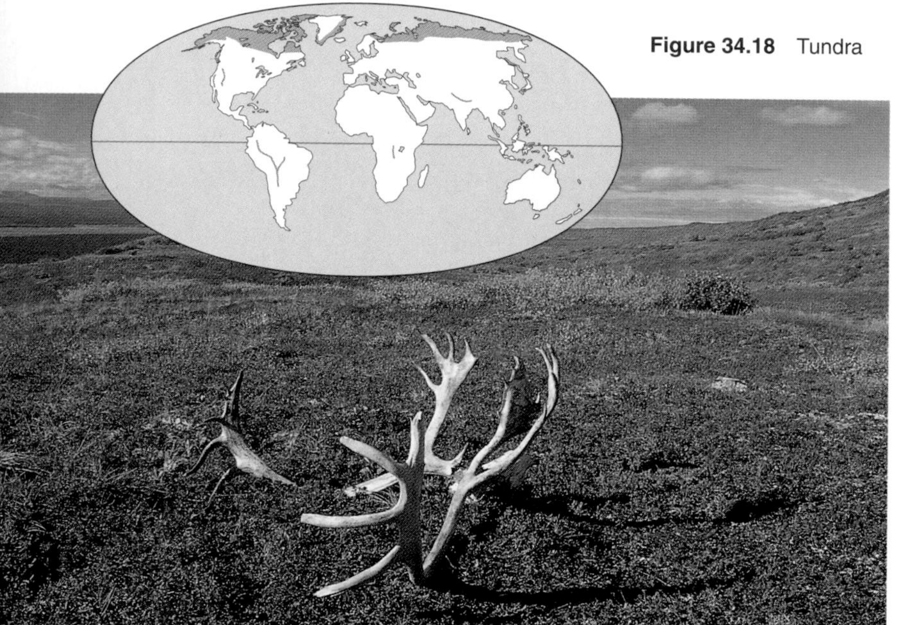

Figure 34.18 Tundra

Animals of the tundra withstand the cold by having good insulation that retains heat. Large herbivores of the tundra include musk oxen and caribou. The principal smaller animals are rodents called lemmings and a few predators such as the arctic fox and snowy owl. Many animals, especially birds, are migratory, using the tundra as a summer breeding ground. During the brief growing season, clouds of mosquitoes often fill the tundra air. Populations of many tundra insects expand rapidly during the summer, and then decline abruptly when the warm season ends. We'll learn more about the ups and downs of biological populations in Chapter 35. 

? What two abiotic factors account for the rarity of trees in arctic tundra?

Long, very cold winters (short growing season) and permafrost

Chapter Review

Chapter Summary

Ecology is the scientific study of the interactions of organisms with their environments. Most organisms are solar-powered. Those living where there is no light are powered by energy from Earth's interior **(Introduction).** Ecologists study environmental interactions at the organism, population, community, and ecosystem levels. Ecosystem interactions involve living (biotic) communities and nonliving (abiotic) factors, such as energy, nutrients, and water **(34.1).**

The Biosphere (34.2–34.6) The global ecosystem is called the biosphere: all life on Earth and where it lives. Except for energy obtained from the sun and heat lost to space, the biosphere is self-contained. Patchiness characterizes the biosphere, with each habitat, or ecological situation in which organisms live, having a unique community of species **(34.2).** Human activities now affect all parts of the biosphere **(34.3).** Solar energy, water, temperature, soil, oxygen, fire, and wind are among the most important abiotic factors determining the biosphere's structure and dynamics **(34.4).** Natural selection adapts organisms to abiotic factors and to biotic ones such as predation and competition **(34.5).**

Climate often determines the distribution of communities. Most climatic variations are due to the uneven heating of Earth's surface. Warm, moist, rising air produces rain, which waters the tropical rain forests. Descending dry air creates deserts at about 30° north and south of the equator. Earth's rotation deflects moving air, creating the prevailing winds. The winds, the planet's rotation, and the arrangement of the continents direct ocean currents that warm or cool coastal areas. Landforms such as mountains can affect local climate **(34.6).**

Aquatic Ecosystems (34.7–34.8) Oceans cover about 75% of Earth's surface. Light and the availability of nutrients are the major factors shaping communities in the sea. Ocean water is called the pelagic zone, and the ocean bottom is called the benthic zone; the area where the sea meets the land is the intertidal zone. Intertidal wetlands, areas permanently or periodically saturated with water, include estuaries, salt marshes, and tidepools (34.7). Freshwater ecosystems include lakes, rivers, streams, and wetlands such as marshes and ponds. Light, temperature, and the availability of nutrients and dissolved oxygen shape lake and pond communities. Temperature, nutrients, currents, and water clarity vary from the source of a river to its mouth, and river communities vary accordingly (34.8).

Terrestrial Ecosystems (34.9–34.18) Climatic differences, mainly temperature and rainfall, shape the major types of ecosystems, called biomes, that cover Earth's land surface (34.9). Several kinds of tropical forests occur in the warm, moist belt along the equator. The tropical rain forest is the most diverse ecosystem on Earth. Large-scale human destruction of these forests endangers many species and may alter world climate (34.10). Forest management requires compromise among the often conflicting interests of private industry, agriculture, government, conservation groups, and indigenous people (34.11). Drier tropical areas and some nontropical areas are characterized by the savanna, a grassland with scattered trees. Grazing by large herbivores and fire help maintain savannas (34.12). Deserts are the driest biomes. The misuse of surrounding land is contributing to the growth of some deserts (34.13). The chaparral biome is a shrubland with cool, rainy winters and dry, hot summers, when fires often occur (34.14). Temperate grasslands are found in the interiors of the continents, where winters are cold. Drought, fires, and grazing animals prevent trees from growing. Farms have replaced most of North America's temperate grasslands (34.15). Forests of broadleaf deciduous trees grow in some temperate areas. North America's deciduous forests have been drastically altered by agriculture and urban development (34.16). Coniferous forests include pine forests, largely maintained by periodic fires, temperate rain forests, and the taiga, an extensive biome of the far north and high mountains. Taiga occurs where there are short summers and long, snowy winters (34.17). Arctic tundra, a treeless biome where it is too cold for large trees to survive, lies between the taiga and the permanently frozen polar regions. Alpine tundra occurs above the treeline on high mountains. The vegetation of the tundra includes shrubs, grasses, mosses, and lichens (34.18).

Testing Your Knowledge

Multiple Choice

1. Changes in the seasons are caused by
   a. the tilt of Earth's axis toward or away from the sun
   b. annual cycles of temperature and rainfall
   c. variation in the distance between Earth and the sun
   d. an annual cycle in the sun's energy output
   e. the periodic buildup of heat energy at the equator

2. What makes the Gobi Desert of Asia a desert?
   a. The growing season there is very short.
   b. Its vegetation is sparse.
   c. It is hot.
   d. Temperatures vary little from summer to winter.
   e. It is dry.

3. Andrea was a passenger on a plane that flew over temperate deciduous forest, then grassland and desert, finally landing at an airport in chaparral. The route of Andrea's flight was between
   a. New York and Denver
   b. Philadelphia and Los Angeles
   c. Denver and Los Angeles
   d. Washington, D.C. and Phoenix
   e. Seattle and Washington, D.C.

4. Which of the following sea creatures might be described as a pelagic animal of the aphotic zone?
   a. a coral reef fish
   b. a giant clam near a deep-sea hydrothermal vent
   c. an intertidal snail
   d. a deep-sea squid
   e. a harbor seal

Matching

1. Ground permanently frozen	a. tropical rain forest
2. Deciduous trees such as hickory and birch	b. savanna
3. Mediterranean climate	c. temperate forest
4. Spruce, fir, pine, and hemlock trees	d. temperate grassland
5. Home of zebras, baboons, and lions	e. chaparral
6. The steppes, pampas, and plains	f. tundra
7. The most complex and diverse biome	g. taiga

Describing, Comparing, and Explaining

1. Explain how the following factors change from the source of a river to its mouth: nutrient content, current, sediments, temperature, oxygen content, food sources.

2. Choose any animal or plant in your geographic area and write a paragraph describing how it is adapted to abiotic and biotic factors in its environment.

3. What climatic conditions allow tropical rain forests to grow along the equator in places such as Brazil and Southeast Asia, but create deserts like the Sahara 30° north and south of the equator?

Thinking as a Scientist

The North American pronghorn looks and acts like the antelopes of Africa. But the pronghorn is really the only survivor of a family of mammals restricted to North America. Propose a hypothesis to explain how these widely separated animals came to be so much alike.

Science, Technology, and Society

Near Lawrence, Kansas, there was, until 1990, a rare patch of the original North American temperate grassland that had never been plowed. It was home to numerous native grasses, annual plants, and grassland animals. Among the species present were two endangered plants. Environmental activists thought the area should be set aside as a nature preserve, and they started to raise money to save it. In 1990, the owner of the land plowed it, stating that there are no federal laws protecting endangered plants on private grasslands, and that he did not want to be told what he could do with his property. What issues and values are in conflict in this situation? How could this story have had a more satisfactory ending for all concerned?

Media Resources

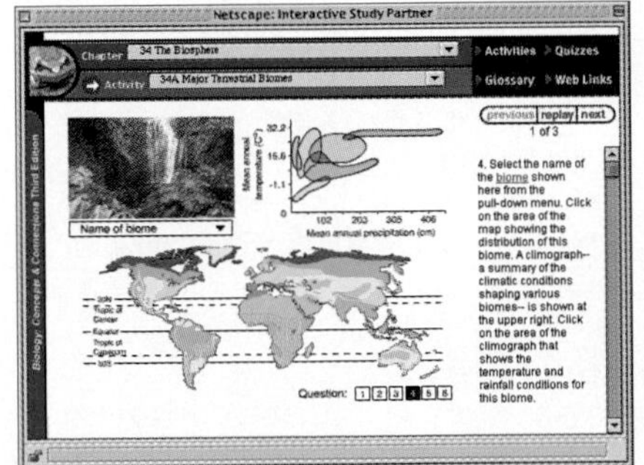

For further review of the concepts in this chapter, use the activities and quizzes on the CD-ROM.

Visit the Special Edition of The Biology Place web site for activities, quizzes, news articles, and web links.

35 Population Dynamics

Outline

City-dwellers often detour around buildings and trees where starlings roost. Ranchers and farmers watch dense flocks of these birds devour grain from fields and feedlots. Yet if you lived in North America a little over a century ago, you would not have seen this bird. Once restricted to Europe and Asia, the European starling is now an abundant and destructive pest in North America, eastern Australia, New Zealand, and South Africa. Omnivorous, aggressive, and tenacious, starlings often replace native species. They oust woodpeckers, bluebirds, and swallows from nesting sites and may pull nestlings of these and other species out of nests to make room for their own offspring.

During the 1800s and early 1900s, introducing foreign species of animals and plants to North America was a popular, unregulated activity. Many people belonged to "acclimatization societies," whose purpose was to bring in species from other countries. Private groups and state game agencies imported the ring-necked pheasant, an Asian native, for sport hunting. Civic authorities in over 100 cities introduced the now-widespread house sparrow for "aesthetic reasons" and to control pest insects. A citizens' group introduced the starling as part of a campaign to bring all the birds mentioned in Shakespeare's works to the New World. In *Henry the Fourth,* a character named Hotspur alludes to the starling's ability to mimic human speech: "I'll have a starling shall be taught to speak nothing but 'Mortimer.'" Who would have thought this line from a play would trigger an environmental calamity 300 years after it was written?

Shakespeare enthusiasts released about 120 starlings in New York's Central Park in 1890. New Yorkers cheered as a breeding pair built a nest under the eaves of the American Museum of Natural History. From that foothold, starlings spread rapidly throughout the U.S. and Canada, as shown on the map at the right. In less than a century, the North American starling population increased to about 100 million. The United States now spends millions of dollars each year trying to control starlings. Mass trappings, hunting, electrified wires on buildings, fireworks, chemical repellents, and poisons have all proved ineffective in controlling the birds.

The starling population in North America has some features in common with the global human population. Both are expanding and are virtually uncontrolled. Both are also harming other species—the starling threatens other bird species; the human population threatens much of the biosphere. In this chapter, we discuss why neither of these populations can grow indefinitely. But when will they stop, and what will stop them?

Population dynamics, the subject of this chapter, is concerned with changes in population size and the factors that regulate populations over time. In the first few modules, we look at how ecologists study populations and some of the major factors that control populations in nature. Then, we return to the topic of uncontrolled populations and see how the principles of population dynamics apply to the global human population.

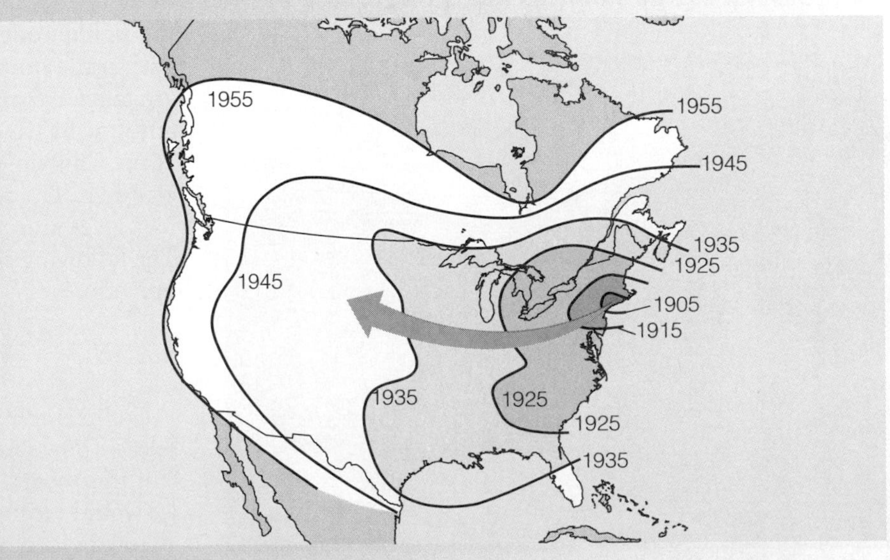

The spread of starlings across North America

35.1 Populations are defined in several ways

The starling population of North America and the global human population are both very large. Most of our knowledge of population dynamics comes from studies of much smaller groups that are confined by more restricted geographical boundaries—for instance, a population of moose in a certain mountain valley, or a population of algae in a lake. Ecologists generally define a **population** as an interacting group of individuals of a species that use common resources and are regulated by the same natural phenomena, such as temperature, water and food supply, and predation.

A researcher must define a population by geographical boundaries appropriate to the questions being asked. For example, a population biologist studying the contribution of asexual reproduction to the population growth of sea anemones might define a population as all the anemones of one species in a tidepool. Another researcher studying the effects of hunting on deer might define a population as all the deer within a particular state. Yet another researcher, attempting to determine which segment of the human population will be most affected by the AIDS epidemic, might study the HIV infection rate of the human population in one nation or throughout the world. Regardless of the scale, two important characteristics of any population are its density and its dispersion patterns. We discuss these characteristics next.

What is the relationship between a population and a species?

A population is a local subset of a species.

Population Structure and Dynamics

35.2 Density and dispersion patterns are important population variables

35A
See Activity 35A on the CD-ROM.

Population density is the number of individuals of a species per unit area or volume—the number of oak trees per km^2 in a forest, for example, or the number of earthworms per m^3 in the forest's soil.

How do we measure population density? In rare cases, it is possible to actually count all individuals within the boundaries of the population. As a hypothetical example, we could count the total number of oak trees (say, 200) in a forest covering 50 km^2. The population density would be the total number of trees divided by the area, or $4/km^2$.

In most cases, it is impractical or impossible to count all individuals in a population. Instead, ecologists use a variety of sampling techniques to estimate population densities. For example, they might base an estimate of the density of alligators in the Florida Everglades on a count of individuals in a few sample plots of 1 km^2 each. The larger the number and size of sample plots, the more accurate the estimates. In some cases, population densities are estimated not by counts of organisms but by indirect indicators, such as number of bird nests or rodent burrows, or signs such as animal droppings or tracks.

Figure 35.2A Researchers using the mark-recapture method

Another sampling technique commonly used to estimate wildlife populations is the **mark-recapture method.** In Figure 35.2A, researchers are sampling a population of mouselike rodents called meadow voles. Box traps are set within the boundaries of the population; captured voles are tagged ("marked") and then released. After a certain amount of time (two weeks, in this study), traps are set again. This time, both marked and unmarked voles will be captured. The proportion of marked to unmarked individuals gives an estimate of the size of the entire population. The following equation gives an estimate of the number of individuals *(N)* in the population:

$$N = \frac{\text{Marked individuals} \times \text{Total catch second time}}{\text{Recaptured marked individuals}}$$

For example, suppose the researcher captures, tags, and releases 50 voles. Two weeks later, 100 voles are captured. If 5 of this second catch are voles that have been recaptured, we would estimate the entire population of voles as $N = 50 \times 100$ divided by $5 = 1000$ voles. This method assumes that a marked individual has the same chance of being trapped as an unmarked individual. This is not always a safe assumption, however. An animal that has been trapped before may be wary of the traps, or, having learned that traps contain food, may have deliberately returned for more. Still, the mark-recapture method is a useful tool for estimating population density.

The **dispersion pattern** of a population refers to the way individuals are spaced within their area. A **clumped** pattern, in which individuals are aggregated in patches, is the most common in nature. Clumping often results from an unequal distribution of resources in the environment. For instance, the cottonwood trees in Figure 35.2B are clumped along a stream channel in patches of moist and sandy soil. Clumping of animals is often associated with uneven food distribution or with mating or other social behavior. For instance, mosquitoes often swarm in great numbers, thereby increasing their chances for mating.

A **uniform,** or even, pattern of dispersion often results from interactions among individuals of a population. For instance, creosote bushes in the desert (Figure 35.2C, left) tend to be uniformly spaced because their roots compete for water and dissolved nutrients. Animals often exhibit uniform dispersion as a result of social interactions. Examples are birds nesting in large numbers on small islands, or people living in a housing development (Figure 35.2C, right).

Some populations exhibit both clumped and uniform dispersion patterns simultaneously. If you look at the entire population, you usually see clumps of individuals. For instance, if you studied dispersion patterns of the human population of the state of Pennsylvania, you would find most of the population clumped in cities. Within each clump, however, individuals or family groups might be more or less uniformly dispersed.

In a third, or **random,** type of dispersion, individuals in a population are spaced in a patternless, unpredictable way. Clams living in a coastal mudflat, for instance, might be randomly dispersed at times of the year when they are not breeding, when resources are plentiful and do not affect their distribution. Clams might also be randomly dispersed if a great number of factors—food, shelter, predators, and dissolved oxygen, for example—are affecting them in conflicting ways with chaotic results. However, environmental conditions and social interactions make random dispersion rare. Researchers often test for clumped or uniform dispersion patterns by comparing a population's actual dispersal with mathematical models of random dispersion.

Figure 35.2B Clumped dispersion of cottonwood trees

Estimates of population density and dispersion patterns are both important in analyzing population dynamics. They enable researchers to compare and contrast the growth or stability of populations occupying different geographic areas. Later in this chapter, we will explore some of the factors that can alter population density and dispersion. In the next module, we turn our attention to how populations grow.

? Using the mark-recapture method, we estimate the population size of a species of small mice in a particular field to be $N = 350$. Later, experiments on the behavior of these mice show that they can locate a baited trap faster if they have already been rewarded with food by visiting that trap once before. Is our original estimate of 350 individuals too low or too high? Explain your answer in terms of the equation for the mark-recapture method.

Too low; because marked individuals are more likely to be captured during the second trapping than unmarked individuals, the numerator (recaptured marked individuals) in the equation was disproportionately high, and thus we underestimated N.

Figure 35.2C Uniform dispersion of creosote bushes (left) and human habitations (right)

35.3 Idealized models help us understand population growth

No other organisms reproduce faster than bacteria. Some bacteria can divide every 20 minutes under ideal laboratory conditions. Thus, starting with a single bacterium, after 20 minutes, there are two bacteria; after 40 minutes, there are four; after 60 minutes, eight—and so on, as shown in the table in Figure 35.3A. If this rate of population growth could continue for only a day and a half—a mere 36 hours—there would be bacteria enough to form a layer a foot deep over the entire planet!

Time	Number of Cells	
0 minutes	1	$= 2^0$
20	2	$= 2^1$
40	4	$= 2^2$
60	8	$= 2^3$
80	16	$= 2^4$
100	32	$= 2^5$
120 (= 2 hours)	64	$= 2^6$
3 hours	512	$= 2^9$
4 hours	4096	$= 2^{12}$
8 hours	16,777,216	$= 2^{24}$
12 hours	68,719,476,736	$= 2^{36}$

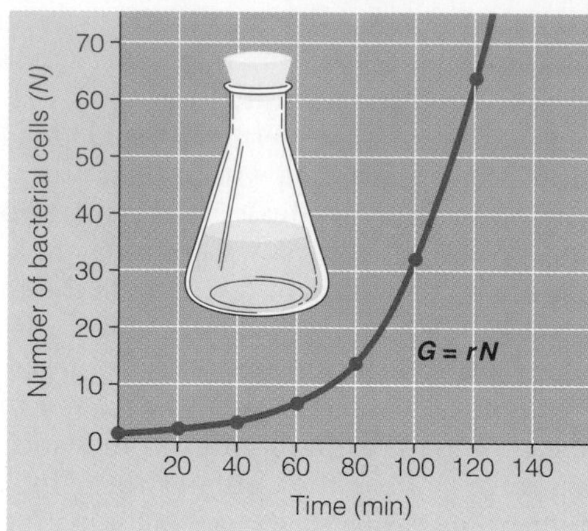

Figure 35.3A Exponential growth of bacteria

The Exponential Growth Model The rate of expansion of a population under ideal conditions is called exponential growth when the whole population multiplies by a constant factor during constant time intervals (the generation time). For the bacterial population in Figure 35.3A, the constant factor is 2 (because each parent cell splits to produce two daughter cells), and the generation time is 20 minutes. The progression for bacterial growth—2, 4, 8, 16, etc.—is the number 2 raised to a successively higher power (exponent) each generation (that is, $2^1 = 2$; $2^2 = 4$; $2^3 = 8$; $2^4 = 16$; etc.).

Suppose you have a summer job working in a microbiology research lab, and you are asked to monitor the growth of a bacterial population. You would sample the population by counting the number of bacterial cells (N) at regular intervals of time (t). You could then plot the numbers you obtained for N against t. The graph in Figure 35.3A shows the type of curve you would obtain if you plotted the number of cells in a bacterial population that was expanding exponentially (and connected the points on the graph with a smooth curve).

The simple equation $G = rN$ describes the J-shaped curve, which is typical of exponential growth. The letter G stands for the growth rate of the population, N stands for the population size (the number of individuals in the population), and r stands for the **intrinsic rate of increase,** an organism's inherent capacity to reproduce. The value for r depends on the kind of organism, but it remains constant for any population expanding without limits. We can obtain a rough estimate of the value of r by subtracting the death rate (the number of individuals dying in a given unit of time) from the birth rate (the number of individuals born in that same unit of time).

Overall, the exponential equation tells us that, with r being constant, the rate at which a population grows depends on the number of individuals already in the population. The bigger the value of N, the faster the population increases. As time goes by, N gets bigger faster and faster. On a graph, the lower part of the J results from the relatively slow growth when the population is small. The steep, upper part of the J results from N being large.

The **exponential growth model** gives an idealized picture of the unregulated growth of a population. For bacteria, unregulated growth means there is no restriction on the abilities of the cells to live, grow, and reproduce. Given a few days of unregulated growth, bacteria would smother every other living thing. Obviously, long periods of exponential increases are not common in the real world, or life could not continue on Earth.

Population-Limiting Factors and the Logistic Growth Model In nature, a population may grow exponentially for a while, but eventually, some environmental factor or factors will limit its growth. Population size then stops growing or may even crash. Environmental factors that restrict population growth are called **population-limiting factors.**

You can see the effect of population-limiting factors in the graph in Figure 35.3B, which illustrates the growth of a population of fur seals on St. Paul Island, off the coast of Alaska. (For simplicity, only the mated bulls were counted. Each has a harem of a number of females, as shown in the photograph.) Before 1925, the seal population on the island remained low because of uncontrolled hunting, although it changed from year to year. After hunting was controlled, the population increased rapidly until about 1935, when it leveled off and began fluctuating around a population size of about 10,000 bull seals. At this point, a number of population-limiting factors, including some hunting and the amount of space suitable for breeding, restricted population growth.

The fur seal growth curve resembles the **logistic growth model,** a description of idealized population growth that is slowed by limiting factors. Figure 35.3C compares the logistic growth model (red) with the exponential growth model (black). As you can see, the logistic curve is J-shaped at first, but gradually levels off to resemble a lazy S more than a J.

The equation for logistic growth is more complicated than the exponential equation, because it describes the effect of population-limiting factors as well as population growth. As shown in Figure 35.3C, the logistic equation is the exponential equation modified by the term $(K - N)/K$. This term represents the overall effect of population-limiting factors.

The logistic equation is actually simpler than it may appear. Notice that the only new letter in the equation is K, which stands for carrying capacity. **Carrying capacity** is the number of individuals in a population that the environment can just maintain ("carry") with no net increase or decrease. For the fur seal population on St. Paul Island, for instance, K is about 10,000 mated males. The value of K varies, depending on species and habitat. K might be considerably less than 10,000 for a fur seal population on an island with fewer breeding sites.

Figure 35.3B Growth of a population of fur seals

Let's see how the term $(K - N)/K$ works in producing the S-shaped logistic curve. When the population first starts growing, N is close to zero—very small compared to the carrying capacity K. At this time, N has little effect on the term $(K - N)/K$; in fact, the term nearly equals K/K, or 1. When this is the case, population growth G is close to $rN(1)$, or just rN—that is, exponential growth. However, as the population increases and N gets close to carrying capacity, it has a large effect on the term $(K - N)/K$. In fact, the term becomes an increasingly small fraction. And the value rN is multiplied by that fraction, slowing down the population growth more and more. At carrying capacity, the population is as big as it can theoretically get in its environment; at this point, $N = K$, $(K - N)/K = 0$, and the population growth rate (G) is zero.

Thus, the term $(K - N)/K$ accounts for the leveling off of the logistic growth curve. To summarize: When the population is still very small, the term has little effect, and the logistic curve is very close to the J-shaped curve. As the population takes off, N makes the term $(K - N)/K$ and the growth rate smaller and smaller. The curve then gets farther and farther away from the J-shaped curve and becomes S shaped.

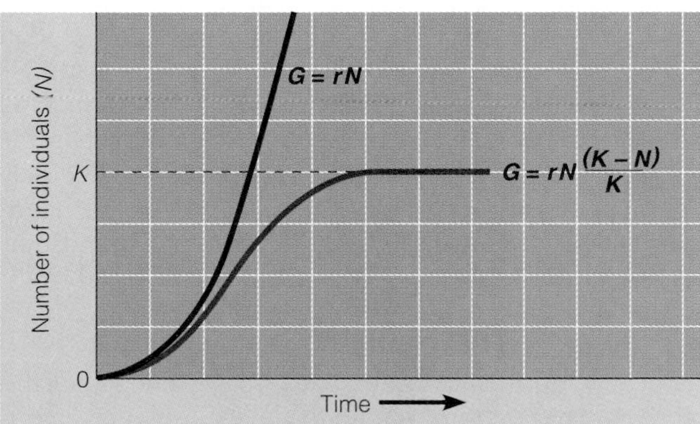

Figure 35.3C Logistic growth and exponential growth compared

What does the logistic growth model suggest to us about real populations in nature? The model predicts that a population's growth rate will be small when the population size is *either* small or large, and highest when the population is at an intermediate level relative to the carrying capacity. At a low population level, resources are abundant, and the population is able to grow nearly exponentially. At this point, however, the increase is small because N is small. In contrast, at a high population level, population-limiting factors strongly oppose the population's potential to increase. In nature, there might be less food available per individual or fewer breeding sites, nest sites, or shelters. What actually happens is that limiting factors make the birth rate decrease, the death rate increase, or both. Eventually, the population stabilizes at the carrying capacity (K), when the birth rate equals the death rate.

It is important to realize that both the logistic growth model and the exponential growth model are mathematical ideals. No natural populations fit either one perfectly. Overall, these models are useful starting points for studying population growth. Ecologists use them to predict how populations will grow in certain environments and as a basis for constructing more complex models. The models have stimulated many experiments and discussions that have led to a greater understanding of populations in nature. In the next module, we take a closer look at some of the factors that limit the growth of natural populations.

? A laboratory jar containing a population of beetle larvae (mealworms) has reached a stable population size. We decide to add twice as much food per day to the jar, but this turns out to have no effect on population size. What is the most likely explanation?

The population was already at carrying capacity before we increased food supply, and the key limiting factor was something other than food availability.

35.4 Density-dependent and density-independent factors limit population growth

Two general types of factors limit population growth. In one case, limits on growth are related to the density of the population itself. In the other, population density has nothing to do with population growth.

Population-limiting factors whose effects depend on population density are called **density-dependent factors.** Such factors affect a greater percentage of individuals in a population as the number of individuals increases. Limited food supply and the buildup of poisonous wastes are examples of density-dependent factors. Such factors depress a population's growth rate by increasing the death rate, decreasing the birth rate, or both. We often see density-dependent regulation in laboratory populations. For example, when a pair of fruit flies is placed in a jar with a limited amount of food, the population grows logistically. After a rapid increase, population growth levels off, as the flies become so numerous that they outstrip their limited food supply (Figure 35.4A). Each individual in a large population has a smaller share of the limited food than it would in a small population. Also, the more flies, the more concentrated poisonous wastes become in the jar.

Laboratory populations are one thing; natural populations are another. We do not often see clear-cut cases of density-dependent factors regulating populations in nature. To test whether such factors are operating, it is necessary to change the density of individuals in the population while keeping other factors constant. This is sometimes done in managing game populations. For instance, state agencies often allow hunters to reduce populations of white-tail deer to levels that keep the animals from permanently damaging the plants they use for food. White-tail deer are browsers, preferring the highly nutritious parts of woody shrubs—young stems, leaves, and buds. When deer populations are kept low and high-quality food is therefore abundant, a high percentage of females become pregnant and bear offspring; in fact, many of them produce twins (Figure 35.4B). On the other hand, when populations are high and food quality is poor, many females fail to ovulate, the spontaneous abortion rate is usually higher, and twinning is rare. These observations support the hypothesis that food supply is a density-dependent factor regulating white-tail deer populations.

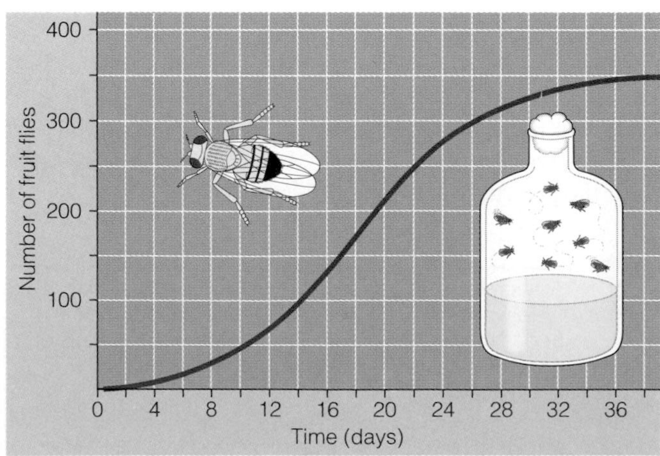

Figure 35.4A Logistic growth of a laboratory population of fruit flies

Figure 35.4B A deer with twin fawns, born when population density was low

Population-limiting factors whose occurrence is not affected by population density are called **density-independent factors.** Included are such abiotic factors as climate and weather. A freeze in the fall, for example, may kill a certain percentage of insects in a population. The date and severity of the first freeze obviously are not affected by the density of the insect population. Density-independent factors such as a killing frost affect the same *percentage* of individuals regardless of population size. (In larger populations, of course, greater numbers will die.)

In many natural populations, density-independent factors limit population size well before resources or other density-dependent factors become important. In such cases, the population may decline suddenly. If we look at the growth curve of such a population, we see something like exponential growth followed by a rapid decline, rather than a leveling off. Figure 35.4C (on the facing page) shows this effect for a population of aphids, insects that feed on the phloem sap of plants. These and many other insects often show virtually exponential growth in the spring and then rapid die-offs when it becomes hot and dry in the summer. A few individuals may remain, and these may allow population growth to resume again if favorable conditions return. Some insect populations—many mosquitoes and grasshoppers, for instance—will die off entirely, leaving only eggs, which will initiate population growth the following year. In addition to seasonal changes in the weather, environmental factors, such as fire, floods, storms, and habitat disruption by human activity, can affect populations in a density-independent manner.

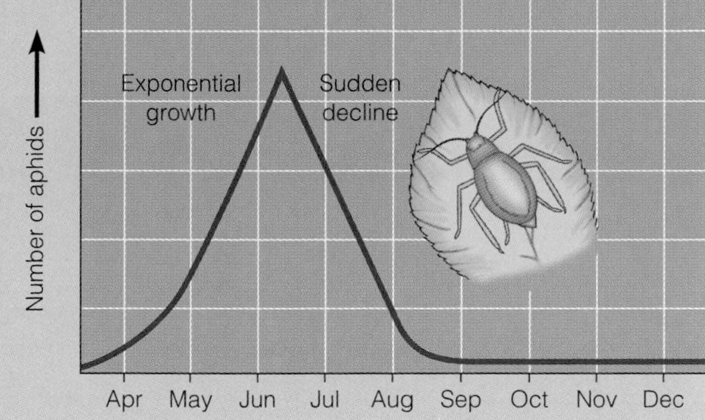

Figure 35.4C The effect of a density-independent factor on aphid population size

Over the long term, most populations are probably regulated by a mixture of density-independent and density-dependent factors. Many populations remain fairly stable in size and are presumably close to a carrying capacity that is determined by density-dependent factors. In addition, however, many show short-term fluctuations due to density-independent factors. In some cases, the distinction between density-dependent and density-independent factors is not clear. In the case of white-tail deer, for example, in very cold, snowy areas, many individuals may starve to death. The severity of this effect is related to the harshness of the winter; cold temperatures increase energy requirements (and therefore the need for food), while deeper snow makes it harder to find food. But the severity of the effect is also density-dependent, because the larger the population, the less food available per individual. 

? Of the following factors with the potential to limit growth of a human population, which one is most density-independent? (a) mass drowning caused by hurricane flood; (b) freezing deaths due to a shortage of housing

(a)

Some populations have "boom-and-bust" cycles 35.5

Figure 35.5 illustrates long-term changes in two populations, the snowshoe hare and the lynx, that fluctuate cyclically. The lynx is one of the main predators of the snowshoe hare in the taiga of North America. About every 10 years, both hare and lynx populations have a rapid increase (a "boom") followed by a sharp decline (a "bust").

What causes boom-and-bust cycles? Since ups and downs in the two populations seem to almost match each other on the graph, does this mean that changes in one directly affect the other? In other words, does predation by the lynx make the hare population fluctuate, and do the ups and downs of the hare population cause the changes in the lynx population?

For the lynx and many other predators that depend heavily on a single species of prey, the availability of prey can influence population changes. Thus, the 10-year cycles in the lynx population probably result at least in part from the 10-year cycles in the hare population. For the hare, however, researchers have discovered that populations will cycle about every 10 years, whether or not lynx are present. Thus, we cannot conclude from the graph that lynx predation alone causes the cyclical changes in the hare population. Another hypothesis is that fluctuations in hare populations are tied to fluctuations in the populations of plants eaten by the hares. Some evidence indicates that when certain plants are damaged by herbivores, their nutrient content decreases. Recent experimental studies performed in the field support the hypothesis that the 10-year cycles of snowshoe hares result from the combined effects of predation and fluctuations in the hare's food sources.

Populations of many rodents, such as lemmings and voles, also exhibit boom-and-bust changes, often cycling every three to five years. For such short-term cycles, some researchers hypothesize that either predation or food supply alone may be the underlying cause. Another hypothesis, based on laboratory studies of mice and other small rodents, is that stress from crowding may alter hormonal balance and reduce fertility. The causes of cycles probably vary among species and maybe even among populations of the same species. 

? State a research question that arises from evidence that the population cycle of snowshoe hares is caused at least partly by a cycle in the availability or nutritional value of the plants eaten by the hares.

What causes the cyclical changes in the plants?

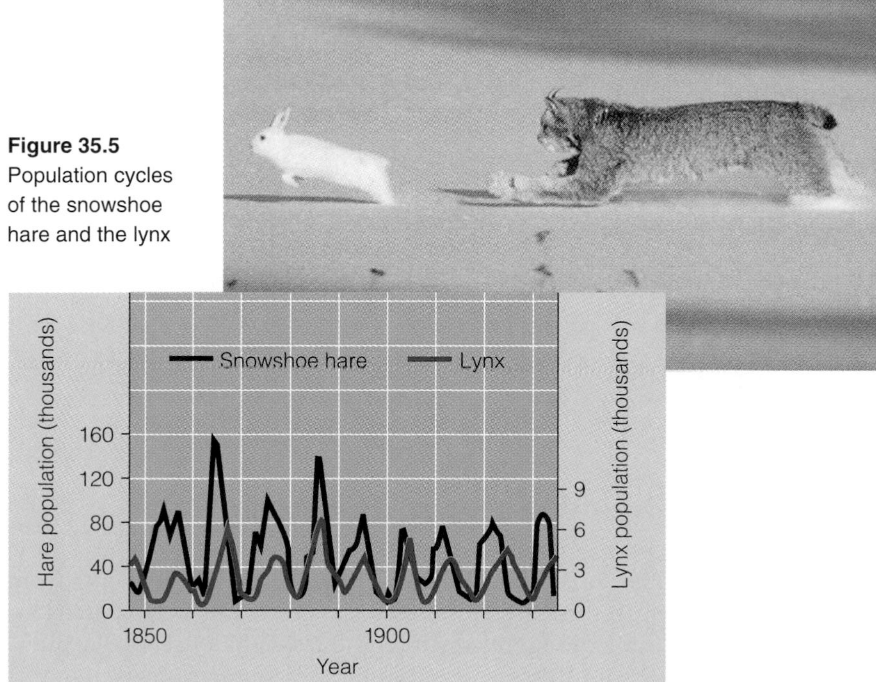

Figure 35.5 Population cycles of the snowshoe hare and the lynx

Life Histories and Their Evolution

35.6 Life tables track mortality and survivorship in populations

The principles of population growth have broad application. When the life insurance industry was established about a century ago, insurance companies took notice of some of the early scientific studies of human populations. Needing to determine how long, on average, an individual of a given age could be expected to live, they began using what are called **life tables.** The life table below was compiled in 1980 from a sample of 10 million U.S. citizens. Using this table, an insurance agent would predict that a 22-year-old has about a 0.982 (98.2%) chance of surviving to age 30. Population ecologists have constructed life tables for populations of many other species of animals, as well as for some plants.

A graph like the one in Figure 35.6 makes the data in a life table easy to comprehend at a glance. These so-called survivorship curves vary with species and often with populations within species. By using a percentage scale instead of actual ages on the horizontal axis, we can compare species with widely varying life spans on the same graph. The curve for the human population tells us that most people die in older age intervals, as we see in the last column of the life table. Species that exhibit this Type I curve—whales and elephants, for example, as well as humans—usually produce few offspring but give them good care, increasing the likelihood that they will survive to maturity.

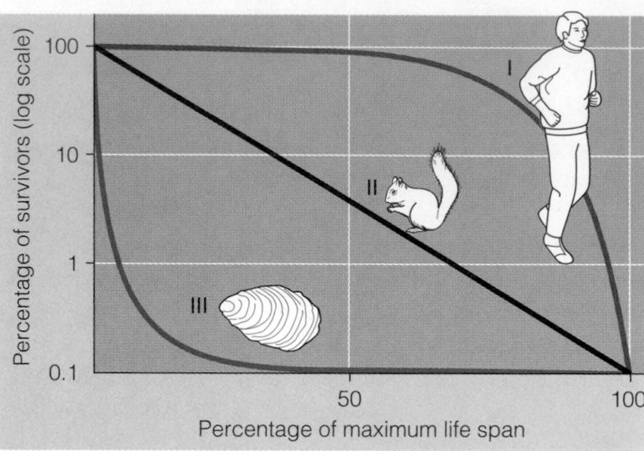

Figure 35.6 Three types of survivorship curves

Life Table for the U.S. Population in 1980

Age Interval	Number Living at Start of Age Interval (*N*)	Number Dying During Interval (*D*)	Mortality (Death Rate) During Interval (*D*/*N*)	Chance of Surviving Interval (1 − *D*/*N*)
0–10	10,000,000	121,678	0.012	0.988
10–20	9,878,322	124,163	0.013	0.987
20–30	9,754,159	174,161	0.018	0.982
30–40	9,579,998	202,773	0.021	0.979
40–50	9,377,225	410,607	0.044	0.956
50–60	8,966,618	882,352	0.098	0.902
60–70	8,084,266	1,810,106	0.224	0.776
70–80	6,274,160	2,999,619	0.478	0.522
80–90	3,274,541	2,628,753	0.803	0.197
90–100	645,788	645,788	1.000	0.000

In contrast, a Type III curve indicates high death rates for the very young, and then a period when death rates are much lower for those few individuals who survive to a certain age. Species with this type of survivorship curve usually produce very large numbers of offspring but provide little or no care for them. An oyster, for instance, may release millions of eggs, but most offspring die as larvae from predation or other causes. A Type II curve is intermediate, with mortality more constant over the life span. This type of survivorship has been observed in several invertebrates, including hydras, and in certain rodents, such as the gray squirrel.

Life tables and survivorship curves enable us to compare populations on the basis of individual life spans. Another way to compare populations is by the way their life history is adapted to the environment, as we see next.

What is the key feature of the "mortality" column in the life table for a population exhibiting a Type II survivorship curve?

The mortality is about the same for every age interval of individuals.

35.7 Evolution shapes life histories

An organism's **life history** is the series of events from birth through reproduction to death. Life history traits include the age at which reproduction first occurs, the number of offspring, the amount of parental care given to offspring, and the energy cost of reproduction. All such traits influence the growth rate of a population. Life history traits, like body features, are shaped by evolution operating through natural selection.

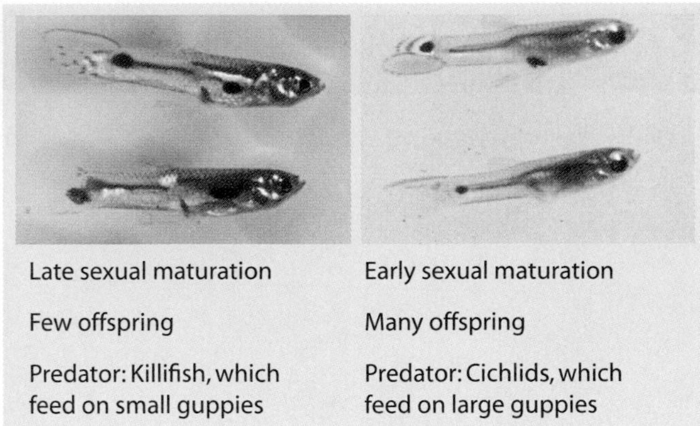

Late sexual maturation
Few offspring
Predator: Killifish, which feed on small guppies
Early sexual maturation
Many offspring
Predator: Cichlids, which feed on large guppies

Figure 35.7A Guppies from two different environments

Figure 35.7B Seeds of a coconut palm, an equilibrial species (left); seeds of a dandelion, an opportunistic species (right)

Let's look at the effect of one environmental factor, predation, in shaping a life history through natural selection. Figure 35.7A illustrates typical mature individuals from two wild populations of the guppy, a popular home-aquarium fish. Guppies vary in color and body form, but they are all one species. On the island of Trinidad in the West Indies, certain guppy populations live where predators called killifish eat mainly small, immature guppies. Other guppy populations live where larger fish, called cichlids, eat mostly mature, large-bodied guppies. Where preyed upon by cichlids, guppies tend to be smaller, mature earlier, and produce more offspring each time they give birth than those in areas without cichlids. Thus, guppy populations differ in certain life history traits depending on the kind of predators in their environment. If the differences between the populations result from natural selection, the life history traits should be heritable. To test for heritability, researchers raised guppies from both populations in the laboratory without predators. The two populations retained their life history differences when followed through several generations, indicating that the differences were indeed inherited.

In a follow-up experiment, researchers introduced guppies from a cichlid habitat into a guppy-free area where there were predators that ate small, immature guppies. The scientists predicted that the new kind of predation would bring about changes in the guppy life history traits. So they were not surprised when, within two years, they began to find female guppies maturing later and producing fewer and larger offspring than those from a control site with cichlids. A laboratory test without predators showed that the changes in the population were heritable. These studies demonstrate not only that life history traits are shaped by natural selection, but that questions about evolution can be tested by field experiments.

In nature, every population has a particular life history adapted to its environment. Some populations, usually of small-bodied species, exhibit an **opportunistic life history.** Individuals reproduce when young, and they produce many offspring; the population tends to grow exponentially when conditions are favorable. Such a population typically lives in an unpredictable environment and is controlled by density-independent factors such as the weather. For example, the dandelion (Figure 35.7B, right) and many other annual weeds grow quickly in open, disturbed areas, producing a large number of seeds in a brief time when the weather is favorable. Although most of the seeds will not produce mature plants, their large number and ability to disperse to new habitats ensure that at least some will grow and eventually produce seeds themselves. For such species, natural selection has fostered an emphasis on the quantity of reproduction rather than on individual survivorship. In general, populations with an opportunistic life history exhibit a Type III survivorship curve (see Figure 35.6).

In contrast, some populations, mostly larger-bodied species, exhibit an **equilibrial life history,** which generally results in a Type I survivorship curve. Individuals usually mature later and produce few offspring but care for their young. The population size may be quite stable, held near the carrying capacity by density-dependent factors. Natural selection has resulted in the production of better-endowed offspring that can become established in the well-adapted population into which they are born. The life histories of many large terrestrial vertebrates fit this model. Among polar bears, for instance, a female has only one or two offspring every three years, but the cubs remain in her protective custody for over two years. In the plant kingdom, the coconut palm tree shown in Figure 35.7B (left) also fits the equilibrial model. Compared to most other trees, it produces relatively few, very large seeds. A large endosperm (the coconut "milk" and "meat") provides nutrients for the embryo; this is a plant's version of parental care.

For most populations, these two life history categories are only hypothetical models, starting points for studying the complex interplay of the forces of natural selection in nature. Most populations probably fall between the opportunistic and equilibrial extremes. Next we begin our look at how the principles of population dynamics apply to the human population. 

? How do the names *opportunistic* and *equilibrial* contrast the key characteristics of these life histories?

Opportunistic life histories are characterized by an ability to produce a large number of offspring very rapidly when the environment affords a temporary opportunity for exponential growth; equilibrial life histories are characterized by a population size that fluctuates only slightly from carrying capacity.

35.8 The human population has been growing very rapidly for centuries

Figure 35.8A shows the stifling, unsanitary conditions in a slum that has arisen within a garbage dump in Manila. One of the world's largest cities, Manila has well over 10 million people. Its problems are not unique, however, and scenes much like this are common in a number of the world's large cities.

What about the human population as a whole? As Figure 35.8B shows, it has grown almost continuously for thousands of years. Growth was relatively slow for most of those years, with a slight dip in the fourteenth century, when the bubonic plague, or Black Death, wiped out about 25% of the population of Europe. By about 1650, approximately 500 million people inhabited the world. Between 1650 and 1850, the population doubled to 1 billion; it doubled again to 2 billion between 1850 and 1930, and doubled still again by 1975 to more than 4 billion. In 1998, the population numbered about 5.9 billion people. At the current rate of increase, it takes less than three years for world population to add the equivalent of another United States (about 270 million). If the present growth persists, scientists estimate that there will be about 8 billion people on Earth by the year 2025.

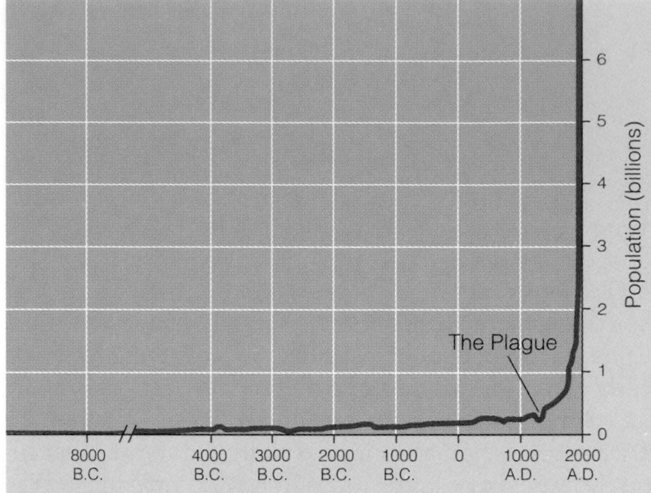

Figure 35.8B The history of human population growth

The growth of human populations, like that of other species, depends on birth rates and death rates. About 10,000 years ago, agricultural societies replaced a lifestyle of hunting, gathering, and scavenging. With this change, human birth rates increased and death rates decreased, leading to the steady, but relatively slow, rate of population growth you see in the graph. Since the mid-1700s, when the Industrial Revolution began in England, the global human population has skyrocketed. This unprecedented growth has resulted mainly from falling death rates, especially infant mortality, throughout most of the world. Improved nutrition, better medical care, and sanitation—all *technological,* rather than biological, attributes of our species—have contributed to an increased percentage of newborns that survive long enough to leave offspring of their own. The effect of decreasing mortality on population growth is compounded in most developing countries, which tend to have relatively high birth rates. For example, the birth rate in Sri Lanka has decreased fairly steadily for nearly 50 years, but it has never declined enough to offset the drop in the death rate (Figure 35.8C). Thus, the overall effect is a growing population (Figures 35.8D and 35.8E).

Figure 35.8A A slum in Manila, Philippines

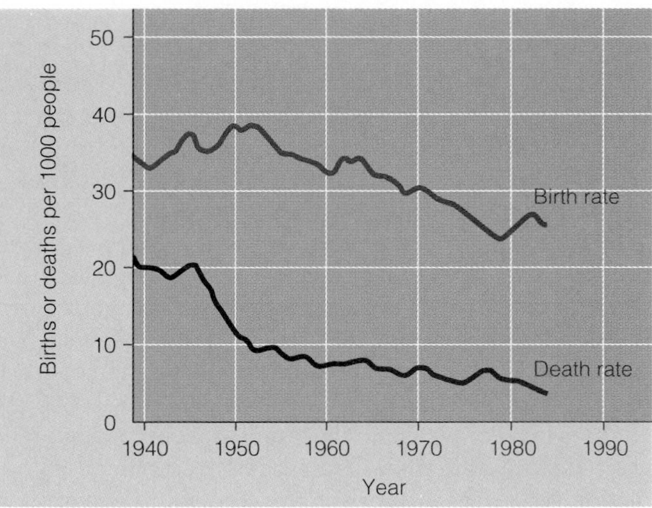

Figure 35.8C Changes in birth and death rates in Sri Lanka

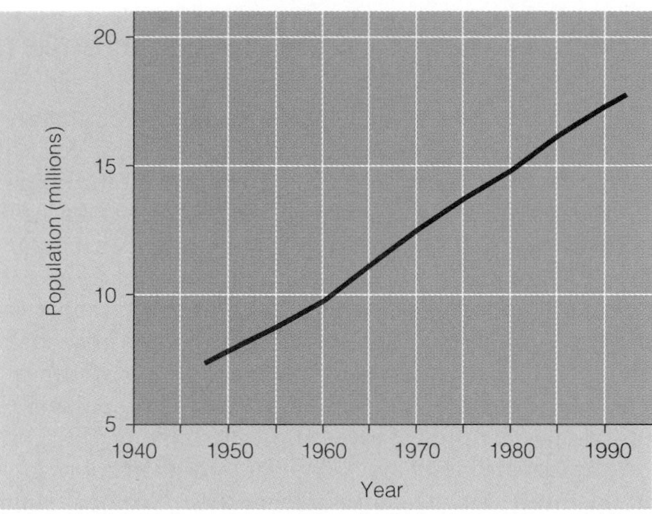

Figure 35.8D Population growth in Sri Lanka

For hundreds of years, agricultural and industrial technology have enabled us to exploit environmental resources more and more efficiently and to inhibit the mechanisms that would normally control our population growth. Put another way, technology has increased Earth's carrying capacity for people, allowing our long-term explosive growth. It is unlikely that any other population of large animals has ever sustained the level of growth depicted in Figure 35.8B for so long a time.

What is the future of the human population? It is difficult to predict the future size of any population accurately, and the human population is no exception. What we do know is that no population can continue to grow indefinitely. Exactly what the world's carrying capacity for humans is and under what circumstances we will approach it are topics of great concern and debate. Has our population already exceeded the total number that Earth can sustain, or have we not yet reached it? Ideally, human populations would reach carrying capacity smoothly and then level off. But if the population has already exceeded carrying capacity, a sharp decline might occur in the future. If the population fluctuates around carrying capacity, it might mean periods of increase would be followed by mass death, as has occurred during plagues and famines.

Barring some worldwide calamity, it is likely that the human population will continue to grow well into the next century. Whatever happens, we know that the human population must eventually stop growing. Unlike other organisms, we have the ability to decide whether this occurs mainly as a result of a decrease in the birth rate or an increase in the death rate. Next we take a closer look at population growth, today and in the future, in some developed and developing nations. 

?

What was the approximate % increase in the Sri Lankan population over the 20-year span from 1960 to 1980?

50%, an increase from about 10 million to 15 million people

Figure 35.8E Schoolchildren in Sri Lanka

35.9 Age-structure diagrams predict the future growth of populations

Worldwide, human population growth is a mosaic of various rates of growth in different countries. Some developed countries, such as Sweden, have virtually stable populations because birth rates and death rates balance. In sharp contrast to Sweden, most developing nations have burgeoning populations in which birth rates greatly exceed death rates. Partly as a result of such unchecked growth, many people in such countries face serious housing, water, and food shortages, as well as severe pollution problems.

The **age structure** of a population is the proportion of individuals in different age groups. By studying age structure, we get an idea of how a population will probably grow in the future. The diagrams in Figure 35.9A compare four different kinds of age structure that are characteristic of human populations in various nations. In each age-structure diagram, the total colored area represents 100%, or the total population. Within this total, each of the three different colors represents the fraction of the population for one of three main age groups (0–14, 15–44, and 45 and older). Within each of these age groups, each horizontal bar represents the population in a 5-year age group. The area to the left of each vertical center line represents the percentage of males in each age group in the population; females are represented on the right side of the line.

As indicated below (left), a population that is expanding rapidly is bottom-heavy, and its age-structure diagram resembles a pyramid. Characteristic of many developing nations, such as Sri Lanka and Nigeria, rapidly expanding populations have a higher percentage of children (age 14 or younger) than individuals in reproductive years. In many developing nations, the death rate, especially for children and young adults, has declined significantly in the last 30 years. A high percentage of young adults and children who are likely to reproduce in the near future sustains explosive population growth.

In contrast to most developing nations, typical developed countries have slow-growing or stable populations. A triangular age-structure diagram (second from left, below) is characteristic of a slowly expanding population, such as that of the United States, Canada, or Australia. In contrast to a rapidly expanding population, the percentage of young adults is higher, relative to the number of children, and the age-structure diagram lacks a wide base. In general, a population whose age groups are nearly balanced will grow at a slower rate than a population with a large percentage of young individuals.

In some developed countries, continued population growth results from a combination of the birth rate and immigration exceeding the death rate. For example, immigration (legal and illegal combined) contributes about 40% of the current growth of the U.S. population. And while the average woman in the United States tends to have only two children, the total number of women having children remains high enough to exceed the death rate. Population researchers predict that the U.S. population will continue to grow well into the next century, perhaps increasing from about 270 million today to about 390 million by the year 2050.

In sharp contrast to the United States, some developed nations, such as Italy, Denmark, and Austria, have stable populations. As indicated in Figure 35.9A (second from right), the lower half of an age-structure diagram of a sta-

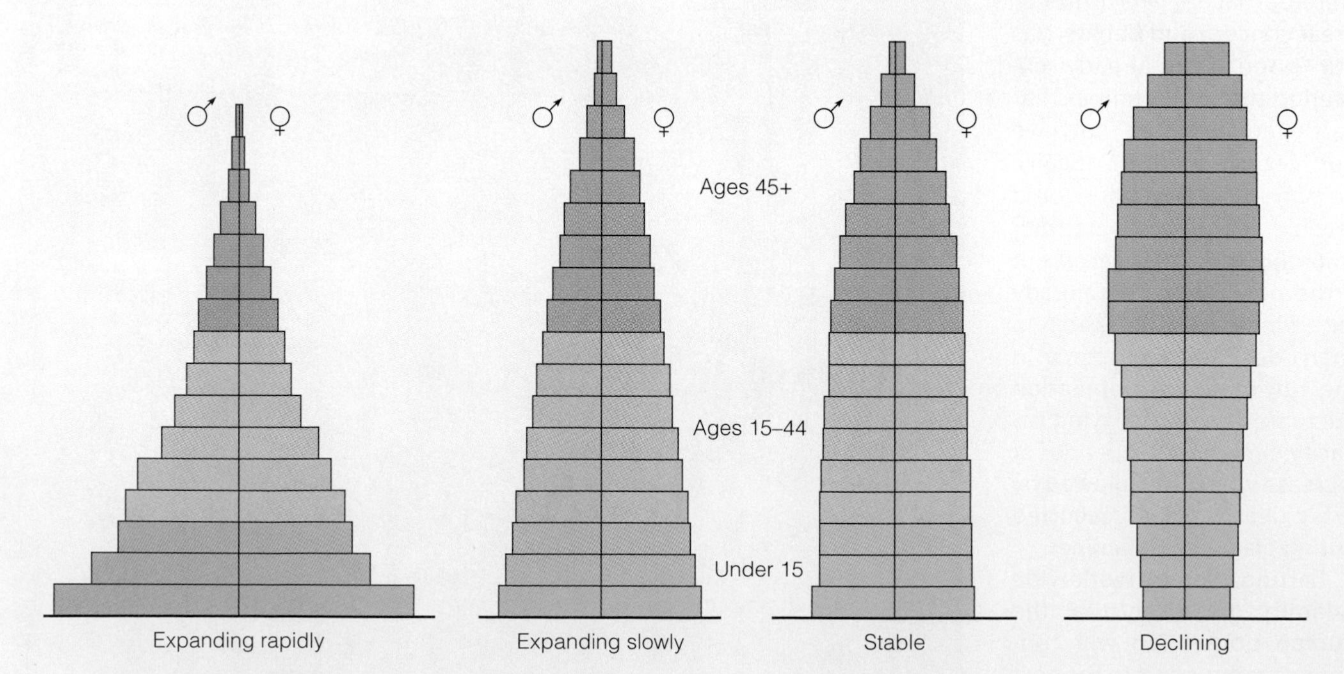

Figure 35.9A Age-structure diagrams of four kinds of populations

ble population has nearly vertical sides. In such a population, the birth rate and death rate are closely matched, and migration in or out of the country has little or no effect on the population as a whole.

The age-structure diagram on the far right in Figure 35.9A is characteristic of nations, such as Hungary and Germany, whose populations are declining. The diagram has a narrow base because the population has relatively few children and young, reproductive adults, and a relatively large number of older individuals.

The diagrams in Figure 35.9B compare the average age structures of developed countries and developing countries worldwide. For the two types of countries, the diagrams indicate the average age structures in 1995 and the projected age structures in the year 2025, using the same three colors as in Figure 35.9A. The age-structure diagrams predict that the population of older people in developed nations will increase slightly in the next 25 years; at the same time, developed nations will see a slight decrease in the number of young people. In marked contrast, the population of developing nations will continue to explode, driven by a large number of young people.

Age-structure diagrams not only reveal a population's growth trends, they also indicate social conditions. Developing nations tend to have very different social problems than do developed nations. For instance, we can predict that most developing nations will have an increasing number of working-age people who are unemployed. They may also have an increasing need for infant-care services and schools as their populations continue to expand. In contrast, many developed nations must plan for more health-care facilities for their aging populations. For many developed nations, the age-structure diagrams tell us that a decreasing number of working-age people will soon be supporting an increasing number of retired people. Financial programs, such as the U.S. Social Security system, which are crucial to many older citizens, will become severely strained as greater numbers of individuals reach retirement age.

A unique feature of human population growth is our ability to control it with voluntary contraception and government-sponsored family planning. Reduced family size contributes to population control. Especially in developed countries, many women choose to delay reproduction, often because of better opportunities for employment and advanced education. Delayed reproduction dramatically decreases population growth rates. You can develop a sense of this by imagining two human populations in which women each produce three children but begin reproduction at different ages. In one population, women first give birth at age 15, and in the other at age 30. If we start with a group of newborn girls, then after 30 years, women in the first population will already begin to have grandchildren, whereas women in the second population will be giving birth to their first children. After 60 years, women in the first population will have a large number of great-great grandchildren (who will themselves begin to reproduce 15 years later), but women in the second population will just begin to see their grandchildren being born.

It seems more desirable for population control to result from a decrease in the birth rate by social changes or individual choice than by an increase in the death rate. As we discuss in Chapter 38, our rapidly expanding global population and the consumption of our planet's limited resources pose serious threats to our future and that of most other species. For better or worse, we have the unique opportunity to decide the fate of our species and the rest of the biosphere. In the next module, we hear from a scientist who studies one of the serious environmental effects of the human population explosion.

(a) How does the age structure of the U.S. population explain the current surplus in the Social Security fund? (b) If the system is not changed, why will the surplus give way to a deficit sometime in the next few decades?

(a) The largest population segment, the "boomers," are currently in the workforce, paying into the system. (b) The "boomers" will retire over the next few decades and begin drawing from the system at a time when there will be fewer employees paying into Social Security.

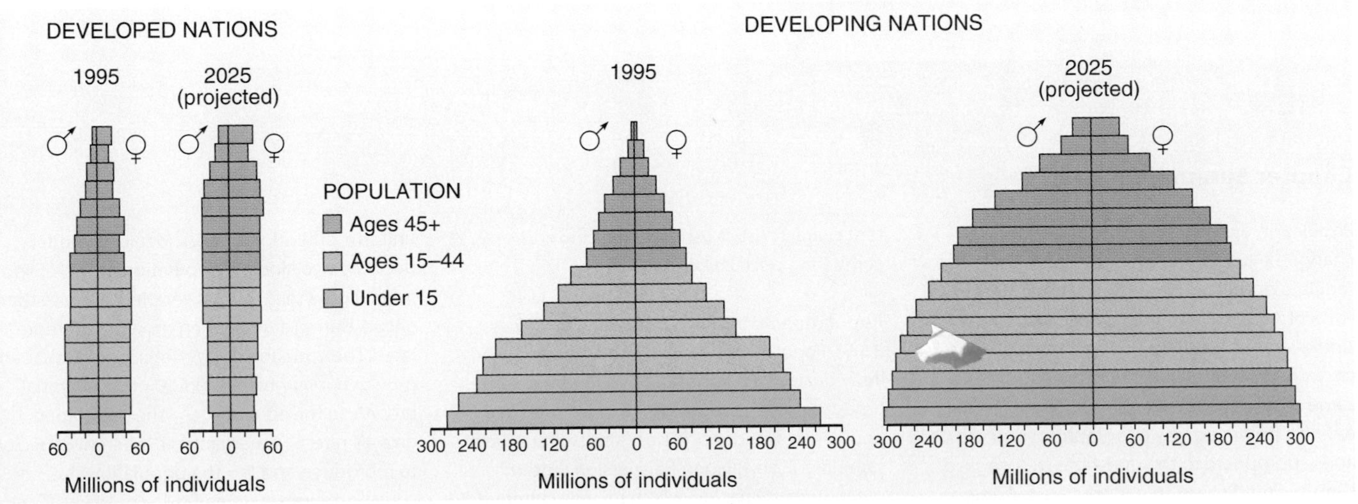

Figure 35.9B Age structures of developed and developing nations in 1995 and 2025 (projected)

TALKING ABOUT SCIENCE

35.10 Global warming may underlie geographic shifts in populations

Figure 35.10 Camille Parmesan studying butterfly populations in northern Spain

Global warming is a controversial and troubling issue (see Module 7.13). Studying butterflies in Europe and western North America, ecologist Camille Parmesan of the University of Texas (Austin) has discovered some of the first evidence of the effects of global warming on animal populations. We asked her about this in a recent interview.

In former geological periods, when there were major climatic changes, the distribution of many species of plants and animals, including many insects, shifted toward or away from the poles, and this is something to look for today. Many butterflies are sensitive to temperature, so they are excellent research models to study the effects of climate change on population dynamics.

Do we know that global warming is occurring, and that human activities are causing it?

Scientists generally agree that there is warming, but we can't expect 100% agreement that this is human-caused. Politically, it's very touchy. We know that pulses of CO_2 are correlated with temperature change, and that the level of CO_2 in the atmosphere is higher now than it's been in about 160,000 years. We are in a natural warm period right now, and many believe we are making it worse by adding CO_2 and other greenhouse gases to the atmosphere. The increases in CO_2 result mainly from energy production and use—burning fossil fuels—in developed nations. Developing countries have much less to do with it than we do.

Tell us more about your butterfly studies.

For over 16 years, I've worked on a species called Edith's Checkerspot *(Euphydryas editha)*. *"Editha"* lives in western North America, and from earlier research and museum records, I learned where populations existed at particular times in the past, back to 1904. Then I did four years of field work, literally going around to each place where the populations had been found before, to determine the current status of populations. I expected to see a subtle population change, due to climate change. But while I was doing the field work, I remember thinking, "They're almost all extinct in the south, and they're doing really well in the north." And the data showed such a *big* difference—that they had actually shifted their range! This had to be the result of some environmental factor that's consistent all the way from Mexico to Canada. Because I had gone to each site, I knew they were still good habitats for *editha*. So it was obvious that the southern populations had not become extinct because of land-use changes—because of destruction of habitat; this population shift did not occur because someone had built houses there.

Earlier research showed that an increase of only 2°C can have a big impact on the fitness of *editha* females. In the western U.S. (where *editha* lives), records indicate that temperatures have gone up about 0.7°C since about 1900. It seems likely that *editha*'s geographic shift is related to this temperature change. Overall, our research indicates that regional warming trends have already affected butterfly populations in western North America and Europe. The ranges of most species have moved north (66%), but only two have moved south.

What does this mean for the human population?

Current models predict that global temperatures will rise by 2°C by 2050. This amount of change means we must worry about flooding of coastal areas, increased heat stress for plants and animals, including people, and an increase in populations in higher latitudes of insects that carry tropical diseases. We have the technology to build dikes, but imagine the massive efforts and costs of the task—to protect continental coastlines and island nations.

Do you have a take-home message for students?

I think we have to stop thinking in terms of everything being stable. We can't just fence off some areas and expect them to be OK. Humans have made Earth a managed system, and we've got to think in terms of stewardship.

We'll discuss these issues and how they relate to communities and ecosystems in Chapter 38.

What is a "geographic shift" in a population?

A change in the population's geographic range relative to a change in some environmental factor, such as temperature

CHAPTER REVIEW

Chapter Summary

Population dynamics is concerned with changes in population size and the factors that regulate populations over time **(Introduction)**. An ecological definition of a population is a single-species group of individuals that use common resources and are regulated by the same environmental factors. Researchers must define a population by geographical boundaries appropriate to the questions being asked. Density and dispersion patterns are two important characteristics used in analyzing and comparing populations **(35.1)**.

Population Structure and Dynamics (35.2–35.5) Population density is the number of individuals in a given area or volume. It is sometimes possible to count all the individuals in a population, but density is usually estimated by sampling. A population's dispersion pattern is the pattern of spacing, which may be clumped, uniform, or random **(35.2)**. Idealized models describe two kinds of population growth. Exponential growth is the accelerating increase that occurs during a time when growth is unregulated. The equation $G = rN$ describes a J-shaped growth curve, which is typical of exponential growth; in the equation, G = the population growth rate, r = an organism's inherent capacity to reproduce, and N = the population size. Logistic growth is slowed by population-limiting factors and tends to level off at carrying capacity, which is the number of individuals the environment can support with no net increase or decrease. The equation $G = rN(K - N)/K$ describes a logistic growth curve, where K = carrying capacity, and the term $(K - N)/K$ accounts for the leveling off of the curve. The logistic growth model predicts that a population's growth rate will be low when the population size is either small or large, and highest when the population is at an intermediate level relative to the carrying capacity (35.3). Density-dependent factors, such as limited food supply and the buildup of toxic wastes, depress a population's growth rate by increasing the death rate, decreasing the birth rate, or both. These factors affect a greater percentage of individuals as the number of individuals increases. Density-independent factors limit many natural populations. These factors, such as weather, fire, and floods, affect the same percentage of individuals regardless of population density. Most populations are probably regulated by a mixture of density-dependent and density-independent factors (35.4). Some populations go through boom-and-bust cycles of growth and decline (35.5).

Life Histories and Their Evolution (35.6–35.7) Life tables and survivorship curves predict an individual's statistical chance of dying or surviving during each interval in its life history. The three types of survivorship—living a full life span and dying of old age, dying young, or maintaining a constant rate of mortality across the life span—reflect important species differences in life history (35.6). Natural selection shapes a species' life history, the series of events from birth through reproduction to death. There are two hypothetical extremes. Opportunistic populations produce many offspring and grow exponentially in unpredictable environments. Equilibrial populations raise few offspring and maintain relatively stable populations in stable environments (35.7).

The Human Population (35.8–35.10) The human population as a whole has doubled three times in the last three centuries. It now stands at about 5.9 billion and may reach 8 billion by the year 2025. Most of the increase is due to improved health and technology, which have decreased the death rate. The human population faces an uncertain future. Eventually, births and deaths will balance—that is, our population will stop growing. Our behavior will probably determine whether this results from a decrease in birth rates or an increase in death rates (35.8). The age structure of a population—its proportion of individuals in different age groups—affects its future growth. A human population with a large percentage of young adults and children (as in developing nations) will grow more rapidly and have different societal needs than a population whose age groups are more nearly balanced. Populations in developed nations are expanding slowly (United States), are stable with approximately equal birth and death rates (Italy), or are declining (Germany) (35.9). Global warming, resulting chiefly from increased output of greenhouse gases (mainly CO_2 and methane) by developed nations, is a serious consequence of population expansion (35.10).

Testing Your Knowledge

Multiple Choice

1. Which of the following shows the effects of a density-dependent limiting factor?
   a. A forest fire kills all the pine trees in a patch of forest.
   b. Early rainfall triggers the explosion of a locust population.
   c. Drought decimates a wheat crop.
   d. Silt from logging kills half the young salmon in a stream.
   e. Voles multiply, and their food supply begins to dwindle.

2. With regard to its percent increase, a population that is growing logistically
   a. grows fastest when density is lowest
   b. has a high intrinsic rate of increase
   c. grows fastest at an intermediate population density
   d. grows fastest as it approaches carrying capacity
   e. is always slowed by density-independent factors

3. Pine trees in a forest tend to shade and kill pine seedlings that sprout nearby. This causes the pine trees to __________. *(Explain your answer.)*
   a. increase exponentially
   b. grow in a clumped pattern
   c. grow in a uniform pattern
   d. exceed their carrying capacity
   e. grow in a random pattern

4. To figure out the human population density of your community, you would need to know the number of people living there and
   a. the land area in which they live
   b. the birth rate of the population
   c. whether population growth is logistic or exponential
   d. the dispersion pattern of the population
   e. the carrying capacity

5. Skyrocketing growth of the human population since the beginning of the Industrial Revolution appears to be mainly a result of
   a. migration to thinly settled regions of the globe
   b. better nutrition boosting the birth rate
   c. a drop in the death rate due to better health care
   d. the concentration of humans in cities
   e. social changes that make it desirable to have more children

Describing, Comparing, and Explaining

1. Compare exponential and logistic population growth. Under what conditions might each occur? What might limit growth?

2. What is survivorship? What does a survivorship curve show? Explain what the three survivorship curves in Figure 35.6 tell us about humans, squirrels, and oysters.

Thinking as a Scientist

The mountain gorilla, spotted owl, giant panda, snow leopard, and grizzly bear are all endangered by human encroachment on their environments. Another thing these animals have in common is that they all have equilibrial life histories. Why might they be more easily endangered than animals with opportunistic life histories? What general type of survivorship curve would you expect these species to exhibit? Explain your answer.

Science, Technology, and Society

Many people regard the rapid population growth of developing countries as our most serious environmental problem. Others feel that the growth of developed countries, though slower, is actually a greater threat to the environment. What kinds of environmental problems result from population growth in (a) developing countries, and (b) developed countries? Which do you think is the greater threat? Why?

Media Resources

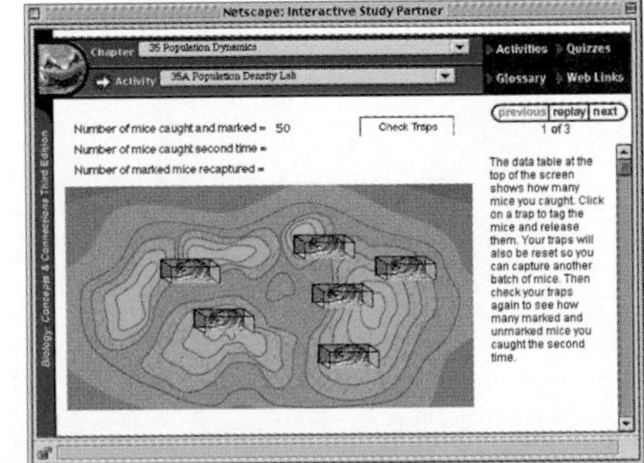

For further review of the concepts in this chapter, use the activities and quizzes on the CD-ROM.

Visit the Special Edition of The Biology Place web site for activities, quizzes, news articles, and web links.

36 Communities and Ecosystems

Outline

A 4-millimeter-long wasp called *Apanteles glomeratus* stabs through the skin of a caterpillar and lays her eggs (photograph at the left). The caterpillar, a larva of the cabbage white butterfly (*Pieris rapae,* below left), is doomed. It will be destroyed from within, as the wasp larvae hatch and nourish themselves on its internal organs. We benefit from the wasp's behavior, for *Pieris* caterpillars eat cabbages and broccoli and are abundant agricultural pests.

Apanteles wasps help control cabbage butterfly populations, but they never come close to eliminating them. There are always fewer *Apanteles* than *Pieris* caterpillars, for *Apanteles* has problems of its own. Other wasps, called ichneumons (below center), can detect when a *Pieris* caterpillar contains *Apanteles* larvae. When a female ichneumon finds one of these caterpillars, she pierces it and deposits her eggs inside the *Apanteles* larvae. And the story may go on. Yet another wasp, a chalcid (below right), may lay its eggs inside the ichneumon larvae. When this happens, the hapless caterpillar houses a three-step food chain: chalcids eating ichneumons eating *Apanteles* eating the caterpillar. Usually, only the chalcids will emerge from the dead husk of the caterpillar.

Though he was unaware of this unusual food chain, eighteenth-century English satirist Jonathan Swift wrote:

> So, Nat'ralists observe, a Flea
> Hath smaller Fleas that on him prey,
> And these have smaller Fleas to bite 'em,
> And so proceed *ad infinitum*.

Swift's flea analogy was meant to ridicule picky literary critics, but his lines paint a good picture of the complex relationships between *Pieris* caterpillars and wasps.

Connections among organisms is the major topic of this chapter. In the first few modules, we see that a biological community derives its structure from the interactions and interdependence of the organisms living within it. Later, we find that ecosystem functioning involves complex interactions between the community and its physical environment. When we get to ecosystems, we take another look at Jonathan Swift's poem; although it works as satire, its last line can't be taken literally for an ecosystem, because no food chain, including the extensive one in a *Pieris* caterpillar, can be infinite.

A cabbage white butterfly *(Pieris rapae)*

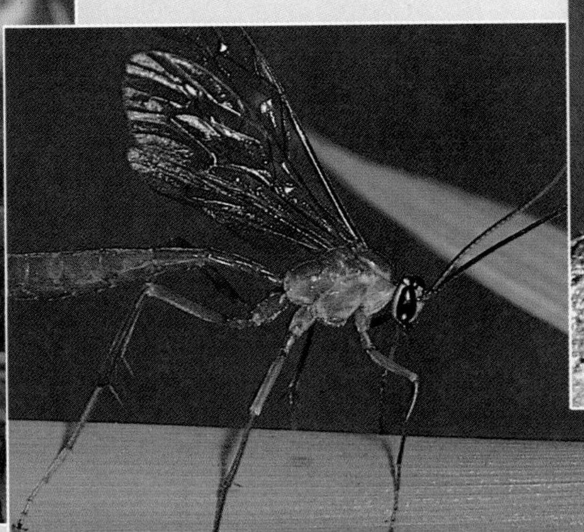

An ichneumon wasp

A chalcid wasp

36.1 A community is all the organisms inhabiting a particular place

In the previous chapter, we saw that a population is a group of interacting individuals of a particular species. We now move one step up the hierarchy of nature to the level of the community. A biological **community** is all the populations of organisms living together and potentially interacting in a particular area. For example, the wild turkeys, grass, shrubs, and trees in Figure 36.1 are components of a community in a deciduous forest. Wild turkeys interact with many other species, such as the oak trees whose acorns they eat.

Just as a population has certain characteristics, such as density and dispersion pattern, a community has its own set of properties. Its defining characteristics are its diversity, its prevalent form of vegetation, its stability, and its trophic structure.

The *diversity* of a community—the variety of different kinds of organisms that make it up—has two components. One is species richness, or the total number of different species in the community. The other is the relative abundance of the different species. For example, imagine two communities, each with 100 individuals distributed among four different species (A, B, C, and D) as follows:

Community 1: 25A, 25B, 25C, 25D
Community 2: 97A, 1B, 1C, 1D

The species richness is the same for both communities, because they both contain four species, but the relative abundance is very different. Suppose these communities were two forests. You would easily notice the four different types of trees in Community 1, but without looking carefully, you might see only the abundant species A in the second forest. You could get a true picture of diversity in these communities only by carefully examining both species richness and relative abundance.

Figure 36.1 Wild turkeys in a forest community

The second property of a community, its *prevalent form of vegetation*, applies mainly to terrestrial situations. For example, deciduous trees are the prevalent components of the community in a temperate deciduous forest biome (see Module 34.16). When we look at the biological community in a biome, we see not only which plants are dominant, but also how the plants are arranged, or "structured." For instance, a deciduous forest has a pronounced vertical structure: The treetops form a top layer, or canopy, under which there is a subcanopy of lower branches, and small shrubs and herbs carpet the forest floor. Thus, the prevalent form of a community's vegetation has two components: the kinds of dominant plants and their particular structure. The types and structural features of plants largely determine the kinds of animals that live in a community.

The third property of a community, *stability*, refers to the community's ability to resist change and return to its original species composition after being disturbed. Stability depends on both the type of community and the nature of disturbances. For example, a forest dominated by cedar and hemlock trees is a highly stable community in that it may last for thousands of years with little change in species composition. Large cedars and hemlocks even withstand most lightning-caused fires, which kill small trees and shrubs growing in forest openings or in the shade of the dominant trees. However, when a fire does kill the dominant trees, a cedar/hemlock forest might seem less stable than, say, a grassland, because it will take much longer for the forest to return to its original species composition.

The fourth property of a community is its **trophic structure** (from the Greek *trophe*, nourishment), the feeding relationships among the various species making up the community. A community's trophic structure determines the passage of energy and nutrients from plants and other photosynthetic organisms to herbivores and then to carnivores. For example, as we saw in the chapter's introduction, the energy in a cabbage leaf eaten by a moth caterpillar may end up supporting one, two, or three kinds of carnivorous wasp larvae.

With the four main properties of a community in mind, we turn next to the forces that tie species populations together into communities. These forces—actually, the interactions among the species themselves—are of three main types: competition, predation, and symbiosis. In discussing these interactions, we'll see that they are all influenced by evolution through natural selection. We begin with a look at competition. 

? How could a community appear to have relatively little diversity even though it is rich in species?

This would be the case if one or a few of the diverse species accounted for almost all the organisms in the community, with the other species being rare.

36.2 Competition may occur when a shared resource is limited

36A
See Activity 36A on the CD-ROM.

When two populations (for example, two species of birds in a forest) both require a limited resource (nest sites, for instance), individuals of the two species compete for the resource. The contest, called **interspecific competition,** can inhibit the growth of both populations. Sometimes the competition even eliminates one of the populations from a community. Interspecific competition may play a major role in structuring a community.

In 1934, Russian ecologist G. F. Gause studied the effects of interspecific competition in laboratory experiments with two closely related species of protists, *Paramecium aurelia* and *P. caudatum.* Gause cultured the protists under stable conditions with a constant amount of food added every day. When he grew the two species in separate cultures, each population grew rapidly and then leveled off at what was apparently the carrying capacity of the culture (see Module 35.3). In contrast, when Gause cultured the two species together, *P. aurelia* apparently had a competitive edge in obtaining food, and *P. caudatum* was driven to extinction in the culture. Gause concluded that two species so similar that they compete for the same limiting resources cannot coexist in the same place. One will use the resources more efficiently and thus reproduce more rapidly. Even a slight reproductive advantage will eventually lead to local elimination of the inferior competitor. Gause's ideas were termed the **competitive exclusion principle.**

Numerous tests of the competitive exclusion principle include field experiments on the interactions between populations of two species of barnacles that attach to intertidal rocks on the North Atlantic coast (Figure 36.2). Both types of barnacles, *Balanus* and *Chthamalus,* grow on rocks that are exposed during low tide; when immersed at high tide, they feed on organic particles suspended in the water. Both types are attached as adults, but have free-swimming larvae that may settle and begin to develop on virtually any rock surface.

Chthamalus (brown in the drawing in Figure 36.2) occupies the upper parts of the rocks, which are out of the water longer during low tides. *Balanus* (gray in the drawing) fails to survive as high on the rocks as *Chthamalus,* apparently because *Balanus* dries out in the air. When experimenters removed *Balanus* from the lower rocks, *Chthamalus* spread lower, colonizing the unoccupied rocks. However, when both species colonize the same rock, *Balanus* eventually displaces *Chthamalus* on the lower part of the rock. Researchers conclude that the upper limit of *Balanus'* distribution is set mainly by the availability of water, whereas the lower limit of *Chthamalus'* distribution is set by competition.

The competitive exclusion principle applies to what is called a population's niche. In ecology, a **niche** is a population's role in its community, or the sum total of its use of the biotic and abiotic resources of its habitat. For example, the temperature range within which a population lives, the time of day individuals feed, and the type of food they consume are all part of its niche. In Figure 36.2, the attachment sites on intertidal rocks, the amount of exposure to seawater and air, and interactions with other species are some of the aspects of each barnacle's niche. Combining the niche concept with the competitive exclusion principle, we might predict that populations of two species cannot coexist in a community if their niches are too similar.

There are actually more questions than answers about the role of competition and competitive exclusion in natural communities. It is hard to determine which facets of the niche are critical when species seem to be competing. And it's even harder to assess what has happened in the evolutionary past. When competitive exclusion has occurred, has it actually determined how many species coexist in a community? Did "outcompeted" populations become locally extinct or migrate out of the area? Or were they altered by natural selection, becoming able to use a different set of resources? These are some of the fascinating but difficult questions confronting researchers in community ecology.

What experiment could you perform using the intertidal rock shown in Figure 36.2 to test the hypothesis that it is mainly susceptibility to drying, not competitive exclusion, that keeps *Balanus* from populating the upper parts of the rock?

You could remove *Chthamalus* from the upper part of the rock to see whether *Balanus* still failed to spread upward even in the absence of a competition.

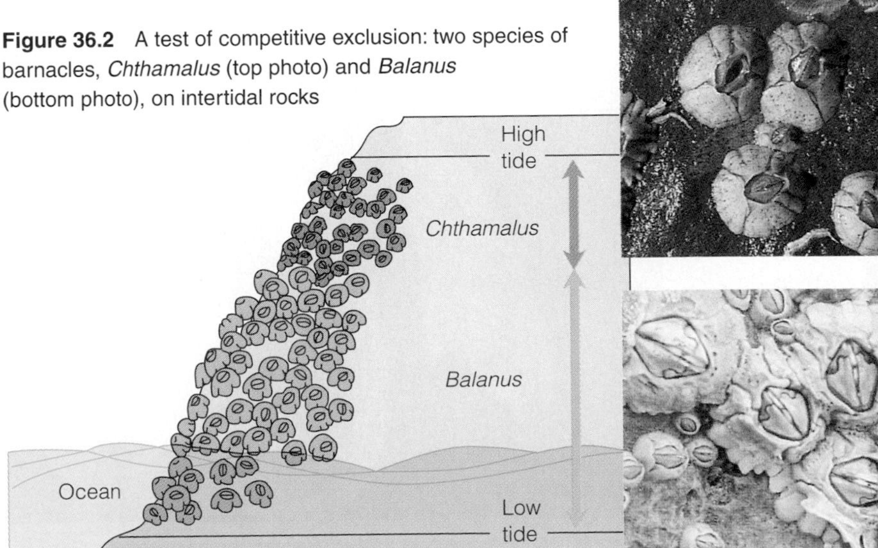

Figure 36.2 A test of competitive exclusion: two species of barnacles, *Chthamalus* (top photo) and *Balanus* (bottom photo), on intertidal rocks

36.3 Predation leads to diverse adaptations in both predator and prey

36A

In **predation,** an interaction where one species eats another, the consumer is called a **predator** and the food species is known as the **prey.** We will use these terms not only for cases of animals eating other animals, but also for plant-herbivore interactions, where the plant is prey.

No species is entirely free from predation, especially when young, and mechanisms of defense against being eaten have evolved in every species. Some species grow too big to be consumed by certain predators; adult elephants, for instance, are generally safe from attack by lions and other large cats. Other species can flee or hide, and many plant and animal species have protective armor, spines, or external glands that produce noxious chemicals. These and other antipredator defense mechanisms evolve through natural selection, as predators and prey interact. Adaptations that favor the predator, such as speed and agility, claws, fangs, and ambush tactics, counter the defense systems of prey. Some predator-prey interactions illustrate the concept of **coevolution,** a series of reciprocal adaptations in two species. Coevolution occurs when a change in one species acts as a new selective force on another species, and counteradaptation of the second species, in turn, affects the selection of individuals in the first species.

Figure 36.3A illustrates an example of coevolution of an herbivorous insect (the caterpillar of the butterfly *Heliconius,* top left) and a plant (the passionflower *Passiflora,* a tropical vine). *Passiflora* produces toxic chemicals that protect its leaves from most herbivorous insects, but *Heliconius* caterpillars have digestive enzymes that break down the toxins. As a result, *Heliconius* gains access to a food source that few other insects can eat. These poison-resistant insects seem to be a strong selection force favoring the survival of *Passiflora* plants that have additional defenses. For instance, the leaves of some species of *Passiflora* produce yellow sugar deposits that look like *Heliconius* eggs. You can see two eggs in the top right photograph of Figure 36.3A and two of the sugar deposits in the bottom photo. Female *Heliconius* butterflies avoid laying their eggs on leaves that already have eggs. Presumably, this is an adaptation that ensures an adequate food supply for offspring, since only a few caterpillars will hatch and feed on any one leaf. Because the butterfly often mistakes the yellow sugar deposits for eggs, *Passiflora* species with the deposits are more likely to escape predation.

Because plants cannot run away from herbivores, chemical toxins, often in combination with various kinds of antipredator spines and thorns, are their main arsenals against being eaten to extinction. Among such chemical weapons are the poison strychnine, produced by a tropical vine called *Strychnos toxifera;* morphine, from the opium poppy; nicotine, produced by the tobacco plant; mescaline, from peyote cactus; tannins, from a variety of plant species; and many substances we use as flavorings but that are toxic to some predators (cinnamon, cloves, and mint, for instance).

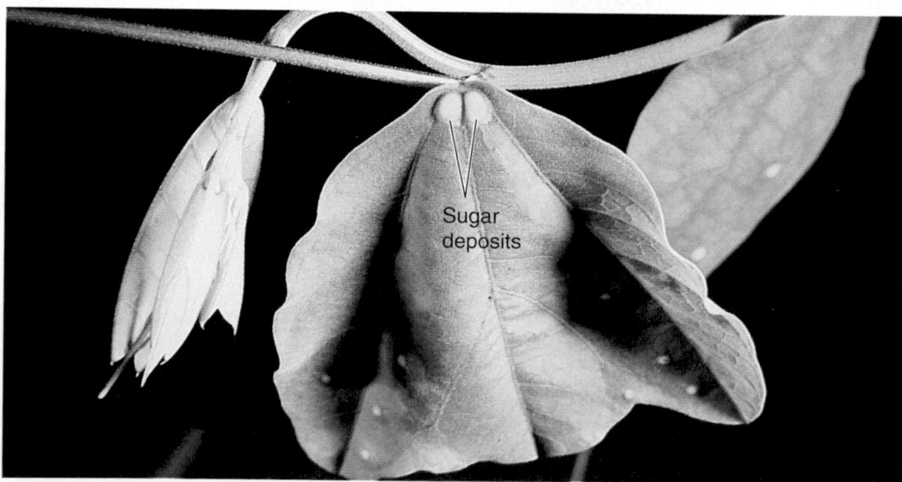

Figure 36.3A Coevolution: *Heliconius* and the passionflower vine

Animal defenses against predators are extremely diverse. Mechanical defenses, such as the porcupine's sharp quills, may be the most obvious, but chemical defenses are also widespread. Animals with effective chemical defenses are often brightly colored, a warning to predators. The vivid markings of the poison-arrow frog (Figure 36.3B), an inhabitant of rain forests in Costa Rica, warn of deadly alkaloids in the frog's skin; predators learn about this as soon as they touch the frog. In some parts of South America, human hunters in the rain forest tip their arrows with poisons from similar frogs to bring down large mammals.

Camouflage is an especially common type of defense in the animal kingdom. Figure 36.3C shows that a canyon tree frog *(Hyla arenicolor),* common in the southwestern U.S., becomes almost invisible on a granite background.

Figure 36.3B Chemical defenses: the poison-arrow frog

Figure 36.3C Camouflage: a canyon tree frog on granite

Mimicry is another effective defense mechanism. The insect on the bottom in Figure 36.3D is a honeybee, which has a stinger armed with a toxic chemical. The insect on the top is a nonstinging flower fly. Many predators learn to avoid the harmful bee and may not attack the harmless fly because it looks like a bee. This type of mimicry, in which a palatable species mimics an unpalatable one, is called **Batesian mimicry.** For this type of mimicry to be effective, the mimic must be considerably less abundant than the species it copies; otherwise, predators would learn that, for example, yellow insects that look like bees are good rather than bad to eat.

Figure 36.3E illustrates another kind of mimicry, called **Müllerian mimicry,** in which two unpalatable species that inhabit the same community mimic each other. On the bottom is a type of wasp called a yellow jacket; on the top is a species of bee called a cuckoo bee. Both species have stingers that release toxic chemicals. Presumably, both gain an adaptive advantage beyond their own defenses because predators will learn more quickly to avoid any prey with this appearance. 

Figure 36.3D Batesian mimicry: a flower fly (top) and a honeybee (bottom)

Figure 36.3E Müllerian mimicry: a cuckoo bee (top) and a yellow jacket (bottom)

? Is the evolution of mimicry, in which a resemblance of one species to another evolves, an example of coevolution? Explain your answer.

No. Coevolution is characterized by adaptive responses of two species to one another. In the case of mimicry, the "model" species is not generally adapting to changes in the mimic.

36.4 Predation can maintain diversity in a community

In simple laboratory experiments where a single predator species is kept with a single prey species having no refuge, the predator may devour all the prey and then itself perish from starvation. In Figure 36.4A, the creature on the left in the micrograph is a predatory protist called *Didinium.* It is shown here eating a *Paramecium.* The graph shows *Didinium's* effect on a population of *Paramecium* when the two are grown together in a test tube. At first, the *Paramecium* population expands rapidly, but *Didinium* soon kills all the *Paramecium* and then dies out itself. If this scenario occurred often in nature, predators would reduce the diversity of species in communities. However, predators rarely drive their prey to extinction, for several reasons. For one thing, natural communities are complex, with many species; predators themselves are often preyed upon, limiting their numbers. Also, a predator may be able to switch to an alternative food source when the population of one prey species dwindles. And the defensive mechanisms of prey play a major role in keeping prey populations from extinction.

Several studies have shown that predator-prey relationships can actually help maintain community diversity rather than reduce it. Experiments by American ecologist Robert Paine in the 1960s were among the first to provide evidence. Paine removed the dominant predator, a sea star of the genus *Pisaster* (Figure 36.4B), from experimental areas within the intertidal zone of the Washington coast. The result was that *Pisaster's* main prey, a mussel of the genus *Mytilus,* outcompeted many of the other shoreline organisms (barnacles and snails, for instance) for the important resource of space on the rocks. The number of species dropped from fifteen to eight.

The experiments of Paine and others have generated the concept of keystone predators. A **keystone predator** is a species, such as *Pisaster,* that reduces the density of the strongest competitors in a community. In doing so, the predator helps maintain species diversity by preventing competitive exclusion of weaker competitors. In Paine's study, predation by *Pisaster* was a key factor in maintaining populations of at least seven other species.

What is the advantage to a keystone predator of being specialized to feed mainly on those prey species that are otherwise the most successful among potential prey species?

The most competitive prey species probably represent the most abundant and dependable food source for the predator.

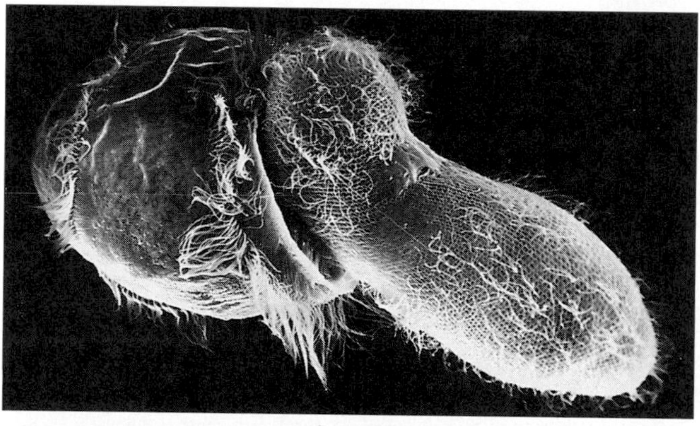

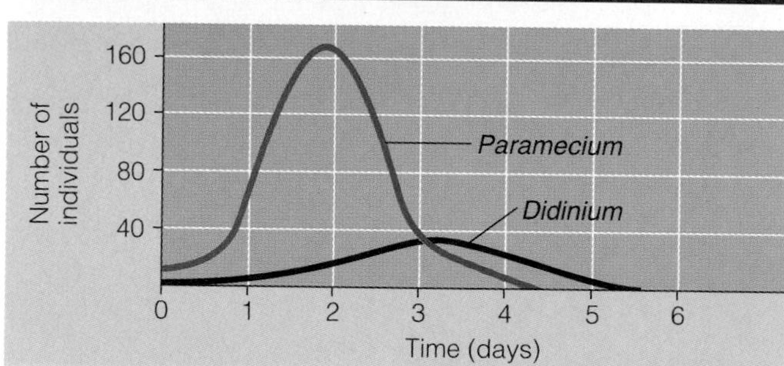

Figure 36.4A Predator-prey dynamics in a laboratory experiment

Figure 36.4B *Pisaster* sea star, a keystone predator, eating a mussel

36.5 Symbiotic relationships help structure communities

36A

A **symbiotic relationship** is an interaction between two or more species in which one species lives in or on another species. There are three main types of symbiotic relationships: parasitism, commensalism, and mutualism. Parastism and mutualism can be key factors in community structure.

Parasitism (Greek, *para,* near, and *sitos,* food) is a kind of predator-prey relationship in which one organism, the parasite, derives its food at the expense of its symbiotic associate, the host. Parasites are usually smaller than their hosts. An example of a parasite is a tapeworm that lives inside the intestines of a larger animal and absorbs nutrients

Figure 36.5A Rabbits in Australia before a parasite was introduced

from its host. Natural selection favors the parasites that are best able to find and feed on hosts. At the same time, defensive abilities of the hosts are also selected for. As an example, plants make chemicals toxic to fungal and bacterial parasites, along with ones toxic to predatory animals (sometimes they are the same chemicals). In vertebrates, the immune system provides a multipronged defense against internal parasites.

At times, it is actually possible to watch the effects of natural selection in host-parasite relationships. Scenes like the one shown in Figure 36.5A were common in Australia during the 1940s. The continent was overrun by hundreds of millions of European rabbits, whose population had exploded from just twelve pairs imported a century earlier. The rabbits destroyed huge expanses of Australia and threatened the sheep and cattle industries. In 1950, myxoma virus, a parasite that infects rabbits, was deliberately introduced into Australia to control the rabbit population. Spread rapidly by mosquitoes, the virus devastated the rabbit population. The virus was less deadly to the offspring of surviving rabbits, however, and it caused less and less harm over the years. Apparently, genotypes in the rabbit population were selected that were better able to resist the parasite. Meanwhile, the deadliest strains of the virus perished with their hosts as natural selection favored strains that could infect hosts but not kill them. Thus, natural selection stabilized this host-parasite relationship, and today, the virus that was originally introduced has only a mild effect on the rabbit population. Pest managers continue to control the rabbits by introducing other deadly strains of the myxoma virus.

In contrast to parasitism, in **commensalism** (Latin, *com*, together, and *mensa*, table), one partner benefits without significantly affecting the other. Few cases of absolute commensalism probably exist, because it is unlikely that one of the partners will be completely unaffected. Commensal associations sometimes involve one species' obtaining food that is inadvertently exposed by another. For instance, several kinds of birds feed on insects flushed out of the grass by grazing cattle. It is difficult to imagine how this could affect the cattle, but the relationship may help or hinder them in some way not yet recognized.

The third type of symbiosis, **mutualism** (Latin, *mutualis*, reciprocal) benefits both partners in the relationship. We have discussed several mutualistic associations in previous chapters—for instance, the legume plants and their nitrogen-fixing bacteria, and the interactions between flowering plants and their pollinators. Figure 36.5B illustrates another case of mutualism. This is part of a branch of a bull's horn acacia tree, which grows in Central and South America. The tree provides room and board for ants of the genus *Pseudomyrmex*. The ants live in large, hollow thorns and eat sugar secreted by the tree. As the photograph suggests, they also eat the yellow structures at the tips of leaflets; these are protein-rich swellings that seem to have no function for the tree except to attract ants. The ants benefit the host tree by attacking virtually anything that touches it. They sting other insects and large herbivores and even clip surrounding vegetation that grows near the tree. When the ants are removed, the trees usually die, probably because herbivores damage them so much that they are unable to compete with surrounding vegetation for light and growing space.

The complex interplay of species in symbiotic relationships highlights an important point about communities: Their structure depends on a web of diverse connections among organisms. In the next module, we see what can happen when community structure is disrupted.

Which type of symbiotic relationship is represented by mycorrhizae, the associations of plant roots and fungi described in Module 32.11?

Mutualism

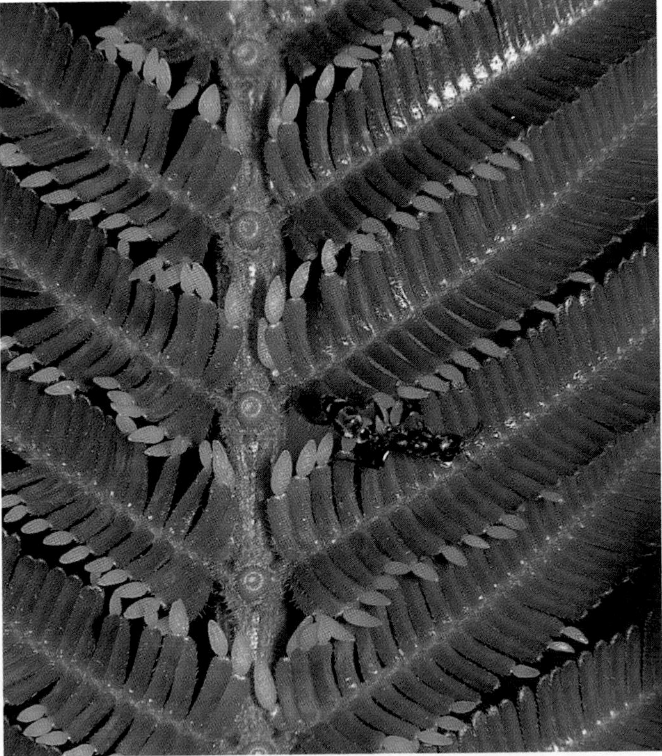

Figure 36.5B Mutualism between an acacia tree and ants

36.6 Disturbance is a prominent feature of most communities

A traditional view of biological communities is that they are characterized by **stability,** the tendency to remain in a more or less constant balance due largely to interactions among organisms. According to this equilibrium view, when a community is disturbed by such factors as storms, fires, floods, and droughts, it tends to return to an original, balanced condition. Disturbances affect all communities, and an increasingly popular view in ecology is that disturbance and change characterize most biological communities more than stability and balance. A **disturbance** is a force that alters a biological community and usually removes organisms from it.

We tend to think of disturbances as having negative impacts, but this is usually only part of the story. Disturbances can affect community structure on virtually any scale. Small-scale disturbances often have positive effects, such as creating new opportunities for species. For example, when a tree falls, it disturbs the immediate surroundings; however, the fallen log fosters new habitats, and the depression left by its roots may fill with water and be used as egg-laying sites by frogs, salamanders, and numerous insects.

Communities may change drastically after a flood, fire, glacial advance or retreat, or volcanic eruption strips away their vegetation. A variety of species may colonize the disturbed area. Later these may be replaced as yet other species colonize the area. This process of community change resulting from disturbance is called **ecological succession.**

When a community arises in a virtually lifeless area with no soil, the process is called **primary succession.** Examples of such areas are new volcanic islands or the rubble left by a retreating glacier. Often the only life forms initially present are autotrophic microorganisms. Lichens and mosses, which grow from windblown spores, are commonly the first large photosynthetic organisms to colonize the barren ground. Soil develops gradually, as organic matter accumulates from the decomposed remains of the early colonizers. Once soil is present, the lichens and mosses are overgrown by grasses, shrubs, and trees that sprout from seeds blown in from nearby areas or carried in by animals. In turn, the area may be colonized by plants that will become the community's prevalent form of vegetation. Primary succession from barren soil to a community, such as a deciduous forest, can take hundreds or thousands of years.

Secondary succession occurs where a disturbance has destroyed an existing community but left the soil intact. For instance, forested areas in the eastern U.S. that are cleared for farming will, if abandoned, undergo secondary succession and may eventually return to forest. The earliest plants to recolonize an area are often species with opportunistic life histories (see Module 35.7). Many are herbaceous (nonwoody) species that grow from windblown or animal-borne seeds. These plants thrive where there is little competition from other plants. If the area is not burned or heavily grazed, woody shrubs may eventually replace most of the herbaceous species. Later yet, trees may replace most of the shrubs.

An early hypothesis viewed succession as a linear sequence that inevitably moves to a permanent final stage—called a climax community—for a particular climate and soil type. Indeed, some communities, such as those in temperate deciduous forests dominated by oak and hickory trees, seem to be mature stages that will persist indefinitely. In many areas, however, what appear to be climax communities may not be stable over long periods, and many communities are routinely disturbed and never reach a climax. For instance, historically, prairie grasslands were often swept by fire. Below are photographs taken ① before, ② during, and ③ after a controlled burn conducted by ecologists studying the long-term effects of fire on a tall-grass prairie in Kansas. Without fire, some grassland communities would eventually become forests. In this case, we might say that forest is the climax community, but that makes little sense if the forest community never develops. In effect, periodic disturbances (fires) stabilize the community at a stage that does not fit the traditional idea of a climax community. (The trees in the photographs below grow along a stream and were not affected by the fire.) Understanding the diverse effects of disturbance in communities is especially important today; as we discuss in Chapter 38, humans are the most widespread and significant agents of disturbance.

What is the main abiotic factor that distinguishes primary from secondary succession?

Absence (primary) versus presence (secondary) of soil at the onset of succession

Figure 36.6 Effect of fire on an ecosystem

TALKING ABOUT SCIENCE

36.7 Ecologist Frank Gilliam discusses the role of fire in ecosystems

Figure 36.7A Plant ecologist Frank Gilliam

Frank S. Gilliam is a professor of biological sciences at Marshall University in Huntington, West Virginia. He has studied temperate deciduous forests in the Appalachian Mountains of Virginia and West Virginia, coastal plain pine forests of the southeastern U.S., and tall-grass prairies in Kansas. One of Dr. Gilliam's special interests is the role of fire in shaping ecosystems. The photos on the facing page were taken during one of his field studies. In a 1996 interview, he summarized his view of fire in nature this way:

> Today we realize that we cannot afford to view fire as the destructive scourge we used to hear about. Like wind, rain, soil, and chemical nutrients, it is an important abiotic factor, even a creative force, in many ecosystems.

We asked him to elaborate.

> Any ecosystem that has a rapid buildup of fuel, periodic droughts, and other conditions that promote burning tends to be strongly influenced by fire. In a typical deciduous forest in North America, fuels do not tend to build up, because leaves and other potential fuel materials decompose fairly rapidly, and because precipitation is relatively high. In contrast, fuel in the form of pine needles tends to accumulate in our southeastern pine forests. The photographs here indicate some of the changes we saw after a fire in the lower coastal plain of South Carolina (Figure 36.7B). The soils there are mainly acidic, and this keeps the decomposition rate on the ground very low. Seasonal droughts are also fairly common, and unless humans prevent fires, these forests tend to burn about every 5–7 years. The conditions of the environment select plants that are fire-adapted, and unless the fuel builds up for many years, fires do not get hot enough to kill the large pine trees; only the fuel on the ground and low-growing plants burn. Following a burn, the numbers and variety of nonwoody plants (the herbs) usually increase dramatically, because fire makes more nutrients, such as nitrogen, phosphorus, and potassium, available to them. In one 100-square-meter plot, for instance, we found that a burn in the winter was correlated with a nearly two-fold increase in the number of herbs growing in the forest.
>
> In another case, the tall-grass prairie I studied in Kansas, essentially all of the plant material is fuel for the next year; and it's ideal fuel—standing dead grass that's well aerated, with no wind buffers. In presettlement times, lightning strikes would start fires that spread for thousands of hectares. Under natural conditions, grasslands tend to burn about every 2–4 years, and because all of a year's plant growth becomes top-quality fuel at the end of the growing season, many grasslands can burn every year. Tall-grass prairies are not just fire-tolerant; they are fire-dependent. A single fire under natural conditions produces essentially no change in the species composition, but these areas typically get enough precipitation that shrubs and trees will replace the grasses unless there is a fire. People tend to think of fire as a type of disturbance in nature. But in a grassland—or, for that matter, in many forests—is fire really a disturbance, or is the absence of fire a disturbance? Whatever the answer, our general attitude about fire has changed drastically. Even in an area like the deciduous forests of the central Appalachians, where fire does not play an essential role, it is usually not terribly destructive. And fire is an integral component of many ecosystems.

We asked Dr. Gilliam how and when he got interested in ecological research and fire in particular.

> I had a childhood interest in the outdoors and was lucky to grow up in Kentucky, which has fairly large areas of state parks and national forests. Actually, I can trace a lot of my interest to my ninth-grade biology course. I loved that course. It had a strong emphasis on ecology and evolution, which are closely intertwined. I like to think of Charles Darwin as one of the earliest ecologists; he predated the word *ecology*, but the work he did—the whole idea of considering the environment and the interactions between organisms and the environment—was ecological. Another thing about that course I'll never forget was a film about grasslands and how fire is important to them, as a constructive, creative force, rather than a destructive one. That was completely different from the Smokey the Bear attitude that forest fires are all bad.
>
> As a freshman in college, I knew I wanted to be a plant biologist, but what that really meant, I had no idea. I try to emphasize that to my students: Don't worry so much about exactly what you will be doing. You can't know anyway until you get farther along. Don't try to make a choice when you have too little information.

Figure 36.7B An experimental burn and regrowth (bottom) in a South Carolina pine forest

As a plant ecologist, Dr. Gilliam does research at both the community level and the ecosystem level. We asked him to explain the distinction between community-level research questions and ecosystem-level research, and to tell us where studies on ecological succession fit.

> Communities and ecosystems are concepts, ways of looking at things. If you're interested in how a particular species interacts with other organisms, that's the community level. If you're asking questions about how organisms change over time, such as succession, you're asking community-level questions. You may not be focusing only on the organisms, but you're emphasizing them. The ecosystem level adds the abiotic environment to the picture as an equal component—you might say an equal partner—with the organisms. The organisms and things like fire or precipitation patterns are part of the interacting system—the ecosystem—you're studying.

We focus on ecosystems for the rest of this chapter. 

? Why is a fire in a pine forest likely to destroy mature trees if it occurs in an area that has been "protected" from fire for many decades?

Because fuel accumulates in the form of overgrown plants on the forest floor, resulting in a fire that is hot enough to ignite and destroy mature trees

Ecosystem Structure and Dynamics

36.8 Energy flow and chemical cycling are the two fundamental processes of ecosystems

The terrarium in Figure 36.8 at the right contains an **ecosystem,** a biotic (living) community and the abiotic environment with which it interacts. The arrows in Figure 36.8 highlight the two fundamental processes—energy flow and chemical cycling—that are the basis of ecosystem interactions. The yellow, orange, and red arrows represent **energy flow,** the passage of energy through the components of the ecosystem. Energy enters the terrarium in the form of sunlight. The plants (which are autotrophs) convert the light energy to chemical energy. Animals (heterotrophs) obtain some of this chemical energy in the form of organic compounds when they eat the plants. Other heterotrophs, such as bacteria and fungi in the soil, obtain much of the chemical energy when they decompose the dead remains of plants and animals. Every use of the chemical energy by the organisms involves a loss of some energy to the surroundings in the form of heat. Eventually, therefore, the ecosystem would run out of energy if it were not powered by a continuous inflow of new energy from an outside source. For most ecosystems, the sun is the outside energy source, but exceptions include several unusual kinds of ecosystems powered by chemical energy obtained from inorganic compounds (see Chapter 34's introduction).

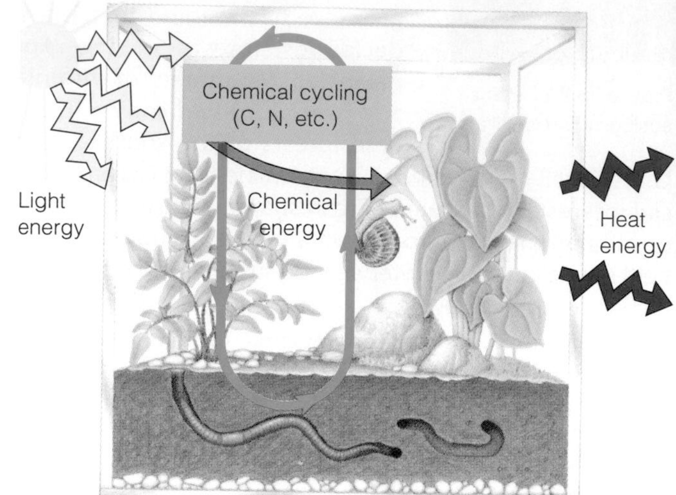

Figure 36.8 A terrarium ecosystem

In contrast to energy flow, **chemical cycling** (blue arrows) involves the circular movement of materials *within* the ecosystem. An ecosystem, especially an artificial one like a terrarium, is more or less self-contained in terms of materials. Chemical elements such as carbon and nitrogen are cycled between abiotic components (air, water, and soil) and biotic components of the ecosystem. The plants acquire these elements in inorganic form from the air and soil and fix them into organic molecules, some of which animals consume. Microorganisms that break down organic wastes and dead organisms return most of the elements in inorganic form to the soil and air. Some elements are also returned as the inorganic by-products of plant and animal metabolism.

In summary, energy flow and chemical cycling both involve the transfer of substances through the feeding levels of the ecosystem. However, energy flows into and out of the ecosystem, whereas chemicals are recycled within the ecosystem. In the rest of this chapter, we discuss energy flow and chemical cycling in ecosystems. 

? Why is the transfer of energy in an ecosystem referred to as energy *flow,* not energy *cycling*?

Because energy passes through an ecosystem, entering as sunlight and leaving as heat. It is not recycled within the ecosystem.

36.9 Trophic structure determines ecosystem dynamics

The community of living organisms within every ecosystem has a trophic structure, a pattern of feeding relationships, consisting of several different levels. The trophic structure determines the route that energy takes in flowing through the ecosystem, as well as the pattern of chemical cycling. The sequence of food transfer from trophic level to trophic level is known as a **food chain.** In Figure 36.9 (facing page), the trophic levels are arranged vertically, and the names of the levels appear in colored boxes. The arrows connecting the organisms point from the food to the consumer.

Figure 36.9 compares a terrestrial food chain and an aquatic food chain. Starting at the bottom, the trophic level that supports all others consists of autotrophs, which ecologists call the **producers** in an ecosystem. Photosynthetic producers are organisms that use light energy to power the synthesis of organic compounds. Plants are the main producers on land. In water, the producers are mainly photosynthetic protists and cyanobacteria, collectively known as phytoplankton. Multicellular algae and aquatic plants are also important producers in shallow waters.

All organisms in trophic levels above the producers are heterotrophs, or consumers, and all consumers are directly or indirectly dependent on the output of producers. Herbivores, which eat plants, algae, or autotrophic bacteria, are the **primary consumers** of autotrophs and their products.

Primary consumers on land include grasshoppers and many other insects, snails, and certain vertebrates, such as grazing mammals and birds that eat seeds and fruit. In aquatic environments, primary consumers include a variety of zooplankton (mainly protists and microscopic animals such as small shrimps) that prey on the phytoplankton.

Above the primary consumers, the trophic levels are made up of carnivores, which eat the consumers from the level below. On land, **secondary consumers** include many small mammals, such as the mouse shown here eating an herbivorous insect, and a great variety of small birds, frogs, and spiders, as well as lions and other large carnivores that eat grazers. In aquatic ecosystems, secondary consumers are mainly small fishes that eat small bottom-dwelling invertebrates and zooplankton.

Higher trophic levels include **tertiary consumers,** such as snakes that eat mice and other secondary consumers. Most ecosystems have secondary and tertiary consumers and, as the figure indicates, some also have a higher level, **quaternary consumers.** These include hawks in terrestrial ecosystems, and killer whales in the marine environment.

Not shown in the figure is another trophic level of consumers called **detritivores,** which derive their energy from **detritus,** the dead material produced by all the trophic levels. Detritus includes animal wastes, plant litter, and all sorts of dead organisms. Most organic matter eventually becomes detritus and is consumed by detritivores. A great variety of animals, often called scavengers, eat detritus. For instance, earthworms, many rodents, and insects eat fallen leaves and other detritus. Other scavengers include crayfish, catfish, and vultures.

An ecosystem's main detritivores are the prokaryotes and fungi. Enormous numbers of these microorganisms in the soil and in mud at the bottom of lakes and oceans convert (recycle) most of the ecosystem's organic materials to inorganic compounds that plants or phytoplankton can use. The breakdown of organic materials to inorganic ones is called **decomposition.** In a sense, all organisms perform decomposition. In cellular metabolism, they all break down organic material and release inorganic products, such as carbon dioxide and ammonia, to the environment.

Food chains provide an overview of ecosystem structure and function, but they are an oversimplification. Natural ecosystems rarely, if ever, have a single, unbranched food chain. We take a more realistic look at trophic relationships in the next module.

I'm eating a cheese pizza. At which trophic level(s) am I feeding?

Primary consumer (flour and tomato sauce) and secondary consumer (cheese, a product from cows, which are primary consumers)

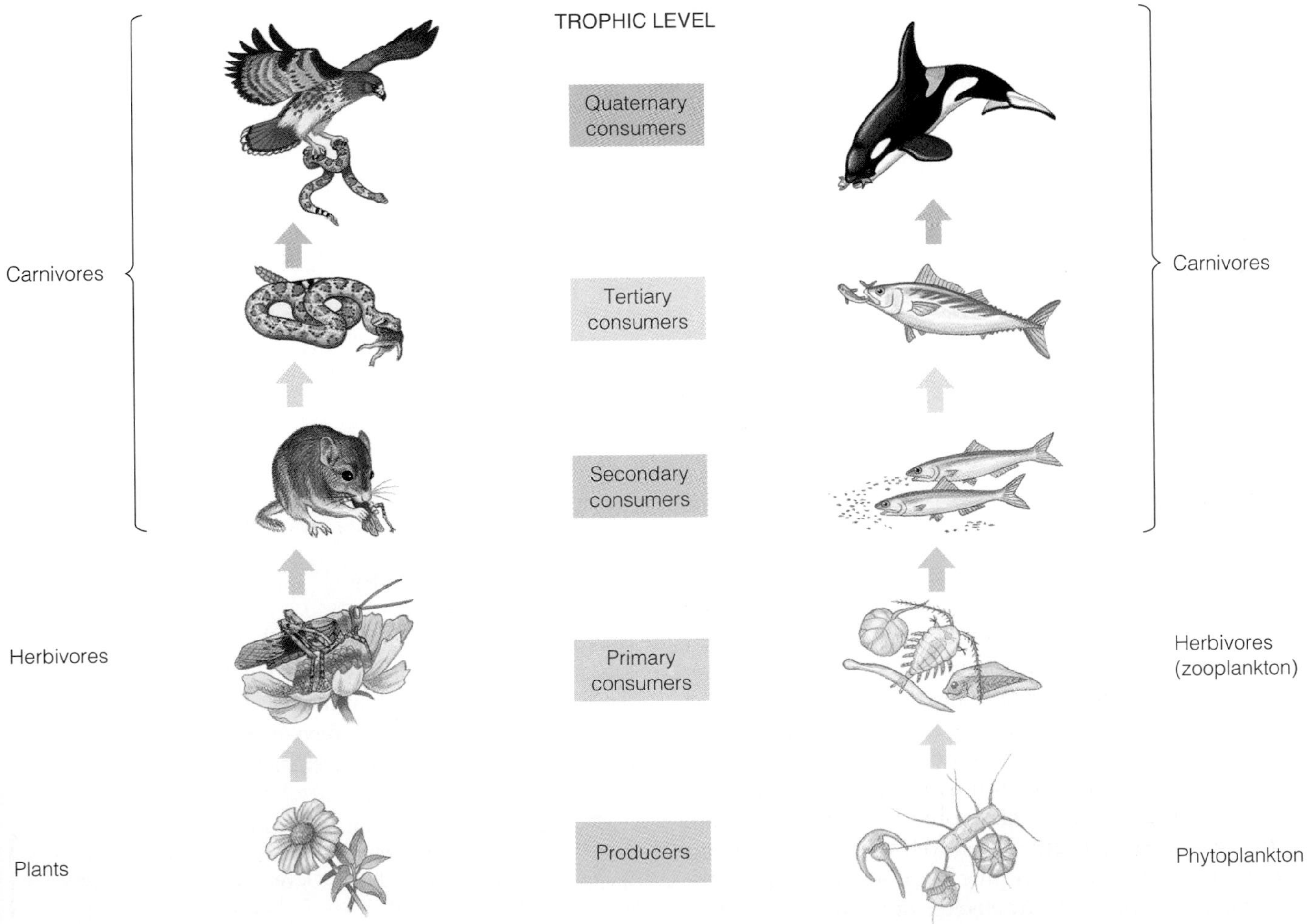

TROPHIC LEVEL

Terrestrial	Trophic level	Aquatic
Carnivores	Quaternary consumers	Carnivores
Carnivores	Tertiary consumers	Carnivores
Carnivores	Secondary consumers	Carnivores
Herbivores	Primary consumers	Herbivores (zooplankton)
Plants	Producers	Phytoplankton

Figure 36.9 **A terrestrial food chain** **An aquatic food chain**

36.10 Food chains interconnect, forming food webs

36B

A more realistic view of the trophic structure of an ecosystem than a food chain is a **food web,** a network of interconnecting food chains. Figure 36.10 shows a simplified example of a food web in a salt marsh. Rooted plants, such as grasses and sedges, are the main producers along the shoreline. In deeper water, a few submerged algae supplement the production of food by phytoplankton.

The tan arrows represent primary consumption. Notice that a consumer may eat more than one type of producer, and several species of primary consumers may feed on the same species of producer. The purple arrows represent secondary consumption. On the far left, the shrew is strictly a secondary consumer, eating insects and earthworms. The duck (to the right of the shrew), however, is both a primary consumer, eating plant seeds, and a secondary consumer, eating insects and other small invertebrates. Likewise, as indicated by the purple and blue arrows pointing to it, the owl is both a secondary and a tertiary consumer, because many of the small mammals and birds it eats are primary as well as secondary consumers. There are no quaternary consumers in this ecosystem, and this is often the case in nature. Finally, the ecosystem also includes detritivores (at the bottom of the figure) that consume dead organic material from all trophic levels.

Figure 36.10 A food web

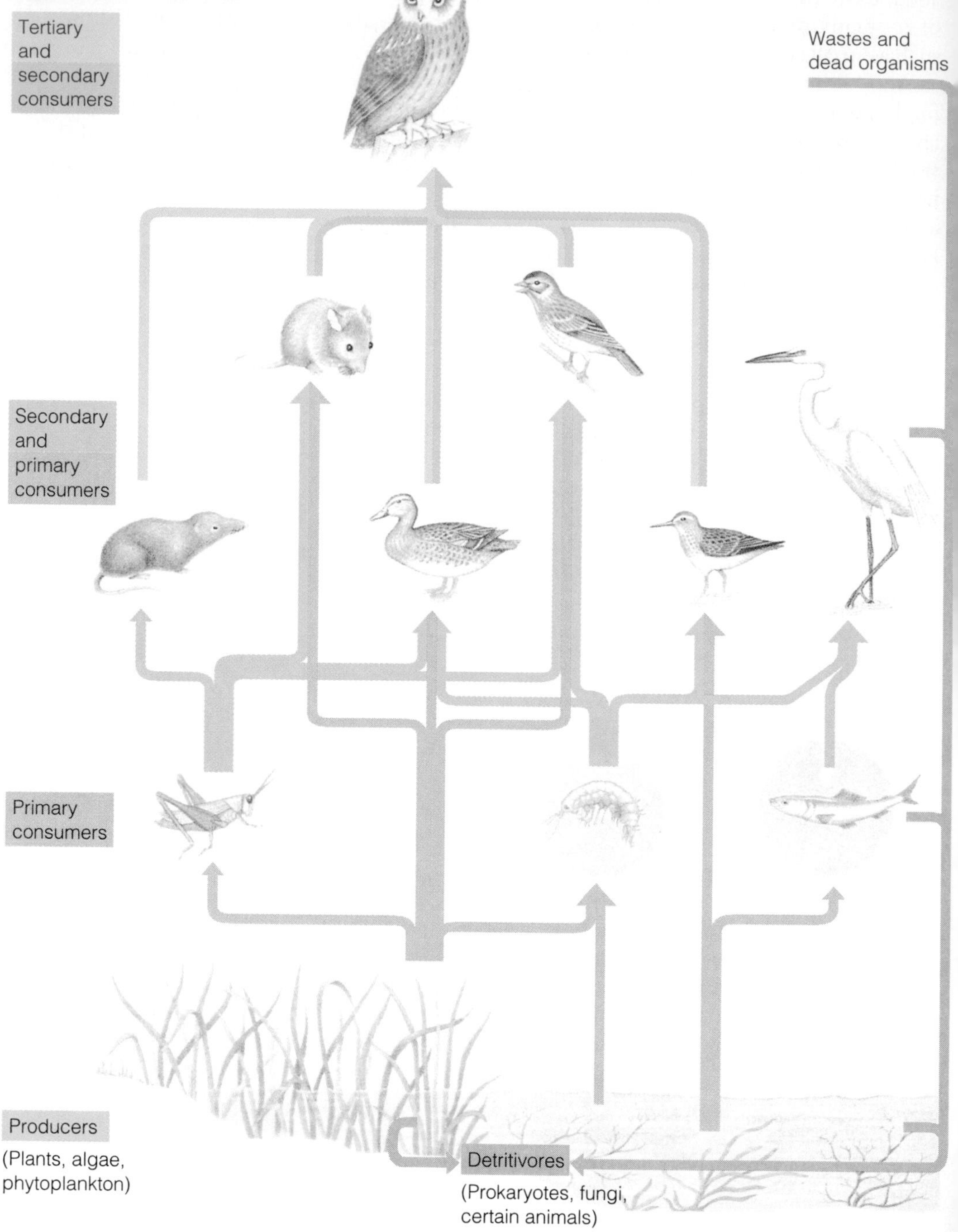

Tertiary and secondary consumers

Wastes and dead organisms

Secondary and primary consumers

Primary consumers

Producers
(Plants, algae, phytoplankton)

Detritivores
(Prokaryotes, fungi, certain animals)

Though more realistic than a food chain, this figure is still a highly simplified model of the feeding relationships in the ecosystem. An actual food web would involve many more organisms at each trophic level, and virtually all the animals would have a more diverse diet than shown here. Indicating "who eats whom," the arrows in a food web diagram outline an ecosystem's overall structure. The interconnecting arrows show the linkages among the organisms, connections based on the movement of chemical nutrients and energy. In the next module, we take a closer look at energy flow in ecosystems.

In the "who eats whom" dynamics of a food web, even consumers of the highest level in the ecosystem eventually become food for ________.

detritivores, or decomposers

Energy supply limits the length of food chains 36.11

Each day, planet Earth receives far more solar energy than is necessary to support the biosphere. Most of this energy is absorbed, scattered, or reflected by the atmosphere or by Earth's surface. Of the visible light that reaches leaves, algae, and cyanobacteria, only about 1% is converted to chemical energy by photosynthesis. But on a global scale, this is enough to produce about 170 billion tons of organic material per year in the biosphere.

Ecologists call the amount, or mass, of living organic material in an ecosystem (or part of one) the **biomass.** They call the rate at which producers convert solar energy to chemical energy (organic compounds) **primary productivity.** The primary productivity in the biosphere is about 170 billion tons of biomass per year. Because the consumers in an ecosystem acquire their organic fuels from producers (either directly or via other consumers), primary productivity sets the biosphere's spending limit for energy. The energy content in 170 billion tons of organic matter is the total annual energy budget for virtually all living organisms.

Let's see how the biosphere's energy budget gets divided up. Figure 36.11, called an energy pyramid, represents an ecosystem with four trophic levels. Each tier of the pyramid represents one trophic level, and the width of each tier indicates about how much chemical energy is actually incorporated into the organic matter of that trophic level in a year. The 10,000 kcal of energy per square meter per year incorporated by producers amounts to 1% of 1 million kcal/m^2/yr of solar energy available to the producers. In this generalized pyramid, 10% of the energy incorporated at each trophic level is built into the next higher level. The 90% decline from one trophic level to the next is a rough average. In actual ecosystems, the decline varies with the species present, generally ranging from as high as 98% to about 80%. For simplicity, we have not included the detritivore trophic level. In many ecosystems, detritivores incorporate more energy than any other trophic level except primary producers, but this does not alter the trend you see in the figure.

The basic trend of energy flow declining significantly with each higher trophic level holds for all ecosystems. This is partly a consequence of the laws of thermodynamics (see Module 5.2): Energy cannot be created or destroyed, though it can be changed from one form to another (first law of thermodynamics); and all energy transformations (including the metabolic reactions in living organisms) involve the conversion of some energy to heat, which is lost from the system (second law). Thus, not all of the chemical energy stored as biomass by plants can be converted to the chemical energy of consumers. Much of that energy is lost from the ecosystem as heat.

Let's look more closely at energy flow by focusing on a single stage, the transfer of organic matter from producers to primary consumers (herbivores). In most ecosystems, herbivores manage to eat only a fraction of the plant material produced, and they can't digest all of what they do consume. For example, a grasshopper eating a blade of grass might digest and absorb only about half the organic material it eats, passing the indigestible wastes as feces (which are available to detritivores). Of the organic compounds it does absorb from its food, the grasshopper typically uses about two-thirds as fuel for cellular respiration. In the process of releasing energy that the grasshopper can put to work, cellular respiration degrades the organic compounds to inorganic waste products and heat. Thus, of the compounds absorbed, only the store of chemical energy left over after respiration can add to the biomass of the grasshopper's trophic level, the primary consumer level. Only this amount of energy is available to the next trophic level. If we tracked the fate of organic materials at higher trophic levels, we would find that energy is lost there in similar ways.

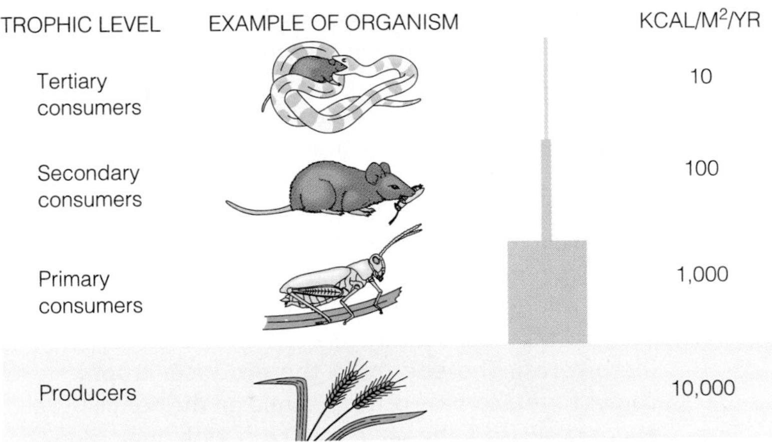

Figure 36.11 An energy pyramid

An important implication of this stepwise decline of energy in a trophic structure is that the amount of energy available to top-level consumers is small compared with that available to lower-level consumers. Only a fraction of the energy fixed by photosynthesis flows through a food chain to a tertiary consumer, such as a snake feeding on a mouse. This explains why most food chains are limited to three to five levels; there is simply not enough energy at the very top of an energy pyramid to support another trophic level.

We can now see why the last line of Jonathan Swift's verse quoted in the chapter's opening essay—"And so proceed *ad infinitum*"—cannot be taken literally for an ecosystem. The number of levels in a trophic structure is anything but infinite. Instead, as the energy pyramid indicates, trophic structures and their component food chains are severely limited by the availability of energy. 

? Why is a pound of bacon so much more expensive than a pound of corn?

Because it took at least 10 pounds of feed corn to produce that pound of bacon

36.12 An energy pyramid explains why meat is a luxury for humans

The dynamics of energy flow apply to the human population as much as to other organisms. Like many other consumers, we depend entirely on productivity by plants for our food. As omnivores, we eat both plant material and meat. When we eat grain or fruit, we are primary consumers; when we eat beef or other meat from herbivores, we are secondary consumers. When we eat fish like trout and salmon (which eat insects and other small animals), we are tertiary or quaternary consumers.

The energy pyramid on the left below indicates energy flow from primary producers to humans as vegetarians (herbivores). The energy in the producer trophic level comes from a corn crop. The pyramid on the right illustrates energy flow from the same corn crop, with humans as secondary consumers, eating cattle. These pyramids are generalized models, based on the rough estimate that about 10% of the energy available in a trophic level appears at the next higher trophic level. Thus, the pyramids indicate that the human population has about ten times more energy available to it when people eat grain than when they process the same amount of grain through another trophic level and eat grain-fed beef. Put another way, the pyramids indicate that it takes about ten times more energy to feed the human population when we eat meat than when we eat plants directly.

Actually, the 10% figure is high for energy flow involving cattle and humans. As endotherms, cattle expend a great deal of the energy they take in on heat production—much more than do ectotherms, such as grasshoppers (see Module 25.4). Accounting for the energy loss in heat production, it may actually take closer to 100 times more energy to feed us on cattle (and other mammals and birds) than on plants directly.

Eating meat of any kind is an expensive luxury, both economically and environmentally. In many countries, people cannot afford to buy much meat or their country cannot afford to produce it, and people are vegetarians by necessity. Whenever meat is eaten, producing it requires that more land be cultivated, more water be used for irrigation, and more chemical fertilizers and pesticides be applied to croplands used for growing grain. It is likely that, as the human population expands, meat consumption will become even more of a luxury than it is today.

The laws of thermodynamics and the fact that energy does not cycle within ecosystems explain why the human population has a limited supply of energy available to it. We turn next to the subject of chemical nutrients, all of which differ from energy in that they cycle within ecosystems.

Why does demand for meat also tend to drive up prices of grains such as wheat and rice, fruits, and vegetables?

Potential supply of plants for direct consumption as food for humans is diminished by the use of agricultural land to grow feed for cattle, chickens, and other meat sources.

TROPHIC LEVEL

Secondary consumers

Primary consumers

Producers

Human vegetarians

Corn

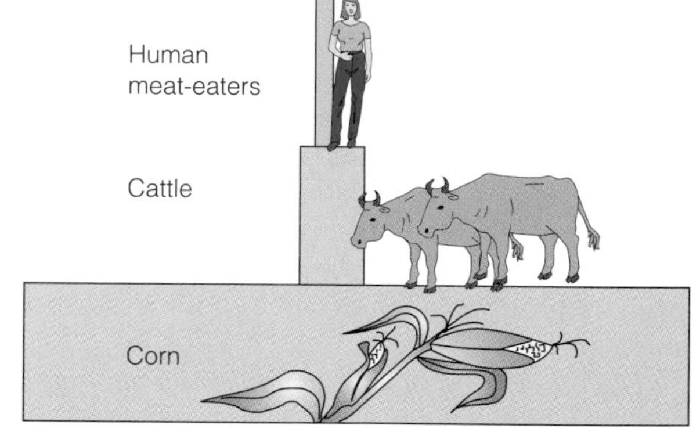

Figure 36.12 Food energy available to the human population at different trophic levels

36.13 Chemicals are recycled between organic matter and abiotic reservoirs

The sun (or in some cases Earth's interior) supplies ecosystems with energy, but there are no extraterrestrial sources of water or the other chemical nutrients essential to life on the planet. Life, therefore, depends on the recycling of chemicals. In the next four modules, we look at the cyclic movement of four substances within the biosphere: water, carbon, nitrogen, and phosphorus. In each case, we see that chemicals pass back and forth between organic matter and the abiotic components of ecosystems. We call the part of the ecosystem where a chemical accumulates or is stockpiled outside of living organisms an **abiotic reservoir.** The main abiotic reservoirs are highlighted in white boxes in the next four figures. Let's begin with water.

What is the main abiotic reservoir for carbon?

The atmospheric stock of CO_2

Figure 36.14 below illustrates the global water cycle, which is driven by heat from the sun. Three major processes driven by solar heat—precipitation, evaporation, and transpiration from plants—continuously move water between the land, oceans, and the atmosphere. The widths of the blue arrows indicate the relative amounts of water that move to and from the various locations each year. The numbers in parentheses indicate actual amounts of water as billion billion (10^{18}) grams per year. Over the oceans (left side of the figure), evaporation exceeds precipitation by 36×10^{18} grams per year. The result is a net movement of this amount of water vapor in clouds that are carried by winds from the oceans across the land. On land (right side of the figure), precipitation exceeds evaporation and transpiration. The excess precipitation forms systems of surface water (such as lakes and rivers) and groundwater, all of which flow back to the sea, completing the water cycle. The water cycle has a global character because there is a large reservoir of water in the atmosphere. Thus, water molecules that have evaporated from the Pacific Ocean, for instance, may appear in a lake or in an animal's body far inland in North America.

Human activity affects the global water cycle in a number of important ways. One of the main sources of atmospheric water is transpiration from the dense vegetation making up tropical rain forests. The destruction of these forests, which is occurring rapidly today, will change the amount of water vapor in the air. This, in turn, will most likely alter local, and perhaps global, weather patterns.

Another change in the water cycle caused by humans results from pumping large amounts of groundwater to the surface to use for irrigation. This practice can increase the rate of evaporation over land, and unless this loss is balanced by increased rainfall over land, groundwater supplies can be depleted. Large areas in the midwestern U.S., the southwestern American desert, parts of California, and areas bordering the Gulf of Mexico currently face this problem.

? What is the main way that life contributes to the water cycle?

Plants move water from the ground to the atmosphere via transpiration.

Solar heat
Water vapor over the sea
Net movement of water vapor by wind (36)
Water vapor over the land
Precipitation over the sea (283)
Evaporation from the sea (319)
Evaporation and transpiration (59)
Precipitation over the land (95)
Oceans
Surface water and groundwater
Flow of water from land to sea (36)

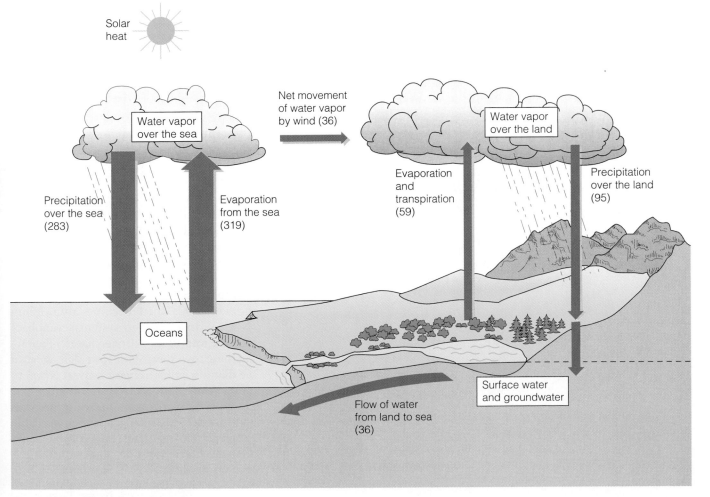

Figure 36.14 The global water cycle

36.15 The carbon cycle depends on photosynthesis and respiration

36C

Like water, the element carbon has an atmospheric reservoir and cycles globally. Moving clockwise from the top of Figure 36.15, you see that carbon dioxide (CO_2) from the atmosphere is converted into organic compounds of plants, algae, and cyanobacteria by photosynthesis. Some of this organic material is then eaten by primary consumers, such as rabbits, and serves as the carbon source for these organisms. Higher-level consumers obtain their carbon by eating lower-level consumers. Meanwhile, carbon compounds in detritus—animal wastes, plant litter, and dead organisms of all kinds—are consumed and decomposed by detritivores. Cellular respiration by plants, animals, soil microbes, and other organisms breaks down organic compounds to CO_2, which returns to the atmosphere.

On a global scale, the return of CO_2 to the atmosphere by respiration closely balances its removal by photosynthesis. However, the increased burning of wood and fossil fuels (coal and petroleum) is steadily raising the level of CO_2 in the atmosphere. This may lead to significant environmental problems, such as global warming, as we will discuss in Module 38.4.

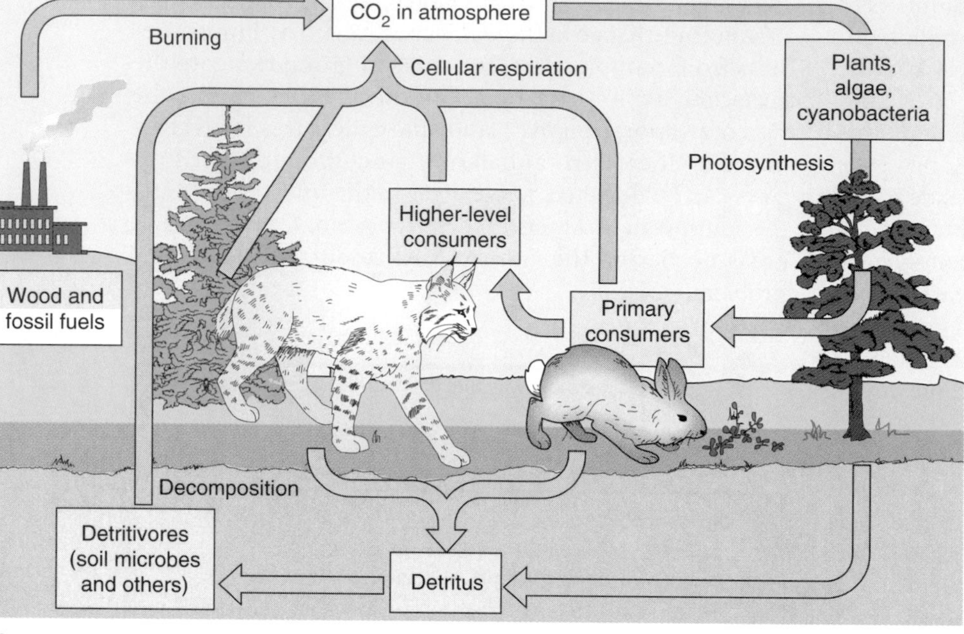

Figure 36.15 The carbon cycle

?

What would happen to the carbon cycle if all the detritivores suddenly went on "strike" and stopped working?

Carbon would accumulate in organic mass, the atmospheric reservoir of carbon would decline, and plants would eventually be starved for CO_2.

36.16 The nitrogen cycle relies heavily on bacteria

36D

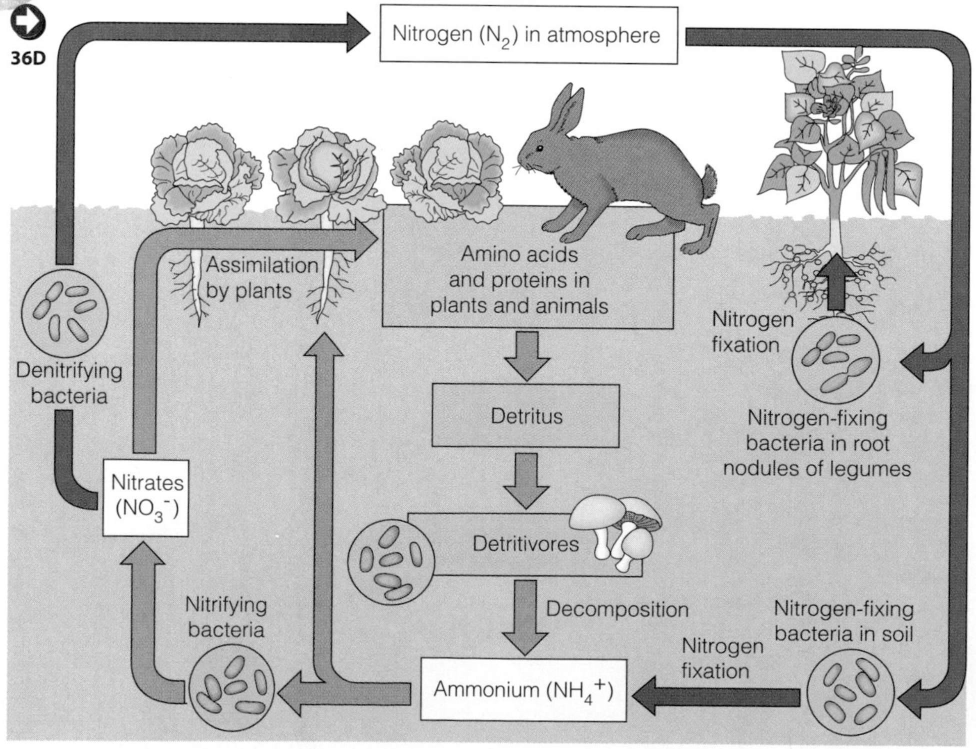

Figure 36.16 The nitrogen cycle

The atmosphere contains a huge reservoir of nitrogen; almost 80% of the atmosphere is N_2. Plants, however, can only use nitrogen in the form of nitrate (NO_3^-) ions or ammonium (NH_4^+) ions. These compounds can be made from N_2 by soil bacteria. Notice the two groups of nitrogen-fixing bacteria on the right side of the figure. Nitrogen fixers in the soil and in some plants (peas, beans, and other legumes, for example) convert atmospheric N_2 to ammonia (NH_3), which becomes ammonium (NH_4^+). (In aquatic ecosystems, cyanobacteria are important nitrogen fixers.) Following the arrows toward the lower left, you can see that nitrifying bacteria convert NH_4^+ to NO_3^-, the main source of nitrogen for plants. Used by plants to make amino acids and proteins, the nitrogen is then available to consumer organisms. Bacteria and fungi acting as detritivores decompose nitrogen-containing detritus back into ammonium, thus keeping the nitrogen available to plants. Another group of soil bacteria, the denitrifiers (far left in the figure), complete the nitrogen cycle by converting soil nitrates to atmospheric N_2. Although not shown in the figure, some NH_4^+ and NO_3^- ions are made in the atmosphere by chemical reactions involving N_2 and NH_3 gas. These ions reach the soil in precipitation and dust and are a crucial source of nitrogen for plants in some ecosystems.

Much of the nitrogen cycling by bacteria in many ecosystems involves the inner cycle in the diagram, the lighter purple arrows linking plants and animals, detritivores, and nitrifying bacteria. The outer cycle (the darker purple arrows) often moves only a tiny fraction of nitrogen into and out of natural ecosystems. However, human activity has altered the nitrogen cycle balance in many areas. Sewage-treatment facilities usually empty large amounts of dissolved inorganic nitrogen compounds into rivers or streams. Farmers routinely apply large amounts of inorganic nitrogen fertilizers, mainly ammonium compounds and nitrates, to croplands. Lawns and golf courses also receive sizable doses of fertilizer. Crop and lawn plants take up some of the nitrogen compounds, and denitrifying bacteria convert some into atmospheric N_2, but chemical fertilizers usually exceed the soil's natural recycling capacity. The excess nitrogen compounds often enter streams, lakes, and groundwater.

In lakes and streams, these nitrogen compounds continue to fertilize, causing heavy growth of algae. Groundwater pollution by nitrogen fertilizers is a serious problem in many agricultural areas. Nitrates in drinking water are converted to nitrites, which can be toxic, in the human digestive tract. In Module 32.10, we discussed some alternatives to the extensive use of agricultural fertilizers.

What is the main abiotic reservoir of nitrogen?

The atmospheric supply of N_2.

The phosphorus cycle depends on the weathering of rock 36.17

In contrast to nitrogen, the element phosphorus has its main abiotic reservoirs in rocks, rather than in the atmosphere. (This is also true of the elements potassium and calcium.) At the center of Figure 36.17, the weathering of rock gradually adds phosphates (compounds containing PO_4^{3-}) to the soil. Plants absorb the dissolved phosphate ions in the soil and build them into organic compounds. Consumers obtain phosphorus in organic form from plants. Decomposers return phosphates to the soil. As indicated in the lower left of Figure 36.17, some phosphates also precipitate out of solution at the bottom of deep lakes and oceans. The phosphates in this form may eventually become part of new rocks and will not cycle back into living organisms until geological processes uplift the rocks and expose them to weathering.

Because weathering is generally a slow process, the amount of phosphates available to plants in natural ecosystems is often quite low. Plant growth can, in fact, be limited by the small amount of soluble phosphates in the soil. And in lakes that have not been altered by human activity, a low level of dissolved phosphates often keeps algal growth to a minimum, thereby helping keep the water clear. In many areas, however, excess, rather than limited, phosphates are a problem. Like nitrogen compounds, phosphates are a major component of sewage outflow. They are also used extensively in agricultural fertilizers and are a common ingredient in pesticides. Phosphate pollution of lakes and rivers, like nitrate pollution, leads to heavy algal growth. We discuss some other effects of human alteration of nutrient cycles in the next two modules.

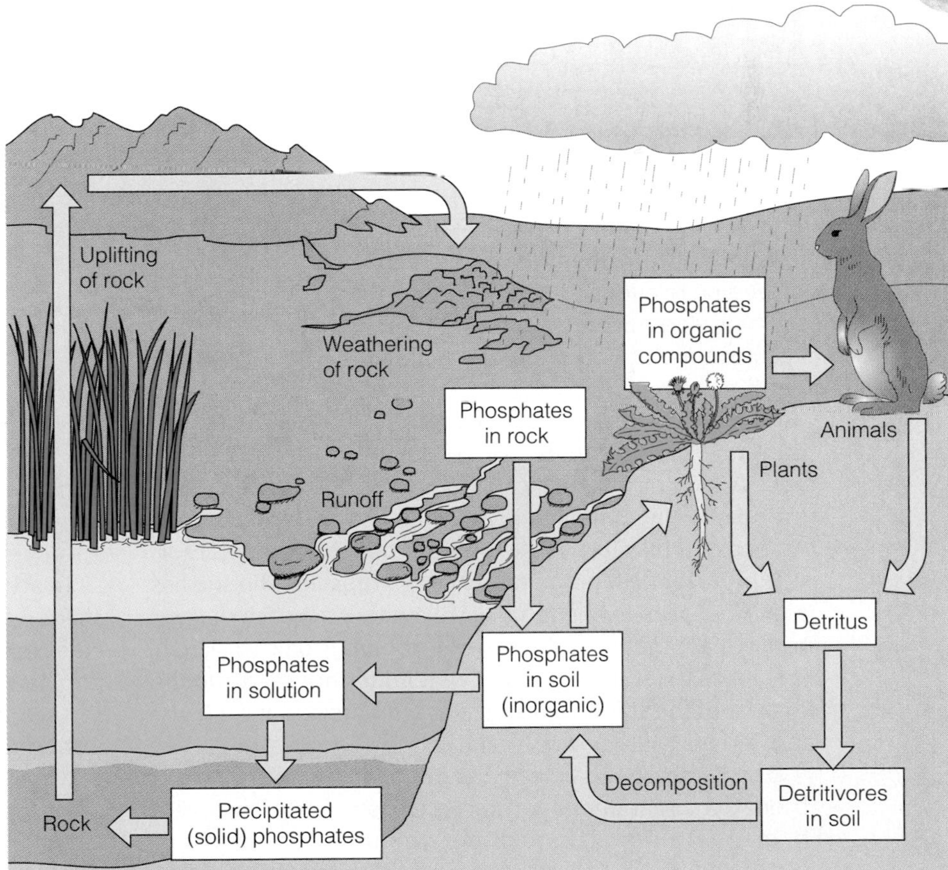

Figure 36.17 The phosphorus cycle

Over the short term, why does phosphorus cycling tend to be more localized than either carbon or nitrogen cycling?

Because phosphorus is cycled almost entirely within the soil rather than transferred over long distances via the atmosphere

Ecosystem Alteration

36.18 Ecosystem alteration can upset chemical cycling

Figure 36.18A A dam at the Hubbard Brook study site

Figure 36.18B Logged watersheds in the Hubbard Brook Forest

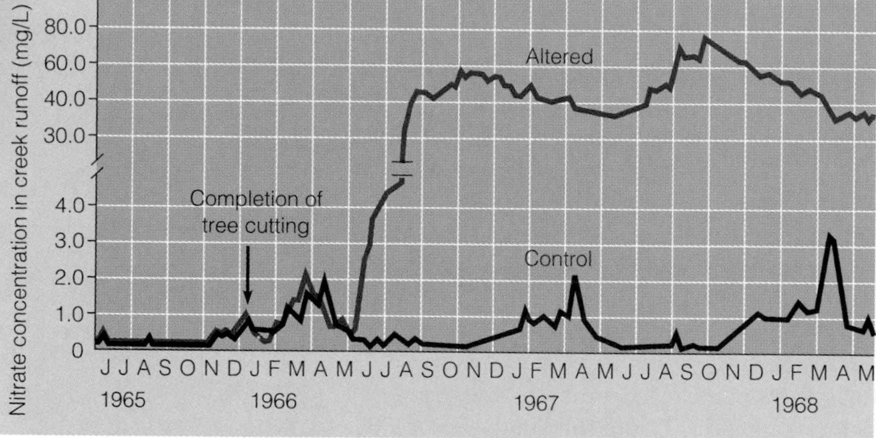

Figure 36.18C The loss of nitrate from a deforested watershed

The cycling of any chemical element in an ecosystem depends on the web of feeding relationships among plants, animals, and microbes. Obtaining an accurate picture of chemical cycling requires long-term study, and a number of research groups have been conducting such studies for several decades. For example, since 1963, one team has been monitoring nutrient dynamics in a nearly mature deciduous forest ecosystem called the Hubbard Brook Experimental Forest, in the White Mountains of New Hampshire. The site has a number of watersheds, small valleys that are each drained by a stream that is a tributary of Hubbard Brook. Thick rock is close to the soil surface, and water does not seep into the rock from the soil. As a result, water drains out of each watershed only via its own stream.

The Hubbard Brook team set out to study water and nutrient dynamics under natural conditions and after severe human intrusion. They first determined the amounts of water and key nutrients—for example, nitrate (NO_3^-) and calcium (Ca^{2+})—that normally move in and out of six of the watersheds. Figure 36.18A shows a small concrete dam with a V-shaped spillway for water. Such a dam was built across the stream at the bottom of each watershed to monitor water and nutrient losses from the stream. When monitoring began, about 60% of the water that fell as rain and snow exited through the streams, and the remaining 40% was lost by transpiration from plants and evaporation from the soil. Preliminary data also indicated that the flow of nutrients into and out of the watersheds was nearly balanced and was relatively small compared with the quantity of nutrients being recycled within the forest itself.

In an experiment begun in 1966, one of the six watersheds was completely logged and then sprayed with herbicides for three years to prevent regrowth of plants. All the original plant material was left in place to decompose. This severely altered watershed is the white (snow-covered) area just right of center in Figure 36.18B. The inflow and outflow of water and minerals for this watershed were compared with a control (unaltered) watershed for three years. Water runoff from the altered system increased by 30–40%, apparently because there were no plants to absorb and transpire water from the soil. Net losses of nutrients were huge, as shown for nitrate in Figure 36.18C. The red line traces the levels of nitrate in runoff from the deforested system from June 1965 through May 1968. The black line indicates nitrate levels in the control watershed. Notice that nitrate levels in the runoff from the deforested area began to rise markedly about five months after the plants were killed. (The break between 4.0 and 30.0 mg/L simplifies the graph; it represents a month when the rate of nitrate loss increased very rapidly.) Within eight months, the nitrate loss was some 60 times greater in the altered watershed than in the control. Without plants to take up and hold nitrate, this nutrient drained out of the ecosystem. In fact, nitrate in the stream reached a level considered unsafe for drinking water.

After 30 years, data from Hubbard Brook point to some other long-term trends. It now appears that none of the watersheds was truly undisturbed even when the study began. For instance, prior to and during the study, acid precipitation seems to have caused an increase in the levels of Ca^{2+} in stream water (see Module 2.16). Apparently, since the 1950s, acid rain and snow have dissolved most of the Ca^{2+} from the forest soil, and the streams have carried it away. As of a report published in 1996, plant growth at Hubbard Brook had virtually ceased because of a lack of Ca^{2+} in the soil. 

? How can clear-cutting a forest (removing all trees) damage the water quality of nearby lakes?

Without the growing trees to assimilate minerals from the soil, more of the minerals run off and end up polluting water resources.

Altered ecosystems trigger changes in other ecosystems 36.19

The Hubbard Brook experiment shows that major change in a terrestrial ecosystem disrupts chemical cycling and moves large amounts of chemical nutrients to other areas. Most of the nutrients lost from deforested lands and from agricultural areas are highly soluble in water and readily pass into aquatic ecosystems.

What do we really mean when we say chemical nutrients can pollute a receiving ecosystem? Let's trace the chain of events that can occur when a freshwater lake, for instance, receives an overload of nutrients from surrounding terrestrial ecosystems. A natural lake typically has a moderate growth of algae and plants and often a rich diversity of fishes and invertebrates. Freshwater lakes change as a result of natural events. Some may gradually undergo **eutrophication,** meaning they become increasingly more productive. As they change, they remain in equilibrium with their surroundings; the input of new nutrients is balanced by the losses from the lake due to outflow and the settling of organic matter to the bottom. The kinds of organisms living in the lake may change, but productivity increases slowly enough that the species diversity usually remains high.

Unfortunately, the natural balance of nutrient cycling in freshwater lakes is easily upset. When a lake receives an excess of mineral nutrients, its entire trophic structure can change very quickly. Today, a lake may receive runoff of inorganic fertilizers from agriculture, as well as sewage, factory wastes, and animal wastes from pastures and stockyards. The water becomes polluted with these materials—actually, overfertilized—and the lake's photosynthetic organisms multiply rapidly. The result is accelerated eutrophication. Shallow areas may become weed-choked, and algal and bacterial populations often grow explosively. The photograph in Figure 36.19 illustrates the effect of experimental eutrophication of part of a freshwater lake. Researchers used a plastic curtain to separate the part of the lake in the background from the portion in the foreground. They then added inorganic phosphorus to the distant part of the lake, keeping the other part as a control. Within two months, a cyanobacterial bloom changed the color of the polluted lake to the pale blue you see here.

Figure 36.19 Experimental eutrophication of part of a lake
(Reprinted with permission from D. W. Schindler, *Science* 184 (1974): 897, Figure 1.49. ©1974 American Association for the Advancement of Science.)

Heavy bacterial and algal growth increases oxygen production during the day but greatly reduces oxygen levels at night, when the photosynthesizers respire. As the organisms die and accumulate at the bottom of the lake, the bacteria decomposing them can use up much of the oxygen dissolved in deep waters. When this happens, the lake may lose most of its species diversity, with only the most tolerant organisms surviving. Human-caused eutrophication, for example, wiped out commercially important fish in Lake Erie during the 1950s and 1960s. Since then, tighter regulations on the dumping of wastes into the lake have enabled some fish populations to rebound, but many of the native species of fishes and invertebrates have not recovered. Today, accelerated eutrophication is the most common problem affecting lakes throughout the world. About 60% of the lakes in the U.S. are affected to some degree.

Human disruption of both aquatic and terrestrial ecosystems is a global problem. In the next module, we take a look at measures that some countries are taking to make human development more compatible with the ecosystems they depend on. 

? How does excessive addition of mineral nutrients to a lake eventually result in the loss of most fish species?

The eutrophication initially causes population explosions of algae and the organisms that feed on them. The respiration of so much life, including the decomposers working on all the organic refuse, consumes most of the lake's oxygen, which the fish require.

36.20 Zoned reserves are an attempt to reverse ecosystem disruption

Today, few, if any, ecosystems remain unaltered by human activities. Accelerated eutrophication reduces species diversity in lakes and rivers because many organisms cannot tolerate the rapid changes in water quality. Large tracts of forest in taiga biomes are still being clear-cut in the U.S., Canada, and Siberia to satisfy demands for lumber and urban development. The current rate of conversion of tropical forests to farmland threatens the survival of thousands of species and may alter global weather patterns.

In an attempt to slow the disruption of ecosystems, a number of countries are setting up what they call zoned reserves. A **zoned reserve** is an extensive region of land that includes one or more areas undisturbed by humans. The undisturbed areas are surrounded by lands that have been changed by human activity and are used for economic gain.

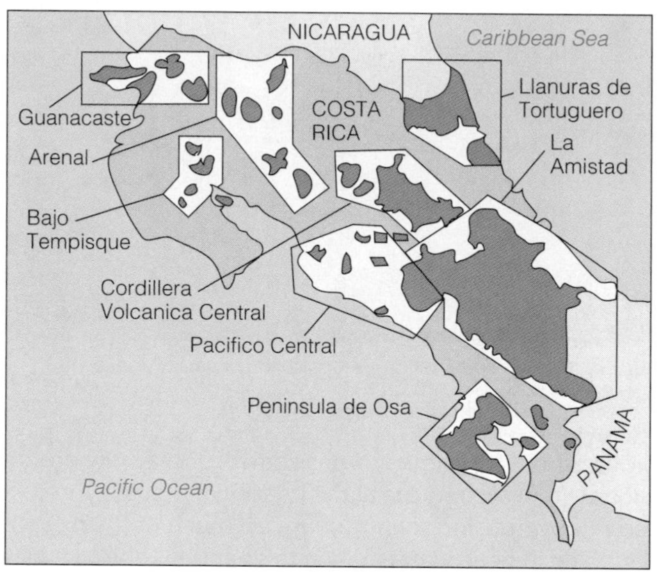

Figure 36.20 Zoned reserves in Costa Rica

The key factor of the zoned reserve concept is the development of a social and economic climate in the surrounding lands that is compatible with ecosystem conservation. These surrounding areas continue to be used to support the human population, but they are protected from extensive alteration. As a result, they serve as a buffer zone, or shield, against further intrusion into the undisturbed areas.

The small Central American nation of Costa Rica has become a world leader in establishing zoned reserves. In exchange for reducing its international debt, the Costa Rican government established eight zoned reserves, called "conservation areas," as shown on the map here. The green areas are national park lands, which remain relatively unchanged by human activity; the yellow areas are buffer zones, privately owned areas where people live and work.

Costa Rica is making progress toward managing its zoned reserves so that the buffer zones provide a steady, lasting supply of forest products, water, and hydroelectric power and also support sustainable agriculture and tourism. An important goal is providing a stable economic base for people living there. Destructive practices that are not compatible with long-term ecosystem stability, and from which there is often little local profit, are gradually being discouraged. Such destructive practices include massive logging, large-scale single-crop agriculture, and extensive mining. Costa Rica looks to its zoned reserve system to maintain at least 80% of its native species. We take a closer look at conservation efforts in the tropics in Chapters 37 and 38.

What philosophy about the relationship of humans to forests appears to be shared by the government of Costa Rica and Ariel Lugo, the scientist you met in Module 34.11?

Both agree that forest preservation is a realistic goal only if the strategy is compatible with an acceptable standard of living for the local people.

Chapter Review

Chapter Summary

A biological community derives its structure from the interactions and interdependence of the organisms that it comprises. The functioning of an ecosystem depends on the complex interactions between its community of organisms and the physical environment **(Introduction).** All the organisms in a particular area make up a community. A number of factors characterize every community: diversity (variety of species), the prevalent form of vegetation (dominant plants and their structure), stability (the ability to resist change and recover after a disturbance), and trophic structure (feeding relationships). Three main types of interactions connect species in communities: interspecific competition, predation, and symbiosis **(36.1).**

Structural Features of Communities (36.2–36.7) Interspecific competition occurs between two populations if they both require the same limited resource. A population's niche is its role in the community, or the sum total of its use of the biotic and abiotic resources of its habitat (where the population lives). If two populations have very similar niches, they may not be able to coexist **(36.2).** In some cases, as predators adapt to prey, natural selection also shapes the prey's defenses. This process of reciprocal adaptation is known as coevolution **(36.3).** A keystone predator may maintain community diversity by controlling the strongest competitors among its prey species **(36.4).** Three types of symbiotic relationships exist in communities. In parasitism, a parasite obtains food at the expense of its host. In commensalism, one species benefits while the other is unaffected. In mutualism, both partners benefit **(36.5).**

Human activities and other disturbances, such as fire, change a community. Ecological succession is a transition in the species composition of a community. Primary succession is the gradual colonization of barren rocks by living organisms. Secondary succession occurs after a disturbance has removed the vegetation but left the soil intact (36.6). Fire is a key abiotic factor in many ecosystems. Grasslands are dependent on fire (36.7).

Ecosystem Structure and Dynamics (36.8–36.17) A community interacts with abiotic factors, forming an ecosystem. Energy flows from the sun, through plants, animals, and decomposers, and is lost as heat. Chemicals are recycled among air, water, soil, and organisms (36.8). A food chain is the stepwise flow of energy and nutrients from plants (producers), to herbivores (primary consumers), to carnivores (secondary and higher-level consumers). The breakdown of organic compounds to inorganic compounds is called decomposition. Detritivores (animal scavengers, fungi, and prokaryotes) decompose waste matter and recycle nutrients (36.9). A food web is a network of interconnecting food chains (36.10).

Primary productivity is the rate at which producers convert sunlight to chemical energy in organic material (biomass). An energy pyramid diagram reveals the flow of energy from producers to primary consumers and to higher trophic levels. Some of the energy consumers gain from their food is used in cellular respiration and lost as heat. Only 10% or less of the energy in food is stored at each trophic level and available to the next level. This stepwise energy loss limits most food chains to three to five levels (36.11). Because the energy pyramid tapers so sharply, a field of corn or other plant crops can support many more vegetarians than meat-eaters (36.12). Ecosystems require daily infusions of energy, but nutrients are recycled between organisms and their abiotic environments (36.13). Heat from the sun drives the global water cycle of precipitation, evaporation, and transpiration (36.14). Carbon is taken from the atmosphere by photosynthesis, used to make organic molecules, and returned to the atmosphere by cellular respiration (36.15). Nitrogen is plentiful in the atmosphere as N_2, but N_2 is unusable by plants. Various bacteria in soil (and legume root nodules) convert N_2 to nitrogen compounds that plants can use, ammonium (NH_4^+) and nitrate (NO_3^-). An additional source of these ions is chemical reactions in the atmosphere. Some soil bacteria break down organic matter and recycle nitrogen, as ammonium or nitrate, to plants; other bacteria return N_2 to the atmosphere (36.16). Phosphorus and other soil minerals are also recycled; they are stored primarily in rock (36.17).

Ecosystem Alteration (36.18–36.20) Experimental studies of ecosystems show that drastic alterations, such as the total removal of vegetation, can increase the runoff of water and loss of soil nutrients. Environmental changes caused by humans, such as acid rain, can unbalance nutrient cycling over the long term (36.18). Nutrient runoff may fertilize a lake and cause excessive growth of algae, which may, in turn, choke other aquatic life (36.19). The alteration of ecosystems by human activities threatens global climate and the existence of thousands of species. To slow the disruption of ecosystems, some nations are establishing zoned reserves, undisturbed wildlands surrounded by buffer zones of compatible economic development (36.20).

Testing Your Knowledge

Multiple Choice

1. Which of the following best illustrates ecological succession?
   a. A mouse eats seeds, and an owl eats the mouse.
   b. Decomposition in soil releases nitrogen that plants can use.
   c. Grass grows on a sand dune, then shrubs, and then trees.
   d. Imported pheasants increase, while local quail disappear.
   e. Overgrazing causes a loss of nutrients from soil.

2. A bat locates insect prey in the dark by bouncing high-pitched sounds off them. One species of moth escapes predation by diving to the ground when it hears "sonar" of a particular bat species. This illustrates ____ between the bat and moth. *(Explain your answer.)*
   a. mutualism
   b. competitive exclusion
   c. ecological succession
   d. commensalism
   e. coevolution

3. Local conditions such as heavy rainfall or the removal of plants may limit the amount of nitrogen, phosphorus, or calcium available to a particular ecosystem, but the amount of carbon available to the system is seldom a problem. Why?
   a. Organisms do not need very much carbon.
   b. Plants can make their own carbon using water and sunlight.
   c. Plants are much better at absorbing carbon from the soil.
   d. Many nutrients come from the soil, but carbon comes from the air.
   e. Symbiotic bacteria help plants capture carbon.

Describing, Comparing, and Explaining

In Southeast Asia, there's an old saying, "There is only one tiger to a hill." In terms of energy flow in ecosystems, explain why big predatory animals such as tigers and sharks are relatively rare.

Thinking as a Scientist

An ecologist studying plants in the desert performed the following experiment. She staked out two identical plots, including a few sagebrush plants and numerous small, annual wildflowers. She found the same five wildflower species in roughly equal numbers on both plots. She then enclosed one of the plots with a fence to keep out kangaroo rats, the most common grain-eaters of the area. After two years, to her surprise, four of the wildflower species were no longer present in the fenced plot, but one species had increased drastically. The control plot had not changed. Using the principles of ecology, propose a hypothesis to explain her results. What additional evidence would support your hypothesis?

Science, Technology, and Society

Sometime in 1986, near Detroit, a freighter pumped out water ballast containing larvae of European zebra mussels. The mollusks multiplied wildly, spreading through Lake Erie and entering Lake Ontario. In some places, they have become so numerous that they have blocked the intake pipes of power plants and water-treatment plants, fouled boat hulls, and sunk buoys. We have seen similar population explosions before, with rabbits in Australia and starlings in North America, for example. What makes this kind of population explosion occur? What might happen to native organisms that suddenly must share the Great Lakes ecosystem with zebra mussels? How would you suggest trying to solve the mussel population problem?

Media Resources

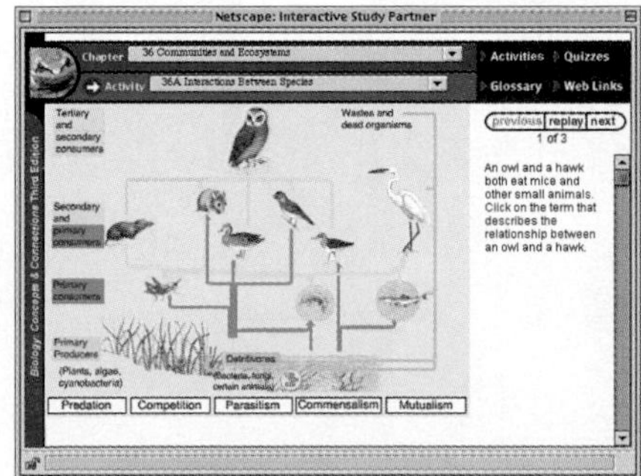

For further review of the concepts in this chapter, use the activities and quizzes on the CD-ROM.

Visit the Special Edition of The Biology Place web site for activities, quizzes, news articles, and web links.

37 Behavioral Adaptations to the Environment

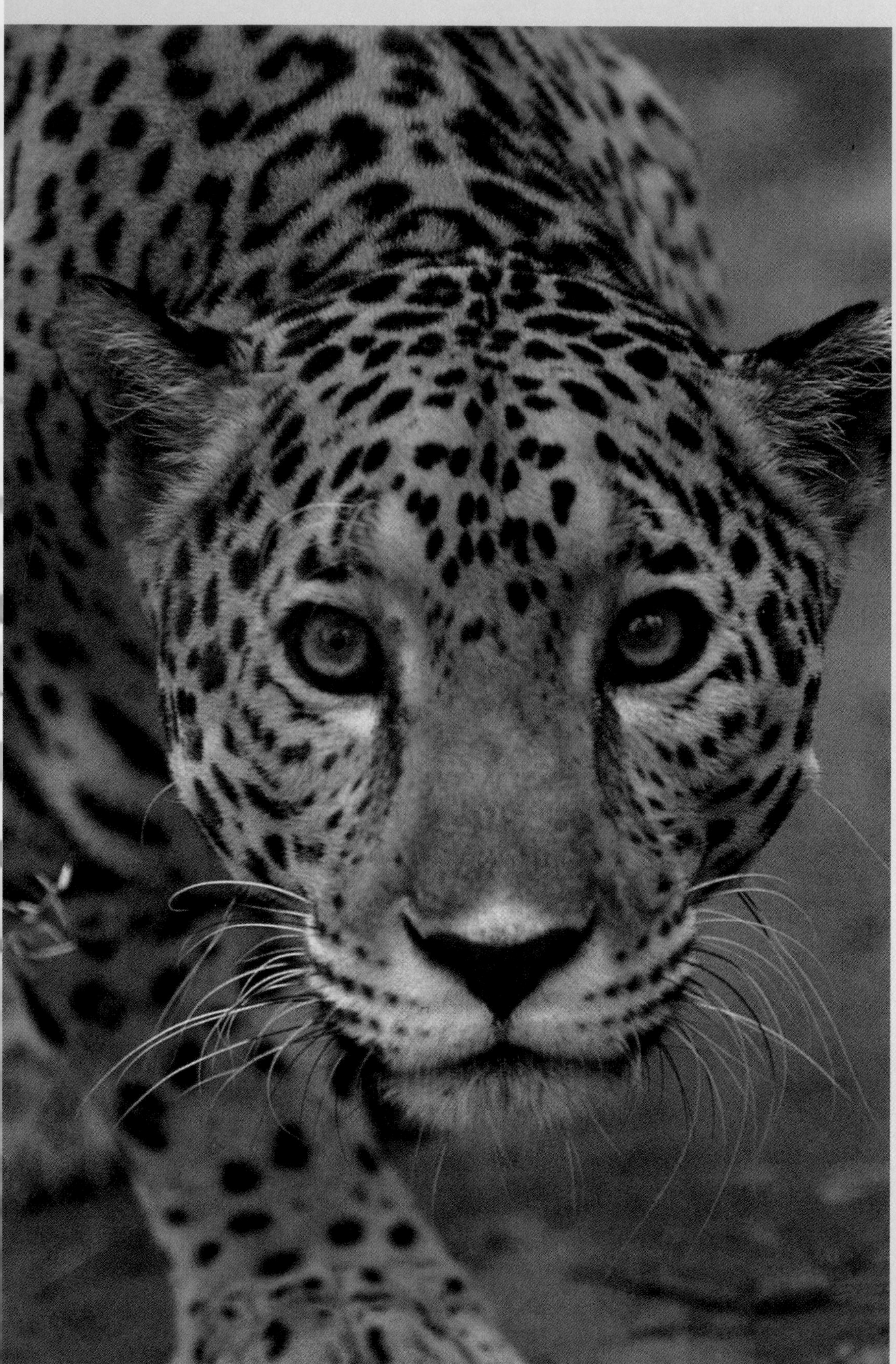

Outline

In his treatise on Mexican wildlife, American naturalist Aldo Leopold wrote:

> The chesty roar of jaguar in the night causes men to edge toward the blaze and draw serapes tighter. It silences the yapping dogs and starts the tethered horses milling. In announcing its mere presence in the blackness of the night, the jaguar puts the animate world on edge.

The jaguar is the largest and strongest member of the cat family in the Americas. An adult male stands nearly 2.5 feet at the shoulder and can weigh up to 300 pounds. Its broad face is shaped by massive jaw muscles that can crush heavy bones.

Jaguars were once common throughout most of South and Central America and ranged through Mexico into the southcentral United States. Today, they are rare almost everywhere except in a few strongholds in the dense rain forests of Central America and Brazil. In 1984, the tiny Caribbean country of Belize established the world's only jaguar refuge, the Cockscomb Forest Jaguar Preserve. The refuge idea grew out of a study of jaguar behavior and ecology carried out in the Cockscomb area in the early 1980s by Alan Rabinowitz, a zoologist with the New York Zoological Society.

Field research in animal behavior is never easy, but Rabinowitz, in his twenties at the time, weathered monumental difficulties. Nearly all the jaguars he studied were shot within a few months by irate ranchers and farmers who believed the cats were killing livestock. Rabinowitz himself contracted amoebic dysentery, hookworm, and fungal infections and barely survived the crash of his plane. One of his assistants died from snakebite. Still, he made some significant discoveries. In just two years, aided by the local Maya, he managed to monitor the activities of six adult jaguars over extended periods. This required trapping the animals, anesthetizing them, and then fitting each one with a radio-transmitter collar (as shown at the right). Using a directional receiver tuned to the transmitter's signals, he could then trace the jaguar's movements in the dense jungle.

Rabinowitz confirmed that jaguars, like most big cats, are solitary hunters that can kill animals twice their size. He also discovered that jaguars shun contact with others of their species, except during the breeding season. Male jaguars in Belize may have overlapping hunting grounds but rarely use the same areas at the same time. A male announces his presence by defecating in open areas, by scratching the ground, and by grunting or low growling. During the breeding season, male and female jaguars pair off, and their behavior changes markedly. The male may lick and caress the female, and after a litter is born, he will often help feed the nursing female and the cubs for days or even weeks.

Studying an animal's behavior is essential to understanding its evolution and ecological interactions. For example, the jaguar's hunting techniques are similar to those of several other species of big cat, such as the mountain lion and leopard. This similarity suggests that these hunting techniques evolved in a common ancestor of the big cats. From an ecological standpoint, the jaguar's signaling behavior prevents direct confrontation between individuals and increases its chances of having nearly exclusive use of part of the limited hunting grounds available to the population. Its breeding season behavior directly affects its reproductive success. We will return to the jaguar several times as we pursue this chapter's main objective—to illustrate the connections between animal behavior, evolution, and ecology.

Fitting a jaguar with a radio-transmitter collar

37.1 Behavioral biology is the study of how animals behave in their natural environments

Animal behavior is broadly defined as externally observable muscular activity triggered by some stimulus, or, more simply, what an animal does when interacting with its natural environment. People have been describing animal activities throughout history, but the science of behavior, or **behavioral biology,** was not set on firm ground until the twentieth century. Nobel laureates Karl von Frisch, Konrad Lorenz, and Niko Tinbergen were among the first experimentalists in behavioral biology. In the early 1900s, Austrian zoologist Karl von Frisch, who pioneered the use of experimental methods in behavior, studied honeybee behavior in detail. Bavarian naturalist Konrad Lorenz, often regarded as the founder of behavioral biology, emphasized the importance of comparing the behavior of various animals. In experiments begun in the 1930s, Lorenz showed that the same stimulus may elicit very different forms of behavior in different animal species. Dutch biologist Niko Tinbergen worked closely with Lorenz, concentrating on experimental studies of innate behavior and on simple forms of learning.

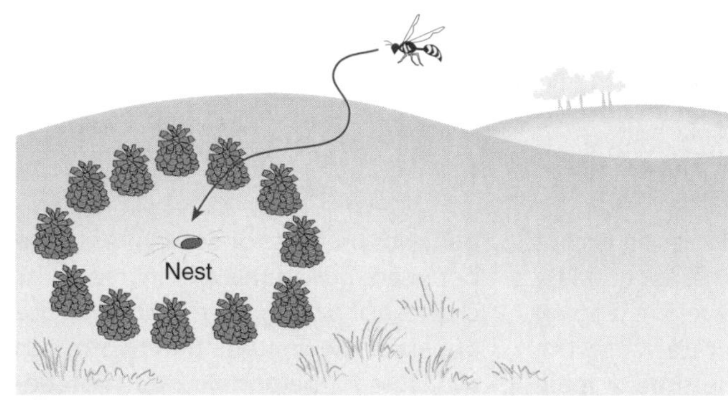

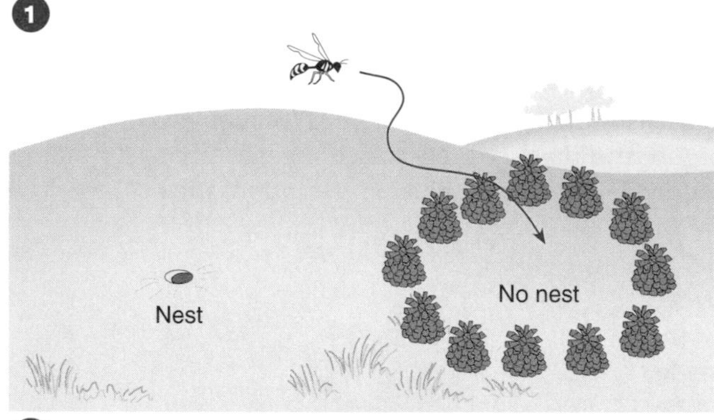

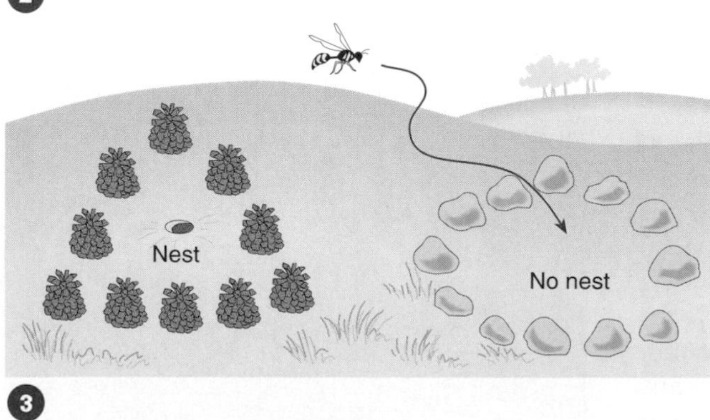

Figure 37.1 Nest-locating behavior of the digger wasp

Figure 37.1 illustrates a classic Tinbergen experiment. It deals with nesting behavior in an insect called the digger wasp, which builds its nest in a small burrow in the ground. A female wasp will often excavate and care for four or five separate nests, flying to each one daily, cleaning it, and bringing food to the single larva in the nest. To test his prediction that the female digger wasp uses landmarks to keep track of her nests, Tinbergen ① placed a circle of pinecones around a nest opening and waited for the mother wasp to return. When she did, he watched as she tended the nest. When she flew away, he ② moved the pinecones a few feet to one side of the nest opening. The next time the wasp returned, she flew to the center of the pinecone circle instead of to the actual nest opening. This experiment indicated that the wasp did use landmarks, and that she could learn new ones to keep track of her nests. But it also raised another question: Did the wasp respond to the pinecones themselves or to their circular arrangement? To answer this question, Tinbergen ③ arranged the pinecones in a triangle around the nest and made a circle of small stones off to one side of the nest opening. This time, the wasp flew to the stones, indicating that she cued in on the *arrangement* of the landmarks rather than the landmarks themselves.

In asking how a digger wasp locates her nests, Tinbergen's studies approached the wasp's behavior at the level that behavioral biologists call proximate (immediate) causes. A **proximate cause** explains behavior in terms of immediate interactions with the environment. For the digger wasp, the proximate cause of nest-locating behavior is the environmental cue: the arrangement of the landmarks. If we determine the proximate cause of a particular behavior, we have an answer to a "how" question—for the wasp, we have learned how she locates her nest.

But there is more to the wasp's behavior than this. As we have seen throughout this book, biologists also pursue "why" questions, those that can only be answered from an evolutionary perspective. Behavioral biologists call the evolutionary causes of behavior **ultimate causes.** The ultimate cause of the wasp's behavior was not addressed by Tinbergen's experiments. It has to do with natural selection acting on genetically based phenotypic differences: Presumably the fitness (reproductive success) of digger wasps has been enhanced by the female's ability to store information about nest location and to use that information to find and service her nests. The search for ultimate causes, an area of biology called **behavioral ecology,** dominates research in behavioral biology today. It is a search for answers to such questions as why natural selection favors a particular behavior, and how a behavior enhances an animal's fitness.

When you touch a hot plate, your arm recoils. What are the proximate and ultimate causes of this behavior?

The proximate cause is a simple reflex, a neural pathway linking stimulation of receptors in your finger to motor response by muscles of your arm and hand; the ultimate cause is the natural selection for a behavior that minimizes damage to the body, thereby contributing to survival and reproductive success.

37.2 Behavior results from both genes and environmental factors

How much is our ability to solve math problems programmed by genes and how much by experience? A virtuoso pianist must have extraordinary hand coordination and the ability to make the mechanical instrument resound with the dynamics and emotions of the music. How much of an artist's virtuosity results from genes (what we call inborn, or innate, talent), and how much comes from study and training? And is the capacity for disciplined training itself mainly learned or inherited?

Many studies indicate that behavior often involves a combination of genetic programming and environmental experiences (learning). For example, much of the digger wasp's nesting behavior is genetically programmed and not changeable by experience. However, as Tinbergen showed, the wasp's nest-locating behavior can be modified by the experience of seeing different landmarks.

Figure 37.2 shows another example of behavior with a strong genetic component. It is the gathering of nest material by two closely related species of African parrots, often called lovebirds. A female lovebird builds her nest with thin strips of vegetation that she cuts with her beak. In captivity, lovebirds will use small sheets of paper in place of vegetation. As shown in the top part of the drawing here, the female of one species, Fischer's lovebird, cuts fairly long strips and carries them back to her nest site one at a time in her beak. In contrast, the peach-faced lovebird cuts shorter strips and usually carries several at a time, by tucking them into the feathers of her lower back.

By interbreeding the two species and recording the behavior of hybrids, researchers have confirmed that these behavior patterns are inherited. A hybrid female cuts strips of intermediate length and attempts to carry them using hybrid behavior. For example, she usually tries tucking the strips under her feathers without releasing them from her beak. Eventually, after failing to transport the strips under her feathers, she will learn to carry a strip in her beak. But she will still act out part of the tucking sequence by turning her head to the rear before flying off.

A myth that is still perpetuated to some extent by popular media is that behavior is due either to genes (nature) or to environmental influences (nurture). In biology, however, the nature versus nurture debate is not about either/or; it is about the degrees to which genes and the environment influence phenotypic traits, including behavioral ones. As we discussed in Chapter 9, phenotype depends on both genes and the environment; behavioral traits have genetic and environmental components, as do all of an animal's structural and functional features.

What is likely the ultimate cause of nest-building behavior by lovebirds?

Behavior that results in effective nest building has evolved via natural selection because it enhances the fitness, or reproductive success, of individuals that inherit and express that behavior.

Single long strip carried in beak
(Fischer's lovebird)

Several short strips tucked under feathers
(peach-faced lovebird)

Figure 37.2 Genetic and environmental components of behavior in lovebirds

37.3 Innate behavior often appears as fixed action patterns

Lorenz and Tinbergen were among the first to demonstrate the importance of **innate behavior,** behavior that appears to be performed in virtually the same way by all individuals of a species. The environment is involved in the sense that the genes whose expression underlies the behavior require an environment to be expressed. Many of Lorenz's and Tinbergen's studies were concerned with essentially unchangeable behavioral sequences called **fixed action patterns (FAPs).** Like someone who has memorized a poem or a piece of music but must start over at the beginning if interrupted, an animal can only perform the FAP as a whole. Once an animal initiates a FAP, it usually carries the sequence to completion, even if it receives different stimuli before it finishes.

Figure 37.3A A graylag goose retrieving an egg—a FAP

Figure 37.3A illustrates one of the fixed action patterns that Lorenz and Tinbergen studied in detail. The bird is the graylag goose, a common European species that nests in shallow depressions on the ground. If the goose happens to bump one of her eggs out of the nest, she always retrieves it in the same manner. As indicated in the figure, she stands up, extends her neck, uses her beak and a side-to-side head motion to nudge the egg back, and then sits down on the nest again. If the egg slips away (or is pulled away by an experimenter) while the goose is retrieving it, she stops her side-to-side head motion but still goes through the other motions as though the egg were there. Only after she sits back down on her eggs does she seem to notice that an egg is still outside the nest. Then she begins another retrieval sequence with the egg. If the egg is again pulled away, the goose will still complete the retrieval motion as before. She even performed the sequence when Lorenz and Tinbergen placed a foreign object, such as a small toy or a ball, near the nest.

Fixed action patterns are important in the lives of many species. Figure 37.3B (on the facing page) illustrates some key events in the life cycle of the European cuckoo, which lays its eggs in the nests of other species of birds. ① When a female cuckoo (upper left) is ready to lay her eggs, she finds a nest of a suitable host species and waits for the host bird to leave the nest unattended. ② She needs only a few seconds to fly to the nest, pick up one of the host's eggs in her beak, and lay one of her own eggs in its place. ③ Mission accomplished, she flies off, abandoning her offspring to the foster parents and eating the stolen egg. When the host bird returns, she usually accepts the cuckoo's egg and incubates it with her own eggs.

The cuckoo's timing is precise, and its egg usually hatches before the host eggs. After a cuckoo hatches, a series of FAPs ensures its survival and dooms the host offspring. First, the hatchling cuckoo, with its eyes not yet open, ejects the unhatched host eggs from the nest (left photograph in Figure 37.3B). The process of ejection, which is innate, is a FAP. After ejecting the host's eggs, the young cuckoo will be fed and nurtured by its foster parents. When a hatchling senses that an adult bird is near, it responds with another FAP: It begs for food by raising its head, opening its mouth, and cheeping. In turn, the foster parent responds with yet another FAP: It stuffs food in the gaping mouth. These innate behaviors are replayed over and over, even after the young cuckoo is much larger than the adults (right photo).

In its simplest form, a FAP can be thought of as an innate response to a certain stimulus. A stimulus that triggers, or "releases," a FAP is called a **sign stimulus.** The sign stimulus for the graylag goose egg retrieval is the presence of an egg, or other object, near the nest. For the hatchling European cuckoo disposing of host eggs, it may be the feel of an unhatched egg. For parent birds, it is the chick's gaping mouth.

Egg-laying behavior

Ejection of host eggs from nest by cuckoo hatchling

Feeding of cuckoo chick by foster mother

Figure 37.3B Fixed action patterns in the life of a European cuckoo

Humans also have FAPs. For instance, when a human infant is touched on the cheek, the infant responds by turning toward the touch and rooting around for a nipple with its mouth. An infant's smile is also a FAP, induced by an adult's face or even something that vaguely resembles a face, such as two dark spots on a white circle.

In all cases, the specific sign stimulus is a proximate cause of the FAP. Experiments with invertebrates suggest that a sign stimulus activates a specific, genetically programmed nervous pathway that makes appropriate muscles in the animal perform the FAP. There is no evidence to suggest that animals integrate the information in sign stimuli in the brain and then make a decision.

What can we say about FAPs in the context of behavioral ecology? In many cases, we can see a clear advantage of FAPs over learned behavior. FAPs enable an animal to perform certain vital activities correctly the first time. No time need be spent in learning. Thus, a graylag goose automatically retrieves its eggs, and a newly hatched cuckoo does not have to learn how to obtain food from its foster parents. Similarly, a newborn jaguar starts nursing without any previous experience in obtaining milk from its mother. Innate behavior is also important for animals, such as many insects, whose life spans are so short that there is little time for learning. In the context of ultimate causes, natural selection seems to have favored innate behavior that enables animals to perform tasks essential to survival without any previous learning. We discuss learning next. 

? What is behavioral ecology?

The investigation of ultimate causes of behaviors, the evolutionary basis for behaviors as mechanisms that enhance reproductive success

37.4 Learning ranges from simple behavioral changes to complex problem solving

Learning is a change in behavior resulting from experience. As this table indicates, there are various forms of learning, ranging from a simple behavioral change in response to a single stimulus, to complex problem solving involving entirely new behavior.

TYPES OF LEARNING

Learning Type	Defining Characteristics
Habituation	Loss of a response to a stimulus after repeated exposure
Imprinting	Learning that is irreversible and limited to a critical time period in an animal's life; often results in a strong bond between new offspring and parents
Association	Behavioral change resulting from a link between a behavior and a reward or punishment; includes classical conditioning and trial-and-error learning
Imitation	Learning by observing and mimicking others
Innovation	Inventive behavior that arises in response to a new situation without trial and error or imitation; problem solving by reasoning

One of the simplest forms of learning is **habituation,** in which an animal learns not to respond to a repeated stimulus that conveys little or no information. There are many examples in both invertebrate and vertebrate animals. The cnidarian *Hydra,* for example, contracts when disturbed by a slight touch; it stops responding, however, if disturbed repeatedly by such a stimulus. Similarly, a scarecrow stimulus will usually make birds avoid a tree with ripe fruit for a few days. But the birds soon become habituated to the scarecrow and may even land on it on their way to the fruit tree. Once habituated to a stimulus, an animal still senses the stimulus—its sensory organs detect it—but the animal has learned not to respond to it.

Habituation is highly adaptive. It allows an animal's nervous system to focus on stimuli that signal food, mates, or real danger and not waste time or energy on a vast number of other stimuli that are irrelevant to its survival and reproduction.

What type of learning enables your brain to ignore the constant sensations of touch from the clothes you're wearing?

Habituation

37.5 Imprinting is learning that involves both innate behavior and experience

Learning often interacts closely with genetically determined, innate behavior. Some of the most interesting cases involve the phenomenon known as imprinting. **Imprinting** is learning that is limited to a specific time period in an animal's life and that is irreversible. One result of imprinting is the formation of a strong bond between two animals, often a hatchling or other newborn animal and its parent. The specific time during which imprinting occurs is called the **critical period.**

In perhaps his most famous study, Konrad Lorenz used the graylag goose to demonstrate imprinting. He divided a batch of eggs from a nest, leaving some with the mother and putting the rest in an incubator. The young reared by the mother served as the control group. They showed normal behavior, following the mother about as goslings and eventually growing up to mate and interact with other geese. The geese from the artificially incubated eggs formed the experimental group. These geese spent their first few hours after hatching with Lorenz, rather than with their mother. From that day on, they steadfastly followed Lorenz (Figure 37.5A, on the facing page) and showed no recognition of their mother or other adults of their own species. This early imprinting lingered into adulthood: The birds continued to prefer the company of Lorenz and other humans to that of their own species. Some of them even tried to mate with humans.

In other experiments, Lorenz demonstrated that the most important imprinting stimulus for graylag geese was movement of an object (normally the parent bird) away from the hatchlings. The effect of movement was increased if the moving object emitted some sound. The sound did not have to be that of a goose, however; Lorenz found that a box with a ticking clock in it was readily and permanently accepted as a "mother."

The critical period for imprinting varies with the species. Lorenz found it to be the first two days after hatching for the graylag goose. During that time, the hatchlings apparently have no innate sense of their species (what we might think of as "mother" or "I am a goose, you are a goose)." Instead, they simply respond to and identify with the first object they encounter that has certain simple characteristics. Imprinting has both innate and learned components. Its innate component is the ability or tendency to imprint during a critical period. The actual imprinting itself is a form of learning.

Since Lorenz's pioneering work, many other examples of imprinting have been discovered, and not all involve parent-offspring bonding. In some species, new offspring

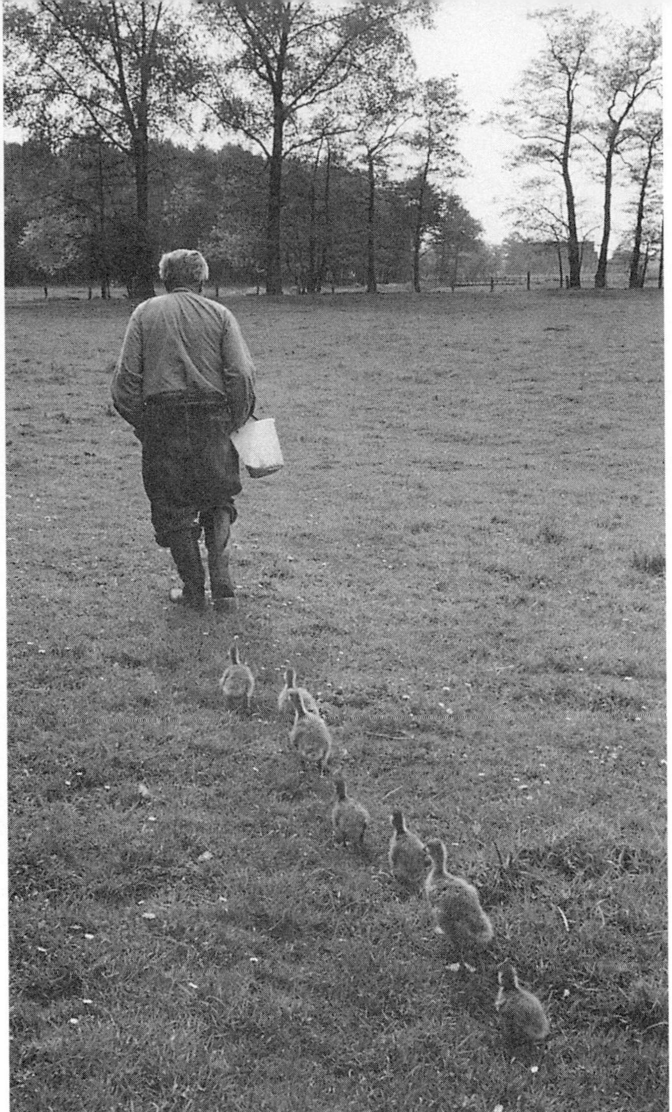

Figure 37.5A Konrad Lorenz with geese imprinted on him

imprint on certain components of their home environment rather than on parents. Newly hatched salmon, for instance, do not receive any parental care but seem to imprint on the complex mixture of odors unique to the freshwater stream where they hatch. They retain their imprinting into adulthood, and it enables them to find their way back to the stream to spawn after spending a year or more at sea. As we saw in the introduction to Chapter 29, adult salmon recognize the odors of their home stream and swim toward their source from great distances.

For many kinds of birds, imprinting plays an important role in song development. In most cases, when we hear a bird sing, it is a male trying to attract a mate or issuing an aggressive warning to would-be competitors. Each species of songbird has its own particular song, although the song varies somewhat among individuals of a population and, in some species, among populations.

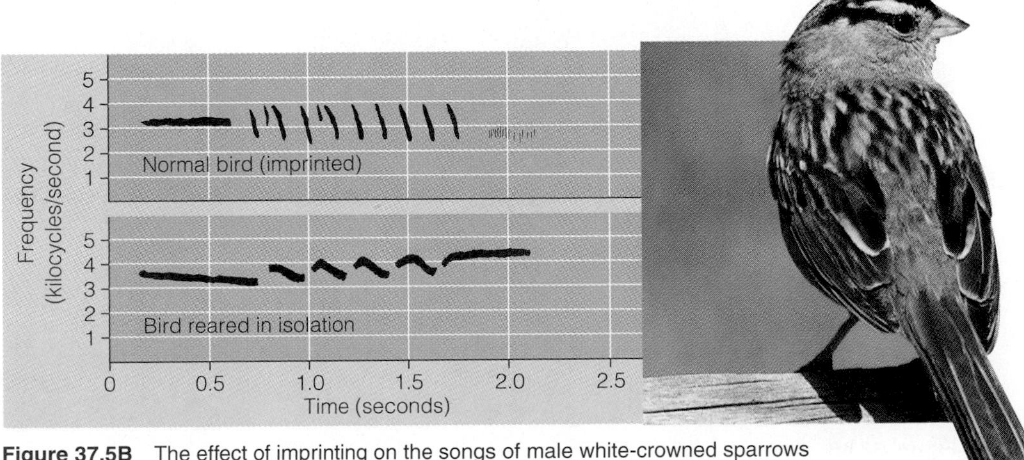

Figure 37.5B The effect of imprinting on the songs of male white-crowned sparrows

Researchers study bird songs using instruments that record individual sounds making up the song. Figure 37.5B shows two sound tracings, illustrating the song patterns of two male white-crowned sparrows. The upper tracing shows the complete song of the species, sung by a normal male (song tracings of other males show only minor differences). Researchers found that when isolated males raised in soundproof chambers heard a taped song of a normal male of their species during a critical period (10–50 days after hatching), the isolated males imprinted on the song. As a result, several months later they learned to sing their species' song normally. Isolated males that did not hear their species' song until after 50 days did not learn to sing normally. Instead, these males sang an abnormal song like the one in the lower tracing. Researchers also discovered a purely genetic component of the white-crowned sparrow song process: Isolated males exposed to the songs of other species during the critical period did not adopt these songs, but sang an abnormal song similar to the lower one in Figure 37.5B.

Imprinting also occurs in many adult animals. For instance, just after giving birth, female sheep and many other mammals imprint on the smell of their own offspring. Afterwards, the females will reject all other young of their species.

The ability of parents and offspring to keep track of each other and the ability of male songbirds to attract mates are examples of capabilities that can have direct and immediate effects on survival and reproduction. Thus, imprinting functions in complex behavior that is strongly connected to evolutionary fitness. Imprinting provides a way for such behavior to become more or less fixed in an animal's nervous system.

How does the research on sparrows support the hypothesis that singing the species' song is a product of both learning and heredity?

To sing its species' song correctly, a male must imprint on the song by hearing it during the critical learning period. However, heredity primes the bird to learn only its species' song, and not that of some other species.

37.6 Many animals learn by association and imitation

Figure 37.6A Associative learning by ducks

Associative learning is learning that a particular stimulus or a particular response is linked to a reward or punishment. If you keep a pet, you probably have observed one type of associative learning first-hand. A dog or cat will learn to associate a particular sound or word with some type of punishment or reward. The ducks in the pond pictured in Figure 37.6A have learned to associate the presence of people with handouts, and they congregate rapidly whenever someone approaches the shoreline. This type of learning, in which an arbitrary stimulus is associated with a reward or a punishment, is called **classical conditioning.** Eventually, the animal learns to respond to the arbitrary stimulus even in the absence of a reward or punishment.

In natural settings, a much more common form of associative learning is **trial-and-error learning.** In this case, an animal learns to associate one of its own behavioral acts with a positive or negative effect. The animal then tends to repeat the response if it is rewarded or avoid the response if it is harmed. For example, predators quickly learn to associate certain kinds of prey with painful experiences. The coyote in Figure 37.6B has a face full of quills obtained when it tried attacking a porcupine (right photograph). The porcupine's sharp quills and ability to roll into a quill-covered ball are strong deterrents against many predators. Coyotes, mountain lions, and domestic dogs often learn the hard way to avoid attacking porcupines nose-first. Trial-and-error learning often provides responses that are important to survival.

Another form of learning is **imitation**—learning by observing and mimicking the behavior of others. Imitation is similar to song imprinting in birds, though imitation is not limited to a critical period. Many predators, including cats, coyotes, and wolves, seem to learn some of their basic hunting tactics by observing and imitating their mother. In many such cases, genetic programming and trial-and-error learning also seem to play key roles. For example, jaguar cubs may develop a sense of which prey are easiest to kill partly by imitating their mother, but trial-and-error learning is also involved, and the underlying basis of stalking behavior is probably genetic. Similarly, a young coyote or mountain lion may learn by a combination of trial and error and imitation to flip a porcupine onto its back, exposing its unprotected belly.

? What type of learning in humans is exemplified by identification with a "role model"?

Imitation

Figure 37.6B Trial-and-error learning by a coyote (left), shown with quills from a porcupine (right)

Innovation is problem solving without prior experience 37.7

Figure 37.7A A dog demonstrating its *lack* of reasoning ability

Dogs often get into situations like the one in Figure 37.7A. Eventually, by trial and error, the animal might make the situation worse by looping its leash around the stump again, or it might actually find a new route to its food bowl. If it did reach the food, it might learn which route was correct and then be able to solve a similar problem more quickly. In any case, a dog does not show an ability to solve problems except by trial and error or by learning from prior experience.

Innovation, sometimes called reasoning, is the ability to respond appropriately to a new situation without prior experience. Innovation involves an ability to analyze problems and to think of possible solutions to them. This type of learning is rare, and we see it mostly in primates. For example, if a chimpanzee is placed in a room with several boxes on the floor and a banana hung high above its head, the chimp will gradually "size up" the situation and then stack the boxes in order to reach the food. In Figure 37.7B, a chimp is using a stick to extend its reach across the water in order to fetch a desired object. Few animals exhibit this kind of problem-solving ability.

The subject of innovation in problem solving raises the larger question of what goes on in the brain of a nonhuman animal. To what extent is an animal aware of the need to solve a problem? To what extent is it aware of itself and its surroundings in general? It is important to recognize, and avoid as much as possible, the pitfall of **anthropomorphism,** ascribing to other animals human motivations or conscious awareness, or assuming that they experience feelings such as pain or pleasure in the same ways we do. This is not to say that our thoughts and those of other animals are completely dissimilar. It's just that we do not know for certain whether, or to what extent, any nonhuman animal has mental experiences similar to ours.

In humans, conscious thinking and self-awareness are generally called **consciousness.** Because of our ignorance of other animals' mental states, many researchers maintain that questions of consciousness in animals other than humans are outside the realm of science. To avoid any trace of anthropomorphism, some researchers focus on behavior patterns that can be described in terms of simple stimuli and responses. In contrast, other researchers, especially those studying chimpanzees and other primates in the wild, find much evidence that some degree of consciousness is not restricted to humans. In a 1992 book entitled *Animal Minds,* Donald Griffin of Princeton University advanced the view (not shared by many other biologists) that conscious thinking is part of the behavior of many animals. Griffin sees animal consciousness arising through the normal process of natural selection and, like so many other major animal functions, having roots extending far back into evolutionary history.

Ultimately, answers to questions about animal thinking may profoundly affect how we interact with other animals and how we view ourselves. Historically, a prevalent view has been that our intellect sets us apart from other animals. There may seem to be a vast difference between human consciousness and that of other animals. But is this difference a fundamental biological one, or is it a matter of degree? Are we simply at one end of a continuum of higher mental functions? The study of consciousness poses unique challenges for scientists.

? Why is it not necessarily anthropomorphic to postulate that some nonhuman animals have consciousness?

The hypothesis that consciousness exists outside our species need not imply that other animals have human-like thoughts.

Figure 37.7B A chimpanzee solving a problem by innovation

37.8 An animal's behavior reflects its evolution

We have stressed that behavior is an evolutionary adaptation that enhances survival and reproductive success (fitness). In fact, behavior is the result of the fine-tuning of an animal to its environment by natural selection. We have seen evidence of this in every type of behavior we have discussed so far—including the hunting and reproductive behavior of jaguars, nest location by digger wasps, and imprinting and other forms of learning in many animals. In all cases, the behavior patterns involve an interplay between the organism and its environment. In the next several modules, we pay special attention to the ecological role of behavior—that is, its role in enabling an animal to survive in its environment. 

? Why is behavior relevant to ecology?

Ecology is the study of organisms' relationships with their environments, and an animal's behavior is part of this organism-environment interface.

37.9 Biological rhythms synchronize behavior with the environment

Animals exhibit a great variety of rhythmic (regularly repeated) behavior patterns. Many mammals—for example, bats, deer, and most cats, including the jaguar—sleep or doze a great deal during the day and feed at dusk and dawn or at night. In contrast, most birds sleep at night and are active during daylight hours. Patterns that are repeated daily, such as sleep/wake cycles in animals and plants, are called **circadian rhythms.** As we saw in plants, internal timers called biological clocks underlie circadian rhythms (see Module 33.10). External cues, especially light/dark cycles, adjust the clocks, keeping body rhythms tightly coordinated with the outside world.

How are circadian rhythms studied? The animal in the photograph in Figure 37.9A is a flying squirrel, an inhabitant of North American forests, which is active at night and usually sleeps in a hollow tree from dawn to dusk (see Chapter 1's opening essay). To study its activity rhythms, a researcher places a flying squirrel in a cage containing a wheel in which the squirrel can run. The cage is connected to a chart recorder, which moves graph paper past a pen at a fixed speed. Whenever the squirrel runs in the wheel, the pen is activated and marks the paper.

The graphs in Figure 37.9A trace the activity patterns of two flying squirrels held under different light conditions for 23 days. The longer black bars indicate periods of extended activity. Graph 1 shows the activity pattern of a squirrel exposed to 12 hours of light alternating with 12 hours of darkness, simulating natural conditions. Graph 2

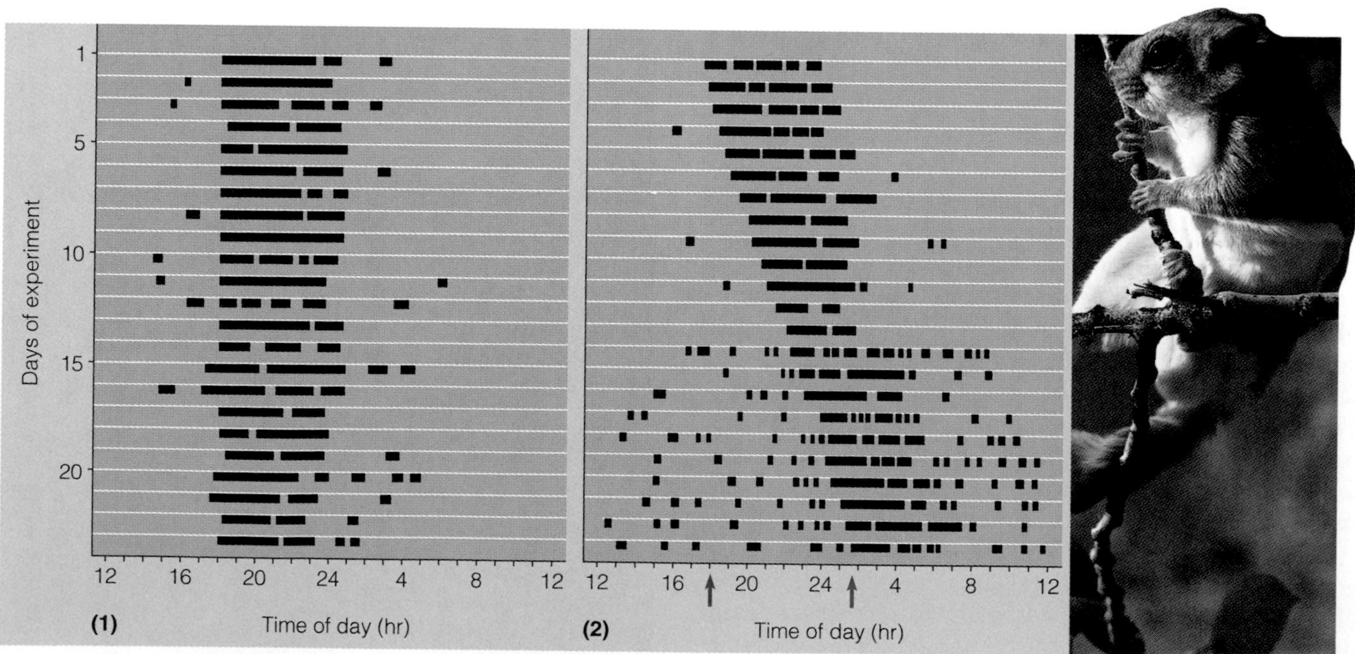

Figure 37.9A Activity rhythms of two flying squirrels under different conditions of light and darkness

shows the activity pattern of a squirrel held in constant darkness for 23 days.

As you can see, the activity of both squirrels remained rhythmic throughout the recording period, with a distinct period of extended activity every day. The activity rhythm of the squirrel exposed to cycles of 12 hours of light and 12 hours of dark remained virtually unchanged for all 23 days. In contrast, the other squirrel's high activity period shifted slightly each day, and after 23 days was nearly 8 hours out of synchronization with the actual time of day. (The small red arrows indicate when the period of greatest activity began on days 1 and 23.) When the squirrel held in the dark was returned to a regular cycle of 12 hours of light and 12 hours of dark, its activity cycle shifted back to that of the other animal in a few days. (This is not shown on the graph.)

Research with many different species has shown that without environmental cues, biological clocks keep time in a free-running way. In the study shown here, the flying squirrel's clock kept time at 24 hours, 21 minutes. Thus, the activity pattern of the squirrel kept in the dark shifted by 21 minutes each day. In contrast, humans have a biological clock that makes circadian rhythms follow an approximately 25-hour cycle. In humans, flying squirrels, and other organisms that have been studied, environmental cues are needed to keep circadian rhythms synchronized with external conditions. Sunrise and sunset are important cues.

How do researchers study circadian rhythms in humans? In one study, conducted in 1989, Italian interior designer Stefania Follini volunteered to test the effects on her body rhythms of long-term isolation below the ground. Follini spent 131 days alone in a plastic, 3.5-m-by-6-m cubicle 3 m underground in a cave near Carlsbad, New Mexico (Figure 37.9B). Temperature in the cubicle was held constant at 21°C. Follini had control of artificial lighting, but the cave itself was totally dark; therefore, without a clock, she had no cues about day or night hours. Researchers monitored a number of physiological factors, including Follini's blood pressure, heart rate, and body temperature. Follini herself kept track of what she thought were days and nights and the passage of time.

During the course of the study, Follini's sleep/wake rhythm followed the typical 25-hour period of the free-running human biological clock. Unexpectedly, however, her blood pressure and heart rate followed a cycle of 48 hours to 7 days. When Follini emerged from the cave in June, at the end of the study, she thought it was March. She had also developed a severe calcium deficiency, lost 24 pounds, and stopped menstruating. She did not regain normal biological rhythms until months after she had emerged, and she began menstruating again only after receiving hormone injections. The reasons for these changes are unknown, but researchers think at least some were related to the emotional stress of being isolated from all human contact and normal environmental stimuli for months. Follini's experiences and those of others who have participated in similar projects lead many researchers to be wary of long-term isolation studies. Most studies of human

Figure 37.9B Subterranean living quarters for studying human circadian rhythms

circadian rhythms are now performed in hospitals or research labs, where subjects have contact with other people and are more closely monitored.

It is important that we learn more about biological rhythms, clocks, and the cues that set them, because body rhythms affect our general well-being, work efficiency, and decision-making ability. Working night shifts, keeping irregular hours, and traveling by jet across several time zones often lead to fatigue, reduced job performance, and depression because our internal rhythms can't adjust instantaneously to a different time frame. Researchers studying ways to help the body reset its internal clocks when necessary have had some encouraging results. Specific light/dark regimes can often be used to reset our biological clocks. Also, some people can minimize the symptoms of jet lag by adjusting when they eat, exercise, and sleep.

Researchers are also making progress toward understanding the basic mechanism of biological clocks. In humans and other mammals, a cluster of neurons in the brain's hypothalamus, called the suprachiasmatic nucleus, functions as the clock. Current hypotheses propose that this and other clocks depend on differential gene expression. Experiments with rodents, for example, indicate that in response to light/dark cycles, certain genes in the cells of the clock turn on, triggering the synthesis of specific proteins. 

? How does research, such as the flying squirrel studies and experiments with humans living below ground, suggest that both an internal clock and environmental cues are important in body rhythms?

Shielded from environmental information about the time of day, rhythms persist, but each on its own free-running cycle. It is the cues of the day/night cycle that synchronize these rhythms with one another and with the environment.

37.10 Kineses and orientation behavior place animals in favorable environments

Many protists and some animals exhibit a very simple response to certain stimuli: When they sense a stimulus, they merely start or stop moving, change their speed of movement, or turn more or less frequently in a random (nondirected) manner. A random movement in response to a stimulus is called a **kinesis** (plural, *kineses;* Greek for movement). Human body lice, for instance, exhibit kineses; they move randomly but become more active in dry areas and less active in moist ones. The more they move, the greater the chance they will find a moist area, which is more favorable to their survival. Once in a suitable place, their decreased activity tends to keep them there.

In contrast to kineses are directed movements, also called **orientation behavior.** Orientation behavior results in an animal's placing its body in a specific position relative to an environmental cue. One form of orientation behavior is a taxis (Greek *tasso,* "put in order"). A **taxis** (plural, *taxes*) is a more or less automatic movement directed toward (positive) or away from (negative) a stimulus. For example, trout exhibit positive *rheo*taxis (Greek *rheos,* current); they automatically orient in an upstream direction, toward the current. Many animals locate mates by *chemo*taxis. Females of many species, especially insects, emit chemical signals that attract males.

A taxis is a more finely tuned response to environmental stimuli than a change in speed or the turning of a random movement. It is, therefore, a more efficient way for an animal to locate a mate or find food. It is not surprising that orientation behavior is much more prevalent than kineses in the animal kingdom.

Which behavior, a kinesis or a taxis (choose one), requires that an animal detect the direction of some stimulus?

Taxis

37.11 Movement from place to place often depends on internal maps

What does an animal's brain do with the information it obtains about the outside world? Watching an animal solve a problem makes us aware that its nervous system has a substantial ability to process information. An expanding research effort on animal cognition seeks to understand information processing.

The term cognition is variously defined. In a narrow sense, it is essentially synonymous with consciousness, or awareness. In a broad sense—the way we use the term in this book—**cognition** is the ability of an animal's nervous system to perceive, store, process, and use information gathered by its sensory receptors. The study of animal cognition, called **cognitive ethology,** attempts to illustrate the connection between data processing by nervous systems and animal behavior. Cognitive ethology includes, but is not limited to, the study of animal awareness.

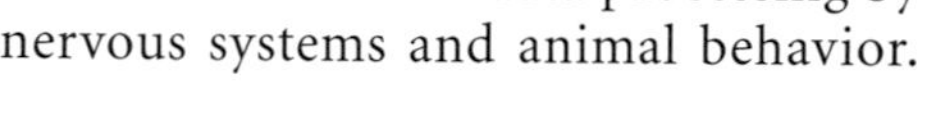

A central hypothesis of cognitive ethology is that many animals formulate **cognitive maps,** internal representations, or codes, of the spatial relationships among objects in their surroundings. In a broad sense, most animals that move through space probably have some internal map of the way external objects are arranged.

The most extensive studies of cognitive maps have been made for animals that exhibit **seasonal migration,** the regular back-and-forth movement of animals between two geographic areas at particular times of the year. It enables many species to access rich food resources throughout the year and to breed or winter in areas that favor survival.

Figure 37.11A shows the migratory route (red line) of one long-distance traveler, the gray whale. During summer, these giant mammals feast on small, bottom-dwelling invertebrates that abound in northern oceans. In the autumn, they leave their northern feeding grounds and begin a long trip south along the North American coastline. Arriving in warm, shallow lagoons off Baja California (Mexico) in the winter months, they breed, and pregnant females give birth to young before migrating back north. The yearly round trip, some 20,000 km, is the longest for any mammal.

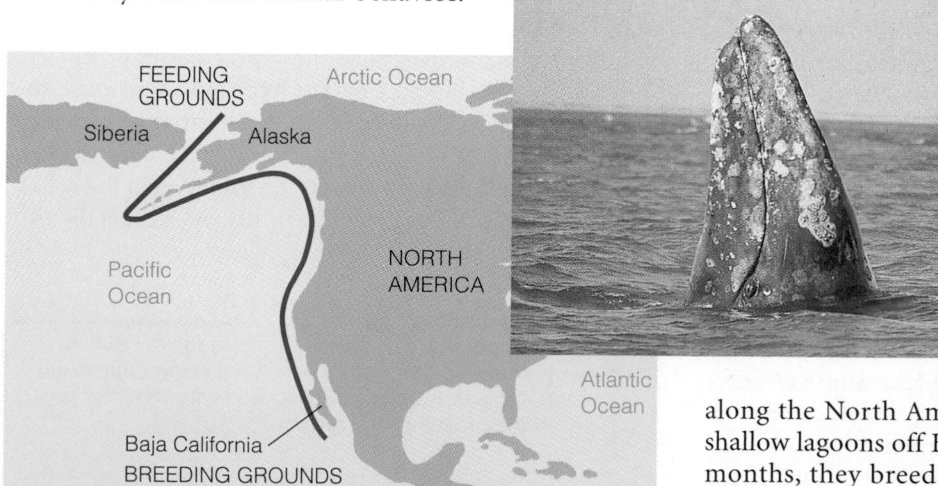

Figure 37.11A The migratory route of the gray whale

Many other species of mammals, as well as numerous birds and some insects, also migrate seasonally. For instance, many insect-eating birds winter in the tropics and breed at high latitudes. Their breeding grounds—such as the tundra and taiga of the Northern Hemisphere—harbor large populations of insects but only in the summer months. The birds migrate to the breeding grounds in the spring and back to their wintering grounds in the late summer or fall.

Among the insects, the monarch butterfly has one of the most remarkable seasonal migrations. During winter, these insects festoon certain trees at the western tip of Cuba, in a few mountain valleys of central Mexico, and at a few sites along the California coast (Figure 38.5C). In the late summer and fall, all of North America's monarchs fly to these wintering sites. They remain on the trees for 4–5 months, not feeding but living off food molecules stored in their tissues. Suitable wintering sites are rare because temperatures must be just cool enough that the monarchs do not metabolize their stored food before spring, and warm enough that they do not freeze. With the onset of spring, monarchs mate at the wintering sites and begin migrating northward. As they arrive at regional destinations, they lay eggs and then die. Two or more generations are produced during the summer, repopulating the United States and southern Canada. With the approach of fall, the summer's last generation of monarchs flies south to the wintering grounds. They migrate as far as 4000 km and end up at a specific site, although they have not flown the route before.

Researchers have found that migrating animals stay on course by using a variety of environmental cues. Gray whales, for instance, seem to use coastal landmarks to pilot their way north and south. Migrating south in the autumn, they orient with the North American coastline on their left. Migrating north in the spring, they keep the coast on their right. Whale watchers sometimes see gray whales stick their heads straight up out of the water (Figure 37.11A, inset), perhaps to obtain a visual fix on land. Many birds migrate at night, navigating by the stars the way ancient human sailors did. In contrast, monarch butterflies migrate during the day, resting in trees and bushes at night; genetic programming may enable them to use the sun as a compass.

Navigating by the sun or by most stars requires an ability to keep track of compass direction. Many migrators also must have an internal timing mechanism that compensates for the continuous daily movement of Earth relative to celestial objects. The timing mechanism must also allow for the apparent change in position of celestial objects as the animal moves over its migration route. Almost nothing is known yet about the nature of these timing mechanisms.

At least one night-migrating bird, the indigo bunting, seems to avoid the need for a timing mechanism by fixing on the North Star, the one bright star in northern skies that appears almost stationary. Figure 37.11B illustrates an experimental setup that was used to study the bunting's navigational mechanism. During the migratory season, wild and laboratory-reared birds were placed in funnel-like cages in a planetarium (see photograph). Each funnel had an ink pad at its base and was lined with blotting paper. When a bird stepped on the ink pad and then tried to fly in a certain direction, it tracked ink on the paper. The researchers found that wild buntings and those raised in the lab and introduced to the northern sky in a planetarium tracked ink in the direction of the North Star. Birds raised under a sky with a different fixed-location star oriented to that star. Apparently, buntings learn a star map and fix on a stationary star when navigating at night.

Another interesting, and more or less open, question about migration is how birds continue navigating when the sun or stars are obscured by clouds. There is strong evidence that some birds can orient to Earth's magnetic field. Magnetite, the iron-containing mineral once used by sailors as a primitive compass, is probably involved in sensing the field. This mineral has been found in the heads of pigeons, in the abdomens of bees, and in certain bacteria that orient to a magnetic field. Future research may show that magnetic sensing is a widespread, important part of a complex navigation mechanism in many animals.

Why is a timekeeping mechanism essential for stellar navigation?

Because the positions of the stars change with time of night and season

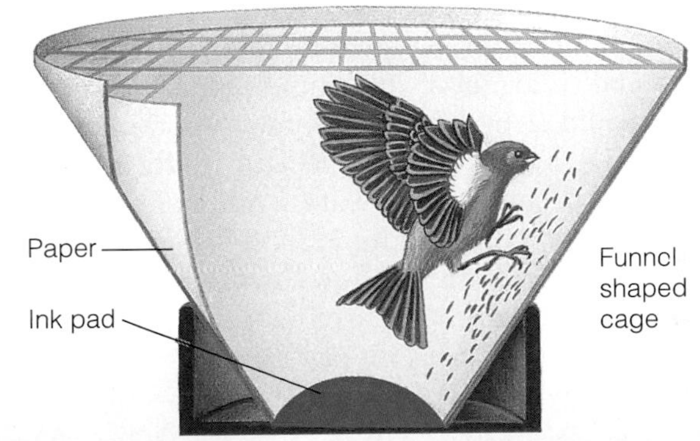

Figure 37.11B An experiment demonstrating star navigation

37.12 Behavioral ecologists use cost/benefit analysis in studying feeding behavior

An animal's ability to obtain energy efficiently is a key factor affecting fitness. Animals feed in a great many ways. Some animals are feeding "generalists," while others are "specialists." The gull in Figure 37.12A is an extreme generalist; it will eat just about anything that is readily available—plant or animal, alive or dead. In sharp contrast, the koala of Australia, an extreme feeding specialist, eats only the leaves of eucalyptus trees (Figure 37.12B).

Most animals have some variety in their diet but are more selective than gulls, even if they are generalists. Often an animal will concentrate on a particular item of food, sometimes to the exclusion of other foods. The mechanism that enables an animal to find particular foods efficiently is called a **search image.** (We often use search images to help us find something more efficiently. For example, when looking for a particular package on a kitchen shelf, you probably scan rapidly, to find packages of a certain size and color rather than reading labels.) If the favored food item becomes scarce, the animal may develop a search image for a different food item.

Why might an animal be selective in its choice of food? Some researchers propose that the explanation lies in what is called **optimal foraging,** feeding behavior that yields the lowest cost/benefit ratio. In other words, optimal foraging provides maximal energy gain with minimal energy expense and minimal time spent foraging (searching for, securing, and eating food). We would expect that natural selection would favor animals that forage optimally, but it is difficult to test this hypothesis in nature. Whenever an animal has food choices, there are a number of trade-offs. Consider the bass in Figure 37.12C, for example. It can readily consume both minnows and crayfish. If it eats a minnow, it will probably get more usable energy per unit of weight (a crayfish has a lot of hard-to-digest exoskeleton), but the minnow is smaller and may be harder to catch. On the other hand, it may take more time to eat a crayfish because of its large claws and tough exoskeleton. Complicating the picture even more, the bass must be alert to other predators while feeding. What would expose it more to a predatory turtle or larger fish, chasing a minnow or mouthing a thrashing crayfish?

In most natural environments, there are so many variables that it's hard to imagine any animal could forage in an absolutely optimal manner. Nonetheless, numerous studies indicate that when prey is plentiful, many different species forage in such a way that their overall energy intake-to-expenditure ratio is high. A bass, for instance, forages efficiently, if not exactly optimally, by switching between minnows, crayfish, aquatic insects, and other invertebrates as conditions, such as a prey's relative sizes, density, and ease of capture, change.

In another case, Alan Rabinowitz found that jaguars in Belize ate mainly small mammals, especially the abundant and slow-moving armadillo (Figure 37.12D, right). An armadillo may weigh only about 5 kg—just a few mouthfuls for a jaguar. It also has hard body armor—no problem for the big cat's jaws, but not digestible. Why would the most powerful cat in North America eat mainly bite-sized prey, especially when it has much larger game available to it? For instance, tapirs (Figure 37.12D, left), which may weigh up to 200 kg and are known to be vulnerable to jaguars, also inhabit the jungles of Belize. But Rabinowitz found no evidence of jaguars eating these large mammals in his study area. Most likely, the jaguars concentrate on armadillos because they are abundant and easy to catch. Tapirs, by contrast, run fast, usually into very dense undergrowth, and are not nearly as abundant.

The kangaroo rat (Figure 37.12E), an herbivorous rodent found in North American deserts, illustrates

Figure 37.12A A feeding generalist, a gull

Figure 37.12B A feeding specialist, a koala

Figure 37.12C A bass eating a crayfish

Figure 37.12D Alternative prey available to jaguars in Belize: the armadillo (right), the preferred prey, and the tapir (left), generally ignored.

the effects of trade-offs in optimal foraging more clearly than either the bass or the jaguar. Foraging at night, the kangaroo rat fills its cheek pouches with high-energy seeds and carries them home to its burrow. Careful studies show that when a choice of seeds is available, the animal picks up seeds that contain more energy than most of those it leaves behind. Later, in the safety of its burrow, the rat may be even more selective, eating only the very richest seeds from its cache. Thus, the kangaroo rat does not forage exactly optimally, for it expends energy gathering seeds it does not consume, but it exhibits a healthy compromise: It selects high-energy food in a manner that reduces time spent above the ground, where it is exposed to predators.

Figure 37.12E A kangaroo rat with a collection of seeds

Early humans were hunter-gatherers, but evidence from anthropology suggests that nutrition was based more on gathering than on hunting. How does this fit in with the hypothesis of optimal foraging?

Meat is nutritious, but hunting also poses relatively high costs in effort and risk compared to the gathering of plant products and dead animals.

Social Behavior and Sociobiology

37.13 Social behavior is an important component of population biology

Of the types of behavior we have examined so far, several involve interactions between two or more individuals of a species. Imprinting, for instance, often involves interaction between a parent and an offspring. Many animals migrate and feed in large groups (flocks, packs, herds, or schools). Wolves, for example, usually hunt in a pack consisting of a tightly knit group of family members. Hunting in packs enables them to kill large animals, such as moose or elk, that would be unavailable to a single individual.

Biologists define **social behavior** broadly as any kind of interaction between two or more animals, usually of the same species. An essential ingredient of all types of social behavior is communication—some means of transferring information between individuals. Because social behavior involves group interactions, it can often affect the growth and regulation of animal populations. For example, the large prey animals that wolf packs can kill provide more food for each wolf than smaller prey; hence, more wolves may survive in a particular area. As we will see in the next several modules, social behavior may also affect fitness directly by actually determining which animals in a population will produce offspring.

What process is required for social behavior within a population?

Communication

37.14 Rituals involving agonistic behavior often resolve confrontations between competitors

Agonistic behavior (from the Greek *agon,* struggle) includes a variety of threats or actual combat that settles disputes between individuals in a population. Conflicts often arise over limited resources, such as food or mates. An agonistic encounter may involve a test of strength or, more commonly, exaggerated posturing and other symbolic displays—rituals—that make the individuals look large or aggressive. Eventually, one individual stops threatening and becomes submissive, exhibiting some type of appeasement display—in effect, surrendering.

Figure 37.14 Ritual wrestling by rattlesnakes

Because violent combat may injure the victor as well as the vanquished in a way that reduces reproductive fitness, we would predict that natural selection would favor ritualized contests. And, in fact, this is what usually happens in nature. The rattlesnakes pictured in Figure 37.14, for example, are rival males wrestling over access to a mate. If they bit each other, both would die from the toxin in their fangs, but this is a pushing, rather than a biting, match. One snake usually tires before the other, and the stronger one pins the loser's head to the ground. In a way, the snakes are like two people who settle a serious argument by arm wrestling instead of resorting to fists or knives. In a typical case, the agonistic ritual inhibits further aggressive activity. Once two individuals have settled a dispute by agonistic behavior, future encounters between them usually involve less dispute, with the original loser giving way to the original victor. Often the victor of an agonistic ritual gains first or exclusive access to mates, and so this form of social behavior can directly affect an individual's evolutionary fitness. 

? Why is "fighting to the death" an unusual form of agonistic behavior among animals?

Because ritualized posturing or nonlethal combat can usually produce a winner without injuries that lower reproductive potential for the winner and eliminate it altogether for the loser

37.15 Dominance hierarchies are maintained by agonistic behavior

Many animals live in social groups maintained by agonistic behavior. Chickens are an example. If several hens unfamiliar to one another are put together, they respond by chasing and pecking one another. Eventually, they establish a clear "peck order." The alpha (top-ranked) hen in the peck order (the one on the left in Figure 37.15) is dominant; she is not pecked by any other hens and can usually drive off all the others by threats rather than actual pecking. The alpha hen also has first access to resources such as food, water, and roosting sites. The beta (second-ranked) hen similarly subdues all others except the alpha, and so on down the line to the omega, or lowest animal.

Figure 37.15 Chickens exhibiting peck order

Peck order in chickens is an example of a **dominance hierarchy,** a ranking of individuals based on social interactions. Once a hierarchy is established, each animal's status in the group is fixed, often for several months or even years. Consequently, rather than fighting with others, group members can concentrate on finding food, watching for predators, locating a mate, or caring for young.

Dominance hierarchies are common, especially in vertebrate populations. In a wolf pack, for example, there is a dominance hierarchy among the females, and the hierarchy may control the pack's size. When food is abundant, the alpha female mates and also allows others to do so. When food is scarce, she usually monopolizes males for herself and keeps other females from mating. Next we hear from a scientist who has conducted long-term studies of dominance hierarchies in nature. 

? How does a dominance hierarchy enhance mean reproductive success for individuals of some animal populations?

By reducing the amount of time and energy that agonistic behavior diverts from feeding and other survival activities

37.16 Behavioral biologist Jane Goodall discusses dominance hierarchies and mental abilities in chimpanzees

Chimpanzees are our closest relatives. Chimps and humans are more like each other genetically than either is like other apes. Dr. Jane Goodall (Figure 37.16A), one of the world's best-known biologists, has studied these remarkable primates in their natural habitat in East Africa since the early 1960s. Many of her discoveries are described in her seven books, her appearances on National Geographic Society television specials, and her frequent lecture tours. In all of her writings and interviews, Dr. Goodall promotes a better understanding of animal behavior, especially that of primates. She also works tirelessly to encourage better living conditions for animals in medical research labs and zoos. Despite her crowded schedule, Goodall still spends as much time as possible studying chimpanzees in the wild.

Dominance hierarchies are an integral part of chimpanzee life, as Goodall relates:

> Some male chimpanzees devote much time and effort to improving or maintaining their position in the hierarchy. For the most part, the male uses the impressive charging display, during which he races across the ground, hurls rocks, drags branches, leaps up and shakes the vegetation [see Figure 37.16B]. In other words, he makes himself look larger and more dangerous than he may actually be. In this way he can often intimidate a rival without having to risk an actual fight, which could be dangerous for him as well as for his rival. The more frequent, the more vigorous, and the more imaginative his charging display, the more likely it is that he will attain a high social position.

Despite the focus male chimpanzees have on social position, the benefits of high position are not clear. In Goodall's words:

> We don't understand why the chimps devote so much time, effort, and risk [to gaining high rank]. What is the advantage of high rank for a male chimpanzee? What are the evolutionary benefits, the reproductive benefits? A high-ranking male has prior access to the best food, but this is not of great benefit since, if the food is short, the chimps typically move about in ones and twos. He can usurp a choice resting place, but he almost never does. He can inhibit other males from copulating with a female in estrus when they are all in a group together, but there is a mechanism in chimp society that enables even a low-ranking male to appropriate a sexually attractive female. All he has to do is persuade her to follow him, away from the other males, to some peripheral area of the community range. It's almost as though humans aren't the only creatures who value high rank for its own sake, and the power that it gives.

And what about female chimpanzees?

> Females have a hierarchy too.... The reproductive advantage to the high-ranking female is clear. She can better appropriate choice food items and thus make her milk richer. In addition, her offspring are likely to become high-ranked since she will support them. In the supportive family group situation, all have a better chance of survival.

Studies of chimpanzee behavior can help us understand certain aspects of our own behavior. They can also make us more aware of what we have in common with other species. Jane Goodall's years of chimpanzee research have convinced her that chimpanzees are truly conscious beings. As she explains:

> Science has been very quick to recognize the incredible similarity in [the anatomy] of the chimpanzee brain and human brain ... and all the other amazing physiological similarities [between the two species]. So it stands to reason that you would find similarities in the emotions ... and in certain kinds of behavior and intellect.

Why is the study of chimpanzee behavior relevant to understanding the origins of certain human behaviors?

Because chimps are our closest relatives

Figure 37.16A Jane Goodall with Goblin, an alpha male

Figure 37.16B Charging display

37.17 Territorial behavior parcels space and resources

No one parceled out space to sunbathers on the beach in Figure 37.17A. The fact is that people tend to space themselves out like this when close to others, establishing what we might call personal territories.

Many animals exhibit territorial behavior. A **territory** is an area, usually fixed in location, that individuals defend and from which other members of the same species are usually excluded. The size of the territory varies with the species, the function, and the resources available. Territories are typically used for feeding, mating, rearing young, or combinations of these activities.

Another beach scene, Figure 37.17B, shows a nesting colony of gannets, seabirds of the Atlantic Ocean. The birds are tightly and evenly spaced on a promontory on the Newfoundland coast. Space is at a premium, and the birds maintain tiny nesting territories by agonistic behavior—calling and pecking at each other. Each gannet is literally only a peck away from its closest neighbors. Such small territories are feasible because the gannets don't use them for feeding; they feed in large flocks at sea, where they display little agonistic behavior.

Figure 37.17A Sunbathers

Figure 37.17B Gannets at a nesting ground

In contrast to the small nesting territory of gannets and many other colonial seabirds, most cats, including jaguars, leopards, cheetahs, and even domestic cats, defend much larger territories (often several square kilometers in area), which they use for foraging as well as breeding.

Figure 37.17C A cheetah spray-urinating

Individuals that have established a territory usually proclaim their territorial rights continually; this is the function of most bird songs, as well as the noisy bellowing of sea lions, the chattering of squirrels, and the defecating of jaguars in open areas along jungle trails. Scent markers are frequently used to signal territories. The male cheetah in Figure 37.17C, a resident of Africa's Serengeti National Park, is spraying urine on a tree. The odor will serve as a chemical "No Trespassing" sign. Other males that approach the area will sniff the marked tree and recognize that the urine is not their own. Usually, the intruder will avoid the marked territory and a potentially deadly confrontation with its proprietor.

Natural selection does not always favor territoriality, and not all species are territorial. However, for those animals that are, the territory can provide exclusive access to food supplies, breeding areas, and places to raise young. Moreover, familiarity with a specific area may help individuals avoid predators. In a territorial species, such benefits outweigh the energy costs of defending territory and thereby increase fitness.

What type of dispersion pattern for an animal population is most often associated with territorial behavior? (*Hint:* Review Module 35.2.)

Uniform pattern

37.18 Mating behavior often involves elaborate courtship rituals

Many animals are strongly programmed to view any organism of the same species as a competitor, to be driven away if possible. Even animals that forage and travel in groups maintain a certain distance from their companions. How, then, is mating accomplished? In many species, prospective mates must perform an elaborate courtship ritual, unique to the species. Courtship rituals confirm that individuals are of the same species but of the opposite sex, that they are physically primed for mating, and perhaps that they are not threats to each other.

Figure 37.18A on the next page depicts the courtship and mating of the common loon, a species that breeds on secluded lakes in the northern United States and Canada. In courtship, a male and female loon swim side by side while performing a series of displays. ① The courting birds frequently turn their heads away from each other. (In sharp contrast, a male loon defending his territory often charges at an intruder with his beak pointed straight ahead.) ② The birds then dip their beaks in the water, and ③ submerge their heads and necks. Prior to copulation,

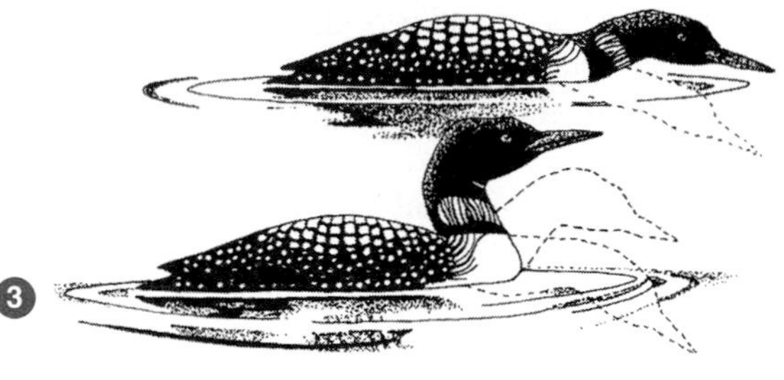

the male invites the female onto land by ④ turning his head backward with his beak held downward. There, ⑤ they copulate.

The common loon and numerous other species exhibit courtship in pairs that are more or less isolated from the rest of the population. In certain other species, courtship is a group activity during which members of one or both sexes choose mates from a group of candidates. For instance, sage grouse, chickenlike birds that inhabit high sagebrush plateaus in the western United States, perform mating rituals in large groups. Each day in the early spring, fifty or more males congregate in an open area called an arena (or lek). The males strut about, erecting their tail feathers in a bright, fanlike display. Dominant males usually defend a small territory near the center of the lek. Females arrive several weeks after the males, and after watching the males perform (Figure 37.18B), a female will select one, and the pair will copulate. Usually, all the females choose dominant males, with the result that only about 10% of the males actually mate.

Is there an advantage to group mating rituals like this? In species that reproduce sexually, an individual's own genes alone do not determine reproductive success; rather, it is the combination of that individual's genes with those of its mate. Researchers hypothesize that there is a connection between a dominant male's display and his evolutionary fitness. We might suppose, therefore, that in choosing a dominant male from several in an arena, a female sage grouse gives her genes the best chance for future survival. Devising ways to test this hypothesis of an ultimate cause is one of the fascinating challenges of research in behavioral ecology.

? How is a female bird's fitness associated with her ability to choose a mate by keying on displays and adornments that "advertise" the healthiness of the male?

She is more likely to have healthy offspring by mating with a healthy male than with a sickly one.

Figure 37.18A Courtship and mating of the common loon

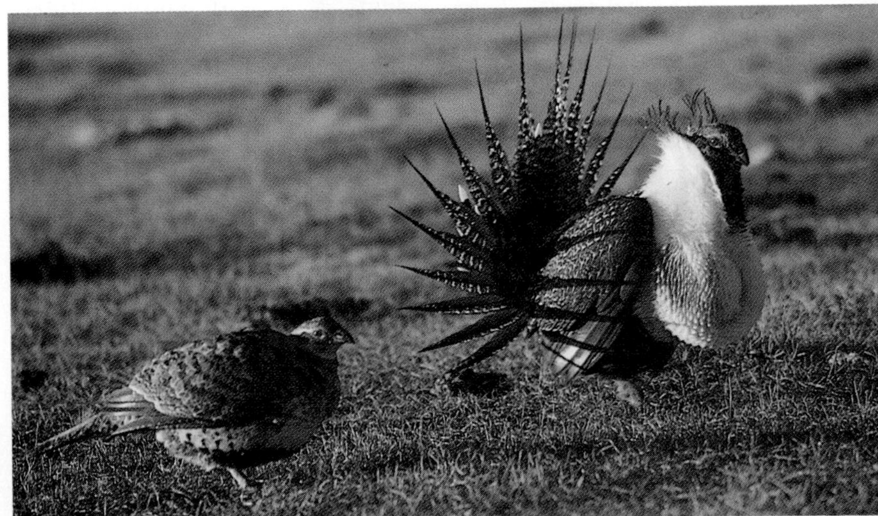

Figure 37.18B Courtship display by a male sage grouse

37.19 Complex social organization hinges on complex signaling

37A
See Activity 37A on the CD-ROM.

So far, two major concepts have emerged from our discussion of social behavior. (1) Social behavior provides organization within populations of animals. For instance, agonistic behavior maintains dominance hierarchies and territoriality, and courtship behavior is often a prerequisite to mating. (2) Social behavior and the social organization it provides depend on some form of signaling, or communication, among the participating animals. As we have seen, animals use a variety of signals, including sounds, such as the growl of a male jaguar announcing his presence at night; odors, such as the urine signs left by many cats to mark their territories; visual displays, such as beak pointing in loons; and touches, such as the caressing of mating jaguars. Signals detected by sight or touch often communicate aggression or nonaggression between individuals at close range.

In general, the more complex the social organization, the more complex the signaling required to sustain it. In fact, animals with a complex social structure often use more than one type of signal simultaneously. Figure 37.19A shows a ring-tailed lemur, a tree-dwelling primate of Madagascar that lives in social groups averaging 15 individuals. Visual displays, scent communication, and vocalizations maintain the dominance hierarchy in the group. The animal shown here is communicating aggression with its prominent tail. Prior to this display, it smeared its tail with odorous secretions from glands on its forelegs. By waving its scented tail over its head, the lemur transmits both visual and chemical signals.

Figure 37.19A A lemur communicating aggression

In addition to primates, many other vertebrates exhibit intricate social behavior. However, many of the most complex social systems are found among the invertebrates. The social system of honeybees is a prime example. Often numbering over 50,000 individuals, a honeybee colony has complicated communication needs. For example, a worker bee that has located a good source of food must communicate this information to other workers in the hive, enabling the food to be harvested in quantity. Intrigued with the question of how bees communicate, Karl von Frisch performed several experiments in the 1940s. He put out dishes of scented sugar water as food sources, varying their distance and direction from hives. In addition, he modified the hives so he could see inside them. Von Frisch and several later researchers discovered that a returning worker bee passes some very complex information to other workers, using a unique signaling system. The figures on the facing page illustrate von Frisch's hypothesis concerning the bees' communication system.

When a worker returns to a hive, others gather around it (Figure 37.19B), and the bee regurgitates some nectar that the others taste. The nectar probably lets the other workers know the type of food that has been found. The worker then performs one of two "dances" that seem to indicate the location of the food. If the source is within 50 m or so of the hive, the bee moves rapidly sideways in tight circles, performing a "round dance" (Figure 37.19C). The round dance translates as "food is near" but does not indicate direction. The other workers then leave the hive and begin foraging nearby; tasting the nectar probably helps them identify a scent to fly toward.

A worker returning from a longer distance performs a "waggle dance," instead of a round dance. As shown in part 1 of Figure 37.19D, the waggle dance involves a half-circle swing in one direction, followed by a straight run, then a half-circle swing in the other direction. During the straight run, the bee waggles its abdomen vigorously. During the waggle dance, the worker also regurgitates some nectar, thus "telling" other workers the type of food to seek.

The waggle dance itself may impart a remarkably detailed code. According to von Frisch's hypothesis, it tells other workers both the distance and the direction of a food source. The speed at which a worker completes the cycles of the dance indicates distance to the food. A certain high rate, for example, would translate as "about 100 m away," whereas a certain slow rate would mean "about 1000 m away," with intermediate rates corresponding to intermediate distances. The code for direction requires that the waggle dance be performed on a vertical surface in the hive. The angle of the straight run in relation to the vertical is the same as the horizontal angle of the food's location in relation to the position of the sun. Thus, if the dancer runs directly upward (Figure 37.19D, part 1), the other workers will fly directly toward the sun when they leave the hive. If the dancer runs directly downward (part 2), the others will fly directly away from the sun. And if the dancer runs at an angle, say 30° to the right of vertical (part 3), the other workers will fly 30° to the right of the horizontal direction of the sun, and so forth. Today, some researchers question the signficance of the pattern of the waggle dance and are exploring other hypotheses for how bees report a source of food.

BEE COMMUNICATION

B. Bees clustering around a recently returned worker

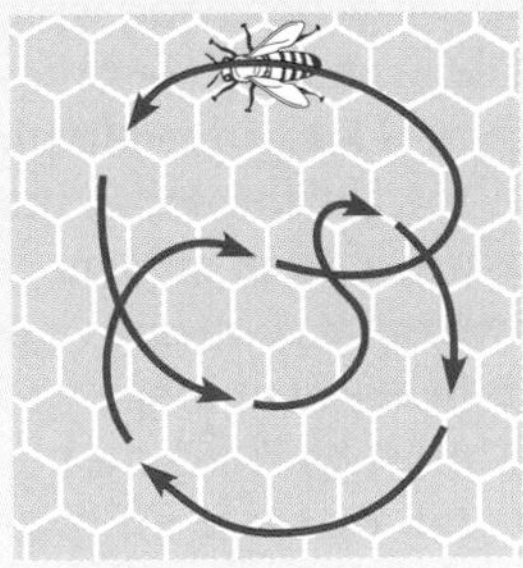

C. Round dance, indicating that food is nearby, in an unspecified direction

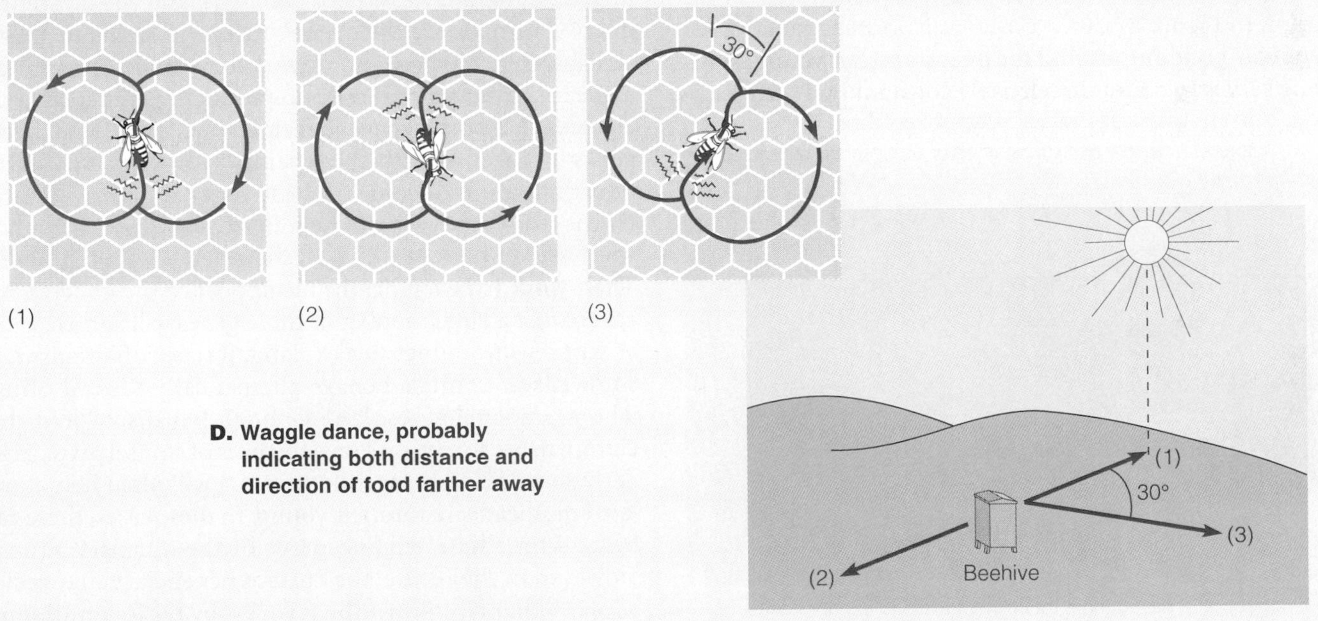

D. Waggle dance, probably indicating both distance and direction of food farther away

Figures 37.19B–D Bee communication

Bees also have some kind of innate timekeeping ability that enables them to adjust for the sun's movement. If a rainstorm prevents them from foraging for several hours, they will still fly in the proper direction toward food, even though the sun's angle has changed.

How do bees inside a dark hive follow the dance movements of their fellow workers? Many questions remain to be answered, but several types of communication, including touch, taste, and sound, seem to be involved. Apparently, as the workers swarm closely around a dancer, they detect dance cues by physical contact. The dancers also seem to make different buzzing sounds during the straight run of the waggle dance; researchers think that these sounds correlate with the distance and direction of the food source. Used together, these different signals make it possible for worker bees to communicate the complex information needed to forage as a group and thus gather enough food to supply their large colony. 

? According to von Frisch's hypothesis, what does a waggle dance "say" to worker bees in a hive?

It indicates the direction and distance of a food source.

37.20 Altruistic acts can be explained by evolution

All the workers in a honeybee hive are sterile females. They never reproduce, but spend their lives laboring on behalf of the one fertile queen that lays all the eggs in the hive. When a worker stings an intruder in defense of the hive, the worker usually dies. Such behavior, which reduces an individual's fitness while increasing the fitness of a recipient (in a beehive, the queen), is known as **altruism.** It is an important component of social behavior in many animal species.

Altruistic behavior is often evident in animals that live in a colony consisting of a cooperative group. The animals in the photographs below are uniquely social rodents called naked mole rats. Almost hairless and nearly blind, they live in colonies of 75–250 or more animals in underground chambers and tunnels in eastern Africa. With a social structure resembling that of honeybees, each colony has only one reproducing female, called the queen (Figure 37.20A). The queen mates with one to three males, called kings. The rest of the colony consists of nonreproductive females and males (Figure 37.20B), which care for and protect the queen, the kings, and new offspring still dependent on the queen. In Figure 37.20A, you can see nonreproductive individuals huddling around the queen and her young. Huddling serves to maintain relatively constant body temperature; naked mole rats are unusual mammals in not being strong thermoregulators.

Figure 37.20A The queen of a naked mole rat colony nursing offspring, while surrounded by other individuals of the colony

Figure 37.20B Nonreproductive members of a mole rat colony

You might wonder about the evolutionary advantage of altruism. While trying to protect the queen or kings from a snake that invades the colony, a nonreproductive naked mole rat may sacrifice its own life and therefore its reproductive success. How can altruistic behavior evolve if it reduces the reproductive success of self-sacrificing individuals? One answer is that the frequency of genes for altruism may increase if individuals that benefit from altruistic acts are themselves also carrying those genes. This is most likely to be the case if the altruists and their beneficiaries are related. According to this idea, known as **kin selection,** altruistic behavior evolves because it increases the number of copies of a gene common to a group, regardless of which individuals in the group transmit the gene. Using DNA analysis (RFLP analysis; see Module 12.11), researchers have found that all the individuals in a naked mole rat colony are closely related. Genetically, the queen appears to be a sibling, daughter, or mother of the kings, and the nonreproductive animals are the queen's direct descendants or her siblings. Hence, when a nonreproductive mole rat enhances a queen's or king's chances of reproducing, it increases the chances that some genes identical to its own (that is, the genes it shares with the beneficiary) will be passed to the next generation. If some of those genes are involved in altruism, this behavior will be inherited by offspring. Likewise, worker bees in a hive all share genes with the queen. Their work (or even death) in support of the queen helps ensure that a large number of those genes will survive.

Kin selection does not explain all types of altruism. In some cases, animals behave altruistically toward others who are not relatives. Jane Goodall has discovered that chimpanzees sometimes save the lives of nonrelatives. Similarly, female dolphins without young will often help unrelated mothers care for their young. In these cases, there can be no immediate enhancement of the altruists' fitness. However, in the future, the current beneficiary may reciprocate—that is, "return the favor"—by performing some other helpful act. Thus, we can explain altruism toward nonrelatives as **reciprocal altruism:** an altruistic act repaid at a later time by the beneficiary (or by another member of the social system).

Altruism and its evolution are controversial topics. The debate has been especially heated over whether kin selection and reciprocal altruism are applicable to human social behavior, an issue we take up in the next module. 

? What is the ultimate cause for altruistic behavior among kin?

Natural selection reinforces altruistic behavior through the reproductive success of closely related individuals that have many genes in common with the altruist, including genes for altruism.

Edward O. Wilson promoted the field of sociobiology and is a leading conservation activist 37.21

In 1975, Edward O. Wilson of Harvard University published a book entitled *Sociobiology: The New Synthesis.* Drawing from numerous studies on vertebrate and invertebrate animals, Wilson promoted a relatively new area of research called **sociobiology,** the study of the evolutionary basis of social behavior. Wilson and other sociobiologists advocate the concept that social behavior evolves, like an animal's anatomical traits, as an expression of genes that have been perpetuated by natural selection. Professor Wilson discussed the social impact of his sociobiology book in a 1995 interview:

> In *Sociobiology: The New Synthesis,* I primarily intended to cover the social insects and vertebrate animals. I then saw that I could not leave out the most familiar vertebrate animal, *Homo sapiens.* I included two chapters on human beings primarily for completeness. I expected the book to have an impact, but I didn't expect it to stir up a hornet's nest of controversy in the social science community, as it did.

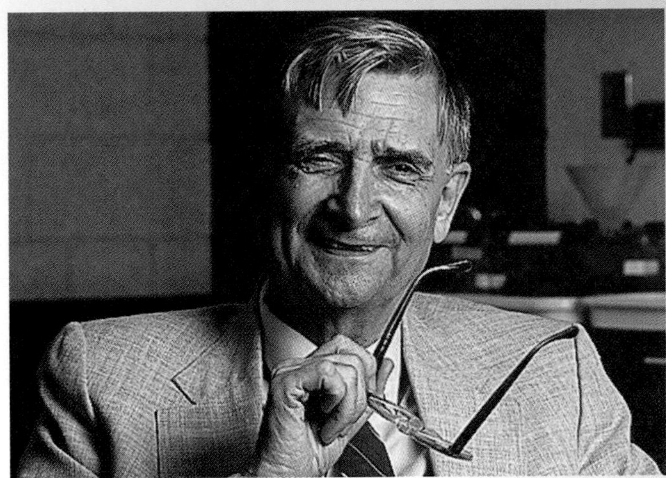

Figure 37.21 Evolutionary biologist and conservation activist Edward O. Wilson

Sociobiology involves a search for evidence that social activities improve an animal's fitness. The concepts of kin selection and reciprocal altruism, with their focus on how the survival of genes underlies social behavior, are key ideas in sociobiology. Professor Wilson elaborated on why these ideas were so controversial.

> *Sociobiology* came out at a time when most scholars in the social sciences believed not only that heredity has no importance in human social behavior, but that it is dangerous to speak of heredity, because to do so might imply that human destiny is fixed, and that there is nothing we can do about social ills. This was a primary reason for resistance both from social scientists, who had already settled on a sociocultural explanation for social behavior, and from some biologists. Since the 1970s, the evidence for genetic influence—certainly not genetic determination in a rigid sense but strong genetic influence on social behavior—has grown very substantially. It is now perhaps a mainstream of thinking.

In *Sociobiology,* Wilson predicted that biology would someday serve as part of the foundation of the social sciences. We asked him if he saw that starting to happen.

> Yes, I believe that sociobiology has already had a profound influence in some social sciences. For example, it has spawned a whole new and quite successful field in psychology called evolutionary psychology, which deals with biological components of human behavior. It has had considerable influence in anthropology, especially in the developing field of comparative studies of nonhuman primate species in relation to human social organization. It has had some influence in economics in the consideration of human choice behavior. And in legal theory, sociobiology has begun to influence thinking about matters such as adoption rights.

We asked Wilson if he thought biology's influence on social studies meant that the impact of science on society is increasing.

> Science is no longer just a fun thing, like landing on the moon or discovering a new species of bird. It is vital—and people know it. They see science as a major part of modern ethics and legislative action. They also see the environment as something that they have got to know about.

In a more recent book, *The Diversity of Life,* Wilson discusses a "biodiversity ethic." As he put it in our interview:

> My own ethic is to make every effort to save all species. This will not be possible in every case everywhere in the world. We are destined to lose a substantial fraction, 10% or more, of the world's biodiversity, no matter what we do. Biologists define **biodiversity** in the broadest sense as all of the variety of life—from the different genes [alleles] at the same chromosome position within populations, up through different species of organisms, on up to different aggregations of species in ecosystems. We should never knowingly allow a species to go extinct if appropriate measures can save it. That, in essence, is the biodiversity ethic.
>
> I realized in the early 1980s that there was growing evidence for a genetic predisposition for many kinds of human behavior. I and others believed that there was a human preference for certain types of natural habitats and an inherent tendency to affiliate with other forms of life. The reason this phenomenon is so important is that it will provide a powerful base for an environmental ethic. In my opinion, we need to value biodiversity for the contribution made to the satisfaction of human nature and not just for the physical welfare it provides us.

At the end of yet another book, an autobiography entitled *Naturalist,* Wilson distills his philosophy down to three truths. As he stated for us,

> Quite simply they are: that the human species is a product of biological evolution; that human beings arose in an arena of natural environments and biodiversity, and that therefore natural environments are a precious part of human heritage; and finally, that neither philosophy nor religion can ever make much sense unless they take the first two points into consideration.

?

What is sociobiology?

The study of the evolutionary basis of social behavior

37.22 Culture may reinforce innate human social behavior

Figure 37.22 Transmission of culture by education: a dance lesson in Indonesia

It is sometimes argued that cultural change is so rapid that it overwhelms any influence of evolution by natural selection on human behavior. However, sociobiologists consider this argument an oversimplification. As an alternative hypothesis, they propose that human culture often enhances or reinforces behavioral traits derived through natural selection. Let's assume, for instance, that incest aversion has an innate component. If most members of a society share the aversion, it is likely to become formalized in laws and taboos. A society that shuns or imprisons members who practice incest lowers the reproductive fitness of incestuous individuals. Thus, the cultural stigma may amplify the evolutionary component of the behavior by acting as an environmental factor in natural selection.

The photograph at the left focuses on one of our most important cultural attributes—education. All cultures are transmitted by the tutoring of the younger generation by the older. Is the passing on of knowledge a purely cultural attribute, or does it have a genetic component? The sociobiologist view is that the positive reinforcements a culture gives to education are proximate causes that amplify a behavior that ultimately evolved because it increased fitness. The challenge in sociobiology is to uncover evidence of the ultimate causes of education and other cultural traits central to humanity. We trace some of the evolutionary threads of human culture in the final chapter. 

? How does the very long childhood of humans, compared to that of most other animals, contribute to culture? (*Hint:* Review Module 15.7.)

It prolongs the opportunity for the young to learn from parents, relatives, teachers, and other mentors.

Chapter Review

Chapter Summary

Classical Concepts in Behavior (Introduction–37.7) The study of an animal's behavior is key to understanding its evolution and ecological roles **(Introduction)**. Behavioral biology is the study of what animals do when interacting with their environment. Behavior can be interpreted in terms of proximate causes, or immediate interaction with the environment. Behavioral ecologists are especially interested in the ultimate causes of behavior, which are evolutionary. Natural selection preserves behaviors that enhance fitness **(37.1)**. Animal behavior, including that of humans, often involves a combination of genetic programming (innate behavior) and environmental experiences (learning) **(37.2)**. Sign stimuli, such as an egg rolling out of a bird's nest or a baby bird's open mouth, trigger innate, unchangeable fixed action patterns (FAPs). The genetic programming underlying FAPs ensures that activities, such as many parent-offspring interactions, are performed correctly without practice **(37.3)**. Learning is a change in behavior resulting from experience. Learning ranges from simple behavioral change in response to a single stimulus to complex problem solving. Habituation is learning to ignore a stimulus **(37.4)**. Imprinting is irreversible learning limited to a critical period in an animal's life. For example, geese imprint on their mother during a short critical period. Imprinting enhances fitness by enabling rapid learning **(37.5)**. Many animals can also learn by association, linking behaviors to positive or negative effects, either by trial and error or by imitation **(37.6)**. Some animals, such as chimpanzees, can "think through" solutions to problems, a type of learning called innovation. Behavioral biologists disagree about conscious thinking in animals. Are humans unique, or are we simply at one end of a thinking continuum? **(37.7)**

Ecological Roles of Behavior (37.8–37.12) Behavior is the result of evolution fine-tuning an animal to its environment **(37.8)**. Many animals, including humans, display rhythmic behavior that is synchronized with the environment. Daily (circadian) rhythms, such as sleep, appear to be timed by an internal biological clock. In the absence of environmental cues, these rhythms continue, but they become out of phase with the environment **(37.9)**. The simplest animal movements are kineses, changes in rate of movement that tend to keep animals in favorable environments. More complex orientation behaviors, such as taxes toward or away from stimuli, involve positioning in reference to a cue **(37.10)**. Cognitive ethology is the study of animal cognition, the ability of an animal's nervous system to perceive, store, process, and use information. Many animals formulate cognitive maps, internal representations of spatial relationships among objects in their surroundings. Some animals, such as birds and monarch butterflies, undertake long-range migrations. Animals may navigate using the sun, stars, landmarks, or Earth's magnetism **(37.11)**. Animals are generally selective and efficient in their food choices. Natural selection seems to have shaped feeding behavior to maximize energy gain and minimize the expenditure of time and energy. This behavior is known as optimal foraging **(37.12)**.

Social Behavior and Sociobiology (37.13–37.22) Social behavior—the interactions among members of a population—influences population dynamics **(37.13)**. Agonistic behavior is social behavior consisting of threats and combat that settles disputes between individuals in a population. Agonistic behavior can directly affect an individual's evolutionary fitness because the victor often gains first or exclusive access to mates **(37.14)**.

Dominance hierarchies, maintained by agonistic behavior, partition resources among members of a population **(37.15)**. Dominance hierarchies are an integral part of the life of many primates, such as chimpanzees **(37.16)**. Territorial behavior, also maintained by agonistic behavior, is another form of social behavior that partitions resources. Territoriality can enhance fitness if the benefits of possessing a territory outweigh the energy costs of defending one **(37.17)**. Courtship behaviors advertise the species, sex, and physical condition of potential mates. Many species court in pairs; some perform courtship rituals in groups assembled in specific areas called arenas, or leks **(37.18)**. Social behavior depends on signaling, in the form of sounds, scents, displays, or touches. Honeybees, for example, perform dances that seem to communicate the direction and distance of nectar to other members of the colony **(37.19)**. Some animals exhibit altruism, behavior that reduces an individual's fitness while increasing the fitness of another individual. Altruism can be explained in terms of kin selection: An animal can increase the survival of genes like its own by helping relatives. In reciprocal altruism, a favor may later be repaid by the beneficiary **(37.20)**. Various forms of altruism and kin selection are key ideas of sociobiology, the study of the evolutionary basis of social behavior. Sociobiologists generally believe that natural selection underlies many human behaviors. Humans may even be innately predisposed to having an environmental ethic **(37.21)**. Human culture may reinforce our genetically programmed innate behavior **(37.22)**.

Testing Your Knowledge

Multiple Choice

1. At various times, a behavioral biologist studied squirrels, sparrows, sharks, deer, and caterpillars. Which of the following behaviors do you think he observed least often? *(Explain your answer.)*
   a. agonistic behavior
   b. habituation
   c. imprinting
   d. innovation
   e. orientation behavior

2. Pheasants do not feed their chicks. Immediately after hatching, a pheasant chick starts pecking at seeds and insects on the ground. How might a behavioral ecologist explain the ultimate cause of this behavior?
   a. Pecking is a fixed action pattern.
   b. Pheasants learned to peck, and their offspring inherited this behavior.
   c. Pheasants that pecked survived and reproduced best.
   d. Pecking is a result of imprinting during a critical period.
   e. Pecking is an example of habituation.

3. Which of the following is true of animals that use the sun to navigate? *(Explain your answer.)*
   a. They cannot travel long distances.
   b. Most live in the sea, where there are few landmarks.
   c. They must have accurate biological clocks.
   d. Most migrate in large schools, flocks, or herds.
   e. They more easily travel east and west than north and south.

4. Ants carry dead ants out of the anthill and dump them on a "trash pile." If a live ant is painted with a chemical from dead ants, other ants repeatedly carry it, kicking and struggling, to the trash pile, until the substance wears off. Which of the following best explains this behavior?
   a. The chemical is a sign stimulus for a fixed action pattern.
   b. The ants have become imprinted on the chemical.
   c. The ants continue the behavior until they become habituated.
   d. The ants can learn only by trial and error.
   e. The chemical triggers a negative taxis.

Describing, Comparing, and Explaining

1. Almost all the behaviors of a housefly are innate. What are some advantages and disadvantages to the fly of innate behaviors compared with behaviors that are mainly learned?

2. A wolf pack has both a dominance hierarchy and a territory defended against other wolf packs. What are the benefits to the pack of a dominance hierarchy? What are the benefits of holding a territory?

3. A chorus of frogs fills the air on a spring evening. The frog calls are courtship signals. What are the functions of courtship behaviors? How might a behavioral biologist explain the proximate cause of this behavior? The ultimate cause?

Thinking as a Scientist

Scientists studying scrub jays found that it is common for "helpers" to assist mated pairs of birds in raising their young. The helpers lack territories and mates of their own. Instead, they help the territory owners gather food for their offspring. Propose a hypothesis to explain what advantage there might be for the helpers to engage in this behavior instead of seeking their own territories and mates. How would you test your hypothesis? (*Hint:* Reread the discussion about naked mole rats in Module 37.20.) If your hypothesis is correct, what kind of results would you expect your tests to yield?

Science, Technology, and Society

1. Researchers are very interested in studying identical twins who were raised apart. Among other things, they hope to answer questions about the roles of inheritance and upbringing in human behavior. So far, data suggest that identical twins raised apart are much more alike than researchers would have predicted. They have similar IQs, personalities, mannerisms, habits, and interests. Why do identical twins make such good subjects for this kind of research? What do the results suggest to you?

2. How aware are animals? Do they think and feel the same kinds of things we do? These questions bear on animal rights, a subject much in the news. Many important biological discoveries have come from experiments performed on animals, yet some animal rights activists believe animal experimentation is cruel and should be stopped. They have harassed researchers, even vandalized laboratories and set animals free. Why are animals used in experiments? Are there uses of animals that should be discontinued? What kinds of guidelines should researchers follow in using animals in experiments, and who should establish and enforce the guidelines?

Media Resources

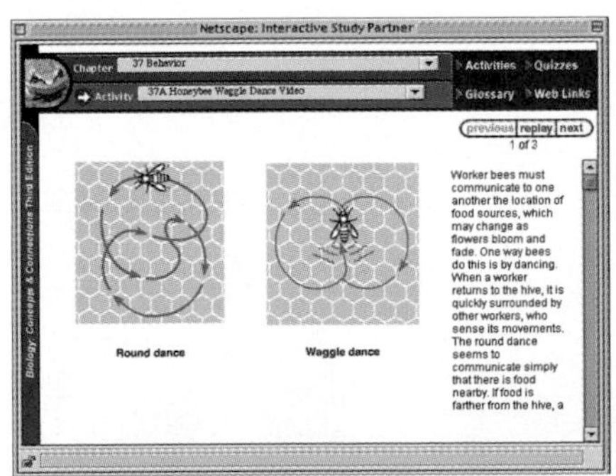

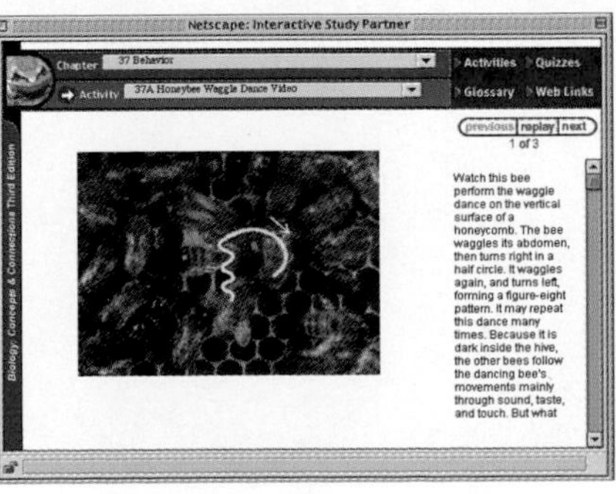

For further review of the concepts in this chapter, use the activities and quizzes on the CD-ROM.

Visit the Special Edition of The Biology Place web site for activities, quizzes, news articles, and web links.

38 Conservation Biology

Outline

The key deer, a miniature subspecies of the white-tail deer, is about the size of a German shepherd dog. It is found only in the Florida Keys and is a population with a unique gene pool. Key deer were cut off from whitetail populations on the mainland when the sea level rose after the last ice age. Confined to a few islands, they have never been numerous and were nearly exterminated by poachers in the early 1900s. Conservation efforts brought the key-deer population back from fewer than 50 animals to about 400 in the early 1970s. Key deer are now declining again as the human population in the Florida Keys mushrooms. Federal law prohibits disturbing the animals, but housing developments are reducing their habitat, and motorists on new highways have become the main threat to the deer's survival. Now down to about 250 individuals, the deer are dying on the roads faster than they can reproduce. Survival of this population and its unique gene pool depends on careful monitoring and the preservation of some natural habitat in the Florida Keys. The U.S. government maintains the National Key Deer Refuge on the island called Big Pine Key.

The Geographic Distribution of Biodiversity

38.5 Some locations in the biosphere are especially rich in biodiversity

Conservation of Species Populations

38.6 Conservation often centers on sustaining fragmented populations

38.7 Identifying critical habitat factors is a central goal in conservation research

38.8 Increased fragmentation threatens many populations: A case study

Managing and Sustaining Ecosystems

38.9 Sustaining ecosystems and landscapes is a conservation priority

38.10 Edges and corridors can strongly influence landscape biodiversity

38.11 Restoring degraded habitats is a developing science

38.12 Sustainable development is an ultimate goal

In many ways, the plight of the key deer illustrates the effect of modern human culture worldwide. We are now presiding over an alarming **biodiversity crisis,** a rapid decrease in Earth's great variety of life. Throughout the biosphere, human activities are altering trophic structure, energy flow, chemical cycling, and natural disturbances—ecosystem processes upon which we and other species depend. By some estimates we are doing more damage to the biosphere and pushing more species toward extinction than the changes that triggered the mass extinctions of dinosaurs about 65 million years ago. At that time, climatic changes resulting from an asteroid collision with Earth, and perhaps increased volcanism, caused extinction rates to increase. Today, a high rate of ecosystem alteration by a single species, *Homo sapiens,* is correlated with a high rate of species loss. Extinction is a natural phenomenon that has been occurring almost since life first evolved. It is the current accelerated rate of extinction that underlies the biodiversity crisis.

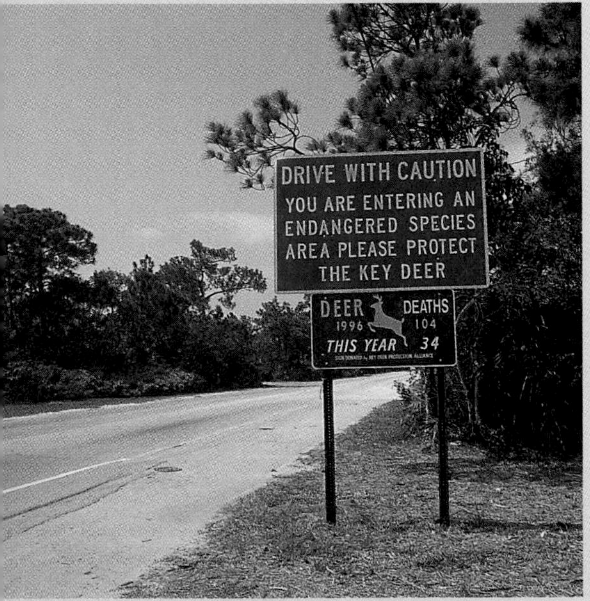

A highway sign on Big Pine Key in Florida

We can only estimate how many species of organisms currently exist. To date, scientists have described and formally named about 1.5 million species. Some biologists believe that the number is about 10 million, but others estimate it to be between 30 million and 80 million. The number of populations with unique gene pools, such as the key deer, adds another huge dimension to Earth's biodiversity.

Because we can only estimate the number of species currently in existence, we cannot determine the actual rate of species loss or the real magnitude of the biodiversity crisis. Some estimates indicate that the global extinction rate may be as much as 50 times higher than at any time in the past 100,000 years.

Biology is the science of life, thus it is fitting that our final chapter be about modern attempts to conserve life. **Conservation biology** is a goal-oriented science that seeks to counter the biodiversity crisis. In this chapter, we examine the biodiversity crisis and some of the major factors that have led to the crisis. We also take a look at some of the research and conservation strategies biologists are using in attempts to slow the rate of species loss. As we proceed, we will see that conservation biology relies on research at all levels of ecology, from populations through ecosystems.

38.1 Habitat destruction, exotic species, and overexploitation are the major threats to biodiversity

38A
See Activity 38A on the CD-ROM.

Human alteration of habitats poses the single greatest threat to biodiversity throughout the biosphere. Major changes result from the expansion of agriculture to feed the burgeoning human population, urban development, forestry, mining, and environmental pollution. Figure 38. 1A shows an all-too-common scene in the tropics, the clearing of a rain forest for lumber, agriculture, or human housing. The amount of human-altered land surface is approaching 50%, and we use over half of all accessible surface fresh water. Some of the most productive aquatic habitats in estuaries and intertidal wetlands are also prime locations for commercial and residential developments. The loss of marine habitats is also severe, especially in coastal areas and coral reefs.

Ranking second behind habitat loss as an important cause of the biodiversity crisis is competition of exotic (nonnative) species with native species. Exotic species are imported in various ways. People inadvertently carry hitchhiking seeds or insects with them when they travel throughout the world, and many foreign plants and animals have been intentionally introduced for agricultural or ornamental purposes. Most transplanted species fail to survive outside their normal range, but there are many cases of viable transplants. If your campus is in an urban setting, there is a good chance that the birds you see most often as you walk between classes are starlings, rock doves (often called "pigeons"), and house sparrows—all exotic species that have replaced native birds in many areas of North America. Some exotics play important roles in their new communities, usually through predation on native species or competition for resources. One of the largest rapid-extinction events yet recorded is the loss of freshwater fishes in Lake Victoria in east Africa. About 200 of the 300 species of native fishes found nowhere else but in this lake have been lost, mainly due to the introduction in the 1960s by Europeans of an exotic predator, the Nile perch (Figure 38.1B). One of the largest freshwater fishes (up to 2 m long and weighing up to 450 kg), the Nile perch was introduced to provide high-protein food for the growing human population. Unfortunately, the perch's main effect has been to wipe out the smaller native species, reducing its own food supply to a critical level.

Figure 38.1B The Nile perch *(Lates niloticus)*, an exotic predator

Figure 38.1A Clearing a rain forest

Other significant threats to biodiversity, such as overexploitation of wildlife, often compound problems of shrinking habitat and exotic species. Animal species whose numbers have been drastically reduced by excessive commercial harvest or sport hunting include whales, the American bison, Galápagos tortoises, and numerous fishes. Many fish stocks in the ocean have been reduced to levels that cannot sustain further human exploitation. In addition to the commercially important species, members of many other species are often killed by harvesting methods; for example, dolphins, marine turtles, and seabirds are caught in fishing nets, and countless numbers of invertebrates are killed by marine trawls. An expanding, often illegal, world trade in wildlife products (such as rhinocerous horns, elephant tusks, and grizzly bear gallbladders) also threatens many species.

What is an "exotic species"?

A species that has been accidentally or purposefully transferred from one location to another, where it did not occur naturally

38.2 Biodiversity is vital to human welfare

38A

Why should we care about the loss of biodiversity? First of all, large-scale extinctions of other species may be a warning to us that we are altering the biosphere so fast that we ourselves are threatened. Secondly, we depend on many other species for food, clothing, shelter, oxygen, soil fertility—the list goes on and on. In the United States, 25% of all prescriptions dispensed from pharmacies contain substances derived from plants. For instance, two substances effective against Hodgkin's disease and certain other forms of cancer come from the rosy periwinkle, a flowering plant native to the island of Madagascar (Figure 38.2). Madagascar alone harbors some 8000 species of flowering plants, 80% of which occur only there. Among these unique plants are several species of wild coffee trees, some of which yield beans lacking caffeine (naturally "decaffeinated"). With an estimated 200,000 species of plants and animals, Madagascar is among the top five most biologically diverse countries in the world. Unfortunately, most of Madagascar's species are in serious trouble. People have lived on the island for only about 2000 years, but in that time, Madagascar has lost 80% of its forests and about 50% of its native species. Madagascar's dilemma represents that of much of the developing world. The island is home to over 10 million people, most of whom are desperately poor and hardly in a position to be concerned with environmental conservation. Yet the people of Madagascar as well as others around the globe could derive vital benefits from the biodiversity that is being destroyed.

Figure 38.2 The rosy periwinkle *(Catharanthus roseus)*, a source of anticancer drugs

Another reason to be concerned about the changes that underlie the biodiversity crisis is that the human population itself is threatened by large-scale alterations in the biosphere. Like all other species, we evolved in Earth's ecosystems, and we are dependent on the living and nonliving components of these systems. By allowing the extinction of species and the degradation of habitats to continue, we are taking a risk with our own species' survival.

In an attempt to counter what they see as a tendency of policymakers and governments to undervalue the biosphere's life-sustaining features, a team of ecologists and economists recently estimated the cost of replacing ecosystem "services" as a measure of the services' value. For example, they estimated part of the value of a wetland from the cost of flood damage that occurred because of the loss of the wetland's ability to hold flood water. For a single year in the late 1990s, the scientists estimated the average annual value of the biosphere at 33 trillion U.S. dollars. In contrast, the global gross national product for the same year was 18 trillion U.S. dollars. Although rough, these estimates help make the important point that we cannot afford to continue to take ecosystems for granted.

The situation is different for those of us in affluent, developed nations than for people in developing nations. We extract and use far more than our share of Earth's resources, and as a result, we are responsible for the greatest amount of environmental degradation. We elaborate on this in the next module.

What are two reasons to be concerned about the relationship of the biodiversity crisis to human welfare?

(1) The environmental degradation threatening other species may also take a toll on human populations; (2) We are dependent on the biodiversity of ecosystems, both directly through use of the organisms and their products and indirectly through the roles diverse organisms play in ecosystem dynamics.

38.3 Technology and the population explosion compound our impact on habitats and other species

38A

Technology and other cultural advancements have produced many benefits, including a significant improvement in human health. But they also fuel our population explosion; and feeding, clothing, and housing billions of people even at a minimal level strains the biosphere. Even more of a strain are the huge amounts of resources consumed by certain segments of the human population.

Because of our technological advances, those of us in developed nations have become mass consumers. The United States, for example, has less than 5% of the global population but consumes far more than 5% of the world's resources. The average U.S. citizen consumes almost 14 times as much energy as the average person in China and 36 times the amount consumed by the average person in India (see the table below). Overall, the United States consumes more energy than the total population of Central and South America, Africa, India, and China combined—in 1993, some 3.5 billion people compared to the U.S.'s 260 million! The high rate of resource use compounds the danger imposed by the human population explosion.

What are the consequences of overpopulation and our penchant for consuming resources? In future years, if more than a small fraction of the people on Earth assume the high standard of living that the developed countries now enjoy, the resources that sustain the human population, such as soil, water, and fossil fuels, could be depleted. We already see harmful effects virtually everywhere today. Figure 38.3A is a scene from a remote ecosystem wounded by human activity. On March 24, 1989, a supertanker ran aground and spilled more than 10 million gallons of crude oil into Prince William Sound in Alaska. This ecological disaster was a shocking reminder that our technological tentacles reach far; as we burn gasoline in Los Angeles, Chicago, or New York, the impact of our demand for oil is felt thousands of miles away.

In a similar way, pollutants emitted into the atmosphere in Seattle, London, Moscow, or just about anywhere may be carried aloft and foul the air thousands of miles away. Acid precipitation, for instance, can destroy lakes far from sources of the sulfur and nitrogen pollutants that cause this problem. (We discussed acid precipitation in detail in Module 2.16.)

Compounds called chlorofluorocarbons (CFCs), such as freons used as refrigerants and as propellants in aerosol cans, release chlorine when they break down. Rising to the upper atmosphere, the chlorine reacts with ozone (O_3) gas,

ENERGY CONSUMPTION PER CAPITA FOR SELECTED COUNTRIES (1991)

Country	Population (millions)	Energy per Person (gigajoules*)
Bangladesh	116.6	2
Nigeria	99.1	6
India	859.2	9
Indonesia	181.4	10
Brazil	163.3	23
China	1,151.3	23
Turkey	58.5	30
Mexico	85.7	56
Japan	123.8	140
Germany	79.5	187
Soviet Union (former)	292.0	193
United States	252.8	320

*1 gigajoule = 239,000 kcal

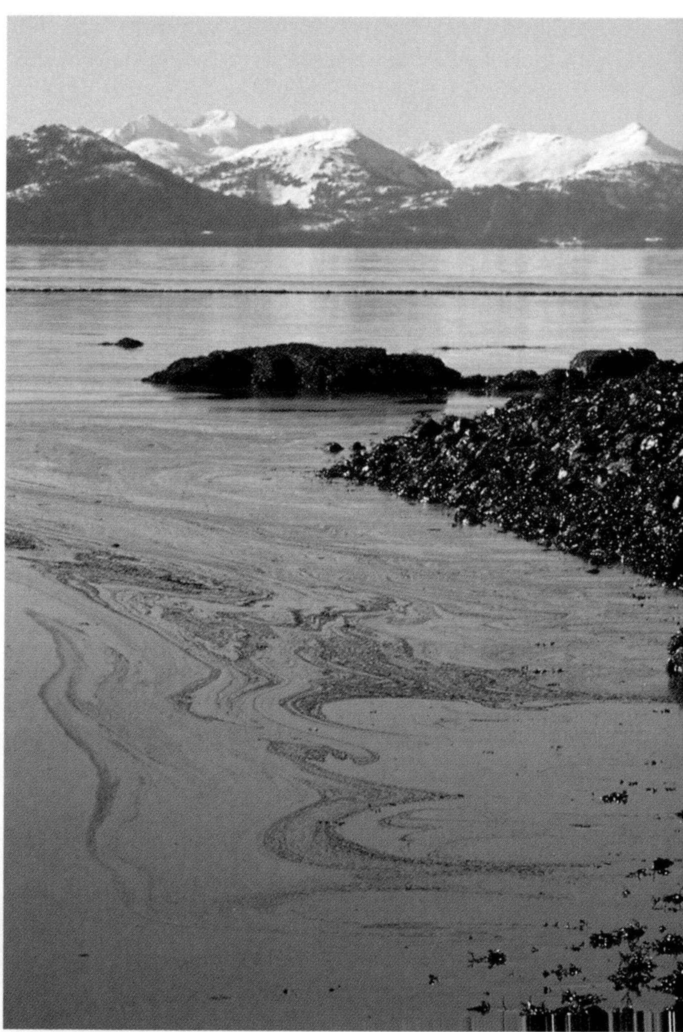

Figure 38.3A An oil spill in a remote ecosystem in Alaska

reducing it to O_2 gas. The **ozone layer** in the upper atmosphere protects life on Earth from the harmful ultraviolet rays in sunlight. Because studies indicated that CFCs were depleting the ozone layer, a multinational agreement was reached in 1987 to phase out the use of these chemicals. However, even if CFC use stopped today, quantities of these chemicals already in the air would continue to deplete the ozone layer—possibly to dangerously low levels—for decades.

Chemical pesticides are another case of the far-reaching effects of human alteration of the biosphere. As we discussed in Module 34.3, chemical pesticides have helped us grow more food and fight infectious diseases. DDT, for example, helped reduce populations of mosquitoes and the diseases, such as malaria and yellow fever, that they carry. Once common in the United States. malaria was virtually eliminated from southern states during the early 1950s by the use of DDT. (Wetland drainage and improved sanitation during the early 1900s eliminated the disease from the northern and midwestern United States.) However, many populations of mosquitoes have become resistant to DDT and other chemical pesticides, and many such chemicals tend to cause harmful side effects in nontarget species. For instance, DDT and related compounds accumulate in the body tissues of many animals, including humans. DDT is long-lived and readily stored in fat. At least in some species, the effects have been deadly.

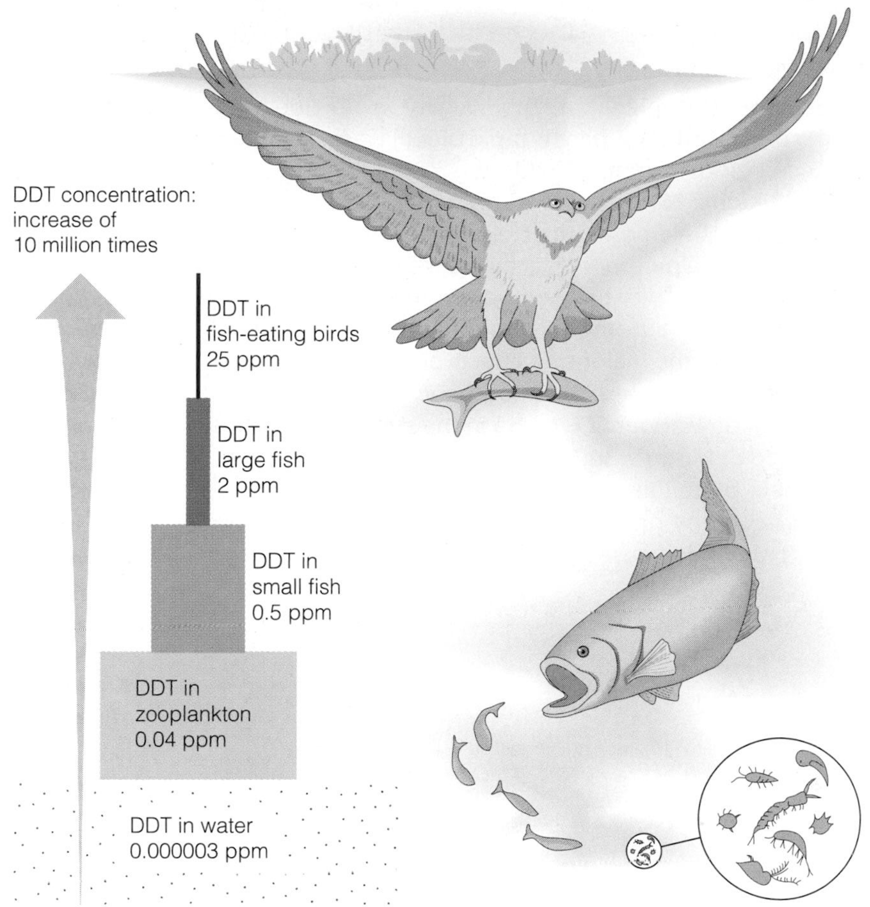

Figure 38.3B Biological magnification of the pesticide DDT in a food chain

Researchers began finding DDT in the fat tissues of a large number of birds and mammals in the early 1960s, after about a decade of widespread use of the chemical. They even found traces in marine mammals in the Arctic, far from any places DDT had been used. The chemical had been transported and concentrated as it passed through food chains. Because each predator consumes many prey, the concentrations of DDT and other substances that persist in living tissue can become increasingly magnified as the substances move through food chains. Because of this process, called **biological magnification,** the top predators in food chains can accumulate DDT and other persistent substances to toxic levels. You can see the effect in Figure 38.3B. The tiny dots in the pyramid represent DDT. In this particular case, the chemical was magnified by a factor of about 10 million, from 0.000003 parts per million (ppm) in the water to 25 ppm in the osprey.

During the 1960s and 1970s, scientists studying predatory birds such as ospreys and eagles found a correlation between high levels of DDT in parent birds and a thinning of eggshells. Young birds died before hatching, producing a marked decline in populations.

The United States banned the use of DDT in 1976, but U.S. companies still manufactured and sold it to other countries until 1984. DDT and several closely related chemicals are still used in many developing nations. Despite this continued threat to the biosphere, the osprey and several other predators have made a comeback in the past several decades because the pesticide is no longer used in their immediate environment.

The DDT story represents a common approach to altering the environment to suit our needs. Technology produced a substance that seemed to hold great promise for helping us feed more people and reduce suffering and death. The substance was used extensively soon after its insecticidal qualities were discovered, before tests of its potential danger to the biosphere were performed. In fact, the biosphere became the testing ground. As we are about to see, this approach may now be putting the life-support systems for all species to the ultimate test.

How is biological magnification directly relevant to the health of most humans in developed countries?

People in developed countries generally eat more meat than do people in developing countries. As secondary or tertiary consumers in a food chain, meat-eaters acquire a greater dose of some toxic chemicals than if they fed exclusively on plants as primary consumers.

38.4 Rapid global warming could alter the entire biosphere

38A

A change that could affect biodiversity worldwide by altering the entire biosphere is rapid global warming. We introduced global warming in Module 7.13 and in the interview in Module 35.10, however, this subject is so timely and pertinent to the biodiversity crisis that we need to paint a more complete picture of its causes and potential effects.

Fossil fuels power most of our industries, agricultural equipment, and automobiles, and they heat most of our homes. One of the major side effects of the combustion of fossil fuels is an increase in atmospheric CO_2. There is nothing odd or new about CO_2 being in the atmosphere. As we discussed in Module 36.15, the addition of CO_2 to the atmosphere by cellular respiration roughly balances the removal of CO_2 by photosynthesis. What is troubling is the current rate of increase in atmospheric CO_2. When a monitoring station in Hawaii began making very accurate measurements in 1958, the CO_2 concentration was about 315 ppm (Figure 38.4A). Today, the concentration of CO_2 in the atmosphere is more than 360 ppm, an increase of about 14% since the measurements began.

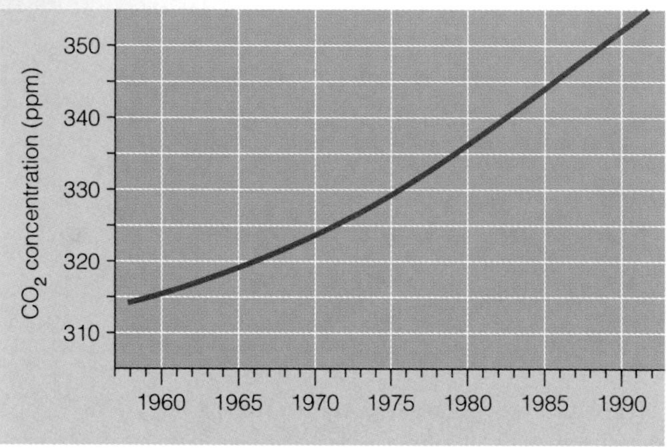

Figure 38.4A The increase of atmospheric CO_2 since 1958

Carbon dioxide is one of several so-called greenhouse gases—molecules that can trap heat and cause atmospheric warming, known as the **greenhouse effect** (Figure 38.4B). CO_2 absorbs infrared radiation and slows its escape from Earth. Other greenhouse gases are methane and nitrous oxide, both of which are also increasing in the atmosphere as a result of fossil-fuel consumption, industry, and agriculture. As a natural phenomenon, the greenhouse effect is essential for virtually all life on Earth; without the CO_2 that respiration puts into the atmosphere, the average air temperature at Earth's surface would be only about minus 18°C.

Our current cause for alarm arises from the potential for too much warming. Studies of climatic changes through geological time and mathematical models lead some climatologists to predict that in the next 50–100 years, at the present rate greenhouse gases are increasing, atmospheric temperatures could rise by about 2°C. It is also possible that as the temperature of the atmosphere increases, populations of soil bacteria will increase and in turn produce even more CO_2 and methane. The upswinging (yellow) curve in Figure 38.4C shows the range of global warming predictions generated by several computer models. The jagged black line shows the actual increase in global temperatures recorded since about 1850.

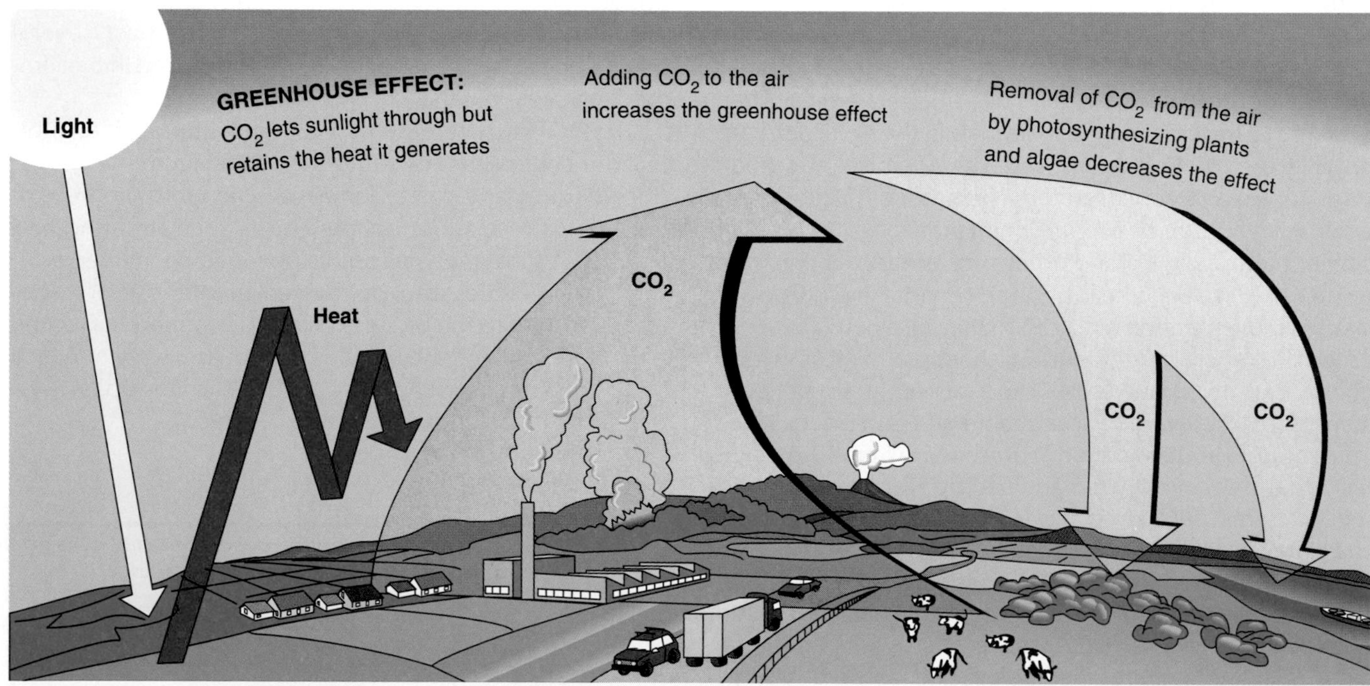

Figure 38.4B The greenhouse effect and factors influencing it

What are the possible consequences of this much global warming? As Figure 38.4C shows, the world has experienced a temperature rise of about 0.5°C in the past century. Half a degree Celsius sounds minor, and on a local or regional basis, it is. However, on a global scale, an increase of less than 2°C would be enough to melt polar ice and raise sea levels significantly. Unless massive dikes were built, the projected rise in sea level would, by the end of the twenty-first century, flood coastal areas, many of which are environmentally sensitive and heavily populated. In addition to flooding, a warming trend might alter patterns of global precipitation. For instance, the grain belts of the central United States and central Asia might become much drier and unable to support the crops currently grown there. Furthermore, forested areas in semiarid zones could lose their trees and become deserts.

Overall, large amounts of habitat would be altered by rapid global warming, and in many cases, these changes could seriously impair efforts to reduce species loss. For instance, warming by as much as 2°C would probably be intolerable for many species found in high mountain areas. One computer model predicts that nearly 60% of the species of small mammals presently inhabiting mountain peaks in the Great Basin in the western U.S. could be lost because they would be unable to migrate to a colder area if warming occurs. Similarly, global warming would probably have a serious impact on biodiversity in reserves and other areas that are currently protected. Warming may alter habitats in these areas, making them uninhabitable for many species whose survival depends on them.

At present, the future of greenhouse warming is uncertain, and most scientists acknowledge some uncertainties about global climate change. However, many researchers emphasize that few if any environmental issues have been as thoroughly studied by as many scientists as this one, and probably most scientists would agree that the balance of evidence points to a measurable human influence on the global climate.

On the other hand, some researchers and government officials call for more data before any conclusions are reached about whether the increases in atmospheric CO_2 and methane are mainly from human sources and whether the increases have actually caused the temperature rise. Skeptics also raise the possibility that smoke and increased cloud cover from fossil-fuel consumption and deforestation may *decrease* warming by reducing the amount of solar heat that reaches Earth's surface. Another possibility is that marine algae, which take up enormous amounts of CO_2, may help slow greenhouse warming. Despite these uncertainties, an increasing number of scientists conclude that immediate steps to slow the warming trend are necessary to prevent catastrophic global change.

What can be done to lessen the chances of a greenhouse disaster? International cooperation and national and individual action are needed to decrease fossil-fuel consumption and to reduce the destruction of forests and replant many areas throughout the world. Several nations have instituted tree-planting programs, and the leaders of many nations have called for efforts to stabilize CO_2 output by the end of the century. Because fossil-fuel combustion currently powers much of our industry and economic growth, meeting the terms of the treaty will require strong individual efforts and acceptance of some major lifestyle changes. There is much to gain by conserving energy at home, recycling, and reducing our use of cars by walking, bicycling, or taking mass transit. Other strategies include developing solar, wind, and geothermal energy sources to reduce our reliance on fossil fuels. Although such measures are expensive, so are building massive dikes to prevent coastal cities from flooding, or changing a food-producing nation such as the United States to a food-dependent one.

Whereas rapid climate change could alter the entire biosphere, its effect on biodiversity would be varied because species are not evenly distributed, as we see next.

What was the approximate % increase in atmospheric CO_2 between 1960 and 1990?

About 12%, from 315 ppm to about 352 ppm (352 − 315 = 37 and 37/315 × 100 = 11.7%)

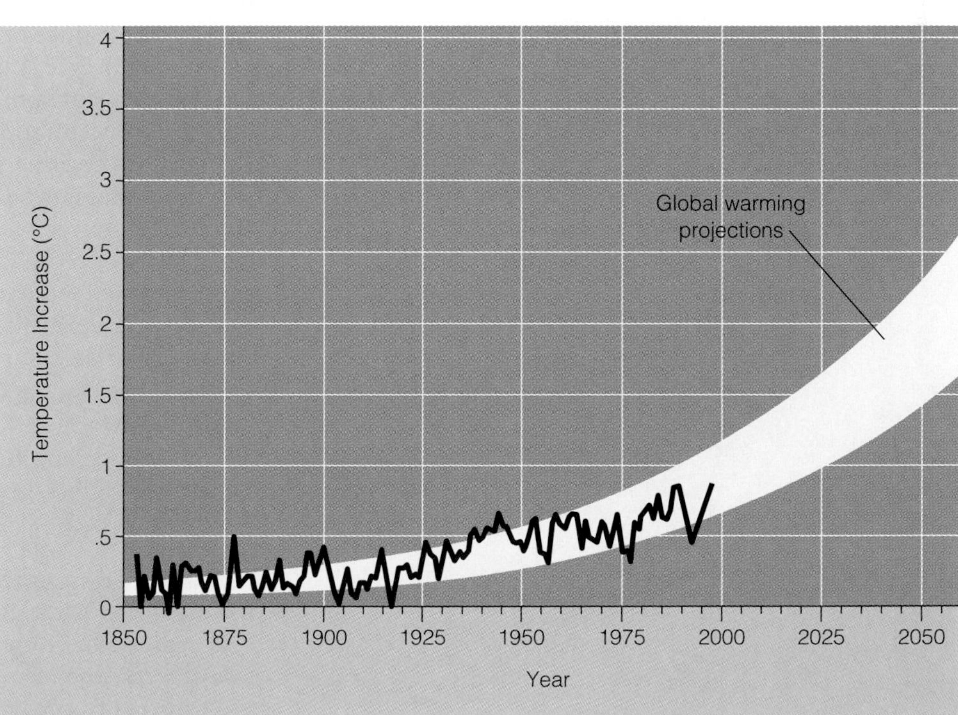

Figure 38.4C Atmospheric temperatures and projections for global warming

The Geographic Distribution of Biodiversity

38.5 Some locations in the biosphere are especially rich in biodiversity

38A

Environments are patchy and so is the distribution of organisms. To better understand the nature of the biodiversity crisis, we need to take a brief look at the geographic distribution of species. One pattern in species distribution, indicated in Figure 38.5A, is related to latitude. In general, species numbers increase steadily from the Arctic to the tropics. For instance, fewer than 100 species of terrestrial birds are found in Arctic areas, and the numbers increase steadily to a high of over 600 in some tropical regions. Many other organisms, such as microbes, flowering plants, reptiles, and mammals, exhibit a similar pattern. Ecologists debate the underlying causes of these patterns. For some organisms, the greater amounts of solar energy that strike Earth at lower latitudes may explain the patterns. Another possible cause may be a greater number of patches that result from the more frequent local disturbances in the tropics. In general, the most diverse areas (those with the greatest number of patches) harbor the greatest number of different plant species. In turn, the varied food resources provided by a diverse plant community generally support a high diversity of animals. Actually, many ecologists consider it likely that a combination of factors underlies the distribution patterns of species.

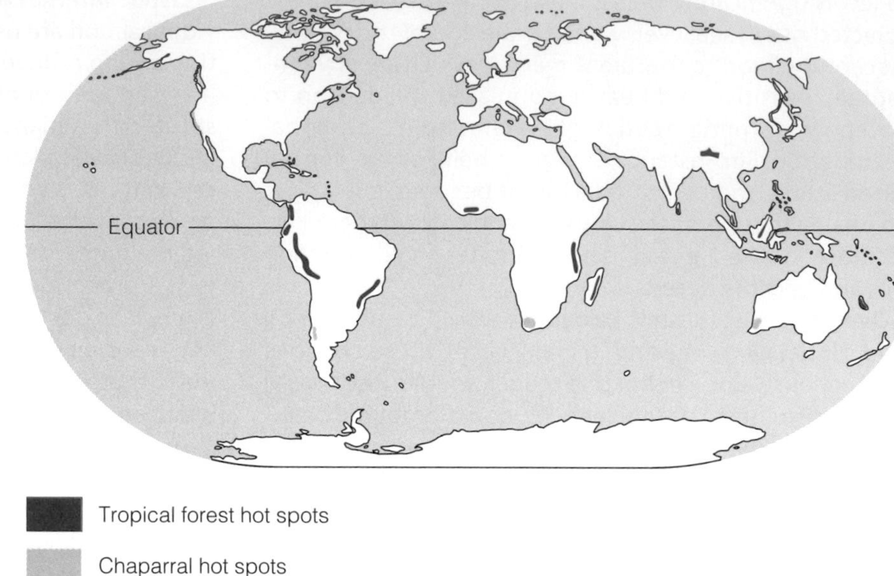

Figure 38.5B Location of some biodiversity hot spots

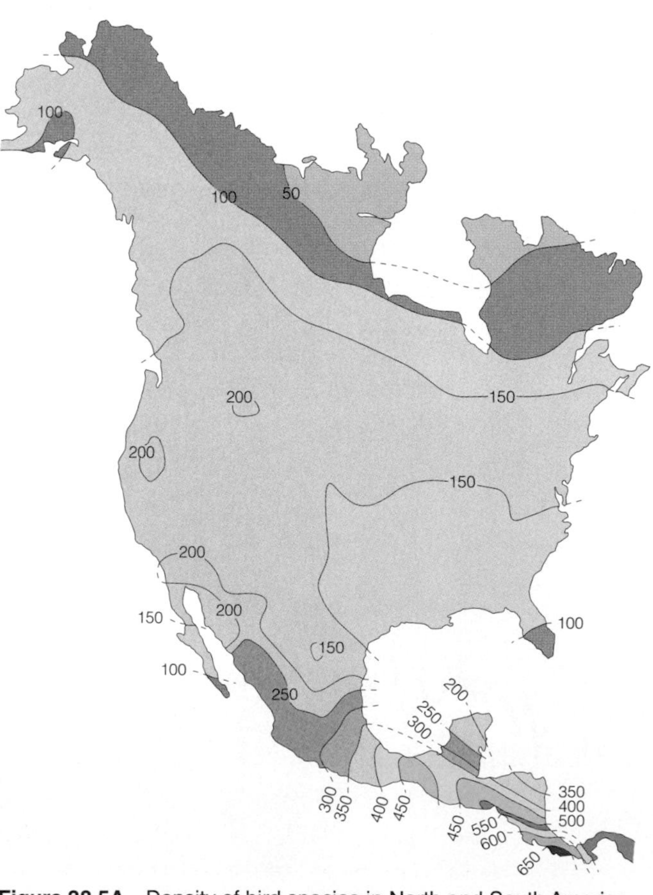

Figure 38.5A Density of bird species in North and South America

Another feature of species distribution is indicated in Figure 38.5B. The small red areas on this map represent **biodiversity hot spots,** relatively small areas with an exceptional concentration of species. Many of the organisms in biodiversity hot spots are **endemic species,** meaning they are found nowhere else. The 18 hot spots on this map account for only about 0.5% of the global land surface, but these areas collectively contain the biosphere's greatest concentrations of endemic vascular plants. About 50,000 species of vascular plants—20% of all those known to science—exist here. Fourteen of the plant hot spots are tropical forest biomes; the other four are dry shrublands (chaparral). In some cases, hot spots for one group of organisms also harbor a large proportion of other endemic species. For instance, the 18 vascular-plant hot spots also contain about 7% of all species of land vertebrates.

Because endemic species are limited to specific areas, they are highly sensitive to habitat degradation. Several of the biodiversity hot spots shown in Figure 38.5B represent only a small fraction of the habitats that were formerly available to the endemic species in those areas; in a third of these hot spots, about 90% of the habitats have been lost to human development. At the current rate of habitat alter-

Figure 38.5C Monarch butterflies *(Danaus plexippus)* overwintering

ation, the rest could lose similar amounts in the next 10–15 years. Current estimates indicate that about half of the species will be lost in areas where 90% of the habitat is seriously altered. Thus, biodiversity hot spots can also be hot spots of extinction, and they rank high on the list of areas demanding strong global conservation efforts. Identifying and acting to save the thousands of endemic species in hot spots is an essential part of modern conservation efforts.

The concentration of species in limited areas provides an opportunity to protect many species in very limited areas. However, species endangerment is a truly global problem, and focusing on hot spots should not detract from efforts to conserve habitats and species diversity in other areas. Another consideration is the spectre of global warming; climatic changes may alter the habitats in biodiversity hot spots such that these areas are no long hospitable to species that currently rely on them.

Figure 38. 5C emphasizes yet another reason for the uneven distribution of species—localized concentrations of species that migrate seasonally. The tree in the photograph is in the highlands of central Mexico. It is covered with overwintering monarch butterflies. Monarchs occupy much of the United States and Canada during the summer months. In the autumn, they migrate to local sites in Mexico and California where temperatures are above freezing but cold enough to slow the insects' metabolism and allow them to survive on stored food reserves until the following spring. Overwintering populations, such as the one shown here, are susceptible to habitat disturbances because they are concentrated in small areas.

The problem of preserving species that migrate between nations and congregate seasonally is especially complex. Habitat preservation in Canada, the United States, or Mexico alone will not remove the threats to the monarch, and the situation is similar for many species of songbirds, marine mammals, sea turtles, and other organisms that congregate in breeding areas or feeding sites seasonally.

Sea turtles, such as the loggerhead turtle (Figure 38.5D), are threatened in their ocean feeding grounds and on land. Loggerheads take about 20 years to reach sexual maturity, and great numbers of juveniles and adults have drowned at sea when caught in fish and shrimp nets. The adults mate at sea, and the females migrate to specific sites on sandy beaches to lay their eggs. Buried in shallow depressions, the eggs are susceptible to predators, especially raccoons, and many egg-laying sites have become housing developments and beachside resorts. An ongoing international effort to conserve sea turtles focuses on protecting egg-laying sites and minimizing the death rate of adults and juveniles at sea. 

?

What is a "biodiversity hot spot"?

A relatively small area with a disproportionate number of species

Figure 38.5D An adult loggerhead turtle *(Caretta caretta)* swimming in the Caribbean Sea

Conservation of Species Populations

38.6 Conservation often centers on sustaining fragmented populations

38A

Much of the popular and political discussion about biodiversity and the biodiversity crisis centers on species and populations. The U.S. Endangered Species Act (ESA) defines an *endangered* species as one that is "in danger of extinction throughout all or a significant portion of its range." Also defined for protection by the ESA, *threatened* species are those that are likely to become endangered in the foreseeable future. Except in cases where only a few individuals of a species remain, it is often difficult to determine if a species actually is on the brink of extinction. Researchers assess such factors as the reproductive success of individuals and the amount of suitable habitat that is likely to remain available. However, we can only estimate what most species and species populations need over the long term. Species viability often triggers heated debate, especially when scientists and politicians must decide whether individual species should be officially listed as endangered or threatened.

The focus of scientific studies concerned with sustaining species is on the dynamics of populations that have been reduced in numbers and fragmented by human activities. Because habitats are patchy, the populations of many species were subdivided into groups with varying degrees of isolation from one another before humans began altering habitats significantly. Gene flow among populations varied according to the degree of isolation. Today, severe **population fragmentation,** the splitting and consequent isolation of portions of populations by habitat degradation, is one of the most harmful effects of habitat loss due to human activities. For convenience, we'll refer to the portions, or fragments, of a population as subpopulations. A decrease in the overall size of populations and a reduction in gene flow among subpopulations usually accompany fragmentation.

Figure 38.6A Fragmentation of a forest ecosystem

The aerial photograph in Figure 38.6A illustrates fragmentation of a coniferous forest ecosystem in the Mt. Hood National Forest in northwestern Oregon. The forest itself was originally contiguous. The open areas in the photo were logged, creating forest fragments, some of which are islands within clear-cut areas. A common result of human activities, this kind of habitat alteration has reduced and fragmented many species populations. Populations that were divided historically have been further fragmented. One of the most controversial endangered species, the northern spotted owl, inhabits coniferous forests of the U.S. Pacific Northwest (Figure 38.6B). Owl populations declined markedly and were fragmented when these forests were logged.

In most cases, subpopulations are separated into habitat patches that vary in quality. For instance, in a fragmented forest, the presence of some large, dead trees can make the difference between a high-quality patch and a low-quality one for squirrel, owls, and many birds that require rot cavities for their nests. Patches with abundant high-quality resources tend to have stable, persistent subpopulations. Reproductive individuals in a high-quality patch tend to produce more offspring than the patch can sustain. Such an

Figure 38.6B The northern spotted owl *(Strix occidentalis caurina)*

area of habitat where a subpopulation's reproductive success exceeds its death rate is called a **source habitat.** Source habitats produce enough new individuals that some disperse to other areas, often in search of food or a place to reproduce. In contrast, a habitat where a subpopulation's death rate exceeds reproductive success is a **sink habitat.** The persistence of many subpopulations in sink habitats depends on individuals dispersing from source habitats. Today, because of habitat loss, more sink habitats exist than were present for most species historically, and the dispersal of individuals to sinks can sometimes threaten the survival of subpopulations in source habitats. For example, the remaining source habitats of the northern spotted owl are relatively small fragments of old-growth rain forests. The owls tend to disperse from small fragments of source habitats into larger, neighboring areas where trees are regrowing after logging. Such areas are sink habitats for the owls, and sustaining the owl populations depends in part on keeping enough reproductive individuals in the source habitats. Some researchers have suggested surrounding old-growth fragments with distinct boundaries—such as clear-cut areas—that the owls will not enter. The spotted owl situation illustrates the importance of identifying source and sink habitats, and of protecting source habitats. 

? Why is the number of fragmented populations increasing?

By destroying habitat, human activities are fragmenting the populations of many species into subpopulations that live in the separate patches of habitat that remain.

38.7 Identifying critical habitat factors is a central goal in conservation research

What factors make some habitats sources and other sinks? Identifying the specific combination of habitat factors that is critical for a species is pivotal in conservation biology. Figures 38.7A–38.7C illustrate several critical factors for the red-cockaded woodpecker *(Picoides borealis),* an endangered, endemic species originally found throughout the southeastern United States. This species requires mature pine forests, preferably ones dominated by the longleaf pine *(Pinus palustris).* Most woodpeckers nest in dead trees, but the red-cockaded woodpecker drills its nest holes in mature, living pine trees (Figure 38.7A). The heartwood of mature longleaf pines is usually rotted and softened by fungi, allowing the woodpeckers adequate space for nesting once they excavate into the heartwood. Red-cockaded woodpeckers also drill small holes around the entrance to their nest cavity, making resin from the tree ooze down the trunk. The resin seems to repel some predators, such as corn snakes that eat bird eggs and nestlings (see the opening essay for Chapter 23).

Figure 38.7B shows another critical factor for this woodpecker—low growth of plants among the mature pine trees. Biologists have found that a habitat becomes a sink, with breeding birds tending to abandon nests, when vegetation among the pines is thick and higher than about 15 feet, as in Figure 38.7C. Apparently, the birds require a clear flight path between their home trees and the neighboring feeding grounds. Historically, periodic fires swept through longleaf pine forests, keeping the undergrowth low. The recent recovery of the red-cockaded woodpecker from near-extinction to sustainable populations is largely due to recognizing the key habitat factors and protecting some longleaf pine forests that support viable numbers of the birds. The use of controlled fires to reduce forest undergrowth helps maintain mature pine trees as well as the woodpeckers. Next, we take a more detailed look at the population dynamics of a freshwater fish species, historically and at present. 

? Are critical habitat factors generally more favorable in a (choose one) *source* habitat or *sink* habitat?

Source habitat

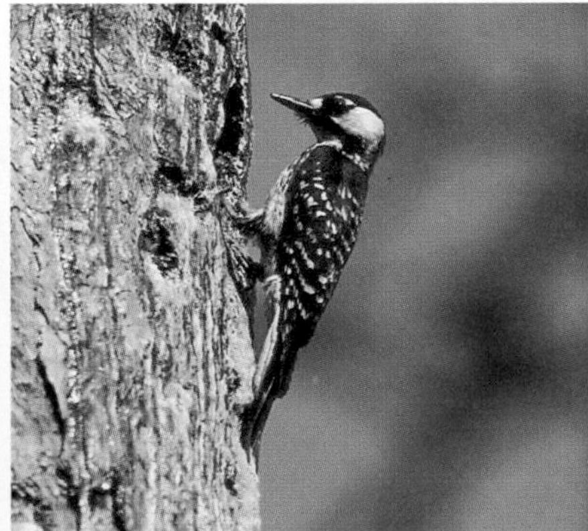

Figure 38.7A The red-cockaded woodpecker *(Picoides borealis)* at its nest site in a longleaf pine tree

Figure 38.7B Forest habitat that can sustain red-cockaded woodpeckers

Figure 38.7C Forest habitat that cannot sustain red-cockaded woodpeckers

38.8 Increased fragmentation threatens many populations: A case study

38A

Human activities have altered habitats and in turn affected the distribution and long-term survival of many species. One example is the bull trout *(Salvelinus confluentus)* shown in Figure 38.8A. An endangered species, the bull trout inhabits lakes, rivers, and mountain streams in northwestern Canada and the United States, including Alaska. It requires cold, fast-flowing streams with pebble-covered bottoms and little or no silt for breeding sites and for egg laying (Figure 38.8B).

Figure 38.8C indicates the general condition of a bull trout population in one area historically (left) and currently (right). The human-caused habitat changes indicated on the right are associated with a decline in numbers of fish and a reduction in the population's chances for long-term survival. The figure is an idealized representation of an actual case.

Historical records indicate that before human intervention, this bull trout population consisted of four subpopulations (S1–S4 within yellow outlines on the drawing). Subpopulations S1 and S2 were resident in high-mountain streams, remaining in these habitats throughout life and reproducing there. The S3 subpopulation inhabited a larger area of the stream system and used the lakes and river for occasional feeding forays. Members of the S4 subpopulation were mainly resident in the lakes and river but ascended the streams to reproduce once each year. The large blue and pink arrows represent gene flow among the subpopulations. Gene flow was maintained chiefly by movements of individuals in S3 and S4 to and from breeding and egg-laying sites in the streams. The S2 subpopulation was the most isolated one, because of its location in the mountains above a portion of the stream that dries up (dashed line) during many summers. However, areas occupied by other subpopulations served as source habitats for the S2 subpopulation. Some adult trout used the S2 breeding and egg-laying sites in wet years, and as a result, some genetic mixing occurred with the S2 fish.

The right side of Figure 38.8C indicates four major alterations in habitat—hydroelectric dams, logging, road building, and mining. Overall, these changes have further fragmented the bull trout population and reduced the size and connectedness of its subpopulations. For example, the s1 subpopulation (historically S1) is virtually isolated by mining operations that release toxic chemicals into the stream. The s2 subpopulation is even more isolated than it was historically; logging and roads have increased the rate at which water leaves the area, and the stream remains dry except during a brief time in the spring when the snow melts and the streams are all flooded. Clear-cut areas and roads have also increased the amount of silt in neighboring portions of the streams, and the increased siltation has destroyed some bull trout breeding and egg-laying sites. The s3 subpopulation is reduced in both size and distribution by the effects of mining, logging, and roads. Two hydroelectric dams on the river fragmented the original river/lake subpopulation (S4, left side of Figure 38.8C), creating a fifth subpopulation (s5). The s4 subpopulation has been reduced in numbers because rel-

Figure 38.8A The bull trout *(Salvelinus confluentus)*, an endangered species of freshwater fish

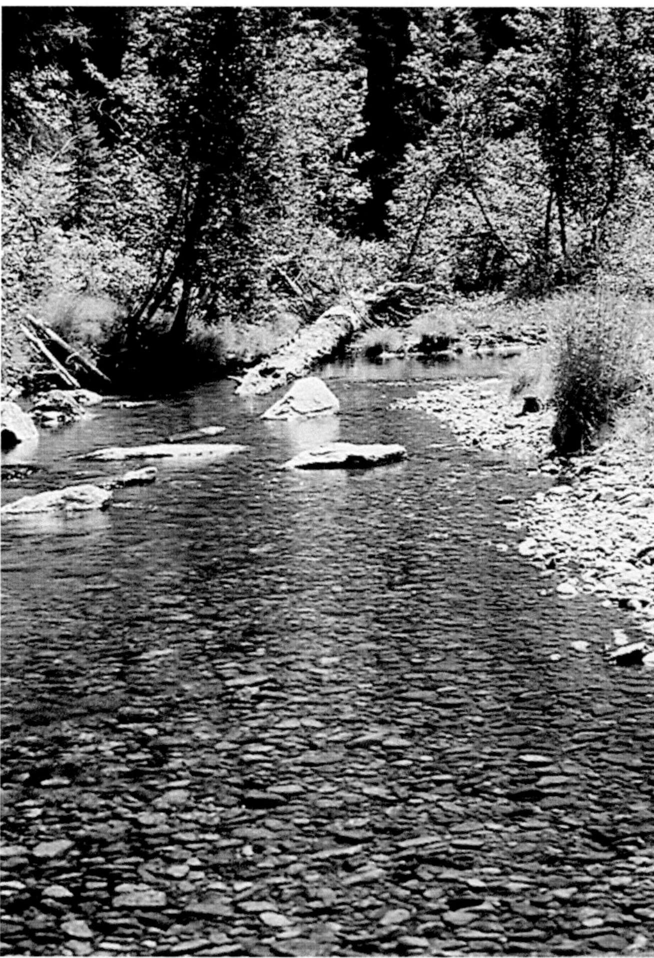

Figure 38.8B A mountain stream in northwestern Montana with the habitat conditions that bull trout use for egg laying

atively few individuals manage to overcome human-caused obstacles and ascend to breeding and egg-laying sites. Finally, the new s5 subpopulation is completely cut off from the other subpopulations and has lost access to many historic breeding and egg-laying sites.

A method called **population viability analysis (PVA)** is an increasingly popular approach to conservation problems, such as those posed by the bull trout. Usually constructed by computer simulations, a PVA incorporates as much information on a population's current status as available and predicts its chances for long-term survival. For instance, useful data for a PVA on a population of a freshwater fish species such as the bull trout would be the size of the breeding population in each subpopulation, the reproductive rates of each subpopulation, the number of other fish (both predators and prey of the threatened or endangered species), the availability of aquatic insects and other invertebrates that fish eat, the amount of suitable habitat (including breeding and egg-laying sites), the amount of gene flow among the subpopulations, and the kinds and frequency of natural disturbances, such as periodic floods. The PVA would also factor in the effects of human activities that may chemically and physically alter portions of the population's habitat.

The conditions of bull trout populations represent the magnitude and complexity of problems that many species and conservation workers trying to sustain populations currently face. Efforts to save the bull trout and many other species require understanding the dynamics of fragmented populations and making the best of the current conditions. Conservation biologists accept the fact that no ecosystems are unaffected by humans. Their efforts to save species that are threatened and endangered often involve estimating how large populations must be to sustain themselves indefinitely and working out ways to provide enough quality habitat to support these populations. Most often it is necessary to weigh a species' biological and ecological needs against other conflicting demands. For example, an ongoing, sometimes bitter debate in the U.S. Pacific Northwest pits saving habitat for populations of the northern spotted owl, timber wolf, grizzly bear, and native trout against demands for jobs in the timber, mining, and other industries that rely on extracting natural resources. Programs to restock wolves and to bolster populations of grizzly bears and other large carnivores are opposed by some recreationists concerned for their safety and by many ranchers concerned with potential losses of livestock. Similar problems confront decision makers in virtually every part of the globe. To become more effective globally, conservation is now moving toward broader planning, focusing more on ecosystems than on individual species, as we see next. 

? Why is it likely that viability analysis of any population will always be incomplete?

It is rarely, if ever, possible to know all the factors that affect a population's viability.

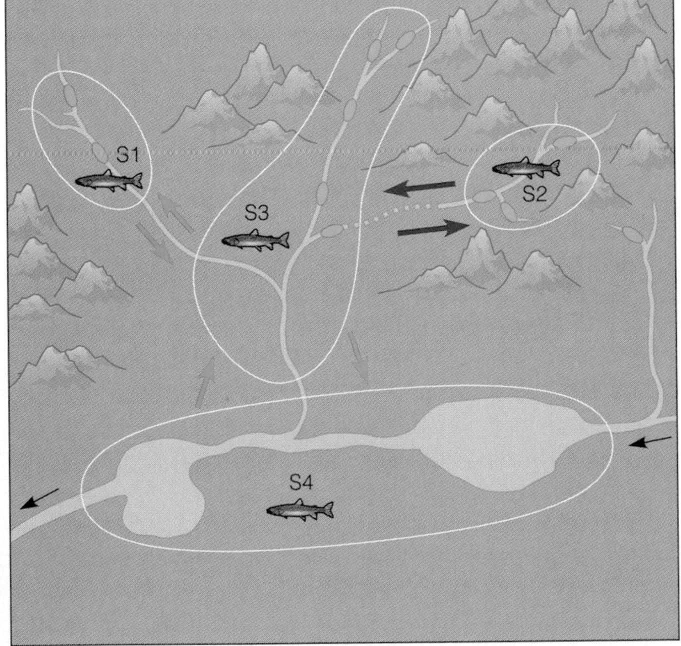

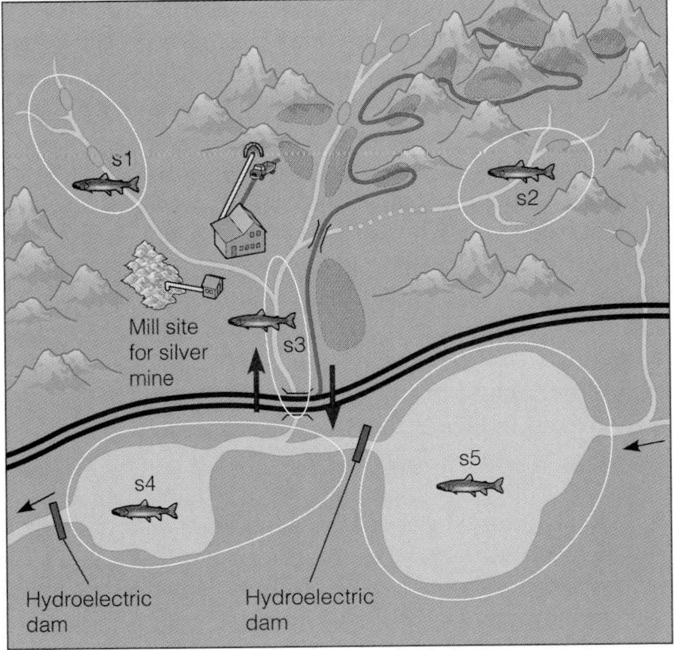

Egg-laying sites in mountain streams

Regular, frequent dispersal and gene flow between subpopulations

Irregular, infrequent dispersal; minimal gene flow between subpopulations

Egg-laying sites in mountain streams

Clear-cut (logged) areas

Roads

Irregular, infrequent dispersal; minimal gene flow between subpopulations

Figure 38.8C The effects of habitat changes on a bull trout population

38.9 Sustaining ecosystems and landscapes is a conservation priority

38A

Most conservation efforts in the past have focused on saving individual species, and this work continues. More and more, however, conservation biology aims at sustaining the biodiversity of entire communities and ecosystems, and on an even broader scale, the biodiversity of whole landscapes. Ecologically, a landscape is a regional assemblage of interacting ecosystems, such as a forest, adjacent fields, wetlands, streams, and streamside habitats. **Landscape ecology** is the application of ecological principles to the study of land-use patterns. Its goal is to understand these patterns and make species conservation a functional part of the picture. Landscape ecology also involves correlating the way species are distributed with the distribution of abiotic factors, such as temperature, precipitation, and elevation.

Figure 38.9A illustrates a method called gap analysis that researchers often use to study the distribution of organisms relative to landscape features and habitat types. **Gap analysis** employs computerized maps, including those generated by satellite sensing and aerial photography, along with information on the distribution of organisms. The most useful analyses include confirmation of satellite and aerial data by sampling on the ground. The uppermost overlay in Figure 38.9A illustrates the distribution of dominant plants and major physical features of the landscape, such as a river. The next overlay (dots) depicts the distribution of a rare, endangered species in the area. An early goal of gap analysis was to locate "gaps" between the distribution of rare, endemic species and protected areas. Thus, the third overlay in Figure 38.9A (gold) shows the position of protected habitats in the area.

Map of vegetation patterns and river course

Distribution of rare, endemic species

Distribution of protected areas

Final overlay map

Figure 38.9A Construction of a gap analysis map

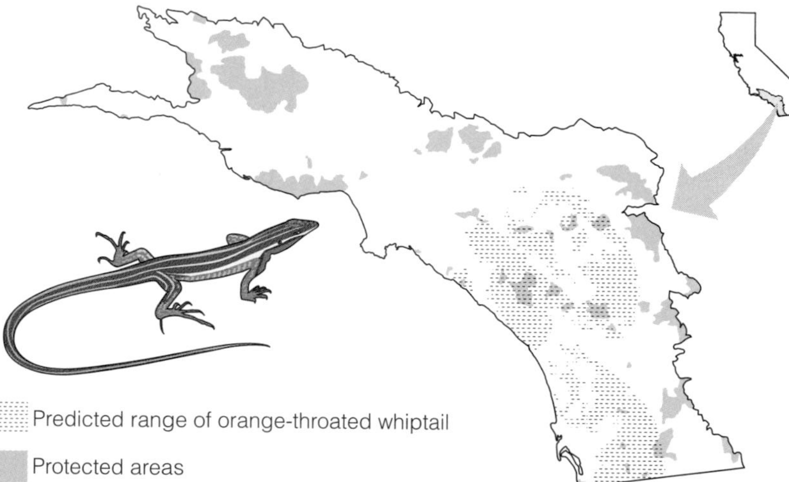

Figure 38.9B Gap analysis map for an endemic lizard, the orange-throated whiptail *(Cnemidophorus hyperythrus)*, in Southern California

Parks, wilderness areas, and other legally protected nature reserves, on land and in aquatic environments, are important components of efforts to sustain biodiversity. However, as indicated by the final overlay maps in Figure 38.9A and in Figure 38.9B, protected areas often lie outside the areas where the greatest concentrations of rare and endangered species are found. Areas that are vital to the survival of endangered species may not be included in protected areas. Gap analysis can highlight these problems, and work can then be focused on sustaining the biodiversity of the whole area. Many researchers now conclude that sustaining biodiversity in most areas will depend on protecting some habitat areas from alteration and surrounding these core habitats with areas where human development occurs but in ways that are compatible with the long-term protection of the protected areas. This is the concept of zoned reserves, which we discussed in Module 36.20.

How is a "landscape" different from an "ecosystem"?

A landscape is more inclusive in that it consists of several interacting ecosystems in the same region.

Edges and corridors can strongly influence landscape biodiversity 38.10

38A

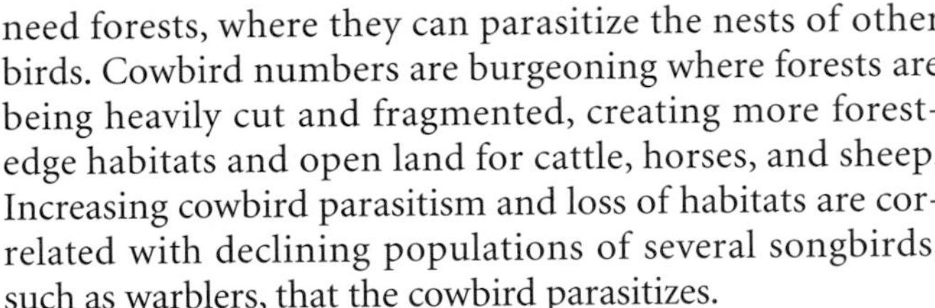

Figure 38.10A A landscape with distinct edges

Edges are prominent features of landscapes. The photograph in Figure 38.10A shows a landscape area in Australia that includes a dry forest, a rocky area with grassy islands, and a flat, grass-covered lakeshore. The boundaries or edges between these distinct areas define the extent of these features and the differences between them. Human activities, such as logging and road building, often create edges that are more abrupt than those delineating natural landscapes.

Edges have their own sets of physical conditions, such as soil type and surface features that differ from either side. Edges also may have their own type and amount of disturbance. For instance, the edge of a forest often has more blown-down trees than a forest interior because the edge is less protected from strong winds.

Because of their specific physical features, edges also have their own communities of organisms. Some organisms thrive in edges because they require resources of the two adjacent areas. For instance, white-tailed deer thrive in edge habitats, where they can browse on woody shrubs, and their populations often expand when forests are logged.

Edges can have positive or negative effects on biodiversity. A recent study in a tropical rain forest in western Africa indicated that edge communities can be important sites of speciation. On the other hand, communities where human activities have generated more edges often have fewer species and are dominated by relatively few species that are adapted to edges. In one example, populations of the brown-headed cowbird (Figure 38.10B), an edge-adapted species that lays its eggs in the nests of other birds, are currently expanding in many areas of North America. Cowbirds forage in open fields on insects disturbed by or attracted to cattle and other large herbivores; the cowbirds also need forests, where they can parasitize the nests of other birds. Cowbird numbers are burgeoning where forests are being heavily cut and fragmented, creating more forest-edge habitats and open land for cattle, horses, and sheep. Increasing cowbird parasitism and loss of habitats are correlated with declining populations of several songbirds, such as warblers, that the cowbird parasitizes.

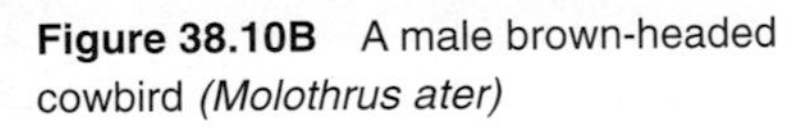

Figure 38.10B A male brown-headed cowbird *(Molothrus ater)*

Another important landscape feature, especially where habitats have been severely fragmented, is a **movement corridor,** a narrow strip or series of small clumps of quality habitat connecting otherwise isolated patches. Streamside habitats often serve as corridors, and government policy in some nations prohibits destruction of these areas. In places where there is extremely heavy human use, government agencies sometimes construct artificial corridors. Figure 38.10C shows a highway underpass that allows movement between protected areas for the few remaining Florida panthers *(Felis concolor coryi)*. High fences along the highway reduce road kills of panthers and other species.

Corridors can promote dispersal and help sustain populations, and they are especially important to species that migrate between different habitats seasonally. On the other hand, a corridor can be harmful—as, for example, in the spread of diseases, especially among small subpopulations in closely situated habitat patches. A movement corridor that connects a source habitat to a sink habitat may also reduce population numbers in the source. The effects of movement corridors have not been thoroughly studied, and researchers tend to evaluate the potential effects of corridors on a case-by-case basis.

How can "living on the edge" be a good thing for some species, such as white-tailed deer and cowbirds?

Because they use a combination of resources from the two ecosystems on either side of the edge

Figure 38.10C Underpass serving as movement corridor for Florida panthers

38.11 Restoring degraded habitats is a developing science

38A

Eventually, some areas that we alter and degrade are abandoned. For instance, mining activities may go on for several decades, and lands are then abandoned in a degraded state. Many ecosystems are damaged inadvertently by such mishaps as oil spills. A new and expanding effort in conservation biology, called **restoration ecology,** uses ecological principles to develop ways to return degraded ecosystems to conditions as similar as possible to their natural, predegraded state.

One of the key strategies in restoration ecology is **bioremediation,** the use of living organisms, usually prokaryotes, fungi, or plants, to detoxify polluted ecosystems. Some success, for instance, has been achieved in using the bacterium *Pseudomonas,* supplied with growth stimulants, to clean up oil spills on beaches (see Module 16.18).

Some other promising cases in bioremediation involve the potential use of lichens and plants to concentrate mining wastes. Researchers in the United Kingdom recently discovered that a lichen species *(Trapelia involuta)* grows on soil polluted with uranium dust left over from mining (Figure 38.11A). The lichen actually concentrates uranium in a dark pigment similar to melanin in human skin. The dark-brown structures in the photograph are uranium-rich fruiting bodies of the lichen fungus. Other researchers are experimenting with plants capable of extracting metals from contaminated soil. Prokaryotes, lichens, and plants may be useful in restoring degraded sites, and studies indicate that some species can concentrate metals in commercially marketable quantities.

A second key strategy in restoration ecology is **augmentation of ecosystem processes.** A researcher must first find out if certain key factors, such as chemical nutrients, that have been removed from an area are limiting its recovery rate. For instance, the soils of many tropical areas become nutrient-deficient and unproductive less than 5 years after being cleared for farming. Encouraging the growth of plants, such as legumes, that thrive on nutrient-poor soils can hasten the rate of recovery of some tropical areas. In the photograph in Figure 38.11B, Dr. Ariel Lugo of the U.S. Forest Service stands next to an exotic plant, a legume called *Albizia* that colonized deforested sites with nitrogen-poor soil in Puerto Rico (see Module 34.11). Prior to a 1997 hurricane, indigenous plants had begun to regrow in these sites. Apparently, the rapid buildup of organic material from dense stands of *Albizia* enabled the indigenous plants to recolonize the area and overgrow the exotic plant.

Figure 38.11A Metal-concentrating lichens (grayish) growing on uranium mineral (green)

Figure 38.11B Ariel Lugo with an *Albizia* plant

Some of the most successful restoration projects have been in wetlands that have been only partially degraded, for instance by dredging and filling. In these cases, restoring the natural water-flow patterns and replanting native plants has led to recolonization by animals. Restoring wetlands that have been extensively degraded and stripped of most native species is much more challenging, as are similar restoration efforts in most ecosystems. As Princeton University biologist Andrew Dobson puts it, "It is a relatively straightforward exercise to take apart an ecosystem or an automobile engine. In contrast, reassembling the engine (or the ecosystem) will reveal a deeper level of understanding of how each of its components functions." Restoration ecologists seek to approximate the original, not to duplicate it. 

? As complementary strategies for restoration ecology, contrast the use of organisms in bioremediation versus augmentation for altering the chemical composition of a degraded ecosystem.

In bioremediation, certain organisms are used to remove harmful chemicals from the environment; in augmentation, certain organisms are used to add essential chemicals to the environment.

Sustainable development is an ultimate goal 38.12

38A

In numbers, geographical range, and capacity to alter the biosphere, our species is clearly one of the most successful ones ever to inhabit planet Earth. Unfortunately, as we have seen in this chapter, our success is linked with many serious problems, and we seem to have set ourselves and the rest of the biosphere on a precarious path into the future. Facing increasing degradation of ecosystems, fragmentation of habitats, and loss of biodiversity, how can we best manage Earth's resources? Among the limited choices, which habitat areas are most practical to protect and manage if we are to save rare species or the greatest number of species? These challenges are pressing, and we must tackle them before scientists complete all the studies that would allow us to plot a surer path.

Many nations, scientific societies, and private foundations have embraced the concept of **sustainable development,** the long-term prosperity of human societies and the ecosystems that support them. Sustainable development will depend on the continued research and application of ecological knowledge. It will also require a commitment that few people or nations have undertaken—to promote ecosystem processes and biodiversity.

Figure 38.12 The gray treefrog *(Hyla versicolor)*

It is unlikely that human nature will suddenly change drastically—that we will abruptly lose our environmental manipulativeness. What we must seek instead are ways to be more accommodating with other species and with the biosphere. Reality demands that we work toward changing some of our values, toward learning to revere the natural processes that sustain us, and toward reducing our orientation toward short-term personal gain. Quoting Michael Dombeck, Chief of the U.S. Forest Service, "We must learn to live within the limits of the land. An old proverb goes something like: 'We haven't inherited the land from our forefathers; we have borrowed it from our children.'"

The photo on this book's cover and in Figure 38.12, a striking image of an amphibian, the gray treefrog *(Hyla versicolor),* symbolizes the biodiversity crisis. As a group, amphibians have been on Earth for over 350 million years. They survived two major extinction events, including the one that decimated the dinosaurs about 65 million years ago. Today, however, many species of frogs, salamanders, and other amphibians are declining at unprecedented rates. Rapid, human-caused habitat loss is most likely a major cause. In some populations of *Hyla versicolor* and other amphibians, researchers are finding abnormal numbers of deformed individuals, probably a result of developmental changes induced by various kinds of pollutants.

An awareness of our unique ability to alter the biosphere and jeopardize the existence of other species, as well as our own, may help us choose a path toward a sustainable future, one in which each generation inherits an adequate supply of natural and economic resources and a relatively stable environment. Despite the uncertainties, now is not the time for gloom and doom, but a time to aggressively pursue more knowledge about life and to work toward long-term sustainability. 

? Why is a concern for the well-being of future generations essential for progress toward sustainable development?

Sustainable development is a long-term goal—longer than a human lifetime. Preoccupation with personal gain in the here-and-now is an obstacle to sustainable development because it discourages behavior that benefits future generations.

CHAPTER REVIEW

Chapter Summary

The Biodiversity Crisis: An Overview (Introduction–38.4) Modern human culture and the rapidly growing global human population have created a biodiversity crisis, a rapid decrease in the variety of species on Earth. Extinction rates are unusually high. The goal of conservation biology is to find ways to counter the rapid loss of species **(Introduction).** Habitat destruction poses the single greatest threat to biodiversity. Competition with exotic species (those that have been transferred to an area where they did not occur naturally) also threatens many species in their native habitats. Overexploitation of wildlife often compounds problems caused by habitat loss and exotic species **(38.1).** Biodiversity is vital to human welfare. We evolved in Earth's ecosystems, and large-scale changes in them threaten us as well as other species **(38.2).** The explosive growth of the human population and of technology continues today. Populations of developing nations are growing the fastest, but the technology and resource consumption of the less-populous developed nations put a much greater strain on the biosphere. Oil spills, acid rain, ozone depletion, and chemical pesticides, which are concentrated in food chains by biological magnification, affect the entire world **(38.3).** Burning of fossil fuels is increasing the amount of CO_2 and other greenhouse gases in the air, which may warm Earth enough to change climate patterns, melt polar ice, and flood coastal regions. Global warming may increase the rate of species loss **(38.4).**

The Geographic Distribution of Biodiversity (38.5) Environments are patchy and as a result species are not evenly distributed. Some areas are unusually rich in biodiversity. Biodiversity hot spots are relatively small areas with a large concentration of species. Many organisms in these hot spots are endemic, found nowhere else. Endemic species are especially prone to extinction. Many migratory species, both terrestrial and aquatic, require international protection.

Conservation of Species Populations (38.6–38.8) Habitat degradation has seriously fragmented many species populations. Conservation efforts often center on sustaining fragmented populations. A source habitat for a portion (subpopulation) of a fragmented population is where reproductive success exceeds the death rate. In a sink habitat, the subpopulation's death rate exceeds reproduction. Protecting source habitats is vital in species conservation **(38.6).** Identifying critical habitat factors in source habitats is a central goal of conservation biology **(38.7).** Increased fragmentation threatens many species, including those whose populations were historically highly fragmented. Conservation biologists often use computer simulations in a population viability analysis (PVA) to assess a species' chances for survival **(38.8).**

Managing and Sustaining Ecosystems (38.9–38.12) Conservation efforts are increasingly aimed at learning how to sustain whole ecosystems and landscapes. Landscape ecology employs ecological principles to study land-use patterns and make species conservation a functional part of the patterns. Landscape researchers often use a computerized mapping technique called gap analysis **(38.9).** Edges (boundaries) between ecosystems, such as forests and grasslands, have their own set of features and assemblages of species. The increased frequency and abruptness of edges caused by human activities can increase species losses. Movement corridors, strips or clumps of quality habitat connecting otherwise isolated habitat patches, may be helpful or harmful to fragmented populations **(38.10).** Restoration of degraded habitats is increasingly important. Two strategies are bioremediation, the use of living organisms to detoxify polluted areas, and augmentation of ecosystem processes, resupplying an area with key factors that were removed **(38.11).** We are the most environmentally manipulative of all animal species. The attitudes and environmental awareness of people are of utmost importance in the search for solutions to the biodiversity crisis and other problems facing humanity. Understanding the biosphere's limits and vulnerability and our own linkages to the natural world may help us make decisions that lead to a sustainable future **(38.12).**

Testing Your Knowledge

Multiple Choice

1. ______________ poses the single greatest threat to biodiversity.
   a. Introduction of exotic species
   b. Overhunting
   c. Movement corridors
   d. Habitat loss
   e. Global warming

2. Which of the following is characteristic of endemic species?
   a. They are often found in biodiversity hot spots.
   b. They are distributed widely in the biosphere.
   c. They all require edges between ecosystems.
   d. Their trophic position makes them very susceptible to the effects of biological magnification.
   e. Researchers have completed PVAs for most of them.

3. Which of the following is least likely to be a sound conservation practice?
   a. identifying critical habitat factors
   b. developing a population viability analysis
   c. identifying and protecting biodiversity hot spots
   d. establishing movement corridors
   e. reducing the effects of population fragmentation

4. Landscape ecology is an indication of what major trend in conservation biology?
   a. moving away from a focus on air and water pollution to a focus on species conservation
   b. switching from an ecosystem approach to a population approach
   c. a trend toward a central focus on fragmented populations
   d. a trend toward development of more PVAs
   e. a trend toward sustaining biodiversity of multiple ecosystems

5. Ospreys and other top predators in food chains are most severely affected by pesticides such as DDT because
   a. their systems are especially sensitive to chemicals
   b. of their rapid reproductive rates
   c. the pesticides become concentrated in their prey
   d. they cannot store the pesticides in their tissues
   e. they are directly exposed to pesticides in the air

6. Some biologists think that half the species on Earth may be exterminated in the next century. Mass extinctions have happened before. How would this one be different?
   a. This time species are threatened by a change in climate.
   b. This extinction is happening at a much faster rate.
   c. A much larger fraction of Earth's species are likely to be affected.
   d. All life on Earth may eventually be extinguished.
   e. This extinction is much more gradual and difficult to notice.

7. Which of the following consumes the most energy each day? *(Explain your answer.)*

a. the United States
b. South America
c. China
d. India
e. Africa

Describing, Comparing, and Explaining

1. What is the greenhouse effect? How is it important to life on Earth?

2. What are the possible causes and consequences of global warming? Why is international cooperation necessary if we are to solve this problem?

Thinking as a Scientist

1. An oil tanker runs aground in a rocky bay, and a local environmental group wants to assess the damage that the spilled oil caused to the bay's marine life. No previous survey of the organisms in this area has ever been done. Two members of the environmental group collect intertidal organisms in both the oil-affected bay and a neighboring bay roughly similar to it. The group then hires a biologist to classify and count the organisms from both bays. Her report states that several important mollusks have significantly lower populations in the oil-affected bay than in the other bay. Concluding that the oil spill killed off the mollusks, the environmental group hires a lawyer and sues the oil company for criminal damage to the local environment. A judge throws the case out for lack of solid evidence. Which of the following arguments do you suppose the oil company's attorney used? Explain your choice.

a. No effort was made to determine why the animals were damaged by the oil spill, and a proposed effect without a proposed mechanism has no scientific credibility.
b. No amount of data can prove a hypothesis. Instead of trying to do so, the group or the biologist should have designed the experiment to falsify the hypothesis that the oil was having an effect.
c. The oil-free bay was not a true control.
d. The group's conclusion was hasty because it was based on only some of the species present.

2. Five years after the oil spill discussed above, most of the oil has disappeared from the intertidal area, but the environmental group rehires the biologist to do some more work on the bay. The new piece of evidence that would most strengthen their conclusion that the oil had depressed the mollusk populations would be for the biologist to find that

a. the two bays now have nearly identical populations of all intertidal animals, including mollusks
b. the mollusk populations in the oil-affected bay are even lower than right after the spill
c. the mollusks in the unpolluted bay have changed in species composition, so they now share few species with the oil-affected bay
d. laboratory tests show that mollusks in both bays can tolerate far higher oil concentrations than those experienced during the spill

3. The oceans are a poorly understood factor in global warming. Some researchers are concerned that warming of the oceans might cause the large amounts of CO_2 dissolved in seawater to come out of solution, the way air bubbles form in a glass of tap water when it warms up. Why might this concern them? On the other hand, how might the oceans help reduce global warming?

4. Biologists in the United States are concerned that populations of many migratory songbirds, such as warblers, are declining. Evidence suggests that some of these birds might be victims of pesticides. Most of the pesticides implicated in songbird mortality have not been used in the U.S. since the 1970s. Suggest a hypothesis to explain the current decline in songbird numbers.

Science, Technology, and Society

1. You may have heard that human activities cause the extinction of one species every hour. Such estimates vary widely, because we do not know how many species exist, nor how fast their habitat is being destroyed. You can make your own estimate of the rate of extinction. Start with the number of species thought to exist. To keep things simple, ignore extinction in the temperate latitudes and focus on the 80% of plants and animals that live in the tropical rain forest. Assume that destruction of the forest continues at a rate of 1% per year, so the forest will be gone in 100 years. Assume (optimistically) that half the rain forest species will survive in preserves, forest remnants, and zoos. How many species will disappear in the next century? How many species is that per year? Per day? Recent studies of the rain forest canopy have led some experts, such as Terry Erwin of the Smithsonian Institution, to suggest that there might be as many as 30 million species on Earth. How does starting with this figure change your estimates?

2. One of the reasons the developed countries consume so much energy is that the price of energy does not reflect its real costs. For example, the monetary cost of the 1991 Gulf War would have doubled the price of every barrel of oil imported into the United States that year. What kinds of hidden environmental costs are not reflected in the prices of fossil fuels? How are these costs paid, and by whom? Do you think these costs could or should be figured into the price of oil? How might that be done?

3. Many atmospheric scientists say that global warming is under way and we need to take action now to avoid drastic environmental change. Some say it is too soon to tell and we should gather more data before we act. What are the advantages and disadvantages of doing something now to slow global warming? What are the advantages and disadvantages of waiting until more data are available? Is there any way to know for sure whether there is a link between CO_2 and global warming?

4. Until recently, response to environmental problems has been fragmented—an antipollution law here, incentives for recycling there. Meanwhile, the problems of the gap between the rich and poor nations, diminishing resources, and pollution continue to grow. Now people and governments are starting to envision a sustainable society. The Worldwatch Institute, a respected environmental monitoring organization, estimates that we must reach sustainability by the year 2030 to avoid economic and environmental disaster. To get there, we must begin shaping a sustainable society during this decade. In what ways is our present system not sustainable? What might a sustainable society be like? Do you think a sustainable society is an achievable goal? Why or why not? What is the alternative? What might we do to work toward sustainability? What are the major roadblocks in the way of achieving sustainability? How would your life be different in a sustainable society?

Media Resources

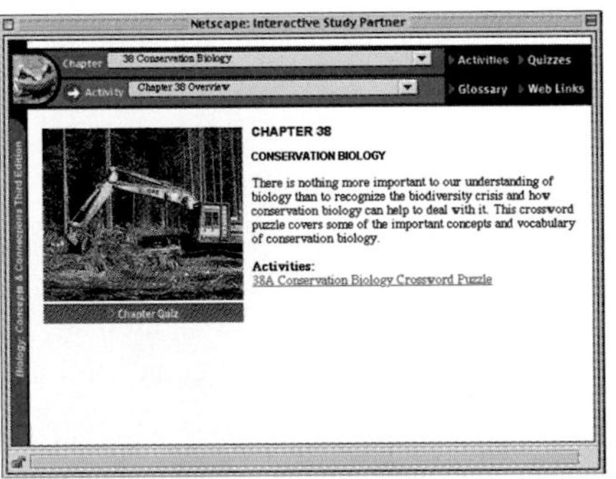

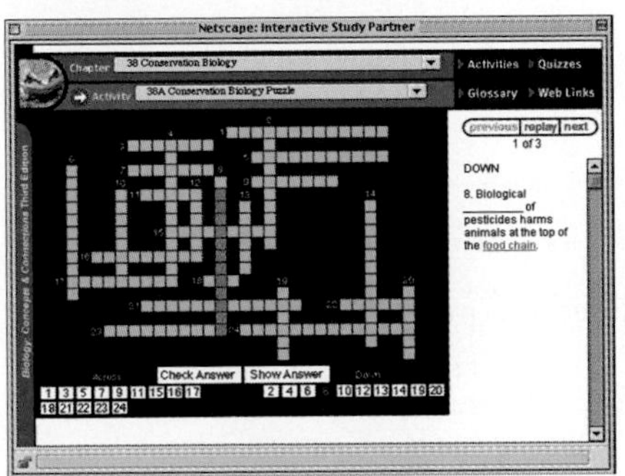

For further review of the concepts in this chapter, use the activities and quizzes on the CD-ROM.

Visit the Special Edition of The Biology Place web site for activities, quizzes, news articles, and web links.

APPENDIX 1

Metric Conversion Table

Measurement	Unit and Abbreviation	Metric Equivalent	Approximate Metric-to-English Conversion Factor	Approximate English-to-Metric Conversion Factor
Length	1 kilometer (km)	= 1000 (10^3) meters	1 km = 0.6 mile	1 mile = 1.6 km
	1 meter (m)	= 100 (10^2) centimeters = 1000 millimeters	1 m = 1.1 yards 1 m = 3.3 feet 1 m = 39.4 inches	1 yard = 0.9 m 1 foot = 0.3 m
	1 centimeter (cm)	= 0.01 (10^{-2}) meter	1 cm = 0.4 inch	1 foot = 30.5 cm 1 inch = 2.5 cm
	1 millimeter (mm)	= 0.001 (10^{-3}) meter	1 mm = 0.04 inch	
	1 micrometer (μm)	= 10^{-6} meter (10^{-3} mm)		
	1 nanometer (nm)	= 10^{-9} meter (10^{-3} μm)		
	1 angstrom (Å)	= 10^{-10} meter (10^{-4} μm)		
Area	1 hectare (ha)	= 10,000 square meters	1 ha = 2.5 acres	1 acre = 0.4 ha
	1 square meter (m^2)	= 10,000 square centimeters	1 m^2 = 1.2 square yards 1 m^2 = 10.8 square feet	1 square yard = 0.8 m^2 1 square foot = 0.09 m^2
	1 square centimeter (cm^2)	= 100 square millimeters	1 cm^2 = 0.16 square inch	1 square inch = 6.5 cm^2
Mass	1 metric ton (t)	= 1000 kilograms	1 t = 1.1 tons	1 ton = 0.91 t
	1 kilogram (kg)	= 1000 grams	1 kg = 2.2 pounds	1 pound = 0.45 kg
	1 gram (g)	= 1000 milligrams	1 g = 0.04 ounce 1 g = 15.4 grains	1 ounce = 28.35 g
	1 milligram (mg)	= 10^{-3} gram	1 mg = 0.02 grain	
	1 microgram (μg)	= 10^{-6} gram		
Volume (Solids)	1 cubic meter (m^3)	= 1,000,000 cubic centimeters	1 m^3 = 1.3 cubic yards 1 m^3 = 35.3 cubic feet	1 cubic yard = 0.8 m^3 1 cubic foot = 0.03 m^3
	1 cubic centimeter (cm^3 or cc)	= 10^{-6} cubic meter	1 cm^3 = 0.06 cubic inch	1 cubic inch = 16.4 cm^3
	1 cubic millimeter (mm^3)	= 10^{-9} cubic meter (10^{-3} cubic centimeter)		
Volume (Liquids and Gases)	1 kiloliter (kL or kl)	= 1000 liters	1 kL = 264.2 gallons	1 gallon = 3.79 L
	1 liter (L)	= 1000 milliliters	1 L = 0.26 gallon 1 L = 1.06 quarts	1 quart = 0.95 L
	1 milliliter (mL or ml)	= 10^{-3} liter = 1 cubic centimeter	1 mL = 0.03 fluid ounce 1 mL = approx. $\frac{1}{4}$ teaspoon 1 mL = approx. 15–16 drops	1 quart = 946 mL 1 pint = 473 mL 1 fluid ounce = 29.6 mL 1 teaspoon = approx. 5 mL
Volume (Liquids and Gases)	1 microliter (μl or μL)	= 10^{-6} liter (10^{-3} milliliters)		
Time	1 second (s)	= $\frac{1}{60}$ minute		
	1 millisecond (ms)	= 10^{-3} second		
Temperature	Degrees Celsius (°C)		°F = $\frac{9}{5}$°C + 32	°C = $\frac{5}{9}$(°F − 32)

The Amino Acids of Proteins

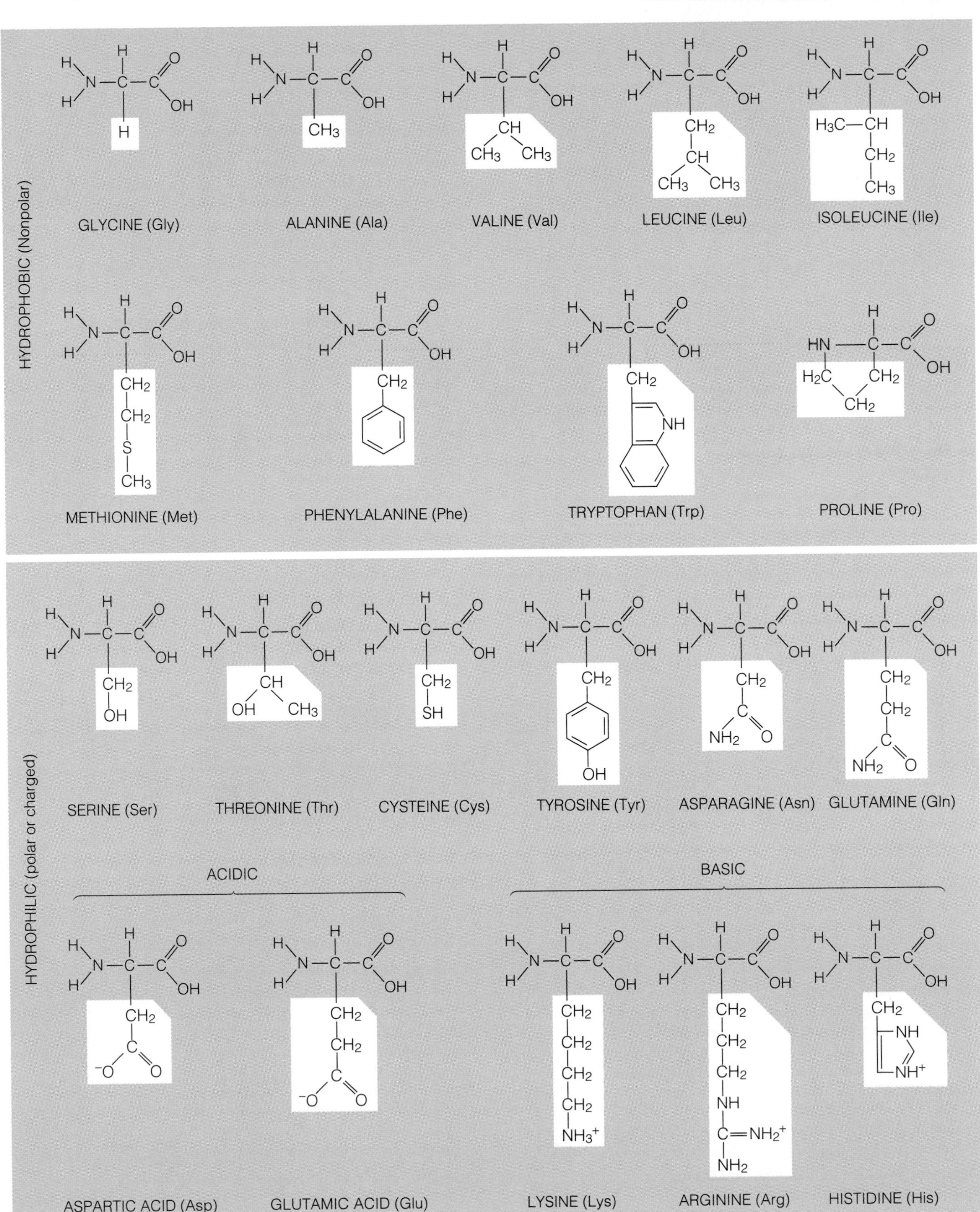
HYDROPHOBIC (Nonpolar)
GLYCINE (Gly)
ALANINE (Ala)
VALINE (Val)
LEUCINE (Leu)
ISOLEUCINE (Ile)
METHIONINE (Met)
PHENYLALANINE (Phe)
TRYPTOPHAN (Trp)
PROLINE (Pro)
HYDROPHILIC (polar or charged)
SERINE (Ser)
THREONINE (Thr)
CYSTEINE (Cys)
TYROSINE (Tyr)
ASPARAGINE (Asn)
GLUTAMINE (Gln)
ACIDIC
BASIC
ASPARTIC ACID (Asp)
GLUTAMIC ACID (Glu)
LYSINE (Lys)
ARGININE (Arg)
HISTIDINE (His)

APPENDIX 3

Further Readings

Chapter 1 Introduction

Krajick, K. "The Secret Life of Backyard Trees." *Discover,* November 1995. Forest canopy research in North America.

Moffett, M.W. *The High Frontier: Exploring the Tropical Rainforest Canopy.* Cambridge, MA: Harvard University Press, 1993. A vividly illustrated and written account of current canopy research in the tropics.

Moore, J. A. *Science as a Way of Knowing: The Foundation of Modern Biology.* Cambridge, MA: Harvard University Press, 1993. A lively account by a famous developmental biologist.

Myers, N. "Pushed to the Edge." *Natural History,* March 1999. A brief essay on the effects of slash-and-burn farming on tropical rain forests.

UNIT I THE LIFE OF THE CELL

Alberts, B., D. Bray, A. Johnson, J. Lewis, M. Raff, K. Roberts, and P. Walter. *Essential Cell Biology.* New York: Garland, 1998. A shorter version of the "bible" of modern cell biology.

Becker, W. M., J. B. Reece, and M. F. Poenie. *The World of the Cell,* 3rd ed. Menlo Park, CA: Benjamin/Cummings, 1996. A student-oriented text for undergraduates, broader in coverage than the short Alberts text.

Mathews, C. K., and K. E. van Holde. *Biochemistry,* 2nd ed. Menlo Park, CA: Benjamin/Cummings, 1996. A well-illustrated biochemistry text.

Chapter 2 The Chemical Basis of Life

Atkins, P. W. *Molecules.* New York: Scientific American Library, 1987. A beautifully illustrated tour of the world of molecules.

Mohnen, V. A. "The Challenge of Acid Rain." *Scientific American,* August 1988. Causes of the acid precipitation problem and possible solutions.

Pennisi, E. "Water: The Power, Promise, and Turmoil of North America's Fresh Water." *National Geographic.* A special 1993 issue examining how we use and abuse our water resources.

Chapter 3 The Molecules of Cells

Asimov, I. *The World of Carbon,* 2nd ed. New York: Macmillan, 1962. A primer on organic chemistry by one of America's most popular science writers.

Horgan, J. "Stubbornly Ahead of His Time." *Scientific American,* March 1993. An essay about Linus Pauling.

Straus, E. "How Proteins Take Shape." *Science News,* September 6, 1997. How chaperone proteins help other proteins fold correctly.

Chapter 4 A Tour of the Cell

De Duve, C. *A Guided Tour of the Living Cell.* New York: Scientific American Books, 1986. A beautifully illustrated introduction to the cell.

Featherstone, C. "Coming to Grips With the Golgi." *Science,* December 18, 1998. Researchers debate just how the Golgi apparatus works.

Rothman, J. E., and L. Orci. "Budding Vesicles in Living Cells." *Scientific American,* March 1996. More about transport vesicles.

Stossel, T. P. "The Machinery of Cell Crawling." *Scientific American,* September 1994. How protein scaffolds enable cells to move.

Strange, C. J. "Biological Ties That Bind." *BioScience,* January 1997. The varied and important roles of the extracellular matrix in animals.

Chapter 5 The Working Cell

Deeth, R. "Chemical Choreography." *New Scientist,* July 5, 1997. Computer modeling is being used to study how enzymes work in the cell.

Horwitz, A. F. "Integrins and Health." *Scientific American,* May 1997. These plasma-membrane proteins are crucial to body function.

MacNeil, P. L. "Cell Wounding and Healing." *American Scientist,* May/June 1991. How cell membranes repair breaks.

Sharon, N., and H. Lis. "Carbohydrates in Cell Recognition." *Scientific American,* January 1993. Sugars on cell surfaces: research and applications.

Chapter 6 How Cells Harvest Chemical Energy

Angier, N. "A Stupid Cell with All the Answers." *Discover,* November 1986. The important roles yeast is playing in basic biological research.

Derr, M. "The End of the Road." *Scientific American,* April 1995. A debilitating disease in which mitochondria fail to make enough ATP for body exercise.

Kearney, J. T. "Training the Olympic Athlete." *Scientific American,* June 1996. Includes discussion of how athletes work to achieve an optimum balance between aerobic and anaerobic energy generation.

Chapter 7 Photosynthesis: Using Light to Make Food

Caldwell, M. "The Amazing All-Natural Light Machine." *Discover,* December 1995. The antenna complexes of a photosynthetic prokaryote.

Culotta, E. "Will Plants Profit from High CO_2?" *Science,* May 5, 1995. An analysis of possible plant responses to increasing atmospheric CO_2.

Govindjee and W. J. Coleman. "How Plants Make Oxygen." *Scientific American,* February 1990.

UNIT II CELLULAR REPRODUCTION AND GENETICS

Mange, E. L., and A. P. Mange. *Basic Human Genetics,* 2nd ed. Sunderland, MA: Sinauer Associates, 1998. A basic text on human inheritance.

P. J. Russell. *Genetics,* 5th ed. Menlo Park, CA: Benjamin/Cummings, 1998. A general genetics textbook.

Chapter 8 The Cellular Basis of Reproduction and Inheritance

"Cancer." *Scientific American,* September 1996. A single-topic issue.

Glover, D. M., C. Gonzalez, and J. W. Raff. "The Centrosome." *Scientific American,* June 1993. Structure and function of the mitotic spindle.

McIntosh, J. R., and K. L. McDonald. "The Mitotic Spindle." *Scientific American,* October 1989.

Murray, A. W., and M. W. Kirschner. "What Controls the Cell Cycle." *Scientific American,* March 1991.

See also the cell biology texts listed under Unit I.

Chapter 9 Patterns of Inheritance

Beardsley, T. "Fetal Checkup." *Scientific American,* June 1997. Fetal cells for testing can be obtained from the mother's blood.

Gillis, A. M. "Turning Off the *X* Chromosome." *BioScience,* March 1994. The inactivation of one *X* chromosome in the cells of females.

Miller, K. R. "Whither the *Y*? *Discover,* February 1995. Commentary on the evolution of the human sex chromosomes.

Orel, V. *Gregor Mendel, the First Geneticist.* New York: Oxford University Press, 1995. A recent biography of a key figure in biology.

Chapter 10 Molecular Biology of the Gene

"Defeating AIDS: What Will It Take?" *Scientific American,* July 1998. A special section consisting of nine articles.

Larsen, E. "The Flu Hunters." *Time,* February 22, 1998. How medical detectives tracked the mysterious and deadly 1997 flu.

Watson, J. D. *The Double Helix.* New York: Atheneum, 1968. The controversial best-seller by the codiscoverer of the double helix.

Chapter 11 The Control of Gene Expression

"Cancer." *Scientific American,* September 1996. A single-topic issue.

De Robertis, E., G. Oliver, and C. Wright. "Homeobox Genes and the Vertebrate Body Plan." *Scientific American,* July 1990.

Nüsslein-Volhard, C. "Gradients That Organize Embryo Development." *Scientific American,* August 1996. A look at early fruit-fly development.

Tjian, R. "Molecular Machines That Control Genes." *Scientific American,* February 1995. How genes are regulated in eukaryotes.

Wilmut, I. "Cloning for Medicine." *Scientific American,* December 1998. An article by the scientist whose cloning research led to "Dolly."

Chapter 12 DNA Technology and the Human Genome

Beardsley, T. "Genes in the Not So Public Domain." *Scientific American,* April 1995. Who owns the base sequences in DNA databases?

Greider, C. W., and E. H. Blackburn. "Telomeres, Telomerase, and Cancer." *Scientific American*, February 1996. An enzyme that seems to keep cancer cells immortal could be a target in the war against cancer.

"Making Gene Therapy Work." *Scientific American*, June 1997. A section consisting of five articles on progress toward gene therapy.

Miller, R. V. "Bacterial Gene Swapping in Nature." *Scientific American*, January 1998. Gene exchange occurs more often than once assumed.

Moxon, E. R., and C. Wills. "DNA Microsatellites: Agents of Evolution?" *Scientific American*, January 1999. Repetitive DNA sequences can influence how organisms adapt to their environments.

Pennisi, E. "Genome Data Shakes Tree of Life." *Science*, May 1, 1998. New genome sequences reveal unexpected evolutionary comnections.

Velander, W. H., H. Lubon, and W. N. Drohan. "Transgenic Livestock as Drug Factories." *Scientific American*, January 1997. How dairy animals can be made to produce therapeutic human proteins in their milk.

Watzman, H. "DNA to Unravel Secrets of the Dead Sea Scrolls." *New Scientist*, January 14, 1995. DNA technology in historical research.

UNIT III CONCEPTS OF EVOLUTION

Darwin, C. *On the Origin of Species by Means of Natural Selection, or the Preservation of Favoured Races in the Struggle for Life.* London: Murray, 1859. The classic that revolutionized biology.

Futuyma, E. J. *Evolutionary Biology*, 3rd ed. Sunderland, MA: Sinauer Associates, 1998. An authoritative undergraduate text.

Chapter 13 How Populations Evolve

Browne, J. *Charles Darwin Voyaging.* Princeton, NJ: Princeton University Press, 1995. An engaging biography.

Dawkins, R. *The Blind Watchmaker.* New York: Norton, 1986. A popular account of how organismal complexity can arise in the absence of design.

Gibbons, A. "The Mystery of Humanity's Missing Mutations." *Science*, January 6, 1995. Did human evolution include a bottleneck?

Hardison, R. "The Evolution of Hemoglobin." *American Scientist*, March-April 1999. Discussion of the role of gene regulation in the history of life.

Levy, S. B. "The Challenge of Antibiotic Resistance." *Scientific American*, March 1998. How resistant prokaryotes arise.

Wise, D.U. "Creationism's Geological Time Scale." *American Scientist*, March-April 1998. A scientist's call for some evidence for the creationists' model.

Chapter 14 The Origin of Species

Goldschmidt, T. *Darwin's Dreampond, Drama in Lake Victoria.* Cambridge, MA: MIT Press, 1996. A researcher's popular account of field work on speciation.

Grant, P. R. "Natural Selection and Darwin's Finches." *Scientific American*, October 1991. How a single drought can change a population.

Knowlton, N. "A Tale of Two Seas." *Natural History*, June 1994. Allopatric speciation resulting from the Panamanian land bridge.

Rennie, J. "Darwin's Current Bulldog." *Scientific American*, August 1994. A profile of Ernst Mayr, Darwin scholar and coauthor of the modern synthesis.

Chapter 15 Tracing Evolutionary History

Benton, M., and D. Harper. *Basic Paleontology.* Essex, England: Addison Wesley Longman, 1997. An introductory text.

Carroll, S. B. "Homeotic Genes and the Evolution of Arthropods and Chordates." *Nature*, August 10, 1995. The duplication of homeotic genes may have been key milestones in animal evolution.

Dawkins, R. "God's Utility Function." *Scientific American*, November 1995. An argument that the diversity of life is the result of a survival contest among selfish genes.

Dalziel, I. W. D. "Earth Before Pangea." *Scientific American*, January 1995. An overview of continental movements.

Gaffney, E. S., L. Dingus, and M. K. Smith. "Why Cladistics?" *Natural History*, June 1995. Dinosaur phylogeny revisited.

Monastersky, R. "Life's Closest Call, What Caused the Spectacular Extinctions at the End of the Permian Period?" *Science News*, February 1, 1997.

Pääbo, S. "Ancient DNA." *Scientific American*, November 1993. Analyzing DNA from fossils.

UNIT IV THE EVOLUTION OF BIOLOGICAL DIVERSITY

Gamlin, L., and G. Vines (eds.). *The Evolution of Life.* New York: Oxford University Press, 1987. A vivid survey of life and life science.

Margulis, L., and K. V. Schwartz. *Five Kingdoms: An Illustrated Guide to the Phyla of Life on Earth*, 2nd ed. New York: W. H. Freeman, 1987.

Monastersky, R. "The Rise of Life on Earth." *National Geographic*, March 1998. A superbly illustrated overview of prokaryotic and eukaryotic evolution.

Smith, J. M., and E. Szathmary. *The Major Transitions in Evolution.* New York: W. H. Freeman, 1995. A panorama of the history of life.

Chapter 16 The Origin and Evolution of Microbial Life: Prokaryotes and Protists

DeDuve, C. "The Beginnings of Life on Earth." *American Scientist*, September 1995. A clear discussion of early molecular evolution.

Hooper, J. "A New Germ Theory." *Atlantic Monthly*, February 1999. Microbes may be responsible for some common "noninfectious" diseases.

Lewin, R. "Bacteria Rule OK?" *New Scientist*, June 3, 1995. Evidence for the endosymbiotic origin of eukaryotes.

Madigan, M. T., J. M. Martinko, and J. Parker. *Brock's Biology of Microorganisms*, 8th ed. Englewood Cliffs, NJ: Prentice-Hall, 1997. A comprehensive text.

Simpson, S. "Life's First Scalding Steps." *Science News*, January 9, 1999. Life may have arisen in hot water at the bottom of the sea.

Chapter 17 Plants, Fungi, and the Colonization of Land

Attenborough, D. *The Private Life of Plants.* Princeton, NJ: Princeton University Press, 1995. A lively, strikingly illustrated popular account.

Berner, R. A. "The Rise of Plants and Their Effect on Weathering and Atmospheric CO_2." *Science*, April 25, 1997. How plants changed Earth.

Cilaghi, C., and P. L. Nimis. "Lichens, Air Pollution, and Lung Cancer." *Nature*, May 29, 1997. Is the diversity of lichens correlated with air quality and cancer incidence?

Diamond, J. "How to Tame a Wild Plant." *Discover*, September 1994. How humans domesticated food plants.

Moore-Landecker, E. *Fundamentals of the Fungi*, 3rd ed. Englewood Cliffs, NJ: Prentice-Hall, 1990. An introductory text.

Wood, W. "Deadly Fungus is Kind to Hearts." *New Scientist*, February 11, 1995. Discusses the significance of drugs made from fungi.

Sterberg, S. "The Emerging Fungal Threat." *Science*, December 9, 1994. The potential rise in fungal diseases in humans.

Chapter 18 The Evolution of Animal Diversity

Brusca, R. G., and G. J. Brusca. *Invertebrates.* Sunderland, MA: Sinauer Associates, 1990. A strong text on invertebrate animals.

Currie, P. J., and K. Padian (eds.). *Encyclopedia of Dinosaurs.* San Diego, CA: Academic Press, 1997. A comprehensive volume.

Forey, P., and P. Janvier. "Evolution of the Early Vertebrates." *American Scientist*, November/December 1994. The earliest fishes.

Morris, S.C. *Crucible of Creation: The Burgess Shale and the Rise of Animals.* New York: Oxford University Press, 1998. A lively text on scientific discovery and animal fossils.

Rice, J. A. (ed.). "The Marvelous Mammalian Parade." *Natural History*, April 1994. Sixteen illustrated articles on extinct mammals.

Wright, K. "When Life Was Odd." *Discover*, March 1997. An illustrated essay on Precambrian animals.

Xerces Society. *Wings, Essays on Invertebrate Conservation.* Fall 1996. Brief essays on the ecological roles of aquatic insects.

Zimmer, C. "Coming onto the Land." *Discover*, June 1995. Vertebrates that first invaded land.

Chapter 19 Human Evolution

Agnew, N., and M. Demas. "Preserving the Laetoli Footprints." *Scientific American*, September 1998. Vividly illustrated discussion of new efforts to save early human imprints.

Ben-Ari, E.T. "Molecular Biographies." *BioScience*, February 1999. Using the human genome to track human history.

Gore, R. "The Dawn of Humans, Expanding Worlds." *National Geographic*, May 1997. A pictorial essay on early human migrations.

Kahn, P., and A. Gibbons. "DNA From an Extinct Human." *Science,* July 11, 1997. A news brief about the analysis of Neanderthal mitochondrial DNA.

Leakey, M, and A. Walker. "Early Hominid Fossils From Africa." *Scientific American,* June 1997. An illustrated essay on some of the earliest hominid fossils.

Shreeve, J. "Sunset on the Savanna." *Discover,* July 1996. Did our ancestors evolve mainly in forests or savannas?

Tattersall, I. *The Human Odyssey: Four Million Years of Human Evolution.* New York: Prentice-Hall General Reference, 1993. A readable and beautifully illustrated book based on the new Hall of Human Biology and Evolution at the American Museum of Natural History.

UNIT V ANIMALS: FORM AND FUNCTION

Louw, G. *Physiological Animal Ecology.* New York: Wiley, 1993. An introduction to how animal form and function correlate with environmental conditions.

Marieb, E. *Human Anatomy and Physiology,* 4th ed. Menlo Park, CA: Benjamin/Cummings, 1998. A basic text, beautifully illustrated.

Schmidt-Nielsen, K. *Animal Physiology: Adaptations and Environment,* 5th ed. New York: Cambridge University Press, 1997. A basic text emphasizing form and function, adaptations, and homeostasis.

Chapter 20 Unifying Concepts of Animal Structure and Function

Ingber, D. E. "The Architecture of Life." *Scientific American,* January 1998. The universal laws that underlie anatomical structure.

Schmidt-Nielsen, K. "How Are Control Systems Controlled?" *American Scientist,* January/February 1994. Many unanswered questions.

Chapter 21 Nutrition and Digestion

Brown, J. L., and E. Pollitt. "Malnutrition, Poverty and Intellectual Development." *Scientific American,* February 1996. The importance of essential nutrients during childhood.

Fackelmann, K. "The Fat Fracas." *Science News,* May 2, 1998. The pros and cons of weight loss.

Sanderson, S., and R. Wasserug. "Suspension-Feeding Vertebrates." *Scientific American,* March 1990. An article emphasizing whales.

Seppa, N. "Ulcer Bacterium's Drug Resistance Unmasked." *Science News,* April 25, 1998. Research on the genetics of resistance and cancer links.

Willett, W. C. "Diet and Health: What Should We Eat?" *Science,* April 22, 1994. A critical analysis of international diets.

Chapter 22 Respiration: The Exchange of Gases

Bartecchi, C. E., T. D. MacKenzie, and R. W. Schrier. "The Global Tobacco Epidemic." *Scientific American,* May 1995. Tobacco use on the rise.

Kooyman, G. L., and P. J. Ponganis. "The Challenges of Diving to Depth." *American Scientist,* November-December 1997. Some adaptations of deep-diving vertebrates.

Lee, J., and L. Manning. "Environmental Lung Disease." *New Scientist,* September 16, 1995. How airborne pollutants affect our lungs.

Seppa, N. "Secondary Smoke Carries High Price." *Science News,* January 17, 1998. A summary of the health hazards of tobacco smoke.

Chapter 23 Circulation

Barinaga, M. "How Much Pain for Cardiac Gain?" *Science,* May 30, 1997. A summary of the debate over how much exercise we need.

Buchanan, M. "Fascinating Rhythm." *New Scientist,* January 3, 1998. The irregular rhythm of the healthy heart.

Coghlan, A. "Blood-clotting Drug Lays Siege to Tumours." *New Scientist,* October 19, 1996. Possible treatment of cancer by inducing clots.

Nucci, M., and A. Abuchowski. "The Search for Blood Substitutes." *Scientific American,* February 1998. Meeting the needs of global blood banks.

Radetsky, P. "The Mother of All Blood Cells." *Discover,* March 1995. Potential uses of research on stem cells.

Chapter 24 The Immune System

Beardsley, T. "Better Than a Cure." *Scientific American,* January 1995. The importance of vaccinating children.

"Life, Death, and the Immune System." *Scientific American,* September 1993. A special issue devoted to the immune system.

Richardson, S. "The End of the Self." *Discover,* April 1996. A radical new theory challenges the traditional self/nonself view of immunity.

See also the AIDS reference listed for Chapter 10.

Chapter 25 Control of the Internal Environment

Line, L. "Staying the Winter." *National Wildlife,* February/March 1995. How chickadees and other songbirds survive northern winters.

Heinrich, B. *The Thermal Warriors.* Cambridge, MA: Harvard University Press, 1996. A lively account of thermoregulation in insects.

McClanahan, L. L., R. Ruibal, and V. H. Shoemaker. "Frogs and Toads in Deserts." *Scientific American,* March 1994. Amphibian adaptations to dry environments.

Travis, J. "Chilled Brains." *Science News,* December 6, 1997. How torpor in ground squirrels may lead to new ways to treat strokes.

Wu, C. "Freeze! Insect Proteins Halt Ice Growth." *Science News,* August 30, 1997. Brief discussion of antifreeze proteins.

Chapter 26 Chemical Regulation

Barinaga, M. "How Jet-Lag Hormone Does Double Duty in the Brain." *Science,* July 25, 1997. Recent research on melatonin.

Lacy, P. E. "Treating Diabetes With Transplanted Cells." *Scientific American,* July 1995. Pancreatic islet cells can be transplanted.

Leutwyler, K. "Don't Stress." *Scientific American,* January 1998. Discusses the long-term effects of stress hormones.

McLachlan, J. A., and S. F. Arnold. "Environmental Estrogens." *American Scientist,* September-October 1996. The controversial hypothesis that environmental chemicals can function as chemical signals in animals.

Snyder, S., and D. Bredt. "Biological Roles of Nitric Oxide." *Scientific American,* May 1992. The varied roles of a gaseous signal.

Chapter 27 Reproduction and Embryonic Development

Diamond, J. M. *Why Is Sex Fun? The Evolution of Human Sexuality.* Basic Books, 1998. Some fascinating hypotheses.

Smith, B. R. "Visualizing Human Embryos." *Scientific American,* March 1999. The use of magnetic resonance microscopy to study early human development.

Travis, J. "Why Do Women Menstruate?" *Science News,* April 12, 1997. Hypotheses and debate.

Wolpert, L. *Principles of Development.* London: Oxford University Press, 1998. An excellent new developmental biology text.

Chapter 28 Nervous Systems

Bear, M. F., Connors, B. W., and M. A Paradiso. *Neuroscience: Exploring the Brain.* Baltimore, MD: Williams & Wilkins. 1996. An accessible text.

Gazzaniga, M.S. "The Split Brain Revisited." *Scientific American,* July 1998. The behavioral and evolutionary implications of our " right brain" and "left brain."

Gillis, A. M. "Why Sleep?" *BioScience,* June 1996. Does sleep restock our brain with energy?

Gould, J. L., and C. G. Gould. *The Animal Mind.* New York: W.H. Freeman/Scientific American Library, 1994. An introduction to the scientific study of cognition in nonhuman animals.

Kempermann, G., and F. H. Gage. "New Nerve Cells for the Adult Brain." *Scientific American,* May 1999. New discoveries that offer hope for future treatments of serious brain disorders.

LeDoux, J. E. "Emotion, Memory and the Brain." *Scientific American,* June 1994. Research on the chemical basis of memory.

Youdim, M. G. H., and P. Riederer. "Understanding Parkinson's Disease." *Scientific American,* January 1997.

Chapter 29 The Senses

Freeman, D. H. "In the Realm of the Chemical." *Discover,* June 1993. On the evolutionary significance of the senses of taste and smell.

Howlett, R. "Flipper's Secret." *New Scientist,* June 28, 1997. A summary of research on sonar use by dolphins.

Konishi, M. "Listening with Two Ears." *Scientific American,* April 1993. How the brain creates aural perceptions.

Lee, J. "Before Your Very Eyes." *New Scientist,* March 15, 1997. Progress toward understanding how the brain creates visual perceptions.

Richardson, S. "The Small Files." *Discover,* August 1995. How odors are sorted.

Chapter 30 How Animals Move

Alexander, R. M. *The Human Machine.* New York: Columbia University Press. 1992. The mechanics of human body movement.

Deyo, R. A. "Low-Back Pain." *Scientific American,* August 1998. The major problems and treatments.

Rome, L. C. "Testing a Muscle's Design." *American Scientist,* July-August, 1997. The molecular basis of muscle action.

Suter, R. B. "Walking on Water." *American Scientist,* March-April 1999. A beautifully illustrated account of how spiders move on water surfaces.

Thomas, A. L., R. "On the Tails of Birds." *BioScience,* April 1997. The variety and aerodynamic functions of bird tails.

Triantafyllou, M. S., and G. S. Triantafyllou. "An Efficient Swimming Machine." *Scientific American,* March 1995. The mechanics of how fish swim.

UNIT VI PLANTS: FORM AND FUNCTION

Moore, R., W. D. Clark, and D.S. Vodopich. *Botany,* 2d ed. Dubuque, IA: W. C. Brown, 1998. An introductory text.

Chapter 31 Plant Structure, Reproduction, and Development

Gillis, A. M. "Using a Mousy, Little Flower to Understand Flamboyant Ones." *BioScience,* May 1995. A plant species as a research organism.

Mestel, R. "Not Doing It, Plant-Style." *Discover,* January, 1995. What prevents certain plants from fertilizing themselves.

Pearce, F. "Lure of the Rings." *New Scientist,* December 14, 1996. How the growth rings of trees may help us understand global climate change.

Perkins, S. "Transgenic Plants Provoke Petition." *Science News,* September 27, 1997. Should we be concerned about the transfer of engineered genes between species?

Vaughan, D. A., and L. A. Sitch. "Gene Flow from the Jungle to Farmers." *BioScience,* January 1991. Problems and methods of maintaining genetic diversity in crop plants.

Chapter 32 Plant Nutrition and Transport

Canny, M.J. "Transporting Water in Plants." *American Scientist,* March-April 1998. Transport can occur even with air bubbles in the xylem.

Gibbons, B. "Do We Treat Our Soil Like Dirt?" *National Geographic,* September 1984. A pictorial essay of soil and its significance.

Chapter 33 Control Systems in Plants

Evans, M. L., R. Moore, and K.-H. Hasenstein. "How Roots Respond to Gravity." *Scientific American,* December 1986.

Mlot, D. "Fine-tuned Plant Response to Insect Attack." *Science News,* December 20, 1997. How plants recruit one animal to kill another.

Vogel, S. "When Leaves Save the Tree." *Natural History,* September 1993. How leaf responses help prevent wind damage to trees.

UNIT VII ECOLOGY

Nybakken, J.W. *Marine Biology, An Ecological Approach.* Menlo Park, CA: Benjamin/Cummings. 1997. An introduction to marine ecosystems.

Ricklefs, R. E. *The Economy of Nature,* 3rd ed. New York: W. H. Freeman, 1993. A classic basic text.

Smith, R. L., and T. M. Smith. *Elements of Ecology.* 4th ed. Menlo Park, CA: Benjamin/Cummings. 1998. A brief introductory text.

Chapter 34 The Biosphere: An Introduction to Earth's Diverse Environments

"The Ecology of Large Rivers." *BioScience,* March 1995. A special issue.

Kleiner, K. "Fanning the Wildfires." *New Scientist,* October 19, 1996. The biological and political aspects of fire in ecosystems.

Nicol, S., and I. Allison. "The Frozen Skin of the Southern Ocean." *American Scientist,* September-October 1997. The biotic and abiotic factors of Antarctica.

Stevens, J. E. "The Antarctic Pack-Ice Ecosystem." *BioScience,* March 1995.

"Unknown Oceans." *New Scientist,* supplement, November 2, 1996. Discussion of some major research questions.

Chapter 35 Population Dynamics

Brown, J. L., and E. Pollitt. "Malnutrition, Poverty, and Intellectual Development." *Scientific American,* February 1996. Malnutrition's global effects on children.

Cohen, J. "Ten Myths of Population." *Discover,* April 1996. An analysis of human population growth.

Kates, R. W. "Sustaining Life on the Earth." *Scientific American,* October 1994. The cultural context of environmental problems.

Olshansky, S. J., B. A. Carnes, and D. Grahn. "Confronting the Boundaries of Human Longevity." *American Scientist,* January-February 1998. The evolution of the human life history.

Chapter 36 Communities and Ecosystems

American Institute of Biological Sciences. "Special Section on Symbiosis." *BioScience,* April 1998. Seven articles on symbiotic associations.

Asner, G. P., T. R. Seastedt, and A. R. Townsend. "The Decoupling of Terrestrial Carbon and Nitrogen Cycles." *BioScience,* April 1997. Human alterations of chemical cycles.

Power, M. E., et al. "Challenges in the Quest for Keystones." *BioScience,* September 1996. The importance and difficulty of identifying keystone species in communities.

Reice, S. R. "Nonequilibrium Determinants of Biological Community Structure." *American Scientist,* September-October 1994. The key role of disturbance in ecosystems.

Smil, V. "Global Population and the Nitrogen Cycle." *Scientific American,* July 1997. Nitrogen fertilizer's role in feeding the human population and potential dangers of its overuse.

Wills, C. "Safety in Diversity." *New Scientist,* March 23, 1996. The key role of parasites in maintaining species richness in communities.

Chapter 37 Behavioral Adaptations to the Environment

Ben-Ari, E. T. "Pheromones: What's in a Name?" *BioScience,* July 1998. External chemical signals and evidence for their roles in human behavior.

Buss, D. M. "The Strategies of Human Mating." *American Scientist,* May-June 1994. A behavioral ecology perspective.

Greenspan, R. J. "Understanding the Genetic Construction of Behavior." *Scientific American,* April 1995.

Queller, D. C., and J. E. Strassmann. "Kin Selection and Social Insects." *BioScience,* March 1998. Illustrates how kin selection explains the evolution of insect societies.

Schoech, S. J., "Physiology of Helping Florida Scrub Jays." *American Scientist,* January-February 1998. Hormonal mechanisms underlying cooperative breeding.

Travis, J. "Timely Surprises." *Science News,* July 11, 1998. A brief summary of research on biological clocks.

Chapter 38 Conservation Biology

American Institute of Biological Science. "Special Section on Belowground Biodiversity." *BioScience,* February 1999. A look at species diversity in subsurface habitats in aquatic and terrestrial environments.

Ashley, M. V. "Molecular Conservation Genetics." *American Scientist,* January-February 1999. Application of molecular biology to problems of species conservation.

Blaustein, A. R., and D. B. Wake. "The Puzzle of Declining Amphibian Populations." *Scientific American,* April 1995. What are the causes of current declines?

Chadwick, D. H. "Blue Refuges, U.S. National Marine Sanctuaries." *National Geographic,* March 1998. An illustrated essay on coastal preserves.

Chapin, F. S. III, et al. "Ecosystem Consequences of Changing Biodiversity." *BioScience,* January 1998. The biodiversity crisis and future research needs.

Doyle, R. "Plants at Risk in the U. S." *Scientific American,* August 1997. A state-by-state tally of loss of plant diversity.

Karl, T. R., N. Nicholls, and J. Gregory. "The Coming Climate." *Scientific American,* May 1997. An intergovernmental panel discusses possible scenarios.

Lubchenco, J. "Entering the Century of the Environment: A New Social Contract for Science." *Science,* January 23, 1998. The challenge for scientists working toward ecosystem sustainability.

Meffe, G. K., and C. R. Carroll. *Principles of Conservation Biology,* 2e. Sunderland, MA: Sinauer Associates. 1997. An authoritative text.

Raloff, J. "Those Old Dioxin Blues." *Science News,* May 17, 1997. An evaluation of toxins as the cause of fish kills in the Great Lakes.

Suplee, C. "Unlocking the Climate Puzzle." *National Geographic,* May 1998. Probable causes and potential consequences of global warming.

Weeks, W. W. *Beyond the Ark: Tools for an Ecosystem Approach to Conservation.* Washington, DC: Island Press. 1997. A brief introduction to modern concepts in conservation biology.

Wilson, E. O. *The Diversity of Life.* Cambridge, MA: Harvard University Press, 1992. Eloquent essays on the importance of biodiversity.

Chapter Review Answers

Chapter 1 Answers

Testing Your Knowledge

Multiple Choice: **1.** d **2.** c **3.** e **4.** d **5.** c **6.** b **7.** d **8.** b **9.** b **10.** d

Describing, Comparing, and Explaining

1. Natural selection screens heritable variations by favoring the reproductive success of some individuals over others. Favored individuals pass more genes to the next generation than individuals that are not favored. As a result, the genetic makeup of a population changes. The change results from a screening (editing) of individuals (consequently their genes), not from the creation of new genes or new individuals.
2. Darwin hypothesized that natural selection operates in populations. A population is a group of interbreeding individuals with varied traits that are inherited. When natural selection favors the reproductive success of certain individuals in a population more than others, it changes the proportions of heritable variations in the population.
3. In pursuit of answers to questions about nature, a scientist uses a logical thought process involving the key elements: observations about natural phenomena, questions derived from observations, hypotheses posed as tentative answers to questions, logical predictions of the outcome of tests if the hypotheses are true, and actual tests of hypotheses. Scientific research is not a rigid method because a scientist must adapt these key components to the set of conditions particular to each study. For example, the kinds of tests that can be performed may be different—more precisely controlled—if a project is conducted in a laboratory setting than if it is conducted in a forest or other outdoor environment.
4. Science deals only with hypotheses that are testable by the scientific process and that are concerned with the natural world. By contrast, religion often seeks answers to hypotheses that cannot be tested by the scientific process. Religious answers may invoke supernatural causes.
5. Technology is the application of scientific knowledge. For example, the use of solar power to run a calculator or heat a home are applications of our knowledge, derived by the scientific process, of the nature of light as a type of energy and how light energy can be converted to other forms of energy. Another example is the use of pieces of DNA removed from bacteria to insert new genes into crop plants. This process, often called genetic engineering, stems from decades of scientific research on the structure and function of DNA from many kinds of organisms.

Thinking as a Scientist

1. A place to start would be your local library, where you would find a field guide on mammals of North America (or of your local area). The guide would indicate the general size and shape of flying squirrel tracks and feces. You could then look for feces and other signs such as partly chewed nuts in the woods, especially around trees that have holes leading into root cavities. (Other species of squirrels leave similar signs.) If you locate a site that you think the squirrels frequent, you could use a method that field researchers often use to detect the presence of nocturnal animals. Set out trays with sheets of paper or wood coated with lampblack. Bait (e.g., peanut butter or corn for flying squirrels) is sometimes placed on or near the trays. If a flying squirrel lands on the lampblack, it will leave identifiable tracks. Another possibility would be to go out just after a snowstorm and look for glide marks the squirrels make when they land in the snow. The field guide would probably show what these look like and indicate how to tell them from similar signs left by the wings of hawks and owls.
2. The headline's conclusion is not supported by the information in the article. Although there were about half as many suicides among caffeinated-coffee-drinking nurses, the study did not rule out the effects of other factors that could affect suicide rates, such as antidepressant drugs and a physician's advice. Also, the amount of caffeine in a cup of coffee may vary; did the researcher take this into account (i.e., determine if the amount of caffeine intake was significantly different between the two groups of nurses)? A headline such as "Preliminary Results May Point to Beneficial Effects of Caffeine" would be more appropriate. The researcher used observations (results of a previous study), posed a question, a hypothesis (2–3 cups of coffee per day reduces the rate of suicide among nurses), a prediction of a test outcome, and a test (comparing the number of suicides in two groups of nurses). However, the study was flawed, and the scientific process compromised, by a failure to consider all possible factors influencing suicides.

Science, Technology, and Society

1. *Some issues and questions to consider:* Have you had other science courses? How were they conducted? Have you had prior exposure to the thought process scientists use to make discoveries? Have your experiences with science been mostly learning about information scientists have discovered or about how they have discovered it? What about your own, perhaps subconscious, use of the scientific thought process? Is there any relationship between scientific thinking and what you do when you push a power button on a washing machine or TV set and nothing happens?
2. Virtually any news report or magazine contains stories that are mainly about biology or at least have biological connections. How about biological connections in advertisements?

Chapter 2 Answers

Testing Your Knowledge

Multiple Choice: **1.** b **2.** c **3.** c **4.** e **5.** a (It needs to share 2 more electrons for a full outer shell of 8.) **6.** c **7.** d

True/False: **1.** F (Only salt and water are compounds.) **2.** T **3.** T **4.** T **5.** F (Molecules are farther apart.) **6.** F (The smallest particle is an atom.) **7.** F (Pure water has a pH of 7.) **8.** F (Most acid precipitation results from burning fossil fuels.) **9.** T **10.** F (Reactants are the starting materials.)

Describing, Comparing, and Explaining

1. For diagram, see Figure 2.10A. Water molecules form hydrogen bonds because they are polar. The unique properties of water that result from hydrogen bonding are cohesion, surface tension, the ability to absorb and store large amounts of heat, a high boiling point, a solid form (ice) that is less dense than liquid water, and solvent properties.
2. First: Because increasing the temperature of water (the average speed of its molecules) requires breaking hydrogen bonds, a process that uses heat, a large amount of heat can be added to water before the water's temperature starts to rise. Conversely, when the surrounding temperature falls, new hydrogen bonds form in water, with the release of heat that slows the cooling process. Second: When the body becomes overheated, water evaporating from its surface decreases the body's temperature (evaporative cooling), because the hotter water molecules leave.
3. A covalent bond forms when atoms complete their outer shells by sharing electrons. Atoms can also complete their outer shells by gaining or losing electrons. This leaves the atoms as ions, with – and + charges, and the oppositely charged ions are attracted to each other, forming an ionic bond.
4. An acid is a compound that donates H^+ ions to a solution. A base is a compound that accepts H^+ ions and removes them from solution. Acidity is described by the pH scale, which measures H^+ concentration on a scale of 0 (most acidic) to 14 (most basic).

Thinking as a Scientist

1. Fluorine needs 1 electron for a full outer shell of 8, and if potassium loses 1 electron, its outer shell will have 8. Potassium will lose an electron (becoming a + ion) and fluorine will pick it up (becoming a – ion). The ions will form an ionic bond.
2. Give a mouse sugar or oxygen gas containing a radioactive isotope of oxygen, then see whether the carbon dioxide it exhales is radioactive.

3. *Some issues and questions to consider:* The situation is complex and the residents' case seems to be weak, at best. Assuming they are downwind from both the city and the power plant, their air might be more polluted and their precipitation more acidic than air sampled at either collection site; however, there seems to be no evidence backing their claim. Evidence from sampling stations suggests that both the city and the power plant are adding pollutants to the air, but there is no indication that the added pollution is causing acid precipitation. If samples collected at a monitoring station in the residents' neighborhood showed an increase in acid precipitation, the residents' case might still be weak (they could not tell if the power plant was the main source of pollutants if their air comes from both the city and the power plant). Also, the courts might not view the residents' air pollution problem as serious unless the pollutant levels in their air exceeded national standards. Where would you place sampling stations to get the best idea of the source of air pollutants? Would the residents' case be strengthened if they discovered that the power-plant monitoring station was *not* downwind from the city or that for three of the five years, the wind actually blew in the other direction most days and nights? A scientific consultant might intitially suggest a study to determine if other air pollutants, infectious diseases, ground-water pollution, or other factors are responsible for the tree deaths. Would eliminating these factors as causative agents strengthen the acid precipitation case?

Science, Technology, and Society

1. *Some issues and questions to consider:* Which is less expensive, power from nuclear power plants or fossil-fuel plants? Does the price of electricity reflect its actual cost, including environmental costs? Which would be more harmful, the environmental effects of acid rain and global warming from fossil-fuel power plants, or the effects of nuclear wastes or potential nuclear accidents? Do you favor development of nuclear energy or fossil-fuel power plants? Which would you prefer to have near you? Do your answers to these last two questions differ? If so, why?
2. *Some issues and questions to consider:* Is it the *kinds* of atoms present in chemical wastes that is important, or the way the atoms are combined to form particular substances? How important is the way the chemical wastes are disposed of? Do chemicals produced by human technology differ from naturally occurring substances? If so, how?

Chapter 3 Answers

Testing Your Knowledge

Multiple Choice: **1.** d (The second kind of molecule is a polymer of the first.) **2.** c **3.** d **4.** e **5.** a **6.** b **7.** d

Describing, Comparing, and Explaining

1. Triglycerides—store energy. Phospholipids—are major components of membranes. Waxes—make up waterproof coatings. Steroids—one kind, cholesterol, is a component of cell membranes; other kinds function as hormones.
2. Weak bonds that stabilize the three-dimensional structure of a protein are disrupted, and the protein unfolds. Function depends on shape, so if the protein is the wrong shape, it won't function properly.
3. Proteins are made of 20 amino acids arranged in many different sequences into chains of many different lengths. Genes, defined stretches of DNA, dictate the amino acid sequences of proteins in the cell.
4. Proteins function in structure, contraction, storage, defense, transport, signaling, and catalyzing chemical reactions (see Module 3.11).
5. This is a hydrolysis reaction, which consumes water. It is essentially the reverse of the diagram in Figure 3.5.

Thinking as a Scientist

1. *Some issues and questions to consider:* How will you choose your test subjects? How many subjects should you have? Will you give them all vitamin C, or just some of them? What criteria will you use to divide the test subjects into groups? What is a control group? Should the subjects know whether they are getting vitamin C or not? Should the experimenters who are giving out the drug and measuring the severity of cold symptoms know which of the subjects are getting vitamin C? What is a "double blind" study? If there is a difference between your groups, how can you be sure it is due to vitamin C?
2. 20 choices for first amino acid, 20 choices for second. 20×20 or $20^2 = 400$ possibilities for 2 amino acids. $20 \times 20 \times 20 \times ... = 20^{129}$ possibilities for a protein 129 amino acids long.
3. A long hydrocarbon chain ending in a carboxyl group is a fatty acid. Fatty acids and glycerol are the molecules that result when fats (triglycerides) are hydrolyzed. The evidence thus suggests that the cake mix does contain fat. To prove this, however, intact fat molecules must be isolated from the mix.
4. b; this sequence is the only one that would give the fragments listed when cut as described with the three enzymes. Answer a is excluded because it lacks some of the fragment sequences (for example, CCC); similarly, answer d lacks fragment LSLSCSL. Answer c would not give the correct fragments when treated with the enzyme that cuts between C and S.

Science, Technology, and Society

1. *Some issues and questions to consider:* How are these chemicals important in agriculture, medicine, and public health? How have they affected humans? Wildlife? Natural vegetation? Are the chemicals themselves harmful, or the way that they are used? What influences have shaped your opinions: The media? Personal experience? Reading? Friends and family? How might the opinion of a villager in a developing country differ from yours?
2. *Some issues and questions to consider:* Is it the responsibility of a corporation to protect the livelihoods of farmers and exporters from whom they buy a product? Or should these individuals look out for themselves? How might producers protect themselves from economic shocks? What are the roles of the governments of developed and developing countries in this kind of trade? How can consumers make their preferences known? Do the concerns of consumers affect manufacturers?

Chapter 4 Answers

Testing Your Knowledge

Multiple Choice: **1.** e **2.** c (Small cells have a greater ratio of surface area to volume.) **3.** b **4.** d **5.** c

Describing, Comparing, and Explaining

1. Tight junctions form leakproof bonds. Anchoring junctions link cells to each other but allow materials to pass between them; they can also connect a cell to an adjacent extracellular matrix. Communicating junctions are channels that allow flow from cell to cell.
2. Both process energy. A chloroplast converts light energy to chemical energy (sugar molecules). A mitochondrion converts chemical energy (food molecules) to another form of chemical energy (ATP).
3. Different conditions and conflicting processes can occur simultaneously within separate, membrane-enclosed compartments. Also, there is increased area for membrane-bound enzymes that carry out metabolic processes.

Thinking as a Scientist

Cell 1: $S = 1256\ \mu m^2$; $V = 4187\ \mu m^3$; $S/V = 0.3$. Cell 2: $S = 5024\ \mu m^2$; $V = 33{,}493\ \mu m^3$; $S/V = 0.15$. The smaller cell has a larger surface area relative to volume for absorbing food and oxygen and excreting waste. Small cells thus perform these activities more efficiently.

Science, Technology, and Society

Some issues and questions to consider: Were the cells Moore's property, a gift, or just surplus? Was Moore asked to donate the cells? Was he informed about how the cells might be used? Is it important to ask permission or inform the patient in such a case? How much did the researchers modify the cells? What did they have to do to them to sell the product? Do the researchers and the university have a right to make money from Moore's cells? Is the fact that they saved Moore's life a factor here? Does Moore have the right to sell his cells? Would Moore have been able to sell the cells without the researchers' help?

Chapter 5 Answers

Testing Your Knowledge

Multiple Choice: **1.** d **2.** b **3.** a **4.** c (Only active transport can move solute against a concentration gradient.)

Describing, Comparing, and Explaining

Heating, pickling, and salting denature enzymes, changing their shapes so they do not fit substrates. Freezing decreases the kinetic energy of molecules, so they lack energy of activation, even in the presence of enzymes.

Thinking as a Scientist

(a) The more enzyme present, the faster the rate of reaction, because it is more likely that enzyme and substrate molecules will meet. (b) The more substrate, the faster the reaction, for the same reason, but only up to a point. An enzyme molecule can work only so fast; once it is saturated (working at top speed), more substrate does not increase the rate.

Science, Technology, and Society

Some issues and questions to consider: Does a woman have a right to work in an unsafe environment, even if it may put her child at risk? What are the rights of the child? Does the company have the right to "protect" women who are not pregnant? Is the company trying to protect the mother and child, or protect itself from a potential lawsuit? Who is responsible for protecting employees and their children? The employees? The company? The government? Suppose a woman risks lead exposure and bears a retarded child? Who is responsible if, 20 years from now, the child decides to sue?

Chapter 6 Answers

Testing Your Knowledge

Multiple Choice: **1.** c **2.** c (NAD^+ and FAD, which are recycled by electron transport, are limited.) **3.** e **4.** d **5.** c **6.** b

Describing, Comparing, and Explaining

1. Disadvantage: Less ATP is produced (only 2 per glucose molecule versus 36 aerobically). Advantage: No oxygen (O_2) is needed.
2. Glycolysis is considered the most ancient because it occurs in all living cells.
3. Oxygen picks up electrons and hydrogen atoms produced by the oxidation of glucose, at the end of the electron transport chain. Carbon dioxide results from the breakdown of glucose molecules in glycolysis and the Krebs cycle.

Thinking as a Scientist

1. 100 kcal per day is 700 kcal per week. On the basis of the table in Module 6.3, walking 4 mi/hr would require 700/231 = about 3 hr; swimming, 1.3 hr; running, 0.8 hr.
2. The amino acids that make up proteins have amino groups containing N atoms. Amino groups must be added to fats or carbohydrates in order to make amino acids for making proteins.
3. 10 NAD^+ and 2 FAD are needed to pick up the electrons and hydrogen atoms from a glucose molecule. NAD^+ and FAD are recycled between electron transport and glycolysis and the Krebs cycle. We need a small additional supply to replace those that are lost or damaged.
4. **a.** No, this shows the blue color getting more intense. The reaction *decolorizes* the blue dye.
   b. No, this shows the dye being decolorized, but it also shows the three mixtures with different initial color intensities. The intensities should have started out the same since all mixtures used the same concentration of dye.
   c. Correct. The mixtures all start out the same, and then the ones with more succinic acid (reactant) decolorize faster. The mixture with the highest concentration of succinic acid decolorizes the fastest. (If we included mixtures with even greater concentrations of succinic acid, we would predict that eventually the mitochondria would be working as fast as possible, and no additional increase in the rate of decolorization would result from further additions of succinic acid.)

Science, Technology, and Society

Some issues and questions to consider: Is your customer aware of the danger? Do you have an obligation to protect the customer, even against her wishes? Does your employer have the right to dismiss you for informing the customer? For refusing to serve the customer? Could you or the restaurant later be held liable for injury to the fetus? Or is the mother responsible for willfully disregarding warnings about drinking?

Chapter 7 Answers

Testing Your Knowledge

Multiple Choice: **1.** d **2.** c **3.** c **4.** b **5.** a **6.** e **7.** c (NADPH and ATP from light-dependent reactions are used by the Calvin cycle.)

Describing, Comparing, and Explaining

1. See Figure 7.11.
2. Young trees take up CO_2 faster than old trees per unit of weight. However, when large old trees are harvested and replaced with seedlings, the total weight of living trees is greatly decreased. Furthermore, harvesting the old trees releases large amounts of CO_2 by the burning and decomposition of wastes or products made from the wood. The overall effect of harvesting and replanting is thus an increase in CO_2 released and a worsening of the greenhouse effect.
3. In photosynthesis, electrons are from chlorophyll; in cellular respiration, electrons are from organic molecules. In photosynthesis, electron energy is from light; in cellular respiration, electron energy is stored in chemical bonds of organic molecules. In photosynthesis, electrons are finally picked up by $NADP^+$; in cellular respiration, electrons are finally picked up by O_2. In both processes, energy is used to transport H^+ through a membrane. As H^+ flows back, the energy is used to make ATP.
4. Plants can break down the sugar for energy in cellular respiration or use the sugar as a raw material for making other organic molecules. Excess sugar is stored as starch.

Thinking as a Scientist

1. The oxygen atoms in glucose come from CO_2. You could give some plants CO_2 containing the isotope ^{18}O and H_2O containing ordinary oxygen (^{16}O) and see whether the ^{18}O ended up in glucose or in O_2 gas. You could then use H_2O labeled with ^{18}O, and unlabeled CO_2. Only ^{18}O from labeled CO_2 would end up in glucose.
2. **a.** No. If this were true, there would be no bubbles from the plant at distances greater than 45 cm.
   b. No. The *Elodea* does increase its rate of photosynthesis at close sunlamp distances. The problem is the high rate of photosynthesis at greater distances.
   c. No. This may be true, but the problem is that the photosynthesis at the low light intensities is too high, not that the photosynthesis at the high light intensities is too low.
   d. Correct. The rate of photosynthesis seen at 45–75 cm seems to be a "background" rate, and the increase in photosynthesis as the sunlamp is moved very close is added to this background rate. One would predict that even if the sunlamp were turned off, the rate of photosynthesis seen at 45–75 cm would persist.
3. When a rain forest tree is eaten by an animal, or when the tree or its parts die and decompose, the organic molecules contained in the tree are consumed in cellular respiration in the cells of the animal or decomposer. The same amount of oxygen produced by the tree when the organic molecules were made is consumed in cellular respiration when they are oxidized. The rain forest could produce a surplus of oxygen if the forest were actually growing in size, storing up more energy in photosynthesis than was consumed in cellular respiration.

Science, Technology, and Society

Some issues and questions to consider: What are the risks that we take and costs we must pay if greenhouse warming continues? How certain do we have to be that warming is caused by human activities before we act? Is it possible that greenhouse warming may not be a threat, that it may actually be beneficial? What can we do to reduce CO_2 emissions? What are the risks and costs if we reduce CO_2 emissions and greenhouse warming turns out to not be a real threat? Is it possible that the costs and sacrifices of reducing CO_2 emissions might actually improve our lifestyle?

Chapter 8 Answers

Testing Your Knowledge

Multiple Choice: **1.** c **2.** a **3.** b **4.** e **5.** e (A diploid cell would have an even number of chromosomes; the odd number suggests meiosis I has been completed. Sister chromatids are together only in prophase and metaphase of meiosis II.) **6.** c **7.** b **8.** c **9.** b **10.** d

Describing, Comparing, and Explaining

1. Various orientations of chromosomes at metaphase I of meiosis lead to different combinations of chromosomes in gametes. Crossing over during prophase I results in an exchange of chromosome segments and new combinations of genes. Random fertilization of eggs by sperm further increases possibilities for variation in offspring.
2. Mitosis is a single division of a cell nucleus that produces two genetic copies of the parent nucleus and thus, when accompanied by cytokinesis, two daughter cells that are genetic copies of the parent cell. Meiosis consists of two consecutive divisions that reduce a diploid parent cell to four haploid cells. In animals and plants, mitosis occurs in most body tissues, meiosis only in the gamete-producing organs (testes and ovaries, in animals). The two sister chromatids of individual chromosomes separate in anaphase of mitosis. In meiosis, homologous chromosomes separate in anaphase I; the sister chromatids of each chromosome don't separate until anaphase II.
3. See the photos in Module 8.7. Interphase: Growth; metabolic activity; DNA synthesis. Prophase: Chromosomes shorten and thicken; mitotic spindle forms. Metaphase: Chromosomes line up on a plane going through the cell's equator. Anaphase: Sister chromatids separate and move to the poles of the cell. Telophase: Daughter nuclei form around chromosomes; cytokinesis occurs.
4. In culture, normal cells usually divide only when they are in contact with a surface but not touching other cells on all sides (the cells usually grow to form only a single layer). The density-dependent inhibition of cell division apparently results from local depletion of substances called growth factors. Growth factors are proteins secreted by certain cells that stimulate other cells to divide; they act via signal-transduction pathways to signal the cell-cycle control system of the affected cell to proceed past its checkpoints. The cell-cycle control systems of cancer cells do not function properly. Cancer cells generally do not require externally supplied growth factors to complete the cell cycle, and they divide indefinitely (in contrast to nomal mammalian cells, which stop dividing after 20–50 generations)—two reasons why they are relatively easy to grow in the lab. Furthermore, cancer cells can often grow without contacting a solid surface, making it possible to culture them in suspension in liquid medium.
5. A ring of microfilaments pinches an animal cell in two, a process called cleavage. In a plant cell, membranous vesicles form a disc called the cell plate at the midline of the parent cell, cell plate membranes fuse with the plasma membrane, and a cell wall grows in the space, separating the daughter cells.
6. See Figures 8.21A and 8.21B.

Thinking as a Scientist

1. **a.** No. In order for this to happen, the chromosomes of the two gametes that fused would have to represent, together, a complete set of the donor's maternal chromosomes (the ones that orignally came from the donor's mother) and a complete set of the donor's paternal chromosomes (from the donor's father). It is much more likely that the zygote would be missing one or more maternal chromosomes and would have an excess of paternal chromosomes, or vice versa.

   b. Correct. Consider what would have to happen to produce a zygote genetically identical to the gamete donor: The zygote would have to have a complete set of the donor's maternal chromosomes and a complete set of the donor's paternal chromosomes The first gamete in this union could contain any mixture of maternal and paternal chromosomes, but once that first gamete was "chosen," the second one would have to have one particular combination of chromosomes—the combination that supplies whatever the first gamete did not supply. So, for example, if the first three chromosomes of the first gamete were maternal, maternal, and paternal, the first three of the second gamete would have to be paternal, paternal, and maternal. The chance that all 23 chromosome pairs would be complementary in this way is only one in 2^{23} (that is, one in 8,388,608). Because of independent assortment, it is much more likely that the zygote would have an unpredictable combination of chromosomes from the donor's father and mother.

   c. No. First, the zygote *could* be genetically identical to the gamete donor (see b). Second, the zygote could not be identical to either of the gamete donor's parents because the donor only has half the genetic material of each of his or her parents. For example, even if the zygote were formed by two gametes containing only paternal chromosomes, the combined set of chromosomes could not be identical to that of the donor's father because it would still be missing half of the father's chromosomes.

   d. No; see answer c.
2. Some possible hypotheses: The replication of the DNA of the bacterial chromosome takes less time than the replication of the DNA in a eukaryotic cell. The time required for a growing bacterium to roughly double its cytoplasm is much less than for a eukaryotic cell. Bacteria have a cell-cycle control system much simpler than that of eukaryotes.
3. 1 cm^3 = 1000 mm^3, so 5000 mm^3 of blood contain $5000 \times 1000 \times 5{,}000{,}000 = 25{,}000{,}000{,}000{,}000$ or 2.5×10^{13} red blood cells. The $\frac{1}{120}$ of the cells that are replaced each day = $2.5 \times 10^{13}/120 = 2.1 \times 10^{11}$ cells. There are $24 \times 60 \times 60 = 86{,}400$ seconds in a day. Therefore, the number of cells replaced each second = $2.1 \times 10^{11}/86{,}400$ = about 2×10^6, or 2 million. Thus, about 2 million cell divisions must occur each second to replace red blood cells that are lost.
4. Each chromosome is on its own in mitosis; chromosome replication and the separation of sister chromatids occur independently for each horse or donkey chromosome. Therefore, mitotic divisions, starting with the zygote, are not impaired. In meiosis, however, homologous chromosomes must pair in prophase I. This process of synapsis cannot occur properly because horse and donkey chromosomes do not match in number or content.
5. To control for the age of the father, compare women of various ages who are all married to men of the same age. To see if the age of the father increases the occurrence of Down syndrome, look at men of various ages whose wives are all the same age.

Science, Technology, and Society

Some issues and questions to consider: Could it be that less money is spent on prevention because effective prevention is so much cheaper? Or because prevention has been tried, and it does not work well? Are lifestyle changes the kind of measures that could benefit from a shift in resources? Is prevention an individual matter of avoiding exposure, or a social matter of preventing exposure? How might the answer to this question shape prevention policy? If more money were devoted to prevention, how could it be used to encourage you or others to make lifestyle changes? Would prevention work better for younger or older people? Might older people, aready exposed to cancer-causing agents, actually be harmed by a shift of resources to prevention?

Chapter 9 Answers

Testing Your Knowledge

Multiple Choice: **1.** c **2.** d **3.** d (Neither parent is ruby-eyed, but some offspring are, so it is recessive. Different ratios among male and female offspring show it is sex-linked.) **4.** e **5.** a **6.** e

Genetics Problems

1. Cross two peas heterozygous for pod color and flower color; both have green pods and purple flowers (genotype *GgPp*). Offspring are produced in the ratio of 9 green purple : 3 green white : 3 yellow purple : 1 yellow white. The ratio from the dihybrid cross is the combined 3:1 ratio of two independent monohybrid crosses. Alleles of the parents do not "stick together" when passed to offspring.
2. Genes on the single *X* chromosome in males are always expressed because there are no corresponding genes on the *Y* chromosome to mask them. A male needs only one recessive color-blindness allele (from his mother) to show the trait; a girl must inherit the allele from both parents, which is less likely.
3. See Figure 9.16A. The parental gametes are *WS* and *ws*. Recombinant gametes are *Ws* and *wS*, produced by crossing over.
4. Height appears to be a quantitative trait, resulting from polygenic inheritance, like human skin color. See Module 9.13.

5. The brown allele appears to be dominant, the white allele recessive. The brown parent appears to be homozygous dominant, *BB*, and the white mouse is homozygous recessive, *bb*. The F_1 mice are all heterozygous, *Bb*. If two of the F_1 mice are mated, $\frac{3}{4}$ of the F_2.
6. The best way to find out whether an F_2 mouse is homozygous dominant or heterozygous is to do a testcross: Mate the brown mouse with a white mouse. If the brown mouse is homozygous, all the offspring will be brown. If the brown mouse is heterozygous, you would expect half the offspring to be brown and half to be white.
7. Freckles is dominant, so Tim and Jan must both be heterozygous. There is a $\frac{3}{4}$ chance that they will produce a child with freckles, a $\frac{1}{4}$ chance they will produce a child without freckles. The probability that the next two children will have freckles is $\frac{3}{4} \times \frac{3}{4} = \frac{9}{16}$.
8. As in problem 7, both Tim and Jan are heterozygous, and Michael is homozygous recessive. The probability of the next child having freckles is $\frac{3}{4}$. The probability of the next child having a straight hairline is $\frac{1}{4}$. The probability that the next child will have freckles and a straight hairline is $\frac{3}{4} \times \frac{1}{4} = \frac{3}{16}$.
9. The genotype of the black short-haired parent rabbit is BBSS. The genotype of the brown long-haired parent rabbit is *bbss*. The F_1 rabbits will all be black and short-haired, BbSs. The F_2 rabbits will be $\frac{9}{16}$ black short-haired, $\frac{3}{16}$ black long-haired, $\frac{3}{16}$ brown short-haired, and $\frac{1}{16}$ brown long-haired.
10. Half their children will be heterozygous and have elevated cholesterol levels. There is a $\frac{1}{4}$ chance that their next child will be homozygous, *hh*, and have an extremely high cholesterol level, like Zoe.
11. If the genes are not linked, the proportions among the offspring will be 25% gray red, 25% gray purple, 25% black red, 25% black purple. The actual percentages show that the genes are linked. The recombination frequency is 6%.
12. The recombination frequencies are: black dumpy 36%, purple dumpy 41%, and black purple, 6% (see question 11). Since these recombination frequencies reflect distances between the genes, the sequence must be purple-black-dumpy (or dumpy-black-purple).
13. The bristle-shape alleles are sex-linked, carried on the *X* chromosome. Normal bristles is dominant (*F*) and forked is recessive (*f*). The genotype of the female parent is X^fX^f. The genotype of the male parent is X^FY. Their female offspring are X^FX^f, their male offspring X^fY.
14. $\frac{1}{4}$ will be boys suffering from hemophilia. $\frac{1}{4}$ will be female carriers. (The mother is a heterozygous carrier, and the father is normal.)
15. In order for a woman to be color-blind, she must inherit *X* chromosomes bearing the color blindness allele from both parents. Her father has only one *X* chromosome, which he passes on to all his daughters, so he must be color-blind. A male only needs to inherit the color blindness allele from a carrier mother; both his parents are usually phenotypically normal.

Thinking as a Scientist

Start out by breeding the cat to get a population to work with. If the curl allele is recessive, two curl cats can have only curl kittens. If the allele is dominant, curl cats can have "normal" kittens. If the curl allele is sex-linked, ratios will differ in male and female offspring of some crosses. If the curl allele is autosomal, the same ratios will be seen among males and females. Once you have established that the curl allele is dominant and autosomal, you can determine if a particular curl cat is true-breeding (homozygous) by doing a testcross with a normal cat. If the curl cat is homozygous, all offspring of the testcross will be curl; if heterozygous, half the offspring will be curl and half normal.

Science, Technology, and Society

Some issues and questions to consider: Do biologists actually see the structures and molecules of cells? What about past evolutionary processes, the origin of life, the physical appearence or behavior of dinosaurs? In other fields of science, what is the evidence for atoms, subatomic particles, the formation of stars, the makeup of the Earth's interior, the past positions of the continents? How clear does evidence have to be before it is acceptable? What prompts a scientist to propose an explanation? What if more than one explanation can account for the observations? Is it possible to be absolutely sure that an explanation is correct? That it is incorrect? What is the place of words like "correct," "incorrect," "fact," and "truth" in science? Are some of the facts in this textbook "wrong"?

Chapter 10 Answers

Testing Your Knowledge

Multiple Choice: **1.** e (Only the phage DNA enters a host cell; T4 DNA determines both DNA and protein.) **2.** e **3.** b **4.** c

Describing, Comparing, and Explaining

1. Ingredients: Original DNA, nucleotides, several enzymes and other proteins, including DNA polymerases and DNA ligase. Steps: Original DNA strands separate at a specific site (origin of replication), nucleotides line up along each strand according to base-pairing rules, DNA polymerase links the nucleotides to form new strands. New nucleotides are added only to the 3′ end of a growing strand. One new strand is made in one continuous piece; the other new strand is made in a series of short pieces that are then joined by DNA ligase. Product: Two identical DNA molecules, each with one old strand and one new strand.
2. A gene is the polynucleotide sequence with information for making one polypeptide. Each codon—a triplet of bases in DNA or RNA—codes for one amino acid. Transcription occurs when RNA polymerase produces mRNA using one strand of DNA as a template. A ribosome is the site of translation, or polypeptide synthesis, and tRNA molecules serve as interpreters of the genetic code. Each tRNA molecule has an amino acid attached at one end, and a three-base anticodon at the other end. Beginning at the start codon, mRNA moves relative to the ribosome a codon at a time. A tRNA with a complementary anticodon pairs with each codon, adding its amino acid to the polypeptide chain. The amino acids are linked by peptide bonds. Translation stops at a stop codon, and the finished polypeptide is released. The polypeptide folds to form a functional protein, sometimes in combination with other polypeptides.

Thinking as a Scientist

1.

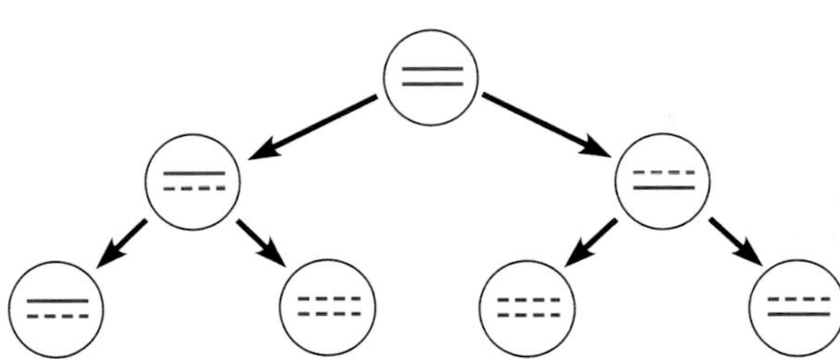

2. mRNA: GAUGCGAUCCGCUAACUGA. Amino acids: Met-Arg-Ser-Ala-Asn.
3. **a.** No. The conservative replication theory would predict that after one cell division, the DNA of half the cells would be entirely radioactive, and the DNA of the other half of the cells would be all new and completely nonradioactive.
b. No. This is the reverse of the correct answer; see answer c.
c. Correct. The conservative replication hypothesis predicts that, after one cell division, half the cells will contain entirely new, nonradioactive DNA, and half the cells will contain entirely old, radioactive DNA. The semiconservative replication hypothesis predicts that each cell will have DNA consisting of one old strand (radioactive) and one new strand (nonradioactive), and therefore that all the DNA molecules in all the cells will contain some radioactivity.
d. No. This experiment could distinguish very well between the hypotheses.

Science, Technology, and Society

Some issues and questions to consider: Is it fair to issue a patent for a gene or gene product that occurs naturally in every human being? Or should a patent be issued only for something new that is invented, rather than found? Suppose another scientist slightly modifies the gene or protein.

How different does the gene or protein have to be in order that the patent is not infringed? Might patents encourage secrecy and interfere with the free flow of scientific information? What are the benefits to the holder of a patent? When research discoveries cannot be patented, what are the scientists' incentives for doing the research? What are the incentives for the institution or company that is providing financial support?

Chapter 11 Answers

Testing Your Knowledge

Multiple Choice: **1.** b (Different genes are active in different kinds of cells.) **2.** c **3.** b **4.** a **5.** b **6.** e

Describing, Comparing, and Explaining

1. The nucleus of a differentiated tadpole intestine cell can shape the development of an entire embryo. Salamander cells can dedifferentiate and regenerate a lost leg. A carrot plant can grow from a single root cell.
2. A mutation in a single gene can influence the actions of many other genes if the mutated gene is a control gene, such as a homeotic gene. A single control gene may encode a protein that affects (activates or represses) the expression of a number of other genes. In addition, some of the affected genes may themselves be control genes that, in turn, affect other batteries of genes. Cascades of gene expression are common in embryonic development.

Thinking as a Scientist

1. **a.** No. The 22K protein does continue to be made for 40 hours, but the 13K protein is last seen at 32 hours. There is no 25K, and the 36K protein is detectable at 32 hours.
   b. No. Both the 22K and the 36K proteins are clearly visible.
   c. Correct. The 22K protein is the series of bands just to the left of the 20K standard, and the 13K protein is the series of bands just to the left of the 10K standard. There are no bands where one would expect to see the 25K protein, and the 36K protein is the three bands about midway between the 30K and 40K standards. Notice that a number of the bands shown are not relevant to the question.
   d. No. The 22K and 13K proteins are visible at 8 hours, there is no 25K protein, and the 36K protein appears only at 32 hours.
2. **a.** If the mutated repressor could still bind to the operator on the DNA, it would continuously repress the operon; enzymes for lactose utilization would not be made whether or not lactose was present.
   b. The *lac* genes would continue to be transcribed and the enzymes made, whether or not lactose was present.
   c. Same predicted result as for b.
   d. RNA polymerase would not be able to transcribe the genes; no proteins would be made whether or not lactose was present.
3. *XXY*

Chapter 12 Answers

Testing Your Knowledge

Multiple Choice: **1.** d **2.** e **3.** c **4.** b **5.** c (Bacteria lack the RNA-splicing machinery needed to delete eukaryotic introns.) **6.** b **7.** d

Describing, Comparing, and Explaining

1. Isolate plasmids from *E. coli.* Cut the plasmids and the DNA containing the GH gene with restriction enzyme to produce molecules with sticky ends. Join the plasmids and the fragments of human DNA with ligase. Allow *E. coli* to take up recombinant plasmids. Bacteria will then replicate plasmids and multiply, producing clones of bacterial cells. Identify a clone carrying and expressing the GH gene. Grow up large amounts of the bacteria and extract and purify GH from the culture. For the purposes of this question, we assume that the GH gene provided does not have introns.
2. Medicine: Genes can be used in human gene therapy, to engineer viruses for vaccines, or to produce transgenic lab animals for AIDS research. Proteins can be hormones, enzymes, blood-clotting factor, etc., or starting material for making vaccines. Agriculture: Foreign genes can be inserted into plant cells or animal eggs to produce transgenic crop plants or farm animals, or inserted into viruses for making animal vaccines. Animal growth hormones are examples of agriculturally useful proteins that can be made using recombinant DNA technology.

Thinking as a Scientist

1. She could start with DNA isolated from liver cells (the entire genome) and carry out the procedure outlined in Module 12.5 to produce a collection of recombinant bacterial clones, each carrying a small piece of liver cell DNA. To find the clone with the desired gene, she could then make a probe of radioactive RNA with a nucleotide sequence complementary to part of the gene: GACCUGACUGU. This probe would bind to the gene, labeling it and identifying the clone that carries it. Alternatively, the biochemist could start with mRNA isolated from liver cells and use it as a template to make DNA (using reverse transcriptase). Cloning this DNA rather than the entire genome would yield a smaller library of genes to be screened—only those active in liver cells. Furthermore, the genes would lack introns, making the desired gene easier to manipulate after isolation.
2. **a.** No. For a marker to be most useful, it must be found in every person with the disorder but not in unaffected ("normal") people. But there is no band that is present for *all* the people with the disorder (whether type 1 or type 2) and *only* for these people.
   b. No. Every band found for type 2 is also found in at least some type 1 cases and in at least some normal people.
   c. Correct. The type 1 marker band is highlighted in the figure below: Notice that it is found in every case of type 1 disease but in no cases of type 2 disease or in any family members without disease.
   d. No. The sample size here is small, and we would feel much better about our conclusions if we saw the same marker in other families with the disease, but these results are definite progress.

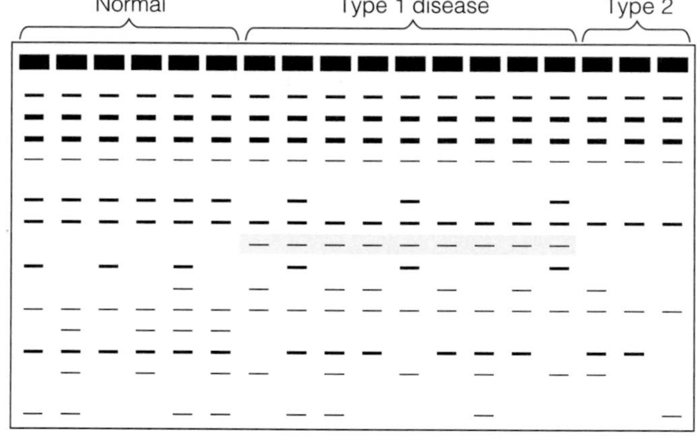

3. Determining the nucleotide sequences is just the first step. Once researchers have written out the DNA "book," they will have to try to figure out what it means—what the nucleotide sequences code for, and how they work.

Science, Technology, and Society

1. *Some issues and questions to consider:* What are some of the unknowns in recombinant DNA experiments? Do we know enough to anticipate and deal with possible unforeseen and negative consequences? Do we want this kind of power over evolution? Who should make these decisions? If scientists doing the research were to make the decisions about guidelines, what factors might shape their judgment? What might shape the judgment of business executives in the decision-making process? Does the public have a right to a voice in the direction of scientific research? Does the public know enough about biology to get involved in this decision-making process? Who represents "the public," anyway?
2. *Some issues and questions to consider:* What kinds of impact will gene therapy have on the individuals who are treated? On society? Who will decide what patients and diseases will be treated? What costs will be involved, and who will pay them? How do we draw the line between treating disorders and "improving" the human species?
3. *Some issues and questions to consider:* Should genetic testing be mandatory or voluntary? Under what circumstances? Why might employers and insurance companies be interested in genetic data? Since genetic characteristics differ among ethnic groups and between the sexes, might it be used to discriminate? Which of these questions do you think is most important? Which issues are likely to be the most serious in the future?

Chapter 13 Answers

Testing Your Knowledge

Multiple Choice: **1.** e **2.** c **3.** b (Erratic rainfall and differential reproductive success would ensure that a mixture of both forms remained in the population.) **4.** b **5.** d

Describing, Comparing, and Explaining

1. Your paragraph should include such evidence as fossils and the fossil record, biogeography, comparative anatomy, comparative embryology, DNA and protein comparisons, artificial selection, and examples of natural selection (peppered moth).
2. If $q^2 = 0.002$, then $q = 0.04$ (approximately). Since $p + q = 1$, $p = 1 - q = 0.96$. The proportion of heterozygotes is $2pq = 2 \times 0.96 \times 0.04 = 0.08$. About 8% of African Americans are carriers.

Thinking as a Scientist

The unstriped snails appear to be better adapted. Striped snails make up 47% of the living population, but 56% of the broken shells. Assuming all the broken shells result from the meals of birds, we would predict that bird predation would reduce the frequency of striped snails, and the frequency of unstriped individuals would increase.

Science, Technology, and Society

Some issues and questions to consider: Who should decide curriculum, "experts" or members of the community? Are the two alternatives both scientific ideas? Who judges what is scientific? If it is more fair to consider alternatives, should the door be open to all alternatives? Are constitutional issues (separation of church and state) involved here? Can a teacher be compelled to teach an idea he or she thinks is wrong? Should a student be required to learn an idea he or she thinks is wrong?

Chapter 14 Answers

Testing Your Knowledge

Multiple Choice: **1.** c **2.** b **3.** d **4.** c

Describing, Comparing, and Explaining

1. Different physical appearance may indicate different species or just differences within a species. Isolated populations may or may not be able to interbreed. Organisms that reproduce only asexually and fossil organisms do not have the potential to interbreed and produce fertile offspring; therefore, the biological species concept does not apply to them.
2. Horses and donkeys are not the same species because neither a mule nor a hinny (the hybrid of a male horse and a female donkey) can interbreed with either a horse or a donkey. This is an example of hybrid sterility, a postzygotic reproductive barrier.

Thinking as a Scientist

A broad hypothesis would be that cultivated American cotton arose from a sequence of hybridization, meiotic failure, and self-fertilization. We can divide this broad statement into at least three testable hypotheses. *Hypothesis 1:* The first step in the origin of cultivated American cotton was hybridization between a wild American cotton plant (with 13 pairs of small chromosomes) and an Old World cotton plant (with 13 pairs of large chromosomes). If this hypothesis is correct, we would predict that the hybrid offspring would have 13 small chromosomes and 13 large chromosomes. *Hypothesis 2:* The second step in the origin of cultivated American cotton was a failure of meiosis in the hybrid offspring. If this hypothesis is true, we would expect the gametes resulting from meiotic failure to each have 13 large chromosomes and 13 small chromosomes. *Hypothesis 3:* The third step in the origin of cultivated American cotton was self-fertilization of the gametes resulting from meiotic failure. If this hypothesis is true, we would expect the outcome of self-fertilization to be a polyploid plant with 52 chromosomes—13 pairs of large ones and 13 pairs of small ones. This is the genetic makeup of cultivated American cotton.

Science, Technology, and Society

Some issues and questions to consider: The rationale behind protecting all endangered groups is the desire to preserve genetic diversity. Each species, subspecies, and hybrid group may represent a unique mix of genes. Studies of the degree of genetic distinctiveness of a subspecies or hybrid group may help decision makers if cost is an issue. If the species as a whole is not at risk, it seems appropriate to determine how distinctive the gene pool of a subspecies or hybrid group is before assigning it a lower priority for saving. A question for society in general is: What is the value of any particular species and its genetically distinct subgroups? And how far are we willing (should we be willing) to go to preserve a genetically distinct group of organisms? How should the costs of preserving genetic diversity compare with the costs of other public projects?

Chapter 15 Answers

Testing Your Knowledge

Multiple Choice: **1.** c **2.** d **3.** e **4.** b **5.** b (The large ground finch and the mangrove finch are on quite separate branches of the tree, indicating that they are more distantly related than any of the other pairs in this question.)

Describing, Comparing, and Explaining

1. The number of differences in amino acid sequence reflects evolutionary relationship. More similarities means that organisms had a more recent common ancestor and are more closely related.
2. An asteroid or comet may have hit Earth, raising dust, blocking sunlight, cooling the climate, and killing plants. Evidence includes a layer of iridium in rocks and a large crater under the Caribbean Sea. Fossils suggest that increased volcanic activity in India may have released dust and cooled the climate. Continental drift may have altered climate and shorelines at the end of the Paleozoic. There is evidence for cooling and a change in sea levels.

Thinking as a Scientist

1. The rock is about 2.6 billion years old. (If the half-life of potassium-40 is 1.3 billion years, there would be 6 grams left after 1.3 billion years and 3 grams left after another 1.3 billion years.)
2. **a.** No, this diagram does not fit the data in the table. As shown in the first horizontal row of data in the table, the melting point of the hybrid DNA is higher for M-A and M-D hybrids than it is for M-B and M-C hybrids. This diagram indicates that M and A are closely related (which the data indicate), but the diagram also indicates that both M and A are closely related to B (apparently not true, because both M-B and A-B hybrid DNAs melted at relatively low temperature). The diagram also indicates that C and D are closely related, but the data do not support this.
   b. Correct. This diagram fits the data in the table. Except for the melting points of the nonhybridized DNA on the diagonal of the table, the highest melting temperatures are M-A, M-D, A-D, and B-C. The data indicate that these fall into two groups: M, A, and D all have high melting points with each other but not with B or C, and B and C have high melting points with each other but not with M, A, or D. Therefore, M, A, and D seem to be closely related (in one lineage) and B and C appear to be in another lineage.
   c. No, this diagram is incorrect. It shows a B-C group and an A-D group, but it predicts that M would be as related to B and C as it is to A and D. The data indicate that M is very closely related to A and D but not closely related to B and C.

Science, Technology, and Society

Some issues and questions to consider: Whereas previous mass extinctions have resulted from catastrophic events, such as asteroid collisions or volcanism, this mass extinction is the result of human-caused environmental alteration. The rate of this mass extinction appears to be much greater than that of the others. Just because life bounces back, does that mean we will? Do we have any ethical responsibility to preserve other species?

Chapter 16 Answers

Testing Your Knowledge

Multiple Choice: **1.** b **2.** c **3.** a (Algae are autotrophs; slime molds are heterotrophs.) **4.**e **5.** c

Describing, Comparing, and Explaining

1. Small, free-living prokaryotes were probably engulfed by a larger cell and took up residence inside. A symbiotic relationship developed between the host cell and engulfed cells. DNA, RNA, ribosomes, and inner membranes of mitochondria and chloroplasts are similar to those of bacteria. These organelles make some of their own proteins, replicate their own DNA, and reproduce by a fissionlike process.

2. *Chlamydomonas* is a eukaryotic cell, much more complex than a prokaryotic bacterium. It is autotrophic, while amoebas are heterotrophic. It is unicellular, unlike multicellular sea lettuce.

Thinking as a Scientist

1. Not a good idea; all life depends on bacteria. You could predict that eliminating all bacteria from an environment would result in a buildup of toxic wastes and dead organisms (both of which bacteria decompose), a shutdown of all chemical cycling, and the consequent death of all organisms except a few kinds of bacteria.
2. Conditions on Earth today are quite different from those on the primitive Earth. The ancient Earth may have had a reducing atmosphere containing chemicals different from today's oxidizing atmosphere, which contains abundant oxygen (O_2). The ozone layer, formed from oxygen, now shields Earth's surface from UV radiation. If organic chemicals were somehow formed today, the organisms already here would probably consume them.

Chapter 17 Answers

Testing Your Knowledge

Multiple Choice: **1.** b **2.** c (It is the only gametophyte among the possible answers.) **3.** a **4.** e **5.** b

Describing, Comparing, and Explaining

1. The alga is surrounded and supported by water, and it has no supporting tissues, vascular system, or special adaptations for obtaining or conserving water. Its whole body is photosynthetic, and its gametes and embryos are dispersed into the water. The seed plant has tissues that support it against gravity; vascular tissues that carry food and water; and special organs (roots, stems, leaves) that absorb, support, and photosynthesize. It is covered by a waterproof cuticle and has stomata for gas exchange. Its gametes are protected by waterproof jackets, sperm are carried by pollen grains, and embryos are protected and provided for by seeds.
2. Animals carry pollen from flower to flower and thus help fertilize eggs. They also disperse seeds by consuming fruit or carrying fruit that clings to their fur. In return, they get food (nectar, pollen, fruit).
3. Plants are autotrophs. They have chlorophyll and make their own food by photosynthesis. Fungi are heterotrophs that digest food externally and absorb nutrient molecules. There are also many structural differences: The threadlike fungal mycelium is different from the plant body, their cell walls are made of different substances, etc.

Thinking as a Scientist

1. Antibiotics probably kill off bacteria that compete with fungi for food. Similarly, bad tastes and odors deter animals that eat or compete with fungi. They are valuable warnings to animals that might eat spoiled food. Those fungi that produce antibiotics and bad-smelling and bad-tasting chemicals would survive and reproduce better than fungi unable to inhibit competitors. Animals that could recognize the smells and tastes also would survive and reproduce better than their competitors. Thus, natural selection would favor fungi that produce the chemicals and, to some extent, the competitors deterred by them.
2. Mosses are haploid plants (the dominant stage in the moss life cycle is the gametophyte, which is haploid). The diploid (sporophyte) generation is dominant in most other plants. Mutations are more likely to show up in a haploid organism because they have only one set of genes. Recessive mutations may not be expressed in a diploid organism. *Some factors to consider in designing your experiment:* What are the advantages/disadvantages of performing the experiment in the laboratory? In the field? What variables would be important to control? How many potted plants should you use? At what distances from the radiation source should you place them? What would serve as a control group for the experiment? What age of plants should you use?

Science, Technology, and Society

Some issues and questions to consider: What are the other functions of forest land? How are other uses affected by logging? Must all the trees in an area be clear-cut? Should a particular area have multiple uses, or should different areas be used for different purposes? How much timber do we need? Could we conserve and recycle more? Are government-managed forests subsidizing private industry? Should we protect habitats, as well as species? Aren't trees a renewable resource? Does the rate of regrowth match the rate of harvest? Will the ancient forests grow back? Are jobs and the economy at least as important as the owl?

Chapter 18 Answers

Testing Your Knowledge

Multiple Choice: **1.** c **2.** d **3.** b **4.** a **5.** c **6.** a (The invertebrates include all animals except the vertebrates.) **7.** c

Matching: **1.** i **2.** f **3.** e **4.** c **5.** a **6.** d **7.** h **8.** b **9.** g

Describing, Comparing, and Explaining

1. Birds share a number of reptile characteristics: amniotic eggs, scales on their legs, beaks and toenails with keratin, and general body form. Their adaptations for flight include feathers, wings, a short tail, bones with air sacs, breast muscles anchored to a keel-like breastbone, a high rate of metabolism, endothermic metabolism, and an efficient circulatory system and lungs.
2. The gastrovascular cavity of a flatworm is an incomplete digestive tract; the worm takes in food and expels waste through the same opening. An earthworm has a complete digestive tract; food travels one way, and different areas are specialized for different functions. The flatworm's body is solid and unsegmented. The earthworm has a coelom, allowing its internal organs to grow and move independently of its outer body wall. Fluid in the coelom cushions internal organs, acts as a skeleton, and aids circulation. Segmentation of the earthworm, including its coelom, allows for greater flexibility and mobility.
3. Cnidarians and most adult echinoderms are radially symmetrical, while most other animals, such as arthropods and chordates, are bilaterally symmetrical. Most radially symmetrical animals stay in one spot or float passively. Most bilateral animals are more active and move head-first through their environment.
4. For example, the legs of a horseshoe crab are used for walking, while the antennae of a grasshopper have a sensory function. Some appendages on the abdomen of a lobster are used for swimming, while the scorpion catches prey with its pincers. (Note that the scorpion stinger and insect wings are not considered jointed appendages.)

Thinking as a Scientist

1. Wet conditions probably increased the survival and reproduction of snails, critical in the fluke life cycle. Possible methods of control include better sanitation so human feces with eggs do not enter water, draining fields to kill snails, poisoning snails, and wearing boots to prevent infection when working in the fields.
2. Important characteristics include symmetry, the presence and type of body cavity, segmentation, type of digestive tract, type of skeleton, and appendages.
3. Agnathans are the trunk, continuing to the present time. Cartilaginous and body fish are offshoots of the agnathans. Bony fishes gave rise to amphibians, and amphibians gave rise to reptiles. Birds and mammals are separate branches from the reptile branch of the vertebrate tree.

Science, Technology, and Society

Some issues and questions to consider: How does the decline of the reefs relate to agriculture? Deforestation? Overfishing? Rapid population growth? What value are the reefs to the local people? What is their value as a world biological resource? What might be the consequences if the reefs disappear? What is likely to make the situation worse? What is likely to improve the situation? In what ways might developed countries contribute to this problem? In what ways might developed countries be able to help the local people preserve the reefs? How might developed countries benefit from helping?

Chapter 19 Answers

Testing Your Knowledge

Multiple Choice: **1.** d. **2.** b **3.** c **4.** b **5.** a **6.** e **7.** c

Describing, Comparing, and Explaining

1. Several primate characteristics make it easy for us to make and use tools—

mobile digits, opposable fingers and thumb, and great sensitivity of touch. Primates also have forward-facing eyes, which enhances depth perception and eye-hand coordination, and a relatively large brain.
2. Chimpanzees make and use simple tools, raid other social groups of their own species, can learn sign language and may use symbolic communication in the wild, and seem self-aware and able to use some form of reasoning to solve problems.
3. Our intelligence and culture—accumulated and transmitted knowledge, beliefs, arts, and products—have enabled us to overcome our physical limitations and alter the environment to fit our needs and desires.
4. The australopithecines roamed the African savanna, probably in small groups. They may have subsisted on nuts and seeds, birds' eggs, and whatever animals they could catch or scavenge from larger predators. They lived mostly on the ground but would climb trees to pick fruits, escape predators, or steal carcasses left by leopards and other large cats.

Thinking as a Scientist
1. Most anthropologists think that humans and apes diverged from a common apelike ancestor 4–8 million years ago. Primate fossils 4–10 million years old might help us understand how humans first evolved.
2. Bipedalism would have allowed early hominids to see farther and run faster in the more open savanna environment, enabling them to find food and escape predators more easily. Bipedalism would have freed the hands to carry objects and use tools. Some anthropologists believe that tool users with larger brains would have been favored by natural selection.

Science, Technology, and Society
Some issues and questions to consider: If you live in an urban area, you probably see commercial and domestic development expanding into areas at the city's edge. In rural areas, agricultural and forestry practices alter land and habitats. News reports often mention global temperature change, acid precipitation, and other widespread changes associated with human activity. Evidence of a decrease in human-caused environmental changes is difficult to find.

Chapter 20 Answers

Testing Your Knowledge
Multiple Choice: **1.** b **2.** c **3.** c (Expelling salt opposes the increases, thereby maintaining a constant internal environment.)
Matching: **1.** d **2.** c **3.** a **4.** d **5.** a **6.** c **7.** b **8.** d **9.** b
Describing, Comparing, and Explaining
1. Fibrous connective tissue cells are embedded in a strong matrix of collagen fibers, which join bones together. Stratified squamous epithelium consists of many cell layers, which protect the body. Neurons are cells with long branches that conduct signals to other cells. Bone cells are surrounded by a matrix containing fibers and calcium salts, forming a hard protective covering around the brain.
2. The surfaces of the intestine, excretory system, and lungs are highly folded and divided, with many blood vessels, increasing their surface area for exchange. Smaller creatures have a greater surface-to-volume ratio, and their cells are closer to the surface, enabling direct exchange between cells and the outside environment.

Thinking as a Scientist
The lizard does not have a constant body temperature, but it does maintain a markedly more constant temperature and a higher temperature than the turtle. While the air is varying over a 10°C range (21–31°C), the turtle varies over an 8°C range (21–29°C), and the lizard varies over only a 4°C range. On the other hand, the lizard's temperature regulation is not nearly as strong as that of the rat, whose temperature range is 36–37°C. Temperature-regulating ability is determined from both temperature constancy and the ability of an animal to maintain a difference between its body temperature and the temperature of the environment.

Chapter 21 Answers

Testing Your Knowledge
Multiple Choice: **1.** a **2.** e **3.** d (They generally serve as coenzymes with catalytic functions.)
Describing, Comparing, and Explaining
You ingest the sandwich one bite at a time. In the oral cavity, chewing begins mechanical digestion, and salivary enzyme action on starch begins chemical digestion. When you swallow, food passes through the pharynx and esophagus to the stomach. Mechanical and chemical digestion continue in the stomach, where pepsin and HCl in gastric juice begin protein digestion. In the small intestine, enzymes from the pancreas and intestinal wall break down starch, protein, and fat to monomers. Bile from the liver and gallbladder emulsifies fat droplets for attack by enzymes. Most nutrients are absorbed into the bloodstream through the walls of the small intestine. In the large intestine, water is absorbed from undigested material, and feces are produced and eliminated.

Thinking as a Scientist
1. A gastrovascular cavity is a sac with one opening. Food must be taken in and wastes expelled through the same opening. In an alimentary canal, food moves in a single direction and is processed sequentially, without backing up. An animal with an alimentary canal can eat continuously, and each part of the canal carries out specific digestive steps as food passes through on a one-way trip.
2. To prove a mineral is essential, you could feed experimental subjects a diet without the mineral and look for ill effects. One problem with this is that it might harm the subjects. Also, it is very difficult to remove all traces of a mineral from the diet. Slight contamination or cheating by subjects might supply sufficient amounts. An additional problem is that animal subjects might not have the same needs as humans.
3. A slice of pizza contains about 450 kcal. Each liter of oxygen consumed liberates 4.83 kcal. It would take $450/4.83 = 93$ liters of oxygen to liberate the energy in the pizza.

Science, Technology, and Society
Some issues and questions to consider: What are the roles of the family, school, advertising, the media, and peers regarding a person's self-image and perception of weight? Are people, especially young people, getting useful, realistic information about diet, self-image, and health? Why or why not? How might the available information be improved? Will better information be enough? How and why do information and environment affect males and females differently?

Chapter 22 Answers

Testing Your Knowledge
Multiple Choice: **1.** c **2.** b **3.** d **4.** a
Describing, Comparing, and Explaining
1. Advantages of breathing air: It can hold a higher concentration of O_2 than water and is easier to move over the respiratory surface. Disadvantage of breathing air: It dries out the respiratory surface; living cells on the respiratory surface must be protected from the drying effect.
2. Nasal cavity, pharynx, larynx, trachea, bronchus, bronchiole, alveolus, through wall of alveolus into blood vessel, blood plasma, into red blood cell, attaches to hemoglobin, carried by blood through heart, blood vessel in muscle, dropped off by hemoglobin, out of red blood cell, through blood vessel wall, through plasma membrane into muscle cell.

Thinking as a Scientist
If gas exchange is to work, the atmosphere must have a higher concentration of oxygen than the organism; thus, in this study, you would expect the air to have the highest partial pressure (159). You would also predict that the rate of oxygen consumption of the exercising insect would be greater than that of the resting insect. You would therefore expect the lowest partial pressure of oxygen (40) in the exercising insect, and the intermediate partial pressure (60) in the resting insect.

Science, Technology, and Society
Some issues and questions to consider: Does expertise in the area necessarily bias a researcher? Would it be more fair to select scientists who know less about the subject? Should objections to Dr. Burns' methods or to his conclusions be more important as selection criteria for the panel? Who selects the panel? Might they be biased? Does the source of research funds

make an expert's opinions suspect? What are criteria for awarding funds? Are there strings attached?

Chapter 23 Answers

Testing Your Knowledge

Multiple Choice: **1.** b **2.** d (The second sound is the closing of the semilunar valves as the ventricles empty.) **3.** c **4.** c **5.** a

Describing, Comparing, and Explaining

1. Pulmonary vein, left atrium, left ventricle, aorta, artery, arteriole, capillary bed (in finger, for example), venule, vein, vena cava, right atrium, right ventricle, pulmonary artery, capillary bed in lung, pulmonary vein.
2. Capillaries are very numerous, producing a large surface area close to body cells. The capillary wall is only one epithelial cell thick. Clefts between epithelial cells allow materials to leak in and out.

Thinking as a Scientist

1. Proteins are important solutes in blood, accounting for much of the osmotic pressure that draws fluid into the blood. If protein concentration is reduced, the inward pull of osmotic pressure will fail to balance the outward push of blood pressure. The net pressure at the arterial end of a capillary forces fluid out into the interstitial fluid. Net pressure at the venous end is not great enough to draw fluid back in.

Science, Technology, and Society

Some issues and questions to consider: Is it ethical to have a child in order to save the life of another? Is it right to conceive a child as a means to an end—to produce a tissue or organ? Is this a less acceptable reason than most reasons parents have for bearing children? Do parents even need a reason for conceiving a child? Do doctors find this acceptable? If they do, why do we seldom hear about it, and why did this story make the front page? Do parents have the right to make decisions like this for their young children? How will the donor (and recipient) feel about this when the donor is old enough to know what happened?

Chapter 24 Answers

Testing Your Knowledge

Multiple Choice: **1.** e **2.** b **3.** b **4.** d **5.** a

Matching: **1.** b **2.** f **3.** d **4.** e **5.** a **6.** g **7.** c

Describing, Comparing, and Explaining

1. AIDS is mainly transmitted in blood and semen. It enters the body through slight wounds during sexual contact or via needles contaminated with infected blood. AIDS is deadly because it infects helper T cells, crippling immunity and leaving the body vulnerable to other infections and cancer.
2. Inflammation is triggered by tissue injury. Damaged cells release histamine and other chemicals, which cause nearby blood vessels to dilate and become leakier. Blood plasma leaves vessels, and phagocytes are attracted to the site of injury. An increase in blood flow, fluid accumulation, and increased cell population cause redness, heat, and swelling. Inflammation disinfects and cleans the area and curtails the spread of infection from the injured area.

Thinking as a Scientist

1. Each time the flu virus changes its antigens, a different population of lymphocytes is activated to defend the body. This primary immune response to a new invader is slower and probably starts with fewer cells than the secondary immune response carried out by memory cells. Since the flu virus can be different each time, memory cells are less able to respond to it. The body is always one step behind, immune to the last flu virus but not the next one.
2. Before you start any tests, you might look for cases in which the self-nonself hypothesis does not explain observed phenomena or cases that have been interpreted as exceptions to the self-nonself concept. Why might this be a useful approach? Promoters of the damage hypothesis argue that immune system failures, such as autoimmune diseases, indicate that the self-nonself hypothesis is flawed. Why do you suppose they make this argument? One case being studied in detail is the immune system in newborn mammals. According to the self-nonself hypothesis, a newborn's immune system is incapable of recognizing foreign antigens, and this ability develops during a critical period soon after birth. (You could read about this in a more advanced general biology textbook or an immunology book.) However, new evidence suggests that the neonatal immune system can respond to at least some foreign antigens. Does this evidence support the damage hypothesis? How so? To test the damage hypothesis, you might try to identify specific danger signals that trigger immune responses. Several research groups are currently focusing on such signals. A possible question is: Could damaged tissue cells signal the antigen-presenting cells (APCs) of the immune system, making the APCs in turn cause changes in T cells? Perhaps the signal is a protein that leaks from a damaged cell. Or perhaps it is simply the loss of physical contact of an APC with a tissue cell that dies suddenly. Can you think of other possibilities? The idea that a body defense system would evolve in response to actual danger—that is, a threat to survival and the perpetuation of genes—is consistent with evolutionary theory. Natural selection would tend to favor those individuals with such a defense system and not favor individuals lacking such a system. Why would natural selection favor individuals whose body defense system distinguishes self from nonself? Could the ability to distinguish self from nonself molecules increase evolutionary fitness?

Science, Technology, and Society

Some issues and questions to consider: How important is it to protect students from AIDS? Is this a function of schools? Do schools serve other such "noneducational" purposes? Should parents or citizens' and church groups—on either side of the issue—have a say in this, or is it a matter between the school and the student? Does the distribution of condoms condone or sanction sexual activity or promiscuity? Is a school legally liable if a school-issued condom fails to protect a student? Are there alternative measures, such as education, that might be as effective for slowing the spread of AIDS?

Chapter 25 Answers

Testing Your Knowledge

Multiple Choice: **1.** c **2.** a **3.** e **4.** d

Matching: **1.** a **2.** c **3.** b **4.** d **5.** b **6.** a

Describing, Comparing, and Explaining

1. In salt water, the fish loses water by osmosis. It drinks salt water and disposes of salts through its gills. Its kidneys conserve water and excrete some salts. In fresh water, it gains water by osmosis. Its kidneys excrete a lot of urine, so it loses some salts. Its gills and digestive tract take up salts to replenish those lost.
2. Countercurrent heat exchangers are sets of parallel blood vessels found in animals that live in cold environments. A countercurrent mechanism in fish gills extracts oxygen from water. In the heat exchanger, cool blood from the skin and warm blood from the body interior flow in opposite directions. In the gill, water and blood flow in opposite directions. In the heat exchanger, as blood loses heat, it comes into contact with progressively colder blood, maximizing heat transfer. In the gill, as blood picks up oxygen, it comes into contact with water containing higher and higher concentrations of oxygen, maximizing the diffusion of oxygen.

Thinking as a Scientist

A countercurrent heat exchanger in the birds' legs reduces the loss of heat from the body. You would expect the temperature of blood flowing from the body into the legs to be warmer than blood flowing back to the body from the legs.

Science, Technology, and Society

Some issues and questions to consider: Could drug use endanger the safety of the employee or others? Is drug testing relevant to jobs where safety is not a factor? Is drug testing an invasion of privacy, interfering in the private life of employees? Is an employer justified in banning drug use off the job if it does not affect safety or ability to do the job? Do the same criteria apply to employers requiring the test? Could an employer use a drug test to regulate other employee behavior that is legal, such as smoking?

Chapter 26 Answers

Testing Your Knowledge

Multiple Choice: **1.** d **2.** e **3.** a **4.** b (Negative feedback: When thyroxine increases, it inhibits TSH, which reduces thyroxine secretion.) **5.** d

Matching: **1.** e, w **2.** a, u **3.** g, r **4.** c, s **5.** f, p **6.** h, t **7.** d, q **8.** b, v

Describing, Comparing, and Explaining

1. The hypothalamus secretes releasing hormones and inhibiting hormones, which are carried by the blood to the anterior pituitary. In response to these signals from the hypothalamus, the anterior pituitary increases or decreases its secretion of a variety of hormones that directly affect body activities or influence other glands. Neurosecretory cells that extend from the hypothalamus into the posterior pituitary secrete hormones that are stored in the posterior pituitary until they are released into the blood.
2. Only cells with the proper receptors will respond to a hormone. For a steroid hormone, the presence (or absence) and types of receptor proteins inside the cell determine the hormone's effect. For a nonsteroid hormone, the types of receptors on the cell's plasma membrane are key, and the second messenger may have different effects inside different cells.

Thinking as a Scientist

a. No. Blood sugar level goes too low. Diabetes would tend to make the blood sugar level go too high after a meal.

b. No. Insulin is working, as seen by the homeostatic blood sugar response to feeding.

c. **Correct.** Without glucagon, exercise and fasting lower blood sugar, the cells cannot mobilize any sugar reserves, and their blood sugar level drops. Insulin (which lowers blood sugar) has no effect.

d. No. If this were true, blood sugar level would increase too much after a meal.

Science, Technology, and Society

Some issues and questions to consider: Why are parents concerned about their children's height? Just because the hormone is available, does this mean that parents should request its use? What is the role of the physician in this situation? Would this kind of use be worth the risk? Is a child able to make an informed decision? How might the child feel in later years if the hormone has unintended side effects? In what ways would usage by an older adult be different? How can people be prevented from misusing a legal drug? How are the distribution and use of these kinds of medications regulated?

Chapter 27 Anwers

Testing Your Knowledge

Multiple Choice: **1.** c **2.** c (The outer layer in a gastrula is the ectoderm; of the choices given, only the brain develops from ectoderm.) **3.** a **4.** d **5.** d **6.** a

Matching: **1.** e **2.** g **3.** d **4.** h **5.** f **6.** a **7.** b **8.** c

Describing, Comparing, and Explaining

1. **A.** FSH. **B.** estrogen. **C.** LH. **D.** progesterone. **P.** menstruation. **Q.** pre-ovulatory phase—follicle develops. **R.** ovulation. **S.** post-ovulatory phase—corpus luteum develops.
2. Asexual reproduction is advantageous in a stable, favorable environment; it allows an isolated animal to produce many offspring quickly and precisely. Sexual reproduction increases variability, which may enhance success in a variable environment.
3. Both produce haploid gametes. Spermatogenesis produces four small sperm; oogenesis produces one large ovum. In humans, the ovary contains all the primary oocytes at birth, while testes can keep making primary spermatocytes throughout life. Oogenesis is not complete until fertilization, but sperm mature without eggs.
4. The extraembryonic membranes provide a moist environment for the embryos of terrestrial vertebrates, and enable the embryos to absorb food and oxygen and dispose of wastes. Such membranes are not needed when an embryo is surrounded by water, as are those of fishes and amphibians.
5. The nerve cells may follow chemical trails to the muscle cells and identify and attach to them by means of specific surface proteins.

Thinking as a Scientist

1. The researcher might find out whether chemicals from the notochord stimulate the nearby ectoderm to become the neural tube, a process called induction. Transplanted notochord tissue might cause ectoderm anywhere in the embryo to become neural tissue. Control: Transplant non-notochord tissue under the ectoderm of the belly area.
2. **a.** No. The endometrium reaches its peak development after ovulation.
b. No. If menstruation started on January 29, it probably stopped around February 2, and the follicle started to grow then. By February 14, the follicle is very well developed.
c. Correct. If the woman has 34-day cycles and we can assume that the post-ovulatory phase is 14 days, she has 20 days in her pre-ovulatory phase. She started menstruating 17 days ago, so on February 14 it's probably 3 days before ovulation. Estrogen is rising rapidly, the follicle is about to ovulate, and (if recent data on the life span of sperm in the female reproductive tract are reliable) intercourse could now likely result in fertilization just after ovulation, in 3 days.
d. No. Menstruation stopped about 5 days after it started, and the LH peak in the middle of the cycle has nothing to do with stopping it.

Science, Technology, and Society

Some issues and questions to consider: If it is technologically possible to save a baby, does that mean we have to do it? What would bring "the greatest good to the greatest number" of babies? If we diverted resources away from premature babies, would they really be used for prenatal care? Who pays for the treatment of premature babies? Prenatal care?

Chapter 28 Answers

Testing Your Knowledge

Multiple Choice: **1.** b **2.** e **3.** a **4.** e (Both I and III would prevent action potentials from occurring; II could actually increase the generation of action potentials.)

Describing, Comparing, and Explaining

1. At the point where the action potential is triggered, sodium ions rush into the neuron. They diffuse laterally and cause sodium gates to open in the adjacent part of the membrane, triggering another action potential. The moving wave of action potentials, each triggering the next, is a moving nerve signal. Behind the action potential, sodium gates are temporarily inactivated, so the action potential can only go forward. At a synapse, the transmitting cell releases a chemical neurotransmitter, which binds to receptors on the receiving cell and triggers a nerve signal in the receiving cell.
2. Fingertips to brain: sensory neuron to spinal cord, interneurons to brain. The motor area of the cerebral cortex makes the decision, and the cerebellum helps coordinate movements. Brain to muscles: interneurons in the brain and spinal cord to motor neurons to muscles.

Thinking as a Scientist

The results show the cumulative effect of all incoming signals on neuron D. Comparing experiments #1 and #2, the more nerve signals D receives from C, the more it sends; C is excitatory. Because neuron A is not varied here, its action is unknown; it may be either excitatory or mildly inhibitory. Comparing experiments #2 and #3, neuron B must be a strongly inhibitory neurotransmitter, because when B is transmitting, D stops.

Science, Technology, and Society

Some issues and questions to consider: What is the role of alcohol in crime? What are its effects on families and in the workplace? In what ways is the individual responsible for alcohol abuse? The family? Society? Who is affected by alcohol abuse? How effective are treatment and punishment in curbing alcohol abuse? Who pays for alcohol abuse and consequent treatment or punishment? Is it possible to enjoy alcohol without abusing it?

Chapter 29 Answers

Testing Your Knowledge

Multiple Choice: **1.** d (He could hear the tuning fork against his skull, so the cochlea, nerve, and brain are OK. Apparently, sounds are not being transmitted to the cochlea; therefore, the bones are the problem.) **2.** b **3.** a **4.** b **5.** e

Describing, Comparing, and Explaining

1. Sound waves strike the eardrum, which moves the bones of the middle ear. The bones vibrate fluid in the cochlea, which moves the basilar membrane. Hair cells on the basilar membrane move against the overlying tissue, causing permeability change and receptor potentials in the hair cells.

The hair cells stimulate sensory neurons, which transmit action potentials to the brain. Louder sounds move hair cells more, generating a greater frequency of action potentials. Different pitches affect different parts of the basilar membrane; different hair cells transmit to different parts of the brain.

2. Sensation is the detection of stimuli (light) by the photoreceptors of the retina and transmission of action potentials to the brain. Perception is the interpretation of these nerve signals—sorting out the patterns of light and dark, and determining their meaning.

Thinking as a Scientist

1. *Some possible hypotheses:* Paired sensory receptors enable an animal to determine the direction from which stimuli come. Paired receptors enable comparison of the intensity of stimuli on either side. Paired receptors enable comparison of slightly different images seen by the eyes or sounds heard by the ears (thus enabling the brain to perceive depth and distance).
2. Do the turtles hear the surf? Plug the ears of some turtles and not others. If turtles without earplugs head for the water and turtles with earplugs get lost, they probably hear the ocean. Or do they smell the water? Plug their nostrils, and follow the same process.
3. **a.** No. Rods and cones detect light, not sound, so it is unlikely that they would have anything to do with this form of synesthesia.
   b. No. The ears are transducing the sound and producing action potentials, but these action potentials are being interpreted incorrectly.
   c. No. If action potentials weren't being produced, there would be no sensation at all.
   d. Correct. The auditory nerve is apparently connected to a visual part of the brain in the cerebral cortex. The ears are producing action potentials, which may be going to the visual cortex.
   e. No. Receptor potentials can only affect action potential generation or frequency; they cannot enter the brain themselves.

Science, Technology, and Society

Some issues and questions to consider: The sound is loud enough to impair hearing, but how long an exposure is necessary for this to occur? Does exposure have to occur all at once, or is damage cumulative? Who is responsible, concert promoters or listeners? Should there be regulations regarding sound exposure at concerts (as there are for job-related noise)? Are young people sufficiently mature and aware to heed such warnings?

Chapter 30 Answers

Testing Your Knowledge

Multiple Choice: **1.** c **2.** e **3.** a (Water supports aquatic animals, reducing the effects of gravity.) **4.** d **5.** c **6.** d **7.** e **8.** d **9.** a (Each neuron controls a smaller number of muscle fibers.) **10.** e

Describing, Comparing, and Explaining

1. The bird's wings are airfoils, with convex upper surfaces and flat or concave lower surfaces. As the wings beat, air passing over them travels farther than air beneath. Air molecules above the wings are more spread out, lowering pressure. Higher pressure beneath the wings pushes them up.
2. Advantages of an exoskeleton include strength, good protection for the body, and flexibility at joints. The major disadvantage is that the exoskeleton must be shed periodically as the insect grows, leaving the insect temporarily weak and vulnerable.
3. Action potentials from the brain travel down the spinal cord and along a motor neuron to the muscle. The neuron releases a neurotransmitter, which triggers action potentials in a muscle fiber membrane. These action potentials initiate the release of calcium from the ER of the cell. Calcium enables myosin heads of the thick filaments to bind with actin of the thin filaments. ATP provides energy for the movement of myosin heads, which causes the thick and thin filaments to slide along one another, shortening the muscle fiber. The shortening of muscle fibers pulls on bones, bending the arm. If more motor units are activated, the contraction is stronger.
4. The human skeleton has a larger skull than the baboon, balanced on top of (instead of in front of) the backbone. The human backbone is S-shaped, not arched like the baboon's. The human pelvic girdle is shorter, rounder, and oriented more vertically. The human hand is adapted for gripping, and our feet are adapted for support of the entire body and bipedal locomotion.

Thinking as a Scientist

1. Chemical A would work better, because acetylcholine triggers contraction. Blocking it would prevent contraction. B would actually increase contraction, because Ca^{2+} signals filaments to interact and slide.
2. Circular muscles in the earthworm body wall decrease the diameter of each segment, squeezing internal fluid and lengthening the segment. Longitudinal muscles shorten and fatten each segment. Different parts of the earthworm can lengthen while others shorten, producing a crawling motion. The whole roundworm body moves at once, because of a lack of segmentation. The body can only shorten or bend, not lengthen, because of a lack of circular muscles. Roundworms simply thrash from side to side.
3. ATP causes the myosin heads of the thick filaments to detach from the thin filaments (Figure 30.9B, Step 1). If there is no ATP present, the myosin heads remain attached to the thin filaments, and the muscle fiber remains fixed in position.

Science, Technology, and Society

1. *Some issues and questions to consider:* Are the places where you live, work, or attend class accessible to a person in a wheelchair? If you were in a wheelchair, would you have trouble with doors, stairs, drinking fountains, toilet facilities, and eating facilities? What kinds of transportation would be available to you, and how convenient would they be? What activities would you have to forgo? How might your disability alter your relationships with your friends and family? How well would you manage on your own?
2. *Some issues and questions to consider:* The major negative consequence is that the bones will become completely hardened before they have grown to their full size. This means that the steroid user will be shorter than he or she would otherwise have been. Do steroid users know this? Do they think the effects of steroids are temporary or reversible? Why is such emphasis placed on appearance and athletic prowess? Will steroid users later regret their use? If steroid use were not against the rules, would it be acceptable?
3. *Some issues and questions to consider:* Do young people look that far ahead? What does it take to get them to think about their later years? Who will they believe? Will scare tactics work? Or are good health habits simply their own responsibility?

Chapter 31 Answers

Testing Your Knowledge

Multiple Choice: **1.** e (Xylem forms just inside cambium; old xylem is pushed inside.) **2.** b **3.** e **4.** e

Matching: **1.** f **2.** b **3.** e **4.** a **5.** c **6.** d

Describing, Comparing, and Explaining

1. Bees deposit the pollen on the stigma of a carpel, and a pollen tube grows to the ovary at the base of the carpel. Sperm travel down the pollen tube and fertilize egg cells in ovules. The ovules grow into seeds, and the ovary grows into the flesh of the fruit (actually, in an apple, just the core). As the seeds mature, the fruit ripens and falls (or is picked).
2. Bulbs, root sprouts, and runners are all examples of vegetative reproduction. Vegetative reproduction is less wasteful and costly than sexual reproduction and less hazardous for young plants.
3. Tomato: Fruit (ripened ovary). Celery stalk: Leaf stalk (petiole). Peanut: Seed (ovule). Strawberry: Fruit (ripened ovary). Lettuce: Leaf blades. Artichoke: Terminal bud. Beet: Root.

Thinking as a Scientist

Modern methods of plant breeding and propagation have increased crop yields but have decreased genetic variability, so plants become more vulnerable to epidemics. Primitive varieties of crop plants could contribute to gene banks and be used for breeding new strains.

Science, Technology, and Society

Some issues and questions to consider: Why are the forests being cut, when preserving them is more beneficial in the long run? Why can't the developing nations create new forest products themselves? Should developing and developed nations share in the profits from the forest? What incentive is there for a company to create a new product if they can't keep the profits? What about companies paying an exploration fee for exclusive rights to take plants from a certain area? What about a contract for the developing country to get a percentage of the profits from new products? Are there reasons to preserve the forest, other than profits and beneficial new products?

Chapter 32 Answers

Testing Your Knowledge

Multiple Choice: **1.** d **2.** d **3.** e **4.** b

Describing, Comparing, and Explaining

If plant starts to dry out, K^+ is pumped out of guard cells. Water follows by osmosis, guard cells become flaccid, and stomata close. This prevents wilting, but it keeps leaves from taking in carbon dioxide, needed for photosynthesis.

Thinking as a Scientist

1. *Hypothesis:* The hydrogen ions in acid precipitation displace positively charged nutrient ions from negatively charged clay particles. *Test:* In the laboratory, place equal amounts and types of soil in separate filters. The pore size of the filter must not allow any undissolved soil particles to pass through. Spray (to simulate rain) soil samples in the filters with solutions of different pH (e.g., pH 5, 6, 7, 8, 9). Determine the concentration of nutrient ions in the solutions. (The only variable in the solutions should be the hydrogen ion concentration. Ideally, the solutions would contain no dissolved nutrient ions.) Collect fluid that drips through soil samples and filters. Determine the hydrogen ion concentration and the nutrient ion concentration in each sample of fluid. *Prediction:* If the hypothesis is correct, the fluid collected from the soil samples exposed to pH lower than 5.6 (acid rain) will contain the highest concentration of positively charged nutrient ions.
2. *Hypothesis:* When fixed nitrogen levels increase to a certain level in the soil, genes coding for nitrogen-fixing enzymes in nitrogen-fixing bacteria are switched off. *Test:* Expose cultures of nitrogen-fixing bacteria (symbiotic and nonsymbiotic ones found in soil) to solutions of different concentrations of fixed nitrogen (that is, NO_3^- and NH_4^+). Determine the concentrations of nitrogen-fixing enzymes produced by the bacteria in each sample. *Prediction:* If your hypothesis is correct, and the fixed nitrogen concentration is high enough in some of the samples, you would expect the enzyme concentration to be measurably lower in samples whose fixed nitrogen concentration is above the level that causes negative feedback (see Figure 32.15).
3. The hypothesis is supported if transpiration varies with light intensity when humidity and temperature are about the same. These conditions are seen at two places in the table; at hours 11 and 12, recordings for temperature and humidity are about the same, but light intensity increased markedly from 11 to 12, as did the transpiration rate. The recordings made at hours 3 and 4 show the same effects. Also, the recordings made at hours 1 and 2 generally support the hypothesis. Here, both temperature and humidity decreased, so you might expect the transpiration rate to stay about the same or perhaps increase, because the temperature decrease is small; however, the transpiration rate dropped, as did the light intensity.

Science, Technology, and Society

1. *Some issues and questions to consider:* How were the farmers assigned or sold "rights" to the water? How is the price established when a farmer buys or sells water rights? Is there enough water for everyone who "owns" it? What kinds of crops are the farmers growing using the water? What will the water be used for in the city? Are there other users with no rights, such as wildlife? Is there any effort being made to curb urban growth and conserve water? Should millions of people be living in what is essentially a desert? What are the reasons for farming desert land?

Chapter 33 Answers

Testing Your Knowledge

Multiple Choice: **1.** b **2.** c (A short-day plant requires a long night. The 13-hour night of answer C is the only uninterrupted night longer than 12 hours.) **3.** b **4.** a **5.** d **6.** b **7.** b

Matching: **1.** g **2.** e **3.** f **4.** a **5.** b **6.** c **7.** h **8.** d

Describing, Comparing, and Explaining

1. Fruit produces ethylene gas, which triggers the ripening and aging of the fruit. Ventilation prevents a buildup of ethylene and delays its effects.
2. The terminal bud produces auxins, which counter the effects of cytokinins from the roots and inhibit the growth of axillary buds. If the terminal bud is removed, the cytokinins predominate, and lateral growth occurs at the axillary buds.

Thinking as a Scientist

1. The red wavelengths in the room lights quickly convert the phytochrome in the mums to the P_{fr} form, which inhibits flowering in a long-night plant. The mums will not flower unless Jon can rig up some far-red lights. Exposure to a burst of far-red light would convert the phytochrome to the P_r form, allowing flowering to occur.
2. He could remove leaves at different stages of being eaten to see how long it took for changes to occur in nearby leaves. He could capture the "hormone" in an agar block, as in the phototropism experiments in Module 33.1, and apply it to an undamaged plant. Or he could block "hormone" movement out of a damaged leaf or into a nearby leaf.
3. The results illustrated in the graph generally support the hypothesis; measured by chlorophyll fluorescence, greater leaf damage occurred in air with no isoprene than in air with isoprene. The experimenters exposed leaves to N_2 to prevent the leaves themselves from emitting isoprene. By doing so, they were able to control the amount of isoprene in the air to which the leaves were exposed.

Science, Technology, and Society

Some issues and questions to consider: Is the hormone safe for human consumption? What are its effects in the environment? Could its production produce impurities or wastes that might be harmful? What kinds of tests need to be done to demonstrate its safety? How much does it cost to make and use? Are the benefits worth the costs and risks? Are there reasons for using an artificial chemical on food simply to improve its appearance? A scientist could seek answers by studying the stability of the hormone in a variety of laboratory simulations of natural conditions. The toxicity of the hormone, of the materials used to produce it, and of its breakdown products could be determined in laboratory tests.

Chapter 34 Answers

Testing Your Knowledge

Multiple Choice: **1.** a **2.** e **3.** b **4.** d

Matching: **1.** f **2.** c **3.** e **4.** g **5.** b **6.** d **7.** a

Describing, Comparing, and Explaining

1. At the source, nutrient content is generally low; the current is stronger; there is usually less sediment, lower temperature, and higher dissolved O_2 concentration; and food is from attached algae or organic material from land. At the mouth, the nutrient content is generally higher, and the current slower; there is usually more sediment, higher temperature, and lower dissolved O_2 concentration; most food is from phytoplankton.
2. Consider the main abiotic factors discussed in Module 34.4. Module 34.5 describes an example of one species' adaptations.
3. The sun's rays strike the equator directly. Near the equator, heated air rises and then cools, and moisture condenses and falls as rain. Dry air descends at about 30° north and south of the equator, forming deserts.

Thinking as a Scientist

Apparently, unrelated animals adapted in similar ways to similar environments, temperate grassland and savanna; convergent evolution.

Science, Technology, and Society

Some issues and questions to consider: What reasons are there for protecting areas like this? Is a public refuge or preserve the only way to protect species and habitats? Is there some way to protect organisms without confiscating land? Does protection take precedence over property rights? Would the prairie have been better off without the publicity? Should the public be able to condemn land to protect it? Should there be state or federal laws to protect endangered plants on private land? Should there be laws to protect endangered habitats, like this patch of prairie?

Chapter 35 Answers

Testing Your Knowledge

Multiple Choice: **1.** e **2.** c **3.** c (Seedlings would tend to grow only in gaps with sufficient light, spacing the trees fairly evenly.) **4.** a **5.** c

Describing, Comparing, and Explaining

1. Exponential growth is accelerating, unlimited population growth described by the equation $G = rN$. It might occur briefly when there are no factors slowing growth. Logistic growth occurs when density-dependent factors slow growth as population density approaches carrying capacity. It is described by $G = rN(K - N)/K$.
2. Survivorship is the fraction of individuals in a given age interval that survive to the next interval. It is a measure of the probability of surviving at any given age. A survivorship curve shows the fraction of individuals in a population surviving at each age interval during the life span. Almost all oysters die young, with a few living a full life span. Few humans die young; most live out a full life span and die of old age. Squirrels have approximately constant mortality and about an equal chance of surviving at all ages.

Thinking as a Scientist

Equilibrial populations tend to live in fairly stable environments, held near carrying capacity by density-dependent limiting factors. They reproduce later than opportunistic species and have fewer offspring. Human intrusion is density-independent, and the lower reproductive rate of equilibrial species makes it hard for them to recover from this kind of disruption. We would expect species with equilibrial life histories to have a Type I survivorship curve (see Module 35.7).

Science, Technology, and Society

Some issues and questions to consider: How does population growth in developing countries relate to food supply, pollution, and the use of natural resources such as fossil fuels? How are these things affected by population growth in developed countries? Which of these factors is most critical to our survival? Are they affected more by the growth of developing or developed countries? What will happen as developing countries become more developed? Will it be possible for everyone to live at the level of the developed world?

Chapter 36 Answers

Testing Your Knowledge

Multiple Choice: **1.** c **2.** e (The bat and moth have adapted to each other.) **3.** d

Describing, Comparing, and Explaining

These animals are secondary or tertiary consumers, at the top of the energy pyramid. Stepwise energy loss means not much energy is left for them; thus, they are rare.

Thinking as a Scientist

Hypothesis: The kangaroo rat is a keystone predator in the desert. (Apparently, predation by the rats kept the one plant from outcompeting the others; removing the rats reduced prey diversity.) *Additional supporting evidence:* Observations of the rats preferentially eating dominant plants or finding that the dominant plant recovers from herbivore damage faster.

Science, Technology, and Society

Some issues and questions to consider: What relationships might exist in the mussels' native habitat that are altered in the Great Lakes? What are possible predators? Competitors? Parasites? How might the mussel compete with Great Lakes organisms? Might competitive exclusion occur? What happens to displaced competitors? Might the Great Lakes species adapt in some way? Might the mussels adapt? Could possible solutions present problems of their own?

Chapter 37 Answers

Testing Your Knowledge

Multiple Choice: **1.** d (Innovation, or reasoning, is most often seen in primates. Most animals do not exhibit this kind of problem-solving ability.) **2.** c **3.** c (Such clocks allow them to compensate for Earth's rotation and the resulting change in the sun's position in the sky.) **4.** a

Describing, Comparing, and Explaining

1. Main advantage: Flies do not live long. Innate behaviors can be performed the first time without learning, enabling flies to find food, mates, etc., without practice. Main disadvantage: Innate behaviors are rigid; flies cannot learn to adapt to specific situations.
2. A dominance hierarchy minimizes energy wasted in fighting, maximizes efficiency in hunting, and ensures that some in the pack will get adequate resources. A territory ensures adequate resources for the pack.
3. Courtship behaviors reduce aggression between potential mates and confirm their species, sex, and physical condition. Environmental changes such as rainfall, temperature, day length, and the presence of females probably lead frogs to start calling, so these would be the proximate causes. The ultimate cause relates to evolution. Fitness (reproductive success) is enhanced for frogs that engage in courtship behaviors.

Thinking as a Scientist

One likely hypothesis is that the helper is closely related to one or both of the birds in the mated pair. Because closely related birds share relatively many genes, the helper bird is indirectly enhancing its own fitness by helping its relatives raise their young. (In other words, this behavior evolved by kin selection.) The easiest way to test the hypothesis would be to determine the relatedness of the birds by comparing tissue samples using molecular methods such as those described in Module 16.12. If birds are closely related, their DNA and proteins should be more similar than those of more distantly related or unrelated birds.

Science, Technology, and Society

1. Identical twins are genetically the same, so any differences between them are due to environment. Thus, the study of identical twins enables researchers to sort out the effects of "nature" and "nurture" on human behavior. The data suggest that many aspects of human behavior are inborn. Some people find these studies frightening because they seem to leave less room for free will and self-improvement than we would like.
2. Animals are used in experiments for various reasons. A particular species of animal may have features that make it well suited to answer an important biological question. Squids, for example, have a giant nerve fiber that made possible the discovery of how all nerve cells function. Animal experiments play a major role in medical research. Many vaccines that protect humans against deadly diseases, and drugs that can cure diseases, have been developed using animal experiments. Animals also benefit, as vaccines and drugs are developed for combating their own pathogens. Some researchers point out that the number of animals used in research is a small fraction of those killed as strays by animal shelters and a minuscule fraction of those killed for human food. They also maintain that modern research facilities are models of responsible and considerate treatment of animals. Whether or not this is true, the possibility that at least some kinds of animals used in research suffer physical pain as a result, and perhaps mental anguish as well, raises serious ethical issues. *Some questions to consider:* What are some medical treatments or products that have undergone testing in animals? Have you benefited from any of them? Are there alternatives to using animals in experiments? Would alternatives put humans at risk? Are all kinds of animal experiments equally valuable? In your opinion, what kinds of experiments are acceptable, and what kinds are unnecessary? What kinds of treatment are humane, and what kinds are inhumane?

Chapter 38 Answers

Testing Your Knowledge

Multiple Choice: **1.** d **2.** a **3.** d **4.** e **5.** c **6.** b **7.** a (The U.S. uses more energy each year than all these other areas combined.)

Describing, Comparing, and Explaining

1. Carbon dioxide and several other gases in the atmosphere absorb infrared radiation and thus slow the escape of heat from Earth. This is called the greenhouse effect. The greenhouse effect is beneficial to life on Earth; without CO_2 in the atmosphere, the temperature at the surface of Earth would be only about $-18°C$, which is too cold for most life to exist.

2. Fossil-fuel consumption, industry, and agriculture are increasing the quantity of greenhouse gases—such as CO_2, methane, and nitrous oxide—in the atmosphere. This could trap more heat and raise atmospheric temperatures 2–5°C over the next century. Logging and the clearing of forests for farming contribute to global warming by reducing the uptake of CO_2 by plants (and adding CO_2 to the air when trees are burned). Global warming could shift patterns of precipitation, turning farmland into deserts. It could also cause polar ice caps to melt, raising ocean levels and flooding coastal areas. Global warming is an international problem; air and climate do not recognize international boundaries. Greenhouse gases produced by industrialized nations add to those resulting from deforestation in less-developed tropical nations. Cooperation and commitment to a less consumerist lifestyle will be needed to slow global warming.

Thinking as a Scientist

1. Choice c is the most convincing. Without samples prior to the oil spill, there is no reason to believe that the oil spill killed off the mollusks. The two bays may have had different mollusk populations before the spill.
2. Choice a would strengthen the conclusion somewhat; however, it may not be strong enough to win a lawsuit. It is circumstantial evidence that prior to the spill, the mollusk populations in the two bays were nearly the same, that the oil killed off the mollusks, and that with the oil mostly gone, the mollusks have repopulated the bay. Why isn't this very strong evidence?
3. If global warming causes CO_2 to bubble out of the ocean into the air, the additional CO_2 in the atmosphere could compound the problem, trapping even more heat and making global warming worse. On the other hand, if the ocean can hold a large amount of CO_2, perhaps it can absorb some of the excess CO_2 we are adding to the atmosphere, and thereby slow global warming.
4. These birds might be affected by pesticides while in their wintering grounds in Central and South America, where such chemicals may still be in use. The birds are also affected by deforestation throughout their range.

Science, Technology, and Society

1. Current counts indicate there are about 1.5 million species of living things. Assume 80% of all living things (not just plants and animals) live in tropical rain forests. This means there are $1.5 \times 0.8 = 1.2$ million species there. If half the species survive, this means 0.6 million species will be extinct in 100 years, or 0.6 million/100 = 6000 per year. This means that 6000/365 = 16 species will disappear per day, or almost one per hour. If there are 30 million species on Earth, 24 million live in the tropics, and 12 million will disappear in the next century. This is 120,000 per year, 329 per day, or 14 per hour.
2. *Some issues and questions to consider:* How does our use of fossil fuels affect the environment? What about oil spills? Disruption of wildlife habitat for construction of oil fields, pipelines, etc.? Burning of fossil fuels and possible climate change and flooding from global warming? Pollution of lakes and destruction of property by acid precipitation? Health effects of polluted air on humans? How are we paying for these "side effects" of fossil-fuel use? In taxes? In health insurance premiums? Do we pay a nonfinancial price in terms of poorer health and quality of life? Could oil companies be required to pick up the tab for environmental effects of fossil-fuel use? Could these costs be covered by an oil tax? How would this change the price of oil? How would a change in the price of oil change our pattern of energy use, our lifestyle, and our environment?
3. *Some issues and questions to consider:* Does the current evidence constitute "proof" that global warming is occurring? Why is it difficult to prove conclusively that human activities have an effect on global warming? How much information do we need before we act? What are the advantages and disadvantages of acting now versus waiting and gathering more data? Which of the following would be more serious, and why? Acting now to curb global warming, and later finding out that we were wrong, that global warming is not a threat? Or failing to take action now, and later finding out that global warming is a real and serious problem?
4. *Some issues and questions to consider:* How do population growth, resource consumption, pollution, and reduction in biodiversity relate to sustainability? How do poverty, economic growth and development, and political issues relate to sustainability? Why might developed and developing nations take different views of a sustainable society? What would life be like in a sustainable society? Have any steps toward sustainability been taken in your community? What are the obstacles to sustainability in your community? What steps have you taken toward a sustainable lifestyle? How old will you be in 2030? What do you think life will be like then?

APPENDIX 5

Photographic Credits

Frontispiece ©Dwight R. Kuhn.

Unit openers: Part I ©Glenn Hoffman. Part II ©CNRI/Science Photo Library/Photo Researchers, Inc. Part III ©Tom and Susan Bean 1986/DRK Photo. Part IV ©Michael Giannechini/Photo Researchers, Inc. Part V ©Art Wolfe. Part VI ©Larry Ulrich. Part VII ©S.R. Maglione/Photo Researchers, Inc.

Chapter 1: **Chapter opener 1** ©Mark Moffett/Minden Pictures. **Chapter opener 2** Mark Moffett©National Geographic Society. **1.00, 1.01** ©Richard Allen Wood/Animals Animals. **1.01** ©Richard Allen Wood/Animals Animals. 1.01 N.L. Max, University of California/BPS. **1.03a** ©Therese Frare/Courtesy of Nalini Nadkarni. **1.03b** ©Mark Moffett/Minden Pictures. **1.04a** ©A.B. Dowsett/Science Source/Photo Researchers, Inc. **1.04b** ©Ralph Robinson/Visuals Unlimited. **1.04c** ©D. P. Wilson/Photo Researchers, Inc. **1.04d** ©David Matherly/Visuals Unlimited. **1.04e** ©Leonard Lee Rue III/Earth Scenes/Animals Animals. **1.04f** ©Michael and Patricia Fogden. **1.05a left** ©Frans Lanting/Minden Pictures. **1.05a right** ©Kjell B. Sandved/Photo Researchers. **1.05a bottom** ©Kjell B. Sandved/Visuals Unlimited. **1.05b** Richard Wagner/UCSF Graphics. **1.06a** Photo Courtesy Department of Library Services/American Museum of Natural History. **1.06c left** ©Frans Lanting/Minden 1991. **1.06c right** ©Fred Felleman/Tony Stone Images. **1.07b left** ©Gregory Dimijian, M.D./Photo Researchers, Inc. **1.07b right** ©P. Henry/ Jacana/Photo Researchers, Inc. **1.08b left** ©Kevin Schafer/Martha Hill. **1.08b right** ©G. Prance/Visuals Unlimited.

Chapter 2: **Chapter opener 1** ©Carr Clifton. **Chapter opener 2** ©Robert Fletterick/UCSF. **2.01b** ©Dr. Jeremy Burgess/Science Photo Library/Photo Researchers, Inc. **2.01c** ©Manfred Kage/Peter Arnold, Inc. **2.01d** ©John Shaw/Tom Stack and Associates. **2.02** Courtesy of James Pushnik, California State University, Chico. **2.05a-b** ©Howard Sochurek. **2.10b** ©Kevin Schafer. **2.11** ©G.I. Bernard/Animals Animals. **2.12** ©David Madison 1999. **2.16** ©Maresa Pryor/Earth Scenes.

Chapter 3: **Chapter opener 1** ©Jane Burton/Bruce Coleman. **Chapter opener 2** Vollrath & Edmunds, Nature 340:305-317. **3.04a** ©Scott Camazine/Photo Researchers, Inc. **3.06** ©Alain McGlaughlin. **3.07 top** ©Biophoto Associates/Photo Researchers, Inc. **3.07 center** ©L.M. Beidler, Florida State University. **3.07 bottom** ©Biophoto Associates/Photo Researchers, Inc. **3.08a** ©Frans Lanting/Minden Pictures. **3.10** ©Mike Neveux. **3.11** ©Evan Agostini/Gamma-Liaison. **3.14b** Courtesy of the Graphics Systems Research Group, IBM U.K. Scientific Centre. **3.19** Courtesy of Linus Pauling.

Chapter 4: **Chapter opener** Courtesy Jan Hinsch/Leica, Inc. **4.01a top** ©Glenn Hoffman. **4.01a bottom** Courtesy of Carl Zeiss, Inc., Thornwood, NY 10594. **4.01b-c top** William Dentler/Biological Photo Service. **4.01b-c bottom** Courtesy of Carl Zeiss, Inc., Thornwood, NY **4.09** ©Barry King/Biological Photo Service. **4.10** ©Garry Cole/Biological Photo Service. **4.11a** ©Richard Rodewald/Biological Photo Service. **4.13a** W.P. Wergin, courtesy of E.H. Newcomb, University of Wisconsin, Madison. **4.13b** ©Roland Birke/Peter Arnold, Inc. **4.15** Courtesy of W.P. Wergin and E.H. Newcomb, University of Wisconsin/Biological Photo Service. **4.16** Courtesy of Dr. Nicolae Simionescu. **4.17a** Courtesy of Dr. B.R. Brinkley, Dept. of Cell Biology, Baylor College of Medicine. **4.17b** Courtesy of Dr. John Heuser, Washington University, St. Louis. **4.18a** W.L. Dentler, University of Kansas/Biological Photo Service. **4.21** ©Dr. Gopal Murti/Phototake.

Chapter 5: **Chapter opener 1** ©Robert and Linda Mitchell. **Chapter opener 2** ©James E. Lloyd, 1977. **5.01a-b, 5.02a** ©Alain McGlaughlin. **5.02b** ©Benjamin/Cummings. **5.10a** Courtesy of David Robertson. **5.10b** ©Joseph Hoffman, Yale University School of Medicine. **5.19c left** ©M. Abbey/Visuals Unlimited. **5.19c center** ©D.W Fawcett/Science Source/Photo Researchers, Inc. **5.19c right** M.M. Perry and A.B. Gilbert, J. Cell Sci. 39(1979): 257. Copyright 1979 The Company of Biologists Ltd.

Chapter 6: **Chapter opener** ©Eric Schabtach and Ira Herskowitz. **Chapter opener 2, 6.01** ©David Madison 1999 **6.06b** ©Stephen Frisch. **6.15c** Obester Winery, Half Moon Bay, CA. Photo by Kevin Schafer. **6.18** Photograph by Jessie Cohen, National Zoological Park, Smithsonian Institution. ©1999 Smithsonian Institution.

Chapter 7: **Chapter opener 1** ©Hans Reinhard/Bruce Coleman. **Chapter opener 2** ©John Sohlden/Visuals Unlimited, 1987. **7.01a-b** ©Larry Ulrich **7.01c** David J. Wrobel, Monterey Bay Aquarium/BPS. **7.01d** ©Patricia Sparling/Visuals Unlimited. **7.02 left** ©M. Eichelberger/Visuals Unlimited **7.02 right** Courtesy of W.P. Wergin and E.H. Newcomb, University of Wisconsin/Biological Photo Service. **7.03a** ©Runk/Schoenberger/Grant Heilman. **7.07a** ©Christine L. Case. **7.13a** ©Tom and Pat Leeson. **7.14** ©Stella Johnson.

Chapter 8: **Chapter opener 1** ©Robert D. Burke, Department of Biology, University of Victoria. **Chapter opener 2** ©Daniel W. Gotshall/Visuals Unlimited. **8.01a** ©Biophoto Associates/Photo Researchers, Inc. **8.01b** ©Dave Parker/Globe Photos. **8.03b** ©Lee D. Simon/Photo Researchers, Inc. **8.04a** ©Andrew Bajer/University of Oregon. **8.04b** ©Biophoto Associates/Photo Researchers, Inc. **8.06** ©Ed Reschke. **8.07a** ©David M. Phillips/Visuals Unlimited. **8.07b** Micrograph by B.A. Palevitz. Courtesy of E.H. Newcomb, University of Wisconsin. **8.11a** ©Carolina Biological Suppy/Phototake. **8.11c** ©Biophoto Associates/Science Source/Photo Researchers, Inc. **8.17b** Courtesy of Stan Short/Jackson Laboratory. **8.18a** ©Cabisco/Visuals Unlimited. **8.19 left** ©SIU/Visuals Unlimited. **8.19 right** ©CNRI/SPL/Photo Researchers, Inc. **8.20a** ©CNRI/Science Photo Library/Photo Researchers, Inc. **8.20b** ©Richard Hutchings/Photo Researchers, Inc. **End of chapter 1** ©Carolina Biological Supply/Phototake.

Chapter 9: **Chapter opener** ©Hans Reinhard/Bruce Coleman Photography. **9.02a** Bettmann Archives. **9.05b** Science Education Resources Pty. Ltd., Victoria, Australia **9.08a** ©Llewellyn/Uniphoto Picture Agency. **9.08a** ©Eliane Sulle/The Image Bank. **9.08a** ©Boisvieux/Explorer/Photo Researchers, Inc. **9.08a** ©Anthony Loveday/Benjamin Cummings. **9.09b** ©Dick Zimmerman/Shooting Star. **9.10c** ©Blair Seitz/Photo Researchers, Inc. **9.10d** ©Howard Sochurek. **9.14** ©Bill Longcore/Photo Researchers, Inc. **9.15a** ©Kevin Keister/Benjamin/Cummings. **9.15b** ©Custom Medical Stock Photos. **9.20a** From Thomas Hunt Morgan: The Man and His Science, Garland Allan (Princeton University Press, 1978). Photo by Dr. Tore Mohr, Fredrikstad, Norway. **9.22a top** ©Jean Claude Revy/Phototake NYC. **9.22a bottom** ©Carolina Biological Supply/PhotoTake NYC. **9.23b** ©Culver Pictures, Inc. **End of chapter 1** ©Norma Jubinville.

Chapter 10: **Chapter opener** ©Oliver Meckes/E.O.S./Photo Researchers, Inc. **10.02d** ©Ken Eward/Science Source/Photo Researchers, Inc. **10.03a** Cold Spring Harbor Archives. **10.03b** Photo Researchers, Inc. **10.03d** ©Richard Wagner, UCSF Graphics. **10.06b** ©Christine Case. **10.10** Courtesy of D.M. Prescott, University of Colorado Medical School; reproduced from Prog. Nucleic Acid Res. III, 35 (1964), with permission. **10.11c** Photo provided by T. A. Steitz. **10.12a** ©Joachim Frank, Howard Hughes Medical Institute. **10.19** ©Visuals Unlimited. **10.20a** ©Science Source/Photo Researchers, Inc. **10.20b** ©Scott Camazine/Science Source/Photo Researchers, Inc. **10.22** ©Lennart Nilsson, Boehringer Ingelheim International.

Chapter 11: **Chapter opener 1** ©David Scharf. **11.01a** ©Biophoto Associates/Science Source/Photo Researchers, Inc. **11.02a-d** ©Ed Reschke. **11.03b** Courtesy of the Roslin Institute, Edinburgh. **11.05a top** ©A.L. Olins, Univ. of Tennessee/ Biological Photo Service. **11.05a bottom** GF Bahr, Armed Forces Institute of Pathology. **11.05b** ©Grant Heilman/Grant Heilman Photography. **11.10a** ©F.R. Turner, Indiana University. **11.16** Courtesy of the University of Washington.

Chapter 12: **Chapter opener 1** ©Dennis Kunkel/Phototake. **Chapter opener 1 (inset)** Leonardo da Vinci, Vitruvian Man, ca. 1492. Photo ©Scala/Art Resource, N.Y. **Chapter opener 2** ©Nita Winter. **12.02c** Courtesy of Huntington Potter, University of South Florida, and David Dressler, Oxford University. **12.14a** ©Associated Press/World Wide Photos. **12.14b** Courtesy of Virginia Walbot, Stanford University. **12.15** Courtesy of Applied Biosystems. **12.16a** ©Steve Miller/AP/Wide World Photos. **12.18** ©Hank Morgan/Photo Researchers, Inc. **12.21** ©Bill Beatty/Visuals Unlimited. **12.22a** Courtesy of University of California, San Francisco. **12.22b** ©Robin Heyden/Benjamin Cummings. **12.22c** Courtesy of Dr. Nancy Wexler, Columbia University. **12.22d** Courtesy of Davis Freeman, University of Washington.

Chapter 13: **Chapter opener 1** ©William Paton/NHPA. **Chapter opener 2** ©D. & M. Plage/Oxford Scientific Films/Animals Animals. **13.01a** ©Frans Lanting/Minden Pictures. **13.01b** Larry Burrows, Life Magazine ©Time Inc. **13.02a** ©John Reader/Science Photo Library/Photo Researchers, Inc. **13.02b left** ©Manfred Kage/Peter Arnold, Inc. **13.02b right** ©Chip Clark, NMNH Smithsonian Institution **13.02c** ©Tom Till. **13.02d** ©Walter Hodge/Peter Arnold, Inc. **13.02e** ©Tom and Susan Bean 1986/DRK Photo. **13.02f** Philip Gingerich ©1991 Discover Magazine. **13.04a** ©Jack Wilburn/Earth Scenes/ Animals Animals. **13.04b.1** ©Gerard Lacz/Peter Arnold. **13.04b.2-3** ©Ron Kimball. **13.04b.4-5** ©Gerard Lacz/Peter Arnold, Inc. **13.04c.1** ©Frans Lanting/Minden Pictures. **13.04c.2-3** ©Tom and Pat Leeson. **13.04c.4** ©D. Robert Franz/The Wildlife Collection. **13.04c.5** ©Joe McDonald/Tom Stack and Associates. **13.05a** ©Animals Animals/Oxford Scientific Films. **13.05b** ©Breck P. Kent/Animals Animals. **13.06** USAF, NOAA/NESDIS at Univ. of CO, CIRES/National Snow and Ice Data Center. **13.11b** ©Frank S. Balthis. **13.13** ©R. Andrew Odum/Peter Arnold, Inc. **13.16** ©Gunter Ziesler/Peter Arnold, Inc. **13.17** ©S. Rhia/Custom Medical Stock Photography.

Chapter 14: **Chapter opener 1**©E. R. Degginger/Animals Animals. **Chapter opener 2** ©Robert E. Hynes / National Geographic Image Collection. **14.01a left** ©Don & Pat Valenti/Tom Stack & Associates. **14.01a right** ©John Shaw/Tom Stack and Associates. **14.01b** ©Photodisc. **14.03b** ©Michael Fogden/Animals Animals. **14.03c** ©John Eastcott/The Image Works. **14.04 top left** ©Larry Ulrich/DRK. **14.04 bottom left** ©John Shaw/Bruce Coleman, Inc. **14.04 right** ©M.P.L. Fogden/Bruce Coleman. **14.05a** ©David Cavagnaro. **14.06b** Courtesy of the University of Amsterdam. **14.07a** ©Marge Lawson. **14.08c top** ©William E. Townsend Jr./Photo Researchers, Inc. **14.08c bottom left** ©T. Kitchin/Tom Stack and Associates. **14.08c bottom right** ©Phillip Rouillard. **14.09** ©Kathleen Cameron/Addison Wesley Longman Inc.

Chapter 15: **Chapter opener 1** ©1989 Donna Braginetz. **Chapter opener 2** ©Bruce Selyem/Museum of the Rockies. **15.03c** ©Zig Leszczynski/Animals Animals. **15.04a** ©Kevin Schafer. **15.04b** Courtesy Tom Simkin/Smithsonian Institution. From "Krakatau 1883: The Volcanic Eruption and its Effects," by Tom Simkin and Richard S. Fiske, Smithsonian Institution Press, 1983. **15.06a** ©Jim Strawser/Grant Heilman. **15.06b** ©Richard Schiel/Earth Scenes. **15.07a** ©Jane Burton/Bruce Coleman. **15.07c** ©Disney Enterprises, Inc. **15.11** ©Tom Hugh/Photo Researchers, Inc. **15.14b** ©A.B. Dowsett/ Science Source/Photo Researchers, Inc. **15.14b** ©Ralph Robinson/Visuals Unlimited.

Chapter 16: **Chapter opener 1** ©Jonathan Blair 1990/Woodfin Camp and Associates. **Chapter opener 2, 16.01b, 16.01d** Stanley Awramik/Biological Photo Service. **16.03a** ©Roger Ressmeyer/Corbis. **16.06b** ©Sidney Fox. **16.07** ©Dr. Tony Brain/David Parker/Science Photo Library/Photo Researchers, Inc. **16.09a-c** ©David M. Phillips/Visuals Unlimited. **16.10, 16.12** ©Christine L. Case. **16.13a** ©Lee D. Simon/Science Source/ Photo Researchers, Inc. **16.13b** Photo from Fran Heyl Assoc.: photo by Dr. David Hasty, computer enhancement by pixelation. **16.13c** ©H.S. Pankratz, T.C. Beaman/Biological Photo Service. **16.13d** ©David M. Phillips/Visuals Unlimited. **16.14a** ©Frederick D. Atwood. **16.14b** ©M. I. Walker/Photo Researchers, Inc. **16.15a** ©Christine L. Case. **16.15b left** ©L. West/Bruce Coleman, Inc. **16.15b right** Centers for Disease Control. **16.16a** The Bettmann Archive. **16.17** Douglas Munnecke/BPS. **16.18a** Courtesy of Exxon Corporation. **16.18b** Courtesy of Susan Barnett. **16.21** ©Roland Birke/Peter Arnold. **16.22a left** ©E. White/Visuals Unlimited. **16.22a right** John Mansfield/University of Wisconsin, Madison. **16.22b** ©Peter Parks/Oxford Scientific Films/Animals Animals. **16.22c** ©Dr. Masamichi Aikawa. **16.22d left** ©M. Abbey/Visuals Unlimited. **16.22d right** ©Eric Grave/Science Source/ Photo Researchers, Inc. **16.23 top** Courtesy of Matt Springer, Stanford University. **16.23 center, bottom** Courtesy of Robert Kay, MRC Cambridge. **16.24a** ©E.R. Degginger/Earth Scenes. **16.24b** ©George Loun/Visuals Unlimited. **16.25a** ©Biophoto Associates/Photo Researchers, Inc. **16.25b** ©Manfred Kage/Peter Arnold, Inc. **16.25c** ©Herb Charles Ohlmeyer/Fran Heyl Associates. **16.25c inset** ©Manfred Kage/Peter Arnold. **16.26a** ©Lewis Trusty/Animals Animals. **16.26b** ©Gary Robinson/Visuals Unlimited. **16.26c** ©D.P. Wilson/Eric and David Hosking/Photo Researchers, Inc.

Chapter 17: **Chapter opener 1** ©Bob Abraham/Stock Market. **Chapter opener 2** ©J.M. Graham/University of Florida. **17.01b** ©Rod Planck/Tom Stack & Associates. **17.02a** ©L. E. Graham. **17.02b** ©Chip Clark, NMNH Smithsonian Institution. **17.03b** ©Stephen J. Krasemann/DRK Photo. **17.03c inset** ©Glenn Oliver/Visuals Unlimited. **17.03c** ©Walter H. Hodge/Peter Arnold, Inc. **17.05** ©Glenn Oliver/Visuals Unlimited. **17.06** ©Milton Rand/ Tom Stack and Associates. **17.07** Field Museum, Neg# CSGEO 75400c, Chicago. **17.08** ©Dr. William M. Harlow / Photo Researchers, Inc. **17.09a** ©F. Stuart Westmorland/Photo Researchers, Inc. **17.09b** ©Linda Bartlett/Addison Wesley Longman. **17.09c** ©Lara Hartley. **17.11a** ©M.E. Warren/Photo Researchers, Inc. **17.13a** ©Jane Burton/Bruce Coleman, Inc. **17.13b** ©Scott Camazine/Photo Researchers, Inc. **17.13c** ©Dwight R. Kuhn. **17.14a** ©Dee Wilder. **17.14b** ©Bob and Clara Calhoun/Bruce Coleman. **17.14c** ©Merlin D. Tuttle/Bat Conservation International. **17.15a** ©John D. Cunningham/Visuals Unlimited. **17.15b** Visuals Unlimited©Cabisco. **17.15c** ©Gregory G. Dimijian/Photo Researchers, Inc. **17.16b** ©Kjell B. Sandved/Visuals Unlimited. **17.16c** ©Kerry T. Givens/Tom Stack & Associates. **17.16d** ©Kerry T. Givens/Tom Stack and Associates. **17.18a** ©Robert Lee/Photo Researchers, Inc. **17.18c** ©Michael Giannechini/Photo Researchers, Inc. **17.19a** ©Stuart Bebb/Oxford Scientific Films/Animals Animals/Earth Scenes. **17.19b** ©Leonard Lee Rue III/Photo Researchers, Inc. **17.19c** ©Camera MD Studio. **17.20a** ©MG3/Vision/Photo Researchers, Inc. **17.20b** ©Christine Case.

Chapter 18: **Chapter opener 1**©A. Witte/C. Mahaney/Tony Stone Images. **Chapter opener 2** ©Chris Huss/Wildlife Collection. **18.01** ©Anthony Mercieca/Photo Researchers, Inc. **18.03a** ©Charles Wyttenbach/BPS. **18.04a** ©Gwen Fidler/Comstock. **18.04b** ©Claudia Mills/Friday Harbor Labs. **18.04c** ©Ken Lucas/Planet Earth Pictures. **18.06b** Centers for Disease Control. **18.06c** ©Drs. Kessel and Shih/Peter Arnold. **18.08a** ©S. Stephanowicz/ Science Photo Library/Photo Researchers, Inc. **18.08b** ©Andrew Syred/Science Photo Library/Photo Researchers, Inc. **18.09b** ©Robert and Linda Mitchell. **18.09c** ©Chris Huss. **18.09d** H.W. Pratt/BPS. **18.09e** ©Carl Roessler/Animals Animals. **18.09f** ©Charles R. Wyttenbach/Biological Photo Service. **18.10b** ©John Gerlach/Tom Stack and Associates. **18.10c** ©Mike Neveux **18.11a** ©Jeff Goodman/NHPA. **18.11b** ©Kjell Sandved. **18.11c** ©Robert and Linda Mitchell. **18.12b** ©Zig Leszczynski/Animals Animals. **18.12c left** ©Robert and Linda Mitchell. **18.12c center** ©Paul Skelcher/Rainbow. **18.12c right** ©David Scharf. **18.12d** ©Tom Stack/Tom Stack and Associates. **18.12e** ©Robert and Linda Mitchell. **18.14b** ©Gary Milburn/Tom Stack and Associates. **18.14c** ©David Wrobel. **18.15a** ©Robert Brons/Biological Photo Service. **18.15b** ©Runk/Schoenberger/Grant Heilman Photography, Inc. **18.17a** ©Tom McHugh, Steinhart Aquarium/Photo Researchers, Inc. **18.18a** ©Tom McHugh, Steinhart Aquarium/Photo Researchers, Inc. **18.18c top** ©Richard Ellis/Photo Researchers, Inc. **18.18c bottom** ©Tom McHugh/Photo Researchers, Inc. **18.19a** ©1990 Suzanne L. & Joseph T. Collins/Photo Researchers, Inc. **18.19b** ©Hans Pfletschinger/Peter Arnold, Inc. **18.20a** ©Robert and Linda Mitchell. **18.20b** ©C. Allan Morgan. **18.20c** The Natural History Museum, London. **18.21b** ©B.K. Wheeler/VIREO. **18.21c** ©David Dare Parker/Auscape International. **18.22a** ©Jean Phillipe Varin/ Jacana/Photo Researchers, Inc. **18.22b** ©Steffan Widstrand/Bruce Coleman. **18.22c** ©Mitch Reardon/Photo Researchers, Inc. **18.24** ©C. Seghers, 2/Photo Researchers, Inc.

Chapter 19: **Chapter opener 1** ©Paul Hanny/Gamma-Liaison. **19.00** Kenneth Garrett©National Geographic Society. **Chapter opener 2** ©E.H. Rao/ Photo Researchers, Inc. **19.01b** ©Frans Lanting/Minden Pictures. **19.01c left** ©Norman Owen/Bruce Coleman, Inc. **19.01c right** ©Wardene Weisser/Bruce Coleman, Inc. **19.01d left** ©Frans Lanting/Minden Pictures. **19.01d right** ©Chicago Zoological Park/Tom McHugh/Photo Researchers, Inc. **19.02a** ©E. R. Degginger/Animals Animals. **19.02b** ©Frans Lanting/Minden Pictures. **19.02c** ©Nancy Adams/Tom Stack and Associates. **19.02d** ©Ken Regan/Camera 5. **19.04a** ©John Reader/Science Photo Library/Photo Researchers, Inc. **19.04b** ©John Gurche. **19.06** ©Special Collections, Department of Library Services, American Museum of Natural History. **19.08** ©Anthony Bannister/Earth Scenes. **19.09** ©Elliot Erwitt/Magnum Photos, Inc. **19.10** ©Mary Evans Picture Library/Photo Researchers, Inc.

Chapter 20: Chapter opener 1 ©Jean-Paul Ferrero/Auscape. **20.01** ©Janice Sheldon. **20.07** ©Ed Reschke. **20.10a inset** ©Howard Sochurek. **20.10a** ©Mehau Kulyk/Science Photo Library/Photo Researchers, Inc. **20.10b-c** ©Howard Sochurek. **20.10d** Courtesy of Marcuse E. Raichle, Washington University School of Medicine. **20.11c** ©Science Photo Library/Photo Researchers, Inc. **20.12a** ©Frans Lanting/Minden Pictures.

Chapter 21: **Chapter opener 1** ©Richard Schlecht / National Geographic Image Collection. **Chapter opener 2** ©Brandon Cole. **21.01a** ©Nigel Dennis/NHPA. **21.01b** ©Jett Britnell/DRK Photos. **21.01c** ©Tom Eisner, Cornell University. **21.01d** ©Hans Pfletschinger/Peter Arnold, Inc. **21.01e** ©J. R. D'Ambrosio/Sygma. **21.12a left** ©Brian Milne/Animals Animals. **21.12a right** ©Hans and Jusy Beste/Animals Animals. **21.14** ©William Thompson.

Chapter 22: **Chapter opener 1** ©Thomas Mangelsen. **Chapter opener 2** ©Dennis and Bo Hansen/Ric Ergenbright Photography. **22.07a** ©Martin Rotker. **22.07b** ©Martin Rotker. **22.08b** ©Dr. Hans Rainer Dunker, Justus Liebig University, Giessen, Germany.

Chapter 23: **Chapter opener 1** ©Kennan Ward 1986/DRK. **Chapter opener 2** ©Malcolm Boulton/Photo Researchers, Inc. **23.01a** ©Lennart Nilsson, The Body Victorious, Dell Publishing Company. **23.08b** ©Custom Medical Stock Photography. **23.12a** ©D.W. Fawcett/Photo Researchers, Inc. **23.14** ©David M. Phillips/Visuals Unlimited. **23.15** ©Victor Eroschenko/Benjamin/Cummings. **23.16b** ©Boehringer Ingelheim International GmbH. Photo: Lennart Nilsson.

Chapter 24: **Chapter opener** ©Jim Bourg/Gamma-Liaison. **24.01a** ©Lennart Nilsson/Boehringer Ingelheim International GmbH. **24.03d** ©Bruce Iverson/Photo Researchers, Inc. **24.10a** ©A. J. Olson, The Scripps Research Institute, 1986. **24.12b** ©Custom Medical Stock Photography. **24.14** ©Boehringer Ingelheim International GmbH. Photo: Lennart Nilsson.

Chapter 25: **Chapter opener 1** ©Ron and Valerie Taylor/Ardea London. Chapter opener 2 ©Art Wolfe. **25.02a** ©Robert and Linda Mitchell. **25.03** ©E. Lyons/Bruce Coleman, Inc. **25.04** Courtesy of J.M. Storey. **25.06** ©David Madison 1999. **25.07** Courtesy of Dr. John Crowe, University of California, Davis.

Chapter 26: **Chapter opener** ©Chris Howes/Planet Earth Pictures. **26.06a** ©Ivan Polunin/Bruce Coleman, Inc. **26.11** ©AP/Wide World Photos. **26.12** ©Kennan Ward.

Chapter 27: **Chapter opener** ©David Crews. **27.01a** ©David Wrobel. **27.01b** ©Jim Solliday/Biological Photo Service. **27.01c** ©Hans Pfletschinger/Peter Arnold, Inc. **27.01d** ©Dwight Kuhn. **27.02b** ©C. Edelman/La Vilette/Photo Researchers, Inc. **27.08** ©Anthony Loveday/Benjamin/Cummings. Contraceptive devices courtesy of Planned Parenthood, Inc. **27.09a** ©David Scharf/Peter Arnold, Inc. **27.12a** ©Cabisco/Visuals Unlimited. **27.12c** Courtesy of Thomas Poole, SUNY Health Science Center. **27.12d** ©Hans Pfletschinger/Peter Arnold, Inc. **27.17a-e** ©Lennart Nilsson/A Child is Born, Dell Publishing. **27.19** ©Andy Walker, Midland Fertility Services/Science Photography Library/Photo Researchers, Inc.

Chapter 28: **Chapter opener 1** ©Jeffrey L. Rotman/Bruce Coleman. **Chapter opener 2** A. L. Hodgkin and R.D. Keynes, J. Phys. 131(1956):615. **28.07** Dr. E.R. Lewis, T. E. Everhart, Y. Y. Zeevi, Science 165(1969):1140-43. **28.17** Courtesy of Dr. Roger Sperry. **28.18b** ©Alexander Tsiarias/Photo Researchers, Inc.

Chapter 29: **Chapter opener 1** ©Henry Ausloos/NHPA. **Chapter opener 2** ©Keith Gunnar/Bruce Coleman, Inc. **29.03c** Courtesy of Dr. R.A. Steinbrecht. **29.03d** ©Joe McDonald/Animals Animals. **29.04** ©Syd Greenberg/Photo Researchers, Inc. **29.04b** ©John L. Pontier/Animals Animals. **29.04c** Woods Hole, Mass./Biological Photo Service.

Chapter 30: **Chapter opener** ©Mark W. Moffett/Minden Pictures. **30.01a** ©Marty Snyderman. **30.01b** ©John Cancalosi/Tom Stack and Associates. **30.01c** ©Gerard Lacz/NHPA. **30.01e** ©Stephen J. Krasemann/DRK Photo. **30.02a** ©Dwight Kuhn. **30.02b** ©Tony Florio/Photo Researchers, Inc. **30.02c** ©E.R. Degginger/Animals Animals. **30.02d** ©Marty Snyderman. **30.06a** ©Carolina Biological Supply/PhotoTake NYC. **30.08** Courtesy of Clara Franzini-Armstrong, University of Pennsylvania. **30.11a** ©Matthew Mulbry. **30.11b** ©Bruce Frisch. **30.12a** ©Al Tielemans 1990/Duomo. **30.12b** ©Norman Owen Tomalin/Bruce Coleman, Inc.

Chapter 31: **Chapter opener 1** National Park Service. **Chapter opener 2** ©Tom McHugh/Photo Researchers, Inc. **31.01a** Courtesy of J. Thorsch. **31.01b** UCSB Photography Department. **31.04c top** ©Larry Mellichamp/Visuals Unlimited. **31.04c botttom** ©John D. Cunningham/Visuals Unlimited. **31.05c** ©Nels Lersten, Iowa State University. **31.05d-e** ©Bruce Iverson. **31.05f** ©Randy Moore/Visuals Unlimited. **31.06b-d, 31.07c** ©Ed Reschke. **31.08b** ©Runk/Schoenberger/Grant Heilman Photography, Inc. **31.12a** ©W.H. Hodge/Peter Arnold, Inc. **31.12c** ©Anthony Loveday/Benjamin/Cummings. **31.13a** ©Barry Runk/Grant Heilman. **31.13b** ©Runk/Schoenberger/Grant Heilman. **31.14a** ©Kevin Schafer. **31.14b** ©Frank Balthis. **31.14c** ©Galen Rowell/Mountain Light. **31.14d** ©Francois Gohier/Photo Researchers, Inc. **31.15a** ©Runk/Schoenberger/Grant Heilman. **31.15b** ©Plantek/Photo Researchers, Inc.

Chapter 32: **Chapter opener 1** ©Jane Grushow/Grant Heilman. **Chapter opener 1 (inset)** ©C. P. Vance/Visuals Unlimited. **Chapter opener 2** The Bettmann Archive. **32.01b** ©Renee Lynn/Photo Researchers, Inc. **32.02a** ©Runk/Schoenberger/Grant Heilman. **32.05a** ©Ray F. Evert/University of Wisconsin. **32.05c** ©M.H. Zimmermann. **32.06b** ©Grant Heilman/Grant Heilman Photography. **32.07a** Courtesy of James Pushnik, California State University, Chico. **32.07b** ©Holt Studios/Earth Scenes. **32.07c-d** Courtesy of James Pushnik, California State University, Chico. **32.08a** ©Steven C. Wilson/Entheos. **32.09a** ©Runk/Schoenberger/Grant Heilman, Inc. **32.09b** ©Kevin Horan/Tony Stone Images. **32.10** Courtesy of Stephen Moore. **32.11** ©R. L. Peterson/Biological Photo Service. **32.12a** ©Kevin Schafer. **32.12b** ©Jim Strawser/Grant Heilman. **32.12c** ©Robert and Linda Mitchell. **32.12d** ©Jeff Lepore/Photo Researchers, Inc. **32.14a** ©Breck P. Kent/Earth Scenes. **32.14b** From E.H. Newcomb and S.R. Tandon/Biological Photo Service. **32.15a** Louisiana State University Public Relations. **32.16** Courtesy of Dr. John C. Sanford, Cornell University.

Chapter 33: **Chapter opener 1** ©Runk/Schoenberger/Grant Heilman Photography. **Chapter opener 2** ©E.R. Degginger/Photo Researchers, Inc. **33.01a** ©John Kaprielan/Photo Researchers, Inc. **33.03a** ©David Newman/Visuals Unlimited. **33.04** ©Walter Chandoha. **33.05a** ©Tugio Sasaki, Institute for Agricultural Research, Japan. **33.05b** Courtesy of Fred Jensen, Kearney Agricultural Center. **33.06** ©Larry Ulrich. **33.07a** ©Runk/Schoenberger/Grant Heilman, Inc. **33.07b** ©Ed Reschke. **33.08** ©John Colwell/Grant Heilman Photography, Inc. **33.09a** ©Michael Evans, Ohio State University. **33.09b** ©Scott Camazine/Photo Researchers, Inc. **33.10** ©Frank B. Salisbury. **33.13** Courtesy of Dr. Coruzzi. **33.14b** Courtesy of Barbara Baker, University of California, Berkeley. **33.15** ©Mary De Chirico/Benjamin/Cummings.

Chapter 34: **Chapter opener** ©Emory Kristof / National Geographic Image Collection. **34.00a** ©J. Edmond/Visuals Unlimited. **34.00b** ©Robert Hessler/Planet Earth. **34.01** ©Carl Wirsen/Woods Hole Oceanographic Institute. **34.02a** Courtesy of NASA. **34.02b** ©Stephen Krasemann/Photo Researchers, Inc. **34.03** The Bettmann Archive. **34.04** ©Arthur C. Smith III/Grant Heilman Photography. **34.05** ©Erwin and Peggy Bauer/Natural Selection. **34.07a** ©M.E. Warren /Photo Researchers, Inc. **34.07b** ©Ron Sefton/Bruce Coleman, Inc. **34.07d** ©Douglas Faulkner/Photo Researchers, Inc. **34.08a** ©Michael Gadomski/Bruce Coleman, Inc. **34.08b** ©Wendell D. Metzen/Bruce Coleman, Inc. **34.10** ©Peter Ward/Bruce Coleman, Inc. **34.11a** ©Alain McLaughlin/Benjamin/Cummings. **34.12** ©S.R. Maglione/Photo Researchers, Inc. **34.13** ©Charlie Ott/Photo Researchers, Inc. **34.14** ©Jack Wilburn/Earth Scenes. **34.15** ©Tom McHugh/Photo Researchers, Inc. **34.16** ©Carr Clifton. **34.17** ©Kirtley Perkins/Photo Researchers, Inc. **34.18** ©J. Warden/Superstock, Inc.

Chapter 35: **Chapter opener** ©Maslowski/Visuals Unlimited. **35.02a** ©Jerry O. Wolff. **35.02b** ©Kirtley Perkins 1983/Visuals Unlimited. **35.02c left** ©Robert P. Carr/Bruce Coleman, Inc. **35.02c right** ©Breck P. Kent/Animals Animals/Earth Scenes. **35.03b** ©Erwin and Peggy Bauer/Natural Selection. **35.04b** ©Runk/Schoenberger/Grant Heilman Photography. **35.05** ©Alan Carey/Photo Researchers, Inc. **35.07a** Courtesy of John Endler, UCSD. **35.07b** ©R. Calentine/Visuals Unlimited. **35.07c** ©Max and Bea Hunn/Visuals Unlimited. **35.08a** ©Alain Evrard/Photo Researchers, Inc. **35.08e** ©Arvind Garg. **35.10** ©Cynthia Johnson/Time Magazine. **35.10** ©Joanna B. Pinneo/Aurora.

Chapter 36: Chapter opener 1 ©Y. Sato. **36.00** ©John Serrao/Visuals Unlimited. **36.00** ©Norm Thomas/Photo Researchers, Inc. **36.00** ©E.R. Degginger/Photo Researchers, Inc. **36.01** ©Marcia W. Griffen/Animals Animals. **36.02b** ©Heather Angel/Biofotos. **36.03a** ©Lawrence E. Gilbert/Biological Photo Service. **36.03b** ©Kevin Schafer/Tom Stack and Associates. **36.03c** ©C. Allan Morgan/Peter Arnold, Inc. **36.03d top** ©J. Alcock/Visuals Unlimited. **36.03d bottom** ©Scott Camazine/Photo Researchers, Inc. **36.03e top** ©E.S. Ross. **36.03e bottom** ©Runk/Schoenberger/Grant Heilman Photography. **36.04a** Dr. Gregory A. Antipa. **36.04b** ©William E. Townsend/Photo Researchers, Inc. **36.05a** Australian Embassy Photo Library. **36.05b** ©Robert and Linda Mitchell. **36.06, 36.07a-b** ©Courtesy of Frank Gilliam, Marshall University. **36.18a** ©John D. Cunningham/Visuals Unlimited. **36.18b** Provided by the Northeastern Forest Experiment Station, Forest Service, United States Department of Agriculture.

Chapter 37: Chapter opener 1 ©Frans Lanting/Minden Pictures. **Chapter opener 2** ©Alan Rabinowitz. **37.03b left** ©Ian Wyllie/Oxford Scientific Films. **37.03b right** ©Michael Leach/OSF/Animals Animals. **37.05a** Thomas McAvoy, Life Magazine.©Time Inc. **37.05b** ©Calvin Larsen/Photo Researchers, Inc. **37.06a** ©David Weintraub 1986/Photo Researchers, Inc. **37.06b left** ©Harry Engels/Animals Animals. **37.06b right** ©Stephen J. Krasemann/Photo Researchers, Inc. **37.07b** Frans de Waal/Primate Center Library. **37.09a** ©Stouffer/Animals Animals. **37.09b** ©Thomas Ives. **37.11a** ©François Gohier/Photo Researchers, Inc. **37.11b** ©Jonathan Blair/Woodfin Camp. **37.12a** ©Kenneth Mallory, New England Aquarium. **37.12b** ©John Cancalosi/Peter Arnold, Inc. **37.12c** ©Runk/Schoenberger/Grant Heilman Photography. **37.12d left** ©Joe McDonald/Animals Animals. **37.12d right** ©Kevin Schafer/Peter Arnold, Inc. **37.12e** ©Joe McDonald/Animals Animals. **37.14** ©Gordon Wiltsie. **37.15** ©Renne Lynn/Benjamin/Cummings. **37.16a** ©Ken Regan/Camera 5. **37.16b** ©Warren and Genny Garst/Tom Stack and Associates. **37.17a** ©Bill Bachman/Photo Researchers, Inc. **37.17b** ©Gohier/Photo Researchers, Inc. **37.17c** ©Animals Animals/J. & B. Photo. **37.18a** ©Peter Roberts. **37.18b** ©Anthony Mercieca/Photo Researchers, Inc. **37.19a** ©J.P. Varin/Jacana/Photo Researchers, Inc. **37.19b** ©Kenneth Lorenzen, UC Davis. **37.20a** Courtesy of Dr. Jennifer Jarvis, University of Cape Town. **37.20b** ©Ray Mendez/Animals Animals. **37.21** ©John Bohn. **37.22** ©Ivan Polunin/Bruce Coleman.

Chapter 38: Chapter opener 1 ©Tom & Pat Leeson/Photo Researchers, Inc. **Chapter opener 2** ©Tess Young/Tom Stack & Associates. **38.01a** ©Juan Manuel Renjifo/Earth Scenes. **38.01b** ©Darrel Plowes. **38.02** ©Richard Shiell/Earth Scenes. **38.03a** ©Al Grillo/Alaska Stock Images. **38.05c** ©Frans Lanting. **38.05d** ©Mark Conlin/Planet Earth Pictures. **38.06a** ©Gary Braasch/Woodfin Camp & Associates. **38.06b** ©Greg Vaughn/Tom Stack & Associates. **38.07a** ©Rob Curtis. **38.07b** ©Raymond K. Gehman/National Geographic Society. **38.07c** ©Blanche Haning. **38.08a** ©William H. Mullins/Photo Researchers, Inc. **38.08b** ©Al Bratkovich. **38.10a** ©David Hosking/Photo Researchers, Inc. **38.10b** ©Calvin Larsen/Photo Researchers, Inc. **38.10c** Courtesy Florida Department of Transportation. **38.11a** ©The Natural History Museum, London. **38.11b** Courtesy of Ariel Lugo/USDA/Forest Service, photo by Mr. Jesus Ayala O'Neill. **38.12a** ©Dwight Kuhn.

APPENDIX 6

Illustration and text credits

The following figures are adapted from Neil Campbell, Jane Reece and Larry Mitchell *Biology,* 5th ed. (Menlo Park, CA: Benjamin/Cummings, 1999) ©1999 Addison Wesley Longman: Campbell 5e: 2.04a-b, 2.07a-b, 2.08, 2.09, 2.10a, 2.13, 2.14, 3.04b, 4.02, 4.04, 4.10, 4.11a-b, 4.15, 4.16, 4.17b, 4.19a-b, 5.04c, 5.06, 5.08, 5.11b, 5.12, 5.13a-c, 5.14a-b, 5.15a-b, 5.17, 6.08, 6.14, 7.02, 7.05, 7.06a-b, 7.07a-c, 7.08, 7.12a-c, 7.11, 8.03a, 8.08a-b, 8.10, 8.13, 8.14, 8.15, 8.16, 8.21a-c, 9.02c-d, 9.03a-b, 9.05a, 9.12a-b, 9.21, 10.1c, 10.2b, 10.03d, 10.05a-b, 10.05c, 10.07, 10.08a, 10.09a, 10.11a-b, 10.12, 10.18a, 11.01b, 11.03a, 12.4, 13.20, 14.02, 15.01, 15.03a-b, 15.07b, 15.08, 15.10, 15.12a, 15.13a, 16.03b, 16.20a-b, 17.03a, 17.17, 17.18b, 18.04d, 18.05, 18.06a-b, 18.09a, 18.12a, 18.15a, 18.21a, 18.23, 19.3b, 20.11a-b, 20.13a, 21.03a-b, 21.04, 21.06, 21.11, 21.12a-b, 22.03, 22.04, 22.05a-b, 23.02a-b, 23.03a-b, 23.04a, 23.05, 23.09a-b, 23.10, 23.11b, 23.13, 23.16a, 23.17, 24.02, 24.03, 24.06, 24.08a, 24.11, 24.12a, 25.05a-b, 25.09a-d, 26.05a-b, 26.07, 26.08, 26.10, 27.03d, 27.04a-c, 27.12c, 27.13a, 27.13b, 28.02, 28.05, 28.07, 28.10, 28.11a, 28.12a, 28.14, 28.18a, 28.18c, 28.19, 29.03a, 29.05, 29.09a, 29.09c, 29.10, 29.11, 29.12a, 30.03a-c, 30.08, 30.10a-b, 31.02,31.07b, 31.08a-b, 31.11a-b, 31.13a-b, 32.02b, 32.03, 32.04, 32.06a, 32.08b, 33.01b, 33.01c, 33.01d, 33.11, 33.12a, 33.14b, 34.06, 34.07c, 34.09, 35.08b, 36.02a, 36.09, 36.14, 37.02, 37.03a, 37.09a, 37.19c-d, 38.05a, 38.05b, 38.08c.

The following figures are adapted from Neil Campbell, *Biology,* 4th ed. (Menlo Park, CA: Benjamin/Cummings, 1997) © 1997 The Benjamin/Cummings Publishing Company: 3.01, 3.02, 3.04c, 5.04a-b, 5.16, 6.16, 7.09, 8.08a, 8.13, 8.17a, 8.19, 8.23a-b, 9.05b, 9.17, 9.19a-b, 9.20b, 10.06a, 10.13b, 10.14, 10.16b, 10.17, 10.18b, 10.21a, 10.21b, 11.7a, 12.2a-b, 12.05, 12.06, 12.07, 12.08b, 12.10, 12.11a-b, 12.11c, 12.19, 12.20, 13.02f, 14.06a, 14.08a-b, 16.01c, 16.05, 16.06a, 16.06c, 17.1a, 18.3c, 18.07a-c, 18.10a, 18.14a, 21.08, 21.16, 24.05, 24.07, 24.13a, 24.13c, 26.04a-b, 26.06b, 27.14, 30.09b, 31.06d, 31.10, 32.01a.

The following figures are adapted from Neil Campbell, *Biology,* 3rd ed. (Redwood City, CA: Benjamin/Cummings, 1993) © 1993 The Benjamin/Cummings Publishing Company: 4.14, 6.07a, 6.11b, 8.05, 8.07b, 8.20c, 9.07, 9.22b-d, 10.1b, 10.09b, 18.3d, 20.01, 22.02a-d, 24.03a-c, 25.11, 26.03, 27.03b, 27.09b, 27.12a, 28.02, 28.03a, 28.16, 29.06, 29.09b, 30.07, 32.13, 36.08, 37.07a.

The following figures are adapted from Neil Campbell, *Biology,* 2nd ed. (Redwood City, CA: Benjamin/Cummings, 1990) © 1990 The Benjamin/Cummings Publishing Company: 8.18b, 9.13, 9.16, 10.04b, 10.16a, 11.7b, 14.03a, 16.27, 20.02, 24.13b, 25.01, 31.08a, 32.03, 33.03b, 36.11.

The following figures are adapted from Neil Campbell, *Biology,* 1st ed. (Menlo Park, CA: Benjamin/Cummings, 1987) © 1987 The Benjamin/Cummings Publishing Company: 5.18, 5.19a-b, 14.05b.

The following figures are adapted from Lawrence G. Mitchell, John A. Mutchmor, and Warren D. Dolphin, *Zoology* (Menlo Park, CA: Benjamin/Cummings, 1988) ©1988 The Benjamin/Cummings Publishing Company: 2.06, 18.01, 18.15b, 18.16, 20.00, 21.10b, 25.08, 28.00, 29.12b, 30.02e, 37.01.

The following figures are adapted from Elaine N. Marieb, *Human Anatomy and Physiology,* 4th ed. (Menlo Park, CA: Benjamin/Cummings, 1998) © 1998 The Benjamin/Cummings Publishing Company: 4.8a, 20.9, 21.5, 21.6, 22.10a, 22.12, 24.9, 27.2a, 27.2c, 27.3a, 27.18b, 28.12b, 29.7a-b, 30.5, 30.6b.

The following figure is adapted from Elaine N. Marieb, *Human Anatomy and Physiology,* 1st ed. (Redwood City, CA: Benjamin/Cummings, 1989) © 1989 The Benjamin/Cummings Publishing Company: 25.12.

The following figures are adapted from Gerard J. Tortora, Berdell R. Funke, and Christine L. Case, *Microbiology: An Introduction,* 5th ed. (Menlo Park, CA: Benjamin/Cummings, 1998) © 1998 The Benjamin/Cummings Publishing Company: 10.01a, 12.3, 16.16b, 16.17, 24.10b.

The following figures are adapted from C. K. Mathews and K. E. van Holde, *Biochemistry* (Menlo Park, CA: Benjamin/Cummings, 1996) © 1996 The Benjamin/Cummings Publishing Company: 6.12, 6.13, 11.5a.

Chapter 1 Introduction: Adapted text quotation from Dr. Jerry Franklin in "The Secret Life of Backyard Trees" by Kevin Krajick, *Discover,* November 1995.

Figure 1.8A: Excerpt from Denise Grady, "A Move to Limit Antibiotic Use in Animal Feed," *New York Times,* March 8, 1999. Copyright ©1999 by The New York Times. Reprinted by permission.

Module 3.19: Talking About Science: Adapted from an interview in Neil Campbell, *Biology,* 1st ed. (Menlo Park, CA: Benjamin/Cummings, 1987).

Figure 4.18B: Adapted from B. Alberts, D. Bray, J. Lewis, M. Raff, K. Roberts, and J. D. Watson, *Molecular Biology of the Cell,* 2nd ed. (New York: Garland, 1989) p. 648.

Table 6.3: Data from C. M. Taylor and G. M. McLeod, *Rose's Laboratory Handbook for Dietetics,* 5th ed. (New York: Macmillan, 1949), p. 18; J. V. G. A. Durnin and R. Passmore, 1967, *Energy and Protein Requirements* in *FAO/WHO Technical Report* No. 522, 1973; W. D. McArdle, F. I. Katch, and V. L. Katch, 1981, *Exercise Physiology* (Philadelphia, PA: Lea & Feibiger, 1981); R. Passmore and J. V. G. A. Durnin, *Physiological Reviews* 35 (1955): 801-840.

Module 7.16: Talking About Science: Adapted from an interview in Neil Campbell, Jane Reece and Larry Mitchell, *Biology,* 5th ed. (Menlo Park, CA: Addison Wesley Longman, 1999).

Figure 8.22a-b: Adapted from F. Vogel and A. G. Motulsky, *Human Genetics* (New York: Springer-Verlag, 1982). ©1982 Springer-Verlag.

Figure 9.08b: Adapted from EVERYONE HERE SPOKE SIGN LANGUAGE by Nora Ellen Groce. Copyright © 1985 by Nora Ellen Groce. Reprinted by permission of Harvard University Press.

Figure 9.14: From *Introduction to Genetics Analysis* 4/E by Suzuki, Griffiths, Miller and Lewontin. Copyright © 1976, 1981, 1986, 1989, 1993, 1996 by W. H. Freeman and Company. Used with permission.

Module 9.15: Text quotation from Sheryl Gay Stolberg, "America's Doctor, David Satcher," *New York Times,* September 13, 1997. Copyright ©1997 by The New York Times.

Figure 10.00: Figure adapted from James Watson, N. H. Hopkins, J. W. Roberts, J. A. Steitz, and A. M. Weiner, *Molecular Biology of the Gene,* 4th ed. (Menlo Park, CA: Benjamin/Cummings, 1987). Copyright © 1987 James Watson.

Module 10.20: Text quotation from Barbara J. Culliton, "Emerging Viruses, Emerging Threat," *Science,* 19 January 1990, vol. 247, p. 279.

Figure 11.3c: Adapted from W. H. Becker, *The World of the Cell,* p. 592 (Redwood City, CA: Benjamin/Cummings, 1986) © 1986 by The Benjamin/Cummings Publishing Company.

Figure 11.12: Adapted from Illustration by William McGinnis, UCSD.

Module 11.16: Talking About Science: Adapted from an interview in Neil Campbell, Jane Reece and Larry Mitchell, *Biology,* 5th ed. (Menlo Park, CA: Addison Wesley Longman, 1999).

Table 11.17: Data from *1999 Facts & Figures: Estimated New Cancer Cases and Deaths,* published by the American Cancer Society.

Module 12.22: Text quotation from Leroy Hood, *Science News,* January 21, 1989. Science Service, Inc.

Module 14.9: Talking About Science: Adapted from an interview in Neil Campbell, Jane Reece and Larry Mitchell, *Biology,* 5th ed. (Menlo Park, CA: Addison Wesley Longman, 1999).

Figure 15.3d: From W. K. Purves and G. H. Orians, *Life: The Science of Biology,* 2nd ed. (New York: Sinauer Associates, 1987).

Figure 15.5: Adapted from an illustration by Patricia J. Wynne.

Figure16.01a: Artist: Peter Sawyer ©NMNH Smithsonian Institution.

Module 16.3: Talking About Science: Adapted from an interview in Neil Campbell, *Biology,* 2nd ed. (Redwood City, CA: Benjamin/Cummings, 1990).

Table 21.15: C.L. Rock and A. M. Coulston, "Weight-Control Approaches: A Review by the California Dietetic Association." Copyright ©1988 by The

American Dietetic Association. Reprinted by permission from the Journal of the American Dietetic Association Vol. 88 (1988): 44-48.

Table 21.17: Data from RDA Subcommittee, *Recommended Dietary Allowances* (Washington, DC: National Academy Press, 1989); M. E. Shils and V. R. Young, *Modern Nutrition in Health and Disease* (Philadelphia, PA: Lea & Feibiger, 1988).

Table 21.18: Data from (1) M. E. Shils, "Magnesium," in M. E. Shils and V. R. Young, eds. *Modern Nutrition in Health and Disease* (Philadelphia, PA: Lea & Feibiger, 1988); (2) V. F. Fairbanks and E. Beutler, "Iron" [same as 1]; (3) N. W. Solomons, "Zinc and Copper" [same as (1)]; (4) RDA Subcommittee, *Recommended Dietary Allowances* (Washington, DC National Academy Press, 1989); (5) E. J. Underwood, *Trace Elements in Human and Animal Nutrition* (New York: Academic Press, 1977).

Fig. 22.10b: Illustration copyright © Irving Geis.

Chapter 24 opening essay: Excerpt reprinted by courtesy of SPORTS ILLUSTRATED November 18, 1991. Copyright © 1991 Time Inc. "I'll Deal With It" by Magic Johnson with Roy Johnson. All rights reserved.

Figure 26.9: Data from C. J. Byrne, D. F. Saxton, et al., *Laboratory Tests: Implication for Nursing Care,* 2nd ed. (Menlo Park, CA: Addison-Wesley, 1986). © 1986 Addison-Wesley Publishing Company.

Figure 27.3c: Adapted from Robert Crooks and Karla Baur, *Our Sexuality,* 5th ed., figure 5.9 (Redwood City, CA: Benjamin/Cummings, 1993). © 1993 The Benjamin/Cummings Publishing Company.

Table 27.8: Data from R. Hatcher et al., *Contraceptive Technology: 1990-1992*, p. 134 (New York, Irvington, 1990).

Figure 28.6: Adapted from Wayne M. Becker and David W. Deamer, *The World of The Cell,* 2nd ed., figure 20.16. (Redwood City, CA: Benjamin/Cummings, 1991). © 1991 The Benjamin/Cummings Publishing Company.

Table 28.11: Data from H. R. Roberts and J. J. Barone, *Food Technology,* 37 (9): 32-39. Copyright ©1983 Institute of Food Technologists.

Figure 28.15A: Adapted from *The Vertebrate Body,* 6th by Alfred Sherwood Romer and Thomas S. Parsons, p. 569 (Philadelphia, PA: Saunders, 1986). Copyright©1986 by Saunders College Publishing, reproduced by permission of the publisher.

Module 28.18: Excerpt from "Left-Brain, Right-Brain" by Roger Sperry, *Saturday Review,* August 9, 1975, pp. 30-33.

Figure 30.3b: Baboon skeleton from Fleagle, J. G., *Primate Adaptations and Evolution* (Orlando, FL: Academic Press,1988). Reprinted by permission.

Module 33.13: Talking About Science: Adapted from an interview in Neil Campbell, Jane Reece and Larry Mitchell, *Biology,* 5th ed. (Menlo Park, CA: Addison Wesley Longman, 1999).

Module 33.15: Talking About Science: Adapted from an interview in Neil Campbell, Jane Reece and Larry Mitchell, *Biology,* 4th ed. (Menlo Park, CA: Addison Wesley Longman, 1996).

Module 34.3: Text quotation from Rachel Carson, *Silent Spring* (New York: Houghton Mifflin, 1962).

Module 34.11: Talking About Science: Adapted from an interview in Neil Campbell, *Biology,* 3rd ed. (Redwood City, CA: Benjamin/Cummings, 1993).

Figure 35.00: From Cleveland Hickman, *Zoology,* 8th ed., (St. Louis, MO: C. V. Mosby, 1988) p. 557.

Table 35.6: Data from F. Black and H. D. Skipper, Jr. *Life Insurance,* 12th ed. (Englewood Cliffs, NJ: Prentice-Hall, 1988).

Figure 35.09a: Adapted from Population and the Environment: The Challenges Ahead (United Nations Population Fund, 1991).

Module 36.07: Talking About Science: Adapted from an interview in Neil Campbell, *Biology,* 4th ed. (Menlo Park, CA: Benjamin/Cummings, 1996).

Figure 36.10: From *Ecology and Field Biology,* 4th ed. by Robert Leo Smith. Copyright ©1990 by Robert Leo Smith. Reprinted by permission of Addison Wesley Longman, Inc.

Figure 36.18c: From G. E. Likens, et al. "Effects of Forest Cutting and Herbicide Treatment on Nutrient Budgets in the Hubbard Brook Watershed Ecosystem," *Ecological Monographs,* 1966, 50: 22-32. Copyright ©1966 by The Ecological Society of America. Reprinted by permission.

Figure 36.20: From *Life: The Science of Biology,* 3rd ed. by Purves and Orians. Copyright © 1992 by Sinauer and W. H. Freeman and Company. Used with permission.

Figure 37.3b: Adapted from an illustration by Patricia J. Wynne.

Figure 37.5b: Courtesy of Masakazu Konishi.

Figure 37.11c: From *Animal Navigation* by Waterman. Copyright © 1989 by Scientific American Books, Inc. Used with permission by W. H. Freeman and Company.

Module 37.16: Talking About Science: Adapted from an interview in Neil Campbell, *Biology,* 3rd ed. (Redwood City, CA: Benjamin/Cummings, 1993).

Module 37.21: Talking About Science: Adapted from an interview in Neil Campbell, *Biology,* 4th ed. (Menlo Park, CA: Benjamin/Cummings, 1996).

Figure 37.18a: From Judith W. McIntyre, *The Common Loon: Spirit of Northern Lakes,* p. 97 (Minneapolis, MN: University of Minnesota Press, 1988). Copyright © 1988 by University of Minnesota Press.

Figure 38.3B: From G. Tyler Miller, *Living in the Environment,* 2nd ed. (Belmont, CA: Wadsworth, 1979), p. 87. ©1979 Wadsworth Publishing Company. All rights reserved.

Figure 38.4c: Data from "Global Warming Trends" by P. D. Jones and T. M. L. Wigley, *Scientific American,* August 1990.

Figure 38.09a: Adapted from an illustration by Tamara Sayre.

Figure 38.09b: USGS-BRD-Gap Analysis.

GLOSSARY

A

abiotic factor (A-by-OT-ik) A nonliving component of an ecosystem, such as air, water, or temperature.

abiotic reservoir The part of an ecosystem where a chemical, such as carbon or nitrogen, accumulates or is stockpiled outside of living organisms.

abscisic acid (ABA) (ab-SIS-ik) A plant hormone that inhibits cell division and promotes dormancy; interacts with gibberellins in regulating seed germination.

absorption The uptake of small nutrient molecules by an organism's own body; the third main stage of food processing, following digestion.

accommodation The automatic changes made by the eye as it focuses on near objects.

acetylcholine (uh-SEE-tul-KOLE-een) A nitrogen-containing neurotransmitter; among other effects, it slows the heart rate and makes skeletal muscles contract.

achondroplasia (a-KON-druh-PLAY-zhuh) A form of human dwarfism caused by a single dominant allele; the homozygous condition is lethal.

acid A substance that increases the hydrogen ion concentration in a solution.

acid precipitation Rain, snow, sleet, hail, drizzle, etc., with a pH below 5.6; can damage or destroy organisms by acidifying lakes, streams, and possibly land habitats.

acrosome A membrane-enclosed sac at the tip of a sperm; contains enzymes that help the sperm penetrate an egg.

ACTH Adrenocorticotropic hormone; a protein hormone secreted by the anterior pituitary that stimulates the adrenal cortex to secrete corticosteroids.

actinomycete (ak-TIN-oh-MY-seat) One of a group of bacteria characterized by a mass of branching cell chains called filaments.

action potential A self-propagating change in the voltage across the plasma membrane of a neuron; a nerve signal.

activator A protein that switches on a gene or group of genes.

active site The part of an enzyme molecule where a substrate molecule attaches (by means of weak chemical bonds); typically, a pocket or groove on the enzyme's surface.

active transport The movement of a substance across a biological membrane against its concentration gradient, aided by specific transport proteins and requiring input of energy (often as ATP).

adaptation *See* evolutionary adaptation.

adaptive radiation The emergence of numerous species from a common ancestor introduced to new and diverse environments.

adhesion The attraction between different kinds of molecules.

adipose tissue A type of connective tissue whose cells contain fat.

adrenal cortex (uh-DREEN-ul) The outer portion of an adrenal gland, controlled by ACTH from the anterior pituitary; secretes hormones called glucocorticoids and mineralocorticoids.

adrenal gland One of a pair of endocrine glands, located adjacent to a kidney in mammals, composed of an outer cortex and a central medulla.

adrenal medulla (uh-DREEN-ul meh-DUL-uh) The central portion of an adrenal gland, controlled by nerve signals, secretes the fight-or-flight hormones epinephrine and norepinephrine.

aerobic (air-OH-bik) Containing or requiring molecular oxygen (O_2).

agonistic behavior (AG-un-NIS-tik) Confrontational behavior involving a contest waged by threats, displays, or actual combat, which settles disputes over limited resources, such as food or mates.

AIDS Acquired immune deficiency syndrome; the late stages of HIV infection; characterized by a reduced number of T cells; usually results in death caused by other diseases.

Agnatha (uh-NATH-uh) A class of vertebrate animals that are superficially fishlike but lack jaws and paired fins.

alga (AL-guh) (plural, algae) One of a great variety of protists, most of which are unicellular or colonial photosynthetic autotrophs with chloroplasts containing the pigment chlorophyll a; heterotrophic and multicellular protists closely related to unicellular autotrophs are also regarded as algae.

alimentary canal (AL-ih-MEN-teh-ree) A digestive tract consisting of a tube running between a mouth and an anus.

allantois (uh-LAN-toe-iss) In animals, an extraembryonic membrane that develops from the yolk sac; helps dispose of the embryo's nitrogenous wastes and forms part of the umbilical cord in mammals.

allele (uh-LEEL) An alternative form of a gene.

allergen (AL-er-jen) An antigen that causes an allergy.

allergy A disorder of the immune system caused by an abnormal sensitivity to an antigen; symptoms are triggered by histamines released from mast cells.

allopatric speciation The formation of a new species as a result of an ancestral population's becoming isolated by a geographical barrier.

alpha helix (AL-fuh HE-lix) The spiral shape resulting from the coiling of a polypeptide in a protein's secondary structure.

alternation of generations A life cycle in which there is both a multicellular diploid form, the sporophyte, and a multicellular haploid form, the gametophyte; a characteristic of plants and multicellular green algae.

altruism (AL-troo-IZ-em) Behavior that reduces an individual's fitness while increasing the fitness of another individual.

alveolus (al-VEE-oh-lus) (plural, alveoli) One of millions of tiny sacs within the vertebrate lungs where gas exchange occurs.

amine (uh-MEEN) An organic compound with one or more amino groups.

amino acid (uh-MEE-noh) An organic molecule containing a carboxyl group and an amino group; serves as the monomer of proteins.

amino group In an organic molecule, a functional group consisting of a nitrogen atom bonded to two hydrogen atoms.

amniocentesis (AM-nee-oh-sen-TEE-sis) A technique for diagnosing genetic defects while a fetus is in the uterus; a sample of amiotic fluid, obtained via a needle inserted into the amnion, is analyzed for telltale chemicals and defective fetal cells.

amnion (AM-nee-on) In vertebrate animals, the extraembryonic membrane that encloses the fluid-filled amniotic sac containing the embryo.

amniotic egg (AM-nee-ah-tik) A shelled egg in which an embryo develops within a fluid-filled amniotic sac and is nourished by yolk; produced by reptiles, birds, and egg-laying mammals, it enables them to complete their life cycles on dry land.

amoeba (uh-ME-ba) A type of protist characterized by great flexibility and the presence of pseudopodia.

amoebocyte (uh-ME-buh-SITE) An amoebalike cell that moves by pseudopodia, found in most animals; depending on the species, may digest and distribute food, dispose of wastes, make skeleton, fight infections, and change into other cell types.

Amphibia A class of vertebrate animals that consists of the amphibians; such as frogs, toads, and salamanders.

amygdala (uh-MIG-deh-la) An integrative center of the cerebrum; functionally the part of the limbic system that seems to label information to be remembered.

anabolic steroid (an-uh-BAH-lic STEHRoid) A synthetic variant of the male hormone testosterone that mimics some of its effects.

anaerobic (an-air-OH-bik) Lacking or not requiring molecular oxygen (O_2).

anaphase The third stage of mitosis, beginning when sister chromatids separate from each other and ending when a complete set of daughter chromosomes have arrived at each of the two poles of the cell.

anaphylactic shock (AN-uh-feh-LAK-tik) A potentially fatal allergic reaction caused by extreme sensitivity to an allergen; involves an abrupt dilation of blood vessels and a sharp drop in blood pressure.

anchoring junction A junction that connects tissue cells to each other (or to an extracellular matrix) and allows materials to pass from cell to cell.

androgen (AN-druh-jen) A steroid sex hormone secreted by the gonads that promotes the development and maintenance of the male reproductive system and male body features.

anemia (uh-NEE-me-ah) A condition in which an abnormally low amount of hemoglobin or a low number of red blood cells results in the body cells not receiving enough oxygen.

angiosperm (AN-jee-oh-spurm) A flowering plant, which forms seeds inside a protective chamber called an ovary.

Animalia (AN-eh-MAL-ee-uh) The kingdom that contains the animals.

Annelida (uh-NEL-ih-duh) The phylum that contains the segmented worms, or annelids; characterized by uniform segmentation; includes earthworms, polychaetes, and leeches.

annual A plant that completes its life cycle in a single year or growing season.

anterior Pertaining to the front, or head, of a bilaterally symmetrical animal.

anterior pituitary (puh-TOO-e-TAIR-ee) An endocrine gland, adjacent to the hypothalamus and the posterior pituitary, that synthesizes several hormones, including some that control the activity of other endocrine glands.

anther A sac in which pollen grains develop, located at the tip of a flower's stamen.

anthropoid (AN-thruh-poid) A member of a primate group made up of the apes (gibbon, orangutan, gorilla, chimpanzee, and bonobo), monkeys, and humans.

anthropomorphism (AN-thruh-puh-MOR-fiz-em) Ascribing human motivation, conscious awareness, or feelings to other animals.

antibody (AN-tih-BOD-ee) A protein dissolved in blood plasma that attaches to a specific kind of antigen and helps counter its effects.

anticodon (AN-tee-KO-dahn) On a tRNA molecule, a specific sequence of three nucleotides that is complementary to a codon triplet on mRNA.

antidiuretic hormone (ADH) (AN-tee-DIE-yoo-RET-ik) A hormone made by the hypothalamus and secreted by the posterior pituitary that promotes water retention by the kidneys.

antigen (AN-tih-jen) A foreign (nonself) molecule that elicits an immune response.

antigenic determinant A region on the surface of an antigen molecule to which an antibody binds.

antigen-presenting cell (APC) One of a family of white blood cells (e.g., a macrophage) that ingests a foreign substance or a microbe and attaches antigenic portions of the ingested material to its own surface, thereby displaying the antigens to a helper T cell.

antihistamine (AN-tee-HISS-tuh-meen) A drug that interferes with the action of histamine, providing temporary relief from an allergic reaction.

aorta (ay-OR-tuh) An artery that conveys blood directly from the heart to other arteries.

aphotic zone (ay-FOE-tik) The region of an aquatic ecosystem beneath the photic zone, where light does not penetrate enough for photosynthesis to take place.

apical dominance (APP-ih-kul) In a plant, the hormonal inhibition of axillary buds by a terminal bud.

apical meristem (APP-ih-kul MAIR-ih-stem) A meristem at the tip of a plant root, or in the terminal bud and axillary bud of a shoot.

apicomplexan (APP-ee-kum-PLEX-un) One of a group of parasitic protozoans, some of which cause human diseases.

appendicular skeleton (APP-en-DIK-yoo-ler) Components of the skeletal system that support the fins of a fish or the arms and legs of a land vertebrate; cartilage and bones of the shoulder girdle, pelvic girdle, and the forelimbs and hind limbs. *See also* axial skeleton.

appendix (uh-PEN-dix) A small, fingerlike extension of the vertebrate cecum; contains a mass of white blood cells that contribute to immunity.

aqueous humor (AY-kwee-us HYOO-mer) Plasmalike liquid in the space between the lens and the cornea in the vertebrate eye; helps maintain the shape of the eye, supplies nutrients and oxygen to its tissues, and disposes of its wastes.

arachnid A member of a major arthropod group that includes spiders, scorpions, ticks, and mites.

Archaea (AR-kee-a) One of two prokaryotic domains of life, the other being Bacteria.

archenteron (ar-KEN-teh-ron) In a developing animal, the endoderm-lined cavity formed during gastrulation; the digestive cavity of a gastrula.

arteriole (ar-TEHR-ee-ole) A vessel that conveys blood between an artery and a capillary bed.

artery A vessel that carries blood away from the heart to other parts of the body.

arthritis (ar-THRY-tiss) A skeletal disorder characterized by inflamed joints and deterioration of the cartilage between bones.

Arthropoda (ar-THROP-uh-duh) The most diverse phylum in the animal kingdom; includes the horseshoe crab, arachnids (e.g., spiders, ticks, scorpions, and mites), crustaceans (e.g., crayfish, lobsters, crabs, barnacles), millipedes, centipedes, and insects. Arthropods are characterized by a chitinous exoskeleton, molting, jointed appendages, and a body formed of distinct groups of segments.

artificial selection Selective breeding of domesticated plants and animals to promote the occurrence of desirable inherited traits in offspring.

asexual reproduction The creation of offspring by a single parent, without the participation of sperm and egg.

associative learning Learning that a particular stimulus or response is linked to a reward or punishment; includes classical conditioning and trial-and-error learning.

astigmatism (uh-STIG-muh-TIZ-em) Blurred vision caused by a misshapen lens or cornea.

atom The smallest unit of matter that retains the properties of an element.

atomic number The number of protons in each atom of a particular element.

atomic weight The approximate total mass of an atom; given as a whole number, the atomic weight approximately equals the mass number.

ATP synthase A complex (cluster) of several proteins found in a cellular membrane (including the inner membrane of mitochondria, the thylakoid membrane of chloroplasts, and the plasma membrane of prokaryotes) that functions in chemiosmosis with adjacent electron transport chains, using the energy of a hydrogen-ion concentration gradient to make ATP. An ATP synthase provides a port through which hydrogen ions (H^+) diffuse.

atrium (AY-tree-um) (plural, atria) A heart chamber that receives blood from the veins.

auditory canal Part of the vertebrate outer ear that channels sound waves from the pinna or outer body surface to the eardrum.

australopithecines (aw-STRAY-loh-PITH-eh-sins) The first hominids; scavenger-gatherer-hunters who lived on African savannas between about 4.4 million years ago and 1.5 million years ago.

autoimmune disease An immunological disorder in which the immune system attacks the body's own molecules.

autonomic nervous system (AW-tuh-NAHM-ik) A subdivision of the motor nervous system of vertebrates that regulates the internal environment; made up of sympathetic and parasympathetic subdivisions.

autosome A chromosome not directly involved in determining the sex of an organism; in mammals, for example, any chromosome other than *X* or *Y*.

autotroph (AW-tuh-trofe) An organism that makes its own food, thereby sustaining itself without eating other organisms or their molecules. Plants, algae, and photosynthetic bacteria are autotrophs.

auxin (AWK-sin) A plant hormone, indoleacetic acid or a related compound, whose chief effect is to promote the growth and development of shoots.

Aves A class of vertebrate animals that consists of the birds.

axial skeleton (AK-see-ul) Components of the skeletal system that support the central trunk of the body; the skull, backbone, and rib cage in a vertebrate. *See also* appendicular skeleton.

axillary bud (AK-seh-LAIR-ee) An embryonic shoot present in the angle formed by a leaf and stem.

axon (AK-son) A neuron fiber that conducts signals to another neuron or to an effector cell.

B

bacillus (buh-SIL-us) (plural, bacilli) A rod-shaped prokaryotic cell.

backbone A series of segmental units called vertebrae, present in all vertebrates.

Bacteria One of two prokaryotic domains of life, the other being Archaea.

bacteriophage (bak-TEER-ee-oh-FAJE) A virus that infects bacteria; also called a phage.

bacterium (plural, bacteria) An organism that is a member of the domain Bacteria *(see).*

bark All the tissues external to the vascular cambium in a plant that is growing in thickness. Bark is made up of secondary phloem, cork cambium, and cork.

Bartholin's glands (BAR-toe-linz) Glands near the vaginal opening in a human female that secrete lubricating fluid during sexual arousal.

basal body (BAY-sul) A eukaryotic cell organelle consisting of a $9 + 0$ arrangement of microtubule triplets; may organize the microtubule assembly of a cilium or flagellum; structurally identical to a centriole.

basal ganglia (GANG-lee-uh) Clusters of nerve cell bodies located deep within the cerebrum that are important in motor coordination.

basal metabolic rate (BMR) The number of kilocalories a resting animal requires to fuel its essential body processes for a given time.

base A substance that decreases the hydrogen ion (H^+) concentration in a solution.

basement membrane The extracellular matrix, consisting of a dense mat of proteins and sticky polysaccharides, that anchors an epithelium to underlying tissues.

Batesian mimicry (BAYTZ-ee-un MIM-ih-kree) A type of mimicry in which a species that a predator can eat looks like a different species that is poisonous or otherwise harmful to the predator.

B cell A type of lymphocyte that matures in the bone marrow and later produces antibodies; responsible for humoral immunity. *See also* T cell.

behavioral ecology The scientific search for evolutionary bases of behavior.

behavioral isolation A type of prezygotic barrier between species; two species remain isolated because individuals of neither species are sexually attracted to individuals of the other species.

benign tumor An abnormal mass of cells that remains at its original site in the body.

benthic zone A seafloor, or the bottom of a freshwater lake, pond, river, or stream.

biennial A plant that completes its life cycle in two years.

bilateral symmetry An arrangement of body parts such that an organism can be divided equally by a single cut passing longitudinally through it. A bilaterally symmetrical organism has mirror-image right and left sides.

bile A solution of bile salts secreted by the liver, which emulsifies fats and aids in their digestion.

binary fission A means of asexual reproduction in which a parent organism, often a single cell, divides into two individuals of about equal size.

binomial A two-part, latinized name of a species; for example, *Homo sapiens.*

biodiversity All of the variety of life; usually refers to the variety of species that make up a community; concerns both species richness (the total number of different species) and the relative abundance of the different species.

biodiversity crisis The current rapid decline in the variety of life on Earth, largely due to the effects of human culture.

biodiversity hot spot A small geographic area with an exceptional concentration of species, especially endemic species (those found nowhere else).

biogeography The geographical distribution of species.

biological clock An internal timekeeper that controls an organism's biological rhythms; marks time with or without environmental cues but often requires signals from the environment to remain tuned to an appropriate period. *See also* circadian rhythm.

biological community *See* community.

biological magnification The accumulation of persistent chemicals in the living tissues of consumers in food chains.

biological species concept The definition of a species as a population or group of populations whose members have the potential in nature to interbreed and produce fertile offspring; a biological species is also called a sexual species.

biomass The amount, or mass, of organic material in an ecosystem.

biome (BY-ome) A terrestrial ecosystem, largely determined by climate, usually classified according to the predominant vegetation, and characterized by organisms adapted to the particular environments.

bioremediation The use of living organisms to detoxify and restore polluted and degraded ecosystems.

biosphere The global ecosystem; that portion of Earth that is alive; all of life and where it lives.

biotechnology The use of living organisms (often microbes) to perform useful tasks; today, usually involves DNA technology.

biotic factor (by-OT-ik) A living component of a biological community; an organism, or a factor pertaining to an organism or organisms.

bivalve A member of a group of mollusks that includes clams, mussels, scallops, and oysters.

blastocoel (BLAS-toe-SEAL) In a developing animal, a central, fluid-filled cavity in a blastula.

blastocyst (BLAS-toe-sist) A mammalian embryo (equivalent to an amphibian blastula) made up of a hollow ball of cells that results from cleavage and that implants in the mother's endometrium.

blastopore (BLAS-toe-pore) A small indentation on one side of a blastula where cells that will form endoderm and mesoderm leave the surface and move inward.

blastula (BLAS-tyoo-luh) An embryonic stage that marks the end of cleavage during animal development; a hollow ball of cells in many species.

blind spot The place on the retina of the vertebrate eye where the optic nerve passes through the eyeball and where there are no photoreceptor cells.

blood A type of connective tissue with a fluid matrix called plasma in which blood cells are suspended.

blood-brain barrier A system of capillaries in the brain that restricts passage of most substances into the brain, thereby preventing large fluctuations in the brain's environment.

blood pressure The force that blood exerts against the walls of blood vessels.

body cavity A fluid-containing space between the digestive tract and the body wall.

bone A type of connective tissue, consisting of living cells held in a rigid matrix of collagen fibers embedded in calcium salts.

bone marrow Blood-cell forming tissue (red bone marrow) or stored fat (yellow bone marrow) found in cavities within bones.

bottleneck effect Genetic drift resulting from a drastic reduction in population size.

Bowman's capsule A cup-shaped swelling at the receiving end of a nephron in the vertebrate kidney; collects the filtrate from the blood.

brainstem A functional unit of the vertebrate brain, composed of the midbrain, medulla oblongata, and the pons; serves mainly as a sensory filter, selecting which information reaches higher brain centers.

breathing The alternation of inhalation and exhalation, supplying a lung or gill with O_2-rich air or water and expelling CO_2-rich air or water.

breathing control center A brain center that directs the activity of organs involved in breathing.

bronchiole (BRONK-ee-ole) A thin breathing tube that branches from a bronchus within a lung.

bronchus (BRONK-us) (plural, bronchi) One of a pair of breathing tubes that branch from the trachea into the lungs.

brown alga One of a group of marine, multicellular, autotrophic protists, the most common and largest type of seaweed. Brown algae include the kelps.

bryophyte (BRYE-eh-FITE) One of a group of plants that lack xylem and phloem; a nonvascular plant. Bryophytes include mosses and their close relatives.

bud In a plant, an undeveloped shoot covered with modified leaves called scales that protect underlying embryonic tissues.

budding A means of asexual reproduction whereby a new individual developed from an outgrowth of a parent splits off and lives independently.

buffer A chemical substance that resists changes in pH by accepting H^+ ions from or donating H^+ ions to solutions.

bulbourethral gland (BUL-bo-yoo-REE-thrul) One of a pair of glands near the base of the penis in the human male that secrete fluid that lubricates and neutralizes acids in the urethra during sexual arousal.

C

calcitonin (KAL-sih-TONE-in) A peptide hormone secreted by the thyroid gland that lowers the blood calcium level.

Calvin cycle The second of two stages of photosynthesis, the Calvin cycle is a cyclic series of chemical reactions that occur in the stroma of a chloroplast, using the carbon in CO_2 and the ATP and NADPH produced by the light reactions to make the energy-rich sugar molecule G3P.

cAMP Cyclic adenosine monophosphate; a molecule synthesized from ATP, which serves as a second messenger, mediating the effect of many steroid hormones on target cells.

cancer cell A cell that is not subject to normal cell-cycle control mechanisms and that will divide continuously, often killing the organism if unchecked.

capillary (KAP-ih-lair-ee) A microscopic blood vessel that conveys blood between an artery and a vein or between an arteriole and a venule; enables the exchange of nutrients and dissolved gases between the blood and interstitial fluid.

capillary bed A network of capillaries that infiltrate every organ and tissue in the body.

capsule A sticky layer that surrounds the bacterial cell wall, protects the cell surface, and sometimes helps glue the cell to surfaces.

carbohydrate (KAR-bo-HI-drate) A class of biological molecules consisting of simple single-monomer sugars (monosaccharides), two-monomer sugars (disaccharides), and other multi-unit sugars (polysaccharides).

carbon fixation The incorporation of carbon from atmospheric CO_2 into the carbon in organic compounds. During photosynthesis in a C_3 plant, carbon is fixed into a three-carbon sugar as it enters the Calvin cycle. In C_4 and CAM plants, carbon is fixed into a four-carbon sugar.

carbon skeleton The chain of carbon atoms that forms the structural backbone of an organic molecule.

carbonyl group (KAR-beh-nil) In an organic molecule, a functional group consisting of a carbon atom linked by a double bond to an oxygen atom.

carboxyl group (kar-BOK-sul) In an organic molecule, a functional group consisting of an oxygen atom double-bonded to a carbon atom that is also bonded to a hydroxyl group.

carboxylic acid An organic compound containing a carboxyl group.

carcinogen (kar-SIN-eh-jen) A cancer-causing agent, either high-energy radiation (such as X rays or UV light) or a chemical.

carcinoma (KAR-sih-NO-muh) Cancer that originates in the coverings of the body, such as skin or the lining of the intestinal tract.

cardiac cycle (KAR-dee-ak) The alternating contractions and relaxations of the heart.

cardiac muscle Striated muscle that forms the contractile tissue of the heart.

cardiac output The volume of blood per minute the left ventricle pumps into the aorta.

cardiovascular system (KAR-dee-oh-VAS-kyuh-ler) A closed circulatory system with a heart and branching network of arteries, capillaries, and veins.

carnivore An animal that eats other animals. *See also* herbivore *and* omnivore.

carpel (KAR-pul) The female part of a flower, consisting of a stalk with an ovary at the base and a stigma, which traps pollen, at the tip.

carrier An individual who is heterozygous for a recessively inherited disorder and who therefore does not show symptoms of that disorder.

carrying capacity In a population, the number of individuals that an environment can sustain.

cartilage (KAR-til-ij) A type of connective tissue, consisting of living cells embedded in a rubbery matrix with collagenous fibers.

Casparian strip (kas-PAIR-ee-un) A waxy barrier in the walls of endodermal cells in a root that prevents water and ions from entering the xylem without crossing one or more cell membranes.

cecum (SEE-kum) (plural, ceca) A blind outpocket of a hollow organ such as an intestine.

cell A basic unit of living matter separated from its environment by a plasma membrane; the fundamental structural unit of life.

cell body The part of a cell, such as a neuron, that houses the nucleus.

cell cycle An orderly sequence of events (including interphase and the mitotic phase) from the time a eukaryotic cell divides to form two daughter cells to the time those daughter cells divide again.

cell-cycle control system A cyclically operating set of proteins that triggers and coordinates events in the eukaryotic cell cycle.

cell division The reproduction of a cell.

cell junction A structure that connects tissue cells to one another.

cell-mediated immunity The type of specific immunity brought about by T cells; fights body cells infected with pathogens. *See also* humoral immunity.

cell plate A double membrane across the midline of a dividing plant cell, between which the new cell wall forms during cytokinesis.

cellular differentiation The specialization in the structure and function of cells that occurs during the development of an organism; results from selective activation and deactivation of the cells' genes.

cellular metabolism (meh-TAB-ul-izm) The chemical activities of cells.

cellular respiration The aerobic harvesting of energy from food molecules; the energy-releasing chemical breakdown of food molecules, such as glucose, and the storage of potential energy in a form that cells can use to perform work; involves glycolysis, the Krebs cycle, the electron transport chain, and chemiosmosis.

cellular slime mold A type of protist that has unicellular amoeboid cells and multicellular reproductive bodies in its life cycle.

cellulose (SEL-yuh-lohs) A large polysaccharide composed of many glucose monomers linked into cablelike fibrils that provide structural support in plant cell walls.

centralization The presence of a central nervous system (CNS), distinct from a peripheral nervous system.

central nervous system (CNS) The integration and command center of the nervous system; the brain and, in vertebrates, the spinal cord.

central vacuole (VAK-yoo-ohl) A membrane-enclosed sac occupying most of the interior of a mature plant cell, having diverse roles in reproduction, growth, and development.

centriole (SEN-tree-ole) A structure in an animal cell, composed of microtubule triplets arranged in a 9+0 pattern. An animal cell usually has a pair of centrioles within each of its centrosomes *(see)*.

centromere (SEN-troh-mere) The region of a chromosome where two sister chromatids are joined and where spindle microtubules attach during mitosis and meiosis. The centromere divides at the onset of anaphase during mitosis and anaphase II of meiosis.

centrosome (SEN-troh-soam) Material in the cytoplasm of a eukaryotic cell that gives rise to microtubules; important in mitosis and meiosis; also called microtubule-organizing center.

cephalization (SEF-uh-le-ZAY-shun) The concentration of a nervous system at the anterior end.

cephalopod A member of a group of mollusks that includes squids and octopuses.

cerebellum (SEH-reh-BELL-um) Part of the vertebrate hindbrain; mainly a planning center that interacts closely with the cerebrum in coordinating body movement.

cerebral cortex (seh-REE-brul KOR-tex) A folded sheet of gray matter forming the surface of the cerebrum. In humans, it contains integrating centers for higher brain functions such as reasoning, speech, language, and imagination.

cerebral hemisphere The right or left half of the vertebrate cerebrum.

cerebrospinal fluid (seh-REE-bro-SPINE-ul) Blood-derived fluid that surrounds, protects against infection, nourishes, and cushions the brain and spinal cord.

cerebrum (seh-REE-brum) The largest, most sophisticated, and dominant part of the vertebrate forebrain, made up of right and left cerebral hemispheres.

cervix (SUR-viks) The neck of the uterus, which opens into the vagina.

chaparral (CHAP-uh-RAL) A biome dominated by spiny evergreen shrubs adapted to periodic drought and fires; found where cold ocean currents circulate offshore, creating mild, rainy winters and long, hot, dry summers.

charophyte A member of a group of green algae closely related to plants. Species of charophytes living today probably represent the ancestors of the plant kingdom.

chemical bond An attraction between two atoms resulting from a sharing of outer-shell electrons or the presence of opposite charges on the atoms; the bonded atoms gain complete outer electron shells.

chemical energy Energy stored in the chemical bonds of molecules; a form of potential energy.

chemical reaction A process leading to chemical changes in matter; involves the making and/or breaking of chemical bonds.

chemiosmosis (KEM-ee-oz-MOH-sis) The production of ATP using the energy of hydrogen-ion (H^+) gradients across membranes to phosphorylate ADP; powers most ATP synthesis in cells.

chemoautotroph An organism that obtains both energy and carbon from inorganic chemicals; makes its own organic compounds from CO_2 without using light energy.

chemoheterotroph (KEM-oh-HET-er-oh-TROFE) An organism that obtains energy and carbon from organic molecules.

chemoreceptor (KEM-oh-reh-SEP-ter) A sensory receptor that detects chemical changes within the body or a specific kind of molecule in the external environment.

chiasma (kye-AZ-muh) (plural, chiasmata) The microscopically visible site where crossing over has occurred between chromatids of homologous chromosomes during prophase I of meiosis.

chloroplast (KLOR-uh-plast) An organelle found in plants and photosynthetic protists. Enclosed by two concentric membranes, a chloroplast absorbs sunlight and uses it to power the synthesis of organic food molecules (sugars).

choanocyte (ko-AN-eh-site) A flagellated feeding cell found in sponges. Also called a collar cell, it has a collarlike ring that traps food particles around the base of its flagellum.

choanoflagellate An ancestral colonial protist from which sponges, and possibly all animals, probably arose.

Chondrichthyes (kon-DRIK-theez) A class of cartilaginous fishes that includes sharks, rays, and skates.

Chordata (kore-DAY-tuh) The phylum of the chordates; characterized by a dorsal hollow nerve cord, a notochord, gill structures, and a post-anal tail; includes lancelets, tunicates, and vertebrates.

chorion (KOR-ee-ON) In animals, the outermost extraembryonic membrane, which becomes the mammalian embryo's part of the placenta.

chorionic villus (KOR-ee-on-ik VILL-us) An outgrowth of the chorion, containing embryonic blood vessels. As part of the placenta, chorionic villi absorb nutrients and oxygen from, and pass wastes into, the mother's bloodstream.

chorionic villus sampling (CVS) A technique for diagnosing genetic defects while the fetus is in the uterus. A small sample of the fetal portion of the placenta is removed and analyzed.

choroid (KOR-oid) A thin, pigmented layer in the vertebrate eye, surrounded by the sclera. The iris is part of the choroid.

chromatin (KRO-muh-tin) The combination of DNA and proteins that constitute eukaryotic chromosomes; often used to refer to the diffuse, very extended form taken by the chromosomes when a eukaryotic cell is not dividing.

chromosome (KRO-muh-soam) A threadlike, gene-carrying structure found in the nucleus of a eukaryotic cell and most visible during mitosis and meiosis. Also, the main gene-carrying structure of a prokaryotic cell. Chromosomes consist of chromatin.

chromosome theory of inheritance A basic principle in biology stating that genes are located on chromosomes and that the behavior of chromosomes during meiosis accounts for inheritance patterns.

ciliate (SIL-ee-it) A type of protozoan that moves by means of cilia.

cilium (SIL-ee-um) (plural, cilia) A short appendage that propels some protists through the water and moves fluids across the surface of many tissue cells in animals. In common with eukaryotic flagella, cilia have a 9 + 2 arrangement of microtubules covered by the cell's plasma membrane.

circadian rhythm (sur-KAY-dee-un) In an organism, a biological cycle of about 24 hours that is controlled by a biological clock, usually under the influence of environmental cues; a pattern of activity that is repeated daily. *See also* biological clock.

circulatory system The organ system that transports materials such as nutrients, O_2, and hormones to body cells and transports CO_2 and other wastes from body cells.

cladistic analysis (kluh-DIS-tik) The study of evolutionary history; specifically, the scientific search for monophyletic taxa (clades), taxonomic groups composed of an ancestor and all its descendants.

cladistic taxonomy An approach to classifying organisms entirely according to when branches arise in a phylogenetic lineage.

classical conditioning A type of learning in which an arbitrary stimulus is associated with a reward or punishment.

cleavage (KLEE-vij) (1) Cytokinesis in animal cells and in some protists, characterized by pinching in of the plasma membrane. (2) In animal development, the succession of rapid cell divisions without cell growth, that converts the animal zygote into a ball of cells.

cleavage furrow The first sign of cytokinesis during cell division in an animal cell; a shallow groove in the cell surface near the old metaphase plate.

cline A gradation in an inherited trait along a geographical continuum; variation in a population's phenotypic features that parallels an environmental gradient.

clitoris (KLIT-er-is) An organ in the female that engorges with blood and becomes erect during sexual arousal.

clonal selection (KLONE-ul) The production of a lineage of genetically identical cells that recognize and attack the specific antigen that stimulated their proliferation. Clonal selection is the mechanism that underlies the immune system's specificity and memory of antigens.

clone As a verb, to produce genetically identical copies of a cell, organism, or DNA molecule. As a noun, the collection of cells, organisms, or molecules resulting from cloning; also (colloquially), a single organism that is genetically identical to another because it arose from the cloning of a somatic cell.

Cnidaria (nigh-DARE-ee-uh) The phylum that contains the hydras, jellyfishes, sea anemones, corals, and related animals characterized by cnidocytes, radial symmetry, a gastrovascular cavity, polyps, and medusae.

cnidocyte (NID-eh-site) A specialized cell for which the phylum Cnidaria is named; consists of a capsule containing a fine coiled thread, which, when discharged, functions in defense and prey capture.

coccus (KOK-us) (plural, cocci) A spherical prokaryotic cell.

cochlea (KOK-lee-uh) A coiled tube in the inner ear of birds and mammals that contains the hearing organ, the organ of Corti.

codominance The expression of two different alleles of a gene in a heterozygote.

codon (KO-dahn) A three-nucleotide sequence in mRNA that specifies a particular amino acid or polypeptide termination signal; the basic unit of the genetic code.

coelom (SEE-loam) A body cavity completely lined with mesoderm.

coenzyme (ko-EN-zime) An organic molecule (usually a vitamin or a compound synthesized from a vitamin) that acts as a cofactor, helping an enzyme catalyze a metabolic reaction.

coevolution Evolutionary change, in which adaptations in one species act as a selective force on a second species, inducing adaptations that in turn act as a selective force on the first species; mutual influence on the evolution of two different interacting species.

cofactor A nonprotein substance (such as a copper, iron, or zinc atom, or an organic molecule) that helps an enzyme catalyze a metabolic reaction. *See also* coenzyme.

cognition The ability of an animal's nervous system to perceive, store, process, and use information obtained by its sensory receptors.

cognitive ethology The scientific study of cognition; the study of the connection between data processing by nervous systems and animal behavior.

cognitive map A representation within the nervous system of spatial relations among objects in an animal's environment.

cohesion (ko-HE-zhun) The attraction between molecules of the same kind.

collecting duct A tube in the vertebrate kidney that concentrates urine while conveying it to the renal pelvis.

collenchyma cell (ko-LEN-kim-uh) In plants, a cell with a thick primary wall and no secondary wall, functioning mainly in supporting growing parts.

colon (KO-lun) Large intestine; the tubular portion of the vertebrate alimentary tract between the small intestine and the anus; functions mainly in water absorption and the formation of feces.

commensalism (keh-MEN-sul-izm) A symbiotic relationship in which one partner benefits without significantly affecting the other.

communicating junction A channel between adjacent tissue cells through which water and other small molecules pass freely.

community An assemblage of all the organisms living together and potentially interacting in a particular area.

companion cell In a plant, a cell connected to a sieve-tube member whose nucleus and ribosomes provide proteins for the sieve-tube member.

competitive exclusion principle The concept that populations of two species cannot coexist in a community if their niches are nearly identical. Using resources more efficiently and having a reproductive advantage, one of the populations will eventually outcompete and eliminate the other.

competitive inhibitor A substance that reduces the activity of an enzyme by binding to the enzyme's active site in place of the substrate; a competitive inhibitor's structure mimics that of the enzyme's substrate.

complement proteins A family of nonspecific defensive blood proteins that cooperate with other components of the vertebrate defense system to protect against microbes; can enhance phagocytosis, directly lyse pathogens, and amplify the inflammatory response.

complete metamorphosis (MET-uh-MOR-fuh-sis) The transformation of a larva into an adult that looks very different, and often functions very differently in its environment, than the larva.

compound A substance containing two or more elements in a fixed ratio; for example, table salt (NaCl) consists of one atom of the element sodium (Na) for every atom of chlorine (Cl).

compound eye The photoreceptor in many invertebrates; made up of many tiny light detectors, each of which detects light from a tiny portion of the field of view.

computed tomography (CT) A technology that uses a computer to create X-ray images of a series of sections through the body.

concentration gradient An increase or decrease in the density of a chemical substance in an area. Cells often maintain concentration gradients of H^+ ions across their membranes. When a gradient exists, the ions or other chemical substances involved tend to move from where they are more concentrated to where they are less concentrated.

cone (1) In vertebrates, a photoreceptor cell in the retina, stimulated by bright light and enabling color vision. (2) In conifers, a reproductive structure bearing pollen or ovules.

conifer A gymnosperm, or naked-seed plant, that produces cones.

conjugation The union (mating) of two bacterial cells or protist cells and the transfer of DNA between the two cells.

conjunctiva A mucous membrane that helps keep the eye moist; lines the inner surface of the eyelids and covers the front of the eyeball, except the cornea.

connective tissue Tissue consisting of cells held in an abundant extracellular matrix, which they produce.

consciousness Awareness; a mental state characterized by conscious thinking and self-awareness.

conservation biology The science of species preservation; the scientific study of ways to slow the current high rate of species loss.

continental drift A change in the position of continents resulting from the incessant slow movement (floating) of the plates of Earth's crust on the underlying molten mantle. It has caused continents to periodically fuse and break apart throughout geological history.

controlled experiment A component of the process of science whereby a scientist carries out two parallel tests, an experimental test and a control test. The experimental test differs from the control by one factor, the variable.

convergent evolution Adaptive change resulting in nonhomologous (analogous) similarities among organisms. Species from different evolutionary lineages come to resemble each other (evolve analogous structures) as a result of living in very similar environments.

copulation Sexual intercourse, necessary for internal fertilization to occur.

cork The outermost protective layer of a plant's bark, produced by the cork cambium.

cork cambium Meristematic tissue that produces cork cells during secondary growth of a plant.

cornea (KOR-nee-uh) The transparent frontal portion of the sclera, which admits light into the vertebrate eye.

coronary artery (KORE-uh-nair-ee) The large blood vessel that conveys blood from the aorta to the tissues of the heart.

corpus callosum (KOR-pus kuh-LO-sum) The thick band of nerve fibers that connect the right and left cerebral hemispheres in placental mammals, enabling the hemispheres to process information together.

corpus luteum (KOR-pus LOO-tee-um) A small body of endocrine tissue that develops from an ovarian follicle after ovulation; secretes progesterone and estrogen during pregnancy.

cortex In plants, the ground tissue system of a root, made up mostly of parenchyma cells, which store food and absorb minerals that have passed through the epidermis. *See also* adrenal cortex, cerebral cortex, *and* renal cortex.

corticosteroid A family of hormones synthesized and secreted by the adrenal cortex, consisting of the mineralocorticoids and the glucocorticoids.

cotyledon (KOT-ih-LEED-un) The first leaf that appears on an embryo of a flowering plant; a seed leaf. Monocot embryos have one cotyledon; dicot embryos have two.

countercurrent exchange The transfer of a substance from a fluid or volume of air moving in one direction to another fluid or volume of air moving in the opposite direction.

countercurrent heat exchanger Parallel blood vessels that convey warm and cold blood in opposite directions, maximizing heat transfer to the cold blood.

covalent bond (ko-VALE-ent) An attraction between atoms that share one or more pairs of outer-shell electrons; symbolized by a single line between the atoms.

crista (KRIS-tuh) (plural, cristae) A fold of the inner membrane of a mitochondrion. Enzyme molecules embedded in cristae make ATP.

critical period In animal behavior, the specific time during which imprinting takes place.

crop A pouchlike organ in a digestive tract where food is softened and may be stored temporarily.

cross *See* hybridization.

cross-fertilization The fusion of sperm and egg derived from two different individuals.

crossing over The exchange of segments between chromatids of homologous chromosomes during synapsis in prophase I of meiosis; also, the exchange of segments between DNA molecules in prokaryotes.

crustacean A member of a major arthropod group that includes lobsters, crayfish, crabs, shrimps, and barnacles.

CT *See* computed tomography.

cuticle (KYOO-tih-kul) (1) In animals, a tough, nonliving outer layer of the skin. (2) In plants, a waxy coating on the surface of stems and leaves that helps retain water.

cyanobacteria (sy-AN-oh-bak-TEER-ee-uh) Photosynthetic, oxygen-producing bacteria, formerly called blue-green algae.

cystic fibrosis (SIS-tik fy-BRO-sis) A genetic disease that occurs in people with two copies of a certain recessive allele; characterized by an excessive secretion of mucus and consequent vulnerability to infection; fatal if untreated.

cytokinesis (SY-toh-kih-NEE-sis) The division of the cytoplasm to form two separate daughter cells. Cytokinesis usually occurs during telophase of mitosis, and the two processes make up the mitotic (M) phase of the cell cycle.

cytokinin (SY-toh-KINE-in) One of a family of plant hormones that promotes cell division, retards aging in flowers and fruits, and may interact antagonistically with auxins in regulating plant growth and development.

cytoplasm (SY-toh-PLAZ-em) Everything inside a cell between the plasma membrane and the nucleus; consists of a semifluid medium and organelles.

cytoskeleton A meshwork of fine fibers that provides structural support for a eukaryotic cell.

cytotoxic T cell (SY-toh-TOK-sik) A type of lymphocyte that attacks body cells infected with pathogens.

D

decomposition The breakdown of organic materials into inorganic ones.

dehydration synthesis (dee-hy-DRAY-shun SIN-thuh-sis) A chemical process in which a polymer forms as monomers are linked by the removal of water molecules. One molecule of water is removed for each pair of monomers linked. Also called condensation.

dehydrogenase (DEE-hy-DRAH-jen-ace) An enzyme that catalyzes a chemical reaction during which one or more hydrogen atoms are removed from a molecule.

deletion The loss of one or more nucleotides from a gene by mutation; the loss of a fragment of a chromosome.

denaturation (dee-NAY-chur-A-shun) A process in which a protein unravels, losing its specific conformation and hence function; can be caused by changes in pH, salt concentration, or high temperature; also refers to the separation of the two strands of the DNA double helix, caused by similar factors.

dendrite (DEN-drite) A neuron fiber that conveys signals from its tip inward, toward the rest of the neuron; in a motor neuron, one of several short, branched extensions that convey nerve signals toward the cell body.

density-dependent factor A population-limiting factor whose effects depend on population density.

density-dependent inhibition The arrest of cell division that occurs when cells grown in a laboratory dish touch one another; generally due to an inadequate supply of growth factors.

density-independent factor A population-limiting factor whose occurrence and effects are not affected by population density.

deoxyribonucleic acid (DNA) (dee-OK-see-ry-boh-noo-KLAY-ik) The genetic material that organisms inherit from their parents; a double-stranded helical macromolecule consisting of nucleotide monomers with deoxyribose sugar and the nitrogenous bases adenine (A), cytosine (C), guanine (G), and thymine (T). *See also* gene.

derived characters Homologous features that have changed from a primitive (ancestral) condition and that are unique to an evolutionary lineage; features found in members of a lineage but not found in ancestors of the lineage.

desert A biome characterized by organisms adapted to sparse rainfall (less than 30 cm per year) and rapid evaporation.

determinate growth Termination of growth after reaching a certain size, as in most animals. *See also* indeterminate growth.

detritivore (deh-TRITE-ih-vore) An organism that derives its energy from organic wastes and dead organisms.

detritus (deh-TRITE-us) Nonliving matter.

deuterostome An animal with a coelom that forms from hollow outgrowths of the digestive tube of the early embryo. The deuterostomes include the echinoderms and the chordates.

diabetes mellitus (DY-uh-BEET-is meh-LITE-is) A human hormonal disease in which body cells cannot absorb enough glucose from the blood and become energy-starved; body fats and proteins are then consumed for their energy. Insulin-dependent diabetes results when the pancreas does not produce insulin; non-insulin-dependent diabetes results when body cells fail to respond to insulin.

dialysis (dy-AL-ih-sis) Separation and disposal of metabolic wastes from the blood by mechanical means; an artificial method of performing the functions of the kidneys.

diaphragm (DY-ah-fram) The sheet of muscle separating the chest cavity from the abdominal cavity in mammals; its contraction expands the chest cavity, and its relaxation reduces it.

diastole (dy-ASS-tuh-lee) The stage of the heart cycle in which the heart muscle is relaxed, allowing the chambers to fill with blood. *See also* systole.

diatom (DY-eh-tom) A unicellular photosynthetic alga with a unique, glassy cell wall containing silica.

dicot (DY-kot) A flowering plant whose embryo has two seed leaves, or cotyledons.

diffusion The spontaneous movement of particles of any kind from where they are more concentrated to where they are less concentrated.

digestion The mechanical and chemical breakdown of food into molecules small enough for the body to absorb; the second main stage of food processing, following ingestion.

digestive system The organ system that ingests food, breaks it down into smaller chemical units, and absorbs the nutrient molecules.

dihybrid cross (DIE-HI-brid) An experimental mating of individuals in which the inheritance of two traits is tracked.

dikaryotic phase (die-KAIR-ee-ot-ik) A series of stages in the life cycle of many fungi in which cells contain two nuclei.

dinoflagellate (DINE-uh-FLAJ-uh-lit) A unicellular photosynthetic alga with two flagella situated in perpendicular grooves in cellulose plates covering the cell.

diploid cell In an organism that reproduces sexually, a cell containing two homologous sets of chromosomes, one set inherited from each parent; a $2n$ cell.

directional selection Natural selection that acts against the relatively rare individuals at one end of a phenotypic range.

disaccharide (di-SAK-uh-ride) A sugar molecule consisting of two monosaccharides linked by dehydration synthesis.

dispersion pattern The manner in which individuals in a population are spaced within their area. Three types of dispersion patterns are clumped (individuals are aggregated in patches), uniform (individuals are evenly distributed), and random (unpredictable distribution).

distal tubule In the vertebrate kidney, the portion of a nephron that helps refine filtrate and empties it into a collecting duct.

disturbance In an ecological sense, a force that changes a biological community and usually removes organisms from it. Disturbances, such as fire and storms, play pivotal roles in structuring many biological communities.

diversifying selection Natural selection that favors extreme over intermediate phenotypes.

DNA *See* deoxyribonucleic acid.

DNA-DNA hybridization A method of determining the relatedness between different species by measuring the extent of hydrogen bonding between single strands of their DNA molecules; also refers more generally to hydrogen bonding between DNA from any two sources (for example, between an organism's DNA and a chemically synthesized DNA probe).

DNA fingerprint An individual's unique collection of DNA restriction fragments, detected by electrophoresis and nucleic acid probes.

DNA ligase (LY-gase) An enzyme, essential for DNA replication, that catalyzes the covalent bonding of adjacent DNA nucleotides; used in genetic engineering to paste a specific piece of DNA containing a gene of interest into a bacterial plasmid or other vector.

DNA polymerase (puh-LIM-eh-rase) An enzyme that assembles DNA nucleotides into polynucleotides using a preexisting strand of DNA as a template.

DNA sequence analysis Determination of the nucleotide sequences of segments of DNA.

doldrums (DOLE-drums) An area of calm or very light winds near the equator, caused by rising warm air.

domain A taxonomic category above the kingdom level; the three domains of life are Archaea, Bacteria, and Eukarya.

dominance hierarchy The ranking of individuals based on social interactions; usually maintained by agonistic behavior.

dominant allele In a heterozygote, the allele that determines the phenotype with respect to a particular gene.

dorsal Pertaining to the back of a bilaterally symmetrical animal.

double bond A type of covalent bond in which two atoms share two pairs of electrons; symbolized by a pair of lines between the bonded atoms.

double fertilization In flowering plants, the formation of both a zygote and a cell with a triploid nucleus, which develops into the endosperm.

Down syndrome A human genetic disorder resulting from the presence of an extra chromosome 21; characterized by heart and respiratory defects, and varying degrees of mental retardation.

Duchenne muscular dystrophy (di-SHEN DIS-truh-fee) A human genetic disease caused by a sex-linked recessive allele; characterized by progressive weakening and a loss of muscle tissue.

duodenum (doo-AH-duh-num) The first portion of the vertebrate small intestine after the stomach, where acid chyme from the stomach is mixed with bile and digestive enzymes.

duplication Repetition of part of a chromosome resulting from fusion with a fragment from a homologous chromosome; can result from an error in meiosis or from mutagenesis.

dynein arm (DY-neen) A protein extension from a microtubule doublet in a cilium or flagellum; involved in energy conversions that drive the bending of cilia and flagella.

E

eardrum A sheet of connective tissue separating the outer ear from the middle ear that vibrates when stimulated by sound waves and passes the waves to the middle ear.

Echinodermata (ee-KIGH-noh-DER-ma-tuh) The phylum of echinoderms, including sea stars, sea urchins, and sand dollars; characterized by a rough or spiny skin, a water vascular system, an endoskeleton, and radial symmetry in adults.

ecological succession The process of biological community change resulting from disturbance; transition in the species composition of a biological community, often following a flood, fire, or volcanic eruption. *See also* primary succession; secondary succession.

ecology The scientific study of how organisms interact with their environments.

ecosystem (EE-koh-sis-tem) All the organisms in a given area, along with the nonliving (abiotic) factors with which they interact; a biological community and its physical environment.

ectoderm (EK-toh-durm) The outer layer of three embryonic cell layers in a gastrula; forms the skin of the gastrula and gives rise to the epidermis and nervous system in the adult.

ectopic pregnancy (ek-TOP-ik) The implantation and development of an embryo outside the uterus.

ectotherm (EK-toh-thurm) An animal that warms itself mainly by absorbing heat from its surroundings.

effector A cell, tissue, or organ capable of carrying out some action in response to a command from the nervous system.

effector cell (1) In the immune system, a T cell or B cell specialized to defend against a specific antigen. (2) A muscle cell or gland cell that carries out a command from the nervous system.

ejaculation (eh-JACK-yoo-LAY-shun) Discharge of semen from the penis.

electroencephalogram (EEG) (eh-LEK-troh-en-SEF-el-o-gram) A graph that shows the patterns of electrical activity in the brain during arousal and sleep.

electromagnetic energy Solar energy, or radiation, which travels in space as rhythmic waves and can be measured in photons.

electromagnetic receptor A sensory receptor that detects energy of different wavelengths, such as electricity, magnetism, and light.

electron A subatomic particle with a single negative electrical charge; one or more electrons move around the nucleus of an atom.

electron carrier A molecule that conveys electrons within a cell; one of several membrane molecules that make up electron transport chains. Electron carriers shuttle electrons during the redox reactions that release energy ultimately used for ATP synthesis.

electronegativity The tendency for an atom to pull electrons toward itself.

electron microscope (EM) An instrument that focuses an electron beam through, or onto the surface of, a specimen. An electron microscope achieves a thousandfold greater resolving power than a light microscope; the most powerful EM can distinguish objects as small as 0.2 nanometer.

electron shell An energy level representing the distance of an electron from the nucleus of an atom.

electron transport chain A series of electron-carrier molecules that shuttle electrons during the redox reactions that release energy used to make ATP; located in the inner membrane of mitochondria, the thylakoid membranes of chloroplasts, and the plasma membranes of prokaryotes.

electrophoresis *See* gel electrophoresis.

element A substance that cannot be broken down to other substances by chemical means; scientists recognize 92 chemical elements occurring in nature.

embryo (EM-bree-oh) A developing stage of a multicellular organism. In humans, the stage in the development of offspring from the first division of the zygote until body structures begin to appear, about the ninth week of gestation.

embryo sac The female gametophyte contained in the ovule of a flowering plant.

emphysema (EM-fuh-SEEM-uh) A respiratory disease caused by smoking, in which the alveoli become brittle and rupture, reducing the lungs' capacity for gas exchange.

endemic species A species of organism whose distribution is limited to a specific geographic area.

endergonic reaction (EN-der-gahn-ik) An energy-requiring chemical reaction, which yields products with more potential energy than the reactants. The amount of energy stored in the products equals the difference between the potential energy in the reactants and that in the products.

endocrine gland (EN-deh-krin) A ductless gland that synthesizes hormone molecules and secretes them directly into the bloodstream.

endocrine system The organ system consisting of ductless glands that secrete hormones and the molecular receptors on or in target cells that respond to the hormones; cooperates with the nervous system in regulating body functions and maintaining homeostasis.

endocytosis (END-oh-sigh-TOE-sis) The movement of materials into the cytoplasm of a cell via membranous vesicles or vacuoles.

endoderm (EN-doh-durm) The innermost of three embryonic cell layers in a gastrula; forms the archenteron in the gastrula, and gives rise to the innermost linings of the digestive tract and other hollow organs in the adult.

endodermis The innermost layer (a one-cell-thick cylinder) of the cortex of a plant root; forms a selective barrier determining which substances pass from the cortex into the vascular tissue.

endomembrane system A network of membranous organelles that partition the cytoplasm of eukaryotic cells into functional compartments. Some of the organelles are structurally connected to each other, whereas others are structurally separate but functionally connected by the traffic of membranous vesicles between them.

endometrium (EN-doh-MEE-tree-um) The inner lining of the uterus in mammals, richly supplied with blood vessels that provide the maternal part of the placenta and nourish the developing embryo.

endoplasmic reticulum (ER) An extensive membranous network in a eukaryotic cell, continuous with the outer nuclear membrane and composed of ribosome-studded (rough) and ribosome-free (smooth) regions. *See also* rough ER; smooth ER.

endorphin (en-DOR-fin) A pain-inhibiting hormone produced by the brain and anterior pituitary; also serves as a neurotransmitter.

endoskeleton A hard skeleton located within the soft tissues of an animal; includes spicules of sponges, the hard plates of echinoderms, and the cartilage and bony skeletons of many vertebrates.

endosperm In flowering plants, a nutrient-rich mass formed by the union of a sperm cell with two polar nuclei during double fertilization; provides nourishment to the developing embryo in the seed.

endospore A thick-coated, protective cell produced within a bacterial cell exposed to harsh conditions.

endosymbiosis (EN-do-SIM-bee-OSE-sis) A process by which the mitochondria and chloroplasts of eukaryotic cells probably evolved, from symbiotic associations between small prokaryotic cells living inside larger ones.

endotherm An animal that derives most of its body heat from its own metabolism.

endotoxin A poisonous component of the cell walls of certain bacteria.

energy The capacity to perform work, or to move matter in a direction it would not move if left alone.

energy coupling In cellular metabolism, the use of energy released from an exergonic reaction to drive an endergonic reaction.

energy flow The passage of energy through the components of an ecosystem.

energy of activation (E_A) The amount of energy that reactants must absorb before a chemical reaction will start.

enhancer A eukaryotic DNA sequence that helps stimulate the transcription of a gene at some distance from it. An enhancer functions by means of a transcription factor called an activator, which binds to it and then to the rest of the transcription apparatus. *See* silencer.

entropy (EN-truh-pee) A measure of disorder; one form of disorder is heat, which is random molecular motion.

enzyme (EN-zime) A protein that serves as a biological catalyst, changing the rate of a chemical reaction without itself being changed into a different molecule in the process.

epidermis (EP-ih-DER-mis) (1) In animals, the living layer or layers of cells forming the protective covering, or outer skin. (2) In plants, the tissue system forming the protective outer covering of leaves, young stems, and young roots.

epididymis (EP-ih-DID-eh-mis) A long coiled tube into which sperm pass from the testis and are stored until mature and ejaculated.

epinephrine (EP-ih-NEF-rin) An amine hormone (also called adrenaline) secreted by the adrenal medulla that prepares body organs for "fight or flight"; also serves as a neurotransmitter.

epithelial tissue (EP-ih-THEEL-ee-ul) A sheet of tightly packed cells lining organs and cavities; also called epithelium.

epithelium (EP-ih-THEEL-ee-um) (plural, epithelia) Epithelial tissue.

erythrocyte (eh-RITH-ruh-site) *See* red blood cell.

esophagus (eh-SOF-eh-gus) The channel through which food passes in a digestive tract; usually receives food from the pharynx.

estivation (ES-tih-VAY-shun) An animal's state of reduced activity (torpor) during periods of high environmental temperatures and reduced food and water supplies.

estrogen (ES-truh-jen) A steroid hormone secreted by the gonads; maintains the female reproductive system and promotes the development of female body features.

estuary (ES-choo-air-ee) An area where fresh water merges with seawater.

ethylene A gas that functions as a hormone in plants, triggering aging responses such as fruit ripening and leaf drop.

eugenics (yoo-JEN-iks) A socially rejected practice, among humans, of attempting to eliminate genetic disorders and "undesirable" inherited traits by selective breeding.

Eukarya (yoo-KAIR-ee-uh) The domain of eukaryotes, organisms made of eukaryotic cells; includes all of the protists, plants, fungi, and animals.

eukaryotic cell (yoo-KAIR-ee-OT-ik) A type of cell that has a membrane-enclosed nucleus and other membrane-enclosed organelles. All organisms except bacteria and archaea are composed of eukaryotic cells.

Eustachian tube (yoo-STAY-shun) An air passage between the middle ear and throat of vertebrates, that equalizes air pressure on either side of the eardrum.

eutherians (YOO-ther-ee-uns) Placental mammals; those whose young complete their embryonic development within the uterus, joined to the mother by the placenta.

eutrophication (YOO-trofe-ih-KAY-shun) An increase in productivity of an aquatic ecosystem.

evolution Genetic change in a population or species over generations; all the changes that transform life on Earth; the heritable changes that have produced Earth's diversity of organisms.

evolutionary adaptation An inherited characteristic that enhances an organism's ability to survive and reproduce in a particular environment.

evolutionary species concept The definition of a species as a cluster of recognizably distinct organisms that represent a lineage of descent; the cluster may be recognized by genotypic and/or phenotypic features.

exaptation(EX-ap-tay-shun) A structure that has evolved in one environmental context and later becomes adapted for a different function in a different environmental context.

excretion (ek-SKREE-shun) The disposal of nitrogen-containing metabolic wastes.

excretory system (EK-skruh-tor-ee) The organ system that disposes of nitrogen-containing waste products of cellular metabolism.

exergonic reaction (EK-ser-gahn-nik) An energy-releasing chemical reaction in which the reactants contain more potential energy than the products. The reaction releases an amount of energy equal to the difference in potential energy between the reactants and the products.

exocytosis (EK-soh-sigh-TOE-sis) The movement of materials out of the cytoplasm of a cell via membranous vesicles or vacuoles.

exon (EK-sahn) In eukaryotes, a coding portion of a gene. *See* intron.

exoskeleton A hard, external skeleton that protects an animal and provides points of attachment for muscles.

exotoxin A poisonous protein secreted by eubacteria.

exponential growth model A mathematical description of idealized, unregulated population growth.

external fertilization The fusion of gametes that parents have discharged into the environment.

extracellular matrix A substance in which the cells of an animal tissue are embedded; consists of protein and polysaccharides.

extraembryonic membranes Four membranes (the yolk sac, amnion, chorion, and allantois) that form a life-support system for the developing embryo of a reptile, bird, or mammal.

eye cup The simplest type of photoreceptor, a cluster of photoreceptor cells shaded by a cuplike cluster of pigmented cells; detects light intensity and direction.

F

facilitated diffusion The passage of a substance across a biological membrane down its concentration gradient, aided by specific transport proteins.

facultative anaerobe (FAK-ul-tay-tiv AN-uh-robe) A microorganism that makes ATP by aerobic respiration if oxygen is present, but that switches to fermentation when oxygen is absent.

farsightedness An inability to focus on close objects; occurs when the eyeball is shorter than normal and the focal point of the lens is behind the retina. Also called hyperopia.

fat A large lipid molecule made from an alcohol called glycerol and three fatty acids; a triglyceride. Most fats function as energy-storage molecules.

fertilization The union of the nucleus of a sperm cell with the nucleus of an egg cell, producing a zygote.

fertilization membrane A barrier that forms around an egg seconds after fertilization, preventing additional sperm from penetrating the egg.

fetal alcohol syndrome (FAS) Birth defects caused by alcohol passed from a mother's bloodstream into the fetal bloodstream via the placenta.

fetoscopy (fee-TOS-cup-ee) A technique for examining a fetus for anatomical deformities. A needle-thin tube containing a viewing scope is inserted into the uterus, giving a direct view of the fetus.

fetus (FEE-tus) A developing human from the ninth week of gestation until birth; has all the major structures of an adult.

F factor A piece of DNA that can exist as a bacterial plasmid; carries genes for making sex pili and other structures needed for conjugation, as well as a site where DNA replication can start. F stands for fertility.

F_1 generation The offspring of two parental (P generation) individuals; F_1 stands for first filial.

F_2 generation The offspring of the F_1 generation; F_2 stands for second filial.

fiber (1) In animals, an elongate, supportive thread in the matrix of connective tissue; an extension of a neuron; a muscle cell. (2) In plants, a long, slender sclerenchyma cell that usually occurs in a bundle.

fibrin (FY-brin) The activated form of the blood-clotting protein fibrinogen, which aggregates into threads that form the fabric of a blood clot.

fibrinogen (FY-brin-uh-jen) The plasma protein that is activated to form a clot when a blood vessel is injured.

filtrate Fluid extracted by the excretory system from the blood or body cavity. The excretory system produces urine from the filtrate after extracting valuable solutes from it and concentrating it.

filtration In the vertebrate kidney, the extraction of water and small solutes, including metabolic wastes, from the blood by the nephrons.

first law of thermodynamics The natural law that the total amount of energy in the universe is constant and that energy can be transferred and transformed, but it can never be destroyed; also called the principle of energy conservation.

fission A means of asexual reproduction whereby a parent separates into two or more genetically identical individuals of about equal size.

fitness The contribution an individual makes to the gene pool of the next generation, relative to the contribution of other individuals in the population.

fixed action pattern (FAP) A genetically programmed, virtually unchangeable behavioral sequence performed in response to a certain stimulus.

flagellate (FLAJ-uh-lit) A protist (protozoan) that moves by means of one or more flagella.

flagellum (fluh-JEL-um) (plural, flagella) A long appendage that propels protists through the water and moves fluids across the surface of many tissue cells in animals. A cell may have one or more flagella. Like cilia, flagella have a 9 + 2 arrangement of microtubules covered by the cell's plasma membrane. *See also* prokaryotic flagellum.

flatworm A member of the phylum Platyhelminthes.

flower In an angiosperm, a short stem with four sets of modified leaves, bearing structures that function in sexual reproduction.

fluid mosaic A description of membrane structure, depicting a cellular membrane as a mosaic of diverse protein molecules embedded in a fluid bilayer made of phospholipid molecules.

fluke One of a group of parasitic flatworms.

follicle (FOL-ih-kul) A cluster of cells that surround, protect, and nourish a developing egg cell in the ovary; also secretes estrogen.

follicle-stimulating hormone (FSH) A protein hormone secreted by the anterior pituitary that stimulates the production of eggs by the ovaries and sperm by the testes.

food chain A sequence of food transfers from producers through several levels of consumers in an ecosystem.

food-conducting cell A specialized, living plant cell with thin primary walls; arranged end-to-end, such cells collectively form phloem tissue. Also called sieve-tube member.

food web A network of interconnecting food chains.

forebrain One of three ancestral and embryonic regions of the vertebrate brain; develops into the thalamus, hypothalamus, and cerebrum.

fossil A preserved remnant or impression of an organism that lived in the past.

fossil fuel An energy deposit formed from the remains of extinct organisms.

fossil record The chronicle of evolution over millions of years of geological time engraved in the order in which fossils appear in rock strata.

founder effect Random change in the gene pool that occurs in a small colony of a population.

fovea (FO-vee-uh) An eye's center of focus and the place on the retina where photoreceptors are highly concentrated.

fragmentation A means of asexual reproduction whereby a single parent breaks into parts that regenerate into whole new individuals.

fruit A ripened, thickened ovary of a flower, which protects dormant seeds and aids in their dispersal.

fruiting body A stage in an organism's life cycle that functions only in reproduction; for example, a mushroom is a fruiting body of many fungi.

functional group An assemblage of atoms that form the chemically reactive part of an organic molecule.

Fungi (FUN-jie) The kingdom that contains the fungi.

fungus (plural, fungi) A heterotrophic eukaryote that digests its food externally and absorbs the resulting small nutrient molecules. Most fungi consist of a netlike mass of filaments called hyphae. Molds, mushrooms, and yeasts are examples of fungi.

G

gallbladder An organ that stores bile and releases it as needed into the small intestine.

gametangium (GAM-eh-TANJ-ee-um) (plural, gametangia) A reproductive organ that houses and protects the gametes of a plant.

gamete (GAM-eet) A sex cell; a haploid egg or sperm. The union of two gametes of opposite sex (fertilization) produces a zygote.

gametic isolation (guh-MEE-tik) A type of prezygotic barrier between species; the species remain isolated because male and female gametes of the different species cannot fuse, or they die before they unite.

gametophyte (guh-MEE-tuh-FITE) The multicellular haploid form in the life cycle of organisms undergoing alternation of generations; mitotically produces haploid gametes that unite and grow into the sporophyte generation.

ganglion (GANG-lee-un) (plural, ganglia) A cluster of nerve cell bodies of the cells whose axons and dendrites make up nerves.

gas exchange *See* respiration.

gastric gland A tubular structure in the vertebrate stomach that secretes gastric juice.

gastric ulcer An open sore in the lining of the stomach, resulting when pepsin and hydrochloric acid destroy the lining tissues faster than they can regenerate.

gastrin A digestive hormone that stimulates the secretion of gastric juice.

gastropod A member of the largest group of mollusks, including snails and slugs.

gastrovascular cavity A digestive compartment with a single opening, the mouth; may function in circulation, body support, waste disposal, and gas exchange, as well as digestion.

gastrula (GAS-troo-luh) The embryonic stage resulting from gastrulation in animal development. Most animals have a gastrula made up of three layers of cells: ectoderm, endoderm, and mesoderm.

gastrulation (GAS-troo-LAY-shun) The phase of embryonic development that transforms the blastula into a gastrula. Gastrulation adds more cells to the embryo and sorts the cells into distinct cell layers.

gel electrophoresis (JEL eh-LEK-tro-for-EE-sis) A technique for separating and purifying macromolecules. A mixture of molecules is placed on a gel between a positively charged electrode and a negatively charged one; negative charges on the molecules are attracted to the positive electrode, and the molecules migrate toward that electrode; the molecules separate in the gel according to their rates of migration.

gene A discrete unit of hereditary information consisting of a specific nucleotide sequence in DNA (or RNA, in some viruses). Most of the genes of a eukaryote are located in its chromosomal DNA; a few are carried by the DNA of mitochondria and chloroplasts.

gene cloning The production of multiple copies of a gene.

gene expression The process whereby genetic information flows from genes to proteins; the flow of genetic information from the genotype to the phenotype.

gene flow The gain or loss of alleles from a population by the movement of individuals or gametes into or out of the population.

gene pool All the genes in a population at any one time.

genetic code The set of rules giving the correspondence between nucleotide triplets (codons) in mRNA and amino acids in protein.

genetic drift A change in the gene pool of a population due to chance.

genetic marker (1) An allele tracked in a genetic study. (2) A specific section of DNA that earmarks a particular allele; may contain specific restriction sites (points where restriction enzymes cut the DNA) that occur only in DNA that contains the allele.

genetic recombination The production, by crossing over and/or independent assortment of chromosomes during meiosis, of offspring with allele combinations different from those in the parents. The term may also be used more specifically to mean the production by crossing over of eukaryotic or prokaryotic *chromosomes* with gene combinations different from those in the original chromosomes.

genetics The science of heredity.

genome (JEE-nome) A complete (haploid) set of an organism's genes; an organism's genetic material.

genomic library (jen-OME-ik) A set of DNA segments from an organism's genome; each segment is usually carried by a plasmid or phage.

genotype (JEE-noh-tipe) The genetic makeup of an organism.

genus (JEE-nus) (plural, genera) In classification, the taxonomic category above species; the first part of a species' binomial; for example, *Homo.*

germination The beginning of growth.

gestation (je-STAY-shun) Pregnancy; the state of carrying developing young within the female reproductive tract.

gibberellin (JIB-er-EL-in) One of a family of plant hormones that trigger the germination of seeds and interact with auxins in regulating growth and fruit development.

gill An extension of the body surface of an animal, specialized for gas exchange and/or suspension feeding.

gizzard A pouchlike organ in a digestive tract, where food is mechanically ground.

glans The rounded, highly sensitive head of the clitoris in females and penis in males.

glomerulus (glum-AIR-yoo-lus) (plural, glomeruli) In the vertebrate kidney, the part of a nephron consisting of the capillaries that are surrounded by Bowman's capsule; together, a glomerulus and Bowman's capsule produce the filtrate from the blood.

glucagon (GLUKE-uh-gahn) A peptide hormone secreted by islet cells in the pancreas that raises the level of glucose in the blood.

glucocorticoid (GLUKE-oh-KOR-tih-koid) A corticosteroid hormone secreted by the adrenal cortex that increases the blood glucose level and helps maintain the body's response to long-term stress.

glycogen (GLY-kuh-jen) A complex, extensively branched polysaccharide of many glucose monomers; serves as an energy-storage molecule in liver and muscle cells.

glycolysis (gly-KOL-eh-sis) The multistep chemical breakdown of a molecule of glucose into two molecules of pyruvic acid; the first stage of cellular respiration in all organisms; occurs in the cytoplasmic fluid.

glycoprotein (gly-koh-PRO-teen) A macromolecule consisting of one or more polypeptides linked to short chains of sugars.

goiter An enlargement of the thyroid gland resulting from a dietary iodine deficiency.

Golgi apparatus (GOLE-jee) An organelle in eukaryotic cells consisting of stacks of membranous sacs that modify, store, and ship products of the endoplasmic reticulum.

gonad A sex organ in an animal; an ovary or a testis.

Gondwana (gahnd-WAH-na) The southern landmass formed during the Mesozoic era when continental drift split Pangaea (all land masses fused); *see also* Laurasia.

gradualist model The view that evolution occurs as a result of populations becoming isolated from common ancestral stock and gradually becoming genetically unique as they are adapted by natural selection to their local environments; Darwin's view of the origin of species.

granum (GRAN-um) (plural, grana) A stack of hollow disks formed of thylakoid membrane in a chloroplast. Grana are the sites where light energy is trapped by chlorophyll and converted to chemical energy during the light reactions of photosynthesis.

gravitropism (greh-VIH-tro-PIZ-em) A plant's growth response to gravity.

green alga One of a group of photosynthetic protists that includes unicellular, colonial, and multicellular species. Green algae are plantlike in having biflagellated cells (gametes in colonial and multicellular species), chloroplasts with chlorophyll a, cellulose cell walls, and starch.

greenhouse effect The warming of the atmosphere caused by CO_2, CH_4, and other gases which absorb infrared radiation and slow its escape from Earth's surface.

ground tissue system A tissue of mostly parenchyma cells that makes up the bulk of a young plant and is continuous throughout its body. The ground tissue system fills the space between the epidermis and the vascular tissue system.

growth factor A protein secreted by certain body cells that stimulates other cells to divide.

growth hormone (GH) A protein hormone secreted by the anterior pituitary that promotes development and growth and stimulates metabolism.

guard cell A specialized epidermal cell in plants that regulates the size of a stoma, allowing gas exchange between the surrounding air and the photosynthetic cells in the leaf.

gymnosperm (JIM-noh-spurm) A naked-seed plant; its seed is said to be naked because it is not enclosed in a fruit.

H

habitat A place where an organism lives; an environmental situation in which an organism lives.

habitat isolation A type of prezygotic barrier between species; the species remain isolated because they breed in different habitats.

habituation Learning not to respond to a repeated stimulus that conveys little or no information.

hair cell A type of mechanoreceptor that detects sound waves and other forms of movement in air or water.

haploid cell In the life cycle of an organism that reproduces sexually, a cell containing a single set of chromosomes; an *n* cell.

Hardy-Weinberg equilibrium The principle that the shuffling of genes that occurs during sexual reproduction, by itself, cannot change the overall genetic makeup of a population.

HDL *See* high-density lipoprotein.

heart attack Death of cardiac muscle cells and the resulting failure of the heart to deliver enough blood to the body.

heartwood In the center of trees, the darkened, older layers of secondary xylem made up of cells that no longer transport water and are clogged with resins. *See also* sapwood.

heat The amount of energy associated with the movement of the atoms and molecules in a body of matter. Heat is energy in its most random form.

helper T cell A type of lymphocyte that helps activate other types of T cells and may help stimulate B cells to produce antibodies.

hemoglobin (HE-moh-glo-bin) An iron-containing protein in red blood cells that reversibly binds O_2 and transports it to body tissues.

hemophilia (HE-muh-FIL-ee-uh) A human genetic disease caused by a sex-linked recessive allele, characterized by excessive bleeding following injury.

hepatic portal vessel A blood vessel that conveys blood from capillaries surrounding the intestine directly to the liver.

herbivore An animal that eats only plants. *See also* carnivore; omnivore.

hermaphroditism (her-MAF-ruh-di-tizm) A condition in which an individual has both female and male gonads and functions as both a male and female in sexual reproduction by producing both sperm and eggs.

heterotroph (HET-ur-oh-TROFE) An organism that cannot make its own organic food molecules and must obtain them by consuming other organisms or their organic products; a consumer or a decomposer in a food chain.

heterozygote advantage Greater reproductive success of heterozygous individuals compared to homozygotes; tends to preserve variation in gene pools.

heterozygous (HET-ur-oh-ZY-gus) Having two different alleles for a given gene.

hibernation Long-term torpor during cold weather; an animal's metabolic rate is reduced, and it survives on energy stored in body fat.

high-density lipoprotein (HDL) A cholesterol-carrying particle in the blood, made up of cholesterol and other lipids surrounded by a single layer of phospholipids in which proteins are embedded. An HDL particle carries less cholesterol than a related lipoprotein, LDL, and may be correlated with a decreased risk of blood vessel blockage.

hindbrain One of three ancestral and embryonic regions of the vertebrate brain; develops into the medulla oblongata, pons, and cerebellum.

hippocampus (HIP-eh-KAM-pus) An integrative center of the cerebrum; functionally, part of the limbic system that plays central roles in memory and learning.

histamine (HISS-tuh-meen) A chemical alarm signal released by injured cells that causes blood vessels to dilate during an inflammatory response.

histone A small basic protein molecule associated with DNA and important in DNA packing in the eukaryotic chromosome.

HIV Human immunodeficiency virus, the retrovirus that attacks the human immune system and causes AIDS.

homeobox (HOME-ee-oh-BOX) A 180-nucleotide sequence within a homeotic gene encoding the part of the protein that binds to the DNA of the genes regulated by the protein.

homeostasis (HOME-ee-oh-STAY-sis) The steady state of body functioning; a state of equilibrium characterized by a dynamic interplay between outside forces that tend to change an organism's internal environment and the internal control mechanisms that oppose such changes.

homeotic gene (HOME-ee-OT-ik) A master control gene that determines the identity of a body structure of a developing organism, presumably by controlling the developmental fate of groups of cells. (In plants, such genes are called organ-identity genes.)

hominid (HAHM-ih-nid) A species on the human branch of the evolutionary tree; a member of the family Hominidae, including *Homo sapiens* and our ancestors.

homologous chromosomes (hoh-MOL-uh-gus) The two chromosomes that make up a matched pair in a diploid cell. Homologous chromosomes are of the same length, centromere position, and staining pattern and possess genes for the same characteristics at corresponding loci. One homologous chromosome is inherited from the organism's father, the other from the mother.

homologous structures Structures that are similar in different species of common ancestry.

homozygous (HO-moh ZY-gus) Having two identical alleles for a given gene.

hormone A regulatory chemical that travels in the blood from its production site, usually an endocrine gland, to other sites, where target cells respond to the regulatory signal.

human chorionic gonadotropin (HCG) (KOR-ee-ahn-ik goh-NAD-uh-TROPE-in) A hormone secreted by the chorion that maintains the corpus luteum of the ovary during the first three months of pregnancy.

humoral immunity The type of specific immunity brought about by antibody-producing B cells; fights bacteria and viruses in body fluids. *See also* cell-mediated immunity.

humus (HYOO-mus) Decomposing organic material found in topsoil.

Huntington's disease A human genetic disease caused by a dominant allele; characterized by uncontrollable body movements and degeneration of the nervous system; usually fatal 10–20 years after the onset of symptoms.

hybrid The offspring of parents of two different species or of two different varieties of one species; the offspring of two parents that differ in one or more inherited traits; an individual that is heterozygous for one or more pair of genes.

hybrid breakdown A type of postzygotic barrier between species; the species remain isolated because the offspring of hybrids are weak or infertile.

hybrid inviability A type of postzygotic barrier between species; the species remain isolated because hybrid zygotes do not develop or hybrids do not become sexually mature.

hybridization The cross-fertilization of two different varieties of an organism or of two different species; also called a cross.

hybrid sterility A type of postzygotic barrier between species; the species remain isolated because hybrids fail to produce functional gametes.

hydrocarbon A chemical compound composed only of the elements carbon and hydrogen.

hydrogen bond A type of weak chemical bond formed when the partially positive hydrogen atom participating in a polar covalent bond in one molecule is attracted to the partially negative atom participating in a polar covalent bond in another molecule (or in another part of the same macromolecule).

hydrolysis (hi-DROL-ih-sis) A chemical process in which macromolecules are broken down by the chemical addition of water molecules to the bonds linking their monomers; an essential part of digestion.

hydrophilic (HI-druh-FIL-ik) "Water-loving"; pertaining to polar or charged molecules (or parts of molecules) that are soluble in water.

hydrophobic (HI-druh-FO-bik) "Water-fearing"; pertaining to nonpolar molecules (or parts of molecules) that do not dissolve in water.

hydrostatic skeleton A skeletal system composed of fluid held under pressure in a closed body compartment; the main skeleton of most cnidarians, flatworms, nematodes, and annelids.

hydroxyl group (hi-DROK-sul) In an organic molecule, a functional group consisting of a hydrogen atom bonded to an oxygen atom.

hymen A thin membrane that partly covers the vaginal opening in the human female; ruptured by sexual intercourse or other vigorous activity.

hypercholesterolemia (HI-per-ko-LES-tur-ah-LEEM-ee-uh) An inherited human disease characterized by an excessively high level of cholesterol in the blood.

hypertension Abnormally high blood pressure; a persistent blood pressure of 140/90 or higher.

hypertonic solution In comparing two solutions, the one with the greater concentration of solutes.

hyperventilating Taking several deep breaths so rapidly that the CO_2 level in the blood is reduced, causing the breathing control centers to temporarily shut down breathing movements.

hypha (HI-fuh) (plural, hyphae) One of many filaments making up the body of a fungus.

hypoglycemia (HI-po-gly-SEE-mee-uh) An abnormally low level of glucose in the blood that results when the pancreas secretes too much insulin into the blood.

hypothalamus (HI-po-THAL-uh-mus) The master control center of the endocrine system, located in the ventral portion of the vertebrate forebrain. The hypothalamus functions in maintaining homeostasis, especially in coordinating the endocrine and nervous systems; secretes hormones of the posterior pituitary and releasing hormones that regulate the anterior pituitary.

hypothesis (hi-POTH-uh-sis) (plural, hypotheses) A tentative explanation a scientist proposes for a specific phenomenon that has been observed.

hypotonic solution In comparing two solutions, the one with the lower concentration of solutes.

I

immune system The organ system that protects the body by recognizing and attacking specific kinds of pathogens and cancer cells.

immunity Resistance to specific body invaders.

immunodeficiency disease An immunological disorder in which the immune system lacks one or more components, making the body susceptible to infectious agents that would ordinarily not be pathogenic.

imprinting Learning that is limited to a specific critical period in an animal's life and that is irreversible.

incomplete dominance A type of inheritance in which the phenotype of a heterozygote *(Aa)* is intermediate between the phenotypes of the two types of homozygotes *(AA* and *aa)*.

incomplete metamorphosis (MET-uh-MOR-fuh-sis) The transformation of young (sometimes called larvae) into adults that are similar to the young except in size and body proportions.

indeterminate growth Growth that continues throughout life, as in most plants. *See also* determinate growth.

indoleacetic acid (IAA) The naturally occurring auxin, a plant hormone.

induction During embryonic development, the influence of one group of cells on another group of cells.

inferior vena cava (VEE-nuh KAY-vuh) A large vein that returns O_2-poor blood to the heart from the lower, or posterior, part of the body. *See also* superior vena cava.

inflammatory response A nonspecific body defense caused by a release of histamine and other chemical alarm signals, which trigger increased blood flow, a local increase in white blood cells, and fluid leakage from the blood. The results include redness, heat, and swelling in the affected tissues.

ingestion The act of eating; the first main stage of food processing.

ingroup In a cladistic study of evolutionary relationships among taxa of organisms, the group of taxa that is actually being analyzed (*see also* outgroup).

innate behavior Behavior that appears to be performed in virtually the same way by all members of a species.

inner ear One of three main regions of the vertebrate ear; includes the cochlea, organ of Corti, and semicircular canals.

insulin A protein hormone, secreted by islet cells in the pancreas, that lowers the level of glucose in the blood.

integration The interpretation of sensory signals within neural processing centers of the central nervous system.

integumentary system (in-TEG-yoo-MEN-ter-ee) The organ system consisting of the skin and its derivatives, such as hair and nails in mammals; helps protect the body from drying out, mechanical injury, and infection.

interferon (IN-ter-FEER-on) A nonspecific defensive protein produced by virus-infected cells and capable of helping other cells resist viruses.

intermediate One of the compounds that form between the initial reactant in a metabolic pathway, such as glucose in glycolysis, and the final product, such as pyruvic acid in glycolysis.

intermediate filament An intermediate-sized protein fiber that is one of the three main kinds of fibers making up the cytoskeleton of eukaryotic cells; ropelike, made of fibrous proteins.

interneuron (IN-ter-NOOR-on) A nerve cell, entirely within the central nervous system, that integrates sensory signals and may relay command signals to motor neurons.

internode The portion of a plant stem between two nodes.

interphase The period in the eukaryotic cell cycle when the cell is not actually dividing. *See* mitosis.

interspecific competition A contest between individuals of two populations that require a limited resource; may inhibit population growth and help structure communities.

interstitial fluid (IN-ter-STISH-yul) An aqueous solution that surrounds body cells, and through which materials pass back and forth between the blood and the body tissues.

intertidal zone (IN-ter-TIDE-ul) A shallow zone where the waters of an estuary or ocean meet land.

intestine The region of a digestive tract between the gizzard or stomach and the anus, where chemical digestion and nutrient absorption usually occur.

intrinsic rate of increase An organism's inherent capacity to reproduce.

intron (IN-trahn) In eukaryotes, a nonexpressed (noncoding) portion of a gene; it is excised from the RNA transcript. *See* exon.

inversion A change in a chromosome resulting from reattachment in a reverse direction of a chromosome fragment to the original chromosome. Mutagens and errors during meiosis can cause inversions.

invertebrate An animal that lacks a backbone.

in vitro fertilization (IVF) (VEE-tro) Uniting sperm and egg in a laboratory container, followed by the placement of a resulting early embryo in the mother's uterus.

ion (EYE-on) An atom or molecule that has gained or lost one or more electrons, thus acquiring an electrical charge.

ionic bond (eye-ON-ik) An attraction between two ions with opposite electrical charges; the electrical attraction of the opposite charges holds the ions together.

iris The colored part of the vertebrate eye, formed by the anterior portion of the choroid.

islet cells (EYE-lit) Clusters of endocrine cells in the pancreas that produce insulin and glucagon.

isomers (EYE-sum-ers) Organic compounds with the same molecular formula but different structures and, therefore, different properties.

isotonic solution (eye-soh-TAHN-ik) A solution having the same solute concentration as another solution.

isotope (EYE-so-tope) A variant form of an atom; isotopes of an element have the same number of protons but different numbers of neutrons.

K

karyotype (KAIR-ee-uh-tipe) A display of micrographs of the metaphase chromosomes of a cell, arranged by size and centromere position.

kelp A giant brown alga, up to 100 m long, that forms extensive undersea forests.

keystone predator A predator species that reduces the density of the strongest competitors in a community, thereby helping maintain species diversity.

kilocalorie (kcal) a quantity of heat equal to 1000 calories. Used to measure the energy content of food, it is usually called a "calorie."

kinesis (kih-NEE-sis) Random movement in response to a stimulus.

kinetic energy (kih-NET-ik) Energy that is actually doing work; the energy of a mass of matter that is moving. Moving matter performs work by transferring its motion to other matter, such as leg muscles pushing bicycle pedals.

kinetochore (kih-NET-oh-core) A specialized protein structure at the centromere region on a sister chromatid. Spindle microtubules attach to the kinetochore during mitosis and meiosis.

kingdom In classification, the broad taxonomic category above phylum or division.

kin selection The concept that altruism evolves because it increases the number of copies of a gene common to a genetically related group of organisms; a hypothesis about the ultimate cause of altruism.

Koch's postulates A set of criteria used to establish that a particular infectious agent causes a disease.

Krebs cycle The metabolic cycle that is fueled by acetyl CoA formed after glycolysis in cellular respiration; chemical reactions in the Krebs cycle complete the metabolic breakdown of glucose molecules to carbon dioxide; occurs in the matrix of mitochondria and supplies most of the NADH molecules that carry energy to the electron transport chains.

L

labia majora (LAY-bee-uh muh-JOR-uh) A pair of outer thickened folds of skin that protect the female genital region.

labia minora (LAY-bee-uh mi-NOR-uh) A pair of inner folds of skin, bordering and protecting the female genital region.

labor A series of strong, rhythmic contractions of the uterus that expel a baby out of the uterus and vagina during childbirth.

lancelet One of a group of invertebrate chordates.

landscape ecology The application of ecological principles to the study of land-use patterns; the scientific study of the biodiversity of interacting ecosystems.

larva (plural, larvae) An immature individual that is structurally and often ecologically very different from an adult.

larynx (LAIR-inks) The voicebox, containing the vocal cords.

lateral Pertaining to the side of a bilaterally symmetrical animal.

lateral line system A row of sensory organs along each side of a fish's body. Sensitive to changes in water pressure, it enables a fish to detect minor vibrations in the water.

Laurasia (lahr-AZE-yuh) The northern landmass formed when continental drift split Pangaea (all land masses fused) during the Mesozoic era; *see also* Gondwana.

LDL *See* low-density lipoprotein.

leaf The main site of photosynthesis in a plant; consists of a flattened blade and a stalk (petiole) that joins the leaf to the stem.

learning A behavioral change resulting from experience.

lens The structure in an eye that focuses light rays onto the retina.

leukemia (loo-KEE-me-ah) A type of cancer of the blood-forming tissues, characterized by an excessive production of white blood cells and an abnormally high number of them in the blood; cancer of the bone marrow cells that produce leukocytes.

leukocyte (LOO-kuh-site) *See* white blood cell.

lichen (LY-ken) A mutualistic association between a fungus and an alga or between a fungus and a cyanobacterium.

life cycle The entire sequence of stages in the life of an organism, from the adults of one generation to the adults of the next.

life history The series of events from birth through reproduction to death.

life table A listing of survivals and deaths in a population in a particular time period and predictions of how long, on average, an individual of a given age will live.

light microscope (LM) An optical instrument with lenses that refract (bend) visible light to magnify images and project them into a viewer's eye or onto photographic film.

light reactions The first of two stages in photosynthesis, the light reactions are the steps in which solar energy is absorbed and converted to chemical energy in the form of ATP and NADPH. The light reactions power the sugar-producing Calvin cycle but produce no sugar themselves.

limbic system (LIM-bik) A functional unit of several integrating and relay centers located deep in the human forebrain; interacts with the cerebral cortex in creating emotions and storing memories.

linked genes Genes located close enough together on a chromosome to be usually inherited together.

lipid An organic compound consisting mainly of carbon and hydrogen atoms linked by nonpolar convalent bonds, and therefore mostly hydrophobic. Lipids include fats, waxes, phospholipids, and steroids that are insoluble in water.

liver The largest organ in the vertebrate body. The liver performs diverse functions such as producing bile, preparing nitrogenous wastes for disposal, and detoxifying poisonous chemicals in the blood.

lobe-finned fish A bony fish with strong, muscular fins supported by bones. Lobefins are extinct except for one species, the coelacanth.

local regulator A chemical messenger that is secreted into the interstitial fluid and causes changes in cells very near the point of secretion. Prostaglandins and neurotransmitters are local regulators.

locomotion Active movement from place to place.

locus (plural, loci) The particular site where a gene is found on a chromosome. Homologous chromosomes have corresponding gene loci.

logistic growth model A mathematical description of idealized population growth that is restricted by limiting factors.

long-day plant A plant that flowers in late spring or early summer when day length is increasing.

long-term depression (LTD) A reduced responsiveness to an action potential (nerve signal) by a receiving neuron.

long-term potentiation (LTP) An enhanced responsiveness to an action potential (nerve signal) by a receiving neuron.

loop of Henle (HEN-lee) In the vertebrate kidney, the portion of a nephron that helps concentrate the filtrate while conveying it between a proximal tubule and a distal tubule.

low-density lipoprotein (LDL) A cholesterol-carrying particle in the blood, made up of cholesterol and other lipids surrounded by a single layer of phospholipids in which proteins are embedded. An LDL particle carries more cholesterol than a related lipoprotein, HDL, and high LDL levels in the blood correlate with a tendency to develop blocked blood vessels and heart disease.

lung An internal sac, lined with moist epithelium, where gases are exchanged between inhaled air and the blood.

luteinizing hormone (LH) (LOO-tee-in-EYE-zing) A protein hormone secreted by the anterior pituitary that stimulates ovulation in females and androgen production in males.

Lyme disease A debilitating human disease caused by the bacterium *Borrelia burgdorferi;* characterized at first by a red rash at the site of tick bite and, if not treated, by heart disease, arthritis, and nervous disorders.

lymph A fluid similar to interstitial fluid that circulates in the lymphatic system.

lymphatic system (lim-FAT-ik) The organ system through which lymph circulates; includes lymph vessels, lymph nodes, and the spleen. The lymphatic system helps remove toxins and pathogens from the blood and interstitial fluid, and returns fluid and solutes from the interstitial fluid to the circulatory system.

lymphocyte (LIM-fuh-site) A type of white blood cell that is chiefly responsible for the immune response; found mostly in the lymphatic system. *See* B cell *and* T cell.

lymphoma (lim-FOH-muh) Cancer of the tissues that form white blood cells.

lysogenic cycle (lie-so-JEN-ik) A type of bacteriophage replication cycle in which the viral genome is incorporated into the bacterial host chromosome as a prophage; new phages are not produced, and the host cell is not killed or lysed unless the viral genome leaves the host chromosome.

lysosome (LYE-suh-soam) A digestive organelle in eukaryotic cells; contains hydrolytic enzymes that digest the cell's food and wastes.

lytic cycle (LIT-ik) A type of viral replication cycle resulting in the release of new viruses by lysis (breaking open) of the host cell.

M

macroevolution The main events in the evolutionary history of life on Earth.

macromolecule A giant molecule in a living organism: a protein, polysaccharide, or nucleic acid.

macronutrient A chemical substance that an organism must obtain in relatively large amounts. *See also* micronutrient.

macrophage (MAK-roh-faje) A large, amoeboid, phagocytic white blood cell that develops from a monocyte.

magnetic resonance imaging (MRI) Imaging technology that uses magnetism and radio waves to induce hydrogen nuclei in water molecules to emit faint radio signals. A computer creates images of the body from the radio signals.

magnification An increase in the apparent size of an object.

malignant tumor An abnormal tissue mass that can spread into neighboring tissue and to other parts of the body; a cancerous tumor.

mantle In a mollusk, the outgrowth of the body surface that drapes over the animal. The mantle produces the shell and forms the mantle cavity.

marsupial (mar-SOO-pee-ul) A pouched mammal, such as a kangaroo, opossum, or koala. Marsupials give birth to embryonic offspring that complete development while housed in a pouch and attached to nipples on the mother's abdomen.

marsupium (mar-SOO-pee-um) (plural, marsupia) The external pouch on the abdomen of a female marsupial.

mass number The sum of the number of protons and neutrons in an atom's nucleus.

mast cell A vertebrate body cell that produces histamine and other molecules that trigger the inflammatory response.

mating type A group of sexually compatible individuals in a population. Certain species of prokaryotes, protists, and fungi have mating types.

matter Anything that occupies space and has mass.

mechanical isolation A type of prezygotic barrier between species; the species remain isolated because structural differences between them prevent fertilization.

mechanoreceptor (meh-KAN-oh-reh-SEP-ter) A sensory receptor that detects physical deformations in the environment, associated with pressure, touch, stretch, motion, and sound.

medulla oblongata (meh-DUL-uh OB-long-GAH-tuh) Part of the vertebrate hindbrain continuous with the spinal cord; passes data between the spinal cord and forebrain and controls autonomic, homeostatic functions, including breathing, heart rate, swallowing, and digestion.

medusa (meh-DOO-suh) (plural, medusae) One of two types of cnidarian body forms; an umbrellalike body form. Also called a jellyfish.

meiosis (my-OH-sis) In a sexually reproducing organism, the division of a single diploid nucleus into four haploid daughter nuclei. Meiosis and cytokinesis produce haploid gametes from diploid cells in the reproductive organs of the parents.

memory cell One of a clone of long-lived lymphocytes formed during the primary immune response; remains in a lymph node until activated by exposure to the same antigen that triggered its formation. When activated, a memory cell forms a large clone that mounts the secondary immune response.

meninges (meh-NIN-jeez) Layers of connective tissue that enwrap and protect the brain and spinal cord.

menstrual cycle (MEN-stroo-ul) The hormonally synchronized cyclic buildup and breakdown of the endometrium of some primates, including humans.

menstruation (MEN-stroo-AY-shun) Uterine bleeding resulting from shedding of the endometrium during a menstrual cycle.

meristem (MAIR-eh-STEM) Plant tissue consisting of undifferentiated cells that divide and generate new cells and tissues.

mesoderm (MEZ-oh-durm) The middle layer of the three embryonic cell layers in a gastrula; gives rise to muscles, bones, the dermis of the skin, and most other organs in the adult.

mesophyll (MEZ-oh-fil) The green tissue in the interior of a leaf; a leaf's ground tissue system; the main site of photosynthesis.

messenger RNA (mRNA) The type of ribonucleic acid that encodes genetic information from DNA and conveys it to ribosomes, where the information is translated into amino acid sequences.

metamorphosis (MET-uh-MOR-fuh-sis) The transformation of a larva into an adult.

metaphase (MET-eh-faze) The second stage of mitosis. During metaphase, all the cell's duplicated chromosomes are lined up at an imaginary plane equidistant between the poles of the mitotic spindle.

metastasis (meh-TAS-teh-sis) The spread of cancer cells beyond their original site.

microevolution A change in a population's gene pool over a succession of generations; evolutionary changes in species over relatively brief periods of geological time.

microfilament The thinnest of the three main kinds of protein fibers making up the cytoskeleton of a eukaryotic cell; a solid, helical rod composed of the globular protein actin. Microfilaments help some cells change shape and move by assembling at one end while disassembling at the other.

micrograph A photograph taken through a microscope.

micronutrient An element that an organism needs in very small amounts and that functions as a component or cofactor of enzymes. *See also* macronutrient.

microtubule The thickest of the three main kinds of fibers making up the cytoskeleton of a eukaryotic cell; a straight, hollow tube made of globular proteins called tubulins. Microtubules form the basis of the structure and movement of cilia and flagella.

microtubule-organizing center (MTOC) A specialized place in the cell where microtubules of the mitotic spindle begin to form.

microvillus (plural, microvilli) A microscopic projection on the surface of a cell. Microvilli increase a cell's surface area.

midbrain One of three ancestral and embryonic regions of the vertebrate brain; develops into sensory integrating and relay centers that send sensory information to the cerebrum.

middle ear One of three main regions of the vertebrate ear; a chamber containing three small bones (the hammer, anvil, and stirrup), which convey vibrations from the eardrum to the oval window.

mineral In nutrition, a chemical element other than carbon, hydrogen, oxygen, or nitrogen that an organism requires for proper body functioning.

mineralocorticoid (min-er-AL-oh-KOR-te-koid) A corticosteroid hormone secreted by the adrenal cortex that helps maintain salt and water homeostasis and may increase blood pressure in response to long-term stress.

mitochondrial matrix (MY-toh-KON-dree-ul MAY-triks) The fluid contained within the inner membrane of a mitochondrion.

mitochondrion (MY-toh-KON-dree-on) (plural, mitochondria) An organelle in eukaryotic cells where cellular respiration occurs. Enclosed by two concentric membranes, it is where most of the cell's ATP is made.

mitosis (MY-toh-sis) The division of a single nucleus into two genetically identical daughter nuclei. Mitosis and cytokinesis make up the mitotic (M) phase of the cell cycle.

mitotic spindle A spindle-shaped structure formed of microtubules and associated proteins that is involved in the movements of chromosomes during mitosis and meiosis. (A spindle is shaped roughly like a football.)

modern synthesis A comprehensive theory of evolution that incorporates genetics and includes most of Darwin's ideas, focusing on populations as the fundamental units of evolution.

molecular biology The study of the molecular basis of genes and gene expression; molecular genetics.

molecule A group of two or more atoms held together by covalent bonds.

Mollusca (mol-LUS-kuh) The phylum that contains the mollusks; characterized by a muscular foot, mantle, mantle cavity, and radula; includes gastropods (snails and slugs), bivalves (clams, oysters, scallops), and cephalopods (squids and octopuses).

molting In arthropods, the process of shedding an old exoskeleton and secreting a new, larger one.

monoclonal antibody (MON-oh-KLONE-ul) An antibody secreted by a clone of cells and, consequently, specific for the one antigen that triggered the development of the clone.

monocot (MON-uh-kot) A flowering plant whose embryos have a single seed leaf, or cotyledon.

monoculture The cultivation of a single plant variety in a large land area.

monocyte (MAHN-uh-site) A phagocytic white blood cell that can engulf bacteria and viruses in infected tissue; has a large oval or horseshoe-shaped nucleus.

monoecious (mon-EE-shus) Having individuals that produce both sperm and eggs (usually refers to plants).

monohybrid cross An experimental mating of individuals in which the inheritance of a single characteristic is tracked.

monomer (MON-uh-mer) A chemical subunit that serves as a building block of a polymer.

monophyletic taxon (MON-oh-fy-leh-tik) A taxonomic group composed of an ancestor and all of its descendants.

monosaccharide (MON-oh-SAK-uh-ride) The smallest kind of sugar molecule; a single-unit sugar. Monosaccharides are the building blocks of more complex sugars and polysaccharides.

monotreme (MON-uh-treem) An egg-laying mammal, such as the duck-billed platypus.

morphs Two or more different kinds of individuals or forms of a phenotypic characteristic in a population.

motor neuron A nerve cell that conveys command signals from the central nervous system to effector cells, such as muscle cells or gland cells.

motor output The conduction of signals from a processing center in a central nervous system to effector cells.

motor unit A motor neuron and all the muscle fibers it controls.

mouth An opening through which food is taken into an animal's body.

movement corridor A series of small clumps or a narrow strip of quality habitat (usable by organisms) that connects otherwise isolated patches of quality habitat.

mRNA *See* messenger RNA.

mucous membrane (MYOO-kus) Smooth, moist epithelium that lines the digestive tract and air tubes leading to the lungs.

Müllerian mimicry (myoo-LER-ee-un MIM-ih-kree) A mutual mimicry by two species, both of which are poisonous or otherwise harmful to a predator.

mutagen (MYOO-tuh-jen) A chemical or physical agent that interacts with DNA and causes a mutation.

mutagenesis (MYOO-tuh-jen-ih-sis) The creation of a mutation.

mutation A change in the nucleotide sequence of DNA; the ultimate source of genetic diversity.

mutualism A symbiotic relationship in which both partners benefit.

mycelium (my-SEEL-ee-um) (plural, mycelia) The densely branched network of hyphae in a fungus.

mycorrhiza (MY-ko-RY-za) (plural, mycorrhizae) A mutualistic association of plant roots and fungi.

myelin sheath (MY-eh-lin) A series of cells, each wound around, and thus insulating, the axon of a nerve cell in vertebrates. Each pair of cells in the sheath is separated by a space called a node of Ranvier.

myofibril (MY-oh-FY-brul) A contractile thread in a muscle cell (fiber) made up of many sarcomeres. Longitudinal bundles of myofibrils make up a muscle fiber.

N

NAD^+ Nicotinamide adenine dinucleotide; a coenzyme that assists enzymes by conveying electrons (from hydrogen atoms) during the redox reactions of cellular metabolism. The plus sign indicates that the molecule is oxidized and ready to pick up hydrogens; the reduced, hydrogen (electron)-carrying form is $NADH_2$.

natural killer cell A nonspecific defensive cell that attacks cancer cells and infected body cells, especially those harboring viruses.

natural selection Differential success in reproduction by different phenotypes resulting from interactions with the environment. Evolution occurs when natural selection produces changes in the relative frequencies of alleles in a population's gene pool.

nearsightedness An inability to focus on distant objects; occurs when the eyeball is longer than normal and the lens focuses distant objects in front of the retina. Also called myopia.

negative feedback A control mechanism in which a chemical reaction, metabolic pathway, or hormone-secreting gland is inhibited by the products of the reaction, pathway, or gland. As the concentration of the products builds up, the product molecules themselves inhibit the process that produced them.

Nematoda (NEM-eh-TOAD-uh) The phylum that contains the roundworms, or nematodes; characterized by a pseudocoelom, a cylindrical, wormlike body form, and a tough cuticle.

nephron The tubular excretory unit and associated blood vessels of the vertebrate kidney; extracts filtrate from the blood and refines it into urine.

nerve A ropelike bundle of neuron fibers (axons and dendrites) tightly wrapped in connective tissue.

nerve cord An elongated bundle of axons and dendrites, usually extending longitudinally from the brain or anterior ganglia. One or more nerve cords and the brain make up the central nervous system in many animals.

nerve net A weblike system of neurons, characteristic of radially symmetrical animals such as Hydra.

nervous system The organ system that forms a communication and coordination network among all parts of an animal's body.

nervous tissue Tissue made up of neurons and supportive cells.

neural tube (NOOR-ul) An embryonic cylinder that develops from ectoderm after gastrulation; gives rise to the brain and spinal cord.

neuromuscular junction A synapse between an axon of a motor neuron and a muscle fiber.

neuron (NOOR-on) A nerve cell; the fundamental structural and functional unit of the nervous system, specialized for carrying signals from one location in the body to another.

neurosecretory cell A nerve cell that synthesizes hormones and secretes them into the blood, as well as conducting nerve signals.

neurotransmitter A chemical messenger that carries information from a transmitting neuron to a receiving cell, either another neuron or an effector cell.

neutral variation Genetic variation that provides no apparent selective advantage for some individuals over others.

neutron An electrically neutral particle (a particle having no electrical charge), found in the nucleus of an atom.

neutrophil (NOO-truh-fil) A nonspecific defensive, phagocytic white blood cell that can engulf bacteria and viruses in infected tissue; has a multilobed nucleus.

niche (nich) A population's role in its community; the sum total of a population's use of the biotic and abiotic resources of its habitat.

nitrogen fixation The conversion of atmospheric nitrogen (N_2) into nitrogen compounds (NH_4^+, NO_3^-) that plants can absorb and use.

nitrogenous base (nigh-TRAH-jen-us) An organic molecule that is a base and contains the element nitrogen.

node The point of attachment of a leaf on a stem.

node of Ranvier (RAHN-vee-ay) An unmyelinated region on a myelinated axon of a nerve cell, where signal transmission occurs.

noncompetitive inhibitor A substance that impedes the activity of an enzyme without entering an active site—and thus without competing directly with the normal substrate. By binding elsewhere on the enzyme, a noncompetitive inhibitor changes the shape of the enzyme so that the active site no longer functions.

nondisjunction An accident of meiosis or mitosis in which a pair of homologous chromosomes or a pair of sister chromatids fail to separate at anaphase.

nonpolar covalent bond An attraction between atoms that share one or more pairs of electrons equally because the atoms have similar electronegativity.

nonrandom mating The selection of mates other than by chance.

nonself molecule A foreign antigen; a protein or other macromolecule that is not part of an organism's body. *See also* self protein.

nonsteroid hormone A regulatory chemical synthesized from amino acids that acts on target cells via a second messenger such as cAMP. Amine hormones, peptide hormones, and protein hormones are nonsteroids.

norepinephrine (NOR-ep-ih-NEF-rin) An amine hormone (also called noradrenaline) secreted by the adrenal medulla that prepares body organs for fight or flight; also serves as a neurotransmitter.

notochord (NO-tuh-kord) A flexible, cartilagelike, longitudinal rod located between the digestive tract and nerve cord in chordate animals; present only in embryos in many species.

nuclear envelope A double membrane, perforated with pores, that encloses the nucleus and separates it from the rest of the eukaryotic cell.

nucleic acid (noo-KLAY-ik) A polymer consisting of many nucleotide monomers; serves as a blueprint for proteins and, through the actions of proteins, for all cellular structures and activities. The two types of nucleic acids are DNA and RNA.

nucleoid region (NOO-klee-oid) The region in a prokaryotic cell consisting of a concentrated mass of DNA.

nucleolus (noo-KLEE-uh-lus) A structure within the nucleus of a eukaryotic cell where ribosomal RNA is made and assembled with proteins to make ribosomal subunits; consists of parts of the chromatin DNA, RNA transcribed from the DNA, and proteins imported from the cytoplasm.

nucleosome (NOO-klee-uh-soam) The beadlike unit of DNA packaging in a eukaryotic cell; consists of DNA wound around a protein core made up of eight histone molecules.

nucleotide (NOO-klee-oh-tide) An organic monomer consisting of a five-carbon sugar covalently bonded to a nitrogenous base and a phosphate group. Nucleotides are the building blocks of nucleic acids.

nucleus (plural, nuclei) (1) An atom's central core, containing protons and neutrons. (2) The genetic control center of a eukaryotic cell.

O

old-growth forest An ancient forest that has never been seriously disturbed by humans. Trees dominating an old-growth forest may be thousands of years old.

ommatidia (ome-uh-TID-ee-uh) The functional units of a compound eye; each ommatidium includes a cornea, lens, and photoreceptor cells.

omnivore An animal that eats both plants and animals. *See also* carnivore; herbivore.

oncogene (ONK-oh-jean) A cancer-causing gene; usually contributes to malignancy by abnormally enhancing the amount or activity of a growth factor made by the cell.

oogenesis (OH-uh-JEN-eh-sis) The formation of ova (egg cells).

open circulatory system A circulatory system in which blood is pumped through open-ended vessels and out among the body cells. In an animal with an open circulatory system, blood and interstitial fluid are one and the same.

operator In prokaryotic DNA, a sequence of nucleotides near the start of an operon to which an active repressor can attach. The binding of repressor prevents RNA polymerase from attaching to the promoter and transcribing the genes of the operon.

operculum (oh-PUR-kyoo-lum) (plural, opercula) A protective flap on each side of a fish's head that covers a chamber housing the gills.

operon (OP-ur-on) A unit of genetic regulation common in prokaryotes; a cluster of genes with related functions, along with the promoter and operator that control their transcription.

optimal foraging Feeding behavior that provides maximal energy gain with minimal energy expense and minimal time spent searching for, securing, and eating food.

organ A structure consisting of several tissues adapted as a group to perform specific functions.

organelle (or-gan-EL) A structure with a specialized function within a cell.

organic compound A chemical compound containing the element carbon and usually synthesized by cells.

organism An individual living thing, such as a bacterium, fungus, protist, plant, or animal.

organ of Corti (KOR-tee) The hearing organ in birds and mammals, located within the cochlea.

organ system A group of organs that work together in performing vital body functions.

orgasm Rhythmic contractions of the reproductive structures, accompanied by extreme pleasure, at the peak of sexual excitement in both sexes; includes ejaculation by the male.

orientation behavior Directed movement resulting in an organism's placing itself in a specific position relative to an environmental cue.

osmoconformer (OZ-moh-con-FORM-er) An organism whose body fluids have a solute concentration equal to that of its surroundings. Osmoconformers do not have a net gain or loss of water by osmosis.

osmoregulation The control of the gain and loss of water and dissolved solutes in an organism.

osmoregulator An organism whose body fluids have a solute concentration different from that of its environment and that must use energy in controlling water loss or gain.

osmosis (oz-MOH-sis) The movement of water across a selectively permeable membrane.

Osteichthyes (os-tee-IK-theez) The vertebrate class of bony fishes; for example, trout and goldfish.

osteoporosis (OS-tee-oh-puh-RO-sis) A skeletal disorder characterized by thinning, porous, and easily broken bones; most common among women after menopause and often related to low estrogen levels.

outer ear One of three main regions of the ear in reptiles, birds, and mammals; made up of the auditory canal and, in many birds and mammals, the pinna.

outgroup In a cladistic study of evolutionary relationships among taxa of organisms, a taxon or group of taxa with a known relationship to, but not a member of, the taxa being studied (*see also* ingroup).

oval window In the vertebrate ear, a membrane-covered gap in the skull bone, through which sound waves pass from the middle ear into the inner ear.

ovarian cycle (oh-VAIR-ee-un) Hormonally synchronized cyclic events in the mammalian ovary, culminating in ovulation.

ovary (1) In animals, the female gonad, which produces egg cells and reproductive hormones. (2) In flowering plants, the basal portion of a carpel in which the egg-containing ovules develop.

oviduct (OH-vih-dukt) The tube that conveys egg cells away from an ovary; also called a Fallopian tube.

ovulation (AH-vyoo-LAY-shun) The release of an egg cell from an ovarian follicle.

ovule (OH-vyool) A reproductive structure in a seed plant; contains the female gametophyte and the developing egg. An ovule develops into a seed.

ovum (OH-vum) (plural, ova) An unfertilized egg, or female gamete.

oxidation The loss of electrons from a substance involved in a redox reaction; always accompanies reduction.

oxytocin (OK-sih-TOE-sin) A peptide hormone, made by the hypothalamus and secreted by the posterior pituitary, that stimulates contraction of the uterus and mammary gland cells.

ozone layer The layer of O_3 in the upper atmosphere that protects life on Earth from the harmful ultraviolet rays in sunlight.

P

pacemaker The SA (sinoatrial) node; a specialized region of cardiac muscle that maintains the heart's pumping rhythm (heartbeat) by setting the rate at which the heart contracts.

paedomorphosis (PED-uh-MOR-fuh-sis) The retention of juvenile body features in an adult.

pancreas (PAN-kree-us) A gland with dual functions: The nonendocrine portion secretes digestive enzymes and an alkaline solution into the small intestine via a duct; the endocrine portion secretes the hormones insulin and glucagon into the blood.

Pangaea (pan-JEE-uh) The supercontinent consisting of all the major landmasses of Earth fused together. Continental drift formed Pangaea near the end of the Paleozoic era.

parasitism (PAIR-eh-SIH-tiz-em) A symbiotic relationship in which the parasite, a type of predator, lives within or on the surface of a host, from which it derives its food.

parasympathetic division One of two sets of neurons in the autonomic nervous system; generally promotes body activities that gain and conserve energy, such as digestion and reduced heart rate. *See also* sympathetic division.

parathyroid glands (PAIR-uh-THY-roid) Four endocrine glands embedded in the surface of the thyroid gland that secrete parathyroid hormone.

parathyroid hormone (PTH) A peptide hormone secreted by the parathyroid glands that raises blood calcium level.

parenchyma cell (pur-EN-kim-uh) In plants, a relatively unspecialized cell with a thin primary wall and no secondary wall; functions in photosynthesis, food storage, and aerobic respiration, and may differentiate into other cell types.

parsimony (PAR-seh-mo-nee) In scientific studies, the search for the least complex explanation for an observed phenomenon.

partial pressure A measure of the relative amount of gas in a mixture.

passive immunity Temporary immunity obtained by acquiring ready-made antibodies or immune cells; lasts only a few weeks or months because the immune system has not been stimulated by antigens.

passive transport The diffusion of a substance across a biological membrane, without any input of energy.

pathogen A disease-causing organism.

pattern formation During embryonic development, the emergence of a spatial organization in which the tissues and organs of the organism are all in their correct places.

pedigree A family tree representing the occurrence of heritable traits in parents and offspring across a number of generations.

pelagic zone (pel-AJ-ik) The region of an ocean occupied by seawater.

peptide bond The covalent linkage between two amino acid units in a polypeptide; formed by dehydration synthesis.

peptidoglycan (PEP-tid-oh-GLY-kan) A polymer of complex sugars cross-linked by short polypeptides; a material unique to eubacterial cell walls.

perception The brain's meaningful interpretation, or conscious understanding, of sensory information.

perennial (peh-REN-ee-ul) A plant that lives for many years.

perforin (PUR-fer-in) A protein secreted by a cytotoxic T cell that lyses (ruptures) an infected cell by perforating its membrane.

peripheral nervous system (PNS) The network of nerves and ganglia carrying signals into and out of the central nervous system.

peristalsis (PER-ih-STAHL-sis) Rhythmic waves of contraction of smooth muscles. Peristalsis propels food through a digestive tract and also enables many animals, such as earthworms, to crawl.

permafrost Continuously frozen ground found in the tundra.

PET *See* positron-emission tomography.

petal A modified leaf of a flowering plant. Petals are the often colorful parts of a flower that advertise it to insects and other pollinators.

P generation The parent individuals from which offspring are derived in studies of inheritance; P stands for parental.

phage (FAJE) *See* bacteriophage.

phagocyte (FAG-geh-site) A white blood cell (e.g., a neutrophil or a monocyte) that engulfs bacteria, foreign proteins, and the remains of dead body cells.

phagocytosis (FAG-oh-sigh-TOE-sis) Cellular "eating;" a type of endocytosis whereby a cell engulfs macromolecules, other cells, or particles into its cytoplasm.

pharynx (FAIR-inks) The organ in a digestive tract that receives food from the oral cavity; in terrestrial vertebrates, the throat region where the air and food passages cross.

phenotype (FEE-noh-tipe) The expressed traits of an organism.

phenylketonuria (PKU) (FEN-ul-KEE-tuh-NOOR-ee-uh) A recessive genetic disorder characterized by an inability to properly break down the amino acid phenylalanine, an accumulation of toxins in the blood, and mental retardation.

phloem (FLO-um) The portion of a plant's vascular system that conveys phloem sap throughout a plant; made up of sieve-tube members.

phloem sap The solution of sugars, other nutrients, and hormones conveyed throughout a plant via phloem tissue.

phosphate group (FOS-fate) A functional group consisting of a phosphorus atom covalently bonded to four oxygen atoms.

phospholipid (FOS-fo-LIP-id) A type of lipid molecule containing the element phosphorus in a hydrophilic "head," to which are attached two fatty acids that form two hydrophobic "tails." A bilayer (double layer) of phospholipids makes up the basic fabric of biological membranes.

phosphorylation (fos-for-uh-LAY-shun) The transfer of a phosphate group, usually from ATP, to a molecule. Nearly all cellular work depends on ATP energizing other molecules by phosphorylation.

photic zone (FOE-tik) The region of an aquatic ecosystem into which light penetrates and where photosynthesis occurs.

photoautotroph An organism that obtains energy from sunlight and carbon from CO_2 by photosynthesis.

photoheterotroph An organism that obtains energy from sunlight and carbon from organic sources.

photon (FOE-tahn) A fixed quantity of light energy. The shorter the wavelength of light, the greater the energy of a photon.

photoperiod The length of the day relative to the length of the night; an environmental stimulus plants use to detect the time of year.

photophosphorylation (FOE-toe-fos-for-uh-LAY-shun) The production of ATP by chemiosmosis during the light reactions of photosynthesis.

photopsin (foe-TOP-sin) One of a family of visual pigments in the cones of the vertebrate eye that absorb bright, colored light.

photoreceptor A type of electromagnetic receptor that detects light.

photorespiration In a plant cell, the breakdown of a two-carbon compound produced by the Calvin cycle. The Calvin cycle produces the two-carbon compound, instead of its usual three-carbon product G3P, when leaf cells fix O_2, instead of CO_2. Photorespiration produces no sugar molecules or ATP.

photosynthesis (FOE-toe-SIN-thuh-sis) The process by which plants, autotrophic protists, and some bacteria use light energy to make sugars and other organic food molecules from carbon dioxide and water.

photosystem A light-harvesting unit of a chloroplast's thylakoid membrane; consists of several hundred antenna molecules, a reaction-center chlorophyll, and a primary electron acceptor.

phototropism (foh-TOT-treh-PIZ-em) The growth of a plant shoot toward or away from light.

pH scale A measure of the relative acidity of a solution, ranging in value from 0 (most acidic) to 14 (most basic); pH stands for potential hydrogen and refers to the concentration of hydrogen ions (H^+).

phylogenetic tree (FYE-lo-jeh-NET-ik) A branching diagram that represents a hypothesis about evolutionary relationships among organisms.

phylogeny (fye-LAHJ-en-ee) The evolutionary history of a species or group of related species.

phylum (FYE-lum) (plural, phyla) In classification, the taxonomic category above class and below kingdom; members of a phylum all have a similar general body plan.

phytochrome (FYE-teh-krome) A colored protein in plants that contains a special set of atoms that absorbs light.

phytoplankton (FYE-toh-PLANK-tun) Algae and photosynthetic bacteria that drift passively in aquatic environments.

pili (PIL-eye) (singular, pilus) Short projections on the surface of prokaryotic cells that help prokaryotes attach to other surfaces; specialized sex pili are used in conjugation to hold the mating cells together.

pineal gland (PIN-ee-ul) An outgrowth of the vertebrate brain that secretes the hormone melatonin, which coordinates daily and seasonal body activities, such as reproductive activity, with environmental light conditions.

pinna (PIN-uh) The flaplike part of the outer ear, projecting from the body surface of many birds and mammals; collects sound waves and channels them to the auditory canal.

pinocytosis (PIN-oh-sigh-TOE-sis) Cellular "drinking"; a type of endocytosis in which the cell takes fluid and dissolved solutes into small membranous vesicles.

pith Part of the ground tissue system of a dicot plant. Pith fills the center of a stem and may store food.

placenta (pluh-SEN-tuh) In most mammals, the organ that provides nutrients and oxygen to the embryo and helps dispose of its metabolic wastes; formed of the embryo's chorion and the mother's endometrial blood vessels.

Placentals (pluh-SEN-tuls) Mammals whose young complete their embryonic development in the uterus, nourished via the mother's blood vessels in the placenta; also called placental mammals or eutherians.

Plantae (PLANT-ay) The kingdom that contains the plants.

plasma The liquid matrix of the blood in which the blood cells are suspended.

plasma cell An antibody-secreting B cell.

plasma membrane The thin layer of lipids and proteins that sets a cell off from its surroundings and acts as a selective barrier to the passage of ions and molecules into and out of the cell; consists of a phospholipid bilayer in which are embedded molecules of protein and cholesterol.

plasmid A small ring of DNA separate from the chromosome(s); plasmids are found in prokayotes and yeast.

plasmodesma (PLAZ-moh-DEZ-muh) (plural, plasmodesmata) An open channel in a plant cell wall, through which strands of cytoplasm connect from adjacent walls.

plasmodial slime mold (plaz-MO-dee-ul) A type of protist that has amoeboid cells, flagellated cells, and an amoeboid plasmodial feeding stage in its life cycle.

plasmodium (1) A single mass of cytoplasm containing many nuclei. (2) The amoeboid feeding stage in the life cycle of a plasmodial slime mold.

platelet A piece of membrane-enclosed cytoplasm from a large cell in the bone marrow of a mammal; a blood-clotting element.

plate tectonics Forces within planet Earth that cause movements of the crust, resulting in continental drift, volcanoes, and earthquakes.

Platyhelminthes (PLAT-ee-hel-MIN-theez) The phylum that contains the flatworms, the bilateral animals with a thin, flat body form, gastrovascular cavity or no digestive system, and no body cavity; the free-living flatworms, flukes, and tapeworms.

pleated sheet The folded arrangement of a polypeptide in a protein's secondary structure.

pleiotropy (PLY-ih-tro-pee) The control of more than one phenotypic characteristic by a single gene.

polar covalent bond An attraction between atoms that share electrons unequally because the atoms differ in electronegativity. The shared electrons are pulled closer to the more electronegative atom, making it partially negative and the other atom partially positive.

polar molecule A molecule containing polar covalent bonds.

pollen grain In a seed plant, the male gametophyte that develops within the anthers of stamens.

pollination In seed plants, the delivery, by wind or animals, of pollen from the male parts of a plant to the stigma of a carpel on the female.

polygenic inheritance (POL-ee-JEN-ik) The additive effect of two or more gene loci on a single phenotypic characteristic.

polymer (POL-uh-mer) A large molecule consisting of many identical or similar molecular units, called monomers, covalently joined together in a chain.

polymerase chain reaction (PCR) (puh-LIM-eh-rase) A techique used to obtain many copies of a DNA molecule or part of a DNA molecule. A small amount of DNA mixed with the enzyme DNA polymerase, DNA nucleotides, and a few other ingredients replicates repeatedly in a test tube.

polymorphism The existence of different forms; may pertain to a population in which two or more morphs are present in readily noticeable frequencies.

polynucleotide (POL-ee-NOO-klee-oh-tide) A polymer made up of many nucleotides covalently bonded together.

polyp (POL-ip) One of two types of cnidarian body forms; a columnar, hydralike body.

polypeptide A chain of amino acids linked by peptide bonds.

polyploid cell (POL-eh-ploid) A cell with more than two complete sets of chromosomes.

polysaccharide (POL-ee-SAK-uh-ride) A carbohydrate polymer consisting of hundreds to thousands of monosaccharides (sugars) linked by covalent bonds.

pons (PAHNZ) Part of the vertebrate hindbrain that functions with the medulla oblongata in passing data between the spinal cord and forebrain and in controlling autonomic, homeostatic functions.

population A group of interacting individuals belonging to one species and living in the same geographic area.

population dynamics Changes in population size resulting from various forces that control and regulate populations over time.

population fragmentation The splitting and consequent isolation of a biological population, usually by human-caused habitat degradation.

population genetics The study of genetic changes in populations; the science of microevolutionary changes in populations.

population viability analysis (PVA) Scientific analysis of the current status of a population or species and predictions of its chances for long-term survival.

Porifera (por-IF-er-uh) The phylum that contains the sponges, characterized by choanocytes, a porous body wall, and no true tissues.

positive feedback A control mechanism in which the products of a process stimulate the process that produced them.

positron-emission tomography (PET) Imaging technology that uses radioactively labeled biological molecules, such as glucose, to obtain information about metabolic processes at specific locations in the body. The labeled molecules are injected into the bloodstream, and a PET scan for radioactive emissions determines which tissues have taken up the molecules.

posterior Pertaining to the rear, or tail, of a bilaterally symmetrical animal.

posterior pituitary An extension of the hypothalamus composed of nervous tissue that secretes hormones made in the hypothalamus; a temporary storage site for hypothalamic hormones.

potential energy Stored energy; the capacity to perform work that matter possesses because of its location or arrangement. Water behind a dam or chemical bonds both possess potential energy.

predation An interaction between species in which one species, the predator, eats the other, the prey.

predator A consumer in a biological community.

prepuce (PRE-pyoos) A fold of skin covering the head of the clitoris and penis.

prey An organism eaten by a predator.

primary consumer An organism that eats only autotrophs.

primary growth Growth in the length of a plant root or shoot produced by an apical meristem.

primary oocyte (OH-eh-SITE) A diploid cell, in prophase I of meiosis, that can be hormonally triggered to develop into an ovum.

primary productivity The rate at which an ecosystem's producers convert solar energy to chemical energy.

primary spermatocyte (spur-MAT-eh-SITE) A diploid cell in the testis that undergoes meiosis I.

primary structure The first level of protein structure; the specific sequence of amino acids making up a polypeptide chain.

primary succession A type of ecological succession in which a biological community arises in an area without soil. *See also* secondary succession.

primitive characters Homologous features found in members of a lineage and also in the ancestors of the lineage; ancestral features.

principle of independent assortment A general rule in inheritance that when gametes form during meiosis, each pair of alleles for a particular characteristic segregate independently; also known as Mendel's second law of inheritance.

principle of segregation A general rule in inheritance that individuals have two alleles for each gene, and that when gametes form by meiosis, the two alleles separate, and each resulting gamete ends up with only one allele of each gene; also known as Mendel's first law of inheritance.

probe In DNA technology, a labeled single-stranded nucleic acid molecule used to find a specific gene, or other nucleotide sequence, within a mass of DNA. The probe hydrogen-bonds to the complementary sequence in the targeted DNA.

producer An organism that makes organic food molecules from CO_2, H_2O, and other inorganic raw materials: a plant, alga, or autotrophic bacterium.

progesterone (pro-JES-teh-roan) A steroid hormone secreted by the corpus luteum of the ovary; maintains the uterine lining during pregnancy.

progestin (pro-JES-tin) One of a family of steroid hormones, including progesterone, produced by the mammalian ovary; progestins prepare the uterus for pregnancy.

programmed cell death The timely death (and disposal of the remains) of certain cells, triggered by certain genes; an essential process in normal development; also called apoptosis.

prokaryotic cell (pro-KAIR-ee-OT-ik) A type of cell lacking a membrane-enclosed nucleus and other membrane-enclosed organelles; found only in the domains Bacteria and Archaea.

prokaryotic cell wall A fairly rigid, chemically complex wall that protects the prokaryotic cell and helps maintain its shape.

prokaryotic flagellum A long surface projection that propels a prokaryotic cell through its liquid environment; totally different from the flagellum of a eukaryotic cell.

prolactin (PRL) (pro-LAK-tin) A protein hormone secreted by the anterior pituitary that stimulates milk production in mammals.

promoter A specific nucleotide sequence in DNA, located at the start of a gene, that is the binding site for RNA polymerase and the place where transcription begins.

prophage (PRO-faje) Phage DNA that has inserted by genetic recombination into the DNA of a prokaryotic chromosome.

prophase The first stage of mitosis, during which duplicated chromosomes condense to form structures visible with a light microscope and the mitotic spindle forms and begins moving the chromosomes toward the center of the cell.

prosimian (pro-SIM-ee-un) A member of the primate group comprised of lorises, bushbabies, tarsiers, and lemurs.

prostaglandin (PROS-tuh-GLAN-din) One of a large family of local regulators secreted by virtually all tissues and performing a wide variety of regulatory functions.

prostate gland (PROS-tate) A gland in human males that secretes an acid-neutralizing component of semen.

protein A biological macromolecule made of one or more polypeptides.

protist (PRO-tist) A member of the kingdom Protista.

Protista (pro-TIS-tuh) In the standard five-kingdom classification system, the kingdom that contains the unicellular eukaryotes (and closely related multicellular organisms) called protists.

proton A subatomic particle with a single positive electrical charge, found in the nucleus of an atom.

proto-oncogene (PRO-toe-ONK-oh-jean) A normal gene that can be converted to a cancer-causing gene.

protostome An animal with a coelom that develops from solid masses of cells that arise between the digestive tube and the body wall of the embryo. The protostomes include the mollusks, annelids, and arthropods.

protozoan (PRO-teh-ZO-un) A protist that lives primarily by ingesting food; a heterotrophic, animal-like protist.

proximal tubule In the vertebrate kidney, the portion of a nephron immediately downstream from Bowman's capsule that conveys and helps refine filtrate.

proximate cause In behavioral biology, the immediate explanation for an organism's behavior; the interactions of an organism with the environment or the particular environmental stimuli that trigger a behavioral response in the organism.

pseudocoelom (SOO-doe-SEE-loam) A body cavity that is in direct contact with the wall of the digestive tract.

pseudopodium (SOO-doe-PODE-ee-um) (plural, pseudopodia) A temporary extension of an amoeboid cell. Pseudopodia function in moving cells and engulfing food.

pulmonary artery A large blood vessel that conveys blood from the heart to a lung.

pulmonary circuit One of two main blood circuits in terrestrial vertebrates; conveys blood between the heart and the lungs. *See also* systemic circuit.

pulmonary vein A blood vessel that conveys blood from a lung to the heart.

pulse The rhythmic stretching of the arteries caused by the pressure of blood forced through the arteries by contractions of the ventricles during systole.

punctuated equilibrium The idea that speciation occurs in spurts followed by long periods of little change.

Punnett square A diagram used in the study of inheritance to show the results of random fertilization.

pupil The opening in the iris, which admits light into the interior of the vertebrate eye; muscles in the iris regulate its size.

purine (PYOOR-een) One of two families of nitrogenous bases found in nucleotides. Adenine (A) and guanine (G) are purines.

pyloric sphincter (PIE-lore-ik SFINK-ter) In the vertebrate digestive tract, a muscular ring that regulates the passage of food out of the stomach and into the small intestine.

pyrimidine (pir-IM-ih-deen) One of two families of nitrogenous bases found in nucleotides. Cytosine (C), thymine (T), and uracil (U) are pyrimidines.

Q

quaternary consumer (KWA-tur-ner-ee) An organism that eats tertiary consumers.

quaternary structure The fourth level of protein structure; the shape resulting from the association of two or more polypeptide subunits.

R

radial symmetry An arrangement of the body parts of an organism like pieces of a pie around an imaginary central axis. Any slice passing longitudinally through a radially symmetrical organism's central axis divides it into mirror-image halves.

radioactive isotope An isotope whose nucleus decays spontaneously, giving off particles and energy.

radiometric dating A method for determining the age of fossils and rocks from the ratio of a radioactive isotope to the nonradioactive istope(s) of the same element in the sample.

radula (RAD-yoo-luh) A toothed, rasping organ used to scrape up or shred food; found in many mollusks.

ray-finned fish A bony fish having fins supported by thin, flexible skeletal rays. All but one living species of bony fishes are rayfins. *See* lobe-finned fish.

reabsorption In the vertebrate kidney, the reclaiming of water and valuable solutes from the filtrate.

reactant A starting material in a chemical reaction.

reaction center In a photosystem in a chloroplast, the chlorophyll *a* molecule and the primary electron acceptor that trigger the light reactions of photosynthesis. The chlorophyll donates an electron excited by light energy to the primary electron acceptor, which passes an electron to an electron transport chain.

reading frame The way in which a cell's mRNA-translating machinery groups the mRNA nucleotides into codons.

receptor On or in a cell, a specific protein molecule whose shape fits that of a specific molecular messenger, such as a hormone.

receptor-mediated endocytosis (END-oh-sigh-TOE-sis) The movement of specific molecules into a cell by the inward budding of membranous vesicles. The vesicles contain proteins with receptor sites specific to the molecules being taken in.

receptor potential The electrical signal produced by sensory transduction.

recessive allele In a heterozygous individual, the allele that has no noticeable effect on the phenotype.

reciprocal altruism (AL-troo-IZ-em) In animal behavior, a selfless act repaid at a later time by the beneficiary or by another member of the beneficiary's social system.

recombinant DNA A DNA molecule carrying genes derived from two or more sources.

recombinant DNA technology Techniques for synthesizing recombinant DNA in vitro and transferring it into cells, where it can be replicated and may be expressed; also known as genetic engineering.

recombination frequency With respect to two given genes, the number of recombinant progeny from a mating divided by the total number of progeny. Recombinant progeny carry combinations of alleles different from those in either of the parents as a result of independent assortment of chromosomes or crossing over.

Recommended Dietary Allowance (RDA) A recommendation for daily nutrient intake established by nutritionists.

rectum The terminal portion of the large intestine where the feces are stored until they are eliminated.

red alga One of a group of marine, mostly multicellular, autotrophic protists, which includes the reef-building coralline algae.

red blood cell A blood cell containing hemoglobin, which transports O_2. Also called erythrocyte.

red-green color blindness A category of common, sex-linked human disorders involving several genes on the *X* chromosome; characterized by a malfunction of light-sensitive cells in the eyes; affects mostly males but also homozygous females.

redox reaction Short for oxidation-reduction; a chemical reaction in which electrons are lost from one substance (oxidation) and added to another (reduction). Oxidation and reduction always occur together.

reduction The gain of electrons by a substance involved in a redox reaction; always accompanies oxidation.

referred pain Pain that issues from an internal organ but that is felt on the body surface.

reflex An unconscious response to a specific stimulus.

regeneration The regrowth of body parts from pieces of an organism.

regulatory gene A gene that codes for a protein, such as a repressor, that controls the transcription of another gene or group of genes.

releasing hormone A hormone, secreted by the hypothalamus, that makes the anterior pituitary secrete hormones.

REM sleep Rapid-eye-movement sleep; a period of sleep when the brain is highly active, brain waves are fairly rapid and regular, and the eyes move rapidly under the closed eyelids. Most dreams occur during REM sleep. *See also* slow-wave (SW) sleep.

renal cortex The outer portion of the vertebrate kidney.

renal medulla The inner portion of the vertebrate kidney, beneath the renal cortex.

repetitive DNA Nucleotide sequences that are present in many copies in the DNA of a genome. The repeated sequences may be long or short and may be located next to each other or dispersed in the DNA.

repressor A protein that blocks the transcription of a gene or operon.

reproductive barrier A biological feature of a species that prevents it from interbreeding with other species even when populations of the two species live together.

Reptilia A class of vertebrate animals that consists of the reptiles, including snakes, lizards, turtles, crocodiles, and alligators.

resolving power A measure of the clarity of an image; the ability of an optical instrument to show two objects as separate.

respiration (1) Gas exchange, or breathing; the exchange of O_2 and CO_2 between an organism and its environment. An aerobic organism takes up O_2 and gives off CO_2. (2) Cellular respiration; the aerobic harvest of energy from food molecules by cells.

respiratory system The organ system that functions in exchanging gases with the environment; it supplies the blood with O_2 and disposes of CO_2.

resting potential The voltage across the plasma membrane of a resting neuron.

restriction enzyme A bacterial enzyme that cuts up foreign DNA, thus protecting bacteria against intruding DNA from phages and other organisms. Restriction enzymes are used in DNA technology to cut DNA molecules in reproducible ways.

restriction fragments Molecules of DNA produced from a longer DNA molecule cut up by a restriction enzyme; used in RFLP (restriction fragment length polymorphism) analysis, genome mapping, and other applications.

retina (RET-ih-nuh) The light-sensitive layer in an eye, made up of photoreceptor cells and sensory neurons.

retrovirus An RNA virus that reproduces by means of a DNA molecule; it reverse-transcribes its RNA into DNA, inserts the DNA into a cellular chromosome, and then transcribes more copies of the RNA from the viral DNA. HIV and a number of cancer-causing viruses are retroviruses.

reverse transcriptase (tran-SKRIP-tase) An enzyme that catalyzes the synthesis of DNA on an RNA template.

RFLPs (RIF-lips) Restriction fragment length polymorphisms; the differences in homologous DNA sequences that are reflected in different lengths of restriction fragments produced when the DNA is cut up with restriction enzymes.

rhizome (RY-zoam) A horizontal stem that grows below the ground.

rhodopsin (ro-DOP-sin) Visual pigment, in the rods of the vertebrate eye, that absorbs dim light.

ribonucleic acid (RNA) (ry-boh-noo-KLAY-ik) A type of nucleic acid consisting of nucleotide monomers with a ribose sugar and the nitrogenous bases adenine (A), cytosine (C), guanine (G), and uracil (U); usually single-stranded; functions in protein synthesis and as the genome of some viruses.

ribosomal RNA (rRNA) (RY-beh-so-mul) The type of ribonucleic acid that, together with proteins, makes up ribosomes; the most abundant type of RNA.

ribosomal RNA (rRNA) sequence analysis Determination of the nucleotide sequence in ribosomal RNA molecules.

ribosome (RY-beh-soam) A cell organelle consisting of RNA and protein organized into two subunits and functioning as the site of protein synthesis in the cytoplasm. The ribosomal subunits are constructed in the nucleolus.

ribozyme (RY-beh-zime) An enzymatic RNA molecule that catalyzes chemical reactions.

RNA *See* ribonucleic acid.

RNA polymerase (puh-LIM-eh-rase) An enzyme that links together the growing chain of RNA nucleotides during transcription, using a DNA strand as a template.

RNA splicing The removal of introns and joining of exons in eukaryotic RNA, forming an mRNA molecule with a continuous coding sequence; occurs before mRNA leaves the nucleus.

RNA world A hypothetical period in the evolution of life when RNA served as rudimentary genes and the sole catalytic molecules.

rod A photoreceptor cell in the vertebrate retina, enabling vision in dim light.

root cap A cone of cells at the tip of a plant root that protects the root's apical meristem.

root hair An outgrowth of an epidermal cell on a root, which increases the root's absorptive surface area.

root nodule A swelling on a plant root consisting of plant cells that contain nitrogen-fixing bacteria.

root pressure The upward push of xylem sap in a vascular plant, caused by the active pumping of minerals into the xylem by root cells.

root system All of a plant's roots that anchor it in the soil, absorb and transport minerals and water, and store food.

rough endoplasmic reticulum (rough ER) (reh-TIK-yuh-lum) A network of interconnected membranous sacs in a eukaryotic cell's cytoplasm. Rough ER membranes are studded with ribosomes that make membrane proteins and secretory proteins. The rough ER constructs membrane from phospholipids and proteins.

roundworm A member of the phylum Nematoda.

R plasmid A bacterial plasmid that carries genes for enzymes that destroy particular antibiotics, thus making the bacterium resistant to the antibiotics.

rRNA *See* ribosomal RNA.

ruminant mammal (ROO-min-ent) A mammal with a four-chambered stomach housing microorganisms that can digest cellulose; examples are cattle, deer, and sheep.

S

SA (sinoatrial) node (SY-noh-A-tree-ul) *See* pacemaker.

saccule (SAK-yool) A fluid-filled inner ear chamber containing hair cells that detect the position of the head relative to gravity.

sapwood Light-colored, water-conducting secondary xylem in a tree. *See also* heartwood.

sarcoma (sar-KO-muh) Cancer of the supportive tissues, such as bone, cartilage, and muscle.

sarcomere (SAR-ko-meer) The fundamental unit of muscle contraction, composed of thin filaments of actin and thick filaments of myosin; the region between two Z lines in a myofibril.

saturated Pertaining to fats and fatty acids whose hydrocarbon chains contain the maximum number of hydrogens and therefore have no double covalent bonds. Saturated fats and fatty acids solidify at room temperature.

savanna A biome dominated by grasses and scattered trees.

scanning electron microscope (SEM) A microscope that uses an electron beam to study the surface architecture of a cell or other specimen.

sclera (SKLER-uh) A layer of connective tissue forming the outer surface of the vertebrate eye. The cornea is the frontal part of the sclera.

sclereid (SKLER-id) In plants, a very hard, dead sclerenchyma cell found in nutshells and seed coats; a stone cell.

sclerenchyma cell (skler-EN-kim-uh) In plants, a supportive cell with rigid secondary walls hardened with lignin.

scrotum A pouch of skin outside the abdomen that houses a testis; functions in cooling sperm, thereby keeping them viable.

search image The mechanism that enables an animal to find a particular kind of food efficiently.

secondary consumer An organism that eats primary consumers.

secondary growth An increase in a plant's girth, involving cell division in the vascular cambium and the cork cambium.

secondary oocyte (OH-eh-SITE) A haploid cell resulting from meiosis I in oogenesis, which will become an ovum after meiosis II.

secondary spermatocyte (spur-MAT-eh-SITE) A haploid cell resulting from meiosis I in spermatogenesis, which will become a sperm cell after meiosis II.

secondary structure The second level of protein structure; the regular patterns of coils or folds of a polypeptide chain.

secondary succession A type of ecological succession that occurs where a disturbance has destroyed an existing biological community but left the soil intact. *See also* primary succession.

second law of thermodynamics The natural law that energy conversions reduce the order of the universe. Energy changes are always accompanied by entropy, an increase in disorder, such as heat.

second messenger A molecule such as cAMP that in response to a hormone triggers a change in enzymatic activity in the hormone's target cell.

secretion (1) The discharge of molecules synthesized by a cell. (2) In the vertebrate kidney, the discharge of wastes from the blood into the filtrate from the nephron tubules.

secretory protein A protein that is secreted by a cell, such as an antibody.

seed A plant embryo packaged with a food supply within a protective covering.

seed coat A tough outer covering of a seed, formed from the outer coat of an ovule; in a flowering plant, it encloses and protects the embryo and endosperm.

seed dormancy The temporary suspension of growth and development of a seed.

segmentation Subdivision along the length of an animal body into a series of repeated parts called segments.

selective permeability (PUR-mee-uh-BIL-ih-tee) A property of biological membranes that allows some substances to cross more easily than others and blocks the passage of other substances altogether.

self-fertilization The fusion of sperm and egg that are produced by the same individual organism.

self protein A protein on the surface of an antigen-presenting cell that can hold a foreign antigen and display it to helper T cells. Each individual has a unique set of self proteins that serve as molecular markers for the body. Lymphocytes do not attack self proteins unless the proteins are displaying foreign antigens; therefore, self proteins mark normal body cells as off-limits to the immune system. The technical name for self proteins is *major histocompatibility complex (MHC) proteins. See also* nonself molecule.

semen (SEE-men) The sperm-containing fluid that is ejaculated by the male during orgasm.

semicircular canals Fluid-filled channels in the inner ear that detect changes in the head's rate of rotation or angular movement.

seminal vesicle (SEM-ih-nul VES-ih-cul) A gland in males that secretes a fluid component of semen that lubricates and nourishes sperm.

seminiferous tubule (SEM-ih-NIF-er-us) A coiled sperm-producing tube in a testis.

sensation A feeling, or general awareness, of stimuli resulting from sensory information reaching the central nervous system.

sensory adaptation The tendency of sensory neurons to become less sensitive when they are stimulated repeatedly.

sensory input The conduction of signals from sensory receptors to processing centers in the central nervous system.

sensory neuron A nerve cell that receives information from sensory receptors and conveys signals into the central nervous system.

sensory transduction The conversion of a stimulus signal into an electrical signal by a sensory receptor cell.

sepal (SEE-pul) A modified leaf of a flowering plant. A whorl of sepals encloses and protects the flower bud before it opens.

sex chromosome A chromosome that determines whether an individual is male or female.

sex-linked gene A gene located on a sex chromosome.

sexual reproduction The creation of offspring by the fusion of two haploid sex cells (gametes), forming a diploid zygote.

sexual species A biological species (*see* biological species concept).

shoot system All of a plant's stems, leaves, and reproductive structures.

short-day plant A plant that flowers in late summer, fall, or winter, when day length is shortening.

sieve plate In a plant, a pore in the end wall of a sieve-tube member through which phloem sap flows.

sieve-tube member A food-conducting cell in a plant; chains of sieve-tube members make up phloem tissue.

signal transduction In cell biology, a series of molecular changes that converts a signal on a target cell's surface into a specific response inside the cell.

sign stimulus In animal behavior, a stimulus that triggers a fixed action pattern.

silencer A eukaryotic DNA sequence that functions to inhibit the start of gene transcription; may act analogously to an enhancer, by binding a repressor.

sink habitat An area of habitat where a species' death rate exceeds its reproductive success.

sister chromatid (KRO-muh-tid) One of the two identical parts of a duplicated chromosome in a eukaryotic cell.

skeletal muscle Striated muscle attached to the skeleton. The contraction of striated muscles provides voluntary movements of the body.

skeletal system The organ system that provides body support and protects body organs such as the brain, heart, and lungs.

slime mold *See* cellular slime mold *and* plasmodial slime mold.

slow-wave (SW) sleep A period of sleep characterized by delta waves, which are fairly regular, and strong bursts of brain activity. *See also* REM sleep.

smooth endoplasmic reticulum (smooth ER) A network of interconnected membranous tubules in a eukaryotic cell's cytoplasm. Smooth ER lacks ribosomes. Enzymes embedded in the smooth ER membrane function in the synthesis of certain kinds of molecules, such as lipids.

smooth muscle Muscle made up of cells without striations, found in the walls of organs such as the digestive tract, urinary bladder, and arteries.

social behavior Any kind of interaction between two or more animals, usually of the same species.

sociobiology The study of the evolutionary basis of social behavior.

sodium-potassium (Na^+-K^+) pump A membrane protein that transports sodium ions out of, and potassium ions into, a cell against their concentration gradients. The process is powered by ATP.

soil horizon A distinct layer of soil.

solute (SOL-yoot) A substance that is dissolved in a solution.

solution A fluid mixture of two or more substances, consisting of a dissolving agent, the solvent, and a substance that is dissolved, the solute.

solvent The dissolving agent in a solution. Water is the most versatile known solvent.

somatic cell (so-MAT-ik) Any cell in a multicellular organism except a sperm or egg cell or a cell that develops into a sperm or egg.

somatic nervous system The division of the motor nervous system of vertebrates composed of neurons that carry signals to skeletal muscles.

somite (SO-mite) A block of mesoderm in a chordate embryo that gives rise to vertebrae and other segmental structures.

source habitat An area of habitat where a species' reproductive success exceeds its death rate and from which new individuals often disperse to other areas.

Southern blotting A technique that reveals the presence of certain nucleotide sequences in a sample of DNA. After gel electrophoresis *(see)* of the cut-up DNA separates the fragments into bands, the DNA fragments are blotted onto a sheet of paper and then tested with a radioactively labeled nucleic acid probe.

speciation (SPEE-shee-A-shun) The evolution of new species.

species *See* biological species concept.

sperm A male gamete.

spermatogenesis (spur-MAT-oh-JEN-eh-sis) The formation of sperm cells.

spermicide A sperm-killing chemical, in the form of cream, jelly, or foam, that works with a barrier device as a method of contraception.

spinal cord The dorsal hollow nerve cord in vertebrates, located within the vertebral column; with the brain, makes up the central nervous system.

spinal nerve In the vertebrate peripheral nervous system, a nerve that carries signals to or from the spinal cord.

sporangium (spuh-RANJ-ee-um) (plural, sporangia) A bulbous structure at the tips of some branches.

spore (1) In plants and algae, a haploid cell that can develop into a multicellular individual without fusing with another cell. (2) In prokaryotes, protists, and fungi, any of a variety of thick-walled life cycle stages capable of surviving unfavorable environmental conditions.

sporophyte (SPOR-uh-FITE) The multicellular diploid form in the life cycle of organisms undergoing alternation of generations; results from a union of gametes and meiotically produces haploid spores that grow into the gametophyte generation.

stability In an ecological sense, the tendency of a biological community to remain in more-or-less constant balance due largely to interactions among organisms.

stabilizing selection Natural selection that favors intermediate variants by acting against extreme phenotypes.

stamen (STAY-men) A pollen-producing male reproductive part of a flower, consisting of a stalk and an anther.

starch A storage polysaccharide found in the roots of plants and certain other cells; a polymer of glucose.

start codon (KO-dahn) On mRNA, the specific three-nucleotide sequence (AUG) to which an initiator tRNA molecule binds, starting translation of genetic information.

stem Part of a plant's shoot system that supports the leaves and reproductive structures.

stem cell In the bone marrow, a type of cell that gives rise to all the types of blood cells.

steroid (STEHR-oid) A type of lipid whose carbon skeleton is bent to form four fused rings: three 6-sided rings and one 5-sided ring; examples are cholesterol, testosterone, and estrogen.

steroid hormone A regulatory chemical, a lipid made from cholesterol, that activates the transcription of specific genes in target cells.

stigma (STIG-muh) (plural, stigmata) The sticky tip of a flower's carpel, which traps pollen grains.

stimulus (plural, stimuli) (1) In the context of a nervous system, a factor that triggers sensory transduction. (2) In behavioral biology, a factor that triggers a specific response.

stoma (STO-muh) (plural, stomata) A pore surrounded by guard cells in the epidermis of a leaf. When stomata are open, CO_2 enters a leaf, and water and O_2 exit. A plant conserves water when its stomata are closed.

stomach A pouchlike organ in a digestive tract, which grinds and churns food and may store it temporarily.

stop codon In mRNA, one of three triplets (UAG, UAA, UGA) that signal gene translation to stop.

stretch receptor A type of mechanoreceptor sensitive to changes in muscle length; detects the position of body parts.

stroma (STRO-muh) A thick fluid enclosed by the inner membrane of a chloroplast. Sugars are made in the stroma by the enzymes of the Calvin cycle.

stromatolite (stro-MAT-ul-lite) Rock formed of layered, fossilized bacterial mats.

substrate (1) A specific substance (reactant) on which an enzyme acts. Each enzyme recognizes only the specific substrate or substrates of the reaction it catalyzes. (2) A surface in or on which an organism lives.

substrate-level phosphorylation The formation of ATP occurring when an enzyme transfers a phosphate group from an organic molecule (e.g., one of the intermediates in glycolysis or the Krebs cycle) to ADP.

succession *See* ecological succession, primary succession, *and* secondary succession.

summation (suh-MAY-shun) The overall effect of all the information a neuron receives at a particular instant.

superior vena cava (VEE-nuh KAY-vuh) A large vein that returns O_2-poor blood to the heart from the upper body and head. *See also* inferior vena cava.

supporting cell In the nervous system, a cell that protects, insulates, and reinforces a neuron.

surface tension A measure of how difficult it is to stretch or break the surface of a liquid.

suspension feeder An animal that extracts food particles suspended in the surrounding water.

sustainable development The long-term prosperity of human societies and the ecosystems that support them.

swim bladder A gas-filled internal sac that helps bony fish maintain buoyancy.

symbiosis A close association between organisms of two or more species.

sympathetic division One of two sets of neurons in the autonomic nervous system; generally prepares the body for energy-consuming activities, such as fleeing or fighting. *See also* parasympathetic division.

sympatric speciation The formation of a new species as a result of a genetic change that produces a reproductive barrier between the changed population (mutants) and the parent population.

synapse (SIN-aps) A junction, or relay point, between two neurons, or between a neuron and an effector cell. Electrical or chemical signals are relayed from one cell to another at a synapse.

synapsis (seh-NAP-sis) The pairing of duplicated homologous chromosomes, forming a tetrad, during prophase I of meiosis. During synapsis, chromatids of homologous chromosomes can exchange segments by crossing over.

synaptic cleft (seh-NAP-tik) A narrow gap separating the synaptic knob of a transmitting neuron from a receiving neuron or an effector cell.

synaptic knob The relay point at the tip of a transmitting neuron's axon, where signals are sent to another neuron or to an effector.

systematics The scientific study of biological diversity and its classification.

systemic acquired resistance A defensive response in plants infected with a pathogenic microbe; helps protect healthy tissue from the microbe.

systemic circuit One of two main blood circuits in terrestrial vertebrates; conveys blood between the heart and the rest of the body. *See also* pulmonary circuit.

systole (SIS-tuh-lee) The contraction stage of the heart cycle, when the heart chambers actively pump blood. *See also* diastole.

T

T_3 *See* triiodothyronine.

T_4 *See* thyroxine.

taiga (TIE-guh) The northern (boreal) coniferous forest, which extends across North America and Eurasia, to the southern border of the arctic tundra; also found just below alpine tundra on mountainsides in temperate zones.

tapeworm A parasitic flatworm characterized by the absence of a digestive tract.

target cell A cell that responds to a regulatory signal, such as a hormone.

taxis (TAK-sis) (plural, taxes) Virtually automatic orientation toward or away from a stimulus.

taxon (TAK-sahn) (plural, taxa) A proper name, such as Phylum Chordata, Class Mammalia, or *Homo sapiens,* in the taxonomic hierarchy used to classify organisms.

taxonomy The branch of biology concerned with identifiying, naming, and classifying species.

T cell A type of lymphocyte that matures in the thymus and is responsible for cell-mediated immunity; also involved in humoral immunity. *See also* B cell.

telophase The fourth and final stage of mitosis, during which daughter nuclei form at the two poles of a cell. Telophase usually occurs together with cytokinesis.

telomere The repetitive DNA *(see)* at each end of a eukaryotic chromosome.

temperate zones Latitudes between the tropics and the Arctic Circle in the north and the Antarctic Circle in the south; regions with milder climates than the tropics or polar regions.

temperature A measure of the intensity of heat, reflecting the average kinetic energy or speed of molecules.

temporal isolation A type of prezygotic barrier between species; the species remain isolated because they breed at different times.

tendon Fibrous connective tissue connecting a muscle to a bone.

terminal bud Embryonic tissue at the tip of a shoot, made up of developing leaves and a compact series of nodes and internodes.

terminator A special sequence of nucleotides in DNA that marks the end of a gene; it signals RNA polymerase to release the newly made RNA molecule, which then departs from the gene.

territory An area that an individual or individuals defend and from which other members of the same species are usually excluded.

tertiary consumer (TER-she-air-ee) An organism that eats secondary consumers.

tertiary structure The third level of protein structure; the overall, three-dimensional shape of a polypeptide in a protein.

testcross The mating between an individual of unknown genotype for a particular characteristic and an individual that is homozygous recessive for that same characteristic.

testis (plural, testes) The male gonad in an animal; produces sperm and, in many species, reproductive hormones.

testosterone (tes-TOS-teh-roan) An androgen hormone that stimulates an embryo to develop into a male and promotes male body features.

tetrad A paired set of homologous chromosomes, each composed of two sister chromatids. Tetrads form during prophase I of meiosis.

thalamus (THAL-uh-mus) An integrating and relay center of the vertebrate forebrain; sorts and relays selected information to specific areas in the cerebral cortex.

theory A widely accepted explanatory idea that is broad in scope and supported by a large body of evidence.

thermodynamics (THER-moh-dy-NAM-iks) The study of energy transformations that occur in a collection of matter. *See* first law of thermodynamics *and* second law of thermodynamics.

thermoreceptor A sensor (sensory receptor) that detects heat or cold.

thermoregulation The maintenance of internal temperature within a range that allows cells to function efficiently.

thigmotropism (thig-MOH-treh-PIZ-em) Growth movement of a plant in response to touch.

threshold potential The minimum change in a membrane's voltage that must occur to generate a nerve signal (action potential).

thylakoid (THY-luh-koid) One of a number of disk-shaped membranous sacs inside a chloroplast. Thylakoid membranes contain chlorophyll and the enzymes of the light reactions of photosynthesis. A stack of thylakoids is called a granum.

thymus gland (THY-mus) An endocrine gland in the neck region of mammals that is active in establishing the immune system; secretes several hormones that promote the development and differentiation of T cells.

thyroid gland (THY-roid) An endocrine gland that secretes thyroxine (T_4), triiodothyronine (T_3), and calcitonin.

thyroid-stimulating hormone (TSH) A protein hormone secreted by the anterior pituitary that stimulates the thyroid gland to secrete its hormones.

thyroxine (T_4) (thy-ROK-sin) An amine hormone secreted by the thyroid gland that stimulates metabolism in virtually all body tissues.

tight junction A junction that binds tissue cells together in a leakproof sheet.

Ti plasmid A bacterial plasmid that induces tumors in plants cells that it infects; often used as a vector to introduce new genes into plant cells. Ti stands for tumor-inducing.

tissue A cooperative unit of many similar cells that perform a specific function within a multicellular organism.

torpor (TOR-per) A state of reduced activity by an endotherm. Torpor reduces energy consumption because the metabolic rate, body temperature, heart rate, and breathing rate decrease.

trace element An element that is essential for the survival of an organism but only in minute quantities.

trachea (TRAY-kee-uh) (plural, tracheae) (1) The windpipe; the portion of the respiratory tube between the larynx and the bronchi. (2) One of many tiny tubes that branch throughout an insect's body, enabling gas exchange between outside air and body cells.

tracheid (TRAY-kee-id) A tapered, porous, water-conducting and supportive cell in plants. Chains of tracheids or vessel elements make up the water-conducting, supportive tubes in xylem.

tracheole The narrowest tube in an insect's tracheal system. *See also* trachea.

transcription The synthesis of RNA on a DNA template.

transcription factor In the eukaryotic cell, a protein that functions in initiating or regulating transcription. Transcription factors bind to DNA or to other proteins that bind to DNA.

transduction (1) The transfer of bacterial genes from one bacterial cell to another by a phage. (2) *See* sensory transduction. (3) *See* signal transduction.

transfer RNA (tRNA) A type of ribonucleic acid that functions as an interpreter in translation. Each tRNA molecule has a specific anticodon, picks up a specific amino acid, and conveys the amino acid to the appropriate codon on mRNA.

transformation The incorporation of new genes into a cell from DNA that the cell takes up from the fluid around it.

transgenic organism An organism that contains genes from another species.

translation The synthesis of a polypeptide using the genetic information encoded in an mRNA molecule. There is a change of "language" from nucleotides to amino acids.

translocation (1) During protein synthesis, the movement of a tRNA molecule carrying a growing polypeptide chain from the A site to the P site on a ribosome. (2) A change in a chromosome resulting from a chromosomal fragment attaching to a nonhomologous chromosome; can occur as a result of an error in meiosis or from mutagenesis.

transmission electron microscope (TEM) A microscope that uses an electron beam to study the internal structure of thinly sectioned specimens.

transpiration The evaporative loss of water from a plant.

transport vesicle A tiny membranous sac in a cell's cytoplasm carrying molecules produced by the cell. The vesicle buds form the endoplasmic reticulum or Golgi and eventually fuses with another membranous organelle or the plasma membranes, releasing its contents.

transposon (trans-POZE-ahn) A transposable genetic element, or "jumping gene"; a segment of DNA that can move from one site to another within a cell and serve as an agent of genetic change.

trial-and-error learning Learning to associate a particular behavioral act with a positive or negative effect.

triglyceride (try-GLIS-eh-ride) A fat, which consists of a molecule of glycerol linked to three molecules of fatty acid.

triiodothyronine (T_3) (TRY-eye-o-doe-THY-ruh-neen) An amine hormone secreted by the thyroid gland that stimulates metabolism in virtually all body tissues.

trimester In human development, one of three 3-month-long periods of pregnancy.

triplet code *See* genetic code.

tRNA *See* transfer RNA.

trophic structure (TRO-fik) The feeding relationships in an ecosystem; determines the route of energy flow and the pattern of chemical cycling in an ecosystem.

trophoblast (TROFE-uh-blast) In mammalian development, the outer portion of a blastocyst. Cells of the trophoblast secrete enzymes that enable the blastocyst to implant in the endometrium of the mother's uterus.

tropical deciduous forest A biome found in tropical areas with distinct wet and dry seasons; characterized by trees and shrubs that drop their leaves seasonally.

tropical rain forest A biome found in humid tropical areas where annual rainfall exceeds 250 cm and a season of reduced rainfall lasts no longer than a few months; among the most complex and species-rich ecosystems. Numerous species of large trees (especially evergreen angiosperms) are predominant.

tropical thorn forest A biome common in dry equatorial lowlands; characterized by thorny shrubs and trees and nonwoody plants that retain water for long periods.

tropics Latitudes between 23.5° north and south.

tropism (TRO-piz-em) A growth response that makes a plant grow toward or away from a stimulus.

true-breeding variety Organisms for which sexual reproduction produces offspring with inherited trait(s) identical to those of the parents; the organisms are homozygous for the characteristic(s) under consideration.

tubal ligation A means of sterilization in which a woman's two oviducts (Fallopian tubes) are tied closed to prevent eggs from reaching the uterus; a segment of each oviduct is removed.

tuber An enlargement at the end of a rhizome, in which food is stored.

tumor An abnormal mass of cells that forms within otherwise normal tissue.

tumor-suppressor gene A gene whose product inhibits cell division, thereby preventing uncontrolled cell growth.

tundra A biome at the northernmost limits of plant growth and at high altitudes, characterized by dwarf woody shrubs, grasses, mosses, and lichens.

tunicate One of a group of invertebrate chordates.

U

ultimate cause In behavioral biology, the evolutionary explanation for an organism's behavior.

ultrasound imaging A technique for examining a fetus in the uterus. High-frequency sound waves echoing off the fetus are used to produce an image of the fetus.

unsaturated Pertaining to fats and fatty acids whose hydrocarbon chains lack the maximum number of hydrogen atoms and therefore have one or more double covalent bonds. Unsaturated fats and fatty acids do not solidify at room temperature.

ureter (yoo-REE-ter or YOO-reh-ter) A duct that conveys urine from the kidney to the urinary bladder.

urethra (yoo-REE-thruh) A duct that conveys urine from the urinary bladder to the outside. In the male, the urethra also conveys semen out of the body during ejaculation.

uterus (YOO-ter-us) In the reproductive system of a mammalian female, the organ where the development of young occurs; the womb.

utricle (YOO-trih-kul) A fluid-filled inner ear chamber containing hair cells that detect the position of the head relative to gravity.

V

vaccination (VAK-sih-NAY-shun) A procedure that presents the immune system with a harmless variant or derivative of a pathogen, thereby stimulating the immune system to mount a long-term defense against the pathogen.

vaccine (VAK-seen) A harmless variant or derivative of a pathogen used to stimulate a host organism's immune system to mount a long-term defense against the pathogen.

vacuole (VAK-yoo-ohl) A membrane-enclosed sac, part of the endomembrane system of a eukaryotic cell, having diverse functions.

vagina (vu-JI-nuh) Part of the female reproductive system between the uterus and the outside opening; the birth canal in mammals; also accommodates the male's penis and receives sperm during copulation.

vascular bundle (VAS-kyu-ler) A strand of vascular tissues (both xylem and phloem) in a plant stem.

vascular cambium (VAS-kyu-ler KAM-bee-um) During secondary growth of a plant, the cylinder of meristematic cells, surrounding the xylem and pith, which produces secondary xylem and phloem.

vascular plant A plant with xylem and phloem.

vascular system *See* vascular tissue.

vascular tissue A network of cells forming narrow tubes that extend throughout a plant. The two types of vascular tissue in a plant are xylem, which channels water and minerals upward from the roots, and phloem, which channels sugars produced in the leaves throughout the plant.

vas deferens (vas DEF-er-enz) (plural, vasa deferentia) Part of the male reproductive system that conveys sperm away from the testis; the sperm duct; in humans, the tube that conveys sperm between the epididymis and the common duct that leads to the urethra.

vasectomy (veh-SEK-tum-ee) Surgical removal of a section of the two sperm ducts (vasa deferentia) to prevent sperm from reaching the urethra; a means of sterilization in the male.

vector In molecular biology, a piece of DNA, usually a plasmid or a viral genome, that is used to move genes from one cell to another.

vegetative reproduction Asexual reproduction by a plant.

vein (1) In animals, a vessel that returns blood to the heart. (2) In plants, a vascular bundle in a leaf, composed of xylem and phloem.

ventilation A mechanism that provides contact between an animal's respiratory surface and the air or water to which it is exposed. Contact between a respiratory surface and air or water enables gas exchange to occur.

ventral Pertaining to the underside, or bottom, of a bilaterally symmetrical animal.

ventricle (VEN-trih-kul) (1) A heart chamber that pumps blood out of a heart. (2) A space in the vertebrate brain, filled with cerebrospinal fluid.

venule (VEN-yool) A vessel that conveys blood between a capillary bed and a vein.

vertebra (VER-teh-bruh) (plural, vertebrae) One of a series of segmented units making up the backbone of a vertebrate animal.

vertebrate (VER-teh-brate) A chordate animal with a backbone; includes agnathans, cartilaginous fishes, bony fishes, amphibians, reptiles, birds, and mammals.

vessel element A short, open-ended, water-conducting and supportive cell in plants. Chains of vessel elements or tracheids make up the water-conducting, supportive tubes in xylem.

villus (VIL-us) (plural, villi) (1) A fingerlike projection of the inner surface of the small intestine. (2) A fingerlike projection of the chorion of the mammalian placenta. Large numbers of villi increase the surface areas of these organs.

vital capacity The maximum volume of air that a respiratory system can inhale and exhale.

vitamin An organic nutrient that an organism requires in very small quantities. Vitamins generally function as coenzymes.

vitreous humor (VIH-tree-us HYOO-mer) A jellylike substance filling the space behind the lens in the vertebrate eye; helps maintain the shape of the eye.

vocal cord One of a pair of stringlike tissues in the larynx. Air rushing past the tensed vocal cords makes them vibrate, producing sounds.

W

water-conducting cells Specialized, dead plant cells with lignin-containing secondary walls, arranged end-to-end, forming xylem tissue. *See also* tracheid; vessel element.

water vascular system In echinoderms, a radially arranged system of water-filled canals that branch into extensions called tube feet. The system provides movement and circulates water, facilitating gas exchange and waste disposal.

wavelength The distance between crests of adjacent waves, such as those of the electromagnetic spectrum.

wax A type of lipid molecule consisting of one fatty acid linked to an alcohol; functions as a waterproof coating on many biological surfaces, such as apples and other fruits.

wetland An ecosystem intermediate between an aquatic one and a terrestrial one. Wetland soil is saturated with water permanently or periodically.

white blood cell A blood cell that functions in defending the body against infections and cancer cells. Also called leukocyte.

wild type The phenotype most commonly found in nature.

wood Secondary xylem of a plant. *See also* heartwood; sapwood.

wood ray A column of parenchyma cells that radiates from the center of a log and transports water to its outer living tissues.

X

xylem (ZY-lum) The nonliving portion of a plant's vascular system that provides support and conveys xylem sap from the roots to the rest of the plant. Xylem is made up of vessel elements and/or tracheids, water-conducting cells.

xylem sap The solution of inorganic nutrients conveyed in xylem tissue from a plant's roots to its shoots.

Y

yolk plug A cluster of endodermal cells at the surface of an amphibian gastrula. The yolk plug marks the position of the blastopore and the site of the future anus.

yolk sac An extraembryonic membrane that develops from endoderm; produces the embryo's first blood cells, germ cells, and gives rise to the allantois.

Z

zoned reserve An extensive region of land that includes one or more areas that are undisturbed by humans. The undisturbed areas are surrounded by lands that have been altered by human activity.

zooplankton (ZO-oh-PLANK-tun) Animals that drift in aquatic environments.

zygote (ZY-goat) The fertilized egg, which is diploid, that results from the union of a sperm cell nucleus and an egg cell nucleus.

INDEX

B

C

E

G

I

N

O

P

Q

R

T

U

V

W

X

Y

Z